Chemistry

ATOMS FIRST

THIRD EDITION

Julia Burdge
COLLEGE OF WESTERN IDAHO

Jason Overby
COLLEGE OF CHARLESTON

Mc
Graw
Hill
Education

CHEMISTRY: ATOMS FIRST, THIRD EDITION

ISBN 978-1-259-63813-8
MHID 1-259-63813-8

Chief Product Officer, SVP Products & Markets: *G. Scott Virkler*
Vice President, General Manager, Products & Markets: *Marty Lange*
Vice President, Content Design & Delivery: *Betsy Whalen*
Managing Director: *Thomas Timp*
Director: *David Spurgeon, Ph.D.*
Director, Product Development: *Rose Koos*
Director of Digital Content: *Robin Reed*
Digital Product Analyst: *Patrick Diller*
Marketing Manager: *Matthew Garcia*
Director of Digital Content: *Shirley Hino, Ph.D.*
Digital Product Developer: *Joan Weber*
Director, Content Design & Delivery: *Linda Avenarius*
Program Manager: *Lora Neyens*
Content Project Managers: *Sherry Kane/Rachael Hillebrand*
Buyer: *Laura M. Fuller*
Design: *David Hash*
Content Licensing Specialists: *Carrie Burger/Lorraine Buczek*
Cover Image: @XYZ/Shutterstock.com
Compositor: *Aptara, Inc.*
Printer: *LSC Communications*

Library of Congress Cataloging-in-Publication Data

Names: Burdge, Julia. | Overby, Jason, 1970-
Title: Chemistry : atoms first / Julia Burdge, College of Western Idaho,
    Jason Overby, College of Charleston.
Other titles: Atoms first
Description: Third edition. | New York, NY : McGraw-Hill Education, [2017] |
    Includes index.
Identifiers: LCCN 2016033779 | ISBN 9781259638138 (alk. paper) | ISBN
    1259638138 (alk. paper)
Subjects: LCSH: Chemistry—Textbooks.
Classification: LCC QD31.3 .B87 2017 | DDC 540—dc23 LC record available at https://lccn.loc.gov/2016033779

mheducation.com/highered

About the Authors

Julia Burdge received her Ph.D. (1994) from the University of Idaho in Moscow, Idaho. Her research and dissertation focused on instrument development for analysis of trace sulfur compounds in air and the statistical evaluation of data near the detection limit.

In 1994 she accepted a position at The University of Akron in Akron, Ohio, as an assistant professor and director of the Introductory Chemistry program. In the year 2000, she was tenured and promoted to associate professor at The University of Akron on the merits of her teaching, service, and research in chemistry education. In addition to directing the general chemistry program and supervising the teaching activities of graduate students, she helped establish a future-faculty development program and served as a mentor for graduate students and post-doctoral associates. Julia has recently relocated back to the northwest to be near family. She lives in Boise, Idaho; and she holds an affiliate faculty position as associate professor in the Chemistry Department at the University of Idaho and teaches general chemistry at the College of Western Idaho.

In her free time, Julia enjoys horseback riding, precious time with her three children, and quiet time at home with Erik Nelson, her partner and best friend.

Jason Overby received his B.S. degree in chemistry and political science from the University of Tennessee at Martin. He then received his Ph.D. in inorganic chemistry from Vanderbilt University (1997) studying main group and transition metal metallocenes and related compounds. Afterwards, Jason conducted postdoctoral research in transition metal organometallic chemistry at Dartmouth College.

Jason began his academic career at the College of Charleston in 1999 as an assistant professor. Currently, he is an associate professor with teaching interests in general and inorganic chemistry. He is also interested in the integration of technology into the classroom, with a particular focus on adaptive learning. Additionally, he conducts research with undergraduates in inorganic and organic synthetic chemistry as well as computational organometallic chemistry.

In his free time, he enjoys boating, exercising, and cooking. He is also involved with USA Swimming as a nationally-certified starter and stroke-and-turn official. He lives in South Carolina with his wife Robin and two daughters, Emma and Sarah.

Brief Contents

Contents

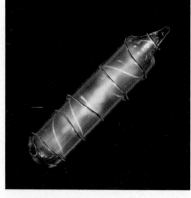

3 QUANTUM THEORY AND THE ELECTRONIC STRUCTURE OF ATOMS 66

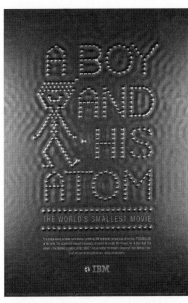

4 PERIODIC TRENDS OF THE ELEMENTS 124

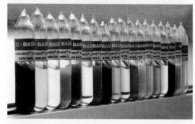
© BASF

5   IONIC AND COVALENT COMPOUNDS   162

© Brand X Pictures/PunchStock

6   REPRESENTING MOLECULES   210

7 MOLECULAR GEOMETRY, INTERMOLECULAR FORCES, AND BONDING THEORIES   246

7.1 Molecular Geometry 247
• The VSEPR Model 248 • Electron-Domain Geometry and Molecular Geometry 249 • Deviation from Ideal Bond Angles 253 • Geometry of Molecules with More Than One Central Atom 253

7.2 Molecular Geometry and Polarity 255

7.3 Intermolecular Forces 259
• Dipole-Dipole Interactions 259 • Hydrogen Bonding 260 • Dispersion Forces 261 • Ion-Dipole Interactions 263

7.4 Valence Bond Theory 264

7.5 Hybridization of Atomic Orbitals 267
• Hybridization of s and p Orbitals 268 • Hybridization of s, p, and d Orbitals 271

7.6 Hybridization in Molecules Containing Multiple Bonds 275

7.7 Molecular Orbital Theory 282
• Bonding and Antibonding Molecular Orbitals 283 • σ Molecular Orbitals 283 • Thinking Outside the Box: Phases 284 • Bond Order 285 • π Molecular Orbitals 285 • Molecular Orbital Diagrams 287 • Thinking Outside the Box: Molecular Orbitals in Heteronuclear Diatomic Species 288

7.8 Bonding Theories and Descriptions of Molecules with Delocalized Bonding 290

© Carol and Mike Werner/Science Source

8 CHEMICAL REACTIONS   308

8.1 Chemical Equations 309
• Interpreting and Writing Chemical Equations 309 • Balancing Chemical Equations 311 • Patterns of Chemical Reactivity 314

8.2 Combustion Analysis 317
• Determination of Empirical Formula 318

8.3 Calculations with Balanced Chemical Equations 320
• Moles of Reactants and Products 320 • Mass of Reactants and Products 321

8.4 Limiting Reactants 323
• Determining the Limiting Reactant 324 • Reaction Yield 326 • Thinking Outside the Box: Atom Economy 330

8.5 Periodic Trends in Reactivity of the Main Group Elements 331
• General Trends in Reactivity 332 • Hydrogen ($1s^1$) 332 • Reactions of the Active Metals 333 • Reactions of Other Main Group Elements 334 • Comparison of Group 1A and Group 1B Elements 337

© LWA/Photodisc/Getty Images

© Dirk Wiersma/Science Source

© Syracuse Newspapers/J. Berry/The Image Works

© Francisco Negroni/Alamy Stock Photo

US Department of Energy/Science Source

© Shawn Knol/Getty Images

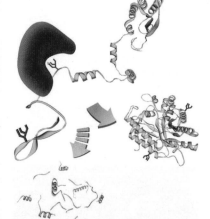

© Kenneth Eward/Science Source

© Richard Megna/Fundamental Photographs

16 ACIDS, BASES, AND SALTS 716

© Purestock/Alamy Stock Photo

© Lisa Stokes/Moment Open/Getty Images

© Friedrich Saurer/Alamy Stock Photo

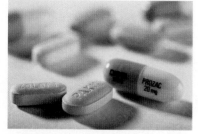

© Jonathan Nourok/Getty Images

20 NUCLEAR CHEMISTRY 940

© Pallava Bagla/Corbis

21 ENVIRONMENTAL CHEMISTRY 974

© Digital Vision/Getty Images

© Imaginechina via AP Images

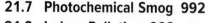
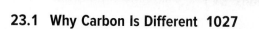

© Digital Vision/Getty Images

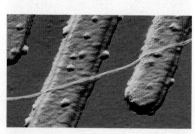
© Delft University of Technology/Science Source

25 NONMETALLIC ELEMENTS AND THEIR COMPOUNDS   (ONLINE ONLY)

© Craig Ruttle/AP Images

26 METALLURGY AND THE CHEMISTRY OF METALS   (ONLINE ONLY)

© Javier Larrea/Getty Images

List of Applications

Preface

The third edition of *Atoms First* by Burdge and Overby continues to build on the innovative success of the first and second editions. Changes to this edition include specific refinements intended to augment the student-centered pedagogical features that continue to make this book effective and popular both with professors, and with their students.

NEW! Student Hot Spot and Student-Centered Refinements using Heat Maps

Using heat maps from the adaptive reading tool SmartBook®, and the detailed analysis of student performance it provides, we were able to target specific learning objectives for minor re-wording, further explanation, or better illustration. Because SmartBook is a dynamic learning tool, we have a multitude of live data that show us exactly where students have been struggling with content; and we have direct insight into student learning that may not always be evident through other assessment methods. The data, such as average time spent answering each question and the percentage of students who correctly answered the question on the first attempt, revealed the learning objectives that students found particularly difficult.

> All properties of matter are either *extensive* or *intensive*. The measured value of an ***extensive property*** depends on the amount of matter. *Mass is an extensive property. More matter means more mass. Values of the same extensive property can be added together.* For example, two copper coins

This has allowed our revisions to be truly student-centered. For example, given specific known topics where students are struggling, we are able to clarify concepts or provide visual interpretations such as the below figure.

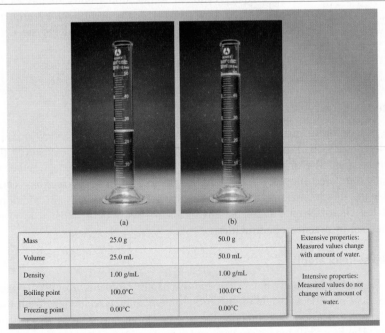

Figure 1.12 Some extensive properties (mass and volume) and intensive properties (density, boiling point, and freezing point) of water. The measured values of the extensive properties depend on the amount of water. The measured values of the intensive properties are independent of the amount of water.

(Photos): © H.S. Photos/Alamy Stock Photo

Further, armed with this powerful insight into the places many students struggle with content, we are able to provide strategically-timed access to additional learning resources. In the text, we have identified areas of particularly difficult content as "Student Hot Spots"—and use them to direct students to a variety of learning resources specific to that content. Students will be able to access over 1,000 digital learning resources throughout this text's SmartBook. These learning resources present summaries of concepts and worked examples, including over 200 videos of chemistry faculty solving problems or modeling concepts which students can view over and over again.

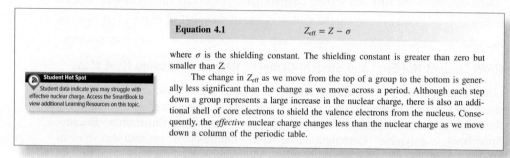

In the SmartBook version of the text, learning resources for these Student Hot Spots are embedded with the content for immediate access.

Guided by these direct student results of content understanding, we have edited the content in most of the chapters. Many of the changes are subtle, although some are more extensive. Our ability to employ live student-assessment data for revisions to address areas of common misunderstanding is unprecedented and has afforded us the opportunity to forever change how we provide the best possible learning materials to ensure that our students are optimally equipped to *engage* in chemistry.

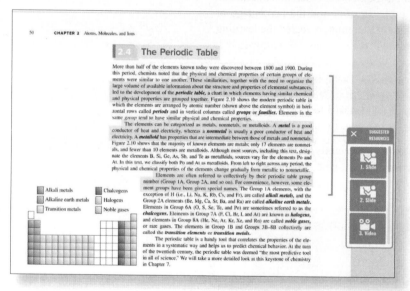

Updated Pedagogy

At the suggestion of many users, we have changed the Section Review questions to multiple choice. This provides an inviting opportunity for self-assessment at the end of each section. Students report using these questions to determine whether or not they have mastered the necessary skills to proceed to the next section—and most consider the multiple-choice format to be especially user-friendly. In addition, over 125 of the end-of-chapter problems have been revised and/or updated to provide a refreshed set of practice opportunities.

Key Skills–Relocated!

Newly located immediately before the end-of-chapter problems, Key Skills pages are modules that provide a review of specific problem-solving techniques from that particular chapter. These are techniques the authors know are vital to success in later chapters. The Key Skills pages are designed to be easy for students to find touchstones to hone specific skills from earlier chapters—in the context of later chapters. The answers to the Key Skills Problems can be found in the Answer Appendix in the back of the book.

New and Updated Chapter Content

Chapter 1—To continue providing the best flow of atoms first content, we have reorganized Chapter 1, placing classification and properties of matter at the end of the chapter. The benefit of this change is two-fold: It puts all of the numerical introduction to measurement and units together at the beginning; and it makes the transition from Chapter 1 (concluding with matter) to Chapter 2 (atoms) a little more seamless. Additionally, we have expanded coverage of dimensional analysis especially concerning units raised to powers and added a new figure illustrating intensive and extensive properties.

Chapter 3—Refreshed with a new introduction and opening image, our chapter on Quantum Theory and the Electronic Structure of Atoms has been updated for clarity in the introduction to energy and energy changes, discussion of the uncertainty principle, and the examination of electron configurations.

Chapter 6—We have refined discussion around several topics in the chapter on Representing Molecules, including multiple bonds, formal charge, and an introduction to resonance. Additionally, we've reordered the steps to building Lewis structures and reworked Worked Example 6.4 that demonstrates how to draw Lewis structures.

Chapter 12—We have included a new, atoms-first introduction to the packing of spheres in crystalline solids—providing a better foundation for understanding the origin of cubic packing in solid-state structures. Additional content has also been added to our section on phase changes.

Chapter 13—In this chapter, Physical Properties of Solutions, we've reworded sections 13.2 (A Molecular View of the Solution Process) and 13.3 (Concentration Units). We also have a new photo illustrating the Tyndall effect (Figure 13.13) as well as new computational end-of-chapter questions for section 13.3.

Chapter 15—In response to student data from SmartBook, we have made changes to some of the key figures in the introduction to equilibrium—improving the visual presentation in ways we believe will resonate with students. We've also updated the introduction to equilibrium constants & reaction quotients as well as the introduction to Le Châtelier's principle.

1.1	**The Study of Chemistry**
	• Chemistry You May Already Know
	• The Scientific Method
1.2	**Scientific Measurement**
	• SI Base Units • Mass • Temperature
	• Derived Units: Volume and Density
1.3	**Uncertainty in Measurement**
	• Significant Figures
	• Calculations with Measured Numbers
	• Accuracy and Precision
1.4	**Using Units and Solving Problems**
	• Conversion Factors
	• Dimensional Analysis—Tracking Units
1.5	**Classification of Matter**
	• States of Matter • Mixtures
1.6	**The Properties of Matter**
	• Physical Properties • Chemical Properties • Extensive and Intensive Properties

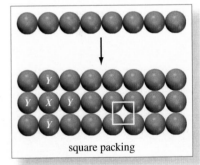

square packing

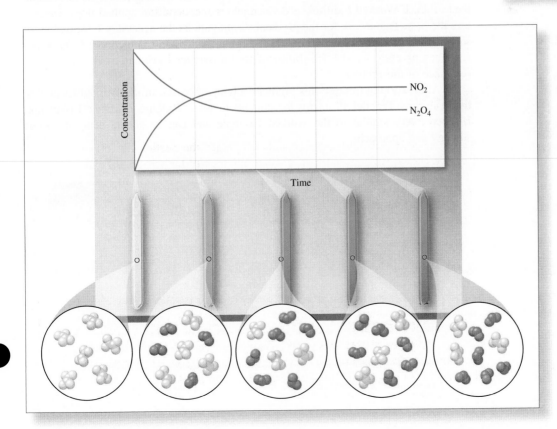

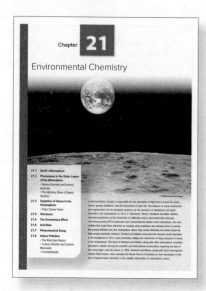

Chapter 21—Based on numerous requests, we have added a new chapter on environmental chemistry, a timely and relevant subdiscipline of chemistry. The topics in this chapter have proven to be of interest to students and instructors alike.

Chapter 26—In response to feedback from professors and to accommodate the inclusion of a dedicated chapter on environmental chemistry, we have moved the chapter on metallurgy and the chemistry of the metals to the online material. Therefore, what was Chapter 21 in the second edition has been renumbered Chapter 26, Metallurgy and the Chemistry of Metals. Both Chapter 25 (Nonmetallic Elements and Their Compounds) and Chapter 26 are available as a free digital download via the Instructor Resources in Connect and for text customization in McGraw-Hill Create.

The Construction of a Learning System

Writing a textbook and its supporting learning tools is a multifaceted process. McGraw-Hill's 360° Development Process is an ongoing, market-oriented approach to building accurate and innovative learning systems. It is dedicated to continual large scale and incremental improvement, driven by multiple customer feedback loops and checkpoints.

This is initiated during the early planning stages of new products and intensifies during the development and production stages. The 360° Development Process then begins again upon publication, in anticipation of the next version of each print and digital product. This process is designed to provide a broad, comprehensive spectrum of feedback for refinement and innovation of learning tools for both student and instructor. The 360° Development Process includes market research, content reviews, faculty and student focus groups, course- and product-specific symposia, accuracy checks, and art reviews, all guided by carefully selected Content Advisors.

The Learning System Used in *Chemistry: Atoms First*

Building Problem-Solving Skills. The entirety of the text emphasizes the importance of problem solving as a crucial element in the study of chemistry. Beginning with Chapter 1, a basic guide fosters a consistent approach to solving problems throughout the text. Each **Worked Example** is divided into four consistently applied steps: *Strategy* lays the basic framework for the problem; *Setup* gathers the necessary information for solving the problem; *Solution* takes us through the steps and calculations; *Think About It* makes us consider the feasibility of the answer or information illustrating the relevance of the problem.

After working through this problem-solving approach in the Worked Examples, there are three Practice Problems for students to solve. *Practice Problem A* (Attempt) is always very similar to the Worked Example and can be solved using the same strategy and approach.

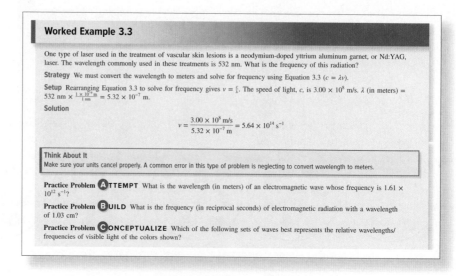

Although *Practice Problem B* (Build) probes comprehension of the same concept as Practice Problem A, it generally is sufficiently different in that it cannot be solved using the exact approach used in the Worked Example. Practice Problem B takes problem solving to another level by requiring students to develop a strategy independently. *Practice Problem C* (Conceptualize) provides an exercise that further probes the student's conceptual understanding of the material and many employ concept and molecular art. The regular use of the Worked Example and Practice Problems in this text will help students develop a robust and versatile set of problem-solving skills.

Section Review. Every section of the book that contains Worked Examples and Practice Problems ends with a Section Review. The Section Review enables the student to evaluate whether they understand the concepts presented in the section.

Key Skills. Newly located immediately before end-of-chapter problems, Key Skills are easy to find review modules where students can return to refresh and hone specific skills that the authors know are vital to success in later chapters. The answers to the Key Skills can be found in the Answer Appendix in the back of the book.

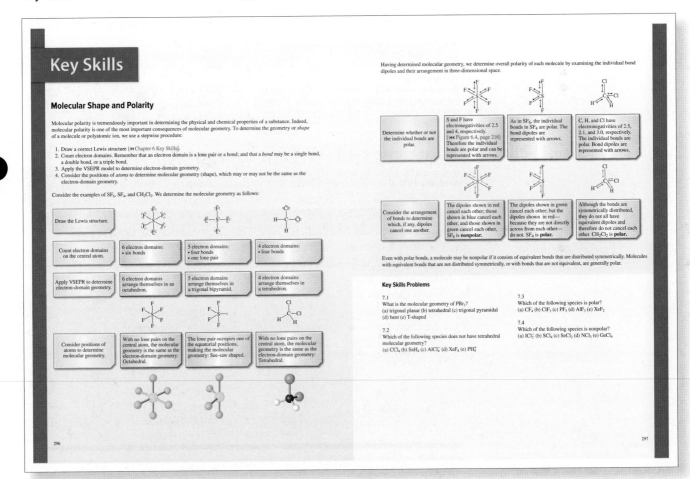

Student Hot Spots. In the text, we have identified areas of particularly difficult content as "Student Hot Spots"—and use them to direct students to a variety of learning resources specific to that content. Students will be able to access over 1,000 digital learning resources throughout this text's SmartBook. These learning resources present summaries of concepts and worked examples, including over 200 videos of chemistry faculty solving problems or modeling concepts which students can view over and over again.

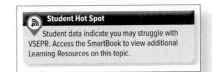

Student Hot Spot

Student data indicate you may struggle with VSEPR. Access the SmartBook to view additional Learning Resources on this topic.

Applications. Each chapter offers a variety of tools designed to help facilitate learning. *Student Annotations* provide helpful hints and simple suggestions to the student.

The nomenclature of molecular compounds follows in a similar manner to that of ionic compounds. Most molecular compounds are composed of two nonmetals (see [◄◄ Section 2.6, Figure 2.10]). To name such a compound, we first name the element that appears first in the formula. For HCl that would be hydrogen. We then name the second element, changing the ending of its name to –*ide*. For HCl, the second element is chlorine, so we would change chlorine to chloride. Thus, the systematic name of HCl is *hydrogen chloride.* Similarly, HI is hydrogen iodide (iod*ine* ⟶ iod*ide*) and SiC is silicon carbide (carb*on* ⟶ carb*ide*).

Student Annotation: Recall that compounds composed of two elements are called *binary* compounds.

Thinking Outside the Box is an application providing a more in-depth look into a specific topic. *Learning Outcomes* provide a brief overview of the concepts the student should understand after reading the chapter. It's an opportunity to review areas that the student does not feel confident about upon reflection.

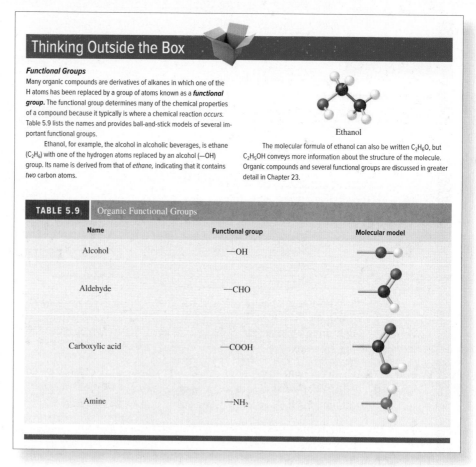

Thinking Outside the Box

Functional Groups

Many organic compounds are derivatives of alkanes in which one of the H atoms has been replaced by a group of atoms known as a **functional group.** The functional group determines many of the chemical properties of a compound because it typically is where a chemical reaction *occurs.* Table 5.9 lists the names and provides ball-and-stick models of several important functional groups.

Ethanol, for example, the alcohol in alcoholic beverages, is ethane (C_2H_6) with one of the hydrogen atoms replaced by an alcohol (—OH) group. Its name is derived from that of *ethane,* indicating that it contains *two* carbon atoms.

Ethanol

The molecular formula of ethanol can also be written C_2H_6O, but C_2H_5OH conveys more information about the structure of the molecule. Organic compounds and several functional groups are discussed in greater detail in Chapter 23.

TABLE 5.9	Organic Functional Groups	
Name	**Functional group**	**Molecular model**
Alcohol	—OH	
Aldehyde	—CHO	
Carboxylic acid	—COOH	
Amine	—NH₂	

Visualization. This text seeks to enhance student understanding through a variety of both unique and conventional visual techniques. A truly unique element in this text is the inclusion of a distinctive feature entitled **Visualizing Chemistry.** These two-page spreads appear as needed to emphasize fundamental, vitally important principles of chemistry. Setting them apart visually makes them easier to find and revisit as needed throughout the course term. Each Visualizing Chemistry feature concludes with a "What's the Point?" box that emphasizes the correct take-away message.

There is a series of conceptual end-of-chapter problems for each *Visualizing Chemistry* piece. The answers to the *Visualizing Chemistry* problems, *Key Skills* problems, and all odd-numbered end of chapter *Problems* can be found in the Answer Appendix at the end of the text.

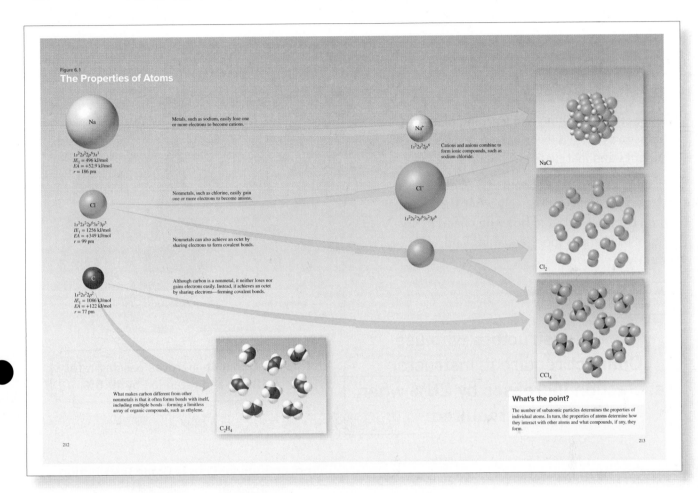

Flow Charts and a variety of inter-textual materials such as *Rewind* and *Fast Forward Buttons* and *Section Review* are meant to enhance student understanding and comprehension by reinforcing current concepts and connecting new concepts to those covered in other parts of the text.

Media. Many *Visualizing Chemistry* pieces have been made into captivating and pedagogically-effective *animations* for additional reinforcement of subject matter first encountered in the textbook. Each *Visualizing Chemistry* animation is noted by an icon.

Integration of Electronic Homework. You will find the *electronic homework* integrated into the text in numerous places. All Practice Problem B's are available in our electronic homework program for practice or assignments. A large number of the end-of-chapter problems are in the electronic homework system ready to assign to students.

For us, this text will always remain a work in progress. We encourage you to contact us with any comments or questions.

Julia Burdge
juliaburdge@hotmail.com

Jason Overby
overbyj@cofc.edu

Video 7.8
Chemical bonding—formation of molecular orbitals.

 **connect®**

©Getty Images/iStockphoto

McGraw-Hill Connect®
Learn Without Limits

Connect is a teaching and learning platform that is proven to deliver better results for students and instructors.

Connect empowers students by continually adapting to deliver precisely what they need, when they need it, and how they need it, so your class time is more engaging and effective.

73% of instructors who use **Connect** require it; instructor satisfaction **increases** by 28% when **Connect** is required.

Connect's Impact on Retention Rates, Pass Rates, and Average Exam Scores

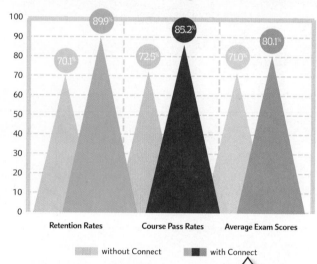

Using **Connect** improves passing rates by **12.7%** and retention by **19.8%**.

Analytics

Connect Insight®

Connect Insight is Connect's new one-of-a-kind visual analytics dashboard—now available for both instructors and students—that provides at-a-glance information regarding student performance, which is immediately actionable. By presenting assignment, assessment, and topical performance results together with a time metric that is easily visible for aggregate or individual results, Connect Insight gives the user the ability to take a just-in-time approach to teaching and learning, which was never before available. Connect Insight presents data that empowers students and helps instructors improve class performance in a way that is efficient and effective.

Impact on Final Course Grade Distribution

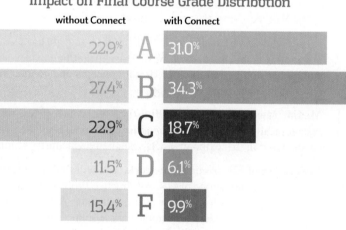

Students can view their results for any **Connect** course.

Adaptive

©Getty Images/iStockphoto

THE **ADAPTIVE** **READING EXPERIENCE** DESIGNED TO TRANSFORM THE WAY STUDENTS READ

> More students earn **A's** and **B's** when they use McGraw-Hill Education **Adaptive** products.

SmartBook®

Proven to help students improve grades and study more efficiently, SmartBook contains the same content within the print book, but actively tailors that content to the needs of the individual. SmartBook's adaptive technology provides precise, personalized instruction on what the student should do next, guiding the student to master and remember key concepts, targeting gaps in knowledge and offering customized feedback, and driving the student toward comprehension and retention of the subject matter. Available on smartphones and tablets, SmartBook puts learning at the student's fingertips—anywhere, anytime.

> Over **5.7 billion questions** have been answered, making McGraw-Hill Education products more intelligent, reliable, and precise.

STUDENTS WANT

Mc Graw Hill Education SMARTBOOK®

95% of students reported **SmartBook** to be a more effective way of reading material

100% of students want to use the Practice Quiz feature available within **SmartBook** to help them study

100% of students reported having reliable access to off-campus wifi

90% of students say they would purchase **SmartBook** over print alone

95% reported that **SmartBook** would impact their study skills in a positive way

Mc Graw Hill Education

*Findings based on a 2015 focus group survey at Pellissippi State Community College administered by McGraw-Hill Education

www.mheducation.com

Instructor and Student Resources

Learn without Limits

A robust set of questions, problems, and interactive figures are presented and aligned with the textbook's learning goals. The integration of **ChemDraw by PerkinElmer,** the industry standard in chemical drawing software, allows students to create accurate chemical structures in their online homework assignments. As an instructor, you can edit existing questions and write entirely new problems. Track individual student performance—by question, assignment, or in relation to the class overall—with detailed grade reports. Integrate grade reports easily with Learning Management Systems (LMS), such as WebCT and Blackboard—and much more. Also available within Connect, our adaptive SmartBook has been supplemented with additional learning resources tied to each learning objective to provide point-in-time help to students who need it. To learn more, visit www.mheducation.com.

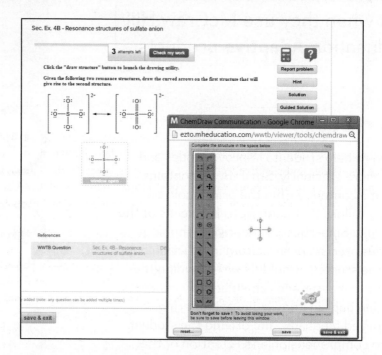

Instructors have access to the following instructor resources through Connect.

- **Art** Full-color digital files of all illustrations, photos, and tables in the book can be readily incorporated into lecture presentations, exams, or custom-made classroom materials. In addition, all files have been inserted into PowerPoint slides for ease of lecture preparation.
- **Animations** Numerous full-color animations illustrating important processes are also provided. Harness the visual impact of concepts in motion by importing these files into classroom presentations or online course materials.
- **PowerPoint Lecture Outlines** Ready-made presentations that combine art and lecture notes are provided for each chapter of the text.
- **Computerized Test Bank** Over 3,000 test questions that accompany *Chemistry: Atoms First* are available utilizing the industry-leading test generation software TestGen. These same questions are also available and assignable through Connect for online tests.
- **Instructor's Solutions Manual** This supplement contains complete, worked-out solutions for the Practice Problem C questions, Key Skills questions, and *all* the end-of-chapter problems in the text.

Fueled by LearnSmart—the most widely used and intelligent adaptive learning resource—**LearnSmart Prep** is designed to get students ready for a forthcoming course by quickly and effectively addressing prerequisite knowledge gaps that may cause problems down the road. By distinguishing what students know from what they don't, and honing in on concepts they are most likely to forget, LearnSmart Prep maintains a continuously adapting learning path individualized for each student, and tailors content to focus on what the student needs to master in order to have a successful start in the new class.

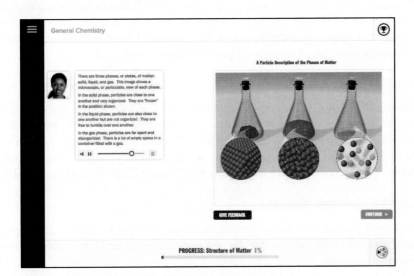

Based on the same world-class, superbly adaptive technology as LearnSmart, **McGraw-Hill LearnSmart Labs** is a must-see, outcomes-based lab simulation. It assesses a student's knowledge and adaptively corrects deficiencies, allowing the student to learn faster and retain more knowledge with greater success. First, a student's knowledge is adaptively leveled on core learning outcomes: Questioning reveals knowledge deficiencies that are corrected by the delivery of content that is conditional on a student's response. Then, a simulated lab experience requires the student to think and act like a scientist: Recording, interpreting, and analyzing data using simulated equipment found in labs and clinics. The student is allowed to make mistakes—a powerful part of the learning experience! A virtual coach provides subtle hints when needed, asks questions about the student's choices, and allows the student to reflect on and correct those mistakes. Whether your need is to overcome the logistical challenges of a traditional lab, provide better lab prep, improve student performance, or make your online experience one that rivals the real world, LearnSmart Labs accomplishes it all.

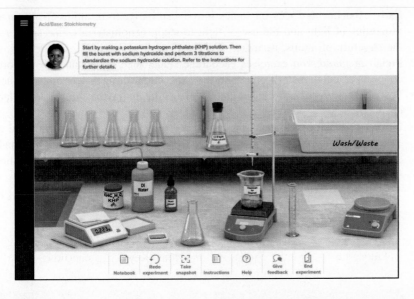

McGraw-Hill Create™

With **McGraw-Hill Create,** you can easily rearrange chapters, combine material from other content sources, and quickly upload content you have written, like your course syllabus or teaching notes. Find the content you need in Create by searching through thousands of leading McGraw-Hill textbooks. Arrange your book to fit your teaching style. Create even allows you to personalize your book's appearance by selecting the cover and adding your name, school, and course information. Order a Create book and you'll receive a complimentary print review copy in 3–5 business days or a complimentary electronic review copy (eComp) via email in minutes. Go to ww.mcgrawhillcreate.com today and register to experience how McGraw-Hill Create empowers you to teach *your* students *your* way. www.mcgrawhillcreate.com

My Lectures—Tegrity®

McGraw-Hill Tegrity records and distributes your class lecture with just a click of a button. Students can view anytime/anywhere via computer, iPod, or mobile device. It indexes as it records your PowerPoint® presentations and anything shown on your computer so students can use keywords to find exactly what they want to study. Tegrity is available as an integrated feature of McGraw-Hill Connect Chemistry and as a standalone.

Student Solutions Manual

Students will find answers to the Visualizing Chemistry and Key Skills questions and detailed solutions and explanations for the odd-numbered problems from the text in the solutions manual.

Laboratory Manual

Laboratory Manual to Accompany Chemistry: Atoms First by Gregg Dieckmann and John Sibert from the University of Texas at Dallas. This laboratory manual presents a lab curriculum that is organized around an atoms-first approach to general chemistry. The philosophy behind this manual is to (1) provide engaging experiments that tap into student curiosity, (2) emphasize topics that students find challenging in the general chemistry lecture course, and (3) create a laboratory environment that encourages students to "solve puzzles" or "play" with course content and not just "follow recipes." The laboratory manual represents a terrific opportunity to get students turned on to science while creating an environment that connects the relevance of the experiments to a greater understanding of their world. This manual has been written to provide instructors with tools that engage students, while providing important connections to the material covered in an atoms-first lecture course.

Important features of this laboratory manual:

- Early experiments focus on topics introduced early in an atoms-first course—properties of light and the use of light to study nanomaterials, line spectra and the structure of atoms, periodic trends, etc.
- Prelab or *foundation* exercises encourage students to understand the important concepts/calculations/procedures in the experiment through working together.
- Postlab or *reflection* exercises put the lab content in the context of a larger chemistry/science picture.
- Instructor's resources (found in the Instructor Resources on Connect®) provided with each experiment outline variations that can be incorporated to enrich the student experience or tailor the lab to the resources/equipment available at the institution.

Acknowledgments

We wish to thank the many people—past and present—who have contributed to the development of this new text.

Titus Vasile Albu, *University of Tennessee–Chattanooga*
Mohd Asim Ansari, *Fullerton College*
Andrew Axup, *St. Ambrose University*
Mary Fran Barber, *Wayne State University*
David L. Boatright, *University of West Georgia*
Michael Bukowski, *Penn State University–Altoona*
Jerry Burns, *Pellissippi State Community College*
Tara Carpenter, *University of Maryland–BC*
David Carter, *Angelo State University*
David Carter, *Angelo State*
Gezahegn Chaka, *Collin County Community College*
Ngee Sing Chong, *Middle Tennessee State University*
Allen Clabo, *Francis Marion University*
Colleen Craig, *University of Washington*
Guy Dadson, *Fullerton College*
David Dearden, *Brigham Young University*
Mark Dibben, *USAFA Preparatory School*
Gregg Dieckmann, *University of Texas–Dallas*
Stephen Drucker, *University of Wisconsin–Eau Claire*
Ronald Duchovic, *Indiana University Purdue University–Fort Wayne*
Jack Eichler, *University of California–Riverside*
Anthony Fernandez, *Merrimack College*
Lee Friedman, *University of Maryland–College Park*
Rachel Garcia, *San Jacinto College*
Kate Graham, *College of St. Benedict/St. John's University*
Patrick Greco, *Sinclair Community College*
Tracy Hamilton, *University of Alabama at Birmingham*
Susan Hendrickson, *University of Colorado–Boulder*
Christine Hrycyna, *Purdue University*
James Jeitler, *Marietta College*
Scott Kennedy, *Anderson University*
Farooq A. Khan, *University of West Georgia*
William Kuhn, *USAFA Preparatory School*
Joseph Langat, *Florida State College at Jacksonville*
John Lee, *University of Tennessee–Chattanooga*
Debbie Leedy, *Glendale Community College*

Yinfa Ma, *Missouri University of Science and Technology*
Helene Maire-Afeli, *University of South Carolina–Union*
John Marvin, *Brescia University*
Roy McClean, *United States Naval Academy*
Anna McKenna, *College of St. Benedict/St. John's University*
Jack McKenna, *St. Cloud State University*
Jeremy Mitchell-Koch, *Emporia State University*
Matt Morgan, *Hamline University*
Douglas Mulford, *Emory University*
Patricia Muisener, *University of South Florida*
Chip Nataro, *Lafayette College*
Anne-Marie Nickel, *Milwaukee School of Engineering*
Delana Nivens, *Armstrong Atlantic State University*
Edith Osborne, *Angelo State University*
Hansa Pandya, *Richland College*
Katherine Parks, *Motlow College*
Mike Rennekamp, *Columbus State Community College*
Dawn Richardson, *Collin College–Frisco*
John Richardson, *Austin College*
Dawn Rickey, *Colorado State University*
Raymond Sadeghi, *University of Texas at San Antonio*
Nicholas Schlotter, *Hamline University*
Sarah Schmidtke, *The College of Wooster*
Jacob Schroeder, *Clemson University*
Stephen Schvaneveldt, *Clemson University*
John Sibert, *University of Texas–Dallas*
Regina Stevens-Truss, *Kalamazoo College*
John Stubbs, *University of New England*
Katherine Stumpo, *University of Tennessee–Martin*
Steve Theberge, *Merrimack College*
Lori Van Der Sluys, *Penn State University*
Jason Vohs, *St. Vincent College*
Stan Whittingham, *Binghamton University*
Nathan Winter, *St. Cloud State University*
Kimberly Woznack, *California University of Pennsylvania*

Raymond Chang's contributions have been invaluable. His unfaltering diligence and legendary attention to detail have added immeasurably to the quality of this book.

The following individuals helped write and review learning goal-oriented content for LearnSmart: David G Jones, Vistamar School and Adam I. Keller, Columbus State Community College.

We both thank and acknowledge our families for their continued and devoted support.

Finally, we must acknowledge our McGraw-Hill family for their inspiration, excitement, and support of this project: Managing Director Thomas Timp; Director of Chemistry David Spurgeon, PhD; Associate Director of Digital Content Robin Reed; Content Project Manager Sherry Kane; Senior Designer David Hash; Senior Director of Digital Content Shirley Hino and Senior Marketing Manager Matthew Garcia.

Chemistry

ATOMS FIRST

THIRD EDITION

Julia Burdge
COLLEGE OF WESTERN IDAHO

Jason Overby
COLLEGE OF CHARLESTON

Chapter

1

Chemistry: The Science of Change

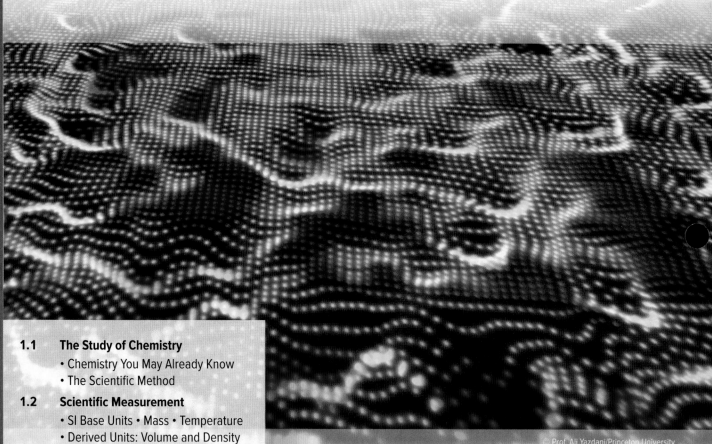

© Prof. Ali Yazdani/Princeton University

RECENT STUDIES of interactions involving nanoparticles of noble metals, including gold, silver, and platinum, have enabled scientists to explain and exploit something known as *localized surface plasmon resonances,* depicted here. Among other things, this work has led to the development of photothermal ablation—a novel treatment for certain cancers. Specially designed gold nanoshells are injected into the patient and preferentially attach themselves to the target tumor cells. Near-infrared radiation (light of slightly longer wavelength than can be detected by the human eye) is then directed at the tumor, causing the gold nanoshells to emit heat. This heat destroys the tumor cells to which the nanoshells are attached, leaving the surrounding and nearby healthy cells unharmed.

Before You Begin, Review These Skills

- Basic algebra
- Scientific notation [◄◄ Appendix 1]

1.1   THE STUDY OF CHEMISTRY

Chemistry often is called the *central science* because knowledge of the principles of chemistry can facilitate understanding of other sciences, including physics, biology, geology, astronomy, oceanography, engineering, and medicine. ***Chemistry*** is the study of *matter* and the *changes* that matter undergoes. Matter is what makes up our bodies, our belongings, our physical environment, and in fact our entire universe. ***Matter*** is anything that has mass and occupies space.

Chemistry You May Already Know

You may already be familiar with some of the terms used in chemistry. Even if this is your first chemistry course, you may have heard of *molecules* and know them to be tiny pieces of a substance—much too tiny to see. Further, you may know that molecules are made up of *atoms,* even smaller pieces of matter. And even if you don't know what a *chemical formula* is, you probably know that H_2O is water. You may have used, or at least heard, the term *chemical reaction;* and you are undoubtedly familiar with a variety of common processes that are chemical reactions, such as those shown in Figure 1.1. Don't worry if you are not familiar with these terms; they will be defined in the early chapters of this book.

The processes in Figure 1.1 are all things that you can observe at the *macroscopic level.* In other words, these processes and their results are visible to the human eye. In studying chemistry, you will learn to visualize and understand these same processes at the *submicroscopic* or *molecular level.*

Student Annotation: Macroscopic means *large enough to be seen with the unaided eye.*

Student Annotation: Submicroscopic means *too small to be seen, even with a microscope.* Atoms and molecules are *submicroscopic.*

The Scientific Method

Advances in our understanding of chemistry (and other sciences) are the result of scientific experiments. Although scientists do not all take the same approach to experimentation, they must follow a set of guidelines known as the ***scientific method*** to have their results added to the larger body of knowledge within a given field. The flowchart in Figure 1.2 illustrates this basic process. The method begins with the gathering of data via observations and experiments. Scientists study these data and try to identify *patterns* or *trends.* When they find a pattern or trend, they may summarize their findings with a ***law,*** a concise verbal or mathematical statement of a reliable relationship between phenomena. Scientists may then formulate a ***hypothesis,*** a tentative explanation for their observations. Further experiments are designed to test the hypothesis. If experiments indicate that the hypothesis is incorrect, the scientists go back to the drawing board, try to come up with a different interpretation of their data, and formulate a new hypothesis. The new hypothesis will then be tested by experiment. When a hypothesis stands the test of extensive experimentation, it may evolve into a theory. A ***theory*** is a unifying principle that explains a body of experimental observations and the laws that are based on them. Theories can also be used to predict related phenomena, so theories are constantly being tested. If a theory is disproved by experiment, then it must be discarded or modified so that it becomes consistent with experimental observations.

Figure 1.1 Many familiar processes are chemical reactions: (a) The flame of a gas stove is the combustion of natural gas, which is primarily methane. (b) The bubbles produced when Alka-Seltzer dissolves in water are carbon dioxide, produced by a chemical reaction between two ingredients in the tablets. (c) The formation of rust is a chemical reaction that occurs when iron, water, and oxygen are all present. (d) Many baked goods "rise" as the result of a chemical reaction that produces carbon dioxide.

(a): © fStop/PunchStock; (b): © Brand X Pictures/PunchStock; (c): © Image Source/Corbis; (d): © Sharon Dominick/Getty

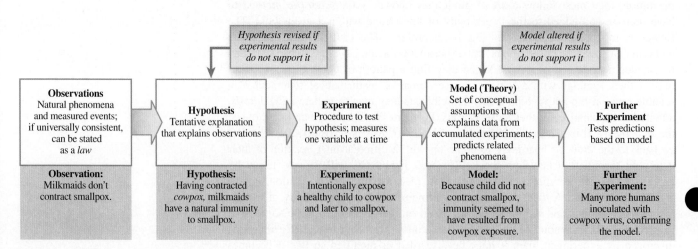

Figure 1.2 Flowchart of the scientific method.

A fascinating example of the use of the scientific method is the story of how smallpox was eradicated. Late in the eighteenth century, an English doctor named Edward Jenner observed that even during outbreaks of smallpox in Europe, milkmaids seldom contracted the disease. He reasoned that when people who had frequent contact with cows contracted *cowpox,* a similar but far less harmful disease, they developed a natural immunity to smallpox. He predicted that intentional exposure to the cowpox virus would produce the same immunity. In 1796, Jenner exposed an 8-year-old boy to the cowpox virus using pus from the cowpox lesions of an infected milkmaid. Six weeks later, he exposed the boy to the *smallpox* virus and, as Jenner had predicted, the boy did *not* contract the disease. Subsequent experiments using the same technique (later dubbed *vaccination* from the Latin *vacca* meaning *cow*) confirmed that immunity to smallpox could be induced.

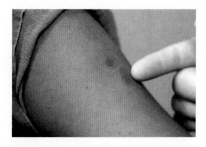

Until recently, almost everyone had a smallpox vaccine scar—usually on the upper arm.

© Chris Livingston/Getty Images

A superbly coordinated international effort on the part of healthcare workers was successful in eliminating smallpox worldwide. In 1980, the World Health Organization declared smallpox officially eradicated. This historic triumph over a dreadful disease, one of the greatest medical advances of the twentieth century, began with Jenner's astute observations, inductive reasoning, and careful experimentation—the essential elements of the *scientific method.*

Student Annotation: The last naturally occurring case was in 1977 in Somalia.

1.2 SCIENTIFIC MEASUREMENT

Scientists use a variety of devices to measure the properties of matter. A meterstick is used to measure length; a burette, pipette, graduated cylinder, and volumetric flask are used to measure volume (Figure 1.3); a balance is used to measure mass; and a thermometer is used to measure temperature. Properties that can be measured are called *quantitative* properties because they are expressed using numbers. When we express a measured quantity with a number, though, we must always include the appropriate unit; otherwise, the measurement is meaningless. For example, to say that the depth of a swimming pool is 3 is insufficient to distinguish between one that is 3 *feet* (0.9 meter) and one that is 3 *meters* (9.8 feet) deep. Units are essential to reporting measurements correctly.

The two systems of units with which you are probably most familiar are the *English system* (foot, gallon, pound, etc.) and the *metric system* (meter, liter, kilogram, etc.). Although there has been an increase in the use of metric units in the United States in recent years, English units still are used commonly. For many years, scientists recorded measurements in metric units, but in 1960, the General Conference on Weights and Measures, the international authority on units, proposed a revised metric system for universal use by scientists. We will use both metric and revised metric (SI) units in this book.

Student Annotation: According to the U.S. Metric Association (USMA), the United States is "the only significant holdout" with regard to adoption of the metric system. The other countries that continue to use traditional units are Myanmar (formerly Burma) and Liberia.

SI Base Units

The revised metric system is called the ***International System of Units*** (abbreviated SI, from the French *Système Internationale d'Unités*). Table 1.1 lists the seven SI base units. All other units of measurement can be derived from these base units. The ***SI unit*** for *volume,* for instance, is derived by cubing (raising to the power 3) the SI base unit for *length*. The prefixes listed in Table 1.2 are used to denote decimal fractions and decimal multiples of SI units. The use of these prefixes enables scientists to tailor the magnitude of a unit to a particular application. For example, the meter (m) is appropriate for describing the dimensions of a classroom, but the kilometer (km), 1000 m, is more appropriate for describing the distance between two cities. Units that you will encounter frequently in the study of chemistry include those for mass, temperature, volume, and density.

Mass

Although the terms *mass* and *weight* often are used interchangeably, they do not mean the same thing. Strictly speaking, weight is the force exerted by an object or sample due to gravity. **Mass** is a measure of the amount of matter in an object or sample. Because gravity varies from location to location (gravity on the moon is only about one-sixth that on Earth), the weight of an object varies depending on where it is measured. The mass of an object remains the same regardless of where it is measured. The SI base unit of mass is the kilogram (kg), but in chemistry the smaller gram (g) often is more convenient and is more commonly used:

$$1 \text{ kg} = 1000 \text{ g} = 1 \times 10^3 \text{ g}$$

Occasionally, the most convenient and/or commonly used unit for a particular application is not an SI unit. One such example is the atomic mass unit. The **atomic mass unit (amu),** as the name suggests, is used to express the masses of atoms—and other objects of similar size. In terms of SI units, the amu is equal to $1.6605378 \times 10^{-24}$ g or $1.6605378 \times 10^{-27}$ kg. Another example is the **angstrom (Å),** a measure of length that is equal to 1×10^{-10} m.

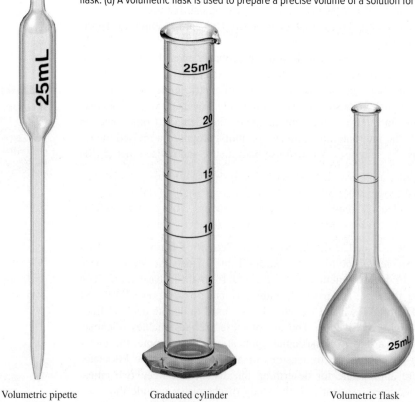

Figure 1.3 (a) A burette is used to measure the volume of a liquid that has been added to a container. A reading is taken before and after the liquid is delivered, and the volume delivered is determined by subtracting the first reading from the second. (b) A volumetric pipette is used to deliver a precise amount of liquid. (c) A graduated cylinder is used to measure a volume of liquid. It is less precise than the volumetric flask. (d) A volumetric flask is used to prepare a precise volume of a solution for use in the laboratory.

25mL

25mL

25mL

Burette
(a)

Volumetric pipette
(b)

Graduated cylinder
(c)

Volumetric flask
(d)

TABLE 1.1	Base SI Units	
Base quantity	**Name of unit**	**Symbol**
Length	meter	m
Mass	kilogram	kg
Time	second	s
Electric current	ampere	A
Temperature	kelvin	K
Amount of substance	mole	mol
Luminous intensity	candela	cd

TABLE 1.2		Prefixes Used with SI Units	
Prefix	**Symbol**	**Meaning**	**Example**
Tera-	T	1×10^{12} (1,000,000,000,000)	1 teragram (Tg) = 1×10^{12} g
Giga-	G	1×10^{9} (1,000,000,000)	1 gigawatt (GW) = 1×10^{9}
Mega-	M	1×10^{6} (1,000,000)	1 megahertz (MHz) = 1×10^{6}
Kilo-	k	1×10^{3} (1,000)	1 kilometer (km) = 1×10^{3} m
Deci-	d	1×10^{-1} (0.1)	1 deciliter (dL) = 1×10^{-1} L
Centi-	c	1×10^{-2} (0.01)	1 centimeter (cm) = 1×10^{-2} m
Milli-	m	1×10^{-3} (0.001)	1 millimeter (mm) = 1×10^{-3} m
Micro-	μ	1×10^{-6} (0.000001)	1 microliter (μL) = 1×10^{-6} L
Nano-	n	1×10^{-9} (0.000000001)	1 nanosecond (ns) = 1×10^{-9} s
Pico-	p	1×10^{-12} (0.000000000001)	1 picogram (pg) = 1×10^{-12} g

Temperature

There are two temperature scales used in chemistry: the *Celsius* scale and the *absolute* or *Kelvin* scale. Their units are the *degree Celsius* (°C) and the *kelvin* (K), respectively. The **Celsius** scale [named after Swedish physicist Ander Celsius (1701–1744)] was originally defined using the freezing point (0°C) and the boiling point (100°C) of pure water at sea level. As Table 1.1 shows, the SI base unit of temperature is the **kelvin.** Kelvin is also known as the *absolute* temperature scale because the lowest temperature theoretically possible is 0 K, a temperature referred to as *absolute zero*. No *degree* sign (°) is used to represent a temperature on the Kelvin scale.

Units of the Celsius and Kelvin scales are equal in magnitude, so *a degree Celsius* is equivalent to a *kelvin*. Thus, if the temperature of an object increases by 5°C, it also increases by 5 K. Absolute zero on the Kelvin scale is equivalent to −273.15°C on the Celsius scale. We use the following equation to convert a temperature from units of degrees Celsius to kelvins:

$$K = {}^{\circ}C + 273.15 \qquad \textbf{Equation 1.1}$$

Student Annotation: There is no such thing as a negative temperature on the Kelvin scale.

Student Annotation: The theoretical basis of the Kelvin scale has to do with the behavior of gases. [Chapter 11]

Student Annotation: Depending on the precision required, the conversion from degrees Celsius to kelvins often is done simply by adding 273, rather than 273.15.

Worked Example 1.1 illustrates conversions between these two temperature scales.

Worked Example 1.1

Normal human body temperature can range over the course of the day from about 36°C in the early morning to about 37°C in the afternoon. Express these two temperatures and the range that they span using the Kelvin scale.

Strategy Use Equation 1.1 to convert temperatures from the Celsius scale to the Kelvin scale. Then convert the range of temperatures from degrees Celsius to kelvins, keeping in mind that 1°C is equivalent to 1 K.

Setup Equation 1.1 is already set up to convert the two temperatures from degrees Celsius to kelvins. No further manipulation of the equation is needed. The range in kelvins will be the same as the range in degrees Celsius.

Solution 36°C + 273 = 309 K, 37°C + 273 = 310 K, and the range of 1°C is equal to a range of 1 K.

Think About It

Check your math and remember that converting a temperature from degrees Celsius to kelvins is different from converting a *difference* in temperature from degrees Celsius to kelvins.

Practice Problem Ⓐ**TTEMPT** Express the freezing point of water (0°C), the boiling point of water (100°C), and the range spanned by the two temperatures using the Kelvin scale.

Practice Problem Ⓑ**UILD** According to the website of the National Aeronautics and Space Administration (NASA), the average temperature of the universe is 2.7 K. Convert this temperature to degrees Celsius.

Practice Problem Ⓒ**ONCEPTUALIZE** If a single degree on the Celsius scale is represented by the rectangle on the left, which of the rectangles on the right best represents a single kelvin?

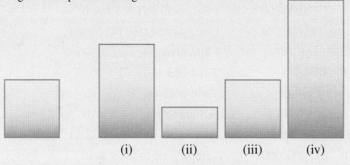

(i)   (ii)   (iii)   (iv)

Outside of scientific circles, the Fahrenheit temperature scale is the one most used in the United States. Before the work of Daniel Gabriel Fahrenheit (German physicist, 1686–1736), there were numerous different, arbitrarily defined temperature scales, none of which gave consistent measurements. Accounts of exactly how Fahrenheit devised his temperature scale vary from source to source. In one account, in 1724, Fahrenheit labeled as 0° the lowest artificially attainable temperature at the time (the temperature of a mixture of ice, water, and a substance called *ammonium chloride*). Using a traditional scale consisting of 12 degrees, he labeled the temperature of a healthy human body as the twelfth degree. On this scale, the freezing point of water occurred at the fourth degree. For better resolution, each degree was further divided into eight smaller degrees. This convention makes the freezing point of water 32°F and normal body temperature 96°F. (Today we consider normal body temperature to be somewhat higher than 96°F.)

The boiling point of water on the Fahrenheit scale is 212°, meaning that there are 180 degrees (212°F minus 32°F) between the freezing and boiling points. This separation is considerably more degrees than the 100 between the freezing point and boiling point of water on the Celsius scale. Thus, the size of a degree on the Fahrenheit scale is only 100/180 or five-ninths of a degree on the Celsius scale. Equation 1.2 gives the relationship between temperatures on the Fahrenheit and Celsius scales.

Equation 1.2 $\text{temperature in } °F = \dfrac{9°F}{5°C} \times (\text{temperature in } °C) + 32°F$

Worked Example 1.2 lets you practice converting from Celsius to Fahrenheit.

Worked Example 1.2

A body temperature above 39°C constitutes a high fever. Convert this temperature to the Fahrenheit scale.

Strategy We are given a temperature in degrees Celsius and are asked to convert it to degrees Fahrenheit.

Setup We use Equation 1.2:

$$\text{temperature in Fahrenheit} = \frac{9°\text{F}}{5°\text{C}} \times (\text{temperature in degrees Celsius}) + 32°\text{F}$$

Solution

$$\text{temperature in Fahrenheit} = \frac{9°\text{F}}{5°\text{C}} \times (39°\text{C}) + 32°\text{F} = 102°\text{F}$$

> **Think About It**
> Knowing that "normal" body temperature on the Fahrenheit scale is approximately 99°F (98.6°F is the number most often cited), 102°F seems like a reasonable answer.

Practice Problem 🅐TTEMPT The average temperature at the summit of Mt. Everest ranges from −36°C during the coldest month (January) and −19°C during the warmest month (July). Convert these temperatures and the range they span to Fahrenheit.

Practice Problem 🅑UILD The average surface temperatures of planets in our solar system range from 867°F on Venus to −330°F on Neptune. Convert these temperatures and the range they span to Celsius.

Student Annotation: The average surface temperature of Pluto is −375°F, but Pluto is no longer classified as a planet.

Practice Problem 🅒ONCEPTUALIZE If a single degree on the Fahrenheit scale is represented by the rectangle on the left, which of the rectangles on the right best represents a single degree on the Celsius scale? Which best represents a single kelvin?

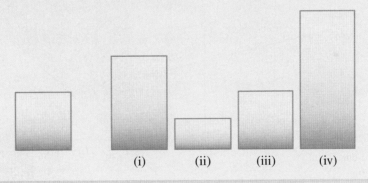

(i)   (ii)   (iii)   (iv)

Derived Units: Volume and Density

There are many quantities, such as volume and density, that require units not included in the base SI units. In these cases, we must combine base units to *derive* appropriate units.

The derived SI unit for volume, the meter cubed (m^3), is a much larger volume than is usually convenient. The more commonly used metric unit, the *liter* (L), is derived by cubing the *decimeter* (one-tenth of a meter) and is therefore also referred to as the cubic decimeter (dm^3). Another commonly used metric unit of volume is the *milliliter* (mL), which is derived by cubing the centimeter (1/100 of a meter). The milliliter is also referred to as the cubic centimeter (cm^3). Figure 1.4 illustrates the relationship between the liter (or dm^3) and the milliliter (or cm^3).

Density is the ratio of mass to volume. A familiar demonstration of density is the attempt to mix water and oil. Oil floats on water because, in addition to not *mixing* with water, oil has a lower *density* than water. That is, given *equal volumes*

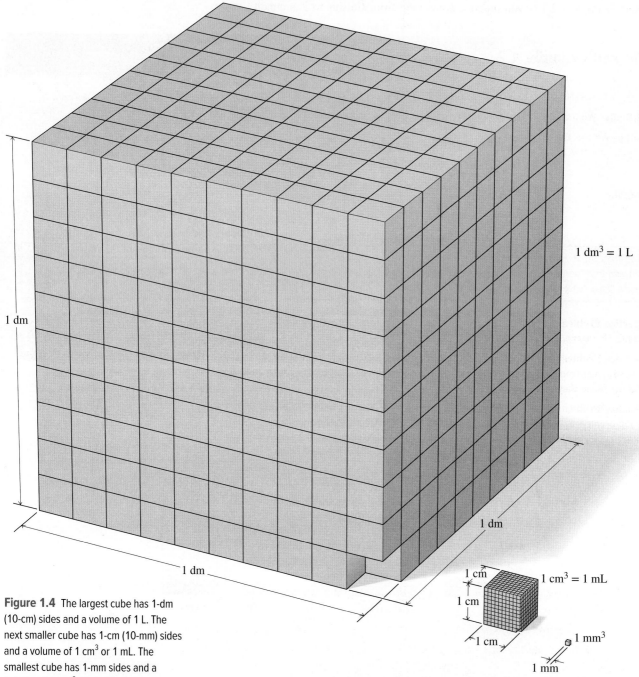

$1\ dm^3 = 1\ L$

$1\ cm^3 = 1\ mL$

$1\ mm^3$

Figure 1.4 The largest cube has 1-dm (10-cm) sides and a volume of 1 L. The next smaller cube has 1-cm (10-mm) sides and a volume of 1 cm^3 or 1 mL. The smallest cube has 1-mm sides and a volume of 1 mm^3. Note that although there are 10 cm in a decimeter, there are 1000 cm^3 in a cubic decimeter.

of the two liquids, the oil will have a *smaller mass* than the water. Density is calculated using the following equation:

Equation 1.3	$d = \dfrac{m}{V}$

where *d*, *m,* and *V* denote density, mass, and volume, respectively. The SI-derived unit for density is the kilogram per cubic meter (kg/m^3). This unit is too large for most common uses, however, so grams per cubic centimeter (g/cm^3) and its equivalent, grams per milliliter (g/mL), are used to express the densities of most solids and liquids. Water, for example, has a density of 1.00 g/cm^3 at 4°C. Because gas densities generally are very low, we typically express them in units of grams per liter (g/L):

$$1\ g/cm^3 = 1\ g/mL = 1000\ kg/m^3$$

$$1\ g/L = 0.001\ g/mL$$

Worked Example 1.3 shows how to calculate density, and how to use density in the calculation of volume.

Worked Example 1.3

Ice cubes float in a glass of water because solid water is less dense than liquid water. (a) Calculate the density of ice given that, at 0°C, a cube that is 2.0 cm on each side has a mass of 7.36 g, and (b) determine the volume occupied by 23 g of ice at 0°C.

Strategy (a) Determine density by dividing mass by volume (Equation 1.3), and (b) use the calculated density to determine the volume occupied by the given mass.

Setup (a) We are given the mass of the ice cube, but we must calculate its volume from the dimensions given. The volume of the ice cube is $(2.0 \text{ cm})^3$, or 8.0 cm^3. (b) Rearranging Equation 1.3 to solve for volume gives $V = m/d$.

Solution

(a) $d = \dfrac{7.36 \text{ g}}{8.0 \text{ cm}^3} = 0.92 \text{ g/cm}^3$ or 0.92 g/mL (b) $V = \dfrac{23 \text{ g}}{0.92 \text{ g/cm}^3} = 25 \text{ cm}^3$ or 25 mL

> **Think About It**
> For a sample with a density *less* than 1 g/cm^3, the number of cubic centimeters should be *greater* than the number of grams. In this case, $25 \text{ (cm}^3) > 23 \text{ (g)}$.

Practice Problem **A**TTEMPT Given that 20.0 mL of mercury has a mass of 272 g, calculate (a) the density of mercury and (b) the volume of 748 g of mercury.

Practice Problem **B**UILD Calculate (a) the density of a solid substance if a cube measuring 2.33 cm on one side has a mass of 117 g and (b) the mass of a cube of the same substance measuring 7.41 cm on one side.

Practice Problem **C**ONCEPTUALIZE Using the picture of the graduated cylinder and its contents, arrange the following in order of increasing density: blue liquid, pink liquid, yellow liquid, grey solid, blue solid, green solid.

Section 1.2 Review

Scientific Measurement

1.2.1 The coldest temperature ever recorded on Earth was −128.6°F (recorded at Vostok Station, Antarctica, on July 21, 1983). Express this temperature in degrees Celsius and in kelvins.
   (a) −89.2°C, −89.2 K   (d) −173.9°C, 99.3 K
   (b) −289.1°C, −15.9 K   (e) −7.0°C, 266.2 K
   (c) −89.2°C, 183.9 K

1.2.2 What is the density of an object that has a volume of 34.2 cm^3 and a mass of 19.6 g?
   (a) 0.573 g/cm^3 (d) 53.8 g/cm^3
   (b) 1.74 g/cm^3 (e) 14.6 g/cm^3
   (c) 670 g/cm^3

1.2.3 A sample of water is heated from room temperature to just below the boiling point. The overall change in temperature is 72°C. Express this temperature change in kelvins.
   (a) 345 K   (d) 201 K
   (b) 72 K   (e) 273 K
   (c) 0 K

1.2.4 Given that the density of gold is 19.3 g/cm^3, calculate the volume (in cm^3) of a gold nugget with a mass of 5.98 g.
   (a) 3.23 cm^3 (d) 0.310 cm^3
   (b) 5.98 cm^3 (e) 13.3 cm^3
   (c) 115 cm^3

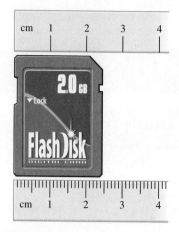

Figure 1.5 The width we report for the memory card depends on which ruler we use to measure it.

1.3   UNCERTAINTY IN MEASUREMENT

Science makes use of two types of numbers: exact and inexact. *Exact* numbers include numbers with defined values, such as 2.54 in the definition 1 inch (in) = 2.54 cm, 1000 in the definition 1 kg = 1000 g, and 12 in the definition 1 dozen = 12 objects. (The number 1 in each of these definitions is also an exact number.) Exact numbers also include those that are obtained by counting. Numbers measured by any method other than counting are *inexact*.

Measured numbers are inexact because of the measuring devices that are used, the individuals who use them, or both. For example, a ruler that is poorly calibrated will result in measurements that are in error—no matter how carefully it is used. Another ruler may be calibrated properly but have insufficient resolution for the necessary measurement. Finally, whether or not an instrument is properly calibrated or has sufficient resolution, there are unavoidable differences in how different people see and interpret measurements.

Significant Figures

An inexact number must be reported in such a way as to indicate the uncertainty in its value. This is done using significant figures. **Significant figures** are the *meaningful digits* in a reported number. Consider the measurement of the memory card in Figure 1.5 using the ruler above it. The card's width is between 2 and 3 cm. We may record the width as 2.5 cm, but because there are no gradations between 2 and 3 cm on this ruler, we are *estimating* the second digit. Although we are certain about the 2 in 2.5, we are *not* certain about the 5. The last digit in a measured number is referred to as the *uncertain digit;* and the uncertainty associated with a measured number is generally considered to be ± 1 in the place of the last digit. Thus, when we report the width of the memory card to be 2.5 cm, we are implying that its width is 2.5 ± 0.1 cm—and that its actual width may be as low as 2.4 cm or as high as 2.6 cm. Each of the digits in a measured number, including the uncertain digit, is a significant figure. The reported width of the memory card, 2.5 cm, contains *two* significant figures.

A ruler with millimeter gradations would enable us to be certain about the second digit in this measurement and to estimate a third digit. Now consider the measurement of the memory card using the ruler below it. We may record the width as 2.45 cm. Again, we estimate one digit beyond those we can read. The reported width of 2.45 cm contains three significant figures. Reporting the width as 2.45 cm implies that the width is 2.45 ± 0.01 cm.

The number of significant figures in any number can be determined using the following guidelines:

1. Any digit that is not zero is significant (112.1 has four significant figures).
2. Zeros located between nonzero digits are significant (305 has three significant figures, and 50.08 has four significant figures).
3. Zeros to the left of the first nonzero digit are not significant (0.0023 has two significant figures, and 0.000001 has one significant figure).
4. Zeros to the right of the last nonzero digit are significant if the number contains a decimal point (1.200 has four significant figures).
5. Zeros to the right of the last nonzero digit in a number that does not contain a decimal point may or may not be significant (100 may have one, two, or three significant figures—it is impossible to tell without additional information). To avoid ambiguity in such cases, it is best to express such numbers using scientific notation. If the intended number of significant figures is one, the number is written as 1×10^2; if the intended number of significant figures is two, the number is written as 1.0×10^2; and if the intended number of significant figures is three, the number is written as 1.00×10^2.

Worked Example 1.4 lets you practice determining the number of significant figures in a number.

Worked Example 1.4

Determine the number of significant figures in the following measurements: (a) 443 cm, (b) 15.03 g, (c) 0.0356 kg, (d) 3.000×10^{-7} L, (e) 50 mL, (f) 0.9550 m.

Strategy All nonzero digits are significant, so the goal will be to determine which of the zeros is significant.

Setup Zeros are significant if they appear between nonzero digits or if they appear after a nonzero digit in a number that contains a decimal point. Zeros may or may not be significant if they appear to the right of the last nonzero digit in a number that does not contain a decimal point.

Solution (a) 3; (b) 4; (c) 3; (d) 4; (e) 1 or 2, an ambiguous case; (f) 4.

> **Think About It**
> Be sure that you have identified zeros correctly as either significant or not significant. They are significant in (b) and (d); they are not significant in (c); it is not possible to tell in (e); and the number in (f) contains one zero that is significant, and one that is not.

Practice Problem Ⓐ**TTEMPT** Determine the number of significant figures in the following measurements: (a) 1129 m, (b) 0.0003 kg, (c) 1.094 cm, (d) 3.5×10^{12} atoms, (e) 150 mL, (f) 9.550 km.

Practice Problem Ⓑ**UILD** Using scientific notation, express the number *one million* to (a) two significant figures, (b) four significant figures, (c) seven significant figures; and using decimal notation, express the number *one tenth* to (d) three significant figures, (e) five significant figures, and (f) one significant figure.

Practice Problem Ⓒ**ONCEPTUALIZE** Report the number of colored objects contained within each square and, in each case, indicate the number of significant figures in the number you report.

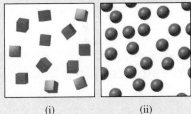

(i)   (ii)

Calculations with Measured Numbers

Because we often use one or more measured numbers to calculate a desired result, a second set of guidelines specifies how to handle significant figures in calculations.

1. In addition and subtraction, the answer cannot have more digits to the right of the decimal point than any of the original numbers. For example:

$$
\begin{array}{ll}
102.50 & \longleftarrow \text{two digits after the decimal point} \\
+\,0.231 & \longleftarrow \text{three digits after the decimal point} \\
\hline
102.731 & \longleftarrow \text{round to } 102.73
\end{array}
$$

$$
\begin{array}{ll}
143.29 & \longleftarrow \text{two digits after the decimal point} \\
-20.1 & \longleftarrow \text{one digit after the decimal point} \\
\hline
123.19 & \longleftarrow \text{round to } 123.2
\end{array}
$$

The rounding procedure works as follows. Suppose we want to round 102.13 and 54.86 each to one digit to the right of the decimal point. To begin, we look at the digit(s) that will be dropped. If the leftmost digit to be dropped is less than 5, as in 102.13, we *round down* (to 102.1), meaning that we simply drop the digit(s). If the leftmost digit to be dropped is equal to or greater than 5, as in 54.86, we *round up* (to 54.9), meaning that we add 1 to the preceding digit.

2. In multiplication and division, the number of significant figures in the final product or quotient is determined by the original number that has the smallest number of significant figures. The following examples illustrate this rule:

$$1.4 \times 8.011 = 11.2154 \qquad \longleftarrow \text{round to } 11$$
$$\text{(limited by 1.4 to } two \text{ significant figures)}$$

$$11.57/305.88 = 0.037825290964 \qquad \longleftarrow \text{round to } 0.03783$$
$$\text{(limited by 11.57 to } four \text{ significant figures)}$$

3. *Exact numbers* can be considered to have an infinite number of significant figures and do not limit the number of significant figures in a calculated result. For example, a penny minted after 1982 has a mass of 2.5 g. If we have three such pennies, the total mass is

$$3 \times 2.5 \text{ g} = 7.5 \text{ g}$$

The answer should *not* be rounded to one significant figure because 3, having been determined by counting, is an *exact* number.

4. In calculations with multiple steps, rounding the result of each step can result in "rounding error." Consider the following two-step calculation:

First step: $A \times B = C$

Second step: $C \times D = E$

Suppose that $A = 3.66$, $B = 8.45$, and $D = 2.11$. The value of E depends on whether we round off C prior to using it in the second step of the calculation.

Method 1	Method 2
$C = 3.66 \times 8.45 = 30.9$	$C = 3.66 \times 8.45 = 30.93$
$E = 30.9 \times 2.11 = 65.2$	$E = 30.93 \times 2.11 = 65.3$

In general, it is best to retain at least one extra digit until the end of a multistep calculation, as shown by method 2, to minimize rounding error.

Worked Examples 1.5 and 1.6 show how significant figures are handled in arithmetic operations.

Worked Example 1.5

Perform the following arithmetic operations and report the result to the proper number of significant figures: (a) 317.5 mL + 0.675 mL, (b) 47.80 L − 2.075 L, (c) 13.5 g ÷ 45.18 L, (d) 6.25 cm × 1.175 cm, (e) 5.46×10^2 g + 4.991×10^3 g.

Strategy Apply the rules for significant figures in calculations, and round each answer to the appropriate number of digits.

Setup (a) The answer will contain one digit to the right of the decimal point to match 317.5, which has the fewest digits to the right of the decimal point. (b) The answer will contain two digits to the right of the decimal point to match 47.80. (c) The answer will contain three significant figures to match 13.5, which has the fewest number of significant figures in the calculation. (d) The answer will contain three significant figures to match 6.25. (e) To add numbers expressed in scientific notation, first write both numbers to the same power of 10. That is, $4.991 \times 10^3 = 49.91 \times 10^2$, so the answer will contain two digits to the right of the decimal point (when multiplied by 10^2) to match both 5.46 and 49.91.

Solution

(a) 317.5 mL
$\underline{+\ 0.675 \text{ mL}}$
318.175 mL   ⟵ round to 318.2 mL

(b) 47.80 L
$\underline{-\ 2.075 \text{ L}}$
45.725 L   ⟵ round to 45.73 L

(c) $\dfrac{13.5 \text{ g}}{45.18 \text{ L}} = 0.298804781$ g/L   ⟵ round to 0.299 g/L

(d) 6.25 cm × 1.175 cm = 7.34375 cm^2 ⟵ round to 7.34 cm^2

(e) 5.46×10^2 g
$\underline{+\ 49.91 \times 10^2 \text{ g}}$
55.37×10^2 g = 5.537×10^3 g

Think About It
It may look as though the rule of addition has been violated in part (e) because the final answer (5.537×10^3 g) has three places past the decimal point, not two. However, the rule was applied to get the answer 55.37×10^2 g, which has *four* significant figures. Changing the answer to correct scientific notation doesn't change the number of significant figures, but in this case it changes the number of places past the decimal point.

Practice Problem Ⓐ**TTEMPT** Perform the following arithmetic operations, and report the result to the proper number of significant figures: (a) 105.5 L + 10.65 L, (b) 81.058 m − 0.35 m, (c) 3.801×10^{21} atoms + 1.228×10^{19} atoms, (d) 1.255 dm × 25 dm, (e) 139 g ÷ 275.55 mL.

Practice Problem Ⓑ**UILD** Perform the following arithmetic operations, and report the result to the proper number of significant figures: (a) 1.0267 cm × 2.508 cm × 12.599 cm, (b) 15.0 kg ÷ 0.036 m^3, (c) 1.113×10^{10} kg − 1.050×10^9 kg, (d) 25.75 mL + 15.00 mL, (e) 46 cm^3 + 180.5 cm^3.

Practice Problem Ⓒ**ONCEPTUALIZE** A citrus dealer in Florida sells boxes of 100 oranges at a roadside stand. The boxes routinely are packed with one to three extra oranges to help ensure that customers are happy with their purchases. The average weight of an orange is 7.2 ounces, and the average weight of the boxes in which the oranges are packed is 3.2 pounds. Determine the total weight of five of these 100-orange boxes.

Worked Example 1.6

An empty container with a volume of 9.850×10^2 cm^3 is weighed and found to have a mass of 124.6 g. The container is filled with a gas and reweighed. The mass of the container and the gas is 126.5 g. Determine the density of the gas to the appropriate number of significant figures.

Strategy This problem requires two steps: subtraction to determine the mass of the gas, and division to determine its density. Apply the corresponding rule regarding significant figures to each step.

Setup In the subtraction of the container mass from the combined mass of the container and the gas, the result can have only one place past the decimal point: 126.5 g − 124.6 g = 1.9 g. Thus, in the division of the mass of the gas by the volume of the container, the result can have only two significant figures.

Solution

$$\begin{array}{r} 126.5 \text{ g} \\ -124.6 \text{ g} \\ \hline \end{array}$$

mass of gas = 1.9 g ◄——— one place past the decimal point (two significant figures)

density = $\dfrac{1.9 \text{ g}}{9.850 \times 10^2 \text{ cm}^3}$ = 0.00193 g/cm^3 ◄——— round to 0.0019 g/cm^3

The density of the gas is 1.9×10^{-3} g/cm^3.

> **Think About It**
> In this case, although each of the three numbers we started with has *four* significant figures, the solution has only *two* significant figures.

Practice Problem Ⓐ**TTEMPT** An empty container with a volume of 150.0 cm^3 is weighed and found to have a mass of 72.5 g. The container is filled with a liquid and reweighed. The mass of the container and the liquid is 194.3 g. Determine the density of the liquid to the appropriate number of significant figures.

Practice Problem Ⓑ**UILD** Another empty container with an unknown volume is weighed and found to have a mass of 81.2 g. The container is then filled with a liquid with a density of 1.015 g/cm^3 and reweighed. The mass of the container and the liquid is 177.9 g. Determine the volume of the container to the appropriate number of significant figures.

Practice Problem Ⓒ**ONCEPTUALIZE** Several pieces of aluminum metal with a total mass of 11.63 g are dropped into a graduated cylinder of water to determine their combined volume. The graduated cylinder is shown before and after the metal has been added. Use the information shown here to determine the density of aluminum. Be sure to report your answer to the appropriate number of significant figures.

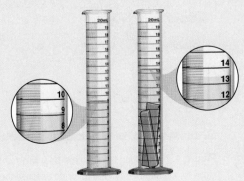

Accuracy and Precision

Accuracy and precision are two ways to gauge the quality of a set of measured numbers. Although the difference between the two terms may be subtle, it is important. ***Accuracy*** tells us how close a measurement is to the *true* value. ***Precision*** tells us how close a series of replicate measurements (measurements of the same thing) are to one another (Figure 1.6).

Suppose that three students are asked to determine the mass of an aspirin tablet. Each student weighs the aspirin tablet three times. The results (in grams) are tabulated here.

	Student A	**Student B**	**Student C**
	0.335	0.357	0.369
	0.331	0.375	0.373
	0.333	0.338	0.371
Average value	0.333	0.357	0.371

The true mass of the tablet is 0.370 g. Student A's results are more precise than those of student B, but neither set of results is very accurate. Student C's results are both precise (very small deviation of individual masses from the average mass) and accurate (average value very close to the true value). Figure 1.7 shows all three students' results in relation to the true mass of the tablet. Highly accurate measurements are usually precise, as well, although highly precise measurements do not necessarily guarantee *accurate* results. For example, an improperly calibrated meterstick or a faulty balance may give precise readings that are significantly different from the correct value.

Student Annotation: Even properly calibrated measuring devices can give varied results. Replicate measurements, such as those represented in the table, are used to determine the variability in the value of a measured quantity.

(a) (b) (c)

Figure 1.6 The distribution of papers shows the difference between accuracy and precision. (a) Good accuracy and good precision. (b) Poor accuracy but good precision. (c) Poor accuracy and poor precision.

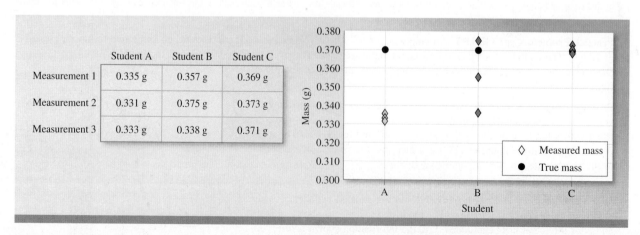

	Student A	Student B	Student C
Measurement 1	0.335 g	0.357 g	0.369 g
Measurement 2	0.331 g	0.375 g	0.373 g
Measurement 3	0.333 g	0.338 g	0.371 g

Figure 1.7 Graphing the students' data illustrates the difference between precision and accuracy. Student A's results are precise (values are close to one another) but not accurate because the average value is far from the true value. Student B's results are neither precise nor accurate. Student C's results are both precise and accurate.

Section 1.3 Review

Uncertainty in Measurement

1.3.1 To the proper number of significant figures, what volume of water does the graduated cylinder contain? Note: The volume of water in a graduated cylinder should be read at the *bottom* of the meniscus (the curved surface at the top).
(a) 32.2 mL
(b) 30.25 mL
(c) 32.5 mL
(d) 32.50 mL
(e) 32.500 mL

1.3.2 The true dependence of *y* on *x* is represented by the black line. Three students measured *y* as a function of *x* and plotted their data on the graph. Which set of data (red, green, or purple) has the best accuracy? Which has the best precision?
(a) red, green
(b) green, green
(c) green, purple
(d) purple, purple
(e) purple, green

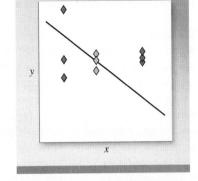

1.3.3 Specify the number of significant figures in each of the following numbers and determine the result of the following calculation to the correct number of significant figures.

$$63.102 \times 10.18 =$$

(a) 642.4
(b) 642.38
(c) 640
(d) 642
(e) 642.378

1.3.4 Specify the number of significant figures in each of the following numbers and determine the result of the following calculation to the correct number of significant figures.

$$3.115 + 0.2281 + 712.5 + 45 =$$

(a) 760.8431
(b) 760.843
(c) 760.84
(d) 760.8
(e) 761

1.3.5 What is the result of the following calculation to the correct number of significant figures?

$$153.1 \div 5.3 =$$

(a) 28.868
(b) 28.87
(c) 28.9
(d) 29
(e) 30

1.3.6 What is the result of the following calculation to the correct number of significant figures?

$$(6.266 - 6.261) \div 522.0 =$$

(a) 9.5785×10^{-6}
(b) 9.579×10^{-6}
(c) 9.58×10^{-6}
(d) 9.6×10^{-6}
(e) 1×10^{-5}

Thinking Outside the Box

Tips for Success in Chemistry Class

Success in a chemistry class depends largely on problem-solving ability. The Worked Examples throughout the text are designed to help you develop problem-solving skills. Each is divided into four steps: Strategy, Setup, Solution, and Think About It.

Strategy: Read the problem carefully and determine what is being asked and what information is provided. The Strategy step is where you should think about what skills are required and lay out a plan for solving the problem. Give some thought to what you expect the result to be. If you are asked to determine the number of atoms in a sample of matter, for example, you should expect the answer to be a whole number. Determine what, if any, units should be associated with the result. When possible, make a ballpark estimate of the magnitude of the correct result, and make a note of your estimate.

Setup: Next, gather the information necessary to solve the problem. Some of the information will have been given in the problem itself. Other information, such as equations, constants, and tabulated data (including atomic masses) should also be brought together in this step. Write down and label clearly all of the information you will use to solve the problem. Be sure to write appropriate units with each piece of information.

Solution: Using the necessary equations, constants, and other information, calculate the answer to the problem. Pay particular attention to the units associated with each number, tracking and canceling units carefully throughout the calculation. In the event that multiple calculations are required, label any intermediate results, but don't round to the necessary

number of significant figures until the final calculation. Always carry at least one extra digit in intermediate calculations. Make sure that the final answer has the correct number of significant figures.

Think About It: Consider your calculated result and ask yourself whether or not it makes sense. Compare the units and the magnitude of your result with your ballpark estimate from the Strategy step. If your result does not have the appropriate units, or if its magnitude or sign is not reasonable, check your solution for possible errors. A very important part of problem solving is being able to judge whether the answer is reasonable. It is relatively easy to spot a wrong sign or incorrect units, but you should also develop a sense of *magnitude* and be able to tell when an answer is either way too big or way too small. For example, if a problem asks you to determine the length of a sheet of paper in millimeters and you calculate a number that is less than 1, you should know that it cannot be correct.

Finally, each worked example is followed by three practice problems. The first, "Attempt," typically is a very similar problem that can be solved using the same strategy. The second and third, "Build" and "Conceptualize" generally test the same skills, but require approaches slightly different from the one used to solve the preceding sample and practice problems.

Regular use of the worked examples and practice problems in this text can help you develop an effective set of problem-solving skills. They can also help you assess whether you are ready to move on to the next new concepts. If you struggle with the practice problems, then you probably need to review the corresponding worked example and the concepts that led up to it.

1.4   USING UNITS AND SOLVING PROBLEMS

Solving problems correctly in chemistry requires careful manipulation of both numbers and units. Paying attention to the units will benefit you greatly as you proceed through this, or any other, chemistry course.

Conversion Factors

A ***conversion factor*** is a fraction in which the same quantity is expressed one way in the numerator and another way in the denominator. By definition, for example, 1 in = 2.54 cm. We can derive a conversion factor from this equality by writing it as the following fraction:

$$\frac{1 \text{ in}}{2.54 \text{ cm}}$$

Because the numerator and denominator express the same length, this fraction is equal to 1; as a result, we can equally well write the conversion factor as:

$$\frac{2.54 \text{ cm}}{1 \text{ in}}$$

Further, because both forms of this conversion factor are equal to 1, we can multiply a quantity by either form without changing the value of that quantity. This is useful for changing the units in which a given quantity is expressed—something you will do often throughout this text. For instance, if we need to convert a length from inches to centimeters, we multiply the length in inches by the form of the conversion factor with the unit inches in the *denominator*:

$$12.00 \text{ in} \times \frac{2.54 \text{ cm}}{1 \text{ in}} = 30.48 \text{ cm}$$

Our choice of this form of the conversion factor allows us to cancel the unit inches—and gives us the desired unit, centimeters. The result contains four significant figures because exact numbers, such as those obtained from definitions, do not limit the number of significant figures in the result of a calculation. Thus, the number of significant figures in the answer to this calculation is based on the number 12.00, not the number 2.54.

Dimensional Analysis—Tracking Units

The use of conversion factors in problem solving is called **dimensional analysis** or the *factor-label method*. Many problems require the use of more than one conversion factor. The conversion of 12.00 into meters, for example, takes two steps: one to convert inches to centimeters, which we have already demonstrated; and one to convert centimeters to meters. The additional conversion factor required is derived from the equality:

$$1 \text{ m} = 100 \text{ cm}$$

and is expressed as either:

$$\frac{100 \text{ cm}}{1 \text{ m}} \quad \text{or} \quad \frac{1 \text{ m}}{100 \text{ cm}}$$

We choose the conversion factor that will introduce the unit meter and cancel the unit centimeter (i.e., the one on the right). We can set up a problem of this type as the following series of unit conversions so that it is unnecessary to calculate an intermediate answer at each step:

$$12.00 \text{ in} \times \frac{2.54 \text{ cm}}{1 \text{ in}} \times \frac{1 \text{ m}}{100 \text{ cm}} = 0.3048 \text{ m}$$

Careful tracking of units and their cancellation can be a valuable tool in checking your work. If we had accidentally used the *reciprocal* of one of the conversion factors, the resulting units would have been something other than meters. Unexpected or nonsensical units can reveal an error in your problem-solving strategy.

Worked Example 1.7 shows how to derive conversion factors and use them to do unit conversions.

Student Annotation: If we had accidentally used the reciprocal of the conversion from centimeters to meters, the result would have been 3048 cm²/m, which would make no sense—both because the units are nonsensical and because the numerical result is not reasonable. You know that 12 inches is a foot and that a foot is not equal to *thousands* of meters!

Worked Example 1.7

The Food and Drug Administration (FDA) recommends that dietary sodium intake be no more than 2400 mg per day. What is this mass in pounds (lb), if 1 lb = 453.6 g?

Strategy This problem requires a two-step dimensional analysis, because we must convert milligrams to grams and then grams to pounds. Assume the number 2400 has four significant figures.

Setup The necessary conversion factors are derived from the equalities 1 g = 1000 mg and 1 lb = 453.6 g:

$$\frac{1 \text{ g}}{100 \text{ mg}} \quad \text{or} \quad \frac{1000 \text{ mg}}{1 \text{ g}} \quad \text{and} \quad \frac{1 \text{ lb}}{453.6 \text{ g}} \quad \text{or} \quad \frac{453.6 \text{ g}}{1 \text{ lb}}$$

From each pair of conversion factors, we select the one that will result in the proper unit cancellation.

Solution

$$2400 \text{ mg} \times \frac{1 \text{ g}}{1000 \text{ mg}} \times \frac{1 \text{ lb}}{453.6 \text{ g}} = 0.005291 \text{ lb}$$

(Continued on next page)

Think About It
Make sure that the magnitude of the result is reasonable and that the units have canceled properly. If we had mistakenly multiplied by 1000 and 453.6 instead of dividing by them, the result ($2400 \text{ mg} \times 1000 \text{ mg/g} \times 453.6 \text{ g/lb} = 1.089 \times 10^9 \text{ mg}^2\text{/lb}$) would be unreasonably large—and the units would not have canceled properly.

Practice Problem A TTEMPT The American Heart Association recommends that healthy adults limit dietary cholesterol to no more than 300 mg per day. Convert this mass of cholesterol to ounces (1 oz = 28.3459 g). Assume 300 mg has just one significant figure.

Practice Problem B UILD A gold nugget has a mass of 0.9347 oz. What is its mass in milligrams?

Practice Problem C ONCEPTUALIZE The diagram contains several objects that are constructed using colored blocks and grey connectors. Note that each of the objects is essentially identical, consisting of the same number and arrangement of blocks and connectors. Give the appropriate conversion factor for each of the specified operations.

(a) We know the number of objects and wish to determine the number of red blocks.

(b) We know the number of yellow blocks and wish to determine the number of objects.

(c) We know the number of yellow blocks and wish to determine the number of white blocks.

(d) We know the number of grey connectors and wish to determine the number of yellow blocks.

Many familiar quantities require units raised to specific powers. For example, an area may be expressed in units of *length* squared (e.g., square meters, m^2; or square inches, in^2). Volumes sometimes are expressed in units of length *cubed* (e.g., cubic feet, ft^3; or cubic centimeters, cm^3). More often, though, volumes are expressed in liters (L) or milliliters (mL). It's important to remember that these are the common names given to specific units of length cubed. The liter is defined as a decimeter (dm) cubed: $1 \text{ L} = 1 \text{ dm}^3$; and the milliliter is defined as the centimeter cubed: $1 \text{ mL} = 1 \text{ cm}^3$. (See Figure 1.6.) When units are squared or cubed, special care must be taken when using them in dimensional analysis. For example, converting from cubic meters to cubic centimeters requires the following operation:

$$2.75 \text{ m}^3 \times \frac{100 \text{ cm}}{1 \text{ m}} \times \frac{100 \text{ cm}}{1 \text{ m}} \times \frac{100 \text{ cm}}{1 \text{ m}} = 2.75 \times 10^6 \text{ cm}^3$$

or

$$2.75 \text{ m}^3 \times \left(\frac{100 \text{ cm}}{1 \text{ m}}\right)^3 = 2.75 \times 10^6 \text{ cm}^3$$

Failing to raise the conversion factor to the same power as the unit itself is a common error—and one that can happen easily when the units L or mL appear—because they do not explicitly show the power 3.

Worked Example 1.8 shows how to handle problems in which conversion factors are squared or cubed in dimensional analysis.

Worked Example 1.8

An average adult has 5.2 L of blood. What is the volume of blood in cubic meters?

Strategy There are several ways to solve a problem such as this. One way is to convert liters to cubic centimeters and then cubic centimeters to cubic meters.

Setup $1 \text{ L} = 1000 \text{ cm}^3$ and $1 \text{ cm} = 1 \times 10^{-2} \text{ m}$. When a unit is raised to a power, the corresponding conversion factor must also be raised to that power in order for the units to cancel appropriately.

Solution

$$5.2 \text{ L} \times \frac{1000 \text{ cm}^3}{1 \text{ L}} \times \left(\frac{1 \times 10^{-2} \text{ m}}{1 \text{ cm}}\right)^3 = 5.2 \times 10^{-3} \text{ m}^3$$

Think About It
Based on the preceding conversion factors, $1 \text{ L} = 1 \times 10^{-3} \text{ m}^3$. Therefore, 5 L of blood would be equal to $5 \times 10^{-3} \text{ m}^3$, which is close to the calculated answer.

Practice Problem **A**TTEMPT The density of silver is 10.5 g/cm^3. What is its density in kg/m^3?

Practice Problem **B**UILD The density of mercury is 13.6 g/cm^3. What is its density in pounds per cubic foot (lb/ft^3)? ($1 \text{ lb} = 453.6 \text{ g}$, $1 \text{ in} = 2.54 \text{ cm}$)

Practice Problem **C**ONCEPTUALIZE Each diagram [(i) or (ii)] shows the objects contained within a cubical space. In each case, determine to the appropriate number of significant figures the number of objects that would be contained within a cubical space in which the length of the cube's edge is exactly five times that of the cube shown in the diagram.

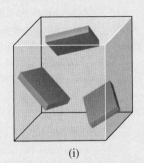
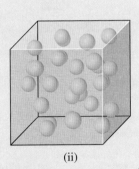

(i)   (ii)

Section 1.4 Review

Using Units and Solving Problems

1.4.1   Convert 43.1 cm^3 to liters.
(a) 43.1 L
(b) 43,100 L
(c) 0.0431 L
(d) 4310 L
(e) 0.043 L

1.4.2   What is the volume of a 5.75-g object that has a density of 3.97 g/cm^3?
(a) 1.45 cm^3
(b) 0.690 cm^3
(c) 22.8 cm^3
(d) 0.0438 cm^3
(e) 5.75 cm^3

1.4.3   The density of lithium metal is 535 kg/m^3. What is this density in g/cm^3?
(a) 0.000535 g/cm^3
(b) 0.535 g/cm^3
(c) 0.0535 g/cm^3
(d) 0.54 g/cm^3
(e) 53.5 g/cm^3

1.4.4   How many cubic centimeters are there in a cubic meter?
(a) 10
(b) 100
(c) 1000
(d) 1×10^4
(e) 1×10^6

© Ingram Publishing/Fototsearch

Video 1.1
Matter—three states of matter

1.5 CLASSIFICATION OF MATTER

Chemists classify all matter either as a *pure substance* or as a *mixture* of substances. A **substance** is a form of matter that has a specific chemical composition and distinct, observable properties such as color, state of matter (Is it a solid, a liquid, or a gas?), and solubility (Does it dissolve in water or not?). Familiar examples of substances and some of their properties include:

- salt (sodium chloride): white, crystalline solid; dissolves in water
- iron: greyish metal; rusts when left exposed to air and water; does not dissolve in water
- mercury: silvery liquid; does not dissolve in water
- carbon dioxide: colorless gas; does not support combustion (can be used to extinguish flames)
- oxygen: colorless gas; supports combustion (will accelerate a fire)

We know that these substances are distinct from one another because they have different properties. Further, we know that these are substances rather than mixtures of substances because none of them consists of more than one substance; that is, sodium chloride consists of only sodium chloride, iron consists of only iron, and so forth.

By contrast, if we were to dissolve a sample of sodium chloride in a glass of water, the result would be a mixture of the two substances: sodium chloride and water. Salt water is a mixture, *not* a pure substance.

States of Matter

All substances can, in principle, exist as a solid, a liquid, and a gas, the three physical states depicted in Figure 1.8. In a solid, particles are held close together in an orderly fashion with little freedom of motion. As a result, a solid does not conform

Figure 1.8 Molecular-level illustrations of a solid, a liquid, and a gas.

to the shape of its container. Particles in a liquid are close together but are not held rigidly in position; they are free to move past one another. Thus, a liquid conforms to the shape of the part of the container it fills. In a gas, the particles are separated by distances that are very large compared to the size of the particles. A sample of gas assumes both the shape and the volume of its container.

We can convert a substance from one state to another without changing the identity of the substance. For example, if solid water (ice) is heated, it will melt to form liquid water. If the liquid water is heated further, it will vaporize to form a gas (water vapor). Conversely, cooling water vapor will cause it to condense into liquid water. When the liquid water is cooled further, it will freeze into ice. Figure 1.9 shows the three physical states of water.

Mixtures

A *mixture* is a combination of two or more substances in which each substance retains its distinct identity. Like pure substances, mixtures can be solids, liquids, or gases. Some familiar examples are trail mix, sterling silver, apple juice, seawater, and air. Mixtures do not have a universal constant composition. Therefore, samples of air collected in different locations will differ in composition because of differences in altitude, pollution, and other factors. Various brands of apple juice may differ in composition because of the use of different varieties of apples, or there may be differences in processing, packaging, and so on.

Mixtures are either *homogeneous* or *heterogeneous*. The mixture we get when we dissolve sodium chloride in water is a **homogeneous mixture** because the composition of the mixture is uniform throughout. We cannot distinguish the components of a homogeneous mixture such as salt water—any sample we examine will have the same composition. If we mix sand with iron filings, however, the sand and the iron filings remain distinct and discernible from each other (Figure 1.10). This type of mixture is called a **heterogeneous mixture** because the composition is *not* uniform. It is possible, indeed *probable* that any two samples of such a mixture will differ in composition.

A mixture, whether homogeneous or heterogeneous, can be separated into the substances it contains without changing the identities of the individual substances. Thus, sugar can be recovered from sugar-water by evaporating the mixture to dryness. The solid sugar will be left behind, and the water component can be recovered by condensing the water vapor that evaporates. To separate the sand-iron mixture, we can use a magnet to remove the iron filings from the sand, because sand is not attracted to the magnet [see Figure 1.10(b)]. After separation, the components of the mixture will have the same composition and properties as they did prior to being mixed.

Figure 1.9 Water as a solid (ice), liquid, and gas. (We can't actually see water vapor, any more than we can see the nitrogen and oxygen that make up most of the air we breathe. When we see steam or clouds, what we are actually seeing is water vapor that has condensed upon encountering cold air.)
© McGraw-Hill Education./Charles D. Winters, photographer

Student Annotation: Homogeneous mixtures are also known as *solutions*. [Chapter 9]

© Nathan Griffith/Alamy

Student Annotation: Condensation refers to the change from gas to liquid.

(a)   (b)

Figure 1.10 (a) A heterogeneous mixture contains iron filings and sand. (b) A magnet is used to separate the iron filings from the mixture.

(a,b): © McGraw-Hill Education/Charles D. Winters, photographer

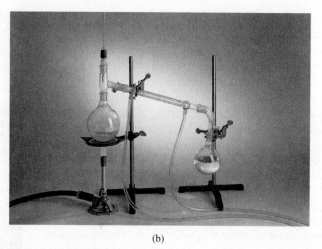

(a)   (b)   (c)

Figure 1.11 (a) Filtration can be used to separate a heterogeneous mixture of a liquid and a solid, such as coffee and coffee grounds. The filter, in this case a coffee filter, allows only the liquid coffee to pass through. (b) Distillation can be used to separate components with different boiling points. The component with the lowest boiling point is vaporized first, leaving behind components with higher boiling points—including dissolved solids. The vapor can be condensed and recovered by cooling. (c) A variety of chromatographic techniques, including paper chromatography, can be used to separate mixtures. Here, paper chromatography is used to separate the dye in candy coatings into its components.

(a): © Bloomimage/Corbis; (b,c): © Richard Megna/Fundamental Photographs

The processes used to separate mixtures are called physical processes. A ***physical process*** is one that does not change the identity of any substance. For example, melting ice causes a change in the physical state of water (solid to liquid), but it does not change the identity of the substance (water). Examples of physical processes that can be used to separate mixtures are shown in Figure 1.11.

1.6   THE PROPERTIES OF MATTER

Substances are identified by their properties as well as by their composition. Properties of a substance may be ***quantitative*** (measured and described using numbers) or ***qualitative*** (not requiring explicit measurement and described without the use of numbers). For example, the *mass* of a sample of matter must be measured and expressed using a number. Mass is a *quantitative* property. The *color* of a substance does not require a measurement or a number to describe. Color is a *qualitative* property.

Physical Properties

Color, melting point, boiling point, and physical state are all physical properties. A ***physical property*** is one that can be observed and measured without changing the *identity* of a substance. For example, we can determine the melting point of ice by heating a block of ice and measuring the temperature at which the ice is converted to water. Liquid water differs from ice in appearance but not in composition. Melting is a ***physical change***—one in which the state of matter changes, but the identity of the matter does not change. We can recover the original ice by cooling the water until it freezes. Therefore, the melting point of a substance is a *physical* property. Similarly, when we say that oil is less dense than water, we are referring to the physical property of density.

Chemical Properties

The statement "iron rusts when it is exposed to water and air" describes a ***chemical property*** of iron, because for us to observe this property, a ***chemical change*** or chemi-cal process must occur. In this case, the chemical change is *corrosion* or *oxidation* of iron. After a chemical change, the original substance (iron metal, in this case) no longer exists. What remains is a different substance (*rust,* in this case). There is no *physical* process by which we can recover the iron from the rust.

Every time we bake cookies, we bring about a chemical change. When heated, the leavening agent (typically baking soda) in cookie dough undergoes a chemical change that produces a gas. The gas forms numerous little bubbles in the dough during the baking process, causing the cookies to "rise." Once the cookies are baked, we cannot recover the baking soda by cooling the cookies, or by *any* physical process. When we eat the cookies, we initiate further chemical changes that occur during digestion and metabolism.

Extensive and Intensive Properties

All properties of matter are either *extensive* or *intensive*. The measured value of an **extensive property** depends on the amount of matter. Values of the same extensive property can be added together. For example, two copper coins will have a combined mass that is the sum of the individual masses of each coin, and the volume occupied by two copper coins is the sum of their individual volumes. Both *mass* and *volume* are *extensive* properties.

The value of an **intensive property** does *not* depend on the amount of matter. Consider again the example of copper coins. The density of copper is the same regardless of how much copper we have; and the same is true regarding the melting point of copper. *Density* and *melting point* are *intensive* properties. Unlike mass and volume, which are additive, density, melting point, and other *intensive* properties are not additive. Figure 1.12 illustrates some of the extensive and intensive properties of water.

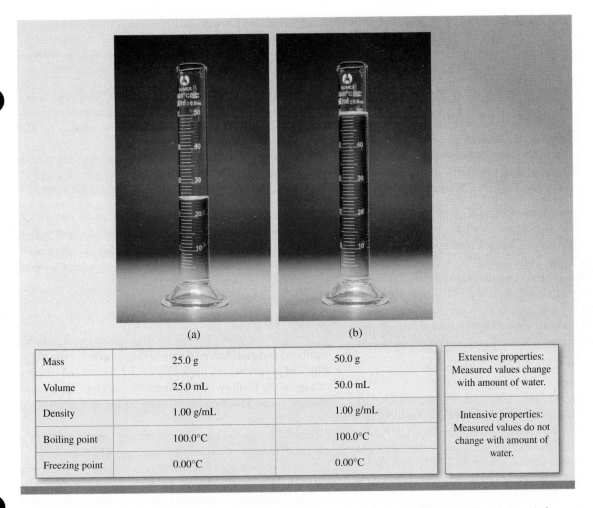

	(a)	(b)	
Mass	25.0 g	50.0 g	Extensive properties: Measured values change with amount of water.
Volume	25.0 mL	50.0 mL	
Density	1.00 g/mL	1.00 g/mL	Intensive properties: Measured values do not change with amount of water.
Boiling point	100.0°C	100.0°C	
Freezing point	0.00°C	0.00°C	

Figure 1.12 Some extensive properties (mass and volume) and intensive properties (density, boiling point, and freezing point) of water. The measured values of the extensive properties depend on the amount of water. The measured values of the intensive properties are independent of the amount of water.

(Photos): © H.S. Photos/Alamy Stock Photo

Learning Outcomes

- Identify the key components of the scientific method.
- Recall the common base SI units of measurement and their associated symbols.
- Utilize SI unit prefixes.
- Perform conversions between different temperature scales.
- Apply derived units, such as volume and density, to perform calculations.
- Apply significant figure rules in calculations.
- Distinguish between accuracy and precision.

- Utilize conversion factors to conduct unit conversions.
- Apply dimensional analysis toward solving problems with multiple steps or conversions.
- Differentiate between states of matter.
- Determine whether a mixture is heterogeneous or homogeneous.
- Categorize properties of matter as being quantitative or qualitative; physical or chemical; extensive or intensive.

Chapter Summary

SECTION 1.1

- *Chemistry* is the study of *matter* and the changes matter undergoes.
- Chemists do research using a set of guidelines and practices known as the *scientific method,* in which observations give rise to *laws,* data give rise to *hypotheses,* hypotheses are tested with experiments, and successful hypotheses give rise to *theories,* which are further tested by experiment.

SECTION 1.2

- Scientists use a system of units referred to as the *International System of Units,* or *SI units.*
- There are seven *base* SI units including the kilogram (for *mass*) and the *kelvin* (for temperature). SI units for such quantities as volume and *density* are derived from the base units. Some commonly used units are not SI units, such as the degree *Celsius,* the *atomic mass unit (amu),* and the *angstrom (Å).*

SECTION 1.3

- Measured numbers are *inexact.* Numbers that are obtained by counting or that are part of a definition are *exact* numbers.
- *Significant figures* are used to specify the uncertainty in a measured number or in a number calculated using measured numbers. Significant figures must be carried through calculations so that the implied uncertainty in the final answer is reasonable.
- *Accuracy* refers to how close measured numbers are to a *true* value. *Precision* refers to how close measured numbers are to *one another.*

SECTION 1.4

- A *conversion factor* is a fraction in which the numerator and denominator are the same quantity expressed in different units. Multiplying by a conversion factor is *unit conversion.*
- *Dimensional analysis* is a series of unit conversions used in the solution of a multistep problem.

SECTION 1.5

- All matter exists either as a *substance* or as a mixture of substances. A *mixture* may be *homogeneous* (uniform composition throughout) or *heterogeneous.* Mixtures may be separated using *physical processes.*

SECTION 1.6

- Substances are identified by their *quantitative* (involving numbers) and *qualitative* (not involving numbers) properties.
- *Physical properties* are those that can be determined without changing the identity of the matter in question. A *physical change* is one in which the identity of the matter involved does not change.
- *Chemical properties* are determined only as the result of a *chemical change* or *chemical process,* in which the original substance is converted to a different substance. Physical and chemical properties may be *extensive* (dependent on the amount of matter) or *intensive* (independent of the amount of matter).

Key Words

Key Equations

1.1	$K = {}^\circ C + 273.15$	Temperature in kelvins is determined by adding 273.15 to the temperature in Celsius. Often we simply add 273, depending on the precision with which the Celsius temperature is known.
1.2	$\text{temp in } {}^\circ F = \dfrac{9\,{}^\circ F}{5\,{}^\circ C} \times (\text{temp in } {}^\circ C) + 32\,{}^\circ F$	Temperature in Celsius is used to determine temperature in Fahrenheit.
1.3	$d = \dfrac{m}{V}$	Density is the ratio of mass to volume. For liquids and solids, densities are typically expressed in g/cm^3.

Key Skills

Dimensional Analysis

Solving problems in chemistry often involves mathematical combinations of measured values and constants. A conversion factor is a fraction (equal to one) derived from an equality. For example, 1 inch is, by definition, equal to 2.54 centimeters:

$$\boxed{1 \text{ in}} \quad = \quad \boxed{2.54 \text{ cm}}$$

We can derive two different conversion factors from this equality:

$$\boxed{\dfrac{1 \text{ in}}{2.54 \text{ cm}}} \quad \text{or} \quad \boxed{\dfrac{2.54 \text{ cm}}{1 \text{ in}}}$$

Which fraction we use depends on what units we start with, and what units we expect our result to have. If we are converting a distance given in centimeters to inches, we multiply by the first fraction:

$$\boxed{37.6 \text{ cm}} \quad \times \quad \boxed{\dfrac{1 \text{ in}}{2.54 \text{ cm}}} \quad = \quad \boxed{14.8 \text{ in}}$$

If we are converting a distance given in inches to centimeters, we multiply by the second fraction:

$$\boxed{5.23 \text{ in}} \quad \times \quad \boxed{\dfrac{2.54 \text{ cm}}{1 \text{ in}}} \quad = \quad \boxed{13.3 \text{ cm}}$$

In each case, the units cancel to give the desired units in the result.

When a unit is raised to a power to express, for example, an area (cm^2) or a volume (cm^3), the conversion factor must be raised to the same power. For example, converting an area expressed in square centimeters to square inches requires that we square the conversion factor; converting a volume expressed in cubic centimeters to cubic meters requires that we cube the conversion factor. The following individual flowcharts converting an area in cm^2 to m^2 show why this is so:

$$\boxed{48.5 \text{ cm}^2} \; = \; \boxed{48.5 \text{ cm}} \; \times \; \boxed{\text{cm}} \qquad \boxed{\left(\dfrac{1 \text{ in}}{2.54 \text{ cm}}\right)^2} \; = \; \boxed{\dfrac{1 \text{ in}}{2.54 \text{ cm}}} \; \times \; \boxed{\dfrac{1 \text{ in}}{2.54 \text{ cm}}}$$

$$\boxed{48.5 \text{ cm}} \; \times \; \boxed{\text{cm}} \; \times \; \boxed{\dfrac{1 \text{ in}}{2.54 \text{ cm}}} \; \times \; \boxed{\dfrac{1 \text{ in}}{2.54 \text{ cm}}} \; = \; \boxed{7.52 \text{ in}^2}$$

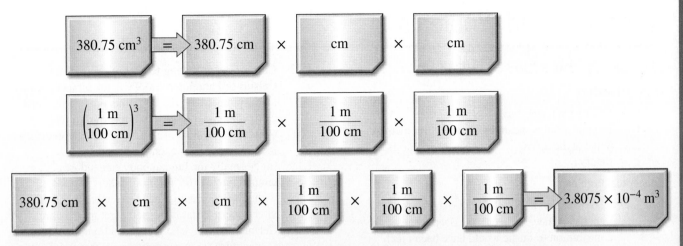

Failure to raise the conversion factor to the appropriate power would result in units not canceling properly.

Often the solution to a problem requires several different conversions, which can be combined on a single line. For example: If we know that a 157-lb athlete running at 7.09 miles per hour consumes 55.8 cm³ of oxygen per kilogram of body weight for every minute spent running, we can calculate how many liters of oxygen this athlete consumes by running 10.5 miles (1 kg = 2.2046 lb, 1 L = 1 dm³):

$$157 \text{ lb} \times \frac{1 \text{ kg}}{2.2046 \text{ lb}} \times \frac{1 \text{ h}}{7.09 \text{ mi}} \times \frac{55.8 \text{ cm}^3}{\text{kg} \cdot \text{min}} \times \frac{60 \text{ min}}{1 \text{ h}} \times \left(\frac{1 \text{ dm}}{10 \text{ cm}}\right)^3 \times 10.5 \text{ mi} = 353 \text{ dm}^3$$

$$353 \text{ dm}^3 = 353 \text{ L}$$

Key Skills Problems

1.1
Given that the density of gold is 19.3 g/cm³, calculate the volume (in cm³) of a gold nugget with a mass of 5.98 g.
(a) 3.23 cm³ (b) 5.98 cm³ (c) 115 cm³ (d) 0.310 cm³
(e) 13.3 cm³

1.2
The SI unit for energy is the joule (J), which is equal to the kinetic energy possessed by a 2.00-kg mass moving at 1.00 m/s. Convert this velocity to mph (1 mi = 1.609 km).
(a) 4.47×10^{-7} mph (b) 5.79×10^6 mph (c) 5.79 mph
(d) 0.0373 mph (e) 2.24 mph

1.3
Determine the density of the following object in g/cm³. A cube with edge length = 0.750 m and mass = 14.56 kg.
(a) 0.0345 g/cm³ (b) 1.74 g/cm³ (c) 670 g/cm³ (d) 53.8 g/cm³
(e) 14.6 g/cm³

1.4
A 28-kg child can consume a maximum of 23 children's acetaminophen tablets in an 8-h period without exceeding the safety-limit maximum allowable dose. Given that each children's tablet contains 80 mg of acetaminophen, determine the maximum allowable dose in mg per pound of body weight for one day.
(a) 80 mg/lb (b) 90 mg/lb (c) 430 mg/lb (d) 720 mg/lb
(e) 3.7 mg/lb

Questions and Problems

SECTION 1.1: THE STUDY OF CHEMISTRY

Review Questions

1.1   Define the terms *chemistry* and *matter*.
1.2   Explain what is meant by the scientific method.
1.3   What is the difference between a hypothesis and a theory?

Conceptual Problems

1.4   Classify each of the following statements as a hypothesis, law, or theory. (a) Beethoven's contribution to music would have been much greater if he had married. (b) An autumn leaf gravitates toward the ground because there is an attractive force between the leaf and Earth. (c) All matter is composed of very small particles.

1.5   Classify each of the following statements as a hypothesis, law, or theory. (a) The force acting on an object is equal to its mass times its acceleration. (b) The universe as we know it started with a big bang. (c) There are many civilizations more advanced than ours on other planets.

SECTION 1.2: SCIENTIFIC MEASUREMENT

Review Questions

1.6   Name the SI base units that are important in chemistry, and give the SI units for expressing the following: (a) length, (b) volume, (c) mass, (d) time, (e) temperature.
1.7   Write the numbers represented by the following prefixes: (a) mega-, (b) kilo-, (c) deci-, (d) centi-, (e) milli-, (f) micro-, (g) nano-, (h) pico-.
1.8   What units do chemists normally use for the density of liquids and solids? For the density of gas? Explain the differences.
1.9   What is the difference between mass and weight? If a person weighs 168 lb on Earth, about how much would the person weigh on the moon?
1.10   Describe the three temperature scales used in everyday life and in the laboratory: the Fahrenheit, Celsius, and Kelvin scales.

Computational Problems

1.11   Bromine is a reddish-brown liquid. Calculate its density (in g/mL) if 586 g of the substance occupies 188 mL.
1.12   The density of ethanol, a colorless liquid that is commonly known as grain alcohol, is 0.798 g/mL. Calculate the mass of 17.4 mL of the liquid.

1.13   Convert the following temperatures to degrees Celsius or Fahrenheit: (a) 95°F, the temperature on a hot summer day; (b) 12°F, the temperature on a cold winter day; (c) a 103°F fever; (d) a furnace operating at 1852°F; (e) −273.15°C (theoretically the lowest attainable temperature).
1.14   (a) Normally the human body can endure a temperature of 105°F for only short periods of time without permanent damage to the brain and other vital organs. What is this temperature in degrees Celsius? (b) Ethylene glycol is a liquid that is used as an antifreeze in car radiators. It freezes at −11.5°C. Calculate its freezing temperature in degrees Fahrenheit. (c) The temperature on the surface of the sun is about 6300°C. What is this temperature in degrees Fahrenheit?
1.15   The density of water at 40°C is 0.992 g/mL. What is the volume of 2.50 g of water at this temperature?
1.16   The density of platinum is 21.5 g/cm^3 at 25°C. What is the volume of 87.6 g of Pt at this temperature?
1.17   Convert the following temperatures to kelvins: (a) 113°C, the melting point of sulfur; (b) 37°C, the normal body temperature; (c) 357°C, the boiling point of mercury.
1.18   Convert the following temperatures to degrees Celsius: (a) 77 K, the boiling point of liquid nitrogen; (b) 4.2 K, the boiling point of liquid helium; (c) 601 K, the melting point of lead.
1.19   A 18.5-g sample of lead pellets at 20°C is mixed with a 45.8-g sample of lead pellets at the same temperature. What are the final mass, temperature, and density of the combined sample? (The density of lead at 20°C is 11.35 g/cm^3. Assume no heat is lost to the surroundings.)
1.20   A student pours 61.1 g of water at 10°C into a beaker containing 95.3 g of water at 10°C. What are the final mass, temperature, and density of the combined water? (The density of water at 10°C is 1.00 g/mL. Assume no heat is lost to the surroundings.)

SECTION 1.3: UNCERTAINTY IN MEASUREMENT

Review Questions

1.21   Indicate which of the following numbers is an exact number: (a) 50,247 tickets were sold at a sporting event; (b) 750 mL of water was used to make a birthday cake; (c) 10 eggs were used to make a breakfast; (d) 0.41 g of oxygen was inhaled in each breath; (e) Earth orbits the sun every 365.24 days.

1.22 Define *significant figure*. Discuss the importance of using the proper number of significant figures in measurements and calculations.

1.23 Distinguish between the terms *accuracy* and *precision*. In general, explain why a precise measurement does not always guarantee an accurate result.

Computational Problems

1.24 Express the following numbers in scientific notation:
(a) 0.000000027
(b) 356
(c) 47,764
(d) 0.096

1.25 Express the following numbers as decimals:
(a) 1.52×10^{-2}
(b) 7.78×10^{-8}
(c) 3.29×10^{-6}
(d) 8.41×10^{-1}

1.26 Express the answers to the following calculations in scientific notation:
(a) $145.75 + (2.3 \times 10^{-1})$
(b) $79,500 \div (2.5 \times 10^{2})$
(c) $(7.0 \times 10^{-3}) - (8.0 \times 10^{-4})$
(d) $(1.0 \times 10^{4}) \times (9.9 \times 10^{6})$

1.27 Express the answers to the following calculations in scientific notation:
(a) $0.0095 + (8.5 \times 10^{-3})$
(b) $653 \div (5.75 \times 10^{-8})$
(c) $850,000 - (9.0 \times 10^{5})$
(d) $(3.6 \times 10^{-4}) \times (3.6 \times 10^{6})$

1.28 Determine the number of significant figures in each of the following measurements:
(a) 4867 mi, (b) 56 mL, (c) 60,104 tons, (d) 2900 g, (e) 40.2 g/cm³, (f) 0.0000003 cm, (g) 0.7 min, (h) 46 amu.

1.29 Determine the number of significant figures in each of the following measurements:
(a) 0.006 L, (b) 0.0605 dm, (c) 60.5 mg, (d) 605.5 cm², (e) 9.60×10^{3} g, (f) 6 kg, (g) 60 m, (h) 1.42 Å.

1.30 Carry out the following operations as if they were calculations of experimental results, and express each answer in the correct units with the correct number of significant figures:
(a) 5.6792 m + 0.6 m + 4.33 m,
(b) 3.70 g − 2.9133 g,
(c) 4.51 cm × 3.6666 cm.

1.31 Carry out the following operations as if they were calculations of experimental results, and express

each answer in the correct units with the correct number of significant figures:
(a) 7.310 km ÷ 5.70 km,
(b) $(3.26 \times 10^{-3} \text{ mg}) - (7.88 \times 10^{-5} \text{ mg})$,
(c) $(4.02 \times 10^{6} \text{ dm}) + (7.74 \times 10^{7} \text{ dm})$.

1.32 Three students (A, B, and C) are asked to determine the volume of a sample of water. Each student measures the volume three times with a graduated cylinder. The results in milliliters are: A (87.1, 88.2, 87.6); B (86.9, 87.1, 87.2); C (87.6, 87.8, 87.9). The true volume is 87.0 mL. Comment on the precision and the accuracy of each student's results.

1.33 Three apprentice carpenters (X, Y, and Z) are assigned the task of measuring the width of a doorway. Each one makes three measurements. The results in inches are X (31.5, 31.6, 31.4); Y (32.8, 32.3, 32.7); Z (31.9, 32.2, 32.1). The true width is 32.0 in. Comment on the precision and the accuracy of each carpenter's measurements.

1.34 Report the quantity being measured to the appropriate number of significant figures.

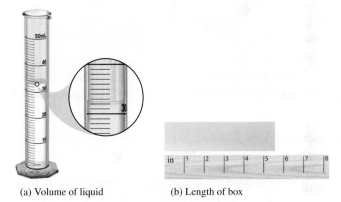

(a) Volume of liquid   (b) Length of box

1.35 Report each temperature to the appropriate number of significant figures.

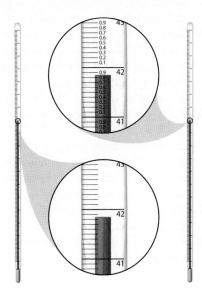

1.36   The density of the metal bar shown is 8.16 g/cm^3. Determine its mass to the appropriate number of significant figures.

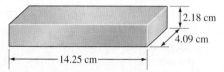

2.18 cm
4.09 cm
14.25 cm

1.37   The following shows an experiment used to determine the density of a gas. The evacuated bulb has a volume of 135.6 mL. It was weighed, filled with the gas, and weighed again. Determine the density of the gas to the appropriate number of significant figures.

243.07 g   243.22 g

SECTION 1.4: USING UNITS AND SOLVING PROBLEMS

Computational Problems

1.38   Carry out the following conversions: (a) 22.6 m to decimeters, (b) 25.4 mg to kilograms, (c) 556 mL to liters, (d) 10.6 kg/m^3 to g/cm^3.

1.39   Carry out the following conversions: (a) 242 lb to milligrams, (b) 68.3 cm^3 to cubic meters, (c) 7.2 m^3 to liters, (d) 28.3 µg to pounds.

1.40   Carry out the following conversions: (a) 242 amu to grams, (b) 87 amu to kilograms, (c) 2.21 Å to meters, (d) 1.73 Å to nanometers. [Conversion factors for atomic mass units (amu) and angstroms (Å) were introduced in Section 1.2.]

1.41   Carry out the following conversions: (a) 1.1×10^{-22} g to atomic mass units, (b) 1.08×10^{-29} kg to atomic mass units, (c) 8.3×10^{-9} m to angstroms, (d) 132 pm to angstroms. [Conversion factors for atomic mass units (amu) and angstroms (Å) were introduced in Section 1.2.]

1.42   The average speed of helium at 25°C is 1255 m/s. Convert this speed to miles per hour (mph).

1.43   How many seconds are there in a solar year (365.24 days)?

1.44   How many minutes does it take light from the sun to reach Earth? (The distance from the sun to Earth is 93 million mi; the speed of light is 3.00×10^8 m/s.)

1.45   A slow jogger runs a mile in 13 min. Calculate the speed in (a) in/s, (b) m/min, (c) km/h (1 mi = 1609 m; 1 in = 2.54 cm).

1.46   A 6.0-ft person weighs 183 lb. Express this person's height in meters and weight in kilograms (1 lb = 453.6 g; 1 m = 3.28 ft).

1.47   The speed limit in many school zones in the United States is 20 mph. What is the speed limit in kilometers per hour (1 mi = 1609 m)?

1.48   For a fighter jet to take off from the deck of an aircraft carrier, it must reach a speed of 62 m/s. Calculate the speed in miles per hour.

1.49   The "normal" lead content in human blood is about 0.40 part per million (that is, 0.40 g of lead per million grams of blood). A value of 0.80 part per million (ppm) is considered to be dangerous. How many grams of lead are contained in 6.0×10^3 g of blood (the amount in an average adult) if the lead content is 0.62 ppm?

1.50   Carry out the following conversions: (a) 1.42 km to miles, (b) 32.4 yd to centimeters, (c) 3.0×10^{10} cm/s to ft/s.

1.51   Carry out the following conversions: (a) 185 nm to meters, (b) 4.5 billion years (roughly the age of Earth) to seconds (assume exactly 365 days in a year), (c) 71.2 cm^3 to cubic meters, (d) 88.6 m^3 to liters.

1.52   Aluminum is a lightweight metal (density = 2.70 g/cm^3) used in aircraft construction, high-voltage transmission lines, beverage cans, and foils. What is its density in kg/m^3?

1.53   The density of ammonia gas under certain conditions is 0.625 g/L. Calculate its density in g/cm^3.

1.54   A human brain weighs about 1 kg and contains about 10^{11} cells. Assuming that each cell is completely filled with water (density = 1 g/mL), calculate the length of one side of such a cell if it were a cube. If the cells were spread out into a thin layer that was a single cell thick, what would be the total surface area (in square meters) for one side of the cell layer?

SECTION 1.5: CLASSIFICATION OF MATTER

Review Questions

1.55   Give an example for each of the following terms: (a) matter, (b) substance, (c) mixture.

1.56   Give an example of a homogeneous mixture and an example of a heterogeneous mixture.

Conceptual Problems

1.57 Identify each of the diagrams shown here as a solid, liquid, gas, or mixture of two substances.

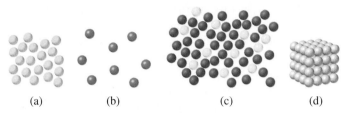

(a)   (b)   (c)   (d)

1.58 Classify each of the following as a pure substance, a homogeneous mixture, or a heterogeneous mixture: (a) seawater, (b) helium gas, (c) salt, (d) diet cola, (e) a milkshake, (f) bottled water, (g) concrete, (h) 24K gold, (i) liquid nitrogen.

SECTION 1.6: THE PROPERTIES OF MATTER

Review Questions

1.59 What is the difference between a qualitative property and a quantitative property?

1.60 Using examples, explain the difference between a physical property and a chemical property.

1.61 How does an intensive property differ from an extensive property?

1.62 Determine which of the following properties are intensive and which are extensive: (a) length, (b) volume, (c) temperature, (d) mass.

Conceptual Problems

1.63 Classify the following as qualitative or quantitative statements, giving your reasons. (a) The sun is approximately 93 million miles from Earth. (b) Leonard da Vinci was a better painter than Michelangelo. (c) Ice is less dense than water. (d) Butter tastes better than margarine. (e) A stitch in time saves nine.

1.64 Determine whether the following statements describe chemical or physical properties. (a) Oxygen gas supports combustion. (b) Ingredients in antacids reduce acid reflux. (c) Water boils above 100°C in a pressure cooker. (d) Carbon dioxide is denser than air. (e) Uranium combines with fluorine to form a gas.

1.65 Determine whether each of the following describes a physical change or a chemical change: (a) The helium gas inside a balloon tends to leak out after a few hours. (b) A flashlight beam slowly gets dimmer and finally goes out. (c) Frozen orange juice is reconstituted by adding water to it. (d) The growth of plants depends on the sun's energy in a process called photosynthesis. (e) A spoonful of sugar dissolves in a cup of coffee.

1.66 Determine whether each of the following describes a physical change or a chemical change: (a) A soda loses its fizz and goes flat. (b) A bruise develops on a football player's arm and gradually changes color. (c) A pile of leaves is burned. (d) Frost forms on a windshield after a cold night. (e) Wet clothes are hung out to dry in the sun.

ADDITIONAL PROBLEMS

1.67 Using the appropriate number of significant figures, report the length of the blue rectangle (a) using the ruler shown above the rectangle and (b) using the ruler shown below the rectangle.

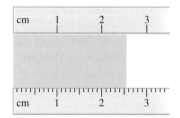

1.68 A piece of metal with a mass of 13.2 g was dropped into a graduated cylinder containing 17.00 mL of water. The graduated cylinder after the addition of the metal is shown. Determine the density of the metal to the appropriate number of significant figures. [Note: The volume of water in a graduated cylinder should be read at the *bottom* of the meniscus (the curved surface at the top).]

1.69 Which of the following statements describe physical properties and which describe chemical properties? (a) Iron has a tendency to rust. (b) Rainwater in industrialized regions tends to be acidic. (c) Hemoglobin molecules are red. (d) When a glass of water is left out in the sun, the water gradually disappears. (e) Carbon dioxide in air is consumed by plants during photosynthesis.

1.70 Give one qualitative and one quantitative statement about each of the following: (a) water, (b) carbon, (c) iron, (d) hydrogen gas, (e) sucrose (cane sugar), (f) table salt, (g) mercury, (h) gold, (i) air.

1.71 In 2004, about 95.0 billion pounds of sulfuric acid were produced in the United States. Convert this quantity to tons.

1.72 In determining the density of a rectangular metal bar, a student made the following measurements: length, 8.53 cm; width, 2.4 cm; height, 1.0 cm; mass, 52.7064 g. Calculate the density of the metal to the correct number of significant figures.

1.73 Calculate the mass of each of the following: (a) a sphere of gold with a radius of 10.0 cm (volume of a sphere with a radius r is $V = \frac{4}{3}\pi r^3$; density of gold = 19.3 g/cm^3); (b) a cube of platinum of edge length 0.040 mm (density = 21.4 g/cm^3); (c) 50.0 mL of ethanol (density = 0.798 g/mL).

1.74 A cylindrical glass tube 12.7 cm in length is filled with mercury (density = 13.6 g/mL). The mass of mercury needed to fill the tube is 105.5 g. Calculate the inner diameter of the tube (volume of a cylinder of radius r and length h is $V = \pi r^2 h$).

1.75 The following procedure was used to determine the volume of a flask. The flask was weighed dry and then filled with water. If the masses of the empty flask and filled flask were 56.12 g and 87.39 g, respectively, and the density of water is 0.9976 g/cm^3, calculate the volume of the flask in cubic centimeters.

1.76 The speed of sound in air at room temperature is about 343 m/s. Calculate this speed in miles per hour (1 mi = 1609 m).

1.77 A piece of platinum metal weighing 234.0 g is placed in a graduated cylinder containing 187.1 mL of water. The volume of water now reads 198.0 mL. From these data, calculate the density of platinum.

1.78 The experiment described in Problem 1.77 is a crude but convenient way to determine the density of some solids. Describe a similar experiment that would enable you to measure the density of ice. Specifically, what would be the requirements for the liquid used in your experiment?

1.79 A copper sphere has a mass of 2.17×10^3 g, and its volume is 242.2 cm^3. Calculate the density of copper.

1.80 Lithium has a very low density (density = 0.53 g/cm^3). What is the volume occupied by 1.20×10^3 g of lithium?

1.81 The medicinal thermometer commonly used in homes can be read to $\pm 0.1°F$, whereas those in the doctor's office may be accurate to $\pm 0.1°C$. Percent error is often expressed as the absolute value of the difference between the true value and the experimental value, divided by the true value:

$$\text{percent error} = \frac{|\text{true value} - \text{experimental value}|}{\text{true value}} \times 100\%$$

The vertical lines indicate absolute value. In degrees Celsius, express the percent error expected from each of these thermometers in measuring a person's body temperature of 38.9°C.

1.82 Vanillin (used to flavor vanilla ice cream and other foods) is the substance whose aroma the human nose detects in the smallest amount. The threshold limit is 2.0×10^{-11} g per liter of air. If the current price of 50 g of vanillin is \$112, determine the cost to supply enough vanillin so that the aroma could be detected in a large aircraft hangar with a volume of 5.0×10^7 ft^3.

1.83 Suppose that a new temperature scale has been devised on which the melting point of ethanol ($-117.3°C$) and the boiling point of ethanol ($78.3°C$) are taken as 0°S and 100°S, respectively, where S is the symbol for the new temperature scale. Derive an equation relating a reading on this scale to a reading on the Celsius scale. What would this thermometer read at 25°C?

1.84 At what temperature does the numerical value on the Celsius scale equal that on the temperature scale described in Problem 1.83?

1.85 A resting adult requires about 240 mL of pure oxygen per minute and breathes about 12 times every minute. If inhaled air contains 20 percent oxygen by volume and exhaled air 16 percent, what is the volume of air per breath? (Assume that the volume of inhaled air is equal to that of exhaled air.)

1.86 (a) Referring to Problem 1.85, calculate the total volume (in liters) of air an adult breathes in a day. (b) In a city with heavy traffic, the air contains 2.1×10^{-6} L of carbon monoxide (a poisonous gas) per liter. Calculate the average daily intake of carbon monoxide in liters by a person.

1.87 The total volume of seawater is 1.5×10^{21} L. Assume that seawater contains 3.1 percent sodium chloride by mass and that its density is 1.03 g/mL. Calculate the total mass of sodium chloride in kilograms and in tons (1 ton = 2000 lb; 1 lb = 453.6 g).

1.88 Magnesium is used in alloys, in batteries, and in the manufacture of chemicals. It is obtained mostly from seawater, which contains about 1.3 g of magnesium for every kilogram of seawater. Referring to Problem 1.87, calculate the volume of seawater (in liters) needed to extract 8.0×10^4 tons of magnesium, which is roughly the annual production in the United States.

1.89 The unit "troy ounce" is often used for precious metals such as gold and platinum (1 troy ounce = 31.103 g). (a) A gold coin weighs 2.41 troy ounces. Calculate its mass in grams. (b) Is a troy ounce heavier or lighter than an ounce (1 lb = 16 oz; 1 lb = 453.6 g)?

1.90 The surface area and average depth of the Pacific Ocean are 1.8×10^8 km^2 and 3.9×10^3 m, respectively. Calculate the volume of water in the ocean in liters.

1.91 Calculate the percent error for the following measurements: (a) The density of alcohol (ethanol) is found to be 0.802 g/mL (true value = 0.798 g/mL). (b) The mass of gold in an earring is analyzed to be 0.837 g (true value = 0.864 g).

1.92 Venus, the second closest planet to the sun, has a surface temperature of 7.3×10^2 K. Convert this temperature to degrees Celsius and degrees Fahrenheit.

1.93 Chalcopyrite contains 34.63 percent copper by mass. How many grams of copper can be obtained from 5.11×10^3 kg of chalcopyrite?

1.94 It has been estimated that 8.0×10^4 tons of gold have been mined. Assume gold costs $1100 per ounce. What is the total value of this quantity of gold?

1.95 One gallon of gasoline in an automobile's engine produces on the average 9.5 kg of carbon dioxide, which is a greenhouse gas; that is, it promotes the warming of Earth's atmosphere. Calculate the annual production of carbon dioxide in kilograms if there are 40 million cars in the United States and each car covers a distance of 5000 mi at an average consumption rate of 20 miles per gallon.

1.96 A sheet of aluminum foil has a total area of 1.000 ft^2 and a mass of 3.636 g. What is the thickness of the foil in millimeters (density of aluminum = 2.699 g/cm^3)?

1.97 The world's total petroleum reserve is estimated at 2.0×10^{22} joules [a joule (J) is the unit of energy where $1 \text{ J} = 1 \text{ kg} \cdot \text{m}^2/\text{s}^2$]. At the present rate of consumption, 1.8×10^{20} joules per year (J/yr), how long would it take to exhaust the supply?

1.98 A sample of DNA, the genetic material of life, was estimated to have a mass of 308,859 amu. What is this mass in grams? The average width of a DNA double strand is approximately 22 Å to 26 Å. Express this range of widths in meters.

1.99 Pheromones are substances secreted by females of many insect species to attract mates. Typically, 1×10^{-8} g of a pheromone is sufficient to reach all targeted males within a radius of 0.50 mi. Calculate the density of the pheromone (in grams per liter) in a cylindrical air space having a radius of 0.50 mi and a height of 40 ft. (Volume of a cylinder of radius r and height h is $\pi r^2 h$.)

1.100 Chlorine is used to disinfect swimming pools. The recommended concentration for this purpose is 1 ppm chlorine, or 1 g of chlorine per million grams of water. Calculate the volume of a chlorine solution (in milliliters) a homeowner should add to her swimming pool if the solution contains 6.0 percent chlorine by mass and there are 2.0×10^4 gallons (gal) of water in the pool (1 gal = 3.79 L; assume the density of both the water and the chlorine solution to be = 1.0 g/mL).

1.101 A graduated cylinder is filled to the 40.00-mL mark with a mineral oil. The masses of the cylinder before and after the addition of the mineral oil are 124.966 g and 159.446 g, respectively. In a separate experiment, a metal ball bearing of mass 18.713 g is placed in the cylinder and the cylinder is again filled to the 40.00-mL mark with the mineral oil. The combined mass of the ball bearing and mineral oil is 50.952 g. Calculate the density and radius of the ball bearing (volume of a sphere of radius r is $\frac{4}{3}\pi r^3$).

1.102 In water conservation, chemists spread a thin film of a certain inert material over the surface of water to cut down on the rate of evaporation of water in reservoirs. This technique was pioneered by Benjamin Franklin three centuries ago. Franklin found that 0.10 mL of oil could spread over the surface of water about 40 m^2 in area. Assuming that the oil forms a *monolayer,* that is, a layer that is only one molecule thick, estimate the length of each oil molecule in nanometers (1 nm = 1×10^{-9} m).

1.103 A chemist in the nineteenth century collected a sample of unknown matter. In general, do you think it would be more difficult to prove that it is a pure substance or a mixture? Explain.

1.104 A gas company in Massachusetts charges $1.30 for 15.0 ft^3 of natural gas. (a) Convert this rate to dollars per liter of gas. (b) If it takes 0.304 ft^3 of gas to boil a liter of water, starting at room temperature (25°C), how much would it cost to boil a 2.1-L kettle of water?

1.105 You are given a liquid. Briefly describe the steps you would take to show whether it is a pure substance or a homogeneous mixture.

1.106 A bank teller is asked to assemble $1 sets of coins for his clients. Each set is made up of three quarters, one nickel, and two dimes. The masses of the coins are quarter, 5.645 g; nickel, 4.967 g; and dime, 2.316 g. What is the maximum number of complete sets that can be assembled from 33.871 kg of quarters, 10.432 kg of nickels, and 7.990 kg of dimes? What is the total mass (in grams) of the assembled sets of coins?

1.107 A 250-mL glass bottle was filled with 242 mL of water at 20°C and tightly capped. It was then left outdoors overnight, where the average temperature was −5°C. Predict what would happen. The density of water at 20°C is 0.998 g/cm^3 and that of ice at −5°C is 0.916 g/cm^3.

1.108 Bronze is an alloy made of copper and tin. Calculate the mass of a bronze cylinder of radius 6.44 cm and length 44.37 cm. The composition of the bronze is 79.42 percent copper and 20.58 percent tin and the densities of copper and tin are 8.94 g/cm^3 and 7.31 g/cm^3, respectively. What assumption should you make in this calculation?

1.109 Lead poisoning affects nearly every system in the body and can occur with no symptoms, potentially causing it to go undiagnosed. The primary source of lead exposure is the deteriorating lead-based paint in older homes, where young children are particularly vulnerable to exposure because of their tendency to put things in their mouths. In addition to the prevention of exposure, the Centers for Disease Control and Prevention (CDC) guidelines recommend that public health actions be initiated when lead levels exceed 10 micrograms of lead per deciliter of blood. Determine whether or not the following lead levels would exceed the CDC's threshold level: (a) 3.0×10^{-4} grams per liter of blood, (b) 2.0×10^{-5} milligrams per milliliter of blood, (c) 6.5×10^{-8} grams per cubic centimeter.

1.110 A chemist mixes two liquids A and B to form a homogeneous mixture. The densities of the liquids are 2.0514 g/mL for A and 2.6678 g/mL for B. When she drops a small object into the mixture, she finds that the object becomes suspended in the liquid; that is, it neither sinks nor floats. If the mixture is made of 41.37 percent A and 58.63 percent B by volume, what is the density of the object? Can this procedure be used in general to determine the densities of solids? What assumptions must be made in applying this method?

1.111 In January 2009, the National Aeronautics and Space Administration (NASA) reported that a planet in our galaxy, known as HD 80606b, underwent a temperature change from 980°F to 2240°F over the course of six hours. (a) Convert these temperatures and the range they span to degrees Celsius, and to kelvins. (b) Determine the rate of temperature change per second in degrees Fahrenheit, degrees Celsius, and kelvins.

1.112 TUMS is a popular remedy for acid indigestion. A typical TUMS tablet contains calcium carbonate plus some inert substances. When ingested, it combines with the gastric juice (hydrochloric acid) in the stomach to give off carbon dioxide gas. When a 1.328-g tablet reacted with 40.00 mL of hydrochloric acid (density = 1.140 g/mL), carbon dioxide gas was given off and the resulting solution weighed 46.699 g. Calculate the number of liters of carbon dioxide gas released if its density is 1.81 g/L.

1.113 Determine (a) the temperature at which the Celsius and Fahrenheit values are numerically equal, and (b) the temperature at which the Kelvin and Fahrenheit values are numerically equal. (c) Is there a temperature at which the Celsius and Kelvin values are numerically equal? Explain.

1.114 The hottest temperature ever recorded on Earth was 136°F (recorded at Al 'Aziziyah, Libya, on September 13, 1922). Express this temperature in degrees Celsius and in kelvins.

1.115 The drug *cidofovir* is approved by the Federal Drug Administration (FDA) for the treatment of certain viral infections of the eye in patients with compromised immune systems. It is distributed in vials containing 375 mg of the drug dissolved in 5 mL of water. The manufacturer specifies that the drug should be kept at room temperature (68°F–77°F). The vial contents are first diluted with saline and then administered intravenously with a recommended dosage of 5 mg cidofovir per kilogram of body weight. (a) Convert cidofovir's recommended storage-temperature range to the Celsius scale. (b) If the fluid in a single vial of cidofovir has a volume of 5.00 mL and a mass of 5.89 g, what is the density of the fluid, in g/mL, to the appropriate number of significant figures? (c) Convert the density in part (b) to g/L and to kg/m^3. (d) What mass of cidofovir should be administered to a 185-lb man?

1.116 The composition of pennies has changed over the years, depending on a number of factors, including the availability of various metals. A penny minted in 1825 was pure copper; a penny minted in 1860 was 88 percent copper and 12 percent nickel; a penny minted in 1965 was 95 percent copper and 5 percent zinc; and a penny minted today is 97.5 percent zinc and 2.5 percent copper. Given that the densities of copper, nickel, and zinc are 8.92 g/cm^3, 8.91 g/cm^3, and 7.14 g/cm^3, respectively, determine the density of each penny.

Answers to In-Chapter Materials

PRACTICE PROBLEMS

1.1A 273 K and 373 K, range = 100 K. **1.1B** −270.5°C.
1.2A −33°F, −2.2°F, range 32°F. **1.2B** 464°C, −201°C, range 665°C.
1.3A (a) 13.6 g/mL, (b) 55 mL. **1.3B** (a) 9.25 g/cm^3, (b) 3.76 × 10^3 g.
1.4A (a) 4, (b) 1, (c) 4, (d) 2, (e) 2 or 3, (f) 4. **1.4B** (a) 1.0 × 10^6,
(b) 1.000 × 10^6, (c) 1.000000 × 10^6, (d) 0.100, (e) 0.10000, (f) 0.1.
1.5A (a) 116.2 L, (b) 80.71 m, (c) 3.813 × 10^{21} atoms, (d) 31 dm^2,
(e) 0.504 g/mL. **1.5B** (a) 32.44 cm^3, (b) 4.2 × 10^2 kg/m^3,
(c) 1.008 × 10^{10} kg, (d) 40.75 mL, (e) 227 cm^3. **1.6A** 0.8120 g/cm^3.
1.6B 95.3 cm^3. **1.7A** 0.01 oz. **1.7B** 2.649 × 10^4 mg.
1.8A 1.05 × 10^4 kg/m^3. **1.8B** 849 lb/ft^3.

SECTION REVIEW

1.2.1 c. **1.2.2** a. **1.2.3** b. **1.2.4** d. **1.3.1** c. **1.3.2** c. **1.3.3** a. **1.3.4** e.
1.3.5 d. **1.3.6** e. **1.4.1** c. **1.4.2** a. **1.4.3** b. **1.4.4** e.

Chapter 2

Atoms and the Periodic Table

© Science Photo Library/Science Source

OF MATTER that exists as isolated atoms, helium is probably the most familiar. Sightings of helium-filled balloons are fairly common. Helium is the product of radioactive decay of the elements uranium (U) and thorium (Th) and is found in and around natural gas deposits. In 1925, the U.S. government officially recognized the strategic value of helium and began an initiative to stockpile it. Estimates vary as to how long Earth's supply of helium will last. Some scientists believe that the shortage could become critical in less than a decade. Because our supply is finite and nonrenewable, efforts by chemical engineers to capture and recycle helium are vitally important as demand increases. In addition to being used to fill balloons, helium is used as a coolant for nuclear reactors and superconducting magnets, to pressurize liquid fuel rockets, and for a host of other applications.

Before You Begin, Review These Skills

- Significant figures [◄◄ Section 1.3]
- Dimensional analysis [◄◄ Section 1.4]

2.1 ATOMS FIRST

Even if you have never studied chemistry before, you probably know already that atoms are extraordinarily small building blocks of matter. Specifically, an **atom** is the smallest quantity of matter that still retains the properties of matter. Further, an **element** is a substance that cannot be broken down into two or more simpler substances by any means. Familiar examples of elements include gold, oxygen, and helium. Consider the example of helium. If we were to divide the helium in a balloon in half, and then divide one of the halves in half, and so on, we would eventually (after a large number of these hypothetical divisions) be left with a sample of helium consisting of just one helium atom. This atom could not be further divided to give two smaller samples of helium. If this is difficult to imagine, think of a collection of DVDs. You could separate the collection into smaller and smaller numbers of DVDs. But when you were down to a single DVD, although you could break it into smaller pieces, none of the pieces would be a DVD. Similarly, an *atom* is the smallest quantity of helium. If we were, somehow, to break this quantity apart, the resulting pieces would not be helium. Figure 2.1 illustrates this concept.

The concept of the atom was first proposed by the philosopher Democritus in the fifth century B.C. Although the notion of matter consisting of atoms endured for many centuries, it was first formalized by the English scientist and

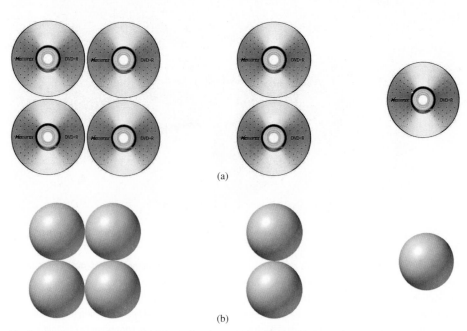

(a)

(b)

Figure 2.1 (a) A collection of DVDs can be separated into smaller and smaller numbers of DVDs; but a single DVD cannot be separated into smaller pieces that are still DVDs. (b) A collection of atoms can be separated into smaller and smaller numbers of atoms; but if we were to separate an atom into smaller pieces, none of the pieces would be an atom.

(a) © David A. Tietz/Editorial Image, LLC

schoolteacher *John Dalton.*[1] Dalton said that atoms, of which all matter consists, are tiny, *indivisible* particles. We now know that, in fact, although atoms are tiny, they are not indivisible. Rather, they are made up of still smaller *sub*atomic particles. The nature, number, and arrangement of subatomic particles determine the properties of atoms, which in turn determine the properties of all things material.

Our goal in this book will be to understand how the nature of atoms gives rise to the properties of all matter. To accomplish this, we will take a somewhat untraditional approach. Rather than beginning with observations on the macroscopic scale and working our way backward to the atomic nature of matter to explain these observations, we start by examining the structure of atoms and the tiny subatomic particles that atoms contain.

2.2   SUBATOMIC PARTICLES AND ATOMIC STRUCTURE

Contrary to Dalton's description of the atom as *indivisible,* a series of investigations beginning in the middle of the nineteenth century indicated that atoms are made up of even smaller things. These investigations led to the discovery of electrons, protons, and neutrons, and contributed to our understanding of the structure of atoms.

Discovery of the Electron

In the late 1800s, many scientists were doing research involving **radiation,** the emission and transmission of energy in the form of waves. One device that was commonly used to investigate this phenomenon was the *cathode ray tube.* A cathode ray tube consists of two metal plates sealed inside a glass tube from which most of the air has been evacuated (Figure 2.2). When the metal plates are connected to a high-voltage

Video 2.1
Cathode ray tube experiment

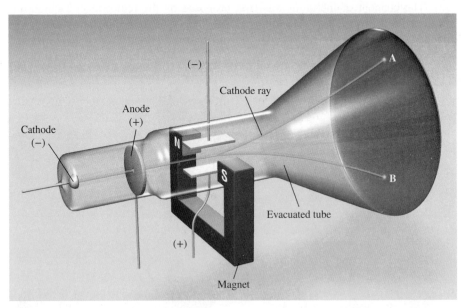

Figure 2.2 A cathode ray tube with an electric field perpendicular to the direction of the cathode rays and an external magnetic field. The symbols N and S denote the north and south poles of the magnet. The cathode rays will strike the end of the tube at point A in the presence of a magnetic field, and at point B in the presence of an electric field. (In the absence of any external field—or when the effects of the electric field and magnetic field cancel each other—the cathode rays will not be deflected but will travel in a straight line and strike the middle of the circular screen.)

---

[1]John Dalton (1766–1844). English chemist, mathematician, and philosopher. In addition to his atomic theory, Dalton also formulated several laws governing the behavior of gases, and gave the first detailed description of a particular type of color blindness, from which he suffered.

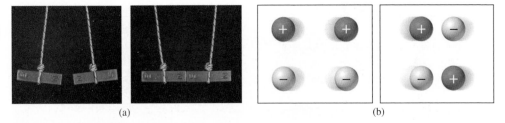

(a)   (b)

Figure 2.3 (a) When the same poles of two magnets are brought together, the magnets repel each other. When opposite poles are brought together, the magnets attract each other. (b) Like charges, either both positive or both negative, repel each other. Opposite charges attract each other.

(a - photos): © H.S. Photos/Science Source

source, the negatively charged plate, called the *cathode,* emits radiation known as **cathode rays.** A cathode ray moves toward the positively charged plate, called the *anode,* where it passes through a hole and continues traveling to the other end of the tube. Although a cathode ray itself is invisible, its path is revealed when the ray strikes a phosphor-coated surface, *producing a bright light.* Most striking about the cathode ray tube experiments was that the results did not depend on the material from which the cathode was made. Cathode rays appeared to be a component even more *fundamental* than atoms themselves, and common to *all* matter.

> **Student Annotation:** Cathode ray tubes were the forerunners of the tubes used in old televisions and computer monitors.

One of the things researchers used to interpret their results was the fact that *like* charges *repel* one another, and *opposite* charges *attract* one another. This is, in essence, *Coulomb's law* and is similar to a phenomenon you may have experienced while playing with magnets as a child. When two magnets are brought close together with the same poles facing each other, there is a repulsive force between them. When opposite poles are brought together, there is an attractive force between the magnets. Figure 2.3 summarizes the repulsive and attractive forces between magnets, and between electrical charges.

> **Student Annotation:** Coulomb's law is extraordinarily important in the understanding of chemistry. We will examine it in more detail in the next few chapters.

In addition to originating at the negatively charged cathode and traveling toward the positively charged anode, cathode rays can be deflected by electric or magnetic fields (Figure 2.4). The rays are repelled by a plate bearing a negative charge, and attracted to a plate bearing a positive charge. This prompted English physicist J. J. Thomson (1856–1940) to propose that the "rays" are actually a stream of negatively charged *particles.* Thomson performed a series of experiments in which he applied a variable electric field and measured the degree of deflection of cathode rays. This enabled him to determine the charge-to-mass *ratio* of the tiny, negatively charged particles that we now know as **electrons.** The charge-to-mass ratio he calculated was 1.76×10^8 C/g, where C stands for *coulomb,* the derived SI unit of electric charge.

Early in the twentieth century, American physicist R. A. Millikan (1868–1953) determined the charge on an electron by examining the motion of tiny oil drops in an

Video 2.2
Millikan's oil-drop experiment

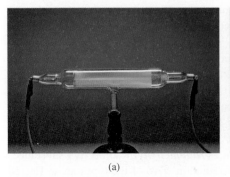

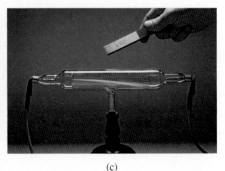

(a)   (b)   (c)

Figure 2.4 (a) A cathode ray produced in a discharge tube. The ray itself is invisible, but a fluorescent coating on the glass causes it to appear green. (b) The cathode ray bends toward one pole of a magnet and (c) bends away from the opposite pole.

© McGraw-Hill Education/Charles D. Winters, photographer

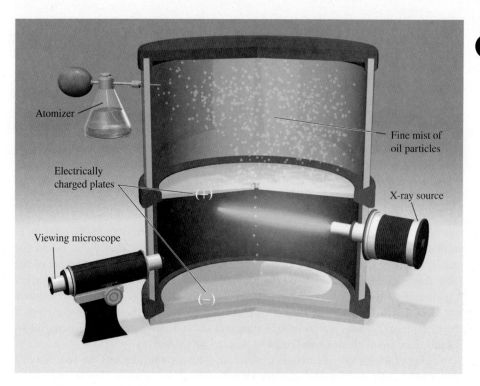

Figure 2.5 Schematic diagram of Millikan's oil-drop experiment.

ingenious experiment (Figure 2.5). The oil drops picked up static charge from particles in the air and were suspended in an electric field. The magnitude of electric field necessary to keep a drop suspended depended on the amount of charge the drop had acquired. Millikan found that the charge on any particular drop was always a multiple of -1.6022×10^{-19} C, which he deduced was the charge on a *single* electron. He then used the charge he had determined, and the charge-to-mass ratio that Thomson had determined, to calculate the *mass* of an electron:

$$\text{mass of an electron} = \frac{\text{charge}}{\text{charge/mass}} = \frac{-1.6022 \times 10^{-19} \text{ C}}{-1.76 \times 10^{8} \text{ C/g}} = 9.10 \times 10^{-28} \text{ g}$$

Radioactivity

Another scientist who worked with cathode ray tubes was German physicist Wilhelm Röntgen (1845–1923). Röntgen noticed that pointing cathode rays at glass or metal caused the emission of another type of ray, which he called *X rays*. X rays were sufficiently energetic to penetrate matter and caused a variety of materials to *fluoresce* (give off light). Unlike cathode rays, X rays were not deflected by magnetic or electric fields, which indicated they could not consist of charged particles.

Soon after Röntgen's discovery of X rays, a French physicist Antoine Becquerel (1852–1908), while investigating fluorescence, discovered that uranium caused the darkening of photographic plates that had been wrapped carefully to prevent their exposure to light. Like X rays, the radiation from the uranium was highly energetic and could not be deflected by magnetic or electric fields. Unlike X rays, these rays were emitted spontaneously, and not as the result of exposure to cathode rays. This spontaneous emission of radiation is known as *radioactivity.* Radioactive substances such as uranium can produce three types of radiation. Two of the types are deflected by charged plates, as shown in Figure 2.6. *Alpha (α) rays* consist of positively charged particles, called *α particles,* that are deflected *away* from a positively charged plate. *Beta (β) rays,* or *β particles,* are *electrons,* so they are deflected away from

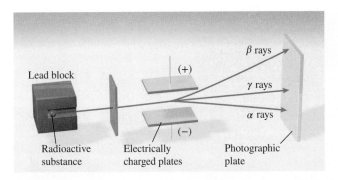

Figure 2.6 Three types of rays emitted by radioactive elements. So-called β rays actually consist of negatively charged particles (electrons) and are therefore attracted by the positively charged plate. The opposite is true for so-called α rays—they are actually positively charged particles and are drawn to the negatively charged plate. Because γ rays are not particles and have no charge, their path is unaffected by an external electric field.

the *negatively* charged plate. The third type of radioactive emission consists of high-energy **gamma (γ) rays.** Like X rays, γ rays have no charge and are unaffected by external electric or magnetic fields.

The Proton and the Nuclear Model of the Atom

By the early 1900s, scientists knew that atoms contained negatively charged electrons and that atoms themselves were electrically neutral. In order to be neutral, an atom must therefore contain something positively charged. Thomson proposed that an atom could be thought of as a sphere of positively charged matter in which negatively charged electrons were embedded uniformly, like the raisins in a scoop of rum raisin ice cream. This so-called plum-pudding model was the accepted theory for a number of years—until it was challenged by the work of Ernest Rutherford (New Zealand physicist, 1871–1937), one of Thomson's own students.

Rutherford and his associates, Hans Geiger and Ernest Marsden, carried out a series of experiments in which they directed α particles from a radioactive source at very thin foils of gold and other metals. They observed that, as expected, the majority of particles penetrated the foil either completely undeflected or with only a small angle of deflection. Every now and then, however, an α particle was scattered (or deflected) at a large angle. In some instances, the α particle actually bounced back in the direction of the radioactive source. This was an extraordinarily surprising finding. In Thomson's model, the positive charge of the atom was so diffuse that the relatively massive, positively charged α particles should all have passed through the foil with little or no deflection. To quote Rutherford's initial reaction to these results: "It was as incredible as if you had fired a 15-in shell at a piece of tissue paper and it came back and hit you." Figure 2.7 illustrates the *expected* and *actual* results of the α-scattering experiment.

According to Rutherford, the results of the α-scattering experiment meant that most of the atom must be empty space. This would explain why the majority of α particles passed through the gold foil with little or no deflection. The atom's positive charges, Rutherford proposed, were all concentrated in an extremely dense central *core* within the atom, which he called the **nucleus.** Whenever an α particle came close to a nucleus in the scattering experiment, it experienced a large repulsive force and therefore a large deflection. Moreover, an α particle traveling directly toward a nucleus would be completely repelled and its direction would be reversed.

The positively charged particles in the nucleus are called **protons.** In separate experiments, it was found that each proton carried the same *magnitude* of charge as an electron (just opposite in sign) but had a mass of 1.67262×10^{-24} g. Although this is an extremely small mass, it is nearly 2000 times the mass of an electron.

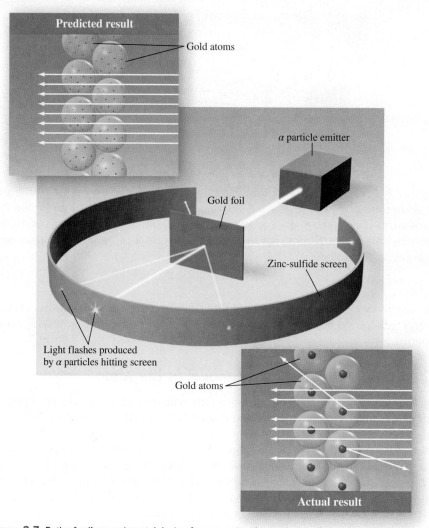

Figure 2.7 Rutherford's experimental design for measuring the scattering of α particles by a piece of gold foil. The plum-pudding model predicted that the α particles would all pass through the gold foil undeflected. The actual result: Most of the α particles do pass through the gold foil with little or no deflection, but a few are deflected at large angles. Occasionally, an α particle bounces off the foil back toward the source. The nuclear model explains the results of Rutherford's experiments.

Based on these data, the atom was believed to consist of a nucleus that accounted for most of the mass of the atom, but which occupied only a tiny fraction of its volume. Atomic (and molecular) dimensions often are expressed using the SI unit *picometer* (*pm*), where

$$1 \text{ pm} = 1 \times 10^{-12} \text{ m}$$

Student Annotation: Although the concept of atomic radius can be useful, you should not get the impression that atoms have well-defined boundaries or surfaces. As we will learn in Chapter 3, the outer edges of atoms are not sharply defined but rather, are relatively "fuzzy."

A typical *atomic radius* is on the order of about 100 pm, whereas the radius of an atomic nucleus is only about 5×10^{-3} pm. You can appreciate the relative sizes of an atom and its nucleus by imagining that if an atom were the size of the New Orleans Superdome, the volume of its nucleus would be comparable to that of a marble. While the protons are confined to the nucleus of the atom, the electrons are distributed *around* the nucleus at relatively large distances from it.

The Neutron

Although Rutherford's nuclear model of the atoms was a marked improvement over Thomson's plum-pudding model, protons and electrons alone could not account for

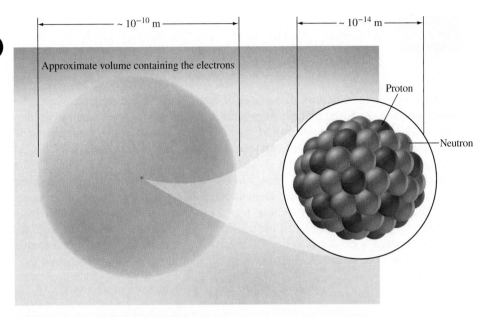

Figure 2.8 The protons and neutrons in an atom are contained in the tiny volume of the nucleus. Electrons are distributed within the sphere surrounding the nucleus.

all of the mass that most atoms were known to possess. It was known that hydrogen, the simplest atom, contained only one proton and that the helium atom contained two protons. Therefore, the ratio of the mass of a helium atom to that of a hydrogen atom should be 2:1. (Because electrons are much lighter than protons, their contribution to atomic mass can be ignored in an analysis such as this.) In reality, however, the mass ratio is 4:1. Rutherford and others postulated that there must be another type of subatomic particle in the atomic nucleus, the proof of which was provided in 1932 by the English physicist James Chadwick (1891–1974). When Chadwick bombarded a thin sheet of beryllium with α particles, a very high-energy radiation was emitted by the metal that was not deflected by either electric or magnetic fields. Later experiments showed that the rays—although similar to γ rays—actually consisted of a third type of subatomic *particle,* which Chadwick named **neutrons** because they were electrically neutral particles having a mass slightly greater than that of protons. The mystery of the mass ratio could now be explained. A typical helium nucleus consists of two protons and two neutrons, whereas a typical hydrogen nucleus contains only a proton; the mass ratio, therefore, is 4:1.

Figure 2.8 shows the location of the elementary particles (protons, neutrons, and electrons) in an atom. There are other subatomic particles, but the electron, the proton, and the neutron are the three fundamental components of the atom that are important in chemistry. Table 2.1 lists the masses and charges of these three elementary particles.

TABLE 2.1	Masses and Charges of Subatomic Particles			
Particle	Mass (g)	Mass (amu)	Charge (C)	Charge unit
Electron*	9.10938×10^{-28}	5.4858×10^{-4}	-1.6022×10^{-19}	-1
Proton	1.67262×10^{-24}	1.0073	$+1.6022 \times 10^{-19}$	$+1$
Neutron	1.67493×10^{-24}	1.0086	0	0

*More refined measurements have resulted in a small change to Millikan's original value.

2.3 ATOMIC NUMBER, MASS NUMBER, AND ISOTOPES

All atoms can be identified by the number of protons and neutrons they contain. The **atomic number (Z)** is the number of protons in the nucleus of each atom of an element. It also indicates the number of *electrons* in the atom—because for an atom to be *neutral,* it must contain the same number of protons and electrons. The elemental identity of an atom can be determined solely from its atomic number. For example, the atomic number of nitrogen is 7. Thus, each atom of the element nitrogen has seven protons. Or, viewed another way, any atom that contains seven protons is a nitrogen atom.

The **mass number (A)** is the total number of neutrons *and* protons, collectively referred to as **nucleons,** present in the nucleus of an atom of an element. Except for the most common form of hydrogen, which has one proton and no neutrons, all atomic nuclei contain both protons and neutrons. In general, the mass number is given by:

mass number (A) = number of protons (Z) + number of neutrons

The number of neutrons in an atom equals the difference between the mass number and the atomic number, or (A − Z). For example, the mass number of fluorine is 19 and the atomic number is 9 (indicating 9 protons in the nucleus). Thus, the number of neutrons in an atom of fluorine is 19 − 9 = 10. The atomic number, number of neutrons, and mass number are all positive integers (whole numbers).

The accepted way to denote the atomic number and mass number of an atom of an element (X) is as follows:

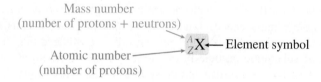

Once again, contrary to Dalton's atomic theory, atoms of a given element are *not* all identical. Instead, most elements have two or more **isotopes,** atoms that have the same atomic number (Z) but different mass numbers (A). For example, there are three isotopes of hydrogen, called *hydrogen* (or *protium*), *deuterium,* and *tritium.* Hydrogen has one proton and no neutrons in its nucleus, deuterium has one proton and *one* neutron, and tritium has one proton and *two* neutrons. Thus, to represent the isotopes of hydrogen, we write

Similarly, the two common isotopes of uranium (Z = 92), which have mass numbers of 235 and 238, respectively, can be represented as follows:

$$^{235}_{92}\text{U} \qquad ^{238}_{92}\text{U}$$

The first isotope, with 235 − 92 = 143 neutrons in its nucleus, is used in nuclear reactors and atomic bombs, whereas the second isotope, with 146 neutrons, lacks the properties necessary for these applications. With the exception of hydrogen, which has different names for each of its isotopes, the isotopes of other elements are identified by their mass numbers. The two isotopes of uranium are called uranium-235 (pronounced "uranium two thirty-five") and uranium-238 (pronounced "uranium two thirty-eight"). Because the atomic number subscript can be deduced from the elemental symbol, it may be omitted from these representations without the loss of any information. The symbols ^{3}H and ^{235}U are sufficient to specify the isotopes tritium and uranium-235, respectively.

The chemical properties of an element are determined primarily by the number of protons and electrons in its atoms; not by the number of neutrons. Therefore, isotopes of the same element typically exhibit very similar chemical properties.

Worked Example 2.1 shows how to calculate the number of protons, neutrons, and electrons using atomic numbers and mass numbers.

Worked Example 2.1

Determine the numbers of protons, neutrons, and electrons in each of the following species:

(a) $^{35}_{17}$Cl, (b) ^{37}Cl, (c) ^{41}K, and (d) carbon-14.

Strategy Recall that the superscript denotes the mass number (A), and the subscript denotes the atomic number (Z). In cases where no subscript is shown, as in parts (b), (c), and (d), the atomic number can be deduced from the elemental symbol or name. For the purpose of determining the number of electrons, remember that atoms are neutral, so the number of electrons is equal to the number of protons.

Setup Number of protons = Z, number of neutrons = $A - Z$, and number of electrons = number of protons. Recall that the 14 in carbon-14 is the *mass number*.

Solution

(a) The atomic number is 17, so there are 17 protons. The mass number is 35, so the number of neutrons is $35 - 17 = 18$. The number of electrons equals the number of protons, so there are 17 electrons.

(b) Because the element is again Cl (chlorine), the atomic number is again 17, so there are 17 protons. The mass number is 37, so the number of neutrons is $37 - 17 = 20$. The number of electrons equals the number of protons, so there are 17 electrons, too.

(c) The atomic number of K (potassium) is 19, so there are 19 protons. The mass number is 41, so there are $41 - 19 = 22$ neutrons. There are 19 electrons.

(d) Carbon-14 can also be represented as ^{14}C. The atomic number of carbon is 6, so there are 6 protons and 6 electrons. There are $14 - 6 = 8$ neutrons.

> **Think About It**
> Verify that the number of protons and the number of neutrons for each example sum to the mass number that is given. In part (a), for example, there are 17 protons and 18 neutrons, which sum to give a mass number of 35, the value given in the problem. In part (b), 17 protons + 20 neutrons = 37. In part (c), 19 protons + 22 neutrons = 41. In part (d), 6 protons + 8 neutrons = 14.

Practice Problem A TTEMPT How many protons, neutrons, and electrons are there in an atom of (a) $^{10}_{5}$B, (b) ^{36}Ar, (c) $^{85}_{38}$Sr, and (d) carbon-11?

Practice Problem B UILD Give the correct symbols to identify an atom that contains (a) 4 protons, 4 electrons, and 5 neutrons; (b) 23 protons, 23 electrons, and 28 neutrons; (c) 54 protons, 54 electrons, and 70 neutrons; and (d) 31 protons, 31 electrons, and 38 neutrons.

Practice Problem C ONCEPTUALIZE Based on the numbers of nucleons, write the nuclear symbol for each of the following diagrams:

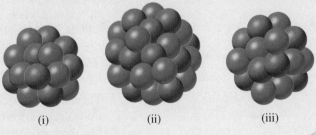

(i)   (ii)   (iii)

Section 2.3 Review

Atomic Number, Mass Number, and Isotopes

2.3.1 How many neutrons are there in an atom of ^{60}Ni?
  (a) 60   (b) 30   (c) 28   (d) 32   (e) 29

2.3.2 An atom with a mass number of 114 has 66 neutrons in its nucleus. What isotope is it?
  (a) ^{114}Cd   (b) ^{114}Dy   (c) ^{66}Cd   (d) ^{66}Dy   (e) ^{180}Ta

2.4   NUCLEAR STABILITY

The nucleus occupies a very small portion of the total volume of an atom, but it contains most of the atom's mass because both the protons and the neutrons reside there. In studying the stability of the atomic nucleus, it is helpful to know something about its density, because it tells us how tightly the particles are packed together. As a sample calculation, let us assume that a nucleus has a radius of 5×10^{-3} pm and a mass of 1×10^{-22} g. These figures correspond roughly to a nucleus containing 30 protons and 30 neutrons. Density is mass/volume, and we can calculate the volume from the known radius (the volume of a sphere is $\frac{4}{3}\pi r^3$, where r is the radius of the sphere). First we convert the picometer units to centimeters. Then, we calculate the density in g/cm^3:

$$r = (5 \times 10^{-3} \text{ pm})\left(\frac{1 \times 10^{-12} \text{ m}}{1 \text{ pm}}\right)\left(\frac{100 \text{ cm}}{1 \text{ m}}\right) = 5 \times 10^{-13} \text{ cm}$$

$$\text{density} = \frac{\text{mass}}{\text{volume}} = \frac{1 \times 10^{-22} \text{ g}}{\frac{4}{3}\pi r^3} = \frac{1 \times 10^{-22} \text{ g}}{\frac{4}{3}\pi(5 \times 10^{-13} \text{ cm})^3}$$

$$= 2 \times 10^{14} \text{ g/cm}^3$$

This is an exceedingly high density. The highest density known for an element is 22.6 g/cm^3, for iridium (Ir). Thus, the average atomic nucleus is roughly 9×10^{12} (or 9 *trillion*) times as dense as the densest element known!

The enormously high density of the nucleus means that some very strong force is needed to hold the particles together so tightly. From *Coulomb's law* we know that like charges repel and unlike charges attract one another. We would thus expect the protons to repel one another strongly, particularly when we consider how close they must be to each other. This indeed is so. However, in addition to the repulsion, there are also short-range attractions between proton and proton, proton and neutron, and neutron and neutron. The stability of any nucleus is determined by the difference between coulombic repulsion and the short-range attraction. If repulsion outweighs attraction, the nucleus disintegrates, emitting particles and/or radiation. If attractive forces prevail, the nucleus is stable.

Patterns of Nuclear Stability

The principal factor that determines whether a nucleus is stable is the *neutron-to-proton ratio (n/p)*. For stable atoms of elements having low atomic number ($\leq$20), the n/p value is close to 1. As the atomic number increases, the neutron-to-proton ratios of the stable nuclei also increase. This deviation at higher atomic numbers arises because more neutrons are needed to counteract the strong repulsion among the protons and stabilize the nucleus. The following rules are useful in gauging whether or not a particular nucleus is expected to be stable:

1. There are more stable nuclei containing 2, 8, 20, 50, 82, or 126 protons or neutrons than there are containing other numbers of protons or neutrons. For example, there are 10 stable isotopes of tin (Sn) with the atomic number 50 and only 2 stable isotopes of antimony (Sb) with the atomic number 51. The numbers 2, 8, 20, 50, 82, and 126 are called *magic numbers*.

2. There are many more stable nuclei with even numbers of both protons and neutrons than with *odd numbers* of these particles (Table 2.2).

3. All isotopes of the elements with atomic numbers higher than 82 are radioactive.

4. All isotopes of technetium (Tc, $Z = 43$) and promethium (Pm, $Z = 61$) are radioactive.

Student Annotation: Of the two stable isotopes of antimony mentioned in rule 1, both have even numbers of neutrons: $^{121}_{51}\text{Sb}$ and $^{123}_{51}\text{Sb}$.

Figure 2.9 shows a plot of the number of neutrons versus the number of protons in various isotopes. The stable nuclei are located in an area of the graph known as the *belt of stability*. Most radioactive nuclei lie outside this belt. Above the belt of stability, the nuclei have higher neutron-to-proton ratios than those within the belt (for the same number of protons).

TABLE 2.2	Number of Stable Isotopes with Even and Odd Numbers of Protons and Neutrons	
Protons	**Neutrons**	**Number of stable isotopes**
Odd	Odd	4
Odd	Even	50
Even	Odd	53
Even	Even	164

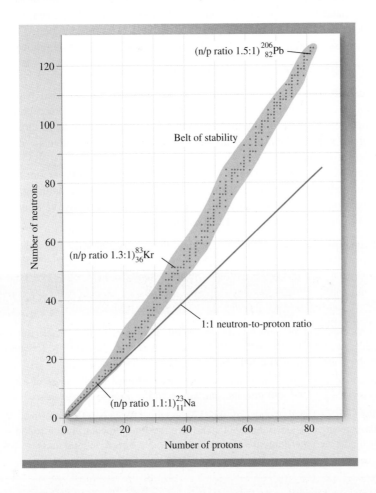

Figure 2.9 Plot of neutrons versus protons for various stable isotopes, represented by dots. The straight line represents the points at which the neutron-to-proton ratio is 1. The shaded area represents the belt of stability. Note that as atomic number increases, the optimal number of neutrons per proton also increases.

Section 2.4 Review

Nuclear Stability

2.4.1   What is the density of the nucleus of an oxygen-16 atom with a nuclear radius of 6.05×10^{-3} pm and a mass of 16 amu?
(a) 1.72×10^{37} g/cm^3 (d) 2.86×10^{13} g/cm^3
(b) 2.86×10^{-17} g/cm^3 (e) 1.05×10^{-11} g/cm^3
(c) 2.86×10^{19} g/cm^3

2.4.2   Which of the following isotopes are predicted to be unstable?
(a) ^{46}Ti   (d) ^{63}Cu
(b) ^{72}Rb   (e) ^{29}Si
(c) ^{20}Ne

2.5   AVERAGE ATOMIC MASS

Atomic mass is the mass of an atom in atomic mass units. Recall that an atomic mass unit is equal to $1.6605378 \times 10^{-24}$ g [◄◄ Section 1.2]. In fact, the amu is defined as one-twelfth the mass of a carbon-12 atom. However, when you look up the atomic mass of carbon in a table such as the one on the inside front cover of this book, you will find that its value is 12.01 amu, not 12.00 amu. The difference arises because most naturally occurring elements (including carbon) have more than one isotope. This means that generally, when we determine the atomic mass of an element, what we are actually measuring is the ***average atomic mass*** of the naturally occurring *mixture* of isotopes. For example, the natural abundances of carbon-12 and carbon-13 are 98.93 percent and 1.07 percent, respectively. The atomic mass of carbon-13 has been determined to be 13.003355 amu. Thus, the average atomic mass of natural carbon can be calculated as follows:

$$(0.9893)(12.00000 \text{ amu}) + (0.0107)(13.003355 \text{ amu}) = 12.01 \text{ amu}$$

In calculations involving percentages, we need to convert each percent abundance to a fractional abundance. For example, 98.93 percent becomes 98.93/100, or 0.9893. Because there are many more carbon-12 atoms than carbon-13 atoms in naturally occurring carbon, the average atomic mass is much closer to the mass of carbon-12 than to that of carbon-13.

When we say that the atomic mass of carbon is 12.01 amu, we are referring to the average value. If we could examine an individual atom of naturally occurring carbon, we would find either an atom of atomic mass of exactly 12 amu or one of 13.003355 amu, but never one of 12.01 amu.

Worked Example 2.2 shows how to calculate the average atomic mass of oxygen.

> **Student Annotation:** Average atomic mass is also referred to as *atomic weight*.

Worked Example 2.2

Oxygen is the most abundant element in both Earth's crust and the human body. The atomic masses of its three stable isotopes, $^{16}_{8}O$ (99.757 percent), $^{17}_{8}O$ (0.038 percent), $^{18}_{8}O$ (0.205 percent), are 15.9949, 16.9991, and 17.9992 amu, respectively. Calculate the average atomic mass of oxygen using the relative abundances given in parentheses.

Strategy Each isotope contributes to the average atomic mass based on its relative abundance. Multiplying the mass of each isotope by its fractional abundance (percent value divided by 100) will give its contribution to the average atomic mass.

Setup Each percent abundance must be converted to a fractional abundance: 99.757 percent to 99.757/100 or 0.99757, 0.038 percent to 0.038/100 or 0.00038, and 0.205 percent to 0.205/100 or 0.00205. Once we find the contribution to the average atomic mass for each isotope, we can then add the contributions together to obtain the average atomic mass.

Solution $(0.99757)(15.9949 \text{ amu}) + (0.00038)(16.9991 \text{ amu}) + (0.00205)(17.9992) = 15.9994 \text{ amu}$

> **Think About It**
> The average atomic mass should be closest to the atomic mass of the most abundant isotope (in this case, oxygen-16) and, to four significant figures, should be the same number that appears in the periodic table on the inside front cover of this book (in this case, 16.00 amu).

Practice Problem ▲TTEMPT The atomic masses of the two stable isotopes of copper, $^{63}_{29}Cu$ (69.17 percent) and $^{65}_{29}Cu$ (30.83 percent), are 62.929599 and 64.927793 amu, respectively. Calculate the average atomic mass of copper.

Practice Problem ⒷUILD The average atomic mass of nitrogen is 14.0067. The atomic masses of the two stable isotopes of nitrogen, ^{14}N and ^{15}N, are 14.003074002 and 15.000108898 amu, respectively. Use this information to determine the percent abundance of each nitrogen isotope.

Practice Problem ⒸONCEPTUALIZE The following diagrams show collections of metal spheres. Each collection consists of two or more different types of metal—represented here by different colors. The masses of metal spheres are as follows: white = 2.3575 g, black = 3.4778 g, and blue = 5.1112 g. For each diagram, determine the average mass of a single metal sphere.

(i)

(ii)

(iii)

Thinking Outside the Box

Measuring Atomic Mass

The most direct and most accurate method for determining atomic and molecular masses is mass spectrometry. In a *mass spectrometer,* such as that depicted, a gaseous sample is bombarded by a stream of high-energy electrons. Collisions between the electrons and the gaseous atoms (or molecules) produce positively charged species, called *ions,* by dislodging an electron from the atoms or molecules. These positive ions (of mass m and charge e) are accelerated as they pass through two oppositely charged plates. The emerging ions are deflected into a circular path by a magnet. The radius of the path depends on the charge-to-mass ratio (i.e., e/m). Ions with a small e/m ratio trace a wider arc than those having a larger e/m ratio, so ions with equal charges but different masses are separated from one another. The mass of each ion (and hence its parent atom or molecule) is determined from the magnitude of its deflection. Eventually, the ions arrive at the detector, which registers a current for each type of ion. The amount of current generated is directly proportional to the number of ions, so it enables us to determine the relative abundance of isotopes.

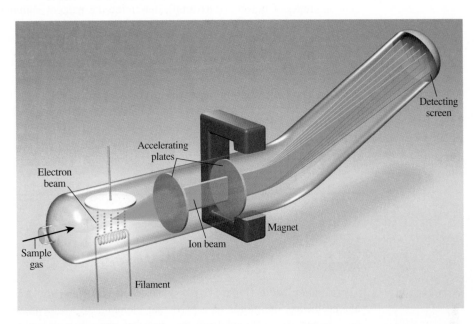

Schematic diagram of one type of mass spectrometer.

The first mass spectrometer, developed in the 1920s by the English physicist F. W. Aston, was crude by today's standards. Nevertheless, it provided indisputable evidence of the existence of isotopes, such as neon-20 (natural abundance 90.48 percent) and neon-22 (natural abundance 9.25 percent). When more sophisticated and sensitive mass spectrometers became available, scientists were surprised to discover that neon has a *third* stable isotope (neon-21) with natural abundance 0.27 percent. This example illustrates how very important experimental accuracy is to a quantitative science like chemistry. Early experiments failed to detect neon-21 because its natural abundance was so small. Only 27 in 10,000 Ne atoms are neon-21.

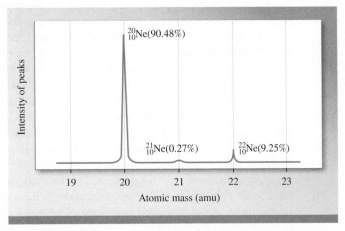

Mass spectrum of neon.

The atomic masses of many isotopes have been accurately determined to five or six significant figures. For most purposes, though, we will use average atomic masses, which are generally given to four significant figures (see the table of atomic masses on the inside front cover). For simplicity, we will usually omit the word *average* when we discuss the atomic masses of the elements.

Section 2.5 Review

Average Atomic Mass

2.5.1   Boron has two naturally occurring isotopes, ^{10}B and ^{11}B, which have masses 10.0129 and 11.0093 amu, respectively. Given the average atomic mass of boron (10.81 amu), determine the percent abundance of each isotope.

(a) 50% ^{10}B, 50% ^{11}B (d) 93% ^{10}B, 7% ^{11}B
(b) 20% ^{10}B, 80% ^{11}B (e) 22% ^{10}B, 78% ^{11}B
(c) 98% ^{10}B, 2% ^{11}B

2.5.2   The two naturally occurring isotopes of antimony, ^{121}Sb (57.21 percent) and ^{123}Sb (42.79 percent), have masses of 120.904 and 122.904 amu, respectively. What is the average atomic mass of Sb?

(a) 121.90 amu   (d) 121.34 amu
(b) 122.05 amu   (e) 122.18 amu
(c) 121.76 amu

2.6   THE PERIODIC TABLE

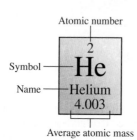

Atomic number
Symbol
Name
Average atomic mass

The ***periodic table*** (Figure 2.10) consists of 118 *elements*. According to Dalton's atomic theory, the atoms of a particular element are all identical; and the atoms of one element are different from the atoms of any other element. Thus, according to Dalton, what makes the elements helium and gold different is that they consist of different kinds of atoms. Further, what makes helium *helium* is that it consists entirely of *identical* helium *atoms*.

More than half of the elements known today were discovered in the nineteenth century. During this time, some elements were found to share similar physical and chemical properties. These similarities prompted chemists to arrange the known elements in vertical columns called "groups" or "families," which ultimately led to the development of the periodic table. Placing the elements in groups according to their properties enabled chemists to organize a growing body of scientific information.

Each element in the periodic table is represented by a tile containing the element's name; the chemical symbol consisting of one or, more commonly, *two* letters; the atomic number, an integer; and the average atomic mass, commonly given to four significant figures. The elements are arranged in ***periods***, horizontal rows, in order of increasing atomic number. The first period contains just two elements, hydrogen (H) and helium (He). The second and third periods each contain eight elements: lithium (Li) through neon (Ne), and sodium (Na) through argon (Ar), respectively. The fourth and fifth periods each contain 18 elements: potassium (K) through krypton (Kr), and rubidium (Rb) through xenon (Xe), respectively.

Most elements can be categorized as metals or nonmetals. One of the properties used to distinguish metals from nonmetals is their ability to conduct heat and/or electricity. A ***metal*** is a good conductor of heat and electricity, whereas a ***nonmetal*** is usually a poor conductor of heat and electricity. A ***metalloid*** is an element with properties that are intermediate between those of metals and nonmetals. Figure 2.10 shows that the majority of known elements are metals; only 17 elements are nonmetals, and fewer than 10 elements are metalloids. Although most sources, including this text,

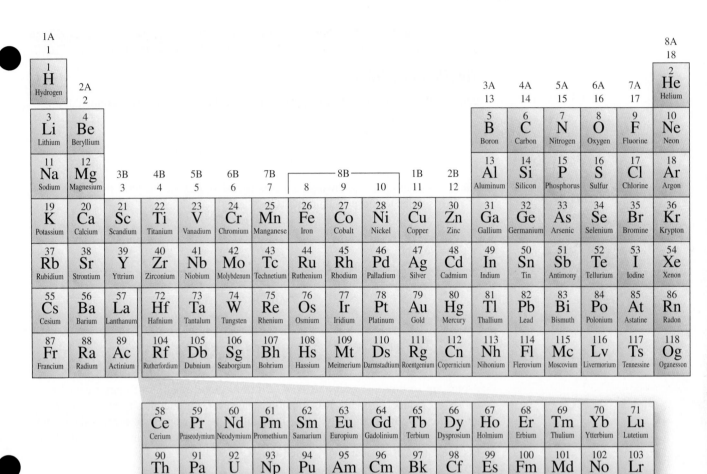

Figure 2.10 The modern periodic table. The elements are arranged according to atomic number (see Section 2.3), which is shown above each element's symbol. With the exception of hydrogen (H), nonmetals appear at the far right of the table. The two rows of metals beneath the main body of the table are set apart to keep the table from being too wide. Actually, cerium (58) should follow lanthanum (57), and thorium (90) should follow actinium (89). The 1–18 group designation has been recommended by the International Union of Pure and Applied Chemistry (IUPAC) but is not yet in wide use. In this text, we generally use the standard U.S. notation for group numbers (1A–8A and 1B–8B). As of this writing, the names of elements 113, 115, 117, and 118 have been proposed and recommended by IUPAC.

designate the elements B, Si, Ge, As, Sb, and Te as metalloids, sources vary for the elements Po and At. In this text, we classify both Po and At as metalloids. From left to right across any period, the physical and chemical properties of the elements change gradually from metallic to nonmetallic.

A vertical column of elements in the periodic table is known as a *group.* Groups of elements sometimes are referred to collectively by their group number (Group 1A, Group 2A, and so on). For convenience, however, some element groups have been given special names. The Group 1A elements (Li, Na, K, Rb, Cs, and Fr) are called the *alkali metals,* and the Group 2A elements (Be, Mg, Ca, Sr, Ba, and Ra) are called the *alkaline earth metals.* Elements in Group 6A (O, S, Se, Te, and Po) are sometimes referred to as the *chalcogens.* Elements in Group 7A (F, Cl, Br, I, and At) are known as the *halogens,* and elements in Group 8A (He, Ne, Ar, Kr, Xe, and Rn) are called the *noble gases.* The elements in Group 1B and Groups 3B–8B collectively are called the *transition elements* or the *transition metals.*

The periodic table is a remarkably useful tool that correlates the properties of elements in a systematic way and helps us to predict an element's behavior. We will look at the periodic table and its development in more detail in Chapter 4.

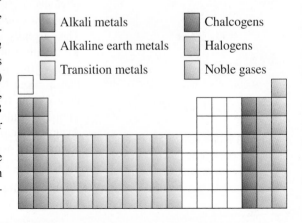

Section 2.6 Review

The Periodic Table

2.6.1   Which of the following series of elemental symbols lists a nonmetal, a metal, and a metalloid?

(a) Ca, Cu, Si   (d) O, Na, S

(b) K, Mg, B   (e) Ag, Cr, As

(c) Br, Ba, Ge

2.6.2   Which of the following elements would you expect to have properties most similar to those of chlorine (Cl)?

(a) Cu   (d) Cr

(b) F   (e) S

(c) Na

2.7   THE MOLE AND MOLAR MASS

Atoms are so tiny that even the *smallest* macroscopic quantity of matter contains an *enormous* number of them. It is important, however, for chemists to know how many atoms there are in a sample. Clearly, it would not be convenient to express the quantities used in the laboratory in terms of the numbers of atoms involved. Instead, chemists use a unit of measurement called the *mole*.

The Mole

If you're buying bagels for yourself, you probably buy them individually, but if you're buying them for everyone in your chemistry class, you will probably want to buy them by the dozen. One dozen bagels contains exactly 12 bagels. In fact, a dozen of anything contains exactly 12 of that thing. Pencils typically come 12 to a box, and 12 such boxes may be shipped to the campus bookstore in a bigger box. The bigger box contains one gross of pencils (144). Whether a dozen or a gross, each is a convenient quantity that contains a reasonable, specific, exact number of items.

Chemists, too, have adopted such a number to make it easy to express the number of atoms in a typical macroscopic sample of matter. Atoms are *so* much smaller than bagels or pencils, though, that the number used by chemists is *significantly* bigger than a dozen or a gross. The quantity used by chemists is the ***mole*** (mol), which is defined as the amount of a substance that contains as many elementary entities (atoms, for example) as there are atoms in exactly 0.012 kg (12 g) of carbon-12. The number of atoms in exactly 12 g of carbon-12, which is determined experimentally, is known as ***Avogadro's number*** (N_A), in honor of Italian scientist Amedeo Avogadro (1776–1856). The currently accepted value of Avogadro's number is 6.0221418×10^{23}, although we usually round it to 6.022×10^{23}. Thus, a dozen doughnuts contains 12 doughnuts; a gross of pencils contains 144 pencils; and a mole of helium gas, an amount that at room temperature and ordinary pressure would fill slightly more than half of a 10-gal fish tank, contains 6.0221418×10^{23} He atoms. Figure 2.11 shows samples containing one mole each of several familiar substances. Note that Avogadro's number has no units. When we convert between moles and numbers of atoms, what we actually use is Avogadro's *constant*, which has units of reciprocal moles: 6.0221418×10^{23} mol^{-1}.

Worked Example 2.3 shows how to convert between moles and atoms.

Figure 2.11 One mole each of several substances: (left to right) aluminum, copper, water, salt, and sugar; and (in balloon) helium.

© McGraw-Hill Education/Charles D. Winters, photographer

Worked Example 2.3

Calcium is the most abundant metal in the human body. A typical human body contains roughly 30 moles of calcium. Determine (a) the number of Ca atoms in 30.00 moles of calcium and (b) the number of moles of calcium in a sample containing 1.00×10^{20} Ca atoms.

Strategy Use Avogadro's constant to convert from moles to atoms and from atoms to moles.

Setup When the number of moles is known, we multiply by Avogadro's constant to convert to atoms. When the number of atoms is known, we divide by Avogadro's constant to convert to moles.

Solution (a) $30.00 \text{ mol Ca} \times \dfrac{6.022 \times 10^{23} \text{ Ca atoms}}{1 \text{ mol Ca}} = 1.807 \times 10^{25}$ Ca atoms

(b) $1.00 \times 10^{20} \text{ Ca atoms} \times \dfrac{1 \text{ mol Ca}}{6.022 \times 10^{23} \text{ Ca atoms}} = 1.66 \times 10^{-4}$ mol Ca

Think About It
Make sure that units cancel properly in each solution and that the result makes sense. In part (a), for example, the number of moles (30) is greater than one, so the number of atoms is greater than Avogadro's number. In part (b), the number of atoms (1×10^{20}) is less than Avogadro's number, so there is less than a mole of substance.

Practice Problem (A)TTEMPT Potassium is the second most abundant metal in the human body. Calculate (a) the number of atoms in 7.31 moles of potassium and (b) the number of moles of potassium that contains 8.91×10^{25} atoms.

Practice Problem (B)UILD Calculate (a) the number of atoms in 1.05×10^{-6} mole of helium and (b) the number of moles of helium that contains 2.33×10^{21} atoms.

Practice Problem (C)ONCEPTUALIZE These diagrams show collections of objects. For each diagram, express the number of objects using units of *dozen* and using units of *gross*. (Report each answer to four significant figures but explain why the answers to this problem actually have more than four significant figures.)

(i)   (ii)   (iii)

Molar Mass

Like the dozen and the gross, which are used to quantify specific, convenient numbers of bagels and pencils, respectively, the mole is used to quantify a specific, convenient number of atoms. Unlike the dozen and the gross, which are arrived at by *counting* the objects they quantify, the mole is not a number that can be arrived at by counting. The number of atoms in a macroscopic sample of matter is simply too big to be counted. Instead, the number of atoms in a quantity of substance is determined by weighing—like nails in a hardware store (Figure 2.12). The number of nails needed to build a fence or a house is fairly large. Therefore, when nails are purchased for such a project, they are not *counted;* but rather, they are *weighed* to determine their number. How many nails there are to a pound depends on the size and type of nail.

For example, if we want to buy 1000 1.5-in 4d common nails, we would not count out a thousand nails. Instead, we would weigh out an amount just over three pounds. Using data from the table in Figure 2.12:

$$1000 \text{ 4d common nails} \times \frac{1 \text{ lb}}{316 \text{ 4d common nails}} = 3.16 \text{ lb}$$

If we were to weigh out something *other* than the calculated amount, we could determine the number of nails using the same conversion factor:

$$5.00 \text{ lb} \times \frac{316 \text{ 4d common nails}}{1 \text{ lb}} = 1580 \text{ 4d common nails}$$

Figure 2.12 Bulk nails are sold by the pound. How many nails there are in a pound depends on the size and type of nail.

Photo: © Ryan McVay/Getty Images

Type of nail	Size (in)	Nails/lb
3d box	1.5	635
6d box	2	236
10d box	3	94
4d casing	1.5	473
8d casing	2.5	145
2d common	1	876
4d common	1.5	316
6d common	2	181
8d common	2.5	106

Note that the same mass of a different type of nail would contain a different number of nails. For example, we might weigh out 5.00 lb of 3-in 10d box nails:

$$5.00 \text{ lb} \times \frac{94 \text{ 10d box nails}}{1 \text{ lb}} = 470 \text{ 10d box nails}$$

The number of nails in each case is determined by weighing a sample of nails, and the number of nails in a given mass depends on the type of nail. Numbers of atoms, too, are determined by *weighing* a sample; and the number of atoms per unit mass also depends on the *type* of atom in the sample.

The ***molar mass*** (***ℳ***) of a substance is the mass in grams of one mole of the substance. By definition, the mass of a mole of carbon-12 is exactly 12 g. Note that the molar mass of carbon-12 is numerically equal to its atomic mass.

Mass of 1 carbon-12 atom: exactly 12 amu

Mass of 1 mole of carbon-12: exactly 12 g

The molar mass of *naturally occurring* carbon, which contains a small amount of carbon-13, is numerically equal to the *average* atomic mass of carbon: 12.01 g. Likewise, the average atomic mass of helium is 4.003 amu and its molar mass is 4.003 g; the average atomic mass of calcium is 40.08 amu and its molar mass is 40.08 g. In fact, any element's molar mass in *grams* is numerically equal to its average atomic mass in *atomic mass units*.

Although the term *molar mass* specifies the mass of one mole of a substance, making the appropriate units simply grams (g), we usually express molar masses in units of grams per mole (g/mol) to facilitate cancellation of units in calculations.

Worked Example 2.4 shows how to use molar mass to convert between mass and moles.

Student Hot Spot

Student data indicate you may struggle with Avogadro's number and conversions of moles and atoms. Log in to Connect to view additional Learning Resources on this topic.

Worked Example 2.4

Determine (a) the number of moles of C in 25.00 g of carbon, (b) the number of moles of He in 10.50 g of helium, and (c) the number of moles of Na in 15.75 g of sodium.

Strategy Molar mass of an element is numerically equal to its average atomic mass. Use the molar mass for each element to convert from mass to moles.

Setup (a) The molar mass of carbon is 12.01 g/mol. (b) The molar mass of helium is 4.003 g/mol. (c) The molar mass of sodium is 22.99 g/mol.

Solution

(a) $25.00 \text{ g C} \times \dfrac{1 \text{ mol C}}{12.01 \text{ g C}} = 2.082 \text{ mol C}$ (c) $15.75 \text{ g Na} \times \dfrac{1 \text{ mol Na}}{22.99 \text{ g Na}} = 0.6851 \text{ mol Na}$

(b) $10.50 \text{ g He} \times \dfrac{1 \text{ mol He}}{4.003 \text{ g He}} = 2.623 \text{ mol He}$

Think About It
Always double-check unit cancellations in problems such as these—errors are common when molar mass is used as a conversion factor. Also make sure that the results make sense. For example, in the case of part (c), a mass smaller than the molar mass corresponds to *less* than a mole.

Practice Problem ATTEMPT Determine the number of moles in (a) 12.25 g of argon (Ar), (b) 0.338 g of gold (Au), and (c) 59.8 g of mercury (Hg).

Practice Problem BUILD Determine the mass in grams of (a) 2.75 moles of calcium, (b) 0.075 mole of helium, and (c) 1.055×10^{-4} mole of potassium.

Practice Problem CONCEPTUALIZE Plain doughnuts from a particular bakery have an average mass of 32.6 g, whereas jam-filled doughnuts from the same bakery have an average mass of 40.0 g. (a) Determine the mass of a dozen plain doughnuts and the mass of a dozen jam-filled doughnuts. (b) Determine the number of doughnuts in a kilogram of plain and the number in a kilogram of jam-filled. (c) Determine the mass of plain doughnuts that contains the same number of doughnuts as a kilogram of jam-filled. (d) Determine the total mass of a dozen doughnuts consisting of three times as many plain as jam-filled.

Interconverting Mass, Moles, and Numbers of Atoms

Molar mass is the conversion factor that we use to convert from mass (*m*) to moles (*n*), and vice versa. We use Avogadro's constant to convert from number of moles (*n*) to number of atoms (*N*), and vice versa. The flowchart in Figure 2.13 summarizes the operations involved in these conversions.

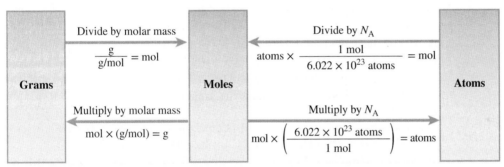

Figure 2.13 Flowchart for conversions among mass, moles, and number of atoms.

Worked Example 2.5 lets you practice these conversions.

Worked Example 2.5

Determine (a) the number of C atoms in 0.515 g of carbon, and (b) the mass of helium that contains 6.89×10^{18} He atoms.

Strategy Use the conversions depicted in Figure 2.13 to convert (a) from grams to moles to atoms, and (b) from atoms to moles to grams.

Setup (a) The molar mass of carbon is 12.01 g/mol. (b) The molar mass of helium is 4.003 g/mol. $N_A = 6.022 \times 10^{23}$.

Solution

(a) $0.515 \text{ g C} \times \dfrac{1 \text{ mol C}}{12.01 \text{ g C}} \times \dfrac{6.022 \times 10^{23} \text{ C atoms}}{1 \text{ mol C}} = 2.58 \times 10^{22} \text{ C atoms}$

(b) $6.89 \times 10^{18} \text{ He atoms} \times \dfrac{1 \text{ mol He}}{6.022 \times 10^{23} \text{ He atoms}} \times \dfrac{4.003 \text{ g He}}{1 \text{ mol He}} = 4.58 \times 10^{-5} \text{ g He}$

Think About It
A ballpark estimate of your result can help you prevent common errors. For example, the mass in part (a) is smaller than the molar mass of carbon. Therefore, you should expect a number of atoms *smaller* than Avogadro's number. Likewise, the number of atoms in part (b) is smaller than Avogadro's number. Therefore, you should expect a mass of helium *smaller* than the molar mass of helium.

(Continued on next page)

Practice Problem **A**TTEMPT   Determine (a) the number of atoms in 105.5 g of gold, and (b) the mass of calcium that contains 8.075×10^{12} Ca atoms.

Practice Problem **B**UILD   Determine the mass of calcium that contains the same number of atoms as 81.06 g of helium, and (b) the number of gold atoms that has the same mass as 7.095×10^{31} argon atoms.

Practice Problem **C**ONCEPTUALIZE   A particular commemorative set of coins contains two 1.00-oz silver coins and three 0.500-oz gold coins. (a) How many gold coins are there in 49.0 lb of coin sets? (b) How many silver coins are there in a collection of sets that has a total mass of 63.0 lb?
(c) What is the total mass (in lb) of a collection of sets that contains 93 gold coins? (d) What is the mass of silver coins (in lb) in a collection of coin sets that contains 9.00 lb of gold coins?

Section 2.7 Review

The Mole and Molar Mass

2.7.1   How many moles of arsenic are there in 6.50 g of arsenic?
  (a) 487 mol   (c) 11.5 mol   (e) 0.163 mol
  (b) 8.68×10^{-2} mol   (d) 2.05×10^{-3} mol

2.7.2   How many atoms are there in 3.559×10^{-6} mol of krypton?
  (a) 1.691×10^{29} atoms   (c) 5.912×10^{-30} atoms   (e) 4.666×10^{-19} atoms
  (b) 2.143×10^{17} atoms   (d) 2.143×10^{18} atoms

2.7.3   How many atoms are there in 30.1 g of magnesium?
  (a) 4.41×10^{26} atoms   (c) 2.06×10^{-24} atoms   (e) 7.46×10^{23} atoms
  (b) 1.21×10^{-21} atoms   (d) 7.46×10^{22} atoms

2.7.4   What mass of mercury contains the same number of atoms as 90.15 g of helium?
  (a) 8.91 g Hg   (c) 0.112 g Hg   (e) 1.80 g Hg
  (b) 4.518×10^{3} g Hg   (d) 7.24×10^{4} g Hg

Learning Outcomes

- Understand the concept of the atom and the nature of an element.
- Recognize the importance of experiments conducted by Thomson, Millikan, Röntgen, and Rutherford in regard to understanding the nature and structure of atoms.
- Understand the different types of radiation that radioactive substances can produce.
- Identify the location and physical properties of electrons, protons, and neutrons in atoms.
- Understand the nature and importance of isotopes.
- Calculate the mass number of an isotope.
- Utilize the mass number of an isotope to determine the number of electrons, protons, or neutrons, given other relevant information.

- List the rules for nuclear stability and use them to predict whether a particular nucleus is stable.
- Understand the nature of the atomic mass scale.
- Calculate the average atomic mass of an element given the atomic mass and relative abundance of each of its naturally occurring isotopes.
- Understand the concept of the mole.
- Use the relationships between Avogadro's number, moles, molar mass, and grams to conduct calculations.
- Interconvert between mass, moles, and number of atoms.

Chapter Summary

SECTION 2.1
- *Atoms* are the tiny building blocks of all matter. *Elements* are substances that cannot be broken down into simpler substances.

SECTION 2.2
- Studies with *radiation* and *cathode ray tubes* indicated that atoms contained subatomic particles, one of which was the *electron.*
- Experiments with *radioactivity* have shown that some atoms give off different types of radiation, called *α rays, β rays, X rays,* and *γ rays.* Alpha rays are composed of *α particles.* Beta rays are composed of *β particles,* which are actually electrons. X rays and γ rays (gamma rays) are high-energy radiation.
- Most of the mass of an atom resides in a tiny, dense region known as the *nucleus.* The nucleus contains positively charged particles called *protons* and electrically neutral particles called *neutrons.*

SECTION 2.3
- The *atomic number (Z)* is the number of protons in the nucleus of an atom. Atomic number determines the identity of the atom.
- The *mass number (A)* is the sum of the protons and neutrons in the nucleus. Protons and neutrons are referred to collectively as *nucleons.*
- Atoms with the same atomic number but different mass numbers are called *isotopes.*

SECTION 2.4
- Stable nuclei with low atomic numbers have neutron-to-proton ratios close to 1. Heavier stable nuclei have higher ratios. Nuclear stability is favored by certain numbers of nucleons including even numbers and "magic" numbers.

SECTION 2.5
- *Atomic mass* is the mass of an atom in atomic mass units. The periodic table contains the *average atomic mass* (sometimes called the *atomic weight*) of each element.

SECTION 2.6
- The *periodic table* arranges the elements in rows (*periods*) and columns (*groups*). Elements in the same group exhibit similar properties.
- *Metals* are good conductors of electricity; *nonmetals* are not good conductors of electricity. *Metalloids* have properties intermediate between those of metals and nonmetals.
- Some of the groups have special names including *alkali metals* (Group 1A, except hydrogen), *alkaline earth metals* (Group 2A), *chalcogens* (Group 6A), *halogens* (Group 7A), *noble gases* (Group 8A), and *transition metals,* also known as *transition elements* (Group 1B and Groups 3B–8B).

SECTION 2.7
- A *mole* is the amount of a substance that contains 6.0221418×10^{23} [*Avogadro's number (N_A)*] of elementary particles (such as atoms). Avogadro's constant, with units of reciprocal moles, is used to convert between moles and numbers of atoms.
- *Molar mass ($\mathcal{M}$)* is the mass of one mole of a substance, usually expressed in *grams.* The molar mass of an element in grams is numerically equal to the *atomic* mass of the element in *amu.*
- Molar mass and Avogadro's constant can be used to interconvert among *mass, moles,* and *number of atoms.*

Key Words

Alkali metal, 53
Alkaline earth metal, 53
Alpha (α) particle, 42
Alpha (α) rays, 42
Atom, 39
Atomic mass, 50
Atomic number (Z), 46
Average atomic mass, 50
Avogadro's number (N_A), 54
Beta (β) particle, 42
Beta (β) rays, 42
Cathode rays, 41
Chalcogens, 53

Electron, 41
Element, 39
Gamma (γ) rays, 43
Group, 53
Halogens, 53
Isotope, 46
Mass number (A), 46
Metal, 52
Metalloid, 52
Molar mass ($\mathcal{M}$), 56
Mole, 54
Neutron, 45
Noble gases, 53

Nonmetal, 52
Nucleons, 46
Nucleus, 43
Period, 52
Periodic table, 52
Proton, 43
Radiation, 40
Radioactivity, 42
Transition elements, 53
Transition metal, 53
X rays, 42

Key Skills

Interconversion Among Mass, Moles, and Numbers of Atoms

Many problems require you to convert a mass to a number of *moles* or to a number of *atoms*. Conversely, you may need to convert from a number of moles or a number of atoms to a *mass*. The conversion factors necessary for these operations are molar mass (in grams per mole) and Avogadro's constant, N_A (in number of atoms per mole). It's easy to make the mistake of multiplying when you should divide, or vice versa. Track your units carefully to make sure that you have not made this common error. Individual flowcharts for each operation are shown in the examples below.

One of the most common tasks you will encounter is the conversion of mass to moles:

How many moles of helium does a 71.3-g sample of helium contain?

$$\text{mass (g)} \times \frac{1}{\text{molar mass}}\left(\frac{\text{mol}}{\text{g}}\right) = \text{moles}$$

From the periodic table, we get the molar mass of helium: 4.003 g/mol:

$$71.3\ \text{g} \times \left(\frac{1\ \text{mol}}{4.003\ \text{g}}\right) = 17.8\ \text{mol}$$

You may also be asked to convert from the mass of a sample to the number of atoms it contains:

How many helium atoms does a 197-g sample of helium contain?

$$\text{mass (g)} \times \frac{1}{\text{molar mass}}\left(\frac{\text{mol}}{\text{g}}\right) \times \text{moles} \times N_A\left(\frac{\text{atoms}}{\text{mol}}\right) = \text{atoms}$$

Again, we use the molar mass of helium, 4.003 g/amu. Avogadro's constant is 6.0221418×10^{23} atoms/mol. (Because the original mass has only three significant figures, we can use 6.022×10^{23} atoms/mol in this calculation.)

$$197\ \text{g} \times \left(\frac{1\ \text{mol}}{4.003\ \text{g}}\right) = 49.2\ \text{mol} \times \left(\frac{6.022 \times 10^{23}\ \text{atoms}}{\text{mol}}\right) = 2.96 \times 10^{25}\ \text{atoms}$$

What number of moles of helium contains 8.80×10^{27} helium atoms?

| atoms | $\times$ | $\dfrac{1}{N_A}\left(\dfrac{mol}{atoms}\right)$ | $=$ | moles |

In this case also, with only three significant figures in the number of atoms, we can use 6.022×10^{23} atoms/mol for Avogadro's number.

| 8.80×10^{27} atoms | $\times$ | $\dfrac{1\ mol}{6.022 \times 10^{23}\ atoms}$ | $=$ | 1.46×10^4 mol |

What is the mass of 8.80×10^{27} helium atoms?

| moles | $\times$ | molar mass $\left(\dfrac{g}{mol}\right)$ | $=$ | mass (g) |

Again, we use the molar mass of helium:

| 1.46×10^4 mol | $\times$ | $\dfrac{4.003\ g}{1\ mol}$ | $=$ | 5.85×10^4 g |

Key Skills Problems

2.1
Convert 0.189 g Cu to moles.
(a) 12.0 mol (b) 0.19 mol (c) 2.97×10^{-3} mol
(d) 1.14×10^{-25} mol (e) 1.14×10^{23} mol

2.2
How many Zn atoms are in 545 g Zn?
(a) 8.33 atoms (b) 8 atoms (c) 1.38×10^{-23} atoms
(d) 5.02×10^{24} atoms (e) 3.28×10^{26} atoms

2.3
One container holds a mixture of metals containing 72.09 g of sodium and 6.99 g of strontium. Another container holds 47.87 g of titanium. What mass of calcium would have to be added to the second container for it to hold the same number of atoms as the first container?
(a) 40.08 g (b) 113.6 g (c) 2.834 g (d) 3.215 g
(e) 88.80 g

2.4
Arrange the following in order of increasing mass:
  i. 1.85×10^{24} Fe atoms
  ii. 7.68×10^{23} Ba atoms
  iii. 9.68×10^{23} Pd atoms
  iv. 1.22×10^{24} Kr atoms
(a) i < iii < ii < iv (b) ii < i < iv < iii (c) i < iv < iii < ii
(d) iv < iii < i < ii (e) ii < iv < iii < i

Questions and Problems

SECTION 2.1: ATOMS FIRST

Review Questions

2.1 Define the terms *atom* and *element*.
2.2 Use a familiar macroscopic example as an analogy to describe Dalton's atomic theory. Explain why a substance such as clay or oil is not useful for an analogy of this type.

SECTION 2.2: SUBATOMIC PARTICLES AND ATOMIC STRUCTURE

Review Questions

2.3 Define the following terms: (a) α particle, (b) β particle, (c) γ ray, (d) X ray.
2.4 Name the types of radiation known to be emitted by radioactive elements.
2.5 Compare the properties of the following: α particles, cathode rays, protons, neutrons, and electrons.
2.6 Describe the contributions of the following scientists to our knowledge of atomic structure: J. J. Thomson, R. A. Millikan, Ernest Rutherford, and James Chadwick.
2.7 Describe the experimental basis for believing that the nucleus occupies a very small fraction of the volume of the atom.

Computational Problems

2.8 The diameter of a neon atom is about 1.5×10^2 pm. If we could line up neon atoms side by side in contact with one another, how many atoms would it take to span a distance of 1 cm?
2.9 The radius of an atom is on the order of 10,000 times as large as its nucleus. If the atom were enlarged such that the radius of its nucleus were 1.0 in, what would be the radius of the atom in miles, and in meters? (5280 ft = 1 mi = 1609 m)

SECTION 2.3: ATOMIC NUMBER, MASS NUMBER, AND ISOTOPES

Review Questions

2.10 Use the argon-40 isotope to define atomic number and mass number. Why does knowledge of the atomic number enable us to deduce the number of electrons present in an atom?
2.11 Why do all atoms of an element have the same atomic number, although they may have different mass numbers?
2.12 What do we call atoms of the same elements with different mass numbers?
2.13 Explain the meaning of each term in the symbol $_Z^A X$.

Computational Problems

2.14 What is the mass number of an iron atom that has 31 neutrons?
2.15 Calculate the number of neutrons of ^{243}Pu.
2.16 For each of the following species, determine the number of protons and the number of neutrons in the nucleus: $_3^6$Li, $_{13}^{28}$Al, $_{13}^{29}$Al, $_{23}^{50}$V, $_{34}^{77}$Se, $_{77}^{193}$Ir.
2.17 Indicate the number of protons, neutrons, and electrons in each of the following species: $_7^{15}$N, $_{25}^{55}$Mn, $_{35}^{81}$Br, $_{82}^{207}$Pb, $_{49}^{115}$In, $_{40}^{94}$Zr
2.18 Write the appropriate symbol for each of the following isotopes: (a) $Z = 19$, $A = 41$; (b) $Z = 46$, $A = 106$; (c) $Z = 52$, $A = 125$; (d) $Z = 38$, $A = 88$.
2.19 Write the appropriate symbol for each of the following isotopes: (a) $Z = 75$, $A = 187$; (b) $Z = 83$, $A = 209$; (c) $Z = 33$, $A = 75$; (d) $Z = 93$, $A = 236$.
2.20 Determine the mass number of (a) a beryllium atom with 5 neutrons, (b) a sodium atom with 12 neutrons, (c) a selenium atom with 44 neutrons, and (d) a gold atom with 118 neutrons.
2.21 Determine the mass number of (a) a chlorine atom with 18 neutrons, (b) a phosphorus atom with 17 neutrons, (c) an antimony atom with 70 neutrons, and (d) a palladium atom with 59 neutrons.
2.22 The following radioactive isotopes are used in medicine for things such as imaging organs, studying blood circulation, and treating cancer. Give the number of neutrons present in each isotope: ^{198}Au, ^{47}Ca, ^{60}Co, ^{18}F, ^{125}I, ^{131}I, ^{42}K, ^{43}K, ^{24}Na, ^{32}P, ^{85}Sr, ^{99}Tc.

SECTION 2.4: NUCLEAR STABILITY

Review Questions

2.23 State the general rules for predicting nuclear stability.
2.24 What is the belt of stability?
2.25 Why is it impossible for the isotope $_2^2$He to exist?

Computational Problems

2.26 The radius of a uranium-235 nucleus is about 7.0×10^{-3} pm. Calculate the density of the nucleus in g/cm^3. (Assume the atomic mass is 235 amu.)
2.27 For each pair of isotopes listed, predict which one is less stable: (a) $_3^6$Li or $_3^9$Li, (b) $_{11}^{23}$Na or $_{11}^{25}$Na, (c) $_{20}^{48}$Ca or $_{21}^{48}$Sc.
2.28 For each pair of elements listed, predict which one has more stable isotopes: (a) Pd or Ag, (b) Na or Mg, (c) I or Xe.
2.29 In each pair of isotopes shown, indicate which one you would expect to be radioactive: (a) $_{10}^{20}$Ne or $_{10}^{17}$Ne, (b) $_{20}^{40}$Ca or $_{20}^{45}$Ca, (c) $_{42}^{95}$Mo or $_{43}^{92}$Tc.

2.30   In each pair of isotopes shown, indicate which one you would expect to be radioactive: (a) $^{195}_{80}$Hg or $^{196}_{80}$Hg, (b) $^{209}_{83}$Bi or $^{242}_{96}$Cm, (c) $^{24}_{13}$Al or $^{88}_{38}$Sr.

SECTION 2.5: AVERAGE ATOMIC MASS

Review Questions

2.31   What is the mass (in amu) of a carbon-12 atom? Why is the atomic mass of carbon listed as 12.01 amu in the table on the inside front cover of this book?

2.32   Explain clearly what is meant by the statement, "The atomic mass of gold is 197.0 amu."

2.33   What information would you need to calculate the average atomic mass of an element?

Computational Problems

2.34   The atomic masses of $^{35}_{17}$Cl (75.78 percent) and $^{37}_{17}$Cl (24.22 percent) are 34.969 and 36.966 amu, respectively. Calculate the average atomic mass of chlorine. The percentages in parentheses denote the relative abundances.

2.35   The atomic masses of ^{20}Ne (90.48 percent), ^{21}Ne (0.27 percent), and ^{22}Ne (9.25 percent) are 19.9924356, 20.9938428, and 21.9913831 amu, respectively. Calculate the average atomic mass of neon. The percentages in parentheses denote the relative abundances.

2.36   The atomic masses of ^{138}La and ^{139}La are 137.907105 and 138.906347 amu, respectively. Calculate the natural abundances of these two isotopes. To eight significant figures, the average atomic mass of lanthanum is 138.90545 amu.

2.37   The atomic masses of ^{6}Li and ^{7}Li are 6.0151 amu and 7.0160 amu, respectively. Calculate the natural abundances of these two isotopes. The average atomic mass of Li is 6.941 amu.

2.38   The element rubidium has two naturally occurring isotopes. The atomic mass of ^{85}Rb (72.17 percent abundant) is 84.911794 amu. Determine the atomic mass of ^{87}Rb (27.83 percent abundant). The average atomic mass of Rb is 85.4678 amu.

2.39   There are three naturally occurring isotopes of the element magnesium. The atomic masses of ^{25}Mg (10.00 percent abundant) and ^{26}Mg (11.01 percent abundant) are 24.9858374 and 25.9825937 amu, respectively. What is the atomic mass of ^{24}Mg (78.99 percent abundant) given the average atomic mass of Mg is 24.3050 amu?

SECTION 2.6: THE PERIODIC TABLE

Review Questions

2.40   What is the periodic table, and what is its significance in the study of chemistry?

2.41   Write the names and symbols for four elements in each of the following categories: (a) nonmetal, (b) metal, (c) metalloid.

2.42   Give two examples of each of the following: (a) alkali metals, (b) alkaline earth metals, (c) halogens, (d) noble gases, (e) chalcogens, (f) transition metals.

2.43   The explosion of an atomic bomb in the atmosphere releases many radioactive isotopes into the environment. One of the isotopes is ^{90}Sr. Via a relatively short food chain, it can enter the human body. Considering the position of strontium in the periodic table, explain why it is particularly harmful to humans.

Conceptual Problems

2.44   Elements whose names end with -ium are usually metals; sodium is one example. Identify a nonmetal whose name also ends with -ium.

2.45   Describe the changes in properties (from metals to nonmetals or from nonmetals to metals) as we move (a) down a periodic group and (b) across the periodic table from left to right.

2.46   Consult the WebElements Periodic Table of the Elements (http://www.webelements.com) to find (a) two metals less dense than water, (b) two metals more dense than mercury, (c) the densest known solid metallic element, and (d) the densest known solid nonmetallic element.

2.47   Group the following elements in pairs that you would expect to show similar properties: K, F, P, Na, Cl, and N.

2.48   Group the following elements in pairs that you would expect to show similar chemical properties: I, Ba, O, Br, S, and Ca.

2.49   Write the symbol for each of the following biologically important elements in the given periodic table: iron (present in hemoglobin for transporting oxygen), iodine (present in the thyroid gland), sodium (present in intracellular and extracellular fluids), phosphorus (present in bones and teeth), sulfur (present in proteins), and magnesium (present in chlorophyll).

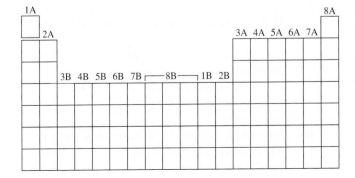

SECTION 2.7: THE MOLE AND MOLAR MASS

Review Questions

2.50   Define the term *mole*. What is the unit for mole in calculations? What does the mole have in common with the dozen and the gross? What does Avogadro's number represent?

2.51   What is the molar mass of an element? What units are commonly used for molar mass?

Computational Problems

2.52   Earth's population is about 7.0 billion. Suppose that every person on Earth participates in a process of counting identical particles at the rate of two particles per second. How many years would it take to count 6.0×10^{23} particles? Assume that there are 365 days in a year.

2.53   The diameter of a human hair is 25.4 μm. If atoms with a diameter of 121 pm were aligned side by side across the hair, how many atoms would be required?

2.54   How many atoms are there in 6.30 moles of selenium (Se)?

2.55   How many moles of nickel (Ni) atoms are there in 9.00×10^9 (9 billion) Ni atoms?

2.56   How many moles of strontium (Sr) atoms are in 93.7 g of Sr?

2.57   How many grams of platinum (Pt) are there in 26.4 moles of Pt?

2.58   What is the mass in grams of a single atom of each of the following elements: (a) Os, (b) Kr?

2.59   What is the mass in grams of a single atom of each of the following elements: (a) Sb, (b) Pd?

2.60   What is the mass in grams of 2.00×10^{12} tin (Sn) atoms?

2.61   How many atoms are present in 4.09 g of scandium (Sc)?

2.62   Which of the following has more atoms: 4.56 g of helium atoms or 2.36 g of manganese atoms?

2.63   Which of the following has a greater mass: 173 atoms of gold or 7.5×10^{-22} mole of silver?

ADDITIONAL PROBLEMS

2.64   A sample of uranium is found to be losing mass gradually. Explain what is happening to the sample.

2.65   The element francium (Fr) was the last element of the periodic table discovered in nature. Because of its high radioactivity, it is estimated that no more than 30 g of francium exists at any given time throughout the Earth's crust. Assuming a molar mass of 223 g/mol for francium, what is the approximate number of francium atoms in the Earth's crust?

2.66   One isotope of a metallic element has 35 neutrons in the nucleus. The neutral atom has 30 electrons. Write the symbol for this atom.

2.67   An isotope of a nonmetallic element has mass number 131. An atom of this isotope contains 54 electrons. Write the symbol for this atom.

2.68   Using the information provided in the table, write a symbol for each atom. If there is not enough information, state what additional information is needed to write a correct symbol for that atom.

	Atom of element			
	A	B	C	D
Number of electrons	6	11	29	
Number of protons	6		29	36
Number of neutrons	6	7		47

2.69   Using the information provided in the table, write a symbol for each atom. If there is not enough information, state what additional information is needed to write a correct symbol for that atom.

	Atom of element			
	A	B	C	D
Number of electrons	10		21	50
Number of protons		75	21	50
Number of neutrons	12	110	21	

2.70   Which of the following symbols provides more information about the atom: ^{23}Na or $_{11}$Na? Explain.

2.71   Discuss the significance of assigning an atomic mass of exactly 12 amu to the carbon-12 isotope.

2.72   List the elements that exist as gases at room temperature. (*Hint:* Most of these elements can be found in Groups 5A, 6A, 7A, and 8A.)

2.73   For the noble gases (the Group 8A elements) $^4_2$He, $^{20}_{10}$Ne, $^{40}_{18}$Ar, $^{84}_{36}$Kr, $^{132}_{54}$Xe (a) determine the number of protons and neutrons in the nucleus of each atom, and (b) determine the ratio of neutrons to protons in the nucleus of each atom. Describe any general trend you discover in the way this ratio changes with increasing atomic number.

2.74   The carat is the unit of mass used by jewelers. One carat is exactly 200 mg. How many carbon atoms are present in a 1.5-carat diamond?

2.75   In the geologic record of Earth, the disappearance of the dinosaurs roughly 65 million years ago is marked by a thin line known as the K-T boundary. Because the K-T boundary contains an unusually high concentration of the element *iridium,* which is ordinarily quite rare in Earth's crust, geologists have deduced that the mass extinction was caused by the catastrophic impact of an asteroid. The two naturally occurring isotopes of iridium are ^{191}Ir and ^{193}Ir, with atomic masses of 190.960584 and 192.962917 amu, respectively; and abundances of 37.3 percent and 62.7 percent, respectively. Determine the average atomic mass of iridium and compare your result with the atomic mass given in the periodic table.

2.76   One atom of a particular element with only one naturally occurring isotope has a mass of 3.818×10^{-23} g. What element is this?

2.77   Identify each of the following elements: (a) a halogen containing 53 electrons, (b) a radioactive noble gas with 86 protons, (c) a Group 6A element with 34 electrons, (d) an alkali metal that contains 11 electrons, (e) a Group 4A element that contains 82 electrons.

2.78   Show the locations of (a) alkali metals, (b) alkaline earth metals, (c) the halogens, and (d) the noble gases in the given outline of a periodic table. Also draw dividing lines between metals and metalloids and between metalloids and nonmetals.

2.79   While most isotopes of light elements such as oxygen and phosphorus contain relatively equal amounts of protons and neutrons in the nucleus, recent results indicate that a new class of isotopes called neutron-rich isotopes can be prepared. These neutron-rich isotopes push the limits of nuclear stability as the large numbers of neutrons approach the "neutron drip line." Neutron-rich isotopes may play a critical role in the nuclear processes of stars. Determine the number of neutrons in the following neutron-rich isotopes: (a) ^{40}Mg, (b) ^{44}Si, (c) ^{48}Ca, (d) ^{43}Al.

2.80   Fill in the blanks in the table:

Symbol		$^{121}_{51}$Sb	
Protons	14		
Neutrons	15		117
Electrons			79

2.81   Fill in the blanks in the table:

Symbol	^{101}Ru		
Protons			62
Neutrons		108	88
Electrons		73	

2.82   (a) Describe Rutherford's experiment and how the results revealed the nuclear structure of the atom. (b) Consider the ^{23}Na atom. Given that the radius and mass of the nucleus are 3.04×10^{-15} m and 3.82×10^{-23} g, respectively, calculate the density of the nucleus in g/cm^3. The radius of a ^{23}Na atom is 186 pm. Calculate the density of the space occupied by the electrons outside the nucleus in the sodium atom. Do your results support Rutherford's model of an atom? (The volume of a sphere of radius r is $\frac{4}{3}\pi r^3$.)

2.83   A cube made of platinum (Pt) has an edge length of 1.0 cm. (a) Calculate the number of Pt atoms in the cube. (b) Atoms are spherical in shape. Therefore, the Pt atoms in the cube cannot fill all the available space. If only 74 percent of the space inside the cube is taken up by Pt atoms, calculate the radius in picometers of a Pt atom. The density of Pt is 21.45 g/cm^3, and the mass of a single Pt atom is 3.240×10^{-22} g. (The volume of a sphere of radius r is $\frac{4}{3}\pi r^3$.)

2.84   Compare the number of significant figures given for the atomic masses of aluminum, bismuth, lead, and molybdenum on webelements.com. Explain why their atomic masses are reported to such different numbers of significant figures.

Answers to In-Chapter Materials

PRACTICE PROBLEMS

2.1A (a) p = 5, n = 5, e = 5. (b) p = 18, n = 18, e = 18. (c) p = 38, n = 47, e = 38. (d) p = 6, n = 5, e = 6. **2.1B** (a) $^9_4$Be, (b) $^{51}_{23}$V, (c) $^{124}_{54}$Xe, (d) $^{69}_{31}$Ga. **2.2A** 63.55 amu. **2.2B** 99.64% ^{14}N, 0.36% ^{15}N. **2.3A** (a) 4.40×10^{24} atoms K, (b) 1.48×10^2 mol K. **2.3B** (a) 6.32×10^{17} atoms He, (b) 3.87×10^{-3} mol He. **2.4A** (a) 0.3066 mol Ar, (b) 1.72×10^{-3} mol Au, (c) 0.298 mol Hg. **2.4B** (a) 1.10×10^2 g Ca, (b) 0.30 g He, (c) 4.125×10^{-3} g K. **2.5A** (a) 3.225×10^{23} atoms Au, (b) 5.374×10^{-10} g Ca. **2.5B** (a) 811.6 g Ca, (b) 1.439×10^{31} atoms Au.

SECTION REVIEW

2.3.1 d. **2.3.2** a. **2.4.1** d. **2.4.2** b. **2.5.1** b. **2.5.2** c. **2.6.1** c. **2.6.2** b. **2.7.1** b. **2.7.2** b. **2.7.3** e. **2.7.4** b.

Chapter 3

Quantum Theory and the Electronic Structure of Atoms

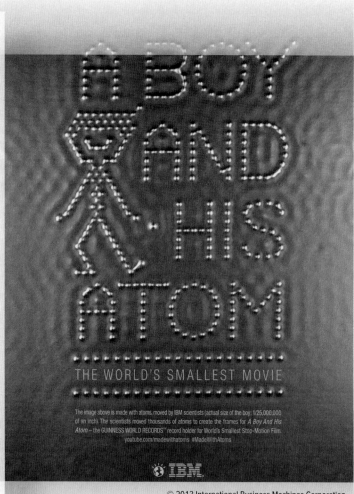

THE WORLD'S SMALLEST MOVIE

The image above is made with atoms, moved by IBM scientists (actual size of the boy: 1/25,000,000 of an inch). The scientists moved thousands of atoms to create the frames for *A Boy And His Atom* – the GUINNESS WORLD RECORDS™ record holder for World's Smallest Stop-Motion Film. youtube.com/madewithatoms #MadeWithAtoms

© 2013 International Business Machines Corporation

THE VERY principles of quantum mechanics make visualizing atoms a difficult and daunting task. However, scientists have harnessed the unique behavior of electrons in atoms to develop such advanced imaging techniques as *atomic force microscopy* (AFM) and *scanning tunneling microscopy* (STM). Shown here is an image created using STM. This image is one of 242 made in the production of a one-minute stop-motion film. Each dot in the image is an oxygen atom magnified over a hundred million times. The ability to image at the atomic level allows scientists to investigate chemical phenomena on *very* small scales.

Before You Begin, Review These Skills

- Tracking units [◄◄ Section 1.4]
- Nuclear model of the atom [◄◄ Section 2.2]

3.1   ENERGY AND ENERGY CHANGES

Before beginning any description of the electronic structure of the atom, it is necessary to define and discuss energy. ***Energy*** is the capacity to do work or transfer heat.

Forms of Energy

All forms of energy are either kinetic or potential. ***Kinetic energy*** is the energy that results from *motion*. It is calculated with the equation

$$E_k = \frac{1}{2}mu^2 \qquad \textbf{Equation 3.1}$$

where m is the mass of the object and u is its velocity. One form of kinetic energy of particular interest to chemists is ***thermal energy,*** which is the energy associated with the random motion of atoms and molecules. We can monitor changes in thermal energy by measuring temperature changes.

Student Annotation: Some textbooks use the letter *v* to denote velocity.

 Potential energy is the energy possessed by an object by virtue of its position. The two forms of potential energy of greatest interest to chemists are *chemical* and *electrostatic*. ***Chemical energy*** is potential energy stored within the structural units of chemical substances. The amount of chemical energy in a sample of matter depends on the types and arrangements of atoms in the structural units that make up the sample.

Student Annotation: "Structural units" may be *ions* [►► Section 5.3] or *molecules* [►► Section 5.5].

 Electrostatic energy is potential energy that results from the interaction of charged particles. Oppositely charged particles *attract* each other, and particles of like charges *repel* each other. The magnitude of the resulting electrostatic potential energy is proportional to the product of the two charges (Q_1 and Q_2) divided by the distance between them (d).

$$E_{el} \propto \frac{Q_1 Q_2}{d} \qquad \textbf{Equation 3.2}$$

Student Annotation: The symbol ∝ can be read as "is proportional to." Equation 3.2 can also be written as an equality with a proportionality constant, *c*:

$$E_{el} = c\frac{Q_1 Q_2}{d}$$

If the charges Q_1 and Q_2 are *opposite* (i.e., one positive and one negative), the result is a *negative* value for E_{el}, which indicates *attraction*. Like charges (i.e., both positive or both negative) result in a *positive* value for E_{el}, indicating *repulsion*.

 Kinetic and potential energy are interconvertible—that is, one form can be converted to the other. For example, dropping an object and allowing it to fall converts potential energy to kinetic energy. Although energy can assume many different forms that are interconvertible, the total amount of energy in the universe is constant—that is, energy can be neither created nor destroyed. When energy of one form disappears, the same amount of energy must appear in another form or forms. This principle is known as the ***law of conservation of energy.***

Units of Energy

The SI unit of energy is the *joule (J),* named for the English physicist James Joule.[1] The joule is a fairly small quantity of energy. It is the amount of kinetic energy possessed by a 2-kg mass moving at a speed of 1 m/s.

$$E_k = \frac{1}{2}mu^2 = \frac{1}{2}(2 \text{ kg})(1 \text{ m/s})^2 = 1 \text{ kg} \cdot \text{m}^2/\text{s}^2 = 1 \text{ J}$$

The joule can also be defined as the amount of energy exerted when a force of 1 newton (N) is applied over a distance of 1 meter.

$$1 \text{ J} = 1 \text{ N} \cdot \text{m}$$

where

$$1 \text{ N} = 1 \text{ kg} \cdot \text{m/s}^2$$

Because the magnitude of a joule is so small, we very often express large amounts of energy using the unit kilojoule (kJ).

$$1 \text{ kJ} = 1000 \text{ J}$$

Worked Examples 3.1 and 3.2 show how to calculate kinetic energy, and how to compare electrostatic energies between charged particles.

Worked Example 3.1

Calculate the kinetic energy of a helium atom moving at a speed of 125 m/s.

Strategy Use Equation 3.1 $\left(E_k = \frac{1}{2}mu^2\right)$ to calculate the kinetic energy of an atom. Note that for units to cancel properly, giving E_k in *joules,* the mass of the atom must be expressed in *kilograms.*

Setup The mass of a helium atom is 4.003 amu. The factor for conversion of amu to g is 1.661×10^{-24} g/1 amu [◄◄ Section 1.2]. Therefore, the mass of a helium atom in kilograms is

$$4.003 \text{ amu} \times \frac{1.661 \times 10^{-24} \text{ g}}{1 \text{ amu}} \times \frac{1 \text{ kg}}{1 \times 10^3 \text{ g}} = 6.649 \times 10^{-27} \text{ kg}$$

Solution

$$E_k = \frac{1}{2}mu^2$$

$$= \frac{1}{2}(6.649 \times 10^{-27} \text{ kg})(125 \text{ m/s})^2$$

> **Student Annotation:** Remember that the base units of the joule are kg · m²/s².

$$= (5.19 \times 10^{-23} \text{ kg} \cdot \text{m}^2/\text{s}^2 = 5.19 \times 10^{-23} \text{ J})$$

Think About It
We expect the energy of a single atom, even a fast-moving one, to be extremely small.

Practice Problem Ⓐ**TTEMPT** Calculate the energy in joules of a 5.25-g object moving at a speed of 655 m/s.

Practice Problem Ⓑ**UILD** Calculate the velocity (in m/s) of a 0.345-g object that has $E_k = 23.5$ J.

Practice Problem Ⓒ**ONCEPTUALIZE** By what factor does the kinetic energy of a particle change if the speed is doubled while the mass is reduced by half?

---

[1]James Prescott Joule (1818–1889). As a young man, Joule was tutored by John Dalton. He is most famous for determining the mechanical equivalent of heat, the conversion between mechanical energy and thermal energy.

Worked Example 3.2

How much greater is the electrostatic potential energy between charges of +2 and −2 than that between charges of +1 and −1 if the opposite charges in each case are separated by the same distance?

Strategy Use Equation 3.2 $\left(E_{el} \propto \frac{Q_1Q_2}{d}\right)$ to compare the magnitudes of the two E_{el} values.

Because the distance between charges is the same in both cases, we can solve for the ratio of E_{el} values without actually knowing the distance. Both the distance and the proportionality constant cancel in the solution.

Setup

$$E_{el(+2,-2)} = c\frac{Q_1Q_2}{d}, \ Q_1 = +2 \text{ and } Q_2 = -2$$

$$E_{el(+1,-1)} = c\frac{Q_1Q_2}{d}, \ Q_1 = +1 \text{ and } Q_2 = -1$$

Solution

$$\frac{c \times \left(\dfrac{2 \times (-2)}{d}\right)}{c \times \left(\dfrac{1 \times (-1)}{d}\right)} = 4$$

The electrostatic potential energy between charges of +2 and −2 is four times as large as that between charges of +1 and −1.

> **Think About It**
> Doubling both charges causes a *fourfold* increase in the magnitude of the electrostatic potential energy between charged particles.

Practice Problem **A**TTEMPT How much greater is the electrostatic potential energy between charges of +3 and −2 than that between charges of +2 and −2 if the opposite charges in each case are separated by the same distance?

Practice Problem **B**UILD What must the separation between charges of +2 and −2 be for the electrostatic potential energy between them to be the same as that between charges of +2 and −3 separated by a distance of 1.00 mm?

Practice Problem **C**ONCEPTUALIZE For charges of +2 and −2, separated by a distance of d, the electrostatic potential energy is E. In terms of E, determine the electrostatic potential energy between each of the pairs of charges shown.

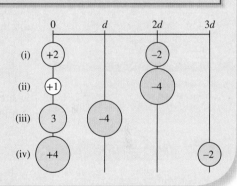

Section 3.1 Review

Energy and Energy Changes

3.1.1 Calculate the kinetic energy of a 5.0-kg mass moving at 26 m/s.
   (a) 1.7×10^3 J   (d) 65 J
   (b) 3.4×10^3 J   (e) 13×10^3 J
   (c) 130 J

3.1.2 Calculate the velocity (in m/s) of a 2.34-g object that has $E_k = 46.2$ J.
   (a) 6.28 m/s   (d) 199 m/s
   (b) 3.95×10^4 m/s   (e) 141 m/s
   (c) 0.199 m/s

3.1.3 Arrange the following pairs of charged particles in order of increasing magnitude of electrostatic attraction (E_{el}).
   (a) i < ii < iii < iv   (d) ii < i = iii < iv
   (b) iv < iii < ii < i   (e) iv < i = ii < iii
   (c) i = iii < ii < iv

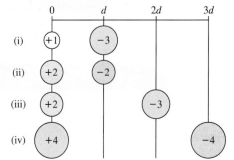

3.2 THE NATURE OF LIGHT

The term *light* usually refers to *visible* light, which is the light we can detect with our eyes. Visible light, however, is only a small part of the continuum of radiation known as the **electromagnetic spectrum.** In addition to visible light, the electromagnetic spectrum includes radio waves, microwave radiation, infrared and ultraviolet radiation, X rays, and gamma rays, as shown in Figure 3.1. You are probably familiar with some of these terms. Ultraviolet radiation, for instance, can cause sunburn; microwave radiation is commonly used to cook and/or reheat food; X rays are used for certain medical diagnoses and during routine dental checkups; and you may recall from Chapter 2 that gamma rays emitted from some radioactive materials. Although these phenomena may seem very different from visible light—and different from one another—they all constitute the transmission of energy in the form of *waves*.

Properties of Waves

The fundamental properties of waves are illustrated in Figure 3.2. Waves are characterized by their wavelength, frequency, and amplitude. **Wavelength** λ (lambda) is the distance between identical points on successive waves (e.g., successive peaks or successive troughs). The **frequency** ν (nu) is the *number* of waves that pass through a particular point in 1 second. **Amplitude** is the vertical distance from the midline of a wave to the top of the peak or the bottom of the trough.

The speed of a wave depends on the type of wave and the nature of the medium through which the wave is traveling (e.g., air, water, or a vacuum). The speed of light through a vacuum, *c*, is 2.99792458×10^8 m/s. The speed, wavelength, and frequency of a wave are related by the equation

Equation 3.3	$c = \lambda v$

where λ and v are expressed in meters (m) and reciprocal seconds (s^{-1}), respectively. While wavelength in meters is convenient for this equation, the units customarily used

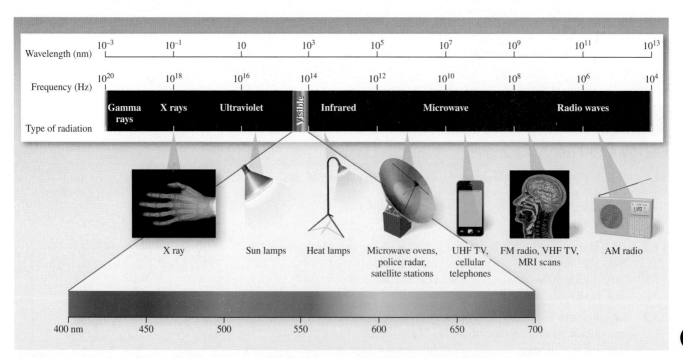

Figure 3.1 Electromagnetic spectrum. Each type of radiation is spread over a specific range of wavelengths (and frequencies). Visible light ranges from 400 nm (violet) to 700 nm (red).

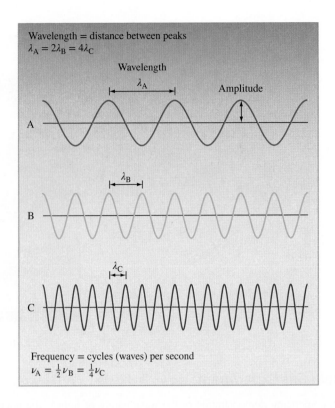

Figure 3.2 Characteristics of waves: wavelength, amplitude, and frequency.

to express the wavelength of electromagnetic radiation depend on the type of radiation and the magnitude of the corresponding wavelength. Visible wavelengths are usually expressed in nanometers (nm, or 10^{-9} m). Microwave and X-ray wavelengths are usually expressed in centimeters (cm, or 10^{-2} m) and angstroms (Å, or 10^{-10} m), respectively.

The Electromagnetic Spectrum

In 1873 James Clerk Maxwell,[2] a Scottish physicist and mathematician, proposed that visible light consisted of electromagnetic waves. According to Maxwell's theory, an **_electromagnetic wave_** has an electric field component and a magnetic field component. These two components have the same wavelength and frequency, and hence the same speed, but they travel in mutually perpendicular planes (Figure 3.3). Maxwell's theory provides a mathematical description of the general behavior of light; and describes accurately how energy in the form of radiation can be propagated through space as oscillating electric and magnetic fields.

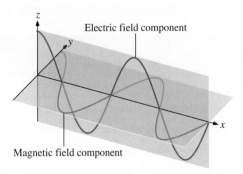

Figure 3.3 Electric field and magnetic field components of an electromagnetic wave. These two components have the same wavelength, frequency, and amplitude, but they vibrate in two mutually perpendicular planes.

---

[2]James Clerk Maxwell (1831–1879). Maxwell is considered by many to be one of the most influential physicists in history. His demonstration that electricity, magnetism, and light are all manifestations of the electromagnetic field is referred to as "the second great unification of physics."

Figure 3.4 Double-slit experiment.
(a) Red lines correspond to the maximum intensity resulting from constructive interference. Dashed blue lines correspond to the minimum intensity resulting from destructive interference. (b) Interference pattern with alternating bright and dark lines.

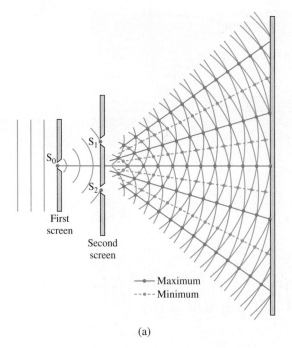

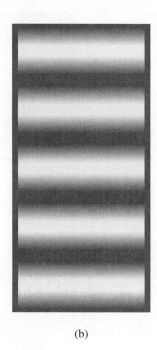

(a)   (b)

The Double-Slit Experiment

A simple yet convincing demonstration of the wave nature of light is the phenomenon of *interference*. When a light source passes through a narrow opening called a *slit,* a bright line is generated in the path of the light through the slit. When the same light source passes through two closely spaced slits, however, as shown in Figure 3.4, the result is not two bright lines, one in the path of each slit, but rather a series of light and dark lines known as an *interference pattern.* When the light sources recombine after passing through the slits, they do so *constructively* where the two waves are *in phase* (giving rise to the light lines) and *destructively* where the waves are *out of phase* (giving rise to the dark lines). Constructive interference and destructive interference are properties of waves.

The various types of electromagnetic radiation in Figure 3.1 differ from one another in wavelength and frequency. Radio waves, which have long wavelengths and low frequencies, are emitted by large antennas, such as those used by broadcasting stations. The shorter, visible light waves are produced by the motions of electrons within atoms. The shortest waves, which also have the highest frequency, are γ (gamma) rays, which result from nuclear processes [◀◀ Section 2.2]. As we will see shortly, the higher the frequency, the more energetic the radiation. Thus, ultraviolet radiation, X rays, and γ rays are high-energy radiation, whereas infrared radiation, microwave radiation, and radio waves are low-energy radiation.

Worked Example 3.3 illustrates the conversion between wavelength and frequency.

Worked Example 3.3

One type of laser used in the treatment of vascular skin lesions is a neodymium-doped yttrium aluminum garnet, or Nd:YAG, laser. The wavelength commonly used in these treatments is 532 nm. What is the frequency of this radiation?

Strategy We must convert the wavelength to meters and solve for frequency using Equation 3.3 ($c = \lambda v$).

Setup Rearranging Equation 3.3 to solve for frequency gives $v = \frac{c}{\lambda}$. The speed of light, c, is 3.00×10^8 m/s. λ (in meters) = 532 nm $\times \frac{1 \times 10^{-9}\,\text{m}}{1\,\text{nm}} = 5.32 \times 10^{-7}$ m.

Solution

$$v = \frac{3.00 \times 10^8 \text{ m/s}}{5.32 \times 10^{-7} \text{ m}} = 5.64 \times 10^{14}\text{ s}^{-1}$$

Think About It
Make sure your units cancel properly. A common error in this type of problem is neglecting to convert wavelength to meters.

Practice Problem **A**TTEMPT What is the wavelength (in meters) of an electromagnetic wave whose frequency is 1.61×10^{12} s^{-1}?

Practice Problem **B**UILD What is the frequency (in reciprocal seconds) of electromagnetic radiation with a wavelength of 1.03 cm?

Practice Problem **C**ONCEPTUALIZE Which of the following sets of waves best represents the relative wavelengths/ frequencies of visible light of the colors shown?

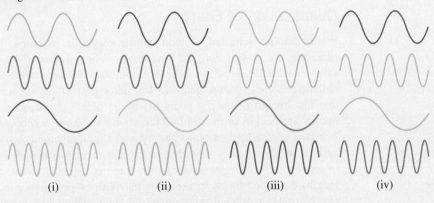

(i)   (ii)   (iii)   (iv)

Section 3.2 Review

The Nature of Light

3.2.1 Calculate the wavelength (in nanometers) of light with frequency 3.45×10^{14} s^{-1}.
   (a) 1.15×10^{-6} nm
   (b) 1.04×10^{23} nm
   (c) 8.70×10^{2} nm
   (d) 115 nm
   (e) 9.66×10^{-24} nm

3.2.2 Calculate the frequency of light with wavelength 126 nm.
   (a) 2.38×10^{15} s^{-1}
   (b) 4.20×10^{-16} s^{-1}
   (c) 37.8 s^{-1}
   (d) 2.65×10^{-2} s^{-1}
   (e) 3.51×10^{19} s^{-1}

3.2.3 Of the waves pictured, which has the greatest frequency, which has the greatest wavelength, and which has the greatest amplitude?
   (a) i, ii, iii
   (b) i, iii, ii
   (c) ii, i, ii
   (d) ii, i, iii
   (e) ii, iii, ii

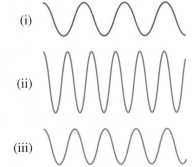

(i)

(ii)

(iii)

3.2.4 When traveling through a translucent medium, such as glass, light moves more slowly than it does when traveling through a vacuum. Red light with a wavelength of 684 nm travels through Pyrex glass with a frequency of 2.92×10^{14} s^{-1}. Calculate the speed of this light.
   (a) 3.00×10^{8} m/s
   (b) 2.00×10^{8} m/s
   (c) 2.92×10^{6} m/s
   (d) 4.23×10^{7} m/s
   (e) 2.23×10^{8} m/s

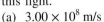

3.3   QUANTUM THEORY

Early attempts by nineteenth-century physicists to figure out the structure of the atom met with only limited success. This was largely because they were attempting to understand the inner workings of atoms using the laws of classical physics, which describe the behavior of *macroscopic* objects. It took a long time to realize—and an even longer time to accept—that the behavior of subatomic particles is *not* governed by the same physical laws as larger objects.

Quantization of Energy

When a solid is heated, it emits electromagnetic radiation, known as **blackbody radiation,** over a wide range of wavelengths. The red glow of the element of an electric stove and the bright white light of a tungsten lightbulb are examples of blackbody radiation. Measurements taken in the latter part of the nineteenth century showed that the amount of energy given off by an object at a certain temperature depends on the wavelength of the emitted radiation. Attempts to account for this dependence in terms of established wave theory and thermodynamic laws were only partially successful. One theory was able to explain short-wavelength dependence but failed to account for the longer wavelengths. Another theory accounted for the longer wavelengths but failed for short wavelengths. With no *one* theory that could explain both observations, it seemed that something fundamental was missing from the laws of classical physics.

In 1900, German physicist Max Planck[3] provided the solution and launched a new era in physics with an idea that departed drastically from accepted concepts. Classical physics assumed that radiant energy was continuous; that is, it could be emitted or absorbed in any amount. Based on data from blackbody radiation experiments, Planck proposed that radiant energy could be emitted or absorbed only in discrete quantities, like small packages or bundles. Planck gave the name **quantum** to the smallest quantity of energy that can be emitted (or absorbed) in the form of electromagnetic radiation. The energy *E* of a single quantum of energy is given by

Equation 3.4	$E = h\nu$

where *h* is called *Planck's constant* and *v* is the *frequency* of the radiation. The value of Planck's constant is 6.63×10^{-34} J · s.

According to quantum theory, energy is always emitted in whole-number multiples of *hv*. At the time Planck presented his theory, he could not explain why energies should be fixed or quantized in this manner. Starting with this hypothesis, however, he had no difficulty correlating the experimental data for the emission by solids over the entire range of wavelengths; the experimental data supported his new *quantum theory.*

The idea that energy is *quantized* rather than *continuous* may seem strange, but the concept of quantization has many everyday analogies. For example, as you walk up a staircase, you can stop only at elevations corresponding to the steps—not at an elevation *between* steps. On a piano, you can play only the notes for which there are keys—not notes between keys. Even processes in living systems involve quantized phenomena. The eggs laid by hens are quanta (hens lay only whole eggs). Similarly, when a dog gives birth to a litter, the number of puppies is always an

---

[3]Max Karl Ernst Ludwig Planck (1858–1947). Planck received the Nobel Prize in Physics in 1918 for his quantum theory. He also made significant contributions in thermodynamics and other areas of physics.

integer. Each puppy is a "quantum" of that animal. Planck's quantum theory revolutionized physics; and the flurry of research that ensued altered our concept of nature forever.

Photons and the Photoelectric Effect

In 1905, only five years after Planck presented his quantum theory, Albert Einstein[4] used the theory to explain another well-known but mysterious physical phenomenon, the *photoelectric effect*. In the photoelectric effect, electrons are ejected from the surface of a metal exposed to light. However, the light must be of a certain minimum frequency, called the *threshold frequency* (Figure 3.5). The *number* of electrons ejected was proportional to the intensity (or brightness) of the light, but the *energies* of the ejected electrons were not. Below the threshold frequency no electrons were ejected no matter how intense the light.

The photoelectric effect could not be explained by the wave theory of light, which associated the energy of light with its intensity. Einstein, however, made an extraordinary assumption. He suggested that a beam of light is really a stream of *particles*. These "particles" of light are now called *photons*. Using Planck's quantum theory of radiation as a starting point, Einstein deduced that each photon must possess energy E given by the equation

$$E_{\text{photon}} = h\nu$$

where h is Planck's constant and ν is the frequency of the light. Electrons are held in a metal by attractive forces, and so removing them from the metal requires light of a sufficiently high frequency (which corresponds to a sufficiently high *energy*) to break them free. Shining a beam of light onto a metal surface can be thought of as shooting a beam of particles—*photons*—at the metal atoms. If the frequency of the photons is such that $h\nu$ exactly equals the energy that binds the electrons in the metal, then the light will have just enough energy to knock the electrons loose. If we use light of a *higher* frequency, then not only will the electrons be knocked loose, but they will also acquire some kinetic energy. This situation is summarized by the equation

$$h\nu = \text{KE} + W \qquad \textbf{Equation 3.5}$$

where KE is the kinetic energy of the ejected electron and W is the binding energy of the electron in the metal. Rewriting Equation 3.5 as

$$\text{KE} = h\nu - W$$

shows that the more energetic the photon (i.e., the higher its frequency), the greater the kinetic energy of the ejected electron. If the frequency of light is below the threshold frequency, the photon will simply bounce off the surface and no electrons will be ejected. If the frequency is equal to the threshold frequency, it will dislodge the most loosely held electron. Above the threshold frequency, it will not only dislodge the electron, but also impart certain kinetic energy to the ejected electron.

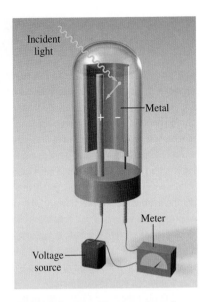

Figure 3.5 Apparatus for studying the photoelectric effect. Light of a certain frequency falls on a clean metal surface. Ejected electrons are attracted toward the positive electrode. The flow of electrons is registered by a detecting meter.

Student Annotation: Note that this is simply Equation 3.4 in which the "quantum" has been specified as a *photon of light.*

---

[4]Albert Einstein (1879–1955). A German-born American physicist, Einstein is regarded by many as one of the two greatest physicists the world has known (the other being Newton). The three papers (on special relativity, Brownian motion, and the photoelectric effect) that he published in 1905 while employed as a technical assistant in the Swiss patent office in Berne have profoundly influenced the development of physics. He received the Nobel Prize in Physics in 1921 for his explanation of the photoelectric effect.

Thinking Outside the Box

Everyday Occurrences of the Photoelectric Effect

Chances are good that you encounter the photoelectric effect regularly. Some everyday applications include the type of device that prevents a garage door from closing when something is in the door's path and motion-detection systems used in museums and other high-security environments. These sorts of devices work simply by responding to an interruption in a beam of light. In each case, a beam of light normally shines on a *photocathode,* a surface that emits photoelectrons, and the photoelectrons, accelerated toward an anode by high voltage, constitute a current. When the beam of light is interrupted, the flow of electrons stops and the current is cut off. In the case of the garage-door safety device, when the current stops, the movement of the door stops. In the case of the motion-detection systems, when the current stops, an alarm may sound or a light may turn on.

One of the more exotic uses of the photoelectric effect is night-vision goggles. Although you may never have looked through such goggles, you have probably seen night-vision images on the news or in a suspense-filled movie such as *Silence of the Lambs.* Typically, the night-vision images we see are from what are known as third-generation night-vision devices. These devices use a photocathode material that emits photoelectrons when struck by photons in the *infrared* region of the electromagnetic spectrum. (First- and second-generation devices used different photocathode materials and relied more on the amplification of low-level visible light.) The photoelectrons emitted by the photocathode enter a microchannel plate (MCP), an array of tiny parallel tubes, where each strikes the internal surface of a tube causing many more electrons to be ejected—a process called *secondary emission.* This effectively amplifies the current generated by each photoelectron. The amplified current is then accelerated by high voltage toward a phosphorus screen, where incident electrons cause the emission of visible light, generating the familiar green glow of night vision.

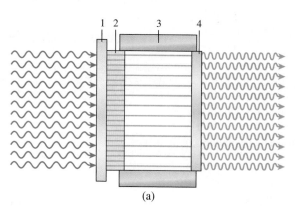

(a)

(b)

(a) Night-vision goggles schematic. 1. Photocathode (gallium arsenide). 2. Microchannel plate (MCP). 3. High-voltage source. 4. Phosphorus screen.
(b) The green glow of night vision.

(b): Source: Official Marine Corps photo by Sgt. Brian A. Tuthill

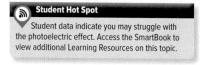

Student Hot Spot

Student data indicate you may struggle with the photoelectric effect. Access the SmartBook to view additional Learning Resources on this topic.

Now consider two beams of light having the same frequency (greater than the threshold frequency) but different intensities. The more intense beam of light consists of a larger number of photons, so it ejects more electrons from the metal's surface than the less-intense beam of light. Thus, the more intense the light, the greater the number of electrons emitted by the target metal; the higher the frequency of the light, the greater the kinetic energy of the ejected electrons.

Worked Example 3.4 shows how to determine the energy of a single photon of light of a given wavelength, and how to determine the kinetic energy of an electron ejected from a metal via the photoelectric effect.

Worked Example 3.4

Calculate the energy (in joules) of (a) a photon with a wavelength of 5.00×10^4 nm (infrared region) and (b) a photon with a wavelength of 52 nm (ultraviolet region). (c) Calculate the maximum kinetic energy of an electron ejected by the photon in part (b) from a metal with a binding energy of 3.7 eV.

Strategy In parts (a) and (b), we are given the wavelength of light. Use Equation 3.3 ($c = \lambda v$) to convert wavelength to frequency, then use Equation 3.4 ($E = hv$) to determine the energy of the photon for each wavelength. In part (c), we are asked to determine the kinetic energy of an ejected electron. For this we use Equation 3.5 ($hv = KE + W$). The binding energy, given in electron volts, must be converted to joules in order for units to cancel.

Setup The wavelengths must be converted from nanometers to meters:

(a) $5.00 \times 10^4 \ \text{nm} \times \dfrac{1 \times 10^{-9} \ \text{m}}{1 \ \text{nm}} = 5.00 \times 10^{-5} \ \text{m}$

(b) $52 \ \text{nm} \times \dfrac{1 \times 10^{-9} \ \text{m}}{1 \ \text{nm}} = 5.2 \times 10^{-8} \ \text{m}$

Planck's constant, h, is 6.63×10^{-34} J · s.

(c) $W = 3.7 \ \text{eV} \times \dfrac{1.602 \times 10^{-19} \ \text{J}}{1 \ \text{eV}} = 5.9 \times 10^{-19} \ \text{J}$

Solution

(a) $v = \dfrac{c}{\lambda} = \dfrac{3.00 \times 10^8 \ \text{m/s}}{5.00 \times 10^{-5} \ \text{m}} = 6.00 \times 10^{12} \ \text{s}^{-1}$ and

$$E = hv = (6.63 \times 10^{-34} \ \text{J} \cdot \text{s})(6.00 \times 10^{12} \ \text{s}^{-1}) = 3.98 \times 10^{-21} \ \text{J}$$

This is the energy of a single photon with wavelength 5.00×10^4 nm.

(b) Following the same procedure as in part (a), the energy of a photon of wavelength 52 nm is 3.8×10^{-18} J.

(c) $KE = hv - W = 3.8 \times 10^{-18} \ \text{J} - 5.9 \times 10^{-19} \ \text{J} = 3.2 \times 10^{-18} \ \text{J}$

Think About It
Remember that frequency and wavelength are *inversely* proportional (Equation 3.3). Thus, as wavelength *decreases*, frequency and energy *increase*. Note that in part (c), subtracting the binding energy made a relatively small change to the energy of the incident photon. If the incident photon had been in the X-ray region of the spectrum, the difference between its energy and the kinetic energy of the ejected electron would have been negligible.

Practice Problem **A**TTEMPT Calculate the energy (in joules) of (a) a photon with wavelength 2.11×10^2 nm, and (b) a photon with wavelength 1.69×10^3 mm. (c) Calculate the maximum kinetic energy of an electron ejected by the photon in part (a) from a metal with a binding energy of 4.66 eV.

Practice Problem **B**UILD (a) Calculate the wavelength (in nanometers) of light with energy 1.89×10^{-20} J per photon. (b) For light of wavelength 410 nm, calculate the number of photons per joule. (c) Determine the binding energy (in electron volts) of a metal if the maximum kinetic energy possessed by an electron ejected from it [using one of the photons in part (b)] is 2.93×10^{-19} J.

Practice Problem **C**ONCEPTUALIZE A blue billiard ball with a mass of 165 g rests in a shallow well on an otherwise flat surface. When a red billiard ball with the same mass moving at any velocity less than 1.20 m/s strikes the blue ball, the blue ball does not move (i). When the red ball strikes the blue ball moving at exactly 1.20 m/s, the blue ball is just barely dislodged from the well (ii). What will be the velocity (iii) of the blue ball when it is struck by the red ball moving at 1.75 m/s?

(i)

(ii)

(iii)

Einstein's theory of light posed a dilemma for scientists. On the one hand, it explains the photoelectric effect. On the other hand, the particle theory of light is inconsistent with the known wavelike properties of light. The only way to resolve the dilemma is to accept the idea that light possesses properties characteristic of both particles *and* waves. Depending on the experiment, light behaves either as a wave or as a stream of particles. This concept was totally alien to the way physicists had thought about matter and radiation, and it took a long time for them to accept it. We will see in Section 3.5 that possessing properties of both particles and waves is not unique to light but ultimately is characteristic of all matter, including electrons.

Section 3.3 Review

Quantum Theory

3.3.1   Calculate the energy per photon of light with wavelength 650 nm.

(a) 1.29×10^{-31} J
(b) 4.31×10^{-40} J
(c) 1.02×10^{-27} J
(d) 1.44×10^{-48} J
(e) 3.06×10^{-19} J

3.3.2   Calculate the wavelength (in centimeters) of light that has energy 1.32×10^{-23} J/photon.

(a) 5.02×10^{-9} cm
(b) 6.64×10^{3} cm
(c) 2.92×10^{-63} cm
(d) 1.51 cm
(e) 66.4 cm

3.3.3   Calculate the maximum kinetic energy of an electron ejected by a photon of the light in 3.3.1 from a metal with a binding energy of 1.56 eV.

(a) 5.61×10^{-20} J
(b) 684 J
(c) 2.50×10^{-19} J
(d) 5.56×10^{-19} J
(e) 6.50×10^{-7} J

3.3.4   A clean metal surface is irradiated with light of three different wavelengths: λ_1, λ_2, and λ_3. The kinetic energies of the ejected electrons are as follows: λ_1: 2.9×10^{-20} J; λ_2: approximately zero; λ_3: 4.2×10^{-19} J. Arrange the light in order of increasing wavelength.

(a) $\lambda_1 < \lambda_2 < \lambda_3$
(b) $\lambda_2 < \lambda_1 < \lambda_3$
(c) $\lambda_3 < \lambda_2 < \lambda_1$
(d) $\lambda_3 < \lambda_1 < \lambda_2$
(e) $\lambda_2 < \lambda_3 < \lambda_1$

3.3.5   Shown here are waves of electromagnetic radiation of two different frequencies and two different amplitudes. Assume that intensity of radiation (photons per second) is directly proportional to amplitude. Which of the waves is made up of photons of greater energy? Which wave delivers more photons during a given period of time? Which wave delivers more total energy during a given time period?

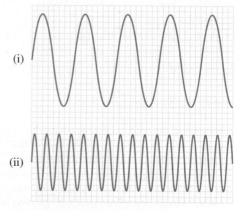

(a) i, i, i
(b) i, ii, i
(c) ii, i, ii
(d) ii, ii, i
(e) ii, ii, ii

BOHR'S THEORY OF THE HYDROGEN ATOM

In addition to explaining the photoelectric effect, Planck's quantum theory and Einstein's ideas made it possible for scientists to unravel another nineteenth-century mystery in physics: atomic line spectra.

In the seventeenth century, Newton had shown that sunlight is composed of various color components that can be recombined to produce white light. Since that time, chemists and physicists have studied the characteristics of such *emission spectra.* The emission spectrum of a substance can be seen by energizing a sample of material with either thermal energy or some other form of energy (such as a high-voltage electrical discharge if the substance is a gas). A "red-hot" or "white-hot" iron bar freshly removed from a fire produces a characteristic glow. The glow is the visible portion of its emission spectrum. The heat given off by the same iron bar is another portion of its emission spectrum—the infrared region. A feature common to the emission spectrum of the sun and that of a heated solid is that both are continuous; that is, all wavelengths of visible light are present in each spectrum (Figure 3.6).

> **Student Annotation:** If you have ever seen a rainbow, you are familiar with this phenomenon. The rainbow is the visible portion of the sun's emission spectrum.

Atomic Line Spectra

Unlike those of the sun or a white-hot iron bar, the emission spectra of atoms in the gas phase do not show a continuous spread of wavelengths from red to violet; rather, the atoms produce bright lines in distinct parts of the visible spectrum. A *line spectrum* is the emission of light only at *specific wavelengths.* Figure 3.7 is a schematic diagram of a discharge tube that is used to study emission spectra.

Every element has a unique emission spectrum, so the characteristic lines in atomic spectra can be used in chemical analysis to identify elements, much as fingerprints are used to identify people. When the lines of the emission spectrum of a known element exactly match the lines of the emission spectrum of an unknown sample, the identity of the element in the sample is established. Although the procedure of identifying elements by their line spectra had been used for many years in chemical analysis, the origin of the spectral lines was not

(a)

(b)

Figure 3.6 The visible white light emitted by (a) the sun and (b) a white-hot iron bar. In each case, the white light is the combination of all visible wavelengths (see Figure 3.1).

(a) © Doug Menuez/Getty Images; (b) © McGraw-Hill Education/Charles D. Winters, photographer

Figure 3.7 (a) Experimental arrangement for studying the emission spectra of atoms and molecules. The gas being studied is in a discharge tube containing two electrodes. As electrons flow from the negative electrode to the positive electrode, they collide with the gas particles. The collisions lead to the emission of light by the atoms (or molecules). The emitted light is separated into its components by a prism. (b) Visible line emission spectrum of hydrogen.

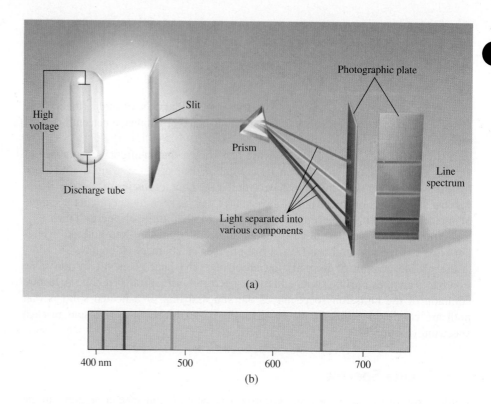

(a)

400 nm   500   600   700

(b)

understood until early in the twentieth century. Figure 3.8 shows the emission spectra of several elements.

In 1885, Johann Balmer[5] developed a simple equation that could be used to calculate the wavelengths of the four visible lines in the emission spectrum of hydrogen. Johannes Rydberg[6] developed Balmer's equation further, yielding an equation that could calculate not only the *visible* wavelengths, but those of *all* hydrogen's spectral lines:

Equation 3.6

$$\frac{1}{\lambda} = R_\infty \left(\frac{1}{n_1^2} - \frac{1}{n_2^2} \right)$$

Student Annotation: The Rydberg equation is a mathematical relationship that was derived from experimental data. Although it predates quantum theory by decades, it agrees remarkably well with it for one-electron systems such as the hydrogen atom.

In Equation 3.6, now known as the *Rydberg equation*, λ is the wavelength of a line in the spectrum; R_∞ is the Rydberg constant (1.09737316×10^7 m^{-1}); and n_1 and n_2 are positive integers, where $n_2 > n_1$.

The Line Spectrum of Hydrogen

In 1913, not long after Planck's and Einstein's discoveries, a theoretical explanation of the emission spectrum of the hydrogen atom was presented by Danish physicist Niels Bohr.[7] Bohr's treatment is very complex and is no longer considered to be correct in all its details. We will concentrate only on his important assumptions and final results, which account for the observed spectral lines and which provide an important step toward the understanding of quantum theory.

---

[5]Johann Jakob Balmer (1825–1898). From 1859 until his death in 1898, Balmer, a Swiss mathematician, taught math at a secondary school for girls in Basel, Switzerland. Although physicists did not understand why his equation worked until long after his death, the visible series of lines in the spectrum of hydrogen is named for him.

[6]Johannes Robert Rydberg (1854–1919). A Swedish mathematician and physicist, Rydberg analyzed many atomic spectra in an effort to understand the periodic properties of elements. Although he was nominated twice for the Nobel Prize in Physics, he never received it.

[7]Niels Henrik David Bohr (1885–1962). One of the founders of modern physics, he received the Nobel Prize in Physics in 1922 for his theory explaining the line spectrum of hydrogen.

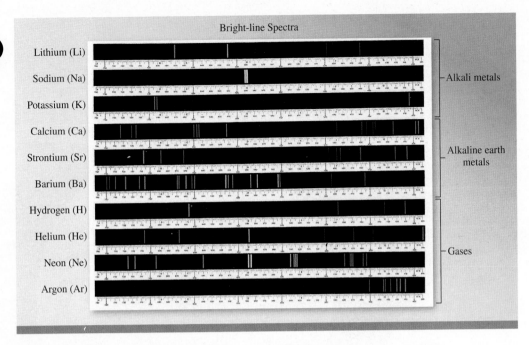

Figure 3.8 Emission spectra of several elements.

Source: © McGraw-Hill Education

When Bohr first approached this problem, physicists already knew that the atom contains electrons and protons. They thought of an atom as an entity in which electrons whirled around the nucleus in circular orbits at high velocities. This was an appealing description because it resembled the familiar model of planetary motion around the sun. However, according to the laws of classical physics, an electron moving in an orbit of a hydrogen atom would experience an acceleration toward the nucleus by radiating away energy in the form of electromagnetic waves. Thus, such an electron would quickly spiral into the nucleus and annihilate itself with the proton. To explain why this does not happen, Bohr postulated that the electron is allowed to occupy only certain orbits of specific energies. In other words, the energies of the electron are quantized. An electron in any of the allowed orbits will not radiate energy and therefore will not spiral into the nucleus.

Bohr attributed the emission of radiation by an energized hydrogen atom to the electron dropping from a higher-energy orbit to a lower one and giving up a quantum of energy (a photon) in the form of light (Figure 3.9). Using arguments based on electrostatic interaction and Newton's laws of motion, Bohr showed that the energies that the electron in the hydrogen atom can possess are given by

$$E_n = -2.18 \times 10^{-18}\ \text{J} \left(\frac{1}{n^2}\right) \qquad \textbf{Equation 3.7}$$

where n is an integer with values $n = 1, 2, 3$, and so on. The negative sign in Equation 3.7 is an arbitrary convention, signifying that the energy of the electron in the atom is *lower* than the energy of a *free electron,* which is an electron that is infinitely far from the nucleus. The energy of a free electron is arbitrarily assigned a value of zero. Mathematically, this corresponds to setting n equal to infinity in Equation 3.7:

$$E_n = -2.18 \times 10^{-18}\ \text{J} \left(\frac{1}{\infty^2}\right) = 0$$

As the electron gets closer to the nucleus (as n decreases), E_n becomes larger in absolute value, but also more negative. The most negative value, then, is reached when $n = 1$,

Video 3.1

Emission spectrum of hydrogen

Figure 3.9

Emission spectrum of hydrogen.

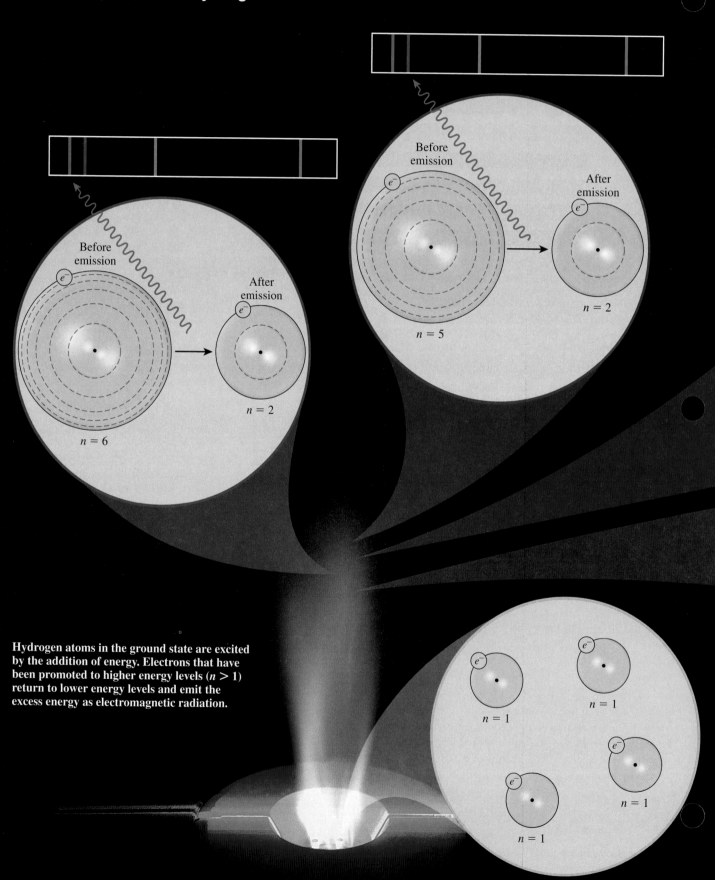

Hydrogen atoms in the ground state are excited by the addition of energy. Electrons that have been promoted to higher energy levels ($n > 1$) return to lower energy levels and emit the excess energy as electromagnetic radiation.

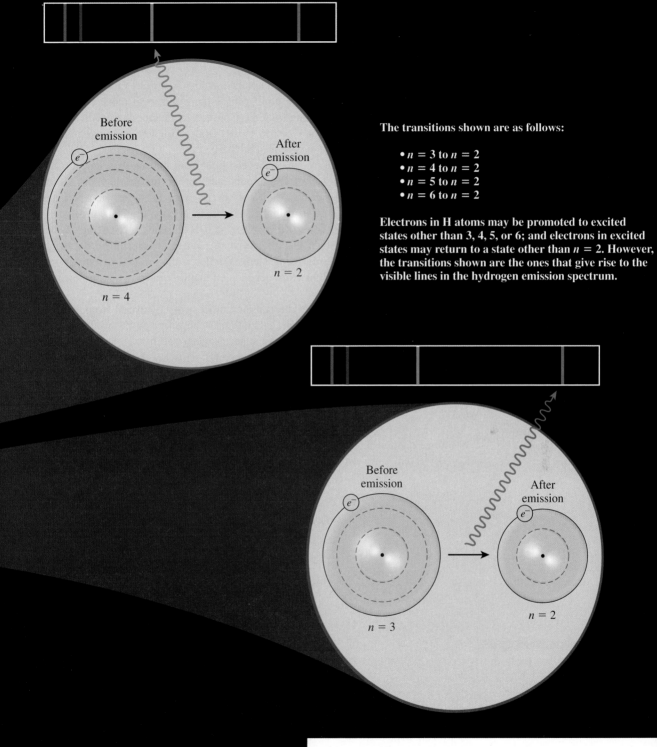

The transitions shown are as follows:

- $n = 3$ to $n = 2$
- $n = 4$ to $n = 2$
- $n = 5$ to $n = 2$
- $n = 6$ to $n = 2$

Electrons in H atoms may be promoted to excited states other than 3, 4, 5, or 6; and electrons in excited states may return to a state other than $n = 2$. However, the transitions shown are the ones that give rise to the visible lines in the hydrogen emission spectrum.

What's the point?

Each line in the visible emission spectrum of hydrogen is the result of an electronic transition from a higher excited state ($n = 3, 4, 5,$ or 6) to a lower excited state ($n = 2$). The energy gap between the initial and final states determines the wavelength of the light emitted.

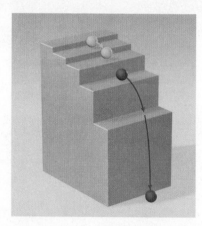

Figure 3.10 Mechanical analogy for the emission processes. The ball can rest on any step but not between steps.

Student Annotation: It is important to recognize that the electron in a hydrogen atom can move from a higher-energy state to *any* lower-energy state. It does not necessarily move from a higher-energy state to the ground state.

Student Annotation: The symbol delta Δ is commonly used to denote *final* minus *initial*.

Student Annotation: When $n_i > n_f$, ΔE is *negative* indicating energy is *emitted*. When $n_f > n_i$, ΔE is *positive* indicating energy is *absorbed*.

which corresponds to the most stable energy state. We call this the **ground state,** the lowest energy state of an atom. The stability of the electron diminishes as n increases. Each energy state in which $n > 1$ is called an **excited state.** Each excited state is higher in energy than the ground state. In the hydrogen atom, an electron for which n is greater than 1 is said to be in an excited state.

The radius of each circular orbit in Bohr's model depends on n^2. Thus, as n increases from 1 to 2 to 3, the orbit radius increases very rapidly. The higher the excited state, the farther away the electron is from the nucleus (and the less tightly held it is by the nucleus).

Bohr's theory enables us to explain the line spectrum of the hydrogen atom. Radiant energy absorbed by the atom causes the electron to move from the ground state ($n = 1$) to an excited state ($n > 1$). Conversely, radiant energy (in the form of a photon) is *emitted* when the electron moves from a higher-energy excited state to a lower-energy excited state or the ground state.

The quantized movement of the electron from one energy state to another is analogous to the movement of a tennis ball either up or down a set of stairs (Figure 3.10). The ball can be on any of several steps but never between steps. The journey from a lower step to a higher one is an energy-requiring process, whereas movement from a higher step to a lower step is an energy-releasing process. The quantity of energy involved in either type of change is determined by the distance between the beginning and ending steps. Similarly, the amount of energy needed to move an electron in the Bohr atom depends on the difference in energy levels between the initial and final states.

To apply Equation 3.7 to the emission process in a hydrogen atom, let us suppose that the electron is initially in an excited state characterized by n_i. During emission, the electron drops to a lower energy state characterized by n_f (the subscripts i and f denote the *initial* and *final* states, respectively). This lower energy state may be the ground state, but it can be any state lower than the initial excited state. The difference between the energies of the initial and final states is

$$\Delta E = E_f - E_i$$

From Equation 3.7,

$$E_n = -2.18 \times 10^{-18} \text{ J} \left(\frac{1}{n_f^2} \right)$$

and

$$E_n = -2.18 \times 10^{-18} \text{ J} \left(\frac{1}{n_i^2} \right)$$

Therefore,

$$\Delta E = \left(\frac{-2.18 \times 10^{-18} \text{ J}}{n_f^2} \right) - \left(\frac{-2.18 \times 10^{-18} \text{ J}}{n_i^2} \right)$$

$$= -2.18 \times 10^{-18} \text{ J} \left(\frac{1}{n_f^2} - \frac{1}{n_i^2} \right)$$

Because this transition results in the emission of a photon of frequency v and energy hv, we can write

Equation 3.8 $$\Delta E = hv = -2.18 \times 10^{-18} \text{ J} \left(\frac{1}{n_f^2} - \frac{1}{n_i^2} \right)$$

A photon is emitted when $n_i > n_f$. Consequently, the term in parentheses is *positive*, making ΔE *negative* (energy is lost to the surroundings). A photon is absorbed when

TABLE 3.1	Emission Series in the Hydrogen Spectrum		
Series	n_f	n_i	**Spectrum region**
Lyman	1	2, 3, 4, . . .	Ultraviolet
Balmer	2	3, 4, 5, . . .	Visible and ultraviolet
Paschen	3	4, 5, 6, . . .	Infrared
Brackett	4	5, 6, 7, . . .	Infrared

$n_f > n_i$, making the term in parentheses *negative*, so ΔE is *positive*. Each spectral line in the emission spectrum of hydrogen corresponds to a particular transition in a hydrogen atom. When we study a large number of hydrogen atoms, we observe all possible transitions and hence the corresponding spectral lines. The brightness of a spectral line depends on how many photons of the same wavelength are emitted.

To calculate the wavelength of an emission line, we substitute c/λ for v and then divide both sides of Equation 3.8 by hc. In addition, because wavelength can have only positive values, we take the absolute value of the right side of the *equation*. (In this case, we do so simply by eliminating the negative sign.)

Student Annotation: Because 2.18×10^{-18} J/hc = 1.096×10^7 m^{-1}, which to three significant figures is equal to R_∞, this equation is essentially the same as the Rydberg equation (Equation 3.6).

$$\frac{1}{\lambda} = \frac{2.18 \times 10^{-18} \text{ J}}{hc}\left(\frac{1}{n_f^2} - \frac{1}{n_i^2}\right)$$ **Equation 3.9**

The emission spectrum of hydrogen includes a wide range of wavelengths from the infrared to the ultraviolet. Table 3.1 lists the series of transitions in the hydrogen spectrum, each with a different value of n_f. The series are named after their discoverers (Lyman, Balmer, Paschen, and Brackett). The Balmer series was the first to be studied because some of its lines occur in the visible region.

Figure 3.11 shows transitions associated with spectral lines in each of the emission series. Each horizontal line represents one of the allowed energy levels for the electron in a hydrogen atom. The energy levels are labeled with their n values.

Student Hot Spot

Student data indicate you may struggle with calculating energy levels in the Bohr hydrogen atom. Access the SmartBook to view additional Learning Resources on this topic.

Figure 3.11 Energy levels in the hydrogen atom and the various emission series. Each series terminates at a different value of n.

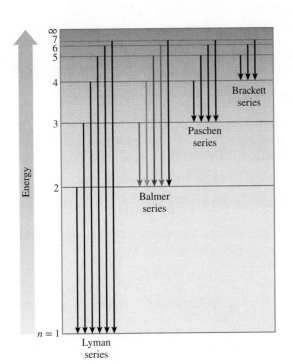

Worked Example 3.5 illustrates the use of Equation 3.9.

Worked Example 3.5

Calculate the wavelength (in nanometers) of the photon emitted when an electron transitions from the $n = 4$ state to the $n = 2$ state in a hydrogen atom.

Strategy Use Equation 3.9 to calculate λ.

Setup According to the problem, the transition is from $n = 4$ to $n = 2$, so $n_i = 4$ and $n_f = 2$. The required constants are $h = 6.63 \times 10^{-34}$ J $\cdot$ s and $c = 3.00 \times 10^8$ m/s.

Solution

$$\frac{1}{\lambda} = \frac{2.18 \times 10^{-18} \text{ J}}{hc} \left(\frac{1}{n_f^2} - \frac{1}{n_i^2} \right)$$

$$= \frac{2.18 \times 10^{-18} \text{ J}}{(6.63 \times 10^{-34} \text{ J} \cdot \text{s})(3.00 \times 10^8 \text{ m/s})} \left(\frac{1}{2^2} - \frac{1}{4^2} \right)$$

$$= 2.055 \times 10^6 \text{ m}^{-1}$$

Student Annotation: Remember to keep at least one extra digit in intermediate answers to avoid rounding error in the final result [Section 1.5].

Therefore,

$$\lambda = 4.87 \times 10^{-7} \text{ m} = 487 \text{ nm}$$

Think About It

Look again at the line spectrum of hydrogen in Figure 3.7 and make sure that your result matches one of them. Note that for an emission, n_i is always greater than n_f, and Equation 3.9 gives a positive result.

Practice Problem (A)TTEMPT What is the wavelength (in nanometers) of a photon emitted during a transition from the $n = 3$ state to the $n = 1$ state in the H atom?

Practice Problem (B)UILD What is the value of n_i for an electron that emits a photon of wavelength 93.14 nm when it returns to the ground state in the H atom?

Practice Problem (C)ONCEPTUALIZE For each pair of transitions, determine which one results in emission of the larger amount of energy.

(a) $n = 6$ to $n = 3$ $n = 3$ to $n = 2$

(b) $n = 3$ to $n = 1$ $n = 10$ to $n = 2$

(c) $n = 2$ to $n = 1$ $n = 99$ to $n = 2$

Section 3.4 Review

Bohr's Theory of the Hydrogen Atom

3.4.1 Calculate the energy of an electron in the $n = 3$ state in a hydrogen atom.
(a) 2.42×10^{-19} J   (c) 7.27×10^{-19} J   (e) -6.54×10^{-18} J
(b) -2.42×10^{-19} J   (d) -7.27×10^{-19} J

3.4.2 Calculate ΔE of an electron that goes from $n = 1$ to $n = 5$.
(a) 8.72×10^{-20} J   (c) 5.45×10^{-17} J   (e) -2.09×10^{-18} J
(b) -8.72×10^{-20} J   (d) 2.09×10^{-18} J

3.4.3 What is the wavelength (in meters) of light emitted when an electron in a hydrogen atom goes from $n = 5$ to $n = 3$?
(a) 4.87×10^{-7} m   (c) 1.28×10^{-6} m   (e) 1.02×10^{-7} m
(b) 6.84×10^{-7} m   (d) 3.65×10^{-7} m

3.4.4 What wavelength (in nanometers) corresponds to the transition of an electron in a hydrogen atom from $n = 2$ to $n = 1$?
(a) 182 nm   (c) 724 nm   (e) 122 nm
(b) 91.2 nm   (d) 812 nm

WAVE PROPERTIES OF MATTER

Bohr's theory was both fascinating and puzzling. It fit the experimental data, but physicists did not understand the underlying principle. Why, for example, was an electron restricted to orbiting the nucleus at certain fixed distances? For a decade, no one, not even Bohr himself, could offer a logical explanation. In 1924, French physicist Louis de Broglie[8] provided a solution to this puzzle. De Broglie reasoned that if energy (light) can, under certain circumstances, behave like a stream of particles (photons), then perhaps particles such as electrons can, under certain circumstances, exhibit wavelike properties.

The de Broglie Hypothesis

In developing his revolutionary theory, de Broglie incorporated his observations of macroscopic phenomena that exhibited quantized behavior. For example, a guitar string has certain discrete frequencies of vibration, like those shown in Figure 3.12(a). The waves generated by plucking a guitar string are *standing* or *stationary waves* because they do not travel along the string. Some points on the string, called **nodes,** do not move at all; that is, the amplitude of the wave at these points is *zero.* There is a node at each end, and there may be one or more nodes between the ends. The greater the frequency of vibration, the shorter the wavelength of the standing wave and the greater the number of nodes. According to de Broglie, an electron in an atom behaves like a *standing wave;* however, as Figure 3.12(a) shows, only certain wavelengths are possible or *allowed.*

De Broglie argued that if an electron does behave like a standing wave in the hydrogen atom, the wavelength must fit the circumference of the orbit exactly; that is, the circumference of the orbit must be an integral multiple of the wavelength, as shown in Figure 3.12(b). Otherwise, the wave would partially cancel itself by destructive interference on each successive orbit, quickly reducing its amplitude to zero.

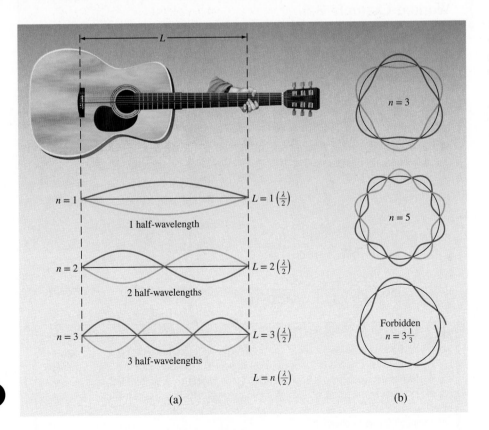

Figure 3.12 (a) Standing waves of a vibrating guitar string. The length of the string must be equal to a whole number times one-half the wavelength ($\lambda/2$). (b) In a circular orbit, only whole number multiples of wavelengths are allowed. Any fractional number of wavelengths would result in cancellation of the wave due to destructive interference.

[8]Louis Victor Pierre Raymond Duc de Broglie (1892–1987). A member of an old and noble family in France, he held the title of a prince. In his doctoral dissertation, he proposed that matter and radiation have the properties of both wave and particle. For this work, de Broglie was awarded the Nobel Prize in Physics in 1929.

The relationship between the circumference of an allowed orbit $(2\pi r)$ and the wavelength (λ) of the electron is given by

Equation 3.10	$2\pi r = n\lambda$

where r is the radius of the orbit, λ is the wavelength of the electron wave, and n is a positive integer (1, 2, 3, ...). Because n is an integer, r can have only certain values (integral multiples of λ) as n increases from 1 to 2 to 3 and so on. And, because the energy of the electron depends on the size of the orbit (or the value of r), the energy can have only certain values, too. Thus, the energy of the electron in a hydrogen atom, if it behaves like a standing wave, must be quantized.

De Broglie's reasoning led to the conclusion that waves can behave like particles and particles can exhibit wavelike properties. De Broglie deduced that the particle and wave properties are related by the following expression:

Equation 3.11	$\lambda = \dfrac{h}{mu}$

where λ, m, and u are the wavelength associated with a moving particle, its mass, and its velocity, respectively. Equation 3.11 implies that a particle in motion can be treated as a wave, and a wave can exhibit the properties of a particle. To help you remember this important point, notice that the left side of Equation 3.11 involves the wavelike property of wavelength, whereas the right side involves mass, a property of particles. A wavelength calculated using Equation 3.11 is typically referred to as a *de Broglie wavelength.*

Worked Example 3.6 illustrates how de Broglie's theory and Equation 3.11 can be applied.

Student Hot Spot

Student data indicate you may struggle with the de Broglie wavelength. Access the SmartBook to view additional Learning Resources on this topic.

Worked Example 3.6

Calculate the de Broglie wavelength of the "particle" in the following two cases: (a) a 25-g bullet traveling at 612 m/s and (b) an electron ($m = 9.109 \times 10^{-31}$ kg) moving at 63.0 m/s.

Strategy Use Equation 3.11 $\left(\lambda = \frac{h}{mu}\right)$ to calculate the de Broglie wavelengths. Remember that the mass in Equation 3.11 must be expressed in kilograms in order for the units to cancel properly.

Setup Planck's constant, h, is 6.63×10^{-34} J · s or, for the purpose of making the unit cancellation obvious, 6.63×10^{-34} kg · m²/s.

Student Annotation: 1 J = 1 kg · m²/s².

Solution

(a) $25 \text{ g} \times \dfrac{1 \text{ kg}}{1000 \text{ g}} = 0.025 \text{ kg}$

$$\lambda = \frac{h}{mu} = \frac{6.63 \times 10^{-34} \text{ kg} \cdot \text{m}^2/\text{s}}{(0.025 \text{ kg})(612 \text{ m/s})} = 4.3 \times 10^{-35} \text{ m}$$

(b) $\lambda = \dfrac{h}{mu} = \dfrac{6.63 \times 10^{-34} \text{ kg} \cdot \text{m}^2/\text{s}}{(9.109 \times 10^{-31} \text{ kg})(63.0 \text{ m/s})} = 1.16 \times 10^{-5} \text{ m}$

Think About It

While you are new at solving these problems, always write out the units of Planck's constant (J · s) as kg · m²/s. This will enable you to check your unit cancellations and detect common errors such as expressing mass in grams rather than kilograms. Note that the calculated wavelength of a macroscopic object, even one as small as a bullet, is extremely small. An object must be at least as small as a subatomic particle in order for its wavelength to be large enough for us to observe.

Practice Problem Ⓐ**TTEMPT** Calculate the de Broglie wavelength (in nanometers) of a hydrogen atom ($m = 1.674 \times 10^{-27}$ kg) moving at 1500 cm/s.

Practice Problem Ⓑ**UILD** Use Equation 3.11 to calculate the *momentum, p* (defined as mass times velocity, $m \times u$) associated with a photon of radiation of wavelength 810 nm. The velocity of a photon is the speed of light, *c*.

Practice Problem Ⓒ**ONCEPTUALIZE** Consider the impact of early electron diffraction experiments on scientists' understanding of the behavior of matter. Which of the following imaginary macroscopic experiments most closely corresponds to the remarkable outcome of electron diffraction?

(a) Combining one marble with another by one method yields two marbles; but combining the two marbles by another method yields four marbles.

(b) Combining one marble with another by one method yields two marbles; but combining the two marbles by another method yields zero marbles.

(c) Combining one marble with another by any method yields two marbles.

> **Student Annotation:** Momentum has units of kg · m/s or N · s, where N is the *newton*, the SI unit of *force*. The newton is a derived SI unit: $1\,N = 1\,kg \cdot m/s^2$.

Diffraction of Electrons

Shortly after de Broglie introduced his equation and predicted that electrons should exhibit wave properties, successful electron diffraction experiments were carried out by American physicists Clinton Davisson[9] and Lester Germer[10] in the United States and physicist G.P. Thomson[11] in England. These experiments demonstrated that electrons do indeed possess wavelike properties. By directing a beam of electrons (which are most definitely particles) through a thin piece of gold foil, Thomson obtained a set of concentric rings on a screen, similar to the diffraction pattern observed when X rays (which are most definitely waves) were used. Figure 3.13 shows X-ray and electron diffraction patterns for aluminum.

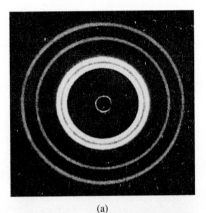

(a)

Section 3.5 Review

Wave Properties of Matter

3.5.1 Calculate the de Broglie wavelength associated with a helium-4 atom (4.00 amu) moving at 3.0×10^6 m/s.
  (a) 2.0×10^{-20} m (d) 1.8×10^{-19} m
  (b) 3.3×10^{-11} m (e) 6.6×10^{-27} m
  (c) 3.3×10^{-14} m

3.5.2 At what speed must a helium-4 atom be traveling to exhibit a de Broglie wavelength of 1.50 Å?
  (a) 1.10×10^{-24} m/s (d) 665 m/s
  (b) 6.65×10^{-2} m/s (e) 2650 m/s
  (c) 1.50×10^{-3} m/s

3.5.3 Determine the minimum speed required for a hydrogen atom to have a de Broglie wavelength in the ultraviolet region (10–400 nm) of the electromagnetic spectrum.
  (a) 3.96 m/s (d) 9.90×10^{-4} m/s
  (b) 0.990 m/s (e) 52.4 m/s
  (c) 1.64×10^{-27} m/s

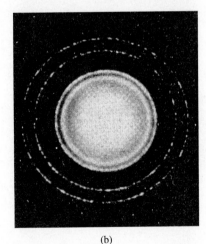

(b)

Figure 3.13 (a) X-ray diffraction pattern of aluminum foil. (b) Electron diffraction of aluminum foil. The similarity of these two patterns shows that electrons can behave like X rays and display wave properties.

(a,b): Source: Erwin Schrodinger, Wave Mechanical Model of the Atom, 1925.

---

[9]Clinton Joseph Davisson (1881–1958). He and G.P. Thomson shared the Nobel Prize in Physics in 1937 for demonstrating the wave properties of electrons.

[10]Lester Halbert Germer (1896–1972). Discoverer (with Davisson) of the wave properties of electrons.

[11]George Paget Thomson (1892–1975). Son of J.J. Thomson, he received the Nobel Prize in Physics in 1937, along with Clinton Davisson, for demonstrating the wave properties of electrons.

3.6   QUANTUM MECHANICS

The discovery that waves could have matterlike properties and that matter could have wavelike properties was revolutionary. Although scientists had long believed that energy and matter were distinct entities, the distinction between them, at least at the atomic level, was no longer clear. Bohr's theory was tremendously successful in explaining the line spectrum of hydrogen, but it failed to explain the spectra of atoms with more than one electron. The electron appeared to behave as a particle in some circumstances and as a wave in others. Neither description could completely explain the behavior of electrons in atoms. This left scientists frustrated in their quest to understand exactly where the electrons in an atom are.

The Uncertainty Principle

To describe the problem of trying to locate a subatomic particle that behaves like a wave, Werner Heisenberg[12] formulated what is now known as the ***Heisenberg uncertainty principle:*** It is impossible to know simultaneously both the *momentum p* (defined as mass times velocity, $m \times u$) and the *position x* of a particle with certainty. Stated mathematically,

Student Annotation: Like the de Broglie wavelength equation, Equation 3.12 requires that mass be expressed in kilograms. Unit cancellation will be more obvious if you express Planck's constant in kg · m²/s rather than J · s.

Equation 3.12	$$\Delta x \cdot \Delta p \geq \frac{h}{4\pi}$$

For a particle of mass *m*,

Equation 3.13	$$\Delta x \cdot m\Delta u \geq \frac{h}{4\pi}$$

where Δx and Δu are the uncertainties in measuring the position and velocity of the particle, respectively. The significance of Equation 3.13 is this: If the measured uncertainties of position and velocity are large (say, in a crude experiment), then their product can be substantially greater than $h/4\pi$ (hence the > part of the ≥ sign). But even with the most favorable conditions for measuring position and velocity, the product of their uncertainties can never be *less* than $h/4\pi$ (hence the = part of the ≥ sign). Thus, making measurement of the velocity of a particle *more* precise (i.e., making Δu a *small* quantity) means that the position must become correspondingly *less* precise (i.e., Δx will become *larger*). Similarly, if the position of the particle is known more precisely, its velocity measurement must become less precise.

If the Heisenberg uncertainty principle is applied to the hydrogen atom, we find that the electron cannot orbit the nucleus in a well-defined path, as Bohr thought. If it did, we could determine precisely both the position of the electron (from the radius of the orbit) and its speed (from its kinetic energy) at the same time. This would violate the uncertainty principle.

Worked Example 3.7 shows how to use the Heisenberg uncertainty principle.

Worked Example 3.7

An electron in a hydrogen atom is known to have a velocity of 5×10^6 m/s ± 1 percent. Using the uncertainty principle, calculate the minimum uncertainty in the position of the electron and, given that the diameter of the hydrogen atom is less than 1 angstrom (Å), comment on the magnitude of this uncertainty compared to the size of the atom.

Strategy The uncertainty in the velocity, 1 percent of 5×10^6 m/s, is Δu. Using Equation 3.13 $\left(\Delta x \cdot m\Delta u \geq \frac{h}{4\pi}\right)$, calculate Δx and compare it with the diameter of the hydrogen atom.

---

[12]Werner Karl Heisenberg (1901–1976). One of the founders of modern quantum theory, Heisenberg, a German physicist, received the Nobel Prize in Physics in 1932.

Setup The mass of an electron (from Table 2.1, rounded to three significant figures and converted to kilograms) is 9.11×10^{-31} kg. Planck's constant, h, is 6.63×10^{-34} kg · m²/s.

Solution

$$\Delta u = 0.01 \times 5 \times 10^6 \text{ m/s} = 5 \times 10^4 \text{ m/s}$$

$$\Delta x = \frac{h}{4\pi \cdot m\Delta u}$$

Therefore,

$$\Delta x = \frac{6.63 \times 10^{-34} \text{ kg} \cdot \text{m}^2/\text{s}}{4\pi(9.11 \times 10^{-31} \text{ kg})(5 \times 10^4 \text{ m/s})} \geq 1 \times 10^{-9} \text{ m}$$

> **Think About It**
> A common error is expressing the mass of the particle in grams instead of kilograms, but you should discover this inconsistency if you check your unit cancellation carefully. Remember that if one uncertainty is small, the other must be large. The uncertainty principle applies in a practical way only to submicroscopic particles. In the case of a macroscopic object, where the mass is much larger than that of an electron, small uncertainties, relative to the size of the object, are possible for both position and velocity.

The *minimum* uncertainty in the position x is 1×10^{-9} m = 10 Å. The uncertainty in the electron's position is 10 times larger than the atom!

Practice Problem (A)**TTEMPT** Calculate the minimum uncertainty in the position of the 25-g bullet from Worked Example 3.6 given that the uncertainty in its velocity is ±1 percent.

Practice Problem (B)**UILD** (a) Calculate the minimum uncertainty in the momentum of an object for which the uncertainty in position is 3 Å. (b) To what minimum uncertainty in velocity does this correspond if the particle is a neutron (mass = 1.0087 amu)? (c) To what minimum uncertainty in velocity does it correspond if the particle is an electron (mass = 5.486×10^{-4} amu)?

Practice Problem (C)**ONCEPTUALIZE** Using Equation 3.13, we can calculate the minimum uncertainty in the position or the velocity of any moving particle, including a macroscopic object such as a marble. Calculate the uncertainty in position of a 10-g marble moving at 2.5 m/s (±5 percent) and comment on the significance of your result.

The Schrödinger Equation

Bohr made a significant contribution to our understanding of atoms, and his suggestion that the energy of an electron in an atom is quantized remains unchallenged, but his theory did not provide a complete description of the behavior of electrons in atoms. In 1926, Austrian physicist Erwin Schrödinger,[13] using a complicated mathematical technique, formulated an equation that describes the behavior and energies of submicroscopic particles in general, an equation analogous to Newton's laws of motion for macroscopic objects. The *Schrödinger equation* requires advanced calculus to solve, and we will not discuss it here. The equation, however, incorporates both particle behavior, in terms of mass m, and wave behavior, in terms of a *wave function* ψ (psi), which depends on the location in space of the system (such as an electron in an atom).

The wave function itself has no direct physical meaning; however, the probability of finding the electron in a certain region in space is proportional to the square of the wave function, ψ^2. The idea of relating ψ^2 to probability stemmed from a wave theory analogy. According to wave theory, the intensity of light is proportional to the square of the amplitude of the wave, or ψ^2. The most likely place to find a photon is where the intensity is greatest—that is, where the value of ψ^2 is greatest. A similar argument associates ψ^2 with the likelihood of finding an electron in regions surrounding the nucleus.

Schrödinger's equation launched an entirely new field called *quantum mechanics* (or *wave mechanics*), and began a new era in physics and chemistry. We now refer to the developments in quantum theory from 1913—when Bohr presented his model of the hydrogen atom—to 1926 as "old quantum theory."

---

[13]Erwin Schrödinger (1887–1961). Schrödinger formulated wave mechanics, which laid the foundation for modern quantum theory. He received the Nobel Prize in Physics in 1933.

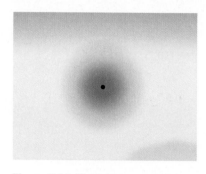

Figure 3.14 Representation of the electron density distribution surrounding the nucleus in the hydrogen atom. It shows a higher probability of finding the electron closer to the nucleus.

The Quantum Mechanical Description of the Hydrogen Atom

The Schrödinger equation specifies the possible energy states the electron can occupy in a hydrogen atom and identifies the corresponding wave functions (ψ). These energy states and wave functions are characterized by a set of *quantum numbers* (to be discussed shortly) with which we can construct a comprehensive model of the hydrogen atom.

Although quantum mechanics does not allow us to specify the exact location of an electron in an atom, it does define the region where the electron is most likely to be at a given time. The concept of ***electron density*** gives the probability that an electron will be found in a particular region of an atom. The square of the wave function, ψ^2, defines the distribution of electron density in three-dimensional space around the nucleus. Regions of high electron density represent a high probability of locating the electron (Figure 3.14).

To distinguish the quantum mechanical description of an atom from Bohr's model, we speak of an atomic *orbital,* rather than an orbit. An ***atomic orbital*** can be thought of as the wave function of an electron in an atom. When we say that an electron is in a certain orbital, we mean that the distribution of the electron density or the probability of locating the electron in space is described by the square of the wave function associated with that orbital. An atomic orbital, therefore, has a characteristic energy, as well as a characteristic distribution of electron density.

Section 3.6 Review

Quantum Mechanics

3.6.1 What is the minimum uncertainty in the position of an electron moving at a speed of 4×10^6 m/s $\pm$ 1 percent? (The mass of an electron is 9.11×10^{-31} kg.)

  (a) 2×10^{-8} m   (d) 6×10^{-9} m
  (b) 1×10^{-12} m   (e) 7×10^{-8} m
  (c) 1×10^{-9} m

3.6.2 What is the minimum uncertainty in the position of a proton moving at a speed of 4×10^6 m/s $\pm$ 1 percent? (The mass of a proton is 1.67×10^{-27} kg.)

  (a) 8×10^{-13} m   (d) 1×10^{-13} m
  (b) 3×10^{-12} m   (e) 8×10^{-10} m
  (c) 4×10^{-11} m

3.7 QUANTUM NUMBERS

In Bohr's model of the hydrogen atom, only one number, *n,* was necessary to describe the location of the electron. In quantum mechanics, three ***quantum numbers*** are required to describe the *distribution of electron density* in an atom. These numbers are derived from the mathematical solution of Schrödinger's equation for the hydrogen atom. They are called the *principal* quantum number, the *angular momentum* quantum number, and the *magnetic* quantum number. Each atomic orbital in an atom is characterized by a unique set of these three quantum numbers.

Student Annotation: The three quantum numbers n, ℓ, and m_ℓ specify the *size, shape,* and *orientation* of an orbital, respectively.

Principal Quantum Number (*n*)

The ***principal quantum number (n)*** designates the *size* of the orbital. The larger n is, the greater the average distance of an electron in the orbital from the nucleus and therefore the larger the orbital. The principal quantum number can have integral values of 1, 2, 3, and so forth, and it corresponds to the quantum number in Bohr's model of the hydrogen atom. Recall from Equation 3.7 that in a hydrogen atom, the value of n determines the energy of an orbital. (As we will see shortly, this is *not* the case for an atom that contains more than one electron.)

Angular Momentum Quantum Number (ℓ)

The *angular momentum quantum number* (ℓ) describes the *shape* of the atomic orbital [▶I Section 3.8]. The values of ℓ are integers that depend on the value of the principal quantum number, n. For a given value of n, the possible values of ℓ range from 0 to $n - 1$. If $n = 1$, there is only one possible value of ℓ; that is, 0 ($n - 1$ where $n = 1$). If $n = 2$, there are two values of ℓ: 0 and 1. If $n = 3$, there are three values of ℓ: 0, 1, and 2. The value of ℓ is designated by the letters s, p, d, and f as follows:[14]

ℓ	0	1	2	3
Orbital designation	s	p	d	f

Thus, if $\ell = 0$, we have an s orbital; if $\ell = 1$, we have a p orbital; and so on.

A collection of orbitals with the same value of n is frequently called a *shell*. One or more orbitals with the same n and ℓ values are referred to as a *subshell*. For example, the shell designated by $n = 2$ is composed of two subshells: $\ell = 0$ and $\ell = 1$ (the allowed ℓ values for $n = 2$). These subshells are called the $2s$ and $2p$ subshells where 2 denotes the value of n, and s and p denote the values of ℓ.

Magnetic Quantum Number (m_ℓ)

The *magnetic quantum number* (m_ℓ) describes the orientation of the orbital in space (Section 3.8). Within a subshell, the value of m_ℓ depends on the value of ℓ. For a certain value of ℓ, there are ($2\ell + 1$) integral values of m_ℓ as follows:

$$-\ell, \, ... \, 0, \, ... + \ell$$

If $\ell = 0$, there is only one possible value of m_ℓ: 0. If $\ell = 1$, then there are *three* values of m_ℓ: −1, 0, and +1. If $\ell = 2$, there are *five* values of m_ℓ, namely, −2, −1, 0, +1, and +2, and so on. The number of m_ℓ values indicates the number of *orbitals* in a subshell with a particular ℓ value; that is, each m_ℓ value refers to a different orbital.

Table 3.2 summarizes the allowed values of the three quantum numbers, n, ℓ, and m_ℓ, and Figure 3.15 illustrates schematically how the allowed values of quantum numbers give rise to the number of subshells and orbitals in each shell of an atom. The number of subshells in a shell is equal to n, and the number of orbitals in a shell is equal to n^2.

> **Student Hot Spot**
>
> Student data indicate you may struggle with describing energy levels with quantum numbers. Access the SmartBook to view additional Learning Resources on this topic.

TABLE 3.2	Allowed Values of the Quantum Numbers n, ℓ, and m_ℓ		
When n is	**ℓ can be**	**When ℓ is**	**m_ℓ can be**
1	only 0	0	only 0
2	0 or 1	0	only 0
		1	−1, 0, or +1
3	0, 1, or 2	0	only 0
		1	−1, 0, or +1
		2	−2, −1, 0, +1, or +2
4	0, 1, 2, or 3	0	only 0
		1	−1, 0, or +1
		2	−2, −1, 0, +1, or +2
		3	−3, −2, −1, 0, +1, +2, or +3
.	.	.	.
.	.	.	.
.	.	.	.

---

[14]The unusual sequence of letters (s, p, d, and f) has an historical origin. Physicists who studied atomic emission spectra tried to correlate their observations of spectral lines with the energy states involved in the transitions. They described the emission lines as *s*harp, *p*rincipal, *d*iffuse, and *f*undamental.

Figure 3.15 Illustration of how quantum numbers designate shells, subshells, and orbitals.

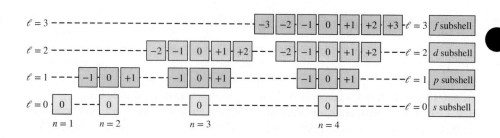

Worked Example 3.8 gives you some practice with the allowed values of quantum numbers.

Worked Example 3.8

What are the possible values for the magnetic quantum number (m_ℓ) when the principal quantum number (n) is 3 and the angular momentum quantum number (ℓ) is 1?

Strategy Use the rules governing the allowed values of m_ℓ. Recall that the possible values of m_ℓ depend on the value of ℓ, not on the value of n.

Setup The possible values of m_ℓ are $-\ell, ... 0, ... +\ell$.

Solution The possible values of m_ℓ are -1, 0, and $+1$.

> **Think About It**
> Consult Table 3.2 to make sure your answer is correct. Table 3.2 confirms that it is the value of ℓ, not the value of n, that determines the possible values of m_ℓ.

Practice Problem (A)TTEMPT What are the possible values for m_ℓ when the principal quantum number (n) is 2 and the angular momentum quantum number (ℓ) is 0?

Practice Problem (B)UILD What are the possible values for m_ℓ when the principal quantum number (n) is 3 and the angular momentum quantum number (ℓ) is 2?

Practice Problem (C)ONCEPTUALIZE Imagine that a shoe cobbler's place of business has trendy, V-shaped cabinets, four of which are shown here. When a customer brings in a pair of shoes for repair, the cobbler keeps the pair in a shoebox in one of these cabinets. The location of each pair of shoes is recorded using a set of numbers that designate the cabinet (C), the shelf (S), and the specific box (B).

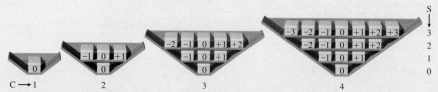

Each cabinet has a number, corresponding to its number of shelves. Thus, for the cabinets shown here, the value of C can be 1, 2, 3, or 4. Shelves within each cabinet are numbered sequentially from the bottom, starting with zero. For the smallest cabinet, with just one shelf, 0 is the only shelf designation. For the other cabinets, shelf designations can have integer values of 0 through C − 1. In addition, each individual box has a number on it. Boxes in the bottom row (row 0) all have the number 0 on them. Any box that resides directly above a box labeled 0 is also labeled 0. Boxes to the right or the left of the zero box on each shelf are numbered sequentially, starting with +1 (for boxes on the right), and starting with −1 (for boxes on the left). Using this numbering system, the cobbler can specify the location of a pair of shoes by designating three numbers: C, S, and B. For each of the following sets of numbers (C, S, B) determine whether or not they designate a box in one of the cabinets. For a set of numbers that does *not* designate a box in one of the cabinets, explain why.

(a) (1, 0, 0); (b) (0, 0, 0); (c) (3, 2, −2); (d) (2, 0, 0); (e) (4, 3, +1); (f) (2, 2, +2).

Electron Spin Quantum Number (m_s)

Whereas three quantum numbers are sufficient to describe an atomic orbital, an additional quantum number becomes necessary to describe an *electron* that *occupies* the orbital.

Experiments on the emission spectra of hydrogen and sodium atoms indicated that each line in the emission spectra could be split into two lines by the application of an external magnetic field. The only way physicists could explain these results was to assume that electrons act like tiny magnets. If electrons are thought of as spinning on their own axes, as Earth does, their magnetic properties can be accounted for. According to electromagnetic theory, a spinning charge generates a magnetic field, and it is this motion that causes an electron to behave like a magnet. Figure 3.16 shows the two possible "spins" of an electron. To specify the electron's spin, we use the ***electron spin quantum number (m_s).*** Because there are two possible directions of spin, opposite each other, m_s has two possible values: $+\frac{1}{2}$ and $-\frac{1}{2}$.

Conclusive proof of electron spin was established by German physicists Otto Stern[15] and Walther Gerlach[16] in 1924. Figure 3.17 shows the basic experimental arrangement. A beam of gaseous atoms generated in a hot furnace passes through a nonuniform magnetic field. The interaction between an electron and the magnetic field causes the atom to be deflected from its straight-line path. Because the direction of spin is random, the electrons in *half* of the atoms will be spinning in one direction. Those atoms will be deflected in one way. The electrons in the other half of the atoms will be spinning in the *opposite* direction. Those atoms will be deflected in the other direction. Thus, two spots of equal intensity are observed on the detecting screen.

To summarize, we can designate an *orbital* in an atom with a set of *three* quantum numbers. These three quantum numbers indicate the size (n), shape (ℓ), and orientation (m_ℓ) of the orbital. A fourth quantum number (m_s) is necessary to designate the spin of an electron in the orbital.

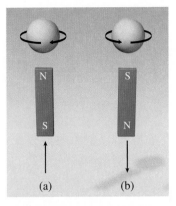

Figure 3.16 (a) Clockwise and (b) counterclockwise spins of an electron. The magnetic fields generated by these two spinning motions are analogous to those from the two magnets. The upward and downward arrows are used to denote the direction of spin.

Student Annotation: Two electrons in the same orbital with opposite spins are referred to as "*paired.*"

Figure 3.17 Experimental arrangement for demonstrating electron spin. A beam of atoms is directed through a magnetic field. When a hydrogen atom, with a single electron, passes through the field, it is deflected in one direction or the other, depending on the direction of the electron's spin. In a stream consisting of many atoms, there will be equal distributions of the two kinds of spins, so two spots of equal intensity are detected on the screen.

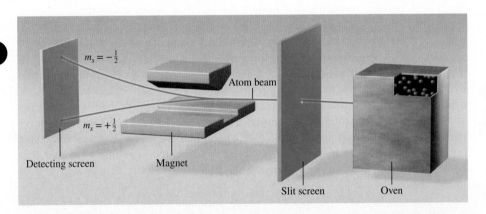

$m_s = -\frac{1}{2}$

$m_s = +\frac{1}{2}$

Atom beam

Detecting screen   Magnet   Slit screen   Oven

Section 3.7 Review

Quantum Numbers

3.7.1 How many orbitals are there in a subshell designated by the quantum numbers $n = 3$, $\ell = 2$?
(a) 2   (b) 3   (c) 5   (d) 7   (e) 10

3.7.2 How many subshells are there in the shell designated by $n = 3$?
(a) 1   (b) 2   (c) 3   (d) 6   (e) 9

3.7.3 What is the total number of orbitals in the shell designated by $n = 3$?
(a) 1   (b) 2   (c) 3   (d) 6   (e) 9

3.7.4 What is the minimum value of the principal quantum number for an orbital in which $m_\ell = +2$?
(a) 1   (b) 2   (c) 3   (d) 4   (e) 5

---

[15]Otto Stern (1888–1969). He made important contributions to the study of the magnetic properties of atoms and the kinetic theory of gases. Stern was awarded the Nobel Prize in Physics in 1943.

[16]Walther Gerlach (1889–1979). Gerlach's main area of research was in quantum theory.

3.8   ATOMIC ORBITALS

Strictly speaking, an atomic orbital does not have a well-defined shape because the wave function characterizing the orbital extends from the nucleus to infinity. In that sense, it is difficult to say what an orbital looks like. On the other hand, it is certainly useful to think of orbitals as having specific shapes. Being able to visualize atomic orbitals is essential to understanding the formation of chemical bonds and molecular geometry, which are discussed in Chapters 5 and 7. In this section, we will look at each type of orbital separately.

s Orbitals

For any value of the principal quantum number (n), the value 0 is possible for the angular momentum quantum number (ℓ), corresponding to an s subshell. Furthermore, when $\ell = 0$, the magnetic quantum number (m_ℓ) has only one possible value, 0, corresponding to an s orbital. Therefore, there is an s subshell in every shell, and each s subshell contains just one orbital, an *s orbital.*

Figure 3.18 illustrates three ways to represent the distribution of electrons: the probability density, the spherical distribution of electron density, and the radial probability distribution (the probability of finding the electron as a function of distance from the nucleus) for the 1s, 2s, and 3s orbitals of hydrogen. The boundary surface (the outermost surface of the spherical representation) is a common way to represent atomic orbitals, incorporating the volume in which there is about a 90 percent probability of finding the electron at any given time.

All s orbitals are spherical in shape but differ in size, which increases as the principal quantum number increases. The radial probability distribution for the 1s orbital exhibits a maximum at 52.9 pm (0.529 Å) from the nucleus. Interestingly, this distance is equal to the radius of the $n = 1$ orbit in the Bohr model of the hydrogen atom. The radial probability distribution plots for the 2s and 3s orbitals exhibit two and three maxima, respectively, with the greatest probability occurring at a greater distance from the nucleus as n increases. Between the two maxima for the 2s orbital, there is a point on the plot where the probability drops to zero. This corresponds to a *node* in the electron density, where the standing wave has zero amplitude. There are two such nodes in the radial probability distribution plot of the 3s orbital.

Although the details of electron density variation within each boundary surface are lost, the most important features of atomic orbitals are their overall shapes and relative sizes, which are adequately represented by boundary surface diagrams.

p Orbitals

When the principal quantum number (n) is 2 or greater, the value 1 is possible for the angular momentum quantum number (ℓ), corresponding to a p subshell. And, when $\ell = 1$, the magnetic quantum number (m_ℓ) has three possible values: -1, 0, and $+1$, each corresponding to a different *p orbital.* Therefore, there is a p subshell in every shell for which $n \geq 2$, and each p subshell contains three p orbitals. These three p orbitals are labeled p_x, p_y, and p_z (Figure 3.19), with the subscripted letters indicating the axis along which each orbital is oriented. These three p orbitals are identical in size, shape, and energy; they differ from one another only in orientation. Note, however, that there is no simple relation between the values of m_ℓ and the x, y, and z directions. For our purpose, you need only remember that because there are three possible values of m_ℓ, there are three p orbitals with different orientations.

The boundary surface diagrams of p orbitals in Figure 3.19 show that each p orbital can be thought of as two lobes on opposite sides of the nucleus. Like s orbitals, p orbitals increase in size from 2p to 3p to 4p orbital and so on.

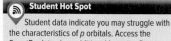

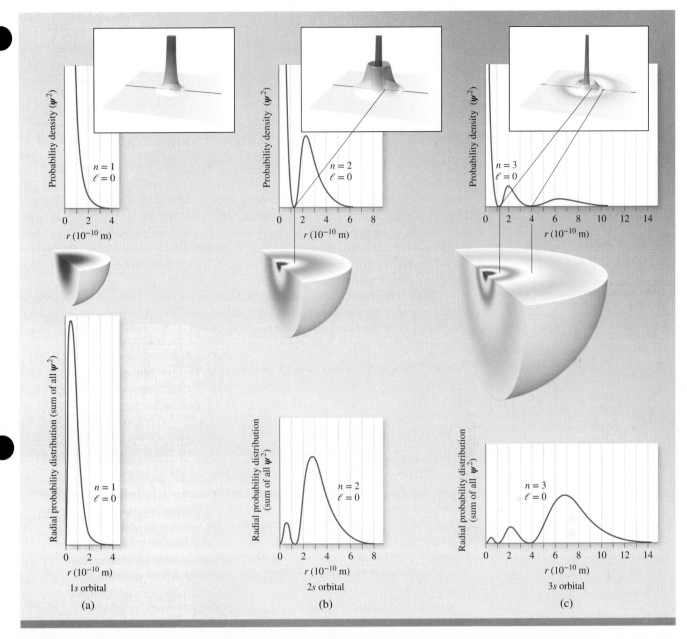

Figure 3.18 From top to bottom, the *probability density* and corresponding relief map, the distribution of electron density represented spherically with shading that corresponds to the relief map above, and the *radial probability distribution* for (a) the 1*s*, (b) the 2*s*, and (c) the 3*s* orbitals of hydrogen.

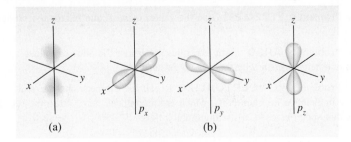

Figure 3.19 (a) Electron distribution in a *p* orbital. (b) Boundary surfaces for the p_x, p_y, and p_z orbitals.

d Orbitals and Other Higher-Energy Orbitals

When the principal quantum number (n) is 3 or greater, the value 2 is possible for the angular momentum quantum number (ℓ), corresponding to a *d* subshell. When $\ell = 2$, the magnetic quantum number (m_ℓ) has *five* possible values, −2, −1, 0, +1, and +2,

Figure 3.20 Boundary surfaces for the *d* orbitals.

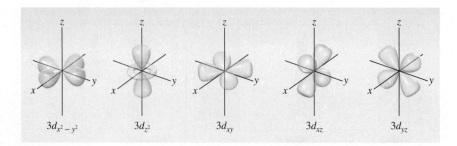

$3d_{x^2-y^2}$ $3d_{z^2}$ $3d_{xy}$ $3d_{xz}$ $3d_{yz}$

each corresponding to a different ***d orbital.*** Again, there is no direct correspondence between a given orientation and a particular m_ℓ value. All the 3*d* orbitals in an atom are identical in energy and are labeled with subscripts denoting their orientation with respect to the *x, y,* and *z* axes and to the planes defined by them. The *d* orbitals that have higher principal quantum numbers (4*d*, 5*d*, etc.) have shapes similar to those shown for the 3*d* orbitals in Figure 3.20.

The ***f orbitals*** are important when accounting for the behavior of elements with atomic numbers greater than 57, but their shapes are difficult to represent. In general chemistry, we will not concern ourselves with the shapes of orbitals having ℓ values greater than 2.

Worked Example 3.9 shows how to label orbitals with quantum numbers.

Worked Example 3.9

List the values of n, ℓ, and m_ℓ for each of the orbitals in a 4*d* subshell.

Strategy Consider the significance of the number and the letter in the 4*d* designation and determine the values of n and ℓ. There are multiple possible values for m_ℓ, which will have to be deduced from the value of ℓ.

Setup The integer at the beginning of an orbital designation is the principal quantum number (n). The letter in an orbital designation gives the value of the angular momentum quantum number (ℓ). The magnetic quantum number (m_ℓ) can have integral values of $-\ell$, ... 0, ... $+\ell$.

Solution The values of n and ℓ are 4 and 2, respectively, so the possible values of m_ℓ are -2, -1, 0, $+1$, and $+2$.

Think About It
Consult Figure 3.15 to verify your answers.

Practice Problem ATTEMPT Give the values of n, ℓ, and m_ℓ for the orbitals in a 3*d* subshell.

Practice Problem BUILD Using the rules governing the allowed values of quantum numbers, explain why there is no 2*d* subshell.

Practice Problem CONCEPTUALIZE Recall the cabinets, shelves, and shoeboxes in Practice Problem 3.8C, which are reproduced here. Write the set of three numbers that specifies each of the highlighted boxes.

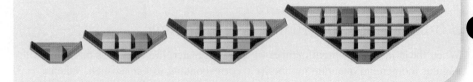

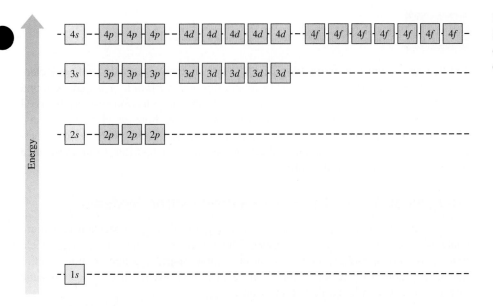

Figure 3.21 Orbital energy levels in the hydrogen atom. Each box represents one orbital. Orbitals with the same principal quantum number (n) all have the same energy.

Energies of Orbitals

The energies of orbitals in the hydrogen atom depend only on the value of the principal quantum number (n), and energy increases as n increases. For this reason, orbitals in the same shell have the same energy regardless of their subshell (Figure 3.21).

$$1s < 2s = 2p < 3s = 3p = 3d < 4s = 4p = 4d = 4f$$

Thus, all four orbitals (one $2s$ and three $2p$) in the second shell have the same energy; all nine orbitals (one $3s$, three $3p$, and five $3d$) in the third shell have the same energy; and all sixteen orbitals (one $4s$, three $4p$, five $4d$, and seven $4f$) in the fourth shell have the same energy. The energy picture is more complex for many-electron atoms than it is for hydrogen, as is discussed in Section 3.9.

Section 3.8 Review

Atomic Orbitals

3.8.1 How many orbitals are there in the $5f$ subshell?
(a) 5
(b) 7
(c) 14
(d) 16
(e) 28

3.8.2 The energy of an orbital in the hydrogen atom depends on _____.
(a) n, ℓ, and m_ℓ
(b) n and ℓ
(c) n only
(d) ℓ only
(e) m_ℓ only

3.8.3 In a hydrogen atom, which orbitals are higher in energy than a $3s$ orbital? (Select all that apply.)
(a) $3p$
(b) $4s$
(c) $2p$
(d) $3d$
(e) $4p$

3.8.4 Which of the following sets of quantum numbers, n, ℓ, and m_ℓ, corresponds to a $3p$ orbital?
(a) $3, 0, 0$
(b) $3, 1, 0$
(c) $3, 2, -1$
(d) $1, 1, -2$
(e) $1, 3, 1$

3.9 ELECTRON CONFIGURATIONS

The hydrogen atom is a particularly simple system because it contains only one electron. The electron may reside in the 1s orbital (the *ground state*), or it may be found in some higher-energy orbital (an *excited state*). With many-electron systems, we need to know the ground-state ***electron configuration***—that is, how the electrons are distributed in the various atomic orbitals. To do this, we need to know the relative energies of atomic orbitals in a many-electron system, which differ from those in a one-electron system such as hydrogen.

Energies of Atomic Orbitals in Many-Electron Systems

Student Annotation: "Splitting" of energy levels refers to the splitting of a shell into subshells of different energies, as shown in Figure 3.23.

Consider the two emission spectra shown in Figure 3.22. The spectrum of helium contains more lines than that of hydrogen. This indicates that there are more possible transitions, corresponding to emission in the visible range, in a helium atom than in a hydrogen atom. This is due to the *splitting* of energy levels caused by electrostatic interactions between helium's two electrons.

Figure 3.23 shows the general order of orbital energies in a many-electron atom. In contrast to the hydrogen atom, in which the energy of an orbital depends only on the value of *n* (see Figure 3.21), the energy of an orbital in a many-electron system

Figure 3.22 Comparison of the emission spectra of H and He.

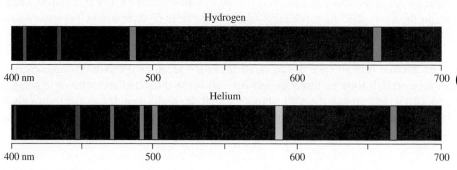

Figure 3.23 Orbital energy levels in many-electron atoms. For a given value of *n*, orbital energy increases with the value of *ℓ*.

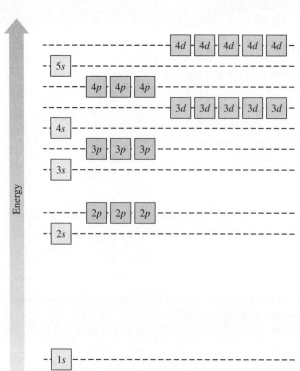

depends on both the value of n and the value of ℓ. For example, $3p$ orbitals all have the same energy, but they are higher in energy than the $3s$ orbital and lower in energy than the $3d$ orbitals. In a many-electron atom, for a given value of n, the energy of an orbital increases with increasing value of ℓ. One important consequence of the splitting of energy levels is the relative energies of d orbitals in one shell and the s orbital in the next higher shell. As Figure 3.23 shows, the $4s$ orbital is lower in energy than the $3d$ orbitals. Likewise, the $5s$ orbital is lower in energy than the $4d$ orbital, and so on. This fact becomes important when we determine the order in which electrons in an atom populate the atomic orbitals.

The Pauli Exclusion Principle

According to the ***Pauli exclusion principle,***[17] no two electrons in an atom can have the same four quantum numbers. If two electrons in an atom have the same n, ℓ, and m_ℓ values (meaning that they occupy the same *orbital*), then they must have different values of m_s; that is, one must have $m_s = +\frac{1}{2}$ and the other must have $m_s = -\frac{1}{2}$. Because there are only two possible values for m_s, and no two electrons in the same orbital may have the same value for m_s, a maximum of *two* electrons may occupy an atomic orbital, and these two electrons must have opposite spins.

We can indicate the arrangement of electrons in atomic orbitals with labels that identify each orbital (or subshell) and the number of electrons in it. Thus, we could describe a hydrogen atom in the ground state using $1s^1$.

Student Annotation: $1s^1$ is read as "one s one."

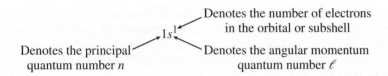

Denotes the principal quantum number n ·········· $1s^1$ ·········· Denotes the number of electrons in the orbital or subshell / Denotes the angular momentum quantum number ℓ

We can also represent the arrangement of electrons in an atom using *orbital diagrams* in which each orbital is represented by a labeled box. The orbital diagram for a hydrogen atom in the ground state is

$$\text{H} \quad \boxed{\uparrow}$$
$$1s^1$$

The upward arrow denotes one of the two possible spins (one of the two possible m_s values) of the electron in the hydrogen atom (the other possible spin is indicated with a downward arrow). Under certain circumstances, as we will see shortly, it is useful to indicate the explicit locations of electrons.

The orbital diagram for a helium atom in the ground state is

Student Annotation: The *ground state* for a many-electron atom is the one in which all the electrons occupy orbitals of the lowest possible energy.

$$\text{He} \quad \boxed{\uparrow\downarrow}$$
$$1s^2$$

The label $1s^2$ indicates there are *two* electrons in the $1s$ orbital. Note also that the arrows in the box point in opposite directions, representing opposite electron spins. Generally, when an orbital diagram includes an orbital with a single electron, we represent it with an upward arrow—although we could represent it equally well with a downward arrow. The choice is arbitrary and has no effect on the energy of the electron.

Student Annotation: $1s^2$ is read as "one s two," *not as* "one s squared."

The Aufbau Principle

We can continue the process of writing electron configurations for elements based on the order of orbital energies and the Pauli exclusion principle. This process is based

---

[17]Wolfgang Pauli (1900–1958). One of the founders of quantum mechanics, Austrian physicist Pauli was awarded the Nobel Prize in Physics in 1945.

on the **Aufbau principle,** which makes it possible to "build" the periodic table of the elements and determine their electron configurations by steps. Each step involves adding one proton to the nucleus and one electron to the appropriate atomic orbital. Through this process we gain a detailed knowledge of the electron configurations of the elements. As we will see in later chapters, knowledge of electron configurations helps us understand and predict the properties of the elements. It also explains why the elements fit into the periodic table the way they do.

After helium, the next element in the periodic table is lithium, which has three electrons. Because of the restrictions imposed by the Pauli exclusion principle, an orbital can accommodate no more than two electrons. Thus, the third electron cannot reside in the $1s$ orbital. Instead, it must reside in the next available orbital with the lowest possible energy. According to Figure 3.23, this is the $2s$ orbital. Therefore, the electron configuration of lithium is $1s^2 2s^1$, and the orbital diagram is

$$\text{Li} \quad \boxed{\uparrow\downarrow} \quad \boxed{\uparrow}$$
$$\qquad\quad 1s^2 \qquad 2s^1$$

Similarly, we can write the electron configuration of beryllium as $1s^2 2s^2$ and represent it with the orbital diagram

$$\text{Be} \quad \boxed{\uparrow\downarrow} \quad \boxed{\uparrow\downarrow}$$
$$\qquad\quad 1s^2 \qquad 2s^2$$

With both the $1s$ and the $2s$ orbitals filled to capacity, the next electron, which is needed for the electron configuration of boron, must reside in the $2p$ subshell. Because all three $2p$ orbitals are of equal energy, or **degenerate,** the electron can occupy any one of them. By convention, we usually show the first electron to occupy the p subshell in the first empty box in the orbital diagram.

$$\text{B} \quad \boxed{\uparrow\downarrow} \quad \boxed{\uparrow\downarrow} \quad \boxed{\uparrow\ \ |\ \ |\ \ }$$
$$\qquad\ 1s^2 \qquad 2s^2 \qquad\ 2p^1$$

Hund's Rule

Will the sixth electron, which is needed to represent the electron configuration of carbon, reside in the $2p$ orbital that is already half occupied, or will it reside in one of the other, empty $2p$ orbitals? According to **Hund's rule,**[18] the most stable arrangement of electrons in orbitals of equal energy is the one in which the number of electrons with the same spin is maximized. As we have seen, no two electrons in any orbital may have the same spin, so maximizing the number of electrons with the same spin requires putting the electrons in separate orbitals. Accordingly, in any subshell, an electron will occupy an empty orbital rather than one that already contains an electron.

The electron configuration of carbon is, therefore, $1s^2 2s^2 2p^2$, and its orbital diagram is

$$\text{C} \quad \boxed{\uparrow\downarrow} \quad \boxed{\uparrow\downarrow} \quad \boxed{\uparrow\ |\ \uparrow\ |\ \ }$$
$$\qquad\ 1s^2 \qquad 2s^2 \qquad\ 2p^2$$

Similarly, the electron configuration of nitrogen is $1s^2 2s^2 2p^3$, and its orbital diagram is

$$\text{N} \quad \boxed{\uparrow\downarrow} \quad \boxed{\uparrow\downarrow} \quad \boxed{\uparrow\ |\ \uparrow\ |\ \uparrow}$$
$$\qquad\ 1s^2 \qquad 2s^2 \qquad\ 2p^3$$

[18]Frederick Hund (1896–1997). A German physicist, Hund worked mainly in quantum mechanics. He also helped to develop the molecular orbital theory of chemical bonding.

Once all the 2p orbitals are singly occupied, additional electrons will have to pair with those already in the orbitals. Thus, the electron configurations and orbital diagrams for O, F, and Ne are

O	$1s^2 2s^2 2p^4$	$\boxed{\uparrow\downarrow}$	$\boxed{\uparrow\downarrow}$	$\boxed{\uparrow\downarrow \mid \uparrow \mid \uparrow}$	
		$1s^2$	$2s^2$	$2p^4$	
F	$1s^2 2s^2 2p^5$	$\boxed{\uparrow\downarrow}$	$\boxed{\uparrow\downarrow}$	$\boxed{\uparrow\downarrow \mid \uparrow\downarrow \mid \uparrow}$	
		$1s^2$	$2s^2$	$2p^5$	
Ne	$1s^2 2s^2 2p^6$	$\boxed{\uparrow\downarrow}$	$\boxed{\uparrow\downarrow}$	$\boxed{\uparrow\downarrow \mid \uparrow\downarrow \mid \uparrow\downarrow}$	
		$1s^2$	$2s^2$	$2p^6$	

By examining the ground-state electron configurations of O, F, and Ne, we can see that some atoms have configurations in which all of the electrons are paired (Ne), and some have configurations in which one or more electrons are unpaired (O and F). These two conditions give rise to different magnetic properties of atoms. Those with all paired electrons are called ***diamagnetic;*** and those with one or more unpaired electrons are called ***paramagnetic.*** Diamagnetic substances are repelled by a magnetic field. For example, neon atoms, with electron configuration $1s^2 2s^2 2p^6$ (no unpaired electrons), are weakly repelled by a magnet. Paramagnetic substances are attracted to a magnetic field. Oxygen atoms, with electron configuration $1s^2 2s^2 2p^4$ (two unpaired electrons), are weakly attracted to a magnet. One way that the magnetic behavior of a substance can be determined is using a magnetic balance like the one shown in Figure 3.24.

The mass of a substance suspended in an electromagnet is determined with the magnet off, and again with the magnet on. If the substance is diamagnetic, it will be repelled by the magnetic field and will have a lower apparent mass with the magnet on. If a substance is paramagnetic, it will be attracted by the magnetic field and will have a greater apparent mass with the magnet on. The magnetic behavior of a substance can provide information about the arrangement of its electrons.

General Rules for Writing Electron Configurations

Based on the preceding examples, we can formulate the following general rules for determining the electron configuration of an element in the ground state:

1. Electrons will reside in the available orbitals of the lowest possible energy.
2. Each orbital can accommodate a maximum of two electrons.
3. Electrons will not pair in degenerate orbitals if an empty orbital is available.
4. Orbitals will fill in the order indicated in Figure 3.23. Figure 3.25 provides a simple way for you to remember the proper order.

Video 3.2
Atomic structure—electron configuration

Student Annotation: Remember that in this context, *degenerate* means "of equal energy." Orbitals in the same subshell are degenerate.

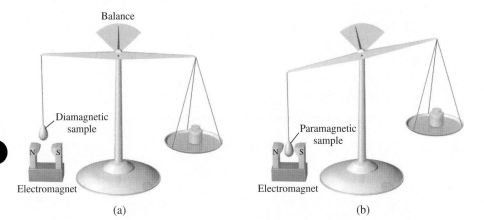

Balance

Diamagnetic sample

N S

Electromagnet

(a)

Paramagnetic sample

N S

Electromagnet

(b)

Figure 3.24 The sample is suspended in an electromagnet and is weighed twice: once with the electromagnet off and once with it on. For a diamagnetic sample (a), the apparent mass will be slightly lower when the magnet is on. For a paramagnetic sample (b), the apparent mass will be greater when the magnet is on because the sample is attracted by the magnetic field. This technique is of particular use for species containing transition metals while other methods are typically employed for other elements on the periodic table.

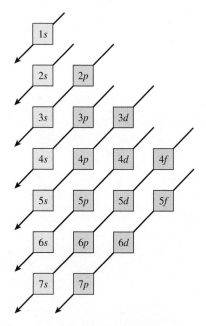

Figure 3.25 A simple way to remember the order in which orbitals fill with electrons.

Worked Example 3.10 illustrates the procedure for determining the ground-state electron configuration of an atom.

Worked Example 3.10

Write the electron configuration and give the orbital diagram of a calcium (Ca) atom ($Z = 20$).

Strategy Use the general rules given and the Aufbau principle to "build" the electron configuration of a calcium atom and represent it with an orbital diagram.

Setup Because $Z = 20$, we know that a Ca atom has 20 electrons. They will fill orbitals in the order designated in Figure 3.23, obeying the Pauli exclusion principle and Hund's rule. Orbitals will fill in the following order: $1s$, $2s$, $2p$, $3s$, $3p$, $4s$. Each s subshell can contain a maximum of two electrons, whereas each p subshell can contain a maximum of six electrons.

Solution

Ca $1s^2 2s^2 2p^6 3s^2 3p^6 4s^2$

⇅	⇅	⇅ ⇅ ⇅	⇅	⇅ ⇅ ⇅	⇅
$1s^2$	$2s^2$	$2p^6$	$3s^2$	$3p^6$	$4s^2$

> **Think About It**
> Review Figure 3.23 to make sure you have filled the orbitals in the correct order and the sum of electrons is 20. Remember that the $4s$ orbital fills before the $3d$ orbitals.

Practice Problem 🅐 TTEMPT Write the electron configuration and give the orbital diagram of a rubidium (Rb) atom ($Z = 37$).

Practice Problem 🅑 UILD Write the electron configuration and give the orbital diagram of a bromine (Br) atom ($Z = 35$).

Practice Problem 🅒 ONCEPTUALIZE Imagine an alternate universe in which the allowed values of the magnetic quantum number, m_ℓ, can have values of $-(\ell + 1)$ ... 0 ... $+ (\ell + 1)$. In this alternate universe, what would be the maximum number of electrons that could have the principal quantum number 3 in a given atom?

Section 3.9 Review

Electron Configurations

3.9.1   Which of the following electron configurations correctly represents the Ti atom?
(a) $1s^2 2s^2 2p^6 3s^2 3p^6 3d^4$
(b) $1s^2 2s^2 2p^6 3s^2 3p^6 4s^2 3d^2$
(c) $1s^2 2s^2 2p^6 3s^2 3p^6 4s^2 3d^{10}$
(d) $1s^2 2s^2 2p^6 3s^2 3p^6 3d^{10}$
(e) $1s^2 2s^2 2p^6 3s^2 3p^6 4s^4$

3.9.2   What element is represented by the following electron configuration?

$$1s^2 2s^2 2p^6 3s^2 3p^6 4s^2 3d^{10} 4p^4$$

(a) Br   (d) Se
(b) As   (e) Te
(c) S

3.9.3 Which orbital diagram is correct for the ground-state S atom?

$$\boxed{\uparrow\downarrow}\quad \boxed{\uparrow\downarrow}\quad \boxed{\uparrow\downarrow}\,\boxed{\uparrow\downarrow}\,\boxed{\uparrow\downarrow}\quad \boxed{\uparrow\downarrow}\quad \boxed{\uparrow\downarrow}\,\boxed{\uparrow\downarrow}\,\boxed{\;}$$
$$1s^2\qquad 2s^2\qquad\quad 2p^6\qquad\quad 3s^2\qquad\quad 3p^4$$

$$\boxed{\uparrow\downarrow}\quad \boxed{\uparrow\downarrow}\quad \boxed{\uparrow\downarrow}\,\boxed{\uparrow\downarrow}\,\boxed{\uparrow\downarrow}\quad \boxed{\uparrow\downarrow}\quad \boxed{\uparrow\downarrow}\,\boxed{\uparrow}\,\boxed{\uparrow}$$
$$1s^2\qquad 2s^2\qquad\quad 2p^6\qquad\quad 3s^2\qquad\quad 3p^4$$

$$\boxed{\uparrow\downarrow}\quad \boxed{\uparrow\downarrow}\quad \boxed{\uparrow\downarrow}\,\boxed{\uparrow\downarrow}\,\boxed{\;}$$
$$1s^2\qquad 2s^2\qquad\quad 2p^4$$

$$\boxed{\uparrow\downarrow}\quad \boxed{\uparrow\downarrow}\quad \boxed{\uparrow\downarrow}\,\boxed{\uparrow}\,\boxed{\uparrow}$$
$$1s^2\qquad 2s^2\qquad\quad 2p^4$$

$$\boxed{\uparrow\downarrow}\quad \boxed{\uparrow\downarrow}\quad \boxed{\uparrow\downarrow}\,\boxed{\uparrow\downarrow}\,\boxed{\uparrow\downarrow}$$
$$1s^2\qquad 2s^2\qquad\quad 2p^6$$

3.10 ELECTRON CONFIGURATIONS AND THE PERIODIC TABLE

The electron configurations of all elements except hydrogen and helium can be represented using a **noble gas core,** which shows in brackets the completed-shell electron configuration of the noble gas element that most recently precedes the element in question, followed by the electron configuration in the outermost occupied subshells. Figure 3.26 gives the ground-state electron configurations of elements from H ($Z = 1$) through element 118. Notice the similar pattern of electron configurations in the elements lithium ($Z = 3$) through neon ($Z = 10$) and those of sodium ($Z = 11$) through argon ($Z = 18$). Both Li and Na, for example, have the configuration ns^1 in their outermost occupied subshells. For Li, $n = 2$; for Na, $n = 3$. Both F and Cl have electron configuration ns^2np^5, where $n = 2$ for F and $n = 3$ for Cl, and so on.

As mentioned in Section 3.9, the $4s$ subshell is filled before the $3d$ subshell in a many-electron atom (see Figure 3.23). Thus, the electron configuration of potassium ($Z = 19$) is $1s^22s^22p^63s^23p^64s^1$. Because $1s^22s^22p^63s^23p^6$ is the electron configuration of argon, we can simplify the electron configuration of potassium by writing $[Ar]4s^1$, where [Ar] denotes the "argon core."

$$\mathrm{K}\qquad \underbrace{1s^22s^22p^63s^23p^6}_{[Ar]}4s^1 \qquad\longrightarrow\qquad [Ar]4s^1$$

The placement of the outermost electron in the $4s$ orbital (rather than in the $3d$ orbital) of potassium is strongly supported by experimental evidence. The physical and chemical properties of potassium are very similar to those of lithium and sodium, the first two alkali metals. In both lithium and sodium, the outermost electron is in an s orbital (there is no doubt that their outermost electrons occupy s orbitals because there is no $1d$ or $2d$ subshell). Based on its similarities to the other alkali metals, we expect potassium to have an analogous electron configuration; that is, we expect the last electron in potassium to occupy the $4s$ rather than the $3d$ orbital.

The elements from Group 3B through Group 1B are *transition metals* [◄◄ Section 2.6]. Transition metals either have incompletely filled d subshells or readily give rise to cations that have incompletely filled d subshells. In the first transition metal series, from scandium ($Z = 21$) through copper ($Z = 29$), additional electrons

Student Annotation: Although zinc and the other elements in Group 2B sometimes are included under the heading "transition metals," they neither have nor readily acquire partially filled d subshells. Strictly speaking, they are *not* transition metals.

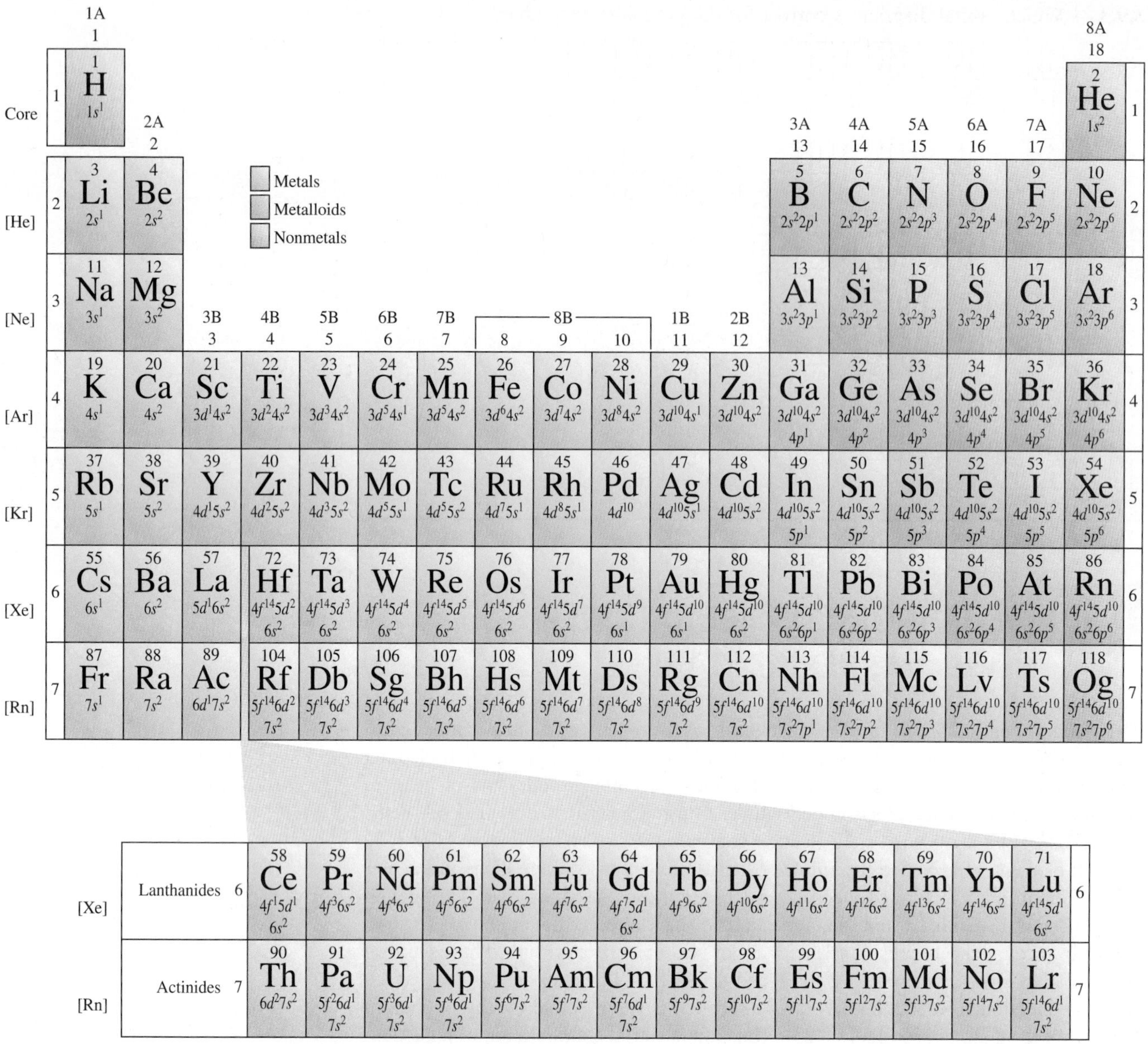

Figure 3.26 Outermost ground-state electron configurations for the known elements.

are placed in the $3d$ orbitals according to Hund's rule. However, there are two anomalies. The electron configuration of chromium ($Z = 24$) is $[Ar]4s^13d^5$ and not $[Ar]4s^23d^4$, as we might expect. A similar break in the pattern is observed for copper, whose electron configuration is $[Ar]4s^13d^{10}$ rather than $[Ar]4s^23d^9$. The reason for these anomalies is that a slightly greater stability is associated with d subshells that are either half filled ($3d^5$) or completely filled ($3d^{10}$).

For elements Zn ($Z = 30$) through Kr ($Z = 36$), the $3d$, $4s$, and $4p$ subshells fill in a straightforward manner. With rubidium ($Z = 37$), electrons begin to enter the $n = 5$ energy level.

Some of the electron configurations in the second transition metal series [yttrium ($Z = 39$) through silver ($Z = 47$)] are also irregular, but the details of many of these irregularities are beyond the scope of this text and we will not be concerned with them.

The sixth period of the periodic table begins with cesium ($Z = 55$), barium ($Z = 56$), and lanthanum ($Z = 57$), whose electron configurations are $[Xe]6s^1$, $[Xe]6s^2$, and $[Xe]6s^25d^1$, respectively. Following lanthanum, there is a gap in the periodic table where the ***lanthanide (rare earth) series*** belongs. The lanthanides are a series of 14 elements that have incompletely filled $4f$ subshells or that readily give rise to *cations*

Student Annotation: When $n = 4$, ℓ can equal 3, corresponding to an f subshell. There are seven possible values for m_ℓ when $\ell = 3$: -3, -2, -1, 0, $+1$, $+2$, and $+3$. Therefore, there are seven f orbitals.

that have incompletely filled $4f$ subshells. The lanthanides (and the actinides, to be discussed next) are shown at the bottom of the periodic table to keep the table from being too wide.

Student Annotation: In theory, the lanthanides arise from the filling of the seven degenerate $4f$ orbitals. In reality, however, the energies of the $5d$ and $4f$ orbitals are very close and the electron configurations of these elements sometimes involve $5d$ electrons.

In theory, the lanthanides arise from the filling of the seven degenerate $4f$ orbitals. In reality, however, the energies of the $5d$ and $4f$ orbitals are very close and the electron configurations of these elements sometimes involve $5d$ electrons. For example, in lanthanum itself ($Z = 57$) the $4f$ orbital is slightly higher in energy than the $5d$ orbital. Thus, lanthanum's electron configuration is $[Xe]6s^2 5d^1$ rather than $[Xe]6s^2 4f^1$.

After the $4f$ subshell is completely filled, the next electron enters the $5d$ subshell of lutetium ($Z = 71$). This series of elements, including lutetium and hafnium ($Z = 72$) and extending through mercury ($Z = 80$), is characterized by the filling of the $5d$ subshell. The $6p$ subshells are filled next, which takes us to radon ($Z = 86$).

The last row of elements begins with francium ($Z = 87$; electron configuration $[Rn]7s^1$) and radium ($Z = 88$; electron configuration $[Rn]7s^2$), and then continues with the *actinide series,* which starts at thorium ($Z = 90$) and ends with lawrencium ($Z = 103$). Most of these elements are not found in nature but have been synthesized in nuclear reactions, which are the subject of Chapter 20. The actinide series has partially filled $5f$ and/or $6d$ subshells. The elements lawrencium ($Z = 103$) through copernicium ($Z = 112$) have a filled $5f$ subshell and are characterized by the filling of the $6d$ subshell.

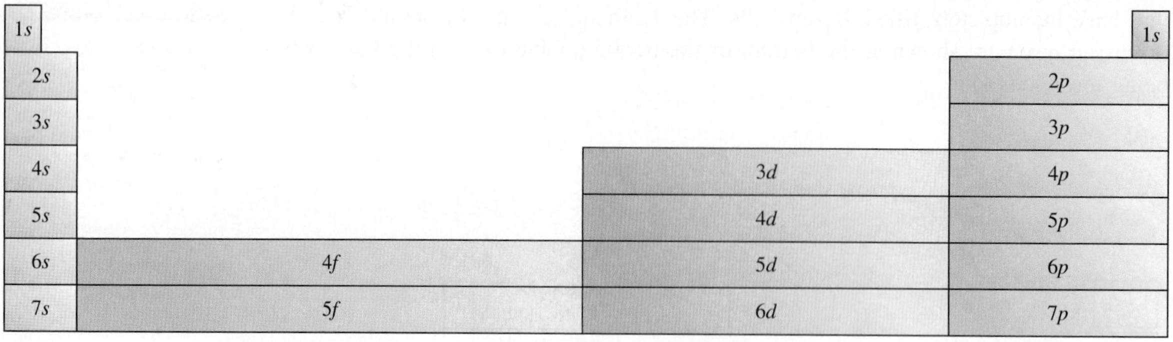

Figure 3.27 Classification of groups of elements in the periodic table according to the type of subshell being filled with electrons.

With few exceptions, you should be able to write the electron configuration of any element, using Figure 3.23 (or Figure 3.25) as a guide. Elements that require particular care are the transition metals, the lanthanides, and the actinides. You may notice from looking at the electron configurations of gadolinium ($Z = 64$) and curium ($Z = 96$) that half-filled f subshells also appear to exhibit slightly enhanced stability. As we noted earlier, at larger values of the principal quantum number n, the order of subshell filling may be irregular due to the closeness of the energy levels.

In Figure 3.27 we group the elements according to the type of subshell in which the outermost electrons are placed. Elements whose outermost electrons are in an s subshell are referred to as s-block elements, those whose outermost electrons are in a p subshell are referred to as p-block elements, and so on.

Worked Example 3.11 shows how to write electron configurations.

Worked Example 3.11

Without referring to Figure 3.26, write the electron configuration for an arsenic atom ($Z = 33$) in the ground state.

Strategy Use Figure 3.23 or Figure 3.25 to determine the order in which the subshells will fill, and then assign electrons to the appropriate subshells.

Setup The noble gas core for As is [Ar], where $Z = 18$ for Ar. The order of filling beyond the noble gas core is $4s$, $3d$, and $4p$. Fifteen electrons must go into these subshells because there are $33 - 18 = 15$ electrons in As beyond its noble gas core.

Solution

$$[Ar]4s^2 3d^{10} 4p^3$$

> **Think About It**
> Arsenic is a p-block element; therefore, we should expect its outermost electrons to reside in a p subshell.

Practice Problem (A)**TTEMPT** Without referring to Figure 3.26, write the electron configuration for a radium atom ($Z = 88$) in the ground state.

Practice Problem (B)**UILD** Without referring to Figure 3.26, determine the identity of the element with the following electron configuration:

$$[Xe]6s^2 4f^{14} 5d^{10} 6p^5$$

Practice Problem (C)**ONCEPTUALIZE** Consider again the alternate universe and its allowed values of m_ℓ from Worked Example 3.10C. At what atomic numbers (in the first four rows of the alternate universe's periodic table) would you expect the ground-state electron configuration to differ from that predicted by the Aufbau principle? (*Hint:* In *our* periodic table, in the first four rows, the ground-state electron configurations differ from those predicted by the Aufbau principle at atomic numbers 24 and 29.)

Section 3.10 Review

Electron Configurations and the Periodic Table

3.10.1 Which of the following electron configurations correctly represents the Ag atom?
(a) $[Kr]5s^24d^9$
(c) $[Kr]5s^14d^{10}$
(e) $[Kr]5s^14d^{10}$
(b) $[Kr]5s^24d^{10}$
(d) $[Kr]5s^24d^9$

3.10.2 What element is represented by the following electron configuration: $[Kr]5s^24d^{10}5p^5$?
(a) Tc
(c) I
(e) Te
(b) Br
(d) Xe

3.10.3 Which of the following is a *d*-block element? (Select all that apply.)
(a) Sb
(c) Ca
(e) U
(b) Au
(d) Zn

3.10.4 Which of the following is a *p*-block element? (Select all that apply.)
(a) Pb
(c) Sr
(e) Na
(b) C
(d) Xe

Learning Outcomes

- Identify energy as being kinetic or potential and, if potential, as being chemical or electrostatic.
- Understand the law of conservation of energy.
- Describe properties of waves including wavelength, frequency, and amplitude.
- Calculate the wavelength or frequency of light given one of these values.
- Cite examples of the electromagnetic spectrum in regard to wavelength and type of radiation.
- Understand the basis of quantum theory and its relationship to the frequency of radiation.
- Describe the basis of the photoelectric effect in terms of photon energy and frequency of radiation.
- Identify the relevance of the de Broglie hypothesis in regard to how electrons may behave as waves as opposed to particles.
- Give in your own words the relevance of the Heisenberg uncertainty principle to the theoretical structure of atomic orbitals.

- Understand how electron density helps to define the shape of atomic orbitals.
- Provide the meaning of each type of quantum number (principal, angular momentum, magnetic, and electron spin).
- Apply quantum number rules to determine allowable values for each type of quantum number.
- Understand the basis of atomic orbitals.
- Arrange atomic orbitals based upon energy levels.
- Understand the meaning of the Pauli exclusion principle and how it relates to electron configurations.
- Apply Hund's rule in drawing electron orbital diagrams.
- Determine the electron configuration of an atom using the Aufbau principle.
- Use the periodic table to determine the electron configuration of an atom.

Chapter Summary

SECTION 3.1

- *Energy* is the capacity to do work or transfer heat. Energy may be *kinetic energy* (the energy associated with *motion*) or *potential energy* (energy possessed by virtue of *position*). *Thermal energy* is a form of kinetic energy.

Chemical energy and *electrostatic energy* are forms of potential energy.
- The *law of conservation of energy* states that energy can neither be created nor destroyed. The SI unit of energy is the *joule (J)*.

SECTION 3.2

- What we commonly refer to as "light" is actually the visible portion of the *electromagnetic spectrum.* All light has certain common characteristics including wavelength, frequency, and amplitude.
- *Wavelength* (λ) is the distance between two crests or two troughs of a wave. *Frequency (v)* is the number of waves that pass a point per unit time. *Amplitude* is the distance between the midpoint and crest or trough of a wave.
- *Electromagnetic waves* have electric and magnetic components that are mutually perpendicular and in phase.

SECTION 3.3

- *Blackbody radiation* is the electromagnetic radiation given off by a solid when it is heated.
- Max Planck proposed that energy, like matter, was composed of tiny, indivisible "packages" called *quanta. Quanta* is the plural of *quantum.*
- Albert Einstein used Planck's revolutionary quantum theory to explain the *photoelectric effect,* in which electrons are emitted when light of a certain minimum frequency, the *threshold frequency,* shines on a metal surface.
- A *quantum* of light is referred to as a *photon.*

SECTION 3.4

- An *emission spectrum* is the light given off by an object when it is excited thermally. An emission spectrum may be *continuous,* including all the wavelengths within a particular range, or it may be a *line spectrum,* consisting only of certain discrete wavelengths.
- The *ground state* is the lowest possible energy state for an atom. An *excited state* is any energy level higher than the ground state.

SECTION 3.5

- A *node* is a point at which a standing wave has zero amplitude.
- Having observed that light could exhibit particle-like behavior, de Broglie proposed that matter might also exhibit wavelike behavior. The *de Broglie wavelength* is the wavelength associated with a particle of very small mass. Soon after de Broglie's proposal, experiments showed that electrons could exhibit diffraction—a property of waves.

SECTION 3.6

- According to the *Heisenberg uncertainty principle,* the product of the uncertainty of the *location* and the uncertainty of the *momentum* of a very small particle must have a certain minimum value. It is thus impossible to know simultaneously both the location and momentum of an electron.

- The *electron density* gives the probability of finding an electron in a particular region in an atom. An *atomic orbital* is the region of three-dimensional space, defined by ψ^2 (the square of the wave function, ψ), where the probability of finding an electron is high. An atomic orbital can accommodate a maximum of *two* electrons.

SECTION 3.7

- An atomic orbital is defined by three *quantum numbers:* the *principal quantum number (n),* the *angular momentum quantum number* (ℓ), and the *magnetic quantum number* (m_ℓ).
- The principal quantum number (n) indicates distance from the nucleus. Possible values of n are (1, 2, 3, ...). The angular momentum quantum number (ℓ) indicates the shape of the orbital. Possible values of ℓ are (0, 1, ..., n − 1). The magnetic quantum number (m_ℓ) indicates the orbital's orientation in space. Possible values of m_ℓ are ($-\ell$, ... 0, ... $+\ell$).
- Two electrons that occupy the same atomic orbital in the ground state must have different *electron spin quantum numbers (m_s),* either $+\frac{1}{2}$ or $-\frac{1}{2}$.

SECTION 3.8

- The value of the angular momentum quantum number (ℓ) determines the type of the atomic orbital: $\ell = 0$ corresponds to an *s orbital,* $\ell = 1$ corresponds to a *p orbital,* $\ell = 2$ corresponds to a *d orbital,* and $\ell = 3$ corresponds to an *f orbital.*

SECTION 3.9

- The *electron configuration* specifies the arrangement of electrons in the atomic orbitals of an atom.
- According to the *Pauli exclusion principle,* no two electrons in an atom in the ground state can have the same four quantum numbers, n, ℓ, m_ℓ, and m_s.
- The *Aufbau principle* describes the theoretical, sequential building up of the elements in the periodic table by the stepwise addition of protons and electrons.
- Atomic orbitals that have the same energy are called *degenerate.* According to *Hund's rule,* degenerate orbitals must all contain one electron before any can contain two electrons.
- Atoms with electron configurations in which there are one or more unpaired electrons are *paramagnetic* and are weakly attracted to a magnetic field. Atoms with electron configurations in which there are no unpaired electrons are *diamagnetic* and are weakly repelled by a magnetic field.

SECTION 3.10

- The *noble gas core* makes it possible to abbreviate the writing of electron configurations.
- The *lanthanide (rare earth) series* and *actinide series* appear at the bottom of the periodic table. They represent the filling of f orbitals.

Key Words

Actinide series, 107
Amplitude, 70
Angular momentum
  quantum number (ℓ), 93
Atomic orbital, 92
Aufbau principle, 102
Blackbody radiation, 74
Chemical energy, 67
d orbital, 98
de Broglie wavelength, 88
Degenerate, 102
Diamagnetic, 103

Electromagnetic spectrum, 70
Electromagnetic wave, 71
Electron configuration, 100
Electron density, 92
Electron spin quantum
  number (m_s), 95
Electrostatic energy, 67
Emission spectrum, 79
Energy, 67
Excited state, 84
f orbital, 98
Frequency (v), 70

Ground state, 84
Heisenberg uncertainty principle, 90
Hund's rule, 102
Joule (J), 68
Kinetic energy, 67
Lanthanide (rare earth) series, 106
Law of conservation of energy, 67
Line spectrum, 79
Magnetic quantum number (m_ℓ), 93
Noble gas core, 105
Node, 87
p orbital, 96

Paramagnetic, 103
Pauli exclusion principle, 101
Photoelectric effect, 75
Photon, 75
Potential energy, 67
Principal quantum number (n), 92
Quantum, 74
Quantum numbers, 92
s orbital, 96
Thermal energy, 67
Threshold frequency, 75
Wavelength (λ), 70

Key Equations

3.1 $E_k = \dfrac{1}{2}mu^2$	The kinetic energy of a moving object is calculated using the mass (m) and velocity (u) of the object.
3.2 $E_{el} \propto \dfrac{Q_1 Q_2}{d}$	The electrostatic potential energy (E_{el}) between two charged objects is calculated using the magnitudes of charge (Q_1 and Q_2) and the distance (d) between the charges.
3.3 $c = \lambda v$	The wavelength (λ) and frequency (v) of electromagnetic radiation are related to one another through the speed of light (c). If wavelength is known, frequency can be determined, and vice versa.
3.4 $E = hv$	The energy of a photon (E) is equal to the product of Planck's constant (h) and frequency (v) of the photon.
3.5 $hv = KE + W$	The energy (hv) of a photon used to eject electrons from a metal surface via the photoelectric effect is equal to the sum of kinetic energy of the ejected electron (KE) and the binding energy (W).
3.6 $\dfrac{1}{\lambda} = R_\infty \left(\dfrac{1}{n_f^2} - \dfrac{1}{n_i^2} \right)$	When an electron transitions from one quantum state to another (n_i to n_f), the difference in energy between the two states is emitted (or absorbed) in the form of light. The wavelength of the emitted/absorbed light can be calculated using Equation 3.6.
3.7 $E_n = -2.18 \times 10^{-18} \text{ J} \left(\dfrac{1}{n^2} \right)$	The energy of an electron for a given value of n (E_n) is inversely proportional to the square of n—and is by convention a negative number.
3.8 $\Delta E = hv = -2.18 \times 10^{-18} \text{ J} \left(\dfrac{1}{n_f^2} - \dfrac{1}{n_i^2} \right)$	The difference in energy between two quantum states (n_i to n_f) is calculated using Equation 3.8.
3.9 $\dfrac{1}{\lambda} = \dfrac{2.18 \times 10^{-18} \text{ J}}{hc} \left(\dfrac{1}{n_f^2} - \dfrac{1}{n_i^2} \right)$	Similar to Equation 3.6, Equation 3.9 allows calculation of the wavelength of emitted/absorbed light when n_i and n_f are known.
3.10 $2\pi r = n\lambda$	This is the relationship between the allowed orbit ($2\pi r$) and wavelength (λ) of an electron behaving as a standing wave.
3.11 $\lambda = \dfrac{h}{mu}$	The de Broglie wavelength (λ) of a particle can be calculated using Planck's constant (h), the mass of the particle in kilograms (m), and velocity (u) of the particle.
3.12 $\Delta x \cdot \Delta p \geq \dfrac{h}{4\pi}$	The Heisenberg uncertainty principle states that the product of uncertainties in position (Δx) and momentum (Δp) of a particle cannot be less than Planck's constant (h) over 4π. Knowing the uncertainty in one (position or momentum) allows us to calculate the minimum uncertainty in the other.
3.13 $\Delta x \cdot m\Delta u \geq \dfrac{h}{4\pi}$	Similar to Equation 3.12, when mass of the particle is known, knowing the uncertainty in position (Δx) allows us to calculate the minimum uncertainty in its velocity (Δu).

Determining Ground-State Valence Electron Configurations Using the Periodic Table

An easy way to determine the electron configuration of an element is by using the periodic table. Although the table is arranged by atomic number, it is also divided into blocks that indicate the type of orbital occupied by an element's outermost electrons. Outermost valence electrons of elements in the s block (shown in yellow), reside in s orbitals; those of elements in the p block (blue) reside in p orbitals; and so on.

To determine the ground-state electron configuration of any element, we start with the most recently completed noble gas core, and count across the following period to determine the valence electron configuration. Consider the example of Cl, which has atomic number 17. The noble gas that precedes Cl is Ne, with atomic number 10. Therefore, we begin by writing [Ne]. The noble gas symbol in square brackets represents the core electrons—with a completed p subshell. To complete the electron configuration, we count from the left of period 3 as shown by the red arrow, adding the last (rightmost) configuration label from each block the arrow touches:

There are seven electrons in addition to the noble gas core. Two of them reside in an s subshell, and five of them reside in a p subshell. By simply counting across the third period, we can determine the specific subshells that contain the valence electrons, and arrive at the correct ground-state electron configuration: $[Ne]3s^2 3p^5$.

For Ga, with atomic number 31, the preceding noble gas is Ar, with atomic number 18. Counting across the fourth period (green arrow) gives the ground-state electron configuration: $[Ar]4s^2 3d^{10} 4p^1$.

There are a few elements for which this method will not give the correct configuration. For example, there is no element with a ground-state valence electron configuration ending in $3d^4$ or $3d^9$. Instead, Cr and Cu are $[Ar]4s^1 3d^5$ and $[Ar]4s^1 3d^{10}$, respectively. Remember that this is the result of the unusual stability of either a *half-filled* or a *filled d* subshell [◄◄ see Section 3.10].

We can also use the periodic table to determine the identity of an element, given its ground-state electron configuration. For example, given the configuration $[Ne]3s^2 3p^4$, we focus on the last entry in the configuration: $3p^4$. This tells us that the element is in the *third* period (3), in the p block (p), and that it has *four* electrons in its p subshell (superscript 4). This corresponds to atomic number 16, which is the element sulfur (S).

Key Skills Problems

3.1
What is the noble gas core for Mo?
(a) Ar (b) Kr (c) Xe (d) Ne (e) Rn

3.2
Which of the following electron configurations correctly represents the V atom?
(a) $[Ar]3d^5$ (b) $[Ar]4s^2 3d^2$ (c) $[Ar]4s^2 3d^4$ (d) $[Ar]4s^2 3d^3$
(e) $[Kr]4s^2 3d^5$

3.3
What element is represented by the electron configuration $[Kr]5s^2 4d^{10} 5p^1$?
(a) Sn (b) Ga (c) In (d) Tl (e) Zr

3.4
What is the electron configuration of the Lu atom?
(a) $[Xe]6s^2 4f^{14}$ (b) $[Xe]6s^2 5d^1$ (c) $[Xe]6s^2 4f^{13}$
(d) $[Xe]6s^2 4f^{14} 5d^1$ (e) $[Xe]4f^{14}$

Questions and Problems

SECTION 3.1: ENERGY AND ENERGY CHANGES

Review Questions

3.1   Define these terms: *potential energy, kinetic energy, law of conservation of energy.*

3.2   What are the units for energy commonly employed in chemistry?

3.3   A truck initially traveling at 60 km/h is brought to a complete stop at a traffic light. Does this change violate the law of conservation of energy? Explain.

3.4   Describe the interconversions of forms of energy occurring in these processes: (a) You throw a softball up into the air and catch it. (b) You switch on a flashlight. (c) You ride the ski lift to the top of the hill and then ski down. (d) You strike a match and let it burn completely.

Computational Problems

3.5   Determine the kinetic energy of (a) a 1.25-kg mass moving at 5.75 m/s, (b) a car weighing 3250 lb moving at 35 mph, (c) an electron moving at 475 m/s, (d) a helium atom moving at 725 m/s.

3.6   Determine the kinetic energy of (a) a 29-kg mass moving at 122 m/s, (b) a tennis ball weighing 58.5 g moving at 71.3 mph, (c) a beryllium atom moving at 355 m/s, (d) a neutron moving at 3.000×10^3 m/s.

3.7   Determine (a) the velocity of a Ne atom that has $E_k = 1.86 \times 10^{-20}$ J, (b) the velocity of a Kr atom that has $E_k = 7.50 \times 10^{-21}$ J, (c) the mass and identity of an atom moving at 385 m/s that has $E_k = 4.812 \times 10^{-21}$ J.

3.8   Determine (a) the velocity of an electron that has $E_k = 6.11 \times 10^{-21}$ J, (b) the velocity of a neutron that has $E_k = 8.03 \times 10^{-22}$ J, (c) the mass and identity of a subatomic particle moving at 1.447×10^3 m/s that has $E_k = 9.5367 \times 10^{-25}$ J.

3.9   (a) How much greater is the electrostatic energy between charges of +3 and −3 than charges of +2 and −3 if the charges in each case are separated by the same distance? (b) By what distance would charges of +2 and −3 have to be separated for the electrostatic energy between them to be the same as that between charges of +3 and −3 separated by a distance of *d?*

3.10   (a) How much greater is the electrostatic energy between charges of +4 and −3 than charges of +2 and −3 if the charges in each case are separated by the same distance? (b) By what distance would charges of +4 and −3 have to be separated for the electrostatic energy between them to be the same as that between charges of +2 and −3 separated by a distance of *d?*

SECTION 3.2: THE NATURE OF LIGHT

Review Questions

3.11   What is a wave? Using a diagram, define the following terms associated with waves: wavelength, frequency, amplitude.

3.12   What are the units for wavelength and frequency of electromagnetic waves? What is the speed of light in meters per second and miles per hour?

3.13   List the types of electromagnetic radiation, starting with the radiation having the longest wavelength and ending with the radiation having the shortest wavelength.

3.14   Give the high and low wavelength values that define the visible region of the electromagnetic spectrum.

Computational Problems

3.15   (a) What is the wavelength (in nanometers) of light having a frequency of 1.90×10^{13} Hz? (b) What is the frequency (in hertz) of light having a wavelength of 235 nm?

3.16   (a) What is the frequency of light having a wavelength of 97 nm? (b) What is the wavelength (in meters) of radiation having a frequency of 9.55×10^7 Hz? (This is the type of radiation used by FM radio stations.)

3.17   The SI unit of time is the second, which is defined as 9,192,631,770 cycles of radiation associated with a certain emission process in the cesium atom. Calculate the wavelength of this radiation (to three significant figures). In which region of the electromagnetic spectrum is this wavelength found?

Conceptual Problems

3.18   The average distance between Mars and Earth is about 1.3×10^8 mi. How long would it take TV pictures transmitted from the Mars rover robot geologist on the Martian surface to reach Earth (1 mi = 1.61 km)?

3.19   How many minutes would it take a radio wave to travel from the planet Venus to Earth? (The average distance from Venus to Earth = 28 million mi.)

3.20   Four waves represent light in four different regions of the electromagnetic spectrum: visible, microwave, infrared, and ultraviolet. Determine the best match of regions to the waves shown here. Explain your choices.

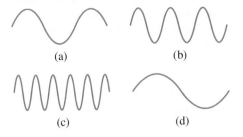

(a)   (b)

(c)   (d)

SECTION 3.3: QUANTUM THEORY

Review Questions

3.21 Briefly explain Planck's quantum theory and explain what a quantum is. What are the units for Planck's constant?

3.22 Give two everyday examples that illustrate the concept of quantization.

3.23 Explain what is meant by the photoelectric effect.

3.24 What is a photon? What role did Einstein's explanation of the photoelectric effect play in the development of the particle-wave interpretation of the nature of electromagnetic radiation?

Computational Problems

3.25 A photon has a wavelength of 705 nm. Calculate the energy of the photon in joules.

3.26 The blue color of the sky results from the scattering of sunlight by one of the components of air. The blue light has a frequency of about 7.5×10^{14} Hz. (a) Calculate the wavelength (in nanometers) associated with this radiation, and (b) calculate the energy (in joules) of a single photon associated with this frequency.

3.27 A photon has a frequency of 6.5×10^9 Hz.
(a) Convert this frequency into wavelength (nm). Does this frequency fall in the visible region?
(b) Calculate the energy (in joules) of this photon.
(c) Calculate the energy (in joules) of 1 mole of photons all with this frequency.

3.28 What is the wavelength (in nanometers) of radiation that has an energy content of 2.13×10^3 kJ/mol? In which region of the electromagnetic spectrum is this radiation found?

3.29 Calculate the difference in energy (in joules) between a photon with $\lambda = 610$ nm (yellow) and a photon with $\lambda = 550$ nm (green).

3.30 How much more energy per photon is there in green light of wavelength 552 nm than in red light of wavelength 675 nm?

3.31 When copper is bombarded with high-energy electrons, X rays are emitted. Calculate the energy (in joules) associated with the photons if the wavelength of the X rays is 0.154 nm.

3.32 A particular form of electromagnetic radiation has a frequency of 9.87×10^{15} Hz. (a) What is its wavelength in nanometers? In meters? (b) To what region of the electromagnetic spectrum would you assign it? (c) What is the energy (in joules) of one quantum of this radiation?

3.33 Photosynthesis makes use of visible light to bring about chemical changes. Explain why heat energy in the form of infrared radiation is ineffective for photosynthesis.

3.34 The retina of a human eye can detect light when radiant energy incident on it is at least 4.0×10^{-17} J. For light of 585-nm wavelength, how many photons does this energy correspond to?

3.35 The radioactive ^{60}Co isotope is used in nuclear medicine to treat certain types of cancer. Calculate the wavelength and frequency of an emitted gamma photon having the energy of 1.29×10^{11} J/mol.

3.36 The binding energy of magnesium metal is 5.86×10^{-19} J. Calculate the minimum frequency of light required to release electrons from the metal.

3.37 What is the kinetic energy of the ejected electron if light of frequency 2.00×10^{15} s^{-1} is used to irradiate the magnesium metal in Problem 3.36?

3.38 A red light was shined onto a metal sample and the result shown in (i) was observed. When the light source was changed to a blue light, the result shown in (ii) was observed. Explain how these results can be interpreted with respect to the photoelectric effect. Describe the result you would expect for each of the following: (a) The intensity of red light is increased. (b) The intensity of blue light is increased. (c) Violet light is used.

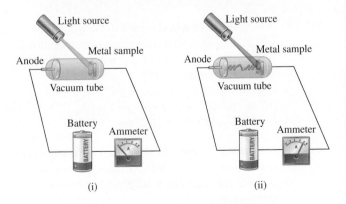

(i)   (ii)

3.39 A photoelectric experiment was performed by separately shining a laser at 450 nm (blue light) and a laser at 560 nm (yellow light) on a clean metal surface and measuring the number and kinetic energy of the ejected electrons. Which light would generate more electrons? Which light would eject electrons with greater kinetic energy? Assume that the same total amount of energy is delivered to the metal surface by each laser and that the frequencies of the laser lights exceed the threshold frequency.

SECTION 3.4: BOHR'S THEORY OF THE HYDROGEN ATOM

 Visualizing Chemistry
Figure 3.9

VC 3.1 Which of the following best explains why we see only four lines in the emission spectrum of hydrogen?
(a) Hydrogen has only four different electronic transitions.
(b) Only four of hydrogen's electronic transitions correspond to visible wavelengths.
(c) The other lines in hydrogen's emission spectrum can't be seen easily against the black background.

VC 3.2 One way to see the emission spectrum of hydrogen is to view the hydrogen in an electric discharge tube through a *spectroscope,* a device that separates the wavelengths. Why can we not view the emission spectrum simply by pointing the spectroscope at a sample of hydrogen confined in a glass tube or flask?
(a) Without the electrons being in excited states, there would be no emission of light.
(b) The glass would make it impossible to see the emission spectrum.
(c) Hydrogen alone does not exhibit an emission spectrum—it must be combined with oxygen.

VC 3.3 How many lines would we see in the emission spectrum of hydrogen if the downward transitions from excited states all ended at $n = 1$ and no transitions ended at $n = 2$?
(a) We would still see four lines.
(b) We would see five lines.
(c) We would not see any lines.

VC 3.4 For a hydrogen atom in which the electron has been excited to $n = 4$, how many different transitions can occur as the electron eventually returns to the ground state?
(a) 1
(b) 3
(c) 6

Review Questions

3.40 What are emission spectra? How do line spectra differ from continuous spectra?

3.41 What is an energy level? Explain the difference between ground state and excited state.

3.42 Briefly describe Bohr's theory of the hydrogen atom and how it explains the appearance of an emission spectrum. How does Bohr's theory differ from concepts of classical physics?

3.43 Explain the meaning of the negative sign in Equation 3.8.

Computational Problems

3.44 Consider the following energy levels of a hypothetical atom:

E_4: -1.0×10^{-19} J
E_3: -5.0×10^{-19} J
E_2: -10×10^{-19} J
E_1: -15×10^{-19} J

(a) What is the wavelength of the photon needed to excite an electron from E_1 to E_4? (b) What is the energy (in joules) a photon must have in order to excite an electron from E_2 to E_3? (c) When an electron drops from the E_3 level to the E_1 level, the atom is said to undergo emission. Calculate the wavelength of the photon emitted in this process.

3.45 The first line of the Balmer series occurs at a wavelength of 656.3 nm. What is the energy difference between the two energy levels involved in the emission that results in this spectral line?

3.46 Calculate the wavelength (in nanometers) of a photon emitted by a hydrogen atom when its electron drops from the $n = 5$ state to the $n = 2$ state.

3.47 Calculate the frequency (hertz) and wavelength (nanometers) of the emitted photon when an electron drops from the $n = 4$ to the $n = 3$ level in a hydrogen atom.

3.48 What wavelength of light is needed to excite the electron in the hydrogen atom from the ground state to $n = 10$? In what region of the electromagnetic spectrum does a photon of this wavelength lie?

3.49 An electron in the hydrogen atom makes a transition from an energy state of principal quantum number n_i to the $n = 2$ state. If the photon emitted has a wavelength of 434 nm, what is the value of n_i?

Conceptual Problems

3.50 Explain why elements produce their own characteristic colors when they emit photons.

3.51 Some copper-containing substances emit green light when they are heated in a flame. How would you determine whether the light is of one wavelength or a mixture of two or more wavelengths?

3.52 Is it possible for a fluorescent material (one that absorbs and then reemits light) to emit radiation in the ultraviolet region after absorbing visible light? Explain your answer.

3.53 Explain how astronomers are able to tell which elements are present in distant stars by analyzing the electromagnetic radiation emitted by the stars.

SECTION 3.5: WAVE PROPERTIES OF MATTER

Review Questions

3.54 How does de Broglie's hypothesis account for the fact that the energies of the electron in a hydrogen atom are quantized?

3.55 Why is Equation 3.11 meaningful only for submicroscopic particles, such as electrons and atoms, and not for macroscopic objects?

3.56 Explain why we cannot observe the wave behavior of a macroscopic moving object such as a baseball.

Computational Problems

3.57 Thermal neutrons are neutrons that move at speeds comparable to those of particles in air at room temperature. These neutrons are most effective in initiating a nuclear chain reaction among ^{235}U isotopes. Calculate the wavelength (in nanometers) associated with a beam of neutrons moving at 7.00×10^2 m/s (mass of a neutron = 1.675×10^{-27} kg).

3.58 Protons can be accelerated to speeds near that of light in particle accelerators. Estimate the wavelength (in nanometers) of such a proton moving at 2.90×10^8 m/s (mass of a proton = 1.673×10^{-27} kg).

3.59 What is the de Broglie wavelength (in centimeters) of a 10.2-g honeybee flying at 3.51 mph (1 mi = 1.61 km)?

3.60 What is the de Broglie wavelength (in nanometers) associated with a 2.7-g Ping-Pong ball traveling at 14 mph?

3.61 How fast must a neutron be traveling to have a de Broglie wavelength of 10.5 Å?

3.62 What is the minimum speed an electron must be traveling to have a *visible* de Broglie wavelength? (Assume the range of visible wavelengths is 400–700 nm.)

SECTION 3.6: QUANTUM MECHANICS

Review Questions

3.63 What are the inadequacies of Bohr's theory?

3.64 What is the Heisenberg uncertainty principle? What is the Schrödinger equation?

3.65 What is the physical significance of the wave function?

3.66 How is the concept of electron density used to describe the position of an electron in the quantum mechanical treatment of an atom?

3.67 What is an atomic orbital? How does an atomic orbital differ from an orbit?

Computational Problems

3.68 The speed of a thermal neutron (see Problem 3.57) is known to within 2.0 km/s. What is the minimum uncertainty in the position of the thermal neutron?

3.69 Alveoli are tiny sacs of air in the lungs. Their average diameter is 5.0×10^{-5} m. Calculate the uncertainty in the velocity of an oxygen molecule (5.3×10^{-26} kg) trapped within a sac. (*Hint:* The maximum uncertainty in the position of the molecule is given by the diameter of the sac.)

Conceptual Problems

3.70 In the beginning of the twentieth century, some scientists thought that a nucleus may contain both electrons and protons. Use the Heisenberg uncertainty principle to show that an electron cannot be confined within a nucleus. Repeat the calculation for a proton. Comment on your results. Assume the radius of a nucleus to be 1.0×10^{-15} m. The masses of an electron and a proton are 9.109×10^{-31} kg and 1.673×10^{-27} kg, respectively. (*Hint:* Treat the radius of the nucleus as the uncertainty in position.)

3.71 Suppose that photons of blue light (430 nm) are used to locate the position of a 2.80-g Ping-Pong ball in flight and that the uncertainty in the position is equal to one wavelength. What is the minimum uncertainty in the speed of the Ping-Pong ball? Comment on the magnitude of your result.

SECTION 3.7: QUANTUM NUMBERS

Review Questions

3.72 Describe the four quantum numbers used to characterize an electron in an atom.

3.73 Which quantum number defines a shell? Which quantum numbers define a subshell?

3.74 Which of the four quantum numbers (n, ℓ, m_ℓ, m_s) determine (a) the energy of an electron in a hydrogen atom and in a many-electron atom, (b) the size of an orbital, (c) the shape of an orbital, (d) the orientation of an orbital in space?

Conceptual Problems

3.75 An electron in a certain atom is in the $n = 2$ quantum level. List the possible values of ℓ and m_ℓ that it can have.

3.76 An electron in an atom is in the $n = 3$ quantum level. List the possible values of ℓ and m_ℓ that it can have.

3.77 Indicate which of the following sets of three quantum numbers (n, ℓ, m_ℓ) are wrong and explain why they are wrong: (a) 4, 1, 0; (b) 3, 3, +2; (c) 2, 1, 0; (d) 2, 1, +2; (e) 1, 0, 0.

3.78 Each of the following sets of quantum numbers (n, ℓ, m_ℓ) is missing one quantum number. For each set, determine all possible values for the missing number: (a) 1, ?, 0; (b) 3, 2, ?; (c) 2, ?, +1; (d) 4, 1, ?; (e) 2, 0, ?.

SECTION 3.8: ATOMIC ORBITALS

Review Questions

3.79 Describe the shapes of s, p, and d orbitals. How are these orbitals related to the quantum numbers n, ℓ, and m_ℓ?

3.80 List the hydrogen orbitals in increasing order of energy.

3.81 Describe the characteristics of an s orbital, p orbital, and d orbital. Which of the following orbitals do not exist: $1p$, $2s$, $2d$, $3p$, $3d$, $3f$, $4g$?

3.82 Why is a boundary surface diagram useful in representing an atomic orbital?

Conceptual Problems

3.83 Give the values of the quantum numbers associated with the following orbitals: (a) $2p$, (b) $3s$, (c) $5d$.

3.84 Give the values of the four quantum numbers of an electron in the following orbitals: (a) $3s$, (b) $4p$, (c) $3d$.

3.85 Describe how a $1s$ orbital and a $2s$ orbital are similar. Describe how they are different.

3.86 What is the difference between a $2p_x$ and a $2p_y$ orbital?

3.87 Why do the $3s$, $3p$, and $3d$ orbitals have the same energy in a hydrogen atom but different energies in a many-electron atom?

3.88 Make a chart of all allowable orbitals in the first four principal energy levels of the hydrogen atom. Designate each by type (for example, s, p), and indicate how many orbitals of each type there are.

3.89 For each of the following pairs of hydrogen orbitals, indicate which is higher in energy: (a) $1s$, $2s$; (b) $2p$, $3p$; (c) $3d_{xy}$, $3d_{yz}$; (d) $3s$, $3d$; (e) $4f$, $5s$.

3.90 Which orbital in each of the following pairs is lower in energy in a many-electron atom: (a) $2s$, $2p$; (b) $3p$, $3d$; (c) $3s$, $4s$; (d) $4d$, $5f$?

3.91 A $3s$ orbital is illustrated here. Using this as a reference to show the relative size of the other four orbitals, answer the following questions.

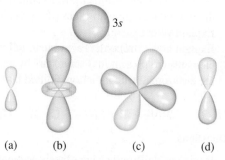

(a)   (b)   (c)   (d)

(a) Which orbital has the greatest value of n?
(b) How many orbitals have a value of $\ell = 1$?
(c) How many other orbitals with the same value of n would have the same general shape as orbital (b)?

SECTION 3.9: ELECTRON CONFIGURATIONS

Review Questions

3.92 What is electron configuration? Describe the roles that the Pauli exclusion principle and Hund's rule play in writing the electron configuration of elements.

3.93 Explain the meaning of the symbol $4d^6$.

3.94 State the Aufbau principle, and explain the role it plays in classifying the elements in the periodic table.

Computational Problems

3.95 Indicate the total number of (a) p electrons in N $(Z = 7)$; (b) s electrons in Si $(Z = 14)$; and (c) $3d$ electrons in S $(Z = 16)$.

3.96 Calculate the total number of electrons that can occupy (a) one s orbital, (b) three p orbitals, (c) five d orbitals, (d) seven f orbitals.

3.97 Determine the total number of electrons that can be held in all orbitals having the same principal quantum number n when (a) $n = 1$, (b) $n = 2$, (c) $n = 3$, and (d) $n = 4$.

3.98 Determine the maximum number of electrons that can be found in each of the following subshells: $3s$, $3d$, $4p$, $4f$, $5f$.

Conceptual Problems

3.99 The ground-state electron configurations listed here are incorrect. Explain what mistakes have been made in each and write the correct electron configurations.

Al: $1s^2 2s^2 2p^4 3s^2 3p^3$
B: $1s^2 2s^2 2p^5$
F: $1s^2 2s^2 2p^6$

3.100 The electron configuration of an atom in the ground state is $1s^2 2s^2 2p^6 3s^2$. Write a complete set of quantum numbers for each of the electrons. Name the element.

3.101 List the following atoms in order of increasing number of unpaired electrons: B, C, N, O, F.

3.102 Determine the number of unpaired electrons in each atom: K, Ca, Sc, Ti, V, Cr, Mn.

3.103 Determine the number of unpaired electrons in each of the following atoms in the ground state and identify each as diamagnetic or paramagnetic: (a) Rb, (b) As, (c) I, (d) Cr, (e) Zn.

3.104 Determine the number of unpaired electrons in each of the following atoms in the ground state and identify each as diamagnetic or paramagnetic: (a) C, (b) S, (c) Cu, (d) Pb, (e) Ti.

3.105 Indicate which of the following sets of quantum numbers in an atom are unacceptable and explain why: (a) $\left(1, 1, +\frac{1}{2}, -\frac{1}{2}\right)$; (b) $\left(3, 0, -1, +\frac{1}{2}\right)$; (c) $\left(2, 0, +1, +\frac{1}{2}\right)$; (d) $\left(4, 3, -2, +\frac{1}{2}\right)$; (e) $(3, 2, +1, 1)$.

3.106 Portions of orbital diagrams representing the ground-state electron configurations of certain elements are shown here. Which of them violate the Pauli exclusion principle? Which violate Hund's rule?

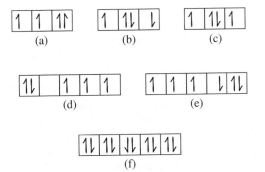

SECTION 3.10: ELECTRON CONFIGURATIONS AND THE PERIODIC TABLE

Review Questions

3.107 Describe the characteristics of transition metals.

3.108 What is the noble gas core? How does it simplify the writing of electron configurations?

3.109 What are the group and period of the element tantalum?

3.110 Define the following terms and give an example of each: *lanthanides, actinides.*

3.111 Explain why the ground-state electron configurations of Cr and Cu are different from what we might expect.

3.112 Write the electron configuration of a xenon core.

3.113 Comment on the correctness of the following statement: The probability of finding two electrons with the same four quantum numbers in an atom is zero.

Conceptual Problems

3.114 Use the Aufbau principle to obtain the ground-state electron configuration of tellurium.

3.115 Use the Aufbau principle to obtain the ground-state electron configuration of iridium.

3.116 Write the ground-state electron configurations for the following elements: B, V, C, As, I, Au.

3.117 Write the ground-state electron configurations for the following elements: Ge, Fe, Zn, Ni, W, Tl.

3.118 What is the symbol of the element with the following ground-state electron configurations? (a) $[Ar]4s^2 3d^8$; (b) $[Ne]3s^2 3p^2$; (c) $[Kr]5s^2 4d^{10} 5p^4$; (d) $[Ar]4s^2 3d^{10} 4p^3$.

3.119 What is the symbol of the element with the following ground-state electron configurations? (a) $[Ar]4s^2 3d^{10} 4p^1$; (b) $[Kr]5s^2 4d^1$; (c) $[Ar]4s^1 3d^5$; (d) $[Rn]7s^2 6d^1$.

ADDITIONAL PROBLEMS

3.120 When a substance containing cesium is heated in a Bunsen burner flame, photons with an energy of 4.30×10^{-19} J are emitted. What color is the cesium flame?

3.121 Discuss the current view of the correctness of the following statements. (a) The electron in the hydrogen atom is in an orbit that never brings it closer than 100 pm to the nucleus. (b) Atomic absorption spectra result from transitions of electrons from lower to higher energy levels. (c) A many-electron atom behaves somewhat like a solar system that has a number of planets.

3.122 Distinguish carefully between the following terms: (a) *wavelength* and *frequency,* (b) *wave properties* and *particle properties,* (c) *quantization of energy* and *continuous variation in energy.*

3.123 What is the maximum number of electrons in an atom that can have the following quantum numbers? Specify the orbitals in which the electrons would be found. (a) $n = 2, m_s = +\frac{1}{2}$; (b) $n = 4, m_\ell = +1$; (c) $n = 3, \ell = 2$; (d) $n = 2, \ell = 0, m_s = -\frac{1}{2}$; (e) $n = 4, \ell = 3, m_\ell = -2$.

3.124 Identify the following individuals and their contributions to the development of quantum theory: Bohr, de Broglie, Einstein, Planck, Heisenberg, Schrödinger.

3.125 Calculate the wavelength and frequency of an emitted photon of gamma radiation that has energy of 3.14×10^{11} J/mol.

3.126 A baseball pitcher's fastball has been clocked at 101 mph. (a) Calculate the wavelength of a 0.141-kg baseball (in nanometers) at this speed. (b) What is the wavelength of a hydrogen atom at the same speed (1 mi = 1609 m)?

3.127 A ruby laser produces radiation of wavelength 633 nm in pulses with a duration of 1.00×10^{-9} s. (a) If the laser produces 0.376 J of energy per pulse, how many photons are produced in each pulse? (b) Calculate the power (in watts) delivered by the laser per pulse (1 W = 1 J/s).

3.128   Four atomic energy levels of an atom are shown here. When an electron in an excited atom moves from (d), (d) $\longrightarrow$ (c), (c) $\longrightarrow$ (b), and (b) $\longrightarrow$ (a), a photon of light is emitted each time. The wavelengths of the various photons are 575 nm, 162 nm, and 131 nm. Determine the energies of each photon emitted and match each emission to the appropriate wavelength knowing the energy levels are drawn to scale.

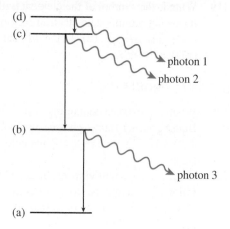

3.129   A single atom of an unknown element travels at 15 percent of the speed of light. The de Broglie wavelength of this atom is 1.06×10^{-16} m. What element is this?

3.130   Spectral lines of the Lyman and Balmer series do not overlap. Verify this statement by calculating the longest wavelength associated with the Lyman series and the shortest wavelength associated with the Balmer series (in nanometers).

3.131   Only a fraction of the electric energy supplied to a tungsten lightbulb is converted to visible light. The rest of the energy shows up as infrared radiation (i.e., heat). A 75-W lightbulb converts 15.0 percent of the energy supplied to it into visible light (assume the wavelength to be 550 nm). How many photons are emitted by the lightbulb per second (1 W = 1 J/s)?

3.132   The figure here illustrates a series of transitions that occur in a hydrogen atom.

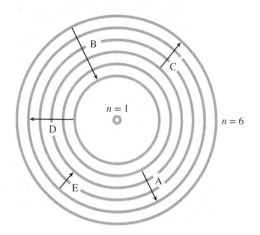

(a) Which transitions are absorptions and which are emissions?

(b) Rank the emissions in order of increasing energy.

(c) Rank the emissions in order of increasing wavelength of light emitted.

3.133   When one of helium's electrons is removed, the resulting species is the helium ion, He^+. The He^+ ion contains only one electron and is therefore a "hydrogen-like ion." Calculate the wavelengths, in increasing order, of the first four transitions in the Balmer series of the He^+ ion. Compare these wavelengths with the same transitions in an H atom. Comment on the differences. (The Rydberg constant for He is 4.39×10^7 m^{-1}.)

3.134   The retina of a human eye can detect light when radiant energy incident on it is at least 4.0×10^{-17} J. For light of 575-nm wavelength, how many photons does this correspond to?

3.135   An electron in an excited state in a hydrogen atom can return to the ground state in two different ways: (a) via a direct transition in which a photon of wavelength λ_1 is emitted and (b) via an intermediate excited state reached by the emission of a photon of wavelength λ_2. This intermediate excited state then decays to the ground state by emitting another photon of wavelength λ_3. Derive an equation that relates λ_1 to λ_2 and λ_3.

3.136   In 1996, physicists created an anti-atom of hydrogen. In such an atom, which is the antimatter equivalent of an ordinary atom, the electric charges of all the component particles are reversed. Thus, the nucleus of an anti-atom is made of an antiproton, which has the same mass as a proton but bears a negative charge, and the electron is replaced by an anti-electron (also called a positron) with the same mass as an electron but bearing a positive charge. Would you expect the energy levels, emission spectra, and atomic orbitals of an antihydrogen atom to be different from those of a hydrogen atom? What would happen if an anti-atom of hydrogen collided with a hydrogen atom?

3.137   The electron configurations described in this chapter all refer to gaseous atoms in their ground states. An atom may absorb a quantum of energy and promote one of its electrons to a higher-energy orbital. When this happens, we say that the atom is in an excited state. The electron configurations of some excited atoms are given. Identify these atoms and write their ground-state configurations:

(a) $1s^1 2s^1$

(b) $1s^1 2s^2 2p^2 3d^1$

(c) $1s^2 2s^2 2p^6 4s^1$

(d) [Ar]$4s^1 3d^{10} 4p^4$

(e) [Ne]$3s^2 3p^4 3d^1$

3.138   Draw the shapes (boundary surfaces) of the following orbitals: (a) $2p_y$, (b) $3dz^2$, (c) $3d_{x^2-y^2}$. (Show coordinate axes in your sketches.)

3.139   Draw orbital diagrams for atoms with the following electron configurations:
(a) $1s^2 2s^2 2p^5$
(b) $1s^2 2s^2 2p^6 3s^2 3p^3$
(c) $1s^2 2s^2 2p^6 3s^2 3p^6 4s^2 3d^7$

3.140   Consider the graph here.

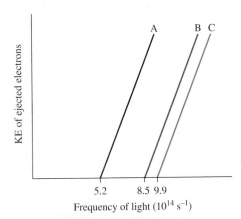

(a) Calculate the binding energy (W) of each metal. Which metal has the highest binding energy? (b) A photon with a wavelength of 333 nm is fired at the three metals. Which, if any, of the metals will eject an electron?

3.141   Scientists have found interstellar hydrogen atoms with quantum number n in the hundreds. Calculate the wavelength of light emitted when a hydrogen atom undergoes a transition from $n = 236$ to $n = 235$. In what region of the electromagnetic spectrum does this wavelength fall?

3.142   Ionization energy is the minimum energy required to remove an electron from an atom. It is usually expressed in units of kJ/mol, that is, the energy in kilojoules required to remove one mole of electrons from one mole of atoms. (a) Calculate the ionization energy for the hydrogen atom. (b) Repeat the calculation, assuming in this second case that the electrons are removed from the $n = 2$ state instead of from the ground state.

3.143   An electron in a hydrogen atom is excited from the ground state to the $n = 4$ state. Comment on the correctness of the following statements (true or false).
(a) $n = 4$ is the first excited state.
(b) It takes more energy to ionize (remove) the electron from $n = 4$ than from the ground state.
(c) The electron is farther from the nucleus (on average) in $n = 4$ than in the ground state.

(d) The wavelength of light emitted when the electron drops from $n = 4$ to $n = 1$ is longer than that from $n = 4$ to $n = 2$.
(e) The wavelength the atom absorbs in going from $n = 1$ to $n = 4$ is the same as that emitted as it goes from $n = 4$ to $n = 1$.

3.144   The ionization energy of a certain element is 412 kJ/mol (see Problem 3.142). However, when the atoms of this element are in the first excited state, the ionization energy is only 126 kJ/mol. Based on this information, calculate the wavelength of light emitted in a transition from the first excited state to the ground state.

3.145   The cone cells of the human eye are sensitive to three wavelength ranges, which the eye interprets as blue (419 nm), green (531 nm), and red (558 nm). If the optic nerve in the eye requires 2.0×10^{-17} J of energy to initiate the sight impulses to the brain, how many photons of blue light (419 nm), green light (531 nm), and red light (558 nm) are needed to see these colors?

3.146   (a) An electron in the ground state of the hydrogen atom moves at an average speed of 5×10^6 m/s. If the speed is known to an uncertainty of 20 percent, what is the minimum uncertainty in its position? Given that the radius of the hydrogen atom in the ground state is 5.29×10^{-11} m, comment on your result. The mass of an electron is 9.1094×10^{-31} kg. (b) A 0.15-kg baseball thrown at 99 mph has a momentum of 6.7 kg · m/s. If the uncertainty in measuring the momentum is 1.0×10^{-7} of the momentum, calculate the uncertainty in the baseball's position.

3.147   The UV light that is responsible for tanning the skin falls in the 320-nm to 400-nm region. Calculate the total energy (in joules) absorbed by a person exposed to this radiation for 2.5 h, given that there are 2.0×10^{16} photons hitting Earth's surface per square centimeter per second over a 80-nm (320-nm to 400-nm) range and that the exposed body area is 0.45 m². Assume that only half of the radiation is absorbed and the other half is reflected by the body. (*Hint:* Use an average wavelength of 360 nm in calculating the energy of a photon.)

3.148   The sun is surrounded by a white circle of gaseous material called the corona, which becomes visible during a total eclipse of the sun. The temperature of the corona is in the millions of degrees Celsius, which is high enough to remove some or all of the electrons from atoms. One way astronomers have been able to estimate the temperature of the corona is by studying the emission lines of ions of certain elements. For example, the emission spectrum of Fe^{14+} ions (iron atoms from which 14 electrons have been removed) has been recorded and analyzed. Knowing that it takes 3.5×10^4 kJ/mol to convert Fe^{13+} to Fe^{14+}, estimate the temperature of the sun's corona. (*Hint:* The average kinetic energy of one mole of a gas is $\frac{3}{2} RT$.)

3.149   When an electron makes a transition between energy levels of a hydrogen atom, there are no restrictions on the initial and final values of the principal quantum number n. However, there is a quantum mechanical rule that restricts the initial and final values of the orbital angular momentum ℓ. This is the *selection rule,* which states that $\Delta\ell = \pm 1$; that is, in a transition, the value of ℓ can increase or decrease by only 1. According to this rule, which of the following transitions are allowed:
(a) $1s \longrightarrow 2s$ (b) $2p \longrightarrow 1s$ (c) $1s \longrightarrow 3d$
(d) $3d \longrightarrow 4f$ (e) $4d \longrightarrow 3s$?

3.150   Blackbody radiation is the term used to describe the dependence of the radiation energy emitted by an object on wavelength at a certain temperature. Planck proposed the quantum theory to account for the dependence. Shown in the figure is a plot of the radiation energy emitted by our sun versus wavelength. This curve is characteristic of objects at about 6000 K, which is the temperature at the surface of the sun. At a higher temperature, the curve has a similar shape but the maximum will shift to a shorter wavelength. (a) What does this curve reveal about two consequences of great biological significance on Earth? (b) How are astronomers able to determine the temperature at the surface of stars in general?

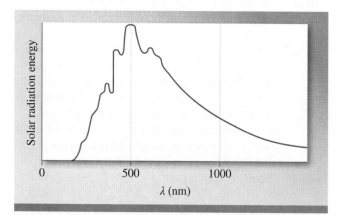

3.151   Suppose that photons of red light (675 nm) are used to locate the position of a 2.80-g Ping-Pong ball in flight and that the uncertainty in the position is equal to one wavelength. What is the minimum uncertainty in the speed of the Ping-Pong ball? Comment on the magnitude of your result.

3.152   In an electron microscope, electrons are accelerated by passing them through a voltage difference. The kinetic energy thus acquired by the electrons is equal to the voltage times the charge on the electron. Thus, a voltage difference of 1 volt imparts a kinetic energy of 1.602×10^{-19} volt-coulomb or 1.602×10^{-19} J. Calculate the wavelength associated with electrons accelerated by 5.00×10^3 volts.

3.153   According to Einstein's special theory of relativity, the mass of a moving particle, m_{moving}, is related to its mass at rest, m_{rest}, by the equation

$$m_{moving} = \frac{m_{rest}}{\sqrt{1 - (u/c)^2}}$$

where u and c are the speeds of the particle and light, respectively. (a) In particle accelerators, protons, electrons, and other charged particles are often accelerated to speeds close to the speed of light. Calculate the wavelength (in nanometers) of a proton moving at 50.0 percent the speed of light. The mass of a proton is 1.67×10^{-27} kg. (b) Calculate the mass of a 6.0×10^{-2}-kg tennis ball moving at 63 m/s. Comment on your results.

3.154   The mathematical equation for studying the photoelectric effect is

$$h\nu = W + \frac{1}{2} m_e u^2$$

where ν is the frequency of light shining on the metal; W is the energy needed to remove an electron from the metal; and m_e and u are the mass and speed of the ejected electron, respectively. In an experiment, a student found that a maximum wavelength of 351 nm is needed to just dislodge electrons from a zinc metal surface. Calculate the velocity (in m/s) of an ejected electron when the student employed light with a wavelength of 313 nm.

Answers to In-Chapter Materials

PRACTICE PROBLEMS

3.1A 1.13×10^3 J. **3.1B** 3.70×10^2 m/s. **3.2A** 1.5 times greater. **3.2B** 0.667 mm. **3.3A** 1.86×10^{-4} m. **3.3B** 2.91×10^{10} s^{-1}. **3.4A** (a) 9.43×10^{-19} J, (b) 1.18×10^{-25} J, (c) 1.96×10^{-19} J. **3.4B** (a) 1.05×10^4 nm, (b) 2.1×10^{18}, (c) 1.2 eV. **3.5A** 103 nm. **3.5B** 7. **3.6A** 26 nm. **3.6B** 8.2×10^{-28} kg · m/s. **3.7A** (a) 3×10^{-34} m. **3.7B** (a) $\pm 2 \times 10^{-25}$ kg · m/s, (b) $\pm 1 \times 10^2$ m/s, (c) $\pm 2 \times 10^5$ m/s. **3.8A** Only 0. **3.8B** $-2, -1, 0, +1, +2$. **3.9A** $n = 3$, $\ell = 2$, $m_\ell = -2$, $-1, 0, +1, +2$. **3.9B** For a d orbital, $\ell = 2$, but when $n = 2$, ℓ cannot be 2. **3.10A** $1s^2 2s^2 2p^6 3s^2 3p^6 4s^2 3d^{10} 4p^6 5s^1$. **3.10B** $1s^2 2s^2 2p^6 3s^2 3p^6 4s^2 3d^{10} 4p^5$. **3.11A** [Rn]$7s^2$. **3.11B** At.

SECTION REVIEW

3.1.1 a. **3.1.2** d. **3.1.3** c. **3.2.1** c. **3.2.2** a. **3.2.3** c. **3.2.4** b. **3.3.1** e. **3.3.2** d. **3.3.3** a. **3.3.4** d. **3.3.5** c. **3.4.1** b. **3.4.2** d. **3.4.3** c. **3.4.4** e. **3.5.1** c. **3.5.2** d. **3.5.3** b. **3.6.1** c. **3.6.2** a. **3.7.1** c. **3.7.2** c. **3.7.3** e. **3.7.4** c. **3.8.1** b. **3.8.2** c. **3.8.3** b,e. **3.8.4** b. **3.9.1** b. **3.9.2** d. **3.9.3** d. **3.10.1** c. **3.10.2** c. **3.10.3** b,d. **3.10.4** a,b,d.

Periodic Trends of the Elements

© Dzhavakhadze Zurab Itar-Tass Photos/Newscom

THE PERIODIC recurrence of the properties of the elements—a phenomenon known as *periodicity*—provides the foundation of the modern periodic table. Based on the work of Dmitri Mendeleev and Lothar Meyer, this careful tabulation of the elements allowed chemists to predict the existence and properties of elements that had not yet been discovered. Mendeleev himself predicted the eventual discovery of an element similar to aluminum, which he dubbed eka-aluminum. Indeed, when the element *gallium* was subsequently discovered, it had the very properties that Mendeleev had predicted for eka-aluminum. Since that time, chemists and physicists have succeeded in completing the current periodic table by producing elements that are not found in nature. In May 2012, flerovium (114) was named officially in honor of the Flerov Laboratory of Nuclear Reactions in Dubna, Russia, where it was produced by a team of scientists. The other element to be named officially in May 2012 is livermorium (116), named in honor of the Lawrence Livermore National Laboratory in Livermore, California.

- Electrostatic energy [◄◄ Section 3.1]
- Electron configurations and the periodic table [◄◄ Section 3.10]

4.1 DEVELOPMENT OF THE PERIODIC TABLE

Although chemists in the nineteenth century generally agreed that matter consisted of atoms, they knew nothing of the *structure* of atoms. Nevertheless, the need to organize a growing body of scientific information about the known elements inspired a number of attempts to arrange the elements in a table. Because accurate measurements of the atomic masses of many elements had already been made, one seemingly logical approach was to organize the elements in order of increasing atomic mass.

In 1864, English chemist John Newlands[1] noticed that when the elements were arranged in order of increasing atomic mass, every eighth element had similar properties. Newlands referred to this peculiar relationship as the *law of octaves.* However, this "law" turned out to be inadequate for elements beyond calcium, and Newlands's work was not accepted by the scientific community.

In 1869, Russian chemist Dmitri Mendeleev[2] and German chemist Lothar Meyer[3] independently proposed a much more extensive tabulation of the elements based on the regular, periodic recurrence of properties—a phenomenon known as *periodicity.*

Mendeleev's classification system was a great improvement over Newlands's for two reasons. First, it grouped the elements together more accurately, according to their properties. Second, and equally important, it made it possible to predict the properties of several elements that had not yet been discovered. For example, Mendeleev proposed the existence of an unknown element that he called eka-aluminum and predicted a number of its properties. (*Eka* is a Sanskrit word meaning "first"; thus, eka-aluminum would be the first element under aluminum in the same group.) When gallium was discovered four years later, its properties matched the predicted properties of eka-aluminum remarkably well:

	Eka-Aluminum (Ea)	Gallium (Ga)
Atomic mass	68 amu	69.9 amu
Melting point	Low	30.15°C
Density	5.9 g/cm^3	5.94 g/cm^3
Formula of oxide	Ea_2O_3	Ga_2O_3

Mendeleev's periodic table included 66 known elements. By 1900, 30 more had been added to the list, filling in some of the empty spaces. Figure 4.1 gives the time period during which each element was discovered.

Although this periodic table was remarkably successful, the early versions had some inconsistencies that were impossible to overlook. For example, the atomic mass of argon (39.95 amu) is greater than that of potassium (39.10 amu), but argon comes before potassium in the periodic table. If elements were arranged solely according to increasing atomic mass, argon would appear in the position occupied by potassium in our modern periodic table. It simply would not make sense to put sodium (a metal) in a group where every other member is a gas—or to put argon (a gas) in a group

---

[1]John Alexander Reina Newlands (1838–1898). Newlands's work was a step in the right direction in the classification of the elements. Unfortunately, because of its shortcomings, he was subjected to much criticism and even ridicule. At one meeting he was asked if he had ever examined the elements according to the order of their initial letters! Nevertheless, in 1887 Newlands was honored by the Royal Society of London for his contribution.

[2]Dmitri Ivanovich Mendeleev (1836–1907). His work on the periodic classification of elements is regarded by many as the most significant achievement in chemistry in the nineteenth century.

[3]Julius Lothar Meyer (1830–1895). In addition to his contribution to the periodic table, Meyer also discovered the chemical affinity of hemoglobin for oxygen.

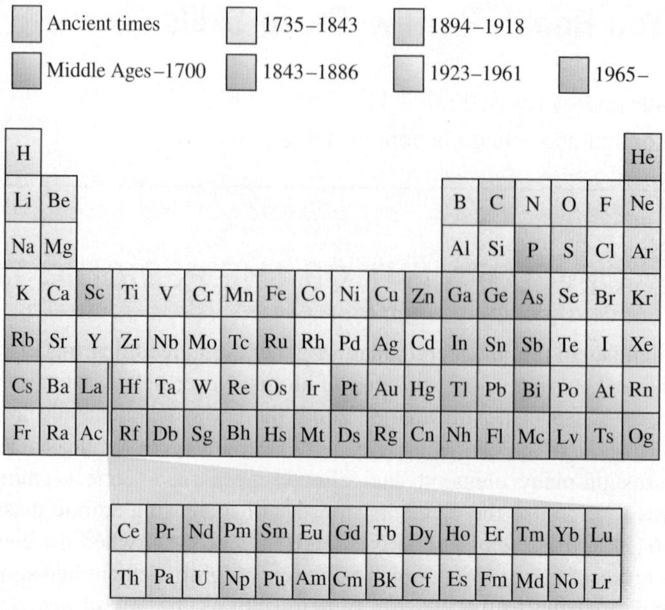

Figure 4.1 Periodic table of elements classified by dates of discovery.

where every other member is a metal. This and other discrepancies suggested that some fundamental property other than atomic mass must be the basis of periodicity. The fundamental property turned out to be the number of protons in an atom's nucleus, something that could not have been known by Mendeleev and his contemporaries.

In 1913 a young English physicist, Henry Moseley,[4] discovered a correlation between what he called *atomic number* and the frequency of X rays generated by bombarding an element with high-energy electrons. Moseley noticed that, in general, the frequencies of X rays emitted from the elements increased with increasing atomic mass. Among the few exceptions he found were argon and potassium. Although argon has a greater atomic mass than potassium, the X-ray emission from potassium indicated that it has the greater atomic number. Ordering the periodic table using atomic number enabled scientists to make sense out of the discrepancies that had puzzled them earlier. Moseley concluded that the atomic number was equal to the number of protons in the nucleus and to the number of electrons in an atom.

Entries in modern periodic tables usually include an element's atomic number along with its symbol. Electron configurations of elements help to explain the periodic recurrence of physical and chemical properties. The importance and usefulness of the periodic table lie in the fact that we can use our understanding of the general properties and trends within a group or a period to predict with considerable accuracy the properties of any element, even though that element may be unfamiliar to us.

Worked Example 4.1 shows how the periodic table can be used to predict similarities in the properties of elements.

Worked Example 4.1

What elements would you expect to exhibit properties most similar to those of chlorine?

Strategy Because elements in the same group tend to have similar properties, you should identify elements in the same group as chlorine.

Setup Chlorine is a member of Group 7A.

Solution Fluorine, bromine, and iodine, the other nonmetals in Group 7A, should have properties most similar to those of chlorine.

---

[4]Henry Gwyn-Jeffreys Moseley (1887–1915). Moseley discovered the relationship between X-ray spectra and atomic number. A lieutenant in the Royal Engineers, he was killed in action at the age of 28 during the British campaign in Gallipoli, Turkey.

Think About It
Astatine (At) is also in Group 7A. Astatine, though, is classified as a metalloid, and we have to be careful comparing nonmetals to *metalloids* (or to *metals*). As a metalloid, the properties of astatine should be less similar to those of chlorine than the other members of Group 7A.

Student Annotation: Actually, astatine is radioactive and very little is known about its properties.

Practice Problem ATTEMPT (a) What element(s) would you expect to exhibit properties most similar to those of silicon (Si)? (b) Which of the elements in Group 5A would you expect to exhibit properties *least* similar to those of nitrogen?

Practice Problem BUILD Arrange the following Group 5A elements in order of increasing similarity of properties to N: As, Bi, and P.

Practice Problem CONCEPTUALIZE Three different groups are highlighted in the periodic table shown here. Which of the three highlighted groups contains the largest number of elements with similar properties? Explain.

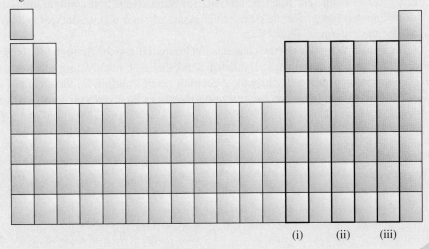

(i)   (ii)   (iii)

Section 4.1 Review

Development of the Periodic Table

4.1.1   Which of the following elements would you expect to have chemical properties most similar to those of S?
(a) P   (c) Se   (e) Sr
(b) Cl   (d) Na

4.1.2   Which of the following elements would you expect to have properties similar to those of Ba?
(a) Sr   (c) Na   (e) B
(b) Rb   (d) K

4.1.3   The first synthesis of element 117 was reported in 2010. Based on its position in the periodic table, which of the following properties would you expect it to exhibit?
(i)   gaseous at room temperature
(ii) unreactive
(iii) properties similar to At
(iv) properties similar to Ra

(a) i and iv   (c) iii only   (e) i and iii
(b) i, ii, and iv   (d) iv only

4.2 THE MODERN PERIODIC TABLE

Figure 4.2 shows the modern periodic table together with the outermost ground-state electron configurations of the elements. (The outermost electron configurations of the elements are also given in Figure 3.26.) Starting with hydrogen, the electronic subshells are filled in the order shown in Figure 3.25 [⏮ Section 3.9].

Classification of Elements

Student Annotation: In this context, outermost electrons refers to those that are placed in orbitals last using the Aufbau principle [Section 3.9].

Based on the type of subshell containing the outermost electrons, the elements can be divided into categories—the main group elements, the noble gases, the transition elements (or transition metals), the lanthanides, and the actinides. The ***main group elements*** (also called the *representative elements*) are the elements in Groups 1A through 7A. With the exception of helium, each of the *noble gases* (the Group 8A elements) has a completely filled *p* subshell. The valence-electron configurations are $1s^2$ for helium and ns^2np^6 for the other noble gases, where *n* is the principal quantum number for the outermost shell.

Student Annotation: Even when an atom from Group 2B loses one or two electrons, as is common, its electron configuration includes a completed *d* subshell [Section 3.9]. Thus, strictly speaking, Zn, Cd, and Hg are *not* transition metals.

The transition metals are the elements in Groups 1B and 3B through 8B. Transition metals either have incompletely filled *d* subshells or readily lose electrons to *achieve* incompletely filled *d* subshells. According to this definition, the elements of Group 2B (Zn, Cd, and Hg) are *not* transition metals as their electron configurations

	1A 1	2A 2	3B 3	4B 4	5B 5	6B 6	7B 7	8B 8	8B 9	8B 10	1B 11	2B 12	3A 13	4A 14	5A 15	6A 16	7A 17	8A 18	
1	1 **H** $1s^1$																	2 **He** $1s^2$	1
2	3 **Li** $2s^1$	4 **Be** $2s^2$											5 **B** $2s^22p^1$	6 **C** $2s^22p^2$	7 **N** $2s^22p^3$	8 **O** $2s^22p^4$	9 **F** $2s^22p^5$	10 **Ne** $2s^22p^6$	2
3	11 **Na** $3s^1$	12 **Mg** $3s^2$											13 **Al** $3s^23p^1$	14 **Si** $3s^23p^2$	15 **P** $3s^23p^3$	16 **S** $3s^23p^4$	17 **Cl** $3s^23p^5$	18 **Ar** $3s^23p^6$	3
4	19 **K** $4s^1$	20 **Ca** $4s^2$	21 **Sc** $4s^23d^1$	22 **Ti** $4s^23d^2$	23 **V** $4s^23d^3$	24 **Cr** $4s^13d^5$	25 **Mn** $4s^23d^5$	26 **Fe** $4s^23d^6$	27 **Co** $4s^23d^7$	28 **Ni** $4s^23d^8$	29 **Cu** $4s^13d^{10}$	30 **Zn** $4s^23d^{10}$	31 **Ga** $4s^24p^1$	32 **Ge** $4s^24p^2$	33 **As** $4s^24p^3$	34 **Se** $4s^24p^4$	35 **Br** $4s^24p^5$	36 **Kr** $4s^24p^6$	4
5	37 **Rb** $5s^1$	38 **Sr** $5s^2$	39 **Y** $5s^24d^1$	40 **Zr** $5s^24d^2$	41 **Nb** $5s^14d^4$	42 **Mo** $5s^14d^5$	43 **Tc** $5s^24d^5$	44 **Ru** $5s^14d^7$	45 **Rh** $5s^14d^8$	46 **Pd** $4d^{10}$	47 **Ag** $5s^14d^{10}$	48 **Cd** $5s^24d^{10}$	49 **In** $5s^25p^1$	50 **Sn** $5s^25p^2$	51 **Sb** $5s^25p^3$	52 **Te** $5s^25p^4$	53 **I** $5s^25p^5$	54 **Xe** $5s^25p^6$	5
6	55 **Cs** $6s^1$	56 **Ba** $6s^2$	57 **La** $6s^25d^1$	72 **Hf** $6s^25d^2$	73 **Ta** $6s^25d^3$	74 **W** $6s^25d^4$	75 **Re** $6s^25d^5$	76 **Os** $6s^25d^6$	77 **Ir** $6s^25d^7$	78 **Pt** $6s^15d^9$	79 **Au** $6s^15d^{10}$	80 **Hg** $6s^25d^{10}$	81 **Tl** $6s^26p^1$	82 **Pb** $6s^26p^2$	83 **Bi** $6s^26p^3$	84 **Po** $6s^26p^4$	85 **At** $6s^26p^5$	86 **Rn** $6s^26p^6$	6
7	87 **Fr** $7s^1$	88 **Ra** $7s^2$	89 **Ac** $7s^26d^1$	104 **Rf** $7s^26d^2$	105 **Db** $7s^26d^3$	106 **Sg** $7s^26d^4$	107 **Bh** $7s^26d^5$	108 **Hs** $7s^26d^6$	109 **Mt** $7s^26d^7$	110 **Ds** $7s^26d^8$	111 **Rg** $7s^26d^9$	112 **Cn** $7s^26d^{10}$	113 **Nh** $7s^27p^1$	114 **Fl** $7s^27p^2$	115 **Mc** $7s^27p^3$	116 **Lv** $7s^27p^4$	117 **Ts** $7s^27p^5$	118 **Og** $7s^27p^6$	7

58 **Ce** $6s^24f^15d^1$	59 **Pr** $6s^24f^3$	60 **Nd** $6s^24f^4$	61 **Pm** $6s^24f^5$	62 **Sm** $6s^24f^6$	63 **Eu** $6s^24f^7$	64 **Gd** $6s^24f^75d^1$	65 **Tb** $6s^24f^9$	66 **Dy** $6s^24f^{10}$	67 **Ho** $6s^24f^{11}$	68 **Er** $6s^24f^{12}$	69 **Tm** $6s^24f^{13}$	70 **Yb** $6s^24f^{14}$	71 **Lu** $6s^24f^{14}5d^1$
90 **Th** $7s^26d^2$	91 **Pa** $7s^25f^26d^1$	92 **U** $7s^25f^36d^1$	93 **Np** $7s^25f^46d^1$	94 **Pu** $7s^25f^6$	95 **Am** $7s^25f^7$	96 **Cm** $7s^25f^76d^1$	97 **Bk** $7s^25f^9$	98 **Cf** $7s^25f^{10}$	99 **Es** $7s^25f^{11}$	100 **Fm** $7s^25f^{12}$	101 **Md** $7s^25f^{13}$	102 **No** $7s^25f^{14}$	103 **Lr** $7s^25f^{14}6d^1$

Figure 4.2 Valence-electron configurations of the elements. For simplicity, the filled *f* subshells are not shown in elements 72 through 86, 104 through 118.

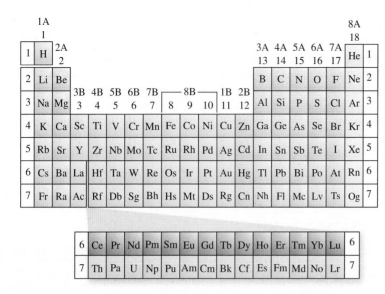

Figure 4.3 Periodic table with color coding of main group elements (grey), noble gases (yellow), transition metals (blue), Group 2B metals (light green), lanthanides (dark green), and actinides (pink).

always include a completed *d* subshell. They are *d*-block elements, though, so they generally are included in the discussion of transition metals.

The lanthanides and actinides are sometimes called *f*-block transition elements because they have incompletely filled *f* subshells. Figure 4.3 distinguishes the groups of elements discussed here.

There is a distinct pattern to the electron configurations of the elements in a particular group. See, for example, the electron configurations of Groups 1A and 2A in Table 4.1. Each member of Group 1A has a noble gas core plus one additional electron, giving each alkali metal the general electron configuration of [noble gas]ns^1. Similarly, the Group 2A alkaline earth metals have a noble gas core and an outer electron configuration of ns^2.

The outermost electrons of an atom are called **valence electrons.** Valence electrons determine how atoms interact with one another. Having the same valence-electron configuration is what causes the elements in the same group, such as those in Group 2A, to exhibit similar chemical properties. This observation holds true for the other main group elements as well. For example, the halogens (Group 7A) all have valence-electron configurations of ns^2np^5, and they exhibit similar properties.

In predicting properties for Groups 3A through 7A, we must take into account that each of these groups contains both *metals* and *nonmetals*. For example, the elements in Group 4A all have the same outer electron configuration, ns^2np^2, but there is considerable variation in chemical properties among these elements because carbon is a *nonmetal*, silicon and germanium are *metalloids*, and tin and lead are *metals*.

TABLE 4.1	Electron Configurations of Group 1A and Group 2A Elements		
Group 1A		**Group 2A**	
Li	[He]$2s^1$	Be	[He]$2s^2$
Na	[Ne]$3s^1$	Mg	[Ne]$3s^2$
K	[Ar]$4s^1$	Ca	[Ar]$4s^2$
Rb	[Kr]$5s^1$	Sr	[Kr]$5s^2$
Cs	[Xe]$6s^1$	Ba	[Xe]$6s^2$
Fr	[Rn]$7s^1$	Ra	[Rn]$7s^2$

As a group, the noble gases behave very similarly. The noble gases are chemically inert because they all have completely filled outer ns and np subshells, a condition that imparts unusual stability. With the exception of the heavier members of the group, krypton and xenon, they do not generally interact with other elements.

Although the outer electron configuration of the transition metals is not always the same within a group and there is often no regular pattern in the way the electron configuration changes from one metal to the next in the same period, all transition metals share many characteristics, such as magnetic properties, that set them apart from other elements. These properties are similar because all these metals have incompletely filled d subshells. Likewise, the lanthanide and actinide elements resemble one another because they have incompletely filled f subshells.

Worked Example 4.2 shows how to determine the electron configuration from the number of electrons in an atom.

Worked Example 4.2

Without using a periodic table, give the ground-state electron configuration and block designation (s, p, d, or f block) of an atom with (a) 17 electrons, (b) 37 electrons, and (c) 22 electrons. Classify each atom as a main group element or transition metal.

Strategy Use the Aufbau principle discussed in Section 3.9. Start writing each electron configuration with principal quantum number $n = 1$, and then continue to assign electrons to orbitals in the order presented in Figure 3.23 until all the electrons have been accounted for.

Setup According to Figure 3.23, orbitals fill in the following order: $1s$, $2s$, $2p$, $3s$, $3p$, $4s$, $3d$, $4p$, $5s$, $4d$, $5p$, $6s$, and so on. Recall that an s subshell contains one orbital, a p subshell contains three orbitals, and a d subshell contains five orbitals. Remember, too, that each orbital can accommodate a maximum of two electrons. The block designation of an element corresponds to the type of subshell occupied by the last electrons added to the configuration according to the Aufbau principle.

Solution

(a) $1s^2 2s^2 2p^6 3s^2 3p^5$, p block, main group

(b) $1s^2 2s^2 2p^6 3s^2 3p^6 4s^2 3d^{10} 4p^6 5s^1$, s block, main group

(c) $1s^2 2s^2 2p^6 3s^2 3p^6 4s^2 3d^2$, d block, transition metal

> **Think About It**
> Consult Figure 3.26 to confirm your answers.

Practice Problem Ⓐ**TTEMPT** Without using a periodic table, give the ground-state electron configuration and block designation (s, p, d, or f block) of an atom with (a) 15 electrons, (b) 20 electrons, and (c) 35 electrons.

Practice Problem Ⓑ**UILD** Identify the elements represented by (a) $1s^2 2s^2 2p^6 3s^2 3p^1$, (b) $1s^2 2s^2 2p^6 3s^2 3p^6 4s^2 3d^{10}$, and (c) $1s^2 2s^2 2p^6 3s^2 3p^6 4s^2 3d^{10} 4p^6 5s^2$.

Practice Problem Ⓒ**ONCEPTUALIZE** Determine the *total* number of electrons and the number of *valence* electrons for each of the indicated elements.

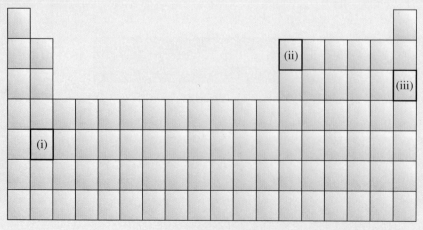

Section 4.2 Review

The Modern Periodic Table

4.2.1 What is the correct electron configuration for a germanium (Ge) atom in the ground state?

(a) $1s^2 2s^2 2p^6 3s^2 3p^6 4s^2 4p^2$ (d) $1s^2 2s^2 2p^6 3s^2 3p^6 4s^2 4p^2 4d^{10}$

(b) $1s^2 2s^2 2p^6 3s^2 3p^6 4s^2 3d^{10} 4p^2$ (e) $1s^2 2s^2 2p^6 3s^2 3p^2 3d^{10}$

(c) $1s^2 2s^2 2p^6 3s^2 3p^6$

4.2.2 What element is represented by the ground-state electron configuration $1s^2 2s^2 2p^6 3s^2 3p^6 4s^2 3d^{10} 4p^6 5s^2 4d^{10} 5p^2$?

(a) Pb   (d) Sn

(b) Ge   (e) Sb

(c) In

4.3 EFFECTIVE NUCLEAR CHARGE

As we have seen, the electron configurations of the elements show a periodic variation with increasing atomic number. In this and the next few sections, we will examine how electron configuration explains the periodic variation of physical and chemical properties of the elements. We begin by introducing the concept of *effective nuclear charge.*

Nuclear charge (Z) is simply the number of protons in the nucleus of an atom. *Effective nuclear charge (Z_{eff})* is the actual magnitude of positive charge that is "experienced" by an electron in the atom. The only atom in which the nuclear charge and effective nuclear charge are the same is hydrogen, which has only one electron. In all other atoms, the electrons are simultaneously attracted to the nucleus and repelled by one another. This results in a phenomenon known as **shielding.** An electron in a many-electron atom is partially shielded from the positive charge of the nucleus by the other electrons in the atom.

One way to illustrate how electrons in an atom shield one another is to consider the amounts of energy required to remove the two electrons from a helium atom, shown in Figure 4.4. Experiments show that it takes 3.94×10^{-18} J to remove the first electron but 8.72×10^{-18} J to remove the second one. There is no shielding once the first electron is removed, so the second electron feels the full effect of the +2 nuclear charge and is more difficult to remove.

Although all the electrons in an atom shield one another to some extent, those that are most effective at shielding are the *core* electrons. As a result, the value of Z_{eff} increases steadily from left to right across a period of the periodic table because the number of core electrons remains the same (only the number of protons, Z, and the number of *valence* electrons increases).

Student Annotation: *Shielding* is also known as *screening*.

Student Annotation: Core electrons are those in the completed inner shells.

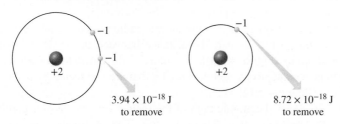

Figure 4.4 Removal of the first electron in He requires less energy than removal of the second electron because of shielding.

As we move to the right across period 2, the nuclear charge increases by 1 with each new element, but the *effective* nuclear charge increases only by an average of 0.64. (If the valence electrons did *not* shield one another, the effective nuclear charge would also increase by 1 each time a proton was added to the nucleus.)

	Li	**Be**	**B**	**C**	**N**	**O**	**F**
Z	3	4	5	6	7	8	9
Z_{eff} (felt by valence electrons)	1.28	1.91	2.42	3.14	3.83	4.45	5.10

In general, the effective nuclear charge is given by

Equation 4.1 $$Z_{\text{eff}} = Z - \sigma$$

where σ is the shielding constant. The shielding constant is greater than zero but smaller than Z.

The change in Z_{eff} as we move from the top of a group to the bottom is generally less significant than the change as we move across a period. Although each step down a group represents a large increase in the nuclear charge, there is also an additional shell of core electrons to shield the valence electrons from the nucleus. Consequently, the *effective* nuclear charge changes less than the nuclear charge as we move down a column of the periodic table.

4.4   PERIODIC TRENDS IN PROPERTIES OF ELEMENTS

Several physical and chemical properties of the elements depend on effective nuclear charge. To understand the trends in these properties, it is helpful to visualize the electrons of an atom in *shells*. Recall that the value of the principal quantum number (n) increases as the distance from the nucleus increases [◄◄ Section 3.8]. If we take this statement literally, and picture all the electrons in a shell at the same distance from the nucleus, the result is a sphere of uniformly distributed negative charge, with its distance from the nucleus depending on the value of n. With this as a starting point, we will examine the periodic trends in atomic radius, ionization energy, electron affinity, and metallic character.

Atomic Radius

Intuitively, we think of the ***atomic radius*** as the distance between the nucleus of an atom and its valence shell (i.e., the outermost shell that is occupied by one or more electrons), because we usually envision atoms as spheres with discrete boundaries. According to the quantum mechanical model of the atom, though, there is no specific distance from the nucleus beyond which an electron may not be found [◄◄ Section 3.8]. Therefore, the atomic radius requires a specific definition.

There are two ways in which the atomic radius is commonly defined. One is the ***metallic radius,*** which is half the distance between the nuclei of two adjacent, identical metal atoms [Figure 4.5(a)]. The other is the ***covalent radius,*** which is half the distance between adjacent, identical nuclei that are connected by a *chemical bond* [►► Chapter 5] [Figure 4.5(b)].

Figure 4.6 shows the atomic radii of the main group elements according to their positions in the periodic table. There are two distinct trends. The atomic radius

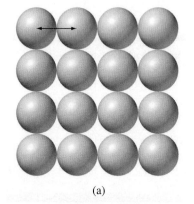

(a)

(b)

Figure 4.5 (a) Atomic radius in metals is defined as half the distance between adjacent metal atoms. (b) Atomic radius in nonmetals is defined as half the distance between bonded identical atoms in a molecule.

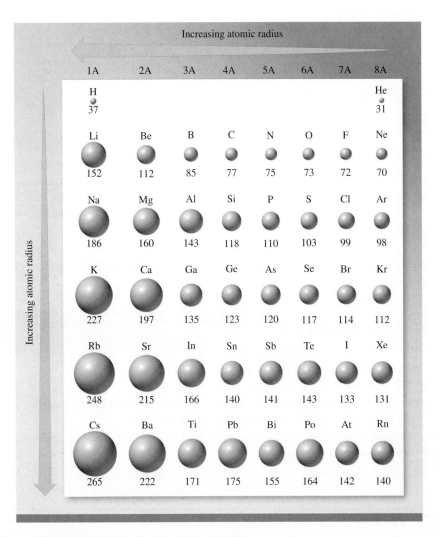

Figure 4.6 Atomic radii of the elements (in picometers).

decreases as we move from left to right across a period and *increases* from top to bottom as we move down within a group.

The increase down a group is fairly easily explained. As we step down a column, the outermost occupied shell has an ever-increasing value of *n,* so it lies farther from the nucleus, making the radius bigger.

Now let's try to understand the decrease in radius from left to right across a period. Although this trend may at first seem counterintuitive, given that the number of valence electrons is increasing with each new element, consider the shell model in which all the electrons in a shell form a uniform sphere of negative charge around the nucleus at a distance specified by the value of *n*. As we move from left to right across a period, the effective nuclear charge increases and each step to the right adds another electron to the valence shell. Coulomb's law (Equation 4.2) dictates that there will be a more powerful attraction between the nucleus and the valence shell when the magnitudes of both charges increase. The result is that as we step across a period, the valence shell is drawn closer to the nucleus, making the atomic radius smaller. Figure 4.7 shows how the effective nuclear charge, charge on the valence shell, and atomic radius vary across period 2. We can picture the valence shells in all the atoms as being initially at the same distance (determined by *n*) from the nuclei, but being pulled closer by a larger attractive force resulting from increases in both Z_{eff} and the number of valence electrons.

Video 4.1
Periodic table—atomic radius

Student Annotation: Although the overall trend in atomic size for transition elements is also to decrease from left to right and increase from top to bottom, the observed radii do not vary in as regular a way as do those of the main group elements.

Student Hot Spot

Student data indicate you may struggle with atomic radii. Access the SmartBook to view additional Learning Resources on this topic.

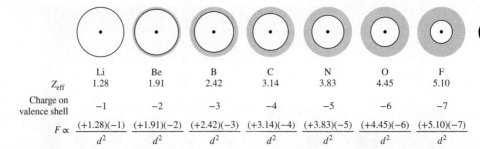

	Li	Be	B	C	N	O	F
Z_{eff}	1.28	1.91	2.42	3.14	3.83	4.45	5.10
Charge on valence shell	−1	−2	−3	−4	−5	−6	−7
$F \propto$	$\dfrac{(+1.28)(-1)}{d^2}$	$\dfrac{(+1.91)(-2)}{d^2}$	$\dfrac{(+2.42)(-3)}{d^2}$	$\dfrac{(+3.14)(-4)}{d^2}$	$\dfrac{(+3.83)(-5)}{d^2}$	$\dfrac{(+4.45)(-6)}{d^2}$	$\dfrac{(+5.10)(-7)}{d^2}$

Figure 4.7 Atomic radius decreases from left to right across a period because of the increased electrostatic attraction between the effective nuclear charge and the charge on the valence shell. The white circle shows the atomic size in each case. The comparison of attractive forces between the nuclei and valence shells is done using Coulomb's law (Equation 4.2).

Worked Example 4.3 shows how to use these trends to compare the atomic radii of different elements.

Worked Example 4.3

Referring only to a periodic table, arrange the elements P, S, and O in order of increasing atomic radius.

Strategy Use the left-to-right (decreasing) and top-to-bottom (increasing) trends to compare the atomic radii of two of the three elements at a time.

Setup Sulfur is to the right of phosphorus in the third row, so sulfur should be smaller than phosphorus. Oxygen is above sulfur in Group 6A, so oxygen should be smaller than sulfur.

Solution

O < S < P.

> **Think About It**
> Consult Figure 4.6 to confirm the order. Note that there are circumstances under which the trends alone will be insufficient to compare the radii of two elements. Using only a periodic table, for example, it would not be possible to determine that bromine ($r = 114$ pm) has a smaller radius than silicon ($r = 118$ pm).

Practice Problem ATTEMPT Referring only to a periodic table, arrange the elements Ge, Se, and F in order of increasing atomic radius.

Practice Problem BUILD For which of the following pairs of elements can the atomic radii *not* be compared using the periodic table alone: Si and Se, Se and Cl, or P and O?

Practice Problem CONCEPTUALIZE Based on size and using only a periodic table, identify the colored spheres as Al, B, Mg, and Sr.

| (i) | (ii) | (iii) | (iv) |

Ionization Energy

Ionization energy (IE) is the minimum energy required to remove an electron from an atom in the gas phase. The result is an *ion,* a chemical species with a net charge. We will represent this by writing the initial state, a sodium atom in the gas phase, on the left; and the final state, a sodium ion (also in the gas phase) and the electron that has been removed, on the right. The initial and final states are separated by an arrow, pointing to the right, which represents the ionization process.

$$Na(g) \longrightarrow Na^+(g) + e^-$$

> **Student Annotation:** This sort of representation is known as a *chemical equation*. We will discuss chemical equations in more detail in Chapter 8. Note that we can indicate the physical state of a substance with an italicized letter in parentheses: (*s*), (*l*), or (*g*).

Typically, we express ionization energy in kJ/mol, the number of kilojoules required to remove a mole of electrons from a mole of gaseous atoms. Sodium, for example, has an ionization energy of 495.8 kJ/mol, meaning that the energy input required to remove one mole of electrons from one mole of gaseous sodium atoms is 495.8 kJ. Specifically, 495.8 kJ/mol is the *first* ionization energy of sodium, $IE_1(Na)$, which corresponds to the removal of the most loosely held electron from each sodium atom. Figure 4.8(a) shows the first ionization energies of the main group elements according to their positions in the periodic table. Figure 4.8(b) shows a graph of IE_1 as a function of atomic number.

> **Student Annotation:** As with atomic radius, ionization energy changes in a similar but somewhat less regular way among the transition elements.

The *loss* of one or more electrons from an atom yields a ***cation,*** an ion with a net *positive* charge. For example, a sodium atom (Na) can readily lose an electron to become a sodium cation, which is represented by Na^+:

Na Atom	**Na$^+$ Ion**
11 protons	11 protons
11 electrons	10 electrons

In general, as effective nuclear charge increases, ionization energy also increases. Thus, IE_1 increases from left to right across a period. Despite this trend, the graph in Figure 4.8(b) shows that IE_1 for a Group 3A element is smaller than that for the corresponding Group 2A element. Likewise, IE_1 for a Group 6A element is smaller than that for the corresponding Group 5A element. Both of these *interruptions* of the upward trend in IE_1 can be explained by using electron configuration.

Recall that the energy of an electron in a many-electron system depends not only on the principal quantum number (n), but also on the angular momentum quantum number (ℓ) [◂◂ Section 3.9, Figure 3.23]. Within a given shell, electrons with the

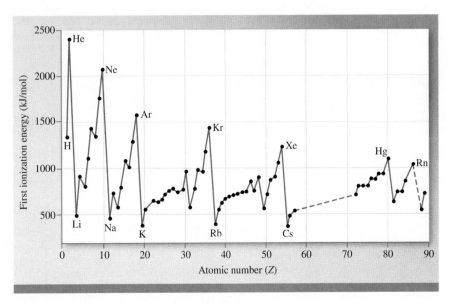

(a)

IE_1 values for main group elements (kJ/mol)

1A 1							8A 18
H 1312	2A 2	3A 13	4A 14	5A 15	6A 16	7A 17	**He** 2372
Li 520	**Be** 899	**B** 800	**C** 1086	**N** 1402	**O** 1314	**F** 1681	**Ne** 2080
Na 496	**Mg** 738	**Al** 577	**Si** 786	**P** 1012	**S** 999	**Cl** 1256	**Ar** 1520
K 419	**Ca** 590	**Ga** 579	**Ge** 761	**As** 947	**Se** 941	**Br** 1143	**Kr** 1351
Rb 403	**Sr** 549	**In** 558	**Sn** 708	**Sb** 834	**Te** 869	**I** 1009	**Xe** 1170
Cs 376	**Ba** 503	**Tl** 589	**Pb** 715	**Bi** 703	**Po** 813	**At** (926)	**Rn** 1037

(b)

Figure 4.8 (a) First ionization energies (in kJ/mol) of the main group elements. (b) First ionization energy as a function of atomic number.

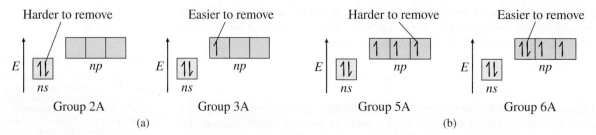

Figure 4.9 (a) It is harder to remove an electron from an *s* orbital than it is to remove an electron from a *p* orbital with the same principal quantum number. (b) Within a *p* subshell, it is easier to remove an electron from a doubly-occupied orbital than from a singly-occupied orbital.

higher value of ℓ have a higher energy (are less tightly held by the nucleus) and are therefore *easier* to remove. Figure 4.9(a) shows the relative energies of an *s* subshell ($\ell = 0$) and a *p* subshell ($\ell = 1$). Ionization of an element in Group 2A requires the removal of an electron from an *s* orbital, whereas ionization of an element in Group 3A requires the removal of an electron from a *p* orbital; therefore, the element in Group 3A has a lower ionization energy than the element in Group 2A.

As for the decrease in ionization energy in elements of Group 6A compared to those in Group 5A, both ionizations involve the removal of a *p* electron, but the ionization of an atom in Group 6A involves the removal of a *paired* electron. The repulsive force between two electrons in the same orbital makes it easier to remove one of them, making the ionization energy for the Group 6A element actually lower than that for the Group 5A element. [See Figure 4.9(b).]

The first ionization energy IE_1 decreases as we move from top to bottom within a group due to the increasing atomic radius. Although the effective nuclear charge does not change significantly as we step down a group, the atomic radius increases because the value of *n* for the valence shell increases. According to Coulomb's law, the attractive force between a valence electron and the effective nuclear charge gets *weaker* as the distance between them increases. This makes it easier to remove an electron, and so IE_1 decreases.

It is possible to remove additional electrons in subsequent ionizations, giving IE_2, IE_3, and so on. The second and third ionizations of sodium, for example, can be represented, respectively, as

$$Na^+(g) \longrightarrow Na^{2+}(g) + e^- \quad \text{and} \quad Na^{2+}(g) \longrightarrow Na^{3+}(g) + e^-$$

However, the removal of successive electrons requires ever-increasing amounts of energy because it is harder to remove an electron from a cation than from an atom (and it gets even harder as the charge on the cation increases). Table 4.2 lists the

TABLE 4.2		Ionization Energies (in kJ/mol) for Elements 3 through 11*								
	Z	**IE₁**	**IE₂**	**IE₃**	**IE₄**	**IE₅**	**IE₆**	**IE₇**	**IE₈**	**IE₉**	**IE₁₀**
Li	3	520	7,298	11,815							
Be	4	899	1,757	14,848	21,007						
B	5	800	2,427	3,660	25,026	32,827					
C	6	1,086	2,353	4,621	6,223	37,831	47,277				
N	7	1,402	2,856	4,578	7,475	9,445	53,267	64,360			
O	8	1,314	3,388	5,301	7,469	10,990	13,327	71,330	84,078		
F	9	1,681	3,374	6,050	8,408	11,023	15,164	17,868	92,038	106,434	
Ne	10	2,080	3,952	6,122	9,371	12,177	15,238	19,999	23,069	115,380	131,432
Na	11	496	4,562	6,910	9,543	13,354	16,613	20,117	25,496	28,932	141,362

*Shaded cells represent the removal of core electrons.

ionization energies of the elements in period 2 and of sodium. These data show that it takes much more energy to remove core electrons than to remove valence electrons. There are two reasons for this. First, core electrons are closer to the nucleus, and second, core electrons experience a greater effective nuclear charge because there are fewer filled shells shielding them from the nucleus. Both of these factors contribute to a greater attractive force between the electrons and the nucleus, which must be overcome to remove the electrons.

Ionization energy is an example of a *chemical property* [◄◄ Section 1.6] and the process of ionization is a **chemical process,** one in which the identity of the matter involved *changes.*

Worked Example 4.4 shows how to use these trends to compare first ionization energies, and subsequent ionization energies, of specific atoms.

Student Annotation: Although the symbols Na and Na$^+$ are nearly identical, the sodium *atom* and the sodium *ion* are chemically distinct and have very different properties.

Worked Example 4.4

Would you expect Na or Mg to have the greater first ionization energy (IE_1)? Which should have the greater second ionization energy (IE_2)?

Strategy Consider effective nuclear charge and electron configuration to compare the ionization energies. Effective nuclear charge increases from left to right in a period (thus increasing IE), and it is more difficult to remove a paired core electron than an unpaired valence electron.

Setup Na is in Group 1A, and Mg is beside it in Group 2A. Na has one valence electron, and Mg has two valence electrons.

Solution $IE_1(Mg) > IE_1(Na)$ because Mg is to the right of Na in the periodic table (i.e., Mg has the greater effective nuclear charge, so it is more difficult to remove its electron). $IE_2(Na) > IE_2(Mg)$ because the second ionization of Mg removes a valence electron, whereas the second ionization of Na removes a core electron.

> **Think About It**
> The first ionization energies of Na and Mg are 496 and 738 kJ/mol, respectively. The second ionization energies of Na and Mg are 4562 and 1451 kJ/mol, respectively.

Practice Problem (A)TTEMPT Which element, Mg or Al, will have the higher first ionization energy and which will have the higher third ionization energy?

Practice Problem (B)UILD Explain why Rb has a lower IE_1 than Sr, but Sr has a lower IE_2 than Rb.

Practice Problem (C)ONCEPTUALIZE Imagine an arrangement of atomic orbitals in an alternate universe, in which the *s* subshell contains *two* orbitals instead of one, and the *p* subshell contains *four* orbitals rather than three. Under these circumstances, in which groups would you expect the anomalously low first ionization energies to occur?

Electron Affinity

Electron affinity (EA) is the energy released when an atom in the gas phase accepts an electron. Consider the process in which a gaseous chlorine atom accepts an electron:

$$Cl(g) + e^- \longrightarrow Cl^-(g)$$

When a mole of gaseous chlorine atoms accepts a mole of electrons, 349.0 kJ/mol of energy are released (the definition of electron affinity). A positive electron affinity indicates a process that is energetically favorable. In general, the larger and more positive the *EA* value, the more favorable the process and the more apt it is to occur. Figure 4.10 shows electron affinities for the main group elements.

Student Annotation: Some books define electron affinity as the energy absorbed, rather than the energy released. The numeric values are the same, only the sign is different.

Figure 4.10 (a) Electron affinities (kJ/mol) of the main group elements. (b) Electron affinity as a function of atomic number.

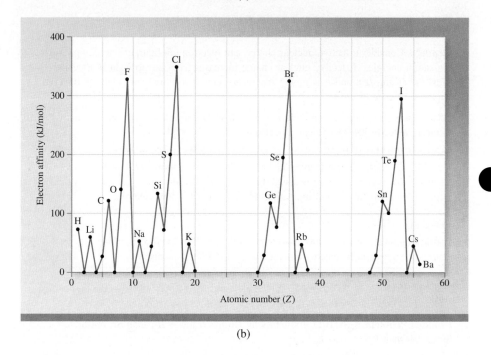

(a)

(b)

The species formed by the addition of an electron to a neutral atom is an anion. An ***anion*** is an ion whose net charge is *negative* due to an *increase* in the number of electrons. The Cl atom, for example, gains an electron to become the Cl^- anion:

Cl Atom	Cl^- Ion
17 protons	17 protons
17 electrons	18 electrons

Like ionization energy, electron affinity increases from left to right across a period. This trend in *EA* is due to the increase in effective nuclear charge from left to right (i.e., it becomes progressively easier to add a negatively charged electron as the positive charge of the element's nucleus increases). There are also periodic interruptions of the upward trend of *EA* from left to right, similar to those observed for IE_1, although they do *not* occur for the same elements. For example, the *EA* of a Group 2A element is lower than that for the corresponding Group 1A element, and the *EA* of a Group 5A element is lower than that for the corresponding Group 4A element. These exceptions to the trend are due to the electron configurations of the elements involved.

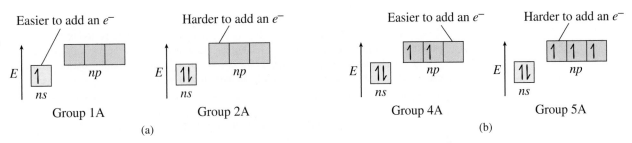

Figure 4.11 (a) It is easier to add an electron to an *s* orbital than to add one to a *p* orbital with the same principal quantum number. (b) Within a *p* subshell, it is easier to add an electron to an empty orbital than to add one to an orbital that already contains an electron.

It is harder to add an electron to a Group 2A element (ns^2) than to the Group 1A element (ns^1) in the same period because the electron added to the Group 2A element is placed in an orbital of higher energy (a *p* orbital versus an *s* orbital). Likewise, it is harder to add an electron to a Group 5A element (ns^2np^3) than to the corresponding Group 4A element (ns^2np^2) because the electron added to the Group 5A element must be placed in an orbital that already contains an electron. Figure 4.11 illustrates these points.

Student Annotation: There is a much less significant and less regular variation in electron affinities from top to bottom within a group. See Figure 4.11(a).

Just as more than one electron can be removed from an atom, more than one electron can also be added to an atom. While many first electron affinities are positive, subsequent electron affinities are always negative. Considerable energy is required to overcome the repulsive forces between the electron and the negatively charged ion. The addition of two electrons to a gaseous oxygen atom can be represented as:

Process	Electron affinity
$O(g) + e^- \longrightarrow O^-(g)$	$EA_1 = 141$ kJ/mol
$O^-(g) + e^- \longrightarrow O^{2-}(g)$	$EA_2 = -744$ kJ/mol

The term *second electron affinity* may seem like something of a misnomer, because an anion in the gas phase has no real "affinity" for an electron. A process that must absorb a significant amount of energy such as the addition of an electron to a gaseous O^- ion happens only in concert with one or more processes that more than compensate for the required energy input.

Worked Example 4.5 lets you practice using the periodic table to compare the electron affinities of elements.

Worked Example 4.5

For each pair of elements, indicate which one you would expect to have the greater first electron affinity, EA_1: (a) Al or Si, (b) Si or P.

Strategy Consider the effective nuclear charge and electron configuration to compare the electron affinities. The effective nuclear charge increases from left to right in a period (thus generally increasing EA), and it is more difficult to add an electron to a partially occupied orbital than to an empty one. Writing out orbital diagrams for the valence electrons is helpful for this type of problem.

Setup

(a) Al is in Group 3A and Si is beside it in Group 4A. Al has three valence electrons ([Ne]$3s^23p^1$), and Si has four valence electrons ([Ne]$3s^23p^2$).

(b) P is in Group 5A (to the right of Si), so it has five valence electrons ([Ne]$3s^23p^3$).

Solution

(a) $EA_1(Si) > EA_1(Al)$ because Si is to the right of Al and therefore has a greater effective nuclear charge.

(Continued on next page)

(b) $EA_1(\text{Si}) > EA_1(\text{P})$ because although P is to the right of Si in the third period of the periodic table (giving P the larger Z_{eff}), adding an electron to a P atom requires placing it in a $3p$ orbital that is partially occupied. The energy cost of *pairing* electrons outweighs the energy advantage of adding an electron to an atom with a larger effective nuclear charge.

Think About It

The first electron affinities of Al, Si, and P are 42.5, 134, and 72.0 kJ/mol, respectively.

$$\boxed{\uparrow\downarrow}\quad \boxed{\uparrow\ \ \ \ }$$
$$3s^2 \qquad 3p^1$$
Valence orbital diagram for Al

$$\boxed{\uparrow\downarrow}\quad \boxed{\uparrow\ \ \uparrow\ \ }$$
$$3s^2 \qquad 3p^2$$
Valence orbital diagram for Si

$$\boxed{\uparrow\downarrow}\quad \boxed{\uparrow\ \ \uparrow\ \ \uparrow}$$
$$3s^2 \qquad 3p^3$$
Valence orbital diagram for P

Practice Problem **A**TTEMPT Would you expect Mg or Al to have the greater EA_1?

Practice Problem **B**UILD Explain why the EA_1 for Ge is greater than the EA_1 for As.

Practice Problem **C**ONCEPTUALIZE In the same hypothetical arrangement described in Practice Problem 4.4C, in which groups would you expect the anomalously low electron affinities to occur?

Metallic Character

Metals tend to

Student Annotation: *Malleability* is the property that allows metals to be pounded into thin sheets. *Ductility* is the capacity to be drawn out into wires.

- Be shiny, lustrous, malleable, and ductile
- Be good conductors of both heat and electricity
- Have low ionization energies (so they commonly form *cations*)

Nonmetals, on the other hand, tend to

- Vary in color and lack the shiny appearance associated with metals
- Be brittle, rather than malleable
- Be poor conductors of both heat and electricity
- Have high electron affinities (so they commonly form *anions*)

Student Annotation: Metallic character is actually defined using a combination of physical and chemical properties. Apart from ionization energy and electron affinity, the chemical properties of metals and nonmetals will be discussed in Chapter 8.

Metallic character increases from top to bottom in a group and decreases from left to right within a period.

Metalloids are elements with properties intermediate between those of metals and nonmetals. Because the definition of metallic character depends on a combination of properties, there may be some variation in the elements identified as metalloids in different sources. Astatine (At), for example, is listed as a metalloid in some sources and a nonmetal in others.

Many of the periodic trends in properties of the elements can be explained using Coulomb's law, which states that the force (F) between two charged objects (Q_1 and Q_2) is directly proportional to the product of the two charges and *inversely* proportional to the distance (d) between the objects squared. Recall that the *energy* between two oppositely charged particles, E_{el}, is inversely proportional to d [◄◄ Section 1.3, Equation 3.2]. The SI unit of force is the newton (1 N = 1 kg · m/s^2) and the SI unit of energy is the joule (1 J = 1 kg · m^2/s^2).

TABLE 4.3	Attractive Force Between Oppositely Charged Objects at a Fixed Distance ($d = 1$) from Each Other		
Q_1	Q_2		Attractive force is proportional to
+1	−1		1
+2	−2		4
+3	−3		9

$$F \propto \frac{Q_1 \times Q_2}{d^2}$$ **Equation 4.2**

When the charges have opposite signs, F is negative—indicating an *attractive* force between the objects. When the charges have the same sign, F is positive—indicating a *repulsive* force. Table 4.3 shows how the magnitude of the attractive force between two oppositely charged objects at a fixed distance from each other varies with changes in the magnitudes of the charges.

Worked Example 4.6 illustrates how Coulomb's law can be used to compare the magnitudes of attractive forces between charged objects.

Worked Example 4.6

For carbon and nitrogen, use the effective nuclear charges given in Figure 4.7 and the atomic radii given in Figure 4.6 to compare the attractive force between the nucleus in each atom and the valence electron that would be removed by the first ionization.

Strategy Use Coulomb's law to calculate a number to which the attractive force will be proportional in each case.

Setup From Figure 4.7, the effective nuclear charges of C and N are 3.14 and 3.83, respectively; and the radii of C and N are 77 pm and 75 pm, respectively. The first ionization energies are 1086 kJ/mol (C) and 1402 kJ/mol (N). The charge on the valence electron in each case is −1.

Solution

For C: $F \propto \dfrac{(3.14) \times (-1)}{(77 \text{ pm})^2} = -5.3 \times 10^{-4}$

For N: $F \propto \dfrac{(3.83) \times (-1)}{(75 \text{ pm})^2} = -6.8 \times 10^{-4}$

Note that in this type of comparison, it doesn't matter what units we use for the distance between the charges. We are not trying to calculate a particular attractive force, only to compare the magnitudes of these two attractive forces.

> **Think About It**
> Remember that the negative sign indicates simply that the force is attractive rather than repulsive. The calculated number for nitrogen is about 28 percent larger than that for carbon.

Practice Problem ATTEMPT Between which two charges is the attractive force larger: +3.26 and −1.15 separated by a distance of 1.5 pm or +2.84 and −3.63 separated by a distance of 2.5 pm?

Practice Problem BUILD What must the distance be between charges of +2.25 and −1.86 in order for the attractive force between them to be the same as that between charges of +4.06 and −2.11 separated by a distance of 2.16 pm?

(Continued on next page)

Practice Problem **C** **ONCEPTUALIZE** Rank these pairs of charged objects in order of increasing magnitude of the attractive force between them.

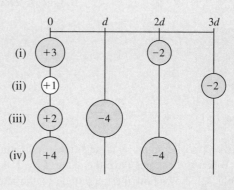

The product of a positive number and a negative number is a negative number. When we are simply comparing the magnitudes of attractive forces, however, it is unnecessary to include the sign. Worked Example 4.6 illustrates the use of Coulomb's law to compare the attractive forces between pairs of opposite charges.

Another trend in the chemical behavior of main group elements is the diagonal relationship. A ***diagonal relationship*** refers to similarities between pairs of elements in different groups and periods of the periodic table. Specifically, the first three members of the second period (Li, Be, and B) exhibit many similarities to the elements located diagonally below them in the periodic table (Mg, Al, and Si). The reason for this phenomenon is the similarity of charge densities of their cations. (Charge density is the charge on an ion divided by its volume.) Cations with comparable charge densities behave similarly and therefore exhibit some of the same chemical properties. Thus, the chemical properties of lithium resemble those of magnesium in some ways; the same holds for beryllium and aluminum and for boron and silicon. Each of these pairs is said to exhibit a diagonal relationship.

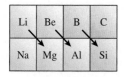

Diagonal relationships

Section 4.4 Review

Periodic Trends in Properties of Atoms

4.4.1   Arrange the elements Ca, Sr, and Ba in order of increasing IE_1.
   (a) Ca < Sr < Ba
   (b) Ba < Sr < Ca
   (c) Ba < Ca < Sr
   (d) Sr < Ba < Ca
   (e) Sr < Ca < Ba

4.4.2   Arrange the elements Li, Be, and B in order of increasing IE_2.
   (a) Li < Be < B
   (b) Li < B < Be
   (c) Be < B < Li
   (d) Be < Li < B
   (e) B < Be < Li

4.4.3   For each of the following pairs of elements, indicate which will have the greater EA_1: Rb or Sr, C or N, O or F.
   (a) Rb, C, O
   (b) Sr, N, F
   (c) Sr, C, F
   (d) Sr, N, O
   (e) Rb, C, F

4.4.4   Which element, K or Ca, will have the greater IE_1, which will have the greater IE_2, and which will have the greater EA_1?

(a) Ca, K, K   (d) Ca, Ca, K

(b) K, K, Ca   (e) Ca, Ca, Ca

(c) K, Ca, K

4.4.5   Which pair of opposite charges has the greatest attractive force?

(a) $+1$ and -1 separated by 1.0 pm

(b) $+1$ and -3 separated by 2.0 pm

(c) $+2$ and -2 separated by 1.5 pm

(d) $+3$ and -3 separated by 2.5 pm

(e) $+3$ and -3 separated by 3.0 pm

4.4.6   What must the separation between charges of $+2$ and -3 be for the attractive force between them to be the same as that between charges of $+3$ and -3 that are separated by a distance of 1.5 pm?

(a) 1.0 pm   (d) 2.0 pm

(b) 1.2 pm   (e) 2.3 pm

(c) 1.5 pm

4.5   ELECTRON CONFIGURATION OF IONS

Because many substances are made up of monatomic anions and cations, it is helpful to know how to write the electron configurations of these ionic species. Just as for atoms, we use the Pauli exclusion principle and Hund's rule to write the ground-state electron configurations of cations and anions.

> **Student Annotation:** "Monatomic" ions are those consisting of a single atom that has either gained (anion) or lost (cation) one or more electrons. We will encounter *polyatomic* ions in Chapter 5.

We can use the periodic table to predict the charges on many of the ions formed by main group elements. With very few exceptions, metals tend to form cations and nonmetals form anions. Elements in Groups 1A and 2A, for example, form ions with charges of $+1$ and $+2$, respectively. Most of the anions that form from elements of Groups 4A through 7A have charges equal to the corresponding group number minus 8. Elements in Groups 6A and 7A form ions with charges of -2 and -1, respectively. For example, the monatomic anion formed by oxygen (Group 6A) has a charge of $6 - 8 = -2$. Knowing something about electron configurations enables us to explain these charges.

Ions of Main Group Elements

In Section 4.4, we learned about the tendencies of atoms to lose or gain electrons. In every period of the periodic table, the element with the highest IE_1 is the Group 8A element, the noble gas. [See Figure 4.8(b).] Also, Group 8A is the only group in which none of the members has *any* tendency to accept an electron; that is, they all have negative EA values. [See Figure 4.10(b).] The $1s^2$ configuration of He and the ns^2np^6 ($n \geq 2$) valence-electron configurations of the other noble gases are extraordinarily stable. In fact, other main group elements tend to either lose or gain the number of electrons needed to achieve the same number of electrons as the nearest noble gas. Species with identical electron configurations, such as a Group 7A atom that has gained one electron and the noble gas immediately to its right in the periodic table, are called *isoelectronic*. Figure 4.12 shows the charges on the most common monatomic ions.

> **Student Annotation:** It is a common error to mistake species with the same valence-electron configuration for isoelectronic species. For example, F^- and Ne are isoelectronic. F^- and Cl^- are not.

To write the electron configuration of an ion formed by a main group element, we first write the configuration for the atom and either add or remove the

Figure 4.12 Common monatomic ions arranged by their positions in the periodic table. Note that mercury(I), Hg_2^{2+}, is actually a *poly*atomic ion.

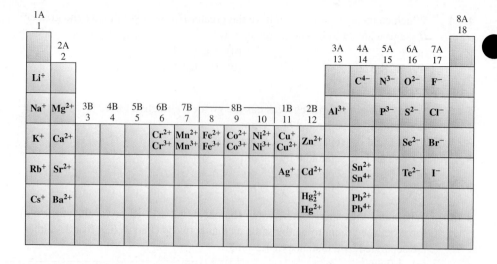

appropriate number of electrons. Electron configurations for the sodium and chloride ions are

$$Na:\ 1s^22s^22p^63s^1 \longrightarrow Na^+:\ 1s^22s^22p^6 \text{ (10 electrons total, isoelectronic with Ne)}$$

$$Cl:\ 1s^22s^22p^63s^23p^5 \longrightarrow Cl^-:\ 1s^22s^22p^63s^23p^6 \text{ (18 electrons total, isoelectronic with Ar)}$$

We can also write electron configurations for ions using the noble gas core.

$$Na:[Ne]3s^1 \longrightarrow Na^+:[Ne]$$

$$Cl:[Ne]3s^23p^5 \longrightarrow Cl^-:[Ne]3s^23p^6 \text{ or } [Ar]$$

Worked Example 4.7 gives you some practice writing electron configurations for the ions of main group elements.

Worked Example 4.7

Write electron configurations for the following ions of main group elements: (a) N^{3-}, (b) Ba^{2+}, and (c) Be^{2+}.

Strategy First write electron configurations for the atoms. Then add electrons (for anions) or remove electrons (for cations) to account for the charge.

Setup

(a) N^{3-} forms when N ($1s^22s^22p^3$ or $[He]2s^22p^3$), a main group nonmetal, gains three electrons.

(b) Ba^{2+} forms when Ba ($1s^22s^22p^63s^23p^64s^23d^{10}4p^65s^24d^{10}5p^66s^2$ or $[Xe]6s^2$) loses two electrons.

(c) Be^{2+} forms when Be ($1s^22s^2$ or $[He]2s^2$) loses two electrons.

Solution

(a) $[He]2s^22p^6$ or $[Ne]$

(b) $[Kr]5s^24d^{10}5p^6$ or $[Xe]$

(c) $1s^2$ or $[He]$

Think About It
Be sure to add electrons to form an anion, and remove electrons to form a cation.

Practice Problem ATTEMPT Write electron configurations for (a) O^{2-}, (b) Ca^{2+}, and (c) Se^{2-}.

Practice Problem BUILD List all the species (atoms and/or ions) that are likely to have the following electron configuration: $1s^2 2s^2 2p^6$.

Practice Problem CONCEPTUALIZE Select the correct valence orbital diagram for the Mg^{2+} ion and for the S^{2-} ion.

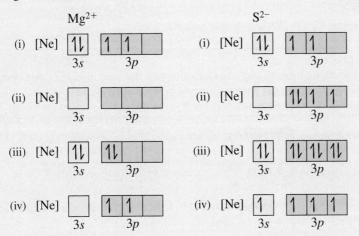

Ions of *d*-Block Elements

Recall that the 4s orbital fills before the 3d orbitals for the elements in the first row of the *d* block (Sc to Zn) [◄◄ Section 3.9]. Following the pattern for writing electron configurations for main group ions, then, we might expect the two electrons lost in the formation of the Fe^{2+} ion to come from the 3d subshell. It turns out, though, that an atom always loses electrons first from the shell with the *highest* value of *n*. In the case of Fe, that would be the 4s subshell.

$$Fe:[Ar]4s^2 3d^6 \longrightarrow Fe^{2+}:[Ar]3d^6$$

Iron can also form the Fe^{3+} ion, in which case the third electron is removed from the 3d subshell.

$$Fe:[Ar]4s^2 3d^6 \longrightarrow Fe^{3+}:[Ar]3d^5$$

In general, when a *d*-block element becomes an ion, it loses electrons first from the *ns* subshell and then from the $(n-1)d$ subshell.

Worked Example 4.8 gives you some practice writing electron configurations for the ions of *d*-block elements.

Student Annotation: This explains, in part, why many of the transition metals can form ions with a +2 charge.

Worked Example 4.8

Write electron configurations for the following ions of *d*-block elements: (a) Zn^{2+}, (b) Mn^{2+}, and (c) Cr^{3+}.

Strategy First write electron configurations for the atoms. Then add electrons (for anions) or remove electrons (for cations) to account for the charge. The electrons removed from a *d*-block element must come first from the outermost *s* subshell, not the partially filled *d* subshell.

(Continued on next page)

Setup

(a) Zn^{2+} forms when Zn ($1s^2 2s^2 2p^6 3s^2 3p^6 4s^2 3d^{10}$ or $[Ar]4s^2 3d^{10}$) loses two electrons.

(b) Mn^{2+} forms when Mn ($1s^2 2s^2 2p^6 3s^2 3p^6 4s^2 3d^5$ or $[Ar]4s^2 3d^5$) loses two electrons.

(c) Cr^{3+} forms when Cr ($1s^2 2s^2 2p^6 3s^2 3p^6 4s^1 3d^5$ or $[Ar]4s^1 3d^5$) loses three electrons—one from the 4s subshell and two from the 3d subshell.

Student Annotation: Remember that the electron configuration of Cr is anomalous in that it has only one 4s electron, making its d subshell half filled [◄◄ Section 3.10].

Solution

(a) $[Ar]3d^{10}$ (b) $[Ar]3d^5$ (c) $[Ar]3d^3$

> **Think About It**
> Be sure to *add* electrons to form an anion and *remove* electrons to form a cation. Also, double-check to ensure that electrons removed from a d-block element come first from the *ns* subshell and then, if necessary, from the $(n - 1)d$ subshell.

Practice Problem Ⓐ**TTEMPT** Write electron configurations for (a) Co^{3+}, (b) Cu^{2+}, and (c) Ag^+.

Practice Problem Ⓑ**UILD** What common d-block ion (see Figure 4.12) is isoelectronic with Zn^{2+}?

Practice Problem Ⓒ**ONCEPTUALIZE** Select the correct valence orbital diagram for the Fe^{2+} ion and for the Fe^{3+} ion.

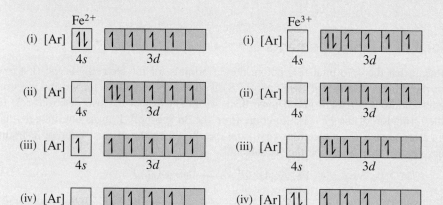

Section 4.5 Review

Electron Configuration of Ions

4.5.1   What is the charge on a titanium ion that is isoelectronic with argon?

(a) 1+   (d) 4+

(b) 2+   (e) 5+

(c) 3+

4.5.2   What ions of As and Sr are isoelectronic with krypton?

(a) 3+, 2−   (d) 3−, 2+

(b) 3−, 2−   (e) 2+, 3−

(c) 2−, 1+

4.5.3   Select the correct ground-state electron configuration for Ti^{2+}.
   (a) $[Ar]4s^2 3d^2$
   (b) $[Ar]4s^2 3d^4$
   (c) $[Ar]4s^2$
   (d) $[Ar]3d^2$
   (e) $[Ar]4s^1 3d^1$

4.5.4   Select the correct ground-state electron configuration for S^{2-}.
   (a) $[Ne]3p^4$
   (b) $[Ne]3s^2 3p^6$
   (c) $[Ne]3s^2 3p^2$
   (d) $[Ne]3p^6$
   (e) $[Ne]$

4.5.5   Which of the following ions is diamagnetic?
   (a) Co^{2+}
   (b) Co^{3+}
   (c) Cu^+
   (d) Cu^{2+}
   (e) Ni^{2+}

4.6   IONIC RADIUS

When an atom gains or loses one or more electrons to become an ion, its radius changes. The *ionic radius,* the radius of a cation or an anion, affects the physical and chemical properties of the substance of which it is a part. The three-dimensional structure of such a substance, for example, depends on the relative sizes of its cations and anions.

Comparing Ionic Radius with Atomic Radius

When an atom loses an electron and becomes a cation, its radius decreases due in part to a reduction in electron-electron repulsions (and consequently a reduction in shielding) in the valence shell. A significant decrease in radius occurs when *all* of an atom's valence electrons are removed. This is the case with ions of most main group elements, which are isoelectronic with the noble gases preceding them. Consider Na, which loses its $3s$ electron to become Na^+.

Video 4.2
Periodic table—atomic and ionic radii

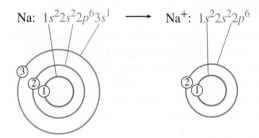

$$Na: \ 1s^2 2s^2 2p^6 3s^1 \longrightarrow Na^+: \ 1s^2 2s^2 2p^6$$

The valence electron of Na has a principal quantum number of $n = 3$. When it has been removed, the resulting Na^+ ion no longer has any electrons in the $n = 3$ shell. The outermost electrons of the Na^+ ion have a principal quantum number of $n = 2$. Because the value of n determines the distance from the nucleus, this corresponds to a smaller radius.

When an atom gains one or more electrons and becomes an anion, its radius increases due to increased electron-electron repulsions. Adding an electron causes the rest of the electrons in the valence shell to spread out and take up more space to maximize the distance between them.

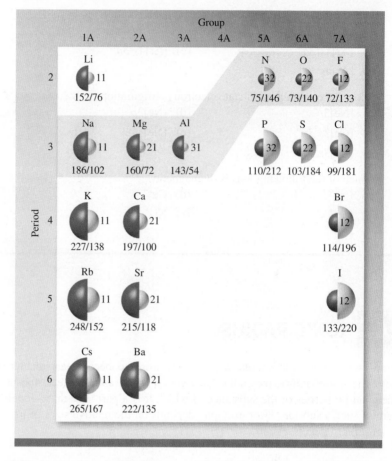

Figure 4.13 A comparison of atomic and ionic radii (in picometers) for main group elements and their common ions (those that are isoelectronic with noble gases).

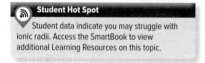
Figure 4.13 shows the ionic radii for those ions of main group elements that are isoelectronic with noble gases and compares them to the radii of the parent atoms. Note that the ionic radius, like the atomic radius, increases from top to bottom in a group.

Isoelectronic Series

An *isoelectronic series* is a series of two or more species that have identical electron configurations, but different nuclear charges. For example, O^{2-}, F^-, and Ne constitute an isoelectronic series. Although these three species have identical electron configurations, they have different radii. In an isoelectronic series, the species with the smallest nuclear charge (i.e., the smallest atomic number, Z) will have the largest radius. The species with the largest nuclear charge (i.e., the largest Z) will have the smallest radius.

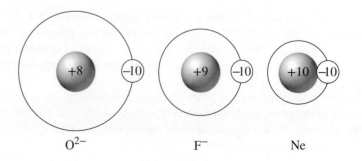

Worked Example 4.9 shows how to identify members of an isoelectronic series and how to arrange them according to radius.

Worked Example 4.9

Identify the isoelectronic series in the following group of species, and arrange the ions in the series in order of increasing radius: K^+, Ne, Ar, Kr, P^{3-}, S^{2-}, and Cl^-.

Strategy Isoelectronic series are species with identical electron configurations but different nuclear charges. Determine the number of electrons in each species. The radii of ions in an isoelectronic series decrease with increasing nuclear charge.

Setup The number of electrons in each species is as follows: 18 (K^+), 10 (Ne), 18 (Ar), 36 (Kr), 18 (P^{3-}), 18 (S^{2-}), and 18 (Cl^-). The nuclear charges of the species with 18 electrons are +19 (K^+), +18 (Ar), +15 (P^{3-}), +16 (S^{2-}), and +17 (Cl^-).

Solution The ions in this isoelectronic series are K^+, P^{3-}, S^{2-}, and Cl^-. In order of increasing radius: $K^+ < Cl^- < S^{2-} < P^{3-}$.

> **Think About It**
> Consult Figure 4.13 to check your result. With identical electron configurations, the attractive force between the valence electrons and the nucleus will be strongest for the largest nuclear charge. Thus, the larger the nuclear charge, the closer in the valence electrons will be pulled and the smaller the radius will be.

Practice Problem **A**TTEMPT Arrange the following isoelectronic series in order of increasing radius: Se^{2-}, Br^-, and Rb^+.

Practice Problem **B**UILD List all the common ions that are isoelectronic with Ne.

Practice Problem **C**ONCEPTUALIZE Which periodic table's highlighted portion includes elements that can form an isoelectronic series? (Select all that apply.)

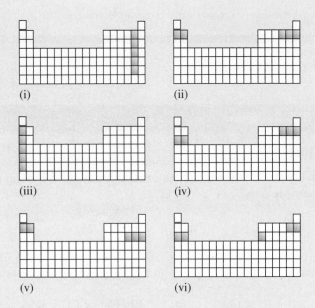

(i)   (ii)

(iii)   (iv)

(v)   (vi)

Thinking Outside the Box

Mistaking Strontium for Calcium

One of the consequences of two chemical species having similar properties is that human physiology sometimes mistakes one species for another. Healthy bones require constant replenishment of calcium. Our dietary calcium comes primarily from dairy products but can also be found in some vegetables—especially dark, leafy greens such as spinach and kale. Strontium-90, a radioactive isotope, is found in the fallout of atomic bomb explosions and is a component in the waste generated by nuclear power facilities. Strontium-90 released into the atmosphere will eventually settle on land and water, and it can reach our bodies through ingestion—especially of vegetation—and through the inhalation of airborne particles. Because calcium and strontium are chemically similar, Sr^{2+} ions are mistaken by the body for Ca^{2+} ions and are incorporated into the bones. Constant exposure to the radiation emitted by strontium-90 affects not only the bone and surrounding soft tissue, but also the bone marrow, damaging and destroying stem cells vital to the immune system. Such long-term exposure leads to an increased risk of leukemia and other cancers.

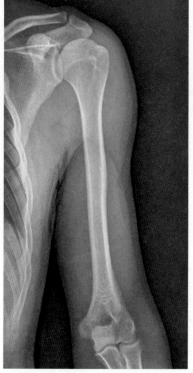

X-ray image of human bones

© Zephyr/Science Source

Ca and Sr, being in the same group, have similar chemical properties.

Section 4.6 Review

Ionic Radius

4.6.1 Identify the largest ion and the smallest ion in the following isoelectronic series: Mg^{2+}, N^{3-}, O^{2-}, Na^+, C^{4-}.

(a) Mg^{2+}, C^{4-}

(b) O^{2-}, Na^+

(c) Na^+, N^{3-}

(d) C^{4-}, Mg^{2+}

(e) N^{3-}, Mg^{2+}

4.6.2 Arrange the following species in the correct order of increasing radius: K^+, Cl^-, Ca^{2+}.

(a) $Cl^- < K^+ < Ca^{2+}$

(b) $Ca^{2+} < K^+ < Cl^-$

(c) $K^+ < Ca^{2+} < Cl^-$

(d) $Ca^{2+} < Cl^- < K^+$

(e) $Cl^- < Ca^{2+} < K^+$

Learning Outcomes

- Explain how elements are arranged in the periodic table.
- Use the location of an element in the periodic table to predict some of its characteristics.
- Utilize the periodic table to determine the electron configuration of an element.
- Describe the importance of valence electrons to chemical characteristics.
- Define effective nuclear charge, atomic radius, ionization energy, and electron affinity.
- Predict differences in effective nuclear charge, atomic radius, ionization energy, and electron affinity between elements using periodic trends.

- Provide some of the characteristics of metals, nonmetals, and metalloids.
- Predict the charge of an ion formed from a main group element.
- Determine the electron configurations of ions of main group and *d*-block elements.
- Predict the sizes of ions relative to atoms of the same element.
- Define isoelectronic species and arrange a series of isoelectronic ions according to radius.

Chapter Summary

SECTION 4.1

- The modern periodic table was devised independently by Dmitri Mendeleev and Lothar Meyer in the nineteenth century. The elements that were known at the time were grouped based on their physical and chemical properties. Using his arrangements of the elements, Mendeleev successfully predicted the existence of elements that had not yet been discovered.
- Early in the twentieth century, Henry Moseley refined the periodic table with the concept of the *atomic number,* thus resolving a few inconsistencies in the tables proposed by Mendeleev and Meyer.
- Elements in the same group of the periodic table tend to have similar physical and chemical properties.

SECTION 4.2

- The periodic table can be divided into the **main group elements** (also known as the *representative elements*) and the *transition metals.* It is further divided into smaller groups or *columns* of elements that all have the same configuration of **valence electrons.**
- The 18 columns of the periodic table are labeled 1A through 8A (*s*- and *p*-block elements) and 1B through 8B (*d*-block elements), or by the numbers 1 through 18.

SECTION 4.3

- **Effective nuclear charge (Z_{eff})** is the nuclear charge that is "felt" by the valence electrons. It is usually lower than the nuclear charge due to **shielding** by the core electrons.

SECTION 4.4

- *Atomic radius* is the distance between an atom's nucleus and its valence shell. The atomic radius of a metal atom is

defined as the **metallic radius,** which is one-half the distance between adjacent, identical nuclei in a metal solid. The atomic radius of a nonmetal is defined as the **covalent radius,** which is one-half the distance between adjacent, identical nuclei in a substance. In general, atomic radii *decrease* from left to right across a *period* of the periodic table and *increase* from top to bottom down a *group.*

- *Ionization energy (IE)* is the energy required to remove an electron from an atom. The first ionization energy (IE_1) is smaller than subsequent ionization energies [e.g., second (IE_2), third (IE_3), and so on]. The first ionization of any atom removes a valence electron. An *ion* with a net positive charge is a **cation.** Ionization energies increase dramatically when core electrons are being removed. Ionization energy is a chemical property, and ionization is a **chemical process.**
- First ionization energies (IE_1 values) tend to increase across a period and decrease down a group. Exceptions to this trend can be explained based upon the electron configuration of the element.
- *Electron affinity (EA)* is the energy released when an atom in the gas phase accepts an electron [$A(g) + e^- \longrightarrow A^-(g)$]. An ion with a net negative charge is an *anion.*
- Electron affinities tend to increase across a period. As with first ionization energies, exceptions to the trend can be explained based on the electron configuration of the element. Electron affinity is also a chemical property.
- Metals tend to be shiny, lustrous, malleable, ductile, and conducting (for both heat and electricity). Metals typically lose electrons to form cations.
- Nonmetals vary in color, tend to be brittle, and are generally not good conductors (for either heat or electricity). They can gain electrons to form anions.

- In general, metallic character decreases across a period and increases down a group of the periodic table. *Metalloids* are elements with properties intermediate between metals and nonmetals.
- A *diagonal relationship* refers to the similar properties shared by elements in different groups and periods in the periodic table.

SECTION 4.5

- Ions of main group elements are *isoelectronic* with noble gases. When a *d*-block element loses one or more electrons, it loses them first from the shell with the highest principal quantum number (e.g., electrons in the $4s$ subshell are lost before electrons in the $3d$ subshell).

SECTION 4.6

- *Ionic radius* is the distance between the nucleus and valence shell of a cation or an anion. A cation is smaller than its parent atom. An anion is larger than its parent atom.
- An *isoelectronic series* consists of one or more ions and a noble gas, all of which have identical electron configurations. Within an isoelectronic series of ions, the greater the nuclear charge, the smaller the radius.

Key Words

Anion, 138	Effective nuclear charge (Z_{eff}), 131	Isoelectronic series, 148
Atomic radius, 132	Electron affinity (*EA*), 137	Main group elements, 128
Cation, 135	Ion, 135	Metallic radius, 132
Chemical process, 137	Ionic radius, 147	Metalloid, 140
Covalent radius, 132	Ionization energy (*IE*), 135	Shielding, 131
Diagonal relationship, 142	Isoelectronic, 143	Valence electrons, 129

Key Equations

4.1 $Z_{eff} = Z - \sigma$

Effective nuclear charge (Z_{eff}) is equal to *nuclear* charge (Z) minus the shielding constant (σ).

4.2 $F \propto \dfrac{Q_1 \times Q_2}{d^2}$

The force (F) between two charged objects (Q_1 and Q_2) is proportional to the product of the two charges divided by the distance (d) squared.

Periodic Trends in Atomic Radius, Ionization Energy, and Electron Affinity

The properties of an element are determined in large part by the *size* (radius) and the valence-shell *electron configuration* of its atoms and ions. Together, principal quantum number (n), effective nuclear charge (Z_{eff}), and charge on the valence shell (number of valence electrons) determine atomic radius.

From left to right across a period, Z_{eff} and charge on the valence shell both *increase*. Each step to the right involves the addition of a proton, which increases Z, and the addition of an electron. Each additional electron resides in the same shell. Remember that electrons in the same shell do not shield one another well. The result is that Z_{eff} increases by nearly as much as Z. As the magnitude of opposite charges increases, coulombic attraction between them increases and they are drawn closer together, thus reducing the atomic radius.

From top to bottom within a group, valence electrons reside in shells with increasingly larger values of n, putting them farther away from the nucleus—thereby increasing the atomic radius. Valence-shell electron configuration and, to a large extent, Z_{eff} remain the same from top to bottom in a group. Variations in atomic radii of the *d*-block elements vary less regularly than the main group elements.

From left to right across a period, Z_{eff} *increases* and radius *decreases*. Both factors contribute to stronger coulombic attraction between the positively charged nucleus and the valence electrons—making it harder to remove an electron; therefore, ionization energy (IE_1) *increases*. From top to bottom within a group, although Z_{eff} remains fairly constant, radius increases. The increased distance between the nucleus and the valence electrons corresponds to weaker coulombic attraction, making it easier to remove an electron; therefore, IE_1 *decreases*.

Because Z_{eff} *increases* and radius *decreases* from left to right across a period, the coulombic attraction between the positively charged nucleus and an added electron gets stronger, making it easier to add an electron; therefore, electron affinity (*EA*) *increases*. And because Z_{eff} remains fairly constant and radius *increases* from top to bottom within a group, there is a smaller coulombic attraction between the nucleus and an added electron—making it harder to add an electron; therefore, *EA decreases*.

Specific anomalies in the general trends of IE_1 and EA are determined by valence-electron configuration.

Ionization energy (IE_1) increases → (Ionization energy (IE_1) decreases ↓)

1							2
1312							2372
3	4	5	6	7	8	9	10
520	899	800	1086	1402	1314	1681	2080
11	12	13	14	15	16	17	18
496	738	577	786	1012	999	1256	1520
19	20	31	32	33	34	35	36
419	590	579	761	947	941	1143	1351
21	22	49	50	51	52	53	54
403	549	558	708	834	869	1009	1170
55	56	81	82	83	84	85	86
376	503	589	715	703	813	(926)	1037

Electron affinity (EA) increases → (Electron affinity (EA) decreases ↓)

1							2
+72.8							(0.0)
3	4	5	6	7	8	9	10
+59.6	≤0	+26.7	+122	−7	+141	+328	(−29)
11	12	13	14	15	16	17	18
+52.9	≤0	+42.5	+134	+72.0	+200	+349	(−35)
19	20	31	32	33	34	35	36
+48.4	+2.37	+28.9	+119	+78.2	+195	+325	(−39)
21	22	49	50	51	52	53	54
+46.9	+5.03	+28.9	+107	+103	+190	+295	(−41)
55	56	81	82	83	84	85	86
+45.5	+13.95	+19.3	+35.1	+91.3	+183	+270	(−41)

For example, although the trend is an increase in IE_1 from left to right, IE_1 for Group 3A is lower than IE_1 for Group 2A—within a period. This is because the electron removed by ionization comes from the p subshell, which is higher in energy than the s subshell [Figure 4.9(a)]. Likewise, IE_1 for Group 6A is lower than IE_1 for Group 5A. In this case, the electron removed by ionization is one of a pair of electrons in a p orbital. Because paired electrons in a single orbital repel one another, removing one of them is relatively easy [Figure 4.9(b)].

Similarly, despite the general trend, EA for Group 2A is lower than EA for Group 1A. In this case, the electron added by electron affinity goes into the p subshell, which is higher in energy than the s subshell [Figure 4.11(a)]. And EA for Group 5A is lower than that for Group 4A because the added electron must go into an already occupied p orbital [Figure 4.11(b)].

Each of these periodic properties can be explained and understood using Coulomb's law: $F \propto \dfrac{Q_1 \times Q_2}{d^2}$.

Key Skills Problems

4.1
Group 8A exhibits the highest first ionization energy. Which group do you expect to exhibit the highest *second* ionization energy (EA_2)?
(a) 7A   (b) 6A   (c) 5A   (d) 2A   (e) 1A

4.2
Which of the following best describes why Z_{eff} does not change significantly from top to bottom within a group?
(a) There are fewer completed electron shells.
(b) There are fewer valence electrons.
(c) There are more valence electrons.
(d) There are more completed electron shells.
(e) The number of protons in the nucleus does not change significantly.

4.3
Often we can compare properties of two elements based solely on periodic trends. For which pair of elements is the periodic trend not sufficient to determine which has the higher first ionization energy?
(a) C and Si   (b) Al and Ga   (c) Ga and Si
(d) Tl and Sn   (e) B and Si

4.4
The colored spheres represent the ions Ca^{2+}, Cl^-, K^+, P^{3-}, and S^{2-}. Based on size and using only a periodic table, determine which of the following has the ions in the same order as the diagram.
(a) Ca^{2+}, Cl^-, K^+, P^{3-}, S^{2-}
(b) Ca^{2+}, K^+, P^{3-}, S^{2-}, Cl^-
(c) P^{3-}, S^{2-}, Cl^-, K^+, Ca^{2+}
(d) Ca^{2+}, K^+, Cl^-, S^{2-}, P^{3-}
(e) P^{3-}, S^{2-}, Cl^-, Ca^{2+}, K^+

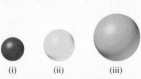

(i)   (ii)   (iii)   (iv)   (v)

Questions and Problems

SECTION 4.1: DEVELOPMENT OF THE PERIODIC TABLE

Review Questions

4.1 Briefly describe the significance of Mendeleev's periodic table.

4.2 What is Moseley's contribution to the modern periodic table?

4.3 Describe the general layout of a modern periodic table.

4.4 What is the most important relationship among elements in the same group in the periodic table?

SECTION 4.2: THE MODERN PERIODIC TABLE

Review Questions

4.5 Classify each of the following elements as a metal, a nonmetal, or a metalloid: Sb, Kr, Co, Na, Al, F, Sr, As, Br, Ge.

4.6 List the properties of metals.

4.7 List the properties of nonmetals.

4.8 What is a main group element? Give names and symbols of six main group elements: two metals, two nonmetals, and two metalloids.

4.9 Without referring to a periodic table, write the name and give the symbol for one element in each of the following groups: 1A, 2A, 3A, 4A, 5A, 6A, 7A, 8A, transition metals.

4.10 What are valence electrons? For atoms of main group elements, the number of valence electrons is equal to the group number. Write electron configurations to show that this is true for the following elements: Na, Ca, Li, I, N, Se, Si.

4.11 Write the outer electron configurations for the (a) alkali metals, (b) alkaline earth metals, (c) halogens, (d) noble gases.

4.12 Use the first-row transition metals (Sc to Cu) as an example to illustrate the characteristics of the electron configurations of transition metals.

4.13 For centuries, arsenic has been the poison of choice for murders and murder mysteries alike. Based on its position in the periodic table, suggest a possible reason for its toxicity.

Conceptual Problems

4.14 In the periodic table, the element hydrogen is sometimes grouped with the alkali metals and sometimes with the halogens. Explain how hydrogen can resemble the Group 1A *and* the Group 7A elements.

4.15 An atom of a certain element has 16 electrons. Consulting only the periodic table, identify the element and write its ground-state electron configuration.

4.16 Group the following electron configurations in pairs that would represent elements with similar properties:
(a) $1s^2 2s^2 2p^6 3s^2$
(b) $1s^2 2s^2 2p^3$
(c) $1s^2 2s^2 2p^6 3s^2 3p^6 4s^2 3d^{10} 4p^6$
(d) $1s^2 2s^2$
(e) $1s^2 2s^2 2p^6$
(f) $1s^2 2s^2 2p^6 3s^2 3p^3$

4.17 Group the following electron configurations in pairs that would represent elements with similar properties:
(a) $1s^2 2s^2 2p^5$
(b) $1s^2 2s^1$
(c) $1s^2 2s^2 2p^6$
(d) $1s^2 2s^2 2p^6 3s^2 3p^5$
(e) $1s^2 2s^2 2p^6 3s^2 3p^6 4s^1$
(f) $1s^2 2s^2 2p^6 3s^2 3p^6 4s^2 3d^{10} 4p^6$

4.18 Without referring to a periodic table, write the electron configuration of elements with the following atomic numbers: (a) 10, (b) 22, (c) 28, (d) 35.

4.19 Specify the group on the periodic table in which each of the following elements is found: (a) $[Ar]4s^1$; (b) $[Ar]4s^2 3d^{10} 4p^3$; (c) $[Ne]3s^2 3p^3$; (d) $[Ar]4s^2 3d^6$.

4.20 For each of the following ground-state electron configurations, determine the corresponding group number on the periodic table: (a) $ns^2 (n-1)d^8$; (b) $ns^2 (n-1)d^{10} np^3$; (c) $ns^2 (n-1)d^3$.

4.21 Determine what element is designated by each of the following: (a) fifth period, $ns^2 (n-1)d^{10} np^2$; (b) fourth period, $ns^2 (n-1)d^3$; (c) third period, $ns^2 np^5$; (d) sixth period, ns^2.

SECTION 4.3: EFFECTIVE NUCLEAR CHARGE

Review Questions

4.22 Explain the term *effective nuclear charge.*

4.23 Explain why there is a greater increase in effective nuclear charge from left to right across a period than there is from top to bottom in a group.

Computational Problems

4.24 The electron configuration of B is $1s^2 2s^2 2p^1$.
(a) If each core electron (i.e., the $1s$ electrons) were totally effective in shielding the valence electrons (i.e., the $2s$ and $2p$ electrons) from the nucleus and the valence electrons did not shield one another, what would be the shielding constant (σ) and the effective nuclear charge (Z_{eff}) for the $2s$ and $2p$ electrons? (b) In reality, the shielding constants for the $2s$ and $2p$ electrons in B are slightly different. They are 2.42 and 2.58, respectively. Calculate Z_{eff} for these electrons, and explain the differences from the values you determined in part (a).

4.25 The electron configuration of C is $1s^2 2s^2 2p^2$. (a) If each core electron (i.e., the $1s$ electrons) were totally effective in screening the valence electrons (i.e., the $2s$ and $2p$ electrons) from the nucleus and the valence electrons did not shield one another, what would be the shielding constant (σ) and the effective nuclear charge (Z_{eff}) for the $2s$ and $2p$ electrons? (b) In reality, the shielding constants for the $2s$ and $2p$ electrons in C are slightly different. They are 2.78 and 2.86, respectively. Calculate Z_{eff} for these electrons, and explain the differences from the values you determined in part (a).

SECTION 4.4: PERIODIC TRENDS IN PROPERTIES OF ELEMENTS

Review Questions

4.26 Define *atomic radius* and explain why such a definition is necessary.

4.27 How does atomic radius change (a) from left to right across a period and (b) from top to bottom in a group?

4.28 Equation 4.2 is used to calculate the force between charged particles. Explain the significance of the sign of the result.

4.29 Use the second period of the periodic table as an example to show that the size of atoms decreases as we move from left to right. Explain the trend.

4.30 Why is the radius of the lithium atom considerably larger than the radius of the hydrogen atom?

4.31 Explain why the atomic radius of S is larger than that of O, but smaller than that of P.

4.32 Define *ionization energy*. Explain why ionization energy measurements are usually made when atoms are in the gaseous state. Why is the second ionization energy always greater than the first ionization energy for any element? What types of elements have the highest ionization energies and what types have the lowest ionization energies?

4.33 Define the terms *cation* and *anion*.

4.34 (a) Define the term *electron affinity*. (b) Explain why electron affinity measurements are made with gaseous atoms. (c) Ionization energy is always a positive quantity, whereas electron affinity may be either positive or negative. Explain.

4.35 Explain the trend in electron affinity from aluminum to chlorine (see Figure 4.10).

4.36 Define *diagonal relationship* and explain what causes it.

4.37 According to the information in Section 4.4, there are two alkaline earth metals (Group 2A) that exhibit diagonal relationships with other elements. Identify the alkaline earth metals and the element with which each has a diagonal relationship.

Computational Problems

4.38 A hydrogen-like ion is an ion containing only one electron. The energies of the electron in a hydrogen-like ion are given by

$$E_n = -(2.18 \times 10^{-18} \text{ J}) Z^2 \left(\frac{1}{n^2} \right)$$

where n is the principal quantum number and Z is the atomic number of the element. Calculate the ionization energy (in kJ/mol) of the He^+ ion.

4.39 Plasma is a state of matter consisting of positive gaseous ions and electrons. In the plasma state, a mercury atom could be stripped of its 80 electrons and therefore would exist as Hg^{80+}. Use the equation in Problem 4.38 to calculate the energy required for the last ionization step, that is,

$$Hg^{79+}(g) \longrightarrow Hg^{80+}(g) + e^-$$

Conceptual Problems

4.40 Consider two ions with opposite charges separated by a distance d. What effect does each of the following changes have on the force between the ions? (a) The positive charge is doubled. (b) The positive charge is doubled and the negative charge is halved. (c) The distance is doubled. (d) The distance is halved. (e) Both charges and the distance are doubled.

4.41 Consider two ions with opposite charges separated by a distance d. What effect does each of the following changes have on the force between the ions? (a) Both charges are doubled. (b) The positive charge is doubled and the negative charge is tripled. (c) The negative charge and the distance are doubled. (d) The positive charge is doubled and the distance is halved. (e) Both charges and the distance are halved.

4.42 Arrange the following pairs of charged particles in order of increasing magnitude of attractive force.

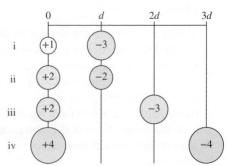

4.43 Arrange the following pairs of charged particles in order of increasing magnitude of attractive force.

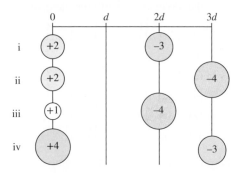

4.44 On the basis of their positions in the periodic table, select the atom with the larger atomic radius in each of the following pairs:
(a) Mg, P; (b) Sr, Be; (c) As, Br; (d) Cl, I; (e) Xe, Kr.

4.45 Arrange the following atoms in order of increasing atomic radius: Si, Mg, Cl, P, Al.

4.46 Which is the largest atom in the third period of the periodic table?

4.47 Which is the smallest atom in Group 6A?

4.48 Based on size, identify the spheres shown as Na, Mg, O, and S.

4.49 Based on size, identify the spheres shown as K, Ca, S, and Se.

4.50 Arrange the following in order of increasing first ionization energy: Na, Cl, Al, S, and Cs.

4.51 Arrange the following in order of increasing first ionization energy: F, K, P, Ca, and Ne.

4.52 Use the third period of the periodic table as an example to illustrate the change in first ionization energies of the elements as we move from left to right. Explain the trend.

4.53 In general, the first ionization energy increases from left to right across a given period. Aluminum, however, has a lower first ionization energy than magnesium. Explain.

4.54 The first and second ionization energies of K are 419 and 3052 kJ/mol, and those of Ca are 590 and 1145 kJ/mol, respectively. Compare their values and comment on the differences.

4.55 Two atoms have the electron configurations $1s^2 2s^2 2p^6$ and $1s^2 2s^2 2p^6 3s^1$. The first ionization energy of one is 2080 kJ/mol, and that of the other is 496 kJ/mol. Match each ionization energy with one of the given electron configurations. Justify your choice.

4.56 The graph shows IE_1, IE_2, IE_3, and IE_4 for three main group elements: A, B, and C. Using information from the graph, determine the group number for each element.

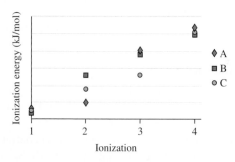

4.57 The graph shows IE_1, IE_2, IE_3, and IE_4 for three main group elements: A, B, and C. Using information from the graph, determine the group number for each element.

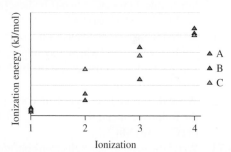

4.58 Arrange the elements in each of the following groups in order of increasing electron affinity:
(a) Li, Na, K; (b) F, Cl, Br, I.

4.59 Specify which of the following elements you would expect to have the greatest electron affinity: He, K, Co, S, Cl.

4.60 Considering their electron affinities, do you think it is possible for the alkali metals to form an anion like M^-, where M represents an alkali metal?

4.61 Explain why alkali metals have a greater affinity for electrons than alkaline earth metals.

4.62 The effective nuclear charges for oxygen and sulfur are 4.45 and 5.48, respectively. Use this information and the atomic radii given in Figure 4.6 to compare the attractive force between the nucleus in each atom and the valence electron that would be removed by the first ionization.

4.63 The effective nuclear charges for sodium and magnesium are 2.51 and 3.31, respectively. Use this information and the atomic radii given in Figure 4.6 to compare the attractive force between the nucleus in each atom and the valence electron that would be removed by the first ionization.

SECTION 4.5: ELECTRON CONFIGURATION OF IONS

Review Questions

4.64 How does the electron configuration of ions derived from main group elements give them stability?

4.65   What do we mean when we say that two ions or an atom and an ion are *isoelectronic?*

4.66   Is it possible for the atoms of one element to be isoelectronic with the atoms of another element? Explain.

4.67   Explain why an isoelectronic series cannot include more than one member of the same group.

Conceptual Problems

4.68   Write the ground-state electron configurations of the following ions: (a) Li^+, (b) H^-, (c) N^{3-}, (d) F^-, (e) S^{2-}, (f) Al^{3+}, (g) Se^{2-}, (h) Br^-.

4.69   Write the ground-state electron configurations of the following ions: (a) Rb^+, (b) Sr^{2+}, (c) Sn^{2+}, (d) Te^{2-}, (e) Ba^{2+}, (f) In^{3+}, (g) Tl^+, (h) Tl^{3+}.

4.70   Write the ground-state electron configurations of the following ions, which play important roles in biochemical processes in our bodies: (a) Na^+, (b) Mg^{2+}, (c) Cl^-, (d) Ca^{2+}.

4.71   Write the ground-state electron configurations of the following ions, which play important roles in various biological processes: (a) Fe^{2+}, (b) Cu^{2+}, (c) Co^{2+}, (d) Mn^{2+}.

4.72   Write the ground-state electron configurations of the following metal ions: (a) Sc^{3+}, (b) Ti^{4+}, (c) V^{5+}, (d) Cr^{3+}, (e) Mn^{3+}.

4.73   Write the ground-state electron configurations of the following metal ions: (a) Ni^{2+}, (b) Cu^+, (c) Ag^+, (d) Au^+, (e) Au^{3+}.

4.74   Identify the ions, each with a net charge of $+1$, that have the following electron configurations: (a) [Ar], (b) $[Ar]3d^{10}$, (c) $[Kr]5s^24d^{10}$, (d) [Xe].

4.75   Identify the ions, each with a net charge of $+3$, that have the following electron configurations: (a) $[Ar]3d^3$, (b) [Ar], (c) $[Kr]4d^6$, (d) $[Xe]4f^{14}5d^6$.

4.76   Which of the following species are isoelectronic with each other: C, Cl^-, Mn^{2+}, B^-, Ar, Zn, Fe^{3+}, Ge^{2+}?

4.77   Group the species that are isoelectronic: Be^{2+}, F^-, Fe^{2+}, N^{3-}, He, S^{2-}, Co^{3+}, Ar.

4.78   For each pair of ions, determine which will have the greater number of unpaired electrons: (a) Fe^{2+}, Fe^{3+}; (b) P^{3+}, P^{5+}; (c) Cr^{2+}, Cr^{3+}.

4.79   Rank the following ions in order of increasing number of unpaired electrons: Ti^{3+}, Fe^{2+}, V^{3+}, Cu^+, Mn^{4+}.

4.80   A cation with a net $+3$ charge is shown to have a total of 10 electrons. Identify the element from which the cation is derived.

4.81   An anion with a net -2 charge has a total of 36 electrons. Identify the element from which the anion is derived.

4.82   A M^{2+} ion derived from a metal in the first transition metal series has four electrons in the $3d$ subshell. What element might M be?

4.83   A metal ion with a net $+3$ charge has five electrons in the $3d$ subshell. Identify the metal.

4.84   Identify the atomic ground-state electron configurations that do not exist. For those that do exist, identify the element.

(a) [Ar] ⇅   ↑ ↑ ↑ __ __
(b) [Ne] ⇅   ⇅ __ __ __
(c) [Kr] ⇅   ↑ ↑ ↑
(d) [Ne] ⇅   ↑ ↑ ↑
(e) [Ar] ⇅   ↑ ↑ ↑ ↑ __
(f) [Kr] ↑   ⇅ ⇅ ⇅ ⇅ ⇅

4.85   Each of the following ground-state electron configurations represents one or more of the transition metal ions in Figure 4.12. Identify the ion or ions represented by each.

(a) [Ar] ___   ⇅ ↑ ↑ ↑ ↑
(b) [Kr] ___   ⇅ ⇅ ⇅ ⇅ ⇅
(c) [Ar] ___   ⇅ ⇅ ⇅ ⇅ ↑
(d) [Ar] ___   ⇅ ⇅ ⇅ ⇅ ⇅
(e) [Ar] ___   ↑ ↑ ↑ __ __
(f) [Ar] ___   ⇅ ⇅ ↑ ↑ ↑

SECTION 4.6: IONIC RADIUS

Review Questions

4.86   Define *ionic radius.* How does the size of an atom change when it becomes (a) an anion and (b) a cation?

4.87   Explain why, for a cation and an anion that are isoelectronic, the anion is larger than the cation.

Conceptual Problems

4.88   List the following ions in order of increasing ionic radius: N^{3-}, Na^+, F^-, Mg^{2+}, O^{2-}.

4.89   Indicate which one of the two species in each of the following pairs is smaller: (a) Cl or Cl^-; (b) Na or Na^+; (c) O^{2-} or S^{2-}; (d) Mg^{2+} or Al^{3+}; (e) Au or Au^{3+}.

4.90   Explain which of the following anions is larger and why: Se^{2-} or Te^{2-}.

4.91   Explain which of the following cations is larger and why: Cu^+ or Cu^{2+}.

4.92   Which of the following is the most realistic representation of an atom from Group 7A becoming an ion?

(a)

(b)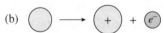

(c)

(d)

(e)

4.93 Which of the following is the most realistic representation of an atom from Group 2A becoming an ion?

(a) ◯ + (e⁻) + (e⁻) ⟶ (2⁺)

(b) ◯ ⟶ (2⁺) + (e⁻) + (e⁻)

(c) ◯ + (e⁻) + (e⁻) ⟶ (2⁻)

(d) ◯ ⟶ (2⁺) + (e⁻) + (e⁻)

(e) ◯ + (e⁻) + (e⁻) ⟶ (2⁻)

4.94 Although sodium and potassium occur in similar amounts in the Earth's crust, living cells accumulate potassium ions almost exclusively. K^+ cannot be replaced by Na^+ in cells, but it is possible for other ions to replace K^+. Based on the similarity of their chemical properties, which of the following ions might be suitable replacements for K^+ in cells: Rb^+, Al^{3+}, Ga^+, Cs^+, Fe^{2+}?

4.95 Ga^{3+} ions are commonly used in a variety of solid state devices that emit light. Assuming ions with sufficiently similar properties could successfully be used in place of Ga^{3+} in such a device, which one of the following ions would be most suitable: Fe^{3+}, Ca^{2+}, Ti^{3+}, In^{3+}, Tl^+?

4.96 State whether each of the following properties of the main group elements generally increases or decreases (a) from left to right across a period and (b) from top to bottom within a group: metallic character, atomic size, ionization energy, acidity of oxides.

4.97 Referring to the periodic table, name (a) the halogen in the fourth period, (b) an element similar to phosphorus in chemical properties, (c) the metal in the fifth period with the lowest ionization energy, (d) an element that has an atomic number smaller than 20 and is similar to strontium.

ADDITIONAL PROBLEMS

4.98 Write equations representing the following processes:
(a) the electron affinity of S^-
(b) the third ionization energy of titanium
(c) the electron affinity of Mg^{2+}
(d) the ionization energy of O^{2-}

4.99 Arrange the following isoelectronic species in order of increasing ionization energy: O^{2-}, F^-, Na^+, Mg^{2+}.

4.100 The following table gives numbers of electrons and protons in ions of several elements.
(a) Which species are cations? (b) Which

species are anions? (c) Write the symbols for all of the species.

	Ion of element					
	A	B	C	D	E	F
Number of electrons	10	18	46	54	46	36
Number of protons	7	20	48	52	49	41

4.101 Arrange the following species in isoelectronic pairs: O^+, Ar, S^{2-}, Ne, Zn, Cs^+, N^{3-}, As^{3+}, N, Xe.

4.102 In which of the following are the species written in decreasing order by size of radius: (a) Be, Mg, Ba; (b) N^{3-}, O^{2-}, F^-; (c) Tl^{3+}, Tl^{2+}, Tl^+?

4.103 Which of the following properties show a clear periodic variation: (a) first ionization energy, (b) molar mass of the elements, (c) number of isotopes of an element, (d) atomic radius?

4.104 Rb and Br atoms are shown here.

Rb   Br

What isoelectronic ions are formed by Rb and Br? Which scene most accurately reflects the relative sizes of the ions once a pair of isoelectronic ions is formed by Rb and Br?

(a)   (b)   (c)   (d)

4.105 Both Mg^{2+} and Ca^{2+} are important biological ions. One of their functions is to bind to the phosphate group of ATP molecules or amino acids of proteins. For Group 2A metals in general, the tendency for binding to the anions increases in the order $Ba^{2+} < Sr^{2+} < Ca^{2+} < Mg^{2+}$. Explain this trend.

4.106 For each pair of elements listed, give three properties that show their chemical similarity: (a) sodium and potassium; (b) chlorine and bromine.

4.107 Given the following valence orbital diagrams, rank these elements in order of increasing (a) atomic size and (b) ionization energy.

A [⬆⬇] [⬆][⬆][ ]   C [⬆⬇] [ ][ ][ ]
  3s   3p   5s   5p

B [⬆⬇] [⬆⬇][⬆⬇][⬆]   D [⬆⬇] [⬆⬇][⬆][⬆]
  2s   2p   3s   3p

4.108 Fill in the blanks in the table:

Symbol		$^{54}_{26}Fe^{3+}$	
Protons	5		46
Neutrons	6	16	59
Electrons	2	18	44
Net charge		−3	

4.109 Contrary to the generalized trend that atomic radius increases down a group, the atomic radius of Zr (159 pm) is actually larger than that of Hf (156 pm) even though the number of electrons increases from 40 (Zr) to 72 (Hf). Suggest a reason for this anomaly. (*Hint:* Many elements of the sixth period after Hf exhibit this same phenomenon in which their atomic radius is not significantly larger or is even smaller than that of the fifth period element above it.)

4.110 Explain why the first electron affinity of sulfur is 200 kJ/mol but the second electron affinity is −649 kJ/mol.

4.111 The H^- ion and the He atom each have two $1s$ electrons. Which of the two species is larger? Explain.

4.112 Given the following electron configurations, rank these elements in order of increasing atomic radius: $[Kr]5s^2$, $[Ne]3s^23p^3$, $[Ar]4s^23d^{10}4p^3$, $[Kr]5s^1$, $[Kr]5s^24d^{10}5p^4$.

4.113 The formula for calculating the energies of an electron in a hydrogen-like ion is given in Problem 4.38. This equation can be applied only to one-electron atoms or ions. One way to modify it for more complex species is to replace Z with $Z − \sigma$ or Z_{eff}. Calculate the value of σ if the first ionization energy of helium is 3.94×10^{-18} J per atom. (Disregard the minus sign in the given equation in your calculation.)

4.114 Why do noble gases have negative electron affinity values?

4.115 The effective nuclear charge felt by the $3p$ valence electron in aluminum is 4.07. Explain how the analogous effective nuclear charge felt by the $4p$ valence electron in gallium can be 6.22, despite the presence of 18 additional electrons in gallium.

4.116 What is the percent change in the force of attraction when two equal but oppositely charged ions are moved from a distance of 125 pm to a distance of 145 pm?

4.117 The atomic radius of K is 227 pm and that of K^+ is 138 pm. Calculate the percent decrease in volume that occurs when $K(g)$ is converted to $K^+(g)$. (The volume of a sphere is $\frac{4}{3}\pi r^3$, where r is the radius of the sphere.)

4.118 The atomic radius of F is 72 pm and that of F^- is 133 pm. Calculate the percent increase in volume that occurs when $F(g)$ is converted to $F^-(g)$. (See Problem 4.117 for the volume of a sphere.)

4.119 A technique called photoelectron spectroscopy is used to measure the ionization energy of atoms. A gaseous sample is irradiated with UV light, and electrons are ejected from the valence shell. The kinetic energies of the ejected electrons are measured. Because the energy of the UV photon

and the kinetic energy of the ejected electron are known, we can write

$$h\nu = IE + \frac{1}{2}mu^2$$

where h is Planck's constant, ν is the frequency of the UV light, and m and u are the mass and velocity of the electron, respectively. In one experiment, the kinetic energy of the ejected electron from potassium is found to be 5.34×10^{-19} J using a UV source of wavelength 162 nm. Calculate the ionization energy of potassium. How can you be sure that this ionization energy corresponds to the electron in the valence shell (i.e., the most loosely held electron)?

4.120 The energy needed for the following process is 1.96×10^4 kJ/mol:

$$Li(g) \longrightarrow Li^{3+}(g) + 3e^-$$

If the first ionization energy of lithium is 520 kJ/mol, calculate the second ionization energy of lithium, that is, the energy required for the process

$$Li^+(g) \longrightarrow Li^{2+}(g) + e^-$$

(*Hint:* You need the equation in Problem 4.38.)

4.121 Using your knowledge of the periodic trends with size, identify these spheres based on their relative sizes as Rb, Rb^+, Mg, F^-, and F.

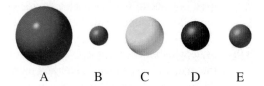

A   B   C   D   E

4.122 What is the electron affinity of the Na^+ ion?

4.123 The ionization energies of sodium (in kJ/mol), starting with the first and ending with the eleventh, are 496, 4562, 6910, 9543, 13,354, 16,613, 20,117, 25,496, 28,932, 141,362, 159,075. Plot the log of ionization energy (y axis) versus the number of ionization (x axis); for example, log 496 is plotted versus 1 (labeled IE_1, the first ionization energy), log 4562 is plotted versus 2 (labeled IE_2, the second ionization energy), and so on. (a) Label IE_1 through IE_{11} with the electrons in orbitals such as $1s$, $2s$, $2p$, and $3s$. (b) What can you deduce about electron shells from the breaks in the curve?

4.124 Experimentally, the electron affinity of an element can be determined by using a laser light to ionize the anion of the element in the gas phase:

$$X^-(g) + h\nu \longrightarrow X(g) + e^-$$

Referring to Figure 4.10, calculate the photon wavelength (in nanometers) corresponding to the electron affinity for chlorine. In what region of the electromagnetic spectrum does this wavelength fall?

4.125 Explain, in terms of their electron configurations, why Fe^{2+} is more easily converted to Fe^{3+} than Mn^{2+} is converted to Mn^{3+}.

4.126 Explain why, in general, atomic radius and ionization energy have opposite periodic trends.

4.127 Explain why the electron affinity of nitrogen is approximately zero, while the elements on either side, carbon and oxygen, have substantial positive electron affinities.

4.128 This graph charts the first six ionization energies of an unknown gaseous atom.

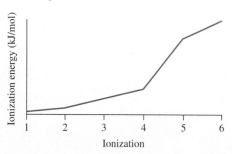

(a) If this element is in the third period, what is its electron configuration?

(b) If this element is in the first row of transition metals, what is its electron configuration?

(c) For each of the elements in parts (a) and (b), explain why there is such a large gap between IE_4 and IE_5.

4.129 Although it is possible to determine the second, third, and higher ionization energies of an element, the same cannot usually be done with the electron affinities of an element. Explain.

4.130 Calculate the maximum wavelength of light (in nanometers) required to ionize a single lithium atom.

4.131 As discussed in the chapter, the atomic mass of argon is greater than that of potassium. This observation created a problem in the early development of the periodic table because it meant that argon should be placed after potassium. (a) How was this difficulty resolved? (b) From the following data, calculate the average atomic masses of argon and potassium: Ar-36 (35.9675 amu, 0.337 percent), Ar-38 (37.9627 amu, 0.063 percent), Ar-40

(39.9624 amu, 99.60 percent), K-39 (38.9637 amu, 93.258 percent), K-40 (39.9640 amu, 0.0117 percent), K-41 (40.9618 amu, 6.730 percent).

4.132 Why do elements that have high ionization energies also have more positive electron affinities? Which group of elements would be an exception to this generalization?

4.133 Predict the atomic number and ground-state electron configuration of the next member of the alkali metals after francium.

4.134 Thallium (Tl) is a neurotoxin and mostly forms only the Tl^+ ion. Aluminum (Al), which causes anemia and dementia, normally only forms the Al^{3+} ion. The first, second, and third ionization energies of Tl are 589, 1971, and 2878 kJ/mol, respectively. The first, second, and third ionization energies of Al are 577.5, 1817, and 2745 kJ/mol, respectively. Plot the ionization energies of Al and Tl versus atomic number and explain the trends.

4.135 The first four ionization energies of an element are approximately 738, 1450, 7.7×10^3, and 1.1×10^4 kJ/mol. To which periodic group does this element belong? Explain your answer.

4.136 What must the charges be on the following elements for them to be isoelectronic with a noble gas? (a) Zr, (b) Cr, (c) Nb, (d) Y.

4.137 The first six ionizations of a gaseous atom can be achieved with light of the following wavelengths, respectively: 218 nm, 113 nm, 29 nm, 22 nm, 17 nm, and 14 nm. To what group of the periodic table does the atom belong?

4.138 A monatomic ion has a charge of +2. The nucleus of the parent atom has a mass number of 55 and the number of neutrons in the nucleus is 1.2 times that of the number of protons. Give the name and symbol of the element.

4.139 An enzyme that requires a metal ion to exhibit biological activity is called a *metalloenzyme*. For a metalloenzyme that requires Zn^{2+} to function, which of the following ions might successfully replace Zn^{2+} while maintaining the activity of the metalloenzyme: Cu^{2+}, Sn^{4+}, Sc^{3+}, Pb^{2+}, Co^{2+}?

Answers to In-Chapter Materials

PRACTICE PROBLEMS

4.1A (a) Ge, (b) Bi. **4.1B** Bi < As < P. **4.2A** (a) $1s^22s^22p^63s^23p^3$, p block, (b) $1s^22s^22p^63s^23p^64s^2$, s block, (c) $1s^22s^22p^63s^23p^64s^23d^{10}4p^5$, p block. **4.2B** (a) Al, (b) Zn, (c) Sr. **4.3A** F < Se < Ge. **4.3B** Si and Se. **4.4A** Mg, Mg. **4.4B** Rb has a smaller Z_{eff} but IE_2 for Rb corresponds to removal of a core electron. **4.5A** Al. **4.5B** Adding an electron to As involves pairing. **4.6A** The attractive force is slightly larger between +3.26 and −1.15 separated by 1.5 pm. **4.6B** 1.51 pm. **4.7A** (a) [Ne],

(b) [Ar], (c) [Kr]. **4.7B** N^{3-}, O^{2-}, F^-, Ne, Na^+, Mg^{2+}, Al^{3+}. **4.8A** (a) $[Ar]3d^6$, (b) $[Ar]3d^9$, (c) $[Kr]4d^{10}$. **4.8B** Cu^+. **4.9A** $Rb^+ < Kr < Br^- < Se^{2-}$. **4.9B** F^-, O^{2-}, N^{3-}, Na^+, Mg^{2+}, Al^{3+}.

SECTION REVIEW

4.1.1 c. **4.1.2** a. **4.1.3** e. **4.2.1** b. **4.2.2** d. **4.4.1** b. **4.4.2** c. **4.4.3** e. **4.4.4** a. **4.4.5** c. **4.4.6** b. **4.5.1** d. **4.5.2** d. **4.5.3** d. **4.5.4** b. **4.5.5** c. **4.6.1** d. **4.6.2** b.

Chapter 5

Ionic and Covalent Compounds

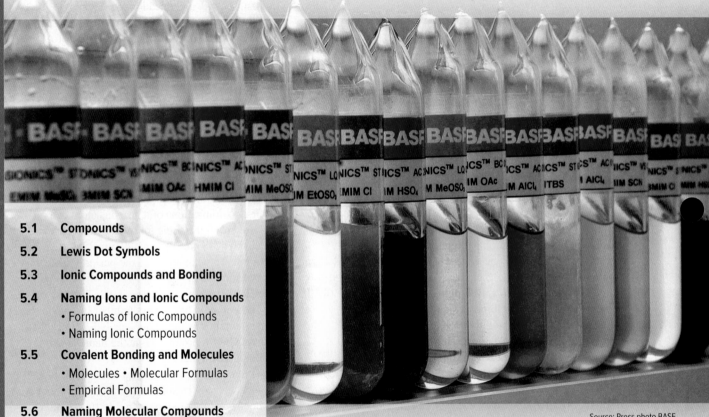

Source: Press photo BASF

MANY COMPOUNDS exhibit predominantly either *ionic* or *covalent* bonding. Recently, though, chemists have succeeded in preparing a new class of compounds called *ionic liquids*. These unusual compounds demonstrate significant concurrent covalent and ionic bonding. Many of these substances possess unusual and useful properties, including being *liquid* at room temperature—a property not normally associated with ionic compounds. The unusual properties of ionic liquids, such as those shown here, allow chemists to use them for cellulose processing, storage media for solar-thermal energy systems, waste recycling, and as solvents for green chemistry applications.

Before You Begin, Review These Skills

- Atomic radius [◄◄ Section 4.4]
- Ions of main group elements [◄◄ Section 4.5]
- The mole [◄◄ Section 2.7]

5.1 COMPOUNDS

Most of the substances we encounter every day are not elements. Rather, they are *compounds*. A **compound** is a substance composed of two or more elements combined in a specific ratio and held together by *chemical bonds* [▶▶ Sections 5.3 and 5.5]. Unlike a mixture [◄◄ Section 1.5], a compound cannot be separated into simpler substances by a physical process. Familiar examples of compounds include water, which is a combination of the elements oxygen and hydrogen; and salt, which is a combination of the elements sodium and chlorine. The discovery of the correlation between electron configuration and periodic properties of the elements [◄◄ Section 4.2] gave chemists a way to explain both the *existence* and the *formation* of compounds.

© H.S. Photos/Alamy Stock Photo

5.2 LEWIS DOT SYMBOLS

Early in the twentieth century, American chemist Gilbert Lewis[1] suggested that atoms combine to achieve a more stable electron configuration. In fact, we have already seen that common monatomic ions result when atoms of main group elements either gain or lose enough electrons to become isoelectronic with noble gases—thereby achieving maximum stability.

> **Student Annotation:** Remember that for two species to be *isoelectronic*, they must have exactly the same electron configuration [◄◄ Section 4.5].

When atoms interact to form compounds, it is their *valence electrons* that actually interact. Therefore, it is helpful to have a method for depicting the valence electrons of the atoms involved. This is done using Lewis dot symbols. A **Lewis dot symbol** consists of the element's symbol surrounded by dots, where each dot represents a valence electron. For the main group elements, the number of dots in the Lewis dot symbol is the same as the group number, as shown in Figure 5.1. (Because they have incompletely filled inner shells, transition metals typically are not represented with Lewis dot symbols.)

Figure 5.1 shows that dots are placed above and below as well as to the left and right of the symbol. The exact order in which the dots are placed around the element symbol is not important, but the *number* of dots is. For example, boron is in Group 3A and therefore has three valence electrons. Thus, any of the following would be correct for the Lewis dot symbol for boron:

$$\dot{\text{B}}\cdot \qquad \overset{\centerdot}{\text{B}}\cdot \qquad \cdot\overset{\centerdot}{\text{B}} \qquad \cdot\dot{\text{B}}$$

When writing Lewis dot symbols, although each side can have a maximum of two dots, we do not "pair" dots until absolutely necessary. Thus, we would *not* represent boron with a pair of dots on one side and a single dot on another.

[1] Gilbert Newton Lewis (1875–1946). Lewis made many important contributions in the areas of chemical bonding, thermodynamics, acids and bases, and spectroscopy. Despite the significance of Lewis's work, he was never awarded a Nobel Prize.

Figure 5.1 Lewis dot symbols of the main group elements.

1A 1																	8A 18
·H	2A 2											3A 13	4A 14	5A 15	6A 16	7A 17	He:
·Li	·Be·											·Ḃ·	·Ċ·	·Ṅ·	·Ö·	:Ḟ·	:Ṅe:
·Na	·Mg·	3B 3	4B 4	5B 5	6B 6	7B 7	┌ 8	8B 9	┐ 10	1B 11	2B 12	·Ȧl·	·Ṡi·	·Ṗ·	·Ṡ·	:Ċl·	:Ȧr:
·K	·Ca·											·Ġa·	·Ġe·	·Ȧs·	·Ṡe·	:Ḃr·	:Ḳr:
·Rb	·Sr·											·İn·	·Ṡn·	·Ṡb·	·Ṫe·	:İ·	:Ẋe:
·Cs	·Ba·											·Ṫl·	·Ṗb·	·Ḃi·	·Ṗo·	:Ȧt·	:Ṙn:
·Fr	·Ra·																

For main group metals such as Na or Mg, the number of dots in the Lewis dot symbol for the atom is equal to the number of electrons lost when the atom forms a cation that is isoelectronic with the preceding noble gas. For nonmetals of the second period (B through F), the number of unpaired dots is the number of bonds the atom can form.

In addition to atoms, we can also represent atomic *ions* with Lewis dot symbols. To do so, we simply add (for anions) or subtract (for cations) the appropriate number of dots from the Lewis dot symbol of the *atom* and include the ion's charge.

Worked Example 5.1 shows how to use Lewis dot symbols to represent atomic ions.

Worked Example 5.1

Write Lewis dot symbols for (a) fluoride ion (F^-), (b) potassium ion (K^+), and (c) sulfide ion (S^{2-}).

Strategy Starting with the Lewis dot symbol for each element, add dots (for anions) or remove dots (for cations) as needed to achieve the correct charge on each ion. Don't forget to include the appropriate charge on the Lewis dot symbol.

Setup The Lewis dot symbols for F, K, and S are :Ḟ·, ·K, and ·Ṡ·, respectively.

Solution

(a) $\left[:\ddot{F}: \right]^-$ (b) K^+ (c) $\left[:\ddot{S}: \right]^{2-}$

Think About It

For ions that are isoelectronic with noble gases, cations should have no dots remaining around the element symbol, whereas anions should have eight dots around the element symbol. Note, too, that for anions, we put square brackets around the Lewis dot symbol and place the negative charge outside the brackets. Because the symbol for a common cation such as the potassium ion has no remaining dots, square brackets are not necessary.

Practice Problem (A)**TTEMPT** Write Lewis dot symbols for (a) Ca^{2+}, (b) N^{3-}, and (c) I^-.

Practice Problem (B)**UILD** Indicate the charge on each of the ions represented by the following Lewis dot symbols: (a) $\left[:\ddot{O}: \right]^?$, (b) $H^?$, and (c) $\left[:\ddot{P}: \right]^?$.

Practice Problem (C)**ONCEPTUALIZE** For each of the highlighted positions on the periodic table, write a Lewis structure for an atom of the element and for the common ion that it forms. Use the generic symbol X for each element. For example, rather than writing ·Na and [Na]$^+$ for sodium and sodium ion, respectively, you should write ·X and [X]$^+$.

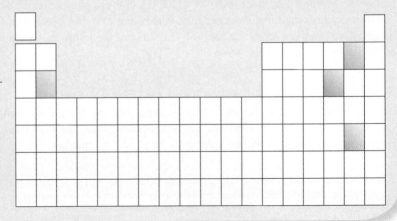

Section 5.2 Review

Lewis Dot Symbols

5.2.1   Using only a periodic table, determine the correct Lewis dot symbol for a silicon (Si) atom.

(a) $:\overset{}{Si}:$ (c) $:\overset{\cdot}{Si}\cdot$ (e) $.:Si$

(b) $\cdot\overset{\cdot}{Si}\cdot$ (d) $\cdot Si\cdot$

5.2.2   Using only a periodic table, determine the correct Lewis dot symbol for the bromide ion (Br^-).

(a) $\left[:\overset{..}{Br}\cdot\right]^-$ (c) $\left[:\overset{..}{Br}\cdot\right]^-$ (e) $\left[:\overset{..}{Br}:\right]^-$

(b) $\left[:\overset{..}{Br}\right]^-$ (d) $\left[Br\cdot\right]^-$

5.2.3   To what group of the periodic table does element X belong if its Lewis dot symbol is $\cdot\overset{\cdot}{X}\cdot$?

(a) 2A   (c) 4A   (e) It is not possible to tell.

(b) 3A   (d) 5A

5.2.4   To which group does the element Y belong if the Lewis dot symbol for its anion is $\left[:\overset{..}{Y}:\right]^{m-}$, where m represents the charge?

(a) 4A   (c) 6A   (e) It is not possible to tell.

(b) 5A   (d) 7A.

5.3   IONIC COMPOUNDS AND BONDING

Recall from Chapter 4 that atoms of elements with low ionization energies, typically *metals,* tend to form *cations;* and those with high positive electron affinities, typically *nonmetals,* tend to form *anions.* When cations and anions are brought together, they can combine to form an ***ionic compound.*** Most ionic compounds are ***binary compounds,*** substances that consist of just two different ions derived from two different elements—one *metal* and one *nonmetal.*

Typically, ionic compounds consist of metal cations and nonmetal anions, held together by ***ionic bonding,*** which simply refers to the electrostatic attraction between oppositely charged particles. As an example, let us consider the ionic compound sodium chloride, known commonly as *salt.*

The electron configuration of sodium is $[Ne]3s^1$ and that of chlorine is $[Ne]3s^23p^5$. When sodium and chlorine atoms come into contact with each other, the valence electron of sodium is transferred to the chlorine atom. We can imagine these processes taking place separately and represent each process using Lewis dot symbols.

$$Na\cdot \longrightarrow Na^+ + e^-$$

$$:\overset{..}{Cl}\cdot + e^- \longrightarrow \left[:\overset{..}{Cl}:\right]^-$$

The sum of these two equations is

$$Na\cdot + :\overset{..}{Cl}\cdot \longrightarrow Na^+ + \left[:\overset{..}{Cl}:\right]^-$$

Student Annotation: Because the electron produced by ionization of the sodium atom is immediately added to the chlorine atom, it does not appear in the overall equation.

The electrostatic attraction, or *ionic bonding,* between the resulting cation (Na^+) and anion (Cl^-) draws them together to form the electrically neutral compound, sodium chloride, which we represent with the chemical formula NaCl. The ***chemical formula,*** or simply ***formula,*** of an ionic compound denotes the constituent elements of the compound and the ratio in which they combine. Sodium chloride, for example, consists of equal numbers of sodium ions and chloride ions. Note that although an ionic compound consists of oppositely charged ions, we do not show the charges on the ions in the chemical formula.

Student Annotation: A compound in which the ratio of combination is something other than 1:1 would indicate the ratio of combination with subscript numbers, as we will see shortly.

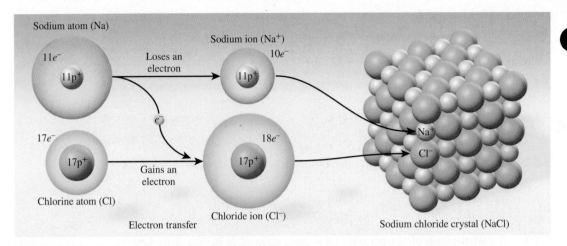

Figure 5.2 An electron is transferred from the sodium atom to the chlorine atom, giving a sodium ion and a chloride ion. The oppositely charged ions are attracted to each other and form a solid lattice.

Student Annotation: Remember that the electron affinity, *EA*, is the energy *released* when a gaseous atom accepts an electron. A *positive EA* corresponds to a process that is energetically favorable. [◄◄ Section 4.4].

Video 5.1
Electron transfer between sodium and chlorine

The structure of an ionic compound consists of a vast array of interspersed cations and anions called a *lattice.* For example, solid sodium chloride is a three-dimensional network of alternating cations and anions (Figure 5.2). As Figure 5.2 shows, no Na^+ ion in NaCl is associated with just one particular Cl^- ion. In fact, each Na^+ ion is surrounded by six Cl^- ions and vice versa. In other ionic compounds, the actual structure may be different, but the arrangement of cations and anions is such that the compounds are all electrically neutral.

The net energy change associated with the formation of Na^+ and Cl^- ions indicates that energy must be supplied in order for the overall transfer of an electron from Na to Cl to take place. The ionization of sodium requires the input of 496 kJ/mol, and the electron affinity of chlorine is 349 kJ/mol [◄◄ Section 4.4]. If the transfer of electrons *from* sodium and *to* chlorine were the only steps in the formation of NaCl, an input of [496 + (−349)] = 147 kJ of energy would be required for the formation of one mole of NaCl. In reality, the formation of ionic compounds typically gives *off* energy—in many cases, a great deal of energy. In fact, the formation of ionic bonds *releases* a large amount of energy that more than compensates for the energy input required to transfer electrons from metal atoms to nonmetal atoms. We can quantify the energy change associated with the formation of ionic bonds with *lattice energy.*

Lattice energy is the amount of energy required to convert a mole of ionic solid to its constituent ions in the gas phase. For example, the lattice energy of sodium chloride is 788 kJ/mol. Thus, it takes 788 kJ of energy to convert one mole of NaCl(*s*) to one mole each of $Na^+(g)$ and $Cl^-(g)$. The magnitude of lattice energy is a measure of an ionic compound's stability. The greater the lattice energy, the more stable the compound. Table 5.1 lists the lattice energies for some ionic compounds.

Lattice energy depends on the magnitudes of the charges and on the distance between them. For example, LiI, NaI, and KI all have the same anion (I^-) and all have cations with the same charge (+1). The trend in their lattice energies (LiI > NaI > KI) can be explained on the basis of ionic radius. The radii of alkali metal ions increase as we move down a group in the periodic table ($r_{Li^+} < r_{Na^+} < r_{K^+}$) [◄◄ Section 4.6]. Knowing the radius of each ion, we can use Coulomb's law to compare the attractive forces between the ions in these three compounds as shown in Figure 5.3(a).

LiI, with the smallest distance between ions, has the strongest attractive forces. It should therefore have the largest lattice energy. KI, with the largest distance between ions, has the weakest attractive forces, and should have the smallest lattice energy. NaI has an intermediate distance between ions and should have an intermediate lattice energy. Thus, Coulomb's law correctly predicts the relative magnitudes of the lattice energies of LiI, NaI, and KI.

TABLE 5.1	Lattice Energies of Selected Ionic Compounds	
Compound	**Lattice energy (kJ/mol)**	**Melting point (°C)**
LiF	1017	845
LiCl	860	610
LiBr	787	550
LiI	732	450
NaCl	788	801
NaBr	736	750
NaI	686	662
KCl	699	772
KBr	689	735
KI	632	680
$MgCl_2$	2527	714
Na_2O	2570	Sub*
MgO	3890	2800
ScN	7547	> 3000

*Na_2O sublimes at 1275°C.

Now consider the compounds LiF, MgO, and ScN, shown in Figure 5.3(b). With similar distances between ions (0.76 + 1.33 = 2.09 Å for LiF; 0.72 + 1.40 = 2.12 Å for MgO; and 0.88 + 1.46 = 2.34 Å for ScN), we can see that the charges play a major role in determining the magnitude of lattice energy.

Worked Example 5.2 shows how to use ionic radii and Coulomb's law to compare lattice energies for ionic compounds.

Worked Example 5.2

Arrange MgO, CaO, and SrO in order of increasing lattice energy.

Strategy Consider the charges on the ions and the distances between them. Apply Coulomb's law to determine the relative lattice energies.

Setup MgO, CaO, and SrO all contain the same anion (O^{2-}) and all contain cations with the same charge (+2). In this case, then, the distance between ions will determine the relative lattice energies. Recall that lattice energy increases as the distance between ions decreases (because the force between oppositely charged particles increases as the distance between them decreases). Because all three compounds contain the same anion, we need only consider the radii of the cations when determining the distance between ions. From Figure 4.13, the ionic radii are 0.72 Å (Mg^{2+}), 1.00 Å (Ca^{2+}), and 1.18 Å (Sr^{2+}).

Solution MgO has the smallest distance between ions, whereas SrO has the largest distance between ions. Therefore, in order of increasing lattice energy: SrO < CaO < MgO.

> **Think About It**
> Mg, Ca, and Sr are all Group 2A metals, so we could have predicted this result without knowing their radii. Recall that ionic radii increase as we move down a column in the periodic table, and charges that are farther apart are more easily separated (meaning the lattice energy will be smaller). The lattice energies of SrO, CaO, and MgO are 3217, 3414, and 3890 kJ/mol, respectively.

Practice Problem ATTEMPT Determine which compound has the larger lattice energy: $MgCl_2$ or $SrCl_2$.

Practice Problem BUILD Arrange the compounds NaF, MgO, and AlN in order of increasing lattice energy.

Practice Problem CONCEPTUALIZE Common ions of four hypothetical elements are shown along with their radii in nanometers. Arrange all the binary ionic compounds that could form from these ions in order of increasing lattice energy.

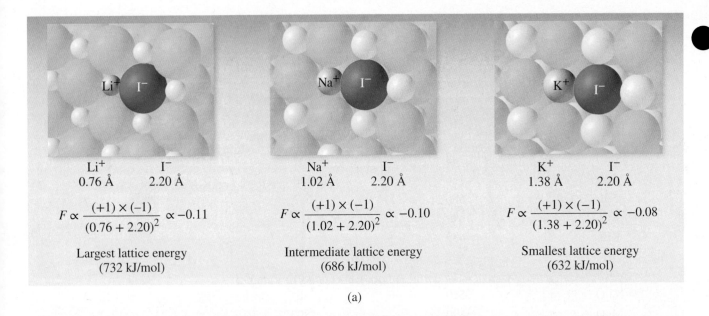

(a)

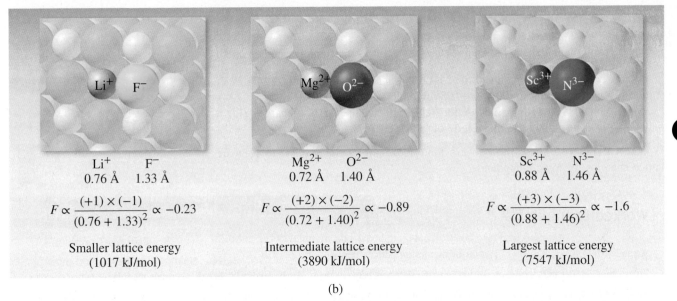

(b)

Figure 5.3 (a) As the distance between two oppositely charged ions increases, the attractive force between them is diminished and the lattice energy of the compound *decreases*. (b) As the magnitudes of charges increase, the attractive force between oppositely charged ions gets larger and lattice energy *increases*. Note that the effect of charge magnitude is more dramatic than that of distance between ions.

Section 5.3 Review

Ionic Compounds and Bonding

5.3.1   Will the lattice energy of KF be larger or smaller than that of LiF, larger or smaller than that of KCl, and larger or smaller than that of KI?

    (a) larger, larger, and smaller

    (b) smaller, larger, and smaller

    (c) smaller, larger, and larger

    (d) smaller, smaller, and smaller

    (e) larger, smaller, and larger

5.3.2   Lattice energies are graphed for three series of compounds in which the ion charges are +2, −2; +2, −1; and +1, −1. The ions in each series of compounds are separated by different distances. Identify the series.

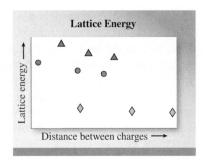

(a)   red +2, −2; blue +1, −1; green +2, −1

(b)   red +2, −2; blue +2, −1; green +1, −1

(c)   red +2, −1; blue +2, −2; green +1, −1

(d)   red +1, −1; blue +2, −1; green +2, −2

(e)   red +2, −1; blue +1, −1; green +2, −2

5.4   NAMING IONS AND IONIC COMPOUNDS

When chemistry was a relatively new science, it was possible for practitioners to memorize the names of the relatively small number of known compounds. Often a compound's name was derived from its physical appearance, properties, origin, or application—for example, milk of magnesia, laughing gas, formic acid (*formica* is the Latin word for ant, and *formic acid* is the compound responsible for the sting of an ant bite), and baking soda.

Today there are many millions of known compounds, and many more being made every year, so it would be impossible to memorize all their names. Fortunately, it is unnecessary, because over the years chemists have devised a system for naming chemical substances. The rules are the same worldwide, facilitating communication among scientists and providing a useful way of labeling an overwhelming variety of substances. Mastering these rules now will benefit you tremendously as you progress through your chemistry course. You must be able to name a compound, given its chemical formula, and you must be able to write the chemical formula of a compound, given its name.

We begin our discussion of chemical *nomenclature* with monatomic ions [◄◄ Section 4.4]. A monatomic ion is one that consists of a single atom with *more* electrons than protons, in the case of an anion, or *fewer* electrons than protons, in the case of a cation. What charges are possible for the monatomic ions of elements in Groups 1A through 7A? The cations that form from elements of Groups 1A, 2A, and 3A have charges equal to their respective group numbers. Most of the anions that form from elements of Groups 4A through 7A have charges equal to the corresponding group number minus 8. For example, the monatomic anion formed by oxygen (Group 6A) has a charge of 6 − 8 = −2. You should be able to determine the charges on ions of elements in Groups 1A through 7A (for any element that forms only one common ion) using only a periodic table. [◄◄ Figure 4.12] shows the charges of many of the monatomic ions from across the periodic table.

A monatomic cation is named simply by adding the word *ion* to the name of the element. Thus, the ion of potassium, K^+, is known as potassium ion. Similarly, the cations formed by the elements magnesium (Mg^{2+}) and aluminum (Al^{3+}) are called magnesium ion and aluminum ion, respectively. It is not necessary for the name to specify the charge on these ions because their charges are equal to their group numbers.

Certain metals, especially the *transition metals,* can form cations of more than one possible charge. Iron, for example, can form Fe^{2+} and Fe^{3+}. An older nomenclature system that is still in limited use assigns the ending *–ous* to the cation with the *smaller* positive charge and the ending *–ic* to the cation with the *greater* positive charge:

$$Fe^{2+}: \text{ferrous ion}$$

$$Fe^{3+}: \text{ferric ion}$$

TABLE 5.2	Names and Formulas of Some Common Monatomic Ions

Name	Formula
Cations	
aluminum	Al^{3+}
barium	Ba^{2+}
cadmium	Cd^{2+}
calcium	Ca^{2+}
cesium	Cs^+
chromium(III)	Cr^{3+}
cobalt(II)	Co^{2+}
copper(I)	Cu^+
copper(II)	Cu^{2+}
hydrogen	H^+
iron(II)	Fe^{2+}
iron(III)	Fe^{3+}
lead(II)	Pb^{2+}
lithium	Li^+
magnesium	Mg^{2+}
manganese(II)	Mn^{2+}
mercury(II)	Hg^{2+}
potassium	K^+
silver	Ag^+
sodium	Na^+
strontium	Sr^{2+}
tin(II)	Sn^{2+}
zinc	Zn^{2+}
Anions	
bromide	Br^-
chloride	Cl^-
fluoride	F^-
hydride	H^-
iodide	I^-
nitride	N^{3-}
oxide	O^{2-}
sulfide	S^{2-}

This method of naming ions has some distinct limitations. First, the *–ous* and *–ic* suffixes indicate the *relative* charges of the two cations involved, not the *actual* charges. Thus, Fe^{3+} is the ferric ion, but Cu^{2+} is the cupric ion. In addition, the *–ous* and *–ic* endings make it possible to name only two cations with different charges. Some metals, such as manganese (Mn), can form cations with three or more different charges.

Therefore, it has become increasingly common to designate different cations with Roman numerals, using the Stock system.[2] In this system, the Roman numeral I indicates a positive charge of one, II means a positive charge of two, and so on, as shown for manganese:

$$Mn^{2+}: \text{manganese(II) ion}$$
$$Mn^{3+}: \text{manganese(III) ion}$$
$$Mn^{4+}: \text{manganese(IV) ion}$$

These names are pronounced "manganese-two ion," "manganese-three ion," and "manganese-four ion," respectively. Using the Stock system, the ferrous and ferric ions are iron(II) and iron(III), respectively. To avoid confusion, and in keeping with modern practice, we will use the Stock system to name compounds in this textbook.

A monatomic anion is named by changing the ending of the element's name to *–ide,* and adding the word *ion.* Thus, the anion of chlorine, Cl^-, is called *chloride ion.* The anions of carbon, nitrogen, and oxygen (C^{4-}, N^{3-}, and O^{2-}) are called *carbide, nitride,* and *oxide.* Because there is only one possible charge for an ion formed from a nonmetal, it is unnecessary for the ion's name to specify its charge. The common anions formed by nonmetal elements are shown in Figure 4.12. Table 5.2 lists alphabetically a number of common monatomic ions.

Formulas of Ionic Compounds

As we have seen, the formulas of ionic compounds give the ratio of combination of ions—generally in the smallest possible whole numbers. In order for ionic compounds to be electrically neutral, the sum of the charges on the cation and anion in the formula must be zero. When the charges on the cation and anion are numerically equal, it is easy to see that they will combine in a 1:1 ratio, as in the case of sodium chloride (NaCl). If the charges on the cations and anions are numerically different, you can apply the following guideline to make the formula electrically neutral (and thus obtain the simplest formula): Write a subscript for the cation that is numerically equal to the charge on the anion and a subscript for the anion that is numerically equal to the charge on the cation. Let's consider some examples.

Potassium Bromide The potassium ion (K^+) and the bromide ion (Br^-) combine to form the ionic compound *potassium bromide.* The sum of the charges is $1 + (-1) = 0$, so no subscripts are necessary. The formula is KBr.

Zinc Iodide The zinc ion (Zn^{2+}) and the iodide ion (I^-) combine to form *zinc iodide.* The sum of the charges of one Zn^{2+} ion and one I^- ion is $+2 + (-1) = +1$. To make the charges add up to zero, we multiply the -1 charge of the anion by 2 and add the subscript "2" to the symbol for iodine. Thus, the formula for zinc iodide is ZnI_2.

Aluminum Oxide The cation is Al^{3+} and the anion is O^{2-}. The following diagram can be used to determine the subscripts for this compound:

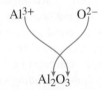

$$Al^{3+} \qquad O^{2-}$$

$$Al_2O_3$$

---

[2]Alfred E. Stock (1876–1946). German chemist, Stock did most of his research in the synthesis and characterization of boron, beryllium, and silicon compounds. He was a pioneer in the study of mercury poisoning.

The sum of the charges for aluminum oxide is $2(+3) + 3(-2) = 0$. Thus, the formula is Al_2O_3.

Naming Ionic Compounds

An ionic compound is named using the name of the cation followed by the name of the anion, eliminating the word *ion* from each. Several examples were given in the Formulas of Ionic Compounds section. Other examples are sodium bromide (NaBr), calcium fluoride (CaF_2), lithium nitride (Li_3N), and sodium sulfide (Na_2S).

In cases where a metal cation may have more than one possible charge, recall that the charge is indicated in the name of the ion with a Roman numeral in parentheses. Thus, the compounds $FeCl_2$ and $FeCl_3$ are named *iron(II) chloride* and *iron(III) chloride,* respectively. (These are pronounced "iron-two chloride" and "iron-three chloride.") Figure 5.4 summarizes the steps for identifying and naming ionic compounds.

Worked Examples 5.3 and 5.4 illustrate how to name ionic compounds and write formulas for ionic compounds based on the information given in Figure 5.4 and Table 5.2.

Worked Example 5.3

Name the following ionic compounds: (a) CaO, (b) Mg_3N_2, and (c) Fe_2S_3.

Strategy Begin by identifying the cation and the anion in each compound, and then combine the names for each, eliminating the word *ion*.

Setup CaO contains Ca^{2+} and O^{2-}, the calcium ion and the oxide ion; Mg_3N_2 contains Mg^{2+} and N^{3-}, the magnesium ion and the nitride ion; and $FeCl_2$ contains Fe^{2+} and Cl^-, the iron(III) ion and the chloride ion. We know that the iron in Fe_2S_3 is Fe^{3+} because it combines with the S^{2-} ion in a 2:3 ratio to give a neutral formula.

Solution (a) Combining the cation and anion names, and eliminating the word *ion* from each of the individual ions' names, we get *calcium oxide* as the name of CaO; (b) Mg_3N_2 is *magnesium nitride;* and (c) Fe_2S_3 is *iron(III) sulfide.*

> **Think About It**
> Be careful not to confuse the subscript in a formula with the charge on the metal ion. In part (c), for example, the subscript on Fe is 2, but this is an iron(III) compound.

Practice Problem Ⓐ**TTEMPT** Name the following ionic compounds: (a) BaS, (b) Cu_2S, (c) $CrCl_3$.

Practice Problem Ⓑ**UILD** Name the following ionic compounds: (a) $AlBr_3$, (b) Rb_2Se, (c) Cu_3N_2.

Practice Problem Ⓒ**ONCEPTUALIZE** The diagram represents a small sample of an ionic compound where red spheres represent oxide ions and grey spheres represent lead ions. Deduce the correct formula and name of the compound.

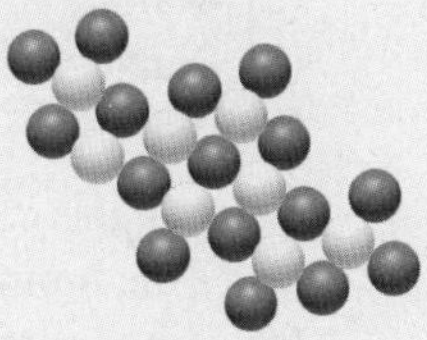

Worked Example 5.4

Deduce the formulas of the following ionic compounds: (a) mercury(II) chloride, (b) lead(II) bromide, and (c) potassium nitride.

Strategy Identify the ions in each compound, and determine their ratios of combination using the charges on the cation and anion in each.

If the compound contains a metal and a nonmetal, the compound is most likely

Ionic

Metal cation has only one possible charge.

Metal cation has more than one possible charge.

- Alkali metal cations
- Alkaline earth metal cations
- Ag^+, Al^{3+}, Cd^{2+}, Zn^{2+}

- Other metal cations

Naming

Naming

- Name metal first.
- Add *-ide* ending to root of nonmetal name.

- Name metal first.
- Specify charge of metal cation with Roman numeral in parentheses.
- Add *-ide* to root of nonmetal name.

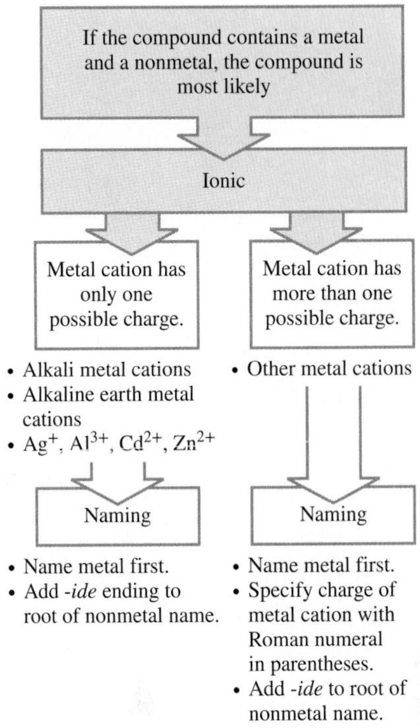

Figure 5.4 Flowchart for naming binary ionic compounds.

(Continued on next page)

Setup (a) Mercury(II) chloride is a combination of Hg^{2+} and Cl^-. To produce a neutral compound, these two ions must combine in a 1:2 ratio. (b) Lead(II) bromide is a combination of Pb^{2+} and Br^-. These ions also combine in a 1:2 ratio. (c) Potassium nitride is a combination of K^+ and N^{3-}. These ions combine in a 3:1 ratio.

Solution The formulas are (a) $HgCl_2$, (b) $PbBr_2$, and (c) K_3N.

> **Think About It**
> Make sure that the charges sum to zero in each compound formula. In part (a), for example, $Hg^{2+} + 2Cl^- = (+2) + 2(-1) = 0$; in part (b), $(+2) + 2(-1) = 0$; and in part (c), $3(+1) + (-3) = 0$.

Practice Problem **A**TTEMPT Deduce the formulas of the following ionic compounds: (a) lead(II) chloride, (b) cobalt(III) oxide, and (c) sodium selenide.

Practice Problem **B**UILD Deduce the formulas of the following ionic compounds: (a) iron(III) sulfide, (b) mercury(II) iodide, and (c) potassium sulfide.

Practice Problem **C**ONCEPTUALIZE The diagram represents a small sample of an ionic compound where yellow spheres represent sulfide ions and blue spheres represent lead ions. Deduce the correct formula and name of the compound.

Section 5.4 Review

Naming Ions and Ionic Compounds

5.4.1 What is the correct name of the compound PbS?
(a) Lead(II) sulfur
(b) Lead(II) sulfide
(c) Lead(IV) sulfide
(d) Lead(II) sulfuride
(e) Lead(IV) sulfuride

5.4.2 What is the correct formula for the compound iron(III) nitride?
(a) Fe_2N_3
(b) FeN_3
(c) Fe_3N_2
(d) FeN
(e) Fe_3N

5.4.3 What is the correct formula for sodium nitride?
(a) Na_3N
(b) NaN
(c) NaN_3
(d) Na_3N_2
(e) Na_2N_3

5.4.4 What is the correct name for the compound $ZnBr_2$?
(a) Zinc brominide
(b) Zinc bromide
(c) Zinc bromine
(d) Zinc boride
(e) Zinc dibrominide

5.5   COVALENT BONDING AND MOLECULES

We learned in Section 5.4 that ionic compounds tend to form between metals and nonmetals when electrons are transferred from an element with a low ionization energy (the metal) to one with a high electron affinity (the nonmetal). When compounds form between elements with more similar properties, electrons are not transferred from one element to another but instead are *shared* to give each atom a noble gas electron configuration. It was Gilbert Lewis who first suggested that a chemical bond involves atoms sharing electrons, and this approach is known as the ***Lewis theory of bonding.***

Lewis theory depicts the formation of the bond in H_2 as

$$H\cdot + \cdot H \longrightarrow H\!:\!H$$

In essence, two H atoms move close enough to each other to *share* the electron pair. Although there are still two atoms and just two electrons, this arrangement allows each H atom to "count" both electrons as its own and to "feel" as though it has the noble gas electron configuration of helium. This type of arrangement, where two atoms share a pair of electrons, is known as ***covalent bonding,*** and the shared pair of electrons constitutes the ***covalent bond.*** In a covalent bond, each electron in a shared pair is attracted to the nuclei of both atoms. It is this attraction that holds the two atoms together.

Student Annotation:
For the sake of simplicity, the shared pair of electrons can be represented by a dash, rather than by two dots: H—H.

Molecules

The formation of a covalent bond between two H atoms produces a species known as a *molecule.* A ***molecule*** is a neutral combination of at least two atoms in a specific arrangement held together by chemical forces (also called *chemical bonds*). A molecule may contain two or more atoms of a single element, or it may contain atoms of two or more elements joined in a fixed ratio.

Thus, a molecule can be of an *element* or it can be of a *compound,* which, by definition, is made up of two or more elements. This definition stems from Dalton's atomic theory, which we first encountered in Chapter 2. Dalton's theory actually consisted of three hypotheses, the first of which was that matter is composed of atoms [◄◄ Section 2.1]. The second hypothesis was that *compounds* are composed of atoms of more than one element; and in any given compound, the same types of atoms are always present in the same relative numbers. Hydrogen gas, for example, is an element, but it consists of molecules, each of which is made up of two H atoms. Water, on the other hand, is a *compound* that consists of molecules, each of which always contains two H atoms and one O atom. We will revisit Dalton's theory and learn about his third hypothesis when we discuss chemical reactions [►► Chapter 8].

Dalton's second hypothesis suggests that to form a certain compound, we need not only atoms of the right *kinds* of elements, but specific *numbers* of these atoms as well. This idea is an extension of a law published in 1799 by Joseph Proust, a French chemist. According to Proust's ***law of definite proportions,*** different samples of a given compound always contain the same elements in the same mass *ratio.* Thus, if we were to analyze samples of carbon dioxide gas obtained from different sources, such as the exhaust from a car in Mexico City or the air above a pine forest in northern Maine, each sample would contain the same ratio by mass of oxygen to carbon. Consider the following results of the analysis of three samples of carbon dioxide, each from a different source:

Sample	Mass of O (g)	Mass of C (g)	Ratio (g O : g C)
123 g carbon dioxide	89.4	33.6	2.66:1
50.5 g carbon dioxide	36.7	13.8	2.66:1
88.6 g carbon dioxide	64.4	24.2	2.66:1

In any sample of pure carbon dioxide, there are 2.66 g of oxygen for every gram of carbon present. This constant mass ratio can be explained by assuming that the elements exist in tiny particles of fixed mass (atoms), and that compounds are formed by the combination of fixed numbers of each type of particle.

Dalton's second hypothesis also supports the ***law of multiple proportions.*** According to this law, if two elements can combine with each other to form two or more *different* compounds, the ratio of masses of one element that combine with a fixed mass of the other element can be expressed in small whole numbers. For example, carbon combines with oxygen to form two different compounds: carbon dioxide and carbon monoxide. In any sample of pure carbon monoxide, there are 1.33 g of oxygen for every gram of carbon.

Sample	Mass of O (g)	Mass of C (g)	Ratio (g O : g C)
16.3 g carbon monoxide	9.31	6.99	1.33:1
25.9 g carbon monoxide	14.8	11.1	1.33:1
88.4 g carbon monoxide	50.5	37.9	1.33:1

Thus, the mass ratio of oxygen to carbon in carbon *di*oxide is 2.66; and the ratio of oxygen to carbon in carbon *mon*oxide is 1.33. According to the law of multiple proportions, the *ratio* of two such *mass ratios* can be expressed as small whole numbers.

$$\frac{\text{mass ratio of O to C in carbon dioxide}}{\text{mass ratio of O to C in carbon monoxide}} = \frac{2.66}{1.33} = 2:1$$

For samples containing equal masses of carbon, the mass ratio of oxygen in carbon dioxide to oxygen in carbon monoxide is 2:1. Modern measurement techniques indicate that one atom of carbon combines with two atoms of oxygen in carbon dioxide and with one atom of oxygen in carbon monoxide. This result is consistent with the law of multiple proportions (Figure 5.5).

The hydrogen molecule, symbolized as H_2, is called a **_diatomic molecule_** because it contains *two* atoms. Other elements that normally exist as diatomic molecules are nitrogen (N_2), oxygen (O_2), and the Group 7A elements—fluorine (F_2), chlorine (Cl_2), bromine (Br_2), and iodine (I_2). These are known as **_homonuclear_** diatomic molecules because both atoms in each molecule are of the same element. A diatomic molecule can also contain atoms of different elements. Examples of these **_heteronuclear_** diatomic molecules include hydrogen chloride (HCl) and carbon monoxide (CO).

Most molecules contain more than two atoms. They can all be atoms of the same element, as in ozone (O_3) and white phosphorus (P_4), or they can be combinations of two or more different elements, as in water (H_2O) and methane (CH_4). Molecules containing more than two atoms are called **_polyatomic molecules._**

Molecules are far too small for us to observe them directly. An effective means of visualizing them is by the use of molecular models. Throughout this book, we will represent matter at the molecular level using *molecular art,* the two-dimensional equivalent of molecular models. In these pictures, atoms are represented as spheres, and atoms of particular elements are represented using specific colors. Table 5.3

Homonuclear diatomic

Heteronuclear diatomic

Polyatomic

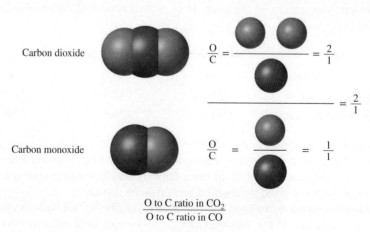

Carbon dioxide $\quad \dfrac{O}{C} = \dfrac{\quad}{\quad} = \dfrac{2}{1}$

$$\frac{\quad}{\quad} = \frac{2}{1}$$

Carbon monoxide $\quad \dfrac{O}{C} = \dfrac{\quad}{\quad} = \dfrac{1}{1}$

$$\frac{\text{O to C ratio in } CO_2}{\text{O to C ratio in CO}}$$

Figure 5.5 An illustration of the law of multiple proportions.

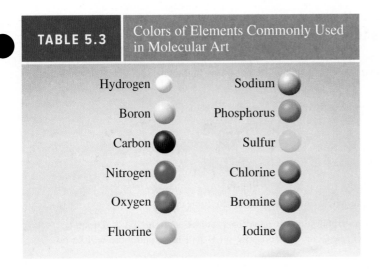

TABLE 5.3	Colors of Elements Commonly Used in Molecular Art

Hydrogen   Sodium

Boron   Phosphorus

Carbon   Sulfur

Nitrogen   Chlorine

Oxygen   Bromine

Fluorine   Iodine

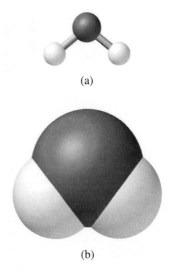

(a)

(b)

Figure 5.6 Water represented with (a) the ball-and-stick model, and (b) the space-filling model.

lists some of the elements that you will encounter most often and the colors used to represent them in this book. Molecular art can be of *ball-and-stick* models, in which the bonds connecting atoms appear as sticks [Figure 5.6(a)], or of *space-filling* models, in which the atoms appear to overlap one another [Figure 5.6(b)]. Ball-and-stick and space-filling models illustrate the specific, three-dimensional arrangement of the atoms. The ball-and-stick models generally do a good job of illustrating the arrangement of atoms, but they exaggerate the distances between atoms, relative to their sizes. By contrast, the space-filling models give a more accurate picture of the interatomic distances but can sometimes obscure the details of the three-dimensional arrangement.

Molecular Formulas

A *chemical formula* (see [◄◄Section 5.3]) can be used to denote the composition of any substance—ionic or molecular. A **molecular formula** shows the exact number of atoms of each element in a *molecule*. In our discussion of molecules, each example was given with its molecular formula in parentheses. Thus, H_2 is the molecular formula for hydrogen, O_2 is oxygen, O_3 is ozone, and H_2O is water. The subscript number indicates the number of atoms of an element present in the molecule. There is no subscript for O in H_2O because there is only one oxygen atom in a molecule of water. The number "one" is never used as a subscript in a chemical formula. Oxygen (O_2) and ozone (O_3) are allotropes of oxygen. An **allotrope** is one of two or more distinct forms of an element. Two of the allotropic forms of the element carbon—diamond and graphite—have dramatically different properties (and *prices*).

We can also represent molecules with *structural formulas*. The **structural formula** shows not only the elemental composition, but also the general arrangement of atoms within the molecule. In the case of water, each of the hydrogen atoms is connected to the oxygen atom. Thus, water can be represented with the structural formula HOH.

In Chapter 6, you will learn how to use the molecular formula to deduce the structural formula and the three-dimensional arrangement of atoms in a molecule. We will use all of these methods for representing molecules throughout the book, so you should be familiar with each method and with the information it provides. Worked Example 5.5 shows how to write a molecular formula from the corresponding molecular model.

Worked Example 5.5

Write the molecular formula of ethanol based on its ball-and-stick model, shown here.

Ethanol

Strategy Refer to the labels on the atoms (or see Table 5.3).

Setup There are *two* carbon atoms, *six* hydrogen atoms, and *one* oxygen atom, so the subscript on C will be 2 and the subscript on H will be 6, and there will be no subscript on O.

Solution C_2H_6O

> **Think About It**
> Often the molecular formula for a compound such as ethanol (consisting of carbon, hydrogen, and oxygen) is written so that the formula more closely resembles the actual arrangement of atoms in the molecule. Thus, the molecular formula for ethanol is commonly written as C_2H_5OH.

Practice Problem **A**TTEMPT Chloroform was used as an anesthetic for childbirth and surgery during the nineteenth century. Write the molecular formulas for (a) chloroform and (b) acetone, based on the molecular models shown here.

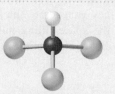

Chloroform   Acetone

Practice Problem **B**UILD Draw (a) a space-filling molecular model of carbon disulfide (CS_2), and (b) a ball-and-stick molecular model for xenon tetrafluoride (XeF_4).

Practice Problem **C**ONCEPTUALIZE How many ball-and-stick models of ethanol molecules can be constructed using the collection of balls shown here? How many of each color ball will be left over?

Empirical Formulas

Student Annotation: Note that acetone contains a double bond between the oxygen atom and one of the carbon atoms. Multiple bonds between atoms will be discussed in Chapter 6.

Student Annotation: The formulas of ionic compounds are usually *empirical* formulas.

In addition to the methods we have learned so far, molecular substances can also be represented using *empirical formulas*. The word *empirical* means "from experience" or, in the context of chemical formulas, "from experiment." The empirical formula tells what elements are present in a molecule and in what whole-number ratio they are combined. For example, the molecular formula of hydrogen peroxide is H_2O_2, but its empirical formula is simply HO. Hydrazine, which has been used as a rocket fuel, has the molecular formula N_2H_4, so its empirical formula is NH_2. Although the ratio of nitrogen to hydrogen is 1:2 in both the molecular formula (N_2H_4) and the empirical formula (NH_2), only the molecular formula tells us the actual number of N atoms (two) and H atoms (four) present in a hydrazine molecule.

TABLE 5.4	Molecular and Empirical Formulas			
Compound	Molecular formula	Model	Empirical formula	Molecular model
Water	H_2O		H_2O	
Hydrogen peroxide	H_2O_2		HO	
Ethane	C_2H_6		CH_3	
Propane	C_3H_8		C_3H_8	
Acetylene	C_2H_2		CH	
Benzene	C_6H_6		CH	

In many cases, the empirical and molecular formulas are identical. In the case of water, for example, there is no combination of smaller whole numbers that can convey the ratio of two H atoms for every one O atom, so the empirical formula is the same as the molecular formula: H_2O. Table 5.4 lists the molecular and empirical formulas for several compounds and shows molecular models for each.

Empirical formulas are the *simplest* chemical formulas; they are written by reducing the subscripts in molecular formulas to the smallest possible whole numbers (without altering the relative numbers of atoms). Molecular formulas are the *true* formulas of molecules. As we will see in Chapter 8, when chemists analyze an unknown compound, the first step is usually the determination of the compound's empirical formula. Worked Example 5.6 lets you practice determining empirical formulas from molecular formulas.

Worked Example 5.6

Write the empirical formulas for the following molecules: (a) glucose ($C_6H_{12}O_6$), a substance known as blood sugar; (b) adenine ($C_5H_5N_5$), also known as vitamin B_4; and (c) nitrous oxide (N_2O), a gas that is used as an anesthetic ("laughing gas") and as an aerosol propellant for whipped cream.

(Continued on next page)

Strategy To write the empirical formula, the subscripts in the molecular formula must be reduced to the smallest possible whole numbers (without altering the relative numbers of atoms).

Setup The molecular formulas in parts (a) and (b) each contain subscripts that are divisible by common numbers. Therefore, we will be able to express the formulas with smaller whole numbers than those in the molecular formulas. In part (c), the molecule has only one O atom, so it is impossible to simplify this formula further.

Solution

(a) Dividing each of the subscripts in the molecular formula for glucose by 6, we obtain the empirical formula CH_2O. If we had divided the subscripts by 2 or 3, we would have obtained the formulas $C_3H_6O_3$ and $C_2H_4O_2$, respectively. Although the ratio of carbon to hydrogen to oxygen atoms in each of these formulas is correct (1:2:1), neither is the simplest formula because the subscripts are not in the smallest possible whole-number ratio.

(b) Dividing each subscript in the molecular formula of adenine by 5, we get the empirical formula CHN.

(c) Because the subscripts in the formula for nitrous oxide are already the smallest possible whole numbers, its empirical formula is the same as its molecular formula N_2O.

Think About It

Make sure that the *ratio* in each empirical formula is the same as that in the corresponding molecular formula and that the subscripts are the smallest possible whole numbers. In part (a), for example, the ratio of C:H:O in the molecular formula is 6:12:6, which is equal to 1:2:1, the ratio expressed in the empirical formula.

Practice Problem **A**TTEMPT Write empirical formulas for the following molecules: (a) caffeine ($C_8H_{10}N_4O_2$), a stimulant found in tea and coffee; (b) butane (C_4H_{10}), used in cigarette lighters; and (c) glycine ($C_2H_5NO_2$), an amino acid.

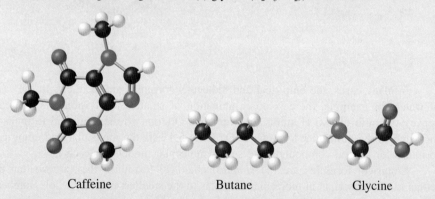

Caffeine   Butane   Glycine

Practice Problem **B**UILD For which of the following molecular formulas is the formula shown in parentheses the correct empirical formula? (a) $C_{12}H_{22}O_{12}$ ($C_6H_{11}O_6$), (b) $C_8H_{12}O_4$ ($C_4H_6O_2$), (c) Na_2O_2 (Na_2O).

Practice Problem **C**ONCEPTUALIZE Which of the following molecules has/have the same empirical formula as acetic acid ($HC_2O_2H_3$)?

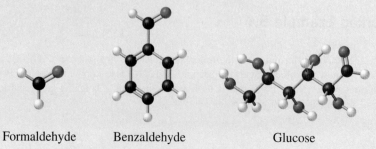

Formaldehyde   Benzaldehyde   Glucose

Section 5.5 Review

Covalent Bonding and Molecules

5.5.1   What is the correct formula for the compound carbon tetrachloride?
(a) C_4Cl
(b) CCl_4
(c) C_4Cl_4
(d) CCl_2
(e) CCl

5.5.2   Give the correct molecular formula and the correct empirical formula for the compound shown.
(a) C_6H_6, C_6H_6
(b) CH, C_6H_6
(c) C_4H_4, CH
(d) C_6H_6, CH
(e) C_6H_6, C_2H_2

5.6   NAMING MOLECULAR COMPOUNDS

The nomenclature of molecular compounds follows in a similar manner to that of ionic compounds. Most molecular compounds are composed of two nonmetals (see [◄◄ Section 2.6, Figure 2.10]). To name such a compound, we first name the element that appears first in the formula. For HCl that would be hydrogen. We then name the second element, changing the ending of its name to –*ide*. For HCl, the second element is chlorine, so we would change chlorine to chloride. Thus, the systematic name of HCl is *hydrogen chloride*. Similarly, HI is hydrogen iodide (iod*ine* ⟶ iod*ide*) and SiC is silicon carbide (carb*on* ⟶ carb*ide*).

Student Annotation: Recall that compounds composed of two elements are called *binary* compounds.

Specifying Numbers of Atoms

It is quite common for one pair of elements to form several different binary molecular compounds. In these cases, confusion in naming the compounds is avoided by the use of Greek prefixes to denote the number of atoms of each element present. Some of the Greek prefixes are listed in Table 5.5, and several compounds named using these prefixes are listed in Table 5.6.

The prefix *mono–* is generally omitted for the first element. SO_2, for example, is named *sulfur dioxide,* not *monosulfur dioxide.* Thus, the absence of a prefix for the first element usually means there is only one atom of that element present in the molecule. In addition, for ease of pronunciation, we usually eliminate the last letter of a prefix that ends in *o* or *a* when naming an oxide. Thus, N_2O_5 is *dinitrogen pentoxide,* rather than *dinitrogen pentaoxide.* Worked Example 5.7 gives you some practice naming binary molecular compounds from their formulas.

Student Hot Spot

Student data indicate you may struggle with naming covalent compounds. Access the SmartBook to view additional Learning Resources on this topic.

TABLE 5.5	Greek Prefixes		
Prefix	**Meaning**	**Prefix**	**Meaning**
Mono–	1	Hexa–	6
Di–	2	Hepta–	7
Tri–	3	Octa–	8
Tetra–	4	Nona–	9
Penta–	5	Deca–	10

TABLE 5.6	Some Compounds Named Using Greek Prefixes		
Compound	**Name**	**Compound**	**Name**
CO	Carbon monoxide	SO_3	Sulfur trioxide
CO_2	Carbon dioxide	NO_2	Nitrogen dioxide
SO_2	Sulfur dioxide	N_2O_5	Dinitrogen pentoxide

Worked Example 5.7

Name the following binary molecular compounds: (a) NF_3, (b) N_2O_4.

Strategy Each compound will be named using the systematic nomenclature including, where necessary, appropriate Greek prefixes.

Setup With binary compounds, we start with the name of the element that appears *first* in the formula, and we change the ending of the *second* element's name to *–ide*. We use prefixes, where appropriate, to indicate the number of atoms of each element. In part (a), the molecule contains one nitrogen atom and three fluorine atoms. We will omit the prefix *mono–* for nitrogen because it is the first element listed in the formula, and we will use the prefix *tri–* to denote the number of fluorine atoms. In part (b), the molecule contains two nitrogen atoms and four oxygen atoms, so we will use the prefixes *di–* and *tetra–* in naming the compound. Recall that in naming an oxide, we omit the last letter of a prefix that ends in *a* or *o*.

Solution (a) nitrogen trifluoride and (b) dinitrogen tetroxide

Think About It
Make sure that the prefixes match the subscripts in the molecular formulas and that the word *oxide* is not preceded immediately by *a* or *o*.

Practice Problem Ⓐ**TTEMPT** Name the following binary molecular compounds:
(a) Cl_2O, (b) $SiCl_4$.

Practice Problem Ⓑ**UILD** Name the following binary molecular compounds:
(a) ClO_2, (b) CBr_4.

Practice Problem Ⓒ**ONCEPTUALIZE** Name the molecular compound shown.

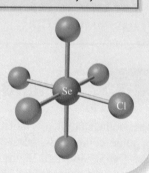

Writing the formula for a molecular compound, given its systematic name, is usually straightforward. For instance, the name *phosphorus pentachloride* indicates the presence of one phosphorus atom (no prefix) and five chlorine atoms (penta–), so the corresponding molecular formula is PCl_5. Note once again that the order of the elements is the same in both the name and the formula. Worked Example 5.8 gives you some practice determining the formulas of binary molecular compounds from their names.

Worked Example 5.8

Write the chemical formulas for the following binary molecular compounds: (a) sulfur tetrafluoride and (b) tetraphosphorus decasulfide.

Strategy The formula for each compound will be deduced using the systematic nomenclature guidelines.

Setup In part (a), there is no prefix for sulfur, so there is only one sulfur atom in a molecule of the compound. Therefore, we will use no prefix for the S in the formula. The prefix *tetra–* means that there are four fluorine atoms. In part (b), the prefixes *tetra–* and *deca–* denote four and ten, respectively.

Solution (a) SF_4 and (b) P_4S_{10}

Think About It
Double-check that the subscripts in the formulas match the prefixes in the compound names: (a) 4 = *tetra*, (b) 4 = *tetra* and 10 = *deca*.

Practice Problem Ⓐ**TTEMPT** Give the molecular formula for each of the following compounds: (a) carbon disulfide and (b) dinitrogen trioxide.

Practice Problem Ⓑ**UILD** Give the molecular formula for each of the following compounds: (a) sulfur trioxide and (b) disulfur decafluoride.

Practice Problem Ⓒ**ONCEPTUALIZE** Draw a molecular model of sulfur trioxide.

There is another group of substances for which we can use Greek prefixes to denote numbers of atoms. It includes such compounds as $SnCl_4$ and $PbCl_4$, which each consist of a metal and a nonmetal—but in which the apparent charge on the metal is unusually high. For example, knowing that Cl forms the chloride ion (Cl^-), we would deduce a charge of +4 on Sn in $SnCl_4$. Given that a great deal of energy would be required to remove four electrons from a tin atom [◄◄ Section 4.4], it seems unlikely that a compound containing a metal with such a high "apparent charge" is actually ionic. In fact, the bonding in $SnCl_4$ is covalent, rather than ionic. Part of the evidence of covalent bonding is that $SnCl_4$ is a liquid at room temperature. (Ionic compounds typically are solids at room temperature.) Although such compounds consist of a metal and a nonmetal but are not truly ionic in nature, we can use either the ionic or the molecular system of nomenclature for binary compounds. Thus, $SnCl_4$ can be named either tin(IV) chloride or tin tetrachloride.

Student Annotation: The concept of "apparent charge" or *oxidation state* [▶▶| Section 9.4] will be discussed in detail in Chapter 9.

Compounds Containing Hydrogen

The names of molecular compounds containing hydrogen do not usually conform to the systematic nomenclature guidelines. Traditionally, many of these compounds are called either by their common, nonsystematic names or by names that do not indicate explicitly the number of H atoms present:

Student Annotation: Binary compounds containing carbon and hydrogen are *organic* compounds and do not follow the same naming conventions as other molecular compounds. Organic compounds and their nomenclature are discussed in detail in Chapter 23.

B_2H_6	Diborane	PH_3	Phosphine
SiH_4	Silane	H_2O	Water
NH_3	Ammonia	H_2S	Hydrogen sulfide

Even the order in which the elements are written in these hydrogen-containing compounds is irregular. In water and hydrogen sulfide, H is written first, whereas it is written last in the other compounds.

Acids make up another important class of molecular compounds that contain hydrogen. One definition of an *acid* is a substance that produces hydrogen ions (H^+) when dissolved in water. Several binary molecular compounds produce hydrogen ions when dissolved in water and are, therefore, acids. In these cases, two different names can be assigned to the same chemical formula. For example, HCl, *hydrogen chloride,* is a gaseous compound. When it is dissolved in water, however, we call it *hydrochloric acid.* The rules for naming binary acids such as these are as follows: remove the *–gen* ending from hydrogen (leaving *hydro–*), change the *–ide* ending on the second element to *–ic*, combine the two words, and add the word *acid.*

Student Annotation: In Chapter 16 we will explore acids and bases in greater detail; and we will see that there are other ways to define the terms *acid* and *base.*

hydrogen (−*gen*) becomes hydro
+
chloride (−*ide,* + *ic*) becomes chloric
+
acid ⟶ hydrochloric acid

Likewise, hydrogen fluoride (HF) becomes *hydrofluoric acid.* Table 5.7 lists these and other examples.

In order for a compound to produce hydrogen ions upon dissolving, it must contain at least one *ionizable hydrogen atom.* An ionizable hydrogen atom is one that separates from the molecule upon dissolving and becomes a hydrogen ion, H^+.

TABLE 5.7	Some Simple Acids	
Formula	**Binary compound name**	**Acid name**
HF	Hydrogen fluoride	Hydrofluoric acid
HCl	Hydrogen chloride	Hydrochloric acid
HBr	Hydrogen bromide	Hydrobromic acid
HI	Hydrogen iodide	Hydroiodic acid
HCN*	Hydrogen cyanide	Hydrocyanic acid

*Although HCN is not a *binary* compound, it is included in this table because it is similar chemically to HF, HCl, HBr, and HI.

Organic Compounds

So far our discussion of nomenclature has focused on *inorganic compounds,* which are generally defined as compounds that do not contain carbon. Another important class of molecular substances is *organic* compounds, which have their own system of nomenclature. *Organic compounds* contain carbon and hydrogen, sometimes in combination with other elements such as oxygen, nitrogen, sulfur, and the halogens. The simplest organic compounds are those that contain only carbon and hydrogen and are known as *hydrocarbons.* Among hydrocarbons, the simplest examples are compounds known as *alkanes.* The name of an alkane depends on the number of carbon atoms in the molecule. Table 5.8 gives the molecular formulas, systematic names of some of the simplest alkanes, and ball-and-stick models.

TABLE 5.8	Formulas, Names, and Models of Some Simple Alkanes	
Formula	**Name**	**Molecular model**
CH_4	Methane	
C_2H_6	Ethane	
C_3H_8	Propane	
C_4H_{10}	Butane	
C_5H_{12}	Pentane	
C_6H_{14}	Hexane	
C_7H_{16}	Heptane	
C_8H_{18}	Octane	
C_9H_{20}	Nonane	
$C_{10}H_{22}$	Decane	

Thinking Outside the Box

Functional Groups

Many organic compounds are derivatives of alkanes in which one of the H atoms has been replaced by a group of atoms known as a **functional group.** The functional group determines many of the chemical properties of a compound because it typically is where a chemical reaction *occurs.* Table 5.9 lists the names and provides ball-and-stick models of several important functional groups.

Ethanol, for example, the alcohol in alcoholic beverages, is ethane (C_2H_6) with one of the hydrogen atoms replaced by an alcohol (—OH) group. Its name is derived from that of *ethane,* indicating that it contains *two* carbon atoms.

Ethanol

The molecular formula of ethanol can also be written C_2H_6O, but C_2H_5OH conveys more information about the structure of the molecule. Organic compounds and several functional groups are discussed in greater detail in Chapter 23.

TABLE 5.9	Organic Functional Groups	
Name	**Functional group**	**Molecular model**
Alcohol	—OH	
Aldehyde	—CHO	
Carboxylic acid	—COOH	
Amine	—NH₂	

Section 5.6 Review

Naming Molecular Compounds

5.6.1 What is the correct systematic name of SF_6?
(a) Sulfur fluoride
(b) Sulfur hexafluoride
(c) Sulfur hexafluorine
(d) Sulfur fluorine
(e) Sulfur hexafluorinide

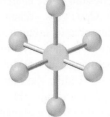

5.6.2 What is the correct systematic name of P_2I_4?
(a) Phosphorus iodide
(b) Phosphorus tetraiodide
(c) Diphosphorus iodide
(d) Diphosphorus tetraiodide
(e) Diphosphorus tetraiodinide

(Continued on next page)

5.6.3   What is the correct systematic name of BrO_2?
 (a) Bromine oxide   (d) Bromine oxygenide
 (b) Bromine dioxygen   (e) Bromine dioxygenide
 (c) Bromine dioxide

5.6.4   What is the name of the compound shown?
 (a) Methane monoxide
 (b) Methanol
 (c) Carbon tetrahydrogen monoxide
 (d) Methane
 (e) Tetrahydrogen carbon monoxide

5.7   COVALENT BONDING IN IONIC SPECIES

So far, the compounds we have encountered have been either *ionic,* held together by ionic bonding, or *molecular,* held together by covalent bonding; and the ions we have encountered have all been *monatomic* ions. However, many common ionic substances contain *poly*atomic ions, which are held together by covalent bonding. Moreover, some ionic substances, known as *hydrates,* actually contain water molecules within their formulas. In this section, we examine the chemical species that are held together by a combination of ionic and covalent bonding.

Polyatomic Ions

Ions that consist of a combination of two or more atoms are called **polyatomic ions.** Because these ions are commonly encountered in general chemistry, you must know the names, formulas, and charges of the polyatomic ions listed in Table 5.10. Although most of the common polyatomic ions are anions, a few are cations. For compounds containing polyatomic ions, formulas are determined following the same rule as for ionic compounds containing only monatomic ions: ions must be combined in a ratio that gives a neutral formula overall. The following examples illustrate how this is done.

 Ammonium Chloride The cation is NH_4^+ and the anion is Cl^-. The sum of the charges is $1 + (-1) = 0$, so the ions combine in a 1:1 ratio and the resulting formula is NH_4Cl.

 Calcium Phosphate The cation is Ca^{2+} and the anion is PO_4^{3-}. The following diagram can be used to determine the subscripts:

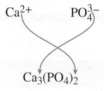

$$Ca^{2+} \qquad PO_4^{3-}$$

$$Ca_3(PO_4)_2$$

The sum of the charges is $3(+2) + 2(-3) = 0$. Thus, the formula for calcium phosphate is $Ca_3(PO_4)_2$. When we add a subscript to a polyatomic ion, we must first put parentheses around the ion's formula to indicate that the subscript applies to *all* the atoms in the polyatomic ion. Other examples are sodium cyanide ($NaCN$), potassium permanganate ($KMnO_4$), and ammonium sulfate [$(NH_4)_2SO_4$].

 Unlike the naming of *molecular* compounds, the naming of ionic compounds does not require the use of Greek prefixes. For example, Li_2CO_3 is lithium carbonate, not dilithium carbonate, even though there are two lithium ions for every one carbonate ion. Prefixes are unnecessary because each of the ions has a specific, *known* charge. Lithium ion always has a charge of $+1$, and carbonate ion always has a charge of -2. The only ratio in which they *can* combine to form a neutral compound is two Li^+ ions for every one CO_3^{2-} ion. Therefore, the name *lithium carbonate* is sufficient to convey the compound's empirical formula.

TABLE 5.10	Common Polyatomic Ions

Name	Formula/Charge
Cations	
ammonium	NH_4^+
hydronium	H_3O^+
mercury(I)	Hg_2^{2+}
Anions	
acetate	$C_2H_3O_2^-$
azide	NH_3^-
carbonate	CO_3^{2-}
chlorate	ClO_3^-
chlorite	ClO_2^-
chromate	CrO_4^{2-}
cyanide	CN^-
dichromate	$Cr_2O_7^{2-}$
dihydrogen phosphate	$H_2PO_4^-$
hydrogen carbonate or bicarbonate	HCO_3^-
hydrogen phosphate	HPO_4^{2-}
hydrogen sulfate or bisulfate	HSO_4^-
hydroxide	OH^-
hypochlorite	ClO^-
nitrate	NO_3^-
nitrite	NO_2^-
oxalate	$C_2O_4^{2-}$
perchlorate	ClO_4^-
permanganate	MnO_4^-
peroxide	O_2^{2-}
phosphate	PO_4^{3-}
phosphite	PO_3^{3-}
sulfate	SO_4^{2-}
sulfite	SO_3^{2-}
thiocyanate	SCN^-

Student Annotation: Some oxoanions occur in series of ions that contain the same central atom and have the same charge, but contain different numbers of oxygen atoms.

perchlorate	ClO_4^-
chlorate	ClO_3^-
chlorite	ClO_2^-
hypochlorite	ClO^-
nitrate	NO_3^-
nitrite	NO_2^-
phosphate	PO_4^{3-}
phosphite	PO_3^{3-}
sulfate	SO_4^{2-}
sulfite	SO_3^{2-}

Worked Example 5.9 lets you practice naming compounds that contain polyatomic ions.

Worked Example 5.9

Name the following ionic compounds: (a) $Fe_2(SO_4)_3$, (b) $Al(OH)_3$, (c) Hg_2O.

Strategy Begin by identifying the cation and the anion in each compound, and then combine the names for each, eliminating the word *ion*.

(Continued on next page)

Setup $Fe_2(SO_4)_3$ contains Fe^{3+} and SO_4^{2-}, the iron(III) ion, and the sulfate ion. We know that the iron in $Fe_2(SO_4)_3$ is iron(III), Fe^{3+}, because it is combined with the sulfate ion in a 2:3 ratio; $Al(OH)_3$ contains Al^{3+} and OH^-, the aluminum ion and the hydroxide ion; and Hg_2O contains Hg_2^{2+} and O^{2-}, the mercury(I) ion and the oxide ion.

Solution (a) Combining the cation and anion names, and eliminating the word *ion* from each of the individual ions' names, we get *iron(III) sulfate* as the name of $Fe_2(SO_4)_3$; (b) $Al(OH)_3$ is *aluminum hydroxide;* and (c) Hg_2O is *mercury(I) oxide.*

> **Think About It**
> Be careful not to confuse the subscript in a formula with the charge on the metal ion. In part (a), for example, the subscript on Fe is 2, but this is an iron(III) compound.

Practice Problem Ⓐ**TTEMPT** Name the following ionic compounds: (a) Na_2SO_4, (b) $Cu(NO_3)_2$, (c) $Fe_2(CO_3)_3$.

Practice Problem Ⓑ**UILD** Write formulas for the following ionic compounds: (a) potassium dichromate, (b) lithium oxalate, (c) copper(II) nitrate.

Practice Problem Ⓒ**ONCEPTUALIZE** The diagram represents a small sample of an ionic compound where red spheres represent nitrate ions and grey spheres represent iron ions. Deduce the correct formula and name of the compound.

Oxoanions are polyatomic anions that contain one or more oxygen atoms and one atom (the "central atom") of another element. Examples include the chlorate (ClO_3^-), nitrate (NO_3^-), and sulfate (SO_4^{2-}) ions. Often, oxoanions occur in series of two or more ions that have the same central atom but different numbers of O atoms (e.g., NO_3^- and NO_2^-). Starting with the oxoanions whose names end in *–ate* (see Table 5.10), we can name these ions as follows:

1. The ion with one *more* O atom than the *–ate* ion is called the *per …ate* ion. Thus, ClO_3^- is the chlorate ion, so ClO_4^- is the *perchlorate ion.*
2. The ion with one *less* O atom than the *–ate* anion is called the *–ite* ion. Thus, ClO_2^- is the *chlorite ion.*
3. The ion with *two* fewer O atoms than the *–ate* ion is called the *hypo …ite* ion. Thus, ClO^- is the *hypochlorite ion.*

At a minimum, you must commit to memory the formulas and charges of the oxoanions whose names end in *–ate* so that you can apply these guidelines when necessary.

Oxoacids

In addition to the binary acids discussed in Section 5.6, there is another important class of acids known as *oxoacids* that, when dissolved in water, produce hydrogen ions and the corresponding oxoanions. The formula of an oxoacid can be determined by adding enough H^+ ions to the corresponding oxoanion to yield a formula with no net charge. For example, the formulas of oxoacids based on the nitrate (NO_3^-) and sulfate (SO_4^{2-}) ions are HNO_3 and H_2SO_4, respectively. The names of oxoacids are derived from the names of the corresponding oxoanions using the following guidelines:

1. An acid based on an *–ate* ion is called *…ic* acid; thus, $HClO_3$ is called *chloric acid.*
2. An acid based on an *–ite* ion is called *…ous* acid; thus, $HClO_2$ is called *chlorous acid.*
3. Prefixes in oxoanion names are retained in the names of the corresponding oxoacids; thus, $HClO_4$ and $HClO$ are called *perchloric acid* and *hypochlorous acid,* respectively.

Most of the acids we have encountered thus far have been ***monoprotic***—meaning they each have just *one* ionizable hydrogen atom. Many oxoacids, such as H_2SO_4 and H_3PO_4, are ***polyprotic***—meaning they have *more* than one ionizable hydrogen atom. In these cases, the names of anions in which one or more (but not all) of the hydrogen ions have been removed must indicate the number of H ions that remain, as shown for the anions derived from phosphoric acid:

$$H_3PO_4 \qquad \text{Phosphoric acid}$$
$$H_2PO_4^- \qquad \text{Dihydrogen phosphate ion}$$
$$HPO_4^{2-} \qquad \text{Hydrogen phosphate ion}$$
$$PO_4^{3-} \qquad \text{Phosphate ion}$$

Worked Examples 5.10 and 5.11 let you practice naming oxoacids and oxoanions.

Worked Example 5.10

Name the following species: (a) BrO_4^-, (b) HCO_3^-, (c) H_2CO_3.

Strategy Each species is either an oxoanion or an oxoacid. Identify the "reference oxoanion" (the one with the *–ate* ending) for each, and apply the rules to determine appropriate names.

Setup (a) Chlorine, bromine, and iodine (members of Group 7A) all form analogous series of oxoanions with one to four oxygen atoms. Thus, the reference oxoanion is bromate (BrO_3^-), which is analogous to chlorate (ClO_3^-). In parts (b) and (c), HCO_3^- and H_2CO_3 have one and two more hydrogens, respectively, than the carbonate ion (CO_3^{2-}).

Solution (a) BrO_4^- has one more O atom than the bromate ion (BrO_3^-), so BrO_4^- is the *perbromate* ion. (b) CO_3^{2-} is the carbonate ion. Because HCO_3^- has one ionizable hydrogen atom, it is called the *hydrogen carbonate ion*. (c) With two ionizable hydrogen atoms and no charge on the compound, H_2CO_3 is *carbonic acid*.

Think About It

Remembering all these names and formulas is greatly facilitated by memorizing the common ions that end in *–ate:*

chlorate	ClO_3^-		nitrate	NO_3^-
iodate	IO_3^-		carbonate	CO_3^{2-}
bromate	BrO_3^-		oxalate	$C_2O_4^{2-}$
sulfate	SO_4^{2-}		chromate	CrO_4^{2-}
phosphate	PO_4^{3-}		permanganate	MnO_4^-

Practice Problem ATTEMPT Name the following species: (a) HBrO, (b) HSO_4^-, (c) $H_2C_2O_4$.

Practice Problem BUILD Name the following species: (a) HIO_3, (b) $HCrO_4^-$, (c) $HC_2O_4^-$.

Practice Problem CONCEPTUALIZE The diagrams show models of a series of oxoanions. Which of the models represents an anion whose name ends in *–ate?*

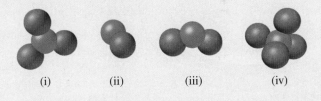

(i)   (ii)   (iii)   (iv)

Worked Example 5.11

Determine the formula of sulfurous acid.

Strategy The *–ous* ending in the name of an acid indicates that the acid is derived from an oxoanion ending in *–ite*. Determine the formula and charge of the oxoanion, and add enough hydrogens to make a neutral formula.

Setup The sulfite ion is SO_3^{2-}.

Solution The formula of sulfurous acid is H_2SO_3.

Think About It
Remembering all these names and formulas is greatly facilitated by memorizing the common ions that end in *–ate*. (See Think About It box in Worked Example 5.10.)

Practice Problem **ATTEMPT** Determine the formula of perbromic acid. (Refer to the information in Worked Example 5.10.)

Practice Problem **BUILD** Determine the formula of chromic acid.

Practice Problem **CONCEPTUALIZE** Referring to the diagrams in Practice Problem 5.10C, which of the ions shown would be part of an acid whose name begins with a prefix?

Hydrates

A *hydrate* is a compound that has a specific number of water molecules within its solid structure. In its normal state, for example, each unit of copper(II) sulfate has five water molecules associated with it. The systematic name for this compound is copper(II) sulfate pentahydrate, and its formula is written as $CuSO_4 \cdot 5H_2O$. The water molecules can be driven off by heating. When this occurs, the resulting compound is $CuSO_4$, which is sometimes called anhydrous copper(II) sulfate; *anhydrous* means that the compound no longer has water molecules associated with it. Hydrates and the corresponding anhydrous compounds often have distinctly different physical and chemical properties (Figure 5.7).

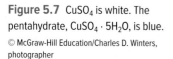
Figure 5.7 $CuSO_4$ is white. The pentahydrate, $CuSO_4 \cdot 5H_2O$, is blue.
© McGraw-Hill Education/Charles D. Winters, photographer

Formula	Common name	Systematic name
H_2O	Water	Dihydrogen monoxide
NH_3	Ammonia	Trihydrogen nitride
CO_2	Dry ice	Solid carbon dioxide
NaCl	Salt	Sodium chloride
N_2O	Nitrous oxide, laughing gas	Dinitrogen monoxide
$CaCO_3$	Marble, chalk, limestone	Calcium carbonate
$NaHCO_3$	Baking soda	Sodium hydrogen carbonate
$MgSO_4 \cdot 7H_2O$	Epsom salt	Magnesium sulfate heptahydrate
$Mg(OH)_2$	Milk of magnesia	Magnesium hydroxide

TABLE 5.11 Common and Systematic Names of Some Familiar Inorganic Compounds

Some other hydrates are

$BaCl_2 \cdot 2H_2O$	Barium chloride dihydrate
$LiCl \cdot H_2O$	Lithium chloride monohydrate
$MgSO_4 \cdot 7H_2O$	Magnesium sulfate heptahydrate
$Sr(NO_3)_2 \cdot 4H_2O$	Strontium nitrate tetrahydrate

Familiar Inorganic Compounds

Some compounds are better known by their common names than by their systematic chemical names. Familiar examples are listed in Table 5.11.

Section 5.7 Review

Covalent Bonding in Ionic Species

5.7.1 What is the correct formula for nitrous acid?
(a) HNO
(b) HNO_2
(c) HNO_3
(d) H_2NO_2
(e) H_2NO_3

5.7.2 What is the formula of nickel(II) nitrate hexahydrate?
(a) $NiNO_3 \cdot 6H_2O$
(b) $Ni_2NO_3 \cdot 6H_2O$
(c) $Ni(NO_3)_2 \cdot 6H_2O$
(d) $NiNO_3 \cdot 12H_2O$
(e) $Ni(NO_3)_2 \cdot 12H_2O$

5.7.3 What is the correct name of the compound Hg_2CrO_4?
(a) Mercury(I) chromate
(b) Mercury(II) chromate
(c) Mercury dichromate
(d) Dimercury chromate
(e) Monomercury chromate

5.7.4 What is the formula of the compound iron(III) carbonate?
(a) $FeCO_3$
(b) Fe_3CO_3
(c) Fe_2CO_3
(d) $Fe_2(CO_3)_3$
(e) $Fe_3(CO_3)_2$

5.8 **MOLECULAR AND FORMULA MASSES**

Using a compound's molecular formula and atomic masses from the periodic table, we can determine the mass (in atomic mass units [◄◄ Section 1.2]) of an individual molecule—a quantity known as the *molecular mass.* The molecular mass is simply the sum of the atomic masses of the atoms that make up the molecule. We multiply the atomic mass of each element by the number of atoms of that element in the molecule and then sum the masses for each element present. For example,

$$\text{molecular mass of } H_2O = 2(\text{atomic mass of H}) + \text{atomic mass of O}$$
$$= 2(1.008 \text{ amu}) + 16.00 \text{ amu} = 18.02 \text{ amu}$$

Because the atomic masses on the periodic table are *average* atomic masses, the result of such a determination is an *average* molecular mass, sometimes referred to as the *molecular weight.*

Although an ionic compound does not have a *molecular* mass, we can use its empirical formula to determine its *formula mass* (the mass of a "formula unit"), sometimes called the *formula weight.* Worked Example 5.12 illustrates how to determine molecular mass and formula mass.

Worked Example 5.12

Calculate the molecular mass or the formula mass, as appropriate, for each of the following compounds: (a) propane, C_3H_8; (b) lithium hydroxide, LiOH; (c) barium acetate, $Ba(C_2H_3O_2)_2$.

Strategy Determine the molecular mass (for each molecular compound) or formula mass (for each ionic compound) by summing all the atomic masses.

Setup Using the formula for each compound, determine the number of atoms of each element present. A molecule of propane contains three C atoms and eight H atoms. The compounds in parts (b) and (c) are ionic and will therefore have formula masses rather than molecular masses. A formula unit of lithium hydroxide contains one Li atom, one O atom, and one H atom. A formula unit of barium acetate contains one Ba atom, four C atoms, six H atoms, and four O atoms. (Remember that the subscript after the parentheses means that there are two acetate ions, each of which contains two C atoms, three H atoms, and two O atoms.)

Solution For each compound, multiply the number of atoms by the atomic mass of each element and then sum the calculated values.

(a) The molecular mass of propane is $3(12.01 \text{ amu}) + 8(1.008 \text{ amu}) = 44.09 \text{ amu}$.

(b) The formula mass of lithium hydroxide is $6.941 \text{ amu} + 16.00 \text{ amu} + 1.008 \text{ amu} = 23.95 \text{ amu}$.

(c) The formula mass of barium acetate is $137.3 \text{ amu} + 4(12.01 \text{ amu}) + 6(1.008 \text{ amu}) + 4(16.00 \text{ amu}) = 255.4 \text{ amu}$.

> **Think About It**
> Double-check that you have counted the number of atoms correctly for each compound and that you have used the proper atomic masses from the periodic table.

Practice Problem **A**TTEMPT Calculate the molecular or formula mass of each of the following compounds: (a) magnesium chloride ($MgCl_2$), (b) sulfuric acid (H_2SO_4), and (c) oxalic acid ($H_2C_2O_4$).

Practice Problem **B**UILD Calculate the molecular or formula mass of each of the following compounds: (a) calcium carbonate ($CaCO_3$), (b) nitrous acid (HNO_2), and (c) ammonium sulfide [$(NH_4)_2S$].

Practice Problem **C** **ONCEPTUALIZE** Determine the molecular masses of caffeine and ibuprofen from the molecular models shown here.

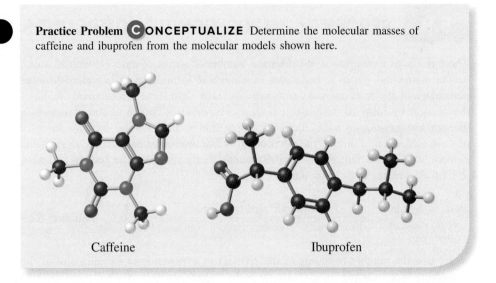

Caffeine   Ibuprofen

Section 5.8 Review

Molecular and Formula Masses

5.8.1   Determine the molecular mass of citric acid ($H_3C_6H_5O_7$).

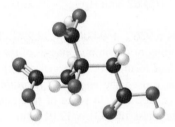

Citric acid

(a) 192.12 amu   (d) 29.02 amu
(b) 189.10 amu   (e) 89.07 amu
(c) 132.07 amu

5.8.2   Determine the formula mass of calcium citrate [$Ca_3(C_6H_5O_7)_2$].
(a) 309.34 amu   (d) 498.44 amu
(b) 69.10 amu   (e) 418.28 amu
(c) 229.18 amu

5.8.3   Determine the molecular mass of perchloric acid.
(a) 52.46 amu   (d) 100.46 amu
(b) 68.46 amu   (e) 116.46 amu
(c) 84.46 amu

5.8.4   Determine the formula mass of iron(III) chromate.
(a) 171.84 amu   (d) 403.83 amu
(b) 459.67 amu   (e) 759.65 amu
(c) 399.52 amu

5.8.5   Determine the formula mass of manganese(II) chloride tetrahydrate.
(a) 125.84 amu   (d) 197.90 amu
(b) 179.89 amu   (e) 215.92 amu
(c) 143.86 amu

5.9   PERCENT COMPOSITION OF COMPOUNDS

The formula of a compound indicates the number of atoms of each element in a unit of the compound. From a molecular or empirical formula, we can calculate what percentage of the total mass is contributed by each element in a compound. A list of the percent by mass of each element in a compound is known as the compound's ***percent composition by mass.*** Percent composition is calculated by dividing the mass of each element in a unit of the compound by the molecular or formula mass of the compound and then multiplying by 100 percent. Mathematically, the percent composition of an element in a compound is expressed as

Student Annotation: Often the terms *percent composition by mass* and *percent composition* are used interchangeably. In this book, unless otherwise specified, *percent composition* means percent composition *by mass*.

$$\begin{matrix} \text{percent by mass} \\ \text{of an element} \end{matrix} = \frac{n \times \text{atomic mass of element}}{\text{molecular or formula mass of compound}} \times 100\% \quad \textbf{Equation 5.1}$$

where n is the number of atoms of the element in a molecule or formula unit of the compound. For example, in a molecule of hydrogen peroxide (H_2O_2), there are two H atoms and two O atoms. The atomic masses of H and O are 1.008 and 16.00 amu, respectively, so the molecular mass of H_2O_2 is 34.02 amu. Therefore, the percent composition of H_2O_2 is calculated as follows:

$$\%H = \frac{2 \times 1.008 \text{ amu H}}{34.02 \text{ amu H}_2O_2} \times 100\% = 5.926\%$$

$$\%O = \frac{2 \times 16.00 \text{ amu O}}{34.02 \text{ amu H}_2O_2} \times 100\% = 94.06\%$$

The sum of percentages is 5.926% + 94.06% = 99.99%. The small discrepancy from 100 percent is due to rounding of the atomic masses of the elements. We could equally well have used the *empirical* formula of hydrogen peroxide (HO) for the calculation. In this case, we would have used the ***empirical formula mass,*** the mass in atomic mass units of one empirical formula, in place of the molecular mass. The empirical formula mass of H_2O_2 (the mass of HO) is 17.01 amu.

$$\%H = \frac{1.008 \text{ amu H}}{17.01 \text{ amu H}_2O_2} \times 100\% = 5.926\%$$

$$\%O = \frac{16.00 \text{ amu O}}{17.01 \text{ amu H}_2O_2} \times 100\% = 94.06\%$$

Student Hot Spot

Student data indicate you may struggle with percent composition by mass. Access the SmartBook to view additional Learning Resources on this topic.

Because both the molecular formula and the empirical formula tell us the composition of the compound, they both give the same percent composition by mass. Worked Example 5.13 shows how to calculate percent composition by mass.

Worked Example 5.13

Lithium carbonate, Li_2CO_3, was the first "mood-stabilizing" drug approved by the FDA for the treatment of mania and manic-depressive illness, also known as bipolar disorder. Calculate the percent composition by mass of lithium carbonate.

Strategy   Use Equation 5.1 to determine the percent by mass contributed by each element in the compound.

Setup   Lithium carbonate is an ionic compound that contains Li, C, and O. In a formula unit, there are two Li atoms, one C atom, and three O atoms with atomic masses 6.941, 12.01, and 16.00 amu, respectively. The formula mass of Li_2CO_3 is 2(6.941 amu) + 12.01 amu + 3(16.00 amu) = 73.89 amu.

Solution For each element, multiply the number of atoms by the atomic mass, divide by the formula mass, and multiply by 100 percent.

$$\%Li = \frac{2 \times 6.941 \text{ amu Li}}{73.89 \text{ amu Li}_2\text{CO}_3} \times 100\% = 18.79\%$$

$$\%C = \frac{12.01 \text{ amu C}}{73.89 \text{ amu Li}_2\text{CO}_3} \times 100\% = 16.25\%$$

$$\%O = \frac{3 \times 16.00 \text{ amu O}}{73.89 \text{ amu Li}_2\text{CO}_3} \times 100\% = 64.96\%$$

Think About It

Make sure that the percent composition results for a compound sum to approximately 100. (In this case, the results sum to exactly 100 percent—18.79% + 16.25% + 64.96% = 100.00%—but remember that because of rounding, the percentages may sum to very slightly more or very slightly less.)

Practice Problem **A**TTEMPT Determine the percent composition by mass of the artificial sweetener aspartame ($C_{14}H_{18}N_2O_5$).

Practice Problem **B**UILD Determine the simplest molecular formula for a compound that is 62.04 percent carbon, 10.41 percent hydrogen, and 27.55 percent oxygen by mass if the molecular mass of the compound is 58.08 amu.

Practice Problem **C**ONCEPTUALIZE Determine the percent composition by mass of acetaminophen, the active ingredient in over-the-counter pain relievers such as Tylenol.

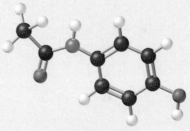

Acetaminophen

Section 5.9 Review

Percent Composition of Compounds

5.9.1 What is the percent composition by mass of aspirin ($C_9H_8O_4$)?
(a) 44.26% C, 3.28% H, 52.46% O
(b) 60.00% C, 4.47% H, 35.53% O
(c) 41.39% C, 3.47% H, 55.14% O
(d) 42.86% C, 6.35% H, 50.79% O
(e) 42.86% C, 38.09% H, 19.05% O

Aspirin

5.9.2 What is the percent composition by mass of sodium bicarbonate ($NaHCO_3$)?
(a) 20.89% Na, 2.75% H, 32.74% C, 43.62% O
(b) 44.20% Na, 1.94% H, 23.09% C, 30.76% O
(c) 21.28% Na, 0.93% H, 33.35% C, 44.43% O
(d) 24.20% Na, 2.94% H, 22.74% C, 50.12% O
(e) 27.37% Na, 1.20% H, 14.30% C, 57.14% O

5.10   MOLAR MASS

Although chemists often wish to combine substances in specific mole ratios, there is no direct way to measure the number of moles in a sample of matter. Instead, chemists determine how many moles there are of a substance by measuring its mass (usually in grams). The molar mass of the substance is then used to convert from grams to moles.

Student Annotation: Because the number of moles specifies the number of particles (atoms, molecules, or ions), using molar mass as a conversion factor, in effect, allows us to count the particles in a sample of matter by weighing the sample.

The ***molar mass*** (*$\mathcal{M}$*) of a substance is the mass in grams of one mole of the substance. By definition, the mass of a mole of carbon-12 is exactly 12 g. Note that the molar mass of carbon is numerically equal to its atomic mass. Likewise, the atomic mass of calcium is 40.08 amu and its molar mass is 40.08 g, the atomic mass of sodium is 22.99 amu and its molar mass is 22.99 g, and so on. In general, an element's molar mass in grams is numerically equal to its atomic mass in atomic mass units. The molar mass (in grams) of any compound is numerically equal to its molecular or formula mass (in amu). The molar mass of water, for example, is 18.02 g, and the molar mass of sodium chloride (NaCl) is 58.44 g.

Recall from [◄◄ Section 1.2] that

$$1 \text{ amu} = 1.661 \times 10^{-24} \text{ g}$$

This is the reciprocal of Avogadro's number. Expressed another way:

$$1 \text{ g} = 6.022 \times 10^{23} \text{ amu}$$

In effect, there is one mole of atomic mass units in a gram.

When it comes to expressing the molar mass of elements such as oxygen and hydrogen, we have to be careful to specify what form of the element we mean. For instance, the element oxygen exists predominantly as diatomic molecules (O_2). Thus, if we say one mole of oxygen and by *oxygen* we mean O_2, the molecular mass is 32.00 amu and the molar mass is 32.00 g. If, on the other hand, we mean a mole of atomic oxygen (O), then the molar mass is only 16.00 g, which is numerically equal to the atomic mass of O (16.00 amu). You should be able to tell from the context which form of an element is intended, as the following examples illustrate:

Context	*Oxygen* means	Molar mass
How many moles of oxygen react with 2 moles of hydrogen to produce water?	O_2	32.00 g
How many moles of oxygen are there in 1 mole of water?	O	16.00 g
Air is approximately 21 percent oxygen.	O_2	32.00 g
Many organic compounds contain oxygen.	O	16.00 g

Although the term *molar mass* specifies the mass of one mole of a substance, making the appropriate units simply grams (g), we usually express molar masses in units of grams per mole (g/mol) to facilitate calculations involving moles.

Interconverting Mass, Moles, and Numbers of Particles

Molar mass is the conversion factor that we use to convert from mass (*m*) to moles (*n*), and vice versa. We use Avogadro's number N_A to convert from number of moles to number of particles (*N*), and vice versa. *Particles* in this context may refer to atoms, molecules, ions, or formula units. Figure 5.8 summarizes the operations involved in these conversions.

Figure 5.8 Flowchart for conversions among mass, moles, and number of particles.

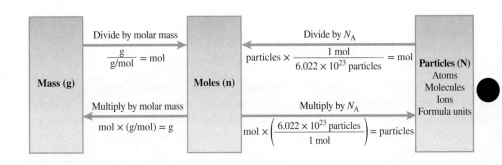

Worked Examples 5.14 and 5.15 illustrate how the conversions are done.

Worked Example 5.14

Determine (a) the number of moles of CO_2 in 10.00 g of carbon dioxide and (b) the mass of 0.905 mole of sodium chloride.

Strategy Use molar mass to convert from mass to moles and to convert from moles to mass.

Setup The molar mass of carbon dioxide is 44.01 g/mol. The molar mass of a compound is numerically equal to its formula mass. The molar mass of sodium chloride (NaCl) is 58.44 g/mol.

Solution

(a) $10.00 \text{ g } CO_2 \times \dfrac{1 \text{ mol } CO_2}{44.01 \text{ g } CO_2} = 0.2272 \text{ mol } CO_2$

(b) $0.905 \text{ mol } NaCl \times \dfrac{58.44 \text{ g NaCl}}{1 \text{ mol } NaCl} = 52.9 \text{ g NaCl}$

> **Think About It**
> Always double-check unit cancellations in problems such as these—errors are common when molar mass is used as a conversion factor. Also make sure that the results make sense. In both cases, a mass smaller than the molar mass corresponds to less than a mole of substance.

Practice Problem **A**TTEMPT Determine the mass in grams of 2.75 moles of glucose ($C_6H_{12}O_6$).

Practice Problem **B**UILD Determine the number of moles in 59.8 g of sodium nitrate ($NaNO_3$).

Practice Problem **C**ONCEPTUALIZE Consider 50-g samples of three different compounds: A ($\mathcal{M} = 43$ g/mol), B ($\mathcal{M} = 75$ g/mol), and C ($\mathcal{M} = 58$ g/mol). Which sample contains the largest number of moles? Which sample contains the smallest number of moles?

Worked Example 5.15

(a) Determine the number of water molecules and the numbers of H and O atoms in 3.26 g of water.

(b) Determine the mass of 7.92×10^{19} carbon dioxide molecules.

Strategy Use molar mass and Avogadro's number to convert from mass to molecules, and vice versa. Use the molecular formula of water to determine the numbers of H and O atoms.

Setup (a) Starting with mass (3.26 g of water), we use molar mass (18.02 g/mol) to convert to moles of water. From moles, we use Avogadro's number to convert to number of water molecules. In part (b), we reverse the process in part (a) to go from number of molecules to mass of carbon dioxide.

Solution

(a) $3.26 \text{ g } H_2O \times \dfrac{1 \text{ mol } H_2O}{18.02 \text{ g } H_2O} \times \dfrac{6.022 \times 10^{23} \text{ } H_2O \text{ molecules}}{1 \text{ mol } H_2O} = 1.09 \times 10^{23} \text{ } H_2O \text{ molecules}$

Using the molecular formula, we can determine the number of H and O atoms in 3.26 g of H_2O as follows:

$$1.09 \times 10^{23} \text{ } H_2O \text{ molecules} \times \dfrac{2 \text{ H atoms}}{1 \text{ } H_2O \text{ molecule}} = 2.18 \times 10^{23} \text{ H atoms}$$

$$1.09 \times 10^{23} \text{ } H_2O \text{ molecules} \times \dfrac{1 \text{ O atom}}{1 \text{ } H_2O \text{ molecule}} = 1.09 \times 10^{23} \text{ O atoms}$$

(b) $7.92 \times 10^{19} \text{ } CO_2 \text{ molecules} \times \dfrac{1 \text{ mol } CO_2}{6.022 \times 10^{23} \text{ } CO_2 \text{ molecules}} \times \dfrac{44.01 \text{ g } CO_2}{1 \text{ mol } CO_2} = 5.79 \times 10^{-3} \text{ g } CO_2$

(Continued on next page)

Think About It
Again, check the cancellation of units carefully and make sure that the magnitudes of your results are reasonable.

Practice Problem **A**TTEMPT Calculate the number of oxygen molecules in 35.5 g of O_2.

Practice Problem **B**UILD Calculate the mass of 12.3 moles of SO_3 molecules.

Practice Problem **C**ONCEPTUALIZE Determine the number of C, H, and O atoms in 3.82 moles of vanillin, the primary component of the extract of vanilla beans.

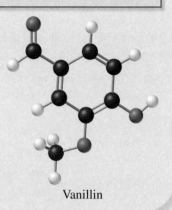

Vanillin

Determination of Empirical Formula and Molecular Formula from Percent Composition

Student Annotation: Comparison of calculated and experimentally determined percent composition can be used to verify the purity of a compound. We will explore one of the experimental methods used to determine percent composition in Chapter 8.

In Section 5.9, we learned how to use the chemical formula (either molecular or empirical) to determine the percent composition by mass. With the concepts of the mole and molar mass, we can now use percent composition that has been determined experimentally to determine the empirical formula and/or the molecular formula of a compound.

The empirical formula gives only the ratio of combination of the atoms in a molecule, so there may be numerous compounds with the same empirical formula. If we know the approximate molar mass of the compound, though, we can determine the molecular formula from the empirical formula. For instance, the molar mass of glucose is about 180 g. The *empirical formula mass* of CH_2O is about 30 g [12.01 g + 2(1.008 g) + 16.00 g]. To determine the molecular formula, we first divide the molar mass by the empirical formula mass: 180 g/30 g = 6. This tells us that there are six empirical-formula units per molecule in glucose. Multiplying each subscript by 6 (recall that when none is shown, the subscript is understood to be a 1) gives the molecular formula, $C_6H_{12}O_6$.

Student Hot Spot

Student data indicate you may struggle with determining empirical formulas. Access the SmartBook to view additional Learning Resources on this topic.

Worked Example 5.16 shows how to determine the empirical formula and molecular formula of a compound from its percent composition and approximate molar mass.

Worked Example 5.16

Determine the empirical formula of a compound that is 30.45 percent nitrogen and 69.55 percent oxygen by mass. Given that the molar mass of the compound is approximately 92 g/mol, determine the molecular formula of the compound.

Strategy Assume a 100-g sample so that the mass percentages of nitrogen and oxygen given in the problem statement correspond to the masses of N and O in the compound. Then, using the appropriate molar masses, convert the grams of each element to moles. Use the resulting numbers as subscripts in the empirical formula, reducing them to the lowest possible whole numbers for the final answer. To calculate the molecular formula, first divide the molar mass given in the problem statement by the empirical formula mass. Then, multiply the subscripts in the empirical formula by the resulting number to obtain the subscripts in the molecular formula.

Setup The empirical formula of a compound consisting of N and O is N_xO_y. The molar masses of N and O are 14.01 and 16.00 g/mol, respectively. One hundred grams of a compound that is 30.45 percent nitrogen and 69.55 percent oxygen by mass contains 30.45 g N and 69.55 g O.

Solution

$$30.45 \text{ g N} \times \frac{1 \text{ mol N}}{14.01 \text{ g N}} = 2.173 \text{ mol N}$$

$$69.55 \text{ g O} \times \frac{1 \text{ mol O}}{16.00 \text{ g O}} = 4.347 \text{ mol O}$$

This gives a formula of $N_{2.173}O_{4.347}$. Dividing both subscripts by the smaller of the two to get the smallest possible whole numbers ($2.173/2.173 = 1$; $4.347/2.173 \approx 2$) gives an empirical formula of NO_2.

Finally, dividing the approximate molar mass (92 g/mol) by the empirical formula mass [14.01 g/mol + 2(16.00 g/mol) = 46.01 g/mol] gives $92/46.01 \approx 2$. Then, multiplying both subscripts in the empirical formula by 2 gives the molecular formula, N_2O_4.

> **Think About It**
> Use the method described in Worked Example 5.13 to calculate the percent composition of the molecular formula N_2O_4 and verify that it is the same as that given in this problem.

Practice Problem ATTEMPT Determine the empirical formula of a compound that is 53.3 percent C, 11.2 percent H, and 35.5 percent O by mass. What is the molecular formula if the molar mass is approximately 90 g/mol?

Practice Problem BUILD Determine the empirical formula of a compound that is 89.9 percent C and 10.1 percent H by mass. What is the molecular formula if the molar mass is approximately 120 g/mol?

Practice Problem CONCEPTUALIZE The percent composition by mass was determined for a compound containing only C and H; and the percent C was found to be four times that of H. Further, the compound's molecular mass is twice its empirical formula mass. Determine the molecular formula of the compound.

Section 5.10 Review

Molar Mass

5.10.1 How many molecules are in 30.1 g of sulfur dioxide (SO_2)?
(a) 1.81×10^{25} (c) 6.02×10^{23} (e) 5.00×10^{-23}
(b) 2.83×10^{23} (d) 1.02×10^{24}

5.10.2 How many moles of hydrogen are there in 6.50 g of ammonia (NH_3)?
(a) 0.382 mol   (c) 0.215 mol   (e) 2.66 mol
(b) 1.39 mol   (d) 1.14 mol

5.10.3 Determine the empirical formula of a compound that has the following composition: 92.3 percent C and 7.7 percent H.
(a) CH   (c) C_4H_6 (e) C_4H_3
(b) C_2H_3 (d) C_6H_7

5.10.4 Determine the molecular formula of a compound that has the following composition: 48.6 percent C, 8.2 percent H, and 43.2 percent O, and the molar mass is approximately 148 g/mol.
(a) $C_3H_6O_2$ (c) $C_8H_4O_3$ (e) $C_4H_4O_6$
(b) $C_6H_{12}O_4$ (d) $C_5H_8O_5$

Learning Outcomes

- Understand the term *compound* and be able to distinguish compounds, elements, and mixtures.
- Prepare Lewis dot symbols of elements and ions.
- Define ionic bonding and provide examples of compounds that contain ionic bonds.
- Use Coulomb's law and distance between ions to rank lattice energies of ionic compounds.
- Describe the Lewis theory of bonding.
- Utilize rules of nomenclature to name the different types of compounds including: ionic compounds, covalent compounds, oxoacids, hydrates, and simple alkanes.
- Recognize covalent bonding in ionic species.
- Name polyatomic ions and know their formulas and charges.

- Calculate the percent composition by mass and molecular/formula/molar mass of a compound.
- Use the relationships between Avogadro's number, moles, molar mass, and grams to perform calculations involving compounds.
- Interconvert between mass, moles, and number of particles.
- Determine the empirical formula of a compound from percent composition or from combustion analysis data.
- Utilize the empirical formula and molar mass to determine the molecular formula of a compound.
- Compare properties of ionic and covalent compounds.

Chapter Summary

SECTION 5.1

- A ***compound*** is a substance made up of two or more elements.

SECTION 5.2

- A ***Lewis dot symbol*** depicts an atom or an atomic ion of a main group element with dots (representing the valence electrons) arranged around the element's symbol. Main group atoms lose or gain one or more electrons to become isoelectronic with noble gases.

SECTION 5.3

- An ***ionic compound*** is one that consists of cations and anions in an electrically neutral combination that is held together by electrostatic attraction known as ***ionic bonding.***
- ***Binary compounds*** consist of two elements. Ionic compounds typically are binary and consist of a metal and a nonmetal.
- A ***chemical formula*** (or simply ***formula***) denotes the composition of a substance.
- A three-dimensional array of alternating cations and anions is called a ***lattice. Lattice energy*** is the energy required to separate a mole of an ionic compound into its constituent ions in the gas phase.

SECTION 5.4

- Cations are named using the name of the element from which they are derived. When a cation's charge is known unambiguously, it is not specified in the name. When an element can form cations of different charges, the charge is specified with a Roman numeral in parentheses. Anions are named similarly,

but the element's ending is changed to *–ide.* Ionic compounds are named systematically using the rules of nomenclature.

SECTION 5.5

- According to the ***Lewis theory of bonding, covalent bonding*** results when atoms *share* valence electrons. The atoms in molecules and those in polyatomic ions are held together by ***covalent bonds.***
- A ***molecule*** is an electrically neutral group of two or more atoms.
- According to the ***law of definite proportions,*** any sample of a given compound will always contain the same elements in the same mass ratio. The ***law of multiple proportions*** states that if two elements can form more than one compound with one another, the mass ratio of one will be related to the mass ratio of the other by a small whole number.
- Molecules consisting of just two atoms are called ***diatomic.*** Diatomic molecules may be ***homonuclear*** (just one kind of atom) or ***heteronuclear*** (two kinds of atoms). In general, molecules containing more than two atoms are called ***polyatomic.***
- A ***molecular formula*** specifies the exact numbers of atoms in a molecule of a compound. A ***structural formula*** shows the arrangement of atoms in a substance. Molecules are depicted using molecular art that may be ball-and-stick models or space-filling models.
- An ***allotrope*** is one of two or more different forms of an element.
- An ***empirical formula*** expresses, in the smallest possible whole numbers, the ratio of the combination of atoms of the elements in a compound. The empirical and molecular formulas of a compound may or may not be identical

SECTION 5.6

- Molecular compounds are usually composed of two or more nonmetals. The nomenclature of molecular compounds is similar to that of ionic compounds, but because there are often multiple possible combinations, Greek prefixes are used to specify the number of each kind of atom in the molecule.
- An *acid* is a substance that generates hydrogen ions when it dissolves in water. An *ionizable hydrogen atom* is one that can be removed in water to become a hydrogen ion, H^+.
- *Inorganic compounds* are generally those that do not contain carbon. *Organic compounds* contain carbon and hydrogen, sometimes in combination with other elements. *Hydrocarbons* contain only carbon and hydrogen. The simplest hydrocarbons are the *alkanes*. A *functional group* is a group of atoms that determines the chemical properties of an organic compound.

SECTION 5.7

- Many chemical species contain both *ionic* and *covalent* bonds. *Polyatomic ions* are those that contain more than one atom held together by covalent bonding. *Oxoanions* are polyatomic ions that contain one or more oxygen atoms.
- *Oxoacids* are acids based on oxoanions. Acids with one ionizable hydrogen atom are called *monoprotic*. Acids with more than one ionizable hydrogen atom are called *polyprotic*.

- *Hydrates* are compounds whose formulas include a specific number of water molecules. Some familiar compounds are known by common or nonsystematic names.

SECTION 5.8

- *Molecular mass* is calculated by summing the masses of all atoms in a molecule. *Molecular weight* is another term for molecular mass.
- For ionic compounds, we use the analogous terms *formula mass* and *formula weight*.
- Molecular masses, molecular weights, formula masses, and formula weights are expressed in atomic mass units (amu).

SECTION 5.9

- Molecular mass, formula mass, or *empirical formula mass* can be used to determine *percent composition by mass* of a compound. Empirical formula mass is expressed in atomic mass units (amu).

SECTION 5.10

- *Molar mass* ($\mathcal{M}$) is the mass of one mole of a substance, usually expressed in grams. The molar mass of a substance in grams is numerically equal to the *atomic, molecular,* or *formula* mass of the substance in atomic mass units.
- Molar mass and Avogadro's number can be used to interconvert among *mass, moles,* and *number of particles* (atoms, molecules, ions, formula units, etc.).

Key Words

Acid, 181
Alkane, 182
Allotrope, 175
Binary compound, 165
Chemical formula, 165
Compound, 163
Covalent bond, 173
Covalent bonding, 173
Diatomic molecule, 174
Empirical formula, 176
Empirical formula mass, 192
Formula, 165
Formula mass, 190
Formula weight, 190
Functional group, 183

Heteronuclear, 174
Homonuclear, 174
Hydrate, 188
Hydrocarbon, 182
Inorganic compound, 182
Ionic bonding, 165
Ionic compound, 165
Ionizable hydrogen atom, 181
Lattice, 166
Lattice energy, 166
Law of definite proportions, 173
Law of multiple proportions, 173
Lewis dot symbol, 163
Lewis theory of bonding, 173
Molar mass ($\mathcal{M}$), 194

Molecular formula, 175
Molecular mass, 190
Molecular weight, 190
Molecule, 173
Monoprotic acid, 187
Organic compound, 182
Oxoacids, 186
Oxoanions, 186
Percent composition by mass, 192
Polyatomic ion, 184
Polyatomic molecule, 174
Polyprotic acid, 187
Structural formula, 175

Key Equation

5.1   percent by mass of an element =
$$\frac{n \times \text{atomic mass of element}}{\text{molecular or formula mass of compound}} \times 100\%$$

Using a compound's formula (molecular or empirical), we can calculate its percent composition by mass.

Key Skills

Ionic Compounds: Nomenclature and Molar Mass Determination

The process of naming binary ionic compounds follows the simple procedure outlined in Section 5.4. Naming compounds that contain polyatomic ions follows essentially the same procedure; but it does require you to recognize the common polyatomic ions (see [◄◄ Table 5.10]). Because many ionic compounds contain polyatomic ions, it is important that you know their names, formulas, and charges—well enough that you can identify them readily.

In ionic compounds with ratios of combination other than 1:1, subscript numbers are used to denote the number of each ion in the formula.

Examples: $CaBr_2$, Na_2S, $AlCl_3$, Al_2O_3, FeO, Fe_2O_3

Recall that because the common ions of main group elements have predictable charges, it is unnecessary to use prefixes to denote their numbers when naming compounds that contain them. Thus, the names of the first four examples shown are calcium bromide, sodium sulfide, aluminum chloride, and aluminum oxide. The last two contain transition metal ions, many of which have more than one possible charge. In these cases, to avoid ambiguity, the charge on the metal ion is designated with a Roman numeral in parentheses. The names of these two compounds are iron(II) oxide and iron(III) oxide, respectively.

When a subscript number is required for a polyatomic ion, the ion's formula must first be enclosed in parentheses.

Examples: $Ca(NO_3)_2$, $(NH_4)_2S$, $Ba(C_2H_3O_2)_2$, $(NH_4)_3(SO_4)_2$, $Fe_3(PO_4)_2$, $Co_2(CO_3)_3$

Names: calcium nitrate, ammonium sulfide, barium acetate, ammonium sulfate, iron(II) phosphate, cobalt(III) carbonate

The process of naming ionic compounds given their formulas can be summarized with the following flowchart:

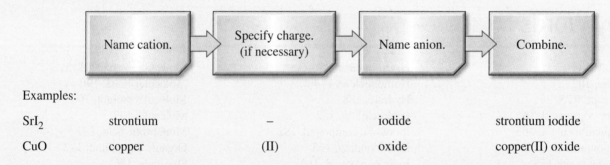

Examples:

| SrI_2 | strontium | – | iodide | strontium iodide |
| CuO | copper | (II) | oxide | copper(II) oxide |

It is equally important that you be able to write the formula of an ionic compound given its name. Again, knowledge of the common polyatomic ions is critical. The process of writing a compound's formula given its name is summarized as follows:

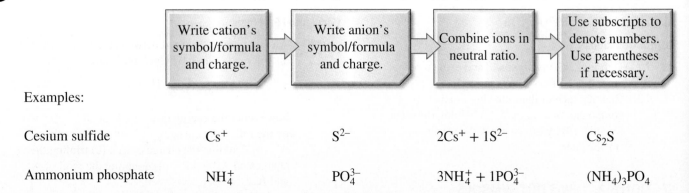

Examples:

Cesium sulfide	Cs^+	S^{2-}	$2Cs^+ + 1S^{2-}$	Cs_2S
Ammonium phosphate	NH_4^+	PO_4^{3-}	$3NH_4^+ + 1PO_4^{3-}$	$(NH_4)_3PO_4$

Knowing the formula of an ionic compound, or any compound, enables us to determine the compound's molar mass. To do this, it is important to count atoms carefully—especially in compounds with polyatomic ions enclosed in parentheses. For example, the formula $Ba(C_2H_3O_2)_2$ contains *one* barium, *four* carbons, *six* hydrogens, and *four* oxygens. Its molar mass is calculated as:

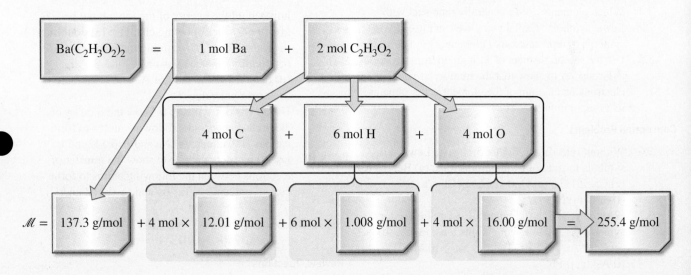

Key Skills Problems

5.1
What is the correct name for $CaSO_4$?
(a) calcium sulfoxide (b) calcium sulfite (c) calcium sulfur oxide (d) calcium sulfate (e) calcium sulfide tetroxide

5.2
What is the correct formula for nickel(II) perchlorate?
(a) $NiClO_4$ (b) Ni_2ClO_4 (c) $Ni(ClO_4)_2$ (d) $NiClO_3$
(e) $Ni(ClO_3)_2$

5.3
Determine the molar mass of $Co(NO_2)_2$.
(a) 151.0 g/mol (b) 209.9 g/mol (c) 163.9 g/mol
(d) 119.0 g/mol (e) 104.0 g/mol

5.4
Determine the molar mass of iron(III) sulfate.
(a) 151.9 g/mol (b) 344.1 g/mol (c) 399.9 g/mol
(d) 271.9 g/mol (e) 359.7 g/mol

Questions and Problems

SECTION 5.1: COMPOUNDS

Review Questions

5.1 Define the term *compound* and explain how a compound differs from a mixture.

5.2 Classify each of the following as an element, a compound, or a mixture: (a) helium, (b) sugar, (c) gold, (d) hydrogen peroxide, (e) air, (f) seawater.

SECTION 5.2: LEWIS DOT SYMBOLS

Review Questions

5.3 What is a Lewis dot symbol? What information does a Lewis dot symbol provide?

5.4 Which elements do we generally represent with Lewis symbols? Explain why we can't use Lewis symbols to represent every element.

5.5 Use the second member of each group from Group 1A to Group 7A to show that the number of valence electrons on an atom of the element is the same as its group number.

Conceptual Problems

5.6 Without referring to Figure 5.1, write Lewis dot symbols for the following atoms/ions: (a) Be, (b) K, (c) Ca, (d) Ga, (e) O, (f) Li^+, (g) Cl^-, (h) S^{2-}, (i) Sr^{2+}, (j) N^{3-}.

5.7 Without referring to Figure 5.1, write Lewis dot symbols for the following atoms/ions: (a) Br, (b) N, (c) I, (d) As, (e) F, (f) P^{3-}, (g) Na^+, (h) Mg^{2+}, (i) As^{3+}, (j) Pb^{2+}.

5.8 Indicate the charge on each of the ions represented by the following Lewis dot symbols: (a) $[:\ddot{Se}:]^?$, (b) $K^?$, and (c) $[:\ddot{As}:]^?$.

5.9 To what group of the periodic table does element X belong if its Lewis dot symbol is (a) $\cdot\dot{X}\cdot$, (b) $:\dot{X}:$, (c) $[:\ddot{X}:]^-$?

SECTION 5.3: IONIC COMPOUNDS AND BONDING

Review Questions

5.10 Explain what *ionic bonding* is.

5.11 Explain how ionization energy and electron affinity determine whether atoms of elements will combine to form ionic compounds.

5.12 Explain why ions with charges greater than ±3 are seldom found in ionic compounds.

5.13 What is *lattice energy* and what does it indicate about the stability of an ionic compound?

Conceptual Problems

5.14 An ionic bond is formed between a cation A^+ and an anion B^-. Based on Coulomb's law

$$E \propto \frac{Q_1 \times Q_2}{d}$$

how would the energy of the ionic bond be affected by the following changes: (a) tripling the radius of A^+, (b) doubling the charge on A^+, (c) tripling the charges on A^+ and B^-, (d) doubling the radii of A^+ and B^-?

5.15 An ionic bond is formed between a cation A^+ and an anion B^-. Based on Coulomb's law

$$E \propto \frac{Q_1 \times Q_2}{d}$$

how would the energy of the ionic bond be affected by the following changes: (a) doubling the radius of A^+, (b) tripling the charge on A^+, (c) doubling the charges on A^+ and B^-, and (d) decreasing the radii of A^+ and B^- to half their original values?

5.16 Use Lewis dot symbols to show the transfer of electrons between the following atoms to form cations and anions: (a) Na and F, (b) K and I.

5.17 Use Lewis dot symbols to show the transfer of electrons between the following atoms to form cations and anions: (a) Ba and O, (b) Al and N.

SECTION 5.4: NAMING IONS AND IONIC COMPOUNDS

Review Questions

5.18 Give an example of each of the following: (a) a monatomic cation, (b) a monatomic anion.

5.19 What is an ionic compound? How is electrical neutrality maintained in an ionic compound?

5.20 Explain why the chemical formulas of ionic compounds are usually the same as their empirical formulas.

5.21 What is the Stock system? What are its advantages over the older system of naming cations?

Conceptual Problems

5.22 Give the formulas and names of the compounds formed from the following pairs of ions: (a) Rb^+ and S^{2-}, (b) Cs^+ and I^-, (c) Ca^{2+} and N^{3-}, (d) Al^{3+} and Se^{2-}.

5.23 Give the formulas and names of the compounds formed from the following pairs of ions: (a) K^+ and P^{3-}, (b) Fe^{3+} and Se^{2-}, (c) Ba^{2+} and N^{3-}, (d) Ga^{3+} and S^{2-}.

5.24 Write the formulas for the following ionic compounds:
(a) copper bromide (containing the Cu^+ ion),
(b) manganese oxide (containing the Mn^{3+} ion),
(c) mercury iodide (containing the Hg_2^{2+} ion),
(d) magnesium phosphide.

5.25 Write the formulas for the following ionic compounds: (a) sodium oxide, (b) iron sulfide (containing the Fe^{2+} ion), (c) cobalt telluride (containing the Co^{3+} and Te^{2-} ions), (d) barium fluoride.

5.26 Name the following compounds: (a) $CdCl_2$, (b) FeI_3, (c) FeS, (d) $SnCl_4$.

5.27 Name the following compounds: (a) BaH_2, (b) K_3N, (c) ZnO, (d) BaO_2.

5.28 In the diagrams shown here, match each of the drawings with the following ionic compounds: Al_2O_3, LiH, Na_2S, $MgCl_2$. (Green spheres represent cations and red spheres represent anions.)

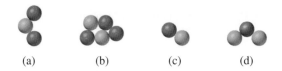

(a)   (b)   (c)   (d)

5.29 Given the formulas for the ionic compounds, draw the correct ratio of cations to anions as shown in Problem 5.28: (a) BaS, (b) CaF_2, (c) Mg_3N_2, (d) K_2O.

SECTION 5.5: COVALENT BONDING AND MOLECULES

Review Questions

5.30 Describe Lewis's contribution to our understanding of the covalent bond.

5.31 State the laws of definite proportions and multiple proportions. Illustrate each with an example.

5.32 The elements nitrogen and oxygen can form a variety of different compounds. Two such compounds, NO and N_2O_4, were decomposed into their constituent elements. One produced 0.8756 g N for every gram of O; the other produced 0.4378 g N for every gram of O. Show that these results are consistent with the law of multiple proportions.

5.33 Two different compounds, each containing only phosphorus and chlorine, were decomposed into their constituent elements. One produced 0.2912 g P for every gram of Cl; the other produced 0.1747 g P for every gram of Cl. Show that these results are consistent with the law of multiple proportions.

5.34 Sulfur reacts with fluorine to produce three different compounds. The mass ratio of fluorine to sulfur for each compound is given in the following table:

Compound	Mass F: Mass S
S_2F_{10}	2.962
SF_4	2.370
SF_6	3.555

Show that these data are consistent with the law of multiple proportions.

5.35 Both FeO and Fe_2O_3 contain only iron and oxygen. The mass ratio of oxygen to iron for each compound is given in the following table:

Compound	Mass O: Mass Fe
FeO	0.2865
Fe_2O_3	0.4297

Show that these data are consistent with the law of multiple proportions.

5.36 Describe the two commonly used molecular models.

5.37 What does a chemical formula represent? Determine the ratio of the atoms in the following molecular formulas: (a) NO, (b) NCl_3, (c) N_2O_4, (d) P_4O_6.

5.38 Define molecular formula and empirical formula. What are the similarities and differences between the empirical formula and molecular formula of a compound?

5.39 Give an example of a case in which two molecules have different molecular formulas but the same empirical formula.

Conceptual Problems

5.40 For each of the following diagrams, determine whether it represents molecules of a compound or molecules of an element.

(a)   (b)   (c)

5.41 For each of the following diagrams, determine whether it represents molecules of a compound or molecules of an element.

(a)   (b)   (c)

5.42   Identify the following as elements or compounds: P_4, PH_3, N_2O, C_6H_6, Hg, Cl_2, N_2, PCl_5.

5.43   Identify the following as elements or compounds: NH_3, N_2, S_8, NO, CO, CO_2, H_2, SO_2.

5.44   For the two compounds pictured, evaluate the following ratio:

$$\frac{\text{g green: 1.00 g yellow (right)}}{\text{g green: 1.00 g yellow (left)}}$$

5.45   For the two compounds pictured, evaluate the following ratio:

$$\frac{\text{g blue: 1.00 g red (right)}}{\text{g blue: 1.00 g red (left)}}$$

5.46   Write the empirical formulas of the following compounds: (a) Al_2Br_6, (b) $Na_2S_2O_4$, (c) N_2O_5, (d) $K_2Cr_2O_7$.

5.47   Write the empirical formulas of the following compounds: (a) C_2H_6, (b) C_6H_{12}, (c) C_5H_{10}, (d) P_4S_6, (e) C_2H_6O.

5.48   Write the molecular formula of ethanol from the model shown.

5.49   Write the molecular formula of alanine, an amino acid used in protein synthesis, from the model shown.

SECTION 5.6: NAMING MOLECULAR COMPOUNDS

Review Questions

5.50   Describe how the naming of *molecular* binary compounds is different from the naming of *ionic* binary compounds. Explain why the two approaches are different.

5.51   Define the term *acid*.

5.52   Describe the difference between naming HF when it is in the gas phase and HF when it is dissolved in water. Give another example of a compound that has one name when in the gas phase and another name when dissolved in water.

5.53   What is the difference between an *inorganic* compound and an *organic* compound?

Conceptual Problems

5.54   Write chemical formulas for the following molecular compounds: (a) nitrogen triiodide, (b) tetranitrogen decoxide, (c) xenon trioxide, (d) diiodine pentoxide.

5.55   Name the following binary molecular compounds: (a) S_2Br_2, (b) IF_5, (c) P_4O_7, (d) Br_2O_3.

5.56   Write the molecular formulas and names of the following compounds.

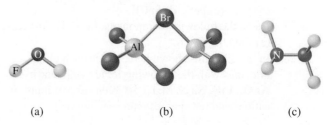

(a)   (b)   (c)

5.57   Write the molecular formulas and names of the following compounds.

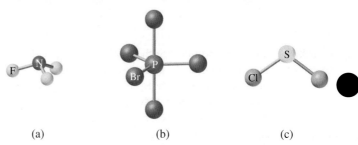

(a)   (b)   (c)

5.58   The GABA molecule, shown here, is a naturally produced inhibitory neurotransmitter in mammals. GABA is also marketed as a supplement that is said, among other things, to decrease anxiety, promote REM sleep, and improve mental clarity. Identify the functional group(s) in the GABA molecule.

5.59   The molecule shown here is one of the intermediates in a particular metabolic pathway in some prokaryotes. It is also a precursor to one of the compounds responsible for the characteristic flavor of strawberries. Identify the functional group(s) in this molecule.

SECTION 5.7: COVALENT BONDING IN IONIC SPECIES

Conceptual Problems

5.60 Name the following compounds: (a) K_3PO_4, (b) CoC_2O_4, (c) Li_2CO_3, (d) $K_2Cr_2O_7$, (e) NH_4NO_2, (f) HIO_3, (g) $SrSO_4$, (h) $Al(OH)_3$.

5.61 Name the following compounds: (a) $KClO$, (b) Ag_2CO_3, (c) HNO_2, (d) $KMnO_4$, (e) $CsClO_3$, (f) KNH_4SO_4, (g) $Fe(BrO_4)_2$, (h) K_2HPO_4.

5.62 Write the formulas for the following compounds: (a) rubidium nitrite, (b) potassium sulfate, (c) sodium hydrogen sulfide, (d) magnesium phosphate, (e) calcium hydrogen phosphate, (f) potassium dihydrogen phosphate, (g) ammonium sulfate, (h) silver perchlorate.

5.63 Write the formulas for the following compounds: (a) copper(I) cyanide, (b) strontium chlorite, (c) perbromic acid, (d) hydroiodic acid, (e) disodium ammonium phosphate, (f) lead(II) carbonate, (g) tin(II) sulfite, (h) cadmium thiocyanate.

SECTION 5.8: MOLECULAR AND FORMULA MASSES

Review Questions

5.64 What is meant by the term *molecular mass,* and why is the molecular mass that we calculate generally an *average* molecular mass?

5.65 Explain the difference between the terms *molecular mass* and *formula mass.* To what type of compound does each term refer?

Computational Problems

5.66 Calculate the molecular mass (in amu) of each of the following substances: (a) C_6H_6O, (b) H_2SO_4, (c) C_6H_6, (d) $C_6H_{12}O_6$, (e) BCl_3, (f) N_2O_5, (g) H_3PO_4.

5.67 Calculate the molecular mass (in amu) of each of the following substances: (a) CH_3Cl, (b) N_2O_4, (c) SO_2, (d) C_6H_{12}, (e) H_2O_2, (f) $C_{12}H_{22}O_{11}$, (g) NH_3.

5.68 Calculate the molecular mass or formula mass (in amu) of each of the following substances: (a) BrN_3, (b) C_4H_{10}, (c) NI_3, (d) Al_2S_3, (e) $Fe(NO_3)_2$, (f) PCl_3, (g) $(NH_4)_2CO_3$.

5.69 Calculate the molecular mass or formula mass (in amu) of each of the following substances: (a) CH_3Cl, (b) N_2O, (c) SO_2, (d) C_6H_{12}, (e) $NaBr$, (f) Cs_2SO_4, (g) $Ba_3(PO_4)_2$.

Conceptual Problems

5.70 Calculate the molecular mass (in amu) of each of the following substances.

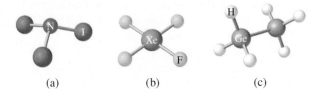

(a)   (b)   (c)

5.71 Calculate the formula mass (in amu) of each of the following substances.

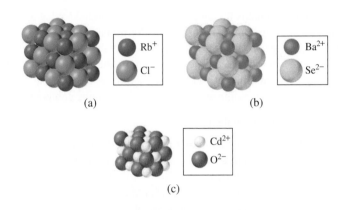

(a)   (b)

(c)

SECTION 5.9: PERCENT COMPOSITION OF COMPOUNDS

Review Questions

5.72 Use dinitrogen monoxide (N_2O) to explain what is meant by the percent composition by mass of a compound.

5.73 Describe how the knowledge of the percent composition by mass of an unknown compound can help us identify the compound.

Computational Problems

5.74 For many years chloroform ($CHCl_3$) was used as an inhalation anesthetic in spite of the fact that it is also a toxic substance that may cause severe liver, kidney, and heart damage. Calculate the percent composition by mass of this compound.

5.75 Tin (Sn) exists in Earth's crust as SnO_2. Calculate the percent composition by mass of Sn and O in SnO_2.

5.76 The anticaking agent added to Morton salt is calcium silicate ($CaSiO_3$). This compound can absorb up to 2.5 times its mass of water and still remain a free-flowing powder. Calculate the percent composition of $CaSiO_3$.

5.77 Tooth enamel is $Ca_5(PO_4)_3(OH)$. Calculate the percent composition of the elements present.

Conceptual Problems

5.78 Interhalogens are compounds that contain two different halogens. Arrange these interhalogen compounds in order of increasing percentage fluorine by mass: BrF_3, IF_7, IF_3, ClF.

5.79 All the substances listed here are fertilizers that contribute nitrogen to the soil. Based on mass percentage, which of these is the richest source of nitrogen?
(a) Urea [$(NH_2)_2CO$]
(b) Ammonium nitrate (NH_4NO_3)
(c) Guanidine [$HNC(NH_2)_2$]
(d) Ammonia (NH_3)

SECTION 5.10: MOLAR MASS

Review Questions

5.80   What is the molar mass of a compound? What are the commonly used units for molar mass?

5.81   *Molar* mass is numerically equivalent to *molecular* mass for a covalent compound, although the units are different. What is the advantage of using the term *molar mass* when discussing ionic compounds?

Computational Problems

5.82   Calculate the molar mass of the following substances: (a) Li_2CO_3, (b) CS_2, (c) $CHCl_3$ (chloroform), (d) $C_6H_8O_6$ (ascorbic acid, or vitamin C), (e) KNO_3, (f) Mg_3N_2.

5.83   Calculate the molar mass of a compound if 0.372 mole of it has a mass of 152 g.

5.84   How many molecules of ethane (C_2H_6) are present in 0.334 g of C_2H_6?

5.85   Calculate the number of C, H, and O atoms in 1.50 g of glucose ($C_6H_{12}O_6$), a sugar.

5.86   Urea [$(NH_2)_2CO$] is used for fertilizer and many other things. Calculate the number of N, C, O, and H atoms in 1.68×10^4 g of urea.

5.87   Pheromones are a special type of compound secreted by the females of many insect species to attract the males for mating. One pheromone has the molecular formula $C_{10}H_{16}O$. Normally, the amount of this pheromone secreted by a female insect is about 1.0×10^{-12} g. How many molecules does this quantity contain?

5.88   The density of water is 1.00 g/mL at 4°C. How many water molecules are present in 5.81 mL of water at this temperature?

5.89   Cinnamic alcohol is used to add a pleasant scent to soaps, perfumes, and some cosmetics. Its molecular formula is $C_9H_{10}O$. (a) Calculate the percent composition by mass of C, H, and O in cinnamic alcohol. (b) How many molecules of cinnamic alcohol are contained in a sample of mass 1.028 g?

5.90   The chemical formula for rust can be represented by Fe_2O_3. How many moles of Fe are present in 1.00 kg of the compound?

5.91   Tin(II) fluoride (SnF_2) is added to some toothpastes to aid in the prevention of tooth decay. What mass of F (in grams) does 7.10 g of the compound contain?

5.92   In pharmacology, the term *bioavailability* refers to the fraction of an administered dose of drug that reaches the "site of action" via the circulatory system. A particular drug with the chemical formula $C_{16}H_{16}ClNO_2S$ has a bioavailability of 55 percent. How many molecules of this drug reach the site of action in the patient's body if a 75.0-mg dose is administered?

5.93   Toxicologists use the term LD_{50} to describe the number of grams of a substance per kilogram of body weight that constitutes a lethal dose for 50 percent of test animals. Calculate the number of arsenic(III) oxide formula units corresponding to an LD_{50} value of 0.018 for a 212-lb man, assuming that the test animals and humans have the same LD_{50}.

5.94   The amino acid cysteine plays an important role in the three-dimensional structure of proteins by forming "disulfide bridges." The percent composition of cysteine is 29.74% C, 5.82% H, 26.41% O, 11.56% N, and 26.47% S. What is its molecular formula if its molar mass is approximately 121 g/mol?

5.95   Equilin is an estrogen isolated from the urine of pregnant mares. The percent composition of equilin is 80.56% C, 7.51% H, and 11.92% O. What is its molecular formula if its molar mass is approximately 268 g/mol?

Conceptual Problems

5.96   The organic compound squaric acid, shown here, is useful for preparing a variety of dyes. Determine the number of moles of squaric acid in 210.5 g.

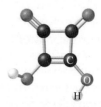

5.97   The arsenic-containing compound cacodylic acid, shown here, is one of the components of the herbicide "Agent Blue." Determine the mass (in grams) of 7.11 moles of cacodylic acid?

ADDITIONAL PROBLEMS

5.98   Predict the formula and name of a binary compound formed from the following elements: (a) Na and H, (b) B and O, (c) Na and S, (d) Al and F, (e) F and O, (f) Sr and Cl.

5.99   The formula for calcium oxide is CaO. What are the formulas for sodium oxide and strontium oxide?

5.100   What is wrong with the name (given in parentheses or brackets) for each of the following compounds: (a) $BaCl_2$ (barium dichloride), (b) Fe_2O_3 [iron(II) oxide], (c) $CsNO_2$ (cesium nitrate), (d) $Mg(HCO_3)_2$ [magnesium(II) bicarbonate]?

5.101 What is wrong with or ambiguous about the phrase "four molecules of NaCl"?

5.102 Rank the following oxohalide compounds in order of increasing percentage of oxygen by mass: $KBrO_3$, $KClO_3$, KIO_3.

5.103 Which of the following are elements, which are molecules but not compounds, which are compounds but not molecules, and which are both compounds and molecules? (a) SO_2, (b) S_8, (c) Cs, (d) N_2O_5, (e) O, (f) O_2, (g) O_3, (h) CH_4, (i) KBr, (j) S, (k) P_4, (l) LiF.

5.104 The following phosphorus sulfides are known: P_4S_3, P_4S_7, and P_4S_{10}. Do these compounds obey the law of multiple proportions? Explain.

5.105 Ethane and acetylene are two gaseous hydrocarbons. Chemical analyses show that in one sample of ethane, 2.65 g of carbon are combined with 0.665 g of hydrogen, and in one sample of acetylene, 4.56 g of carbon are combined with 0.383 g of hydrogen. (a) Are these results consistent with the law of multiple proportions? Explain. (b) Write reasonable molecular formulas for these compounds.

5.106 Which of the following compounds are likely to be ionic? Which are likely to be molecular? CH_4, NaBr, BaF_2, CCl_4, ICl, CsCl, NF_3.

5.107 Which of the following compounds are likely to be ionic? Which are likely to be molecular? $SiCl_4$, LiF, $BaCl_2$, B_2H_6, KCl, C_2H_4.

5.108 The mineral *tschermigite* is essentially pure $NH_4Al(SO_4)_2 \cdot 12H_2O$. Determine the percent oxygen by mass in the mineral.

5.109 The formulas for the fluorides of the third-period elements are NaF, MgF_2, AlF_3, SiF_4, PF_5, SF_6, and ClF_3. Classify each of these compounds as covalent or ionic.

5.110 Complete the table by filling in the blanks with appropriate information.

Cation	Anion	Formula	Name
Co^{2+}	PO_4^{3-}		
Hg_2^{2+}	I^-		
		Cu_2CO_3	
			Lithium nitride
Al^{3+}	S^{2-}		

5.111 Complete the table by filling in the blanks with appropriate information.

Cation	Anion	Formula	Name
			Magnesium bicarbonate
		$SrCl_2$	
Fe^{3+}	NO_2^-		
			Manganese(II) chlorate
		$SnBr_4$	

5.112 Determine the molecular and empirical formulas of the compounds shown here. (Black spheres are carbon, and white spheres are hydrogen.)

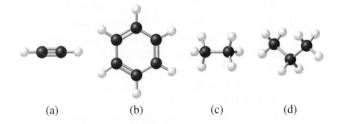

(a)   (b)   (c)   (d)

5.113 Rank the four compounds shown in Problem 5.112 according to increasing percent carbon by mass.

5.114 For each of the following pairs of elements, state whether the binary compound they form is likely to be ionic or covalent. Write the empirical formula and name of the compound: (a) I and Cl, (b) Mg and F.

5.115 For each of the following pairs of elements, state whether the binary compound they form is likely to be ionic or covalent. Write the empirical formula and name of the compound: (a) B and F, (b) K and Br.

5.116 Cysteine, shown here, is one of the 20 amino acids found in proteins in humans. Write the molecular formula of cysteine, and calculate its molar mass. What is the mass of 0.153 mole of cysteine?

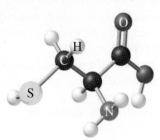

5.117 Isoflurane, shown here, is a common inhalation anesthetic. Write its molecular formula, and calculate its molar mass. How many moles of isoflurane are in 3.82 g?

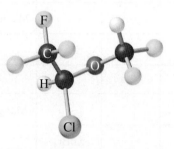

5.118 The atomic mass of element X is 33.42 amu. A 27.22-g sample of X combines with 84.10 g of another element Y to form a compound XY. Calculate the atomic mass of Y.

5.119 How many moles of O are needed to combine with 0.212 mole of C to form (a) CO and (b) CO_2?

5.120 A sample of ammonium dichromate contains 1.81×10^{24} hydrogen atoms. How many grams of nitrogen does the sample contain?

5.121 Which of the following are ionic compounds and which are covalent compounds: $RbCl$, PF_5, BrF_3, KO_2, CI_4?

5.122 Classify the following substances as ionic compounds or covalent compounds containing discrete molecules: CH_4, KF, CO, $SiCl_4$, $BaCl_2$.

5.123 The formula of aluminum molybdate is $Al_2(MoO_4)_3$. What is the formula of magnesium molybdate?

5.124 Determine what is wrong with the chemical formula and write the correct chemical formula for each of the following compounds: (a) $(NH_3)_2CO_3$ (ammonium carbonate), (b) $CaOH$ (calcium hydroxide), (c) $CdSO_3$ (cadmium sulfide), (d) $ZnCrO_4$ (zinc dichromate).

5.125 Which of the following has the greater mass: 0.72 g of O_2 or 0.0011 mole of chlorophyll ($C_{55}H_{72}MgN_4O_5$)?

5.126 Chemical analysis shows that the oxygen-carrying protein hemoglobin is 0.34 percent Fe by mass. What is the minimum possible molar mass of hemoglobin? The actual molar mass of hemoglobin is about 65,000 g. How would you account for the discrepancy between your minimum value and the experimental value?

5.127 One mole of a particular compound is shown to contain 6.02×10^{23} atoms of hydrogen, 35.45 g of chlorine, and 64.0 g of oxygen. Determine the formula of this compound.

5.128 The compound 2,3-dimercaptopropanol ($HSCH_2CHSHCH_2OH$), commonly known as British Anti-Lewisite (BAL), was developed during World War I as an antidote to arsenic-containing poison gas. (a) If each BAL molecule binds one arsenic (As) atom, how many As atoms can be removed by 1.0 g of BAL? (b) BAL can also be used to remove poisonous heavy metals like mercury (Hg) and lead (Pb). If each BAL binds one Hg atom, calculate the mass percent of Hg in a BAL-Hg complex. (An H atom is removed when a BAL molecule binds an Hg atom.)

5.129 Mustard gas ($C_4H_8Cl_2S$) is a poisonous gas that was used in World War I and banned afterward. It causes general destruction of body tissues, resulting in the formation of large water blisters. There is no effective antidote. Calculate the percent composition by mass of the elements in mustard gas.

5.130 Myoglobin stores oxygen for metabolic processes in muscle. Chemical analysis shows that it contains 0.34 percent Fe by mass. What is the molar mass of myoglobin? (There is one Fe atom per molecule.)

5.131 Hemoglobin ($C_{2952}H_{4664}N_{812}O_{832}S_8Fe_4$) is the oxygen carrier in blood. (a) Calculate its molar mass. (b) An average adult has about 5.0 L of blood. Every milliliter of blood has approximately 5.0×10^9 erythrocytes, or red blood cells, and every red blood cell has about 2.8×10^8 hemoglobin molecules. Calculate the mass of hemoglobin molecules in grams in an average adult.

5.132 A mixture of NaBr and Na_2SO_4 contains 29.96 percent Na by mass. Calculate the percent by mass of each compound in the mixture.

5.133 Calculate the number of cations and anions in each of the following compounds: (a) 8.38 g of KBr, (b) 5.40 g of Na_2SO_4, (c) 7.45 g of $Ca_3(PO_4)_2$.

5.134 Calculate the percent composition by mass of all the elements in calcium phosphate [$Ca_3(PO_4)_2$], a major component of bone.

5.135 Which of the following substances contains the greatest mass of chlorine: (a) 5.0 g Cl_2, (b) 60.0 g $NaClO_3$, (c) 0.10 mol KCl, (d) 30.0 g $MgCl_2$, (e) 0.50 mol Cl_2?

5.136 The natural abundances of the two stable isotopes of hydrogen (hydrogen and deuterium) are 99.99 percent ^1_1H and 0.01 percent ^2_1H. Assume that water exists as either H_2O or D_2O. Calculate the number of D_2O molecules in exactly 400 mL of water (density 1.00 g/mL).

5.137 Air is a mixture of many gases and as such, it does not have a molar mass. However, we can calculate an *effective molar mass* as a weighted average of the individual components. To do so, we only need to consider the three major components: nitrogen, oxygen, and argon. Given that one mole of air at sea level is made up of 78.08 percent nitrogen, 20.95 percent oxygen, and 0.97 percent argon, what is the effective molar mass of air?

5.138 Serine, shown here, is a biologically important molecule involved in the function of RNA and

DNA, fat and fatty acid metabolism, muscle formation, and the maintenance of a healthy immune system. Identify the functional group(s) in the serine molecule.

Answers to In-Chapter Materials

PRACTICE PROBLEMS

5.1A (a) Ca^{2+}, (b) $\left[:\ddot{N}:\right]^{3-}$, (c) $\left[:\ddot{I}:\right]^{-}$. **5.1B** (a) 2−, (b) +, (c) 3−. **5.2A** $MgCl_2$. **5.2B** NaF < MgO < AlN. **5.3A** (a) barium sulfide, (b) copper(I) sulfide, (c) chromium(III) chloride. **5.3B** (a) aluminum bromide, (b) rubidium selenide, (c) copper(II) nitride. **5.4A** (a) $PbCl_2$, (b) Co_2O_3, (c) Na_2Se. **5.4B** (a) Fe_2S_3,

(b) HgI_2, (c) K_2S. **5.5A** (a) $CHCl_3$, (b) C_3H_6O. **5.5B** (a)

(b)   **5.6A** (a) $C_4H_5N_2O$, (b) C_2H_5, (c) $C_2H_5NO_2$.

5.6B (a). **5.7A** (a) dichlorine monoxide, (b) silicon tetrachloride. **5.7B** (a) chlorine dioxide, (b) carbon tetrachloride. **5.8A** (a) CS_2, (b) N_2O_3. **5.8B** (a) SO_3, (b) S_2F_{10}. **5.9A** (a) sodium sulfate,

(b) copper(II) nitrate, (c) iron(III) carbonate. **5.9B** (a) $K_2Cr_2O_7$, (b) $Li_2C_2O_4$, (c) $Cu(NO_3)_2$. **5.10A** (a) hypobromous acid, (b) hydrogen sulfate ion, (c) oxalic acid. **5.10B** (a) iodic acid, (b) hydrogen chromate ion, (c) hydrogen oxalate ion. **5.11A** $HBrO_4$. **5.11B** H_2CrO_4. **5.12A** (a) 95.21 amu, (b) 98.09 amu, (c) 90.04 amu. **5.12B** (a) 100.09 amu, (b) 47.02 amu, (c) 68.15 amu. **5.13A** 57.13% C, 6.165% H, 9.521% N, 27.18% O. **5.13B** C_3H_6O. **5.14A** 495 g. **5.14B** 0.704 mol. **5.15A** 6.68×10^{23} O_2 molecules. **5.15B** 985 g. **5.16A** C_2H_5O; $C_4H_{10}O_2$. **5.16B** C_3H_4; C_9H_{12}.

SECTION REVIEW

5.2.1 $\cdot\dot{S}\dot{i}\cdot$. **5.2.2** $\left[:\ddot{B}r:\right]^{-}$. **5.2.3** d. **5.2.4** e. **5.3.1** c. **5.3.2** b. **5.4.1** b. **5.4.2** d. **5.4.3** a. **5.4.4** b. **5.5.1** b. **5.5.2** d. **5.6.1** b. **5.6.2** d. **5.6.3** c. **5.6.4** b. **5.7.1** b. **5.7.2** c. **5.7.3** a. **5.7.4** d. **5.8.1** a. **5.8.2** d. **5.8.3** d. **5.8.4** b. **5.8.5** d. **5.9.1** b. **5.9.2** e. **5.10.1** b. **5.10.2** d. **5.10.3** a. **5.10.4** b.

Chapter **6**

Representing Molecules

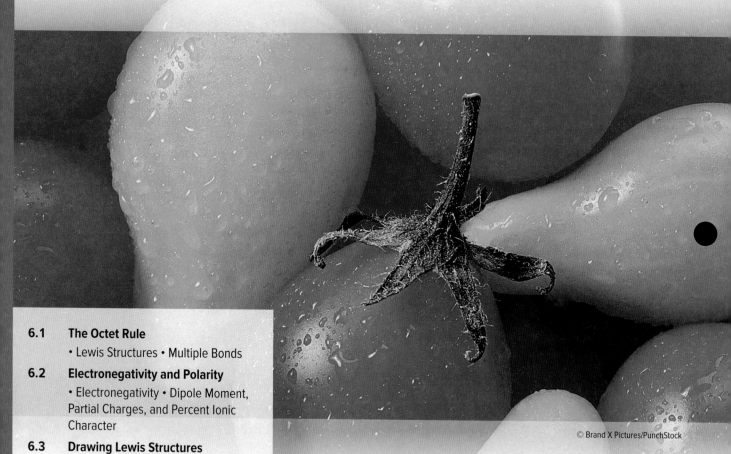

© Brand X Pictures/PunchStock

IN MOST chemical species, all of the electrons are paired. One class of molecules, known as *radicals* or *free* radicals, have one or more *unpaired* electrons. Free radicals are important in the preparation of polymers, and are critical in many biological processes; however, free radicals also are responsible for some undesirable processes that cause cell damage. Fortunately, the body has mechanisms to minimize radical damage, and biochemists have identified molecules known as *antioxidants* that can prevent or slow this damage. Fruits and vegetables (including tomatoes, shown here) are good sources of important antioxidants such as vitamins C and E, and the phytochemical *lycopene.*

Before You Begin, Review These Skills

- Lewis dot symbols [◄◄ Section 5.2]
- Coulombs [◄◄ Section 2.2]

6.1 THE OCTET RULE

Lewis summarized much of his theory of chemical bonding with the octet rule. According to the *octet rule,* atoms will lose, gain, or share electrons to achieve a noble gas electron configuration. This rule enables us to predict many of the formulas for compounds consisting of specific elements. The octet rule holds for nearly all the compounds made up of second period elements and is therefore especially important in the study of organic compounds, which contain mostly C, N, and O atoms. Whether a particular atom achieves an octet by losing, gaining, or sharing electrons depends on the properties of the atom—which are determined by the number of subatomic particles within the atom. Figure 6.1 (pages 212–213) summarizes how atomic properties determine what sort of bonds an atom will form.

As with ionic bonding, covalent bonding of many-electron atoms involves only the valence electrons. Consider the fluorine molecule (F_2). The electron configuration of F is $1s^2 2s^2 2p^5$. The $1s$ electrons are low in energy and stay near the nucleus most of the time, so they do not participate in bond formation. Thus, each F atom has seven valence electrons (two $2s$ electrons and five $2p$ electrons). The Lewis symbol of fluorine indicates that only one of its valence electrons is unpaired [◄◄ Figure 5.1, page 164], so the formation of the F_2 molecule can be represented as follows:

$$:\!\ddot{F}\cdot \ + \ \cdot\ddot{F}\!: \ \longrightarrow \ :\!\ddot{F}\!\!:\!\!\ddot{F}\!: \ \text{ or } \ :\!\ddot{F}\!-\!\ddot{F}\!:$$

Only two valence electrons participate in the formation of the F_2 bond. The other, nonbonding, electrons are called *lone pairs*—pairs of valence electrons that are not involved in covalent bond formation. Thus, each F in F_2 has three lone pairs of electrons.

$$\longrightarrow :\!\ddot{F}\!-\!\ddot{F}\!: \longleftarrow \text{lone pair}$$

Lewis Structures

The structures used to represent molecules held together by covalent bonds, such as H_2 and F_2, are called *Lewis structures.* A **Lewis structure** is a representation of covalent bonding in which shared electron pairs are shown either as dashes or as pairs of dots between two atoms, and lone pairs are shown as pairs of dots on individual atoms. Only valence electrons are shown in a Lewis structure.

To draw the Lewis structure of the water molecule, recall that the Lewis dot symbol for oxygen has two unpaired dots [◄◄ Figure 5.1, page 164], meaning that it has two unpaired electrons and can form two bonds. Because hydrogen has only one electron, it can form only one covalent bond. Thus, the Lewis structure for water is

$$H\!:\!\ddot{O}\!:\!H \quad \text{ or } \quad H\!-\!\ddot{O}\!-\!H$$

In this case, the O atom has two lone pairs. The hydrogen atom has no lone pairs because its only valence electron is used to form a covalent bond.

Student Annotation: For nearly all elements, achieving a noble gas electron configuration results in eight valence electrons around each atom—hence the name *octet* rule.

Student Annotation: Recall that a pair of shared electrons can be represented either with two dots or with a dash. [◄◄ Section 5.5]

Student Annotation: Lewis structures are also referred to as Lewis dot structures. In this text, we will use the term *Lewis structure* to avoid confusion with the term *Lewis dot symbol.*

Figure 6.1

The Properties of Atoms

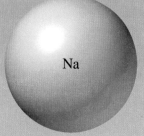

Na

$1s^2 2s^2 2p^6 3s^1$
$IE_1 = 496$ kJ/mol
$EA = +52.9$ kJ/mol
$r = 186$ pm

Metals, such as sodium, easily lose one
or more electrons to become cations.

Cl

$1s^2 2s^2 2p^6 3s^2 3p^5$
$IE_1 = 1256$ kJ/mol
$EA = +349$ kJ/mol
$r = 99$ pm

Nonmetals, such as chlorine, easily gain
one or more electrons to become anions.

Nonmetals can also achieve an octet by
sharing electrons to form covalent bonds.

C

$1s^2 2s^2 2p^2$
$IE_1 = 1086$ kJ/mol
$EA = +122$ kJ/mol
$r = 77$ pm

Although carbon is a nonmetal, it neither loses nor
gains electrons easily. Instead, it achieves an octet
by sharing electrons—forming covalent bonds.

What makes carbon different from other
nonmetals is that it often forms bonds with itself,
including multiple bonds—forming a limitless
array of organic compounds, such as ethylene.

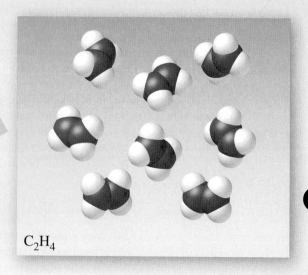

C_2H_4

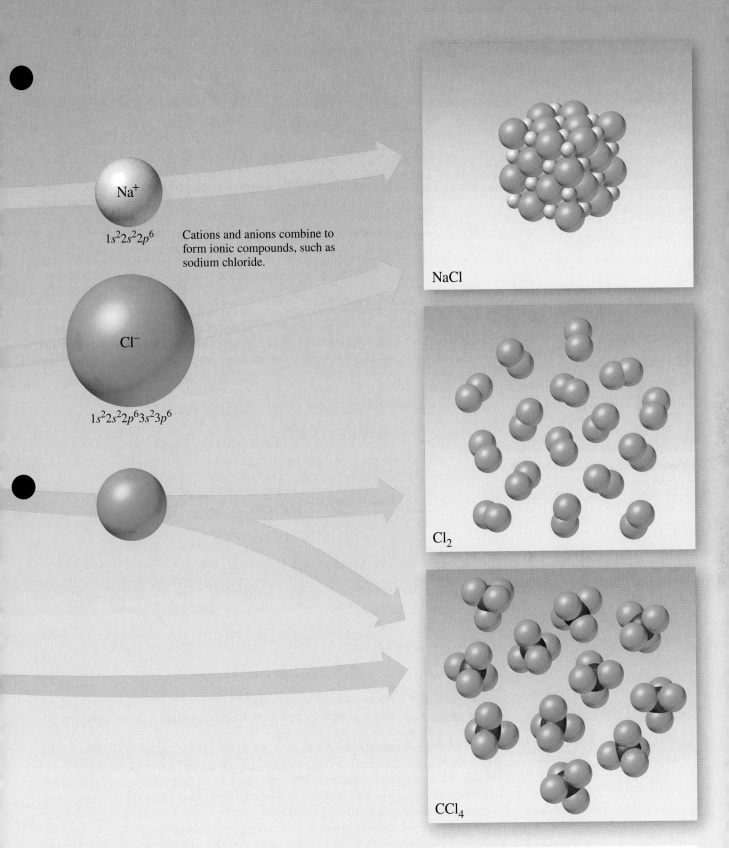

Na⁺

$1s^22s^22p^6$

Cl⁻

$1s^22s^22p^63s^23p^6$

Cations and anions combine to form ionic compounds, such as sodium chloride.

NaCl

Cl₂

CCl₄

What's the point?

The number of subatomic particles determines the properties of individual atoms. In turn, the properties of atoms determine how they interact with other atoms and what compounds, if any, they form.

In the F_2 and H_2O molecules, the F, H, and O atoms each achieve a stable noble gas configuration by sharing electrons, thus illustrating the octet rule:

$$\text{F with } 8\ e^- \quad \text{:F:F:} \quad \text{F with } 8\ e^-$$

$$\text{H with } 2\ e^- \quad \text{H:O:H} \quad \text{H with } 2\ e^-$$

$$\text{O with } 8\ e^-$$

The octet rule works best for elements in the second period of the periodic table. These elements have only $2s$ and $2p$ valence subshells, which can hold a total of eight electrons. When an atom of one of these elements forms a covalent compound, it can attain the noble gas electron configuration [Ne] by sharing electrons with other atoms in the same compound. In Section 6.6, we will discuss some important exceptions to the octet rule.

Multiple Bonds

When the bond between two atoms consists of *one* shared pair of electrons, the bond is said to be a ***single bond.*** When the bond between two atoms consists of *two* pairs of electrons, the bond is said to be a ***double bond;*** and when the bond between two atoms consists of *three* pairs of electrons, the bond is said to be a ***triple bond.*** Collectively, double and triple bonds are known as ***multiple bonds.*** Double bonds are found in molecules such as carbon dioxide (CO_2) and ethylene (C_2H_4):

$$\text{Each O has } 8\ e^- \quad \text{:O::C::O:} \quad \text{or} \quad \text{:O=C=O:}$$
$$\text{C has } 8\ e^-$$

$$\text{Each H has } 2\ e^- \quad \begin{array}{c} \text{H } \text{H} \\ \text{H:C::C:H} \end{array} \quad \text{or} \quad \begin{array}{cc} \text{H} & \text{H} \\ \diagdown & \diagup \\ \text{C} = \text{C} \\ \diagup & \diagdown \\ \text{H} & \text{H} \end{array}$$
$$\text{Each C has } 8\ e^-$$

Triple bonds occur in molecules such as nitrogen (N_2) and acetylene (C_2H_2):

$$\text{Each N has } 8\ e^-$$
$$\text{:N:::N:} \quad \text{or} \quad \text{:N}\equiv\text{N:}$$

$$\text{Each H has } 2\ e^- \quad \text{H:C:::C:H} \quad \text{or} \quad \text{H}-\text{C}\equiv\text{C}-\text{H}$$
$$\text{Each C has } 8\ e^-$$

In ethylene and acetylene, all the valence electrons are used in bonding; there are no lone pairs on the carbon atoms. In fact, most stable molecules containing carbon do not have lone pairs on the carbon atoms.

Multiple bonds are shorter than single bonds. ***Bond length*** is defined as the distance between the nuclei of two covalently bonded atoms in a molecule (Figure 6.2). Table 6.1 shows some experimentally determined bond lengths. For a given pair of atoms, such as carbon and nitrogen, triple bonds are shorter than double bonds, and double bonds are shorter than single bonds.

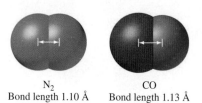

N₂
Bond length 1.10 Å

CO
Bond length 1.13 Å

Figure 6.2 Bond length is the distance between the nuclei of two bonded atoms.

The shorter multiple bonds are also stronger than single bonds. We can quantify the strength of a bond by measuring the quantity of energy required to break it. Specifically, we measure the energy change associated with breaking one mole of bonds of a particular type. For example, the bond energy of the diatomic hydrogen molecule is

$$H_2(g) \longrightarrow H(g) + H(g) \qquad \text{Bond energy} = 436.4 \text{ kJ/mol}$$

According to this equation, breaking all the covalent bonds in one mole of gaseous H_2 molecules (one mole of H—H bonds) requires 436.4 kJ of energy. Because multiple

TABLE 6.1	Average Bond Lengths of Common Single, Double, and Triple Bonds		
Bond type	**Bond length (pm)**	**Bond type**	**Bond length (pm)**
C–H	107	C=N	138
O–H	96	C≡N	116
C–O	143	N–N	147
C=O	121	N=N	124
C≡O	113	N≡N	110
C–C	154	N–O	136
C=C	133	N=O	122
C≡C	120	O–O	148
C–N	143	O=O	121

bonds involve the sharing of multiple pairs of electrons, the bond energies of double and triple bonds between two atoms are higher than a single bond between the same atoms. We will examine the energy changes associated with breaking chemical bonds in more detail in Chapter 10.

6.2 ELECTRONEGATIVITY AND POLARITY

So far, we have described chemical bonds as either *ionic,* when they occur between a metal and a nonmetal, or *covalent,* when they occur between nonmetals. In fact, ionic and covalent bonds are simply the extremes in a spectrum of bonding. Bonds that fall between these two extremes are *polar,* meaning that electrons are shared but are not shared equally. Such bonds are referred to as *polar covalent bonds.* The following shows a comparison of the different types of bonds, where M and X represent two atoms.

M:X	$M^{\delta+}X^{\delta-}$	M^+X^-
Pure covalent bond	**Polar covalent bond**	**Ionic bond**
Neutral atoms held together by *equally* shared electrons	Partially charged atoms held together by *unequally* shared electrons	Oppositely charged ions held together by electrostatic attraction

To illustrate the spectrum of bonding, let's consider three substances: H_2, HF, and NaF. In the H_2 molecule, where the two bonding atoms are identical, the electrons are shared equally. That is, the electrons in the covalent bond spend roughly the same amount of time in the vicinity of each H atom. In the HF molecule, on the other hand, where the two bonding atoms are different, the electrons are *not* shared equally. They spend more time in the vicinity of the F atom than in the vicinity of the H atom. (The δ symbol is used to denote partial charges on the atoms.) In NaF, the electrons are not shared at all but rather are transferred from sodium to fluorine.

One way to visualize the distribution of electrons in species such as H_2, HF, and NaF is to use electrostatic potential models (Figure 6.3). These models show regions where electrons spend a lot of time in red, and regions where electrons spend very little time in blue. (Regions where electrons spend a moderate amount of time appear green.)

Student Annotation: In fact, the transfer of electrons in NaF is *nearly* complete. Even in an ionic bond, the electrons in question spend a small amount of time near the cation.

Figure 6.3 Electron density maps show the distribution of charge in a covalent species (H_2), a polar covalent species (HF), and an ionic species (NaF). The most electron-rich regions are red; the most electron-poor regions are blue.

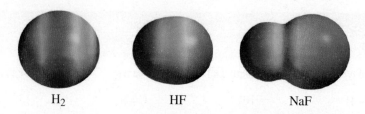

H_2 HF   NaF

Electronegativity

Student Annotation: Electrons spend *more* time around atoms with *high* electronegativities, and *less* time around atoms with *low* electronegativities.

Video 6.1
Periodic table—electronegativity

Student Annotation: The trend in electronegativities can also be related to atomic radius. The smaller an atom, the closer the shared electrons will be to its nucleus; therefore, the more powerfully they will be attracted to it.

Electronegativity is the ability of an atom to draw shared electrons (the electrons in a covalent bond) toward itself. It determines how the electron density [◄◄ Section 3.6] in a molecule or a polyatomic ion is distributed—what we have previously referred to as "where the electrons spend most of their time." An element with a high electronegativity has a greater tendency to attract electron density than an element with low electronegativity. We can make a qualitative prediction of an atom's electronegativity based on its electron affinity (*EA*) and its first ionization energy (IE_1) [◄◄ Section 4.4]. An atom such as fluorine, which has a high electron affinity (tends to *accept* electrons) and a high ionization energy (has little or no tendency to *lose* electrons), has a *high* electronegativity. Sodium, on the other hand, has a low electron affinity, a low ionization energy, and therefore a *low* electronegativity.

Electronegativity is a relative concept, meaning that an element's electronegativity can be measured only in relation to the electronegativity of other elements. Linus Pauling[1] devised one method for calculating the relative electronegativities of most elements. These values are shown in Figure 6.4. In general, electronegativity increases from left to right across a period in the periodic table, as the metallic character of the elements decreases. Within each group, electronegativity decreases with increasing atomic number and increasing metallic character. The trend in transition metals is less regular. The most electronegative elements (the halogens, oxygen, nitrogen, and sulfur) are found in the upper right-hand corner of the periodic table, and the least electronegative elements (the alkali and alkaline earth metals) are clustered near the lower left-hand corner. These trends are readily apparent in the graph in Figure 6.5.

Figure 6.4 Electronegativities of common elements.

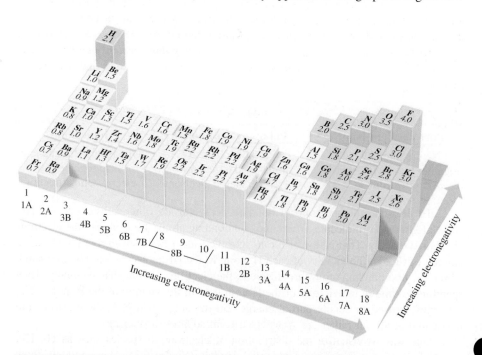

---

[1] Linus Carl Pauling (1901–1994). This American chemist is regarded by many as the most influential chemist of the twentieth century. Pauling received the Nobel Prize in Chemistry in 1954 for his work on protein structure, and the Nobel Peace Prize in 1962 for his tireless campaign against the testing and proliferation of nuclear arms. He is the only person ever to have received two unshared Nobel Prizes.

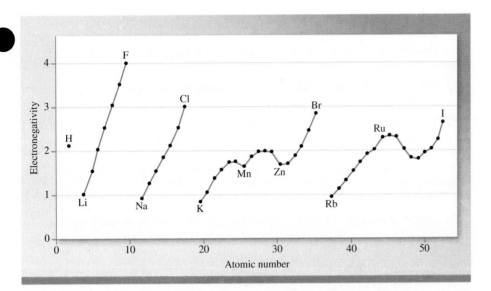

Figure 6.5 Variation of electronegativity with atomic number.

Electronegativity and electron affinity are related but distinct concepts. Both indicate the tendency of an atom to attract electrons. Electron affinity, however, refers to an isolated atom's ability to attract an additional electron in the gas phase, whereas electronegativity refers to the ability of an atom in a chemical bond (with another atom) to attract the *shared* electrons. Electron affinity, moreover, is an experimentally measurable quantity, whereas electronegativity is an estimated number that cannot be measured directly.

Atoms of elements with widely different electronegativities tend to form ionic compounds with each other because the atom of the less electronegative element gives up its electrons to the atom of the more electronegative element. Atoms of elements with comparable electronegativities tend to form polar covalent bonds, or simply polar bonds, with each other because the shift in electron density from one atom to the other is usually small. Only atoms of the same element, which have the same electronegativity, can be joined by a *pure* covalent bond.

There is no sharp distinction between *nonpolar covalent* and *polar covalent* or between *polar covalent* and *ionic,* but the following guidelines can help distinguish among them:

- A bond between atoms whose electronegativities differ by less than 0.5 is generally considered purely covalent or **nonpolar.**
- A bond between atoms whose electronegativities differ by the range of 0.5 to 2.0 is generally considered *polar covalent.*
- A bond between atoms whose electronegativities differ by 2.0 or more is generally considered *ionic.*

Worked Example 6.1 shows how to use electronegativities to determine whether a chemical bond is nonpolar, polar, or ionic.

Video 6.2
Chemical bonding—ionic covalent and polar covalent bonds

Worked Example 6.1

Classify the following bonds as nonpolar, polar, or ionic: (a) the bond in ClF, (b) the bond in CsBr, and (c) the carbon-carbon double bond in C_2H_4.

Strategy Using the information in Figure 6.4, determine which bonds have identical, similar, and widely different electronegativities.

Setup Electronegativity values from Figure 6.4 are Cl (3.0), F (4.0), Cs (0.7), Br (2.8), and C (2.5).

(Continued on next page)

Solution

(a) The difference between the electronegativities of F and Cl is $4.0 - 3.0 = 1.0$, making the bond in ClF polar.

(b) In CsBr, the difference is $2.8 - 0.7 = 2.1$, making the bond ionic.

(c) In C_2H_4, the two atoms are identical. (Not only are they the same element, but each C atom is bonded to two H atoms.) The carbon-carbon double bond in C_2H_4 is nonpolar.

> ### Think About It
> By convention, the difference in electronegativity is always calculated by subtracting the smaller number from the larger one, so the result is always positive.

Practice Problem A TTEMPT   Classify the following bonds as nonpolar, polar, or ionic: (a) the bonds in H_2S, (b) the H—O bonds in H_2O_2, and (c) the O—O bond in H_2O_2.

Practice Problem B UILD   Using data from Figure 6.4, list all the main group elements that can form ionic compounds with N.

Practice Problem C ONCEPTUALIZE
Electrostatic potential maps are shown for HCl and LiH. Determine which diagram is which. (The H atom is shown on the left in both.)

Dipole Moment, Partial Charges, and Percent Ionic Character

The shift of electron density in a polar bond is symbolized by placing a crossed arrow (a dipole arrow) above the Lewis structure to indicate the direction of the shift. For example,

$$\overset{\longmapsto}{\text{H--}\overset{..}{\underset{..}{\text{F}}}\text{:}}$$

The consequent charge separation can be represented as

$$\overset{\delta+}{\text{H}}\text{--}\overset{..}{\underset{..}{\text{F}}}\text{:}^{\delta-}$$

where the deltas denote partial positive and negative charges. A quantitative measure of the polarity of a bond is its **_dipole moment_ (μ),** which is calculated as the product of the charge (Q) and the distance (r) between the charges:

Equation 6.1 $\qquad\qquad\qquad \mu = Q \times r$

> **Student Annotation:** The distance, r, between partial charges in a polar diatomic molecule is the bond length expressed in meters. Bond lengths are usually given in angstroms (Å) or picometers (pm), so it is generally necessary to convert to meters.

In order for a diatomic molecule containing a polar bond to be electrically neutral, the partial positive and partial negative charges must have the same magnitude. Therefore, the Q term in Equation 6.1 refers to the magnitude of the partial charges and the calculated value of μ is always positive. Dipole moments are usually expressed in debye units (D), named for Peter Debye.[2] In terms of more familiar SI units,

$$1 \text{ D} = 3.336 \times 10^{-30} \text{ C} \cdot \text{m}$$

> **Student Annotation:** We usually express the charge on an electron as -1. This refers to *units of electronic charge*. However, remember that the charge on an electron can also be expressed in *coulombs* [Section 2.2]. The conversion factor between the two is necessary to calculate dipole moments: $1 \text{ e}^- = 1.6022 \times 10^{-19}$ C.

where C is coulombs and m is meters. Table 6.2 lists several polar diatomic molecules, their bond lengths, and their experimentally measured dipole moments.

---

[2]Peter Joseph William Debye (1884–1966). An American chemist and physicist of Dutch origin, Debye made many significant contributions to the study of molecular structure, polymer chemistry, X-ray analysis, and electrolyte solutions. He was awarded the Nobel Prize in Chemistry in 1936.

TABLE 6.2	Bond Lengths and Dipole Moments of the Hydrogen Halides	
Molecule	**Bond length (Å)**	**Dipole moment (D)**
HF	0.92	1.82
HCl	1.27	1.08
HBr	1.41	0.82
HI	1.61	0.44

Worked Example 6.2 shows how to use bond lengths and dipole moments to determine the magnitude of the partial charges in a polar diatomic molecule.

Worked Example 6.2

Burns caused by hydrofluoric acid [HF(aq)] are unlike any other acid burns and present unique medical complications. HF solutions typically penetrate the skin and damage internal tissues, including bone, often with minimal surface damage. Less concentrated solutions actually can cause greater injury than more concentrated ones by penetrating more deeply before causing injury, thus delaying the onset of symptoms and preventing timely treatment. Determine the magnitude of the partial positive and partial negative charges in the HF molecule.

Student Annotation: Hydrofluoric acid has several important industrial applications, including the etching of glass and the manufacture of electronic components.

Strategy Rearrange Equation 6.1 to solve for Q. Convert the resulting charge in coulombs to charge in units of electronic charge.

Setup According to Table 6.2, $\mu = 1.82$ D and $r = 0.92$ Å for HF. The dipole moment must be converted from debye to C · m and the distance between the ions must be converted to meters.

$$\mu = 1.82 \text{ D} \times \frac{3.336 \times 10^{-30} \text{ C} \cdot \text{m}}{1 \text{ D}} = 6.07 \times 10^{-30} \text{ C} \cdot \text{m}$$

$$r = 0.92 \text{ Å} \times \frac{1 \times 10^{-10} \text{ m}}{1 \text{ Å}} = 9.2 \times 10^{-11} \text{ m}$$

Solution In coulombs:

$$Q = \frac{\mu}{r} = \frac{6.07 \times 10^{-30} \text{ C} \cdot \text{m}}{9.2 \times 10^{-11} \text{ m}} = 6.598 \times 10^{-20} \text{ C}$$

In units of electronic charge:

$$6.598 \times 10^{-20} \text{ C} \times \frac{1 \; e^-}{1.6022 \times 10^{-19} \text{ C}} = 0.41 \; e^-$$

Therefore, the partial charges in HF are +0.41 and −0.41 on H and F, respectively.

$$^{+0.41}\text{H}-\ddot{\underset{\cdot\cdot}{\text{F}}}\text{:}\;^{-0.41}$$

Think About It
Calculated partial charges should always be less than 1. If a "partial" charge were 1 or greater, it would indicate that at least one electron had been transferred from one atom to the other. Remember that polar bonds involve unequal *sharing* of electrons, not a complete *transfer* of electrons.

Practice Problem Ⓐ**TTEMPT** Using data from Table 6.2, determine the magnitude of the partial charges in HBr.

Practice Problem Ⓑ**UILD** Given that the partial charges on C and O in carbon monoxide are +0.020 and −0.020, respectively, calculate the dipole moment of CO. (The distance between the partial charges, r, is 113 pm.)

(Continued on next page)

Practice Problem Ⓒ**ONCEPTUALIZE** Two pairs of elements are highlighted in the periodic table shown here. Consider two binary compounds, one consisting of the two elements highlighted in yellow and one consisting of the two elements highlighted in blue. For which pair of elements will the partial charges be largest? Explain.

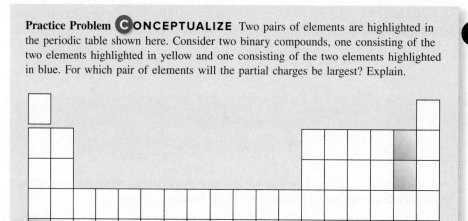

Although the designations "covalent," "polar covalent," and "ionic" can be useful, sometimes chemists wish to describe and compare chemical bonds with more precision. For this purpose, we can use Equation 6.1 to calculate the dipole moment we would *expect* if the charges on the atoms were discrete instead of partial—that is, if an electron had actually been transferred from one atom to the other. Comparing this calculated dipole moment with the measured value gives us a quantitative way to describe the nature of a bond using the term *percent ionic character*. **Percent ionic character** is defined as the ratio of observed μ to calculated μ, multiplied by 100%.

Equation 6.2 $\dfrac{\text{percent ionic}}{\text{character}} = \dfrac{\mu \text{ (observed)}}{\mu \text{ (calculated assuming discrete charges)}} \times 100\%$

Figure 6.6 illustrates the relationship between percent ionic character and the electronegativity difference in a heteronuclear diatomic molecule.

Figure 6.6 Relationship between percent ionic character and electronegativity difference.

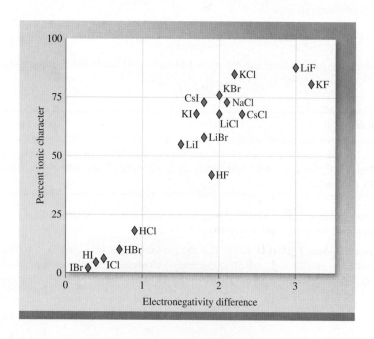

Worked Example 6.3 shows how to calculate percent ionic character using Equation 6.2.

Worked Example 6.3

Using data from Table 6.2, calculate the percent ionic character of the bond in HI.

Strategy Use Equation 6.1 to calculate the dipole moment in HI assuming that the charges on H and I are +1 and −1, respectively; and use Equation 6.2 to calculate percent ionic character. The magnitude of the charges must be expressed as coulombs (1 $e^- = 1.6022 \times 10^{-19}$ C); the bond length (r) must be expressed as meters (1 Å = 1×10^{-10} m); and the calculated dipole moment should be expressed as debyes (1 D = 3.336×10^{-30} C · m).

Setup From Table 6.2, the bond length in HI is 1.61 Å (1.61×10^{-10} m) and the measured dipole moment of HI is 0.44 D.

Solution The dipole we would expect if the magnitude of charges were 1.6022×10^{-19} C is

$$\mu = Q \times r = (1.6022 \times 10^{-19}\ \text{C}) \times (1.61 \times 10^{-10}\ \text{m}) = 2.58 \times 10^{-29}\ \text{C} \cdot \text{m}$$

Converting to debyes gives

$$2.58 \times 10^{-29}\ \text{C} \cdot \text{m} \times \frac{1\ \text{D}}{3.336 \times 10^{-30}\ \text{C} \cdot \text{m}} = 7.73\ \text{D}$$

The percent ionic character of the H—I bond is $\dfrac{0.44\ \text{D}}{7.73\ \text{D}} \times 100\% = 5.7\%$.

> **Think About It**
> A purely covalent bond (in a homonuclear diatomic molecule such as H_2) would have 0 percent ionic character. In theory, a purely ionic bond would be expected to have 100 percent ionic character, although no *purely* ionic bond is known to exist.

Practice Problem ATTEMPT Using data from Table 6.2, calculate the percent ionic character of the bond in HF.

Practice Problem BUILD Using information from Figure 4.13 (page 148), and given that the NaI bond has 59.7 percent ionic character, determine the measured dipole moment of NaI.

Practice Problem CONCEPTUALIZE One metal and three nonmetals are highlighted on the periodic table shown here. List the nonmetal elements in order of increasing ionic character of the bond each might form with the highlighted metal.

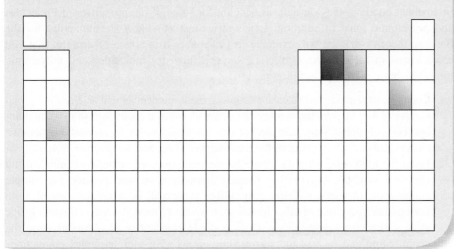

Section 6.2 Review

Electronegativity and Polarity

6.2.1 In which of the following molecules are the bonds *most* polar?

(a) H_2Se
(b) H_2O
(c) CO_2

(d) BCl_3
(e) PCl_5

6.2.2 Arrange molecules A through E in order of increasing percent ionic character.

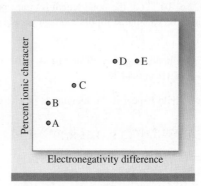

(a) $A < B < C < D < E$
(b) $A = B < C < D < E$
(c) $A = B < C < D = E$

(d) $A < B < C < D = E$
(e) $A < B = C = D < E$

6.2.3 Using data from Table 6.2, calculate the magnitude of the partial charges in HI.

(a) 0.39
(b) 1.8
(c) 0.057

(d) 0.60
(e) 0.15

6.2.4 Using data from Table 6.2, calculate the percent ionic character of HCl.

(a) 85.0
(b) 6.10
(c) 17.7

(d) 2.03
(e) 20.3

6.3   DRAWING LEWIS STRUCTURES

Although the octet rule and Lewis structures alone do not present a complete picture of covalent bonding, they do help us account for some of the properties of molecules and polyatomic ions. In addition, Lewis structures provide a starting point for the bonding theories that we will examine in Chapter 7. It is crucial, therefore, that you learn a system for drawing correct Lewis structures for molecules and polyatomic ions. The basic steps are as follows:

1. Count the total number of valence electrons present. Remember that an element's *group number* (1A–8A) gives the number of valence electrons it contributes to the total number. For polyatomic ions, add electrons to the total number to account for negative charges; subtract electrons from the total number to account for positive charges.

2. From the molecular formula, draw the skeletal structure of the compound, using chemical symbols and placing bonded atoms next to one another. For simple compounds, this step is fairly easy. Often there will be a unique central atom surrounded by a group of other identical atoms. In general, the *least* electronegative atom will be the central atom. Draw a single covalent bond (dash) between the central atom and each of the surrounding atoms. (For more complex compounds

Student Annotation: H cannot be a central atom because it only forms *one* covalent bond.

whose structures might not be obvious, you may need to be given information in addition to the molecular formula.)

3. For each bond (dash) in the skeletal structure, subtract two electrons from the total valence electrons (determined in step 2) to determine the number of remaining electrons.

4. Use the remaining electrons to complete the octets of the terminal atoms (those bonded to the central atom) by placing pairs of electrons on each atom. (Remember that an H atom only requires two electrons to complete its valence shell.) If there is more than one type of terminal atom, complete the octets of the most electronegative atoms first.

5. If any electrons remain after step 4, place them in pairs on the central atom.

6. If the central atom has fewer than eight electrons after completing steps 1 to 5, move one or more pairs from the terminal atoms to form multiple bonds between the central atom and the terminal atoms.

Steps for Drawing Lewis Structures

Step	CH_4	CCl_4	H_2O	O_2	CN^-
1	8	32	8	12	10
2	H—C—H with H above and H below	Cl—C—Cl with Cl above and Cl below	H—O—H	O—O	C—N
3	$8 - 8 = 0$	$32 - 8 = 24$	$8 - 4 = 4$	$12 - 2 = 10$	$10 - 2 = 8$
4	H—C—H with H above and H below	:Cl̈—C—Cl̈: with :Cl̈ above and :Cl̈: below	H—O—H	:Ö—Ö:	:C—N̈:
5	—	—	H—Ö—H	—	—
6	—	—	—	:Ö=Ö:	[:C≡N:]⁻

Worked Example 6.4 shows how to draw a Lewis structure.

Worked Example 6.4

Draw the Lewis structure for carbon disulfide (CS_2).

Strategy Use the procedure just described in steps 1 through 6 for drawing Lewis structures.

Setup

Step 1: The total number of valence electrons is 16: 6 from each S atom and 4 from the C atom [$2(6) + 4 = 16$].

Step 2: C and S have identical electronegativities. We will draw the skeletal structure with the unique atom, C, at the center.

S—C—S

(Continued on next page)

Step 3: Subtract 4 electrons to account for the bonds in the skeletal structure, leaving us 12 electrons to distribute.

Step 4: Distribute the 12 remaining electrons as 3 lone pairs on each S atom.

$$:\ddot{S}-C-\ddot{S}:$$

Step 5: There are no electrons remaining after step 4, so step 5 does not apply.

Step 6: To complete carbon's octet, use one lone pair from each S atom to make a double bond to the C atom.

Solution

$$\ddot{S}=C=\ddot{S}:$$

> **Think About It**
> Counting the total number of valence electrons should be relatively simple to do, but it is often done hastily and is therefore a potential source of error in this type of problem. Remember that the number of valence electrons for each element is equal to the group number of that element.

Practice Problem Ⓐ**TTEMPT** Draw the Lewis structure for NF_3.

Practice Problem Ⓑ**UILD** Draw the Lewis structure for ClO_3^-.

Practice Problem Ⓒ**ONCEPTUALIZE** Of the three Lewis structures shown here, identify any that are not correct and specify what is wrong.

$$[:C\!\!=\!\!N:]^-\qquad :O\!\!=\!\!\ddot{C}\!\!=\!\!O:\qquad :\ddot{O}-\ddot{S}-\ddot{O}:$$
(i)   (ii)   (iii)

Section 6.3 Review

Drawing Lewis Structures

6.3.1 Identify the correct Lewis structure for formic acid (HCOOH).

(a) $H-\ddot{C}-\ddot{O}-\ddot{O}-H$

(b) $H-\ddot{O}\!\!=\!\!C\!\!=\!\!\ddot{O}-H$

(c) $H-\ddot{O}-\ddot{C}-\ddot{O}-H$

(d)
$$\overset{\displaystyle :O:}{\underset{\displaystyle}{\overset{\displaystyle \|}{H-C-\ddot{O}-H}}}$$

(e)
$$\overset{\displaystyle :\ddot{O}:}{\underset{\displaystyle H}{\overset{\displaystyle |}{:\ddot{O}-C-H}}}$$

6.3.2 Identify the correct Lewis structure for hydrogen peroxide (H_2O_2).

(a) $H-\ddot{O}\!\!=\!\!\ddot{O}-H$

(b) $H-O\!\!\equiv\!\!O-H$

(c) $H-\ddot{O}-\ddot{O}-H$

(d) $H\!\!=\!\!O\!\!=\!\!O\!\!=\!\!H$

(e) $H-H-\ddot{O}-\ddot{O}:$

6.4   LEWIS STRUCTURES AND FORMAL CHARGE

So far you have learned one method of electron "bookkeeping." In Section 6.2, you learned how to calculate *partial charges*. Another commonly used method of electron bookkeeping is ***formal charge,*** which can be used to determine the most plausible Lewis structures when more than one possibility exists for a compound. Formal charge is determined by comparing the number of electrons associated with an atom in a

Video 6.3
Chemical Bonding—formal charge calculations

Lewis structure with the number of electrons that would be associated with the isolated atom. In an isolated atom, the number of electrons associated with the atom is simply the number of valence electrons. (As usual, we need not be concerned with the core electrons.)

To determine the number of electrons associated with an atom in a Lewis structure, keep in mind the following:

- *All* the atom's *nonbonding* electrons are associated with the atom.
- *Half* of the atom's *bonding* electrons are associated with the atom.

An atom's formal charge is calculated as follows:

formal charge = valence electrons − associated electrons   **Equation 6.3**

We can illustrate the concept of formal charge using the ozone molecule (O_3). Use the step-by-step method for drawing Lewis structures to draw the Lewis structure for ozone, and then determine the formal charge on each O atom by subtracting the number of associated electrons from the number of valence electrons.

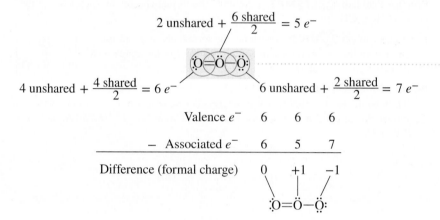

Student Annotation: While you are new at determining formal charges, it may be helpful to draw Lewis structures with all dots, rather than dashes. This can make it easier to see how many electrons are associated with each atom.

Remember that for the purpose of counting associated electrons, those shared by two atoms are evenly split between them.

The sum of the formal charges must equal the overall charge on the species. Because O_3 is a molecule, its formal charges must sum to zero. For ions, the formal charges must sum to the overall charge on the ion.

Formal charges do *not* represent actual charges on atoms in a molecule. In the O_3 molecule, for example, there is no evidence that the central atom bears a net +1 charge or that one of the terminal atoms bears a −1 charge. Assigning formal charges to the atoms in the Lewis structure merely helps us keep track of the electrons involved in bonding in the molecule.

Worked Example 6.5 lets you practice determining formal charges.

Worked Example 6.5

The widespread use of fertilizers has resulted in the contamination of some groundwater with nitrates, which are potentially harmful. Nitrate toxicity is due primarily to its conversion in the body to nitrite (NO_2^-), which interferes with the ability of hemoglobin to transport oxygen. Determine the formal charges on each atom in the nitrate ion (NO_3^-).

Strategy Use steps 1 through 6 on pages 222–224 for drawing Lewis structures to draw the Lewis structure of (NO_3^-). For each atom, subtract the associated electrons from the valence electrons.

(Continued on next page)

Setup

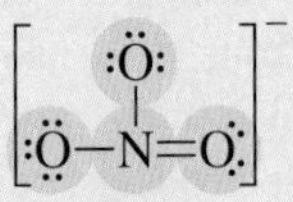

$$\left[\begin{array}{c} :\ddot{O}: \\ | \\ :\ddot{O}-N=\ddot{O}: \end{array}\right]^{-}$$

The N atom has five valence electrons and four associated electrons (one from each single bond and two from the double bond). Each singly bonded O atom has six valence electrons and seven associated electrons (six in three lone pairs and one from the single bond). The doubly bonded O atom has six valence electrons and six associated electrons (four in two lone pairs and two from the double bond).

Solution The formal charges are as follows: +1 (N atom), −1 (singly bonded O atoms), and 0 (doubly bonded O atom).

> **Think About It**
> The sum of formal charges $(+1) + (-1) + (-1) + (0) = -1$ is equal to the overall charge on the nitrate ion.

Practice Problem (A)**TTEMPT** Determine the formal charges on each atom in the carbonate ion (CO_3^{2-}).

Practice Problem (B)**UILD** Determine the formal charges and use them to determine the overall charge, if any, on the species represented by the Lewis structure shown here.

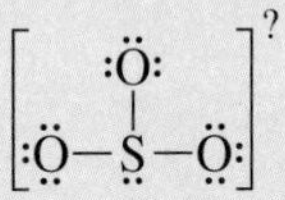

$$\left[\begin{array}{c} :\ddot{O}: \\ | \\ :\ddot{O}-S-\ddot{O}: \end{array}\right]^{?}$$

Practice Problem (C)**ONCEPTUALIZE** The hypothetical element A is shown here in three different partial Lewis structures. For each structure, determine what the formal charge is on A if it is a member of (a) Group 7A, (b) Group 5A, and (c) Group 3A.

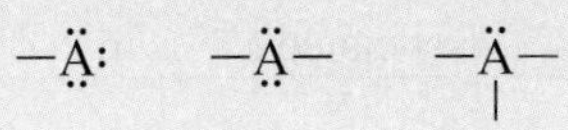

$$-\ddot{A}: \qquad -\ddot{A}- \qquad -\underset{|}{\overset{}{\ddot{A}}}-$$
$$\text{(i)} \qquad \text{(ii)} \qquad \text{(iii)}$$

Sometimes, there is more than one possible skeletal arrangement of atoms for the Lewis structure for a given species. In such cases, we often can select the best skeletal arrangement by using formal charges and the following guidelines:

- For molecules, a Lewis structure in which all formal charges are zero is preferred to one in which there are nonzero formal charges.
- Lewis structures with small formal charges (0 and ±1) are preferred to those with large formal charges (±2, ±3, and so on).
- The best skeletal arrangement of atoms will give rise to Lewis structures in which the formal charges are consistent with electronegativities. For example, the more electronegative atoms should have the more negative formal charges.

Worked Example 6.6 shows how formal charge can be used to determine the best skeletal arrangement of atoms for the Lewis structure of a molecule or polyatomic ion.

Worked Example 6.6

Formaldehyde (CH_2O), which can be used to preserve biological specimens, is commonly sold as a 37 percent aqueous solution. Use formal charges to determine which skeletal arrangement of atoms shown here is the best choice for the Lewis structure of CH_2O.

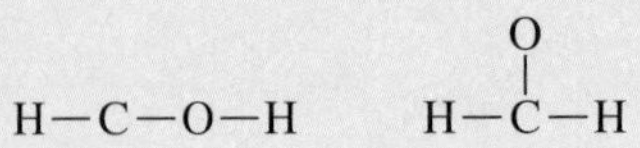

$$\text{H}-\text{C}-\text{O}-\text{H} \qquad \overset{\overset{\displaystyle O}{\displaystyle |}}{\text{H}-\text{C}-\text{H}}$$

Strategy Complete the Lewis structures for each of the CH_2O skeletons shown and determine the formal charges on the atoms in each one.

Setup The completed Lewis structures for the skeletons shown are

$$H-\ddot{C}=\ddot{O}-H \qquad H-\overset{\overset{\displaystyle \ddot{O}}{\|}}{C}-H$$

In the structure on the left, the formal charges are as follows:

Both H atoms: 1 valence e^- − 1 associated e^- (from single bond) = 0

C atom: 4 valence e^- − 5 associated e^- (two in the lone pair, one from the single bond, and two from the double bond) = −1

O atom: 6 valence e^- − 5 associated e^- (two from the lone pair, one from the single bond, and two from the double bond) = +1

$$H-\ddot{C}=\ddot{O}-H$$
Formal charges   0   −1   +1   0

In the structure on the right, the formal charges are as follows:

Both H atoms: 1 valence e^- − 1 associated e^- (from single bond) = 0

C atom: 4 valence e^- − 4 associated e^- (one from each single bond, and two from the double bond) = 0

O atom: 6 valence e^- − 6 associated e^- (four from the two lone pairs and two from the double bond) = 0

$$H-\overset{\overset{\displaystyle \ddot{O}}{\|}}{C}-H$$
Formal charges all zero

Solution Of the two possible arrangements, the structure on the left has an O atom with a positive formal charge, which is inconsistent with oxygen's high electronegativity. Therefore, the structure on the right, in which both H atoms are attached directly to the C atom and all atoms have a formal charge of zero, is the better choice for the Lewis structure of CH_2O.

Think About It

For a molecule, formal charges of zero are preferred. When there are nonzero formal charges, they should be consistent with the electronegativities of the atoms in the molecule. A positive formal charge on oxygen, for example, is inconsistent with oxygen's high electronegativity.

Practice Problem **A**TTEMPT Two possible arrangements are shown for the Lewis structure of a carboxyl group, —COOH. Use formal charges to determine which of the two arrangements is better.

$$-\overset{\overset{\displaystyle \ddot{O}}{\|}}{C}-\ddot{O}-H \qquad -\ddot{C}-\ddot{O}=\ddot{O}-H$$

Practice Problem **B**UILD Use Lewis structures and formal charges to determine the best skeletal arrangement of atoms in (NCl_2^-).

Practice Problem **C**ONCEPTUALIZE For each partial Lewis structure shown here, determine what group element A must belong to in order for its formal charge to be zero.

$$-\overset{\|}{A}- \qquad -\ddot{A}- \qquad -\overset{\overset{\displaystyle \cdot\cdot}{}}{\underset{|}{\ddot{A}}}- \qquad =A=$$

    (i)   (ii)   (iii)   (iv)

Section 6.4 Review

Lewis Structures and Formal Charge

6.4.1   Determine the formal charges on H, C, and N, respectively, in HCN.

(a) 0, +1, and −1   (d) 0, +1, and +1

(b) −1, +1, and 0   (e) 0, 0, and 0

(c) 0, −1, and +1

6.4.2   Which of the Lewis structures shown is most likely preferred for NCO⁻?

(a) $\left[\ddot{N}=C=\ddot{O} \right]^{-}$

(b) $\left[:\ddot{N}-C\equiv O: \right]^{-}$

(c) $\left[:N\equiv C-\ddot{O}: \right]^{-}$

(d) $\left[:\ddot{N}-C=\ddot{O} \right]^{-}$

(e) $\left[\ddot{N}=C-\ddot{O}: \right]^{-}$

6.5   RESONANCE

Our drawing of the Lewis structure for ozone (O_3) satisfied the octet rule for the central O atom because we placed a double bond between it and one of the two terminal O atoms. In fact, we can put the double bond at either end of the molecule, as shown by the following two equivalent Lewis structures:

$$\ddot{O}=\ddot{O}-\ddot{O}: \quad \longleftrightarrow \quad :\ddot{O}-\ddot{O}=\ddot{O}$$

A single bond between O atoms should be longer than a double bond between O atoms, but experimental evidence indicates that both of the bonds in O_3 are equal in length (128 pm). Because neither one of these two Lewis structures accounts for the known bond lengths in O_3, we use both Lewis structures to represent the ozone molecule.

Each of the Lewis structures is called a resonance structure. A ***resonance structure*** is one of two or more equally valid Lewis structures for a single species that cannot be represented accurately by a single Lewis structure. The double-headed arrow (⟷) indicates that the structures shown are resonance structures. Like the medieval European traveler to Africa who described a rhinoceros as a cross between a griffin and a unicorn (two familiar but imaginary animals), we describe ozone, a real molecule, in terms of two familiar but nonexistent structures.

A common misconception about resonance is that a molecule such as ozone exists as one resonance structure at one time, and as another resonance structure the other times. In fact, the molecule never actually exists as either resonance structure. Neither of the resonance structures adequately represents the actual molecule, which has its own unique, stable structure. "Resonance" is a human invention, designed to address the limitations of a simple bonding model. To extend the animal analogy, a rhinoceros is not a *griffin* part of the time and a *unicorn* part of the time. It is always a rhinoceros—that is, a distinct, *real* creature.

The carbonate ion provides another example of resonance:

$$\left[\ddot{O}=C-\ddot{O}: \atop :\ddot{O} \right]^{2-} \longleftrightarrow \left[:\ddot{O}-C-\ddot{O}: \atop \ddot{O} \right]^{2-} \longleftrightarrow \left[:\ddot{O}-C=\ddot{O} \atop :\ddot{O}: \right]^{2-}$$

According to experimental evidence, all three carbon-oxygen bonds in CO_3^{2-} are equivalent. Therefore, the properties of the carbonate ion are best explained by considering its resonance structures together.

The concept of resonance applies equally well to organic systems. A good example is the benzene molecule (C_6H_6):

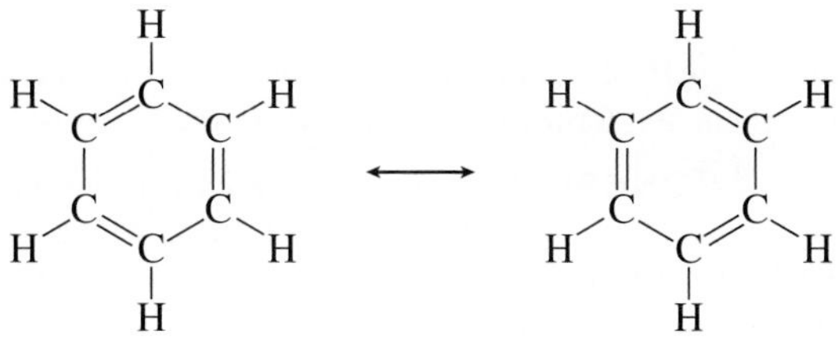

If one of these resonance structures corresponded to the actual structure of benzene, there would be two different bond lengths between adjacent C atoms, one with the properties of a single bond and the other with the properties of a double bond. In fact, the distance between all adjacent C atoms in benzene is 140 pm, which is shorter than a C—C bond (154 pm) and longer than a C=C bond (133 pm).

A simpler way of drawing the structure of the benzene molecule and other compounds containing the benzene ring is to show only the skeleton and not the carbon and hydrogen atoms. By this convention, the resonance structures are represented by

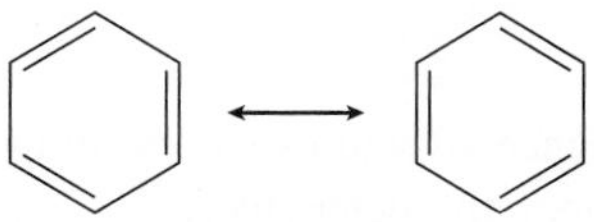

Note that the C atoms at the corners of the hexagon and the H atoms are not shown, although they are understood to be there. Only the bonds between the C atoms are shown.

Resonance structures differ only in the positions of their *electrons*—not in the positions of their atoms. Thus, :N=N=Ö: and :N≡N—Ö: are resonance structures of each other, whereas :N=N=Ö: and :N=O=N: are not.

Worked Example 6.7 shows how to draw resonance structures.

> **Student Hot Spot**
>
> Student data indicate you may struggle with resonance structures. Access the SmartBook to view additional Learning Resources on this topic.

> **Student Annotation:** The representation of organic compounds is discussed in more detail in Chapter 23.

Worked Example 6.7

High oil and gasoline prices have renewed interest in alternative methods of producing energy, including the "clean" burning of coal. Part of what makes "dirty" coal *dirty* is its high sulfur content. Burning dirty coal produces sulfur dioxide (SO_2), among other pollutants. Sulfur dioxide is oxidized in the atmosphere to form sulfur trioxide (SO_3), which subsequently combines with water to produce sulfuric acid—a major component of acid rain. Draw all possible resonance structures of sulfur trioxide.

Strategy Draw two or more Lewis structures for SO_3 in which the atoms are arranged the same way but the electrons are arranged differently.

Setup Following the steps for drawing Lewis structures, we determine that a correct Lewis structure for SO_3 contains two sulfur-oxygen single bonds and one sulfur-oxygen double bond. But the double bond can be put in any one of three positions in the molecule.

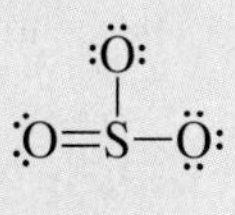

Solution

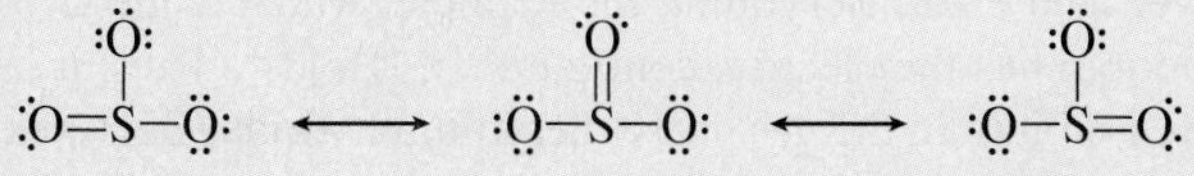

Think About It
Always make sure that resonance structures differ only in the positions of the electrons, *not* in the positions of the atoms.

(Continued on next page)

Practice Problem **A**TTEMPT Draw all possible resonance structures for the nitrate ion (NO_3^-).

Practice Problem **B**UILD Draw three resonance structures for the thiocyanate ion (NCS^-), and determine the formal charges in each resonance structure.

Practice Problem **C**ONCEPTUALIZE The Lewis structure of a molecule consisting of the hypothetical elements A, B, and C is shown here. Of the four other structures, identify any that is *not* a resonance structure of the original and explain why it is not a resonance structure.

$$:\ddot{C}-A=\ddot{C}:\qquad :C=\ddot{B}-\ddot{C}:\qquad :\ddot{C}-\ddot{A}-\ddot{C}:\qquad :\ddot{C}-\ddot{A}-\ddot{C}:\qquad :\ddot{C}-\ddot{A}-\ddot{C}:$$

(i)   (ii)   (iii)   (iv)

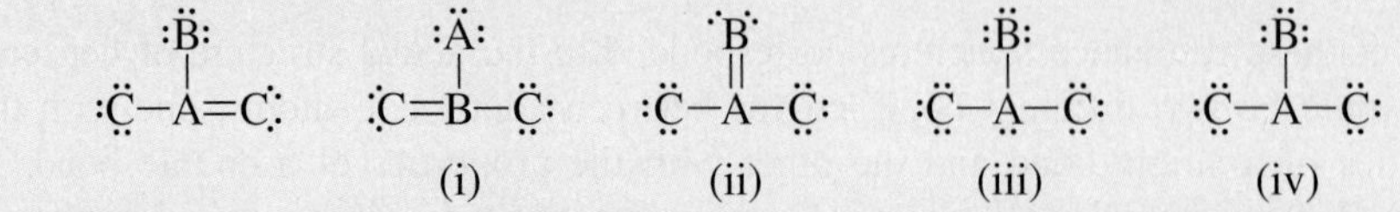

Section 6.5 Review

Resonance

6.5.1   How many resonance structures can be drawn for the nitrite ion (NO_2^-)? (N and O must obey the octet rule.)

(a) 1   (d) 4
(b) 2   (e) 5
(c) 3

6.5.2   Indicate which of the following are resonance structures of $:\ddot{C}l-Be-\ddot{C}l:$ (select all that apply).

(a) $\ddot{C}l=Be=\ddot{C}l$ (d) $\ddot{B}e=Cl=\ddot{C}l$
(b) $:Cl\equiv Be-\ddot{C}l:$ (e) $:\ddot{C}l-Be\equiv Cl:$
(c) $:Be\equiv Cl-\ddot{C}l:$

6.6   EXCEPTIONS TO THE OCTET RULE

The octet rule almost always holds for second period elements. Exceptions to the octet rule fall into three categories:

1. The central atom has fewer than eight electrons due to a shortage of electrons.
2. The central atom has fewer than eight electrons due to an odd number of electrons.
3. The central atom has more than eight electrons.

Incomplete Octets

In some compounds, the number of electrons surrounding the central atom in a stable molecule is fewer than eight. Beryllium, for example, which is the Group 2A element in the second period, has the electron configuration $[He]2s^2$. Thus, it has two valence electrons in the $2s$ orbital. In the gas phase, beryllium hydride (BeH_2) exists as discrete molecules. The Lewis structure of BeH_2 is

$$H-Be-H$$

Only four electrons surround the Be atom, so there is no way to satisfy the octet rule for beryllium in this molecule.

Elements in Group 3A also tend to form compounds in which they are surrounded by fewer than eight electrons. Boron, for example, has the electron configuration

$[He]2s^22p^1$, so it has only three valence electrons. Boron reacts with the halogens to form a class of compounds having the general formula BX_3, where X is a halogen atom. Thus, there are only six electrons around the boron atom in boron trifluoride:

$$\ddot{\underset{\ddot{F}}{\overset{:\ddot{F}:}{\vert}}} - B - \ddot{F}:$$

We actually *can* satisfy the octet rule for boron in BF_3 by using a lone pair on one of the F atoms to form a double bond between the F atom and boron. This gives rise to three additional resonance structures:

$$:\ddot{F} = B - \ddot{F}: \quad \longleftrightarrow \quad :\ddot{F} - B - \ddot{F}: \quad \longleftrightarrow \quad :\ddot{F} - B = F:$$

Although these resonance structures result in boron carrying a negative formal charge while fluorine carries a positive formal charge, a situation that is inconsistent with the electronegativities of the atoms involved, the experimentally determined bond length in BF_3 (130.9 pm) is shorter than a single bond (137.3 pm). The shorter bond length would appear to support the inclusion of these three resonance structures in the representation of BF_3.

On the other hand, boron trifluoride combines with ammonia to give a species in which boron's octet is complete. This is easier to picture starting with the resonance structure in which boron has only six valence electrons around it:

$$\ddot{\underset{:\ddot{F}:}{\overset{:\ddot{F}:}{\vert}}} - B \quad + \quad :N - H \quad \longrightarrow \quad :\ddot{F} - B - N - H$$

No *one* of the resonance structures is sufficient to describe the bonding in BF_3. Rather, *all* of them are necessary to account for the physical and chemical properties that we observe.

The B—N bond in F_3B—NH_3 is different from the covalent bonds discussed so far in the sense that both electrons are contributed by the N atom. This type of bond is called a ***coordinate covalent bond*** (also referred to as a ***dative bond***), which is defined as a covalent bond in which one of the atoms donates both electrons. Although the properties of a coordinate covalent bond do not differ from those of a normal covalent bond (i.e., the electrons are shared in both cases), the distinction is useful for keeping track of valence electrons and assigning formal charges.

The formation of a coordinate covalent bond is an example of a *Lewis acid-base process*. A ***Lewis base*** is a species that has a lone pair of electrons that it can *donate*; and a ***Lewis acid*** is a species that can *accept* a pair of electrons. In this example, NH_3 is the Lewis base. (It has a lone pair, which it donates to BF_3.) BF_3 is the Lewis acid. (Its central atom has only six electrons around it, enabling it to accept the lone pair—thereby completing boron's octet.) The electrons donated by the base and accepted by the acid constitute the bond that forms between the two species. The Lewis definition of acids and bases is one of several we will encounter; and it will be described in more detail in Chapter 16.

Student Annotation: When experimental evidence supports the "correctness" of a resonance structure, we say that the resonance structure is "important." If the experimental evidence supports a particular resonance structure more strongly than the others, we call that resonance structure "more important."

Odd Numbers of Electrons

Some molecules, such as nitrogen dioxide (NO_2), contain an odd number of electrons.

$$:\ddot{O} = \dot{N} - \ddot{O}:$$

Because we need an even number of electrons for every atom in a molecule to have a complete octet, the octet rule cannot be obeyed for all the atoms in these molecules. Molecules with an odd number of electrons are sometimes referred to as ***free radicals*** (or just *radicals*). Many radicals are highly reactive, because there is a tendency for the unpaired electron to form a covalent bond with an unpaired electron on another molecule. When two nitrogen dioxide molecules collide, for example, they form dinitrogen tetroxide, a molecule in which the octet rule *is* satisfied for both the N and O atoms.

$$\ddot{\underset{\ddot{O}}{\overset{\ddot{O}}{N}}} \cdot \quad + \quad \cdot \underset{\ddot{O}}{\overset{\ddot{O}}{N}} \quad \longrightarrow \quad \underset{\ddot{O}}{\overset{\ddot{O}}{N}} - \underset{\ddot{O}}{\overset{\ddot{O}}{N}}$$

Worked Example 6.8 lets you practice drawing Lewis structures for a species with an odd number of electrons.

Worked Example 6.8

Draw the Lewis structure of chlorine dioxide (ClO_2).

Strategy The skeletal structure is

$$O{-}Cl{-}O$$

This puts the unique atom, Cl, in the center and puts the more electronegative O atoms in terminal positions.

Setup There are a total of 19 valence electrons (7 from the Cl and 6 from each of the two O atoms). We subtract 4 electrons to account for the two bonds in the skeleton, leaving us with 15 electrons to distribute as follows: three lone pairs on each O atom, one lone pair on the Cl atom, and the last remaining electron also on the Cl atom.

Solution

$$\ddot{\underset{..}{O}} {-} \dot{\underset{..}{Cl}} {-} \ddot{\underset{..}{O}}{:}$$

> **Think About It**
> ClO_2 is used primarily to bleach wood pulp in the manufacture of paper, but it is also used to bleach flour, disinfect drinking water, and deodorize certain industrial facilities. It was critical in the eradication of toxic mold in homes in New Orleans that were damaged by the devastating floodwaters of Hurricane Katrina in 2005.

Practice Problem **A**TTEMPT Draw the Lewis structure for the OH species.

Practice Problem **B**UILD Draw the Lewis structure for the NS_2 molecule.

Practice Problem **C**ONCEPTUALIZE Hypothetical elements A and B combine to form a number of molecules and polyatomic ions. (Element A is a member of Group 5A; element B is a member of Group 6A.) Using the chemical formulas, determine which of the following species must be represented by a Lewis structure with an unpaired electron.

$$AB \quad AB_2 \quad A_2B \quad AB_3 \quad AB_3^- \quad AB_2^-$$

Severe flooding in New Orleans after Hurricane Katrina in 2005.

© Wesley Bocxe/Science Source

Student Annotation: The OH species is a *radical*, not to be confused with the hydroxide *ion* (OH^-).

Thinking Outside the Box

Species with Unpaired Electrons

Beginning about a week after the September 11, 2001, attacks, letters containing anthrax bacteria were mailed to several news media offices and to two U.S. senators. Of the 22 people who subsequently contracted anthrax, 5 died. Anthrax is a spore-forming bacterium (*Bacillus anthracis*) and, like smallpox, is classified by the CDC as a *Category A bioterrorism agent*. Spore-forming bacteria are notoriously difficult to kill, making the cleanup of the buildings contaminated by anthrax costly and time consuming. The American Media Inc. (AMI) building in Boca Raton, Florida, was not deemed safe to enter until July 2004, after it had been treated with chlorine dioxide (ClO_2), the only structural fumigant approved by the Environmental Protection Agency (EPA) for anthrax decontamination. The effectiveness of ClO_2 in killing anthrax and other hardy biological agents stems in part from its being a *radical,* meaning that it contains an odd number of electrons.

While ClO_2 is normally produced at a required site using a chlorine dioxide generator, radicals occur naturally in a number of places ranging from interstellar space to the human body. The hydroxyl radical (OH·) was first detected in interstellar space in 1963 but is also found in the atmosphere and mammalian cells. In the troposphere, hydroxyl radical interacts with various air pollutants; in cells, this same species attacks macromolecules such as carbohydrates, amino acids, and DNA—potentially leading to cancer. The superoxide radical (O_2^-) is generated in humans as part of the immune response, but excess amounts are harmful. Fortunately, the enzyme *superoxide dismutase* can provide some protection by "scavenging" this biologically toxic radical. Finally, nitric oxide (NO) is an important radical species involved in a number of signaling roles in cells.

The American Media Inc. building in Boca Raton, Florida.
© Eliot J Schechter/Stringer/AFP/Getty Images

Expanded Octets

Atoms of the second period elements cannot have more than eight valence electrons around them, but atoms of elements in and beyond the third period of the periodic table can. In addition to the $3s$ and $3p$ orbitals, elements in the third period also have $3d$ orbitals that can be used in bonding. These orbitals enable an atom to form an *expanded octet*. One compound in which there is an expanded octet is sulfur hexafluoride, a very stable compound. The electron configuration of sulfur is $[Ne]3s^23p^4$. In SF_6, each of sulfur's 6 valence electrons forms a covalent bond with a fluorine atom, so there are 12 electrons around the central sulfur atom:

$$
\begin{array}{ccc}
 & :\!\ddot{F}\!: & \\
:\!\dot{F} & |\ \ & \ddot{F}\!: \\
 & S & \\
:\!\ddot{F} & |\ \ & \ddot{F}\!: \\
 & :\!\ddot{F}\!: &
\end{array}
$$

In Chapter 7, we will see that these 12 electrons, or six bonding pairs, are accommodated in six orbitals that originate from the one $3s$, the three $3p$, and two of the five $3d$ orbitals. Sulfur also forms many compounds in which it does obey the octet rule. In sulfur dichloride, for instance, S is surrounded by only 8 electrons:

$$:\!\ddot{C}l\!-\!\ddot{S}\!-\!\ddot{C}l\!:$$

When drawing Lewis structures of compounds containing a central atom from the third period and beyond, occasionally you may find that the octet rule is satisfied for all the atoms before all the valence electrons have been distributed. When this happens, the extra electrons should be placed as lone pairs on the central atom. Worked Example 6.9 illustrates this approach.

Worked Example 6.9 involves compounds that do not obey the octet rule.

Worked Example 6.9

Draw the Lewis structures of (a) boron triiodide (BI_3), (b) arsenic pentafluoride (AsF_5), and (c) xenon tetrafluoride (XeF_4).

Strategy Follow the step-by-step procedure for drawing Lewis structures. The skeletal structures are

$$
\text{(a)} \ \ \begin{array}{c} I \\ | \\ I\!-\!B\!-\!I \end{array}
\qquad
\text{(b)} \ \ \begin{array}{c} F \\ | \\ F\!\diagdown \\ F\!\diagup As\!-\!F \\ | \\ F \end{array}
\qquad
\text{(c)} \ \ \begin{array}{c} F \\ | \\ F\!-\!Xe\!-\!F \\ | \\ F \end{array}
$$

Note that the skeletal structure already has more than an octet around the As atom.

(Continued on next page)

Setup (a) There are a total of 24 valence electrons in BI_3 (3 from the B and 7 from each of the three I atoms). We subtract 6 electrons to account for the three bonds in the skeleton, leaving 18 electrons to distribute as three lone pairs on each I atom. (b) There are 40 total valence electrons [5 from As (Group 5A) and 7 from each of the five F atoms (Group 7A)]. We subtract 10 electrons to account for the five bonds in the skeleton, leaving 30 to be distributed. Next, place three lone pairs on each F atom, thereby completing all their octets and using up all the electrons. (c) There are 36 total valence electrons (8 from Xe and 7 from each of the four F atoms). We subtract 8 electrons to account for the bonds in the skeleton, leaving 28 to distribute. We first complete the octets of all four F atoms. When this is done, four electrons remain, so we place two lone pairs on the Xe atom.

Solution

$$\text{(a)} \ \ddot{\underset{..}{I}}\!-\!\overset{\displaystyle :\ddot{I}:}{\underset{\displaystyle :\ddot{I}:}{B}}\!-\!\ddot{\underset{..}{I}}: \qquad \text{(b)} \ \overset{\displaystyle :\ddot{F}:}{\underset{\displaystyle :\ddot{F}:}{\overset{:\ddot{F}:}{\underset{:\ddot{F}:}{As}}}}\!-\!\ddot{\underset{..}{F}}: \qquad \text{(c)} \ \overset{\displaystyle :\ddot{F}:}{\underset{\displaystyle :\ddot{F}:}{:\ddot{F}\!-\!Xe\!-\!\ddot{F}:}}$$

Think About It

Boron is one of the elements that does not always follow the octet rule. Like BF_3, however, BI_3 can be drawn with a double bond to satisfy the octet of boron. This gives rise to a total of four resonance structures:

$$\overset{\displaystyle :\ddot{I}:}{:\ddot{I}\!-\!B\!-\!\ddot{I}:} \ \longleftrightarrow \ \overset{\displaystyle :\ddot{I}:}{:\ddot{I}\!=\!B\!-\!\ddot{I}:} \ \longleftrightarrow \ \overset{\displaystyle :\ddot{I}}{:\ddot{I}\!-\!B\!-\!\ddot{I}:} \ \longleftrightarrow \ \overset{\displaystyle :\ddot{I}:}{:\ddot{I}\!-\!B\!=\!\ddot{I}}$$

Atoms beyond the second period can accommodate more than an octet of electrons, whether those electrons are used in bonds or reside on the central atom as lone pairs.

Practice Problem Ⓐ**TTEMPT** Draw the Lewis structures of (a) beryllium fluoride (BeF_2), (b) phosphorus pentachloride (PCl_5), and (c) the iodine tetrachloride ion ICl_4^-.

Practice Problem Ⓑ**UILD** Draw the Lewis structures of (a) boron trichloride (BCl_3), (b) antimony pentafluoride (SbF_5), and (c) krypton difluoride (KrF_2).

Practice Problem Ⓒ**ONCEPTUALIZE** Elements in the same group exhibit similar chemistry and sometimes form analogous species. For example, nitrogen and phosphorus (both members of Group 5A) can combine with chlorine in a 1:3 ratio to form NCl_3 and PCl_3, respectively. Phosphorus can also combine with chlorine in a 1:5 ratio. Explain why nitrogen cannot.

In some cases, a species can be represented by a Lewis structure in which the octet rule is obeyed, or by one in which the central atom has an expanded octet. The sulfate ion (SO_4^{2-}), for example, can be represented by Structure I, which obeys the octet rule, or by Structure II, which does not.

$$\left[\ \overset{\displaystyle :\ddot{O}:^{-1}}{\underset{\displaystyle :\ddot{O}:_{-1}}{^{-1}:\ddot{O}\!-\!\overset{+2}{S}\!-\!\ddot{O}:^{-1}}} \ \right]^{2-} \qquad\qquad \left[\ \overset{\displaystyle :\ddot{O}:^{-1}}{\underset{\displaystyle :\ddot{O}:_{-1}}{:\ddot{O}\!=\!S\!=\!\ddot{O}:}} \ \right]^{2-}$$

Structure I   Structure II

In Structure I, although the octet rule is obeyed for the central atom, there are nonzero formal charges (shown in blue) on all of the atoms. In Structure II, by relocating a lone pair from each of two oxygen atoms and creating two double bonds, we change three of the formal charges to zero. Although some chemists have a strong preference for one or the other (structures in which the octet rule is obeyed or structures in which formal charges are minimized), both are valid Lewis structures.

Worked Example 6.10 gives you some practice drawing resonance structures for species that can obey the octet rule, but that also can be drawn with expanded octets to minimize formal charges.

Worked Example 6.10

Draw two resonance structures for sulfurous acid (H_2SO_3): one that obeys the octet rule for the central atom and one that minimizes the formal charges. Determine the formal charge on each atom in both structures.

Strategy Begin by drawing the skeletal structure and counting the total number of valence electrons. Use the steps outlined in Section 6.3 to draw the first structure and reposition one or more lone pairs to adjust the formal charges for the second structure.

Setup The skeletal structure is

$$\begin{matrix} & & O & & \\ & & | & & \\ H-O-&S&-O-H \end{matrix}$$

Note that each hydrogen in an oxoacid is attached to an oxygen atom, and *not* directly to the central atom. The total number of valence electrons is 26 (6 from S, 6 from each O, and 1 from each H).

Solution Following the steps outlined on pages 222–224, we get the first structure:

$$\begin{matrix} & & :\overset{..}{\underset{..}{O}}:^{-1} & & \\ & & | & & \\ H-\overset{..}{\underset{..}{O}}-&\overset{+1}{S}&-\overset{..}{\underset{..}{O}}-H \end{matrix}$$

From the top O atom, which has three lone pairs, we reposition one lone pair to create a double bond between O and S to get the second structure:

$$\begin{matrix} & & \overset{..}{O} & & \\ & & \| & & \\ H-\overset{..}{\underset{..}{O}}-&S&-\overset{..}{\underset{..}{O}}-H \end{matrix}$$

Incorporating the double bond results in every atom having a formal charge of zero.

Think About It

In some species, such as the sulfate ion, it is possible to incorporate too many double bonds. Structures with *three* and *four* double bonds to sulfur would give formal charges on S and O that are inconsistent with the electronegativities of these elements. In general, if you are trying to minimize formal charges by expanding the central atom's octet, only add enough double bonds to make the formal charge on the central atom zero.

Practice Problem (A)TTEMPT Draw three resonance structures for the hydrogen sulfite ion (HSO_3^-), one that obeys the octet rule for the central atom and two that expand the octet of the central atom. Calculate the formal charges on all atoms in each structure and determine which, if any, of the resonance structures has formal charges that are inconsistent with the elements' electronegativities.

Practice Problem (B)UILD Draw two resonance structures for each species—one that obeys the octet rule and one in which the formal charge on the central atom is zero: SO_3, SO_2.

Practice Problem (C)ONCEPTUALIZE Three resonance structures are shown for a polyatomic ion, in which the central atom (X) is a member of Group 5A. List the structures in order of decreasing formal charge on the central atom.

$$\left[\begin{matrix} :O: \\ \| \\ \overset{..}{O}=A-\overset{..}{O}: \\ \| \\ :O: \end{matrix}\right]^{3-} \qquad \left[\begin{matrix} :O: \\ \| \\ :\overset{..}{O}-A-\overset{..}{O}: \\ | \\ :O: \end{matrix}\right]^{3-} \qquad \left[\begin{matrix} :O: \\ \| \\ :\overset{..}{O}-A-\overset{..}{O}: \\ | \\ :O: \end{matrix}\right]^{3-}$$

$$\text{(i)} \qquad\qquad\qquad \text{(ii)} \qquad\qquad\qquad \text{(iii)}$$

Section 6.6 Review

Exceptions to the Octet Rule

6.6.1 In which of the following species does the central atom not obey the octet rule?

(a) BrO_3^- (d) HCN

(b) ClO_2^- (e) ICl_4^-

(c) CO_2

(Continued on next page)

6.6.2   Which elements cannot have more than an octet of electrons? (Select all that apply.)
(a) N   (d) Br
(b) C   (e) O
(c) S

6.6.3   How many electrons are around the central atom in BBr$_3$?
(a) 4   (d) 10
(b) 6   (e) 12
(c) 8

6.6.4   How many lone pairs are there on the central atom in the Lewis structure of ICl$_2^-$?
(a) 0   (d) 3
(b) 1   (e) 4
(c) 2

Learning Outcomes

- Define the octet rule as it relates to Lewis structures of compounds.
- Apply rules for drawing Lewis structures to determining the Lewis structures of compounds.
- Determine the polarity of a bond using differences in electronegativity.
- Define electronegativity, dipole moment, partial charge, and percent ionic character.
- Determine the formal charge on the atoms in a Lewis structure.
- Use formal charges to identify the most likely structure of a compound when more than one Lewis structure can be drawn.

- Define resonance and determine the resonance structures of a species.
- Determine Lewis structures of species that do not follow the octet rule, including radicals.
- Explain why period one and period two atoms cannot exceed the octet rule.
- Explain why period three atoms and beyond can exceed the octet rule.
- Draw Lewis structures with and without expanded octets for species in which both are possible.

Chapter Summary

SECTION 6.1

- According to the *octet rule,* atoms will lose, gain, or share electrons to achieve a noble gas configuration. Pairs of valence electrons that are *not* involved in the covalent bonding in a molecule or polyatomic ion (i.e., valence electrons that are not shared) are called *lone pairs*.
- *Lewis structures* are drawn to represent molecules and polyatomic ions, showing the arrangement of atoms and the positions of all valence electrons. Lewis structures represent the shared pairs of valence electrons either as two dots, ··, or as a single dash, —. Any *unshared* electrons are represented as dots.
- One shared pair of electrons between atoms constitutes a *single bond. Multiple bonds* form between atoms that share

more than one pair of electrons. Two shared pairs constitute a *double bond,* and three shared pairs constitute a *triple bond. Bond length* is determined in part by the number of shared electron pairs between two atoms.

SECTION 6.2

- Bonds in which electrons are not shared equally are *polar* and are referred to as *polar covalent bonds*.
- *Electronegativity* is an atom's ability to draw shared electrons toward itself. Bonds between elements of widely different electronegativities ($\Delta \geq 2.0$) are *ionic*. Covalent bonds between atoms with significantly different electronegativities ($0.5 \leq \Delta < 2.0$) are *polar*. Bonds between atoms with very similar electronegativities ($\Delta < 0.5$) are *nonpolar*.

- *Percent ionic character* quantifies the polarity of a bond, and is determined by comparing the measured dipole moment to the one predicted by assuming that the bonded atoms have discrete charges.
- The *dipole moment (μ)* is a quantitative measure of the polarity of a bond.

SECTION 6.3

- Lewis structures of molecules or polyatomic ions can be drawn using the following step-by-step procedure:

  1. Count the total number of valence electrons, adding electrons to account for a negative charge and subtracting electrons to account for a positive charge.
  2. Use the molecular formula to draw the skeletal structure.
  3. Subtract two electrons for each bond in the skeletal structure.
  4. Distribute the remaining valence electrons to complete octets, completing the octets of the more electronegative atoms first.
  5. Place any remaining electrons on the central atom.
  6. Include double or triple bonds, if necessary, to complete the octets of all atoms.

SECTION 6.4

- *Formal charge* is a way of keeping track of the valence electrons in a species. Formal charges should be consistent with electronegativities and can be used to determine the best arrangement of atoms and electrons for a Lewis structure.

SECTION 6.5

- *Resonance structures* are two or more equally correct Lewis structures that differ in the positions of the electrons but *not* in the positions of the atoms. Different resonance structures of a compound can be separated by a resonance arrow, $\longleftrightarrow$.

SECTION 6.6

- A *Lewis base* is a species that can donate a lone pair of electrons to form a bond. A *Lewis acid* is a species that can accept a pair of electrons from a Lewis base.
- In an ordinary covalent bond, each atom contributes one electron to the shared pair of electrons. In cases where just one of the atoms contributes *both* of the electrons, the bond is called a *coordinate covalent bond* or a *dative bond*.
- A species that contains an odd number of electrons is called a *free radical.*

Key Words

Bond length, 214	Free radical, 231	Octet rule, 211
Coordinate covalent bond, 231	Lewis acid, 231	Percent ionic character, 220
Dative bond, 231	Lewis base, 231	Polar, 215
Dipole moment (μ), 218	Lewis structure, 211	Polar covalent bond, 215
Double bond, 214	Lone pair, 211	Resonance structure, 228
Electronegativity, 216	Multiple bond, 214	Single bond, 214
Formal charge, 224	Nonpolar, 217	Triple bond, 214

Key Equations

6.1	$\mu = Q \times r$	Dipole moment (μ) is calculated as the product of charge magnitude (Q) and distance between the charges (bond length, r) in a diatomic molecule. Because molecules are neutral, the partial charges in a heteronuclear diatomic molecule are equal in magnitude and opposite in sign. Equation 6.1 can be used to calculate the dipole moment when the magnitude of partial charges is known—or it can be used to determine the magnitude of partial charges when the experimentally determined dipole moment is known.
6.2	percent ionic character $= \dfrac{\mu \text{ (observed)}}{\mu \text{ (calculated assuming discrete charges)}} \times 100\%$	Percent ionic character of a bond is equal to the ratio of the observed dipole moment to the dipole moment calculated, assuming discrete charges on the atoms.
6.3	formal charge = valence electrons − associated electrons	Formal charge on an atom in a Lewis structure is equal to the number of valence electrons (group number) minus half of the electrons it shares with other atoms in the structure.

Key Skills

Drawing Lewis Structures

The first step in solving many problems is drawing a correct Lewis structure. The process of drawing a Lewis structure was first described in pages 222–224. The steps are summarized in the following flowchart:

1. Count the total valence electrons. Recall that each atom contributes a number equal to its group number; and remember to add or subtract valence electrons to account for charge on a polyatomic ion.*

2. Use the chemical formula to draw a skeletal structure. Usually the central atom is less electronegative than the terminal atoms, although hydrogen cannot be a central atom because it can form only one bond.

3. For each bond in the skeletal formula, subtract two from the total number of valence electrons.

4. Distribute the remaining electrons, satisfying first the octets of the more electronegative (usually terminal) atoms.

5. If all terminal atoms have complete octets, and there are valence electrons still to be distributed, place them on the central atom as lone pairs.

6. If the valence electrons run out before all octets are satisfied, use multiple bonds to complete the octets of all atoms.

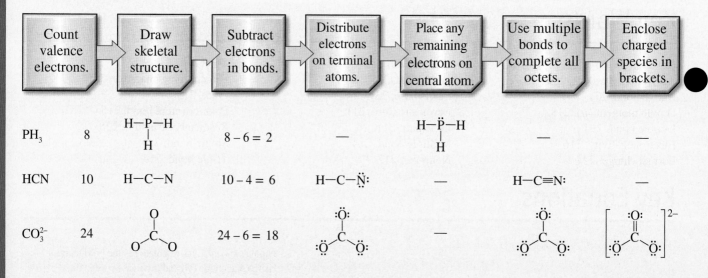

*Remember to enclose structures of charged species in square brackets with the superscript charge.

There are exceptions to the octet rule:

- Be and B, small atoms with low electronegativity, need not obey the octet rule.
- Elements in the third period and beyond need not obey the octet rule.
- Species with an odd number of valence electrons cannot obey the octet rule.
- A larger central atom (from the third period or beyond) can accommodate more than eight electrons and can have an "expanded" octet.

	Count valence electrons.	Draw skeletal structure.	Subtract electrons in bonds.	Distribute electrons on terminal atoms.	Place any remaining electrons on central atom.	Enclose charged species in brackets.
BeI_2	16	I—Be—I	$16 - 4 = 12$	:Ï—Be—Ï:	—	—
$XeCl_2$	22	Cl—Xe—Cl	$22 - 4 = 18$	:C̈l—Xe—C̈l:	:C̈l—Ẍe—C̈l:	—
PCl_5	40	Cl—P(—Cl)(—Cl)(—Cl)—Cl	$40 - 10 = 30$	structure with Cl lone pairs	—	—
SF_6	48	F—S(—F)...	$48 - 12 = 36$	structure with F lone pairs	—	—
ICl_4	36	Cl—I(—Cl)(—Cl)—Cl	$36 - 8 = 28$	:C̈l—I(—C̈l:)(—C̈l:)—C̈l:	structure with I lone pairs	$\left[\text{:C̈l—Ï—C̈l:}\right]^-$

Key Skills Problems

6.1
Which of the following atoms must always obey the octet rule? (Select all that apply.)
(a) C   (b) N   (c) S   (d) Br   (e) Xe

6.2
Which of the following species has an odd number of electrons? (Select all that apply.)
(a) N_2O (b) NO_2 (c) NO_2^- (d) NO_3^- (e) NS

6.3
How many lone pairs are on the central atom in $XeOF_2$?
(a) 0   (b) 1   (c) 2   (d) 3   (e) 4

6.4
How many lone pairs are on the central atom in the perchlorate ion?
(a) 0   (b) 1   (c) 2   (d) 3   (e) 4

Questions and Problems

SECTION 6.1: THE OCTET RULE

Review Questions

6.1 Explain Lewis's contribution to our understanding of the covalent bond.

6.2 Explain the difference between a *Lewis symbol* and a *Lewis structure*.

6.3 Use examples to illustrate each of the following terms: *lone pair, Lewis structure, octet rule, bond length.*

6.4 Explain the octet rule and why it applies mainly to second period elements.

6.5 Explain how the octet rule applies to hydrogen.

6.6 Compare single, double, and triple bonds in a molecule, and give an example of each. For the same bonding atoms, how does the bond length change from single bond to double bond to triple bond?

6.7 For a given pair of bonded atoms, explain how bond *length* relates to bond *strength*.

Conceptual Problems

6.8 For each of the following pairs of elements, state whether the binary compound they form is likely to be ionic or covalent. Write the empirical formula and name of the compound: (a) Al and Cl, (b) S and F.

6.9 For each of the following pairs of elements, state whether the binary compound they form is likely to be ionic or covalent. Write the empirical formula and name of the compound: (a) Ba and F, (b) C and Br.

SECTION 6.2: ELECTRONEGATIVITY AND POLARITY

Review Questions

6.10 Define *electronegativity* and explain the difference between electronegativity and electron affinity. Describe in general how the electronegativities of the elements change according to their position in the periodic table.

6.11 What is a *polar covalent bond?* Name two compounds that contain one or more polar covalent bonds.

Computational Problems

6.12 The radical species ClO has a dipole moment of 1.24 D and the Cl—O bond distance is 1.57 Å. Determine the magnitude of the partial charges in ClO.

6.13 Given that the partial charges on Al and F in aluminum monofluoride are +0.019 and −0.019, respectively, calculate the dipole moment of AlF. (The distance between the partial charges, r, is 165 pm.)

6.14 The measured dipole moment of bromine monofluoride, BrF, is 1.42 D and the Br—F bond distance is 1.76 Å. Determine the percent ionic character of the bond in BrF.

6.15 Given that the BaO bond has 85.3 percent ionic character and the Ba—O bond length is 194 pm, determine the measured dipole moment of BaO.

6.16 Four different atoms are represented by colored spheres: 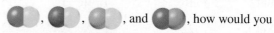, and . Their electronegativities are as follows: = 3.6, = 2.1, = 1.8, and = 1.6. If the atoms of these elements form the molecules , , , and , how would you arrange these molecules in order of increasing covalent bond character?

Conceptual Problems

6.17 Arrange the following bonds in order of increasing ionic character: carbon to hydrogen, fluorine to hydrogen, bromine to hydrogen, sodium to chlorine, potassium to fluorine, lithium to chlorine.

6.18 List the following bonds in order of increasing ionic character: the lithium-to-fluorine bond in LiF, the potassium-to-oxygen bond in K_2O, the nitrogen-to-nitrogen bond in N_2, the sulfur-to-oxygen bond in SO_2, the chlorine-to-fluorine bond in ClF_3.

6.19 Classify the following bonds as ionic, polar covalent, or nonpolar covalent, and explain: (a) the PCl bond in PCl_3, (b) the CS bond in CS_2, (c) the CH bond in CH_4, (d) the AlF bond in AlF_3.

6.20 Classify the following bonds as ionic, polar covalent, or nonpolar covalent, and explain: (a) the CaO bond in CaO, (b) the CC bond in Cl_3CCCl_3, (c) the CCl bond in Cl_3CCCl_3, (d) the SeCl bond in $SeCl_2$.

6.21 List the following bonds in order of increasing ionic character: potassium to iodine, carbon to oxygen, lithium to fluorine, boron to fluorine.

SECTION 6.3: DRAWING LEWIS STRUCTURES

Conceptual Problems

6.22 Draw Lewis structures for the following molecules and ions: (a) NCl_3, (b) OCS, (c) H_2O_2, (d) CH_3COO^-, (e) CN^-, (f) $CH_3CH_2NH_3^+$.

6.23 Draw Lewis structures for the following molecules and ions: (a) OF_2, (b) N_2F_2, (c) Si_2H_6, (d) OH^-, (e) CH_2ClCOO^-, (f) $CH_3NH_3^+$.

6.24 Draw Lewis structures for the following molecules: (a) ICl, (b) PH_3, (c) P_4 (each P is bonded to three other P atoms), (d) H_2S, (e) N_2H_4, (f) $HClO_3$, (g) $COBr_2$ (C is bonded to O and Br atoms).

6.25 Draw Lewis structures for the following molecules: (a) ClF_3, (b) H_2Se, (c) NH_2OH, (d) $POCl_3$ (P is bonded to O and Cl atoms), (e) CH_3CH_2Br, (f) NCl_3, (g) CH_3NH_2.

SECTION 6.4: LEWIS STRUCTURES AND FORMAL CHARGE

Review Questions

6.26 Explain the concept of *formal charge.*

6.27 Do formal charges represent an actual separation of charges?

Conceptual Problems

6.28 Draw Lewis structures for the following ions: (a) O_2^{2-}, (b) C_2^{2-}, (c) NO^+, (d) NH_4^+.

6.29 Draw Lewis structures for the following ions: (a) NO_2^+, (b) SCN^-, (c) S_2^{2-}, (d) ClF_2^+. Show formal charges.

6.30 The following Lewis structures are incorrect. Explain what is wrong with each one, and give a correct Lewis structure for the molecule. (Relative positions of atoms are shown correctly.)

(a) $H-\ddot{C}=\ddot{N}$

(b) $H=C=C=H$

(c) $\ddot{O}-Sn-\ddot{O}$

(d) F B F with F below

(e) $H-\ddot{O}=\ddot{F}$:

(f) C–F with H and O

(g) F N F with F below

6.31 The skeletal structure of acetic acid shown here is correct but some of the bonds are wrong. (a) Identify the incorrect bonds and explain what is wrong with them. (b) Draw the correct Lewis structure for acetic acid.

$$H=\overset{\overset{\displaystyle H}{|}}{C}-\overset{\overset{\displaystyle :O:}{||}}{C}-O-H$$

SECTION 6.5: RESONANCE

Review Questions

6.32 What are *resonance structures?* Is it possible to isolate one resonance structure of a compound for analysis? Explain.

6.33 What are the rules for drawing resonance structures?

Conceptual Problems

6.34 Draw three resonance structures for the chlorate ion (ClO_3^-). Show formal charges.

6.35 Draw all of the resonance structures for the following species and show formal charges: (a) HCO_2^-, (b) $CH_2NO_2^-$. The relative positions of the atoms are as follows:

O   H   O

H   C   C   N

O   H   O

6.36 Draw two resonance structures for diazomethane (CH_2N_2). Show formal charges. The skeletal structure of the molecule is

H

C   N   N

H

6.37 Draw three resonance structures for hydrazoic acid (HN_3). The atomic arrangement is HNNN. Show formal charges.

6.38 Draw three resonance structures for the molecule N_2O in which the atoms are arranged in the order NNO. Indicate formal charges.

6.39 Draw three reasonable resonance structures for the OCN$^-$ ion. Show formal charges.

6.40 Indicate which of the following are resonance structures of $:\ddot{C}l-Be-\ddot{C}l:$. For the others, explain why they are *not* resonance structures of $:\ddot{C}l-Be-\ddot{C}l:$.

(a) $:\ddot{C}l=Be=\ddot{C}l:$

(b) $:Cl\equiv Be-\ddot{C}l:$

(c) $:Be\equiv Cl-\ddot{C}l:$

(d) $:\ddot{C}l-Be\equiv Cl:$

(e) $:Be=Cl=\ddot{C}l:$

6.41 Indicate which of the following are resonance structures of

$$\overset{:O:}{\underset{:\ddot{C}l\qquad\ddot{C}l:}{||}}C$$ (select all that apply).

(a) :O: with Cl C Cl

(b) :O: with Cl C Cl:

(c) :O: with Cl C Cl:

(d) :O: with Cl C Cl

(e) :O: with Cl C Cl

6.42 Draw one additional resonance structure of the adenine molecule shown here, which is part of the DNA structure.

6.43 Draw a resonance structure of the guanine molecule shown here, which is part of the DNA structure.

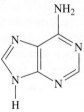

SECTION 6.6: EXCEPTIONS TO THE OCTET RULE

Review Questions

6.44   Why does the octet rule not hold for many compounds containing elements in the third period of the periodic table and beyond?

6.45   Give three examples of compounds that do not satisfy the octet rule. Draw a Lewis structure for each.

6.46   Because fluorine has seven valence electrons $(2s^2 2p^5)$, in principle, seven covalent bonds could form around the atom as in FH_7 or FCl_7. Explain why these compounds have never been prepared.

6.47   What is a *coordinate covalent bond?* How is it different from an ordinary covalent bond? How is it the same?

Conceptual Problems

6.48   The AlI_3 molecule has an incomplete octet around Al. Draw three resonance structures of the molecule in which the octet rule is satisfied for both the Al and the I atoms. Show formal charges.

6.49   In the vapor phase, beryllium chloride consists of discrete $BeCl_2$ molecules. Is the octet rule satisfied for Be in this compound? If not, can you form an octet around Be by drawing another resonance structure? Is this structure plausible? Explain.

6.50   Until the early 1960s, the noble gases were not known to form any compounds. Since then, a few compounds of Kr, Xe, and Rn—and one compound of Ar—have been made in laboratories. Draw Lewis structures for the following molecules:
(a) XeF_2, (b) XeF_4, (c) XeF_6, (d) $XeOF_4$, (e) XeO_2F_2. In each case Xe is the central atom.

6.51   Draw a Lewis structure for the tetrachloraluminate ion, $AlCl_4^-$. This ion can be considered as a combination of $AlCl_3$ and Cl^-. What kind of bond joins $AlCl_3$ and Cl^- in $AlCl_4^-$?

6.52   Draw Lewis structures for SeF_4 and SeF_6. Is the octet rule satisfied for Se?

6.53   Draw a Lewis structure for $SbCl_5$. Does this molecule obey the octet rule?

6.54   Draw Lewis structures for the radical species ClF_2 and BrO_2.

6.55   Draw Lewis structures for the radical species HO_2 and ClO.

6.56   Draw two resonance structures for the bromate ion (BrO_3^-), one that obeys the octet rule and one in which the formal charge on the central atom is zero.

6.57   Draw two resonance structures for the sulfite ion (SO_3^{2-}), one that obeys the octet rule and one in which the formal charge on the central atom is zero.

ADDITIONAL PROBLEMS

6.58   Draw Lewis structures for BrF_3, ClF_5, and IF_7. Identify those in which the octet rule is not obeyed.

6.59   Draw three reasonable resonance structures for the azide ion N_3^- in which the atoms are arranged as NNN. Show formal charges.

6.60   The amide group plays an important role in determining the structure of proteins:

$$-\overset{..}{N}-\overset{\overset{\displaystyle \overset{..}{O}}{\|}}{C}-$$
$$\underset{H}{|}$$

Draw another resonance structure for this group. Show formal charges.

6.61   Give an example of an ion or molecule containing Al that (a) obeys the octet rule, (b) has an expanded octet, and (c) has an incomplete octet.

6.62   Draw four reasonable resonance structures for the PO_3F^{2-} ion. The central P atom is bonded to the three O atoms and to the F atom. Show formal charges.

6.63   Attempts to prepare the compounds CF_2, LiO_2, $CsCl_2$, and PI_5 as stable species under atmospheric conditions have failed. Suggest possible reasons for the failure.

6.64   Draw reasonable resonance structures for the following ions: (a) HSO_4^-, (b) PO_4^{3-}, (c) HSO_3^-, (d) SO_3^{2-}.

6.65   Are the following statements true or false?
(a) Formal charges represent an actual separation of charges. (b) All second period elements obey the octet rule in their compounds. (c) The resonance structures of a molecule can be separated from one another in the laboratory.

6.66   A rule for drawing plausible Lewis structures is that the central atom is generally less electronegative than the surrounding atoms. Explain why this is so.

6.67   Which of the following molecules has the longest nitrogen-to-nitrogen bond: N_2H_4, N_2O, N_2, N_2O_4? Explain.

6.68   Most organic acids can be represented as RCOOH, where COOH is the carboxyl group and R is the rest of the molecule. [For example, R is CH_3 in acetic acid (CH_3COOH).] (a) Draw a Lewis structure for the carboxyl group. (b) Upon ionization, the carboxyl group is converted to the carboxylate group (COO^-). Draw resonance structures for the carboxylate group.

6.69   Which of the following species are isoelectronic: NH_4^+, C_6H_6, CO, CH_4, N_2, $B_3N_3H_6$? (The term *isoelectronic* means the same thing in the context of molecules and polyatomic ions as it does with regard to atoms and monatomic ions: isoelectronic species have the same number and arrangement of electrons.)

6.70 The following is a simplified (skeletal) structure of the amino acid histidine. Draw a complete Lewis structure of the molecule.

6.71 The following is a simplified (skeletal) structure of the amino acid tryptophan. Draw a complete Lewis structure of the molecule.

6.72 The following species have been detected in interstellar space: (a) CH, (b) OH, (c) C_2, (d) HNC, (e) HCO. Draw Lewis structures for these species.

6.73 The sulfur oxides SO_2 and SO_3 are significant pollutants in the air and contributors to the formation of acid rain. It is thought that these compounds can be converted into radical anions inside of cells. Draw plausible Lewis structures of the radical anions SO_2^- and SO_3^-.

6.74 Draw Lewis structures for the following organic molecules: (a) tetrafluoroethylene (C_2F_4), (b) propane (C_3H_8), (c) butadiene ($CH_2CHCHCH_2$), (d) propyne (CH_3CCH), (e) benzoic acid (C_6H_5COOH). (To draw C_6H_5COOH, replace an H atom in benzene with a COOH group.)

6.75 The triiodide ion (I_3^-) in which the I atoms are arranged in a straight line is stable, but the corresponding F_3^- ion does not exist. Explain.

6.76 Draw Lewis structures for the following organic molecules: C_2H_3F, C_3H_6, C_4H_8. In each there is one C=C bond, and the rest of the carbon atoms are joined by C—C bonds.

6.77 Methyl isocyanate (CH_3NCO) is used to make certain pesticides. In December 1984, water leaked into a tank containing this substance at a chemical plant, producing a toxic cloud that killed thousands of people in Bhopal, India. Draw Lewis structures for CH_3NCO, showing formal charges.

6.78 The chlorine nitrate ($ClONO_2$) molecule is believed to be involved in the destruction of ozone in the Antarctic stratosphere. Draw a plausible Lewis structure for this molecule.

6.79 Several resonance structures for the molecule CO_2 are shown here. Which of these species is/are most important in describing the bonding in this molecule? Explain why the others are less important.

(a) :Ö=C=Ö: (c) :O≡C̈—Ö:
(b) :O≡C—Ö: (d) :Ö—C—Ö:

6.80 For each of the following organic molecules draw a Lewis structure in which the carbon atoms are bonded to each other by single bonds: (a) C_2H_6, (b) C_4H_{10}, (c) C_5H_{12}. For parts (b) and (c) show only structures in which each C atom is bonded to no more than two other C atoms.

6.81 Draw Lewis structures for the following chlorofluorocarbons (CFCs), which are partly responsible for the depletion of ozone in the stratosphere: (a) $CFCl_3$, (b) CF_2Cl_2, (c) CHF_2Cl, (d) CF_3CHF_2.

6.82 Draw Lewis structures for the following organic molecules: (a) methanol (CH_3OH); (b) ethanol (CH_3CH_2OH); (c) tetraethyl lead [$Pb(CH_2CH_3)_4$], which is used in "leaded gasoline"; (d) methylamine (CH_3NH_2), which is used in tanning; (e) mustard gas ($ClCH_2CH_2SCH_2CH_2Cl$), a poisonous gas used in World War I; (f) urea [$(NH_2)_2CO$], a fertilizer; and (g) glycine (NH_2CH_2COOH), an amino acid.

6.83 Draw Lewis structures for the following four isoelectronic species: (a) CO, (b) NO^+, (c) CN^-, (d) N_2. Show formal charges. (See Problem 6.69.)

6.84 Oxygen forms three types of ionic compounds in which the anions are oxide (O^{2-}), peroxide (O_2^{2-}), and superoxide (O_2^-). Draw Lewis structures of these ions.

6.85 Is there a group of elements that always violates the octet rule? Explain.

6.86 Draw three resonance structures for (a) the cyanate ion (NCO^-) and (b) the isocyanate ion (CNO^-). In each case, rank the resonance structures in order of increasing importance.

6.87 The N—O bond distance in nitric oxide is 115 pm, which is intermediate between a triple bond (106 pm) and a double bond (120 pm). (a) Draw two resonance structures for NO, and comment on their relative importance. (b) Is it possible to draw a resonance structure having a triple bond between the atoms?

6.88 Within the comet Hale-Bopp, a number of sulfur-containing diatomic molecules have been detected including sulfur mononitride (SN) and sulfur monoxide (SO). (a) Determine the partial charges in SN given the dipole moment (1.81 D) and S—N bond distance (149 pm). (b) If the partial charges on S and O in sulfur monoxide are +0.022 and −0.022, respectively, calculate the dipole moment of SO. (The distance between the partial charges, r, is 1.48 Å.)

6.89   In the gas phase, aluminum chloride exists as a dimer (a unit of two) with the formula Al_2Cl_6. Its skeletal structure is given by

$$\begin{array}{ccccc}
Cl & & Cl & & Cl \\
 & \diagdown & | & \diagdown & \\
 & Al & & Al & \\
 & \diagup & | & \diagup & \\
Cl & & Cl & & Cl
\end{array}$$

Complete the Lewis structure and identify the coordinate covalent bonds in the molecules. Would you expect this molecule to be polar? Explain.

6.90   Draw a Lewis structure for nitrogen pentoxide (N_2O_5) in which each N is bonded to three O atoms.

6.91   Determine the percent ionic character of RbF and RbI. The dipole moments of RbF and RbI are 8.55 D and 11.48 D, respectively. [Use information from Figure 4.13 (page 148) to calculate the bond distances in each ionic compound.] Comment on the difference between your two answers.

6.92   Nitrogen dioxide (NO_2) is a stable compound. Explain why there is a tendency for two such molecules to combine to form dinitrogen tetroxide (N_2O_4). Draw four resonance structures of N_2O_4, showing formal charges.

6.93   Electrostatic potential maps for three compounds A, B, and C are shown here.

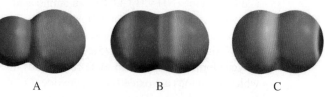

A   B   C

Using the data in the following table, determine which compound corresponds to which electrostatic potential map and determine the dipole moment of each compound. Fill in the missing data in the table.

Compound	Partial charges	Distance between charges (pm)	Dipole moment
	±0.19	213	
	±0.051	214	
	±0.68	315	

6.94   Vinyl chloride (C_2H_3Cl) differs from ethylene (C_2H_4) in that one of the H atoms is replaced with a Cl atom. Vinyl chloride is used to prepare poly(vinyl chloride), which is an important polymer used in pipes. (a) Draw the Lewis structure of vinyl chloride. (b) The repeating unit in poly(vinyl chloride) is $-CH_2-CHCl-$. Draw a portion of the molecule showing three such repeating units.

6.95   Pyridine has a structure similar to that of benzene, but one of the carbon atoms (and the hydrogen bonded to it) are replaced by a nitrogen atom. Draw the two resonance structures of pyridine and using information from Table 6.1, estimate the length of the C—N bonds.

6.96   Among the common inhaled anesthetics are:
Halothane $(CF_3CHClBr)$
Isoflurane $(CF_3CHClOCHF_2)$
Enflurane $(CHFClCF_2OCHF_2)$
Methoxyflurane $(CHCl_2CF_2OCH_3)$
Draw Lewis structures of these molecules.

6.97   In 1999 an unusual cation containing only nitrogen (N_5^+) was prepared. Draw three resonance structures of the ion, showing formal charges. (*Hint:* The N atoms are joined in a linear fashion.)

6.98   In each of the following Lewis structures, Z and X represent different third period main group elements.

$$\left[\begin{array}{c} \ddot{O} \\ | \\ H-\ddot{O}-Z-\ddot{O}: \\ | \\ :\ddot{O}: \end{array} \right]^{-} \qquad \begin{array}{c} :\ddot{F}: \\ | \\ :\ddot{X}-\ddot{F}: \\ | \\ :\ddot{F}: \end{array}$$

Draw the Lewis structure for the compound ZX_3.

6.99   The American chemist Robert S. Mulliken suggested a different definition for the electronegativity (EN) of an element, given by

$$EN = \frac{IE_1 + EA}{2}$$

where IE_1 is the first ionization energy and EA is the electron affinity of the element. Calculate the electronegativities of O, F, and Cl using the preceding equation and data from the Chapter 4 Key Skills (pages 153–154). Compare the electronegativities of these elements on the Mulliken and Pauling scales. (To convert to the Pauling scale, divide each EN value by 230 kJ/mol.)

6.100 Electrostatic potential maps for three compounds A, B, and C are shown here.

A   B   C

Using the data in the following table, determine which compound corresponds to which electrostatic potential map and fill in the missing data in the table.

Compound	Partial charges	Distance between charges (pm)	Dipole moment
	±0.16	157	
	±0.43		3.10
		258	8.89

Answers to In-Chapter Materials

PRACTICE PROBLEMS

6.1A (a) polar, (b) polar, (c) nonpolar. **6.1B** Li through Fr in Group 1A, and Ca through Ra in Group 2A. **6.2A** 0.12. **6.2B** 0.11 D. **6.3A** 41%. **6.3B** 9.23 D.

6.4A $:\!\ddot{F}\!-\!\overset{\displaystyle :\ddot{F}:}{\underset{|}{N}}\!-\!\ddot{F}\!:$ **6.4B** $\left[:\!\ddot{O}\!-\!\overset{\displaystyle :\ddot{O}:}{Cl}\!-\!\ddot{O}\!:\right]^{-}$ **6.5A** C atom = 0, double-bonded O atom = 0, single-bonded O atoms = −1. **6.5B** S atom = +1, O atoms = −1, overall charge = −2.

6.6A $-\overset{\displaystyle \overset{\ddot{O}}{\|}}{C}\!-\!\ddot{O}\!-\!H$. **6.6B** Cl−N−Cl.

6.7A $\left[:\!O\!=\!\overset{\displaystyle :\ddot{O}:}{\underset{|}{N}}\!-\!\ddot{O}\!:\right]^{-}\!\!\longleftrightarrow\left[:\!\ddot{O}\!-\!\overset{\displaystyle \overset{\ddot{O}}{\|}}{N}\!-\!\ddot{O}\!:\right]^{-}\!\!\longleftrightarrow\left[:\!\ddot{O}\!-\!\overset{\displaystyle :\ddot{O}:}{\underset{|}{N}}\!=\!O\!:\right]^{-}$

6.7B $\left[\overset{-2\quad 0\quad +1}{:\!\ddot{N}\!-\!C\!\equiv\!S\!:}\right]^{-}\!\!\longleftrightarrow\left[\overset{-1\quad 0\quad\ \ 0}{:\!\ddot{N}\!=\!C\!=\!\ddot{S}\!:}\right]^{-}\!\!\longleftrightarrow\left[\overset{0\quad\ 0\quad -1}{:\!N\!\equiv\!C\!-\!\ddot{S}\!:}\right]^{-}$

6.8A $\cdot\ddot{O}\!-\!H$ **6.8B** $:\!\ddot{S}\!=\!\dot{N}\!-\!\ddot{S}\!:$ **6.9A** (a) $:\!\ddot{F}\!-\!Be\!-\!\ddot{F}\!:$ (b) $\overset{\displaystyle \ddot{Cl}}{\underset{\displaystyle \ddot{Cl}}{\overset{\displaystyle |}{\underset{\displaystyle |}{\overset{:\ddot{Cl}:}{P}\!-\!\ddot{Cl}\!:}}}}$

(c) $\left[\overset{\displaystyle :\ddot{Cl}:}{\underset{\displaystyle :\ddot{Cl}:}{:\ddot{Cl}\!-\!\overset{|}{\underset{|}{I}}\!-\!\ddot{Cl}\!:}}\right]^{-}$ **6.9B** (a) $:\!\ddot{Cl}\!-\!\overset{\displaystyle :\ddot{Cl}:}{B}\!-\!\ddot{Cl}\!:$ (b) $\overset{\displaystyle \ddot{F}}{\underset{\displaystyle \ddot{F}}{\overset{:\ddot{F}:}{\underset{:\ddot{F}:}{Sb}}\!-\!\ddot{F}\!:}}$

(c) $:\!\ddot{F}\!-\!Kr\!-\!\ddot{F}\!:$

6.10A

$\left[H\!-\!\ddot{O}\!-\!\overset{\displaystyle \overset{\ddot{O}^{-1}}{|}}{\underset{\displaystyle \ddot{O}^{-1}}{S^{+1}}}\!-\!\ddot{O}^{-1}\right]^{-}$ $\left[H\!-\!\ddot{O}\!-\!\overset{\displaystyle \overset{\ddot{O}}{\|}}{\underset{\displaystyle \ddot{O}^{-1}}{S}}\!-\!\ddot{O}^{-1}\right]^{-}$ $\left[H\!-\!\ddot{O}\!-\!\overset{\displaystyle \overset{\ddot{O}}{\|}}{\underset{\displaystyle \ddot{O}^{-1}}{S}}\!=\!O\!:\right]^{-}$

$\left[\overset{\displaystyle :\ddot{O}:}{\underset{\displaystyle :\ddot{O}:}{:\ddot{O}\!-\!\overset{|}{\underset{|}{P}}\!-\!\ddot{O}\!:}}\right]^{3-}$ $\left[\overset{\displaystyle \overset{\ddot{O}}{\|}}{\underset{\displaystyle :\ddot{O}:}{:\ddot{O}\!-\!\overset{}{\underset{|}{P}}\!-\!\ddot{O}\!:}}\right]^{3-}$ $H\!-\!\ddot{O}\!-\!\overset{\displaystyle :\ddot{O}:}{\underset{|}{Cl}}\!-\!\ddot{O}\!:$ $\quad H\!-\!\ddot{O}\!-\!\overset{\displaystyle \overset{\ddot{O}}{\|}}{Cl}\!=\!O\!:$

6.10B

$\left[:\!\ddot{O}\!-\!\overset{\displaystyle \overset{\ddot{O}}{\|}}{\underset{\displaystyle \ddot{O}:}{S}}\!-\!\ddot{O}\!:\longleftrightarrow:\!\ddot{O}\!-\!\overset{\displaystyle \overset{\ddot{O}}{\|}}{S}\!=\!O\!:\right]$ $\left[:\!\ddot{O}\!=\!\ddot{S}\!-\!\ddot{O}\!:\longleftrightarrow:\!\ddot{O}\!=\!\ddot{S}\!=\!O\!:\right]$

SECTION REVIEW

6.2.1 b. **6.2.2** d. **6.2.3** c. **6.2.4** c. **6.3.1** $H\!-\!\overset{\displaystyle \overset{:\ddot{O}:}{\|}}{C}\!-\!\ddot{O}\!-\!H$

6.3.2 $H\!-\!\ddot{O}\!-\!\ddot{O}\!-\!H$ **6.4.1** e. **6.4.2** $\left[:\!\ddot{N}\!=\!C\!=\!\ddot{O}\!:\right]^{-},\left[:\!\ddot{N}\!-\!C\!\equiv\!O\!:\right]^{-},$
$\left[:\!N\!\equiv\!C\!-\!\ddot{O}\!:\right]^{-}$, best structure $\left[:\!N\!\equiv\!C\!-\!\ddot{O}\!:\right]^{-}$ **6.5.1** b.

6.5.2 $\ddot{O}\!=\!\ddot{S}\!-\!\ddot{O}\!:\longleftrightarrow:\!\ddot{O}\!-\!\ddot{S}\!=\!\ddot{O}$ **6.6.1** e. **6.6.2** a, b, e. **6.6.3** b. **6.6.4** d.

Chapter 7

Molecular Geometry, Intermolecular Forces, and Bonding Theories

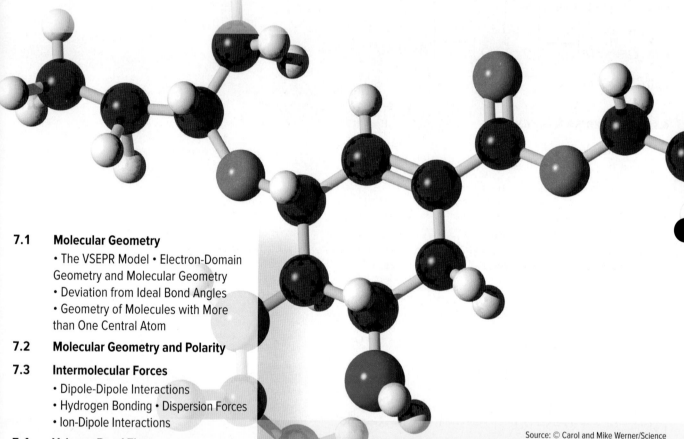

Source: © Carol and Mike Werner/Science

MOLECULAR STRUCTURE profoundly affects the properties of a substance and determines how it interacts with other chemical species. By paying particular attention to the structure and shape of molecules, medicinal chemists can design drugs to inhibit the action of specific pathogenic viruses and bacteria. One example of this is the antiviral drug Tamiflu, a model of which is shown here. Tamiflu prevents reproductive budding of the influenza virus by binding effectively in the active site of an important enzyme. By understanding the importance of molecular structure, synthetic chemists can now design better and more effective drugs for a multitude of diseases and illnesses.

- Shapes of atomic orbitals [◄◄ Section 3.8]
- Electron configurations of atoms [◄◄ Section 3.9]
- Drawing Lewis structures [◄◄ Section 6.3]

7.1 MOLECULAR GEOMETRY

Many familiar chemical and biochemical processes depend heavily on the three-dimensional shapes of the molecules and/or ions involved. Our sense of smell is one example; the effectiveness of a particular drug is another. Although the shape of a molecule or polyatomic ion must be determined experimentally, we can predict their shapes reasonably well using Lewis structures [◄◄ Section 6.3] and the *valence-shell electron-pair repulsion* *(VSEPR)* model. In this section, we will focus primarily on determining the shapes of molecules of the general type AB_x, where A is a central atom surrounded by x B atoms and x can have integer values of 2 to 6. For example, NH_3 is an AB_3 molecule in which A is nitrogen, B is hydrogen, and $x = 3$. The Lewis structure of NH_3 is shown here.

Student Annotation: Any atom that is bonded to two or more other atoms can be considered a "central" atom.

$$H-\ddot{N}-H$$
$$|$$
$$H$$

Lewis structure of NH_3

Table 7.1 lists examples of each type of AB_x molecule and polyatomic ion that we will consider. Throughout this chapter, we will discuss concepts that apply both to molecules and to polyatomic ions, but we will usually refer to them collectively as "molecules."

Having the molecular formula alone is insufficient to predict the shape of a molecule. For instance, AB_2 molecules may be linear or bent:

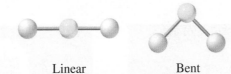

Linear   Bent

Moreover, AB_3 molecules may be planar, pyramidal, or T-shaped:

Trigonal planar   Trigonal pyramidal   T-shaped

To determine shape, we must start with a correct Lewis structure and apply the VSEPR model.

TABLE 7.1	Examples of AB_x Molecules and Polyatomic Ions
AB_2	$BeCl_2$, SO_2, H_2O, NO_2^-
AB_3	BF_3, NH_3, ClF_3, SO_3^{2-}
AB_4	CCl_4, NH_4^+, SF_4, XeF_4, ClO_4^-
AB_5	PCl_5, IF_5, SbF_5, BrF_5
AB_6	SF_6, UF_6, $TiCl_6^{3-}$

The VSEPR Model

The basis of the VSEPR model is that electron pairs in the valence shell of an atom *repel* one another. As we learned in Chapter 6, there are two types of electron pairs: bonding pairs and nonbonding pairs (also known as lone pairs). Furthermore, bonding pairs may be found in single bonds or in multiple bonds. For clarity, we will refer to electron *domains* instead of electron pairs when we use the VSEPR model. An ***electron domain*** in this context is a lone pair or a bond, regardless of whether the bond is single, double, or triple. Consider the following examples:

	CO_2	O_3	NH_3	PCl_5	XeF_4
	$\ddot{O}=C=\ddot{O}$	$\ddot{O}=\ddot{O}-\ddot{O}$	H—N̈—H | H	see structure	see structure
	2 double bonds	1 single bond 1 double bond + 1 lone pair	3 single bonds + 1 lone pair	5 single bonds	4 single bonds + 2 lone pairs
Total number of electron domains on central atom	2 electron domains	3 electron domains	4 electron domains	5 electron domains	6 electron domains

Note the number of electron domains on the central atom in each molecule. The VSEPR model predicts that because these electron domains repel one another, they will arrange themselves to be as far apart as possible, thus minimizing the repulsive interactions between them.

We can visualize the arrangement of electron domains using balloons, as shown in Figure 7.1. Like the B atoms in our AB_x molecules, the balloons are all connected

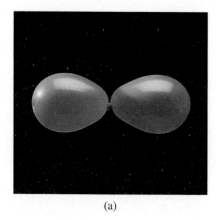

(a)

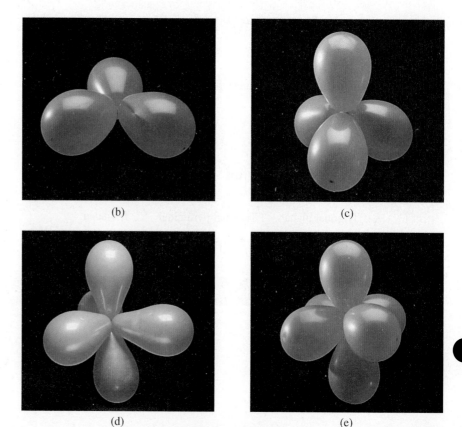

(b) (c) (d) (e)

Figure 7.1 The arrangements adopted by (a) two, (b) three, (c) four, (d) five, and (e) six balloons.

Source: (all): © McGraw-Hill Education/Stephen Frisch, photographer

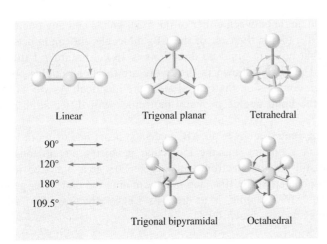

to a central, fixed point, which represents the central atom (A). When they are as far apart as possible, they adopt the five geometries shown in the figure. When there are only two balloons, they orient themselves to point in opposite directions [Figure 7.1(a)]. With three balloons, the arrangement is a trigonal plane [Figure 7.1(b)]. With four balloons, the arrangement adopted is a tetrahedron [Figure 7.1(c)]. With five balloons, three of them adopt positions in a trigonal plane whereas the other two point opposite to each other, forming an axis that is perpendicular to the trigonal plane [Figure 7.1(d)]. This geometry is called a trigonal bipyramid. Finally, with six balloons, the arrangement is an octahedron, which is essentially a square bipyramid [Figure 7.1(e)]. Each of the AB_x molecules we consider will have one of these five electron-domain geometries: linear, trigonal planar, tetrahedral, trigonal bipyramidal, or octahedral.

Electron-Domain Geometry and Molecular Geometry

It is important to distinguish between the ***electron-domain geometry,*** which is the arrangement of electron domains (bonds and lone pairs) around the central atom, and the ***molecular geometry,*** which is the arrangement of bonded *atoms.* Figure 7.2 illustrates the molecular geometries of AB_x molecules in which all the electron domains are bonds—that is, there are no lone pairs on any of the central atoms. In these cases, the molecular geometry is the same as the electron-domain geometry.

In an AB_x molecule, a ***bond angle*** is the angle between two adjacent A—B bonds. In an AB_2 molecule, there are only two bonds and therefore only one bond angle, and, provided that there are no lone pairs on the central atom, the bond angle is 180°. AB_3 and AB_4 molecules have three and four bonds, respectively. However, in each case there is only one bond angle possible between any two A—B bonds. In an AB_3 molecule, the bond angle is 120°, and in an AB_4 molecule, the bond angle is 109.5°—again, provided that there are no lone pairs on the central atoms. Similarly, in an AB_6 molecule, the bond angles between adjacent bonds are all 90°. (The angle between any two A—B bonds that point in opposite directions is 180°.)

AB_5 molecules contain two different bond angles between adjacent bonds. The reason for this is, unlike those in the other AB_x molecules, the positions occupied by bonds in a trigonal bipyramid are not all equivalent. The three bonds arranged in a trigonal plane are referred to as ***equatorial.*** The bond angle between any two of the three equatorial bonds is 120°. The two bonds that form an axis perpendicular to the trigonal plane are referred to as ***axial.***

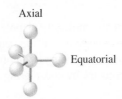

Video 7.1
Chemical bonding—VSEPR and
molecular geometry

Student Annotation: It would be impossible
to overstate the importance of being able to
draw Lewis structures correctly, especially for
students who will go on to study organic
chemistry.

The bond angle between either of the axial bonds and any one of the equatorial bonds is 90°. (As in the case of the AB_6 molecule, the angle between any two A—B bonds that point in opposite directions is 180°.) Figure 7.2 illustrates these bond angles. The angles shown in the figure are the bond angles observed when all the electron domains on the central atom are identical. As we will see later in this section, the bond angles in many molecules will differ slightly from these *ideal* values.

When the central atom in an AB_x molecule bears one or more lone pairs, the electron-domain geometry and the molecular geometry are no longer the same. However, we still use the electron-domain geometry as a first step in determining the molecular geometry. The first step in determining the molecular geometry of O_3 (or any species), for example, is to draw its Lewis structure. Two different resonance structures can be drawn for O_3:

$$:\!\ddot{O}\!=\!\ddot{O}\!-\!\ddot{O}\!: \quad\longleftrightarrow\quad :\!\ddot{O}\!-\!\ddot{O}\!=\!\ddot{O}\!:$$

Either one can be used to determine its geometry.

The next step is to count the electron domains on the central atom. In this case, there are three: one single bond, one double bond, and one lone pair. Using the VSEPR model, we first determine the electron-domain geometry. According to the information in Figure 7.2, three electron domains on the central atom will be arranged in a trigonal plane. Molecular geometry, however, is dictated by the arrangement of *atoms*. If we consider only the positions of the three atoms in this molecule, the molecular geometry (the molecule's shape) is *bent*.

Electron-domain geometry: Molecular geometry:
trigonal planar   bent

In addition to the five basic geometries depicted in Figure 7.2, you must be familiar with how molecular geometry can differ from electron-domain geometry. Table 7.2 shows the common molecular geometries where there are one or more lone pairs on the central atom. Note the positions occupied by the lone pairs in the trigonal bipyramidal electron-domain geometry. When there are lone pairs on the central atom in a trigonal bipyramid, they preferentially occupy equatorial positions, because repulsion is greater when the angle between electron domains is 90° or less. Placing a lone pair in an *axial* position would put it at 90° to three other electron domains. Placing it in an *equatorial* position puts it at 90° to only two other domains, thus minimizing the number of strong repulsive interactions.

All positions are equivalent in the octahedral geometry, so one lone pair on the central atom can occupy any of the positions. If there is a second lone pair in this geometry, though, it must occupy the position opposite the first. This arrangement minimizes the repulsive forces between the two lone pairs (they are 180° apart instead of 90° apart).

In summary, the steps to determine the electron-domain and molecular geometries are as follows:

1. Draw the Lewis structure of the molecule or polyatomic ion.
2. Count the number of electron domains on the central atom.
3. Determine the electron-domain geometry by applying the VSEPR model.
4. Determine the molecular geometry by considering the positions of the atoms only.

Student Hot Spot

Student data indicate you may struggle with
VSEPR. Access the SmartBook to view additional
Learning Resources on this topic.

TABLE 7.2	Electron-Domain and Molecular Geometries of Molecules with Lone Pairs on the Central Atom					
Total number of electron domains	Number of lone pairs	Type of molecule	Electron-domain geometry	Placement of lone pairs	Molecular geometry	Example
3	1	AB_2	Trigonal planar		Bent	SO_2
4	1	AB_3	Tetrahedral		Trigonal pyramidal	NH_3
4	2	AB_2	Tetrahedral		Bent	H_2O
5	1	AB_4	Trigonal bipyramidal		Seesaw-shaped	SF_4
5	2	AB_3	Trigonal bipyramidal		T-shaped	ClF_3
5	3	AB_2	Trigonal bipyramidal		Linear	IF_2^-
6	1	AB_5	Octahedral		Square pyramidal	BrF_5
6	2	AB_4	Octahedral		Square planar	XeF_4

Worked Example 7.1 shows how to determine the shape of a molecule or polyatomic ion.

Worked Example 7.1

Determine the shapes of (a) SO_3 and (b) ICl_4^-.

Strategy Use Lewis structures and the VSEPR model to determine first the electron-domain geometry and then the molecular geometry (shape).

Setup (a) The Lewis structure of SO_3 is:

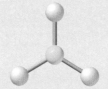

There are three electron domains on the central atom: one double bond and two single bonds.

(b) The Lewis structure of ICl_4^- is:

$$\left[\begin{array}{c} :\ddot{C}l: \\ | \\ :\ddot{C}l-\ddot{I}-\ddot{C}l: \\ | \\ :\ddot{C}l: \end{array} \right]^-$$

There are six electron domains on the central atom in ICl_4^-: four single bonds and two lone pairs.

Solution

(a) According to the VSEPR model, three electron domains will be arranged in a trigonal plane. Because there are no lone pairs on the central atom in SO_3, the molecular geometry is the same as the electron-domain geometry. Therefore, the shape of SO_3 is trigonal planar.

Electron-domain geometry: trigonal planar ⟶ Molecular geometry: trigonal planar

(b) Six electron domains will be arranged in an octahedron. Two lone pairs on an octahedron will be located on opposite sides of the central atom, making the shape of ICl_4^- square planar.

Electron-domain geometry: octahedral ⟶ Molecular geometry: square planar

---

Think About It
Compare these results with the information in Figure 7.2 and Table 7.2. Make sure that you can draw Lewis structures correctly. Without a correct Lewis structure, you will be unable to determine the shape of a molecule.

Practice Problem ⒶTTEMPT Determine the shapes of (a) CO_2 and (b) SCl_2.

Practice Problem ⒷUILD (a) From what group must the terminal atoms come in an AB_x molecule where the central atom is from Group 6A, for the electron-domain geometry and the molecular geometry both to be trigonal planar? (b) From what group must the terminal atoms come in an AB_x molecule where the central atom is from Group 7A, for the electron-domain geometry to be octahedral and the molecular geometry to be square pyramidal?

Practice Problem ⒸONCEPTUALIZE These four models may represent molecules or polyatomic ions. Lone pairs on the central atom, if any, are not shown. Which of these could represent a species in which there are lone pairs on the central atom? Which could represent a species in which there are no lone pairs on the central atom?

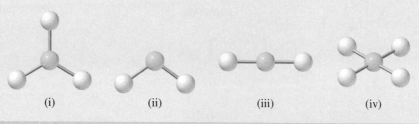

(i)   (ii)   (iii)   (iv)

Deviation from Ideal Bond Angles

Some electron domains are better than others at repelling neighboring domains. As a result, the bond angles may be slightly different from those shown in Figure 7.2. For example, the electron-domain geometry of ammonia (NH_3) is tetrahedral, so we might predict the H—N—H bond angles to be 109.5°. In fact, the bond angles are about 107°, slightly smaller than predicted. The lone pair on the nitrogen atom repels the N—H bonds more strongly than the bonds repel one another. It therefore "squeezes" them closer together than the ideal tetrahedral angle of 109.5°.

In effect, a lone pair takes up more *space* than the bonding pairs. This can be understood by considering the attractive forces involved in determining the location of the electron pairs. A lone pair on a central atom is attracted only to the nucleus of that atom. A bonding pair of electrons, on the other hand, is simultaneously attracted by the nuclei of both of the bonding atoms. As a result, the lone pair has more freedom to spread out and greater capacity to repel other electron domains. Also, because they contain more electron density, multiple bonds repel more strongly than single bonds. Consider the bond angles in each of the following examples:

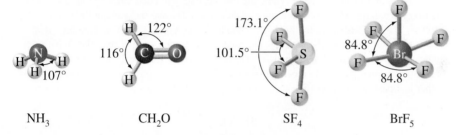

Geometry of Molecules with More Than One Central Atom

Thus far we have considered the geometries of molecules having only one central atom. We can determine the overall geometry of more complex molecules by treating them as though they have multiple central atoms. Methanol (CH_3OH), for example, has a central C atom and a central O atom, as shown in the following Lewis structure:

$$\text{H}-\overset{\displaystyle \text{H}}{\underset{\displaystyle \text{H}}{\text{C}}}-\ddot{\text{O}}-\text{H}$$

Both the C and the O atoms are surrounded by four electron domains. In the case of C, they are three C—H bonds and one C—O bond. In the case of O, they are one O—C bond, one O—H bond, and two lone pairs. In each case, the electron-domain geometry is tetrahedral. However, the molecular geometry of the C part of the molecule is *tetrahedral,* whereas the molecular geometry of the O part of the molecule is *bent.* Note that although the Lewis structure makes it appear as though there is a 180° angle between the O—C and O—H bonds, the angle is actually approximately 109.5°, the angle in a tetrahedral arrangement of electron domains.

Worked Example 7.2 shows how to determine when bond angles differ from ideal values.

> **Student Annotation:** When we specify the geometry of a particular portion of a molecule, we refer to it as the geometry "about" a particular atom. In methanol, for example, we say that the geometry is *tetrahedral about the C atom and bent about the O atom.*

Worked Example 7.2

Acetic acid, the substance that gives vinegar its characteristic smell and sour taste, is sometimes used in combination with corticosteroids to treat certain types of ear infections. Its Lewis structure is

$$\text{H}-\overset{\displaystyle \text{H}}{\underset{\displaystyle \text{H}}{\text{C}}}-\overset{\displaystyle \ddot{\text{O}}}{\text{C}}-\ddot{\text{O}}-\text{H}$$

(Continued on next page)

Determine the molecular geometry about each of the central atoms, and determine the approximate value of each of the bond angles in the molecule. Which if any of the bond angles would you expect to be smaller than the ideal values?

Strategy Identify the central atoms and count the number of electron domains around each of them. Use the VSEPR model to determine each electron-domain geometry, and the information in Table 7.2 to determine the molecular geometry about each central atom.

Setup The leftmost C atom is surrounded by four electron domains: one C—C bond and three C—H bonds. The middle C atom is surrounded by three electron domains: one C—C bond, one C—O bond, and one C=O (double) bond. The O atom is surrounded by four electron domains: one O—C bond, one O—H bond, and two lone pairs.

Solution The electron-domain geometry of the leftmost C is tetrahedral. Because all four electron domains are bonds, the molecular geometry of this part of the molecule is also tetrahedral. The electron-domain geometry of the middle C is trigonal planar. Again, because all the domains are bonds, the molecular geometry is also trigonal planar. The electron-domain geometry of the O atom is tetrahedral. Because two of the domains are lone pairs, the molecular geometry about the O atom is bent. Bond angles are determined using electron-domain geometry. Therefore, the approximate bond angles about the leftmost C are 109.5°, those about the middle C are 120°, and those about the O are 109.5°. The angle between the two single bonds on the middle carbon will be *less* than 120°, because the double bond repels the single bonds more strongly than they repel each other. Likewise, the bond angle between the two bonds on the O will be less than 109.5°, because the lone pairs on O repel the single bonds more strongly than they repel each other and push the two bonding pairs closer together. The angles are labeled as follows:

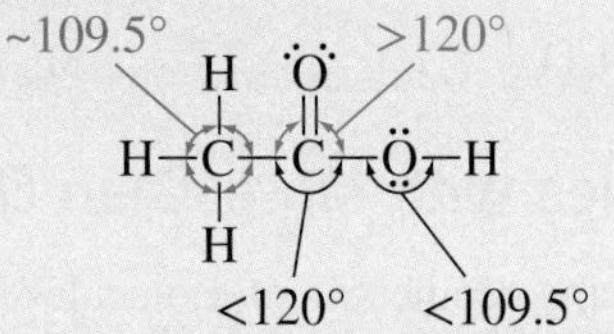

$$\sim 109.5° \qquad > 120°$$

$$< 120° \qquad < 109.5°$$

Think About It
Compare these answers with the information in Figure 7.2 and Table 7.2.

Practice Problem Ⓐ**TTEMPT** Ethanolamine ($HOCH_2CH_2NH_2$) has a smell similar to ammonia and is commonly found in biological tissues. Its Lewis structure is

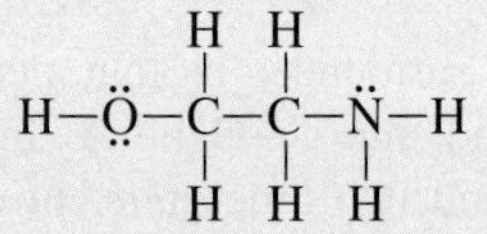

Determine the molecular geometry about each central atom and label all the bond angles. Cite any expected deviations from ideal bond angles.

Practice Problem Ⓑ**UILD** The bond angle in NH_3 is significantly smaller than the ideal bond angle of 109.5° because of the lone pair on the central atom. Explain why the bond angle in SO_2 is very close to 120° despite there being a lone pair on the central atom.

Practice Problem Ⓒ**ONCEPTUALIZE** Which of these models represents a species in which there is deviation from ideal bond angles?

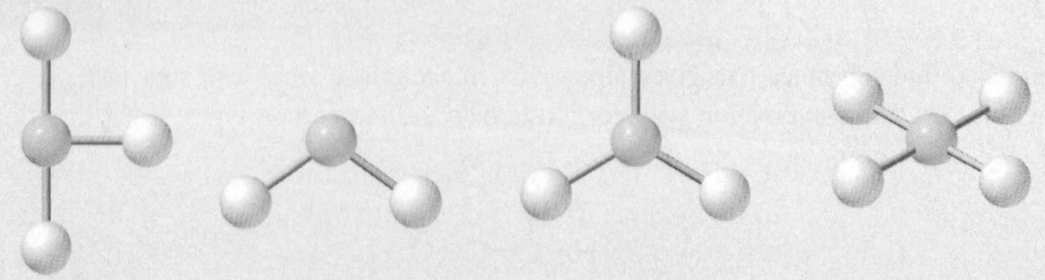

(i)   (ii)   (iii)   (iv)

Section 7.1 Review

Molecular Geometry

7.1.1 What are the electron-domain geometry and molecular geometry of CO_3^{2-}?
    (a) tetrahedral, trigonal planar
    (b) tetrahedral, trigonal pyramidal
    (c) trigonal pyramidal, trigonal pyramidal
    (d) trigonal planar, trigonal planar
    (e) tetrahedral, tetrahedral

7.1.2 What are the electron-domain geometry and molecular geometry of ClO_3^-?
    (a) tetrahedral, trigonal planar
    (b) tetrahedral, trigonal pyramidal
    (c) trigonal pyramidal, trigonal pyramidal
    (d) trigonal planar, trigonal planar
    (e) tetrahedral, tetrahedral

7.1.3 What is the approximate value of the bond angle indicated?

    (a) $< 90°$ (d) $> 120°$
    (b) $< 109.5°$ (e) $< 120°$
    (c) $> 109.5°$

7.1.4 What is the approximate value of the bond angle indicated?

    (a) $< 180°$ (d) $> 109.5°$
    (b) $> 180°$ (e) $< 90°$
    (c) $< 109.5°$

7.2 MOLECULAR GEOMETRY AND POLARITY

Molecular geometry is tremendously important in understanding the physical and chemical behavior of a substance. *Molecular polarity,* for example, is one of the most important consequences of molecular geometry, because molecular polarity influences physical, chemical, and biological properties. Recall from Section 6.2 that a bond between two atoms of different electronegativities is polar and that a diatomic molecule containing a polar bond is a *polar molecule*. Whether a molecule made up of three or more atoms is polar depends not only on the polarity of the individual bonds but also on its molecular geometry.

Each of the CO_2 and H_2O molecules contains two identical atoms bonded to a central atom and two polar bonds. However, only one of these molecules is polar. To understand why, think of each individual bond dipole as a vector. The overall dipole moment of the molecule is determined by vector addition of the individual bond dipoles.

In the case of CO_2, we have two identical vectors pointing in opposite directions. When the vectors are placed on a Cartesian coordinate system, they have no *y* component

Video 7.2
Chemical bonding—molecular geometry and polarity

and their x components are equal in magnitude but opposite in sign. The sum of these two vectors is zero in both the x and y directions. Thus, although the *bonds* in CO_2 are polar, the *molecule* is nonpolar.

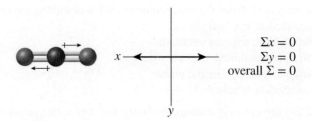

The vectors representing the bond dipoles in water, although equal in magnitude and opposite in the x direction, are not opposite in the y direction. Therefore, although their x components sum to zero, their y components do not. This means that there is a net resultant dipole and H_2O is *polar.*

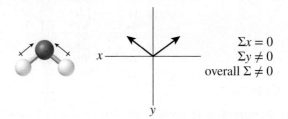

In AB_x molecules where $x \geq 3$, it may be less obvious whether the individual bond dipoles cancel one another. Consider the molecule BF_3, for example, which has a trigonal planar geometry:

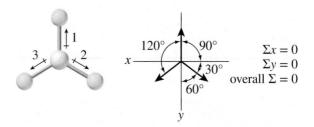

We will simplify the math in this analysis by assigning the vectors representing the three identical B–F bonds an arbitrary magnitude of 1.00. The x, y coordinates for the end of arrow 1 are (0, 1.00). Determining the coordinate for the ends of arrows 2 and 3 requires the use of trigonometric functions. You may have learned the mnemonic SOH CAH TOA, where the letters stand for

$$Sin = Opposite \ over \ Hypotenuse$$
$$Cos = Adjacent \ over \ Hypotenuse$$
$$Tan = Opposite \ over \ Adjacent$$

The x coordinate for the end of arrow 2 corresponds to the length of the line opposite the 60° angle. The hypotenuse of the triangle has a length of 1.00 (the arbitrarily assigned value). Therefore, using SOH,

$$\sin 60° = 0.866 = \frac{opposite}{hypotenuse} = \frac{opposite}{1}$$

so the x coordinate for the end of arrow 2 is 0.866.

The magnitude of the y coordinate corresponds to the length of the line adjacent to the 60° angle. Using TOA,

$$\tan 60° = 1.73 = \frac{\text{opposite}}{\text{adjacent}} = \frac{0.866}{\text{adjacent}}$$

$$\text{adjacent} = \frac{0.866}{1.73} = 0.500$$

so the y coordinate for the end of arrow 2 is -0.500. (The trigonometric formula gives us the length of the side. We know from the diagram that the sign of this y component is negative.)

Arrow 3 is similar to arrow 2. Its x component is equal in magnitude but opposite in sign, and its y component is the same magnitude and sign as that for arrow 2. Therefore, the x and y coordinates for all three vectors are

	x	y
Arrow 1	0	1
Arrow 2	0.866	-0.500
Arrow 3	-0.866	-0.500
Sum =	0	0

Because the individual bond dipoles (represented here as the vectors) sum to zero, the molecule is nonpolar overall.

Although it is somewhat more complicated, a similar analysis can be done to show that all x, y, and z coordinates sum to zero when there are four identical polar bonds arranged in a tetrahedron about a central atom. In fact, any time there are identical bonds symmetrically distributed around a central atom, with no lone pairs on the central atom, the molecule will be nonpolar overall, even if the bonds themselves are polar.

In cases where *non*-identical bonds are distributed symmetrically around the central atom, the nature of the atoms surrounding the central atom determines whether the molecule is polar overall. For example, CCl_4 and $CHCl_3$ have the same molecular geometry (tetrahedral), but CCl_4 is nonpolar because the bond dipoles cancel one another. In $CHCl_3$, however, the bonds are not all identical, and therefore the bond dipoles do not sum to zero. The $CHCl_3$ molecule is polar.

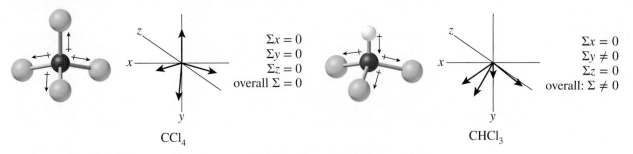

Worked Example 7.3 shows you how to determine whether a molecule is polar.

Worked Example 7.3

Determine whether (a) PCl_5 and (b) H_2CO (C double bonded to O) are polar.

Strategy For each molecule, draw the Lewis structure, use the VSEPR model to determine its molecular geometry, and then determine whether the individual bond dipoles cancel.

(Continued on next page)

Setup

(a) The Lewis structure of PCl_5 is

With five identical electron domains around the central atom, the electron-domain and molecular geometries are trigonal bipyramidal. The equatorial bond dipoles will cancel one another, just as in the case of BF_3, and the axial bond dipoles will also cancel each other.

(b) The Lewis structure of H_2CO is

The bond dipoles, although symmetrically distributed around the C atom, are not identical and therefore will not sum to zero.

Solution

(a) PCl_5 is nonpolar.   (b) H_2CO is polar.

Think About It

Make sure that your Lewis structures are correct and that you count electron domains on the central atom carefully. This will give you the correct electron-domain and molecular geometries. Molecular polarity depends both on individual bond dipoles and molecular geometry.

Practice Problem Ⓐ**TTEMPT** Determine whether (a) CH_2Cl_2 and (b) XeF_4 are polar.

Practice Problem Ⓑ**UILD** For each of the following hypothetical molecules, draw one structure that is polar and one structure that is nonpolar: (a) H_2SF_4 and (b) H_2PCl_3.

Practice Problem Ⓒ**ONCEPTUALIZE** Which of these models could represent a polar molecule?

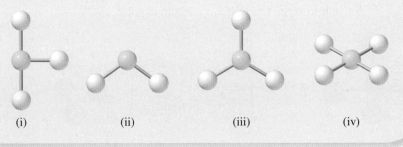

(i)   (ii)   (iii)   (iv)

Dipole moments can be used to distinguish between molecules that have the same chemical formula but different arrangements of atoms. Such compounds are called **structural isomers.** For example, there are two structural isomers of dichloro-ethylene ($C_2H_2Cl_2$). Because the individual bond dipoles sum to zero in *trans*-dichloroethylene, the *trans* isomer is nonpolar.

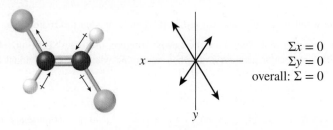

$\Sigma x = 0$
$\Sigma y = 0$
overall: $\Sigma = 0$

The bond dipoles in the *cis* isomer do not cancel one another, so *cis*-dichloroethylene is polar.

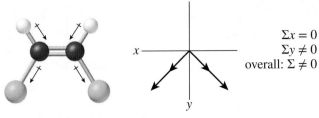

$$\Sigma x = 0$$
$$\Sigma y \neq 0$$
overall: $\Sigma \neq 0$

Because of the difference in polarity, these two isomers can be distinguished experimentally by measuring the dipole moment.

Student Annotation: Polarity is an important property that determines, in part, the properties of a molecule or polyatomic ion.

Section 7.2 Review

Molecular Geometry and Polarity

7.2.1 Identify the polar molecules in the following group: HBr, CH_4, CS_2.
(a) HBr only
(b) HBr and CS_2
(c) HBr, CH_4, and CS_2
(d) CH_4 and CS_2
(e) CH_4 only

7.2.2 Identify the nonpolar molecules in the following group: SO_2, NH_3, XeF_2.
(a) SO_2, NH_3, and XeF_2
(b) SO_2 only
(c) XeF_2 only
(d) SO_2 and XeF_2
(e) SO_2 and NH_3

7.3 INTERMOLECULAR FORCES

An important consequence of molecular polarity is the existence of attractive forces between neighboring molecules, which we refer to as ***intermolecular forces.*** We have already encountered an example of intermolecular forces in the form of ionic bonding [◄◄ Section 5.3], where the magnitude of attraction between oppositely charged particles is governed by Coulomb's law [◄◄ Section 4.4]. Because the particles that make up an ionic compound have discrete (*full*) charges, the attractive forces that hold them together are especially powerful. In fact, that's the reason they're solids at room temperature. The intermolecular forces we will encounter in this chapter are also the result of Coulombic attractions, but the attractions involve only partial charges [◄◄ Section 6.2] rather than discrete charges and are therefore weaker than the forces involved in ionic bonding. Nevertheless, the magnitudes of such intermolecular forces can be sufficient to hold the molecules of a substance together in a solid (e.g., sugar), whereas substances with relatively weaker intermolecular forces typically are liquids (e.g., water) and substances with no apparent intermolecular attractions are gases (e.g., oxygen).

Student Annotation: Gases actually do exhibit intermolecular attractions; but the attractions are so weak compared to the kinetic energies of the individual molecules, that they are negligible, and do not hold the molecules together [►► Chapter 11].

We will begin our discussion of intermolecular forces with the attractive forces that act between atoms or molecules in a pure substance. These forces are known collectively as ***van der Waals forces,*** and they include *dipole-dipole interactions,* including *hydrogen bonding* and *dispersion forces.*

Dipole-Dipole Interactions

Dipole-dipole interactions are attractive forces that act between *polar molecules.* Recall that a diatomic molecule containing elements of significantly different electronegativities, such as HCl, has an unequal distribution of electron density and therefore has partial charges, positive ($\delta+$) at one end and negative ($\delta-$) at the other. The partial positive charge on one molecule is attracted to the partial negative charge on a neighboring

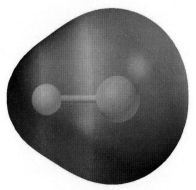

HCl

Figure 7.3 Arrangement of polar molecules in a liquid (left) and in a solid (right).

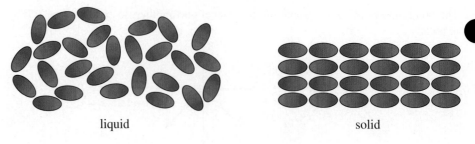

liquid   solid

molecule. Figure 7.3 shows the orientation of polar molecules in a solid and in a liquid. The arrangement is somewhat less orderly in the liquid than it is in the solid.

Because this attractive force between polar molecules is Coulombic, the magnitude of the attractive forces depends on the magnitude of the dipole. In general, the larger the dipole, the larger the attractive force. Certain physical properties such as boiling point *reflect* the magnitude of intermolecular forces. A substance in which the particles are held together by larger intermolecular attractions will require more energy to *separate* the particles and will therefore boil at a higher temperature. Table 7.3 lists several compounds with similar molar masses along with their dipole moments and boiling points.

Hydrogen Bonding

Hydrogen bonding is a special type of dipole-dipole interaction. But, whereas dipole-dipole interactions act between any polar molecules, hydrogen bonding occurs only in molecules that contain H bonded to a small, highly electronegative atom, such as N, O, or F. In a molecule such as HF, which is shown in Figure 7.4, the F atom to which H is bonded draws electron density toward itself. Being small and highly electronegative, the F atom draws electron density away from H quite effectively. Because H has only one electron, this leaves the hydrogen nucleus practically unshielded—giving H a large partial positive charge. This large partial positive charge is powerfully attracted to the large partial negative charge (lone pairs) on the small, highly electronegative F atom of a neighboring HF molecule. The result is an especially strong dipole-dipole attraction.

In Worked Example 6.2 we calculated the partial charges on H and F as +0.41 and −0.41, respectively. The partial charges on H and Cl, by contrast, are only +0.18 and −0.18,

> **Student Annotation:** It is a common error to assume that hydrogen bonding occurs in any molecule that contains hydrogen. It *only* occurs to a significant degree in molecules with N—H, O—H, or F—H bonds.

TABLE 7.3	Dipole Moments and Boiling Points of Compounds with Similar Molecular Masses		
Compound	**Structural formula**	**Dipole moment (D)**	**Boiling point (°C)**
Propane	$CH_3CH_2CH_3$	0.1	−42
Dimethyl ether	CH_3OCH_3	1.3	−25
Methyl chloride	CH_3Cl	1.9	−24
Acetaldehyde	CH_3CHO	2.7	21
Acetonitrile	CH_3CN	2.9	82

Figure 7.4 Hydrogen bonds between HF molecules.

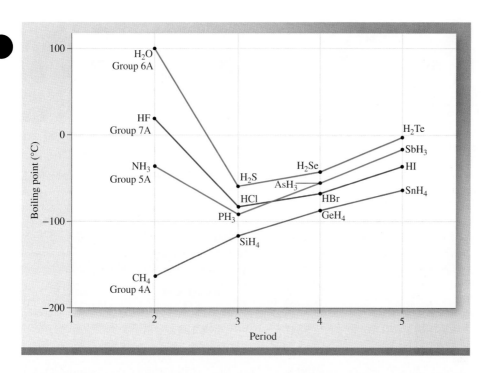

Figure 7.5 Boiling points of the hydrogen compounds of elements from Groups 4A through 7A. Although normally the boiling point increases with increasing mass within a group, the *lightest* compound has the *highest* boiling point in Groups 5A through 7A. This departure from the observed trend is due to hydrogen bonding.

respectively. They are smaller in HCl because Cl is larger and less electronegative than F. Because of the high electronegativity of F, and the high partial positive charges that result, HF exhibits significant hydrogen bonding, whereas HCl does not.

Figure 7.5 shows the boiling points of the binary hydrogen compounds of Groups 4A through 7A. Within the series of hydrogen compounds of Group 4A, the boiling point increases with increasing molar mass. For Groups 5A through 7A, the same trend is observed for all but the smallest member of each series, which has what appears to be an unexpectedly high boiling point. This departure from the trend in boiling points illustrates how powerful hydrogen bonding can be. NH_3 (Group 5A), HF (Group 7A), and H_2O (Group 6A) have anomalously high boiling points because they exhibit strong hydrogen bonding. CH_4 (the smallest member of the Group 4A series) does not exhibit hydrogen bonding because it does not contain an N—H, O—H, or F—H bond; hence, it has the lowest boiling point of that group.

Student Annotation: Strictly speaking, HCl does exhibit hydrogen bonding to a slight degree. But only molecules containing an N—H, O—H, or F—H bond exhibit significant hydrogen bonding.

Student Annotation: Hydrogen bonding also occurs in mixtures, between solute and solvent molecules that contain N—H, F—H, or O—H bonds.

Dispersion Forces

Nonpolar gases, such as N_2 and O_2, can be liquefied under the right conditions of pressure and temperature, so nonpolar molecules must also exhibit attractive intermolecular forces. These intermolecular forces are Coulombic in nature (as are all intermolecular forces), but they differ from other intermolecular forces because they arise from the movement of electrons in nonpolar molecules.

On average, the distribution of electron density in a nonpolar molecule is uniform and symmetrical (which is why a nonpolar molecule has no dipole moment). However, because electrons in a molecule have some freedom to move about, at any given point in time the molecule may have a nonuniform distribution of electron density, giving it a fleeting, temporary dipole—called an ***instantaneous dipole.*** An instantaneous dipole in one molecule can induce dipoles in neighboring molecules. For example, the temporary partial negative charge on a molecule repels the electrons in a molecule next to it. This repulsion polarizes the second molecule, which then acquires a temporary dipole. It in turn polarizes the next molecule and so on, leaving a collection of ordinarily nonpolar molecules with partial positive and negative charges and Coulombic attractions between them as shown in Figure 7.6. The resulting attractive forces are called ***London***[1] ***dispersion forces*** or simply ***dispersion forces.***

---

[1]Fritz London (1900–1954), a German physicist, was a theoretical physicist whose major work was on superconductivity in liquid helium.

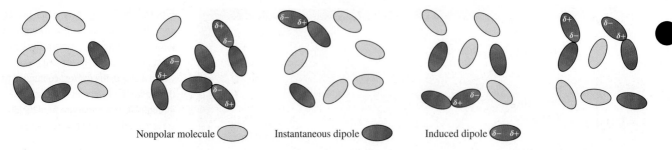

Figure 7.6 Instantaneous dipoles in ordinarily nonpolar molecules can induce temporary dipoles in neighboring molecules, causing the molecules to be attracted to one another. This type of interaction is responsible for the condensation of nonpolar gases.

TABLE 7.4	Molar Masses, Boiling Points, and States of the Halogens at Room Temperature		
Molecule	**Molar mass (g/mol)**	**Boiling point (°C)**	**State (room temp.)**
F_2	38.0	−188	Gas
Cl_2	70.9	−34	Gas
Br_2	159.8	59	Liquid
I_2	253.8	184	Solid

The magnitude of dispersion forces depends on how mobile the electrons in the molecule are. In small molecules, such as F_2, the electrons are relatively close to the nuclei and cannot move about very freely; thus, the electron distribution in F_2 is not easily *polarized*. In larger molecules, such as Cl_2, the electrons are somewhat farther away from the nuclei and therefore move about more freely. The electron density in Cl_2 is more easily polarized than that in F_2, resulting in larger instantaneous dipoles, larger induced dipoles, and larger intermolecular attractions overall. The trend in magnitude of dispersion forces with increasing molecular size is illustrated by the correlation between molar mass and boiling point in Table 7.4. (It is also illustrated by the data in Table 12.2 in Section 12.3.) Dispersion forces act not only between nonpolar molecules, but between *all* molecules.

Worked Example 7.4 lets you practice determining what kinds of forces exist between particles in liquids and solids.

Student Annotation: Molecules that more readily acquire an instantaneous dipole are said to be *polarizable*. In general, larger molecules have greater *polarizability* than small ones.

 Student Hot Spot

Student data indicate you may struggle with intermolecular forces. Access the SmartBook to view additional Learning Resources on this topic.

Worked Example 7.4

What kind(s) of intermolecular forces exist in (a) $CCl_4(l)$, (b) $CH_3COOH(l)$, (c) $CH_3COCH_3(l)$, and (d) $H_2S(l)$?

Strategy Draw Lewis dot structures and apply VSEPR theory [◄◄ Section 7.1] to determine whether each molecule is polar or nonpolar. Nonpolar molecules exhibit dispersion forces only. Polar molecules exhibit both dipole-dipole interactions and dispersion forces. Polar molecules with N—H, F—H, or O—H bonds exhibit dipole-dipole interactions (including hydrogen bonding) and dispersion forces.

Setup The Lewis dot structures for molecules (a) to (d) are

```
      :Cl:                H  :O:              H  :O:  H
       |                  |   ||              |   ||  |
  :Cl—C—Cl:          H—C—C—O—H          H—C—C—C—H          H—S—H
       |                  |                   |       |
      :Cl:                H                   H       H

      (a)                 (b)                 (c)                 (d)
```

Solution

(a) CCl_4 is nonpolar, so the only intermolecular forces are dispersion forces.

(b) CH_3COOH is polar and contains an $O-H$ bond, so it exhibits dipole-dipole interactions (including hydrogen bonding) and dispersion forces.

(c) CH_3COCH_3 is polar but does *not* contain $N-H$, $O-H$, or $F-H$ bonds, so it exhibits dipole-dipole interactions and dispersion forces.

(d) H_2S is polar but does not contain $N-H$, $O-H$, or $F-H$ bonds, so it exhibits dipole-dipole interactions and dispersion forces.

> **Think About It**
> Being able to draw correct Lewis structures is, once again, vitally important. Review, if you need to, the procedure for drawing them [◄◄ Section 6.3].

Practice Problem **A**TTEMPT What kind(s) of intermolecular forces exist in (a) $CH_3CH_2CH_2CH_2CH_3(l)$, (b) $CH_3CH_2OH(l)$, (c) $H_2CO(l)$, and (d) $O_2(l)$?

Practice Problem **B**UILD What kind(s) of intermolecular forces exist in (a) $CH_2Cl_2(l)$, (b) $CH_3CH_2CH_2OH(l)$, (c) $H_2O_2(l)$, and (d) $N_2(l)$?

Practice Problem **C**ONCEPTUALIZE Using the molecular formula C_2H_6O, draw two Lewis structures that exhibit different intermolecular forces. List the types of intermolecular forces exhibited by each structure.

Ion-Dipole Interactions

Ion-dipole interactions are Coulombic attractions between ions (either positive or negative) and polar molecules. These interactions occur in mixtures of ionic and polar species such as an aqueous solution of sodium chloride. The magnitude of ion-dipole interactions depends on the charge and the size of the ion, and on the dipole moment and size of the polar molecule. Cations generally interact more strongly with dipoles than anions (of the same magnitude charge) because they tend to be smaller.

> **Student Annotation:** Hydration [▶▶ Section 9.2] is one example of an ion-dipole interaction.

Figure 7.7 shows the ion-dipole interaction between the Na^+ and Mg^{2+} ions with a water molecule, which has a large dipole moment (1.87 D). Because the Mg^{2+} ion has a *higher charge* and a *smaller size* [◄◄ Section 4.6] than the Na^+ ion (the ionic radii of Mg^{2+} and Na^+ are 72 and 102 pm, respectively), Mg^{2+} interacts more strongly with water molecules. The properties of solutions are discussed in Chapter 13.

Although the intermolecular forces discussed so far are all *attractive* forces, molecules also exert repulsive forces on one another (when *like* charges approach one another). When two molecules approach each other closely, the repulsions between electrons and between nuclei in the molecules become significant. The magnitude of these repulsive forces rises steeply as the distance separating the molecules in a condensed phase decreases. This is the reason liquids and solids are so hard to compress. In these phases, the molecules are already in close contact with one another, and so they greatly resist being compressed further.

Section 7.3 Review

Intermolecular Forces

7.3.1 What kind(s) of intermolecular forces exist between benzene molecules (C_6H_6)?
    (a) Dispersion forces only
    (b) Dipole-dipole interactions only
    (c) Hydrogen bonding only
    (d) Dispersion forces and dipole-dipole interactions
    (e) Dispersion forces and hydrogen bonding

7.3.2 Which of the following exhibits significant hydrogen bonding? (Select all that apply.)
    (a) HBr   (c) H_2 (e) CH_3CN
    (b) H_2CF_2 (d) H_2O_2

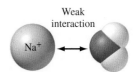

Weak interaction

Na^+

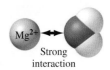

Mg^{2+}

Strong interaction

Figure 7.7 Ion-dipole interactions between ions and water molecules. In each case the positive charge on the ion is attracted to the partial negative charge on the oxygen atom in the water molecule. The attraction is stronger when the distance between the two species is smaller.

7.4   VALENCE BOND THEORY

In Chapter 6, we encountered our first chemical bonding theory, that proposed by Gilbert Lewis, which describes the bonds in molecules using the octet rule [◄◄ Section 6.1]. Although the Lewis theory of bonding and Lewis structures have provided a relatively simple way for us to visualize arrangements of electrons in molecules, they are insufficient to explain the differences between the covalent bonds in molecules such as H_2, F_2, and HF. Although Lewis theory describes the bonds in these three molecules in exactly the same way, they really are quite different from one another, as evidenced by their bond lengths and bond energies listed in Table 7.5. Understanding these differences and why covalent bonds form in the first place requires a bonding model that combines Lewis's notion of atoms sharing electron pairs and the quantum mechanical descriptions of atomic orbitals.

According to **valence bond theory,** atoms share electrons when an atomic orbital on one atom overlaps with an atomic orbital on the other. Each of the overlapping atomic orbitals must contain a single, unpaired electron. Furthermore, the two electrons shared by the bonded atoms must have opposite spins [◄◄ Section 3.7]. The nuclei of both atoms are attracted to the shared pair of electrons. It is this mutual attraction for the shared electrons that holds the atoms together.

The quantum mechanical model of the atom treats electrons in atoms as waves, rather than particles. Therefore, rather than use arrows to denote the locations, the convention used here shows a singly occupied atomic s orbital as light yellow and a doubly occupied orbital as a darker version of yellow. The H—H bond in H_2 forms when the singly occupied $1s$ orbitals of the two H atoms overlap.

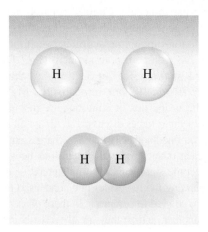

Note that in this case, although there are still just two electrons, each H atom "thinks" it owns them both. So when the two singly occupied orbitals overlap, *both* orbitals end up doubly occupied.

> **Student Annotation:** Recall that *bond energy* is the amount of energy required to break a mole of identical bonds [◄◄ Section 6.1].

H—H

:F̈—F̈:

H—F̈:

Lewis dot structures
of H_2, F_2, and HF

TABLE 7.5	Bond Lengths and Bond Energies of H_2, F_2, and HF	
	Bond length (Å)	Bond energy (kJ/mol)
H_2	0.74	436.4
F_2	1.42	150.6
HF	0.92	568.2

Similarly, the F—F bond in F_2 forms when the singly occupied 2p orbitals of the two F atoms overlap.

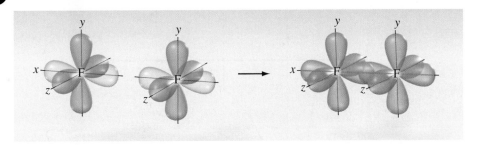

Following the convention introduced earlier, atomic p orbitals will be represented as light blue if singly occupied and darker blue if doubly occupied. Empty p orbitals will appear white. Recall that the ground-state electron configuration of the F atom is $[He]2s^22p^5$ [◄◄ Section 3.9]. (The ground-state orbital diagram of F is shown above.)

We can also use the valence bond model to depict the formation of an H—F bond. In this case, the singly occupied 1s orbital of the H atom overlaps with the singly occupied 2p orbital of the F atom.

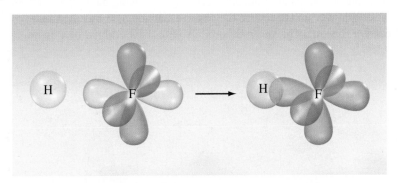

According to the quantum mechanical model, the sizes, shapes, and energies of the 1s orbital of H and the 2p orbital of F are different. Therefore, it is not surprising that the bonds in H_2, F_2, and HF vary in strength and length.

One of the advantages of valence bond theory over Lewis theory is that it explains *why* covalent bonds form. According to valence bond theory, a covalent bond will form between two atoms if the potential energy of the resulting molecule is lower than the combined potential energies of the isolated atoms. Simply put, this means that the formation of covalent bonds gives off energy. While this may not seem intuitively obvious, you know that energy must be *supplied* to a molecule to *break* covalent bonds. It should make sense, then, that the reverse process—the *formation* of bonds—must give *off* energy. Figure 7.8 illustrates how the total potential energy of two hydrogen atoms varies with distance between the nuclei.

Valence bond theory also introduces the concept of *directionality* to chemical bonds. For example, we expect the bond formed by the overlap of a p orbital to coincide with the axis along which the p orbital lies. Consider the molecule H_2S. Unlike the other molecules that we have encountered, H_2S does not have the bond angle that Lewis theory and the VSEPR model would lead us to predict. (With four electron domains on the central atom, we would expect the bond angle to be on the order of 109.5°.) In fact, the H—S—H bond angle is 92°. Looking at this in terms of valence bond theory, the central atom (S) has two unpaired electrons, each of which resides in a 3p orbital. The orbital diagram for the ground-state electron configuration of the S atom is

$2s^2$	$2p^5$
⇅	⇅ ⇅ ↑

Orbital diagram for F

Student Annotation: For you to understand the material in this section and [▶▶ Section 7.5], you must be able to draw orbital diagrams for ground-state electron configurations [◄◄ Section 3.9].

S [Ne]

$3s^2$ $3p^4$

Figure 7.8 The change in potential energy of two hydrogen atoms as a function of internuclear distance. The minimum potential energy (−436 kJ/mol) occurs when the distance between the nuclei is 74 pm. The yellow spheres represent the 1s orbitals of hydrogen.

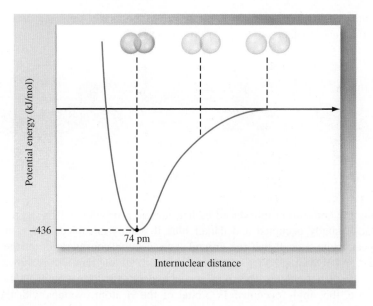

Remember that p orbitals are mutually perpendicular, lying along the x, y, and z axes [◀◀ Section 3.8]. We can rationalize the observed bond angle by envisioning the overlap of each of the singly occupied 3p orbitals with the 1s orbital of a hydrogen atom.

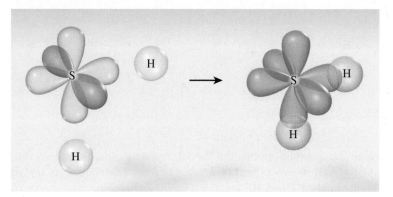

In summary, the important features of valence bond theory are as follows:

- A bond forms when singly occupied atomic orbitals on two atoms overlap.
- The two electrons shared in the region of orbital overlap must be of opposite spin.
- Formation of a bond results in a lower potential energy for the system.

Worked Example 7.5 shows how to use valence bond theory to explain the bonding in a molecule.

Worked Example 7.5

Hydrogen selenide (H$_2$Se) is a foul-smelling gas that can cause eye and respiratory tract inflammation. The H—Se—H bond angle in H$_2$Se is approximately 92°. Use valence bond theory to describe the bonding in this molecule.

Strategy Consider the central atom's ground-state electron configuration and determine what orbitals are available for bond formation.

Setup The ground-state electron configuration of Se is [Ar]4$s^2$3d^{10}4p^4. Its orbital diagram (showing only the 4p orbitals) is

$4p^4$

Solution Two of the $4p$ orbitals are singly occupied and therefore available for bonding. The bonds in H_2Se form as the result of the overlap of a hydrogen $1s$ orbital with each of these orbitals on the Se atom.

Think About It
Because the $4p$ orbitals on the Se atom are all mutually perpendicular, we should expect the angles between bonds formed by their overlap to be approximately 90°.

Practice Problem **A**TTEMPT Use valence bond theory to describe the bonding in phosphine (PH_3), which has H—P—H bond angles of approximately 94°.

Practice Problem **B**UILD For which molecule(s) can we not use valence bond theory alone to explain the bonding: SO_2 (O—S—O bond angle ~120°), CH_4 (H—C—H bond angles = 109.5°), AsH_3 (H—As—H bond angles = 92°)? Explain.

Practice Problem **C**ONCEPTUALIZE Which of these models could represent a species for which valence bond theory alone is sufficient to explain the observed bond angle? Explain.

~90°	~110°	~120°	~180°
(i)	(ii)	(iii)	(iv)

Section 7.4 Review

Valence Bond Theory

7.4.1   Which of the following atoms, in its ground state, does not have unpaired electrons? (Select all that apply.)
   (a) O   (c) B   (e) Ne
   (b) C   (d) F

7.4.2   According to valence bond theory, how many bonds would you expect a nitrogen atom (in its ground state) to form?
   (a) 2   (c) 4   (e) 6
   (b) 3   (d) 5

7.5   HYBRIDIZATION OF ATOMIC ORBITALS

Although valence bond theory is useful and can explain more of our experimental observations than Lewis bond theory, it fails to explain the bonding in many of the molecules that we encounter. According to valence bond theory, for example, an atom must have a singly occupied atomic orbital to form a bond with another atom. How then do we explain the bonding in $BeCl_2$? The central atom, Be, has a ground-state electron configuration of $[He]2s^2$, so it has no unpaired electrons. With no singly occupied atomic orbitals in its ground state, how does Be form two bonds?

   Furthermore, in cases where the ground-state electron configuration of the central atom *does* have the required number of unpaired electrons, how do we explain the observed bond angles? Carbon, like sulfur, has two unpaired electrons in its ground state. Using valence bond theory as our guide, we might envision the formation of two covalent bonds with oxygen, as in CO_2. If the two unpaired electrons on C

Video 7.3
Hybrid orbitals—orbital hybridization and valence bond theory

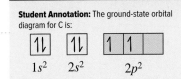

Student Annotation: The ground-state orbital diagram for C is:

$$\underset{1s^2}{\boxed{\uparrow\downarrow}}\quad \underset{2s^2}{\boxed{\uparrow\downarrow}}\quad \underset{2p^2}{\boxed{\uparrow}\,\boxed{\uparrow}\,\boxed{}}$$

(each residing in a $2p$ orbital) were to form bonds, however, the O—C—O bond angle should be on the order of 90°, like the bond angle in H_2S. In fact, the bond angle in CO_2 is 180°.

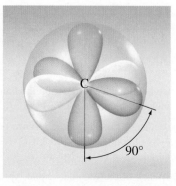

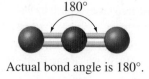

Actual bond angle is 180°.

Bond angle should be 90°.

:C̈l—Be—C̈l:

$BeCl_2$

To explain these and other observations, we need to extend our discussion of orbital overlap to include the concept of **hybridization,** or *mixing* of atomic orbitals.

The idea of hybridization of atomic orbitals begins with the molecular geometry and works backward to explain the bonds and the observed bond angles in a molecule. To extend our discussion of orbital overlap and introduce the concept of hybridization of atomic orbitals, we first consider beryllium chloride ($BeCl_2$), which has two electron domains on the central atom. Using its Lewis structure (shown above) and the VSEPR model, we predict that $BeCl_2$ will have a Cl—Be—Cl bond angle of 180°. If this is true, though, how does Be form two bonds with no unpaired electrons, and why is the angle between the two bonds 180°?

To answer the first part of the question, we envision the *promotion* of one of the electrons in the $2s$ orbital to an empty $2p$ orbital. Recall that electrons can be promoted from a lower atomic orbital to a higher one [◄◄ Section 3.4]. The ground-state electron configuration is the one in which all the electrons occupy orbitals of the lowest possible energy. A configuration in which one or more electrons occupy a *higher* energy orbital is called an *excited* state. An excited state generally is denoted with a star (e.g., Be* for an excited-state Be atom). Showing only the valence orbitals, we can represent the promotion of one of the valence electrons of beryllium as

$$\text{Be} \quad \boxed{\uparrow\downarrow} \quad \boxed{} \quad \xrightarrow{\text{promotion}} \quad \text{Be*} \quad \boxed{\uparrow} \quad \boxed{\uparrow}$$
$$\qquad\qquad 2s^2 \qquad\quad 2p \qquad\qquad\qquad\qquad\qquad 2s^1 \qquad 2p^1$$

$3s^2 \qquad 3p^5$

Orbital diagram for Cl

With one of its valence electrons promoted to the $2p$ subshell, the Be atom now has two unpaired electrons and therefore can form two bonds. However, the orbitals in which the two unpaired electrons reside are different from each other, so we would expect bonds formed as a result of the overlap of these two orbitals (each with a $3p$ orbital on a Cl atom) to be different.

$2s \qquad 3p \qquad\qquad \ne \qquad\qquad 2p \qquad 3p$

Experimentally, though, the bonds in $BeCl_2$ are identical in length and strength.

Hybridization of s and p Orbitals

To explain how beryllium forms two identical bonds, we must mix the orbitals in which the unpaired electrons reside, thus yielding two equivalent orbitals. The mixing of beryllium's 2s orbital with one of its 2p orbitals, a process known as *hybridization,* yields two **hybrid orbitals** that are neither s nor p, but have some character of each. The hybrid orbitals are designated 2sp or simply sp.

Student Annotation: Hybrid orbitals are another type of *electron domain.*

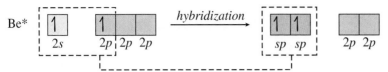

Mixing of one s orbital and one p orbital to yield two sp orbitals

The mathematical combination of the quantum mechanical wave functions for an s orbital and a p orbital gives rise to two new, equivalent wave functions. As shown in Figure 7.9(a), each sp hybrid orbital has one small lobe and one large lobe and, like any two electron domains on an atom, they are oriented in opposite directions with a 180° angle between them. The figure shows the atomic and hybrid orbitals separately for clarity. Note that the hybrid orbitals are shown in two ways: the first is a more realistic shape, whereas the second is a simplified shape that we use to keep the figures clear and make the visualization of orbitals easier. Note also that the representations of hybrid orbitals are green. Figures 7.9(b) and (c) show the locations of the atomic orbitals and the hybrid orbitals, respectively, relative to the beryllium nucleus.

Student Annotation: In picturing the shapes of the resulting hybrid orbitals, it may help to remember that orbitals, like waves, can combine constructively or destructively. We can think of the large lobe of each sp hybrid orbital as the result of a *constructive* combination and the small lobe of each as the result of a *destructive* combination.

With two sp hybrid orbitals, each containing a single unpaired electron, we can see how the Be atom is able to form two identical bonds with two Cl atoms [Figure 7.4(c)]. Each of the singly occupied sp hybrid orbitals on the Be atom overlaps with the singly occupied 3p atomic orbital on a Cl atom.

We can do a similar analysis of the bonds and the trigonal-planar geometry of boron trifluoride (BF_3). The ground-state electron configuration of the B atom is

Student Annotation: The energy required to promote an electron in an atom is more than compensated for by the energy given off when a bond forms.

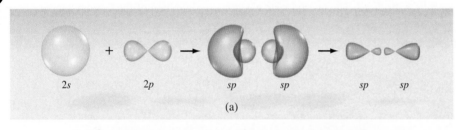

(a)

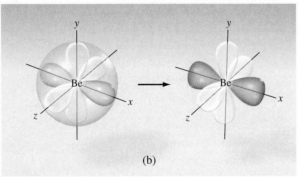

(b)

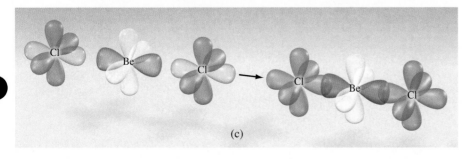

(c)

Figure 7.9 (a) An atomic s orbital (yellow) and one atomic p orbital (blue) combine to form two sp hybrid orbitals (green). The realistic hybrid orbital shapes are shown first. The thinner representations are used to keep diagrams clear. (b) The 2s orbital and one of the 2p orbitals on Be combine to form two sp hybrid orbitals. Unoccupied orbitals are shown in white. Like any two electron domains, the hybrid orbitals on Be are 180° apart. (c) The hybrid orbitals on Be each overlap with a singly occupied 3p orbital on a Cl atom.

$[He]2s^22p^1$, containing just one unpaired electron. Promotion of one of the $2s$ electrons to an empty $2p$ orbital gives the three unpaired electrons needed to explain the formation of *three* bonds. The ground-state and excited-state electron configurations can be represented by

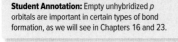

Because the three bonds in BF_3 are identical, we must hybridize the three singly occupied atomic orbitals (the one s and two p orbitals) to give three singly occupied hybrid orbitals.

Mixing of one s orbital and two p orbitals to yield three sp^2 orbitals

Student Annotation: Empty unhybridized p orbitals are important in certain types of bond formation, as we will see in Chapters 16 and 23.

$$\begin{array}{c} H \\ | \\ H-C-H \\ | \\ H \end{array}$$

CH_4

Figure 7.10 illustrates the hybridization and bond formation in BF_3.

In both cases (i.e., for $BeCl_2$ and BF_3), some but not *all* of the p orbitals are hybridized. When the remaining unhybridized atomic p orbitals do not contain electrons, as in the case of BF_3, they will not be part of the discussion of bonding in this chapter. As we will see in Section 7.6, though, unhybridized atomic orbitals that *do* contain electrons are important in our description of the bonding in a molecule.

We can now apply the same kind of analysis to the methane molecule (CH_4). The Lewis structure of CH_4 has four electron domains around the central carbon atom. This means that we need four hybrid orbitals, which in turn means that four atomic orbitals must be hybridized. The ground-state electron configuration of the C atom contains two

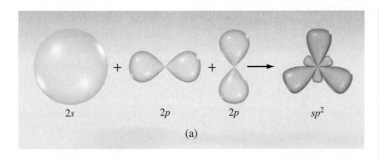

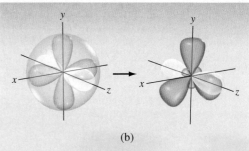

Figure 7.10 (a) An s atomic orbital and two p atomic orbitals combine to form three sp^2 hybrid orbitals. (b) The three sp^2 hybrid orbitals on B are arranged in a trigonal plane. (Empty atomic orbitals are shown in white.) (c) Hybrid orbitals on B overlap with $2p$ orbitals on F.

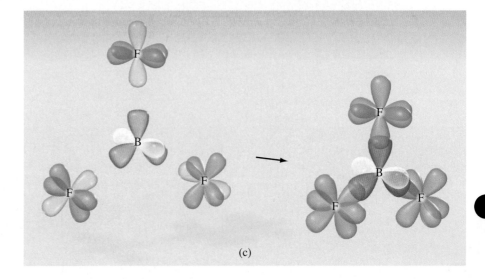

unpaired electrons. Promotion of one electron from the $2s$ orbital to the empty $2p$ orbital yields the four unpaired electrons needed for the formation of four bonds.

$$C \quad \boxed{1\!\downarrow}_{2s^2} \quad \boxed{1}\,\boxed{1}\,\boxed{\;}_{2p^2} \xrightarrow{\text{promotion}} C^* \quad \boxed{1}_{2s^1} \quad \boxed{1}\,\boxed{1}\,\boxed{1}_{2p^3}$$

Hybridization of the s orbital and the three p orbitals yields four hybrid orbitals designated sp^3. We can then place the electrons that were originally in the s and p atomic orbitals into the sp^3 hybrid orbitals.

$$C^* \quad \boxed{1}_{2s} \quad \boxed{1}\,\boxed{1}\,\boxed{1}_{2p\;2p\;2p} \xrightarrow{\text{hybridization}} \boxed{1}\,\boxed{1}\,\boxed{1}\,\boxed{1}_{sp^3\;sp^3\;sp^3\;sp^3}$$

Mixing of one s orbital and
three p orbitals to yield four sp^3 orbitals

The set of four sp^3 hybrid orbitals on carbon, like any four electron domains on a central atom, assumes a tetrahedral arrangement. Figure 7.11 illustrates how the hybridization of the C atom results in the formation of the four bonds and the 109.5° bond angles observed in CH_4.

Hybridization of *s, p,* and *d* Orbitals

Recall that elements in the third period of the periodic table and beyond do not necessarily obey the octet rule because they have d orbitals that can hold additional electrons [◀◀ Section 6.6]. In order to explain the bonding and geometry of molecules in which there are more than four electron domains on the central atom, we must include d orbitals in our hybridization scheme. PCl_5, for example, has five electron domains around the P atom. To explain the five bonds in this molecule, we will need five singly occupied hybrid orbitals. The ground-state electron configuration of the P atom is $[Ne]3s^23p^3$, which contains three unpaired electrons. In this case, though, because all three of the p orbitals are occupied, promotion of an electron from the $3s$ orbital to a $3p$ orbital would *not* result in additional unpaired electrons. However, we can

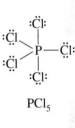

PCl_5

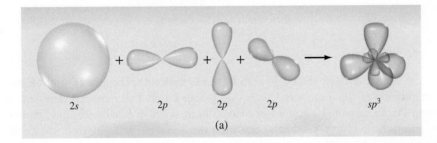

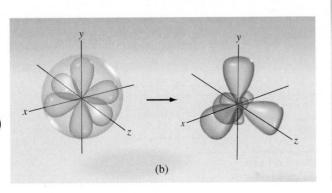

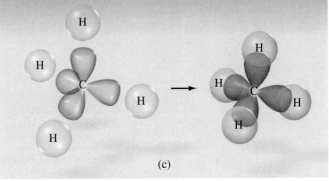

Figure 7.11 (a) An *s* atomic orbital and three *p* atomic orbitals combine to form four sp^3 hybrid orbitals. (b) The four sp^3 hybrid orbitals on C are arranged in a tetrahedron. (c) Hybrid orbitals on C overlap with 1*s* orbitals on H. For clarity, the small lobes of the hybrid orbitals are not shown.

envision the promotion of an electron from the $3s$ orbital to an empty $3d$ orbital resulting in the five unpaired electrons needed.

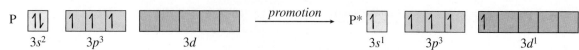

Hybridization of the s orbital, the three p orbitals, and one of the d orbitals yields hybrid orbitals that are designated sp^3d. After placing the five electrons in the five hybrid orbitals, we can rationalize the formation of five bonds in the molecule.

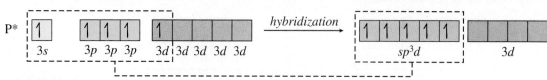

Mixing of one s orbital, three p orbitals,
and one d orbital to yield five sp^3d orbitals

Student Annotation: Note that the superscript numbers in hybrid orbital notation are used to designate the number of atomic orbitals that have undergone hybridization. When the superscript is 1, it is not shown (analogous to the subscripts in chemical formulas).

The sp^3d orbitals have shapes similar to those we have seen for the sp, sp^2, and sp^3 hybrid orbitals—that is, one large lobe and one small lobe. In addition, the five hybrid orbitals adopt a trigonal bipyramidal arrangement, enabling us to explain the geometry and bond angles in PCl_5.

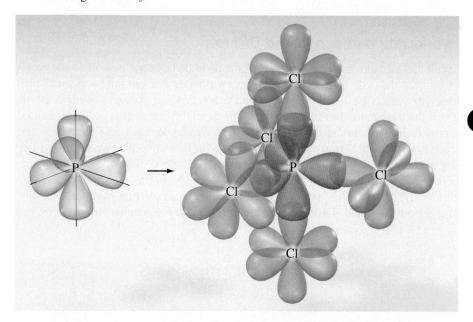

A similar analysis can be done with the SF_6 molecule. The ground-state electron configuration of the S atom is $[Ne]3s^23p^4$, giving it only two unpaired electrons. To obtain the six unpaired electrons needed to form six S—F bonds, we must promote *two* electrons to empty d orbitals: one from the $3s$ orbital and one from the doubly occupied $3p$ orbital. The resulting hybrid orbitals are designated sp^3d^2.

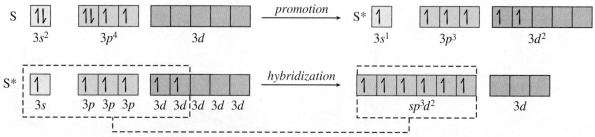

Mixing of one s orbital, three p orbitals,
and two d orbitals to yield six sp^3d^2 orbitals

The six bonds in SF_6 form, therefore, when each sp^3d^2 hybrid orbital on the S atom overlaps with a singly occupied $2p$ orbital on an F atom.

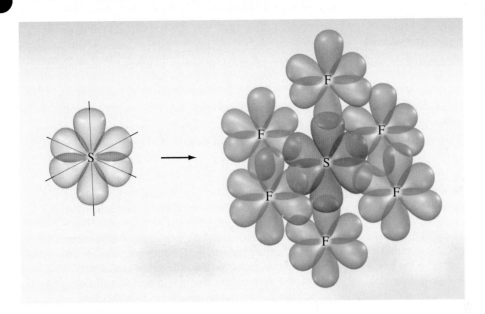

Table 7.6 shows how the number of electron domains on a central atom corresponds to a set of hybrid orbitals. In general, the hybridized bonding in a molecule can be described using the following steps:

1. Draw the Lewis structure.
2. Count the number of electron domains on the central atom. This is the number of hybrid orbitals necessary to account for the molecule's geometry. (This is also the number of *atomic* orbitals that must undergo hybridization.)
3. Draw the ground-state orbital diagram for the central atom.
4. Maximize the number of unpaired valence electrons by promotion.
5. Combine the necessary number of atomic orbitals to generate the required number of hybrid orbitals.
6. Place electrons in the hybrid orbitals, putting one electron in each orbital before pairing any electrons.

It is important to recognize that we do not use hybrid orbitals to *predict* molecular geometries, but rather to *explain* geometries that are already known. As we saw in Section 7.3, the bonding in many molecules can be explained without the use of hybrid orbitals. Hydrogen sulfide (H_2S), for example, has a bond angle of 92°. This bond angle is best explained without the use of hybrid orbitals.

Worked Example 7.6 shows how to use hybridization to explain the bonding and geometry in a molecule.

TABLE 7.6	Number of Electron Domains and Hybrid Orbitals on Central Atom				
Number of electron domains on central atom	2	3	4	5	6
Hybrid orbitals	sp	sp^2	sp^3	sp^3d	sp^3d^2
Geometry	Linear	Trigonal planar	Tetrahedral	Trigonal bipyramidal	Octahedral

Worked Example 7.6

Ammonia (NH_3) is a trigonal pyramidal molecule with H—N—H bond angles of about 107°. Describe the formation of three equivalent N—H bonds, and explain the angles between them.

Strategy Starting with a Lewis structure, determine the number and type of hybrid orbitals necessary to rationalize the bonding in NH_3.

Setup The Lewis structure of NH_3 is H—N̈—H .
107°
H

The ground-state electron configuration of the N atom is $[He]2s^2 2p^3$. Its valence orbital diagram is

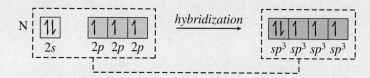

$2s^2$ $2p^3$

Solution Although the N atom has the three unpaired electrons needed to form three N—H bonds, we would expect bond angles of ~90° (not 107°) to form from the overlap of the three mutually perpendicular 2p orbitals. Hybridization, therefore, is necessary to explain the bonding in NH_3. (Recall from Section 7.3 that valence bond theory alone could be used to explain the bonding in H_2S, where the bond angles are ~90°—hybridization was unnecessary.) Although we often need to promote an electron to maximize the number of unpaired electrons, no promotion is necessary for the nitrogen in NH_3. We already have the three unpaired electrons necessary, and the promotion of an electron from the 2s orbital to one of the 2p orbitals would not result in any additional unpaired electrons. Furthermore, there are no empty d orbitals in the second shell. According to the Lewis structure, there are four electron domains on the central atom (three bonds and a lone pair of electrons). Four electron domains on the central atom require four hybrid orbitals, and four hybrid orbitals require the hybridization of four atomic orbitals: one s and three p. This corresponds to sp^3 hybridization. Because the atomic orbitals involved in the hybridization contain a total of five electrons, we place five electrons in the resulting hybrid orbitals. This means that one of the hybrid orbitals will contain a lone pair of electrons.

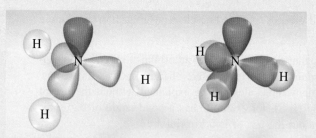

Each N—H bond is formed by the overlap between an sp^3 hybrid orbital on the N atom and the 1s atomic orbital on an H atom. Because there are four electron domains on the central atom, we expect them to be arranged in a tetrahedron. In addition, because one of the electron domains is a lone pair, we expect the H—N—H bond angles to be slightly smaller than the ideal tetrahedral bond angle of 109.5°.

Think About It
This analysis agrees with the experimentally observed geometry and bond angles of 107° in NH_3.

Practice Problem **A**TTEMPT Use hybrid orbital theory to describe the bonding and explain the bond angles in bromine pentafluoride (BrF_5).

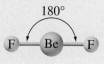

Practice Problem **B**UILD Use hybrid orbital theory to describe the bonding and explain the bond angles in BeF_2.

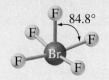

Practice Problem **C**ONCEPTUALIZE Indicate for which of the following species hybrid orbitals must be used to explain the geometry: CCl_4, Cl_2, SO_3^{2-}, ClF.

Hybridization of Atomic Orbitals

7.5.1 How many orbitals does a set of sp^2 hybrid orbitals contain?

    (a) 2   (d) 5

    (b) 3   (e) 6

    (c) 4

7.5.2 How many p atomic orbitals are required to generate a set of sp^3 hybrid orbitals?

    (a) 0   (d) 3

    (b) 1   (e) 4

    (c) 2

7.6 HYBRIDIZATION IN MOLECULES CONTAINING MULTIPLE BONDS

The concept of valence bond theory and hybridization can also be used to describe the bonding in molecules containing double and triple bonds, such as ethylene (C_2H_4) and acetylene (C_2H_2). The Lewis structure of ethylene is

$$\begin{array}{ccc} H & & H \\ \diagdown & & \diagup \\ & C = C & \\ \diagup & & \diagdown \\ H & & H \end{array}$$

Each carbon atom is surrounded by three electron domains (two single bonds and one double bond). Thus, we expect the hybridization about each C atom to be sp^2, just like the B atom in BF_3. Applying the procedure described in Section 7.5, we first maximize the number of unpaired electrons by promoting an electron from the $2s$ orbital to the empty $2p$ orbital.

We then hybridize the required number of atomic orbitals, which in this case is three (one for each electron domain on the C atom).

The three equivalent sp^2 hybrid orbitals, arranged in a trigonal plane, enable us to explain the three bonds about each C atom. In this case, however, each C atom is left with a singly occupied, *unhybridized* atomic orbital. As we will see, it is the singly occupied p orbitals *not* involved in hybridization that give rise to multiple bonds in molecules.

    In the bonding schemes that we have described thus far, the overlap of atomic orbitals or hybrid orbitals occurs directly between the two nuclei involved in bonding. Such bonds, in which the shared electron density is concentrated directly along the internuclear axis, are called **sigma (σ) bonds.** The ethylene molecule (also known as *ethene*) contains five sigma bonds: one between the two C atoms (the result of the overlap of one of the sp^2 hybrid orbitals on each C atom) and four between the C and H atoms (each the result of the overlap of an sp^2 hybrid orbital on a C atom and the $1s$ orbital on an H atom). The leftover unhybridized p orbital is perpendicular to the plane in which the atoms of the molecule lie. Figure 7.12(a) illustrates the formation and the overlap of sp^2 hybrid orbitals and shows the positions of the remaining p orbital on each C atom.

Video 7.4
Chemical bonding—sigma-pi bonding

Figure 7.12 (a) A sigma bond forms when sp^2 hybrid orbitals on the C atoms overlap. Each C atom has one remaining unhybridized p orbital. (b) The remaining p orbitals overlap to form a pi bond.

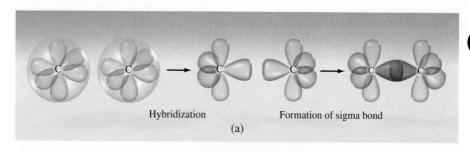

Hybridization   Formation of sigma bond

(a)

Remaining p orbitals
shown as simplified shape

Actual shape of
remaining p orbitals

Two lobes of a pi bond

(b)

Video 7.5
Formation of pi bonds in ethylene.

Student Annotation: Bond energy was introduced in [◄◄ Section 6.1].

Remember that the shapes used to represent the atomic and hybrid orbitals are simplified to make visualization of the molecules easier. The actual shapes of both atomic and hybrid orbitals are such that when the sp^2 hybrid orbitals overlap to form a sigma bond between the two C atoms, the remaining unhybridized p orbitals also overlap, although to a smaller extent. Figure 7.12(b) shows the overlap of the unhybridized p orbitals on the two C atoms in ethylene. The resulting regions of electron density are concentrated above and below the plane of the molecule, in contrast to a sigma bond in which the electron density is concentrated directly along the internuclear axis. Bonds that form from the sideways overlap of p orbitals are called *pi (π) bonds*. The two regions of overlap shown in Figure 7.12(b) together make up one pi bond. It is the formation of the pi bond that makes the ethylene molecule planar.

A sigma bond and a pi bond together constitute a *double* bond. Because the sideways overlap of p orbitals is not as effective as the overlap of hybrid orbitals that point directly toward each other, the contribution of the pi bond to the overall strength of the bond is less than that of the sigma bond. The bond energy of a carbon-carbon double bond (620 kJ/mol) is greater than that of a carbon-carbon single bond (347 kJ/mol), but not twice as large.

Worked Example 7.7 shows how to use a Lewis structure to determine the number of sigma and pi bonds in a molecule.

Worked Example 7.7

Thalidomide ($C_{13}H_{10}N_2O_4$) is a sedative and antiemetic that was widely prescribed during the 1950s, although not in the United States, for pregnant women suffering from morning sickness. Its use was largely discontinued when it was determined to be responsible for thousands of devastating birth defects. Determine the number of carbon-carbon sigma bonds and the total number of pi bonds in thalidomide.

Thalidomide

Strategy Use the Lewis structure to determine the number of single and double bonds. Then, to convert the number of single and double bonds to the number of sigma and pi bonds, remember that a single bond is composed of a sigma bond, whereas a double bond is usually composed of one sigma bond and one pi bond.

Setup There are nine carbon-carbon single bonds and three carbon-carbon double bonds. Overall there are seven double bonds in the molecule (three C=C and four C=O).

Solution Thalidomide contains 12 carbon-carbon sigma bonds and a total of seven pi bonds (three in carbon-carbon double bonds and four in carbon-oxygen double bonds).

Think About It
The Lewis structure given for thalidomide is one of two possible resonance structures. Draw the other resonance structure, and count sigma and pi bonds again. Make sure you get the same answer.

Practice Problem ATTEMPT The active ingredient in Tylenol and a host of other over-the-counter pain relievers is acetaminophen ($C_8H_9NO_2$). Determine the total number of sigma and pi bonds in the acetaminophen molecule.

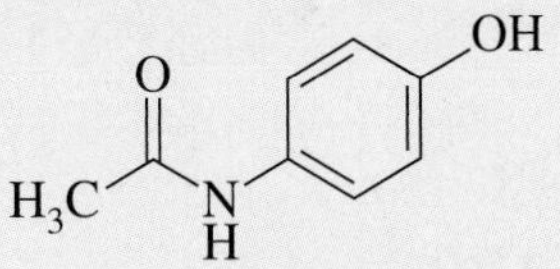

Acetaminophen

Practice Problem BUILD Determine the total number of sigma and pi bonds in a molecule of aspirin ($C_9H_8O_4$).

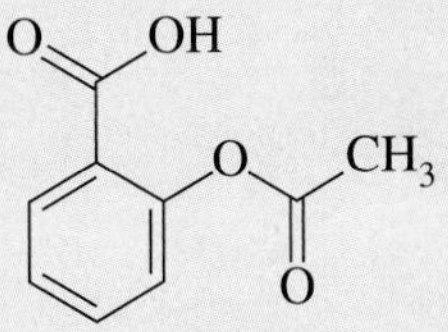

Aspirin

Practice Problem CONCEPTUALIZE In terms of valence bond theory and hybrid orbitals, explain why C_2H_2 and C_2H_4 contain pi bonds, whereas C_2H_6 does not.

Because the p orbitals that form pi bonds must be parallel to each other, pi bonds restrict the rotation of a molecule in a way that sigma bonds do not. For example, the molecule 1,2-dichloroethane exists as a single isomer. Although we can draw the molecule in several different ways, including the two shown in Figure 7.13(a), all of them are equivalent because the molecule can rotate freely about the sigma bond between the two carbon atoms.

On the other hand, 1,2-dichloroethylene exists as two distinct isomers—*cis* and *trans*—as shown in Figure 7.13(b). The double bond between the carbon atoms consists of one sigma bond and one pi bond. The pi bond restricts rotation about the sigma bond, making the molecules rigid, planar, and not interchangeable. To change one isomer into the other, the pi bond would have to be broken and rotation would have to occur about the sigma bond and the pi bond. This process would require a significant input of energy.

The acetylene molecule (C_2H_2) is linear. Because each carbon atom has two electron domains around it in the Lewis structure, the carbon atoms are sp hybridized. As before, promotion of an electron first maximizes the number of unpaired electrons.

$$H-C\equiv C-H$$
$$C_2H_2$$

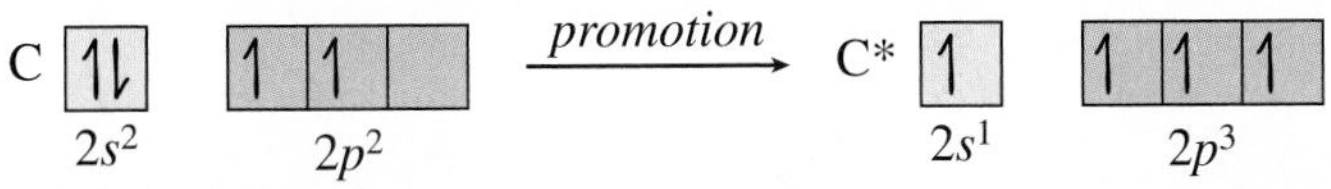

The $2s$ orbital and one of the $2p$ orbitals then mix to form two sp hybrid orbitals.

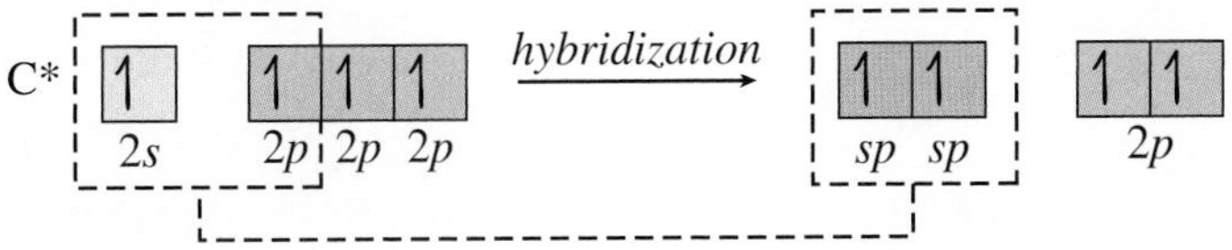

This leaves two unhybridized p orbitals (each containing an electron) on each C atom. Figure 7.14 shows the sigma and pi bonds in the acetylene molecule (also known as

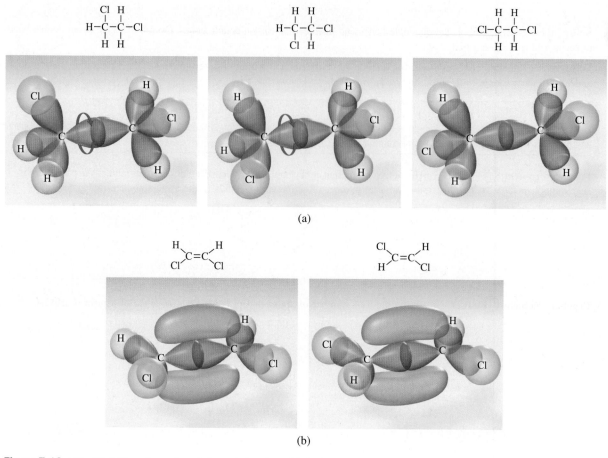

(a)

(b)

Figure 7.13 (a) In 1,2-dichloroethane, there is free rotation about the C—C single bond. All three Lewis structures represent the same molecule. (b) In 1,2-dichloroethylene, there is no rotation about the C=C double bond. The two Lewis structures represent two different molecules.

Video 7.6
Formation of pi bonds in acetylene.

Figure 7.14 (a) Formation of the sigma bond in acetylene. (b) Formation of the pi bonds in acetylene.

ethyne). Just as one sigma bond and one pi bond make up a *double* bond, one sigma bond and two pi bonds make up a *triple* bond. Figure 7.15 summarizes the formation of bonds in ethane, ethylene, and acetylene.

Worked Example 7.8 shows how hybrid orbitals and pi bonds can be used to explain the bonding in formaldehyde, a molecule with a carbon-oxygen double bond.

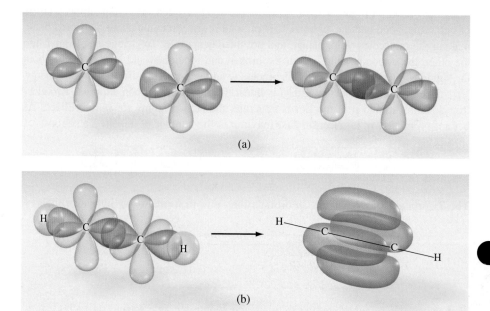

(a)

(b)

Worked Example 7.8

In addition to its use in aqueous solution as a preservative for laboratory specimens, formaldehyde gas is used as an antibacterial fumigant. Use hybridization to explain the bonding in formaldehyde (CH_2O).

Strategy Draw the Lewis structure of formaldehyde, determine the hybridization of the C and O atoms, and describe the formation of the sigma and pi bonds in the molecule.

Setup The Lewis structure of formaldehyde is

$$\overset{\displaystyle \overset{\displaystyle \cdot\cdot}{\ddot{O}}}{\underset{}{H-C-H}}$$

The C and O atoms each have three electron domains around them. [Carbon has two single bonds (C—H) and a double bond (C=O); oxygen has a double bond (O=C) and two lone pairs.]

Solution Three electron domains correspond to sp^2 hybridization. For carbon, promotion of an electron from the $2s$ orbital to the empty $2p$ orbital is necessary to maximize the number of unpaired electrons. For oxygen, no promotion is necessary. Each undergoes hybridization to produce sp^2 hybrid orbitals; and each is left with a singly occupied, unhybridized p orbital.

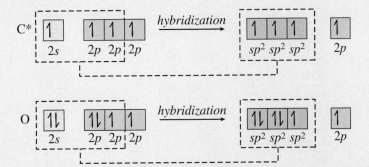

A sigma bond is formed between the C and O atoms by the overlap of one of the sp^2 hybrid orbitals from each of them. Two more sigma bonds form between the C atom and the H atoms by the overlap of carbon's remaining sp^2 hybrid orbitals with the $1s$ orbital on each H atom. Finally, the remaining p orbitals on C and O overlap to form a pi bond.

Student Annotation: The two lone pairs on the O atom are the electrons in the doubly occupied sp^2 hybrid orbitals.

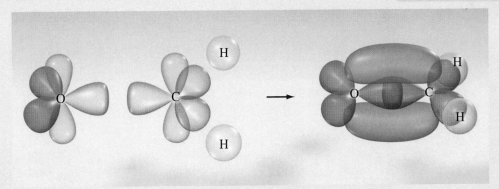

Think About It
Our analysis describes the formation of both a sigma bond and a pi bond between the C and O atoms. This corresponds correctly to the double bond predicted by the Lewis structure.

Practice Problem Ⓐ**TTEMPT** Use valence bond theory and hybrid orbitals to explain the bonding in hydrogen cyanide (HCN).

Practice Problem Ⓑ**UILD** Use valence bond theory and hybrid orbitals to explain the bonding in diatomic nitrogen (N_2).

Practice Problem Ⓒ**ONCEPTUALIZE** Explain why hybrid orbitals are necessary to explain the bonding in N_2 and O_2, but not to explain the bonding in H_2 or Br_2.

Figure 7.15

Formation of Pi Bonds in Ethylene and Acetylene

2s and 2p atomic orbitals on two C atoms

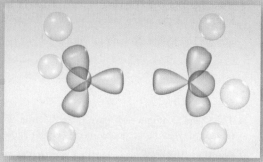

In ethane (C_2H_6) the C atoms are sp^3-hybridized.

In ethylene (C_2H_4) the C atoms are sp^2-hybridized.

Overlap of sp^2 hybrid orbitals forms a sigma bond between the two C atoms.

In acetylene (C_2H_2) the C atoms are sp-hybridized.

Overlap of sp hybrid orbitals forms a sigma bond between the two C atoms.

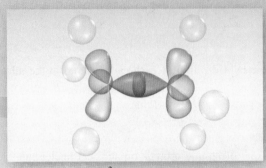

Overlap of sp^3 hybrid orbitals forms a sigma bond between the two C atoms.

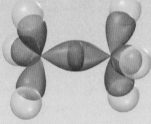

Each C atom forms three sigma bonds with H atoms.

Each C atom forms two sigma bonds with H atoms. Each C atom has one leftover unhybridized p orbital.

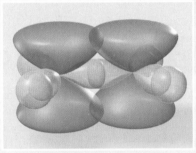

Using the more realistic shape shows how p orbitals overlap.

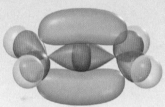

The parallel, unhybridized p orbitals overlap to form a pi bond with two lobes.

Each C atom forms one sigma bond with an H atom. Each C atom has two leftover unhybridized p orbitals.

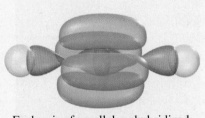

Each pair of parallel, unhybridized p orbitals overlaps to form a pi bond with two lobes.

What's the point?

When carbon is sp^2- or sp-hybridized, parallel, unhybridized p orbitals interact to form pi bonds. Each pi bond consists of two lobes as a result of the overlap.

Section 7.6 Review

Hybridization in Molecules Containing Multiple Bonds

7.6.1   Which of the following molecules contains one or more pi bonds? (Select all that apply.)

(a) N_2

(b) Cl_2

(c) CO_2

(d) CH_3OH

(e) CCl_4

7.6.2   From left to right, give the hybridization of each carbon atom in the allene molecule ($H_2C{=}C{=}CH_2$).

(a) sp^2, sp^2, sp^2

(b) sp^3, sp^2, sp^3

(c) sp^2, sp, sp^2

(d) sp^3, sp, sp^3

(e) sp^3, sp^3, sp^3

7.6.3   Which of the following pairs of atomic orbitals on adjacent nuclei can overlap to form a sigma bond? Assume that the x axis is the internuclear axis. (Select all that apply.)

(a) $1s$ and $2s$

(b) $1s$ and $2p_x$

(c) $2p_y$ and $2p_y$

(d) $3p_y$ and $3p_z$

(e) $2p_x$ and $2p_x$

7.6.4   Which of the following pairs of atomic orbitals on adjacent nuclei can overlap to form a pi bond? Assume that the x axis is the internuclear axis. (Select all that apply.)

(a) $1s$ and $2s$

(b) $1s$ and $2p_x$

(c) $2p_y$ and $2p_y$

(d) $3p_y$ and $3p_z$

(e) $2p_x$ and $2p_x$

Liquid oxygen is attracted to the poles of a magnet because O_2 is paramagnetic.

Liquid nitrogen is not attracted to the poles of a magnet because N_2 is diamagnetic.

Source: (both): © McGraw-Hill Education/Charles D. Winters, photographer

O_2

Video 7.7

Chemical bonding—paramagnetic liquid oxygen.

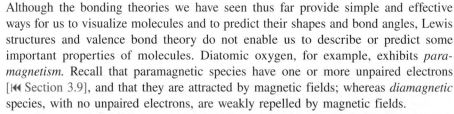

7.7   MOLECULAR ORBITAL THEORY

Although the bonding theories we have seen thus far provide simple and effective ways for us to visualize molecules and to predict their shapes and bond angles, Lewis structures and valence bond theory do not enable us to describe or predict some important properties of molecules. Diatomic oxygen, for example, exhibits *paramagnetism.* Recall that paramagnetic species have one or more unpaired electrons [◄◄ Section 3.9], and that they are attracted by magnetic fields; whereas *diamagnetic* species, with no unpaired electrons, are weakly repelled by magnetic fields.

Because O_2 exhibits paramagnetism, it must contain unpaired electrons. According to the Lewis structure of O_2 (shown below) and the valence bond theory description of O_2, however, all the electrons in O_2 are paired. To account for the paramagnetism of O_2, we need a new bonding theory. The theory that correctly describes this and other important molecular properties is *molecular orbital theory*.

According to *molecular orbital theory,* the atomic orbitals involved in bonding actually combine to form new orbitals that are the "property" of the entire molecule, rather than of the individual atoms forming the bonds. These new orbitals are called *molecular orbitals.* In molecular orbital theory, electrons shared by atoms in a molecule reside in the molecular orbitals.

Molecular orbitals are like atomic orbitals in several ways: they have specific shapes and specific energies, and they can each accommodate a maximum of two electrons. As was the case with atomic orbitals, two electrons residing in the same molecular orbital must have opposite spins, as required by the Pauli exclusion principle. And, like hybrid orbitals, the number of molecular orbitals we get is equal to the number of atomic orbitals we combine.

Our treatment of molecular orbital theory in this text will be limited to descriptions of bonding in diatomic molecules consisting of elements from the first two periods of the periodic table (H through Ne).

Bonding and Antibonding Molecular Orbitals

To begin our discussion, we consider H_2, the simplest homonuclear diatomic molecule. According to *valence* bond theory, an H_2 molecule forms when two H atoms are close enough for their $1s$ atomic orbitals to overlap. According to *molecular* orbital theory, two H atoms come together to form H_2 when their $1s$ atomic orbitals combine to give molecular orbitals. Figure 7.16 shows the $1s$ atomic orbitals of the isolated H atoms and the molecular orbitals that result from their constructive and destructive combinations. The *constructive* combination of the two $1s$ orbitals gives rise to a molecular orbital [Figure 7.16(b)] that lies along the internuclear axis directly between the two H nuclei. Just as electron density shared between two nuclei in overlapping atomic orbitals drew the nuclei together, electron density in a molecular orbital that lies between two nuclei will draw them together, too. Thus, this molecular orbital is referred to as a ***bonding molecular orbital.***

The *destructive* combination of the $1s$ atomic orbitals also gives rise to a molecular orbital that lies along the internuclear axis, but, as Figure 7.16(c) shows, this molecular orbital, which consists of two lobes, does not lie between the two nuclei. Electron density in this molecular orbital would actually pull the two nuclei in opposite directions, rather than toward each other. This is referred to as an ***antibonding molecular orbital.***

σ Molecular Orbitals

Molecular orbitals that lie along the internuclear axis (such as the bonding and antibonding molecular orbitals in H_2) are referred to as σ molecular orbitals. Specifically, the *bonding* molecular orbital formed by the combination of two $1s$ atomic orbitals is designated σ_{1s} and the *antibonding* orbital is designated σ_{1s}^*, where the asterisk distinguishes an antibonding orbital from a bonding orbital. Figure 7.16(d) summarizes the combination of two $1s$ atomic orbitals to yield two molecular orbitals: one bonding and one antibonding.

Like atomic orbitals, molecular orbitals have specific energies. The combination of two atomic orbitals of equal energy, such as two $1s$ orbitals on two H atoms, yields one molecular orbital that is lower in energy (bonding) and one molecular orbital that is higher in energy (antibonding) than the original atomic orbitals. The bonding molecular orbital in H_2 is concentrated between the nuclei, along the internuclear axis. Electron density in this molecular orbital both attracts the nuclei and shields them from each other, stabilizing the molecule. Thus, the bonding molecular orbital is lower in energy than the isolated atomic orbitals. In contrast, the antibonding molecular

Student Annotation: Recall that a homonuclear diatomic molecule is one in which both atoms are the same element [◄◄ Section 5.5].

Student Annotation: Remember that the quantum mechanical approach treats atomic orbitals as *wave functions* [◄◄ Section 3.6], and that one of the properties of waves is their capacity for both constructive combination and destructive combination [◄◄ Section 3.2].

Video 7.8
Chemical bonding—formation of molecular orbitals.

Student Annotation: The designations σ and π are used in molecular orbital theory just as they are in valence bond theory: σ refers to electron density along the internuclear axis, and π refers to electron density that influences both nuclei but that does not lie directly along the internuclear axis.

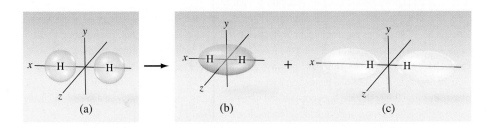

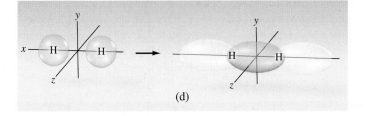

Figure 7.16 (a) Two *s* atomic orbitals combine to give two sigma molecular orbitals. (b) One of the molecular orbitals is lower in energy than the original atomic orbitals (darker), and (c) one is higher in energy (lighter). The two light yellow lobes make up one molecular orbital. (d) Atomic and molecular orbitals are shown relative to the H nuclei.

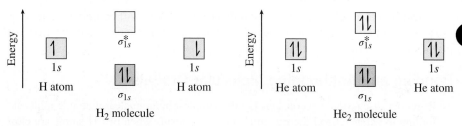

Figure 7.17 Relative energies of atomic orbitals in H and molecular orbitals in H_2.

Figure 7.18 Relative energies of atomic orbitals in He and molecular orbitals in He_2.

orbital has most of its electron density outside the internuclear region. Electron density in this orbital does not shield one nucleus from the other, which increases the nuclear repulsions and makes the antibonding molecular orbital *higher* in energy than the isolated atomic orbitals.

Showing all the molecular orbitals in a molecule can make for a complicated picture. Rather than represent molecules with pictures of their molecular orbitals, we generally use diagrams in which molecular orbitals are represented with boxes placed at the appropriate relative energy levels. Figure 7.17 shows the energies of the σ_{1s} and σ_{1s}^* molecular orbitals in H_2, relative to the energy of the original $1s$ orbitals on two isolated H atoms. Like atomic orbitals, molecular orbitals fill in order of increasing energy. Note that the electrons that originally resided in the atomic orbitals both occupy the lowest-energy molecular orbital, σ_{1s}, with opposite spins.

We can construct a similar molecular orbital diagram for the hypothetical molecule He_2. Like the H atom, the He atom has a $1s$ orbital (unlike H, though, the $1s$ orbital on He has two electrons, not one). The combination of $1s$ orbitals to form molecular orbitals in He_2 is essentially the same as what we have described for H_2. The placement of electrons is shown in Figure 7.18.

Student Annotation: The molecular orbital diagram shows the two H atoms having electrons with paired spins (↑ and ↓). H atoms whose electrons have parallel spin (↑ and ↑, or ↓ and ↓) actually repel one another and will not bond to form H_2.

Thinking Outside the Box

In phase   Out of phase

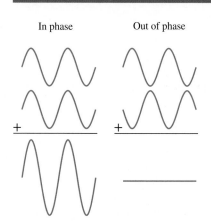

Phases

In the double-slit experiment [◀◀ Section 3.2, Figure 3.4], bright lines result from constructive interference of light waves and dark lines result from destructive interference of light waves. To picture how interference works, imagine combining two sets of one-dimensional waves shown here. Note that the amplitude of each wave can be positive, zero, or negative. The sign of a wave's amplitude, positive or negative, is referred to as its **phase.** On the left, the waves are in phase—meaning that their peaks and troughs coincide exactly. In this case, addition yields a wave in which the amplitude is twice that of the individual waves. On the right, the waves are out of phase. Addition causes the amplitudes to cancel one another at each point along the *x* axis, resulting in a flat line.

Recall that according to the quantum mechanical model of the atom, it is the *square* of the wave function (Ψ^2) that defines the distribution of electron density in three-dimensional space around a nucleus. Because it is obtained by squaring the wave function, Ψ^2 can be greater than or equal to zero—it cannot have *negative* values. However, the wave function itself (Ψ) can have values that are positive, zero, or negative. Thus, like one-dimensional waves, wave functions can have positive or negative *phase.* Because they can have both positive and negative phases, atomic orbitals can add *con*structively (constructive interference) or *de*structively (destructive interference).

The different phases of wave functions can be represented by the use of different colors in our depictions of atomic orbitals. For example, rather than representing a *p* orbital with two blue lobes, we can represent it with one blue lobe (positive phase) and one red lobe (negative) as shown here. Although pictures of atomic orbitals without phases are adequate for showing the regions of electron density in an atom, the phases help us account for the formation of bonding and antibonding molecular orbitals. Constructive interference of atomic orbitals produces *bonding* molecular orbitals. Destructive interference produces *antibonding* molecular orbitals.

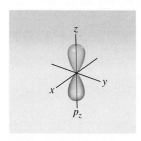

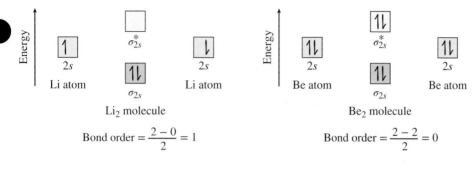

Figure 7.19 Bond order determination for Li_2 and Be_2.

Bond order $= \dfrac{2-0}{2} = 1$

Bond order $= \dfrac{2-2}{2} = 0$

Bond Order

With molecular orbital diagrams such as those for H_2 and He_2, we can begin to see the power of molecular orbital theory. For a diatomic molecule described using molecular orbital theory, we can calculate the **bond order**. The value of the bond order indicates, qualitatively, how *stable* a molecule is. The higher the bond order, the more stable the molecule. Bond order is calculated in the following way:

$$\text{bond order} = \frac{\text{\# of electrons in bonding molecular orbitals} - \text{\# of electrons in antibonding molecular orbitals}}{2}$$

Equation 7.1

In the case of H_2, where both electrons reside in the σ_{1s} orbital, the bond order is $[(2-0)/2] = 1$. In the case of He_2, where the two additional electrons reside in the σ_{1s}^* orbital, the bond order is $[(2-2)/2] = 0$. Molecular orbital theory predicts that a molecule with a bond order of zero will not exist and He_2, in fact, does *not* exist.

We can do similar analyses of the molecules Li_2 and Be_2. (The Li and Be atoms have ground-state electron configurations of $[He]2s^1$ and $[He]2s^2$, respectively.) The $2s$ atomic orbitals also combine to form the corresponding σ and σ^* molecular orbitals. Figure 7.19 shows the molecular orbital diagrams and bond orders for Li_2 and Be_2.

As predicted by molecular orbital theory, Li_2, with a bond order of 1, is a stable molecule, whereas Be_2, with a bond order of 0, does not exist.

Student Annotation: When we draw molecular orbital diagrams, we need only show the valence orbitals and electrons.

π Molecular Orbitals

To consider diatomic molecules beyond Be_2, we must also consider the combination of p atomic orbitals. Like s orbitals, p orbitals combine both constructively, to give bonding molecular orbitals that are lower in energy than the original atomic orbitals, and destructively, to give antibonding molecular orbitals that are higher in energy than the original atomic orbitals. However, the orientations of p_x, p_y, and p_z orbitals give rise to two different types of molecular orbitals: σ molecular orbitals, in which the regions of electron density in the bonding and antibonding molecular orbitals lie along the internuclear axis; and π molecular orbitals, in which the regions of electron density affect both nuclei but do *not* lie along the internuclear axis.

Orbitals that lie along the internuclear axis, as the $2p_x$ orbitals do in Figure 7.20(a), point directly toward each other and combine to form σ molecular orbitals. Figure 7.20(b) shows the combination of two $2p_x$ atomic orbitals to give two molecular orbitals designated σ_{2p_x} and $\sigma_{2p_x}^*$. Figure 7.20(c) shows the relative energies of these molecular orbitals.

Orbitals that are aligned parallel to each other, like the $2p_y$ and $2p_z$ orbitals shown in Figure 7.21, combine to form π molecular orbitals. These bonding molecular orbitals are designated π_{2p_y} and π_{2p_z}; the corresponding antibonding molecular orbitals are designated $\pi_{2p_y}^*$ and $\pi_{2p_z}^*$. Often we refer to the molecular orbitals collectively using the designations $\pi_{2p_{y,z}}$ and $\pi_{2p_{y,z}}^*$. Figure 7.21(a) shows the constructive and destructive combination of parallel p orbitals. Figure 7.21(b) shows the locations of the molecular orbitals resulting from the combination of p_y and p_z orbitals relative to the two atomic nuclei. Again, electron density in the resulting *bonding* molecular orbitals serves to hold the nuclei together, whereas electron density in the *antibonding* molecular orbitals does not.

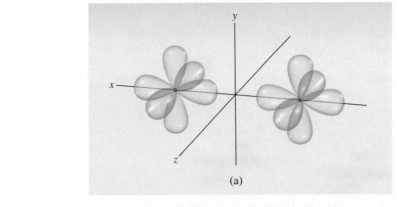

Figure 7.20 (a) Two sets of 2p orbitals. (b) The p atomic orbitals that point toward each other (p_x) combine to give bonding and antibonding σ molecular orbitals. (c) The antibonding σ molecular orbital is higher in energy than the corresponding bonding σ molecular orbital.

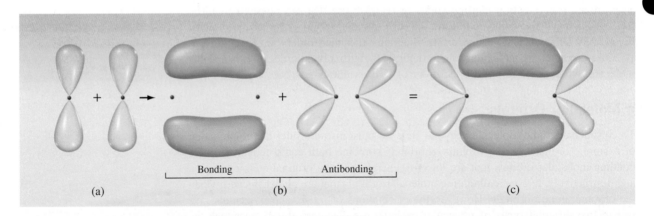

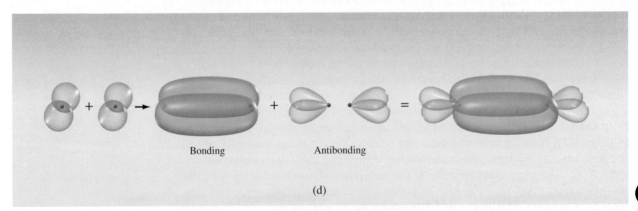

Figure 7.21 Formation of molecular orbitals from the interaction of parallel p orbitals on adjacent atoms (a). Bonding and antibonding molecular orbitals shown separately (b), and together relative to the two nuclei (c). (d) Formation of molecular orbitals from interaction of p orbitals perpendicular to those shown in part (a).

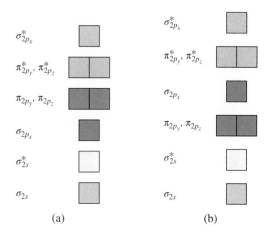

Figure 7.22 (a) Ordering of molecular orbital energies for O_2 and F_2. (b) Ordering of molecular orbital energies for Li_2, B_2, C_2, and N_2. Bonding orbitals are darker; antibonding orbitals are lighter.

Just as the p atomic orbitals within a particular shell are higher in energy than the s orbital in the same shell, all the molecular orbitals resulting from the combination of p atomic orbitals are higher in energy than the molecular orbitals resulting from the combination of s atomic orbitals. To understand better the relative energy levels of the molecular orbitals resulting from p orbital combinations, consider the fluorine molecule (F_2).

In general, molecular orbital theory predicts that the more effective the interaction or overlap of the atomic orbitals, the lower in energy the resulting bonding molecular orbital will be and the higher in energy the resulting antibonding molecular orbital will be. Thus, the relative energy levels of molecular orbitals in F_2 can be represented by the diagram in Figure 7.22(a). The p_x orbitals, which lie along the internuclear axis, overlap most effectively, giving the lowest-energy bonding molecular orbital and the highest-energy antibonding molecular orbital.

The order of orbital energies shown in Figure 7.22(a) assumes that p orbitals interact only with other p orbitals and s orbitals interact only with other s orbitals—that there is no significant interaction between s and p orbitals. In fact, the relatively smaller nuclear charges of boron, carbon, and nitrogen atoms cause their atomic orbitals to be held less tightly than those of atoms with larger nuclear charges, and some s-p interaction does take place. This results in a change in the relative energies of the σ_{2p_x} and $\pi_{2p_{y,z}}$ molecular orbitals. Although energies of several of the resulting molecular orbitals change, the most important of these changes is the energy of the σ_{2p} orbital, making it higher than the $\pi_{2p_{y,z}}$ orbitals. The relative energy levels of molecular orbitals in the Li_2, B_2, C_2, and N_2 molecules can be represented by the diagram in Figure 7.22(b).

Molecular Orbital Diagrams

Beginning with oxygen, the nuclear charge is sufficiently large to prevent the interaction of s and p orbitals. Thus, for O_2 and Ne_2, the order of molecular orbital energies is the same as that for F_2, which is shown in Figure 7.22(a). Figure 7.23 gives the molecular orbital diagrams, magnetic properties, bond orders, and bond energies for Li_2, B_2, C_2, N_2, O_2, F_2, and Ne_2. Note that the filling of molecular orbitals follows the same rules as the filling of atomic orbitals [◄◄ Section 3.9]:

- Lower energy orbitals fill first.
- Each orbital can accommodate a maximum of two electrons with opposite spins.
- Hund's rule is obeyed.

There are several important predictions made by the molecular orbital diagrams in Figure 7.23. First, molecular orbital theory correctly predicts that Ne_2, with a bond

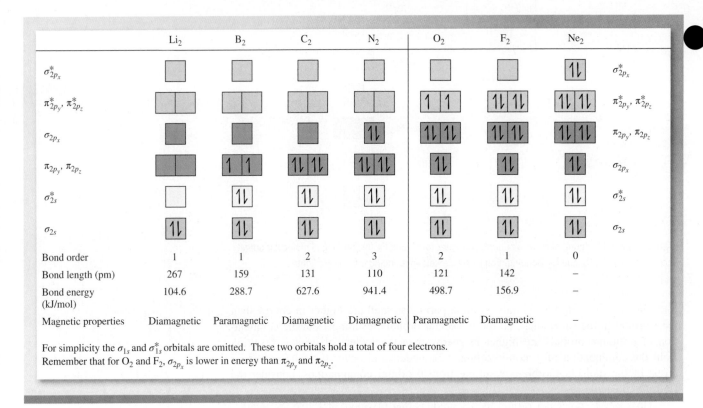

	Li$_2$	B$_2$	C$_2$	N$_2$	O$_2$	F$_2$	Ne$_2$	
$\sigma^*_{2p_x}$	□	□	□	□	□	□	⇅	$\sigma^*_{2p_x}$
$\pi^*_{2p_y}$, $\pi^*_{2p_z}$	□ □	□ □	□ □	□ □	↑ ↑	⇅ ⇅	⇅ ⇅	$\pi^*_{2p_y}$, $\pi^*_{2p_z}$
σ_{2p_x}	■	■	■	⇅	⇅ ⇅	⇅ ⇅	⇅ ⇅	π_{2p_y}, π_{2p_z}
π_{2p_y}, π_{2p_z}	■ ■	↑ ↑	⇅ ⇅	⇅ ⇅	⇅	⇅	⇅	σ_{2p_x}
σ^*_{2s}	□	⇅	⇅	⇅	⇅	⇅	⇅	σ^*_{2s}
σ_{2s}	⇅	⇅	⇅	⇅	⇅	⇅	⇅	σ_{2s}
Bond order	1	1	2	3	2	1	0	
Bond length (pm)	267	159	131	110	121	142	–	
Bond energy (kJ/mol)	104.6	288.7	627.6	941.4	498.7	156.9	–	
Magnetic properties	Diamagnetic	Paramagnetic	Diamagnetic	Diamagnetic	Paramagnetic	Diamagnetic	–	

For simplicity the σ_{1s} and σ^*_{1s} orbitals are omitted. These two orbitals hold a total of four electrons.
Remember that for O$_2$ and F$_2$, σ_{2p_x} is lower in energy than π_{2p_y} and π_{2p_z}.

Figure 7.23 Molecular orbital diagrams for second-period homonuclear diatomic molecules.

Thinking Outside the Box

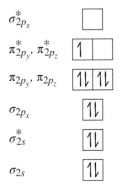

$\sigma^*_{2p_x}$	□
$\pi^*_{2p_y}$, $\pi^*_{2p_z}$	↑ □
π_{2p_y}, π_{2p_z}	⇅ ⇅
σ_{2p_x}	⇅
σ^*_{2s}	⇅
σ_{2s}	⇅

Molecular orbital diagram for NO

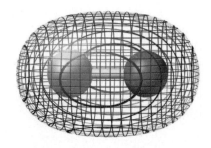

Molecular Orbitals in Heteronuclear Diatomic Species

Our description of molecular orbital theory can also be applied to heteronuclear diatomic species, in which the two atoms are different, such as NO. In a case such as this, our description of the molecular orbitals involved in bonding must be modified slightly.

The atomic orbitals of a more electronegative atom are lower in energy than the corresponding atomic orbitals of a less electronegative atom. The 2*s* and 2*p* atomic orbitals are lower in energy for oxygen, which is more electronegative than nitrogen. (See Figure 6.4 on page 216.) When atomic orbitals of different energies interact to form molecular orbitals, the lower-energy atomic orbital contributes more to the *bonding* molecular orbital; and the higher-energy atomic orbital contributes more to the *antibonding* molecular orbital. The result is that the bonding molecular orbital more closely resembles the atomic orbital of the *more* electronegative atom, and the antibonding molecular orbital more closely resembles the atomic orbital of the *less* electronegative atom. The result of this is that electron density in the resulting bonding molecular orbital is *greater* in the vicinity of the *more* electronegative atom.

Recall that the order of energies of molecular orbitals in second-period homonuclear diatomic molecules is different for B$_2$, C$_2$, and N$_2$, than it is for O$_2$, F$_2$, and Ne$_2$ (Figure 7.23). For second-period heteronuclear diatomic species containing one atom from each group, such as NO and CO, there is no simple rule that can tell us which order the molecular orbitals follow. Note that the molecular orbitals in NO follow the same order as those in O$_2$.

Often the bond order determined from a molecular orbital diagram corresponds to the number of bonds in the Lewis structure of the molecule. In the case of NO, though, the Lewis structure contains a double bond whereas the molecular orbital approach gives a bond order of 2.5. In fact, the molecular orbital approach gives a bond order that is more consistent with experimental data. The experimentally determined strength of the bond in NO (631 kJ/mol) is greater than that of the average nitrogen-oxygen double bond (607 kJ/mol) [▶▶ Section 10.7].

order of 0, does not exist. Second, it correctly predicts the magnetic properties of the molecules that do exist. Both B_2 and O_2 are known to be paramagnetic. Third, although bond order is only a qualitative measure of bond strength, the calculated bond orders of the molecules correlate well with the measured bond energies. The N_2 molecule, with a bond order of 3, has the largest bond energy of the five molecules. The B_2 and F_2 molecules, each with a bond order of 1, have the smallest bond enthalpies. Its ability to predict correctly the properties of molecules makes molecular orbital theory a powerful tool in the study of chemical bonding.

Worked Example 7.9 shows how to use molecular orbital diagrams to determine the magnetic properties and bond order of a molecule or polyatomic ion.

Worked Example 7.9

The superoxide ion (O_2^-) has been implicated in a number of degenerative conditions, including aging and Alzheimer's disease. Using molecular orbital theory, determine whether (O_2^-) is paramagnetic or diamagnetic, and then calculate its bond order.

Strategy Start with the molecular orbital diagram for O_2, add an electron, and then use the resulting diagram to determine the magnetic properties and bond order.

Setup The molecular orbital diagram for O_2 is shown in Figure 7.23. The additional electron must be added to the lowest-energy molecular orbital available.

Solution In this case, either of the two singly occupied π_{2p}^* orbitals can accommodate an additional electron. This gives a molecular orbital diagram in which there is one unpaired electron, making (O_2^-) paramagnetic. The new diagram has six electrons in bonding molecular orbitals and three in antibonding molecular orbitals. We can ignore the electrons in the σ_{2s} and σ_{2s}^* orbitals because their contributions to the bond order cancel each other. The bond order is $(6 - 3)/2 = 1.5$.

$\sigma_{2p_x}^*$ ☐

$\pi_{2p_y}^*,\ \pi_{2p_z}^*$ | ↑↓ | ↑ |

$\pi_{2p_y},\ \pi_{2p_z}$ | ↑↓ | ↑↓ |

σ_{2p_x} | ↑↓ |

σ_{2s}^* | ↑↓ |

σ_{2s} | ↑↓ |

Molecular orbital diagram for O_2^-

Think About It

Experiments confirm that the superoxide ion is paramagnetic. Also, any time we add one or more electrons to an antibonding molecular orbital, as we did in this problem, we should expect the bond order to decrease. Electrons in antibonding orbitals cause a bond to be less stable.

Practice Problem Ⓐ**TTEMPT** Use molecular orbital theory to determine whether N_2^{2-} is paramagnetic or diamagnetic, and then calculate its bond order.

Practice Problem Ⓑ**UILD** Use molecular orbital theory to determine whether F_2^{2+} is paramagnetic or diamagnetic, and then calculate its bond order.

Practice Problem Ⓒ**ONCEPTUALIZE** For most of the homonuclear diatomic species shown in Figure 7.23, *addition* and *removal* of one or more electrons (to form polyatomic ions) have opposite effects on the bond order. For some species, addition and removal of electrons have the *same* effect on bond order. Identify the species for which this is true and explain how it can be so.

Section 7.7 Review

Molecular Orbital Theory

7.7.1 Calculate the bond order of N_2^{2+}, and determine whether it is paramagnetic or diamagnetic.

(a) 2, paramagnetic

(b) 2, diamagnetic

(c) 3, paramagnetic

(d) 3, diamagnetic

(e) 1, paramagnetic

(Continued on next page)

7.7.2   Which of the following species is paramagnetic? (Select all that apply.)

(a) C_2^{2-}

(b) O_2^{2+}

(c) F_2^{2+}

(d) F_2^{2-}

(e) C_2^{2+}

7.7.3   Calculate the bond order of He_2^+.

(a) 0

(b) 0.5

(c) 1.0

(d) 1.5

(e) 2

7.7.4   Which if any of the following species has a bond order of 0? (Select all that apply.)

(a) B_2^{2+}

(b) Ne_2^{2+}

(c) F_2^{2-}

(d) He_2^{2+}

(e) H_2^{2-}

7.8   BONDING THEORIES AND DESCRIPTIONS OF MOLECULES WITH DELOCALIZED BONDING

The progression of bonding theories in this chapter illustrates the importance of model development. Scientists use models to understand experimental results and to predict future observations. A model is useful as long as it agrees with observation. When it fails to do so, it must be replaced with a new model. What follows is a synopsis of the strengths and weaknesses of the bonding theories presented in Chapters 6 and 7.

Lewis Theory

Strength: The Lewis theory of bonding enables us to make qualitative predictions about bond strengths and bond lengths. Lewis structures are easy to draw and are widely used by chemists.

Weakness: Lewis structures are two dimensional, whereas molecules are three dimensional. In addition, Lewis theory fails to account for the differences between bonds in compounds such as H_2, F_2, and HF. It also fails to explain *why* bonds form.

The Valence-Shell Electron-Pair Repulsion Model

Strength: The VSEPR model enables us to predict the shapes of many molecules and polyatomic ions.

Weakness: Because the VSEPR model is based on the Lewis theory of bonding, it also fails to explain why bonds form.

Valence Bond Theory

Strength: Valence bond theory describes the formation of covalent bonds as the overlap of atomic orbitals. It further explains that bonds form because the resulting molecule has a lower potential energy than the original, isolated atoms.

Weakness: Valence bond theory alone fails to explain the bonding in many molecules such as $BeCl_2$, BF_3, and CH_4, in which the central atom in its ground state does not have enough unpaired electrons to form the observed number of bonds.

Hybridization of Atomic Orbitals

Strength: The hybridization of atomic orbitals is not a separate bonding theory; rather, it is an *extension* of valence bond theory. Using hybrid orbitals, we can understand the bonding and geometry of more molecules, including $BeCl_2$, BF_3, and CH_4.

Weakness: Valence bond theory and hybrid orbitals fail to predict some of the important properties of molecules, such as the paramagnetism of O_2.

Molecular Orbital Theory

Strength: Molecular orbital theory enables us to predict accurately the magnetic and other properties of molecules and ions.

Weakness: Pictures of molecular orbitals can be very complex.

Although molecular orbital theory is in many ways the most powerful of the bonding models, it is also the most complex, so we continue to use the other models when they do an adequate job of explaining or predicting the properties of a molecule. For example, if you need to predict the three-dimensional shape of an AB_x molecule on an exam, you should draw its Lewis structure and apply the VSEPR model. On the other hand, if you need to determine the bond order of a diatomic molecule or ion, you should draw a molecular orbital diagram. In general chemistry, it is best to use the *simplest* theory that can answer a particular question.

Because they remain useful, we don't discard the old models when we develop new ones. In fact, the bonding in some molecules, such as benzene (C_6H_6), is best described using a combination of models. Benzene can be represented with two resonance structures [◄◄ Section 6.5].

Video 7.9
Chemical bonding—sigma and pi bonding in benzene.

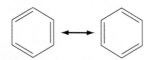

According to its Lewis structure and valence bond theory, the benzene molecule contains 12 σ bonds (6 carbon-carbon and 6 carbon-hydrogen) and 3 π bonds. From experimental evidence, however, we know that benzene does not have 3 single bonds and 3 double bonds between carbon atoms. Rather, there are 6 equivalent carbon-carbon bonds. This is precisely the reason that two different Lewis structures are necessary to represent the molecule. Neither one alone accurately depicts the nature of the carbon-carbon bonds. In fact, the π bonds in benzene are ***delocalized,*** meaning that they are spread out over the entire molecule, rather than confined between two specific atoms. (Bonds that are confined between two specific atoms are called ***localized*** bonds.) Valence bond theory does a good job of describing the localized σ bonds in benzene, but molecular orbital theory does a better job of using delocalized π bonds to describe the bonding scheme in benzene.

To describe the σ bonds in benzene, begin with a Lewis structure and count the electron domains on the carbon atoms. (Either resonance structure will give the same result.) Each C atom has three electron domains around it (two single bonds and one double bond). Recall from Table 7.6 that an atom that has three electron domains is sp^2 hybridized. To obtain the three unpaired electrons necessary on each C atom, one electron from each C atom must be promoted from the doubly occupied $2s$ orbital to an empty $2p$ orbital.

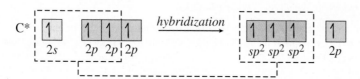

This actually creates four unpaired electrons. Next, the orbitals are sp^2 hybridized, leaving one singly occupied, unhybridized $2p$ orbital on each C atom.

The sp^2 hybrid orbitals adopt a trigonal planar arrangement and overlap with one another (and with $1s$ orbitals on H atoms) to form the σ bonds in the molecule.

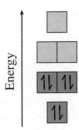

The remaining unhybridized $2p$ orbitals (one on each C atom) combine to form molecular orbitals. Because the p orbitals are all parallel to one another, only π_{2p} and π_{2p}^* molecular orbitals form. The combination of these six $2p$ atomic orbitals forms six molecular orbitals: three bonding and three antibonding. These molecular orbitals are delocalized over the entire benzene molecule.

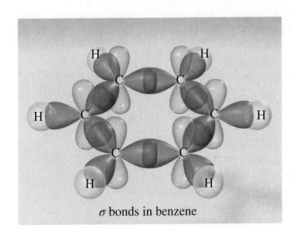

σ bonds in benzene

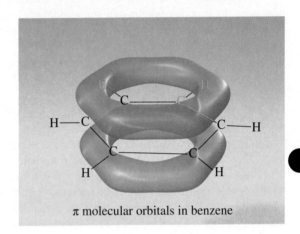

π molecular orbitals in benzene

In the ground state, the lower energy bonding molecular orbitals contain all six electrons. The electron density in the delocalized π molecular orbitals lies above and below the plane that contains all the atoms and the σ bonds in the molecule.

Worked Example 7.10 lets you practice using multiple bonding theories to describe the bonding in a species with delocalized π bonds.

Worked Example 7.10

It takes three resonance structures to represent the carbonate ion CO_3^{2-}.

$$\left[\ddot{\underset{\ddot{O}}{\overset{:\ddot{O}:}{C}}}\ddot{O}:\right]^{2-} \longleftrightarrow \left[\underset{:\ddot{O}}{\overset{:O}{C}}\ddot{O}:\right]^{2-} \longleftrightarrow \left[\underset{:\ddot{O}}{\overset{:\ddot{O}:}{C}}\ddot{O}:\right]^{2-}$$

None of the three, though, is a completely accurate depiction. As with benzene, the bonds that are shown in the Lewis structure as one double and two single are actually three equivalent bonds. Use a combination of valence bond theory and molecular orbital theory to explain the bonding in CO_3^{2-}.

Strategy Starting with the Lewis structure, use valence bond theory and hybrid orbitals to describe the σ bonds. Then use molecular orbital theory to describe the delocalized π bonding.

Setup The Lewis structure of the carbonate ion shows three electron domains around the central C atom, so the carbon must be sp^2 hybridized.

Solution Each of the sp^2 hybrid orbitals on the C atom overlaps with a singly occupied p orbital on an O atom, forming the three σ bonds. Each O atom has an additional, singly occupied p orbital perpendicular to the one involved in σ bonding. The unhybridized p orbital on C overlaps with the p orbitals on O to form π bonds, which have electron densities above and below the plane of the molecule. Because the species can be represented with resonance structures, we know that the π bonds are delocalized.

Think About It

Although the Lewis structure of CO_3^{2-} shows three electron domains on one of the O atoms, we generally do not treat *terminal* atoms (those with single bonds to only one other atom) as though they are hybridized because it is unnecessary to do so.

Practice Problem **A**TTEMPT Use a combination of valence bond theory and molecular orbital theory to describe the bonding in ozone (O_3).

Practice Problem **B**UILD Use a combination of valence bond theory and molecular orbital theory to describe the bonding in the nitrite ion (NO_2^-).

Practice Problem **C**ONCEPTUALIZE For which of the following species is the bonding best described as delocalized?

$$SO_3 \quad SO_3^{2-} \quad S_8 \quad O_2$$

Section 7.8 Review

Bonding Theories and Descriptions of Molecules with Delocalized Bonding

7.8.1 Which of the following contain one or more delocalized π bonds? (Select all that apply.)

(a) O_2
(b) CO_2
(c) NO_2^-
(d) CH_4
(e) CH_2Cl_2

7.8.2 Which of the atoms in BCl_3 need hybrid orbitals to describe the bonding in the molecule?

(a) all four atoms
(b) only the B atom
(c) only the three Cl atoms
(d) only the B atom and one Cl atom
(e) only the B atom and two Cl atoms

7.8.3 Which of the following can hybrid orbitals be used for? (Select all that apply.)

(a) to explain the geometry of a molecule
(b) to explain how a central atom can form more bonds than the number of unpaired electrons in its ground-state configuration
(c) to predict the geometry of a molecule
(d) to explain the magnetic properties of a molecule
(e) to predict the magnetic properties of a molecule

7.8.4 Which of the following enables us to explain the paramagnetism of O_2?

(a) Lewis theory
(b) valence bond theory
(c) valence-shell electron-pair repulsion
(d) hybridization of atomic orbitals
(e) molecular orbital theory

Learning Outcomes

- Use the valence-shell electron-pair repulsion (VSEPR) model to determine the shape of a molecule.
- Define electron domain.
- Describe the difference between electron-domain geometry and molecular geometry.
- Recognize the difference between axial and equatorial positions in AB_5 molecules.
- Understand why deviations from ideal bond angles occur.
- Determine when molecules will be polar or nonpolar.
- Identify structural isomers of compounds.
- List and briefly describe the different types of intermolecular forces: dipole-dipole, hydrogen bonding, dispersion forces, and ion-dipole.

- Identify the intermolecular forces present in a given substance.
- Use valence bond theory to describe the bonding in molecules.
- Predict the hybridization of molecules to explain bonding in molecules.
- Describe the bonding in molecules containing double and triple bonds.
- Understand how atomic orbitals combine to form molecular orbitals according to molecular orbital theory.
- Identify bonding and antibonding molecular orbitals.
- Calculate the bond order of diatomic species.
- Describe localized and delocalized bonds in molecules.

Chapter Summary

SECTION 7.1

- According to the *valence-shell electron-pair repulsion (VSEPR)* model, electron pairs in the valence shell of an atom repel one another. An *electron domain* is a lone pair or a bond. Any bond (single, double, or triple) constitutes one electron domain.
- The arrangement of electron domains about a central atom, determined using the VSEPR model, is called the *electron-domain geometry*. The arrangement of *atoms* in a molecule is called the *molecular geometry*. The basic molecular geometries are linear, bent, trigonal planar, tetrahedral, trigonal pyramidal, trigonal bipyramidal, seesaw-shaped, T-shaped, octahedral, square pyramidal, and square planar.
- The *bond angle* is the angle between two adjacent bonds in a molecule or polyatomic ion. A trigonal bipyramid contains two types of bonds: *axial* and *equatorial*.

SECTION 7.2

- The polarity of a molecule depends on the polarity of its individual bonds and on its molecular geometry. Even a molecule containing polar bonds may be nonpolar overall if the bonds are distributed symmetrically.
- *Structural isomers* are molecules with the same chemical formula, but different structural arrangements.

SECTION 7.3

- The particles (atoms, molecules, or ions) in the condensed phases (solids and liquids) are held together by *intermolecular forces.* Intermolecular forces are electrostatic attractions between opposite charges or partial charges.

- Intermolecular forces acting between atoms or molecules in a pure substance are called *van der Waals forces* and include *dipole-dipole interactions* (including *hydrogen bonding*) and *London dispersion forces* (also called simply *dispersion forces*).
- Dipole-dipole interactions exist between *polar* molecules, whereas *nonpolar* molecules are held together by *dispersion* forces alone. Dispersion forces are those between *instantaneous dipoles* and *induced* dipoles. When a nonpolar molecule acquires an instantaneous dipole, it is said to be *polarized.* Dispersion forces act between all molecules, nonpolar and polar.
- Hydrogen bonding is an especially strong type of dipole-dipole interaction that occurs in molecules that contain H—N, H—O, or H—F bonds.
- *Ion-dipole interactions* are those that occur (in solutions) between ions and polar molecules.

SECTION 7.4

- According to *valence bond theory,* bonds form between atoms when atomic orbitals overlap, thus allowing the atoms to share valence electrons. A bond forms when the resulting molecule is lower in energy than the original, isolated atoms.

SECTION 7.5

- To explain the bonding in some molecules, we need to employ the concept of *hybridization,* in which atomic orbitals mix to form *hybrid orbitals.*
- In order to use hybrid-orbital analysis, we must already know the molecular geometry and bond angles in a molecule. Hybrid orbitals are not used to predict molecular geometries.

SECTION 7.6

- *Sigma (σ) bonds* form when the region of orbital overlap lies directly between the two atoms. *Pi (π) bonds* form when parallel, unhybridized *p* orbitals interact. Usually, a *double* bond consists of one sigma bond and one pi bond; and a *triple* bond consists of one sigma bond and two pi bonds.

SECTION 7.7

- A paramagnetic species is one that contains unpaired electrons. A diamagnetic species is one in which there are no unpaired electrons. Paramagnetic species are weakly attracted by a magnetic field, whereas diamagnetic species are weakly repelled by a magnetic field.
- According to *molecular orbital theory,* atomic orbitals combine to form new *molecular orbitals* that are associated with the molecule rather than with individual atoms. Molecular orbitals may be *sigma,* if the orbital lies directly along the internuclear axis, or *pi,* if the orbital does not lie directly along the internuclear axis.
- Molecular orbitals may be *bonding* or *antibonding.* A *bonding molecular orbital* is lower in energy than the isolated atomic orbitals that combined to form it. The corresponding *antibonding molecular orbital* is higher in energy than the isolated atomic orbitals.
- Bonding molecular orbitals result from the interaction of atomic orbitals that are *in phase;* antibonding molecular orbitals result from the interaction of atomic orbitals that are out *of phase.* Electrons in bonding molecular orbitals contribute to the strength of a bond. Electrons in antibonding molecular orbitals detract from the strength of a bond. *Bond order* is a measure of the *strength* of a bond and can be determined using a molecular orbital diagram.

SECTION 7.8

- It is generally best to use the bonding theory that most easily describes the bonding in a particular molecule or polyatomic ion. In species that can be represented by two or more resonance structures, the pi bonds are *delocalized,* meaning that they are spread out over the molecule and not constrained to just two atoms. *Localized* bonds are those constrained to two atoms. Many species are best described using a combination of valence bond theory and molecular orbital theory.

Key Words

Key Equation

| 7.1 | $\text{bond order} = \dfrac{\begin{array}{c}\text{number of electrons} \\ \text{in bonding} \\ \text{molecular orbitals}\end{array} - \begin{array}{c}\text{number of electrons} \\ \text{in antibonding} \\ \text{molecular orbitals}\end{array}}{2}$ | Bond order is calculated by subtracting the number of electrons in antibonding molecular orbitals from the number in bonding molecular orbitals, and dividing the result by 2. |

Key Skills

Molecular Shape and Polarity

Molecular polarity is tremendously important in determining the physical and chemical properties of a substance. Indeed, molecular polarity is one of the most important consequences of molecular geometry. To determine the geometry or *shape* of a molecule or polyatomic ion, we use a stepwise procedure:

1. Draw a correct Lewis structure [◀◀ Chapter 6 Key Skills].
2. Count electron domains. Remember that an electron domain is a lone pair or a bond; and that a *bond* may be a single bond, a double bond, or a triple bond.
3. Apply the VSEPR model to determine electron-domain geometry.
4. Consider the positions of *atoms* to determine molecular geometry (shape), which may or may not be the same as the electron-domain geometry.

Consider the examples of SF_6, SF_4, and CH_2Cl_2. We determine the molecular geometry as follows:

Draw the Lewis structure.	SF_6 Lewis structure	SF_4 Lewis structure	CH_2Cl_2 Lewis structure
Count electron domains on the central atom.	6 electron domains: • six bonds	5 electron domains: • four bonds • one lone pair	4 electron domains: • four bonds
Apply VSEPR to determine electron-domain geometry.	6 electron domains arrange themselves in an octahedron.	5 electron domains arrange themselves in a trigonal bipyramid.	4 electron domains arrange themselves in a tetrahedron.
Consider positions of atoms to determine molecular geometry.	With no lone pairs on the central atom, the molecular geometry is the same as the electron-domain geometry: Octahedral.	The lone pair occupies one of the equatorial positions, making the molecular geometry: See-saw shaped.	With no lone pairs on the central atom, the molecular geometry is the same as the electron-domain geometry: Tetrahedral.

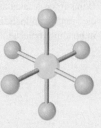  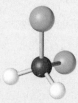

Having determined molecular geometry, we determine overall polarity of each molecule by examining the individual bond dipoles and their arrangement in three-dimensional space.

| Determine whether or not the individual bonds are polar. | S and F have electronegativities of 2.5 and 4, respectively. [◄◄ Figure 6.4, page 216] Therefore the individual bonds are polar and can be represented with arrows. | As in SF$_6$, the individual bonds in SF$_4$ are polar. The bond dipoles are represented with arrows. | C, H, and Cl have electronegativities of 2.5, 2.1, and 3.0, respectively. The individual bonds are polar. Bond dipoles are represented with arrows. |

| Consider the arrangement of bonds to determine which, if any, dipoles cancel one another. | The dipoles shown in red cancel each other; those shown in blue cancel each other; and those shown in green cancel each other, SF$_6$ is **nonpolar**. | The dipoles shown in green cancel each other; but the dipoles shown in red— because they are not directly across from each other— do not. SF$_4$ is **polar**. | Although the bonds are symmetrically distributed, they do not all have equivalent dipoles and therefore do not cancel each other. CH$_2$Cl$_2$ is **polar**. |

Even with polar bonds, a molecule may be nonpolar if it consists of equivalent bonds that are distributed symmetrically. Molecules with equivalent bonds that are not distributed symmetrically, or with bonds that are not equivalent, are generally polar.

Key Skills Problems

7.1
What is the molecular geometry of PBr$_3$?
(a) trigonal planar (b) tetrahedral (c) trigonal pyramidal
(d) bent (e) T-shaped

7.2
Which of the following species does not have tetrahedral molecular geometry?
(a) CCl$_4$ (b) SnH$_4$ (c) AlCl$_4^-$ (d) XeF$_4$ (e) PH$_4^+$

7.3
Which of the following species is polar?
(a) CF$_4$ (b) ClF$_3$ (c) PF$_5$ (d) AlF$_3$ (e) XeF$_2$

7.4
Which of the following species is nonpolar?
(a) ICl$_2^-$ (b) SCl$_4$ (c) SeCl$_2$ (d) NCl$_3$ (e) GeCl$_4$

Questions and Problems

SECTION 7.1: MOLECULAR GEOMETRY

Review Questions

7.1 How is the geometry of a molecule defined, and why is the study of molecular geometry important?

7.2 Sketch the shape of a linear triatomic molecule, a trigonal planar molecule containing four atoms, a tetrahedral molecule, a trigonal bipyramidal molecule, and an octahedral molecule. Give the bond angles in each case.

7.3 How many atoms are directly bonded to the central atom in a tetrahedral molecule, a trigonal bipyramidal molecule, and an octahedral molecule?

7.4 Discuss the basic features of the VSEPR model. Explain why the magnitude of repulsion decreases in the following order: lone pair–lone pair > lone pair–bonding pair > bonding pair–bonding pair.

7.5 In the trigonal bipyramidal arrangement, why does a lone pair occupy an equatorial position rather than an axial position?

Conceptual Problems

7.6 Predict the geometry of the following molecules and ion using the VSEPR model: (a) CH_3I, (b) ClF_3, (c) H_2S, (d) SO_3, (e) SO_4^{2-}.

7.7 Predict the geometry of the following molecules and ion using the VSEPR model: (a) CBr_4, (b) BCl_3, (c) NF_3, (d) H_2Se, (e) NO_2^-.

7.8 Predict the geometries of the following species: (a) $AlCl_3$, (b) $AlCl_4^-$, (c) $ZnCl_2$, (d) $ZnCl_4^{2-}$.

7.9 Predict the geometries of the following species using the VSEPR method: (a) PCl_3, (b) $CHCl_3$, (c) SiH_4, (d) $TeCl_4$.

7.10 Predict the geometries of the following ions: (a) NH_4^+, (b) NH_2^-, (c) CO_3^{2-}, (d) ICl_2^-, (e) ICl_4^-, (f) AlH_4^-, (g) $SnCl_5^-$, (h) H_3O^+, (i) BeF_4^{2-}.

7.11 Predict the geometry of the following molecules using the VSEPR method: (a) $HgBr_2$, (b) N_2O (arrangement of atoms is NNO), (c) SCN^- (arrangement of atoms is SCN).

7.12 Which of the following species are tetrahedral: $SiCl_4$, SeF_4, XeF_4, CI_4, $CdCl_4^{2-}$?

7.13 Describe the geometry around each of the three central atoms in the CH_3COOH molecule.

7.14 Describe the geometry about each of the central atoms in ethanolamine, shown here.

7.15 Which of the following shows a deviation from ideal bond angles that is not possible for an AB_x molecule? Explain.

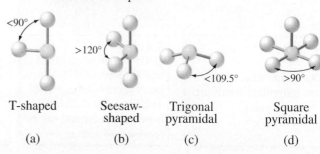

T-shaped	Seesaw-shaped	Trigonal pyramidal	Square pyramidal
(a)	(b)	(c)	(d)

SECTION 7.2: MOLECULAR GEOMETRY AND POLARITY

Review Questions

7.16 Explain why an atom cannot have a permanent dipole moment.

7.17 The bonds in beryllium hydride (BeH_2) molecules are polar, and yet the dipole moment of the molecule is zero. Explain.

Conceptual Problems

7.18 Determine whether (a) NBr_3, (b) OCS, and (c) XeF_4 are polar.

7.19 Determine whether (a) BrF_5, (b) ClF_3, and (c) BCl_3 are polar.

7.20 Each of the molecules shown contains polar bonds. Which of the *molecules* is polar?

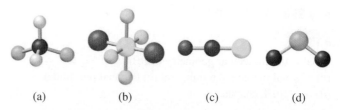

(a)	(b)	(c)	(d)

7.21 Each of the molecules shown contains polar bonds. Which of the *molecules* is polar?

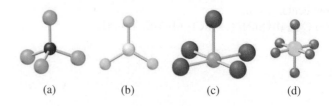

(a)	(b)	(c)	(d)

SECTION 7.3: INTERMOLECULAR FORCES

Review Questions

7.22 Give an example for each type of intermolecular force: (a) dipole-dipole interaction, (b) ion-dipole interaction, (c) dispersion forces, (d) van der Waals forces.

7.23 Explain the term *polarizability*. What kind of molecules tend to have high polarizabilities? What is the relationship between polarizability and intermolecular forces?

7.24 Explain the difference between a temporary dipole moment and a permanent dipole moment.

7.25 What physical properties are determined by the strength of intermolecular forces in solids and in liquids?

7.26 Explain why hydrogen bonding is exhibited by some hydrogen-containing compounds and not by others.

7.27 Describe the types of intermolecular forces that govern the folding of a protein molecule into its physiologically functioning three-dimensional state.

Conceptual Problems

7.28 The compounds Br_2 and ICl are isoelectronic (have the same number of electrons) and have similar molar masses, yet Br_2 melts at $-7.2°C$ and ICl melts at $27.2°C$. Explain.

7.29 If you lived in Alaska, which of the following natural gases could you keep in an outdoor storage tank in winter: methane (CH_4), propane (C_3H_8), or butane (C_4H_{10})? Explain why.

7.30 The binary hydrogen compounds of the Group 4A elements and their boiling points are: CH_4, $-162°C$; SiH_4, $-112°C$; GeH_4, $-88°C$; and SnH_4, $-52°C$. Explain the increase in boiling points from CH_4 to SnH_4.

7.31 List the types of intermolecular forces that exist between molecules (or atoms or ions) in each of the following species: (a) benzene (C_6H_6), (b) CH_3Cl, (c) PF_3, (d) NaCl, (e) CS_2.

7.32 Ammonia is both a donor and an acceptor of hydrogen in hydrogen-bond formation. Draw a diagram showing the hydrogen bonding of an ammonia molecule with two other ammonia molecules.

7.33 Which of the following species are capable of hydrogen bonding among themselves: (a) C_2H_6, (b) HI, (c) KF, (d) BeH_2, (e) CH_3COOH?

7.34 Arrange the following in order of increasing boiling point: RbF, CO_2, CH_3OH, CH_3Br. Explain your reasoning.

7.35 Diethyl ether has a boiling point of $34.5°C$, and 1-butanol has a boiling point of $117°C$.

$$\underset{\text{Diethyl ether}}{H-\overset{\overset{\displaystyle H}{|}}{\underset{\underset{\displaystyle H}{|}}{C}}-\overset{\overset{\displaystyle H}{|}}{\underset{\underset{\displaystyle H}{|}}{C}}-O-\overset{\overset{\displaystyle H}{|}}{\underset{\underset{\displaystyle H}{|}}{C}}-\overset{\overset{\displaystyle H}{|}}{\underset{\underset{\displaystyle H}{|}}{C}}-H} \qquad \underset{\text{1-Butanol}}{H-\overset{\overset{\displaystyle H}{|}}{\underset{\underset{\displaystyle H}{|}}{C}}-\overset{\overset{\displaystyle H}{|}}{\underset{\underset{\displaystyle H}{|}}{C}}-\overset{\overset{\displaystyle H}{|}}{\underset{\underset{\displaystyle H}{|}}{C}}-\overset{\overset{\displaystyle H}{|}}{\underset{\underset{\displaystyle H}{|}}{C}}-OH}$$

Both of these compounds have the same numbers and types of atoms. Explain the difference in their boiling points.

7.36 Which member of each of the following pairs of substances would you expect to have a higher boiling point: (a) O_2 and Cl_2, (b) SO_2 and CO_2, (c) HF and HI?

7.37 Which substance in each of the following pairs would you expect to have the higher boiling point: (a) Ne or Xe, (b) CO_2 or CS_2, (c) CH_4 or Cl_2, (d) F_2 or LiF, (e) NH_3 or PH_3? Explain why.

7.38 Explain in terms of intermolecular forces why (a) NH_3 has a higher boiling point than CH_4 and (b) KCl has a higher melting point than I_2.

7.39 What kind of attractive forces must be overcome to (a) melt ice, (b) boil molecular bromine, (c) melt solid iodine, and (d) dissociate F_2 into F atoms?

7.40 Determine which molecule will condense to a liquid at the higher temperature and explain why: NH_3 or CH_4.

7.41 Determine which molecule will condense to a liquid at the higher temperature and explain why: SiH_4 or PH_3.

7.42 The following compounds have the same molecular formulas (C_4H_{10}). Which one would you expect to have a higher boiling point?

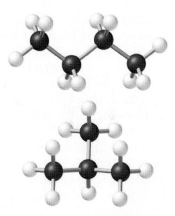

7.43 Explain the difference in the melting points of the following compounds.

m.p. 45°C	m.p. 115°C

(*Hint:* One of the two can form *intra*molecular hydrogen bonds.)

SECTION 7.4: VALENCE BOND THEORY

Review Questions

7.44 What is valence bond theory? How does it differ from the Lewis concept of chemical bonding?

7.45   Use valence bond theory to explain the bonding in Cl_2 and HCl. Show how the atomic orbitals overlap when a bond is formed.

7.46   According to valence bond theory, how many bonds would you expect each of the following atoms (in the ground state) to form: Be, C?

7.47   According to valence bond theory, how many bonds would you expect each of the following atoms (in the ground state) to form: P, S?

Conceptual Problems

7.48   For which molecule(s) can we *not* use valence bond theory to explain the bonding: $BeCl_2$ (Cl—Be—Cl bond angle = 180°), Br_2, SF_6 (F—S—F bond angles = 90° and 180°)? Explain.

7.49   For which molecule(s) can we *not* use valence bond theory to explain the bonding: N_2, BF_3 (F—B—F bond angles = 120°), HI? Explain.

SECTION 7.5:   HYBRIDIZATION OF ATOMIC ORBITALS

Review Questions

7.50   What is the hybridization of atomic orbitals? Why do we never refer to isolated atoms as hybridized?

7.51   Is it possible for two $2p$ orbitals on an atom to combine to give two hybrid orbitals? Explain.

7.52   Determine the hybridization of the central atom in a molecule with the following molecular geometries: (a) tetrahedral, (b) trigonal planar, (c) trigonal bipyramidal, (d) linear, (e) octahedral.

7.53   What is the angle between any two of the following two hybrid orbitals on an atom: (a) sp, (b) sp^2, (c) sp^3? Explain why the answer to this question would be more complicated for sp^3d and sp^3d^2 hybrid orbitals.

Conceptual Problems

7.54   Describe the bonding scheme of the AsH_3 molecule in terms of hybridization.

7.55   What is the hybridization of Si in SiH_4 and in H_3Si—SiH_3?

7.56   What is the hybridization of the nitrogen atoms in the following species: (a) NH_3, (b) H_2N—NH_2, (c) NO_3^-?

7.57   Describe the hybridization of phosphorus in PF_5.

7.58   Describe the change in hybridization (if any) of the Al atom when aluminum trichloride, $AlCl_3$, interacts with a chloride ion to form the tetrachloroaluminate ion, $AlCl_4^-$. (*Hint:* Start by drawing Lewis structures.)

7.59   Describe the changes in hybridization (if any) of the B and N atoms as a result of the formation of F_3B—NH_3 from BF_3 and NH_3. (*Hint:* Start by drawing Lewis structures.)

SECTION 7.6:   HYBRIDIZATION IN MOLECULES CONTAINING MULTIPLE BONDS

   Visualizing Chemistry
Figure 7.15

VC 7.1   How is a sigma bond different from a pi bond?
(a) A sigma bond is a bonding molecular orbital; a pi bond is an antibonding molecular orbital.
(b) A sigma bond is a single bond, whereas a pi bond is a double bond.
(c) The electron density in a sigma bond lies along the internuclear axis; that of a pi bond does not.

VC 7.2   Pi bonds form when _____ atomic orbitals on _____ atom(s) overlap.
(a) perpendicular, adjacent
(b) parallel, adjacent
(c) parallel, the same

VC 7.3   Formation of two pi bonds requires the combination of _____ atomic orbitals.
(a) two
(b) four
(c) six

VC 7.4   Why are there no pi bonds in ethane (C_2H_6)?
(a) The remaining unhybridized p orbitals do not contain any electrons.
(b) There are no unhybridized p orbitals remaining on either C atom.
(c) The remaining unhybridized p orbitals are not parallel to each other.

Review Questions

7.60   Describe the difference between a sigma bond and a pi bond.

7.61   Which of the following pairs of atomic orbitals of adjacent nuclei can overlap to form a sigma bond? Which overlap to form a pi bond? Which cannot overlap (no bond)? Consider the x axis to be the internuclear axis, that is, the line joining the nuclei of the two atoms. (a) $1s$ and $1s$, (b) $1s$ and $2p_x$, (c) $2p_x$ and $2p_y$, (d) $3p_y$ and $3p_y$, (e) $2p_x$ and $2p_x$, (f) $1s$ and $2s$.

Conceptual Problems

7.62   Determine the hybridization of each carbon atom in the following molecules:
(a) H_3C—CH_3
(b) H_3C—CH=CH_2
(c) CH_3—$C{\equiv}C$—CH_2OH
(d) CH_3CH=O
(e) CH_3COOH

7.63   Determine the hybridization of the carbon atoms in the following species: (a) CO, (b) CO_2, (c) CN^-.

7.64 The allene molecule ($H_2C=C=CH_2$) is linear (the three C atoms lie on a straight line). Determine the hybridization of each carbon atom. Draw diagrams to show the formation of sigma bonds and pi bonds in allene.

7.65 What is the hybridization of the central N atom in the azide ion (N_3^-)? (The arrangement of atoms is NNN.)

7.66 How many sigma bonds and pi bonds are there in each of the following molecules?

(a)　　　　　(b)　　　　　(c)

7.67 How many pi bonds and sigma bonds are there in the tetracyanoethylene molecule?

7.68 Tryptophan is one of the 20 amino acids in the human body. Describe the hybridization of the C atoms and the N atoms, and determine the number of sigma and pi bonds in the molecule.

7.69 Benzo[*a*]pyrene is a potent carcinogen found in coal and cigarette smoke. Determine the number of sigma and pi bonds in the molecule.

SECTION 7.7: MOLECULAR ORBITAL THEORY

Review Questions

7.70 What is molecular orbital theory? How does it differ from valence bond theory?

7.71 Define the following terms: *bonding molecular orbital, antibonding molecular orbital, pi molecular orbital,* and *sigma molecular orbital.*

7.72 Sketch the shapes of the following molecular orbitals: σ_{1s}, σ_{1s}^*, π_{2p}, π_{2p}^*. How do their energies compare?

7.73 Explain the significance of bond order. How can bond order be used for quantitative comparisons of the strengths of chemical bonds?

Conceptual Problems

7.74 Using molecular orbital theory, explain the changes that occur in the H—H internuclear distance as the molecule H_2 is ionized first to H_2^+ and then to H_2^{2+}.

7.75 The formation of H_2 from two H atoms is an energetically favorable process. Yet, statistically there is less than a 100 percent chance that any two H atoms will undergo the reaction. Apart from energy considerations, how would you account for this observation based on the electron spins in the two H atoms?

7.76 Draw a molecular orbital energy level diagram for each of the following species: He_2, HHe, He_2^+. Compare their relative stabilities in terms of bond orders. (Treat HHe as a diatomic molecule with three electrons.)

7.77 Arrange the following species in order of increasing stability: Li_2, Li_2^+, Li_2^-. Justify your choice with a molecular orbital energy level diagram.

7.78 It has long been predicted that Be_2 could not exist because the molecule would not be stable; however, in 2009, the existence of Be_2 was confirmed by spectroscopy. Use molecular orbital theory to explain why the Be_2 molecule was predicted not to exist.

7.79 Which of these species has a longer bond, B_2 or B_2^+? Explain in terms of molecular orbital theory.

7.80 Acetylene (C_2H_2) has a tendency to lose two protons (H^+) and form the carbide ion (C_2^{2-}), which is present in a number of ionic compounds, such as CaC_2 and MgC_2. Describe the bonding in the C_2^{2-} ion in terms of molecular orbital theory. Compare the bond order in C_2^{2-} with that in C_2.

7.81 Compare the Lewis and molecular orbital treatments of the oxygen molecule.

7.82 Explain why the bond order of N_2 is greater than that of N_2^+, but the bond order of O_2 is less than that of O_2^+.

7.83 Compare the relative bond orders of the following species and indicate their magnetic properties (i.e., diamagnetic or paramagnetic): O_2, O_2^+, O_2^- (superoxide ion), O_2^{2-} (peroxide ion).

7.84 Use molecular orbital theory to compare the relative stabilities of F_2 and F_2^+.

7.85 A single bond is almost always a sigma bond, and a double bond is almost always made up of a sigma bond and a pi bond. There are very few exceptions to this rule. Show that the B_2 and C_2 molecules are examples of the exceptions.

7.86 Draw the molecular orbital diagram for the cyanide ion (CN^-). (Assume the ordering of molecular orbitals to be like that in N_2.) Write the electron configuration of the cyanide ion (CN^-).

7.87 Given that BeO is diamagnetic, use a molecular orbital diagram to determine whether the orbitals are ordered like those of Be_2 or those of O_2.

SECTION 7.8: BONDING THEORIES AND DESCRIPTIONS OF MOLECULES WITH DELOCALIZED BONDING

Review Questions

7.88   How does a delocalized molecular orbital differ from a molecular orbital such as that found in H_2 or C_2H_4? What do you think are the minimum conditions (e.g., number of atoms and types of orbitals) for forming a delocalized molecular orbital?

7.89   In Chapter 6 we saw that the resonance concept is useful for dealing with species such as the benzene molecule and the carbonate ion. How does molecular orbital theory deal with these species?

Conceptual Problems

7.90   Both ethylene (C_2H_4) and benzene (C_6H_6) contain the C=C bond, yet benzene is more stable in that it does not combine readily with other molecules— whereas ethylene *does*. Explain this difference in the chemical properties of these two species.

7.91   Chemists often represent benzene with the structure shown on the left. Explain why the structure on the left is a better representation of the benzene molecule than the one on the right.

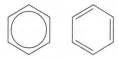

7.92   Determine which of these molecules has a more delocalized orbital, and justify your choice. (*Hint:* Both molecules contain two benzene rings. In naphthalene, the two rings are fused together. In biphenyl, the two rings are joined by a single bond around which the two rings can rotate.)

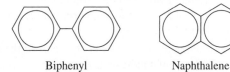

Biphenyl   Naphthalene

7.93   Nitryl fluoride (FNO_2) is used in rocket propellants. The fluorine and oxygen atoms are bonded to the nitrogen atom. (a) Draw a Lewis structure for FNO_2. (b) Indicate the hybridization of the nitrogen atom. (c) Describe the bonding in terms of molecular orbital theory. Where would you expect delocalized molecular orbitals to form?

7.94   Describe the bonding in the nitrate ion NO_3^- in terms of delocalized molecular orbitals.

7.95   What is the hybridization of the central O atom in O_3? Describe the bonding in O_3 in terms of delocalized molecular orbitals.

ADDITIONAL PROBLEMS

7.96   Which of the following species is not likely to have a tetrahedral shape: (a) $SiBr_4$, (b) NF_4^+, (c) SF_4, (d) $BeCl_4^{2-}$, (e) BF_4^-, (f) $AlCl_4^-$?

7.97   Draw the Lewis structure of mercury(II) bromide. Is this molecule linear or bent? How would you establish its geometry?

7.98   Liquid bromine (Br_2) and the interhalogen compound ICl have nearly identical molecular masses, yet their boiling points differ by 38°C. Which one has the higher boiling point and why?

7.99   Predict the geometry of sulfur dichloride (SCl_2) and the hybridization of the sulfur atom.

7.100   Antimony pentafluoride (SbF_5) combines with XeF_4 and XeF_6 to form ionic compounds, $XeF_3^+SbF_6^-$ and $XeF_5^+SbF_6^-$. Describe the geometries of the cations and the anion in these two compounds.

7.101   The molecular model of vitamin C is shown here. (a) Write the molecular formula of the compound. (b) What is the hybridization of each C and O atom? (c) Describe the geometry about each C and O atom.

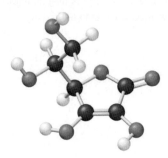

7.102   The molecular model of nicotine (a stimulant) is shown here. (a) Write the molecular formula of the compound. (b) What is the hybridization of each C and N atom? (c) Describe the geometry about each C and N atom.

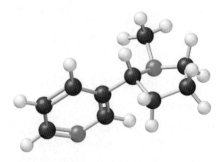

7.103   Predict the bond angles for the following molecules: (a) $BeCl_2$, (b) BCl_3, (c) CCl_4, (d) CH_3Cl, (e) Hg_2Cl_2 (arrangement of atoms: $ClHgHgCl$), (f) $SnCl_2$, (g) H_2O_2, (h) SnH_4.

7.104 The germanium pentafluoride anion (GeF_5^-) has been observed as part of an ionic compound with cesium at low temperatures. The anion is formed by the interaction of GeF_4 with F^-. Which figure best illustrates the change in molecular geometry around Ge in the transformation from GeF_4 to GeF_5^-?

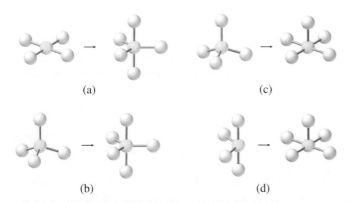

(a)   (c)

(b)   (d)

7.105 Draw Lewis structures and give the other information requested for the following molecules: (a) BF_3. Shape: planar or nonplanar? (b) ClO_3^-. Shape: planar or nonplanar? (c) HCN. Polar or nonpolar? (d) OF_2. Polar or nonpolar? (e) NO_2. Estimate the ONO bond angle.

7.106 Which figure best illustrates the hybridization of arsenic in (a) arsenic pentafluoride (AsF_5) and (b) arsenic trifluoride (AsF_3)?

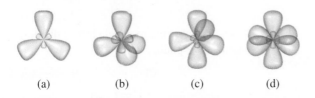

(a)   (b)   (c)   (d)

7.107 Which of the following substances has the highest polarizability: CH_4, H_2, CCl_4, SF_6, H_2S?

7.108 Draw Lewis structures and give the other information requested for the following: (a) SO_3. Polar or nonpolar molecule? (b) PF_3. Polar or nonpolar? (c) F_3SiH. Polar or nonpolar? (d) SiH_3^-. Shape: planar or pyramidal? (e) Br_2CH_2. Polar or nonpolar molecule?

7.109 Which of the following molecules are linear: ICl_2^-, IF_2^+, OF_2, SnI_2, $CdBr_2$?

7.110 Semiconducting materials such as indium phosphide (InP) can be prepared from a single-source precursor like $(CH_3)_3In-P(CH_3)_3$. Which pair of figures best illustrates the hybridization of In and P, respectively, in $(CH_3)_3In-P(CH_3)_3$?

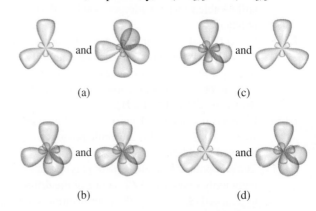

(a)   (c)

(b)   (d)

7.111 The N_2F_2 molecule can exist in either of the following two forms.

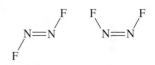

(a) What is the hybridization of N in each case?
(b) Identify each structure as polar or nonpolar.

7.112 Cyclopropane (C_3H_6) has the shape of a triangle in which a C atom is bonded to two H atoms and two other C atoms at each corner. Cubane (C_8H_8) has the shape of a cube in which a C atom is bonded to one H atom and three other C atoms at each corner. (a) Draw Lewis structures of these molecules. (b) Compare the CCC angles in these molecules with those predicted for an sp^3-hybridized C atom. (c) Would you expect these molecules to be easy to make? Explain.

7.113 The compound 1,2-dichloroethane ($C_2H_4Cl_2$) is nonpolar, while *cis*-dichloroethylene ($C_2H_2Cl_2$) has a dipole moment. The reason for the difference is that groups connected by a single bond can rotate with respect to each other, but no rotation occurs when a double bond connects the groups. On the basis of bonding considerations, explain why rotation occurs in 1,2-dichloroethane but not in *cis*-dichloroethylene.

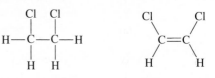

1,2-dichloroethane   *cis*-dichloroethylene

7.114   If water were a linear molecule, (a) would it still be polar, and (b) would the water molecules still be able to form hydrogen bonds with one another?

7.115   Select the substance in each pair that should have the higher boiling point. In each case identify the principal intermolecular forces involved and account briefly for your choice: (a) K_2S or $(CH_3)_3N$, (b) Br_2 or $CH_3CH_2CH_2CH_3$.

7.116   Carbon suboxide (C_3O_2) is a colorless pungent-smelling gas. Does this molecule possess a dipole moment? Explain.

7.117   The compound 3′-azido-3′-deoxythymidine, commonly known as AZT, is one of the drugs used to treat AIDS. What are the hybridization states of the C and N atoms in this molecule?

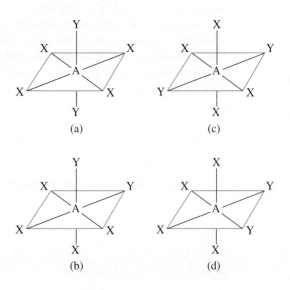

7.118   The following molecules (AX_4Y_2) all have an octahedral geometry. Group the molecules that are equivalent to each other.

7.119   The amino acid selenocysteine is one of the components of selenoproteins, more than 20 of which have been identified so far in human cells.

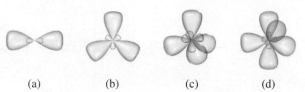

In the structure of this interesting amino acid, how many of each type of hybrid orbital illustrated here are present?

(a)   (b)   (c)   (d)

7.120   Write the ground-state electron configuration for B_2. Is the molecule diamagnetic or paramagnetic?

7.121   What is the hybridization of C and of N in this molecule?

7.122   Does the following molecule have a dipole moment? Explain.

7.123   Gaseous or highly volatile liquid anesthetics are often preferred in surgical procedures because once inhaled, these vapors can quickly enter the bloodstream through the alveoli and then enter the brain. Several common gaseous anesthetics are shown here with their boiling points.

Halothane
50°C

Isoflurane
48.5°C

Enflurane
56.5°C

Based on intermolecular force considerations, explain the advantages of using these anesthetics. (*Hint:* The brain barrier is made of membranes that have a nonpolar interior region.)

7.124   Although both carbon and silicon are in Group 4A, very few Si=Si bonds are known. Account for the instability of silicon-to-silicon double bonds in general. (*Hint:* Compare the atomic radii of C and Si in Figure 4.6 on page 133. What effect would the larger size have on pi bond formation?)

7.125   So-called greenhouse gases, which contribute to global warming, have a dipole moment or can be bent or distorted during molecular vibration into shapes that have a *temporary* dipole moment. Based on this, which of the following gases are greenhouse gases: N_2, O_2, O_3, CO, CO_2, NO_2, N_2O, CH_4, $CFCl_3$?

7.126   Two of the drugs that are prescribed for the treatment and prevention of kidney stones are hydrochlorothiazide and chlorthalidone, shown here. What types of intermolecular forces exist between chlorthalidone molecules and between hydrochlorothiazide molecules?

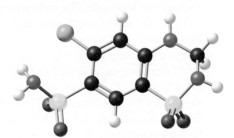

Hydrochlorothiazide

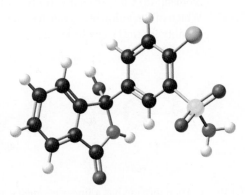

Chlorthalidone

7.127   The disulfide bond, −S−S−, plays an important role in determining the three-dimensional structure of proteins. Describe the nature of the bond and the hybridization of the S atoms.

7.128   The stable allotropic form of phosphorus is P_4, in which each P atom is bonded to three other P atoms. Draw a Lewis structure of this molecule and describe its geometry. At high temperatures, P_4 dissociates to form P_2 molecules containing a P=P bond. Explain why P_4 is more stable than P_2.

7.129   The BO^+ ion is paramagnetic. Determine (a) whether the order of molecular-orbital energies is like that in B_2 or in O_2; (b) the bond order; and (c) the number of unpaired electrons in the ion.

7.130   Use molecular orbital theory to explain the bonding in the azide ion (N_3^-). (See Problem 7.65.)

7.131   Which best illustrates the change in geometry about the central S atom when SO_3 gains two electrons to become the SO_3^{2-} ion?

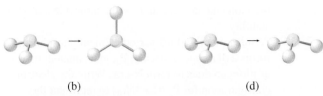

(a)   (c)

(b)   (d)

7.132   Draw three Lewis structures for compounds with the formula $C_2H_2F_2$. Indicate which of the compounds is/are polar.

7.133   Consider three of the vibrational modes of carbon dioxide

$$\overset{\leftarrow}{O}=C=\vec{O} \qquad \vec{O}=\overset{\leftarrow}{C}=\vec{O} \qquad \overset{\uparrow}{O}=\overset{\uparrow}{C}=\underset{\downarrow}{O}$$

where the arrows indicate the movement of the atoms. (During a complete cycle of vibration, the atoms move toward one extreme position and then reverse their direction to the other extreme position.) Which of the preceding vibrations are responsible for CO_2 behaving as a greenhouse gas? (See Problem 7.125.)

7.134   Aluminum trichloride ($AlCl_3$) is an electron-deficient molecule. It has a tendency to form a dimer (a molecule made up of two $AlCl_3$ units), Al_2Cl_6. (a) Draw a Lewis structure for the dimer. (b) Describe the hybridization of Al in $AlCl_3$ and in Al_2Cl_6. (c) Sketch the geometry of the dimer. (d) Do either of these molecules possess a dipole moment?

7.135 Progesterone is a hormone responsible for female sex characteristics. In the usual shorthand structure, each point where lines meet represents a C atom, and most H atoms are not shown. Draw the complete structure of the molecule, showing all C and H atoms. Indicate which C atoms are sp^2- and sp^3-hybridized.

7.136 The molecule benzyne (C_6H_4) resembles benzene in that it has a six-membered ring of carbon atoms, but it is far less stable. Draw a Lewis structure of the molecule and account for the molecule's lack of stability.

7.137 Assume that the third-period element phosphorus forms a diatomic molecule, P_2, in an analogous way as nitrogen does to form N_2. (a) Write the electron configuration for P_2. Use [Ne_2] to represent the electron configuration for the first two periods. (b) Calculate its bond order. (c) What are its magnetic properties (diamagnetic or paramagnetic)?

7.138 Consider an N_2 molecule in its first excited electronic state, that is, when an electron in the highest occupied molecular orbital is promoted to the lowest empty molecular orbital. (a) Identify the molecular orbitals involved, and sketch a diagram to show the transition. (b) Compare the bond order and bond length of N_2^* with N_2, where the asterisk denotes the excited molecule. (c) Is N_2^* diamagnetic or paramagnetic? (d) When N_2^* loses its excess energy and converts to the ground state N_2, it emits a photon of wavelength 470 nm, which makes up part of the auroras' lights. Calculate the energy difference between these levels.

7.139 The Lewis structure for O_2 is

$$\ddot{O}=\ddot{O}$$

Use molecular orbital theory to show that the structure actually corresponds to an excited state of the oxygen molecule.

7.140 Draw the Lewis structure of ketene (C_2H_2O) and describe the hybridization of the C atoms. (The molecule does not contain O—H bonds.) On separate diagrams, sketch the formation of the sigma and pi bonds.

7.141 The compound TCDD, or 2,3,7,8-tetrachlorodibenzo-*p*-dioxin, is highly toxic.

It gained considerable notoriety in 2004 when it was implicated in the attempted murder of a Ukrainian politician. (a) Describe its geometry, and state whether the molecule has a dipole moment. (b) How many pi bonds and sigma bonds are there in the molecule?

7.142 Name the kinds of attractive forces that must be overcome to (a) boil liquid ammonia, (b) melt solid phosphorus (P_4), (c) dissolve CsI in liquid HF, (d) melt potassium metal.

7.143 Carbon monoxide (CO) is a poisonous compound due to its ability to bind strongly to Fe^{2+} in the hemoglobin molecule. The molecular orbitals of CO have the same energy order as those of the N_2 molecule. (a) Draw a Lewis structure of CO and assign formal charges. Explain why CO has a rather small dipole moment of 0.12 D. (b) Compare the bond order of CO with that from molecular orbital theory. (c) Which of the atoms (C or O) is more likely to form bonds with the Fe^{2+} ion in hemoglobin?

7.144 Which of the following ions possess a dipole moment: (a) ClF_2^+, (b) ClF_2^-, (c) IF_4^+, (d) IF_4^-?

7.145 Which of the following geometries has a greater stability for tin(IV) hydride (SnH_4)?

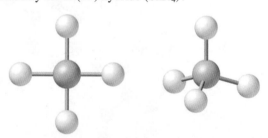

7.146 Carbon dioxide has a linear geometry and is nonpolar. Yet we know that the molecule exhibits bending and stretching motions that create a dipole moment. How would you reconcile these seemingly conflicting descriptions of CO_2?

7.147 The bond length of the transient diatomic molecule CF is 129.1 pm and that of the molecular ion CF^+ is 117.3 pm. Using molecular orbital theory, explain why the CF bond shortens with the loss of an electron.

7.148 Which of the following compounds is most likely to exist as a liquid at room temperature: ethane (C_2H_6), hydrazine (N_2H_4), fluoromethane (CH_3F)? Explain.

Answers to In-Chapter Materials

PRACTICE PROBLEMS

7.1A (a) linear, (b) bent. **7.1B** (a) Group 6A, (b) Group 7A.
7.2A Bent about O, tetrahedral about each C, trigonal pyramidal about N.
All bond angles are ~109.5°. Angles labeled in blue are <109.5°.

7.2B SO_2 contains double bonds, which are not pushed together by the central atom's lone pair as easily as the single bonds in NH_3.
7.3A (a) polar, (b) nonpolar.
7.3B

polar	nonpolar	polar	nonpolar
(a)		(b)	

7.4A (a) Dispersion forces, (b) dipole-dipole forces, H bonding, and dispersion forces, (c) dipole-dipole and dispersion forces, (d) dispersion forces. **7.4B** (a) Dipole-dipole and dispersion forces, (b) dipole-dipole forces, H bonding, and dispersion forces, (c) dipole-dipole forces, H bonding, and dispersion forces, (d) dispersion forces. **7.5A** Singly occupied $3p$ orbitals from the P atom overlap with s orbitals from H atoms. **7.5B** We cannot use valence bond theory to explain the bonding in SO_2 or CH_4. In the case of SO_2, although the central atom has two unpaired electrons and can form two bonds, the unpaired electrons on S are in $3p$ orbitals. Formation of two bonds by the overlap of two $3p$ orbitals on S would be expected to result in a bond angle of approximately 90°. In the case of CH_4, the central atom does not have enough unpaired electrons to form four bonds. **7.6A** Two of the $4p$ electrons in Br are promoted to empty d orbitals. The s orbital, all three p orbitals, and two of the d orbitals hybridize to form six sp^3d^2 hybrid orbitals. One of the hybrid

orbitals contains the lone pair. Each of the remaining hybrid orbitals contains one electron and overlaps with a singly occupied $2p$ orbital on an F atom. The arrangement of hybrid orbitals is octahedral and the bond angles are ~90°. **7.6B** One of the $2s$ electrons in Be is promoted to an empty p orbital. The s orbital and one p orbital hybridize to form two sp hybrid orbitals. Each hybrid orbital contains one electron and overlaps with a singly occupied $2p$ orbital on an F atom. The arrangement is linear with a bond angle of ~180°. **7.7A** 16 σ bonds and 4 π bonds. **7.7B** 17 σ bonds and 5 π bonds. **7.8A** C and N atoms are sp-hybridized. The triple bond between C and N is composed of one sigma bond (from overlap of hybrid orbitals) and two pi bonds (from interaction of remaining p orbitals). The single bond between H and C is the result of an sp orbital from C overlapping with an s orbital from H. **7.8B** Each N atom is sp-hybridized. One sp orbital on each is singly occupied and one contains a lone pair. The singly occupied sp orbitals overlap to form a sigma bond between the N atoms. The remaining unhybridized p orbitals interact to form two pi bonds. **7.9A** Paramagnetic; bond order = 2. **7.9B** Paramagnetic; bond order = 2. **7.10A** Two different resonance structures are possible; therefore we consider all three atoms to be sp^2-hybridized. One of the hybrid orbitals on the central O atom contains the lone pair, the other two form sigma bonds to the terminal O atoms. Each atom has one remaining unhybridized p orbital. The p orbitals combine to form π molecular orbitals. **7.10B** Two different resonance structures are possible; therefore we consider all three atoms to be sp^2-hybridized. One of the hybrid orbitals on the central N atom contains the lone pair, the other two form sigma bonds to the terminal O atoms. Each atom has one remaining unhybridized p orbital. The p orbitals combine to form π molecular orbitals.

SECTION REVIEW
7.1.1 d. **7.1.2** b. **7.1.3** e. **7.1.4** c. **7.2.1** a. **7.2.2** c. **7.3.1** a. **7.3.2** d. **7.4.1** b, e. **7.4.2** b. **7.5.1** b. **7.5.2** d. **7.6.1** a, c. **7.6.2** c. **7.6.3** a, b, e. **7.6.4** c, e. **7.7.1** b. **7.7.2** c, e. **7.7.3** b. **7.7.4** a, c, e. **7.8.1** c. **7.8.2** b. **7.8.3** a, b. **7.8.4** e.

Chapter 8

Chemical Reactions

A DELICACY in Japanese cuisine, the fugu fish is featured in many upscale restaurants. Parts of the fish, most notably the eyes, liver, and ovaries, contain *tetrodoxin*—a potentially deadly neurotoxin. As recently as the early 1980s, dozens of deaths in Japan were attributed to fugu consumption every year. In 1984, serving the liver of the fugu was banned, which helped to reduce the average annual number of fatalities. Although the fish remains popular, the sale of fugu is highly regulated by the Japanese government and only specially trained, licensed chefs are permitted to prepare it. To facilitate study of the chemistry and biochemistry of tetrodoxin, chemists have synthesized it in the laboratory. One such synthesis requires a sequence of 32 steps, resulting in an overall reaction yield of just 0.52 percent. Another synthesis requires 67 steps, but has a greater overall yield of 1.22 percent.

Before You Begin, Review These Skills

- The mole and molar masses of elements [◄◄ Section 2.7]
- Formulas of ionic compounds [◄◄ Section 5.4]
- Formulas of molecular compounds [◄◄ Section 5.5]
- The mole and molar masses of compounds [◄◄ Section 5.10]

8.1 CHEMICAL EQUATIONS

Thus far, we have examined the structure of atoms and how the number and arrangement of subatomic particles give rise to the properties of atoms. The *properties* of atoms determine how they interact with one another to become the matter that we encounter every day. In this chapter, we investigate the chemical changes that matter can undergo, how we represent these changes with *chemical equations,* and how we can *use* chemical equations to solve quantitative problems.

The third hypothesis of John Dalton's atomic theory described a ***chemical reaction*** as a process that neither creates nor destroys atoms, but that rearranges atoms in chemical compounds. Examples of chemical reactions include the rusting of iron and the explosive combination of hydrogen and oxygen gases to produce water. Like the other chemical processes that we have encountered (electron transitions, ionization, electron affinity, and formation of chemical bonds), chemical reactions involve changes in energy.

A ***chemical equation*** uses chemical symbols to denote what occurs in a chemical reaction. We have seen how chemists represent elements and compounds using chemical symbols. Now we will look at how chemists represent chemical reactions using chemical equations.

Student Annotation: Recall that chemical changes can also be referred to as *chemical processes* [◄◄ Section 1.6]. The terms *chemical change, chemical process,* and *chemical reaction* are roughly equivalent.

Student Annotation: The three hypotheses that make up Dalton's atomic theory are:

1. Matter is composed of tiny, indivisible particles called atoms; and all atoms of a given element are identical [◄◄ Section 2.1].
2. *Compounds* are made up of specific combinations of atoms of two or more different elements [◄◄ Section 5.5].
3. Chemical reactions cause the *rearrangement* of atoms, but do not cause either the creation or the destruction of atoms.

Interpreting and Writing Chemical Equations

A chemical equation represents a *chemical statement.* When you encounter a chemical equation, you may find it useful to read it as though it were a sentence. Read

$$NH_3 + HCl \longrightarrow NH_4Cl$$

as "Ammonia and hydrogen chloride react to produce ammonium chloride." Read

$$CaCO_3 \longrightarrow CaO + CO_2$$

as "Calcium carbonate reacts to produce calcium oxide and carbon dioxide." Thus, the plus signs can be interpreted simply as the word *and,* and the arrows can be interpreted as the phrase "react(s) to produce."

In addition to interpreting chemical equations, you must be able to write chemical equations to represent reactions. For example, the equation for the process by which sulfur and oxygen react to produce sulfur dioxide is written as

$$S + O_2 \longrightarrow SO_2$$

Likewise, we write the equation for the reaction of sulfur trioxide and water to produce sulfuric acid as

$$SO_3 + H_2O \longrightarrow H_2SO_4$$

Each chemical species that appears to the left of the arrow is called a ***reactant.*** Reactants are those substances that are *consumed* in the course of a chemical reaction. Each species that appears to the right of the arrow is called a ***product.*** Products are the substances that *form* during the course of a chemical reaction.

Chemists usually indicate the physical states of reactants and products with italicized letters in parentheses following each species in the equation. Gases, liquids, and solids are labeled with (g), (l), and (s), respectively. Chemical species that are dissolved in water are said to be **aqueous** and are labeled (aq). The equation examples given previously can be written as follows:

$$NH_3(g) + HCl(g) \longrightarrow NH_4Cl(s) \qquad S(s) + O_2(g) \longrightarrow SO_2(g)$$

$$CaCO_3(s) \longrightarrow CaO(s) + CO_2(g) \qquad SO_3(g) + H_2O(l) \longrightarrow H_2SO_4(l)$$

At this point, it is useful to discuss how substances are represented in chemical equations. Compounds, whether they are ionic or molecular, are represented with their chemical formulas. However, because *free* elements (those that are *uncombined*) exist in a variety of forms, and because some elements can exist as different allotropes [◄◄ Section 5.5], there are two different ways to represent them.

Metals

Because metals usually do not exist in discrete molecular units but rather in complex, three-dimensional networks of atoms [►► Chapter 12], we always use their empirical formulas in chemical equations. The empirical formulas are the same as the symbols that represent the elements. For example, the empirical formula for iron is Fe, the same as the symbol for the element.

Nonmetals

There is no single rule regarding the representation of nonmetals in chemical equations. Carbon, for example, exists in several allotropic forms. Regardless of the allotrope, we use its empirical formula C to represent elemental carbon in chemical equations. Often the symbol C will be followed by the specific allotrope in parentheses. Thus, we represent two of carbon's allotropic forms as C(graphite) and C(diamond).

For nonmetals that exist as polyatomic molecules, we generally use the molecular formula in equations: H_2, N_2, O_2, F_2, Cl_2, Br_2, I_2, and P_4, for example. In the case of sulfur, however, we primarily use the empirical formula S rather than the molecular formula S_8—although the molecular formula S_8 is equally correct and sometimes used.

Noble Gases

All the noble gases exist as isolated atoms, so we use their symbols: He, Ne, Ar, Kr, Xe, and Rn.

Metalloids

The metalloids, like the metals, generally have complex three-dimensional networks, so we also represent them with their empirical formulas—that is, their symbols: B, Si, Ge, and so on.

Chemical equations are also used to represent *physical* processes. Sucrose $(C_{12}H_{22}O_{11})$ dissolving in water, for example, is a physical process [◄◄ Section 1.6] that can be represented with the following chemical equation.

$$C_{12}H_{22}O_{11}(s) \xrightarrow{\text{H}_2\text{O}} C_{12}H_{22}O_{11}(aq)$$

The H_2O over the arrow in the equation denotes the process of dissolving a substance in water. Although formulas or symbols are sometimes omitted for simplicity, they can be written over an arrow in a chemical equation to indicate the conditions under which the reaction takes place. For example, in the chemical equation

$$2KClO_3(s) \xrightarrow{\Delta} 2KCl(s) + 3O_2(g)$$

the symbol Δ indicates that the addition of heat energy is necessary to make $KClO_3$ react to form KCl and O_2.

Balancing Chemical Equations

Based on what you have learned so far, the chemical equation for the explosive reaction of hydrogen gas with oxygen gas to form liquid water would be

$$H_2(g) \;+\; O_2(g) \longrightarrow H_2O(l)$$

This equation as it is written, however, violates the *law of conservation of mass,* because four atoms (two H and two O) react to produce only three atoms (two H and one O). The law of conservation of mass is another way of stating Dalton's third hypothesis—that atoms can be neither created nor destroyed.

The equation must be *balanced* so that the same number of each kind of atom appears on both sides of the reaction arrow. Balancing is achieved by writing appropriate *stoichiometric coefficients* (often referred to simply as *coefficients*) to the left of the chemical formulas. In the following case, we write a coefficient of 2 to the left of both the $H_2(g)$ and the $H_2O(l)$.

$$2H_2(g) \;+\; O_2(g) \longrightarrow 2H_2O(l)$$

There are now four H atoms and two O atoms on each side of the arrow. When balancing a chemical equation, we can change only the coefficients that precede the chemical formulas, *not* the subscripts within the chemical formulas. Changing the subscripts would change the formulas for the species involved in the reaction. For example, changing the product from H_2O to H_2O_2 would result in equal numbers of each kind of atom on both sides of the equation, but the equation we set out to balance represented the combination of hydrogen gas and oxygen gas to form water (H_2O), not hydrogen peroxide (H_2O_2). Additionally, we cannot add reactants or products to the chemical equation for the purpose of balancing it. To do so would result in an equation that represents the wrong reaction. The chemical equation must be made quantitatively correct without changing its qualitative chemical statement.

Balancing a chemical equation requires something of a trial-and-error approach. You may find that you change the coefficient for a particular reactant or product, only to have to change it again later in the process. In general, it will facilitate the balancing process if you do the following:

1. Change the coefficients of compounds (e.g., CO_2) before changing the coefficients of elements (e.g., O_2).
2. Treat polyatomic ions that appear on both sides of the equation (e.g., CO_3^{2-}) as units, rather than counting their constituent atoms individually.
3. Count atoms and/or polyatomic ions carefully, and track their numbers each time you change a coefficient.

To balance the chemical equation for the combustion of butane, we first take an inventory of the numbers of each type of atom on each side of the arrow.

$$C_4H_{10}(g) + O_2(g) \longrightarrow CO_2(g) + H_2O(l)$$

$$4 - C - 1$$

$$10 - H - 2$$

$$2 - O - 3$$

Student Annotation: When we study electrochemistry in detail, we will learn a method for balancing certain equations that does allow addition of H_2O, H^+, and OH^- [▶▶ Section 18.1].

Student Annotation: *Combustion* refers to burning in the presence of oxygen. Combustion of a hydrocarbon such as butane produces carbon dioxide and water.

Initially, there are 4 C atoms on the left and 1 on the right; 10 H atoms on the left and 2 on the right; and 2 O atoms on the left and 3 on the right. As a first step, we will place a coefficient of 4 in front of $CO_2(g)$ on the product side.

$$C_4H_{10}(g) + O_2(g) \longrightarrow \mathbf{4}CO_2(g) + H_2O(l)$$

$$4 - C - 4$$
$$10 - H - 2$$
$$2 - O - 9$$

This changes the tally of atoms as shown. Thus, the equation is balanced for carbon, but not for hydrogen or oxygen. Next, we place a coefficient of 5 in front of $H_2O(l)$ on the product side and tally the atoms on both sides again.

$$C_4H_{10}(g) + O_2(g) \longrightarrow \mathbf{4}CO_2(g) + \mathbf{5}H_2O(l)$$

$$4 - C - 4$$
$$10 - H - 10$$
$$2 - O - 13$$

Now the equation is balanced for carbon and hydrogen. Only oxygen remains to be balanced. There are 13 O atoms on the product side of the equation (8 in CO_2 molecules and another 5 in H_2O molecules), so we need 13 O atoms on the reactant side. Because each oxygen molecule contains 2 O atoms, we will have to place a coefficient of $\frac{13}{2}$ in front of $O_2(g)$.

$$C_4H_{10}(g) + \frac{13}{2}O_2(g) \longrightarrow \mathbf{4}CO_2(g) + \mathbf{5}H_2O(l)$$

$$4 - C - 4$$
$$10 - H - 10$$
$$13 - O - 13$$

With equal numbers of each kind of atom on both sides of the equation, this equation is now balanced. For now, however, you should practice balancing equations with the smallest possible *whole* number coefficients. Multiplying each coefficient by 2 gives all whole numbers and a final balanced equation.

$$\mathbf{2}C_4H_{10}(g) + \mathbf{13}O_2(g) \longrightarrow \mathbf{8}CO_2(g) + \mathbf{10}H_2O(l)$$

$$8 - C - 8$$
$$20 - H - 20$$
$$26 - O - 26$$

Student Annotation: A balanced equation is, in a sense, a mathematical equality. We can multiply or divide through by any number, and the equality will still be valid.

Student Annotation: You may wish to review how to deduce a compound's formula from its name [|◄◄ Sections 5.4 and 5.6].

Worked Examples 8.1 and 8.2 let you practice writing and balancing chemical equations.

Worked Example 8.1

Write and balance the chemical equation for the aqueous reaction of barium hydroxide and perchloric acid to produce aqueous barium perchlorate and water.

Strategy Determine the formulas and physical states of all reactants and products, and use them to write a chemical equation that makes the correct chemical statement. Finally, adjust coefficients in the resulting chemical equation to ensure that there are identical numbers of each type of atom on both sides of the reaction arrow.

Setup The reactants are $Ba(OH)_2$ and $HClO_4$, and the products are $Ba(ClO_4)_2$ and H_2O. Because the reaction is aqueous, all species except H_2O will be labeled (*aq*) in the equation. Being a liquid, H_2O will be labeled (*l*).

Solution The chemical statement "barium hydroxide and perchloric acid react to produce barium perchlorate and water" can be represented with the following unbalanced equation.

$$Ba(OH)_2(aq) + HClO_4(aq) \longrightarrow Ba(ClO_4)_2(aq) + H_2O(l)$$

Perchlorate ions (ClO_4^-) appear on both sides of the equation, so count them as units, rather than count the individual atoms they contain. Thus, the tally of atoms and polyatomic ions is

$$1 — Ba — 1$$
$$2 — O — 1 \text{ (not including O atoms in } ClO_4^- \text{ ions)}$$
$$3 — H — 2$$
$$1 — ClO_4^- — 2$$

The barium atoms are already balanced, and placing a coefficient of 2 in front of $HClO_4(aq)$ balances the number of perchlorate ions.

$$Ba(OH)_2(aq) + 2HClO_4(aq) \longrightarrow Ba(ClO_4)_2(aq) + H_2O(l)$$

$$1 — Ba — 1$$
$$2 — O — 1 \text{ (not including O atoms in } ClO_4^- \text{ ions)}$$
$$4 — H — 2$$
$$2 — ClO_4^- — 2$$

Placing a coefficient of 2 in front of $H_2O(l)$ balances both the O and H atoms, giving us the final balanced equation.

$$Ba(OH)_2(aq) + 2HClO_4(aq) \longrightarrow Ba(ClO_4)_2(aq) + 2H_2O(l)$$

$$1 — Ba — 1$$
$$2 — O — 2 \text{ (not including O atoms in } ClO_4^- \text{ ions)}$$
$$4 — H — 4$$
$$2 — ClO_4^- — 2$$

Think About It
Check to be sure the equation is balanced by counting all the atoms individually.

$$1 — Ba — 1$$
$$10 — O — 10$$
$$4 — H — 4$$
$$2 — Cl — 2$$

Practice Problem **A**TTEMPT Write and balance the chemical equation that represents the combustion of propane (i.e., the reaction of propane gas with oxygen gas to produce carbon dioxide gas and liquid water).

Practice Problem **B**UILD Write and balance the chemical equation that represents the reaction of sulfuric acid with sodium hydroxide to form water and sodium sulfate.

Practice Problem **C**ONCEPTUALIZE Write a balanced equation for the reaction shown here.

Worked Example 8.2

Butyric acid (also known as butanoic acid, $C_4H_8O_2$) is one of many compounds found in milk fat. First isolated from rancid butter in 1869, butyric acid has received a great deal of attention in recent years as a potential anticancer agent. Write a balanced equation for the metabolism of butyric acid. Assume that the overall process of metabolism and combustion are the same (i.e., reaction with oxygen to produce carbon dioxide and water).

Strategy Begin by writing an unbalanced equation to represent the combination of reactants and formation of products as stated in the problem, and then balance the equation.

Butyric acid

(Continued on next page)

Setup Metabolism in this context refers to the combination of $C_4H_8O_2$ with O_2 to produce CO_2 and H_2O.

Solution

$$C_4H_8O_2(aq) + O_2(g) \longrightarrow CO_2(g) + H_2O(l)$$

Balance the number of C atoms by changing the coefficient for CO_2 from 1 to 4.

$$C_4H_8O_2(aq) + O_2(g) \longrightarrow 4CO_2(g) + H_2O(l)$$

Balance the number of H atoms by changing the coefficient for H_2O from 1 to 4.

$$C_4H_8O_2(aq) + O_2(g) \longrightarrow 4CO_2(g) + 4H_2O(l)$$

Finally, balance the number of O atoms by changing the coefficient for O_2 from 1 to 5.

$$C_4H_8O_2(aq) + 5O_2(g) \longrightarrow 4CO_2(g) + 4H_2O(l)$$

Think About It

Count the number of each type of atom on each side of the reaction arrow to verify that the equation is properly balanced. There are 4 C, 8 H, and 12 O in the reactants and in the products, so the equation is balanced.

Practice Problem **A**TTEMPT Another compound found in milk fat that appears to have anticancer and anti-obesity properties is conjugated linoleic acid (CLA; $C_{18}H_{32}O_2$). Assuming again that the only products are CO_2 and H_2O, write a balanced equation for the metabolism of CLA.

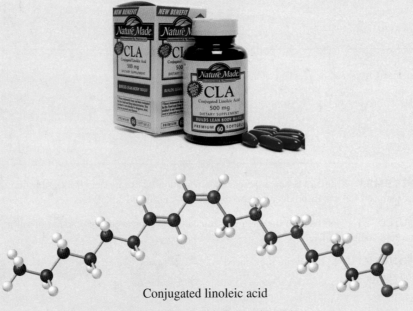

Conjugated linoleic acid

© McGraw-Hill Education/John Flournoy, photographer

Practice Problem **B**UILD Write a balanced equation for the combination of ammonia gas with solid copper(II) oxide to form copper metal, nitrogen gas, and liquid water.

Practice Problem **C**ONCEPTUALIZE The compound shown on the left reacts with nitrogen dioxide to form the compound shown on the right and iodine. Write a balanced equation for the reaction.

Methyl iodide   Nitromethane

Patterns of Chemical Reactivity

As you continue to study chemistry, you will encounter a wide variety of chemical reactions. The sheer number of different reactions can seem daunting at times, but most of them fall into a relatively small number of categories. Becoming familiar with several reaction types and learning to recognize *patterns* of reactivity will help you make sense out of the reactions in this book. Three of the most commonly encountered reaction types are *combination, decomposition,* and *combustion.*

Combination

A reaction in which two or more reactants combine to form a single product is known as a ***combination reaction.*** Examples include the reaction of ammonia and hydrogen chloride to form ammonium chloride,

$$NH_3(g) + HCl(g) \longrightarrow NH_4Cl(s)$$

and the reaction of nitrogen and hydrogen gases to form ammonia,

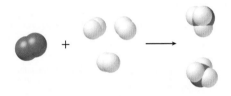

$$N_2(g) + 3H_2(g) \longrightarrow 2NH_3(g)$$

Decomposition

A reaction in which two or more products form from a single reactant is known as a ***decomposition reaction.*** A decomposition reaction is essentially the opposite of a combination reaction. Examples of this type of reaction include the decomposition of calcium carbonate to produce calcium oxide and carbon dioxide gas,

$$CaCO_3(s) \xrightarrow{\Delta} CaO(s) + CO_2(g)$$

and the decomposition of hydrogen peroxide to produce water and oxygen gas,

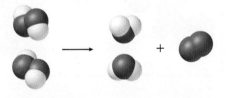

$$2H_2O_2(aq) \longrightarrow 2H_2O(l) + O_2(g)$$

Combustion

A ***combustion reaction*** is one in which a substance burns in the presence of oxygen. Combustion of a compound that contains C and H (or C, H, and O) produces carbon dioxide gas and water. By convention, we will consider the water produced in a combustion reaction to be *liquid* water. Examples of this type of combustion are the combustion of formaldehyde,

$$CH_2O(l) + O_2(g) \longrightarrow CO_2(g) + H_2O(l)$$

and the combustion of methane,

$$CH_4(g) + 2O_2(g) \longrightarrow CO_2(g) + 2H_2O(l)$$

Although these combustion reactions are shown here as balanced equations, oxygen is generally supplied in excess in such processes to ensure complete combustion. Worked Example 8.3 lets you practice classifying reactions.

Worked Example 8.3

Determine whether each of the following equations represents a combination reaction, a decomposition reaction, or a combustion reaction: (a) $H_2(g) + Br_2(g) \longrightarrow 2HBr(g)$, (b) $2HCO_2H(l) + O_2(g) \longrightarrow 2CO_2(g) + 2H_2O(l)$, (c) $2KClO_3(s) \longrightarrow 2KCl(s) + 3O_2(g)$.

Strategy Look at the reactants and products in each balanced equation to see if two or more reactants combine into one product (a combination reaction), if one reactant splits into two or more products (a decomposition reaction), or if the main products formed are carbon dioxide gas and water (a combustion reaction).

Setup The equation in part (a) depicts *two reactants* and *one product*. The equation in part (b) represents a combination of a compound containing C, H, and O—with O_2—to produce CO_2 and H_2O. The equation in part (c) represents *two products* being formed from a *single reactant*.

Solution These equations represent (a) a combination reaction, (b) a combustion reaction, and (c) a decomposition reaction.

Think About It

Make sure that a reaction identified as a combination has only one product [as in part (a)], a reaction identified as a combustion consumes O_2 and produces CO_2 and H_2O [as in part (b)], and a reaction identified as a decomposition has only one reactant [as in part (c)].

Student Annotation: Combustion of a compound that contains elements other than C, H, and O will produce other products. For example, combustion of a compound containing sulfur will produce SO_2.

Practice Problem **A**TTEMPT Identify each of the following as a combination, decomposition, or combustion reaction: (a) $C_2H_4O_2(l) + 2O_2(g) \longrightarrow 2CO_2(g) + 2H_2O(l)$, (b) $2Na(s) + Cl_2(g) \longrightarrow 2NaCl(s)$, (c) $2NaH(s) \longrightarrow 2Na(s) + H_2(g)$.

Practice Problem **B**UILD Using the chemical species A_2, B, and AB, write a balanced equation for a combination reaction.

Practice Problem **C**ONCEPTUALIZE Each of the diagrams represents a reaction mixture before and after a chemical reaction. Identify each of the reactions shown as combination, decomposition, or combustion.

before	after
(i)	

before	after
(ii)	

Section 8.1 Review

Chemical Equations

8.1.1   What are the stoichiometric coefficients in the following equation when it is balanced?

$$CH_4(g) + H_2O(g) \longrightarrow H_2(g) + CO_2(g)$$

(a) 1, 2, 2, 2   (d) 2, 2, 2, 1

(b) 2, 1, 1, 2   (e) 1, 2, 4, 1

(c) 1, 2, 2, 1

8.1.2   Which is the correctly balanced form of the given equation?

$$S(s) + O_3(g) \longrightarrow SO_2(g)$$

(a) $S(s) + O_3(g) \rightarrow SO_2(g)$ (d) $3S(s) + 2O_3(g) \rightarrow SO_3(g)$

(b) $3S(s) + 6O_3(g) \rightarrow 3SO_3(g)$ (e) $3S(s) + 2O_3(g) \rightarrow 3SO_3(g)$

(c) $3S(s) + O_3(g) \rightarrow 3SO_3(g)$

8.1.3   Write a balanced chemical equation for the reaction represented here.

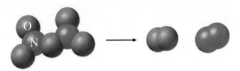

    (a) $N_2O_5(g) \rightarrow N_2(g) + O_2(g)$ (d) $2N_2O_5(g) \rightarrow 2N_2(g) + O_2(g)$
    (b) $2N_2O_5(g) \rightarrow 2N_2(g) + 5O_2(g)$ (e) $N_2O_5(g) \rightarrow 2N_2(g) + 5O_2(g)$
    (c) $2N_2O_5(g) \rightarrow N_2(g) + 5O_2(g)$

8.1.4   Which chemical equation represents the reaction shown?

    (a) $6N(g) + 18H(g) \rightarrow 6NH_3(g)$ (d) $N_2(g) + 3H_2(g) \rightarrow 2NH_3(g)$
    (b) $3N_2(g) + 3H_2(g) \rightarrow 2NH_3(g)$ (e) $3N_2(g) + 6H_2(g) \rightarrow 6NH_3(g)$
    (c) $2N_2(g) + 3H_2(g) \rightarrow 2NH_3(g)$

8.1.5   Identify the type of reaction shown.

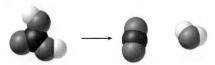

    (a) combination   (d) ionization
    (b) decomposition   (e) hybridization
    (c) combustion

8.2   COMBUSTION ANALYSIS

As we saw in Section 5.5, knowing the mass of each element contained in a sample of a substance enables us to determine the empirical formula of the substance. One common, practical use of this ability is the experimental determination of empirical formula by *combustion analysis.*

    Combustion analysis of organic compounds (containing carbon, hydrogen, and sometimes oxygen) is carried out using an apparatus like the one shown in Figure 8.1. A sample of known mass is placed in the furnace and heated in the presence of oxygen.

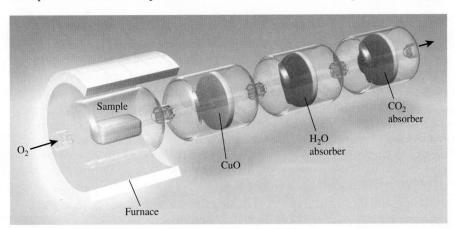

Figure 8.1 Schematic of a combustion analysis apparatus. CO_2 and H_2O produced in combustion are trapped and weighed. The amounts of these products are used to determine how much carbon and hydrogen the combusted sample contained. (CuO is used to ensure complete combustion of all carbon to CO_2.)

The carbon dioxide and water produced from carbon and hydrogen, respectively, in the combustion reaction are collected in "traps," which are weighed before and after the combustion. The difference in mass of each trap before and after the reaction is the mass of the collected product. Knowing the mass of each product, we can determine the percent composition of the compound. And, from percent composition, we can determine the empirical formula.

Determination of Empirical Formula

When a compound such as glucose is burned in a combustion analysis apparatus, carbon dioxide (CO_2) and water (H_2O) are produced. Because only oxygen gas is added to the reaction, the carbon and hydrogen present in the products must have come from the glucose. The oxygen in the products may have come from the glucose, but it may also have come from the added oxygen. Suppose that in one such experiment the combustion of 18.8 g of glucose produced 27.6 g of CO_2 and 11.3 g of H_2O. We can calculate the mass of carbon and hydrogen in the original 18.8-g sample of glucose as follows:

$$\text{mass of C} = 27.6 \text{ g } CO_2 \times \frac{1 \text{ mol } CO_2}{44.01 \text{ g } CO_2} \times \frac{1 \text{ mol } C}{1 \text{ mol } CO_2} \times \frac{12.01 \text{ g C}}{1 \text{ mol } C} = 7.53 \text{ g C}$$

$$\text{mass of H} = 11.3 \text{ g } H_2O \times \frac{1 \text{ mol } H_2O}{18.02 \text{ g } H_2O} \times \frac{2 \text{ mol } H}{1 \text{ mol } H_2O} \times \frac{1.008 \text{ g H}}{1 \text{ mol } H} = 1.26 \text{ g H}$$

Thus, 18.8 g of glucose contains 7.53 g of carbon and 1.26 g of hydrogen. The remaining mass [18.8 g − (7.53 g + 1.26 g) = 10.0 g] is oxygen.

The number of moles of each element present in 18.8 g of glucose is

$$\text{moles of C} = 7.53 \text{ g } C \times \frac{1 \text{ mol C}}{12.01 \text{ g } C} = 0.627 \text{ mol C}$$

$$\text{moles of H} = 1.26 \text{ g } H \times \frac{1 \text{ mol H}}{1.008 \text{ g } H} = 1.25 \text{ mol H}$$

$$\text{moles of O} = 10.0 \text{ g } O \times \frac{1 \text{ mol O}}{16.00 \text{ g } O} = 0.626 \text{ mol O}$$

> **Student Hot Spot**
>
> Student data indicate you may struggle with combustion analysis. Access the SmartBook to view additional Learning Resources on this topic.

> **Student Annotation:** Determination of an empirical formula from combustion data can be especially sensitive to rounding error. When solving problems such as these, don't round until the very end.

The empirical formula of glucose can therefore be written $C_{0.627}H_{1.25}O_{0.626}$. Because the numbers in an empirical formula must be integers, we divide each of the subscripts by the smallest subscript, 0.626. Thus, we have $0.627/0.626 \approx 1$, $1.25/0.626 \approx 2$, and $0.626/0.626 = 1$; and we obtain CH_2O for the empirical formula. Using the method discussed in Chapter 5 [◀◀ Section 5.10], we can determine the molecular formula of this compound if we know the molar mass.

Worked Example 8.4 shows how to determine the molecular formula of a compound from its combustion data and molar mass.

Worked Example 8.4

Combustion of a 5.50-g sample of benzene produces 18.59 g CO_2 and 3.81 g H_2O. Determine the empirical formula and the molecular formula of benzene, given that its molar mass is approximately 78 g/mol.

Strategy From the product masses, determine the mass of C and the mass of H in the 5.50-g sample of benzene. Sum the masses of C and H; the difference between this sum and the original sample mass is the mass of O in the sample (if O is in fact present in benzene). Convert the mass of each element to moles, and use the results as subscripts in a chemical formula. Convert the subscripts to whole numbers by dividing each by the smallest subscript. This gives the empirical formula. To calculate the molecular formula, first divide the molar mass given in the problem statement by the empirical formula mass. Then, multiply the subscripts in the empirical formula by the resulting number to obtain the subscripts in the molecular formula.

Setup The necessary molar masses are CO_2, 44.01 g/mol; H_2O, 18.02 g/mol; C, 12.01 g/mol; H, 1.008 g/mol; and O, 16.00 g/mol.

Solution We calculate the mass of carbon and the mass of hydrogen in the products (and therefore in the original 5.50-g sample) as follows:

$$\text{mass of C} = 18.59 \text{ g CO}_2 \times \frac{1 \text{ mol CO}_2}{44.01 \text{ g CO}_2} \times \frac{1 \text{ mol C}}{1 \text{ mol CO}_2} \times \frac{12.01 \text{ g C}}{1 \text{ mol C}} = 5.073 \text{ g C}$$

$$\text{mass of H} = 3.81 \text{ g H}_2\text{O} \times \frac{1 \text{ mol H}_2\text{O}}{18.02 \text{ g H}_2\text{O}} \times \frac{2 \text{ mol H}}{1 \text{ mol H}_2\text{O}} \times \frac{1.008 \text{ g H}}{1 \text{ mol H}} = 0.426 \text{ g H}$$

The total mass of products is 5.073 g + 0.426 g = 5.499 g. Because the combined masses of C and H account for the entire mass of the original sample (5.499 g ≈ 5.50 g), this compound must not contain O.

Converting mass to moles for each element present in the compound,

$$\text{moles of C} = 5.073 \text{ g C} \times \frac{1 \text{ mol C}}{12.01 \text{ g C}} = 0.4224 \text{ mol C}$$

$$\text{moles of H} = 0.426 \text{ g H} \times \frac{1 \text{ mol H}}{1.008 \text{ g H}} = 0.423 \text{ mol H}$$

gives the formula $C_{0.4224}H_{0.423}$. Converting the subscripts to whole numbers (0.4224/0.4224 = 1; 0.423/0.4224 ≈ 1) gives the empirical formula CH.

Finally, dividing the approximate molar mass (78 g/mol) by the empirical formula mass (12.01 g/mol + 1.008 g/mol = 13.02 g/mol) gives 78/13.02 ≈ 6. Then, multiplying both subscripts in the empirical formula by 6 gives the molecular formula C_6H_6.

Think About It
Use the molecular formula to determine the molar mass and make sure that the result agrees with the molar mass given in the problem. For C_6H_6, the molar mass is 6(12.01 g/mol) + 6(1.008 g/mol) = 78.11 g/mol, which agrees with the 78 g/mol given in the problem statement.

Practice Problem **A**TTEMPT The combustion of a 28.1-g sample of ascorbic acid (vitamin C) produces 42.1 g CO_2 and 11.5 g H_2O. Determine the empirical and molecular formulas of ascorbic acid. The molar mass of ascorbic acid is approximately 176 g/mol.

Practice Problem **B**UILD Determine the masses of products in the combustion of 11.85 g of a compound with empirical formula $C_4H_{10}O$.

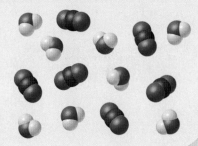

Practice Problem **C**ONCEPTUALIZE The models here represent the products of a combustion analysis experiment. Determine the empirical formula of the compound being analyzed if it (a) contains only carbon and hydrogen and (b) if it contains carbon, hydrogen, and oxygen and has a molar mass of approximately 60 g/mol. (c) Explain how combustion analysis of two different compounds can produce the same products in the same amounts.

Section 8.2 Review

Combustion Analysis

8.2.1 Determine the masses of CO_2 and H_2O produced by the combustion of 0.986 g of a compound with empirical formula $C_3H_6O_2$.
(a) 1.76 g CO_2, 0.719 g H_2O (d) 0.329 g CO_2, 0.657 g H_2O
(b) 0.480 g CO_2, 0.081 g H_2O (e) 0.657 g CO_2, 0.329 g H_2O
(c) 1.76 g CO_2, 1.44 g H_2O

8.2.2 What is the empirical formula of a compound containing C, H, and O if combustion of 1.23 g of the compound yields 1.8 g CO_2 and 0.74 g H_2O?
(a) CH_3O (c) CHO   (e) CH_2O
(b) C_2H_3O (d) $C_2H_3O_2$

8.2.3 What are the empirical and molecular formulas of a hydrocarbon if combustion of 2.10 g of the compound yields 6.59 g CO_2 and 2.70 g H_2O and its molar mass is about 84 g/mol?
(a) CH, C_5H_5 (c) CH_2, C_6H_{12} (e) CH_2, C_4H_8
(b) CH, C_6H_6 (d) CH_2, C_5H_{10}

8.3   CALCULATIONS WITH BALANCED CHEMICAL EQUATIONS

Often we would like to predict how much of a particular product will form from a given amount of a reactant. Other times, we perform an experiment, measure the amount of product formed, and use this information to deduce the quantity or composition of a reactant. Balanced chemical equations can be powerful tools for this type of problem solving.

Moles of Reactants and Products

Based on the equation for the reaction of carbon monoxide with oxygen to produce carbon dioxide,

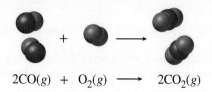

$$2CO(g) + O_2(g) \longrightarrow 2CO_2(g)$$

Student Annotation: When reactants are combined in exactly the mole ratio specified by the balanced chemical equation, they are said to be combined *in stoichiometric amounts*.

2 moles of CO combine with 1 mole of O_2 to produce 2 moles of CO_2. In stoichiometric calculations, we say that 2 moles of CO are *equivalent* to 2 moles of CO_2, which can be represented as

$$2 \text{ mol CO} \simeq 2 \text{ mol } CO_2$$

where the symbol $\simeq$ means "is stoichiometrically equivalent to" or simply "is equivalent to." The ratio of moles of CO consumed to moles of CO_2 produced is 2:2 or 1:1. Regardless of the number of moles of CO consumed in the reaction, the same number of moles of CO_2 will be produced. We can use this constant ratio as a conversion factor that can be written as

$$\frac{2 \text{ mol CO}}{2 \text{ mol } CO_2} \quad \text{or} \quad \frac{1 \text{ mol CO}}{1 \text{ mol } CO_2}$$

The ratio can also be written as the reciprocal,

$$\frac{2 \text{ mol } CO_2}{2 \text{ mol CO}} \quad \text{or} \quad \frac{1 \text{ mol } CO_2}{1 \text{ mol CO}}$$

These conversion factors enable us to determine how many moles of CO_2 will be produced upon reaction of a given amount of CO, or how much CO is necessary to produce a specific amount of CO_2. Consider the complete reaction of 3.82 moles of CO to form CO_2. To calculate the number of moles of CO_2 produced, we use the conversion factor with moles of CO_2 in the numerator and moles of CO in the denominator.

$$\text{moles } CO_2 \text{ produced} = 3.82 \text{ mol CO} \times \frac{1 \text{ mol } CO_2}{1 \text{ mol CO}} = 3.82 \text{ mol } CO_2$$

Similarly, we can use other ratios represented in the balanced equation as conversion factors. For example, we have 1 mol $O_2 \simeq 2$ mol CO_2 and 2 mol CO $\simeq 1$ mol O_2. The corresponding conversion factors allow us to calculate the amount of CO_2 produced upon reaction of a given amount of O_2, and the amount of one reactant necessary to react completely with a given amount of the other. Using the preceding example, we can determine the ***stoichiometric amount*** of O_2 (the exact number of moles of O_2 needed to react with 3.82 moles of CO).

$$\text{moles } O_2 \text{ needed} = 3.82 \text{ mol CO} \times \frac{1 \text{ mol } O_2}{2 \text{ mol CO}} = 1.91 \text{ mol } O_2$$

Worked Example 8.5 illustrates how to determine reactant and product amounts using a balanced chemical equation.

Worked Example 8.5

Urea [$(NH_2)_2CO$] is a by-product of protein metabolism. This waste product is formed in the liver and then filtered from the blood and excreted in the urine by the kidneys. Urea can be synthesized in the laboratory by the combination of ammonia and carbon dioxide according to the equation

(a) Calculate the amount of urea that will be produced by the complete reaction of 5.25 moles of ammonia. (b) Determine the stoichiometric amount of carbon dioxide required to react with 5.25 moles of ammonia.

$$2NH_3(g) + CO_2(g) \longrightarrow (NH_2)_2CO(aq) + H_2O(l)$$

Strategy Use the balanced chemical equation to determine the correct stoichiometric conversion factors, and then multiply by the number of moles of ammonia given.

Setup According to the balanced chemical equation, the conversion factor for ammonia and urea is either

$$\frac{2 \text{ mol } NH_3}{1 \text{ mol } (NH_2)_2CO} \quad \text{or} \quad \frac{1 \text{ mol } (NH_2)_2CO}{2 \text{ mol } NH_3}$$

To multiply by moles of NH_3 and have the units cancel properly, we use the conversion factor with moles of NH_3 in the denominator. Similarly, the conversion factor for ammonia and carbon dioxide can be written as

$$\frac{2 \text{ mol } NH_3}{1 \text{ mol } CO_2} \quad \text{or} \quad \frac{1 \text{ mol } CO_2}{2 \text{ mol } NH_3}$$

Again, we select the conversion factor with ammonia in the denominator so that moles of NH_3 will cancel in the calculation.

Solution

(a) moles $(NH_2)_2CO$ produced $= 5.25 \text{ mol } NH_3 \times \dfrac{1 \text{ mol } (NH_2)_2CO}{2 \text{ mol } NH_3} = 2.63 \text{ mol } (NH_2)_2CO$

(b) moles CO_2 produced $= 5.25 \text{ mol } NH_3 \times \dfrac{1 \text{ mol } CO_2}{2 \text{ mol } NH_3} = 2.63 \text{ mol } CO_2$

Think About It

As always, check to be sure that units cancel properly in the calculation. Also, the balanced equation indicates that there will be *fewer* moles of urea produced than ammonia consumed. Therefore, your calculated number of moles of urea (2.63) should be *smaller* than the number of moles given in the problem (5.25). Similarly, the stoichiometric coefficients in the balanced equation are the same for carbon dioxide and urea, so your answers to this problem should also be the same for both species.

Practice Problem (A)TTEMPT Nitrogen and hydrogen react to form ammonia according to the following balanced equation: $N_2(g) + 3H_2(g) \longrightarrow 2NH_3(g)$. Calculate the number of moles of hydrogen required to react with 0.0880 mole of nitrogen, and the number of moles of ammonia that will form.

Practice Problem (B)UILD Tetraphosphorus decoxide (P_4O_{10}) reacts with water to produce phosphoric acid. Write a balanced equation for this reaction and determine the number of moles of each reactant required to produce 5.80 moles of phosphoric acid.

Practice Problem (C)ONCEPTUALIZE The models represent the reaction of nitric acid with tin metal to form metastannic acid (H_2SnO_3), water, and nitrogen dioxide. Determine how many moles of nitric acid must react to produce 8.75 mol H_2SnO_3. (Don't forget to balance the equation.)

Mass of Reactants and Products

Balanced chemical equations give us the relative amounts of reactants and products in terms of moles. However, because we measure reactants and products in the laboratory by weighing them, most often such calculations start with mass rather than the number of moles.

Worked Example 8.6 illustrates how to determine amounts of reactants and products in terms of grams.

Worked Example 8.6

Dinitrogen monoxide (N_2O), also known as *nitrous oxide* or "laughing gas," is used as an anesthetic in dentistry. It is manufactured by heating ammonium nitrate. The balanced equation is

$$NH_4NO_3(s) \xrightarrow{\Delta} N_2O(g) + 2H_2O(l)$$

(a) Calculate the mass of ammonium nitrate that must be heated in order to produce 10.0 g of nitrous oxide. (b) Determine the corresponding mass of water produced in the reaction.

Strategy For part (a), use the molar mass of nitrous oxide to convert the given mass of nitrous oxide to moles, use the appropriate stoichiometric conversion factor to convert to moles of ammonium nitrate, and then use the molar mass of ammonium nitrate to convert to grams of ammonium nitrate. For part (b), use the molar mass of nitrous oxide to convert the given mass of nitrous oxide to moles, use the stoichiometric conversion factor to convert from moles of nitrous oxide to moles of water, and then use the molar mass of water to convert to grams of water.

Setup The molar masses are as follows: 80.05 g/mol for NH_4NO_3, 44.02 g/mol for N_2O, and 18.02 g/mol for H_2O. The conversion factors from nitrous oxide to ammonium nitrate and from nitrous oxide to water are, respectively,

$$\frac{1 \text{ mol } NH_4NO_3}{1 \text{ mol } N_2O} \quad \text{and} \quad \frac{2 \text{ mol } H_2O}{1 \text{ mol } N_2O}$$

Solution

(a) $10.0 \text{ g } N_2O \times \dfrac{1 \text{ mol } N_2O}{44.02 \text{ g } N_2O} = 0.227 \text{ mol } N_2O$

$0.227 \text{ mol } N_2O \times \dfrac{1 \text{ mol } NH_4NO_3}{1 \text{ mol } N_2O} = 0.227 \text{ mol } NH_4NO_3$

$0.227 \text{ mol } NH_4NO_3 \times \dfrac{80.05 \text{ g } NH_4NO_3}{1 \text{ mol } NH_4NO_3} = 18.2 \text{ g } NH_4NO_3$

Thus, 18.2 g of ammonium nitrate must be heated in order to produce 10.0 g of nitrous oxide.

(b) Starting with the number of moles of nitrous oxide determined in the first step of part (a),

$$0.227 \text{ mol } N_2O \times \frac{2 \text{ mol } H_2O}{1 \text{ mol } N_2O} = 0.454 \text{ mol } H_2O$$

$$0.454 \text{ mol } H_2O \times \frac{18.02 \text{ g } H_2O}{1 \text{ mol } H_2O} = 8.18 \text{ g } H_2O$$

Therefore, 8.18 g of water will also be produced in the reaction.

Think About It
Use the law of conservation of mass to check your answers. Make sure that the combined mass of both products is equal to the mass of reactant you determined in part (a). In this case (rounded to the appropriate number of significant figures), 10.0 g + 8.18 g = 18.2 g. Remember that small differences may arise as the result of rounding.

Practice Problem Ａ TTEMPT Calculate the mass of water produced by the metabolism of 56.8 g of glucose ($C_6H_{12}O_6$). (Remember that the chemical equation for metabolism is the same as the equation for combustion.)

Practice Problem Ｂ UILD What mass of glucose must be metabolized in order to produce 175 g of water?

Practice Problem Ｃ ONCEPTUALIZE The models here represent the reaction of nitrogen dioxide with water to form nitrogen monoxide and nitric acid. What mass of nitrogen dioxide must react for 100.0 g HNO_3 to be produced? (Don't forget to balance the equation.)

Section 8.3 Review

Calculations with Balanced Chemical Equations

8.3.1 How many moles of LiOH will be produced if 0.550 mol Li reacts according to the following equation?

$$2Li(s) + 2H_2O(l) \longrightarrow 2LiOH(aq) + H_2(g)$$

(a) 0.550 mol
(b) 1.10 mol
(c) 0.275 mol

(d) 2.20 mol
(e) 2.00 mol

8.3.2 What mass of lithium nitride is produced when 75.0 g of lithium metal react with nitrogen according to the following equation?

$$6Li(s) + N_2(g) \longrightarrow 2Li_3N(s)$$

(a) 63 g
(b) 125 g
(c) 376 g

(d) 753 g
(e) 1130 g

8.3.3 What mass of iron(III) oxide must react with carbon monoxide to produce 124.5 kg iron metal according to the following equation?

$$Fe_2O_3(s) + 3CO(g) \longrightarrow 2Fe(s) + 3CO_2(g)$$

(a) 356.0 kg
(b) 222.4 kg
(c) 178.0 kg

(d) 712.0 kg
(e) 0.1780 kg

8.3.4 Determine the stoichiometric amount (in grams) of O_2 necessary to react with 5.71 g Al according to the following equation:

$$4Al(s) + 3O_2(g) \longrightarrow 2Al_2O_3(s)$$

(a) 5.08 g
(b) 9.03 g
(c) 2.54 g

(d) 4.28 g
(e) 7.61 g

8.4 LIMITING REACTANTS

When a chemist carries out a reaction, the reactants usually are not present in stoichiometric amounts. Because the goal of a reaction is usually to produce the maximum quantity of a useful compound from the starting materials, an excess of one reactant is commonly supplied to ensure that the more expensive or more important reactant is converted completely to the desired product. Consequently, some of the reactant supplied in excess will be left over at the end of the reaction. The reactant used up first in a reaction is called the ***limiting reactant,*** because the amount of this reactant *limits* the amount of product that can form. When all the limiting reactant has been consumed, no more product can be formed. ***Excess reactants*** are those present in quantities *greater* than necessary to react with the quantity of the limiting reactant.

The concept of a limiting reactant applies to everyday tasks, too, such as making beef and mushroom kabobs. Suppose you want to make the maximum number of kabobs possible, each of which will consist of 1 skewer, 4 mushrooms, and 3 pieces of beef. If you have 4 skewers, 12 mushrooms, and 15 pieces of beef, how many kabobs can you make? The answer, as Figure 8.2 illustrates, is 3. After making 3 kabobs, your supply of mushrooms will be exhausted. Although you will still have 1 skewer and 6 pieces of beef remaining, you will not have the necessary ingredients to make any more kabobs according to the recipe. In this example, because the mushrooms will run out

Video 8.1
Limiting reactant in reaction of NO and O_2.

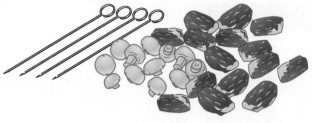

4 skewers, 12 mushrooms, 15 pieces of beef
(a)

3 assembled kabobs, 1 skewer, and 6 pieces of beef
(b)

Figure 8.2 Part (a) 4 skewers, 12 mushrooms, 15 pieces of beef, (b) 3 assembled kabobs, 1 skewer, and 6 pieces of beef. The number of mushrooms limits the number of kabobs that can be assembled according to the recipe.

first, mushrooms are the limiting "reactant." The total amount of product, in this case the total number of kabobs, is limited by the amount of one ingredient.

Determining the Limiting Reactant

In problems involving limiting reactants, the first step is to determine which is the limiting reactant. After the limiting reactant has been identified, the rest of the problem can be solved using the approach outlined in Section 8.3. Consider the formation of methanol (CH_3OH) from carbon monoxide and hydrogen.

$$CO(g) + 2H_2(g) \longrightarrow CH_3OH(l)$$

Suppose that initially we have 5 moles of CO and 8 moles of H_2, the ratio shown in Figure 8.3(a). We can use the stoichiometric conversion factors to determine how many moles of H_2 are necessary for all the CO to react. From the balanced equation, we have 1 mol CO ≏ 2 mol H_2. Therefore, the amount of H_2 necessary to react with 5 mol CO is

$$\text{moles of } H_2 = 5 \text{ mol CO} \times \frac{2 \text{ mol } H_2}{1 \text{ mol CO}} = 10 \text{ mol } H_2$$

Because there are only 8 moles of H_2 available, there is insufficient H_2 to react with all the CO. Therefore, H_2 is the limiting reactant and CO is the excess reactant. H_2 will be used up first, and when it is gone, the formation of methanol will cease and there will be some CO left over, as shown in Figure 8.3(b). To determine how much CO will be left over when the reaction is complete, we must first calculate the amount of CO that will react with all 8 moles of H_2.

$$\text{moles of } CO = 8 \text{ mol } H_2 \times \frac{1 \text{ mol CO}}{2 \text{ mol } H_2} = 4 \text{ mol CO}$$

Thus, there will be 4 moles of CO consumed and 1 mole (5 mol − 4 mol) left over.

Worked Example 8.7 illustrates how to combine the concept of a limiting reactant with the conversion between mass and moles. Figure 8.4 (pp. 328–329) illustrates the steps for this type of calculation.

Figure 8.3 The reaction of (a) H_2 and CO to form (b) CH_3OH. Each molecule represents 1 mole of substance. In this case, H_2 is the limiting reactant and there is 1 mole of CO remaining when the reaction is complete.

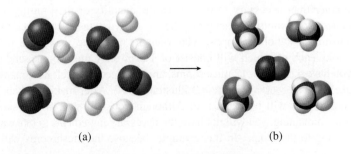

(a)   (b)

Worked Example 8.7

Alka-Seltzer® tablets contain aspirin, sodium bicarbonate, and citric acid. When they come into contact with water, the sodium bicarbonate ($NaHCO_3$) and citric acid ($H_3C_6H_5O_7$) react to form carbon dioxide gas, among other products.

$$3NaHCO_3(aq) + H_3C_6H_5O_7(aq) \longrightarrow 3CO_2(g) + 3H_2O(l) + Na_3C_6H_5O_7(aq)$$

The formation of CO_2 causes the trademark fizzing when the tablets are dropped into a glass of water. An Alka-Seltzer tablet contains 1.700 g of sodium bicarbonate and 1.000 g of citric acid. Determine, for a single tablet dissolved in water, (a) which ingredient is the limiting reactant, (b) what mass of the excess reactant is left over when the reaction is complete, and (c) what mass of CO_2 forms.

Strategy Convert each of the reactant masses to moles. Use the balanced equation to write the necessary stoichiometric conversion factor and determine which reactant is limiting. Again, using the balanced equation, write the stoichiometric conversion factors to determine the number of moles of excess reactant remaining and the number of moles of CO_2 produced. Finally, use the appropriate molar masses to convert moles of excess reactant and moles of CO_2 to grams.

Setup The required molar masses are 84.01 g/mol for $NaHCO_3$, 192.12 g/mol for $H_3C_6H_5O_7$, and 44.01 g/mol for CO_2. From the balanced equation we have 3 mol $NaHCO_3 \simeq 1$ mol $H_3C_6H_5O_7$, 3 mol $NaHCO_3 \simeq 3$ mol CO_2, and 1 mol $H_3C_6H_5O_7 \simeq 3$ mol CO_2. The necessary stoichiometric conversion factors are, therefore,

$$\frac{3 \text{ mol } NaHCO_3}{1 \text{ mol } H_3C_6H_5O_7} \quad \frac{1 \text{ mol } H_3C_6H_5O_7}{3 \text{ mol } NaHCO_3} \quad \frac{3 \text{ mol } CO_2}{3 \text{ mol } NaHCO_3} \quad \frac{3 \text{ mol } CO_2}{1 \text{ mol } NaHCO_3}$$

Solution

$$1.700 \text{ g } NaHCO_3 \times \frac{1 \text{ mol } NaHCO_3}{84.01 \text{ g } NaHCO_3} = 0.02024 \text{ mol } NaHCO_3$$

$$1.000 \text{ g } H_3C_6H_5O_7 \times \frac{1 \text{ mol } H_3C_6H_5O_7}{192.12 \text{ g } H_3C_6H_5O_7} = 0.005205 \text{ mol } H_3C_6H_5O_7$$

(a) To determine which reactant is limiting, calculate the amount of citric acid necessary to react completely with 0.02024 mol sodium bicarbonate.

$$0.02024 \text{ mol } NaHCO_3 \times \frac{1 \text{ mol } H_3C_6H_5O_7}{3 \text{ mol } NaHCO_3} = 0.006745 \text{ mol } H_3C_6H_5O_7$$

The amount of $H_3C_6H_5O_7$ required to react with 0.02024 mol of $NaHCO_3$ is more than a tablet contains. Therefore, citric acid is the limiting reactant and sodium bicarbonate is the excess reactant.

(b) To determine the mass of excess reactant ($NaHCO_3$) left over, first calculate the amount of $NaHCO_3$ that will react.

$$0.005205 \text{ mol } H_3C_6H_5O_7 \times \frac{3 \text{ mol } NaHCO_3}{1 \text{ mol } H_3C_6H_5O_7} = 0.01562 \text{ mol } NaHCO_3$$

Thus, 0.01562 mol of $NaHCO_3$ will be consumed, leaving 0.00462 mol unreacted. Convert the unreacted amount to grams.

$$0.00462 \text{ mol } NaHCO_3 \times \frac{84.01 \text{ g } NaHCO_3}{1 \text{ mol } NaHCO_3} = 0.388 \text{ g } NaHCO_3$$

(c) To determine the mass of CO_2 produced, first calculate the number of moles of CO_2 produced from the number of moles of limiting reactant ($H_3C_6H_5O_7$) consumed.

$$0.005205 \text{ mol } H_3C_6H_5O_7 \times \frac{3 \text{ mol } CO_2}{1 \text{ mol } H_3C_6H_5O_7} = 0.01562 \text{ mol } CO_2$$

Convert this amount to grams.

$$0.01562 \text{ mol } CO_2 \times \frac{44.01 \text{ g } CO_2}{1 \text{ mol } CO_2} = 0.6874 \text{ g } CO_2$$

To summarize the results: (a) citric acid is the limiting reactant, (b) 0.388 g sodium bicarbonate remains unreacted, and (c) 0.6874 g carbon dioxide is produced.

(Continued on next page)

Think About It

In a problem such as this, it is a good idea to check your work by calculating the amounts of the other products in the reaction. According to the law of conservation of mass, the combined starting mass of the two reactants (1.700 g + 1.000 g = 2.700 g) should equal the sum of the masses of products and leftover excess reactant. In this case, the masses of H_2O and $Na_3C_6H_5O_7$ produced are 0.2815 g and 1.343 g, respectively. The mass of CO_2 produced is 0.6874 g [from part (c)] and the amount of excess $NaHCO_3$ is 0.388 g [from part (b)]. The total, 0.2815 g + 1.343 g + 0.6874 g + 0.388 g, is 2.700 g, identical to the total mass of the reactants.

Practice Problem **A**TTEMPT Ammonia is produced by the reaction of nitrogen and hydrogen according to the equation $N_2(g) + 3H_2(g) \longrightarrow 2NH_3(g)$. Calculate the mass of ammonia produced when 35.0 g of nitrogen react with 12.5 g of hydrogen. Which is the excess reactant and how much of it will be left over when the reaction is complete?

Practice Problem **B**UILD Potassium hydroxide and phosphoric acid react to form potassium phosphate and water according to the equation $3KOH(aq) + H_3PO_4(aq) \longrightarrow K_3PO_4(aq) + 3H_2O(l)$. Determine the starting mass of each reactant if 55.7 g K_3PO_4 are produced and 89.8 g H_3PO_4 remain unreacted.

Practice Problem **C**ONCEPTUALIZE The diagrams show a reaction mixture before and after a chemical reaction. Write the balanced equation for the reaction and identify the limiting reactant.

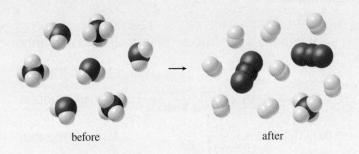

before   after

Reaction Yield

When you use stoichiometry to calculate the amount of product formed in a reaction, you are calculating the *theoretical yield* of the reaction. The theoretical yield is the amount of product that forms when *all* the limiting reactant reacts to form the desired product. It is the *maximum* obtainable yield, predicted by the balanced equation. In practice, the *actual yield*—the amount of product actually obtained from a reaction— is almost always less than the theoretical yield. There are many reasons for the difference between the actual and theoretical yields. For instance, some of the reactants may not react to form the desired product. They may react to form different products, in something known as *side reactions,* or they may simply remain unreacted. In addition, it may be difficult to isolate and recover all the product at the end of the reaction. Chemists often determine the efficiency of a chemical reaction by calculating its *percent yield,* which tells *what percentage the actual yield is of the theoretical yield.* It is calculated as follows:

Equation 8.1 $$\% \text{ yield} = \frac{\text{actual yield}}{\text{theoretical yield}} \times 100\%$$

Percent yields may range from a tiny fraction to 100 percent. (They cannot exceed 100 percent.) Chemists try to maximize percent yield in a variety of ways. Factors that can affect percent yield, including temperature and pressure, are discussed in Chapter 15.

Worked Example 8.8 shows how to calculate the percent yield of a pharmaceutical manufacturing process.

Worked Example 8.8

Aspirin, acetylsalicylic acid ($C_9H_8O_4$), is the most commonly used pain reliever in the world. It is produced by the reaction of salicylic acid ($C_7H_6O_3$) and acetic anhydride ($C_4H_6O_3$) according to the following equation.

$$C_7H_6O_3 + C_4H_6O_3 \longrightarrow C_9H_8O_4 + HC_2H_3O_2$$

salicylic acid   acetic anhydride   acetylsalicylic acid   acetic acid

In a certain aspirin synthesis, 104.8 g of salicylic acid and 110.9 g of acetic anhydride are combined. Calculate the percent yield of the reaction if 105.6 g of aspirin are produced.

Strategy Convert reactant grams to moles, and determine which is the limiting reactant. Use the balanced equation to determine the number of moles of aspirin that can be produced, and convert this number of moles to grams for the theoretical yield. Use the actual yield (given in the problem) and the calculated theoretical yield to calculate the percent yield.

Setup The necessary molar masses are 138.12 g/mol for salicylic acid, 102.09 g/mol for acetic anhydride, and 180.15 g/mol for aspirin.

Solution

$$104.8 \text{ g } C_7H_6O_3 \times \frac{1 \text{ mol } C_7H_6O_3}{138.12 \text{ g } C_7H_6O_3} = 0.7588 \text{ mol } C_7H_6O_3$$

$$110.9 \text{ g } C_4H_6O_3 \times \frac{1 \text{ mol } C_4H_6O_3}{102.09 \text{ g } C_4H_6O_3} = 1.086 \text{ mol } C_4H_6O_3$$

Because the two reactants combine in a 1:1 mole ratio, the reactant present in the smallest number of moles (in this case, salicylic acid) is the limiting reactant. According to the balanced equation, one mole of aspirin is produced for every mole of salicylic acid consumed.

$$1 \text{ mol salicylic acid } (C_7H_6O_3) \simeq 1 \text{ mol aspirin } (C_9H_8O_4)$$

Therefore, the theoretical yield of aspirin is 0.7588 mol. We convert this to grams using the molar mass of aspirin:

$$0.7588 \text{ mol } C_9H_8O_4 \times \frac{180.15 \text{ g } C_9H_8O_4}{1 \text{ mol } C_9H_8O_4} = 136.7 \text{ g } C_9H_8O_4$$

Thus, the theoretical yield is 136.7 g. If the actual yield is 105.6 g, the percent yield is

$$\% \text{ yield} = \frac{105.6 \text{ g}}{136.7 \text{ g}} \times 100\% = 77.25\% \text{ yield}$$

Think About It
Make sure you have used the proper molar masses and remember that percent yield can never exceed 100 percent.

Practice Problem ATTEMPT Diethyl ether is produced from ethanol according to the following equation:

$$2CH_3CH_2OH(l) \longrightarrow CH_3CH_2OCH_2CH_3(l) + H_2O(l)$$

Calculate the percent yield if 68.6 g of ethanol react to produce 16.1 g of ether.

Practice Problem BUILD What mass of ether will be produced if 207 g of ethanol react with a 73.2 percent yield?

Practice Problem CONCEPTUALIZE The diagrams show a mixture of reactants and the mixture of recovered products for an experiment using the chemical reaction introduced in Practice Problem 8.7C. Identify the limiting reactant and determine the percent yield of carbon dioxide.

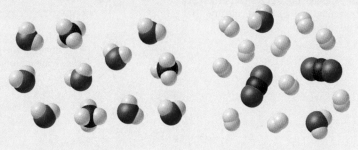

starting material   recovered products

Figure 8.4

Limiting Reactant Problems

 START

Determine what mass of NH_3 forms when 84.06 g N_2 and 22.18 g H_2 react according to the equation:

$$N_2 + 3H_2 \longrightarrow 2NH_3$$

Convert to moles.

$$\frac{84.06 \text{ g } N_2}{28.02 \text{ g/mol}} = 3.000 \text{ mol } N_2$$

$$\frac{22.18 \text{ g } H_2}{2.016 \text{ g/mol}} = 11.00 \text{ mol } H_2$$

Determine moles NH_3.

Total mass before reaction:

$$84.06 \text{ g } N_2 + 22.18 \text{ g } H_2 = 106.24 \text{ g}$$

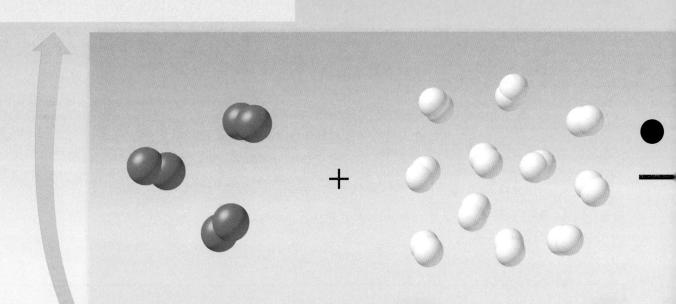

Compare the total mass *after* the reaction with the total mass *before* the reaction. The small difference between the masses before and after is due to rounding.

$$\sum \text{after reaction} = 102.2 \text{ g } NH_3 + 4.03 \text{ g } H_2 = 106.2 \text{ g}$$

Add the mass of the product and the mass of leftover excess reactant to get the total mass after reaction.

Use coefficients as conversion factors.

Method 1

$$3.000 \; \cancel{\text{mol N}_2} \times \frac{2 \; \text{mol NH}_3}{1 \; \cancel{\text{mol N}_2}} = 6.000 \; \text{mol NH}_3$$

$$11.00 \; \cancel{\text{mol H}_2} \times \frac{2 \; \text{mol NH}_3}{3 \; \cancel{\text{mol H}_2}} = 7.333 \; \text{mol NH}_3$$

or

Method 2

$$3.000 \; \text{N}_2 + 9.000 \; \text{H}_2 \longrightarrow 6.000 \; \text{NH}_3$$

$$3.667 \; \text{N}_2 + 11.00 \; \text{H}_2 \longrightarrow 7.333 \; \text{NH}_3$$

Rewrite the balanced equation using actual amounts. According to the balanced equation, 3.667 mol N_2 are required to react with 11.00 mol H_2.

Either way, the *smaller* amount of product is correct.

6.000 mol NH_3

Convert to grams.

$$6.000 \; \cancel{\text{mol NH}_3} \times \frac{17.03 \; \text{g NH}_3}{1 \; \cancel{\text{mol NH}_3}} = 102.2 \; \text{g NH}_3$$

CHECK

N_2 was the limiting reactant. Calculate how much H_2 is left over.

$$2.00 \; \cancel{\text{mol H}_2} \times \frac{2.016 \; \text{g H}_2}{1 \; \cancel{\text{mol H}_2}} = 4.03 \; \text{g H}_2$$

Convert to grams.

11.00 mol initially
− 9.00 mol consumed
2.00 mol H_2 remaining

What's the point?

There is more than one correct method for solving many types of problems. This limiting reactant problem shows two different routes to the correct answer, and shows how the result can be compared to the information given in the problem to determine whether or not it is reasonable and correct.

Section 8.4 Review

Limiting Reactants

8.4.1 How many moles of NH_3 can be produced by the combination of 3.0 mol N_2 and 1.5 mol H_2?

(a) 2.0 mol   (d) 6.0 mol

(b) 1.5 mol   (e) 1.0 mol

(c) 0.50 mol

8.4.2 What mass of $CaSO_4$ is produced according to the given equation when 5.00 g of each reactant are combined?

$$CaF_2(s) + H_2SO_4(aq) \longrightarrow CaSO_4(s) + 2HF(g)$$

(a) 10.0 g   (d) 8.72 g

(b) 11.6 g   (e) 5.02 g

(c) 6.94 g

8.4.3 What is the percent yield for a process in which 10.4 g CH_3OH react and 10.1 g CO_2 form according to the following equation?

$$2CH_3OH(l) + 3O_2(g) \longrightarrow 2CO_2(g) + 4H_2O(l)$$

(a) 97.1%   (d) 103%

(b) 70.7%   (e) 37.9%

(c) 52.1%

8.4.4 What mass of water is produced by the reaction of 50.0 g CH_3OH with an excess of O_2 when the yield is 53.2 percent?

$$2CH_3OH(l) + 3O_2(g) \longrightarrow 2CO_2(g) + 4H_2O(l)$$

(a) 28.1 g   (d) 15.0 g

(b) 56.2 g   (e) 26.6 g

(c) 29.9 g

8.4.5 Reactants A (red) and B (blue) combine to form a single product C according to the equation $2A + B \longrightarrow C$. What is the limiting reactant in the reaction vessel shown?

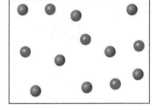

(a) A

(b) B

(c) C

(d) None. Reactants are present in stoichiometric amounts.

Thinking Outside the Box

Atom Economy

Many of the reactions carried out by chemists produce multiple products—not all of which are useful. The consequence is that even a reaction that proceeds quantitatively and without side reactions can generate a significant amount of chemical waste. Often the processes necessary to dispose of such waste add significantly to the cost of production. Although percent yield is an excellent measure of the overall efficiency of a chemical reaction, it does not indicate how efficiently the reactants are converted into *useful* products.

    In the 1990s, as a part of the focus on green chemistry, the concept of *atom economy* was developed by Barry Trost, a professor of chemistry at Stanford University. Atom economy is a measure of how much of the starting mass of reactants ends up in the final mass of desired products in a chemical reaction. The atom economy of a reaction can be determined according to the equation

Student Annotation: A reaction is said to *proceed quantitatively* if its percent yield is very close to 100 percent.

Student Annotation: In green chemistry, also called *sustainable* chemistry, chemical processes are designed to minimize the use and production of hazardous substances. [http://www.epa.gov/greenchemistry/]

Equation 8.2   Atom economy $= \dfrac{\text{Mass of atoms in desired product(s)}}{\text{Mass of atoms in all reactants}} \times 100\%$

Note how this approach is different from a percent yield analysis. In fact, according to this formula, the atom economy of a reaction can be low even if the percent yield is very high. Consider the following example. Historically, phenol (C_6H_5OH) has been manufactured by combining benzene (C_6H_6) with sulfuric acid (H_2SO_4) and sodium hydroxide (NaOH).

The atom economy of this process is

$$\frac{\text{Mass of } C_6H_5OH}{\text{Mass of } C_6H_6, H_2SO_4, \text{ and NaOH}} \times 100\% = \frac{94.11 \text{ g}}{256.19 \text{ g}} \times 100\% = 36.73\%$$

Today, the manufacture of phenol is done by combining benzene with propene (C_3H_6) and oxygen.

In terms of phenol production alone, the atom economy of this process is

$$\frac{\text{Mass of } C_6H_5OH}{\text{Mass of } C_6H_6, C_3H_6, \text{ and } O_2} \times 100\% = \frac{94.11 \text{ g}}{152.19 \text{ g}} \times 100\% = 61.84\%$$

But when we consider that the other product of the reaction, acetone, is a substance with commercial value and is easily recovered from this process, the atom economy rises to 100 percent. Atom economy encourages chemists to find better, less expensive, and more sustainable methods of synthesizing important compounds.

Another striking example is the nonsteroidal anti-inflammatory drug ibuprofen. Beginning in the 1960s, ibuprofen was manufactured using a six-step industrial process that generated a considerable amount of chemical waste and had an atom economy of just over 40 percent. By the 1990s, one manufacturer, Hoechst Celanese, had developed a more efficient synthesis of ibuprofen. The new method, known as the BHC method, costs less, has only three steps, produces far less waste, and has an atom economy of over 77 percent. Moreover, if we consider that one of the other products in the BHC method, acetic acid, can be recovered and ultimately used in other industrial processes, the atom economy is over 99 percent.

8.5 PERIODIC TRENDS IN REACTIVITY OF THE MAIN GROUP ELEMENTS

Ionization energy and electron affinity [◄◄ Section 4.4] enable us to understand the types of reactions that elements undergo and the types of compounds they are likely to form. These two parameters actually measure similar things. Ionization energy is a measure of how powerfully an atom attracts its own electrons, while electron affinity is a measure of how powerfully an atom can attract electrons from another source. As a very simple example of how this helps us to understand a chemical reaction, consider the combination of a sodium atom and a chlorine atom, shown in Figure 8.5.

Student Annotation: Recall that ionization energy and electron affinity refer specifically to atoms and ions in the *gas phase* [◄◄ Section 4.4].

Figure 8.5 Formation of NaCl from its constituent elements. Note that although the charges are not all shown, the solid consists of a three-dimensional array of alternating oppositely charged ions.

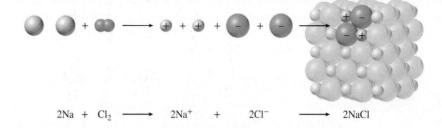

$$2Na \;+\; Cl_2 \;\longrightarrow\; 2Na^+ \;+\; 2Cl^- \;\longrightarrow\; 2NaCl$$

Video 8.2
Periodic table—properties of alkali and alkaline earth metals.

Sodium, with its low ionization energy, has a relatively weak attraction for its one valence electron. Chlorine, with its energetically favorable electron affinity, has the ability to attract electrons from another source. In this case, the electron that is loosely held by the Na atom, and powerfully attracted by the Cl atom, is transferred from Na to Cl, thus producing a sodium ion (Na^+) and a chloride ion (Cl^-). According to Coulomb's law, oppositely charged objects attract each other. The positively charged sodium ion and the negatively charged chloride ion are drawn together by electrostatic attraction, and the result is the formation of the solid ionic compound sodium chloride (NaCl).

General Trends in Reactivity

Before we examine the reactivity of elements in individual groups, let's identify some overall trends. We have said that elements in the same group resemble one another in chemical behavior because they have similar valence electron configurations. This statement, although correct in the general sense, must be applied with caution. Chemists have long known that the properties of the first member of each group (Li, Be, B, C, N, O, and F) are different from those of the rest of the members of the same group. Lithium, for example, exhibits many, but not all, of the properties characteristic of the Group 1A (alkali) metals. For example, unlike the other Group 1A elements, Li reacts with the O_2 and N_2 in air to form a simple oxide (Li_2O) and nitride (Li_3N), respectively. Similarly, beryllium is a somewhat atypical member of Group 2A (alkaline earth metals) in that it forms covalent compounds, and so on. The differences can be attributed to the unusually small size of the first element in each group [◄◄ Figure 4.6, page 133].

Another trend in the chemical reactivity of main group elements is the *diagonal relationship*. Diagonal relationships refer to similarities between pairs of elements in different groups and periods of the periodic table [◄◄ Section 4.4]. Not only do the first three members of the second period (Li, Be, and B) exhibit chemical properties similar to those of the elements located diagonally below them in the periodic table (Mg, Al, and Si), they also show similar patterns in reactivity. For example, as noted in Section Review Question 8.3.2 (page 323), lithium reacts with N_2 to form lithium nitride (Li_3N). Magnesium reacts in an analogous way to form magnesium nitride (Mg_3N_2).

$$6Li(s) \;+\; N_2(g) \;\longrightarrow\; 2Li_3N(s)$$
$$3Mg(s) \;+\; N_2(g) \;\xrightarrow{\Delta}\; Mg_3N_2(s)$$

The reaction of magnesium with nitrogen is shown in Figure 8.6. Lithium is the only member of Group 1A to exhibit this reactivity toward nitrogen.

A comparison of the properties of elements in the same group is most valid if the elements in question have a similar metallic (or nonmetallic) character. The elements in Groups 1A and 2A, for example, are all metals, whereas those in Groups 7A and 8A are all nonmetals. We have to be more careful when comparing the elements of Groups 3A through 6A, though, because a single group may contain metals, metalloids, and nonmetals. In these groups, we should expect a greater variation in chemical properties even though all group members have similar valence electron configurations.

Figure 8.6 Magnesium burning in pure nitrogen.
© Richard Megna/Fundamental Photographs

Hydrogen ($1s^1$)

Although it is the simplest and most abundant element in the universe, there is no completely suitable position for hydrogen in the periodic table (it really belongs in a group by itself). Traditionally hydrogen is shown at the top of Group 1A, because, like the alkali metals, it has a single s valence electron and forms a cation with a charge of +1 (H^+). On the other hand, hydrogen also forms the *hydride* ion (H^-) in ionic compounds such as NaH and CaH_2. In this respect, hydrogen resembles the members of Group 7A (halogens), all of which form −1 anions (F^-, Cl^-, Br^-, and I^-) in ionic compounds. Ionic hydrides react with water to produce hydrogen gas and the corresponding metal hydroxides.

$$2NaH(s) + 2H_2O(l) \longrightarrow 2NaOH(aq) + 2H_2(g)$$

$$CaH_2(s) + 2H_2O(l) \longrightarrow Ca(OH)_2(aq) + 2H_2(g)$$

Arguably, the most important compound of hydrogen is water, which forms when hydrogen burns in air.

$$2H_2(g) + O_2(g) \longrightarrow 2H_2O(l)$$

Reactions of the Active Metals

Group 1A Elements (ns^1, $n \geq 2$)

These elements all have low ionization energies, making it easy for them to become M^+ ions. In fact, these metals are so reactive that they are never found in nature in the pure elemental state. They react with water to produce hydrogen gas and the corresponding metal hydroxide (Figure 8.7),

$$2M(s) + 2H_2O(l) \longrightarrow 2MOH(aq) + H_2(g)$$

where M denotes an alkali metal. When exposed to air, they gradually lose their shiny appearance as they react with oxygen to form metal oxides. Lithium forms lithium oxide (containing the oxide ion, O^{2-}).

$$4Li(s) + O_2(g) \longrightarrow 2Li_2O(s)$$

(a)　　　　　　　　　　　　　(b)　　　　　　　　　　　　　(c)

Figure 8.7 Alkali metals reacting with water: (a) sodium, (b) potassium, and (c) cesium. The reaction becomes increasingly violent as ionization energy of the metal decreases down the group.

Figure 8.8 Barium reacting with water.

© McGraw-Hill Education/Charles D. Winters, photographer

Student Annotation: Because they have less metallic character than the other Group 2A elements, beryllium and magnesium form some molecular compounds such as BeH$_2$ and MgH$_2$.

The other alkali metals all form oxides or *peroxides* (containing the peroxide O$_2^{2-}$).

$$2Na(s) + O_2(g) \longrightarrow Na_2O_2(s)$$

Potassium, rubidium, and cesium also form *superoxides* (containing the superoxide ion, O$_2^-$).

$$K(s) + O_2(g) \longrightarrow KO_2(s)$$

The type of oxide that forms when an alkali metal reacts with oxygen has to do with the stability of the various oxides. Because these oxides are all ionic compounds, their stability depends on how strongly the cations and anions attract one another. Lithium tends to form predominantly the oxide because lithium oxide is more stable than lithium peroxide.

Group 2A Elements (ns^2, $n \geq 2$)

As a group, the alkaline earth metals are somewhat less reactive than the alkali metals. Both the first and the second ionization energies decrease (and metallic character increases) from beryllium to barium. Group 2A elements tend to form M^{2+} ions, where M denotes an alkaline earth metal atom.

The reactions of alkaline earth metals with water vary considerably. Beryllium does not react with water; magnesium reacts slowly with steam; and calcium, strontium, and barium react vigorously with cold water (Figure 8.8).

$$Ca(s) + 2H_2O(l) \longrightarrow Ca(OH)_2(aq) + H_2(g)$$

$$Sr(s) + 2H_2O(l) \longrightarrow Sr(OH)_2(aq) + H_2(g)$$

$$Ba(s) + 2H_2O(l) \longrightarrow Ba(OH)_2(aq) + H_2(g)$$

The reactivity of the alkaline earth metals toward oxygen also increases from Be to Ba. Beryllium and magnesium form oxides (BeO and MgO) only at elevated temperatures, whereas CaO, SrO, and BaO form at room temperature.

Magnesium reacts with aqueous acid [represented by H$^+$(aq)] to produce hydrogen gas.

$$Mg(s) + 2H^+(aq) \longrightarrow Mg^{2+}(aq) + H_2(g)$$

Calcium, strontium, and barium also react with aqueous acid solutions to produce hydrogen gas; however, because the metals also react with water, the two different reactions (with H$^+$ and with H$_2$O) occur simultaneously.

Reactions of Other Main Group Elements

Group 3A Elements (ns^2np^1, $n \geq 2$)

Boron, the first member of the group, is a metalloid; the others (Al, Ga, In, and Tl) are metals. Boron does not form binary ionic compounds and is unreactive toward both oxygen and water. Aluminum, the next element in the group, readily forms aluminum oxide when exposed to air.

$$4Al(s) + 3O_2(g) \longrightarrow 2Al_2O_3(s)$$

The aluminum oxide forms a protective coating, preventing the underlying metal from reacting further (Figure 8.9). This fact makes it possible to use aluminum for structural materials, such as aluminum siding and the shells of airplanes. Without the protective coating, layer after layer of Al atoms would become oxidized, and the structure would eventually crumble.

Aluminum forms the Al^{3+} ion. It reacts with acid according to the equation

$$2Al(s) + 6H^+(aq) \longrightarrow 2Al^{3+}(aq) + 3H_2(g)$$

The other Group 3A metals (Ga, In, and Tl) can form both M^+ and M^{3+} ions. As we move down the group, the M^+ ion becomes the more stable of the two.

The metallic elements in Group 3A also form many molecular compounds. For example, aluminum reacts with hydrogen to form AlH_3, which has properties similar to those of BeH_2. The progression of properties across the second row of the periodic table illustrates the gradual shift from metallic to nonmetallic character in the main group elements.

Group 4A Elements (ns^2np^2, $n \geq 2$)

Carbon, the first member of the group, is a nonmetal, whereas silicon and germanium, the next two members, are metalloids. Tin and lead, the last two members of the group, are metals. They do not react with water, but they do react with aqueous acid to produce hydrogen gas.

$$Sn(s) + 2H^+(aq) \longrightarrow Sn^{2+}(aq) + H_2(g)$$

$$Pb(s) + 2H^+(aq) \longrightarrow Pb^{2+}(aq) + H_2(g)$$

The Group 4A elements form compounds containing both +2 and +4 cations. For carbon and silicon, the +4 ion is the more stable one. For example, CO_2 is more stable than CO—although the C in CO_2 is not really a C^{4+} ion. Similarly, SiO_2 is a stable compound, but SiO does not exist under ordinary conditions. As we move down the group, however, the relative stability of the two ions is reversed. In tin compounds, the +4 ion is only slightly more stable than the +2 ion. In lead compounds, the +2 ion is the more stable one. The outer electron configuration of lead is $6s^26p^2$, and lead tends to lose only the $6p$ electrons to form Pb^{2+} rather than both the $6p$ and $6s$ electrons to form Pb^{4+}.

Group 5A Elements (ns^2np^3, $n \geq 2$)

Nitrogen and phosphorus are *nonmetals,* arsenic and antimony are *metalloids,* and bismuth is a *metal.* Like Group 4A, Group 5A contains elements in all three categories. Thus, we expect significant variation in chemical properties among the group members.

Elemental nitrogen is a diatomic gas (N_2). It forms a variety of oxides (NO, N_2O, NO_2, N_2O_4, and N_2O_5), all of which are gases except for N_2O_5, which is a solid at room temperature. Nitrogen has a tendency to accept three electrons to form the nitride ion (N^{3-}). Most metal nitrides, such as Li_3N and Mg_3N_2, are ionic compounds. Phosphorus exists as individual P_4 molecules (white phosphorus) or chains of P_4 molecules (red phosphorus). It forms two solid oxides with the formulas P_4O_6 and P_4O_{10}. The industrially important oxoacids—nitric acid and phosphoric acid—form when N_2O_5 and P_4O_{10}, respectively, react with water (Figure 8.10).

$$N_2O_5(s) + H_2O(l) \longrightarrow 2HNO_3(aq)$$

$$P_4O_{10}(s) + 6H_2O(l) \longrightarrow 4H_3PO_4(aq)$$

Arsenic, antimony, and bismuth have extensive three-dimensional structures. Bismuth is far less reactive than metals in the preceding groups.

Group 6A Elements (ns^2np^4, $n \geq 2$)

The first three members of the group (oxygen, sulfur, and selenium) are nonmetals, whereas the last two (tellurium and polonium) are metalloids. Oxygen is a colorless, odorless, diatomic gas; elemental sulfur and selenium exist as the molecules S_8 and Se_8, respectively; and tellurium and polonium have more extensive three-dimensional structures. (Polonium is a radioactive element that is difficult to study in the laboratory.) Oxygen has a tendency to accept two electrons to form the oxide ion (O^{2-}) in many compounds. Sulfur, selenium, and tellurium also form ions by accepting two electrons: S^{2-}, Se^{2-}, and Te^{2-}. The elements in Group 6A (especially oxygen) form a large number of molecular compounds with nonmetals. Some of the important

(a)

(b)

Figure 8.9 (a) Finely divided aluminum metal being sprinkled into a flame to form aluminum oxide (Al_2O_3). (b) A thin layer of aluminum oxide prevents the underlying metal from further oxidation.

(a) © McGraw-Hill Education/Charles D. Winters, photographer; (b) © McGraw-Hill Education/Ken Cavanagh, photographer

Student Annotation: Often in cases where the apparent charge on a cation is especially high (≥ 4), the "ion" is not really an *ion;* that is, it has achieved a stable electron configuration by *sharing* rather than by *losing* electrons [◄◄ Section 5.5]. In Chapter 9, we will learn a method for keeping track of the electrons in covalent species.

(a)   (b)

Figure 8.10 Addition of nonmetal oxides to water containing an acid-base indicator. In each case, the color change indicates that an acid is produced by the reaction. (a) $N_2O_5(s) + H_2O(l) \longrightarrow 2HNO_3(aq)$, (b) $P_4O_{10}(s) + 6H_2O(l) \longrightarrow 4H_3PO_4(aq)$.
© Richard Megna/Fundamental Photographs

compounds of sulfur are SO_2, SO_3, and H_2S. Sulfuric acid, an oxoacid, forms when sulfur trioxide reacts with water (Figure 8.11).

$$SO_3(g) + H_2O(l) \longrightarrow H_2SO_4(aq)$$

Group 7A Elements (ns^2np^5, $n \geq 2$)

All the halogens are nonmetals with the general formula X_2, where X denotes a halogen element. Like the Group 1A metals, the Group 7A nonmetals are too reactive to be found in nature in the elemental form. (Astatine, the last member of Group 7A, is radioactive. Very little is known about its properties.)

The halogens have high ionization energies and large, energetically favorable electron affinities. Anions derived from the halogens (F^-, Cl^-, Br^-, and I^-) are called *halides*. The vast majority of alkali metal halides are ionic compounds. The halogens also form many molecular compounds among themselves, such as ICl and BrF_3, and with nonmetals in other groups, such as NF_3, PCl_5, and SF_6. The halogens react with hydrogen to form hydrogen halides (Figure 8.12).

$$H_2(g) + X_2(g) \longrightarrow 2HX(g)$$

Figure 8.11 A forest damaged by acid rain.
© Will & Deni McIntyre/Corbis

Figure 8.12 Colorless H_2 gas reacts with purple I_2 to form colorless HI gas.

© David A. Tietz/Editorial Image, LLC

$$H_2(g) \quad + \quad I_2(g) \quad \longrightarrow \quad 2HI(g)$$

This reaction is explosive when it involves fluorine, but it becomes less and less violent as we substitute chlorine, bromine, and iodine. The hydrogen halides dissolve in water to form hydrohalic acids.

Group 8A Elements (ns^2np^6, $n \geq 2$)

All the noble gases exist as monatomic species. With the exception of helium, which has the electron configuration $1s^2$, their atoms have completely filled outer ns and np subshells. Their electron configurations give the noble gases their great stability. The Group 8A ionization energies are among the highest of all the elements; their electron affinities are all less than zero, so they have no tendency to accept extra electrons [◄◄ Chapter 4 Key Skills].

For years, the noble gases were called *inert gases* because they were not known to react with anything. Beginning in 1963, however, compounds were prepared from xenon and krypton by exposing them to very strong oxidizing agents such as fluorine and oxygen. In 2000, the first successful preparation of an argon compound, HArF, was reported—although it is stable only at temperatures below $-256°C$. Some of the more stable compounds that have been prepared are XeF_4, XeO_3, $XeOF_4$, and KrF_2 (Figure 8.13). Although the chemistry of the noble gases is interesting, the compounds of noble gases are not involved in any natural biological processes or major commercial applications.

Comparison of Group 1A and Group 1B Elements

Although the outer electron configurations of Groups 1A and 1B are similar (members of both groups have a single valence electron in an s orbital), their chemical properties are very different.

The first ionization energies of Cu, Ag, and Au are 745, 731 and 890 kJ/mol, respectively. Because these values are considerably larger than those of the alkali metals, the Group 1B elements are much less reactive. The higher ionization energies of the Group 1B elements result from incomplete shielding of the nucleus by the inner d electrons (compared with the more effective shielding by the completely filled noble gas cores). Consequently, the outer s electrons of the Group 1B elements are more strongly attracted by the nucleus. In fact, copper, silver, and gold are so unreactive that they are usually found in the uncombined state in nature. The inertness, rarity, and attractive appearance of these metals make them valuable in the manufacture of coins and jewelry. For this reason, these metals are also known as "coinage metals." Similar reasoning can be used to explain the differences in the chemistry of the elements in Group 2A from that of the elements in Group 2B.

Student Annotation: Note that the common ions formed by the other main group elements are those that make them isoelectronic with a noble gas. In Group 1A elements, for example, each atom *loses* one electron to become isoelectronic with the noble gas that immediately precedes it; in Group 7A elements, each atom *gains* one electron to become isoelectronic with the noble gas that immediately follows it; and so on.

Student Annotation: Because of the trend in ionization energy within a group [◄◄ Section 4.4], you might expect radon to react with fluorine more readily than krypton or xenon and, in fact, it *does*. However, primarily because of radon's radioactivity, very little is known about the compound that forms.

Figure 8.13 Xenon tetrafluoride (XeF_4) crystals.

Courtesy Argonne National Laboratory

Learning Outcomes

- Define chemical reaction, chemical equation, reactant, and product.
- Recognize physical states of reactants and products in a chemical equation.
- Balance chemical equations by changing stoichiometric coefficients.
- Describe the three commonly encountered reaction types (combustion, combination, and decomposition).
- Use combustion analysis to determine the empirical formula of a compound.

- Determine amounts of reactant required or product formed using stoichiometry.
- Identify the limiting reactant in a reaction.
- Define theoretical yield, actual yield, and percent yield.
- Predict the theoretical yield of a reaction.
- Find the percent yield of a reaction.
- Use the periodic table to predict general trends in reactivity of main group metals.

Chapter Summary

SECTION 8.1

- A *chemical equation* is a written representation of a *chemical reaction* or a physical process. Chemical species on the left side of the equation are called *reactants,* and those on the right side of the equation are called *products.*
- The physical state of each reactant and product is specified in parentheses as (*s*), (*l*), (*g*), or (*aq*) for *solid, liquid, gas,* and *aqueous* (dissolved in water), respectively.

- Chemical equations must be balanced to obey the *law of conservation of mass,* which is a restatement of the third hypothesis of Dalton's atomic theory. Matter is neither created nor destroyed by chemical reactions.
- Chemical equations are balanced only by changing the *stoichiometric coefficients* or simply *coefficients* of the reactants and/or products, and never by changing the formulas of the reactants and/or products (i.e., by changing their subscript numbers).

- Three commonly encountered reaction types are **combustion** (in which a substance burns in the presence of oxygen), **combination** (in which two or more reactants combine to form a single product), and **decomposition** (in which a reactant splits apart to form two or more products).

SECTION 8.2

- **Combustion analysis** is used to determine the empirical formula of a compound. The *empirical formula* can be used to calculate percent composition.

SECTION 8.3

- A balanced chemical equation can be used to determine how much product will form from given amounts of reactants, how much of one reactant is necessary to react with a given amount of another, or how much reactant is required to produce a specified amount of product. Reactants that are combined in exactly the ratio specified by the balanced equation are said to be "combined in **stoichiometric amounts.**"

SECTION 8.4

- The **limiting reactant** is the reactant that is consumed completely in a chemical reaction. An **excess reactant** is the reactant that is not consumed completely. The maximum amount of product that can form depends on the amount of limiting reactant.
- The **theoretical yield** of a reaction is the amount of product that will form if all the limiting reactant is consumed by the desired reaction. The **actual yield** is the amount of product actually recovered. **Percent yield** [(actual/theoretical) × 100%] is a measure of the efficiency of a chemical reaction.

SECTION 8.5

- Although members of a group in the periodic table exhibit similar chemical and physical properties, the first member of each group tends to be significantly different from the other members. Hydrogen is essentially a group unto itself.
- The alkali metals (Group 1A) tend to be highly reactive toward oxygen, water, and acid. Group 2A metals are less reactive than Group 1A metals, but the heavier members all react with water to produce metal hydroxides and hydrogen gas. Groups that contain both metals and nonmetals (e.g., Groups 4A, 5A, and 6A) tend to show greater variability in their physical and chemical properties.

Key Words

Key Equations

8.1	$\% \text{ yield} = \dfrac{\text{actual yield}}{\text{theoretical yield}} \times 100\%$	The amount of product actually produced in a reaction will nearly always be less than that predicted by the balanced equation. We use the actual (measured) amount of product and the calculated amount of product to determine the percent yield of a reaction.
8.2	$\text{Atom economy} = \dfrac{\text{Mass of atoms in desired product(s)}}{\text{Mass of atoms in all reactants}} \times 100\%$	The atom economy of a reaction measures how efficiently reactants are converted into a useful product (or products). It is determined by using the total mass of the desired product(s) divided by the total mass of all reactants multiplied by 100 percent.

Key Skills

Limiting Reactant

The amount of product that can be produced in a chemical reaction typically is limited by the amount of *one* of the reactants—known as the *limiting* reactant. The practice of identifying the limiting reactant, calculating the maximum possible amount of product, and determining the percent yield and remaining amount of an excess reactant requires several skills:

- Balancing chemical equations [◄◄ Section 8.1]
- Determining molar mass [◄◄ Key Skills Chapter 5]
- Converting mass to moles [◄◄ Section 5.10]
- Using stoichiometric conversion factors [◄◄ Section 8.3]
- Converting moles to mass [◄◄ Section 5.10]

Consider the following example. Hydrazine (N_2H_4) reacts with dinitrogen tetroxide (N_2O_4) to form nitrogen monoxide (NO) and water. Determine the mass of NO that can be produced when 10.45 g of N_2H_4 and 53.68 g of N_2O_4 are combined. The unbalanced equation is

$$N_2H_4 + N_2O_4 \longrightarrow NO + H_2O$$

We first balance the equation.

$$N_2H_4 + 2N_2O_4 \longrightarrow 6\,NO + 2H_2O$$

Next, we determine the necessary molar masses.

$$N_2H_4: 2(14.01) + 4(1.008) = \boxed{\dfrac{32.05\text{ g}}{\text{mol}}} \qquad N_2O_4: 2(14.01) + 4(16.00) = \boxed{\dfrac{92.02\text{ g}}{\text{mol}}} \qquad NO: 14.01 + 16.00 = \boxed{\dfrac{30.01\text{ g}}{\text{mol}}}$$

We convert the reactant masses given in the problem to moles. Then we determine the mole amount of NO that could be produced from the mole amount of each reactant by multiplying each of the *reactant* mole amounts by the appropriate stoichiometric conversion factor, which we derive from the balanced equation. According to the balanced equation,

$$1\text{ mol }N_2H_4 \simeq 6\text{ mol NO} \qquad \text{and} \qquad 2\text{ mol }N_2O_4 \simeq 6\text{ mol NO}$$

$$\dfrac{10.45\text{ g }N_2H_4}{32.05\text{ g/mol}} = 0.32605\text{ mol }N_2H_4 \qquad 0.32605\text{ mol }N_2H_4 \times \dfrac{6\text{ mol NO}}{1\text{ mol }N_2H_4} = 1.9563\text{ mol NO}$$

$$\dfrac{53.68\text{ g }N_2O_4}{92.02\text{ g/mol}} = 0.58335\text{ mol }N_2O_4 \qquad 0.58335\text{ mol }N_2O_4 \times \dfrac{6\text{ mol NO}}{2\text{ mol }N_2O_4} = 1.7501\text{ mol NO}$$

The reactant that produces the smaller amount of product is the limiting reactant; in this case, N_2O_4.

We continue the problem using the mole amount of NO produced by reaction of the given amount of N_2O_4. To convert from moles to mass (grams), we multiply the number of moles NO by the molar mass of NO.

$$1.7501 \text{ mol NO} \quad \times \quad \frac{30.01 \text{ g}}{\text{mol}} \quad = \quad 52.52 \text{ g NO}$$

Thus, 52.52 g NO can be produced by the reaction. Note that we retained an extra significant figure until the end of the calculation.

To determine the mass of remaining excess reactant, we must first determine what amount was consumed in the reaction. To do this, we multiply the mole amount of limiting reactant (N_2O_4) by the appropriate stoichiometric conversion factor. According to the balanced equation,

$$1 \text{ mol } N_2H_4 \rightleftharpoons 2 \text{ mol } N_2O_4$$

$$0.58335 \text{ mol } N_2O_4 \quad \times \quad \frac{1 \text{ mol } N_2H_4}{2 \text{ mol } N_2O_4} \quad = \quad 0.29168 \text{ mol } N_2H_4$$

This is the amount of N_2H_4 consumed. The amount remaining is the difference between this and the original amount. We convert the remaining mole amount to grams using the molar mass of N_2H_4.

$$0.32605 \text{ mol } N_2H_4 \quad - \quad 0.29168 \text{ mol } N_2H_4 \quad = \quad 0.03437 \text{ mol } N_2H_4$$

$$0.03437 \text{ mol } N_2H_4 \quad \times \quad \frac{32.05 \text{ g}}{\text{mol}} \quad = \quad 1.102 \text{ g } N_2H_4$$

Thus, 1.102 g N_2H_4 remain when the reaction is complete.

We can check our work in a problem such as this by also calculating the mass of the other product, in this case water. The mass of all products plus the mass of any remaining reactant must equal the sum of starting reactant masses.

Key Skills Problems

8.1
Calculate the mass of water produced in the previous example.

(a) 21.02 g (b) 10.51 g (c) 11.61 g (d) 11.75 g (e) 5.400 g

Use the following information to answer questions 8.2, 8.3, and 8.4.

Calcium phosphide (Ca_3P_2) and water react to form calcium hydroxide and phosphine (PH_3). In a particular experiment, 225.0 g Ca_3P_2 and 125.0 g water are combined.

$$Ca_3P_2(s) + H_2O(l) \longrightarrow Ca(OH)_2(aq) + PH_3(g)$$

(Don't forget to balance the equation.)

8.2
How much PH_3 can be produced?

(a) 350.0 g (b) 235.0 g (c) 78.59 g (d) 83.96 g (e) 41.98 g

8.3
How much $Ca(OH)_2$ can be produced?

(a) 91.51 g (b) 274.5 g (c) 513.8 g (d) 85.63 g (e) 257.0 g

8.4
How much of the excess reactant remains when the reaction is complete?

(a) 14.37 g (b) 235.0 g (c) 78.56 g (d) 83.96 g (e) 41.98 g

Questions and Problems

SECTION 8.1: CHEMICAL EQUATIONS

Review Questions

8.1 Use the formation of water from hydrogen and oxygen to explain the following terms: *chemical reaction, reactant,* and *product.*

8.2 What is the difference between a chemical reaction and a chemical equation?

8.3 Why must a chemical equation be balanced? What law is not obeyed by an unbalanced chemical equation?

Conceptual Problems

8.4 Write an unbalanced equation to represent each of the following reactions: (a) nitrogen and oxygen react to form nitrogen dioxide, (b) dinitrogen pentoxide reacts to form dinitrogen tetroxide and oxygen, (c) ozone reacts to form oxygen, (d) chlorine and sodium iodide react to form iodine and sodium chloride, and (e) magnesium and oxygen react to form magnesium oxide.

8.5 Write an unbalanced equation to represent each of the following reactions: (a) potassium hydroxide and phosphoric acid react to form potassium phosphate and water; (b) zinc and silver chloride react to form zinc chloride and silver; (c) sodium hydrogen carbonate reacts to form sodium carbonate, water, and carbon dioxide; (d) ammonium nitrite reacts to form nitrogen and water; and (e) carbon dioxide and potassium hydroxide react to form potassium carbonate and water.

8.6 For each of the following unbalanced chemical equations, write the corresponding chemical statement.
(a) $S_8 + O_2 \longrightarrow SO_2$
(b) $CH_4 + O_2 \longrightarrow CO_2 + H_2O$
(c) $N_2 + H_2 \longrightarrow NH_3$
(d) $P_4O_{10} + H_2O \longrightarrow H_3PO_4$
(e) $S + HNO_3 \longrightarrow H_2SO_4 + NO_2 + H_2O$

8.7 For each of the following unbalanced chemical equations, write the corresponding chemical statement.
(a) $K + H_2O \longrightarrow KOH + H_2$
(b) $Ba(OH)_2 + HCl \longrightarrow BaCl_2 + H_2O$
(c) $Cu + HNO_3 \longrightarrow Cu(NO_3)_2 + NO + H_2O$
(d) $Al + H_2SO_4 \longrightarrow Al_2(SO_4)_3 + H_2$
(e) $HI \longrightarrow H_2 + I_2$

8.8 Balance the following equations using the method outlined in Section 8.1.
(a) $C + O_2 \longrightarrow CO$
(b) $CO + O_2 \longrightarrow CO_2$
(c) $H_2 + Br_2 \longrightarrow HBr$
(d) $Rb + H_2O \longrightarrow RbOH + H_2$
(e) $Mg + O_2 \longrightarrow MgO$
(f) $O_3 \longrightarrow O_2$
(g) $H_2O_2 \longrightarrow H_2O + O_2$
(h) $N_2 + H_2 \longrightarrow NH_3$

(i) $Zn + AgCl \longrightarrow ZnCl_2 + Ag$
(j) $S_8 + O_2 \longrightarrow SO_2$
(k) $NaOH + H_2SO_4 \longrightarrow Na_2SO_4 + H_2O$
(l) $Cl_2 + NaI \longrightarrow NaCl + I_2$
(m) $KOH + H_3PO_4 \longrightarrow K_3PO_4 + H_2O$
(n) $CH_4 + Br_2 \longrightarrow CBr_4 + HBr$

8.9 Balance the following equations using the method outlined in Section 8.1.
(a) $N_2O_5 \longrightarrow N_2O_4 + O_2$
(b) $KNO_3 \longrightarrow KNO_2 + O_2$
(c) $NH_4NO_3 \longrightarrow N_2O + H_2O$
(d) $NH_4NO_2 \longrightarrow N_2 + H_2O$
(e) $NaHCO_3 \longrightarrow Na_2CO_3 + H_2O + CO_2$
(f) $P_4O_{10} + H_2O \longrightarrow H_3PO_4$
(g) $HCl + CaCO_3 \longrightarrow CaCl_2 + H_2O + CO_2$
(h) $Al + H_2SO_4 \longrightarrow Al_2(SO_4)_3 + H_2$
(i) $CO_2 + KOH \longrightarrow K_2CO_3 + H_2O$
(j) $CH_4 + O_2 \longrightarrow CO_2 + H_2O$
(k) $Be_2C + H_2O \longrightarrow Be(OH)_2 + CH_4$
(l) $Cu + HNO_3 \longrightarrow Cu(NO_3)_2 + NO + H_2O$
(m) $S + HNO_3 \longrightarrow H_2SO_4 + NO_2 + H_2O$
(n) $NH_3 + CuO \longrightarrow Cu + N_2 + H_2O$

8.10 Which of the following equations best represents the reaction shown in the diagram?
(a) $8A + 4B \longrightarrow C + D$
(b) $4A + 8B \longrightarrow 4C + 4D$
(c) $2A + B \longrightarrow C + D$
(d) $4A + 2B \longrightarrow 4C + 4D$
(e) $2A + 4B \longrightarrow C + D$

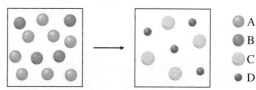

○ A
● B
● C
● D

8.11 Which of the following equations best represents the reaction shown in the diagram?
(a) $A + B \longrightarrow C + D$
(b) $6A + 4B \longrightarrow C + D$
(c) $A + 2B \longrightarrow 2C + D$
(d) $3A + 2B \longrightarrow 2C + D$
(e) $3A + 2B \longrightarrow 4C + 2D$

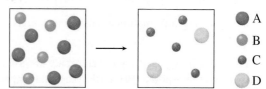

● A
● B
● C
○ D

8.12 Determine whether each of the following equations represents a combination reaction, a decomposition reaction, or a combustion reaction.
(a) $2NaHCO_3 \longrightarrow Na_2CO_3 + CO_2 + H_2O$
(b) $NH_3 + HCl \longrightarrow NH_4Cl$
(c) $2CH_3OH + 3O_2 \longrightarrow 2CO_2 + 4H_2O$

8.13 Determine whether each of the following equations represents a combination reaction, a decomposition reaction, or a combustion reaction.
(a) $C_3H_8 + 5O_2 \longrightarrow 3CO_2 + 4H_2O$
(b) $2NF_2 \longrightarrow N_2F_4$
(c) $CuSO_4 \cdot 5H_2O \longrightarrow CuSO_4 + 5H_2O$

SECTION 8.2: COMBUSTION ANALYSIS

Review Questions

8.14 Explain how the combined mass of CO_2 and H_2O produced in combustion analysis can be greater than the mass of the sample being analyzed.

8.15 Explain why, in combustion analysis, we cannot determine the amount of oxygen in the sample directly from the amount of oxygen in the products H_2O and CO_2.

Conceptual Problems

8.16 The diagram shows the products of a combustion analysis. Determine the empirical formula of the compound being analyzed if (a) it is a hydrocarbon and (b) it is a compound containing C, H, and O, and has a formula weight of approximately 92.

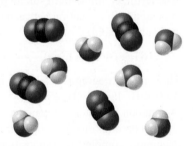

8.17 Which of the following diagrams could represent the products of combustion of a sample of
(a) acetylene (C_2H_2) and (b) ethylene (C_2H_4)?

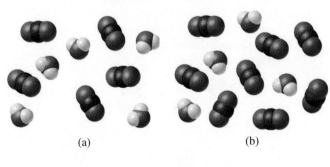

(a)   (b)

(c)

Computational Problems

8.18 Determine the empirical formula of an organic compound containing only C and H given that combustion of a 1.50-g sample of the compound produces 4.71 g CO_2 and 1.93 g H_2O.

8.19 Menthol is a flavoring agent extracted from peppermint oil. It contains C, H, and O. In one analysis, combustion of 10.00 mg of the substance yields 11.53 mg H_2O and 28.16 mg CO_2. What is the empirical formula of menthol?

8.20 Perchloroethylene (C_2Cl_4), also known as "perc," is the solvent used in most dry cleaning. It was originally synthesized from another compound consisting of C and Cl. Determine the empirical formula of the precursor compound given that a 5.01-g sample yields 1.86 g CO_2 in a combustion analysis experiment.

8.21 Another method for synthesizing perchloroethylene (see Problem 8.20) starts with a compound that contains C, H, and Cl. Determine the empirical formula of this precursor given that combustion of a 2.90-g sample produces 2.58 g CO_2 and 1.06 g H_2O.

8.22 Butyl acetate is a compound containing C, H, and O that occurs naturally in many fruits. The synthetic form is used to impart fruity flavors to candy, ice cream, and other sweets. Determine the empirical formula and the molecular formula of butyl acetate given that its molar mass is about 116 g/mol and the combustion of a 9.73-g sample yields 22.12 g CO_2 and 9.06 g H_2O.

8.23 Succinic acid, a substance used by the food and beverage industries to control acidity, contains only C, H, and O and has a molar mass of about 118 g/mol. Given that combustion of 1.99 g of this compound produces 2.97 g CO_2 and 0.911 g H_2O, determine its empirical and molecular formulas.

SECTION 8.3: CALCULATIONS WITH BALANCED CHEMICAL EQUATIONS

Review Questions

8.24 On what law is stoichiometry based? Why is it essential to use balanced equations in solving stoichiometric problems?

8.25 Describe the steps involved in balancing a chemical equation.

Computational Problems

8.26 Consider the combustion of carbon monoxide (CO) in oxygen gas.

$$2CO(g) + O_2(g) \longrightarrow 2CO_2(g)$$

Starting with 3.60 moles of CO, calculate the number of moles of CO_2 produced if there is enough oxygen gas to react with all the CO.

8.27 Silicon tetrachloride ($SiCl_4$) can be prepared by heating Si in chlorine gas.

$$Si(s) + 2Cl_2(g) \longrightarrow SiCl_4(l)$$

In one reaction, 0.507 mole of $SiCl_4$ is produced. How many moles of molecular chlorine were used in the reaction?

8.28 Ammonia is a colorless gas with a pungent, characteristic odor. It is prepared by the reaction between hydrogen and nitrogen.

$$3H_2(g) + N_2(g) \longrightarrow 2NH_3(g)$$

In a particular reaction, 6.0 moles of NH_3 were produced. How many moles of H_2 and how many moles of N_2 were consumed to produce this amount of NH_3?

8.29 Consider the combustion of butane (C_4H_{10}).

$$2C_4H_{10}(g) + 13O_2(g) \longrightarrow 8CO_2(g) + 10H_2O(l)$$

In a particular reaction, 5.0 moles of C_4H_{10} react with an excess of O_2. Calculate the number of moles of CO_2 formed.

8.30 The annual production of sulfur dioxide from burning coal and fossil fuels, auto exhaust, and other sources is about 26 million tons. The equation for the reaction is

$$S(s) + O_2(g) \longrightarrow SO_2(g)$$

How many tons of sulfur would it take to produce that quantity of SO_2?

8.31 When baking soda (sodium bicarbonate or sodium hydrogen carbonate, $NaHCO_3$) is heated, it releases carbon dioxide gas, which is responsible for the rising of cookies, cakes, and other baked goods. (a) Write a balanced equation for the decomposition of the compound (one of the products is Na_2CO_3). (b) Calculate the mass of $NaHCO_3$ required to produce 20.5 g of CO_2.

8.32 Potassium cyanide (KCN) reacts with acids to form a deadly poisonous gas, hydrogen cyanide (HCN). The reaction is represented by the equation

$$KCN(aq) + HCl(aq) \longrightarrow KCl(aq) + HCN(g)$$

If a sample of 0.140 g of KCN is treated with an excess of HCl, calculate the amount of HCN produced, in grams.

8.33 Part of wine making is fermentation, a complex chemical process that converts glucose to ethanol and carbon dioxide.

$$\underset{\text{glucose}}{C_6H_{12}O_6} \longrightarrow \underset{\text{ethanol}}{2C_2H_5OH} + 2CO_2$$

What is the maximum quantity of ethanol (in grams and in liters) that can be produced from 500.4 g of glucose (density of ethanol = 0.789 g/mL)?

8.34 When copper(II) sulfate pentahydrate ($CuSO_4 \cdot 5H_2O$) is heated in air above 100°C, it loses the water molecules and its blue color.

$$CuSO_4 \cdot 5H_2O \longrightarrow CuSO_4 + 5H_2O$$

If 9.60 g of $CuSO_4$ are left after heating 15.01 g of the blue compound, calculate the number of moles of H_2O originally present in the compound.

8.35 For many years, the extraction of gold from other materials involved the use of potassium cyanide.

$$4Au + 8KCN + O_2 + 2H_2O \longrightarrow 4KAu(CN)_2 + 4KOH$$

What is the minimum mass of KCN needed to extract 29.0 g (about an ounce) of gold?

8.36 Limestone ($CaCO_3$) is decomposed to quicklime (CaO) and carbon dioxide by heating. Calculate how many grams of quicklime can be produced from 1.0 kg of limestone.

8.37 Nitrous oxide (N_2O) is also called "laughing gas." It can be prepared by the thermal decomposition of ammonium nitrate (NH_4NO_3). The other product is H_2O. (a) Write a balanced equation for this reaction. (b) How many grams of N_2O are formed if 0.46 mole of NH_4NO_3 is used in the reaction?

8.38 The fertilizer ammonium sulfate [$(NH_4)_2SO_4$] is prepared by the reaction between ammonia (NH_3) and sulfuric acid.

$$2NH_3(g) + H_2SO_4(aq) \longrightarrow (NH_4)_2SO_4(aq)$$

How many kilograms of NH_3 are needed to produce 1.00×10^5 kg of $(NH_4)_2SO_4$?

8.39 A common laboratory preparation of oxygen gas is the thermal decomposition of potassium chlorate ($KClO_3$). Assuming complete decomposition, calculate the number of grams of O_2 gas that can be obtained from 46.0 g of $KClO_3$. (The products are KCl and O_2.)

SECTION 8.4: LIMITING REACTANTS

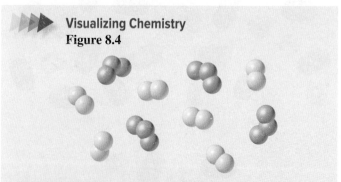

▶▶▶ **Visualizing Chemistry**
Figure 8.4

The diagram shows reactants A_3 and B_2 prior to reaction.

VC 8.1 For which of these products would A_3 be the limiting reactant?

(a)   (b)   (c)

VC 8.2   For which of these products would B_2 be the limiting reactant?

(a)   (b)   (c)

VC 8.3   For which product are the reactants present in stoichiometric amounts?

(a)   (b)   (c)

VC 8.4   Assuming that B_2 is the limiting reactant, which of the following best represents the remaining A_3 when the reaction is complete?

(a)   (b)

(c)

Review Questions

8.40   Define *limiting reactant* and *excess reactant*. What is the significance of the limiting reactant in predicting the amount of the product obtained in a reaction? Can there be a limiting reactant if only one reactant is present? Explain.

8.41   Give an everyday example that illustrates the limiting reactant concept.

8.42   Why is the theoretical yield of a reaction determined only by the amount of the limiting reactant?

8.43   Why is the actual yield of a reaction almost always smaller than the theoretical yield?

Conceptual Problems

8.44   Consider the reaction

$$2A + B \longrightarrow C$$

(a) In the diagram here that represents the reaction, which reactant, A or B, is the limiting reactant?

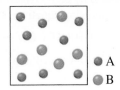

(b) Assuming a complete reaction, draw a molecular-model representation of the amounts of reactants and products left after the reaction. The atomic arrangement in C is ABA.

8.45   Consider the reaction

$$N_2 + 3H_2 \longrightarrow 2NH_3$$

Assuming each model represents one mole of the substance, show the number of moles of the product

and the excess reactant left after the complete reaction.

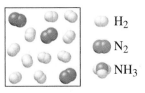

8.46   Reactants A (red) and B (blue) combine in the reaction vessel shown to form a single product C (purple) according to the equation $2A + B \longrightarrow C$.

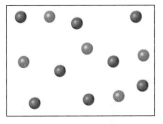

Which of the following scenes represents the contents of the reaction vessel after the reaction is complete?

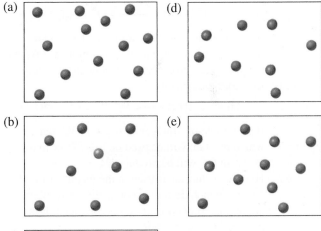

8.47   Consider this reaction, where each red sphere represents an oxygen atom and each blue sphere represents a nitrogen atom. Write the balanced equation and identify the limiting reactant.

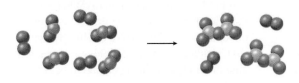

Computational Problems

8.48 Nitric oxide (NO) reacts with oxygen gas to form nitrogen dioxide (NO_2), a dark brown gas.

$$2NO(g) + O_2(g) \longrightarrow 2NO_2(g)$$

In one experiment, 0.886 mole of NO is mixed with 0.503 mole of O_2. Determine which of the two reactants is the limiting reactant. Calculate also the number of moles of NO_2 produced.

8.49 The depletion of ozone (O_3) in the stratosphere has been a matter of great concern among scientists in recent years. It is believed that ozone can react with nitric oxide (NO) that is discharged from high-altitude jet planes. The reaction is

$$O_3 + NO \longrightarrow O_2 + NO_2$$

If 0.740 g of O_3 reacts with 0.670 g of NO, how many grams of NO_2 will be produced? Which compound is the limiting reactant? Calculate the number of moles of the excess reactant remaining at the end of the reaction.

8.50 Propane (C_3H_8) is a minor component of natural gas and is used in domestic cooking and heating. (a) Balance the following equation representing the combustion of propane in air.

$$C_3H_8 + O_2 \longrightarrow CO_2 + H_2O$$

(b) How many grams of carbon dioxide can be produced by burning 3.65 moles of propane? Assume that oxygen is the excess reactant in this reaction.

8.51 Consider the reaction

$$MnO_2 + 4HCl \longrightarrow MnCl_2 + Cl_2 + 2H_2O$$

If 0.86 mole of MnO_2 and 48.2 g of HCl react, which reactant will be used up first? How many grams of Cl_2 will be produced?

8.52 Hydrogen fluoride is used in the manufacture of Freons (which destroy ozone in the stratosphere) and in the production of aluminum metal. It is prepared by the reaction

$$CaF_2 + H_2SO_4 \longrightarrow CaSO_4 + 2HF$$

In one process, 6.00 kg of CaF_2 are treated with an excess of H_2SO_4 to yield 2.86 kg of HF. Calculate the percent yield of HF.

8.53 Nitroglycerin ($C_3H_5N_3O_9$) is a powerful explosive. Its decomposition may be represented by

$$4C_3H_5N_3O_9 \longrightarrow 6N_2 + 12CO_2 + 10H_2O + O_2$$

This reaction generates a large amount of heat and gaseous products. It is the sudden formation of these gases, together with their rapid expansion, that produces the explosion. (a) What is the maximum amount of O_2 in grams that can be obtained from 2.00×10^2 g of nitroglycerin? (b) Calculate the percent yield in this reaction if the amount of O_2 generated is found to be 6.55 g.

8.54 Titanium(IV) oxide (TiO_2) is a white substance produced by the action of sulfuric acid on the mineral ilmenite ($FeTiO_3$).

$$FeTiO_3 + H_2SO_4 \longrightarrow TiO_2 + FeSO_4 + H_2O$$

Its opaque and nontoxic properties make it suitable as a pigment in plastics and paints. In one process, 8.00×10^3 kg of $FeTiO_3$ yielded 3.67×10^3 kg of TiO_2. What is the percent yield of the reaction?

8.55 Ethylene (C_2H_4), an important industrial organic chemical, can be prepared by heating hexane (C_6H_{14}) at 800°C.

$$C_6H_{14} \longrightarrow C_2H_4 + \text{other products}$$

If the yield of ethylene production is 42.5 percent, what mass of hexane must be used to produce 481 g of ethylene?

8.56 When heated, lithium reacts with nitrogen to form lithium nitride.

$$6Li(s) + N_2(g) \longrightarrow 2Li_3N(s)$$

What is the theoretical yield of Li_3N in grams when 12.3 g of Li are heated with 33.6 g of N_2? If the actual yield of Li_3N is 5.89 g, what is the percent yield of the reaction?

8.57 Disulfur dichloride (S_2Cl_2) is used in the vulcanization of rubber, a process that prevents the slippage of rubber molecules past one another when stretched. It is prepared by heating sulfur in an atmosphere of chlorine.

$$S_8(l) + 4Cl_2(g) \longrightarrow 4S_2Cl_2(l)$$

What is the theoretical yield of S_2Cl_2 in grams when 4.06 g of S_8 are heated with 6.24 g of Cl_2? If the actual yield of S_2Cl_2 is 6.55 g, what is the percent yield?

8.58 Maleic anhydride ($C_4H_3O_2$) is an important industrial intermediate chemical prepared from benzene according to the following reaction.

$$4C_6H_6 + 15O_2 \longrightarrow 4C_4H_3O_2 + 8CO_2 + 6H_2O$$

What is the atom economy of this reaction?

8.59 The Cativa process is used industrially to prepare acetic acid by carbonylation of methanol according to the following equation.

$$CH_3OH + CO \longrightarrow CH_3CO_2H$$

What is the atom economy of this process?

SECTION 8.5: PERIODIC TRENDS IN REACTIVITY OF THE MAIN GROUP ELEMENTS

Review Questions

8.60 Why do members of a group exhibit similar chemical properties?

8.61 Why are Group 1B elements more stable than Group 1A elements even though they seem to have the same outer electron configuration, ns^1, where n is the principal quantum number of the outermost shell?

Conceptual Problems

8.62 Use the alkali metals and alkaline earth metals as examples to show how we can predict the chemical properties of elements simply from their electron configurations.

8.63 Explain why the noble gases as a group are highly stable chemically.

Additional Problems

8.64 The diagram represents the products (CO_2 and H_2O) formed by the combustion of a hydrocarbon (a compound containing only C and H atoms). Write an equation for the reaction. (*Hint:* The molar mass of the hydrocarbon is about 30 g.)

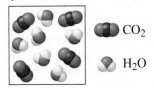

8.65 Consider the reaction of hydrogen gas with oxygen gas.

$$2H_2(g) + O_2(g) \longrightarrow 2H_2O(g)$$

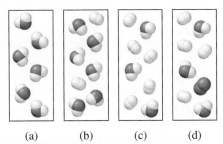

Assuming a complete reaction, which of the diagrams (a–d) represents the amounts of reactants and products left after the reaction?

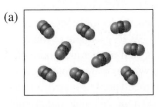

(a)   (b)   (c)   (d)

8.66 Industrially, nitric acid is produced by the Ostwald process represented by the following equations.

$$4NH_3(g) + 5O_2(g) \longrightarrow 4NO(g) + 6H_2O(l)$$

$$2NO(g) + O_2(g) \longrightarrow 2NO_2(g)$$

$$2NO_2(g) + H_2O(l) \longrightarrow HNO_3(aq) + HNO_2(aq)$$

What mass of NH_3 (in grams) must be used to produce 1.00 ton of HNO_3 by the Ostwald process, assuming an 80 percent yield in each step (1 ton = 2000 lb; 1 lb = 453.6 g)?

8.67 A sample of a compound of Cl and O reacts with an excess of H_2 to give 0.233 g of HCl and 0.403 g of H_2O. Determine the empirical formula of the compound.

8.68 What is the maximum mass of P_2I_4 that can be prepared from 7.25 g of P_4O_6 and 12.00 g of iodine according to the reaction

$$5P_4O_6 + 8I_2 \longrightarrow 4P_2I_4 + 3P_4O_{10}$$

8.69 How many moles of O are needed to combine with 0.212 mole of C to form (a) CO and (b) CO_2?

8.70 A mixture of 8.00 moles of H_2 and 6.00 moles of O_2 is ignited, forming water. What is the composition of the mixture by mass after the reaction is completed?

8.71 The explosive nitroglycerin ($C_3H_5N_3O_9$) has also been used as a drug to treat heart patients to relieve pain (angina pectoris). We now know that nitroglycerin produces nitric oxide (NO), which causes muscles to relax and allows the arteries to dilate. If each nitroglycerin molecule releases one NO per atom of N, calculate the mass percent of NO in nitroglycerin.

8.72 Carbon monoxide reacts with oxygen to produce carbon dioxide according to the following balanced equation.

$$2CO(g) + O_2(g) \longrightarrow 2CO_2(g)$$

Shown is a reaction vessel containing the reactants.

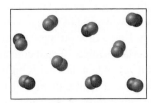

Which of the following represents the contents of the reaction vessel when the reaction is complete?

(a)

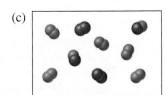

(c)

(b)

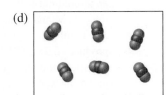

(d)

8.73 An iron bar weighed 664 g. After the bar had been standing in moist air for a month, exactly one-eighth of the iron turned to rust (Fe_2O_3). Calculate the final mass of the iron bar and rust.

8.74 One of the reactions that occurs in a blast furnace, where iron ore is converted to cast iron, is

$$Fe_2O_3 + 3CO \longrightarrow 2Fe + 3CO_2$$

Suppose that 1.64×10^3 kg of Fe are obtained from a 2.62×10^3-kg sample of Fe_2O_3. Assuming that the reaction goes to completion, what is the percent purity of Fe_2O_3 in the original sample?

8.75 An impure sample of zinc (Zn) is treated with an excess of sulfuric acid (H_2SO_4) to form zinc sulfate ($ZnSO_4$) and molecular hydrogen (H_2). (a) Write a balanced equation for the reaction. (b) If 0.0764 g of H_2 is obtained from 3.86 g of the sample, calculate the percent purity of the sample. (c) What assumptions must you make in part (b)?

8.76 A mixture of methane (CH_4) and ethane (C_2H_6) of mass 13.43 g is completely burned in oxygen. If the total mass of CO_2 and H_2O produced is 64.84 g, calculate the fraction of CH_4 in the mixture.

8.77 Aspirin or acetylsalicylic acid is synthesized by combining salicylic acid with acetic anhydride.

$$C_7H_6O_3 + C_4H_6O_3 \longrightarrow C_9H_8O_4 + HC_2H_3O_2$$

salicylic   acetic   aspirin   acetic acid
acid   anhydride

(a) How much salicylic acid is required to produce 0.400 g of aspirin (about the content in a tablet), assuming acetic anhydride is present in excess? (b) Calculate the amount of salicylic acid needed if only 74.9 percent of salicylic is converted to aspirin. (c) In one experiment, 9.26 g of salicylic acid react with 8.54 g of acetic anhydride. Calculate the theoretical yield of aspirin and the percent yield if only 10.9 g of aspirin are produced.

8.78 A certain sample of coal contains 1.6 percent sulfur by mass. When the coal is burned, the sulfur is converted to sulfur dioxide. To prevent air pollution, this sulfur dioxide is treated with calcium oxide (CaO) to form calcium sulfite ($CaSO_3$). Calculate the daily mass (in kilograms) of CaO needed by a power plant that uses 6.60×10^6 kg of coal per day.

8.79 A certain metal oxide has the formula MO where M denotes the metal. A 39.46-g sample of the compound is strongly heated in an atmosphere of hydrogen to remove oxygen as water molecules. At the end, 31.70 g of the metal are left over. If O has an atomic mass of 16.00 amu, calculate the atomic mass of M and identify the element.

8.80 The combustion of a 5.50-g sample of oxalic acid produces 5.38 g CO_2 and 1.10 g H_2O. Determine the empirical and molecular formulas of oxalic acid. The molar mass of oxalic acid is approximately 90 g/mol.

8.81 When 0.273 g of Mg is heated strongly in a nitrogen (N_2) atmosphere, a chemical reaction occurs. The product of the reaction weighs 0.378 g. Calculate the empirical formula of the compound containing Mg and N. Name the compound.

8.82 Octane (C_8H_{18}) is a component of gasoline. Complete combustion of octane yields H_2O and CO_2. Incomplete combustion produces H_2O and CO, which not only reduces the efficiency of the engine using the fuel but is also toxic. In a certain test run, 1.000 gallon (gal) of octane is burned in an engine. The total mass of CO, CO_2, and H_2O produced is 11.53 kg. Calculate the efficiency of the process; that is, calculate the fraction of octane converted to CO_2. The density of octane is 2.650 kg/gal.

8.83 Leaded gasoline contains an additive to prevent engine "knocking." On analysis, the additive compound is found to contain carbon, hydrogen, and lead (Pb) (hence, "leaded gasoline"). When 51.36 g of this compound are burned in an apparatus such as that shown in Figure 8.1, 55.90 g of CO_2 and 28.61 g of H_2O are produced. Determine the empirical formula of the gasoline additive.

8.84 Because of its detrimental effect on the environment, the lead compound described in Problem 8.83 has been replaced in recent years by methyl *tert*-butyl ether (a compound of C, H, and O) to enhance the performance of gasoline. (As of 1999, this compound is also being phased out because of its contamination of drinking water.) When a 12.1-g sample of the compound is burned in an apparatus like the one shown in Figure 8.1, 30.2 g of CO_2 and 14.8 g of H_2O are formed. What is the empirical formula of the compound?

8.85 Industrially, hydrogen gas can be prepared by combining propane gas (C_3H_8) with steam at about 400°C. The products are carbon monoxide (CO) and hydrogen gas (H_2). (a) Write a balanced equation for the reaction. (b) How many kilograms of H_2 can be obtained from 2.84×10^3 kg of propane?

8.86 A reaction having a 90 percent yield may be considered a successful experiment. However, in the synthesis of complex molecules such as chlorophyll and many anticancer drugs, a chemist often has to carry out a multiple-step synthesis. What is the overall percent yield for such a synthesis, assuming it is a 30-step reaction with a 90 percent yield at each step?

8.87 Potash is any potassium mineral that is used for its potassium content. Most of the potash produced in the United States goes into fertilizer. The major sources of potash are potassium chloride (KCl) and potassium sulfate (K_2SO_4). Potash production is often reported as the potassium oxide (K_2O) equivalent or the amount of K_2O that could be made from a given mineral. (a) If KCl costs $0.55 per kilogram, for what price (dollar per kilogram) must K_2SO_4 be sold to supply the same amount of potassium on a per dollar basis? (b) What mass (in kilograms) of K_2O contains the same number of moles of K atoms as 1.00 kg of KCl?

8.88 A 21.496-g sample of magnesium is burned in air to form magnesium oxide and magnesium nitride. When the products are treated with water, 2.813 g of gaseous ammonia is generated. Calculate the amounts of magnesium nitride and magnesium oxide formed.

8.89 What is wrong or ambiguous with each of the following statements? (a) NH_4NO_2 is the limiting reactant in the reaction

$$NH_4NO_2(s) \longrightarrow N_2(g) + 2H_2O(l)$$

(b) The limiting reactants for the reaction shown here are NH_3 and $NaCl$.

$$NH_3(aq) + NaCl(aq) + H_2CO_3(aq) \longrightarrow NaHCO_3(aq) + NH_4Cl(aq)$$

8.90 A sample of iron weighing 15.0 g was heated with potassium chlorate ($KClO_3$) in an evacuated container. The oxygen generated from the decomposition of $KClO_3$ converted some of the Fe to Fe_2O_3. If the combined mass of Fe and Fe_2O_3 was 17.9 g, calculate the mass of Fe_2O_3 formed and the mass of $KClO_3$ decomposed.

8.91 What mass of water is produced by the combustion of the oxygen in 30.0 g of air (20.0 percent oxygen by mass) with C_4H_{10}?

8.92 A sample of 10.0 g of sodium reacts with oxygen to form 13.87 g of sodium oxide (Na_2O) and sodium peroxide (Na_2O_2). Calculate the percent composition of the product mixture.

8.93 Write a balanced equation for the reaction shown here.

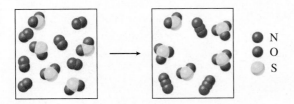

N
O
S

8.94 Write a balanced equation for the reaction shown here.

N
H
Cl
F

8.95 The compound $NH_4V_3O_8$ can be prepared according to the following sequence of reactions.

$$N_2 + 3H_2 \longrightarrow 2NH_3$$
$$2NH_3 + V_2O_5 + H_2O \longrightarrow 2NH_4VO_3$$
$$3NH_4VO_3 + 2HCl \longrightarrow NH_4V_3O_8 + 2NH_4Cl + H_2O$$

What is the maximum number of moles of $NH_4V_3O_8$ that can be prepared starting from 2.5 moles each of N_2 and H_2. Assume all other reactants are not limiting reagents.

8.96 The interhalogen compound ClF_3 reacts with NH_3 to produce three products according to this figure.

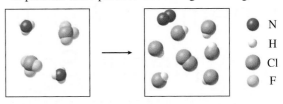

N
H
Cl
F

If 30.0 g of NH_3 and 175 g of ClF_3 are allowed to react, what masses of each product would be recovered assuming complete reaction?

8.97 The organic chemical ethylene oxide, C_2H_4O, is commonly used to sterilize the majority of medical supplies such as bandages, sutures, and surgical implements. It has traditionally been prepared according to the following reaction.

$$H_2C{=}CH_2 + Cl_2 + Ca(OH)_2 \longrightarrow$$

$$\overset{O}{H_2C{-}CH_2} + CaCl_2 + H_2O$$

A more recent synthesis for ethylene oxide is

$$H_2C{=}CH_2 + \frac{1}{2}O_2 \longrightarrow \overset{O}{H_2C{-}CH_2}$$

Compare the atom economies of these two methods for preparing ethylene oxide.

8.98 Heating 2.40 g of the oxide of metal X (molar mass of X = 55.9 g/mol) in carbon monoxide (CO) yields the pure metal and carbon dioxide. The mass of the metal product is 1.68 g. From the data given, show that the simplest formula of the oxide is X_2O_3 and write a balanced equation for the reaction.

8.99 A compound X contains 63.3 percent manganese (Mn) and 36.7 percent O by mass. When X is heated, oxygen gas is evolved and a new compound Y containing 72.0 percent Mn and 28.0 percent O is formed. (a) Determine the empirical formulas of X and Y. (b) Write a balanced equation for the conversion of X to Y.

Answers to In-Chapter Materials

PRACTICE PROBLEMS

8.1A $C_3H_8(g) + 5O_2(g) \longrightarrow 3CO_2(g) + 4H_2O(l)$. **8.1B** $H_2SO_4(aq) + 2NaOH(aq) \longrightarrow Na_2SO_4(aq) + 2H_2O(l)$. **8.2A** $C_{18}H_{32}O_2(aq) + 25O_2(g) \longrightarrow 18CO_2(g) + 16H_2O(l)$. **8.2B** $2NH_3(g) + 3CuO(s) \longrightarrow 3Cu(s) + N_2(g) + 3H_2O(l)$. **8.3A** (a) combustion, (b) combination, (c) decomposition. **8.3B** $A_2 + 2B \longrightarrow 2AB$. **8.4A** $C_3H_4O_3$ and $C_6H_8O_6$. **8.4B** 28.14 g CO_2, 14.40 H_2O. **8.5A** 0.264 mol H_2 and 0.176 mol NH_3. **8.5B** $P_4O_{10} +$

$6H_2O \longrightarrow 4H_3PO_4$; 1.45 mol P_4O_{10}, 8.70 mol H_2O. **8.6A** 34.1 g. **8.6B** 292 g. **8.7A** 42.6 g ammonia; nitrogen is the limiting reactant, 4.95 g hydrogen left over. **8.7B** 44.2 g KOH, 115.5 g H_3PO_4. **8.8A** 29.2%. **8.8B** 122 g.

SECTION REVIEW

8.1.1 e. **8.1.2** e. **8.1.3** b. **8.1.4** d. **8.1.5** b. **8.2.1** a. **8.2.2** e. **8.2.3** d. **8.3.1** a. **8.3.2** b. **8.3.3** c. **8.3.4** a. **8.4.1** e. **8.4.2** c. **8.4.3** b. **8.4.4** c. **8.4.5** a.

Chapter 9

Chemical Reactions in Aqueous Solutions

© Dirk Wiersma/Science Source

"HARD WATER" is the result of dissolved metal ions such as Ca^{2+} and Mg^{2+}. The presence of these dissolved ions can prevent soap from lathering properly; and can result in the formation of hard-water deposits, such as bathtub rings. Dissolved ions also contribute to the corrosion of metals that are exposed to the water. To combat these problems, chemists have devised "water-softening" methods to selectively replace or remove aqueous ions from solution. Typically this is achieved by passing hard water over water pellets (shown here) where insoluble salts such as calcium carbonate ($CaCO_3$) form around the pellets—thereby removing calcium ions from the water. These methods are also useful in the removal of other more harmful metal ions such as lead and arsenic.

Before You Begin, Review These Skills

- Identifying compounds as either molecular or ionic [◄◄ Sections 5.3 and 5.5]
- Names, formulas, and charges of the common polyatomic ions [◄◄ Table 5.10]
- Stoichiometric calculations [◄◄ Section 8.3]

9.1 GENERAL PROPERTIES OF AQUEOUS SOLUTIONS

Video 9.1
Solutions—strong, weak, and nonelectrolytes.

A *solution* is a homogeneous mixture [◄◄ Section 1.5] of two or more substances. Solutions may be gaseous (such as air), solid (such as brass), or liquid (such as salt-water). Usually, the substance present in the largest amount is referred to as the *solvent* and any substance present in a smaller amount is called a *solute.* For example, if we dissolve a teaspoon of sugar in a glass of water, water is the solvent and sugar is the solute. In this chapter, we will focus on the properties of *aqueous* solutions—those in which *water* is the solvent. Throughout the remainder of this chapter, unless otherwise noted, *solution* will refer specifically to an *aqueous* solution.

Electrolytes and Nonelectrolytes

You have probably heard of electrolytes in the context of sports drinks such as Gatorade. Electrolytes in body fluids are necessary for the transmission of electrical impulses, which are critical to physiological processes such as nerve impulses and muscle contractions. In general, an *electrolyte* is a substance that dissolves in water to yield a solution that conducts electricity. By contrast, a *nonelectrolyte* is a substance that dissolves in water to yield a solution that does *not* conduct electricity. Every water-soluble substance fits into one of these two categories.

> **Student Annotation:** A substance that *dissolves* in a particular solvent is said to be "soluble" in that solvent. In this chapter, we will use the word *soluble* to mean "water soluble."

The difference between an aqueous solution that conducts electricity and one that does not is the presence or absence of *ions*. As an illustration, consider solutions of sugar and salt. The physical processes of sugar (sucrose, $C_{12}H_{22}O_{11}$) dissolving in water and salt (sodium chloride, NaCl) dissolving in water can be represented with the following chemical equations.

$$C_{12}H_{22}O_{11}(s) \xrightarrow{\text{H}_2\text{O}} C_{12}H_{22}O_{11}(aq) \quad \text{and} \quad NaCl(s) \xrightarrow{\text{H}_2\text{O}} Na^+(aq) + Cl^-(aq)$$

Note that the sucrose molecules remain intact upon dissolving, becoming aqueous sucrose *molecules.* Conversely, when sodium chloride is dissolved, it *dissociates,* producing separate aqueous *ions*—namely, sodium ions and chloride ions. **Dissociation** is the process by which an *ionic* compound, upon dissolution, breaks apart into its *constituent* ions. In the case of sodium chloride, the constituent ions are Na^+ and Cl^-. It is the presence of these ions that allows an aqueous solution of sodium chloride to conduct electricity. A solution of sucrose does not conduct electricity because it does not contain ions. Thus, sodium chloride is an *electrolyte* and sucrose is a *nonelectrolyte.*

Like sucrose, which is a molecular compound [◄◄ Section 5.5], *most* water-soluble molecular compounds are nonelectrolytes. The molecular compounds that *are* electrolytes include *acids* and *molecular bases.* Acids and molecular bases are electrolytes because they undergo **ionization** in solution. Ionization is the process by which a molecular compound forms aqueous ions when it dissolves. HCl, for example, ionizes to produce aqueous H^+ ions and Cl^- ions.

$$HCl(g) \xrightarrow{\text{H}_2\text{O}} H^+(aq) + Cl^-(aq)$$

Recall from Chapter 5 that we defined an *acid* as a substance that produces hydrogen ions (H^+) when dissolved in water. We can similarly define a ***base*** as a compound that produces *hydroxide* ions (OH^-) when dissolved in water. Ammonia (NH_3), for example, ionizes in water to produce hydroxide (OH^-) and ammonium (NH_4^+) ions.

$$NH_3(g) + H_2O(l) \rightleftharpoons NH_4^+(aq) + OH^-(aq)$$

Both HCl and NH_3 are *molecular* compounds. Unlike NaCl, they themselves do not *consist* of ions. Nevertheless, they are electrolytes because they undergo *ionization* when dissolved in water—resulting in solutions that contain ions and therefore can conduct electricity.

Strong Electrolytes and Weak Electrolytes

In a solution of sodium chloride, all of the dissolved compound exists in the form of ions. Thus, NaCl, which is an ionic compound [◄◄ Section 5.3], is said to have *dissociated completely*. An electrolyte that dissociates completely is known as a ***strong electrolyte***. All water-soluble ionic compounds dissociate completely upon dissolving, so all water-soluble ionic compounds are strong electrolytes.

The list of molecular compounds that are strong electrolytes is fairly short. It comprises the seven strong acids, which are listed in Table 9.1. A strong acid ionizes completely, resulting in a solution that contains hydrogen ions and the corresponding anions but essentially no acid molecules.

Apart from the strong acids, molecular compounds that are electrolytes are *weak* electrolytes. A ***weak electrolyte*** is a compound that produces ions upon dissolving but exists in solution *predominantly* as molecules that are *not* ionized. Consider the following example. Acetic acid ($HC_2H_3O_2$) is not one of the strong acids listed in Table 9.1, so like most acids, it is a *weak* acid. Its ionization in water is represented by the following chemical equation.

$$HC_2H_3O_2(l) \rightleftharpoons H^+(aq) + C_2H_3O_2^-(aq)$$

Note the use of the double arrow, $\rightleftharpoons$, in this equation and in two earlier equations, including the ionization of ammonia (NH_3) and one of those in Table 9.1. The double "equilibrium" arrow denotes a reaction that does not result in *all* of the reactant(s) (e.g., acetic acid) being converted permanently to product(s) (e.g., hydrogen ions and acetate ions). Instead, the *forward* reaction (ionization of $HC_2H_3O_2$ to produce H^+ and

TABLE 9.1	The Strong Acids
Acid	**Ionization equation**
Hydrochloric acid	$HCl(aq) \longrightarrow H^+(aq) + Cl^-(aq)$
Hydrobromic acid	$HBr(aq) \longrightarrow H^+(aq) + Br^-(aq)$
Hydroiodic acid	$HI(aq) \longrightarrow H^+(aq) + I^-(aq)$
Nitric acid	$HNO_3(aq) \longrightarrow H^+(aq) + NO_3^-(aq)$
Chloric acid	$HClO_3(aq) \longrightarrow H^+(aq) + ClO_3^-(aq)$
Perchloric acid	$HClO_4(aq) \longrightarrow H^+(aq) + ClO_4^-(aq)$
Sulfuric acid*	$H_2SO_4(aq) \longrightarrow H^+(aq) + HSO_4^-(aq)$
	$HSO_4^-(aq) \rightleftharpoons H^+(aq) + SO_4^{2-}(aq)$

*Note that although each sulfuric acid molecule has two ionizable hydrogen atoms, it only undergoes the first ionization completely, effectively producing one H^+ ion and one HSO_4^- ion per H_2SO_4 molecule. The second ionization happens only to a very small extent.

$C_2H_3O_2^-$) and *reverse* reaction (recombination of H^+ and $C_2H_3O_2^-$ to produce $HC_2H_3O_2$) both occur, and a state of *dynamic chemical equilibrium* is established. Because there is a stronger tendency for the ions to recombine than for the molecules to ionize, most of the $HC_2H_3O_2$ molecules remain intact in solution—with only a very small percentage existing as separate aqueous ions.

The ionization of a weak base, while similar in many ways to the ionization of a weak acid, requires some additional explanation. Ammonia (NH_3) is a common weak base. Recall that the ionization of ammonia in water is represented by the equation

$$NH_3(g) + H_2O(l) \rightleftharpoons NH_4^+(aq) + OH^-(aq)$$

Note that the ammonia molecule does not ionize by breaking apart into ions. Rather, it does so by ionizing a *water* molecule. The H^+ ion from a water molecule attaches to an ammonia molecule, producing an ammonium ion (NH_4^+) and leaving what remains of the water molecule, the OH^- ion, in solution.

$$NH_3(g) \quad + \quad H_2O(l) \quad \rightleftharpoons \quad NH_4^+(aq) \quad + \quad OH^-(aq)$$

As with the ionization of a weak acid, the reverse process predominates and at any given point in time, there will be far more NH_3 molecules present than there will be NH_4^+ and OH^- ions. The simultaneous occurrence of both forward and reverse reactions is indicated by the use of the double arrow.

We can distinguish between electrolytes and nonelectrolytes experimentally using an apparatus like the one pictured in Figure 9.1. A lightbulb is connected to a battery using a circuit that includes the contents of the beaker. For the bulb to light, electric current must flow from one electrode to the other. Pure water is a very poor conductor of electricity because H_2O ionizes to only an extremely small extent. There are virtually no ions in pure water to conduct the current, so H_2O is considered a nonelectrolyte. If we add a small amount of salt (sodium chloride), however, the lightbulb will begin to glow as soon as the salt dissolves in the water. Sodium chloride dissociates completely in water to give Na^+ and Cl^- ions. Because the NaCl solution conducts electricity, we say that NaCl is an electrolyte.

If the solution contains a nonelectrolyte, as it does in Figure 9.1(a), the bulb will not light. If the solution contains an electrolyte, as it does in Figure 9.1(b) and (c), the bulb will light. The cations in solution are attracted to the negative electrode, and the anions are attracted to the positive electrode. This movement sets up an electric current that is equivalent to the flow of electrons along a metal wire. How brightly the bulb burns depends upon the number of ions in solution. In Figure 9.1(b), the solution contains a *weak* electrolyte and therefore a relatively small number of ions, so the bulb lights only weakly. The solution in Figure 9.1(c) contains a *strong* electrolyte, which produces a relatively large number of ions, so the bulb lights brightly.

Although the experimental method illustrated in Figure 9.1 can be useful, often you will have to characterize a compound as a nonelectrolyte, a weak electrolyte, or a strong electrolyte just by looking at its formula. A good first step is to determine whether the compound is *ionic* or *molecular*.

An ionic compound contains a *cation* (which is either a metal ion or the ammonium ion) and an *anion* (which may be atomic or polyatomic). A binary compound that contains a metal and a nonmetal is almost always ionic. This is a good time to review the polyatomic anions in Table 5.10 [◄◄ Section 5.7]. You will need to be able to recognize them in the formulas of compounds. Any ionic compound that dissolves in water is a strong electrolyte.

Student Annotation: In a state of *dynamic chemical equilibrium*, or simply *equilibrium*, both forward and reverse reactions continue to occur. However, because they are occurring at the same *rate*, no net change is observed over time in the amounts of reactants or products. Chemical equilibrium is the subject of Chapters 15 through 17.

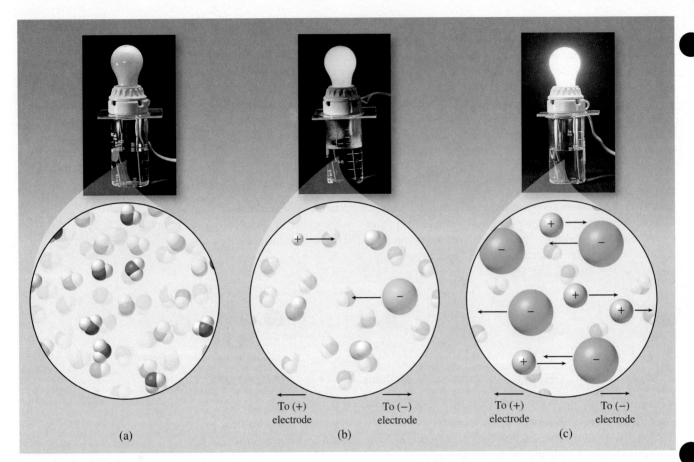

To (+)   To (−)   To (+)   To (−)
electrode electrode   electrode electrode
    (a)   (b)   (c)

Figure 9.1 An apparatus for distinguishing between electrolytes and nonelectrolytes and between weak electrolytes and strong electrolytes. A solution's ability to conduct electricity depends on the number of ions it contains. (a) Pure water contains almost no ions and does not conduct electricity; therefore, the lightbulb is not lit. (b) A weak electrolyte solution such as HF(aq) contains a small number of ions, so the lightbulb is dimly lit. (c) A strong electrolyte solution such as NaCl(aq) contains a large number of ions, so the lightbulb is brightly lit. The molar amounts of dissolved substances in the beakers in (b) and (c) are equal.

© McGraw-Hill Education/Stephen Frisch, photographer

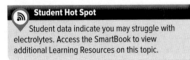

Student Hot Spot

Student data indicate you may struggle with electrolytes. Access the SmartBook to view additional Learning Resources on this topic.

Student Annotation: For "insoluble" ionic compounds, which are actually very *slightly* soluble, the small amount that dissolves does dissociate completely—making them strong electrolytes.

If a compound does not contain a metal cation or the ammonium cation, it is molecular. In this case, you will need to determine whether or not the compound is an acid. Acids generally can be recognized by the way their formulas are written, with the ionizable hydrogens written first. $HC_2H_3O_2$, H_2CO_3, and H_3PO_4 are acetic acid, carbonic acid, and phosphoric acid, respectively. Formulas of carboxylic acids, such as acetic acid, often are written with their ionizable hydrogen atoms *last* in order to keep the atoms of the functional group together in the formula. Thus, either $HC_2H_3O_2$ or CH_3COOH is correct for acetic acid. To make it easier to identify compounds as acids, in this chapter we will write all acid formulas with the ionizable H atom(s) first. If a compound is an acid, it is an electrolyte. If it is one of the acids listed in Table 9.1, it is a *strong* acid and therefore a *strong* electrolyte. Any acid not listed in Table 9.1 is a *weak* acid and therefore a *weak* electrolyte.

If a molecular compound is not an acid, you must then consider whether or not it is a weak base. Many weak bases are related to ammonia in that they consist of a nitrogen atom bonded to hydrogen and/or carbon atoms. Examples include methylamine (CH_3NH_2), pyridine (C_5H_5N), and hydroxylamine (NH_2OH). *Weak* bases are *weak* electrolytes.

If a molecular compound is neither an acid nor a weak base, it is a nonelectrolyte. The flowchart in Figure 9.2 can be useful for classification of water-soluble compounds.

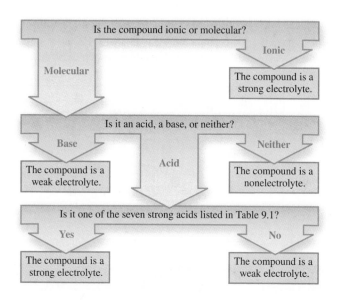

Figure 9.2 Flowchart for determining if a compound is a strong electrolyte, a weak electrolyte, or a nonelectrolyte.

Worked Example 9.1 shows how to classify compounds as nonelectrolytes, weak electrolytes, and strong electrolytes.

Worked Example 9.1

Sports drinks typically contain sucrose ($C_{12}H_{22}O_{11}$), fructose ($C_6H_{12}O_6$), sodium citrate ($Na_3C_6H_5O_7$), potassium citrate ($K_3C_6H_5O_7$), and ascorbic acid ($H_2C_6H_6O_6$), among other ingredients. Classify each of these ingredients as a nonelectrolyte, a weak electrolyte, or a strong electrolyte.

Strategy Identify each compound as ionic or molecular; identify each molecular compound as acid, base, or neither; and identify each acid as strong or weak.

Setup Sucrose and fructose contain no cations and are therefore molecular compounds—neither is an acid or a base. Sodium citrate and potassium citrate contain metal cations and are therefore ionic compounds. Ascorbic acid does not appear on the list of strong acids in Table 9.1, so it is a weak acid.

Solution Sucrose and fructose are nonelectrolytes. Sodium citrate and potassium citrate are strong electrolytes. Ascorbic acid is a weak electrolyte.

Think About It
Remember that any soluble ionic compound is a strong electrolyte, whereas most molecular compounds are nonelectrolytes or weak electrolytes. The only molecular compounds that are strong electrolytes are the strong acids listed in Table 9.1.

Practice Problem A TTEMPT A so-called "enhanced water" contains citric acid ($H_3C_6H_5O_7$), magnesium lactate [$Mg(C_3H_5O_3)_2$], calcium lactate [$Ca(C_3H_5O_3)_2$], and potassium phosphate (K_3PO_4). Classify each of these compounds as a nonelectrolyte, a weak electrolyte, or a strong electrolyte.

Practice Problem B UILD For each of the following solutions, classify the solute as a nonelectrolyte, a weak electrolyte, or a strong electrolyte.

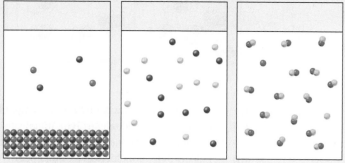

(Continued on next page)

Practice Problem **C**ONCEPTUALIZE Determine which diagram, if any, could represent an aqueous solution of each of the following compounds: LiCl, $CuSO_4$, K_2SO_4, H_2CO_3, $Al_2(SO_4)_3$, $AlCl_3$, Na_3PO_4. (Red and blue spheres represent different chemical species.)

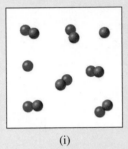

(i)

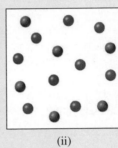

(ii)

(iii)

(iv)

Section 9.1 Review

General Properties of Aqueous Solutions

9.1.1   Soluble ionic compounds are _____.
  (a) always nonelectrolytes
  (b) always weak electrolytes
  (c) always strong electrolytes
  (d) never strong electrolytes
  (e) sometimes nonelectrolytes

9.1.2   Soluble molecular compounds are _____.
  (a) always nonelectrolytes
  (b) always weak electrolytes
  (c) always strong electrolytes
  (d) never strong electrolytes
  (e) sometimes nonelectrolytes

9.1.3   Which of the following compounds is a weak electrolyte?
  (a) LiCl
  (b) $(C_2H_5)_2NH$
  (c) KNO_3
  (d) NaI
  (e) HNO_3

9.1.4   Which of the following compounds is a strong electrolyte?
  (a) HF
  (b) HCN
  (c) NaF
  (d) NH_3
  (e) H_2O

9.1.5   Each of the following figures represents an aqueous solution. Classify each solute as a nonelectrolyte, a weak electrolyte, or a strong electrolyte.

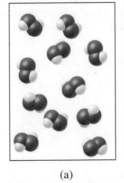

(a)

(b)

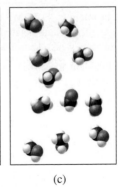

(c)

  (a) strong electrolyte, weak electrolyte, weak electrolyte
  (b) nonelectrolyte, weak electrolyte, nonelectrolyte
  (c) weak electrolyte, nonelectrolyte, weak electrolyte
  (d) nonelectrolyte, nonelectrolyte, nonelectrolyte
  (e) strong electrolyte, weak electrolyte, nonelectrolyte

PRECIPITATION REACTIONS

When an aqueous solution of lead(II) nitrate [$Pb(NO_3)_2$] is added to an aqueous solution of sodium iodide (NaI), a yellow insoluble solid—lead(II) iodide (PbI_2)—forms. Sodium nitrate ($NaNO_3$), the other reaction product, remains in solution. Figure 9.3 shows this reaction in progress. An insoluble solid product that separates from a solution is called a ***precipitate,*** and a chemical reaction in which a precipitate forms is called a ***precipitation reaction.***

   Precipitation reactions usually involve ionic compounds, but only certain combinations of electrolyte solutions result in the formation of a precipitate. Whether or not a precipitate forms when two solutions are mixed depends on the solubility of the products.

Solubility Guidelines for Ionic Compounds in Water

When an ionic substance such as sodium chloride dissolves in water, the water molecules remove individual ions from the three-dimensional solid structure and surround them. This process, called ***hydration,*** is shown in Figure 9.4. Water is an excellent solvent for ionic compounds because H_2O is a *polar* molecule [◄◄ Section 6.2]. There is a partial negative charge on the oxygen atom, denoted by the δ− symbol, and partial positive charges, denoted by the δ+ symbol, on each of the hydrogen atoms. The oxygen atoms in the surrounding water molecules are attracted to the cations, while the hydrogen atoms are attracted to the anions. These attractions explain the orientation of water molecules around each of the ions in solution. The surrounding water molecules prevent the cations and anions from recombining.

Video 9.2
Precipitation of $BaSO_4$.

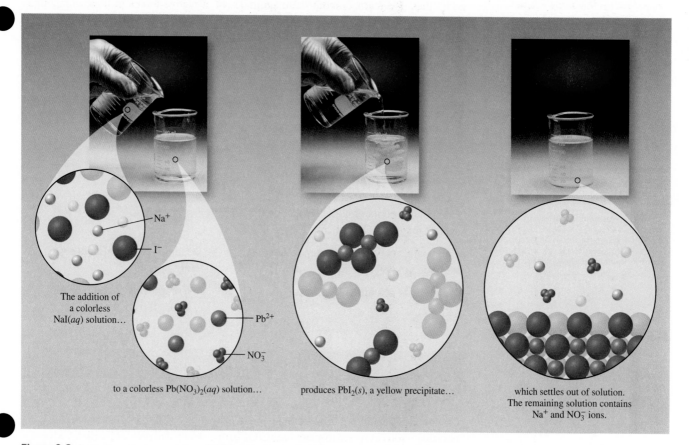

The addition of a colorless NaI(*aq*) solution...

to a colorless $Pb(NO_3)_2$(*aq*) solution...

produces PbI_2(*s*), a yellow precipitate...

which settles out of solution. The remaining solution contains Na^+ and NO_3^- ions.

Figure 9.3 A colorless aqueous solution of NaI is added to a colorless aqueous solution of $Pb(NO_3)_2$. A yellow precipitate, PbI_2, forms. Na^+ and NO_3^- ions remain in solution.

Figure 9.4 Hydration of anions and cations of a soluble ionic compound. Water molecules surround each anion with their partial positive charges (H atoms) oriented toward the negatively charged anion; and they surround each cation with their partial negative charges (O atoms) oriented toward the positively charged cation.

Solubility is defined as the maximum amount of solute that will dissolve in a given quantity of solvent at a specific temperature. Not all ionic compounds dissolve in water. Whether or not an ionic compound is water soluble depends on the relative magnitudes of the water molecules' attraction to the ions, and the ions' attraction for each other. We learned about the magnitudes of attractive forces in ionic compounds in Chapter 5; now it is useful to learn some guidelines that enable us to predict the solubility of ionic compounds. Table 9.2 lists groups of compounds that are *soluble* and shows the *insoluble* exceptions. Table 9.3 lists groups of compounds that are *insoluble* and shows the *soluble* exceptions.

TABLE 9.2	Solubility Guidelines: Soluble Compounds
Water-soluble compounds	**Insoluble exceptions**
Compounds containing an alkali metal cation (Li^+, Na^+, K^+, Rb^+, Cs^+) or the ammonium ion (NH_4^+)	
Compounds containing the nitrate ion (NO_3^-), acetate ion ($C_2H_3O_2^-$), or chlorate ion (ClO_3^-)	
Compounds containing the chloride ion (Cl^-), bromide ion (Br^-), or iodide ion (I^-)	Compounds containing Ag^+, Hg_2^{2+}, or Pb^{2+}
Compounds containing the sulfate ion (SO_4^{2-})	Compounds containing Ag^+, Hg_2^{2+}, Pb^{2+}, Ca^{2+}, Sr^{2+}, or Ba^{2+}

TABLE 9.3	Solubility Guidelines: Insoluble Compounds
Water-insoluble compounds	**Soluble exceptions**
Compounds containing the carbonate ion (CO_3^{2-}), phosphate ion (PO_4^{3-}), chromate ion (CrO_4^{2-}), or sulfide ion (S^{2-})	Compounds containing Li^+, Na^+, K^+, Rb^+, Cs^+, or NH_4^+
Compounds containing the hydroxide ion (OH^-)	Compounds containing Li^+, Na^+, K^+, Rb^+, Cs^+, or Ba^{2+}

Worked Example 9.2 gives you some practice applying the solubility guidelines.

Worked Example 9.2

Classify each of the following compounds as soluble or insoluble in water: (a) $AgNO_3$, (b) $CaSO_4$, (c) K_2CO_3.

Strategy Use the guidelines in Tables 9.2 and 9.3 to determine whether or not each compound is expected to be water soluble.

Setup

(a) $AgNO_3$ contains the nitrate ion (NO_3^-). According to Table 9.2, *all* compounds containing the nitrate ion are soluble.

(b) $CaSO_4$ contains the sulfate ion (SO_4^{2-}). According to Table 9.2, compounds containing the sulfate ion are soluble unless the cation is Ag^+, Hg_2^{2+}, Pb^{2+}, Ca^{2+}, Sr^{2+}, or Ba^{2+}. Thus, the Ca^{2+} ion is one of the insoluble exceptions.

(c) K_2CO_3 contains an alkali metal cation (K^+) for which, according to Table 9.2, there are no insoluble exceptions. Alternatively, Table 9.3 shows that most compounds containing the carbonate ion (CO_3^{2-}) are insoluble—but compounds containing a Group 1A cation such as K^+ are soluble exceptions.

Solution (a) Soluble, (b) Insoluble, (c) Soluble.

> **Think About It**
> Check the ions in each compound against the information in Tables 9.2 and 9.3 to confirm that you have drawn the right conclusions.

Practice Problem Ⓐ**TTEMPT** Classify each of the following compounds as soluble or insoluble in water:

(a) $PbCl_2$, (b) $(NH_4)_3PO_4$, (c) $Fe(OH)_3$.

Practice Problem Ⓑ**UILD** Classify each of the following compounds as soluble or insoluble in water:

(a) $MgBr_2$, (b) $Ca_3(PO_4)_2$, (c) $KClO_3$.

Practice Problem Ⓒ**ONCEPTUALIZE** Using Tables 9.2 and 9.3, identify a compound that will cause precipitation of two different insoluble ionic compounds when an aqueous solution of it is added to an aqueous solution of iron(III) sulfate.

Molecular Equations

The reaction shown in Figure 9.3 can be represented with the chemical equation

$$Pb(NO_3)_2(aq) + 2NaI(aq) \longrightarrow 2NaNO_3(aq) + PbI_2(s)$$

Based on this chemical equation, the metal cations seem to exchange anions. That is, the Pb^{2+} ion, originally paired with NO_3^- ions, ends up paired with I^- ions; similarly, each Na^+ ion, originally paired with an I^- ion, ends up paired with an NO_3^- ion. This equation, as written, is called a *molecular equation,* which is a chemical equation written with all compounds represented by their chemical formulas, making it look as though they exist in solution as molecules or formula units.

You now know enough chemistry to predict the products of this type of chemical reaction! Simply write the formulas for the reactants, and then write formulas for the compounds that would form if the cations in the reactants were to trade anions. For example, if you want to write the equation for the reaction that occurs when solutions of sodium sulfate and barium hydroxide are combined, you would first write the formulas of the reactants [◄◄ Chapter 5]:

$$Na_2SO_4(aq) + Ba(OH)_2(aq) \longrightarrow$$

Then you would write the formula for one product by combining the cation from the first reactant (Na^+), with the anion from the second reactant (OH^-); and write the formula for the other product by combining the cation from the second reactant (Ba^{2+}) with the anion from the first (SO_4^{2-}). Thus, the equation is

$$Na_2SO_4(aq) + Ba(OH)_2(aq) \longrightarrow 2NaOH + BaSO_4$$

Although we have balanced the equation [◄◄ Section 8.1], we have not yet put phases in parentheses for the products.

The final step in predicting the outcome of such a reaction is to determine which of the products, if any, will precipitate from solution. We do this using the solubility guidelines for ionic compounds (Tables 9.2 and 9.3). The first product (NaOH) contains a Group 1A cation (Na^+) and will therefore be soluble. We indicate its phase as (aq). The second product ($BaSO_4$) contains the sulfate ion (SO_4^{2-}). Sulfate compounds are soluble unless the cation is Ag^+, Hg_2^{2+}, Pb^{2+}, Ca^{2+}, Sr^{2+}, or Ba^{2+}. $BaSO_4$ is therefore insoluble and will precipitate. We indicate its phase as (s):

$$Na_2SO_4(aq) + Ba(OH)_2(aq) \longrightarrow 2NaOH(aq) + BaSO_4(s)$$

Reactions in which compounds exchange ions are sometimes called **metathesis** or **double replacement** reactions.

Ionic Equations

Although molecular equations are useful, especially from the standpoint of knowing which solutions to combine in the laboratory, they are in a sense unrealistic. As we saw in Section 9.1, soluble ionic compounds are *strong electrolytes*. As such, they exist in solution as hydrated *ions,* rather than as formula units. Thus, it would be more realistic to represent the aqueous species in the reaction of $Na_2SO_4(aq)$ with $Ba(OH)_2(aq)$ as follows:

$$Na_2SO_4(aq) \longrightarrow 2Na^+(aq) + SO_4^{2-}(aq)$$

$$Ba(OH)_2(aq) \longrightarrow Ba^{2+}(aq) + 2OH^-(aq)$$

$$NaOH(aq) \longrightarrow Na^+(aq) + OH^-(aq)$$

If we were to rewrite the equation, representing the dissolved compounds as hydrated ions, it would be

$$2Na^+(aq) + SO_4^{2-}(aq) + Ba^{2+}(aq) + 2OH^-(aq) \longrightarrow 2Na^+(aq) + 2OH^-(aq) + BaSO_4(s)$$

This version of the equation is called an **ionic equation,** a chemical equation in which any compound that exists completely or predominantly as *ions* in solution is represented as those ions. Species that are *insoluble* or that exist in solution completely or predominantly as *molecules* are represented with their chemical formulas, as they were in the molecular equation.

Net Ionic Equations

$Na^+(aq)$ and $OH^-(aq)$ both appear as reactants and products in the ionic equation for the reaction of $Na_2SO_4(aq)$ with $Ba(OH)_2(aq)$. Ions that appear on both sides of the equation arrow are called **spectator ions** because they do not participate in the reaction. Spectator ions cancel one another, just as identical terms on both sides of an algebraic equation cancel one another, so we need not show spectator ions in chemical equations.

Spectator ions

$$\overgroup{2Na^+(aq)} + SO_4^{2-}(aq) + Ba^{2+}(aq) + \overgroup{2OH^-(aq)} \longrightarrow \overgroup{2Na^+(aq)} + \overgroup{2OH^-(aq)} + BaSO_4(s)$$

Eliminating the spectator ions yields the following equation:

$$Ba^{2+}(aq) + SO_4^{2-}(aq) \longrightarrow BaSO_4(s)$$

This version of the equation is called a **net ionic equation,** which is a chemical equation that includes only the species that are actually involved in the reaction. The net ionic equation, in effect, tells us what actually happens when we combine solutions of sodium sulfate and barium hydroxide.

The steps necessary to determine the molecular, ionic, and net ionic equations for a precipitation reaction are as follows:

1. Write and balance the molecular equation, predicting the products by assuming that the cations trade anions.
2. Write the ionic equation by separating strong electrolytes into their constituent ions.
3. Write the net ionic equation by identifying and canceling spectator ions on both sides of the equation.

If both products of a reaction are strong electrolytes, then all the ions in solution are spectator ions. In this case, there is no net ionic equation and no reaction takes place.

Worked Example 9.3 illustrates the stepwise determination of molecular, ionic, and net ionic equations.

> **Student Hot Spot**
>
> Student data indicate you may struggle with precipitation reactions. Access the SmartBook to view additional Learning Resources on this topic.

Worked Example 9.3

Write the molecular, ionic, and net ionic equations for the reaction that occurs when aqueous solutions of lead acetate $[Pb(C_2H_3O_2)_2]$ and calcium chloride $(CaCl_2)$ are combined.

Strategy Predict the products by exchanging ions and balance the equation. Determine which product will precipitate based on the solubility guidelines in Tables 9.2 and 9.3. Rewrite the equation showing strong electrolytes as ions. Identify and cancel spectator ions.

Solution

Molecular equation:

$$Pb(C_2H_3O_2)_2(aq) + CaCl_2(aq) \longrightarrow PbCl_2(s) + Ca(C_2H_3O_2)_2(aq)$$

Ionic equation:

$$Pb^{2+}(aq) + 2C_2H_3O_2^-(aq) + Ca^{2+}(aq) + 2Cl^-(aq) \longrightarrow PbCl_2(s) + Ca^{2+}(aq) + 2C_2H_3O_2^-(aq)$$

Net ionic equation:

$$Pb^{2+}(aq) + 2Cl^-(aq) \longrightarrow PbCl_2(s)$$

> **Think About It**
> Remember that the charges on ions in a compound must sum to zero. Make sure that you have written correct formulas for the products and that each of the equations you have written is balanced. If you find that you are having trouble balancing an equation, check to make sure you have correct formulas for the products.

Practice Problem **A**TTEMPT Write the molecular, ionic, and net ionic equations for the combination of $Sr(NO_3)_2(aq)$ and $Li_2SO_4(aq)$.

Practice Problem **B**UILD Determine the reactants and write the balanced molecular, ionic, and net ionic equations for the combination that results in the formation of $NH_4NO_3(aq)$ and $BaCO_3(s)$.

Practice Problem **C**ONCEPTUALIZE Which diagram best represents the result when equal volumes of equal-concentration aqueous solutions of barium nitrate and potassium phosphate are combined?

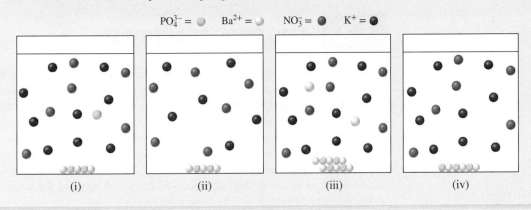

PO_4^{3-} = ⬤ Ba^{2+} = ⬤ NO_3^- = ⬤ K^+ = ⬤

(i)   (ii)   (iii)   (iv)

Section 9.2 Review

Precipitation Reactions

9.2.1 Which of the following are water soluble? (Select all that apply.)
(a) Na_2S
(d) $CuBr_2$
(b) $Ba(C_2H_3O_2)_2$
(e) Hg_2Cl_2
(c) $CaCO_3$

9.2.2 Which of the following are water insoluble? (Select all that apply.)
(a) Ag_2CrO_4
(d) $BaSO_4$
(b) Li_2CO_3
(e) $ZnCl_2$
(c) $Ca_3(PO_4)_2$

9.2.3 What are the spectator ions in the ionic equation for the combination of $Li_2CO_3(aq)$ and $Ba(OH)_2(aq)$?
(a) CO_3^{2-} and OH^-
(d) Ba^{2+} and OH^-
(b) Li^+ and OH^-
(e) Ba^{2+} and CO_3^{2-}
(c) Li^+ and Ba^{2+}

9.2.4 Write the balanced net ionic equation for the combination of $Fe(NO_3)_2(aq)$ and $Na_2CO_3(aq)$.
(a) $Na^+(aq) + CO_3^{2-}(aq) \rightarrow NaCO_3(s)$
(b) $Fe^{2+}(aq) + CO_3^{2-}(aq) \rightarrow FeCO_3(s)$
(c) $2Na^+(aq) + CO_3^{2-}(aq) \rightarrow Na_2CO_3(s)$
(d) $Fe^{2+}(aq) + 2NO_3^-(aq) \rightarrow Fe(NO_3)_2(s)$
(e) $Na^+(aq) + NO_3^-(aq) \rightarrow NaNO_3(s)$

9.2.5 Write the balanced net ionic equation for the combination shown. (Note that the reaction represented here is not balanced.)

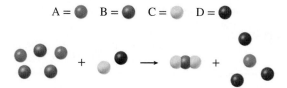

(a) $2A + 3B + C + D \rightarrow BC_2 + A + 3D$
(b) $3B + 6C \rightarrow 3BC_2$
(c) $A + 3B + C \rightarrow BC_2 + 2D$
(d) $3B + C \rightarrow BC_2$
(e) $B + 2C \rightarrow BC_2$

9.3   ACID-BASE REACTIONS

Another type of reaction occurs when two solutions, one containing an acid and one containing a base, are combined. We frequently encounter acids and bases in everyday life (Figure 9.5). Ascorbic acid, for instance, is also known as vitamin C, acetic acid is the component responsible for the sour taste and characteristic smell of vinegar, and hydrochloric acid is the acid in muriatic acid and is also the principal ingredient in gastric juice (stomach acid). Ammonia, found in many cleaning products, and sodium hydroxide, found in drain cleaner, are common bases. Acid-base chemistry is extremely important to biological processes. Let's look again at the properties of acids and bases, and then look at acid-base reactions.

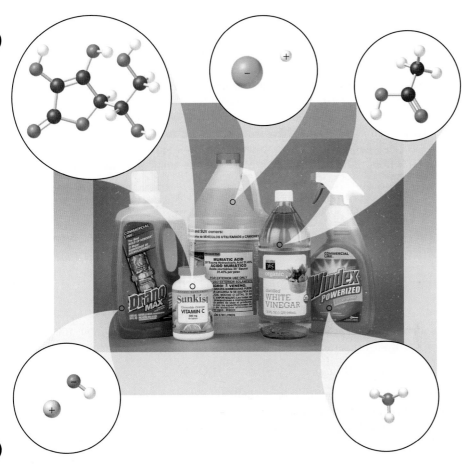

Figure 9.5 Some common acids and bases. From left to right: Sodium hydroxide (NaOH), ascorbic acid ($C_6H_8O_6$ or, with its ionizable hydrogens written first, $H_2C_6H_6O_6$), hydrochloric acid (HCl), acetic acid ($HC_2H_3O_2$), and ammonia (NH_3). HCl and NaOH are both strong electrolytes and exist in solution entirely as ions. Water molecules are not shown.

© David A. Tietz/Editorial Image, LLC

Strong Acids and Bases

As we saw in Section 9.1, the seven strong acids ionize completely in solution. (See Table 9.1.) All other acids are weak acids. The strong bases are the hydroxides of Group 1A and heavy Group 2A metals. These are soluble ionic compounds, which dissociate completely and exist entirely as ions in solution. Thus, both strong acids and strong bases are strong electrolytes. Table 9.4 lists the strong acids and strong bases. It is important that you know these compounds.

Video 9.3
Acids and bases—dissociation of strong and weak acids.

Brønsted Acids and Bases

In Section 5.6 we defined an acid as a substance that ionizes in water to produce H^+ ions, and a base as a substance that ionizes (or dissociates, in the case of an ionic base)

TABLE 9.4	Strong Acids and Strong Bases	
Strong acids	**Strong bases**
HCl	LiOH
HBr	NaOH
HI	KOH
HNO_3	RbOH
$HClO_3$	CsOH
$HClO_4$	$Ca(OH)_2$
H_2SO_4	$Sr(OH)_2$
	$Ba(OH)_2$

Student Annotation: Although three of the Group 2A hydroxides [$Ca(OH)_2$, $Sr(OH)_2$, and $Ba(OH)_2$] are typically classified as strong bases, only $Ba(OH)_2$ is sufficiently soluble to be used commonly in the laboratory. For any ionic compound, what does dissolve—even if it is only a tiny amount—dissociates completely.

in water to produce OH^- ions. These definitions are attributed to the Swedish chemist Svante Arrhenius.[1] Although the **Arrhenius acid** and **Arrhenius base** definitions are useful, they are restricted to the behavior of compounds in aqueous solution. More inclusive definitions were proposed by the Danish chemist Johannes Brønsted[2] in 1932. A **Brønsted acid** is a proton *donor,* and a **Brønsted base** is a proton *acceptor.* In this context, the word *proton* refers to a hydrogen atom that has lost its electron—also known as a *hydrogen ion* (H^+). Consider the ionization of the weak base ammonia (NH_3).

This equation shows that NH_3 is a base in the Arrhenius sense; that is, it produces OH^- in solution. It is also a base in the Brønsted sense because it accepts a proton (H^+) from the water molecule to become the ammonium ion (NH_4^+).

Student Annotation: The H atom consists of a proton and an electron. When the electron is lost, all that remains is the proton—hence the use of the term proton in this context.

$$NH_3(aq) \quad + \quad H_2O(l) \quad \rightleftharpoons \quad NH_4^+(aq) \quad + \quad OH^-(aq)$$

Now consider the ionization of hydrofluoric acid (HF), a weak acid.

$$HF(aq) \rightleftharpoons H^+(aq) + F^-(aq)$$

HF is an acid in the Arrhenius sense because it produces H^+ in solution. It is also an acid in the Brønsted sense because it donates a proton to water.

$$HF(aq) + H_2O(l) \rightleftharpoons H_3O^+(aq) + F^-(aq)$$

These two chemical equations mean essentially the same thing, but it is important to recognize that an aqueous proton (H^+) does not exist as an isolated species in solution. Rather, it is hydrated just as other aqueous ions are. (See Figure 9.4.) The proton, being positively charged, is strongly attracted to the partial negative charge on the oxygen atom in a water molecule. Thus, it is convenient and more *realistic* for us to represent the ionization of HF with the equation

Student Annotation: In reality, aqueous protons are surrounded by water molecules, just as other ions are. We use H_3O^+ in chemical equations, showing just one of the water molecules involved, to emphasize that the proton is hydrated in solution.

$$HF(aq) + H_2O(l) \rightleftharpoons H_3O^+(aq) + F^-(aq)$$

where we include the water molecule to which the proton becomes attached, both before and after the ionization. (We could show H_3O^+ as $H_2O \cdot H^+$ to emphasize that it is a water molecule attached to a proton.) With the equation written this way, we show that HF donates a proton to H_2O, thus converting the H_2O molecule to the **hydronium ion (H_3O^+)**.

Student Annotation: The terms and symbols *hydrogen ion, proton, hydronium ion,* H^+, and H_3O^+ all refer to the same aqueous species and are used interchangeably.

$$HF(aq) \quad + \quad H_2O(l) \quad \rightleftharpoons \quad H_3O^+(aq) \quad + \quad F^-(aq)$$

Both H^+ and H_3O^+ will be used in chemical equations throughout the text. You should be aware that they refer to the same aqueous species.

The Brønsted definitions of acids and bases are not restricted to species in aqueous solution. In fact, Brønsted acid-base reactions sometimes take place in the gas phase. For example, in the reaction between HCl and NH_3 gases, HCl acts as the Brønsted acid, donating its proton to NH_3, which, by accepting the proton, acts as a Brønsted base. The products of this proton transfer are the chloride ion (Cl^-) and the ammonium ion (NH_4^+), which subsequently combine to form the ionic solid ammonium chloride.

$$HCl(g) + NH_3(g) \longrightarrow NH_4Cl(s)$$

---

[1]Svante August Arrhenius (1859–1927) made important contributions to the study of chemical kinetics and electrolyte solutions. (He also speculated that life had come to Earth from other planets.) Arrhenius was awarded the Nobel Prize in Chemistry in 1903.

[2]Johannes Nicolaus Brønsted (1879–1947). In addition to his theory of acids and bases, Brønsted worked on thermodynamics and the separation of mercury into its isotopes. In some books, Brønsted acids and bases are called Brønsted-Lowry acids and bases. Brønsted and Thomas Martin Lowry (1874–1936) developed essentially the same acid-base theory independently in 1923.

Most of the strong acids are *monoprotic acids,* meaning that each acid molecule has one proton to donate. One of the strong acids, H_2SO_4, is a *diprotic acid,* meaning that each acid molecule has two protons that it can donate. Other diprotic acids include oxalic acid ($H_2C_2O_4$) and carbonic acid (H_2CO_3). There are also *triprotic acids,* those with three protons, although they are relatively less common than mono- or diprotic acids. Examples include phosphoric acid (H_3PO_4) and citric acid ($H_3C_6H_5O_7$).

In general, acids with more than one proton are called *polyprotic.* Of the polyprotic acids, only sulfuric acid (H_2SO_4) is a strong acid—although it is strong only in its *first* ionization in water. H_2SO_4 ionizes completely to yield H^+ and HSO_4^-, the subsequent ionization of the hydrogen sulfate ion (HSO_4^-) happens only to a very small extent. (See Table 9.1.) Note the single and double arrows in the following two equations.

$$H_2SO_4(aq) \longrightarrow H^+(aq) + HSO_4^-(aq)$$

$$HSO_4^-(aq) \rightleftharpoons H^+(aq) + SO_4^{2-}(aq)$$

For all other polyprotic acids, each ionization is incomplete and is represented by an equation with a double arrow. The first, second, and third ionizations of phosphoric acid are represented as

$$H_3PO_4(aq) \rightleftharpoons H^+(aq) + H_2PO_4^-(aq)$$

$$H_2PO_4^-(aq) \rightleftharpoons H^+(aq) + HPO_4^{2-}(aq)$$

$$HPO_4^{2-}(aq) \rightleftharpoons H^+(aq) + PO_4^{3-}(aq)$$

Some of each of the species shown is present in a solution of phosphoric acid. Because each successive ionization happens to a smaller and smaller extent, the relative concentrations of species in solution are as follows:

$$[H_3PO_4] > [H^+] \approx [H_2PO_4^-] > [HPO_4^{2-}] > [PO_4^{3-}]$$

Student Annotation: These relative concentrations are true only in an aqueous solution of phosphoric acid that contains no other dissolved compounds. We will look in detail at how to determine concentrations in aqueous solutions of polyprotic acids in Chapter 16.

Just as some acids produce more than one H^+ ion, some strong bases produce more than one OH^- ion. Barium hydroxide, for example, dissociates to produce 2 moles of hydroxide ion for every mole of $Ba(OH)_2$ dissolved.

$$Ba(OH)_2(s) \xrightarrow{H_2O} Ba^{2+}(aq) + 2OH^-(aq)$$

Compounds such as this are referred to as *dibasic* bases, indicating that they produce 2 moles of hydroxide per mole of compound. Those that produce only 1 mole of hydroxide per mole of compound, such as NaOH, are called *monobasic* bases.

Acid-Base Neutralization

A *neutralization reaction* is a reaction between an acid and a base. In general, an aqueous acid-base reaction produces water and a *salt,* which is an ionic compound made up of the cation from a base and the anion from an acid. The substance we know as table salt, NaCl, is a familiar example. It is a product of the following acid-base reaction.

Student Annotation: A compound in which the anion is oxide (O^{2-}) or hydroxide (OH^-) is not considered a salt.

$$HCl(aq) + NaOH(aq) \longrightarrow H_2O(l) + NaCl(aq)$$

However, because the acid, base, and salt are all strong electrolytes, they exist entirely as ions in solution. The ionic equation is

$$H^+(aq) + Cl^-(aq) + Na^+(aq) + OH^-(aq) \longrightarrow H_2O(l) + Na^+(aq) + Cl^-(aq)$$

The net ionic equation is

$$H^+(aq) + OH^-(aq) \longrightarrow H_2O(l)$$

Video 9.4
Acids and bases—neutralization reaction of NaOH and HCl.

Both Na^+ and Cl^- are spectator ions. If we were to carry out the preceding reaction using stoichiometric amounts [◄◄ Section 8.3] of HCl and NaOH, the result would be neutral saltwater with no leftover acid or base.

The following are also examples of acid-base neutralization reactions, represented by molecular equations.

$$HNO_3(aq) + KOH(aq) \longrightarrow H_2O(l) + KNO_3(aq)$$

$$H_2SO_4(aq) + 2NaOH(aq) \longrightarrow 2H_2O(l) + Na_2SO_4(aq)$$

$$2HC_2H_3O_2(aq) + Ba(OH)_2(aq) \longrightarrow 2H_2O(l) + Ba(C_2H_3O_2)_2(aq)$$

$$HCl(aq) + NH_3(aq) \longrightarrow NH_4Cl(aq)$$

> **Student Annotation:** Acid-base neutralization reactions, like precipitation reactions [◄◄ Section 9.2], are *metathesis* reactions, where two species exchange ions.

The last equation looks different because it does not show water as a product. Recall, however, that $NH_3(aq)$ ionizes to give $NH_4^+(aq)$ and $OH^-(aq)$. If we include these two species as reactants in place of $NH_3(aq)$, the equation becomes

$$HCl(aq) + NH_4^+(aq) + OH^-(aq) \longrightarrow H_2O(l) + NH_4Cl(aq)$$

Worked Example 9.4 involves an acid-base neutralization reaction.

Worked Example 9.4

Milk of magnesia, an over-the-counter laxative, is a mixture of magnesium hydroxide [$Mg(OH)_2$] and water. Because $Mg(OH)_2$ is insoluble in water (see Table 9.3), milk of magnesia is a *suspension* rather than a solution. The undissolved solid is responsible for the milky appearance of the product. When acid such as HCl is added to milk of magnesia, the suspended $Mg(OH)_2$ dissolves, and the result is a clear, colorless solution. Write balanced molecular, ionic, and net ionic equations for this reaction.

> **Student Annotation:** Most suspended solids will settle to the bottom of the bottle, making it necessary to "shake well before using." Shaking redistributes the solid throughout the liquid.

| (a) Milk of magnesia | (b) Addition of HCl | (c) Resulting clear, colorless solution |

© McGraw-Hill Education/Charles D. Winters, photographer

Strategy Determine the products of the reaction; then write and balance the equation. Remember that one of the reactants, $Mg(OH)_2$, is a solid. Identify any strong electrolytes and rewrite the equation showing strong electrolytes as ions. Identify and cancel the spectator ions.

Setup Because this is an acid-base neutralization reaction, one of the products is water. The other product is a salt comprising the cation from the base, Mg^{2+}, and the anion from the acid, Cl^-. In order for the formula to be neutral, these ions combine in a 1:2 ratio, giving $MgCl_2$ as the formula of the salt.

Solution

$$Mg(OH)_2(s) + 2HCl(aq) \longrightarrow 2H_2O(l) + MgCl_2(aq)$$

Of the species in the molecular equation, only HCl and $MgCl_2$ are strong electrolytes. Therefore, the ionic equation is

$$Mg(OH)_2(s) + 2H^+(aq) + 2Cl^-(aq) \longrightarrow 2H_2O(l) + Mg^{2+}(aq) + 2Cl^-(aq)$$

Cl^- is the only spectator ion. The net ionic equation is

$$Mg(OH)_2(s) + 2H^+(aq) \longrightarrow 2H_2O(l) + Mg^{2+}(aq)$$

Think About It

Make sure your equation is balanced and that you only show strong electrolytes as ions. $Mg(OH)_2$ is *not* shown as aqueous ions because it is insoluble.

Practice Problem ❶**TTEMPT**   Write balanced molecular, ionic, and net ionic equations for the neutralization reaction between $Ba(OH)_2(aq)$ and $HF(aq)$.

Practice Problem ❷**UILD**   Write balanced molecular, ionic, and net ionic equations for the neutralization reaction between $NH_3(aq)$ and $H_2SO_4(aq)$.

Practice Problem ❸**ONCEPTUALIZE**   Which diagram best represents the ions remaining in solution after stoichiometric amounts of aqueous barium hydroxide and hydrobromic acid are combined?

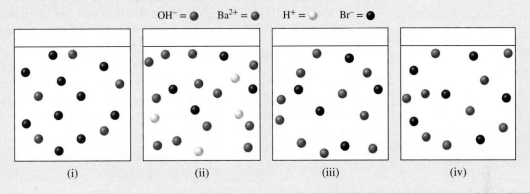

$OH^- =$ ● $Ba^{2+} =$ ● $H^+ =$ ○ $Br^- =$ ●

(i)   (ii)   (iii)   (iv)

Section 9.3 Review

Acid-Base Reactions

9.3.1   Identify the Brønsted acid in the following equation:

$$H_2SO_4(aq) + NH_3(aq) \longrightarrow (NH_4)_2SO_4(aq)$$

  (a) $H_2SO_4(aq)$ (d) $(NH_4)_2SO_4(aq)$
  (b) $NH_3(aq)$ (e) This equation does not contain a Brønsted acid.
  (c) $H_2O(l)$

9.3.2   Identify the Brønsted base in the following equation:

$$HCl(aq) + NO_2^-(aq) \longrightarrow HNO_2(aq) + Cl^-(aq)$$

  (a) $HCl(aq)$ (d) $Cl^-(aq)$
  (b) $NO_2^-(aq)$ (e) $H_2O(l)$
  (c) $HNO_2(aq)$

9.3.3   Which of the following is the correct net ionic equation for the reaction of H_2SO_4 and KOH?
  (a) $H_2(aq) + 2OH^-(aq) \longrightarrow 2H_2O(l)$
  (b) $2H^+(aq) + 2OH^-(aq) \longrightarrow 2H_2O(l)$
  (c) $2H^+(aq) + OH^-(aq) \longrightarrow H_2O(l)$
  (d) $H_2SO_4(aq) + 2OH^-(aq) \longrightarrow 2H_2O(l) + SO_4^{2-}(aq)$
  (e) $H^+(aq) + HSO_4^-(aq) + 2OH^-(aq) \longrightarrow 2H_2O(l) + SO_4^{2-}(aq)$

9.3.4   Which of the following is the correct net ionic equation for the reaction of HF and LiOH?
  (a) $HF(aq) + LiOH(aq) \longrightarrow H_2O(l) + LiF(aq)$
  (b) $H^+(aq) + F^-(aq) + Li^+(aq) + OH^-(aq) \longrightarrow H_2O(l) + Li^+(aq) + F^-(aq)$
  (c) $H^+(aq) + OH^-(aq) \longrightarrow H_2O(l)$
  (d) $HF(aq) + OH^-(aq) \longrightarrow H_2O(l) + F^-(aq)$
  (e) $H^+(aq) + Li^+(aq) + OH^-(aq) \longrightarrow H_2O(l) + Li^+(aq)$

9.4 OXIDATION-REDUCTION REACTIONS

In Sections 9.2 and 9.3, we encountered two types of chemical reactions that can occur when two electrolyte solutions are combined: *precipitation,* in which ionic compounds exchange ions, and *acid-base neutralization,* in which a proton is transferred from an acid to a base. In this section, we will learn about **oxidation-reduction reactions,** commonly called *redox* reactions. A **redox reaction** is a chemical reaction in which *electrons* are transferred from one reactant to another. For example, if we place a piece of zinc metal into a solution that contains copper ions, the following reaction will occur.

$$Zn(s) + Cu^{2+}(aq) \longrightarrow Zn^{2+}(aq) + Cu(s)$$

This reaction is shown in Figure 9.6. In this process, zinc atoms are *oxidized* (they lose electrons) and copper ions are *reduced* (they gain electrons). Each zinc atom loses two electrons to become a zinc ion, and each copper ion gains two electrons to become a copper atom.

$$Zn(s) \longrightarrow Zn^{2+}(aq) + 2e^-$$
$$Cu^{2+}(aq) + 2e^- \longrightarrow Cu(s)$$

These two equations show electrons as a product in the zinc reaction and as a reactant in the copper reaction. Each of these two equations represents a **half-reaction,** the oxidation or the reduction reaction in a redox reaction. The sum of the two half-reaction equations is the overall equation for the redox reaction.

$$Zn(s) \longrightarrow Zn^{2+}(aq) + 2e^-$$
$$\underline{Cu^{2+}(aq) + 2e^- \longrightarrow Cu(s)}$$
$$Zn(s) + Cu^{2+}(aq) + 2e^- \longrightarrow Zn^{2+}(aq) + Cu(s) + 2e^-$$

Although these two processes can be represented by separate equations, they cannot occur separately. In order for one species to gain electrons, another must lose them, and vice versa.

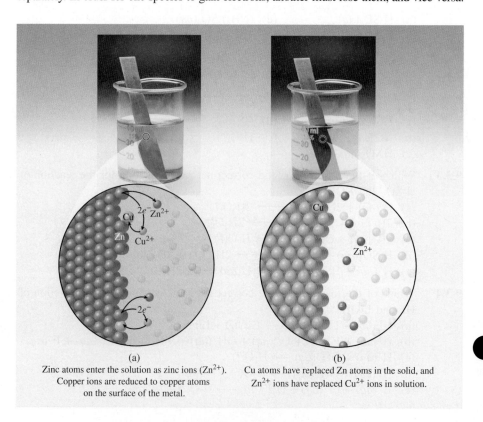

Figure 9.6 Oxidation of zinc in a solution of copper(II) sulfate.

© McGraw-Hill Education/Charles D. Winters, photographer

(a)
Zinc atoms enter the solution as zinc ions (Zn^{2+}). Copper ions are reduced to copper atoms on the surface of the metal.

(b)
Cu atoms have replaced Zn atoms in the solid, and Zn^{2+} ions have replaced Cu^{2+} ions in solution.

Oxidation is the *loss* of electrons. The opposite process, the *gain* of electrons, is called **reduction.** In the reaction of Zn with Cu^{2+}, Zn is called the **reducing agent** because it donates electrons, causing Cu^{2+} to be reduced. Cu^{2+} is called the **oxidizing agent,** on the other hand, because it accepts electrons, causing Zn to be oxidized.

Another example of a redox reaction is the formation of calcium oxide (CaO) from its constituent elements.

$$2Ca(s) + O_2(g) \longrightarrow 2CaO(s)$$

In this reaction, each calcium atom loses two electrons (is oxidized) and each oxygen atom gains two electrons (is reduced). The corresponding half-reactions are

$$2Ca \longrightarrow 2Ca^{2+} + 4e^-$$
$$O_2(g) + 4e^- \longrightarrow 2O^{2-}$$

The resulting Ca^{2+} and O^{2-} ions combine to form CaO.

Redox reactions take place because atoms of different elements have different tendencies to gain electrons. Oxygen, for instance, has a much greater tendency to gain electrons than calcium. Calcium, being a metal, has a significant tendency to lose electrons. Compounds that form between elements with significantly different tendencies to gain electrons generally are ionic. By knowing the charges on the monatomic ions in such compounds, we can keep track of the electrons that have been lost and gained.

Oxidation Numbers

When elements of similar abilities to gain electrons combine, they tend to form molecular compounds, as in the formation of HF and NH_3 from their respective elements.

$$H_2(g) + F_2(g) \longrightarrow 2HF(g)$$
$$N_2(g) + 3H_2(g) \longrightarrow 2NH_3(g)$$

In the formation of hydrogen fluoride (HF), therefore, fluorine does not gain an electron per se—and hydrogen does not lose one. Experimental evidence shows, however, that there is a partial transfer of electrons from H to F. Oxidation numbers provide us with a way to "balance the books" with regard to electrons in a chemical equation. The **oxidation number,** also called the **oxidation state,** is the charge an atom would have *if* electrons were transferred completely. For example, we can rewrite the preceding equations for the formation of HF and NH_3 as follows:

$$H_2(g) \quad + \quad F_2(g) \quad \longrightarrow \quad 2HF(g)$$

$$\begin{array}{ccc} \boxed{0} & \boxed{0} & \boxed{+1}\,\boxed{-1} \\ \boxed{0} & \boxed{0} & \boxed{+1}\,\boxed{-1} \end{array}$$

$$N_2(g) \quad + \quad 3H_2(g) \quad \longrightarrow \quad 2NH_3(g)$$

$$\begin{array}{ccc} \boxed{0} & \boxed{0} & \boxed{-3}\,\boxed{+1} \\ \boxed{0} & \boxed{0} & \boxed{-3}\,\boxed{+3} \end{array}$$

The numbers in circles below each element are the oxidation numbers, and the numbers in the boxes are each atom's contribution to overall charge. In both of the reactions shown, the reactants are all homonuclear diatomic molecules. Thus, we would expect *no* transfer of electrons from one atom to the other and the oxidation number of each is zero. For the product molecules, however, for the sake of determining oxidation numbers, we assume that complete electron transfer has taken place and that each atom has either gained or lost one or more electrons. The oxidation numbers reflect the number of electrons assumed to have been transferred.

Oxidation numbers enable us to identify elements that are oxidized and reduced at a glance. The elements that show an *increase* in oxidation number—hydrogen in the preceding examples—are oxidized, whereas the elements that show a *decrease* in oxidation number—fluorine and nitrogen—are reduced.

Video 9.5
Chemical reactions—formation of Ag_2S by oxidation reduction.

The following guidelines will help you assign oxidation numbers. There are essentially two rules:

1. The oxidation number of any element, in its elemental form, is zero.
2. The oxidation numbers in any chemical species must sum to the overall charge on the species. That is, oxidation numbers must sum to zero for any molecule and must sum to the charge on any polyatomic ion. The oxidation number of a monatomic ion is equal to the charge on the ion.

In addition to these two rules, it is necessary to know the elements that always, or nearly always, have the same oxidation number. Table 9.5 lists elements whose oxidation numbers are "reliable," in order of decreasing reliability.

To determine oxidation numbers in a compound or a polyatomic ion, you must use a stepwise, systematic approach. Draw a circle under each element's symbol in the chemical formula. Then draw a square under each circle. In the circle, write the oxidation number of the element; in the square, write the total contribution to charge by that element. Start with the oxidation numbers you know, and use them to figure out the ones you don't know. Here is an example:

$$KMnO_4$$

Oxidation number
Total contribution to charge

Fill in the oxidation number first for the element that appears highest on the list in Table 9.5. Potassium (K) is a Group 1A metal. In its compounds, it always has the oxidation number +1. We write +1 in the circle beneath the K. Because there is only one K atom in this formula, the total contribution to charge is also +1, so we also write +1 in the square beneath the K.

$$KMnO_4$$

Oxidation number
Total contribution to charge

Next on the list is oxygen (O). In compounds, O usually has the oxidation number −2, so we assign it −2. Because there are four O atoms in the formula, the total contribution to charge by O atoms is 4(−2) = −8.

$$KMnO_4$$

Oxidation number
Total contribution to charge

TABLE 9.5	Elements with Reliable Oxidation Numbers in Compounds or Polyatomic Ions	
Element	**Oxidation number**	**Exceptions**
Fluorine	−1	
Group 1A or 2A metal	+1 or +2, respectively	
Hydrogen	+1	Any combination with a Group 1A or 2A metal to form a metal hydride. Examples: LiH and CaH$_2$—the oxidation number of H is −1 in both examples.
Oxygen	−2	Any combination with something higher on the list that necessitates its having a different oxidation number (see rule 2 for assigning oxidation numbers). Examples: H$_2$O$_2$ and KO$_2$—the oxidation number of O for H$_2$O$_2$ is −1 and for KO$_2$ is −$\frac{1}{2}$.
Group 7A (other than fluorine)	−1	Any combination with something higher on the list that necessitates its having a different oxidation number (see rule 2 for assigning oxidation numbers). Examples: ClF, BrO$_4^-$, and IO$_3^-$—the oxidation numbers of Cl, Br, and I are +1, +7, and +5, respectively. Remember that these exceptions do not apply to fluorine, which *always* has an oxidation state of −1 when it is part of a compound.

The numbers in the squares, all the contributions to overall charge, must sum to zero. This requires putting $+7$ in the box beneath the Mn atom. Because there is just one Mn atom in this formula, the contribution to charge is the same as the oxidation number. Thus, $(+1) + (+7) + (-8) = 0$.

$$KMnO_4$$

Oxidation number ⓐ +1 +7 −2
Total contribution to charge ▢ +1 +7 −8

Worked Example 9.5 lets you determine oxidation numbers in three more compounds and a polyatomic ion.

Worked Example 9.5

Determine the oxidation number of each atom in the following compounds and ion: (a) SO_2, (b) NaH, (c) CO_3^{2-}, (d) N_2O_5.

Strategy For each compound, assign an oxidation number first to the element that appears higher in Table 9.5. Then use rule 2 to determine the oxidation number of the other element.

Setup

(a) O appears in Table 9.5 but S does not, so we assign oxidation number -2 to O. Because there are two O atoms in the molecule, the total contribution to charge by O is $2(-2) = -4$. The lone S atom must therefore contribute $+4$ to the overall charge.

(b) Both Na and H appear in Table 9.5, but Na appears higher in the table, so we assign the oxidation number $+1$ to Na. This means that H must contribute -1 to the overall charge. .... **Student Annotation:** H^- is the hydride ion.

(c) We assign the oxidation number -2 to O. Because there are three O atoms in the carbonate ion, the total contribution to charge by O is -6. To have the contributions to charge sum to the charge on the ion (-2), the C atom must contribute $+4$.

(d) We assign the oxidation number -2 to O. Because there are five O atoms in the N_2O_5 molecule, the total contribution to charge by O is -10. To have the contributions to charge sum to zero, the contribution by N must be $+10$, and because there are two N atoms, each one must contribute $+5$. Therefore, the oxidation number of N is $+5$.

Solution

(a) In SO_2, the oxidation numbers of S and O are $+4$ and -2, respectively. SO_2

+4 −2
+4 −4

(b) In NaH, the oxidation numbers of Na and H are $+1$ and -1, respectively. NaH

+1 −1
+1 −1

(c) In CO_3^{2-}, the oxidation numbers of C and O are $+4$ and -2, respectively. CO_3^{2-}

+4 −2
+4 −6

(d) In N_2O_5, the oxidation numbers of N and O are $+5$ and -2, respectively. N_2O_5

+5 −2
+10 −10

Think About It
Use the circle and square system to verify that the oxidation numbers you have assigned do indeed sum to the overall charge on each species.

Practice Problem Ⓐ**TTEMPT** Assign oxidation numbers to each atom in the following compounds and ion: H_2O_2, ClO_3^-, H_2SO_4.

Practice Problem Ⓑ**UILD** Assign oxidation numbers to each atom in the following species: O_2^{2-}, ClO_3^-, MnO_2.

Practice Problem Ⓒ**ONCEPTUALIZE** Write the balanced equation for the reaction represented by the models and determine oxidation states for each element before and after the reaction.

As with formal charges [◄◄ Section 6.4], each atom's oxidation number makes a contribution to the overall charge on the species. Note that the oxidation numbers in both HF $[(+1) + (-1) = 0]$ and NH_3 $[(-3) + 3(+1) = 0]$ sum to zero. Because compounds are electrically neutral, the oxidation numbers in any compound will sum to zero. For a polyatomic ion, oxidation numbers must sum to the charge on the ion. (The oxidation number of a monatomic ion is equal to its charge.)

Oxidation of Metals in Aqueous Solutions

Recall from the beginning of this section that zinc metal reacts with aqueous copper ions to form aqueous zinc ions and copper metal. One way that this reaction might be carried out is for zinc metal to be immersed in a solution of copper(II) chloride $(CuCl_2)$, as depicted in Figure 9.6. The molecular equation for this reaction is

$$Zn(s) \quad + \quad CuCl_2(aq) \quad \longrightarrow \quad ZnCl_2(aq) \quad + \quad Cu(s)$$

This is an example of a ***displacement reaction.*** Zinc *displaces,* or *replaces,* copper in the dissolved salt by being oxidized from Zn to Zn^{2+}. Copper is displaced from the salt (and removed from solution) by being reduced from Cu^{2+} to Cu. Chloride (Cl^-), which is neither oxidized nor reduced, is a spectator ion in this reaction.

What would happen, then, if we placed copper metal into a solution containing zinc chloride $(ZnCl_2)$? Would $Cu(s)$ be oxidized to $Cu^{2+}(aq)$ by $Zn^{2+}(aq)$ the way $Zn(s)$ is oxidized to $Zn^{2+}(aq)$ by $Cu^{2+}(aq)$? The answer is no. In fact, no reaction would occur if we were to immerse copper metal into an aqueous solution of $ZnCl_2$.

$$Cu(s) + ZnCl_2(aq) \longrightarrow \text{no reaction}$$

No reaction occurs between $Cu(s)$ and $Zn^{2+}(aq)$, whereas a reaction does occur between $Zn(s)$ and $Cu^{2+}(aq)$ because zinc is more easily oxidized than copper.

The ***activity series*** (Table 9.6) is a list of metals (and hydrogen) arranged from top to bottom in order of decreasing ease of oxidation. The second column shows the oxidation half-reaction corresponding to each element in the first column. Note the positions of zinc and copper in the table. Zinc appears higher in the table and is therefore oxidized more easily. In fact, an element in the series will be oxidized by the ions of any element that appears below it. According to Table 9.6, therefore, zinc metal will be oxidized by a solution containing any of the following ions: Cr^{3+}, Fe^{2+}, Cd^{2+}, Co^{2+}, Ni^{2+}, Sn^{2+}, H^+, Cu^{2+}, Ag^+, Hg^{2+}, Pt^{2+}, or Au^{2+}. On the other hand, zinc will not be oxidized by a solution containing Mn^{2+}, Al^{2+}, Mg^{2+}, Na^+, Ca^{2+}, Ba^{2+}, K^+, or Li^+ ions.

Metals listed at the top of the activity series are called the *active metals.* These include the alkali and alkaline earth metals. These metals are so reactive that they are not found in nature in their elemental forms. Metals at the bottom of the series, such as copper, silver, platinum, and gold, are called the *noble metals* because they have very little tendency to react. These are the metals most often used for jewelry and coins.

Student Annotation: In this context, *activity* and *ease of oxidation* mean the same thing.

Student Annotation: When a metal is *oxidized* by an aqueous solution, it becomes an aqueous ion.

Balancing Simple Redox Equations

To learn how to balance redox equations, let's revisit the practice of balancing equations. In Chapter 8, you learned to balance equations by counting the number of each kind of atom on each side of the equation arrow. For the purpose of balancing redox equations, it is also necessary to count electrons. For example, consider the net ionic equation for the reaction of chromium metal with nickel ion.

$$Cr(s) + Ni^{2+}(aq) \longrightarrow Cr^{3+}(aq) + Ni(s)$$

TABLE 9.6	Activity Series	
	Element	**Oxidation half-reaction**
	Lithium	$Li \longrightarrow Li^+ + e^-$
	Potassium	$K \longrightarrow K^+ + e^-$
	Barium	$Ba \longrightarrow Ba^{2+} + 2e^-$
	Calcium	$Ca \longrightarrow Ca^{2+} + 2e^-$
	Sodium	$Na \longrightarrow Na^+ + e^-$
	Magnesium	$Mg \longrightarrow Mg^{2+} + 2e^-$
	Aluminum	$Al \longrightarrow Al^{3+} + 3e^-$
	Manganese	$Mn \longrightarrow Mn^{2+} + 2e^-$
	Zinc	$Zn \longrightarrow Zn^{2+} + 2e^-$
	Chromium	$Cr \longrightarrow Cr^{3+} + 3e^-$
	Iron	$Fe \longrightarrow Fe^{2+} + 2e^-$
	Cadmium	$Cd \longrightarrow Cd^{2+} + 2e^-$
	Cobalt	$Co \longrightarrow Co^{2+} + 2e^-$
	Nickel	$Ni \longrightarrow Ni^{2+} + 2e^-$
	Tin	$Sn \longrightarrow Sn^{2+} + 2e^-$
	Lead	$Pb \longrightarrow Pb^{2+} + 2e^-$
	Hydrogen	$H_2 \longrightarrow 2H^+ + 2e^-$
	Copper	$Cu \longrightarrow Cu^{2+} + 2e^-$
	Silver	$Ag \longrightarrow Ag^+ + e^-$
	Mercury	$Hg \longrightarrow Hg^{2+} + 2e^-$
	Platinum	$Pt \longrightarrow Pt^{2+} + 2e^-$
	Gold	$Au \longrightarrow Au^{3+} + 3e^-$

Increasing ease of oxidation (indicated by upward arrow along the left side of the table)

Although this equation has equal numbers of each type of atom on both sides, it is not balanced because there is a charge of $+2$ on the reactant side and a charge of $+3$ on the product side. To balance it, we can separate it into its half-reactions.

$$Cr(s) \longrightarrow Cr^{3+}(aq) + 3e^-$$

$$Ni^{2+}(aq) + 2e^- \longrightarrow Ni(s)$$

When we add half-reactions to get the overall reaction, the electrons must cancel. Because any electrons lost by one species must be gained by the other, electrons may *not* appear in an overall chemical equation. Therefore, prior to adding these two half-reactions, we must multiply the chromium half-reaction by 2,

$$2[Cr(s) \longrightarrow Cr^{3+}(aq) + 3e^-]$$

and the nickel half-reaction by 3.

$$3[Ni^{2+}(aq) + 2e^- \longrightarrow Ni(s)]$$

Then when we add the half-reactions, the electrons cancel and we get the balanced overall equation.

$$2Cr(s) \longrightarrow 2Cr^{3+}(aq) + 6e^-$$
$$+ \; 3Ni^{2+}(aq) + 6e^- \longrightarrow 3Ni(s)$$
$$\overline{}$$
$$2Cr(s) + 3Ni^{2+}(aq) \longrightarrow 2Cr^{3+}(aq) + 3Ni(s)$$

This is known as the ***half-reaction method*** of balancing redox equations. We will use this method extensively when we examine more complex redox reactions in Chapter 18.

The activity series enables us to predict whether or not a metal will be oxidized by a solution containing a particular salt or by an acid. Worked Examples 9.6 and 9.7 give you more practice making such predictions and balancing redox equations.

Worked Example 9.6

Using the activity series, predict which of the following reactions will occur, and for those that will occur, write the net ionic equation and indicate which element is oxidized and which is reduced.

(a) $Fe(s) + PtCl_2(aq) \longrightarrow$? (b) $Cr(s) + AuCl_3(aq) \longrightarrow$? (c) $Pb(s) + Zn(NO_3)_2(aq) \longrightarrow$?

Strategy Recognize that the salt in each equation (the compound on the reactant side) is a strong electrolyte. What is important is the identity of the metal cation *in* the salt. For each equation, compare the positions in Table 9.6 of the solid metal and the metal cation from the salt to determine whether or not the solid metal will be oxidized. If the cation appears lower in the table, the solid metal will be oxidized (i.e., the reaction will occur). If the cation appears higher in the table, the solid metal will not be oxidized (i.e., no reaction will occur).

Setup

(a) The cation in $PtCl_2$ is Pt^{2+}. Platinum appears lower in Table 9.6 than iron, so $Pt^{2+}(aq)$ will oxidize $Fe(s)$.

(b) The cation in $AuCl_3$ is Au^{3+}. Gold appears lower in Table 9.6 than chromium, so $Au^{3+}(aq)$ will oxidize $Cr(s)$.

(c) The cation in $Zn(NO_3)_2$ is Zn^{2+}. Zinc appears higher in Table 9.6 than lead, so $Zn^{2+}(aq)$ will not oxidize $Pb(s)$.

Solution

(a) $Fe(s) + Pt^{2+}(aq) \longrightarrow Fe^{2+}(aq) + Pt(s)$; iron is oxidized (0 to +2) and platinum is reduced (+2 to 0).

(b) $Cr(s) + Au^{3+}(aq) \longrightarrow Cr^{3+}(aq) + Au(s)$; chromium is oxidized (0 to +3) and gold is reduced (+3 to 0).

(c) No reaction.

Think About It

Check your conclusions by working each problem backward. For part (b), for example, write the net ionic equation in reverse, using the products as the reactants: $Au(s) + Cr^{3+}(aq) \longrightarrow$? Now compare the positions of gold and chromium in Table 9.6 again. Chromium is higher, so chromium(III) ions cannot oxidize gold. This confirms your conclusion that the forward reaction (the oxidation of chromium by gold ions) will occur.

Practice Problem (A)TTEMPT Using the activity series, predict which of the following reactions will occur, and for those that will occur, write the net ionic equation and indicate which element is oxidized and which is reduced:

(a) $Co(s) + BaI_2(aq) \longrightarrow$? (b) $Sn(s) + CuBr_2(aq) \longrightarrow$? (c) $Ag(s) + NaCl(aq) \longrightarrow$?

Practice Problem (B)UILD Using the activity series, predict which of the following reactions will occur, and for those that will occur, write the net ionic equation and indicate which element is oxidized and which is reduced:

(a) $Ni(s) + Cu(NO_3)_2(aq) \longrightarrow$? (b) $Ag(s) + KCl(aq) \longrightarrow$? (c) $Al(s) + AuCl_3(aq) \longrightarrow$?

Practice Problem (C)ONCEPTUALIZE Given the following data, construct an activity series similar to Table 9.6 for five metals: A, B, C, D, and E. The data indicate the results of specific combinations of metals and metal ions.

Experiment 1: $A(s) + D^+(aq) \longrightarrow A^+(aq) + D(s)$ Experiment 4: $C(s) + A^+(aq) \longrightarrow$ no reaction

Experiment 2: $C(s) + B^+(aq) \longrightarrow C^+(aq) + B(s)$ Experiment 5: $B(s) + E^+(aq) \longrightarrow B^+(aq) + E(s)$

Experiment 3: $D(s) + B^+(aq) \longrightarrow$ no reaction   Experiment 6: $D(s) + E^+(aq) \longrightarrow$ no reaction

Worked Example 9.7

Predict which of the following reactions will occur, and for those that will occur, balance the equation and indicate which element is oxidized and which is reduced.

(a) $Al(s) + CaCl_2(aq) \longrightarrow$? (b) $Cr(s) + Pb(C_2H_3O_2)_2(aq) \longrightarrow$? (c) $Sn(s) + HI(aq) \longrightarrow$?

Strategy As in Worked Example 9.6, identify the cation in the aqueous species and for each equation, compare the positions in Table 9.6 of the solid metal and the cation to determine whether or not the solid metal will be oxidized. If the cation appears lower in the table, the reaction will occur.

Setup

(a) The cation in $CaCl_2$ is Ca^{2+}. Calcium appears higher in Table 9.6 than aluminum, so $Ca^{2+}(aq)$ will not oxidize $Al(s)$.

(b) The cation in $Pb(C_2H_3O_2)_2$ is Pb^{2+}. Lead appears lower in Table 9.6 than chromium, so $Pb^{2+}(aq)$ will oxidize $Cr(s)$.

(c) The cation in HI is H^+. Hydrogen appears lower in Table 9.6 than tin, so $H^+(aq)$ will oxidize $Sn(s)$.

Solution

(a) No reaction.

(b) The two half-reactions are represented by the following:

$$\text{Oxidation: } Cr(s) \longrightarrow Cr^{3+}(aq) + 3e^-$$
$$\text{Reduction: } Pb^{2+}(aq) + 2e^- \longrightarrow Pb(s)$$

To balance the charges, we must multiply the oxidation half-reaction by 2 and the reduction half-reaction by 3.

$$2 \times [Cr(s) \longrightarrow Cr^{3+}(aq) + 3e^-] = 2Cr(s) \longrightarrow 2Cr^{3+}(aq) + 6e^-$$
$$3 \times [Pb^{2+}(aq) + 2e^- \longrightarrow Pb(s)] = 3Pb^{2+}(aq) + 6e^- \longrightarrow 3Pb(s)$$

We can then add the two half-reactions, canceling the electrons on both sides to get

$$2Cr(s) + 3Pb^{2+}(aq) \longrightarrow 2Cr^{3+}(aq) + 3Pb(s)$$

The overall, balanced molecular equation is

$$2Cr(s) + 3Pb(C_2H_3O_2)_2(aq) \longrightarrow 2Cr(C_2H_3O_2)_3(aq) + 3Pb(s)$$

Chromium is oxidized (0 to +3) and lead is reduced (+2 to 0).

(c) The two half-reactions are as follows:

$$\text{Oxidation: } Sn(s) \longrightarrow Sn^{2+}(aq) + 2e^-$$
$$\text{Reduction: } 2H^{2+}(aq) + 2e^- \longrightarrow H_2(g)$$

Adding the two half-reactions and canceling the electrons on both sides yields

$$Sn(s) + 2H^+(aq) \longrightarrow Sn^{2+}(aq) + H_2(g)$$

The overall, balanced molecular equation is

$$Sn(s) + 2HI(aq) \longrightarrow SnI_2(aq) + H_2(g)$$

Tin is oxidized (0 to +2) and hydrogen is reduced (+1 to 0). Reactions in which hydrogen ion is reduced to hydrogen gas are known as *hydrogen displacement* reactions.

Think About It
Check your conclusions by working each problem backward. Write each equation in reverse and compare the positions of the elements in the activity series.

Practice Problem ⒶTTEMPT Predict which of the following reactions will occur, and for those that will occur, give the overall, balanced molecular equation and indicate which element is oxidized and which is reduced. (a) $Mg(s) + Cr(C_2H_3O_2)_3(aq) \longrightarrow$? (b) $Cu(s) + HBr(aq) \longrightarrow$? (c) $Cd(s) + AgNO_3(aq) \longrightarrow$?

Practice Problem ⒷUILD Predict which of the following reactions will occur, and for those that will occur, indicate which element is oxidized and which is reduced. (a) $Pt(s) + Cu(NO_3)_2(aq) \longrightarrow$? (b) $Ag(s) + AuCl_3(aq) \longrightarrow$? (c) $Sn(s) + HNO_3(aq) \longrightarrow$?

Practice Problem ⒸONCEPTUALIZE Metals M and N are represented by yellow and white spheres, respectively. Based on the diagrams before and after the reaction, write the corresponding balanced equation and assign oxidation numbers to the metals and their ions.

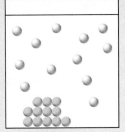

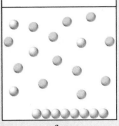

before   after

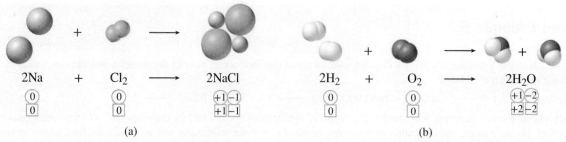

$$2Na \quad + \quad Cl_2 \quad \longrightarrow \quad 2NaCl$$

$$2H_2 \quad + \quad O_2 \quad \longrightarrow \quad 2H_2O$$

(a)   (b)

Figure 9.7 (a) Reaction between sodium and chlorine to form sodium chloride, and (b) reaction between hydrogen and oxygen to form water. For each element, the oxidation number appears in the circle and the total contribution to charge appears in the square below it. Remember that ionic compounds such as NaCl do not exist as discrete formula units but rather as three-dimensional networks or *lattices* [|◀◀ Section 5.3] of alternating oppositely charged ions.

Other Types of Redox Reactions

Several of the reaction types that you have already encountered are also redox reactions.

Combination Reactions

Combination reactions such as the formation of ammonia from its constituent elements can involve oxidation and reduction.

$$N_2(g) \quad + \quad 3H_2(g) \quad \longrightarrow \quad 2NH_3(g)$$

In this reaction, nitrogen is reduced from 0 to −3, while hydrogen is oxidized from 0 to +1. Other examples of combination reactions include those shown in Figure 9.7.

Decomposition

Decomposition can also be a redox reaction, as illustrated by the following examples.

$$2NaH(s) \quad \longrightarrow \quad 2Na(s) \quad + \quad H_2(g)$$

$$2KClO_3(s) \quad \longrightarrow \quad 2KCl(s) \quad + \quad 3O_2(g)$$

$$2H_2O_2(aq) \quad \longrightarrow \quad 2H_2O(l) \quad + \quad O_2(g)$$

The decomposition of hydrogen peroxide, shown in the preceding equation, is an example of a ***disproportionation reaction,*** in which one element undergoes both oxidation and reduction. In the case of H_2O_2, the oxidation number of O is initially −1. In the products of the decomposition, O has an oxidation number of −2 in H_2O and of 0 in O_2.

oxidation
reduction

$$2H_2O_2(aq) \quad \longrightarrow \quad 2H_2O(l) \quad + \quad O_2(g)$$

Combustion

Finally, *combustion* [|◀◀ Section 8.1] is a redox process.

$$CH_4(g) \quad + \quad 2O_2(g) \quad \longrightarrow \quad CO_2(g) \quad + \quad 2H_2O(l)$$

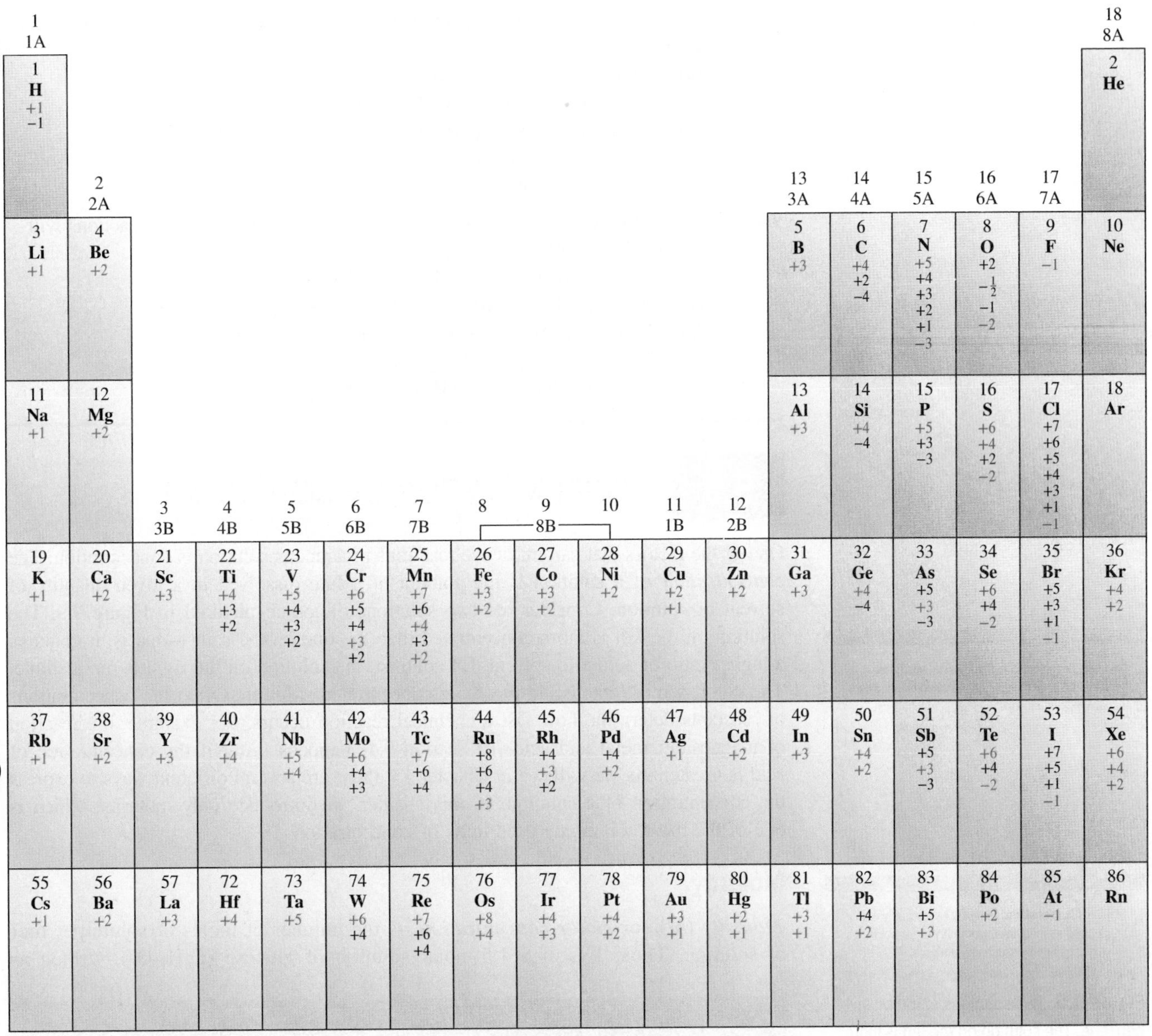

Figure 9.8 Periodic table showing oxidation numbers for each element. The most common oxidation numbers are shown in red.

Figure 9.8 shows the known oxidation numbers of elements in compounds—arranged according to their positions in the periodic table.

Section 9.4 Review

Oxidation-Reduction Reactions

9.4.1 Determine the oxidation number of sulfur in each of the following species: H_2S, HSO_3^-, SCl_2, and S_8.

   (a) $+2, +6, -2, +\frac{1}{4}$ (d) $-1, +4, +2, 0$

   (b) $-2, +3, +2, 0$ (e) $-2, +4, +2, 0$

   (c) $-2, +5, +2, -\frac{1}{4}$

9.4.2 What species is the reducing agent in the following equation?

$$Mg(s) + 2HCl(aq) \longrightarrow MgCl_2(aq) + H_2(g)$$

   (a) $Mg(s)$ (d) $Mg^{2+}(aq)$

   (b) $H^+(aq)$ (e) $H_2(g)$

   (c) $Cl^-(aq)$

(Continued on next page)

9.4.3   Which of the following equations represents a redox reaction? (Select all that apply.)

(a) $2Mg(s) + O_2(g) \rightarrow 2MgO(s)$

(b) $Cu(s) + PtCl_2(aq) \rightarrow CuCl_2(aq) + Pt(s)$

(c) $NH_4Cl(aq) + AgNO_3(aq) \rightarrow NH_4NO_3(aq) + AgCl(s)$

(d) $2NaN_3(s) \rightarrow 2Na(s) + 3N_2(g)$

(e) $CaCO_3(s) \rightarrow CaO(s) + CO_2(g)$

9.4.4   According to the activity series, which of the following redox reactions will occur? (Select all that apply.)

(a) $Fe(s) + NiBr_2(aq) \rightarrow FeBr_2(aq) + Ni(s)$

(b) $Sn(s) + Pb(NO_3)_2(aq) \rightarrow Sn(NO_3)_2(aq) + Pb(s)$

(c) $Mg(s) + BaI_2(aq) \rightarrow MgI_2(aq) + Ba(s)$

(d) $Pb(s) + PtCl_2(aq) \rightarrow PbCl_2(aq) + Pt(s)$

(e) $Zn(s) + CaBr_2(aq) \rightarrow ZnBr_2(aq) + Ca(s)$

9.5   CONCENTRATION OF SOLUTIONS

One of the factors that can influence reactions in aqueous solution is concentration. The *concentration* of a solution is the amount of solute dissolved in a given quantity of solvent or solution. Consider the two solutions of iodine pictured in Figure 9.9. The solution on the left is more concentrated than the one on the right—that is, it contains a higher ratio of solute to solvent. By contrast, the solution on the right is more dilute. The color is more intense in the more concentrated solution. Often the concentrations of reactants determine how fast a chemical reaction occurs. For example, the reaction of magnesium metal and acid [◄◄ Section 9.4] happens faster if the concentration of acid is greater. As we will see in Chapter 13, there are several different ways to express the concentration of a solution. In this chapter, we introduce only molarity, which is one of the most commonly used units of concentration.

Molarity

Molarity (M), or ***molar concentration,*** is the number of moles of solute per liter of solution. Thus, 1 L of a 1.5-molar solution of glucose ($C_6H_{12}O_6$), written as

Student Annotation: The qualitative terms *concentrated* and *dilute* are relative terms, like *expensive* and *cheap*.

Student Annotation: Molarity can equally well be defined as millimoles per milliliter (mmol/mL), which can simplify some calculations.

Figure 9.9 Two solutions of iodine in benzene. The solution on the left is more concentrated. The solution on the right is more dilute.

© McGraw-Hill Education/Charles D. Winters, photographer

Concentrated solution:
More solute particles per unit volume

Dilute solution:
Fewer solute particles per unit volume

1.5 M $C_6H_{12}O_6$, contains 1.5 moles of dissolved glucose. Half a liter of the same solution would contain 0.75 mole of dissolved glucose, a milliliter of the solution would contain 1.5×10^{-3} mole of dissolved glucose, and so on.

$$\text{molarity} = \frac{\text{moles solute}}{\text{liters solution}} \qquad \textbf{Equation 9.1}$$

To calculate the molarity of a solution, we divide the number of moles of solute by the volume of the solution in liters.

Equation 9.1 can be rearranged in three ways to solve for any of the three variables: molarity (M), moles of solute (mol), or volume of solution in liters (L).

$$(1) \; M = \frac{\text{mol}}{L} \qquad (2) \; L = \frac{\text{mol}}{M} \qquad (3) \; mol = M \times L$$

Student Annotation: Students sometimes have difficulty seeing how units cancel in these equations. It may help to write M as mol/L until you become completely comfortable with these equations.

Worked Example 9.8 illustrates how to use these equations to solve for molarity, volume of solution, and moles of solute.

Worked Example 9.8

For an aqueous solution of glucose ($C_6H_{12}O_6$), determine (a) the molarity of 2.00 L of a solution that contains 50.0 g of glucose, (b) the volume of this solution that would contain 0.250 mole of glucose, and (c) the number of moles of glucose in 0.500 L of this solution.

Strategy Convert the mass of glucose given to moles, and use the equations for interconversions of M, liters, and moles to calculate the answers.

Setup The molar mass of glucose is 180.2 g.

$$\text{Moles of glucose} = \frac{50.0 \text{ g}}{180.2 \text{ g/mol}} = 0.277 \text{ mol}$$

Solution

(a) $\text{molarity} = \dfrac{0.277 \text{ mol } C_6H_{12}O_6}{2.00 \text{ L solution}} = 0.139 \; M$

Student Annotation: A common way to state the concentration of this solution is to say, "This solution is 0.139 M in glucose."

(b) $\text{volume} = \dfrac{0.250 \text{ mol } C_6H_{12}O_6}{0.139\text{-}M \text{ solution}} = 1.80 \text{ L}$

(c) moles of $C_6H_{12}O_6$ in 0.500 L = 0.500 L × 0.139 M = 0.0695 mol

Think About It
Check to see that the magnitudes of your answers are logical. For example, the mass given in the problem corresponds to 0.277 mole of solute. If you are asked, as in part (b), for the volume that contains a number of moles smaller than 0.277, make sure your answer is smaller than the original volume.

Practice Problem Ⓐ**TTEMPT** For an aqueous solution of sucrose ($C_{12}H_{22}O_{11}$), determine (a) the molarity of 5.00 L of a solution that contains 235 g of sucrose, (b) the volume of this solution that would contain 1.26 mole of sucrose, and (c) the number of moles of sucrose in 1.89 L of this solution.

Practice Problem Ⓑ**UILD** For an aqueous solution of sodium chloride (NaCl), determine (a) the molarity of 3.75 L of a solution that contains 155 g of sodium chloride, (b) the volume of this solution that would contain 4.58 moles of sodium chloride, and (c) the number of moles of sodium chloride in 22.75 L of this solution.

Practice Problem Ⓒ**ONCEPTUALIZE** The diagrams represent solutions of two different concentrations. What volume of solution 2 contains the same amount of solute as 5.00 mL of solution 1? What volume of solution 1 contains the same amount of solute as 30.0 mL of solution 2?

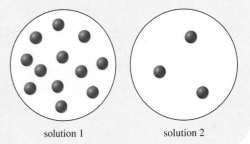

solution 1   solution 2

The procedure for preparing a solution of known molarity is shown in Figure 9.10. First, the solute is weighed accurately and transferred, often with a funnel, to a volumetric flask of the desired volume. Next, water is added to the flask, which is then swirled to dissolve the solid. After all the solid has dissolved, more water is added

Video 9.6
Figure 9.10 Preparing a solution from a solid.

Figure 9.10

Preparing a Solution from a Solid

The mass likely will not be exactly the calculated number.

3.896 g

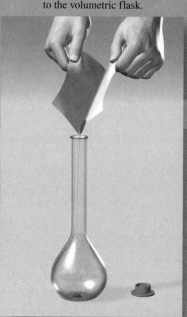

Transfer the weighed KMnO₄ to the volumetric flask.

Weigh out the solid KMnO₄. (The tare function on a digital balance automatically subtracts the mass of the weighing paper.)

2.003 g

Calculate the mass of KMnO₄ necessary for the target concentration of 0.1 *M*.

$$\frac{0.1 \text{ mol}}{\text{L}} \times 0.2500 \text{ L} = 0.02500 \text{ mol}$$

$$0.02500 \text{ mol} \times \frac{158.04 \text{ g}}{\text{mol}} = 3.951 \text{ g KMnO}_4$$

KMnO₄

500 mL

0.000 g

250ml

Add water sufficient to dissolve the $KMnO_4$.

Swirl the flask to dissolve the solid.

Add more water.

Fill exactly to the calibration mark using a wash bottle or eye dropper.

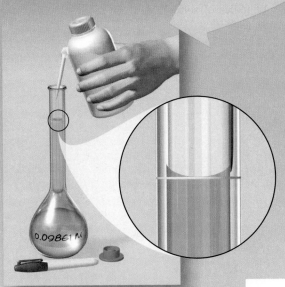

0.09861 M

After capping and inverting the flask to ensure complete mixing, we calculate the actual concentration of the prepared solution.

$$3.896 \text{ g } KMnO_4 \times \frac{1 \text{ mol}}{158.04 \text{ g}} = 0.024652 \text{ mol}$$

$$\frac{0.024652 \text{ mol}}{0.2500 \text{ L}} = 0.09861 \ M$$

What's the point?

The goal is to prepare a solution of precisely known concentration, with that concentration being very close to the target concentration of 0.1 M. Note that because 0.1 is a *specified* number, it does not limit the number of significant figures in our calculations.

slowly to bring the level of solution exactly to the volume mark. Knowing the volume of the solution in the flask and the quantity of compound dissolved, we can determine the molarity of the solution using Equation 9.1. Note that this procedure does not require that we know the exact amount of water added. Because of the way molarity is defined, it is important only that we know the final volume of the *solution.*

Dilution

Concentrated "stock" solutions of commonly used substances typically are kept in the laboratory stockroom. Often we need to dilute these stock solutions before using them. **_Dilution_** is the process of preparing a less concentrated solution from a more concentrated one. Suppose that we want to prepare 1.00 L of a 0.400 M $KMnO_4$ solution from a solution of 1.00 M $KMnO_4$. For this purpose we need 0.400 mole of $KMnO_4$. Because there is 1.00 mole of $KMnO_4$ in 1.00 L of a 1.00 M $KMnO_4$ solution, there is 0.400 mole of $KMnO_4$ in 0.400 L of the same solution.

$$\frac{1.00 \text{ mol } KMnO_4}{1.00 \text{ L of solution}} = \frac{0.400 \text{ mol } KMnO_4}{0.400 \text{ L of solution}}$$

Therefore, we must withdraw 400 mL from the 1.00 M $KMnO_4$ solution and dilute it to 1.00 L by adding water (in a 1.00-L volumetric flask). This method gives us 1.00 L of the desired 0.400 M $KMnO_4$.

In carrying out a dilution process, it is useful to remember that adding more solvent to a given amount of the stock solution changes (decreases) the concentration of the solution without changing the number of moles of solute present in the solution (Figure 9.11).

Video 9.7
Solution—dilution.

Equation 9.2	moles of solute before dilution = moles of solute after dilution

Using arrangement (3) of Equation 9.1, we can calculate the number of moles of solute.

$$\text{moles of solute} = \frac{\text{moles of solute}}{\text{liters of solution}} \times \text{liters of solution}$$

Because the number of moles of solute before the dilution is the same as that after dilution, we can write

Equation 9.3	$M_c \times L_c = M_d \times L_d$

Figure 9.11 Dilution changes the concentration of a solution; it does not change the number of moles of solute in the solution.

© David A. Tietz/Editorial Image, LLC

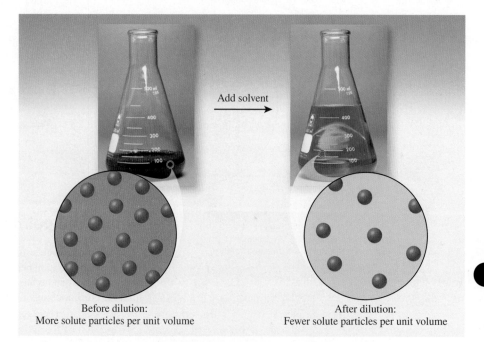

Add solvent

Before dilution:
More solute particles per unit volume

After dilution:
Fewer solute particles per unit volume

where the subscripts c and d stand for *concentrated* and *dilute,* respectively. Thus, by knowing the molarity of the concentrated stock solution (M_c) and the desired final molarity (M_d) and volume (L_d) of the dilute solution, we can calculate the volume of stock solution required for the dilution (L_c).

Because most volumes measured in the laboratory are in milliliters rather than liters, it is worth pointing out that Equation 9.3 can be written with volumes of the concentrated and dilute solutions in milliliters.

$$M_c \times mL_c = M_d \times mL_d \qquad \textbf{Equation 9.4}$$

In this form of the equation, the product of each side is in millimoles (mmol) rather than moles. We apply Equation 9.4 in Worked Example 9.9.

Student Hot Spot

Student data indicate you may struggle with dilutions. Access the SmartBook to view additional Learning Resources on this topic.

Student Annotation: Remember: molarity × mL = millimoles. This saves steps in titration problems.

Worked Example 9.9

What volume of 12.0 *M* HCl, a common laboratory stock solution, must be used to prepare 250.0 mL of 0.125 *M* HCl?

Strategy Because the desired final volume is given in milliliters, we use Equation 9.4 to determine the volume of 12.0 *M* HCl required for the dilution.

Setup M_c = 12.0 *M*, M_d = 0.125 *M*, mL_d = 250.0 mL

Solution

$$12.0 \ M \times mL_c = 0.125 \ M \times 250.0 \ mL$$

$$mL_c = \frac{0.125 \ M \times 250.0 \ mL}{12.0 \ M} = 2.60 \ mL$$

Student Annotation: It is very important to note that, for safety, when diluting a concentrated acid, the acid must be added to the water, and *not* the other way around.

Think About It
Plug the answer into Equation 9.4, and make sure that the product of concentration and volume on both sides of the equation give the same result.

Practice Problem **A**TTEMPT What volume of 6.0 *M* H_2SO_4 is needed to prepare 500.0 mL of a solution that is 0.25 *M* in H_2SO_4?

Practice Problem **B**UILD What volume of 0.20 *M* H_2SO_4 can be prepared by diluting 125 mL of 6.0 *M* H_2SO_4?

Practice Problem **C**ONCEPTUALIZE The diagrams represent a concentrated stock solution (left) and a dilute solution (right) that can be prepared by dilution of the stock solution. How many milliliters of the concentrated stock solution are needed to prepare solutions of the same concentration as the dilute solution of each of the following final volumes? (a) 50.0 mL, (b) 100.0 mL, (c) 250.0 mL

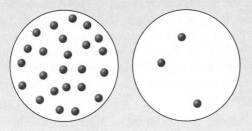

Serial Dilution

A series of dilutions may be used in the laboratory to prepare a number of increasingly dilute solutions from a stock solution. The method involves preparing a solution as described in Figure 9.10 on pages 380-381, and diluting a portion of the prepared solution to make a *more* dilute solution. For example, we could use a 0.400 *M* $KMnO_4$

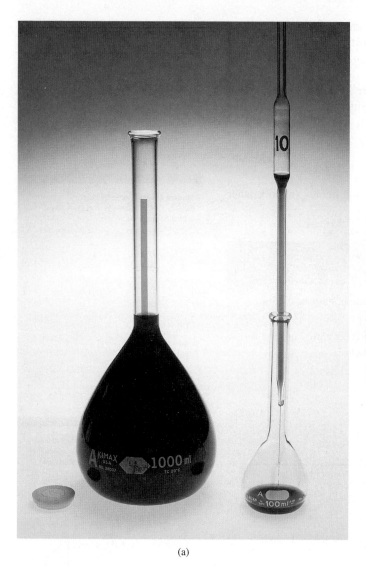

(a)

(b)

Figure 9.12 Serial dilution. (a) A solution of precisely known concentration is prepared in a volumetric flask. A precise volume of the solution is transferred to a second volumetric flask and subsequently diluted. (b) A precise volume of the second solution is transferred to a third volumetric flask and diluted. The process is repeated several times, each time producing more dilute solution. In this example, the concentration is reduced by a factor of 10 at each stage.
© McGraw-Hill Education/Charles D. Winters, photographer.

solution to prepare a series of five increasingly dilute solutions, with the concentration decreasing by a factor of 10 at each stage. Using a volumetric pipette, we withdraw 10.00 mL of the 0.400-M solution and deliver it into a 100.00-mL volumetric flask as shown in Figure 9.12(a). We then dilute to the volumetric mark and cap and invert the flask to ensure complete mixing. The concentration of the newly prepared solution is determined using Equation 9.4, where M_c is 0.400 M, and mL_c and mL_d are 10.00 mL and 100.00 mL, respectively.

$$0.400\ M \times 10.00\ mL = M_d \times 100.00\ mL$$

$$M_d = 0.0400\ M \text{ or } 4.00 \times 10^{-2}\ M$$

Repeating this process four more times, each time using the most recently prepared solution as the "concentrated" solution and diluting 10.00 mL to 100.00 mL, we get five $KMnO_4$ solutions with concentrations $4.00 \times 10^{-2}\ M$, $4.00 \times 10^{-3}\ M$, $4.00 \times 10^{-4}\ M$, $4.00 \times 10^{-5}\ M$, and $4.00 \times 10^{-6}\ M$ [Figure 9.12(b)]. This type of serial dilution is commonly used to prepare "standard" solutions with precisely known concentrations, for quantitative analysis.

Thinking Outside the Box

Visible Spectrophotometry

As we saw in Chapter 3, white light is composed of all the colors of the rainbow. In fact, a rainbow results from the separation of white light by water droplets into the wavelengths that make up the visible spectrum [◄◄ Section 3.2]. Selective absorption of visible light is what makes some solutions appear colored; and, for a solution that is colored, the intensity of color is related to the solution's concentration (see Figure 9.9). This effect gives rise to a type of analysis known as *visible spectrophotometry*. A visible spectrophotometer compares the intensity of light that enters a sample (called the incident light) I_0, with the intensity of the light that is transmitted through the sample, I. *Transmittance* (T) is the ratio of I to I_0.

$$T = \frac{I}{I_0}$$

Absorbance (A) measures how much light is absorbed by the solution and is defined as the negative logarithm of transmittance.

Student Annotation: Both transmittance and absorbance are *unitless* quantities.

$$A = -\log T = -\log\frac{I}{I_0}$$

Plotting absorbance as a function of wavelength gives an absorption spectrum. The absorption spectrum—that is, the characteristic absorption over a range of wavelengths—can serve as a sort of fingerprint for the identification of a compound in solution.

The quantitative relationship between absorbance and a solution's concentration is called the Beer-Lambert law and is expressed as

$$A = \varepsilon bc$$

where ε = proprity constant called the molar absorptivity

b = path length of solution (in centimeters) through which light travels

c = molar concentration of solution

The *molar absorptivity* is specific to a chemical species and is a measure of how strongly the species absorbs light of a particular wavelength. The diagram here shows how absorbance depends on path length and concentration. Quantitative analysis using visible spectrophotometry generally requires selection of the appropriate wavelength for analysis (usually the wavelength at which absorbance is highest), determination of absorbance for a series of solutions of known concentration (the standards), construction of a calibration curve, and calculation of an unknown concentration using the calibration curve.

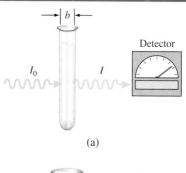

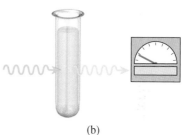

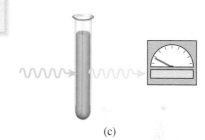

(a) A colored solution absorbs some of the incident visible light, diminishing the light's intensity from I_0 to I. (b) The intensity is reduced more when the light travels through a longer path length of the same solution or (c) when the light travels through the same path length of a more concentrated solution.

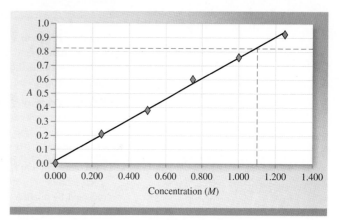

A calibration curve with absorbance (A) on the y axis and molar concentration on the x axis. Linear regression is done using a spreadsheet or graphing calculator to generate the line that best fits all the calibration data. An unknown concentration can be determined by drawing a dashed line from the point on the calibration line corresponding to the measured absorbance to the x axis, as shown. In this case, a measured absorbance of 0.83 corresponds to a concentration of 1.1 M.

Worked Example 9.10 illustrates the method of serial dilution to prepare a series of standard HCl solutions.

Worked Example 9.10

Starting with a 2.0-M stock solution of hydrochloric acid, four standard solutions (1 to 4) are prepared by sequential diluting 10.00 mL of each solution to 250.00 mL. Determine (a) the concentrations of all four standard solutions and (b) the number of moles of HCl in each solution.

Strategy In part (a), because the volumes are all given in milliliters, we will use Equation 9.4, rearranged to solve for M_d, to determine the molar concentration of each standard solution. In part (b), Equation 9.1, rearranged to solve for moles, can be used to calculate the number of moles in each. We must remember to convert each solution's volume to liters so that units will cancel properly.

Setup (a) $M_d = \dfrac{M_c \times mL_c}{mL_d}$; (b) mol $= M \times$ L, 250.00 mL $= 2.500 \times 10^{-1}$ L

Solution (a) $M_{d1} = \dfrac{2.00\ M \times 10.00\ \text{mL}}{250.00\ \text{mL}} = 8.00 \times 10^{-2}\ M$

$M_{d2} = \dfrac{8.00 \times 10^{-2}\ M \times 10.00\ \text{mL}}{250.00\ \text{mL}} = 3.20 \times 10^{-3}\ M$

$M_{d3} = \dfrac{3.20 \times 10^{-3}\ M \times 10.00\ \text{mL}}{250.00\ \text{mL}} = 1.28 \times 10^{-4}\ M$

$M_{d4} = \dfrac{1.28 \times 10^{-4}\ M \times 10.00\ \text{mL}}{250.00\ \text{mL}} = 5.12 \times 10^{-6}\ M$

(b) $\text{mol}_1 = 8.00 \times 10^{-2}\ M \times 2.500 \times 10^{-1}\ L = 2.00 \times 10^{-2}$ mol
$\text{mol}_2 = 3.20 \times 10^{-3}\ M \times 2.500 \times 10^{-1}\ L = 8.00 \times 10^{-4}$ mol
$\text{mol}_3 = 1.28 \times 10^{-4}\ M \times 2.500 \times 10^{-1}\ L = 3.20 \times 10^{-5}$ mol
$\text{mol}_4 = 5.12 \times 10^{-6}\ M \times 2.500 \times 10^{-1}\ L = 1.28 \times 10^{-6}$ mol

Think About It
Serial dilution is one of the fundamental practices of homeopathy. Some remedies undergo so many serial dilutions that very few (if any) molecules of the original substance still exist in the final preparation.

Practice Problem **A** **TTEMPT** Starting with a 6.552-M stock solution of HNO_3, five standard solutions are prepared via serial dilution. At each stage, 25.00 mL of solution is diluted to 100.00 mL. Determine (a) the concentration of and (b) the number of moles of HNO_3 in each standard solution.

Practice Problem **B** **UILD** Five standard solutions of HBr are prepared by serial dilution in which, at each stage, 10.00 mL is diluted to 150.00 mL. Given that the concentration of the most dilute solution is $3.22 \times 10^{-6}\ M$, determine the concentration of the original HBr stock solution.

Practice Problem **C** **ONCEPTUALIZE** The lower diagram represents a concentrated stock solution of a strong electrolyte. Which of the solutions represented here could be prepared by diluting a sample of the stock solution? Select all that apply.

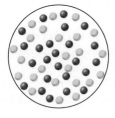

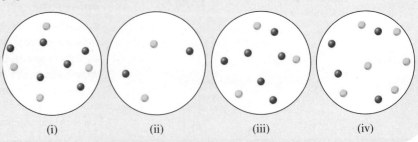

(i)   (ii)   (iii)   (iv)

The pH Scale

The acidity of an aqueous solution depends on the concentration of hydronium ions $[H_3O^+]$. This concentration can range over many orders of magnitude, which can make reporting the numbers cumbersome. To describe the acidity of a solution, rather than report the molar concentration of hydronium ions, we typically use the more convenient pH scale. The *pH* of a solution is defined as the negative base-10 logarithm of the hydronium ion concentration (in mol/L).

$$pH = -\log [H_3O^+] \text{ or } pH = -\log [H^+] \qquad \textbf{Equation 9.5}$$

The pH of a solution is a dimensionless quantity, so the units of concentration must be removed from $[H_3O^+]$ before taking the logarithm. Because $[H_3O^+] = [OH^-] = 1.0 \times 10^{-7}$ M in pure water at 25°C, the pH of pure water at 25°C is

$$-\log (1.0 \times 10^{-7}) = 7.00$$

At 25°C, therefore, a neutral solution has pH 7.00. An acidic solution has pH < 7.00; a basic solution has pH > 7.00. Table 9.7 shows the calculation of pH for solutions ranging from 0.10 M to 1.0×10^{-14} M.

In the laboratory, pH is measured with a pH meter (Figure 9.13). Table 9.8 lists the pH values of a number of common fluids. Note that the pH of body fluids varies greatly, depending on the location and function of the fluid. The low pH (high acidity) of gastric juices is vital for digestion of food, whereas the higher pH of blood is required to facilitate the transport of oxygen.

A measured pH can be used to determine experimentally the concentration of hydronium ion in solution. Solving Equation 9.5 for $[H_3O^+]$ gives

$$[H_3O^+] = 10^{-pH} \qquad \textbf{Equation 9.6}$$

Student Annotation: Equation 9.5 converts numbers that can span an enormous range (~10^{-1} to 10^{-14}) to numbers generally ranging from ~1 to 14.

Student Annotation: A word about significant figures: When we take the log of a number with two significant figures, we report the result to two places past the decimal point. Thus, pH 7.00 has *two* significant figures, not three.

Student Annotation: 10^x is the inverse function of log. (It is the second function on the same key on most calculators.) You must be comfortable performing these operations on your calculator.

TABLE 9.7	Benchmark pH Values for a Range of Hydronium Ion Concentrations at 25°C		
$[H_3O^+]$ (*M*)	$-\log [H_3O^+]$	pH	
0.10	$-\log (1.0 \times 10^{-1})$	1.00	
0.010	$-\log (1.0 \times 10^{-2})$	2.00	
1.0×10^{-3}	$-\log (1.0 \times 10^{-3})$	3.00	
1.0×10^{-4}	$-\log (1.0 \times 10^{-4})$	4.00	
1.0×10^{-5}	$-\log (1.0 \times 10^{-5})$	5.00	
1.0×10^{-6}	$-\log (1.0 \times 10^{-6})$	6.00	Acidic
1.0×10^{-7}	$-\log (1.0 \times 10^{-7})$	7.00	Neutral
1.0×10^{-8}	$-\log (1.0 \times 10^{-8})$	8.00	Basic
1.0×10^{-9}	$-\log (1.0 \times 10^{-9})$	9.00	
1.0×10^{-10}	$-\log (1.0 \times 10^{-10})$	10.00	
1.0×10^{-11}	$-\log (1.0 \times 10^{-11})$	11.00	
1.0×10^{-12}	$-\log (1.0 \times 10^{-12})$	12.00	
1.0×10^{-13}	$-\log (1.0 \times 10^{-13})$	13.00	
1.0×10^{-14}	$-\log (1.0 \times 10^{-14})$	14.00	

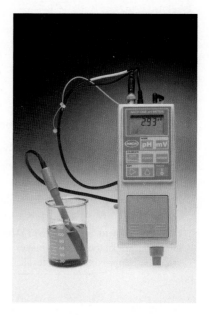

Figure 9.13 A pH meter is commonly used in the laboratory to determine the pH of a solution. Although many pH meters have a range of 1 to 14, pH values can actually be less than 1 and greater than 14.

© McGraw-Hill Education/Charles D. Winters, photographer.

TABLE 9.8	Typical pH Values of Some Common Fluids		
Fluid	**pH**	**Fluid**	**pH**
Stomach acid	1.0	Saliva	6.4–6.9
Lemon juice	2.0	Milk	6.5
Vinegar	3.0	Pure water	7.0
Grapefruit juice	3.2	Blood	7.35–7.45
Orange juice	3.5	Tears	7.4
Urine	4.8–7.5	Milk of magnesia	10.6
Rainwater (in clean air)	5.5	Household ammonia	11.5

Worked Examples 9.11 and 9.12 illustrate calculations involving pH.

Worked Example 9.11

Determine the pH of a solution at 25°C in which the hydronium ion concentration is (a) 3.5×10^{-4} M, (b) 1.7×10^{-7} M, and (c) 8.8×10^{-11} M.

Strategy Given $[H_3O^+]$, use Equation 9.5 to solve for pH.

Setup

(a) pH = $-\log(3.5 \times 10^{-4})$ (b) pH = $-\log(1.7 \times 10^{-7})$ (c) pH = $-\log(8.8 \times 10^{-11})$

Solution

(a) pH = 3.46   (b) pH = 6.77   (c) pH = 10.06

Think About It

When a hydronium ion concentration falls between two "benchmark" concentrations in Table 9.7, the pH falls between the two corresponding pH values. In part (c), for example, the hydronium ion concentration (8.8×10^{-11} M) is greater than 1.0×10^{-11} M but less than 1.0×10^{-10} M. Therefore, we expect the pH to be between 11.00 and 10.00.

$[H_3O^+]$ (M)	$-\log[H_3O^+]$	pH
1.0×10^{-10}	$-\log(1.0 \times 10^{-10})$	10.00
$8.8 \times 10^{-11*}$	$-\log(8.8 \times 10^{-11})$	10.06[†]
1.0×10^{-11}	$-\log(1.0 \times 10^{-11})$	11.00

*$[H_3O^+]$ between two benchmark values
[†]pH between two benchmark values

Recognizing the benchmark concentrations and corresponding pH values is a good way to determine whether or not your calculated result is reasonable.

Practice Problem ATTEMPT Determine the pH of a solution at 25°C in which the hydronium ion concentration is (a) 3.2×10^{-9} M, (b) 4.0×10^{-8} M, and (c) 5.6×10^{-2} M.

Practice Problem BUILD Determine the pH of a solution at 25°C in which the hydronium ion concentration is (a) 1.2 M, (b) 3.0×10^{-11} M, and (c) 8.6×10^{-12} M.

Practice Problem CONCEPTUALIZE Strong acid is added in 1-mL increments to a liter of water at 25°C. Which of the following graphs best approximates the result of plotting hydronium ion concentration as a function of mL acid added? Which graph best approximates the result of plotting pH as a function of mL acid added?

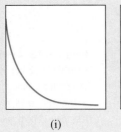
(i)

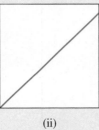
(ii)

(iii)

(iv)

Worked Example 9.12

Calculate the hydronium ion concentration in a solution at 25°C in which the pH is (a) 4.76, (b) 11.95, and (c) 8.01.

Strategy Given pH, use Equation 9.6 to calculate $[H_3O^+]$.

Setup	Solution
(a) $[H_3O^+] = 10^{-4.76}$	(a) $[H_3O^+] = 1.7 \times 10^{-5}\ M$
(b) $[H_3O^+] = 10^{-11.95}$	(b) $[H_3O^+] = 1.1 \times 10^{-12}\ M$
(c) $[H_3O^+] = 10^{-8.01}$	(c) $[H_3O^+] = 9.8 \times 10^{-9}\ M$

Think About It

If you use the calculated hydronium ion concentrations to recalculate pH, you will get numbers slightly different from those given in the problem. In part (a), for example, $-\log (1.7 \times 10^{-5}) = 4.77$. The small difference between this and 4.76 (the pH given in the problem) is due to a rounding error. Remember that a concentration derived from a pH with two digits to the right of the decimal point can have only two significant figures. Note also that the benchmarks can be used equally well in this circumstance. A pH between 4 and 5 corresponds to a hydronium ion concentration between $1 \times 10^{-4}\ M$ and $1 \times 10^{-5}\ M$.

Practice Problem (A)TTEMPT Calculate the hydronium ion concentration in a solution at 25°C in which the pH is (a) 9.90, (b) 1.45, and (c) 7.01.

Practice Problem (B)UILD Calculate the hydronium ion concentration in a solution at 25°C in which the pH is (a) 2.11, (b) 11.59, and (c) 6.87.

Practice Problem (C)ONCEPTUALIZE What is the value of the exponent in the hydronium ion concentration for solutions with pH values of 5.90, 10.11, and 1.25?

Solution Stoichiometry

Soluble ionic compounds such as $KMnO_4$ are strong electrolytes, so they undergo complete dissociation upon dissolution and exist in solution entirely as ions. $KMnO_4$ dissociates, for example, to give 1 mole of potassium ion and 1 mole of permanganate ion for every mole of potassium permanganate. Thus, a 0.400-M solution of $KMnO_4$ will be 0.400 M in K^+ and 0.400 M in MnO_4^-.

In the case of a soluble ionic compound with other than a 1:1 combination of constituent ions, we must use the subscripts in the chemical formula to determine the concentration of each ion in solution. Sodium sulfate (Na_2SO_4) dissociates, for example, to give twice as many sodium ions as sulfate ions.

$$Na_2SO_4(s) \xrightarrow{H_2O} 2Na^+(aq) + SO_4^{2-}(aq)$$

Therefore, a solution that is 0.35 M in Na_2SO_4 is actually 0.70 M in Na^+ and 0.35 M in SO_4^{2-}. Frequently, molar concentrations of dissolved species are expressed using square brackets. Thus, the concentrations of species in a 0.35-M solution of Na_2SO_4 can be expressed as follows: $[Na^+] = 0.70\ M$ and $[SO_4^{2-}] = 0.35\ M$. Worked Example 9.13 lets you practice relating concentrations of compounds and concentrations of individual ions using solution stoichiometry.

Student Annotation: If we only need to express the concentration of the compound, rather than the concentrations of the individual ions, we could express the concentration of this solution as $[Na_2SO_4] = 0.35\ M$.

Student Annotation: Square brackets around a chemical species can be read as "the concentration of" that species. For example, $[Na^+]$ is read as "the concentration of sodium ion."

Worked Example 9.13

Using square-bracket notation, express the concentration of (a) chloride ion in a solution that is 1.02 M in $AlCl_3$, (b) nitrate ion in a solution that is 0.451 M in $Ca(NO_3)_2$, and (c) Na_2CO_3 in a solution in which $[Na^+] = 0.124$ M.

Strategy Use the concentration given in each case and the stoichiometry indicated in the corresponding chemical formula to determine the concentration of the specified ion or compound.

Setup

(a) There are 3 moles of Cl^- ion for every 1 mole of $AlCl_3$,

$$AlCl_3(s) \xrightarrow{H_2O} Al^{3+}(aq) + 3Cl^-(aq)$$

so the concentration of Cl^- will be three times the concentration of $AlCl_3$.

(b) There are 2 moles of nitrate ion for every 1 mole of $Ca(NO_3)_2$,

$$Ca(NO_3)_2(s) \xrightarrow{H_2O} Ca^{2+}(aq) + 2NO_3^-(aq)$$

so $[NO_3^-]$ will be twice $[Ca(NO_3)_2]$.

(c) There is 1 mole of Na_2CO_3 for every 2 moles of sodium ion,

$$Na_2CO_3(s) \xrightarrow{H_2O} 2Na^+(aq) + CO_3^{2-}(aq)$$

so $[Na_2CO_3]$ will be half of $[Na^+]$. (Assume that Na_2CO_3 is the only source of Na^+ ions in this solution.)

Solution

(a) $[Cl^-] = [AlCl_3] \times \dfrac{3 \text{ mol } Cl^-}{1 \text{ mol } AlCl_3} = \dfrac{1.02 \text{ mol } AlCl_3}{L} \times \dfrac{3 \text{ mol } Cl^-}{1 \text{ mol } AlCl_3} = \dfrac{3.06 \text{ mol } Cl^-}{L} = 3.06 \text{ } M$

(b) $[NO_3^-] = [Ca(NO_3)_2] \times \dfrac{2 \text{ mol } NO_3^-}{1 \text{ mol } Ca(NO_3)_2} = \dfrac{0.451 \text{ mol } Ca(NO_3)_2}{L} \times \dfrac{2 \text{ mol } NO_3^-}{1 \text{ mol } Ca(NO_3)_2} = \dfrac{0.902 \text{ mol } NO_3^-}{L} = 0.902 \text{ } M$

(c) $[Na_2CO_3] = [Na^+] \times \dfrac{1 \text{ mol } Na_2CO_3}{2 \text{ mol } Na^+} = \dfrac{0.124 \text{ mol } Na^+}{L} \times \dfrac{1 \text{ mol } Na_2CO_3}{2 \text{ mol } Na^+} = \dfrac{0.0620 \text{ mol } Na_2CO_3}{L} = 0.0620 \text{ } M$

> **Think About It**
> Make sure that units cancel properly. Remember that the concentration of an ion can never be less than the concentration of its dissolved parent compound. It will always be the concentration of the parent compound times its stoichiometric subscript in the chemical formula.

Practice Problem **A**TTEMPT Using the square-bracket notation, express the concentrations of ions in a solution that is 0.750 M in aluminum sulfate $[Al_2(SO_4)_3]$.

Practice Problem **B**UILD Determine the concentration of the compound for a solution in which $[SO_4^{2-}] = 0.36$ M, if the compound is (a) Na_2SO_4, (b) $MgSO_4$, and (c) $Al_2(SO_4)_3$.

Practice Problem **C**ONCEPTUALIZE Which of the diagrams could represent an aqueous solution that contains both NaCl and $BaCl_2$? Select all that apply.

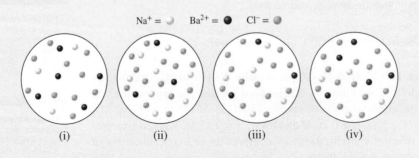

$Na^+ = $ ⚪ $Ba^{2+} = $ ⚫ $Cl^- = $ ◍

(i)   (ii)   (iii)   (iv)

Section 9.5 Review

Concentration of Solutions

9.5.1 Calculate the molar concentration of a solution prepared by dissolving
58.5 g NaOH in enough water to yield 1.25 L of solution.
(a) 1.46 *M*   (d) 1.17 *M*
(b) 46.8 *M*   (e) 0.855 *M*
(c) 2.14×10^{-2} *M*

9.5.2 What mass of glucose ($C_6H_{12}O_6$) in grams must be used to prepare 500 mL
of a solution that is 2.50 *M* in glucose?
(a) 225 g   (d) 1.25 g
(b) 125 g   (e) 625 g
(c) 200 g

9.5.3 What volume in mL of a 1.20 *M* HCl solution must be diluted to prepare
1.00 L of 0.0150 *M* HCl?
(a) 15.0 mL   (d) 85.0 mL
(b) 12.5 mL   (e) 115 mL
(c) 12.0 mL

9.5.4 Determine the pH of a solution at 25°C in which $[H_3O^+] = 6.35 \times 10^{-8}$ *M*.
(a) 7.65   (d) 6.35
(b) 6.80   (e) 8.00
(c) 7.20

9.5.5 Determine $[H_3O^+]$ in a solution at 25°C if pH = 5.75.
(a) 1.8×10^{-6} *M*   (d) 2.4×10^{-9} *M*
(b) 5.6×10^{-9} *M*   (e) 1.0×10^{-6} *M*
(c) 5.8×10^{-6} *M*

9.5.6 For a solution that is 0.18 *M* in Na_2CO_3, express the concentration in
terms of each of the individual ions.
(a) 0.18 *M* in Na^+, 0.18 *M* in CO_3^{2-}
(b) 0.18 *M* in Na^+, 0.090 *M* in CO_3^{2-}
(c) 0.090 *M* in Na^+, 0.18 *M* in CO_3^{2-}
(d) 0.36 *M* in Na^+, 0.18 *M* in CO_3^{2-}
(e) 0.18 *M* in Na^+, 0.36 *M* in CO_3^{2-}

9.6   AQUEOUS REACTIONS AND CHEMICAL ANALYSIS

Certain aqueous reactions are useful for determining how much of a particular
substance is present in a sample. For example, if we want to know the concentration
of lead in a sample of water, or if we need to know the concentration of an acid,
knowledge of precipitation reactions, acid-base reactions, and solution stoichiometry
will be useful. Two common types of such quantitative analyses are *gravimetric
analysis* and *acid-base titration*.

Student Annotation: Experiments that measure the amount of a substance present are called *quantitative analysis*.

Gravimetric Analysis

Gravimetric analysis is an analytical technique based on the measurement of mass.
One type of gravimetric analysis experiment involves the formation and isolation of
a precipitate, such as AgCl(*s*):

Student Annotation: According to the information in Table 9.2, AgCl is an insoluble exception to the chlorides, which typically are soluble.

$$AgNO_3(aq) + NaCl(aq) \longrightarrow NaNO_3(aq) + AgCl(s)$$

This reaction is often used in gravimetric analysis because the reactants can be obtained in pure form. The net ionic equation is

$$Ag^+(aq) + Cl^-(aq) \longrightarrow AgCl(s)$$

Suppose, for example, that we wanted to test the purity of a sample of NaCl by determining the percent by mass of Cl. First, we would accurately weigh out some NaCl and dissolve it in water. To this mixture, we would add enough $AgNO_3$ solution to cause the precipitation of all the Cl^- ions present in solution as AgCl. (In this procedure, NaCl is the limiting reagent and $AgNO_3$ is the excess reagent.) We would then separate, dry, and weigh the AgCl precipitate. From the measured mass of AgCl, we would be able to calculate the mass of Cl using the percent by mass of Cl in AgCl. Because all the Cl in the precipitate came from the dissolved NaCl, the amount of Cl that we calculate is the amount that was present in the original NaCl sample. We could then calculate the percent by mass of Cl in the NaCl and compare it to the known composition of NaCl to determine its purity.

Gravimetric analysis is a highly accurate technique, because the mass of a sample can be measured accurately. However, this procedure is applicable only to reactions that go to completion or have nearly 100 percent yield. In addition, if AgCl were soluble to any significant degree, it would not be possible to remove all the Cl^- ions from the original solution, and the subsequent calculation would be in error. Worked Example 9.14 shows the calculations involved in a gravimetric experiment.

Gravimetric analysis is a quantitative method, not a qualitative one, so it does not establish the identity of the unknown substance. Thus, the results in Worked Example 9.14 do *not* identify the cation. However, knowing the percent by mass of Cl greatly helps us narrow the possibilities. Because no two compounds containing the same anion (or cation) have the same percent composition by mass, comparison of the percent by mass obtained from gravimetric analysis with that calculated from a series of known compounds could reveal the identity of the unknown compounds.

Worked Example 9.14

A 0.8633-g sample of an ionic compound containing chloride ions and an unknown metal cation is dissolved in water and treated with an excess of $AgNO_3$. If 1.5615 g of AgCl precipitate forms, what is the percent by mass of Cl in the original compound?

Strategy Using the mass of AgCl precipitate and the percent composition of AgCl, determine what mass of chloride the precipitate contains. The chloride in the precipitate was originally in the unknown compound. Using the mass of chloride and the mass of the original sample, determine the percent Cl in the compound.

Setup To determine the percent Cl in AgCl, divide the molar mass of Cl by the molar mass of AgCl:

$$\frac{34.45 \text{ g}}{(34.45 \text{ g} + 107.9 \text{ g})} \times 100\% = 24.72\%$$

The mass of Cl in the precipitate is 0.2472×1.5615 g = 0.3860 g.

Solution The percent Cl in the unknown compound is the mass of Cl in the precipitate divided by the mass of the original sample.

$$\frac{0.3860 \text{ g}}{0.8633 \text{ g}} \times 100\% = 44.71\%$$

Think About It
Pay close attention to which numbers correspond to which quantities. It is easy in this type of problem to lose track of which mass is the precipitate and which is the original sample. Dividing by the wrong mass at the end will result in an incorrect answer.

Practice Problem Ⓐ**TTEMPT** A 0.5620-g sample of an ionic compound containing the bromide ion (Br^-) is dissolved in water and treated with an excess of $AgNO_3$. If the mass of the AgBr precipitate that forms is 0.8868 g, what is the percent by mass of Br in the original compound?

Practice Problem Ⓑ**UILD** A sample that is 63.9 percent chloride by mass is dissolved in water and treated with an excess of $AgNO_3$. If the mass of the AgCl precipitate that forms is 1.085 g, what was the mass of the original sample?

Practice Problem **C**ONCEPTUALIZE Which diagram best represents the solution (originally containing sodium chloride) from which the chloride has been removed by the addition of excess silver nitrate?

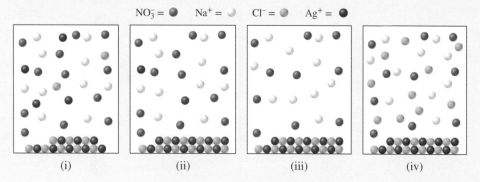

$NO_3^- = $ ● $\quad$ $Na^+ = $ ◐ $\quad$ $Cl^- = $ ● $\quad$ $Ag^+ = $ ●

(i) $\qquad$ (ii) $\qquad$ (iii) $\qquad$ (iv)

Acid-Base Titrations

Quantitative studies of acid-base neutralization reactions are most conveniently carried out using a technique known as titration. In **titration,** a solution of accurately known concentration, called a **standard solution,** is added gradually to another solution of unknown concentration, until the chemical reaction between the two solutions is complete, as shown in Figure 9.14. If we know the volumes of the standard and unknown solutions used in the titration, along with the concentration of the standard solution, we can calculate the concentration of the unknown solution.

A solution of the strong base sodium hydroxide can be used as the standard solution in a titration, but it must first be *standardized,* because sodium hydroxide in solution reacts with carbon dioxide in the air, making its concentration unstable over time. We can standardize the sodium hydroxide solution by titrating it against an acid solution of accurately known concentration. The acid often chosen for this task is a monoprotic acid called potassium hydrogen phthalate (KHP), for which the molecular formula is

Video 9.8
Chemical reactions—titrations.

Student Annotation: *Standardization* in this context is the meticulous determination of concentration.

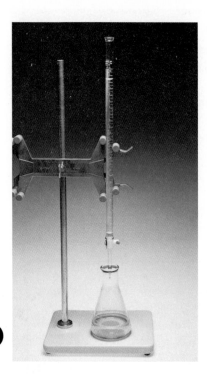

Figure 9.14 Apparatus for titration.
© McGraw-Hill Education/Charles D. Winters, photographer.

$KHC_8H_4O_4$. KHP is a white, soluble solid that is commercially available in highly pure form. The reaction between KHP and sodium hydroxide is

$$KHC_8H_4O_4(aq) + NaOH(aq) \longrightarrow KNaC_8H_4O_4(aq) + H_2O(l)$$

Student Annotation: Note that KHP is a *monoprotic* acid, so it reacts in a 1:1 ratio with hydroxide ion.

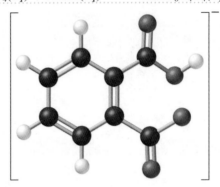

$$HC_8H_4O_4^-$$

and the net ionic equation is

$$HC_8H_4O_4^-(aq) + OH(aq) \longrightarrow C_8H_4O_4^{2-}(aq) + H_2O(l)$$

Student Annotation: The endpoint in a titration is used to approximate the equivalence point. A careful choice of indicators, which we will discuss in Chapter 16, helps make this approximation reasonable. Phenolphthalein, although very common, is not appropriate for every acid-base titration.

To standardize a solution of NaOH with KHP, a known amount of KHP is transferred to an Erlenmeyer flask and some distilled water is added to make up a solution. Next, NaOH solution is carefully added to the KHP solution from a burette until all the acid has reacted with base. This point in the titration, where the acid has been completely neutralized, is called the ***equivalence point.*** It is usually signaled by the ***endpoint,*** where an indicator causes a sharp change in the color of the solution. In acid-base titrations, ***indicators*** are substances that have distinctly different colors in acidic and basic media. One commonly used indicator is phenolphthalein, which is colorless in acidic and neutral solutions but reddish pink in basic solutions. At the equivalence point, all the KHP present has been neutralized by the added NaOH and the solution is still colorless. However, if we add just one more drop of NaOH solution from the burette, the solution will be basic and will immediately turn pink. Worked Example 9.15 illustrates just such a titration.

Worked Example 9.15

In a titration experiment, a student finds that 25.49 mL of an NaOH solution is needed to neutralize 0.7137 g of KHP. What is the concentration (in *M*) of the NaOH solution?

Strategy Using the mass given and the molar mass of KHP, determine the number of moles of KHP. Recognize that the number of moles of NaOH in the volume given is equal to the number of moles of KHP. Divide moles of NaOH by volume (in liters) to get molarity.

Setup The molar mass of KHP ($KHC_8H_4O_4$) = [39.1 g + 5(1.008 g) + 8(12.01 g) + 4(16.00 g)] = 204.2 g/mol.

Solution

$$\text{moles of KHP} = \frac{0.7137 \text{ g}}{204.2 \text{ g/mol}} = 0.003495 \text{ mol}$$

Because moles of KHP = moles of NaOH, then moles of NaOH = 0.003495 mol.

$$\text{molarity of NaOH} = \frac{0.003495 \text{ mol}}{0.02549 \text{ L}} = 0.1371 \text{ } M$$

Think About It
Remember that molarity can also be defined as mmol/mL. Try solving the problem again using millimoles and make sure you get the same answer.

$$0.003495 \text{ mol} = 3.495 \times 10^{-3} \text{ mol}$$

$$= 3.495 \text{ mmol}$$

and

$$\frac{3.495 \text{ mmol}}{25.49 \text{ mL}} = 0.1371 \text{ } M$$

Practice Problem ATTEMPT How many grams of KHP are needed to neutralize 22.36 mL of a 0.1205 *M* NaOH solution?

Practice Problem BUILD What volume (in mL) of a 0.2550 *M* NaOH solution can be neutralized by 10.75 g of KHP?

Practice Problem CONCEPTUALIZE Which diagram best represents a solution (originally containing KHP for standardization of NaOH titrant) at the *equivalence point,* and which best represents the solution at the *endpoint?*

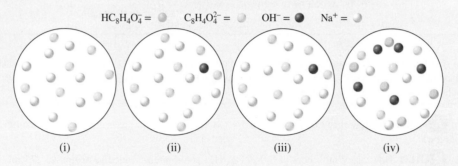

$HC_8H_4O_4^- =$ $C_8H_4O_4^{2-} =$ $OH^- =$ $Na^+ =$

(i)   (ii)   (iii)   (iv)

The reaction between NaOH and KHP is a relatively simple acid-base neutralization. Suppose, though, that instead of KHP, we wanted to use a diprotic acid such as H_2SO_4 for the titration. The reaction is represented by

$$2NaOH(aq) + H_2SO_4(aq) \longrightarrow Na_2SO_4(aq) + 2H_2O(l)$$

Because 2 mol NaOH ≏ 1 mol H_2SO_4, we need twice as much NaOH to react completely with an H_2SO_4 solution of the *same* molar concentration and volume as a monoprotic acid such as HCl. On the other hand, we would need twice the amount of HCl to neutralize a $Ba(OH)_2$ solution compared to an NaOH solution having the same concentration and volume because 1 mole of $Ba(OH)_2$ yields 2 moles of OH^- ions.

$$2HCl(aq) + Ba(OH)_2(aq) \longrightarrow BaCl_2(aq) + 2H_2O(l)$$

In any acid-base titration, regardless of what acid and base are reacting, the total number of moles of H^+ ions that have reacted at the equivalence point must be equal to the total number of moles of OH^- ions that have reacted. Worked Example 9.16 explores the titration of an NaOH solution with a diprotic acid.

Student Hot Spot

Student data indicate you may struggle with titrations. Access the SmartBook to view additional Learning Resources on this topic.

Worked Example 9.16

What volume (in mL) of a 0.203 *M* NaOH solution is needed to neutralize 25.0 mL of a 0.188 *M* H_2SO_4 solution?

Strategy First, write and balance the chemical equation that corresponds to the neutralization reaction.

$$2NaOH(aq) + H_2SO_4(aq) \longrightarrow Na_2SO_4(aq) + 2H_2O(l)$$

The base and the diprotic acid combine in a 2:1 ratio: $2NaOH ≏ H_2SO_4$. Use the molarity and the volume given to determine the number of millimoles of H_2SO_4. Use the number of millimoles of H_2SO_4 to determine the number of millimoles of NaOH. Using millimoles of NaOH and the concentration given, determine the volume of NaOH that will contain the correct number of millimoles.

Student Annotation: Remember: molarity × mL = millimoles. This saves steps in titration problems.

Setup The necessary conversion factors are:

From the balanced equation: $\dfrac{2 \text{ mmol NaOH}}{1 \text{ mmol } H_2SO_4}$

From the molarity of the NaOH given: $\dfrac{1 \text{ mL NaOH}}{0.203 \text{ mmol NaOH}}$

(Continued on next page)

Solution

$$\text{millimoles of } H_2SO_4 = 0.188 \ M \times 25.0 \ \text{mL} = 4.70 \ \text{mmol}$$

$$\text{millimoles of NaOH required} = 4.70 \ \text{mmol } H_2SO_4 \times \frac{2 \ \text{mmol NaOH}}{1 \ \text{mmol } H_2SO_4} = 9.40 \ \text{mmol NaOH}$$

$$\text{volume of 0.203 } M \text{ NaOH} = 9.40 \ \text{mmol NaOH} \times \frac{1 \ \text{mL NaOH}}{0.203 \ \text{mmol NaOH}} = 46.3 \ \text{mL}$$

Think About It
Notice that the two concentrations 0.203 M and 0.188 M are similar. Both round to the same value (~0.20 M) to two significant figures. Therefore, the titration of a diprotic acid with a monobasic base of roughly equal concentration should require roughly twice as much base as the beginning volume of acid: $2 \times 25.0 \ \text{mL} \approx 46.3 \ \text{mL}$.

Practice Problem Ⓐ**TTEMPT** How many milliliters of a 1.42 M H_2SO_4 solution are needed to neutralize 95.5 mL of a 0.336 M KOH solution?

Practice Problem Ⓑ**UILD** How many milliliters of a 0.211 M HCl solution are needed to neutralize 275 mL of a 0.0350 M $Ba(OH)_2$ solution?

Practice Problem Ⓒ**ONCEPTUALIZE** Which diagram best represents the ions in solution at the equivalence point in the titration of $Ba(OH)_2$ with HCl?

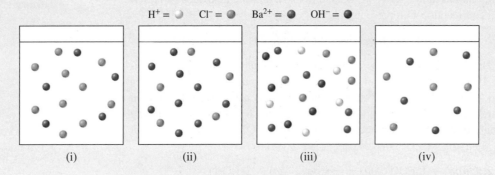

$$H^+ = \bigcirc \qquad Cl^- = \bullet \qquad Ba^{2+} = \bullet \qquad OH^- = \bullet$$

(i)   (ii)   (iii)   (iv)

Worked Example 9.17 shows how titration with a standard base can be used to determine the molar mass of an unknown acid.

Worked Example 9.17

A 0.1216-g sample of a monoprotic acid is dissolved in 25 mL water, and the resulting solution is titrated with 0.1104 M NaOH solution. A 12.5-mL volume of the base is required to neutralize the acid. Calculate the molar mass of the acid.

Strategy Using the concentration and volume of the base, we can determine the number of moles of base required to neutralize the acid. We then determine the number of moles of acid and divide the mass of the acid by the number of moles to get molar mass.

Setup Because the acid is monoprotic, it will react in a 1:1 ratio with the base; therefore, the number of moles of acid will be equal to the number of moles of base. The volume of base in liters is 0.0125 L.

Solution

$$\text{moles of base} = 0.0125 \ \text{L} \times \frac{0.1104 \ \text{mol}}{\text{L}} = 0.00138 \ \text{mol}$$

Because moles of base = moles of acid, the moles of acid = 0.00138 mol. Therefore,

$$\text{molar mass of the acid} = \frac{0.1216 \ \text{g}}{0.00138 \ \text{mol}} = 88.1 \ \text{g/mol}$$

Think About It
In order for this technique to work, we must know whether the acid is monoprotic, diprotic, or polyprotic. A diprotic acid, for example, would combine in a 1:2 ratio with the base, and the result would have been a molar mass twice as large.

Practice Problem (A)TTEMPT What is the molar mass of a monoprotic acid if 28.1 mL of 0.0788 M NaOH is required to neutralize a 0.205-g sample?

Practice Problem (B)UILD What is the molar mass of a diprotic acid if 30.5 mL of 0.1112 M NaOH is required to neutralize a 0.1365-g sample?

Practice Problem (C)ONCEPTUALIZE Consider aqueous solutions of two different acids. Each contains the same mass of acid, and each requires the same volume of 0.10 M NaOH for complete neutralization—and yet the two acids do not have the same molar mass. Explain how this is possible.

Section 9.6 Review

Aqueous Reactions and Chemical Analysis

9.6.1 What mass of AgCl will be recovered if a solution containing 5.00 g of NaCl is treated with enough $AgNO_3$ to precipitate all the chloride ion?
(a) 12.3 g
(d) 9.23 g
(b) 5.00 g
(e) 10.0 g
(c) 3.03 g

9.6.2 A 10.0-g sample of an unknown ionic compound is dissolved, and the solution is treated with enough $AgNO_3$ to precipitate all the chloride ion and 30.1 g of AgCl are recovered. Which of the following compounds could be the unknown?
(a) NaCl
(d) $MgCl_2$
(b) $NaNO_3$
(e) KCl
(c) $BaCl_2$

9.6.3 If 25.0 mL of an H_2SO_4 solution requires 39.9 mL of 0.228 M NaOH to neutralize, what is the concentration of the H_2SO_4 solution?
(a) 0.728 M
(d) 0.228 M
(b) 0.364 M
(e) 0.910 M
(c) 0.182 M

9.6.4 What volume of 0.144 M H_2SO_4 is required to neutralize 25.0 mL of 0.0415 M $Ba(OH)_2$?
(a) 7.20 mL
(d) 50.0 mL
(b) 3.60 mL
(e) 12.5 mL
(c) 14.4 mL

Learning Outcomes

- Categorize compounds as nonelectrolytes, weak electrolytes, or strong electrolytes.
- Identify weak and strong acids and bases.
- Apply solubility guidelines toward determining whether a reaction will produce a precipitate.
- Produce the molecular, ionic, and net ionic equations for a reaction.
- Identify the spectator ions in a reaction.
- Identify the driving force for a reaction in aqueous solution.
- Understand the definition of Arrhenius and Brønsted acids and bases.

- Identify an acid as monoprotic, diprotic, or triprotic.
- Predict the neutralization reaction between an acid and a base.
- Identify the various components of an oxidation-reduction reaction including reducing/oxidizing agents and half-reactions.
- Apply oxidation number rules toward determining the oxidation number of each element in a compound or polyatomic ion.
- Utilize the activity series to determine whether a reaction occurs between a metal and ions of another metal.

- Predict the balanced equation for an oxidation-reduction reaction.
- Identify the different types of reactions that may be oxidation-reduction reactions, including combustion, decomposition, and disproportionation.
- Calculate the molarity of a solution.
- Determine the concentration of a solution that has been diluted in addition to applying dilution principles toward serial dilutions.

- Use the pH scale to classify a solution as being acidic, basic, or neutral.
- Use pH, or concentration of hydronium ion, to conduct calculations asking for one of these terms.
- Apply concepts of stoichiometry toward reactions in solution and their associated problems, including gravimetric analysis and titrations.

Chapter Summary

SECTION 9.1

- A *solution* is a homogeneous mixture consisting of a *solvent* and one or more dissolved species called *solutes.*
- An *electrolyte* is a compound that dissolves in water to give an electrically conducting solution. *Nonelectrolytes* dissolve to give nonconducting solutions. Acids and *bases* are electrolytes.
- Electrolytes may be ionic or molecular. Ionic electrolytes undergo *dissociation* in solution; molecular electrolytes undergo *ionization.* *Strong electrolytes* dissociate (or ionize) completely. *Weak electrolytes* ionize only partially.

SECTION 9.2

- A *precipitation reaction* results in the formation of an insoluble product called a *precipitate.* From general guidelines about solubilities of ionic compounds, we can predict whether a precipitate will form in a reaction.
- *Hydration* is the process in which water molecules surround solute particles.
- *Solubility* is the amount of solute that will dissolve in a specified amount of a given solvent at a specified temperature.
- A *molecular equation* represents a reaction as though none of the reactants or products has dissociated or ionized.
- An *ionic equation* represents the strong electrolytes in a reaction as ions.
- A *spectator ion* is one that is *not* involved in the reaction. Spectator ions appear on both sides of the ionic equation. A *net ionic equation* is an ionic equation from which spectator ions have been eliminated. Reactions in which cations in two ionic compounds exchange anions are called *metathesis* or *double replacement* reactions.

SECTION 9.3

- The hydrogen ion in solution is more realistically represented as the *hydronium ion* (H_3O^+). The terms *hydrogen ion, hydronium ion,* and *proton* are used interchangeably in the context of acid-base reactions.
- *Arrhenius acids* ionize in water to give H_3O^+ ions, whereas *Arrhenius bases* ionize (or *dissociate*) in water to give

OH⁻ ions. *Brønsted acids* donate protons (H^+ ions), whereas *Brønsted bases* accept protons.
- Brønsted acids may be *monoprotic, diprotic,* or *triprotic,* depending on the number of protons they have to donate (the number of ionizable hydrogen atoms they have). In general, an acid with more than one ionizable hydrogen atom is called *polyprotic.* Bases may be *monobasic* or *dibasic.*
- The reaction of an acid and a base is a *neutralization reaction.* The products of a neutralization reaction are water and a *salt.*

SECTION 9.4

- *Oxidation-reduction,* or *redox, reactions* are those in which *electrons* are exchanged. Oxidation and reduction always occur *simultaneously.* You cannot have one without the other.
- *Oxidation* is the loss of electrons; *reduction* is the gain of electrons. In a redox reaction, the *oxidizing agent* is the reactant that gets reduced and the *reducing agent* is the reactant that gets oxidized.
- *Oxidation numbers* or *oxidation states* help us keep track of charge distribution and are assigned to all atoms in a compound or ion according to specific rules.
- Many redox reactions can be further classified as *combination, decomposition, displacement, hydrogen displacement, combustion,* or *disproportionation* reactions. The *activity series* can be used to determine whether or not a displacement reaction will occur.
- A *half-reaction* is a chemical equation representing only the oxidation or only the reduction of an oxidation-reduction reaction. Redox equations, which must be balanced for both mass and charge, can be balanced using the *half-reaction method.*

SECTION 9.5

- The *concentration* of a solution is the amount of solute dissolved in a given amount of solution. *Molarity* (*M*) or *molar concentration* expresses concentration as the number of moles of solute in 1 L of solution.
- Adding a solvent to a solution, a process known as *dilution,* decreases the concentration (molarity) of the solution without changing the total number of moles of solute present in the solution.

- The *pH* scale measures acidity: pH = −log [H⁺].
- pH = 7.00 is neutral, pH < 7.00 is acidic, and pH > 7.00 is basic.

SECTION 9.6

- *Gravimetric analysis* often involves a precipitation reaction.
- Acid-base *titration* involves an acid-base reaction. Typically, a solution of known concentration (a *standard solution*) is added gradually to a solution of unknown concentration with the goal of determining the unknown concentration.
- The point at which the reaction in the titration is complete is called the *equivalence point*. An *indicator* is a substance that changes color at or near the equivalence point of a titration. The point at which the indicator changes color is called the *endpoint* of the titration.

Key Words

Key Equations

9.1	$\text{molarity} = \dfrac{\text{moles solute}}{\text{liters solution}}$	One common expression of concentration is molarity, which is determined by dividing moles of solute by volume of solution in liters.
9.2	moles of solute before dilution = moles of solute after dilution	When a given volume of concentrated solution is diluted, the concentration changes but the number of moles of solute does not change.
9.3	$M_c \times L_c = M_d \times L_d$	When a given volume of concentrated solution is diluted, the molarity multiplied by liters before a dilution is equal to molarity multiplied by liters after a dilution. This enables us to calculate the final molarity after dilution, the number of liters of concentrated stock solution required to perform a desired dilution, and so forth.
9.4	$M_c \times mL_c = M_d \times mL_d$	Often it is more convenient to multiply molarity by milliliters rather than liters. Because the units will cancel, we can use any units of volume in this equation.
9.5	$pH = -\log [H_3O^+]$ or $pH = -\log [H^+]$	The pH of an aqueous solution is calculated as minus the base-10 log of hydronium ion concentration.
9.6	$[H_3O^+] = 10^{-pH}$ or $[H^+] = 10^{-pH}$	Hydronium ion concentration can also be calculated from pH.

Key Skills

Net Ionic Equations

A molecular equation is necessary to do stoichiometric calculations [◀◀ Section 8.3] but molecular equations often misrepresent the species in a solution.

Net ionic equations are preferable in many instances because they indicate more succinctly the species in solution and the actual chemical process that a chemical equation represents. Writing net ionic equations is an important part of solving a variety of problems including those involving precipitation reactions, redox reactions, and acid-base neutralization reactions. To write net ionic equations, you must draw on several skills from earlier chapters:

- Recognition of the common polyatomic ions [◀◀ Section 5.7]
- Balancing chemical equations and labeling species with (s), (l), (g), or (aq) [◀◀ Section 8.1]
- Identification of strong electrolytes, weak electrolytes, and nonelectrolytes [◀◀ Section 9.1]

Writing a net ionic equation begins with writing and balancing the molecular equation. For example, consider the precipitation reaction that occurs when aqueous solutions of sodium iodide and lead(II) nitrate are combined.

$$\boxed{Pb(NO_3)_2(aq)} \; + \; \boxed{NaI(aq)} \longrightarrow$$

Exchanging the ions of the two aqueous reactants gives us the formulas of the products. The phases of the products are determined by considering the solubility guidelines [◀◀ Tables 9.2 and 9.3].

$$\boxed{Pb(NO_3)_2(aq)} \; + \; \boxed{NaI(aq)} \longrightarrow \boxed{PbI_2(s)} \; + \; \boxed{NaNO_3(aq)}$$

We balance the equation and separate the soluble strong electrolytes to get the ionic equation.

$$\boxed{Pb(NO_3)_2(aq)} \; + \; 2\,\boxed{NaI(aq)} \longrightarrow \boxed{PbI_2(s)} \; + \; 2\,\boxed{NaNO_3(aq)}$$

$$\boxed{Pb^{2+}(aq)} + 2\,\boxed{NO_3^-(aq)} + 2\,\boxed{Na^+(aq)} + 2\,\boxed{I^-(aq)} \longrightarrow \boxed{PbI_2(s)} + 2\,\boxed{Na^+(aq)} + 2\,\boxed{NO_3^-(aq)}$$

We then identify the spectator ions, those that are identical on both sides of the equation, and eliminate them.

$$\boxed{Pb^{2+}(aq)} + 2\,\boxed{NO_3^-(aq)} + 2\,\boxed{Na^+(aq)} + 2\,\boxed{I^-(aq)} \longrightarrow \boxed{PbI_2(s)} + 2\,\boxed{Na^+(aq)} + 2\,\boxed{NO_3^-(aq)}$$

What remains is the net ionic equation.

$$\boxed{Pb^{2+}(aq)} \; + \; \boxed{2I^-(aq)} \longrightarrow \boxed{PbI_2(s)}$$

Consider now the reaction that occurs when aqueous solutions of hydrochloric acid and potassium fluoride are combined.

$$HCl(aq) + KF(aq) \longrightarrow$$

Again, exchanging the ions of the two aqueous reactants gives us the formulas of the products.

$$HCl(aq) + KF(aq) \longrightarrow HF(aq) + KCl(aq)$$

This equation is already balanced. We separate soluble strong electrolytes into their constituent ions. In this case, although the products are both aqueous, only one is a strong electrolyte. The other, HF, is a *weak* electrolyte.

$$H^+(aq) + Cl^-(aq) + K^+(aq) + F^-(aq) \longrightarrow HF(aq) + K^+(aq) + Cl^-(aq)$$

We identify the spectator ions and eliminate them.

$$H^+(aq) + \boxed{Cl^-(aq)} + \boxed{K^+(aq)} + F^-(aq) \longrightarrow HF(aq) + \boxed{K^+(aq)} + \boxed{Cl^-(aq)}$$

What remains is the net ionic equation.

$$H^+(aq) + F^-(aq) \longrightarrow HF(aq)$$

You must be able to identify the species in solution as strong, weak, or nonelectrolytes so that you know which should be separated into ions and which should be left as molecular or formula units.

Key Skills Problems

9.1
What is the balanced net ionic equation for the precipitation of $FeSO_4(s)$ when aqueous solutions of K_2SO_4 and $FeCl_2$ are combined?
(a) $2K^+(aq) + SO_4^{2-}(aq) + Fe^{2+}(aq) + 2Cl^-(aq) \longrightarrow$
$FeSO_4(s) + 2K^+(aq) + 2Cl^-(aq)$
(b) $Fe^{2+}(aq) + SO_4^{2-}(aq) \longrightarrow FeSO_4(s)$
(c) $K_2SO_4(aq) + FeCl_2(aq) \longrightarrow FeSO_4(s) + 2KCl(aq)$
(d) $Fe^{2+}(aq) + 2SO_4^{2-}(aq) \longrightarrow FeSO_4(s)$
(e) $2K^+(aq) + SO_4^{2-}(aq) + Fe^{2+}(aq) + 2Cl^-(aq) \longrightarrow FeSO_4(s)$

9.2
Consider the following net ionic equation: $Cd^{2+}(aq) + 2OH^-(aq) \longrightarrow Cd(OH)_2(s)$. If the spectator ions in the ionic equation are $NO_3^-(aq) + K^+(aq)$, what is the molecular equation for this reaction?
(a) $CdNO_3(aq) + KOH(aq) \longrightarrow Cd(OH)_2(s) + KNO_3(aq)$
(b) $Cd^{2+}(aq) + NO_3^-(aq) + 2K^+(aq) + OH^-(aq) \longrightarrow$
$Cd(OH)_2(s) + 2K^+(aq) + NO_3^-(aq)$
(c) $Cd(NO_3)_2(aq) + 2KOH(aq) \longrightarrow Cd(OH)_2(s) + 2KNO_3(aq)$
(d) $Cd(OH)_2(s) + 2KNO_3(aq) \longrightarrow Cd(NO_3)_2(aq) + 2KOH(aq)$
(e) $Cd^{2+}(aq) + NO_3^-(aq) + K^+(aq) + OH^-(aq) \longrightarrow$
$Cd(OH)_2(s) + K^+(aq) + NO_3^-(aq)$

9.3
The net ionic equation for the neutralization of acetic acid $(HC_2H_3O_2)$ with lithium hydroxide $[LiOH(aq)]$ is
(a) $H^+(aq) + OH^-(aq) \longrightarrow H_2O(l)$
(b) $H^+(aq) + C_2H_3O_2^-(aq) \longrightarrow HC_2H_3O_2(aq)$
(c) $HC_2H_3O_2(aq) + OH^-(aq) \longrightarrow H_2O(l) + C_2H_3O_2^-(aq)$
(d) $HC_2H_3O_2(aq) + Li^+(aq) + OH^-(aq) \longrightarrow$
$H_2O(l) + LiC_2H_3O_2(aq)$
(e) $H^+(aq) + C_2H_3O_2^-(aq) + OH^-(aq) \longrightarrow H_2O(l) + C_2H_3O_2^-(aq)$

9.4
When steel wool $[Fe(s)]$ is placed in a solution of $CuSO_4(aq)$, the steel becomes coated with copper metal and the characteristic blue color of the solution fades. What is the net ionic equation for this reaction?
(a) $Fe(s) + CuSO_4(aq) \longrightarrow FeSO_4(aq) + Cu(s)$
(b) $Fe^{2+}(aq) + Cu(s) \longrightarrow Fe(s) + Cu^{2+}(aq)$
(c) $FeSO_4(aq) + Cu(s) \longrightarrow Fe(s) + CuSO_4(aq)$
(d) $Fe(s) + Cu^{2+}(aq) \longrightarrow Fe^{2+}(aq) + Cu(s)$
(e) $Fe(s) + Cu(aq) \longrightarrow Fe(aq) + Cu(s)$

Questions and Problems

SECTION 9.1: GENERAL PROPERTIES OF AQUEOUS SOLUTIONS

Review Questions

9.1   Define *solute, solvent,* and *solution* by describing the process of dissolving a solid in a liquid.

9.2   What is the difference between a nonelectrolyte and an electrolyte? Between a weak electrolyte and a strong electrolyte?

9.3   What is the difference between the symbols $\longrightarrow$ and $\rightleftharpoons$ in chemical equations?

9.4   Although water is classified as a nonelectrolyte, in reality it is an *extremely weak* electrolyte and therefore cannot conduct electricity. Nevertheless, it can be dangerous to operate electrical appliances with wet hands. Explain.

9.5   Identify the species present in aqueous solutions of the following strong electrolytes: (a) LiF, (b) NH_4NO_3, (c) $CaBr_2$, (d) Na_2CO_3.

9.6   The passage of electricity through an electrolyte solution is caused by the movement of (a) electrons only, (b) cations only, (c) anions only, (d) both cations and anions.

9.7   You are given a water-soluble compound X. Describe how you would determine whether it is an electrolyte or a nonelectrolyte. If it is an electrolyte, how would you determine whether it is strong or weak?

Conceptual Problems

9.8   These diagrams show aqueous solutions of three different compounds: (a) AB, (b) AC, and (c) AD. Arrange the compounds in order of increasing electrolyte strength.

(a)   (b)   (c)

9.9   These diagrams show aqueous solutions of three different compounds: (a) A_2E, (b) A_2F, and (c) A_2G. Arrange the compounds in order of increasing electrolyte strength.

(a)   (b)   (c)

9.10   The aqueous solutions of three compounds are shown in the diagram. Identify each compound as a nonelectrolyte, a weak electrolyte, or a strong electrolyte.

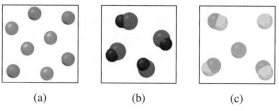

(a)   (b)   (c)

9.11   Which of the following diagrams best represents the hydration of NaCl when dissolved in water? The Cl^- ion is larger in size than the Na^+ ion.

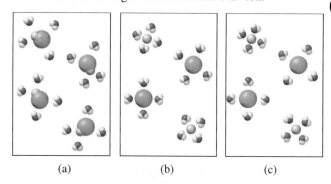

(a)   (b)   (c)

9.12   Identify each of the following substances as a strong electrolyte, weak electrolyte, or nonelectrolyte: (a) H_2O, (b) KCl, (c) HNO_3, (d) $HC_2H_3O_2$, (e) $C_{12}H_{22}O_{11}$.

9.13   Identify each of the following substances as a strong electrolyte, weak electrolyte, or nonelectrolyte: (a) $Ba(NO_3)_2$, (b) Ne, (c) NH_3, (d) NaOH, (e) HF.

SECTION 9.2: PRECIPITATION REACTIONS

Review Questions

9.14   Describe hydration. What properties of water enable its molecules to interact with ions in solution?

9.15 What is the difference between a *molecular equation* and an *ionic equation*? Between an *ionic equation* and a *net ionic equation*? What is the advantage of writing net ionic equations?

Conceptual Problems

9.16 Two aqueous solutions of KOH and MgCl₂ are mixed. Which diagram best represents the resulting combination?

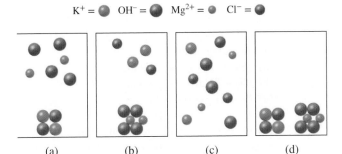

$K^+ = $ $OH^- = $ $Mg^{2+} = $ $Cl^- = $

(a)   (b)   (c)   (d)

9.17 Two aqueous solutions of AgNO₃ and NaCl are mixed. Which diagram best represents the mixture?

$Ag^+ = $ $Cl^- = $ $Na^+ = $ $NO_3^- = $

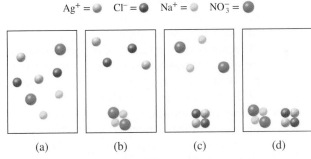

(a)   (b)   (c)   (d)

9.18 Which reaction is represented by the net ionic equation for the combination of aqueous solutions of LiNO₃ and NaC₂H₃O₂?

$Li^+ = $ $NO_3^- = $ $Na^+ = $ $C_2H_3O_2^- = $

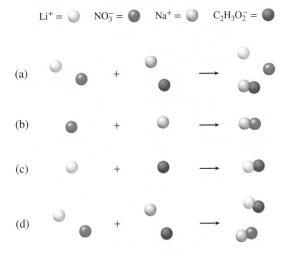

(a)   +   ⟶

(b)   +   ⟶

(c)   +   ⟶

(d)   +   ⟶

(e) There is no net ionic equation. No reaction occurs.

9.19 Which reaction is represented by the net ionic equation for the combination of aqueous solutions of LiOH and Cu(NO₃)₂?

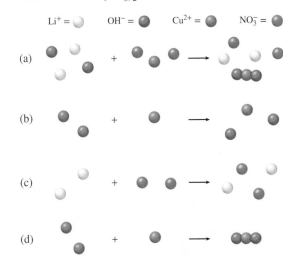

$Li^+ = $ $OH^- = $ $Cu^{2+} = $ $NO_3^- = $

(a)   +   ⟶

(b)   +   ⟶

(c)   +   ⟶

(d)   +   ⟶

(e) There is no net ionic equation. No reaction occurs.

9.20 Characterize the following compounds as soluble or insoluble in water:
(a) CaCO₃, (b) ZnSO₄, (c) Hg(NO₃)₂, (d) HgSO₄, (e) NH₄ClO₄.

9.21 Characterize the following compounds as soluble or insoluble in water: (a) Ca₃(PO₄)₂, (b) Mn(OH)₂, (c) AgClO₃, (d) K₂S, (e) Pb₃(PO₄)₂.

9.22 Write ionic and net ionic equations for the following reactions:
(a) $Na_2S(aq) + ZnCl_2(aq) \longrightarrow$
(b) $K_3PO_4(aq) + 3Sr(NO_3)_2(aq) \longrightarrow$
(c) $Mg(NO_3)_2(aq) + 2NaOH(aq) \longrightarrow$

9.23 Write ionic and net ionic equations for the following reactions:
(a) $AgNO_3(aq) + Na_2SO_4(aq) \longrightarrow$
(b) $BaCl_2(aq) + ZnSO_4(aq) \longrightarrow$
(c) $(NH_4)_2CO_3(aq) + CaCl_2(aq) \longrightarrow$

9.24 Which of the following processes will likely result in a precipitation reaction? (a) Mixing a FeCl₃ solution with a LiOH solution; (b) mixing a KNO₃ solution with a CaCl₂ solution; (c) mixing a (NH₄)₂CO₃ solution with a Na₂CO₃ solution; (d) mixing a Sr(NO₃)₂ solution with a Ca(OH)₂ solution. For mixtures that result in precipitation, write the formula of the precipitate.

9.25 Which of the following processes will likely result in a precipitation reaction? (a) Mixing a NaNO₃ solution with a CuSO₄ solution; (b) mixing a BaCl₂ solution with a K₂SO₄ solution; (c) mixing a AgNO₃ solution with a LiC₂H₃O₂ solution. Write a net ionic equation for each combination.

SECTION 9.3: ACID-BASE REACTIONS

Review Questions

9.26 List the general properties of acids and bases.

9.27 Give Arrhenius's and Brønsted's definitions of an acid and a base. Why are Brønsted's definitions more useful in describing acid-base properties?

9.28 Give an example of a monoprotic acid, a diprotic acid, and a triprotic acid.

9.29 What are the products of an acid-base neutralization reaction?

9.30 What factors qualify a compound as a salt? Specify which of the following compounds are salts: CH_4, NaF, $NaOH$, CaO, $BaSO_4$, HNO_3, NH_3, KBr.

9.31 Identify the following as a weak or strong acid or base: (a) NH_3, (b) H_3PO_4, (c) $LiOH$, (d) $HCOOH$ (formic acid), (e) H_2SO_4, (f) HF, (g) $Ba(OH)_2$.

Conceptual Problems

9.32 Identify each of the following species as a Brønsted acid, base, or both: (a) HI, (b) $C_2H_3O_2^-$, (c) $H_2PO_4^-$, (d) HSO_4^-.

9.33 Identify each of the following species as a Brønsted acid, base, or both: (a) PO_4^{3-}, (b) ClO_2^-, (c) NH_4^+, (d) HCO_3^-.

9.34 Balance the following equations and write the corresponding ionic and net ionic equations (if appropriate):
(a) $HBr(aq) + NH_3(aq) \longrightarrow$
(b) $Ba(OH)_2(aq) + H_3PO_4(aq) \longrightarrow$
(c) $HClO_4(aq) + Mg(OH)_2(s) \longrightarrow$

9.35 Balance the following equations and write the corresponding ionic and net ionic equations (if appropriate):
(a) $HC_2H_3O_2(aq) + KOH(aq) \longrightarrow$
(b) $H_2CO_3(aq) + NaOH(aq) \longrightarrow$
(c) $HNO_3(aq) + Ba(OH)_2(aq) \longrightarrow$

SECTION 9.4: OXIDATION-REDUCTION REACTIONS

Review Questions

9.36 Give an example of a redox reaction that is (a) a combination reaction, (b) a decomposition reaction, (c) a displacement reaction, and (d) a disproportionation reaction.

9.37 What is an oxidation number? How is it used to identify redox reactions?

9.38 Explain how oxidation number is different from *charge* or *partial charge*. Explain under what circumstances it is the same.

9.39 Describe how the activity series is organized, and how it is used to make predictions about redox reactions.

9.40 Use the following reaction to define the terms *redox reaction, half-reaction, oxidizing agent,* and *reducing agent.*

$$4Na(s) + O_2(g) \longrightarrow 2Na_2O(s)$$

9.41 Is it possible to have a reaction in which oxidation occurs and reduction does not? Explain.

Conceptual Problems

9.42 For the complete redox reactions represented here, write the half-reactions and identify the oxidizing and reducing agents.
(a) $4Fe + 3O_2 \longrightarrow 2Fe_2O_3$
(b) $Cl_2 + 2NaBr \longrightarrow 2NaCl + Br_2$
(c) $Si + 2F_2 \longrightarrow SiF_4$
(d) $H_2 + Cl_2 \longrightarrow 2HCl$

9.43 For the complete redox reactions represented here, break down each reaction into its half-reactions, identify the oxidizing agent, and identify the reducing agent.
(a) $2Sr + O_2 \longrightarrow 2SrO$
(b) $2Li + H_2 \longrightarrow 2LiH$
(c) $2Cs + Br_2 \longrightarrow 2CsBr$
(d) $3Mg + N_2 \longrightarrow Mg_3N_2$

9.44 Phosphorus forms a variety of oxoacids. Indicate the oxidation number of phosphorus in each of the following acids: (a) HPO_3, (b) H_3PO_2, (c) H_3PO_3, (d) H_3PO_4, (e) $H_4P_2O_7$, (f) $H_5P_3O_{10}$.

9.45 Arrange the following species in order of increasing oxidation number of the sulfur atom: (a) H_2S, (b) S_8, (c) H_2SO_4, (d) S^{2-}, (e) HS^-, (f) SO_2, (g) SO_3.

9.46 Give the oxidation number for the following species: H_2, Se_8, P_4, O, U, As_4, B_{12}.

9.47 Give the oxidation numbers for the underlined atoms in the following molecules and ions: (a) $\underline{Cl}F$, (b) $\underline{I}F_7$, (c) $\underline{C}H_4$, (d) $\underline{C}_2H_2$, (e) $\underline{C}_2H_4$, (f) $K_2\underline{Cr}O_4$, (g) $K_2\underline{Cr}_2O_7$, (h) $K\underline{Mn}O_4$, (i) $Na\underline{H}CO_3$, (j) $\underline{Li}_2$, (k) $Na\underline{I}O_3$, (l) $K\underline{O}_2$, (m) $\underline{P}F_6^-$, (n) $K\underline{Au}Cl_4$.

9.48 Give the oxidation numbers for the underlined atoms in the following molecules and ions: (a) $Mg_3\underline{N}_2$, (b) $Cs\underline{O}_2$, (c) $Ca\underline{C}_2$, (d) $\underline{C}O_3^{2-}$, (e) $\underline{C}_2O_4^{2-}$, (f) $Zn\underline{O}_2^{2-}$, (g) $Na\underline{B}H_4$, (h) $\underline{W}O_4^{2-}$.

9.49 Give the oxidation numbers for the underlined atoms in the following molecules and ions: (a) $\underline{Cs}_2O$, (b) $Ca\underline{I}_2$, (c) $\underline{Al}_2O_3$, (d) $H_3\underline{As}O_3$, (e) $\underline{Ti}O_2$, (f) $\underline{Mo}O_4^{2-}$, (g) $\underline{Pt}Cl_4^{2-}$, (h) $\underline{Pt}Cl_6^{2-}$, (i) $\underline{Sn}F_2$, (j) $\underline{Cl}F_3$, (k) $\underline{Sb}F_6^-$.

9.50 Determine which of the following metals can react with water: (a) Au, (b) Li, (c) Hg, (d) Ca, (e) Pt.

9.51 Nitric acid is a strong oxidizing agent. State which of the following species is *least* likely to be produced when nitric acid reacts with a strong reducing agent such as zinc metal, and explain why: N_2O, NO, NO_2, N_2O_4, N_2O_5, NH_4^+.

9.52   Predict the outcome of the reactions represented by the following equations by using the activity series, and balance the equations.
(a) $Cu(s) + HCl(aq) \longrightarrow$
(b) $Au(s) + NaBr(aq) \longrightarrow$
(c) $Mg(s) + CuSO_4(aq) \longrightarrow$
(d) $Zn(s) + KBr(aq) \longrightarrow$

9.53   One of the following oxides does not react with molecular oxygen: NO, N_2O, SO_2, SO_3, P_4O_6. Based on oxidation numbers, which one is it? Explain.

9.54   Classify the following redox reactions as *combination, decomposition, displacement,* and/or *disproportionation.*
(a) $P_4 + 10Cl_2 \longrightarrow 4PCl_5$
(b) $2NO \longrightarrow N_2 + O_2$
(c) $Cl_2 + 2KI \longrightarrow 2KCl + I_2$
(d) $3HNO_2 \longrightarrow HNO_3 + H_2O + 2NO$

9.55   Classify the following redox reactions as *combination, decomposition, displacement,* and/or *disproportionation.*
(a) $2H_2O_2 \longrightarrow 2H_2O + O_2$
(b) $Mg + 2AgNO_3 \longrightarrow Mg(NO_3)_2 + 2Ag$
(c) $NH_4NO_2 \longrightarrow N_2 + 2H_2O$
(d) $H_2 + Br_2 \longrightarrow 2HBr$

SECTION 9.5: CONCENTRATION OF SOLUTIONS

Visualizing Chemistry
Figure 9.10

VC 9.1   Which of the following would result in the actual concentration of the prepared solution being higher than the final, calculated value?
(a) Loss of some of the solid during transfer to the volumetric flask.
(b) Neglecting to add the last bit of water with the wash bottle to fill to the volumetric mark.
(c) Neglecting to tare the balance with the weigh paper on the pan.

VC 9.2   Why can't we prepare the solution by first filling the volumetric flask to the mark and then adding the solid?
(a) The solid would not all dissolve.
(b) The solid would not all fit into the flask.
(c) The final volume would not be correct.

VC 9.3   What causes the concentration of the prepared solution not to be exactly 0.1 M?
(a) Rounding error in the calculations.
(b) The volume of the flask is not exactly 250 mL.
(c) The amount of solid weighed out is not exactly the calculated mass.

VC 9.4   The volumetric flask used to prepare a solution from a solid is shown before and after the last of the water has been added.

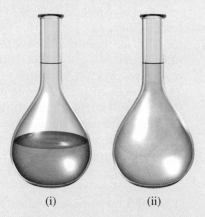

(i)   (ii)

Which of the following statements is true?
(a) The concentration of solute is greater in (i) than in (ii).
(b) The concentration of solute is smaller in (i) than in (ii).
(c) The concentration of solute in (i) is equal to the concentration of solute in (ii).

Review Questions

9.56   Write the equation for calculating molarity. Why is molarity a convenient concentration unit in chemistry?

9.57   Describe the steps involved in preparing a solution of known molar concentration using a volumetric flask.

9.58   Describe the basic steps involved in diluting a solution of known concentration.

9.59   Write the equation that enables us to calculate the concentration of a diluted solution. Give units for all the terms.

9.60   Define pH. Why do chemists normally choose to discuss the acidity of a solution in terms of pH rather than hydrogen ion concentration $[H^+]$?

9.61   The pH of a solution is 6.7. From this statement alone, can you conclude that the solution is acidic? If not, what additional information would you need? Can the pH of a solution be zero or negative? If so, give examples to illustrate these values.

Computational Problems

9.62   Describe how you would prepare 275 mL of a 1.05 M $NaNO_3$ solution.

9.63   Calculate the mass of CaI_2 in grams required to prepare 5.00×10^2 mL of a 2.80-M solution.

9.64   How many grams of $RbOH$ are present in 35.0 mL of a 5.50 M $RbOH$ solution?

9.65   How many grams of $MgCl_2$ are present in 60.0 mL of a 0.100 M $MgCl_2$ solution?

9.66   Calculate the molarity of each of the following solutions: (a) 6.57 g of methanol (CH_3OH) in 1.50×10^2 mL of solution, (b) 10.4 g of calcium chloride ($CaCl_2$) in 2.20×10^2 mL of solution, (c) 7.82 g of naphthalene ($C_{10}H_8$) in 85.2 mL of benzene solution.

9.67   Calculate the molarity of each of the following solutions: (a) 29.0 g of ethanol (C_2H_5OH) in 545 mL of solution, (b) 15.4 g of sucrose ($C_{12}H_{22}O_{11}$) in 74.0 mL of solution, (c) 9.00 g of sodium chloride (NaCl) in 86.4 mL of solution.

9.68   Determine how many grams of each of the following solutes would be needed to make 2.50×10^2 mL of a 0.100-M solution: (a) cesium bromide (CsBr), (b) calcium sulfate ($CaSO_4$), (c) sodium phosphate (Na_3PO_4), (d) lithium dichromate ($Li_2Cr_2O_7$), (e) potassium oxalate ($K_2C_2O_4$).

9.69   Calculate the volume in milliliters of a solution required to provide the following: (a) 2.14 g of sodium chloride from a 0.270-M solution, (b) 4.30 g of ethanol from a 1.50-M solution, (c) 0.85 g of acetic acid ($HC_2H_3O_2$) from a 0.30-M solution.

9.70   Water is added to 75.0 mL of a 0.992 M KNO_3 solution until the volume of the solution is exactly 250 mL. What is the concentration of the final solution?

9.71   Describe how to prepare 1.00 L of a 0.646 M HCl solution, starting with a 2.00 M HCl solution.

9.72   You have 505 mL of a 0.125 M HCl solution and you want to dilute it to exactly 0.100 M. How much water should you add?

9.73   How would you prepare 60.0 mL of 0.200 M HNO_3 from a stock solution of 4.00 M HNO_3?

9.74   A volume of 46.2 mL of a 0.568 M calcium nitrate [$Ca(NO_3)_2$] solution is mixed with 80.5 mL of a 1.396 M calcium nitrate solution. Calculate the concentration of the final solution.

9.75   A volume of 35.2 mL of a 1.66 M $KMnO_4$ solution is mixed with 16.7 mL of a 0.892 M $KMnO_4$ solution. Calculate the concentration of the final solution.

9.76   The maximum level of fluoride that the EPA allows in U.S. drinking water is 4 mg/L. Convert this concentration to molarity.

9.77   According to the website labtestsonline.org, a fasting blood sugar (glucose, $C_6H_{12}O_6$) level of at least 126 milligrams per deciliter (mg/dL) can indicate type 2 diabetes. Convert this concentration to molarity.

9.78   Determine the pH of a solution at 25°C in which the hydronium ion concentration is (a) 5.3×10^{-4} M, (b) 7.1×10^{-6} M, and (c) 4.8×10^{-12} M.

9.79   Determine the pH of a solution at 25°C in which the hydronium ion concentration is (a) 2.7×10^{-3} M, (b) 1.9×10^{-9} M, and (c) 5.6×10^{-2} M.

9.80   Calculate the hydronium ion concentration in mol/L for each of the following solutions: (a) a solution whose pH is 5.20, (b) a solution whose pH is 16.00, (c) a solution whose hydroxide concentration is 3.7×10^{-9} M.

9.81   Calculate the hydronium ion concentration in mol/L for solutions with the following pH values: (a) 2.42, (b) 11.21, (c) 6.96, (d) 15.00.

9.82   Fill in the word *acidic, basic,* or *neutral* for the following solutions at 25°C:
(a) pH > 7; solution is _____.
(b) pH = 7; solution is _____.
(c) pH < 7; solution is _____.

9.83   Complete the following table for a solution at 25°C:

pH	[H^+]	Solution is
<7		
	$<1.0 \times 10^{-7}$ M	
		Neutral

9.84   (a) What is the Na^+ concentration in each of the following solutions: 3.25 M sodium sulfate, 1.78 M sodium carbonate, 0.585 M sodium bicarbonate? (b) What is the concentration of a lithium carbonate solution that is 0.595 M in Li^+?

9.85   (a) Determine the chloride ion concentration in each of the following solutions: 0.150 M $BaCl_2$, 0.566 M NaCl, 1.202 M $AlCl_3$. (b) What is the concentration of a $Sr(NO_3)_2$ solution that is 2.55 M in nitrate ion?

9.86   What volume of 0.112 M ammonium sulfate contains 5.75 g of ammonium ion?

9.87   Determine the resulting nitrate ion concentration when 95.0 mL of 0.992 M potassium nitrate and 155.5 mL of 1.570 M calcium nitrate are combined.

9.88   Absorbance values for five standard solutions of a colored solute were determined at 410 nm with a 1.00-cm path length, giving the following table of data:

Solute concentration (M)	A
0.250	0.165
0.500	0.317
0.750	0.510
1.000	0.650
1.250	0.837

The absorbance of a solution of unknown concentration containing the same solute was 0.400. What is the concentration of the unknown solution (page 385)?

9.89   Referring to Problem 9.88, (a) determine the absorbance values you would expect for solutions with the following concentrations: 0.4 M, 0.6 M, 0.8 M, 1.1 M. (b) Using data in the table in Problem 9.88, calculate the average molar absorptivity of the compound and determine the units of molar absorptivity (page 385).

Conceptual Problems

9.90   Which best represents the before-and-after molecular-level view of the dilution of a concentrated stock solution?

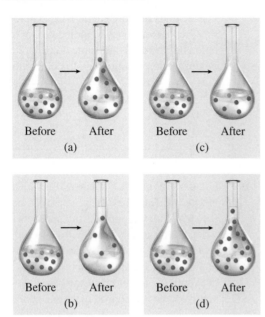

9.91   Which best represents an aqueous solution of sodium sulfate?

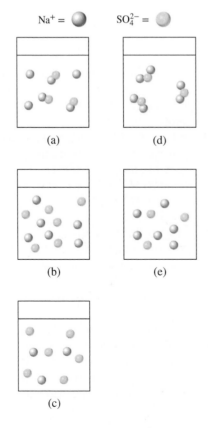

SECTION 9.6: AQUEOUS REACTIONS AND CHEMICAL ANALYSIS

Review Questions

9.92   Describe the basic steps involved in gravimetric analysis. How does this procedure help us determine the identity of a compound or the purity of a compound if its formula is known?

9.93   Explain why distilled water must be used in the gravimetric analysis of chlorides.

9.94   Describe the basic steps involved in an acid-base titration. Why is this technique of great practical value?

9.95   How does an acid-base indicator work?

9.96   A student carried out two titrations using an NaOH solution of unknown concentration in the burette. In one titration she weighed out 0.2458 g of KHP (page 394) and transferred it to an Erlenmeyer flask. She then added 20.00 mL of distilled water to dissolve the acid. In the other titration she weighed out 0.2507 g of KHP but added 40.00 mL of distilled water to dissolve the acid. Assuming no experimental error, would she obtain the same result for the concentration of the NaOH solution?

9.97   Would the volume of a 0.10 M NaOH solution needed to titrate 25.0 mL of a 0.10 M HNO_2 (a weak acid) solution be different from that needed to titrate 25.0 mL of a 0.10 M HCl (a strong acid) solution?

Computational Problems

9.98   A sample of 0.6760 g of an unknown compound containing barium ions (Ba^{2+}) is dissolved in water and treated with an excess of Na_2CO_3. If the mass of the $BaCO_3$ precipitate formed is 0.4105 g, what is the percent by mass of Ba in the original unknown compound?

9.99   If 45.0 mL of 0.250 M $CaCl_2$ is added to 25.0 mL of 1.00 M $AgNO_3$, what is the mass in grams of AgCl precipitate?

9.100   The concentration of Cu^{2+} ions in the water (which also contains sulfate ions) discharged from a certain industrial plant is determined by adding excess sodium sulfide (Na_2S) solution to 0.800 L of the water. The molecular equation is

$$Na_2S(aq) + CuSO_4(aq) \longrightarrow Na_2SO_4(aq) + CuS(aq)$$

Write the net ionic equation and calculate the molar concentration of Cu^{2+} in the water sample if 0.0177 g of solid CuS is formed.

9.101   How many grams of NaCl are required to precipitate most of the Ag ions from 2.50×10^2 mL of a 0.0113 M $AgNO_3$ solution? Write the net ionic equation for the reaction.

9.102   Calculate the concentration (in molarity) of an NaOH solution if 25.0 mL of the solution is needed to neutralize 17.4 mL of a 0.312 M HCl solution.

9.103 A quantity of 18.68 mL of a KOH solution is needed to neutralize 0.4218 g of KHP. What is the concentration (in molarity) of the KOH solution?

9.104 Determine the volume of 1.025 M HCl needed to neutralize each of the following:
(a) 100.0 mL of a 0.300 M NaOH solution
(b) 100.0 mL of a 0.910 M Ba(OH)$_2$ solution

9.105 Calculate the volume in milliliters of a 1.015 M NaOH solution required to titrate the following solutions.
(a) 25.00 mL of a 2.430 M HCl solution
(b) 25.00 mL of a 4.500 M H$_2$SO$_4$ solution
(c) 25.00 mL of a 1.500 M H$_3$PO$_4$ solution

Conceptual Problems

9.106 Which of the following best represents the contents of a beaker in which equal volumes of 0.10 M BaCl$_2$ and 0.10 M AgNO$_3$ were combined?

Ag$^+$ = ◯   NO$_3^-$ = ⬤   Ba^{2+} = ⬤   Cl$^-$ = ◯

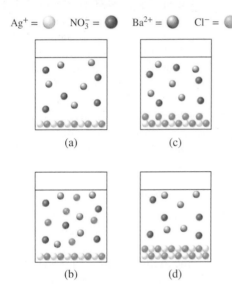

(a)   (c)

(b)   (d)

9.107 Which of the following best represents the contents of a beaker in which equal volumes of 0.10 M NaCl and 0.10 M Pb(NO$_3$)$_2$ were combined?

Na$^+$ = ◯   NO$_3^-$ = ⬤   Pb^{2+} = ⬤   Cl$^-$ = ◯

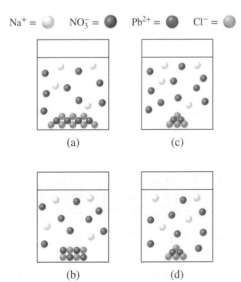

(a)   (c)

(b)   (d)

9.108 Diagram (a) shows a solution of a base and an acid before the neutralization reaction. Each of the after-reaction diagrams, (b)−(d), shows the products of reaction with one of the following acids: HCl, H$_2$SO$_4$, H$_3$PO$_4$. Determine which diagram corresponds to which acid. Blue spheres = OH$^-$ ions, red spheres = acid molecules, green spheres = anions of the acids. Assume all the acid-base neutralization reactions go to completion.

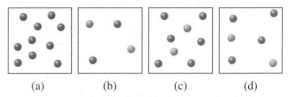

(a)   (b)   (c)   (d)

9.109 Diagram (a) shows a solution of HCl and a base before the neutralization reaction. Of diagrams (b)−(d), which represents the products of reaction when the base is sodium hydroxide and which represents the products when the base is barium hydroxide? Blue spheres = base, red spheres = H$^+$, grey spheres = cations of the bases. Assume all the acid-base neutralization reactions go to completion.

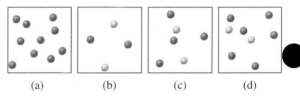

(a)   (b)   (c)   (d)

ADDITIONAL PROBLEMS

9.110 Oxygen (O$_2$) and carbon dioxide (CO$_2$) are colorless and odorless gases. Suggest two chemical tests that would allow you to distinguish between these two gases.

9.111 Which of the following aqueous solutions would you expect to be the best conductor of electricity at 25°C? Explain your answer.
(a) 0.20 M NaCl
(b) 0.60 M HC$_2$H$_3$O$_2$
(c) 0.25 M HCl
(d) 0.20 M Mg(NO$_3$)$_2$

9.112 A 5.00×10^2 mL sample of 2.00 M HCl solution is treated with 4.47 g of magnesium. Calculate the concentration of the acid solution after all the metal has reacted. Assume that the volume remains unchanged.

9.113 Calculate the volume of a 0.156 M CuSO$_4$ solution that would react with 7.89 g of zinc.

9.114 Sodium carbonate (Na$_2$CO$_3$) is available in very pure form and can be used to standardize acid solutions. What is the molarity of an HCl solution if 28.3 mL of the solution is required to react with 0.256 g of Na$_2$CO$_3$?

9.115 Identify each of the following compounds as a nonelectrolyte, a weak electrolyte, or a strong electrolyte: (a) ethanolamine (H$_2$NC$_2$H$_4$OH),

(b) potassium fluoride (KF), (c) ammonium nitrate (NH_4NO_3), (d) isopropanol (C_3H_7OH).

9.116 Identify each of the following compounds as a nonelectrolyte, a weak electrolyte, or a strong electrolyte: (a) lactose ($C_{12}H_{22}O_{11}$), (b) lactic acid ($HC_3H_5O_3$), (c) dimethylamine [$(CH_3)_2NH$], (d) barium hydroxide [$Ba(OH)_2$].

9.117 Determine the predominant species (there may be more than one) in an aqueous solution for each of the compounds in Problem 9.115.

9.118 Determine the predominant species (there may be more than one) in an aqueous solution for each of the compounds in Problem 9.116.

9.119 A 3.664-g sample of a monoprotic acid was dissolved in water. It took 20.27 mL of a 0.1578 M NaOH solution to neutralize the acid. Calculate the molar mass of the acid.

9.120 Acetic acid ($HC_2H_3O_2$) is an important ingredient of vinegar. A sample of 50.0 mL of a commercial vinegar is titrated against a 1.00 M NaOH solution. What is the concentration (in M) of acetic acid present in the vinegar if 5.75 mL of the base is needed for the titration?

9.121 A 15.00-mL solution of potassium nitrate (KNO_3) was diluted to 125.0 mL, and 25.00 mL of this solution was then diluted to 1.000×10^3 mL. The concentration of the final solution is 0.00383 M. Calculate the concentration of the original solution.

9.122 When 2.50 g of a zinc strip was placed in an $AgNO_3$ solution, silver metal formed on the surface of the strip. After some time had passed, the strip was removed from the solution, dried, and weighed. If the mass of the strip was 3.37 g, calculate the mass of Ag and Zn metals present.

9.123 Calculate the mass of the precipitate formed when 2.27 L of 0.0820 M $Ba(OH)_2$ is mixed with 3.06 L of 0.0664 M Na_2SO_4.

9.124 What is the oxidation number of O in HFO?

9.125 Classify the following reactions according to the types discussed in the chapter:
(a) $Cl_2 + 2OH^- \longrightarrow Cl^- + ClO^- + H_2O$
(b) $Ca^{2+} + CO_3^{2-} \longrightarrow CaCO_3$
(c) $NH_3 + H^+ \longrightarrow NH_4^+$
(d) $2CCl_4 + CrO_4^{2-} \longrightarrow 2COCl_2 + CrO_2Cl_2 + 2Cl^-$
(e) $Ca + F_2 \longrightarrow CaF_2$
(f) $2Li + H_2 \longrightarrow 2LiH$
(g) $Ba(NO_3)_2 + Na_2SO_4 \longrightarrow 2NaNO_3 + BaSO_4$
(h) $CuO + H_2 \longrightarrow Cu + H_2O$
(i) $Zn + 2HCl \longrightarrow ZnCl_2 + H_2$
(j) $2FeCl_2 + Cl_2 \longrightarrow 2FeCl_3$

9.126 Calculate the concentration of the acid (or base) remaining in solution when 10.7 mL of 0.211 M HNO_3 is added to 16.3 mL of 0.258 M NaOH.

9.127 (a) Describe a preparation for magnesium hydroxide [$Mg(OH)_2$] and predict its solubility. (b) Milk of magnesia contains mostly $Mg(OH)_2$ and is effective in treating acid (mostly hydrochloric acid)

indigestion. Calculate the volume of a 0.035 M HCl solution (a typical acid concentration in an upset stomach) needed to react with two spoonfuls (approximately 10 mL) of milk of magnesia [at 0.080 g $Mg(OH)_2$/mL].

9.128 A quantitative definition of solubility is the number of grams of a solute that will dissolve in a given volume of water at a particular temperature. Describe an experiment that would enable you to determine the solubility of a soluble compound.

9.129 A 60.0-mL 0.513 M glucose ($C_6H_{12}O_6$) solution is mixed with 120.0 mL of a 2.33 M glucose solution. What is the concentration of the final solution? Assume the volumes are additive.

9.130 An ionic compound X is only slightly soluble in water. What test would you employ to show that the compound does indeed dissolve in water to a certain extent?

9.131 You are given a colorless liquid. Describe three chemical tests you would perform on the liquid to show that it is water.

9.132 Using the apparatus shown in Figure 9.1, a student found that a sulfuric acid solution caused the lightbulb to glow brightly. However, after the addition of a certain amount of a barium hydroxide [$Ba(OH)_2$] solution, the light began to dim even though $Ba(OH)_2$ is also a strong electrolyte. Explain.

9.133 Which of the diagrams shown corresponds to the reaction between AgOH(s) and HNO_3(aq)? Write a balanced equation for the reaction. (For simplicity, water molecules are not shown.)

(a)　　　　　(b)　　　　　(c)

● Ag^+
● NO_3^-

9.134 Which of the diagrams shown corresponds to the reaction between $Ba(OH)_2$(aq) and H_2SO_4(aq)? Write a balanced equation for the reaction. (For simplicity, water molecules are not shown.)

(a)　　　　　(b)　　　　　(c)

● Ba^{2+}
● SO_4^{2-}

9.135 You are given a soluble compound of an unknown molecular formula.

(a) Describe three tests that would show that the compound is an acid.

(b) Once you have established that the compound is an acid, describe how you would determine its molar mass using an NaOH solution of known concentration. (Assume the acid is monoprotic.)

(c) How would you find out whether the acid is weak or strong? You are provided with a sample of NaCl and an apparatus like that shown in Figure 9.1 for comparison.

9.136 You are given two colorless solutions, one containing NaCl and the other sucrose ($C_{12}H_{22}O_{11}$). Suggest a chemical and a physical test that would allow you to distinguish between these two solutions.

9.137 The concentration of lead ions (Pb^{2+}) in a sample of polluted water that also contains nitrate ions (NO_3^-) is determined by adding solid sodium sulfate (Na_2SO_4) to exactly 500 mL of the water. (a) Write the molecular and net ionic equations for the reaction. (b) Calculate the molar concentration of Pb^{2+} if 0.00450 g of Na_2SO_4 was needed for the complete precipitation of Pb^{2+} ions as $PbSO_4$.

9.138 Hydrochloric acid is not an oxidizing agent in the sense that sulfuric acid and nitric acid are. Explain why the chloride ion is not a strong oxidizing agent like SO_4^{2-} and NO_3^-.

9.139 Explain how you would prepare potassium iodide (KI) by means of (a) an acid-base reaction and (b) a reaction between an acid and a carbonate compound.

9.140 Sodium reacts with water to yield hydrogen gas. Why is this reaction not used in the laboratory preparation of hydrogen?

9.141 Describe how you would prepare the following compounds: (a) $Mg(OH)_2$, (b) AgI, (c) $Ba_3(PO_4)_2$.

9.142 Someone spilled concentrated sulfuric acid on the floor of a chemistry laboratory. To neutralize the acid, would it be preferable to pour concentrated sodium hydroxide solution or spray solid sodium bicarbonate over the acid? Explain your choice and the chemical basis for the action.

9.143 Describe in each case how you would separate the cations or anions in the following aqueous solutions: (a) $NaNO_3$ and $Ba(NO_3)_2$, (b) $Mg(NO_3)_2$ and KNO_3, (c) KBr and KNO_3, (d) K_3PO_4 and KNO_3, (e) Na_2CO_3 and $NaNO_3$.

9.144 The following are common household compounds: salt (NaCl), sugar (sucrose), vinegar (contains acetic acid, $HC_2H_3O_2$), baking soda ($NaHCO_3$), washing soda ($Na_2CO_3 \cdot 10H_2O$), boric acid (H_3BO_3, used in eyewash), Epsom salts ($MgSO_4 \cdot 7H_2O$),

sodium hydroxide (used in drain openers), ammonia, milk of magnesia [$Mg(OH)_2$], and calcium carbonate. Based on what you have learned in this chapter, describe tests that would allow you to identify each of these compounds.

9.145 Sulfites (compounds containing the SO_3^{2-} ions) are used as preservatives in dried fruits and vegetables and in wine making. In an experiment to test for the presence of sulfite in fruit, a student first soaked several dried apricots in water overnight and then filtered the solution to remove all solid particles. She then treated the solution with hydrogen peroxide (H_2O_2) to oxidize the sulfite ions to sulfate ions. Finally, the sulfate ions were precipitated by treating the solution with a few drops of a barium chloride ($BaCl_2$) solution. Write a balanced equation for each of the preceding steps.

9.146 A 0.8870-g sample of a mixture of NaCl and KCl is dissolved in water, and the solution is then treated with an excess of $AgNO_3$ to yield 1.913 g of AgCl. Calculate the percent by mass of each compound in the mixture.

9.147 Chlorine forms a number of oxides with the following oxidation numbers: +1, +3, +4, +6, and +7. Write a formula for each of these compounds.

9.148 One of the uses of oxalic ($H_2C_2O_4$) acid is rust removal. It reacts with rust (Fe_2O_3) according to the equation

$$Fe_2O_3(s) + 6H_2C_2O_4(aq) \longrightarrow$$
$$2Fe(C_2O_4)_3^{3-}(aq) + 3H_2O(l) + 6H^+(aq)$$

Calculate the number of grams of rust that can be removed by 5.00×10^2 mL of a 0.100-M solution of oxalic acid.

9.149 Acetylsalicylic acid ($HC_9H_7O_4$) is a monoprotic acid commonly known as "aspirin." A typical aspirin tablet, however, contains only a small amount of the acid. In an experiment to determine its composition, an aspirin tablet was crushed and dissolved in water. It took 12.25 mL of 0.1466 M NaOH to neutralize the solution. Calculate the number of grains of aspirin in the tablet (one grain = 0.0648 g).

9.150 A 0.9157-g mixture of $CaBr_2$ and NaBr is dissolved in water, and $AgNO_3$ is added to the solution to form AgBr precipitate. If the mass of the precipitate is 1.6930 g, what is the percent by mass of NaBr in the original mixture?

9.151 Hydrogen halides (HF, HCl, HBr, HI) are highly reactive compounds that have many industrial and laboratory uses. (a) In the laboratory, HF and HCl can be generated by reacting CaF_2 and NaCl with concentrated sulfuric acid. Write appropriate equations for the reactions. (*Hint:* These are not redox reactions.) (b) Why is it that HBr and HI cannot be prepared similarly, that is, by reacting NaBr and NaI with concentrated sulfuric acid? (*Hint:* H_2SO_4 is a stronger oxidizing agent than both Br_2 and I_2.) (c) HBr can be prepared by combining phosphorus tribromide (PBr_3) with water. Write an equation for this reaction.

9.152 A 325-mL sample of solution contains 25.3 g of $CaCl_2$. (a) Calculate the molar concentration of Cl^- in this solution. (b) How many grams of Cl^- are in 0.100 L of this solution?

9.153 Phosphoric acid (H_3PO_4) is an important industrial chemical used in fertilizers, detergents, and the food industry. It is produced by two different methods. In the *electric furnace method,* elemental phosphorus (P_4) is burned in air to form P_4O_{10}, which is then combined with water to give H_3PO_4. In the *wet process,* the mineral phosphate rock [$Ca_5(PO_4)_3F$] is combined with sulfuric acid to give H_3PO_4 (and HF and $CaSO_4$). Write equations for these processes, and classify each step as precipitation, acid-base, or redox reaction.

9.154 Ammonium nitrate (NH_4NO_3) is one of the most important nitrogen-containing fertilizers. Its purity can be analyzed by titrating a solution of NH_4NO_3 with a standard NaOH solution. In one experiment, a 0.2041-g sample of industrially prepared NH_4NO_3 required 24.42 mL of 0.1023 M NaOH for neutralization. (a) Write a net ionic equation for the reaction. (b) What is the percent purity of the sample?

9.155 Potassium superoxide (KO_2) is used in some self-containing breathing equipment by firefighters. It reacts with carbon dioxide in respired (exhaled) air to form potassium carbonate and oxygen gas. (a) Write an equation for the reaction. (b) What is the oxidation number of oxygen in the O_2^{2-} ion? (c) How many liters of respired air can react with 7.00 g of KO_2 if each liter of respired air contains 0.063 g of CO_2?

9.156 Barium sulfate ($BaSO_4$) has important medical uses. The dense salt absorbs X rays and acts as an opaque barrier. Thus, X-ray examination of a patient who has swallowed an aqueous suspension of $BaSO_4$ particles allows the radiologist to diagnose an ailment of the patient's digestive tract. Given the following starting compounds, describe how you would prepare $BaSO_4$ by neutralization and by precipitation: $Ba(OH)_2$, $BaCl_2$, $BaCO_3$, H_2SO_4, and K_2SO_4.

9.157 Is the following reaction a redox reaction? Explain.
$$3O_2(g) \longrightarrow 2O_3(g)$$

9.158 Absorbance values for five standard $KMnO_4$ solutions were determined at a wavelength of 528 nm. The data are tabulated here.

Solute concentration (M)	A
0.0500	0.172
0.100	0.310
0.150	0.502
0.200	0.680
0.250	0.822

The absorbance of a $KMnO_4$ solution with unknown concentration, analyzed under the same conditions, was 0.550. Determine the concentration of the unknown solution. (See page 385.)

9.159 Draw molecular models to represent the following acid-base reactions.
(a) $OH^- + H_3O^+ \longrightarrow 2H_2O$
(b) $NH_4^+ + NH_2^- \longrightarrow 2NH_3$
Identify the Brønsted acid and base in each case.

9.160 On standing, a concentrated nitric acid gradually turns yellow. Explain. (*Hint:* Nitric acid slowly decomposes. Nitrogen dioxide is a colored gas.)

9.161 When preparing a solution of known concentration, explain why one must first dissolve the solid completely before adding enough solvent to fill the volumetric flask to the mark.

9.162 Can the following decomposition reaction be characterized as an acid-base reaction? Explain.
$$NH_4Cl(s) \longrightarrow NH_3(g) + HCl(g)$$

9.163 Give a chemical explanation for each of the following: (a) When calcium metal is added to a sulfuric acid solution, hydrogen gas is generated. After a few minutes, the reaction slows down and eventually stops even though none of the reactants is used up. Explain. (b) In the activity series, aluminum is above hydrogen, yet the metal appears to be unreactive toward hydrochloric acid. Why? (*Hint:* Al forms an oxide, Al_2O_3, on the surface.) (c) Sodium and potassium lie above copper in the activity series. Explain why Cu^{2+} ions in a $CuSO_4$ solution are not converted to metallic copper upon the addition of these metals. (d) A metal M reacts slowly with steam. There is no visible change when it is placed in a pale green iron(II) sulfate solution. Where should we place M in the activity series? (e) Before aluminum metal was obtained by electrolysis, it was produced by reducing its chloride ($AlCl_3$) with an active metal. What metals would you use to produce aluminum in that way?

9.164 The recommended procedure for preparing a very dilute solution is not to weigh out a very small mass

412   **CHAPTER 9**   Chemical Reactions in Aqueous Solutions

or measure a very small volume of a stock solution. Instead, it is done by serial dilution. A sample of 0.8214 g of $KMnO_4$ was dissolved in water and made up to the volume in a 500-mL volumetric flask. A 2.000-mL sample of this solution was transferred to a 1000-mL volumetric flask and diluted to the mark with water. Next, 10.00 mL of the diluted solution was transferred to a 250-mL flask and diluted to the mark with water. (a) Calculate the concentration (in molarity) of the final solution. (b) Calculate the mass of $KMnO_4$ needed to directly prepare the final solution and comment on why dilute solutions are not typically prepared directly.

9.165 The following "cycle of copper" experiment is performed in some general chemistry laboratories. The series of reactions starts with copper and ends with metallic copper. The steps are as follows: (1) A piece of copper wire of known mass is allowed to react with concentrated nitric acid [the products are copper(II) nitrate, nitrogen dioxide, and water]. (2) The copper(II) nitrate is treated with a sodium hydroxide solution to form copper(II) hydroxide precipitate. (3) On heating, copper(II) hydroxide decomposes to yield copper(II) oxide. (4) The copper(II) oxide is combined with concentrated sulfuric acid to yield copper(II) sulfate. (5) Copper(II) sulfate is treated with an excess of zinc metal to form metallic copper. (6) The remaining zinc metal is removed by treatment with hydrochloric acid, and metallic copper is filtered, dried, and weighed. (a) Write a balanced equation for each step and classify the reactions. (b) Assuming that a student started with 65.6 g of copper, calculate the theoretical yield at each step. (c) Considering the nature of the steps, comment on why it is possible to recover most of the copper used at the start.

9.166 Use the periodic table framework given here to show the names and positions of two metals that can (a) displace hydrogen from cold water and (b) displace hydrogen from acid. (c) Also show two metals that do not react with either water or acid.

Use trends in periodic properties to explain the differences in your answers to parts (a) and (b).

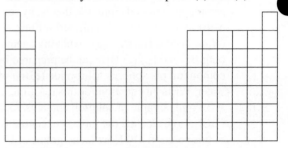

9.167 A 22.02-mL solution containing 1.615 g $Mg(NO_3)_2$ is mixed with a 28.64-mL solution containing 1.073 g NaOH. Calculate the concentrations of the ions remaining in solution after the reaction is complete. Assume volumes are additive.

9.168 Because the acid-base and precipitation reactions discussed in this chapter all involve ionic species, their progress can be monitored by measuring the electrical conductance of the solution. Match each of the following reactions with one of the diagrams shown here. The electrical conductance is shown in arbitrary units. Explain the significance of the point at which the slope changes in each diagram.
(1) A 1.0 M KOH solution is added to 1.0 L of 1.0 M $HC_2H_3O_2$.
(2) A 1.0 M NaOH solution is added to 1.0 L of 1.0 M HCl.
(3) A 1.0 M $BaCl_2$ solution is added to 1.0 L of 1.0 M K_2SO_4.
(4) A 1.0 M NaCl solution is added to 1.0 L of 1.0 M $AgNO_3$.
(5) A 1.0 M $HC_2H_3O_2$ solution is added to 1.0 L of 1.0 M NH_3.

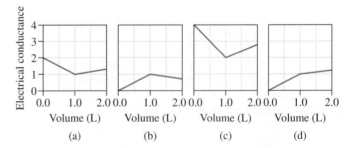

Answers to In-Chapter Materials

PRACTICE PROBLEMS

9.1A Weak electrolyte, strong electrolyte, strong electrolyte, strong electrolyte. **9.1B** Strong electrolyte, strong electrolyte, weak electrolyte. **9.2A** (a) insoluble, (b) soluble, (c) insoluble. **9.2B** (a) soluble, (b) insoluble, (c) soluble. **9.3A** $Sr(NO_3)_2(aq) + Li_2SO_4(aq) \longrightarrow$ $SrSO_4(s) + 2LiNO_3(aq)$; $Sr^{2+}(aq) + 2NO_3^-(aq) + 2Li^+(aq) +$ $SO_4^{2-}(aq) \longrightarrow SrSO_4(s) + 2Li^+(aq) + 2NO_3^-(aq)$; $Sr^{2+}(aq) +$ $SO_4^{2-}(aq) \longrightarrow SrSO_4(s)$. **9.3B** $(NH_4)_2CO_3(aq) + Ba(NO_3)_2(aq)$; $(NH_4)_2CO_3(aq) + Ba(NO_3)_2(aq) \longrightarrow BaCO_3(s) + 2NH_4NO_3(aq)$; $2NH_4^+(aq) + CO_3^{2-}(aq) + Ba^{2+}(aq) + 2NO_3^-(aq) \longrightarrow BaCO_3(s) +$ $2NH_4^+(aq) + 2NO_3^-(aq)$; $Ba^{2+}(aq) + CO_3^{2-}(aq) \longrightarrow BaCO_3(s)$. **9.4A** $Ba(OH)_2(aq) + 2HF(aq) \longrightarrow BaF_2(s) + H_2O(l)$; $Ba^{2+}(aq)$ $+ 2OH^-(aq) + 2HF(aq) \longrightarrow BaF_2(s) + 2H_2O(l)$; $Ba^{2+}(aq) +$ $2OH^-(aq) + 2HF(aq) \longrightarrow BaF_2(s) + 2H_2O(l)$. **9.4B** $2NH_3(aq) +$ $H_2SO_4(aq) \longrightarrow (NH_4)_2SO_4(aq)$; $2NH_3(aq) + H^+(aq) + HSO_4^-(aq)$ $\longrightarrow 2NH_4^+(aq) + SO_4^{2-}(aq)$; $2NH_3(aq) + H^+(aq) + HSO_4^-(aq)$ $\longrightarrow 2NH_4^+(aq) + SO_4^{2-}(aq)$. **9.5A** $H = +1$, $O = -1$; $Cl = +5$, $O = -2$; $H = +1$, $S = +6$, $O = -2$. **9.5B** $O = -1$; $Cl = +1$, $O = -2$; $Mn = +4$, $O = -2$. **9.6A** (a) no reaction, (b) $Sn(s) + Cu^{2+}(aq)$ $\longrightarrow Sn^{2+}(aq) + Cu(s)$, (c) no reaction. **9.6B** (a) $Ni(s) + Cu^{2+}(aq)$ $\longrightarrow Ni^{2+}(aq) + Cu(s)$, (b) no reaction, (c) $Al(s) + Au^{3+}(aq)$ $\longrightarrow Al^{3+}(aq) + Au(s)$. **9.7A** (a) $3Mg(s) + 2Cr(C_2H_3O_2)_3(aq)$ $\longrightarrow 3Mg(C_2H_3O_2)_2(aq) + 2Cr(s)$, Mg is oxidized and Cr is reduced; (b) no reaction; (c) $Cd(s) + 2AgNO_3(aq) \longrightarrow 2Ag(s) +$ $Cd(NO_3)_2(aq)$, Cd is oxidized and Ag is reduced. **9.7B** (a) no reaction, (b) Au is reduced and Ag is oxidized, (c) H is reduced and Sn is oxidized. **9.8A** (a) 0.137 M, (b) 9.18 L, (c) 0.259 mol. **9.8B** (a) 0.707 M, (b) 6.48 L, (c) 16.1 mol. **9.9A** 21 mL. **9.9B** 3.8 mL. **9.10A** (a) 1.638 M, 0.4095 M, 0.1024 M, 2.559×10^{-2} M, 6.398 $\times 10^{-3}$ M; (b) 0.1638 mol, 4.095×10^{-2} mol, 1.024×10^{-2} mol, 2.559×10^{-3} mol, 6.398×10^{-4} mol. **9.10B** 2.45 M. **9.11A** (a) 8.49, (b) 7.40, (c) 1.25. **9.11B** (a) −0.08, (b) 10.52, (c) 11.07. **9.12A** 1.3×10^{-10} M, (b) 3.5×10^{-2} M, (c) 9.8×10^{-8} M. **9.12B** (a) 7.8×10^{-3} M, (b) 2.6×10^{-12} M, (c) 1.3×10^{-7} M. **9.13A** $[Al^{3+}] = 1.50$ M, $[SO_4^{2-}] = 2.25$ M. **9.13B** (a) 0.36 M, (b) 0.36 M, (c) 0.12 M. **9.14A** 67.13%. **9.14B** 0.420 g. **9.15A** 0.5502 g KHP. **9.15B** 206.4 mL. **9.16A** 11.3 mL. **9.16B** 91.2 mL. **9.17A** 92.6 g/mol. **9.17B** 80.5 g/mol.

SECTION REVIEW

9.1.1 c. **9.1.2** e. **9.1.3** b. **9.1.4** c. **9.1.5** b. **9.2.1** a, b, d. **9.2.2** a, c, d. **9.2.3** b. **9.2.4** b. **9.2.5** b. **9.3.1** a. **9.3.2** b. **9.3.3** e. **9.3.4** d. **9.4.1** e. **9.4.2** a. **9.4.3** a, b, d. **9.4.4** a, d. **9.5.1** d. **9.5.2** a. **9.5.3** b. **9.5.4** c. **9.5.5** a. **9.5.6** d. **9.6.1** a. **9.6.2** d. **9.6.3** c. **9.6.4** a.

Chapter 10

Energy Changes in Chemical Reactions

© Syracuse Newspapers/J.Berry/The Image Works

FOSSIL FUELS are a nonrenewable source of energy. According to some estimates, the world's supply of petroleum fuels will be exhausted within the next 50 years—prompting chemists to begin the search for alternative fuels. Of particular interest are biofuels—fuels derived from renewable sources such as corn, soybeans, rapeseeds, and even algae (shown here). Biodiesel is one such biofuel that is readily produced from a variety of oils and fats, and that may have potential as an alternative to petroleum for use in automobiles.

Before You Begin, Review These Skills

- Tracking units [◄◄ Section 1.4]
- Balancing chemical equations [◄◄ Section 8.1]
- Drawing Lewis structures [◄◄ Section 6.3]

10.1 ENERGY AND ENERGY CHANGES

You have already learned that matter can undergo physical changes and chemical changes [◄◄ Section 1.6]. The melting of ice, for example, is a physical change that can be represented by the following equation.

$$H_2O(s) \longrightarrow H_2O(l)$$

The formation of water from its constituent elements, represented by the following equation, is an example of a chemical change.

$$2H_2(g) + O_2(g) \longrightarrow 2H_2O(l)$$

In each case, there is energy involved in the change. Energy (in the form of heat) must be *supplied* to melt ice, whereas energy (in the form of heat and light) is *produced* by the explosive combination of hydrogen and oxygen gases. In fact, every change that matter undergoes is accompanied by either the absorption or the release of energy [◄◄ Section 3.1]. In this chapter, we will focus on the energy changes associated with physical and chemical processes.

To analyze energy changes associated with physical processes and chemical reactions, we must first define the **system,** the specific part of the universe that is of interest to us. For chemists, systems usually include the substances involved in physical and chemical changes. In an experiment involving the melting of an ice cube, for example, the system would be the ice, and the surroundings would be the rest of the universe. In an acid-base neutralization experiment, the system may be the reactants HCl and NaOH; and the rest of the universe, including the container itself and the water in which the reactants are dissolved, may constitute the **surroundings.**

Many chemical reactions are carried out for the purpose of exploiting the associated energy change, rather than for the purpose of obtaining the products of the reactions. For example, combustion reactions involving fossil fuels are carried out for the thermal energy they produce, not for their products, which are carbon dioxide and water.

It is important to distinguish between thermal energy and heat. **Heat** is the transfer of thermal energy between two bodies that are at different temperatures. Although the term *heat* by itself implies the transfer of energy, we customarily talk of "heat flow," meaning "energy absorbed" or "energy released," when describing the energy changes that occur during a process. **Thermochemistry** is the study of the heat (the transfer of thermal energy) associated with chemical reactions.

The combustion of hydrogen gas in oxygen is one of many chemical reactions that release considerable quantities of energy (Figure 10.1).

$$2H_2(g) + O_2(g) \longrightarrow 2H_2O(l) + energy$$

In this case, we label the mixture of reactants and product (hydrogen, oxygen, and water molecules) the *system*. Because energy cannot be created or destroyed, any energy released by the system must be gained by the surroundings. Thus, the heat

Student Annotation: Although "surroundings" includes everything in the universe apart from the system, when chemists refer to the *surroundings* they generally mean the *immediate* surroundings—the part of the universe that is closest to the system.

Figure 10.1 The *Hindenburg*, a German airship filled with hydrogen gas, was destroyed in a horrific fire as it landed at Lakehurst, New Jersey, in 1937.

© Bettmann/Corbis

415

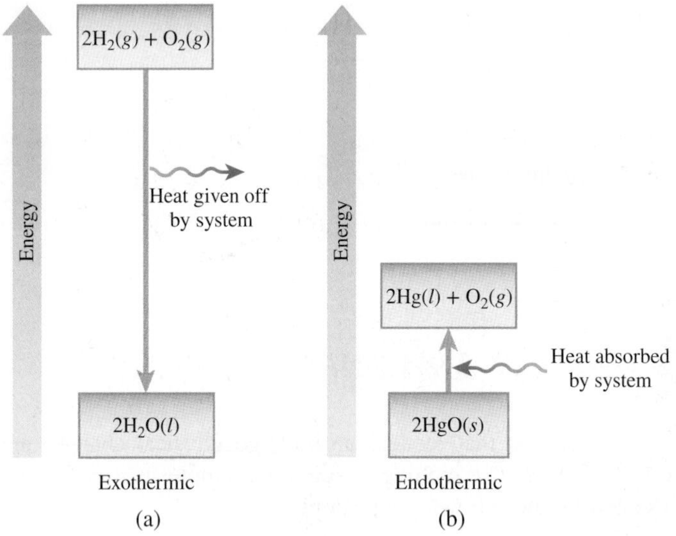

Figure 10.2 (a) An exothermic reaction. Heat is given off by the system. (b) An endothermic reaction. Heat is absorbed by the system.

$2H_2(g) + O_2(g)$

Heat given off by system

$2H_2O(l)$

Exothermic

(a)

$2Hg(l) + O_2(g)$

Heat absorbed by system

$2HgO(s)$

Endothermic

(b)

Student Annotation: Emission of a photon from an atom when an electron falls from an excited state to a lower state [◄◄ Section 3.4] and most first electron affinities [◄◄ Section 4.4] are examples of exothermic processes that we have encountered in previous chapters.

Video 10.1
Thermochemistry—heat flow in endothermic and exothermic reactions.

Student Annotation: In these examples, all the energy released or absorbed by the system is *thermal* energy, which is commonly referred to as heat. In Section 10.2, we will consider examples in which some of the energy is in the form of *work*.

Student Annotation: The triple equal sign is used to denote a definition. Recall that numbers with defined values are exact numbers, which do not limit the number of significant figures in a calculation [◄◄ Section 1.5].

generated by the combustion process is transferred from the system to its surroundings. This reaction is an example of an ***exothermic process,*** which is any process that gives off heat—that is, transfers thermal energy *from* the system *to* the surroundings. Figure 10.2 shows the energy change for the combustion of hydrogen gas.

Next, consider the decomposition of mercury(II) oxide (HgO) at high temperatures.

$$energy + 2HgO(s) \longrightarrow 2Hg(l) + O_2(g)$$

This reaction is an ***endothermic process*** because heat has to be supplied to the system (i.e., to HgO) by the surroundings [see Figure 10.2(b)] in order for the reaction to occur. Thus, thermal energy is transferred *from* the surroundings *to* the system in an endothermic process.

According to Figure 10.2, the energy of the products of an exothermic reaction is lower than the energy of the reactants. The difference in energy between the reactants H_2 and O_2 and the product H_2O is the heat released by the system to the surroundings. In an endothermic reaction, on the other hand, the energy of the products is higher than the energy of the reactants. Here, the difference between the energy of the reactant HgO and the products Hg and O_2 is the heat absorbed by the system from the surroundings.

Recall that the SI unit for energy is the joule, which is the amount of energy possessed by a 2-kg object moving at 1 m/s [◄◄ Section 3.1]. Another unit used to express energy is the calorie (cal), which was originally defined as the amount of energy required to raise the temperature of one gram of water by one degree (°C). Although the calorie is not an SI unit, its use is still quite common. The calorie is now defined in terms of the joule.

$$1 \text{ cal} \equiv 4.184 \text{ J}$$

Because this is a definition, the number 4.184 is an *exact* number, which does not limit the number of significant figures in a calculation [◄◄ Section 1.3]. You may be familiar with the term *calorie* from nutrition labels. In fact, the "calories" listed on food packaging are really *kilocalories*. Often the distinction is made by capitalizing the "C" in "calorie" when it refers to the energy content of food.

$$1 \text{ Cal} \equiv 1000 \text{ cal}$$

and

$$1 \text{ Cal} \equiv 4184 \text{ J}$$

Section 10.1 Review

Energy and Energy Changes

10.1.1 Calculate the number of calories in 723.01 J.
(a) 172.80 cal
(d) 3025 cal
(b) 172.8 cal
(e) 0.173 cal
(c) 3025.1 cal

10.1.2 The label on packaged food indicates that it contains 215 Cal per serving. Convert this amount of energy to joules.
(a) 51.4 J
(d) 9.00×10^2 J
(b) 5.14×10^4 J
(e) 9.00×10^5 J
(c) 5.14×10^{-2} J

10.1.3 From the figure shown here, which of the following equations represents an endothermic process?
(a) $2AX(g) + B_2(g) \longrightarrow A_2(g) + B_2X_2(g)$
(b) $A_2(g) + B_2X_2(g) \longrightarrow 2AX(g) + B_2(g)$
(c) $2AX(g) + B_2X_2(g) \longrightarrow A_2(g) + B_2(g)$
(d) $A_2(g) + B_2(g) \longrightarrow 2AX(g) + B_2X_2(g)$
(e) $B_2X_2(g) + B_2(g) \longrightarrow A_2(g) + 2AX(g)$

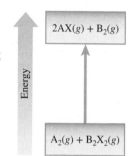

10.2 INTRODUCTION TO THERMODYNAMICS

Thermochemistry is part of a broader subject called **thermodynamics,** which is the scientific study of the interconversion of heat and other kinds of energy. The laws of thermodynamics provide useful guidelines for understanding the energetics and directions of processes. In this section, we will introduce the first law of thermodynamics, which is particularly relevant to the study of thermochemistry. We will continue our discussion of thermodynamics in [▶▶| Chapter 14].

We have defined a system as the part of the universe we are studying. There are three types of systems. An **open system** can exchange mass and energy with its surroundings. For example, an open system may consist of a quantity of water in an open container, as shown in Figure 10.3(a). If we close the flask, as in Figure 10.3(b), so that no water vapor can escape from or condense into the container, we create a **closed system,** which allows the transfer of *energy* but not *mass*. By placing the water in an insulated container, as shown in Figure 10.3(c), we can construct an **isolated system,** which does not exchange either mass or energy with its surroundings.

Student Annotation: The energy exchanged between open systems or closed systems and their surrounding is usually in the form of heat.

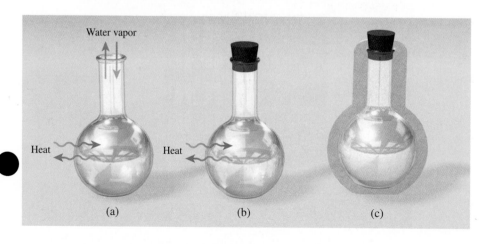

Figure 10.3 (a) An open system allows exchange of both energy and matter with the surroundings. (b) A closed system allows exchange of energy but not matter. (c) An isolated system does not allow exchange of energy or matter. (This flask is enclosed by an insulating vacuum jacket.)

States and State Functions

In thermodynamics, we study changes in the *state of a system,* which is defined by the values of all relevant macroscopic properties, such as composition, energy, temperature, pressure, and volume. Energy, pressure, volume, and temperature are said to be *state functions*—properties that are determined by the state of the system, regardless of how that condition was achieved. In other words, when the state of a system changes, the magnitude of change in any state function depends only on the initial and final states of the system and not on how the change is accomplished.

Consider, for example, your position in a six-story building. Your elevation depends upon which floor you are on. If you change your elevation by taking the stairs from the ground floor up to the fourth floor, the change in your elevation depends only upon your initial state (the ground floor—the floor you started on) and your final state (the fourth floor—the floor you went to). It does not depend on whether you went directly to the fourth floor or up to the sixth and then down to the fourth floor. Your overall change in elevation is the same either way because it depends only on your initial and final elevations. Thus, elevation is a state function.

The amount of effort it takes to get from the ground floor to the fourth floor, on the other hand, depends on how you get there. More effort has to be exerted to go from the ground floor to the sixth floor and back down to the fourth floor than to go from the ground floor to the fourth floor directly. The effort required for this change in elevation is *not* a state function. Furthermore, if you subsequently return to the ground floor, your overall change in elevation will be zero, because your initial and final states are the same, but the amount of effort you exerted going from the ground floor to the fourth floor and back to the ground floor is *not* zero. Even though your initial and final states are the same, you do not get back the effort that went into climbing up and down the stairs.

Energy is a state function, too. Using potential energy as an example, your net increase in gravitational potential energy is always the same, regardless of how you get from the ground floor to the fourth floor of a building (Figure 10.4).

The First Law of Thermodynamics

The *first law of thermodynamics,* which is based on the law of conservation of energy, states that energy can be converted from one form to another but cannot be created or destroyed. It would be impossible to demonstrate this by measuring the total amount of energy in the universe; in fact, just determining the total energy content of a small sample of matter would be extremely difficult. Fortunately, because energy is a state function, we can demonstrate the first law by measuring the change

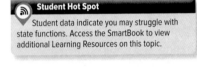

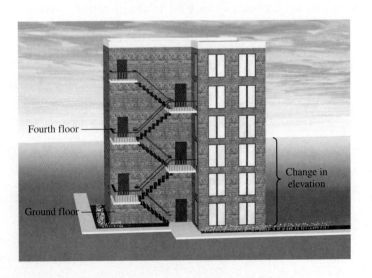

Figure 10.4 The change in elevation that occurs when a person goes from the ground floor to the fourth floor in a building does not depend on the path taken.

in the energy of a system between its initial state and its final state in a process. The change in internal energy, ΔU, is given by

$$\Delta U = U_f - U_i$$

Student Annotation: The symbol U is used to denote the internal energy of a *system*, as opposed to the symbol E, which is used to denote the energy possessed by a particle or associated with a particular process.

where U_i and U_f are the internal energies of the system in the initial and final states, respectively; and the symbol Δ means *final* minus *initial*.

The internal energy of a system has two components: kinetic energy and potential energy. The kinetic energy component consists of various types of molecular motion and the movement of electrons within molecules. Potential energy is determined by the attractive interactions between electrons and nuclei and by repulsive interactions between electrons and between nuclei in individual molecules, as well as by interactions between molecules. It is impossible to measure all these contributions accurately, so we cannot calculate the total energy of a system with any certainty. *Changes* in energy, on the other hand, can be determined experimentally.

Student Annotation: Recall that the symbol Δ is commonly used to mean *final* minus *initial*.

Consider the reaction between 1 mole of sulfur and 1 mole of oxygen gas to produce 1 mole of sulfur dioxide.

$$S(s) + O_2(g) \longrightarrow SO_2(g)$$

Student Annotation: Elemental sulfur exists as S_8 molecules but we typically represent it simply as S to simplify chemical equations [◄◄ Section 8.1].

In this case our system is composed of the reactant molecules and the product molecules. We do not know the internal energy content of either the reactants or the product, but we can accurately measure the *change* in energy content ΔU given by

$$\Delta U = U(\text{products}) - U(\text{reactants})$$
$$= \text{energy content of 1 mol } SO_2(g) - \text{energy content of 1 mol } S(s)$$
$$\text{and 1 mol } O_2(g)$$

This reaction gives off heat: 296.4 kJ, to be exact. Therefore, the energy of the product is less than that of the reactants, and ΔU is negative (-296.4 kJ/mol).

Student Annotation: Later in this chapter, we will learn about the experimental techniques used to measure energy changes [►► Section 10.4].

The release of heat that accompanies this reaction indicates that some of the chemical energy contained in the system has been converted to thermal energy. Furthermore, the thermal energy released by the system is absorbed by the surroundings. The transfer of energy from the system to the surroundings does not change the total energy of the universe. That is, the sum of the energy changes is zero,

$$\Delta U_{\text{sys}} + \Delta U_{\text{surr}} = 0$$

where the subscripts "sys" and "surr" denote system and surroundings, respectively. Thus, if a system undergoes an energy change ΔU_{sys}, the rest of the universe, or the surroundings, must undergo a change in energy that is equal in magnitude but opposite in sign.

$$\Delta U_{\text{sys}} = -\Delta U_{\text{surr}}$$

Energy released in one place must be gained elsewhere. Furthermore, because energy can be changed from one form to another, the energy lost by one system can be gained by another system in a different form. For example, the energy released by burning coal in a power plant may ultimately turn up in our homes as electric energy, heat, light, and so on.

Work and Heat

Recall from Section 3.1 that energy is defined as the capacity to do work or transfer heat. When a system releases or absorbs heat, its internal energy changes. Likewise, when a system does work on its surroundings, or when the surroundings do work on the system, the system's internal energy also changes. The overall change in the system's internal energy is given by

Student Annotation: The units for heat and work are the same as those for energy: joules, kilojoules, or calories.

$$\Delta U = q + w \qquad \textbf{Equation 10.1}$$

where q is heat (released or absorbed by the system) and w is work (done *on* the system or done *by* the system). Note that it is possible for the heat and work components to cancel each other out and for there to be no change in the system's internal energy.

TABLE 10.1	Sign Conventions for Heat (q) and Work (w)	
Process		**Sign**
Heat absorbed by the system (endothermic process)		q is positive
Heat released by the system (exothermic process)		q is negative
Work done on the system by the surroundings (e.g., a volume decrease)		w is positive
Work done by the system on the surroundings (e.g., a volume increase)		w is negative

Figure 10.5 (a) When heat is released by the system (to the surroundings), q is negative. When work is done by the system (on the surroundings), w is negative. (b) When heat is absorbed by the system (from the surroundings), q is positive. When work is done on the system (by the surroundings), w is positive.

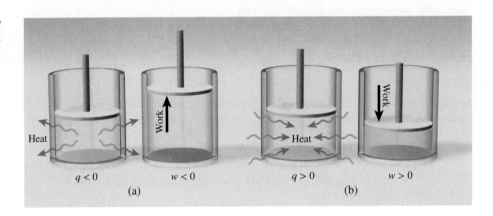

Student Annotation: Interestingly, although neither q nor w is a state function (each depends on the path between the initial and final states of the system), their sum, ΔU, does *not* depend on the path between initial and final states because U *is* a state function.

In chemistry, we are normally interested in the energy changes associated with the system rather than the surroundings. Therefore, unless otherwise indicated, ΔU will refer specifically to ΔU_{sys}. The sign conventions for q and w are as follows: q is positive for an endothermic process and negative for an exothermic process, and w is positive for work done on the system by the surroundings and negative for work done by the system on the surroundings. Table 10.1 summarizes the sign conventions for q and w.

The drawings in Figure 10.5 illustrate the logic behind the sign conventions for q and w. If a system releases heat to the surroundings or does work on the surroundings [Figure 10.5(a)], we would expect its internal energy to decrease because they are energy-depleting processes. For this reason, both q and w are negative. Conversely, if heat is added to the system or if work is done on the system [Figure 10.5(b)] then the internal energy of the system increases. In this case, both q and w are positive.

Worked Example 10.1 shows how to determine the overall change in the internal energy of a system.

Worked Example 10.1

Calculate the overall change in internal energy, ΔU, in joules, for a system that absorbs 188 J of heat and does 141 J of work on its surroundings.

Strategy Combine the two contributions to internal energy using Equation 10.1 and the sign conventions for q and w.

Setup The system absorbs heat, so q is *positive*. The system does work on the surroundings, so w is negative.

Solution

$$\Delta U = q + w = 188 \text{ J} + (-141 \text{ J}) = 47 \text{ J}$$

Think About It
Consult Table 10.1 to make sure you have used the proper sign conventions for q and w.

Practice Problem **ATTEMPT** Calculate the change in total internal energy for a system that releases 1.34×10^4 kJ of heat and does 2.98×10^4 kJ of work on the surroundings.

Practice Problem **BUILD** Calculate the magnitude of q for a system that does 7.05×10^5 kJ of work on its surroundings and for which the change in total internal energy is -9.55×10^3 kJ. Indicate whether heat is absorbed or released by the system.

Practice Problem **CONCEPTUALIZE** The diagram on the left shows a system before a process. Which of the diagrams on the right could represent the system after it undergoes a process in which the system absorbs heat and ΔU is positive?

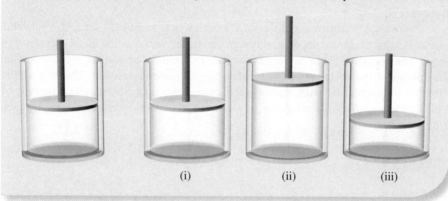

(i)   (ii)   (iii)

Section 10.2 Review

Introduction to Thermodynamics

10.2.1   Calculate the overall change in internal energy for a system that releases 43 J in heat in a process in which no work is done.
(a) 43 J
(b) -2.3×10^{-2} J
(c) 0 J
(d) 2.3×10^{-2} J
(e) -43 J

10.2.2   Calculate w, and determine whether work is done by the system or on the system when 928 kJ of heat is released and $\Delta U = -1.47 \times 10^3$ kJ.
(a) $w = -1.36 \times 10^6$ kJ, done by the system
(b) $w = 1.36 \times 10^6$ kJ, done on the system
(c) $w = -5.4 \times 10^2$ kJ, done by the system
(d) $w = 2.4 \times 10^3$ kJ, done on the system
(e) $w = -2.4 \times 10^3$ kJ, done by the system

10.3   ENTHALPY

To calculate ΔU, we must know the values and signs of both q and w. As we will see in Section 10.4, we determine q by measuring temperature changes. To determine w, we need to know whether the reaction occurs under constant-volume conditions, or under constant-pressure conditions.

Reactions Carried Out at Constant Volume or at Constant Pressure

Imagine carrying out the decomposition of sodium azide (NaN_3) in two different experiments. In the first experiment, the reactant is placed in a metal cylinder with a fixed volume. When detonated, the NaN_3 reacts, generating a large quantity of N_2 gas inside the closed, fixed-volume container.

$$2NaN_3(s) \longrightarrow 2Na(s) + 3N_2(g)$$

The effect of this reaction will be an increase in the pressure inside the container, similar to what happens if you shake a bottle of soda vigorously prior to opening it. (The concept of pressure will be examined in detail in Chapter 11. However, if you have ever put air in the tire of an automobile or a bicycle, you are familiar with the concept.)

Now imagine carrying out the same reaction in a metal cylinder with a movable piston. As this explosive decomposition proceeds, the piston in the metal cylinder will move. The gas produced in the reaction pushes the cylinder upward, thereby increasing the volume of the container and preventing any increase in pressure. This is a simple example of mechanical work done by a chemical reaction. Specifically, this type of work is known as *pressure-volume,* or *PV,* work. The amount of work done by such a process is given by

Equation 10.2	$w = -P\Delta V$

where P is the external, opposing pressure and ΔV is the change in the volume of the container as the result of the piston being pushed upward. In keeping with the sign conventions in Table 10.1, an increase in volume results in a negative value for w, whereas a decrease in volume results in a positive value for w. Figure 10.6 illustrates this reaction (a) being carried out at a constant volume, and (b) at a constant pressure.

Worked Example 10.2 shows how to calculate the work associated with a volume change.

Figure 10.6 (a) The explosive decomposition of NaN_3 at constant volume results in an increase in pressure inside the vessel. (b) The decomposition at constant pressure, in a vessel with a movable piston, results in an increase in volume. The resulting change in volume, ΔV, can be used to calculate the work done by the system.

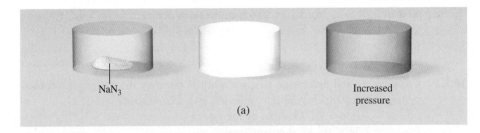

(a)

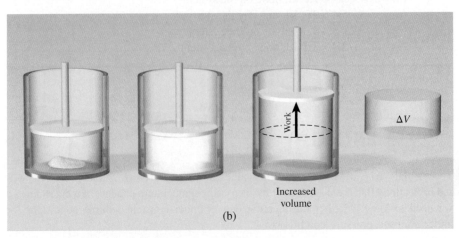

(b)

Worked Example 10.2

Determine the work done (in joules) when a sample of gas expands from 552 mL to 891 mL at constant temperature (a) against a constant pressure of 1.25 atm, (b) against a constant pressure of 1.00 atm, and (c) against a vacuum (1 L · atm = 101.3 J).

Strategy Determine change in volume (ΔV), identify the external pressure (P), and use Equation 10.2 to calculate w. The result of Equation 10.2 will be in L · atm; use the equality 1 L · atm = 101.3 J to convert to joules.

Setup $\Delta V = (891 - 552)$ mL = 339 mL. (a) $P = 1.25$ atm, (b) $P = 1.00$ atm, (c) $P = 0$ atm.

Solution

(a) $w = -(1.25 \text{ atm})(339 \text{ mL}) \left(\dfrac{1 \text{ L}}{1000 \text{ mL}} \right) \left(\dfrac{101.3 \text{ J}}{1 \text{ L} \cdot \text{atm}} \right) = -42.9$ J   (c) $w = -(0 \text{ atm})(339 \text{ mL}) \left(\dfrac{1 \text{ L}}{1000 \text{ mL}} \right) \left(\dfrac{101.3 \text{ J}}{1 \text{ L} \cdot \text{atm}} \right) = 0$ J

(b) $w = -(1.00 \text{ atm})(339 \text{ mL}) \left(\dfrac{1 \text{ L}}{1000 \text{ mL}} \right) \left(\dfrac{101.3 \text{ J}}{1 \text{ L} \cdot \text{atm}} \right) = -34.3$ J

Think About It

Remember that the negative sign in the answers to parts (a) and (b) indicate that the system does work on the surroundings. When an expansion happens against a vacuum, no work is done. This example illustrates that work is *not* a state function. For an equivalent change in volume, the work varies depending on the external pressure against which the expansion must occur.

Practice Problem A TTEMPT Calculate the work done by or on the system during the following processes: (a) a sample of gas expands from 1.2 L to 3.8 L against an external pressure of 1.01 atm; (b) a sample of gas is compressed from 87.5 mL to 72.9 mL by a pressure of 2.72 atm; (c) the volume of a 2.00-L gas sample increases by a factor of 2 against an external pressure of 0.998 atm.

Practice Problem B UILD (a) Against what external pressure must a gas sample expand from 50.0 mL to 219 mL to do 8.34 J of work? (b) To what volume will a 200.0-mL gas sample be compressed by an external pressure of 7.8 atm if 100.0 J of work are done on the system? (c) To what volume must a 200.0-mL gas sample expand against an external pressure of 7.8 atm to do 100.0 J of work?

Practice Problem C ONCEPTUALIZE The diagram on the left shows a sample of gas contained in a cylinder with a movable piston. Diagrams (i), (ii), and (iii) show the sample after a process has occurred. Fill in the table with the sign of w for each instance.

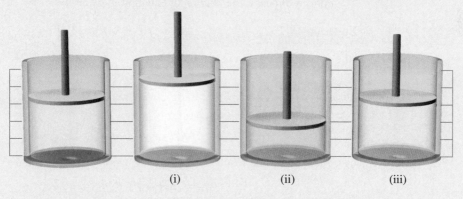

	(i)	(ii)	(iii)
	\multicolumn{3}{c}{w (positive, negative, or zero)}
External $P = 5$ atm			
External $P = 0$ atm			
	(i)	(ii)	(iii)

When a chemical reaction is carried out at constant volume, then no PV work can be done because $\Delta V = 0$ in Equation 10.2. From Equation 10.1 it follows that

$$\Delta U = q - P\Delta V \qquad \textbf{Equation 10.3}$$

and, because $P\Delta V = 0$ at constant volume,

$$q_V = \Delta U \qquad \textbf{Equation 10.4}$$

We add the subscript "V" to indicate that this is a constant-volume process. This equality may seem strange at first. We said earlier that q is *not* a state function. However, for a process carried out under constant-volume conditions, q can have only one specific value, which is equal to ΔU. In other words, although q is *not* a state function, q_V *is* one.

Constant-volume conditions are often inconvenient and sometimes impossible to achieve. Most reactions occur in open containers, under conditions of constant pressure (usually at whatever the atmospheric pressure happens to be where the experiments are conducted). In general, for a constant-pressure process, we write

$$\Delta U = q + w$$
$$= q_P - P\Delta V$$

or

Equation 10.5 $$q_P = \Delta U + P\Delta V$$

where the subscript "P" denotes constant pressure.

Enthalpy and Enthalpy Changes

There is a thermodynamic function of a system called ***enthalpy (H),*** which is defined by Equation 10.6,

Equation 10.6 $$H = U + PV$$

where U is the internal energy of the system and P and V are the pressure and volume of the system, respectively. Because U and PV have energy units, enthalpy also has energy units. Furthermore, U, P, and V are all state functions—that is, the changes in $(U + PV)$ depend only on the initial and final states. It follows, therefore, that the change in H, or ΔH, also depends only on the initial and final states. Thus, H is a state function.

For any process, the *change* in enthalpy is given by

Equation 10.7 $$\Delta H = \Delta U + \Delta(PV)$$

If the pressure is held constant, then

Equation 10.8 $$\Delta H = \Delta U + P\Delta V$$

If we solve Equation 10.8 for ΔU,

$$\Delta U = \Delta H - P\Delta V$$

Then, substituting the result for ΔU into Equation 10.5, we obtain

$$q_P = (\Delta H - P\Delta V) + P\Delta V$$

The $P\Delta V$ terms cancel, and for a constant-pressure process, the heat exchanged between the system and the surroundings is equal to the enthalpy change.

Equation 10.9 $$q_P = \Delta H$$

Again, q is *not* a state function, but q_P *is* one; that is, the heat change at constant pressure can have only one specific value and is equal to ΔH.

We now have two quantities—ΔU and ΔH—that can be associated with a reaction. If the reaction occurs under constant-volume conditions, then the heat change, q_V, is equal to ΔU. If the reaction is carried out at constant pressure, on the other hand, the heat change, q_P, is equal to ΔH.

Because most laboratory reactions are constant-pressure processes, the heat exchanged between the system and surroundings is equal to the change in enthalpy for the process. For any reaction, we define the change in enthalpy, called the ***enthalpy of reaction (ΔH),*** as the difference between the enthalpies of the products and the enthalpies of the reactants.

Student Annotation: The enthalpy of reaction is often symbolized by ΔH_{rxn}. The subscript can be changed to denote a specific type of reaction or physical process: ΔH_{vap} can be used for the enthalpy of *vaporization*, for example.

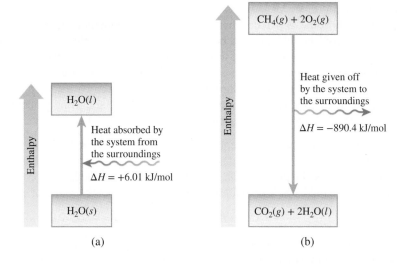

$$\Delta H = H(\text{products}) - H(\text{reactants}) \qquad \textbf{Equation 10.10}$$

The enthalpy of reaction can be positive or negative, depending on the process. For an endothermic process (where heat is absorbed by the system from the surroundings), ΔH is positive (i.e., $\Delta H > 0$). For an exothermic process (where heat is released by the system to the surroundings), ΔH is negative (i.e., $\Delta H < 0$).

We will now apply the idea of enthalpy changes to two common processes, the first involving a physical change and the second involving a chemical change.

Thermochemical Equations

Under ordinary atmospheric conditions at sea level, ice melts to form liquid water at temperatures above 0°C. Measurements show that for every mole of ice converted to liquid water under these conditions, 6.01 kJ of heat energy is absorbed by the system (the ice). Because the pressure is constant, the heat change is equal to the enthalpy change, ΔH. This is an *endothermic* process ($\Delta H > 0$), because heat is absorbed by the ice from its surroundings (Figure 10.7a). The equation for this physical change is

$$H_2O(s) \longrightarrow H_2O(l) \qquad\qquad \Delta H = +6.01 \text{ kJ/mol}$$

The "per mole" in the unit for ΔH means that this is the enthalpy change *per mole of the reaction (or process) as it is written*—that is, when 1 mole of ice is converted to 1 mole of liquid water.

Now consider the combustion of methane (CH_4), the principal component of natural gas.

$$CH_4(g) + 2O_2(g) \longrightarrow CO_2(g) + 2H_2O(l) \qquad \Delta H = -890.4 \text{ kJ/mol}$$

From experience we know that burning natural gas releases heat to the surroundings, so it is an exothermic process. Under constant-pressure conditions, this heat change is equal to the enthalpy change and ΔH must have a negative sign [Figure 10.7(b)]. Again, the "per mole" in the units for ΔH means that when 1 mole of CH_4 reacts with 2 moles of O_2 to yield 1 mole of CO_2 and 2 moles of liquid H_2O, 890.4 kJ of heat is released to the surroundings.

It is important to keep in mind that the ΔH value in kJ/mol does not mean *per mole* of a particular reactant or product. It refers to all the species in a reaction in the molar amounts specified by the coefficients in the balanced equation. Thus, for the combustion of methane, the ΔH value of -890.4 kJ/mol can be expressed in any of the following ways.

$$\frac{-890.4 \text{ kJ}}{1 \text{ mol CH}_4} \qquad \frac{-890.4 \text{ kJ}}{2 \text{ mol O}_2} \qquad \frac{-890.4 \text{ kJ}}{1 \text{ mol CO}_2} \qquad \frac{-890.4 \text{ kJ}}{2 \text{ mol H}_2\text{O}}$$

Student Annotation: Although, strictly speaking, it is unnecessary to include the sign of a positive number, we will include the sign of all positive ΔH values to emphasize the thermochemical sign convention.

Student Annotation: When you specify that a particular amount of heat is released, it is not necessary to include a negative sign.

Although the importance of expressing ΔH in units of kJ/mol (rather than just kilojoules) will become apparent when we study thermodynamics in greater detail [▶| Chapter 14], you should learn and become comfortable with this convention now.

The equations for the melting of ice and the combustion of methane are examples of *thermochemical equations,* which are chemical equations that show the enthalpy changes as well as the mass relationships. It is essential to specify a balanced chemical equation when quoting the enthalpy change of a reaction. The following guidelines are helpful in interpreting, writing, and manipulating thermochemical equations.

1. When writing thermochemical equations, we must always specify the physical states of all reactants and products, because they help determine the actual enthalpy changes. In the equation for the combustion of methane, for example, changing the liquid water product to water vapor changes the value of ΔH.

$$CH_4(g) + 2O_2(g) \longrightarrow CO_2(g) + 2H_2O(g) \qquad \Delta H = -802.4 \text{ kJ/mol}$$

The enthalpy change is -802.4 kJ rather than -890.4 kJ (page 425) because 88.0 kJ are needed to convert 2 moles of liquid water to 2 moles of water vapor; that is,

$$2H_2O(l) \longrightarrow 2H_2O(g) \qquad \Delta H = +88.0 \text{ kJ/mol}$$

2. If we multiply both sides of a thermochemical equation by a factor n, then ΔH must also be multiplied by the same factor. Thus, for the melting of ice, if $n = 2$, we have

$$2H_2O(s) \longrightarrow 2H_2O(l) \qquad \Delta H = 2(+6.01 \text{ kJ/mol}) = +12.02 \text{ kJ/mol}$$

3. When we reverse a chemical equation, we change the roles of reactants and products. Consequently, the magnitude of ΔH for the equation remains the same, but its sign changes. For example, if a reaction consumes thermal energy from its surroundings (i.e., if it is endothermic), then the reverse reaction must release thermal energy back to its surroundings (i.e., it must be exothermic) and the enthalpy change expression must also change its sign. Thus, reversing the melting of ice and the combustion of methane, the thermochemical equations become

$$H_2O(l) \longrightarrow H_2O(s) \qquad \Delta H = -6.01 \text{ kJ/mol}$$

$$CO_2(g) + 2H_2O(l) \longrightarrow CH_4(g) + 2O_2(g) \qquad \Delta H = +890.4 \text{ kJ/mol}$$

What was an endothermic process becomes an exothermic process when reversed, and vice versa.

Worked Example 10.3 illustrates the use of a thermochemical equation to relate the mass of a product to the energy consumed in the reaction.

Worked Example 10.3

Given the thermochemical equation for photosynthesis,

$$6H_2O(l) + 6CO_2(g) \longrightarrow C_6H_{12}O_6(s) + 6O_2(g) \qquad \Delta H = +2803 \text{ kJ/mol}$$

calculate the solar energy required to produce 75.0 g of $C_6H_{12}O_6$.

Strategy The thermochemical equation shows that for every mole of $C_6H_{12}O_6$ produced, 2803 kJ is absorbed. We need to find out how much energy is absorbed for the production of 75.0 g of $C_6H_{12}O_6$. We must first find out how many moles there are in 75.0 g of $C_6H_{12}O_6$.

Setup The molar mass of $C_6H_{12}O_6$ is 180.2 g/mol, so 75.0 g of $C_6H_{12}O_6$ is

$$75.0 \text{ g } C_6H_{12}O_6 \times \frac{1 \text{ mol } C_6H_{12}O_6}{180.2 \text{ g } C_6H_{12}O_6} = 0.416 \text{ mol } C_6H_{12}O_6$$

We will multiply the thermochemical equation, including the enthalpy change, by 0.416, in order to write the equation in terms of the appropriate amount of $C_6H_{12}O_6$.

Solution

$$(0.416 \text{ mol})[6H_2O(l) + 6CO_2(g) \longrightarrow C_6H_{12}O_6(s) + 6O_2(g)]$$

and $\qquad$ $(0.416 \text{ mol})(\Delta H) = (0.416 \text{ mol})(2803 \text{ kJ/mol})$ gives

$$2.50H_2O(l) + 2.50CO_2(g) \longrightarrow 0.416C_6H_{12}O_6(s) + 2.50O_2(g) \qquad \Delta H = +1.17 \times 10^3 \text{ kJ}$$

Therefore, 1.17×10^3 kJ of energy in the form of sunlight is consumed in the production of 75.0 g of $C_6H_{12}O_6$. Note that the "per mole" units in ΔH are canceled when we multiply the thermochemical equation by the number of moles of $C_6H_{12}O_6$.

Think About It
The specified amount of $C_6H_{12}O_6$ is less than half a mole. Therefore, we should expect the associated enthalpy change to be less than half that specified in the thermochemical equation for the production of 1 mole of $C_6H_{12}O_6$.

Practice Problem **A**TTEMPT Calculate the solar energy required to produce 5255 g of $C_6H_{12}O_6$.

Practice Problem **B**UILD Calculate the mass (in grams) of O_2 that is produced by photosynthesis when 2.49×10^4 kJ of solar energy is consumed.

Practice Problem **C**ONCEPTUALIZE The diagrams represent systems before and after reaction for two related chemical processes. ΔH for the first reaction is 1755.0 kJ/mol. Determine the value of ΔH for the second reaction.

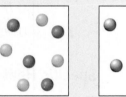

before $\qquad$ after $\qquad$ $\Delta H = 1755.0$ kJ/mol

before $\qquad$ after $\qquad$ $\Delta H = ?$

Section 10.3 Review

Enthalpy

10.3.1 Given the thermochemical equation, $H_2(g) + Br_2(l) \longrightarrow 2HBr(g)$, $\Delta H = -72.4$ kJ/mol, calculate the amount of heat released when a kilogram of $Br_2(l)$ is consumed in this reaction.
(a) 7.24×10^4 kJ $\qquad$ (d) 227 kJ
(b) 453 kJ $\qquad$ (e) 724 kJ
(c) 906 kJ

10.3.2 Given the thermochemical equation, $2Cu_2O(s) \longrightarrow 4Cu(s) + O_2(g)$, $\Delta H = +333.8$ kJ/mol, calculate the mass of copper produced when 1.47×10^4 kJ is consumed in this reaction.
(a) 11.2 kg $\qquad$ (d) 334 kg
(b) 176 kg $\qquad$ (e) 782 kg
(c) 44.0 kg

10.4 CALORIMETRY

In the study of thermochemistry, heat changes that accompany physical and chemical processes are measured with a *calorimeter*, a closed container designed specifically for this purpose. We begin our discussion of **calorimetry,** the measurement of heat changes, by defining two important terms: *specific heat* and *heat capacity*.

TABLE 10.2	Specific Heat Values of Some Common Substances
Substance	Specific heat (J/g · °C)
Al(s)	0.900
Au(s)	0.129
C (graphite)	0.720
C (diamond)	0.502
Cu(s)	0.385
Fe(s)	0.444
Hg(l)	0.139
H₂O(l)	4.184
C₂H₅OH(l) (ethanol)	2.46

Specific Heat and Heat Capacity

The *specific heat (s)* of a substance is the amount of heat required to raise the temperature of 1 g of the substance by 1°C. The *heat capacity (C)* is the amount of heat required to raise the temperature of an *object* by 1°C. We can use the specific heat of a substance to determine the heat capacity of a specified quantity of that substance. For example, we can use the specific heat of water, 4.184 J/(g · °C), to determine the heat capacity of a kilogram of water.

Student Annotation: Although heat capacity is typically given for an object rather than for a substance—the "object" may be a given quantity of a particular substance.

$$\text{Heat capacity of 1 kg of water} = \frac{4.184 \text{ J}}{1 \text{ g} \cdot {}^{\circ}\text{C}} \times 1000 \text{ g} = 4184 \text{ or } 4.184 \times 10^3 \text{ J/}{}^{\circ}\text{C}$$

Note that specific heat has the units J/(g · °C) and heat capacity has the units J/°C. Table 10.2 shows the specific heat values of some common substances. If we know the specific heat and the amount of a substance, then the change in the sample's temperature (ΔT) will tell us the amount of heat (q) that has been absorbed or released in a particular process. One equation for calculating the heat associated with a temperature change is given by

Student Hot Spot

Student data indicate you may struggle with specific heat capacity. Access the SmartBook to view additional Learning Resources on this topic.

Equation 10.11	$q = sm\Delta T$

where s is the specific heat, m is the mass of the substance undergoing the temperature change, and ΔT is the temperature change: $\Delta T = T_{\text{final}} - T_{\text{initial}}$. Another equation for calculating the heat associated with a temperature change is given by

Student Annotation: Note that $C = sm$. Although specific heat is a property of substances, and heat capacity is a property of objects, we can define a specified quantity of a substance as an "object" and determine its heat capacity using its mass and its specific heat.

Equation 10.12	$q = C\Delta T$

where C is the heat capacity and ΔT is the temperature change: The sign convention for q is the same as that for an enthalpy change: q is positive for endothermic processes and negative for exothermic processes. Worked Example 10.4 shows how to use the specific heat of a substance to calculate the amount of heat needed to raise the temperature of the substance by a particular amount.

Worked Example 10.4

Calculate the amount of heat (in kilojoules) required to heat 255 g of water from 25.2°C to 90.5°C.

Strategy Use Equation 10.11 ($q = sm\Delta T$) to calculate q.

Setup $s = 4.184$ J/g · °C, $m = 255$ g, and $\Delta T = 90.5$°C − 25.2°C = 65.3°C

Solution

$$q = \frac{4.184 \text{ J}}{\text{g} \cdot {}^\circ\text{C}} \times 255 \text{ g} \times 65.3{}^\circ\text{C} = 6.97 \times 10^4 \text{ J or } 69.7 \text{ kJ}$$

> **Think About It**
> Look carefully at the cancellation of units and make sure that the number of kilojoules is smaller than the number of joules. It is a common error to multiply by 1000 instead of dividing in conversions of this kind.

Practice Problem ATTEMPT Calculate the amount of heat (in kilojoules) required to heat 1.01 kg of water from 0.05°C to 35.81°C.

Practice Problem BUILD What will be the final temperature of a 514-g sample of water, initially at 10.0°C, after the addition of 90.8 kJ?

Practice Problem CONCEPTUALIZE Shown here are two samples of the same substance. When equal amounts of heat are added to both samples, the temperature of the sample on the left increases by 15.3°C. Determine the increase in temperature of the sample on the right.

Constant-Pressure Calorimetry

A crude constant-pressure calorimeter can be constructed from two Styrofoam coffee cups, as shown in Figure 10.8. This device, called a coffee-cup calorimeter, can be used to measure the heat exchanged between the system and surroundings for a variety of reactions, such as acid-base neutralization, heat of solution, and heat of dilution. Because the pressure is constant, the heat change for the process (q) is equal to the enthalpy change (ΔH). In such experiments, we consider the reactants and products to be the system, and the water in the calorimeter to be the surroundings. We neglect the small heat capacity of the Styrofoam cups in our calculations. In the case of an exothermic reaction, the heat released by the system is absorbed by the water (surroundings), thereby increasing its temperature. Knowing the mass of the water in the calorimeter, the specific heat of water, and the change in temperature, we can calculate q_P of the system using the equation

$$q_{\text{sys}} = -sm\Delta T \qquad \textbf{Equation 10.13}$$

Note that the minus sign makes q_{sys} a negative number if ΔT is a positive number (i.e., if the temperature goes up). This is in keeping with the sign conventions listed in Table 10.1. A negative ΔH or a negative q indicates an exothermic process, whereas a positive ΔH or a positive q indicates an endothermic process. Table 10.3 lists some of the reactions that can be studied with a constant-pressure calorimeter. Figure 10.9 (pages 430–431) shows how constant-pressure calorimetry can be used to determine ΔH for a reaction.

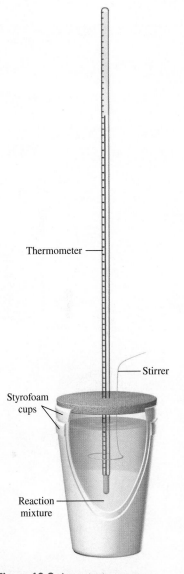

Figure 10.8 A constant-pressure coffee-cup calorimeter made of two Styrofoam cups. The nested cups help to insulate the reaction mixture from the surroundings. Two solutions of known volume containing the reactants at the same temperature are carefully mixed in the calorimeter. The heat produced or absorbed by the reaction can be determined by measuring the temperature change.

TABLE 10.3	Heats of Some Typical Reactions and Physical Processes Measured at Constant Pressure	
Type of reaction	**Example**	**ΔH (kJ/mol)**
Heat of neutralization	$HCl(aq) + NaOH(aq) \longrightarrow H_2O(l) + NaCl(aq)$	−56.2
Heat of ionization	$H_2O(l) \longrightarrow H^+(aq) + OH^-(aq)$	+56.2
Heat of fusion	$H_2O(s) \longrightarrow H_2O(l)$	+6.01
Heat of vaporization	$H_2O(l) \longrightarrow H_2O(g)$	+44.0*

*Measured at 25°C. At 100°C, the value is +40.79 kJ.

Video 10.2
Figure 10.9 Determination of $\Delta H_{\text{rxn}}^\circ$ by constant-pressure calorimetry.

 Student Hot Spot
Student data indicate you may struggle with constant-pressure calorimetry. Log in to Connect to view additional Learning Resources on this topic.

Figure 10.9

Determination of ΔH°_{rxn}
by Constant-Pressure Calorimetry

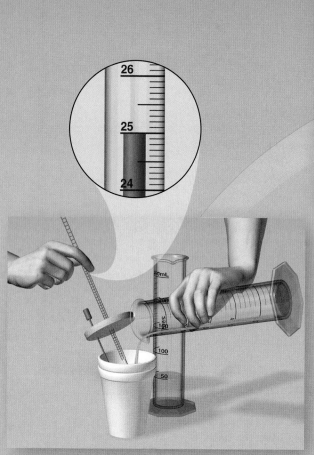

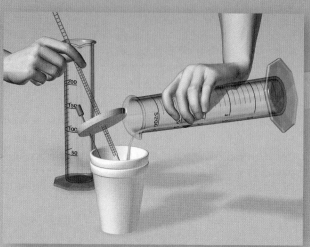

We pour the solutions into the calorimeter one at a time.

We start with 50.0 mL each of 1.00 *M* HCl and
1.00 *M* NaOH. Both solutions are at room temperature,
which in this example is 25.0°C. The net ionic equation
that represents the reaction is

$$H^+(aq) + OH^-(aq) \longrightarrow H_2O(l)$$

We have 0.0500 L × 1.00 *M* = 0.0500 mol of each reactant.

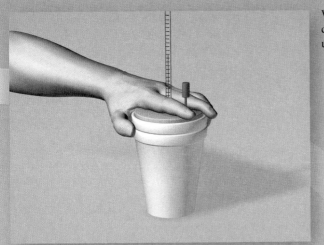

When both solutions have been added, we cap the calorimeter to prevent loss of energy to the environment and use the stirrer to ensure that the solutions are mixed thoroughly.

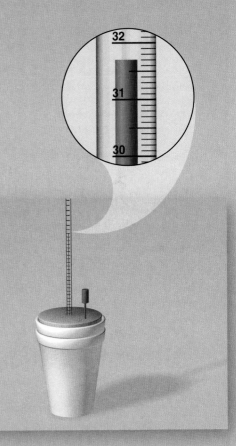

As the reaction proceeds, the temperature of the water increases as it absorbs the energy given off by the reaction. We record the maximum water temperature as 31.7°C.

Assuming the density and specific heat of the solution to be the same as those of water (1 g/mL, 4.184 J/g · °C), we calculate q_{soln} as follows:

q_{soln} = specific heat of water × mass of water × temperature change

$$= \frac{4.184 \text{ J}}{1 \text{ g} \cdot {}^\circ\text{C}} \times 100.0 \text{ g} \times (31.7 - 25.0)^\circ\text{C} = 2803 \text{ J}$$

Assuming that the heat capacity of the calorimeter is negligible, we know that $q_{soln} = -q_{rxn}$, and we can write

$$q_{rxn} = -2803 \text{ J}$$

This is the heat of reaction when 0.0500 mol H^+ reacts with 0.0500 mol OH^-. To determine ΔH_{rxn}, we divide q_{rxn} by the number of moles. (These reactants are present in stoichiometric amounts. If there were a *limiting* reactant, we would divide by the number of moles of limiting reactant.)

$$\Delta H_{rxn} = \frac{-2803 \text{ J}}{0.0500 \text{ mol}} = -5.61 \times 10^4 \text{ J/mol or } -56.1 \text{ kJ/mol}$$

This result is very close to the number we get using Equation 10.17 [▶▶ Section 10.6] and the data in Appendix 2.

What's the point?

Constant-pressure calorimetry can be used to determine ΔH_{rxn}—the heat of reaction for the reactant quantities specified by the balanced equation. However, when we carry out calorimetry experiments in the laboratory, we typically use much smaller quantities of reactants than those represented in a chemical equation. By measuring the temperature change of the surroundings (a known quantity of water in which the reactants are dissolved), we can determine q_{rxn} for the reactant quantities in the experiment. We can then divide q_{rxn} by the number of moles of reactant to determine ΔH_{rxn}.

Constant-pressure calorimetry can also be used to determine the heat capacity of an object or the specific heat of a substance. Suppose, for example, that we have a lead pellet with a mass of 26.47 g originally at 89.98°C. We drop the pellet into a constant-pressure calorimeter containing 100.0 g of water at 22.50°C. The temperature of the water increases to 23.17°C. In this case, we consider the pellet to be the system and the water to be the surroundings. Because it is the temperature of the surroundings that we measure, and because $q_{sys} = -q_{surr}$, we calculate q_P of the surroundings as

Student Annotation: The temperature of the water stops changing when it and the temperature of the lead pellet are equal. Therefore, the final temperature of the pellet is also 23.17°C.

Equation 10.14 $$q_{surr} = sm\Delta T$$

Thus, q_{water} of the water is

$$q = \frac{4.184\text{ J}}{\text{g}\cdot{}^\circ\text{C}} \times 1000.0\text{ g} \times (23.17^\circ\text{C} - 22.50^\circ\text{C}) = 280\text{ J}$$

and q_{pellet} is −280 J. The negative sign indicates that heat is *released* by the pellet. Dividing q_{pellet} by the temperature change (ΔT) gives us the heat capacity of the pellet (C_{pellet}).

$$C_{pellet} = \frac{-280\text{ J}}{23.17^\circ\text{C} - 89.98^\circ\text{C}} = 4.19\text{ J/}^\circ\text{C}$$

Furthermore, because we know the *mass* of the pellet, we can determine the specific heat of lead (s_{Pb}).

$$s_{Pb} = \frac{C_{pellet}}{m_{pellet}} = \frac{4.19\text{ J/}^\circ\text{C}}{26.47\text{ g}} = 0.158\text{ J/g}\cdot{}^\circ\text{C or } 0.16\text{ J/g}\cdot{}^\circ\text{C}$$

Video 10.3
Figure 10.10 Determination of specific heat by constant-pressure calorimetry.

Worked Example 10.5 shows how to use constant-pressure calorimetry to calculate the heat capacity (C) of a substance. Figure 10.10 (pages 434–435) illustrates the process of determining the specific heat of a metal using constant-pressure calorimetry.

Worked Example 10.5

A metal pellet with a mass of 100.0 g, originally at 88.4°C, is dropped into 125 g of water originally at 25.1°C. The final temperature of both the pellet and the water is 31.3°C. Calculate the heat capacity C (in J/°C) of the pellet.

Strategy Water constitutes the *surroundings;* the pellet is the *system*. Use Equation 10.14 ($q_{surr} = sm\Delta T$) to determine the heat absorbed by the water; then use Equation 10.12 ($q = C\Delta T$) to determine the heat capacity of the metal pellet.

Setup m_{water} = 125 g, s_{water} = 4.184 J/g · °C, and ΔT_{water} = 31.3°C − 25.1°C = 6.2°C. The heat absorbed by the water must be released by the pellet: $q_{water} = -q_{pellet}$, m_{pellet} = 100.0 g, and ΔT_{pellet} = 31.3°C − 88.4°C = −57.1°C.

Solution From Equation 10.14, we have

$$q_{water} = \frac{4.184\text{ J}}{\text{g}\cdot{}^\circ\text{C}} \times 125\text{ g} \times 6.2^\circ\text{C} = 3242.6\text{ J}$$

Thus,

$$q_{pellet} = -3242.6\text{ J}$$

From Equation 10.12, we have

$$-3242.6\text{ J} = C_{pellet} \times (-57.1^\circ\text{C})$$

Thus,

$$C_{pellet} = 57\text{ J/}^\circ\text{C}$$

Think About It
The units cancel properly to give appropriate units for heat capacity. Moreover, ΔT_{pellet} is a negative number because the temperature of the pellet decreases.

Practice Problem **TTEMPT** What would the final temperature be if the pellet from Worked Example 10.5, initially at 95°C, were dropped into a 218-g sample of water, initially at 23.8°C?

Practice Problem **B**UILD What mass of water could be warmed from 23.8°C to 46.3°C by the pellet in Worked Example 10.5 initially at 116°C?

Practice Problem **C**ONCEPTUALIZE Two samples of the same substance are shown. The temperatures of the two samples are indicated on the thermometer. Which of the final temperatures shown [(i), (ii), or (iii)] best represents the final temperature when the two samples are combined?

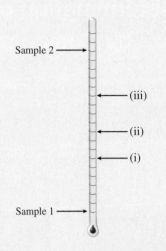

Sample 1   Sample 2

Constant-Volume Calorimetry

The heat of combustion is usually measured using constant-volume calorimetry. Typically, a known mass of the compound to be analyzed is placed in a steel container called a *constant-volume bomb,* or simply a *bomb,* which is pressurized with oxygen. The closed bomb is then immersed in a known amount of water in an insulated container, as shown in Figure 10.11 on page 436. (Together, the steel bomb and the water in which it is submerged constitute the *calorimeter.*) The sample is ignited electrically, and the heat released by the combustion of the sample is absorbed by the bomb and the water and can be determined by measuring the increase in temperature of the water. The special design of this type of calorimeter allows us to assume that no heat (or mass) is lost to the surroundings during the time it takes to carry out the reaction and measure the temperature change. Therefore, we can call the bomb and the water in which it is submerged an *isolated* system. Because no heat enters or leaves the system during the process, the heat change of the system overall (q_{system}) is zero and we can write

$$q_{cal} = -q_{rxn}$$

where q_{cal} and q_{rxn} are the heat changes for the calorimeter and the reaction, respectively. Thus,

$$q_{rxn} = -q_{cal}$$

To calculate q_{cal}, we need to know the heat capacity of the calorimeter (C_{cal}) and the change in temperature, that is,

$$q_{cal} = C_{cal}\Delta T \qquad \textbf{Equation 10.15}$$

And, because $q_{rxn} = -q_{cal}$,

$$q_{rxn} = -C_{cal}\Delta T \qquad \textbf{Equation 10.16}$$

The heat capacity of the calorimeter (C_{cal}) is determined by burning a substance with an accurately known heat of combustion. For example, it is known that the combustion of a 1.000-g sample of benzoic acid (C_6H_5COOH) releases 26.38 kJ of heat. If the measured temperature increase is 4.673°C, then the heat capacity of the calorimeter is given by

$$C_{cal} = \frac{q_{cal}}{\Delta T} = \frac{26.38 \text{ kJ}}{4.673°C} = 5.645 \text{ kJ/}°C$$

Figure 10.10

Determination of Specific Heat
by Constant-Pressure Calorimetry

We add metal shot (125.0 g at 100.0°C) to the water, and we cap the calorimeter to prevent loss of energy to the environment.

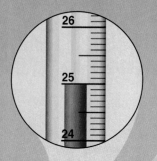

We place 100.0 mL (100.0 g) of water in the calorimeter. The temperature of this water is 25.0°C.

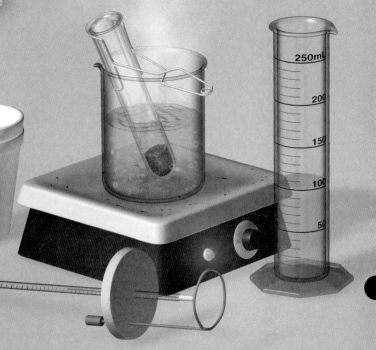

We place 125.0 g of metal shot in a test tube and immerse it in boiling water long enough to heat all of the metal to the boiling point of water (100.0°C).

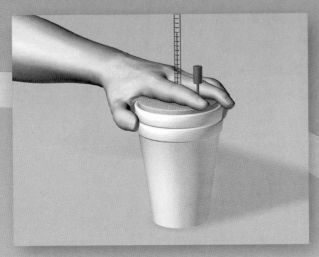

As energy is transferred from the metal shot to the water, the temperature of the water increases and the temperature of the metal shot decreases. We use the stirrer to ensure thorough mixing. The thermometer measures the temperature of the water.

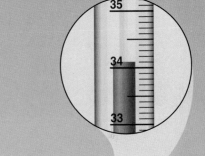

When the temperature of the metal shot and the water are equal, the temperature of the water has reached a maximum value. We record this temperature as 34.1°C.

We know that $q_{water} = -q_{metal}$.
Substituting in the information given we write:

q_{water} = specific heat of water × mass of water × temperature change

$$= \frac{4.184 \text{ J}}{1 \text{ g} \cdot °\text{C}} \times 100.0 \text{ g} \times (34.1 - 25.0)°\text{C} = 3807 \text{ J}$$

q_{metal} = specific heat of metal × mass of metal × temperature change

$$= x \times 125.0 \text{ g} \times (34.1 - 100.0)°\text{C} = -8238 \, x \text{ g} \cdot °\text{C}$$

and

$$3807 \text{ J} = -(-8238 \, x) \text{ g} \cdot °\text{C}$$

$$x = \frac{3807 \text{ J}}{8238 \text{ g} \cdot °\text{C}} = 0.46 \text{ J/g} \cdot °\text{C}$$

The specific heat of the metal is therefore 0.46 J/g · °C.

What's the point?

We can determine the specific heat of a metal by combining a known mass of the metal at a known temperature with a known mass of water at a known temperature. Assuming the calorimeter has a negligible heat capacity, the amount of energy lost by the hotter metal is equal to the amount of energy gained by the cooler water.

Figure 10.11 A constant-volume bomb calorimeter. The calorimeter is filled with oxygen gas at high pressure before it is placed in the bucket. The sample is ignited electrically, and the heat produced by the reaction is determined by measuring the temperature increase in the known amount of water surrounding the bomb.

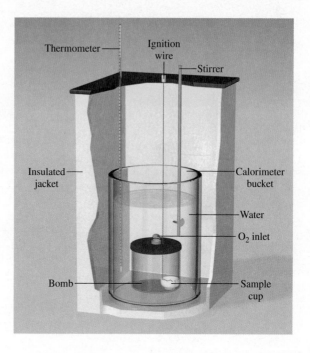

Once C_{cal} has been determined, the calorimeter can be used to measure the heat of combustion of other substances. Because a reaction in a bomb calorimeter occurs under constant-volume rather than constant-pressure conditions, the measured heat change corresponds to the *internal energy* change (ΔU) rather than to the *enthalpy* change (ΔH) (see Equations 10.4 and 10.9). It is possible to correct the measured heat changes so that they correspond to ΔH values, but the corrections usually are quite small, so we will not concern ourselves with the details here.

Thinking Outside the Box

Heat Capacity of Calorimeters

Although we usually assume that no heat is absorbed by the Styrofoam cups we use in the laboratory for constant-pressure calorimetry, in reality, the calorimeter generally does absorb a small portion of the heat produced by a chemical reaction. We can determine the heat capacity of a coffee-cup calorimeter by combining reactant solutions with precisely known concentrations and masses. Once we have determined the heat capacity, we can correct for the heat absorbed by the calorimeter when we carry out other experiments.

Consider an experiment in which we combine 50.0 mL of 0.250 *M* HCl(*aq*) with 50.0 mL of 0.250 *M* NaOH. The calorimeter and both solutions are initially at 23.50°C, and the density and specific heat of the combined solution are the same as that of water (1.000 g/mL and 4.184 J/g · °C, respectively). We can determine the amount of heat the reaction will generate using moles of reactants and the heat of neutralization value from Table 10.3. We have 0.0500 L × 0.250 *M* = 0.0125 mole of each reactant. According to Table 10.3, the heat of neutralization is −56.2 kJ/mol. Thus, we expect the enthalpy change of the system to be 0.0125 mol × (−56.2 kJ/mol) = −0.703 kJ. Converting this to joules and using Equation 10.13, we calculate the temperature change we expect from the combination of these reactants.

$$q_{sys} = -703 \text{ J} = -\frac{4.184 \text{ J}}{\text{g} \cdot °\text{C}} \times 100.0 \text{ g} \times \Delta T$$

$$\Delta T = \frac{-703 \text{ J}}{-(4.184 \text{ J/g} \cdot °\text{C})(100.0 \text{ g})} = 6.18°\text{C}$$

Thus, we expect the temperature to increase by 1.68°C. However, the measured final temperature is 25.09°C, an increase of only 1.59°C. The water temperature increased by less than expected because the calorimeter absorbed part of the heat produced by the neutralization reaction. We can determine how much heat the calorimeter absorbed using Equation 10.14. This time, we use the measured temperature change to calculate the amount of heat absorbed by the water.

$$q_{surr} = \frac{4.184 \text{ J}}{\text{g} \cdot °\text{C}} \times 100.0 \text{ g} \times 1.59°\text{C} = 655 \text{ J}$$

We know that the reaction produced 703 J but the water absorbed only 665 J. The remaining energy, 703 − 665 = 38 J, is $q_{calorimeter}$, the heat absorbed by the calorimeter itself. We use Equation 10.12 to calculate the heat capacity of the calorimeter.

$$38 \text{ J} = C_{calorimeter} \times 1.59°\text{C}$$

$$C_{calorimeter} = 23.9 \text{ J/°C}$$

Student Annotation: The calorimeter had the same initial temperature and the same final temperature as the solutions.

Worked Example 10.6 shows how to use constant-volume calorimetry to determine the energy content per gram of a substance.

Worked Example 10.6

A Famous Amos bite-sized chocolate chip cookie weighing 7.25 g is burned in a bomb calorimeter to determine its energy content. The heat capacity of the calorimeter is 39.97 kJ/°C. During the combustion, the temperature of the water in the calorimeter increases by 3.90°C. Calculate the energy content (in kilojoules per gram) of the cookie.

Strategy Use Equation 10.16 ($q_{rxn} = -C_{cal}\Delta T$) to calculate the heat released by the combustion of the cookie. Divide the heat released by the mass of the cookie to determine its energy content per gram.

Setup C_{cal} = 39.97 kJ/°C and ΔT = 3.90°C.

Solution From Equation 10.16 we have

$$q_{rxn} = -C_{cal}\Delta T = -(39.97 \text{ kJ/°C})(3.90°C) = -1.559 \times 10^2 \text{ kJ}$$

Student Annotation: The negative sign in the result indicates that heat is released by the combustion.

Because energy content is a positive quantity, we write

$$\text{energy content per gram} = \frac{1.559 \times 10^2 \text{ kJ}}{7.25 \text{ g}} = 21.5 \text{ kJ/g}$$

Think About It
According to the label on the cookie package, a serving size is four cookies, or 29 g, and each serving contains 150 Cal. Convert the energy per gram to Calories per serving to verify the result.

$$\frac{21.5 \text{ kJ}}{g} \times \frac{1 \text{ Cal}}{4.184 \text{ kJ}} \times \frac{29 \text{ g}}{\text{serving}} = 1.5 \times 10^2 \text{ Cal/serving}$$

Practice Problem A **TTEMPT** A serving of Grape-Nuts cereal (5.80 g) is burned in a bomb calorimeter with a heat capacity of 43.7 kJ/°C. During the combustion, the temperature of the water in the calorimeter increased by 1.92°C. Calculate the energy content (in kilojoules per gram) of Grape-Nuts.

Practice Problem B **UILD** The energy content of raisin bread is 13.1 kJ/g. Calculate the temperature increase when a slice of raisin bread (32.0 g) is burned in the calorimeter in Worked Example 10.6.

Practice Problem C **ONCEPTUALIZE** Suppose an experiment to determine the energy content of food used a calorimeter that contained less water than it did when it was calibrated. Explain how this would affect the result of the experiment.

Section 10.4 Review

Calorimetry

10.4.1 A 1.000-g sample of benzoic acid is burned in a calorimeter to determine its heat capacity, C_{cal}. The reaction gives off 26.42 kJ of heat and the temperature of the water in the calorimeter increases from 23.40°C to 27.20°C. What is the heat capacity of the calorimeter?
   (a) 3.80 kJ/°C   (d) 0.144 kJ/°C
   (b) 6.95 kJ/°C   (e) 7.81 kJ/°C
   (c) 100 kJ/°C

10.4.2 In a constant-pressure calorimetry experiment, a reaction gives off 21.8 kJ of heat. The calorimeter contains 150 g of water, initially at 23.4°C. What is the final temperature of the water? The heat capacity of the calorimeter is negligibly small.
   (a) 11.3°C   (d) 58.1°C
   (b) 26.9°C   (e) 23.4°C
   (c) 37.1°C

(Continued on next page)

10.4.3　A reaction, carried out in a bomb calorimeter with $C_{cal} = 5.01$ kJ/°C, gives off 318 kJ of heat. The initial temperature of the water is 24.8°C. What is the final temperature of the water in the calorimeter?

(a)　88.3°C

(b)　63.5°C

(c)　29.8°C

(d)　162°C

(e)　76.7°C

10.4.4　Quantities of 50.0 mL of 1.00 *M* HCl and 50.0 mL of 1.00 *M* NaOH are combined in a constant-pressure calorimeter. Both solutions are initially at 24.4°C. Calculate the final temperature of the combined solutions. (Use the data from Table 10.3. Assume that the mass of the combined solutions is 100.0 g and that the solution's specific heat is the same as that for water, 4.184 J/g · °C.) The heat capacity of the calorimeter is negligibly small.

(a)　31.1°C

(b)　29.0°C

(c)　44.2°C

(d)　91.8°C

(e)　35.7°C

10.5　HESS'S LAW

Because enthalpy is a state function, the change in enthalpy that occurs when reactants are converted to products in a reaction is the same whether the reaction takes place in one step or in a series of steps. This observation is called ***Hess's law***.[1] An analogy for Hess's law can be made to the floors in a building. Suppose, for example, that you take the elevator from the first floor to the sixth floor of the building. The net gain in your gravitational potential energy (which is analogous to the enthalpy change for the overall process) is the same whether you go directly there or stop at each floor on your way up (breaking the trip into a series of steps).

　　Recall from Section 10.3 that the enthalpy change for the combustion of a mole of methane depends on whether the product water is liquid or gas. More heat is given off by the reaction that produces liquid water. We can use this example to illustrate Hess's law by envisioning the first of these reactions happening in two steps. In step 1, methane and oxygen are converted to carbon dioxide and liquid water, releasing heat.

$$CH_4(g) + 2O_2(g) \longrightarrow CO_2(g) + 2H_2O(l) \qquad \Delta H = -890.4 \text{ kJ/mol}$$

In step 2, the liquid water is vaporized, which requires an input of heat.

$$2H_2O(l) \longrightarrow 2H_2O(g) \qquad \Delta H = +88.0 \text{ kJ/mol}$$

We can add balanced chemical equations just as we can add algebraic equalities, canceling identical items on opposite sides of the equation arrow.

$$CH_4(g) + 2O_2(g) \longrightarrow CO_2(g) + 2\cancel{H_2O(l)} \qquad \Delta H = -890.4 \text{ kJ/mol}$$
$$+ 2\cancel{H_2O(l)} \longrightarrow 2H_2O(g) \qquad \Delta H = +88.0 \text{ kJ/mol}$$
$$\overline{CH_4(g) + 2O_2(g) \longrightarrow CO_2(g) + 2H_2O(g) \qquad \Delta H = -802.4 \text{ kJ/mol}}$$

When we add thermochemical equations, we add the ΔH values as well. This gives us the overall enthalpy change for the net reaction. Using this method, we can deduce the enthalpy changes for many reactions, some of which may not be possible to carry out directly. In general, we apply Hess's law by arranging a series of chemical equations (corresponding to a series of steps) in such a way that they sum to the desired overall equation. Often, in applying Hess's law, we must manipulate the equations

---

[1]Germain Henri Hess (1802–1850), a Swiss chemist, was born in Switzerland but spent most of his life in Russia. For formulating Hess's law, he is called the father of thermochemistry.

involved, multiplying by appropriate coefficients, reversing equations, or both. It is important to follow the guidelines [◄◄ Section 10.3] for the manipulation of thermochemical equations and to make the corresponding change to the enthalpy change of each step.

Worked Example 10.7 illustrates the use of this method for determining ΔH.

Worked Example 10.7

Given the following thermochemical equations,

$$NO(g) + O_3(g) \longrightarrow NO_2(g) + O_2(g) \qquad O_3(g) \longrightarrow \tfrac{3}{2}O_2(g) \qquad O_2(g) \longrightarrow 2O(g)$$
$$\Delta H = -198.9 \text{ kJ/mol} \qquad\qquad \Delta H = -142.3 \text{ kJ/mol} \qquad \Delta H = +495 \text{ kJ/mol}$$

determine the enthalpy change for the reaction

$$NO(g) + O(g) \longrightarrow NO_2(g)$$

Strategy Arrange the given thermochemical equations so that they sum to the desired equation. Make the corresponding changes to the enthalpy changes, and add them to get the desired enthalpy change.

Setup The first equation has NO as a reactant with the correct coefficient, so we use it as is.

$$NO(g) + O_3(g) \longrightarrow NO_2(g) + O_2(g) \qquad \Delta H = -198.9 \text{ kJ/mol}$$

The second equation must be reversed so that the O_3 introduced by the first equation will cancel (O_3 is not part of the overall chemical equation). We also must change the sign on the corresponding ΔH value.

$$\tfrac{3}{2}O_2(g) \longrightarrow O_3(g) \qquad \Delta H = +142.3 \text{ kJ/mol}$$

These two steps sum to give the following:

$$
\begin{aligned}
NO(g) + O_3(g) &\longrightarrow NO_2(g) + O_2(g) & \Delta H &= -198.9 \text{ kJ/mol} \\
+ \ \tfrac{1}{2}O_2(g)\, \tfrac{3}{2}O_2(g) &\longrightarrow O_3(g) & \Delta H &= +142.3 \text{ kJ/mol} \\
\hline
NO(g) + \tfrac{1}{2}O_2(g) &\longrightarrow NO_2(g) & \Delta H &= -56.6 \text{ kJ/mol}
\end{aligned}
$$

We then replace the $\tfrac{1}{2}O_2$ on the left with O by incorporating the last equation. To do so, we divide the third equation by 2 and reverse its direction. As a result, we must also divide its ΔH value by 2 and change its sign.

$$O(g) \longrightarrow \tfrac{1}{2}O_2(g) \qquad \Delta H = -247.5 \text{ kJ/mol}$$

Finally, we sum all the steps and add their enthalpy changes.

Solution

$$
\begin{aligned}
NO(g) + O_3(g) &\longrightarrow NO_2(g) + O_2(g) & \Delta H &= -198.9 \text{ kJ/mol} \\
\tfrac{3}{2}O_2(g) &\longrightarrow O_3(g) & \Delta H &= +142.3 \text{ kJ/mol} \\
+ \ O(g) &\longrightarrow \tfrac{1}{2}O_2(g) & \Delta H &= -247.5 \text{ kJ/mol} \\
\hline
NO(g) + O(g) &\longrightarrow NO_2(g) & \Delta H &= -304 \text{ kJ/mol}
\end{aligned}
$$

Think About It
Double-check the cancellation of identical items.

Practice Problem ⒶTTEMPT Use the thermochemical equations provided in Worked Example 10.7 to determine the enthalpy change for the reaction $2NO(g) + 4O(g) \longrightarrow 2NO_2(g) + O_2(g)$.

Practice Problem ⒷUILD Use the thermochemical equations provided in Worked Example 10.7 to determine the enthalpy change for the reaction $2NO_2(g) \longrightarrow 2NO(g) + O_2(g)$.

Practice Problem ⒸONCEPTUALIZE The diagrams shown are representations of four systems before and after reactions involving five different chemical species—each represented by a different color sphere. The ΔH values are given for the first three. Determine ΔH for the last reaction.

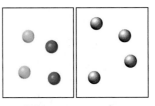

before   after
$\Delta H = 25 \text{ kJ/mol}$

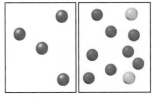

before   after
$\Delta H = 100 \text{ kJ/mol}$

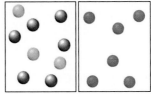

before   after
$\Delta H = -60 \text{ kJ/mol}$

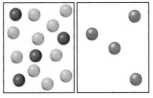

before   after
$\Delta H = ?$

Section 10.5 Review

Hess's Law

10.5.1 Given the following information, determine ΔH for $3H_2(g) + O_3(g) \longrightarrow 3H_2O(g)$.

$$2H_2(g) + O_2(g) \longrightarrow 2H_2O(g) \qquad \Delta H = -483.6 \text{ kJ/mol}$$

$$3O_2(g) \longrightarrow 2O_3(g) \qquad \Delta H = +284.6 \text{ kJ/mol}$$

(a) -199 kJ/mol   (d) $+768.2$ kJ/mol

(b) -1010 kJ/mol   (e) -440.8 kJ/mol

(c) -867.7 kJ/mol

10.5.2 Given the following information, determine ΔH_{rxn} for $P_4O_6(s) + 2O_2(g) \longrightarrow P_4O_{10}(s)$.

$$P_4(s) + 3O_2(g) \longrightarrow P_4O_6(s) \qquad \Delta H = -1640.1 \text{ kJ/mol}$$

$$P_4(s) + 5O_2(g) \longrightarrow P_4O_{10}(s) \qquad \Delta H = -2940.1 \text{ kJ/mol}$$

(a) -1300.0 kJ/mol   (d) $+982.6$ kJ/mol

(b) $+4580.2$ kJ/mol   (e) -982.6 kJ/mol

(c) -4580.2 kJ/mol

10.6 STANDARD ENTHALPIES OF FORMATION

So far we have learned that we can determine the enthalpy change that accompanies a reaction by measuring the heat absorbed or released (at constant pressure). According to Equation 10.10, ΔH can also be calculated if we know the enthalpies of all reactants and products. However, there is no way to measure the *absolute* value of the enthalpy of a substance. Only values *relative* to an arbitrary reference can be determined. This problem is similar to the one geographers face in expressing the elevations of specific mountains or valleys. Rather than trying to devise some type of "absolute" elevation scale (perhaps based on the distance from the center of Earth), by common agreement all geographical heights and depths are expressed relative to sea level, an arbitrary reference with a defined elevation of "zero" meters or feet. Similarly, chemists have agreed on an arbitrary reference point for enthalpy.

The "sea level" reference point for all enthalpy expressions is called the **standard enthalpy of formation (ΔH_f°),** which is defined as the heat change that results when 1 mole of a compound is formed from its constituent elements in their standard states. The superscript degree sign denotes standard-state conditions, and the subscript f stands for *formation*. The phrase "in their standard states" refers to the most stable form of an element under standard conditions, meaning at ordinary atmospheric pressure. The element oxygen, for example, can exist as atomic oxygen (O), diatomic oxygen (O_2), or ozone (O_3). By far the most stable form at ordinary atmospheric pressure, though, is diatomic oxygen. Thus, the standard state of oxygen is O_2. Although the standard state does not specify a temperature, we will always use (ΔH_f°) values measured at 25°C.

Appendix 2 lists the standard enthalpies of formation for a number of elements and compounds. By convention, the standard enthalpy of formation of any element in its most stable form is zero. Again, using the element oxygen as an example, we can write $\Delta H_f^\circ (O_2) = 0$, but $\Delta H_f^\circ (O_3) \neq 0$ and $\Delta H_f^\circ (O) \neq 0$. Similarly, graphite is a more stable allotropic form of carbon than diamond under standard conditions and 25°C, so we have $\Delta H_f^\circ (\text{graphite}) = 0$ and $\Delta H_f^\circ (\text{diamond}) \neq 0$.

The importance of the standard enthalpies of formation is that once we know their values, we can readily calculate the *standard enthalpy of reaction* (ΔH°_{rxn}), defined as the enthalpy of a reaction carried out under standard conditions. For example, consider the hypothetical reaction

$$a\text{A} + b\text{B} \longrightarrow c\text{C} + d\text{D}$$

where *a*, *b*, *c*, and *d* are stoichiometric coefficients. For this reaction, ΔH°_{rxn} is given by

$$\Delta H^\circ_{rxn} = [c\Delta H^\circ_f(\text{C}) + d\Delta H^\circ_f(\text{D})] - [a\Delta H^\circ_f(\text{A}) + b\Delta H^\circ_f(\text{B})] \qquad \textbf{Equation 10.17}$$

We can generalize Equation 10.17 as

$$\Delta H^\circ_{rxn} = \Sigma n\Delta H^\circ_f(\text{products}) - \Sigma m\Delta H^\circ_f(\text{reactants}) \qquad \textbf{Equation 10.18}$$

where *m* and *n* are the stoichiometric coefficients for the reactants and products, respectively, and Σ (sigma) means "the sum of." In these calculations, the stoichiometric coefficients are treated as numbers without units. Thus, the result has units of kJ/mol, where again, "per mole" means per mole of reaction as written. To use Equation 10.18 to calculate ΔH°_{rxn}, we must know the ΔH°_f values of the compounds that take part in the reaction. These values, tabulated in Appendix 2, are determined either by the direct method or the indirect method.

The *direct* method of measuring ΔH°_f works for compounds that can be synthesized from their elements easily and safely. Suppose we want to know the enthalpy of formation of carbon dioxide. We must measure the enthalpy of the reaction when carbon (graphite) and molecular oxygen in their standard states are converted to carbon dioxide in its standard state.

$$\text{C(graphite)} + \text{O}_2(g) \longrightarrow \text{CO}_2(g) \qquad \Delta H^\circ_{rxn} = -393.5 \text{ kJ/mol}$$

We know from experience that this combustion goes to completion. Thus, from Equation 10.18 we can write

$$\Delta H^\circ_{rxn} = \Delta H^\circ_f(\text{CO}_2) - [\Delta H^\circ_f(\text{graphite}) + \Delta H^\circ_f(\text{O}_2)] = -393.5 \text{ kJ/mol}$$

Because graphite and O_2 are the most stable allotropic forms of their respective elements, ΔH°_f (graphite) and ΔH°_f (O_2) are both zero. Therefore,

$$\Delta H^\circ_{rxn} = \Delta H^\circ_f(\text{CO}_2) = -393.5 \text{ kJ/mol}$$

or

$$\Delta H^\circ_f(\text{CO}_2) = -393.5 \text{ kJ/mol}$$

Arbitrarily assigning a value of zero to ΔH°_f for each element in its standard state does *not* affect the outcome of these calculations. Remember, in thermochemistry we are interested only in enthalpy *changes* because they can be determined experimentally, whereas the absolute enthalpy values cannot. The choice of a zero "reference level" for enthalpy is intended to simplify the calculations. Referring again to the terrestrial altitude analogy, we find that Mt. Everest (the highest peak in the world) is 8708 feet higher than Mt. Denali (the highest peak in North America). This difference in altitude would be the same whether we had chosen sea level or the center of Earth as our reference elevation.

Other compounds that can be studied by the direct method are SF_6, P_4O_{10}, and CS_2. The equations representing their syntheses are

$$\text{S(rhombic)} + 3\text{F}_2(g) \longrightarrow \text{SF}_6(g)$$

$$\text{P}_4(\text{white}) + 5\text{O}_2(g) \longrightarrow \text{P}_4\text{O}_{10}(s)$$

$$\text{C(graphite)} + 2\text{S(rhombic)} \longrightarrow \text{CS}_2(l)$$

S(rhombic) and P(white) are the most stable allotropes of sulfur and phosphorus, respectively, at 1 atm and 25°C, so their ΔH°_f values are zero.

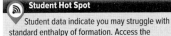

Worked Example 10.8 shows how ΔH_f° values can be used to determine ΔH_{rxn}°.

Worked Example 10.8

Using data from Appendix 2, calculate ΔH_{rxn}° for $Ag^+(aq) + Cl^-(aq) \longrightarrow AgCl(s)$.

Strategy Use Equation 10.18 $[\Delta H_{rxn}^\circ = \Sigma n\Delta H_f^\circ(\text{products}) - \Sigma m\Delta H_f^\circ(\text{reactants})]$ and ΔH_f° values from Appendix 2 to calculate ΔH_{rxn}°.

Setup The ΔH_f° values for $Ag^+(aq)$, $Cl^-(aq)$, and $AgCl(s)$ are $+105.9$, -167.2, and -127.0 kJ/mol, respectively.

Solution Using Equation 10.18,

$$\Delta H_{rxn}^\circ = \Delta H_f^\circ(AgCl) - [\Delta H_f^\circ(Ag^+) + \Delta H_f^\circ(Cl^-)]$$

$$= -127.0 \text{ kJ/mol} - [(+105.9 \text{ kJ/mol}) + (-167.2 \text{ kJ/mol})]$$

$$= -127. \text{ kJ/mol} - (-61.3 \text{ kJ/mol}) = -65.7 \text{ kJ/mol}$$

Think About It
Watch out for misplaced or missing minus signs. This is an easy place to lose track of them.

Practice Problem **A**TTEMPT Using data from Appendix 2, calculate ΔH_{rxn}° for $CaCO_3(s) \longrightarrow CaO(s) + CO_2(g)$.

Practice Problem **B**UILD Using data from Appendix 2, calculate ΔH_{rxn}° for $2SO(g) + \frac{2}{3}O_3(g) \longrightarrow 2SO_2(g)$.

Practice Problem **C**ONCEPTUALIZE The diagrams represent a system before and after a chemical reaction. Using the table of ΔH_f° values for the species involved in the reaction, determine ΔH_{rxn}° for the process represented by the diagrams.

before   after

Species	ΔH_f° (kJ/mol)
	78.0
	−188.5
	106.5

Many compounds cannot be synthesized from their elements directly. In some cases, the reaction proceeds too slowly, or side reactions produce substances other than the desired compound. In these cases, ΔH_f° can be determined by an *indirect* approach, using Hess's law. If we know a series of reactions for which ΔH_{rxn}° can be measured, and we can arrange them in such a way as to have them sum to the equation corresponding to the formation of the compound of interest, we can calculate ΔH_f° for the compound.

Worked Example 10.9 shows how to use Hess's law to calculate the ΔH_f° value by the indirect method for a compound that cannot be produced easily from its constituent elements.

Worked Example 10.9

Given the following information, calculate the standard enthalpy of formation of acetylene (C_2H_2) from its constituent elements.

$$C(\text{graphite}) + O_2(g) \longrightarrow CO_2(g) \qquad \Delta H_{rxn}^\circ = -393.5 \text{ kJ/mol} \quad (1)$$

$$H_2(g) + \tfrac{1}{2}O_2(g) \longrightarrow H_2O(l) \qquad \Delta H_{rxn}^\circ = -285.8 \text{ kJ/mol} \quad (2)$$

$$2C_2H_2(g) + 5O_2(g) \longrightarrow 4CO_2(g) + 2H_2O(l) \qquad \Delta H_{rxn}^\circ = -2598.8 \text{ kJ/mol} \quad (3)$$

Strategy Arrange the equations that are provided so that they will sum to the desired equation. This may require reversing or multiplying one or more of the equations. For any such change, the corresponding change must also be made to the ΔH_{rxn}° value.

Setup The desired equation, corresponding to the standard enthalpy of formation of acetylene, is

$$2C(\text{graphite}) + H_2(g) \longrightarrow C_2H_2(g)$$

We multiply Equation (1) and its ΔH°_{rxn} value by 2.

$$2C(graphite) + 2O_2(g) \longrightarrow 2CO_2(g) \qquad\qquad \Delta H^\circ_{rxn} = -787.0 \text{ kJ/mol}$$

We include Equation (2) and its ΔH°_{rxn} value as is.

$$H_2(g) + \tfrac{1}{2}O_2(g) \longrightarrow H_2O(l) \qquad\qquad \Delta H^\circ_{rxn} = -285.8 \text{ kJ/mol}$$

We reverse Equation (3) and divide it by 2 (i.e., multiply through by 1/2).

$$2CO_2(g) + H_2O(l) \longrightarrow C_2H_2(g) + \tfrac{5}{2}O_2(g) \qquad\qquad \Delta H^\circ_{rxn} = +1299.4 \text{ kJ/mol}$$

Student Annotation: The original ΔH°_{rxn} value of Equation (3) has its sign reversed and it is divided by 2.

Solution Summing the resulting equations and the corresponding ΔH°_{rxn} values.

$$2C(graphite) + 2O_2(g) \longrightarrow 2CO_2(g) \qquad\qquad \Delta H^\circ_{rxn} = -787.0 \text{ kJ/mol}$$

$$H_2(g) + \tfrac{1}{2}O_2(g) \longrightarrow H_2O(l) \qquad\qquad \Delta H^\circ_{rxn} = -285.8 \text{ kJ/mol}$$

$$2CO_2(g) + H_2O(l) \longrightarrow C_2H_2(g) + \tfrac{5}{2}O_2(g) \qquad\qquad \Delta H^\circ_{rxn} = +1299.4 \text{ kJ/mol}$$

$$\overline{2C(graphite) + H_2(g) \longrightarrow C_2H_2(g) \qquad\qquad \Delta H^\circ_f = +226.6 \text{ kJ/mol}}$$

> **Think About It**
> Remember that a ΔH°_{rxn} is only a ΔH°_f when there is just *one product*, just *one mole* is produced, and all the reactants are *elements in their standard states*.

Practice Problem **A**TTEMPT Use the following data to calculate ΔH°_f for $CS_2(l)$:

$$C(graphite) + O_2(g) \longrightarrow CO_2(g) \qquad\qquad \Delta H^\circ_{rxn} = -393.5 \text{ kJ/mol}$$

$$S(rhombic) + O_2(g) \longrightarrow SO_2(g) \qquad\qquad \Delta H^\circ_{rxn} = -296.4 \text{ kJ/mol}$$

$$CS_2(l) + 3O_2(g) \longrightarrow CO_2(g) + 2SO_2(g) \qquad\qquad \Delta H^\circ_{rxn} = -1073.6 \text{ kJ/mol}$$

Practice Problem **B**UILD ΔH°_f of hydrogen chloride [HCl(g)] is -92.3 kJ/mol. Given the following data, determine the identity of the two missing products and calculate ΔH°_{rxn} for Equation (3). [*Hint:* Start by writing the chemical equation that corresponds to ΔH°_f for HCl(g).]

$$N_2(g) + 4H_2(g) + Cl_2(g) \longrightarrow 2NH_4Cl(s) \qquad\qquad \Delta H^\circ_{rxn} = -603.78 \text{ kJ/mol} \quad (1)$$

$$N_2(g) + 3H_2(g) \longrightarrow 2NH_3(g) \qquad\qquad \Delta H^\circ_{rxn} = -92.6 \text{ kJ/mol} \quad (2)$$

$$NH_4Cl(s) \longrightarrow \qquad\qquad\qquad\qquad\qquad\qquad\qquad\qquad (3)$$

Practice Problem **C**ONCEPTUALIZE The diagrams represent a system before and after a chemical reaction for which ΔH°_{rxn} is -2624.9 kJ/mol. Use this information to complete the table of ΔH°_f values for the species involved in the reaction.

before	after

Species	ΔH°_f (kJ/mol)
●	-148.7
●	?
●	255.1

Section 10.6 Review

Standard Enthalpies of Formation

10.6.1 Using data from Appendix 2, calculate ΔH°_{rxn} for $H_2(g) + F_2(g) \longrightarrow 2HF(g)$.
  (a) -271.6 kJ/mol
  (d) 0 kJ/mol
  (b) -543.2 kJ/mol
  (e) -135.8 kJ/mol
  (c) $+271.6$ kJ/mol

(Continued on next page)

10.6.2   Using data from Appendix 2, calculate ΔH°_{rxn} for $2NO_2(g) \longrightarrow N_2O_4(g)$.

(a) −24.19 kJ/mol   (d) +67.7 kJ/mol
(b) −33.85 kJ/mol   (e) −58.04 kJ/mol
(c) +9.66 kJ/mol

10.6.3   Using the following data, calculate ΔH°_f for $CO(g)$.

$$C(graphite) + O_2(g) \longrightarrow CO_2(g) \qquad\qquad \Delta H^\circ_{rxn} = -393.5 \text{ kJ/mol}$$

$$CO(g) + \tfrac{1}{2}O_2(g) \longrightarrow CO_2(g) \qquad\qquad \Delta H^\circ_{rxn} = -283.0 \text{ kJ/mol}$$

(a) −393.5 kJ/mol   (d) +110.5 kJ/mol
(b) −676.5 kJ/mol   (e) −110.5 kJ/mol
(c) +676.5 kJ/mol

10.7   BOND ENTHALPY AND THE STABILITY OF COVALENT MOLECULES

Why do covalent bonds form? According to valence bond theory, a covalent bond will form between two atoms if the potential energy of the resulting molecule is lower than that of the isolated atoms. Simply put, this means that the formation of covalent bonds is exothermic. While this fact may not seem intuitively obvious, you know that energy must be supplied to a molecule to *break* covalent bonds. Because the formation of a bond is the *reverse* process, we should expect energy to be given off when a bond forms.

Student Annotation: Recall that we previously used the term *bond energy* to refer to this property, before the term *enthalpy* had been introduced [◄◄ Section 6.1].

One measure of the stability of a molecule is its ***bond enthalpy,*** which is the enthalpy change associated with breaking a particular bond in 1 mole of gaseous molecules. (Bond enthalpies in solids and liquids are affected by neighboring molecules.) The experimentally determined bond enthalpy of the diatomic hydrogen molecule, for example, is

$$H_2(g) \longrightarrow H(g) + H(g) \qquad\qquad \Delta H^\circ = 436.4 \text{ kJ/mol}$$

According to this equation, breaking the covalent bonds in 1 mole of gaseous H_2 molecules requires 436.4 kJ of energy. For the less stable chlorine molecule,

$$Cl_2(g) \longrightarrow Cl(g) + Cl(g) \qquad\qquad \Delta H^\circ = 242.7 \text{ kJ/mol}$$

Bond enthalpies can also be directly measured for heteronuclear diatomic molecules, such as HCl,

$$HCl(g) \longrightarrow H(g) + Cl(g) \qquad\qquad \Delta H^\circ = 431.9 \text{ kJ/mol}$$

as well as for molecules containing multiple bonds.

$$O_2(g) \longrightarrow O(g) + O(g) \qquad\qquad \Delta H^\circ = 498.7 \text{ kJ/mol}$$

$$N_2(g) \longrightarrow N(g) + N(g) \qquad\qquad \Delta H^\circ = 941.4 \text{ kJ/mol}$$

Measuring the strength of covalent bonds in polyatomic molecules is more complicated. For example, measurements show that the energy needed to break the first O−H bond in H_2O is different from that needed to break the second O−H bond.

$$H_2O(g) \longrightarrow H(g) + OH(g) \qquad\qquad \Delta H^\circ = 502 \text{ kJ/mol}$$

$$OH(g) \longrightarrow H(g) + O(g) \qquad\qquad \Delta H^\circ = 427 \text{ kJ/mol}$$

In each case, an O−H bond is broken, but the first step requires the input of more energy than the second. The difference between the two ΔH° values suggests that the second O−H bond itself undergoes change, because of the changes in its chemical environment.

TABLE 10.4	Bond Enthalpies		
Bond	Bond enthalpy (kJ/mol)	Bond	Bond enthalpy (kJ/mol)
H—H*	436.4	N—N	193
H—N	393	N=N	418
H—O	460	N≡N	941.4
H—S	368	N—O	176
H—P	326	N=O	607
H—F	568.2	N—F	272
H—Cl	431.9	N—Cl	200
H—Br	366.1	N—Br	243
H—I	298.3	N—I	159
C—H	414	O—O	142
C—C	347	O=O	498.7
C=C	620	O—P	502
C≡C	812	O=S	469
C—N	276	O—F	190
C=N	615	O—Cl	203
C≡N	891	O—Br	234
C—O	351	O—I	234
C=O†	745	P—P	197
C≡O	1070	P=P	489
C—P	263	S—S	268
C—S	255	S=S	352
C=S	477	F—F	156.9
C—F	453	Cl—Cl	242.7
C—Cl	339	Cl—F	193
C—Br	276	Br—Br	192.5
C—I	216	I—I	151.0

*Bond enthalpies shown in red are for diatomic molecules.

†The C=O bond enthalpy in CO_2 is 799 kJ/mol.

Student Annotation: Bond enthalpies for diatomic molecules have more significant figures than those for polyatomic molecules. Those for polyatomic molecules are *average* values based on the bonds in more than one compound.

We can now understand why the bond enthalpy of the same O—H bond in two different molecules, such as methanol (CH_3OH) and water (H_2O), will not be the same: their environments are different. For polyatomic molecules, therefore, we speak of the *average* bond enthalpy of a particular bond. For example, we can measure the enthalpy of the O—H bond in 10 different polyatomic molecules and obtain the average O—H bond enthalpy by dividing the sum of the bond enthalpies by 10. Table 10.4 lists the average bond enthalpies of a number of diatomic and polyatomic molecules. As we noted earlier, triple bonds are stronger than double bonds, and double bonds are stronger than single bonds.

A comparison of the thermochemical changes that take place during a number of reactions reveals a strikingly wide variation in the enthalpies of different reactions. For example, the combustion of hydrogen gas in oxygen gas is fairly *exothermic*.

$$H_2(g) + \tfrac{1}{2}O_2(g) \longrightarrow H_2O(l) \qquad \Delta H^\circ = -285.8 \text{ kJ/mol}$$

Figure 10.12 Enthalpy changes in (a) an exothermic reaction and (b) an endothermic reaction. The $\Delta H°$ values are calculated using Equation 10.18 and tabulated $\Delta H_f°$ values from Appendix 2.

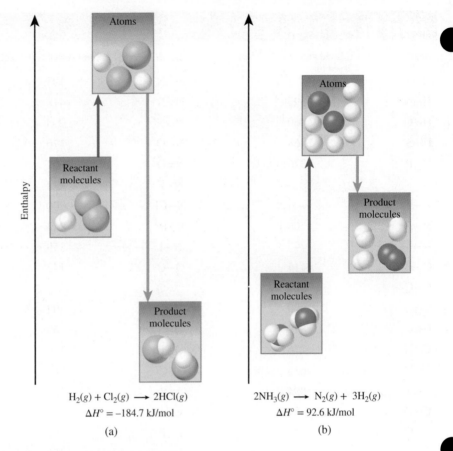

$$H_2(g) + Cl_2(g) \longrightarrow 2HCl(g)$$
$$\Delta H° = -184.7 \text{ kJ/mol}$$
(a)

$$2NH_3(g) \longrightarrow N_2(g) + 3H_2(g)$$
$$\Delta H° = 92.6 \text{ kJ/mol}$$
(b)

The formation of glucose from carbon dioxide and water, on the other hand, best achieved by photosynthesis, is highly *endothermic*.

$$6CO_2(g) + 6H_2O(l) \longrightarrow C_6H_{12}O_6(s) + 6O_2(g) \qquad \Delta H° = 2801 \text{ kJ/mol}$$

We can account for such variations by looking at the stability of individual reactant and product molecules. After all, most chemical reactions involve the making and breaking of bonds. Therefore, knowing the bond enthalpies and hence the stability of molecules reveals something about the thermochemical nature of the reactions that molecules undergo.

In many cases, it is possible to predict the approximate enthalpy of a reaction by using the average bond enthalpies. Because energy is always required to break chemical bonds and chemical bond formation is always accompanied by a release of energy, we can estimate the enthalpy of a reaction by counting the total number of bonds broken and formed in the reaction and recording all the corresponding enthalpy changes. The enthalpy of reaction in the gas phase is given by

Equation 10.19
$$\begin{aligned} \Delta H° &= \Sigma \text{BE (reactants)} - \Sigma \text{BE (products)} \\ &= \text{total energy } \textit{input} - \text{total energy } \textit{released} \\ &\quad\; (\textit{to break} \text{ bonds}) \qquad (\text{by bond } \textit{formation}) \end{aligned}$$

Student Annotation: To many students, Equation 10.19 appears to be backward. Ordinarily you calculate as *final minus initial*. Here we are determining the difference between the amount of heat we have to add to break reactant bonds and the amount of heat released when product bonds form. The sign of the final answer tells us if the process is endothermic (+) or exothermic (−).

where BE stands for average bond enthalpy and Σ is the summation sign. As written, Equation 10.19 takes care of the sign convention for $\Delta H°$. Thus, if the total energy input needed to break bonds in the reactants is less than the total energy released when bonds are formed in the products, then $\Delta H°$ is negative and the reaction is exothermic [Figure 10.12(a)]. On the other hand, if less energy is released (bond making) than absorbed (bond breaking), $\Delta H°$ is positive and the reaction is endothermic [Figure 10.12(b)].

If all the reactants and products are diatomic molecules, then the equation for the enthalpy of reaction will yield accurate results because the bond enthalpies of diatomic molecules are accurately known. If some or all of the reactants and products are polyatomic molecules, the equation will yield only approximate results because the bond enthalpies used will be averages.

Worked Example 10.10 shows how to estimate enthalpies of reaction using bond enthalpies.

Worked Example 10.10

Use bond enthalpies from Table 10.4 to estimate the enthalpy of reaction for the combustion of methane.

$$CH_4(g) + 2O_2(g) \longrightarrow CO_2(g) + 2H_2O(l)$$

Strategy Draw Lewis structures to determine what bonds are to be broken and what bonds are to be formed. (Don't skip the step of drawing Lewis structures. This is the only way to know for certain what types and numbers of bonds must be broken and formed.)

Setup

$$
\begin{array}{ccc}
\underset{\displaystyle\overset{\displaystyle H}{|}}{\overset{\displaystyle H}{H-C-H}} \; + \;
\begin{array}{c} :O=O: \\ :O=O: \end{array}
& \longrightarrow &
:O=C=O: \; + \;
\begin{array}{c} H-\ddot{O}-H \\ H-\ddot{O}-H \end{array}
\end{array}
$$

Bonds to break: 4 C—H and 2 O=O

Bonds to form: 2 C=O and 4 H—O

Bond enthalpies from Table 10.4: 414 kJ/mol (C—H), 498.7 kJ/mol (O=O), 799 kJ/mol (C=O in CO_2), and 460 kJ/mol (H—O)

Solution

$$[4(414 \text{ kJ/mol}) + 2(498.7 \text{ kJ/mol})] - [2(799 \text{ kJ/mol}) + 4(460 \text{ kJ/mol})] = -785 \text{ kJ/mol}$$

Remember that heats of reaction are expressed in kJ/mol, where the "per mole" refers to *per mole of reaction as written* [◄◄ Section 10.3].

Think About It
Use Equation 10.18 [◄◄ Section 10.6] and data from Appendix 2 to calculate this enthalpy of reaction again; then compare your results using the two approaches. The difference in this case is due to two things: Most tabulated bond enthalpies are averages and, by convention, we show the product of combustion as liquid water—but average bond enthalpies apply to species in the gas phase, where there is little or no influence exerted by neighboring molecules.

Practice Problem **A**TTEMPT Use bond enthalpies from Table 10.4 to estimate the enthalpy of reaction for the combination of carbon monoxide and oxygen to produce carbon dioxide.

$$2CO(g) + O_2(g) \longrightarrow 2CO_2(g)$$

Practice Problem **B**UILD Using the following chemical equation, data from Table 10.4, and data from Appendix 2, determine the P—Cl bond enthalpy.

$$PH_3(g) + 3HCl(g) \longrightarrow PCl_3(g) + 3H_2(g)$$

Practice Problem **C**ONCEPTUALIZE Four different chemical reactions are represented here. For each reaction, indicate whether it is endothermic or exothermic—or if there is not enough information to determine.

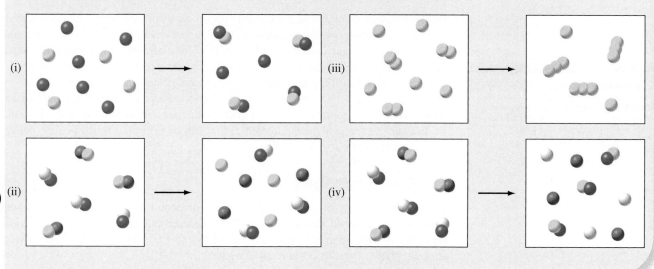

Section 10.7 Review

Bond Enthalpy and the Stability of Covalent Molecules

10.7.1   Use data from Table 10.4 to estimate ΔH_{rxn} for the reaction of ethylene with hydrogen to produce ethane.

$$C_2H_4(g) + H_2(g) \longrightarrow C_2H_6(g)$$

(a) -391 kJ/mol   (d) -118 kJ/mol
(b) $+710$ kJ/mol   (e) $+118$ kJ/mol
(c) -710 kJ/mol

10.7.2   Use data from Table 10.4 to estimate ΔH_{rxn} for the reaction of fluorine and chlorine to produce ClF.

$$F_2(g) + Cl_2(g) \longrightarrow 2ClF(g)$$

(a) -77.5 kJ/mol   (d) -13.6 kJ/mol
(b) -206.6 kJ/mol   (e) $+13.6$ kJ/mol
(c) $+206.6$ kJ/mol

10.7.3   Use bond enthalpies to determine ΔH_{rxn} for the reaction shown.

(a) -1139 kJ/mol   (d) $+114$ kJ/mol
(b) $+1139$ kJ/mol   (e) -114 kJ/mol
(c) -346 kJ/mol

10.7.4   Use bond enthalpies to determine ΔH_{rxn} for the reaction shown.

(a) -1028 kJ/mol   (d) $+392$ kJ/mol
(b) -200 kJ/mol   (e) $+200$ kJ/mol
(c) -392 kJ/mol

10.8 LATTICE ENERGY AND THE STABILITY OF IONIC SOLIDS

Unlike covalent compounds, ionic compounds consist of vast arrays of interspersed cations and anions, not discrete molecular units. For example, solid sodium chloride (NaCl) consists of equal numbers of Na^+ and Cl^- ions arranged in a three-dimensional network, the lattice, of alternating cations and anions (see Figure 5.2, page 166).

We learned in Chapter 5 that the energy change associated with the formation of a lattice is the lattice energy. Although lattice energy is a useful measure of an ionic compound's stability, it is *not* a quantity that we can measure directly. Instead, we use various thermodynamic quantities that can be measured, and calculate lattice energy using Hess's law [◄◄ Section 10.5].

Student Annotation: Recall that *lattice energy* is the amount of energy required to convert 1 mole of ionic solid to its constituent ions in the gas phase [◄◄ Section 5.3]. For NaCl, the equation representing the process is:

$$NaCl(s) \longrightarrow Na^+(g) + Cl^-(g)$$

The Born-Haber Cycle

We have described the formation of an ionic compound as though it happens when gaseous ions coalesce into a solid. In fact, the reactions that produce ionic solids generally do not occur this way. Figure 10.13 illustrates the formation of sodium chloride (NaCl) from its constituent elements.

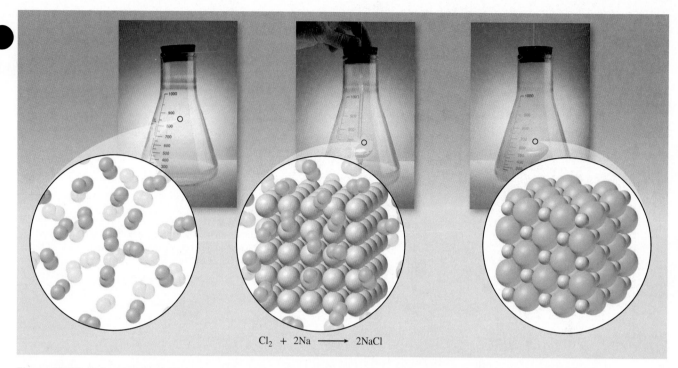

$$Cl_2 + 2Na \longrightarrow 2NaCl$$

Figure 10.13 Sodium metal and chlorine gas combine to produce sodium chloride in a highly exothermic reaction.
(all): © Richard Megna/Fundamental Photographs

We can imagine the reaction of $Na(s)$ and $Cl_2(g)$ to form $NaCl(s)$ as taking place in a series of steps for which the energy changes can be measured. This method of determining the lattice energy is known as the ***Born-Haber cycle.*** Table 10.5 lists the energy changes associated with each step.

The net reaction resulting from the series of steps in Table 10.5 is

$$Na(s) + \tfrac{1}{2}Cl_2(g) \longrightarrow Na^+(g) + Cl^-(g)$$

The final step in the formation of $NaCl(g)$ would be the coalescence of $Na^+(g) + Cl^-(g)$. This is the step for which we cannot measure the energy change directly. However, we *can* measure the standard heat of formation of $NaCl(s)$. (It is tabulated in Appendix 2 as −410.9 kJ/mol.) Although the formation of $NaCl(s)$ from its constituent elements is not actually a step in our imaginary process, knowing its value enables us to calculate the lattice energy of NaCl. Figure 10.14 (pages 450–451) illustrates how this is done using all of these thermodynamic data and Hess's law. The numbered steps correspond to the steps in Table 10.5. The Born-Haber cycle enables us to calculate the lattice energy, which we cannot measure directly, using the quantities that we can measure.

Video 10.4
Figure 10.14 Born-Haber cycle.

TABLE 10.5	Hypothetical Steps in the Formation of $Na^+(g)$ and $Cl^-(g)$ from $Na(s)$ and $Cl_2(g)$	

Chemical equation	Energy change (kJ/mol)
$Na(s) \longrightarrow Na(g)$	107.7*
$\tfrac{1}{2}Cl_2(g) \longrightarrow Cl(g)$	121.7†
$Na(g) \longrightarrow Na^+(g) + e^-$	495.9‡
$Cl(g) + e^- \longrightarrow Cl^-(g)$	−349§

*Standard heat of formation (ΔH_f°) of $Na(g)$ from Appendix 2.
†Standard heat of formation (ΔH_f°) of $Cl(g)$ from Appendix 2.
‡First ionization energy (IE_1) of Na from Figure 4.8.
§ΔH° for this process is negative. (Recall that by definition, *EA* is the amount of energy *released* [◄◄ Section 4.4]. This ΔH is equal to −*EA*.)

Figure 10.14

Born-Haber Cycle

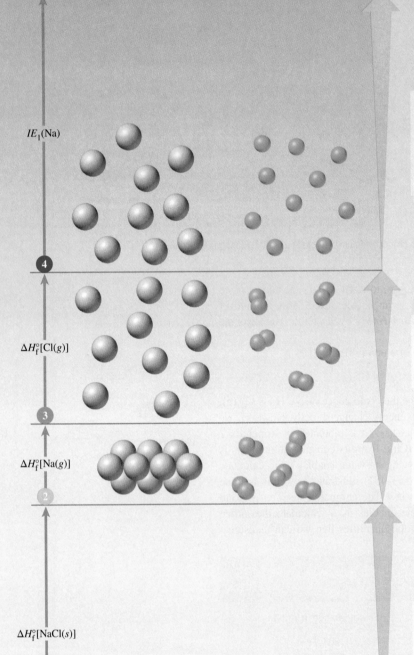

$IE_1(\text{Na})$

4

$\Delta H_f^\circ[\text{Cl}(g)]$

3

$\Delta H_f^\circ[\text{Na}(g)]$

2

$\Delta H_f^\circ[\text{NaCl}(s)]$

1

The first ionization energy of sodium, $IE_1(\text{Na})$, gives the amount of energy required to convert 1 mole of $\text{Na}(g)$ to 1 mole of $\text{Na}^+(g)$:

$$\text{Na}(g) \longrightarrow \text{Na}^+(g) + e^-$$

Step 3
The tabulated value of ΔH_f° for $\text{Cl}(g)$ gives the amount of energy needed to convert $\frac{1}{2}$ mole of $\text{Cl}_2(g)$ to 1 mole of $\text{Cl}(g)$:

$$\tfrac{1}{2}\text{Cl}_2(g) \longrightarrow \text{Cl}(g)$$

Step 2
The tabulated value of ΔH_f° for $\text{Na}(g)$ gives the amount of energy needed to convert 1 mole of $\text{Na}(s)$ to 1 mole of $\text{Na}(g)$:

$$\text{Na}(s) \longrightarrow \text{Na}(g)$$

Step 1
The tabulated value of ΔH_f° for $\text{NaCl}(s)$ gives us the energy produced when 1 mole of Na and $\frac{1}{2}$ mole of Cl_2 combine to form 1 mole of NaCl:

$$\text{Na}(s) + \tfrac{1}{2}\text{Cl}_2(g) \longrightarrow \text{NaCl}(s)$$

Step 1 in the Born-Haber cycle involves converting 1 mole of NaCl into 1 mole of Na and $\frac{1}{2}$ mole of Cl_2 (the reverse of the ΔH_f° reaction):

$$\text{NaCl}(s) \longrightarrow \text{Na}(s) + \tfrac{1}{2}\text{Cl}_2(g)$$

Therefore, ΔH for step 1 is $-\Delta H_f^\circ[\text{NaCl}(s)]$

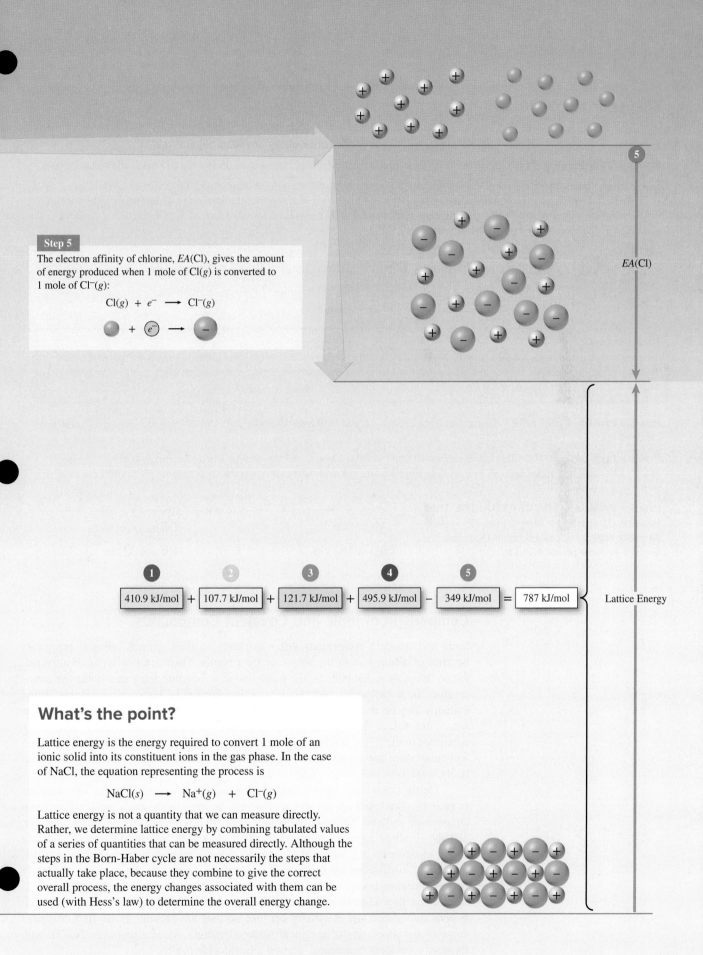

The electron affinity of chlorine, EA(Cl), gives the amount of energy produced when 1 mole of Cl(g) is converted to 1 mole of Cl$^-$(g):

$$Cl(g) + e^- \longrightarrow Cl^-(g)$$

410.9 kJ/mol + 107.7 kJ/mol + 121.7 kJ/mol + 495.9 kJ/mol − 349 kJ/mol = 787 kJ/mol

EA(Cl)

Lattice Energy

What's the point?

Lattice energy is the energy required to convert 1 mole of an ionic solid into its constituent ions in the gas phase. In the case of NaCl, the equation representing the process is

$$NaCl(s) \longrightarrow Na^+(g) + Cl^-(g)$$

Lattice energy is not a quantity that we can measure directly. Rather, we determine lattice energy by combining tabulated values of a series of quantities that can be measured directly. Although the steps in the Born-Haber cycle are not necessarily the steps that actually take place, because they combine to give the correct overall process, the energy changes associated with them can be used (with Hess's law) to determine the overall energy change.

Worked Example 10.11 shows how to use the Born-Haber cycle to calculate the lattice energy.

Worked Example 10.11

Using data from Figures 4.8 and 4.10 and Appendix 2, calculate the lattice energy of cesium chloride (CsCl).

Strategy Using Figure 10.14 as a guide, combine the pertinent thermodynamic data and use Hess's law to calculate the lattice energy.

Setup From Figure 4.8, $IE_1(Cs) = 376$ kJ/mol. From Figure 4.10, $EA_1(Cl) = 349.0$ kJ/mol. From Appendix 2, $\Delta H_f^\circ[Cs(g)] = 76.50$ kJ/mol, $\Delta H_f^\circ[Cl(g)] = 121.7$ kJ/mol, $\Delta H_f^\circ[CsCl(s)] = -442.8$ kJ/mol. Because we are interested in magnitudes only, we can use the absolute values of the thermodynamic data. And, because only the standard heat of formation of CsCl(s) is a negative number, it is the only one for which the sign changes.

Solution

$$\{\Delta H_f^\circ[Cs(g)] + \Delta H_f^\circ[Cl(g)] + IE_1(Cs) + |\Delta H_f^\circ[CsCl(s)]|\} - EA_1(Cl) = \text{lattice energy}$$

$$= (76.50 \text{ kJ/mol} + 121.7 \text{ kJ/mol} + 376 \text{ kJ/mol} + 442.8 \text{ kJ/mol}) - 349.0 \text{ kJ/mol}$$

$$= 668 \text{ kJ/mol}$$

> **Think About It**
> Compare this value to that for NaCl in Figure 10.14 (787 kJ/mol). Both compounds contain the same anion (Cl^-) and both have cations with the same charge (+1), so the relative sizes of the cations will determine the relative strengths of their lattice energies. Because Cs^+ is larger than Na^+, the lattice energy of CsCl is smaller than the lattice energy of NaCl.

Practice Problem (A)**TTEMPT** Using data from Figures 4.8 and 4.10 and Appendix 2, calculate the lattice energy of rubidium iodide (RbI).

Practice Problem (B)**UILD** The lattice energy of MgO is 3890 kJ/mol, and the second ionization energy (IE_2) of Mg is 1450.6 kJ/mol. Using these data, as well as data from Figures 4.8 and 4.10 and Appendix 2, determine the second electron affinity for oxygen, $EA_2(O)$.

Practice Problem (C)**ONCEPTUALIZE** Five points (A through E) lie along a line. The known distances between points are given. Determine the distance between points A and C.

A _____ B _____ C _ D _____ E

AB = 5.05 cm   DE = 4.65 cm

BD = 7.65 cm   CE = 6.27 cm

Comparison of Ionic and Covalent Compounds

Ionic and covalent compounds differ markedly in their general physical properties because of differences in the nature of their bonds. There are two types of attractive forces in covalent compounds, the *intramolecular* bonding force that holds the atoms together in a molecule, and the *intermolecular* forces between molecules. Bond enthalpy can be used to quantify the in*tra*molecular bonding force. In*ter*molecular forces, the forces that hold a group of molecules together, are usually quite weak compared to the forces holding atoms together *within* a molecule. Consequently, covalent compounds are usually gases, liquids, or low-melting solids—reflecting their relatively weak intermolecular attractive forces.

On the other hand, the electrostatic forces [◀◀ Section 5.3] holding ions together in an ionic compound are usually very strong, so ionic compounds are solids at room temperature and have high melting points. Many ionic compounds are soluble in water, and the resulting aqueous solutions conduct electricity because the compounds are strong electrolytes [◀◀ Section 9.1]. Most covalent compounds are insoluble in water, or if they do dissolve, their aqueous solutions generally do not conduct electricity because the compounds are nonelectrolytes. Molten ionic compounds conduct electricity because they contain mobile cations and anions; liquid or molten covalent compounds do *not* conduct electricity because no ions are present. Table 10.6 compares some of the properties of a typical ionic compound, sodium chloride (NaCl), with those of a covalent compound, carbon tetrachloride (CCl$_4$).

	Comparison of Some Properties of an Ionic Compound	
TABLE 10.6	(NaCl) and a Covalent Compound (CCl₄)	

Property	NaCl	CCl$_4$
Appearance	White solid	Colorless liquid
Melting point (°C)	801	−23
Molar heat of fusion* (kJ/mol)	30.2	2.5
Boiling point (°C)	1413	76.5
Molar heat of vaporization* (kJ/mol)	600	30
Density (g/cm³)	2.17	1.59
Solubility in water	High	Very low
Electrical conductivity		
Solid	Poor	Poor
Liquid	Good	Poor
Aqueous	Good	Poor

*The *molar heat of fusion* and *molar heat of vaporization* are the amounts of heat needed to melt 1 mole of the solid and to vaporize 1 mole of the liquid, respectively.

Section 10.8 Review

Lattice Energy and the Stability of Ionic Solids

10.8.1   Will the lattice energy of KF be larger or smaller than that of LiF; larger or smaller than that of KCl; larger or smaller than that of KI?
(a) larger, larger, and smaller
(b) smaller, larger, and smaller
(c) smaller, larger, and larger
(d) smaller, smaller, and smaller
(e) larger, smaller, and larger

10.8.2   Using the following data, calculate the lattice energy of KF.
(a) 808 kJ/mol
(b) −286 kJ/mol
(c) 261 kJ/mol
(d) 1355 kJ/mol
(e) −261 kJ/mol

$\Delta H_f^\circ[K(g)] = 89.99$ kJ/mol, $\Delta H_f^\circ[F(g)] = 80.0$ kJ/mol,

$IE_1(K) = 419$ kJ/mol, $\Delta H_f^\circ[KF(s)] = -574$ kJ/mol, and $EA_1(F) = 328$ kJ/mol

Learning Outcomes

- Identify the system and surroundings for a given experiment.
- Identify a process as endothermic or exothermic.
- Understand the various units used to measure energy including joules and calories.
- Identify a system as being open, closed, or isolated.
- Understand the concept of state functions and be able to identify state functions and nonstate functions.
- Define work and heat and understand the sign conventions associated with these terms.
- Understand the key differences between internal energy measurements using constant-pressure and constant-volume calorimetry.
- Understand enthalpy and enthalpy changes.

- Calculate the enthalpy change of a reaction and understand how it depends upon stoichiometric amounts of products and reactants.
- Perform calorimetric calculations involving specific heat or heat capacity.
- Apply Hess's law in the determination of the heat of reaction of a multistep process.
- Understand the nature of the standard state, particularly as it applies to standard heat of formation.
- Use the standard heats of formation of products and reactants to calculate the enthalpy change of a reaction.
- Use bond enthalpies to calculate an estimate of the enthalpy change of a reaction.
- Calculate the lattice energy of a compound using the Born-Haber cycle.

Chapter Summary

SECTION 10.1

- The *system* is the particular part of the universe that we are interested in studying—such as the reactants and products in a chemical reaction. The term *surroundings* refers to the rest of the universe. System + surroundings = universe.
- *Heat* refers to the flow of thermal energy between two bodies at different temperatures. *Thermochemistry* is the study of the heat associated with chemical reactions and physical processes.
- In an *exothermic process,* heat is released to the surroundings, so the energy of the system decreases. In an *endothermic process,* heat is absorbed from the surroundings, so the energy of the system increases.

SECTION 10.2

- *Thermodynamics* is the study of the conversions among different types of energy. Thermochemistry is a branch of thermodynamics.
- An *open system* is one that can exchange both matter and energy with its surroundings. A *closed system* is one that can exchange energy but not matter with its surroundings. An *isolated system* is one that cannot exchange either energy or matter with its surroundings.
- The *state of a system* is defined by the values of all relevant macroscopic properties, such as temperature, volume, and pressure. A *state function* is one whose value depends only on the state of the system and not on how that state was achieved. State functions include energy, pressure, volume, and temperature.
- The *first law of thermodynamics* states that energy cannot be created or destroyed, but it can be changed from one form to another. The first law of thermodynamics is based on the law of conservation of energy.

SECTION 10.3

- *Enthalpy (H)* is the heat exchanged between the system and surroundings at constant pressure. It is a state function. *Enthalpy of reaction* (ΔH_{rxn}) is the heat exchanged at constant pressure for a specific reaction.

- A *thermochemical equation* is a balanced chemical equation for which the enthalpy change (ΔH_{rxn}) is given.

SECTION 10.4

- *Calorimetry* is the science of measuring temperature changes to determine heats associated with chemical reactions. Calorimetry may be carried out at constant pressure (in a coffee-cup calorimeter) or at constant volume (in a bomb calorimeter).
- The *specific heat (s)* of a substance is the amount of heat required to increase the temperature of 1 g of the substance by 1°C. The *heat capacity (C)* of an object is the amount of heat required to increase the temperature of the object by 1°C.

SECTION 10.5

- *Hess's law* states that the enthalpy change for a reaction that occurs in a series of steps is equal to the sum of the enthalpy changes of the individual steps. Hess's law is valid because enthalpy is a state function.

SECTION 10.6

- The *standard enthalpy of formation* (ΔH_f°) is the enthalpy change associated with the formation of 1 mole of a substance from its constituent elements, each in its standard state. The *standard enthalpy of reaction* (ΔH_{rxn}°) can be calculated for any reaction using tabulated standard enthalpies of formation (ΔH_f°) of the products and reactants.

SECTION 10.7

- *Bond enthalpy* is the energy required to break 1 mole of a particular type of bond. Bond enthalpies are a measure of the stability of covalent bonds and can be used to estimate the enthalpy change for a reaction.

SECTION 10.8

- *Lattice energy* is the amount of energy required to convert a mole of ionic solid to its constituent ions in the gas phase. Lattice energy cannot be measured directly, but is determined using the *Born-Haber cycle* and thermodynamic quantities that can be measured directly.

Key Words

Key Equations

10.1	$\Delta U = q + w$	The change in internal energy of a system (ΔU) is the sum of heat (q) and the work (w) associated with a process. Proper sign conventions must be used for heat and work (Table 10.1).
10.2	$w = -P\Delta V$	Pressure-volume work done by (or on) a system is calculated using the external pressure (P) and the change in volume (ΔV).
10.3	$\Delta U = q - P\Delta V$	The change in internal energy of a system (ΔU) is equal to heat (q) minus pressure-volume work ($P\Delta V$).
10.4	$q_V = \Delta U$	Heat given off (or absorbed) by a system at constant volume (q_V) is equal to the change in internal energy (ΔU).
10.5	$q_P = \Delta U + P\Delta V$	Heat given off (or absorbed) by a system at constant pressure (q_P) is equal to the sum of change in internal energy (ΔU) and pressure-volume work ($P\Delta V$).
10.6	$H = U + PV$	Enthalpy (H) is equal to the sum of internal energy (U) and pressure-volume work (PV).
10.7	$\Delta H = \Delta U + \Delta(PV)$	The change in enthalpy (ΔH) is equal to the sum of change in internal energy (ΔU) and change in the product of pressure and volume [$\Delta(PV)$].
10.8	$\Delta H = \Delta U + P\Delta V$	The change in enthalpy (ΔH) is equal to the sum of change in internal energy (ΔU) and the product of external pressure (P) and change in volume (ΔV).
10.9	$q_P = \Delta H$	Heat given off (or absorbed) by a process at constant pressure (q_P) is equal to change in enthalpy (ΔH).
10.10	$\Delta H = H(\text{products}) - H(\text{reactants})$	Enthalpy change for a reaction (ΔH) is the difference between the enthalpy of products [$H(\text{products})$] and enthalpy of reactants [$H(\text{reactants})$], although this is not the equation generally used to calculate enthalpy changes because the absolute values of enthalpy are not known.
10.11	$q = sm\Delta T$	Heat given off (or absorbed) by a substance (q) is equal to the product of specific heat of the substance (s), mass of the substance (m), and the change in temperature (ΔT).
10.12	$q = C\Delta T$	Heat given off (or absorbed) by an object (q) is equal to the product of heat capacity of the object (C) and the change in temperature (ΔT).
10.13	$q_{\text{sys}} = -sm\Delta T$	Heat given off (or absorbed) by a system (q_{sys}) is equal in magnitude and opposite in sign to the heat given off or absorbed by the surroundings.
10.14	$q_{\text{surr}} = sm\Delta T$	Heat given off (or absorbed) by the surroundings (q_{surr}) is equal in magnitude and opposite in sign to the heat given off or absorbed by the system.
10.15	$q_{\text{cal}} = C_{\text{cal}}\Delta T$	Heat given off (or absorbed) by a calorimeter (q_{cal}) is equal to the product of heat capacity of the calorimeter (C_{cal}) and change in temperature (ΔT).
10.16	$q_{\text{rxn}} = -C_{\text{cal}}\Delta T$	Heat of reaction (q_{rxn}) is equal in magnitude and opposite in sign to heat of calorimeter (q_{cal}).
10.17	$\Delta H^{\circ}_{\text{rxn}} = \begin{aligned}&[c\Delta H^{\circ}_{\text{f}}(\text{C}) + d\Delta H^{\circ}_{\text{f}}(\text{D})] \\ &- [a\Delta H^{\circ}_{\text{f}}(\text{A}) + b\Delta H^{\circ}_{\text{f}}(\text{B})]\end{aligned}$	Standard enthalpy change for a reaction ($\Delta H^{\circ}_{\text{rxn}}$) can be calculated by multiplying the coefficient of each species in the reaction by the corresponding standard enthalpy of formation ($\Delta H^{\circ}_{\text{f}}$).
10.18	$\Delta H^{\circ}_{\text{rxn}} = \begin{aligned}&\Sigma n\Delta H^{\circ}_{\text{f}}(\text{products}) \\ &- \Sigma m\Delta H^{\circ}_{\text{f}}(\text{reactants})\end{aligned}$	Standard enthalpy change for a reaction ($\Delta H^{\circ}_{\text{rxn}}$) is the difference between the sum of standard enthalpies of formation of products $\Sigma\Delta H^{\circ}_{\text{f}}$ (products) and the sum of standard enthalpies of formation of reactants $\Sigma\Delta H^{\circ}_{\text{f}}$ (reactants).
10.19	$\Delta H^{\circ} = \begin{aligned}&\Sigma\text{BE}(\text{reactants}) \\ &- \Sigma\text{BE}(\text{products})\end{aligned}$	The enthalpy change of a reaction can be estimated by subtracting the sum of bond enthalpies in products from the sum of bond enthalpies in reactants.

Key Skills

Enthalpy of Reaction

Using tabulated ΔH_f° values, we can calculate the standard enthalpy of reaction (ΔH_{rxn}°) using Equation 10.18.

$$\Delta H_{rxn}^\circ = \Sigma n \Delta H_f^\circ \text{ (products)} - \Sigma m \Delta H_f^\circ \text{ (reactants)}$$

This method of calculating thermodynamic quantities such as enthalpy of reaction is important not only in this chapter, but also in Chapters 14 and 18. The following examples illustrate the use of Equation 10.18 and data from Appendix 2. Each example provides a specific reminder of one of the important facets of this approach.

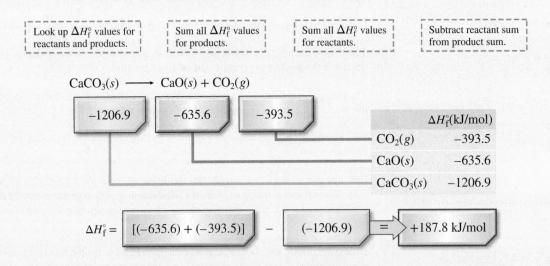

Each ΔH_f° value must be multiplied by the corresponding stoichiometric coefficient in the balanced equation.

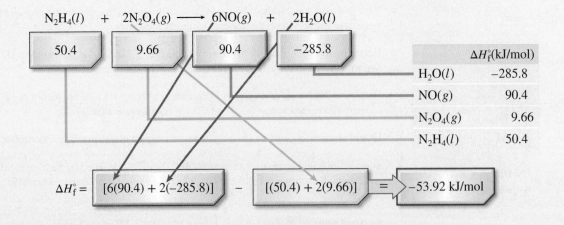

$$Ba(s) + 2H_2O(l) \longrightarrow Ba^{2+}(aq) + 2OH^-(aq) + H_2(g)$$

By definition, the standard enthalpy of formation for an element in its standard state is zero. In addition, many tables of thermodynamic data, including Appendix 2, do not contain values for aqueous strong electrolytes such as barium hydroxide.

However, the tables do include values for the individual aqueous ions. Therefore, determination of this enthalpy of reaction is facilitated by rewriting the equation with $Ba(OH)_2$ written as separate ions.

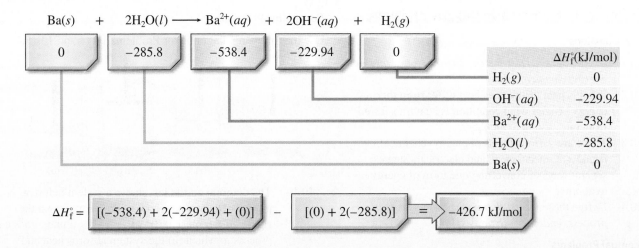

$$\Delta H_f^\circ = [(-538.4) + 2(-229.94) + (0)] - [(0) + 2(-285.8)] = -426.7 \text{ kJ/mol}$$

You will find more than one tabulated ΔH_f° value for some substances, such as water. It is important to select the value that corresponds to the phase of matter represented in the chemical equation. In previous examples, water has appeared in the balanced equations as a liquid. It can also appear as a gas.

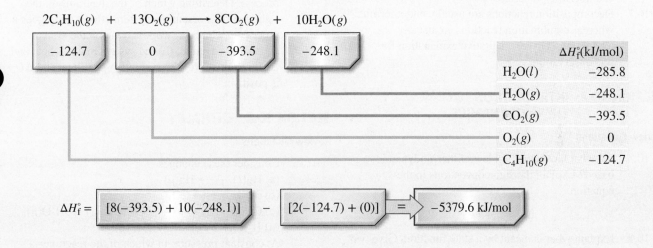

$$\Delta H_f^\circ = [8(-393.5) + 10(-248.1)] - [2(-124.7) + (0)] = -5379.6 \text{ kJ/mol}$$

Key Skills Problems

10.1
Using data from Appendix 2, calculate the standard enthalpy of the following reaction.

$$Mg(OH)_2(s) \longrightarrow MgO(s) + H_2O(l)$$

(a) −608.7 kJ/mol (b) −81.1 kJ/mol (c) −37.1 kJ/mol
(d) +81.1 kJ/mol (e) +37.1 kJ/mol

10.2
Using data from Appendix 2, calculate the standard enthalpy of the following reaction.

$$4HBr(g) + O_2(g) \longrightarrow 2H_2O(l) + 2Br_2(l)$$

(a) −426.8 kJ/mol (b) −338.8 kJ/mol (c) −249.6 kJ/mol
(d) +426.8 kJ/mol (e) +338.8 kJ/mol

10.3
Using data from Appendix 2, calculate the standard enthalpy of the following reaction (you must first balance the equation).

$$P(red) + Cl_2(g) \longrightarrow PCl_3(g)$$

(a) −576.1 kJ/mol (b) −269.7 kJ/mol (c) −539.3 kJ/mol
(d) −602.6 kJ/mol (e) +639.4 kJ/mol

10.4
Using only whole-number coefficients, the combustion of hexane can be represented as

$$2C_6H_{14}(l) + 19O_2(g) \longrightarrow 12CO_2(g) + 14H_2O(l)$$
$$\Delta H^\circ = -8388.4 \text{ kJ/mol}$$

Using this and data from Appendix 2, determine the standard enthalpy of formation of hexane.
(a) −334.8 kJ/mol (b) −167.4 kJ/mol (c) −669.6 kJ/mol
(d) +334.8 kJ/mol (e) +669.6 kJ/mol

Questions and Problems

SECTION 10.1: ENERGY AND ENERGY CHANGES

Review Questions

10.1   Define these terms: *system, surroundings, thermal energy, chemical energy.*

10.2   What is heat? How does heat differ from thermal energy? Under what condition is heat transferred from one system to another?

10.3   These are various forms of energy: chemical, heat, light, mechanical, and electrical. Suggest several ways of converting one form of energy to another.

10.4   Define these terms: *thermochemistry, exothermic process, endothermic process.*

Conceptual Problems

10.5   Stoichiometry is based on the law of conservation of mass. On what law is thermochemistry based?

10.6   Describe two exothermic processes and two endothermic processes.

10.7   Decomposition reactions are usually endothermic, whereas combination reactions are usually exothermic. Give a qualitative explanation for this observation.

SECTION 10.2: INTRODUCTION TO THERMODYNAMICS

Review Questions

10.8   On what law is the first law of thermodynamics based? Explain the sign conventions in the equation

$$\Delta U = q + w$$

10.9   Explain what is meant by a state function. Give two examples of quantities that are state functions and two that are not state functions.

Computational Problems

10.10   In a gas expansion, 36 J of heat is absorbed from the surroundings and the energy of the system decreases by 182 J. Calculate the work done.

10.11   The work done to compress a gas is 112 J. As a result, 51 J of heat is given off to the surroundings. Calculate the change in energy of the gas.

10.12   Calculate *q*, and determine whether heat is absorbed or released when a system does work on the surroundings equal to 64 J and $\Delta U = -23$ J.

10.13   Calculate *w*, and determine whether work is done *by* the system or *on* the system when 98 J of heat is released and $\Delta U = -215$ J.

Conceptual Problems

Use the following diagrams for Problems 10.14 and 10.15.

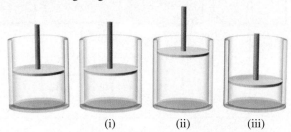

(i)   (ii)   (iii)

10.14   The diagram on the left shows a system before a process. Determine which of the diagrams on the right could represent the system after it undergoes a process in which (a) the system absorbs heat and ΔU is negative; (b) the system absorbs heat and does work on the surroundings; (c) the system releases heat and does work on the surroundings.

10.15   The diagram on the left shows a system before a process. Determine which of the diagrams on the right could represent the system after it undergoes a process in which (a) work is done on the system and ΔU is negative; (b) the system releases heat and ΔU is positive; (c) the system absorbs heat and ΔU is positive.

SECTION 10.3: ENTHALPY

Review Questions

10.16   Consider these changes.
(a) $Hg(l) \longrightarrow Hg(g)$
(b) $3O_2(g) \longrightarrow 2O_3(g)$
(c) $CuSO_4 \cdot 5H_2O(s) \longrightarrow CuSO_4(s) + 5H_2O(g)$
(d) $H_2(g) + F_2(g) \longrightarrow 2HF(g)$
At constant pressure, in which of the reactions is work done by the system on the surroundings? By the surroundings on the system? In which of them is no work done?

10.17   Define these terms: *enthalpy* and *enthalpy of reaction.* Under what condition is the heat of a reaction equal to the enthalpy change of the same reaction?

10.18   In writing thermochemical equations, why is it important to indicate the physical state (i.e., gaseous, liquid, solid, or aqueous) of each substance?

10.19   Consider the reaction:

$$2CH_3OH(g) + 3O_2(g) \longrightarrow 2CO_2(g) + 4H_2O(l)$$
$$\Delta H = -1452.8 \text{ kJ/mol}$$

What is the value of ΔH if (a) the equation is multiplied throughout by 2; (b) the direction of the reaction is reversed so that the products become the reactants, and vice versa; (c) water vapor instead of liquid water is formed as the product?

Computational Problems

10.20 A sample of nitrogen gas expands in volume from 2.5 to 5.8 L at constant temperature. Calculate the work done in joules if the gas expands (a) against a vacuum, (b) against a constant pressure of 1.35 atm, and (c) against a constant pressure of 6.9 atm. (See Equation 10.2.)

10.21 A gas expands in volume from 26.7 to 119.3 mL at constant temperature. Calculate the work done (in joules) if the gas expands (a) against a vacuum, (b) against a constant pressure of 3.5 atm, and (c) against a constant pressure of 10.1 atm. (See Equation 10.2.)

10.22 A gas expands and does PV work on the surroundings equal to 325 J. At the same time, it absorbs 127 J of heat from the surroundings. Calculate the change in energy of the gas.

10.23 The first step in the industrial recovery of zinc from the zinc sulfide ore is roasting, that is, the conversion of ZnS to ZnO by heating

$$2ZnS(s) + 3O_2(g) \longrightarrow 2ZnO(s) + 2SO_2(g)$$
$$\Delta H = -879 \text{ kJ/mol}$$

Calculate the heat evolved (in kilojoules) per gram of ZnS roasted.

10.24 Determine the amount of heat (in kilojoules) given off when 1.26×10^4 g of NO_2 are produced according to the equation

$$2NO(g) + O_2(g) \longrightarrow 2NO_2(g)$$
$$\Delta H = -114.6 \text{ kJ/mol}$$

10.25 Consider the reaction

$$2H_2O(g) \longrightarrow 2H_2(g) + O_2(g)$$
$$\Delta H = +483.6 \text{ kJ/mol}$$

at a certain temperature. If the increase in volume is 32.7 L against an external pressure of 1.00 atm, calculate ΔU for this reaction. (The conversion factor is $1 \text{ L} \cdot \text{atm} = 101.3 \text{ J}$.)

10.26 Consider the reaction

$$H_2(g) + Cl_2(g) \longrightarrow 2HCl(g)$$
$$\Delta H = -184.6 \text{ kJ/mol}$$

If 3 moles of H_2 react with 3 moles of Cl_2 to form HCl, calculate the work done (in joules) against a pressure of 1.0 atm. What is ΔU for this reaction? Assume the reaction goes to completion and that $\Delta V = 0$. (The conversion factor is $1 \text{ L} \cdot \text{atm} = 101.3 \text{ J}$.)

Conceptual Problems

10.27 The diagrams represent systems before and after reaction for two related chemical processes. ΔH for the first reaction is -595.8 kJ/mol. Determine the value of ΔH for the second reaction.

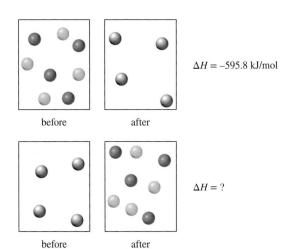

$\Delta H = -595.8$ kJ/mol

before   after

$\Delta H = ?$

before   after

10.28 For most biological processes, the changes in internal energy are approximately equal to the changes in enthalpy. Explain.

SECTION 10.4: CALORIMETRY

▶▶▶ Visualizing Chemistry
Figure 10.9 and Figure 10.10

VC 10.1 Referring to Figure 10.9, which of the following would result in the calculated value of ΔH_{rxn} being too high?
(a) Spilling some of one of the reactant solutions before adding it to the calorimeter
(b) Reading the final temperature before it reached its maximum value
(c) Misreading the thermometer at the beginning of the experiment and recording too low an initial temperature

VC 10.2 How would the ΔH_{rxn} calculated in Figure 10.9 be affected if the concentration of one of the reactant solutions were twice as high as it was supposed to be?
(a) The calculated ΔH_{rxn} would not be affected.
(b) The calculated ΔH_{rxn} would be too low.
(c) The calculated ΔH_{rxn} would be too high.

VC 10.3 For an exothermic reaction like the one depicted in Figure 10.9, if the heat capacity of the calorimeter is not negligibly small, the heat absorbed by the water will be _____ the heat given off by the reaction.
(a) greater than
(b) less than
(c) equal to

VC 10.4 Referring to Figure 10.9, how would the results of the experiment have been different if the reaction had been endothermic?
(a) The results would have been the same.
(b) There would have been a smaller temperature increase.
(c) There would have been a temperature decrease.

VC 10.5 What would happen to the specific heat calculated in Figure 10.10 if some of the warm metal shot were lost during the transfer to the calorimeter?
  (a) It would not affect the calculated value of specific heat.
  (b) It would cause the calculated value of specific heat to be too high.
  (c) It would cause the calculated value of specific heat to be too low.

VC 10.6 What would happen to the specific heat calculated in Figure 10.10 if the test tube containing the metal shot were left in the boiling water for longer than the recommended time?
  (a) It would not affect the calculated value of specific heat.
  (b) It would cause the calculated value of specific heat to be too high.
  (c) It would cause the calculated value of specific heat to be too low.

VC 10.7 What would happen to the specific heat calculated in Figure 10.10 if some of the water were spilled prior to being added to the calorimeter?
  (a) It would not affect the calculated value of specific heat.
  (b) It would cause the calculated value of specific heat to be too high.
  (c) It would cause the calculated value of specific heat to be too low.

VC 10.8 Referring to the process depicted in Figure 10.10, which of the following must be known precisely for the calculated specific heat to be accurate.
  (a) The mass of the boiling water
  (b) The temperature of the metal shot before it is immersed in the boiling water
  (c) The mass of the water that is added to the calorimeter

Review Questions

10.29 What is the difference between specific heat and heat capacity? What are the units for these two quantities? Which is the *intensive* property and which is the *extensive* property?

10.30 Define *calorimetry* and describe two commonly used calorimeters. In a calorimetric measurement, why is it important that we know the heat capacity of the calorimeter? How is this value determined?

Computational Problems

10.31 A 2.21-kg piece of copper metal is heated from 20.5°C to 126.4°C. Calculate the heat absorbed (in kilojoules) by the metal.

10.32 Calculate the amount of heat liberated (in kilojoules) from 187 g of mercury when it cools from 96.5°C to 12.5°C.

10.33 A sheet of gold weighing 10.0 g and at a temperature of 18.0°C is placed flat on a sheet of iron weighing 20.0 g and at a temperature of 55.6°C. What is the final temperature of the combined metals? Assume that no heat is lost to the surroundings. (*Hint:* The heat gained by the gold must be equal to the heat lost by the iron. The specific heats of the metals are given in Table 10.2.)

10.34 A 0.1375-g sample of solid magnesium is burned in a constant-volume bomb calorimeter that has a heat capacity of 3024 J/°C. The temperature increases by 1.126°C. Calculate the heat given off by the burning Mg, in kJ/g and in kJ/mol.

10.35 A quantity of 2.00×10^2 mL of 0.862 *M* HCl is mixed with 2.00×10^2 mL of 0.431 *M* Ba(OH)$_2$ in a constant-pressure calorimeter of negligible heat capacity. The initial temperature of the HCl and Ba(OH)$_2$ solutions is the same at 20.48°C. For the process

$$H^+(aq) + OH^-(aq) \longrightarrow H_2O(l)$$

the heat of neutralization is −56.2 kJ/mol. What is the final temperature of the mixed solution? Assume the specific heat of the solution is the same as that for pure water.

10.36 The fuel value of hamburger is approximately 3.6 kcal/g. If a man eats 1 lb of hamburger for lunch and if none of the energy is stored in his body, estimate the amount of water that would have to be lost in perspiration to keep his body temperature constant. The heat of vaporization of water may be taken as 2.41 kJ/g (1 lb = 453.6 g).

10.37 Metabolic activity in the human body releases approximately 1.0×10^4 kJ of heat per day. Assuming the body contains 50 kg of water, how fast would the body temperature rise if it were an isolated system? How much water must the body eliminate as perspiration to maintain the normal body temperature (98.6°F)? The heat of vaporization of water may be taken as 2.41 kJ/g. Comment on your results.

10.38 A piece of silver with a mass 362 g has a heat capacity of 85.7 J/°C. What is the specific heat of silver?

10.39 A 25.95-g sample of methanol at 35.6°C is added to a 38.65-g sample of ethanol at 24.7°C in a constant-pressure calorimeter. If the final temperature of the combined liquids is 28.5°C and the heat capacity of the calorimeter is 19.3 J/°C, determine the specific heat of methanol.

10.40 A 50.75-g sample of water at 75.6°C is added to a sample of water at 24.1°C in a constant-pressure calorimeter. If the final temperature of the combined water is 39.4°C and the heat capacity of the calorimeter is 26.3 J/°C, calculate the mass of the water originally in the calorimeter.

Conceptual Problems

10.41 Consider the following data:

Metal	Al	Cu
Mass (g)	10	30
Specific heat [J/(g · °C)]	0.900	0.385
Temperature (°C)	40	60

When these two metals are placed in contact, which of the following will take place?
(a) Heat will flow from Al to Cu because Al has a larger specific heat.
(b) Heat will flow from Cu to Al because Cu has a larger mass.
(c) Heat will flow from Cu to Al because Cu has a larger heat capacity.
(d) Heat will flow from Cu to Al because Cu is at a higher temperature.
(e) No heat will flow in either direction.

10.42 Consider two metals A and B, each having a mass of 100 g and an initial temperature of 20°C. The specific heat of A is larger than that of B. Under the same heating conditions, which metal would take longer to reach a temperature of 21°C?

SECTION 10.5: HESS'S LAW

Review Questions

10.43 State Hess's law. Explain, with one example, the usefulness of Hess's law in thermochemistry.

10.44 Describe how chemists use Hess's law to determine the ΔH_f° of a compound by measuring its heat (enthalpy) of combustion.

Computational Problems

10.45 From the following heats of combustion,

$$CH_3OH(l) + \tfrac{3}{2}O_2(g) \longrightarrow CO_2(g) + 2H_2O(l)$$
$$\Delta H_{rxn}^\circ = -726.4 \text{ kJ/mol}$$

$$C(graphite) + O_2(g) \longrightarrow CO_2(g)$$
$$\Delta H_{rxn}^\circ = -393.5 \text{ kJ/mol}$$

$$H_2(g) + \tfrac{1}{2}O_2(g) \longrightarrow H_2O(l)$$
$$\Delta H_{rxn}^\circ = -285.8 \text{ kJ/mol}$$

calculate the enthalpy of formation of methanol (CH₃OH) from its elements.

$$C(graphite) + 2H_2(g) + \tfrac{1}{2}O_2(g) \longrightarrow CH_3OH(l)$$

10.46 Calculate the standard enthalpy change for the reaction

$$2Al(s) + Fe_2O_3(s) \longrightarrow 2Fe(s) + Al_2O_3(s)$$

given that

$$2Al(s) + \tfrac{3}{2}O_2(g) \longrightarrow Al_2O_3(s)$$
$$\Delta H_{rxn}^\circ = -1601 \text{ kJ/mol}$$

$$2Fe(s) + \tfrac{3}{2}O_2(g) \longrightarrow Fe_2O_3(s)$$
$$\Delta H_{rxn}^\circ = -821 \text{ kJ/mol}$$

10.47 From these data,

$$S(rhombic) + O_2(g) \longrightarrow SO_2(g)$$
$$\Delta H_{rxn}^\circ = -296.06 \text{ kJ/mol}$$

$$S(monoclinic) + O_2(g) \longrightarrow SO_2(g)$$
$$\Delta H_{rxn}^\circ = -296.36 \text{ kJ/mol}$$

calculate the enthalpy change for the transformation

$$S(rhombic) \longrightarrow S(monoclinic)$$

(Monoclinic and rhombic are different allotropic forms of elemental sulfur.)

10.48 From the following data,

$$C(graphite) + O_2(g) \longrightarrow CO_2(g)$$
$$\Delta H_{rxn}^\circ = -393.5 \text{ kJ/mol}$$

$$H_2(g) + \tfrac{1}{2}O_2(g) \longrightarrow H_2O(l)$$
$$\Delta H_{rxn}^\circ = -285.8 \text{ kJ/mol}$$

$$2C_2H_6(g) + 7O_2(g) \longrightarrow 4CO_2(g) + 6H_2O(l)$$
$$\Delta H_{rxn}^\circ = -3119.6 \text{ kJ/mol}$$

calculate the enthalpy change for the reaction

$$2C(graphite) + 3H_2(g) \longrightarrow C_2H_6(g)$$

Conceptual Problems

The following diagrams depict three chemical reactions involving five different chemical species—each represented by a different color sphere. Use this information to solve Problems 10.49 and 10.50.

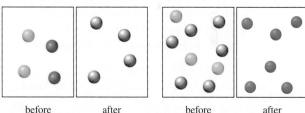

before	after
	$\Delta H = 25$ kJ/mol

before	after
	$\Delta H = -60$ kJ/mol

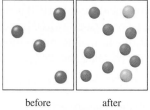

before   after
$\Delta H = 100$ kJ/mol

10.49 Determine the value of ΔH for the following reaction.

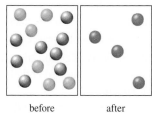

before   after
$\Delta H = ?$

10.50 Determine the value of ΔH for the following reaction.

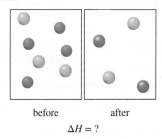

before   after

$\Delta H = ?$

SECTION 10.6: STANDARD ENTHALPIES OF FORMATION

Review Questions

10.51 What is meant by the standard-state condition?

10.52 How are the standard enthalpies of an element and of a compound determined?

10.53 What is meant by the standard enthalpy of a reaction?

10.54 Write the equation for calculating the enthalpy of a reaction. Define all the terms.

Computational Problems

10.55 The ΔH_f° values of the two allotropes of oxygen, O_2 and O_3, are 0 kJ/mol and 142.2 kJ/mol, respectively, at 25°C. Which is the more stable form at this temperature?

10.56 The standard enthalpies of formation of ions in aqueous solutions are obtained by arbitrarily assigning a value of zero to H^+ ions; that is, $\Delta H_f^\circ = [H^+(aq)] = 0$. (a) For the following reaction

$$HCl(g) \xrightarrow{\text{H}_2\text{O}} H^+(aq) + Cl^-(aq)$$
$$\Delta H_f^\circ = -74.9 \text{ kJ/mol}$$

calculate ΔH_f° for the Cl^- ion. (b) Given that ΔH_f° for OH^- ion is -229.6 kJ/mol, calculate the enthalpy of neutralization when 1 mole of a strong monoprotic acid (such as HCl) is titrated by 1 mole of a strong base (such as KOH) at 25°C.

10.57 Calculate the heat of decomposition for this process at constant pressure and 25°C.

$$CaCO_3(s) \longrightarrow CaO(s) + CO_2(g)$$

(Look up the standard enthalpy of formation of the reactant and products in Appendix 2.)

10.58 Calculate the heats of combustion for the following reactions from the standard enthalpies of formation listed in Appendix 2.
(a) $C_2H_4(g) + 3O_2(g) \longrightarrow 2CO_2(g) + 2H_2O(l)$
(b) $2H_2S(g) + 3O_2(g) \longrightarrow 2H_2O(l) + 2SO_2(g)$

10.59 Calculate the heats of combustion for the following reactions from the standard enthalpies of formation listed in Appendix 2.
(a) $2H_2(g) + O_2(g) \longrightarrow 2H_2O(l)$
(b) $2C_2H_2(g) + 5O_2(g) \longrightarrow 4CO_2(g) + 2H_2O(l)$

10.60 The standard enthalpy change for the following reaction is 436.4 kJ/mol.

$$H_2(g) \longrightarrow H(g) + H(g)$$

Calculate the standard enthalpy of formation of atomic hydrogen (H).

10.61 Methanol, ethanol, and n-propanol are three common alcohols. When 1.00 g of each of these alcohols is burned in air, heat is liberated as follows: (a) methanol (CH_3OH), -22.6 kJ; (b) ethanol (C_2H_5OH), -29.7 kJ; (c) n-propanol (C_3H_7OH), -33.4 kJ. Calculate the heats of combustion of these alcohols in kJ/mol.

10.62 Pentaborane-9 (B_5H_9) is a colorless, highly reactive liquid that will burst into flames when exposed to oxygen. The reaction is

$$2B_5H_9(l) + 12O_2(g) \longrightarrow 5B_2O_3(s) + 9H_2O(l)$$

Calculate the kilojoules of heat released per gram of the compound reacted with oxygen. The standard enthalpy of formation of B_5H_9 is 73.2 kJ/mol.

10.63 From the standard enthalpies of formation, calculate ΔH_{rxn}° for the reaction

$$C_6H_{12}(l) + 9O_2(g) \longrightarrow 6CO_2(g) + 6H_2O(l)$$

For $C_6H_{12}(l)$, $\Delta H_f^\circ = -151.9$ kJ/mol.

10.64 At 850°C, $CaCO_3$ undergoes substantial decomposition to yield CaO and CO_2. Assuming that the ΔH_f° values of the reactant and products are the same at 850°C as they are at 25°C, calculate the enthalpy change (in kilojoules) if 66.8 g of CO_2 is produced in one reaction.

10.65 Determine the amount of heat (in kilojoules) given off when 1.26×10^4 g of ammonia is produced according to the equation

$$N_2(g) + 3H_2(g) \longrightarrow 2NH_3(g)$$
$$\Delta H_f^\circ = -92.6 \text{ kJ/mol}$$

Assume that the reaction takes place under standard-state conditions at 25°C.

10.66 Which of the following standard enthalpy of formation values is not zero at 25°C: Na(monoclinic), Ne(g), $CH_4(g)$, S_8(monoclinic), Hg(l), H(g)?

10.67 Which is the more negative quantity at 25°C: ΔH_f° for $H_2O(l)$ or ΔH_f° for $H_2O(g)$?

Conceptual Problems

10.68 Predict the value of ΔH_f° (greater than, less than, or equal to zero) for these elements at 25°C: (a) $Br_2(g)$, $Br_2(l)$; (b) $I_2(g)$, $I_2(s)$.

10.69 In general, compounds with negative ΔH_f° values are more stable than those with positive ΔH_f° values. $H_2O_2(l)$ has a negative ΔH_f° (see Appendix 2). Why, then, does $H_2O_2(l)$ have a tendency to decompose to $H_2O(l)$ and $O_2(g)$?

10.70 Suggest ways (with appropriate equations) that would allow you to measure the ΔH_f° values of $Ag_2O(s)$ and $CaCl_2(s)$ from their elements. No calculations are necessary.

10.71 Using the data in Appendix 2, calculate the enthalpy change for the gaseous reaction shown here. (*Hint:* First determine the limiting reactant.)

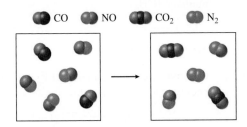

SECTION 10.7: BOND ENTHALPY AND THE STABILITY OF COVALENT MOLECULES

Review Questions

10.72 What is *bond enthalpy?* Bond enthalpies of polyatomic molecules are average values, whereas those of diatomic molecules can be accurately determined. Why?

10.73 Explain why the bond enthalpy of a molecule is usually defined in terms of a gas-phase reaction. Why are bond-breaking processes always endothermic and bond-forming processes always exothermic?

Computational Problems

10.74 From the following data, calculate the average bond enthalpy for the N—H bond.

$$NH_3(g) \longrightarrow NH_2(g) + H(g) \qquad \Delta H° = 435 \text{ kJ/mol}$$

$$NH_2(g) \longrightarrow NH(g) + H(g) \qquad \Delta H° = 381 \text{ kJ/mol}$$

$$NH(g) \longrightarrow N(g) + H(g) \qquad \Delta H° = 360 \text{ kJ/mol}$$

10.75 For the reaction

$$O(g) + O_2(g) \longrightarrow O_3(g) \qquad \Delta H° = -107.2 \text{ kJ/mol}$$

Calculate the average bond enthalpy in O_3.

10.76 The bond enthalpy of $F_2(g)$ is 156.9 kJ/mol. Calculate $\Delta H_f°$ for F(g).

10.77 For the reaction

$$2C_2H_6(g) + 7O_2(g) \longrightarrow 4CO_2(g) + 6H_2O(g)$$

(a) Predict the enthalpy of reaction from the average bond enthalpies in Table 10.4.

(b) Calculate the enthalpy of reaction from the standard enthalpies of formation (see Appendix 2) of the reactant and product molecules, and compare the result with your answer for part (a).

Conceptual Problems

10.78 Use average bond enthalpies from Table 10.4 to estimate ΔH_{rxn} for the following reaction.

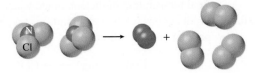

10.79 Use average bond enthalpies from Table 10.4 to estimate ΔH_{rxn} for the following reaction.

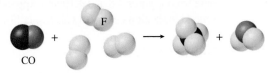

SECTION 10.8: LATTICE ENERGY AND THE STABILITY OF IONIC SOLIDS

▶▶▶ Visualizing Chemistry
Figure 10.14

VC 10.9 What additional information would you need to calculate the lattice energy for a compound if the charges on the cation and anion were +2 and −1, respectively, rather than +1 and −1?
 (a) No additional information is needed.
 (b) IE_2 of the cation.
 (c) IE_2 of the cation and EA_2 of the anion.

VC 10.10 What additional information would you need to calculate the lattice energy for a compound if the charges on the cation and anion were +2 and −2, respectively, rather than +1 and −1?
 (a) No additional information is needed.
 (b) IE_2 of the cation.
 (c) IE_2 of the cation and EA_2 of the anion.

VC 10.11 How would the magnitude of the lattice energy calculated using the Born-Haber cycle change if the charges on the cation and anion were +2 and −2, respectively, rather than +1 and −1?
 (a) Lattice energy would increase.
 (b) Lattice energy would decrease.
 (c) Whether lattice energy would increase or decrease depends on the relative magnitudes of IE_2 of the cation and EA_2 of the anion.

VC 10.12 What law enables us to use the Born-Haber cycle to calculate lattice energy?
 (a) Coulomb's law
 (b) Hess's law
 (c) Law of multiple proportions

Review Questions

10.80 Explain how the lattice energy of an ionic compound such as KCl can be determined using the Born-Haber cycle. On what law is this procedure based?

10.81 Specify which compound in each of the following pairs of ionic compounds should have the higher lattice energy: (a) KCl or MgO, (b) LiF or LiBr, (c) Mg_3N_2 or NaCl, (d) AlN or CaO, (e) NaF or CsF, (f) $MgCl_2$ or MgF_2. Explain your choice.

Computational Problems

10.82 Calculate the lattice energy of LiCl. Use data from Figures 4.8 and 4.10 and Appendix 2.

10.83 Use the Born-Haber cycle outlined in Section 10.8 for NaCl to calculate the lattice energy of $CaCl_2$. Use data from Figures 4.8 and 4.10 and Appendix 2. (The second ionization energy of Ca, IE_2, is 1145 kJ/mol.)

ADDITIONAL PROBLEMS

10.84 Consider the following two reactions.

$$A \longrightarrow 2B \qquad \Delta H^\circ_{rxn} = \Delta H_1$$

$$A \longrightarrow C \qquad \Delta H^\circ_{rxn} = \Delta H_2$$

Determine the enthalpy change for the process

$$2B \longrightarrow C$$

10.85 Consider the reaction

$$NH_3(g) + 3F_2(g) \longrightarrow NF_3(g) + 3HF(g)$$
$$\Delta H^\circ_{rxn} = -881.2 \text{ kJ/mol}$$

Using data from Appendix 2, determine the standard enthalpy of formation of NF_3.

10.86 Based on changes in enthalpy, which of the following reactions will occur more readily?
(a) $Cl(g) + CH_4(g) \longrightarrow CH_3Cl(g) + H(g)$
(b) $Cl(g) + CH_4(g) \longrightarrow CH_3(g) + HCl(g)$

10.87 The standard enthalpy change ΔH° for the thermal decomposition of silver nitrate according to the following equation is +78.67 kJ.

$$AgNO_3(s) \longrightarrow AgNO_2(s) + \tfrac{1}{2}O_2(g)$$

The standard enthalpy of formation of $AgNO_3(s)$ is −123.02 kJ/mol. Calculate the standard enthalpy of formation of $AgNO_2(s)$.

10.88 Hydrazine (N_2H_4) decomposes according to the following reaction.

$$3N_2H_4(l) \longrightarrow 4NH_3(g) + N_2(g)$$

(a) Given that the standard enthalpy of formation of hydrazine is 50.42 kJ/mol, calculate ΔH° for its decomposition. (b) Both hydrazine and ammonia burn in oxygen to produce $H_2O(l)$ and $N_2(g)$. Write balanced equations for each of these processes, and calculate ΔH° for each of them. On a mass basis (per kilogram), which would be the better fuel: hydrazine or ammonia?

10.89 Consider the reaction

$$N_2(g) + 3H_2(g) \longrightarrow 2NH_3(g)$$
$$\Delta H^\circ_f = -92.6 \text{ kJ/mol}$$

When 2 moles of N_2 react with 6 moles of H_2 to form 4 moles of NH_3 at 1 atm and a certain temperature, there is a decrease in volume equal to 98 L. Calculate ΔU for this reaction. (The conversion factor is $1 \text{ L} \cdot \text{atm} = 101.3 \text{ J}$.)

10.90 Calculate the heat released when 2.00 L of $Cl_2(g)$ with a density of 1.88 g/L reacts with an excess of sodium metal at 25°C and 1 atm to form sodium chloride.

10.91 Using the following information and the fact that the average C−H bond enthalpy is 414 kJ/mol, estimate the standard enthalpy of formation of methane (CH_4).

$$C(s) \longrightarrow C(g) \qquad \Delta H^\circ_{rxn} = 716 \text{ kJ/mol}$$

$$2H_2(g) \longrightarrow 4H(g) \qquad \Delta H^\circ_{rxn} = 872.8 \text{ kJ/mol}$$

10.92 Compare the bond enthalpy of F_2 with the overall energy change for the following process.

$$F_2(g) \longrightarrow F^+(g) + F^-(g)$$

Which is the preferred dissociation for F_2, energetically speaking? (*Hint:* You will need data from Figures 4.8 and 4.10.)

10.93 Calculate ΔH° for the reaction

$$H_2(g) + I_2(g) \longrightarrow 2HI(g)$$

(a) using Equation 10.18 and data from Appendix 2, and (b) using Equation 10.19 and data from Table 10.4.

10.94 Consider the reaction

$$2Na(s) + 2H_2O(l) \longrightarrow 2NaOH(aq) + H_2(g)$$

When 2 moles of Na react with water at 25°C and 1 atm, the volume of H_2 formed is 24.5 L. Calculate the work done in joules when 0.34 g of Na reacts with water under the same conditions. (The conversion factor is $1 \text{ L} \cdot \text{atm} = 101.3 \text{ J}$.)

10.95 You are given the following data.

$$H_2(g) \longrightarrow 2H(g) \qquad \Delta H^\circ = 436.4 \text{ kJ/mol}$$

$$Br_2(g) \longrightarrow 2Br(g) \qquad \Delta H^\circ = 192.5 \text{ kJ/mol}$$

$$H_2(g) + Br_2(g) \longrightarrow 2HBr(g) \quad \Delta H^\circ = -72.4 \text{ kJ/mol}$$

Calculate ΔH° for the reaction

$$H(g) + Br(g) \longrightarrow HBr(g)$$

10.96 Methanol (CH_3OH) is an organic solvent and is also used as a fuel in some automobile engines. From the following data, calculate the standard enthalpy of formation of methanol.

$$2CH_3OH(l) + 3O_2(g) \longrightarrow 2CO_2(g) + 4H_2O(l)$$
$$\Delta H^\circ_{rxn} = -1452.8 \text{ kJ/mol}$$

10.97 A 44.0-g sample of an unknown metal at 99.0°C was placed in a constant-pressure calorimeter containing 80.0 g of water at 24.0°C. The final temperature of the system was found to be 28.4°C. Calculate the specific heat of the metal. (The heat capacity of the calorimeter is 12.4 J/°C.)

10.98 A student mixes 88.6 g of water at 74.3°C with 57.9 g of water at 24.8°C in an insulated flask. What is the final temperature of the combined water?

10.99 Producer gas (carbon monoxide) is prepared by passing air over red-hot coke.

$$C(s) + \tfrac{1}{2}O_2(g) \longrightarrow CO(g)$$

Water gas (a mixture of carbon monoxide and hydrogen) is prepared by passing steam over red-hot coke.

$$C(s) + H_2O(g) \longrightarrow CO(g) + H_2(g)$$

For many years, both producer gas and water gas were used as fuels in industry and for domestic cooking. The large-scale preparation of these gases was carried out alternately, that is, first producer gas, then water gas, and so on. Using thermochemistry, explain why it might be advantageous to alternate these two processes.

10.100 Compare the heat produced by the complete combustion of 1 mole of methane (CH_4) with a mole of water gas (0.50 mole H_2 and 0.50 mole CO) under the same conditions. On the basis of your answer, would you prefer methane over water gas as a fuel? Can you suggest two other reasons why methane is preferable to water gas as a fuel?

10.101 The so-called hydrogen economy is based on hydrogen produced from water using solar energy. The gas is then burned as a fuel.

$$2H_2(g) + O_2(g) \longrightarrow 2H_2O(l)$$

A primary advantage of hydrogen as a fuel is that it is nonpolluting. A major disadvantage is that it is a gas and therefore is harder to store than liquids or solids. Calculate the number of moles of H_2 required to produce an amount of energy equivalent to that produced by the combustion of a gallon of octane (C_8H_{18}). The density of octane is 2.66 kg/gal, and its standard enthalpy of formation is −249.9 kJ/mol.

10.102 Ethanol (C_2H_5OH) and gasoline (assumed to be all octane, C_8H_{18}) are both used as automobile fuel. If gasoline is selling for $2.79/gal, what would the price of ethanol have to be to provide the same amount of heat per dollar? The density and ΔH_f° of octane are 0.7025 g/mL and −249.9 kJ/mol, respectively, and of ethanol are 0.7894 g/mL and −277.0 kJ/mol, respectively (1 gal = 3.785 L).

10.103 How many moles of ethane (C_2H_6) would have to be burned for the combustion to heat 855 g of water from 25.0°C to 98.0°C?

10.104 Calculate the heat of reaction for the following process. (The heat of formation for ONF_3 is −163.18 kJ/mol.)

$$2NF_3(g) + O_2(g) \longrightarrow 2ONF_3(g)$$

10.105 The heat of vaporization of a liquid (ΔH_{vap}) is the energy required to vaporize 1.00 g of the liquid at its boiling point. In one experiment, 60.0 g of liquid nitrogen (boiling point = −196°C) is poured into a Styrofoam cup containing 2.00×10^2 g of water at 55.3°C. Calculate the molar heat of vaporization of liquid nitrogen if the final temperature of the water is 41.0°C.

10.106 Explain the cooling effect experienced when ethanol is rubbed on your skin, given that

$$C_2H_5OH(l) \longrightarrow C_2H_5OH(g) \quad \Delta H^{\circ} = 42.2 \text{ kJ/mol}$$

10.107 For which of the following reactions does $\Delta H_{rxn}^{\circ} = \Delta H_f^{\circ}$?
(a) $H_2(g) + S(\text{rhombic}) \longrightarrow H_2S(g)$
(b) $C(\text{diamond}) + O_2(g) \longrightarrow CO_2(g)$
(c) $H_2(g) + CuO(s) \longrightarrow H_2O(l) + Cu(s)$
(d) $O(g) + O_2(g) \longrightarrow O_3(g)$

10.108 Calculate the work done (in joules) when 1.0 mole of water is frozen at 0°C and 1.0 atm. The volumes of 1 mole of water and ice at 0°C are 0.0180 and 0.0196 L, respectively. (The conversion factor is $1 \text{ L} \cdot \text{atm} = 101.3 \text{ J}$.)

10.109 A certain gas initially at 0.050 L undergoes expansion until its volume is 0.50 L. Calculate the work done (in joules) by the gas if it expands (a) against a vacuum and (b) against a constant pressure of 0.20 atm. (The conversion factor is $1 \text{ L} \cdot \text{atm} = 101.3 \text{ J}$.)

10.110 Calculate the standard enthalpy of formation for diamond, given that

$$C(\text{graphite}) + O_2(g) \longrightarrow CO_2(g)$$
$$\Delta H^{\circ} = -393.5 \text{ kJ/mol}$$

$$C(\text{diamond}) + O_2(g) \longrightarrow CO_2(g)$$
$$\Delta H^{\circ} = -395.4 \text{ kJ/mol}$$

10.111 The first step in the industrial recovery of zinc from zinc sulfide ore is known as *roasting*. Roasting is the conversion of ZnS to ZnO by heating according to the following *unbalanced* equation.

$$ZnS(s) + O_2(g) \longrightarrow ZnO(s) + SO_2(g)$$

Determine the amount of heat involved when 1.26×10^4 g of SO_2 are produced in the roasting process.

10.112 Calculate the standard enthalpy change for the fermentation process, in which glucose ($C_6H_{12}O_6$) is converted to ethanol (C_2H_5OH) and carbon dioxide.

10.113 Portable hot packs are available for skiers and people engaged in other outdoor activities in a cold climate. The air-permeable paper packet contains a mixture of powdered iron, sodium chloride, and other components, all moistened by a little water. The exothermic reaction that produces the heat is a very common one—the rusting of iron.

$$4Fe(s) + 3O_2(g) \longrightarrow 2Fe_2O_3(s)$$

When the outside plastic envelope is removed, O_2 molecules penetrate the paper, causing the reaction to begin. A typical packet contains 250 g of iron to warm your hands or feet for up to 4 hours. Using data from Appendix 2, determine how much heat (in kilojoules) is produced by this reaction.

10.114 The bond enthalpy of the C–N bond in the amide group of proteins (see Problem 6.60) can be treated as an average of C–N and C≡N bonds. Calculate the maximum wavelength of light needed to break the bond.

10.115 The total volume of the Pacific Ocean is estimated to be 7.2×10^8 km^3. A medium-sized atomic bomb produces 1.0×10^{15} J of energy upon explosion. Calculate the number of atomic bombs needed to release enough energy to raise the temperature of the water in the Pacific Ocean by 1°C.

10.116 A woman expends 95 kJ of energy in walking a kilometer. The energy is supplied by the metabolic breakdown of food intake and has a 35% efficiency. How much energy does she save by walking the kilometer instead of driving a car that gets 8.2 km per liter of gasoline (approximately 20 mi/gal)? The density of gasoline is 0.71 g/mL, and its enthalpy of combustion is −49 kJ/g.

10.117 The carbon dioxide exhaled by sailors in a submarine is often removed by reaction with an aqueous lithium hydroxide solution. (a) Write a balanced equation for this process. (*Hint:* The products are water and a soluble salt.) (b) If every sailor consumes 1.2×10^4 kJ of energy every day and assuming that this energy is totally supplied by the metabolism of glucose ($C_6H_{12}O_6$), calculate the amounts of CO_2 produced and LiOH required to purify the air.

10.118 The enthalpy of combustion of benzoic acid (C_6H_5COOH) is commonly used as the standard for calibrating constant-volume bomb calorimeters; its value has been accurately determined to be −3226.7 kJ/mol. When 1.9862 g of benzoic acid are burned in a calorimeter, the temperature rises from 21.84°C to 25.67°C. What is the heat capacity of the bomb? (Assume that the quantity of water surrounding the bomb is exactly 2000 g.)

10.119 Calcium oxide (CaO) is used to remove sulfur dioxide generated by coal-burning power stations.

$$2CaO(s) + 2SO_2(g) + O_2(g) \longrightarrow 2CaSO_4(s)$$

Calculate the enthalpy change if 6.6×10^5 g of SO_2 is removed by this process.

10.120 Glauber's salt, sodium sulfate decahydrate ($Na_2SO_4 \cdot 10H_2O$), undergoes a phase transition (i.e., melting or freezing) at a convenient temperature of about 32°C.

$$Na_2SO_4 \cdot 10H_2O(s) \longrightarrow Na_2SO_4 + 10H_2O(l)$$
$$\Delta H° = 74.4 \text{ kJ/mol}$$

As a result, this compound is used to regulate the temperature in homes. It is placed in plastic bags in the ceiling of a room. During the day, the endothermic melting process absorbs heat from the surroundings, cooling the room. At night, it gives off heat as it freezes. Calculate the mass of Glauber's salt in kilograms needed to lower the temperature of air in a room by 8.2°C. The mass of air in the room is 605.4 kg; the specific heat of air is 1.2 J/g · °C.

10.121 An excess of zinc metal is added to 50.0 mL of a 0.100 *M* AgNO$_3$ solution in a constant-pressure calorimeter like the one pictured in Figure 10.8. As a result of the reaction

$$Zn(s) + 2Ag^+(aq) \longrightarrow Zn^{2+}(aq) + 2Ag(s)$$

the temperature rises from 19.25°C to 22.17°C. If the heat capacity of the calorimeter is 98.6 J/°C, calculate the enthalpy change for the given reaction on a molar basis. Assume that the density and specific heat of the solution are the same as those for water, and ignore the specific heats of the metals.

10.122 (a) A person drinks four glasses of cold water (3.0°C) every day. The volume of each glass is 2.5×10^2 mL. How much heat (in kilojoules) does the body have to supply to raise the temperature of the water to 37°C, the body temperature? (b) How much heat would your body lose if you were to ingest 8.0×10^2 g of snow at 0°C to quench your thirst? (The amount of heat necessary to melt snow is 6.01 kJ/mol.)

10.123 Which of the following $\Delta H°_{rxn}$ values is a $\Delta H°_f$ value? For any that is not a $\Delta H°_f$ value, explain why it is not.

(a) $H_2(g) + Br_2(l) \longrightarrow 2HBr(g)$
$$\Delta H°_{rxn} = -72.4 \text{ kJ/mol}$$

(b) $4Al(s) + 3O_2(g) \longrightarrow 2Al_2O_3(s)$
$$\Delta H°_{rxn} = -3339.6 \text{ kJ/mol}$$

(c) $Ag(s) + \frac{1}{2}Cl_2(g) \longrightarrow AgCl(s)$
$$\Delta H°_{rxn} = -127.0 \text{ kJ/mol}$$

(d) $Cu^{2+}(aq) + SO_4^{2-}(aq) \longrightarrow CuSO_4(s)$
$$\Delta H°_{rxn} = +73.25 \text{ kJ/mol}$$

(e) $\frac{1}{2}H_2(g) + \frac{1}{2}N_2(g) + \frac{3}{2}O_2(g) \longrightarrow HNO_3(l)$
$$\Delta H°_{rxn} = -173.2 \text{ kJ/mol}$$

10.124 At 25°C the standard enthalpy of formation of HF(aq) is −320.1 kJ/mol; of OH$^-$(aq), it is −229.6 kJ/mol; of F$^-$(aq), it is −329.1 kJ/mol; and of $H_2O(l)$, it is −285.8 kJ/mol.
(a) Calculate the standard enthalpy of neutralization of HF(aq).

$$HF(aq) + OH^-(aq) \longrightarrow F^-(aq) + H_2O(l)$$

(b) Using the value of −56.2 kJ as the standard enthalpy change for the reaction

$$H^+(aq) + OH^-(aq) \longrightarrow H_2O(l)$$

calculate the standard enthalpy change for the reaction

$$HF(aq) \longrightarrow H^+(aq) + F^+(aq)$$

10.125 Why are cold, damp air and hot, humid air more uncomfortable than dry air at the same temperatures? [The specific heats of water vapor and air are approximately 1.9 J/(g · °C) and 1.0 J/(g · °C), respectively.]

10.126 Vinyl chloride (C_2H_3Cl) differs from ethylene (C_2H_4) in that one of the H atoms is replaced with a Cl atom. Vinyl chloride is used to prepare poly(vinyl chloride), which is an important polymer used in pipes. Calculate the enthalpy change when 1.0×10^3 kg of vinyl chloride forms poly(vinyl chloride).

10.127 A 46-kg person drinks 500 g of milk, which has a "caloric" value of approximately 3.0 kJ/g. If only 17% of the energy in milk is converted to mechanical work, how high (in meters) can the person climb based on this energy intake? [*Hint:* The work done in ascending is given by *mgh*, where *m* is the mass (in kg), *g* is the gravitational acceleration (9.8 m/s²), and *h* is the height (in meters).]

10.128 The height of Niagara Falls on the American side is 51 m. (a) Calculate the potential energy of 1.0 g of water at the top of the falls relative to the ground level. (b) What is the speed of the falling water if all the potential energy is converted to kinetic energy? (c) What would be the increase in temperature of the water if all the kinetic energy were converted to heat? (See Problem 10.127 for information.)

10.129 In the nineteenth century, two scientists named Dulong and Petit noticed that for a solid element, the product of its molar mass and its specific heat is approximately 25 J/°C. This observation, now called Dulong and Petit's law, was used to estimate the specific heat of metals. Verify the law for the metals listed in Table 10.2. The law does not apply to one of the metals. Which one is it? Why?

10.130 Determine the standard enthalpy of formation of ethanol (C_2H_5OH) from its standard enthalpy of combustion (−1367.4 kJ/mol).

10.131 Acetylene (C_2H_2) and benzene (C_6H_6) have the same empirical formula. In fact, benzene can be made from acetylene as follows:

$$3C_2H_2(g) \longrightarrow C_6H_6(l)$$

The enthalpies of combustion for C_2H_2 and C_6H_6 are −1299.4 and −3267.4 kJ/mol, respectively. Calculate the standard enthalpies of formation of C_2H_2 and C_6H_6 and hence the enthalpy change for the formation of C_6H_6 from C_2H_2.

10.132 From the lattice energy of KCl in Table 5.1, and the first ionization energy of K and electron affinity of Cl in Figures 4.8 and 4.10, calculate the $\Delta H°$ for the reaction

$$K(g) + Cl(g) \longrightarrow KCl(s)$$

10.133 The hydroxyl radical (OH) plays an important role in atmospheric chemistry. It is highly reactive and

has a tendency to combine with an H atom from other compounds, causing them to break up. Thus, OH is sometimes called a "detergent" radical because it helps to clean up the atmosphere.
(a) Estimate the enthalpy change for the following reaction.

$$OH(g) + CH_4(g) \longrightarrow CH_3(g) + H_2O(g)$$

(b) The radical is generated when sunlight hits water vapor. Calculate the maximum wavelength [in nanometers (nm)] required to break an O—H bond in H_2O.

10.134 How much metabolic energy must a 5.2-g hummingbird expend to fly to a height of 12 m? (See the hint in Problem 10.127.)

10.135 (a) From the following data, calculate the bond enthalpy of the F_2^- ion.

$F_2(g) \longrightarrow 2F(g)$	$\Delta H°_{rxn} = 156.9$ kJ/mol
$F^-(g) \longrightarrow F(g) + e^-$	$\Delta H°_{rxn} = 333$ kJ/mol
$F_2^-(g) \longrightarrow F_2(g) + e^-$	$\Delta H°_{rxn} = 290$ kJ/mol

(b) Explain the difference between the bond enthalpies of F_2 and F_2^-.

10.136 The average temperature in deserts is high during the day but quite cool at night, whereas that in regions along the coastline is more moderate. Explain.

10.137 Both glucose and fructose are simple sugars with the same molecular formula of $C_6H_{12}O_6$. Sucrose ($C_{12}H_{22}O_{11}$), or table sugar, consists of a glucose molecule bonded to a fructose molecule (a water molecule is eliminated in the formation of sucrose). (a) Calculate the energy released when a 2.0-g glucose tablet is burned in air. (b) To what height can a 65-kg person climb after ingesting such a tablet, assuming only 30% of the energy released is available for work. (See the hint for Problem 10.127.) Repeat the calculations for a 2.0-g sucrose tablet.

10.138 About 6.0×10^{13} kg of CO_2 is fixed (converted to more complex organic molecules) by photosynthesis every year. (a) Assuming all the CO_2 ends up as glucose ($C_6H_{12}O_6$), calculate the energy (in kilojoules) stored by photosynthesis per year. (b) A typical coal-burning electric power station generates about 2.0×10^6 W per year. How many such stations are needed to generate the same amount of energy as that captured by photosynthesis (1 W = 1 J/s)?

10.139 Experiments show that it takes 1656 kJ/mol to break all the bonds in methane (CH_4) and 4006 kJ/mol to break all the bonds in propane (C_3H_8). Based on these data, calculate the average bond enthalpy of the C—C bond.

10.140 From a thermochemical point of view, explain why a carbon dioxide fire extinguisher or water should not be used on a magnesium fire.

10.141   Consider the reaction

$$2H_2(g) + O_2(g) \longrightarrow 2H_2O(l)$$

Under atmospheric conditions (1.00 atm), it was found that the formation of water resulted in a decrease in volume equal to 73.4 L. Calculate ΔU for the process if $\Delta H = -571.6$ kJ/mol. (The conversion factor is $1 \, L \cdot atm = 101.3$ J.)

10.142   *Lime* is a term that includes calcium oxide (CaO, also called quicklime) and calcium hydroxide [Ca(OH)$_2$, also called slaked lime]. It is used in the steel industry to remove acidic impurities, in air-pollution control to remove acidic oxides such as SO_2, and in water treatment. Quicklime is made industrially by heating limestone ($CaCO_3$) above 2000°C.

$$CaCO_3(s) \longrightarrow CaO(s) + CO_2(g)$$
$$\Delta H° = 177.8 \text{ kJ/mol}$$

Slaked lime is produced by treating quicklime with water.

$$CaO(s) + H_2O(l) \longrightarrow Ca(OH)_2(s)$$
$$\Delta H° = -62.5 \text{ kJ/mol}$$

The exothermic reaction of quicklime with water and the rather small specific heats of both quicklime [0.946 J/(g · °C)] and slaked lime [1.20 J/(g · °C)] make it hazardous to store and transport lime in vessels made of wood. Wooden sailing ships carrying lime would occasionally catch fire when water leaked into the hold. (a) If a 500.0-g sample of water reacts with an equimolar amount of CaO (both at an initial temperature of 25°C), what is the final temperature of the product, Ca(OH)$_2$? Assume that the product absorbs all the heat released in the reaction. (b) Given that the standard enthalpies of formation of CaO and H$_2$O are -635.6 and -285.8 kJ/mol, respectively, calculate the standard enthalpy of formation of Ca(OH)$_2$.

10.143   A 4.117-g impure sample of glucose ($C_6H_{12}O_6$) was burned in a constant-volume calorimeter having a heat capacity of 19.65 kJ/°C. If the rise in temperature is 3.134°C, calculate the percent by mass of the glucose in the sample. Assume that the impurities are unaffected by the combustion process and that $\Delta U = \Delta H$. See Appendix 2 for thermodynamic data.

10.144   The combustion of 0.4196 g of a hydrocarbon releases 17.55 kJ of heat. The masses of the products are $CO_2 = 1.419$ g and $H_2O = 0.290$ g. (a) What is the empirical formula of the compound? (b) If the approximate molar mass of the compound is 76 g/mol, calculate its standard enthalpy of formation.

10.145   Photosynthesis produces glucose ($C_6H_{12}O_6$) and oxygen from carbon dioxide and water.

$$6CO_2(g) + 6H_2O(l) \longrightarrow C_6H_{12}O_6(s) + 6O_2(g)$$

(a) How would you determine experimentally the $\Delta H_f°$ value for this reaction? (b) Solar radiation produces about 7.0×10^{14} kg of glucose a year on Earth. What is the corresponding $\Delta H°$ change?

10.146   Ice at 0°C is placed in a Styrofoam cup containing 361 g of a soft drink at 23°C. The specific heat of the drink is about the same as that of water. Some ice remains after the ice and soft drink reach an equilibrium temperature of 0°C. Determine the mass of ice that has melted. Ignore the heat capacity of the cup. (*Hint:* It takes 334 J to melt 1 g of ice at 0°C.)

10.147   Acetylene (C_2H_2) can be made by combining calcium carbide (CaC_2) with water. (a) Write an equation for the reaction. (b) What is the maximum amount of heat (in joules) that can be obtained from the combustion of acetylene, starting with 74.6 g of CaC_2?

10.148   In 1998, scientists using a special type of electron microscope were able to measure the force needed to break a *single* chemical bond. If 2.0×10^{-9} N was needed to break a C—Si bond, estimate the bond enthalpy in kJ/mol. Assume that the bond has to be stretched by a distance of 2 Å (2×10^{-10} m) before it is broken.

10.149   A driver's manual states that the stopping distance quadruples as the speed doubles; that is, if it takes 30 feet to stop a car moving at 25 mph, then it would take 120 feet to stop a car moving at 50 mph. Justify this statement by using mechanics and the first law of thermodynamics. [Assume that when a car is stopped, its kinetic energy ($\frac{1}{2} \, mu^2$) is totally converted to heat.]

10.150   From the enthalpy of formation for CO_2 and the following information, calculate the standard enthalpy of formation for carbon monoxide (CO).

$$CO(g) + \tfrac{1}{2}O_2(g) \longrightarrow CO_2(g)$$
$$\Delta H° = -283.3 \text{ kJ/mol}$$

Why can't we obtain the standard enthalpy of formation directly by measuring the enthalpy of the following reaction?

$$C(\text{graphite}) + \tfrac{1}{2}O_2(g) \longrightarrow CO(g)$$

10.151   The atoms in the H_3^+ ion are arranged in an equilateral triangle. Given the following information

$$2H + H^+ \longrightarrow H_3^+ \qquad \Delta H° = -849 \text{ kJ/mol}$$
$$H_2 \longrightarrow 2H \qquad \Delta H° = 436.4 \text{ kJ/mol}$$

calculate $\Delta H°$ for the reaction

$$H^+ + H_2 \longrightarrow H_3^+$$

10.152   A man ate 0.50 lb of cheese (an energy intake of 4×10^3 kJ). Suppose that none of the energy was stored in his body. What mass (in grams) of water would he need to perspire to maintain his original temperature? (It takes 44.0 kJ to vaporize 1 mole of water.)

10.153 When 1.034 g of naphthalene ($C_{10}H_8$) is burned in a constant-volume bomb calorimeter at 298 K, 41.56 kJ of heat is evolved. Calculate ΔU and w for the reaction on a molar basis.

10.154 A hemoglobin molecule (molar mass = 65,000 g) can bind up to four oxygen molecules. In a certain experiment, a 0.085-L solution containing 6.0 g of deoxyhemoglobin (hemoglobin without oxygen molecules bound to it) was combined with an excess of oxygen in a constant-pressure calorimeter of negligible heat capacity. Calculate the enthalpy of reaction per mole of oxygen bound if the temperature rose by 0.044°C. Assume the solution is dilute so that the specific heat of the solution is equal to that of water.

10.155 A gas company in Massachusetts charges 27 cents for a mole of natural gas (CH_4). Calculate the cost of heating 200 mL of water (enough to make a cup of coffee or tea) from 20°C to 100°C. Assume that only 50% of the heat generated by the combustion is used to heat the water; the rest of the heat is lost to the surroundings.

10.156 A 12.1-g piece of aluminum at 81.7°C is added to a sample of water at 23.4°C in a constant-pressure calorimeter. If the final temperature of the water is 24.9°C, and the heat capacity of the calorimeter is 19.8 J/°C, calculate the mass of the water in the calorimeter.

10.157 Using Table 10.4, compare the following bond enthalpies: C—C in C_2H_6, N—N in N_2H_4, and O—O in H_2O_2. What effect do lone pairs on adjacent atoms appear to have on bond enthalpy? (*Hint:* Start by drawing Lewis structures of each molecule.)

10.158 According to information obtained from www. krispykreme.com, a Krispy Kreme original glazed doughnut weighs 52 g and contains 200 Cal and 12 g of fat. (a) Assuming that the fat in the doughnut is metabolized according to the given equation for tristearin

$$C_{57}H_{110}O_6(s) + 81.5O_2(g) \longrightarrow 57CO_2(g) + 55H_2O(l)$$
$$\Delta H° = -37,760 \text{ kJ/mol}$$

calculate the number of Calories in the reported 12 g of fat in each doughnut. (b) If all the energy contained in a Krispy Kreme doughnut (and just in the fat) were transferred to 6.00 kg of water originally at 25.5°C, what would be the final temperature of the water? (c) When a Krispy Kreme apple fritter weighing 101 g is burned in a bomb calorimeter with $C_{cal} = 95.3$ kJ/°C, the measured temperature increase is 16.7°C. Calculate the number of Calories in a Krispy Kreme apple fritter. (d) What would the $\Delta H°$ value be for the metabolism of 1 mole of the fat tristearin if the water produced by the reaction were gaseous instead of liquid? [*Hint:* Use data from Appendix 2 to determine the $\Delta H°$ value for the reaction $H_2O(l) \longrightarrow H_2O(g)$.]

Answers to In-Chapter Materials

PRACTICE PROBLEMS

10.1A -4.32×10^4 kJ. **10.1B** 6.95×10^5 kJ, heat is absorbed.
10.2A (a) -270 J, (b) 4.02 J, (c) -202 J. **10.2B** (a) 0.487 atm, (b) 73.4 mL, (c) 326.6 mL. **10.3A** 8.174×10^4 kJ. **10.3B** 1.71×10^3 g.
10.4A 151 kJ. **10.4B** 52.2°C. **10.5A** 28°C. **10.5B** 42 g. **10.6A** 14.5 kJ/g.
10.6B 10.5°C rise. **10.7A** -1103 kJ/mol. **10.7B** 113.2 kJ/mol.
10.8A 177.8 kJ/mol. **10.8B** -697.6 kJ/mol. **10.9A** 87.3 kJ/mol.
10.9B $NH_3(g) + HCl(g)$, 176.8 kJ/mol. **10.10A** -557 kJ/mol. (See footnote to Table 10.4 regarding the C=O bond enthalpy in CO_2.)

10.10B 328 kJ/mol. **10.11A** 629 kJ/mol. **10.11B** -841 kJ/mol. (Remember that we define electron affinity as the energy *given off* [◄ Section 4.4]. A negative *EA* value indicates that energy is *absorbed*, as we would expect to be the case with any *second EA*.)

SECTION REVIEW

10.1.1 a. **10.1.2** e. **10.1.3** b. **10.2.1** e. **10.2.2** c. **10.3.1** b. **10.3.2** a. **10.4.1** b.
10.4.2 d. **10.4.3** a. **10.4.4** a. **10.5.1** c. **10.5.2** a. **10.6.1** b. **10.6.2** e. **10.6.3** e.
10.7.1 d. **10.7.2** e. **10.7.3** e. **10.7.4** b. **10.8.1** c. **10.8.2** a.

Chapter 11

Gases

© Francisco Negroni/Alamy Stock Photo

VOLCANIC ERUPTIONS release enormous quantities of the gases dissolved in magma, including water vapor, carbon dioxide, sulfur dioxide, hydrogen sulfide, and several hydrogen halides. Even without eruption, volcanic gases escape continuously into the atmosphere from such geological features as geysers, hot springs, and fumaroles. Knowledge of the physical and chemical properties of gases enables chemists and geologists to understand and to mitigate the hazards that volcanic gases pose to life in the immediate surroundings—and to long term atmospheric conditions on a global scale.

11.1 PROPERTIES OF GASES

Recall from Chapter 1 that matter exists in one of three states: solid, liquid, or gas. In fact, most substances that are solid or liquid at room temperature (25°C) *can* exist as gases under appropriate conditions. Water, for instance, *evaporates* under the right conditions. Water vapor is a gas. In this chapter, we will explore the nature of gases and how their properties at the molecular level give rise to the macroscopic properties that we observe. Figure 11.1 illustrates the three states of matter at the macroscopic and molecular levels.

Relatively few elements exist as gases at room temperature. Those that do are hydrogen, nitrogen, oxygen, fluorine, chlorine, and the noble gases. Of these, the noble gases exist as isolated atoms, whereas the others exist as diatomic molecules [◄◄ Section 5.5]. Figure 11.2 shows where the gaseous elements appear in the periodic table.

Many molecular compounds, most often those with low molar masses, exist as gases at room temperature. Table 11.1 lists some gaseous compounds that may be familiar to you.

Student Annotation: In general, the term *vapor* is used to refer to the gaseous state of a substance that is a liquid or solid at room temperature.

Student Annotation: Recall that oxygen also exists as the triatomic molecule ozone (O_3) [Section 5.5]. Diatomic O_2, however, is the more stable allotrope at room temperature.

Figure 11.1 Solid, liquid, and gaseous states of a substance.

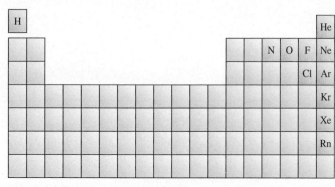

Figure 11.2 Elements that exist as gases at room temperature.

TABLE 11.1	Molecular Compounds That Are Gases at Room Temperature
Molecular formula	**Compound name**
HCl	Hydrogen chloride
NH_3	Ammonia
CO_2	Carbon dioxide
N_2O	Dinitrogen monoxide or nitrous oxide
CH_4	Methane
HCN	Hydrogen cyanide

Gases differ from the condensed phases (solids and liquids) in the following important ways:

1. *A sample of gas assumes both the shape and volume of its container.* Like a liquid, a gas consists of particles (molecules or atoms) that do not have fixed positions in the sample [◂◂ Section 1.5]. As a result, both liquids and gases are able to *flow*. (Recall from Chapter 1 that we refer to liquids and gases collectively as *fluids*.) While a sample of liquid will assume the shape of the part of its container that it occupies, a sample of gas will expand to fill the entire *volume* of its container.

2. *Gases are compressible.* Unlike a solid or a liquid, a gas consists of particles with relatively large distances between them; that is, the distance between any two particles in a gas is much larger than the size of a molecule or atom. Because gas particles are far apart, it is possible to move them closer together by confining them to a smaller volume.

3. *The densities of gases are much smaller than those of liquids and solids; and the density of a gaseous substance is highly variable depending on temperature and pressure.* The densities of gases are typically expressed in g/L, whereas those of liquids and solids are typically expressed in g/mL or g/cm^3.

   When we compress a sample of gas, we decrease its volume. Because its mass remains the same, the ratio of mass to volume (density) increases. Conversely, if we increase the volume to which a sample of gas is confined, we decrease its density.

   If you have ever seen a hot air balloon aloft, you have seen a demonstration of how the density of a gas varies with temperature. Hot air is less dense than cold air, so hot air "floats" on cold air, much like oil floats on water.

4. *Gases form homogeneous mixtures (solutions) with one another in any proportion.* Some liquids (e.g., oil and water) do not mix with one another. Gases, on the other hand, because their particles are so far apart, do not interact with one another to any significant degree unless a chemical reaction takes place between them. This allows molecules of different gases to mix uniformly. That is, gases that don't react with each other are mutually *miscible*.

Each of these four characteristics is the result of the properties of gases at the molecular level.

11.2   THE KINETIC MOLECULAR THEORY OF GASES

The *kinetic molecular theory,* which was put forth in the nineteenth century by a number of physicists, notably Ludwig Boltzmann[1] and James Maxwell,[2] explains how the molecular nature of gases gives rise to their macroscopic

---

[1]Ludwig Eduard Boltzmann (1844–1906). Although Boltzmann, an Austrian physicist, was one of the greatest theoretical physicists of all time, his work was not recognized by other scientists in his own lifetime. He suffered from poor health and severe depression and committed suicide in 1906.

[2]James Clerk Maxwell (1831–1879), a Scottish physicist, was one of the great theoretical physicists of the nineteenth century. His work covered many areas in physics, including the kinetic theory of gases, thermodynamics, and electricity and magnetism.

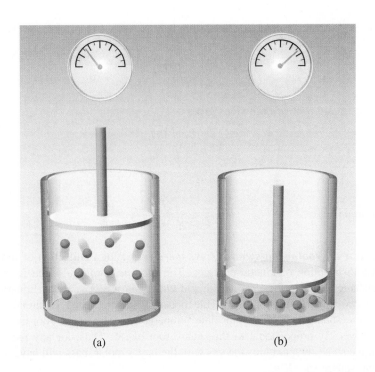

Figure 11.3 Gases can be compressed by decreasing their volume. (a) Before volume decrease. (b) After volume decrease, the increased frequency of collisions between molecules and the walls of their container constitutes higher pressure.

(a)   (b)

properties. The four basic assumptions of the kinetic molecular theory are as follows:

1. A gas is composed of particles that are separated by relatively large distances. The volume occupied by individual molecules is negligible. Gases are compressible because molecules in the gas phase are separated by large distances and can be moved closer together by decreasing the volume occupied by a sample of gas (Figure 11.3).

2. Gas molecules are constantly in random motion, moving in straight paths, colliding with the walls of their container and with one another in perfectly elastic collisions. (Energy is *transferred* but not *lost* in the collisions.)
3. Gas molecules do not exert attractive or repulsive forces on one another.
4. The average kinetic energy, $\overline{E_k}$, of gas molecules in a sample is proportional to the absolute temperature.

$$\overline{E_k} \propto T$$

Recall that kinetic energy is the energy associated with motion [|◀◀ Section 3.1].

$$E_k = \frac{1}{2}mu^2$$

Thus, the kinetic energy of an individual gas molecule is proportional to its mass and its velocity squared. When we talk about a group of gas molecules, we determine the average kinetic energy using the *mean square speed, $\overline{u^2}$*, which is the average of the speed squared for all the molecules in the sample,

$$\overline{u^2} = \frac{u_1^2 + u_2^2 + u_3^2 + \cdots u_N^2}{N}$$

where N is the number of molecules in the sample.

Molecular Speed

One of the important outcomes of the kinetic molecular theory is that the total kinetic energy of a mole of gas (any gas) is equal to $\frac{3}{2}RT$, where R is the ***gas constant (R),*** 8.314 J/K · mol. With assumption 4, we saw that the average kinetic energy of one molecule is $\frac{1}{2}mu^2$. For 1 mole of the gas, we write

$$N_A\left(\frac{1}{2}m\overline{u^2}\right) = \frac{3}{2}RT$$

where N_A is Avogadro's number. Because $m \times N_A = \mathcal{M}$ we can rearrange the preceding equation as follows:

$$\overline{u^2} = \frac{3RT}{\mathcal{M}}$$

Taking the square root of both sides gives

$$\sqrt{\overline{u^2}} = \sqrt{\frac{3RT}{\mathcal{M}}}$$

or

Equation 11.1	$u_{rms} = \sqrt{\dfrac{3RT}{\mathcal{M}}}$

where u_{rms} is the **root-mean-square (rms) speed (u_{rms})**. In a collection of molecules, there is a distribution of kinetic energies. The *root-mean-square speed* is the speed of a molecule that has the *average* kinetic energy in a gas sample. Equation 11.1 indicates two important things: (1) The root-mean-square speed is directly proportional to the square root of the absolute temperature, and (2) the root-mean-square speed is inversely proportional to the square root of $\mathcal{M}$. Thus, for any two samples of gas at the same temperature, the gas with the larger molar mass will have the lower root-mean-square speed, u_{rms}.

Keep in mind that most molecules will have speeds either higher or lower than u_{rms}—and that u_{rms} is temperature dependent. James Maxwell studied extensively the behavior of gas molecules at various temperatures. Figure 11.4(a) shows typical Maxwell speed distribution curves for nitrogen gas at three different temperatures. At a given temperature, the distribution curve tells us the number of molecules moving at a certain speed. The maximum of each curve represents the most probable speed—that is, the speed of the largest number of molecules. Note that the most probable speed increases as temperature increases [the maximum shifts toward the right in Figure 11.4(a)]. Furthermore, the curve also begins to flatten out with increasing temperature, indicating that larger numbers of molecules are moving faster.

Figure 11.4(b) shows the speed distributions of three different gases (Cl_2, N_2, and He) at the same temperature (300 K). The difference in these curves can be explained by noting that lighter molecules, on average, move faster than heavier ones.

Although we can use Equation 11.1 to calculate u_{rms} of a molecule in a particular sample, we will generally find it more useful to compare the u_{rms} values of

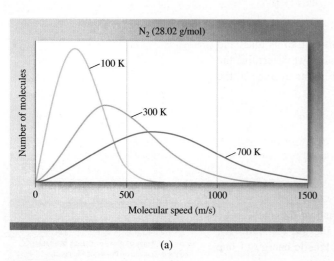

(a)

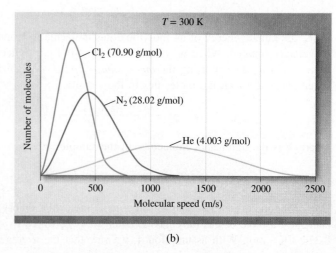

(b)

Figure 11.4 (a) The distribution of speeds for nitrogen gas at three different temperatures. At higher temperatures, more molecules are moving faster. (b) The distribution of speeds for three different gases at the same temperature. On average, lighter molecules move faster than heavier molecules.

molecules in different gas samples. For example, we can write Equation 11.1 for two different gases.

$$u_{rms}(1) = \sqrt{\frac{3RT}{\mathcal{M}_1}} \quad \text{and} \quad u_{rms}(2) = \sqrt{\frac{3RT}{\mathcal{M}_2}}$$

We can then determine the u_{rms} of a molecule in one gas relative to that in the other gas.

$$\frac{u_{rms}(1)}{u_{rms}(2)} = \frac{\sqrt{\dfrac{3RT}{\mathcal{M}_1}}}{\sqrt{\dfrac{3RT}{\mathcal{M}_2}}}$$

Canceling identical terms, when both gases are at the same temperature, we can write

$$\frac{u_{rms}(1)}{u_{rms}(2)} = \sqrt{\frac{\mathcal{M}_2}{\mathcal{M}_1}} \qquad \textbf{Equation 11.2}$$

Student Annotation: Note that because Equation 11.2 contains the ratio of two molar masses, we can express the molar masses as g/mol or kg/mol. (In Equation 11.1, we had to express molar mass as kg/mol for the units to cancel properly.)

Using Equation 11.2 we can compare u_{rms} values of molecules with different molar masses (at a given temperature). Worked Example 11.1 shows how this is done.

Worked Example 11.1

Determine how much faster a helium atom moves, on average, than a carbon dioxide molecule at the same temperature.

Strategy Use Equation 11.2 and the molar masses of He and CO_2 to determine the ratio of their root-mean-square speeds. When solving a problem such as this, it is generally best to label the lighter of the two molecules as molecule 1 and the heavier molecule as molecule 2. This ensures that the result will be greater than 1, which is relatively easy to interpret.

Setup The molar masses of He and CO_2 are 4.003 and 44.01 g/mol, respectively.

Solution

$$\frac{u_{rms}(He)}{u_{rms}(CO_2)} = \sqrt{\frac{44.01 \dfrac{g}{mol}}{4.003 \dfrac{g}{mol}}} = 3.316$$

On average, He atoms move 3.316 times as fast as CO_2 molecules at the same temperature.

Think About It
Remember that the relationship between molar mass and molecular speed (Equation 11.2) is reciprocal. A CO_2 molecule has approximately 10 times the mass of an He atom. Therefore, we should expect an He atom, on average, to be moving approximately $\sqrt{10}$ times (~3.2 times) as fast as a CO_2 molecule.

Practice Problem ATTEMPT Determine the relative root-mean-square speeds of O_2 and SF_6 at a given temperature.

Practice Problem BUILD Determine the molar mass of a gas that moves 4.67 times as fast as CO_2.

Practice Problem CONCEPTUALIZE The diagram on the top represents an equimolar mixture of two gases prior to escaping into the adjoining evacuated chamber. The molar mass of the brown gas is significantly larger than the molar mass of the yellow gas. Which of the diagrams [(i)–(iii)] best represents the contents of the two chambers after a period of time has passed?

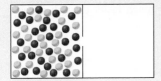
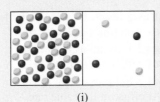
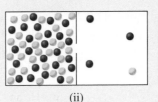
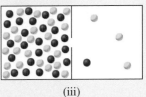

(i)   (ii)   (iii)

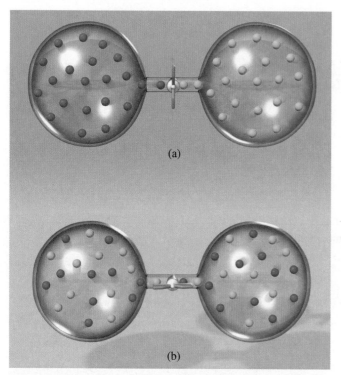

(a)

(b)

Figure 11.5 Diffusion is the mixing of gases. (a) Two different gases in separate containers. (b) When the stopcock is opened, the gases mix by diffusion.

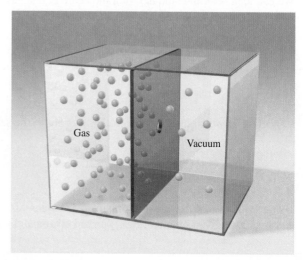

Gas

Vacuum

Figure 11.6 Effusion is the escape of a gas into a vacuum.

Video 11.1
Diffusion of gases.

Student Hot Spot
Student data indicate you may struggle with rates of diffusion and molar masses. Access the SmartBook to view additional Learning Resources on this topic.

Diffusion and Effusion

The random motion of gas molecules gives rise to two readily observable phenomena: diffusion and effusion. *Diffusion* is the mixing of gases as the result of random motion and frequent collisions (Figure 11.5); *effusion* is the escape of gas molecules from a container to a region of vacuum (Figure 11.6). One of the earliest successes of the kinetic molecular theory was its ability to explain diffusion and effusion.

Graham's law states that the rate of diffusion or effusion of a gas is inversely proportional to the square root of its molar mass.

$$\text{Rate} \propto \frac{1}{\sqrt{\mathcal{M}}}$$

This is essentially a restatement of Equation 11.1. Thus, lighter gases diffuse and effuse more rapidly than heavier gases.

Section 11.2 Review

The Kinetic Molecular Theory of Gases

11.2.1 Methane (CH_4) diffuses 2.433 times as fast as a certain closely related gaseous compound. Which of the following could be the unknown gas?
(a) O_2 (d) CH_3I
(b) C_2H_6 (e) F_2
(c) CH_3Br

11.2.2 Which gas effuses faster, He or Ar, and how much faster does it effuse?
(a) He effuses 9.98 times as fast as Ar.
(b) He effuses 3.16 times as fast as Ar.
(c) Ar effuses 9.98 times as fast as He.
(d) Ar effuses 3.16 times as fast as He.
(e) He and Ar effuse at the same rate.

11.3 GAS PRESSURE

According to kinetic molecular theory, gas molecules in a sample are in constant motion, colliding with one another and with the walls of their container. The collisions of gas molecules with the walls of a container constitute the *pressure* exerted by a sample of gas. For example, the air in the tires of your car exerts pressure by virtue of the molecules in the air colliding with the inside walls of the tires. In fact, gases exert pressure on everything they touch. Thus, while you may add enough air to increase the pressure inside your tire to 32 pounds per square inch (psi), there is also a pressure of approximately 14.7 psi, called *atmospheric pressure,* acting on the outside of the tire—and on everything else, including your body. The reason you don't feel the pressure of the atmosphere pushing on the outside of your body is that an equal pressure exists inside your body so that there is no net pressure on you.

Atmospheric pressure, the pressure exerted by Earth's atmosphere, can be demonstrated using the empty metal container shown in Figure 11.7(a). Because the container is open to the atmosphere, atmospheric pressure acts on both the internal and external walls of the container. When we attach a vacuum pump to the opening of the container and draw air out of it, however, we reduce the pressure inside the container. When the pressure against the interior walls is reduced, atmospheric pressure crushes the container [Figure 11.7(b)].

Student Annotation: Common pencil-type pressure gauges actually measure the difference between internal and external pressure. Thus, if the tire is completely flat, the reading of 0 psi means that the pressure *inside* the tire is the same as that *outside* the tire.

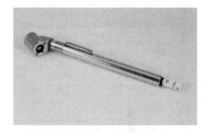

Pencil-type tire gauge.
© David A. Tietz/Editorial Image, LLC

Definition and Units of Pressure

Pressure is defined as the force applied per unit area.

$$\text{pressure} = \frac{\text{force}}{\text{area}}$$

The SI unit of force is the **newton (N),** where

$$1 \text{ N} = 1 \text{ kg} \cdot \text{m/s}^2$$

The SI unit of pressure is the **pascal (Pa),** defined as 1 newton per square meter.

$$1 \text{ Pa} = 1 \text{ N/m}^2$$

Although the pascal is the SI unit of pressure, there are other units of pressure that are more commonly used. Table 11.2 lists the units in which pressure is most commonly expressed in chemistry and their definitions in terms of pascals. Which of these units you encounter most often will depend on your specific field of study. Use of *atmospheres* (atm) is common in chemistry, although use of the *bar* is becoming increasingly common. Use of *millimeters mercury* (mmHg) is common in medicine and meteorology. We will use all of these units in this text.

(a)

(b)

Figure 11.7 (a) An empty metal can. (b) When the air is removed by a vacuum pump, atmospheric pressure crushes the can.

© McGraw-Hill Education/Charles D. Winters, photographer

TABLE 11.2	Units of Pressure Commonly Used in Chemistry	
Unit	**Origin**	**Definition**
standard atmosphere (atm)	Pressure at sea level	1 atm = 101,325 Pa
mmHg	Barometer measurement	1 mmHg = 133.322 Pa
torr	Name given to mmHg in honor of Torricelli, the inventor of the barometer	1 torr = 133.322 Pa
bar	Same order of magnitude as atm, but a decimal multiple of Pa	1 bar = 1 × 10⁵ Pa

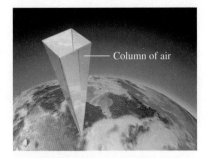

Figure 11.8 A column of air 1 cm × 1 cm from Earth's surface to the top of the atmosphere weighs approximately 1 kg.

Student Annotation: Remember that when a unit is raised to a power, any conversion factor you use must also be raised to that power [Section 1.4].

Calculation of Pressure

The force experienced by an area exposed to Earth's atmosphere is equal to the weight of the column of air above it. For example, the mass of air above a spot on the ground near sea level, with an area of 1 cm^2, is approximately 1 kg (Figure 11.8).

The weight of an object that is subject to Earth's gravitational pull is equal to its mass times the gravitational constant, 9.80665 m/s^2. Thus, the force exerted by this column of air is

$$1 \text{ kg} \times \frac{9.80665 \text{ m}}{\text{s}^2} \approx 10 \text{ kg} \cdot \text{m/s}^2 = 10 \text{ N}$$

Pressure, though, is force per unit area. Specifically, pressure in *pascals* is equal to force in *newtons* per *square meter*. We must first convert area from cm^2 to m^2,

$$1 \text{ cm}^2 \times \left(\frac{1 \text{ m}}{100 \text{ cm}}\right)^2 = 0.0001 \text{ m}^2$$

and then divide force by area.

$$\frac{10 \text{ N}}{0.0001 \text{ m}^2} = 1 \times 10^5 \text{ Pa}$$

This pressure is roughly equal to 1 atm (~1 × 10^5 Pa), which we would expect at sea level.

We can calculate the pressure exerted by a column of any fluid (gas or liquid) in the same way. In fact, this is how atmospheric pressure is commonly measured—by determining the height of a column of mercury it can support.

Measurement of Pressure

A simple ***barometer,*** an instrument used to measure atmospheric pressure, consists of a long glass tube, closed at one end and filled with mercury. The tube is carefully inverted in a container of mercury so that no air enters the tube. When the tube is inverted, and the open end is submerged in the mercury in the container, some of the mercury in the tube will flow out into the container, creating an empty space at the top (closed end) of the tube (Figure 11.9). The weight of the mercury remaining in the tube is supported by atmospheric pressure pushing down on the surface of the mercury in the container. In other words, the pressure exerted by the column of mercury is *equal* to the pressure exerted by the atmosphere. ***Standard atmospheric pressure*** (1 atm) was originally defined as the pressure that would support a column of mercury exactly 760 mm high at 0°C at sea level. The mmHg unit is also called the torr, after the Italian scientist Evangelista Torricelli,[3] who invented the barometer.

Student Annotation: Standard atmospheric pressure expressed in different units:

1 atm*
101,325 Pa
760 mmHg*
760 torr*
1.01325 bar
14.7 psi

*These are exact numbers.

A ***manometer*** is a device used to measure pressures other than atmospheric pressure. The principle of operation of a manometer is similar to that of a barometer. There are two types of manometers, both of which are shown in Figure 11.10. The closed-tube manometer [Figure 11.10(a)] is normally used to measure pressures below atmospheric pressure, whereas the open-tube manometer [Figure 11.10(b)] is generally used to measure pressures equal to or greater than atmospheric pressure.

The pressure exerted by a column of fluid, such as that in a barometer (Figure 11.9), is given by Equation 11.3,

Equation 11.3 $P = hdg$

where h is the height of the column in meters, d is the density of the fluid in kg/m^3, and g is the gravitational constant equal to 9.80665 m/s^2. This equation explains why barometers historically have been constructed using mercury. The height of a column of fluid supported by a given pressure is inversely proportional to the density of the fluid. (At a given P, as d goes down, h must come up—and vice versa.) Mercury's very high density made it possible to construct barometers and manometers of manageable size. For example, a barometer filled with mercury that stands 1 m tall would have to be over 13 m tall if it were filled with water.

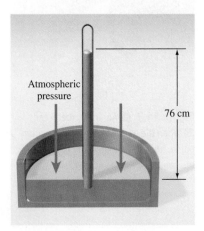

Atmospheric pressure

76 cm

Figure 11.9 Barometer.

[3]Evangelista Torricelli (1608–1674) was supposedly the first person to recognize the existence of atmospheric pressure.

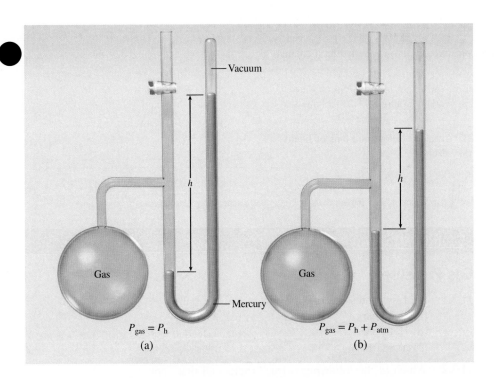

Figure 11.10 (a) Closed-tube manometer. The space labeled "vacuum" actually contains a small amount of mercury vapor. (b) Open-tube manometer.

Worked Example 11.2 shows how to calculate the pressure exerted by a column of fluid.

Worked Example 11.2

Calculate the pressure exerted by a column of mercury 70.0 cm high. Express the pressure in pascals, in atmospheres, and in bars. The density of mercury is 13.5951 g/cm^3.

Strategy Use Equation 11.3 to calculate pressure. Remember that the height must be expressed in meters and density must be expressed in kg/m^3.

Setup

$$h = 70.0 \text{ cm} \times \frac{1 \text{ m}}{100 \text{ cm}} = 0.700 \text{ m}$$

$$d = \frac{13.5951 \text{ g}}{\text{cm}^3} \times \frac{1 \text{ kg}}{1000 \text{ g}} \times \left(\frac{100 \text{ cm}}{1 \text{ m}}\right)^3 = 1.35951 \times 10^4 \text{ kg/m}^3$$

$$g = 9.80665 \text{ m/s}^2$$

Solution

$$\text{pressure} = 0.700 \text{ m} \times \frac{1.35951 \times 10^4 \text{ kg}}{\text{m}^3} \times \frac{9.80665 \text{ m}}{\text{s}^2} = 9.33 \times 10^4 \text{ kg/m} \cdot \text{s}^2 = 9.33 \times 10^4 \text{ Pa}$$

$$9.33 \times 10^4 \text{ Pa} \times \frac{1 \text{ atm}}{101,325 \text{ Pa}} = 0.921 \text{ atm}$$

$$0.921 \text{ atm} \times \frac{1.01325 \text{ bar}}{1 \text{ atm}} = 0.933 \text{ bar}$$

Think About It
Make sure your units cancel properly in this type of problem. Common errors include forgetting to express height in meters and density in kg/m^3. You can avoid these errors by becoming familiar with the value of atmospheric pressure in the various units. A column of mercury slightly less than 760 mm is equivalent to slightly less than 101,325 Pa, slightly less than 1 atm, and slightly less than 1 bar.

(Continued on next page)

Practice Problem **A**TTEMPT   What pressure (in atm) is exerted by a column of mercury exactly 1 m high?

Practice Problem **B**UILD   What would be the height of a column of water supported by the pressure you calculated in Worked Example 11.2? Assume that the density of the water is 1.00 g/cm³.

Practice Problem **C**ONCEPTUALIZE   Arrange the four columns of liquid [(i)–(iv)] in order of increasing pressure they exert.

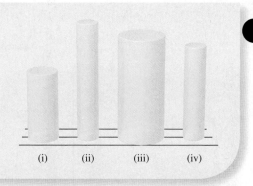

(i)   (ii)   (iii)   (iv)

Section 11.3 Review

Gas Pressure

11.3.1   Express a pressure of 1.15 atm in units of bar.
  (a) 1.17 bar   (d) 874 bar
  (b) 1.13 bar   (e) 1.51×10^{-3} bar
  (c) 0.881 bar

11.3.2   Which of the following is true? (Select all that apply.)
  (a) 0.80 atm = 0.80 torr   (d) 2300 atm = 1748 torr
  (b) 4180 mmHg = 5.573×10^5 Pa   (e) 5.5 atm = 1.0×10^5 Pa
  (c) 433 torr = 433 mmHg

11.3.3   Calculate the height of a column of ethanol that would be supported by atmospheric pressure (1 atm). The density of ethanol is 0.789 g/cm³.
  (a) 1.31×10^4 m   (d) 13.1 m
  (b) 0.789 m   (e) 780 mm
  (c) 600 mm

11.3.4   What pressure (in atm) is exerted by a column of water 50.0 m high? Assume the density of the water is 1.00 g/cm³.
  (a) 490 atm   (d) 50 atm
  (b) 4.84 atm   (e) 1.62 atm
  (c) 0.087 atm

11.3.5   Rank the pressures exerted by these columns of water from lowest to highest.
  (a) i = ii < iii   (d) ii < i = iii
  (b) i < iii < ii   (e) iii < i = ii
  (c) iii < i < ii

(i)   (ii)   (iii)

Video 11.2
Gas laws.

11.4   THE GAS LAWS

In contrast to the condensed phases, all gases, even those with vastly different chemical compositions, exhibit remarkably similar physical behavior. Numerous experiments carried out in the seventeenth and eighteenth centuries showed that the physical state of a sample of gas can be described completely with just four parameters: temperature (T), pressure (P), volume (V), and number of moles (n). Knowing any three of these parameters enables us to calculate the fourth. The relationships between these parameters are known as the ***gas laws***.

Student Annotation: Because they arise from experiment, these laws are referred to as the *empirical* gas laws.

Boyle's Law: The Pressure-Volume Relationship

Imagine that you have a plastic syringe filled with air. If you hold your finger tightly against the tip of the syringe and push the plunger with your other hand, decreasing the

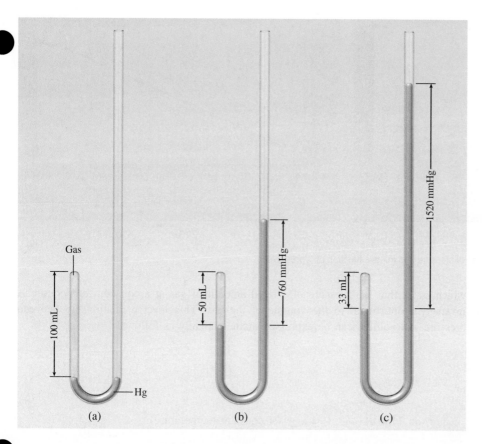

Figure 11.11 Demonstration of Boyle's law. The volume of a sample of gas is inversely proportional to its pressure. (a) $P = 760$ mmHg, $V = 100$ mL. (b) $P = 1520$ mmHg, $V = 50$ mL. (c) $P = 2280$ mmHg, $V = 33$ mL. Note that the total pressure exerted on the gas is the sum of atmospheric pressure (760 mmHg) and the difference in height of the mercury.

volume of the air, you will increase the pressure in the syringe. During the seventeenth century, Robert Boyle[4] conducted systematic studies of the relationship between gas volume and pressure using a simple apparatus like the one shown in Figure 11.11. The J-shaped tube contains a sample of gas confined by a column of mercury. The apparatus functions as an open-end manometer. When the mercury levels on both sides are equal [Figure 11.11(a)], the pressure of the confined gas is equal to atmospheric pressure. When more mercury is added through the open end, the pressure of the confined gas is increased by an amount proportional to the height of the added mercury—and the volume of the gas decreases. If, for example, as shown in Figure 11.11(b), we *double* the pressure on the confined gas by adding enough mercury to make the difference in mercury levels on the left and right 760 mm (the height of a mercury column that exerts a pressure equal to 1 atm), the volume of the gas is reduced by *half*. If we triple the original pressure on the confined gas by adding more mercury, the volume of the gas is reduced to one-third of its original volume [Figure 11.11(c)].

Table 11.3 gives a set of data typical of Boyle's experiments. Figure 11.12(a) and (b) shows some of the volume data plotted as a function of pressure and as a function of the inverse of pressure, respectively. These data illustrate **Boyle's law,**

TABLE 11.3	Typical Data from Experiments with the Apparatus of Figure 11.11								
P (mmHg)	760	855	950	1045	1140	1235	1330	1425	1520	2280
V (mL)	100	89	78	72	66	59	55	54	50	33
	Shown in Figure 11.11(a)								Shown in Figure 11.11(b)	Shown in Figure 11.11(c)

---

[4]Robert Boyle (1627–1691), British chemist and natural philosopher, is best known for the gas law that bears his name. He also made many other significant contributions to the fields of chemistry and physics.

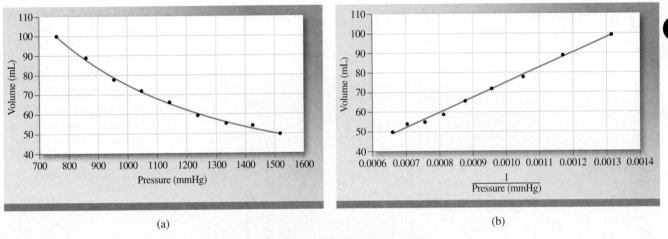

Figure 11.12 Plots of volume (a) as a function of pressure and (b) as a function of 1/pressure.

which states that the pressure of a fixed amount of gas at a constant temperature is inversely proportional to the volume of the gas. This inverse relationship between pressure and volume can be expressed mathematically as follows:

$$V \propto \frac{1}{P}$$

or

Equation 11.4(a) $V = k_1 \dfrac{1}{P}$ (at constant temperature)

where k_1 is a *proportionality constant.* We can rearrange Equation 11.4(a) to get

Equation 11.4(b) $PV = k_1$ **(at constant temperature)**

According to this form of Boyle's law, the *product* of the pressure and the volume of a given sample of gas (at constant temperature) is a *constant.*

Although the individual values of pressure and volume can vary greatly for a given sample of gas, the product of P and V is always equal to the same constant as long as the temperature is held constant and the amount of gas does not change. Therefore, for a given sample of gas under two different sets of conditions at constant temperature, we can write

$$P_1V_1 = k_1 = P_2V_2$$

Equation 11.5 $P_1V_1 = P_2V_2$ **(at constant temperature)**

where V_1 is the volume at pressure P_1 and V_2 is the volume at pressure P_2.

Worked Example 11.3 illustrates the use of Boyle's law.

Worked Example 11.3

If a skin diver takes a breath at the surface, filling his lungs with 5.82 L of air, what volume will the air in his lungs occupy when he dives to a depth where the pressure is 1.92 atm? (Assume constant temperature and that the pressure at the surface is exactly 1 atm.)

Strategy Use Equation 11.5 to solve for V_2.

Setup $P_1 = 1.00$ atm, $V_1 = 5.82$ L, and $P_2 = 1.92$ atm.

Solution

$$V_2 = \frac{P_1 \times V_1}{P_2} = \frac{1.00 \text{ atm} \times 5.82 \text{ L}}{1.92 \text{ atm}} = 3.03 \text{ L}$$

Think About It
At higher pressure, the volume should be smaller. Therefore, the answer makes sense.

Practice Problem ATTEMPT Calculate the volume of a sample of gas at 5.75 atm if it occupies 5.14 L at 2.49 atm. (Assume constant temperature.)

Practice Problem BUILD At what pressure would a sample of gas occupy 7.86 L if it occupies 3.44 L at 4.11 atm? (Assume constant temperature.)

Practice Problem CONCEPTUALIZE Which of the following diagrams could represent a gas sample in a balloon at constant temperature before and after an increase in external pressure?

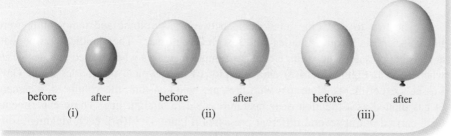

| before | after | before | after | before | after |
| (i) | | (ii) | | (iii) | |

Charles's and Gay-Lussac's Law: The Temperature-Volume Relationship

Video 11.3
Gas laws—Charles's law.

If you took a helium-filled Mylar balloon outdoors on a cold day, the balloon would shrink somewhat when it came into contact with the cold air. This would occur because the volume of a sample of gas depends on the temperature. A more dramatic illustration is shown in Figure 11.13, where liquid nitrogen is being poured over a balloon. The large drop in temperature of the air in the balloon (the boiling liquid nitrogen has a temperature of −196°C) results in a significant decrease in its volume, causing the balloon to shrink.

Figure 11.13 (a) Air-filled balloon. (b) Lowering the temperature with liquid nitrogen causes a volume decrease. The pressure inside the balloon, which is roughly equal to the external pressure, remains constant in this process.

© McGraw-Hill Education/Charles D. Winters, photographer

(a)

(b)

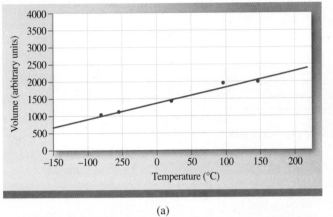

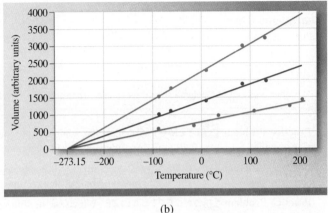

(a)   (b)

Figure 11.14 (a) Plot of the volume of a sample of gas as a function of temperature. (b) Plot of the volume of a sample of gas as a function of temperature at three different pressures.

The first to study the relationship between gas volume and temperature were French scientists Jacques Charles[5] and Joseph Gay-Lussac.[6] Their studies showed that, at constant pressure, the volume of a gas sample increases when heated and decreases when cooled. Figure 11.14(a) shows a plot of data typical of Charles's and Gay-Lussac's experiments. Note that with pressure held constant, the volume of a sample of gas plotted as a function of temperature yields a straight line. These experiments were carried out at several different pressures [Figure 11.14(b)], each yielding a different straight line. Interestingly, if the lines are extrapolated to zero volume, they all meet at the *x* axis at the temperature −273.15°C. The implication is that a gas sample occupies zero volume at −273.15°C. This is not observed in practice, however, because all gases condense to form liquids or solids before −273.15°C is reached.

In 1848, Lord Kelvin[7] realized the significance of the extrapolated lines all meeting at −273.15°C. He identified −273.15°C as ***absolute zero,*** theoretically the lowest attainable temperature. Then he set up an ***absolute temperature scale,*** now called the ***Kelvin temperature scale,*** with absolute zero as the lowest point [◄◄ Section 1.4]. On the Kelvin scale, 1 kelvin (K) is equal in magnitude to 1 degree Celsius. The difference is simply an offset of 273.15. We obtain the absolute temperature by adding 273.15 to the temperature expressed in Celsius, although we often use simply 273 instead of 273.15. Several important points on the two scales match up as follows:

> **Student Annotation:** Remember that a kelvin and a degree Celsius have the same magnitude. Thus, while we add 273.15 to the temperature in °C to get the temperature in K, a *change* in temperature in Celsius is *equal* to the change in temperature in K. A temperature of 20°C is the same as 293.15 K. A *change* in temperature of 20°C, however, is the same as a *change* in temperature of 20 K.

	Kelvin scale (K)	Celsius scale (°C)
Absolute zero	0 K	−273.15°C
Freezing point of water	273.15 K	0°C
Boiling point of water	373.15 K	100°C

The dependence of the volume of a sample of gas on temperature is given by

> **Student Annotation:** Don't forget that volume is proportional to absolute temperature. The volume of a sample of gas at constant pressure doubles if the temperature increases from 100 K to 200 K—but *not* if the temperature increases from 100°C to 200°C!

Equation 11.6(a) $V = k_2T$ **(at constant pressure)**

$$V \propto T$$

or

Equation 11.6(b) $\dfrac{V}{T} = k_2$ (at constant pressure)

---

[5]Jacques Alexandre Cesar Charles (1746–1823) was a gifted lecturer, an inventor of scientific apparatus, and the first person to use hydrogen to inflate balloons.

[6]Joseph Louis Gay-Lussac (1778–1850), French chemist and physicist, like Charles, was a balloon enthusiast. Once he ascended to an altitude of 20,000 ft to collect air samples for analysis.

[7]William Thomson, Lord Kelvin (1824–1907), was a Scottish mathematician and physicist. Kelvin did important work in many branches of physics.

where k_2 is the proportionality constant. Equations 11.6(a) and (b) are expressions of **Charles's and Gay-Lussac's law,** often referred to simply as **Charles's law,** which states that the volume of a fixed amount of gas maintained at constant pressure is directly proportional to the absolute temperature of the gas.

Just as we did with the pressure-volume relationship at constant temperature, we can compare two sets of volume-temperature conditions for a given sample of gas at constant pressure. From Equation 11.6, we can write

$$\frac{V_1}{T_1} = k_2 = \frac{V_2}{T_2}$$

or

$$\frac{V_1}{T_1} = \frac{V_2}{T_2} \text{ (at constant pressure)} \qquad \textbf{Equation 11.7}$$

where V_1 is the volume of the gas at T_1 and V_2 is the volume of the gas at T_2.

Worked Example 11.4 shows how to use Charles's law.

Worked Example 11.4

A sample of argon gas that originally occupied 14.6 L at 25.0°C was heated to 50.0°C at constant pressure. What is its new volume?

Strategy Use Equation 11.7 to solve for V_2. Remember that temperatures must be expressed in kelvin.

Setup $T_1 = 298.15$ K, $V_1 = 14.6$ L, and $T_2 = 323.15$ K.

Solution

$$V_2 = \frac{V_1 \times T_2}{T_1} = \frac{14.6 \text{ L} \times 323.15 \text{ K}}{298.15 \text{ K}} = 15.8 \text{ L}$$

> **Think About It**
> When temperature increases at constant pressure, the volume of a gas sample increases.

Practice Problem Ⓐ**TTEMPT** A sample of gas originally occupies 29.1 L at 0.0°C. What is its new volume when it is heated to 15.0°C? (Assume constant pressure.)

Practice Problem Ⓑ**UILD** At what temperature (in °C) will a sample of gas occupy 82.3 L if it occupies 50.0 L at 75.0°C? (Assume constant pressure.)

Practice Problem Ⓒ**ONCEPTUALIZE** A sample of gas at 50°C is contained in a cylinder with a movable piston shown on the right. Which of the diagrams [(i)–(iv)] best represents the system when the temperature of the sample has been increased to 100°C?

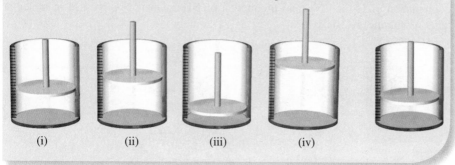

(i)   (ii)   (iii)   (iv)

Avogadro's Law: The Amount-Volume Relationship

In 1811, the Italian scientist Amedeo Avogadro proposed that equal volumes of different gases contain the same number of particles (molecules or atoms) at the same temperature and pressure. This hypothesis gave rise to **Avogadro's law,** which states that the volume of a sample of gas is directly proportional to the number of moles in the sample at constant temperature and pressure.

$$V \propto n$$

or

Equation 11.8(a) $V = k_3n$ **(at constant temperature and pressure)**

Rearranging 11.8(a) gives

Equation 11.8(b) $\dfrac{V}{n} = k_3$

Equations 11.8(a) and (b) are expressions of Avogadro's law.

As with other gas laws, we can compare two sets of conditions using Avogadro's law when n and V both change at constant pressure and temperature and write

$$\frac{V_1}{n_1} = k_3 = \frac{V_2}{n_2}$$

or

Equation 11.9 $\dfrac{V_1}{n_1} = \dfrac{V_2}{n_2}$

where V_1 is the volume of a sample of gas consisting of n_1 moles and V_2 is the volume of a sample consisting of n_2 moles—under conditions of constant temperature and pressure. Coupled with a balanced chemical equation, Avogadro's law enables us to predict the volumes of gaseous reactants and products. Consider the reaction of H_2 and N_2 to form NH_3:

$$3H_2(g) + N_2(g) \longrightarrow 2NH_3(g)$$

The balanced equation reveals the ratio of combination of reactants in terms of *moles* [◂◂ Section 8.3]. However, because the volume of a gas (at a given temperature and pressure) is directly *proportional* to the number of moles, the balanced equation also reveals the ratio of combination in terms of *volume*. Thus, if we were to combine three volumes (liters, milliliters, etc.) of hydrogen gas with one volume of nitrogen gas, assuming they react completely according to the balanced equation, we would expect two volumes of ammonia gas to be produced (Figure 11.15). The ratio of combination of H_2 and N_2 (and production of NH_3), whether expressed in moles or units of volume, is 3:1:2.

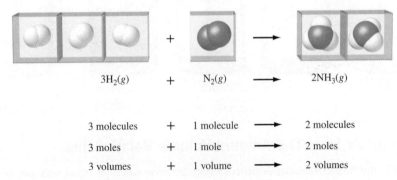

$3H_2(g)$	+	$N_2(g)$	$\longrightarrow$ $2NH_3(g)$
3 molecules	+	1 molecule	$\longrightarrow$ 2 molecules
3 moles	+	1 mole	$\longrightarrow$ 2 moles
3 volumes	+	1 volume	$\longrightarrow$ 2 volumes

Figure 11.15 Illustration of Avogadro's law. The volume of a sample of gas is directly proportional to the number of moles.

Worked Example 11.5 shows how to apply Avogadro's law.

Worked Example 11.5

If we combine 3.0 L of NO and 1.5 L of O_2, and they react according to the balanced equation $2NO(g) + O_2(g) \longrightarrow 2NO_2(g)$, what volume of NO_2 will be produced? (Assume that the reactants and product are all at the same temperature and pressure.)

Strategy Apply Avogadro's law to determine the volume of a gaseous product.

Setup Because volume is proportional to the number of moles, the balanced equation determines in what volume ratio the reactants combine and the ratio of product volume to reactant volume. The amounts of reactants given are stoichiometric amounts [◀◀ Section 8.3].

Solution According to the balanced equation, the volume of NO_2 formed will be equal to the volume of NO that reacts. Therefore, 3.0 L of NO_2 will form.

Think About It
Remember that the coefficients in balanced chemical equations indicate ratios in molecules or moles. Under conditions of constant temperature and pressure, the volume of a gas is proportional to the number of moles. Therefore, the coefficients in balanced equations containing only gases also indicate ratios in liters, provided the reactions occur at constant temperature and pressure.

Student Annotation: It is important to recognize that coefficients indicate ratios in liters only in balanced equations in which all of the reactants and products are gases. We cannot apply the same approach to reactions in which there are solid, liquid, or aqueous species.

Practice Problem ATTEMPT What volume (in liters) of water vapor will be produced when 34 L of H_2 and 17 L of O_2 react according to the equation $2H_2(g) + O_2(g) \longrightarrow 2H_2O(g)$?

Practice Problem BUILD What volumes (in liters) of carbon monoxide and oxygen gas must react according to the equation $2CO(g) + O_2(g) \longrightarrow 2CO_2(g)$ to form 3.16 L of carbon dioxide?

Practice Problem CONCEPTUALIZE A hypothetical gaseous reaction is depicted here with molecular models. Imagine that this reaction takes place in a cylinder with a movable piston at constant temperature and pressure; and that the diagram shown on the right represents the reaction mixture before the reaction. Which of the diagrams [(i)–(iv)] best represents the system when the reaction is complete? (Assume that reactants are combined in stoichiometric amounts. Note that the reaction depicted is not balanced.)

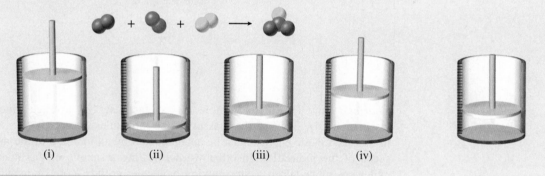

(i)   (ii)   (iii)   (iv)

The Gas Laws and Kinetic Molecular Theory

Each of the empirical gas laws describes a facet of the behavior of macroscopic samples of gas; and each can be explained and understood at the *molecular* level using the kinetic molecular theory of gases.

Boyle's law: The pressure exerted by a gas is the result of the collisions of gas molecules with the walls of their container (assumption 2). The magnitude of the pressure depends on both the frequency of collision and the speed of molecules when they collide with the walls. Decreasing the volume occupied by a sample of gas increases the frequency of these collisions, thus increasing the pressure (see Figure 11.3).

Figure 11.16 Charles's law. (a) The volume of a sample of gas at constant pressure is proportional to its absolute temperature. (b) The pressure of a sample of gas at constant volume is proportional to its absolute temperature.

Charles's Law

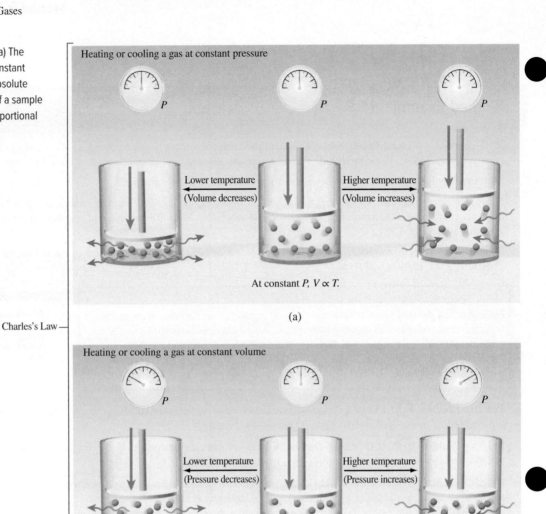

Heating or cooling a gas at constant pressure

Lower temperature (Volume decreases)

Higher temperature (Volume increases)

At constant P, $V \propto T$.

(a)

Heating or cooling a gas at constant volume

Lower temperature (Pressure decreases)

Higher temperature (Pressure increases)

At constant V, $P \propto T$.

(b)

Charles's law: Heating a sample of gas increases its average kinetic energy (assumption 4). Because the masses of the molecules do not change, an increase in average kinetic energy must be accompanied by an increase in the mean square speed of the molecules. In other words, heating a sample of gas makes the gas molecules move faster. Faster-moving molecules collide more frequently and with greater speed at impact, thus increasing the pressure. If the container can expand (as is the case with a balloon or a cylinder with a movable piston), the volume of the gas sample will increase, thereby decreasing the frequency of collisions until the pressure inside the container and the pressure outside are again equal (Figure 11.16).

Avogadro's law: Because the magnitude of the pressure exerted by a sample of gas depends on the frequency of the collisions with the container wall, the presence of more molecules would cause an increase in pressure. Again, the container will expand if it can. Expansion of the container will decrease the frequency of collisions until the pressures inside and outside the container are once again equal (Figure 11.17).

As we will see, the kinetic molecular theory can also be used to explain the behavior of gas *mixtures* [◄◄ Section 11.7].

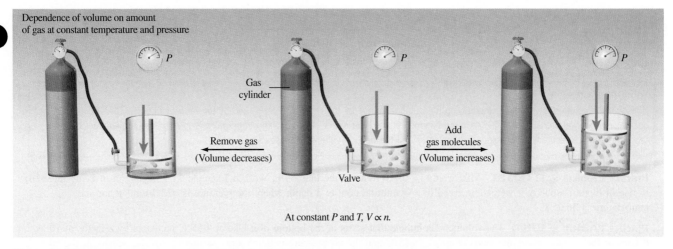

Dependence of volume on amount
of gas at constant temperature and pressure

Gas
cylinder

Remove gas
(Volume decreases)

Add
gas molecules
(Volume increases)

Valve

At constant P and T, $V \propto n$.

Figure 11.17 Avogadro's law. The volume of a gas at constant temperature and pressure is proportional to the number of moles.

The Combined Gas Law: The Pressure-Temperature-Amount-Volume Relationship

Although the gas laws we have discussed so far are useful, each requires that two of the system's parameters be held constant.

Problem type	Relates	Requires constant
Boyle's law	P and V	n and T
Charles's law	T and V	n and P
Avogadro's law	n and V	P and T

Many of the problems we will encounter involve changes to P, T, and V, and, in some cases, also to n. To solve such problems, we need a gas law that relates all the variables. By combining Equations 11.5, 11.7, and 11.9, we obtain the *combined gas law.*

$$\frac{P_1 V_1}{n_1 T_1} = \frac{P_2 V_2}{n_2 T_2}$$ **Equation 11.10(a)**

The combined gas law can be used to solve problems where any or all of the variables change. Note that when a problem involves a fixed quantity of gas, Equation 11.10(a) reduces to the more common form of the combined gas law.

$$\frac{P_1 V_1}{T_1} = \frac{P_2 V_2}{T_2}$$ **Equation 11.10(b)**

Worked Example 11.6 illustrates the use of the combined gas law.

Worked Example 11.6

If a child releases a 6.25-L helium balloon in the parking lot of an amusement park where the temperature is 28.50°C and the air pressure is 757.2 mmHg, what will be the volume of the balloon when it has risen to an altitude where the temperature is −34.35°C and the air pressure is 366.4 mmHg?

Strategy In this case, because there is a fixed amount of gas, we use Equation 11.10(b). The only value we don't know is V_2. Temperatures must be expressed in kelvins. We can use any units of pressure, as long as we are consistent.

Setup $T_1 = 301.65$ K, $T_2 = 238.80$ K. Solving Equation 11.10(b) for V_2 gives

$$V_2 = \frac{P_1 T_2 V_1}{P_2 T_1}$$

(Continued on next page)

Solution

$$V_2 = \frac{757.2 \text{ mmHg} \times 238.80 \text{ K} \times 6.25 \text{ L}}{366.4 \text{ mmHg} \times 301.65 \text{ K}} = 10.2 \text{ L}$$

Think About It

Note that the solution is essentially multiplying the original volume by the ratio of P_1 to P_2, and by the ratio of T_2 to T_1. The effect of decreasing external pressure is to increase the balloon volume. The effect of decreasing *temperature* is to *decrease* the volume. In this case, the effect of decreasing pressure predominates and the balloon volume increases significantly.

Practice Problem **A**TTEMPT What would be the volume of the balloon in Worked Example 11.6 if, instead of being released to rise in the atmosphere, it were submerged in a swimming pool to a depth where the pressure is 922.3 mmHg and the temperature is 26.35°C?

Practice Problem **B**UILD The volume of a bubble that starts at the bottom of a lake at 4.55°C increases by a factor of 10 as it rises to the surface where the temperature is 18.45°C and the air pressure is 0.965 atm. Assuming that the density of the lake water is 1.00 g/cm³, determine the depth of the lake. (*Hint:* You will need to use Equation 11.3.)

Practice Problem **C**ONCEPTUALIZE Which of the following diagrams could represent a gas sample in a balloon before and after an increase in temperature and an increase in external pressure?

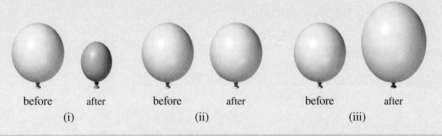

before	after	before	after	before	after
(i)		(ii)		(iii)	

Section 11.4 Review

The Gas Laws

11.4.1   Given $P_1 = 1.50$ atm, $V_1 = 37.3$ mL, and $P_2 = 1.18$ atm, calculate V_2. Assume that n and T are constant.
(a) 0.0211 mL   (d) 12.7 mL
(b) 0.0341 mL   (e) 47.4 mL
(c) 29.3 mL

11.4.2   Given $T_1 = 21.5$°C, $V_1 = 25.0$ mL, and $T_2 = 316$°C, calculate V_2. Assume that n and P are constant.
(a) 100 mL   (d) 3.40 mL
(b) 73.5 mL   (e) 26.5 mL
(c) 25.0 mL

11.4.3   At what temperature will a gas sample occupy 100.0 L if it originally occupies 76.1 L at 89.5°C? Assume constant P.
(a) 276°C   (d) 68.1°C
(b) 118°C   (e) 99.6°C
(c) 203°C

11.4.4   What volume of NH_3 will be produced when 180 mL of H_2 reacts with 60.0 mL of N_2 according to the following equation?

$$3H_2(g) + N_2(g) \longrightarrow 2NH_3(g)$$

Assume constant T and P for reactants and products.
(a) 120 mL   (d) 240 mL
(b) 60 mL   (e) 220 mL
(c) 180 mL

11.4.5   Which diagram could represent the result of increasing the temperature and decreasing the external pressure on a fixed amount of gas in a balloon?

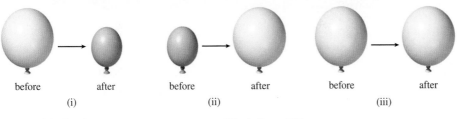

before	after	before	after	before	after
(i)		(ii)		(iii)	

(a) i only
(b) ii only
(c) i and ii

(d) i, ii, and iii
(e) iii only

11.4.6   Which diagram in question 11.4.5 could represent the result of decreasing both the temperature and the pressure?

(a) i only
(b) ii only
(c) i and ii

(d) i, ii, and iii
(e) iii only

11.5   THE IDEAL GAS EQUATION

Recall that the state of a sample of gas is described completely using the four variables T, P, V, and n. Each of the gas laws introduced in Section 11.4 relates one variable of a sample of gas to another while the *other* two variables are held constant. In experiments with gases, however, there are usually changes in more than just two of the variables. Therefore, it is useful for us to combine the equations representing the gas laws into a single equation that will enable us to account for changes in any or all of the four variables.

Summarizing the gas law equations from Section 11.4,

$$\text{Boyle's law: } V \propto \frac{1}{P}$$

$$\text{Charles's law: } V \propto T$$

$$\text{Avogadro's law: } V \propto n$$

we can combine these equations into the following general equation that describes the physical behavior of all gases, namely

$$V \propto \frac{nT}{P}$$

or

$$V = R\frac{nT}{P}$$

where R is the proportionality constant. This equation can be rearranged to give

$$PV = nRT \qquad\qquad \textbf{Equation 11.11}$$

Equation 11.11 is the most commonly used form of the **ideal gas equation,** which describes the relationship among the four variables P, V, n, and T. An **ideal gas** is a hypothetical sample of gas whose pressure-volume-temperature behavior is predicted accurately by the ideal gas equation. Although the behavior of *real* gases generally differs slightly from that predicted by Equation 11.11, in most of the cases we will encounter, the differences are usually small enough for us to use the ideal gas equation to make reasonably good predictions about the behavior of gases.

Student Annotation: We will discuss the conditions that result in deviation from ideal behavior in Section 11.6.

TABLE 11.4	Various Equivalent Expressions of the Gas Constant, R

Numerical value	Unit
0.08206	$L \cdot atm/K \cdot mol$
62.36	$L \cdot torr/K \cdot mol$
0.08314	$L \cdot bar/K \cdot mol$
8.314	$m^3 \cdot Pa/K \cdot mol$
8.314	$J/K \cdot mol$
1.987	$cal/K \cdot mol$

Note that the product of volume and pressure gives units of *energy* (i.e., joules and calories).

The proportionality constant, R, in Equation 11.11 is the gas constant [◄◄ Section 11.2]. The value and units of R depend on the units in which P and V are expressed. (The variables n and T are always expressed in mol and K, respectively.) Recall from Section 11.3 that pressure is commonly expressed in atmospheres, mmHg (torr), pascals, or bar. Volume is typically expressed in liters or milliliters, but can also be expressed in other units, such as m^3. Table 11.4 lists several different expressions of the gas constant, R.

Keep in mind that despite having different numerical values, all these expressions of R are *equal* to one another; just as 1 yard, 3 feet, and 36 inches are all equal to one another. They are simply expressed in different units.

One of the simplest uses of the ideal gas equation is the calculation of one of the variables when the other three are already known. For example, we can calculate the volume of 1 mole of an ideal gas at 0°C and 1 atm, conditions known as *standard temperature and pressure (STP)*. In this case, n, T, and P are given. R is a constant, leaving V as the only unknown. We can rearrange Equation 11.11 to solve for V,

$$V = \frac{nRT}{P}$$

enter the information that is given, and calculate V. Remember that in calculations using the ideal gas equation, temperature must *always* be expressed in kelvins.

$$V = \frac{(1\ mol)(0.08206\ L \cdot atm/K \cdot mol)(273.15\ K)}{1\ atm} = 22.41\ L$$

Thus, the volume occupied by 1 mole of an ideal gas at STP is 22.41 L, a volume slightly less than 6 gal.

Worked Example 11.7 shows how to calculate the molar volume of a gas at a temperature other than 0°C.

Student Annotation: When you use R in a calculation, use the version that facilitates proper cancellation of units.

Student Annotation: In thermochemistry, we often used 25°C as the "standard" temperature—although temperature is *not* actually part of the definition of the *standard state* [Section 10.6]. The standard temperature for gases is defined specifically as 0°C.

Student Annotation: In this problem, because they are specified rather than measured, 0°C, 1 mole, and 1 atm are *exact* numbers and do not affect the number of significant figures in the result [Section 1.3].

Student Hot Spot

Student data indicate you may struggle with the ideal gas law. Access the SmartBook to view additional Learning Resources on this topic.

Worked Example 11.7

Calculate the volume of a mole of ideal gas at room temperature (25°C) and 1 atm.

Strategy Convert the temperature in °C to temperature in kelvins, and use the ideal gas equation to solve for the unknown volume.

Setup The data given are $n = 1$ mol, $T = 298.15$ K, and $P = 1.00$ atm. Because the pressure is expressed in atmospheres, we use $R = 0.08206\ L \cdot atm/K \cdot mol$ to solve for volume in liters.

Solution

$$V = \frac{(1\ mol)(0.08206\ L \cdot atm/K \cdot mol)(298.15\ K)}{1\ atm} = 24.5\ L$$

Think About It

With the pressure held constant, we should expect the volume to increase with increased temperature. Room temperature is higher than the standard temperature for gases (0°C), so the molar volume at room temperature (25°C) *should* be higher than the molar volume at 0°C—and it is.

Practice Problem A **TTEMPT** What is the volume of 5.12 moles of an ideal gas at 32°C and 1.00 atm?

Practice Problem B **UILD** At what temperature (in °C) would 1 mole of ideal gas occupy 50.0 L ($P = 1.00$ atm)?

Practice Problem C **ONCEPTUALIZE** The diagram shown on the right represents a sample of an ideal gas at STP in a container whose volume is not fixed. Which of the diagrams [(i)–(iv)] best represents the sample after the absolute temperature has been doubled and the external pressure has been increased by a factor of 3?

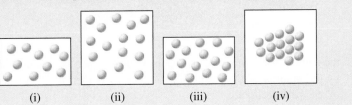

(i)   (ii)   (iii)   (iv)

Applications of the Ideal Gas Equation

Using some simple algebraic manipulation, we can solve for variables other than those that appear explicitly in the ideal gas equation. For example, if we know the molar mass of a gas (g/mol), we can determine its density at a given temperature and pressure. Recall from Section 11.1 that the density of a gas is generally expressed in units of g/L. We can rearrange the ideal gas equation to solve for mol/L.

$$\frac{n}{V} = \frac{P}{RT}$$

If we then multiply both sides by the molar mass, $\mathcal{M}$, we get

$$\mathcal{M} \times \frac{n}{V} = \frac{P}{RT} \times \mathcal{M}$$

where $\mathcal{M} \times n/V$ gives g/L or density, d. Therefore,

$$d = \frac{P\mathcal{M}}{RT} \qquad \textbf{Equation 11.12}$$

Conversely, if we know the density of a gas, we can determine its molar mass.

$$\mathcal{M} = \frac{dRT}{P} \qquad \textbf{Equation 11.13}$$

In a typical experiment, in which the molar mass of a gas is determined, a flask of known volume is evacuated and weighed [Figure 11.18(a)]. It is then filled (to a known pressure) with the gas of unknown molar mass and reweighed [Figure 11.18(b)]. The difference in mass is the mass of the gas sample. Dividing by the known volume of the flask gives the density of the gas, and the molar mass can then be determined using Equation 11.13.

Similarly, the molar mass of a volatile liquid can be determined by placing a small volume of it in the bottom of a flask, the mass and volume of which are known. The flask is then immersed in a hot-water bath, causing the volatile liquid to completely evaporate and its vapor to fill the flask. Because the flask is open, some of the excess vapor escapes. When no more vapor escapes, the flask is capped and removed from the water bath. The flask is then weighed to determine the mass of the vapor. (At this point, some or all of the vapor has condensed but the mass remains

Student Annotation: $n \times \mathcal{M} = m$, where m is mass in grams.

Student Annotation: Another way to arrive at Equation 11.12 is to substitute $m/\mathcal{M}$ for n in the ideal gas equation and rearrange to solve for m/V (density):

$$PV = \frac{m}{\mathcal{M}}RT \quad \text{and} \quad \frac{m}{V} = d = \frac{P\mathcal{M}}{RT}$$

Student Annotation: Because the flask is open to the atmosphere while the volatile liquid vaporizes, we can use atmospheric pressure as P. Also, because the flask is capped at the water bath temperature, we can use the water bath temperature as T.

Figure 11.18 (a) Evacuated flask. (b) Flask filled with gas. The mass of the gas is the difference between the two masses. The density of the gas is determined by dividing mass by volume.

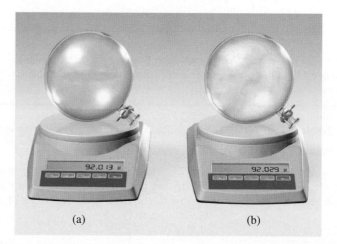

(a)   (b)

the same.) The density of the vapor is determined by dividing the mass of the vapor by the volume of the flask. Equation 11.13 is then used to calculate the molar mass of the volatile liquid.

Worked Examples 11.8 and 11.9 illustrate the use of Equations 11.12 and 11.13.

Worked Example 11.8

Carbon dioxide is effective in fire extinguishers partly because its density is greater than that of air, so CO_2 can smother the flames by depriving them of oxygen. (Air has a density of approximately 1.2 g/L at room temperature and 1 atm.) Calculate the density of CO_2 at room temperature (25°C) and 1.0 atm.

Strategy Use Equation 11.12 to solve for density. Because the pressure is expressed in atm, we should use $R = 0.08206$ L · atm/K · mol. Remember to express temperature in kelvins.

Setup The molar mass of CO_2 is 44.01 g/mol.

Solution

$$d = \frac{P\mathcal{M}}{RT} = \frac{(1 \text{ atm})\left(\dfrac{44.01 \text{ g}}{\text{mol}}\right)}{\left(\dfrac{0.08206 \text{ L} \cdot \text{atm}}{\text{K} \cdot \text{mol}}\right)(298.15 \text{ K})} = 1.8 \text{ g/L}$$

Think About It

The calculated density of CO_2 is greater than that of air under the same conditions (as expected). Although it may seem tedious, it is a good idea to write units for each and every entry in a problem such as this. Unit cancellation is very useful for detecting errors in your reasoning or your solution setup.

Practice Problem (A)TTEMPT Calculate the density of helium in a helium balloon at 25.0°C. (Assume that the pressure inside the balloon is 1.10 atm.)

Practice Problem (B)UILD Calculate the density of air at 0°C and 1 atm. (Assume that air is 80 percent N_2 and 20 percent O_2.)

Practice Problem (C)ONCEPTUALIZE Two samples of gas are shown at the same temperature and pressure. Which sample has the greater density? Which exerts the greater pressure?

(i)   (ii)

Worked Example 11.9

A company has just patented a new synthetic alcohol for alcoholic beverages. The new product is said to have all the pleasant properties associated with ethanol but none of the undesirable effects such as hangover, impairment of motor skills, and risk of addiction. The chemical formula is proprietary. You analyze a sample of the new product by placing a small volume of it in a round-bottomed flask with a volume of 511.0 mL and an evacuated mass of 131.918 g. You submerge the flask in a water bath at 100.0°C and allow the volatile liquid to vaporize. You then cap the flask and remove it from the water bath. You weigh it and determine the mass of the vapor in the flask to be 0.768 g. What is the molar mass of the volatile liquid, and what does it mean with regard to the new product? (Assume the pressure in the laboratory is 1 atm.)

Strategy Use the measured mass of the vapor and the given volume of the flask to determine the density of the vapor at 1 atm and 100.0°C, and then use Equation 11.13 to determine molar mass.

Setup $P = 1$ atm, $V = 0.5110$ L, $R = 0.08206$ L · atm/K · mol, and $T = 373.15$ K.

Solution

$$d = \frac{0.768 \text{ g}}{0.5110 \text{ L}} = 1.5029 \text{ g/L}$$

$$\mathcal{M} = \frac{\left(\dfrac{1.5029 \text{ g}}{\cancel{L}}\right)\left(\dfrac{0.08206 \; \cancel{L} \cdot \cancel{\text{atm}}}{\cancel{K} \cdot \text{mol}}\right)(373.15 \; \cancel{K})}{1 \; \cancel{\text{atm}}} = 46.02 \text{ g/mol}$$

The result is a molar mass suspiciously close to that of ethanol!

Think About It

Because more than one compound can have a particular molar mass, this method is not definitive for identification. However, in this circumstance, further testing of the proprietary formula certainly would be warranted.

Practice Problem Ⓐ**TTEMPT** Determine the molar mass of a gas with a density of 1.905 g/L at 80.0°C and 1.00 atm.

Practice Problem Ⓑ**UILD** Determine the molar mass of a gas with a density of 5.14 g/L at 73.0°C and 1.00 atm.

Practice Problem Ⓒ**ONCEPTUALIZE** These models represent two compounds that contain different amounts of the same two elements. Both compounds are liquids at room temperature. If the compound represented on the left is analyzed by the method described in Worked Example 11.9 and adds 0.412 g to the mass of the evacuated flask, what mass will be added to the flask when the compound represented on the right is analyzed under the same experimental conditions?

Section 11.5 Review

The Ideal Gas Equation

11.5.1 Calculate the volume occupied by 8.75 moles of an ideal gas at STP.
  (a) 196 L
  (b) 268 L
  (c) 0.718 L
  (d) 18.0 L
  (e) 2.56 L

11.5.2 Calculate the pressure exerted by 10.2 moles of an ideal gas in a 7.5-L vessel at 150°C.
  (a) 17 atm
  (b) 31 atm
  (c) 0.72 atm
  (d) 1.3 atm
  (e) 47 atm

11.5.3 Determine the density of a gas with $\mathcal{M} = 146.07$ g/mol at 1.00 atm and 100.0°C.
  (a) 6.85×10^{-3} g/L
  (b) 4.77 g/L
  (c) 146 g/L
  (d) 30.6 g/L
  (e) 17.8 g/L

11.5.4 Determine the molar mass of a gas with $d = 1.963$ g/L at 1.00 atm and 100.0°C.
  (a) 0.0166 g/mol
  (b) 60.1 g/mol
  (c) 16.1 g/mol
  (d) 6.09×10^3 g/mol
  (e) 1.63×10^3 g/mol

Figure 11.19 Molar volumes of some common gases at STP.

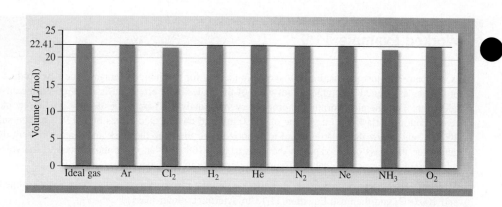

11.6   REAL GASES

The gas laws and the kinetic molecular theory assume that molecules in the gas phase occupy negligible volume (assumption 1) and that they do not exert any force on one another, either attractive or repulsive (assumption 3). Gases that behave as though these assumptions were strictly true are said to exhibit *ideal behavior*. Many gases do exhibit ideal or nearly ideal behavior under ordinary conditions. Figure 11.19 shows the molar volumes of some common gases at STP. All are remarkably close to the ideal value of 22.41 L. Although we generally assume that real gases behave ideally, there are conditions—namely, high pressure and low temperature—under which the behavior of a real gas deviates from ideal.

Factors That Cause Deviation from Ideal Behavior

At high pressures, gas molecules are relatively close together. We can assume that gas molecules occupy no volume only when the distances between molecules are large. When the distances between molecules are reduced, the volume occupied by each individual molecule becomes more significant.

   At low temperatures, gas molecules are moving more slowly. We can assume that there are no intermolecular forces between gas molecules, either attractive or repulsive, when the gas molecules are moving very fast and the magnitude of their kinetic energies is much larger than the magnitude of any intermolecular forces [◀◀ Section 7.3]. When molecules move more slowly, they have lower kinetic energies and the magnitude of the forces between them becomes more significant.

The van der Waals Equation

Because there are conditions under which use of the ideal gas equation would result in large errors (i.e., high pressure and/or low temperature), we must use a slightly different approach when gases do not behave ideally. Analyses of real gases that took into account nonzero molecular volumes and intermolecular forces were first carried out by J. D. van der Waals[8] in 1873. Van der Waals's treatment provides us with an interpretation of the behavior of real gases at the molecular level.

   Consider the approach of a particular molecule toward the wall of its container (Figure 11.20). The intermolecular attractions exerted by neighboring molecules prevent the molecule from hitting the wall as hard as it otherwise would. This results in

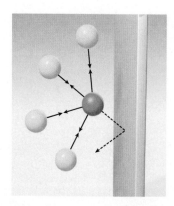

Figure 11.20 The effect of intermolecular attractions on the pressure exerted by a gas.

[8]Johannes Diderik van der Waals (1837–1923), a Dutch physicist, received the Nobel Prize in Physics in 1910 for his work on the properties of gases and liquids.

the pressure exerted by a real gas being lower than that predicted by the ideal gas equation. Van der Waals suggested that the pressure exerted by an ideal gas, P_{ideal}, is related to the experimentally measured pressure, P_{real}, by the equation

$$P_{ideal} = P_{real} + \frac{an^2}{V^2}$$

where a is a constant and n and V are the number of moles and volume of the gas, respectively. The correction term for pressure (an^2/V^2) can be understood as follows. The intermolecular interaction that gives rise to nonideal behavior depends on how frequently any two molecules encounter each other. The number of such encounters increases with the square of the number of molecules per unit volume (n/V), and a is a proportionality constant. The quantity P_{ideal} is the pressure we would measure if there were no intermolecular attractions.

The other correction concerns the volume occupied by the gas molecules. In the ideal gas equation, V represents the volume of the container. However, each molecule actually occupies a very small but nonzero volume. We can correct for the volume occupied by the gas molecules by subtracting a term, nb, from the volume of the container:

$$V_{real} = V_{ideal} - nb$$

where n and b are the number of moles and the proportionality constant, respectively.

Incorporating both corrections into the ideal gas equation gives us the ***van der Waals equation,*** with which we can analyze gases under conditions where ideal behavior is not expected.

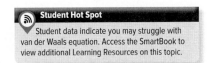

Student Hot Spot

Student data indicate you may struggle with van der Waals equation. Access the SmartBook to view additional Learning Resources on this topic.

experimentally measured pressure   container volume

$$\left(P + \frac{an^2}{V^2}\right)(V - nb) = nRT$$ **Equation 11.14**

corrected   corrected
pressure term   volume term

The van der Waals constants a and b for a number of gases are listed in Table 11.5. The magnitude of a indicates how strongly molecules of a particular type of gas attract one another. The magnitude of b is related to molecular (or atomic) size, although the relationship is not a simple one.

TABLE 11.5	Van der Waals Constants of Some Common Gases				
Gas	$a\left(\dfrac{atm \cdot L^2}{mol^2}\right)$	$b\left(\dfrac{L}{mol}\right)$	Gas	$a\left(\dfrac{atm \cdot L^2}{mol^2}\right)$	$b\left(\dfrac{L}{mol}\right)$
He	0.034	0.0237	O_2	1.36	0.0318
Ne	0.211	0.0171	Cl_2	6.49	0.0562
Ar	1.34	0.0322	CO_2	3.59	0.0427
Kr	2.32	0.0398	CH_4	2.25	0.0428
Xe	4.19	0.0510	CCl_4	20.4	0.138
H_2	0.244	0.0266	NH_3	4.17	0.0371
N_2	1.39	0.0391	H_2O	5.46	0.0305

Worked Example 11.10 shows how to use the van der Waals equation.

Worked Example 11.10

A sample of 3.50 moles of NH_3 gas occupies 5.20 L at 47°C. Calculate the pressure of the gas (in atm) using (a) the ideal gas equation and (b) the van der Waals equation.

Strategy (a) Use the ideal gas equation, $PV = nRT$. (b) Use Equation 11.14 and a and b values for NH_3 from Table 11.5.

Setup $T = 320.15$ K, $a = 4.17$ atm · L/mol², and $b = 0.0371$ L/mol.

Solution

(a) $P = \dfrac{nRT}{V} = \dfrac{(3.50 \text{ mol})\left(\dfrac{0.08206 \text{ L} \cdot \text{atm}}{\text{K} \cdot \text{mol}}\right)(320.15 \text{ K})}{5.20 \text{ L}} = 17.7$ atm

(b) Evaluating the correction terms in the van der Waals equation, we get

$$\frac{an^2}{V^2} = \frac{\left(\dfrac{4.17 \text{ atm} \cdot \text{L}^2}{\text{mol}^2}\right)(3.50 \text{ mol})^2}{(5.20 \text{ L})^2} = 1.89 \text{ atm} \qquad nb = (3.50 \text{ mol})\left(\frac{0.0371 \text{ L}}{\text{mol}}\right) = 0.130 \text{ L}$$

Finally, substituting these results into Equation 11.14, we have

$$(P + 1.89 \text{ atm})(5.20 \text{ L} - 0.130 \text{ L}) = (3.50 \text{ mol})\left(\frac{0.08206 \text{ L} \cdot \text{atm}}{\text{K} \cdot \text{mol}}\right)(320.15 \text{ K})$$

$$P = 16.2 \text{ atm}$$

Think About It
As is often the case, the pressure exerted by the real gas sample is lower than predicted by the ideal gas equation.

Practice Problem Ⓐ**TTEMPT** Using data from Table 11.5, calculate the pressure exerted by 11.9 moles of neon gas in a volume of 5.75 L at 25°C using (a) the ideal gas equation and (b) the van der Waals equation (Equation 11.14). Compare your results.

Practice Problem Ⓑ**UILD** Calculate the pressure exerted by 0.35 mole of oxygen gas in a volume of 6.50 L at 32°C using (a) the ideal gas equation and (b) the van der Waals equation.

Practice Problem Ⓒ**ONCEPTUALIZE** What properties of real gases prevent them from exhibiting ideal behavior? Explain why gases exhibit a greater degree of ideal behavior at very high temperatures and/or at very low pressures.

One way to measure a gas's deviation from ideal behavior is to determine its compressibility factor, Z, where $Z = PV/RT$. For one mole of an ideal gas, Z is equal to 1 at all pressures and temperatures. For real gases, the factors that contribute to nonideal behavior cause the value of Z to deviate from 1. Intermolecular forces help to account for the plots in Figure 11.21(a). Molecules exert both attraction and repulsion on one another. At large separations (low pressures), attraction predominates. In this region, the gas is more compressible than an ideal gas and the curve dips below the horizontal line ($Z < 1$). As molecules are brought closer together under pressure, repulsion begins to play an important role. If the pressure continues to increase, a point is reached when the gas becomes less compressible than an ideal gas because the molecules repel one another and the curve rises above the horizontal line ($Z > 1$). Figure 11.21(b) shows the plots of Z versus pressure at different temperatures. We see that as the temperature increases, the gas behaves more like an ideal gas (the curves become closer to the horizontal line). The increase in the molecules' kinetic energy makes molecular attraction less important.

van der Waals Constants

The constants used in the van der Waals equation allow us to correct for the nonideal behavior of real gases. For example, contrary to assumption 3 of the kinetic molecular theory of gases, there *are* attractive forces between the molecules in a sample of gas. Evidence of this includes the fact that when a gas is cooled and/or compressed, it condenses

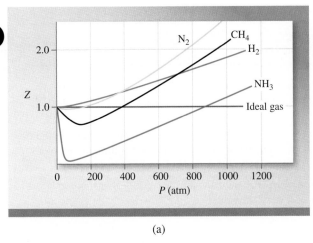

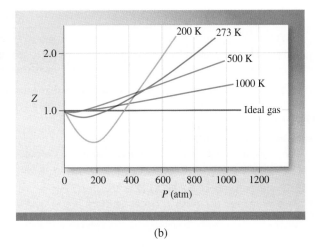

Figure 11.21 (a) Compressibility factor (Z) as a function of pressure for several gases at 0°C. (b) Compressibility factor as a function of pressure for N_2 gas at several different temperatures.

to a liquid. Even in the gas phase, the attractive forces between molecules can impact the observed behavior of a substance. As illustrated in Figure 11.20, a molecule that is attracted to other molecules in a sample of gas will not strike the wall of the container with as high velocity as it would if there were no such intermolecular attractions. The pressure term in the van der Waals equation, $P + a(n/V)^2$, is the experimentally determined pressure, P, plus a correction for the pressure that we do *not* observe because of attractive forces between the gas molecules. Note that the correction factor depends on the moles-per-unit-volume (n/V) squared. The value of the constant a is specific to a particular gas.

Also contrary to one of the assumptions of kinetic molecular theory (assumption 1), gas molecules do not actually have negligible volumes. When two gas molecules (assumed to be spherical) approach each other, the distance of closest approach is the sum of the radii ($2r$). The volume around each molecule into which the center of another molecule cannot penetrate is called the *excluded volume*. The effect of the excluded volume is to limit the fraction of the container volume actually available for molecules to move about in a gas sample. Thus, the volume term in the van der Waals equation, $V - nb$, is the container volume V minus the correction for the excluded volume, nb, where n is the number of moles of the gas and b is the excluded volume per mole of the gas.

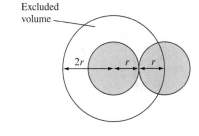

Worked Example 11.11 shows how to use the van der Waals equation to calculate excluded volume.

Worked Example 11.11

Consider a gas sample consisting of molecules with radius r. (a) Determine the excluded volume defined by two molecules and (b) calculate the excluded volume per mole (b) for the gas. Compare the excluded volume per mole with the volume actually occupied by a mole of the molecules.

Strategy (a) Use the formula for volume of a sphere to calculate the volume of a sphere of radius $2r$. This is the excluded volume defined by two molecules. (b) Multiply the excluded volume per molecule by Avogadro's number to determine excluded volume per mole.

Setup The equation for volume of a sphere is $\frac{4}{3}\pi r^3$; $N_A = 6.022 \times 10^{23}$.

Solution (a) Excluded volume defined by two molecules $= \frac{4}{3}\pi(2r)^3 = 8(\frac{4}{3}\pi r^3)$.

(b) Excluded volume per mole $= \dfrac{8(\frac{4}{3}\pi r^3)}{2 \text{ molecules}} \times \dfrac{N_A \text{ molecules}}{1 \text{ mole}} = \dfrac{4N_A(\frac{4}{3}\pi r^3)}{1 \text{ mole}}$.

The volume actually occupied by a mole of molecules with radius r is $N_A(\frac{4}{3}\pi r^3)$.

(Continued on next page)

Think About It
The excluded volume per mole is four times the volume actually occupied by a mole of molecules.

Practice Problem **(A)TTEMPT** Determine the excluded volume per mole and the total volume of the molecules in a mole for a gas consisting of molecules with radius 165 pm. [*Note:* To obtain the volume in liters, we must express the radius in decimeters (dm).]

Practice Problem **(B)UILD** Because the van der Waals constant *b* is the excluded volume per mole of a gas, we can use the value of *b* to estimate the radius of a molecule or atom. Consider a gas that consists of molecules, for which the van der Waals constant *b* is 0.0315 L/mol. Estimate the molecular radius in picometers (pm). Assume that the molecules are spherical.

Practice Problem **(C)ONCEPTUALIZE** Imagine a tiny sealed cubical vessel that contains just one spherical gas molecule. Would both of the van der Waals correction factors, $a(n/V)^2$ and nb, be necessary? Explain.

Section 11.6 Review

Real Gases

11.6.1 Using the van der Waals equation, calculate the pressure exerted by 1.5 moles of carbon dioxide in a 3.75-L vessel at 10°C.
(a) 9.3 atm   (d) 8.6 atm
(b) 8.9 atm   (e) −2.4 atm
(c) 10 atm

11.6.2 Using the van der Waals equation, calculate the temperature (in °C) of a 3.74-mol sample of methane gas in a 1.00-L vessel when the measured pressure is 1.30 atm.
(a) −166.4°C   (d) −269.6°C
(b) −183.6°C   (e) 89.7°C
(c) −268.9°C

11.7   GAS MIXTURES

So far our discussion of the physical properties of gases has focused on the behavior of *pure* gaseous substances, even though the gas laws were all developed based on observations of samples of air, which is a *mixture* of gases. In this section, we will consider gas mixtures and their physical behavior. We will restrict our discussion in this section to gases that do not react with one another—that is, ideal gases.

Dalton's Law of Partial Pressures

When two or more gaseous substances are placed in a container, each gas behaves as though it occupies the container alone. For example, if we place 1.00 mole of N_2 gas in a 5.00-L container at 0°C, it exerts a pressure of

$$P = \frac{(1 \text{ mol})(0.08206 \text{ L} \cdot \text{atm/K} \cdot \text{mol})(273.15 \text{ K})}{5.00 \text{ L}} = 4.48 \text{ atm}$$

If we then add a mole of another gas, such as O_2, the pressure exerted by N_2 does not change. It remains at 4.48 atm. The O_2 gas exerts its own pressure, also 4.48 atm.

Neither gas is affected by the presence of the other. In a mixture of gases, the pressure exerted by each gas is known as the ***partial pressure (P_i)*** of the gas. We use subscripts to denote partial pressures,

$$P_{N_2} = \frac{(1 \text{ mol})(0.08206 \text{ L} \cdot \text{atm/K} \cdot \text{mol})(273.15 \text{ K})}{5.00 \text{ L}} = 4.48 \text{ atm}$$

$$P_{O_2} = \frac{(1 \text{ mol})(0.08206 \text{ L} \cdot \text{atm/K} \cdot \text{mol})(273.15 \text{ K})}{5.00 \text{ L}} = 4.48 \text{ atm}$$

and we can solve the ideal gas equation for each component of any gas mixture.

$$P_i = \frac{n_i RT}{V}$$

Dalton's law of partial pressures states that the total pressure exerted by a gas mixture is the sum of the partial pressures exerted by each component of the mixture.

$$P_{\text{total}} = \Sigma P_i$$

Thus, the total pressure exerted by a mixture of 1.00 mol N_2 and 1.00 mol O_2 in a 5.00-L vessel at 0°C is

$$P_{\text{total}} = P_{N_2} + P_{O_2} = 4.48 \text{ atm} + 4.48 \text{ atm} = 8.96 \text{ atm}$$

Figure 11.22 illustrates Dalton's law of partial pressures.

As with the other gas laws, Dalton's law of partial pressures can be explained using the kinetic molecular theory. Gas molecules do not attract or repel one another (assumption 3), so the pressure exerted by one gas is unaffected by the presence of another gas. Consequently, the total pressure exerted by a mixture of gases is simply the sum of the partial pressures of the individual components in the mixture.

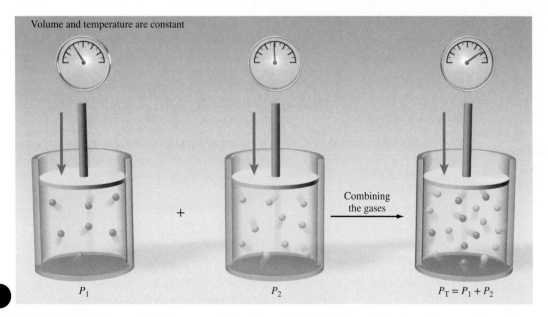

Figure 11.22 Each component of a gas mixture exerts a pressure independent of the other components. The total pressure is the sum of the individual components' partial pressures.

Worked Example 11.12 shows how to apply Dalton's law of partial pressures.

Worked Example 11.12

A 1.00-L vessel contains 0.215 mole of N_2 gas and 0.0118 mole of H_2 gas at 25.5°C. Determine the partial pressure of each component and the total pressure in the vessel.

Strategy Use the ideal gas equation to find the partial pressure of each component of the mixture, and sum the two partial pressures to find the total pressure.

Setup $T = 298.65$ K

Solution

$$P_{N_2} = \frac{(0.215 \text{ mol})\left(\dfrac{0.08206 \text{ L} \cdot \text{atm}}{\text{K} \cdot \text{mol}}\right)(298.65 \text{ K})}{1.00 \text{ L}} = 5.27 \text{ atm}$$

$$P_{H_2} = \frac{(0.0118 \text{ mol})\left(\dfrac{0.08206 \text{ L} \cdot \text{atm}}{\text{K} \cdot \text{mol}}\right)(298.65 \text{ K})}{1.00 \text{ L}} = 0.289 \text{ atm}$$

$$P_{total} = P_{N_2} + P_{H_2} = 5.27 \text{ atm} + 0.289 \text{ atm} = 5.56 \text{ atm}$$

Think About It
The total pressure in the vessel can also be determined by summing the number of moles of mixture components (0.215 + 0.0118 = 0.227 mol) and solving the ideal gas equation for P_{total}:

$$P_{total} = \frac{(0.227 \text{ mol})\left(\dfrac{0.08206 \text{ L} \cdot \text{atm}}{\text{K} \cdot \text{mol}}\right)(298.65 \text{ K})}{1.00 \text{ L}} = 5.56 \text{ atm}$$

Practice Problem (A)TTEMPT Determine the partial pressures and the total pressure in a 2.50-L vessel containing the following mixture of gases at 15.8°C: 0.0194 mol He, 0.0411 mol H_2, and 0.169 mol Ne.

Practice Problem (B)UILD Determine the number of moles of each gas present in a mixture of CH_4 and C_2H_6 in a 2.00-L vessel at 25.0°C and 1.50 atm, given that the partial pressure of CH_4 is 0.39 atm.

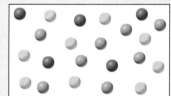

Practice Problem (C)ONCEPTUALIZE The diagram represents a mixture of three different gases. The partial pressure of the gas represented by red spheres is 1.25 atm. Determine the partial pressures of the other gases; and determine the total pressure.

Mole Fractions

The relative amounts of the components of a gas mixture can be specified using *mole fractions*. The **mole fraction** (χ_i) of a component of a mixture is the number of moles of the component divided by the total number of moles in the mixture.

Student Annotation: Mole fractions do not refer only to gas mixtures. They can be used to specify the concentrations of components of mixtures in any phase. Mole fractions are used extensively in Chapter 13.

Equation 11.15 $$\chi_i = \frac{n_i}{n_{total}}$$

There are three things to remember about mole fractions:

1. The mole fraction of a mixture component is always less than 1.
2. The sum of mole fractions for all components of a mixture is always 1.
3. Mole fraction is dimensionless.

In addition, n and P are proportional $\left[n = P \times \left(\frac{V}{RT}\right)\right]$ at a specified T and V, so we can determine mole fraction by dividing the partial pressure of a component by the total pressure.

$$\chi_i = \frac{P_i}{P_{total}}$$

Equation 11.16

Rearranging Equations 11.15 and 11.16 gives

$$\chi_i \times n_{total} = n_i$$

Equation 11.17

and

$$\chi_i \times P_{total} = P_i$$

Equation 11.18

Worked Example 11.13 lets you practice calculations involving mole fractions, partial pressures, and total pressure.

Worked Example 11.13

In 1999, the FDA approved the use of nitric oxide (NO) to treat and prevent lung disease, which occurs commonly in premature infants. The nitric oxide used in this therapy is supplied to hospitals in the form of a N_2/NO mixture. Calculate the mole fraction of NO in a 10.00-L gas cylinder at room temperature (25°C) that contains 6.022 mol N_2 and in which the total pressure is 14.75 atm.

Strategy Use the ideal gas equation to calculate the total number of moles in the cylinder. Subtract moles of N_2 from the total to determine moles of NO. Divide moles NO by total moles to get mole fraction (Equation 11.15).

Setup The temperature is 298.15 K.

Solution

$$\text{total moles} = \frac{PV}{RT} = \frac{(14.75\ \text{atm})(10.00\ \text{L})}{\left(\dfrac{0.08206\ \text{L} \cdot \text{atm}}{\text{K} \cdot \text{mol}}\right)(298.15\ \text{K})} = 6.029\ \text{mol}$$

$$\text{mol NO} = \text{total moles} - \text{mol N}_2 = 6.029 - 6.022 = 0.007\ \text{mol NO}$$

$$\chi_{NO} = \frac{n_{NO}}{n_{total}} = \frac{0.007\ \text{mol NO}}{6.029\ \text{mol}} = 0.001$$

Think About It
To check your work, determine χ_{N_2} by subtracting χ_{NO} from 1. Using each mole fraction and the total pressure, calculate the partial pressure of each component using Equation 11.16 and verify that they sum to the total pressure.

Practice Problem **A**TTEMPT Determine the mole fractions and partial pressures of CO_2, CH_4, and He in a sample of gas that contains 0.250 mole of CO_2, 1.29 moles of CH_4, and 3.51 moles of He, and in which the total pressure is 5.78 atm.

Practice Problem **B**UILD Determine the partial pressure and number of moles of each gas in a 15.75-L vessel at 30.0°C containing a mixture of xenon and neon gases only. The total pressure in the vessel is 6.50 atm, and the mole fraction of xenon is 0.761.

Practice Problem **C**ONCEPTUALIZE A mixture of gases can be represented with red, yellow, and green spheres. The diagram shows such a mixture, but the green spheres are missing. Determine the number of green spheres missing, the mole fraction of yellow, and the mole fraction of green, given that the mole fraction of red is 0.28.

Thinking Outside the Box

Decompression Injury

One of the first lessons taught in SCUBA certification is that divers must never hold their breath during ascent to the surface. Failure to heed this warning can result in serious injury or death. During underwater ascent, the air in a diver's lungs expands. If the air is not expelled, it causes overexpansion and rupture of alveoli—the tiny sacks that normally fill with air on inhalation. This condition, known as "burst lung," can cause air to escape into the chest cavity, where further expansion can collapse the ruptured lung. The potential for these catastrophic injuries is not limited to deep-sea divers, though, as burst lung can occur during a rapid ascent of as little as 3 meters, a depth common in public swimming pools. Furthermore, *spontaneous pneumothorax,* the medical term for this injury, can be caused even

(Continued on next page)

without the failure to exhale if there is an air-filled cyst in the diver's lung, or if there is a region of lung tissue blocked by phlegm due to a respiratory tract infection. In addition to the risk of spontaneous pneumothorax, burst lung can result in gas bubbles entering the bloodstream. Expansion of these bubbles during rapid underwater ascent can block circulation, a condition known as *gas embolism,* which can lead to heart attack or stroke.

 Because the cause of burst lung in divers is rapid *de*compression, treatment for the resulting conditions, especially gas embolism, usually includes *re*compression in a hyperbaric chamber. Hyperbaric chambers

are cylindrical enclosures built to withstand pressures significantly above atmospheric pressure. With the victim of a rapid decompression and, in some cases, medical personnel inside the enclosure, the door is sealed and high-pressure air is pumped in until the interior pressure is high enough to compress the gas bubbles in the victim's system. The victim usually breathes pure oxygen through a mask to help purge the undesirable gases during treatment. The pressure in a typical medical hyperbaric chamber can be increased to as many as six times atmospheric pressure.

Student Annotation: "Monoplace" hyperbaric chambers, which are large enough to accommodate only one person, typically are pressurized with pure oxygen.

Student Annotation: One unfortunate group of patients was undergoing treatment when the power to the chamber was shut off accidentally. All the patients died. At the time, their deaths were attributed to influenza, but they almost certainly died as the result of the unintended rapid decompression.

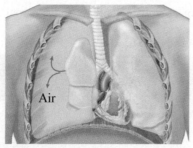

Pneumothorax

Monoplace hyperbaric chamber
© Kike Calvo/VW/The Image Works.

 In 1918, during the Spanish flu epidemic that claimed tens of millions of lives worldwide, physician Orville Cunningham noted that people living at lower elevations appeared to have a greater chance of surviving the flu than those living at higher elevations. Believing this to be the result of increased air pressure, he developed a hyperbaric chamber to treat flu victims. One of Cunningham's earliest and most notable successes was the recovery of a flu-stricken colleague who had been near death. Cunningham subsequently built a hyperbaric chamber large enough to accommodate dozens of patients and treated numerous flu victims, most with success.

 In the decades following the Spanish flu epidemic, hyperbaric therapy fell out of favor with the medical community and was largely discontinued.

Interest in it was revived when the U.S. military ramped up its underwater activities in the 1940s and hyperbaric chambers were constructed to treat military divers suffering from decompression sickness (DCS), also known as "the bends." Significant advancement in hyperbaric methods began in the 1970s when the Undersea Medical Society (renamed the Undersea and Hyperbaric Medical Society in 1976) became involved in the clinical use of hyperbaric chambers. Today, hyperbaric oxygen therapy (HBOT) is used to treat a wide variety of conditions, including carbon monoxide poisoning, anemia caused by critical blood loss, severe burns, and life-threatening bacterial infections. Once considered an "alternative" therapy and viewed with skepticism, HBOT is now covered by most insurance plans.

Section 11.7 Review

Gas Mixtures

11.7.1 What is the partial pressure of He in a 5.00-L vessel at 25°C that contains 0.0410 mole of He, 0.121 mole of Ne, and 0.0922 mole of Ar?
 (a) 1.24 atm
 (b) 0.248 atm
 (c) 0.117 atm
 (d) 2.87 atm
 (e) 0.201 atm

11.7.2 What is the mole fraction of CO_2 in a mixture of 0.756 mole of N_2, 0.189 mole of O_2, and 0.0132 mole of CO_2?
 (a) 0.789
 (b) 0.0138
 (c) 0.0140
 (d) 1.003
 (e) 0.798

11.7.3 What is the partial pressure of oxygen in a gas mixture that contains 4.10 moles of oxygen, 2.38 moles of nitrogen, and 0.917 mole of carbon dioxide and that has a total pressure of 2.89 atm?
 (a) 1.60 atm
 (b) 3.59 atm
 (c) 0.391 atm
 (d) 0.705 atm
 (e) 0.624 atm

11.7.4 In the diagram, each color represents a different gas molecule. Calculate the mole fraction of each gas.
- (a) $\chi_{red} = 0.5$, $\chi_{blue} = 0.4$, $\chi_{green} = 0.7$
- (b) $\chi_{red} = 0.05$, $\chi_{blue} = 0.07$, $\chi_{green} = 0.07$
- (c) $\chi_{red} = 0.3125$, $\chi_{blue} = 0.25$, $\chi_{green} = 0.4375$
- (d) $\chi_{red} = 0.4167$, $\chi_{blue} = 0.333$, $\chi_{green} = 0.5833$
- (e) $\chi_{red} = 0.333$, $\chi_{blue} = 0.333$, $\chi_{green} = 0.333$

11.7.5 Calculate the partial pressure of each gas in the diagram in question 11.7.4 if the total pressure is 8.21 atm.
- (a) $P_{red} = 0.5$ atm, $P_{blue} = 0.25$ atm, $P_{green} = 4.5$ atm
- (b) $P_{red} = 2.57$ atm, $P_{blue} = 2.05$ atm, $P_{green} = 3.59$ atm
- (c) $P_{red} = 3.13$ atm, $P_{blue} = 2.50$ atm, $P_{green} = 2.58$ atm
- (d) $P_{red} = 2.74$ atm, $P_{blue} = 2.74$ atm, $P_{green} = 2.74$ atm
- (e) $P_{red} = 3.125$ atm, $P_{blue} = 2.500$ atm, $P_{green} = 4.375$ atm

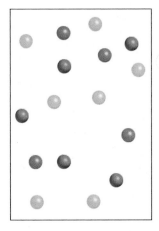

11.8   REACTIONS WITH GASEOUS REACTANTS AND PRODUCTS

In Chapter 8, we used balanced chemical equations to calculate amounts of reactants and/or products in chemical reactions—expressing those amounts in mass (usually grams). However, in the case of reactants and products that are gases, it is more practical to measure and express amounts in volume (liters or milliliters). This makes the ideal gas equation useful in the stoichiometric analysis of chemical reactions that involve gases.

Calculating the Required Volume of a Gaseous Reactant

According to Avogadro's law, the volume of a gas at a given temperature and pressure is proportional to the number of moles. Moreover, balanced chemical equations give the ratio of combination of gaseous reactants in both moles and volume (see Figure 11.15). Therefore, if we know the volume of one reactant in a gaseous reaction, we can determine the required amount of another reactant (at the same temperature and pressure). For example, consider the reaction of carbon monoxide and oxygen to yield carbon dioxide:

$$2CO(g) + O_2(g) \longrightarrow 2CO_2(g)$$

The ratio of combination of CO and O_2 is 2:1, whether we are talking about moles or units of volume. Thus, if we want to determine the stoichiometric amount [◄◄ Section 8.3] of O_2 required to combine with a particular volume of CO, we simply use the conversion factor provided by the balanced equation, which can be expressed as any of the following:

$$\frac{1 \text{ mol } O_2}{2 \text{ mol } CO} \quad \text{or} \quad \frac{1 \text{ L } O_2}{2 \text{ L } CO} \quad \text{or} \quad \frac{1 \text{ mL } O_2}{2 \text{ mL } CO}$$

Let's say we want to determine what volume of O_2 is required to react completely with 65.8 mL of CO at STP. We could use the ideal gas equation to convert the volume of CO to moles, use the stoichiometric conversion factor to convert to moles O_2, and then use the ideal gas equation again to convert moles O_2 to volume. But this method involves several unnecessary steps. We get the same result simply by using the conversion factor expressed in milliliters.

$$65.8 \text{ mL CO} \times \frac{1 \text{ mL } O_2}{2 \text{ mL CO}} = 32.9 \text{ mL } O_2$$

In cases where only one of the reactants is a gas, we *do* need to use the ideal gas equation in our analysis. Recall, for example, the reaction of sodium metal and chlorine gas used to illustrate the Born-Haber cycle [◄◄ Section 10.8].

$$2Na(s) + Cl_2(g) \longrightarrow 2NaCl(s)$$

Given moles (or more commonly the *mass*) of Na, and information regarding temperature and pressure, we can determine the volume of Cl_2 required to react completely.

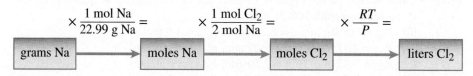

Worked Example 11.14 shows how to use the ideal gas equation in a stoichiometric analysis.

Worked Example 11.14

Sodium peroxide (Na_2O_2) is used to remove carbon dioxide from (and add oxygen to) the air supply in spacecrafts. It works by reacting with CO_2 in the air to produce sodium carbonate (Na_2CO_3) and O_2.

$$2Na_2O_2(g) + 2CO_2(g) \longrightarrow 2Na_2CO_3(s) + O_2(g)$$

What volume (in liters) of CO_2 (at STP) will react with a kilogram of Na_2O_2?

Strategy Convert the given mass of Na_2O_2 to moles, use the balanced equation to determine the stoichiometric amount of CO_2, and then use the ideal gas equation to convert moles of CO_2 to liters.

Setup The molar mass of Na_2O_2 is 77.98 g/mol (1 kg = 1000 g). (Treat the specified mass of Na_2O_2 as an exact number.)

Solution

$$1000 \text{ g } Na_2O_2 \times \frac{1 \text{ mol } Na_2O_2}{77.98 \text{ g } Na_2O_2} = 12.82 \text{ mol } Na_2O_2$$

$$12.82 \text{ mol } Na_2O_2 \times \frac{2 \text{ mol } CO_2}{2 \text{ mol } Na_2O_2} = 12.82 \text{ mol } CO_2$$

$$V_{CO_2} = \frac{(12.82 \text{ mol } CO_2)(0.08206 \text{ L} \cdot \text{atm/K} \cdot \text{mol})(273.15 \text{ K})}{1 \text{ atm}} = 287.4 \text{ L } CO_2$$

Think About It
The answer seems like an enormous volume of CO_2. If you check the cancellation of units carefully in ideal gas equation problems, however, with practice you will develop a sense of whether such a calculated volume is reasonable.

Practice Problem **A**TTEMPT What volume (in liters) of CO_2 can be consumed at STP by 525 g Na_2O_2?

Practice Problem **B**UILD What mass (in grams) of Na_2O_2 is necessary to consume 1.00 L CO_2 at STP?

Practice Problem **C**ONCEPTUALIZE The decomposition reactions of two solid compounds are represented here.

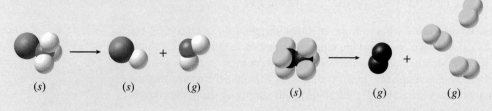

If an equal number of moles of each solid reactant were to decompose, how would the volume of products of the second decomposition compare to the volume of products of the first decomposition?

Determining the Amount of Reactant Consumed Using Change in Pressure

Although none of the empirical gas laws focuses on the relationship between n and P explicitly, we can rearrange the ideal gas equation to show that n is directly proportional to P at constant V and T.

$$n = P \times \left(\frac{V}{RT}\right) \quad \text{(at constant } V \text{ and } T) \qquad \textbf{Equation 11.19(a)}$$

Therefore, we can use the change in pressure in a reaction vessel to determine how many moles of a gaseous reactant are consumed in a chemical reaction.

$$\Delta n = \Delta P \times \left(\frac{V}{RT}\right) \quad \text{(at constant } V \text{ and } T) \qquad \textbf{Equation 11.19(b)}$$

where Δn is the number of moles of gas consumed and ΔP is the change in pressure in the reaction vessel. Worked Example 11.15 shows how to use Equation 11.19(b).

Student Annotation: This refers to a reaction in which there is only *one* gaseous reactant and in which none of the products is a gas, such as the reaction described in Worked Example 11.15. In reactions involving multiple gaseous species, Δn refers to the *net* change in number of moles of gas—and the analysis gets somewhat more complicated.

Worked Example 11.15

Another air-purification method for enclosed spaces involves the use of "scrubbers" containing aqueous lithium hydroxide, which react with carbon dioxide to produce lithium carbonate and water.

$$2LiOH(aq) + CO_2(g) \longrightarrow Li_2CO_3(s) + H_2O(l)$$

Consider the air supply in a submarine with a total volume of 2.5×10^5 L. The pressure is 0.9970 atm, and the temperature is 25°C. If the pressure in the submarine drops to 0.9891 atm as the result of carbon dioxide being consumed by an aqueous lithium hydroxide scrubber, how many moles of CO_2 are consumed?

Strategy Use Equation 11.19(b) to determine Δn, the number of moles of CO_2 consumed.

Setup $\Delta P = 0.9970$ atm $- 0.9891$ atm $= 7.9 \times 10^{-3}$ atm. According to the problem statement, $V = 2.5 \times 10^5$ L and $T = 298.15$ K. For problems in which P is expressed in atmospheres and V in liters, use $R = 0.08206$ L · atm/K · mol.

Solution

$$\Delta n_{CO_2} = 7.9 \times 10^{-3} \text{ atm} \times \frac{2.5 \times 10^5 \text{ L}}{(0.08206 \text{ L} \cdot \text{atm/K} \cdot \text{mol}) \times (298.15 \text{ K})} = 81 \text{ moles } CO_2 \text{ consumed}$$

Think About It
Careful cancellation of units is *essential*. Note that this amount of CO_2 corresponds to 162 moles or 3.9 kg of LiOH. (It's a good idea to verify this yourself.)

Practice Problem A **TTEMPT** Using all the same conditions as those described in Worked Example 11.15, calculate the number of moles of CO_2 consumed if the pressure drops by 0.010 atm.

Practice Problem B **UILD** By how much would the pressure in the submarine drop if 2.55 kg of LiOH were completely consumed by reaction with CO_2? (Assume the same starting P, V, and T as in Worked Example 11.15.)

Practice Problem C **ONCEPTUALIZE** The diagrams represent a reaction in which all of the species (reactants and products) are gases. If stoichiometric amounts of reactants are combined in a reaction vessel of fixed volume, how will the pressure after the reaction compare to the pressure before the reaction? (Assume that temperature is constant.)

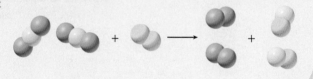

Using Partial Pressures to Solve Problems

The volume of gas produced by a chemical reaction can be measured using an apparatus like the one shown in Figure 11.23. Dalton's law of partial pressures is useful

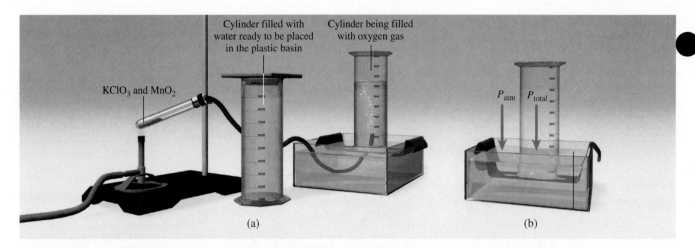

Figure 11.23 (a) Apparatus for measuring the amount of gas produced in a chemical reaction. (b) When the water levels inside and outside the collection vessel are the same, the pressure inside the vessel is equal to atmospheric pressure.

Video 11.4
Gas laws—collecting a gas over water.

in the analysis of these kinds of experimental results. For example, the decomposition of potassium chlorate ($KClO_3$), the reaction used to generate emergency oxygen supplies on airplanes, produces potassium chloride and oxygen.

$$2KClO_3(s) \longrightarrow 2KCl(s) + 3O_2(s)$$

The oxygen gas is collected over water, as shown in Figure 11.23(a). The volume of water displaced by the gas is equal to the volume of gas produced. (Prior to reading the volume of the gas, the level of the graduated cylinder must be adjusted such that the water levels inside and outside the cylinder are the *same*. This ensures that the *pressure* inside the graduated cylinder is the same as atmospheric pressure [Figure 11.23(b)].) However, because the measured volume contains both the oxygen produced by the reaction *and* water vapor, the pressure exerted inside the graduated cylinder is the sum of the two partial pressures.

$$P_{total} = P_{O_2} + P_{H_2O}$$

Video 11.5
Figure 11.24—Molar volume of a gas.

By subtracting the partial pressure of water from the total pressure, which is equal to atmospheric pressure, we can determine the partial pressure of oxygen—and thereby determine how many moles are produced by the reaction. We get the partial pressure of water, which depends on temperature, from a table of values. Table 11.6 lists the partial pressure (also known as the *vapor pressure*) of water at various temperatures. Figure 11.24 illustrates an experimental technique that involves collection of a gas over water.

TABLE 11.6	Vapor Pressure of Water (P_{H_2O}) as a Function of Temperature				
T(°C)	*P*(torr)	*T*(°C)	*P*(torr)	*T*(°C)	*P*(torr)
0	4.6	35	42.2	70	233.7
5	6.5	40	55.3	75	289.1
10	9.2	45	71.9	80	355.1
15	12.8	50	92.5	85	433.6
20	17.5	55	118.0	90	525.8
25	23.8	60	149.4	95	633.9
30	31.8	65	187.5	100	760.0

Worked Example 11.16 shows how to use Dalton's law of partial pressures to determine the amount of gas produced in a chemical reaction and collected over water.

Worked Example 11.16

Calcium metal reacts with water to produce hydrogen gas [◄◄ Section 8.5].

$$Ca(s) + 2H_2O(l) \longrightarrow Ca(OH)_2(aq) + H_2(g)$$

Determine the mass of H_2 produced at 25°C and 0.967 atm when 525 mL of the gas is collected over water as shown in Figure 11.23.

Strategy Use Dalton's law of partial pressures to determine the partial pressure of H_2, use the ideal gas equation to determine moles of H_2, and then use the molar mass of H_2 to convert to mass. (Pay careful attention to units. Atmospheric pressure is given in atmospheres, whereas the vapor pressure of water is tabulated in torr.)

Setup $V = 0.525$ L and $T = 298.15$ K. The partial pressure of water at 25°C is 23.8 torr (see Table 11.6) or (23.8 torr 1 atm/760 torr) = 0.0313 atm. The molar mass of H_2 is 2.016 g/mol.

Solution

$$P_{H_2} = P_{total} - P_{H_2O} = 0.967 \text{ atm} - 0.0313 \text{ atm} = 0.936 \text{ atm}$$

$$\text{moles of } H_2 = \frac{(0.9357 \text{ atm})(0.525 \text{ L})}{\left(\dfrac{0.08206 \text{ L} \cdot \text{atm}}{\text{K} \cdot \text{mol}}\right)(298.15 \text{ K})} = 2.01 \times 10^{-2} \text{ mol}$$

$$\text{moles of } H_2 = (2.008 \times 10^{-2} \text{ mol})(2.016 \text{ g/mol}) = 0.0405 \text{ g } H_2$$

Think About It
Check unit cancellation carefully, and remember that the densities of gases are relatively low. The mass of approximately half a liter of hydrogen at or near room temperature and 1 atm should be a very small number.

Practice Problem Ⓐ TTEMPT Calculate the mass of O_2 produced by the decomposition of $KClO_3$ when 821 mL of O_2 is collected over water at 30.0°C and 1.015 atm.

Practice Problem Ⓑ UILD Determine the volume of gas collected over water when 0.501 g O_2 is produced by the decomposition of $KClO_3$ at 35.0°C and 1.08 atm.

Practice Problem Ⓒ ONCEPTUALIZE The diagram on the top represents the result of an experiment in which the oxygen gas produced by a chemical reaction is collected over water at typical room temperature. Which of the diagrams [(i)–(iv)] best represents the result of the same experiment on a day when the temperature in the laboratory is significantly warmer?

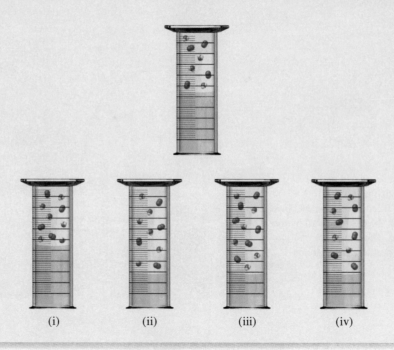

| (i) | (ii) | (iii) | (iv) |

Figure 11.24

Molar Volume of a Gas

The reaction between zinc and hydrochloric acid produces hydrogen gas. The net ionic equation for the reaction is

$$Zn(s) + 2H^+(aq) \longrightarrow Zn^{2+}(aq) + H_2(g)$$

The H_2 gas evolved in the reaction is collected in the inverted graduated cylinder. Using the balanced equation, we determine how much H_2 will be produced.

$$0.072 \text{ g Zn} \times \frac{1 \text{ mol Zn}}{65.41 \text{ g Zn}} = 0.0011 \text{ mol Zn}$$

$$0.0011 \text{ mol Zn} \times \frac{1 \text{ mol } H_2}{1 \text{ mol Zn}} = 0.0011 \text{ mol } H_2$$

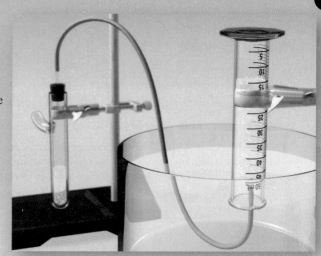

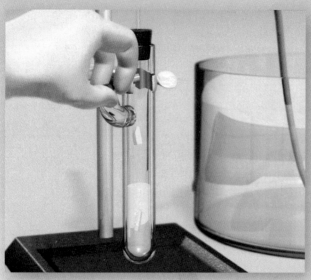

The reservoir is inverted and the zinc drops into the 1.0 *M* HCl.

START

A 0.072-g sample of zinc is placed in the reservoir. The vessel contains approximately 5 mL of 1.0 *M* HCl. A graduated cylinder is filled with water and inverted over the tubing to collect the gas produced by the reaction of zinc and acid. The temperature of the water is 25.0°C and the pressure in the room is 748.0 torr.

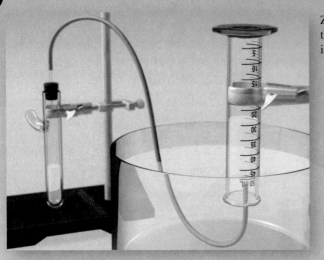

Zn is the limiting reactant. When all of the zinc has been consumed, the reaction is complete and no more gas is evolved.

We adjust the level of the graduated cylinder to make the water level the same inside and outside of the cylinder. This lets us know that the pressure inside the cylinder is the same as the pressure in the room. When the water levels are the same, we can read the volume of gas collected, which is 26.5 mL.

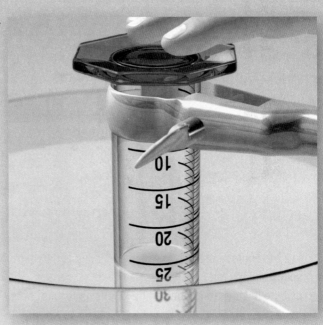

The pressure inside the cylinder is the sum of two partial pressures: that of the collected H_2, and that of water vapor. To determine the partial pressure of H_2, which is what we want, we must subtract the vapor pressure of water from the total pressure. Table 11.6 gives the vapor pressure of water at 25.0°C as 23.8 torr. Therefore, the pressure of H_2 is $748.0 - 23.8 = 724.2$ torr. Using Equation 11.10(b),

$$\frac{P_1V_1}{T_1} = \frac{P_2V_2}{T_2}$$

we calculate the volume the H_2 gas would occupy at STP.

$$\frac{(724.2 \text{ torr})(26.5 \text{ mL})}{298.15 \text{ K}} = \frac{(760 \text{ torr})V_2}{273.15 \text{ K}}$$

$$V_2 = \frac{(724.2 \text{ torr})(26.5 \text{ mL})(273.15 \text{ K})}{(760 \text{ torr})(298.15 \text{ K})} = 23.13 \text{ mL}$$

This is the volume of 0.0011 mol H_2. The molar volume is

$$\frac{23.13 \text{ mL}}{0.0011 \text{ mol}} = 2.1 \times 10^4 \text{ mL/mol or } 21 \text{ L/mol}$$

What's the point?

The gas collected in the graduated cylinder is a mixture of the gas produced by the reaction and water vapor. We determine the pressure of the gas produced by the reaction by subtracting the tabulated partial pressure of water from the total pressure. Knowing the volume, pressure, and temperature of a sample of gas, in this case H_2, we can determine what volume the same sample of gas would occupy at STP. This enables us to determine experimentally the molar volume of H_2 at STP, which turns out to be fairly close to the accepted value of 22.4 L. Several sources of error, including uncertainty in the volume of the collected gas, contribute to the result not being exactly 22.4 L.

Section 11.8 Review

Reactions with Gaseous Reactants and Products

11.8.1   Determine the volume of Cl_2 gas at STP that will react with 1.00 mole of Na solid to produce NaCl according to the equation

$$2Na(s) + Cl_2(g) \longrightarrow 2NaCl(s)$$

(a) 22.4 L   (d) 11.2 L
(b) 44.8 L   (e) 30.6 L
(c) 15.3 L

11.8.2   Determine the mass of NaN_3 required for an air bag to produce 100.0 L of N_2 gas at 85.0°C and 1.00 atm according to the equation

$$2NaN_3(s) \longrightarrow 2Na(s) + 3N_2(g)$$

(a) 332 g   (d) 664 g
(b) 148 g   (e) 442 g
(c) 221 g

11.8.3   What mass of acetylene (C_2H_2) is produced by the reaction of calcium carbide (CaC_2) and water,

$$CaC_2(s) + 2H_2O(l) \longrightarrow C_2H_2(g) + Ca(OH)_2(aq),$$

if 425 mL of the gas is collected over water at 30°C and a pressure of 0.996 atm?

(a) 0.016 g   (d) 0.424 g
(b) 16.3 g   (e) 0.443 g
(c) 0.019 g

Learning Outcomes

- List the characteristics of gases that distinguish them from solids and liquids.
- List the basic assumptions of the kinetic molecular theory of gases.
- Define effusion and diffusion.
- Give a definition for pressure and provide examples of some units for pressure.
- Demonstrate the ability to interconvert units of pressure.
- Describe the major characteristics of the various gas laws: Boyle's, Charles's, Avogadro's, and combined.
- Use the combined gas law to interconvert measurements of pressure, volume, or temperature for a given gas.
- Use the ideal gas equation to determine the pressure, volume, moles, or temperature of a gas given all of the other values.

- Identify the values for standard temperature and standard pressure of gases.
- Apply the ideal gas equation to determine characteristics of a gas including density and molecular weight.
- Identify factors that may cause gases to deviate from ideal gas behavior.
- Use the van der Waals equation to determine the pressure of a real gas.
- Define and be able to calculate mole fraction.
- Use Dalton's law of partial pressures to determine the mole fraction or partial pressure of gases in a mixture of gases.
- Use the ideal gas equation in stoichiometric calculations.

Chapter Summary

SECTION 11.1

- A gas assumes the volume and shape of its container and is compressible. Gases generally have low densities (expressed in g/L) and will mix in any proportions to give homogeneous solutions.

SECTION 11.2

- According to the **kinetic molecular theory,** gases are composed of particles with negligible volume that are separated by large distances; the particles are in constant, random motion, and collisions between the particles and between the particles and their container walls are perfectly elastic; there are no attractive or repulsive forces between the particles; and the average kinetic energy of particles in a sample is proportional to the absolute temperature of the sample.
- R is the **gas constant.**
- The **root-mean-square (rms) speed (u_{rms})** of gas molecules in a sample at a given temperature is inversely proportional to the molecular mass.
- According to **Graham's law,** the rates of **diffusion** (mixing of gases) and **effusion** (escape of a gas from a container into a vacuum) are inversely proportional to the square root of the molar mass of the gas.

SECTION 11.3

- Gases exert **pressure,** which is the force per unit area. The SI units of force and pressure are the **newton (N)** and the **pascal (Pa),** respectively. Other commonly used units of pressure are atmosphere (atm), mmHg, torr, and bar.
- Pressure can be measured using a **barometer** or a **manometer.**
- **Standard atmospheric pressure** (1 atm) is the pressure exerted by the atmosphere at sea level.

SECTION 11.4

- The physical state of a sample of gas can be described using four parameters: temperature (T), pressure (P), volume (V), and number of moles (n). Equations relating these parameters are called the **gas laws.**
- **Boyle's law** states that the volume of a sample of gas at constant temperature is inversely proportional to pressure.
- Experiments done by Charles and Gay-Lussac showed that the volume of a gas at constant pressure is directly proportional to temperature. Lord Kelvin used Charles's and Gay-Lussac's data to propose that **absolute zero** is the lowest theoretically attainable temperature. The **absolute temperature scale,** also known as the **Kelvin temperature scale,** is used for all calculations involving gases.

- **Charles's and Gay-Lussac's law,** commonly known as **Charles's law,** states that the volume of a sample of gas at constant pressure is directly proportional to its absolute temperature.
- **Avogadro's law** states that the volume of a sample of gas at constant temperature and pressure is directly proportional to the number of moles.
- The **combined gas law** combines the laws of Boyle, Charles, and Avogadro and relates pressure, volume, temperature, and number of moles without assuming that any of the parameters is constant.

SECTION 11.5

- The **ideal gas equation,** $PV = nRT$, makes it possible to predict the behavior of gases. An **ideal gas** is one that behaves in a way predicted by the ideal gas equation. The gas constant, R, may be expressed in a variety of units. The units used to express R depend on the units used to express P and V.
- **Standard temperature and pressure (STP)** is defined as 0°C and 1 atm.
- The ideal gas equation can be used to calculate the density of a gas and to interconvert between density and molar mass.

SECTION 11.6

- Deviation from ideal behavior is observed at high pressure and/or low temperature. The **van der Waals equation** makes corrections for the nonzero volume of gas molecules and the attractive forces between molecules.

SECTION 11.7

- Each component in a mixture of gases exerts a **partial pressure (P_i)** independent of the other mixture components. **Dalton's law of partial pressures** states that the total pressure exerted by a gas *mixture* is the sum of the partial pressures of the components.
- **Mole fraction (χ_i)** is the unitless quotient of the number of moles of a mixture component and the total number of moles in the mixture, n_i/n_{total}.

SECTION 11.8

- For a reaction occurring at constant temperature and pressure, and involving only gases, the coefficients in the balanced chemical equation apply to units of volume, as well as to numbers of molecules or moles.
- A balanced chemical equation and the ideal gas equation can be used to determine volumes of gaseous reactants and/or products in a reaction.

Key Words

Key Equations

11.1 $u_{rms} = \sqrt{\dfrac{3RT}{\mathcal{M}}}$	The root-mean-square speed of gas molecules in a sample is inversely proportional to the square root of the molar mass of the gas.
11.2 $\dfrac{u_{rms}(1)}{u_{rms}(2)} = \sqrt{\dfrac{\mathcal{M}_2}{\mathcal{M}_1}}$	Rates of effusion/diffusion of gases of different molar masses can be compared using the square root of the ratio of their molar masses, with the rate of each gas being inversely proportional to the square root of its molar mass.
11.3 $P = hdg$	The pressure exerted by a column of fluid is calculated as the product of the column height (in m), the density of the fluid (in kg/m^3), and the gravitational constant (9.80665 m/s^2).
11.4(a) $V = k_1 \dfrac{1}{P}$ (at constant temperature) **11.4(b)** $PV = k_1$ (at constant temperature)	Boyle's law. At constant temperature, (a) the volume of a gas is inversely proportional to pressure; and (b) the product of volume and pressure for a sample of gas is constant.
11.5 $P_1 V_1 = P_2 V_2$ (at constant temperature)	For a sample of gas at constant temperature, because the product of pressure and volume is constant, we can calculate the change in volume for a given change in pressure—or vice versa.
11.6(a) $V = k_2 T$ (at constant pressure) **11.6(b)** $\dfrac{V}{T} = k_2$ (at constant pressure)	Charles's law. At constant pressure, (a) the volume of a sample of gas is directly proportional to absolute temperature; and (b) the ratio of volume to absolute temperature is constant.
11.7 $\dfrac{V_1}{T_1} = \dfrac{V_2}{T_2}$ (at constant pressure)	For a sample of gas at constant pressure, because the ratio of volume to absolute temperature is constant, we can calculate the change in volume for a given change in temperature.
11.8(a) $V = k_3 n$ (at constant temperature and pressure) **11.8(b)** $\dfrac{V}{n} = k_3$	Avogadro's law. At constant temperature and pressure, (a) the volume of a sample of gas is directly proportional to the number of moles; and (b) the ratio of volume to number of moles is constant.
11.9 $\dfrac{V_1}{n_1} = \dfrac{V_2}{n_2}$	For a sample of gas at constant temperature and pressure, because the ratio of volume to number of moles is constant, we can calculate the change in volume for a given change in number of moles.

11.10(a) $\dfrac{P_1 V_1}{n_1 T_1} = \dfrac{P_2 V_2}{n_2 T_2}$

11.10(b) $\dfrac{P_1 V_1}{T_1} = \dfrac{P_2 V_2}{T_2}$

Combined gas law. (a) The quantity $\frac{PV}{nT}$ is constant; and (b) because the quantity $\frac{PV}{T}$ is constant for a given amount of gas, we can calculate the changes in pressure, volume, and/or temperature as the other parameters change.

11.11 $PV = nRT$

Ideal gas equation. The product of pressure and volume is directly proportional to the product of number of moles and absolute temperature. R is the gas constant. Units of R depend on the units used for the other parameters.

11.12 $d = \dfrac{p\mathcal{M}}{RT}$

By rearranging the ideal gas equation, we can calculate the density of a gas using its molar mass.

11.13 $\mathcal{M} = \dfrac{dRT}{P}$

By rearranging the ideal gas equation, we can calculate the molar mass of a gas using its density.

11.14 $\left(P + \dfrac{an^2}{V^2} \right)(V - nb) = nRT$

The ideal gas equation is modified for real gases by applying a correction to both the pressure term and the volume term. The constants a and b depend on the identity of the gas.

11.15 $\chi_i = \dfrac{n_i}{n_{\text{total}}}$

The mole fraction of a component in a mixture is the ratio of number of moles of the component to the total number of moles.

11.16 $\chi_i = \dfrac{P_i}{P_{\text{total}}}$

Mole fractions in a mixture of gases can also be calculated as the ratio of partial pressure of a component to total pressure.

11.17 $\chi_i \times n_{\text{total}} = n_i$

The number of moles of a component in a gaseous mixture is the product of the component's mole fraction and the total number of moles.

11.18 $\chi_i \times P_{\text{total}} = P_i$

Partial pressure of a component of a gaseous mixture is the product of the component's mole fraction and the total pressure.

11.19(a) $n = P \times \left(\dfrac{V}{RT} \right)$ (at constant V and T)

11.19(b) $\Delta n = \Delta P \times \left(\dfrac{V}{RT} \right)$ (at constant V and T)

The net number of gaseous moles consumed or produced in a reaction can be calculated using the measured change in pressure.

Key Skills

Mole Fractions

Most of the gases that we encounter are mixtures of two or more different gases. The concentrations of gases in a mixture are typically expressed using mole fractions, which are calculated using Equation 11.15.

$$\chi_i = \frac{n_i}{n_{total}}$$

Depending on the information given in a problem, calculating mole fractions may require you to determine molar masses and carry out mass-to-mole conversions [◄◄ Section 5.10].

For example, consider a mixture that consists of known masses of three different gases: 5.50 g He, 7.75 g N_2O, and 10.00 g SF_6. Molar masses of the components are

He: 4.003 = $\dfrac{4.003\ g}{mol}$ N_2O: 2(14.01) + (16.00) = $\dfrac{44.02\ g}{mol}$ SF_6: 32.07 + 6(19.00) = $\dfrac{146.1\ g}{mol}$

We convert each of the masses given in the problem to moles by dividing each by the corresponding molar mass.

$\dfrac{5.50\ g\ He}{4.003\ g/mol}$ ⟹ 1.374 mol He $\dfrac{7.75\ g\ N_2O}{44.02\ g/mol}$ ⟹ 0.1761 mol N_2O $\dfrac{10.00\ g\ SF_6}{146.1\ g/mol}$ ⟹ 0.06846 mol SF_6

We then determine the total number of moles in the mixture.

1.374 mol He + 0.1761 mol N_2O + 0.06846 mol SF_6 = ⟹ 1.619 moles

We divide the number of moles of each component by the total number of moles to get each component's mole fraction.

$\chi_{He} = \dfrac{1.374\ mol\ He}{1.619\ moles}$ ⟹ 0.849 $\chi_{N_2O} = \dfrac{0.1761\ mol\ N_2O}{1.619\ moles}$ ⟹ 0.109 $\chi_{SF_6} = \dfrac{0.06846\ mol\ SF_6}{1.619\ moles}$ = 0.0423

The resulting mole fractions have no units; and for any mixture, the sum of mole fractions of all components is 1. Rounding error may result in the overall sum of mole fractions not being exactly 1. In this case, to the appropriate number of significant figures [◄◄ Section 1.3], the sum is 1.00. (Note that we kept an extra digit throughout the calculations.)

Because at a given temperature, pressure is proportional to the number of moles, mole fractions can also be calculated using the partial pressures of the gaseous components using Equation 11.16.

$$\chi_i = \frac{P_i}{P_{total}}$$

Because gases that do not react with one another are all mutually miscible [◄◄ Section 11.1], gas mixtures are homogeneous— and can also be referred to as *solutions*. And although we first encounter mole fractions in the context of gases, they are also used extensively in the context of other solutions—including *aqueous* solutions [►► Section 13.5]. Determination of mole fraction is done the same way, regardless of the nature of the solution. When liquids are involved, it is sometimes necessary to convert from volume to mass using the liquid's *density* [◄◄ Section 1.2].

volume of liquid (mL)   ×   density of liquid (g/mL)   ⟹   mass of liquid (g)

Consider the following example: 5.75 g of sugar (sucrose, $C_{12}H_{22}O_{11}$) is dissolved in 100.0 mL of water at 25°C. We first determine the molar masses of sucrose and water.

$$H_2O: 2(1.008) + 16.00 = \boxed{\frac{18.02 \text{ g}}{\text{mol}}} \qquad C_{12}H_{22}O_{11}: 12(12.01) + 22(1.008) + 11(16.00) = \boxed{\frac{342.3 \text{ g}}{\text{mol}}}$$

Then we use the density of water to convert the volume given to a mass. The density of water at 25°C is 0.9970 g/mL.

$$\boxed{100.0 \text{ mL } H_2O} \times \boxed{\frac{0.9970 \text{ g}}{\text{mL}}} = \boxed{99.70 \text{ g } H_2O}$$

We convert the masses of both solution components to moles.

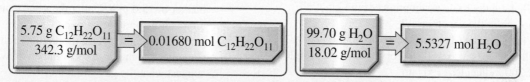

$$\boxed{\frac{5.75 \text{ g } C_{12}H_{22}O_{11}}{342.3 \text{ g/mol}}} = 0.01680 \text{ mol } C_{12}H_{22}O_{11} \qquad \boxed{\frac{99.70 \text{ g } H_2O}{18.02 \text{ g/mol}}} = 5.5327 \text{ mol } H_2O$$

We then sum the number of moles and divide moles of each individual component by the total.

$$\boxed{0.01680 \text{ mol } C_{12}H_{22}O_{11}} + \boxed{5.5327 \text{ mol } H_2O} = 5.5495 \text{ moles}$$

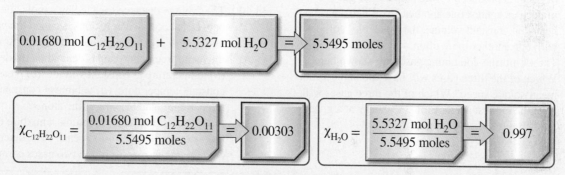

$$\chi_{C_{12}H_{22}O_{11}} = \boxed{\frac{0.01680 \text{ mol } C_{12}H_{22}O_{11}}{5.5495 \text{ moles}}} = 0.00303 \qquad \chi_{H_2O} = \boxed{\frac{5.5327 \text{ mol } H_2O}{5.5495 \text{ moles}}} = 0.997$$

To the appropriate number of significant figures, the mole fractions sum to 1.

Key Skills Problems

11.1
Determine the mole fraction of helium in a gaseous mixture consisting of 0.524 g He, 0.275 g Ar, and 2.05 g CH_4.
(a) 0.0069   (d) 0.493
(b) 0.0259   (e) 0.131
(c) 0.481

11.2
Determine the mole fraction of argon in a gaseous mixture in which the partial pressures of H_2, N_2, and Ar are 0.01887 atm, 0.3105 atm, and 1.027 atm, respectively.
(a) 0.01391   (d) 0.01887
(b) 0.2289   (e) 1.027
(c) 0.7572

11.3
Determine the mole fraction of *water* in a solution consisting of 5.00 g glucose ($C_6H_{12}O_6$) and 250.0 g water.
(a) 0.00200   (d) 1.00
(b) 0.998   (e) 0.907
(c) 0.0278

11.4
Determine the mole fraction of ethanol in a solution containing 15.50 mL ethanol (C_2H_5OH) and 110.0 mL water. (The density of ethanol is 0.789 g/mL; the density of water is 0.997 g/mL.)
(a) 0.0436   (d) 0.958
(b) 6.08   (e) 0.0418
(c) 0.265

Questions and Problems

SECTION 11.1: PROPERTIES OF GASES

Review Questions

11.1   Name five elements and five compounds that exist as gases at room temperature.

11.2   List the physical characteristics of gases.

SECTION 11.2: THE KINETIC MOLECULAR THEORY OF GASES

Review Questions

11.3   What are the basic assumptions of the kinetic molecular theory of gases?

11.4   What does the Maxwell speed distribution curve tell us? Does Maxwell's theory work for a sample of 200 molecules? Explain.

11.5   Which of the following statements is correct? (a) Heat is produced by the collision of gas molecules against one another. (b) When a gas is heated at constant volume, the molecules collide with one another more often.

11.6   Three fluorine-containing gases are shown here. Which of the three gases will have the highest root-mean-square speed? Which of the three gases will have the highest average kinetic energy at a given temperature?

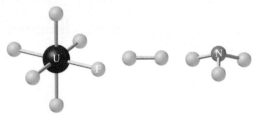

11.7   What is the difference between gas diffusion and effusion?

Computational Problems

11.8   Compare the root-mean-square speeds at 85°C of (a) O_2 and O_3, (b) F_2 and Cl_2, and (c) CH_4 and CCl_4.

11.9   The average temperature at the top of the stratosphere is −3.0°C. Calculate the root-mean-square speeds of N_2, O_2, and O_3 molecules in this region.

11.10   Nickel forms a gaseous compound of the formula $Ni(CO)_x$. Given the fact that under the same conditions of temperature and pressure, methane (CH_4) effuses 3.3 times faster than $Ni(CO)_x$, what is the value of x?

11.11   At a certain temperature, the speeds of six gaseous molecules in a container are 2.0, 2.2, 2.6, 2.7, 3.3, and 3.5 m/s. Calculate the root-mean-square speed and the average speed of the molecules. These two average values are close to each other, but the root-mean-square value is always the larger of the two. Why?

11.12   The ^{235}U isotope undergoes fission when bombarded with neutrons. However, its natural abundance is only 0.72 percent. To separate it from the more abundant ^{238}U isotope, uranium is first converted to UF_6, which is easily vaporized above room temperature. The mixture of the $^{235}UF_6$ and $^{238}UF_6$ gases is then subjected to many stages of effusion. Calculate how much faster $^{235}UF_6$ effuses than $^{238}UF_6$.

11.13   An unknown gas evolved from the fermentation of glucose is found to effuse through a porous barrier in 15.0 min. Under the same conditions of temperature and pressure, it takes an equal volume of N_2 12.0 min to effuse through the same barrier. Calculate the molar mass of the unknown gas, and suggest what the gas might be.

Conceptual Problems

11.14   The average distance traveled by a molecule between successive collisions is called the *mean free path*. For a given amount of a gas, how does the mean free path of a gas depend on (a) density, (b) temperature at constant volume, (c) pressure at constant temperature, (d) volume at constant temperature, and (e) size of the atoms?

11.15   Each pair of diagrams represents a mixture of gases before and after effusion. In each case, determine how the molar masses of the two gases compare.

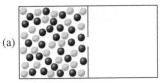

(a)

before   after

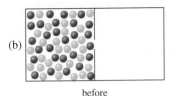

(b)

before   after

SECTION 11.3: GAS PRESSURE

Review Questions

11.16   Define *pressure* and give the common units for pressure.

11.17   Describe how a barometer and a manometer are used to measure gas pressure.

11.18   Explain why the height of mercury in a barometer is independent of the cross-sectional area of the tube.

11.19   Is the atmospheric pressure in a mine that is 500 m below sea level greater or less than 1 atm?

11.20   If the maximum distance that water may be brought up a well by a suction pump is 34 ft (10.3 m), how is it possible to obtain water and oil from hundreds of feet below the surface of Earth?

11.21   Why is it that if the barometer reading falls in one part of the world, it must rise somewhere else?

Computational Problems

11.22   The atmospheric pressure at the summit of Denali is 581 mmHg on a certain day. What is the pressure in atmospheres, in kilopascals, and in bars?

11.23   Convert 375 mmHg to atmospheres, bars, torr, and pascals.

11.24   Calculate the height of a column of isopropanol (C_3H_7OH) that would be supported by atmospheric pressure (1 atm). The density of isopropanol is 0.785 g/cm^3.

11.25   Calculate the height of a column of toluene (C_7H_8) that would be supported by atmospheric pressure. The density of toluene is 0.867 g/cm^3.

11.26   What pressure (in atm and in bars) is exerted by a column of methanol (CH_3OH) 183 m high? The density of methanol is 0.787 g/cm^3.

11.27   What pressure (in atm and in bars) is exerted by a column of ethylene glycol [$CH_2(OH)CH_2(OH)$] 53 m high? The density of ethylene glycol is 1.12 g/cm^3.

SECTION 11.4: THE GAS LAWS

Review Questions

11.28   State each of the following gas laws in words and also in the form of an equation: Boyle's law, Charles's law, Avogadro's law. In each case, indicate the conditions under which the law is applicable, and give the units for each quantity in the equation.

11.29   Explain why a helium weather balloon expands as it rises in the air. Assume that the temperature remains constant.

Computational Problems

11.30   At −11°C, a sample of carbon monoxide gas exerts a pressure of 0.37 atm. What is the pressure when the volume of the gas is reduced to one-third of the original value at the same temperature?

11.31   A methane sample occupying a volume of 2.15 L at a pressure of 5.25 atm is allowed to expand at constant temperature until its pressure reaches 1.85 atm. What is its final volume?

11.32   A sample of air occupies 3.8 L when the pressure is 1.2 atm. (a) What volume does it occupy at 6.6 atm? (b) What pressure is required in order to compress it to 0.075 L? (The temperature is kept constant.)

11.33   The volume of a gas is 7.15 L, measured at 1.00 atm. What is the pressure of the gas in mmHg if the volume is changed to 9.25 L? (The temperature remains constant.)

11.34   Under constant-pressure conditions, a sample of hydrogen gas initially at 88°C and 9.6 L is cooled until its final volume is 3.4 L. What is its final temperature?

11.35   A 28.4-L volume of methane gas is heated from 35°C to 72°C at constant pressure. What is the final volume of the gas?

Conceptual Problems

11.36   Molecular chlorine and molecular fluorine combine to form a gaseous product. Under the same conditions of temperature and pressure it is found that one volume of Cl_2 reacts with three volumes of F_2 to yield two volumes of the product. What is the formula of the product?

11.37   Ammonia burns in oxygen gas to form nitric oxide (NO) and water vapor. How many volumes of NO are obtained from one volume of ammonia at the same temperature and pressure?

11.38   Consider the following gaseous sample in a cylinder fitted with a movable piston. Initially there are n moles of the gas at temperature T, pressure P, and volume V.

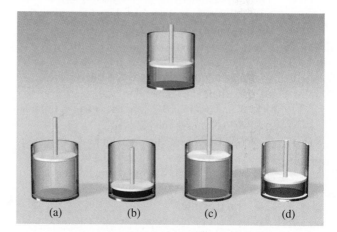

(a)   (b)   (c)   (d)

Choose the cylinder that correctly represents the gas after each of the following changes. (1) The pressure on the piston is tripled at constant n and T. (2) The absolute temperature is doubled at constant n and P. (3) n more moles of the gas are added at constant T and P. (4) Absolute temperature is halved at constant P.

11.39   A gaseous sample of a substance is cooled at constant pressure. Which of the following diagrams best represents the situation if the final temperature is (a) above the boiling point of the substance and (b) below the boiling point but above the freezing point of the substance?

(a)   (b)   (c)   (d)

SECTION 11.5: THE IDEAL GAS EQUATION

Review Questions

11.40   What are standard temperature and pressure (STP)? What is the significance of STP in relation to the volume of 1 mole of an ideal gas?

11.41   Why is the density of a gas much lower than that of a liquid or solid under atmospheric conditions? What units are normally used to express the density of gases?

Computational Problems

11.42   A sample of nitrogen gas in a 4.5-L container at a temperature of 27°C exerts a pressure of 4.1 atm. Calculate the number of moles of gas in the sample.

11.43   Given that 6.9 moles of carbon monoxide gas are present in a container of volume 30.4 L, what is the pressure of the gas (in atm) if the temperature is 82°C?

11.44   What volume will 9.8 moles of sulfur hexafluoride (SF_6) gas occupy if the temperature and pressure of the gas are 105°C and 9.4 atm, respectively?

11.45   The temperature of 2.5 L of a gas initially at STP is raised to 210°C at constant volume. Calculate the final pressure of the gas in atmospheres.

11.46   A gas-filled balloon having a volume of 2.50 L at 1.2 atm and 20°C is allowed to rise to the stratosphere (about 30 km above the surface of Earth), where the temperature and pressure are −23°C and 3.00×10^{-3} atm, respectively. Calculate the final volume of the balloon.

11.47   A gas evolved during the fermentation of glucose (wine making) has a volume of 0.67 L at 22.5°C and 1.00 atm. What was the volume of this gas at the fermentation temperature of 36.5°C and 1.00 atm pressure?

11.48   An ideal gas originally at 0.85 atm and 66°C was allowed to expand until its final volume, pressure, and temperature were 94 mL, 0.60 atm, and 45°C, respectively. What was its initial volume?

11.49   Calculate the volume (in liters) of 124.3 g of CO_2 at STP.

11.50   A gas at 572 mmHg and 35.0°C occupies a volume of 6.15 L. Calculate its volume at STP.

11.51   Dry ice is solid carbon dioxide. A 0.050-g sample of dry ice is placed in an evacuated 4.6-L vessel at 30°C. Calculate the pressure inside the vessel after all the dry ice has been converted to CO_2 gas.

11.52   At STP, 0.280 L of a gas weighs 0.400 g. Calculate the molar mass of the gas.

11.53   At 741 torr and 44°C, 7.10 g of a gas occupies a volume of 5.40 L. What is the molar mass of the gas?

11.54   Ozone molecules in the stratosphere absorb much of the harmful radiation from the sun. Typically, the temperature and pressure of ozone in the stratosphere are 250 K and 1.0×10^{-3} atm, respectively. How many ozone molecules are present in 1.0 L of air under these conditions?

11.55   Assuming that air contains 78 percent N_2, 21 percent O_2, and 1.0 percent Ar, all by volume, how many particles of each type of gas are present in 1.0 L of air at STP?

11.56   A 2.10-L vessel contains 4.65 g of a gas at 1.00 atm and 27.0°C. (a) Calculate the density of the gas in g/L. (b) What is the molar mass of the gas?

11.57   Calculate the density of hydrogen bromide (HBr) gas in g/L at 733 mmHg and 46°C.

11.58   A certain anesthetic contains 64.9 percent C, 13.5 percent H, and 21.6 percent O by mass. At 120°C and 750 mmHg, the mass of 1.00 L of the gaseous compound is 2.30 g. What is the molecular formula of the compound?

11.59   A compound has the empirical formula SF_4. At 20°C, 0.100 g of the gaseous compound occupies a volume of 22.1 mL and exerts a pressure of 1.02 atm. What is the molecular formula of the gas?

Conceptual Problems

11.60   The pressure of 6.0 L of an ideal gas in a flexible container is decreased to one-third of its original pressure, and its absolute temperature is decreased by one-half. What is the final volume of the gas?

11.61   A certain amount of gas at 25°C and at a pressure of 0.800 atm is contained in a vessel. Suppose that the vessel can withstand a pressure no higher than 5.00 atm. How high can you raise the temperature of the gas without bursting the vessel?

SECTION 11.6: REAL GASES

Review Questions

11.62   Cite two pieces of evidence to show that gases do not behave ideally under all conditions. Under which set of conditions would a gas be expected to behave most ideally: (a) high temperature and low pressure, (b) high temperature and high pressure, (c) low temperature and high pressure, or (d) low temperature and low pressure?

11.63   Figure 11.21(a) shows that at 0°C, with the exception of H_2, each of the gases has a pressure at which its compressibility factor is equal to 1—the point at which the curve crosses the ideal gas line. What is the significance of this point? Does each of these gases have a pressure at which the assumptions of ideal behavior (negligible molecular volume and no intermolecular attractions) are valid? Explain.

11.64 Write the van der Waals equation for a real gas. Explain the corrective terms for pressure and volume.

11.65 (a) A real gas is introduced into a flask of volume V. Is the corrected volume of the gas greater or less than V? (b) Ammonia has a larger a value than neon does (see Table 11.5). What can you conclude about the relative strength of the attractive forces between molecules of ammonia and between atoms of neon?

Computational Problems

11.66 At 27°C, 10.0 moles of a gas in a 1.50-L container exert a pressure of 130 atm. Is this an ideal gas?

11.67 Using the data shown in Table 11.5, calculate the pressure exerted by 2.50 moles of CO_2 confined in a volume of 5.00 L at 450 K. Compare the pressure with that predicted by the ideal gas equation.

SECTION 11.7: GAS MIXTURES

Review Questions

11.68 State Dalton's law of partial pressures and explain what *mole fraction* is. Does mole fraction have units?

11.69 What are the approximate partial pressures of N_2 and O_2 in air at the top of a mountain where atmospheric pressure is 0.8 atm? (See Problem 11.55.)

Computational Problems

11.70 A mixture of gases contains 0.31 mol CH_4, 0.25 mol C_2H_6, and 0.29 mol C_3H_8. The total pressure is 1.50 atm. Calculate the partial pressures of the gases.

11.71 A 2.5-L flask at 15°C contains a mixture of N_2, He, and Ne at partial pressures of 0.32 atm for N_2, 0.15 atm for He, and 0.42 atm for Ne. (a) Calculate the total pressure of the mixture. (b) Calculate the volume in liters at STP occupied by He and Ne if the N_2 is removed selectively.

11.72 Dry air near sea level, where atmospheric pressure is 1.00 atm, has the following composition by volume: N_2, 78.08 percent; O_2, 20.94 percent; Ar, 0.93 percent; CO_2, 0.05 percent. Calculate (a) the partial pressure of each gas in atmospheres and (b) the concentration of each gas in mol/L at 0°C. (*Hint:* Because volume is proportional to the number of moles present, mole fractions of gases can be expressed as ratios of volumes at the same temperature and pressure.)

11.73 A mixture of helium and neon gases is collected over water at 28.0°C and 745 mmHg. If the partial pressure of helium is 368 mmHg, what is the partial pressure of neon? (Vapor pressure of water at 28°C = 28.3 mmHg.)

11.74 Helium is mixed with oxygen gas for deep-sea divers. Calculate the percent by volume of oxygen gas in the mixture if the diver has to submerge to a depth where the total pressure is 5.2 atm. The partial pressure of oxygen is maintained at 0.20 atm at this depth.

11.75 A sample of ammonia (NH_3) gas is completely decomposed to nitrogen and hydrogen gases over heated iron wool. If the total pressure is 866 mmHg after the reaction, calculate the partial pressures of N_2 and H_2.

11.76 The Catalina hyperbaric chamber at the University of Southern California's Wrigley Marine Science Center treats mostly victims of diving accidents. In one treatment protocol, the chamber is pressurized to 4.6 atm with compressed air and the patient breathes a mixture of gases that contains oxygen. If the partial pressure of oxygen is 2.8 atm, what is the mole fraction of oxygen?

11.77 Considering the hyperbaric chamber in Problem 11.76, what chamber pressure would be required for a patient to receive the therapeutic partial pressure of O_2 (2.8 atm) without breathing a special mixture of gases through a mask? Assume that the air used to pressurize the chamber is 21 percent O_2 by volume.

Conceptual Problems

11.78 Consider the three containers shown, all of which have the same volume and are at the same temperature. (a) Which container has the smallest mole fraction of gas A (red)? (b) Which container has the highest partial pressure of gas B (green)? (c) Which container has the highest total pressure?

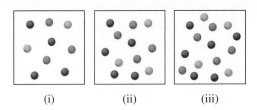

(i)   (ii)   (iii)

11.79 The volume of the box on the right is twice that of the box on the left. The boxes contain helium atoms (red) and hydrogen molecules (green) at the same temperature. (a) Which box has a higher total pressure? (b) Which box has a higher partial pressure of helium?

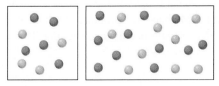

SECTION 11.8: REACTIONS WITH GASEOUS REACTANTS AND PRODUCTS

 Visualizing Chemistry
Figure 11.24

VC 11.1 The molar volume of hydrogen can be determined using the reaction of zinc metal and acid,

$$Zn(s) + 2H^+(aq) \longrightarrow Zn^{2+}(aq) + H_2(g)$$

as shown in Figure 11.24. When the reaction is complete, what does the space above the water in the graduated cylinder contain?
(a) $H_2(g)$, $Zn^{2+}(aq)$, and $H_2O(g)$
(b) $H_2(g)$ and $H_2O(g)$
(c) $H_2(g)$, $H_2O(g)$, and air

VC 11.2 How would the calculated molar volume be affected if we neglected to subtract the partial pressure of water vapor from the total pressure?
(a) It would be greater.
(b) It would be smaller.
(c) It would not change.

VC 11.3 How would the calculated molar volume be affected if we neglected to adjust the level of the graduated cylinder prior to reading the volume of gas collected? Assume that the level of water inside the graduated cylinder is higher than the level outside.
(a) It would be greater.
(b) It would be smaller.
(c) It would not change.

VC 11.4 How would the calculated molar volume be affected if some of the zinc metal failed to drop into the aqueous acid?
(a) It would be greater.
(b) It would be smaller.
(c) It would not change.

Computational Problems

11.80 Consider the formation of nitrogen dioxide from nitric oxide and oxygen:

$$2NO(g) + O_2(g) \longrightarrow 2NO_2(g)$$

If 9.0 L of NO is combined with excess O_2 at STP, what is the volume in liters of the NO_2 produced?

11.81 Methane, the principal component of natural gas, is used for heating and cooking. The combustion process is

$$CH_4(g) + 2O_2(g) \longrightarrow CO_2(g) + 2H_2O(l)$$

If 15.0 moles of CH_4 react with oxygen, what is the volume of CO_2 (in liters) produced at 23.0°C and 0.985 atm?

11.82 When coal is burned, the sulfur present in coal is converted to sulfur dioxide (SO_2), which is largely responsible for the production of acid rain:

$$S(s) + O_2(g) \longrightarrow SO_2(g)$$

Calculate the volume of SO_2 gas (in mL) formed at 30.5°C and 1.04 atm when 3.15 kg of S reacts with excess oxygen.

11.83 In alcohol fermentation, yeast converts glucose to ethanol and carbon dioxide:

$$C_6H_{12}O_6(s) \longrightarrow 2C_2H_5OH(l) + 2CO_2(g)$$

If 5.97 g of glucose reacts and 1.44 L of CO_2 gas is collected at 293 K and 0.984 atm, what is the percent yield of the reaction?

11.84 A compound of P and F was analyzed as follows: Heating 0.2324 g of the compound in a 378-cm^3 container turned all of it to gas, which had a pressure of 97.3 mmHg at 77°C. Then the gas was mixed with calcium chloride solution, which converted all the F to 0.2631 g of CaF_2. Determine the molecular formula of the compound.

11.85 A quantity of 0.225 g of a metal M (molar mass = 27.0 g/mol) reacted with an excess of hydrochloric acid to produce 0.303 L of molecular hydrogen (measured at 17°C and 741 mmHg). Deduce from these data the corresponding equation, and write formulas for the oxide and the sulfate of the metal M.

11.86 What is the mass of the solid NH_4Cl formed when 73.0 g of NH_3 is mixed with an equal mass of HCl? What is the volume of the gas remaining, measured at 14.0°C and 752 mmHg? What gas is it?

11.87 Dissolving 3.00 g of an impure sample of calcium carbonate in hydrochloric acid produced 0.656 L of carbon dioxide (measured at 20.0°C and 792 mmHg). Calculate the percent by mass of calcium carbonate in the sample. State any assumptions.

11.88 Calculate the mass in grams of hydrogen chloride produced when 5.6 L of molecular hydrogen measured at STP react with an excess of molecular chlorine gas.

11.89 Ethanol (C_2H_5OH) burns in air:

$$C_2H_5OH(l) + O_2(g) \longrightarrow CO_2(g) + H_2O(l)$$

Balance the equation and determine the volume of air in liters at 45.0°C and 793 mmHg required to burn 185 g of ethanol. Assume that air is 21.0 percent O_2 by volume.

11.90 A piece of sodium metal reacts completely with water as follows:

$$2Na(s) + 2H_2O(l) \longrightarrow 2NaOH(aq) + H_2(g)$$

The hydrogen gas generated is collected over water at 25.0°C. The volume of the gas is 246 mL measured at 1.00 atm. Calculate the number of grams of sodium used in the reaction. (Vapor pressure of water at 25°C = 0.0313 atm.)

11.91 A sample of zinc metal reacts completely with an excess of hydrochloric acid:

$$Zn(s) + 2HCl(aq) \longrightarrow ZnCl_2(aq) + H_2(g)$$

The hydrogen gas produced is collected over water at 25.0°C using an arrangement similar to that shown in Figure 11.23(a). The volume of the gas is 7.80 L, and the pressure is 0.980 atm. Calculate the amount of zinc metal in grams consumed in the reaction. (Vapor pressure of water at 25°C = 23.8 mmHg.)

ADDITIONAL PROBLEMS

11.92 Under the same conditions of temperature and pressure, which of the following gases would behave most ideally: Ne, N_2, or CH_4? Explain.

11.93 Nitroglycerin, an explosive compound, decomposes according to the equation

$$4C_3H_5(NO_3)_3(s) \longrightarrow 12CO_2(g) + 10H_2O(g) + 6N_2(g) + O_2(g)$$

Calculate the total volume of gases when collected at 1.2 atm and 25°C from 2.6×10^2 g of nitroglycerin. What are the partial pressures of the gases under these conditions?

11.94 The empirical formula of a compound is CH. At 200°C, 0.145 g of this compound occupies 97.2 mL at a pressure of 0.74 atm. What is the molecular formula of the compound?

11.95 When ammonium nitrite (NH_4NO_2) is heated, it decomposes to give nitrogen gas. This property is used to inflate some tennis balls. (a) Write a balanced equation for the reaction. (b) Calculate the quantity (in grams) of NH_4NO_2 needed to inflate a tennis ball to a volume of 86.2 mL at 1.20 atm and 22°C.

11.96 The percent by mass of bicarbonate (HCO_3^-) in a certain Alka-Seltzer product is 32.5 percent. Calculate the volume of CO_2 generated (in mL) at 37°C and 1.00 atm when a person ingests a 3.29-g tablet. (*Hint:* The reaction is between HCO_3^- and HCl acid in the stomach.)

11.97 Three flasks containing gases A (red) and B (blue) are shown here. (a) If the total pressure in (i) is 2.0 atm, what are the pressures in (ii) and (iii)? (b) Calculate the total pressure and the partial pressure of each gas after the valves are opened.

The volumes of (i) and (iii) are 2.0 L each, and the volume of (ii) is 1.0 L. The temperature is the same throughout.

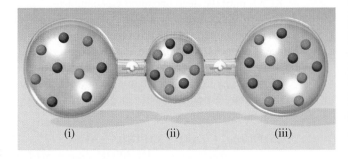

(i)   (ii)   (iii)

11.98 Referring to the hyperbaric chamber in Problem 11.76, in one treatment protocol, the chamber is pressurized to 6.0 atm with compressed air and the patient breathes a mixture of gases that is 47 percent oxygen by volume. In another protocol, the chamber is pressurized with compressed air to 2.8 atm and the patient breathes pure O_2. Determine the partial pressure of O_2 in each treatment protocol and compare the results.

11.99 In the metallurgical process of refining nickel, the metal is first combined with carbon monoxide to form tetracarbonylnickel, which is a gas at 43°C:

$$Ni(s) + 4CO(g) \longrightarrow Ni(CO)_4(g)$$

This reaction separates nickel from other solid impurities. (a) Starting with 86.4 g of Ni, calculate the pressure of $Ni(CO)_4$ in a container of volume 4.00 L. (Assume the preceding reaction goes to completion.) (b) At temperatures above 43°C, the pressure of the gas is observed to increase much more rapidly than predicted by the ideal gas equation. Explain.

11.100 The partial pressure of carbon dioxide varies with seasons. Would you expect the partial pressure in the Northern Hemisphere to be higher in the summer or winter? Explain.

11.101 A healthy adult exhales about 5.0×10^2 mL of a gaseous mixture with each breath. Calculate the number of molecules present in this volume at 37°C and 1.1 atm. List the major components of this gaseous mixture.

11.102 Sodium bicarbonate ($NaHCO_3$) is called baking soda because when heated, it releases carbon dioxide gas, which is responsible for the rising of cookies, some doughnuts, and cakes. (a) Calculate the volume (in liters) of CO_2 produced by heating 5.0 g of $NaHCO_3$ at 180°C and 1.3 atm. (b) Ammonium bicarbonate (NH_4HCO_3) has also been used for the same purpose. Suggest one advantage and one disadvantage of using NH_4HCO_3 instead of $NaHCO_3$ for baking.

11.103 On heating, potassium chlorate ($KClO_3$) decomposes to yield potassium chloride and oxygen gas. In one experiment, a student heated 20.4 g of $KClO_3$ until the decomposition was complete.
(a) Write a balanced equation for the reaction.
(b) Calculate the volume of oxygen (in liters) if it was collected at 0.962 atm and 18.3°C.

11.104 Some commercial drain cleaners contain a mixture of sodium hydroxide and aluminum powder. When the mixture is poured down a clogged drain, the following reaction occurs:

$$2NaOH(aq) + 2Al(s) + 6H_2O(l) \longrightarrow$$
$$2NaAl(OH)_4(aq) + 3H_2(g)$$

The heat generated in this reaction helps melt away obstructions such as grease, and the hydrogen gas released stirs up the solids clogging the drain. Calculate the volume of H_2 formed at 23°C and 1.00 atm if 3.12 g of Al are treated with an excess of NaOH.

11.105 The volume of a sample of pure HCl gas was 189 mL at 25°C and 108 mmHg. It was completely dissolved in about 60 mL of water and titrated with an NaOH solution; 15.7 mL of the NaOH solution was required to neutralize the HCl. Calculate the molarity of the NaOH solution.

11.106 Propane (C_3H_8) burns in oxygen to produce carbon dioxide gas and water vapor. (a) Write a balanced equation for this reaction. (b) Calculate the number of liters of carbon dioxide measured at STP that could be produced from 7.45 g of propane.

11.107 Consider the following apparatus. Calculate the partial pressures of helium and neon after the stopcock is open. The temperature remains constant at 16°C.

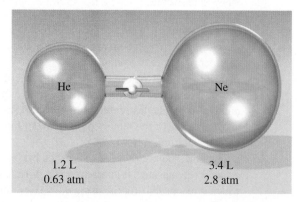

1.2 L   3.4 L
0.63 atm   2.8 atm

11.108 Nitric oxide (NO) reacts with molecular oxygen as follows:

$$2NO(g) + O_2(g) \longrightarrow 2NO_2(g)$$

Initially NO and O_2 are separated as shown here. When the valve is opened, the reaction quickly goes to completion. Determine what gases remain at the end and calculate their partial pressures. Assume that the temperature remains constant at 25°C.

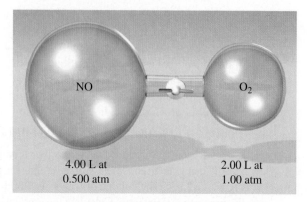

4.00 L at   2.00 L at
0.500 atm   1.00 atm

11.109 Consider the apparatus shown here. When a small amount of water is introduced into the flask by squeezing the bulb of the medicine dropper, water is squirted upward out of the long glass tubing. Explain this observation. (*Hint:* Hydrogen chloride gas is soluble in water.)

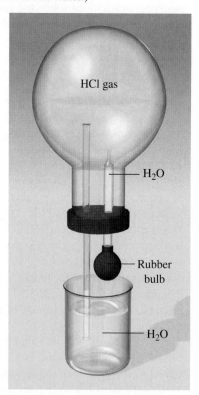

HCl gas

H₂O

Rubber bulb

H₂O

11.110 A 180.0-mg sample of an alloy of iron and metal X is treated with dilute sulfuric acid, liberating hydrogen and yielding Fe^{2+} and X^{3+} ions in solution. It is known that the alloy contains 20.0 percent iron by mass. The alloy yields 50.9 mL of hydrogen collected over water at 22°C and a total pressure of 750.0 torr. What is element X?

11.111 A certain hydrate has the formula $MgSO_4 \cdot xH_2O$. A quantity of 54.2 g of the compound is heated in an oven to drive off the water. If the steam generated exerts a pressure of 24.8 atm in a 2.00-L container at 120°C, calculate x.

11.112 A mixture of Na_2CO_3 and $MgCO_3$ of mass 7.63 g is combined with an excess of hydrochloric acid. The CO_2 gas generated occupies a volume of 1.67 L at 1.24 atm and 26°C. From these data, calculate the percent composition by mass of Na_2CO_3 in the mixture.

11.113 The apparatus shown below can be used to measure atomic and molecular speeds. Suppose that a beam of metal atoms is directed at a rotating cylinder in a vacuum. A small opening in the cylinder allows the atoms to strike a target area. Because the cylinder is rotating, atoms traveling at different speeds will strike the target at different positions. In time, a layer of the metal will deposit on the target area, and the variation in its thickness is found to correspond to Maxwell's speed distribution. In one experiment, it is found that at 850°C some bismuth (Bi) atoms struck the target at a point 2.80 cm from the spot directly opposite the slit. The diameter of the cylinder is 15.0 cm, and it is rotating at 130 revolutions per second. (a) Calculate the speed (in m/s) at which the target is moving. (*Hint:* The circumference of a circle is given by $2\pi r$, where r is the radius.) (b) Calculate the time (in seconds) it takes for the target to travel 2.80 cm. (c) Determine the speed of the Bi atoms. Compare your result in part (c) with the u_{rms} of Bi at 850°C. Comment on the difference.

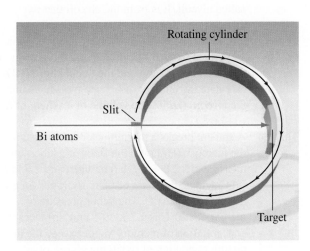

Rotating cylinder

Slit

Bi atoms

Target

11.114 If 10.00 g of water is introduced into an evacuated flask of volume 2.500 L at 65°C, calculate the mass of water vaporized. (*Hint:* Assume that the volume of the remaining liquid water is negligible; the vapor pressure of water at 65°C is 187.5 mmHg.)

11.115 Commercially, compressed oxygen is sold in metal cylinders. If a 120-L cylinder is filled with oxygen to a pressure of 132 atm at 22°C, what is the mass of O_2 present? How many liters of O_2 gas at 1.00 atm and 22°C could the cylinder produce? (Assume ideal behavior.)

11.116 The shells of hard-boiled eggs sometimes crack due to the rapid thermal expansion of the shells at high temperatures. Suggest another reason why the shells may crack.

11.117 Ethylene gas (C_2H_4) is emitted by fruits and is known to be responsible for their ripening. Based on this information, explain why a bunch of bananas ripens faster in a closed paper bag than in an open bowl.

11.118 About 8.0×10^6 tons of urea [$(NH_2)_2CO$] are used annually as a fertilizer. The urea is prepared at 200°C and under high-pressure conditions from carbon dioxide and ammonia (the products are urea and steam). Calculate the volume of ammonia (in liters) measured at 150 atm needed to prepare 1.0 ton of urea.

11.119 Some ballpoint pens have a small hole in the main body of the pen. What is the purpose of this hole?

11.120 The gas laws are vitally important to scuba divers. The pressure exerted by 33 ft of seawater is equivalent to 1 atm pressure. (a) A diver ascends quickly to the surface of the water from a depth of 36 ft without exhaling gas from his lungs. By what factor will the volume of his lungs increase by the time he reaches the surface? Assume that the temperature is constant. (b) The partial pressure of oxygen in air is about 0.20 atm. (Air is 20 percent oxygen by volume.) In deep-sea diving, the composition of air the diver breathes must be changed to maintain this partial pressure. What must the oxygen content (in percent by volume) be when the total pressure exerted on the diver is 4.0 atm? (At constant temperature and pressure, the volume of a gas is directly proportional to the number of moles of gases.)

11.121 Nitrous oxide (N_2O) can be obtained by the thermal decomposition of ammonium nitrate (NH_4NO_3). (a) Write a balanced equation for the reaction. (b) In a certain experiment, a student obtains 0.340 L of the gas at 718 mmHg and 24°C. If the gas weighs 0.580 g, calculate the value of the gas constant.

11.122 Two vessels are labeled A and B. Vessel A contains NH_3 gas at 70°C, and vessel B contains Ne gas at the same temperature. If the average kinetic energy of NH_3 is 7.1×10^{-21} J/molecule, calculate the root-mean-square speed of Ne atoms in m/s.

11.123 Which of the following molecules has the largest a value: CH_4, F_2, C_6H_6, Ne?

11.124   The following procedure is a simple though somewhat crude way to measure the molar mass of a gas. A liquid of mass 0.0184 g is introduced into a syringe like the one shown here by injection through the rubber tip using a hypodermic needle. The syringe is then transferred to a temperature bath heated to 45°C, and the liquid vaporizes. The final volume of the vapor (measured by the outward movement of the plunger) is 5.58 mL, and the atmospheric pressure is 760 mmHg. Given that the compound's empirical formula is CH_2, determine the molar mass of the compound.

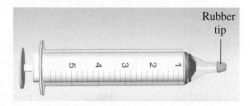

Rubber tip

11.125   In 1995, a man suffocated as he walked by an abandoned mine in England. At that moment there was a sharp drop in atmospheric pressure due to a change in the weather. Suggest what might have caused the man's death.

11.126   Acidic oxides such as carbon dioxide react with basic oxides like calcium oxide (CaO) and barium oxide (BaO) to form salts (metal carbonates).
(a) Write equations representing these two reactions.
(b) A student placed a mixture of BaO and CaO of combined mass 4.88 g in a 1.46-L flask containing carbon dioxide gas at 35°C and 746 mmHg. After the reactions were complete, she found that the CO_2 pressure had dropped to 252 mmHg. Calculate the percent composition by mass of the mixture. Assume that the volumes of the solids are negligible.

11.127   (a) What volume of air at 1.0 atm and 22°C is needed to fill a 0.98-L bicycle tire to a pressure of 5.0 atm at the same temperature? (Note that the 5.0 atm is the gauge pressure, which is the difference between the pressure in the tire and atmospheric pressure. Before filling, the pressure in the tire was 1.0 atm.) (b) What is the total pressure in the tire when the gauge pressure reads 5.0 atm? (c) The tire is pumped by filling the cylinder of a hand pump with air at 1.0 atm and then, by compressing the gas in the cylinder, adding all the air in the pump to the air in the tire. If the volume of the pump is 33 percent of the tire's volume, what is the gauge pressure in the tire after three full strokes of the pump? Assume constant temperature.

11.128   The running engine of an automobile produces carbon monoxide (CO), a toxic gas, at the rate of about 188 g CO per hour. A car is left idling in a poorly ventilated garage that is 6.0 m long, 4.0 m wide, and 2.2 m high at 20°C. (a) Calculate the rate of CO production in mol/min. (b) How long would it take to build up a lethal concentration of CO of 1000 ppmv (parts per million by volume)?

11.129   Interstellar space contains mostly hydrogen atoms at a concentration of about 1 atom/cm³.
(a) Calculate the pressure of the H atoms.
(b) Calculate the volume (in liters) that contains 1.0 g of H atoms. The temperature is 3 K.

11.130   Atop Mt. Everest, the atmospheric pressure is 210 mmHg and the air density is 0.426 kg/m³.
(a) Calculate the air temperature, given that the molar mass of air is 29.0 g/mol. (b) Assuming no change in air composition, calculate the percent decrease in oxygen gas from sea level to the top of Mt. Everest.

11.131   Relative humidity is defined as the ratio (expressed as a percentage) of the partial pressure of water vapor in the air to the equilibrium vapor pressure (see Table 11.6) at a given temperature. On a certain summer day in North Carolina, the partial pressure of water vapor in the air is 3.9×10^3 Pa at 30°C. Calculate the relative humidity.

11.132   Under the same conditions of temperature and pressure, why does 1 L of moist air weigh less than 1 L of dry air? In weather forecasts, an oncoming low-pressure front usually means imminent rainfall. Explain.

11.133   Air entering the lungs ends up in tiny sacs called alveoli. It is from the alveoli that oxygen diffuses into the blood. The average radius of the alveoli is 0.0050 cm, and the air inside contains 14 percent oxygen. Assuming that the pressure in the alveoli is 1.0 atm and the temperature is 37°C, calculate the number of oxygen molecules in one of the alveoli. (*Hint:* The volume of a sphere of radius r is $\frac{4}{3}\pi r^3$.)

11.134   A student breaks a thermometer and spills most of the mercury (Hg) onto the floor of a laboratory that measures 15.2 m long, 6.6 m wide, and 2.4 m high. (a) Calculate the mass of mercury vapor (in grams) in the room at 20°C. The vapor pressure of mercury at 20°C is 1.7×10^{-6} atm. (b) Does the concentration of mercury vapor exceed the air quality regulation of 0.050 mg Hg/m³ of air? (c) One way to deal with small quantities of spilled mercury is to spray sulfur powder over the metal. Suggest a physical and a chemical reason for this action.

11.135 Nitrogen forms several gaseous oxides. One of them has a density of 1.33 g/L measured at 764 mmHg and 150°C. Write the formula of the compound.

11.136 Nitrogen dioxide (NO_2) cannot be obtained in a pure form in the gas phase because it exists as a mixture of NO_2 and N_2O_4. At 25°C and 0.98 atm, the density of this gas mixture is 2.7 g/L. What is the partial pressure of each gas?

11.137 Lithium hydride reacts with water as follows:

$$LiH(s) + H_2O(l) \longrightarrow LiOH(aq) + H_2(g)$$

During World War II, U.S. pilots carried LiH tablets. In the event of a crash landing at sea, the LiH would react with the seawater and fill their life jackets and lifeboats with hydrogen gas. How many grams of LiH are needed to fill a 4.1-L life jacket at 0.97 atm and 12°C?

11.138 The atmosphere on Mars is composed mainly of carbon dioxide. The surface temperature is 220 K, and the atmospheric pressure is about 6.0 mmHg. Taking these values as Martian "STP," calculate the molar volume in liters of an ideal gas on Mars.

11.139 Venus's atmosphere is composed of 96.5 percent CO_2, 3.5 percent N_2, and 0.015 percent SO_2 by volume. Its standard atmospheric pressure is 9.0×10^6 Pa. Calculate the partial pressures of the gases in pascals.

11.140 A student tries to determine the volume of a flask like the one shown in Figure 11.18. These are her results: mass of the flask filled with dry air at 23°C and 744 mmHg = 91.6843 g; mass of evacuated flask = 91.4715 g. Assume the composition of air is 78 percent N_2, 21 percent O_2, and 1 percent argon by volume. What is the volume (in mL) of the flask? (*Hint:* First calculate the average molar mass of air, as shown in Problem 5.137.)

11.141 Apply your knowledge of the kinetic theory of gases to the following situations. (a) Two flasks of volumes V_1 and V_2 ($V_2 > V_1$) contain the same number of helium atoms at the same temperature. (i) Compare the root-mean-square (rms) speeds and average kinetic energies of the helium (He) atoms in the flasks. (ii) Compare the frequency and the force with which the He atoms collide with the walls of their containers. (b) Equal numbers of He atoms are placed in two flasks of the same volume at temperatures T_1 and T_2 ($T_2 > T_1$). (i) Compare the rms speeds of the atoms in the two flasks. (ii) Compare the frequency and the force with which the He atoms collide with the walls of their containers. (c) Equal numbers of He and neon (Ne) atoms are placed in two flasks of the same volume,

and the temperature of both gases is 74°C. Comment on the validity of the following statements: (i) The rms speed of He is equal to that of Ne. (ii) The average kinetic energies of the two gases are equal. (iii) The rms speed of each He atom is 1.47×10^3 m/s.

11.142 At what temperature will He atoms have the same u_{rms} value as N_2 molecules at 25°C?

11.143 Estimate the distance (in nm) between molecules of water vapor at 100°C and 1.0 atm. Assume ideal behavior. Repeat the calculation for liquid water at 100°C, given that the density of water is 0.96 g/cm³ at that temperature. Comment on your results. (Assume each water molecule to be a sphere with a diameter of 0.3 nm.) (*Hint:* First calculate the number density of water molecules. Next, convert the number density to linear density, that is, the number of molecules in one direction.)

11.144 Which of the noble gases would not behave ideally under any conditions? Explain.

11.145 A 5.72-g sample of graphite was heated with 68.4 g of O_2 in a 8.00-L flask. The reaction that took place was

$$C(graphite) + O_2(g) \longrightarrow CO_2(g)$$

After the reaction was complete, the temperature in the flask was 182°C. What was the total pressure inside the flask?

11.146 A 6.11-g sample of a Cu-Zn alloy reacts with HCl acid to produce hydrogen gas. If the hydrogen gas has a volume of 1.26 L at 22°C and 728 mmHg, what is the percent of Zn in the alloy? (*Hint:* Cu does not react with HCl.)

11.147 A stockroom supervisor measured the contents of a 25.0-gal drum partially filled with acetone on a day when the temperature was 18.0°C and atmospheric pressure was 750 mmHg, and found that 15.4 gal of the solvent remained. After tightly sealing the drum, an assistant dropped the drum while carrying it upstairs to the organic laboratory. The drum was dented, and its internal volume was decreased to 20.4 gal. What is the total pressure inside the drum after the accident? The vapor pressure of acetone at 18.0°C is 400 mmHg. (*Hint:* At the time the drum was sealed, the pressure inside the drum, which is equal to the sum of the pressures of air and acetone, was equal to the atmospheric pressure.)

11.148 In 2.00 min., 29.7 mL of He effuses through a small hole. Under the same conditions of pressure and temperature, 10.0 mL of a mixture of CO and CO_2 effuses through the hole in the same amount of time. Calculate the percent composition by volume of the mixture.

11.149 A mixture of methane (CH_4) and ethane (C_2H_6) is stored in a container at 294 mmHg. The gases are burned in air to form CO_2 and H_2O. If the pressure of CO_2 is 356 mmHg measured at the same temperature and volume as the original mixture, calculate the mole fractions of the gases.

11.150 Use the kinetic theory of gases to explain why hot air rises.

11.151 One way to gain a physical understanding of *b* in the van der Waals equation is to calculate the "excluded volume." Assume that the distance of closest approach between two similar atoms is the sum of their radii (2*r*). (a) Calculate the volume around each atom into which the center of another atom cannot penetrate. (b) From your result in part (a), calculate the excluded volume for 1 mole of the atoms, which is the constant *b*. How does this volume compare with the sum of the volumes of 1 mole of the atoms?

11.152 A 5.00-mole sample of NH_3 gas is kept in a 1.92-L container at 300 K. If the van der Waals equation is assumed to give the correct answer for the pressure of the gas, calculate the percent error made in using the ideal gas equation to calculate the pressure.

11.153 The root-mean-square speed of a certain gaseous oxide is 493 m/s at 20°C. What is the molecular formula of the compound?

11.154 Referring to Figure 11.4, we see that the maximum of each speed distribution plot is called the most probable speed (u_{mp}) because it is the speed possessed by the largest number of molecules. It is given by $u_{mp} = \sqrt{2RT/\mathcal{M}}$. (a) Compare u_{mp} with u_{rms} for nitrogen at 25°C. (b) The following diagram shows the Maxwell speed distribution curves for an ideal gas at two different temperatures T_1 and T_2. Calculate the value of T_2.

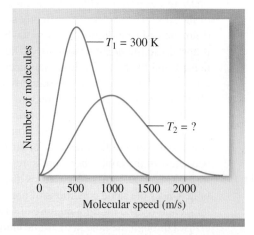

11.155 A gaseous reaction takes place at constant volume and constant pressure in a cylinder as shown here.

Which of the following equations best describes the reaction? The initial temperature (T_1) is twice that of the final temperature (T_2).
(a) $A + B \longrightarrow C$
(b) $AB \longrightarrow C + D$
(c) $A + B \longrightarrow C + D$
(d) $A + B \longrightarrow 2C + D$

11.156 Assuming ideal behavior, which of the following gases will have the greatest volume at STP? (a) 0.82 mole of He, (b) 24 g of N_2, or (c) 5.0×10^{23} molecules of Cl_2.

11.157 Calculate the density of helium in a helium balloon at 25.0°C. (Assume that the pressure inside the balloon is 1.10 atm.)

11.158 Helium atoms in a closed container at room temperature are constantly colliding with one another and with the walls of their container. Does this "perpetual motion" violate the law of conservation of energy? Explain.

11.159 Sulfur hexafluoride (SF_6) has a molar mass of 146 g/mol and neon (Ne) has a molar mass of 20 g/mol. However, the average kinetic energies of these two gases at the same temperature are the same. Explain.

11.160 Consider the molar volumes shown in Figure 11.19. (a) Explain why Cl_2 and NH_3 have molar volumes significantly smaller from that of an ideal gas. (b) Explain why H_2, He, and Ne have molar volumes greater than that of an ideal gas. (*Hint:* Look up the boiling points of the gases shown in the figure.)

11.161 The plot of *Z* versus *P* for a gas at 0°C is shown. Explain the causes of the negative deviation from ideal behavior at lower pressures and the positive deviation from ideal behavior at higher pressures.

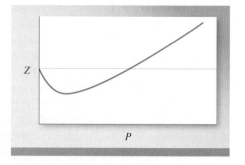

Answers to In-Chapter Materials

PRACTICE PROBLEMS

11.1A 2.137. **11.1B** 2.02 g/mol, H_2. **11.2A** 1.32 atm. **11.2B** 9.52 m.
11.3A 2.23 L. **11.3B** 1.80 atm. **11.4A** 30.7 L. **11.4B** 300°C.
11.5A 34 L. **11.5B** 3.16 L CO, 1.58 L O_2. **11.6A** 5.09 L. **11.6B** 85.0 m.
A common mistake in this problem is failure to subtract the
atmospheric pressure (0.965 atm) from the total pressure at the bottom
of the lake (9.19 atm). The pressure due to the *water* is only 8.23 atm.
11.7A 128 L. **11.7B** 336°C. **11.8A** 0.180 g/L. **11.8B** 1.29 g/L.
11.9A 55.2 g/mol. **11.9B** 146 g/mol. **11.10A** 50.6 atm, 51.6 atm.
11.10B 1.3 atm, 1.3 atm. **11.11A** 0.0453 L/mol, 0.0113 L.
11.11B 146 pm. **11.12A** $P_{He} = 0.184$ atm, $P_{H_2} = 0.390$ atm, $P_{Ne} =$
1.60 atm, $P_{total} = 2.18$ atm. **11.12B** 0.032 mol CH_4, 0.091 mol C_2H_6,
0.123 mol total. **11.13A** $\chi_{CO_2} = 0.0495$, $\chi_{CH_4} = 0.255$, $\chi_{He} = 0.695$,
$P_{CO_2} = 0.286$ atm, $P_{CH_4} = 1.47$ atm, $P_{He} = 4.02$ atm. **11.13B** $P_{Xe} =$
4.95 atm, $P_{Ne} = 1.55$ atm, $n_{Xe} = 3.13$, $n_{Ne} = 0.984$. **11.14A** 151 mL.
11.14B 3.48 g. **11.15A** 1.0×10^2 mol. **11.15B** 0.0052 atm.
11.16A 1.03 g. **11.16B** 0.386 L.

SECTION REVIEW

11.2.1 c. **11.2.2** b. **11.3.1** a. **11.3.2** b, c, d. **11.3.3** d. **11.3.4** b.
11.3.5 e. **11.4.1** e. **11.4.2** a. **11.4.3** c. **11.4.4** a. **11.4.5** b.
11.4.6 d. **11.5.1** a. **11.5.2** e. **11.5.3** b. **11.5.4** b. **11.6.1** b.
11.6.2 b. **11.7.1** e. **11.7.2** b. **11.7.3** a. **11.7.4** c. **11.7.5** b.
11.8.1 d. **11.8.2** b. **11.8.3** e.

Chapter 12

Liquids and Solids

© US Department of Energy/Science Source

DISCOVERED IN 1931, aerogels were originally thought to be nothing more than chemical anomalies—due to their extremely low densities. (The lowest known density of an aerogel is actually less than the density of *air*.) After recognizing their unusual properties, chemists and physicists have applied aerogels to a number of important uses, including as chemical adsorbers for cleaning up chemical and oil spills, thermal insulation for the Mars rover and space suits, biocompatible drug delivery systems, and particle traps for space dust aboard spacecraft.

Before You Begin, Review These Skills

- Intermolecular forces [◀◀ Section 7.3]
- Basic trigonometry [▶▶ Appendix 1]

12.1 THE CONDENSED PHASES

In Chapter 7, we learned that intermolecular forces arise as a consequence of molecular polarity; and that inter*molecular* can actually refer to forces between *molecules, atoms, or ions.* At a given temperature, the magnitude of intermolecular forces determines whether a substance is a solid, a liquid, or a gas. Gases, which consist of rapidly moving particles (molecules or atoms) separated by relatively large distances, were the subject of Chapter 11. Intermolecular attractions are essentially negligible in gases—one of the assumptions of the kinetic molecular theory of gases [◀◀ Section 11.2]. In liquids and solids, the *condensed phases* [◀◀ Section 1.5], intermolecular forces are large enough to keep the particles of a substance in contact with one another. Substances in which the attractive forces are strong enough to keep the particles together, but weak enough to allow flow (molecules moving past one another), are *liquids*. Those in which the attractive forces are strong enough to keep the particles in fixed positions relative to one another are *solids*. The gas, liquid, and solid phases are illustrated at the molecular level in Figure 12.1.

Among liquids and solids, varying magnitudes of intermolecular forces are responsible for several physical properties. These properties include surface tension (liquids), viscosity (liquids), vapor pressure (liquids and solids), and phase-transition temperatures (boiling point for liquids and melting point for solids). In the following sections, we will examine these properties, the relationships between them, and some of their consequences.

Student Annotation: As we saw in Chapter 7, intermolecular forces also occur between ions and polar molecules—most notably in aqueous solutions. However, the focus of this chapter will be the intermolecular forces in pure substances.

(a) Gas

(b) Liquid

(c) Solid

Figure 12.1 (a) Particles in a gas are separated by large distances and are free to move entirely independently of one another. (b) Particles in a liquid are very close together but are free to move about. (c) Particles in a solid are essentially locked in place with respect to one another.

Figure 12.2 Intermolecular forces acting on a molecule in both the surface layer and the interior region of a liquid.

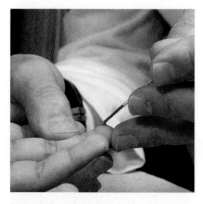

Blood being drawn from a finger into a capillary tube.

© Yoav Levy/Phototake

Figure 12.3 Surface tension causes water to bead up on a just-washed car.

(a) © Purestock/PunchStock; (b) © Ingram Publishing/SuperStock

12.2 PROPERTIES OF LIQUIDS

Several of the physical properties of a liquid depend on the magnitude of its intermolecular forces. In this section we consider three such properties: surface tension, viscosity, and vapor pressure.

Surface Tension

A molecule within a liquid is pulled in all directions by the intermolecular forces between it and the other molecules that surround it. There is no *net* pull in any one direction. A molecule at the surface of the liquid is similarly pulled down and to the sides by neighboring molecules, but Figure 12.2 shows there is no upward pull to balance the downward or inward pull (into the bulk of the liquid). This results in a net pull inward on molecules at the surface, causing the surface of a liquid to tighten like an elastic film, thus minimizing its surface area. The "beading" of water on the surface of a freshly washed car, shown in Figure 12.3, illustrates this phenomenon.

A quantitative measure of the elastic force in the surface of a liquid is the *surface tension,* the amount of energy required to stretch or increase the surface of a liquid by a unit area (e.g., by 1 cm^2). A liquid with strong intermolecular forces has a high surface tension. Water, for instance, with its strong hydrogen bonds, has a very high surface tension.

Another illustration of surface tension is the *meniscus,* the curved surface of a liquid contained in a narrow tube. Figure 12.4(a) shows the concave surface of water in a graduated cylinder. (You probably know from your laboratory class that you are to read the volume level with the bottom of the meniscus.) This is caused by a thin film of water adhering to the wall of the glass cylinder. The surface tension of water causes this film to contract, and as it does, it pulls the water up the cylinder. This effect, known as *capillary action,* is more pronounced in a cylinder with a very small diameter, such as a capillary tube used to draw a small amount of blood. Two types of forces bring about capillary action. One is *cohesion,* the attractions between *like* molecules (in this case, between water molecules). The other is *adhesion,* the attractions between *unlike* molecules (in this case, between water molecules and the molecules that make up the interior surface of the graduated cylinder). If adhesion is stronger than cohesion, as it is in Figure 12.4(a), the contents of the tube will be pulled upward. The upward movement is limited by the weight of the liquid in the tube. In the case of mercury, shown in Figure 12.4(b), the cohesive forces are stronger than the adhesive forces, resulting in a convex meniscus in which the liquid level at the glass wall is lower than that in the middle.

Viscosity

Another property determined by the magnitude of intermolecular forces in a liquid is viscosity. *Viscosity,* with units of N · s/m^2, is a measure of a fluid's resistance to flow. The higher the viscosity, the more slowly a liquid flows. The viscosity of a liquid typically decreases with increasing temperature. The phrase "slow as molasses in winter" refers to the fact that molasses pours more slowly (has a higher viscosity) in cold weather. You may also have noticed that honey and maple syrup seem thinner when they are heated.

(a)

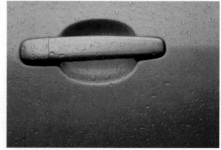

(b)

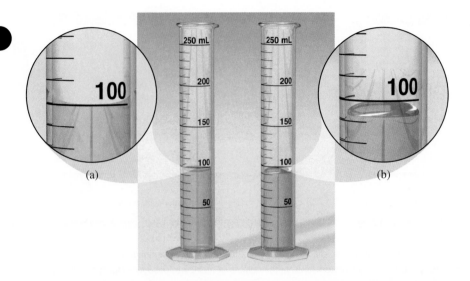

TABLE 12.1	Viscosities of Some Familiar Liquids at 20°C
Liquid	**Viscosity ($N \cdot s/m^2$)**
Acetone (C_3H_6O)	3.16×10^{-4}
Water (H_2O)	1.01×10^{-3}
Ethanol (C_2H_5OH)	1.20×10^{-3}
Mercury (Hg)	1.55×10^{-3}
Blood	4×10^{-3}
Glycerol ($C_3H_8O_3$)	1.49

Liquids that have strong intermolecular forces have higher viscosities than those that have weaker intermolecular forces. Table 12.1 lists the viscosities of some liquids that may be familiar to you. Water's high viscosity, like its high surface tension, is the result of hydrogen bonding. Note how large the viscosity of glycerol is compared to the other liquids listed in Table 12.1. The structure of glycerol, which is a sweet tasting, syrupy liquid used for a wide variety of things including the manufacture of candy and antibiotics, is

$$
\begin{array}{c}
\text{H} \\
| \\
\text{H--C--O--H} \\
| \\
\text{H--C--O--H} \\
| \\
\text{H--C--O--H} \\
| \\
\text{H}
\end{array}
$$

Glycerol

Like water, glycerol can form hydrogen bonds. Each glycerol molecule has three —OH groups that can participate in hydrogen bonding with other glycerol molecules. Furthermore, because of their shape, the molecules have a great tendency to become entangled rather than to slip past one another as the molecules of less viscous liquids do. These interactions contribute to its high viscosity.

Vapor Pressure of Liquids

In Chapter 11 we encountered the term *vapor pressure,* referring to the temperature-dependent partial pressure of water [◄◄ Section 11.8]. In fact, vapor pressure is another property of liquids that depends on the magnitude of intermolecular forces. The molecules in a liquid are in constant motion, and, like the molecules in a gas, they have a distribution of kinetic energies. The most probable kinetic energy for molecules in a

Student Annotation: Substances that have *high* vapor pressures at room temperature are said to be *volatile.* Note that this does *not* mean that a substance is explosive—only that it has a high vapor pressure.

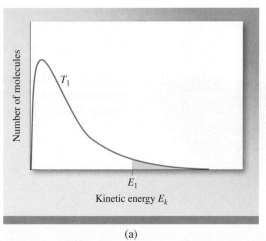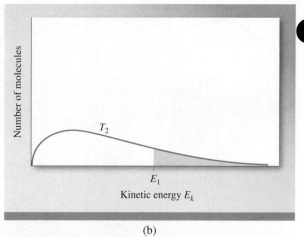

(a)   (b)

Figure 12.5 Kinetic energy distribution curves for molecules in a liquid (a) at temperature T_1 and (b) at a higher temperature T_2. Note that at the higher temperature, the curve flattens out. The shaded areas represent the number of molecules possessing kinetic energy equal to or greater than a certain kinetic energy E_1. The higher the temperature, the greater the number of molecules with high kinetic energy.

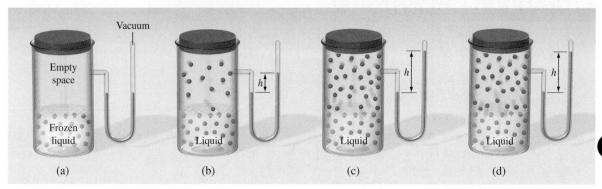

Figure 12.6 Establishment of equilibrium vapor pressure. (a) Initially, there are no molecules in the gas phase. (b) Molecules enter the gas phase, increasing the total pressure above the liquid. (c) The partial pressure of the liquid continues to increase until the rates of vaporization and condensation are equal. (d) Vaporization and condensation continue to occur at the same rate, and there is no further net change in pressure.

Student Annotation: The processes of vaporization and condensation are examples of *phase changes.* These and other phase changes are discussed in detail in Section 12.5.

Student Annotation: We often use the simpler term *vapor pressure* to mean *equilibrium vapor pressure.*

Student Hot Spot

Student data indicate you may struggle with this content. Log in to Connect to view additional Learning Resources on this topic.

sample of liquid increases with increasing temperature, as shown in Figure 12.5. If a molecule at the surface of a liquid has sufficient kinetic energy, it can escape from the liquid phase into the gas phase. This phenomenon is known as **evaporation** or **vaporization.** Consider the apparatus shown in Figure 12.6. As a liquid begins to evaporate, molecules leave the liquid phase and become part of the gas phase in the space above the liquid. Molecules in the gas phase can return to the liquid phase if they strike the liquid surface and again become trapped by intermolecular forces, a process known as **condensation.** Initially, evaporation occurs more rapidly than condensation. As the number of molecules in the gas phase increases, however, so does the rate of condensation. The vapor pressure over the liquid increases until the rate of condensation is equal to the rate of evaporation, which is constant at any given temperature (Figure 12.7). This (or any other) situation, wherein a forward process and reverse process are occurring at the same rate, is called a **dynamic equilibrium.** Although both processes are ongoing (*dynamic*), the number of molecules in the gas phase at any given point in time does not change (*equilibrium*). The pressure exerted by the molecules that have escaped to the gas phase, once the pressure has stopped increasing, is the **equilibrium vapor pressure.**

The average kinetic energy of molecules in a liquid increases with increasing temperature (see Figure 12.5). At a higher temperature, therefore, a greater percentage of molecules at the liquid surface will possess sufficient kinetic energy to escape into the gas phase. Consequently, as we have already seen with water in Table 11.6, vapor pressure increases with increasing temperature. Figure 12.8 shows the plots of vapor pressure versus temperature for three different liquids.

The plots in Figure 12.8 of vapor pressure as a function of temperature are not linear. However, a linear relationship does exist between the natural log of vapor

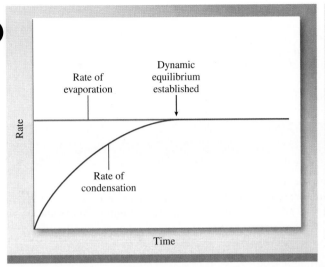

Figure 12.7 Comparison of the rates of vaporization and condensation at constant temperature.

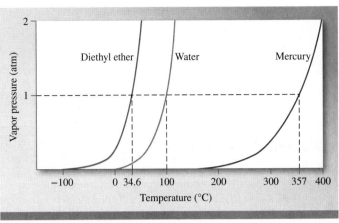

Figure 12.8 The increase in vapor pressure with temperature for three liquids. The normal boiling points of the liquids (at 1 atm) are shown on the horizontal axis. The strong metallic bonding in mercury results in its having a very low vapor pressure at room temperature.

pressure and the reciprocal of absolute temperature. This relationship is called the *Clausius*[1]*-Clapeyron*[2] *equation,*

$$\ln P = -\frac{\Delta H_{vap}}{RT} + C \qquad \textbf{Equation 12.1}$$

where $\ln P$ is the natural logarithm of the vapor pressure, ΔH_{vap} is the molar heat of vaporization (kJ/mol), R is the gas constant (8.314 J/K · mol), and C is a constant that must be determined experimentally for each different compound. The Clausius-Clapeyron equation has the form of the general linear equation $y = mx + b$.

Student Annotation: The units of R are those that enable us to cancel the units of J/mol (or kJ/mol) associated with ΔH_{vap} [◄◄ Section 11.5, Table 11.4].

$$\ln P = \left(-\frac{\Delta H_{vap}}{R}\right)\left(\frac{1}{T}\right) + C$$
$$y = \qquad mx \qquad + b$$

By measuring the vapor pressure of a liquid at several different temperatures and plotting $\ln P$ versus $1/T$, we can determine the slope of the line, which is equal to $-\Delta H_{vap}/R$. (ΔH_{vap} is assumed to be independent of temperature.)

If we know the value of ΔH_{vap} and the vapor pressure of a liquid at one temperature, we can use the Clausius-Clapeyron equation to calculate the vapor pressure of the liquid at a different temperature. At temperatures T_1 and T_2, the vapor pressures are P_1 and P_2. From Equation 12.1 we can write

$$\ln P_1 = -\frac{\Delta H_{vap}}{RT_1} + C \qquad \textbf{Equation 12.2}$$

$$\ln P_2 = -\frac{\Delta H_{vap}}{RT_2} + C \qquad \textbf{Equation 12.3}$$

Subtracting Equation 12.3 from Equation 12.2 we get

$$\ln P_1 - \ln P_2 = -\frac{\Delta H_{vap}}{RT_1} - \left(-\frac{\Delta H_{vap}}{RT_2}\right)$$

$$= \frac{\Delta H_{vap}}{R}\left(\frac{1}{T_2} - \frac{1}{T_1}\right)$$

---

[1]Rudolf Julius Emanuel Clausius (1822–1888), a German physicist, worked mainly in electricity, kinetic theory of gases, and thermodynamics.

[2]Benoit Paul Emile Clapeyron (1799–1864), a French engineer, made contributions to the thermodynamics of steam engines.

and finally

Equation 12.4

$$\ln\frac{P_1}{P_2} = \frac{\Delta H_{vap}}{R}\left(\frac{1}{T_2} - \frac{1}{T_1}\right)$$

Worked Example 12.1 shows how to use Equation 12.4.

Worked Example 12.1

Diethyl ether is a volatile, highly flammable organic liquid that today is used mainly as a solvent. (It was used as an anesthetic during the nineteenth century and as a recreational intoxicant early in the twentieth century during prohibition, when ethanol was difficult to obtain.) The vapor pressure of diethyl ether is 401 mmHg at 18°C, and its molar heat of vaporization is 26 kJ/mol. Calculate its vapor pressure at 32°C.

Strategy Given the vapor pressure at one temperature, P_1, use Equation 12.4 to calculate the vapor pressure at a second temperature, P_2.

$$\ln\frac{P_1}{P_2} = \frac{\Delta H_{vap}}{R}\left(\frac{1}{T_2} - \frac{1}{T_1}\right)$$

Setup Temperature must be expressed in kelvins, so $T_1 = 291.15$ K and $T_2 = 305.15$ K. Because the molar heat of vaporization is given in kJ/mol, we will have to convert it to J/mol for the units of R to cancel properly: $\Delta H_{vap} = 2.6 \times 10^4$ J/mol. The inverse function of ln x is e^x.

Solution

$$\ln\frac{P_1}{P_2} = \frac{2.6 \times 10^4 \text{ J/mol}}{8.314 \text{ J/K} \cdot \text{mol}}\left(\frac{1}{305.15 \text{ K}} - \frac{1}{291.15 \text{ K}}\right)$$

$$= -0.4928$$

$$\frac{P_1}{P_2} = e^{-0.4928} = 0.6109$$

$$\frac{P_1}{0.6109} = P_2$$

$$P_2 = \frac{401 \text{ mmHg}}{0.6109} = 6.6 \times 10^2 \text{ mmHg}$$

Think About It

It is easy to switch P_1 and P_2 or T_1 and T_2 accidentally and get the wrong answer to a problem such as this. One way to help safeguard against this common error is to verify that the vapor pressure is *higher* at the higher temperature.

Practice Problem ATTEMPT The vapor pressure of ethanol is 1.00×10^2 mmHg at 34.9°C. What is its vapor pressure at 55.8°C? (ΔH_{vap} for ethanol is 39.3 kJ/mol.)

Practice Problem BUILD Estimate the molar heat of vaporization of a liquid whose vapor pressure doubles when the temperature is raised from 85°C to 95°C. At what temperature will the vapor pressure be five times the value at 85°?

Practice Problem CONCEPTUALIZE The diagram on the left depicts a system at room temperature. (The space above the liquid contains air and the vapor of the liquid in the container, and is sealed with a movable piston, making the pressure inside the vessel equal to atmospheric pressure.) Which of the diagrams [(i)–(v)] could represent the same system at a higher temperature?

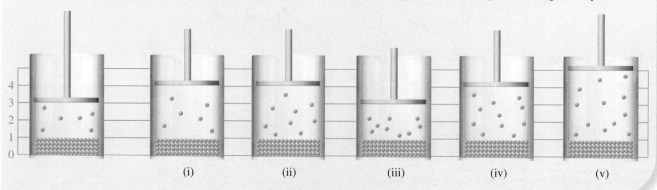

Boiling Point

We have learned that the vapor pressure of a liquid increases with increasing temperature. When the vapor pressure reaches the external pressure, the liquid boils. In fact, the **boiling point** of a substance is defined as the temperature at which its vapor pressure equals the external, atmospheric pressure. As a result, the boiling point of a substance varies with the external pressure. At the top of a mountain, for example, where the atmospheric pressure is lower than that at sea level, the vapor pressure of water (or any liquid) reaches the external pressure at a lower temperature. Thus, the boiling point is lower than it would be at sea level. Like the other properties of liquids described in this section, boiling point varies with the magnitude of intermolecular forces exhibited by a substance. In general, the stronger the intermolecular attractions, the higher the boiling point.

Student Annotation: The temperature at which the vapor pressure of a liquid is equal to 1 atm is called the **normal boiling point.**

Student Hot Spot

Student data indicate you may struggle with boiling point. Access the SmartBook to view additional Learning Resources on this topic.

Section 12.2 Review

Properties of Liquids

12.2.1   At what temperature would diethyl ether have a vapor pressure of 250 mmHg? Use the vapor pressure at 18°C and ΔH_{vap} given in Worked Example 12.1.
   (a) 5.6°C
   (b) 280°C
   (c) 17°C
   (d) 6.5°C
   (e) −270°C

12.2.2   Given the following information for C_6F_6, calculate its ΔH_{vap}. At 300 K, the vapor pressure is 92.47 mmHg, and at 320 K, the vapor pressure is 225.1 mmHg.
   (a) 208 kJ/mol
   (b) 411 kJ/mol
   (c) 16.4 kJ/mol
   (d) 10.3 kJ/mol
   (e) 35.5 kJ/mol

12.2.3   Using the graph, estimate the vapor pressure of the liquid at 100°C.

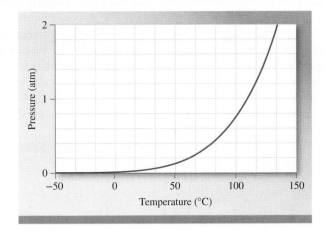

   (a) 100 atm
   (b) 0.75 atm
   (c) 1.00 atm
   (d) 110 atm
   (e) 1.50 atm

12.2.4   Using the result from question 12.2.3 and another point from the graph, estimate ΔH_{vap} for the liquid.
   (a) 15 kJ/mol
   (b) 35 kJ/mol
   (c) 55 kJ/mol
   (d) 75 kJ/mol
   (e) 95 kJ/mol

12.3   THE PROPERTIES OF SOLIDS

As with liquids, the magnitudes of intermolecular forces in *solids* are responsible for some of their important physical properties, including melting point and vapor pressure. Moreover, solids fall generally into two categories: amorphous and crystalline, properties that also are impacted by intermolecular forces.

Melting Point

The ***melting point*** of a solid is the temperature at which the energies of individual particles (molecules, atoms, or ions) enable them to break free of their fixed positions in the solid—allowing them to flow past one another. At the melting point, the solid and liquid phases of a substance coexist in equilibrium. Solid substances with strong intermolecular forces melt at higher temperatures than those with weaker intermolecular forces. Figure 12.9 shows the melting points of the binary hydrogen compounds of Groups 4A through 7A (the same compounds used in Figure 7.5 (page 261) to illustrate the effect of hydrogen bonding on *boiling* point). The trends in melting point are nearly identical to the trends in boiling point—because both properties depend on the magnitude of attractive forces exhibited by a substance.

In general, attractive forces are greater between larger molecules than between smaller molecules; greater between polar molecules than between nonpolar molecules (of roughly equal size); and greater between molecules that can form hydrogen bonds than between molecules of similar size that cannot form hydrogen bonds. And because ion-ion interactions are typically much stronger than other types of intermolecular forces, the melting points of ionic solids tend to be very high. Table 12.2 shows how molar mass and types of intermolecular forces influence melting points of a variety of substances.

Vapor Pressure of Solids

Like liquids, solids have characteristic vapor pressures that depend on the magnitude of intermolecular forces. Because molecules are more tightly held in a solid, the vapor pressure of a solid is generally much lower than that of the corresponding liquid.

The vapor pressures of solids typically are very low at room temperature—often too low to measure easily. However, there are exceptions. Several examples of solids with discernible vapor pressures are shown in Figure 12.10. In the case of naphthalene [Figure 12.10(a)], we can detect its vapor pressure (on the order of 0.1 mmHg at room temperature) by its tarlike acrid smell. Naphthalene ($C_{10}H_8$) was once widely used as a fumigant and pesticide—although it has largely been replaced by other compounds that are less toxic to humans and pets. Iodine [Figure 12.10(b)] produces a visible vapor. Solid

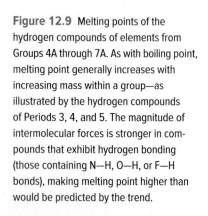

Figure 12.9 Melting points of the hydrogen compounds of elements from Groups 4A through 7A. As with boiling point, melting point generally increases with increasing mass within a group—as illustrated by the hydrogen compounds of Periods 3, 4, and 5. The magnitude of intermolecular forces is stronger in compounds that exhibit hydrogen bonding (those containing N—H, O—H, or F—H bonds), making melting point higher than would be predicted by the trend.

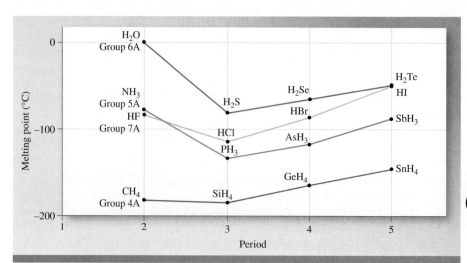

TABLE 12.2	Intermolecular Forces and Melting Points		
Substance	$\mathcal{M}$ **(g/mol)**	**Types of IM Forces**	**Melting Point (°C)**
He	4.003	Dispersion	−272.2
H_2O	18.02	Dispersion, dipole-dipole, hydrogen bonding	0
HF	20.01	Dispersion, dipole-dipole, hydrogen bonding	−83.6
LiF	25.94	Dispersion, ion-ion	845
HCl	36.46	Dispersion, dipole-dipole	−114.2
F_2	38.00	Dispersion	−219.6
Ar	39.95	Dispersion	−189.3
NaCl	58.44	Dispersion, ion-ion	801
KI	166.0	Dispersion, ion-ion	680
$C_6H_{12}O_6$ (fructose)	180.2	Dispersion, dipole-dipole, hydrogen bonding	103
$C_{12}H_{22}O_{11}$ (sucrose)	342.3	Dispersion, dipole-dipole, hydrogen bonding	186

carbon dioxide, commonly known as dry ice, is perhaps the most familiar—and most remarkable example [Figure 12.10(c)]. The vapor pressure of solid CO_2 is more than 56 atm at 20°C—a pressure great enough to cause most sealed containers to explode.

Amorphous Solids

In most solids, the atoms, molecules, or ions occupy positions in a regular three-dimensional arrangement. These substances are referred to as *crystalline;* and they constitute the majority of solids because of their inherent stability. (We will learn more about crystalline solids shortly.) However, if a solid forms under certain extraordinary conditions (e.g., when a liquid is cooled quickly), there may not be sufficient time for the atoms or molecules to move into the positions of a regular crystal before they become locked in place. The resulting substances are known as ***amorphous solids.***

The best-known example of an amorphous solid is glass, which is one of civilization's most valuable and versatile materials. It is also one of the oldest—glass articles date back as far as 1000 B.C. ***Glass*** commonly refers to an optically transparent fusion product of inorganic materials that has cooled to a rigid state without becoming crystalline. In this context, fusion refers to the process by which glass is formed:

(a)   (b)   (c)

Figure 12.10 (a) Traditionally made of naphthalene, mothballs were used in closets and trunks to kill moths and their larvae—thereby protecting clothing, blankets, and other fabrics from being consumed by the pests. (b) Iodine solid, with its relatively high vapor pressure, readily becomes a visible gas. (c) Solid carbon dioxide has an especially high vapor pressure—greater than can be contained by conventional vessels.

(a) © H.S. Photos/Alamy Stock Photo; (b) © Charles D. Winters/Science Source; (c) © McGraw-Hill Education/Charles D. Winters, photographer

Figure 12.11 Two-dimensional representation of (a) crystalline quartz and (b) noncrystalline (amorphous) quartz glass. The small spheres represent silicon. In reality, the structure of quartz is three dimensional. Each Si atom is bonded in a tetrahedral arrangement to four O atoms.

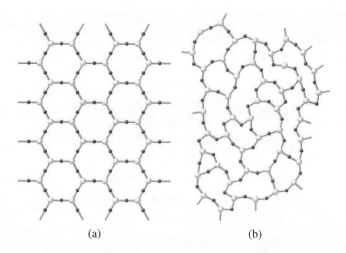

(a)   (b)

mixing molten silicon dioxide (SiO_2), the major component, with compounds such as sodium oxide (Na_2O), boron oxide (B_2O_3), and certain transition metal oxides for color and other properties.

There are hundreds of different types of glass in common use today. Figure 12.11 shows two-dimensional schematic representations of crystalline quartz and amorphous quartz glass. Table 12.3 lists the composition and properties of quartz, Pyrex, and soda-lime glass. The color of glass is due largely to the presence of oxides of metal ions—mostly transition metal ions. For example, green glass contains iron(III) oxide (Fe_2O_3) or copper(II) oxide (CuO), yellow glass contains uranium(IV) oxide (UO_2), blue glass contains cobalt(II) and copper(II) oxides (CoO and CuO), and red glass contains small particles of gold and copper.

Crystalline Solids

A *crystalline solid* possesses rigid and long-range order; its atoms, molecules, or ions occupy specific positions. The arrangement of the particles in a crystalline solid, which we call the *lattice structure*, depends on the nature and size of the particles involved. The forces responsible for the stability of a crystal can be ionic forces, covalent bonds, van der Waals forces, hydrogen bonds, or a combination of some of these forces.

Unit Cells

A *unit cell* is the basic repeating structural unit of a crystalline solid. Figure 12.12 shows a unit cell and its extension in three dimensions. Each sphere represents an atom, ion, or molecule and is called a *lattice point*. For the purpose of clarity, we will limit our discussion in this section to metal crystals in which each lattice point is occupied by an atom.

Every crystalline solid can be described in terms of one of the seven types of unit cells shown in Figure 12.13. The geometry of the cubic unit cell is particularly simple because all sides and all angles are equal. Any of the unit cells, when repeated in space in all three dimensions, forms the lattice structure characteristic of a crystalline solid.

TABLE 12.3	Composition and Properties of Three Types of Glass	
Pure quartz glass	100% SiO_2	Low thermal expansion, transparent to a wide range of wavelengths. Used in optical research.
Pyrex glass	60%–80% SiO_2, 10%–25% B_2O_3, some Al_2O_3	Low thermal expansion; transparent to visible and infrared, but not to ultraviolet light. Used in cookware and laboratory glassware.
Soda-lime glass	75% SiO_2, 15% Na_2O, 10% CaO	Easily attacked by chemicals and sensitive to thermal shocks. Transmits visible light but absorbs ultraviolet light. Used in windows and bottles.

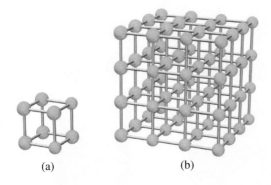

Figure 12.12 (a) A single unit cell and (b) a lattice (three-dimensional array) made up of many unit cells. Each sphere represents a lattice point, which may be an atom, a molecule, or an ion.

(a)   (b)

Packing Spheres

We can understand the geometric requirements for crystal formation by considering the different ways of packing a number of identical atoms to form an ordered three-dimensional structure. The way the atoms are arranged in layers determines the type of unit cell.

To understand the three-dimensional structure of most solids, we begin by considering a one-dimensional *row* of atoms. To make a two-dimensional *plane* from this row of atoms, the simplest method is to duplicate the row and place the duplicate rows directly below the original, such that the atoms align to form columns. This two-dimensional packing arrangement is known as *square* packing. In a square-packed array, the unit cell is a square. The atom labeled *x* has four neighbor atoms (each labeled *y*) touching it directly.

In the simplest case, the layer of atoms in a square-packed array can form a three-dimensional structure by placing a layer above and below the original in such a way that atoms in one layer lie directly over the atoms in the layer below [Figure 12.14(a)]. This procedure can be extended to generate many layers, as in the case of a crystal. With square packing in three dimensions, the atom labeled *x* is now in immediate contact with *six* atoms: *four* atoms in its own layer, *one* atom in the layer above it, and *one* atom in the layer below it. Each atom in this arrangement is said to have a coordination number of 6 because it has *six* immediate neighbors. The **coordination number** is the number of atoms surrounding an atom in a crystal lattice. The value of the coordination number indicates how tightly the atoms are packed together—the larger the coordination number, the more tightly packed the atoms are. The basic repeating unit in this array of atoms is called a **simple cubic cell** (scc) [Figure 12.14(b)].

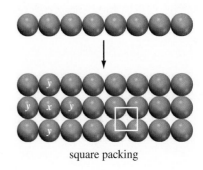

square packing

Student Annotation: The simple cubic cell is also called the *primitive* cubic cell.

Figure 12.13 The seven types of unit cells. Angle α is defined by edges *b* and *c*, angle β by edges *a* and *c*, and angle γ by edges *a* and *b*.

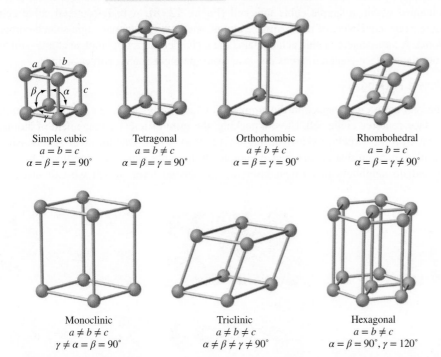

Simple cubic
$a = b = c$
$\alpha = \beta = \gamma = 90°$

Tetragonal
$a = b \neq c$
$\alpha = \beta = \gamma = 90°$

Orthorhombic
$a \neq b \neq c$
$\alpha = \beta = \gamma = 90°$

Rhombohedral
$a = b = c$
$\alpha = \beta = \gamma \neq 90°$

Monoclinic
$a \neq b \neq c$
$\gamma \neq \alpha = \beta = 90°$

Triclinic
$a \neq b \neq c$
$\alpha \neq \beta \neq \gamma \neq 90°$

Hexagonal
$a = b \neq c$
$\alpha = \beta = 90°, \gamma = 120°$

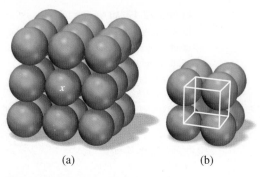

Figure 12.14 Arrangement of identical spheres in a simple cubic cell. (a) Three layers of square-packed spheres. (b) Definition of a simple cubic cell.

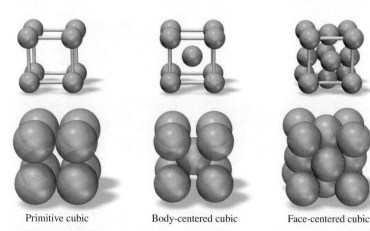

Primitive cubic   Body-centered cubic   Face-centered cubic

Figure 12.15 Three types of cubic cells. The top view makes it easier to see the locations of the lattice points, but the bottom view is more realistic, with the spheres touching one another.

The other types of cubic cells, shown in Figure 12.15, are the ***body-centered cubic cell*** (bcc) and the ***face-centered cubic cell*** (fcc). Unlike the simple cube, the second layer of atoms in the body-centered cubic arrangement fits into the depressions of the first layer and the third layer fits into the depressions of the second layer (Figure 12.16).

The coordination number of each atom in the bcc structure is 8 (each sphere is in contact with four others in the layer above and four others in the layer below). In the face-centered cubic cell, there are atoms at the center of each of the six faces of the cube, in addition to the eight corner atoms. The coordination number in the face-centered cubic cell is 12 (each sphere is in contact with four others in its own layer, four others in the layer above, and four others in the layer below).

Because every unit cell in a crystalline solid is adjacent to other unit cells, most of a cell's atoms are shared by neighboring cells. (The atom at the center of the body-centered cubic cell is an exception.) In all types of cubic cells, for example, each corner atom belongs to eight unit cells whose corners all touch [Figure 12.17(a)]. An atom that lies on an edge, on the other hand, is shared by four unit cells [Figure 12.17(b)], and a face-centered atom is shared by two unit cells [Figure 12.17(c)]. Because a simple cubic cell has lattice points only at each of the eight corners, and because each corner atom is shared by eight unit cells, there will be the equivalent of only *one* complete atom contained within a simple cubic unit cell (Figure 12.18). A body-centered cubic cell contains the equivalent of two complete atoms, one in the center and eight shared corner atoms. A face-centered cubic cell contains the equivalent of four complete atoms—three from the six face-centered atoms and one from the eight shared corner atoms.

Closest Packing

There is more empty space in the simple cubic and body-centered cubic cells than in the face-centered cubic cell. Closest packing, the most efficient arrangement of atoms, starts with the structure shown in Figure 12.19(a), which we call layer A. Focusing on the only atom that is surrounded completely by other atoms, we see that it has six immediate neighbors in its own layer. In the second layer, which we call layer B,

Figure 12.16 In the body-centered cubic arrangement, the spheres in each layer rest in the depressions between spheres in the previous layer.

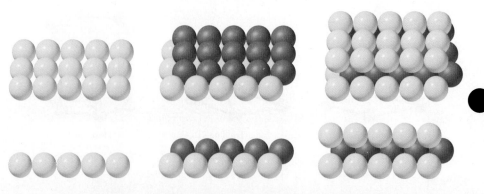

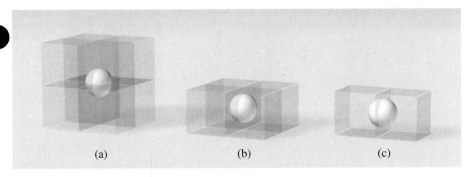

Figure 12.17 (a) A corner atom in any cell is shared by eight unit cells. (b) An edge atom is shared by four unit cells. (c) A face-centered atom in a cubic cell is shared by two unit cells.

Figure 12.18 Because each sphere is shared by eight unit cells and there are eight corners in a cube, there is the equivalent of one complete sphere inside a simple cubic unit cell.

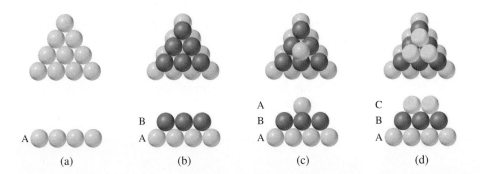

Figure 12.19 (a) In a close-packed layer, each sphere is in contact with six others. (b) Spheres in the second layer fit into the depressions between the first-layer spheres. (c) In the hexagonal close-packed structure, each third-layer sphere is directly over a first-layer sphere. (d) In the cubic close-packed structure, each third-layer sphere fits into a depression that is directly over a depression in the first layer.

atoms are packed into the depressions between the atoms in the first layer so that all the atoms are as close together as possible [Figure 12.19(b)].

There are two ways that a third layer of atoms can be arranged. They may sit in the depressions between second-layer atoms such that the third-layer atoms lie directly over atoms in the first layer [Figure 12.19(c)]. In this case, the third layer is also labeled A. Alternatively, atoms in the third layer may sit in a *different* set of depressions such that they do not lie directly over atoms in the first layer [Figure 12.19(d)]. In this case, we label the third layer C.

Figure 12.20 shows the exploded views and the structures resulting from these two arrangements. The ABA arrangement [Figure 12.20(a)] is known as the *hexagonal close-packed (hcp) structure,* and the ABC arrangement [Figure 12.20(b)] is the *cubic close-packed (ccp) structure,* which corresponds to the face-centered cube already described. In the hcp structure, the spheres in every other layer occupy the same verti-cal position (ABABAB . . .), while in the ccp structure, the spheres in every fourth layer occupy the same vertical position (ABCABCA . . .). In both structures, each sphere has a coordination number of 12 (each sphere is in contact with six spheres in its own layer, three spheres in the layer above, and three spheres in the layer below). Both the hcp and ccp structures represent the most efficient way of packing identical spheres in a unit cell, and the coordination number cannot exceed 12.

Many metals form crystals with hcp or ccp structures. For example, magnesium, titanium, and zinc crystallize with their atoms in an hcp array, while aluminum, nickel,

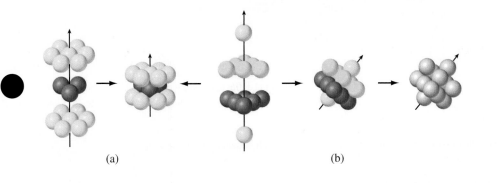

Figure 12.20 Exploded views of (a) a hexagonal close-packed structure and (b) a cubic close-packed structure. This view is tilted to show the face-centered cubic unit cell more clearly. Note that this arrangement is the same as the face-centered unit cell.

Figure 12.21 The relationship between the edge length (a) and radius (r) of atoms in the (a) simple cubic cell, (b) body-centered cubic cell, and (c) face-centered cubic cell.

scc

$a = 2r$

bcc

$b^2 = a^2 + a^2$
$c^2 = a^2 + b^2$
$= 3a^2$
$c = \sqrt{3}a = 4r$
$a = \dfrac{4r}{\sqrt{3}}$

fcc

$b = 4r$
$b^2 = a^2 + a^2$
$16r^2 = 2a^2$
$a = \sqrt{8}r$

(a)   (b)   (c)

Student Annotation: The noble gases, which are monatomic, crystallize in the ccp structure, with the exception of helium, which crystallizes in the hcp structure.

and silver crystallize in the ccp arrangement. A substance will crystallize with the arrangement that maximizes the stability of the solid.

Figure 12.21 summarizes the relationship between the atomic radius r and the edge length a of a simple cubic cell, a body-centered cubic cell, and a face-centered cubic cell. This relationship can be used to determine the atomic radius of a sphere in which the density of the crystal is known.

Thinking Outside the Box

X-Ray Diffraction

Virtually all we know about crystal structure has been learned from X-ray diffraction studies. X-ray diffraction is the scattering of X rays by the units of a crystalline solid. The scattering, or *diffraction patterns,* produced are used to deduce the arrangement of particles in the solid lattice.

In Section 3.2 we discussed the interference phenomenon associated with waves (see Figure 3.4). Because X rays are a form of electromagnetic radiation (i.e., they are *waves*), they exhibit interference phenomena under suitable conditions. In 1912 Max von Laue[3] correctly suggested that, because the wavelength of X rays is comparable in magnitude to the distances between lattice points in a crystal, the lattice should be able to diffract X rays. An X-ray diffraction pattern is the result of interference in the waves associated with X rays.

The figure here shows a typical X-ray diffraction setup. A beam of X rays is directed at a mounted crystal. When X-ray photons encounter the

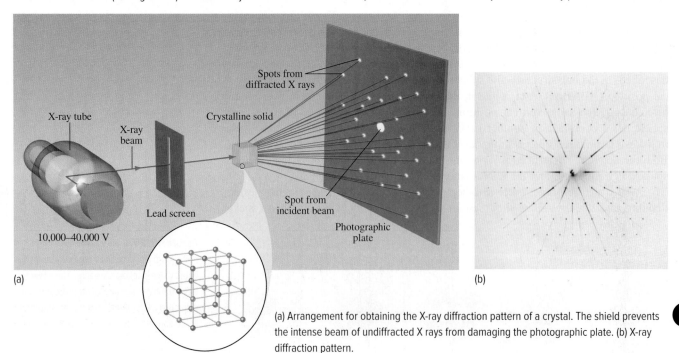

Spots from diffracted X rays

X-ray tube

X-ray beam

Crystalline solid

Lead screen

10,000–40,000 V

Spot from incident beam

Photographic plate

(a)

(b)

(a) Arrangement for obtaining the X-ray diffraction pattern of a crystal. The shield prevents the intense beam of undiffracted X rays from damaging the photographic plate. (b) X-ray diffraction pattern.

[3]Max Theodor Felix von Laue (1879–1960), a German physicist, received the Nobel Prize in Physics in 1914 for his discovery of X-ray diffraction.

electrons in the atoms of a crystalline solid, some of the incoming radiation is reflected, much as visible light is reflected by a mirror; the process is called the scattering of X rays.

To understand how a diffraction pattern arises, consider the scattering of X rays by the atoms in two parallel planes (as shown). Initially, the two incident rays are in phase with each other (their maxima and minima occur at the same positions). The upper wave is scattered, or reflected, by an atom in the first layer, while the lower wave is scattered by an atom in the second layer. For these two scattered waves to be in phase again, the extra distance traveled by the lower wave (the sum of the distance between points B and C and the distance between points C and D) must be an integral multiple of the wavelength (λ) of the X ray; that is,

$$BC + CD = 2d \sin \theta = n\lambda \qquad n = 1, 2, 3, \ldots$$

where θ is the angle between the X rays and the plane of the crystal and d is the distance between adjacent planes. This is known as the Bragg equation, after William H. Bragg and Sir William L. Bragg.[4] The reinforced waves produce a dark spot on a photographic film for each value of θ that satisfies the Bragg equation.

The X-ray diffraction technique offers the most accurate method for determining bond lengths and bond angles in molecules in solids. Because X rays are scattered by electrons, chemists can construct an electron-density contour map from the diffraction patterns by using a complex mathematical procedure. Basically, an electron-density contour map tells us the relative electron densities at various locations in a molecule. The densities reach a maximum near the center of each atom. In this manner, we can determine the positions of the nuclei and hence the geometric parameters of the molecule.

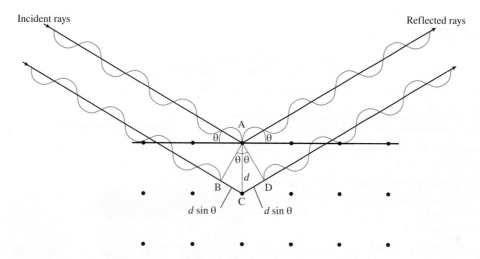

Reflection of X rays from two layers of atoms. The lower wave travels a distance $2d \sin \theta$ longer than the upper wave does. For the two waves to be in phase again after reflection, it must be true that $2d \sin \theta = n\lambda$, where λ is the wavelength of the X ray and $n = 1, 2, 3, \ldots$ The sharply defined spots in the X-ray diffraction figure are observed only if the crystal is large enough to consist of hundreds of parallel layers.

---

[4]William Henry Bragg (1862–1942) and Sir William Lawrence Bragg (1890–1972) were English physicists, father and son. Both worked on X-ray crystallography. The younger Bragg formulated the fundamental equation for X-ray diffraction. The two shared the Nobel Prize in Physics in 1915.

Worked Example 12.2 illustrates the relationships between the unit cell type, cell dimensions, and density of a metal.

Worked Example 12.2

Gold crystallizes in a cubic close-packed structure (face-centered cubic unit cell) and has a density of 19.3 g/cm³. Calculate the atomic radius of an Au atom in angstroms (Å).

Strategy Using the given density and the mass of gold contained within a face-centered cubic unit cell, determine the volume of the unit cell. Then, use the volume to determine the value of a, and use the equation supplied in Figure 12.21(c) to find r. Be sure to use consistent units for mass, length, and volume.

Setup The face-centered cubic unit cell contains a total of four atoms of gold [six faces, each shared by two unit cells, and eight corners, each shared by eight unit cells—Figure 12.21(c)]. $d = m/V$ and $V = a^3$.

(Continued on next page)

Solution First, we determine the mass of gold (in grams) contained within a unit cell.

$$m = \frac{4 \text{ atoms}}{\text{unit cell}} \times \frac{1 \text{ mol}}{6.022 \times 10^{23} \text{ atoms}} \times \frac{197.0 \text{ g Au}}{1 \text{ mol Au}} = 1.31 \times 10^{-21} \text{ g/unit cell}$$

Then we calculate the volume of the unit cell in cm^3.

$$V = \frac{m}{d} = \frac{1.31 \times 10^{-21} \text{ g}}{19.3 \text{ g/cm}^3} = 6.78 \times 10^{-23} \text{ cm}^3$$

Using the calculated volume and the relationship $V = a^3$ (rearranged to solve for a), we determine the length of a side of a unit cell.

$$a = \sqrt[3]{V} = \sqrt[3]{6.78 \times 10^{-23} \text{ cm}^3} = 4.08 \times 10^{-8} \text{ cm}$$

Using the relationship provided in Figure 12.21(c) (rearranged to solve for r), we determine the radius of a gold atom in centimeters.

$$r = \frac{a}{\sqrt{8}} = \frac{4.08 \times 10^{-8} \text{ cm}}{\sqrt{8}} = 1.44 \times 10^{-8} \text{ cm}$$

Finally, we convert centimeters to angstroms.

$$1.44 \times 10^{-8} \text{ cm} \times \frac{1 \times 10^{-2} \text{ m}}{1 \text{ cm}} \times \frac{1 \text{ Å}}{1 \times 10^{-10} \text{ m}} = 1.44 \text{ Å}$$

Think About It
Atomic radii tend to be on the order of 1 Å, so this answer is reasonable.

Practice Problem Ⓐ**TTEMPT** When silver crystallizes, it forms face-centered cubic cells. The unit cell edge length is 4.087 Å. Calculate the density of silver.

Practice Problem Ⓑ**UILD** The density of sodium metal is 0.971 g/cm^3, and the unit cell edge length is 4.285 Å. Determine the unit cell (simple, body-centered, or face-centered cubic) of sodium metal.

Practice Problem Ⓒ**ONCEPTUALIZE** The diagram shows two different arrangements of circles. Using the areas defined by the red rectangles, determine the two-dimensional "density" (ratio of area occupied by circles to total area) for each arrangement; and determine the ratio of densities for the two arrangements. (Report the ratio of densities to two significant figures.)

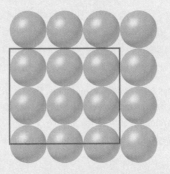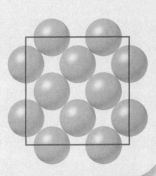

Section 12.3 Review

Crystal Structure

12.3.1   Nickel has a face-centered cubic unit cell with an edge length of 352.4 pm. Calculate the density of nickel.

(a) 2.227 g/cm^3 (d) 8.908 g/cm^3
(b) 4.455 g/cm^3 (e) 11.14 g/cm^3
(c) 38.99 g/cm^3

12.3.2   A metal crystallizes in a body-centered cubic unit cell with an edge length of 3.09×10^{-8} cm and has a density of 5.74 g/cm^3. Determine the approximate molar mass and the identity of the metal.

(a) 51 g/mol, V   (d) 12 g/mol, C
(b) 101 g/mol, Ru   (e) 204 g/mol, Tl
(c) 27 g/mol, Al

12.4 TYPES OF CRYSTALLINE SOLIDS

The structures and properties of crystalline solids, such as melting point, density, and hardness, are determined by the kinds of forces that hold the particles together. We can classify any crystal as one of four types: ionic, covalent, molecular, or metallic.

Ionic Crystals

Ionic crystals are composed of charged spheres (cations and anions) that are held together by Coulombic attraction. Anions typically are considerably larger than cations [◄◄ Section 4.6], and the relative sizes and numbers of the ions in a compound determine how the ions are arranged in the solid lattice. NaCl adopts a face-centered cubic arrangement as shown in Figure 12.22. Note the positions of ions within the unit cell, and within the lattice overall. Both the Na^+ ions and the Cl^- ions adopt face-centered cubic arrangements, and the unit cell defined by the arrangement of cations overlaps with the unit cell defined by the arrangement of anions. Look closely at the unit cell shown in Figure 12.23(a). It is defined as fcc by the positions of the Cl^- ions. Recall that there is the equivalent of four spheres contained in the fcc unit cell (half a sphere at each of six faces and one-eighth of a sphere at each of eight corners). In this case the spheres are Cl^- ions, so the unit cell of NaCl contains four Cl^- ions. Now look at the positions of the Na^+ ions. There are Na^+ ions centered on each edge of the cube, in addition to one Na^+ at the center. Each sphere on the cube's edge is shared by four unit cells, and there are 12 such edges. Thus, the unit cell in Figure 12.22(a) also

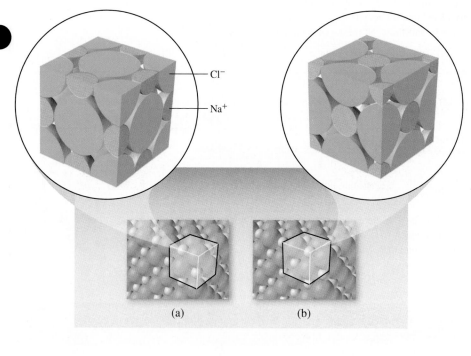

Cl^-

Na^+

(a) (b)

Figure 12.22 The unit cell of an ionic compound can be defined either by (a) the positions of anions or (b) the positions of cations.

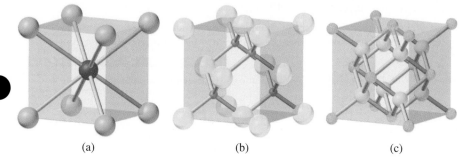

(a) (b) (c)

Figure 12.23 Crystal structures of (a) CsCl, (b) ZnS, and (c) CaF_2. In each case, the smaller sphere represents the cation.

Student Annotation: It is a common mistake to identify the CsCl structure as body-centered cubic. Remember that the lattice points used to define a unit cell must all be identical. In this case, they are all Cl⁻ ions. CsCl has a simple cubic unit cell.

contains four Na^+ ions (one-quarter sphere at each of 12 edges, giving three spheres, and one sphere at the center). The unit cell of an ionic compound always contains the same ratio of cations to anions as the empirical formula of the compound.

Figure 12.23 shows the crystal structures of three ionic compounds: CsCl, ZnS, and CaF_2. Cesium chloride [Figure 12.23(a)] has the simple cubic lattice. Despite the apparent similarity of the formulas of CsCl and NaCl, CsCl adopts a different arrangement because the Cs^+ ion is much larger than the Na^+ ion. Zinc sulfide [Figure 12.23(b)] has the *zincblende* structure, which is based on the face-centered cubic lattice. If the S^{2-} ions occupy the lattice points, the smaller Zn^{2+} ions are arranged tetrahedrally about each S^{2-} ion. Other ionic compounds that have the zincblende structure include CuCl, BeS, CdS, and HgS. Calcium fluoride [Figure 12.23(c)] has the *fluorite* structure. The unit cell in Figure 12.23(c) is defined based on the positions of the cations, rather than the positions of the anions. The Ca^{2+} ions occupy the lattice points, and each F^- ion is surrounded tetrahedrally by four Ca^{2+} ions. The compounds SrF_2, BaF_2, $BaCl_2$, and PbF_2 also have the fluorite structure.

Worked Examples 12.3 and 12.4 show how to determine the number of ions in a unit cell and the density of an ionic crystal, respectively.

Worked Example 12.3

How many of each ion are contained within a unit cell of ZnS?

Strategy Determine the contribution of each ion in the unit cell based on its position.

Setup Referring to Figure 12.23, the unit cell has four Zn^{2+} ions completely contained within the unit cell, and S^{2-} ions at each of the eight corners and at each of the six faces. Interior ions (those completely contained within the unit cell) contribute one, those at the corners each contribute one-eighth, and those on the faces each contribute one-half.

Solution The ZnS unit cell contains four Zn^{2+} ions (interior) and four S^{2-} ions [$8 \times \frac{1}{8}$(corners) and $6 \times \frac{1}{2}$(faces)].

> **Think About It**
> Make sure that the ratio of cations to anions that you determine for a unit cell matches the ratio expressed in the compound's empirical formula.

Practice Problem Ⓐ**TTEMPT** Referring to Figure 12.23, determine how many of each ion are contained within a unit cell of CaF_2.

Practice Problem Ⓑ**UILD** Referring to Figure 12.23, determine how many of each ion are contained within a unit cell of CsCl.

Practice Problem Ⓒ**ONCEPTUALIZE** The diagram here shows the anions in the hypothetical edge-centered unit cell of an ionic compound in which the ions combine in a 1:1 ratio. Referring to the diagram, determine the total number of ions contained within the unit cell.

Worked Example 12.4

The edge length of the NaCl unit cell is 564 pm. Determine the density of NaCl in g/cm^3.

Strategy Use the number of Na^+ and Cl^- ions in a unit cell (four of each) to determine the mass of a unit cell. Calculate volume using the edge length given in the problem statement. Density is mass divided by volume ($d = m/V$). Be careful to use units consistently.

Setup The masses of Na^+ and Cl^- ions are 22.99 amu and 35.45 amu, respectively. The conversion factor from amu to grams is

$$\frac{1 \text{ g}}{6.022 \times 10^{23} \text{ amu}}$$

Student Annotation: Note that the mass of an atomic *ion* is treated the same as the mass of the parent *atom*. In these cases, the mass of an electron is not significant [◄◄ Section 2.2, Table 2.1].

so the masses of the Na^+ and Cl^- ions are 3.818×10^{-23} g and 5.887×10^{-23} g, respectively. The unit cell length is

$$564 \text{ pm} \times \frac{1 \times 10^{-12} \text{ m}}{1 \text{ pm}} \times \frac{1 \text{ cm}}{1 \times 10^{-2} \text{ m}} = 5.64 \times 10^{-8} \text{ cm}$$

Solution The mass of a unit cell is 3.882×10^{-22} g ($4 \times 3.818 \times 10^{-23}$ g $+ 4 \times 5.887 \times 10^{-23}$ g). The volume of a unit cell is 1.794×10^{-22} cm^3 [$(5.64 \times 10^{-8}$ cm$)^3$]. Therefore, the density is given by

$$d = \frac{3.882 \times 10^{-22} \text{ g}}{1.794 \times 10^{-22} \text{ cm}^3} = 2.16 \text{ g/cm}^3$$

Think About It

If you were to hold a cubic centimeter (1 cm^3) of salt in your hand, how heavy would you expect it to be? Common errors in this type of problem include errors of unit conversion—especially with regard to length and volume. Such errors can lead to results that are off by many orders of magnitude. Often you can use common sense to gauge whether a calculated answer is reasonable. For instance, simply getting the centimeter-meter conversion upside down would result in a calculated density of 2.16×10^{12} g/cm^3! You *know* that a cubic centimeter of salt doesn't have a mass that large. (That's billions of kilograms!) If the magnitude of a result is not reasonable, go back and check your work.

Practice Problem ATTEMPT LiF has the same unit cell as NaCl (fcc). The edge length of the LiF unit cell is 402 pm. Determine the density of LiF in g/cm^3.

Practice Problem BUILD NiO also adopts the face-centered cubic arrangement. Given that the density of NiO is 6.67 g/cm^3, calculate the length of the edge of its unit cell (in pm).

Practice Problem CONCEPTUALIZE Referring to the diagram in Practice Problem 12.3C, determine the density of the hypothetical ionic compound given that the radii of the anions and cations are 150 pm and 92 pm, respectively; and that their average masses are 98 amu and 192 amu, respectively.

Most ionic crystals have high melting points, which is an indication of the strong cohesive forces holding the ions together. A measure of the stability of ionic crystals is the lattice energy [◄◄ Section 5.3]; the higher the lattice energy, the more stable the compound. Ionic solids do not conduct electricity well because the ions are fixed in position. In the molten (melted) state or when dissolved in water, however, the compound's ions are free to move and the resulting liquid conducts electricity.

Covalent Crystals

In covalent crystals, atoms are held together in an extensive three-dimensional network entirely by covalent bonds. Well-known examples are two of carbon's allotropes: diamond and graphite. In diamond, each carbon atom is sp^3-hybridized and bonded to four other carbon atoms [Figure 12.24(a)]. The strong covalent bonds in three dimensions contribute to diamond's unusual hardness (it is the hardest material known) and very high melting point (3550°C). In graphite, carbon atoms are arranged in six-membered rings [Figure 12.24(b)]. The atoms are all sp^2-hybridized, and each atom is bonded to three other atoms. The remaining unhybridized $2p$ orbital on each carbon atom is used in pi bonding. In fact, each layer of graphite has the kind of delocalized molecular orbital that is present in benzene [◄◄ Section 7.8]. Because electrons are free to move around in this extensively delocalized molecular orbital, graphite is a good conductor of electricity in directions along the planes of the carbon atoms. The layers are held together by weak van der Waals forces. The covalent bonds in graphite account for its hardness; however, because the layers can slide past one another, graphite is slippery to the touch and is effective as a lubricant. It is also used as the "lead" in pencils.

Another covalent crystal is quartz (SiO_2). The arrangement of silicon atoms in quartz is similar to that of carbon in diamond, but in quartz there is an oxygen atom between each pair of Si atoms. Because Si and O have different electronegativities, the Si—O bond is polar. Nevertheless, SiO_2 is similar to diamond in many respects, such as being very hard and having a high melting point (1610°C).

Figure 12.24 Structures of (a) diamond and (b) graphite. Note that in diamond, each carbon atom is bonded in a tetrahedral arrangement to four other carbon atoms. In graphite, each carbon atom is bonded in a trigonal planar arrangement to three other carbon atoms. The distance between layers in graphite is 335 pm.

© McGraw-Hill Education/Charles D. Winters, photographer

Molecular Crystals

In a molecular crystal, the lattice points are occupied by molecules, so the attractive forces between them are van der Waals forces and/or hydrogen bonding. An example of a molecular crystal is solid sulfur dioxide (SO_2), in which the predominant attractive force is a dipole-dipole interaction. Intermolecular hydrogen bonding is mainly responsible for maintaining the three-dimensional lattice of ice (Figure 12.25). Other examples of molecular crystals are I_2, P_4, and S_8.

Figure 12.25 The three-dimensional structure of ice. The covalent bonds are shown by short solid lines and the weaker hydrogen bonds by long dotted lines between O and H. The empty space in the structure accounts for the low density of ice, relative to liquid water.

© McGraw-Hill Education/Charles D. Winters, photographer

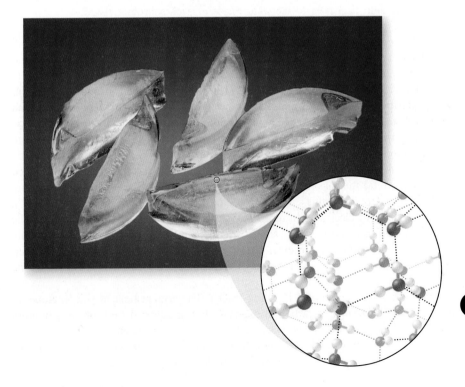

TABLE 12.4	Types of Crystals and Their General Properties		
Type of crystal	**Cohesive forces**	**General properties**	**Examples**
Ionic	Coulombic attraction and dispersion forces	Hard, brittle, high melting point, poor conductor of heat and electricity	NaCl, LiF, MgO, CaCO$_3$
Covalent	Covalent bonds	Hard, brittle, high melting point, poor conductor of heat and electricity	C (diamond),* SiO$_2$ (quartz)
Molecular[†]	Dispersion and dipole-dipole forces, hydrogen bonds	Soft, low melting point, poor conductor of heat and electricity	Ar, CO$_2$, I$_2$, H$_2$O, C$_{12}$H$_{22}$O$_{11}$
Metallic	Metallic bonds	Variable hardness and melting point, good conductor of heat and electricity	All metallic elements, such as Na, Mg, Fe, Cu

*Diamond is a good conductor of heat.
[†]Included in this category are crystals made up of individual atoms.

Except in ice, molecules in molecular crystals are generally packed together as closely as their size and shape allow. Because van der Waals forces and hydrogen bonding are usually quite weak compared with covalent and ionic bonds, molecular crystals are more easily broken apart than ionic and covalent crystals. Indeed, most molecular crystals melt at temperatures below 100°C.

Metallic Crystals

Every lattice point in a metallic crystal is occupied by an atom of the same metal. Metallic crystals are generally body-centered cubic, face-centered cubic, or hexagonal close-packed. Consequently, metallic elements are usually very dense.

The bonding in metals is quite different from that in other types of crystals. In a metal, the bonding electrons are delocalized over the entire crystal. In fact, metal atoms in a crystal can be imagined as an array of positive ions immersed in a sea of delocalized valence electrons (Figure 12.26). The great cohesive force resulting from delocalization is responsible for a metal's strength, whereas the mobility of the delocalized electrons makes metals good conductors of heat and electricity. Table 12.4 summarizes the properties of the four different types of crystals discussed. Note that the data in Table 12.4 refer to the solid phase of each substance listed.

Worked Example 12.5 lets you practice using unit-cell data to determine the density of a metal.

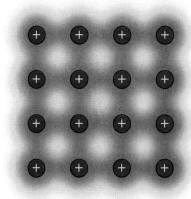

Figure 12.26 A cross section of a metallic crystal. Each circled positive charge represents the nucleus and inner electrons of a metal atom. The grey area surrounding the positive metal ions indicates the mobile "sea" of electrons.

Worked Example 12.5

The metal iridium (Ir) crystallizes with a face-centered cubic unit cell. Given that the length of the edge of a unit cell is 383 pm, determine the density of iridium in g/cm^3.

Strategy A face-centered metallic crystal contains four atoms per unit cell [8 × $\frac{1}{8}$(corners) and 6 × $\frac{1}{2}$(faces)]. Use the number of atoms per cell and the atomic mass to determine the mass of a unit cell. Calculate volume using the edge length given in the problem statement. Density is then mass divided by volume ($d = m/V$). Be sure to make all necessary unit conversions.

Setup The mass of an Ir atom is 192.2 amu. The conversion factor from amu to grams is

$$\frac{1 \text{ g}}{6.022 \times 10^{23} \text{ amu}}$$

so the mass of an Ir atom is 3.192×10^{-22} g. The unit cell length is

$$383 \text{ pm} \times \frac{1 \times 10^{-12} \text{ m}}{1 \text{ pm}} \times \frac{1 \text{ cm}}{1 \times 10^{-2} \text{ m}} = 3.83 \times 10^{-8} \text{ cm}$$

(Continued on next page)

Solution The mass of a unit cell is 1.277×10^{-21} g ($4 \times 3.192 \times 10^{-22}$ g). The volume of a unit cell is 5.618×10^{-23} cm^3 [$(3.83 \times 10^{-8}$ cm)^3]. Therefore, the density is given by

$$d = \frac{1.277 \times 10^{-21} \text{ g}}{5.62 \times 10^{-23} \text{ cm}^3} = 22.7 \text{ g/cm}^3$$

> **Think About It**
> Metals typically have high densities, so common sense can help you decide whether your calculated answer is reasonable.

Practice Problem Ⓐ**TTEMPT** Aluminum metal crystallizes in a face-centered cubic unit cell. If the length of the cell edge is 404 pm, what is the density of aluminum in g/cm^3?

Practice Problem Ⓑ**UILD** Copper crystallizes in a face-centered cubic lattice. If the density of the metal is 8.96 g/cm^3, what is the length of the unit cell edge in picometers?

Practice Problem Ⓒ**ONCEPTUALIZE** Given that the diameter and average mass of a billiard ball are 5.72 cm and 165 g, respectively, determine the density of a billiard ball. Assuming that they can be packed like atoms in a metal, determine the density of a collection of billiard balls packed with a simple cubic unit cell, and those packed with a face-centered unit cell. Explain why the three densities are different despite all referring to the same objects.

12.5   PHASE CHANGES

A *phase* is a homogeneous part of a system that is separated from the rest of the system by a well-defined *boundary*. When an ice cube floats in a glass of water, for example, the liquid water is one phase and the solid water (the ice cube) is another. Although the chemical properties of water are the same in both phases, the physical properties of a solid are different from those of a liquid—and the boundary between the solid (ice) and the liquid (water) is the *phase boundary*.

When a substance goes from one phase to another phase, we say that it has undergone a ***phase change.*** Phase changes in a system are generally caused by the addition or removal of energy, usually in the form of heat. Familiar examples of phase changes include the following:

> **Student Annotation:** Phase changes are *physical* changes [◄◄ Section 1.6].

Example	Phase Change
Freezing of water	$H_2O(l) \longrightarrow H_2O(s)$
Evaporation (or vaporization) of water	$H_2O(l) \longrightarrow H_2O(g)$
Melting (fusion) of ice	$H_2O(s) \longrightarrow H_2O(l)$
Condensation of water vapor	$H_2O(g) \longrightarrow H_2O(l)$
Sublimation of dry ice	$CO_2(s) \longrightarrow CO_2(g)$

The establishment of an equilibrium vapor pressure, as described in Section 12.2, involved two of these phase changes: vaporization and condensation. Figure 12.27 summarizes the various types of phase changes.

Liquid-Vapor

Because boiling point is defined in terms of the vapor pressure of the liquid, the boiling point is related to the ***molar heat of vaporization*** (ΔH_{vap}), the amount of heat required to vaporize a mole of substance at its boiling point. Indeed, the data in Table 12.5 show that the boiling point generally increases as ΔH_{vap} increases. Ultimately, both the boiling point and ΔH_{vap} are determined by the strength of intermolecular forces. For example, argon (Ar) and methane (CH_4), which have only relatively weak dispersion forces, have low boiling points and small molar heats of vaporization. Diethyl ether ($C_2H_5OC_2H_5$)

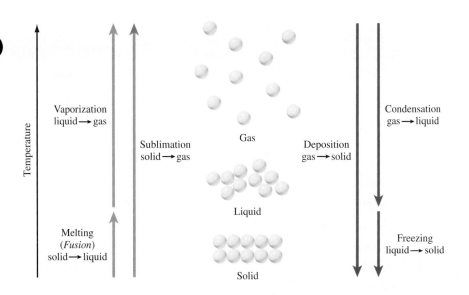

Figure 12.27 The six possible phase changes: melting (fusion), vaporization, sublimation, deposition, condensation, and freezing.

has a dipole moment, and the dipole-dipole forces account for its moderately high boiling point and ΔH_{vap}. Both ethanol (C_2H_5OH) and water have strong hydrogen bonding, which accounts for their high boiling points and large ΔH_{vap} values. Strong metallic bonding causes mercury to have the highest boiling point and ΔH_{vap} of the liquids in Table 12.5. Interestingly, benzene (C_6H_6), although nonpolar, has a high *polarizability* due to the distribution of its electrons in delocalized p molecular orbitals [◄◄ Section 7.8]. The dispersion forces that result can be as strong as (or even stronger than) dipole-dipole forces and/or hydrogen bonds.

The opposite of vaporization is condensation. In principle, a gas can be liquefied (made to condense) either by cooling or by applying pressure. Cooling a sample of gas decreases the kinetic energy of its molecules, so eventually the molecules aggregate to form small drops of liquid. Applying pressure to the gas (compression), on the other hand, reduces the distance between molecules, so they can be pulled together by intermolecular attractions. Many liquefication processes use a combination of reduced temperature and increased pressure.

Every substance has a ***critical temperature*** ($\boldsymbol{T_c}$) above which its gas phase cannot be liquefied, no matter how great the applied pressure. ***Critical pressure*** ($\boldsymbol{P_c}$) is the minimum pressure that must be applied to liquefy a substance *at* its critical temperature. At temperatures above the critical temperature, there is no fundamental distinction between a liquid and a gas—we simply have a fluid. A fluid at a temperature and pressure that exceed T_c and P_c, respectively, is called a ***supercritical fluid.*** Supercritical fluids have some remarkable properties and are used as solvents in a wide variety of industrial applications. The first such large-scale industrial use was the decaffeination of coffee with supercritical CO_2.

Student Annotation: T_c is the *highest* temperature at which a substance can exist as a liquid.

TABLE 12.5	Molar Heats of Vaporization for Selected Liquids	
Substance	**Boiling point (°C)**	**ΔH_{vap} (kJ/mol)**
Argon (Ar)	−186	6.3
Benzene (C_6H_6)	80.1	31.0
Ethanol (C_2H_5OH)	78.3	39.3
Diethyl ether ($C_2H_5OC_2H_5$)	34.6	26.0
Mercury (Hg)	357	59.0
Methane (CH_4)	−164	9.2
Water (H_2O)	100	40.79

TABLE 12.6	Critical Temperatures and Critical Pressures of Selected Substances		
Substance	T_c (°C)	P_c (atm)
Ammonia (NH_3)	132.4	111.5
Argon (Ar)	−122.2	6.3
Benzene (C_6H_6)	288.9	47.9
Carbon dioxide (CO_2)	31.0	73.0
Ethanol (C_2H_5OH)	243	63.0
Diethyl ether ($C_2H_5OC_2H_5$)	192.6	35.6
Mercury (Hg)	1462	1036
Methane (CH_4)	−83.0	45.6
Molecular hydrogen (H_2)	−239.9	12.8
Molecular nitrogen (N_2)	−147.1	33.5
Molecular oxygen (O_2)	−118.8	49.7
Sulfur hexafluoride (SF_6)	45.5	37.6
Water (H_2O)	374.4	219.5

Table 12.6 lists the critical temperatures and critical pressures of a number of common substances. The critical temperature of a substance reflects the strength of its intermolecular forces. Benzene, ethanol, mercury, and water, which have strong intermolecular forces, also have high critical temperatures compared with the other substances listed in the table.

Solid-Liquid

Student Annotation: We first encountered the term *melting point* in Section 12.3. Note that the melting point and the freezing point of a substance are the same temperature.

The transformation of a solid to a liquid is called *melting,* or **fusion.** The opposite process (transformation of a liquid to a solid) is called *freezing*. For a given substance, the melting point and *freezing point* are the temperature at which solid and liquid phases coexist in equilibrium. The melting point and freezing point at 1 atm specifically are called the **normal melting point** and **normal freezing point,** respectively.

Student Annotation: We refer to the *melting point* when we are increasing the temperature of a solid—and to the *freezing point* when we are decreasing the temperature of a liquid. For any substance, these two terms refer to the same temperature.

A familiar liquid-solid equilibrium is that of water and ice. At 0°C and 1 atm, the dynamic equilibrium is represented by

$$\text{ice} \rightleftharpoons \text{water}$$

or

$$H_2O(s) \rightleftharpoons H_2O(l)$$

Student Annotation: In most cases a glass of ice water would not be a *true* example of a dynamic equilibrium because it would not be kept at 0°C. At room temperature, all the ice eventually melts.

A glass of ice water at 0°C provides a practical illustration of this dynamic equilibrium. As the ice cubes melt to form water, some of the water between ice cubes may freeze, thus joining the cubes together. Remember that in a dynamic equilibrium, forward and reverse processes are occurring at the same rate [◀◀ Section 9.1].

Because molecules are more strongly held in the solid phase than in the liquid phase, heat is required to melt a solid into a liquid. The heating curve in Figure 12.28 shows that when a solid is heated, its temperature increases gradually until point A is reached. At this point, the solid begins to melt. During the melting period (A ⟶ B), the first flat portion of the curve in Figure 12.28, heat is being absorbed by the system, yet its temperature remains constant. The heat helps the molecules overcome the attractive forces in the solid. Once the sample has melted completely (point B), the heat absorbed increases the average kinetic energy of the liquid molecules and the liquid temperature rises (B ⟶ C). The vaporization process (C ⟶ D) can be explained similarly. The temperature remains constant during the period when the increased kinetic

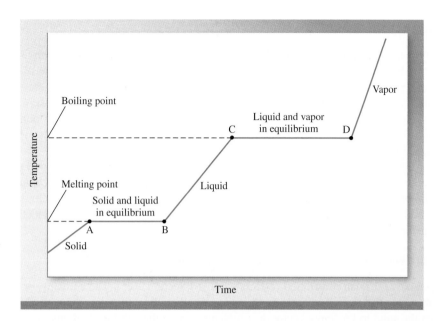

Figure 12.28 A typical heating curve, from the solid phase through the liquid phase to the gas phase of a substance. Because ΔH_{fus} is smaller than ΔH_{vap}, a substance melts in less time than it takes to vaporize. This explains why AB is shorter than CD. The steepness of the solid, liquid, and vapor heating lines is determined by the specific heat of the substance in each state.

energy is used to overcome the cohesive forces in the liquid. When all molecules are in the gas phase, the temperature rises again.

The ***molar heat of fusion*** (ΔH_{fus}) is the energy, usually expressed in kJ/mol, required to melt 1 mole of a solid. Table 12.7 lists the molar heats of fusion for the substances in Table 12.5. A comparison of the data in the two tables shows that ΔH_{fus} is smaller than ΔH_{vap} for each substance. This is consistent with the fact that molecules in a liquid are still rather closely packed together, so some energy (but not a lot of energy, relatively speaking) is needed to bring about the rearrangement from solid to liquid. When a liquid is vaporized, on the other hand, its molecules become completely separated from one another, so considerably more energy is required to overcome the intermolecular attractive forces.

Cooling a substance has the opposite effect of heating it. If we remove heat from a gas sample at a steady rate, its temperature decreases. As the liquid is being formed, heat is given off by the system, because its potential energy is decreasing. For this reason, the temperature of the system remains constant over the condensation period (D ⟶ C). After all the vapor has condensed, the temperature of the liquid begins to drop again. Continued cooling of the liquid finally leads to freezing (B ⟶ A).

Supercooling is a phenomenon in which a liquid can be temporarily cooled to below its freezing point. Supercooling occurs when heat is removed from a liquid so rapidly that the molecules literally have no time to assume the ordered structure of a solid. A supercooled liquid is unstable. Gentle stirring or the addition to it of a small "seed" crystal of the same substance will cause it to solidify quickly.

TABLE 12.7	Molar Heats of Fusion for Selected Substances	
Substance	**Melting point (°C)**	**ΔH_{fus} (kJ/mol)**
Argon (Ar)	−190	1.3
Benzene (C_6H_6)	5.5	10.9
Ethanol (C_2H_5OH)	−117.3	7.61
Diethyl ether ($C_2H_5OC_2H_5$)	−116.2	6.90
Mercury (Hg)	−39	23.4
Methane (CH_4)	−183	0.84
Water (H_2O)	0	6.01

Solid iodine in equilibrium with its vapor.

© McGraw-Hill Education/Ken Karp, photographer

Student Annotation: Equation 12.5 is generally used to approximate ΔH_{sub}. It only holds strictly when all the phase changes occur at the same temperature.

Student Hot Spot

Student data indicate you may struggle with enthalpy changes for phase changes. Access the SmartBook to view additional Learning Resources on this topic.

Solid-Vapor

Solids can be vaporized, so solids, too, have a vapor pressure. *Sublimation* is the process by which molecules go directly from the solid phase to the vapor phase. The reverse process, in which molecules go directly from the vapor phase to the solid phase, is called *deposition.* Naphthalene, which is the substance used to make moth-balls, has a fairly high vapor pressure for a solid (1 mmHg at 53°C); thus, its pungent vapor quickly permeates an enclosed space. Iodine also sublimes. At room temperature, the violet color of iodine vapor is easily visible in a closed container.

Because molecules are more tightly held in a solid, the vapor pressure of a solid is generally much less than that of the corresponding liquid. The *molar enthalpy of sublimation* (ΔH_{sub}) of a substance is the energy, usually expressed in kilojoules, required to sublime 1 mole of a solid. It is equal to the sum of the molar enthalpies of fusion and vaporization.

Equation 12.5 $$\Delta H_{sub} = \Delta H_{fus} + \Delta H_{vap}$$

Equation 12.5 is an illustration of Hess's law [◄◄ Section 10.5]. The enthalpy, or heat change, for the overall process is the same whether the substance changes directly from the solid to the vapor phase or if it changes from the solid to the liquid and then to the vapor phase.

Phase changes can be of significant consequence in biological systems. If you have ever suffered a steam burn, you know that it can be far more serious than a burn caused simply by boiling water—even though steam and boiling water are both at the same temperature. A heating curve helps explain why this is so (see Figure 12.29). When boiling water touches your skin, it is cooled to body temperature because it deposits the heat it contains on your skin. The heat deposited on your skin by a sample of boiling water at 100°C can be represented by the orange line under the curve. When an equivalent mass of steam contacts your skin, it first deposits heat as it condenses and *then* cools to body temperature. The heat deposited on your skin by a sample of steam is represented by the red line under the curve. Notice how much more heat is deposited by steam than by liquid water at the same temperature. The steam contains more heat because it has been heated *and* vaporized. The additional heat that was absorbed by the water to vaporize it is what makes a steam burn worse than a burn from boiling water.

A heating curve can also be used to explain why hikers stranded by blizzards are warned not to consume snow in an effort to stay hydrated. When you drink cold water, your body expends energy to warm the water you consume to body temperature. If you consume snow, your body must first expend the energy necessary to melt the snow, and then to warm it. Because a phase change is involved, the amount of energy required to assimilate snow is much greater than the amount necessary to assimilate an equal mass of water—even if the water is ice-cold. This can contribute to *hypothermia,* a potentially dangerous drop in body temperature.

Figure 12.29 Heating curve of water.

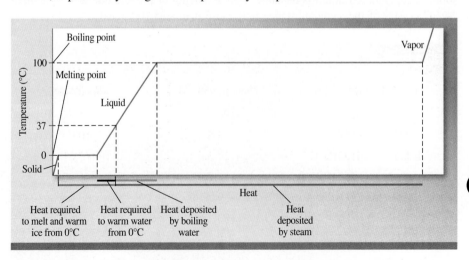

Worked Example 12.6 explores the influence of phase changes on the amount of energy transferred between water and a human body.

Worked Example 12.6

(a) Calculate the amount of heat deposited on the skin of a person burned by 1.00 g of liquid water at 100.0°C and (b) the amount of heat deposited by 1.00 g of steam at 100.0°C. (c) Calculate the amount of energy necessary to warm 100.0 g of water from 0.0°C to body temperature and (d) the amount of heat required to melt 100.0 g of ice at 0.0°C and then warm it to body temperature. (Assume that body temperature is 37.0°C.)

Student Annotation: You may want to review the calculation of the heat exchanged between the system and surroundings for temperature changes and phase changes [◀◀Sections 10.3 and 10.4].

Strategy For the purpose of following the sign conventions, we can designate the water as the *system* and the body as the *surroundings*. (a) Heat is transferred from hot water to the skin in a single step: a temperature change. (b) The transfer of heat from steam to the skin takes place in two steps: a phase change and a temperature change. (c) Cold water is warmed to body temperature in a single step: a temperature change. (d) The melting of ice and the subsequent warming of the resulting liquid water takes place in two steps: a phase change and a temperature change. In each case, the heat transferred during a temperature change depends on the mass of the water, the specific heat of water, and the change in temperature. For the phase changes, the heat transferred depends on the amount of water (in moles) and the molar heat of vaporization (ΔH_{vap}) or molar heat of fusion (ΔH_{fus}). In each case, the total energy transferred or required is the sum of the energy changes for the individual steps.

Setup The required specific heats (s) are 4.184 J/g · °C for water and 1.99 J/g · °C for steam. (Assume that the specific heat values do not change over the range of temperatures in the problem.) From Table 12.5, the molar heat of vaporization (ΔH_{vap}) of water is 40.79 kJ/mol, and from Table 12.7, the molar heat of fusion (ΔH_{fus}) of water is 6.01 kJ/mol. The molar mass of water is 18.02 g/mol. *Note:* The ΔH_{vap} of water is the amount of heat required to vaporize a mole of water. In this problem, however, we want to know how much heat is deposited when water vapor *condenses*, so we must use the negative, −40.79 kJ/mol.

Solution

(a) $\Delta T = 37.0°C - 100.0°C = -63.0°C$

From Equation 10.13, we write

$$q = ms\Delta T = 1.00 \text{ g} \times \frac{4.184 \text{ J}}{\text{g} \cdot °C} \times -63.0°C = -2.64 \times 10^2 \text{ J} = -0.264 \text{ kJ}$$

Thus, 1.00 g of water at 100.0°C deposits 0.264 kJ of heat on the skin. (The negative sign indicates that heat is given off by the system and absorbed by the surroundings.)

(b) $\dfrac{1.00 \text{ g}}{18.02 \text{ g/mol}} = 0.0555$ mol water

$$q_1 = n\Delta H_{vap} = 0.0555 \text{ mol} \times \frac{-40.79 \text{ kJ}}{\text{mol}} = -2.26 \text{ kJ}$$

$$q_2 = ms\Delta T = 1.00 \text{ g} \times \frac{4.184 \text{ J}}{\text{g} \cdot °C} \times -63.0°C = -2.64 \times 10^2 \text{ J} = -0.264 \text{ kJ}$$

The overall energy deposited on the skin by 1.00 g of steam is the sum of q_1 and q_2.

$$-2.26 \text{ kJ} + (-0.264 \text{ kJ}) = -2.53 \text{ kJ}$$

The negative sign indicates that the system (steam) gives off the energy.

(c) $\Delta T = 37.0°C - 0.0°C = 37.0°C$

$$q = ms\Delta T = 100.0 \text{ g} \times \frac{4.184 \text{ J}}{\text{g} \cdot °C} \times 37.0 \text{ }°C = 1.55 \times 10^4 \text{ J} = 15.5 \text{ kJ}$$

The energy required to warm 100.0 g of water from 0.0°C to 37.0°C is 15.5 kJ.

(d) $\dfrac{100.0 \text{ g}}{18.02 \text{ g/mol}} = 5.55$ mol

$$q_1 = n\Delta H_{fus} = 5.55 \text{ mol} \times \frac{6.01 \text{ kJ}}{\text{mol}} = 33.4 \text{ kJ}$$

$$q_2 = ms\Delta T = 100.0 \text{ g} \times \frac{4.184 \text{ J}}{\text{g} \cdot °C} \times 37.0°C = 1.55 \times 10^4 \text{ J} = 15.5 \text{ kJ}$$

The energy required to melt 100.0 g of ice at 0.0°C and warm it to 37.0°C is the sum of q_1 and q_2.

$$33.4 \text{ kJ} + 15.5 \text{ kJ} = 48.9 \text{ kJ}$$

(Continued on next page)

Think About It

In problems that include phase changes, the q values corresponding to the phase-change steps will be the largest contributions to the total. If you find that this is not the case in your solution, check to see if you have made the common error of neglecting to convert the q values corresponding to temperature changes from joules to kilojoules.

Practice Problem **A**TTEMPT   Calculate the amount of energy (in kilojoules) necessary to convert 346 g of liquid water from 0°C to water vapor at 182°C.

Practice Problem **B**UILD   Determine the final state and temperature of 100 g of water originally at 25.0°C after 50.0 kJ of heat have been added to it.

Practice Problem **C**ONCEPTUALIZE   Two samples of the same pure liquid are represented here. If adding 753 J to the sample on the left causes its temperature to increase by 41.2°C, by how much will the temperature of the sample on the right increase when an equal amount of energy is added to it?

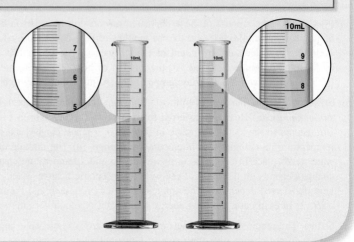

Section 12.5 Review

Phase Changes

12.5.1   How much energy (in kilojoules) is required to convert 25.0 g of liquid water at room temperature (25°C) to steam at 110°C?
   (a) 64.9 kJ   (c) 1339 kJ   (e) 26.9 kJ
   (b) 562 kJ   (d) 1.34 kJ

12.5.2   How much energy (in kilojoules) is given off when 1.00 g of steam at 100.0°C cools to room temperature (25.0°C)?
   (a) 0.326 kJ   (c) 2.58 kJ   (e) 22.1 kJ
   (b) 316 kJ   (d) 48.9 kJ

12.6   PHASE DIAGRAMS

The relationships between the phases of a substance can be represented in a single graph known as a phase diagram. A ***phase diagram*** summarizes the conditions (temperature and pressure) at which a substance exists as a solid, liquid, or gas. Figure 12.30(a) shows the phase diagram of CO_2, which is typical of many substances. The graph is divided

Figure 12.30 (a) The phase diagram of carbon dioxide. Note that the solid-liquid boundary line has a positive slope. There is no liquid phase below 5.2 atm, so only the solid and vapor phases can exist under ordinary atmospheric conditions. (b) Heating solid CO_2 initially at −100°C and 1 atm (point 1) causes it to sublime when it reaches −78°C (point 2). At −25°C, increasing the pressure from 1 atm (point 3) to about 70 (point 4) will cause CO_2 to condense to a liquid.

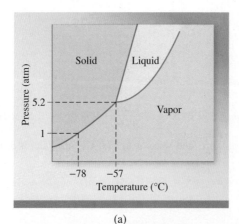

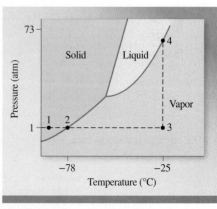

(a)   (b)

into three regions, each of which represents a pure phase. The line separating any two regions, called a *phase boundary line,* indicates conditions under which these two phases can exist in equilibrium. The point at which all three phase boundary lines meet is called the **triple point.** The triple point is the only combination of temperature and pressure at which all three phases of a substance can be in equilibrium with one another. The point at which the liquid-vapor phase boundary line abruptly ends is the critical point, corresponding to the critical temperature (T_c) and the critical pressure (P_c).

To understand the information in a phase diagram, consider the dashed lines between numbered points in Figure 12.30(b). If we start with a sample of CO_2 at 1 atm and $-100°C$, the sample is initially a solid (point 1). If we then add heat to the sample at 1 atm, its temperature increases until it reaches $-78°C$, the sublimation point of CO_2 at 1 atm (point 2). When the entire sample has sublimed at $-78°C$, the temperature of the resulting vapor will begin to increase. We continue adding heat until the temperature of the vapor is $-25°C$ (point 3). At this point, we maintain the temperature at $-25°C$ and begin increasing the pressure until the vapor condenses, which would occur at a pressure of about 70 atm (point 4). With a phase diagram, we can tell in what phase a substance will exist at any given temperature and pressure. Furthermore, we can tell what phase changes will occur as the result of increases or decreases in temperature, pressure, or both.

One of the interesting things about the phase diagram of CO_2 is that its triple point occurs above atmospheric pressure $-57°C$, 5.2 atm. This means that there is no liquid phase at atmospheric pressure, the condition that makes dry ice "dry." At elevated pressure, solid CO_2 does melt. In fact, liquid CO_2 is used as the solvent in many dry-cleaning operations.

The phase diagram of water (Figure 12.31) is unusual because the solid-liquid phase-boundary line has a negative slope. (Compare this with the solid-liquid phase boundary line of CO_2 in Figure 12.30.) As a result, ice can be liquefied within a narrow temperature range by applying pressure.

Worked Example 12.7 lets you practice interpreting the information in a phase diagram.

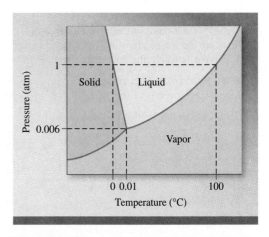

Figure 12.31 The phase diagram of water. Each solid line between two phases specifies the conditions of pressure and temperature under which the two phases can exist in equilibrium. The point at which all three phases can exist in equilibrium (0.006 atm and 0.01°C) is called the triple point.

Student Hot Spot

Student data indicate you may struggle with phase diagrams. Access the SmartBook to view additional Learning Resources on this topic.

Student Annotation: Bismuth is another example of a substance whose solid-liquid phase boundary (under ordinary atmospheric conditions) has a negative slope. Like water, bismuth has a liquid density (1.005 g/cm³ at room temperature) that is higher than its solid density (0.9780 g/cm³ at its melting point of 271°C).

Worked Example 12.7

Using the following phase diagram, (a) determine the normal boiling point and the normal melting point of the substance, (b) determine the physical state of the substance at 2 atm and 110°C, and (c) determine the pressure and temperature that correspond to the triple point of the substance.

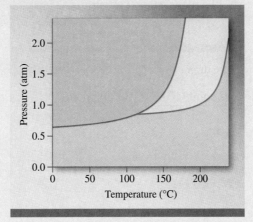

(Continued on next page)

Strategy Each point on the phase diagram corresponds to a pressure-temperature combination. The normal boiling and melting points are the temperatures at which the substance undergoes phase changes. These points fall on the phase boundary lines. The triple point is where the three phase boundaries meet.

Setup By drawing lines corresponding to a given pressure and/or temperature, we can determine the temperature at which a phase change occurs, or the physical state of the substance under specified conditions.

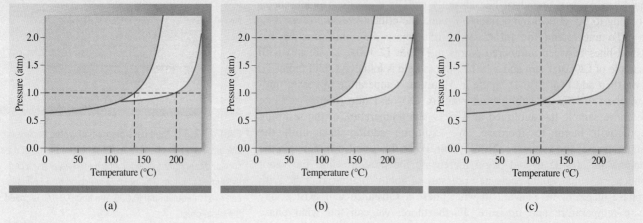

(a)   (b)   (c)

Solution

(a) The normal boiling and melting points are ~205°C and ~140°C, respectively.

(b) At 2 atm and 110°C the substance is a solid.

(c) The triple point occurs at ~0.8 atm and ~115°C.

Think About It

The triple point of this substance occurs at a pressure below atmospheric pressure. Therefore, it will melt rather than sublime when it is heated under ordinary conditions.

Practice Problem **A**TTEMPT Use the following phase diagram to (a) determine the normal boiling point and melting point of the substance, and (b) the physical state of the substance at 1.2 atm and 100°C.

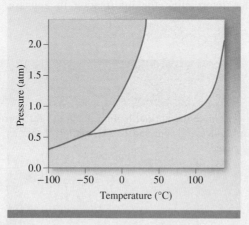

Practice Problem **B**UILD Sketch the phase diagram of a substance using the following data.

Pressure (atm)	Melting point (°C)	Boiling point (°C)	Sublimation point (°C)
0.5	—	—	0
1.0	60	110	—
1.5	75	200	—
2.0	105	250	—
2.5	125	275	—

The triple point is at 0.75 atm and 45°C.

Practice Problem **C**ONCEPTUALIZE Which of the phase diagrams [(i)–(iii)] has an arrow that traces a path including the following changes? For the other two phase diagrams, write a numbered list of changes in temperature, pressure, and phase.

1. Temperature increase with no phase change

2. Pressure decrease causing a solid-to-vapor phase change

3. Temperature increase with no phase change

4. Pressure increase with vapor-to-liquid phase change and liquid-to-solid phase change

5. Temperature increase with solid-to-liquid phase change

6. Pressure decrease with liquid-to-vapor phase change

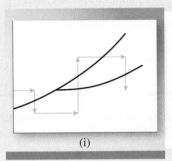
(i)

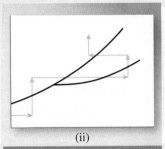
(ii)

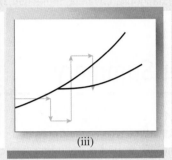
(iii)

Section 12.6 Review

Phase Diagrams

Refer to the following phase diagrams to answer questions 12.6.1 and 12.6.2.

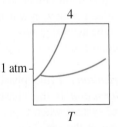

12.6.1 Which phase diagram corresponds to a substance that will sublime rather than melt as it is heated at 1 atm?
(a) 1
(b) 1 and 2
(c) 1, 2, and 4
(d) 3
(e) none

12.6.2 Which phase diagram corresponds to a substance that will liquefy when pressure is increased at a temperature below its freezing point?
(a) 1
(b) 1 and 2
(c) 1, 2, and 4
(d) 3
(e) none

Learning Outcomes

- Use the Clausius-Clapeyron equation to calculate the vapor pressure of a liquid at a given temperature.
- Define unit cell and lattice point.
- Calculate the atomic radius of an atom given its density and type of crystal.
- Identify key characteristics and examples of the major types of crystal: ionic, covalent, molecular, or metallic.

- Calculate the amount of heat lost or gained when a substance undergoes a series of phase and/or temperature changes.
- Define phase boundary, triple point, and critical point.
- Use phase diagrams to determine the phase of a substance at a given temperature and pressure.

Chapter Summary

SECTION 12.1
- The magnitude of intermolecular forces in liquids and solids influences various properties including surface tension, vapor pressure, boiling point, and melting point.

SECTION 12.2
- *Surface tension* is the net pull inward on molecules at the surface of a liquid. Surface tension is related to *cohesion,* the attractive forces between molecules within a substance, and *adhesion,* the attractive forces between molecules in a substance and their container. The balance between cohesion and adhesion determines whether a liquid meniscus is concave or convex. It also gives rise to *capillary action,* in which liquid is drawn upward into a narrow tube against gravity.
- *Viscosity* is resistance to flow, reflecting how easily molecules move past one another.
- *Evaporation* (also known as *vaporization*) is the phase change from liquid to vapor. *Condensation* is the phase change from vapor to liquid. In a closed system, when vaporization and condensation are occurring at the same rate, a state of *dynamic equilibrium* exists and the *vapor pressure* is equal to the *equilibrium vapor pressure.* Vapor pressure measures how easily molecules escape to the vapor phase. A *volatile* substance has a high vapor pressure.
- Surface tension, viscosity, and vapor pressure are all temperature dependent. The *Clausius-Clapeyron equation* relates the vapor pressure of a substance to its absolute temperature.
- The *boiling point* of a substance is the temperature at which its vapor pressure equals the external pressure. The *normal boiling point* is the temperature at which its vapor pressure equals 1 atm.

SECTION 12.3
- The *melting point* or *freezing point* is the temperature at which the solid and liquid phases are in equilibrium.
- The melting point of a solid is strongly related to the strength of the intermolecular forces in the solid. Ionic solids typically have a high melting point due to strong ion-ion interactions.

- Like liquids, solids have vapor pressures that are dependent on the magnitude of the intermolecular forces present. The vapor pressures of solids are usually very low with few exceptions.
- *Amorphous solids* such as *glass* lack regular three-dimensional structure.
- The spheres (molecules, atoms, or ions) in a *crystalline solid* are arranged in a three-dimensional *lattice structure* consisting of a repeating pattern of *unit cells.* The type of unit cell is determined by the positions of the *lattice points.*
- Cubic unit cells may be *simple cubic, body-centered cubic,* or *face-centered cubic,* containing a total of one, two, or four spheres, respectively. The *coordination number* is the number of spheres surrounding each sphere.
- *Closest packing* is the most efficient arrangement of spheres in a solid. It may be *hexagonal* or *cubic.* Each has a coordination number of 12.

SECTION 12.4
- Crystals may be ionic, covalent, molecular, or metallic. Each type of crystalline solid has characteristics determined in part by the types of interactions holding it together.

SECTION 12.5
- The possible *phase changes* are melting or *fusion* ($s \longrightarrow l$), *freezing* ($l \longrightarrow s$), vaporization ($l \longrightarrow g$), *condensation* ($g \longrightarrow l$), *sublimation* ($s \longrightarrow g$), and *deposition* ($g \longrightarrow s$).
- Boiling point is pressure dependent. The *molar heat of vaporization* (ΔH_{vap}) is the amount of heat required to vaporize 1 mole of a substance at its boiling point.
- The terms *normal melting point* and *normal freezing point* refer to the temperature at which the solid-liquid phase transition takes place at 1 atm pressure.
- The *critical temperature* (T_c) is the temperature above which a gas cannot be liquefied by applying pressure. The *critical pressure* (P_c) is the pressure necessary to liquefy a gas at its critical temperature. A substance above its critical temperature and pressure is a *supercritical fluid.*

- The *molar heat of fusion* (ΔH_{fus}) is the amount of heat required to melt 1 mole of a substance at its melting point. *Supercooling* is the process of rapidly lowering a liquid's temperature below its freezing point.
- The *molar enthalpy of sublimation* (ΔH_{sub}) is equal to the sum of the molar heats of fusion and vaporization: $\Delta H_{sub} = \Delta H_{fus} + \Delta H_{vap}$.

SECTION 12.6

- A *phase diagram* indicates the phase of a substance under any combination of temperature and pressure. Lines between phases are called *phase boundaries*.
- The *triple point* is where all three phase boundaries meet. This is the temperature and pressure combination at which all three phases are in equilibrium.

Key Words

Key Equations

12.1 $\ln P = -\dfrac{\Delta H_{vap}}{RT} + C$

The Clausius-Clapeyron equation relates the natural logarithm of the vapor pressure of a substance to its heat of vaporization (ΔH_{vap}) and the absolute temperature.

12.2 $\ln P_1 = -\dfrac{\Delta H_{vap}}{RT_1} + C$

Equations 12.2 and 12.3 are the Clausius-Clapeyron equation written for one substance at two different temperatures, T_1 and T_2.

12.3 $\ln P_2 = -\dfrac{\Delta H_{vap}}{RT_2} + C$

12.4 $\ln \dfrac{P_1}{P_2} = \dfrac{\Delta H_{vap}}{R} \left(\dfrac{1}{T_2} - \dfrac{1}{T_1} \right)$

Subtracting Equation 12.3 from Equation 12.2 gives an equation that can be used to determine vapor pressure at a new temperature—provided that vapor pressure at one temperature and ΔH_{vap} for the substance are known. This equation can also be rearranged to solve for ΔH_{vap} if the vapor pressure is known at two different temperatures.

12.5 $\Delta H_{sub} = \Delta H_{fus} + \Delta H_{vap}$

The heat of sublimation (ΔH_{sub}) is the sum of heat of fusion (ΔH_{fus}) and heat of vaporization (ΔH_{vap}). This is a consequence of Hess's law. Sublimation (the phase change from solid to vapor) can be thought of as a two-step process in which a solid becomes a liquid, and the liquid becomes a vapor. Because the two steps sum to the overall process, their ΔH values sum to the overall ΔH.

Key Skills

Intermolecular Forces

Most of the intermolecular forces discussed in Chapter 7 are those between particles (atoms, molecules, or ions) in a pure substance. However, our ability to predict how easily a substance can be dissolved in a particular solvent relies on our understanding of the forces between the particles of two *different* substances. The axiom "like dissolves like" refers to *polar* (or ionic) substances being more soluble in *polar* solvents, and *nonpolar* substances being more soluble in *nonpolar* solvents. To assess the solubility of a substance, we must identify it as ionic, polar, or nonpolar. The following flowchart illustrates this identification process and the conclusions we can draw about solubility.

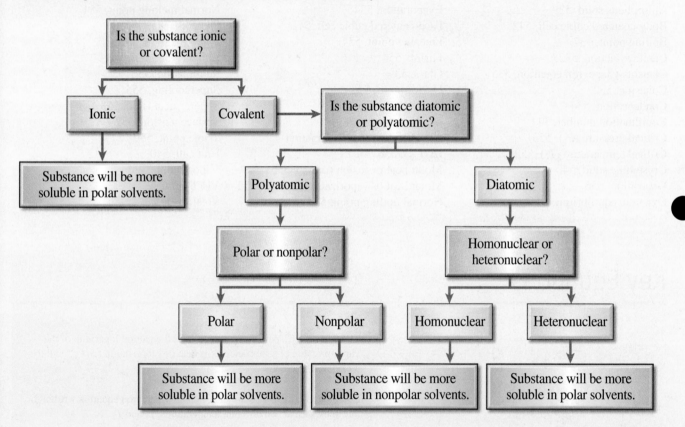

To determine whether or not a substance is ionic, we examine its chemical formula. If the formula contains either a metal cation or the ammonium ion (NH_4^+), it is ionic. Recall that there are some ionic compounds that are considered insoluble in water [◄◄ Table 9.3, page 358]. However, even these compounds are more soluble in *water* than they are in *nonpolar* solvents.

If a substance is not ionic, it is covalent. Covalent substances may be polar or nonpolar, depending on the electronegativities of the atoms involved—and on molecular geometry [◄◄ Chapter 7 Key Skills]. Polar species will be more soluble in polar solvents, such as water; and nonpolar species will be more soluble in nonpolar solvents, such as benzene.

In addition to deciding whether a substance will be more soluble in polar or nonpolar solvents, it is possible to assess *relative* solubilities of two different solutes in the same solvent. For example, we may be asked to compare the solubilities (in water and in benzene) of two different molecules. If one of the molecules is polar and the other is nonpolar, we should expect the *polar* molecule to be more soluble in *water*—and the *nonpolar* one to be more soluble in *benzene*.

Further, some nonpolar substances do dissolve in water because of dispersion forces, which are stronger in larger molecules with greater *polarizability* [◄◄ Section 7.3]. Therefore, we should expect the larger of two nonpolar molecules to be more soluble in water.

Key Skills Problems

12.1
Which of the following would you expect to be more soluble in water than in benzene? (Select all that apply.)

(a) CH_3OH (b) CCl_4 (c) $\begin{array}{c} H \\ \diagdown \\ C=C \\ \diagup \\ Cl \end{array} \begin{array}{c} Cl \\ \diagup \\ \diagdown \\ H \end{array}$ (d) $\begin{array}{c} Cl \\ \diagdown \\ C=C \\ \diagup \\ H \end{array} \begin{array}{c} Cl \\ \diagup \\ \diagdown \\ H \end{array}$ (e) KI

12.2
Which of the following would you expect to be more soluble in benzene than in water? (Select all that apply.)

(a) Br_2 (b) KBr (c) NH_3 (d) $\begin{array}{c} H \\ \diagdown \\ C=C \\ \diagup \\ Cl \end{array} \begin{array}{c} Cl \\ \diagup \\ \diagdown \\ H \end{array}$ (e) $\begin{array}{c} H \\ \diagdown \\ C=C \\ \diagup \\ H \end{array} \begin{array}{c} Cl \\ \diagup \\ \diagdown \\ Cl \end{array}$

12.3
Arrange the following substances in order of *decreasing* solubility in water: Kr, O_2, N_2.
(a) $Kr \approx O_2 > N_2$
(b) $Kr > O_2 \approx N_2$
(c) $Kr \approx N_2 > O_2$
(d) $Kr > N_2 > O_2$
(e) $Kr \approx N_2 \approx O_2$

12.4
Arrange the following substances in order of *increasing* solubility in water: C_2H_5OH, CO_2, N_2O.
(a) $C_2H_5OH < CO_2 < N_2O$
(b) $CO_2 < N_2O < C_2H_5OH$
(c) $N_2O < C_2H_5OH < CO_2$
(d) $CO_2 \approx N_2O < C_2H_5OH$
(e) $CO_2 < C_2H_5OH < N_2O$

Questions and Problems

SECTION 12.2: PROPERTIES OF LIQUIDS

Review Questions

12.1   Explain why liquids, unlike gases, are virtually incompressible.

12.2   What is surface tension? What is the relationship between intermolecular forces and surface tension? How does surface tension change with temperature?

12.3   Despite the fact that stainless steel is much denser than water, a stainless-steel razor blade can be made to float on water. Why?

12.4   Use water and mercury as examples to explain adhesion and cohesion.

12.5   A glass can be filled slightly above the rim with water. Explain why the water does not overflow.

12.6   Draw diagrams showing the capillary action of (a) water and (b) mercury in three tubes of different radii.

12.7   What is viscosity? What is the relationship between intermolecular forces and viscosity?

12.8   Why does the viscosity of a liquid decrease with increasing temperature?

12.9   Why is ice less dense than water?

12.10   Define boiling point. How does the boiling point of a liquid depend on external pressure? Referring to Table 11.6, what is the boiling point of water when the external pressure is 187.5 mmHg?

12.11   As a liquid is heated at constant pressure, its temperature rises. This trend continues until the boiling point of the liquid is reached. No further rise in temperature of the liquid can be induced by heating. Explain.

Computational Problems

12.12   The vapor pressure of benzene (C_6H_6) is 40.1 mmHg at 7.6°C. What is its vapor pressure at 60.6°C? The molar heat of vaporization of benzene is 31.0 kJ/mol.

12.13   Estimate the molar heat of vaporization of a liquid whose vapor pressure doubles when the temperature is raised from 75°C to 100°C.

Conceptual Problems

12.14   Predict which of the following liquids has greater surface tension: ethanol (C_2H_5OH) or dimethyl ether (CH_3OCH_3).

12.15   Predict the viscosity of ethylene glycol relative to that of ethanol and glycerol (see Table 12.1).

$$CH_2{-}OH$$
$$|$$
$$CH_2{-}OH$$

Ethylene glycol

12.16   Vapor pressure measurements at several different temperatures are shown for mercury. Determine graphically the molar heat of vaporization for mercury.

T(°C)	200	250	300	320	340
P(mmHg)	17.3	74.4	246.8	376.3	557.9

12.17   The vapor pressure of liquid X is lower than that of liquid Y at 20°C, but higher at 60°C. What can you deduce about the relative magnitude of the molar heats of vaporization of X and Y?

SECTION 12.3: PROPERTIES OF SOLIDS

Review Questions

12.18   What is an amorphous solid? How does it differ from a crystalline solid?

12.19   Define glass. What is the chief component of glass? Name three types of glass.

12.20   Define the following terms: *crystalline solid, lattice point, unit cell, coordination number, closest packing.*

12.21   Describe the geometries of the following cubic cells: simple cubic, body-centered cubic, face-centered cubic. Which of these structures would give the highest density for the same type of atoms? Which the lowest?

12.22   Classify the solid states in terms of crystal types of the elements in the third period of the periodic table. Predict the trends in their melting points and boiling points.

12.23   The melting points of the oxides of the third-period elements are given in parentheses: Na_2O (1275°C), MgO (2800°C), Al_2O_3 (2045°C), SiO_2 (1610°C), P_4O_{10} (580°C), SO_3 (16.8°C), Cl_2O_7 (−91.5°C). Classify these solids in terms of crystal types.

12.24   Define X-ray diffraction. What are the typical wavelengths (in nanometers) of X rays? (See Figure 3.1.)

12.25   Write the Bragg equation. Define every term and describe how this equation can be used to measure interatomic distances.

Computational Problems

12.26   What is the coordination number of each sphere in (a) a simple cubic cell, (b) a body-centered cubic cell, and (c) a face-centered cubic cell? Assume the spheres are all the same.

12.27   Calculate the number of spheres that would be found within a simple cubic cell, body-centered cubic cell, and face-centered cubic cell. Assume that the spheres are the same.

12.28 Metallic iron crystallizes in a cubic lattice. The unit cell edge length is 287 pm. The density of iron is 7.87 g/cm³. How many iron atoms are within a unit cell?

12.29 Barium metal crystallizes in a body-centered cubic lattice (the Ba atoms are at the lattice points only). The unit cell edge length is 502 pm, and the density of the metal is 3.50 g/cm³. Using this information, calculate Avogadro's number. [*Hint:* First calculate the volume (in cm³) occupied by 1 mole of Ba atoms in the unit cells. Next calculate the volume (in cm³) occupied by one Ba atom in the unit cell. Assume that 68 percent of the unit cell is occupied by Ba atoms.]

12.30 Vanadium crystallizes in a body-centered cubic lattice (the V atoms occupy only the lattice points). How many V atoms are present in a unit cell?

12.31 Europium crystallizes in a body-centered cubic lattice (the Eu atoms occupy only the lattice points). The density of Eu is 5.26 g/cm³. Calculate the unit cell edge length in picometers.

12.32 Crystalline silicon has a cubic structure. The unit cell edge length is 543 pm. The density of the solid is 2.33 g/cm³. Calculate the number of Si atoms in one unit cell.

12.33 A face-centered cubic cell contains 8 X atoms at the corners of the cell and 6 Y atoms at the faces. What is the empirical formula of the solid?

12.34 When X rays of wavelength 0.090 nm are diffracted by a metallic crystal, the angle of first-order diffraction ($n = 1$) is measured to be 15.2°. What is the distance (in picometers) between the layers of atoms responsible for the diffraction?

12.35 The distance between layers in an NaCl crystal is 282 pm. X rays are diffracted from these layers at an angle of 23.0°. Assuming that $n = 1$, calculate the wavelength of the X rays in nanometers.

Conceptual Problems

12.36 Identify the unit cell of molecular iodine (I_2) shown here. (*Hint:* Consider the position of iodine molecules, not individual iodine atoms.)

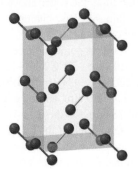

12.37 Shown here is a zinc oxide unit cell. What is the formula of zinc oxide?

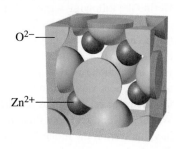

SECTION 12.4: TYPES OF CRYSTALLINE SOLIDS

Review Questions

12.38 Describe and give examples of the following types of crystals: (a) ionic crystals, (b) covalent crystals, (c) molecular crystals, (d) metallic crystals.

12.39 Why are metals good conductors of heat and electricity? Why does the ability of a metal to conduct electricity decrease with increasing temperature?

Conceptual Problems

12.40 A solid is hard, brittle, and electrically nonconducting. Its melt (the liquid form of the substance) and an aqueous solution containing the substance conduct electricity. Classify the solid.

12.41 A solid is soft and has a low melting point (below 100°C). The solid, its melt, and an aqueous solution containing the substance are all nonconductors of electricity. Classify the solid.

12.42 A solid is very hard and has a high melting point. Neither the solid nor its melt conducts electricity. Classify the solid.

12.43 Which of the following are molecular solids and which are covalent solids: Se_8, HBr, Si, CO_2, C, P_4O_6, SiH_4?

12.44 Classify the solid state of the following substances as ionic crystals, covalent crystals, molecular crystals, or metallic crystals: (a) CO_2, (b) B_{12}, (c) S_8, (d) KBr, (e) Mg, (f) SiO_2, (g) LiCl, (h) Cr.

12.45 Explain why diamond is harder than graphite. Why is graphite an electrical conductor but diamond is not?

SECTION 12.5: PHASE CHANGES

Review Questions

12.46 What is a phase change? Name all possible changes that can occur among the vapor, liquid, and solid phases of a substance.

12.47 What is the equilibrium vapor pressure of a liquid? How is it measured, and how does it change with temperature?

12.48   Use any one of the phase changes to explain what is meant by dynamic equilibrium.

12.49   Define the following terms: (a) molar heat of vaporization, (b) molar heat of fusion, (c) molar heat of sublimation. What are their typical units?

12.50   How is the molar heat of sublimation related to the molar heats of vaporization and fusion? On what law are these relationships based?

12.51   What can we learn about the intermolecular forces in a liquid from the molar heat of vaporization?

12.52   The greater the molar heat of vaporization of a liquid, the greater its vapor pressure. True or false?

12.53   Using Table 11.6 as a reference, what is the boiling point of water when the external pressure is 118.0 mmHg?

12.54   A closed container of liquid pentane (bp = 36.1°C) is at room temperature. Why does the vapor pressure initially increase but eventually stop changing?

12.55   What is critical temperature? What is the significance of critical temperature in condensation of gases?

12.56   What is the relationship between intermolecular forces in a liquid and the liquid's boiling point and critical temperature? Why is the critical temperature of water greater than that of most other substances?

12.57   How do the boiling points and melting points of water and carbon tetrachloride vary with pressure? Explain any difference in behavior of these two substances.

12.58   Why is solid carbon dioxide called dry ice?

12.59   The vapor pressure of a liquid in a closed container depends on which of the following: (a) the volume above the liquid, (b) the amount of liquid present, (c) temperature, (d) intermolecular forces between the molecules in the liquid?

12.60   Wet clothes dry more quickly on a hot, dry day than on a hot, humid day. Explain.

12.61   Which of the following phase transitions gives off more heat: (a) 1 mole of steam to 1 mole of water at 100°C, or (b) 1 mole of water to 1 mole of ice at 0°C?

12.62   A beaker of water is heated to boiling by a Bunsen burner. Would adding another burner raise the temperature of the boiling water? Explain.

12.63   Explain why splashing a small amount of liquid nitrogen (b.p. 77 K) is not as harmful as splashing boiling water on your skin.

Computational Problems

12.64   Calculate the amount of heat (in kilojoules) required to convert 25.97 g of water to steam at 100°C.

12.65   How much heat (in kilojoules) is needed to convert 212.8 g of ice at −15°C to steam at 138°C? (The specific heats of ice and steam are 2.03 and 1.99 J/g · °C, respectively.)

12.66   The molar heats of fusion and sublimation of lead are 4.77 and 182.8 kJ/mol, respectively. Estimate the molar heat of vaporization of molten lead.

Conceptual Problems

12.67   Freeze-dried coffee is prepared by freezing brewed coffee and then removing the ice component with a vacuum pump. Describe the phase changes taking place during these processes.

12.68   How is the rate of evaporation of a liquid affected by (a) temperature, (b) the surface area of a liquid exposed to air, (c) intermolecular forces?

12.69   Explain why steam at 100°C causes more serious burns than water at 100°C.

12.70   The following compounds, listed with their boiling points, are liquid at −10°C: butane, −0.5°C; ethanol, 78.3°C; toluene, 110.6°C. At −10°C, which of these liquids would you expect to have the highest vapor pressure? Which the lowest? Explain.

12.71   A student hangs wet clothes outdoors on a winter day when the temperature is −15°C. After a few hours, the clothes are found to be fairly dry. Describe the phase changes in this drying process.

SECTION 12.6: PHASE DIAGRAMS

Review Questions

12.72   What is a phase diagram? What useful information can be obtained from studying a phase diagram?

12.73   Explain how water's phase diagram differs from those of most substances. What property of water causes the difference?

Conceptual Problems

12.74   The blades of ice skates are quite thin, so the pressure exerted on ice by a skater can be substantial. Explain how this facilitates skating on ice.

12.75   A length of wire is placed on top of a block of ice. The ends of the wire extend over the edges of the ice, and a heavy weight is attached to each end. It is found that the ice under the wire gradually melts, so the wire slowly moves through the ice block. At the same time, the water above the wire refreezes. Explain the phase changes that accompany this phenomenon.

12.76   The boiling point and freezing point of sulfur dioxide are −10°C and −72.7°C (at 1 atm), respectively. The triple point is −75.5°C and 1.65×10^{-3} atm, and its critical point is at 157°C and 78 atm. On the basis of this information, draw a rough sketch of the phase diagram of SO_2.

12.77   A phase diagram of water is shown. Label the regions. Predict what would happen as a result of the following changes: (a) Starting at A, we raise the temperature at constant pressure. (b) Starting at B,

we lower the pressure at constant temperature.
(c) Starting at C, we lower the temperature at constant pressure.

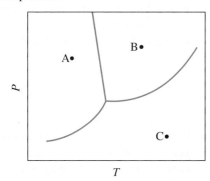

ADDITIONAL PROBLEMS

12.78   At −35°C, liquid HI has a higher vapor pressure than liquid HF. Explain.

12.79   Based on the following properties of elemental boron, classify it as one of the crystalline solids discussed in Section 12.4: high melting point (2300°C), poor conductor of heat and electricity, insoluble in water, very hard substance.

12.80   Referring to Figure 12.30, determine the stable phase of CO_2 at (a) 4 atm and −60°C and (b) 0.5 atm and −20°C.

12.81   Which of the following properties indicates very strong intermolecular forces in a liquid: (a) very low surface tension, (b) very low critical temperature, (c) very low boiling point, (d) very low vapor pressure?

12.82   Given two complementary strands of DNA containing 100 base pairs each, calculate the ratio of two separate strands to hydrogen-bonded double helix in solution at 300 K. (*Hint:* The formula for calculating this ratio is $e^{-\Delta E/RT}$, where ΔE is the energy difference between hydrogen-bonded double-strand DNAs and single-strand DNAs and R is the gas constant.) Assume the energy of hydrogen bonds per base pair to be 10 kJ/mol.

12.83   The average distance between base pairs measured parallel to the axis of a DNA molecule is 3.4 Å. The average molar mass of a pair of nucleotides is 650 g/mol. Estimate the length in centimeters of a DNA molecule of molar mass 5.0×10^9 g/mol. Roughly how many base pairs are contained in this molecule?

12.84   A CO_2 fire extinguisher is located on the outside of a building in Massachusetts. During the winter months, one can hear a sloshing sound when the extinguisher is gently shaken. In the summertime there is often no sound when it is shaken. Explain. Assume that the extinguisher has no leaks and that it has not been used.

12.85   What is the vapor pressure of mercury at its normal boiling point (357°C)?

12.86   A flask of water is connected to a powerful vacuum pump. When the pump is turned on, the water begins to boil. After a few minutes, the same water begins to freeze. Eventually, the ice disappears. Explain what happens at each step.

12.87   The liquid-vapor boundary line in the phase diagram of any substance always stops abruptly at a certain point. Why?

12.88   The interionic distances of several alkali halide crystals are as follows:

Crystal	NaCl	NaBr	NaI	KCl	KBr	KI
Interionic distance (pm)	282	299	324	315	330	353

Plot lattice energy versus the reciprocal interionic distance. How would you explain the plot in terms of the dependence of lattice energy on the distance of separation between ions? What law governs this interaction? (For lattice energies, see Table 5.1.)

12.89   Which has a greater density, crystalline SiO_2 or amorphous SiO_2? Why?

12.90   A student is given four solid samples labeled W, X, Y, and Z. All have a metallic luster. She is told that the solids could be gold, lead sulfide, mica (which is quartz, or SiO_2), and iodine. The results of her investigations are: (a) W is a good electrical conductor; X, Y, and Z are poor electrical conductors. (b) When the solids are hit with a hammer, W flattens out, X shatters into many pieces, Y is smashed into a powder, and Z is not affected. (c) When the solids are heated with a Bunsen burner, Y melts with some sublimation, but X, W, and Z do not melt. (d) In treatment with 6 *M* HNO_3, X dissolves; there is no effect on W, Y, or Z. On the basis of these test results, identify the solids.

12.91   Which of the following statements are false? (a) Dipole-dipole interactions between molecules are greatest if the molecules possess only temporary dipole moments. (b) All compounds containing hydrogen atoms can participate in hydrogen-bond formation. (c) Dispersion forces exist between all atoms, molecules, and ions.

12.92   The diagram shows a kettle of boiling water. Identify the phases in regions A and B.

© Simon Murrell/OJO Images/Getty.

12.93 The south pole of Mars is covered with solid carbon dioxide, which partly sublimes during the summer. The CO_2 vapor recondenses in the winter when the temperature drops to 150 K. Given that the heat of sublimation of CO_2 is 25.9 kJ/mol, calculate the atmospheric pressure on the surface of Mars. [*Hint:* Use Figure 12.30 to determine the normal sublimation temperature of dry ice and Equation 12.4, which also applies to sublimations.]

12.94 The properties of gases, liquids, and solids differ in a number of respects. How would you use the kinetic molecular theory (see Section 11.2) to explain the following observations? (a) Ease of compressibility decreases from gas to liquid to solid. (b) Solids retain a definite shape, but gases and liquids do not. (c) For most substances, the volume of a given amount of material increases as it changes from solid to liquid to gas.

12.95 The standard enthalpy of formation of gaseous molecular iodine is 62.4 kJ/mol. Use this information to calculate the molar heat of sublimation of molecular iodine at 25°C.

12.96 A small drop of oil in water assumes a spherical shape. Explain. (*Hint:* Oil is made up of nonpolar molecules, which tend to avoid contact with water.)

12.97 Under the same conditions of temperature and density, which of the following gases would you expect to behave less ideally: CH_4 or SO_2? Explain.

12.98 The distance between Li^+ and Cl^- is 257 pm in solid LiCl and 203 pm in an LiCl unit in the gas phase. Explain the difference in the bond lengths.

12.99 Heat of hydration, that is, the heat change that occurs when ions become hydrated in solution, is largely due to ion-dipole interactions. The heats of hydration for the alkali metal ions are Li^+, −520 kJ/ mol; Na^+, −405 kJ/mol; K^+, −321 kJ/mol. Account for the trend in these values.

12.100 The fluorides of the second period elements and their melting points are: LiF, 845°C; BeF_2, 800°C; BF_3, −126.7°C; CF_4, −184°C; NF_3, −206.6°C; OF_2, −223.8°C; F_2, −219.6°C. Classify the type(s) of intermolecular forces present in each compound.

12.101 Calculate the $\Delta H°$ for the following processes at 25°C: (a) $Br_2(l) \longrightarrow Br_2(g)$, (b) $Br_2(g) \longrightarrow 2Br(g)$. Comment on the relative magnitudes of these $\Delta H°$ values in terms of the forces involved in each case. (*Hint:* See Table 10.4.)

12.102 Which liquid would you expect to have a greater viscosity, water or diethyl ether? The structure of diethyl ether is shown in Problem 7.35.

12.103 A beaker of water is placed in a closed container. Predict the effect on the vapor pressure of the water when (a) its temperature is lowered, (b) the volume of the container is doubled, (c) more water is added to the beaker.

12.104 Ozone (O_3) is a strong oxidizing agent that can oxidize all the common metals except gold and platinum. A convenient test for ozone is based on its action on mercury. When exposed to ozone, mercury becomes dull looking and sticks to glass tubing (instead of flowing freely through it). Write a balanced equation for the reaction. What property of mercury is altered by its interaction with ozone?

12.105 A sample of limestone ($CaCO_3$) is heated in a closed vessel until it is partially decomposed. Write an equation for the reaction, and state how many phases are present.

12.106 Carbon and silicon belong to Group 4A of the periodic table and have the same valence electron configuration (ns^2np^2). Why does silicon dioxide (SiO_2) have a much higher melting point than carbon dioxide (CO_2)?

12.107 A pressure cooker is a sealed container that allows steam to escape when it exceeds a predetermined pressure. How does this device reduce the time needed for cooking?

12.108 A 1.20-g sample of water is injected into an evacuated 5.00-L flask at 65°C. What percentage of the water will be vapor when the system reaches equilibrium? Assume ideal behavior of water vapor and that the volume of liquid water is negligible. The vapor pressure of water at 65°C is 187.5 mmHg.

12.109 What are the advantages of cooking the vegetable broccoli with steam instead of boiling it in water?

12.110 A quantitative measure of how efficiently spheres pack into unit cells is called *packing efficiency*, which is the percentage of the cell space occupied by the spheres. Calculate the packing efficiencies of a simple cubic cell, a body-centered cubic cell, and a face-centered cubic cell. (*Hint:* Refer to Figure 12.21 and use the relationship that the volume of a sphere is $\frac{4}{3}\pi r^3$, where r is the radius of the sphere.)

12.111 The phase diagram of helium is shown. Helium is the only known substance that has two different liquid phases: helium-I and helium-II. (a) What is the maximum temperature at which helium-II can exist? (b) What is the minimum pressure at which solid helium can exist? (c) What is the normal boiling point of helium-I? (d) Can solid helium sublime?

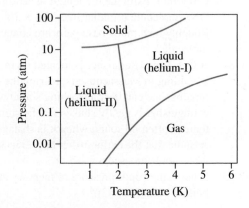

12.112 The phase diagram of sulfur is shown. (a) How many triple points are there? (b) Which is the more stable allotrope under ordinary atmospheric conditions? (c) Describe what happens when sulfur at 1 atm is heated from 80°C to 200°C.

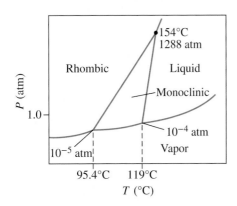

12.113 Provide an explanation for each of the following phenomena: (a) Solid argon (m.p. −189.2°C; b.p. −185.7°C) can be prepared by immersing a flask containing argon gas in liquid nitrogen (b.p. −195.8°C) until it liquefies and then connecting the flask to a vacuum pump. (b) The melting point of cyclohexane (C_6H_{12}) increases with increasing pressure exerted on the solid cyclohexane. (c) Certain high-altitude clouds contain water droplets at −10°C. (d) When a piece of dry ice is added to a beaker of water, fog forms above the water.

12.114 Argon crystallizes in the face-centered cubic arrangement at 40 K. Given that the atomic radius of argon is 191 pm, calculate the density of solid argon.

12.115 Given the phase diagram of carbon, answer the following questions: (a) How many triple points are there and what are the phases that can coexist at each triple point? (b) Which has a higher density, graphite or diamond? (c) Synthetic diamond can be made from graphite. Using the phase diagram, how would you go about making diamond?

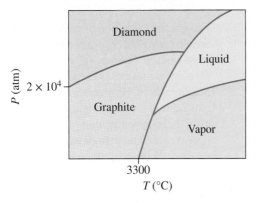

12.116 A chemistry instructor performed the following mystery demonstration. Just before the students arrived in class, she heated some water to boiling in an Erlenmeyer flask. She then removed the flask

from the flame and closed the flask with a rubber stopper. After the class commenced, she held the flask in front of the students and announced that she could make the water boil simply by rubbing an ice cube on the outside walls of the flask. To the amazement of everyone, it worked. Give an explanation for this phenomenon.

12.117 Swimming coaches sometimes suggest that a drop of alcohol (ethanol) placed in an ear plugged with water "draws out the water." Explain this action from a molecular point of view.

12.118 Given the general properties of water and ammonia, comment on the problems that a biological system (as we know it) would have developing in an ammonia medium.

	H_2O	NH_3
Boiling point	373.15 K	239.65 K
Melting point	273.15 K	195.3 K
Molar heat capacity	75.3 J/K · mol	8.53 J/K · mol
Molar heat of vaporization	40.79 kJ/mol	23.3 kJ/mol
Molar heat of fusion	6.0 kJ/mol	5.9 kJ/mol
Viscosity	0.001 N · s/m^2	0.0254 N · s/m^2 (at 240 K)
Dipole moment	1.82 D	1.46 D
Phase at 300 K	Liquid	Gas

12.119 Why do citrus growers spray their trees with water to protect them from freezing?

12.120 Calcium metal crystallizes in a face-centered cubic unit cell with a cell edge length of 558.84 pm. Calculate (a) the radius of a calcium atom in angstroms (Å) and (b) the density of calcium metal in g/cm^3.

12.121 A student heated a beaker of cold water (on a tripod) with a Bunsen burner. When the gas was ignited, she noticed that there was water condensed on the outside of the beaker. Explain what happened.

12.122 The compound dichlorodifluoromethane (CCl_2F_2) has a normal boiling point of −30°C, a critical temperature of 112°C, and a corresponding critical pressure of 40 atm. If the gas is compressed to 18 atm at 20°C, will the gas condense? Your answer should be based on a graphical interpretation.

12.123 Iron crystallizes in a body-centered cubic lattice. The cell length as determined by X-ray diffraction is 286.7 pm. Given that the density of iron is 7.874 g/cm^3, calculate Avogadro's number.

12.124 Sketch the cooling curves of water from about 110°C to about −10°C. How would you also show the formation of supercooled liquid below 0°C that then freezes to ice? The pressure is at 1 atm throughout the process. The curves need not be drawn quantitatively.

12.125   The boiling point of methanol is 65.0°C, and the standard enthalpy of formation of methanol vapor is −201.2 kJ/mol. Calculate the vapor pressure of methanol (in mmHg) at 25°C. (*Hint:* See Appendix 2 for other thermodynamic data of methanol.)

12.126   A sample of water shows the following behavior as it is heated at a constant rate.

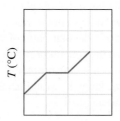

If twice the mass of water has the same amount of heat transferred to it, which of the graphs [(a)–(d)] best describes the temperature variation? Note that the scales for all the graphs are the same.

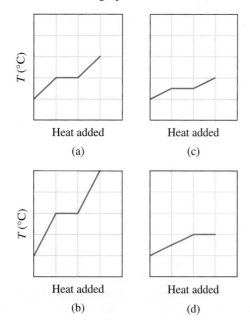

Heat added   Heat added
   (a)   (c)

Heat added   Heat added
   (b)   (d)

12.127   A closed vessel of volume 9.6 L contains 2.0 g of water. Calculate the temperature (in °C) at which only half of the water remains in the liquid phase. (See Table 11.6 for vapor pressures of water at different temperatures.)

12.128   The electrical conductance of copper metal decreases with increasing temperature, but that of a $CuSO_4$ solution increases with increasing temperature. Explain.

12.129   Assuming ideal behavior, calculate the density of gaseous HF at its normal boiling point (19.5°C). The experimentally measured density under the same conditions is 3.10 g/L. Account for the difference between your calculated result and the experimental value.

12.130   Explain why drivers are advised to use motor oil with lower viscosity in the winter and higher viscosity in the summer.

12.131   At what angle would you expect X rays of wavelength 0.154 nm to be reflected from a crystal in which the distance between layers is 312 pm? (Assume $n = 1$.)

12.132   Silicon used in computer chips must have an impurity level below 10^{-9} (i.e., fewer than one impurity atom for every 10^9 Si atoms). Silicon is prepared by the reduction of quartz (SiO_2) with coke (a form of carbon made by the destructive distillation of coal) at about 2000°C.

$$SiO_2(s) + 2C(s) \longrightarrow Si(l) + 2CO(g)$$

Next, solid silicon is separated from other solid impurities by treatment with hydrogen chloride at 350°C to form gaseous trichlorosilane ($SiCl_3H$).

$$Si(s) + 3HCl(g) \longrightarrow SiCl_3H(g) + H_2(g)$$

Finally, ultrapure Si can be obtained by reversing the above reaction at 1000°C.

$$SiCl_3H(g) + H_2(g) \longrightarrow Si(s) + 3HCl(g)$$

The molar heat of vaporization of trichlorosilane is 28.8 kJ/mol and its vapor pressure at 2°C is 0.258 atm. (a) Using this information and the equation

$$\ln \frac{P_1}{P_2} = \frac{\Delta H_{vap}}{R}\left(\frac{1}{T_2} - \frac{1}{T_1}\right)$$

determine the normal boiling point of trichlorosilane. (b) What kind(s) of intermolecular forces exist between trichlorosilane molecules? (c) Each cubic unit cell (edge length $a = 543$ pm) contains eight Si atoms. If there are 1.0×10^{13} boron atoms per cubic centimeter in a sample of pure silicon, how many Si atoms are there for every B atom in the sample? (d) Calculate the density of pure silicon.

12.133   Patients who have suffered from kidney stones often are advised to drink extra water to help prevent the formation of additional stones. An article on WebMD.com recommends drinking at least 3 quarts (2.84 L) of water every day—nearly 50 percent more than the amount recommended for healthy adults. How much energy must the body expend to warm this amount of water consumed at 10°C to body temperature (37°C)? How much *more* energy would have to be expended if the same quantity of water were consumed as *ice* at 0°C? ΔH_{fus} for water is 6.01 kJ/mol. Assume the density and specific heat of water are 1.00 g/cm^3 and 4.184 J/g · °C, respectively, and that both quantities are independent of temperature.

Answers to In-Chapter Materials

PRACTICE PROBLEMS

12.1A 265 mmHg. **12.1B** 75.9 kJ/mol, 109°C. **12.2A** 10.5 g/cm^3.
12.2B Body-centered cubic. **12.3A** 4 Ca, 8 F. **12.3B** 1 Cs, 1 Cl.
12.4A 2.65 g/cm^3. **12.4B** 421 pm. **12.5A** 2.72 g/cm^3. **12.5B** 361 pm.
12.6A 984 kJ. **12.6B** 100°C, liquid and vapor in equilibrium.
12.7A (a) ~110°C, ~−10°C; (b) liquid.
12.7B

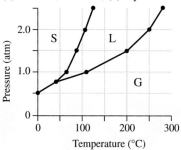

SECTION REVIEW

12.2.1 a. **12.2.2** e. **12.2.3** b. **12.2.4** b. **12.3.1** d.
12.3.2 a. **12.5.1** a. **12.5.2** c. **12.6.1** a. **12.6.2** e.

Chapter 13

Physical Properties of Solutions

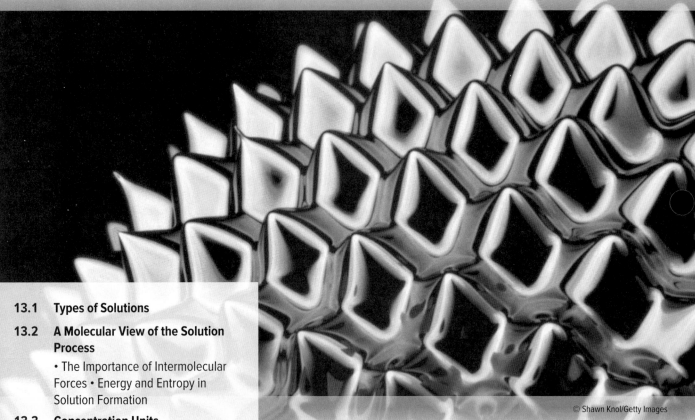

© Shawn Knol/Getty Images

A COLLOID is a uniform dispersion of one substance in another substance. Liquid magnets or *ferrofluids* represent an unusual type of colloid wherein nanoscale ferromagnetic particles are suspended in a carrier fluid, such as an organic solvent or water. When no external magnetic field is present, the fluid is not magnetic. However, when an external magnetic field is applied, the paramagnetic nanoparticles align with the magnet. Depending on the strength of the magnetic field applied, the density and shape of the ferrofluid can change. Chemists and materials scientists have found uses for ferrofluids in magnetic liquid sealants, low-friction seals for rotating shafts, stereo speakers and computer hard drives.

13.1 TYPES OF SOLUTIONS

As we noted in Section 1.5, a solution is a homogeneous mixture of two or more substances. Recall that a solution consists of a *solvent* and one or more *solutes* [◄◄ Section 9.1]. Although many of the most familiar solutions are those in which a solid is dissolved in a liquid (e.g., saltwater or sugar water), the components of a solution may be solid, liquid, or gas. The possible combinations give rise to seven distinct types of solutions, which we classify by the original states of the solution components. Table 13.1 gives an example of each type.

In this chapter, we will focus on solutions in which the solvent is a liquid; and the liquid solvent we will encounter most often is water. Recall that solutions in which water is the solvent are called *aqueous* solutions [◄◄ Section 9.1].

Solutions can also be classified by the amount of solute dissolved relative to the maximum amount that can be dissolved. A *saturated solution* is one that contains the maximum amount of a solute that will dissolve in a solvent at a specific temperature. The amount of solute dissolved in a given volume of a saturated solution is called the *solubility.* It is important to realize that solubility refers to a specific solute, a specific solvent, and a specific *temperature.* For example, the solubility of NaCl in water at 20°C is 36 g per 100 mL. The solubility of NaCl at another temperature, or in another solvent, would be different. An *unsaturated solution* is one that contains less solute than it has the capacity to dissolve. A *supersaturated solution,* on the other hand, contains *more* dissolved solute than is present in a saturated solution (Figure 13.1). It is generally not stable, and eventually the dissolved solute will come out of solution. An example of this phenomenon is shown in Figure 13.2.

Student Annotation: The term *solubility* was also defined in Section 9.2.

TABLE 13.1	Types of Solutions		
Solute	**Solvent**	**State of resulting solution**	**Example**
Gas	Gas	Gas*	Air
Gas	Liquid	Liquid	Carbonated water
Gas	Solid	Solid	H_2 gas in palladium
Liquid	Liquid	Liquid	Ethanol in water
Liquid	Solid	Solid	Mercury in silver
Solid	Liquid	Liquid	Saltwater
Solid	Solid	Solid	Brass (Cu/Zn)

*Gaseous solutions can only contain gaseous solutes.

Figure 13.1 (a) Many solutions consist of a solid dissolved in water. (b) When all the solid dissolves, the solution is unsaturated. (c) If more solid is added than will dissolve, the solution is saturated. (d) A saturated solution is, by definition, in contact with undissolved solid. (e) Some saturated solutions can be made into supersaturated solutions by heating to dissolve more solid, and cooling carefully to prevent crystallization.

© McGraw-Hill Education/Charles D. Winters, photographer

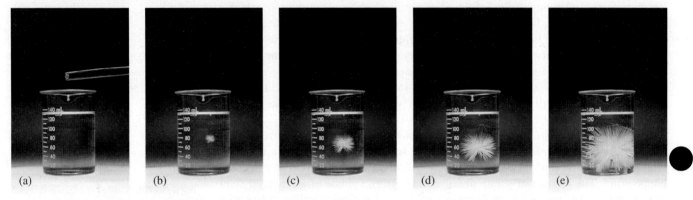

Figure 13.2 In a supersaturated solution, (a) addition of a tiny seed crystal initiates crystallization of excess solute. (b)–(e) Crystallization proceeds rapidly to give a *saturated* solution and the crystallized solid.

© McGraw-Hill Education/Charles D. Winters, photographer

13.2 A MOLECULAR VIEW OF THE SOLUTION PROCESS

In Chapter 9, we learned guidelines that helped us predict whether an ionic solid is soluble in water. We now take a more general look at the factors that determine solubility at the molecular level. This discussion will enable us to understand why so many ionic substances are soluble in water, which is a polar solvent; and it will help us to predict the solubility of ionic and molecular compounds in both polar and nonpolar solvents.

The Importance of Intermolecular Forces

The intermolecular forces that hold molecules together in liquids and solids play a central role in the solution process. When the solute dissolves in the solvent, molecules of the solute disperse throughout the solvent. They are, in effect, *separated* from one another and each solute molecule is *surrounded* by solvent molecules. The process by which solute molecules are surrounded by solvent molecules is called *solvation.* The ease with which solute molecules are separated from one another and surrounded by solvent molecules depends on the relative strengths of the solute-solute attractive forces, the solvent-solvent attractive forces, and the solute-solvent attractive forces.

The solute-solute attractive forces and solvent-solvent attractive forces may be any of those that were covered in Chapter 7, and occur in *pure* substances:

- Dispersion forces—present in *all* substances
- Dipole-dipole forces—present in *polar* substances
- Hydrogen bonding—especially strong dipole-dipole forces exhibited by molecules with O—H, N—H, or F—H bonds
- Ion-ion forces—present in *ionic* substances

Because solutions consist of at least two *different* substances, each of which may have different *properties*, there is a greater variety of intermolecular forces to consider. In addition to those just listed, solutions can also exhibit the following solute-solvent attractive forces.

Intermolecular forces	Example	Model
Ion-dipole. The charge of an ion is attracted to the partial charge on a polar molecule.	NaCl or KI in H_2O	Na^+ - - - - δ^- δ^+ +
Dipole-induced dipole. The partial charge on a polar molecule induces a temporary partial charge on a neighboring nonpolar molecule or atom.	He or CO_2 in H_2O	+ δ^- δ^+ - - - + δ^+ δ^-
Ion-induced dipole. The charge of an ion induces a temporary partial charge on a neighboring nonpolar molecule or atom.	Fe^{2+} and O_2	Fe^{2+} - - - δ^- δ^+ +

Student Annotation: A hemoglobin molecule contains four Fe^{2+} ions. In the early stages of O_2 binding, oxygen molecules are attracted to the Fe^{2+} ions by an ion-induced dipole interaction.

For simplicity, we can imagine the solution process taking place in the three distinct steps shown in Figure 13.3. Step 1 is the separation of solute molecules from one another, and step 2 is the separation of solvent molecules from one another. Both of these steps require an input of energy to overcome intermolecular attractions, so they are *endothermic*. In step 3 the solvent and solute molecules

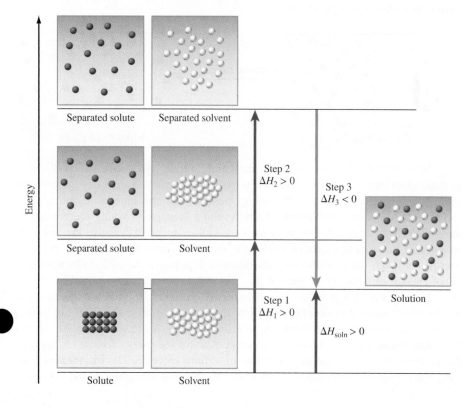

Figure 13.3 A molecular view of the solution process portrayed as taking place in three steps: First the solute and solvent molecules are separated (steps 1 and 2, respectively—both endothermic). Then the solvent and solute molecules mix (step 3—exothermic).

Student Annotation: Like the formation of chemical bonds [◄◄ Section 10.7], the formation of intermolecular attractions is exothermic. If that isn't intuitively obvious, think of it this way: It would require energy to *separate* molecules that are attracted to each other. The reverse process, the *combination* of molecules that attract each other, would give off an equal amount of energy [◄◄ Section 10.3].

mix. This process is usually *exothermic*. The enthalpy change for the overall process, ΔH_{soln}, is given by

$$\Delta H_{soln} = \Delta H_1 + \Delta H_2 + \Delta H_3$$

The overall solution-formation process is exothermic ($\Delta H_{soln} < 0$) when the energy given off in step 3 is greater than the sum of energy required for steps 1 and 2. The overall process is endothermic ($\Delta H_{soln} > 0$) when the energy given off in step 3 is less than the total required for steps 1 and 2. (Figure 13.3 depicts a solution formation that is endothermic overall.)

Energy and Entropy in Solution Formation

Previously, we learned that the driving force behind some processes is the lowering of the system's potential energy. Recall the minimization of potential energy when two hydrogen atoms are 74 pm apart [◄◄ Section 7.4]. However, because there are substances that dissolve endothermically, meaning that the process increases the system's potential energy, something else must be involved in determining whether a substance will dissolve. That something else is *entropy*.

The **entropy** of a system is a measure of how *dispersed* or *spread out* its energy is. Consider two samples of different gases separated by a physical barrier. When we remove the barrier, the gases mix, forming a solution. Under ordinary conditions, we can treat the gases as ideal, meaning we can assume that there are no attractive forces between the molecules in either sample before they mix (no solute-solute or solvent-solvent attractions to break)—and no attractive forces between the molecules in the mixture (no solute-solvent attractions form). The energy of the system does not change—and yet the gases mix spontaneously. The reason such a solution forms is that, although there is no change in the energy of either of the original samples of gas, the energy possessed by each sample of gas spreads out into a larger volume. This increased dispersal of the system's energy is an increase in the *entropy* of the system. There is a natural tendency for entropy to increase—that is, for the energy of a system to become more dispersed—unless there is something preventing that dispersal. Initially, the physical barrier between the two gases prevented their energy from spreading out into the larger volume. It is the increase in entropy that drives the formation of this solution, and many others.

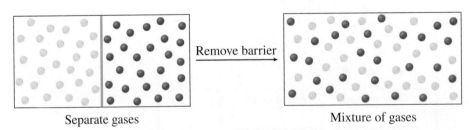

Separate gases   Mixture of gases

Now consider the case of solid ammonium nitrate (NH_4NO_3), which dissolves in water in an *endothermic* process. In this case, the dissolution *increases* the potential energy of the system. However, the energy possessed by the ammonium nitrate solid spreads out to occupy the volume of the resulting solution—causing the entropy of the system to increase. Ammonium nitrate dissolves in water because the favorable increase in the system's entropy outweighs the unfavorable increase in its potential energy. Although the process being endothermic is a barrier to solution formation, it is not enough of a barrier to prevent it.

In some cases, a process is so endothermic that even an increase in entropy is not enough to allow it to happen spontaneously. Sodium chloride (NaCl) is not soluble in a nonpolar solvent such as benzene, for example, because the solvent-solute interactions that would result are too weak to compensate for the energy required to separate the network of positive and negative ions in sodium chloride. The magnitude of the exothermic step in solution formation (ΔH_3) is so small compared to the combined magnitude of the endothermic steps (ΔH_1 and ΔH_2) that the overall process is *highly* endothermic and does not happen to any significant degree despite the increase in entropy that would result.

The saying "like dissolves like" is useful in predicting the solubility of a substance in a given solvent. In short, it means that *polar* substances (including ionic substances) will be more soluble in *polar* solvents; and *nonpolar* substances will be more soluble in *nonpolar* solvents. Put another way, substances with intermolecular forces of similar *type* and *magnitude* are likely to be soluble in each other. For example, both carbon tetrachloride (CCl_4) and benzene (C_6H_6) are nonpolar liquids. The only intermolecular forces present in these substances are dispersion forces [◄◄ Section 7.3]. When these two liquids are mixed, they readily dissolve in each other, because the attraction between CCl_4 and C_6H_6 molecules is comparable in magnitude to the forces between molecules in pure CCl_4 and to those between molecules in pure C_6H_6. Two liquids are said to be ***miscible*** if they are completely soluble in each other in all proportions. Alcohols such as methanol, ethanol, and 1,2-ethylene glycol are miscible with water because they can form hydrogen bonds with water molecules.

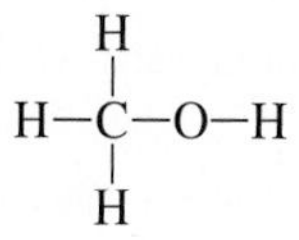

Methanol	Ethanol	1,2-Ethylene glycol

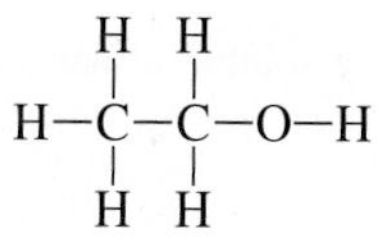

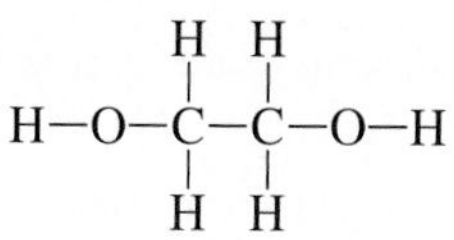

The guidelines listed in Tables 9.2 and 9.3 enable us to predict the solubility of a particular ionic compound in water. When sodium chloride dissolves in water, the ions are stabilized in solution by *hydration,* which involves ion-dipole interactions. In general, ionic compounds are much more soluble in polar solvents, such as water, liquid ammonia, and liquid hydrogen fluoride, than in nonpolar solvents. Because the molecules of nonpolar solvents, such as benzene and carbon tetrachloride, do not have a dipole moment, they cannot effectively solvate the Na^+ and Cl^- ions. The predominant intermolecular interaction between ions and nonpolar compounds is an ion-induced dipole interaction, which typically is much weaker than ion-dipole interactions. Consequently, ionic compounds usually have extremely low solubility in nonpolar solvents.

Student Annotation: *Solvation* refers in a general way to solute particles being surrounded by solvent molecules. When the solvent is water, we use the more specific term *hydration* [◄◄ Section 9.2].

Worked Example 13.1 lets you practice predicting solubility based on intermolecular forces.

Worked Example 13.1

Determine for each solute whether the solubility will be greater in water, which is polar, or in benzene (C_6H_6), which is nonpolar: (a) bromine (Br_2), (b) sodium iodide (NaI), (c) carbon tetrachloride (CCl_4), and (d) formaldehyde (CH_2O).

Strategy Consider the structure of each solute to determine whether it is polar. For molecular solutes, start with a Lewis structure and apply the VSEPR theory [◄◄ Section 7.1]. We expect polar solutes, including ionic compounds, to be more soluble in water. Nonpolar solutes will be more soluble in benzene.

Setup

(a) Bromine is a homonuclear diatomic molecule and is nonpolar.

(b) Sodium iodide is ionic.

(c) Carbon tetrachloride has the Lewis structure shown.

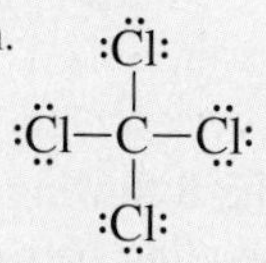

With four electron domains around the central atom, we expect a tetrahedral arrangement. A symmetrical arrangement of identical bonds results in a nonpolar molecule.

(Continued on next page)

(d) Formaldehyde has the Lewis structure shown.

Crossed arrows can be used to represent the individual bond dipoles [◀◀ Section 7.2]. This molecule is polar and can form hydrogen bonds with water.

Solution

(a) Bromine is more soluble in benzene.

(b) Sodium iodide is more soluble in water.

(c) Carbon tetrachloride is more soluble in benzene.

(d) Formaldehyde is more soluble in water.

Think About It

Remember that molecular formula alone is not sufficient to determine the shape or the polarity of a polyatomic molecule. It must be determined by starting with a correct Lewis structure and applying the VSEPR theory [◀◀ Section 7.1].

Practice Problem ATTEMPT Predict whether iodine (I_2) is more soluble in liquid ammonia (NH_3) or in carbon disulfide (CS_2).

Practice Problem BUILD Which of the following should you expect to be more soluble in benzene than in water: C_3H_8, HCl, I_2, CS_2?

Practice Problem CONCEPTUALIZE The first diagram represents a closed system consisting of water and a water-soluble gas in a container fitted with a movable piston. Which of the diagrams [(i)–(iii)] best represents the system when the piston is moved downward, decreasing the volume of the gas over the water? (Each diagram includes a thermometer indicating the temperature of the system.)

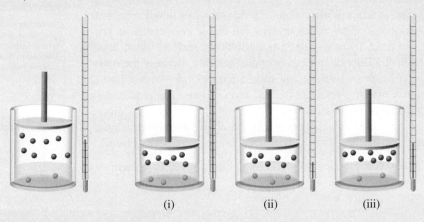

(i)   (ii)   (iii)

Section 13.2 Review

A Molecular View of the Solution Process

13.2.1 Which of the following compounds do you expect to be more soluble in benzene than in water? (Select all that apply.)

(a) SO_2

(b) CO_2

(c) Na_2SO_4

(d) C_2H_6

(e) Br_2

13.2.2 Which of the following compounds dissolved in water would exhibit hydrogen bonding between the solute and solvent?

(a) $H_2(g)$ in $H_2O(l)$

(b) $CH_3OH(l)$ in $H_2O(l)$

(c) $CO_2(g)$ in $H_2O(l)$

(d) $NH_3(g)$ in $H_2O(l)$

(e) $NaCl(s)$ in $H_2O(l)$

13.3 CONCENTRATION UNITS

We learned in Chapter 9 that chemists often express concentration of solutions in units of *molarity*. Recall that molarity, *M*, is defined as the number of moles of solute divided by the number of liters of solution [◄◄ Section 9.5].

$$\text{molarity} = M = \frac{\text{moles of solute}}{\text{liters of solution}}$$

Mole fraction, χ, which is defined as the number of moles of solute divided by the total number of moles, is also an expression of concentration [◄◄ Section 11.7].

$$\text{mole fraction of component A} = \chi_A = \frac{\text{moles of A}}{\text{sum of moles of all components}}$$

In this section, we will learn about *molality* and *percent by mass,* two additional ways to express the concentration of a mixture component. How a chemist expresses concentration depends on the type of problem being solved.

Student Annotation: We have already used percent by mass to describe the composition of a pure substance [◄◄ Section 5.9]. In this chapter, we will use percent by mass to describe *solutions*.

Molality

Molality (m) is the number of moles of solute dissolved in 1 kg (1000 g) of solvent.

$$\text{molality} = m = \frac{\text{moles of solute}}{\text{mass of solvent (in kg)}} \qquad \textbf{Equation 13.1}$$

For example, to prepare a 1 molal (1-*m*) aqueous sodium sulfate solution, we must dissolve 1 mole (142.0 g) of Na_2SO_4 in 1 kg of water.

Student Annotation: It is a common mistake to confuse molarity and molality. Molarity depends on the *volume* of the *solution*. Molality depends on the *mass* of the *solvent*.

Student Hot Spot

Student data indicate you may struggle with molarity and molality. Access the SmartBook to view additional Learning Resources on this topic.

Percent by Mass

The *percent by mass* (also called *percent by weight*) is the ratio of the mass of a solute to the mass of the solution, multiplied by 100 percent. Because the units of mass cancel on the top and bottom of the fraction, any units of mass can be used—provided they are used consistently.

$$\text{percent by mass} = \frac{\text{mass of solute}}{\text{mass of solute} + \text{mass of solvent}} \times 100\% \qquad \textbf{Equation 13.2}$$

For example, we can express the concentration of the sodium sulfate solution used to illustrate molality as follows. (Recall that the solution consists of 142.0 g Na_2SO_4 and 1 kg water.)

$$\text{percent by mass } Na_2SO_4 = \frac{\text{mass of } Na_2SO_4}{\text{mass of } Na_2SO_4 + \text{mass of water}} \times 100\%$$

$$= \frac{142.0 \text{ g}}{1142.0 \text{ g}} \times 100\% = 12.4\%$$

The term *percent* literally means "parts per hundred." If we were to use Equation 13.2 but multiply by 1000 instead of 100, we would get "parts per thousand"; multiplying by 1,000,000 would give "parts per million" or *ppm;* and so on. Parts per million, parts per billion, parts per trillion, and so forth, are often used to express very low concentrations, such as those of some pollutants in the atmosphere or a body of water. For example, if a 1-kg sample of water is found to contain 3 μg (3×10^{-6} g) of arsenic, its concentration can be expressed in parts per billion (ppb) as follows:

$$\frac{3 \times 10^{-6} \text{ g}}{1000 \text{ g}} \times 10^9 = 3 \text{ ppb}$$

Student Hot Spot

Student data indicate you may struggle with calculations involving mass percent and volume percent. Log in to Connect to view additional Learning Resources on this topic.

Worked Example 13.2 shows how to calculate the concentration in molality and in percent by mass.

Worked Example 13.2

A solution is made by dissolving 170.1 g of glucose ($C_6H_{12}O_6$) in enough water to make a liter of solution. The density of the solution is 1.062 g/mL. Express the concentration of the solution in (a) molality, (b) percent by mass, and (c) parts per million.

Strategy Use the molar mass of glucose to determine the number of moles of glucose in a liter of solution. Use the density (in g/L) to calculate the mass of a liter of solution. Subtract the mass of glucose from the mass of solution to determine the mass of water. Use Equation 13.1 to determine the molality. Knowing the mass of glucose and the total mass of solution in a liter, use Equation 13.2 to calculate the percent concentration by mass.

Setup The molar mass of glucose is 180.2 g/mol; the density of the solution is 1.062 g/mL.

Solution

(a)
$$\frac{170.1 \text{ g}}{180.2 \text{ g/mol}} = 0.9440 \text{ mol glucose per liter of solution}$$

$$1 \text{ liter of solution } \times \frac{1062 \text{ g}}{\text{L}} = 1062 \text{ g}$$

$$1062 \text{ g solution} - 170.1 \text{ g glucose} = 892 \text{ g water} = 0.892 \text{ kg water}$$

$$\frac{0.9440 \text{ mol glucose}}{0.892 \text{ kg water}} = 1.06 \ m$$

(b) $\dfrac{170.1 \text{ g glucose}}{1062 \text{ g solution}} \times 100\% = 16.02\%$ glucose by mass

(c) $\dfrac{170.1 \text{ g glucose}}{1062 \text{ g solution}} \times 1{,}000{,}000 = 1.602 \times 10^5$ ppm glucose

Think About It
Pay careful attention to units in problems such as this. Most require conversions between grams and kilograms and/or liters and milliliters.

Practice Problem **A**TTEMPT Determine (a) the molality and (b) the percent by mass of urea for a solution prepared by dissolving 5.46 g urea [$(NH_2)_2CO$] in 215 g of water.

Practice Problem **B**UILD Determine the molality of an aqueous solution that is 4.5 percent urea by mass.

Practice Problem **C**ONCEPTUALIZE For a given solute/solvent pair at a given temperature, which graph [(i)–(iv)] best depicts the relationship between percent composition by mass and molality of the solute?

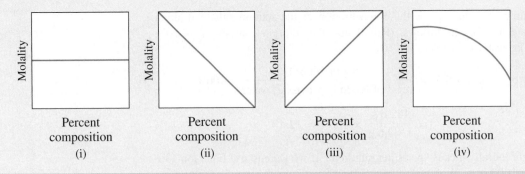

| Percent composition (i) | Percent composition (ii) | Percent composition (iii) | Percent composition (iv) |

Comparison of Concentration Units

The choice of a concentration unit is based on the purpose of the experiment. For instance, we typically use molarity to express the concentrations of solutions for titrations and gravimetric analyses. *Mole fractions* are used to express the concentrations of gases—and of solutions when we are working with vapor pressures, which we will discuss in Section 13.5.

The advantage of molarity is that it is generally easier to measure the volume of a solution, using precisely calibrated volumetric flasks, than to weigh the solvent. Molality, on the other hand, has the advantage of being temperature independent. The volume of a solution typically increases slightly with increasing temperature, which would change the molarity. The mass of solvent in a solution, however, does *not* change with temperature.

Percent by mass is similar to molality in that it is independent of temperature. Furthermore, because it is defined in terms of the ratio of the mass of solute to the mass of solution, we do not need to know the molar mass of the solute to calculate the percent by mass.

Often it is necessary to convert the concentration of a solution from one unit to another. For example, the same solution may be used for different experiments that require different concentration units for calculations. Suppose we want to express the concentration of a 0.396-*m* aqueous glucose ($C_6H_{12}O_6$) solution (at 25°C) in molarity. We know there is 0.396 mole of glucose in 1000 g of the solvent. We need to determine the *volume* of this solution to calculate molarity. To determine volume, we must first calculate its mass.

$$0.396 \text{ mol } C_6H_{12}O_6 \times \frac{180.2 \text{ g}}{1 \text{ mol } C_6H_{12}O_6} = 71.4 \text{ g } C_6H_{12}O_6$$

$$71.4 \text{ g } C_6H_{12}O_6 + 1000 \text{ g } H_2O = 1071 \text{ g solution}$$

Once we have determined the mass of the solution, we use the *density* of the solution to determine its volume. The density of a 0.396 *m* glucose solution is 1.16 g/mL at 25°C. Therefore, its volume is

$$\text{volume} = \frac{\text{mass}}{\text{density}}$$

$$= \frac{1071 \text{ g}}{1.16 \text{ g/mL}} \times \frac{1 \text{ L}}{1000 \text{ mL}}$$

$$= 0.923 \text{ L}$$

Having determined the volume of the solution, the molarity is given by

$$\text{molarity} = \frac{\text{moles of solute}}{\text{liters of solution}}$$

$$= \frac{0.396 \text{ mol}}{0.923 \text{ L}}$$

$$= 0.429 \text{ mol/L} = 0.429 \text{ } M$$

Worked Example 13.3 shows how to convert from one unit of concentration to another.

Worked Example 13.3

"Rubbing alcohol" is a mixture of isopropyl alcohol (C_3H_7OH) and water that is 70 percent isopropyl alcohol by mass (density = 0.79 g/mL at 20°C). Express the concentration of rubbing alcohol in (a) molarity and (b) molality.

Strategy

(a) Use density to determine the total mass of a liter of solution, and use percent by mass to determine the mass of isopropyl alcohol in a liter of solution. Convert the mass of isopropyl alcohol to moles, and divide moles by liters of solution to get molarity.

(b) Subtract the mass of C_3H_7OH from the mass of solution to get the mass of water. Divide moles of C_3H_7OH by the mass of water (in kilograms) to get molality.

Setup The mass of a liter of rubbing alcohol is 790 g, and the molar mass of isopropyl alcohol is 60.09 g/mol.

(Continued on next page)

Solution

(a)
$$\frac{790 \text{ g solution}}{\text{L solution}} \times \frac{70 \text{ g C}_3\text{H}_7\text{OH}}{100 \text{ g solution}} = \frac{553 \text{ g C}_3\text{H}_7\text{OH}}{\text{L solution}}$$

$$\frac{553 \text{ g C}_3\text{H}_7\text{OH}}{\text{L solution}} \times \frac{1 \text{ mol}}{60.09 \text{ g C}_3\text{H}_7\text{OH}} = \frac{9.20 \text{ mol C}_3\text{H}_7\text{OH}}{\text{L solution}} = 9.2 \text{ } M$$

(b) 790 g solution − 553 g C$_3$H$_7$OH = 237 g water = 0.237 kg water

$$\frac{9.20 \text{ mol C}_3\text{H}_7\text{OH}}{0.237 \text{ kg water}} = 39 \text{ } m$$

Rubbing alcohol is 9.2 M and 39 m in isopropyl alcohol.

Think About It

Note the large difference between molarity and molality in this case. Molarity and molality are the same (or similar) only for very dilute aqueous solutions.

Practice Problem **A**TTEMPT An aqueous solution that is 16 percent sulfuric acid (H$_2$SO$_4$) by mass has a density of 1.109 g/mL at 25°C. Determine (a) the molarity and (b) the molality of the solution at 25°C.

Practice Problem **B**UILD Determine the percent sulfuric acid by mass of a 1.49-m aqueous solution of H$_2$SO$_4$.

Practice Problem **C**ONCEPTUALIZE The diagrams represent solutions of a solid substance that is soluble in both water (density 1 g/cm^3) and chloroform (density 1.5 g/cm^3). For which of these solutions will the numerical value of molarity be closest to that of the molality? For which will the values of molarity and molality be most different?

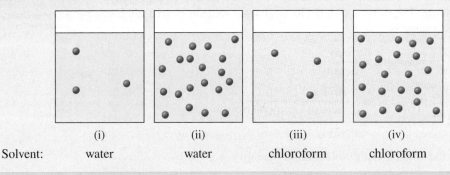

	(i)	(ii)	(iii)	(iv)
Solvent:	water	water	chloroform	chloroform

Section 13.3 Review

Concentration Units

13.3.1 Determine the percent by mass of KCl in a solution prepared by dissolving 1.18 g of KCl in 86.3 g of water.
(a) 1.35%   (c) 1.39%   (e) 2.12%
(b) 1.37%   (d) 2.05%

13.3.2 What is the molality of a solution prepared by dissolving 6.44 g of naphthalene (C$_{10}$H$_8$) in 80.1 g benzene?
(a) 1.13 m (c) 0.804 m (e) 11.7 m
(b) 80.4 m (d) 0.627 m

13.3.3 At 20.0°C, a 0.258-m aqueous solution of glucose (C$_6$H$_{12}$O$_6$) has a density of 1.0173 g/mL. Calculate the molarity of this solution.
(a) 0.258 M (c) 0.456 M (e) 0.448 M
(b) 0.300 M (d) 0.251 M

13.3.4 At 25.0°C, an aqueous solution that is 25.0 percent H$_2$SO$_4$ by mass has a density of 1.178 g/mL. Calculate the molarity and the molality of this solution.
(a) 3.00 M and 3.40 m (c) 3.00 M and 3.00 m (e) 3.44 M and 3.14 m
(b) 3.40 M and 3.40 m (d) 3.00 M and 2.98 m

13.4 FACTORS THAT AFFECT SOLUBILITY

Recall that solubility is defined as the maximum amount of solute that will dissolve in a given quantity of solvent at a specific *temperature*. Temperature affects the solubility of most substances. In this section we will consider the effects of temperature on the aqueous solubility of solids and gases, and the effect of *pressure* on the aqueous solubility of gases.

Temperature

More sugar dissolves in hot tea than in iced tea because the aqueous solubility of sugar, like that of most solid substances, increases as the temperature increases. Figure 13.4 shows the solubility of some common solids in water as a function of temperature. Note how the solubility of a solid and the *change* in solubility over a particular temperature range vary considerably. The relationship between temperature and solubility is complex and often nonlinear.

The relationship between temperature and the aqueous solubility of gases is somewhat simpler than that of solids. Most gaseous solutes become less soluble in water as temperature increases. If you get a glass of water from your faucet and leave it on the kitchen counter for a while, you will see bubbles forming in the water as it warms to room temperature. As the temperature of the water increases, dissolved gases become less soluble and come out of solution—resulting in the formation of bubbles.

One of the more important consequences of the reduced solubility of gases in water at elevated temperature is *thermal pollution*. Hundreds of billions of gallons of water are used every year for industrial cooling, mostly in electric power and nuclear power production. This process heats the water, which is then returned to the rivers and lakes from which it was taken. The increased water temperature has a twofold impact on aquatic life. The rate of metabolism of cold-blooded species such as fish increases with increasing temperature, thereby increasing their need for oxygen. At the same time, the increased water temperature causes a decrease in the solubility of oxygen—making less oxygen available. The result can be disastrous for fish populations.

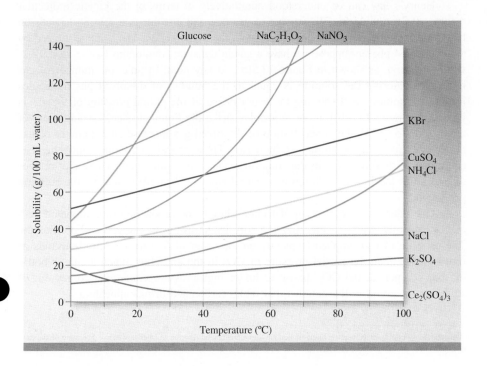

Figure 13.4 Temperature dependence of the solubility of glucose and several ionic compounds in water.

Figure 13.5 A molecular view of Henry's law. When the partial pressure of the gas over the solution increases from (a) to (b), the concentration of the dissolved gas also increases according to Equation 13.3.

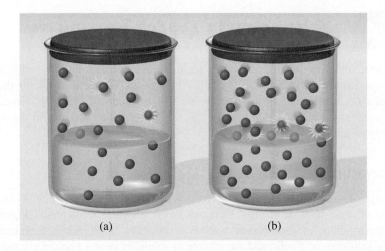

(a)   (b)

Pressure

Although pressure does not influence the solubility of a liquid or a solid significantly, it does greatly affect the solubility of a gas. The quantitative relationship between gas solubility and pressure is given by ***Henry's***[1] ***law,*** which states that the solubility of a gas in a liquid is proportional to the pressure of the gas over the solution,

$$c \propto P$$

and is expressed as

Equation 13.3	$c = kP$

> **Student Annotation:** Henry's law means that if we double the pressure of a gas over a solution (at constant temperature), we double the concentration of gas dissolved in the solution; triple the pressure, triple the concentration; and so on.

where c is the molar concentration (mol/L) of the dissolved gas, P is the pressure (in atm) of the gas over the solution, and k is a proportionality constant called the ***Henry's law constant (k).*** Henry's law constants are specific to the gas-solvent combination and vary with temperature. The units of k are mol/L · atm. If there is a mixture of gases over the solution, then P in Equation 13.3 is the partial pressure of the gas in question.

Henry's law can be understood qualitatively in terms of the kinetic molecular theory [◄◄ Section 11.2]. The amount of gas that will dissolve in a solvent depends on how frequently the gas molecules collide with the liquid surface and become trapped by the condensed phase. Suppose we have a gas in dynamic equilibrium [◄◄ Section 12.2] with a solution, as shown in Figure 13.5(a). At any point in time, the number of gas molecules entering the solution is equal to the number of dissolved gas molecules leaving the solution and entering the vapor phase. If the partial pressure of the gas is increased [Figure 13.5(b)], more molecules strike the liquid surface, causing more of them to dissolve. As the concentration of dissolved gas increases, the number of gas molecules leaving the solution also increases. These processes continue until the concentration of dissolved gas in the solution reaches the point again where the number of molecules leaving the solution per second equals the number entering the solution per second.

One interesting application of Henry's law is the production of carbonated beverages. Manufacturers put the "fizz" in soft drinks using pressurized carbon dioxide. The pressure of CO_2 applied (typically on the order of 5 atm) is many thousands of times greater than the partial pressure of CO_2 in the air. Thus, when a can or bottle of soda is opened, the CO_2 dissolved under high-pressure conditions comes out of solution—resulting in the bubbles that make carbonated drinks appealing.

---

[1]William Henry (1775–1836) was an English chemist whose major contribution to science was his discovery of the law describing the solubility of gases, which now bears his name.

Worked Example 13.4 illustrates the use of Henry's law.

Worked Example 13.4

Calculate the concentration of carbon dioxide in a soft drink that was bottled under a partial pressure of 5.0 atm CO_2 at 25°C (a) before the bottle is opened and (b) after the soda has gone "flat" at 25°C. The Henry's law constant for CO_2 in water at this temperature is 3.1×10^{-2} mol/L · atm. Assume that the partial pressure of CO_2 in air is 0.0003 atm and that the Henry's law constant for the soft drink is the same as that for water.

Strategy Use Equation 13.3 and the given Henry's law constant to solve for the molar concentration (mol/L) of CO_2 at 25°C and the two CO_2 pressures given.

Setup At 25°C, the Henry's law constant for CO_2 in water is 3.1×10^{-2} mol/L · atm.

Solution (a) $c = (3.1 \times 10^{-2}$ mol/L · atm$)(5.0$ atm$) = 1.6 \times 10^{-1}$ mol/L

(b) $c = (3.1 \times 10^{-2}$ mol/L · atm$)(0.0003$ atm$) = 9 \times 10^{-6}$ mol/L

> **Think About It**
> With a pressure approximately 15,000 times smaller in part (b) than in part (a), we expect the concentration of CO_2 to be approximately 15,000 times smaller—and it is.

Practice Problem Ⓐ**TTEMPT** Calculate the concentration of CO_2 in water at 25°C when the pressure of CO_2 over the solution is 4.0 atm.

Practice Problem Ⓑ**UILD** Calculate the pressure of O_2 necessary to generate an aqueous solution that is 3.4×10^{-2} *M* in O_2 at 25°C. The Henry's law constant for O_2 in water at 25°C is 1.3×10^{-3} mol/L · atm.

Practice Problem Ⓒ**ONCEPTUALIZE** The first diagram represents a closed system with two different gases dissolved in water. Which of the diagrams [(i)–(iv)] could represent a closed system consisting of the same two gases at the same temperature?

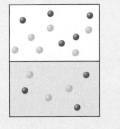 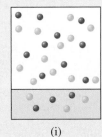 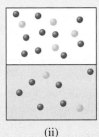 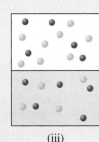 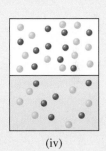

    (i)   (ii)   (iii)   (iv)

Section 13.4 Review

Factors That Affect Solubility

13.4.1 The solubility of N_2 in water at 25°C and an N_2 pressure of 1 atm is 6.8×10^{-4} mol/L. Calculate the concentration of dissolved N_2 in water under atmospheric conditions where the partial pressure of N_2 is 0.78 atm.
(a) 6.8×10^{-4} *M*   (d) 1.5×10^{-4} *M*
(b) 8.7×10^{-4} *M*   (e) 3.1×10^{-4} *M*
(c) 5.3×10^{-4} *M*

13.4.2 Calculate the molar concentration of O_2 in water at 25°C under atmospheric conditions where the partial pressure of O_2 is 0.22 atm. The Henry's law constant for O_2 is 1.3×10^{-3} mol/L · atm.
(a) 2.9×10^{-4} *M*   (d) 1.0×10^{-3} *M*
(b) 5.9×10^{-3} *M*   (e) 1.3×10^{-3} *M*
(c) 1.7×10^{-3} *M*

13.5   COLLIGATIVE PROPERTIES

Colligative properties are properties that depend on the number of solute particles in solution but do not depend on the nature of the solute particles. That is, colligative properties depend on the concentration of solute particles regardless of whether those particles are atoms, molecules, or ions. The colligative properties are vapor-pressure lowering, boiling-point elevation, freezing-point depression, and osmotic pressure. We begin by considering the colligative properties of relatively dilute solutions ($\leq 0.2\ M$) of nonelectrolytes.

Vapor-Pressure Lowering

We have seen that a liquid exerts a characteristic vapor pressure [◄◄ Section 12.2]. When a *nonvolatile* solute (one that does *not* exert a vapor pressure) is dissolved in a liquid, the vapor pressure exerted by the liquid decreases. The difference between the vapor pressure of a pure solvent and that of the corresponding solution depends on the concentration of the solute in the solution. This relationship is expressed by *Raoult's*[2] *law,* which states that the partial pressure of a solvent over a solution, P_1, is given by the vapor pressure of the pure solvent, P_1°, times the mole fraction of the solvent in the solution, χ_1.

Student Annotation: Table 11.6 (page 508) gives the vapor pressure of water at various temperatures.

Equation 13.4	$P_1 = \chi_1 P_1^\circ$

In a solution containing only one solute, $\chi_1 = 1 - \chi_2$, where χ_2 is the mole fraction of the solute. Equation 13.4 can therefore be rewritten as

$$P_1 = (1 - \chi_2)\ P_1^\circ$$

or

$$P_1 = P_1^\circ - \chi_2 P_1^\circ$$

so that

Equation 13.5	$P_1^\circ - P_1 = \Delta P = \chi_2 P_1^\circ$

Thus, the decrease in vapor pressure, ΔP, is directly proportional to the solute concentration expressed as a *mole fraction.*

To understand the phenomenon of vapor-pressure lowering, we must understand the degree of *order* associated with the states of matter involved. As we saw in Section 13.2, molecules in the liquid state are rather highly ordered; that is, they have low entropy. Molecules in the gas phase have significantly less order—they have high entropy. Because there is a natural tendency toward increased entropy, molecules have a certain tendency to leave the region of lower entropy and enter the region of higher entropy. This corresponds to molecules leaving the liquid and entering the gas phase. As we have seen, when a solute is added to a liquid, the liquid's order is disrupted. Thus, the solution has greater entropy than the pure liquid. Because there is a smaller difference in entropy between the solution and the gas phase than there was between the pure liquid and the gas phase, there is a decreased tendency for molecules to leave the solution and enter the gas phase—resulting in a lower vapor pressure exerted by the solvent. This qualitative explanation of vapor-pressure lowering is illustrated in Figure 13.6. The smaller difference in entropy between the solution and gas phases, relative to that between the pure liquid and gas phases, results in a decreased tendency for solvent molecules to enter the gas phase. This results in a lowering of vapor pressure. The solvent in a solution will always exert a lower vapor pressure than the pure solvent.

[2]François Marie Raoult (1839–1901), a French chemist, worked mainly in solution properties and electrochemistry.

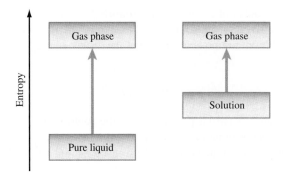

Figure 13.6 The smaller difference in entropy between the solution and gas phases, relative to that between the pure liquid and gas phases, results in a decreased tendency for solvent molecules to enter the gas phase. This results in a lowering of vapor pressure. The solvent in a solution always exerts a lower vapor pressure than the pure solvent.

Worked Example 13.5 shows how to use Raoult's law.

Worked Example 13.5

Calculate the vapor pressure of water over a solution made by dissolving 225 g of glucose in 575 g of water at 35°C. (At 35°C, $P_{H_2O}^{\circ} = 42.2$ mmHg.)

Strategy Convert the masses of glucose and water to moles, determine the mole fraction of water, and use Equation 13.4 to find the vapor pressure over the solution.

Setup The molar masses of glucose and water are 180.2 and 18.02 g/mol, respectively.

Solution

$$\frac{225 \text{ g glucose}}{180.2 \text{ g/mol}} = 1.25 \text{ mol glucose} \quad \text{and} \quad \frac{575 \text{ g water}}{18.02 \text{ g/mol}} = 31.9 \text{ mol water}$$

$$\chi_{water} = \frac{31.9 \text{ mol water}}{1.25 \text{ mol glucose} + 31.9 \text{ mol water}} = 0.962$$

$$P_{H_2O} = \chi_{water} \times P_{H_2O}^{\circ} = 0.962 \times 42.2 \text{ mmHg} = 40.6 \text{ mmHg}$$

The vapor pressure of water over the solution is 40.6 mmHg.

Think About It
This problem can also be solved using Equation 13.5 to calculate the vapor-pressure lowering, ΔP.

Practice Problem **A**TTEMPT Calculate the vapor pressure of a solution made by dissolving 115 g of urea [$(NH_2)_2CO$; molar mass = 60.06 g/mol] in 485 g of water at 25°C. (At 25°C, $P_{H_2O}^{\circ} = 23.8$ mmHg.)

Practice Problem **B**UILD Calculate the mass of urea that should be dissolved in 225 g of water at 35°C to produce a solution with a vapor pressure of 37.1 mmHg. (At 35°C, $P_{H_2O}^{\circ} = 42.2$ mmHg.)

Practice Problem **C**ONCEPTUALIZE The diagrams [(i)–(iv)] represent four closed systems containing aqueous solutions of the same nonvolatile solute at the same temperature. Over which solution is the vapor pressure of water the highest? Over which solution is it the lowest? Over which two solutions is the vapor pressure the same?

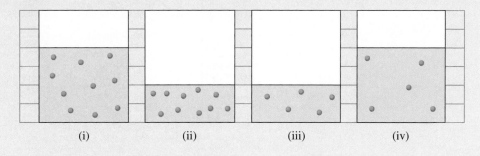

(i)   (ii)   (iii)   (iv)

If both components of a solution are *volatile* (i.e., have measurable vapor pressure), the vapor pressure of the solution is the sum of the individual partial pressures exerted by the solution components. Raoult's law holds equally well in this case:

$$P_A = \chi_A P_A^\circ$$
$$P_B = \chi_B P_B^\circ$$

where P_A and P_B are the partial pressures over the solution for components A and B, P_A° and P_B° are the vapor pressures of the pure substances A and B, and χ_A and χ_B are their mole fractions. The total pressure is given by Dalton's law of partial pressures [◄◄ Section 11.7]:

$$P_T = P_A + P_B$$

or

$$P_T = \chi_A P_A^\circ + \chi_B P_B^\circ$$

For example, benzene and toluene are volatile components that have similar structures and therefore similar intermolecular forces.

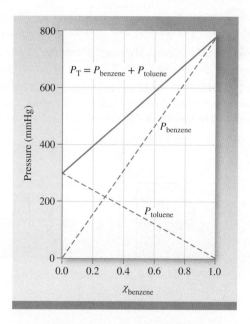

Benzene   Toluene

In a solution of benzene and toluene, the vapor pressure of each component obeys Raoult's law. Figure 13.7 shows the dependence of the total vapor pressure (P_T) in a benzene-toluene solution on the composition of the solution. Because there are only two components in the solution, we need only express the composition of the solution in terms of the mole fraction of *one* component. For any value of $\chi_{benzene}$, the mole fraction of toluene, $\chi_{toluene}$, is given by the equation $(1 - \chi_{benzene})$. The benzene-toluene solution is an example of an ***ideal solution,*** which is simply a solution that *obeys* Raoult's law.

Note that for a mixture in which the mole fractions of benzene and toluene are both 0.5, although the liquid mixture is equimolar, the vapor above the solution is not. Because pure benzene has a higher vapor pressure (75 mmHg at 20°C) than pure toluene (22 mmHg at 20°C), the vapor phase over the mixture will contain a higher concentration of the more volatile benzene molecules than it will the less volatile toluene molecules.

Figure 13.7 The dependence of partial pressures of benzene and toluene on their mole fractions in a benzene-toluene solution ($\chi_{toluene} = 1 - \chi_{benzene}$) at 80°C. This solution is said to be ideal because the vapor pressures obey Raoult's law.

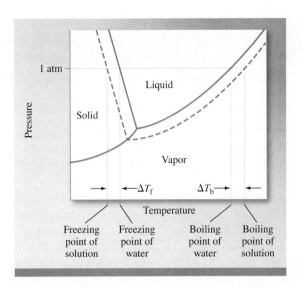

Figure 13.8 Phase diagram illustrating the boiling-point elevation and freezing-point depression of aqueous solutions. The dashed curves pertain to the solution, and the solid curves to the pure solvent. As this diagram shows, the boiling point of the solution is higher than that of water, and the freezing point of the solution is lower than that of water.

Boiling-Point Elevation

Recall that the *boiling point* of a substance is the temperature at which its vapor pressure equals the external atmospheric pressure [◄◄ Section 12.5]. Because the presence of a nonvolatile solute lowers the vapor pressure of a solution, it also affects the boiling point of the solution—relative to that of the pure liquid. Figure 13.8 shows the phase diagram of water and the changes that occur when a nonvolatile solute is added to it. At any temperature the vapor pressure over a solution is lower than that over the pure liquid, so the liquid-vapor curve for the solution lies below that for the pure solvent. Consequently, the dashed solution curve intersects the horizontal line that marks $P = 1$ atm at a higher temperature than the normal boiling point of the pure solvent; that is, a higher temperature is needed to make the solvent's vapor pressure equal to atmospheric pressure.

The boiling-point elevation (ΔT_b) is defined as the difference between the boiling point of the solution (T_b) and the boiling point of the pure solvent (T_b°).

$$\Delta T_b = T_b - T_b^\circ$$

Student Annotation: Rearranging this equation, we get the boiling point of the solution by adding ΔT_b to the boiling point of the pure solvent: $T_b = T_b^\circ + \Delta T_b$.

Because $T_b > T_b^\circ$, ΔT_b is a positive quantity.

The value of ΔT_b is proportional to the concentration, expressed in molality, of the solute in the solution:

$$\Delta T_b \propto m$$

or

$$\Delta T_b = K_b m \qquad \textbf{Equation 13.6}$$

where m is the molality of the solution and K_b is the molal boiling-point elevation constant. The units of K_b are °C/m. Table 13.2 lists values of K_b for several common solvents. Using the boiling-point elevation constant for water and Equation 13.6, you can show that the boiling point of a 1.00-m aqueous solution of a nonvolatile, nonelectrolyte would be 100.5°C.

$$\Delta T_b = K_b m = (0.52°C/m)(1.00\ m) = 0.52°C$$

$$T_b = T_b^\circ + \Delta T_b = 100.0°C + 0.52°C = 100.5°C$$

Student Annotation: The practice of adding salt to water in which food is cooked is for the purpose of increasing the boiling point, thereby cooking the food at a higher temperature.

Freezing-Point Depression

If you have ever lived in a cold climate, you may have seen roads and sidewalks that were "salted" in the winter. The application of a salt such as NaCl or $CaCl_2$ thaws ice (or prevents its formation) by lowering the freezing point of water.

TABLE 13.2	Molal Boiling-Point Elevation and Freezing-Point Depression Constants of Several Common Solvents			
Solvent	Normal boiling point (°C)	K_b (°C/m)	Normal freezing point (°C)	K_f (°C/m)
Water	100.0	0.52	0.0	1.86
Benzene	80.1	2.53	5.5	5.12
Ethanol	78.4	1.22	−117.3	1.99
Acetic acid	117.9	2.93	16.6	3.90
Cyclohexane	80.7	2.79	6.6	20.0

The phase diagram in Figure 13.8 shows that in addition to shifting the liquid-vapor phase boundary down, the addition of a nonvolatile solute also shifts the solid-liquid phase boundary to the left. Consequently, this dashed line intersects the solid horizontal line at 1 atm at a temperature lower than the freezing point of pure water. The freezing-point depression (ΔT_f) is defined as the difference between the freezing point of the pure solvent and the freezing point of the solution.

$$\Delta T_f = T_f^\circ - T_f$$

Because $T_f^\circ > T_f$, ΔT_f is a positive quantity. Again, the change in temperature is proportional to the molal concentration of the solution:

$$\Delta T_f \propto m$$

or

Equation 13.7 $\qquad\qquad\qquad\qquad \Delta T_f = K_f m$

where m is the concentration of solute expressed in molality and K_f is the molal freezing-point depression constant (see Table 13.2). Like K_b, K_f has units of °C/m.

Like boiling-point elevation, freezing-point depression can be explained in terms of differences in entropy. Freezing involves a transition from the more disordered liquid state to the more ordered solid state. For this to happen, energy must be removed from the system. Because a solution has greater disorder than the solvent, there is a bigger difference in entropy between the solution and the solid than there is between the pure solvent and the solid (Figure 13.9). The larger difference in entropy means that more energy must be removed for the liquid-solid transition to happen. Thus, the solution freezes at a lower temperature than does the pure solvent. Boiling-point elevation occurs only when the solute is nonvolatile. Freezing-point depression occurs regardless of the solute's volatility.

Figure 13.9 The solution has greater entropy than the pure solvent. The bigger difference in entropy between the solution and the solid means that more energy must be removed from the solution for it to freeze. Thus, the solution freezes at a lower temperature than the pure solvent.

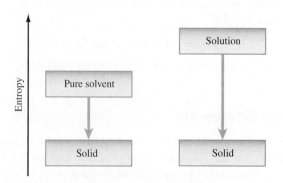

Worked Example 13.6 demonstrates a practical application of freezing-point depression and boiling-point elevation.

Worked Example 13.6

Ethylene glycol [CH$_2$(OH)CH$_2$(OH)] is a common automobile antifreeze. It is water soluble and fairly nonvolatile (b.p. 197°C). Calculate (a) the freezing point and (b) the boiling point of a solution containing 685 g of ethylene glycol in 2075 g of water.

Strategy Convert grams of ethylene glycol to moles, and divide by the mass of water in kilograms to get molal concentration. Use molal concentration in Equations 13.7 and 13.6 to determine ΔT_f and ΔT_b, respectively.

Setup The molar mass of ethylene glycol (C$_2$H$_6$O$_2$) is 62.07 g/mol. K_f and K_b for water are 1.86°C/m and 0.52°C/m, respectively.

Solution

$$\frac{685 \text{ g C}_2\text{H}_6\text{O}_2}{62.07 \text{ g/mol}} = 11.04 \text{ mol C}_2\text{H}_6\text{O}_2 \quad \text{and} \quad \frac{11.04 \text{ mol C}_2\text{H}_6\text{O}_2}{2.075 \text{ kg H}_2\text{O}} = 5.32 \ m \text{ C}_2\text{H}_6\text{O}_2$$

(a) $\Delta T_f = K_f m = (1.86°C/m)(5.32 \ m) = 9.89°C$

The freezing point of the solution is $(0 - 9.89)°C = -9.89°C$.

(b) $\Delta T_b = K_b m = (0.52°C/m)(5.32 \ m) = 2.8°C$

The boiling point of the solution is $(100.0 + 2.8)°C = 102.8°C$.

Think About It
Because it both lowers the freezing point and raises the boiling point, antifreeze is useful at both temperature extremes.

Practice Problem 🅐TTEMPT Calculate the freezing point and boiling point of a solution containing 268 g of ethylene glycol and 1015 g of water.

Practice Problem 🅑UILD What mass of ethylene glycol must be added to 1525 g of water to raise the boiling point to 103.9°C?

Practice Problem 🅒ONCEPTUALIZE The diagrams [(i)–(iv)] represent four different aqueous solutions of the same solute. Which of the solutions has the lowest freezing point? Which has the highest boiling point?

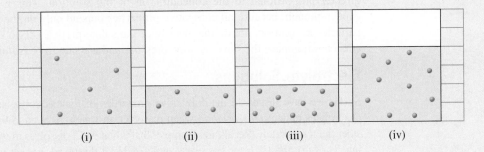

(i)   (ii)   (iii)   (iv)

Osmotic Pressure

Many chemical and biological processes depend on *osmosis,* the selective passage of solvent molecules through a porous membrane from a more dilute solution to a more concentrated one. Figure 13.10 illustrates osmosis. The left compartment of the apparatus contains pure solvent; the right compartment contains a solution made with the same solvent. The two compartments are separated by a *semipermeable membrane,* which allows the passage of solvent molecules but blocks the passage of solute molecules. At the beginning, the liquid levels in the two tubes are equal [Figure 13.10(a)]. As time passes, the level in the right tube rises. It continues to rise until equilibrium is reached, after which no further net change in levels is observed. The *osmotic pressure (π)* of a solution is the pressure required to stop osmosis. As shown in Figure 13.10(b), this pressure can be measured directly from the difference in the final liquid levels.

Video 13.1
Osmosis.

Figure 13.10 Osmotic pressure. (a) The levels of the pure solvent (left) and of the solution (right) are equal at the start. (b) During osmosis, the level on the solution side rises as a result of the net flow of solvent from left to right.

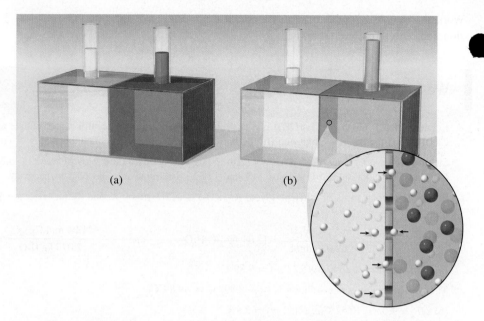

(a)   (b)

The osmotic pressure of a solution is directly proportional to the concentration, expressed in *molarity,* of the solute in solution:

$$\pi \propto M$$

and is given by

Equation 13.8	$\pi = MRT$

where M is the molarity of the solution, R is the gas constant (0.08206 L · atm/K · mol), and T is the absolute temperature. The osmotic pressure (π) is typically expressed in atmospheres.

Like boiling-point elevation and freezing-point depression, osmotic pressure is directly proportional to the concentration of the solution. This is what we would expect, though, because all colligative properties depend only on the number of solute particles in solution, not on the identity of the solute particles. Two solutions of equal concentration have the same osmotic pressure and are said to be ***isotonic*** to each other.

Electrolyte Solutions

So far we have discussed the colligative properties of nonelectrolyte solutions. Because electrolytes undergo *dissociation* when dissolved in water [◀◀ Section 9.1], we must consider them separately. Recall, for example, that when NaCl dissolves in water, it dissociates into $Na^+(aq)$ and $Cl^-(aq)$. For every mole of NaCl dissolved, we get two moles of ions in solution. Similarly, when a formula unit of $CaCl_2$ dissolves, we get three ions: one Ca^{2+} ion and two Cl^- ions. Thus, for every mole of $CaCl_2$ dissolved, we get *three* moles of ions in solution. Colligative properties depend only on the *number* of dissolved particles—not on the *type* of particles. This means that a 0.1-*m* solution of NaCl will exhibit a freezing-point depression twice that of a 0.1-*m* solution of a nonelectrolyte, such as sucrose. Similarly, we expect a 0.1-*m* solution of $CaCl_2$ to depress the freezing point of water three times as much as a 0.1 *m* sucrose solution. To account for this effect, we introduce and define a quantity called the ***van't Hoff***[3] ***factor (i),*** which is given by

$$i = \frac{\text{actual number of particles in solution after dissociation}}{\text{number of formula units initially dissolved in solution}}$$

Thus, i is 1 for all nonelectrolytes. For strong electrolytes such as NaCl and KNO_3, i is 2, and for strong electrolytes such as Na_2SO_4 and $CaCl_2$, i is 3. The van't Hoff

---

[3]Jacobus Henricus van't Hoff (1852–1911). One of the most prominent chemists of his time, van't Hoff, a Dutch chemist, did significant work in thermodynamics, molecular structure and optical activity, and solution chemistry. In 1901 he received the first Nobel Prize in Chemistry.

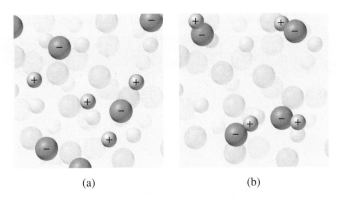

(a) (b)

Figure 13.11 (a) Free ions and (b) ion pairs in solution. Ion pairing reduces the number of dissolved particles in a solution, causing a decrease in the observed colligative properties. Furthermore, an ion pair bears no net charge and therefore cannot conduct electricity in solution.

factor can be thought of as the number of particles a substance breaks into (by *dissociation* or *ionization*) when it dissolves. The increased number of particles has a significant impact on the magnitudes of colligative properties. Consequently, the equations for colligative properties must be modified as follows:

Student Annotation: The amount of vapor-pressure lowering would also be affected by dissociation of an electrolyte. In calculating the mole fraction of solute and solvent, the number of moles of solute would have to be multiplied by the appropriate van't Hoff factor.

$$\Delta T_f = iK_f m \qquad \textbf{Equation 13.9}$$

$$\Delta T_b = iK_b m \qquad \textbf{Equation 13.10}$$

$$\pi = iMRT \qquad \textbf{Equation 13.11}$$

In reality, the colligative properties of electrolyte solutions are usually smaller than predicted by Equations 13.9 through 13.11, especially at higher concentrations, because of the formation of *ion pairs*. An **ion pair** is made up of one or more cations and one or more anions held together by electrostatic forces (Figure 13.11). The presence of an ion pair reduces the number of particles in solution, thus reducing the observed colligative properties. Tables 13.3 and 13.4 list van't Hoff factors calculated from balanced equations and those measured experimentally.

Student Annotation: A van't Hoff factor calculated using the coefficients in a balanced equation is an exact number [◄◄ Section 1.3].

TABLE 13.3	Calculated and Measured van't Hoff Factors of 0.0500 *M* Electrolyte Solutions at 25°C	
Electrolyte	***i* (Calculated)**	***i* (Measured)**
Sucrose*	1	1.0
HCl	2	1.9
NaCl	2	1.9
MgSO$_4$	2	1.3
MgCl$_2$	3	2.7
FeCl$_3$	4	3.4

*Sucrose is a nonelectrolyte. It is listed here for comparison only.

TABLE 13.4	Experimentally Measured van't Hoff Factors of Sucrose and NaCl Solutions at 25°C		
	Concentration		
Compound	**0.100 *m***	**0.00100 *m***	**0.000100 *m***
Sucrose	1.00	1.00	1.00
NaCl	1.87	1.94	1.97

Worked Example 13.7 demonstrates the experimental determination of a van't Hoff factor.

Worked Example 13.7

The osmotic pressure of a 0.0100 M potassium iodide (KI) solution at 25°C is 0.465 atm. Determine the experimental van't Hoff factor for KI at this concentration.

Strategy Use osmotic pressure to calculate the molar concentration of KI, and divide by the *nominal* concentration of 0.0100 M.

Setup $R = 0.08206$ L · atm/K · mol, and $T = 298$ K.

Solution Solving Equation 13.8 for M,

$$M = \frac{\pi}{RT} = \frac{0.465 \text{ atm}}{(0.08206 \text{ L} \cdot \text{atm/K} \cdot \text{mol})(298 \text{ K})} = 0.0190 \ M$$

$$i = \frac{0.0190 \ M}{0.0100 \ M} = 1.90$$

The experimental van't Hoff factor for KI at this concentration is 1.90.

Think About It
The calculated van't Hoff factor for KI is 2. The experimentally determined van't Hoff factor must be less than or equal to the calculated value.

Practice Problem **A**TTEMPT The freezing-point depression of a 0.100 m MgSO$_4$ solution is 0.225°C. Determine the experimental van't Hoff factor of MgSO$_4$ at this concentration.

Practice Problem **B**UILD Using the experimental van't Hoff factor from Table 13.4, determine the freezing point of a 0.100-m aqueous solution of NaCl. (Assume that the van't Hoff factors do not change with temperature.)

Practice Problem **C**ONCEPTUALIZE The diagram represents an aqueous solution of an electrolyte. Determine the experimental van't Hoff factor for the solute.

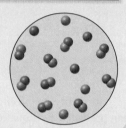

Section 13.5 Review

Colligative Properties

13.5.1 A solution contains 75.0 g of glucose (molar mass 180.2 g/mol) in 425 g of water. Determine the vapor pressure of water over the solution at 35°C. ($P^\circ_{\text{H}_2\text{O}} = 42.2$ mmHg at 35°C.)
   (a) 0.732 mmHg   (d) 41.5 mmHg
   (b) 42.9 mmHg   (e) 42.2 mmHg
   (c) 243 mmHg

13.5.2 Determine the boiling point and the freezing point of a solution prepared by dissolving 678 g of glucose in 2.0 kg of water. For water, $K_b = 0.52$°C/m and $K_f = 1.86$°C/m.
   (a) 101°C and 3.5°C   (d) 112°C and 6.2°C
   (b) 99°C and −3.5°C   (e) 88°C and −6.2°C
   (c) 101°C and −3.5°C

13.5.3 Calculate the osmotic pressure of a solution prepared by dissolving 65.0 g of Na$_2$SO$_4$ in enough water to make 500 mL of solution at 20°C. (Assume no ion pairing.)
   (a) 0.75 atm   (d) 1×10^{-2} atm
   (b) 66 atm   (e) 22 atm
   (c) 44 atm

13.5.4 A 1.00-m solution of HCl has a freezing point of −3.30°C. Determine the experimental van't Hoff factor for HCl at this concentration.
   (a) 1.77   (d) 2
   (b) 2.01   (e) 1
   (c) 1.90

Thinking Outside the Box

Intravenous Fluids

Human blood consists of red blood cells (*erythrocytes*), white blood cells (*leukocytes*), and platelets (*thrombocytes*) suspended in plasma, an aqueous solution containing a variety of solutes including salts and proteins. Each red blood cell is surrounded by a protective semipermeable membrane. Inside this membrane, the concentration of dissolved substances is about 0.3 *M*. Likewise, the concentration of dissolved substances in plasma is also about 0.3 *M*. Having the same concentration (and therefore the same osmotic pressure of ~7.6 atm at 37°C) inside and outside the red blood cell prevents a net movement of water into or out of the cell through the protective semipermeable membrane. To maintain this balance of osmotic pressure, fluids that are given intravenously must be *isotonic* to plasma. Five percent dextrose (sugar) and normal saline, which is 0.9% sodium chloride, are two of the most commonly used isotonic intravenous fluids.

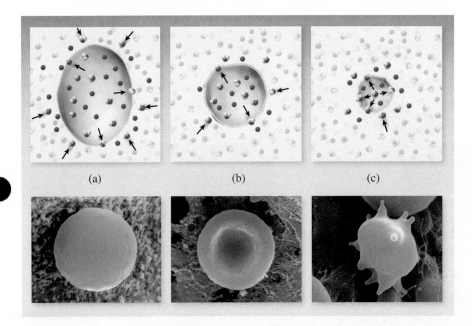

(a)　　　　　　　　(b)　　　　　　　　(c)

A cell in (a) a hypotonic solution, (b) an isotonic solution, and (c) a hypertonic solution. The cell swells and may eventually burst in (a); it shrinks in (c).

© David M. Phillips/Science Source

A solution that has a lower concentration of dissolved substances than plasma is said to be **hypotonic** to plasma. If a significant volume of pure water were administered intravenously, it would dilute the plasma, lowering its concentration and making it hypotonic to the solution inside the red blood cells. If this were to happen, water would enter the red blood cells via osmosis. The cells would swell and could potentially burst, a process called *hemolysis*. On the other hand, if red blood cells were placed in a solution with a higher concentration of dissolved substances than plasma, a solution said to be **hypertonic** to plasma, water would leave the cells via osmosis. The cells would shrink, a process called *crenation,* which is also potentially dangerous. The osmotic pressure of human plasma must be maintained within a very narrow range to prevent damage to red blood cells. Hypotonic and hypertonic solutions can be administered intravenously to treat specific medical conditions, but the patient must be carefully monitored throughout the treatment.

Interestingly, humans have for centuries exploited the sensitivity of cells to osmotic pressure. The process of "curing" meat with salt or with sugar causes crenation of the bacteria cells that would otherwise cause spoilage.

Thinking Outside the Box

Fluoride Poisoning

In 1993, nine patients who had undergone routine hemodialysis treatment at the University of Chicago Hospitals became seriously ill, and three of them died. The illnesses and deaths were attributed to fluoride poisoning, which occurred when the equipment meant to remove fluoride from the water failed. Although hemodialysis is supposed to remove impurities from the blood, the inadvertent use of fluoridated water in the process actually added a toxin to the patients' blood.

Hemodialysis, often called simply *dialysis,* is the cleansing of toxins from the blood of patients whose kidneys have failed. It works by routing a patient's blood temporarily through a special filter called a *dialyzer.* Inside the dialyzer, the blood is separated (by an artificial porous membrane) from an aqueous solution called the *dialysate.* The dialysate contains a variety of

dissolved substances, typically including sodium chloride, sodium bicarbonate or sodium acetate, calcium chloride, potassium chloride, magnesium chloride, and sometimes glucose. Its composition mimics that of blood plasma. When the two solutions, blood and dialysate, are separated by a porous membrane, the smallest of the dissolved solutes pass through the membrane from the side where the concentration is high to the side where the concentration is low. Because the dialysate contains vital components of blood in concentrations equal to those in blood, no net passage of these substances occurs through the membrane. However, the harmful substances that accumulate in the blood of patients whose kidneys do not function properly pass through the membrane into the dialysate, in which their concentration is initially zero, and are thereby removed from the blood. Properly done, hemodialysis therapy can add years to the life of a patient with kidney failure.

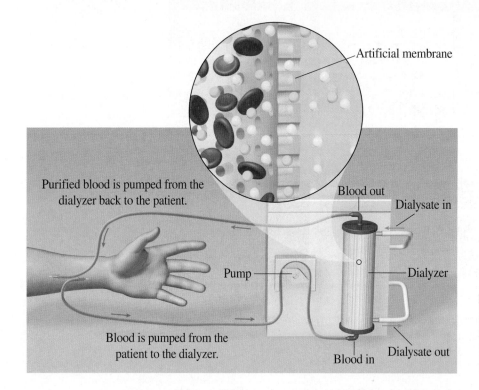

Artificial membrane

Purified blood is pumped from the dialyzer back to the patient.

Blood out

Dialysate in

Pump

Dialyzer

Blood is pumped from the patient to the dialyzer.

Blood in   Dialysate out

Student Annotation: Elevated levels of fluoride are associated with *osteomalacia,* a condition marked by debilitating bone pain and muscle weakness.

Osmosis refers to the movement of solvent through a membrane from the side where the solute concentration is lower to the side where the solute concentration is higher. Hemodialysis involves a more porous membrane, through which both solvent (water) and small solute particles can pass. The size of the membrane pores is such that only small waste products such as excess potassium ion, creatinine, urea, and extra fluid can pass through. Larger components in blood, such as blood cells and proteins, are too large to pass through the membrane. A solute will pass through the membrane from the side where its concentration is higher to the side where its concentration is lower. The composition of the dialysate ensures that the necessary solutes in the blood (e.g., sodium and calcium ions) are not removed. Because it is not normally found in blood, fluoride ion, if present in the dialysate, will flow across the membrane into the blood. In fact, this is true of any sufficiently small solute that is not normally found in

blood—necessitating requirements for the purity of water used to prepare dialysate solutions that far exceed those for drinking water.

Despite the use of fluoride in municipal water supplies and many topical dental products, acute fluoride poisoning is relatively rare. Fluoride is now routinely removed from the water used to prepare dialysate solutions. However, water-supply fluoridation became common during the 1960s and 1970s—just when hemodialysis was first being made widely available to patients. This unfortunate coincidence resulted in large numbers of early dialysis patients suffering the effects of fluoride poisoning, before the danger of introducing fluoride via dialysis was recognized. The level of fluoride considered safe for the drinking water supply is based on the presumed ingestion by a healthy person of 14 L of water per week. Many dialysis patients routinely are exposed to as much as *50 times* that volume, putting them at significantly increased risk of absorbing toxic amounts of fluoride.

13.6 CALCULATIONS USING COLLIGATIVE PROPERTIES

The colligative properties of nonelectrolyte solutions provide a means of determining the molar mass of a solute. Although any of the four colligative properties can be used in theory for this purpose, only freezing-point depression and osmotic pressure are used in practice because they show the most pronounced, and therefore the most easily measured, changes. From the experimentally determined freezing-point depression or osmotic pressure, we can calculate the solution's molality or molarity, respectively. Knowing the mass of dissolved solute, we can readily determine its molar mass. Worked Examples 13.8 and 13.9 illustrate this technique.

Student Hot Spot

Student data indicate you may struggle with determining molar mass from colligative properties. Access the SmartBook to view additional Learning Resources on this topic.

Student Annotation: These calculations require Equations 13.7 and 13.8, respectively.

Worked Example 13.8

Quinine was the first drug widely used to treat malaria, and it remains the treatment of choice for severe cases. A solution prepared by dissolving 10.0 g of quinine in 50.0 mL of ethanol has a freezing point 1.55°C below that of pure ethanol. Determine the molar mass of quinine. (The density of ethanol is 0.789 g/mL.) Assume that quinine is a nonelectrolyte.

Strategy Use Equation 13.7 to determine the molal concentration of the solution. Use the density of ethanol to determine the mass of solvent. The molal concentration of quinine multiplied by the mass of ethanol (in kilograms) gives moles of quinine. The mass of quinine (in grams) divided by moles of quinine gives the molar mass.

Setup mass of ethanol = 50.0 mL × 0.789 g/mL = 39.5 g or 3.95 × 10⁻² kg

K_f for ethanol (from Table 13.2) is 1.99°C/m.

Solution Solving Equation 13.7 for molal concentration,

$$m = \frac{\Delta T_f}{K_f} = \frac{1.55°C}{1.99°C/m} = 0.779\ m$$

The solution is 0.779 m in quinine (i.e., 0.779 mol quinine/kg ethanol solvent).

$$\left(\frac{0.779\ \text{mol quinine}}{\text{kg ethanol}}\right)(3.95 \times 10^{-2}\ \text{kg ethanol}) = 0.0308\ \text{mol quinine}$$

$$\text{molar mass of quinine} = \frac{10.0\ \text{g quinine}}{0.0308\ \text{mol quinine}} = 325\ \text{g/mol}$$

Think About It
Check the result using the molecular formula of quinine: $C_{20}H_{24}N_2O_2$ (324.4 g/mol). Multistep problems such as this one require careful tracking of units at each step.

Practice Problem ATTEMPT Calculate the molar mass of naphthalene, the organic compound in "mothballs," if a solution prepared by dissolving 5.00 g of naphthalene in exactly 100 g of benzene has a freezing point 2.00°C below that of pure benzene.

Practice Problem BUILD What mass of naphthalene must be dissolved in 2.00 × 10² g of benzene to give a solution with a freezing point 2.50°C below that of pure benzene?

Practice Problem CONCEPTUALIZE The first diagram represents an aqueous solution. Which of the diagrams [(i)–(iv)] represents a solution that is isotonic with the first?

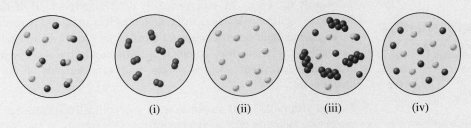

| (i) | (ii) | (iii) | (iv) |

Worked Example 13.9

A solution is prepared by dissolving 50.0 g of hemoglobin (Hb) in enough water to make 1.00 L of solution. The osmotic pressure of the solution is measured and found to be 14.3 mmHg at 25°C. Calculate the molar mass of hemoglobin. (Assume that there is no change in volume when the hemoglobin is added to the water.)

Strategy Use Equation 13.8 to calculate the molarity of the solution. Because the solution volume is 1 L, the molarity is equal to the number of moles of hemoglobin. Dividing the mass of hemoglobin, which is given in the problem statement, by the number of moles gives the molar mass.

Setup $R = 0.08206 \ \text{L} \cdot \text{atm/K} \cdot \text{mol}$, $T = 298$ K, and $\pi = 14.3 \ \text{mmHg}/(760 \ \text{mmHg/atm}) = 1.88 \times 10^{-2}$ atm.

Solution Rearranging Equation 13.8 to solve for molarity, we get

$$M = \frac{\pi}{RT} = \frac{1.88 \times 10^{-2} \ \text{atm}}{(0.08206 \ \text{L} \cdot \text{atm/K} \cdot \text{mol})(298 \ \text{K})} = 7.69 \times 10^{-4} \ M$$

Thus, the solution contains 7.69×10^{-4} mole of hemoglobin.

$$\text{molar mass of hemoglobin} = \frac{50.0 \ \text{g}}{7.69 \times 10^{-4} \ \text{mol}} = 6.50 \times 10^{4} \ \text{g/mol}$$

Think About It
Biological molecules can have *very* high molar masses.

Practice Problem 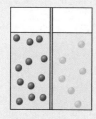 **ATTEMPT** A solution made by dissolving 25 mg of insulin in 5.0 mL of water has an osmotic pressure of 15.5 mmHg at 25°C. Calculate the molar mass of insulin. (Assume that there is no change in volume when the insulin is added to the water.)

Practice Problem **B**UILD What mass of insulin must be dissolved in 50.0 mL of water to produce a solution with an osmotic pressure of 16.8 mmHg at 25°C?

Practice Problem **C**ONCEPTUALIZE The first diagram represents one aqueous solution separated from another by a semipermeable membrane. Which of the diagrams [(i)–(iv)] could represent the same system after the passage of some time?

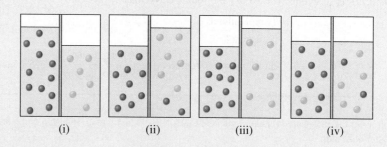

(i)   (ii)   (iii)   (iv)

The colligative properties of an electrolyte solution can be used to determine percent dissociation. *Percent dissociation* is the percentage of dissolved molecules (or formula units, in the case of an ionic compound) that separate into ions in solution. For a strong electrolyte such as NaCl, there should be *complete,* or 100 percent, dissociation. However, the data in Table 13.4 indicate that this is not necessarily the case. An experimentally determined van't Hoff factor smaller than the corresponding calculated value indicates less than 100 percent dissociation. As the experimentally determined van't Hoff factors for NaCl indicate, dissociation of a strong electrolyte is more complete at lower concentration. The *percent ionization* of a weak electrolyte, such as a weak acid, also depends on the concentration of the solution.

Worked Example 13.10 shows how to use colligative properties to determine the percent dissociation of a weak electrolyte.

Worked Example 13.10

A solution that is 0.100 M in hydrofluoric acid (HF) has an osmotic pressure of 2.64 atm at 25°C. Calculate the percent ionization of HF at this concentration.

Strategy Use the osmotic pressure and Equation 13.8 to determine the molar concentration of the particles in solution. Compare the concentration of particles to the *nominal* concentration (0.100 M) to determine what percentage of the original HF molecules are ionized.

Setup $R = 0.08206$ L · atm/K · mol, and $T = 298$ K.

Solution Rearranging Equation 13.8 to solve for molarity,

$$M = \frac{\pi}{RT} = \frac{2.64 \text{ atm}}{(0.08206 \text{ L} \cdot \text{atm/K} \cdot \text{mol})(298 \text{ K})} = 0.108 \ M$$

The concentration of dissolved particles is 0.108 M. Consider the ionization of HF [◄◄ Section 9.3]:

$$HF(aq) \rightleftharpoons H^+(aq) + F^-(aq)$$

According to this equation, if x HF molecules ionize, we get x H^+ ions and x F^- ions. Thus, the total concentration of particles in solution will be the original concentration of HF minus x, which gives the concentration of intact HF molecules, plus $2x$, which is the concentration of ions (H^+ and F^-).

$$(0.100 - x) + 2x = 0.100 + x$$

Therefore, $0.108 = 0.100 + x$ and $x = 0.008$. Because we earlier defined x as the amount of HF ionized, the percent ionization is given by

$$\text{percent ionization} = \frac{0.008 \ M}{0.100 \ M} \times 100\% = 8\%$$

At this concentration HF is 8 percent ionized.

Think About It
For weak acids, the lower the concentration, the greater the percent ionization. A 0.010-M solution of HF has an osmotic pressure of 0.30 atm, corresponding to 23 percent ionization. A 0.0010-M solution of HF has an osmotic pressure of 3.8×10^{-2} atm, corresponding to 56 percent ionization.

Practice Problem ATTEMPT An aqueous solution that is 0.0100 M in acetic acid ($HC_2H_3O_2$) has an osmotic pressure of 0.255 atm at 25°C. Calculate the percent ionization of acetic acid at this concentration.

Practice Problem BUILD An aqueous solution that is 0.015 M in acetic acid ($HC_2H_3O_2$) is 3.5 percent ionized at 25°C. Calculate the osmotic pressure of this solution.

Practice Problem CONCEPTUALIZE The diagrams [(i)–(iv)] represent aqueous solutions of weak electrolytes. List the solutions in order of increasing percent ionization.

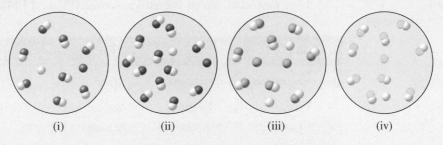

| (i) | (ii) | (iii) | (iv) |

Section 13.6 Review

Calculations Using Colligative Properties

13.6.1 A solution made by dissolving 14.2 g of sucrose in 100 g of water exhibits a freezing-point depression of 0.77°C. Calculate the molar mass of sucrose.
(a) 34 g/mol
(c) 2.4 g/mol
(e) 68 g/mol
(b) 3.4×10^2 g/mol
(d) 1.8×10^2 g/mol

(Continued on next page)

13.6.2 A 0.010-M solution of the weak electrolyte HA has an osmotic pressure of 0.27 atm at 25°C. What is the percent ionization of the electrolyte at this concentration?

(a) 27%   (c) 15%   (e) 90%

(b) 10%   (d) 81%

13.7   COLLOIDS

The solutions discussed so far in this chapter are true homogeneous mixtures. Now consider what happens if we add fine sand to a beaker of water and stir. The sand particles are suspended at first but gradually settle to the bottom of the beaker. This is an example of a heterogeneous mixture. Between the two extremes of homogeneous and heterogeneous mixtures is an intermediate state called a colloidal suspension, or simply, a colloid. A *colloid* is a dispersion of particles of one substance throughout another substance. Colloidal particles are much larger than the normal solute molecules; they range from 1×10^3 pm to 1×10^6 pm. Also, a colloidal suspension lacks the homogeneity of a true solution.

Colloids can be further categorized as aerosols (liquid or solid dispersed in gas), foams (gas dispersed in liquid or solid), emulsions (liquid dispersed in another liquid), sols (solid dispersed in liquid or in another solid), and gels (liquid dispersed in a solid). Table 13.5 lists the different types of colloids and gives one or more examples of each.

One way to distinguish a solution from a colloid is by the *Tyndall*[4] *effect.* When a beam of light passes through a colloid, it is scattered by the dispersed phase (Figure 13.12). No such scattering is observed with true solutions because the solute molecules are too small to interact with visible light. Another demonstration of the Tyndall effect is the scattering of light from automobile headlights in fog (Figure 13.13).

Among the most important colloids are those in which the dispersing medium is water. Such colloids can be categorized as *hydrophilic* (*water loving*) or *hydrophobic* (*water fearing*). Hydrophilic colloids contain extremely large molecules such as proteins. In the aqueous phase, a protein like hemoglobin folds in such a way that the hydrophilic parts of the molecule, the parts that can interact favorably with water molecules by ion-dipole forces or hydrogen-bond formation, are on the outside surface (Figure 13.14).

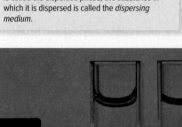

Student Annotation: The substance dispersed is called the *dispersed phase;* the substance in which it is dispersed is called the *dispersing medium.*

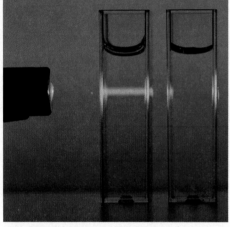

Figure 13.12 The Tyndall effect. Light is scattered by colloidal particles (left) but not by dissolved particles (right).

© McGraw-Hill Education/Charles D. Winters, photographer

Student Annotation: Styrofoam is a registered trademark of the Dow Chemical Company. It refers specifically to extruded polystyrene used for insulation in home construction. "Styrofoam" cups, coolers, and packing peanuts are not really made of Styrofoam.

TABLE 13.5	Types of Colloids		
Dispersing medium	**Dispersed phase**	**Name**	**Example**
Gas	Liquid	Aerosol	Fog, mist
Gas	Solid	Aerosol	Smoke
Liquid	Gas	Foam	Whipped cream, meringue
Liquid	Liquid	Emulsion	Mayonnaise
Liquid	Solid	Sol	Milk of magnesia
Solid	Gas	Foam	Styrofoam
Solid	Liquid	Gel	Jelly, butter
Solid	Solid	Solid sol	Alloys such as steel, gemstones (glass with dispersed metal)

[4]John Tyndall (1820–1893), an Irish physicist, did important work in magnetism and also explained glacier motion.

Figure 13.13 A familiar example of the Tyndall effect: headlights illuminating fog.
© seebes/iStock/Getty Images

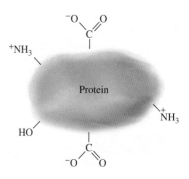

Figure 13.14 Hydrophilic groups on the surface of a large molecule such as a protein stabilize the molecule in water. Note that all the hydrophilic groups can form hydrogen bonds with water.

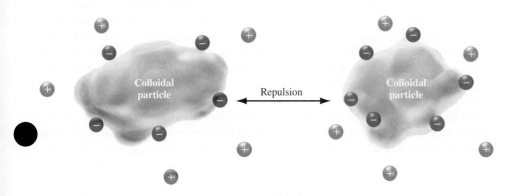

Figure 13.15 Diagram showing the stabilization of hydrophobic colloids. Negative ions are adsorbed onto the surface, and the repulsion between like charges prevents aggregation of the particles.

A hydrophobic colloid normally would not be stable in water, and the particles would clump together, like droplets of oil in water merging to form a film at the water's surface. They can be stabilized, however, by the adsorption of ions on their surface (Figure 13.15). Material that collects on the surface is *adsorbed,* whereas material that passes to the interior is *absorbed.* The adsorbed ions are hydrophilic and can interact with water to stabilize the colloid. In addition, because adsorption of ions leaves the colloid particles *charged,* electrostatic repulsion prevents them from clumping together. Soil particles in rivers and streams are hydrophobic particles that are stabilized in this way. When river water enters the sea, the charges on the dispersed particles are neutralized by the high-salt medium. With the charges on their surfaces neutralized, the particles no longer repel one another and they clump together to form the silt that is seen at the mouth of the river.

Another way hydrophobic colloids can be stabilized is by the presence of other hydrophilic groups on their surfaces. Consider sodium stearate, a soap molecule that has a polar group at one end, often called the "head," and a long hydrocarbon "tail" that is nonpolar (Figure 13.16). The cleansing action of soap is due to the dual nature of the hydrophobic tail and the hydrophilic head. The hydrocarbon tail is readily soluble in oily substances, which are also nonpolar, while the ionic —COO⁻ group remains outside the oily surface. When enough soap molecules have surrounded an oil droplet, as shown in Figure 13.17, the entire system becomes stabilized in water because the exterior portion is now largely hydrophilic. This is how greasy

Student Annotation: A hydrophobic colloid must be *stabilized* to remain suspended in water.

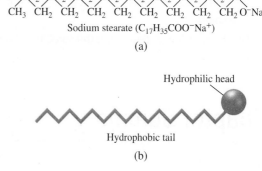

Sodium stearate ($C_{17}H_{35}COO^-Na^+$)

(a)

Hydrophilic head

Hydrophobic tail

(b)

Figure 13.16 (a) A sodium stearate molecule. (b) The simplified representation of the molecule that shows a hydrophilic head and a hydrophobic tail.

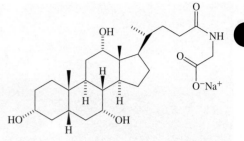

Figure 13.18 Structure of sodium glycocholate. The hydrophobic tail of sodium glycocholate dissolves in ingested fats, stabilizing them on the aqueous medium of the digestive system.

Figure 13.17 The mechanism by which soap removes grease. (a) Grease (oily substance) is not soluble in water. (b) When soap is added to water, the nonpolar tails of soap molecules dissolve in grease. (c) The grease can be washed away when the polar heads of the soap molecules stabilize it in water.

substances are removed by the action of soap. In general, the process of stabilizing a colloid that would otherwise not stay dispersed is called *emulsification,* and a substance used for such stabilization is called an *emulsifier* or *emulsifying agent.*

A mechanism similar to that involving sodium stearate makes it possible for us to digest dietary fat. When we ingest fat, the gallbladder excretes a substance known as bile. Bile contains a variety of substances including bile salts. A *bile salt* is a derivative of cholesterol with an attached amino acid. Like sodium stearate, a bile salt has both a hydrophobic end and a hydrophilic end. (Figure 13.18 shows the bile salt sodium glycocholate.) The bile salts surround fat particles with their hydrophobic ends oriented toward the fat and their hydrophilic ends facing the water, emulsifying the fat in the aqueous medium of the digestive system. This process allows fats to be digested and other nonpolar substances such as fat-soluble vitamins to be absorbed through the wall of the small intestine.

Student Annotation: It is being nonpolar that makes some vitamins soluble in fat. Remember the axiom "like dissolves like."

Learning Outcomes

- Define solubility.
- Define saturated, unsaturated, and supersaturated solution.
- Describe the three types of interactions that determine the extent to which a solute is dissolved in solution.
- Use concentration units to express the concentration of a given solution or to interchange concentration units.
- List and describe the factors that affect the solubility of a solute.
- Use Henry's law to determine the solubility of a gas in solution.

- List, describe, and perform calculations involving colligative properties: vapor-pressure lowering, boiling-point elevation, freezing-point depression, and osmotic pressure.
- Use van't Hoff factors to determine the colligative properties of electrolyte solutions.
- Define isotonic, hypertonic, and hypotonic.
- Use colligative properties to determine the percent dissociation (or percent ionization) of an electrolyte in solution.
- Define colloid and provide examples.

Chapter Summary

SECTION 13.1

- Solutions are homogeneous mixtures of two or more substances, which may be solids, liquids, or gases.
- *Saturated solutions* contain the maximum possible amount of dissolved solute.

- The amount of solute dissolved in a saturated solution is the *solubility* of the solute in the specified solvent at the specified temperature.
- *Unsaturated solutions* contain less than the maximum possible amount of solute.

- *Supersaturated solutions* contain more solute than specified by the solubility.

SECTION 13.2

- Substances with similar intermolecular forces tend to be soluble in one another. "Like dissolves like." Two liquids that are soluble in each other are called *miscible.*
- Solution formation may be endothermic or exothermic overall. An increase in *entropy* is the driving force for solution formation. Solute particles are surrounded by solvent molecules in a process called *solvation.*

SECTION 13.3

- In addition to molarity (M) and mole fraction (χ), *molality* (*m*) and *percent by mass* are used to express the concentrations of solutions.
- *Molality* is defined as the number of moles of solute per kilogram of solvent. *Percent by mass* is defined as the mass of solute divided by the total mass of the solution, all multiplied by 100 percent.
- Molality and percent by mass have the advantage of being temperature independent. Conversion among molarity, molality, and percent by mass requires solution *density.*
- The units of concentration used depend on the type of problem to be solved.

SECTION 13.4

- Increasing the temperature *increases* the solubility of most solids in water and *decreases* the solubility of most gases in water.
- Increasing the pressure increases the solubility of gases in water but does not affect the solubility of solids.
- According to *Henry's law,* the solubility of a gas in a liquid is directly proportional to the partial pressure of the gas over the solution: $c = kP.$
- The proportionality constant k is the *Henry's law constant (k).* Henry's law constants are specific to the gas and solvent, and they are temperature dependent.

SECTION 13.5

- *Colligative properties* depend on the *number* (but not on the *type*) of dissolved particles. The colligative properties are *vapor-pressure lowering, boiling-point elevation, freezing-point depression,* and *osmotic pressure.*

- A *volatile* substance is one that has a measurable vapor pressure. A *nonvolatile* substance is one that does not have a measurable vapor pressure.
- According to *Raoult's law,* the partial pressure of a substance over a solution is equal to the *mole fraction* (χ) of the substance times its *pure vapor pressure* ($P°$). An *ideal solution* is one that obeys Raoult's law.
- *Osmosis* is the flow of solvent through a *semipermeable membrane,* one that allows solvent molecules but not solute particles to pass, from a more dilute solution to a more concentrated one.
- *Osmotic pressure (π)* is the pressure required to prevent osmosis from occurring.
- Two solutions with the same osmotic pressure are called *isotonic. Hypotonic* refers to a solution with a *lower* osmotic pressure. *Hypertonic* refers to a solution with a *higher* osmotic pressure. These terms are often used in reference to human plasma, which has an osmotic pressure of 7.6 atm.
- In electrolyte solutions, the number of dissolved particles is increased by dissociation or ionization. The magnitudes of colligative properties are increased by the *van't Hoff factor (i),* which indicates the degree of dissociation or ionization.
- The experimentally determined van't Hoff factor is generally smaller than the calculated value due to the formation of *ion pairs*—especially at high concentrations. Ion pairs are oppositely charged ions that are attracted to each other and effectively become a single "particle" in solution.

SECTION 13.6

- Experimentally determined colligative properties can be used to calculate the molar mass of a nonelectrolyte or the *percent dissociation* (or *percent ionization*) of a weak electrolyte.

SECTION 13.7

- A *colloid* is a dispersion of particles (about 1×10^3 pm to 1×10^6 pm) of one substance in another substance.
- Colloids can be distinguished from true solutions by the *Tyndall effect,* which is the scattering of visible light by colloidal particles.
- Colloids are classified either as *hydrophilic* (water loving) or *hydrophobic* (water fearing).
- Hydrophobic colloids can be stabilized in water by surface interactions with ions or polar molecules.

Key Words

Key Equations

13.1 $\text{molality} = m = \dfrac{\text{moles of solute}}{\text{mass of solvent (in kg)}}$	Molality (m) of a particular solute in a solution is calculated by dividing the number of moles of that solute by the number of kilograms of solvent.
13.2 $\text{percent by mass} = \dfrac{\text{mass of solute}}{\text{mass of solute} + \text{mass of solvent}} \times 100\%$	Percent mass of a particular solute in a solution is calculated by dividing the mass of that solute by the total mass of solvent and solute.
13.3 $c = kP$	The concentration of gas dissolved in a liquid at a particular temperature is equal to the product of that gas's partial pressure over the solution (P) and the Henry's law constant (k). The value of the Henry's law constant is specific to the gas-solvent-temperature combination.
13.4 $P_1 = \chi_1 P_1^{\circ}$	The pressure exerted by a solvent over a solution (P_1) is equal to the product of that solvent's mole fraction (χ_1) and the pressure exerted by the pure solvent (P_1°).
13.5 $P_1^{\circ} - P_1 = \Delta P = \chi_2 P_1^{\circ}$	The difference between the pressure exerted by a solvent over a solution (P_1) and the pressure exerted by the pure solvent (P_1°) is the vapor-pressure lowering (ΔP) caused by the presence of solute. In the case where only one solute is present, the mole fraction of the solute can be calculated from the value of ΔP. ΔP is the product of the solute's mole fraction (χ_2) and the pressure exerted by the pure solvent (P_1°).
13.6 $\Delta T_b = K_b m$	The amount by which a solvent's boiling point is increased by the presence of a solute (ΔT_b) is calculated as the product of the boiling-point elevation constant (K_b) and the molality of the solute (m). The value of the boiling-point elevation constant depends on the identity of the solvent.
13.7 $\Delta T_f = K_f m$	The amount by which a solvent's boiling point is decreased by the presence of a solute (ΔT_f) is calculated as the product of the freezing-point depression constant (K_f) and the molality of the solute (m). The value of the freezing-point depression constant depends on the identity of the solvent.
13.8 $\pi = MRT$	Osmotic pressure of a solution (π) is the product of solute molarity (M), the gas constant (R) expressed in unit of pressure—typically $0.0823 \ \text{L} \cdot \text{atm/K} \cdot \text{mol}$, and absolute temperature (T).
13.9 $\Delta T_f = i K_f m$	For the purpose of calculating the freezing-point depression of a solution in which the solute is an electrolyte, the solute's molality (m) is multiplied by its van't Hoff factor (i). The value of i depends on the percent dissociation (or ionization) of the solute.
13.10 $\Delta T_b = i K_b m$	For the purpose of calculating the boiling-point elevation of a solution in which the solute is an electrolyte, the solute's molality (m) is multiplied by its van't Hoff factor (i). The value of i depends on the percent dissociation (or ionization) of the solute.
13.11 $\pi = i MRT$	For the purpose of calculating the osmotic pressure of a solution in which the solute is an electrolyte, the solute's molarity (M) is multiplied by its van't Hoff factor (i). The value of i depends on the percent dissociation (or ionization) of the solute.

Entropy as a Driving Force

We have seen that although a decrease in system energy can be the driving force for a process [◀◀ Section 7.4], entropy also plays a role in determining whether a process will occur. Recall that entropy is a measure of how spread out a system's energy is. The simplest way to interpret this is to consider how spread out a system's energy is in *space*. Consider the example of a compressed gas in one side of a divided container. If the barrier between the two compartments is removed, the compressed gas will expand to fill the new, larger volume.

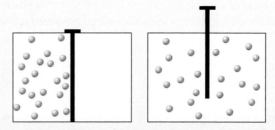

The energy possessed by the gas molecules was originally contained within a smaller volume. After the expansion, the energy possessed by the molecules occupies a larger volume, meaning that the energy is more spread out in space. This spreading out in space of the system's energy is an increase in entropy.

In addition to applying this interpretation of entropy to gas expansion and solution formation, we can apply it to phase changes, such as the sublimation of dry ice [$CO_2(s)$].

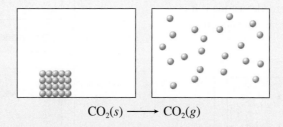

$$CO_2(s) \longrightarrow CO_2(g)$$

The CO_2 molecules in a sample of dry ice possess energy that is confined to the volume of the solid. Sublimation of the sample results in the molecules (and the energy they possess) occupying a much larger volume—corresponding to an increase in entropy. Because of this increase in entropy, although it is endothermic, the sublimation of dry ice does happen spontaneously.

For any process to happen, it must be exothermic, or be accompanied by an entropy increase, or both. An endothermic process (one in which system energy increases) may occur if there is a sufficient increase in the system's entropy. For example, although the dissolution of sucrose ($C_{12}H_{22}O_{11}$) is endothermic, sucrose dissolves in water because of the resulting increase in entropy.

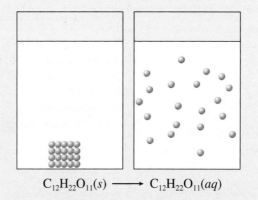

$$C_{12}H_{22}O_{11}(s) \longrightarrow C_{12}H_{22}O_{11}(aq)$$

Likewise, a process that results in an entropy *decrease* may occur if it is sufficiently exothermic. An example of this is the condensation of water vapor on a cool surface. Although the energy of the water molecules is less spread out when it condenses, the condensation process is exothermic enough to compensate for the entropy decrease—and it does happen.

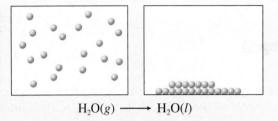

$$H_2O(g) \longrightarrow H_2O(l)$$

Processes that are neither exothermic nor accompanied by an entropy increase do not occur.

Being able to make a qualitative assessment of the entropy change associated with a process, and being able to determine whether a process is exothermic or endothermic, can help us predict which processes are likely to happen, and which are not.

Key Skills Problems

13.1

Which of the following processes is accompanied by an increase in entropy? (Select all that apply.)

(a) $Br_2(l) \longrightarrow Br_2(g)$
(b) $NH_3(g) + HCl(g) \longrightarrow NH_4Cl(s)$
(c) $NaCl(s) \longrightarrow Na^+(g) + Cl^-(g)$
(d) $H_2(g) + O_2(g) \longrightarrow 2H_2O(l)$

13.2

For each of the processes depicted here, determine if it is endothermic or exothermic, or if there is not enough information to determine.

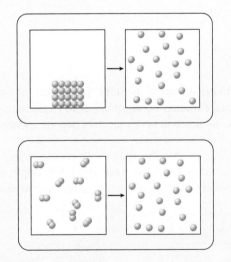

(a) Endothermic, exothermic
(b) Exothermic, endothermic
(c) Endothermic, not enough information to determine
(d) Endothermic, endothermic
(e) Exothermic, not enough information to determine

13.3

For each of the processes depicted here, determine if it is endothermic or exothermic, or if there is not enough information to determine.

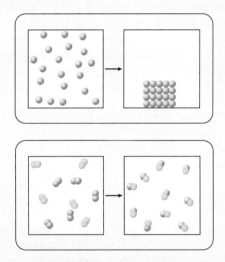

(a) Endothermic, exothermic
(b) Exothermic, endothermic
(c) Endothermic, not enough information to determine
(d) Endothermic, endothermic
(e) Exothermic, not enough information to determine

13.4

For some ionic solutes, dissolution actually causes an overall decrease in entropy. In order for such a dissolution to occur spontaneously, it must be _____.
(a) Exothermic
(b) Endothermic
(c) There is not enough information to determine this.

Questions and Problems

SECTION 13.1: TYPES OF SOLUTIONS

Review Questions

13.1 Distinguish between an unsaturated solution, a saturated solution, and a supersaturated solution.

13.2 Describe the different types of solutions that can be formed by the combination of solids, liquids, and gases. Give examples of each type of solution.

SECTION 13.2: A MOLECULAR VIEW OF THE SOLUTION PROCESS

Review Questions

13.3 Briefly describe the solution process at the molecular level. Use the dissolution of a solid in a liquid as an example.

13.4 Basing your answer on intermolecular force considerations, explain what "like dissolves like" means.

13.5 What is solvation? What factors influence the extent to which solvation occurs? Give two examples of solvation; include one that involves ion-dipole interaction and one in which dispersion forces come into play.

13.6 As you know, some solution processes are endothermic and others are exothermic. Provide a molecular interpretation for the difference.

13.7 Explain why dissolving a solid almost always leads to an increase in disorder.

13.8 Describe the factors that affect the solubility of a solid in a liquid. What does it mean to say that two liquids are miscible?

Conceptual Problems

13.9 Why is naphthalene ($C_{10}H_8$) more soluble than CsF in benzene?

13.10 Explain why ethanol (C_2H_5OH) is not soluble in cyclohexane (C_6H_{12}).

13.11 Arrange the following compounds in order of increasing solubility in water: O_2, LiCl, Br_2, methanol (CH_3OH).

13.12 Explain the variations in solubility in water of the listed alcohols:

Compound	Solubility in water (g/100 g) at 20°C
CH_3OH	∞
CH_3CH_2OH	∞
$CH_3CH_2CH_2OH$	∞
$CH_3CH_2CH_2CH_2OH$	9
$CH_3CH_2CH_2CH_2CH_2OH$	2.7

Note: ∞ means that the alcohol and water are completely miscible in all proportions.

SECTION 13.3: CONCENTRATION UNITS

Review Questions

13.13 Define the following concentration terms and give their units: *percent by mass, mole fraction, molarity, molality.* Compare their advantages and disadvantages.

13.14 Outline the steps required for conversion between molarity, molality, and percent by mass.

13.15 A solution of a particular concentration is prepared at 20°C. Of *molarity, molality, percent by mass,* and *mole fraction,* which will change when the solution is heated to 75°C? Explain.

Computational Problems

13.16 Calculate the amount of water (in grams) that must be added to (a) 5.00 g of urea [$(NH_2)_2CO$] in the preparation of a 8.15 percent by mass solution, and (b) 61.2 g of $MgBr_2$ in the preparation of a 2.5 percent by mass solution.

13.17 Calculate the molality of each of the following solutions: (a) 30.2 g of sucrose ($C_{12}H_{22}O_{11}$) in 285 g of water, (b) 8.35 moles of ethylene glycol ($C_2H_6O_2$) in 1905 g of water.

13.18 Calculate the molality of each of the following aqueous solutions: (a) 3.75 M KCl solution (density of solution = 1.09 g/mL), (b) 36.1 percent by mass NaBr solution.

13.19 Calculate the molalities of the following aqueous solutions: (a) 0.988 M sugar ($C_{12}H_{22}O_{11}$) solution (density of solution = 1.095 g/mL), (b) 1.12 M NaOH solution (density of solution = 1.082 g/mL), (c) 2.98 M $NaHCO_3$ solution (density of solution = 1.077 g/mL).

13.20 For dilute aqueous solutions in which the density of the solution is roughly equal to that of the pure solvent, the molarity of the solution is equal to its molality. Show that this statement is correct for a 0.0045-M aqueous copper(II) chloride ($CuCl_2$) solution.

13.21 The alcohol content of hard liquor is normally given in terms of the "proof," which is defined as twice the percentage by volume of ethanol (C_2H_5OH) present. Calculate the grams of alcohol present in a generous serving (45 mL) of 80-proof cognac. The density of ethanol is 0.798 g/mL.

13.22 PepsiCo, maker of Pepsi, announced in April 2015 that it was replacing the artificial sweetener aspartame ($C_{14}H_{18}N_2O_5$) with sucralose ($C_{12}H_{19}Cl_3O_8$) in Diet Pepsi. Determine the molarity of a 12-oz can of Diet Pepsi that contains 40 mg of sucralose.

13.23 After sales of "new" Diet Pepsi proved disappointing, PepsiCo announced in June 2016 that it would resume production of a version of Diet Pepsi with aspartame ($C_{14}H_{18}N_2O_5$). Determine the molarity of a 12-oz can of Diet Pepsi that contains 111 mg of aspartame.

13.24 The density of an aqueous solution containing 25.0 percent of ethanol (C_2H_5OH) by mass is

0.950 g/mL. (a) Calculate the molality of this solution. (b) Calculate its molarity. (c) What volume of the solution would contain 0.275 mole of ethanol?

SECTION 13.4: FACTORS THAT AFFECT SOLUBILITY

Review Questions

13.25 How do the solubilities of most ionic compounds in water change with temperature? With pressure?

13.26 Discuss the factors that influence the solubility of a gas in a liquid.

13.27 What is thermal pollution? Why is it harmful to aquatic life?

13.28 What is Henry's law? Define each term in the equation, and give its units. How would you account for the law in terms of the kinetic molecular theory of gases? Give two exceptions to Henry's law.

13.29 A student is observing two beakers of water. One beaker is heated to 30°C, and the other is heated to 100°C. In each case, bubbles form in the water. Are these bubbles of the same origin? Explain.

13.30 A man bought a goldfish in a pet shop. Upon returning home, he put the goldfish in a bowl of recently boiled water that had been cooled quickly. A few minutes later the fish was found dead. Explain what happened to the fish.

Computational Problems

13.31 The solubility of KNO_3 is 155 g per 100 g of water at 75°C and 38.0 g at 25°C. What mass (in grams) of KNO_3 will crystallize out of solution if exactly 100 g of its saturated solution at 75°C is cooled to 25°C?

13.32 A 3.20-g sample of a salt dissolves in 9.10 g of water to give a saturated solution at 25°C. What is the solubility (in g salt/100 g of H_2O) of the salt?

13.33 The solubility of CO_2 in water at 25°C and 1 atm is 0.034 mol/L. What is its solubility under atmospheric conditions? (The partial pressure of CO_2 in air is 0.0003 atm.) Assume that CO_2 obeys Henry's law.

13.34 A 50-g sample of impure $KClO_3$ (solubility = 7.1 g per 100 g H_2O at 20°C) is contaminated with 10 percent of KCl (solubility = 25.5 g per 100 g of H_2O at 20°C). Calculate the minimum quantity of 20°C water needed to dissolve all the KCl from the sample. How much $KClO_3$ will be left after this treatment? (Assume that the solubilities are unaffected by the presence of the other compound.)

13.35 Fish breathe the dissolved air in water through their gills. Assuming the partial pressures of oxygen and nitrogen in air to be 0.20 and 0.80 atm, respectively, calculate the mole fractions of oxygen and nitrogen in the air dissolved in water at 298 K. The solubilities of O_2 and N_2 in water at 298 K are 1.3×10^{-3} mol/L · atm and 6.8×10^{-4} mol/L · atm, respectively. Comment on your results.

13.36 The solubility of N_2 in blood at 37°C and at a partial pressure of 0.80 atm is 5.6×10^{-4} mol/L. A deep-sea diver breathes compressed air with the partial pressure of N_2 equal to 4.0 atm. Assume that the total volume of blood in the body is 5.0 L. Calculate the amount of N_2 gas released (in liters at 37°C and 1 atm) when the diver returns to the surface of the water, where the partial pressure of N_2 is 0.80 atm.

Conceptual Problems

13.37 The difference between water-soluble and fat-soluble vitamins is their molecular structures. Water-soluble vitamins tend to have multiple polar groups that can interact with water to form hydrogen bonds, whereas fat-soluble vitamins tend to be nonpolar molecules consisting predominately of hydrocarbon chains. This is an example of what is meant by the expression "like dissolves like." Predict whether vitamin C and vitamin E will be water soluble or fat soluble.

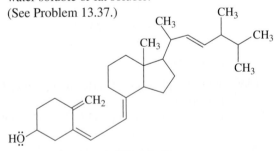

Vitamin C

Vitamin E

13.38 Predict whether each vitamin will be water soluble or fat soluble. (See Problem 13.37.)

Vitamin D

Vitamin B_2 (riboflavin)

13.39 A student carried out the following experiment to measure the pressure of carbon dioxide in the space above the carbonated soft drink in a bottle. First, she weighed the bottle (853.5 g). Next, she carefully removed the cap to let the CO_2 gas escape. She then reweighed the bottle with the cap (851.3 g). Finally, she measured the volume of the soft drink (452.4 mL). Given that the Henry's law constant for CO_2 in water at 25°C is 3.4×10^{-2} mol/L · atm, calculate the pressure of CO_2 over the soft drink in the bottle before it was opened. Explain why this pressure is only an estimate of the true value.

13.40 The first diagram represents an open system with two different gases dissolved in water. Which of the diagrams [(a)–(c)] could represent the same system at a higher temperature?

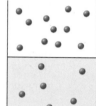

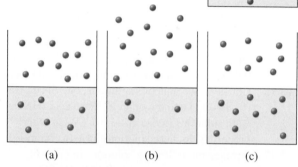

(a)   (b)   (c)

13.41 The diagrams represent an aqueous solution at two different temperatures. In both diagrams, the solution is saturated in two different solutes, each of which is represented by a different color. Using the diagrams, determine for each solute whether the dissolution process is endothermic or exothermic. For which of the solutes do you think the numerical value of ΔH_{soln} is greater? Explain.

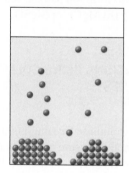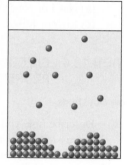

Lower temperature   Higher temperature

SECTION 13.5: COLLIGATIVE PROPERTIES

Review Questions

13.42 What are colligative properties? What is the meaning of the word *colligative* in this context?

13.43 Give two examples of (a) a volatile liquid and (b) a nonvolatile liquid.

13.44 Write the equation representing Raoult's law, and express it in words.

13.45 Use a solution of benzene in toluene to explain what is meant by an ideal solution.

13.46 Write the equations relating boiling-point elevation and freezing-point depression to the concentration of the solution. Define all the terms, and give their units.

13.47 How is vapor-pressure lowering related to a rise in the boiling point of a solution?

13.48 Use a phase diagram to show the difference in freezing points and boiling points between an aqueous urea solution and pure water.

13.49 What is osmosis? What is a semipermeable membrane?

13.50 Write the equation relating osmotic pressure to the concentration of a solution. Define all the terms, and specify their units.

13.51 Explain why molality is used for boiling-point elevation and freezing-point depression calculations and molarity is used in osmotic pressure calculations.

13.52 Why is the discussion of the colligative properties of electrolyte solutions more involved than that of nonelectrolyte solutions?

13.53 What are ion pairs? What effect does ion-pair formation have on the colligative properties of a solution? How does the ease of ion-pair formation depend on (a) charges on the ions, (b) size of the ions, (c) nature of the solvent (polar versus nonpolar), (d) concentration?

13.54 What is the van't Hoff factor? What information does it provide?

13.55 For most intravenous injections, great care is taken to ensure that the concentration of solutions to be injected is comparable to that of blood plasma. Explain.

Computational Problems

13.56 A solution is prepared by dissolving 396 g of sucrose ($C_{12}H_{22}O_{11}$) in 624 g of water. What is the vapor pressure of this solution at 30°C? (The vapor pressure of water is 31.8 mmHg at 30°C.)

13.57 How many grams of sucrose ($C_{12}H_{22}O_{11}$) must be added to 552 g of water to give a solution with a vapor pressure 2.0 mmHg less than that of pure water at 20°C? (The vapor pressure of water at 20°C is 17.5 mmHg.)

13.58 The vapor pressure of benzene is 100.0 mmHg at 26.1°C. Calculate the vapor pressure of a solution containing 24.6 g of camphor ($C_{10}H_{16}O$) dissolved in 98.5 g of benzene. (Camphor is a low-volatility solid.)

13.59 The vapor pressures of ethanol (C_2H_5OH) and 1-propanol (C_3H_7OH) at 35°C are 100 and 37.6 mmHg, respectively. Assume ideal behavior and calculate the partial pressures of ethanol and 1-propanol at 35°C over a solution of ethanol in 1-propanol, in which the mole fraction of ethanol is 0.300.

13.60   The vapor pressure of ethanol (C_2H_5OH) at 20°C is 44 mmHg, and the vapor pressure of methanol (CH_3OH) at the same temperature is 94 mmHg. A mixture of 30.0 g of methanol and 45.0 g of ethanol is prepared (and can be assumed to behave as an ideal solution). (a) Calculate the vapor pressure of methanol and ethanol above this solution at 20°C. (b) Calculate the mole fraction of methanol and ethanol in the vapor above this solution at 20°C. (c) Suggest a method for separating the two components of the solution.

13.61   How many grams of urea [$(NH_2)_2CO$] must be added to 658 g of water to give a solution with a vapor pressure 2.50 mmHg lower than that of pure water at 30°C? (The vapor pressure of water at 30°C is 31.8 mmHg.)

13.62   What are the boiling point and freezing point of a 3.12-m solution of naphthalene in benzene? (The boiling point and freezing point of benzene are 80.1°C and 5.5°C, respectively.)

13.63   An aqueous solution contains the amino acid glycine (NH_2CH_2COOH). Assuming that the acid does not ionize in water, calculate the molality of the solution if it freezes at −1.1°C.

13.64   How many liters of the antifreeze ethylene glycol [$CH_2(OH)CH_2(OH)$] would you add to a car radiator containing 6.50 L of water if the coldest winter temperature in your area is −20°C? Calculate the boiling point of this water-ethylene glycol mixture. (The density of ethylene glycol is 1.11 g/mL.)

13.65   A solution is prepared by condensing 4.00 L of a gas, measured at 27°C and 748 mmHg pressure, into 75.0 g of benzene. Calculate the freezing point of this solution.

13.66   What is the osmotic pressure (in atm) of a 1.57-M aqueous solution of urea [$(NH_2)_2CO$] at 27.0°C?

13.67   What are the normal freezing points and boiling points of the following solutions: (a) 21.2 g NaCl in 135 mL of water and (b) 15.4 g of urea in 66.7 mL of water?

13.68   At 25°C, the vapor pressure of pure water is 23.76 mmHg and that of seawater is 22.98 mmHg. Assuming that seawater contains only NaCl, estimate its molal concentration.

13.69   Both NaCl and $CaCl_2$ are used to melt ice on roads and sidewalks in winter. What advantages do these substances have over sucrose or urea in lowering the freezing point of water?

13.70   A 0.86 percent by mass solution of NaCl is called "physiological saline" because its osmotic pressure is equal to that of the solution in blood cells. Calculate the osmotic pressure of this solution at normal body temperature (37°C). Note that the density of the saline solution is 1.005 g/mL.

13.71   The osmotic pressure of 0.010-M solutions of $CaCl_2$ and urea at 25°C are 0.605 and 0.245 atm, respectively. Calculate the van't Hoff factor for the $CaCl_2$ solution.

13.72   Calculate the osmotic pressure of a 0.0500 M $MgSO_4$ solution at 25°C. (*Hint:* See Table 13.3.)

13.73   The tallest trees known are the redwoods in California. Assuming the height of a redwood to be 105 m (about 350 ft), estimate the osmotic pressure required to push water up to the treetop.

13.74   Calculate the difference in osmotic pressure (in atm) at the normal body temperature between the blood plasma of a diabetic patient and that of a healthy adult. Assume that the sole difference between the two people is due to the higher glucose level in the diabetic patient. The glucose levels are 1.75 and 0.84 g/L, respectively. Based on your result, explain why such a patient frequently feels thirsty.

Conceptual Problems

13.75   Which of the following aqueous solutions has (a) the higher boiling point, (b) the higher freezing point, and (c) the lower vapor pressure: 0.35 m $CaCl_2$ or 0.90 m urea? Explain. Assume complete dissociation.

13.76   Consider two aqueous solutions, one of sucrose ($C_{12}H_{22}O_{11}$) and the other of nitric acid (HNO_3). Both solutions freeze at −1.5°C. What other properties do these solutions have in common?

13.77   Arrange the following solutions in order of decreasing freezing point: 0.10 m Na_3PO_4, 0.35 m NaCl, 0.20 m $MgCl_2$, 0.15 m $C_6H_{12}O_6$, 0.15 m CH_3COOH.

13.78   Arrange the following aqueous solutions in order of decreasing freezing point, and explain your reasoning: 0.50 m HCl, 0.50 m glucose, 0.50 m acetic acid.

13.79   Indicate which compound in each of the following pairs is more likely to form ion pairs in water: (a) NaCl or Na_2SO_4, (b) $MgCl_2$ or $MgSO_4$, (c) LiBr or KBr.

SECTION 13.6: CALCULATIONS USING COLLIGATIVE PROPERTIES

Review Questions

13.80   Describe how you would use freezing-point depression and osmotic pressure measurements to determine the molar mass of a compound. Why are boiling-point elevation and vapor-pressure lowering normally not used for this purpose?

13.81   Describe how you would use the osmotic pressure to determine the percent ionization of a weak, monoprotic acid.

Computational Problems

13.82   The elemental analysis of an organic solid extracted from gum arabic (a gummy substance used in adhesives, inks, and pharmaceuticals) showed that it contained 40.0 percent C, 6.7 percent H, and 53.3 percent O. A solution of 0.650 g of the solid in 27.8 g of the solvent diphenyl gave a freezing-point depression of 1.56°C. Calculate the molar mass and molecular formula of the solid. (K_f for diphenyl is 8.00°C/m.)

13.83 A solution of 2.50 g of a compound having the empirical formula C_6H_5P in 25.0 g of benzene is observed to freeze at 4.3°C. Calculate the molar mass of the solute and its molecular formula.

13.84 The molar mass of benzoic acid (C_6H_5COOH) determined by measuring the freezing-point depression in benzene is twice what we would expect for the molecular formula, $C_7H_6O_2$. Explain this apparent anomaly.

13.85 A solution containing 0.8330 g of a polymer of unknown structure in 170.0 mL of an organic solvent was found to have an osmotic pressure of 5.20 mmHg at 25°C. Determine the molar mass of the polymer.

13.86 A quantity of 7.480 g of an organic compound is dissolved in water to make 300.0 mL of solution. The solution has an osmotic pressure of 1.43 atm at 27°C. The analysis of this compound shows that it contains 41.8 percent C, 4.7 percent H, 37.3 percent O, and 16.3 percent N. Calculate the molecular formula of the compound.

13.87 A solution of 6.85 g of a carbohydrate in 100.0 g of water has a density of 1.024 g/mL and an osmotic pressure of 4.61 atm at 20.0°C. Calculate the molar mass of the carbohydrate.

13.88 A 0.036-M aqueous nitrous acid (HNO_2) solution has an osmotic pressure of 0.93 atm at 25°C. Calculate the percent ionization of the acid.

13.89 A 0.100-M aqueous solution of the base HB has an osmotic pressure of 2.83 atm at 25°C. Calculate the percent ionization of the base.

13.90 Hydrofluoric acid (HF) can be used in the fluoridation of water. HF is a weak acid that only partially ionizes in solution. If an aqueous solution that is 0.15 M in HF has an osmotic pressure of 3.9 atm at 25°C, what is the percent ionization of HF at this concentration?

SECTION 13.7: COLLOIDS

Review Questions

13.91 What are colloids? Referring to Table 13.5, why is there no colloid in which both the dispersed phase and the dispersing medium are gases?

13.92 Describe how hydrophilic and hydrophobic colloids are stabilized in water.

13.93 Describe and give an everyday example of the Tyndall effect.

ADDITIONAL PROBLEMS

13.94 Lysozyme is an enzyme that cleaves bacterial cell walls. A sample of lysozyme extracted from egg white has a molar mass of 13,930 g. A quantity of 0.100 g of this enzyme is dissolved in 150 g of water at 25°C. Calculate the vapor-pressure lowering, the depression in freezing point, the elevation in boiling point, and the osmotic pressure of this solution. (The vapor pressure of water at 25°C is 23.76 mmHg.)

13.95 The blood sugar (glucose) level of a diabetic patient is approximately 0.140 g of glucose/100 mL of blood. Every time the patient ingests 40 g of glucose, her blood glucose level rises to approximately 0.240 g/100 mL of blood. Calculate the number of moles of glucose per milliliter of blood and the total number of moles and grams of glucose in the blood before and after consumption of glucose. (Assume that the total volume of blood in her body is 5.0 L.)

13.96 Trees in cold climates may be subjected to temperatures as low as −60°C. Estimate the concentration of an aqueous solution in the body of the tree that would remain unfrozen at this temperature. Is this a reasonable concentration? Comment on your result.

13.97 A cucumber placed in concentrated brine (saltwater) shrivels into a pickle. Explain.

13.98 Two liquids A and B have vapor pressures of 76 and 132 mmHg, respectively, at 25°C. What is the total vapor pressure of the ideal solution made up of (a) 1.00 mole of A and 1.00 mole of B and (b) 2.00 moles of A and 5.00 moles of B?

13.99 Determine the van't Hoff factor of Na_3PO_4 in a 0.40-m solution whose freezing point is −2.6°C.

13.100 A 262-mL sample of a sugar solution containing 1.22 g of the sugar has an osmotic pressure of 30.3 mmHg at 35°C. What is the molar mass of the sugar?

13.101 Consider the three mercury manometers shown in the diagram. One of them has 1 mL of water on top of the mercury, another has 1 mL of a 1 m urea solution on top of the mercury, and the third one has 1 mL of a 1 m NaCl solution placed on top of the mercury. Which of these solutions is in the tube labeled X, which is in Y, and which is in Z?

13.102 A forensic chemist is given a white powder for analysis. She dissolves 0.50 g of the substance in 8.0 g of benzene. The solution freezes at 3.9°C. Can the chemist conclude that the compound is cocaine ($C_{17}H_{21}NO_4$)? What assumptions are made in the analysis?

13.103 "Time-release" drugs have the advantage of releasing the drug to the body at a constant rate so that the drug concentration at any time is not too high as to have harmful side effects or too low as to be ineffective. A schematic diagram of a pill that works on this basis is shown. Explain how it works.

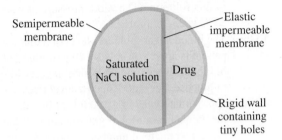

13.104 A solution of 1.00 g of anhydrous aluminum chloride ($AlCl_3$) in 50.0 g of water freezes at $-1.11°C$. Does the molar mass determined from this freezing point agree with that calculated from the formula? Why?

13.105 Explain why reverse osmosis is (theoretically) more desirable as a desalination method than distillation or freezing. What minimum pressure must be applied to seawater at 25°C for reverse osmosis to occur? (Treat seawater as a 0.70 M NaCl solution.)

13.106 What masses of sodium chloride, magnesium chloride, sodium sulfate, calcium chloride, potassium chloride, and sodium bicarbonate are needed to produce 1 L of artificial seawater for an aquarium? The required ionic concentrations are $[Na^+] = 2.56\ M$, $[K^+] = 0.0090\ M$, $[Mg^{2+}] = 0.054\ M$, $[Ca^{2+}] = 0.010\ M$, $[HCO_3^-] = 0.0020\ M$, $[Cl^-] = 2.60\ M$, $[SO_4^{2-}] = 0.051\ M$.

13.107 The osmotic pressure of blood plasma is approximately 7.5 atm at 37°C. Estimate the total concentration of dissolved species and the freezing point of blood.

13.108 The antibiotic gramicidin A can transport Na^+ ions into a certain cell at the rate of 5.0×10^7 Na^+ ions/channel · s. Calculate the time in seconds to transport enough Na^+ ions to increase its concentration by 8.0×10^{-3} M in a cell whose intracellular volume is 2.0×10^{-10} mL.

13.109 A protein has been isolated as a salt with the formula $Na_{20}P$ (this notation means that there are 20 Na^+ ions associated with a negatively charged protein P^{20-}). The osmotic pressure of a 10.0-mL solution containing 0.225 g of the protein is 0.257 atm at 25.0°C. (a) Calculate the molar mass of the protein from these data. (b) Calculate the actual molar mass of the protein.

13.110 A nonvolatile organic compound Z was used to make up two solutions. Solution A contains 5.00 g of Z dissolved in 100 g of water, and solution B contains 2.31 g of Z dissolved in 100 g of benzene. Solution A has a vapor pressure of 754.5 mmHg at the normal boiling point of water, and solution B has the same vapor pressure at the normal boiling point of benzene. Calculate the molar mass of Z in solutions A and B, and account for the difference.

13.111 Hydrogen peroxide with a concentration of 3.0 percent (3.0 g of H_2O_2 in 100 mL of solution) is sold in drugstores for use as an antiseptic. For a 10.0-mL 3.0 percent H_2O_2 solution, calculate (a) the oxygen gas produced (in liters) at STP when the compound undergoes complete decomposition and (b) the ratio of the volume of O_2 collected to the initial volume of the H_2O_2 solution.

13.112 State which of the alcohols listed in Problem 13.12 you would expect to be the best solvent for each of the following substances, and explain why: (a) I_2, (b) KBr, (c) $CH_3CH_2CH_2CH_2CH_3$.

13.113 Before a carbonated beverage bottle is sealed, it is pressurized with a mixture of air and carbon dioxide. (a) Explain the effervescence that occurs when the cap of the bottle is removed. (b) What causes the fog to form near the mouth of the bottle right after the cap is removed?

13.114 Iodine (I_2) is only sparingly soluble in water (left photo). Yet upon the addition of iodide ions (e.g., from KI), iodine is converted to the triiodide ion, which readily dissolves (right photo).

$$I_2(s) + I^-(aq) \rightleftharpoons I_3^-(aq)$$

Describe the change in solubility of I_2 in terms of the change in intermolecular forces.

© David A. Tietz/Editorial Image, LLC

13.115 (a) The root cells of plants contain a solution that is hypertonic in relation to water in the soil. Thus, water can move into the roots by osmosis. Explain why salts such as NaCl and $CaCl_2$ spread on roads to melt ice can be harmful to nearby trees. (b) Just before urine leaves the human body, the collecting ducts in the kidney (which contain the urine) pass through a fluid whose salt concentration is considerably greater than is found in the blood and tissues. Explain how this action helps conserve water in the body.

13.116 Hemoglobin, the oxygen-transport protein, binds about 1.35 mL of oxygen per gram of the protein. The concentration of hemoglobin in normal blood is 150 g/L blood. Hemoglobin is about 95 percent saturated with O_2 in the lungs and only 74 percent saturated with O_2 in the capillaries. Calculate the volume of O_2 released by hemoglobin when 100 mL of blood flows from the lungs to the capillaries.

13.117 Two beakers, one containing a 50-mL aqueous 1.0 M glucose solution and the other a 50-mL aqueous 2.0 M glucose solution, are placed under a tightly sealed bell jar at room temperature. What are the volumes in these two beakers at equilibrium?

13.118 In the apparatus shown, what will happen if the membrane is (a) permeable to both water and the Na^+ and Cl^- ions, (b) permeable to water and the Na^+ ions but not to the Cl^- ions, (c) permeable to water but not to the Na^+ and Cl^- ions?

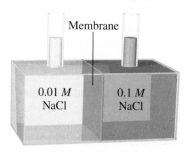

13.119 Concentrated hydrochloric acid is usually available at a concentration of 37.7 percent by mass. What is its molar concentration? (The density of the solution is 1.19 g/mL.)

13.120 Explain each of the following statements: (a) The boiling point of seawater is higher than that of pure water. (b) Carbon dioxide escapes from the solution when the cap is removed from a carbonated soft drink bottle. (c) Molal and molar concentrations of dilute aqueous solutions are approximately equal. (d) In discussing the colligative properties of a solution (other than osmotic pressure), it is preferable to express the concentration in units of molality rather than in molarity. (e) Methanol (b.p. 65°C) is useful as an antifreeze, but it should be removed from the car radiator during the summer season.

13.121 A mixture of NaCl and sucrose ($C_{12}H_{22}O_{12}$) of combined mass 10.2 g is dissolved in enough water to make up a 250-mL solution. The osmotic pressure of the solution is 7.32 atm at 23°C. Calculate the mass percent of NaCl in the mixture.

13.122 A 1.32-g sample of a mixture of cyclohexane (C_6H_{12}) and naphthalene ($C_{10}H_8$) is dissolved in 18.9 g of benzene (C_6H_6). The freezing point of the solution is 2.2°C. Calculate the mass percent of the mixture. (See Table 13.2 for constants.)

13.123 How does each of the following affect the solubility of an ionic compound: (a) lattice energy, (b) solvent (polar versus nonpolar), (c) enthalpies of hydration of cation and anion?

13.124 A solution contains two volatile liquids A and B. Complete the following table, in which the symbol ⟷ indicates attractive intermolecular forces.

Attractive forces	Deviation from Raoult's law	ΔH_{soln}
A ⟷ A, B ⟷ B > A ⟷ B		
	Negative	
		Zero

13.125 The concentration of commercially available concentrated nitric acid is 70.0 percent by mass, or 15.9 M. Calculate the density and the molality of the solution.

13.126 A mixture of ethanol and 1-propanol behaves ideally at 36°C and is in equilibrium with its vapor. If the mole fraction of ethanol in the solution is 0.62, calculate its mole fraction in the vapor phase at this temperature. (The vapor pressures of pure ethanol and 1-propanol at 36°C are 108 and 40.0 mmHg, respectively.)

13.127 Ammonia (NH_3) is very soluble in water, but nitrogen trichloride (NCl_3) is not. Explain.

13.128 For ideal solutions, the volumes are additive. This means that if 5 mL of A and 5 mL of B form an ideal solution, the volume of the solution is 10 mL. Provide a molecular interpretation for this observation. When 500 mL of ethanol (C_2H_5OH) is mixed with 500 mL of water, the final volume is less than 1000 mL. Why?

13.129 Acetic acid is a weak acid that ionizes in solution as follows:

If the freezing point of a 0.106 m CH_3COOH solution is −0.203°C, calculate the percent of the acid that has undergone ionization.

$$CH_3COOH(aq) \rightleftharpoons CH_3COO^-(aq) + H^+(aq)$$

13.130 Which vitamins (see the given structures) do you expect to be water soluble? (See Problem 13.37.)

13.131 Calculate the percent by mass of the solute in each of the following aqueous solutions: (a) 5.75 g of NaBr in 67.9 g of solution, (b) 24.6 g of KCl in 114 g of water, (c) 4.8 g of toluene in 39 g of benzene.

13.132 Acetic acid is a polar molecule and can form hydrogen bonds with water molecules. Therefore, it has a high solubility in water. Yet acetic acid is also soluble in benzene (C_6H_6), a nonpolar solvent that lacks the ability to form hydrogen bonds. A solution of 3.8 g of CH_3COOH in 80 g C_6H_6 has a freezing point of 3.5°C. Calculate the molar mass of the solute, and suggest what its structure might be. (*Hint:* Acetic acid molecules can form hydrogen bonds between themselves.)

13.133   A 2.6-L sample of water contains 192 μg of lead. Does this concentration of lead exceed the safety limit of 0.050 ppm of lead per liter of drinking water?

13.134   Fish in the Antarctic Ocean swim in water at about $-2°C$. (a) To prevent their blood from freezing, what must be the concentration (in molality) of the blood? Is this a reasonable physiological concentration? (b) In recent years, scientists have discovered a special type of protein in the blood of these fish that, although present in quite low concentrations ($\leq 0.001\ m$), has the ability to prevent the blood from freezing. Suggest a mechanism for its action.

© Bill Curtsinger/National Geographic/Getty Images

13.135   Why are ice cubes (e.g., those you see in the trays in the freezer of a refrigerator) cloudy inside?

13.136   If a soft drink can is shaken and then opened, the drink escapes violently. However, if after shaking the can, we tap it several times with a metal spoon, no such "explosion" of the drink occurs. Why?

13.137   Two beakers are placed in a closed container. Beaker A initially contains 0.15 mole of naphthalene ($C_{10}H_8$) in 100 g of benzene (C_6H_6), and beaker B initially contains 31 g of an unknown compound dissolved in 100 g of benzene. At equilibrium, beaker A is found to have lost 7.0 g of benzene. Assuming ideal behavior, calculate the molar mass of the unknown compound. State any assumptions made.

13.138   (a) Derive the equation relating the molality (m) of a solution to its molarity (M)

$$m = \frac{M}{d - \frac{M\mathcal{M}}{1000}}$$

where d is the density of the solution (g/mL) and $\mathcal{M}$ is the molar mass of the solute (g/mol). (*Hint:* Start by expressing the solvent in kilograms in terms of the difference between the mass of the solution and the mass of the solute.) (b) Show that, for dilute aqueous solutions, m is approximately equal to M.

13.139   At 27°C, the vapor pressure of pure water is 23.76 mmHg and that of an aqueous solution of urea is 22.98 mmHg. Calculate the molality of urea in the solution.

13.140   A very long pipe is capped at one end with a semipermeable membrane. How deep (in meters) must the pipe be immersed into the sea for

freshwater to begin to pass through the membrane? Assume the water to be at 20°C, and treat it as a 0.70 M NaCl solution. The density of seawater is 1.03 g/cm³, and the acceleration due to gravity is 9.81 m/s².

13.141   A mixture of liquids A and B exhibits ideal behavior. At 84°C, the total vapor pressure of a solution containing 1.2 moles of A and 2.3 moles of B is 331 mmHg. Upon the addition of another mole of B to the solution, the vapor pressure increases to 347 mmHg. Calculate the vapor pressure of pure A and B at 84°C.

13.142   Use Henry's law and the ideal gas equation to prove the statement that the volume of a gas that dissolves in a given amount of solvent is *independent* of the pressure of the gas. (*Hint:* Henry's law can be modified as $n = kP$, where n is the number of moles of the gas dissolved in the solvent.)

13.143   At 298 K, the osmotic pressure of a glucose solution is 10.50 atm. Calculate the freezing point of the solution. The density of the solution is 1.16 g/mL.

13.144   *Ringer's lactate,* a solution containing several different salts, is often administered intravenously for the initial treatment of trauma patients. One liter of Ringer's lactate contains 0.102 mole of sodium chloride, 4×10^{-3} mole of potassium chloride, 1.5×10^{-3} mole of calcium chloride, and 2.8×10^{-2} mole of sodium lactate. Determine the osmotic pressure of this solution at normal body temperature (37°C). Assume no ion pairing. (The formula of the lactate ion is $CH_3CH_2COO^-$.)

13.145   The diagram here shows vapor pressure curves for pure benzene and a solution of a nonvolatile solute in benzene. Estimate the molality of the solution.

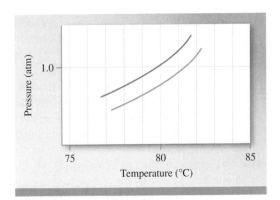

13.146   Valinomycin is an antibiotic. It functions by binding K^+ ions and transporting them across the membrane into cells to offset the ionic balance. The molecule is represented here by its skeletal structure in which the end of each straight line corresponds to a carbon atom (unless a different atom is shown at the end of the line). There are as many H atoms attached to each C atom as necessary to give each C atom a total of four bonds. Using the "like dissolves like"

principle, explain how the molecule functions.
(*Hint:* The —CH₃ groups at the two ends of each Y
shape are nonpolar.)

13.147 Early fluoridation of municipal water supplies was
done by dissolving enough sodium fluoride to
achieve a 1-ppm concentration of fluoride ion.
Convert 1.0 ppm F⁻ to *percent by mass* F⁻ and
molality of NaF.

13.148 Many fluoridation facilities now use fluorosilicic
acid instead of sodium fluoride. Fluorosilicic acid

typically is distributed as a 23 percent (1.596-*m*)
aqueous solution. (a) Calculate the van't Hoff factor
of fluorosilicic acid given that a 23 percent solution
has a freezing point of −15.5°C. (b) The density of
23 percent fluorosilicic acid is 1.19 g/mL. Given
that the osmotic pressure of the solution at 25°C is
242 atm, calculate the molar mass of fluorosilicic
acid.

13.149 A mixture of two volatile liquids is said to be ideal
if each component obeys Raoult's law:

$$P_i = \chi_i P_i^\circ$$

Two volatile liquids A (molar mass 100 g/mol) and B
(molar mass 110 g/mol) form an ideal solution.
At 55°C, A has a vapor pressure of 98 mmHg and B
has a vapor pressure of 42 mmHg. A solution
is prepared by mixing equal masses of A and B.
(a) Calculate the mole fraction of each component
in the solution. (b) Calculate the partial pressures
of A and B over the solution at 55°C. (c) Suppose
that some of the vapor over the solution at 55°C
is condensed to a liquid. Calculate the mole
fraction of each component in the condensed
liquid. (d) Calculate the partial pressures of the
components above the condensed liquid at 55°C.

13.150 Explain why we cannot use osmotic pressure to
determine both molar mass and percent ionization
for an unknown monoprotic acid.

Answers to In-Chapter Problems

PRACTICE PROBLEMS

13.1A CS₂. **13.1B** C₃H₈, I₂, CS₂. **13.2A** (a) 0.423 *m*, (b) 2.48%.
13.2B 0.78 *m*. **13.3A** (a) 1.8 *M*, (b) 1.9 *m*. **13.3B** 12.8%.
13.4A 0.12 *M*. **13.4B** 26 atm. **13.5A** 22.2 mmHg. **13.5B** 103 g.
13.6A f.p. = −7.91°C, b.p. = 102.2°C. **13.6B** 710 g. **13.7A** *i* = 1.21.
13.7B f.p. = −0.35°C. **13.8A** 128 g/mol. **13.8B** 12.5 g. **13.9A** 6.0 ×
10³ g/mol. **13.9B** 0.27 g. **13.10A** 4%. **13.10B** 0.38 atm.

SECTION REVIEW

13.2.1 b, d, e. **13.2.2** b, d. **13.3.1** a. **13.3.2** d. **13.3.3** d. **13.3.4** a.
13.4.1 c. **13.4.2** a. **13.5.1** d. **13.5.2** c. **13.5.3** b. **13.5.4** a. **13.6.1** b.
13.6.2 b.

Chapter **14**

Entropy and Free Energy

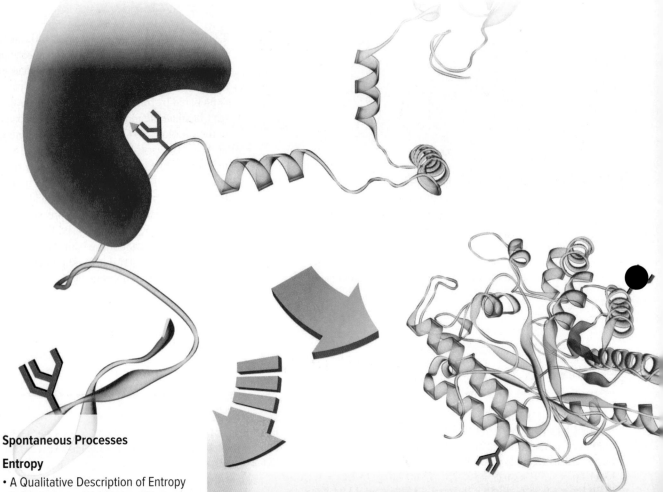

© Kenneth Eward/Science Source.

THE ILLUSTRATION shows the process by which a polypeptide chain in a biological cell folds into a three-dimensional structure called a protein. Although the original polypeptide chain can assume any number of configurations, the protein, if it is to function properly, must adopt a specific arrangement. Thus, the folding of a protein is accompanied by a decrease in system entropy. According to the second law of thermodynamics, any spontaneous process must result in an increase in the entropy of the universe. It follows, therefore, that there must be an increase in the entropy of the surroundings greater in magnitude than the decrease in entropy associated with the protein folding. In fact, the folding of a protein is exothermic and does indeed cause an increase in entropy of the surroundings via the spreading out of the energy produced by the process.

- System and surroundings [◄◄ Section 10.1]
- Hess's law [◄◄ Section 10.5]
- Standard enthalpies of formation [◄◄ Section 10.6]

14.1 SPONTANEOUS PROCESSES

An understanding of thermodynamics enables us to predict whether a reaction will occur when reactants are combined. This is important in the synthesis of new compounds in the laboratory, the manufacturing of chemicals on an industrial scale, and the understanding of natural processes such as cell function. A process that *does* occur under a specific set of conditions is called a ***spontaneous process.*** One that does *not* occur under a specific set of conditions is called ***nonspontaneous process.*** Table 14.1 lists examples of familiar spontaneous processes and their nonspontaneous counterparts. These examples illustrate what we know intuitively: Under a given set of conditions, a process that occurs spontaneously in one direction does not also occur spontaneously in the opposite direction.

Student Annotation: The conditions that most often are specified are *temperature, pressure,* and in the case of a solution, *concentration.*

Processes that result in a decrease in the energy of a system often are spontaneous. For example, the combustion of methane is exothermic.

$$CH_4(g) + 2O_2(g) \longrightarrow CO_2(g) + 2H_2O(l) \qquad \Delta H° = -890.4 \text{ kJ/mol}$$

Thus, the energy of the system is lowered because heat is given off during the course of the reaction. Likewise, in the acid-base neutralization reaction,

$$H^+(aq) + OH^-(aq) \longrightarrow H_2O(l) \qquad \Delta H° = -56.2 \text{ kJ/mol}$$

heat is given off, lowering the energy of the system. Each of these processes is spontaneous, and each results in a lowering of the system's energy.

Now consider the melting of ice:

$$H_2O(s) \longrightarrow H_2O(l) \qquad \Delta H° = 6.01 \text{ kJ/mol}$$

In this case, the process is endothermic and yet it is also spontaneous at temperatures above 0°C. Conversely, the freezing of water is an *exothermic* process.

$$H_2O(l) \longrightarrow H_2O(s) \qquad \Delta H° = -6.01 \text{ kJ/mol}$$

Yet it is *not* spontaneous at temperatures above 0°C.

TABLE 14.1	Familiar Spontaneous and Nonspontaneous Processes
Spontaneous	**Nonspontaneous**
Ice melting at room temperature	Water freezing at room temperature
Sodium metal reacting violently with water to produce sodium hydroxide and hydrogen gas [◄◄ Section 8.5]	Sodium hydroxide reacting with hydrogen gas to produce sodium metal and water
A ball rolling downhill	A ball rolling uphill
The rusting of iron at room temperature	The conversion of rust back to iron metal at room temperature
Water freezing at −10°C	Ice melting at −10°C

Based on the first two examples, and many others like them, we might conclude that exothermic processes tend to be spontaneous and, indeed, a negative ΔH does *favor* spontaneity. The last two examples, however, make it clear that the sign of ΔH *alone* is insufficient to predict spontaneity in every circumstance. For the remainder of this chapter, we will examine the *two* factors that determine whether a process is spontaneous under a given set of conditions.

14.2   ENTROPY

To predict the spontaneity of a chemical or physical process, we need to know both the change in *enthalpy* [◄◄ Section 10.3] and the change in *entropy* associated with the process. We first encountered the concept of entropy in our discussion of solution formation [◄◄ Section 13.2]. We will now look in more detail at what entropy is, and why it matters.

A Qualitative Description of Entropy

Qualitatively, the **entropy (S)** of a system is a measure of how *spread out* or how *dispersed* the system's energy is. The simplest interpretation of this is how spread out a system's energy is in *space*. In other words, for a given system, the greater the volume it occupies, the greater its entropy. This interpretation explains how the process in Figure 14.1 occurs spontaneously despite there being no enthalpy change. Because they are moving, the gas molecules that were originally confined to one side of the container possess *motional energy*. In the absence of a barrier preventing it, the motional energy of molecules will spread out to occupy a larger volume. The dispersal of a system's motional energy to occupy a larger volume when the barrier is removed constitutes an *increase* in the system's entropy. Just as spontaneity is favored by a process being exothermic, spontaneity is also favored by an increase in the system's entropy. Whether it is the enthalpy change, the entropy change, or both, for a process to be spontaneous, *something* must favor spontaneity.

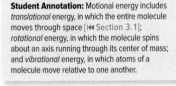

Student Annotation: Motional energy includes *translational* energy, in which the entire molecule moves through space [◄◄ Section 3.1]; *rotational* energy, in which the molecule spins about an axis running through its center of mass; and *vibrational* energy, in which atoms of a molecule move relative to one another.

A Quantitative Definition of Entropy

At this point, it is useful to introduce the mathematical definition of entropy proposed by Ludwig Boltzmann.

Equation 14.1	$S = k \ln W$

Student Annotation: The Boltzmann constant is equal to the gas constant, R (in J/K · mol), divided by Avogadro's constant, N_A.

where k is the Boltzmann constant $(1.38 \times 10^{-23}$ J/K) and W is the number of energetically equivalent different ways the molecules in a system can be arranged. To illustrate what this means, let's consider a simplified version of the process shown in Figure 14.1. Prior to the removal of the barrier between the left and right sides of the container, at any

Figure 14.1 A spontaneous process. The rapidly moving gas molecules originally confined to one side of a container spread out to fill the whole container when the barrier is removed.

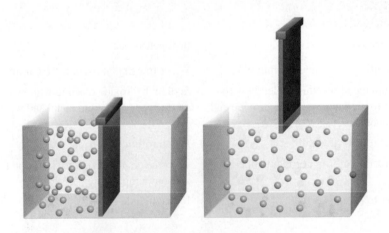

given instant, each molecule has a particular location, somewhere in the left side of the container. To narrow down the possible locations of the molecules, we imagine that each side of the container is divided into a number of equal smaller volumes called *cells*. In the simplest scenario, with just one molecule in the system, the number of possible locations of the molecule is equal to the number of cells. If the system contains *two* molecules, the number of possible arrangements is equal to the number of cells *squared*. (Note that a cell may contain more than one molecule.) Each time we increase the number of molecules by one, the number of possible arrangements increases by a factor equal to the number of cells. In general, for a volume consisting of X cells, and containing N molecules, the number of possible arrangements, W, is given by the equation

$$W = X^N$$ **Equation 14.2**

Student Annotation: The number of possible arrangements is sometimes called the number of *microstates*.

Figure 14.2 illustrates this for a simple case involving just two molecules. We imagine the container is divided into four cells each with volume v. Initially, both molecules are confined to the left side, which consists of two cells. With two molecules in two cells, there are $2^2 = 4$ possible arrangements of the molecules [Figure 14.2(a)]. When the barrier is removed, doubling the volume available to the molecules, the number of cells also doubles. With four cells available, there are $4^2 = 16$ possible arrangements of the molecules. Eight of the 16 arrangements have the molecules on opposite sides of the container [Figure 14.2(b)]. Of the other 8 arrangements, 4 have both molecules on the left side [as shown in Figure 14.2(a)], and 4 have both molecules on the right side (not shown). There are three different states possible for this system.

1. One molecule on each side (eight possible arrangements)
2. Both molecules on the left (four possible arrangements)
3. Both molecules on the right (four possible arrangements)

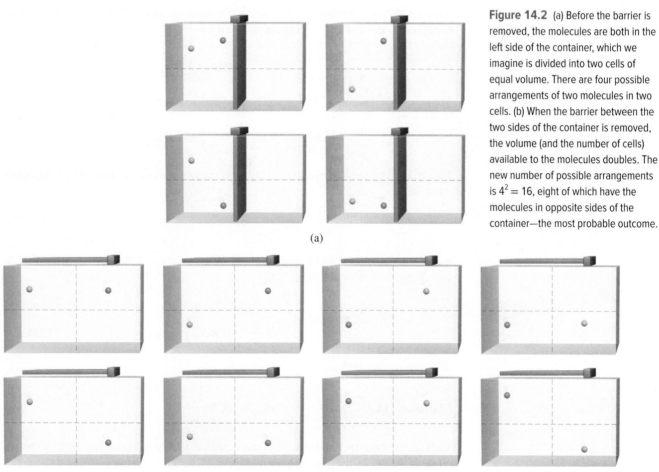

Figure 14.2 (a) Before the barrier is removed, the molecules are both in the left side of the container, which we imagine is divided into two cells of equal volume. There are four possible arrangements of two molecules in two cells. (b) When the barrier between the two sides of the container is removed, the volume (and the number of cells) available to the molecules doubles. The new number of possible arrangements is $4^2 = 16$, eight of which have the molecules in opposite sides of the container—the most probable outcome.

(a)

(b)

The most *probable* state is the one with the *largest* number of possible arrangements. In this case, the most probable state is the one with one molecule on each side of the container. The same principle applies to systems with larger numbers of molecules. Increasing the number of molecules increases the number of possible arrangements, but the most probable state will be the one in which the gas molecules are divided evenly between the two sides of the container.

14.3   ENTROPY CHANGES IN A SYSTEM

Calculating ΔS_{sys}

The change in entropy of a system is the difference between the entropy of the final state and the entropy of the initial state.

Equation 14.3	$\Delta S_{sys} = S_{final} - S_{initial}$

Using Equation 14.1, we can write an expression for the entropy of each state,

$$\Delta S_{sys} = k \ln W_{final} - k \ln W_{initial} = k \ln \frac{W_{final}}{W_{initial}}$$

Combining this result with Equation 14.2 gives

$$\Delta S_{sys} = k \ln \frac{(X_{final})^N}{(X_{initial})^N} = k \ln \left(\frac{X_{final}}{X_{initial}} \right)^N = k N \ln \left(\frac{X_{final}}{X_{initial}} \right)$$

Because X is the number of cells, and the volume of each cell is v, the total volume is related to the number of cells by

$$V = Xv \qquad \text{or} \qquad X = \frac{V}{v}$$

We substitute V_{final}/v for X_{final} and $V_{initial}/v$ for $X_{initial}$ to get

$$\Delta S_{sys} = k N \ln \left(\frac{V_{final}/v}{V_{initial}/v} \right) = k N \ln \frac{V_{final}}{V_{initial}}$$

Finally, because the Boltzmann constant, k, is the gas constant, R, divided by Avogadro's constant,

$$k = \frac{R}{N_A}$$

and because the number of molecules, N, is the product of the number of moles, n, and Avogadro's constant, N_A,

$$N = n \times N_A$$

$$kN = \left(\frac{R}{N_A} \times n \times N_A \right) = nR$$

the equation becomes

Equation 14.4	$\Delta S_{sys} = nR \ln \frac{V_{final}}{V_{initial}}$

Worked Example 14.1 shows how to use Equation 14.4 to calculate the entropy change for a process like the one shown in Figure 14.1, the expansion of an ideal gas at constant temperature.

Worked Example 14.1

Determine the change in entropy for 1.0 mole of an ideal gas originally confined to one-half of a 5.0-L container when the gas is allowed to expand to fill the entire container at constant temperature.

Strategy This is the isothermal expansion of an ideal gas. Because the molecules spread out to occupy a greater volume, we expect there to be an increase in the entropy of the system. Use Equation 14.3 to solve for ΔS_{sys}.

Setup $R = 8.314$ J/K · mol, $n = 1.0$ mole, $V_{final} = 5.0$ L, and $V_{initial} = 2.5$ L.

Solution

$$\Delta S_{sys} = nR \ln \frac{V_{final}}{V_{initial}} = 1.0 \text{ mol} \times \frac{8.314 \text{ J}}{\text{K} \cdot \text{mol}} \times \ln \frac{5.0 \text{ L}}{2.5 \text{ L}} = 5.8 \text{ J/K}$$

Think About It
Remember that for a process to be spontaneous, *something* must favor spontaneity. If the process is spontaneous but not exothermic (in this case, there is no enthalpy change), then we should expect ΔS_{sys} to be positive.

Practice Problem **A**TTEMPT Determine the change in entropy (ΔS_{sys}) for the expansion of 0.10 mole of an ideal gas from 2.0 L to 3.0 L at constant temperature.

Practice Problem **B**UILD To what fraction of its original volume must a 0.50-mole sample of ideal gas be compressed at constant temperature for ΔS_{sys} to be −6.7 J/K?

Practice Problem **C**ONCEPTUALIZE Which equation is correct for calculating ΔS_{sys} for a gaseous reaction that occurs at constant volume?

(i) $\Delta S_{sys} = nR$ (ii) $\Delta S_{sys} = nRT \ln \frac{P_{initial}}{P_{final}}$ (iii) $\Delta S_{sys} = nR \ln \frac{P_{initial}}{P_{final}}$ (iv) $\Delta S_{sys} = \frac{nR}{T} \ln \frac{P_{final}}{P_{initial}}$

Standard Entropy, $S°$

Although Equation 14.1 provides a quantitative definition of entropy, we seldom use it or Equation 14.3 to calculate the entropy change for a real process because of the difficulty involved in determining W, the number of different possible arrangements (Equation 14.2) in a macroscopic system. Instead, for processes other than isothermal expansion or compression of an ideal gas (for which we can use Equation 14.4), we routinely determine entropy changes using tabulated values.

Using calorimetry [◀◀ Section 10.4], it is possible to determine the *absolute* value of the entropy of a substance, S, something we cannot do with either energy or enthalpy. (Recall that we can determine ΔU and ΔH for a process that a system undergoes, but we cannot determine the absolute values of either U or H for a system [◀◀ Sections 10.2 and 10.3].) *Standard entropy* is the absolute entropy of a substance at 1 atm. (Tables of standard entropy values typically are the values at 25°C because so many processes are carried out at room temperature—although temperature is *not* part of the standard state definition and therefore must be specified.) Table 14.2 lists standard entropies of a few elements and compounds. Appendix 2 provides a more extensive listing. The units of entropy are J/K · mol. We use joules rather than kilojoules because entropy values typically are quite small. The entropies of substances (elements and compounds) are always positive (i.e., $S > 0$), even for elements in their standard states. (Recall that the standard *enthalpy* of formation, $\Delta H_f°$, for elements in their standard states is arbitrarily defined as zero, and for compounds it may be either positive or negative [◀◀ Section 10.6].)

Referring to Table 14.2, we can identify several important trends:

- For a given substance, the standard entropy is greater in the liquid phase than in the solid phase. [Compare the standard entropies of Na(*s*) and Na(*l*).] This results from there being greater molecular motion in a liquid, resulting in many possible arrangements of atoms in the liquid phase, whereas the positions of atoms in the solid are fixed.

Student Annotation: For even the simplest of hypothetical systems, where there are only two possible positions for molecules ($X = 2$), most calculators cannot display a number as large as the result of Equation 14.2 for even as few as 500 molecules—much less for the *enormous* number of molecules present in any *real* sample. (If your calculator is like most, with $X = 2$, you can calculate the number of possible arrangements for $N \leq 332$ molecules. Try it: 2 ^ 332 = ? and 2 ^ 333 = ?)

Student Annotation: You will find that tables, including Appendix 2, contain *negative* absolute entropies for some aqueous ions. Unlike a substance, an individual ion cannot be studied experimentally. Therefore, standard entropies of ions are actually *relative* values, where a standard entropy of zero is arbitrarily assigned to the hydrated hydrogen ion. Depending on an ion's extent of hydration, its standard entropy may be positive or negative, relative to that of hydrogen ion.

TABLE 14.2	Standard Entropy Values ($S°$) for Some Substances at 25°C		
Substance	**$S°$ (J/K · mol)**	**Substance**	**$S°$ (J/K · mol)**
$H_2O(l)$	69.9	C(diamond)	2.4
$H_2O(g)$	188.7	C(graphite)	5.69
Na(s)	51.05	$O_2(g)$	205.0
Na(l)	57.56	$O_3(g)$	237.6
Na(g)	153.7	$F_2(g)$	203.34
He(g)	126.1	Au(s)	47.7
Ne(g)	146.2	Hg(l)	77.4

- For a given substance, the standard entropy is greater in the gas phase than in the liquid phase. [Compare the standard entropies of Na(l) and Na(g) and those of $H_2O(l)$ and $H_2O(g)$.] This results from there being much greater molecular motion in a gas, resulting in many more possible arrangements of atoms in the gas phase than in the liquid phase—in part because the gas phase occupies a much greater volume than either of the condensed phases.
- For two monatomic species, the one with the larger molar mass has the greater standard entropy. [Compare the standard entropies of He(g) and Ne(g).]
- For two substances in the same phase, and with similar molar masses, the substance with the more complex molecular structure has the greater standard entropy. [Compare the standard entropies of $O_3(g)$ and $F_2(g)$.] The more complex a molecular structure, the more different types of motion the molecule can exhibit. A diatomic molecule such as F_2, for example, exhibits only one type of vibration, whereas a bent triatomic molecule such as O_3 exhibits three different types of vibrations. Each mode of motion contributes to the total number of available energy levels within which a system's energy can be dispersed. Figure 14.3 illustrates the ways in which the F_2 and O_3 molecules can rotate and vibrate.
- In cases where an element exists in two or more allotropic forms, the form in which the atoms are more mobile has the greater entropy. [Compare the standard entropies of C(diamond) and C(graphite). In diamond, the carbon atoms occupy fixed positions in a three-dimensional array. In graphite, although the carbon atoms occupy fixed positions within the two-dimensional sheets (see Figure 12.24, page 550), the sheets are free to move with respect to one another, which increases the mobility and, therefore, total number of possible arrangements of atoms within the solid.]

Now let's consider a process represented by the following chemical equation.

$$a\text{A} + b\text{B} \longrightarrow c\text{C} + d\text{D}$$

Just as the enthalpy change of a reaction is the difference between the enthalpies of the products and reactants (Equation 10.10), the entropy change is the difference between the entropies of the products and reactants.

Equation 14.5 $\Delta S°_{rxn} = [cS°(\text{C}) + dS°(\text{D})] - [aS°(\text{A}) + bS°(\text{B})]$

Or, using Σ to represent summation and m and n to represent the stoichiometric coefficients of the reactants and products, respectively, Equation 14.5 can be generalized as follows:

Equation 14.6 $\Delta S°_{rxn} = \Sigma n S°(\text{products}) - \Sigma m S°(\text{reactants})$

The standard entropy values of a large number of substances have been measured in J/K · mol. To calculate the standard entropy change for a reaction ($\Delta S°_{rxn}$), we look up the standard entropies of the products and reactants and use Equation 14.5. Worked Example 14.2 demonstrates this approach.

Student Annotation: The *reaction* generally is the *system*. Therefore, $\Delta S°_{rxn}$ is $\Delta S°_{sys}$.

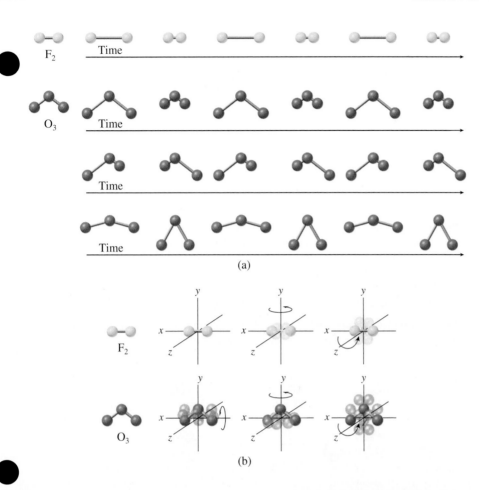

Figure 14.3 In addition to translational motion, molecules exhibit both *vibrations*, in which the atoms' positions relative to one another change, and *rotations*, in which the molecule rotates about its center of mass. (a) A diatomic molecule such as fluorine only exhibits one type of vibration. A bent, triatomic molecule such as ozone exhibits three types of vibration. (b) A diatomic molecule exhibits two different rotations, whereas a bent, triatomic molecule exhibits three different rotations. (Note that rotation of F_2 about the x axis would cause no change in the positions of either atom in the molecule.)

Student Hot Spot

Student data indicate you may struggle with entropy changes. Access the SmartBook to view additional Learning Resources on this topic.

Worked Example 14.2

From the standard entropy values in Appendix 2, calculate the standard entropy changes for the following reactions at 25°C.

(a) $CaCO_3(s) \longrightarrow CaO(s) + CO_2(g)$ (c) $H_2(g) + Cl_2(g) \longrightarrow 2HCl(g)$

(b) $N_2(g) + 3H_2(g) \longrightarrow 2NH_3(g)$

Strategy Look up standard entropy values and use Equation 14.5 to calculate ΔS°_{rxn}. Just as we did when we calculated standard enthalpies of reaction, we consider stoichiometric coefficients to be dimensionless—giving ΔS°_{rxn} units of J/K · mol.

Student Annotation: Recall that here *per mole* means per mole of reaction as written [◄◄ Section 10.3].

Setup From Appendix 2, $S^\circ[CaCO_3(s)] = 92.9$ J/K · mol, $S^\circ[CaO(s)] = 39.8$ J/K · mol, $S^\circ[CO_2(g)] = 213.6$ J/K · mol, $S^\circ[N_2(g)] = 191.5$ J/K · mol, $S^\circ[H_2(g)] = 131.0$ J/K · mol, $S^\circ[NH_3(g)] = 193.0$ J/K · mol, $S^\circ[Cl_2(g)] = 223.0$ J/K · mol, and $S^\circ[HCl(g)] = 187.0$ J/K · mol.

Solution

(a) $\Delta S^\circ_{rxn} = [S^\circ(CaO) + S^\circ(CO_2)] - [S^\circ(CaCO_3)]$
$= [(39.8 \text{ J/K} \cdot \text{mol}) + (213.6 \text{ J/K} \cdot \text{mol})] - (92.9 \text{ J/K} \cdot \text{mol})$
$= 160.5 \text{ J/K} \cdot \text{mol}$

(b) $\Delta S^\circ_{rxn} = [2S^\circ(NH_3)] - [S^\circ(N_2) + 3S^\circ(H_2)]$
$= (2)(193.0 \text{ J/K} \cdot \text{mol}) - [(191.5 \text{ J/K} \cdot \text{mol}) + (3)(131.0 \text{ J/K} \cdot \text{mol})]$
$= -198.5 \text{ J/K} \cdot \text{mol}$

(c) $\Delta S^\circ_{rxn} = [2S^\circ(HCl)] - [S^\circ(H_2) + S^\circ(Cl_2)]$
$= (2)(187.0 \text{ J/K} \cdot \text{mol}) - [(131.0 \text{ J/K} \cdot \text{mol}) + (223.0 \text{ J/K} \cdot \text{mol})]$
$= 20.0 \text{ J/K} \cdot \text{mol}$

(Continued on next page)

> **Think About It**
> Remember to multiply each standard entropy value by the correct stoichiometric coefficient. Like Equation 10.18, Equation 14.5 can only be used with a *balanced* chemical equation.

Practice Problem Ⓐ**TTEMPT** Calculate the standard entropy change for the following reactions at 25°C. Predict first whether each one will be positive, negative, or too close to call.

1. $2CO_2(g) \longrightarrow 2CO(g) + O_2(g)$
2. $3O_2(g) \longrightarrow 2O_3(g)$
3. $2NaHCO_3(s) \longrightarrow Na_2CO_3(s) + H_2O(l) + CO_2(g)$

Practice Problem Ⓑ**UILD** In each of the following reactions, there is one species for which the standard entropy is not listed in Appendix 2. In each case, using the values that *are* in Appendix 2 and the ΔS°_{rxn} that is given, determine the value of the missing standard entropy at 25°C: (a) $K(s) \longrightarrow K(l)$, $\Delta S^\circ_{rxn} = 7.9$ J/K · mol, (b) $2S(\text{rhombic}) + Cl_2(g) \longrightarrow S_2Cl_2(g)$, $\Delta S^\circ_{rxn} = 44.74$ J/K · mol, (c) $O_2(g) + 2MgF_2(s) \longrightarrow 2MgO(s) + 2F_2(g)$, $\Delta S^\circ_{rxn} = 140.76$ J/K · mol.

Practice Problem Ⓒ**ONCEPTUALIZE** For each reaction shown in the diagrams, indicate whether ΔS°_{rxn} is positive, negative, or too close to call.

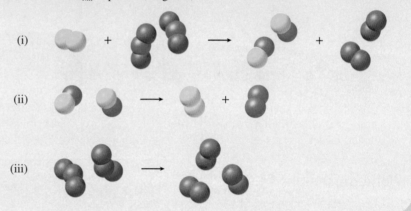

Qualitatively Predicting the Sign of ΔS°_{sys}

Equation 14.5 enables us to calculate ΔS°_{rxn} for a process when the standard entropies of the products and reactants are known. However, sometimes it's useful just to know the *sign* of ΔS°_{rxn}. Although multiple factors can influence the sign of ΔS°_{rxn}, the outcome is often dominated by a single factor, which can be used to make a qualitative prediction. Several processes that lead to an increase in entropy are

- Melting
- Vaporization or sublimation
- Temperature increase
- Reaction resulting in a greater number of gas molecules

When a solid is melted, the molecules have greater energy and are more mobile. They go from being in fixed positions in the solid, to being free to move about in the liquid. As we saw in the discussion of standard entropy, this leads to many more possible arrangements of the molecules and, therefore, greater entropy. The same rationale holds for the vaporization or sublimation of a substance. There is a dramatic increase in energy/mobility, and in the number of possible arrangements of a system's molecules when the molecules go from a condensed phase to the gas phase. Therefore, there is a much larger increase in the system's entropy, relative to the solid-to-liquid transition.

When the temperature of a system is increased, the energy of the system's molecules increases. To visualize this, recall from the discussion of kinetic molecular theory that increasing the temperature of a gas increases its average kinetic energy. This corresponds to an increase in the average speed of the gas molecules and a spreading out of the range of molecular speeds. [See Figure 11.4(a).] If we think of each of the possible molecular speeds within the range as a discrete energy level, we can see that at higher temperatures, there is a greater number of possible molecular speeds and, therefore, a greater number of energy states available to the molecules in the system. With a greater number of available energy states, there is a greater number of possible arrangements of molecules *within* those states and, therefore, a greater entropy.

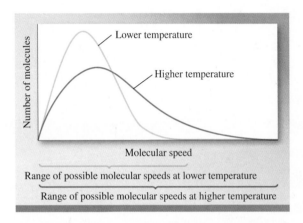

Because the entropy of a substance in the gas phase is always significantly greater than its entropy in either the liquid or solid phase, a reaction that results in an increase in the number of gas molecules causes an increase in the system's entropy. For reactions that do not involve gases, an increase in the number of solid, liquid, or aqueous molecules also usually causes an entropy increase.

By considering these factors, we can usually make a reasonably good prediction of the sign of ΔS°_{rxn} for a physical or chemical process, without having to look up the absolute entropy values for the species involved. Figure 14.4 (pages 628–629) summarizes the factors that can be used to compare entropies and illustrates several comparisons.

In addition to melting, vaporization/sublimation, temperature increase, and reactions that increase the number of gas molecules, which can always be counted upon to result in an entropy increase, the process of *dissolving* a substance often leads to an increase in entropy. In the case of a molecular solute, such as sucrose (sugar), dissolving causes dispersal of the molecules (and consequently, of the system's *energy*) into a larger volume—resulting in an increase in entropy. In the case of an ionic solute, the analysis is slightly more complicated. We saw in our discussion of solution formation [◄◄ Section 13.2] that the dissolution of ammonium nitrate (NH_4NO_3) is spontaneous, even though it is endothermic, because the system's entropy increases when the ionic solid dissociates and is dispersed throughout the solution. In general, this is the case for ionic solutes in which the charges on ions are small. (In the case of NH_4NO_3, they are +1 and −1.) However, when ions are dispersed in water, they become hydrated (surrounded by water molecules in a specific arrangement [◄◄ Figure 9.4, page 358]). This leads to a *decrease* in the entropy of the *water,* as hydration reduces the mobility of some of the water molecules by fixing them in positions around the dissolved ions. When the charges on ions are low, the increase in entropy of the solute typically outweighs the decrease in entropy of the water—resulting in an overall *increase* in the entropy of the system—as is the case with NH_4NO_3. By contrast, when highly charged ions such as Al^{3+} and Fe^{3+} are hydrated, the decrease in entropy of the water can actually outweigh the increase in entropy of the solute, leading to an overall *decrease*

Video 14.1
Factors that influence the entropy of a system.

Figure 14.4

Factors That Influence the Entropy of a System

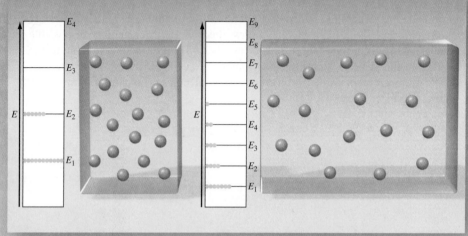

Volume Change

Quantum mechanical analysis shows that the spacing between translational energy levels is inversely proportional to the volume of the container. Thus, when the volume is increased, more energy levels become available within which the system's energy can be dispersed.

Temperature Change

At higher temperatures, molecules have greater kinetic energy—making more energy levels accessible. This increases the number of energy levels within which the system's energy can be dispersed, causing entropy to increase.

Molecular Complexity

Unlike atoms, which exhibit only translational motion, molecules can also exhibit rotational and vibrational motions. The greater a molecule's complexity, the greater the number of possible ways it can rotate and vibrate. The ozone molecule (O_3), for example, is more complex than the fluorine molecule (F_2) and exhibits more different kinds of vibrations and rotations. (See Figure 14.3.) This results in more energy levels within which the system's energy can be dispersed. The number and spacing of additional energy levels have been simplified to keep the illustration clear.

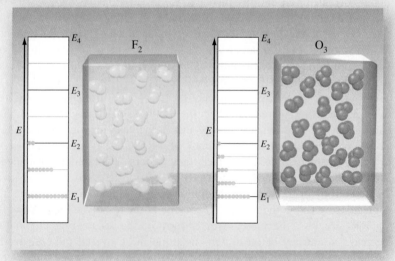

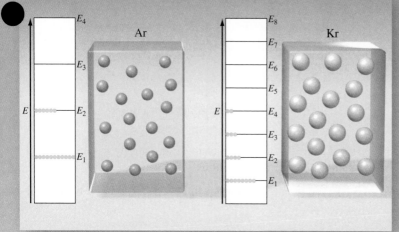

Molar Mass

The energy levels for a substance with a larger molar mass are more closely spaced. Kr, for example, has roughly twice the molar mass of Ar. Thus, Kr has roughly twice as many energy levels within which the system's energy can be dispersed.

Phase Change

Because of greater mobility, there are many more different possible arrangements (W) of molecules in the liquid phase than there are in the solid phase; and there are many, *many* more different possible arrangements of molecules in the gas phase than there are in the liquid phase. Entropy of a substance increases when it is melted ($s \longrightarrow l$), vaporized ($l \longrightarrow g$), or sublimed ($s \longrightarrow g$).

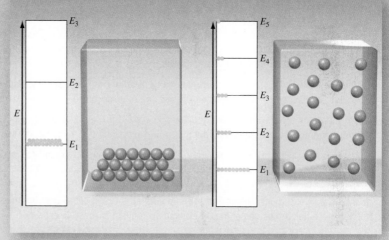

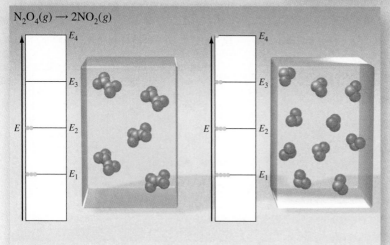

$N_2O_4(g) \longrightarrow 2NO_2(g)$

Chemical Reaction

When a chemical reaction produces more gas molecules than it consumes, the number of different possible arrangements of molecules (W) increases and entropy increases.

What's the point?

Although several factors can influence the entropy of a system or the entropy change associated with a process, often one factor dominates the outcome. Each of these comparisons shows a qualitative illustration of one of the important factors.

TABLE 14.3	Entropy Changes for the Dissolution (ΔS°_{soln}) of Some Ionic Solids at 25°C	
Dissolution equation		ΔS°_{soln} **(J/k · mol)**
$NH_4NO_3(s) \longrightarrow NH_4^+(aq) + NO_3^-(aq)$		108.1
$AlCl_3(s) \longrightarrow Al^{3+}(aq) + 3Cl^-(aq)$		−253.2
$FeCl_3(s) \longrightarrow Fe^{3+}(aq) + 3Cl^-(aq)$		−266.1

in entropy of the system. Table 14.3 lists the changes in entropy associated with the spontaneous dissolution of several ionic solids.

Worked Example 14.3 lets you practice making qualitative predictions of the sign of ΔS°_{rxn}.

Worked Example 14.3

For each process, determine the sign of ΔS for the system: (a) decomposition of $CaCO_3(s)$ to give $CaO(s)$ and $CO_2(g)$, (b) heating bromine vapor from 45°C to 80°C, (c) condensation of water vapor on a cold surface, (d) reaction of $NH_3(g)$ and $HCl(g)$ to give $NH_4Cl(s)$, and (e) dissolution of sugar in water.

Strategy Consider the change in energy/mobility of atoms and the resulting change in number of possible positions that each particle can occupy in each case. An increase in the number of arrangements corresponds to an increase in entropy and therefore a positive ΔS.

Setup Increases in entropy generally accompany solid-to-liquid, liquid-to-gas, and solid-to-gas transitions; the dissolving of one substance in another; a temperature increase; and reactions that increase the net number of moles of gas.

Solution ΔS is (a) positive, (b) positive, (c) negative, (d) negative, and (e) positive.

> **Think About It**
> For reactions involving only liquids and solids, predicting the sign of ΔS° can be more difficult, but in many such cases an increase in the total number of molecules and/or ions is accompanied by an increase in entropy.

Practice Problem **ATTEMPT** For each of the following processes, determine the sign of ΔS: (a) crystallization of sucrose from a supersaturated solution, (b) cooling water vapor from 150°C to 110°C, (c) sublimation of dry ice.

Practice Problem **BUILD** Make a qualitative prediction of the sign of ΔH°_{soln} for $AlCl_3(s)$ and the dissolution of $FeCl_3(s)$. See Table 14.3. Explain your reasoning.

Practice Problem **CONCEPTUALIZE** Consider the gas-phase reaction of A_2 (blue) and B_2 (orange) to form AB_3. What are the correct balanced equation and the sign of ΔS for the reaction?

(a) $A_2 + B_2 \longrightarrow AB_3$, negative

(b) $2A_2 + 3B_2 \longrightarrow 4AB_3$, positive

(c) $2A_2 + 3B_2 \longrightarrow 4AB_3$, negative

(d) $A_2 + 3B_2 \longrightarrow 2AB_3$, negative

(e) $A_2 + 3B_2 \longrightarrow 2AB_3$, positive

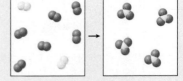

Section 14.3 Review

Entropy Changes in a System

14.3.1 For which of the following physical processes is ΔS negative? (Select all that apply.)

(a) Freezing ethanol

(b) Evaporating water

(c) Mixing carbon tetrachloride and benzene

(d) Heating water

(e) Condensing bromine vapor

14.3.2   For which of the following chemical reactions is ΔS negative?
(Select all that apply.)

(a) $2O_3(g) \longrightarrow 3O_2(g)$

(b) $4Fe(s) + 3O_2(g) \longrightarrow 2Fe_2O_3(s)$

(c) $2H_2O_2(aq) \longrightarrow 2H_2O(l) + O_2(g)$

(d) $2Li(s) + 2H_2O(l) \longrightarrow 2LiOH(aq) + H_2(g)$

(e) $2NH_3(g) \longrightarrow N_2(g) + 3H_2(g)$

14.3.3   Identify the correct balanced equation and the
sign of ΔS for the reaction shown here.

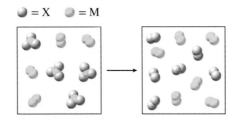

(a) $4MX_3 + 4M_2 \longrightarrow 6X_2 + 6M_2$, ΔS positive

(b) $4MX_3 \longrightarrow 6X_2 + 2M_2$, ΔS negative

(c) $2MX_3 + 2M_2 \longrightarrow 3X_2 + 3M_2$, ΔS negative

(d) $2MX_3 + 2M_2 \longrightarrow 3X_2 + 3M_2$, ΔS positive

(e) $2MX_3 \longrightarrow 3X_2$, ΔS positive

14.4   ENTROPY CHANGES IN THE UNIVERSE

Recall that the *system* typically is the part of the universe we are investigating (e.g.,
the reactants and products in a chemical reaction). The *surroundings* are everything
else [◄◄ Section 10.1]. Together, the system and surroundings make up the *universe*.
We have seen that the *dispersal* or *spreading out* of a system's energy corresponds to
an increase in the system's entropy. Moreover, an increase in the system's entropy is
one of the factors that determines whether a process is spontaneous. However, correctly predicting the spontaneity of a process requires us to consider entropy changes
in both the system and the surroundings.

Consider the following processes:

- An ice cube spontaneously melts in a room where the temperature is 25°C. In
  this case, the motional energy of the air molecules at 25°C is transferred to the
  ice cube (at 0°C), causing the ice to melt. There is no temperature change during a phase change; however, because the molecules are more mobile and there
  are many more different possible arrangements in *liquid* water than there are in
  ice, there is an increase in the entropy of the system. In this case, because the
  process of melting is endothermic, heat is transferred *from* the surroundings *to*
  the system and the temperature of the surroundings decreases. The slight decrease
  in temperature causes a small decrease in molecular motion and a decrease in
  the entropy of the surroundings.

  ΔS_{sys} is positive.

  ΔS_{surr} is negative.

- A cup of hot water spontaneously cools to room temperature as the motional
  energy of the water molecules spreads out to the cooler surrounding air. Although
  the loss of energy from the system and corresponding temperature decrease
  cause a *decrease* in the entropy of the *system,* the increased temperature of the
  surrounding air causes an *increase* in the entropy of the *surroundings.*

  ΔS_{sys} is negative.

  ΔS_{surr} is positive.

Thus, it is not just the entropy of the *system* that determines if a process is spontaneous, the entropy of the *surroundings* is also important. There are also examples of

spontaneous processes in which ΔS_{sys} and ΔS_{surr} are *both* positive. The decomposition of hydrogen peroxide produces water and oxygen gas, $2H_2O_2(l) \longrightarrow 2H_2O(l) + O_2(g)$. Because the reaction results in an increase in the number of gas molecules, we know that there is an increase in the entropy of the system. However, this is an exothermic reaction, meaning that it also gives off heat to the surroundings. An increase in temperature of the surroundings causes an increase in the entropy of the surroundings as well. (Note that there are no spontaneous processes in which ΔS_{sys} and ΔS_{surr} are both negative, which will become clear shortly.)

Calculating ΔS_{surr}

When an exothermic process takes place, the heat transferred from the system to the surroundings increases the temperature of the molecules in the surroundings. Consequently, there is an increase in the number of energy levels accessible to the molecules in the surroundings and the entropy of the surroundings increases. Conversely, in an endothermic process, heat is transferred from the surroundings to the system, decreasing the entropy of the surroundings. Remember that for constant-pressure processes, the heat released or absorbed, q, is equal to the enthalpy change of the system, ΔH_{sys} [◀◀ Section 10.3]. The change in entropy for the surroundings, ΔS_{surr}, is directly proportional to ΔH_{sys}.

$$\Delta S_{surr} \propto -\Delta H_{sys}$$

The minus sign indicates that a negative enthalpy change in the system (an *exothermic* process) corresponds to a positive entropy change in the surroundings. For an *endothermic* process, the enthalpy change in the system is a positive number and corresponds to a negative entropy change in the surroundings.

In addition to being directly proportional to ΔH_{sys}, ΔS_{surr} is inversely proportional to temperature.

$$\Delta S_{surr} \propto \frac{1}{T}$$

Combining the two expressions gives

Equation 14.7	$\Delta S_{surr} = \dfrac{-\Delta H_{sys}}{T}$

The Second Law of Thermodynamics

We have seen that both the system and surroundings can undergo changes in entropy during a process. The sum of the entropy changes for the system and the surroundings is the entropy change for the universe overall.

Equation 14.8	$\Delta S_{univ} = \Delta S_{sys} + \Delta S_{surr}$

The *second law of thermodynamics* says that for a process to be spontaneous as written (in the forward direction), ΔS_{univ} must be positive. Therefore, the system may undergo a *decrease* in entropy, as long as the surroundings undergoes a larger *increase* in entropy, and vice versa. A process for which ΔS_{univ} is *negative* is not spontaneous as written.

In some cases, ΔS_{univ} is neither positive nor negative but is equal to zero. This happens when the entropy changes of the system and surroundings are equal in magnitude and opposite in sign and describes a specific type of process known as an *equilibrium* process. An *equilibrium process* is one that does not occur spontaneously in either the net forward or net reverse direction but can be made to occur by the addition or removal of energy to a system at equilibrium. An example of an equilibrium process is the melting of ice at 0°C. (Remember that at 0°C, ice and liquid water are in equilibrium with each other [◀◀ Section 12.5].)

With Equations 14.6 and 14.7, we can calculate the entropy changes for both the system and surroundings in a process. We can then use the second law of thermodynamics (Equation 14.8) to determine if the process is spontaneous or nonspontaneous as written or if it is an equilibrium process.

Student Annotation: The concept of equilibrium will be examined in detail in Chapters 15, 16, and 17.

Consider the synthesis of ammonia at 25°C.

$$N_2(g) + 3H_2(g) \longrightarrow 2NH_3(g) \qquad \Delta H^{\circ}_{rxn} = -92.6 \text{ kJ/mol}$$

From Worked Example 14.2(b), we have $\Delta S^{\circ}_{sys} = -199$ J/K · mol, and substituting ΔH°_{sys} (−92.6 kJ/mol) into Equation 14.7, we get

$$\Delta S_{surr} = \frac{-(-92.6 \times 1000)\text{J/mol}}{298 \text{ K}} = 311 \text{ J/K} \cdot \text{mol}$$

The entropy change for the universe is

$$\Delta S^{\circ}_{univ} = \Delta S^{\circ}_{sys} + \Delta S^{\circ}_{surr}$$

$$= -199 \text{ J/K} \cdot \text{mol} + 311 \text{ J/K} \cdot \text{mol}$$

$$= 112 \text{ J/K} \cdot \text{mol}$$

Because ΔS°_{univ} is positive, the reaction will be spontaneous at 25°C. Keep in mind, though, that just because a reaction is spontaneous does not mean that it will occur at an observable rate. The synthesis of ammonia is, in fact, extremely slow at room temperature. Thermodynamics can tell us whether a reaction will occur spontaneously under specific conditions, but it does not tell us how fast it will occur.

Worked Example 14.4 lets you practice identifying spontaneous, nonspontaneous, and equilibrium processes.

> **Student Annotation:** The spontaneity that we have seen as favored by a process being exothermic is due to the spreading out of energy from the system to the surroundings; thus, the negative ΔH_{sys} corresponds to a positive ΔS_{surr}. It is this positive contribution to the overall ΔS_{univ} that actually favors spontaneity.

Worked Example 14.4

Determine if each of the following is a spontaneous process, a nonspontaneous process, or an equilibrium process at the specified temperature: (a) $H_2(g) + I_2(g) \longrightarrow 2HI(g)$ at 0°C, (b) $CaCO_3(s) \longrightarrow CaO(s) + CO_2(g)$ at 200°C, (c) $CaCO_3(s) \longrightarrow CaO(s) + CO_2(g)$ at 1000°C, (d) $Na(s) \longrightarrow Na(l)$ at 98°C. (Assume that the thermodynamic data in Appendix 2 do not vary with temperature.)

Strategy For each process, use Equation 14.6 to determine ΔS°_{sys} and Equations 10.18 and 14.7 to determine ΔH°_{sys} and ΔS°_{surr}. At the specified temperature, the process is *spontaneous* if ΔS_{sys} and ΔS_{surr} sum to a positive number, *nonspontaneous* if they sum to a negative number, and an *equilibrium process* if they sum to zero. Note that because the *reaction* is the *system*, ΔS_{rxn} and ΔS_{sys} are used interchangeably.

Setup From Appendix 2,

(a) $S^{\circ}[H_2(g)] = 131.0$ J/K · mol, $S^{\circ}[I_2(g)] = 260.57$ J/K · mol, $S^{\circ}[HI(g)] = 206.3$ J/K · mol; $\Delta H^{\circ}_f [H_2(g)] = 0$ kJ/mol, $\Delta H^{\circ}_f [I_2(g)] = 62.25$ kJ/mol, $\Delta H^{\circ}_f [HI(g)] = 25.9$ kJ/mol

(b), (c) In Worked Example 14.2(a), we determined that for this reaction, $\Delta S^{\circ}_{rxn} = 160.5$ J/K · mol, $\Delta H^{\circ}_f [CaCO_3(s)] = -1206.9$ kJ/mol, $\Delta H^{\circ}_f [CaO(s)] = -635.6$ kJ/mol, $\Delta H^{\circ}_f [CO_2(g)] = -393.5$ kJ/mol.

(d) $S^{\circ}[Na(s)] = 51.05$ J/K · mol, $S^{\circ}[Na(l)] = 57.56$ J/K · mol; $\Delta H^{\circ}_f [Na(s)] = 0$ kJ/mol, $\Delta H^{\circ}_f [Na(l)] = 2.41$ kJ/mol.

Solution

(a) $\Delta S^{\circ}_{rxn} = [2S^{\circ}(HI)] - [S^{\circ}(H_2) + S^{\circ}(I_2)]$

$\qquad = (2)(206.3 \text{ J/K} \cdot \text{mol}) - (131.0 \text{ J/K} \cdot \text{mol} + 260.57 \text{ J/K} \cdot \text{mol}) = 21.03 \text{ J/K} \cdot \text{mol}$

$\Delta H^{\circ}_{rxn} = [2\Delta H^{\circ}_f(HI)] - [\Delta H^{\circ}_f(H_2) + \Delta H^{\circ}_f(I_2)]$

$\qquad = (2)(25.9 \text{ kJ/mol}) - (0 \text{ kJ/mol} + 62.25 \text{ kJ/mol}) = -10.5 \text{ kJ/mol}$

$\Delta S_{surr} = \dfrac{-\Delta H_{rxn}}{T} = \dfrac{-(-10.5 \text{ kJ/mol})}{273 \text{ K}} = 0.0385 \text{ kJ/K} \cdot \text{mol} = 38.5 \text{ J/K} \cdot \text{mol}$

$\Delta S_{univ} = \Delta S_{sys} + \Delta S_{surr} = 21.03 \text{ J/K} \cdot \text{mol} + 38.5 \text{ J/K} \cdot \text{mol} = 59.5 \text{ J/K} \cdot \text{mol}$

ΔS_{univ} is positive, so the reaction is spontaneous at 0°C.

(Continued on next page)

(b), (c) $\Delta S_{rxn}^{\circ} = 160.5$ J/K $\cdot$ mol

$$\Delta H_{rxn}^{\circ} = [\Delta H_f^{\circ}(CaO) + \Delta H_f^{\circ}(CO_2)] - [\Delta H_f^{\circ}(CaCO_3)]$$

$$= [-635.6 \text{ kJ/mol} + (-393.5 \text{ kJ/mol})] - (-1206.9 \text{ kJ/mol}) = 177.8 \text{ kJ/mol}$$

(b) $T = 200°C$ and

$$\Delta S_{surr} = \frac{-\Delta H_{sys}}{T} = \frac{-(177.8 \text{ kJ/mol})}{473 \text{ K}} = -0.376 \text{ kJ/K} \cdot \text{mol} = -376 \text{ J/K} \cdot \text{mol}$$

$$\Delta S_{univ} = \Delta S_{sys} + \Delta S_{surr} = 160.5 \text{ J/K} \cdot \text{mol} + (-376 \text{ J/K} \cdot \text{mol}) = -216 \text{ J/K} \cdot \text{mol}$$

ΔS_{univ} is negative, so the reaction is nonspontaneous at 200°C.

(c) $T = 1000°C$ and

$$\Delta S_{surr} = \frac{-\Delta H_{sys}}{T} = \frac{-(177.8 \text{ kJ/mol})}{1273 \text{ K}} = -0.1397 \text{ kJ/K} \cdot \text{mol} = -139.7 \text{ J/K} \cdot \text{mol}$$

$$\Delta S_{univ} = \Delta S_{sys} + \Delta S_{surr} = 160.5 \text{ J/K} \cdot \text{mol} + (-139.7 \text{ J/K} \cdot \text{mol}) = -20.8 \text{ J/K} \cdot \text{mol}$$

In this case, ΔS_{univ} is positive; therefore, the reaction is spontaneous at 1000°C.

(d) $\Delta S_{rxn}^{\circ} = S°[Na(l)] - S°[Na(s)] = 57.56$ J/K $\cdot$ mol $- 51.05$ J/K $\cdot$ mol $= 6.51$ J/K $\cdot$ mol

$$\Delta H_{rxn}^{\circ} = \Delta H_f^{\circ}[Na(l)] - \Delta H_f^{\circ}[Na(s)] = 2.41 \text{ kJ/mol} - 0 \text{ kJ/mol} = 2.41 \text{ kJ/mol}$$

$$\Delta S_{surr} = \frac{-\Delta H_{rxn}}{T} = \frac{-(2.41 \text{ kJ/mol})}{371 \text{ K}} = -0.0650 \text{ kJ/K} \cdot \text{mol} = -6.50 \text{ J/K} \cdot \text{mol}$$

$$\Delta S_{univ} = \Delta S_{sys} + \Delta S_{surr} = 6.51 \text{ J/K} \cdot \text{mol} + (-6.50 \text{ J/K} \cdot \text{mol}) = 0.01 \text{ J/K} \cdot \text{mol} \approx 0$$

ΔS_{univ} is zero, so the reaction is an equilibrium process at 98°C. In fact, this is the melting point of sodium.

Student Annotation: The small difference between the magnitudes of ΔS_{sys} and ΔS_{surr} results from thermodynamic values not being entirely independent of temperature. The tabulated values of $S°$ and ΔH_f° are for 25°C.

Think About It

Remember that standard enthalpies of formation have units of kJ/mol, whereas standard absolute entropies have units of J/K $\cdot$ mol. Make sure that you convert kilojoules to joules, or vice versa, before combining the terms.

Practice Problem **A**TTEMPT   For each of the following, calculate ΔS_{univ} and identify the process as a spontaneous process, a nonspontaneous process, or an equilibrium process at the specified temperature: (a) $CO_2(g) \longrightarrow CO_2(aq)$ at 25°C, (b) $N_2O_4(g) \longrightarrow 2NO_2(g)$ at 10.4°C, (c) $PCl_3(l) \longrightarrow PCl_3(g)$ at 61.2°C. (Assume that the thermodynamic data in Appendix 2 do not vary with temperature.)

Practice Problem **B**UILD   (a) Calculate ΔS_{univ} and determine if the reaction $H_2O_2(l) \longrightarrow H_2O_2(g)$ is spontaneous, nonspontaneous, or an equilibrium process at 163°C. (b) The reaction $NH_3(g) + HCl(g) \longrightarrow NH_4Cl(s)$ is spontaneous in the forward direction at room temperature but, because it is exothermic, becomes less spontaneous with increasing temperature. Determine the temperature at which it is no longer spontaneous in the forward direction. (c) Determine the boiling point of Br_2. (Assume that the thermodynamic data in Appendix 2 do not vary with temperature.)

Practice Problem **C**ONCEPTUALIZE   The following table shows the signs of ΔS_{sys}, ΔS_{surr}, and ΔS_{univ} for four processes. Where possible, fill in the missing table entries. Indicate where it is not possible to determine the missing sign and explain.

Process	ΔS_{sys}	ΔS_{surr}	ΔS_{univ}
1	−	−	
2	+		+
3	−	+	
4		−	+

Thinking Outside the Box

Thermodynamics and Living Systems

Under normal physiological conditions, polypeptides spontaneously fold into unique three-dimensional structures called native proteins, which can perform various functions. Because the original chain can assume many possible configurations while the native protein can have only one specific arrangement, the folding process is accompanied by a decrease in entropy of the system. (Note that solvent molecules, water in this case, can also play a role in affecting the entropy change.) In accord with the second law of thermodynamics, any spontaneous process must result in an increase in the entropy of the universe. It follows, therefore, that there must be an increase in the entropy of the surroundings that outweighs the decrease in the entropy of the system. The intramolecular attractions between amino acid residues cause the folding of the polypeptide chain to be exothermic. The energy produced by the process spreads out, increasing molecular motion in the surroundings—thereby increasing the entropy of the surroundings.

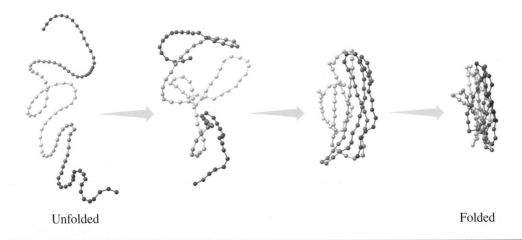

Unfolded   Folded

The Third Law of Thermodynamics

Finally, we consider the third law of thermodynamics briefly in connection with the determination of standard entropy. We have related the entropy of a system to the number of possible arrangements of the system's molecules. The larger the number of possible arrangements, the larger the entropy. Imagine a pure, perfect crystalline substance at absolute zero (0 K). Under these conditions, there is essentially no molecular motion and, because the molecules occupy fixed positions in the solid, there is only one way to arrange the molecules. From Equation 14.1, we write

$$S = k \ln W = k \ln 1 = 0$$

According to the *third law of thermodynamics,* the entropy of a perfect crystalline substance is *zero* at absolute zero. As temperature increases, molecular motion increases, causing an increase in the number of possible arrangements of the molecules and in the number of accessible energy states, among which the system's energy can be dispersed. (See Figure 14.4.) This results in an increase in the system's entropy. Thus, the entropy of any substance at any temperature above 0 K is greater than zero. If the crystalline substance is impure or imperfect in any way, then its entropy is greater than zero even at 0 K because without perfect crystalline order, there is more than one possible arrangement of molecules.

The significance of the third law of thermodynamics is that it enables us to determine experimentally the *absolute* entropies of substances. Starting with the knowledge that the entropy of a pure crystalline substance is zero at 0 K, we can measure the increase in entropy of the substance as it is heated. The change in entropy of a substance, ΔS, is the difference between the final and initial entropy values,

$$\Delta S = S_{final} - S_{initial}$$

Figure 14.5 Entropy increases in a substance as temperature increases from absolute zero.

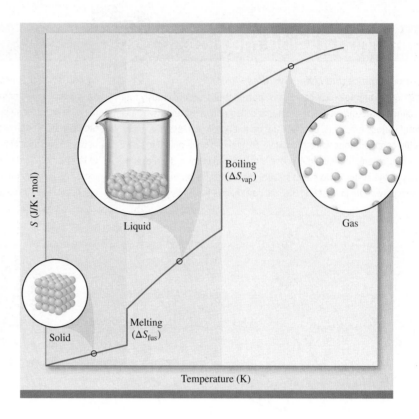

where $S_{initial}$ is zero if the substance starts at 0 K. Therefore, the measured *change* in entropy is equal to the *absolute* entropy of the substance at the final temperature.

$$\Delta S = S_{final}$$

The entropy values arrived at in this way are called *absolute* entropies because they are *true* values—unlike standard enthalpies of formation, which are derived using an arbitrary reference. Because the tabulated values are determined at 1 atm, we usually refer to absolute entropies as *standard* entropies, $S°$. Figure 14.5 shows the increase in entropy of a substance as temperature increases from absolute zero. At 0 K, it has a zero entropy value (assuming that it is a perfect crystalline substance). As it is heated, its entropy increases gradually at first because of greater molecular motion within the crystal. At the melting point, there is a large increase in entropy as the solid is transformed into the liquid. Further heating increases the entropy of the liquid again due to increased molecular motion. At the boiling point, there is a large increase in entropy as a result of the liquid-to-vapor transition. Beyond that temperature, the entropy of the gas continues to increase with increasing temperature.

Section 14.4 Review

Entropy Changes in the Universe

14.4.1   Using data from Appendix 2, calculate $\Delta S°$ (in J/K · mol) for the following reaction.

$$2NO(g) + O_2(g) \longrightarrow 2NO_2(g)$$

(a)   145.3 J/K · mol   (d)   −59.7 J/K · mol
(b)   −145.3 J/K · mol   (e)   −421.2 J/K · mol
(c)   59.7 J/K · mol

14.4.2 Using data from Appendix 2, calculate $\Delta S°$ (in J/K · mol) for the following reaction.

$$CH_4(g) + 2O_2(g) \longrightarrow CO_2(g) + 2H_2O(l)$$

(a) 107.7 J/K · mol   (d) 242.8 J/K · mol
(b) −107.7 J/K · mol   (e) −242.8 J/K · mol
(c) 2.6 J/K · mol

14.4.3 The diagrams show a spontaneous chemical reaction. What can we deduce about ΔS_{surr} for this process?

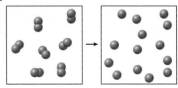

(a) ΔS_{surr} is positive.
(b) ΔS_{surr} is negative.
(c) ΔS_{surr} is zero.
(d) There is not enough information to deduce the sign of ΔS_{surr}.

14.4.4 The diagrams show a spontaneous chemical reaction. What can we deduce about ΔS_{surr} for this process?

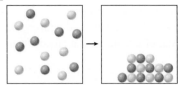

(a) ΔS_{surr} is positive.
(b) ΔS_{surr} is negative.
(c) ΔS_{surr} is zero.
(d) There is not enough information to deduce the sign of ΔS_{surr}.

14.4.5 For the process shown, state whether ΔS_{surr} is positive, negative, or if there is not enough information to determine the sign of ΔS_{surr}.

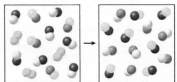

(a) ΔS_{surr} is positive.
(b) ΔS_{surr} is negative.
(c) ΔS_{surr} is zero.
(d) There is not enough information to deduce the sign of ΔS_{surr}.

14.5   PREDICTING SPONTANEITY

Gibbs Free-Energy Change, ΔG

According to the second law of thermodynamics, $\Delta S_{univ} > 0$ for a spontaneous process. What we are usually concerned with and usually *measure,* however, are the properties of the *system* rather than those of the surroundings or those of the universe overall. Therefore, it is convenient to have a thermodynamic function that enables us to determine whether a process is spontaneous by considering the system alone.

We begin with Equation 14.8. For a spontaneous process,

$$\Delta S_{univ} = \Delta S_{sys} + \Delta S_{surr} > 0$$

Substituting $-\Delta H_{sys}/T$ for ΔS_{surr}, we write

$$\Delta S_{univ} = \Delta S_{sys} + \left(-\frac{\Delta H_{sys}}{T}\right) > 0$$

Multiplying both sides of the equation by T gives

$$T\Delta S_{univ} = T\Delta S_{sys} - \Delta H_{sys} > 0$$

Now we have an equation that expresses the second law of thermodynamics (and predicts whether a process is spontaneous) in terms of only the *system*. We no longer need to consider the surroundings. For convenience, we can rearrange the preceding equation, multiply through by -1, and replace the $>$ sign with a $<$ sign.

$$-T\Delta S_{univ} = \Delta H_{sys} - T\Delta S_{sys} < 0$$

According to this equation, a process carried out at constant pressure and temperature is spontaneous if the changes in enthalpy and entropy of the system are such that $\Delta H_{sys} - T\Delta S_{sys}$ is less than zero.

To express the spontaneity of a process more directly, we introduce another thermodynamic function called the **Gibbs**[1] **free energy (G),** or simply **free energy.**

Equation 14.9	$G = H - TS$

Each of the terms in Equation 14.9 pertains to the system. G has units of energy just as H and TS do. Furthermore, like enthalpy and entropy, free energy is a state function. The change in free energy, ΔG, of a system for a process that occurs at constant temperature is

Equation 14.10	$\Delta G = \Delta H - T\Delta S$

Equation 14.10 enables us to predict the spontaneity of a process using the change in enthalpy, the change in entropy, and the absolute temperature. At constant temperature and pressure, for processes that are spontaneous as written (in the forward direction), ΔG is negative. For processes that are not spontaneous as written but that are spontaneous in the reverse direction, ΔG is positive. For systems at equilibrium, ΔG is zero.

> **Student Annotation:** In this context, *free energy* is the energy available to do work. Thus, if a particular process is accompanied by a release of usable energy (i.e., if ΔG is negative), this fact alone guarantees that it is spontaneous, and there is no need to consider what happens to the rest of the universe.

- $\Delta G < 0$ The reaction is spontaneous in the forward direction (and nonspontaneous in the reverse direction).
- $\Delta G > 0$ The reaction is nonspontaneous in the forward direction (and spontaneous in the reverse direction).
- $\Delta G = 0$ The system is at equilibrium.

Often we can predict the sign of ΔG for a process if we know the signs of ΔH and ΔS. Table 14.4 shows how we can use Equation 14.10 to make such predictions.

Based on the information in Table 14.4, you may wonder what constitutes a "low" or a "high" temperature. For the freezing of water, 0°C is the temperature that divides high from low. Water freezes spontaneously at temperatures below 0°C, and ice melts spontaneously at temperatures above 0°C. At 0°C, a system of ice and water is at equilibrium. The temperature that divides "high" from "low" depends, though, on the individual reaction. To determine that temperature, we must set ΔG equal to 0 in Equation 14.10 (i.e., the equilibrium condition).

$$0 = \Delta H - T\Delta S$$

Rearranging to solve for T yields

$$T = \frac{\Delta H}{\Delta S}$$

The temperature that divides high from low for a particular reaction can now be calculated if the values of ΔH and ΔS are known.

---

[1]Josiah Willard Gibbs (1839–1903), an American physicist, was one of the founders of thermodynamics. Gibbs was a modest and private individual who spent almost all his professional life at Yale University. Because he published most of his work in obscure journals, Gibbs never gained the eminence that his contemporary and admirer James Maxwell did. Even today, very few people outside of chemistry and physics have ever heard of Gibbs.

TABLE 14.4		Predicting the Sign of ΔG Using Equation 14.10 and the Signs of ΔH and ΔS		
When ΔH is	**And ΔS is**	**ΔG will be**	**And the process is**	**Example**
Negative	Positive	Negative	Always spontaneous	$2H_2O_2(aq) \longrightarrow 2H_2O(l) + O_2(g)$
Positive	Negative	Positive	Always nonspontaneous	$3O_2(g) \longrightarrow 2O_3(g)$
Negative	Negative	Negative when $T\Delta S < \Delta H$	Spontaneous at low T	$H_2O(l) \longrightarrow H_2O(s)$
		Positive when $T\Delta S > \Delta H$	Nonspontaneous at high T	(freezing of water)
Positive	Positive	Negative when $T\Delta S > \Delta H$	Spontaneous at high T	$2HgO(s) \longrightarrow 2Hg(l) + O_2(g)$
		Positive when $T\Delta S < \Delta H$	Nonspontaneous at low T	

Worked Example 14.5 demonstrates the use of this approach.

Worked Example 14.5

According to Table 14.4, a reaction will be spontaneous only at high temperatures if both ΔH and ΔS are positive. For a reaction in which $\Delta H = 199.5$ kJ/mol and $\Delta S = 476$ J/K · mol, determine the temperature (in °C) above which the reaction is spontaneous.

Strategy The temperature that divides high from low is the temperature at which $\Delta H = T\Delta S$ ($\Delta G = 0$). Therefore, we use Equation 14.10, substituting 0 for ΔG and solving for T to determine temperature in kelvins; we then convert to degrees Celsius.

Setup

$$\Delta S = \left(\frac{476 \text{ J}}{\text{K} \cdot \text{mol}} \right) \left(\frac{1 \text{ kJ}}{1000 \text{ J}} \right) = 0.476 \text{ kJ/K} \cdot \text{mol}$$

Solution

$$T = \frac{\Delta H}{\Delta S} = \frac{199.5 \text{ kJ/mol}}{0.476 \text{ kJ/K} \cdot \text{mol}} = 419 \text{ K}$$

$$= (419 - 273) = 146°C$$

Think About It
Spontaneity is favored by a release of energy (ΔH being negative) and by an increase in entropy (ΔS being positive). When both quantities are positive, as in this case, only the entropy change favors spontaneity. For an endothermic process such as this, which requires the input of heat, it should make sense that adding more heat by increasing the temperature will shift the equilibrium to the right, thus making it "more spontaneous."

Practice Problem **A**TTEMPT A reaction will be spontaneous only at low temperatures if both ΔH and ΔS are negative. For a reaction in which $\Delta H = -380.1$ kJ/mol and $\Delta S = -95.00$ J/K · mol, determine the temperature (in °C) below which the reaction is spontaneous.

Practice Problem **B**UILD Given that the reaction $4Fe(s) + 3O_2(g) + 6H_2O(l) \longrightarrow 4Fe(OH)_3(s)$ is spontaneous at temperatures below 1950°C, estimate the standard entropy of $Fe(OH)_3(s)$.

Practice Problem **C**ONCEPTUALIZE Which of the following graphs best represents the relationship between ΔG and temperature for a process that is exothermic and for which ΔS is negative?

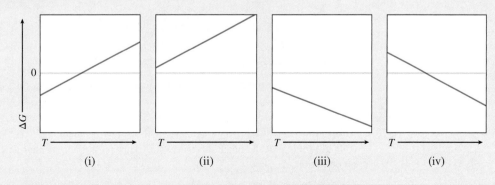

(i)   (ii)   (iii)   (iv)

Student Annotation: The introduction of the term $\Delta G°$ enables us to write Equation 14.10 as

$$\Delta G° = \Delta H° - T\Delta S°$$

Standard Free-Energy Changes, $\Delta G°$

The **standard free energy of reaction** ($\Delta G°_{\text{rxn}}$) is the free-energy change for a reaction when it occurs under standard-state conditions—that is, when reactants in their standard states are converted to products in *their* standard states. The conventions used by chemists to define the standard states of pure substances and solutions are

- Gases   1   atm pressure
- Liquids   Pure liquid
- Solids   Pure solid
- Elements   The most stable allotropic form at 1 atm and 25°C
- Solutions   1 molar concentration

To calculate $\Delta G°_{\text{rxn}}$, we start with the general equation

$$a\text{A} + b\text{B} \longrightarrow c\text{C} + d\text{D}$$

The standard free-energy change for this reaction is given by

Equation 14.11 $\Delta G°_{\text{rxn}} = [c\Delta G°_{\text{f}}(\text{C}) + d\Delta G°_{\text{f}}(\text{D})] - [a\Delta G°_{\text{f}}(\text{A}) + b\Delta G°_{\text{f}}(\text{B})]$

Equation 14.11 can be generalized as follows:

Equation 14.12 $\Delta G°_{\text{rxn}} = \Sigma n\Delta G°_{\text{f}}(\text{products}) - \Sigma m\Delta G°_{\text{f}}(\text{reactants})$

where m and n are stoichiometric coefficients. The term $\Delta G°_{\text{f}}$ is the **standard free energy of formation** of a compound—that is, the free-energy change that occurs when 1 mole of the compound is synthesized from its constituent elements, each in its standard state. For the combustion of graphite,

$$\text{C(graphite)} + \text{O}_2(g) \longrightarrow \text{CO}_2(g)$$

the standard free-energy change (from Equation 14.12) is

$$\Delta G°_{\text{rxn}} = [\Delta G°_{\text{f}}(\text{CO}_2)] - [\Delta G°_{\text{f}}(\text{C, graphite}) + \Delta G°_{\text{f}}(\text{O}_2)]$$

As with standard enthalpy of formation, the standard free energy of formation of any element (in its most stable allotropic form at 1 atm) is defined as zero. Thus,

$$\Delta G°_{\text{f}}(\text{C, graphite}) = 0 \qquad \text{and} \qquad \Delta G°_{\text{f}}(\text{O}_2) = 0$$

Therefore, the standard free-energy change for the reaction in this case is equal to the standard free energy of formation of CO_2.

$$\Delta G°_{\text{rxn}} = \Delta G°_{\text{f}}(\text{CO}_2)$$

 Student Hot Spot

Student data indicate you may struggle with determining free-energy changes. Access the SmartBook to view additional Learning Resources on this topic.

Appendix 2 lists the values of $\Delta G°_{\text{f}}$ at 25°C for a number of compounds.

Worked Example 14.6 demonstrates the calculation of standard free-energy changes.

Worked Example 14.6

Calculate the standard free-energy changes for the following reactions at 25°C:

(a) $\text{CH}_4(g) + 2\text{O}_2(g) \longrightarrow \text{CO}_2(g) + 2\text{H}_2\text{O}(l)$

(b) $2\text{MgO}(s) \longrightarrow 2\text{Mg}(s) + \text{O}_2(g)$

Strategy Look up the $\Delta G°_{\text{f}}$ values for the reactants and products in each equation, and use Equation 14.12 to solve for $\Delta G°_{\text{rxn}}$.

Setup From Appendix 2, we have the following values: $\Delta G°_{\text{f}}[\text{CH}_4(g)] = -50.8$ kJ/mol, $\Delta G°_{\text{f}}[\text{CO}_2(g)] = -394.4$ kJ/mol, $\Delta G°_{\text{f}}[\text{H}_2\text{O}(l)] = -237.2$ kJ/mol, and $\Delta G°_{\text{f}}[\text{MgO}(s)] = -569.6$ kJ/mol. All the other substances are elements in their standard states and have, by definition, $\Delta G°_{\text{f}} = 0$.

Solution

(a) $\Delta G^{\circ}_{rxn} = (\Delta G^{\circ}_{f}[CO_2(g)] + 2\Delta G^{\circ}_{f}[H_2O(l)]) - (\Delta G^{\circ}_{f}[CH_4(g)] + 2\Delta G^{\circ}_{f}[O_2(g)])$

$\qquad = [(-394.4 \text{ kJ/mol}) + (2)(-237.2 \text{ kJ/mol})] - [(-50.8 \text{ kJ/mol}) + (2)(0 \text{ kJ/mol})]$

$\qquad = -818.0 \text{ kJ/mol}$

(b) $\Delta G^{\circ}_{rxn} = (2\Delta G^{\circ}_{f}[Mg(s)] + \Delta G^{\circ}_{f}[O_2(g)]) - (2\Delta G^{\circ}_{f}[MgO(s)])$

$\qquad = [(2)(0 \text{ kJ/mol}) + (0 \text{ kJ/mol})] - (2)(-569.6 \text{ kJ/mol})]$

$\qquad = 1139 \text{ kJ/mol}$

Think About It

Note that, like standard enthalpies of formation (ΔH°_{f}), standard free energies of formation (ΔG°_{f}) depend on the *state* of matter. Using water as an example, $\Delta G^{\circ}_{f}[H_2O(l)] = 237.2 \text{ kJ/mol}$ and $\Delta G^{\circ}_{f}[H_2O(g)] = -228.6 \text{ kJ/mol}$. Always double-check to make sure you have selected the right value from the table.

Practice Problem **A**TTEMPT Calculate the standard free-energy changes for the following reactions at 25°C.

(a) $H_2(g) + Br_2(l) \longrightarrow 2HBr(g)$

(b) $2C_2H_6(g) + 7O_2(g) \longrightarrow 4CO_2(g) + 6H_2O(l)$

Practice Problem **B**UILD For each reaction, determine the value of ΔG°_{f} that is not listed in Appendix 2.

(a) $Li_2O(s) + 2HCl(g) \longrightarrow 2LiCl(s) + H_2O(g)$ $\quad \Delta G^{\circ}_{rxn} = -244.88 \text{ kJ/mol}$

(b) $Na_2O(s) + 2HI(g) \longrightarrow 2NaI(s) + H_2O(l)$ $\quad \Delta G^{\circ}_{rxn} = -435.44 \text{ kJ/mol}$

Practice Problem **C**ONCEPTUALIZE For which of the following species is $\Delta G^{\circ}_{f} = 0$?

(i) $Br_2(l)$ $\qquad$ (ii) $I_2(g)$ $\qquad$ (iii) $CO_2(g)$ $\qquad$ (iv) $Xe(g)$

Using ΔG° to Solve Problems

The sign of ΔG°, the standard free-energy change, indicates whether a process will occur spontaneously under standard conditions. However, ΔG° values change with temperature. One of the uses of Equation 14.10 is to determine the temperature at which a particular equilibrium will begin to favor a desired product. For example, calcium oxide (CaO), also called quicklime, is an extremely valuable inorganic substance with a variety of industrial uses, including water treatment and pollution control. It is prepared by heating limestone ($CaCO_3$), which decomposes at a high temperature.

$$CaCO_3(s) \rightleftharpoons CaO(s) + CO_2(g)$$

The reaction is reversible, and under the right conditions, CaO and CO_2 readily recombine to form $CaCO_3$ again. To prevent this from happening in the industrial preparation, the CO_2 is removed as it forms and the system is never maintained at equilibrium.

An important piece of information for the chemist responsible for maximizing CaO production is the temperature at which the decomposition equilibrium of $CaCO_3$ begins to favor products. We can make a reliable estimate of that temperature as follows. First we calculate ΔH° and ΔS° for the reaction at 25°C, using the data in Appendix 2. To determine ΔH°, we apply Equation 10.18.

$$\Delta H^{\circ} = [\Delta H^{\circ}_{f}(CaO) + \Delta H^{\circ}_{f}(CO_2)] - [\Delta H^{\circ}_{f}(CaCO_3)]$$

$$= [(-635.6 \text{ kJ/mol}) + (-393.5 \text{ kJ/mol})] - (-1206.9 \text{ kJ/mol})$$

$$= 177.8 \text{ kJ/mol}$$

Next we apply Equation 14.6 to find $\Delta S°$.

$$\Delta S° = [S°(CaO) + S°(CO_2)] - S°(CaCO_3)$$
$$= [(39.8 \text{ J/K} \cdot \text{mol}) + (213.6 \text{ J/K} \cdot \text{mol})] - (92.9 \text{ J/K} \cdot \text{mol})$$
$$= 160.5 \text{ J/K} \cdot \text{mol}$$

From Equation 14.10, we can write

$$\Delta G° = \Delta H° - T\Delta S°$$

and we obtain

$$\Delta G° = (177.8 \text{ kJ/mol}) - (298 \text{ K})(0.1605 \text{ kJ/K} \cdot \text{mol})$$
$$= 130.0 \text{ kJ/mol}$$

Student Annotation: Be careful with units in problems of this type. $S°$ values are tabulated using joules, whereas $\Delta H_f°$ values are tabulated using kilojoules.

Because $\Delta G°$ is a large positive number, the reaction does *not* favor product formation at 25°C (298 K). And, because $\Delta H°$ and $\Delta S°$ are both positive, we know that $\Delta G°$ will be negative (product formation will be favored) at high temperatures. We can determine what constitutes a high temperature for this reaction by calculating the temperature at which $\Delta G°$ is zero.

$$0 = \Delta H° - T\Delta S°$$

or

$$T = \frac{\Delta H°}{\Delta S°}$$
$$= \frac{(177.8 \text{ kJ/mol})}{(0.1605 \text{ kJ/K} \cdot \text{mol})}$$
$$= 1108 \text{ K } (835°C)$$

At temperatures higher than 835°C, $\Delta G°$ becomes negative, indicating that the reaction would then favor the formation of CaO and CO_2. At 840°C (1113 K), for example,

$$\Delta G° = \Delta H° - T\Delta S°$$
$$= 177.8 \text{ kJ/mol} - (1113 \text{ K})(0.1605 \text{ kJ/K} \cdot \text{mol})\frac{(1 \text{ kJ})}{(1000 \text{ J})}$$
$$= -0.8 \text{ kJ/mol}$$

Student Hot Spot

Student data indicate you may struggle with the effect of temperature on free energy changes. Access the SmartBook to view additional Learning Resources on this topic.

At still higher temperatures, $\Delta G°$ becomes increasingly negative, thus favoring product formation even more. Note that in this example we used the $\Delta H°$ and $\Delta S°$ values at 25°C to calculate changes to $\Delta G°$ at much higher temperatures. Because both $\Delta H°$ and $\Delta S°$ actually change with temperature, this approach does not give us a truly accurate value for $\Delta G°$, but it does give us a reasonably good estimate.

Equation 14.10 can also be used to calculate the change in entropy that accompanies a phase change. Recall that at the temperature at which a phase change occurs both phases of a substance are present. For example, at the freezing point of water, both liquid water and solid ice coexist in a state of equilibrium [◄◄ Section 12.5], where ΔG is zero. Therefore, Equation 14.10 becomes

$$0 = \Delta H - T\Delta S$$

or

$$\Delta S = \frac{\Delta H}{T}$$

Consider the ice-water equilibrium. For the ice-to-water transition, ΔH is the molar heat of fusion (see Table 12.7) and T is the melting point. The entropy change is therefore

$$\Delta S_{\text{ice} \longrightarrow \text{water}} = \frac{6010 \text{ J/mol}}{273 \text{ K}} = 22.0 \text{ J/K} \cdot \text{mol}$$

Thus, when 1 mole of ice melts at 0°C, there is an increase in entropy of 22.0 J/K · mol. The increase in entropy is consistent with the increase in the number of possible arrangements of molecules from solid to liquid. Conversely, for the water-to-ice transition, the decrease in entropy is given by

$$\Delta S_{\text{water} \longrightarrow \text{ice}} = \frac{-6010 \text{ J/mol}}{273 \text{ K}} = -22.0 \text{ J/K} \cdot \text{mol}$$

The same approach can be applied to the water-to-steam transition. In this case, ΔH is the heat of vaporization and T is the boiling point of water.

Worked Example 14.7 examines the phase transitions in benzene.

Worked Example 14.7

The molar heats of fusion and vaporization of benzene are 10.9 and 31.0 kJ/mol, respectively. Calculate the entropy changes for the solid-to-liquid and liquid-to-vapor transitions for benzene. At 1 atm pressure, benzene melts at 5.5°C and boils at 80.1°C.

Strategy The solid-liquid transition at the melting point and the liquid-vapor transition at the boiling point are *equilibrium* processes. Therefore, because ΔG is zero at equilibrium, in each case we can use Equation 14.10, substituting 0 for ΔG and solving for ΔS, to determine the entropy change associated with the process.

Setup The melting point of benzene is $5.5 + 273.15 = 278.7$ K and the boiling point is $80.1 + 273.15 = 353.3$ K.

Solution

$$\Delta S_{\text{fus}} = \frac{\Delta H_{\text{fus}}}{T_{\text{melting}}} = \frac{10.9 \text{ kJ/mol}}{278.7 \text{ K}}$$

$$= 0.0391 \text{ kJ/K} \cdot \text{mol} \quad \text{or} \quad 39.1 \text{ J/K} \cdot \text{mol}$$

$$\Delta S_{\text{vap}} = \frac{\Delta H_{\text{vap}}}{T_{\text{boiling}}} = \frac{31.0 \text{ kJ/mol}}{353.3 \text{ K}}$$

$$= 0.0877 \text{ kJ/K} \cdot \text{mol} \quad \text{or} \quad 87.7 \text{ J/K} \cdot \text{mol}$$

Think About It
For the same substance, ΔS_{vap} is always significantly larger than ΔS_{fus}. The change in number of arrangements is always bigger in a liquid-to-gas transition than in a solid-to-liquid transition.

Practice Problem Ⓐ**TTEMPT** The molar heats of fusion and vaporization of argon are 1.3 and 6.3 kJ/mol, respectively, and argon's melting point and boiling point are −190°C and −186°C, respectively. Calculate the entropy changes for the fusion and vaporization of argon.

Practice Problem Ⓑ**UILD** Using data from Appendix 2 and assuming that the tabulated values do not change with temperature, (a) calculate $\Delta H_{\text{fus}}^{\circ}$ and $\Delta S_{\text{fus}}^{\circ}$ for sodium metal and determine the melting temperature of sodium, and (b) calculate $\Delta H_{\text{vap}}^{\circ}$ and $\Delta S_{\text{vap}}^{\circ}$ for sodium metal and determine the boiling temperature of sodium.

Practice Problem Ⓒ**ONCEPTUALIZE** Explain why, in general, we can use the equation $\Delta S = \frac{\Delta H}{T}$ to calculate ΔS for a phase change but not for a chemical reaction.

Section 14.5 Review

Predicting Spontaneity

14.5.1   Using data from Appendix 2, calculate ΔG° (in kJ/mol) at 25°C for the reaction

$$CH_4(g) + 2O_2(g) \longrightarrow CO_2(g) + 2H_2O(g)$$

(a) −580.8 kJ/mol   (d) −800.8 kJ/mol
(b) 580.8 kJ/mol   (e) −818.0 kJ/mol
(c) −572.0 kJ/mol

14.5.2   Calculate ΔS_{sub} (in J/K · mol) for the sublimation of iodine in a closed flask at 45°C.

$$I_2(s) \longrightarrow I_2(g) \qquad \Delta H_{sub} = 62.4 \text{ kJ/mol}$$

(a) 1.4 J/K · mol   (d) 0.196 J/K · mol
(b) 196 J/K · mol   (e) 721 J/K · mol
(c) 1387 J/K · mol

14.5.3   At what temperature (in °C) does a reaction go from being nonspontaneous to spontaneous if it has $\Delta H = 171$ kJ/mol and $\Delta S = 161$ J/K · mol?
(a) 270°C   (d) 790°C
(b) 670°C   (e) 28°C
(c) 1100°C

14.6   THERMODYNAMICS IN LIVING SYSTEMS

Many biochemical reactions have a positive ΔG° value, yet they are essential to the maintenance of life. In living systems, these reactions are coupled to an energetically favorable process, one that has a negative ΔG° value. The principle of coupled reactions is based on a simple concept: we can use a thermodynamically favorable reaction to drive an unfavorable one. Suppose, for example, that we want to extract zinc from a zinc sulfide (ZnS). The following reaction will not work because it has a large positive ΔG° value.

$$ZnS(s) \longrightarrow Zn(s) + S(s) \qquad \Delta G^\circ = 198.3 \text{ kJ/mol}$$

On the other hand, the combustion of sulfur to form sulfur dioxide is favored because of its large negative ΔG° value.

$$S(s) + O_2(g) \longrightarrow SO_2(g) \qquad \Delta G^\circ = -300.1 \text{ kJ/mol}$$

By coupling the two processes, we can bring about the separation of zinc from zinc sulfide. In practice, this means heating ZnS in air so that the tendency of S to form SO_2 will promote the decomposition of ZnS.

$$ZnS(s) \longrightarrow Zn(s) + S(s) \qquad \Delta G^\circ = 198.3 \text{ kJ/mol}$$
$$\underline{S(s) + O_2(g) \longrightarrow SO_2(g) \qquad \Delta G^\circ = -300.1 \text{ kJ/mol}}$$
$$ZnS(s) + O_2(g) \longrightarrow Zn(s) + SO_2(g) \qquad \Delta G^\circ = -101.8 \text{ kJ/mol}$$

Coupled reactions play a crucial role in our survival. In biological systems, enzymes facilitate a wide variety of nonspontaneous reactions. In the human body, for example, food molecules, represented by glucose ($C_6H_{12}O_6$), are converted to carbon dioxide and water during metabolism, resulting in a substantial release of free energy.

$$C_6H_{12}O_6(s) + 6O_2(g) \longrightarrow 6CO_2(g) + 6H_2O(l) \qquad \Delta G^\circ = -2880 \text{ kJ/mol}$$

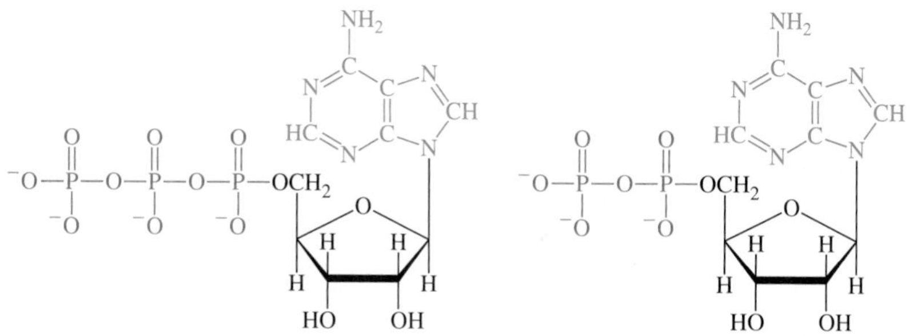

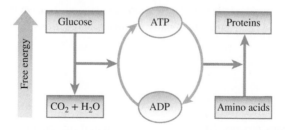

Figure 14.6 Structures of ATP and ADP.

In a living cell, this reaction does not take place in a single step; rather, the glucose molecule is broken down with the aid of enzymes in a series of steps. Much of the free energy released along the way is used to synthesize adenosine triphosphate (ATP) from adenosine diphosphate (ADP) and phosphoric acid (Figure 14.6).

$$ADP + H_3PO_4 \longrightarrow ATP + H_2O \qquad \Delta G^\circ = 31 \text{ kJ/mol}$$

The function of ATP is to store free energy until it is needed by cells. Under appropriate conditions, ATP undergoes hydrolysis to give ADP and phosphoric acid, with a release of 31 kJ/mol of free energy, which can be used to drive energetically unfavorable reactions, such as protein synthesis.

Proteins are polymers made of amino acids. The stepwise synthesis of a protein molecule involves the joining of individual amino acids. Consider the formation of the dipeptide (a unit composed of two amino acids) alanylglycine from alanine and glycine. This reaction represents the first step in the synthesis of a protein molecule.

$$\text{alanine} + \text{glycine} \longrightarrow \text{alanylglycine} \qquad \Delta G^\circ = 29 \text{ kJ/mol}$$

The positive ΔG° value means this reaction does not favor the formation of product, so only a little of the dipeptide would be formed at equilibrium. With the aid of an enzyme, however, the reaction is coupled to the hydrolysis of ATP as follows:

$$\text{ATP} + H_2O + \text{alanine} + \text{glycine} \longrightarrow \text{ADP} + H_3PO_4 + \text{alanylglycine}$$

The overall free-energy change is given by $\Delta G^\circ = -31$ kJ/mol $+ 29$ kJ/mol $= -2$ kJ/mol, which means that the coupled reaction now favors the formation of product and an appreciable amount of alanylglycine will be formed under these conditions. Figure 14.7 shows the ATP-ADP interconversions that act as energy storage (from metabolism) and free-energy release (from ATP hydrolysis) to drive essential reactions.

Figure 14.7 Schematic representation of ATP synthesis and coupled reactions in living systems. The conversion of glucose to carbon dioxide and water during metabolism releases free energy. The released free energy is used to convert ADP to ATP. The ATP molecules are then used as an energy source to drive unfavorable reactions such as protein synthesis from amino acids.

Learning Outcomes

- Distinguish between a spontaneous and nonspontaneous process and cite examples of each.
- Define entropy.
- Calculate the change in entropy of a system given the moles of ideal gas, and initial and final volumes of the gas.
- Describe the conditions for standard entropy.
- Calculate the standard entropy change for a given reaction.
- List key trends in standard entropy of atoms and molecules.
- Predict the sign of ΔS for a process and use the sign to indicate whether the system has undergone an increase or decrease in entropy.
- Calculate ΔS_{surr} given ΔS_{sys} and temperature.

- State in your own words the second law of thermodynamics.
- Determine whether a process is spontaneous given ΔS_{surr} and ΔS_{sys}.
- State in your own words the third law of thermodynamics.
- Define Gibbs free energy.
- Use ΔH and ΔS to calculate ΔG and, in turn, determine whether a process is spontaneous.
- Predict the sign of ΔG given ΔH and ΔS.
- Define standard free energy of formation.
- Calculate the standard free energy of a given reaction.
- Explain, using thermodynamic terms, why energetically unfavored metabolic reactions can occur.

Chapter Summary

SECTION 14.1
- A **spontaneous process** is one that occurs under a specified set of conditions.
- A **nonspontaneous process** is one that does *not* occur under a specified set of conditions.
- Spontaneous processes do not necessarily happen quickly.

SECTION 14.2
- **Entropy** is a thermodynamic state function that measures how dispersed or spread out a system's energy is.

SECTION 14.3
- Entropy change for a process can be calculated using standard entropy values or can be predicted qualitatively based on factors such as temperature, phase, and number of molecules.
- Whether a process is spontaneous depends on the change in *enthalpy* and the change in *entropy* of the system.
- Tabulated **standard entropy** values are absolute values.

SECTION 14.4
- According to the **second law of thermodynamics,** the entropy change for the universe is positive for a *spontaneous* process and zero for an **equilibrium process.**
- According to the **third law of thermodynamics,** the entropy of a perfectly crystalline substance at 0 K is zero.

SECTION 14.5
- The **Gibbs free energy (G)** or simply the **free energy** of a system is the energy available to do work.
- The **standard free energy of reaction (ΔG°_{rxn})** for a reaction tells us whether the equilibrium lies to the right (negative ΔG°_{rxn}) or to the left (positive ΔG°_{rxn}).
- **Standard free energies of formation (ΔG°_{f})** can be used to calculate standard free energies of reaction.

SECTION 14.6
- In living systems, thermodynamically favorable reactions provide the free energy needed to drive necessary but thermodynamically unfavorable reactions.

Key Words

Key Equations

14.1	$S = k \ln W$	The entropy S of a system is equal to the product of the Boltzmann constant (k) and ln of W, the number of possible arrangements of molecules in the system.
14.2	$W = X^N$	The number of possible arrangements W is equal to the number of possible locations of molecules X raised to the number of molecules in the system N.
14.3	$\Delta S_{sys} = S_{final} - S_{initial}$	The entropy change in a system, ΔS_{sys}, is equal to final entropy, S_{final}, minus initial entropy, $S_{initial}$.
14.4	$\Delta S_{sys} = nR \ln \dfrac{V_{final}}{V_{initial}}$	For a gaseous process involving a volume change, entropy change is calculated as the product of the number of moles (n), the gas constant (R), and ln of the ratio of final volume to initial volume [$\ln(V_{final}/V_{initial})$].
14.5	$\Delta S_{rxn}^{\circ} = [cS^{\circ}(C) + dS^{\circ}(D)] - [aS^{\circ}(A) + bS^{\circ}(B)]$	Standard entropy change for a reaction (ΔS_{rxn}°) can be calculated using tabulated values of absolute entropies (S°) for products and reactants.
14.6	$\Delta S_{rxn}^{\circ} = \Sigma nS^{\circ}(\text{products}) - \Sigma mS^{\circ}(\text{reactants})$	ΔS_{rxn}° is calculated as the sum of absolute entropies for products minus the sum of absolute entropies for reactants. Each species in a chemical equation must be multiplied by its coefficient.
14.7	$\Delta S_{surr} = \dfrac{-\Delta H_{sys}}{T}$	Entropy change in the surroundings (ΔS_{surr}) is calculated as the ratio of the negative of the enthalpy change in the system ($-\Delta H_{sys}$) to absolute temperature (T).
14.8	$\Delta S_{univ} = \Delta S_{sys} + \Delta S_{surr}$	Entropy change in the universe (ΔS_{univ}) is equal to the sum of entropy change for the system (ΔS_{sys}) and entropy change for the surroundings (ΔS_{surr}).
14.9	$G = H - TS$	Gibbs free energy (G) is the difference between enthalpy (H) and the product of absolute temperature and entropy (TS).
14.10	$\Delta G = \Delta H - T\Delta S$	The change in free energy (ΔG) is calculated as the difference between change in enthalpy (ΔH) and the product of absolute temperature and change in entropy ($T\Delta S$).
14.11	$\Delta G_{rxn}^{\circ} = [c\Delta G_{f}^{\circ}(C) + d\Delta G_{f}^{\circ}(D)] - [a\Delta G_{f}^{\circ}(A) + b\Delta G_{f}^{\circ}(B)]$	Standard free-energy change for a reaction (ΔG_{rxn}°) can be calculated using tabulated values of free energies of formation (ΔG_{f}°) for products and reactants.
14.12	$\Delta G_{rxn}^{\circ} = \Sigma n\Delta G_{f}^{\circ}(\text{products}) - \Sigma m\Delta G_{f}^{\circ}(\text{reactants})$	ΔG_{rxn}° is calculated as the sum of free energies of formation for products minus the sum of free energies of formation for reactants. Each species in a chemical equation must be multiplied by its coefficient.

Key Skills

Determining $\Delta G°$

$\Delta G°$ indicates whether a chemical reaction or physical process will proceed spontaneously as written under standard conditions. In later chapters, $\Delta G°$ will be necessary for calculations involving chemical equilibrium [▶▶ Chapters 15–17] and electrochemistry [▶▶ Chapter 18]. Using tabulated $\Delta G_f°$ values, we can calculate the standard free-energy change ($\Delta G°$) using Equation 14.11.

$$\Delta G°_{rxn} = \Sigma n\Delta G_f°(\text{products}) - \Sigma m\Delta G_f°(\text{reactants})$$

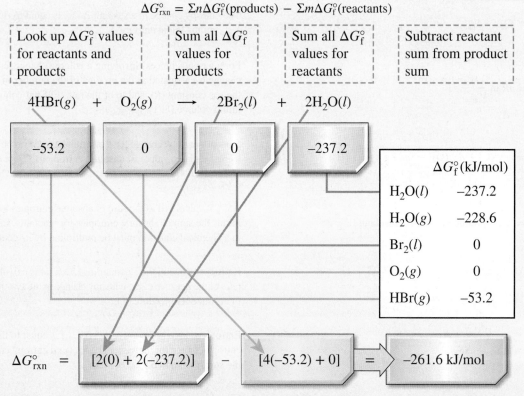

	$\Delta G_f°$(kJ/mol)
$H_2O(l)$	−237.2
$H_2O(g)$	−228.6
$Br_2(l)$	0
$O_2(g)$	0
$HBr(g)$	−53.2

$$\Delta G°_{rxn} = [2(0) + 2(-237.2)] - [4(-53.2) + 0] = -261.6 \text{ kJ/mol}$$

Alternatively, $\Delta G°$ can be calculated using Equation 14.10 ($\Delta G°_{rxn} = \Delta H°_{rxn} - T\Delta S°_{rxn}$), absolute temperature (T), and $\Delta H°/\Delta S°$ values, which themselves typically are calculated from tabulated data. (Remember that the units of $\Delta S°$ must be converted from J/K · mol to kJ/K · mol prior to using Equation 14.10.)

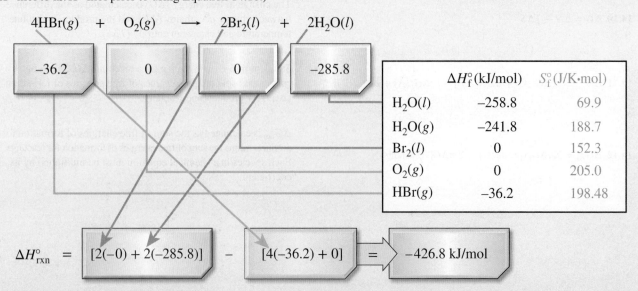

	$\Delta H_f°$(kJ/mol)	$S_f°$(J/K·mol)
$H_2O(l)$	−258.8	69.9
$H_2O(g)$	−241.8	188.7
$Br_2(l)$	0	152.3
$O_2(g)$	0	205.0
$HBr(g)$	−36.2	198.48

$$\Delta H°_{rxn} = [2(-0) + 2(-285.8)] - [4(-36.2) + 0] = -426.8 \text{ kJ/mol}$$

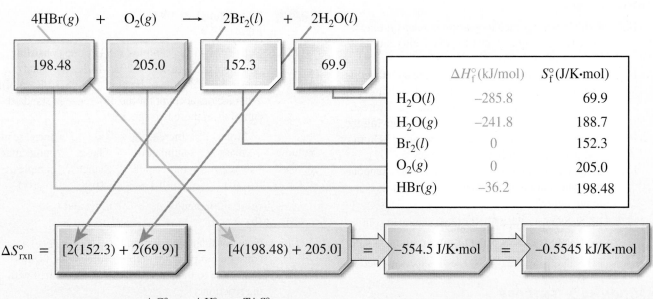

$$4HBr(g) + O_2(g) \longrightarrow 2Br_2(l) + 2H_2O(l)$$

	ΔH_f° (kJ/mol)	S_f° (J/K·mol)
$H_2O(l)$	−285.8	69.9
$H_2O(g)$	−241.8	188.7
$Br_2(l)$	0	152.3
$O_2(g)$	0	205.0
$HBr(g)$	−36.2	198.48

$$\Delta S_{rxn}^\circ = [2(152.3) + 2(69.9)] - [4(198.48) + 205.0] = -554.5 \text{ J/K·mol} = -0.5545 \text{ kJ/K·mol}$$

$$\Delta G_{rxn}^\circ = \Delta H_{rxn}^\circ - T\Delta S_{rxn}^\circ$$
$$= -426.8 \text{ kJ/mol} - (298.15 \text{ K})(-0.5545 \text{ kJ/K} \cdot \text{mol}) = -261.6 \text{ kJ/mol}$$

Key Skills Problems

14.1
Using ΔG_f° values from Appendix 2, calculate the standard free-energy change (ΔG_{rxn}°) of the following reaction at 25.0°C.

$$Mg(OH)_2(s) \longrightarrow MgO(s) + H_2O(l)$$

(a) −35.6 kJ/mol
(b) −1166.2 kJ/mol
(c) +35.6 kJ/mol
(d) +1166.2 kJ/mol
(e) +27.0 kJ/mol

14.2
Calculate ΔG_{rxn}° for the reaction in question 14.1 at 150°C. Use data from Appendix 2 and assume that ΔH_f° and S° values do not change with temperature.
(a) −8098 kJ/mol
(b) −45.1 kJ/mol
(c) +22.9 kJ/mol
(d) +45.1 kJ/mol
(e) −8024 kJ/mol

14.3
Using ΔG_{rxn}° values from Appendix 2, calculate the standard free-energy change of the following reaction at 25.0°C.

$$C_4H_{10}(g) + O_2(g) \longrightarrow CO_2(g) + H_2O(l)$$

(You must first balance the equation.)

(a) +615.9 kJ/mol
(b) −5495.8 kJ/mol
(c) +539.3 kJ/mol
(d) −615.9 kJ/mol
(e) −5511.5 kJ/mol

14.4
Calculate ΔG_{rxn}° for the reaction in question 14.3 at −125.0°C. Use data from Appendix 2 and assume that ΔH_f° and S° values do not change with temperature.
(a) −532.0 kJ/mol
(b) −5626.8 kJ/mol
(c) +536.9 kJ/mol
(d) −5647.0 kJ/mol
(e) −5797.4 kJ/mol

Questions and Problems

SECTION 14.1: SPONTANEOUS PROCESSES

Review Questions

14.1   Explain what is meant by a *spontaneous process*. Give two examples each of spontaneous and nonspontaneous processes.

14.2   Which of the following processes are spontaneous and which are nonspontaneous: (a) dissolving table salt (NaCl) in hot soup, (b) climbing Mt. Everest, (c) spreading fragrance in a room by removing the cap from a perfume bottle, (d) separating helium and neon from a mixture of the gases?

14.3   Which of the following processes are spontaneous and which are nonspontaneous at a given temperature?

(a) $NaNO_3(s) \xrightarrow{H_2O} NaNO_3(aq)$ saturated soln

(b) $NaNO_3(s) \xrightarrow{H_2O} NaNO_3(aq)$ unsaturated soln

(c) $NaNO_3(s) \xrightarrow{H_2O} NaNO_3(aq)$ supersaturated soln

SECTION 14.2: ENTROPY

Review Questions

14.4   Describe what is meant by the term *entropy*. What are the units of entropy?

14.5   What is the relationship between entropy and the number of possible arrangements of molecules in a system?

Conceptual Problems

14.6   Referring to the setup in Figure 14.2, determine the number of possible arrangements, *W,* and calculate the entropy before and after removal of the barrier if the number of molecules is (a) 10, (b) 50, (c) 100.

14.7   In the setup shown, a container is divided into eight cells and contains two molecules. Initially, both molecules are confined to the left side of the container. (a) Determine the number of possible arrangements before and after removal of the central barrier. (b) After the removal of the barrier, how many of the arrangements correspond to the state in which both molecules are in the left side of the container? How many correspond to the state in which both molecules are in the right side of the container? How many correspond to the state in which the molecules are in opposite sides of the container? Calculate the entropy for each state and comment on the most probable state of the system after removal of the barrier.

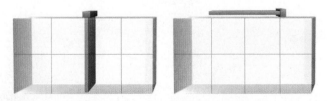

SECTION 14.3: ENTROPY CHANGES IN A SYSTEM

 Visualizing Chemistry
Figure 14.4

VC 14.1   Consider two gas samples at STP: one consisting of a mole of F_2 gas ($S° = 203.34$ J/K · mol) and one consisting of a mole of F gas ($S° = 158.7$ J/K · mol). What factors account for the difference in standard entropies of these two species?

Volume increase	Molar mass increase	Increased number of molecules	Phase change	Increase in molecular complexity
(i)	(ii)	(iii)	(iv)	(iv)

(a) i, ii, iii, and iv   (c) ii, iv, and v

(b) ii and v

VC 14.2   Now consider the reaction $F_2(g) \longrightarrow 2F(g)$ at constant temperature and pressure. What factors contribute to the entropy increase associated with the reaction?

Volume increase	Molar mass increase	Increased number of molecules	Phase change	Increase in molecular complexity
(i)	(ii)	(iii)	(iv)	(iv)

(a) i and iii   (c) i, iv, and v

(b) i, ii, and iii

VC 14.3   Which of the following best describes why entropy always increases with temperature?

(a) As temperature increases, the number of molecules increases.

(b) As temperature increases, energy levels become more closely spaced.

(c) As temperature increases, the molecules become more energetic and can access more energy levels.

VC 14.4   Which of the following best explains why entropy typically increases with molar mass?

(a) As molar mass increases, the number of molecules increases.

(b) As molar mass increases, energy levels become more closely spaced.

(c) As molar mass increases, the molecules become more energetic and can access more energy levels.

Review Questions

14.8   How does the entropy of a system change for each of the following processes?

(a) A solid melts.

(b) A liquid freezes.

(c) A liquid boils.

(d) A vapor is converted to a solid.

(e) A vapor condenses to a liquid.

(f) A solid sublimes.

(g) A solid dissolves in water.

14.9   How does the entropy of a system change for each of the following processes?
(a) Bromine liquid vaporizes.
(b) Water freezes to form ice.
(c) Naphthalene, the key component of mothballs, sublimes.
(d) Sugar crystals form from a supersaturated solution.
(e) A block of lead melts.
(f) Iodine vapor condenses to form solid iodine.
(g) Carbon tetrachloride dissolves in liquid benzene.

14.10   Predict whether the entropy change is positive or negative for each of the following reactions. Give reasons for your predictions.
(a) $2KClO_4(s) \longrightarrow 2KClO_3(s) + O_2(g)$
(b) $H_2O(g) \longrightarrow H_2O(l)$
(c) $2Na(s) + 2H_2O(l) \longrightarrow 2NaOH(aq) + H_2(g)$
(d) $N_2(g) \longrightarrow 2N(g)$

14.11   State whether the sign of the entropy change expected for each of the following processes will be positive or negative, and explain your predictions.
(a) $PCl_3(l) + Cl_2(g) \longrightarrow PCl_5(g)$
(b) $2HgO(s) \longrightarrow 2Hg(l) + O_2(g)$
(c) $H_2(g) \longrightarrow 2H(g)$
(d) $U(s) + 3F_2(g) \longrightarrow UF_6(g)$

Computational Problems

14.12   Calculate ΔS_{sys} for (a) the isothermal expansion of 3.0 moles of an ideal gas from 15.0 L to 20.0 L, (b) the isothermal expansion of 7.5 moles of an ideal gas from 20.0 L to 26.5 L, and (c) the isothermal compression of 2.0 moles of an ideal gas from 75.0 L to 25.0 L.

14.13   Calculate ΔS_{sys} for (a) the isothermal compression of 0.0090 mole of an ideal gas from 152 mL to 80.5 mL, (b) the isothermal compression of 0.045 mole of an ideal gas from 325 mL to 32.5 mL, and (c) the isothermal expansion of 2.83 moles of an ideal gas from 225 L to 385 L.

14.14   Using the data in Appendix 2, calculate the standard entropy changes for the following reactions at 25°C.
(a) $S(rhombic) + O_2(g) \longrightarrow SO_2(g)$
(b) $MgCO_3(s) \longrightarrow MgO(s) + CO_2(g)$
(c) $2C_2H_6(g) + 7O_2(g) \longrightarrow 4CO_2(g) + 6H_2O(l)$

14.15   Using the data in Appendix 2, calculate the standard entropy changes for the following reactions at 25°C.
(a) $H_2(g) + CuO(s) \longrightarrow Cu(s) + H_2O(g)$
(b) $2Al(s) + 3ZnO(s) \longrightarrow Al_2O_3(s) + 3Zn(s)$
(c) $CH_4(g) + 2O_2(g) \longrightarrow CO_2(g) + 2H_2O(l)$

Conceptual Problems

14.16   For each pair of substances listed here, choose the one having the larger standard entropy value at 25°C. The same molar amount is used in the comparison. Explain the basis for your choice.
(a) Li(s) or Li(l), (b) $C_2H_5OH(l)$ or $CH_3OCH_3(l)$ (Hint: Which molecule can hydrogen bond?), (c) Ar(g) or Xe(g), (d) CO(g) or $CO_2(g)$, (e) $O_2(g)$ or $O_3(g)$, (f) $NO_2(g)$ or $N_2O_4(g)$.

14.17   Arrange the following substances (1 mole each) in order of increasing entropy at 25°C: (a) Ne(g),

(b) $SO_2(g)$, (c) Na(s), (d) NaCl(s), (e) $H_2(g)$. Give the reasons for your arrangement.

SECTION 14.4: ENTROPY CHANGES IN THE UNIVERSE

Review Questions

14.18   State the second law of thermodynamics in words, and express it mathematically.

14.19   State the third law of thermodynamics in words, and explain its usefulness in calculating entropy values.

Computational Problems

14.20   Calculate ΔS_{surr} for each of the reactions in Problem 14.14 and determine if each reaction is spontaneous at 25°C.

14.21   Calculate ΔS_{surr} for each of the reactions in Problem 14.15 and determine if each reaction is spontaneous at 25°C.

14.22   Using data from Appendix 2, calculate ΔS_{rxn}° and ΔS_{surr} for each of the reactions in Problem 14.10 and determine if each reaction is spontaneous at 25°C.

14.23   Using data from Appendix 2, calculate ΔS_{rxn}° and ΔS_{surr} for each of the reactions in Problem 14.11 and determine if each reaction is spontaneous at 25°C.

14.24   When a folded protein in solution is heated to a high enough temperature, its polypeptide chain will unfold to become the denatured protein—a process known as "denaturation." The temperature at which most of the protein unfolds is called the "melting" temperature. The melting temperature of a certain protein is found to be 63°C, and the enthalpy of denaturation is 510 kJ/mol. Estimate the entropy of denaturation, assuming that the denaturation is a single-step equilibrium process; that is, folded protein $\rightleftharpoons$ denatured protein. The single polypeptide protein chain has 98 amino acids. Calculate the entropy of denaturation per amino acid.

SECTION 14.5: PREDICTING SPONTANEITY

Review Questions

14.25   Define free energy. What are its units?

14.26   Why is it more convenient to predict the direction of a reaction in terms of ΔG_{sys} instead of ΔS_{univ}? Under what conditions can ΔG_{sys} be used to predict the spontaneity of a reaction?

14.27   What is the significance of the sign of ΔG_{sys}?

14.28   From the following combinations of ΔH and ΔS, predict if a process will be spontaneous at a high or low temperature: (a) both ΔH and ΔS are negative, (b) ΔH is negative and ΔS is positive, (c) both ΔH and ΔS are positive, (d) ΔH is positive and ΔS is negative.

Problems

14.29   Assuming that ΔH and ΔS do not change with temperature, determine ΔG for the denaturation in Problem 14.24 at 20°C.

14.30   Calculate ΔG° for the following reactions at 25°C.
(a) $N_2(g) + O_2(g) \longrightarrow 2NO(g)$
(b) $H_2O(l) \longrightarrow H_2O(g)$
(c) $2C_2H_2(g) + 5O_2(g) \longrightarrow 4CO_2(g) + 2H_2O(l)$
(Hint: Look up the standard free energies of formation of the reactants and products in Appendix 2.)

14.31 Calculate $\Delta G°$ for the following reactions at 25°C.
(a) $2Mg(s) + O_2(g) \longrightarrow 2MgO(s)$
(b) $2SO_2(g) + O_2(g) \longrightarrow 2SO_3(g)$
(c) $2C_2H_6(g) + 7O_2(g) \longrightarrow 4CO_2(g) + 6H_2O(l)$
(See Appendix 2 for thermodynamic data.)

14.32 From the values of ΔH and ΔS, predict which of the following reactions would be spontaneous at 25°C. Reaction A: $\Delta H = 10.5$ kJ/mol, $\Delta S = 30$ J/K · mol; Reaction B: $\Delta H = 1.8$ kJ/mol, $\Delta S = -113$ J/K · mol. If either of the reactions is nonspontaneous at 25°C, at what temperature might it become spontaneous?

14.33 Find the temperatures at which reactions with the following ΔH and ΔS values would become spontaneous.
(a) $\Delta H = -126$ kJ/mol, $\Delta S = 84$ J/K · mol;
(b) $\Delta H = -11.7$ kJ/mol, $\Delta S = -105$ J/K · mol.

14.34 The molar heats of fusion and vaporization of ethanol are 7.61 and 26.0 kJ/mol, respectively. Calculate the molar entropy changes for the solid-liquid and liquid-vapor transitions for ethanol. At 1 atm pressure, ethanol melts at −117.3°C and boils at 78.3°C.

14.35 The molar heats of fusion and vaporization of mercury are 23.4 and 59.0 kJ/mol, respectively. Calculate the molar entropy changes for the solid-liquid and liquid-vapor transitions for mercury. At 1 atm pressure, mercury melts at −38.9°C and boils at 357°C.

14.36 Use the values listed in Appendix 2 to calculate $\Delta G°$ for the following alcohol fermentation.
$$C_6H_{12}O_6(s) \longrightarrow 2C_2H_5OH(l) + 2CO_2(g)$$

14.37 Certain bacteria in the soil obtain the necessary energy for growth by oxidizing nitrites to nitrates.
$$2NO_2^- + O_2 \longrightarrow 2NO_3^-$$
Given that the standard Gibbs free energies of formation of NO_2^- and NO_3^- are −34.6 and −110.5 kJ/mol, respectively, calculate the amount of Gibbs free energy released when 1 mole of NO_2^- is oxidized to 1 mole of NO_3^-.

SECTION 14.6: THERMODYNAMICS IN LIVING SYSTEMS

Review Questions

14.38 What is a coupled reaction? What is its importance in biological reactions?

14.39 What is the role of ATP in biological reactions?

Computational Problem

14.40 Referring to the metabolic process involving glucose in Figure 14.7, calculate the maximum number of moles of ATP that can be synthesized from ADP from the breakdown of 1 mole of glucose.

ADDITIONAL PROBLEMS

14.41 Predict the signs of ΔH, ΔS, and ΔG of the system for the following processes at 1 atm: (a) ammonia melts at −60°C, (b) ammonia melts at −77.7°C, (c) ammonia melts at −100°C. (The normal melting point of ammonia is −77.7°C.)

14.42 A student placed 1 g of each of three compounds A, B, and C in a container and found that after 1 week no change had occurred. Offer some possible

explanations for the fact that no reactions took place. Assume that A, B, and C are totally miscible liquids.

14.43 The enthalpy change in the denaturation of a certain protein is 125 kJ/mol. If the entropy change is 397 J/K · mol, calculate the minimum temperature at which the protein would denature spontaneously.

14.44 Consider the following facts: Water freezes spontaneously at −5°C and 1 atm, and ice has a lower entropy than liquid water. Explain how a spontaneous process can lead to a decrease in entropy.

14.45 Ammonium nitrate (NH_4NO_3) dissolves spontaneously and endothermically in water. What can you deduce about the sign of ΔS for the solution process?

14.46 The standard enthalpy of formation and the standard entropy of gaseous benzene are 82.93 kJ/mol and 269.2 J/K · mol, respectively. Calculate $\Delta H°$, $\Delta S°$, and $\Delta G°$ for the given process at 25°C. Comment on your answers.
$$C_6H_6(l) \longrightarrow C_6H_6(g)$$

14.47 (a) Trouton's rule states that the ratio of the molar heat of vaporization of a liquid (ΔH_{vap}) to its boiling point in kelvins is approximately 90 J/K · mol. Use the following data to show that this is the case and explain why Trouton's rule holds true.

	T_{bp} (°C)	ΔH_{vap} (kJ/mol)
Benzene	80.1	31.0
Hexane	68.7	30.8
Mercury	357	59.0
Toluene	110.6	35.2

(b) Use the values in Table 12.5 to calculate the same ratio for ethanol and water. Explain why Trouton's rule does not apply to these two substances as well as it does to other liquids.

14.48 Referring to Problem 14.47, explain why the ratio is considerably smaller than 90 J/K · mol for liquid HF.

14.49 Predict whether the entropy change is positive or negative for each of these reactions:
(a) $Zn(s) + 2HCl(aq) \rightleftharpoons ZnCl_2(aq) + H_2(g)$
(b) $O(g) + O(g) \rightleftharpoons O_2(g)$
(c) $NH_4NO_3(s) \rightleftharpoons N_2O(g) + 2H_2O(g)$

14.50 A certain reaction is spontaneous at 72°C. If the enthalpy change for the reaction is 19 kJ/mol, what is the *minimum* value of ΔS (in J/K · mol) for the reaction?

14.51 Use the following data to determine the normal boiling point, in kelvins, of mercury. What assumptions must you make to do the calculation?
$Hg(l)$: $\Delta H_f° = 0$ (by definition)
$S° = 77.4$ J/K · mol
$Hg(g)$: $\Delta H_f° = 60.78$ kJ/mol
$S° = 174.7$ J/K · mol

14.52 The reaction $NH_3(g) + HCl(g) \longrightarrow NH_4Cl(s)$ proceeds spontaneously at 25°C even though there is a decrease in entropy in the system (gases are converted to a solid). Explain.

14.53 A certain reaction is known to have a $\Delta G°$ value of −122 kJ/mol. Will the reaction necessarily occur if the reactants are mixed together?

14.54   The molar heat of vaporization of ethanol is 39.3 kJ/mol, and the boiling point of ethanol is 78.3°C. Calculate ΔS for the vaporization of 0.50 mole of ethanol.

14.55   As an approximation, we can assume that proteins exist either in the native (physiologically functioning) state or the denatured state. The standard molar enthalpy and entropy of the denaturation of a certain protein are 512 kJ/mol and 1.60 kJ/K · mol, respectively. Comment on the signs and magnitudes of these quantities, and calculate the temperature at which the denaturation becomes spontaneous.

14.56   When a native protein in solution is heated to a high enough temperature, its polypeptide chain will unfold to become the denatured protein. The temperature at which a large portion of the protein unfolds is called the melting temperature. The melting temperature of a certain protein is found to be 46°C, and the enthalpy of denaturation is 382 kJ/mol. Estimate the entropy of denaturation, assuming that the denaturation is a two-state process; that is, native protein ⟶ denatured protein. The single polypeptide protein chain has 122 amino acids. Calculate the entropy of denaturation per amino acid. Comment on your result.

14.57   A 74.6-g ice cube floats in the Arctic Sea. The pressure and temperature of the system and surroundings are at 1 atm and 0°C, respectively. Calculate ΔS_{sys}°, ΔS_{surr}, and ΔS_{univ} for the melting of the ice cube. What can you conclude about the nature of the process from the value of ΔS_{univ}? (The molar heat of fusion of water is 6.01 kJ/mol.)

14.58   A reaction for which ΔH and ΔS are both negative is
(a) nonspontaneous at all temperatures.
(b) spontaneous at all temperatures.
(c) spontaneous at high temperatures.
(d) spontaneous at low temperatures.
(e) at equilibrium.

14.59   The sublimation of carbon dioxide at −78°C is given by

$$CO_2(s) \longrightarrow CO_2(g) \qquad \Delta H_{sub} = 25.2 \text{ kJ/mol}$$

Calculate ΔS_{sub} when 84.8 g of CO_2 sublimes at this temperature.

14.60   Many hydrocarbons exist as structural isomers, which are compounds that have the same molecular formula but different structures. For example, both butane and isobutane have the same molecular formula of C_4H_{10}

(see Problem 7.42 on page 299). Calculate the mole percent of these molecules in an equilibrium mixture at 25°C, given that the standard free energy of formation of butane is −15.7 kJ/mol and that of isobutane is −18.0 kJ/mol. Does your result support the notion that straight-chain hydrocarbons (i.e., hydrocarbons in which the C atoms are joined along a line) are less stable than branch-chain hydrocarbons?

14.61   Consider the following reaction at 298 K.

$$2H_2(g) + O_2(g) \longrightarrow 2H_2O(l) \quad \Delta H° = -571.6 \text{ kJ/mol}$$

Calculate ΔS_{sys}, ΔS_{surr}, and ΔS_{univ} for the reaction.

14.62   Which of the following is not accompanied by an increase in the entropy of the system: (a) mixing of two gases at the same temperature and pressure, (b) mixing of ethanol and water, (c) discharging a battery, (d) expansion of a gas followed by compression to its original temperature, pressure, and volume?

14.63   Which of the following are not state functions: S, H, q, w, T?

14.64   Give a detailed example of each of the following, with an explanation: (a) a thermodynamically spontaneous process, (b) a process that would violate the first law of thermodynamics, (c) a process that would violate the second law of thermodynamics, (d) an irreversible process, (e) an equilibrium process.

14.65   Hydrogenation reactions (e.g., the process of converting C=C bonds to C—C bonds in the food industry) are facilitated by the use of a transition metal catalyst, such as Ni or Pt. The initial step is the adsorption, or binding, of hydrogen gas onto the metal surface. Predict the signs of ΔH, ΔS, and ΔG when hydrogen gas is adsorbed onto the surface of Ni metal.

14.66   At 0 K, the entropy of carbon monoxide crystal is not zero but has a value of 4.2 J/K · mol, called the residual entropy. According to the third law of thermodynamics, this means that the crystal does not have a perfect arrangement of the CO molecules. (a) What would be the residual entropy if the arrangement were totally random? (b) Comment on the difference between the result in part (a) and 4.2 J/K · mol. (*Hint:* Assume that each CO molecule has two choices for orientation, and use Equation 14.1 to calculate the residual entropy.)

14.67   Which of the following thermodynamic functions are associated only with the first law of thermodynamics: S, U, G, and H?

Answers to In-Chapter Materials

PRACTICE PROBLEMS

14.1A 0.34 J/K. **14.1B** $\frac{1}{5}$. **14.2A** (a) 173.6 J/K · mol, (b) −139.8 J/K · mol, (c) 215.3 J/K · mol. **14.2B** (a) $S°[K(l)] = 71.5$ J/K · mol, (b) $S°[S_2Cl_2(g)]$ = 331.5 J/K · mol, (c) $S°[MgF_2(s)] = 57.24$ J/K · mol. **14.3A** (a) negative, (b) negative, (c) positive. **14.3B** The sign of $\Delta H°$ for both dissolution processes is negative. Something must favor spontaneity; if not entropy change, then enthalpy change. Because these processes both involve decreases in the system's entropy, they must be exothermic, or they could not be spontaneous. **14.4A** (a) $\Delta S_{univ} = -27.2$ J/K · mol, nonspontaneous, (b) $\Delta S_{univ} = -28.1$ J/K · mol, nonspontaneous, (c) $\Delta S_{univ} = 0$ J/K · mol,

equilibrium. **14.4B** (a) $\Delta S_{univ} = 5.2$ J/K · mol, spontaneous, (b) 346°C, (c) 58°C. **14.5A** 3728°C. **14.5B** 108 J/K · mol. **14.6A** (a) −106 kJ/mol, (b) −2935 kJ/mol. **14.6B** (a) $\Delta G_f°$ [Li$_2$O(s)] = −561.2 kJ/mol, (b) $\Delta G_f°$ [NaI(s)] = −286.1 kJ/mol. **14.7A** $\Delta S_{fus} = 16$ J/K · mol, $\Delta S_{vap} = 72$ J/K · mol. **14.7B** (a) $\Delta H_{fus}° = 2.41$ kJ/mol, $\Delta S_{fus}° = 6.51$ J/K · mol, $T_{melting} = 97$°C, (b) $\Delta H_{vap}° = 105.3$ kJ/mol, (c) $\Delta S_{vap}° = 96.1$ J/K · mol, $T_{boiling} = 823$°C.

SECTION REVIEW

14.3.1 a, e. **14.3.2** b. **14.3.3** e. **14.4.1** b. **14.4.2** e. **14.4.3** b. **14.4.4** a. **14.4.5** e. **14.5.1** d. **14.5.2** b. **14.5.3** d.

Chapter 15

Chemical Equilibrium

© Richard Megna/Fundamental Photographs

BIOMAGNIFICATION refers to the accumulation of the greatest levels of toxins in those species that are highest in the food chain. Most often, this involves substances that are more soluble in fatty tissues than in water—making them less likely to be excreted. To study the phenomenon of biomagnification, chemists have developed a laboratory equilibrium system consisting of octanol and water, which are mutually immiscible. In the simplest version of this system, the substance of interest is shaken in a flask containing both octanol and water—and its resulting concentration in each solvent is measured. Using the ratio of octanol solubility to water solubility, a parameter known as the *partition coefficient,* scientists can model the selective absorption of pesticides and other toxins into the fatty tissues of birds, fish, and mammals.

- Gibbs free energy [◀◀ Section 14.5]
- The quadratic equation [▶▶ Appendix 1]

15.1 THE CONCEPT OF EQUILIBRIUM

Up until now, we have treated chemical equations as though they go to completion; that is, we start with only *reactants* and end up with only *products*. In fact, this is not the case with most chemical reactions. Instead, if we start with only reactants, the reaction will proceed, causing reactant concentrations to decrease (as reactants are consumed) and product concentrations to increase (as products are produced). Eventually, though, the concentrations of reactants and products will stop changing—without all of the reactants having been converted to products. The reaction will appear to have stopped, and we will be left with a *mixture* of reactants and products.

As an example, consider the decomposition of dinitrogen tetroxide (N_2O_4) to yield nitrogen dioxide (NO_2).

$$N_2O_4(g) \rightleftharpoons 2NO_2(g)$$

N_2O_4 is a colorless gas, whereas NO_2 is brown. If we begin by placing a sample of pure N_2O_4 in an evacuated flask, the contents of the flask change from colorless to brown as the decomposition produces NO_2 (Figure 15.1). At first, the brown color intensifies as the concentration of NO_2 increases. After some time has passed, though, the intensity of the brown color stops increasing, indicating that the concentration of NO_2 has stopped increasing.

Although the concentration of the product (and of the reactant) has stopped changing, the reaction has not actually stopped. Like most chemical reactions, the decomposition of N_2O_4 is a *reversible process,* meaning that the products of the

Student Annotation: NO_2 is the cause of the brown appearance of some polluted air.

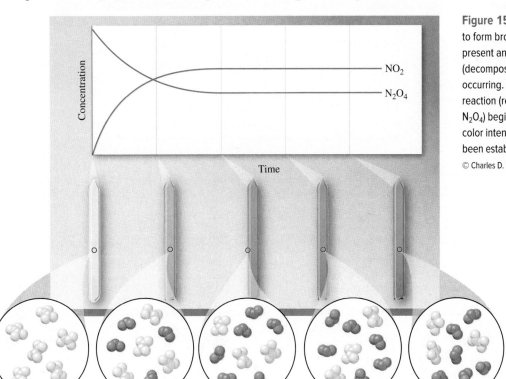

Figure 15.1 Reaction of colorless N_2O_4 to form brown NO_2. Initially, only N_2O_4 is present and only the forward reaction (decomposition of N_2O_4 to give NO_2) is occurring. As NO_2 forms, the reverse reaction (recombination of NO_2 to give N_2O_4) begins to occur. Initially, the brown color intensifies. When equilibrium has been established, the color stops changing.
© Charles D. Winters

reaction can react to form *reactants.* Thus, as the decomposition of N_2O_4 continues, the reverse reaction, the combination of NO_2 molecules to produce N_2O_4, is also occurring. Eventually, the concentrations of both species reach levels where they remain constant because the two processes are occurring at the same rate—and the system is said to have achieved *dynamic equilibrium* or simply **equilibrium.**

In the experiment shown in Figure 15.1, initially:

- N_2O_4 concentration is high.
- NO_2 concentration is zero.

As the reaction proceeds:

- N_2O_4 concentration falls.
- NO_2 concentration rises.

We could equally well have started the N_2O_4/NO_2 experiment with pure NO_2 in a flask. Figure 15.2 shows how the brown color, initially intense, fades as NO_2 combines to form N_2O_4. As before, the intensity stops changing after a period of time. When we start the experiment with pure NO_2, initially:

- NO_2 concentration is high.
- N_2O_4 concentration is zero.

As the reaction proceeds:

- NO_2 concentration falls.
- N_2O_4 concentration rises.

We could also conduct this kind of experiment starting with a mixture of NO_2 and N_2O_4. Again, the forward and reverse reactions would occur, and equilibrium would be reached when the concentrations of both species become constant.

Some important things to remember about equilibrium are:

- Equilibrium is a *dynamic* state—both forward and reverse reactions continue to occur, although there is no net change in reactant and product concentrations over time.
- Equilibrium can be established starting with only *reactants,* with only *products,* or with any *mixture* of reactants and products.

Figure 15.2 N_2O_4/NO_2 equilibrium starting with NO_2. Initially, only NO_2 is present and only the reverse reaction (recombination of NO_2 to give N_2O_4) is occurring. As N_2O_4 forms, the forward reaction (decomposition of N_2O_4) begins to occur. The brown color continues to fade until the forward and reverse reactions are occurring at the same rate.

© Charles D. Winters

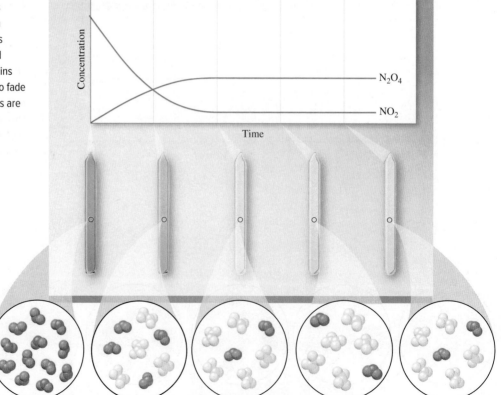

15.2 THE EQUILIBRIUM CONSTANT

To gain a quantitative understanding of chemical equilibrium, it is important for you to first understand the concept of *reaction quotient*. The **reaction quotient (Q_c)** is simply a fraction that we construct using a balanced chemical equation. We multiply the *product* concentrations in the numerator and *reactant* concentrations in the denominator—with *each* concentration raised to a power equal to its stoichiometric coefficient. Consider, for example, the general balanced equation with reactants A and B, and products C and D—with coefficients *a, b, c,* and *d,* respectively.

$$aA + bB \rightleftharpoons cC + dD$$

The reaction quotient, Q_c, for this balanced equation is expressed as Equation 15.1,

$$Q_c = \frac{[C]^c[D]^d}{[A]^a[B]^b} \qquad \textbf{Equation 15.1}$$

where the subscript "c" denotes concentration. Using this method, we can write a reaction quotient for any balanced chemical equation.

Worked Example 15.1 lets you practice writing reaction quotients for a variety of balanced equations.

Worked Example 15.1

Write reaction quotients for the following reactions:

(a) $N_2(g) + 3H_2(g) \rightleftharpoons 2NH_3(g)$

(b) $H_2(g) + I_2(g) \rightleftharpoons 2HI(g)$

(c) $Ag^+(aq) + 2NH_3(aq) \rightleftharpoons Ag(NH_3)_2^+(aq)$

(d) $2O_3(g) \rightleftharpoons 3O_2(g)$

(e) $Cd^{2+}(aq) + 4Br^-(aq) \rightleftharpoons CdBr_4^{2-}(aq)$

(f) $2NO(g) + O_2(g) \rightleftharpoons 2NO_2(g)$

Strategy Use the law of mass action to write reaction quotients.

Setup The reaction quotient for each reaction has the form of the concentrations of *products* over the concentrations of *reactants*, each raised to a power equal to its stoichiometric coefficient in the balanced chemical equation.

Solution

(a) $Q_c = \dfrac{[NH_3]^2}{[N_2][H_2]^3}$

(c) $Q_c = \dfrac{[Ag(NH_3)_2^+]}{[Ag^+][NH_3]^2}$

(e) $Q_c = \dfrac{[CdBr_4^{2-}]}{[Cd^{2+}][Br^-]^4}$

(b) $Q_c = \dfrac{[HI]^2}{[H_2][I_2]}$

(d) $Q_c = \dfrac{[O_2]^3}{[O_3]^2}$

(f) $Q_c = \dfrac{[NO_2]^2}{[NO]^2[O_2]}$

Student Annotation: The molar concentrations in a reaction quotient may be concentrations of *gases* or *aqueous* species.

Think About It
With practice, writing reaction quotients becomes second nature. Without sufficient practice, it will seem inordinately difficult. It is important that you become proficient at this. It is very often the first step in solving equilibrium problems.

Practice Problem (A)TTEMPT Write the reaction quotient for each of the following reactions.

(a) $2N_2O(g) \rightleftharpoons 2N_2(g) + O_2(g)$

(b) $2NOBr(g) \rightleftharpoons 2NO(g) + Br_2(g)$

(c) $HF(aq) \rightleftharpoons H^+(aq) + F^-(aq)$

(d) $CO(g) + H_2O(g) \rightleftharpoons CO_2(g) + H_2(g)$

(e) $CH_4(g) + 2H_2S(g) \rightleftharpoons CS_2(g) + 4H_2(g)$

(f) $H_2C_2O_4(aq) \rightleftharpoons 2H^+(aq) + C_2O_4^{2-}(aq)$

Practice Problem (B)UILD Write the equation for the equilibrium that corresponds to each of the following reaction quotients.

(a) $Q_c = \dfrac{[HCl]^2}{[H_2][Cl_2]}$

(c) $Q_c = \dfrac{[Cr(OH)_4^-]}{[Cr^{3+}][OH^-]^4}$

(e) $Q_c = \dfrac{[H^+][HSO_3^-]}{[H_2SO_3]}$

(b) $Q_c = \dfrac{[HF]}{[H^+][F^-]}$

(d) $Q_c = \dfrac{[H^+][ClO^-]}{[HClO]}$

(f) $Q_c = \dfrac{[NOBr]^2}{[NO]^2[Br_2]}$

(Continued on next page)

Practice Problem **C**ONCEPTUALIZE In principle, in the reaction of A and B to form C, A(g) + B(g) $\rightleftharpoons$ C(g), equilibrium can be established by starting with just a mixture of A and B, starting with just C, or starting with a mixture of A, B, and C. Each of the graphs here depicts the change in the value of the reaction quotient (Q) as equilibrium is established. (The x axis is time.) Identify which graph corresponds to each of the described starting conditions.

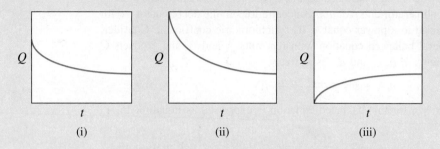

(i)   (ii)   (iii)

In the mid-nineteenth century, Cato Guldberg[1] and Peter Waage[2] studied the equilibrium mixtures of a wide variety of chemical reactions. They made countless observations and found that at constant temperature, the reaction quotient of an equilibrium mixture of reactants and products always has the same value—regardless of the initial concentrations. For a system at equilibrium, the reaction quotient is equal to the *equilibrium constant* (K_c).

Student Annotation: Note that Equation 15.2 is nearly identical to Equation 15.1. The only difference is that Equation 15.2 refers specifically to a system at equilibrium—one in which reactant and product concentrations are constant.

Equation 15.2

$$Q_c = \frac{[C]^c[D]^d}{[A]^a[B]^b} = K_c$$

Equation 15.2 is known as the *law of mass action.* Specifically, for the previous reaction at equilibrium, we write

$$K_c = \frac{[C]^c[D]^d}{[A]^a[B]^b}$$

Student Annotation: The *equilibrium expression* is the equation:

equilibrium expression

$$K_c = \frac{[NO_2]^2_{eq}}{[N_2O_4]_{eq}}$$

equilibrium constant

The equilibrium expression enables us to calculate the equilibrium constant which is simply K_c.

and call it the *equilibrium expression*—a term we will use often. Returning to the N_2O_4/NO_2 system illustrated in Figure 15.3, the equilibrium expression is derived from the balanced chemical equation as

$$K_c = \frac{\left[NO_2\right]^2}{\left[N_2O_4\right]} \qquad N_2O_4(g) \rightleftharpoons 2NO_2(g)$$

Student Hot Spot

Student data indicate you may struggle with reaction quotients. Access the SmartBook to view additional Learning Resources on this topic.

In this equation the coefficient of N_2O_4 is 1, which generally is not written either as a coefficient or as an exponent.

Calculating Equilibrium Constants

Table 15.1 lists the starting and equilibrium concentrations of N_2O_4 and NO_2 in a series of experiments carried out at 25°C. Using the equilibrium concentrations from each of the experiments in the table, we do indeed get a constant value for the reaction coefficient—within the limits of experimental error. (The average value is 4.63×10^{-3}.) Therefore, the equilibrium constant (K_c) for this reaction at 25°C is 4.63×10^{-3}.

---

[1]Cato Maximilian Guldberg (1836–1902), a Norwegian chemist and mathematician, conducted research primarily in the field of thermodynamics.

[2]Peter Waage (1833–1900), a Norwegian chemist, who—like his co-worker, Guldberg—did research in thermodynamics.

TABLE 15.1	Initial and Equilibrium Concentrations of N_2O_4 and NO_2 at 25°C				
	Initial concentrations (M)		**Equilibrium concentrations (M)**		
Experiment	$[N_2O_4]_i$	$[NO_2]_i$	$[N_2O_4]$	$[NO_2]$	$\dfrac{[NO_2]^2}{[N_2O_4]}$
1	0.670	0.00	0.643	0.0547	4.65×10^{-3}
2	0.446	0.0500	0.448	0.0457	4.66×10^{-3}
3	0.500	0.0300	0.491	0.0475	4.60×10^{-3}
4	0.600	0.0400	0.594	0.0523	4.60×10^{-3}
5	0.000	0.200	0.0898	0.0204	4.63×10^{-3}

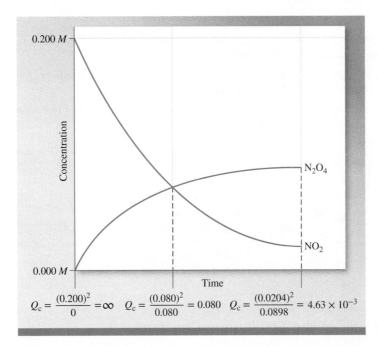

Figure 15.3 The value of the reaction quotient, Q_c, changes as the reaction (experiment 5 from Table 15.1) progresses. When the system reaches equilibrium, Q_c is equal to the equilibrium constant.

$$Q_c = \frac{(0.200)^2}{0} = \infty \qquad Q_c = \frac{(0.080)^2}{0.080} = 0.080 \qquad Q_c = \frac{(0.0204)^2}{0.0898} = 4.63 \times 10^{-3}$$

Like reaction quotients, equilibrium expressions can be written for any reaction for which we know the balanced chemical equation. Further, knowing the equilibrium expression for a reaction, we can use equilibrium concentrations to calculate the value of the corresponding equilibrium constant.

Worked Example 15.2 shows how to use an equilibrium expression and equilibrium concentrations to calculate the value of K_c.

Worked Example 15.2

Carbonyl chloride ($COCl_2$), also called phosgene, is a highly poisonous gas that was used on the battlefield in World War I. It is produced by the reaction of carbon monoxide with chlorine gas.

$$CO(g) + Cl_2(g) \rightleftharpoons COCl_2(g)$$

In an experiment conducted at 74°C, the equilibrium concentrations of the species involved in the reaction were as follows: $[CO] = 1.2 \times 10^{-2}$ M, $[Cl_2] = 0.054$ M, and $[COCl_2] = 0.14$ M. (a) Write the equilibrium expression, and (b) determine the value of the equilibrium constant for this reaction at 74°C.

Strategy Use the law of mass action to write the equilibrium expression and plug in the equilibrium concentrations of all three species to evaluate K_c.

(Continued on next page)

Setup The equilibrium expression has the form of concentrations of products over concentrations of reactants, each raised to the appropriate power—in the case of this reaction, all the coefficients are 1, so all the powers will be 1.

Solution

(a) $K_c = \dfrac{[COCl_2]}{[CO][Cl_2]}$ (b) $K_c = \dfrac{(0.14)}{(1.2 \times 10^{-2})(0.054)} = 216$ or 2.2×10^2

K_c for this reaction at 74°C is 2.2×10^2.

Think About It

When putting the equilibrium concentrations into the equilibrium expression, we leave out the units. It is common practice to express equilibrium constants without units.

Practice Problem **A**TTEMPT In an analysis of the following reaction at 100°C,

$$Br_2(g) + Cl_2(g) \rightleftharpoons 2BrCl(g)$$

the equilibrium concentrations were found to be $[Br_2] = 2.3 \times 10^{-3}$ M, $[Cl_2] = 1.2 \times 10^{-2}$ M, and $[BrCl] = 1.4 \times 10^{-2}$ M. Write the equilibrium expression, and calculate the equilibrium constant for this reaction at 100°C.

Practice Problem **B**UILD In another analysis at 100°C involving the same reaction, the equilibrium concentrations of the reactants were found to be $[Br_2] = 4.1 \times 10^{-3}$ M and $[Cl_2] = 8.3 \times 10^{-3}$ M. Determine the value of $[BrCl]$.

Practice Problem **C**ONCEPTUALIZE Consider the reaction 2A $\rightleftharpoons$ B. The diagram shown on the right represents a system at equilibrium where A = ⚪ and B = ⚫. Which of the following diagrams [(i)–(iv)] also represents a system at equilibrium? Select all that apply.

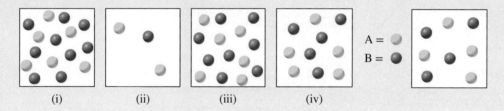

(i)   (ii)   (iii)   (iv)

A = ⚪
B = ⚫

Magnitude of the Equilibrium Constant

One of the things the equilibrium constant tells us is the extent to which a reaction will proceed at a particular temperature if we combine stoichiometric amounts of reactants. To illustrate this, let's consider again the general reaction

$$A + B \rightleftharpoons C$$

If we combine stoichiometric amounts of reactants, 1 mol of A with 1 mol of B, three outcomes are possible:

1. The reaction will go essentially to completion, and the equilibrium mixture will consist predominantly of C (product).
2. The reaction will not occur to any significant degree, and the equilibrium mixture will consist predominantly of A and B (reactants).
3. The reaction will proceed to a significant degree, but will not go to completion; and the equilibrium mixture will contain comparable amounts of A, B, and C.

When the magnitude of K_c is very large, we expect the first outcome. The formation of the $Ag(NH_3)_2^+$ ion is an example of this possibility.

$$Ag^+(aq) + 2NH_3(aq) \rightleftharpoons Ag(NH_3)_2^+(aq) \qquad K_c = 1.5 \times 10^7 \text{ (at 25°C)}$$

If we were to combine aqueous Ag^+ and aqueous NH_3 in a mole ratio of 1:2, the resulting equilibrium mixture would contain mostly $Ag(NH_3)_2^+$ with only very small amounts of reactants remaining. A reaction with a very large equilibrium constant is sometimes said to "lie to the right" or to "favor products."

When the magnitude of K_c is very small, we expect the second outcome. The chemical combination of nitrogen and oxygen gases to give nitrogen monoxide is an example of this possibility.

$$N_2(g) + O_2(g) \rightleftharpoons 2NO(g) \qquad K_c = 4.3 \times 10^{-25} \text{ (at } 25°C\text{)}$$

Nitrogen and oxygen gases do not react to any significant extent at room temperature. If we were to place a mixture of N_2 and O_2 in an evacuated flask and allow the system to reach equilibrium, the resulting mixture would be predominantly N_2 and O_2 with only a *very* small amount of NO. A reaction with a very small equilibrium constant is said to "lie to the left" or to "favor reactants."

The terms *very large* and *very small* are somewhat arbitrary when applied to equilibrium constants. Generally speaking, an equilibrium constant with a magnitude greater than about 1×10^2 can be considered large; one with a magnitude smaller than about 1×10^{-2} can be considered small.

An equilibrium constant that falls between 1×10^2 and 1×10^{-2} indicates that neither products nor reactants are strongly favored. In this case, a system at equilibrium will contain a *mixture* of reactants and products, the exact composition of which will depend on the individual reaction and its stoichiometry.

Unlike the equilibrium constant (K_c), the value of the reaction quotient (Q_c) *changes* as a reaction takes place—because the concentrations of reactants and products change as a reaction progresses (Figure 15.3). Q_c is equal to K_c only when a system is at equilibrium. However, we often find it useful to calculate the value of Q_c for a system that is not at equilibrium (Equation 15.1).

In case it is not already apparent, our discussion of equilibrium is essentially another way of looking at thermodynamics. Saying that a reaction "lies to the right" is the same as saying that it is *spontaneous as written*. A reaction that "lies to the left" is *nonspontaneous* as written. The relationships between reaction quotients, equilibrium constants, and free energy will be discussed in detail in Section 15.4.

Section 15.2 Review

The Equilibrium Constant

15.2.1 Select the equilibrium expression for the reaction

$$2CO(g) + O_2(g) \rightleftharpoons 2CO_2(g)$$

(a) $K_c = \dfrac{2[CO_2]^2}{2[CO]^2[O_2]}$

(d) $K_c = \dfrac{2[CO_2]^2}{[2CO]^2[O_2]}$

(b) $K_c = \dfrac{(2[CO_2])^2}{(2[CO])^2[O_2]}$

(e) $K_c = \dfrac{[CO_2]^2}{[CO]^2[O_2]}$

(c) $K_c = \dfrac{2[CO_2]}{2[CO][O_2]}$

15.2.2 Determine the value of the equilibrium constant, K_c, for the reaction

$$A + 2B \rightleftharpoons C + D$$

if the equilibrium concentrations are [A] = 0.0115 M, [B] = 0.0253 M, [C] = 0.109 M, and [D] = 0.0110 M.

(a) 163

(d) 6.14×10^{-3}

(b) 4.12

(e) 0.243

(c) 2.06

The equilibria we have discussed so far have all been *homogeneous;* that is, the reactants and products have all existed in the same phase—either gaseous or aqueous. In these cases, the equilibrium expression consists of writing the product of the product concentrations at equilibrium over the product of the reactant concentrations at equilibrium, with each concentration raised to a power equal to its stoichiometric coefficient in the balanced chemical equation (Equation 15.1). When the species in a reversible chemical reaction are not all in the same phase, the equilibrium is *heterogeneous.*

Heterogeneous Equilibria

Writing equilibrium expressions for heterogeneous equilibria is also straightforward, but it is slightly different from what we have done so far for homogeneous equilibria. For example, carbon dioxide can combine with elemental carbon to produce carbon monoxide.

$$CO_2(g) + C(s) \rightleftharpoons 2CO(g)$$

The two gases and one solid constitute two separate phases. If we were to write the equilibrium expression for this reaction as we have done previously for homogeneous reactions, including all product concentrations in the numerator and all reactant concentrations in the denominator, we would have

$$K_c^* = \frac{[CO]^2}{[CO_2][C]}$$

(K_c^* is superscripted with an asterisk to distinguish it from the equilibrium constant that we are about to derive.) The "concentration" of a solid, however, is a constant. If we were to double the number of moles of elemental carbon (C) in the preceding reaction, we would also double its volume. The ratio of moles to volume, which is how we define the concentration, remains the same. Because the concentration of solid carbon is a constant, it is incorporated into the value of the equilibrium constant and does not appear explicitly in the equilibrium expression.

$$K_c^* \times [C] = \frac{[CO]^2}{[CO_2]}$$

The product of K_c^* and $[C]_{eq}$ gives the *real* equilibrium constant for this reaction. The corresponding equilibrium expression is

$$K_c = \frac{[CO]^2}{[CO_2]}$$

The same argument applies to the concentrations of pure liquids in heterogeneous equilibria. Only gaseous species and aqueous species appear in equilibrium expressions.

Worked Example 15.3 lets you practice writing equilibrium expressions for heterogeneous equilibria.

Worked Example 15.3

Write equilibrium expressions for each of the following reactions.

(a) $CaCO_3(s) \rightleftharpoons CaO(s) + CO_2(g)$ (c) $2Fe(s) + 3H_2O(l) \rightleftharpoons Fe_2O_3(s) + 2H_2(g)$

(b) $Hg(l) + Hg^{2+}(aq) \rightleftharpoons Hg_2^{2+}(aq)$ (d) $O_2(g) + 2H_2(g) \rightleftharpoons 2H_2O(l)$

Strategy Use the law of mass action to write the equilibrium expression for each reaction. Only gases and aqueous species appear in the expression.

Setup

(a) Only CO_2 will appear in the expression.   (c) Only H_2 will appear in the expression.

(b) Hg^{2+} and Hg_2^{2+} will appear in the expression.   (d) O_2 and H_2 will appear in the expression.

Solution

(a) $K_c = [CO_2]$

(b) $K_c = \dfrac{[Hg_2^{2+}]}{[Hg^{2+}]}$

(c) $K_c = [H_2]^2$

(d) $K_c = \dfrac{1}{[O_2][H_2]^2}$

Think About It

Like writing equilibrium expressions for homogeneous equilibria, writing equilibrium expressions for heterogeneous equilibria becomes second nature if you practice. The importance of developing this skill now cannot be overstated. Your ability to understand the principles and to solve many of the problems in this and Chapters 16 to 18 depends on your ability to write equilibrium expressions *correctly* and *easily*.

Practice Problem **A**TTEMPT Write equilibrium expressions for each of the following reactions.

(a) $SiCl_4(g) + 2H_2(g) \rightleftharpoons Si(s) + 4HCl(g)$

(b) $Hg^{2+}(aq) + 2Cl^-(aq) \rightleftharpoons HgCl_2(s)$

(c) $Ni(s) + 4CO(g) \rightleftharpoons Ni(CO)_4(g)$

(d) $Zn(s) + Fe^{2+}(aq) \rightleftharpoons Zn^{2+}(aq) + Fe(s)$

Practice Problem **B**UILD Which of the following equilibrium expressions corresponds to a heterogeneous equilibrium? How can you tell?

(a) $K_c = [NH_3][HCl]$

(b) $K_c = \dfrac{[H^+][C_2H_3O_2^-]}{[HC_2H_3O_2]}$

(c) $K_c = \dfrac{[Ag(NH_3)_2^+][Cl^-]}{[NH_3]^2}$

(d) $K_c = [Ba^{2+}][F^-]^2$

Practice Problem **C**ONCEPTUALIZE Consider the reaction $A(s) + B(g) \rightleftharpoons C(s)$. Which of the following diagrams [(i)–(iv)] could represent a system at equilibrium? Select all that apply. A = ⚪, B = ⚪, and C = ⚫.

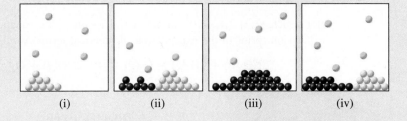

(i)	(ii)	(iii)	(iv)

Manipulating Equilibrium Expressions

In our study of enthalpy, we learned that it is possible to manipulate chemical equations to solve thermochemistry problems [◄◄ Section 10.3]. Recall that when we made a change in a thermochemical equation, we had to make the corresponding change in the ΔH of the reaction. When we reversed a reaction, for example, we had to change the sign of its ΔH. When we change something about how an equilibrium reaction is expressed, we must also make the appropriate changes to the equilibrium expression and the equilibrium constant. Consider the reaction of nitric oxide and oxygen to produce nitrogen dioxide.

$$2NO(g) + O_2(g) \rightleftharpoons 2NO_2(g)$$

The equilibrium expression is

$$K_c = \dfrac{[NO_2]^2}{[NO]^2[O_2]}$$

and the equilibrium constant at 500 K is 6.9×10^5. If we were to reverse this equation, writing the equation for the decomposition of NO_2 to produce NO and O_2,

$$2NO_2(g) \rightleftharpoons 2NO(g) + O_2(g)$$

the new equilibrium expression would be the reciprocal of the original equilibrium expression.

$$K_c' = \frac{[NO]^2[O_2]}{[NO_2]^2}$$

Because the equilibrium expression is the reciprocal of the original, the equilibrium constant is also the reciprocal of the original. At 500 K, therefore, the equilibrium constant for the new equation is $1/(6.9 \times 10^5)$ or 1.5×10^{-6}.

Alternatively, if instead of reversing the original equation, we were to multiply it by 2,

$$4NO(g) + 2O_2(g) \rightleftharpoons 4NO_2(g)$$

the new equilibrium expression would be

$$K_c'' = \frac{[NO_2]^4}{[NO]^4[O_2]^2}$$

which is the original equilibrium expression squared. Again, because the new equilibrium expression is the square of the original, the new equilibrium constant is the square of the original: $K_c = (6.9 \times 10^5)^2$ or 4.8×10^{11}.

Finally, if we were to add the reverse of the original equation

$$2NO_2(g) \rightleftharpoons 2NO(g) + O_2(g)$$

(for which $K_c = 1.6 \times 10^{-6}$ at 500 K) to the equation

$$2H_2(g) + O_2(g) \rightleftharpoons 2H_2O(g)$$

(for which $K_c = 2.4 \times 10^{47}$ at 500 K), the sum of the equations and the equilibrium expression (prior to the cancellation of identical terms) would be

$$2NO_2(g) + 2H_2(g) + O_2(g) \rightleftharpoons 2NO(g) + O_2(g) + 2H_2O(g)$$

and

$$K_c = \frac{[NO]^2[O_2][H_2O]^2}{[NO_2]^2[H_2]^2[O_2]}$$

It may be easier to see before canceling identical terms that for the *sum* of two equations, the equilibrium expression is the *product* of the two corresponding equilibrium expressions.

$$\frac{[NO]^2[O_2][H_2O]^2}{[NO_2]^2[H_2]^2[O_2]} = \frac{[NO]^2[\cancel{O_2}]}{[NO_2]^2} \times \frac{[H_2O]^2}{[H_2]^2[\cancel{O_2}]}$$

Canceling identical terms on the left and right sides of the equation, and on the top and bottom of the equilibrium expression, gives

$$2NO_2(g) + 2H_2(g) \rightleftharpoons 2NO(g) + 2H_2O(g) \qquad \text{and} \qquad K_c = \frac{[NO]^2[H_2O]^2}{[NO_2]^2[H_2]^2}$$

And, because the new equilibrium expression is the product of the individual expressions, the new equilibrium constant is the product of the individual constants. Therefore, for the reaction

$$2NO_2(g) + 2H_2(g) \rightleftharpoons 2NO(g) + 2H_2O(g)$$

at 500 K, $K_c = (1.6 \times 10^{-6})(2.4 \times 10^{47}) = 3.8 \times 10^{41}$.

Using hypothetical equilibria, Table 15.2 summarizes the various ways that chemical equations can be manipulated and the corresponding changes that must be made to the equilibrium expression and equilibrium constant.

TABLE 15.2	Manipulation of Equilibrium Constant Expressions

$$A(g) + B(g) \rightleftharpoons 2C(g) \qquad K_{c_1} = 4.39 \times 10^{-3}$$

$$2C(g) \rightleftharpoons D(g) + E(g) \qquad K_{c_2} = 1.15 \times 10^4$$

Equation	Equilibrium expression	Relationship to original K_c	Equilibrium constant
$2C(g) \rightleftharpoons A(g) + B(g)$ Original equation is reversed.	$K'_{c_1} = \dfrac{[A][B]}{[C]^2}$	$\dfrac{1}{K_{c_1}}$	2.28×10^2 New constant is the reciprocal of the original.
$2A(g) + 2B(g) \rightleftharpoons 4C(g)$ Original equation is multiplied by a number.	$K''_{c_1} = \dfrac{[C]^4}{[A]^2[B]^2}$	$(K_{c_1})^2$	1.93×10^{-5} New constant is the original raised to the same number.
$\frac{1}{2}A(g) + \frac{1}{2}B(g) \rightleftharpoons C(g)$ Original equation is divided by 2.	$K'''_{c_1} = \dfrac{[C]}{[A]^{1/2}[B]^{1/2}}$	$\sqrt{K_{c_1}}$	6.63×10^{-2} New constant is the square root of the original.
$A(g) + B(g) \rightleftharpoons D(g) + E(g)$ Two equations are added.	$K_{c_3} = \dfrac{[D][E]}{[A][B]}$	$K_{c_1} \times K_{c_2}$	50.5 New constant is the product of the two original constants.

*Temperature is the same for both reactions.

Worked Example 15.4 shows how to manipulate chemical equations and make the corresponding changes to their equilibrium constants.

Worked Example 15.4

The following reactions have the indicated equilibrium constants at 100°C.

$$(1) \qquad 2NOBr(g) \rightleftharpoons 2NO(g) + Br_2(g) \qquad K_c = 0.014$$

$$(2) \qquad Br_2(g) + Cl_2(g) \rightleftharpoons 2BrCl(g) \qquad K_c = 7.2$$

Determine the value of K_c for the following reactions at 100°C.

(a) $2NO(g) + Br_2(g) \rightleftharpoons 2NOBr(g)$ (d) $2NOBr(g) + Cl_2(g) \rightleftharpoons 2NO(g) + 2BrCl(g)$

(b) $4NOBr(g) \rightleftharpoons 4NO(g) + 2Br_2(g)$ (e) $NO(g) + BrCl(g) \rightleftharpoons NOBr(g) + \frac{1}{2}Cl_2(g)$

(c) $NOBr(g) \rightleftharpoons NO(g) + \frac{1}{2}Br_2(g)$

Strategy Begin by writing the equilibrium expressions for the reactions that are given. Then, determine the relationship of each equation's equilibrium expression to the equilibrium expression of the original equations, and make the corresponding change to the equilibrium constant for each.

Setup The equilibrium expressions for the reactions that are given are

$$K_c = \frac{[NO]^2[Br_2]}{[NOBr]^2} \qquad \text{and} \qquad K_c = \frac{[BrCl]^2}{[Br_2][Cl_2]}$$

(a) This equation is the reverse of original equation 1. Its equilibrium expression is the reciprocal of that for the original equation.

$$2NO(g) + Br_2(g) \rightleftharpoons 2NOBr(g) \qquad K_c = \frac{[NOBr]^2}{[NO]^2[Br_2]}$$

(b) This is original equation 1 multiplied by a factor of 2. Its equilibrium expression is the original expression squared.

$$4NOBr(g) \rightleftharpoons 4NO(g) + 2Br_2(g) \qquad K_c = \left(\frac{[NO]^2[Br_2]}{[NOBr]^2}\right)^2$$

(c) This is original equation 1 multiplied by $\frac{1}{2}$. Its equilibrium expression is the square root of the original.

$$NOBr(g) \rightleftharpoons NO(g) + \frac{1}{2}Br_2(g) \qquad K_c = \sqrt{\frac{[NO]^2[Br_2]}{[NOBr]^2}} \qquad \text{or} \qquad K_c = \left(\frac{[NO]^2[Br_2]}{[NOBr]^2}\right)^{1/2}$$

(Continued on next page)

(d) This is the sum of original equations 1 and 2. Its equilibrium expression is the product of the two individual expressions.

$$2NOBr(g) + Cl_2(g) \rightleftharpoons 2NO(g) + 2BrCl(g) \qquad K_c = \frac{[NO]^2[BrCl]^2}{[NOBr]^2[Cl_2]}$$

(e) Probably the simplest way to analyze this reaction is to recognize that it is the reverse of the reaction in part (d), multiplied by $\frac{1}{2}$. Its equilibrium expression is the square root of the reciprocal of the expression in part (d).

$$NO(g) + BrCl(g) \rightleftharpoons NOBr(g) + \frac{1}{2}Cl_2(g)$$

$$K_c = \sqrt{\frac{[NOBr]^2[Cl_2]}{[NO]^2[BrCl]^2}} \qquad \text{or} \qquad K_c = \left(\frac{[NOBr]^2[Cl_2]}{[NO]^2[BrCl]^2}\right)^{1/2}$$

Each equilibrium constant will bear the same relationship to the original as the equilibrium expression bears to the original.

Solution

(a) $K_c = 1/0.014 = 71$

(b) $K_c = (0.014)^2 = 2.0 \times 10^{-4}$

(c) $K_c = (0.014)^{1/2} = 0.12$

(d) $K_c = (0.014)(7.2) = 0.10$

(e) $K_c = (1/0.10)^{1/2} = 3.2$

Think About It

The magnitude of an equilibrium constant reveals whether products or reactants are favored, so the reciprocal relationship between K_c values of forward and reverse reactions should make sense. A very large K_c value means that products are favored. In the reaction of hydrogen ion and hydroxide ion to form water, the value of K_c is very large, indicating that the product, water, is favored.

$$H^+(aq) + OH^-(aq) \rightleftharpoons H_2O(l) \qquad K_c = 1.0 \times 10^{14} \text{ (at 25°C)}$$

Simply writing the equation backward doesn't change the fact that water is the predominate species at equilibrium. In the *reverse* reaction, therefore, the favored species is on the *reactant* side.

$$H_2O(l) \rightleftharpoons H^+(aq) + OH^-(aq) \qquad K_c = 1.0 \times 10^{-14} \text{ (at 25°C)}$$

As a result, the magnitude of K_c should correspond to reactants being favored; that is, it should be very small.

Practice Problem ATTEMPT The following reactions have the indicated equilibrium constants at a particular temperature.

$$N_2(g) + O_2(g) \rightleftharpoons 2NO(g) \qquad K_c = 4.3 \times 10^{-25}$$

$$2NO(g) + O_2(g) \rightleftharpoons 2NO_2(g) \qquad K_c = 6.4 \times 10^9$$

Determine the values of the equilibrium constants for the following equations at the same temperature: (a) $2NO(g) \rightleftharpoons N_2(g) + O_2(g)$, (b) $\frac{1}{2}N_2(g) + \frac{1}{2}O_2(g) \rightleftharpoons NO(g)$, and (c) $N_2(g) + 2O_2(g) \rightleftharpoons 2NO_2(g)$.

Practice Problem BUILD Using the data from Practice Problem A, determine the values of the equilibrium constants for the following equations: (a) $4NO(g) \rightleftharpoons N_2(g) + 2NO_2(g)$, (b) $4NO_2(g) \rightleftharpoons 2N_2(g) + 4O_2(g)$, and (c) $2NO(g) + 2NO_2(g) \rightleftharpoons 3O_2(g) + 2N_2(g)$.

Practice Problem CONCEPTUALIZE Consider a chemical reaction represented by the equation $A(g) + B(g) \rightleftharpoons C(g)$, for which the value of K_c is 100. If we multiply the chemical equation by two, $2A(g) + 2B(g) \rightleftharpoons 2C(g)$, the value of K_c becomes 10,000. Does the larger equilibrium constant indicate that more product $[C(g)]$ will form, given the same starting conditions? Explain.

Gaseous Equilibria

When an equilibrium expression contains only gases, we can write an alternate form of the expression in which the concentrations of gases are expressed as partial pressures (atm). Thus, for the equilibrium

$$N_2O_4(g) \rightleftharpoons 2NO_2(g)$$

we can either write the equilibrium expression as

$$K_c = \frac{[NO_2]^2}{[N_2O_4]}$$

or as

$$K_P = \frac{(P_{NO_2})^2}{P_{N_2O_4}}$$

where the subscript "P" in K_P stands for *pressure*, and P_{NO_2} and $P_{N_2O_4}$ are the equilibrium partial pressures of NO_2 and N_2O_4, respectively. In general, K_c is not equal to K_P because the partial pressures of reactants and products expressed in atmospheres are not equal to their concentrations expressed in mol/L. However, a simple relationship between K_P and K_c can be derived using the following equilibrium,

$$a\text{A}(g) \rightleftharpoons b\text{B}(g)$$

where a and b are the stoichiometric coefficients. The equilibrium constant K_c is given by

$$K_c = \frac{[\text{B}]^b}{[\text{A}]^a}$$

and the expression for K_P is

$$K_P = \frac{(P_\text{B})^b}{(P_\text{A})^a}$$

where P_A and P_B are the partial pressures of A and B. Assuming ideal gas behavior,

$$P_\text{A}V = n_\text{A}RT$$

$$P_\text{A} = \frac{n_\text{A}RT}{V} = \left(\frac{n_\text{A}}{V}\right)RT$$

and

$$P_\text{A} = [\text{A}]RT$$

where [A] is the molar concentration of A. Likewise,

$$P_\text{B}V = n_\text{B}RT$$

$$P_\text{B} = \frac{n_\text{B}RT}{V} = \left(\frac{n_\text{B}}{V}\right)RT$$

and

$$P_\text{B} = [\text{B}]RT$$

Substituting the expressions for P_A and P_B into the expression for K_P gives

$$K_P = \frac{(P_\text{B})^b}{(P_\text{A})^a}$$

$$= \frac{[\text{B}]^b}{[\text{A}]^a}(RT)^{b-a}$$

which simplifies to

$$K_P = K_c(RT)^{\Delta n}$$

where $\Delta n = b - a$. In general,

$$\Delta n = \text{moles of gaseous products} - \text{moles of gaseous reactants} \quad \textbf{Equation 15.3}$$

Because pressures are usually expressed in atmospheres, the gas constant R is expressed as 0.08206 L · atm/K · mol, and we can write the relationship between K_P and K_c as

$$K_P = K_c[(0.08206 \text{ L} \cdot \text{atm/K} \cdot \text{mol}) \times T]^{\Delta n} \quad \textbf{Equation 15.4}$$

K_P is equal to K_c only in the special case where $\Delta n = 0$, as in the following equilibrium reaction.

$$\text{H}_2(g) + \text{Br}_2(g) \rightleftharpoons 2\text{HBr}(g)$$

In this case, Equation 15.4 can be written as

$$K_P = K_c[(0.08206 \text{ L} \cdot \text{atm/K} \cdot \text{mol}) \times T]^0$$

$$= K_c$$

Keep in mind that K_P expressions can only be written for reactions in which every species in the equilibrium expression is a gas. (Remember that solids and pure liquids do not appear in the equilibrium expression.)

Worked Examples 15.5 and 15.6 let you practice writing K_P expressions and illustrate the conversion between K_c and K_P.

Worked Example 15.5

Write K_P expressions for (a) $PCl_3(g) + Cl_2(g) \rightleftharpoons PCl_5(g)$, (b) $O_2(g) + 2H_2(g) \rightleftharpoons 2H_2O(l)$, and (c) $F_2(g) + H_2(g) \rightleftharpoons 2HF(g)$.

Strategy Write equilibrium expressions for each equation, using partial pressures for concentrations of the gases.

Setup (a) All the species in this equation are gases, so they will all appear in the K_P expression.
(b) Only the reactants are gases.
(c) All species are gases.

Solution

$$\text{(a) } K_P = \frac{(P_{PCl_5})}{(P_{PCl_3})(P_{Cl_2})} \qquad \text{(b) } K_P = \frac{1}{(P_{O_2})(P_{H_2})^2} \qquad \text{(c) } K_P = \frac{(P_{HF})^2}{(P_{F_2})(P_{H_2})}$$

> **Think About It**
> It isn't necessary for every species in the reaction to be a gas—only those species that appear in the equilibrium expression.

Practice Problem ATTEMPT Write K_P expressions for (a) $2CO(g) + O_2(g) \rightleftharpoons 2CO_2(g)$, (b) $CaCO_3(s) \rightleftharpoons CaO(s) + CO_2(g)$, and (c) $N_2(g) + 3H_2(g) \rightleftharpoons 2NH_3(g)$.

Practice Problem BUILD Write the equation for the gaseous equilibrium corresponding to each of the following K_P expressions.

$$\text{(a) } K_P = \frac{(P_{NO_3})^2}{(P_{NO_2})^2(P_{O_2})} \qquad \text{(b) } K_P = \frac{(P_{CO_2})(P_{H_2})^4}{(P_{CH_4})(P_{H_2O})^2} \qquad \text{(c) } K_P = \frac{(P_{HI})^2}{(P_{I_2})(P_{H_2})}$$

Practice Problem CONCEPTUALIZE These diagrams represent closed systems at equilibrium in which red and yellow spheres represent reactants and/or products. For which system(s) can a K_P expression be written? For which can a K_c expression be written? In each case, select all that apply.

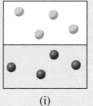
(i)

(ii)

(iii)

(iv)

Worked Example 15.6

The equilibrium constant, K_c, for the reaction

$$N_2O_4(g) \rightleftharpoons 2NO_2(g)$$

is 4.63×10^{-3} at 25°C. What is the value of K_P at this temperature?

Strategy Use Equation 15.4 to convert from K_c to K_P. Be sure to convert temperature in degrees Celsius to kelvins.

Setup Using Equation 15.3,

$$\Delta n = 2(NO_2) - 1(N_2O_4) = 1$$

$T = 298 \, K$.

Solution $K_P = \left[K_c \left(\frac{0.08206 \, L \cdot atm}{K \cdot mol} \right) \times T \right]$

$= (4.63 \times 10^{-3})(0.08206 \times 298)$

$= 0.113$

Think About It

Note that we have essentially disregarded the units of R and T so that the resulting equilibrium constant, K_P, is unitless. Equilibrium constants commonly are treated as unitless quantities.

Practice Problem **A**TTEMPT For the reaction

$$N_2(g) + 3H_2(g) \rightleftharpoons 2NH_3(g)$$

K_c is 2.3×10^{-2} at $375°C$. Calculate K_P for the reaction at this temperature.

Practice Problem **B**UILD $K_P = 2.79 \times 10^{-5}$ for the reaction in Practice Problem A at $472°C$. What is K_c for this reaction at $472°C$?

Practice Problem **C**ONCEPTUALIZE Consider the reaction $2A(l) \rightleftharpoons 2B(g)$ at room temperature. Are the values of K_c and K_P numerically equal? Under what conditions might your answer be different?

Section 15.3 Review

Equilibrium Expressions

15.3.1 Select the equilibrium expression for the reaction

$$H^+(aq) + OH^-(aq) \rightleftharpoons H_2O(l)$$

(a) $K_c = \dfrac{[H_2O]}{[H^+][OH^-]}$ (c) $K_c = [H^+][OH^-]$ (e) $K_c = [H^+][OH^-][H_2O]$

(b) $K_c = \dfrac{[H^+][OH^-]}{[H_2O]}$ (d) $K_c = \dfrac{1}{[H^+][OH^-]}$

15.3.2 Select the equilibrium expression for the reaction

$$CaO(s) + CO_2(g) \rightleftharpoons CaCO_3(s)$$

(a) $K_c = \dfrac{1}{[CO_2]}$ (c) $K_c = [CO_2]$ (e) $K_c = \dfrac{[CO_2]}{[CaO][CaCO_3]}$

(b) $K_c = \dfrac{[CaCO_3]}{[CaO][CO_2]}$ (d) $K_c = \dfrac{[CaO][CO_2]}{[CaCO_3]}$

15.3.3 Given the following information,

$$HF(aq) \rightleftharpoons H^+(aq) + F^-(aq) \qquad K_c = 6.8 \times 10^{-4} \text{ (at } 25°C)$$
$$H_2C_2O_4(aq) \rightleftharpoons 2H^+(aq) + C_2O_4^{2-}(aq) \qquad K_c = 3.8 \times 10^{-6} \text{ (at } 25°C)$$

determine the value of K_c for the following reaction at $25°C$.

$$C_2O_4^{2-}(aq) + 2HF(aq) \rightleftharpoons 2F^-(aq) + H_2C_2O_4(aq)$$

(a) 2.6×10^{-9} (c) 1.2×10^{-1} (e) 6.8×10^{-4}
(b) 1.8×10^{-12} (d) 2.6×10^{5}

15.3.4 K_c for the reaction

$$Br_2(g) \rightleftharpoons 2Br(g)$$

is 1.1×10^{-3} at $1280°C$. Calculate the value of K_P for this reaction at this temperature.

(a) 1.1×10^{-3} (c) 0.14 (e) 8.3×10^{-6}
(b) 18 (d) 9.1×10^{2}

15.4 CHEMICAL EQUILIBRIUM AND FREE ENERGY

Our discussion of equilibrium is really just another way of looking at thermodynamics. The *magnitude* of K gives us the same information as the *sign* of $\Delta G°$ (i.e., whether a significant amount of product will form when reactants are combined).

- A reaction with a very large K ($\gg$ 1) has a negative $\Delta G°$. In this case, we expect reactants to combine spontaneously to form products—and the resulting equilibrium mixture will consist predominantly of products.
- A reaction with a very small K ($\ll$ 1) has a positive $\Delta G°$. In this case, we do not expect reactants to combine spontaneously to form products to any significant degree—and the resulting equilibrium mixture will consist predominantly of reactants.

The magnitude of K and the sign of $\Delta G°$ provide useful information and enable us to solve a wide variety of problems. In fact, we will be using K extensively to solve equilibrium problems in this and the next two chapters. Often, however, an actual experiment will not start with just stoichiometric amounts of reactants. In this section, we will learn how to predict the outcome of a reaction under any starting conditions.

Using Q and K to Predict the Direction of Reaction

If we start an experiment with only reactants, we know that the reactant concentrations will decrease and the product concentrations will increase; that is, the reaction must proceed in the *forward* direction for equilibrium to be established. Likewise, if we start an experiment with only products, we know that the product concentrations will decrease and the reactant concentrations will increase. In this case, the reaction must proceed in the *reverse* direction to achieve equilibrium. But often we must predict the direction in which a reaction will proceed when we start with a mixture of reactants and products. For this situation, we calculate the value of the reaction quotient, Q_c, and compare it to the value of the equilibrium constant, K_c.

The equilibrium constant, K_c, for the gaseous formation of hydrogen iodide from molecular hydrogen and molecular iodine,

$$H_2(g) + I_2(g) \rightleftharpoons 2HI(g)$$

is 54.3 at 430°C. If we were to conduct an experiment starting with a mixture of 0.243 mole of H_2, 0.146 mole of I_2, and 1.98 moles of HI in a 1.00-L container at 430°C, would more HI form or would HI be consumed and more H_2 and I_2 form? Using the starting concentrations, we can calculate the reaction quotient as follows:

$$Q_c = \frac{[HI]_i^2}{[H_2]_i[I_2]_i} = \frac{(1.98)^2}{(0.243)(0.146)} = 111$$

where the subscript "i" indicates *initial* concentration. Because the reaction quotient does not equal K_c ($Q_c = 111$, $K_c = 54.3$), the reaction is not at equilibrium. To establish equilibrium, the reaction will proceed to the left, consuming HI and producing H_2 and I_2, decreasing the value of the numerator and increasing the value in the denominator until the value of the reaction quotient equals that of the equilibrium constant. Thus, the reaction proceeds in the reverse direction (from right to left) to reach equilibrium.

There are three possibilities when we compare Q with K:

$Q < K$ The ratio of initial concentrations of products to reactants is too small. To reach equilibrium, reactants must be converted to products. The system proceeds in the forward direction (from left to right).

$Q = K$ The initial concentrations are equilibrium concentrations. The system is already at equilibrium, and there will be no net reaction in either direction.

$Q > K$ The ratio of initial concentrations of products to reactants is too large. To reach equilibrium, products must be converted to reactants. The system proceeds in the reverse direction (from right to left).

Worked Example 15.7 shows how the value of Q is used to determine the direction of a reaction that is not at equilibrium.

Worked Example 15.7

At 375°C, the equilibrium constant for the reaction

$$N_2(g) + 3H_2(g) \rightleftharpoons 2NH_3(g)$$

is 1.2. At the start of a reaction, the concentrations of N_2, H_2, and NH_3 are 0.071 M, 9.2×10^{-3} M, and 1.83×10^{-4} M, respectively. Determine whether this system is at equilibrium, and if not, determine in which direction it must proceed to establish equilibrium.

Strategy Use the initial concentrations to calculate Q_c, and then compare Q_c with K_c.

Setup

$$Q_c = \frac{[NH_3]_i^2}{[N_2]_i[H_2]_i^3} = \frac{(1.83 \times 10^{-4})^2}{(0.071)(9.2 \times 10^{-3})^3} = 0.61$$

Solution The calculated value of Q_c is less than K_c. Therefore, the reaction is not at equilibrium and must proceed to the right to establish equilibrium.

Think About It

In proceeding to the right, a reaction consumes reactants and produces more products. This increases the numerator in the reaction quotient and decreases the denominator. The result is an increase in Q_c until it is equal to K_c, at which point equilibrium will be established.

Practice Problem **A**TTEMPT The equilibrium constant, K_c, for the formation of nitrosyl chloride from nitric oxide and chlorine

$$2NO(g) + Cl_2(g) \rightleftharpoons 2NOCl(g)$$

is 6.5×10^4 at 35°C. In which direction will the reaction proceed to reach equilibrium if the starting concentrations of NO, Cl_2, and NOCl are 1.1×10^{-3} M, 3.5×10^{-4} M, and 1.9 M, respectively?

Practice Problem **B**UILD Calculate K_P for the formation of nitrosyl chloride from nitric oxide and chlorine at 35°C, and determine whether the reaction will proceed to the right or the left to achieve equilibrium when the starting pressures are $P_{NO} = 1.01$ atm, $P_{Cl_2} = 0.42$ atm, and $P_{NOCl} = 1.76$ atm.

Practice Problem **C**ONCEPTUALIZE Consider the reaction 2A $\rightleftharpoons$ B. The diagram shown on the right represents a system at equilibrium where A = ⚪ and B = ⚫. For each of the following diagrams [(i)–(iv)], indicate whether the reaction will proceed to the right, the left, or neither to achieve equilibrium.

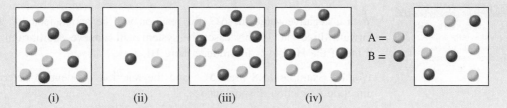

(i)   (ii)   (iii)   (iv)

A = ⚪
B = ⚫

Relationship Between ΔG and $\Delta G°$

Recall that it is the sign of ΔG, not the sign of $\Delta G°$, that indicates whether a reaction will proceed as written under a specified set of conditions [◄◄ Section 14.5]. $\Delta G°$ can be calculated using tabulated thermodynamic data. To determine ΔG, however, we also need the reaction quotient, Q. The relationship between ΔG and $\Delta G°$, which is derived from thermodynamics, is

$$\Delta G = \Delta G° + RT \ln Q \qquad \textbf{Equation 15.5}$$

where R is the gas constant (8.314 J/K · mol or 8.314×10^{-3} kJ/K · mol), T is the absolute temperature at which the reaction takes place, and Q is the reaction quotient [◄◄ Section 15.2]. Thus, ΔG depends on two terms: $\Delta G°$ and $RT \ln Q$. For a given reaction at temperature T, the value of $\Delta G°$ is fixed but that of $RT \ln Q$ can vary because Q varies according to the composition of the reaction mixture.

Student Annotation: The Q used in Equation 15.5 can be either Q_c (for reactions that take place in solution) or Q_P (for reactions that take place in the gas phase).

Consider the following equilibrium:

$$H_2(g) + I_2(s) \rightleftharpoons 2HI(g)$$

Using Equation 14.12 and information from Appendix 2, we find that $\Delta G°$ for this reaction at 25°C is 2.60 kJ/mol. The value of ΔG, however, depends on the pressures of both gaseous species. If we start with a reaction mixture containing solid I_2, in which $P_{H_2} = 4.0$ atm and $P_{HI} = 3.0$ atm, the reaction quotient, Q_P, is

$$Q_P = \frac{(P_{HI})^2}{(P_{H_2})} = \frac{(3.0)^2}{4.0} = \frac{9.0}{4.0}$$

$$= 2.25$$

Using this value in Equation 15.5 gives

$$\Delta G = \frac{2.60 \text{ kJ}}{\text{mol}} + \left(\frac{8.314 \times 10^{-3} \text{ kJ}}{\text{K} \cdot \text{mol}}\right)(298 \text{ K})(\ln 2.25)$$

$$= 4.3 \text{ kJ/mol}$$

Because ΔG is positive, we conclude that, starting with these concentrations, the forward reaction will not occur spontaneously as written. Instead, the *reverse* reaction will occur spontaneously and the system will reach equilibrium by consuming part of the HI initially present and producing more H_2 and I_2.

If, on the other hand, we start with a mixture of gases in which $P_{H_2} = 4.0$ atm and $P_{HI} = 1.0$ atm, the reaction quotient, Q_P, is

$$Q_P = \frac{(P_{HI})^2}{(P_{H_2})} = \frac{(1.0)^2}{(4.0)} = \frac{1}{4}$$

$$= 0.25$$

Using this value in Equation 15.5 gives

$$\Delta G = \frac{2.60 \text{ kJ}}{\text{mol}} + \left(\frac{8.314 \times 10^{-3} \text{ kJ}}{\text{K} \cdot \text{mol}}\right)(298 \text{ K})(\ln 0.25)$$

$$= -0.8 \text{ kJ/mol}$$

Student Hot Spot

Student data indicate you may struggle with the relationship between free energy and equilibrium. Access the SmartBook to view additional Learning Resources on this topic.

With a negative value for ΔG, the reaction will be spontaneous as written—in the forward direction. In this case, the system will achieve equilibrium by consuming some of the H_2 and I_2 to produce more HI.

Worked Example 15.8 uses $\Delta G°$ and the reaction quotient to determine in which direction a reaction is spontaneous.

Worked Example 15.8

The equilibrium constant, K_P, for the reaction

$$N_2O_4(g) \rightleftharpoons 2NO_2(g)$$

is 0.113 at 298 K, which corresponds to a standard free-energy change of 5.4 kJ/mol. In a certain experiment, the initial pressures are $P_{N_2O_4} = 0.453$ atm and $P_{NO_2} = 0.122$ atm. Calculate ΔG for the reaction at these pressures, and predict the direction in which the reaction will proceed spontaneously to establish equilibrium.

Strategy Use the partial pressures of N_2O_4 and NO_2 to calculate the reaction quotient Q_P, and then use Equation 15.5 to calculate ΔG.

Setup The reaction quotient expression is

$$Q_P = \frac{(P_{NO_2})^2}{P_{N_2O_4}} = \frac{(0.122)^2}{0.453} = 0.0329$$

Solution

$$\Delta G = \Delta G^\circ + RT \ln Q_P$$

$$= \frac{5.4\ \text{kJ}}{\text{mol}} + \left(\frac{8.314 \times 10^{-3}\ \text{kJ}}{\text{K} \cdot \text{mol}}\right)(298\ \text{K})(\ln 0.0329)$$

$$= 5.4\ \text{kJ/mol} - 8.46\ \text{kJ/mol}$$

$$= -3.1\ \text{kJ/mol}$$

Because ΔG is negative, the reaction proceeds spontaneously from left to right to reach equilibrium.

Think About It

Remember, a reaction with a positive ΔG° value can be spontaneous if the starting concentrations of reactants and products are such that $Q < K$.

Practice Problem **A**TTEMPT ΔG° for the reaction

$$H_2(g) + I_2(s) \rightleftharpoons 2HI(g)$$

is 2.60 kJ/mol at 25°C. Calculate ΔG, and predict the direction in which the reaction is spontaneous if the starting concentrations are $P_{H_2} = 5.25$ atm and $P_{HI} = 1.75$ atm.

Practice Problem **B**UILD What is the minimum partial pressure of H_2 required for the preceding reaction to be spontaneous in the forward direction at 25°C if the partial pressure of HI is 0.94?

Practice Problem **C**ONCEPTUALIZE Consider the reaction in Worked Example 15.8. Which of the following graphs [(i)–(iv)] best shows what happens to ΔG as the partial pressure of N_2O_4 is increased?

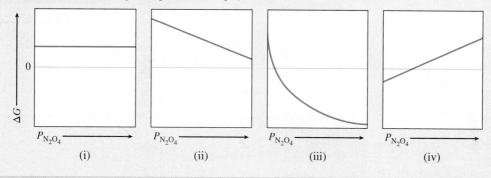

(i)   (ii)   (iii)   (iv)

Relationship Between ΔG° and K

We have defined *equilibrium* as the condition in which forward and reverse reactions are occurring at the same rate—such that no *net* reaction is occurring in either direction, and the concentrations of reactants and products are constant. In this state, $Q = K$ and $\Delta G = 0$. Thus, $\Delta G = \Delta G^\circ + RT \ln Q$ (Equation 15.6) becomes

$$0 = \Delta G^\circ + RT \ln K$$

or

$$\Delta G^\circ = -RT \ln K \qquad \textbf{Equation 15.6}$$

Student Annotation: In this equation and in Equation 15.6, K is K_c for reactions that take place in solution and K_P for reactions that take place in the gas phase.

According to Equation 15.6, then, the larger K is, the more negative ΔG° is. For chemists, Equation 15.6 is one of the most important equations in thermodynamics because it enables us to find the equilibrium constant of a reaction if we know the change in standard free energy, and vice versa.

It is significant that Equation 15.6 relates the equilibrium constant to the standard free-energy change, ΔG°, rather than to the actual free-energy change, ΔG. The actual free-energy change of the system varies as the reaction progresses and becomes zero at equilibrium. On the other hand, ΔG°, like K, is a constant for a particular reaction at a given temperature. Figure 15.4 shows plots of the free energy of a reacting system versus the extent of the reaction for two reactions. Table 15.3 summarizes the relationship

Video 15.1
Chemical equilibrium.

Figure 15.4 (a) $\Delta G° < 0$. At equilibrium, there is a significant conversion of reactants to products. (b) $\Delta G° > 0$. At equilibrium, reactants are favored over products. In both cases, the net reaction toward equilibrium is from left to right (reactants to products) if $Q < K$ and right to left (products to reactants) if $Q > K$. At equilibrium, $Q = K$.

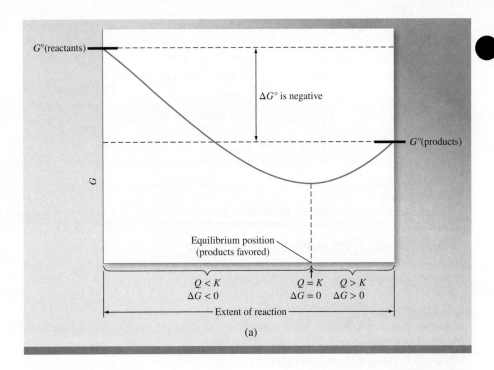

(a)

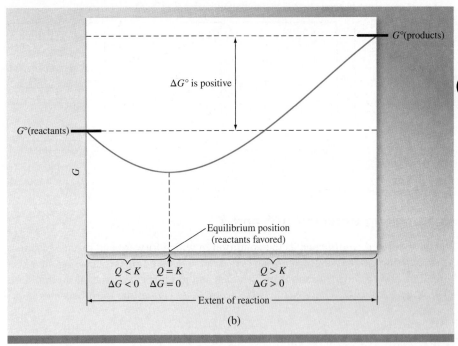

(b)

between the *magnitude* of an equilibrium constant and the *sign* of the corresponding $\Delta G°$. Remember this important distinction: It is the sign of ΔG and not that of $\Delta G°$ that determines the direction of reaction spontaneity. The sign of $\Delta G°$ only tells us the relative amounts of products and reactants when equilibrium is reached, *not* the direction the reaction must go to reach equilibrium.

Student Annotation: The sign of $\Delta G°$ tells us the same thing that the magnitude of K tells us. The sign of ΔG tells us the same thing as the comparison of Q and K values.

TABLE 15.3		Relationship Between K and $\Delta G°$ as Predicted by Equation 15.6	
K	**ln K**	**$\Delta G°$**	**Result at equilibrium**
> 1	Positive	Negative	Products are favored.
= 1	0	0	Neither products nor reactants are favored.
< 1	Negative	Positive	Reactants are favored.

For reactions with very large or very small equilibrium constants, it can be very difficult—sometimes impossible—to determine K values by measuring the concentrations of the reactants and products. Consider, for example, the formation of nitric oxide from molecular nitrogen and molecular oxygen.

$$N_2(g) + O_2(g) \rightleftharpoons 2NO(g)$$

At 25°C, the equilibrium constant, K_P, is

$$K_P = \frac{(P_{NO})^2}{(P_{N_2})(P_{O_2})} = 4.0 \times 10^{-31}$$

The very small value of K_P means that the concentration of NO at equilibrium will be exceedingly low and, for all intents and purposes, impossible to measure directly. In such a case, the equilibrium constant is more conveniently determined using $\Delta G°$, which can be calculated either from tabulated $\Delta G_f°$ values or from $\Delta H°$ and $\Delta S°$.

Worked Examples 15.9 and 15.10 show how to use $\Delta G°$ to calculate K and how to use K to calculate $\Delta G°$, respectively.

Worked Example 15.9

Using data from Appendix 2, calculate the equilibrium constant, K_P, for the following reaction at 25°C.

$$2H_2O(l) \rightleftharpoons 2H_2(g) + O_2(g)$$

Strategy Use data from Appendix 2 and Equation 14.12 to calculate $\Delta G°$ for the reaction. Then use Equation 15.6 to solve for K_P.

Setup

$$\Delta G° = (2\Delta G_f°[H_2(g)] + \Delta G_f°[O_2(g)]) - (2\Delta G_f°[H_2O(l)])$$

$$= [(2)(0 \text{ kJ/mol}) + (0 \text{ kJ/mol})] - [(2)(-237.2 \text{ kJ/mol})]$$

$$= 474.4 \text{ kJ/mol}$$

Solution

$$\Delta G° = -RT \ln K_P$$

$$\frac{474.4 \text{ kJ}}{\text{mol}} = -\left(\frac{8.314 \times 10^{-3} \text{ kJ}}{\text{K} \cdot \text{mol}}\right)(298 \text{ K}) \ln K_P$$

$$-191.5 = \ln K_P$$

$$K_P = e^{-191.5}$$

$$= 7 \times 10^{-84}$$

Think About It
This is an extremely small equilibrium constant, which is consistent with the large, positive value of $\Delta G°$. We know from everyday experience that water does not decompose spontaneously into its constituent elements at 25°C.

Practice Problem **A**TTEMPT Using data from Appendix 2, calculate the equilibrium constant, K_P, for the following reaction at 25°C.

$$2O_3(g) \rightleftharpoons 3O_2(g)$$

Practice Problem **B**UILD K_f for the complex ion $Ag(NH_3)_2^+$ is 1.5×10^7 at 25°C. Using this and data from Appendix 2, calculate the value of $\Delta G_f°$ for $Ag(NH_3)_2^+(aq)$.

Practice Problem **C**ONCEPTUALIZE Which of the following graphs [(i)–(iv)] best shows the relationship between $\Delta G°$ and equilibrium constant (K)?

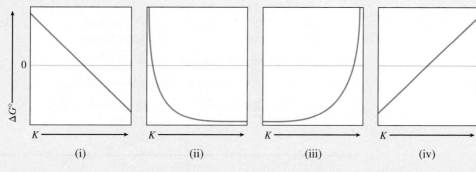

(i)   (ii)   (iii)   (iv)

Worked Example 15.10

The equilibrium constant, K_{sp}, for the dissolution of silver chloride in water at 25°C,

$$AgCl(s) \rightleftharpoons Ag^+(aq) + Cl^-(aq)$$

is 1.6×10^{-10}. Calculate $\Delta G°$ for the process.

Strategy Use Equation 15.6 to calculate $\Delta G°$.

Setup $R = 8.314 \times 10^{-3}$ kJ/K · mol and $T = (25 + 273) = 298$ K.

Solution

$$\Delta G° = -RT \ln K_{sp}$$

$$= -\left(\frac{8.314 \times 10^{-3} \text{ kJ}}{\text{K} \cdot \text{mol}}\right)(298 \text{ K}) \ln (1.6 \times 10^{-10})$$

$$= 55.9 \text{ kJ/mol}$$

> **Think About It**
> The relatively large, positive $\Delta G°$, like the very small K value, corresponds to a process that lies very far to the left. Note that the K in Equation 15.6 can be any type of K_c (K_a, K_b, K_{sp}, etc.) or K_P.

Practice Problem **A**TTEMPT Calculate $\Delta G°$ for the process:

$$BaF_2(s) \rightleftharpoons Ba^{2+}(aq) + 2F^-(aq)$$

The K_{sp} of BaF_2 at 25°C is 1.7×10^{-6}.

Practice Problem **B**UILD K_{sp} for $Co(OH)_2$ at 25°C is 3.3×10^{-16}. Using this and data from Appendix 2, calculate the value of $\Delta G_f°$ for $Co(OH)_2(s)$.

Practice Problem **C**ONCEPTUALIZE Fill in each blank with *positive, negative, zero,* or *impossible to determine from the information given.*

When $Q = K$, ΔG is _____.   When $Q > K$, ΔG is _____.

When $Q < K$, ΔG is _____.

Section 15.4 Review

Free Energy and Chemical Equilibrium

15.4.1 For the reaction

$$A(aq) + B(aq) \rightleftharpoons C(aq)$$

$\Delta G° = -1.95$ kJ/mol at 25°C. What is ΔG (in kJ/mol) at 25°C when the concentrations are $[A] = [B] = 0.315$ M and $[C] = 0.405$ M?
(a) -1.95 kJ/mol   (c) -5.43 kJ/mol   (e) -12.1 kJ/mol
(b) 1.53 kJ/mol   (d) 8.16 kJ/mol

15.4.2 The K_{sp} for iron(III) hydroxide [$Fe(OH)_3$] is 1.1×10^{-36} at 25°C. For the process

$$Fe(OH)_3(s) \rightleftharpoons Fe^{3+}(aq) + 3OH^-(aq)$$

determine $\Delta G°$ (in kJ/mol) at 25°C.
(a) -2.73×10^{-36} kJ/mol   (c) 17.2 kJ/mol   (e) 205 kJ/mol
(b) -17.2 kJ/mol   (d) -205 kJ/mol

15.4.3 The $\Delta G°$ for the reaction

$$N_2(g) + 3H_2(g) \rightleftharpoons 2NH_3(g)$$

is -33.3 kJ/mol at 25°C. What is the value of K_P?
(a) 2×10^{-6} (c) 3×10^{-70} (e) 1×10^{-1}
(b) 7×10^5 (d) 1

15.5 CALCULATING EQUILIBRIUM CONCENTRATIONS

If we know the equilibrium constant for a reaction, we can calculate the concentrations in the equilibrium mixture from the initial reactant concentrations. Consider the following system involving two organic compounds, *cis*- and *trans*-stilbene.

cis-Stilbene   *trans*-Stilbene

The equilibrium constant (K_c) for this system is 24.0 at 200°C. If we know that the starting concentration of *cis*-stilbene is 0.850 *M*, we can use the equilibrium expression to determine the equilibrium concentrations of both species. The stoichiometry of the reaction tells us that for every mole of *cis*-stilbene converted, 1 mole of *trans*-stilbene is produced. We will let x be the equilibrium concentration of *trans*-stilbene in mol/L; therefore, the equilibrium concentration of *cis*-stilbene must be $(0.850 - x)$ mol/L. It is useful to summarize these changes in concentrations in an equilibrium table.

	cis-stilbene $\rightleftharpoons$ *trans*-stilbene	
Initial concentration (M):	0.850	0
Change in concentration (M):	$- x$	$+ x$
Equilibrium concentration (M):	$0.850 - x$	x

We then use the equilibrium concentrations, defined in terms of x, in the equilibrium expression

$$K_c = \frac{[\textit{trans}\text{-stilbene}]}{[\textit{cis}\text{-stilbene}]}$$

$$24.0 = \frac{x}{0.850 - x}$$

$$x = 0.816\ M$$

Having solved for x, we calculate the equilibrium concentrations of *cis*- and *trans*-stilbene as follows:

$$[\textit{cis}\text{-stilbene}] = (0.850 - x)\ M = 0.034\ M$$
$$[\textit{trans}\text{-stilbene}] = x\ M = 0.816\ M$$

A good way to check the answer to a problem such as this is to use the calculated equilibrium concentrations in the equilibrium expression and make sure that we get the correct K_c value.

$$K_c = \frac{0.816}{0.034} = 24$$

Figure 15.5 (pages 678–679) shows in detail how to construct and use an equilibrium table, also known as an "ice" table. Worked Examples 15.11 through 15.13 provide you with some practice building and using ice tables.

Figure 15.5 Equilibrium (ice) tables.

Constructing ice Tables to Solve Equilibrium Problems

To illustrate the use of such a table, let's consider the formation of chloromethane (CH_3Cl), for which $K_c = 4.5 \times 10^3$ at 1500°C. In this experiment, the initial concentrations of reactants are equal:

$$[CH_4] = [Cl_2] = 0.0010\ M$$

In the simplest case, where we start with reactants only, the initial concentrations of the products are zero. We fill in these concentrations, completing the top row of the table.

START

$$CH_4(g) + Cl_2(g) \rightleftarrows CH_3Cl(g) + HCl(g)$$

			0	0
i	0.00100	0.00100		
c				
e				

$$CH_4(g) + Cl_2(g) \rightleftarrows CH_3Cl(g) + HCl(g)$$

i	0.00100	0.00100	0	0
c	-x	-x	+x	+x
e				

The stoichiometry of this reaction indicates that each reactant concentration will decrease by the same unknown amount, x; and that each product concentration will increase by that same amount, x. We enter this information in the middle row.

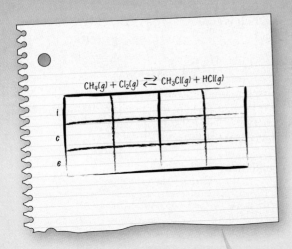

$$CH_4(g) + Cl_2(g) \rightleftarrows CH_3Cl(g) + HCl(g)$$

i				
c				
e				

Begin with a balanced chemical equation. Draw a table under the equation, making a column for each species in the reaction. The table should have three rows: one for *initial* concentrations (labeled "*i*"), one for changes in *concentrations* (labeled "*c*"), and one for final or *equilibrium* concentrations (labeled "*e*").

An ice table can also be used to calculate the value of an equilibrium constant. Consider the formation of chloromethane at a different temperature, 2000°C (2273 K). Using the same starting conditions as before, we find that the equilibrium concentration of chloromethane is $9.7 \times 10^{-4}\ M$. We fill in the information we know and use stoichiometry to determine the missing information. The equilibrium concentration of HCl is equal to that of CH_3Cl, $9.7 \times 10^{-4}\ M$. Further, the amount by which each reactant concentration decreased is also $9.7 \times 10^{-4}\ M$.

START

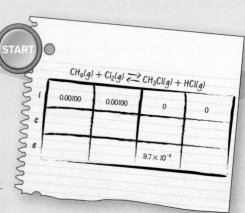

$$CH_4(g) + Cl_2(g) \rightleftarrows CH_3Cl(g) + HCl(g)$$

i	0.00100	0.00100	0	0
c				
e			9.7×10^{-4}	

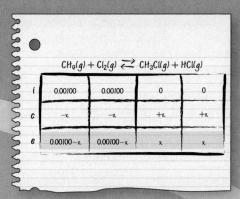

We fill in the bottom row (the equilibrium concentrations) by adding the top- and middle-row entries in each column. We then use the equilibrium expression, in this case

$$K_c = \frac{[CH_3Cl][HCl]}{[CH_4][Cl_2]}$$

to solve for x using the value of K_c and the equilibrium concentrations from the bottom row of the completed table.

CHECK

$$4.50 \times 10^3 = \frac{(x)(x)}{(0.00100 - x)(0.00100 - x)} = \frac{x^2}{(0.00100 - x)^2}$$

Taking the square root of both sides of this equation gives

$$\sqrt{4.50 \times 10^3} = \sqrt{\frac{x^2}{(0.00100 - x)^2}}$$

$$67.08 = \frac{x}{0.00100 - x}$$

and solving for x gives

$$0.06708 - 67.08x = x$$
$$0.06708 = 68.08x$$
$$x = 9.85 \times 10^{-4}$$

We used x to represent the concentrations of both products at equilibrium, and the amount by which each reactant concentration decreases. Therefore, the equilibrium concentrations are

$$[CH_4] = [Cl_2] = (0.00100 - 9.85 \times 10^{-4})\ M = 1.5 \times 10^{-5}\ M$$

and

$$[CH_3Cl] = [HCl] = 9.85 \times 10^{-4}\ M$$

We can verify our results by plugging the equilibrium concentrations back into the equilibrium expression. We should get a number very close to the original equilibrium constant.

$$K_c = \frac{[CH_3Cl][HCl]}{[CH_4][Cl_2]} = \frac{(9.85 \times 10^{-4})^2}{(1.5 \times 10^{-5})^2} = 4.3 \times 10^3$$

In this case, the small difference between the result and the original K_c value (4.5×10^3) is due to rounding.

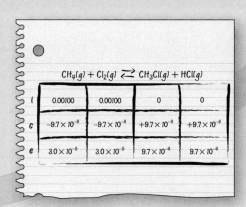

The equilibrium concentration of HCl is equal to that of CH_3Cl, $9.7 \times 10^{-4}\ M$. Further, the amount by which each reactant concentration decreased is also $9.7 \times 10^{-4}\ M$. Therefore, the equilibrium concentrations are $[CH_4] = [Cl_2] = 3.0 \times 10^{-5}\ M$; and $[CH_3Cl] = [HCl] = 9.7 \times 10^{-4}\ M$. Plugging these equilibrium concentrations into the equilibrium expression gives

$$K_c = \frac{[CH_3Cl][HCl]}{[CH_4][Cl_2]} = \frac{(9.7 \times 10^{-4})^2}{(3.0 \times 10^{-5})^2} = 1.0 \times 10^3.$$

Therefore, at 2000°C, K_c for this reaction is 1.0×10^3.

(See Visualizing Chemistry questions VC15.1–VC15.4 on page 705.)

What's the point?

Equilibrium or *ice* tables can be used to solve a variety of equilibrium problems. We begin by entering the known concentrations and then use the stoichiometry of the reaction to complete the table. When the bottom row of the ice table is complete, we can use the equilibrium expression to calculate either equilibrium concentrations, or the value of the equilibrium constant.

Worked Example 15.11

K_c for the reaction of hydrogen and iodine to produce hydrogen iodide,

$$H_2(g) + I_2(g) \rightleftharpoons 2HI(g)$$

is 54.3 at 430°C. What will the concentrations be at equilibrium if we start with 0.240 M concentrations of both H_2 and I_2?

Strategy Construct an equilibrium table to determine the equilibrium concentration of each species in terms of an unknown (x); solve for x, and use it to calculate the equilibrium molar concentrations.

Setup Insert the starting concentrations that we know into the equilibrium table.

	$H_2(g)$	+	$I_2(g)$	$\rightleftharpoons$	$2HI(g)$
Initial concentration (M):	0.240		0.240		0
Change in concentration (M):					
Equilibrium concentration (M):					

Solution We define the change in concentration of one of the reactants as x. Because there is no product at the start of the reaction, the reactant concentration must decrease; that is, this reaction must proceed in the forward direction to reach equilibrium. According to the stoichiometry of the chemical reaction, the reactant concentrations will both decrease by the same amount (x), and the product concentration will increase by twice that amount ($2x$). Combining the initial concentration and the change in concentration for each species, we get expressions (in terms of x) for the equilibrium concentrations.

	$H_2(g)$	+	$I_2(g)$	$\rightleftharpoons$	$2HI(g)$
Initial concentration (M):	0.240		0.240		0
Change in concentration (M):	$-x$		$-x$		$+2x$
Equilibrium concentration (M):	$0.240 - x$		$0.240 - x$		$2x$

Next, we insert these expressions for the equilibrium concentrations into the equilibrium expression and solve for x.

$$K_c = \frac{[HI]^2}{[H_2][I_2]}$$

$$54.3 = \frac{(2x)^2}{(0.240 - x)(0.240 - x)} = \frac{(2x)^2}{(0.240 - x)^2}$$

$$\sqrt{54.3} = \frac{2x}{0.240 - x}$$

$$x = 0.189$$

Using the calculated value of x, we can determine the equilibrium concentration of each species as follows:

$$[H_2] = (0.240 - x)\ M = 0.051\ M$$
$$[I_2] = (0.240 - x)\ M = 0.051\ M$$
$$[HI] = 2x = 0.378\ M$$

Think About It

Always check your answer by inserting the calculated concentrations into the equilibrium expression.

$$\frac{[HI]^2}{[H_2][I_2]} = \frac{(0.378)^2}{(0.051)^2} = 54.9 \approx K_c$$

The small difference between the calculated K_c and the one given in the problem statement is due to rounding.

Practice Problem A TTEMPT Calculate the equilibrium concentrations of H_2, I_2, and HI at 430°C if the initial concentrations are $[H_2] = [I_2] = 0\ M$ and $[HI] = 0.525\ M$.

Practice Problem B UILD Determine the initial concentration of HI if the initial concentrations of H_2 and I_2 are both 0.10 M and their equilibrium concentrations are both 0.043 M at 430°C.

Practice Problem **C**ONCEPTUALIZE   Consider the reaction $A(g) + B(g) \rightleftharpoons C(g)$. The diagram shown on the right depicts the starting condition for a system. Without knowing the value of K_c determine which of the following diagrams [(i)–(iv)] could represent the system at equilibrium. Select all that apply.

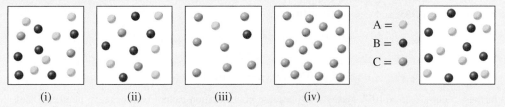

(i)	(ii)	(iii)	(iv)

Worked Example 15.12

For the same reaction and temperature as in Worked Example 15.11, calculate the equilibrium concentrations of all three species if the starting concentrations are as follows: $[H_2] = 0.00623\ M$, $[I_2] = 0.00414\ M$, and $[HI] = 0.0424\ M$.

Strategy Using the initial concentrations, calculate the reaction quotient, Q_c, and compare it to the value of K_c (given in the problem statement of Worked Example 15.11) to determine which direction the reaction will proceed to establish equilibrium. Then, construct an equilibrium table to determine the equilibrium concentrations.

Setup

$$\frac{[HI]^2}{[H_2][I_2]} = \frac{(0.0424)^2}{(0.00623)(0.00414)} = 69.7$$

Therefore, $Q_c > K_c$, so the system will have to proceed to the *left* (reverse) to reach equilibrium. The equilibrium table is

	$H_2(g)$	$+$	$I_2(g)$	$\rightleftharpoons$	$2HI(g)$
Initial concentration (M):	0.00623		0.00414		0.0424
Change in concentration (M):					
Equilibrium concentration (M):					

Solution Because we know the reaction must proceed from right to left, we know that the concentration of HI will decrease and the concentrations of H_2 and I_2 will increase. Therefore, the table should be filled in as follows:

	$H_2(g)$	$+$	$I_2(g)$	$\rightleftharpoons$	$2HI(g)$
Initial concentration (M):	0.00623		0.00414		0.0424
Change in concentration (M):	$+x$		$+x$		$-2x$
Equilibrium concentration (M):	$0.00623 + x$		$0.00414 + x$		$0.0424 - 2x$

Next, we insert these expressions for the equilibrium concentrations into the equilibrium expression and solve for x.

$$K_c = \frac{[HI]^2}{[H_2][I_2]}$$

$$54.3 = \frac{(0.0424 - 2x)^2}{(0.00623 + x)(0.00414 + x)}$$

It isn't possible to solve this equation the way we did in Worked Example 15.11 (by taking the square root of both sides) because the concentrations of H_2 and I_2 are unequal. Instead, we have to carry out the multiplications,

$$54.3(2.58 \times 10^{-5} + 1.04 \times 10^{-2}x + x^2) = 1.80 \times 10^{-3} - 1.70 \times 10^{-1}x + 4x^2$$

Collecting terms we get

$$50.3x^2 + 0.735x - 4.00 \times 10^{-4} = 0$$

(Continued on next page)

This is a quadratic equation of the form $ax^2 + bx + c = 0$. The solution for the quadratic equation (see Appendix 1) is

$$x = \frac{-b \pm \sqrt{b^2 - 4ac}}{2a}$$

Here we have $a = 50.3$, $b = 0.735$, and $c = -4.00 \times 10^{-4}$, so

$$x = \frac{-0.735 \pm \sqrt{(0.735)^2 - 4(50.3)(-4.00 \times 10^{-4}}}{2(50.3)}$$

$$x = 5.25 \times 10^{-4} \quad \text{or} \quad x = -0.0151$$

Only the first of these values, 5.25×10^{-4}, makes sense because concentration cannot be a negative number. Using the calculated value of x, we can determine the equilibrium concentration of each species as follows:

$$[H_2] = (0.00623 + x) \, M = 0.00676 \, M$$
$$[I_2] = (0.00414 + x) \, M = 0.00467 \, M$$
$$[HI] = (0.0424 - 2x) \, M = 0.0414 \, M$$

Think About It
Checking this result gives

$$K_c = \frac{[HI]^2}{[H_2][I_2]} = \frac{(0.0414)^2}{(0.00676)(0.00467)} = 54.3$$

Practice Problem **A**TTEMPT Calculate the equilibrium concentrations of H_2, I_2, and HI at 430°C if the initial concentrations are $[H_2] = [I_2] = 0.378 \, M$ and $[HI] = 0 \, M$.

Practice Problem **B**UILD At 1280°C the equilibrium constant K_c for the reaction

$$Br_2(g) \rightleftharpoons 2Br(g)$$

is 1.1×10^{-3}. If the initial concentrations are $[Br_2] = 6.3 \times 10^{-2} \, M$ and $[Br] = 1.2 \times 10^{-2} \, M$, calculate the concentrations of these two species at equilibrium.

Practice Problem **C**ONCEPTUALIZE For the reaction of hydrogen and iodine gases to form hydrogen iodide, indicate which of these diagrams [(i)–(iv)] represents a starting condition from which equilibrium can be established.

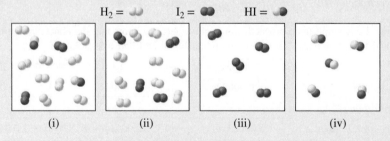

When the magnitude of K (either K_c or K_P) is very small, the solution to an equilibrium problem can be simplified—making it unnecessary to use the quadratic equation. Worked Example 15.13 illustrates this.

Worked Example 15.13

At elevated temperatures, iodine molecules break apart to give iodine atoms according to the equation

$$I_2(g) \rightleftharpoons 2I(g)$$

K_c for this reaction at 205°C is 3.39×10^{-12}. Determine the concentration of atomic iodine when a 1.00-L vessel originally charged with 0.00155 mol of molecular iodine at this temperature is allowed to reach equilibrium.

Strategy Construct an equilibrium table and use initial concentrations and the value of K_c to determine the changes in concentration and the equilibrium concentrations.

Setup The initial concentration of $I_2(g)$ is 0.00155 M and the original concentration of $I(g)$ is zero. $K_c = 3.39 \times 10^{-12}$.

Solution We expect the reactant concentration to decrease by some unknown amount x. The stoichiometry of the reaction indicates that the product concentration will increase by twice that amount, $2x$.

	$I_2(g)$	$\rightleftharpoons$	$2I(g)$
Initial concentration (M):	0.00155		0
Change in concentration (M):	$-x$		$+2x$
Equilibrium concentration (M):	$0.00155 - x$		$2x$

Inserting the equilibrium concentrations into the equilibrium expression for the reaction gives

$$3.39 \times 10^{-12} = \frac{[I]^2}{[I_2]} = \frac{(2x)^2}{(0.00155 - x)}$$

We could solve for x using the quadratic equation, as we did in Worked Example 15.12. However, because the magnitude of K_c is very small, we expect this equilibrium to lie far to the left. The means that very little of the molecular iodine will break apart to give atomic iodine; and the value of x [amount of $I_2(g)$ that reacts] will be very small. In fact, the value of x will be negligible compared to the original concentration of $I_2(g)$. Therefore, $0.00155 - x \approx 0.00155$, and the solution simplifies to

$$3.39 \times 10^{-12} = \frac{(2x)^2}{0.00155} = \frac{4x^2}{0.00155}$$

$$\frac{3.39 \times 10^{-12}(0.00155)}{4} = 1.31 \times 10^{-15} = x^2$$

$$x = \sqrt{1.31 \times 10^{-15}} = 3.62 \times 10^{-8} M$$

According to our ice table, the equilibrium concentration of atomic iodine is $2x$; therefore, $[I(g)] = 2 \times 3.62 \times 10^{-8} = 7.24 \times 10^{-8} M$.

Think About It

Having solved for x by this method, we can see that it is indeed insignificant compared to the original concentration of iodine. To an appropriate number of significant figures, $0.00155 - 3.62 \times 10^{-8} = 0.00155$.

Practice Problem Ⓐ**TTEMPT** Aqueous hydrocyanic acid (HCN) ionizes according to the equation

$$HCN(aq) \rightleftharpoons H^+(aq) + CN^-(aq)$$

At 25°C, K_c for this reaction is 4.9×10^{-10}. Determine the equilibrium concentrations of all species in a 0.100-M solution of aqueous HCN.

Practice Problem Ⓑ**UILD** Consider a weak acid, HA, that ionizes according to the equation

$$HA(aq) \rightleftharpoons H^+(aq) + A^-(aq)$$

At 25°C, a 0.145-M solution of aqueous HA is found to have a hydrogen ion concentration of 2.2×10^{-5} M. Determine the K_c for this reaction at 25°C.

Practice Problem Ⓒ**ONCEPTUALIZE** Each of the diagrams [(i)–(iv)] shows a system before and after equilibrium is established. Indicate which diagram best represents a system in which you can neglect the x in the solution, as you did in Worked Example 15.13.

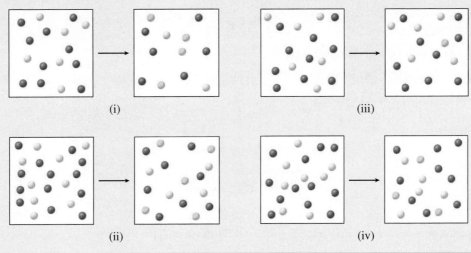

(i)   (iii)

(ii)   (iv)

Here is a summary of the use of initial reactant concentrations to determine equilibrium concentrations.

1. Construct an equilibrium table, and fill in the initial concentrations (including any that are zero).
2. Use initial concentrations to calculate the reaction quotient, Q, and compare Q to K to determine the direction in which the reaction will proceed.
3. Define x as the amount of a particular species consumed, and use the stoichiometry of the reaction to define (in terms of x) the amount of other species consumed or produced.
4. For each species in the equilibrium, add the change in concentration to the initial concentration to get the equilibrium concentration.
5. Use the equilibrium concentrations and the equilibrium expression to solve for x.
6. Using the calculated value of x, determine the concentrations of all species at equilibrium.
7. Check your work by plugging the calculated equilibrium concentrations into the equilibrium expression. The result should be very close to the K_c stated in the problem.

Student Annotation: If $Q < K$, the reaction will occur as written. If $Q > K$, the reverse reaction will occur.

Student Hot Spot

Student data indicate you may struggle with determining K values. Access the SmartBook to view additional Learning Resources on this topic.

The same procedure applies to K_P.

Worked Example 15.14 shows how to solve an equilibrium problem using partial pressures.

Worked Example 15.14

A mixture of 5.75 atm of H_2 and 5.75 atm of I_2 is contained in a 1.0-L vessel at 430°C. The equilibrium constant (K_P) for the reaction

$$H_2(g) + I_2(g) \rightleftharpoons 2HI(g)$$

at this temperature is 54.3. Determine the equilibrium partial pressures of H_2, I_2, and HI.

Strategy Construct an equilibrium table to determine the equilibrium partial pressures.

Setup The equilibrium table is

	$H_2(g)$	+	$I_2(g)$	$\rightleftharpoons$	$2HI(g)$
Initial partial pressure (atm):	5.75		5.75		0
Change in partial pressure (atm):	$-x$		$-x$		$+2x$
Equilibrium partial pressure (atm):	$5.75 - x$		$5.75 - x$		$2x$

Solution Setting the equilibrium expression equal to K_P,

$$54.3 = \frac{(2x)^2}{(5.75 - x)^2}$$

Taking the square root of both sides of the equation gives

$$\sqrt{54.3} = \frac{2x}{5.75 - x}$$

$$7.369 = \frac{2x}{5.75 - x}$$

$$7.369(5.75 - x) = 2x$$

$$42.37 - 7.369x = 2x$$

$$42.37 = 9.369x$$

$$x = 4.52$$

The equilibrium partial pressures are $P_{H_2} = P_{I_2} = 5.75 - 4.52 = 1.23$ atm, and $P_{HI} = 9.04$ atm.

Think About It

Plugging the calculated partial pressures into the equilibrium expression gives

$$\frac{(P_{HI})^2}{(P_{H_2})(P_{I_2})} = \frac{(9.04)^2}{(1.23)^2} = 54.0$$

The small difference between this result and the equilibrium constant given in the problem statement is due to rounding.

Practice Problem A TTEMPT Determine the equilibrium partial pressures of H_2, I_2, and HI if we begin the experiment with 1.75 atm each of H_2 and I_2 at 430°C.

Practice Problem B UILD Determine the equilibrium partial pressures of H_2, I_2, and HI (at 430°C) if we begin the experiment with the following conditions: P_{H_2} = 0.25 atm, P_{I_2} = 0.050 atm, P_{HI} = 2.5 atm.

Practice Problem C ONCEPTUALIZE Consider the reaction $A(g) + B(g) \rightleftharpoons C(s) + D(s)$. The diagram shown on the right represents a system at equilibrium. Each of the following diagrams [(i)–(iv)] is missing spheres of a particular color. Indicate how many spheres of the missing color must be included for each diagram to represent a system at equilibrium.

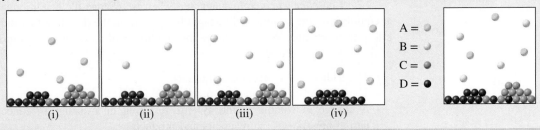

(i)　　(ii)　　(iii)　　(iv)

A = ○
B = ○
C = ●
D = ●

Section 15.5 Review

Using Equilibrium Expressions to Solve Problems

Use the following information to answer questions 15.5.1 and 15.5.2: K_c for the reaction

$$A + B \rightleftharpoons 2C$$

is 1.7×10^{-2} at 250°C.

15.5.1 What will the equilibrium concentrations of A, B, and C be at this temperature if $[A]_i = [B]_i = 0.750\ M$ ($[C]_i = 0$)?
(a) $6.1 \times 10^{-3}\ M$, $6.1 \times 10^{-3}\ M$, 0.092 M　　(d) 0.70 M, 0.70 M, 0.092 M
(b) 0.046 M, 0.046 M, 0.092 M　　(e) 0.087 M, 0.087 M, 0.66 M
(c) 0.70 M, 0.70 M, 0.046 M

15.5.2 What will the equilibrium concentrations of A, B, and C be at this temperature if $[C]_i = 0.875\ M$ ($[A]_i = [B]_i = 0$)?
(a) 0.41 M, 0.41 M, 0.82 M　　(d) 0.43 M, 0.43 M, 0.44 M
(b) 0.41 M, 0.41 M, 0.054 M　　(e) 0.43 M, 0.43 M, 0.43 M
(c) 0.43 M, 0.43 M, 0.0074 M

15.5.3 If $K_c = 3$ for the reaction $X + 2Y \rightleftharpoons Z$ at a certain temperature, then for each of the mixtures of X, Y, and Z shown here, in what direction must the reaction proceed to achieve equilibrium?

X = ●
Y = ○
Z = ●

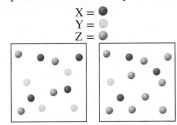

(a) Right, right　　(d) Left, neither
(b) Neither, left　　(e) Right, neither
(c) Neither, right

15.6 LE CHÂTELIER'S PRINCIPLE: FACTORS THAT AFFECT EQUILIBRIUM

One of the interesting and useful features of chemical equilibria is that they can be manipulated in specific ways to maximize production of a desired substance. Consider, for example, the industrial production of ammonia from its constituent elements by the Haber process.

$$N_2(g) + 3H_2(g) \rightleftharpoons 2NH_3(g)$$

More than 100 million tons of ammonia is produced annually by this reaction, with most of the resulting ammonia being used for fertilizers to enhance crop production. Clearly it would be in the best interest of industry to maximize the yield of NH_3. In this section, we will learn about the various ways in which an equilibrium can be manipulated to accomplish this goal.

Le Châtelier's principle states that when we change the conditions of a system at equilibrium, the system will respond by *shifting* in the direction that minimizes the effect of the change. We may change the conditions of a system at equilibrium by any of the following means.

- The addition of a reactant or product
- The removal of a reactant or product
- A change in volume of the system, resulting in a change in concentration or partial pressure of the reactants and products
- A change in temperature

"Shifting" refers to the occurrence of either the forward or reverse reaction such that the effect of the change is partially offset as the system reestablishes equilibrium. An equilibrium that shifts to the right is one in which more products are produced by the forward reaction. An equilibrium that shifts to the left is one in which more reactants are produced by the reverse reaction. Using Le Châtelier's principle, we can predict the direction in which an equilibrium will shift, given the specific change that is made.

Addition or Removal of a Substance

Again using the Haber process as an example,

$$N_2(g) + 3H_2(g) \rightleftharpoons 2NH_3(g)$$

consider a system at 700 K, in which the equilibrium concentrations are as follows:

$$[N_2] = 2.05 \ M \qquad [H_2] = 1.56 \ M \qquad [NH_3] = 1.52 \ M$$

Student Annotation: Remember that at equilibrium, the reaction quotient, Q_c, is equal to the equilibrium constant, K_c.

Using these concentrations in the reaction quotient expression, we can calculate the value of K_c for the reaction at this temperature as follows:

$$Q_c = \frac{[NH_3]^2}{[N_2][H_2]^3} = \frac{(1.52)^2}{(2.05)(1.56)^3} = 0.297 = K_c$$

If we were to change the conditions of this system by adding more N_2, increasing its concentration from 2.05 M to 3.51 M, the system would no longer be at equilibrium. To see that this is true, use the new concentration of nitrogen in the reaction quotient expression. The new calculated value of Q_c (0.173) is no longer equal to the value of K_c (0.297).

$$Q_c = \frac{[NH_3]^2}{[N_2][H_2]^3} = \frac{(1.52)^2}{(3.51)(1.56)^3} = 0.173 \neq K_c$$

For this system to reestablish equilibrium, the net reaction will have to shift in such a way that Q_c is again equal to K_c, which is constant at a given temperature. Recall

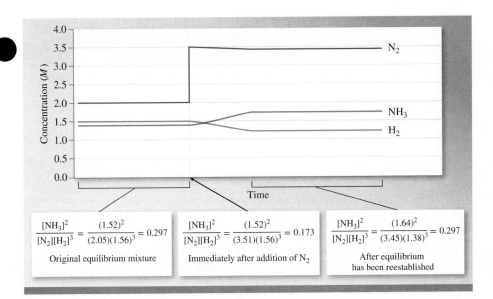

Figure 15.6 Adding more of a reactant to a system at equilibrium causes the equilibrium position to shift toward product. The system responds to the addition of N_2 by consuming some of the added N_2 (and some of the other reactant, H_2) to produce more NH_3.

from Section 15.4 that when Q is less than K, the reaction proceeds to the right to achieve equilibrium. Likewise, an equilibrium that is changed in such a way that Q *becomes* less than K will *shift* to the right to reestablish equilibrium. This means that the forward reaction, the consumption of N_2 and H_2 to produce NH_3, will occur. The result will be a net decrease in the concentrations of N_2 and H_2 (thus making the denominator of the reaction quotient smaller), and a net increase in the concentration of NH_3 (thus making the numerator larger). When the concentrations of all species are such that Q_c is again equal to K_c, the system will have established a new *equilibrium position,* meaning that it will have shifted in one direction or the other, resulting in a new equilibrium concentration for each species. Figure 15.6 shows how the concentrations of N_2, H_2, and NH_3 change when N_2 is added to the original equilibrium mixture.

Conversely, if we were to remove N_2 from the original equilibrium mixture, the lower concentration in the denominator of the reaction quotient would result in Q_c being greater than K_c. In this case the reaction will shift to the left. That is, the reverse reaction will take place, thereby increasing the concentrations of N_2 and H_2 and decreasing the concentration of NH_3 until Q_c is once again equal to K_c.

The addition or removal of NH_3 will cause a shift in the equilibrium, too. The addition of NH_3 will cause a shift to the left; the removal of NH_3 will cause a shift to the right. Figure 15.7(a) shows the additions and removals that cause this equilibrium to shift to the right. Figure 15.7(b) shows those that cause it to shift to the left.

In essence, a system at equilibrium will respond to the addition of a species by consuming some of that species, and it will respond to the removal of a species by producing more of that species. It is important to remember that the addition or removal of a species from an equilibrium mixture does not change the value of the equilibrium constant, K. Rather, it changes temporarily the value of the reaction quotient, Q. Furthermore, to cause a shift in the equilibrium, the species added or removed must be one that appears in the reaction quotient expression. In the case of a heterogeneous equilibrium, altering the amount of a solid or liquid species does not change the position of the equilibrium because doing so does not change the value of Q.

Worked Example 15.15 explores the effects of changing conditions for a system at equilibrium.

Student Annotation: While reestablishing equilibrium causes a decrease in the N_2 concentration, the final concentration will still be higher than that in the original equilibrium mixture. The system responds to the the added reactant by consuming part of it.

Student Hot Spot

Student data indicate you may struggle with how concentration changes affect equilibria. Access the SmartBook to view additional Learning Resources on this topic.

(a)

(b)

Figure 15.7 (a) Addition of a reactant or removal of a product will cause an equilibrium to shift to the right. (b) Addition of a product or removal of a reactant will cause an equilibrium to shift to the left.

Worked Example 15.15

Hydrogen sulfide (H_2S) is a contaminant commonly found in natural gas. It is removed by reaction with oxygen to produce elemental sulfur.

$$2H_2S(g) + O_2(g) \rightleftharpoons 2S(s) + 2H_2O(g)$$

For each of the following scenarios, determine whether the equilibrium will shift to the right, shift to the left, or neither: (a) addition of $O_2(g)$, (b) removal of $H_2S(g)$, (c) removal of $H_2O(g)$, and (d) addition of $S(s)$.

Strategy Use Le Châtelier's principle to predict the direction of shift for each case. Remember that the position of the equilibrium is only changed by the addition or removal of a species that appears in the reaction quotient expression.

Setup Begin by writing the reaction quotient expression,

$$Q_c = \frac{[H_2O]^2}{[H_2S]^2[O_2]}$$

Because sulfur is a solid, it does not appear in the expression. Changes in the concentration of any of the other species will cause a change in the equilibrium position. Addition of a reactant or removal of a product that appears in the expression for Q_c will shift the equilibrium to the right.

addition   addition

$$2H_2S(g) \ + \ O_2(g) \ \xrightarrow{\longleftarrow} \ 2S(s) \ + \ 2H_2O(g)$$

removal

Removal of a reactant or addition of a product that appears in the expression for Q_c will shift the equilibrium to the left.

addition

$$2H_2S(g) \ + \ O_2(g) \ \xleftarrow{\longrightarrow} \ 2S(s) \ + \ 2H_2O(g)$$

removal   removal

Solution

(a) Shift to the right   (b) Shift to the left   (c) Shift to the right   (d) No change

---

Think About It

In each case, analyze the effect the change will have on the value of Q_c. In part (a), for example, O_2 is added, so its concentration increases. Looking at the reaction quotient expression, we can see that a larger concentration of oxygen corresponds to a larger overall denominator—giving the overall fraction a smaller value. Thus, Q will temporarily be smaller than K and the reaction will have to shift to the right, consuming some of the added O_2 (along with some of the H_2S in the mixture) to reestablish equilibrium.

Practice Problem **A**TTEMPT For each change indicated, determine whether the equilibrium

$$PCl_3(g) + Cl_2(g) \rightleftharpoons PCl_5(g)$$

will shift to the right, shift to the left, or neither: (a) addition of $PCl_3(g)$, (b) removal of $PCl_3(g)$, (c) removal of $PCl_5(g)$, and (d) removal of $Cl_2(g)$.

Practice Problem **B**UILD What can be added to the equilibrium that will (a) shift it to the left, (b) shift it to the right, (c) not shift it in either direction?

$$AgCl(s) + 2NH_3(aq) \rightleftharpoons Ag(NH_3)_2^+(aq) + Cl^-(aq)$$

Practice Problem **C**ONCEPTUALIZE Consider the reaction $A(g) + B(g) \rightleftharpoons C(s) + D(s)$. The first of the diagrams on the right represents a system at equilibrium where A = ⬤, B = ⬤, C = ⬤, and D = ⬤. The second diagram on the right represents the system immediately after more A has been added. Which of the following diagrams [(i)–(iv)] could represent the system after equilibrium has been reestablished? Select all that apply.

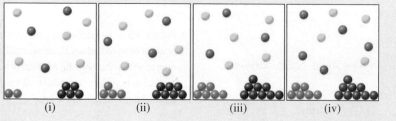

A = ⬤
B = ⬤
C = ⬤
D = ⬤

(i)   (ii)   (iii)   (iv)

Changes in Volume and Pressure

If we were to start with a gaseous system at equilibrium in a cylinder with a movable piston, we could change the volume of the system, thereby changing the concentrations of the reactants and products.

Consider again the equilibrium between N_2O_4 and NO_2.

$$N_2O_4(g) \rightleftharpoons 2NO_2(g)$$

At 25°C the equilibrium constant for this reaction is 4.63×10^{-3}. Suppose we have an equilibrium mixture of 0.643 M N_2O_4 and 0.0547 M NO_2 in a cylinder fitted with a movable piston. If we push down on the piston, the equilibrium will be disturbed and will shift in the direction that minimizes the effect of this disturbance. Consider what happens to the concentrations of both species if we decrease the volume of the cylinder by half. Both concentrations are initially *doubled*: $[N_2O_4] = 1.286$ M and $[NO_2] = 0.1094$ M. If we plug the new concentrations into the reaction quotient expression, we get

$$Q_c = \frac{[NO_2]_{eq}^2}{[N_2O_2]_{eq}} = \frac{(0.1094)^2}{1.286} = 9.31 \times 10^{-3}$$

which is not equal to K_c, so the system is no longer at equilibrium. Because Q_c is greater than K_c, the equilibrium will have to shift to the left for equilibrium to be reestablished (Figure 15.8).

In general, a decrease in volume of a reaction vessel will cause a shift in the equilibrium in the direction that minimizes the total number of moles of gas. Conversely, an increase in volume will cause a shift in the direction that maximizes the total number of moles of gas.

Worked Example 15.16 shows how to predict the equilibrium shift that will be caused by a volume change.

Worked Example 15.16

For each reaction, predict in what direction the equilibrium will shift when the volume of the reaction vessel is decreased.

(a) $PCl_5(g) \rightleftharpoons PCl_3(g) + Cl_2(g)$ (c) $H_2(g) + I_2(g) \rightleftharpoons 2HI(g)$

(b) $2PbS(s) + 3O_2(g) \rightleftharpoons 2PbO(s) + 2SO_2(g)$

Strategy Determine which direction minimized the number of moles of gas in the reaction. Count only moles of *gas*.

Setup We have (a) 1 mole of gas on the reactant side and 2 moles of gas on the product side, (b) 3 moles of gas on the reactant side and 2 moles of gas on the product side, and (c) 2 moles of gas on each side.

Student Annotation: It is a common error to count all the species in a reaction to determine which side has fewer moles. To determine what direction shift a volume change will cause, it is only the number of moles of *gas* that matters.

Solution (a) Shift to the left   (b) Shift to the right   (c) No shift

(Continued on next page)

Think About It
When there is no difference in the number of moles of gas, changing the volume of the reaction vessel will change the concentrations of reactant(s) and product(s)—but the system will remain at equilibrium. (Q will remain equal to K.)

Practice Problem ATTEMPT   For each reaction, predict the direction of shift caused by increasing the volume of the reaction vessel.

(a) $2NOCl(g) \rightleftharpoons 2NO(g) + Cl_2(g)$ (c) $Zn(s) + 2H^+(aq) \rightleftharpoons Zn^{2+}(aq) + H_2(g)$

(b) $CaCO_3(s) \rightleftharpoons CaO(s) + CO_2(g)$

Practice Problem BUILD   For the following equilibrium, give an example of a change that will cause a shift to the right, a change that will cause a shift to the left, and one that will cause no shift.

$$H_2(g) + F_2(g) \rightleftharpoons 2HF(g)$$

Practice Problem CONCEPTUALIZE   Consider the reaction $A(g) + B(g) \rightleftharpoons AB(g)$. The diagram shown on the top right represents a system at equilibrium. Which of the following diagrams [(i)–(iii)] best represents the system when equilibrium has been reestablished following a volume increase of 50 percent?

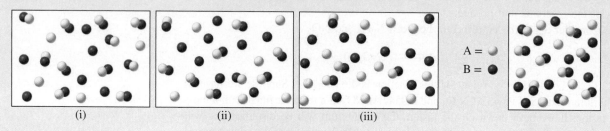

(i)	(ii)	(iii)

A = (light sphere)
B = (dark sphere)

It is possible to change the total pressure of a system without changing its volume—by adding an inert gas such as helium to the reaction vessel. Because the total volume remains the same, the concentrations of reactant and product gases do not change. Therefore, the equilibrium is not disturbed and no shift will occur.

Changes in Temperature

A change in concentration or volume may alter the position of an equilibrium (i.e., the relative amounts of reactants and products), but it does not change the value of the equilibrium constant. Only a change in temperature can alter the value of the equilibrium constant. To understand why, consider the following reaction.

$$N_2O_4(g) \rightleftharpoons 2NO_2(g)$$

The forward reaction is endothermic (absorbs heat, $\Delta H > 0$).

$$\text{Heat} + N_2O_4(g) \rightleftharpoons 2NO_2(g) \qquad \Delta H° = 58.0 \text{ kJ/mol}$$

If we treat heat as though it were a reactant, we can use Le Châtelier's principle to predict what will happen if we add or remove heat. Increasing the temperature (adding heat) will shift the reaction in the forward direction because heat appears on the reactant side. Lowering the temperature (removing heat) will shift the reaction in the reverse direction. Consequently, the equilibrium constant, given by

$$K_c = \frac{[NO_2]^2}{[N_2O_4]}$$

increases when the system is heated and decreases when the system is cooled (Figure 15.9). A similar argument can be made in the case of an exothermic reaction, where heat can be considered to be a product. Increasing the temperature of an exothermic reaction causes the equilibrium constant to decrease, shifting the equilibrium toward reactants.

Another way to understand this effect is to consider what happens to the value of ΔG when temperature changes. Consider Equation 14.10

$$\Delta G = \Delta H - T\Delta S$$

(a)   (b)

Figure 15.9 (a) N_2O_4/NO_2 equilibrium. (b) Because the reaction is endothermic, at higher temperature, the equilibrium $N_2O_4(g) \rightleftarrows 2NO_2(g)$ shifts toward product, making the reaction mixture darker.

© McGraw-Hill Education//Charles D. Winters, photographer.

At equilibrium, because $\Delta G = 0$, we can write

$$\Delta H = T\Delta S$$

For an exothermic reaction, ΔH is negative; and because absolute temperature is a *positive* number, ΔS must also be negative. Increasing the temperature causes the term $T\Delta S$ to be a larger negative number, and subtracting a larger *negative* number from ΔH gives a *positive* value for ΔG. When ΔG is positive, the reaction must proceed to the left to reestablish equilibrium.

Another example of this phenomenon is the equilibrium between the following ions.

$$\underset{\text{blue}}{CoCl_4^{2-}} + 6H_2O \rightleftarrows \underset{\text{pink}}{Co(H_2O)_6^{2+}} + 4Cl^- + \text{heat}$$

The reaction as written, the formation of $Co(H_2O)_6^{2+}$, is exothermic. Thus, the reverse reaction, the formation of $CoCl_4^{2-}$, is endothermic. On heating, the equilibrium shifts to the left and the solution turns blue. Cooling favors the exothermic reaction [the formation of $Co(H_2O)_6^{2+}$] and the solution turns pink (Figure 15.10).

In summary, a temperature increase favors an endothermic reaction, and a temperature decrease favors an exothermic process. Temperature affects the position of an equilibrium by changing the value of the equilibrium constant. Figures 15.11 and 15.12 (pages 692–695) illustrate the effects of various changes to systems at equilibrium.

Videos 15.2 & 15.3
Le Châtelier's principle Figures 15.11 and 15.12

Figure 15.10 (a) An equilibrium mixture of $CoCl_4^{2-}$ ions and $Co(H_2O)_6^{2+}$ ions appears violet. (b) Heating favors the formation of $CoCl_4^{2-}$, making the solution look more blue. (c) Cooling favors the formation of $Co(H_2O)_6^{2+}$, making the solution look more pink.

© McGraw-Hill Education/Charles D. Winters, photographer.

(a)   (b)   (c)

Figure 15.11 Le Châtelier's principle.

Effect of Addition to an Equilibrium Mixture

$$H_2(g) + I_2(g) \rightleftharpoons 2HI(g)$$

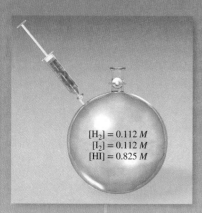

$[H_2] = 0.112\ M$
$[I_2] = 0.112\ M$
$[HI] = 0.825\ M$

$$Q_c = \frac{[HI]^2}{[H_2][I_2]} = \frac{(0.825)^2}{(0.112)(0.112)} = 54.3$$

$$Q_c = K_c$$

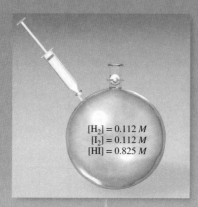

$[H_2] = 0.112\ M$
$[I_2] = 0.112\ M$
$[HI] = 0.825\ M$

$$Q_c = \frac{[HI]^2}{[H_2][I_2]} = \frac{(0.825)^2}{(0.112)(0.112)} = 54.3$$

$$Q_c = K_c$$

$[H_2] = 0.112\ M$
$[I_2] = 0.112\ M$
$[HI] = 0.825\ M$

$$Q_c = \frac{[HI]^2}{[H_2][I_2]} = \frac{(0.825)^2}{(0.112)(0.112)} = 54.3$$

$$Q_c = K_c$$

Add a reactant—$I_2(g)$.

$[H_2] = 0.112\ M$
$[I_2] = 0.499\ M$
$[HI] = 0.825\ M$

$$Q_c = \frac{(0.825)^2}{(0.112)(0.499)} = 12$$

$$Q_c \neq K_c$$

Add a product—$HI(g)$.

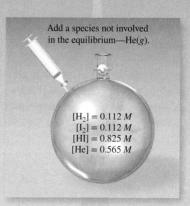

$[H_2] = 0.112\ M$
$[I_2] = 0.112\ M$
$[HI] = 3.17\ M$

$$Q_c = \frac{(3.17)^2}{(0.112)(0.112)} = 801$$

$$Q_c \neq K_c$$

Add a species not involved
in the equilibrium—$He(g)$.

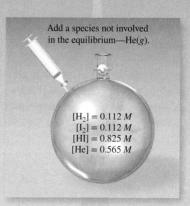

$[H_2] = 0.112\ M$
$[I_2] = 0.112\ M$
$[HI] = 0.825\ M$
$[He] = 0.565\ M$

$$Q_c = \frac{(0.825)^2}{(0.112)(0.112)} = 54.3$$

$$Q_c = K_c$$

Equilibrium shifts toward product.

$[H_2] = 0.0404\ M$
$[I_2] = 0.427\ M$
$[HI] = 0.968\ M$

$$Q_c = \frac{(0.968)^2}{(0.0404)(0.427)} = 54.3$$

$$Q_c = K_c$$

Equilibrium shifts toward reactant.

$[H_2] = 0.362\ M$
$[I_2] = 0.362\ M$
$[HI] = 2.67\ M$

$$Q_c = \frac{(2.67)^2}{(0.362)(0.362)} = 54.4$$

$$Q_c = K_c$$

Equilibrium does not shift in either direction.

$[H_2] = 0.112\ M$
$[I_2] = 0.112\ M$
$[HI] = 0.825\ M$
$[He] = 0.565\ M$

What's the point?

Adding a reactant to an equilibrium mixture shifts the equilibrium toward the product side of the equation. Adding a product shifts the equilibrium toward the reactant side. Adding a species that is neither a reactant nor a product does not cause a shift in the equilibrium.

Effect of Temperature Change

$$N_2O_4(g) \rightleftharpoons 2NO_2(g) \quad \Delta H° = 58.04 \text{ kJ/mol}$$

Temperature decrease drives an
endothermic equilibrium toward reactants.

Temperature increase drives an
endothermic equilibrium toward products.

$$H_2(g) + I_2(g) \rightleftharpoons 2HI(g) \quad \Delta H° = -9.4 \text{ kJ/mol}$$

Temperature increase drives an
exothermic equilibrium toward reactants.

Temperature decrease drives an
exothermic equilibrium toward products.

What's the point?

Increasing the temperature of an equilibrium mixture
causes a shift toward the product side for an endothermic
reaction, and a shift toward the reactant side for an
exothermic reaction.

Figure 15.12 Le Châtelier's principle.

Effect of Volume Change

$[N_2O_4] = 1.08\ M$
$[NO_2] = 0.0707\ M$

$$Q_c = \frac{[NO_2]^2}{[N_2O_4]} = \frac{(0.0707)^2}{1.08} = 4.63 \times 10^{-3}$$

$$Q_c = K_c$$

$[N_2O_4] = 0.540\ M$
$[NO_2] = 0.0354\ M$

$$Q_c = \frac{(0.0354)^2}{0.540} = 2.3 \times 10^{-3}$$

$$Q_c \neq K_c$$

$[N_2O_4] = 2.16\ M$
$[NO_2] = 0.141\ M$

$$Q_c = \frac{(0.141)^2}{2.16} = 9.2 \times 10^{-3}$$

$$Q_c \neq K_c$$

What's the point?

Increasing the volume causes a shift toward the side with the *larger* number of moles of gas. Decreasing the volume of an equilibrium mixture causes a shift toward the side of the equation with the *smaller* number of moles of gas.

$[N_2O_4] = 0.533\ M$
$[NO_2] = 0.0497\ M$

$$Q_c = \frac{(0.0497)^2}{0.533} = 4.6 \times 10^{-3}$$

$$Q_c = K_c$$

$[N_2O_4] = 2.18\ M$
$[NO_2] = 0.100\ M$

$$Q_c = \frac{(0.100)^2}{2.18} = 4.6 \times 10^{-3}$$

$$Q_c = K_c$$

$$[H_2] = 0.112\ M$$
$$[I_2] = 0.112\ M$$
$$[HI] = 0.825\ M$$
$$Q_c = \frac{[HI]^2}{[H_2][I_2]} = \frac{(0.825)^2}{(0.112)(0.112)} = 54.3$$
$$Q_c = K_c$$

$$[H_2] = 0.056\ M$$
$$[I_2] = 0.056\ M$$
$$[HI] = 0.413\ M$$
$$Q_c = \frac{(0.413)^2}{(0.056)(0.056)} = 54.3$$
$$Q_c = K_c$$

$$[H_2] = 0.224\ M$$
$$[I_2] = 0.224\ M$$
$$[HI] = 1.65\ M$$
$$Q_c = \frac{(1.65)^2}{(0.224)(0.224)} = 54.3$$
$$Q_c = K_c$$

What's the point?

For an equilibrium with equal numbers of gaseous moles on both sides, a change in volume does not cause the equilibrium to shift in either direction.

$$[H_2] = 0.056\ M$$
$$[I_2] = 0.056\ M$$
$$[HI] = 0.413\ M$$
$$Q_c = \frac{(0.413)^2}{(0.056)(0.056)} = 54.3$$
$$Q_c = K_c$$

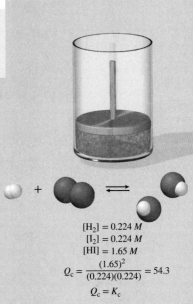

$$[H_2] = 0.224\ M$$
$$[I_2] = 0.224\ M$$
$$[HI] = 1.65\ M$$
$$Q_c = \frac{(1.65)^2}{(0.224)(0.224)} = 54.3$$
$$Q_c = K_c$$

Thinking Outside the Box

Biological Equilibria

In chemistry, and in biology, when a necessary ingredient for a process is in short supply, nature sometimes responds in a way that compensates for the shortage.

An important example of nature responding to a shortage is the production of extra red blood cells in individuals who reside at high elevation—where the concentration of oxygen is lower than at sea level. Extra red blood cells enable the blood to transport adequate oxygen from the lungs to the rest of the body, despite the low oxygen concentration. Enhanced oxygen-carrying capacity of the blood can give an athlete greater aerobic capacity and stamina (at normal oxygen concentrations). This has given rise to attempts by some athletes to increase their red blood count (RBC) artificially—a practice known as "blood doping."

Historically, blood doping has been done with blood infusions or with the drug *erythropoietin*—both of which have been banned by the World Anti-Doping Agency (WADA). In recent years, however, the use of hypoxic sleeping tents has grown in popularity. These tents are kept filled with a mixture of nitrogen and oxygen in which the oxygen concentration is lower than that in natural air. Just as the bodies of people who reside at high elevations naturally produce additional red blood cells to compensate for the lower partial pressure of oxygen in the air, the body of an athlete who spends his or her sleeping hours in a hypoxic tent adjusts in the same way, increasing the athlete's RBC. Those who "live high and train low" are believed to have an advantage over athletes who reside and train at or near sea level.

The combination of oxygen with the hemoglobin (Hb) molecule, which carries oxygen through the blood, is a complex reaction, but for our purposes it can be represented by the following simplified equation:*

$$Hb(aq) + O_2(aq) \rightleftharpoons HbO_2(aq)$$

where HbO_2 is oxyhemoglobin, the hemoglobin-oxygen complex that actually transports oxygen to tissues. The equilibrium expression for this process is

$$K_c = \frac{[HbO_2]}{[Hb][O_2]}$$

At an altitude of 3 km, the partial pressure of oxygen is only about 0.14 atm, compared with 0.20 atm at sea level. According to Le Châtelier's principle, a decrease in oxygen concentration will shift the hemoglobin-oxyhemoglobin equilibrium from right to left. This change depletes the supply of oxyhemoglobin, causing hypoxia. Over time, the body copes with this problem by producing more hemoglobin molecules. As the concentration of Hb increases, the equilibrium gradually shifts back toward the right (toward the formation of oxyhemoglobin). It can take several weeks for the increase in hemoglobin production to meet the body's oxygen needs adequately. A return to full capacity may require several years to occur. Studies show that long-time residents of high-altitude areas have high hemoglobin levels in their blood—sometimes as much as 50 percent more than individuals living at sea level. The production of more hemoglobin and the resulting increased capacity of the blood to deliver oxygen to the body have made high-altitude training and hypoxic tents popular among some athletes.

---

*Biological processes such as this are not true equilibria, but rather they are *steady-state* situations. In a steady state, the constant concentrations of reactants and products are not the result of forward and reverse reactions occurring at the same rate. Instead, reactant concentration is replenished by a previous reaction and product concentration is maintained by a subsequent reaction. Nevertheless, many of the principles of equilibrium, including Le Châtelier's principle, still apply.

Section 15.6 Review

Factors That Affect Chemical Equilibrium

15.6.1 Indicate in which direction the following equilibrium will shift in response to each change.

$$XY_2(s) \rightleftharpoons X(s) + Y_2(g)$$

(i) Addition of $XY_2(s)$, (ii) removal of $Y_2(g)$, (iii) decrease in container volume, (iv) addition of $X(s)$.

(a) Neither, left, right, neither
(b) Left, right, right, neither
(c) Right, neither, neither, left
(d) Neither, right, left, neither
(e) Right, right, left, left

15.6.2 Indicate in which direction the following equilibrium will shift in response to each change.

$$A_2(g) + B_2(g) \rightleftharpoons 2AB(g)$$

(i) Addition of $A_2(g)$, (ii) removal of $B_2(g)$, (iii) increase in container volume, (iv) addition of $AB(g)$.

(a) Left, right, neither, right
(b) Right, left, neither, left
(c) Right, neither, left, right
(d) Neither, left, right, right
(e) Left, neither, neither, left

15.6.3 The diagram shows the gaseous reaction $2A \rightleftharpoons A_2$ at equilibrium. How will the numbers of A and A_2 change if the volume of the container is increased at constant temperature?

(a) A_2 will increase and A will decrease.
(b) A will increase and A_2 will decrease.
(c) A_2 and A will both decrease.
(d) A_2 and A will both increase.
(e) Neither A_2 nor A will change.

15.6.4 The diagrams show equilibrium mixtures of A_2, B_2, and AB at two different temperatures $(T_2 > T_1)$. Is the reaction $A_2 + B_2 \rightleftharpoons 2AB$ endothermic or exothermic?

A = ●
B = ◌

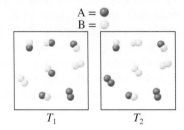

T_1 T_2

(a) Endothermic
(b) Exothermic
(c) Neither
(d) There is not enough information to determine.

Learning Outcomes

- Define reversible process.
- Define equilibrium.
- Differentiate between equilibrium constant and reaction quotient.
- Write the equilibrium constant expression for a given reaction.
- Determine the value of the equilibrium constant given equilibrium concentrations of reactants and products.
- Use the equilibrium constant to predict the relative amounts of reactants and products at equilibrium.
- Differentiate between heterogeneous and homogeneous equilibria.
- Convert between K_c and K_P for a reaction involving gases.
- Calculate ΔG and $\Delta G°$ of a reaction at a specified temperature given Q or K.

- Predict the direction of a reaction given initial concentrations of reactants and products and the value of the equilibrium constant.
- Calculate the equilibrium concentration of reactants or products given initial concentrations.
- Construct an equilibrium table for a reaction and use it to determine equilibrium, initial or final concentrations of a reactant or product.
- Give in your own words the meaning of Le Châtelier's principle.
- Apply Le Châtelier's principle toward determining the shift of a reaction at equilibrium given a change in one of the following: removal or addition of reactant or product, change in volume or pressure, and temperature change.

Chapter Summary

SECTION 15.1

- By our definition of a chemical reaction, reactants are consumed and products are produced. This is known as the *forward* reaction. A *reversible process* is one in which the products can also be consumed to produce reactants, a process known as the *reverse* reaction.
- *Equilibrium* is the condition where the forward and reverse reactions are occurring at the same rate and there is no net change in the reactant and product concentrations over time.
- In theory, equilibrium can be established starting with just reactants, with just products, or with a combination of reactants and products.

SECTION 15.2

- The *reaction quotient* (Q_c) is the product of the *product concentrations* over the product of the *reactant concentrations,* with each concentration raised to a power equal to the corresponding stoichiometric coefficient in the balanced chemical equation. This is known as the *law of mass action.*
- At equilibrium, the reaction quotient is equal to a constant value, the *equilibrium constant* (K_c).
- At equilibrium, concentrations in the reaction quotient are *equilibrium* concentrations and the quotient is called the *equilibrium expression.*
- K_c is constant at constant temperature. The value of K_c can be calculated by plugging equilibrium concentrations into the equilibrium expression.

- A large equilibrium constant $(K_c > 10^2)$ indicates that products are favored at equilibrium. A small equilibrium constant $(K_c < 10^{-2})$ indicates that reactants are favored at equilibrium.

SECTION 15.3

- Solids and liquids do not appear in the equilibrium expression for a heterogeneous reaction.
- When chemical equations that represent equilibria are reversed, multiplied, combined with other equations, or any combination of these processes, the corresponding changes must be made to the equilibrium constants.
- Equilibrium expressions that contain only gases can be written either as K_c expressions or as K_P expressions. K_P expressions have the same form as K_c expressions but contain partial pressures rather than molar concentrations. The reaction quotient, Q, can also be expressed in terms of the pressures of products and reactants. In this case it is labeled Q_P.
- K_c and K_P are not usually equal. The values are the same *only* when the reaction results in no net change in the number of moles of gas.

SECTION 15.4

- The free-energy change (ΔG) is determined using the standard free-energy change $(\Delta G°)$ and the reaction quotient (Q).
- We can predict in which direction a reaction must proceed to achieve equilibrium by comparing the values of Q_c and K_c (or of Q_P and K_P).

- The sign of ΔG tells us whether the reaction is spontaneous under the conditions described.
- $\Delta G°$ is related to the equilibrium constant, K. A negative $\Delta G°$ corresponds to a large K; a positive $\Delta G°$ corresponds to a small K.

SECTION 15.5

- Starting concentrations can be used, along with the equilibrium expression and equilibrium constant, to determine equilibrium concentrations.

SECTION 15.6

- According to *Le Châtelier's principle,* a system at equilibrium will react to stress by shifting in the direction that will partially offset the effect of the stress.
- The stresses that can be applied to a system at equilibrium include the addition or removal of a substance, changes in the volume of the reaction vessel, and changes in the temperature.

Key Words

Equilibrium, 656
Equilibrium constant (K_c or K_P), 658
Equilibrium expression, 658

Law of mass action, 658
Le Châtelier's principle, 686

Reaction quotient (Q_c or Q_P), 657
Reversible process, 655

Key Equations

15.1 $\quad Q_c = \dfrac{[C]^c[D]^d}{[A]^a[B]^b}$	Reaction quotient, Q, can be calculated using starting concentrations or pressures. In cases where both reactants and products are present initially, we can determine which direction the reaction will proceed to reach equilibrium by comparing the value of Q with the value of K. If Q is less K, the reaction will proceed to the right. If Q is greater than K, the reaction will proceed to the left. (If Q is equal to K, the system is already at equilibrium.)
15.2 $\quad Q_c = \dfrac{[C]^c[D]^d}{[A]^a[B]^b} = K_c$ (at equilibrium)	The reaction quotient, Q, is equal to the product of product concentrations (each raised to the appropriate power) divided by the product of reactant concentrations (each raised to the appropriate power). For each chemical species in the reaction quotient, the "appropriate" power is the coefficient of that species in the balanced chemical equation. When the concentrations used to calculate Q are equilibrium concentrations, $Q = K$. Q and K can refer either to Q_c and K_c, or to Q_P and K_P.
15.3 $\quad \Delta n$ = moles of gaseous products − moles of gaseous reactants	For a reaction in which one or more species is a gas, Δn is the number of gaseous moles on the product side of the equation minus the number of gaseous moles on the reactant side: moles gaseous products − moles gaseous reactants.
15.4 $\quad K_P = K_c\,[(0.08206\text{ L}\cdot\text{atm/K}\cdot\text{mol}) \times T]^{\Delta n}$	For a reaction in which all of the species in the equilibrium expression are gases, we can write the expression in terms of pressures (K_P), or we can write the expression in terms of concentrations (K_c). Equation 15.4 is used to convert between the two different equilibrium constants. Note that when there is no change in the number of gaseous moles ($\Delta n = 0$), K_P is equal to K_c.
15.5 $\quad \Delta G = \Delta G° + RT \ln Q$	ΔG is calculated as the sum of $\Delta G°$ and the product of R, T, and ln of the reaction quotient, Q.
15.6 $\quad \Delta G° = -RT \ln K$	At equilibrium, Q is equal to K and ΔG is equal to zero. $\Delta G°$ is equal to minus the product of R, T, and ln of K.

Key Skills

Equilibrium Problems

The principles of equilibrium were introduced in Chapter 15 largely through the use of examples involving gases. However, it is important to recognize that these principles apply to *all* types of equilibria; and that the approach to solving an equilibrium problem is always the same. Many of the important reactions and processes that we will study occur in water (i.e., they are *aqueous equilibria*). Common examples of aqueous equilibria include the ionization of a weak acid such as hydrofluoric acid

$$HF(aq) \rightleftharpoons H^+(aq) + F^-(aq) \qquad\qquad K_a = 7.1 \times 10^{-4}$$

and the dissolution of a slightly soluble salt such as lead(II) chloride

$$PbCl_2(s) \rightleftharpoons Pb^{2+}(aq) + 2Cl^-(aq) \qquad\qquad K_{sp} = 2.4 \times 10^{-4}$$

Each of these equilibria has an equilibrium constant (K_c) associated with it—although they are shown, respectively, as K_a and K_{sp}. The subscripts a and sp simply identify the specific *type* of equilibrium to which the equilibrium constants refer: ionization of a weak <u>a</u>cid, or <u>s</u>olubility <u>p</u>roduct of a slightly soluble salt.

When solving an equilibrium problem, we start with a balanced chemical equation, and use it to write an equilibrium expression. We then construct an equilibrium table and fill in the known concentrations. Often, these are the *initial* concentrations; but they can also be *equilibrium* concentrations, depending on what information is given in the problem. Then, using stoichiometry, we determine the changes in concentrations of reactants and products, and we use the equilibrium *expression* and equilibrium *constant* to solve for the unknown concentrations.

Consider a solution that is 0.15 M in HF. We may wish to know the concentration of H^+ ion at equilibrium. We construct an equilibrium table and write the equilibrium expression.

$$HF(aq) \rightleftharpoons H^+(aq) + F^-(aq)$$

i	0.15	0	0
c			
e			

$$7.1 \times 10^{-4} = \frac{[H^+][F^-]}{[HF]}$$

Student Annotation: The initials i, c, and e stand for *initial, change,* and *equilibrium.*

As we will see in Chapter 16, the initial concentration of H^+ in a problem such as this is not really zero, but it is *very* small and can be neglected in most cases. Using the stoichiometry of the balanced equation, we determine that the concentration of HF will decrease by some unknown amount, which we designate x, and that the concentrations of H^+ and F^- will each increase by the same amount.

$$HF(aq) \rightleftharpoons H^+(aq) + F^-(aq)$$

i	0.15	0	0
c	$-x$	$+x$	$+x$
e	$0.15 - x$	x	x

$$7.1 \times 10^{-4} = \frac{[H^+][F^-]}{[HF]}$$

We enter the equilibrium concentrations (in this case, in terms of the unknown x) into the equilibrium expression and set it equal to the equilibrium constant.

$$7.1 \times 10^{-4} = \frac{x \cdot x}{0.15 - x}$$

In this case, because the equilibrium constant is so small, we neglect the x in the denominator and solving for x gives 0.010 M. This is the concentration of both the H^+ ion and the F^- ion.

Next, consider a saturated solution of $PbCl_2$. We may wish to know the concentrations of Pb^{2+} and Cl^- ions in the solution. In this case, we proceed in the same way by constructing an equilibrium table and writing an equilibrium expression. Recall that solids, such as $PbCl_2$, do not appear in the equilibrium expression. Before the $PbCl_2$ dissolves, the concentrations of lead(II) ion and chloride ion are zero.

$$PbCl_2(s) \rightleftharpoons Pb^{2+}(aq) + 2Cl^-(aq)$$

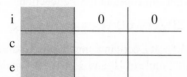

		0	0
i			
c			
e			

$$2.4 \times 10^{-4} = [Pb^{2+}][Cl^-]^2$$

The stoichiometry indicates that $[Pb^{2+}]$ will increase and that $[Cl^-]$ will increase by twice as much. In the case of solubility equilibria, we can use s in place of x, to remind us that the result is the "solubility" of the salt.

$$PbCl_2(s) \rightleftharpoons Pb^{2+}(aq) + 2Cl^-(aq)$$

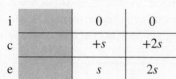

i	0	0
c	$+s$	$+2s$
e	s	$2s$

$$2.4 \times 10^{-4} = [Pb^{2+}][Cl^-]^2$$

We enter the equilibrium concentrations (in this case, in terms of the unknown s) into the equilibrium expression and set it equal to the equilibrium constant.

$$2.4 \times 10^{-4} = (s) \cdot (2s)^2$$

Solving for s gives 0.039 M. This is the concentration Pb^{2+} ion, but the concentration of Cl^- ion is $2s = 0.078\ M$.

Key Skills Problems

15.1

The K_a for hydrocyanic acid (HCN) is 4.9×10^{-10}. Determine the concentration of H^+ in a solution that is 0.25 M in HCN.
(a) $1.2 \times 10^{-10}\ M$
(b) $5.0 \times 10^{-4}\ M$
(c) $2.2 \times 10^{-5}\ M$
(d) $2.5 \times 10^{-5}\ M$
(e) $1.1 \times 10^{-5}\ M$

15.2

Determine the concentrations of Pb^{2+} and I^- in a saturated solution of PbI_2. (K_{sp} for PbI_2 is 1.4×10^{-8}.)
(a) 0.0015 M and 0.0030 M
(b) 0.0015 M and 0.0015 M
(c) 0.0015 M and 0.00075 M
(d) 0.0019 M and 0.0019 M
(e) 0.0019 M and 0.0038 M

15.3

Determine the K_a for a weak acid if a 0.10-M solution of the acid has $[H^+] = 4.6 \times 10^{-4}\ M$.
(a) 2.1×10^{-7}
(b) 2.1×10^{-6}
(c) 0.32
(d) 4.6×10^{-2}
(e) 0.046

15.4

Determine the K_{sp} for the slightly soluble salt A_2X, if $[A^+] = 0.019\ M$.
(a) 9.5×10^{-3}
(b) 3.6×10^{-4}
(c) 1.8×10^{-4}
(d) 3.4×10^{-6}
(e) 6.9×10^{-5}

Questions and Problems

SECTION 15.1: THE CONCEPT OF EQUILIBRIUM

Review Questions

15.1　Define *equilibrium*. Give two examples of a dynamic equilibrium.

15.2　Which of the following statements is correct about a reacting system at equilibrium: (a) the concentrations of reactants are equal to the concentrations of products, (b) the rate of the forward reaction is equal to the rate of the reverse reaction?

15.3　Consider the reversible reaction A $\rightleftharpoons$ B. Explain how equilibrium can be reached by starting with only A, only B, or a mixture of A and B.

SECTION 15.2: THE EQUILIBRIUM CONSTANT

Review Questions

15.4　What is the law of mass action?

15.5　Briefly describe the importance of equilibrium in the study of chemical reactions.

15.6　Define *reaction quotient*. How does it differ from the equilibrium constant?

15.7　Write reaction quotients for the following reactions. (a) $2NO(g) + O_2(g) \rightleftharpoons N_2O_4(g)$, (b) $S(s) + 3F_2(g) \rightleftharpoons SF_6(g)$, (c) $Co^{3+}(aq) + 6NH_3(aq) \rightleftharpoons Co(NH_3)_6^{3+}(aq)$, (d) $HCOOH(aq) \rightleftharpoons HCOO^-(aq) + H^+(aq)$.

15.8　Write the equation for the reaction that corresponds to each of the following reaction quotients.

(a) $Q_c = \dfrac{[H_2]^2[S_2]}{[H_2S]^2}$　　(c) $Q_c = \dfrac{[HgI_4^{2-}]}{[Hg^{2+}][I^-]^4}$

(b) $Q_c = \dfrac{[NO_2]^2[Cl_2]}{[NClO_2]^2}$　　(d) $Q_c = \dfrac{[NO]^2[Br_2]}{[NOBr]^2}$

Computational Problems

15.9　Consider the reaction

$$2NO(g) + 2H_2(g) \rightleftharpoons N_2(g) + 2H_2O(g)$$

At a certain temperature, the equilibrium concentrations are [NO] = 0.50 *M*, [H$_2$] = 0.25 *M*, [N$_2$] = 0.12 *M*, and [H$_2$O] = 2.05 *M*. (a) Write the equilibrium expression for the reaction. (b) Determine the value of the equilibrium constant.

15.10　The equilibrium constant for the reaction

$$2SO_2(g) + O_2(g) \rightleftharpoons 2SO_3(g)$$

is 5.1×10^3 at a certain temperature. If [SO$_2$] = 0.0664 *M* and [O$_2$] = 0.150 *M*, what is [SO$_3$]?

15.11　Consider the following equilibrium process at 700°C.

$$2H_2(g) + S_2(g) \rightleftharpoons 2H_2S(g)$$

Analysis shows that there are 2.50 moles of H$_2$, 1.35×10^{-5} mole of S$_2$, and 8.70 moles of H$_2$S

present in a 12.0-L flask. Calculate the equilibrium constant K_c for the reaction.

15.12　The equilibrium constant for the reaction

$$2H_2(g) + CO(g) \rightleftharpoons CH_3OH(g)$$

is 1.6×10^{-2} at a certain temperature. If there are 1.17×10^{-2} mole of H$_2$ and 3.46×10^{-3} mole of CH$_3$OH at equilibrium in a 5.60-L flask, what is the concentration of CO?

Conceptual Problems

15.13　The first diagram represents a system at equilibrium for the reaction 2A $\rightleftharpoons$ 2B where A = ● and B = ●. How many red spheres must be added to the second diagram for it also to represent a system at equilibrium?

15.14　Referring to the same diagrams as Problem 15.13, indicate how many red spheres must be added to the second diagram for it to represent a system at equilibrium if the reaction involved were A $\rightleftharpoons$ B.

SECTION 15.3: EQUILIBRIUM EXPRESSIONS

Review Questions

15.15　Define *homogeneous* equilibrium and *heterogeneous* equilibrium. Give two examples of each.

15.16　What do the symbols K_c and K_P represent?

15.17　Write the expressions for the equilibrium constants K_P of the following thermal decomposition reactions.
(a) $2NaHCO_3(s) \rightleftharpoons Na_2CO_3(s) + CO_2(g) + H_2O(g)$
(b) $2CaSO_4(s) \rightleftharpoons 2CaO(s) + 2SO_2(g) + O_2(g)$

15.18　Write equilibrium constant expressions for K_c and for K_P, if applicable, for the following processes.
(a) $2CO_2(g) \rightleftharpoons 2CO(g) + O_2(g)$
(b) $3O_2(g) \rightleftharpoons 2O_3(g)$
(c) $CO(g) + Cl_2(g) \rightleftharpoons COCl_2(g)$
(d) $H_2O(g) + C(s) \rightleftharpoons CO(g) + H_2(g)$
(e) $HCOOH(aq) \rightleftharpoons H^+(aq) + HCOO^-(aq)$
(f) $2HgO(s) \rightleftharpoons 2Hg(l) + O_2(g)$

15.19　Write the equilibrium constant expressions for K_c and for K_P, if applicable, for the following reactions.
(a) $2NO_2(g) + 7H_2(g) \rightleftharpoons 2NH_3(g) + 4H_2O(l)$
(b) $2ZnS(s) + 3O_2(g) \rightleftharpoons 2ZnO(s) + 2SO_2(g)$
(c) $C(s) + CO_2(g) \rightleftharpoons 2CO(g)$
(d) $C_6H_5COOH(aq) \rightleftharpoons C_6H_5COO^-(aq) + H^+(aq)$

15.20 Write the equation relating K_c to K_P, and define all the terms.

15.21 What is the rule for writing the equilibrium constant for the overall reaction that is the sum of two or more reactions?

15.22 Give an example of a multiple equilibria reaction.

Computational Problems

15.23 The equilibrium constant (K_c) for the reaction

$$2HCl(g) \rightleftharpoons H_2(g) + Cl_2(g)$$

is 4.17×10^{-34} at 25°C. What is the equilibrium constant for the reaction

$$H_2(g) + Cl_2(g) \rightleftharpoons 2HCl(g)$$

at the same temperature?

15.24 What is K_P at 1273°C for the reaction

$$2CO(g) + O_2(g) \rightleftharpoons 2CO_2(g)$$

if K_c is 2.24×10^{22} at the same temperature?

15.25 The equilibrium constant K_P for the reaction

$$2SO_3(g) \rightleftharpoons 2SO_2(g) + O_2(g)$$

is 1.8×10^{-5} at 350°C. What is K_c for this reaction?

15.26 Consider the reaction

$$N_2(g) + O_2(g) \rightleftharpoons 2NO(g)$$

If the equilibrium partial pressures of N_2, O_2, and NO are 0.15, 0.33, and 0.050 atm, respectively, at 2200°C, what is K_P?

15.27 A reaction vessel contains NH_3, N_2, and H_2 at equilibrium at a certain temperature. The equilibrium concentrations are $[NH_3] = 0.25\ M$, $[N_2] = 0.11\ M$, and $[H_2] = 1.91\ M$. Calculate the equilibrium constant K_c for the synthesis of ammonia if the reaction is represented as
(a) $N_2(g) + 3H_2(g) \rightleftharpoons 2NH_3(g)$

(b) $\frac{1}{2}N_2(g) + \frac{3}{2}H_2(g) \rightleftharpoons NH_3(g)$

15.28 The equilibrium constant K_c for the reaction

$$I_2(g) \rightleftharpoons 2I(g)$$

is 3.8×10^{-5} at 727°C. Calculate K_c and K_P for the equilibrium

$$2I(g) \rightleftharpoons I_2(g)$$

at the same temperature.

15.29 At equilibrium, the pressure of the reacting mixture

$$CaCO_3(s) \rightleftharpoons CaO(s) + CO_2(g)$$

is 0.105 atm at 350°C. Calculate K_P and K_c for this reaction.

15.30 The equilibrium constant K_P for the reaction

$$PCl_5(g) \rightleftharpoons PCl_3(g) + Cl_2(g)$$

is 1.05 at 250°C. The reaction starts with a mixture of PCl_5, PCl_3, and Cl_2 at pressures of 0.177, 0.223,

and 0.111 atm, respectively, at 250°C. When the mixture comes to equilibrium at that temperature, which pressures will have decreased and which will have increased? Explain why.

15.31 Ammonium carbamate ($NH_4CO_2NH_2$) decomposes as follows:

$$NH_4CO_2NH_2(s) \rightleftharpoons 2NH_3(g) + CO_2(g)$$

Starting with only the solid, it is found that when the system reaches equilibrium at 40°C, the total gas pressure (NH_3 and CO_2) is 0.363 atm. Calculate the equilibrium constant K_P.

15.32 Pure phosgene gas ($COCl_2$), 3.00×10^{-2} mol, was placed in a 1.50-L container. It was heated to 800 K, and at equilibrium the pressure of CO was found to be 0.497 atm. Calculate the equilibrium constant K_P for the reaction

$$CO(g) + Cl_2(g) \rightleftharpoons COCl_2(g)$$

15.33 Consider the equilibrium

$$2NOBr(g) \rightleftharpoons 2NO(g) + Br_2(g)$$

If nitrosyl bromide (NOBr) is 34% dissociated at 25°C and the total pressure is 0.25 atm, calculate K_P and K_c for the dissociation at this temperature.

15.34 The following equilibrium constants have been determined for hydrosulfuric acid at 25°C.

$$H_2S(aq) \rightleftharpoons H^+(aq) + HS^-(aq) \qquad K_c' = 9.5 \times 10^{-8}$$
$$HS^-(aq) \rightleftharpoons H^+(aq) + S^{2-}(aq) \qquad K_c'' = 1.0 \times 10^{-19}$$

Calculate the equilibrium constant for the following reaction at the same temperature.

$$H_2S(aq) \rightleftharpoons 2H^+(aq) + S^{2-}(aq)$$

15.35 The following equilibrium constants have been determined for oxalic acid at 25°C.

$$H_2C_2O_4(aq) \rightleftharpoons H^+(aq) + HC_2O_4^-(aq) \qquad K_c' = 6.5 \times 10^{-2}$$
$$HC_2O_4^-(aq) \rightleftharpoons H^+(aq) + C_2O_4^{2-}(aq) \qquad K_c'' = 6.1 \times 10^{-5}$$

Calculate the equilibrium constant for the following reaction at the same temperature.

$$H_2C_2O_4(aq) \rightleftharpoons 2H^+(aq) + C_2O_4^{2-}(aq)$$

15.36 The following equilibrium constants were determined at 1123 K.

$$C(s) + CO_2(g) \rightleftharpoons 2CO(g) \qquad K_P' = 1.3 \times 10^{14}$$
$$CO(g) + Cl_2(g) \rightleftharpoons COCl_2(g) \qquad K_P'' = 6.0 \times 10^{-3}$$

Write the equilibrium constant expression K_P, and calculate the equilibrium constant at 1123 K for

$$C(s) + CO_2(g) + 2Cl_2(g) \rightleftharpoons 2COCl_2(g)$$

15.37 At a certain temperature, the following reactions have the constants shown.

$$S(s) + O_2(g) \rightleftharpoons SO_2(g) \qquad K'_c = 4.2 \times 10^{52}$$

$$2S(s) + 3O_2(g) \rightleftharpoons 2SO_3(g) \qquad K''_c = 9.8 \times 10^{128}$$

Calculate the equilibrium constant K_c for the following reaction at that temperature.

$$2SO_2(g) + O_2(g) \rightleftharpoons 2SO_3(g)$$

Conceptual Problems

15.38 The following diagrams represent the equilibrium state for three different reactions of the type

$$A + X \rightleftharpoons AX \ (X = B, C, or D)$$

$$A + B \rightleftharpoons AB \qquad A + C \rightleftharpoons AC \qquad A + D \rightleftharpoons AD$$

(a) Which reaction has the largest equilibrium constant?

(b) Which reaction has the smallest equilibrium constant?

15.39 The equilibrium constant for the reaction $A \rightleftharpoons B$ is $K_c = 10$ at a certain temperature. (1) Starting with only reactant A, which of the diagrams shown here best represents the system at equilibrium? (2) Which of the diagrams best represents the system at equilibrium if $K_c = 0.10$? Explain why you can calculate K_c in each case without knowing the volume of the container. The grey spheres represent the A molecules, and the green spheres represent the B molecules.

(a)   (b)   (c)   (d)

SECTION 15.4: CHEMICAL EQUILIBRIUM AND FREE ENERGY

Review Questions

15.40 Explain the difference between ΔG and $\Delta G°$.

15.41 Explain why Equation 15.6 is of great importance in chemistry.

15.42 Fill in the missing entries in the following table.

K	$\ln K$	$\Delta G°$	Result at equilibrium
< 1			
		0	
			Products are favored

Computational Problems

15.43 The aqueous reaction

L-glutamate + pyruvate $\rightleftharpoons$ α-ketoglutarate + L-alanine

is catalyzed by the enzyme L-glutamate–pyruvate aminotransferase. At 300 K, the equilibrium constant for the reaction is 1.11. Predict whether the forward reaction will occur if the concentrations of the reactants and products are [L-glutamate] = 3.0×10^{-5} M, [pyruvate] = 3.3×10^{-4} M, [α-ketoglutarate] = 1.6×10^{-2} M, and [L-alanine] = 6.25×10^{-3} M.

15.44 For the autoionization of water at 25°C,

$$H_2O(l) \rightleftharpoons H^+(aq) + OH^-(aq)$$

K_w is 1.0×10^{-14}. What is $\Delta G°$ for the process?

15.45 Consider the following reaction at 25°C.

$$Fe(OH)_2(s) \rightleftharpoons Fe^{2+}(aq) + 2OH^-(aq)$$

Calculate $\Delta G°$ for the reaction. K_{sp} for $Fe(OH)_2$ is 1.6×10^{-14}.

15.46 Calculate $\Delta G°$ and K_P for the following equilibrium reaction at 25°C.

$$2H_2O(g) \rightleftharpoons 2H_2(g) + O_2(g)$$

15.47 (a) Calculate $\Delta G°$ and K_P for the following equilibrium reaction at 25°C.

$$PCl_5(g) \rightleftharpoons PCl_3(g) + Cl_2(g)$$

(b) Calculate ΔG for the reaction if the partial pressures of the initial mixture are $P_{PCl_5} = 0.0029$ atm, $P_{PCl_3} = 0.27$ atm, and $P_{Cl_2} = 0.40$ atm.

15.48 The equilibrium constant (K_P) for the reaction

$$H_2(g) + CO_2(g) \rightleftharpoons H_2O(g) + CO(g)$$

is 4.40 at 2000 K. (a) Calculate $\Delta G°$ for the reaction. (b) Calculate ΔG for the reaction when the partial pressures are $P_{H_2} = 0.25$ atm, $P_{CO_2} = 0.78$ atm, $P_{H_2O} = 0.66$ atm, and $P_{CO} = 1.20$ atm.

15.49 Consider the decomposition of calcium carbonate.

$$CaCO_3(s) \rightleftharpoons CaO(s) + CO_2(g)$$

Calculate the pressure in atm of CO_2 in an equilibrium process (a) at 25°C and (b) at 800°C. Assume that $\Delta H° = 177.8$ kJ/mol and $\Delta S° = 160.5$ J/K · mol for the temperature range.

15.50 The equilibrium constant K_P for the reaction

$$CO(g) + Cl_2(g) \rightleftharpoons COCl_2(g)$$

is 5.62×10^{35} at 25°C. Calculate $\Delta G°_f$ for $COCl_2$ at 25°C.

15.51 At 25°C, $\Delta G°$ for the process

$$H_2O(l) \rightleftharpoons H_2O(g)$$

is 8.6 kJ/mol. Calculate the vapor pressure of water at this temperature.

15.52   Calculate $\Delta G°$ for the process

$$C(\text{diamond}) \rightleftharpoons C(\text{graphite})$$

Is the formation of graphite from diamond favored at 25°C? If so, why is it that diamonds do not become graphite on standing?

15.53   Calculate K_P for the following reaction at 25°C.

$$H_2(g) + I_2(g) \rightleftharpoons 2HI(g) \qquad \Delta G° = 2.60 \text{ kJ/mol}$$

Conceptual Problems

15.54   A and B react to form A_2B according to the following equation.

$$2A + B \rightleftharpoons A_2B$$

All but one of the figures here represent equilibrium mixtures of A, B, and A_2B. Identify the figure in which the mixture is not at equilibrium.

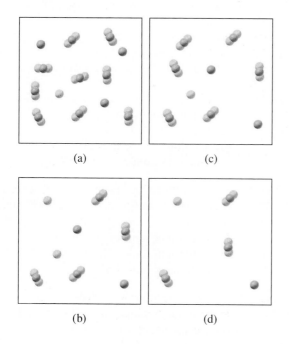

(a)   (c)

(b)   (d)

15.55   If $K_c = 2$ for the reaction $A_2 + B_2 \rightleftharpoons 2AB$ at a certain temperature, which of the following diagrams represents an equilibrium mixture of A_2, B_2, and AB? (Select all that apply.)

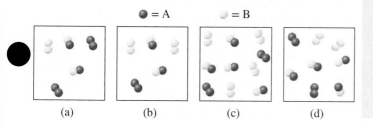

(a)   (b)   (c)   (d)

SECTION 15.5: CALCULATING EQUILIBRIUM CONCENTRATIONS

Visualizing Chemistry
Figure 15.5

VC 15.1 For an equilibrium based on the reaction $A + B \rightleftharpoons C$, where the initial concentrations of A and B are known, and the concentration of C is zero, what is the correct entry for the highlighted cell in the ice table shown here?

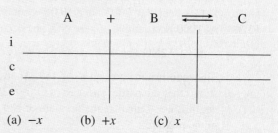

(a) $-x$ (b) $+x$ (c) x

VC 15.2 For an equilibrium based on the reaction $A + B \rightleftharpoons 2C$, where the initial concentrations of A and B are known and the initial concentration of C is zero, what is the correct entry for the highlighted cell in the ice table shown here?

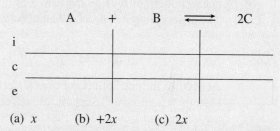

(a) x (b) $+2x$ (c) $2x$

VC 15.3 For an equilibrium based on the reaction $2A \rightleftharpoons B + C$, where the initial concentration of A is known and the initial concentrations of B and C are zero, what is the correct entry for the highlighted cell in the ice table shown here?

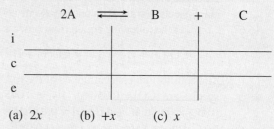

(a) $2x$ (b) $+x$ (c) x

VC 15.4 Of the reactions in questions 15.1–15.3, which has the largest K_c value if this diagram represents the system at equilibrium? $A = \bullet$, $B = \circ$, and $C = \circ$.

(a) $A + B \rightleftharpoons 2C$
(b) $A + B \rightleftharpoons C$
(c) $2A \rightleftharpoons B + C$

Review Questions

15.56 Outline the steps for calculating the concentrations of reacting species in an equilibrium reaction.

Computational Problems

15.57 For the reaction

$$H_2(g) + CO_2(g) \rightleftharpoons H_2O(g) + CO(g)$$

at 700°C, $K_c = 0.534$. Calculate the number of moles of H_2 that are present at equilibrium if a mixture of 0.300 mole of CO and 0.300 mole of H_2O is heated to 700°C in a 10.0-L container.

15.58 At 1000 K, a sample of pure NO_2 gas decomposes.

$$2NO_2(g) \rightleftharpoons 2NO(g) + O_2(g)$$

The equilibrium constant K_P is 158. Analysis shows that the partial pressure of O_2 is 0.25 atm at equilibrium. Calculate the pressure of NO and NO_2 in the mixture.

15.59 The equilibrium constant K_c for the reaction

$$H_2(g) + Br_2(g) \rightleftharpoons 2HBr(g)$$

is 2.18×10^6 at 730°C. Starting with 3.20 moles of HBr in a 12.0-L reaction vessel, calculate the concentrations of H_2, Br_2, and HBr at equilibrium.

15.60 The dissociation of molecular iodine into iodine atoms is represented as

$$I_2(g) \rightleftharpoons 2I(g)$$

At 1000 K, the equilibrium constant K_c for the reaction is 3.80×10^{-5}. Suppose you start with 0.0456 mole of I_2 in a 2.30-L flask at 1000 K. What are the concentrations of the gases at equilibrium?

15.61 The equilibrium constant K_c for the decomposition of phosgene ($COCl_2$) is 4.63×10^{-3} at 527°C.

$$COCl_2(g) \rightleftharpoons CO(g) + Cl_2(g)$$

Calculate the equilibrium partial pressures of all the components, starting with pure phosgene at 0.760 atm.

15.62 Consider the following equilibrium process at 686°C.

$$CO_2(g) + H_2(g) \rightleftharpoons CO(g) + H_2O(g)$$

The equilibrium concentrations of the reacting species are [CO] = 0.050 M, [H_2] = 0.045 M, [CO_2] = 0.086 M, and [H_2O] = 0.040 M.
(a) Calculate K_c for the reaction at 686°C.
(b) If we add CO_2 to increase its concentration to 0.50 mol/L, what will the concentrations of all the gases be when equilibrium is reestablished?

15.63 Consider the heterogeneous equilibrium process.

$$C(s) + CO_2(g) \rightleftharpoons 2CO(g)$$

At 700°C, the total pressure of the system is found to be 4.50 atm. If the equilibrium constant K_P is 1.52, calculate the equilibrium partial pressures of CO_2 and CO.

15.64 The equilibrium constant K_c for the reaction

$$H_2(g) + CO_2(g) \rightleftharpoons H_2O(g) + CO(g)$$

is 4.2 at 1650°C. Initially 0.80 mol H_2 and 0.80 mol CO_2 are injected into a 5.0-L flask. Calculate the concentration of each species at equilibrium.

SECTION 15.6: LE CHÂTELIER'S PRINCIPLE: FACTORS THAT AFFECT CHEMICAL EQUILIBRIUM

 Visualizing Chemistry
Figure 15.11 and Figure 15.12

Solid calcium carbonate decomposes to form solid calcium oxide and carbon dioxide gas.

$$CaCO_3(s) \rightleftharpoons CaO(s) + CO_2(g) \qquad \Delta H° = 177.8 \text{ kJ/mol}$$

Consider an equilibrium mixture of calcium carbonate and its decomposition products in a container.

VC 15.5 Which of the following changes will result in the formation of more CaO(s)?
(a) addition of more $CaCO_3(s)$
(b) addition of more $CO_2(g)$
(c) removal of some of the $CO_2(g)$

VC 15.6 Which of the following changes will result in the formation of more $CaCO_3(s)$?
(a) addition of more CaO(s)
(b) addition of more $CO_2(g)$
(c) removal of some of the $CaCO_3(s)$

VC 15.7 Which of the following changes will not cause the equilibrium to shift in either direction? (Select all that apply.)
(a) removal of some of the $CaCO_3(s)$
(b) removal of some of the $CO_2(g)$
(c) addition of more CaO(s)

VC 15.8 Which of the following will be affected by a change in the system's temperature? (Select all that apply.)
(a) the amount of $CaCO_3(s)$
(b) the amount of CaO(s)
(c) the amount of $CO_2(g)$

A sample of calcium carbonate is in equilibrium with its decomposition products in a closed system:
$CaCO_3(s) \rightleftharpoons CaO(s) + CO_2(g)$.

VC 15.9 If the volume of the system is increased at constant temperature, relative to the original equilibrium mixture, which of the following will be true when equilibrium is reestablished? (Select all that apply.)
(a) There will be a higher concentration of $CO_2(g)$.
(b) There will be a lower concentration of $CO_2(g)$.
(c) There will be more $CO_2(g)$.

VC 15.10 If the volume of the system is decreased at constant temperature, relative to the original equilibrium mixture, which of the following will be true when equilibrium is reestablished? (Select all that apply.)
  (a) There will be more $CaCO_3(s)$.
  (b) There will be more $CaO(s)$.
  (c) There will be more $CO_2(g)$.

A sample of gaseous hydrogen iodide is in equilibrium with its decomposition products in a closed system:
$2HI(g) \rightleftharpoons H_2(g) + I_2(g)$.

VC 15.11 If the volume of the system is increased at constant temperature, relative to the original equilibrium mixture, which of the following will be true when equilibrium is reestablished?
  (a) The concentration of HI will increase and the concentrations of H_2 and I_2 will decrease.
  (b) The concentration of all three species will be lower.
  (c) The concentration of all three species will be unchanged.

VC 15.12 If the volume of the system is decreased at constant temperature, relative to the original equilibrium mixture, which of the following will be true when equilibrium is reestablished?
  (a) There will be more HI.
  (b) There will be less HI.
  (c) The amount of HI will be unchanged.

Review Questions

15.65 Explain Le Châtelier's principle. How does this principle enable us to maximize the yields of desirable reactions and minimize the effect of undesirable ones?

15.66 Use Le Châtelier's principle to explain why the equilibrium vapor pressure of a liquid increases with increasing temperature.

15.67 List four factors that can shift the position of an equilibrium. Only one of these factors can alter the value of the equilibrium constant. Which one is it?

Conceptual Problems

15.68 Which of the following equilibria will shift to the left when the temperature is increased? [ΔH (kJ/mol) values are given in parentheses.] (Select all that apply.)
  (a) $S + H_2 \rightleftharpoons H_2S$ ΔH (−20)
  (b) $C + H_2O \rightleftharpoons CO + H_2$ ΔH (131)
  (c) $H_2 + CO_2 \rightleftharpoons H_2O + CO$ ΔH (41)
  (d) $MgO + CO_2 \rightleftharpoons MgCO_3$ ΔH (−117)
  (e) $2CO + O_2 \rightleftharpoons 2CO_2$ ΔH (−566)

15.69 For which of the following reactions will a change in volume *not* affect the position of the equilibrium? (Select all that apply.)
  (a) $MgO(s) + CO_2(g) \rightleftharpoons MgCO_3(s)$
  (b) $H_2(g) + Cl_2(g) \rightleftharpoons 2HCl(g)$
  (c) $BaCO_3(s) \rightleftharpoons BaO(s) + CO_2(g)$
  (d) $Br_2(l) + H_2(g) \rightleftharpoons 2HBr(g)$
  (e) $C(graphite) + CO_2(g) \rightleftharpoons 2CO(g)$

15.70 Which of the following equilibria will shift to the right when H_2 is added? (Select all that apply.)
  (a) $2H_2 + O_2 \rightleftharpoons 2H_2O$
  (b) $2HI \rightleftharpoons H_2 + I_2$
  (c) $H_2 + CO_2 \rightleftharpoons H_2O + CO$
  (d) $2NaHCO_3 \rightleftharpoons Na_2CO_3 + H_2O + CO_2$
  (e) $2CO + O_2 \rightleftharpoons 2CO_2$

15.71 Which of the following will cause the equilibrium

$$C(graphite) + CO_2(g) \rightleftharpoons 2CO(g)$$

to shift to the right? (Select all that apply.)
  (a) decreasing the volume
  (b) increasing the volume
  (c) adding more C(graphite)
  (d) adding more $CO_2(g)$
  (e) removing $CO(g)$ as it forms

15.72 Consider the following equilibrium system involving SO_2, Cl_2, and SO_2Cl_2 (sulfuryl dichloride).

$$SO_2(g) + Cl_2(g) \rightleftharpoons SO_2Cl_2(g)$$

Predict how the equilibrium position would change if (a) Cl_2 gas were added to the system, (b) SO_2Cl_2 were removed from the system, (c) SO_2 were removed from the system. The temperature remains constant in each case.

15.73 Heating solid sodium bicarbonate in a closed vessel establishes the following equilibrium.

$$2NaHCO_3(s) \rightleftharpoons Na_2CO_3(s) + H_2O(g) + CO_2(g)$$

What would happen to the equilibrium position if (a) some of the CO_2 were removed from the system, (b) some solid Na_2CO_3 were added to the system, (c) some of the solid $NaHCO_3$ were removed from the system? The temperature remains constant.

15.74 Consider the following equilibrium systems.
  (a) $A \rightleftharpoons 2B$ $\Delta H° = 20.0$ kJ/mol
  (b) $A + B \rightleftharpoons C$ $\Delta H° = -5.4$ kJ/mol
  (c) $A \rightleftharpoons B$ $\Delta H° = 0.0$ kJ/mol
Predict the change in the equilibrium constant K_c that would occur in each case if the temperature of the reacting system were raised.

15.75 What effect does an increase in pressure have on each of the following systems at equilibrium? The temperature is kept constant, and, in each case, the reactants are in a cylinder fitted with a movable piston.
  (a) $A(s) \rightleftharpoons 2B(s)$ (d) $A(g) \rightleftharpoons B(g)$
  (b) $2A(l) \rightleftharpoons B(l)$ (e) $A(g) \rightleftharpoons 2B(g)$
  (c) $A(s) \rightleftharpoons B(g)$

15.76   Consider the equilibrium

$$2I(g) \rightleftharpoons I_2(g)$$

What would be the effect on the position of equilibrium of (a) increasing the total pressure on the system by decreasing its volume, (b) adding I_2 to the reaction mixture, and (c) decreasing the temperature?

15.77   Consider the following equilibrium process.

$$PCl_5(g) \rightleftharpoons PCl_3(g) + Cl_2(g) \qquad \Delta H° = 92.5 \text{ kJ/mol}$$

Predict the direction of the shift in equilibrium when (a) the temperature is raised, (b) more chlorine gas is added to the reaction mixture, (c) some PCl_3 is removed from the mixture, (d) the pressure on the gases is increased, (e) a catalyst is added to the reaction mixture.

15.78   Consider the reaction

$$2SO_2(g) + O_2(g) \rightleftharpoons 2SO_3(g) \qquad \Delta H° = -198.2 \text{ kJ/mol}$$

Comment on the changes in the concentrations of SO_2, O_2, and SO_3 at equilibrium if we were to (a) increase the temperature, (b) increase the pressure, (c) increase SO_2, (d) add helium at constant volume.

15.79   Consider the following equilibrium reaction in a closed container.

$$CaCO_3(s) \rightleftharpoons CaO(s) + CO_2(g)$$

What will happen if (a) the volume is increased, (b) some CaO is added to the mixture, (c) some $CaCO_3$ is removed, (d) some CO_2 is added to the mixture, (e) a few drops of an NaOH solution are added to the mixture, (f) a few drops of an HCl solution are added to the mixture (ignore the reaction between CO_2 and water), (g) temperature is increased?

15.80   Consider the gas-phase reaction

$$2CO(g) + O_2(g) \rightleftharpoons 2CO_2(g)$$

Predict the shift in the equilibrium position when helium gas is added to the equilibrium mixture (a) at constant pressure and (b) at constant volume.

15.81   The following diagrams show an equilibrium mixture of O_2 and O_3 at temperatures T_1 and T_2 ($T_2 > T_1$). (a) Write an equilibrium equation showing the forward reaction to be exothermic. (b) Predict how the number of O_2 and O_3 molecules would change if the volume were decreased at constant temperature.

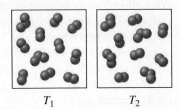

$T_1 \qquad\qquad T_2$

15.82   The following diagrams show the reaction A + B $\rightleftharpoons$ AB at two different temperatures. Is the forward reaction endothermic or exothermic?

600 K   800 K

15.83   The effect of acid (H^+) on the binding of oxygen to hemoglobin can be summarized as follows:

$$HbH^+(aq) + O_2(aq) \rightleftharpoons HbO_2(aq) + H^+(aq)$$

During strenuous exercise, acid can build up in a condition known as *acidosis*. What happens to the concentration of oxyhemoglobin as acidosis occurs?

15.84   The simplified equation representing the binding of oxygen by hemoglobin (Hb) is

$$Hb(aq) + O_2(aq) \rightleftharpoons HbO_2(aq) \qquad \Delta H < 0$$

Determine how each of the following affects the amount of HbO_2: (a) decreasing the temperature; (b) increasing the pressure of O_2; (c) decreasing the amount of hemoglobin.

15.85   Consider the reaction A + B $\rightleftharpoons$ 2C. The following top diagram represents a system at equilibrium where A = ◐, B = ●, and C = ●. Which of the diagrams [(a)–(d)] also represent a system at equilibrium? (Select all that apply.)

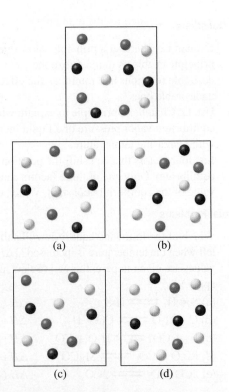

(a)   (b)

(c)   (d)

ADDITIONAL PROBLEMS

15.86 Consider the following reaction at equilibrium.

$$A(g) \rightleftharpoons 2B(g)$$

From the data shown here, calculate the equilibrium constant (both K_P and K_c) at each temperature. Is the reaction endothermic or exothermic?

Temperature (°C)	[A] (*M*)	[B] (*M*)
200	0.0125	0.843
300	0.171	0.764
400	0.250	0.724

15.87 The equilibrium constant K_P for the reaction

$$2H_2O(g) \rightleftharpoons 2H_2(g) + O_2(g)$$

is 2×10^{-42} at 25°C. (a) What is K_c for the reaction at the same temperature? (b) The very small value of K_P (and K_c) indicates that the reaction overwhelmingly favors the formation of water molecules. Explain why, despite this fact, a mixture of hydrogen and oxygen gases can be kept at room temperature without any change.

15.88 For a reaction with a negative $\Delta G°$ value, which of the following statements is false? (a) The equilibrium constant K is greater than one. (b) The reaction is spontaneous when all the reactants and products are in their standard states. (c) The reaction is always exothermic.

15.89 Carbon monoxide (CO) and nitric oxide (NO) are polluting gases contained in automobile exhaust. Under suitable conditions, these gases can be made to react to form nitrogen (N_2) and the less harmful carbon dioxide (CO_2). (a) Write an equation for this reaction. (b) Identify the oxidizing and reducing agents. (c) Calculate the K_P for the reaction at 25°C. (d) Under normal atmospheric conditions, the partial pressures are $P_{N_2} = 0.80$ atm, $P_{CO_2} = 3.0 \times 10^{-4}$ atm, $P_{CO} = 5.0 \times 10^{-5}$ atm, and $P_{NO} = 5.0 \times 10^{-7}$ atm. Calculate Q_P, and predict the direction toward which the reaction will proceed. (e) Will raising the temperature favor the formation of N_2 and CO_2?

15.90 Consider the following reacting system.

$$2NO(g) + Cl_2(g) \rightleftharpoons 2NOCl(g)$$

What combination of temperature and pressure would maximize the yield of nitrosyl chloride (NOCl)? [*Hint:* $\Delta H_f°$(NOCl) = 51.7 kJ/mol.]

15.91 At a certain temperature and a total pressure of 1.2 atm, the partial pressures of an equilibrium mixture

$$2A(g) \rightleftharpoons B(g)$$

are $P_A = 0.60$ atm and $P_B = 0.60$ atm. (a) Calculate the K_P for the reaction at this temperature. (b) If the total pressure were increased to 1.5 atm, what would be the partial pressures of A and B at equilibrium?

15.92 The decomposition of ammonium hydrogen sulfide

$$NH_4HS(s) \rightleftharpoons NH_3(g) + H_2S(g)$$

is an endothermic process. A 6.1589-g sample of the solid is placed in an evacuated 4.000-L vessel at exactly 24°C. After equilibrium has been established, the total pressure inside is 0.709 atm. Some solid NH_4HS remains in the vessel. (a) What is the K_P for the reaction? (b) What percentage of the solid has decomposed? (c) If the volume of the vessel were doubled at constant temperature, what would happen to the amount of solid in the vessel?

15.93 Consider the reaction

$$2NO(g) + O_2(g) \rightleftharpoons 2NO_2(g)$$

At 430°C, an equilibrium mixture consists of 0.020 mole of O_2, 0.040 mole of NO, and 0.96 mole of NO_2. Calculate K_P for the reaction, given that the total pressure is 0.20 atm.

15.94 In the Mond process for the purification of nickel, carbon monoxide is combined with heated nickel to produce $Ni(CO)_4$, which is a gas and can therefore be separated from solid impurities.

$$Ni(s) + 4CO(g) \rightleftharpoons Ni(CO)_4(g)$$

Given that the standard free energies of formation of CO(g) and $Ni(CO)_4(g)$ are −137.3 and −587.4 kJ/mol, respectively, calculate the equilibrium constant of the reaction at 80°C. Assume that $\Delta G_f°$ is temperature independent.

15.95 Consider the reaction

$$N_2(g) + O_2(g) \rightleftharpoons 2NO(g)$$

Given that $\Delta G°$ for the reaction at 25°C is 173.4 kJ/mol, (a) calculate the standard free energy of formation of NO and (b) calculate K_P of the reaction. (c) One of the starting substances in smog formation is NO. Assuming that the temperature in a running automobile engine is 1100°C, estimate K_P for the given reaction. (d) As farmers know, lightning helps to produce a better crop. Why?

15.96 When heated, ammonium carbamate decomposes as follows:

$$NH_4CO_2NH_2(s) \rightleftharpoons 2NH_3(g) + CO_2(g)$$

At a certain temperature, the equilibrium pressure of the system is 0.318 atm. Calculate K_P for the reaction.

15.97 A mixture of 0.47 mole of H_2 and 3.59 moles of HCl is heated to 2800°C. Calculate the equilibrium partial pressures of H_2, Cl_2, and HCl if the total pressure is 2.00 atm. For the reaction

$$H_2(g) + Cl_2(g) \rightleftharpoons 2HCl(g)$$

K_P is 193 at 2800°C.

15.98   When heated at high temperatures, iodine vapor dissociates as follows:

$$I_2(g) \rightleftharpoons 2I(g)$$

In one experiment, a chemist finds that when 0.054 mole of I_2 was placed in a flask of volume 0.48 L at 587 K, the degree of dissociation (i.e., the fraction of I_2 dissociated) was 0.0252. Calculate K_c and K_P for the reaction at this temperature.

15.99   The following reaction represents the removal of ozone in the stratosphere.

$$2O_3(g) \rightleftharpoons 3O_2(g)$$

Calculate the equilibrium constant (K_P) for this reaction. In view of the magnitude of the equilibrium constant, explain why this reaction is not considered a major cause of ozone depletion in the absence of human-made pollutants such as the nitrogen oxides and CFCs. Assume the temperature of the stratosphere is −30°C and ΔG_f° is temperature independent.

15.100   At 1130°C, the equilibrium constant (K_c) for the reaction

$$2H_2S(g) \rightleftharpoons 2H_2(g) + S_2(g)$$

is 2.25×10^{-4}. If $[H_2S] = 4.84 \times 10^{-3}$ M and $[H_2] = 1.50 \times 10^{-3}$ M, calculate $[S_2]$.

15.101   A quantity of 6.75 g of SO_2Cl_2 was placed in a 2.00-L flask. At 648 K, there is 0.0345 mole of SO_2 present. Calculate K_c for the reaction

$$SO_2Cl_2(g) \rightleftharpoons SO_2(g) + Cl_2(g)$$

15.102   Calculate the equilibrium pressure of CO_2 due to the decomposition of barium carbonate ($BaCO_3$) at 25°C.

15.103   One mole of N_2 and three moles of H_2 are placed in a flask at 375°C. Calculate the total pressure of the system at equilibrium if the mole fraction of NH_3 is 0.21. The K_P for the reaction is 4.31×10^{-4}.

15.104   Consider the gas-phase reaction between A_2 (green) and B_2 (red) to form AB at 298 K:

$$A_2(g) + B_2(g) \rightleftharpoons 2AB(g) \qquad \Delta G^\circ = -3.4 \text{ kJ/mol}$$

(a) Which of the following reaction mixtures is at equilibrium?
(b) Which of the following reaction mixtures has a negative ΔG value?
(c) Which of the following reaction mixtures has a positive ΔG value?

The partial pressures of the gases in each frame are equal to the number of A_2, B_2, and AB molecules times 0.10 atm. Round your results to two significant figures.

(i)   (ii)   (iii)

15.105   Consider the following reaction at 1600°C.

$$Br_2(g) \rightleftharpoons 2Br(g)$$

When 1.05 moles of Br_2 are put in a 0.980-L flask, 1.20% of the Br_2 undergoes dissociation. Calculate the equilibrium constant K_c for the reaction.

15.106   The following diagram represents a gas-phase equilibrium mixture for the reaction AB $\rightleftharpoons$ A + B at a certain temperature. Describe what would happen to the system after each of the following changes: (a) the temperature is decreased, (b) the volume is increased, (c) He atoms are added to the mixture at constant volume, (d) a catalyst is added to the mixture.

15.107   The formation of SO_3 from SO_2 and O_2 is an intermediate step in the manufacture of sulfuric acid, and it is also responsible for the acid rain phenomenon. The equilibrium constant K_P for the reaction

$$2SO_2(g) + O_2(g) \rightleftharpoons 2SO_3(g)$$

is 0.13 at 830°C. In one experiment, 2.00 mol SO_2 and 2.00 mol O_2 were initially present in a flask. What must the total pressure at equilibrium be to have an 80.0% yield of SO_3?

15.108   Calculate the pressure of O_2 (in atm) over a sample of NiO at 25°C if $\Delta G^\circ = 212$ kJ/mol for the reaction

$$NiO(s) \rightleftharpoons Ni(s) + \tfrac{1}{2}O_2(g)$$

15.109   The following reaction was described as the cause of sulfur deposits formed at volcanic sites.

$$2H_2S(g) + SO_2(g) \rightleftharpoons 3S(s) + 2H_2O(g)$$

It may also be used to remove SO_2 from power plant stack gases. (a) Identify the type of redox reaction it is. (b) Calculate the equilibrium constant (K_P) at 25°C, and comment on whether this method is feasible for removing SO_2. (c) Would this procedure become more effective or less effective at a higher temperature?

15.110   Eggshells are composed mostly of calcium carbonate ($CaCO_3$) formed by the reaction

$$Ca^{2+}(aq) + CO_3^{2-}(aq) \rightleftharpoons CaCO_3(s)$$

The carbonate ions are supplied by carbon dioxide produced as a result of metabolism. Explain why eggshells are thinner in the summer when the rate of panting by chickens is greater. Suggest a remedy for this situation.

15.111 Calculate $\Delta G°$ and K_P for the following processes at 25°C.

(a) $H_2(g) + Br_2(l) \rightleftharpoons 2HBr(g)$

(b) $\frac{1}{2}H_2(g) + \frac{1}{2}Br_2(l) \rightleftharpoons HBr(g)$

Account for the differences in $\Delta G°$ and K_P obtained for parts (a) and (b).

15.112 Consider the dissociation of iodine.

$$I_2(g) \rightleftharpoons 2I(g)$$

A 1.00-g sample of I_2 is heated to 1200°C in a 500-mL flask. At equilibrium, the total pressure is 1.51 atm. Calculate K_P for the reaction at this temperature.

15.113 The equilibrium constant K_P for the following reaction is 4.31×10^{-4} at 375°C.

$$N_2(g) + 3H_2(g) \rightleftharpoons 2NH_3(g)$$

In a certain experiment a student starts with 0.862 atm of N_2 and 0.373 atm of H_2 in a constant-volume vessel at 375°C. Calculate the partial pressures of all species when equilibrium is reached.

15.114 A quantity of 0.20 mole of carbon dioxide was heated to a certain temperature with an excess of graphite in a closed container until the following equilibrium was reached.

$$C(s) + CO_2(g) \rightleftharpoons 2CO(g)$$

Under these conditions, the average molar mass of the gases was 35 g/mol. (a) Calculate the mole fractions of CO and CO_2. (b) What is K_P if the total pressure is 11 atm? (*Hint:* The average molar mass is the sum of the products of the mole fraction of each gas and its molar mass.)

15.115 When dissolved in water, glucose (corn sugar) and fructose (fruit sugar) exist in equilibrium as follows:

$$\text{fructose} \rightleftharpoons \text{glucose}$$

A chemist prepared a 0.244 M fructose solution at 25°C. At equilibrium, it was found that its concentration had decreased to 0.113 M. (a) Calculate the equilibrium constant for the reaction. (b) At equilibrium, what percentage of fructose was converted to glucose?

15.116 Calculate ΔG for the reaction

$$H_2O(l) \rightleftharpoons H^+(aq) + OH^-(aq)$$

at 25°C for the following initial concentrations.

(a) $[H^+] = 1.0 \times 10^{-7}\ M$, $[OH^-] = 1.0 \times 10^{-7}\ M$

(b) $[H^+] = 1.0 \times 10^{-3}\ M$, $[OH^-] = 1.0 \times 10^{-4}\ M$

(c) $[H^+] = 1.0 \times 10^{-12}\ M$, $[OH^-] = 2.0 \times 10^{-8}\ M$

(d) $[H^+] = 3.5\ M$, $[OH^-] = 4.8 \times 10^{-4}\ M$

15.117 Assuming that the ΔG value from Problem 14.29 is $\Delta G°$ for the denaturation, determine the value of equilibrium constant for the process at 20°C.

15.118 At room temperature, solid iodine is in equilibrium with its vapor through sublimation and deposition. Describe how you would use radioactive iodine, in either solid or vapor form, to show that there is a dynamic equilibrium between these two phases.

15.119 At 1024°C, the pressure of oxygen gas from the decomposition of copper(II) oxide (CuO) is 0.49 atm:

$$4CuO(s) \rightleftharpoons 2Cu_2O(s) + O_2(g)$$

(a) What is K_P for the reaction? (b) Calculate the fraction of CuO that will decompose if 0.16 mole of it is placed in a 2.0-L flask at 1024°C. (c) What would the fraction be if a 1.0-mole sample of CuO were used? (d) What is the smallest amount of CuO (in moles) that would establish the equilibrium?

15.120 A mixture containing 3.9 moles of NO and 0.88 mole of CO_2 was allowed to react in a flask at a certain temperature according to the equation

$$NO(g) + CO_2(g) \rightleftharpoons NO_2(g) + CO(g)$$

At equilibrium, 0.11 mole of CO_2 was present. Calculate the equilibrium constant K_c of this reaction.

15.121 The equilibrium constant K_c for the reaction

$$H_2(g) + I_2(g) \rightleftharpoons 2HI(g)$$

is 54.3 at 430°C. At the start of the reaction, there are 0.714 mole of H_2, 0.984 mole of I_2, and 0.886 mole of HI in a 2.40-L reaction chamber. Calculate the concentrations of the gases at equilibrium.

15.122 For reactions carried out under standard-state conditions, Equation 14.10 takes the form $\Delta G° = \Delta H° - T\Delta S°$. (a) Assuming $\Delta H°$ and $\Delta S°$ are independent of temperature, derive the equation

$$\ln \frac{K_2}{K_1} = \frac{\Delta H°}{R}\left(\frac{T_2 - T_1}{T_1 T_2}\right)$$

where K_1 and K_2 are the equilibrium constants at T_1 and T_2, respectively. (b) Given that at 25°C K_c is 4.63×10^{-3} for the reaction

$$N_2O_4(g) \rightleftharpoons 2NO_2(g) \qquad \Delta H° = 58.0\ \text{kJ/mol}$$

calculate the equilibrium constant at 65°C.

15.123 When a gas was heated under atmospheric conditions, its color deepened. Heating above 150°C caused the color to fade, and at 550°C the color was barely detectable. However, at 550°C, the color was partially restored by increasing the pressure of the system. Which of the following best fits the preceding description: (a) a mixture of hydrogen and bromine, (b) pure bromine, (c) a mixture of nitrogen dioxide and dinitrogen tetroxide. (*Hint:* Bromine has a reddish color, and nitrogen dioxide is a brown gas. The other gases are colorless.) Justify your choice.

15.124 Both Mg^{2+} and Ca^{2+} are important biological ions. One of their functions is to bind to the phosphate group of ATP molecules or amino acids of proteins. For Group 2A metals in general, the equilibrium constant for binding to the anions increases in the order $Ba^{2+} < Sr^{2+} < Ca^{2+} < Mg^{2+}$. What property of the Group 2A metal cations might account for this trend?

15.125   The equilibrium constant K_c for the following reaction is 1.2 at 375°C.

$$N_2(g) + 3H_2(g) \rightleftharpoons 2NH_3(g)$$

(a) What is the value of K_P for this reaction?
(b) What is the value of the equilibrium constant K_c for $2NH_3(g) \rightleftharpoons N_2(g) + 3H_2(g)$? (c) What is K_c for $\frac{1}{2}N_2(g) + \frac{3}{2}H_2(g) \rightleftharpoons NH_3(g)$? (d) What are the values of K_P for the reactions described in parts (b) and (c)?

15.126   The equilibrium constant (K_P) for the formation of the air pollutant nitric oxide (NO) in an automobile engine at 530°C is 2.9×10^{-11}.

$$N_2(g) + O_2(g) \rightleftharpoons 2NO(g)$$

(a) Calculate the partial pressure of NO under these conditions if the partial pressures of nitrogen and oxygen are 3.0 and 0.012 atm, respectively. (b) Repeat the calculation for atmospheric conditions where the partial pressures of nitrogen and oxygen are 0.78 and 0.21 atm and the temperature is 25°C. (The K_P for the reaction is 4.0×10^{-31} at this temperature.) (c) Is the formation of NO endothermic or exothermic? (d) What natural phenomenon promotes the formation of NO? Why?

15.127   A sealed glass bulb contains a mixture of NO_2 and N_2O_4 gases. Describe what happens to the following properties of the gases when the bulb is heated from 20°C to 40°C: (a) color, (b) pressure, (c) average molar mass, (d) degree of dissociation (from N_2O_4 to NO_2), (e) density. Assume that volume remains constant. (*Hint:* NO_2 is a brown gas; N_2O_4 is colorless.)

15.128   At 20°C, the vapor pressure of water is 0.0231 atm. Calculate K_P and K_c for the process

$$H_2O(l) \rightleftharpoons H_2O(g)$$

15.129   Industrially, sodium metal is obtained by electrolyzing molten sodium chloride. The reaction at the cathode is $Na^+ + e^- \longrightarrow Na$. We might expect that potassium metal would also be prepared by electrolyzing molten potassium chloride. However, potassium metal is soluble in molten potassium chloride and therefore is hard to recover. Furthermore, potassium vaporizes readily at the operating temperature, creating hazardous conditions. Instead, potassium is prepared by the distillation of molten potassium chloride in the presence of sodium vapor at 892°C:

$$Na(g) + KCl(l) \rightleftharpoons NaCl(l) + K(g)$$

In view of the fact that potassium is a stronger reducing agent than sodium, explain why this approach works. (The boiling points of sodium and potassium are 892°C and 770°C, respectively.)

15.130   In the gas phase, nitrogen dioxide is actually a mixture of nitrogen dioxide (NO_2) and dinitrogen tetroxide (N_2O_4). If the density of such a mixture is 2.3 g/L at 74°C and 1.3 atm, calculate the partial pressures of the gases and K_P for the dissociation of N_2O_4.

15.131   A 2.50-mole sample of NOCl was initially in a 1.50-L reaction chamber at 400°C. After equilibrium was established, it was found that 28.0% of the NOCl had dissociated.

$$2NOCl(g) \rightleftharpoons 2NO(g) + Cl_2(g)$$

Calculate the equilibrium constant K_c for the reaction.

15.132   About 75% of hydrogen for industrial use is produced by the *steam-reforming* process. This process is carried out in two stages called primary and secondary reforming. In the primary stage, a mixture of steam and methane at about 30 atm is heated over a nickel catalyst at 800°C to give hydrogen and carbon monoxide:

$$CH_4(g) + H_2O(g) \rightleftharpoons CO(g) + 3H_2(g)$$

$$\Delta H° = 206 \text{ kJ/mol}$$

The secondary stage is carried out at about 1000°C, in the presence of air, to convert the remaining methane to hydrogen:

$$CH_4(g) + \frac{1}{2}O_2(g) \rightleftharpoons CO(g) + 2H_2(g)$$

$$\Delta H° = 35.7 \text{ kJ/mol}$$

(a) What conditions of temperature and pressure would favor the formation of products in both the primary and secondary stages? (b) The equilibrium constant K_c for the primary stage is 18 at 800°C. (i) Calculate K_P for the reaction. (ii) If the partial pressures of methane and steam were both 15 atm at the start, what are the pressures of all the gases at equilibrium?

15.133   Photosynthesis can be represented by

$$6CO_2(g) + 6H_2O(l) \rightleftharpoons C_6H_{12}O_6(s) + 6O_2(g)$$

$$\Delta H° = 2801 \text{ kJ/mol}$$

Explain how the equilibrium would be affected by the following changes: (a) partial pressure of CO_2 is increased, (b) O_2 is removed from the mixture, (c) $C_6H_{12}O_6$ (glucose) is removed from the mixture, (d) more water is added, (e) a catalyst is added, (f) temperature is decreased.

15.134   Consider the decomposition of ammonium chloride at a certain temperature:

$$NH_4Cl(s) \rightleftharpoons NH_3(g) + HCl(g)$$

Calculate the equilibrium constant K_P if the total pressure is 2.2 atm at that temperature.

15.135   At 25°C, the equilibrium partial pressures of NO_2 and N_2O_4 are 0.15 atm and 0.20 atm, respectively. If the volume is doubled at constant temperature, calculate the partial pressures of the gases when a new equilibrium is established.

15.136 In 1899 the German chemist Ludwig Mond developed a process for purifying nickel by converting it to the volatile nickel tetracarbonyl [$Ni(CO)_4$] (b.p. = 42.2°C):

$$Ni(s) + 4CO(g) \rightleftharpoons Ni(CO)_4(g)$$

(a) Describe how you can separate nickel and its solid impurities. (b) How would you recover nickel? [ΔH_f° for $Ni(CO)_4$ is −602.9 kJ/mol.]

15.137 Consider the equilibrium reaction described in Problem 15.30. A quantity of 2.50 g of PCl_5 is placed in an evacuated 0.500-L flask and heated to 250°C. (a) Calculate the pressure of PCl_5, assuming it does not dissociate. (b) Calculate the partial pressure of PCl_5 at equilibrium. (c) What is the total pressure at equilibrium? (d) What is the degree of dissociation of PCl_5? (The degree of dissociation is given by the fraction of PCl_5 that has undergone dissociation.)

15.138 Consider the equilibrium system $3A \rightleftharpoons B$. Sketch the changes in the concentrations of A and B over time for the following situations: (a) initially only A is present, (b) initially only B is present, (c) initially both A and B are present (with A in higher concentration). In each case, assume that the concentration of B is higher than that of A at equilibrium.

15.139 The vapor pressure of mercury is 0.0020 mmHg at 26°C. (a) Calculate K_c and K_P for the process $Hg(l) \rightleftharpoons Hg(g)$. (b) A chemist breaks a thermometer and spills mercury onto the floor of a laboratory measuring 6.1 m long, 5.3 m wide, and 3.1 m high. Calculate the mass of mercury (in grams) vaporized at equilibrium and the concentration of mercury vapor (in mg/m³). Does this concentration exceed the safety limit of 0.05 mg/m³? (Ignore the volume of furniture and other objects in the laboratory.)

15.140 Large quantities of hydrogen are needed for the synthesis of ammonia. One preparation of hydrogen involves the reaction between carbon monoxide and steam at 300°C in the presence of a copper-zinc catalyst:

$$CO(g) + H_2O(g) \rightleftharpoons CO_2(g) + H_2(g)$$

Calculate the equilibrium constant (K_P) for the reaction and the temperature at which the reaction favors the formation of CO and H_2O.

15.141 The activity series in Table 9.6 shows that reaction (a) is spontaneous whereas reaction (b) is nonspontaneous at 25°C.
(a) $Fe(s) + 2H^+(aq) \longrightarrow Fe^{2+}(aq) + H_2(g)$
(b) $Cu(s) + 2H^+(aq) \longrightarrow Cu^{2+}(aq) + H_2(g)$
Use the data in Appendix 2 to calculate the equilibrium constant for these reactions and hence confirm that the activity series is correct.

15.142 At 25°C, a mixture of NO_2 and N_2O_4 gases are in equilibrium in a cylinder fitted with a movable piston. The concentrations are [NO_2] = 0.0475 M and [N_2O_4] = 0.487 M. The volume of the gas mixture is halved by pushing down on the piston at constant temperature. Calculate the concentrations of the gases when equilibrium is reestablished. Will the color become darker or lighter after the change? [Hint: K_c for the dissociation of N_2O_4 is 4.63×10^{-3}. $N_2O_4(g)$ is colorless, and $NO_2(g)$ has a brown color.]

15.143 A student placed a few ice cubes in a drinking glass with water. A few minutes later she noticed that some of the ice cubes were fused together. Explain what happened.

15.144 Heating copper(II) oxide at 400°C does not produce any appreciable amount of Cu:

$$CuO(s) \rightleftharpoons Cu(s) + \tfrac{1}{2}O_2(g) \qquad \Delta G^\circ = 127.2 \text{ kJ/mol}$$

However, if this reaction is coupled to the conversion of graphite to carbon monoxide, it becomes spontaneous. Write an equation for the coupled process, and calculate the equilibrium constant for the coupled reaction.

15.145 The equilibrium constant K_c for the reaction

$$2NH_3(g) \rightleftharpoons N_2(g) + 3H_2(g)$$

is 0.83 at 375°C. A 14.6-g sample of ammonia is placed in a 4.00-L flask and heated to 375°C. Calculate the concentrations of all the gases when equilibrium is reached.

15.146 The dependence of the equilibrium constant of a reaction on temperature is given by the van't Hoff equation,

$$\ln K = \frac{-\Delta H^\circ}{RT} + C$$

where C is a constant. The following table gives the equilibrium constant (K_P) for the reaction at various temperatures.

$$2NO(g) + O_2(g) \rightleftharpoons 2NO_2(g)$$

K_P	138	5.12	0.436	0.0626	0.0130
T(K)	600	700	800	900	1000

Determine graphically the ΔH° for the reaction.

15.147 Consider the reaction between NO_2 and N_2O_4 in a closed container.

$$N_2O_4(g) \rightleftharpoons 2NO_2(g)$$

Initially, 1 mole of N_2O_4 is present. At equilibrium, x mole of N_2O_4 has dissociated to form NO_2.
(a) Derive an expression for K_P in terms of x and P, the total pressure. (b) How does the expression in part (a) help you predict the shift in equilibrium due to an increase in P? Does your prediction agree with Le Châtelier's principle?

15.148 The following diagram shows the variation of the equilibrium constant with temperature for the reaction

$$I_2(g) \rightleftharpoons 2I(g)$$

Calculate $\Delta G°$, $\Delta H°$, and $\Delta S°$ for the reaction at 872 K. (*Hint:* See Problem 15.122.)

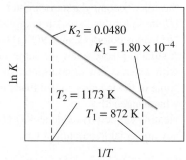

15.149 The K_P for the reaction

$$SO_2Cl_2(g) \rightleftharpoons SO_2(g) + Cl_2(g)$$

is 2.05 at 648 K. A sample of SO_2Cl_2 is placed in a container and heated to 648 K, while the total pressure is kept constant at 9.00 atm. Calculate the partial pressures of the gases at equilibrium.

15.150 Derive the equation

$$\Delta G = RT \ln \frac{Q}{K}$$

where Q is the reaction quotient, and describe how you would use it to predict the spontaneity of a reaction.

15.151 Consider the following reaction at a certain temperature.

$$A_2 + B_2 \rightleftharpoons 2AB$$

The mixing of 1 mole of A_2 with 3 moles of B_2 gives rise to x mole of AB at equilibrium. The addition of two more moles of A_2 produces another x mole of AB. What is the equilibrium constant for the reaction?

15.152 Iodine is sparingly soluble in water but much more so in carbon tetrachloride (CCl_4). The equilibrium constant, also called the partition coefficient, for the distribution of I_2 between these two phases

$$I_2(aq) \rightleftharpoons I_2(CCl_4)$$

is 83 at 20°C. (a) A student adds 0.030 L of CCl_4 to 0.200 L of an aqueous solution containing 0.032 g of I_2. The mixture at 20°C is shaken, and the two phases are then allowed to separate. Calculate the fraction of I_2 remaining in the aqueous phase. (b) The student now repeats the extraction of I_2 with another 0.030 L of CCl_4. Calculate the fraction of the I_2 from the original solution that remains in the aqueous phase. (c) Compare the result in part (b) with a single extraction using 0.060 L of CCl_4. Comment on the difference.

15.153 For the purpose of determining K_P using Equation 15.4, what is Δn for the following equation?

$$C_3H_8(g) + 5O_2(g) \rightleftharpoons 3CO_2(g) + 4H_2O(l)$$

15.154 Industrial production of ammonia from hydrogen and nitrogen gases is done using the Haber process.

$$N_2(g) + 3H_2(g) \rightleftharpoons 2NH_3(g)$$
$$\Delta H° = -92.6 \text{ kJ/mol}$$

Based on your knowledge of the principles of equilibrium, what would the optimal temperature and pressure conditions be for production of ammonia on a large scale? Are the same conditions also optimal from the standpoint of kinetics? Explain.

15.155 For which of the following reactions is K_c equal to K_P? For which can we not write a K_P expression?
(a) $4NH_3(g) + 5O_2(g) \rightleftharpoons 4NO(g) + 6H_2O(g)$
(b) $CaCO_3(s) \rightleftharpoons CaO(s) + CO_2(g)$
(c) $Zn(s) + 2H^+(aq) \rightleftharpoons Zn^{2+}(aq) + H_2(g)$
(d) $PCl_3(g) + 3NH_3(g) \rightleftharpoons 3HCl(g) + P(NH_2)_3(g)$
(e) $NH_3(g) + HCl(g) \rightleftharpoons NH_4Cl(s)$
(f) $NaHCO_3(s) + H^+(aq) \rightleftharpoons H_2O(l) + CO_2(g) +$ $Na^+(aq)$
(g) $H_2(g) + F_2(g) \rightleftharpoons 2HF(g)$
(h) $C(\text{graphite}) + CO_2(g) \rightleftharpoons 2CO(g)$

15.156 At present, the World Anti-Doping Agency has no way to detect the use of hypoxic sleeping tents, other than inspection of an athlete's home. Imagine, however, that a biochemical analysis company is developing a way to determine whether an elevated red blood cell count in an athlete's blood is the result of such a practice. A key substance in the detection process is the proprietary compound OD17X, which is produced by the combination of two other proprietary compounds, OD1A and OF2A. The aqueous reaction is represented by the equation

$$OD1A(aq) + OF2A(aq) \rightleftharpoons OD17X(aq)$$
$$\Delta H = 29 \text{ kJ/mol}$$

(a) Write the equilibrium expression for this reaction. (b) Given the equilibrium concentrations at room temperature of [OD1A] = 2.12 M, [OF2A] = 1.56 M, and [OD17X] = 1.01×10^{-4} M, calculate the value of the equilibrium constant (K_c) at room temperature. (c) The OD17X produced is precipitated with another proprietary substance according to the equation

$$OD17X(aq) + A771A(aq) \rightleftharpoons OD17X\text{-}A77(s)$$

K_c for the precipitation equilibrium is 1.0×10^6 at room temperature. Write the equilibrium expression for the sum of the two reactions and determine the value of the overall equilibrium constant. (d) Determine whether a mixture with

[OD1A] = 3.00 M, [OF2A] = 2.50 M, and [OD17X] = 2.7×10^{-4} M is at equilibrium and, if not, which direction it will have to proceed to achieve equilibrium. (e) Because the production of OD17X is endothermic, it can be enhanced by increasing the temperature. At 250°C, the equilibrium constant for the reaction

$$OD1A(aq) + OF2A(aq) \rightleftharpoons OD17X(aq)$$

is 3.8×10^2. If a synthesis at 250°C begins with 1.0 M of each reactant, what will be the equilibrium concentrations of reactants and products?

15.157 (a) Use the van't Hoff equation in Problem 15.146 to derive the following expression, which relates the equilibrium constants at two different temperatures:

$$\ln \frac{K_1}{K_2} = \frac{\Delta H°}{R}\left(\frac{1}{T_2} - \frac{1}{T_1}\right)$$

How does this equation support the prediction based on Le Châtelier's principle about the shift in equilibrium with temperature? (b) The vapor pressures of water are 31.82 mmHg at 30°C and 92.51 mmHg at 50°C. Calculate the molar heat of vaporization of water.

Answers to In-Chapter Materials

PRACTICE PROBLEMS

15.1A (a) $Q_c = \dfrac{[N_2]^2[O_2]}{[N_2O]^2}$, (b) $Q_c = \dfrac{[NO]^2[Br_2]}{[NOBr]^2}$, (c) $Q_c = \dfrac{[H^+][F^-]}{[HF]}$,

(d) $Q_c = \dfrac{[CO_2][H_2]}{[CO][H_2O]}$, (e) $Q_c = \dfrac{[CS_2][H_2]^4}{[CH_4][H_2O]^2}$, (f) $Q_c = \dfrac{[H^+]^2[C_2O_4^{2-}]}{[H_2C_2O_4]}$.

15.1B (a) $H_2 + Cl_2 \rightleftharpoons 2HCl$, (b) $H^+ + F^- \rightleftharpoons HF$, (c) $Cr^{3+} + 4OH^- \rightleftharpoons Cr(OH)_4^-$, (d) $HClO \rightleftharpoons H^+ + ClO^-$, (e) $H_2SO_3 \rightleftharpoons H^+ + HSO_3^-$, (f) $2NO + Br_2 \rightleftharpoons 2NOBr$.

15.2A $K_c = \dfrac{[BrCl]^2}{[Br_2][Cl_2]}$, $K_c = 7.1$. **15.2B** 0.016 M.

15.3A (a) $K_c = \dfrac{[HCl]^4}{[SiCl_4][H_2]^2}$, (b) $K_c = \dfrac{1}{[Hg^{2+}][Cl^-]^2}$,

(c) $K_c = \dfrac{[Ni(CO)_4]}{[CO]^4}$, (d) $K_c = \dfrac{[Zn^{2+}]}{[Fe^{2+}]}$. **15.3B** The expressions in (a),

(c), and (d) correspond to heterogeneous equilibria. It would be impossible to write a balanced chemical equation using only the species in each of these expressions, indicating that there are species in each reaction (solids or liquids) that do not appear in the equilibrium expressions. **15.4A** (a) 2.3×10^{24}, (b) 6.6×10^{-13}, (c) 2.8×10^{-15}.

15.4B (a) 1.5×10^{34}, (b) 1.3×10^{29}, (c) 8.5×10^{38}.

15.5A (a) $K_P = \dfrac{(P_{CO_2})^2}{(P_{CO})^2(P_{O_2})}$, (b) $K_P = P_{CO_2}$, (c) $K_P = \dfrac{(P_{NH_3})^2}{(P_{N_2})(P_{H_2})^3}$.

15.5B (a) $2NO_2 + O_2 \rightleftharpoons 2NO_3$, (b) $CH_4 + 2H_2O \rightleftharpoons CO_2 + 4H_2$, (c) $I_2 + H_2 \rightleftharpoons 2HI$. **15.6A** 8.1×10^{-6}. **15.6B** 0.104. **15.7A** left. **15.7B** right. **15.8A** 1.3 kJ/mol, reverse reaction is spontaneous. **15.8B** 2.5 atm. **15.9A** 2×10^{57}. **15.9B** −16.8 kJ/mol. **15.10A** 32.9 kJ/mol. **15.10B** −454.4 kJ/mol. **15.11A** $[H_2] = [I_2] = 0.056$ M; $[HI] = 0.413$ M. **15.11B** 0.20 M. **15.12A** $[H_2] = [I_2] = 0.081$ M, $[HI] = 0.594$ M. **15.12B** $[Br] = 8.4 \times 10^{-3}$ M, $[Br_2] = 6.5 \times 10^{-2}$ M. **15.13A** $[H^+] = [CN^-] = 7.0 \times 10^{-6}$ M, $[HCN] \approx 0.100$ M. **15.13B** $K_c = 3.3 \times 10^{-9}$. **15.14A** $P_{H_2} = P_{I_2} = 0.37$ atm, $P_{HI} = 2.75$ atm. **15.14B** $P_{H_2} = 0.41$ atm, $P_{I_2} = 0.21$ atm, $P_{HI} = 2.2$ atm. **15.15A** (a) right, (b) left, (c) right, (d) left. **15.15B** (a) $Ag(NH_3)_2^+$ (aq) or $Cl^-(aq)$, (b) $NH_3(aq)$, (c) $AgCl(s)$. **15.16A** (a) right, (b) right, (c) right. **15.16B** right shift: remove HF; left shift: remove H_2; no shift: reduce volume of container.

SECTION REVIEW
15.2.1 e. **15.2.2** a. **15.3.1** d. **15.3.2** a. **15.3.3** c. **15.3.4** c. **15.4.1** b. **15.4.2** e. **15.4.3** b. **15.5.1** d. **15.5.2** b. **15.5.3** e. **15.6.1** d. **15.6.2** b. **15.6.3** b **15.6.4** b.

Chapter 16

Acids, Bases, and Salts

© Purestock / Alamy Stock Photo

A **SUPERACID,** derived from the reaction of a strong Lewis base with a strong Brønsted acid, may be thousands of times stronger than the original Brønsted acid. For many years, the utility of superacids eluded scientists—who were nonetheless fascinated by their ability to dissolve notoriously insoluble substances, including *candle wax.* In recent years, however, chemists have used superacids in some unusual chemical transformations, generating chemical species that were ordinarily too elusive to be studied easily. Superacids now feature prominently in some important industrial processes, including the manufacture of plastics and production of high-octane gasoline.

Before You Begin, Review These Skills

- The list of strong acids and the list of strong bases [◄◄ Section 9.3]
- Determination of pH [◄◄ Section 9.5]
- How to solve equilibrium problems [◄◄ Section 15.5]

16.1 BRØNSTED ACIDS AND BASES

In Chapter 9, we learned that a Brønsted acid is a substance that can donate a proton and a Brønsted base is a substance that can accept a proton [◄◄ Section 9.3]. In this chapter, we extend our discussion of Brønsted acid-base theory to include conjugate acids and conjugate bases.

When a Brønsted acid donates a proton, what remains of the acid is known as a **conjugate base**. For example, in the ionization of HCl in water,

$$\text{HCl}(aq) + \text{H}_2\text{O}(l) \rightleftharpoons \text{H}_3\text{O}^+(aq) + \text{Cl}^-(aq)$$
$$\text{acid} \qquad\qquad\qquad\qquad\qquad \text{conjugate base}$$

HCl donates a proton to water, producing the hydronium ion (H_3O^+) and the chloride ion (Cl^-), which is the conjugate base of HCl. The two species, HCl and Cl^-, are known as a *conjugate acid-base pair* or simply a **conjugate pair**. Table 16.1 lists the conjugate bases of several familiar species.

Conversely, when a Brønsted base *accepts* a proton, the newly formed *protonated* species is known as a **conjugate acid**. When ammonia (NH_3) ionizes in water,

$$\text{NH}_3(aq) + \text{H}_2\text{O}(l) \rightleftharpoons \text{NH}_4^+(aq) + \text{OH}^-(aq)$$
$$\text{base} \qquad\qquad\qquad \text{conjugate acid}$$

NH_3 accepts a proton from water to become the ammonium ion (NH_4^+). The ammonium ion is the conjugate acid of ammonia. Table 16.2 lists the conjugate acids of several common species.

Any reaction that we describe using Brønsted acid-base theory involves an acid and a base. The acid donates the proton, and the base accepts it. Furthermore, the products of such a reaction are always a conjugate base and a conjugate acid. It is useful to identify and label each species in a Brønsted acid-base reaction. For the ionization of HCl in water, the species are labeled as follows:

loses a proton
gains a proton

$$\text{HCl}(aq) + \text{H}_2\text{O}(l) \rightleftharpoons \text{H}_3\text{O}^+(aq) + \text{Cl}^-(aq)$$
$$\text{acid} \qquad \text{base} \qquad\quad \text{conjugate} \qquad \text{conjugate}$$
$$\text{acid} \qquad\qquad \text{base}$$

And for the ionization of NH_3 in water,

gains a proton
loses a proton

$$\text{NH}_3(aq) + \text{H}_2\text{O}(l) \rightleftharpoons \text{NH}_4^+(aq) + \text{OH}^-(aq)$$
$$\text{base} \qquad \text{acid} \qquad\quad \text{conjugate} \qquad \text{conjugate}$$
$$\text{acid} \qquad\qquad \text{base}$$

Worked Examples 16.1 and 16.2 let you practice identifying conjugate pairs and the species in a Brønsted acid-base reaction.

TABLE 16.1	Conjugate Bases of Some Common Species
Species	**Conjugate base**
CH_3COOH	CH_3COO^-
H_2O	OH^-
NH_3	NH_2^-
H_2SO_4	HSO_4^-

TABLE 16.2	Conjugate Acids of Some Common Species
Species	**Conjugate acid**
NH_3	NH_4^+
H_2O	H_3O^+
OH^-	H_2O
H_2NCONH_2 (urea)	$\text{H}_2\text{NCONH}_3^+$

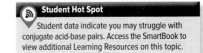

Student Hot Spot

Student data indicate you may struggle with conjugate acid-base pairs. Access the SmartBook to view additional Learning Resources on this topic.

Worked Example 16.1

What is (a) the conjugate base of HNO_3, (b) the conjugate acid of O^{2-}, (c) the conjugate base of HSO_4^-, and (d) the conjugate acid of HCO_3^-?

Strategy To find the conjugate base of a species, *remove* a proton from the formula. To find the conjugate acid of a species, *add* a proton to the formula.

Setup The word *proton*, in this context, refers to H^+. Thus, the formula and the charge will both be affected by the addition or removal of H^+.

Solution (a) NO_3^- (b) OH^- (c) SO_4^{2-} (d) H_2CO_3

> **Think About It**
> A species does not need to be what we think of as an acid in order for it to have a conjugate base. For example, we would not refer to the hydroxide ion (OH^-) as an acid—but it does have a conjugate base, the oxide ion (O^{2-}). Furthermore, a species that can either lose or gain a proton, such as HCO_3^-, has both a conjugate base (CO_3^{2-}) and a conjugate acid (H_2CO_3).

Practice Problem (A)TTEMPT What is (a) the conjugate acid of ClO_4^-, (b) the conjugate acid of S^{2-}, (c) the conjugate base of H_2S, and (d) the conjugate base of $H_2C_2O_4$?

Practice Problem (B)UILD HSO_3^- is the conjugate acid of what species? HSO_3^- is the conjugate base of what species?

Practice Problem (C)ONCEPTUALIZE Which of the models represents a species that has a conjugate base? Which represents a species that is the conjugate base of another species?

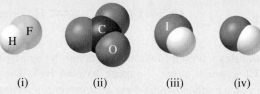

(i)   (ii)   (iii)   (iv)

Worked Example 16.2

Label each of the species in the following equations as an acid, base, conjugate base, or conjugate acid.

(a) $HF(aq) + NH_3(aq) \rightleftharpoons F^-(aq) + NH_4^+(aq)$ (b) $CH_3COO^-(aq) + H_2O(l) \rightleftharpoons CH_3COOH(aq) + OH^-(aq)$

Strategy In each equation, the reactant that loses a proton is the acid and the reactant that gains a proton is the base. Each product is the conjugate of one of the reactants. Two species that differ only by a proton constitute a conjugate pair.

Setup (a) HF loses a proton and becomes F^-; NH_3 gains a proton and becomes NH_4^+.

(b) CH_3COO^- gains a proton to become CH_3COOH; H_2O loses a proton to become OH^-.

Solution

(a) $HF(aq) + NH_3(aq) \rightleftharpoons F^-(aq) + NH_4^+(aq)$ (b) $CH_3COO^-(aq) + H_2O(l) \rightleftharpoons CH_3COOH(aq) + OH^-(aq)$

    acid   base   conjugate base   conjugate acid   base   acid   conjugate acid   conjugate base

> **Think About It**
> In a Brønsted acid-base reaction, there is always an acid and a base, and whether a substance behaves as an acid or a base depends on what it is combined with. Water, for example, behaves as a base when combined with HCl but behaves as an acid when combined with NH_3.

Practice Problem (A)TTEMPT Identify and label the species in each reaction.

(a) $NH_4^+(aq) + H_2O(l) \rightleftharpoons NH_3(aq) + H_3O^+(aq)$ (b) $CN^-(aq) + H_2O(l) \rightleftharpoons HCN(aq) + OH^-(aq)$

Practice Problem (B)UILD (a) Write an equation in which HSO_3^- reacts (with water) to form its conjugate base. (b) Write an equation in which HSO_3^- reacts (with water) to form its conjugate acid.

Practice Problem (C)ONCEPTUALIZE Write the formula and charge for each species in this reaction and identify each as an acid, a base, a conjugate acid, or a conjugate base.

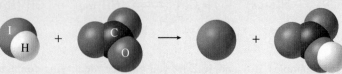

Section 16.1 Review

Brønsted Acids and Bases

16.1.1 Which of the following pairs of species are conjugate pairs? (Select all that apply.)

(a) H_2S and S^{2-}

(b) NH_2^- and NH_3

(c) O_2 and H_2O_2

(d) HBr and Br^-

(e) HCl and OH^-

16.1.2 Which of the following species do *not* have a conjugate base? (Select all that apply.)

(a) $HC_2O_4^-$

(b) OH^-

(c) O^{2-}

(d) CO_3^{2-}

(e) HClO

16.2 MOLECULAR STRUCTURE AND ACID STRENGTH

The strength of an acid is measured by its tendency to ionize.

$$HX \longrightarrow H^+ + X^-$$

Two factors influence the extent to which the acid undergoes ionization. One is the strength of the H—X bond. The stronger the bond, the more difficult it is for the HX molecule to break up and hence the weaker the acid. The other factor is the polarity of the H—X bond. The difference in the electronegativities between H and X results in a polar bond like

$$\overset{\delta+\ \ \ \delta-}{H-X}$$

If the bond is highly polarized (i.e., if there is a large accumulation of positive and negative charges on the H and X atoms, respectively), HX will tend to break up into H^+ and X^- ions. A high degree of polarity, therefore, *generally* gives rise to a stronger acid. In this section, we consider the roles of bond strength and bond polarity in determining the strength of an acid.

Hydrohalic Acids

The halogens form a series of binary acids called the hydrohalic acids (HF, HCl, HBr, and HI). Table 16.3 shows that of this series only HF is a weak acid. The data in the table indicate that the predominant factor in determining the strength of the hydrohalic acids is bond strength. HF has the largest bond enthalpy, making its bond the most difficult to break. In this series of binary acids, acid strength increases as bond strength decreases. The strength of the acids increases as follows:

$$HF \ll HCl < HBr < HI$$

Oxoacids

An oxoacid, as we learned in Chapter 5, contains hydrogen, oxygen, and a central, nonmetal atom [◀◀ Section 5.7]. As the Lewis structures in Figure 16.1 show, oxoacids contain one or more O—H bonds. If the central atom is an electronegative element, or is in a high oxidation state, it will attract electrons, causing the O—H bond to be more polar. This makes it easier for the hydrogen to be lost as H^+, making the acid stronger.

Student Annotation: Remember that protons are *hydrated* in aqueous solution [◀◀ Section 9.3] and that they can also be represented by the formula H_3O^+. Thus, this equation could also be written as:

$$HX + H_2O \longrightarrow H_3O^+ + X^-$$

Student Annotation: The polarity of the H—X bond actually decreases from H—F to H—I, largely because F is the most electronegative element. This would suggest that HF would be the strongest of the hydrohalic acids. Based on the data in Table 16.3, however, bond enthalpy is the more important factor in determining the strengths of these acids.

TABLE 16.3 Bond Enthalpies for Hydrogen Halides and Acid Strengths for Hydrohalic Acids

Bond	Bond enthalpy (kJ/mol)	Acid strength
H—F	562.8	Weak
H—Cl	431.9	Strong
H—Br	366.1	Strong
H—I	298.3	Strong

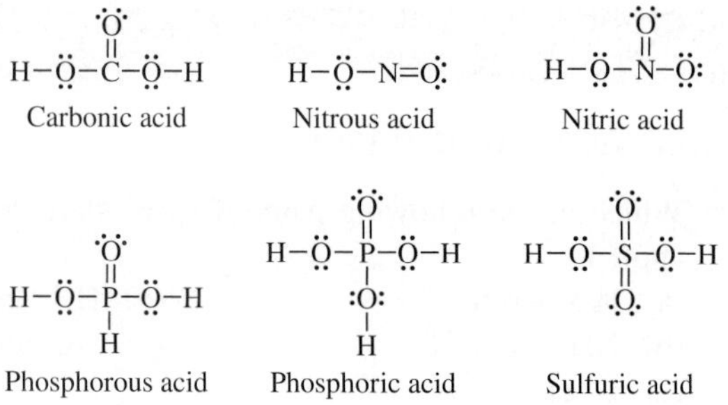

Figure 16.1 Lewis structures of some common oxoacids. Recall that there is more than one possible Lewis structure for oxoacids in which the central atom is from period 3 or below on the periodic table [◀◀ Section 6.6].

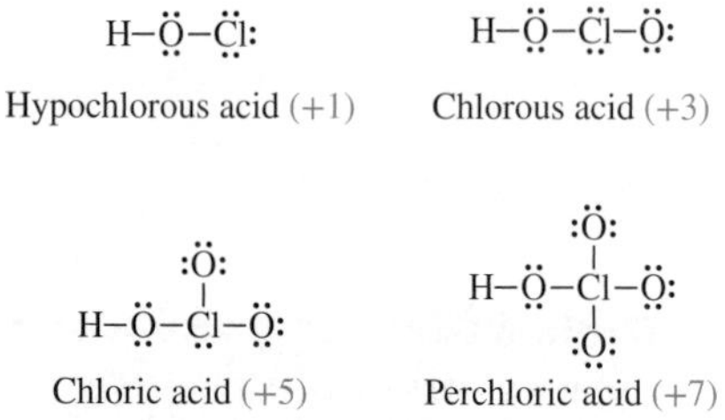

Figure 16.2 Lewis structures of the oxoacids of chlorine. The oxidation number of the Cl atom is shown in parentheses. Note that although hypochlorous acid is written as HClO, the H atom is bonded to the O atom.

Student Annotation: As the number of attached oxygen atoms increases, the oxidation number of the central atom also increases [◀◀ Section 9.4].

To compare their strengths, it is convenient to divide the oxoacids into two groups.

1. *Oxoacids having different central atoms that are from the same group of the periodic table and that have the same oxidation number.* Two examples are

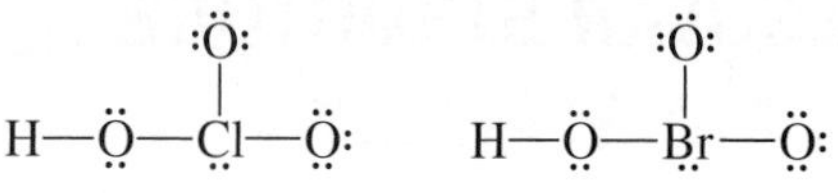

$HClO_3$ $HBrO_3$

Within this group, acid strength increases with increasing electronegativity of the central atom. Cl and Br have the same oxidation number in these acids, +5. However, because Cl is more electronegative than Br, it attracts the electron pair it shares with oxygen (in the Cl—O—H group) to a greater extent than Br does (in the corresponding Br—O—H group). Consequently, the O—H bond is more polar in chloric acid than in bromic acid and ionizes more readily. The relative acid strengths are

$$HClO_3 > HBrO_3$$

2. *Oxoacids having the same central atom but different numbers of oxygen atoms.* Within this group, acid strength increases with increasing oxidation number of the central atom. Consider the oxoacids of chlorine shown in Figure 16.2. In this series, the ability of chlorine to draw electrons away from the OH group (thus making the O—H bond more polar) increases with the number of electronegative O atoms attached to Cl. Thus, $HClO_4$ is the strongest acid because it has the largest number of oxygen atoms attached to Cl. The acid strength decreases as follows:

$$HClO_4 > HClO_3 > HClO_2 > HClO$$

Worked Example 16.3 compares acid strengths based on molecular structure.

Worked Example 16.3

Predict the relative strengths of the oxoacids in each of the following groups.

(a) HClO, HBrO, and HIO; (b) HNO_3 and HNO_2.

Strategy In each group, compare the electronegativities or oxidation numbers of the central atoms to determine which O—H bonds are the most polar. The more polar the O—H bond, the more readily it is broken and the stronger the acid.

Setup

(a) In a group with different central atoms, we must compare electronegativities. The electronegativities of the central atoms in this group decrease as follows: Cl > Br > I.

(b) These two acids have the same central atom but differ in the number of attached oxygen atoms. In a group such as this, the greater the number of attached oxygen atoms, the higher the oxidation number of the central atom and the stronger the acid.

Solution

(a) Acid strength decreases as follows: HClO > HBrO > HIO.

(b) HNO_3 is a stronger acid than HNO_2.

> **Student Annotation:** Another way to compare the strengths of these two is to remember that HNO_3 is one of the seven strong acids. HNO_2 is not.

> **Think About It**
> Four of the strong acids are oxoacids: HNO_3, $HClO_4$, $HClO_3$, and H_2SO_4.

Practice Problem (A)TTEMPT Indicate which is the stronger acid: (a) $HBrO_3$ or $HBrO_4$; (b) H_2SeO_4 or H_2SO_4.

Practice Problem (B)UILD Based on the information in this section, which is the predominant factor in determining the strength of an oxoacid: electronegativity of the central atom or oxidation state of the central atom? Explain.

Practice Problem (C)ONCEPTUALIZE The models show formula units of two compounds with the general formula XOH. In the model on the left, X is a metal. In the model on the right, X is a nonmetal. One of these compounds is a weak acid and the other is a strong base. Identify the acid and the base and explain why their properties are so different despite the similarity in their formulas.

Carboxylic Acids

So far our discussion has focused on inorganic acids. A particularly important group of organic acids is the *carboxylic acids,* whose Lewis structures can be represented by

$$R-\overset{\overset{\ddot{O}}{\|}}{C}-\overset{..}{\underset{..}{O}}-H$$

where R is part of the acid molecule and the shaded portion represents the carboxyl group, —COOH.

The conjugate *base* of a carboxylic acid, called a *carboxylate* anion, $RCOO^-$, can be represented by more than one resonance structure:

$$R-\overset{\overset{\ddot{O}}{\|}}{C}-\overset{..}{\underset{..}{O}}{:}^{-} \quad \longleftrightarrow \quad R-\overset{\overset{:\ddot{O}:^{-}}{|}}{C}=\overset{..}{O}{:}$$

> **Student Annotation:** You learned in Chapter 9 [◄◄ Section 9.3, Figure 9.5] that carboxylic acid formulas are often written with the ionizable H atom last to keep the functional group together. You should recognize the formulas for organic acids written either way. For example, acetic acid may be written as $HC_2H_3O_2$ or as CH_3COOH.

In terms of molecular orbital theory [◄◄ Section 7.7], we attribute the stability of the anion to its ability to spread out or delocalize the electron density over several atoms. The greater the extent of electron delocalization, the more stable the anion and the greater the tendency for the acid to undergo ionization—that is, the stronger the acid.

The strength of carboxylic acids depends on the nature of the R group. Consider, for example, acetic acid and chloroacetic acid.

$$\overset{\displaystyle H}{\underset{\displaystyle H}{H-\overset{|}{\underset{|}{C}}-\overset{\overset{\ddot{O}}{\|}}{C}-\overset{..}{\underset{..}{O}}-H}} \qquad \overset{\displaystyle :\overset{..}{C}l:}{\underset{\displaystyle H}{H-\overset{|}{\underset{|}{C}}-\overset{\overset{\ddot{O}}{\|}}{C}-\overset{..}{\underset{..}{O}}-H}}$$

Acetic acid   Chloroacetic acid

The presence of the electronegative Cl atom in chloroacetic acid shifts the electron density toward the R group, thereby making the O—H bond more polar. Consequently, there is a greater tendency for chloroacetic acid to ionize.

$$\overset{\displaystyle :\overset{..}{C}l:}{\underset{\displaystyle H}{H-\overset{|}{\underset{|}{C}}-\overset{\overset{\ddot{O}}{\|}}{C}-\overset{..}{\underset{..}{O}}-H}} \; \rightleftharpoons \; \left[\overset{\displaystyle :\overset{..}{C}l:}{\underset{\displaystyle H}{H-\overset{|}{\underset{|}{C}}-\overset{\overset{\ddot{O}}{\|}}{C}-\overset{..}{\underset{..}{O}}{:}}} \right]^{-} \; + \; H^{+}$$

Chloroacetic acid is the stronger of the two acids.

Section 16.2 Review

Molecular Structure and Acid Strength

16.2.1   Arrange the following organic acids in order of increasing strength: bromoacetic acid ($CH_2BrCOOH$), chloroacetic acid ($CH_2ClCOOH$), fluoroacetic acid (CH_2FCOOH), iodoacetic acid (CH_2ICOOH).
(a) $CH_2BrCOOH < CH_2ClCOOH < CH_2FCOOH < CH_2ICOOH$
(b) $CH_2ICOOH < CH_2BrCOOH < CH_2ClCOOH < CH_2FCOOH$
(c) $CH_2FCOOH < CH_2ClCOOH < CH_2BrCOOH < CH_2ICOOH$
(d) $CH_2BrCOOH < CH_2ICOOH < CH_2ClCOOH < CH_2FCOOH$
(e) $CH_2ClCOOH < CH_2BrCOOH < CH_2ICOOH < CH_2FCOOH$

16.2.2   Arrange the following acids in order of increasing strength: $HBrO_3$, $HBrO$, $HBrO_4$.
(a) $HBrO_4 < HBrO_3 < HBrO$ (d) $HBrO < HBrO_3 < HBrO_4$
(b) $HBrO < HBrO_4 < HBrO_3$ (e) $HBrO_4 < HBrO < HBrO_3$
(c) $HBrO_3 < HBrO_4 < HBrO$

16.2.3   Consider the following three acids and their bond enthalpies: H_2A (268 kJ/mol), H_2B (344 kJ/mol), H_2C (314 kJ/mol). Which of the three acids is strongest, and which is the weakest?
(a) H_2C; H_2B (d) H_2C; H_2A
(b) H_2B; H_2A (e) H_2B; H_2C
(c) H_2A; H_2B

16.3   THE ACID-BASE PROPERTIES OF WATER

Video 16.1
Chemical equilibrium.

Water is often referred to as the "universal solvent," because it is so common and so important to life on Earth. In addition, most of the acid-base chemistry that you will encounter takes place in aqueous solution. In this section, we take a closer look at water's ability to act as either a Brønsted acid (as in the ionization of NH_3) or a Brønsted base (as in the ionization of HCl). A species that can behave either as a Brønsted acid or a Brønsted base is called **amphoteric.** Water is a very weak electrolyte, but it does undergo ionization to a small extent.

$$H_2O(l) \rightleftharpoons H^+(aq) + OH^-(aq)$$

This reaction is known as the **autoionization of water.** Because we can represent the aqueous proton as either H^+ or H_3O^+ [◄◄ Section 9.3], we can also write the autoionization of water as

$$2H_2O(l) \rightleftharpoons H_3O^+(aq) + OH^-(aq)$$

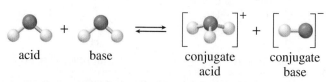

acid   base   conjugate   conjugate
                            acid   base

where one water molecule acts as an acid and the other acts as a base.

As indicated by the double arrow in the equation, the reaction is an equilibrium. The equilibrium expression for the autoionization of water is

Student Annotation: Recall that in a heterogeneous equilibrium such as this, liquids and solids do not appear in the equilibrium expression [◄◄ Section 15.3].

$$K_w = [H_3O^+][OH^-] \text{ or } K_w = [H^+][OH^-]$$

Because the autoionization of water is an important equilibrium that you will encounter frequently in the study of acids and bases, we use the subscript w to indicate that the equilibrium constant is that specifically for the autoionization of *water*. It is important to realize, though, that K_w is simply a K_c for a *specific reaction*. We will frequently replace the c in K_c expressions with a letter or a series of letters to indicate the specific type of reaction to which the K_c refers. For example, K_c for the ionization of a weak acid is called K_a, and K_c for the ionization of a weak base is called K_b. In Chapter 17, we will make extensive use of K_{sp}, where sp stands for "solubility product." Each specially subscripted K is simply a K_c for a specific type of reaction. The constant K_w is sometimes referred to as the ***ion-product constant.***

In pure water, autoionization is the only source of H_3O^+ and OH^-, and the stoichiometry of the reaction tells us that their concentrations are equal. At 25°C, the concentrations of hydronium and hydroxide ions in pure water are $[H_3O^+] = [OH^-] = 1.0 \times 10^{-7}\ M$. Using the equilibrium expression, we can calculate the value of K_w at 25°C as follows:

> **Student Annotation:** Recall that we disregard the units when we substitute concentrations into an equilibrium expression [◄◄ Section 15.2].

$$K_w = [H_3O^+][OH^-] = (1.0 \times 10^{-7})(1.0 \times 10^{-7}) = 1.0 \times 10^{-14}$$

Furthermore, in any aqueous solution at 25°C, the product of H_3O^+ and OH^- concentrations is equal to 1.0×10^{-14}.

$$K_w = [H_3O^+][OH^-] = 1.0 \times 10^{-14} \text{ (at 25°C)} \quad \textbf{Equation 16.1}$$

Although their product is a constant, the individual concentrations of hydronium and hydroxide can be influenced by the addition of an acid or a base. The relative amounts of H_3O^+ and OH^- determine whether a solution is neutral, acidic, or basic.

> **Student Annotation:** Because the product of H_3O^+ and OH^- concentrations is a constant, we cannot alter the concentrations independently. Any change in one also affects the other.

- When $[H_3O^+] = [OH^-]$, the solution is neutral.
- When $[H_3O^+] > [OH^-]$, the solution is acidic.
- When $[H_3O^+] < [OH^-]$, the solution is basic.

Worked Example 16.4 shows how to use Equation 16.1.

Worked Example 16.4

The concentration of hydronium ions in stomach acid is 0.10 *M*. Calculate the concentration of hydroxide ions in stomach acid at 25°C.

Strategy Use the value of K_w to determine $[OH^-]$ when $[H_3O^+] = 0.10\ M$.

Setup $K_w = [H_3O^+][OH^-] = 1.0 \times 10^{-14}$ at 25°C. Rearranging Equation 16.1 to solve for $[OH^-]$,

$$[OH^-] = \frac{1.0 \times 10^{-14}}{[H_3O^+]}$$

Solution

$$[OH^-] = \frac{1.0 \times 10^{-14}}{0.10} = 1.0 \times 10^{-13}\ M$$

> **Think About It**
> Remember that equilibrium constants are temperature dependent. The value of K_w is 1.0×10^{-14} only at 25°C.

Practice Problem (A)TTEMPT The concentration of hydroxide ions in the antacid milk of magnesia is $5.0 \times 10^{-4}\ M$. Calculate the concentration of hydronium ions at 25°C.

Practice Problem (B)UILD The value of K_w at normal body temperature (37°C) is 2.8×10^{-14}. Calculate the concentration of hydroxide ions in stomach acid at body temperature ($[H_3O^+] = 0.10\ M$).

Practice Problem (C)ONCEPTUALIZE The diagram shown on the left represents a system consisting of the weak electrolyte AB(*l*). Like water, liquid AB can autoionize to form A^+ ions (red) and B^- ions (blue). At room temperature, the product of ion concentrations, $[A^+][B^-]$, is always equal to 16 for the volume represented here. Also, because every AB molecule that ionizes produces one A^+ ion and one B^- ion, in a pure sample of AB(*l*), the concentrations $[A^+]$ and $[B^-]$ are equal to each other. Which of the other diagrams [(i)–(iii)] best represents the system after enough of the strong electrolyte NaB has been dissolved to increase the number of B^- ions to 8? (Although AB is a liquid, the molecules are shown far apart to keep the diagrams from being too crowded. Na^+ ions also are not shown, in order to keep the diagrams clear.)

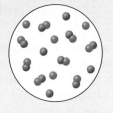

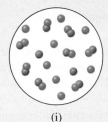

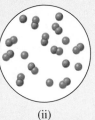

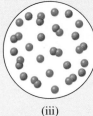

(i)   (ii)   (iii)

Section 16.3 Review

The Acid-Base Properties of Water

16.3.1 Calculate [OH⁻] in a solution in which $[H_3O^+] = 0.0012$ M at 25°C.

(a) 1.2×10^{-3} M   (d) 8.3×10^{-12} M

(b) 8.3×10^{-17} M   (e) 1.2×10^{11} M

(c) 1.0×10^{-14} M

16.3.2 Calculate [H₃O⁺] in a solution in which $[OH^-] = 0.25$ M at 25°C.

(a) 4.0×10^{-14} M   (d) 1.0×10^{-7} M

(b) 1.0×10^{-14} M   (e) 4.0×10^{-7} M

(c) 2.5×10^{13} M

16.4 THE pH AND pOH SCALES

In Chapter 9, we learned about the use of the pH scale to express the acidity of aqueous solutions [◄◄ Section 9.5]. Remember that (1) $pH = -\log[H_3O^+]$ [◄◄ Equation 9.5], (2) pH is a dimensionless quantity, (3) pH is 7 in a neutral solution (where $[H_3O^+] = [OH^-]$) at 25°C, and (4) the pH scale makes it easier to express hydronium ion concentrations that can span many orders of magnitude.

A **pOH** scale analogous to the pH scale can be defined using the negative base-10 logarithm of the *hydroxide* ion concentration of a solution, [OH⁻].

Equation 16.2	$pOH = -\log [OH^-]$

Rearranging Equation 16.2 to solve for hydroxide ion concentration gives

Equation 16.3	$[OH^-] = 10^{-pOH}$

Now consider again the K_w equilibrium expression for water at 25°C.

$$[H_3O^+][OH^-] = 1.0 \times 10^{-14}$$

Taking the negative logarithm of both sides, we obtain

$$-\log([H_3O^+][OH^-]) = -\log(1.0 \times 10^{-14})$$

$$-\log([H_3O^+] + \log[OH^-]) = 14.00$$

$$-\log[H_3O^+] - \log[OH^-] = 14.00$$

$$(-\log[H_3O^+]) + (-\log[OH^-]) = 14.00$$

And from the definitions of pH and pOH we see that at 25°C,

Equation 16.4	$pH + pOH = 14.00$

Equation 16.4 provides another way to express the relationship between the hydronium ion concentration and the hydroxide ion concentration. On the pOH scale, 7.00 is neutral, numbers greater than 7.00 indicate that a solution is acidic, and numbers less than 7.00 indicate that a solution is basic. Table 16.4 lists pOH values for a range of hydroxide ion concentrations at 25°C.

TABLE 16.4	Benchmark pOH Values for a Range of Hydroxide Ion Concentrations at 25°C	

[OH⁻] (M)	pOH	
0.10	1.00	
1.0×10^{-3}	3.00	
1.0×10^{-5}	5.00	Basic
1.0×10^{-7}	7.00	Neutral
1.0×10^{-9}	9.00	Acidic
1.0×10^{-11}	11.00	
1.0×10^{-13}	13.00	

Worked Examples 16.5 and 16.6 illustrate calculations involving pOH.

Worked Example 16.5

Determine the pOH of a solution at 25°C in which the hydroxide ion concentration is (a) 3.7×10^{-5} M, (b) 4.1×10^{-7} M, and (c) 8.3×10^{-2} M.

Strategy Given [OH⁻], use Equation 16.2 to calculate pOH.

Setup

(a) $pOH = -\log (3.7 \times 10^{-5})$

(b) $pOH = -\log (4.1 \times 10^{-7})$

(c) $pOH = -\log (8.3 \times 10^{-2})$

Solution

(a) $pOH = 4.43$

(b) $pOH = 6.39$

(c) $pOH = 1.08$

Think About It
Remember that the pOH scale is, in essence, the *reverse* of the pH scale. On the pOH scale, numbers below 7 indicate a basic solution, whereas numbers above 7 indicate an acidic solution. The pOH benchmarks (abbreviated in Table 16.4) work the same way the pH benchmarks do. In part (a), for example, a hydroxide ion concentration between 1×10^{-4} M and 1×10^{-5} M corresponds to a pOH between 4 and 5.

[OH⁻] (M)	pOH
1.0×10^{-4}	4.00
3.7×10^{-5}*	4.43†
1.0×10^{-5}	5.00

*[OH⁻] between two benchmark values
†pOH between two benchmark values

Practice Problem **A**TTEMPT Determine the pOH of a solution at 25°C in which the hydroxide ion concentration is (a) 5.7×10^{-12} M, (b) 7.3×10^{-3} M, and (c) 8.5×10^{-6} M.

Practice Problem **B**UILD Determine the pOH of a solution at 25°C in which the hydroxide ion concentration is (a) 2.8×10^{-8} M, (b) 9.9×10^{-9} M, and (c) 1.0×10^{-11} M.

Practice Problem **C**ONCEPTUALIZE Without doing any calculations, determine between which two whole numbers the pOH will be for solutions with OH⁻ concentrations of 4.71×10^{-5} M, 2.9×10^{-12} M, and 7.15×10^{-3} M.

Worked Example 16.6

Calculate the hydroxide ion concentration in a solution at 25°C in which the pOH is (a) 4.91, (b) 9.03, and (c) 10.55.

Strategy Given pOH, use Equation 16.3 to calculate [OH⁻].

Setup

(a) $[OH^-] = 10^{-4.91}$

(b) $[OH^-] = 10^{-9.03}$

(c) $[OH^-] = 10^{-10.55}$

Solution

(a) $[OH^-] = 1.2 \times 10^{-5}\ M$

(b) $[OH^-] = 9.3 \times 10^{-10}\ M$

(c) $[OH^-] = 2.8 \times 10^{-11}\ M$

Think About It
Use the benchmark pOH values to determine whether these solutions are reasonable. In part (a), for example, the pOH between 4 and 5 corresponds to [OH⁻] between $1 \times 10^{-4}\ M$ and $1 \times 10^{-5}\ M$.

Practice Problem ATTEMPT Calculate the hydroxide ion concentration in a solution at 25°C in which the pOH is (a) 13.02, (b) 5.14, and (c) 6.98.

Practice Problem BUILD Calculate the hydroxide ion concentration in a solution at 25°C in which the pOH is (a) 11.26, (b) 3.69, and (c) 1.60.

Practice Problem CONCEPTUALIZE What is the value of the exponent in the hydronium ion concentration for solutions with pOH values of 2.90, 8.75, and 11.86?

Section 16.4 Review

The pH and pOH Scales

16.4.1   Determine the pH of a solution at 25°C in which $[H^+] = 6.35 \times 10^{-8}\ M$.
   (a) 7.65   (c) 7.20   (e) 8.00
   (b) 6.80   (d) 6.35

16.4.2   Determine $[H^+]$ in a solution at 25°C if pH = 5.75.
   (a) $1.8 \times 10^{-6}\ M$ (c) $5.8 \times 10^{-6}\ M$ (e) $1.0 \times 10^{-6}\ M$
   (b) $5.6 \times 10^{-9}\ M$ (d) $2.4 \times 10^{-9}\ M$

16.4.3   Determine the pOH of a solution at 25°C in which $[OH^-] = 4.65 \times 10^{-3}\ M$.
   (a) 11.67   (c) 0.32   (e) 2.33
   (b) 13.68   (d) 4.65

16.4.4   Determine $[OH^-]$ in a solution at 25°C if pH = 10.50.
   (a) $3.2 \times 10^{-11}\ M$ (c) $1.1 \times 10^{-2}\ M$ (e) $8.5 \times 10^{-7}\ M$
   (b) $3.2 \times 10^{-4}\ M$ (d) $7.1 \times 10^{-8}\ M$

16.5   STRONG ACIDS AND BASES

Most of this chapter and Chapter 17 deal with equilibrium and the application of the principles of equilibrium to a variety of reaction types. In the context of our discussion of acids and bases, however, it is necessary to review the ionization of strong acids and the dissociation of strong bases. These reactions generally are not treated as *equilibria* but rather as processes that go to completion. This makes the determination of pH for a solution of strong acid or strong base relatively simple.

Student Annotation: We indicate that ionization of a strong acid is *complete* by using a single arrow (⟶) instead of the double, equilibrium arrow (⇌) in the equation.

Strong Acids

There are many different acids, but as we learned in Chapter 9, relatively few qualify as *strong*.

Strong acid	Ionization reaction
Hydrochloric acid	$HCl(aq) + H_2O(l) \longrightarrow H_3O^+(aq) + Cl^-(aq)$
Hydrobromic acid	$HBr(aq) + H_2O(l) \longrightarrow H_3O^+(aq) + Br^-(aq)$
Hydroiodic acid	$HI(aq) + H_2O(l) \longrightarrow H_3O^+(aq) + I^-(aq)$
Nitric acid	$HNO_3(aq) + H_2O(l) \longrightarrow H_3O^+(aq) + NO_3^-(aq)$
Chloric acid	$HClO_3(aq) + H_2O(l) \longrightarrow H_3O^+(aq) + ClO_3^-(aq)$
Perchloric acid	$HClO_4(aq) + H_2O(l) \longrightarrow H_3O^+(aq) + ClO_4^-(aq)$
Sulfuric acid	$H_2SO_4(aq) + H_2O(l) \longrightarrow H_3O^+(aq) + HSO_4^-(aq)$

Video 16.2
Acids and bases—the dissociation of strong and weak acids.

Student Annotation: Remember that although sulfuric acid has two ionizable protons, only the first ionization is complete.

It is a good idea to commit this short list of strong acids to memory.

Because the ionization of a strong acid is complete, the concentration of hydronium ion at equilibrium is equal to the starting concentration of the strong acid. For instance, if we prepare a 0.10-M solution of HCl, the concentration of hydronium ion in the solution is 0.10 M. All the HCl ionizes, and no HCl molecules remain. Thus, at equilibrium (when the ionization is complete), [HCl] = 0 M and $[H_3O^+]$ = $[Cl^-]$ = 0.10 M. Therefore, the pH of the solution (at 25°C) is

$$pH = -\log(0.10) = 1.00$$

This is a very low pH, which is consistent with a relatively concentrated solution of a strong acid.

Student Annotation: As we will see in Section 16.6, a solution of equal concentration but containing a weak acid has a higher pH.

Worked Examples 16.7 and 16.8 let you practice relating the concentration of a strong acid to the pH of an aqueous solution.

Worked Example 16.7

Calculate the pH of an aqueous solution at 25°C that is (a) 0.035 M in HI, (b) 1.2×10^{-4} M in HNO_3, and (c) 6.7×10^{-5} M in $HClO_4$.

Strategy HI, HNO_3, and $HClO_4$ are all strong acids, so the concentration of hydronium ion in each solution is the same as the stated concentration of the acid. Use Equation 9.5 to calculate pH.

Setup

(a) $[H_3O^+]$ = 0.035 M

(b) $[H_3O^+]$ = 1.2×10^{-4} M

(c) $[H_3O^+]$ = 6.7×10^{-5} M

Solution

(a) pH = $-\log(0.035)$ = 1.46

(b) pH = $-\log(1.2 \times 10^{-4})$ = 3.92

(c) pH = $-\log(6.7 \times 10^{-5})$ = 4.17

Think About It
Again, note that when a hydronium ion concentration falls between two of the benchmark concentrations in Table 9.7, the pH falls between the two corresponding pH values. In part (b), for example, the hydronium ion concentration of 1.2×10^{-4} M is greater than 1.0×10^{-4} M and less than 1.0×10^{-3} M. Therefore, we expect the pH to be between 4.00 and 3.00.

$[H_3O^+]$ (M)	$-\log[H_3O^+]$	pH
1.0×10^{-3}	$-\log(1.0 \times 10^{-3})$	3.00
1.2×10^{-4}*	$-\log(1.2 \times 10^{-3})$	3.92[†]
1.0×10^{-4}	$-\log(1.0 \times 10^{-4})$	4.00

*$[H_3O^+]$ between two benchmark values
[†]pH between two benchmark values

Being comfortable with the benchmark hydronium ion concentrations and the corresponding pH values will help you avoid some of the common errors in pH calculations.

Practice Problem Ⓐ**TTEMPT** Calculate the pH of an aqueous solution at 25°C that is (a) 0.081 M in HI, (b) 8.2×10^{-6} M in HNO_3, and (c) 5.4×10^{-3} M in $HClO_4$.

Practice Problem Ⓑ**UILD** Calculate the pH of an aqueous solution at 25°C that is (a) 0.011 M in HNO_3, (b) 3.5×10^{-3} M in HBr, and (c) 9.3×10^{-10} M in HCl.

Practice Problem Ⓒ**ONCEPTUALIZE** Estimate the pH of a solution prepared by dissolving 1.0×10^{-10} mole of a strong acid in a liter of water at 25°C.

Worked Example 16.8

Calculate the concentration of HCl in a solution at 25°C that has pH (a) 4.95, (b) 3.45, and (c) 2.78.

Strategy Use Equation 9.6 to convert from pH to the molar concentration of hydronium ion. In a strong acid solution, the molar concentration of hydronium ion is equal to the acid concentration.

Setup

(a) $[HCl] = [H_3O^+] = 10^{-4.95}$

(b) $[HCl] = [H_3O^+] = 10^{-3.45}$

(c) $[HCl] = [H_3O^+] = 10^{-2.78}$

Solution

(a) $1.1 \times 10^{-5} \, M$

(b) $3.5 \times 10^{-4} \, M$

(c) $1.7 \times 10^{-3} \, M$

Student Annotation: When we take the inverse log of a number with two digits to the right of the decimal point, the result has two significant figures.

Think About It
As pH decreases, acid concentration increases.

Practice Problem A TTEMPT Calculate the concentration of HNO_3 in a solution at 25°C that has pH (a) 2.06, (b) 1.77, and (c) 6.01.

Practice Problem B UILD Calculate the concentration of HBr in a solution at 25°C that has pH (a) 4.81, (b) 1.82, and (c) 3.04.

Practice Problem C ONCEPTUALIZE Which of the plots [(i)–(iv)] best approximates the line that would result if pH were plotted as a function of hydronium ion concentration?

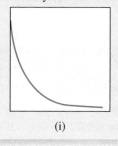

(i)

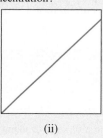

(ii)

(iii)

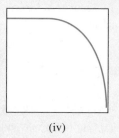

(iv)

Strong Bases

The list of strong bases is also fairly short. It consists of the hydroxides of alkali metals (Group 1A) and the hydroxides of the heaviest alkaline earth metals (Group 2A). The dissociation of a strong base is, for practical purposes, complete. Equations representing dissociations of the strong bases are as follows:

Student Annotation: Recall that $Ca(OH)_2$ and $Sr(OH)_2$ are not very soluble, but what *does* dissolve dissociates completely [◄◄ Section 9.3, Table 9.4].

Group 1A hydroxides	Group 2A hydroxides
$LiOH(aq) \longrightarrow Li^+(aq) + OH^-(aq)$	$Ca(OH)_2(aq) \longrightarrow Ca^{2+}(aq) + 2OH^-(aq)$
$NaOH(aq) \longrightarrow Na^+(aq) + OH^-(aq)$	$Sr(OH)_2(aq) \longrightarrow Sr^{2+}(aq) + 2OH^-(aq)$
$KOH(aq) \longrightarrow K^+(aq) + OH^-(aq)$	$Ba(OH)_2(aq) \longrightarrow Ba^{2+}(aq) + 2OH^-(aq)$
$RbOH(aq) \longrightarrow Rb^+(aq) + OH^-(aq)$	
$CsOH(aq) \longrightarrow Cs^+(aq) + OH^-(aq)$	

Again, because the reaction goes to completion, the pH of such a solution is relatively easy to calculate. In the case of a Group 1A hydroxide, the hydroxide ion concentration is simply the starting concentration of the strong base. In a solution that is 0.018 *M* in NaOH, for example, $[OH^-] = 0.018 \, M$. Its pH can be calculated in two ways. We can either use Equation 16.1 to determine hydronium ion concentration,

$$[H_3O^+][OH^-] = 1.0 \times 10^{-14}$$

$$[H_3O^+] = \frac{1.0 \times 10^{-14}}{[OH^-]} = \frac{1.0 \times 10^{-14}}{0.018} = 5.56 \times 10^{-13} \, M$$

and then Equation 9.5 to determine pH:

$$pH = -\log(5.56 \times 10^{-13} \, M) = 12.25$$

or we can calculate the pOH with Equation 16.2,

$$pOH = -\log (0.018) = 1.75$$

and use Equation 16.4 to convert to pH:

$$pH + pOH = 14.00$$

$$pH = 14.00 - 1.75 = 12.25$$

Both methods give the same result.

In the case of a Group 2A metal hydroxide, we must be careful to account for the reaction stoichiometry. For instance, if we prepare a solution that is 1.9×10^{-4} M in barium hydroxide, the concentration of hydroxide ion at equilibrium (after complete dissociation) is $2(1.9 \times 10^{-4}$ $M)$ or 3.8×10^{-4} M—twice the original concentration of $Ba(OH)_2$. Once we have determined the hydroxide ion concentration, we can determine pH as before:

$$[H_3O^+] = \frac{1.0 \times 10^{-14}}{[OH^-]} = \frac{1.0 \times 10^{-14}}{3.8 \times 10^{-4}} = 2.63 \times 10^{-11}\ M$$

and

$$pH = -\log (2.63 \times 10^{-11}\ M) = 10.58$$

or

$$pOH = -\log (3.8 \times 10^{-4}) = 3.42$$

$$pH + pOH = 14.00$$

$$pH = 14.00 - 3.42 = 10.58$$

Worked Examples 16.9 and 16.10 illustrate calculations involving hydroxide ion concentration, pOH, and pH.

Worked Example 16.9

Calculate the pOH of the following aqueous solutions at 25°C: (a) 0.013 M LiOH, (b) 0.013 M $Ba(OH)_2$, (c) 9.2×10^{-5} M KOH.

Strategy LiOH, $Ba(OH)_2$, and KOH are all strong bases. Use reaction stoichiometry to determine hydroxide ion concentration and Equation 16.2 to determine pOH.

Setup

(a) The hydroxide ion concentration is simply equal to the concentration of the base. Therefore, $[OH^-] = [LiOH] = 0.013$ M.

(b) The hydroxide ion concentration is twice that of the base:

$$Ba(OH)_2(aq) \longrightarrow Ba^{2+}(aq) + 2OH^-(aq)$$

Therefore, $[OH^-] = 2 \times [Ba(OH)_2] = 2(0.013\ M) = 0.026$ M.

(c) The hydroxide ion concentration is equal to the concentration of the base. Therefore, $[OH^-] = [KOH] = 9.2 \times 10^{-5}$ M.

Solution (a) pOH = $-\log (0.013) = 1.89$ (b) pOH = $-\log (0.026) = 1.59$ (c) pOH = $-\log (9.2 \times 10^{-5}) = 4.04$

> **Think About It**
> These are basic pOH values, which is what we should expect for the solutions described in the problem. Note that while the solutions in parts (a) and (b) have the same base concentration, they do not have the same hydroxide concentration and therefore do not have the same pOH.

(Continued on next page)

Practice Problem (A)**TTEMPT** Calculate the pOH of the following aqueous solutions at 25°C: (a) 0.15 *M* NaOH, (b) 8.4×10^{-3} *M* RbOH, (c) 1.7×10^{-5} *M* CsOH.

Practice Problem (B)**UILD** Calculate the pOH of the following aqueous solutions at 25°C: (a) 9.5×10^{-8} *M* NaOH, (b) 6.1×10^{-2} *M* LiOH, (c) 6.1×10^{-2} *M* Ba(OH)$_2$.

Practice Problem (C)**ONCEPTUALIZE** Which of the plots [(i)–(iv)] best represents the relationship between pH and pOH?

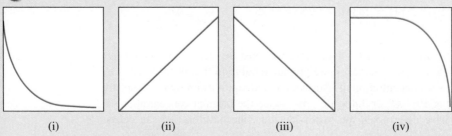

(i)   (ii)   (iii)   (iv)

Worked Example 16.10

An aqueous solution of a strong base has pH 8.15 at 25°C. Calculate the original concentration of base in the solution (a) if the base is NaOH and (b) if the base is Ba(OH)$_2$.

Strategy Use Equation 16.4 to convert from pH to pOH and Equation 16.3 to determine the hydroxide ion concentration. Consider the stoichiometry of dissociation in each case to determine the concentration of the base itself.

Setup $pOH = 14.00 - 8.15 = 5.85$

(a) The dissociation of 1 mole of NaOH produces 1 mole of OH$^-$. Therefore, the concentration of the base is *equal* to the concentration of hydroxide ion.

(b) The dissociation of 1 mole of Ba(OH)$_2$ produces 2 moles of OH$^-$. Therefore, the concentration of the base is only one-half the concentration of hydroxide ion.

Solution $[OH^-] = 10^{-5.85} = 1.41 \times 10^{-6}$ *M*

(a) $[NaOH] = [OH^-] = 1.4 \times 10^{-6}$ *M*

(b) $[Ba(OH)_2] = \frac{1}{2}[OH^-] = 7.1 \times 10^{-7}$ *M*

> **Student Annotation:** Remember to keep an additional significant figure or two until the end of the problem—to avoid *rounding error* [◄◄ Section 1.3].

Think About It
Alternatively, we could determine the hydroxide ion concentration using Equation 9.6, $[H_3O^+] = 10^{-8.15} = 7.1 \times 10^{-9}$ *M* and Equation 16.1,

$$[OH^-] = \frac{1.0 \times 10^{-14}}{7.1 \times 10^{-9} \, M}$$

$$= 1.4 \times 10^{-6} \, M$$

Once [OH$^-$] is known, the solution is the same as shown previously.

Practice Problem (A)**TTEMPT** An aqueous solution of a strong base has pH 8.98 at 25°C. Calculate the concentration of base in the solution (a) if the base is LiOH and (b) if the base is Ba(OH)$_2$.

Practice Problem (B)**UILD** An aqueous solution of a strong base has pH 12.24 at 25°C. Calculate the concentration of base in the solution (a) if the base is NaOH and (b) if the base is Ba(OH)$_2$.

Practice Problem (C)**ONCEPTUALIZE** Which of the plots [(i)–(iv)] best represents pH as a function of concentration for two bases of the same concentration—one monobasic and one dibasic?

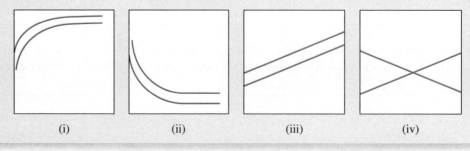

(i)   (ii)   (iii)   (iv)

Section 16.5 Review

Strong Acids and Bases

16.5.1 Calculate the pH of a 0.075-M solution of perchloric acid ($HClO_4$) at 25°C.
(a) 12.88   (c) 6.25   (e) 7.00
(b) 7.75   (d) 1.12

16.5.2 What is the concentration of HBr in a solution with pH 5.89 at 25°C?
(a) $7.8 \times 10^{-9}\ M$ (c) $5.9 \times 10^{-14}\ M$ (e) $1.0 \times 10^{-7}\ M$
(b) $1.3 \times 10^{-6}\ M$ (d) $8.1 \times 10^{-7}\ M$

16.5.3 What is the pOH of a solution at 25°C that is $1.3 \times 10^{-3}\ M$ in $Ba(OH)_2$?
(a) 2.89   (c) 3.19   (e) 11.14
(b) 2.59   (d) 11.11

16.5.4 What is the concentration of KOH in a solution at 25°C that has pOH 3.31?
(a) $2.0 \times 10^{-11}\ M$ (c) $3.3 \times 10^{-7}\ M$ (e) $4.9 \times 10^{-4}\ M$
(b) $3.3 \times 10^{-1}\ M$ (d) $4.5 \times 10^{-4}\ M$

16.5.5 What is the pH of a solution at 25°C that is 0.0095 M in LiOH?
(a) 11.68   (c) 11.98   (e) 12.28
(b) 2.02   (d) 1.72

16.5.6 What is the concentration of $Ca(OH)_2$ in a solution at 25°C if the pH is 9.01?
(a) $1.0 \times 10^{-5}\ M$ (c) $2.0 \times 10^{-5}\ M$ (e) $4.9 \times 10^{-9}\ M$
(b) $5.1 \times 10^{-6}\ M$ (d) $9.8 \times 10^{-9}\ M$

16.5.7 Which diagram best represents a solution of sulfuric acid (H_2SO_4)?

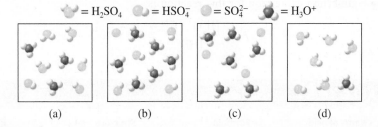

$= H_2SO_4$ $= HSO_4^-$ $= SO_4^{2-}$ $= H_3O^+$

(a)   (b)   (c)   (d)

16.6   WEAK ACIDS AND ACID IONIZATION CONSTANTS

Most acids are **weak acids,** which ionize only to a limited extent in water. At equilibrium, an aqueous solution of a weak acid contains a mixture of aqueous acid molecules, hydronium ions, and the corresponding conjugate base. The degree to which a weak acid ionizes depends on the *concentration* of the acid and the *equilibrium constant* for the ionization.

The Ionization Constant, K_a

Consider a weak monoprotic acid HA. Its ionization in water is represented by

$$HA(aq) + H_2O(l) \rightleftharpoons H_3O^+(aq) + A^-(aq)$$

or by

$$HA(aq) \rightleftharpoons H^+(aq) + A^-(aq)$$

The equilibrium expression for this reaction is

$$K_a = \frac{[H_3O^+][A^-]}{[HA]} \quad \text{or} \quad K_a = \frac{[H^+][A^-]}{[HA]}$$

Student Annotation: Remember that H_3O^+ and H^+ are used interchangeably.

Video 16.3
Using equilibrium tables to solve problems (Figure 16.3).

where K_a is the equilibrium constant for the reaction. More specifically, K_a is called the *acid ionization constant.* Although all weak acids ionize less than 100 percent, they vary in strength. The magnitude of K_a indicates how strong a weak acid is. A large K_a value indicates a stronger acid, whereas a small K_a value indicates a weaker acid. For example, acetic acid (CH_3COOH) and hydrofluoric acid (HF) are both weak acids, but HF is the stronger acid of the two, as evidenced by its larger K_a value. Solutions of equal concentration of the two acids do *not* have the same pH. The pH of the HF solution is lower.

Solution (at 25°C)	K_a	pH
0.10 M HF	7.1×10^{-4}	2.09
0.10 M CH_3COOH	1.8×10^{-5}	2.87

Table 16.5 lists a number of weak acids and their K_a values at 25°C in order of decreasing acid strength.

Calculating pH from K_a

Calculating the pH of a weak acid solution is an equilibrium problem, which we solve using the methods introduced in Chapter 15. Figure 16.3 on pages 734 and 735 shows in detail how an equilibrium table is used to determine the pH of a weak acid. Suppose we want to determine the pH of a 0.50 M HF solution at 25°C. The ionization of HF is represented by

$$HF(aq) + H_2O(l) \rightleftharpoons H_3O^+(aq) + F^-(aq)$$

The equilibrium expression for this reaction is

$$K_a = \frac{[H_3O^+][F^-]}{[HF]} = 7.1 \times 10^{-4}$$

TABLE 16.5	Ionization Constants of Some Weak Acids at 25°C		
Name of acid	**Formula**	**Structure**	**K_a**
Chloroacetic acid	$CH_2ClCOOH$	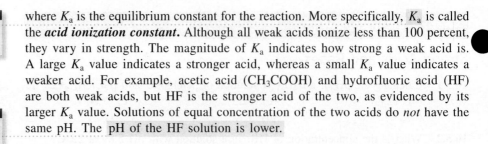 CH₂Cl—C—O—H (with =O)	5.6×10^{-2}
Hydrofluoric acid	HF	H—F	7.1×10^{-4}
Nitrous acid	HNO_2	O=N—O—H	4.5×10^{-4}
Formic acid	HCOOH	H—C—O—H (with =O)	1.7×10^{-4}
Benzoic acid	C_6H_5COOH	⬡—C—O—H (with =O)	6.5×10^{-5}
Acetic acid	CH_3COOH	CH₃—C—O—H (with =O)	1.8×10^{-5}
Hydrocyanic acid	HCN	H—C≡N	4.9×10^{-10}
Phenol	C_6H_5OH	⬡—O—H	1.3×10^{-10}

We construct an equilibrium table and enter the starting concentrations of all species in the equilibrium expression.

$$HF(aq) + H_2O(l) \rightleftharpoons H_3O^+(aq) + F^-(aq)$$

	HF		H_3O^+	F^-
Initial concentration (M):	0.50		0	0
Change in concentration (M):				
Equilibrium concentration (M):				

Using the reaction stoichiometry, we determine the changes in all species.

$$HF(aq) + H_2O(l) \rightleftharpoons H_3O^+(aq) + F^-(aq)$$

	HF		H_3O^+	F^-
Initial concentration (M):	0.50		0	0
Change in concentration (M):	$-x$		$+x$	$+x$
Equilibrium concentration (M):				

Finally, we express the equilibrium concentration of each species in terms of x.

$$HF(aq) + H_2O(l) \rightleftharpoons H_3O^+(aq) + F^-(aq)$$

	HF		H_3O^+	F^-
Initial concentration (M):	0.50		0	0
Change in concentration (M):	$-x$		$+x$	$+x$
Equilibrium concentration (M):	$0.50 - x$		x	x

These equilibrium concentrations are then entered into the equilibrium expression to give

$$K_a = \frac{(x)(x)}{0.50 - x} = 7.1 \times 10^{-4}$$

Rearranging this expression, we get

$$x^2 + 7.1 \times 10^{-4}x - 3.55 \times 10^{-4} = 0$$

This is a quadratic equation, which we can solve using the quadratic formula given in Appendix 1. In the case of a weak acid, however, often we can use a shortcut to simplify the calculation. Because HF is a weak acid, and weak acids ionize only to a slight extent, x must be small compared to 0.50. Therefore, we can make the following approximation.

$$0.50 - x \approx 0.50$$

Now the equilibrium expression becomes

$$\frac{x^2}{0.50 - x} \approx \frac{x^2}{0.50} = 7.1 \times 10^{-4}$$

Rearranging, we get

$$x^2 = (0.50)(7.1 \times 10^{-4}) = 3.55 \times 10^{-4}$$
$$x = \sqrt{3.55 \times 10^{-4}} = 1.9 \times 10^{-2}$$

Thus, we have solved for x without having to use the quadratic equation. At equilibrium we have

$$[HF] = (0.50 - 0.019)\ M = 0.48\ M$$

$$[H_3O^+] = 0.019\ M$$

$$[F^-] = 0.019\ M$$

and the pH of the solution is

$$pH = -\log(0.019) = 1.72$$

Figure 16.3

Using Equilibrium Tables to Solve Problems

Using K_a and Concentration to Determine pH of a Weak Acid

START

Determine the pH of a weak acid with a concentration of 0.10 *M* and a K_a of 2.5×10^{-5}. In this figure, the hydronium ion is represented as H^+.

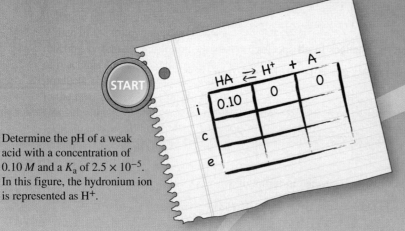

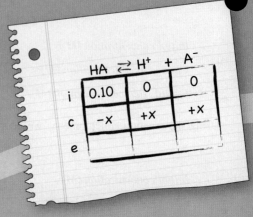

The concentrations will change by an unknown amount,
· [HA] will decrease by x
· $[H^+]$ and $[A^-]$ will increase by x
We enter the anticipated changes in concentrations in the middle row of the table.

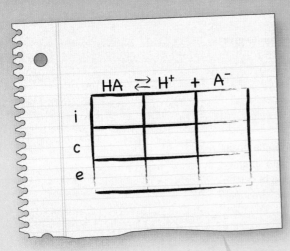

Many equilibrium problems can be solved using an equilibrium table. The table is constructed with a column under each species in the equilibrium equation, and three rows labeled i (initial), c (change), and e (equilibrium). (This is why such a table is sometimes referred to as an "ice" table.)

HA

We use pH to determine the equilibrium concentration of H^+:

$$[H^+] = 10^{-pH} = 10^{-3.82} = 1.5 \times 10^{-4} \, M$$

Because the ionization of HA produces equal amounts of H^+ and A^-, the equilibrium concentration of A^- is also $1.5 \times 10^{-4} \, M$. We enter these concentrations in the bottom row of the table.

Using pH and Concentration to Determine K_a of a Weak Acid

START

Determine the K_a of a weak acid if a 0.12-*M* solution has a pH of 3.82.

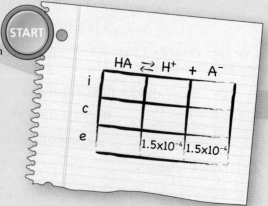

The equilibrium concentrations, which we enter in the last row of the table, are expressed in terms of the unknown x. We write the equilibrium expression for the reaction,

$$K_a = \frac{[H^+][A^-]}{[HA]}$$

and enter the K_a value and the equilibrium concentrations from the last row in the table:

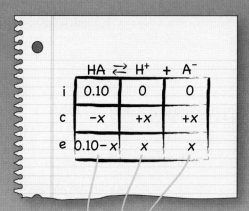

$$2.5 \times 10^{-5} = \frac{(x)(x)}{(0.10 - x)} = \frac{x^2}{(0.10 - x)}$$

Because the magnitude of K_a is small, we expect the amount of HA ionized (x) to be small compared to 0.10. Therefore, we can assume that $0.10 - x \approx 0.10$. This simplifies the solution to

$$2.5 \times 10^{-5} = \frac{x^2}{0.10}$$

Solving for x gives

$$(2.5 \times 10^{-5})(0.10) = x^2$$

$$2.5 \times 10^{-6} = x^2$$

$$\sqrt{2.5 \times 10^{-6}} = x$$

$$x = 1.6 \times 10^{-3} \, M$$

We used x to represent $[H^+]$ at equilibrium. (x is also equal to $[A^-]$ at equilibrium and to the change in $[HA]$.) Therefore, we determine pH by taking the negative log of x.

$$pH = -\log [H^+] = -\log (1.6 \times 10^{-3}) = 2.80$$

CHECK Verify that $\dfrac{1.6 \times 10^{-3} \, M}{0.10 \, M} \times 100\% < 5\%$

and that $\dfrac{(1.6 \times 10^{-3})^2}{(0.10 - 1.6 \times 10^{-3})} \approx K_a$

With the bottom row of the table complete, we can use these equilibrium concentrations to calculate the value of K_a.

$$K_a = \frac{[H^+][A^-]}{[HA]} = \frac{(1.5 \times 10^{-4})(1.5 \times 10^{-4})}{(0.12)} = 1.9 \times 10^{-7}$$

CHECK Verify that when $[HA] = 0.12 \, M$ and $K_a = 1.9 \times 10^{-7}$, $[H^+] = \sqrt{(1.9 \times 10^{-7})(0.12)} = 1.5 \times 10^{-4} \, M$ and pH = 3.82.

What's the point?

· Starting with the molar concentration and K_a of a weak monoprotic acid, we can use an equilibrium table to determine pH.
· Starting with pH and molar concentration, we can use an equilibrium table to determine the K_a of a weak monoprotic acid.

$$HA \rightleftharpoons H^+ + A^-$$

To complete the last row of the table, we also need the equilibrium value of $[HA]$. The amount of weak acid that ionized is equal to the amount of H^+ produced by the ionization. Therefore, at equilibrium,

$$[HA] = 0.12 \, M - 1.5 \times 10^{-4} \, M \approx 0.12 \, M$$

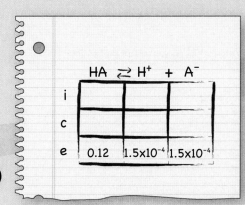

This shortcut gives a good approximation as long as the magnitude of x is significantly smaller than the initial acid concentration. As a rule, it is acceptable to use this shortcut if the calculated value of x is less than 5 percent of the initial acid concentration. In this case, the approximation is acceptable because

$$\frac{0.019\ M}{0.50\ M} \times 100\% = 3.8\%$$

Student Hot Spot

Student data indicate you may struggle with determining the pH of a weak acid solution. Access the SmartBook to view additional Learning Resources on this topic.

This is the formula for the *percent ionization* of the acid [◀◀ Section 13.6]. Recall that the percent ionization of a weak electrolyte, such as a weak acid, depends on concentration. Consider a more dilute solution of HF, one that is 0.050 M. Using the preceding procedure to solve for x, we would get $6.0 \times 10^{-3}\ M$. The following test shows, however, that this answer is *not* a valid approximation because it is greater than 5 percent of 0.050 M.

$$\frac{6.0 \times 10^{-3}\ M}{0.050\ M} \times 100\% = 12\%$$

Student Annotation: In many cases, use of the quadratic equation can be avoided with a method called *successive approximation*, which is presented in Appendix 1.

In this case, we must solve for x using the quadratic equation [◀◀ Worked Example 15.12].

Worked Example 16.11 shows how to use K_a to determine the pH of a weak acid solution.

Worked Example 16.11

The K_a of hypochlorous acid (HClO) is 3.5×10^{-8}. Calculate the pH of a solution at 25°C that is 0.0075 M in HClO.

Strategy Construct an equilibrium table, and express the equilibrium concentration of each species in terms of x. Solve for x using the approximation shortcut, and evaluate whether the approximation is valid. Use Equation 9.5 to determine pH.

Setup

$$HClO(aq) + H_2O(l) \rightleftharpoons H_3O^+(aq) + ClO^-(aq)$$

	HClO	H₂O	H₃O⁺	ClO⁻
Initial concentration (M):	0.0075		0	0
Change in concentration (M):	$-x$		$+x$	$+x$
Equilibrium concentration (M):	$0.0075 - x$		x	x

Solution These equilibrium concentrations are then substituted into the equilibrium expression to give

$$K_a = \frac{(x)(x)}{0.0075 - x} = 3.5 \times 10^{-8}$$

Assuming that $0.0075 - x \approx 0.0075$,

$$\frac{x^2}{0.0075} = 3.5 \times 10^{-8} \qquad x^2 = (3.5 \times 10^{-8})(0.0075)$$

Solving for x, we get

$$x = \sqrt{2.625 \times 10^{-10}} = 1.62 \times 10^{-5}\ M$$

According to the equilibrium table, $x = [H_3O^+]$. Therefore,

$$pH = -\log(1.62 \times 10^{-5}) = 4.79$$

Student Annotation: Applying the 5 percent test indicates that the approximation shortcut is valid in this case: $(1.62 \times 10^{-5}/0.0075) \times 100\% < 5\%$.

Think About It

We learned in Section 16.3 that the concentration of hydronium ion in pure water at 25°C is $1.0 \times 10^{-7}\ M$, yet we use 0 M as the starting concentration to solve for the pH of a solution of weak acid.

$$HA(aq) \rightleftharpoons H^+(aq) + A^-(aq)$$

	HA	H⁺	A⁻
Initial concentration (M):		0	0
Change in concentration (M):			
Equilibrium concentration (M):			

The reason for this is that the *actual* concentration of hydronium ion in pure water is insignificant compared to the amount produced by the ionization of the weak acid. We could use the actual concentration of hydronium as the initial concentration, but doing so would not change the result because $(x + 1.0 \times 10^{-7})\ M \approx x\ M$. In solving problems of this type, we neglect the small concentration of H⁺ due to the autoionization of water.

Practice Problem Ⓐ**TTEMPT** Calculate the pH at 25°C of a 0.18-M solution of a weak acid that has $K_a = 9.2 \times 10^{-6}$.

Practice Problem Ⓑ**UILD** Calculate the pH at 25°C of a 0.065-M solution of a weak acid that has $K_a = 1.2 \times 10^{-5}$.

Practice Problem Ⓒ**ONCEPTUALIZE** The diagrams show solutions of four different weak acids. In which solution is the concentration of the weak acid highest? Which solution has the highest pH value?

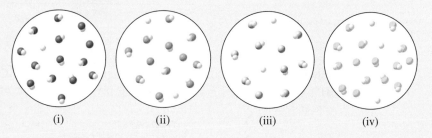

(i)   (ii)   (iii)   (iv)

Percent Ionization

When we encountered acids earlier [◄◄ Section 9.3], we learned that a strong acid is one that ionizes completely. Weak acids ionize only partially. The degree to which a weak acid ionizes is a measure of its strength, just as the magnitude of its ionization constant (K_a) is a measure of its strength. A quantitative measure of the degree of ionization is *percent ionization,* which, for a weak, monoprotic acid (HA) is calculated as follows:

$$\text{percent ionization} = \frac{[H_3O^+]_{eq}}{[HA]_0} \times 100\% \qquad \textbf{Equation 16.5}$$

in which the hydrogen ion concentration, $[H^+]_{eq}$, is the concentration at equilibrium and the weak acid concentration, $[HA]_0$, is the *original* concentration of the acid, which is not necessarily the same as the concentration at equilibrium.

Student Annotation: In fact, a weak acid's original concentration is never *exactly* equal to its equilibrium concentration. However, sometimes the amount that ionizes is so small compared to the original concentration that, to an appropriate number of significant figures, the weak acid concentration does not change and $[HA]_0 \approx [HA]_{eq}$.

Thinking Outside the Box

Acid Rain

Rainwater in unpolluted areas is slightly acidic. The cause is carbon dioxide in the air, which dissolves in raindrops and reacts to form the weak acid, *carbonic* acid (H_2CO_3). In areas where the air is polluted by the burning of fossil fuels, rain can be very acidic. This phenomenon, known as *acid rain,* was first discovered in 1852 by Scottish chemist Robert Angus Smith (1817–1884) in Manchester, England. At that time, the industrial revolution was well under way and the British economy relied heavily on the use of coal to generate steam. The two main atmospheric contributors to the acidity of acid rain are sulfur dioxide (SO_2) and the oxides of nitrogen, NO and NO_2 (collectively referred to as NO_x), which react to produce sulfuric acid (H_2SO_4) and nitric acid (HNO_3)—both *strong* acids. SO_2 and NO_x are both produced by the burning of sulfur-bearing coal.

Although acid rain was first discovered in 1852, scientific and societal efforts to understand and remedy its causes did not emerge for another century. U.S. scientists began to study the causes and effects

of acid rain extensively only in the late 1960s. Public awareness of the phenomenon was heightened in the 1970s when results of studies at the Hubbard Brook Experimental Forest in New Hampshire were published, detailing the devastating impact of acid rain on the ecology of the region.

Despite the dire circumstance described by atmospheric scientists and ecologists in the last half of the twentieth century, there appears to be reason for optimism with regard to acid rain. The Acid Rain Program, part of the 1990 amendment to the Clean Air Act (originally enacted in 1963), has contributed to reducing the acidity of rainwater in industrial areas of the United States to levels below those seen in the 1960s. Reducing sulfur emissions from coal-burning power plants has involved using coal containing less sulfur, fitting smokestacks with chemical scrubbers to remove SO_2, and developing alternative sources of energy. Significant reductions in NO_x emissions have been achieved through the use of catalytic converters, which have been standard equipment on automobiles in the United States since the mid-1970s.

Consider a 0.10-*M* solution of benzoic acid, for which the ionization constant (K_a) is 6.5×10^{-5}. Using the procedure described in Worked Example 16.11, we can determine the equilibrium concentrations of benzoic acid, H^+, and the benzoate ion.

$$C_6H_5COOH(aq) + H_2O(l) \rightleftharpoons H_3O^+(aq) + C_6H_5COO^-(aq)$$

	C_6H_5COOH	H_2O	H_3O^+	$C_6H_5COO^-$
Initial concentration (M):	0.100		0	0
Change in concentration (M):	$-x$		$+x$	$+x$
Equilibrium concentration (M):	$0.100 - x$		x	x

Solving for x gives 0.0025 *M*. Therefore, at equilibrium, $[C_6H_5COOH] = 0.097$ *M* and $[H^+] = [C_6H_5COO^-] = 0.0025$ *M*. The percent ionization of benzoic acid at this concentration is

$$\frac{0.0025\ M}{0.100\ M} \times 100\% = 2.5\%$$

Now consider what happens when we dilute this equilibrium mixture by adding enough water to double the volume. The concentrations of all three species are cut in half; $[C_6H_5COOH] = 0.049$ *M*, $[H^+] = 0.0013$ *M*, and $[C_6H_5COO^-] = 0.0013$ *M*. If we plug these new concentrations into the equilibrium expression and calculate the reaction quotient (Q), we get a number different from K_a. In fact, we get ($0.0013^2/0.049$) = 3.4×10^{-5}, which is *smaller* than K_a. When Q is smaller than K, the reaction must proceed to the right to reestablish equilibrium [◄◄ Section 15.4]. Proceeding to the right, in the case of weak acid ionization, corresponds to more of the acid ionizing— meaning that its percent ionization increases.

Student Annotation: According to Le Châtelier's principle [◄◄ Section 15.6], the reduction in particle concentration caused by dilution stresses the equilibrium. The equilibrium shifts to minimize the effects of the stress by shifting toward the side with more dissolved particles, causing more of the weak acid to ionize.

We can solve again for percent ionization starting with half the original concentration of benzoic acid.

$$C_6H_5COOH(aq) + H_2O(l) \rightleftharpoons H_3O^+(aq) + C_6H_5COO^-(aq)$$

	C_6H_5COOH	H_2O	H_3O^+	$C_6H_5COO^-$
Initial concentration (M):	0.050		0	0
Change in concentration (M):	$-x$		$+x$	$+x$
Equilibrium concentration (M):	$0.050 - x$		x	x

This time, solving for x gives 0.0018 *M*. Therefore, at equilibrium, $[C_6H_5COOH] = 0.048$ *M*, and $[H^+] = [C_6H_5COO^-] = 0.0018$ *M*. The percent ionization of benzoic acid at this concentration is

$$\frac{0.0018\ M}{0.0050\ M} \times 100\% = 3.6\%$$

Figure 16.4 shows the dependence of percent ionization of a weak acid on concentration. Note that as concentration approaches zero, percent ionization approaches 100.

Worked Example 16.12 lets you practice calculating percent ionization of weak acid solutions.

Figure 16.4 Percent ionization of a weak acid depends on the original acid concentration. As concentration approaches zero, ionization approaches 100 percent. The dashed grey line represents 100 percent ionization, which is characteristic of a strong acid.

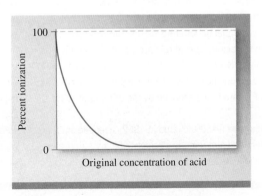

Worked Example 16.12

Determine the pH and percent ionization for acetic acid solutions at 25°C with concentrations (a) 0.15 M, (b) 0.015 M, and (c) 0.0015 M.

Strategy Using the procedure described in Worked Example 16.11, we construct an equilibrium table and for each concentration of acetic acid, we solve for the equilibrium concentration of H$^+$. We use Equation 9.5 to find pH, and Equation 16.5 to find percent ionization.

Setup From Table 16.5, the ionization constant, K_a, for acetic acid is 1.8×10^{-5}.

Solution

(a)

	$CH_3COOH(aq)$ +	$H_2O(l)$	$\rightleftharpoons$ $H_3O^+(aq)$ +	$CH_3COO^-(aq)$
Initial concentration (M):	0.15		0	0
Change in concentration (M):	$-x$		$+x$	$+x$
Equilibrium concentration (M):	$0.15 - x$		x	x

Solving for x gives [H$_3$O$^+$] = 0.0016 M and pH = $-\log (0.0016)$ = 2.78.

$$\text{percent ionization} = \frac{0.0016\ M}{0.15\ M} \times 100\% = 1.1\%$$

(b) Solving in the same way as part (a) gives [H$_3$O$^+$] = 5.2×10^{-4} M and pH = 3.28.

$$\text{percent ionization} = \frac{5.2 \times 10^{-4}\ M}{0.015\ M} \times 100\% = 3.5\%$$

(c) Solving the quadratic equation, or using successive approximation [▶▶ Appendix 1] gives [H$_3$O$^+$] = 1.6×10^{-4} M and pH = 3.78.

$$\text{percent ionization} = \frac{1.6 \times 10^{-4}\ M}{0.0015\ M} \times 100\% = 11\%$$

Think About It
Percent ionization also increases as concentration decreases for weak bases [▶▶ Section 16.7].

Practice Problem A TTEMPT Determine the pH and percent ionization for hydrocyanic acid (HCN) solutions of concentration (a) 0.25 M, (b) 0.0075 M, and (c) 8.3×10^{-5} M.

Practice Problem B UILD At what concentration does hydrocyanic acid exhibit (a) 0.05 percent ionization, (b) 0.10 percent ionization, and (c) 0.15 percent ionization?

Practice Problem C ONCEPTUALIZE Which of the diagrams shows the weak acid with the highest percent ionization? Is it possible for one of the other weak acids shown here ever to have a higher percent ionization than the one you chose? Explain.

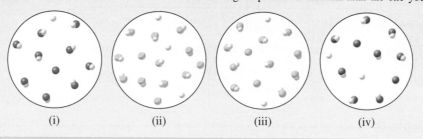

| (i) | (ii) | (iii) | (iv) |

Using pH to Determine K_a

In Chapter 15 we learned that we can determine the value of an equilibrium constant using equilibrium concentrations [◀◀ Section 15.2]. Using a similar approach, we can use the pH of a weak acid solution to determine the value of K_a. Suppose we want to determine the K_a of a weak acid (HA) and we know that a 0.25-M solution of the acid has a pH of 3.47 at 25°C. The first step is to use pH to determine the equilibrium hydronium ion concentration. Using Equation 9.6, we get

Student Annotation: Remember that we keep extra significant figures until the end of a multistep problem to minimize rounding error [◀◀ Section 1.3].

$$[H_3O^+] = 10^{-3.47} = 3.39 \times 10^{-4}\ M$$

We use the starting concentration of the weak acid and the equilibrium concentration of the hydronium ion to construct an equilibrium table and determine the equilibrium concentrations of all three species.

	$HA(aq)$	$+$	$H_2O(l)$	$\rightleftharpoons$	$H_3O^+(aq)$	$+$	$A^-(aq)$
Initial concentration (M):	0.25				0		0
Change in concentration (M):	-3.39×10^{-4}				$+3.39 \times 10^{-4}$		$+3.39 \times 10^{-4}$
Equilibrium concentration (M):	0.2497				3.39×10^{-4}		3.39×10^{-4}

Student Hot Spot

Student data indicate you may struggle with determining the K_a of a weak acid. Access the SmartBook to view additional Learning Resources on this topic.

These equilibrium concentrations are substituted into the equilibrium expression to give

$$K_a = \frac{(3.39 \times 10^{-4})^2}{0.2497} = 4.6 \times 10^{-7}$$

Therefore, the K_a of this weak acid is 4.6×10^{-7}.

Worked Example 16.13 shows how to determine K_a using pH.

Worked Example 16.13

Aspirin (acetylsalicylic acid, $HC_9H_7O_4$) is a weak acid. It ionizes in water according to the equation

$$HC_9H_7O_4(aq) + H_2O(l) \rightleftharpoons H_3O^+(aq) + C_9H_7O_4^-(aq)$$

A 0.10-*M* aqueous solution of aspirin has a pH of 2.27 at 25°C. Determine the K_a of aspirin.

Strategy Determine the hydronium ion concentration from the pH. Use the hydronium ion concentration to determine the equilibrium concentrations of the other species, and plug the equilibrium concentrations into the equilibrium expression to evaluate K_a.

Setup Using Equation 9.6, we have

$$[H_3O^+] = 10^{-2.27} = 5.37 \times 10^{-3} \, M$$

To calculate K_a, though, we also need the equilibrium concentrations of $C_9H_7O_4^-$ and $HC_9H_7O_4$. The stoichiometry of the reaction tells us that $[C_9H_7O_4^-] = [H_3O^+]$. Furthermore, the amount of aspirin that has *ionized* is equal to the amount of hydronium ion in solution. Therefore, the equilibrium concentration of aspirin is $(0.10 - 5.37 \times 10^{-3}) \, M = 0.095 \, M$.

	$HC_9H_7O_4(aq)$	$+$	$H_2O(l)$	$\rightleftharpoons$	$H_3O^+(aq)$	$+$	$C_9H_7O_4^-(aq)$
Initial concentration (M):	0.10				0		0
Change in concentration (M):	-0.005				$+5.37 \times 10^{-3}$		$+5.37 \times 10^{-3}$
Equilibrium concentration (M):	0.095				5.37×10^{-3}		5.37×10^{-3}

Solution Substitute the equilibrium concentrations into the equilibrium expression as follows:

$$K_a = \frac{[H_3O^+][C_9H_7O_4^-]}{[HC_9H_7O_4]} = \frac{(5.37 \times 10^{-3})^2}{0.095} = 3.0 \times 10^{-4}$$

The K_a of aspirin is 3.0×10^{-4}.

Think About It
Check your work by using the calculated value of K_a to solve for the pH of a 0.10-*M* solution of aspirin.

Practice Problem **ATTEMPT** Calculate the K_a of a weak acid if a 0.065-*M* solution of the acid has a pH of 2.96 at 25°C.

Practice Problem **BUILD** Calculate the K_a of a weak acid if a 0.015-*M* solution of the acid has a pH of 5.03 at 25°C.

Practice Problem **CONCEPTUALIZE** Calculate K_a values (to two significant figures) for the weak acids represented in the diagrams.

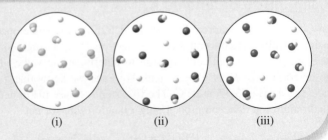

(i)   (ii)   (iii)

Section 16.6 Review

Weak Acids and Acid Ionization Constants

16.6.1 The K_a of a weak acid is 5.5×10^{-4}. What is the pH of a 0.63-M solution of this acid at 25°C?

(a) 3.26   (d) 1.63
(b) 1.83×10^{-2} (e) 0.201
(c) 1.73

16.6.2 A 0.042-M solution of a weak acid has pH 4.01 at 25°C. What is the K_a of this acid?

(a) 9.5×10^{-9} (d) 4.2×10^{-2}
(b) 9.8×10^{-5} (e) 2.3×10^{-7}
(c) 0.91

16.6.3 The diagrams show solutions of three different weak acids with the general formula HA. List the acids in order of increasing K_a value.

(a) i < ii < iii   (d) ii < iii < i
(b) i < iii < ii   (e) iii < ii < i
(c) ii < i < iii

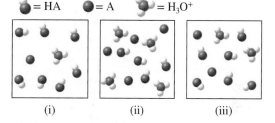

(i)   (ii)   (iii)

16.7 WEAK BASES AND BASE IONIZATION CONSTANTS

Just as most acids are weak, most bases are also weak. The ionization of a *weak base* is incomplete and is treated in the same way as the ionization of a weak acid. In this section, we will see how the ionization constant for a weak base, K_b, is related to the pH of an aqueous solution.

The Ionization Constant, K_b

The ionization of a weak base can be represented by the equation

$$B(aq) + H_2O(l) \rightleftharpoons HB^+(aq) + OH^-(aq)$$

where B is the weak base and HB^+ is its conjugate acid. The equilibrium expression for the ionization is

$$K_b = \frac{[HB^+][OH^-]}{[B]}$$

where K_b is the equilibrium constant—specifically known as the *base ionization constant*. Like weak acids, weak bases vary in strength. Table 16.6 lists a number of common weak bases and their ionization constants. The ability of any one of these substances to act as a base is the result of the lone pair of electrons on the nitrogen atom. The presence of this lone pair is what enables a compound to accept a proton, which is what makes a compound a Brønsted base.

Calculating pH from K_b

Solving problems involving weak bases requires the same approach we used for weak acids. It is important to remember, though, that solving for x in a typical weak base problem gives us the hydroxide ion concentration rather than the hydronium ion concentration.

TABLE 16.6	Ionization Constants of Some Weak Bases at 25°C		
Name of base	**Formula**	**Structure**	K_b
Ethylamine	$C_2H_5NH_2$	$CH_3-CH_2-\overset{\cdot\cdot}{N}-H$ $\quad\quad\quad\quad\quad\mid$ $\quad\quad\quad\quad\quad H$	5.6×10^{-4}
Methylamine	CH_3NH_2	$CH_3-\overset{\cdot\cdot}{N}-H$ $\quad\quad\quad\mid$ $\quad\quad\quad H$	4.4×10^{-4}
Ammonia	NH_3	$H-\overset{\cdot\cdot}{N}-H$ $\quad\quad\mid$ $\quad\quad H$	1.8×10^{-5}
Pyridine	C_5H_5N	N:	1.7×10^{-9}
Aniline	$C_6H_5NH_2$	$\overset{\cdot\cdot}{N}-H$ $\quad\mid$ $\quad H$	3.8×10^{-10}
Urea	H_2NCONH_2	$\quad\quad\quad\quad O$ $\quad\quad\quad\quad\parallel$ $H-\overset{\cdot\cdot}{N}-C-\overset{\cdot\cdot}{N}-H$ $\quad\mid\quad\quad\quad\mid$ $\quad H\quad\quad\quad H$	1.5×10^{-14}

Worked Example 16.14 shows how to use K_b to calculate the pH of a weak base solution.

Worked Example 16.14

What is the pH of a 0.040 M ammonia solution at 25°C?

Strategy Construct an equilibrium table, and express equilibrium concentrations in terms of the unknown x. Plug these equilibrium concentrations into the equilibrium expression, and solve for x. From the value of x, determine the pH.

Setup

$$NH_3(aq) + H_2O(l) \rightleftharpoons NH_4^+(aq) + OH^-(aq)$$

	$NH_3(aq) +$	$H_2O(l) \rightleftharpoons$	$NH_4^+(aq) +$	$OH^-(aq)$
Initial concentration (M):	0.040		0	0
Change in concentration (M):	$-x$		$+x$	$+x$
Equilibrium concentration (M):	$0.040 - x$		x	x

Solution The equilibrium concentrations are substituted into the equilibrium expression to give

$$K_b = \frac{[NH_4^+][OH^-]}{[NH_3]} = \frac{(x)(x)}{0.040 - x} = 1.8 \times 10^{-5}$$

Assuming that $0.040 - x \approx 0.040$ and solving for x gives

$$\frac{(x)(x)}{0.040 - x} \approx \frac{(x)(x)}{0.040} = 1.8 \times 10^{-5}$$

$$x^2 = (1.8 \times 10^{-5})(0.040) = 7.2 \times 10^{-7}$$

$$x = \sqrt{7.2 \times 10^{-7}} = 8.5 \times 10^{-4} \, M$$

Student Annotation: Applying the 5 percent test indicates that the approximation shortcut is valid in this case: $(8.5 \times 10^{-4}/0.040) \times 100\% \approx 2\%$.

According to the equilibrium table, $x = [OH^-]$. Therefore, $pOH = -\log (x)$:

$$-\log (8.5 \times 10^{-4}) = 3.07$$

and $pH = 14.00 - pOH = 14.00 - 3.07 = 10.93$. The pH of a 0.040-$M$ solution of NH_3 at 25°C is 10.93.

Think About It
It is a common error in K_b problems to forget that x is the *hydroxide* ion concentration rather than the *hydronium* ion concentration. Always make sure that the pH you calculate for a solution of base is a basic pH—that is, a pH greater than 7.

Practice Problem **A**TTEMPT Calculate the pH at 25°C of a 0.0028-M solution of a weak base with a K_b of 6.8×10^{-8}.

Practice Problem **B**UILD Calculate the pH at 25°C of a 0.16-M solution of a weak base with a K_b of 2.9×10^{-11}.

Practice Problem **C**ONCEPTUALIZE The diagrams represent solutions of three different weak bases. Arrange the solutions in order of increasing pH. To keep the diagrams from being crowded, hydroxide ions are shown, but water molecules are not.

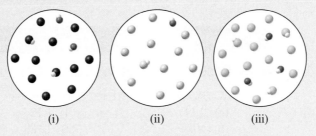

(i)   (ii)   (iii)

Using pH to Determine K_b

Just as we can use pH to determine the K_a of a weak acid, we can also use it to determine the K_b of a weak base. Worked Example 16.15 demonstrates this procedure.

Worked Example 16.15

Caffeine, the stimulant in coffee and tea, is a weak base that ionizes in water according to the equation

$$C_8H_{10}N_4O_2(aq) + H_2O(l) \rightleftharpoons HC_8H_{10}N_4O_2^+(aq) + OH^-(aq)$$

A 0.15-M solution of caffeine at 25°C has a pH of 8.45. Determine the K_b of caffeine.

Strategy Use pH to determine pOH, and pOH to determine the hydroxide ion concentration. From the hydroxide ion concentration, use reaction stoichiometry to determine the other equilibrium concentrations and plug those concentrations into the equilibrium expression to evaluate K_b.

Setup

$$pOH = 14.00 - 8.45 = 5.55$$
$$[OH^-] = 10^{-5.55} = 2.82 \times 10^{-6} \ M$$

Based on the reaction stoichiometry, $[HC_8H_{10}N_4O_2^+] = [OH^-]$, and the amount of hydroxide ion in solution at equilibrium is equal to the amount of caffeine that has ionized. At equilibrium, therefore,

$$[C_8H_{10}N_4O_2] = (0.15 - 2.82 \times 10^{-6}) \ M \approx 0.15 \ M$$

	$C_8H_{10}N_4O_2(aq) +$	$H_2O(l) \rightleftharpoons$	$HC_8H_{10}N_4O_2^+(aq) +$	$OH^-(aq)$
Initial concentration (M):	0.15		0	0
Change in concentration (M):	-2.82×10^{-6}		$+2.82 \times 10^{-6}$	$+2.82 \times 10^{-6}$
Equilibrium concentration (M):	0.15		2.82×10^{-6}	2.82×10^{-6}

(Continued on next page)

Solution Plugging the equilibrium concentrations into the equilibrium expression gives

$$K_b = \frac{[HC_8H_{10}N_4O_2^+][OH^-]}{[C_8H_{10}N_4O_2]} = \frac{(2.82 \times 10^{-6})^2}{0.15} = 5.3 \times 10^{-11}$$

Think About It
Check your answer by using the calculated K_b to determine the pH of a 0.15-M solution.

Practice Problem ATTEMPT Determine the K_b of a weak base if a 0.50-M solution of the base has a pH of 9.59 at 25°C.

Practice Problem BUILD Determine the K_b of a weak base if a 0.35-M solution of the base has a pH of 11.84 at 25°C.

Practice Problem CONCEPTUALIZE Determine the value of K_b (to two significant figures) for each of the bases represented in Practice Problem 16.14C.

Section 16.7 Review

Weak Bases and Base Ionization Constants

16.7.1 What is the pH of a 0.63-M solution of weak base at 25°C if $K_b = 9.5 \times 10^{-7}$?
(a) 3.11   (d) 1.12
(b) 10.89   (e) 12.88
(c) 7.00

16.7.2 A 0.12-M solution of a weak base has a pH of 10.76 at 25°C. Determine K_b.
(a) 2.5×10^{-21} (d) 2.8×10^{-6}
(b) 8.3×10^{8} (e) 1.0×10^{-14}
(c) 4.0×10^{-8}

16.7.3 The diagrams show solutions of three different weak bases. List the bases in order of increasing K_b value.

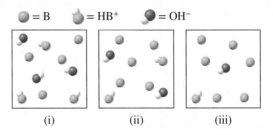

(a) i < ii < iii   (d) ii < iii < i
(b) i < iii < ii   (e) iii < ii < i
(c) ii < i < iii

16.8 CONJUGATE ACID-BASE PAIRS

At the beginning of this chapter, we introduced the concept of conjugate acids and conjugate bases. In this section, we examine the properties of conjugate acids and bases, independent of their parent compounds.

The Strength of a Conjugate Acid or Base

When a strong acid such as HCl dissolves in water, it ionizes completely because its conjugate base (Cl^-) has essentially *no* affinity for the H^+ ion in solution.

$$HCl(aq) \longrightarrow H^+(aq) + Cl^-(aq)$$

Because the chloride ion has no affinity for the H^+ ion, it does not act as a Brønsted base in water. If we dissolve a chloride salt such as NaCl in water, for example, the Cl^- ions in solution would not accept protons from the water.

$$Cl^-(aq) + H_2O(l) \xrightarrow{\hspace{1em}\times\hspace{1em}} HCl(aq) + OH^-(aq)$$

> **Student Annotation:** The products HCl and OH^- do *not* actually form when Cl^- and H_2O are combined.

The chloride ion, which is the conjugate base of a strong acid, is an example of a *weak conjugate base.*

Now consider the case of a weak acid. When HF dissolves in water, the ionization happens only to a *limited* degree because the conjugate base, F^-, has a strong affinity for the H^+ ion.

$$HF(aq) \rightleftharpoons H^+(aq) + F^-(aq)$$

This equilibrium lies far to the left ($K_a = 7.1 \times 10^{-4}$). Because the fluoride ion has a strong affinity for the H^+ ion, it acts as a Brønsted base in water. If we were to dissolve a fluoride salt, such as NaF in water, the F^- ions in solution would, to some extent, accept protons from water.

$$F^-(aq) + H_2O(l) \rightleftharpoons HF(aq) + OH^-(aq)$$

> **Student Annotation:** Note that one of the products of the reaction of a conjugate base with water is always the corresponding weak acid.

The fluoride ion, which is the conjugate base of a weak acid, is an example of a *strong conjugate base.*

Conversely, a strong base has a *weak conjugate acid* and a weak base has a *strong conjugate acid.* For example, H_2O is the weak conjugate acid of the strong base OH^-, whereas the ammonium ion (NH_4^+) is the strong conjugate acid of the weak base ammonia (NH_3). When an ammonium salt is dissolved in water, the ammonium ions donate protons to the water molecules.

$$NH_4^+(aq) + H_2O(l) \rightleftharpoons NH_3(aq) + H_3O^+(aq)$$

> **Student Annotation:** The Group 1A and heavy Group 2A metal hydroxides are classified as *strong bases.* It is the hydroxide ion itself, however, that accepts a proton and is therefore the Brønsted base. Soluble metal hydroxides are simply sources of the hydroxide ion.

In general, there is a reciprocal relationship between the strength of an acid or base and the strength of its conjugate.

Acid	Example	Conjugate base	Formula		Base	Example	Conjugate acid	Formula
strong	HNO_3	weak conjugate	NO_3^-		strong	OH^-	weak conjugate	H_2O
weak	HCN	strong conjugate	CN^-		weak	NH_3	strong conjugate	NH_4^+

It is important to recognize that the words *strong* and *weak* do not mean the same thing in the context of conjugate acids and conjugate bases as they do in the context of acids and bases in general. A strong conjugate reacts with water—either accepting a proton from it or donating a proton to it—to a small but measurable extent. A strong conjugate acid acts as a weak Brønsted acid in water; and a strong conjugate base acts as a weak Brønsted base in water. A *weak* conjugate does not react with water to any measurable extent.

> **Student Annotation:**
> • A strong *conjugate* acid is a *weak* Brønsted acid.
> • A strong *conjugate* base is a *weak* Brønsted base.

The Relationship Between K_a and K_b of a Conjugate Acid-Base Pair

Because it accepts a proton from water to a small extent, what we refer to as a "strong *conjugate* base" is actually a *weak* Brønsted base. Therefore, every strong conjugate base has an ionization constant, K_b. Likewise, every strong conjugate acid, because it acts as a weak Brønsted acid, has an ionization constant, K_a.

A simple relationship between the ionization constant of a weak acid (K_a) and the ionization constant of its conjugate base (K_b) can be derived as follows, using acetic acid as an example.

$$CH_3COOH(aq) \rightleftharpoons H^+(aq) + CH_3COO^-(aq)$$
$$\text{acid} \qquad\qquad\qquad \text{conjugate base}$$

$$K_a = \frac{[H^+][CH_3COO^-]}{[CH_3COOH]}$$

Student Annotation: A conjugate base such as the acetate ion is introduced into a solution by dissolving a soluble salt containing acetate. Sodium acetate, for example, can be used to supply the acetate ion. The sodium ion does not take part in the reaction—it is a *spectator* ion [◄◄ Section 9.2].

The conjugate base, CH_3COO^-, reacts with water according to the equation

$$CH_3COO^-(aq) + H_2O(l) \rightleftharpoons CH_3COOH(aq) + OH^-(aq)$$

and the base ionization equilibrium expression is written as

$$K_b = \frac{[CH_3COOH][OH^-]}{[CH_3COO^-]}$$

As for any chemical equations, we can add these two equilibria and cancel identical terms.

$$\cancel{CH_3COOH(aq)} \rightleftharpoons H^+(aq) + \cancel{CH_3COO^-(aq)}$$
$$\underline{+ \cancel{CH_3COO^-(aq)} + H_2O(l) \rightleftharpoons \cancel{CH_3COOH(aq)} + OH^-(aq)}$$
$$H_2O(l) \rightleftharpoons H^+(aq) + OH^-(aq)$$

The sum is the autoionization of water. In fact, this is the case for any weak acid and its conjugate base:

Student Annotation: Remember that the hydronium ion can be expressed as either H^+ or H_3O^+. Note that when the hydronium ion is used, the equation must start with *two* water molecules.

$$\cancel{HA} \rightleftharpoons H^+ + \cancel{A^-}$$
$$\underline{+ \cancel{A^-} + H_2O \rightleftharpoons \cancel{HA} + OH^-}$$
$$H_2O \rightleftharpoons H^+ + OH^-$$

or for any weak base and its conjugate acid.

$$\cancel{B} + H_2O \rightleftharpoons \cancel{HB^+} + OH^-$$
$$\underline{+ \cancel{HB^+} + H_2O \rightleftharpoons \cancel{B} + H_3O^+}$$
$$2H_2O \rightleftharpoons H_3O^+ + OH^-$$

Recall that when we add two equilibria, the equilibrium constant for the *net* reaction is the product of the equilibrium constants for the individual equations [◄◄ Section 15.3]. Thus, for any conjugate acid-base pair,

Equation 16.6	$K_a \times K_b = K_w$

Equation 16.6 gives the quantitative basis for the reciprocal relationship between the strength of an acid and that of its conjugate base (or between the strength of a base and that of its conjugate acid). Because K_w is a constant, K_b must decrease if K_a increases, and vice versa.

Worked Example 16.16 shows how to determine ionization constants for conjugates.

Worked Example 16.16

Determine (a) K_b of the acetate ion (CH_3COO^-), (b) K_a of the methylammonium ion ($CH_3NH_3^+$), (c) K_b of the fluoride ion (F^-), and (d) K_a of the ammonium ion (NH_4^+).

Strategy Each species listed is either a conjugate base or a conjugate acid. Determine the identity of the acid corresponding to each conjugate base and the identity of the base corresponding to each conjugate acid; then, consult Tables 16.5 and 16.6 for their ionization constants. Use the tabulated ionization constants and Equation 16.6 to calculate each indicated K value.

Setup (a) A K_b value is requested, indicating that the acetate ion is a conjugate base. To identify the corresponding Brønsted acid, add a proton to the formula to get CH_3COOH (acetic acid). The K_a of acetic acid (from Table 16.5) is 1.8×10^{-5}.

(b) A K_a value is requested, indicating that the methylammonium ion is a conjugate acid. Determine the identity of the corresponding Brønsted base by removing a proton from the formula to get CH_3NH_2 (methylamine). The K_b of methylamine (from Table 16.6) is 4.4×10^{-4}.

(c) F^- is the conjugate base of HF; $K_a = 7.1 \times 10^{-4}$.

(d) NH_4^+ is the conjugate acid of NH_3; $K_b = 1.8 \times 10^{-5}$. Solving Equation 16.6 separately for K_a and K_b gives, respectively,

$$K_a = \frac{K_w}{K_b} \quad \text{and} \quad K_b = \frac{K_w}{K_a}$$

Solution

(a) Conjugate base CH_3COO^-: $K_b = \dfrac{1.0 \times 10^{-14}}{1.8 \times 10^{-5}} = 5.6 \times 10^{-10}$ (c) Conjugate base F^-: $K_b = \dfrac{1.0 \times 10^{-14}}{7.1 \times 10^{-4}} = 1.4 \times 10^{-11}$

(b) Conjugate acid $CH_3NH_3^+$: $K_a = \dfrac{1.0 \times 10^{-14}}{4.4 \times 10^{-4}} = 2.3 \times 10^{-11}$ (d) Conjugate acid NH_4^+: $K_a = \dfrac{1.0 \times 10^{-14}}{1.8 \times 10^{-5}} = 5.6 \times 10^{-10}$

Think About It

Because the conjugates of weak acids and bases have ionization constants, salts containing these ions have an effect on the pH of a solution. In Section 16.10 we will use the ionization constants of conjugate acids and conjugate bases to calculate pH for solutions containing dissolved salts.

Practice Problem ATTEMPT Determine (a) K_b of the benzoate ion ($C_6H_5COO^-$), (b) K_b of the chloroacetate ion (CH_2ClCOO^-), and (c) K_a of the ethylammonium ion ($C_2H_5NH_3^+$).

Practice Problem BUILD Determine (a) K_b of the weak base B whose conjugate acid HB^+ has $K_a = 8.9 \times 10^{-4}$ and (b) K_a of the weak acid HA whose conjugate base has $K_b = 2.1 \times 10^{-8}$.

Practice Problem CONCEPTUALIZE For each weak acid solution in diagrams (i)–(iii), identify the corresponding conjugate-base solution in diagrams (iv)–(vi). (Water molecules and hydroxide ions are not shown.)

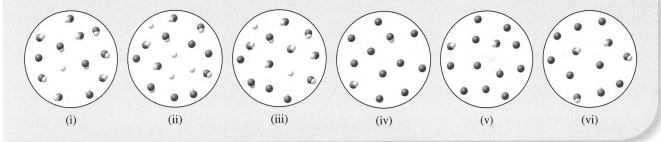

(i)   (ii)   (iii)   (iv)   (v)   (vi)

Section 16.8 Review

Conjugate Acid-Base Pairs

16.8.1 Calculate the K_b of the cyanide ion (CN^-).
- (a) 4.9×10^{-10}
- (b) 2.0×10^{-5}
- (c) 4.9×10^{-24}
- (d) 1.0×10^{-7}
- (e) 2.2×10^{-5}

16.8.2 Which of the following anions is the strongest base? (See Table 16.5.)
- (a) Cyanide ion (CN^-)
- (b) Benzoate ion ($C_6H_5COO^-$)
- (c) Nitrite ion (NO_2^-)
- (d) Phenolate ion ($C_6H_5O^-$)
- (e) Formate ion ($HCOO^-$)

16.8.3 The diagrams show solutions of three different weak acids with formulas HX, HY, and HZ. List the anions X^-, Y^-, and Z^- in order of increasing K_b value.
- (a) $Z^- < Y^- < X^-$
- (b) $Y^- < Z^- < X^-$
- (c) $Z^- < X^- < Y^-$
- (d) $X^- < Y^- < X^-$
- (e) $X^- < Z^- < Y^-$

= HX   = HY   = HZ   = H_3O^+

16.9   DIPROTIC AND POLYPROTIC ACIDS

Diprotic and polyprotic acids undergo successive ionizations, losing one proton at a time [◄◄ Section 9.3], and each ionization has a K_a associated with it. Ionization constants for a diprotic acid are designated K_{a_1} and K_{a_2}. A triprotic acid has K_{a_1}, K_{a_2}, and K_{a_3}. We write a separate equilibrium expression for each ionization, and we may need two or more equilibrium expressions to calculate the concentrations of species in solution at equilibrium. For carbonic acid (H_2CO_3), for example, we write

$$H_2CO_3(aq) \rightleftharpoons H^+(aq) + HCO_3^-(aq) \qquad K_{a_1} = \frac{[H^+][HCO_3^-]}{[H_2CO_3]}$$

$$HCO_3^-(aq) \rightleftharpoons H^+(aq) + CO_3^{2-}(aq) \qquad K_{a_2} = \frac{[H^+][CO_3^{2-}]}{[HCO_3^-]}$$

Note that the conjugate base in the first ionization is the acid in the second ionization. Table 16.7 shows the ionization constants of several diprotic acids and one polyprotic acid. For a given acid, the first ionization constant is much larger than the second ionization constant, and so on. This trend makes sense because it is easier to remove a proton from a neutral species than from one that is negatively charged, and it is easier to remove a proton from a species with a single negative charge than from one with a double negative charge.

TABLE 16.7	Ionization Constants of Some Diprotic and Polyprotic Acids at 25°C					
Name of acid	**Formula**	**Structure**	K_{a_1}	K_{a_2}	K_{a_3}
Sulfuric acid	H_2SO_4		Very large	1.3×10^{-2}	
Oxalic acid	$H_2C_2O_4$		6.5×10^{-2}	6.1×10^{-5}	
Sulfurous acid	H_2SO_3		1.3×10^{-2}	6.3×10^{-8}	
Ascorbic acid (vitamin C)	$H_2C_6H_6O_6$		8.0×10^{-5}	1.6×10^{-12}	
Carbonic acid	H_2CO_3		4.2×10^{-7}	4.8×10^{-11}	
Hydrosulfuric acid*	H_2S	H—S—H	9.5×10^{-8}	1×10^{-19}	
Phosphoric acid	H_3PO_4		7.5×10^{-3}	6.2×10^{-8}	4.8×10^{-13}

*The second ionization constant of H_2S is very low and difficult to measure. The value in this table is an estimate.

Worked Example 16.17 shows how to calculate equilibrium concentrations of all species in solution for an aqueous solution of a diprotic acid.

Worked Example 16.17

Oxalic acid ($H_2C_2O_4$) is a poisonous substance used mainly as a bleaching agent. Calculate the concentrations of all species present at equilibrium in a 0.10-M solution at 25°C.

Strategy Follow the same procedure for each ionization as for the determination of equilibrium concentrations for a monoprotic acid. The conjugate base resulting from the first ionization is the acid for the second ionization, and its starting concentration is the equilibrium concentration from the first ionization.

Setup The ionizations of oxalic acid and the corresponding ionization constants are

$$H_2C_2O_4(aq) \rightleftharpoons H^+(aq) + HC_2O_4^-(aq) \qquad K_{a_1} = 6.5 \times 10^{-2}$$

$$HC_2O_4^-(aq) \rightleftharpoons H^+(aq) + C_2O_4^{2-}(aq) \qquad K_{a_2} = 6.1 \times 10^{-5}$$

Construct an equilibrium table for each ionization, using x as the unknown in the first ionization and y as the unknown in the second ionization.

$$H_2C_2O_4(aq) \rightleftharpoons H^+(aq) + HC_2O_4^-(aq)$$

Initial concentration (M):	0.10	0	0
Change in concentration (M):	$-x$	$+x$	$+x$
Equilibrium concentration (M):	$0.10 - x$	x	x

The equilibrium concentration of the hydrogen oxalate ion ($HC_2O_4^-$) after the first ionization becomes the starting concentration for the second ionization. Additionally, the equilibrium concentration of H^+ is the starting concentration for the second ionization.

$$HC_2O_4^-(aq) \rightleftharpoons H^+(aq) + C_2O_4^{2-}(aq)$$

Initial concentration (M):	x	x	0
Change in concentration (M):	$-y$	$+y$	$+y$
Equilibrium concentration (M):	$x - y$	$x + y$	y

Solution

$$K_{a_1} = \frac{[H^+][HC_2O_4^-]}{[H_2C_2O_4]}$$

$$6.5 \times 10^{-2} = \frac{x^2}{0.10 - x}$$

Applying the approximation and neglecting x in the denominator of the expression gives

$$6.5 \times 10^{-2} \approx \frac{x^2}{0.10}$$

$$x^2 = 6.5 \times 10^{-3}$$

$$x = 8.1 \times 10^{-2} \ M$$

Testing the approximation,

$$\frac{8.1 \times 10^{-2} \ M}{0.10 \ M} \times 100\% = 81\%$$

Clearly the approximation is not valid, so we must solve the following quadratic equation.

$$x^2 + 6.5 \times 10^{-2} \ x - 6.5 \times 10^{-3} = 0$$

The result is $x = 0.054 \ M$. Thus, after the first ionization, the concentrations of species in solution are

$$[H^+] = 0.054 \ M$$

$$[HC_2O_4^-] = 0.054 \ M$$

$$[H_2C_2O_4] = (0.10 - 0.054) \ M = 0.046 \ M$$

(Continued on next page)

Rewriting the equilibrium table for the second ionization, using the calculated value of x, gives the following:

$$HC_2O_4^-(aq) \rightleftharpoons H^+(aq) + C_2O_4^{2-}(aq)$$

	$HC_2O_4^-$	H^+	$C_2O_4^{2-}$
Initial concentration (M):	0.054	0.054	0
Change in concentration (M):	$-y$	$+y$	$+y$
Equilibrium concentration (M):	$0.054 - y$	$0.054 + y$	y

$$K_{a_2} = \frac{[H^+][C_2O_4^{2-}]}{[HC_2O_4^-]}$$

$$6.1 \times 10^{-5} = \frac{(0.054 + y)(y)}{0.054 - y}$$

Assuming that y is very small and applying the approximations $0.054 + y \approx 0.054$ and $0.054 - y \approx 0.054$ gives

$$\frac{(0.054)(y)}{0.054} = y = 6.1 \times 10^{-5}$$

We must test the approximation as follows to see if it is valid.

$$\frac{6.1 \times 10^{-5}\ M}{0.054\ M} \times 100\% = 0.11\%$$

This time, because the ionization constant is much smaller, the approximation is valid. At equilibrium, the concentrations of all species are

$$[H_2C_2O_4] = 0.046\ M$$
$$[HC_2O_4^-] = (0.054 - 6.1 \times 10^{-5})\ M = 0.054\ M$$
$$[H^+] = (0.054 + 6.1 \times 10^{-5})\ M = 0.054\ M$$
$$[C_2O_4^{2-}] = 6.1 \times 10^{-5}\ M$$

Think About It

Note that the second ionization did not contribute significantly to the H^+ concentration. Therefore, we could determine the pH of this solution by considering only the first ionization. This is true in general for polyprotic acids where K_{a_1} is at least $1000 \times K_{a_2}$. [Note that it is necessary to consider the second ionization to determine the concentration of oxalate ion $(C_2O_4^{2-})$.]

Practice Problem **A**TTEMPT Calculate the concentrations of $H_2C_2O_4$, $HC_2O_4^-$, $C_2O_4^{2-}$, and H^+ ions in a 0.20 M oxalic acid solution at 25°C.

Practice Problem **B**UILD Calculate the concentrations of H_2SO_4, HSO_4^-, SO_4^{2-}, and H^+ ions in a 0.14 M sulfuric acid solution at 25°C.

Practice Problem **C**ONCEPTUALIZE Which of the diagrams could represent an aqueous solution of a polyprotic acid? Which could represent a solution of a polybasic base? (Water molecules are not shown.)

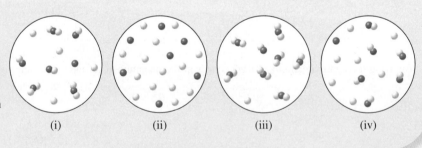

(i)   (ii)   (iii)   (iv)

Section 16.9 Review

Diprotic and Polyprotic Acids

16.9.1 Calculate the equilibrium concentration of CO_3^{2-} in a 0.050-M solution of carbonic acid at 25°C.

(a) $4.2 \times 10^{-7}\ M$ (c) $1.5 \times 10^{-4}\ M$ (e) $0.049\ M$

(b) $4.8 \times 10^{-11}\ M$ (d) $0.050\ M$

16.9.2 What is the pH of a 0.40-M solution of phosphoric acid at 25°C?

(a) 5.48   (c) 1.29   (e) 0.80

(b) 1.26   (d) 12.74

16.9.3 List the molecular and ionic species in order of increasing concentration in a solution of ascorbic acid ($H_2C_6H_6O_6$).

   (a) $[H_2C_6H_6O_6] < [HC_6H_6O_6^-] < [C_6H_6O_6^{2-}] = [H_3O^+]$

   (b) $[H_2C_6H_6O_6] < [HC_6H_6O_6^-] < [C_6H_6O_6^{2-}] < [H_3O^+]$

   (c) $[C_6H_6O_6^{2-}] < [HC_6H_6O_6^-] = [H_3O^+] < [H_2C_6H_6O_6]$

   (d) $[H_3O^+] = [H_2C_6H_6O_6] < [HC_6H_6O_6^-] < [C_6H_6O_6^{2-}]$

   (e) $[C_6H_6O_6^{2-}] = [H_3O^+] < [HC_6H_6O_6^-] < [H_2C_6H_6O_6]$

16.10 ACID-BASE PROPERTIES OF SALT SOLUTIONS

In Section 16.8, we saw that the conjugate base of a weak acid acts as a weak Brønsted base in water. Consider a solution of the salt sodium fluoride (NaF). Because NaF is a strong electrolyte, it dissociates completely in water to give a solution of sodium cations (Na^+) and fluoride anions (F^-). The fluoride ion, which is the conjugate base of hydrofluoric acid, reacts with water to produce hydrofluoric acid and hydroxide ion.

$$F^-(aq) + H_2O(l) \rightleftharpoons HF(aq) + OH^-(aq)$$

This is a specific example of *salt hydrolysis,* in which ions produced by the dissociation of a salt react with water to produce either hydroxide ions or hydronium ions—thus impacting pH. Using our knowledge of how ions from a dissolved salt interact with water, we can determine (based on the identity of the dissolved salt) whether a solution will be neutral, basic, or acidic. Note in the preceding example that sodium ions (Na^+) do not hydrolyze and thus have no impact on the pH of the solution.

Basic Salt Solutions

Sodium fluoride is a salt that dissolves to give a basic solution. In general, an anion that is the conjugate base of a weak acid reacts with water to produce hydroxide ion. Other examples include the acetate ion (CH_3COO^-), the nitrite ion (NO_2^-), the sulfite ion (SO_3^{2-}), and the hydrogen carbonate ion (HCO_3^-). Each of these anions undergoes hydrolysis to produce the corresponding weak acid and hydroxide ion.

$$A^-(aq) + H_2O(l) \rightleftharpoons HA(aq) + OH^-(aq)$$

We can therefore make the qualitative prediction that a solution of a salt in which the anion is the conjugate base of a weak acid will be basic. We calculate the pH of a basic salt solution the same way we calculate the pH of any weak base solution, using the K_b value for the anion. The necessary K_b value is calculated using the tabulated K_a value of the corresponding weak acid (see Table 16.5).

Worked Example 16.18 shows how to calculate the pH of a basic salt solution.

Student Annotation: Recall that a *salt* is an ionic compound formed by the reaction between an acid and a base [◀◀ Section 9.3]. Salts are strong electrolytes that dissociate completely into ions.

Student Annotation: HCO_3^- has an ionizable proton and can also act as a Brønsted acid. However, its tendency to *accept* a proton is stronger than its tendency to *donate* a proton:

$HCO_3^- + H_2O \rightleftharpoons H_2CO_3 + OH^-$
$K_b \approx 10^{-8}$

$HCO_3^- + H_2O \rightleftharpoons CO_3^{2-} + H_3O^+$
$K_a \approx 10^{-11}$

Student Hot Spot

Student data indicate you may struggle with determining the pH of a solution containing a weak acid conjugate ion. Access the SmartBook to view additional Learning Resources on this topic.

Student Annotation: Remember that for any conjugate acid-base pair (Equation 16.8):
$K_a \times K_b = K_w$

Worked Example 16.18

Calculate the pH of a 0.10-*M* solution of sodium hypochlorite (NaOCl) at 25°C.

Strategy A solution of NaOCl contains Na^+ ions and OCl^- ions. The OCl^- ion is the conjugate base of the weak acid, HOCl. Use the K_a value for HOCl (3.5×10^{-8}, from Appendix 3) and Equation 16.6 to determine K_b for OCl^-.

$$K_b = \frac{K_w}{K_a} = \frac{1.0 \times 10^{-14}}{3.5 \times 10^{-8}} = 2.9 \times 10^{-7}$$

Then, solve this pH problem like any equilibrium problem, using an equilibrium table.

(Continued on next page)

Setup It's always a good idea to write the equation corresponding to the reaction that takes place along with the equilibrium expression.

$$OCl^-(aq) + H_2O(l) \rightleftharpoons HOCl(aq) + OH^-(aq) \qquad K_b = \frac{[HOCl][OH^-]}{[OCl^-]}$$

Construct an equilibrium table, and determine, in terms of the unknown x, the equilibrium concentrations of the species in the equilibrium expression.

$$OCl^-(aq) + H_2O(l) \rightleftharpoons HOCl(aq) + OH^-(aq)$$

	OCl^-	H_2O	$HOCl$	OH^-
Initial concentration (M):	0.10		0	0
Change in concentration (M):	$-x$		$+x$	$+x$
Equilibrium concentration (M):	$0.10 - x$		x	x

Solution Substituting the equilibrium concentrations into the equilibrium expression and using the shortcut to solve for x, we get

$$2.9 \times 10^{-7} = \frac{x^2}{0.10 - x} \approx \frac{x^2}{0.10}$$

$$x = \sqrt{(2.9 \times 10^{-7})(0.10)} = 1.7 \times 10^{-4}\ M$$

According to our equilibrium table, $x = [OH^-]$. Therefore, we calculate the pOH first as

$$pOH = -\log (1.7 \times 10^{-4}) = 3.77$$

and then the pH,

$$pH = 14.00 - pOH = 14.00 - 3.77 = 10.23$$

The pH of a 0.10-M solution of NaOCl at 25°C is 10.23.

Think About It

It's easy to mix up pH and pOH in this type of problem. Always make a qualitative prediction regarding the pH of a salt solution first, and then check to make sure that your calculated pH agrees with your prediction. In this case, we would predict a basic pH because the anion in the salt (OCl^-) is the conjugate base of a weak acid ($HOCl$). The calculated pH, 10.23, is indeed basic.

Practice Problem **A**TTEMPT Determine the pH of a 0.15-M solution of sodium acetate (CH_3COONa) at 25°C.

Practice Problem **B**UILD Determine the concentration of a solution of sodium fluoride (NaF) that has pH 8.51 at 25°C.

Practice Problem **C**ONCEPTUALIZE Which of the graphs [(i)–(iv)] best represents the relationship between the pH of a 0.10 M basic salt solution and the K_a of the acid from which the salt is derived?

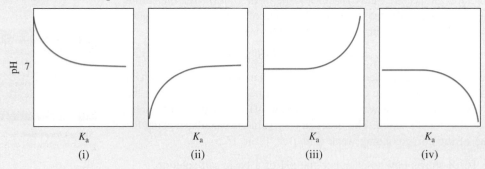

K_a	K_a	K_a	K_a
(i)	(ii)	(iii)	(iv)

Acidic Salt Solutions

When the cation of a salt is the conjugate acid of a weak base, a solution of the salt will be acidic. For example, when ammonium chloride dissolves in water, it dissociates to give a solution of ammonium ions and chloride ions.

$$NH_4Cl(s) \xrightarrow{H_2O} NH_4^+(aq) + Cl^-(aq)$$

The ammonium ion is the conjugate acid of the weak base ammonia (NH_3). It acts as a weak Brønsted acid, reacting with water to produce hydronium ion.

$$NH_4^+(aq) + H_2O(l) \rightleftharpoons NH_3(aq) + H_3O^+(aq)$$

We would therefore predict that a solution containing the ammonium ion is acidic. To calculate the pH, we must determine the K_a for NH_4^+ using the tabulated K_b value for NH_3 and Equation 16.6. Because Cl^- is the weak conjugate base of the strong acid HCl, Cl^- does not hydrolyze and therefore has no impact on the pH of the solution.

Worked Example 16.19 shows how to calculate the pH of an acidic salt solution.

Worked Example 16.19

Calculate the pH of a 0.10-M solution of ammonium chloride (NH_4Cl) at 25°C.

Strategy A solution of NH_4Cl contains NH_4^+ cations and Cl^- anions. The NH_4^+ ion is the conjugate acid of the weak base NH_3. Use the K_b value for NH_3 (1.8×10^{-5} from Table 16.6) and Equation 16.6 to determine K_a for NH_4^+.

$$K_a = \frac{K_w}{K_b} = \frac{1.0 \times 10^{-14}}{1.8 \times 10^{-5}} = 5.6 \times 10^{-10}$$

Setup Again, we write the balanced chemical equation and the equilibrium expression.

$$NH_4^+(aq) + H_2O(l) \rightleftharpoons NH_3(aq) + H_3O^+(aq) \qquad K_a = \frac{[NH_3][H_3O^+]}{[NH_4^+]}$$

Next, construct a table to determine the equilibrium concentrations of the species in the equilibrium expression.

	$NH_4^+(aq)$ +	$H_2O(l)$ ⇌	$NH_3(aq)$ +	$H_3O^+(aq)$
Initial concentration (M):	0.10		0	0
Change in concentration (M):	$-x$		$+x$	$+x$
Equilibrium concentration (M):	$0.10 - x$		x	x

Solution Substituting the equilibrium concentrations into the equilibrium expression and using the shortcut to solve for x, we get

$$5.6 \times 10^{-10} = \frac{x^2}{0.10 - x} \approx \frac{x^2}{0.10}$$

$$x = \sqrt{(5.6 \times 10^{-10})(0.10)} = 7.5 \times 10^{-6} \, M$$

According to the equilibrium table, $x = [H_3O^+]$. The pH can be calculated as follows:

$$pH = -\log(7.5 \times 10^{-6}) = 5.12$$

The pH of a 0.10-M solution of ammonium chloride (at 25°C) is 5.12.

Think About It

In this case, we would predict an acidic pH because the cation in the salt (NH_4^+) is the conjugate acid of a weak base (NH_3). The calculated pH is acidic.

Practice Problem **A**TTEMPT Determine the pH of a 0.25-M solution of pyridinium nitrate ($C_5H_6NNO_3$) at 25°C. [Pyridinium nitrate dissociates in water to give pyridinium ions ($C_5H_6N^+$), the conjugate acid of *pyridinium* (see Table 16.6), and nitrate ions (NO_3^-).]

Practice Problem **B**UILD Determine the concentration of a solution of ammonium chloride (NH_4Cl) that has pH 5.37 at 25°C.

Practice Problem **C**ONCEPTUALIZE Which of the graphs [(i)–(iv)] best represents the relationship between the pH of a 0.10 M acidic salt solution and the K_b of the base from which the salt is derived?

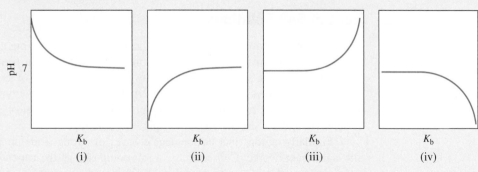

Figure 16.5 The six H_2O molecules surround the Al^{3+} ion in an octahedral arrangement. The attraction of the small Al^{3+} ion for the lone pairs on the oxygen atoms is so great that the O—H bonds in an H_2O molecule attached to the metal cation are weakened, allowing the loss of a proton (H^+) to an incoming H_2O molecule. This hydrolysis of the metal cation makes the solution acidic.

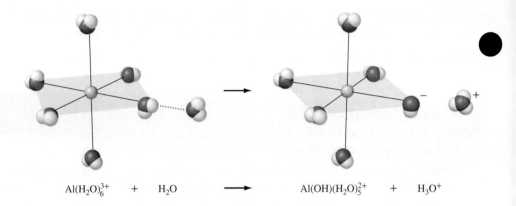

$$Al(H_2O)_6^{3+} + H_2O \longrightarrow Al(OH)(H_2O)_5^{2+} + H_3O^+$$

The metal ion in a dissolved salt can also react with water to produce an acidic solution. The extent of hydrolysis is greatest for the small and highly charged metal cations such as Al^{3+}, Cr^{3+}, Fe^{3+}, Bi^{3+}, and Be^{2+}. For example, when aluminum chloride dissolves in water, each Al^{3+} ion becomes associated with six water molecules (Figure 16.5).

Consider one of the bonds that forms between the metal ion and an oxygen atom from one of the six water molecules in $Al(H_2O)_6^{3+}$.

$$Al \rightleftharpoons O \begin{smallmatrix} H \\ \\ H \end{smallmatrix}$$

The positively charged Al^{3+} ion draws electron density toward itself, increasing the polarity of the O—H bonds. Consequently, the H atoms have a greater tendency to ionize than those in water molecules not associated with the Al^{3+} ion. The resulting ionization process can be written as

$$Al(H_2O)_6^{3+}(aq) + H_2O(l) \rightleftharpoons Al(OH)(H_2O)_5^{2+}(aq) + H_3O^+(aq)$$

or as

$$Al(H_2O)_6^{3+}(aq) \rightleftharpoons Al(OH)(H_2O)_5^{2+}(aq) + H^+(aq)$$

The equilibrium constant for the metal cation hydrolysis is given by

$$K_a = \frac{[Al(OH)(H_2O)_5^{2+}][H^+]}{[Al(H_2O)_6^{3+}]} = 1.3 \times 10^{-5}$$

$Al(OH)(H_2O)_5^{2+}$ can undergo further ionization:

$$Al(OH)(H_2O)_5^{2+}(aq) \rightleftharpoons Al(OH)_2(H_2O)_4^+(aq) + H^+(aq)$$

and so on. It is generally sufficient, however, to take into account only the first stage of hydrolysis when determining the pH of a solution that contains metal ions.

Neutral Salt Solutions

The extent of hydrolysis is greatest for the smallest and most highly charged metal ions because a compact, highly charged ion is more effective in polarizing the O—H bond and facilitating ionization. This is why relatively large ions of low charge, including the metal cations of Groups 1A and 2A (the cations of the strong bases), do not undergo significant hydrolysis (Be^{2+} is an exception). Thus, most metal cations of Groups 1A and 2A do not impact the pH of a solution.

Similarly, anions that are conjugate bases of strong acids do not hydrolyze to any significant degree. Consequently, a salt composed of the cation of a strong base and the anion of a strong acid, such as NaCl, produces a neutral solution.

Student Annotation: The metal cations of the strong bases are those of the alkali metals: (Li^+, Na^+, K^+, Rb^+, and Cs^+) and those of the heavy alkaline earth metals (Sr^{2+} and Ba^{2+}).

To summarize, the pH of a salt solution can be predicted qualitatively by identifying the ions in solution and determining which of them, if any, undergoes significant hydrolysis.

	Examples
A cation that will make a solution acidic is • The conjugate acid of a weak base • A small, highly charged metal ion (other than from Group 1A or 2A)	NH_4^+, CH_3, NH_3^+, C_2H_5, NH_3^+ Al^{3+}, Cr^{3+}, Fe^{3+}, Bi^{3+}
An anion that will make a solution basic is • The conjugate base of a weak acid	CN^-, NO_2^-, CH_3COO^-
A cation that will not affect the pH of a solution is • A Group 1A or heavy Group 2A cation (except Be^{2+})	Li^+, Na^+, Ba^{2+}
An anion that will not affect the pH of a solution is • The conjugate base of a strong acid	Cl^-, NO_3^-, ClO_4^-

Student Hot Spot

Student data indicate you may struggle with determining the pH of a salt solution. Access the SmartBook to view additional Learning Resources on this topic.

Worked Example 16.20 lets you practice predicting the pH of salt solutions.

Worked Example 16.20

Predict whether a 0.10-M solution of each of the following salts will be basic, acidic, or neutral: (a) LiI, (b) NH_4NO_3, (c) $Sr(NO_3)_2$, (d) KNO_2, (e) NaCN.

Strategy Identify the ions present in each solution, and determine which, if any, will impact the pH of the solution.

Setup

(a) Ions in solution: Li^+ and I^-. Li^+ is a Group 1A cation; I^- is the conjugate base of the strong acid HI. Therefore, neither ion hydrolyzes to any significant degree.

(b) Ions in solution: NH_4^+ and NO_3^-. NH_4^+ is the conjugate acid of the weak base NH_3; NO_3^- is the conjugate base of the strong acid HNO_3. In this case, the cation will hydrolyze, making the pH acidic.

$$NH_4^+(aq) + H_2O(l) \rightleftharpoons NH_3(aq) + H_3O^+(aq)$$

(c) Ions in solution: Sr^{2+} and NO_3^-. Sr^{2+} is a heavy Group 2A cation; NO_3^- is the conjugate base of the strong acid HNO_3. Neither ion hydrolyzes to any significant degree.

(d) Ions in solution: K^+ and NO_2^-. K^+ is a Group 1A cation; NO_2^- is the conjugate base of the weak acid HNO_2. In this case, the anion hydrolyzes, thus making the pH basic.

$$NO_2^-(aq) + H_2O(l) \rightleftharpoons HNO_2(aq) + OH^-(aq)$$

(e) Ions in solution: Na^+ and CN^-. Na^+ is a Group 1A cation; CN^- is the conjugate base of the weak acid HCN. In this case, too, the anion hydrolyzes, thus making the pH basic.

$$CN^-(aq) + H_2O(l) \rightleftharpoons HCN(aq) + OH^-(aq)$$

Solution

(a) Neutral   (b) Acidic   (c) Neutral   (d) Basic   (e) Basic

Think About It
It's very important that you be able to identify the ions in solution correctly. If necessary, review the formulas and charges of the common polyatomic ions [◀◀ Section 5.7, Table 5.10].

Practice Problem (A)TTEMPT Predict whether a 0.10-M solution of each of the following salts will be basic, acidic, or neutral.

(a) CH_3COOLi (b) C_5H_5NHCl (c) KF   (d) KNO_3 (e) $KClO_4$

Practice Problem (B)UILD In addition to those given in Worked Example 16.20 and Practice Problem A, identify two salts that will dissolve to give (a) an acidic solution, (b) a basic solution, and (c) a neutral solution.

(Continued on next page)

Practice Problem **C**ONCEPTUALIZE Which of the reactions could correctly illustrate a process by which a salt affects the pH of a solution?

Salts in Which Both the Cation and the Anion Hydrolyze

So far, we have considered salts in which only one ion undergoes hydrolysis. In some salts, both the cation and the anion hydrolyze. Whether a solution of such a salt is basic, acidic, or neutral depends on the relative strengths of the weak acid and the weak base. Although the process of calculating the pH in these cases is more complex than in cases where only one ion hydrolyzes, we can make qualitative predictions regarding pH using the values of K_b (of the salt's anion) and K_a (of the salt's cation).

- When $K_b > K_a$, the solution is basic.
- When $K_b < K_a$, the solution is acidic.
- When $K_b \approx K_a$, the solution is neutral or nearly neutral.

The salt NH_4NO_2, for example, dissociates in solution to give NH_4^+ ($K_a = 5.6 \times 10^{-10}$) and NO_2^- ($K_b = 2.2 \times 10^{-11}$). Because K_a for the ammonium ion is larger than K_b for the nitrite ion, we would expect the pH of an ammonium nitrite solution to be slightly acidic.

Section 16.10 Review

Acid-Base Properties of Salt Solutions

16.10.1 Calculate the pH of a 0.075-M solution of NH_4NO_3 at 25°C.
- (a) 5.19
- (b) 8.81
- (c) 7.00
- (d) 2.93
- (e) 11.07

16.10.2 Calculate the pH of a 0.082-M solution of NaCN at 25°C.
- (a) 5.20
- (b) 8.80
- (c) 7.00
- (d) 2.89
- (e) 11.11

16.10.3 Which of the following salts will produce a basic solution when dissolved in water? (Select all that apply.)
- (a) Sodium hypochlorite (NaClO)
- (b) Potassium fluoride (KF)
- (c) Lithium carbonate ($LiCO_3$)
- (d) Barium chloride ($BaCl_2$)
- (e) Ammonium iodide (NH_4I)

16.10.4 Which of the following salts will produce a neutral solution when dissolved in water?
- (a) Calcium chlorite [$Ca(ClO_2)_2$]
- (b) Potassium iodide (KI)
- (c) Lithium nitrate ($LiNO_3$)
- (d) Barium cyanide [$Ba(CN)_2$]
- (e) Ammonium iodide (NH_4I)

16.10.5 The diagrams represent solutions of three salts NaX (X = A, B, or C). Arrange the three X$^-$ anions in order of increasing base strength. (The Na$^+$ ions are not shown.)

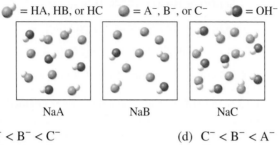

NaA	NaB	NaC

(a) A$^-$ < B$^-$ < C$^-$
(b) A$^-$ < C$^-$ < B$^-$
(c) B$^-$ < A$^-$ < C$^-$
(d) C$^-$ < B$^-$ < A$^-$
(e) C$^-$ < A$^-$ < B$^-$

16.11 ACID-BASE PROPERTIES OF OXIDES AND HYDROXIDES

Oxides of Metals and Nonmetals

Figure 16.6 shows the formulas of a number of oxides of the main group elements in their highest oxidation states. All alkali metal oxides and all alkaline earth metal oxides except BeO are basic. Beryllium oxide and several metallic oxides in Groups 3A and 4A are amphoteric. Nonmetallic oxides in which the oxidation number of the main group element is high are acidic (e.g., N_2O_5, SO_3, and Cl_2O_7), but those in which the oxidation number of the main group element is low (e.g., CO and NO) show no measurable acidic properties. No nonmetallic oxides are known to have basic properties.

The basic metallic oxides react with water to form metal hydroxides.

$$Na_2O(s) + H_2O(l) \longrightarrow 2NaOH(aq)$$

$$BaO(s) + H_2O(l) \longrightarrow Ba(OH)_2(aq)$$

1A 1																	8A 18
	2A 2											3A 13	4A 14	5A 15	6A 16	7A 17	
Li$_2$O	BeO											B$_2$O$_3$	CO$_2$	N$_2$O$_5$		OF$_2$	
Na$_2$O	MgO	3B 3	4B 4	5B 5	6B 6	7B 7	8	—8B— 9	10	1B 11	2B 12	Al$_2$O$_3$	SiO$_2$	P$_4$O$_{10}$	SO$_3$	Cl$_2$O$_7$	
K$_2$O	CaO											Ga$_2$O$_3$	GeO$_2$	As$_2$O$_5$	SeO$_3$	Br$_2$O$_7$	
Rb$_2$O	SrO											In$_2$O$_3$	SnO$_2$	Sb$_2$O$_5$	TeO$_3$	I$_2$O$_7$	
Cs$_2$O	BaO											Tl$_2$O$_3$	PbO$_2$	Bi$_2$O$_5$	PoO$_3$	At$_2$O$_7$	

Basic oxide

Acidic oxide

Amphoteric oxide

Figure 16.6 Oxides of the main group elements in their highest oxidation states.

The reactions between acidic oxides and water are as follows:

$$CO_2(g) + H_2O(l) \rightleftharpoons H_2CO_3(aq)$$

$$SO_3(g) + H_2O(l) \rightleftharpoons H_2SO_4(aq)$$

$$N_2O_5(g) + H_2O(l) \rightleftharpoons 2HNO_3(aq)$$

$$P_4O_{10}(g) + 6H_2O(l) \rightleftharpoons 4H_3PO_4(aq)$$

$$Cl_2O_7(g) + H_2O(l) \rightleftharpoons 2HClO_4(aq)$$

The reaction between CO_2 and H_2O explains why pure water gradually becomes acidic when it is exposed to air, which contains CO_2. The pH of rainwater exposed only to unpolluted air is about 5.5. The reaction between SO_3 and H_2O is largely responsible for acid rain.

Reactions between acidic oxides and bases and those between basic oxides and acids resemble normal acid-base reactions in that the products are a salt and water.

$$CO_2(g) + 2NaOH(aq) \longrightarrow Na_2CO_3(aq) + H_2O(l)$$

$$BaO(s) + 2HNO_3(aq) \longrightarrow Ba(NO_3)_2(aq) + H_2O(l)$$

Aluminum oxide (Al_2O_3) is amphoteric. Depending on the reaction conditions, it can behave either as an acidic oxide or as a basic oxide. For example, Al_2O_3 acts as a base with hydrochloric acid to produce a salt ($AlCl_3$) and water,

$$Al_2O_3(s) + 6HCl(aq) \longrightarrow 2AlCl_3(aq) + 3H_2O(l)$$

and acts as an acid with sodium hydroxide,

$$Al_2O_3(s) + 2NaOH(aq) + 3H_2O(l) \longrightarrow 2NaAl(OH)_4(aq)$$

Only a salt, sodium aluminum hydroxide [$NaAl(OH)_4$, which contains the Na^+ and $Al(OH)_4^-$ ions] is formed in the reaction with sodium hydroxide—no water is produced. Nevertheless, the reaction is still classified as an acid-base reaction because Al_2O_3 neutralizes NaOH. Other amphoteric oxides are ZnO, BeO, and Bi_2O_3.

Silicon dioxide is insoluble and does not react with water. It has acidic properties, however, because it reacts with a very concentrated aqueous base—that is,

$$SiO_2(s) + 2OH^-(aq) \longrightarrow SiO_3^{2-}(aq) + H_2O(l)$$

For this reason, concentrated aqueous, strong bases such as sodium hydroxide (NaOH) should *not* be stored in Pyrex glassware, which is made of SiO_2.

Some transition metal oxides in which the metal has a high oxidation number act as acidic oxides. Two examples are manganese(VII) oxide (Mn_2O_7) and chromium(VI) oxide (CrO_3), both of which react with water to produce acids.

$$Mn_2O_7(s) + H_2O(l) \longrightarrow 2HMnO_4(aq)$$
$$\text{permanganic acid}$$

$$CrO_3(s) + H_2O(l) \longrightarrow H_2CrO_4(aq)$$
$$\text{chromic acid}$$

Basic and Amphoteric Hydroxides

All the alkali and alkaline earth metal hydroxides, except $Be(OH)_2$, are basic. $Be(OH)_2$, as well as $Al(OH)_3$, $Sn(OH)_2$, $Pb(OH)_2$, $Cr(OH)_3$, $Cu(OH)_2$, $Zn(OH)_2$, and $Cd(OH)_2$, is amphoteric. All amphoteric hydroxides are insoluble, but beryllium hydroxide reacts with both acids and bases as follows:

$$Be(OH)_2(s) + 2H^+(aq) \longrightarrow Be^{2+}(aq) + 2H_2O(l)$$

$$Be(OH)_2(s) + 2OH^-(aq) \longrightarrow Be(OH)_4^{2-}(aq)$$

Aluminum hydroxide reacts with both acids and bases in a similar fashion:

$$Al(OH)_3(s) + 3H^+(aq) \longrightarrow Al^{3+}(aq) + 3H_2O(l)$$

$$Al(OH)_3(s) + OH^-(aq) \longrightarrow Al(OH)_4^-(aq)$$

Student Annotation: The similarity in behavior of aluminum hydroxide to that of beryllium hydroxide is a result of the diagonal relationship [◄◄ Section 8.5].

16.12 LEWIS ACIDS AND BASES

So far, we have discussed acid-base properties in terms of the Brønsted theory. For example, a Brønsted base is a substance that must be able to accept protons. By this definition, both the hydroxide ion and ammonia are bases.

$$H^+ \quad + \quad {}^-\!\ddot{O}\!-\!H \quad \longrightarrow \quad H\!-\!\ddot{O}\!-\!H$$

$$H^+ \quad + \quad \underset{\underset{H}{|}}{\overset{\overset{H}{|}}{:\!N\!-\!H}} \quad \longrightarrow \quad \left[\underset{\underset{H}{|}}{\overset{\overset{H}{|}}{H\!-\!N\!-\!H}}\right]^+$$

In each case, the atom to which the proton becomes attached possesses at least one unshared pair of electrons. This characteristic property of OH^-, NH_3, and other Brønsted bases suggests a more general definition of acids and bases.

In 1932, G. N. Lewis defined what we now call a *Lewis base* as a substance that can donate a pair of electrons. A *Lewis acid* is a substance that can accept a pair of electrons. In the protonation of ammonia, for example, NH_3 acts as a Lewis base because it donates a pair of electrons to the proton H^+, which acts as a Lewis acid by accepting the pair of electrons. A Lewis acid-base reaction, therefore, is one that involves the donation of a pair of electrons from one species to another.

The significance of the Lewis concept is that it is more general than other definitions. Lewis acid-base reactions include many reactions that do not involve Brønsted acids. Consider, for example, the reaction between boron trifluoride (BF_3) and ammonia to form an adduct compound.

$$\underset{\text{acid}}{\overset{\overset{F}{|}}{\underset{\underset{F}{|}}{F\!-\!B}}} \quad + \quad \underset{\text{base}}{\overset{\overset{H}{|}}{\underset{\underset{H}{|}}{:\!N\!-\!H}}} \quad \longrightarrow \quad \overset{\overset{F \quad H}{| \quad |}}{\underset{\underset{F \quad H}{| \quad |}}{F\!-\!B\!-\!N\!-\!H}}$$

The B atom in BF_3 is sp^2-hybridized [◄◄ Section 7.5]. The vacant, unhybridized $2p_z$ orbital accepts the pair of electrons from NH_3. Thus, BF_3 functions as an acid according to the Lewis definition, even though it does not contain an ionizable proton. A *coordinate covalent* bond [◄◄ Section 6.6] is formed between the B and N atoms. In fact, every Lewis acid-base reaction results in the formation of a coordinate covalent bond.

Boric acid is another Lewis acid containing boron. Boric acid (a weak acid used in eyewash) is an oxoacid with the following structure.

$$\overset{\overset{H}{|}}{\underset{\underset{H\!-\!\ddot{O}\!-\!B\!-\!\ddot{O}\!-\!H}{}}{:\!O\!:}}$$

Boric acid does not ionize in water to produce H^+. Instead, it produces H^+ in solution by taking a hydroxide ion away from water.

$$B(OH)_3(aq) + H_2O(l) \rightleftharpoons B(OH)_4^-(aq) + H^+(aq)$$

In this Lewis acid-base reaction, boric acid accepts a pair of electrons from the hydroxide ion that is derived from the water molecule, leaving behind the hydrogen ion.

The hydration of carbon dioxide to produce carbonic acid,

$$CO_2(g) + H_2O(l) \rightleftharpoons H_2CO_3(aq)$$

can be explained in terms of Lewis acid-base theory as well. The first step involves the donation of a lone pair on the O atom in H_2O to the C atom in CO_2. An orbital is vacated on the C atom to accommodate the lone pair by relocation of the electron pair in one of the C–O pi bonds, changing the hybridization of the oxygen atom from sp^2 to sp^3.

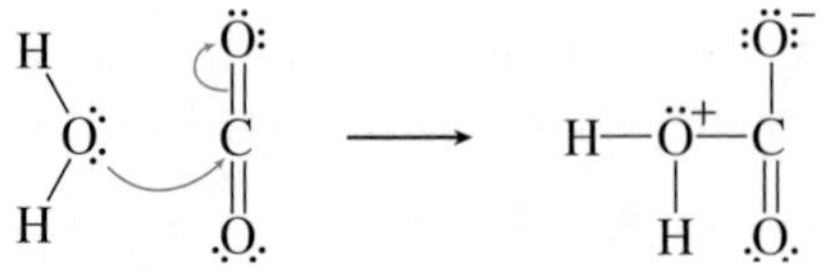

As a result, H_2O is a Lewis base and CO_2 is a Lewis acid. Finally, a proton is transferred onto the O atom bearing the negative charge to form H_2CO_3.

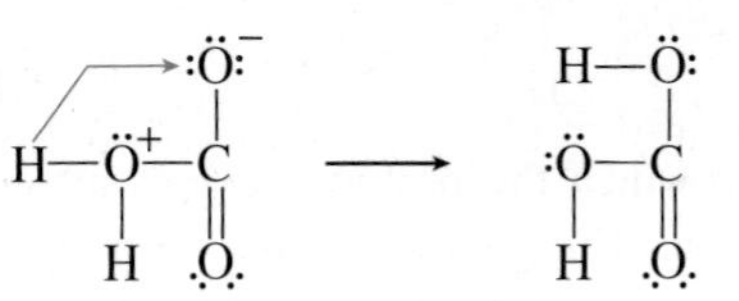

Other examples of Lewis acid-base reactions are

$$Ag^+(aq) + 2NH_3(aq) \rightleftharpoons Ag(NH_3)_2^+(aq)$$
$$Cd^{2+}(aq) + 4I^-(aq) \rightleftharpoons CdI_4^{2-}(aq)$$
$$Ni(s) + 4CO(g) \rightleftharpoons Ni(CO)_4(g)$$

The hydration of metal ions is in itself a Lewis acid-base reaction. When copper(II) sulfate ($CuSO_4$) dissolves in water, each Cu^{2+} ion becomes associated with six water molecules as $Cu(H_2O)_6^{2+}$. In this case, Cu^{2+} acts as the acid, accepting electrons, whereas H_2O acts as the base, donating electrons.

Worked Example 16.21 shows how to classify Lewis acids and bases.

Worked Example 16.21

Identify the Lewis acid and Lewis base in each of the following reactions.

(a) $C_2H_5OC_2H_5 + AlCl_3 \rightleftharpoons (C_2H_5)_2OAlCl_3$

(b) $Hg^{2+}(aq) + 4CN^-(aq) \rightleftharpoons Hg(CN)_4^{2-}(aq)$

Strategy Determine which species in each reaction accepts a pair of electrons (Lewis acid) and which species donates a pair of electrons (Lewis base).

Setup

(a) It can be helpful to draw Lewis structures of the species involved.

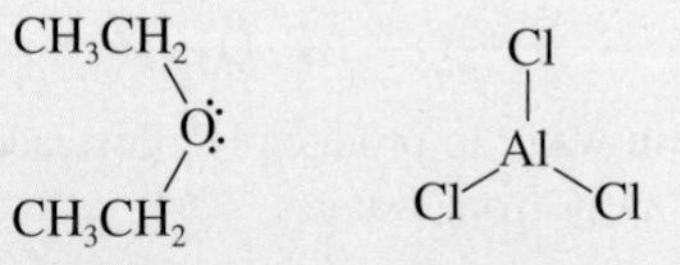

(b) Metal ions act as Lewis acids, accepting electron pairs from anions or molecules with lone pairs.

Solution

(a) The Al is sp^2-hybridized in $AlCl_3$ with an empty $2p_z$ orbital. It is electron deficient, sharing only six electrons. Therefore, the Al atom has the capacity to gain two electrons to complete its octet. This property makes $AlCl_3$ a Lewis acid. On the other hand, the lone pairs on the oxygen atom in $C_2H_5OC_2H_5$ make the compound a Lewis base.

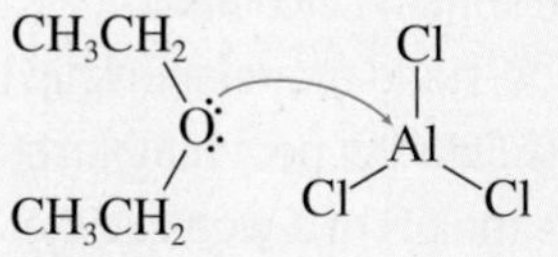

Student Annotation: An electron-deficient molecule is one with less than a complete octet around the central atom.

$$\begin{array}{c} CH_3CH_2 \\ \diagdown \\ \ddot{O}: \\ \diagup \\ CH_3CH_2 \end{array} \quad \begin{array}{c} Cl \\ | \\ Cl \diagdown Al \diagup Cl \end{array}$$

(b) Hg^{2+} accepts four pairs of electrons from the CN^- ions. Therefore, Hg^{2+} is the Lewis acid and CN^- is the Lewis base.

> **Think About It**
>
> In Lewis acid-base reactions, the acid is usually a cation or an electron-deficient molecule, whereas the base is an anion or a molecule containing an atom with lone pairs.

Practice Problem ATTEMPT Identify the Lewis acid and Lewis base in the following reaction.

$$Co^{3+}(aq) + 6NH_3(aq) \rightleftharpoons Co(NH_3)_6^{3+}$$

Practice Problem BUILD Write formulas for the Lewis acid and Lewis base that react to form H_3NAlBr_3.

Practice Problem CONCEPTUALIZE Which of the diagrams best depicts the combination of HCl and water as a Lewis acid-base reaction? Which best depicts the combination as a Brønsted acid-base reaction?

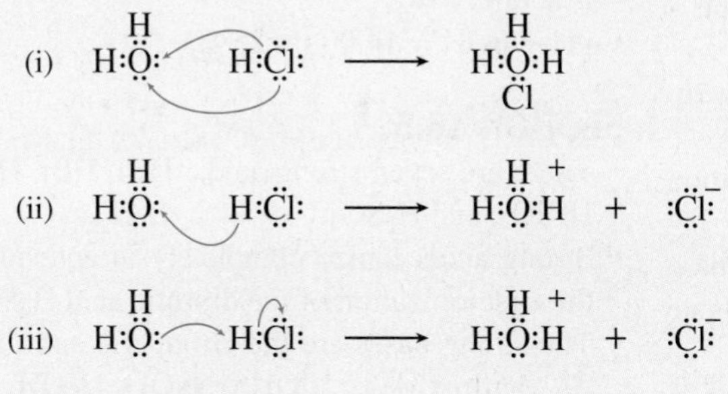

(i) $\quad H:\ddot{O}:\quad H:\ddot{C}l: \quad \longrightarrow \quad H:\ddot{O}:H$ (with H above O, Cl below O)

(ii) $\quad H:\ddot{O}:\quad H:\ddot{C}l: \quad \longrightarrow \quad H:\ddot{O}:H^+ \quad + \quad :\ddot{C}l:^-$ (with H above O)

(iii) $\quad H:\ddot{O}:\quad H:\ddot{C}l: \quad \longrightarrow \quad H:\ddot{O}:H^+ \quad + \quad :\ddot{C}l:^-$ (with H above O)

Section 16.12 Review

Lewis Acids and Bases

16.12.1 Which of the following cannot act as a Lewis base? (Select all that apply.)
  (a) NH_3 (d) Fe^{2+}
  (b) OH^- (e) Al^{3+}
  (c) CH_4

16.12.2 Which of the following is a Lewis acid but not a Brønsted acid? (Select all that apply.)
  (a) H_2O (d) Al^{3+}
  (b) BCl_3 (e) NH_3
  (c) OH^-

Learning Outcomes

- Define conjugate acid and conjugate base and cite examples of conjugate pairs.
- Give in your own words definitions of acids/bases: Brønsted and Lewis.
- Define amphoteric.
- Rank species of each of the following acids based upon relative strength: hydrohalic, oxoacids, and carboxylic.
- Use the equilibrium expression for water and its value to perform calculations and determine whether a solution is acidic, basic, or neutral.
- Use the pH scale to classify a solution as being acidic, basic, or neutral.
- Use pH, pOH, concentration of hydroxide ion, or concentration of protons to conduct calculations asking for one of these terms.

- Identify an acid or base as being strong or weak.
- Use the value of the acid-dissociation constant (K_a) in equilibrium calculations.
- Understand the relationship between K_a, K_b, and K_w.
- Calculate the percent ionization of a weak acid or base.
- Use the pH of a weak acid or base solution to calculate the K_a or K_b, respectively.
- Use K_w to determine the relative strength of conjugate pairs.
- Identify polyprotic acids and explain why the initial ionization is easier than subsequent ones.
- Classify a salt as being basic, acidic, or neutral based upon the acid and base used to form the salt.
- Calculate the pH of a salt solution.
- Classify an oxide as being acidic, basic, or amphoteric.
- Identify a species as a Lewis acid or Lewis base.

Chapter Summary

SECTION 16.1

- A Brønsted acid donates a proton; a Brønsted base accepts a proton.
- When a Brønsted acid donates a proton, the anion that remains is a ***conjugate base***.
- When a Brønsted base accepts a proton, the resulting cation is a ***conjugate acid***.
- The combination of a Brønsted acid and its conjugate base (or the combination of a Brønsted base and its conjugate acid) is called a ***conjugate pair***.

SECTION 16.2

- The strength of an acid is affected by molecular structure.
- Polar and weak bonds to the ionizable hydrogen lead to a stronger acid.
- Resonance stabilization of the conjugate base favors the ionization process, resulting in a stronger acid.

SECTION 16.3

- Water is ***amphoteric,*** meaning it can act both as a Brønsted acid and a Brønsted base.
- Pure water undergoes ***autoionization*** (to a very small extent), resulting in concentrations of H^+ and OH^- of 1.0×10^{-7} M at 25°C.
- K_w is the equilibrium constant for the autoionization of water, also called the ***ion-product constant.***
- $K_w = [H^+][OH^-] = 1.0 \times 10^{-14}$ at 25°C.

SECTION 16.4

- The ***pOH*** scale is analogous to the pH scale, but it measures *basicity:* pOH $= -\log [OH^-]$.

- pOH = 7.00 is neutral, pOH < 7.00 is basic, and pOH > 7.00 is acidic.
- pH + pOH = 14.00 (at 25°C).

SECTION 16.5

- There are seven strong acids: HCl, HBr, HI, HNO_3, $HClO_3$, $HClO_4$, and H_2SO_4.
- Strong acids ionize completely in aqueous solution. (Only the first ionization of the diprotic acid H_2SO_4 is complete.)
- The strong bases are the Group 1A and the heaviest Group 2A hydroxides: LiOH, NaOH, KOH, RbOH, CsOH, $Ca(OH)_2$, $Sr(OH)_2$, and $Ba(OH)_2$.

SECTION 16.6

- A ***weak acid*** ionizes only partially. The ***acid ionization constant, K_a,*** is the equilibrium constant that indicates to what extent a weak acid ionizes.
- We solve for the pH of a solution of weak acid using the concentration of the acid, the K_a value, and an equilibrium table.
- We can also determine the K_a of a weak acid if we know the initial acid concentration and the pH at equilibrium.

SECTION 16.7

- A ***weak base*** ionizes only partially. The ***base ionization constant, K_b,*** is the equilibrium constant that indicates to what extent a weak base ionizes.
- We solve for the pH of a solution of weak base using the concentration of the base, the K_b value, and an equilibrium table.
- We can also determine the K_b of a weak base if we know the concentration and the pH at equilibrium.

SECTION 16.8

- The conjugate base of a strong acid is a *weak conjugate base,* meaning that it does not react with water.
- The conjugate base of a weak acid is a *strong conjugate base,* meaning that it acts as a weak Brønsted base in water.
- The conjugate acid of a strong base is a *weak conjugate acid,* meaning that it does not react with water.
- The conjugate acid of a weak base is a *strong conjugate acid,* meaning that it acts as a weak Brønsted acid in water.
- For any conjugate acid-base pair, $K_a \times K_b = K_w$.

SECTION 16.9

- Diprotic and polyprotic acids have more than one proton to donate. They undergo stepwise ionizations. Each ionization has a K_a value associated with it.
- The K_a values for stepwise ionizations become progressively smaller.
- In most cases, it is only necessary to consider the first ionization of an acid to determine pH. To determine the concentrations of other species at equilibrium, it may be necessary to consider subsequent ionizations.

SECTION 16.10

- Salts dissolve in water to give neutral, acidic, or basic solutions depending on their constituent ions. *Salt hydrolysis* is the reaction of an ion with water to produce hydronium or hydroxide ions.
- Cations that are strong conjugate acids such as NH_4^+ make a solution more acidic.
- Anions that are strong conjugate bases such as F^- make a solution more basic. Anions that are conjugate bases of strong acids have no effect on pH.
- Small, highly charged metal ions hydrolyze to give acidic solutions.

SECTION 16.11

- Oxides of metals generally are basic; oxides of nonmetals generally are acidic.
- Metal hydroxides may be basic or amphoteric.

SECTION 16.12

- Lewis theory provides more general definitions of acids and bases.
- A *Lewis acid* accepts a pair of electrons; a *Lewis base* donates a pair of electrons.
- A Lewis acid is generally electron-poor and need not have a hydrogen atom.
- A Lewis base is an anion or a molecule with one or more lone pairs of electrons.

Key Words

Acid ionization constant (K_a), 732	Conjugate pair, 717	Strong conjugate acid, 745
Amphoteric, 722	Ion-product constant, 723	Strong conjugate base, 745
Autoionization of water, 722	Lewis acid, 759	Weak acid, 731
Base ionization constant (K_b), 741	Lewis base, 759	Weak base, 741
Conjugate acid, 717	pOH, 724	Weak conjugate acid, 745
Conjugate base, 717	Salt hydrolysis, 751	Weak conjugate base, 745

Key Equations

16.1	$K_w = [H_3O^+][OH^-] = 1.0 \times 10^{-14}$ (at 25°C)	The equilibrium constant for autoionization of water is K_w. In any aqueous solution, K_w is equal to the product of hydronium ion and hydroxide ion concentrations. At 25°C, the value of K_w is 1.0×10^{-14}.
16.2	$pOH = -\log [OH^-]$	The pOH of an aqueous solution is calculated as minus the base-10 log of hydroxide ion concentration.
16.3	$[OH^-] = 10^{-pOH}$	Hydroxide ion concentration can also be calculated from pOH.
16.4	$pH + pOH = 14.00$ (at 25°C)	The sum of pH and pOH in any aqueous solution at 25°C is 14.00.
16.5	percentage ionization $= \dfrac{[H_3O^+]_{eq}}{[HA]_0} \times 100\%$	Percent ionization of a weak acid is calculated as the ratio of hydronium ion concentration at equilibrium to original weak acid concentration times 100%.
16.6	$K_a \times K_b = K_w$	For any conjugate acid-base pair, the product of K_a for the acid and K_b for the base is K_w.

Key Skills

Salt Hydrolysis

Salt hydrolysis is critical to the understanding of certain acid-base titrations. When a weak acid is titrated with a strong base, the product of the neutralization is the weak acid's conjugate base.

$$HA(aq) + OH^-(aq) \longrightarrow A^-(aq) + H_2O(l)$$

The conjugate base of a weak acid behaves as a weak Brønsted base in water.

$$A^-(aq) + H_2O(l) \rightleftharpoons HA(aq) + OH^-(aq)$$

If we know the concentrations of the weak acid and the strong base, we can determine pH at the equivalence point of a weak acid–strong base titration as follows:

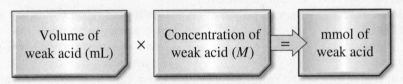

In the case of a monoprotic weak acid such as acetic acid ($HC_2H_2O_2$) and a monobasic strong base such as NaOH, the number of millimoles of base is equal to the number of millimoles of acid at the equivalence point.

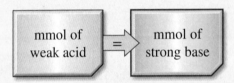

This enables us to determine the volume of strong base necessary to reach the equivalence point.

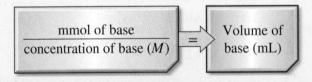

The combination of the original weak acid volume and the volume of base added to reach the equivalence point gives the total volume at the equivalence point. Further, the number of millimoles of conjugate base produced is equal to the number of millimoles of weak acid present at the start of the titration.

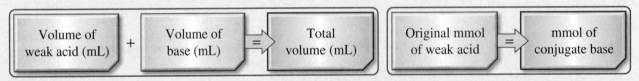

Using this information, we find the concentration of conjugate base; and to get K_b, we use the tabulated K_a for the weak acid and Equation 16.8.

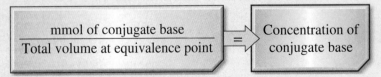

$$K_b = \frac{K_w}{K_a}$$

Once we have both the concentration and the ionization constant for the conjugate base, we construct an ice table and solve for the equilibrium concentration of hydroxide—and for pH. To illustrate the process, let's determine the pH at the equivalence point of the titration of 30.0 mL 0.15 M acetic acid ($HC_2H_3O_2$, $K_a = 1.8 \times 10^{-5}$) with 0.12 M NaOH at 25°C.

The neutralization net ionic equation is $HC_2H_3O_2(aq) + OH^-(aq) \longrightarrow C_2H_3O_2^-(aq) + H_2O(l)$

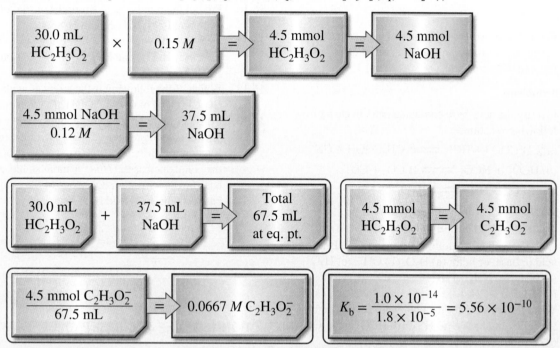

We now construct an equilibrium table and solve for [OH$^-$]. The reaction of acetate ion with water is

$$C_2H_3O_2^-(aq) + H_2O(l) \rightleftharpoons HC_2H_3O_2(aq) + OH^-(aq)$$

	$C_2H_3O_2^-$		$HC_2H_3O_2$	OH^-
i	0.0667		0	0
c	$-x$		$+x$	$+x$
e	$0.0667 - x$		x	x

$$5.56 \times 10^{-10} = \frac{[HC_2H_3O_2][OH^-]}{[C_2H_3O_2^-]}$$

Solving for x gives 6.09×10^{-6} M. This is the concentration of the OH$^-$ ion,

$$pOH = -\log [OH^-] = -\log (6.09 \times 10^{-6}) = 5.22$$

and pH = 14.00 − pOH, which gives

$$pH = 8.78$$

Key Skills Problems

16.1
Calculate the pH of a solution that is 0.22 M in nitrite ion (NO_2^-) at 25°C. K_a for nitrous acid (HNO_2) is 4.5×10^{-4}.
(a) 11.79   (b) 8.34   (c) 5.65   (d) 7.00   (e) 2.21

16.3
Calculate the pH of a solution that is 0.22 M in pyridinium ion ($C_5H_5NH^+$) at 25°C. K_b for pyridine (C_5H_5N) is 1.7×10^{-9}.
(a) 11.06   (b) 7.00   (c) 7.51   (d) 2.94   (e) 4.19

16.2
Determine pH at the equivalence point in the titration of 41.0 mL 0.096 M formic acid with 0.108 M NaOH at 25°C.
(a) 12.94   (b) 7.00   (c) 5.76   (d) 8.24   (e) 1.06

16.4
Determine pH at the equivalence point in the titration of 26.0 mL 1.12 M pyridine with 0.93 M HCl at 25°C.
(a) 7.00   (b) 2.76   (c) 11.24   (d) 1.73   (e) 12.27

Questions and Problems

SECTION 16.1: BRØNSTED ACIDS AND BASES

Review Questions

16.1   For a species to act as a Brønsted base, an atom in the species must possess a lone pair of electrons. Explain why this is so.

Conceptual Problems

16.2   Identify the acid-base conjugate pairs in each of the following reactions.

(a) $CH_3COO^- + HCN \rightleftharpoons CH_3COOH + CN^-$

(b) $HCO_3^- + HCO_3^- \rightleftharpoons H_2CO_3 + CO_3^{2-}$

(c) $H_2PO_4^- + NH_3 \rightleftharpoons HPO_4^{2-} + NH_4^+$

(d) $HClO + CH_3NH_2 \rightleftharpoons CH_3NH_3^+ + ClO^-$

(e) $CO_3^{2-} + H_2O \rightleftharpoons HCO_3^- + OH^-$

16.3   Classify each of the following species as a Brønsted acid or base, or both: (a) H_2O, (b) OH^-, (c) H_3O^+, (d) NH_3, (e) NH_4^+, (f) NH_2^-, (g) NO_3^-, (h) CO_3^{2-}, (i) HBr, (j) HCN.

16.4   Write the formula for the conjugate acid of each of the following bases: (a) HS^-, (b) HCO_3^-, (c) CO_3^{2-}, (d) $H_2PO_4^-$, (e) HPO_4^{2-}, (f) PO_4^{3-}, (g) HSO_4^-, (h) SO_4^{2-}, (i) SO_3^{2-}.

16.5   Write the formulas of the conjugate bases of the following acids: (a) HNO_2, (b) H_2SO_4, (c) H_2S, (d) HCN, (e) HCOOH (formic acid).

16.6   Oxalic acid ($H_2C_2O_4$) has the following structure.

$$O=C-OH$$
$$O=C-OH$$

An oxalic acid solution contains the following species in varying concentrations: $H_2C_2O_4$, $HC_2O_4^-$, $C_2O_4^{2-}$, and H^+.

(a) Draw Lewis structures of $HC_2O_4^-$ and $C_2O_4^{2-}$.

(b) Which of the four species can act only as acids, which can act only as bases, and which can act as both acids and bases?

16.7   Write the formula for the conjugate base of each of the following acids: (a) $CH_2ClCOOH$, (b) HIO_4, (c) H_3PO_4, (d) $H_2PO_4^-$, (e) HPO_4^{2-}, (f) H_2SO_4, (g) HSO_4^-, (h) HIO_3, (i) HSO_3^-, (j) NH_4^+, (k) H_2S, (l) HS^-, (m) HClO.

SECTION 16.2: MOLECULAR STRUCTURE AND ACID STRENGTH

Review Questions

16.8   List four factors that affect the strength of an acid.

16.9   How does the strength of an oxoacid depend on the electronegativity and oxidation number of the central atom?

Conceptual Problems

16.10   Predict the relative acid strengths of the following compounds: H_2O, H_2S, and H_2Se.

16.11   Compare the strengths of the following pairs of acids: (a) H_2SO_4 and H_2SeO_4, (b) H_3PO_4 and H_3AsO_4.

16.12   Which of the following is the stronger acid: $CH_2ClCOOH$ or $CHCl_2COOH$? Explain your choice.

16.13   Consider the following compounds: Experimentally, phenol is found to be a stronger acid than methanol. Explain this difference in terms of the structures of the conjugate bases. (*Hint:* A more stable conjugate base favors ionization. Only one of the conjugate bases can be stabilized by resonance.)

Phenol   Methanol

SECTION 16.3: THE ACID-BASE PROPERTIES OF WATER

Review Questions

16.14   Write the equilibrium expression for the autoionization of water.

16.15   Write an equation relating $[H^+]$ and $[OH^-]$ in solution at 25°C.

16.16   The equilibrium constant for the autoionization of water

$$H_2O(l) \rightleftharpoons H^+(aq) + OH^-(aq)$$

is 1.0×10^{-14} at 25°C and 3.8×10^{-14} at 40°C. Is the forward process endothermic or exothermic?

16.17   Compare the magnitudes of $[H^+]$ and $[OH^-]$ in aqueous solutions that are acidic, basic, and neutral.

Computational Problems

16.18   Calculate the OH^- concentration in an aqueous solution at 25°C with each of the following H_3O^+ concentrations: (a) 1.13×10^{-4} M, (b) 4.55×10^{-8} M, (c) 7.05×10^{-11} M, (d) 3.13×10^{-2} M.

16.19   Calculate the H_3O^+ concentration in an aqueous solution at 25°C with each of the following OH^- concentrations: (a) 2.50×10^{-2} M, (b) 1.67×10^{-5} M, (c) 8.62×10^{-3} M, (d) 1.75×10^{-12} M.

16.20   The value of K_w at 50°C is 5.48×10^{-14}. Calculate the OH^- concentration in each of the aqueous solutions from Problem 16.18 at 50°C.

16.21   The value of K_w at 100°C is 5.13×10^{-13}. Calculate the H_3O^+ concentration in each of the aqueous solutions from Problem 16.19 at 100°C.

SECTION 16.4: THE pH AND pOH SCALES

Review Questions

16.22   Define pOH. Write the equation relating pH and pOH.

16.23   The pOH of a solution is 6.8. From this statement alone, can you conclude that the solution is basic? If not, what additional information would you need?

Computational Problems

16.24   Calculate the concentration of H^+ ions in a 0.62 M NaOH solution.

16.25   Calculate the concentration of OH^- ions in a 1.4×10^{-3} M HCl solution.

16.26   Calculate the pH of each of the following solutions: (a) 2.8×10^{-4} M $Ba(OH)_2$, (b) 5.2×10^{-4} M HNO_3.

16.27   Calculate the pH of each of the following solutions: (a) 0.0010 M HCl, (b) 0.76 M KOH.

16.28   Calculate the number of moles of KOH in 5.50 mL of a 0.360 M KOH solution. What is the pOH of the solution at 25°C?

16.29   The pOH of a solution is 9.40 at 25°C. Calculate the hydrogen ion concentration of the solution.

16.30   A solution is made by dissolving 18.4 g of HCl in enough water to make 662 mL of solution. Calculate the pH of the solution at 25°C.

16.31   How much NaOH (in grams) is needed to prepare 546 mL of solution with a pH of 10.00 at 25°C?

SECTION 16.5: STRONG ACIDS AND BASES

Review Questions

16.32   Which of the following statements are true regarding a 1.0-M solution of a strong acid HA at 25°C? (Choose all that apply.)

(a) $[A^-] > [H^+]$

(b) The pH is 0.00.

(c) $[H^+] = 1.0$ M

(d) $[HA] = 1.0$ M

16.33   Why are ionizations of strong acids and strong bases generally not treated as equilibria?

Computational Problems

16.34   Calculate the pH of an aqueous solution at 25°C that is (a) 0.023 M in HCl, (b) 8.1×10^{-3} M in HNO_3, and (c) 9.7×10^{-6} M in $HClO_4$.

16.35   Calculate the pH of an aqueous solution at 25°C that is (a) 0.075 M in HI, (b) 6.5×10^{-6} M in $HClO_4$, and (c) 1.5×10^{-4} M in HCl.

16.36   Calculate the concentration of HBr in a solution at 25°C that has a pH of (a) −0.21, (b) 5.46, and (c) 1.54.

16.37   Calculate the concentration of HNO_3 in a solution at 25°C that has a pH of (a) 2.91, (b) 5.33, and (c) −0.18.

16.38   Calculate the pOH and pH of the following aqueous solutions at 25°C: (a) 0.016 M KOH, (b) 1.35 M NaOH, (c) 0.094 M $Ba(OH)_2$.

16.39   Calculate the pOH and pH of the following aqueous solutions at 25°C: (a) 0.0715 M LiOH, (b) 0.0441 M $Ba(OH)_2$, (c) 0.17 M NaOH.

16.40   An aqueous solution of a strong base has a pH of 10.41 at 25°C. Calculate the concentration of the base if the base is (a) LiOH and (b) $Ba(OH)_2$.

16.41   An aqueous solution of a strong base has a pH of 8.93 at 25°C. Calculate the concentration of the base if the base is (a) KOH and (b) $Ba(OH)_2$.

16.42   Determine $[H^+]$ and pH for a raindrop in which the carbonic acid concentration is 1.8×10^{-5} M. Assume that carbonic acid is the only acid in the raindrop and that the second ionization is negligible.

16.43   A sample of acid rain collected from a pond was shown to have a pH = 4.65. What concentration of HNO_3 does this pH correspond to?

SECTION 16.6: WEAK ACIDS AND ACID IONIZATION CONSTANTS

> ▶▶▶ **Visualizing Chemistry**
> **Figure 16.3**
> Three weak acid solutions are shown [(i)–(iii)] .

(i)   (ii)   (iii)

VC 16.1 Which weak acid has the largest K_a value?

(a) i   (b) ii   (c) iii

VC 16.2 Which weak acid has the highest pH?

(a) i   (b) ii   (c) iii

VC 16.3 For which weak acid solution can we most likely neglect x in the denominator of the equilibrium expression in the determination of pH?

(a) i   (b) ii   (c) iii

VC 16.4 In the event that we cannot neglect x in the denominator of the equilibrium expression, _____ to solve for pH.

(a) we cannot use an equilibrium table

(b) we must use the quadratic equation

(c) it is unnecessary

Review Questions

16.44   Explain what is meant by the strength of an acid.

16.45   What does the ionization constant tell us about the strength of an acid?

16.46   List the factors on which the K_a of a weak acid depends.

16.47   Why do we normally not quote K_a values for strong acids such as HCl and HNO_3?

16.48 Why is it necessary to specify temperature when giving K_a values?

16.49 Which of the following solutions has the highest pH: (a) 0.40 M HCOOH, (b) 0.40 M HClO$_4$, (c) 0.40 M CH$_3$COOH?

Computational Problems

16.50 The K_a for hydrofluoric acid is 7.1×10^{-4}. Calculate the pH of a 0.15-M aqueous solution of hydrofluoric acid at 25°C.

16.51 The K_a for benzoic acid is 6.5×10^{-5}. Calculate the pH of a 0.10-M aqueous solution of benzoic acid at 25°C.

16.52 Calculate the pH of an aqueous solution at 25°C that is 0.34 M in phenol (C$_6$H$_5$OH). (K_a for phenol = 1.3×10^{-10}.)

16.53 Calculate the pH of an aqueous solution at 25°C that is 0.095 M in hydrocyanic acid (HCN). (K_a for hydrocyanic acid = 4.9×10^{-10}.)

16.54 Determine the percent ionization of the following solutions of phenol: (a) 0.56 M, (b) 0.25 M, (c) 1.8×10^{-6} M.

16.55 Determine the percent ionization of the following solutions of formic acid at 25°C: (a) 0.016 M, (b) 5.7×10^{-4} M, (c) 1.75 M.

16.56 Calculate the concentration at which a monoprotic acid with $K_a = 4.5 \times 10^{-5}$ will be 1.5% ionized.

16.57 A 0.015-M solution of a monoprotic acid is 0.92% ionized. Calculate the ionization constant for the acid.

16.58 The pH of an aqueous acid solution is 6.20 at 25°C. Calculate the K_a for the acid. The initial acid concentration is 0.010 M.

16.59 Calculate the K_a of a weak acid if a 0.19-M aqueous solution of the acid has a pH of 4.52 at 25°C.

16.60 What is the original molarity of a solution of a weak acid whose K_a is 3.5×10^{-5} and whose pH is 5.26 at 25°C?

16.61 What is the original molarity of a solution of formic acid (HCOOH) whose pH is 3.26 at 25°C? (K_a for formic acid = 1.7×10^{-4}.)

16.62 Determine the K_w for water at a particular temperature where the pH of pure water at this temperature is 6.14.

16.63 In biological and medical applications, it is often necessary to study the autoionization of water at 37°C instead of 25°C. Given that K_w for water is 2.5×10^{-14} at 37°C, calculate the pH of pure water at this temperature.

Conceptual Problems

16.64 Classify each of the following species as a weak or strong acid: (a) HNO$_3$, (b) HF, (c) H$_2$SO$_4$, (d) HSO$_4^-$, (e) H$_2$CO$_3$, (f) HCO$_3^-$, (g) HCl, (h) HCN, (i) HNO$_2$.

16.65 Classify each of the following species as a weak or strong base: (a) LiOH, (b) CN$^-$, (c) H$_2$O, (d) ClO$_4^-$, (e) NH$_2^-$.

16.66 Which of the following statements are true for a 0.10-M solution of a weak acid HA? (Choose all that apply.)
(a) The pH is 1.00.
(b) [H$^+$] >> [A$^-$]
(c) [H$^+$] = [A$^-$]
(d) The pH is less than 1.

SECTION 16.7: WEAK BASES AND BASE IONIZATION CONSTANTS

Review Questions

16.67 Compare the pH values for 0.10-M solutions of NaOH and of NH$_3$ to illustrate the difference between a strong base and a weak base.

16.68 Which of the following has a higher pH: (a) 1.0 M NH$_3$, (b) 0.20 M NaOH (K_b for NH$_3$ = 1.8×10^{-5})?

Computational Problems

16.69 The pH of a 0.045-M solution of a weak base is 9.88 at 25°C. What is the K_b of the base?

16.70 Calculate the pH for each of the following solutions at 25°C: (a) 0.10 M NH$_3$, (b) 0.050 M C$_5$H$_5$N (pyridine). (K_b for pyridine = 1.7×10^{-9}.)

16.71 Calculate the pH at 25°C of a 0.61-M aqueous solution of a weak base B with a K_b of 1.5×10^{-4}.

16.72 What is the original molarity of an aqueous solution of ammonia (NH$_3$) whose pH is 11.22 at 25°C? (Use the K_b for ammonia provided in Problem 16.68.)

16.73 What is the pH at 25°C of a 0.045-M aqueous solution of a weak base B with a K_b of 4.2×10^{-10}?

16.74 Determine the K_b of a weak base if a 0.19-M aqueous solution of the base at 25°C has a pH of 10.88.

SECTION 16.8: CONJUGATE ACID-BASE PAIRS

Review Questions

16.75 Write the equation relating K_a for a weak acid and K_b for its conjugate base. Use NH$_3$ and its conjugate acid NH$_4^+$ to derive the relationship between K_a and K_b.

16.76 From the relationship $K_aK_b = K_w$, what can you deduce about the relative strengths of a weak acid and its conjugate base?

Conceptual Problems

16.77 Calculate K_b for each of the following ions: CN$^-$, F$^-$, CH$_3$COO$^-$, HCO$_3^-$.

16.78 Calculate K_a for each of the following ions: NH$_4^+$, C$_6$H$_5$NH$_3^+$, CH$_3$NH$_3^+$, C$_2$H$_5$NH$_3^+$. (See Table 16.6.)

Computational Problems

16.79 The following diagrams represent aqueous solutions of three different monoprotic acids: HA, HB, and HC. (a) Which conjugate base (A$^-$, B$^-$, or C$^-$) has the smallest K_b value? (b) Which anion is the strongest base? The water molecules have been omitted for clarity.

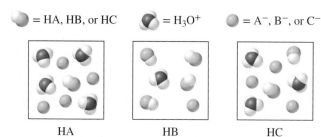

16.80   The following diagrams represent solutions of three salts NaX (X = A, B, or C). (a) Which X⁻ has the weakest conjugate acid? (b) Arrange the three X⁻ anions in order of decreasing base strength. The Na⁺ ion and water molecules have been omitted for clarity.

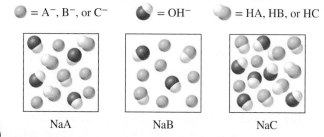

SECTION 16.9: DIPROTIC AND POLYPROTIC ACIDS

Review Questions

16.81   Write all the species (except water) that are present in a phosphoric acid solution. Indicate which species can act as a Brønsted acid, which as a Brønsted base, and which as both a Brønsted acid and a Brønsted base.

16.82   Write the K_{a_1} and K_{a_2} expressions for sulfurous acid, H_2SO_3.

Computational Problems

16.83   Compare the pH of a 0.040 M HCl solution with that of a 0.040 M H_2SO_4 solution. (*Hint:* H_2SO_4 is a strong acid; K_a for HSO_4^- = 1.3×10^{-2}.)

16.84   What are the concentrations of HSO_4^-, SO_4^{2-}, and H^+ in a 0.20 M $KHSO_4$ solution? (*Hint:* H_2SO_4 is a strong acid; K_a for HSO_4^- = 1.3×10^{-2}.)

16.85   Calculate the concentrations of H^+, HCO_3^-, and CO_3^{2-} in a 0.025 M H_2CO_3 solution.

16.86   Calculate the pH at 25°C of a 0.25-M aqueous solution of phosphoric acid (H_3PO_4). (K_{a_1}, K_{a_2}, and K_{a_3} for phosphoric acid are 7.5×10^{-3}, 6.25×10^{-8}, and 4.8×10^{-13}, respectively.)

16.87   Calculate the pH at 25°C of a 0.25-M aqueous solution of oxalic acid ($H_2C_2O_4$). (K_{a_1} and K_{a_2} for oxalic acid are 6.5×10^{-2} and 6.1×10^{-5}, respectively.)

Conceptual Problems

16.88   The first and second ionization constants of a diprotic acid H_2A are K_{a_1} and K_{a_2} at a certain temperature. Under what conditions will $[A^{2-}] = K_{a_1}$?

16.89   (1) Which of the following diagrams represents a solution of a weak diprotic acid? (2) Which diagrams represent chemically implausible situations? (The hydrated proton is shown as a hydronium ion. Water molecules are omitted for clarity.)

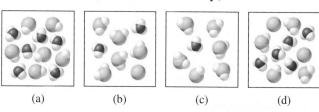

SECTION 16.10: ACID-BASE PROPERTIES OF SALT SOLUTIONS

Review Questions

16.90   Define salt hydrolysis. Categorize salts according to how they affect the pH of a solution.

16.91   Explain why small, highly charged metal ions are able to undergo hydrolysis.

16.92 Al^{3+} is not a Brønsted acid, but $Al(H_2O)_6^{3+}$ is. Explain.

16.93   Specify which of the following salts will undergo hydrolysis: KF, $NaNO_3$, NH_4NO_2, $MgSO_4$, KCN, C_6H_5COONa, RbI, Na_2CO_3, $CaCl_2$, HCOOK.

Computational Problems

16.94   Calculate the pH of a 0.36 M CH_3COONa solution. (K_a for acetic acid = 1.8×10^{-5}.)

16.95   Calculate the pH of a 0.42 M NH_4Cl solution. (K_b for ammonia = 1.8×10^{-5}.)

16.96   Calculate the pH of a 0.082 M NaF solution. (K_a for HF = 7.1×10^{-4}.)

16.97   Calculate the pH of a 0.91 M $C_2H_5NH_3I$ solution. (K_b for $C_2H_5NH_2$ = 5.6×10^{-4}.)

Conceptual Problems

16.98   Predict the pH (>7, <7, or ≈7) of aqueous solutions containing the following salts: (a) KBr, (b) $Al(NO_3)_3$, (c) $BaCl_2$, (d) $Bi(NO_3)_3$.

16.99   Predict whether the following solutions are acidic, basic, or nearly neutral: (a) NaBr, (b) K_2SO_3, (c) NH_4NO_2, (d) $Cr(NO_3)_3$.

16.100   A certain salt, MX (containing the M⁺ and X⁻ ions), is dissolved in water, and the pH of the resulting solution is 7.0. What can you say about the strengths of the acid and the base from which the salt is derived?

16.101   In a certain experiment, a student finds that the pH values of 0.10-M solutions of three potassium salts KX, KY, and KZ are 7.0, 9.0, and 11.0, respectively. Arrange the acids HX, HY, and HZ in order of increasing acid strength.

16.102 Predict whether a solution containing the salt K_2HPO_4 will be acidic, neutral, or basic.

16.103 Predict the pH (>7, <7, or ≈7) of a $NaHCO_3$ solution.

SECTION 16.11: ACID-BASE PROPERTIES OF OXIDES AND HYDROXIDES

Review Questions

16.104 Classify the following oxides as acidic, basic, amphoteric, or neutral: (a) CO_2, (b) K_2O, (c) CaO, (d) N_2O_5, (e) CO, (f) NO, (g) SnO_2, (h) SO_3, (i) Al_2O_3, (j) BaO.

16.105 Write equations for the reactions between (a) CO_2 and $NaOH(aq)$, (b) Na_2O and $HNO_3(aq)$.

Conceptual Problems

16.106 Explain why metal oxides tend to be basic if the oxidation number of the metal is low and tend to be acidic if the oxidation number of the metal is high. (*Hint:* Metallic compounds in which the oxidation numbers of the metals are low are more ionic than those in which the oxidation numbers of the metals are high.)

16.107 Arrange the oxides in each of the following groups in order of increasing basicity: (a) K_2O, Al_2O_3, BaO, (b) CrO_3, CrO, Cr_2O_3.

16.108 $Zn(OH)_2$ is an amphoteric hydroxide. Write balanced ionic equations to show its reaction with (a) HCl, (b) NaOH [the product is $Zn(OH)_4^{2-}$].

16.109 $Al(OH)_3$ is insoluble in water. It dissolves in concentrated NaOH solution. Write a balanced ionic equation for this reaction. What type of reaction is this?

16.110 How do the chemical properties of oxides change from left to right across a period? How do they change from top to bottom within a particular group?

16.111 Write balanced equations for the reactions between each of the following oxides and water: (a) Li_2O, (b) CaO, (c) SO_3.

SECTION 16.12: LEWIS ACIDS AND BASES

Review Questions

16.112 What are the Lewis definitions of an acid and a base? In what way are they more general than the Brønsted definitions?

16.113 In terms of orbitals and electron arrangements, what must be present for a molecule or an ion to act as a Lewis acid (use H^+ and BF_3 as examples)? What must be present for a molecule or ion to act as a Lewis base (use OH^- and NH_3 as examples)?

Conceptual Problems

16.114 Classify each of the following species as a Lewis acid or a Lewis base: (a) CO_2, (b) H_2O, (c) I^-, (d) SO_2, (e) NH_3, (f) OH^-, (g) H^+, (h) BCl_3.

16.115 Describe the following reaction in terms of the Lewis theory of acids and bases.

$$AlCl_3(s) + Cl^-(aq) \longrightarrow AlCl_4^-(aq)$$

16.116 Which would be considered a stronger Lewis acid: (a) BF_3 or BCl_3, (b) Fe^{2+} or Fe^{3+}? Explain.

16.117 All Brønsted acids are Lewis acids, but the reverse is not true. Give two examples of Lewis acids that are not Brønsted acids.

16.118 Identify the Lewis acid and the Lewis base in the following reactions.

(a) $Fe(s) + 5CO(g) \longrightarrow Fe(CO)_5(l)$

(b) $BCl_3(g) + NH_3(g) \longrightarrow Cl_3BNH_3(s)$

(c) $Hg^{2+}(aq) + 4I^-(aq) \longrightarrow HgI_4^{2-}(aq)$

16.119 Identify the Lewis acid and the Lewis base in the following reactions.

(a) $AlBr_3(s) + Br^-(aq) \longrightarrow AlBr_4^-(aq)$

(b) $Cr(s) + 6CO(g) \longrightarrow Cr(CO)_6(s)$

(c) $Cu^{2+}(aq) + 4CN^-(aq) \longrightarrow Cu(CN)_4^{2-}(aq)$

ADDITIONAL PROBLEMS

16.120 H_2SO_4 is a strong acid, but HSO_4^- is a weak acid. Account for the difference in strength of these two related species.

16.121 Which of the following diagrams best represents a strong acid, such as HCl, dissolved in water? Which represents a weak acid? Which represents a very weak acid? (The hydrated proton is shown as a hydronium ion. Water molecules are omitted for clarity.)

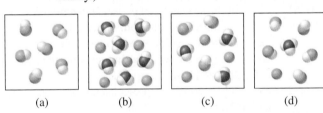

(a)   (b)   (c)   (d)

16.122 Predict the direction that predominates in the reaction

$$F^-(aq) + H_2O(l) \rightleftharpoons HF(aq) + OH^-(aq)$$

16.123 Predict the products and tell whether the following reaction will occur to any measurable extent.

$$CH_3COOH(aq) + Cl^-(aq) \longrightarrow$$

16.124 In a 0.080 M NH_3 solution, what percent of the NH_3 is present as NH_4^+?

16.125 Calculate the pH and percent ionization of a 0.88 M HNO_2 solution at 25°C.

16.126 A typical reaction between an antacid and the hydrochloric acid in gastric juice is

$$NaHCO_3(s) + HCl(aq) \rightleftharpoons NaCl(aq) + H_2O(l) + CO_2(g)$$

Calculate the volume (in liters) of CO_2 generated from 0.350 g of $NaHCO_3$ and excess gastric juice at 1.00 atm and 37.0°C.

16.127 To which of the following would the addition of an equal volume of 0.60 M NaOH lead to a solution having a lower pH: (a) water, (b) 0.30 M HCl, (c) 0.70 M KOH, (d) 0.40 M NaNO$_3$?

16.128 The pH of a 0.0642-M solution of a monoprotic acid is 3.86. Is this a strong acid?

16.129 Like water, liquid ammonia undergoes autoionization:

$$NH_3 + NH_3 \rightleftharpoons NH_4^+ + NH_2^-$$

(a) Identify the Brønsted acids and Brønsted bases in this reaction. (b) What species correspond to H$^+$ and OH$^-$, and what is the condition for a neutral solution?

16.130 HA and HB are both weak acids although HB is the stronger of the two. Will it take a larger volume of a 0.10 M NaOH solution to neutralize 50.0 mL of 0.10 M HB than would be needed to neutralize 50.0 mL of 0.10 M HA?

16.131 A solution contains a weak monoprotic acid HA and its sodium salt NaA both at 0.1 M concentration. Show that [OH$^-$] = K_w/K_a.

16.132 The three common chromium oxides are CrO, Cr$_2$O$_3$, and CrO$_3$. If Cr$_2$O$_3$ is amphoteric, what can you say about the acid-base properties of CrO and CrO$_3$?

16.133 Use the data in Table 16.5 to calculate the equilibrium constant for the following reaction:

$$HCOOH(aq) + OH^-(aq) \rightleftharpoons HCOO^-(aq) + H_2O(l)$$

16.134 Use the data in Table 16.5 to calculate the equilibrium constant for the following reaction:

$$CH_3COOH(aq) + NO_2^- \rightleftharpoons CH_3COO^-(aq) + HNO_2(aq)$$

16.135 Most of the hydrides of Group 1A and Group 2A metals are ionic (the exceptions are BeH$_2$ and MgH$_2$, which are covalent compounds). (a) Describe the reaction between the hydride ion (H$^-$) and water in terms of a Brønsted acid-base reaction. (b) The same reaction can also be classified as a redox reaction. Identify the oxidizing and reducing agents.

16.136 Calculate the pH of a 0.20 M ammonium acetate (CH$_3$COONH$_4$) solution.

16.137 Novocaine, used as a local anesthetic by dentists, is a weak base ($K_b = 8.91 \times 10^{-6}$). What is the ratio of the concentration of the base to that of its acid in the blood plasma (pH = 7.40) of a patient? (As an approximation, use the K_a values at 25°C.)

16.138 Which of the following is the stronger base: NF$_3$ or NH$_3$? (*Hint:* F is more electronegative than H.)

16.139 Which of the following is a stronger base: NH$_3$ or PH$_3$? (*Hint:* The N—H bond is stronger than the P—H bond.)

16.140 The ion product of D$_2$O is 1.35×10^{-15} at 25°C. (a) Calculate pD where pD = $-$log [D$^+$]. (b) For what values of pD will a solution be acidic in D$_2$O? (c) Derive a relation between pD and pOD.

16.141 Give an example of (a) a weak acid that contains oxygen atoms, (b) a weak acid that does not contain oxygen atoms, (c) a neutral molecule that acts as a Lewis acid, (d) a neutral molecule that acts as a Lewis base, (e) a weak acid that contains two ionizable H atoms, (f) a conjugate acid-base pair, both of which react with HCl to give carbon dioxide gas.

16.142 What is the pH of 250.0 mL of an aqueous solution containing 0.616 g of a strong acid?

16.143 (a) Use VSEPR to predict the geometry of the hydronium ion (H$_3$O$^+$). (b) The O atom in H$_2$O has two lone pairs and in principle can accept two H$^+$ ions. Explain why the species H$_4$O^{2+} does not exist. What would be its geometry if it did exist?

16.144 HF is a weak acid, but its strength increases with concentration. Explain. (*Hint:* F$^-$ reacts with HF to form HF$_2^-$. The equilibrium constant for this reaction is 5.2 at 25°C.)

16.145 When chlorine reacts with water, the resulting solution is weakly acidic and reacts with AgNO$_3$ to give a white precipitate. Write balanced equations to represent these reactions. Explain why manufacturers of household bleaches add bases such as NaOH to their products to increase their effectiveness.

16.146 When the concentration of a strong acid is not substantially higher than 1.0×10^{-7} M, the ionization of water must be taken into account in the calculation of the solution's pH. (a) Derive an expression for the pH of a strong acid solution, including the contribution to [H$^+$] from H$_2$O. (b) Calculate the pH of a 1.0×10^{-7} M HCl solution.

16.147 Calculate the pH of a 2.00 M NH$_4$CN solution.

16.148 Calculate the concentrations of all species in a 0.100 M H$_3$PO$_4$ solution.

16.149 In the vapor phase, acetic acid molecules associate to a certain extent to form dimers,

$$2CH_3COOH(g) \rightleftharpoons (CH_3COOH)_2(g)$$

At 51°C, the pressure of a certain acetic acid vapor system is 0.0342 atm in a 360-mL flask. The vapor is condensed and neutralized with 13.8 mL of 0.0568 M NaOH. (a) Calculate the degree of dissociation (α) of the dimer under these conditions:

$$(CH_3COOH)_2 \rightleftharpoons 2CH_3COOH$$

(b) Calculate the equilibrium constant K_P for the reaction in part (a).

16.150 Calculate the concentrations of all the species in a 0.100 M Na$_2$CO$_3$ solution.

16.151 Henry's law constant for CO$_2$ at 38°C is 2.28×10^{-3} mol/L atm. Calculate the pH of a solution of CO$_2$ at 38°C in equilibrium with the gas at a partial pressure of 3.20 atm.

16.152 Hydrocyanic acid (HCN) is a weak acid and a deadly poisonous compound—in the gaseous form (hydrogen cyanide), it is used in gas chambers. Why is it dangerous to treat sodium cyanide with acids (such as HCl) without proper ventilation?

16.153 How many grams of NaCN would you need to dissolve in enough water to make exactly 250 mL of solution with a pH of 10.00?

16.154 A solution of formic acid (HCOOH) has a pH of 2.53. How many grams of formic acid are there in 100.0 mL of the solution?

16.155 Calculate the pH of a 1-L solution containing 0.150 mole of CH_3COOH and 0.100 mole of HCl.

16.156 Predict the products of the following oxides with water: Na_2O, BaO, CO_2, N_2O_5, P_4O_{10}, SO_3. Write an equation for each of the reactions. Specify whether the oxides are acidic, basic, or amphoteric.

16.157 You are given two beakers, one containing an aqueous solution of strong acid (HA) and the other an aqueous solution of weak acid (HB) of the same concentration. Describe how you would compare the strengths of these two acids by (a) measuring the pH, (b) measuring electrical conductance, and (c) studying the rate of hydrogen gas evolution when these solutions are combined with an active metal such as Mg or Zn.

16.158 Use Le Châtelier's principle to predict the effect of the following changes on the extent of hydrolysis of sodium nitrite ($NaNO_2$) solution: (a) HCl is added, (b) NaOH is added, (c) NaCl is added, (d) the solution is diluted.

16.159 A 0.400 *M* formic acid (HCOOH) solution freezes at $-0.758°C$. Calculate the K_a of the acid at that temperature. (*Hint:* Assume that molarity is equal to molality. Carry out your calculations to three significant figures and round off to two for K_a.)

16.160 The disagreeable odor of fish is mainly due to organic compounds (RNH_2) containing an amino group, $-NH_2$, where R is the rest of the molecule. Amines are bases just like ammonia. Explain why putting some lemon juice on fish can greatly reduce the odor.

16.161 A solution of methylamine (CH_3NH_2) has a pH of 10.64. How many grams of methylamine are there in 100.0 mL of the solution?

16.162 Describe the hydration of SO_2 as a Lewis acid-base reaction.

16.163 Both the amide ion (NH_2^-) and the nitride ion (N^{3-}) are stronger bases than the hydroxide ion and hence do not exist in aqueous solutions. (a) Write equations showing the reactions of these ions with water, and identify the Brønsted acid and base in each case. (b) Which of the two is the stronger base?

16.164 When carbon dioxide is bubbled through a clear calcium hydroxide solution, the solution appears milky. Write an equation for the reaction, and explain how this reaction illustrates that CO_2 is an acidic oxide.

16.165 Explain the action of smelling salt, which is ammonium carbonate $[(NH_4)_2CO_3]$. (*Hint:* The thin film of aqueous solution that lines the nasal passage is slightly basic.)

16.166 About half of the hydrochloric acid produced annually in the United States (3.0 billion pounds) is used in metal pickling. This process involves the removal of metal oxide layers from metal surfaces to prepare them for coating. (a) Write the overall and net ionic equations for the reaction between iron(III) oxide, which represents the rust layer over iron, and HCl. Identify the Brønsted acid and base. (b) Hydrochloric acid is also used to remove scale (which is mostly $CaCO_3$) from water pipes. Hydrochloric acid reacts with calcium carbonate in two stages; the first stage forms the bicarbonate ion, which then reacts further to form carbon dioxide. Write equations for these two stages and for the overall reaction. (c) Hydrochloric acid is used to recover oil from the ground. It dissolves rocks (often $CaCO_3$) so that the oil can flow more easily. In one process, a 15 percent (by mass) HCl solution is injected into an oil well to dissolve the rocks. If the density of the acid solution is 1.073 g/mL, what is the pH of the solution?

16.167 Which of the following does not represent a Lewis acid-base reaction?

(a) $H_2O + H^+ \longrightarrow H_3O^+$

(b) $NH_3 + BF_3 \longrightarrow H_3NBF_3$

(c) $PF_3 + F_2 \longrightarrow PF_5$

(d) $Al(OH)_3 + OH^- \longrightarrow Al(OH)_4^-$

16.168 Determine whether each of the following statements is true or false. If false, explain why the statement is wrong.

(a) All Lewis acids are Brønsted acids.

(b) The conjugate base of an acid always carries a negative charge.

(c) The percent ionization of a base increases with its concentration in solution.

(d) A solution of barium fluoride is acidic.

16.169 How many milliliters of a strong monoprotic acid solution at pH = 4.12 must be added to 528 mL of the same acid solution at pH = 5.76 to change its pH to 5.34? Assume that the volumes are additive.

16.170 Hemoglobin (Hb) is a blood protein that is responsible for transporting oxygen. It can exist in the protonated form as HbH^+. The binding of oxygen can be represented by the simplified equation

$$HbH^+ + O_2 \rightleftharpoons HbO_2 + H^+$$

(a) What form of hemoglobin is favored in the lungs where oxygen concentration is highest? (b) In body tissues, where the cells release carbon dioxide produced by metabolism, the blood is more acidic due to the formation of carbonic acid. What form of hemoglobin is favored under this condition? (c) When a person hyperventilates, the concentration of CO_2 in his or her blood decreases. How does this action affect the given equilibrium? Frequently a person who is hyperventilating is advised to breathe into a paper bag. Why does this action help the individual?

16.171 A 20.27-g sample of a metal carbonate (MCO_3) is combined with 500 mL of a 1.00 M HCl solution. The excess HCl acid is then neutralized by 32.80 mL of 0.588 M NaOH. Identify M.

16.172 Calculate the pH of a solution that is 1.00 M HCN and 1.00 M HF. Compare the concentration (in molarity) of the CN^- ion in this solution with that in a 1.00 M HCN solution. Comment on the difference.

16.173 Tooth enamel is largely hydroxyapatite [$Ca_3(PO_4)_3OH$]. When it dissolves in water (a process called *demineralization*), it dissociates as follows:

$$Ca_5(PO_4)_3OH \longrightarrow 5Ca^{2+} + 3PO_4^{3-} + OH^-$$

The reverse process, called *remineralization,* is the body's natural defense against tooth decay. Acids produced from food remove the OH^- ions and thereby weaken the enamel layer. Most toothpastes contain a fluoride compound such as NaF or SnF_2. What is the function of these compounds in preventing tooth decay?

16.174 Use the van't Hoff equation (see Problem 15.146) and the data in Appendix 2 to calculate the pH of water at its normal boiling point.

16.175 Write the formulas and names of the oxides of the second period elements (Li to N). Identify the oxides as acidic, basic, or amphoteric. Use the highest oxidation state of each element.

16.176 The atmospheric sulfur dioxide (SO_2) concentration over a certain region is 0.12 ppm by volume. Calculate the pH of the rainwater due to this pollutant. Assume that the dissolution of SO_2 does not affect its pressure. (K_a for $H_2SO_3 = 1.3 \times 10^{-2}$.)

16.177 Sulfuric acid (H_2SO_4) accounts for as much as 80 percent of the acid in acid rain. Its first ionization is complete, producing H^+ and the hydrogen sulfate ion:

$$H_2SO_4(aq) \rightleftharpoons H^+(aq) + HSO_4^-(aq)$$

Its second ionization, which produces additional H^+ and the sulfate ion, has an ionization constant (K_{a_2}) of 1.3×10^{-2}.

$$HSO_4^-(aq) \rightleftharpoons H^+(aq) + SO_4^{2-}(aq)$$

Calculate the concentration of all species in a raindrop in which the sulfuric acid concentration is 4.00×10^{-5} M. Assume that sulfuric acid is the only acid present.

16.178 A 1.87-g sample of Mg reacts with 80.0 mL of a HCl solution whose pH is −0.544. What is the pH of the solution after all the Mg has reacted? Assume constant volume.

Answers to In-Chapter Materials

PRACTICE PROBLEMS

16.1A (a) $HClO_4$, (b) HS^-, (c) HS^-, (d) $HC_2O_4^-$. **16.1B** SO_3^{2-}, H_2SO_3. **16.2A** (a) NH_4^+ acid, H_2O base, NH_3 conjugate base, H_3O^+ conjugate acid; (b) CN^- base, H_2O acid, HCN conjugate acid, OH^- conjugate base. **16.2B** (a) $HSO_4^- + H_2O \longrightarrow H_3O^+ + SO_4^{2-}$, (b) $HSO_4^- + H_2O \longrightarrow H_2SO_4 + OH^-$. **16.3A** (a) $HBrO_4$, (b) H_2SO_4. **16.3B** Electronegativity. The greater difference in electronegativities of two elements in a bond leads to a weaker bond and thus acidity increases. **16.4A** 2.0×10^{-11} M. **16.4B** 2.8×10^{-13} M. **16.5A** (a) 11.24, (b) 2.14, (c) 5.07. **16.5B** (a) 7.55, (b) 8.00, (c) 11.00. **16.6A** (a) 9.5×10^{-14} M, (b) 7.2×10^{-6} M, (c) 1.0×10^{-7} M. **16.6B** (a) 5.5×10^{-12} M, (b) 2.0×10^{-4} M, (c) 2.5×10^{-2} M. **16.7A** (a) 1.09, (b) 5.09, (c) 2.27. **16.7B** (a) 1.96, (b) 2.46, (c) 7.00. **16.8A** (a) 8.7×10^{-3} M, (b) 1.7×10^{-2} M, (c) 9.8×10^{-7} M. **16.8B** (a) 1.5×10^{-5} M, (b) 1.5×10^{-2} M, (c) 9.1×10^{-4} M. **16.9A** (a) 0.82, (b) 2.08, (c) 4.77. **16.9B** (a) 6.71, (b) 1.21, (c) 0.91. **16.10A** (a) 9.5×10^{-6} M, (b) 4.8×10^{-6} M. **16.10B** (a) 1.7×10^{-2} M, (b) 8.7×10^{-3} M. **16.11A** 2.89. **16.11B** 3.05. **16.12A** (a) 4.96, 0.0044%, (b) 5.72, 0.026%, (c) 6.70, 0.24%. **16.12B** (a) 2.0×10^{-3} M, (b) 4.9×10^{-4} M, (c) 2.2×10^{-4} M. **16.13A** 1.9×10^{-5}. **16.13B** 5.8×10^{-9}. **16.14A** 9.14. **16.14B** 8.33. **16.15A** 3.0×10^{-9}. **16.15B** 1.4×10^{-4}. **16.16A** (a) 1.5×10^{-10}, (b) 1.8×10^{-13}, (c) 1.8×10^{-11}. **16.16B** (a) 1.1×10^{-11}, (b) 4.8×10^{-7}. **16.17A** [$H_2C_2O_4$] = 0.11 M, [$HC_2O_4^-$] = 0.086 M, [$C_2O_4^{2-}$] = 6.1×10^{-5} M, [H^+] = 0.086 M. **16.17B** [H_2SO_4] = 0 M, [HSO_4^-] = 0.13 M, [SO_4^{2-}] = 0.011 M, [H^+] = 0.15 M. **16.18A** 8.96. **16.18B** 0.75 M. **16.19A** 2.92. **16.19B** 0.032 M. **16.20A** (a) basic, (b) acidic, (c) basic, (d) neutral, (e) neutral. **16.20B** (a) NH_4Br and CH_3NH_3I, (b) $NaHPO_4$ and KOBr, (c) NaI and KBr. **16.21A** Lewis acid: Co^{3+}, Lewis base: NH_3. **16.21B** NH_3 and $AlBr_3$.

SECTION REVIEW

16.1.1 b, d. **16.1.2** c, d. **16.2.1** b. **16.2.2** d. **16.2.3** c. **16.3.1** d. **16.3.2** a. **16.4.1** c. **16.4.2** a. **16.4.3** e. **16.4.4** b. **16.5.1** d. **16.5.2** b. **16.5.3** b. **16.5.4** e. **16.5.5** c. **16.5.6** b. **16.5.7** b. **16.6.1** c. **16.6.2** e. **16.6.3** b. **16.7.1** b. **16.7.2** d. **16.7.3** e. **16.8.1** b. **16.8.2** d. **16.8.3** a. **16.9.1** b. **16.9.2** c. **16.9.3** c. **16.10.1** a. **16.10.2** e. **16.10.3** a, b, c. **16.10.4** b, c. **16.10.5** c. **16.12.1** c, d, e. **16.12.2** b, d.

Chapter 17

Acid-Base Equilibria and Solubility Equilibria

© Lisa Stokes/Moment Open/Getty Images

A BUFFER solution is an aqueous mixture of a weak acid and a weak base that is used to maintain the pH of a solution within a desired range. The ability of buffers to minimize the effect of added strong acid or strong base makes them useful in a variety of chemical and biochemical processes. Chemists have developed methods for buffering foods, cosmetics, drugs and drug-delivery systems, and many other familiar products to optimize their performance. Buffers are used in the textile industry to ensure rich, consistent colors in dyed fabrics.

17.1   THE COMMON ION EFFECT

Until now, we have discussed the properties of solutions containing a single solute. In this section, we will examine how the properties of a solution change when a second solute is introduced.

Recall that a system at equilibrium will shift in response to being stressed and that stress can be applied in a variety of ways, including the addition of a reactant or a product [◄◄ Section 15.6]. Consider a liter of solution containing 0.10 mole of acetic acid. Using the K_a for acetic acid (1.8×10^{-5}) and an equilibrium table [◄◄ Section 16.6], the pH of this solution at 25°C can be determined.

$$CH_3COOH(aq) \rightleftharpoons H^+(aq) + CH_3COO^-(aq)$$

	CH_3COOH	H^+	CH_3COO^-
Initial concentration (M):	0.10	0	0
Change in concentration (M):	$-x$	$+x$	$+x$
Equilibrium concentration (M):	$0.10 - x$	x	x

Assuming that $(0.10 - x)\ M \approx 0.10\ M$ and solving for x, we get 1.34×10^{-3}. Therefore, $[CH_3COOH] = 0.09866\ M$, $[H^+] = [CH_3COO^-] = 1.34 \times 10^{-3}\ M$, and pH = 2.87.

> **Student Annotation:** The percent ionization of acetic acid is
> $$\frac{1.34 \times 10^{-3}\ M}{0.10\ M} \times 100\% = 1.3\%$$

Now consider what happens when we add 0.050 mole of sodium acetate (CH_3COONa) to the solution. Sodium acetate dissociates completely in aqueous solution to give sodium ions and acetate ions.

$$CH_3COONa(aq) \xrightarrow{H_2O} Na^+(aq) + CH_3COO^-(aq)$$

Thus, by adding sodium acetate, we have increased the concentration of acetate ion. Because acetate ion is a product in the ionization of acetic acid, the addition of acetate ion causes the equilibrium to shift to the left. The net result is a reduction in the percent ionization of acetic acid.

> **Student Annotation:** By adding sodium acetate, we also add sodium ions to the solution. However, sodium ions do not interact with water or with any of the other species present [◄◄ Section 16.10].

addition

$$CH_3COOH(aq) \longleftarrow H^+(aq) + CH_3COO^-(aq)$$

Equilibrium is driven toward reactant.

Shifting the equilibrium to the left consumes not only some of the added acetate ion, but also some of the hydrogen ion. This causes the pH to change (in this case the pH increases).

Worked Example 17.1 shows how an equilibrium table can be used to calculate the pH of a solution of acetic acid after the addition of sodium acetate.

Worked Example 17.1

Determine the pH at 25°C of a solution prepared by adding 0.050 mole of sodium acetate to 1.0 L of 0.10 M acetic acid. (Assume that the addition of sodium acetate does not change the volume of the solution.)

Strategy Construct a new equilibrium table to solve for the hydrogen ion concentration.

Setup We use the stated concentration of acetic acid, 0.10 M, and $[H^+] \approx 0\ M$ as the initial concentrations in the following table.

> **Student Annotation:** Remember that prior to the ionization of a weak acid, the concentration of hydrogen ion in water at 25°C is $1.0 \times 10^{-7}\ M$. However, because this concentration is insignificant compared to the concentration resulting from the ionization, we can neglect it in our equilibrium table.

$$CH_3COOH(aq) \rightleftharpoons H^+(aq) + CH_3COO^-(aq)$$

	$CH_3COOH(aq)$	$H^+(aq)$	$CH_3COO^-(aq)$
Initial concentration (M):	0.10	0	0.050
Change in concentration (M):	$-x$	$+x$	$+x$
Equilibrium concentration (M):	$0.10 - x$	x	$0.050 + x$

Solution Substituting the equilibrium concentrations, in terms of the unknown x, into the equilibrium expression gives

$$1.8 \times 10^{-5} = \frac{(x)(0.050 + x)}{0.10 - x}$$

Because we expect x to be very small (even smaller than $1.34 \times 10^{-3}\ M$—as discussed), because the ionization of CH_3COOH is suppressed by the presence of CH_3COO^-, we assume

$$(0.10 - x)\ M \approx 0.10\ M \quad \text{and} \quad (0.050 + x)\ M \approx 0.050\ M$$

> **Student Annotation:** In this case, the percent ionization of acetic acid is
> $$\frac{3.6 \times 10^{-5}\ M}{0.10\ M} \times 100\% = 0.036\%$$
> This is considerably smaller than the percent ionization prior to the addition of sodium acetate.

Therefore, the equilibrium expression simplifies to

$$1.8 \times 10^{-5} = \frac{(x)(0.050)}{0.10}$$

and $x = 3.6 \times 10^{-5}\ M$. According to the equilibrium table, $[H^+] = x$, so pH $= -\log(3.6 \times 10^{-5}) = 4.44$.

Think About It

The equilibrium concentrations of CH_3COOH, CH_3COO^-, and H^+ are the same regardless of whether we add sodium acetate to a solution of acetic acid, add acetic acid to a solution of sodium acetate, or dissolve both species at the same time. We could have constructed an equilibrium table starting with the equilibrium concentrations in the 0.10 M acetic acid solution.

	$CH_3COOH(aq)$	$H^+(aq)$	$CH_3COO^-(aq)$
Initial concentration (M):	0.09866	1.34×10^{-3}	5.134×10^{-2}
Change in concentration (M):	$+y$	$-y$	$-y$
Equilibrium concentration (M):	$0.09866 + y$	$1.34 \times 10^{-3} - y$	$5.134 \times 10^{-2} - y$

> **Student Annotation:** This is the sum of the equilibrium concentration of acetate ion in a 0.10-M solution of acetic acid ($1.34 \times 10^{-3}\ M$) and the added acetate ion (0.050 M).

> **Student Annotation:** Treating the problem as though both CH_3COOH and CH_3COO^- are added at the same time and the reaction proceeds to the right simplifies the solution.

In this case, the reaction proceeds to the left. (The acetic acid concentration increases, and the concentrations of hydrogen and acetate ions decrease.) Solving for y gives $1.304 \times 10^{-3}\ M$. $[H^+] = 1.34 \times 10^{-3} - y = 3.6 \times 10^{-5}\ M$ and pH $= 4.44$. We get the same pH either way.

Practice Problem **A**TTEMPT Determine the pH at 25°C of a solution prepared by dissolving 0.075 mole of sodium acetate in 1.0 L of 0.25 M acetic acid. (Assume that the addition of sodium acetate does not change the volume of the solution.)

Practice Problem **B**UILD Determine the pH at 25°C of a solution prepared by dissolving 0.35 mole of ammonium chloride in 1.0 L of 0.25 M aqueous ammonia.

Practice Problem **C**ONCEPTUALIZE Which of the following compounds, when added to an aqueous solution of HF, would cause an increase in the pH? Which would cause a decrease in pH? Which would not have an effect on the pH?

NaF	SnF_2	HCl	NaCl	NaOH	H_2O
(i)	(ii)	(iii)	(iv)	(v)	(vi)

An aqueous solution of a weak electrolyte contains both the weak electrolyte and its ionization products, which are ions. If a soluble salt that contains one of those ions is added, the equilibrium shifts to the left, thereby *suppressing* the ionization of the weak electrolyte. In general, when a compound containing an ion in common with a dissolved substance is added to a solution at equilibrium, the equilibrium shifts to the left. This phenomenon is known as the ***common ion effect.***

> **Student Annotation:** The *common ion* can also be H^+ or OH^-. For example, addition of a strong acid to a solution of a weak acid suppresses ionization of the weak acid. Similarly, addition of a strong base to a solution of weak base suppresses ionization of the weak base.

Section 17.1 Review

The Common Ion Effect

17.1.1 Which of the following would cause a decrease in the percent ionization of a solution of nitrous acid (HNO_2) at equilibrium? (Select all that apply.)
(a) $NaNO_2$
(b) H_2O
(c) $Ca(NO_3)_2$
(d) HNO_3
(e) $NaNO_3$

17.1.2 What is the pH of a solution prepared by adding 0.05 mole of NaF to 1.0 L of 0.1 M HF at 25°C? (Assume that the addition of NaF does not change the volume of the solution.) (K_a for HF = 7.1×10^{-4}.)
(a) 2.1
(b) 2.8
(c) 1.4
(d) 4.6
(e) 7.3

17.2 BUFFER SOLUTIONS

A solution that contains a weak acid and its conjugate base (or a weak base and its conjugate acid) is a *buffer solution* or simply a *buffer.* Buffer solutions, by virtue of their composition, *resist* changes in pH upon addition of small amounts of either an acid or a base. The ability to resist pH change is very important to chemical and biological systems, including the human body. The pH of blood is about 7.4, whereas that of gastric juices is about 1.5. Each of these pH values is crucial for proper enzyme function and the balance of osmotic pressure, and each is maintained within a very narrow pH range by a buffer.

> **Student Annotation:** Any solution of a weak acid contains some conjugate base. In a buffer solution, though, the amounts of weak acid and conjugate base must be *comparable,* meaning that the conjugate base must be supplied by a dissolved salt.

Calculating the pH of a Buffer

Consider a solution that is 1.0 M in acetic acid and 1.0 M sodium acetate. If a small amount of acid is added to this solution, it is consumed completely by the acetate ion,

$$H^+(aq) + CH_3COO^-(aq) \longrightarrow CH_3COOH(aq)$$

thus converting a strong acid (H^+) to a weak acid (CH_3COOH). Addition of a strong acid lowers the pH of a solution. However, a buffer's ability to convert a strong acid to a weak acid minimizes the effect of the addition on the pH.

Similarly, if a small amount of a base is added, it is consumed completely by the acetic acid,

$$CH_3COOH(aq) + OH^-(aq) \longrightarrow CH_3COO^-(aq) + H_2O(l)$$

thus converting a strong base (OH^-) to a weak base (CH_3COO^-). Addition of a strong base increases the pH of a solution. Again, however, a buffer's ability to convert a strong base to a weak base minimizes the effect of the addition on pH.

> **Student Annotation:** Remember that sodium acetate is a strong electrolyte [◀◀ Section 9.1], so it dissociates completely in water to give sodium ions and acetate ions:
> $CH_3COONa(s) \longrightarrow Na^+(aq) + CH_3COO^-(aq)$

Video 17.1
Acids and bases—effect of addition of a strong acid and a strong base on a buffer.

To illustrate the function of a buffer, suppose that we have 1 L of the acetic acid–sodium acetate solution described previously. We can calculate the pH of the buffer using the procedure in Section 17.1.

$$CH_3COOH(aq) \rightleftharpoons H^+(aq) + CH_3COO^-(aq)$$

Initial concentration (M):	1.0	0	1.0
Change in concentration (M):	$-x$	$+x$	$+x$
Equilibrium concentration (M):	$1.0 - x$	x	$1.0 + x$

The equilibrium expression is

$$K_a = \frac{(x)(1.0 + x)}{1.0 - x}$$

> **Student Annotation:** The forward reaction is suppressed by the presence of the common ion, CH_3COO^-, and the reverse process is suppressed by the presence of CH_3COOH.

Because it is reasonable to assume that x will be very small,

$$(1.0 - x)\ M \approx 1.0\ M \qquad \text{and} \qquad (1.0 + x)\ M \approx 1.0\ M$$

Thus, the equilibrium expression simplifies to

$$1.8 \times 10^{-5} = \frac{(x)(1.0)}{1.0} = x$$

At equilibrium, therefore, $[H^+] = 1.8 \times 10^{-5}\ M$ and pH = 4.74.

Now consider what happens when we add 0.10 mole of HCl to the buffer. (We assume that the addition of HCl causes no change in the volume of the solution.) The reaction that takes place when we add a strong acid is the conversion of H^+ to CH_3COOH. The added acid is all consumed, along with an equal amount of acetate ion. We keep track of the amounts of acetic acid and acetate ion when a strong acid (or base) is added by writing the starting amounts above the equation and the final amounts (after the added substance has been consumed) below the equation.

> **Student Annotation:** As long as the amount of strong acid added to the buffer does not exceed the amount of conjugate base originally present, *all* the added acid will be consumed and converted to weak acid.

Upon addition of H^+:	1.0 mol	0.1 mol	1.0 mol
	$CH_3COO^-(aq) + $	$H^+(aq) \longrightarrow$	$CH_3COOH(aq)$
After H^+ has been consumed:	0.9 mol	0 mol	1.1 mol

We can use the resulting amounts of acetic acid and acetate ion to construct a new equilibrium table.

$$CH_3COOH(aq) \rightleftharpoons H^+(aq) + CH_3COO^-(aq)$$

Initial concentration (M):	1.1	0	0.9
Change in concentration (M):	$-x$	$+x$	$+x$
Equilibrium concentration (M):	$1.1 - x$	x	$0.9 + x$

We can solve for pH as we have done before, assuming that x is small enough to be neglected.

$$1.8 \times 10^{-5} = \frac{(x)(0.9 + x)}{1.1 - x} \approx \frac{(x)(0.9)}{1.1}$$

$$x = 2.2 \times 10^{-5}\ M$$

> **Student Annotation:** Had we added 0.10 mole of HCl to 1 L of pure water, the pH would have gone from 7.00 to 1.00!

Thus, when equilibrium is reestablished, $[H^+] = 2.2 \times 10^{-5}\ M$ and pH = 4.66—a change of only 0.08 pH unit.

In the determination of the pH of a buffer such as the one just described, we always neglect the small amount of weak acid that ionizes (x) because ionization is suppressed by the presence of a common ion. Similarly, we ignore the hydrolysis of the acetate ion because of the presence of acetic acid. This enables us to derive an expression for determining the pH of a buffer. We begin with the equilibrium expression

$$K_a = \frac{[H^+][A^-]}{[HA]}$$

Rearranging to solve for $[H^+]$ gives

$$[H^+] = \frac{K_a[HA]}{[A^-]}$$

Taking the negative logarithm of both sides, we obtain

$$-\log [H^+] = -\log K_a - \log \frac{[HA]}{[A^-]}$$

or

$$-\log [H^+] = -\log K_a + \log \frac{[A^-]}{[HA]}$$

Thus,

$$pH = pK_a + \log \frac{[A^-]}{[HA]} \qquad \textbf{Equation 17.1}$$

where

$$pK_a = -\log K_a \qquad \textbf{Equation 17.2}$$

Equation 17.1 is known as the **_Henderson-Hasselbalch equation._** Its more general form is

$$pH = pK_a + \log \frac{[\text{conjugate base}]}{[\text{weak acid}]} \qquad \textbf{Equation 17.3}$$

In the case of our acetic acid (1.0 M) and sodium acetate (1.0 M) buffer, the concentrations of weak acid and conjugate base are _equal_. When this is true, the log term in the Henderson-Hasselbalch equation is zero and the pH is numerically equal to the pK_a. In the case of an acetic acid–acetate ion buffer, $pK_a = -\log 1.8 \times 10^{-5} = 4.74$.

After the addition of 0.10 mole of HCl, we determined that the concentrations of acetic acid and acetate ion were 1.1 M and 0.9 M, respectively. Using these concentrations in the Henderson-Hasselbalch equation gives

$$pH = 4.74 + \log \frac{[CH_3COO^-]}{[CH_3COOH]}$$

$$= 4.74 + \log \frac{0.9\ M}{1.1\ M}$$

$$= 4.74 + (-0.087) = 4.65$$

The small difference between this pH and the 4.66 calculated using an equilibrium table is due to differences in rounding. Figure 17.1 illustrates how a buffer solution resists drastic changes in pH.

Video 17.2
Buffer solutions.

Figure 17.1

Buffer Solutions

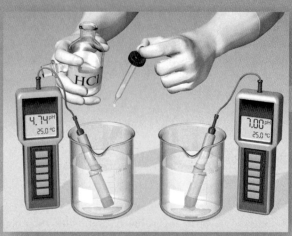

When we add 0.001 mol of strong acid, it is completely consumed by the acetate ion in the buffer.

Before reaction: 0.001 mol   0.010 mol   0.010 mol

$$H^+(aq) + CH_3COO^-(aq) \longrightarrow CH_3COOH(aq)$$

After reaction: 0 mol   0.009 mol   0.011 mol

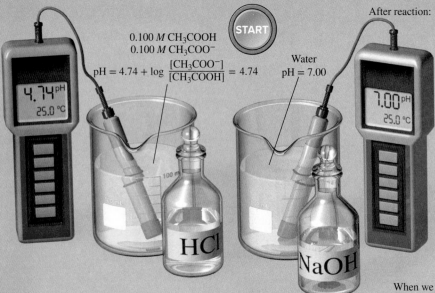

0.100 M CH$_3$COOH
0.100 M CH$_3$COO$^-$

$$\text{pH} = 4.74 + \log \frac{[CH_3COO^-]}{[CH_3COOH]} = 4.74$$

Water
pH = 7.00

START

START

The buffer solution is 0.100 M in acetic acid and 0.100 M in sodium acetate. 100 mL of this buffer contains (0.100 mol/L)(0.10 L) = 0.010 mol each acetic acid and acetate ion.

When we add 0.001 mol of strong base, it is completely consumed by the acetic acid in the buffer.

Before reaction: 0.001 mol   0.010 mol   0.010 mol

$$OH^-(aq) + CH_3COOH(aq) \longrightarrow H_2O(l) + CH_3COO^-(aq)$$

After reaction: 0 mol   0.009 mol   0.011 mol

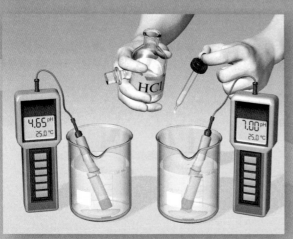

We can calculate the new pH using the Henderson-Hasselbalch equation:

$$pH = 4.74 + \log \frac{0.009}{0.011} = 4.65$$

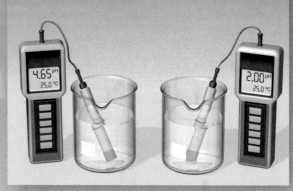

There is nothing in pure water to consume strong acid. Therefore, its pH drops drastically.

$$pH = -\log \frac{0.001 \text{ mol}}{0.10 \text{ L}} = 2.00$$

We can calculate the new pH using the Henderson-Hasselbalch equation:

$$pH = 4.74 + \log \frac{0.011}{0.009} = 4.83$$

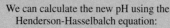

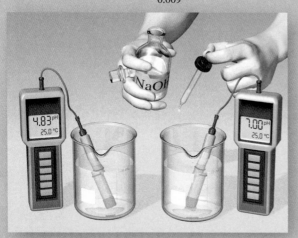

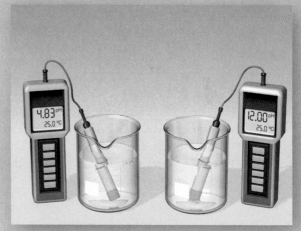

There is nothing in pure water to consume strong base. Therefore, its pH rises drastically.

$$pOH = -\log \frac{0.001 \text{ mol}}{0.10 \text{ L}} = 2.00, pH = 12.00$$

What's the point?

A buffer contains both a weak acid and its conjugate base.*
Small amounts of strong acid or strong base are consumed by
the buffer components, thereby preventing drastic pH changes.
Pure water does not contain species that can consume acid
or base. Even a very small addition of either acid or base causes
a large change in pH.

*A buffer could also be prepared using a weak base and its conjugate acid.

Worked Example 17.2 shows how the Henderson-Hasselbalch equation is used to determine the pH of a buffer after the addition of a strong base.

Worked Example 17.2

Starting with 1.00 L of a buffer that is 1.00 M in acetic acid and 1.00 M in sodium acetate, calculate the pH after the addition of 0.100 mole of NaOH. (Assume that the addition does not change the volume of the solution.)

Strategy Added base will react with the acetic acid component of the buffer, converting OH^- to CH_3COO^-.

$$CH_3COOH(aq) + OH^-(aq) \longrightarrow H_2O(l) + CH_3COO^-(aq)$$

Write the starting amount of each species above the equation and the final amount of each species below the equation. Use the final amounts as concentrations in Equation 17.1.

Setup

Upon addition of OH^-:

1.00 mol	0.10 mol		1.00 mol
$CH_3COOH(aq)$	$+ OH^-(aq)$	$\longrightarrow H_2O(l) +$	$CH_3COO^-(aq)$

After OH^- has been consumed: 0.90 mol   0 mol   1.10 mol

Student Annotation: The volume of the buffer is 1 L in this example, so the number of moles of a substance is equal to the molar concentration. In cases where the buffer volume is something other than 1 L, however, we can still use molar amounts in the Henderson-Hasselbalch equation because the volume would cancel in the top and bottom of the log term.

Solution

$$pH = 4.74 + \log \frac{1.10\ M}{0.90\ M}$$

$$= 4.74 + \log \frac{1.10\ M}{0.90\ M} = 4.83$$

Thus, the pH of the buffer after addition of 0.10 mole of NaOH is 4.83.

Think About It
Always do a "reality check" on a calculated pH. Although a buffer does minimize the effect of added base, the pH does increase. If you find that you've calculated a lower pH after the addition of a base, check for errors like mixing up the weak acid and conjugate base concentrations or losing track of a minus sign.

Practice Problem **A**TTEMPT Calculate the pH of 1 L of a buffer that is 1.0 M in acetic acid and 1.0 M in sodium acetate after the addition of 0.25 mole of NaOH.

Practice Problem **B**UILD How much HCl must be added to a liter of buffer that is 1.5 M in acetic acid and 0.75 M in sodium acetate to result in a buffer pH of 4.10?

Practice Problem **C**ONCEPTUALIZE The first diagram represents a buffer solution. Which of the other diagrams [(i)–(iv)] best represents the buffer after the addition of strong acid?

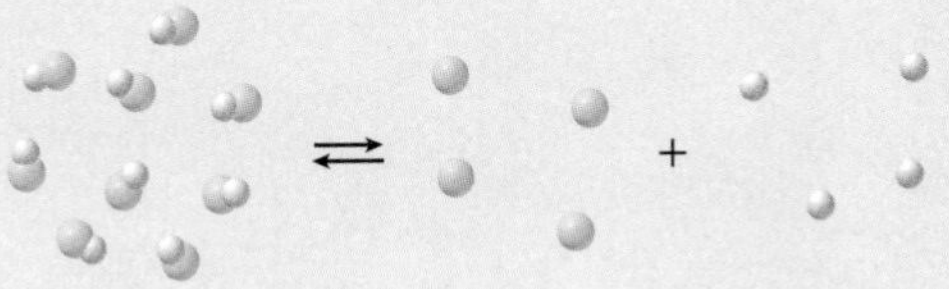

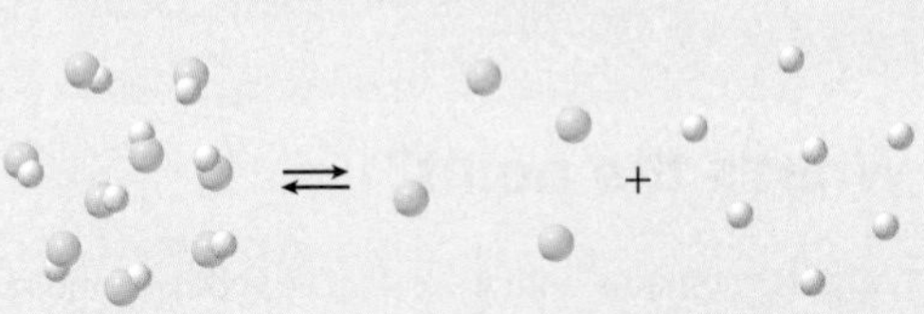

(i)

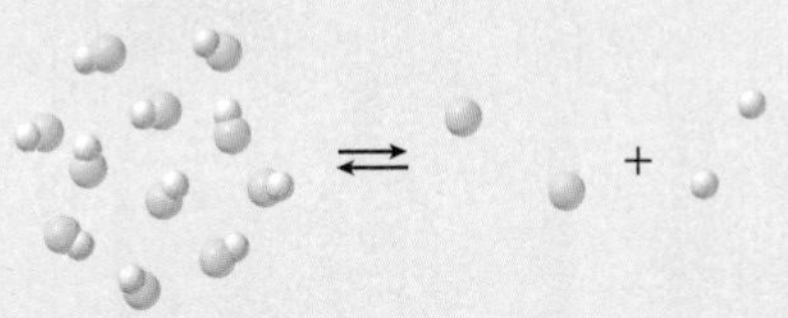

(iii)

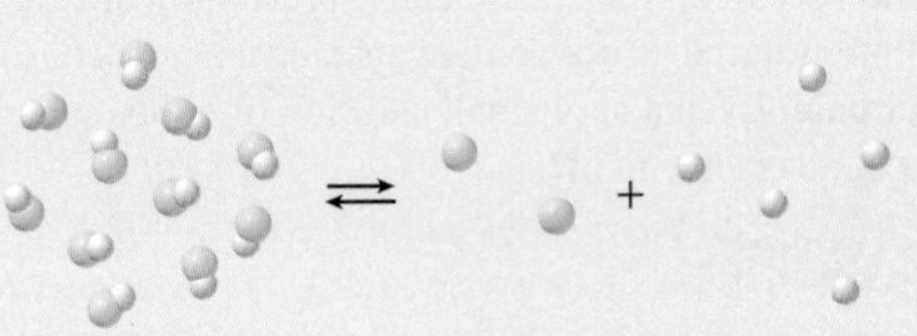

(ii)

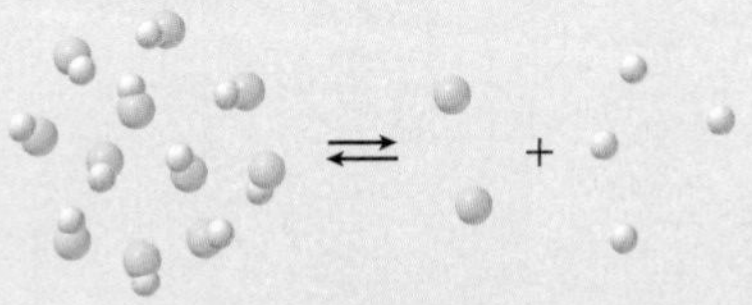

(iv)

Preparing a Buffer Solution with a Specific pH

A solution is only a buffer if it has the capacity to resist pH change when either an acid or a base is added. If the concentrations of a weak acid and conjugate base differ by more than a factor of 10, the solution does not have this capacity. Therefore, we consider a solution a buffer, and can use Equation 17.1 to calculate its pH, only if the following condition is met:

$$10 \geq \frac{[\text{conjugate base}]}{[\text{weak acid}]} \geq 0.1$$

Consequently, the log term in Equation 17.1 can only have values from -1 to 1, and the pH of a buffer cannot be more than one pH unit different from the pK_a of the weak acid it contains. This is known as the *range* of the buffer, where $pH = pK_a \pm 1$. This enables us to select the appropriate conjugate pair to prepare a buffer with a specific, desired pH.

First, we choose a weak acid whose pK_a is close to the desired pH. Next, we substitute the pH and pK_a values into Equation 17.1 to obtain the necessary ratio of [conjugate base]/[weak acid]. This ratio can then be converted to molar quantities for the preparation of the buffer.

Worked Example 17.3 demonstrates this procedure.

Student Hot Spot

Student data indicate you may struggle with preparing buffer solutions. Access the SmartBook to view additional Learning Resources on this topic.

Worked Example 17.3

Select an appropriate weak acid from the table, and describe how you would prepare a buffer with a pH of 9.50.

Strategy Select an acid with a pK_a within one pH unit of 9.50. Use the pK_a of the acid and Equation 17.1 to calculate the necessary ratio of [conjugate base]/[weak acid]. Select concentrations of the buffer components that yield the calculated ratio.

Setup Two of the acids listed in the margin have pK_a values in the desired range: hydrocyanic acid (HCN, $pK_a = 9.31$) and phenol (C_6H_5OH, $pK_a = 9.89$).

Solution Plugging the values for phenol into Equation 17.1 gives

Weak Acid	K_a	pK_a
HF	7.1×10^{-4}	3.15
HNO_2	4.5×10^{-4}	3.35
HCOOH	1.7×10^{-4}	3.77
C_6H_5COOH	6.5×10^{-5}	4.19
CH_3COOH	1.8×10^{-5}	4.74
HCN	4.9×10^{-10}	9.31
C_6H_5OH	1.3×10^{-10}	9.89

$$9.50 = 9.89 + \log \frac{[C_6H_5O^-]}{[C_6H_5OH]}$$

$$9.50 - 9.89 = \log \frac{[C_6H_5O^-]}{[C_6H_5OH]} = -0.39$$

$$\frac{[C_6H_5O^-]}{[C_6H_5OH]} = 10^{-0.39} = 0.41$$

Therefore, the ratio of $[C_6H_5O^-]$ to $[C_6H_5OH]$ must be 0.41 to 1. One way to achieve this would be to dissolve 0.41 mole of C_6H_5ONa and 1.00 mole of C_6H_5OH in 1 L of water.

Think About It

There is an infinite number of combinations of [conjugate base] and [weak acid] that will give the necessary ratio. Note that this pH could also be achieved using HCN and a cyanide salt. For most purposes, it is best to use the least toxic compounds available.

Practice Problem **A**TTEMPT Select an appropriate acid, using pK_a values, and describe how you would prepare a buffer with pH = 4.5.

Practice Problem **B**UILD What range of pH values could be achieved with a buffer consisting of nitrous acid (HNO_2) and sodium nitrite (NO_2^-)?

Practice Problem **C**ONCEPTUALIZE The diagrams represent three different weak acids, each of which can be combined with its conjugate base to prepare a buffer. Which acid can be used to prepare the buffer with the lowest pH?

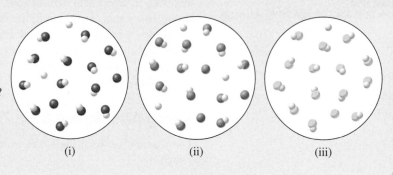

(i)   (ii)   (iii)

Section 17.2 Review

Buffer Solutions

17.2.1 What is the pH of a buffer that is 0.76 M in HF and 0.98 M in NaF?

    (a) 3.26   (d) 10.85
    (b) 3.04   (e) 10.74
    (c) 3.15

17.2.2 Consider 1 L of a buffer that is 0.85 M in formic acid (HCOOH) and 1.4 M in sodium formate (HCOONa). Calculate the pH after the addition of 0.15 mol HCl. (Assume the addition causes no volume change.)

    (a) 4.11   (d) 10.13
    (b) 3.99   (e) 10.01
    (c) 3.87

17.2.3 Consider 1 L of a buffer that is 1.5 M in hydrocyanic acid (HCN) and 1.2 M in sodium cyanide (NaCN). Calculate the pH after the addition of 0.25 mol NaOH. (Assume the addition causes no volume change.)

    (a) 9.21   (d) 4.63
    (b) 9.37   (e) 4.96
    (c) 9.04

17.2.4 The solutions shown contain a combination of the weak acid HA and its sodium salt NaA. (For clarity, the sodium ions and water molecules are not shown.) Arrange the solutions in order of increasing pH.

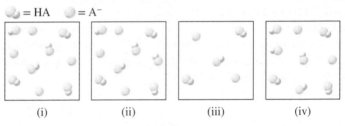

                (i)   (ii)   (iii)   (iv)

    (a) i < iii < iv < ii   (d) iv < ii < i = iii
    (b) iv < ii < iii < i   (e) i = iii < iv < ii
    (c) i < ii < iv < iii

17.3   ACID-BASE TITRATIONS

In Section 9.6 we introduced acid-base titrations as a form of chemical analysis. Having discussed buffer solutions, we can now look in more detail at the quantitative aspects of acid-base titrations. We will consider three types of reactions: (1) titrations involving a strong acid and a strong base, (2) titrations involving a weak acid and a strong base, and (3) titrations involving a strong acid and a weak base. Titrations involving a weak acid and a weak base are complicated by the hydrolysis of both the cation and the anion of the salt formed. These titrations will not be discussed here. Figure 17.2 shows the experimental setup for monitoring the pH over the course of an acid-base titration.

Video 17.3
Acids and bases—titration of HCl with NaOH.

Strong Acid–Strong Base Titrations

The reaction between the strong acid HCl and the strong base NaOH can be represented by

$$\text{NaOH}(aq) + \text{HCl}(aq) \longrightarrow \text{NaCl}(aq) + \text{H}_2\text{O}(l)$$

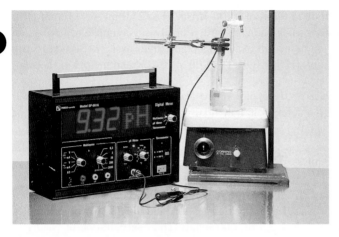

Figure 17.2 A pH meter is used to monitor an acid-base titration.
© McGraw-Hill Education/Ken Karp, photographer

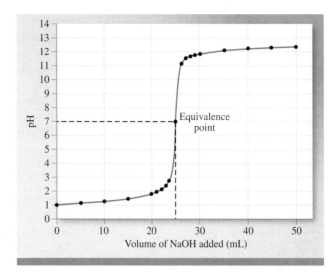

Figure 17.3 Titration curve (pH as a function of volume titrant added) of a strong acid–strong base titration. A 0.100 M NaOH solution, the titrant, is added from a burette to 25.0 mL of a 0.100 M HCl solution in an Erlenmeyer flask.

or by the net ionic equation,

$$OH^-(aq) + H^+(aq) \longrightarrow H_2O(l)$$

Consider the addition of a 0.100 M NaOH solution (from a burette) to a vessel containing 25.0 mL of 0.100 M HCl. For convenience, we will use only three significant figures for volume and concentration and two significant figures for pH. Figure 17.3 shows the titration curve—the plot of pH as a function of titrant volume added.

Before the addition of NaOH begins, the pH of the acid is given by $-\log(0.100)$, or 1.00. When NaOH is added, the pH of the solution increases slowly at first. Near the equivalence point, the pH begins to rise steeply, and at the equivalence point, when equimolar amounts of acid and base have reacted, the curve rises almost vertically. In a strong acid–strong base titration, both the hydrogen ion and hydroxide ion concentrations are very small at the equivalence point (roughly 1×10^{-7} M); consequently, the addition of a single drop of the base causes a large increase in $[OH^-]$ and a steep rise in the pH of the solution. Beyond the equivalence point, the pH again increases slowly with the continued addition of NaOH.

It is possible to calculate the pH of the solution at every stage of titration. Here are three sample calculations.

> **Student Annotation:** Recall that only the digits to the right of the decimal point are significant in a pH value.

> **Student Annotation:** The *titrant* is the solution that is added from the burette.

> **Student Annotation:** Recall that for an acid and base that combine in a 1:1 ratio, the equivalence point is where equal molar amounts of acid and base have been combined [|◀◀ Section 9.6].

1. Consider the addition of 10.0 mL of 0.100 M NaOH to 25.0 mL of 0.100 M HCl: The total volume of the solution is 35.0 mL. The number of millimoles of NaOH in 10.0 mL is

> **Student Annotation:** These calculations could also be done using *moles*, but using millimoles simplifies the calculations. Remember that millimoles = $M \times$ mL [|◀◀ Section 9.5].

$$10.0 \text{ ml} \times \frac{0.100 \text{ mmol NaOH}}{1 \text{ mL}} = 1.00 \text{ mmol}$$

The number of millimoles of HCl originally present in 25.0 mL of solution is

$$25.0 \text{ ml} \times \frac{0.100 \text{ mmol HCl}}{1 \text{ mL}} = 2.50 \text{ mmol}$$

Thus, the amount of HCl left after partial neutralization is $2.50 - 1.00$, or 1.50 mmol. Next, we determine the resulting concentration of H^+. We have 1.50 mmol in 35.0 mL:

$$\frac{1.50 \text{ mmol HCl}}{35.0 \text{ mL}} = 0.0429 \text{ } M$$

Thus $[H^+] = 0.0429$ M, and the pH of the solution is

$$pH = -\log(0.04289) = 1.37$$

2. Consider the addition of 25.0 mL of 0.100 M NaOH to 25.0 mL of 0.100 M HCl: This is a straightforward calculation, because it involves a complete neutralization reaction and neither ion in the salt (NaCl) undergoes hydrolysis [◄◄ Section 16.10]. At the equivalence point, $[H^+] = [OH^-] = 1.00 \times 10^{-7}$ M and the pH of the solution is 7.000.

3. Consider the addition of 35.0 mL of 0.100 M NaOH to 25.0 mL of 0.100 M HCl: The total volume of the solution is now 60.0 mL. The number of millimoles of NaOH added is

$$35.0 \text{ ml} \times \frac{0.100 \text{ mmol NaOH}}{1 \text{ mL}} = 3.50 \text{ mmol}$$

There are 2.50 mmol of HCl in 25.0 mL of solution. After complete neutralization of HCl, 2.50 mmol of NaOH have been consumed, and the number of millimoles of NaOH remaining is 3.5 − 2.5 or 1.00 mmol. The concentration of NaOH in 60.0 mL of solution is

$$\frac{1.00 \text{ mmol NaOH}}{60.0 \text{ mL}} = 0.0167 \text{ } M$$

Thus $[OH^-] = 0.0167$ M and pOH = $-\log (0.0167) = 1.78$. The pH of the solution is 14.00 − 1.78 or 12.22.

Table 17.1 lists the data at eight different points during a strong acid–strong base titration along with the calculated pH at each point.

Weak Acid–Strong Base Titrations

Consider the neutralization reaction between acetic acid (a weak acid) and sodium hydroxide (a strong base).

$$CH_3COOH(aq) + NaOH(aq) \longrightarrow CH_3COONa(aq) + H_2O(l)$$

This equation can be simplified to

$$CH_3COOH(aq) + OH^-(aq) \longrightarrow CH_3COO^-(aq) + H_2O(l)$$

The acetate ion that results from this neutralization undergoes hydrolysis [◄◄ Section 16.10] as follows:

$$CH_3COO^-(aq) + H_2O(l) \rightleftharpoons CH_3COOH(aq) + OH^-(aq)$$

At the equivalence point, therefore, when we only have sodium acetate in solution, the pH will be greater than 7 as a result of the OH^- formed by hydrolysis of the acetate ion.

TABLE 17.1	Determination of pH at Several Different Points in a Strong Acid–Strong Base Titration					
Volume OH⁻ added (mL)	OH⁻ added (mmol)	H⁺ remaining (mmol)	Total volume (mL)	$[H^+]$ (mol/L)		pH
0	0	2.5	25.0	0.100		1.000
5.0	0.50	2.0	30.0	0.0667		1.176
10.0	1.0	1.5	35.0	0.0429		1.364
15.0	1.5	1.0	40.0	0.0250		1.602
20.0	2.0	0.5	45.0	0.0111		1.955
25.0	2.5	0	50.0	1.00×10^{-7}		7.000
Volume OH⁻ added (mL)	OH⁻ added (mmol)	Excess OH⁻ (mmol)	Total volume (mL)	$[OH^-]$ (mol/L)	pOH	pH
30.0	3.0	0.5	55.0	0.0091	2.04	11.96
35.0	3.5	1.0	60.0	0.0167	1.78	12.223

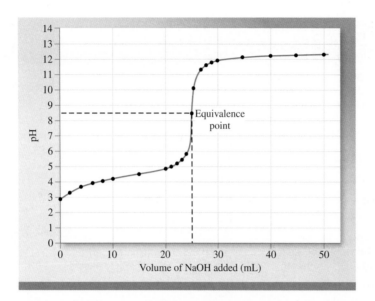

Figure 17.4 Titration curve of a weak acid–strong base titration. A 0.100 M NaOH solution is added from a burette to 25.0 mL of a 0.100 M CH$_3$COOH solution in an Erlenmeyer flask. Because of the hydrolysis of the salt formed, the pH at the equivalence point is greater than 7.

The curve for titration of 25.0 mL of 0.1 M acetic acid with 0.10 M sodium hydroxide is shown in Figure 17.4. Note how the shape of the curve differs from the one in Figure 17.3. Compared to the curve for titration of a strong acid with a strong base, the curve for titration of a weak acid with a strong base has a higher initial pH, a more gradual change in pH as base is added, and a shorter vertical region near the equivalence point.

Again, it is possible to calculate the pH at every stage of the titration. Here are four sample calculations.

1. Prior to the addition of any base, the pH is determined by the ionization of acetic acid. We use its concentration (0.10 M) and its K_a (1.8×10^{-5}) to calculate the H$^+$ concentration using an equilibrium table.

$$CH_3COOH(aq) \rightleftharpoons H^+(aq) + CH_3COO^-(aq)$$

	$CH_3COOH(aq)$	$H^+(aq)$	$CH_3COO^-(aq)$
Initial concentration (M):	0.10	0	0
Change in concentration (M):	$-x$	$+x$	$+x$
Equilibrium concentration (M):	$0.10 - x$	x	x

$$K_a = \frac{[H^+][CH_3COO^-]}{[CH_3COOH]} = \frac{x^2}{0.10 - x} = 1.8 \times 10^{-5}$$

We can neglect x on the bottom of the equation [◄◄ Section 16.6]. Solving for x,

$$\frac{x^2}{0.10} = 1.8 \times 10^{-5}$$

$$x^2 = (1.8 \times 10^{-5})(0.10) = 1.8 \times 10^{-6}$$

$$x = \sqrt{1.8 \times 10^{-6}} = 1.34 \times 10^{-3} \, M$$

gives $[H^+] = 1.34 \times 10^{-3} \, M$ and pH = 2.87.

2. After the first addition of base, some of the acetic acid has been converted to acetate ion via the reaction

$$CH_3COOH(aq) + OH^-(aq) \longrightarrow CH_3COO^-(aq) + H_2O(l)$$

With significant amounts of both acetic acid and acetate ion in solution, we now treat the solution as a buffer and use the Henderson-Hasselbalch equation to calculate the pH.

TABLE 17.2	Determination of pH at Several Different Points in a CH_3COOH-NaOH Titration						
Volume OH⁻ added (mL)	**OH⁻ added (mmol)**	**CH_3COOH remaining**	**CH_3COO^- produced**				**pH**
0	0	2.5	0.0				2.87*
5.0	0.50	2.0	0.50				4.14
10.0	1.0	1.5	1.0				4.56
15.0	1.5	1.0	1.5				4.92
20.0	2.0	0.5	2.0				5.34
25.0	2.5	0.0	2.5				8.72†
Volume OH⁻ added (mL)	**OH⁻ added (mmol)**	**Excess OH⁻ (mmol)**	**Total volume (mL)**	**[OH⁻] (mol/L)**	**pOH**		**pH**
30.0	3.0	0.5	55.0	0.0091	2.04		11.96
35.0	3.5	1.0	60.0	0.017	1.78		12.22

*$[CH_3COOH] = 0.10\ M$, $K_a = 1.8 \times 10^{-5}$.
†$[CH_3COO^-] = 0.050\ M$, $K_b = 5.6 \times 10^{-10}$.

After the addition of 10.0 mL of base, the solution contains 1.5 mmol of acetic acid and 1.0 mmol of acetate ion (see Table 17.2).

$$pH = 4.74 + \log \frac{1.0 \text{ mmol}}{1.5 \text{ mmol}} = 4.56$$

Each of the points between the beginning of the titration and the equivalence point can be calculated in this way.

3. At the equivalence point, all the acetic acid has been neutralized and we are left with acetate ion in solution. (There is also sodium ion, which does not undergo hydrolysis and therefore does not impact the pH of the solution.) At this point, pH is determined by the concentration and the K_b of acetate ion. The equivalence point occurs when 25.0 mL of base has been added, making the total volume 50.0 mL. The 2.5 mmol of acetic acid (see Table 17.2) has all been converted to acetate ion. Therefore, the concentration of acetate ion is

$$[CH_3COO^-] = \frac{2.5 \text{ mmol}}{50.0 \text{ mL}} = 0.050\ M$$

As we did at the beginning of the titration, we construct an equilibrium table.

	$CH_3COO^-(aq)$ + $H_2O(l)$	$\rightleftharpoons$	$OH^-(aq)$ +	$CH_3COOH(aq)$
Initial concentration (M):	0.050		0	0
Change in concentration (M):	$-x$		$+x$	$+x$
Equilibrium concentration (M):	$0.050 - x$		x	x

The K_b for acetate ion is 5.6×10^{-10}.

$$K_b = \frac{[OH^-][CH_3COOH]}{[CH_3COO^-]} = \frac{x^2}{0.050 - x} = 5.6 \times 10^{-10}$$

As before, we can neglect x in the denominator of the equation. Solving for x,

$$\frac{x^2}{0.050} = 5.6 \times 10^{-10}$$

$$x^2 = (5.6 \times 10^{-10})(0.050) = 2.8 \times 10^{-11}$$

$$x = \sqrt{2.8 \times 10^{-11}} = 5.3 \times 10^{-6}\ M$$

gives $[OH^-] = 5.3 \times 10^{-6}\ M$, pOH = 5.28, and pH = 8.72.

4. After the equivalence point, the curve for titration of a weak acid with a strong base is identical to the curve for titration of a strong acid with a strong base. Because all the acetic acid has been consumed, there is nothing in solution to consume the additional added OH^-, and the pH levels off between 12 and 13.

Table 17.2 lists the data for the titration of 25.0 mL of 0.10 M acetic acid with 0.10 M NaOH.

Worked Example 17.4 shows how to calculate the pH for the titration of a weak acid with a strong base.

Worked Example 17.4

Calculate the pH in the titration of 50.0 mL of 0.120 M acetic acid by 0.240 M sodium hydroxide after the addition of (a) 10.0 mL of base, (b) 25.0 mL of base, and (c) 35.0 mL of base.

Strategy The reaction between acetic acid and sodium hydroxide is

$$CH_3COOH(aq) + OH^-(aq) \longrightarrow H_2O(l) + CH_3COO^-(aq)$$

Prior to the equivalence point [part (a)], the solution contains both acetic acid and acetate ion, making the solution a buffer. We can solve part (a) using Equation 17.1, the Henderson-Hasselbalch equation. At the equivalence point [part (b)], all the acetic acid has been neutralized and we have only acetate ion in solution. We must determine the concentration of acetate ion and solve part (b) as an equilibrium problem, using the K_b for acetate ion. After the equivalence point [part (c)], all the acetic acid has been neutralized and there is nothing to consume the additional added base. We must determine the concentration of excess hydroxide ion in the solution and solve for pH using Equations 16.4 and 16.6.

Setup Remember that M can be defined as either mol/L or mmol/mL [◄◄ Section 9.5]. For this type of problem, it simplifies the calculations to use millimoles rather than moles. K_a for acetic acid is 1.8×10^{-5}, so $pK_a = 4.74$. K_b for acetate ion is 5.6×10^{-10}.

(a) The solution originally contains (0.120 mmol/mL)(50.0 mL) = 6.00 mmol of acetic acid. A 10.0-mL amount of base contains (0.240 mmol/mL)(10.0 mL) = 2.40 mmol of base. After the addition of 10.0 mL of base, 2.40 mmol of OH^- has neutralized 2.40 mmol of acetic acid, leaving 3.60 mmol of acetic acid and 2.40 mmol acetate ion in solution.

Upon addition of OH^-	6.00 mmol	2.40 mmol		0 mmol
	$CH_3COOH(aq)$ +	$OH^-(aq)$ ⇌	$H_2O(l)$ +	$CH_3COO^-(aq)$
After OH^- has been consumed:	3.60 mmol	0 mmol		2.40 mmol

(b) After the addition of 25.0 mL of base, the titration is at the equivalence point. We calculate the pH using the concentration and the K_b of acetate ion.

(c) After the addition of 35.0 mL of base, the titration is past the equivalence point and we solve for pH by determining the concentration of excess hydroxide ion.

Solution (a) $pH = pK_a + \log \dfrac{2.40}{3.60} = 4.74 - 0.18 = 4.56$

(b) At the equivalence point, we have 6.0 mmol of acetate ion in the total volume. We determine the total volume by calculating what volume of 0.24 M base contains 6.0 mmol.

$$(\text{volume})(0.240 \text{ mmol/mL}) = 6.00 \text{ mmol}$$

$$\text{volume} = \frac{6.00 \text{ mmol}}{0.240 \text{ mmol/mL}} = 25.0 \text{ mL}$$

Therefore, the equivalence point occurs when 25.0 mL of base has been added, making the total volume 50.0 mL + 25.0 mL = 75.0 mL. The concentration of acetate ion at the equivalence point is therefore

$$\frac{6.00 \text{ mmol } CH_3COO^-}{75.0 \text{ mL}} = 0.0800 \ M$$

We can construct an equilibrium table using this concentration and solve for pH using the ionization constant for CH_3COO^- ($K_b = 5.6 \times 10^{-10}$).

	$CH_3COO^-(aq)$ +	$H_2O(l)$ ⇌	$OH^-(aq)$ +	$CH_3COOH(aq)$
Initial concentration (M):	0.0800		0	0
Change in concentration (M):	$-x$		$+x$	$+x$
Equilibrium concentration (M):	$0.0800 - x$		x	x

(Continued on next page)

Using the equilibrium expression and assuming that x is small enough to be neglected,

$$K_b = \frac{[CH_3COOH][OH^-]}{[CH_3COO^-]} = \frac{(x)(x)}{0.0800 - x} \approx \frac{x^2}{0.0800} = 5.6 \times 10^{-10}$$

$$x = \sqrt{4.48 \times 10^{-11}} = 6.7 \times 10^{-6} \, M$$

According to the equilibrium table, $x = [OH^-]$, so $[OH^-] = 6.7 \times 10^{-6} \, M$. At equilibrium, therefore, $pOH = -\log(6.7 \times 10^{-6}) = 5.17$ and $pH = 14.00 - 5.17 = 8.83$.

(c) After the equivalence point, we must determine the concentration of excess base and calculate pOH and pH using Equations 16.4 and 16.6. A 35.0-mL amount of the base contains $(0.240 \, \text{mmol/mL})(35.0 \, \text{mL}) = 8.40$ mmol of OH^-. After neutralizing the 6.00 mmol of acetic acid originally present in the solution, this leaves $8.40 - 6.00 = 2.40$ mmol of excess OH^-. The total volume is $50.0 + 35.0 = 85.0$ mL. Therefore, $[OH^-] = 2.40 \, \text{mmol}/85.0 \, \text{mL} = 0.0280 \, M$, $pOH = -\log(0.0280) = 1.553$, and $pH = 14.000 - 1.553 = 12.447$.

In summary, (a) $pH = 4.56$, (b) $pH = 8.83$, and (c) $pH = 12.447$.

Think About It

For each point in a titration, decide first what species are in solution and what *type* of problem it is. If the solution contains only a weak acid (or weak base), as is the case before any titrant is added, or if it contains only a conjugate base (or conjugate acid), as is the case at the equivalence point, when pH is determined by salt hydrolysis, it is an *equilibrium* problem that requires a concentration, an ionization constant, and an equilibrium table. If the solution contains comparable concentrations of both members of a conjugate pair, which is the case at points prior to the equivalence point, it is a *buffer* problem and is solved using the Henderson-Hasselbalch equation. If the solution contains excess titrant, either a strong base or strong acid, it is simply a pH problem requiring only a concentration.

Practice Problem Ⓐ**TTEMPT** For the titration of 10.0 mL of 0.15 M acetic acid with 0.10 M sodium hydroxide, determine the pH when (a) 10.0 mL of base has been added, (b) 15.0 mL of base has been added, and (c) 20.0 mL of base has been added.

Practice Problem Ⓑ**UILD** For the titration of 25.0 mL of 0.20 M hydrofluoric acid with 0.20 M sodium hydroxide, determine the volume of base added when pH is (a) 2.85, (b) 3.15, and (c) 11.89. [To solve part (c), you may want to review the approach in Worked Example 9.9 on page 383.]

Practice Problem Ⓒ**ONCEPTUALIZE** Which of the graphs [(i)–(iv)] best represents the plot of pH versus volume of strong base added in the titration of a weak acid?

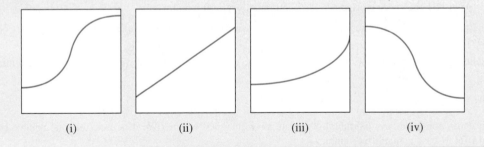

(i)　　　　　(ii)　　　　　(iii)　　　　　(iv)

Strong Acid–Weak Base Titrations

Consider the titration of HCl, a strong acid, with NH_3, a weak base:

$$HCl(aq) + NH_3(aq) \longrightarrow NH_4Cl(aq)$$

or simply

$$H^+(aq) + NH_3(aq) \longrightarrow NH_4^+(aq)$$

The pH at the equivalence point is less than 7 because the ammonium ion acts as a weak Brønsted acid:

$$NH_4^+(aq) + H_2O(l) \rightleftharpoons NH_3(aq) + H_3O^+(aq)$$

or simply

$$NH_4^+(aq) \rightleftharpoons NH_3(aq) + H^+(aq)$$

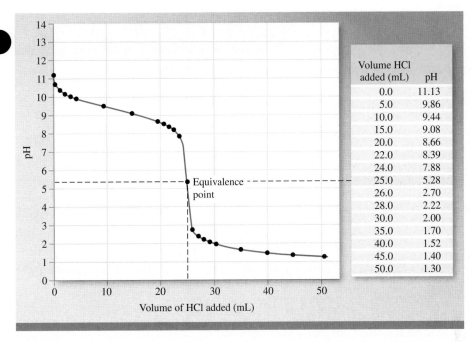

Figure 17.5 Titration curve of a strong acid–weak base titration. A 0.100 M HCl solution is added from a burette to 25.0 mL of a 0.100 M NH$_3$ solution in an Erlenmeyer flask. As a result of salt hydrolysis, the pH at the equivalence point is lower than 7.

Because of the volatility of an aqueous ammonia solution, it is more convenient to use hydrochloric acid as the titrant (i.e., to add HCl solution from the burette). Figure 17.5 shows the titration curve for this experiment.

Analogous to the titration of a weak acid with a strong base, the initial pH is determined by the concentration and the K_b of ammonia.

$$NH_3(aq) + H_2O(l) \rightleftharpoons NH_4^+(aq) + OH^-(aq)$$

Consider the titration of 25.0 mL of 0.10 M NH$_3$ with 0.10 M HCl. We calculate the initial pH by constructing an equilibrium table and solving for x.

	$NH_3(aq)$ +	$H_2O(l)$ ⇌	$NH_4^+(aq)$ +	$OH^-(aq)$
Initial concentration (M):	0.10		0	0
Change in concentration (M):	$-x$		$+x$	$+x$
Equilibrium concentration (M):	$0.10 - x$		x	x

Using the equilibrium expression and assuming that x is small enough to be neglected,

$$K_b = \frac{[NH_4^+][OH^-]}{[NH_3]} = \frac{(x)(x)}{0.10 - x} \approx \frac{x^2}{0.10} = 1.8 \times 10^{-5}$$

$$x^2 = 1.8 \times 10^{-6}$$

$$x = \sqrt{1.8 \times 10^{-6}} = 1.3 \times 10^{-3}$$

$$pH = 11.11$$

The pH at the equivalence point is calculated using the concentration and K_a of the conjugate base of NH$_3$, the NH$_4^+$ ion, and an equilibrium table.

Worked Example 17.5 shows how this is done.

Worked Example 17.5

Calculate the pH at the equivalence point when 25.0 mL of 0.100 M NH_3 is titrated with 0.100 M HCl.

Strategy The reaction between NH_3 and HCl is

$$NH_3(aq) + H^+(aq) \longrightarrow NH_4^+(aq)$$

At the equivalence point, all the NH_3 has been converted to NH_4^+. Therefore, we must determine the concentration of NH_4^+ at the equivalence point and use the K_a for NH_4^+ to solve for pH using an equilibrium table.

Setup The solution originally contains (0.100 mmol/mL)(25.0 mL) = 2.50 mmol NH_4^+. At the equivalence point, 2.50 mmol of HCl has been added. The volume of 0.100 M HCl that contains 2.50 mmol is

$$(\text{volume})(0.100 \text{ mmol/mL}) = 2.50 \text{ mmol}$$

$$\text{volume} = \frac{2.50 \text{ mmol}}{0.100 \text{ mmol/mL}} = 25.0 \text{ mL}$$

It takes 25.0 mL of titrant to reach the equivalence point, so the total solution volume is 25.0 + 25.0 = 50.0 mL. At the equivalence point, all the NH_3 originally present has been converted to NH_4^+. The concentration of NH_4^+ is (2.50 mmol)/(50.0 mL) = 0.0500 M. We must use this concentration as the starting concentration of ammonium ion in our equilibrium table.

Solution

$$NH_4^+(aq) + H_2O(l) \rightleftharpoons NH_3(aq) + H_3O^+(aq)$$

Initial concentration (M):	0.0500	0	0
Change in concentration (M):	−x	+x	+x
Equilibrium concentration (M):	0.0500 − x	x	x

The equilibrium expression is

$$K_a = \frac{[NH_3][H^+]}{[NH_4^+]} = \frac{(x)(x)}{0.0500 - x} \approx \frac{x^2}{0.0500} = 5.6 \times 10^{-10}$$

$$x^2 = 2.8 \times 10^{-11}$$

$$x = \sqrt{2.8 \times 10^{-11}} = 5.3 \times 10^{-6} \, M$$

$[H^+] = x = 5.3 \times 10^{-6} \, M$. At equilibrium, therefore, pH = $-\log (5.3 \times 10^{-6})$ = 5.28.

Think About It

In the titration of a weak base with a strong acid, the species in solution at the equivalence point is the conjugate acid. Therefore, we should expect an *acidic* pH. Once all the NH_3 has been converted to NH_4^+, there is no longer anything in the solution to consume added acid. Thus, the pH after the equivalence point depends on the number of millimoles of H^+ added and not consumed divided by the new total volume.

Practice Problem **A**TTEMPT Calculate the pH at the equivalence point in the titration of 50.0 mL of 0.10 M methylamine (see Table 16.6 on page 742) with 0.20 M HCl.

Practice Problem **B**UILD A 50.0-mL quantity of a 0.20-M solution of one of the weak bases in Table 16.6 is titrated with 0.050 M HCl. At the equivalence point, the pH is 2.99. Identify the weak base.

Practice Problem **C**ONCEPTUALIZE Which of the graphs [(i)–(iv)] best represents the plot of pH versus volume of strong acid added in the titration of a weak base?

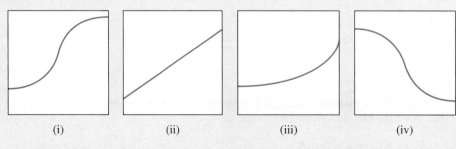

(i)   (ii)   (iii)   (iv)

Acid-Base Indicators

The equivalence point is the point at which the acid has been neutralized completely by the added base. The equivalence point in a titration can be determined by monitoring the pH over the course of the titration, or it can be determined using an *acid-base indicator*. An acid-base indicator is usually a weak organic acid or base for which the ionized and un-ionized forms are different colors.

Consider a weak organic acid that we will refer to as HIn. To be an effective acid-base indicator, HIn and its conjugate base, In$^-$, must have distinctly different colors. In solution, the acid ionizes to a small extent.

$$HIn(aq) \rightleftharpoons H^+(aq) + In^-(aq)$$

In a sufficiently acidic medium, the ionization of HIn is suppressed according to Le Châtelier's principle, and the preceding equilibrium shifts to the left. In this case, the color of the solution will be that of HIn. In a basic medium, on the other hand, the equilibrium shifts to the right and the color of the solution will be that of the conjugate base, In$^-$.

The ***endpoint*** of a titration is the point at which the color of the indicator changes. Not all indicators change color at the same pH, however, so the choice of indicator for a particular titration depends on the strength of the acid (and the base) used in the titration. To use the endpoint to determine the equivalence point of a titration, we must select an appropriate indicator.

The endpoint of an indicator does not occur at a specific pH; rather, there is a range of pH over which the color change occurs. In practice, we select an indicator whose color change occurs over a pH range that coincides with the steepest part of the titration curve. Consider the information in Figure 17.6, which shows the titration curves for hydrochloric acid and acetic acid—each being titrated with sodium hydroxide. Either of the indicators shown can be used for the titration of a strong acid with a strong base because both endpoints coincide with the steepest part of the HCl-NaOH titration curve. However, methyl red changes from red to yellow over the pH range of 4.2 to 6.3. This endpoint occurs significantly *before* the equivalence point in the titration of acetic acid, which occurs at about pH 8.7. Therefore, methyl red is *not* a suitable indicator for use in the titration of acetic acid with sodium hydroxide. Phenolphthalein, on the other hand, *is* a suitable indicator for the CH$_3$COOH-NaOH titration.

Student Annotation: The *endpoint* is where the color changes. The *equivalence point* is where neutralization is complete. Experimentally, we use the endpoint to estimate the equivalence point.

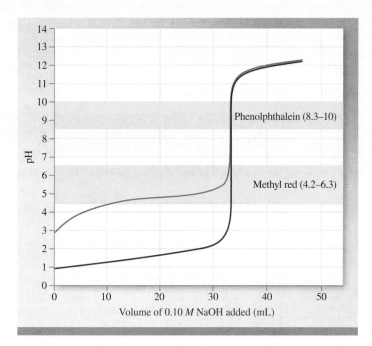

Figure 17.6 Titration curve of a strong acid with a strong base (blue) and titration curve of a weak acid with a strong base (red). The indicator phenolphthalein can be used to determine the equivalence point of either titration. Methyl red can be used for the strong acid–strong base titration but cannot be used for the weak acid–strong base titration because its color change does not coincide with the steepest part of the curve.

Figure 17.7 Solutions containing extracts of red cabbage (obtained by boiling the cabbage in water) produce different colors when treated with an acid and a base. The pH of the solutions increases from left to right.

© McGraw-Hill Education/Ken Karp, photographer

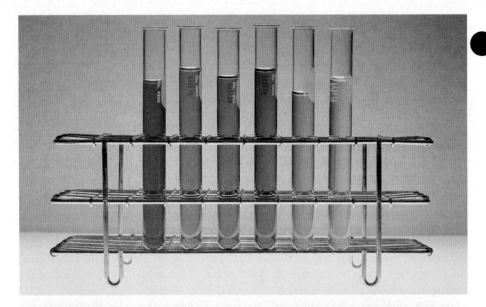

TABLE 17.3	Some Common Acid-Base Indicators		
		Color	
Indicator	In acid	In base	pH range
Thymol blue	Red	Yellow	1.2–2.8
Bromophenol blue	Yellow	Bluish purple	3.0–4.6
Methyl orange	Orange	Yellow	3.1–4.4
Methyl red	Red	Yellow	4.2–6.3
Chlorophenol blue	Yellow	Red	4.8–6.4
Bromothymol blue	Yellow	Blue	6.0–7.6
Cresol red	Yellow	Red	7.2–8.8
Phenolphthalein	Colorless	Reddish pink	8.3–10.0

Many acid-base indicators are plant pigments. For example, boiling red cabbage in water extracts pigments that exhibit a variety of colors at different pH values (Figure 17.7).

Table 17.3 lists a number of indicators commonly used in acid-base titrations. The choice of indicator for a particular titration depends on the strength of the acid and base to be titrated.

Worked Example 17.6 illustrates this point.

Worked Example 17.6

Which indicator (or indicators) listed in Table 17.3 would you use for the acid-base titrations shown in (a) Figure 17.3, (b) Figure 17.4, and (c) Figure 17.5?

Strategy Determine the pH range that corresponds to the steepest part of each titration curve and select an indicator (or indicators) that changes color within that range.

Setup (a) The titration curve in Figure 17.3 is for the titration of a strong acid with a strong base. The steep part of the curve spans a pH range of about 4 to 10.

(b) Figure 17.4 shows the curve for the titration of a weak acid with a strong base. The steep part of the curve spans a pH range of about 7 to 10.

(c) Figure 17.5 shows the titration of a weak base with a strong acid. The steep part of the curve spans a pH range of about 7 to 3.

Solution (a) Most of the indicators listed in Table 17.3, with the exceptions of thymol blue, bromophenol blue, and methyl orange, would work for the titration of a strong acid with a strong base.

(b) Cresol red and phenolphthalein are suitable indicators.

(c) Bromophenol blue, methyl orange, methyl red, and chlorophenol blue are all suitable indicators.

> **Think About It**
> If we don't select an appropriate indicator, the endpoint (color change) will not coincide with the equivalence point.

Practice Problem (A)TTEMPT Referring to Table 17.3, specify at least one indicator that would be suitable for the following titrations: (a) CH_3NH_2 with HBr, (b) HNO_3 with NaOH, (c) HNO_2 with KOH.

Practice Problem (B)UILD For which of the bases in Table 16.6 (page 742) could you titrate a 0.1-M solution of base with 0.1 M nitric acid using the indicator thymol blue?

Practice Problem (C)ONCEPTUALIZE The diagram shows the curve for titration of a particular weak acid with a strong base. Also shown is the region of color change for an acid-base indicator that is not a good choice for this titration. Suppose you were to titrate with a known concentration of base to determine the concentration of a sample of this weak acid. How would the acid concentration you determine from the titration be affected by the use of this indicator? Explain.

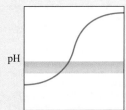

pH

Section 17.3 Review

Acid-Base Titrations

17.3.1 Calculate the pH at the equivalence point in the titration of 30 mL of 0.25 M CH_3COOH with 0.25 M KOH.
  (a) 7.00
  (b) 5.08
  (c) 8.92
  (d) 2.82
  (e) 11.18

17.3.2 Calculate the pH after the addition of 25 mL of 0.10 M NaOH to 50 mL of 0.10 M HF.
  (a) 3.15
  (b) 9.31
  (c) 12.52
  (d) 1.48
  (e) 11.87

17.3.3 Calculate the pH after the addition of 35 mL of 0.10 M NaOH to 30 mL of 0.10 M HCN.
  (a) 11.89
  (b) 2.11
  (c) 12.22
  (d) 1.78
  (e) 13.00

17.4   SOLUBILITY EQUILIBRIA

The solubility of ionic compounds is important in industry, medicine, and everyday life. For example, barium sulfate ($BaSO_4$), an insoluble compound that is opaque to X rays, is used to diagnose ailments of the digestive tract. Tooth decay begins when tooth enamel, which is mainly made of hydroxyapatite [$Ca_5(PO_4)_3OH$], is made more soluble in saliva by the presence of acid.

The general rules for predicting the solubility of ionic compounds in water were introduced in Section 9.2. While these rules are useful, they do not allow us to make quantitative predictions about how much of a given ionic compound will dissolve in water. To develop a quantitative approach, we must start with the principles of chemical equilibrium. Unless otherwise stated, the solvent is water in the following discussion and the temperature is 25°C.

Student Annotation: The compounds described as "insoluble" in Section 9.2, Table 9.3, are actually very *slightly* soluble—each to a different degree.

Solubility Product Expression and K_{sp}

Consider a saturated solution of silver chloride that is in contact with undissolved solid silver chloride. The equilibrium can be represented as

$$AgCl(s) \rightleftharpoons Ag^+(aq) + Cl^-(aq)$$

Although AgCl is not very soluble, *all* the AgCl that does dissolve in water dissociates completely into Ag^+ and Cl^- ions. We can write the equilibrium expression for the dissociation of AgCl as

$$K_{sp} = [Ag^+][Cl^-]$$

Student Annotation: A solubility product equilibrium expression is like any other equilibrium expression: K is equal to the concentrations of products over the concentrations of reactants, each raised to its coefficient from the balanced chemical equation.
Thus, for the process
$$MX_n(s) \rightleftharpoons M^{n+}(aq) + nX^-(aq)$$
the K_{sp} expression is
$$K_{sp} = [M^{n+}][X^-]^n$$
MX_n does not appear in the expression because, as for any heterogeneous equilibrium, the equilibrium expression does not include pure liquids or solids [◄◄ Section 15.3].

where K_{sp} is called the **solubility product constant**. (K_{sp} is just another specially subscripted K_c, where "sp" stands for "solubility product.")

Because each AgCl unit contains only one Ag^+ and one Cl^- ion, its solubility product expression is particularly simple to write. Many ionic compounds dissociate into more than two ions. Table 17.4 and Appendix 4 list a number of slightly soluble ionic compounds along with equations representing their dissolution equilibria and their solubility product constants. (Compounds deemed "soluble" by the solubility rules in Chapter 9 are not listed for the same reason we did not list K_a values for the strong acids in Table 16.5.) In general, the magnitude of K_{sp} indicates the solubility of an ionic compound—the smaller the K_{sp} value, the less soluble the compound. To make a direct comparison of K_{sp} values, however, we must compare salts with similar formulas, such as AgCl with ZnS (one cation, one anion) or CaF_2 with $Fe(OH)_2$ (one cation, two anions).

Calculations Involving K_{sp} and Solubility

There are two ways to express the solubility of a substance: **molar solubility**, which is the number of moles of solute in 1 L of a saturated solution (mol/L), and **solubility**, which is the number of grams of solute in 1 L of a saturated solution (g/L). Both of these expressions refer to concentrations of saturated solutions at a particular temperature (usually 25°C).

Student Annotation: Be careful not to confuse the terms *solubility* and K_{sp}. Solubility is the concentration of a saturated solution. K_{sp} is an equilibrium constant.

Often we know the value of K_{sp} for a compound and are asked to calculate the compound's molar solubility. The procedure for solving such a problem is essentially identical to the procedure for solving weak acid or weak base equilibrium problems.

1. Construct an equilibrium table.
2. Fill in what we know.
3. Figure out what we don't know.

For example, the K_{sp} of silver bromide (AgBr) is 7.7×10^{-13}. We can construct an equilibrium table and fill in the starting concentrations of Ag^+ and Br^- ions.

$AgBr(s) \rightleftharpoons$	$Ag^+(aq)$	$+ Br^-(aq)$
Initial concentration (M):	0	0
Change in concentration (M):		
Equilibrium concentration (M):		

Let s be the molar solubility (in mol/L) of AgBr. Because one unit of AgBr yields one Ag^+ cation and one Br^- anion, both $[Ag^+]$ and $[Br^-]$ are equal to s at equilibrium.

$AgBr(s) \rightleftharpoons$	$Ag^+(aq)$	$+ Br^-(aq)$
Initial concentration (M):	0	0
Change in concentration (M):	$+s$	$+s$
Equilibrium concentration (M):	s	s

TABLE 17.4	Solubility Products of Some Slightly Soluble Ionic Compounds at 25°C	
Compound	**Dissolution equilibrium**	K_{sp}
Aluminum hydroxide	$Al(OH)_3(s) \rightleftharpoons Al^{3+}(aq) + 3OH^-(aq)$	1.8×10^{-33}
Barium carbonate	$BaCO_3(s) \rightleftharpoons Ba^{2+}(aq) + CO_3^{2-}(aq)$	8.1×10^{-9}
Barium fluoride	$BaF_2(s) \rightleftharpoons Ba^{2+}(aq) + 2F^-(aq)$	1.7×10^{-6}
Barium sulfate	$BaSO_4(s) \rightleftharpoons Ba^{2+}(aq) + SO_4^{2-}(aq)$	1.1×10^{-10}
Bismuth sulfide	$Bi_2S_3(s) \rightleftharpoons 2Bi^{3+}(aq) + 3S^{2-}(aq)$	1.6×10^{-72}
Cadmium sulfide	$CdS(s) \rightleftharpoons Cd^{2+}(aq) + S^{2-}(aq)$	8.0×10^{-28}
Calcium carbonate	$CaCO_3(s) \rightleftharpoons Ca^{2+}(aq) + CO_3^{2-}(aq)$	8.7×10^{-9}
Calcium fluoride	$CaF_2(s) \rightleftharpoons Ca^{2+}(aq) + 2F^-(aq)$	4.0×10^{-11}
Calcium hydroxide	$Ca(OH)_2(s) \rightleftharpoons Ca^{2+}(aq) + 2OH^-(aq)$	8.0×10^{-6}
Calcium phosphate	$Ca_3(PO_4)_2(s) \rightleftharpoons 3Ca^{2+}(aq) + 2PO_4^{3-}(aq)$	1.2×10^{-26}
Calcium sulfate	$CaSO_4(s) \rightleftharpoons Ca^{2+}(aq) + SO_4^{2-}(aq)$	2.4×10^{-5}
Chromium(III) hydroxide	$Cr(OH)_3(s) \rightleftharpoons Cr^{3+}(aq) + 3OH^-(aq)$	3.0×10^{-29}
Cobalt(II) sulfide	$CoS(s) \rightleftharpoons Co^{2+}(aq) + S^{2-}(aq)$	4.0×10^{-21}
Copper(I) bromide	$CuBr(s) \rightleftharpoons Cu^+(aq) + Br^-(aq)$	4.2×10^{-8}
Copper(I) iodide	$CuI(s) \rightleftharpoons Cu^+(aq) + I^-(aq)$	5.1×10^{-12}
Copper(II) hydroxide	$Cu(OH)_2(s) \rightleftharpoons Cu^{2+}(aq) + 2OH^-(aq)$	2.2×10^{-20}
Copper(II) sulfide	$CuS(s) \rightleftharpoons Cu^{2+}(aq) + S^{2-}(aq)$	6.0×10^{-37}
Iron(II) hydroxide	$Fe(OH)_2(s) \rightleftharpoons Fe^{2+}(aq) + 2OH^-(aq)$	1.6×10^{-14}
Iron(III) hydroxide	$Fe(OH)_3(s) \rightleftharpoons Fe^{3+}(aq) + 3OH^-(aq)$	1.1×10^{-36}
Iron(III) phosphate	$FePO_4(s) \rightleftharpoons Fe^{3+}(aq) + PO_4^{3-}(aq)$	1.3×10^{-22}
Iron(II) sulfide	$FeS(s) \rightleftharpoons Fe^{2+}(aq) + S^{2-}(aq)$	6.0×10^{-19}
Lead(II) bromide	$PbBr_2(s) \rightleftharpoons Pb^{2+}(aq) + 2Br^-(aq)$	6.6×10^{-6}
Lead(II) carbonate	$PbCO_3(s) \rightleftharpoons Pb^{2+}(aq) + CO_3^{2-}(aq)$	3.3×10^{-14}
Lead(II) chloride	$PbCl_2(s) \rightleftharpoons Pb^{2+}(aq) + 2Cl^-(aq)$	2.4×10^{-4}
Lead(II) chromate	$PbCrO_4(s) \rightleftharpoons Pb^{2+}(aq) + CrO_4^{2-}(aq)$	2.0×10^{-14}
Lead(II) fluoride	$PbF_2(s) \rightleftharpoons Pb^{2+}(aq) + 2F^-(aq)$	4.0×10^{-8}
Lead(II) iodide	$PbI_2(s) \rightleftharpoons Pb^{2+}(aq) + 2I^-(aq)$	1.4×10^{-8}
Lead(II) sulfate	$PbSO_4(s) \rightleftharpoons Pb^{2+}(aq) + SO_4^{2-}(aq)$	1.8×10^{-8}
Lead(II) sulfide	$PbS(s) \rightleftharpoons Pb^{2+}(aq) + S^{2-}(aq)$	3.4×10^{-28}
Magnesium carbonate	$MgCO_3(s) \rightleftharpoons Mg^{2+}(aq) + CO_3^{2-}(aq)$	4.0×10^{-5}
Magnesium hydroxide	$Mg(OH)_2(s) \rightleftharpoons Mg^{2+}(aq) + 2OH^-(aq)$	1.2×10^{-11}
Manganese(II) sulfide	$MnS(s) \rightleftharpoons Mn^{2+}(aq) + S^{2-}(aq)$	3.0×10^{-14}
Mercury(I) bromide	$Hg_2Br_2(s) \rightleftharpoons Hg_2^{2+}(aq) + 2Br^-(aq)$	6.4×10^{-23}
Mercury(I) chloride	$Hg_2Cl_2(s) \rightleftharpoons Hg_2^{2+}(aq) + 2Cl^-(aq)$	3.5×10^{-18}
Mercury(I) sulfate	$Hg_2SO_4(s) \rightleftharpoons Hg_2^{2+}(aq) + SO_4^{2-}(aq)$	6.5×10^{-7}
Mercury(II) sulfide	$HgS(s) \rightleftharpoons Hg^{2+}(aq) + S^{2-}(aq)$	4.0×10^{-54}
Nickel(II) sulfide	$NiS(s) \rightleftharpoons Ni^{2+}(aq) + S^{2-}(aq)$	1.4×10^{-24}
Silver bromide	$AgBr(s) \rightleftharpoons Ag^+(aq) + Br^-(aq)$	7.7×10^{-13}
Silver carbonate	$Ag_2CO_3(s) \rightleftharpoons 2Ag^+(aq) + CO_3^{2-}(aq)$	8.1×10^{-12}
Silver chloride	$AgCl(s) \rightleftharpoons Ag^+(aq) + Cl^-(aq)$	1.6×10^{-10}
Silver chromate	$Ag_2CrO_4(s) \rightleftharpoons 2Ag^+(aq) + CrO_4^{2-}(aq)$	1.2×10^{-12}
Silver iodide	$AgI(s) \rightleftharpoons Ag^+(aq) + I^-(aq)$	8.3×10^{-17}
Silver sulfate	$Ag_2SO_4(s) \rightleftharpoons 2Ag^+(aq) + SO_4^{2-}(aq)$	1.5×10^{-5}
Silver sulfide	$Ag_2S(s) \rightleftharpoons 2Ag^+(aq) + S^{2-}(aq)$	6.0×10^{-51}
Strontium carbonate	$SrCO_3(s) \rightleftharpoons Sr^{2+}(aq) + CO_3^{2-}(aq)$	1.6×10^{-9}
Strontium hydroxide	$Sr(OH)_2(s) \rightleftharpoons Sr^{2+}(aq) + 2OH^-(aq)$	3.2×10^{-4}
Strontium sulfate	$SrSO_4(s) \rightleftharpoons Sr^{2+}(aq) + SO_4^{2-}(aq)$	3.8×10^{-7}
Tin(II) sulfide	$SnS(s) \rightleftharpoons Sn^{2+}(aq) + S^{2-}(aq)$	1.0×10^{-26}
Zinc hydroxide	$Zn(OH)_2(s) \rightleftharpoons Zn^{2+}(aq) + 2OH^-(aq)$	1.8×10^{-14}
Zinc sulfide	$ZnS(s) \rightleftharpoons Zn^{2+}(aq) + S^{2-}(aq)$	3.0×10^{-23}

The equilibrium expression is

$$K_{sp} = [Ag^+][Br^-]$$

Therefore,

$$7.7 \times 10^{-13} = (s)(s)$$

and

$$s = \sqrt{7.7 \times 10^{-13}} = 8.8 \times 10^{-7} \, M$$

Thus, the molar solubility of AgBr is $8.8 \times 10^{-7} \, M$. Furthermore, we can express this solubility in g/L by multiplying the molar solubility by the molar mass of AgBr.

$$\frac{8.8 \times 10^{-7} \, \text{mol AgBr}}{1 \, \text{L}} \times \frac{187.8 \, \text{g}}{1 \, \text{mol AgBr}} = 1.7 \times 10^{-4} \, \text{g/L}$$

Worked Example 17.7 demonstrates this approach.

Worked Example 17.7

Calculate the solubility of copper(II) hydroxide [$Cu(OH)_2$] in g/L.

Strategy Write the dissociation equation for $Cu(OH)_2$, and look up its K_{sp} value in Table 17.4. Solve for molar solubility using the equilibrium expression. Convert molar solubility to solubility in g/L using the molar mass of $Cu(OH)_2$.

Setup The equation for the dissociation of $Cu(OH)_2$ is

$$Cu(OH)_2(s) \rightleftharpoons Cu^{2+}(aq) + 2OH^-(aq)$$

and the equilibrium expression is $K_{sp} = [Cu^{2+}][OH^-]^2$. According to Table 17.4, K_{sp} for $Cu(OH)_2$ is 2.2×10^{-20}. The molar mass of $Cu(OH)_2$ is 97.57 g/mol.

Solution

$$Cu(OH)_2(s) \rightleftharpoons Cu^{2+}(aq) + 2OH^-(aq)$$

	Cu^{2+}	OH^-
Initial concentration (M):	0	0
Change in concentration (M):	$+s$	$+2s$
Equilibrium concentration (M):	s	$2s$

Student Annotation: The stoichiometry of the balanced dissociation equation indicates that the concentration of OH^- increases by twice as much as that of Cu^{2+}.

Therefore,

$$2.2 \times 10^{-20} = (s)(2s)^2 = 4s^3$$

$$s = \sqrt[3]{\frac{2.2 \times 10^{-20}}{4}} = 1.8 \times 10^{-7} \, M$$

The molar solubility of $Cu(OH)_2$ is $1.8 \times 10^{-7} \, M$. Multiplying by its molar mass gives

$$\text{solubility of } Cu(OH)_2 = \frac{1.8 \times 10^{-7} \, \text{mol } Cu(OH)_2}{1 \, \text{L}} \times \frac{97.57 \, \text{g } Cu(OH)_2}{1 \, \text{mol } Cu(OH)_2}$$

$$= 1.7 \times 10^{-5} \, \text{g/L}$$

Think About It
Common errors arise in this type of problem when students neglect to raise an entire term to the appropriate power. For example, $(2s)^2$ is equal to $4s^2$ (not $2s^2$).

Practice Problem (A)TTEMPT Calculate the molar solubility and the solubility in g/L of each salt at 25°C: (a) AgCl, (b) SnS, (c) $SrCO_3$.

Practice Problem (B)UILD Calculate the molar solubility and the solubility in g/L of each salt at 25°C: (a) PbF_2, (b) Ag_2CO_3, (c) Bi_2S_3.

Practice Problem (C)ONCEPTUALIZE The diagrams [(i)–(iii)] represent solutions saturated with three different sparingly soluble ionic compounds. Which compound has the greatest molar solubility?

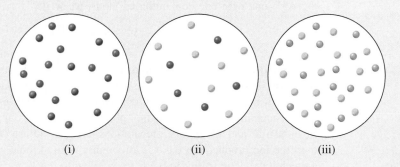

(i)　　　　　　　　(ii)　　　　　　　　(iii)

We can also use molar solubility to determine the value of K_{sp}. Worked Example 17.8 illustrates this procedure.

Worked Example 17.8

The solubility of calcium sulfate ($CaSO_4$) is measured experimentally and found to be 0.67 g/L. Calculate the value of K_{sp} for calcium sulfate.

Strategy Convert solubility to molar solubility using the molar mass of $CaSO_4$, and substitute the molar solubility into the equilibrium expression to determine K_{sp}.

Setup The molar mass of $CaSO_4$ is 136.2 g/mol. The molar solubility of $CaSO_4$ is

$$\text{molar solubility of } CaSO_4 = \frac{0.67 \text{ g } CaSO_4}{1 \text{ L}} \times \frac{1 \text{ mol } CaSO_4}{136.2 \text{ g } CaSO_4}$$

$$s = 4.9 \times 10^{-3} \text{ mol/L}$$

The equation and the equilibrium expression for the dissociation of $CaSO_4$ are

$$CaSO_4(s) \rightleftharpoons Ca^{2+}(aq) + SO_4^{2-}(aq) \quad \text{and} \quad K_{sp} = [Ca^{2+}][SO_4^{2-}]$$

Solution Substituting the molar solubility into the equilibrium expression gives

$$K_{sp} = (s)(s) = (4.9 \times 10^{-3})^2 = 2.4 \times 10^{-5}$$

Think About It
The K_{sp} for $CaSO_4$ is relatively large (compared to many of the K_{sp} values in Table 17.4). In fact, sulfates are listed as soluble compounds in Table 9.2, but calcium sulfate is listed as an insoluble exception. Remember that the term *insoluble* really refers to compounds that are *slightly* soluble, and that different sources may differ with regard to how soluble a compound must be to be considered "soluble."

Practice Problem (A)TTEMPT Given the solubility, calculate the solubility product constant (K_{sp}) of each salt at 25°C: (a) $PbCrO_4$, $s = 4.0 \times 10^{-5}$ g/L; (b) BaC_2O_4, $s = 0.29$ g/L; (c) $MnCO_3$, $s = 4.2 \times 10^{-6}$ g/L.

Practice Problem (B)UILD Given the solubility, calculate the solubility product constant (K_{sp}) of each salt at 25°C: (a) Ag_2SO_3, $s = 4.6 \times 10^{-3}$ g/L; (b) Hg_2I_2, $s = 1.5 \times 10^{-7}$ g/L; (c) $Zn_3(PO_4)_2$, $s = 5.9 \times 10^{-5}$ g/L.

Practice Problem (C)ONCEPTUALIZE Which compound in Practice Problem 17.7C has the largest K_{sp}?

Predicting Precipitation Reactions

For the dissociation of an ionic solid in water, any one of the following conditions may exist: (1) the solution is unsaturated, (2) the solution is saturated, or (3) the solution is supersaturated. For concentrations of ions that do not correspond to equilibrium conditions, we use the reaction quotient (Q) [◂◂ Section 15.2] to predict when a precipitate will form. Note that Q has the same form as K_{sp} except that the concentrations of ions are not equilibrium concentrations. For example, if we mix a solution containing Ag^+ ions with one containing Cl^- ions, we write

$$Q = [Ag^+]_i[Cl^-]_i$$

The subscript "i" denotes that these are *initial* concentrations and do not necessarily correspond to those at equilibrium. If Q is less than or equal to K_{sp}, no precipitate will form. If Q is greater than K_{sp}, AgCl will precipitate out. (Precipitation will continue until the product of the ion concentrations is equal to K_{sp}.)

The ability to predict whether precipitation will occur often has practical value. In industrial and laboratory preparations, we can adjust the concentrations of ions until the ion product exceeds K_{sp} to obtain a desired ionic compound (in the form of a precipitate). The ability to predict precipitation reactions is also useful in medicine. Kidney stones, which can be extremely painful, consist largely of calcium oxalate (CaC_2O_4, $K_{sp} = 2.3 \times 10^{-9}$). The normal physiological concentration of calcium ions in blood plasma is about 5×10^{-3} M. Oxalate ions ($C_2O_4^{2-}$), derived from oxalic acid present in vegetables such as rhubarb and spinach, react with the calcium ions to form insoluble calcium oxalate, which can gradually build up in the kidneys. Proper adjustment of a patient's diet can help to reduce precipitate formation.

Worked Example 17.9 demonstrates the steps involved in predicting precipitation reactions.

Worked Example 17.9

Predict whether a precipitate will form when each of the following is added to 650 mL of 0.0080 M K_2SO_4: (a) 250 mL of 0.0040 M $BaCl_2$; (b) 175 mL of 0.15 M $AgNO_3$; (c) 325 mL of 0.25 M $Sr(NO_3)_2$. (Assume volumes are additive.)

Strategy For each part, identify the compound that might precipitate and look up its K_{sp} value in Table 17.4 or Appendix 4. Determine the concentrations of each compound's constituent ions, and use them to determine the value of the reaction quotient, Q_{sp}; then compare each reaction quotient with the value of the corresponding K_{sp}. If the reaction quotient is greater than K_{sp}, a precipitate will form.

Setup The compounds that might precipitate and their K_{sp} values are (a) $BaSO_4$, $K_{sp} = 1.1 \times 10^{-10}$; (b) Ag_2SO_4, $K_{sp} = 1.5 \times 10^{-5}$; (c) $SrSO_4$, $K_{sp} = 3.8 \times 10^{-7}$.

Solution (a) Concentrations of the constituent ions of $BaSO_4$ are:

$$[Ba^{2+}] = \frac{250 \text{ ml} \times 0.0040 \ M}{650 \text{ mL} + 250 \text{ mL}} = 0.0011 \ M \quad \text{and} \quad [SO_4^{2-}] = \frac{650 \text{ ml} \times 0.0080 \ M}{650 \text{ mL} + 250 \text{ mL}} = 0.0058 \ M$$

Using these concentrations in the equilibrium expression, $[Ba^{2+}][SO_4^{2-}]$, gives a reaction quotient of $(0.0011)(0.0058) = 6.4 \times 10^{-6}$, which is greater than the K_{sp} of $BaSO_4$ (1.1×10^{-10}). Therefore, $BaSO_4$ will precipitate.

(b) Concentrations of the constituent ions of Ag_2SO_4 are

$$[Ag^+] = \frac{175 \text{ mL} \times 0.15 \ M}{650 \text{ mL} + 175 \text{ mL}} = 0.032 \ M \quad \text{and} \quad [SO_4^{2-}] = \frac{650 \text{ mL} \times 0.0080 \ M}{650 \text{ mL} + 175 \text{ mL}} = 0.0063 \ M$$

Using these concentrations in the equilibrium expression, $[Ag^+]^2[SO_4^{2-}]$, gives a reaction quotient of $(0.032)^2(0.0063) = 6.5 \times 10^{-6}$, which is less than the K_{sp} of Ag_2SO_4 (1.5×10^{-5}). Therefore, Ag_2SO_4 will not precipitate.

(c) Concentrations of the constituent ions of $SrSO_4$ are

$$[Sr^{2+}] = \frac{325 \text{ mL} \times 0.25 \ M}{650 \text{ mL} + 325 \text{ mL}} = 0.083 \ M \quad \text{and} \quad [SO_4^{2-}] = \frac{650 \text{ mL} \times 0.0080 \ M}{650 \text{ mL} + 325 \text{ mL}} = 0.0053 \ M$$

Using these concentrations in the equilibrium expression, $[Sr^{2+}][SO_4^{2-}]$, gives a reaction quotient of $(0.083)(0.0053) = 4.4 \times 10^{-4}$, which is greater than the K_{sp} of $SrSO_4$ (3.8×10^{-7}). Therefore, $SrSO_4$ will precipitate.

Think About It

Students sometimes have difficulty deciding what compound might precipitate. Begin by writing down the constituent ions in the two solutions before they are combined. Consider the two possible combinations: the cation from the first solution and the anion from the second, or vice versa. You can consult the information in Tables 9.2 and 9.3 to determine whether one of the combinations is insoluble. Also keep in mind that only an *insoluble* salt will have a tabulated K_{sp} value.

Practice Problem A TTEMPT Predict whether a precipitate will form from each of the following combinations: (a) 25 mL of 1×10^{-5} M Co(NO$_3$)$_2$ and 75 mL of 5×10^{-4} M Na$_2$S; (b) 500 mL of 7.5×10^{-4} M AlCl$_3$ and 100 mL of 1.7×10^{-5} M Hg$_2$(NO$_3$)$_2$; (c) 1.5 L of 0.025 M BaCl$_2$ and 1.25 L of 0.014 M Pb(NO$_3$)$_2$.

Practice Problem B UILD What is the maximum mass (in grams) of each of the following soluble salts that can be added to 150 mL of 0.050 M BaCl$_2$ without causing a precipitate to form: (a) (NH$_4$)$_2$SO$_4$, (b) Pb(NO$_3$)$_2$, (c) NaF? (Assume that the addition of solid causes no change in volume.)

Practice Problem C ONCEPTUALIZE The two sample diagrams represent a saturated solution of the slightly soluble salt MA and a solution of the soluble salt NH$_4$A, respectively. Which of the solutions [(i)–(iv)] of the soluble salt MNO$_3$ can be added to the solution of NH$_4$A without causing a precipitate to form? Assume that the volumes of all solutions are equal, and that they are additive when combined. (For clarity, the water molecules, ammonium ions, and nitrate ions are not shown.)

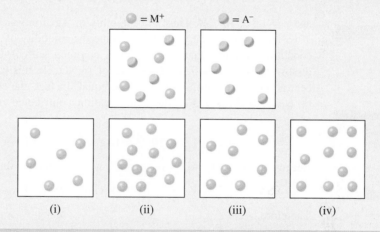

(i)   (ii)   (iii)   (iv)

Section 17.4 Review

Solubility Equilibria

17.4.1 Using the K_{sp} for aluminum hydroxide [Al(OH)$_3$], calculate its molar solubility.
(a) 2.9×10^{-9} M
(d) 8.4×10^{-12} M
(b) 7.7×10^{-12} M
(e) 3.8×10^{-9} M
(c) 4.2×10^{-17} M

17.4.2 What precipitate will form if 0.10-M solutions of Pb(NO$_3$)$_2$ and NaI are mixed?
(a) Pb(NO$_3$)$_2$
(d) NaNO$_3$
(b) NaI
(e) None
(c) PbI$_2$

17.4.3 Diagrams (i) to (iii) represent saturated solutions of MX, MY, and MZ, each in equilibrium with its solid. Each solution also contains the soluble nitrate salt of M^{n+}. Arrange the solutions in order of increasing K_{sp}. (For clarity, the water molecules and nitrate ions are not shown.)
(a) i < iii < ii
(d) iii < ii = i
(b) ii < iii < i
(e) ii < i < iii
(c) i = ii < iii

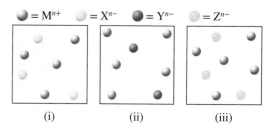

(i)   (ii)   (iii)

17.5   FACTORS AFFECTING SOLUBILITY

In this section we will examine the effect of several factors on solubility including the common ion effect, pH, and the formation of complex ions.

The Common Ion Effect

The solubility product is an equilibrium constant, and precipitation of an ionic compound from solution occurs whenever the ion product exceeds the K_{sp} for that substance. In a saturated solution of AgCl, for example, the ion product $[Ag^+][Cl^-]$ is equal to K_{sp}. The solubility of AgCl in water can be calculated as follows, using the procedure introduced in Section 17.4.

$$1.6 \times 10^{-10} = [Ag^+][Cl^-]$$

In a solution in which AgCl is the only solute, $[Ag^+] = [Cl^-]$. Therefore,

$$1.6 \times 10^{-10} = s^2$$

and $s = 1.3 \times 10^{-5}$ M. Thus, the solubility of AgCl in water at 25°C is 1.3×10^{-5} M.

Now suppose we want to determine the solubility of AgCl in a solution already containing a solute that has an ion in common with AgCl. For example, consider dissolving AgCl in a 0.10-M solution of $AgNO_3$. In this case, the Ag^+ and Cl^- concentrations at equilibrium will not be equal. In fact, the Ag^+ ion concentration will be equal to 0.10 M *plus* the concentration contributed by AgCl. The equilibrium expression is

$$1.6 \times 10^{-10} = [Ag^+][Cl^-] = (0.10 + s)(s)$$

Because we expect s to be very small, we can simplify this calculation as follows:

$$(0.10 + s)\ M \approx 0.10\ M$$

Therefore,

$$1.6 \times 10^{-10} = 0.10s$$

and $s = 1.6 \times 10^{-9}$ M. Thus, AgCl is *significantly* less soluble in 0.10 M $AgNO_3$ than in pure water—due to the common ion effect. Figure 17.8 (pages 804–805) illustrates the common ion effect.

Worked Example 17.10 shows how the common ion effect affects solubility.

Student Annotation: s still represents the concentration of Cl^- at equilibrium and the *solubility* of AgCl.

Student Annotation: The common ion effect is an example of Le Châtelier's principle [Section 15.6].

Video 17.4
Common ion effect.

Worked Example 17.10

Calculate the molar solubility of silver chloride in a solution that is 6.5×10^{-3} M in silver nitrate.

Strategy Silver nitrate is a strong electrolyte that dissociates completely in water. Therefore, the concentration of Ag^+ before any AgCl dissolves is 6.5×10^{-3} M. Use the equilibrium expression, the K_{sp} for AgCl, and an equilibrium table to determine how much AgCl will dissolve.

Setup The dissolution equilibrium and the equilibrium expression are

$$AgCl(s) \rightleftharpoons Ag^+(aq) + Cl^-(aq) \qquad 1.6 \times 10^{-10} = [Ag^+][Cl^-]$$

Solution

	AgCl(s) $\rightleftharpoons$	$Ag^+(aq)$ +	$Cl^-(aq)$
Initial concentration (M):		6.5×10^{-3}	0
Change in concentration (M):		$+s$	$+s$
Equilibrium concentration (M):		$6.5 \times 10^{-3} + s$	s

Substituting these concentrations into the equilibrium expression gives

$$1.6 \times 10^{-10} = (6.5 \times 10^{-3} + s)(s)$$

We expect s to be very small, so

$$6.5 \times 10^{-3} + s \approx 6.5 \times 10^{-3}$$

and

$$1.6 \times 10^{-10} = (6.5 \times 10^{-3})(s)$$

Thus

$$s = \frac{1.6 \times 10^{-10}}{6.5 \times 10^{-3}} = 2.5 \times 10^{-8} \, M$$

Therefore, the molar solubility of AgCl in 6.5×10^{-3} M AgNO$_3$ is 2.5×10^{-8} M.

> **Think About It**
> The molar solubility of AgCl in water is $\sqrt{1.6 \times 10^{-10}} = 1.3 \times 10^{-5}$ M. The presence of 6.5×10^{-3} M AgNO$_3$ reduces the solubility of AgCl by a factor of ~500.

Practice Problem Ⓐ**TTEMPT** Calculate the molar solubility of AgI in (a) pure water and (b) 0.0010 M NaI.

Practice Problem Ⓑ**UILD** Arrange the following salts in order of increasing molar solubility in 0.0010 M AgNO$_3$: AgBr, Ag$_2$CO$_3$, AgCl, AgI, Ag$_2$S.

Practice Problem Ⓒ**ONCEPTUALIZE** The diagram on the left shows a saturated solution of a slightly soluble ionic compound. In the diagram on the right, enough of the nitrate salt of the cation has been added to increase the concentration of cations (yellow). How many anions (blue) must be included in the second diagram for it to correctly represent the solution after the addition?

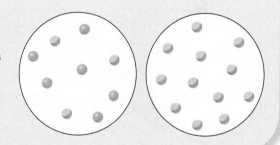

pH

The solubility of a substance can also depend on the pH of the solution. Consider the solubility equilibrium of magnesium hydroxide:

$$\text{Mg(OH)}_2(s) \rightleftharpoons \text{Mg}^{2+}(aq) + 2\text{OH}^-(aq)$$

According to Le Châtelier's principle, adding OH$^-$ ions (increasing the pH) shifts the equilibrium from right to left, thereby decreasing the solubility of Mg(OH)$_2$. (This is actually another example of the common ion effect.) On the other hand, adding H$^+$ ions (decreasing the pH) shifts the equilibrium from left to right, and the solubility of Mg(OH)$_2$ *increases*. Thus, insoluble bases tend to dissolve in acidic solutions. Similarly, insoluble acids tend to dissolve in basic solutions.

To examine the effect of pH on the solubility of Mg(OH)$_2$, we first calculate the pH of a saturated Mg(OH)$_2$ solution:

$$K_{sp} = (s)(2s)^2 = 4s^3$$

$$4s^3 = 1.2 \times 10^{-11}$$

$$s^3 = 3.0 \times 10^{-12}$$

$$s = 1.4 \times 10^{-4} \, M$$

At equilibrium, therefore,

$$[\text{OH}^-] = 2(1.4 \times 10^{-4} \, M) = 2.8 \times 10^{-4} \, M$$

$$\text{pOH} = -\log(2.8 \times 10^{-4}) = 3.55$$

$$\text{pH} = 14.00 - 3.55 = 10.45$$

Figure 17.8

Common Ion Effect

We prepare a saturated solution by adding AgCl to water and stirring.

In the resulting saturated solution, the concentrations of Ag^+ and Cl^- are equal, and the product of their concentrations is equal to K_{sp}.

$$[Ag^+][Cl^-] = 1.6 \times 10^{-10}$$

Therefore, the concentrations are

$$[Ag^+] = 1.3 \times 10^{-5}\ M$$
$$\text{and } [Cl^-] = 1.3 \times 10^{-5}\ M$$

Neither AgCl nor a saturated solution of AgCl is purple. The color has been used in this illustration to clarify the process.

START

After filtering off the solid AgCl, we dissolve enough NaCl to make the concentration of Cl⁻ = 1.0 M.

Because the concentration of Cl⁻ is now larger, the product of Ag⁺ and Cl⁻ concentrations is no longer equal to K_{sp}.

$$[Ag^+][Cl^-] = (1.3 \times 10^{-5}\ M)(1.0\ M) > 1.6 \times 10^{-10}$$

In any solution saturated with AgCl at 25°C, the product of $[Ag^+]$ and $[Cl^-]$ must equal the K_{sp} of AgCl. Therefore, AgCl will precipitate until the product of ion concentrations is again 1.6×10^{-10}.

Note that this causes nearly all the dissolved AgCl to precipitate. With a Cl⁻ concentration of 1.0 M, the highest possible concentration of Ag⁺ is $1.6 \times 10^{-10}\ M$.

$$[Ag^+](1.0\ M) = 1.6 \times 10^{-10}$$

therefore, $[Ag^+] = 1.6 \times 10^{-10}\ M$

The amount of AgCl precipitated is exaggerated for emphasis. The actual amount of AgCl would be extremely small.

What's the point?

When two salts contain the same ion, the ion they both contain is called the "common ion." The solubility of a slightly soluble salt such as AgCl can be decreased by the addition of a *soluble* salt with a common ion. In this example, AgCl is precipitated by adding NaCl. AgCl could also be precipitated by adding a soluble salt containing the Ag⁺ ion, such as $AgNO_3$.

In a solution with a pH of *less* than 10.45, the solubility of $Mg(OH)_2$ would increase. The dissolution process and the effect of additional H^+ ions are summarized as follows:

$$Mg(OH)_2(s) \rightleftharpoons Mg^{2+}(aq) + 2\cancel{OH^-(aq)}$$

$$2H^+(aq) + 2\cancel{OH^-(aq)} \longrightarrow 2H_2O(l)$$

$$Overall: \quad Mg(OH)_2(s) + 2H^+(aq) \rightleftharpoons Mg^{2+}(aq) + 2H_2O(l)$$

If the pH of the medium were higher than 10.45, $[OH^-]$ would be higher and the solubility of $Mg(OH)_2$ would decrease because of the common ion (OH^-) effect.

The pH also influences the solubility of salts that contain a basic anion. For example, the solubility equilibrium for BaF_2 is

$$BaF_2(s) \rightleftharpoons Ba^{2+}(aq) + 2F^-(aq)$$

and

$$K_{sp} = [Ba^{2+}][F^-]^2$$

In an acidic medium, the high $[H^+]$ will shift the following equilibrium to the left, consuming F^-.

$$HF(aq) \rightleftharpoons H^+(aq) + F^-(aq)$$

As the concentration of F^- decreases, the concentration of Ba^{2+} must increase to satisfy the equality $K_{sp} = [Ba^{2+}][F^-]^2$ and maintain the state of equilibrium. Thus, more BaF_2 dissolves. The process and the effect of pH on the solubility of BaF_2 can be summarized as follows:

$$BaF_2(s) \rightleftharpoons Ba^{2+}(aq) + 2\cancel{F^-(aq)}$$

$$2H^+(aq) + 2\cancel{F^-(aq)} \longrightarrow 2HF(aq)$$

$$Overall: \quad BaF_2(s) + 2H^+(aq) \rightleftharpoons Ba^{2+}(aq) + 2HF(aq)$$

The solubilities of salts containing anions that do *not* hydrolyze, such as Cl^-, Br^-, and NO_3^-, are unaffected by pH.

Worked Example 17.11 demonstrates the effect of pH on solubility.

Worked Example 17.11

Which of the following compounds will be more soluble in acidic solution than in water: (a) CuS, (b) AgCl, (c) $PbSO_4$?

Strategy For each salt, write the dissociation equilibrium equation and determine whether it produces an anion that will react with H^+. Only an anion that is the conjugate base of a weak acid will react with H^+.

Setup

(a) $CuS(s) \rightleftharpoons Cu^{2+}(aq) + S^{2-}(aq)$

   S^{2-} is the conjugate base of the weak acid HS^-. S^{2-} reacts with H^+ as follows:

$$S^{2-}(aq) + H^+(aq) \longrightarrow HS^-(aq)$$

(b) $AgCl(s) \rightleftharpoons Ag^+(aq) + Cl^-(aq)$

   Cl^- is the conjugate base of the strong acid HCl. Cl^- does not react with H^+.

(c) $PbSO_4(s) \rightleftharpoons Pb^{2+}(aq) + SO_4^{2-}(aq)$

   SO_4^{2-} is the conjugate base of the weak acid HSO_4^-. It reacts with H^+ as follows:

$$SO_4^{2-}(aq) + H^+(aq) \longrightarrow HSO_4^-(aq)$$

A salt that produces an anion that reacts with H^+ will be more soluble in acid than in water.

Solution CuS and $PbSO_4$ are more soluble in acid than in water. (AgCl is no more or less soluble in acid than in water.)

Think About It

When a salt dissociates to give the conjugate base of a weak acid, H^+ ions in an acidic solution *consume* a product (base) of the dissolution. This drives the equilibrium to the right (more solid dissolves) according to Le Châtelier's principle.

Practice Problem A TTEMPT Determine if the following compounds are more soluble in acidic solution than in pure water: (a) $Ca(OH)_2$, (b) $Mg_3(PO_4)_2$, (c) $PbBr_2$.

Practice Problem B UILD Other than those in Worked Example 17.11 and those in Practice Problem A, list three salts that are more soluble in acidic solution than in pure water.

Practice Problem C ONCEPTUALIZE If an ionic compound's solubility is affected by the presence of acid in solution, will its solubility necessarily also be affected by the presence of base? Explain.

Complex Ion Formation

A *complex ion* is an ion containing a central metal cation bonded to one or more molecules or ions. Complex ions are crucial to many chemical and biological processes. Here we will consider the effect of complex ion formation on solubility. In Chapter 22 we will discuss the chemistry of complex ions in more detail.

    Transition metals have a particular tendency to form complex ions. For example, a solution of cobalt(II) chloride ($CoCl_2$) is pink because of the presence of the $Co(H_2O)_6^{2+}$ ions (Figure 17.9). When HCl is added, the solution turns blue because the complex ion $CoCl_4^{2-}$ forms.

$$Co^{2+}(aq) + 4Cl^-(aq) \rightleftharpoons CoCl_4^{2-}(aq)$$

Copper(II) sulfate ($CuSO_4$) dissolves in water to produce a blue solution. The hydrated copper(II) ions are responsible for this color; many other sulfates (e.g., Na_2SO_4) are colorless. Adding a few drops of concentrated ammonia solution to a $CuSO_4$ solution causes the formation of a light-blue precipitate, copper(II) hydroxide.

$$Cu^{2+}(aq) + 2OH^-(aq) \longrightarrow Cu(OH)_2(s)$$

Student Annotation: Lewis acid-base reactions in which a metal cation combines with a Lewis base result in the formation of complex ions.

Figure 17.9 (Left) An aqueous cobalt(II) chloride solution. The pink color is due to the presence of $Co(H_2O)_6^{2+}$ ions. (Right) After the addition of HCl solution, the solution turns blue because of the formation of the complex $CoCl_4^{2-}$ ions.

© McGraw-Hill Education/Ken Karp, photographer

Thinking Outside the Box

© McGraw-Hill Education

Equilibrium and Tooth Decay

Teeth are protected by a hard enamel layer about 2 mm thick that is composed of a mineral called *hydroxyapatite* [$Ca_5(PO_4)_3OH$]. *Demineralization* is the process by which hydroxyapatite dissolves in the saliva. Because phosphates of alkaline earth metals such as calcium are insoluble [◄◄ Section 9.2, Table 9.3], this process happens only to a very small extent.

$$Ca_5(PO_4)_3OH(s) \xrightleftharpoons{} 5Ca^{2+}(aq) + 3PO_4^{3-}(aq) + OH^-(aq)$$

The reverse process, called *remineralization,* is the body's natural defense against tooth decay.

$$5Ca^{2+}(aq) + 3PO_4^{3-}(aq) + OH^-(aq) \xrightleftharpoons{} Ca_5(PO_4)_3OH(s)$$

Bacteria in the mouth break down some of the food we eat to produce organic acids such as acetic acid and lactic acid. Acid production is greatest from foods that are high in sugar. Thus, after a sugary snack, the H^+ concentration in the mouth increases, causing OH^- ions in the saliva to be consumed.

$$H^+(aq) + OH^-(aq) \longrightarrow H_2O(l)$$

Removal of OH^- from the saliva draws the $Ca_5(PO_4)_3OH$ dissolution equilibrium to the right, promoting demineralization. Once the protective enamel layer is weakened, tooth decay begins. The best way to prevent tooth decay is to eat a diet low in sugar and brush immediately after every meal.

Most toothpastes contain fluoride, which helps to reduce tooth decay. The F^- ions in toothpaste replace some of the OH^- ions during the remineralization process.

$$5Ca^{2+}(aq) + 3PO_4^{3-}(aq) + F^-(aq) \xrightleftharpoons{} Ca_5(PO_4)_3F(s)$$

Because F^- is a weaker base than OH^-, the modified enamel, called *fluoroapatite,* is more resistant to the acid produced by bacteria.

Student Annotation: In children, the growth of the enamel layer (mineralization) occurs faster than demineralization; in adults, demineralization and remineralization occur at roughly equal rates.

The OH^- ions are supplied by the ammonia solution. If more NH_3 is added, the blue precipitate redissolves to produce a beautiful dark-blue solution, this time due to the formation of the complex ion $Cu(NH_3)_4^{2+}$ (Figure 17.10).

$$Cu(OH)_2(s) + 4NH_3(aq) \xrightleftharpoons{} Cu(NH_3)_4^{2+}(aq) + 2OH^-(aq)$$

Thus, the formation of the complex ion $Cu(NH_3)_4^{2+}$ increases the solubility of $Cu(OH)_2$.

A measure of the tendency of a metal ion to form a particular complex ion is given by the ***formation constant*** (K_f) (also called the stability constant), which is the

Figure 17.10 (Left) An aqueous solution of copper(II) sulfate. (Center) After the addition of a few drops of concentrated aqueous ammonia solution, a light-blue precipitate of $Cu(OH)_2$ is formed. (Right) When more concentrated aqueous ammonia solution is added, the $Cu(OH)_2$ precipitate dissolves to form the dark-blue complex ion $Cu(NH_3)_4^{2+}$.

© McGraw-Hill Education/Ken Karp, photographer

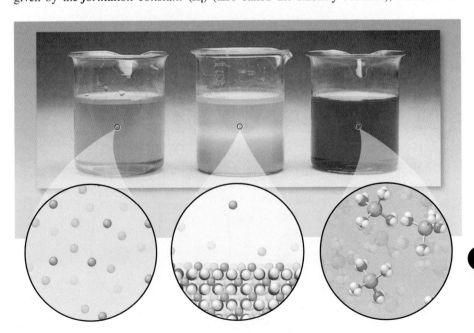

TABLE 17.5	Formation Constants of Selected Complex Ions in Water at 25°C	
Complex ion	**Equilibrium expression**	**Formation constant (K_f)**
$Ag(NH_3)_2^+$	$Ag^+ + 2NH_3 \rightleftharpoons Ag(NH_3)_2^+$	1.5×10^7
$Ag(CN)_2^-$	$Ag^+ + 2CN^- \rightleftharpoons Ag(CN)_2^-$	1.0×10^{21}
$Cu(CN)_4^{2-}$	$Cu^{2+} + 4CN^- \rightleftharpoons Cu(CN)_4^{2-}$	1.0×10^{25}
$Cu(NH_3)_4^{2+}$	$Cu^{2+} + 4NH_3 \rightleftharpoons Cu(NH_3)_4^{2+}$	5.0×10^{13}
$Cd(CN)_4^{2-}$	$Cd^{2+} + 4CN^- \rightleftharpoons Cd(CN)_4^{2-}$	7.1×10^{16}
CdI_4^{2-}	$Cd^{2+} + 4I^- \rightleftharpoons CdI_4^{2-}$	2.0×10^6
$HgCl_4^{2-}$	$Hg^{2+} + 4Cl^- \rightleftharpoons HgCl_4^{2-}$	1.7×10^{16}
HgI_4^{2-}	$Hg^{2+} + 4I^- \rightleftharpoons HgI_4^{2-}$	2.0×10^{30}
$Hg(CN)_4^{2-}$	$Hg^{2+} + 4CN^- \rightleftharpoons Hg(CN)_4^{2-}$	2.5×10^{41}
$Co(NH_3)_6^{3+}$	$Co^{3+} + 6NH_3 \rightleftharpoons Co(NH_3)_6^{3+}$	5.0×10^{31}
$Zn(NH_3)_4^{2+}$	$Zn^{2+} + 4NH_3 \rightleftharpoons Zn(NH_3)_4^{2+}$	2.9×10^9
$Cr(OH)_4^-$	$Cr^{3+} + 4OH^- \rightleftharpoons Cr(OH)_4^-$	8×10^{29}

equilibrium constant for the complex ion formation. The larger K_f is, the more stable the complex ion is. Table 17.5 lists the formation constants of a number of complex ions.

The formation of the $Cu(NH_3)_4^{2+}$ ion can be expressed as

$$Cu^{2+}(aq) + 4NH_3(aq) \rightleftharpoons Cu(NH_3)_4^{2+}(aq)$$

The corresponding formation constant is

$$K_f = \frac{[Cu(NH_3)_4^{2+}]}{[Cu^{2+}][NH_3]^4} = 5.0 \times 10^{13}$$

The large value of K_f in this case indicates that the complex ion is very stable in solution and accounts for the very low concentration of copper(II) ions at equilibrium.

Recall that K for the sum of two reactions is the product of the individual K values [◄◄ Section 15.3]. The dissolution of silver chloride is represented by the equation

$$AgCl(s) \rightleftharpoons Ag^+(aq) + Cl^-(aq)$$

The sum of this equation and the one representing the formation of $Ag(NH_3)_2^+$ is

$$AgCl(s) \rightleftharpoons Ag^+(aq) + Cl^-(aq) \qquad K_{sp} = 1.6 \times 10^{-10}$$
$$\underline{Ag^+(aq) + 2NH_3(aq) \rightleftharpoons Ag(NH_3)_2^+(aq) \qquad K_f = 1.5 \times 10^7}$$
$$AgCl(s) + 2NH_3(aq) \rightleftharpoons Ag(NH_3)_2^+(aq) + Cl^-(aq)$$

and the corresponding equilibrium constant is $(1.6 \times 10^{-10})(1.5 \times 10^7) = 2.4 \times 10^{-3}$. This is significantly larger than the K_{sp} value, indicating that much more AgCl will dissolve in the presence of aqueous ammonia than in pure water. In general, the effect of complex ion formation generally is to *increase* the solubility of a substance.

The solution of an equilibrium problem involving complex ion formation is complicated both by the magnitude of K_f and by the stoichiometry of the reaction. Consider the combination of aqueous copper(II) ions and ammonia to form the complex ion $Cu(NH_3)_4^{2+}$.

$$Cu^{2+}(aq) + 4NH_3(aq) \rightleftharpoons Cu(NH_3)_4^{2+}(aq) \qquad K_f = 5.0 \times 10^{14}$$

Let's say we wish to determine the molar concentration of free copper(II) ion in solution when 0.10 mole of $Cu(NO_3)_2$ is dissolved in a liter of 3.0 M NH_3. We cannot solve this with the same approach we used to determine the pH of a weak acid solution. Not

Student Annotation: Formation of a complex ion *consumes* the metal ion produced by the dissociation of a salt, increasing the salt's solubility simply due to Le Châtelier's principle [◄◄ Section 15.6].

Student Annotation: The term *free* is used to refer to a metal ion that is *not* part of a complex ion.

only can we not neglect x, the amount of copper(II) ion consumed in the reaction, but having to raise the ammonia concentration to the fourth power in the equilibrium expression results in an equation that is not easily solved. Another approach is needed.

Because the magnitude of K_f is so large, we begin by assuming that *all* the copper(II) ion is consumed to form the complex ion. Then we consider the equilibrium in terms of the *reverse* reaction; that is, the dissociation of $Cu(NH_3)_4^{2+}$, for which the equilibrium constant is the reciprocal of K_f.

$$Cu(NH_3)_4^{2+}(aq) \rightleftharpoons Cu^{2+}(aq) + 4NH_3(aq) \qquad K = 2.0 \times 10^{-15}$$

Now we construct an equilibrium table and, because this K is so small, we can expect x (the amount of complex ion that dissociates) to be insignificant compared to the concentration of the complex ion and the concentration of ammonia. [Note that the concentration of ammonia, which had been 3.0 M, has been diminished by $4 \times 0.10 \; M$ due to the amount required to complex 0.10 mole of copper(II) ion.]

$$Cu(NH_3)_4^{2+}(aq) \rightleftharpoons Cu^{2+}(aq) + 4NH_3(aq)$$

	$Cu(NH_3)_4^{2+}$	Cu^{2+}	$4NH_3$
Initial concentration (M):	1.10	0	2.6
Change in concentration (M):	$-x$	$+x$	$+4x$
Equilibrium concentration (M):	$0.10 - x$	x	$2.6 + 4x$

We can neglect x with respect to the concentrations of $Cu(NH_3)_4^{2+}$ and NH_3 ($0.10 - x \approx 0.10$ and $2.6 + 4x \approx 2.6$), and the solution becomes

$$\frac{[Cu^{2+}][NH_3]^4}{[Cu(NH_3)_4^{2+}]} = \frac{x(2.6)^4}{0.10} = 2.0 \times 10^{-15}$$

and $x = 4.4 \times 10^{-18} \; M$. Note that because the formation constant is so large, the amount of copper that remains uncomplexed is extremely small. As always, it is a good idea to check the answer by plugging it into the equilibrium expression.

$$\frac{(4.4 \times 10^{-18})[2.6 + 4(4.4 \times 10^{-18})]^4}{0.10 - 4.4 \times 10^{-18}} = 2.0 \times 10^{-15}$$

Worked Example 17.12 lets you practice applying this approach to a complex ion formation equilibrium problem.

Worked Example 17.12

In the presence of aqueous cyanide, cadmium(II) forms the complex ion $Cd(CN)_4^{2-}$. Determine the molar concentration of free (uncomplexed) cadmium(II) ion in solution when 0.20 mole of $Cd(NO_3)_2$ is dissolved in a liter of 2.0 M sodium cyanide (NaCN).

Strategy Because formation constants are typically very large, we begin by assuming that all the Cd^{2+} ion is consumed and converted to complex ion. We then determine how much Cd^{2+} is produced by the subsequent dissociation of the complex ion, a process for which the equilibrium constant is the reciprocal of K_f.

Setup From Table 17.5, the formation constant (K_f) for the complex ion $Cd(CN)_4^{2-}$ is 7.1×10^{16}. The reverse process,

$$Cd(CN)_4^{2-}(aq) \rightleftharpoons Cd^{2+}(aq) + 4CN^-(aq)$$

has an equilibrium constant of $1/K_f = 1.4 \times 10^{-17}$. The equilibrium expression for the dissociation is

$$1.4 \times 10^{-17} = \frac{[Cd^{2+}][CN^-]^4}{[Cd(CN)_4^{2-}]}$$

The formation of complex ion will consume some of the cyanide originally present. Stoichiometry indicates that four CN^- ions are required to react with one Cd^{2+} ion. Therefore, the concentration of CN^- that we enter in the top row of the equilibrium table will be $[2.0 \; M - 4(0.20 \; M)] = 1.2 \; M$.

Solution We construct an equilibrium table,

$$Cd(CN)_4^{2-}(aq) \rightleftharpoons Cd^{2+}(aq) + 4CN^-(aq)$$

	$Cd(CN)_4^{2-}(aq)$	$Cd^{2+}(aq)$	$4CN^-(aq)$
Initial concentration (M):	0.20	0	1.2
Change in concentration (M):	$-x$	$+x$	$+4x$
Equilibrium concentration (M):	$0.20 - x$	x	$1.2 + 4x$

and, because the magnitude of K is so *small,* we can neglect x with respect to the initial concentrations of $Cd(CN)_4^{2-}$ and $CN^-(0.20 - x \approx 0.20$ and $1.2 + 4x \approx 1.2)$, so the solution becomes

$$\frac{[Cd^{2+}][CN^-]^4}{[Cd(CN)_4^{2-}]} = \frac{x(1.2)^4}{0.20} = 1.4 \times 10^{-17}$$

and $x = 1.4 \times 10^{-18}\ M$.

Think About It

When you assume that all the metal ion is consumed and converted to complex ion, it's important to remember that some of the complexing agent (in this case, CN^- ion) is consumed in the process. Don't forget to adjust its concentration accordingly before entering it in the top row of the equilibrium table.

Practice Problem Ⓐ**TTEMPT** In the presence of aqueous ammonia, cobalt(III) forms the complex ion $Co(NH_3)_6^{3+}$. Determine the molar concentration of free cobalt(III) ion in solution when 0.15 mole of $Co(NO_3)_3$ is dissolved in a liter of 2.5-M aqueous ammonia.

Practice Problem Ⓑ**UILD** Use information from Tables 17.4 and 17.5 to determine the molar solubility of chromium(III) hydroxide in a buffered solution with pH = 11.45.

Practice Problem Ⓒ**ONCEPTUALIZE** Beginning with a saturated solution of AgCl, which of the graphs [(i)–(iv)] best represents how the concentrations of free silver and chloride ions change as NH_3 is added to the solution?

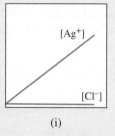
(i)

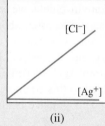
(ii)

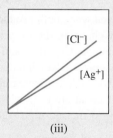
(iii)

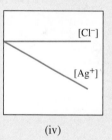
(iv)

Finally, there is a class of hydroxides, called amphoteric hydroxides, which can react with both acids and bases. Examples are $Al(OH)_3$, $Pb(OH)_2$, $Cr(OH)_3$, $Zn(OH)_2$, and $Cd(OH)_2$. $Al(OH)_3$ reacts with acids and bases as follows:

$$Al(OH)_3(s) + 3H^+(aq) \longrightarrow Al^{3+}(aq) + 3H_2O(l)$$

$$Al(OH)_3(s) + OH^-(aq) \rightleftharpoons Al(OH)_4^-(aq)$$

The increase in solubility of $Al(OH)_3$ in a basic medium is the result of the formation of the complex ion $Al(OH)_4^-$ in which $Al(OH)_3$ acts as the Lewis acid and OH^- acts as the Lewis base. Other amphoteric hydroxides react similarly with acids and bases.

Section 17.5 Review

Factors Affecting Solubility

17.5.1 Calculate the molar solubility of AgCl in 0.10 M CaCl$_2$.

(a) $1.6 \times 10^{-10}\ M$
(b) $1.6 \times 10^{-9}\ M$
(c) $8.0 \times 10^{-10}\ M$
(d) $1.3 \times 10^{-5}\ M$
(e) $1.6 \times 10^{-10}\ M$

17.5.2 In which of the solutions would the slightly soluble salt MC_2 be *most* soluble, and in which would it be *least* soluble? (For clarity, water molecules and counter ions in soluble salts are not shown.)

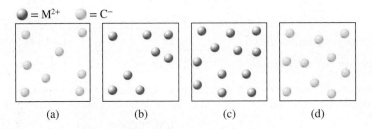

(a)   (b)   (c)   (d)

17.6 SEPARATION OF IONS USING DIFFERENCES IN SOLUBILITY

In chemical analysis, it sometimes is necessary to remove one type of ion from solution by precipitation while leaving other ions in solution. For instance, the addition of sulfate ions to a solution containing both potassium and barium ions causes $BaSO_4$ to precipitate out, thereby removing most of the Ba^{2+} ions from the solution. The other "product," K_2SO_4, is soluble and will remain in solution. The $BaSO_4$ precipitate can be separated from the solution by filtration.

Fractional Precipitation

Compound	K_{sp}
AgCl	1.6×10^{-10}
AgBr	7.7×10^{-13}
AgI	8.3×10^{-17}

Even when both products are insoluble, we can still achieve some degree of separation by choosing the proper reagent to bring about precipitation. Consider a solution that contains Cl^-, Br^-, and I^- ions. One way to separate these ions is to convert them to insoluble silver halides. As the K_{sp} values in the margin show, the solubility of the silver halides decreases from AgCl to AgI. Thus, when a soluble compound such as silver nitrate is slowly added to this solution, AgI begins to precipitate first, followed by AgBr, and then AgCl. This practice is known as *fractional precipitation.*

Video 17.5
Solutions precipitation—fractional crystallization.

Worked Example 17.13 describes the separation of only two ions (Cl^- and Br^-), but the procedure can be applied to a solution containing more than two different types of ions.

Worked Example 17.13

Silver nitrate is added slowly to a solution that is 0.020 *M* in Cl^- ions and 0.020 *M* in Br^- ions. Calculate the concentration of Ag^+ ions (in mol/L) required to initiate the precipitation of AgBr without precipitating AgCl.

Strategy Silver nitrate dissociates in solution to give Ag^+ and NO_3^- ions. Adding Ag^+ ions in sufficient amount will cause the slightly soluble ionic compounds AgCl and AgBr to precipitate from solution. Knowing the K_{sp} values for AgCl and AgBr (and the concentrations of Cl^- and Br^- already in solution), we can use the equilibrium expressions to calculate the maximum concentration of Ag^+ that can exist in solution without exceeding K_{sp} for each compound.

Setup The solubility equilibria, K_{sp} values, and equilibrium expressions for AgCl and AgBr are

$$AgCl(s) \rightleftharpoons Ag^+(aq) + Cl^-(aq) \quad K_{sp} = 1.6 \times 10^{-10} = [Ag^+][Cl^-]$$
$$AgBr(s) \rightleftharpoons Ag^+(aq) + Br^-(aq) \quad K_{sp} = 7.7 \times 10^{-13} = [Ag^+][Br^-]$$

Because the K_{sp} for AgBr is smaller (by a factor of more than 200), AgBr should precipitate first; that is, it will require a lower concentration of added Ag^+ to begin precipitation. Therefore, we first solve for $[Ag^+]$ using the equilibrium expression for AgBr to determine the minimum Ag^+ concentration necessary to initiate precipitation of AgBr. We then solve for $[Ag^+]$ again, using the equilibrium expression for AgCl to determine the *maximum* Ag^+ concentration that can exist in the solution without initiating the precipitation of AgCl.

Solution Solving the AgBr equilibrium expression for Ag^+ concentration, we have

$$[Ag^+] = \frac{K_{sp}}{[Br^-]}$$

and

$$[Ag^+] = \frac{7.7 \times 10^{-13}}{0.020} = 3.9 \times 10^{-11} \, M$$

For AgBr to precipitate from solution, the silver ion concentration must exceed 3.9×10^{-11} M. Solving the AgCl equilibrium expression for the Ag^+ concentration, we have

$$[Ag^+] = \frac{K_{sp}}{[Cl^-]}$$

and

$$[Ag^+] = \frac{1.6 \times 10^{-10}}{0.020} = 8.0 \times 10^{-9} \, M$$

For AgCl *not* to precipitate from solution, the silver ion concentration must stay below 8.0×10^{-9} M. Therefore, to precipitate the Br^- ions without precipitating the Cl^- from this solution, the Ag^+ concentration must be greater than 3.9×10^{-11} M and less than 8.0×10^{-9} M.

Think About It

If we continue adding $AgNO_3$ until the Ag^+ concentration is high enough to begin the precipitation of AgCl, the concentration of Br^- remaining in solution can also be determined using the K_{sp} expression.

$$[Br^-] = \frac{K_{sp}}{[Ag^+]} = \frac{7.7 \times 10^{-13}}{8.0 \times 10^{-9}}$$

$$= 9.6 \times 10^{-5} \, M$$

Thus, by the time AgCl begins to precipitate, $(9.6 \times 10^{-5} \, M) \div (0.020 \, M) = 0.0048$, so less than 0.5 percent of the original bromide ion remains in the solution.

Practice Problem **A**TTEMPT Lead(II) nitrate is added slowly to a solution that is 0.020 M in Cl^- ions. Calculate the concentration of Pb^{2+} ions (in mol/L) required to initiate the precipitation of $PbCl_2$. (K_{sp} for $PbCl_2$ is 2.4×10^{-4}.)

Practice Problem **B**UILD Calculate the concentration of Ag^+ (in mol/L) necessary to initiate the precipitation of (a) AgCl and (b) Ag_3PO_4 from a solution in which $[Cl^-]$ and $[PO_4^{3-}]$ are each 0.10 M. (K_{sp} for Ag_3PO_4 is 1.8×10^{-18}.)

Practice Problem **C**ONCEPTUALIZE The first two diagrams show saturated solutions of the sparingly soluble ionic compounds AX and BX_2. The third diagram shows a solution of soluble salts containing the cations A^+ and B^{2+}. (The anions of the soluble salts are not shown.) Which of the sparingly soluble compounds will precipitate first as NaX is added to the third solution?

Qualitative Analysis of Metal Ions in Solution

The principle of selective precipitation can be used to identify the types of ions present in a solution. This practice is called *qualitative analysis*. There are about 20 common cations that can be analyzed readily in aqueous solution. These cations can be divided into five groups according to the solubility products of their insoluble salts (Table 17.4). Because an unknown solution may contain from 1 to all 20 ions, any analysis must be carried out systematically from group 1 through group 5. The general procedure for separating these 20 ions is as follows:

Student Annotation: Note that these group numbers do *not* correspond to groups in the periodic table.

- *group 1 cations.* When dilute HCl is added to the unknown solution, only the Ag^+, Hg_2^{2+}, and Pb^{2+} ions precipitate as insoluble chlorides. The other ions, whose chlorides are soluble, remain in solution.
- *group 2 cations.* After the chloride precipitates have been removed by filtration, hydrogen sulfide is added to the unknown solution, which is acidic due to the addition of HCl. Metal ions from group 2 react to produce metal sulfides.

$$M^{2+}(aq) + H_2S(aq) \rightleftharpoons MS(s) + 2H^+(aq)$$

In the presence of H^+, this equilibrium shifts to the *left*. Therefore, only the metal sulfides with the *smallest* K_{sp} values precipitate under acidic conditions. These are Bi_2S_3, CdS, CuS, and SnS (see Table 17.4). The solution is then filtered to remove the insoluble sulfides.

• *group 3 cations.* At this stage, sodium hydroxide is added to the solution to make it basic. In a basic solution, the metal sulfide equilibrium shifts to the right and the more soluble sulfides (CoS, FeS, MnS, NiS, ZnS) now precipitate out of solution. The Al^{3+} and Cr^{3+} ions actually precipitate as the hydroxides $Al(OH)_3$ and $Cr(OH)_3$, rather than as the sulfides, because the hydroxides are less soluble. The solution is filtered again to remove the insoluble sulfides and hydroxides.

• *group 4 cations.* After all the group 1, 2, and 3 cations have been removed from solution, sodium carbonate is added to the basic solution to precipitate Ba^{2+}, Ca^{2+}, and Sr^{2+} ions as $BaCO_3$, $CaCO_3$, and $SrCO_3$. These precipitates, too, are removed from solution by filtration.

• *group 5 cations.* At this stage, the only cations possibly remaining in solution are Na^+, K^+, and NH_4^+. The presence of NH_4^+ ions can be determined by adding sodium hydroxide.

$$NaOH(aq) + NH_4^+(aq) \longrightarrow Na^+(aq) + H_2O(l) + NH_3(g)$$

The ammonia gas is detected either by its characteristic odor or by observing a wet piece of red litmus paper turning blue when placed above (not in contact with) the solution. To confirm the presence of Na^+ and K^+ ions, a flame test is often used in which a piece of platinum wire (chosen because platinum is inert) is dipped into the original solution and then held over a Bunsen burner flame. Na^+ ions emit a yellow flame when heated in this manner, whereas K^+ ions emit a violet flame (Figure 17.11). Figure 17.12 summarizes this scheme for separating metal ions.

Video 17.6
Chemical analysis—flame tests of metals.

Figure 17.11 Flame tests for sodium (yellow flame) and potassium (violet flame).

© McGraw-Hill Education/Stephen Frisch, photographer

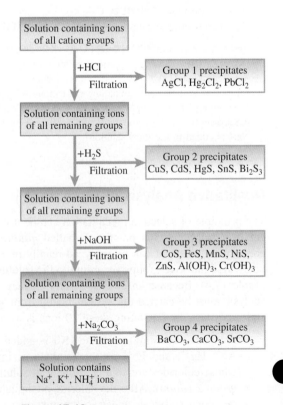

Figure 17.12 A flowchart for the separation of cations in qualitative analysis.

Section 17.6 Review

Separation of Ions Using Differences in Solubility

17.6.1 A solution is 0.10 M in Br^-, CO_3^{2-}, Cl^-, I^-, and SO_4^{2-} ions. Which compound will precipitate first as silver nitrate is added to the solution?
(a) $AgBr$ (c) $AgCl$ (e) Ag_2SO_4
(b) Ag_2CO_3 (d) AgI

17.6.2 Barium nitrate is added slowly to a solution that is 0.10 M in SO_4^{2-} ions and 0.10 M in F^- ions. Calculate the concentration of Ba^{2+} ions (in mol/L) required to initiate the precipitation of $BaSO_4$ without precipitating BaF_2.
(a) $1.7 \times 10^{-6}\ M$ (c) $1.7 \times 10^{-4}\ M$ (e) $1.1 \times 10^{-8}\ M$
(b) $1.1 \times 10^{-9}\ M$ (d) $1.7 \times 10^{-5}\ M$

Learning Outcomes

- Describe the common ion effect.
- Calculate equilibrium concentrations or pH of a solution involving a common ion.
- Calculate the pH of a buffer using equilibrium calculations and the Henderson-Hasselbalch equation.
- Describe how to prepare a buffer with a given pH.
- Select the appropriate weak acid (or weak base) to make a buffer of given pH.
- Predict the pH of a titration given the type of acid and base (strong or weak).
- Calculate the pH throughout a titration between acids and bases: strong plus strong or strong plus weak.
- Use a titration curve to identify what type of titration was employed (e.g., weak acid plus strong base).

- Distinguish between endpoint and equivalence point.
- Select an indicator given the acid and base species used in a titration.
- Define solubility product constant (K_{sp}).
- Use K_{sp} to determine molar solubility and vice versa.
- Use K_{sp} to predict whether a precipitation will occur.
- Cite examples of how the following can affect solubility: common ions, pH, or complex ion formation.
- Use K_f, the formation constant, in equilibrium calculations.
- Explain fractional precipitation.
- Give, in your own words, the meaning and primary basis of qualitative analysis.

Chapter Summary

SECTION 17.1
- The presence of a common ion suppresses the ionization of a weak acid or weak base. This is known as the **common ion effect.**
- A common ion is added to a solution in the form of a *salt.*

SECTION 17.2
- A solution that contains significant concentrations of both members of a conjugate acid-base pair (weak acid–conjugate base or weak base–conjugate acid) is a *buffer solution* or simply a **buffer.**
- Buffer solutions resist pH change upon addition of small amounts of strong acid or strong base. Buffers are important to biological systems.

- The pH of a buffer can be calculated using an equilibrium table or with the **Henderson-Hasselbalch equation.**
- The pK_a of a weak acid is $-\log K_a$. When the weak acid and conjugate base concentrations in a buffer solution are equal, $pH = pK_a$.
- We can prepare a buffer with a specific pH by choosing a weak acid with a pK_a close to the desired pH.

SECTION 17.3
- The titration curve of a strong acid–strong base titration has a long, steep region near the equivalence point.
- Titration curves for weak acid–strong base or weak base–strong acid titrations have a significantly shorter steep region.

- The pH at the equivalence point of a strong acid–strong base titration is 7.00.
- The pH at the equivalence point of a weak acid–strong base titration is above 7.00.
- The pH at the equivalence point of a weak base–strong acid titration is below 7.00.
- Acid-base indicators are usually weak organic acids that exhibit two different colors depending on the pH of the solution. The *endpoint* of a titration is the point at which the color of the indicator changes. It is used to estimate the *equivalence point* of a titration.
- The indicator used for a particular titration should exhibit a color change in the pH range corresponding to the steep region of the titration curve.

SECTION 17.4

- The *solubility product constant* (K_{sp}) is the equilibrium constant that indicates to what extent a slightly soluble ionic compound dissociates in water.
- K_{sp} can be used to determine *molar solubility* or *solubility* in g/L, and vice versa.

- K_{sp} can also be used to predict whether a precipitate will form when two solutions are mixed.

SECTION 17.5

- Solubility is affected by common ions, pH, and complex ion formation. The *formation constant* (K_f) indicates to what extent complex ions form.
- A salt that dissociates to give a strong conjugate base such as fluoride ion will be more soluble in acidic solution than in pure water.
- A salt that dissociates to give hydroxide ion will be more soluble at lower pH and less soluble at higher pH.
- The solubility of an ionic compound increases when the formation of a *complex ion* consumes one of the products of dissociation.

SECTION 17.6

- Ions can be separated using *fractional precipitation.*
- Fractional precipitation schemes can be designed based on K_{sp} values.
- Groups of cations can be identified through the use of selective precipitation. This is the basis of *qualitative analysis.*

Key Words

Buffer, 777
Common ion effect, 777
Complex ion, 807
Endpoint, 793

Formation constant (K_f), 808
Fractional precipitation, 812
Henderson-Hasselbalch equation, 779
Molar solubility, 796

Qualitative analysis, 813
Solubility, 796
Solubility product constant (K_{sp}), 796

Key Equations

17.1	$\text{pH} = \text{p}K_a + \log \dfrac{[\text{A}^-]}{[\text{HA}]}$	The pH of a buffer is calculated as the sum of pK_a of the weak acid component and base-10 log of the ratio of conjugate base to weak acid.
17.2	$\text{p}K_a = -\log K_a$	The pK_a of a weak acid is equal to minus the base-10 log of its K_a value.
17.3	$\text{pH} = \text{p}K_a + \log \dfrac{[\text{conjugate base}]}{[\text{weak acid}]}$	This is the more general form of the Henderson-Hasselbalch equation, which is used to calculate the pH of a buffer. Because the buffer components are dissolved in the same volume, the ratio of conjugate base to weak acid can be calculated using either molar concentrations or absolute molar amounts.

Buffers

The Henderson-Hasselbalch equation (Equation 17.3) enables us to calculate the pH of a buffer if we know the concentrations of the weak acid and conjugate base (or of the weak base and conjugate acid); and to determine the pH of a buffer after the addition of strong acid or base. It can also be used to determine the amounts or *relative* amounts of both members of a conjugate pair necessary to prepare a buffer of specified pH.

The most familiar form of the Henderson-Hasselbalch equation is

$$pH = pK_a + \log \frac{[\text{conjugate base}]}{[\text{weak acid}]}$$

Although the equation contains the ratio of *molar concentrations* of conjugate base and weak acid, because both are contained within the same volume, volume cancels in the numerator and denominator of the log term. Therefore, we can also use *moles* or *millimoles* of conjugate base and weak acid, which is more convenient in many cases.

$$pH = pK_a + \log \frac{\text{mol conjugate base}}{\text{mol weak acid}} \quad \text{or} \quad pH = pK_a + \log \frac{\text{mmol conjugate base}}{\text{mmol weak acid}}$$

If we wish to prepare a buffer in a specific pH range, we must first select a suitable conjugate pair. Because the concentrations of weak acid and conjugate base cannot differ by more than a factor of 10, the log term in the Henderson-Hasselbalch equation can have values only from −1 through 1.

$$pH = pK_a \pm 1$$

Therefore, we must select a weak acid with a pK_a within one pH unit of the desired buffer pH. For example, to prepare a buffer with a pH between 4 and 5, we could use acetic acid ($HC_2H_3O_2$). [From Table 16.5, K_a for acetic acid is 1.8×10^{-5}, therefore its pK_a is $-\log (1.8 \times 10^{-5}) = 4.74$.]

To prepare a buffer with a specific pH, we solve the Henderson-Hasselbalch equation to determine the relative amounts of weak acid and conjugate base. The following flowchart illustrates this process for preparation of a buffer with pH = 4.15 using acetic acid ($pK_a = 4.74$) and sodium acetate ($NaC_2H_3O_2$).

$$4.15 \;\Rightarrow\; 4.74 \;+ \log \left(\frac{[NaC_2H_3O_2]}{[HC_2H_3O_2]} \right)$$

$$-0.59 \;=\; \log \left(\frac{[NaC_2H_3O_2]}{[HC_2H_3O_2]} \right)$$

To eliminate the log term, we must take the antilog (10^x) of both sides of the equation.

$$10^{-0.59} \;=\; \frac{[NaC_2H_3O_2]}{[HC_2H_3O_2]}$$

This gives

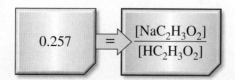

$$0.257 = \frac{[NaC_2H_3O_2]}{[HC_2H_3O_2]}$$

This means that we need 0.257 mole of sodium acetate for every mole of acetic acid. If the amount of one member of the conjugate pair is specified, this ratio enables us to calculate the amount of the other member. For instance, if we know that the buffer must contain 25.0 g acetic acid, we can determine the amount of sodium acetate necessary to achieve the desired pH. We will need the molar masses of both members of the conjugate pair [◄◄ Key Skills Chapter 5].

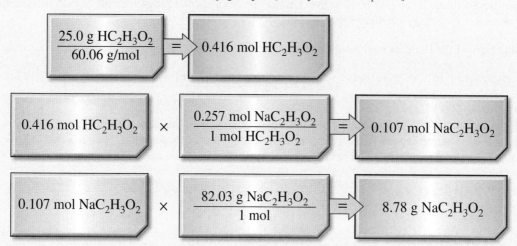

$$\frac{25.0 \text{ g } HC_2H_3O_2}{60.06 \text{ g/mol}} = 0.416 \text{ mol } HC_2H_3O_2$$

$$0.416 \text{ mol } HC_2H_3O_2 \times \frac{0.257 \text{ mol } NaC_2H_3O_2}{1 \text{ mol } HC_2H_3O_2} = 0.107 \text{ mol } NaC_2H_3O_2$$

$$0.107 \text{ mol } NaC_2H_3O_2 \times \frac{82.03 \text{ g } NaC_2H_3O_2}{1 \text{ mol}} = 8.78 \text{ g } NaC_2H_3O_2$$

Therefore, we would need 8.78 g sodium acetate. Because the pH of a buffer does not depend on volume, we could combine these amounts of acetic acid and sodium acetate in any volume that is convenient.

Key Skills Problems

17.1
Which of the acids in Table 16.5 (page 732) can be used to prepare a buffer of pH 6.5? (Select all that apply.)
(a) hydrofluoric acid
(b) benzoic acid
(c) hydrocyanic acid
(d) phenol
(e) none

17.2
What molar ratio of sodium cyanide to hydrocyanic acid is necessary to prepare a buffer with pH = 9.72?
(a) 0.39:1
(b) 0.41:1
(c) 2.3:1
(d) 2.6:1
(e) 1:1

17.3
How many moles of sodium benzoate must be added to 175 mL of 0.955 M benzoic acid to prepare a buffer with pH = 5.05?
(a) 0.18
(b) 0.15
(c) 6.8
(d) 7.2
(e) 1.2

17.4
How much sodium fluoride must be dissolved in 250 mL of 0.98 M HF to prepare a buffer with pH 3.50?
(a) 23 g
(b) 2.2 g
(c) 15 g
(d) 92 g
(e) 0.98 g

Questions and Problems

SECTION 17.1: THE COMMON ION EFFECT

Review Questions

17.1 Use Le Châtelier's principle to explain how the common ion effect affects the pH of a weak acid solution.

17.2 Describe the effect on pH (increase, decrease, or no change) that results from each of the following additions: (a) potassium acetate to an acetic acid solution, (b) ammonium nitrate to an ammonia solution, (c) sodium formate (HCOONa) to a formic acid (HCOOH) solution, (d) potassium chloride to a hydrochloric acid solution, (e) barium iodide to a hydroiodic acid solution.

17.3 Define pK_a for a weak acid. What is the relationship between the value of the pK_a and the strength of the acid?

17.4 The pK_a for each of two monoprotic acids HA and HB is 5.9 and 8.1, respectively. Which of the two is the stronger acid?

Computational Problems

17.5 Determine the pH of (a) a 0.40 M CH_3COOH solution, and (b) a solution that is 0.40 M CH_3COOH and 0.20 M CH_3COONa.

17.6 Determine the pH of (a) a 0.20 M NH_3 solution, and (b) a solution that is 0.20 M NH_3 and 0.30 M NH_4Cl.

SECTION 17.2: BUFFER SOLUTIONS

> **Visualizing Chemistry**
> **Figure 17.1**

VC 17.1 Which pair of substances can be dissolved together to prepare a buffer solution?

CH_3COOH CH_3COONa NaOH HCl NaCl
  (i)   (ii)   (iii)   (iv) (v)

(a) i/ii
(b) i/ii, i/iii, ii/iv
(c) i/ii, i/iii, ii/iii, iv/v

VC 17.2 Consider the buffer shown in Figure 17.1, in which $[CH_3COOH] = 0.10\ M$, $[CH_3COO^-] = 0.10\ M$, and pH = 4.74. How would the pH of the buffer change if we doubled the volume by adding 100 mL of water?
(a) The pH would increase.
(b) The pH would decrease.
(c) The pH would not change.

VC 17.3 Consider a buffer similar to the one shown in Figure 17.1, in which $[CH_3COOH] = 0.50\ M$ and $[CH_3COO^-] = 0.75\ M$. Which substance can we add more of without causing a drastic change in pH?
(a) HCl
(b) NaOH
(c) Neither

VC 17.4 According to Figure 17.1, the reaction that occurs when strong acid is added to the buffer is

$$H^+(aq) + CH_3COO^-(aq) \longrightarrow CH_3COOH(aq)$$

Estimate the value of K for this reaction.
(a) 1×10^{14}
(b) 6×10^{-10}
(c) 6×10^4

Review Questions

17.7 What is a buffer solution? What must a solution contain to be a buffer?

17.8 Using only a pH meter, water, and a graduated cylinder, how would you distinguish between an acid solution and a buffer solution at the same pH?

Computational Problems

17.9 Calculate the pH of the buffer system made up of 0.15 M NH_3/0.35 M NH_4Cl.

17.10 Calculate the pH of the following two buffer solutions: (a) 2.0 M CH_3COONa/2.0 M CH_3COOH, (b) 0.20 M CH_3COONa/0.20 M CH_3COOH. Which is the more effective buffer? Why?

17.11 The pH of a bicarbonate–carbonic acid buffer is 8.00. Calculate the ratio of the concentration of carbonic acid (H_2CO_3) to that of the bicarbonate ion (HCO_3^-).

17.12 What is the pH of the buffer 0.10 M Na_2HPO_4/0.15 M KH_2PO_4?

17.13 The pH of a sodium acetate–acetic acid buffer is 4.50. Calculate the ratio $[CH_3COO^-]/[CH_3COOH]$.

17.14 The pH of blood plasma is 7.40. Assuming the principal buffer system is HCO_3^-/H_2CO_3, calculate the ratio $[HCO_3^-]/[H_2CO_3]$. Is this buffer more effective against an added acid or an added base?

17.15 Calculate the pH of the 0.20 M NH_3/0.20 M NH_4Cl buffer. What is the pH of the buffer after the addition of 10.0 mL of 0.10 M HCl to 65.0 mL of the buffer?

17.16 Calculate the pH of 1.00 L of the buffer 1.00 M CH_3COONa/1.00 M CH_3COOH before and after the addition of (a) 0.080 mol NaOH and (b) 0.12 mol HCl. (Assume that there is no change in volume.)

Conceptual Problems

17.17 Which of the following solutions can act as a buffer: (a) KCl/HCl, (b) KHSO₄/H₂SO₄, (c) Na₂HPO₄/NaH₂PO₄, (d) KNO₂/HNO₂?

17.18 Which of the following solutions can *not* act as a buffer: (a) KCN/HCN, (b) Na₂SO₄/NaHSO₄, (c) NH₃/NH₄NO₃, (d) NaI/HI?

17.19 A diprotic acid, H₂A, has the following ionization constants: $K_{a_1} = 1.1 \times 10^{-3}$ and $K_{a_2} = 2.5 \times 10^{-6}$. To make up a buffer solution of pH 3.25, which combination would you choose: NaHA/H₂A or Na₂A/NaHA?

17.20 A student is asked to prepare a buffer solution at pH 4.95, using one of the following weak acids: HA ($K_a = 2.7 \times 10^{-3}$), HB ($K_a = 4.4 \times 10^{-6}$), HC ($K_a = 2.6 \times 10^{-9}$). Which acid should the student choose? Why?

17.21 The diagrams [(a)–(d)] contain one or more of the compounds: H₂A, NaHA, and Na₂A, where H₂A is a weak diprotic acid. (1) Which of the solutions can act as buffer solutions? (2) Which solution is the most effective buffer solution? Water molecules and Na⁺ ions have been omitted for clarity.

⬤ = H₂A   ◑ = HA⁻   ◔ = A²⁻

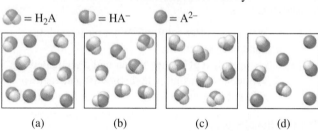

　　(a)　　　　　(b)　　　　　(c)　　　　　(d)

17.22 The diagrams [(a)–(d)] represent solutions containing a weak acid HA (pK_a = 5.0) and its sodium salt NaA. (1) Which solution has the lowest pH? Which has the highest pH? (2) How many different species are present after the addition of two H⁺ ions to solution (a)? (3) How many different species are present after the addition of two OH⁻ ions to solution (b)?

◑ = HA   ⬤ = A⁻

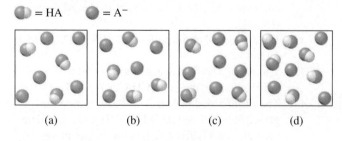

　　(a)　　　　　(b)　　　　　(c)　　　　　(d)

SECTION 17.3: ACID-BASE TITRATIONS

Review Questions

17.23 Briefly describe what happens in an acid-base titration.

17.24 Sketch titration curves for the following acid-base titrations: (a) HCl versus NaOH, (b) HCl versus CH₃NH₂, (c) CH₃COOH versus NaOH. In each case, the base is added to the acid in an Erlenmeyer flask. Your graphs should show the pH on the y axis and the volume of base added on the x axis.

17.25 Explain how an acid-base indicator works in a titration. What are the criteria for choosing an indicator for a particular acid-base titration?

17.26 The amount of indicator used in an acid-base titration must be small. Why?

Computational Problems

17.27 A 0.3115-g sample of a monoprotic acid neutralizes 21.6 mL of 0.08133 M KOH solution. Calculate the molar mass of the acid.

17.28 A 6.50-g quantity of a diprotic acid was dissolved in water and made up to exactly 250 mL. Calculate the molar mass of the acid if 25.0 mL of this solution required 20.3 mL of 1.00 M KOH for neutralization. Assume that both protons of the acid were titrated.

17.29 In a titration experiment, 27.5 mL of 0.500 M H₂SO₄ neutralizes 50.0 mL of NaOH. What is the concentration of the NaOH solution?

17.30 In a titration experiment, 31.4 mL of 1.12 M HCOOH is neutralized by 16.3 mL of Ba(OH)₂. What is the concentration of the Ba(OH)₂ solution?

17.31 A 0.1276-g sample of an unknown monoprotic acid was dissolved in 25.0 mL of water and titrated with a 0.0633 M NaOH solution. The volume of base required to bring the solution to the equivalence point was 18.4 mL. (a) Calculate the molar mass of the acid. (b) After 10.0 mL of base had been added during the titration, the pH was determined to be 5.87. What is the K_a of the unknown acid?

17.32 A solution is made by mixing exactly 500 mL of 0.167 M NaOH with exactly 500 mL of 0.100 M CH₃COOH. Calculate the equilibrium concentrations of H⁺, CH₃COOH, CH₃COO⁻, OH⁻, and Na⁺.

17.33 Calculate the pH at the equivalence point for the following titration: 0.20 M HCl versus 0.20 M methylamine (CH₃NH₂).

17.34 Calculate the pH at the equivalence point for the following titration: 0.10 M HCOOH versus 0.10 M NaOH.

17.35 A 25.0-mL solution of 0.100 M CH₃COOH is titrated with a 0.200 M KOH solution. Calculate the pH after the following additions of the KOH solution: (a) 0.0 mL, (b) 5.0 mL, (c) 10.0 mL, (d) 12.5 mL, (e) 15.0 mL.

17.36 A 10.0-mL solution of 0.300 M NH₃ is titrated with a 0.100 M HCl solution. Calculate the pH after the following additions of the HCl solution: (a) 0.0 mL, (b) 10.0 mL, (c) 20.0 mL, (d) 30.0 mL, (e) 40.0 mL.

Conceptual Problems

17.37 Referring to Table 17.3, specify which indicator or indicators you would use for the following titrations: (a) HCOOH versus NaOH, (b) HCl versus KOH, (c) HNO₃ versus CH₃NH₂.

17.38   A student carried out an acid-base titration by adding NaOH solution from a burette to an Erlenmeyer flask containing an HCl solution and using phenolphthalein as the indicator. At the equivalence point, she observed a faint reddish-pink color. However, after a few minutes, the solution gradually turned colorless. What do you suppose happened?

17.39   The ionization constant K_a of an indicator HIn is 1.0×10^{-6}. The color of the nonionized form is red and that of the ionized form is yellow. What is the color of this indicator in a solution whose pH is 4.00?

17.40   The K_a of a certain indicator is 2.0×10^{-6}. The color of HIn is green and that of In$^-$ is red. A few drops of the indicator are added to an HCl solution, which is then titrated against an NaOH solution. At what pH will the indicator change color?

17.41   Diagrams (a) through (d) represent solutions at various stages in the titration of a weak base B (such as NH$_3$) with HCl. Identify the solution that corresponds to (1) the initial stage before the addition of HCl, (2) halfway to the equivalence point, (3) the equivalence point, (4) beyond the equivalence point. Is the pH greater than, less than, or equal to 7 at the equivalence point? Water and Cl$^-$ ions have been omitted for clarity.

= B   = BH$^+$ = H$_3$O$^+$

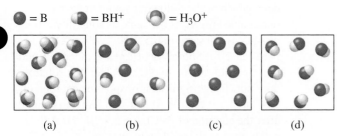

(a)   (b)   (c)   (d)

17.42   Diagrams (a) through (d) represent solutions at various stages in the titration of a weak acid HA with NaOH. Identify the solution that corresponds to (1) the initial stage before the addition of NaOH, (2) halfway to the equivalence point, (3) the equivalence point, (4) beyond the equivalence point. Is the pH greater than, less than, or equal to 7 at the equivalence point? Water and Na$^+$ ions have been omitted for clarity.

= HA   = A$^-$ = OH$^-$

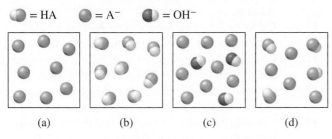

(a)   (b)   (c)   (d)

SECTION 17.4: SOLUBILITY EQUILIBRIA

Review Questions

17.43   Use BaSO$_4$ to distinguish between the terms *solubility* and *solubility product*.

17.44   Why do we usually not quote the K_{sp} values for soluble ionic compounds?

17.45   Write balanced equations and solubility product expressions for the solubility equilibria of the following compounds: (a) CuBr, (b) ZnC$_2$O$_4$, (c) Ag$_2$CrO$_4$, (d) Hg$_2$Cl$_2$, (e) AuCl$_3$, (f) Mn$_3$(PO$_4$)$_2$.

17.46   Write the solubility product expression for the ionic compound A$_x$B$_y$.

17.47   How can we predict whether a precipitate will form when two solutions are mixed?

17.48   Silver chloride has a larger K_{sp} than silver carbonate (see Table 17.4). Does this mean that AgCl also has a larger molar solubility than Ag$_2$CO$_3$?

Computational Problems

17.49   Calculate the concentration of ions in the following saturated solutions: (a) [I$^-$] in AgI solution with [Ag$^+$] = 4.3×10^{-7} M, (b) [Al^{3+}] in Al(OH)$_3$ solution with [OH$^-$] = 9.5×10^{-8} M.

17.50   From the solubility data given, calculate the solubility products for the following compounds: (a) SrF$_2$, 7.3×10^{-2} g/L, (b) Ag$_3$PO$_4$, 6.7×10^{-3} g/L.

17.51   The molar solubility of MnCO$_3$ is 4.2×10^{-6} M. What is K_{sp} for this compound?

17.52   The solubility of an ionic compound MX (molar mass = 346 g) is 4.63×10^{-3} g/L. What is K_{sp} for this compound?

17.53   The solubility of an ionic compound M$_2$X$_3$ (molar mass = 188 g) is 3.6×10^{-17} g/L. What is K_{sp} for this compound?

17.54   Using data from Table 17.4, calculate the molar solubility of CaF$_2$.

17.55   What is the pH of a saturated strontium hydroxide solution?

17.56   The pH of a saturated solution of a metal hydroxide M(OH)$_2$ is 9.68. Calculate the K_{sp} for this compound.

17.57   If 20.0 mL of 0.10 M Ba(NO$_3$)$_2$ is added to 50.0 mL of 0.10 M Na$_2$CO$_3$, will BaCO$_3$ precipitate?

17.58   A volume of 75 mL of 0.060 M NaF is mixed with 25 mL of 0.15 M Sr(NO$_3$)$_2$. Calculate the concentrations in the final solution of NO$_3^-$, Na$^+$, Sr^{2+}, and F$^-$. (K_{sp} for SrF$_2$ = 2.0×10^{-10}.)

Conceptual Problems

17.59   The diagram represents a saturated solution of the salt MA$_2$. Which soluble salt will cause the precipitation of the greatest quantity of MA$_2$ when 0.1 mole is dissolved in the saturated solution? (For clarity, the water molecules are not shown.)
(a) NaA
(b) BaA$_2$
(c) AlA$_3$
(d) M(NO$_3$)$_2$
(e) M$_3$(PO$_4$)$_2$

= M^{2+} = A$^-$

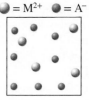

17.60   Diagrams 1 and 2 represent a saturated solution of the slightly soluble salt M$_2$B and a solution of the soluble salt MNO₃, respectively. Which of the solutions (a) through (f) of the soluble salt Na₂B can be added to the solution of MNO₃ without causing a precipitate to form? Select all that apply. Assume that the solution volumes are equal, and that they are additive when combined. (For clarity, the water molecules, nitrate ions, and sodium ions are not shown.)

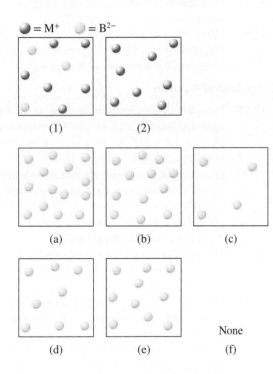

= M⁺   = B²⁻

(1)   (2)

(a)   (b)   (c)

(d)   (e)   (f) None

SECTION 17.5: FACTORS AFFECTING SOLUBILITY

 Visualizing Chemistry
Figure 17.8

VC 17.5 Which of the following would cause precipitation of the largest amount of AgCl when 0.1 mole is dissolved in the saturated solution shown in Figure 17.8?
(a) NaCl
(b) CsCl
(c) Both would cause precipitation of the same amount.

VC 17.6 Which of the following would cause precipitation of the largest amount of AgCl when 0.1 g is dissolved in the saturated solution shown in Figure 17.8?
(a) NaCl
(b) CsCl
(c) Both would cause precipitation of the same amount.

VC 17.7 How would the concentration of silver ion in the last view of Figure 17.8 be different if we dissolved the sodium chloride *before* dissolving the silver chloride?
(a) It would be higher.
(b) It would be lower.
(c) It would not change.

VC 17.8 How would the concentration of silver ion in the last view of Figure 17.8 be different if we added silver nitrate to the saturated solution of AgCl instead of sodium chloride?
(a) It would be higher.
(b) It would be lower.
(c) It would not change.

Review Questions

17.61   How does the common ion effect influence solubility equilibria? Use Le Châtelier's principle to explain the decrease in solubility of CaCO₃ in an Na₂CO₃ solution.

17.62   The molar solubility of AgCl in 6.5×10^{-3} M AgNO₃ is 2.5×10^{-8} M. In deriving K_{sp} from these data, which of the following assumptions are reasonable? (a) K_{sp} is the same as solubility. (b) K_{sp} of AgCl is the same in 6.5×10^{-3} M AgNO₃ as in pure water. (c) Solubility of AgCl is independent of the concentration of AgNO₃. (d) [Ag⁺] in solution does not change significantly upon the addition of AgCl to 6.5×10^{-3} M AgNO₃. (e) [Ag⁺] in solution after the addition of AgCl to 6.5×10^{-3} M AgNO₃ is the same as it would be in pure water.

17.63   Give an example to illustrate the general effect of complex ion formation on solubility.

Computational Problems

17.64   How many grams of CaCO₃ will dissolve in 3.0×10^2 mL of 0.050 M Ca(NO₃)₂?

17.65   The solubility product of PbBr₂ is 8.9×10^{-6}. Determine the molar solubility in (a) pure water, (b) 0.20 M KBr solution, and (c) 0.20 M Pb(NO₃)₂ solution.

17.66   Calculate the molar solubility of AgCl in a 1.00-L solution containing 10.0 g of dissolved CaCl₂.

17.67   Calculate the molar solubility of BaSO₄ in (a) water and (b) a solution containing 1.0 M SO₄²⁻ ions.

17.68   Compare the molar solubility of Mg(OH)₂ in water and in a solution buffered at a pH of 9.0.

17.69   Calculate the molar solubility of Fe(OH)₂ in a solution buffered at (a) a pH of 8.00 and (b) a pH of 10.00.

17.70   The solubility product of Mg(OH)₂ is 1.2×10^{-11}. What minimum OH⁻ concentration must be attained (e.g., by adding NaOH) to decrease the Mg concentration in a solution of Mg(NO₃)₂ to less than 1.0×10^{-10} M?

17.71 Calculate whether a precipitate will form if 2.00 mL of 0.60 M NH_3 is added to 1.0 L of 1.0×10^{-3} M $FeSO_4$.

17.72 If 2.50 g of $CuSO_4$ is dissolved in 9.0×10^2 mL of 0.30 M NH_3, what are the concentrations of Cu^{2+}, $Cu(NH_3)_4^{2+}$, and NH_3 at equilibrium?

17.73 Calculate the concentrations of Cd^{2+}, $Cd(CN)_4^{2-}$, and CN^- at equilibrium when 0.50 g of $Cd(NO_3)_2$ dissolves in 5.0×10^2 mL of 0.50 M NaCN.

17.74 If NaOH is added to 0.010 M Al^{3+}, which will be the predominant species at equilibrium: $Al(OH)_3$ or $Al(OH)_4^-$? The pH of the solution is 14.00. [K_f for $Al(OH)_4^- = 2.0 \times 10^{33}$.]

17.75 Calculate the molar solubility of AgI in a 1.0 M NH_3 solution.

17.76 (a) Calculate the molar solubility of hydroxyapatite given that its K_{sp} is 2×10^{-59}.
(b) Calculate the molar solubility of hydroxyapatite in a buffered aqueous solution with pH = 4.0.

17.77 (a) Calculate the K_{sp} of fluoroapatite given that its molar solubility is 7×10^{-8} M.
(b) Calculate the molar solubility of fluoroapatite in an aqueous solution in which the concentration of fluoride ion is 0.10 M.

Conceptual Problems

17.78 Which of the following ionic compounds will be more soluble in acid solution than in water: (a) $BaSO_4$, (b) $PbCl_2$, (c) $Fe(OH)_3$, (d) $CaCO_3$?

17.79 Which of the following will be more soluble in acid solution than in pure water: (a) CuI, (b) Ag_2SO_4, (c) $Zn(OH)_2$, (d) BaC_2O_4, (e) $Ca_3(PO_4)_2$?

SECTION 17.6: SEPARATION OF IONS USING DIFFERENCES IN SOLUBILITY

Review Questions

17.80 Outline the general procedure of qualitative analysis.

17.81 Give two examples of metal ions in each group (1 through 5) in the qualitative analysis scheme.

Computational Problems

17.82 Solid NaI is slowly added to a solution that is 0.010 M in Cu^+ and 0.010 M in Ag^+. (a) Which compound will begin to precipitate first? (b) Calculate $[Ag^+]$ when CuI just begins to precipitate. (c) What percent of Ag^+ remains in solution at this point?

17.83 Find the approximate pH range suitable for the separation of Fe^{3+} and Zn^{2+} ions by precipitation of $Fe(OH)_3$ from a solution that is initially 0.010 M in both Fe^{3+} and Zn^{2+}.

17.84 In a group 1 analysis, a student obtained a precipitate containing both AgCl and $PbCl_2$. Suggest one reagent that would enable the student to separate AgCl(s) from $PbCl_2$(s).

17.85 In a group 1 analysis, a student adds HCl acid to the unknown solution to make $[Cl^-] = 0.15$ M. Some $PbCl_2$ precipitates. Calculate the concentration of Pb^{2+} remaining in solution.

Conceptual Problems

17.86 Both KCl and NH_4Cl are white solids. Suggest one reagent that would enable you to distinguish between these two compounds.

17.87 Describe a simple test that would allow you to distinguish between $AgNO_3$(s) and $Cu(NO_3)_2$(s).

ADDITIONAL PROBLEMS

17.88 Sketch the titration curve of a weak acid with a strong base like the one shown in Figure 17.4. On your graph, indicate the volume of base used at the equivalence point and also at the half-equivalence point, that is, the point at which half of the acid has been neutralized. Explain how the measured pH at the half-equivalence point can be used to determine K_a of the acid.

17.89 A 200-mL volume of NaOH solution was added to 400 mL of a 2.00 M HNO_2 solution. The pH of the mixed solution was 1.50 units greater than that of the original acid solution. Calculate the molarity of the NaOH solution.

17.90 The pK_a of butyric acid (HBut) is 4.7. Calculate K_b for the butyrate ion (But$^-$).

17.91 A solution is made by mixing exactly 500 mL of 0.167 M NaOH with exactly 500 mL 0.100 M HCOOH. Calculate the equilibrium concentrations of Na^+, $HCOO^-$, OH^-, H^+, and HCOOH.

17.92 Tris [tris(hydroxymethyl)aminomethane] is a common buffer for studying biological systems: (a) Calculate the pH of the tris buffer after mixing 15.0 mL of 0.10 M HCl solution with 25.0 mL of 0.10 M tris. (b) This buffer was used to study an enzyme-catalyzed reaction. As a result of the reaction, 0.00015 mole of H^+ was consumed. What is the pH of the buffer at the end of the reaction? (c) What would be the final pH if no buffer were present?

$$HOCH_2-\underset{\underset{HOCH_2}{|}}{\overset{\overset{HOCH_2}{|}}{C}}-NH_3^+ \underset{}{\overset{pK_a=8.1}{\rightleftharpoons}} HOCH_2-\underset{\underset{HOCH_2}{|}}{\overset{\overset{HOCH_2}{|}}{C}}-NH_2 + H^+$$

17.93 $Cd(OH)_2$ is an insoluble compound. It dissolves in excess NaOH in solution. Write a balanced ionic equation for this reaction. What type of reaction is this?

17.94 A student mixes 50.0 mL of 1.00 M $Ba(OH)_2$ with 86.4 mL of 0.494 M H_2SO_4. Calculate the mass of $BaSO_4$ formed and the pH of the mixed solution.

17.95 For which of the following reactions is the equilibrium constant called a solubility product?
(a) $Zn(OH)_2(s) + 2OH^-(aq) \rightleftharpoons Zn(OH)_4^{2-}(aq)$
(b) $3Ca^{2+}(aq) + 2PO_4^{3-}(aq) \rightleftharpoons Ca_3(PO_4)_2(s)$
(c) $CaCO_3(s) + 2H^+(aq) \rightleftharpoons Ca^{2+}(aq) + H_2O(l) + CO_2(g)$
(d) $PbI_2(s) \rightleftharpoons Pb^{2+}(aq) + 2I^-(aq)$

17.96   A 2.0-L kettle contains 116 g of boiler scale ($CaCO_3$). How many times would the kettle have to be completely filled with distilled water to remove all the deposit at 25°C?

17.97   Equal volumes of 0.12 M $AgNO_3$ and 0.14 M $ZnCl_2$ solution are mixed. Calculate the equilibrium concentrations of Ag^+, Cl^-, Zn^{2+}, and NO_3^-.

17.98   Find the approximate pH range suitable for separating Mg^{2+} and Zn^{2+} by the precipitation of $Zn(OH)_2$ from a solution that is initially 0.010 M in Mg^{2+} and Zn^{2+}.

17.99   Calculate the solubility (in g/L) of Ag_2CO_3.

17.100   A volume of 25.0 mL of 0.100 M HCl is titrated against a 0.100 M CH_3NH_2 solution added to it from a burette. Calculate the pH values of the solution after (a) 10.0 mL of CH_3NH_2 solution has been added, (b) 25.0 mL of CH_3NH_2 solution has been added, (c) 35.0 mL of CH_3NH_2 solution has been added.

17.101   The molar solubility of $Pb(IO_3)_2$ in a 0.10 M $NaIO_3$ solution is 2.4×10^{-11} mol/L. What is K_{sp} for $Pb(IO_3)_2$?

17.102   When a KI solution was added to a solution of mercury(II) chloride, a precipitate [mercury(II) iodide] formed. A student plotted the mass of the precipitate versus the volume of the KI solution added and obtained the following graph. Explain the shape of the graph.

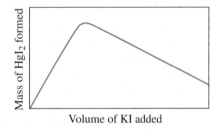

17.103   Barium is a toxic substance that can seriously impair heart function. For an X ray of the gastrointestinal tract, a patient drinks an aqueous suspension of 20 g $BaSO_4$. If this substance were to equilibrate with the 5.0 L of the blood in the patient's body, what would be [Ba^{2+}]? For a good estimate, we may assume that the K_{sp} of $BaSO_4$ at body temperature is the same as at 25°C. Why is $Ba(NO_3)_2$ not chosen for this procedure?

17.104   The pK_a of phenolphthalein is 9.10. Over what pH range does this indicator change from 95 percent HIn to 95 percent In^-?

17.105   Solid NaBr is slowly added to a solution that is 0.010 M in Cu^+ and 0.010 M in Ag^+. (a) Which compound will begin to precipitate first? (b) Calculate [Ag^+] when CuBr just begins to precipitate. (c) What percent of Ag^+ remains in solution at this point?

17.106   Cacodylic acid is $(CH_3)_2AsO_2H$. Its ionization constant is 6.4×10^{-7}. (a) Calculate the pH of 50.0 mL of a 0.10-M solution of the acid. (b) Calculate the pH of 25.0 mL of 0.15 M $(CH_3)_2AsO_2Na$. (c) Mix the solutions in parts (a) and (b). Calculate the pH of the resulting solution.

17.107   Radiochemical techniques are useful in estimating the solubility product of many compounds. In one experiment, 50.0 mL of a 0.010 M $AgNO_3$ solution containing a silver isotope with a radioactivity of 74,025 counts per min per mL was mixed with 100 mL of a 0.030 M $NaIO_3$ solution. The mixed solution was diluted to 500 mL and filtered to remove all the $AgIO_3$ precipitate. The remaining solution was found to have a radioactivity of 44.4 counts per min per mL. What is the K_{sp} of $AgIO_3$?

17.108   The molar mass of a certain metal carbonate, MCO_3, can be determined by adding an excess of HCl acid to react with all the carbonate and then "back-titrating" the remaining acid with NaOH. (a) Write an equation for these reactions. (b) In a certain experiment, 20.00 mL of 0.0800 M HCl was added to a 0.1022-g sample of MCO_3. The excess HCl required 5.64 mL of 0.1000 M NaOH for neutralization. Calculate the molar mass of the carbonate and identify M.

17.109   Acid-base reactions usually go to completion. Confirm this statement by calculating the equilibrium constant for each of the following cases: (a) a strong acid reacting with a strong base; (b) a strong acid reacting with a weak base (NH_3); (c) a weak acid (CH_3COOH) reacting with a strong base; (d) a weak acid (CH_3COOH) reacting with a weak base (NH_3). (*Hint:* Strong acids exist as H^+ ions and strong bases exist as OH^- ions in solution. You need to look up K_a, K_b, and K_w.)

17.110   Calculate x, the number of molecules of water in oxalic acid hydrate ($H_2C_2O_4 \cdot xH_2O$), from the following data: 5.00 g of the compound is made up to exactly 250 mL solution, and 25.0 mL of this solution requires 15.9 mL of 0.500 M NaOH solution for neutralization.

17.111   Describe how you would prepare a 1-L 0.20 M $CH_3COONa/0.20$ M CH_3COOH buffer system by (a) mixing a solution of CH_3COOH with a solution of CH_3COONa, (b) mixing a solution of CH_3COOH with a solution of NaOH, and (c) mixing a solution of CH_3COONa with a solution of HCl.

17.112   Phenolphthalein is the common indicator for the titration of a strong acid with a strong base. (a) If the pK_a of phenolphthalein is 9.10, what is the ratio of the nonionized form of the indicator (colorless) to the ionized form (reddish pink) at pH 8.00? (b) If 2 drops of 0.060 M phenolphthalein are used in a titration involving a 50.0-mL volume, what is the concentration of the ionized form at pH 8.00? (Assume that 1 drop = 0.050 mL.)

17.113 Of the reactions depicted, which best represents (a) what occurs when strong acid is added to a buffer solution, and (b) what occurs when strong base is added to a buffer solution?

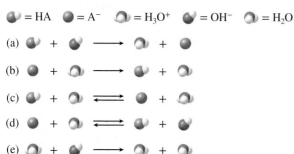

17.114 Oil paintings containing lead(II) compounds as constituents of their pigments darken over the years. Suggest a chemical reason for the color change.

17.115 What reagents would you employ to separate the following pairs of ions in solution: (a) Na^+ and Ba^{2+}, (b) K^+ and Pb^{2+}, (c) Zn^{2+} and Hg^{2+}?

17.116 Look up the K_{sp} values for $BaSO_4$ and $SrSO_4$ in Table 17.4. Calculate the concentrations of Ba^{2+}, Sr^{2+}, and SO_4^{2-} in a solution that is saturated with both compounds.

17.117 In principle, amphoteric oxides, such as Al_2O_3 and BeO, can be used to prepare buffer solutions because they possess both acidic and basic properties (see Section 16.11). Explain why these compounds are of little practical use as buffer components.

17.118 $CaSO_4$ ($K_{sp} = 2.4 \times 10^{-5}$) has a larger K_{sp} value than that of Ag_2SO_4 ($K_{sp} = 1.4 \times 10^{-5}$). Does it necessarily follow that $CaSO_4$ also has greater solubility (g/L)? Explain.

17.119 When lemon juice is added to tea, the color becomes lighter. In part, the color change is due to dilution, but the main reason for the change is an acid-base reaction. What is the reaction? (*Hint:* Tea contains "polyphenols," which are weak acids, and lemon juice contains citric acid.)

17.120 How many milliliters of 1.0 *M* NaOH must be added to 200 mL of 0.10 *M* NaH_2PO_4 to make a buffer solution with a pH of 7.50?

17.121 The maximum allowable concentration of Pb^{2+} ions in drinking water is 0.05 ppm (i.e., 0.05 g of Pb^{2+} in 1 million grams of water). Is this guideline exceeded if an underground water supply is at equilibrium with the mineral anglesite ($PbSO_4$) ($K_{sp} = 1.6 \times 10^{-8}$)?

17.122 The solutions shown contain one or more of the following compounds: H_2A (a weak diprotic acid), NaHA, and Na_2A (the sodium salts of HA^- and A^{2-}). (For clarity, the sodium ions and water molecules are not shown.) Which solutions are buffers? (Select all that apply.)

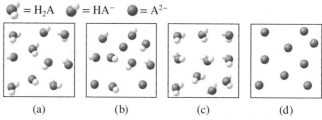

17.123 Which of the following solutions has the highest $[H^+]$: (a) 0.10 *M* HF, (b) 0.10 *M* HF in 0.10 *M* NaF, (c) 0.10 *M* HF in 0.10 *M* SbF_5? (*Hint:* SbF_5 reacts with F^- to form the complex ion SbF_6^-.)

17.124 Distribution curves show how the fractions of a nonionized acid and its conjugate base vary as a function of the pH of the medium. Plot distribution curves for CH_3COOH and its conjugate base CH_3COO^- in solution. Your graph should show fraction as the *y* axis and pH as the *x* axis. What are the fractions and pH at the point where these two curves intersect?

17.125 Calcium oxalate is a major component of kidney stones. Predict whether the formation of kidney stones can be minimized by increasing or decreasing the pH of the fluid present in the kidney. The pH of normal kidney fluid is about 8.2. [The first and second acid ionization constants of oxalic acid ($H_2C_2O_4$) are 6.5×10^{-2} and 6.1×10^{-5}, respectively. The solubility product of calcium oxalate is 3.0×10^{-9}.]

17.126 Water containing Ca^{2+} and Mg^{2+} ions is called *hard water* and is unsuitable for some household and industrial use because these ions react with soap to form insoluble salts, or curds. One way to remove the Ca^{2+} ions from hard water is by adding washing soda ($Na_2CO_3 \cdot 10H_2O$). (a) The molar solubility of $CaCO_3$ is 9.3×10^{-5} *M*. What is its molar solubility in a 0.050 *M* Na_2CO_3 solution? (b) Why are Mg^{2+} ions not removed by this procedure? (c) The Mg^{2+} ions are removed as $Mg(OH)_2$ by adding slaked lime [$Ca(OH)_2$] to the water to produce a saturated solution. Calculate the pH of a saturated $Ca(OH)_2$ solution. (d) What is the concentration of Mg^{2+} ions at this pH? (e) In general, which ion (Ca^{2+} or Mg^{2+}) would you remove first? Why?

17.127 Amino acids are building blocks of proteins. These compounds contain at least one amino group ($-NH_2$) and one carboxyl group ($-COOH$). Consider glycine (NH_2CH_2COOH). Depending on the pH of the solution, glycine can exist in one of three possible forms:

Fully protonated: $^+NH_3-CH_2-COOH$
Dipolar ion: $^+NH_3-CH_2-COO^-$
Fully ionized: $NH_2-CH_2-COO^-$

Predict the predominant form of glycine at pH 1.0, 7.0, and 12.0. The pK_a of the carboxyl group is 2.3 and that of the ammonium group ($-NH_3^+$) is 9.6.

17.128 Consider the ionization of the following acid-base indicator.

$$HIn(aq) \rightleftharpoons H^+(aq) + In^-(aq)$$

The indicator changes color according to the ratios of the concentrations of the acid to its conjugate base. When $[HIn]/[In^-] \geq 10$, color of acid (HIn) predominates. When $[HIn]/[In^-] \leq 0.1$, color of conjugate base (In^-) predominates. Show that the pH range over which the indicator changes from the acid color to the base color is $pH = pK_a \pm 1$, where K_a is the ionization constant of the acid HIn.

17.129 One way to distinguish a buffer solution from an acid solution is by dilution. (a) Consider a buffer solution made of 0.500 M CH_3COOH and 0.500 M CH_3COONa. Calculate its pH and the pH after it has been diluted 10-fold. (b) Compare the result in part (a) with the pH values of a 0.500 M CH_3COOH solution before and after it has been diluted 10-fold.

17.130 (a) Referring to Figure 17.4, describe how you would determine the pK_b of the base. (b) Derive an analogous Henderson-Hasselbalch equation relating pOH to pK_b of a weak base B and its conjugate acid HB^+. Sketch a titration curve showing the variation of the pOH of the base solution versus the volume of a strong acid added from a burette. Describe how you would determine the pK_b from this curve.

17.131 $AgNO_3$ is added slowly to a solution that contains 0.1 M each of Br^-, CO_3^{2-}, and SO_4^{2-} ions. What compound will precipitate first and what compound will precipitate last?

17.132 Diagrams (a) through (d) represent solutions of MX, which may also contain one or both of the soluble salts, MNO_3 and NaX. (Na^+ and NO_3^- ions are not shown.) If (a) represents a saturated solution of MX, classify each of the other solutions as unsaturated, saturated, or supersaturated.

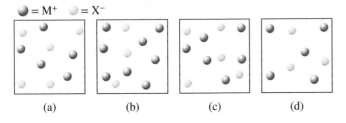

   (a)   (b)   (c)   (d)

17.133 Which of the following compounds, when added to water, will increase the solubility of CdS: (a) $LiNO_3$, (b) Na_2SO_4, (c) KCN, (d) $NaClO_3$?

17.134 The titration curve shown here represents the titration of a weak diprotic acid (H_2A) versus NaOH. (a) Label the major species present at the marked points. (b) Estimate the pK_{a_1} and pK_{a_2} values of the acid. Assume that any salt hydrolysis is negligible.

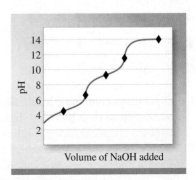

Volume of NaOH added

17.135 A 1.0-L saturated silver carbonate solution at 5°C is filtered to remove undissolved solid and treated with enough hydrochloric acid to decompose the dissolved compound. The carbon dioxide generated is collected in a 19-mL vial and exerts a pressure of 114 mmHg at 25°C. What is the K_{sp} of silver carbonate at 5°C?

17.136 Draw distribution curves for an aqueous carbonic acid solution. Your graph should show fraction of species present as the y axis and pH as the x axis. Note that at any pH, only two of the three species (H_2CO_3, HCO_3^-, and CO_3^{2-}) are present in appreciable concentrations. Use the K_a values in Table 16.7.

17.137 A sample of 0.96 L of HCl gas at 372 mmHg and 22°C is bubbled into 0.034 L of 0.57 M NH_3. What is the pH of the resulting solution? Assume the volume of solution remains constant and that the HCl is totally dissolved in the solution.

17.138 Histidine is one of the 20 amino acids found in proteins. Shown here is a fully protonated histidine molecule, where the numbers denote the pK_a values of the acidic groups.

(a) Show stepwise ionization of histidine in solution. (*Hint:* The H^+ ion will first come off from the strongest acid group followed by the next strongest acid group and so on.) (b) A dipolar ion is one in which the species has an equal number of positive and negative charges. Identify the dipolar ion in part (a). (c) The pH at which the dipolar ion predominates is called the isoelectric point, denoted by pI. The isoelectric point is the average of the pK_a values leading to and following the formation of the dipolar ion. Calculate the pI of histidine. (d) The histidine group plays an important role in buffering blood (the pH of blood is about 7.4). Which conjugate acid-base pair shown in part (a) is responsible for maintaining the pH of blood?

17.139 The solutions (a) through (f) represent various points in the titration of the weak acid HA with NaOH. (For clarity, the sodium ions and water molecules are not shown.)

(a) Which diagram corresponds to the beginning of the titration, prior to the addition of any NaOH?

(b) Which diagram corresponds to the equivalence point of the titration? (Ignore salt hydrolysis.)

(c) Which diagram illustrates the solution *after* the equivalence point, when excess NaOH has been added?

(d) Which diagram corresponds to the point at which the pH is equal to the pK_a of the weak acid?

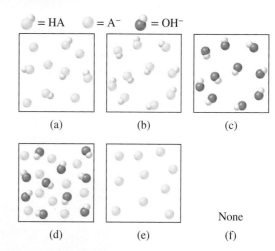

17.140 The buffer range is defined by the equation pH = p$K_a \pm 1$. Calculate the range of the ratio [conjugate base]/[acid] that corresponds to this equation.

17.141 The pK_a of the indicator methyl orange is 3.46. Over what pH range does this indicator change from 90 percent HIn to 90 percent In$^-$?

Answers to In-Chapter Materials

PRACTICE PROBLEMS

17.1A 4.22. **17.1B** 9.11. **17.2A** 5.0. **17.2B** 0.33 mol. **17.3A** Dissolve 0.6 mol CH$_3$COONa and 1 mol CH$_3$COOH in enough water to make 1 L of solution. **17.3B** 2.35–4.35. **17.4A** (a) 5.04, (b) 8.76, (c) 12.2. **17.4B** (a) 8.3 mL, (b) 12.5 mL, (c) 27.0 mL. **17.5A** 5.91. **17.5B** aniline. **17.6A** (a) bromophenol blue, methyl orange, methyl red, or chlorophenol; (b) any but thymol blue, bromophenol blue, or methyl orange; (c) cresol red or phenolphthalein. **17.6B** pyridine or aniline. **17.7A** (a) 1.3×10^{-5} M, 1.8×10^{-3} g/L; (b) 1.0×10^{-3} M, 1.5×10^{-11} g/L; (c) 4.0×10^{-5} M, 5.9×10^{-3} g/L. **17.7B** (a) 2.2×10^{-3} M, 0.53 g/L; (b) 1.3×10^{-4} M, 3.5×10^{-2} g/L; (c) 1.7×10^{-15} M, 8.8×10^{-13} g/L. **17.8A** (a) 1.5×10^{-14}, (b) 1.7×10^{-6}, (c) 1.3×10^{-15}. **17.8B** (a) 1.5×10^{-14}, (b) 4.8×10^{-29}, (c) 9.0×10^{-33}.

17.9A (a) yes, (b) yes, (c) no. **17.9B** (a) 4.4×10^{-8} g, (b) 1.2 g, (c) 0.037 g. **17.10A** (a) 9.1×10^{-9} M, (b) 8.3×10^{-14} M. **17.10B** Ag$_2$S < AgI < AgBr < AgCl < Ag$_2$CO$_3$. **17.11A** (a) yes, (b) yes, (c) no. **17.11B** Any salts containing the CO$_3^{2-}$ ion, the OH$^-$ ion, the S^{2-} ion, or any anion of a weak acid such as the SO$_3^{2-}$ ion or the F$^-$ ion. **17.12A** 1.8×10^{-34} M. **17.12B** 6.8×10^{-2} M. **17.13A** 0.60 M. **17.13B** (a) 1.6×10^{-9} M, (b) 2.6×10^{-6} M.

SECTION REVIEW

17.1.1 a, c, d. **17.1.2** b. **17.2.1** a. **17.2.2** c. **17.2.3** b. **17.2.4** d. **17.3.1** c. **17.3.2** a. **17.3.3** a. **17.4.1** a. **17.4.2** c. **17.4.3** e. **17.5.1** c. **17.5.2** b, d. **17.6.1** d. **17.6.2** b.

Chapter 18

Electrochemistry

© Friedrich Saurer/Alamy

CHEMICAL REACTIONS can be harnessed to release electrical energy; and the resulting energy can be stored in a battery for later use. Eventually, when the stored energy is used up, a battery "runs down," whereas a fuel cell (shown here) operates and produces energy continuously as long as the necessary reactants are supplied. Known in principle since 1838, fuel cells provided both electrical energy and potable water aboard the Apollo space missions. Now fuel cell technology is being used to power automobiles and provide a cleaner alternative to internal combustion engines.

Before You Begin, Review These Skills

- Oxidation numbers [◀◀ Section 9.4]
- Gibbs free energy [◀◀ Section 14.5]
- The reaction quotient, Q [◀◀ Section 15.2]

18.1 BALANCING REDOX REACTIONS

In Chapter 9 we briefly discussed oxidation-reduction or "redox" reactions, those in which electrons are transferred from one species to another. In this section we review how to identify a reaction as a redox reaction and look more closely at how such reactions are balanced.

A redox reaction is one in which there are *changes* in oxidation states, which we identify using the rules introduced in Chapter 9. The following are examples of redox reactions.

$$2KClO_3(s) \longrightarrow 2KCl(s) + 3O_2(g)$$
$$\underset{+1\ +5\ -2}{} \quad \underset{+1\ -1}{} \quad \underset{0}{}$$

$$CH_4(g) + 2O_2(g) \longrightarrow CO_2(g) + 2H_2O(l)$$
$$\underset{-4\ +1}{} \quad \underset{0}{} \quad \underset{+4\ -2}{} \quad \underset{+1\ -2}{}$$

$$Sn(s) + Cu^{2+}(aq) \longrightarrow Cu(s) + Sn^{2+}(aq)$$
$$\underset{0}{} \quad \underset{+2}{\phantom{Cu^{2+}}} \quad \underset{0}{} \quad \underset{+2}{\phantom{Sn^{2+}}}$$

Equations for redox reactions, such as those shown here, can be balanced by inspection, the method of balancing introduced in [◀◀ Section 8.1], but remember that redox equations must be balanced for mass (number of atoms) *and* for charge (number of electrons) [◀◀ Section 9.4]. In this section we introduce the *half-reaction method* to balance equations that cannot be balanced simply by inspection.

Consider the aqueous reaction of the iron(II) ion (Fe^{2+}) with the dichromate ion ($Cr_2O_7^{2-}$).

$$Fe^{2+} + Cr_2O_7^{2-} \longrightarrow Fe^{3+} + Cr^{3+}$$

Because there is no species containing oxygen on the product side of the equation, it would not be possible to balance this equation simply by adjusting the coefficients of reactants and products. However, there are two things about the reaction that make it possible to *add* species to the equation to balance it—without changing the chemical reaction it represents:

- The reaction takes place in aqueous solution, so we can add H_2O as needed to balance the equation.
- This particular reaction takes place in acidic solution, so we can add H^+ as needed to balance the equation. (Some reactions take place in basic solution, enabling us to add OH^- as needed for balancing. We will learn more about this shortly.)

After writing the unbalanced equation, we balance it stepwise as follows:

1. Separate the unbalanced reaction into *half-reactions.* A half-reaction is an oxidation or a reduction that occurs as part of the overall redox reaction.

$$\textit{Oxidation:} \qquad Fe^{2+} \longrightarrow Fe^{3+}$$

$$\textit{Reduction:} \qquad Cr_2O_7^{2-} \longrightarrow Cr^{3+}$$

Video 18.1
Electrochemistry—oxidation-reduction reactions.

Student Annotation: Now is a good time to review how oxidation numbers are assigned [◀◀ Section 9.4].

2. Balance each of the half-reactions with regard to atoms other than O and H. In this case, no change is required for the oxidation half-reaction. We adjust the coefficient of the chromium(III) ion to balance the reduction half-reaction.

Oxidation: $\qquad\qquad\qquad\qquad Fe^{2+} \longrightarrow Fe^3$

Reduction: $\qquad\qquad\qquad\qquad Cr_2O_7^{2-} \longrightarrow 2Cr^{3+}$

3. Balance both half-reactions for O by adding H_2O. Again, the oxidation in this case requires no change, but we must add seven water molecules to the product side of the reduction.

Oxidation: $\qquad\qquad\qquad\qquad Fe^{2+} \longrightarrow Fe^{3+}$

Reduction: $\qquad\qquad\qquad\qquad Cr_2O_7^{2-} \longrightarrow 2Cr^{3+} + 7H_2O$

4. Balance both half-reactions for H by adding H^+. Once again, the oxidation in this case requires no change, but we must add 14 hydrogen ions to the reactant side of the reduction.

Oxidation: $\qquad\qquad\qquad\qquad Fe^{2+} \longrightarrow Fe^{3+}$

Reduction: $\qquad 14H^+ + Cr_2O_7^{2-} \longrightarrow 2Cr^{3+} + 7H_2O$

5. Balance both half-reactions for charge by adding electrons. To do this, we determine the total charge on each side and add electrons to make the total charges equal. In the case of the oxidation, there is a charge of $+2$ on the reactant side and a charge of $+3$ on the product side. Adding one electron to the product side makes the charges equal.

Oxidation: $\qquad Fe^{2+} \longrightarrow Fe^{3+} + e^-$

Total charge: $\qquad +2 \qquad\qquad +2$

In the case of the reduction, there is a total charge of $[(14)(+1) + (1)(-2)] = +12$ on the reactant side and a total charge of $[(2)(+3)] = +6$ on the product side. Adding six electrons to the reactant side makes the charges equal.

Reduction: $\qquad 6e^- + 14H^+ + Cr_2O_7^{2-} \longrightarrow 2Cr^{3+} + 7H_2O$

Total charge: $\qquad\qquad\qquad +6 \qquad\qquad\qquad +6$

6. If the number of electrons in the balanced oxidation half-reaction is not the same as the number of electrons in the balanced reduction half-reaction, multiply one or both of the half-reactions by the number(s) required to make the number of electrons the same in both. In this case, with one electron in the oxidation and six in the reduction, multiplying the oxidation by 6 accomplishes this.

Oxidation: $\qquad\qquad\qquad\qquad 6(Fe^{2+} \longrightarrow Fe^{3+} + e^-)$

$\qquad\qquad\qquad\qquad\qquad\qquad 6Fe^{2+} \longrightarrow 6Fe^{3+} + 6e^-$

Reduction: $\qquad 6e^- + 14H^+ + Cr_2O_7^{2-} \longrightarrow 2Cr^{3+} + 7H_2O$

7. Finally, add the balanced half-reactions back together and cancel the electrons, in addition to any other identical terms that appear on both sides.

$$6Fe^{2+} \longrightarrow 6Fe^{3+} + 6e^-$$
$$\underline{6e^- + 14H^+ + Cr_2O_7^{2-} \longrightarrow 2Cr^{3+} + 7H_2O}$$
$$6Fe^{2+} + 14H^+ + Cr_2O_7^{2-} \longrightarrow 6Fe^{3+} + 2Cr^{3+} + 7H_2O$$

A final check shows that the resulting equation is balanced both for mass and for charge.

Some redox reactions occur in basic solution. When this is the case, balancing by the half-reaction method is done exactly as described for reactions in acidic solution, but it requires two additional steps:

8. For each H^+ ion in the final equation, add one OH^- ion to each side of the equation, combining the H^+ and OH^- ions to produce H_2O.
9. Make any additional cancellations made necessary by the new H_2O molecules.

Worked Example 18.1 shows how to use the half-reaction method to balance a reaction that takes place in basic solution.

Worked Example 18.1

Permanganate ion and iodide ion react in basic solution to produce manganese(IV) oxide and molecular iodine. Use the half-reaction method to balance the equation

$$MnO_4^- + I^- \longrightarrow MnO_2 + I_2$$

Strategy The reaction takes place in basic solution, so apply steps 1 through 9 to balance for mass and for charge.

Setup Identify the oxidation and reduction half-reactions by assigning oxidation numbers.

$$\underset{(+7)(-2)}{MnO_4^-} + \underset{(-1)}{I^-} \longrightarrow \underset{(+4)(-2)}{MnO_2} + \underset{(0)}{I_2}$$

Solution

Step 1. Separate the unbalanced reaction into half-reactions.

$$\textit{Oxidation:} \quad I^- \longrightarrow I_2$$

$$\textit{Reduction:} \quad MnO_4^- \longrightarrow MnO_2$$

Step 2. Balance each half-reaction for mass, excluding O and H.

$$2I^- \longrightarrow I_2$$
$$MnO_4^- \longrightarrow MnO_2$$

Step 3. Balance both half-reactions for O by adding H_2O.

$$2I^- \longrightarrow I_2$$
$$MnO_4^- \longrightarrow MnO_2 + 2H_2O$$

Step 4. Balance both half-reactions for H by adding H^+.

$$2I^- \longrightarrow I_2$$
$$4H^+ + MnO_4^- \longrightarrow MnO_2 + 2H_2O$$

Step 5. Balance the total charge of both half-reactions by adding electrons.

$$2I^- \longrightarrow I_2 + 2e^-$$
$$3e^- + 4H^+ + MnO_4^- \longrightarrow MnO_2 + 2H_2O$$

Step 6. Multiply the half-reactions to make the numbers of electrons the same in both.

$$3(2I^- \longrightarrow I_2 + 2e^-)$$
$$2(3e^- + 4H^+ + MnO_4^- \longrightarrow MnO_2 + 2H_2O)$$

(Continued on next page)

Step 7.   Add the half-reactions back together, canceling electrons.

$$6I^- \longrightarrow 3I_2 + 6e^-$$

$$6e^- + 8H^+ + 2MnO_4^- \longrightarrow 2MnO_2 + 4H_2O$$

$$\overline{8H^+ + 2MnO_4^- + 6I^- \longrightarrow 2MnO_2 + 3I_2 + 4H_2O}$$

Step 8.   For each H^+ ion in the final equation, add one OH^- ion to each side of the equation, combining the H^+ and OH^- ions to produce H_2O.

$$8H^+ + 2MnO_4^- + 6I^- \longrightarrow 2MnO_2 + 3I_2 + 4H_2O$$

$$+ \, 8OH^- \hspace{5cm} + \, 8OH^-$$

$$\overline{\enclose{circle}{8H_2O} + 2MnO_4^- + 6I^- \longrightarrow 2MnO_2 + 3I_2 + \enclose{circle}{4H_2O} + 8OH^-}$$

Step 9.   Carry out any cancellations made necessary by the additional H_2O molecules.

$$4H_2O + 2MnO_4^- + 6I^- \longrightarrow 2MnO_2 + 3I_2 + 8OH^-$$

Think About It

Verify that the final equation is balanced for mass and for charge. Remember that electrons cannot appear in the overall balanced equation.

Practice Problem **A**TTEMPT Use the half-reaction method to balance the following equation in basic solution.

$$CN^- + MnO_4^- \longrightarrow CNO^- + MnO_2$$

Practice Problem **B**UILD Use the half-reaction method to balance the following equation in acidic solution.

$$Fe^{2+} + MnO_4^- \longrightarrow Fe^{3+} + Mn^{2+}$$

Practice Problem **C**ONCEPTUALIZE In Chapter 8, you learned to balance chemical equations by changing coefficients only—it was not permissible to add species to the equation. Explain why it is all right to add water and hydronium (or hydroxide) to the equations in this chapter as part of the balancing process.

Section 18.1 Review

Balancing Redox Reactions

18.1.1   Which of the following equations does not represent a redox reaction? (Select all that apply.)

(a) $NH_3 + HCl \longrightarrow NH_4Cl$ (d) $3NO_2 + H_2O \longrightarrow NO + 2HNO_3$

(b) $2H_2O_2 \longrightarrow 2H_2O + O_2$ (e) $LiCl \longrightarrow Li^+ + Cl^-$

(c) $2O_3 \longrightarrow 3O_2$

18.1.2 MnO_4^- and $C_2O_4^{2-}$ react in basic solution to form MnO_2 and CO_3^{2-}. What are the coefficients of MnO_4^- and $C_2O_4^{2-}$ in the balanced equation?

(a) 1 and 1   (d) 2 and 6

(b) 2 and 1   (e) 2 and 2

(c) 2 and 3

18.2   GALVANIC CELLS

Video 18.2
Electrochemistry—operation of voltaic cell.

When zinc metal is placed in a solution containing copper(II) ions, Zn is oxidized to Zn^{2+} ions whereas Cu^{2+} ions are reduced to Cu [◄◄ Section 9.4].

$$Zn(s) + Cu^{2+}(aq) \longrightarrow Zn^{2+}(aq) + Cu(s)$$

Student Annotation: This reaction is shown in Figure 9.6.

The electrons are transferred directly from the reducing agent, Zn, to the oxidizing agent, Cu^{2+}, in solution. However, if we physically separate two half-reactions from each other, we can arrange it such that the electrons must travel through a wire to pass from the Zn atoms to the Cu^{2+} ions. As the reaction progresses, it generates a flow of electrons through the wire and thereby generates electricity.

The experimental apparatus for generating electricity through the use of a spontaneous reaction is called a **galvanic cell.** Figure 18.1 shows the essential components of a galvanic cell. A zinc bar is immersed in an aqueous $ZnSO_4$ solution in one container, and a copper bar is immersed in an aqueous $CuSO_4$ solution in another container. The cell operates on the principle that the oxidation of Zn to Zn^{2+} and the reduction of Cu^{2+} to Cu can be made to take place simultaneously in separate locations with the transfer of electrons between them occurring through an external wire. The zinc and copper bars are called **electrodes.** By definition, the **anode** in a galvanic cell is the electrode at which *oxidation* occurs and the **cathode** is the electrode at which *reduction* occurs. (Each combination of container, electrode, and solution is called a **half-cell.**) This particular arrangement of electrodes and electrolytes is called a Daniell cell.

Student Annotation: A *galvanic* cell can also be called a *voltaic* cell. Both terms refer to a cell in which a spontaneous chemical reaction generates a flow of electrons.

The half-reactions for the galvanic cell shown in Figure 18.1 are

Oxidation: $Zn(s) \longrightarrow Zn^{2+}(aq) + 2e^-$

Reduction: $Cu^{2+}(aq) + 2e^- \longrightarrow Cu(s)$

To complete the electric circuit, and allow electrons to flow through the external wire, the solutions must be connected by a conducting medium through which the cations and anions can move from one half-cell to the other. This requirement is satisfied by a **salt bridge,** which, in its simplest form, is an inverted U tube containing an inert electrolyte solution, such as KCl or NH_4NO_3. The ions in the salt bridge must not react with the other ions in solution or with the electrodes (see Figure 18.1). During the course of the redox reaction, electrons flow through the external wire from the anode (Zn electrode) to the cathode (Cu electrode). In the solution, the cations (Zn^{2+}, Cu^{2+}, and K^+) move toward the cathode, while the anions (SO_4^{2-} and Cl^-) move toward the anode. Without the salt bridge connecting the two solutions, the buildup of positive charge in the anode compartment (due to the departure of electrons and the resulting formation of Zn^{2+}) and the buildup of negative charge in the cathode compartment (created by the arrival of electrons and the reduction of Cu^{2+} ions to Cu) would quickly prevent the cell from operating.

Student Annotation: This is the origin of the terms *cathode* and *anode*:
• Cations move toward the cathode.
• Anions move toward the anode.

An electric current flows from the anode to the cathode because there is a difference in electrical potential energy between the electrodes. This flow of electric current is analogous to the flow of water down a waterfall, which occurs because there is a difference in the gravitational potential energy, or the flow of gas from a high-pressure region to a low-pressure region. Experimentally the difference in electrical potential between the anode and the cathode is measured by a voltmeter (Figure 18.2, page 836) and the reading (in volts) is called the **cell potential** (E_{cell}). The potential of a cell depends not only on the nature of the electrodes and the ions in solution, but also on the concentrations of the ions and the temperature at which the cell is operated.

Student Annotation: The volt is a derived SI unit: 1 V = 1 J/1 C.

The conventional notation for representing galvanic cells is the cell diagram. For the cell shown in Figure 18.1, if we assume that the concentrations of Zn^{2+} and Cu^{2+} ions are 1 *M*, the cell diagram is

Student Annotation: The terms *cell potential, cell voltage, cell electromotive force,* and *cell emf* are used interchangeably and are all symbolized the same way with E_{cell}.

$$Zn(s) \mid Zn^{2+}(1\ M) \parallel Cu^{2+}(1\ M) \mid Cu(s)$$

Figure 18.1

Construction of a Galvanic Cell

We add a salt bridge, a tube containing a solution of a strong electrolyte—in this case Na_2SO_4. Having this solution in electrical contact with the two solutions in the beakers allows ions to migrate toward the electrodes, ensuring that the two compartments remain electrically neutral.

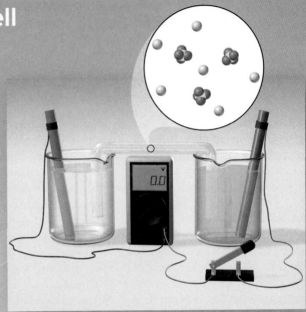

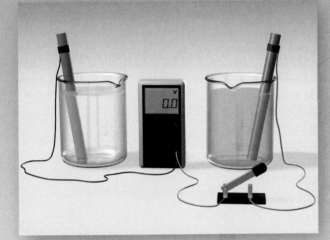

The two metal pieces are the electrodes in the galvanic cell. We connect the electrodes with a length of wire routed through a voltmeter and a switch so that we can complete the circuit when we have completed construction of the cell.

To make the reaction between zinc and copper more useful, we can construct a galvanic cell. In one beaker, we place a piece of zinc metal in a 1.00-M solution of Zn^{2+} ions. In the other, we place a piece of copper metal in a 1.00-M solution of Cu^{2+} ions.

START

As shown in Figure 9.6, when zinc metal (Zn) is immersed in a solution containing copper ions (Cu^{2+}), the zinc is oxidized to zinc ions (Zn^{2+}), and copper ions are reduced to copper metal (Cu).

$$Zn(s) + Cu^{2+}(aq) \longrightarrow Zn^{2+}(aq) + Cu(s)$$

This is an oxidation-reduction reaction in which electrons flow spontaneously from zinc metal to the copper ions in solution. The lightening of the blue color indicates that the concentration of Cu^{2+} has decreased. Copper metal is deposited on the solid zinc surface. Some of the zinc metal has gone into solution as Zn^{2+} ions, which do not impart any color in the solution.

before   after

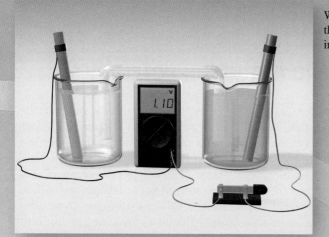

When we close the switch, we complete the circuit; the voltmeter indicates the initial potential of the cell: 1.10 V.

When we replace the voltmeter with a lightbulb, electrons flow from the zinc electrode (the anode) to the copper electrode (the cathode). The flow of electrons lights the bulb. Anions in the salt bridge migrate toward the anode; cations migrate toward the cathode.

As the reaction proceeds, zinc metal from the anode is oxidized, increasing the Zn^{2+} concentration in the beaker on the left; and additional copper metal is deposited on the cathode, decreasing the Cu^{2+} concentration in the beaker on the right. As the concentrations of both ions change, the potential of the cell decreases. After allowing the reaction to proceed for a time, we can reinsert the voltmeter to measure the decreased voltage.

What's the point?

Zinc is a stronger reducing agent than copper, so there is a natural tendency for electrons to flow from zinc metal to copper ions. We can harness this flow of electrons by forcing the half-reactions to occur in separate compartments. Electrons still flow from Zn to Cu^{2+}, but they must flow through the wire connecting the two electrodes. The potential of the cell decreases as the reaction proceeds, as the reading on the voltmeter shows.

The single vertical line represents a phase boundary. For example, the zinc electrode is a solid and the Zn^{2+} ions are in solution. Thus, we draw a line between Zn and Zn^{2+} to show the phase boundary. The double vertical lines denote the salt bridge. By convention, the anode is written first, to the left of the double lines, and the other components appear in the order in which we would encounter them in moving from the anode to the cathode (from left to right in the cell diagram).

18.3 STANDARD REDUCTION POTENTIALS

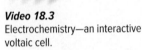

Video 18.3
Electrochemistry—an interactive voltaic cell.

Student Annotation: A cell in which different half-reactions take place will have a different emf.

When the concentrations of the Cu^{2+} and Zn^{2+} ions are both 1.0 M, the cell potential of the cell described in Section 18.2 is 1.10 V at 25°C (see Figure 18.2). This measured potential is related to the half-reactions that take place in the anode and cathode compartments. The overall cell potential is the difference between the electric potentials at the Zn and Cu electrodes—known as the *half-cell potentials*. Just as it is impossible for only one of the half-reactions to occur independently, it is impossible to measure the potential of just a single half-cell. However, if we arbitrarily define the potential value of a particular half-cell as zero, we can use it to determine the *relative* potentials of other half-cells. Then we can use these relative potentials to determine overall *cell* potentials. The hydrogen electrode, shown in Figure 18.3, serves as the reference for this purpose. Hydrogen gas is bubbled into a hydrochloric acid solution at 25°C. The platinum electrode has two functions. First, it provides a surface on which the dissociation (and oxidation) of hydrogen molecules can take place.

$$H_2 \longrightarrow 2H^+ + 2e^-$$

Second, it serves as an electrical conductor to the external circuit.

Under standard state conditions [◄◄ Section 14.5], when the pressure of H_2 is 1 atm and the concentration of HCl solution is 1 M, the potential for the reduction of H^+ at 25°C is defined as *exactly* zero.

$$2H^+(1\ M) + 2e^- \longrightarrow H_2(1\ atm) \qquad E° = 0\ V$$

As before, the ° superscript denotes standard state conditions, and $E°$ is the ***standard reduction potential;*** that is, $E°$ is the potential associated with a reduction half-reaction at an electrode when the ion concentration is 1 M and the gas pressure is 1 atm. Because the hydrogen electrode is used to determine all other electrode potentials, it is called the ***standard hydrogen electrode (SHE).***

Figure 18.2 The galvanic cell described in Figure 18.1. Note the U tube (salt bridge) connecting the two beakers. When the concentrations of Zn^{2+} and Cu^{2+} are 1 molar (1 M) at 25°C, the cell voltage is 1.10 V.
© McGraw-Hill Education/Ken Karp, photographer

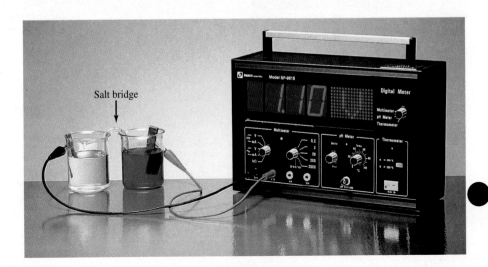

Salt bridge

Figure 18.3 A hydrogen electrode operating under standard state conditions. Hydrogen gas at 1 atm is bubbled through a 1 M HCl solution. The electrode itself is made of platinum.

A half-cell potential is measured by constructing a galvanic cell in which one of the electrodes is the SHE. The measured voltage is then used to determine the potential for the other half-cell. Figure 18.4(a) shows a galvanic cell with a zinc electrode and a SHE. When the circuit is completed in this cell, electrons flow from the zinc electrode to the SHE, thereby oxidizing Zn to Zn^{2+} and reducing H^+ ions to H_2. In this case, therefore, the zinc electrode is the anode (where oxidation takes place) and the SHE is the cathode (where reduction takes place). The cell diagram is

Student Annotation: One indication of the direction of electron flow is the fact that as the reaction proceeds, the mass of the zinc electrode decreases as a result of the oxidation half-reaction:

$$Zn(s) \longrightarrow Zn^{2+}(aq) + 2e^-$$

$$Zn(s) \mid Zn^{2+}(1 \ M) \parallel H^+(1 \ M) \mid H_2(1 \ atm) \mid Pt(s)$$

The measured potential for this cell is 0.76 V at 25°C. We can write the half-cell reactions as follows:

Anode (oxidation): $Zn(s) \longrightarrow Zn^{2+}(1 \ M) + 2e^-$

Cathode (reduction): $2H^+(1 \ M) + 2e^- \longrightarrow H_2(1 \ atm)$

Overall: $Zn(s) + 2H^+(1 \ M) \longrightarrow Zn^{2+}(1 \ M) + H_2(1 \ atm)$

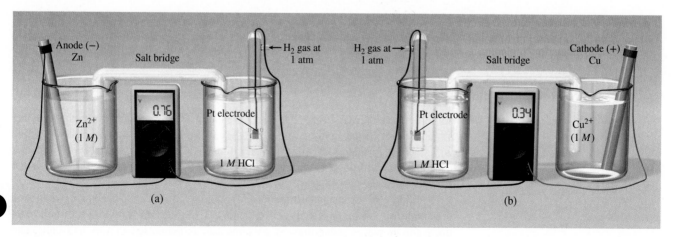

Figure 18.4 (a) A cell consisting of a zinc electrode and a hydrogen electrode. (b) A cell consisting of a copper electrode and a hydrogen electrode. Both cells are operating under standard state conditions. Note that in (a) the SHE is the cathode, but in (b) it is the anode.

By convention, the standard cell potential, E°_{cell}, which is composed of a contribution from the anode and a contribution from the cathode, is given by

Equation 18.1 $$E^\circ_{cell} = E^\circ_{cathode} - E^\circ_{anode}$$

where $E^\circ_{cathode}$ and E°_{anode} are the standard reduction potentials of the cathode and anode, respectively. For the Zn-SHE cell, we write

$$E^\circ_{cell} = E^\circ_{H^+/H_2} - E^\circ_{Zn^{2+}/Zn}$$

$$0.76 \text{ V} = 0 - E^\circ_{Zn^{2+}/Zn}$$

where the subscript "H^+/H_2" means "$2H^+ + 2e^- \longrightarrow H_2$" and the subscript "$Zn^{2+}/Zn$" means "$Zn^{2+} + 2e^- \longrightarrow Zn$." Thus, the standard reduction potential of zinc, $E^\circ_{Zn^{2+}/Zn}$, is -0.76 V.

The standard reduction potential of copper can be determined in a similar fashion, by using a cell with a copper electrode and a SHE [Figure 18.4(b)]. In this case, electrons flow from the SHE to the copper electrode when the circuit is completed; that is, the *copper* electrode is the cathode and the SHE is the *anode*. The cell diagram is

Student Annotation: Reduction of Cu^{2+} ions causes the mass of the Cu electrode to increase.

$$\text{Pt}(s) \mid H_2(1 \text{ atm}) \mid H^+(1 \text{ } M) \parallel Cu^{2+}(1 \text{ } M) \mid Cu(s)$$

and the half-cell reactions are

Anode (oxidation): $\quad H_2(1 \text{ atm}) \longrightarrow 2H^+(1 \text{ } M) + 2e^-$

Cathode (reduction): $\quad \underline{Cu^{2+}(1 \text{ } M) + 2e^- \longrightarrow Cu(s)}$

Overall: $\quad H_2(1 \text{ atm}) + Cu^{2+}(1 \text{ } M) \longrightarrow 2H^+(1 \text{ } M) + Cu(s)$

Under standard state conditions and at 25°C, the measured potential for the cell is 0.34 V, so we write

$$E^\circ_{cell} = E^\circ_{cathode} - E^\circ_{anode}$$
$$= E^\circ_{Cu^{2+}/Cu} - E^\circ_{H^+/H_2}$$
$$0.34 \text{ V} = E^\circ_{Cu^{2+}/Cu} - 0$$

Thus, the standard reduction potential of copper, $E^\circ_{Cu^{2+}/Cu}$, is 0.34 V, where the subscript "Cu^{2+}/Cu" means "$Cu^{2+} + 2e^- \longrightarrow Cu$."

Having determined the standard reduction potentials of Zn and Cu, we can use Equation 18.1 to calculate the cell potential for the Daniell cell described in Section 18.2.

$$E^\circ_{cell} = E^\circ_{cathode} - E^\circ_{anode}$$
$$= E^\circ_{Cu^{2+}/Cu} - E^\circ_{Zn^{2+}/Zn}$$
$$= 0.34 \text{ V} - (-0.76 \text{ V})$$
$$= 1.10 \text{ V}$$

As in the case of ΔG° [◄◄ Section 14.5], we can use the sign of E° to predict whether a reaction lies to the right or to the left. A positive E° means that the redox reaction will favor the formation of products at equilibrium. Conversely, a negative E° indicates that reactants will be favored at equilibrium. We will examine how E°_{cell}, ΔG°, and K are related in Section 18.4.

Table 18.1 lists standard reduction potentials, in order of decreasing reduction potential, for a number of half-cell reactions. To avoid ambiguity, all half-cell reactions are shown as reductions. A galvanic cell is composed of two half-cells, and therefore two half-cell reactions. Whether a particular half-cell reaction occurs as a reduction in a galvanic cell depends on how its reduction potential compares to that

TABLE 18.1 | Standard Reduction Potentials at 25°C[*]

Half-Reaction	$E°$(V)
$F_2(g) + 2e^- \longrightarrow 2F^-(aq)$	+2.87
$O_3(g) + 2H^+(aq) + 2e^- \longrightarrow O_2(g) + H_2O(l)$	+2.07
$Co^{3+}(aq) + e^- \longrightarrow Co^{2+}(aq)$	+1.82
$H_2O_2(aq) + 2H^+(aq) + 2e^- \longrightarrow 2H_2O(l)$	+1.77
$PbO_2(s) + 4H^+(aq) + SO_4^{2-}(aq) + 2e^- \longrightarrow PbSO_4(s) + 2H_2O(l)$	+1.70
$Ce^{4+}(aq) + e^- \longrightarrow Ce^{3+}(aq)$	+1.61
$MnO_4^-(aq) + 8H^+(aq) + 5e^- \longrightarrow Mn^{2+}(aq) + 4H_2O(l)$	+1.51
$Au^{3+}(aq) + 3e^- \longrightarrow Au(s)$	+1.50
$Cl_2(g) + 2e^- \longrightarrow 2Cl^-(aq)$	+1.36
$Cr_2O_7^{2-}(aq) + 14H^+(aq) + 6e^- \longrightarrow 2Cr^{3+}(aq) + 7H_2O(l)$	+1.33
$MnO_2(s) + 4H^+(aq) + 2e^- \longrightarrow Mn^{2+}(aq) + 2H_2O(l)$	+1.23
$O_2(g) + 4H^+(aq) + 4e^- \longrightarrow 2H_2O(l)$	+1.23
$Br_2(l) + 2e^- \longrightarrow 2Br^-(aq)$	+1.07
$NO_3^-(aq) + 4H^+(aq) + 3e^- \longrightarrow NO(g) + 2H_2O(l)$	+0.96
$2Hg^{2+}(aq) + 2e^- \longrightarrow Hg_2^{2+}(aq)$	+0.92
$Hg_2^{2+}(aq) + 2e^- \longrightarrow 2Hg(l)$	+0.85
$Ag^+(aq) + e^- \longrightarrow Ag(s)$	+0.80
$Fe^{3+}(aq) + e^- \longrightarrow Fe^{2+}(aq)$	+0.77
$O_2(g) + 2H^+(aq) + 2e^- \longrightarrow H_2O_2(aq)$	+0.68
$MnO_4^-(aq) + 2H_2O(l) + 3e^- \longrightarrow MnO_2(s) + 4OH^-(aq)$	+0.59
$I_2(s) + 2e^- \longrightarrow 2I^-(aq)$	+0.53
$O_2(g) + 2H_2O(l) + 4e^- \longrightarrow 4OH^-(aq)$	+0.40
$Cu^{2+}(aq) + 2e^- \longrightarrow Cu(s)$	+0.34
$AgCl(s) + e^- \longrightarrow Ag(s) + Cl^-(aq)$	+0.22
$SO_4^{2-}(aq) + 4H^+(aq) + 2e^- \longrightarrow SO_2(g) + 2H_2O(l)$	+0.20
$Cu^{2+}(aq) + e^- \longrightarrow Cu^+(aq)$	+0.15
$Sn^{4+}(aq) + 2e^- \longrightarrow Sn^{2+}(aq)$	+0.13
$2H^+(aq) + 2e^- \longrightarrow H_2(g)$	0.00
$Pb^{2+}(aq) + 2e^- \longrightarrow Pb(s)$	−0.13
$Sn^{2+}(aq) + 2e^- \longrightarrow Sn(s)$	−0.14
$Ni^{2+}(aq) + 2e^- \longrightarrow Ni(s)$	−0.25
$Co^{2+}(aq) + 2e^- \longrightarrow Co(s)$	−0.28
$PbSO_4(s) + 2e^- \longrightarrow Pb(s) + SO_4^{2-}(aq)$	−0.31
$Cd^{2+}(aq) + 2e^- \longrightarrow Cd(s)$	−0.40
$Fe^{2+}(aq) + 2e^- \longrightarrow Fe(s)$	−0.44
$Cr^{3+}(aq) + 3e^- \longrightarrow Cr(s)$	−0.74
$Zn^{2+}(aq) + 2e^- \longrightarrow Zn(s)$	−0.76
$2H_2O(l) + 2e^- \longrightarrow H_2(g) + 2OH^-(aq)$	−0.83
$Mn^{2+}(aq) + 2e^- \longrightarrow Mn(s)$	−1.18
$Al^{3+}(aq) + 3e^- \longrightarrow Al(s)$	−1.66
$Be^{2+}(aq) + 2e^- \longrightarrow Be(s)$	−1.85
$Mg^{2+}(aq) + 2e^- \longrightarrow Mg(s)$	−2.37
$Na^+(aq) + e^- \longrightarrow Na(s)$	−2.71
$Ca^{2+}(aq) + 2e^- \longrightarrow Ca(s)$	−2.87
$Sr^{2+}(aq) + 2e^- \longrightarrow Sr(s)$	−2.89
$Ba^{2+}(aq) + 2e^- \longrightarrow Ba(s)$	−2.90
$K^+(aq) + e^- \longrightarrow K(s)$	−2.93
$Li^+(aq) + e^- \longrightarrow Li(s)$	−3.05

Increasing strength as oxidizing agent

Increasing strength as reducing agent

[*]For all half-reactions the concentration is 1 M for dissolved species and the pressure is 1 atm for gases. These are the standard state values.

of the other half-cell reaction. If it has the greater (or more *positive*) reduction potential of the two, it will occur as a reduction. If it has the smaller (or more *negative*) reduction potential of the two, it will occur in the reverse direction, as an oxidation.

Consider the example of the cell described in Section 18.2. The two half-cell reactions and their standard reduction potentials are

$$Cu^{2+} + 2e^- \longrightarrow Cu \qquad E° = 0.34 \text{ V}$$

$$Zn^{2+} + 2e^- \longrightarrow Zn \qquad E° = -0.76 \text{ V}$$

The Cu half-reaction, having a greater (more positive) reduction potential, is the half-reaction that will occur as a reduction.

$$Cu^{2+} + 2e^- \longrightarrow Cu$$

The Zn half-reaction has a smaller (less positive) reduction potential and will occur, instead, as an oxidation.

$$Zn \longrightarrow Zn^{2+} + 2e^-$$

Adding the two half-reactions gives the overall cell reaction.

$$Zn + Cu^{2+} \longrightarrow Zn^{2+} + Cu$$

Consider, however, what would happen if we were to construct a galvanic cell combining the Zn half-cell with an Mn half-cell. The reduction potential of Mn is −1.18 V.

$$Zn^{2+} + 2e^- \longrightarrow Zn \qquad E° = -0.76 \text{ V}$$

$$Mn^{2+} + 2e^- \longrightarrow Mn \qquad E° = -1.18 \text{ V}$$

In this case, the greater (less negative) reduction potential is that of Zn, so the Zn half-reaction will occur as a reduction and the Zn electrode will be the cathode. The Mn electrode will be the anode. The cell potential is therefore

$$E°_{cell} = E°_{cathode} - E°_{anode}$$

$$= E°_{Zn^{2+}/Zn} - E°_{Mn^{2+}/Mn}$$

$$= (-0.76 \text{ V}) - (-1.18 \text{ V})$$

$$= 0.42 \text{ V}$$

The overall cell reaction is

$$Mn + Zn^{2+} \longrightarrow Mn^{2+} + Zn$$

By using Equation 18.1, we can predict the direction of an overall cell reaction.

It is important to understand that the standard reduction potential is an *intensive* property (like temperature and density), not an *extensive* property (like mass and volume) [◀◀ Section 1.6]. This means that the value of the standard reduction potential does *not* depend on the amount of a substance involved. Therefore, when it is necessary to multiply one of the half-reactions by a coefficient to balance the overall equation, the value of $E°$ for the half-reaction remains the same. Consider a galvanic cell made up of a Zn half-cell and an Ag half-cell.

$$Zn(s) \mid Zn^{2+}(1 \text{ } M) \parallel Ag^+(1 \text{ } M) \mid Ag(s)$$

The half-cell reactions are

$$Ag^+ + e^- \longrightarrow Ag \qquad E° = 0.80 \text{ V}$$

$$Zn^{2+} + 2e^- \longrightarrow Zn \qquad E° = -0.76 \text{ V}$$

The Ag half-reaction, with the more positive standard reduction potential, will occur as a reduction, and the Zn half-reaction will occur as an oxidation. Balancing the equation for the overall cell reaction requires multiplying the reduction (the Ag half-reaction) by 2.

$$2(Ag^+ + e^- \longrightarrow Ag)$$

We can then add the two half-reactions and cancel the electrons to get the overall, balanced equation.

$$2Ag^+ + 2e^- \longrightarrow 2Ag$$
$$\underline{\qquad + Zn \longrightarrow Zn^{2+} + 2e^-}$$
$$2Ag^+ + Zn \longrightarrow 2Ag + Zn^{2+}$$

The standard cell potential can be calculated using Equation 18.1.

$$E^\circ_{cell} = E^\circ_{cathode} - E^\circ_{anode}$$
$$= E^\circ_{Ag^+/Ag} - E^\circ_{Zn^{2+}/Zn}$$
$$= 0.80\ V - (-0.76\ V)$$
$$= 1.56\ V$$

Although we multiplied the Ag half-reaction by 2, we did *not* multiply its standard reduction potential by 2.

Table 18.1 is essentially an extended version of the activity series [◄◄ Section 9.4]. Worked Example 18.2 illustrates how to use standard reduction potentials to predict the direction of the overall reaction in a galvanic cell.

Student Hot Spot

Student data indicate you may struggle with reduction potentials. Access the SmartBook to view additional Learning Resources on this topic.

Student Annotation: It is a common error to multiply E° values by the same number as the half-cell equation. Think of the reduction potential as the height of a waterfall. Just as water falls from a higher elevation to a lower elevation, electrons move from an electrode with a higher potential toward one with a lower potential. The amount of water falling does not affect the net change in its elevation. Likewise, the number of electrons moving from higher potential to lower potential does not affect the size of the change in potential.

Worked Example 18.2

A galvanic cell consists of an Mg electrode in a 1.0 M Mg(NO$_3$)$_2$ solution and a Cd electrode in a 1.0 M Cd(NO$_3$)$_2$ solution. Determine the overall cell reaction, and calculate the standard cell potential at 25°C.

Strategy Use the tabulated values of E° to determine which electrode is the cathode and which is the anode, combine cathode and anode half-cell reactions to get the overall cell reaction, and use Equation 18.1 to calculate E°_{cell}.

Setup The half-cell reactions and their standard reduction potentials are

$$Mg^{2+} + 2e^- \longrightarrow Mg \qquad E^\circ = -2.37\ V$$
$$Cd^{2+} + 2e^- \longrightarrow Cd \qquad E^\circ = -0.40\ V$$

Because the Cd half-cell reaction has the greater (less negative) standard reduction potential, it will occur as the reduction. The Mg half-cell reaction will occur as the oxidation. Therefore, $E^\circ_{cathode} = -0.40\ V$ and $E^\circ_{anode} = -2.37\ V$.

Solution Adding the two half-cell reactions together gives the overall cell reaction.

$$Mg \longrightarrow Mg^{2+} + 2e^-$$
$$\underline{Cd^{2+} + 2e^- \longrightarrow Cd}$$
$$Overall:\quad Mg + Cd^{2+} \longrightarrow Mg^{2+} + Cd$$

The standard cell potential is

$$E^\circ_{cell} = E^\circ_{cathode} - E^\circ_{anode}$$
$$= E^\circ_{Cd^{2+}/Cd} - E^\circ_{Mg^{2+}/Mg}$$
$$= (-0.40\ V) - (-2.37\ V)$$
$$= 1.97\ V$$

(Continued on next page)

Think About It

If you ever calculate a negative voltage for a galvanic cell potential, you have done something wrong—check your work. Under standard state conditions, the overall cell reaction will proceed in the direction that gives a positive $E°_{cell}$.

Practice Problem **A** **TTEMPT** Determine the overall cell reaction and $E°_{cell}$ (at 25°C) of a galvanic cell made of a Cd electrode in a 1.0 M Cd(NO$_3$)$_2$ solution and a Pb electrode in a 1.0 M Pb(NO$_3$)$_2$ solution.

Practice Problem **B** **UILD** A galvanic cell with $E°_{cell}$ = 0.30 V can be constructed using an iron electrode in a 1.0 M Fe(NO$_3$)$_2$ solution, and either a tin electrode in a 1.0 M Sn(NO$_3$)$_2$ solution, or a chromium electrode in a 1.0 M Cr(NO$_3$)$_3$ solution—even though Sn^{2+}/Sn and Cr^{3+}/Cr have different standard reduction potentials. Explain and give the overall balanced reaction for each cell.

Practice Problem **C** **ONCEPTUALIZE** For a galvanic cell consisting of a Zn electrode immersed in a solution that is 0.10 M in Zn^{2+}(aq) and a Cu electrode immersed in a solution that is 0.10 M in Cu^{2+}(aq), which of the graphs [(i)–(iv)] best represents the concentrations of metal ions as a function of time?

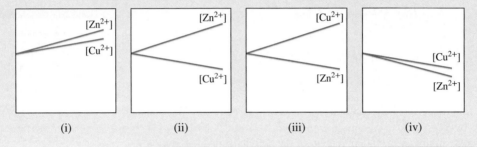

(i)   (ii)   (iii)   (iv)

Standard reduction potentials can also be used to determine what, if any, redox reaction will take place when reactants are combined in the same beaker—rather than divided into half-cells. Unlike a galvanic cell, in which we use reduction potentials to determine if the reaction proceeds in the *forward* or *reverse* direction, when the reaction is not divided into half-cells, we use reduction potentials to determine *if* the reaction will occur as written—or *not*.

Worked Example 18.3 illustrates this technique.

Worked Example 18.3

Predict what redox reaction will take place, if any, when molecular bromine (Br$_2$) is added to (a) a 1-M solution of NaI and (b) a 1-M solution of NaCl. (Assume a temperature of 25°C.)

Strategy In each case, write the equation for the redox reaction that *might* take place and use $E°$ values to determine whether the proposed reaction will actually occur.

Setup From Table 18.1:

$$Br_2(l) + 2e^- \longrightarrow 2Br^-(aq) \qquad E° = 1.07 \text{ V}$$

$$I_2(s) + 2e^- \longrightarrow 2I^-(aq) \qquad E° = 0.53 \text{ V}$$

$$Cl_2(g) + 2e^- \longrightarrow 2Cl^-(aq) \qquad E° = 1.36 \text{ V}$$

Solution (a) If a redox reaction is to occur, it will be the oxidation of I$^-$ ions by Br$_2$.

$$Br_2(l) + 2I^-(aq) \longrightarrow 2Br^-(aq) + I_2(s)$$

Because the reduction potential of Br$_2$ is greater than that of I$_2$, Br$_2$ will be reduced to Br$^-$ and I$^-$ will be oxidized to I$_2$. Thus, the preceding reaction *will* occur.

(b) In this case, the proposed reaction is the reduction of Br$_2$ by Cl$^-$ ions.

$$Br_2(l) + 2Cl^-(aq) \longrightarrow 2Br^-(aq) + Cl_2(g)$$

However, because the reduction potential of Br$_2$ is smaller than that of Cl$_2$, this reaction will *not* occur. Cl$_2$ is more readily reduced than Br$_2$, so Br$_2$ is not reduced by Cl$^-$.

Think About It

We can use Equation 18.1 and treat problems of this type like galvanic cell problems. Write the proposed redox reaction, and identify the "cathode" and the "anode." If the calculated $E°_{cell}$ is positive, the reaction will occur. If the calculated $E°_{cell}$ is negative, the reaction will not occur.

Practice Problem **A**TTEMPT Determine what redox reaction, if any, occurs (at 25°C) when lead metal (Pb) is added to (a) a 1.0-M solution of $NiCl_2$ and (b) a 1.0-M solution of HCl.

Practice Problem **B**UILD Would it be safer to store a cobalt(II) chloride solution in a tin container or an iron container? Explain.

Practice Problem **C**ONCEPTUALIZE A piece of nickel metal is added to a solution that is 1.0 M in three different chloride salts: $CoCl_2$, $NiCl_2$, and $SnCl_2$. Which of the graphs [(i)–(iv)] best represents the concentrations of metal ions in solution as a function of time?

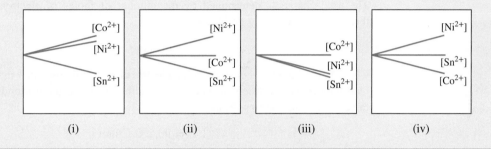

(i)   (ii)   (iii)   (iv)

Section 18.3 Review

Standard Reduction Potentials

18.3.1 Calculate $E°_{cell}$ at 25°C for a galvanic cell made of a Cr electrode in a solution that is 1.0 M in Cr^{3+} and an Au electrode in a solution that is 1.0 M in Au^{3+}.
   (a) −2.03 V   (d) 2.24 V
   (b) 1.11 V   (e) 0.76 V
   (c) −1.11 V

18.3.2 Calculate $E°_{cell}$ at 25°C for a galvanic cell made of a Cr electrode in a solution that is 1.0 M in Cr^{3+} and an Ag electrode in a solution that is 1.0 M in Ag^+.
   (a) 0.06 V   (d) 3.14 V
   (b) −0.06 V   (e) 3.02 V
   (c) 1.54 V

18.3.3 What redox reaction, if any, will occur at 25°C when Al metal is placed in a solution that is 1.0 M in Cr^{3+}?
   (a) $Al^{3+} + Cr \longrightarrow Al + Cr^{3+}$
   (b) $Cr^{3+} + Al \longrightarrow Cr + Al^{3+}$
   (c) $Al^{3+} + Cr^{3+} \longrightarrow Al + Cr$
   (d) $Al + Cr \longrightarrow Al^{3+} + Cr^{3+}$
   (e) No redox reaction will occur.

18.3.4 What redox reaction, if any, will occur at 25°C when Zn metal is placed in a solution that is 1.0 M in Sr^{2+}?
   (a) $Zn^{2+} + Sr \longrightarrow Zn + Sr^{2+}$
   (b) $Sr^{2+} + Zn \longrightarrow Sr + Zn^{2+}$
   (c) $Zn^{2+} + Sr^{2+} \longrightarrow Zn + Sr$
   (d) $Zn + Sr \longrightarrow Zn^{2+} + Sr^{2+}$
   (e) No redox reaction will occur.

18.4   SPONTANEITY OF REDOX REACTIONS UNDER STANDARD STATE CONDITIONS

We now look at how E_{cell}° is related to thermodynamic quantities such as ΔG° and K. In a galvanic cell, chemical energy is converted to electric energy. The electric energy produced in a galvanic cell is the product of the cell potential and the total electric charge (in coulombs) that passes through the cell.

$$\text{electric energy} = \text{volts} \times \text{coulombs}$$

$$= \text{joules}$$

The total charge is determined by the number of moles of electrons (n) that pass through the circuit. By definition,

$$\text{total charge} = nF$$

where F, the Faraday[1] constant, is the electric charge contained in 1 mole of electrons. One faraday is equivalent to 96,485.3 C, although we usually round the number to three significant figures. Thus,

$$1\ F = 96,500\ \text{C/mol } e^{-}$$

Because

$$1\ \text{J} = 1\ \text{C} \times 1\ \text{V}$$

we can also express the units of faraday as

$$1\ F = 96,500\ \text{J/V} \cdot \text{mol } e^{-}$$

The measured cell potential is the maximum voltage that the cell can produce. This value is used to calculate the maximum amount of electric energy that can be obtained from the chemical reaction. This energy is used to do electrical work ($w_{electrical}$), so

$$w_{max} = w_{electrical}$$

$$= -nFE_{cell}$$

where w_{max} is the maximum amount of work that can be done. The negative sign on the right-hand side indicates that the electrical work is done *by* the system *on* the surroundings. In Chapter 14 we defined *free energy* as the energy available to do *work*. Specifically, the change in free energy, ΔG, represents the maximum amount of useful work that can be obtained from a reaction.

$$\Delta G = w_{max}$$

Therefore, we can write

Equation 18.2	$\Delta G = -nFE_{cell}$

Both n and F are positive quantities and ΔG is negative for a spontaneous process, so E_{cell} must be positive for a spontaneous process. For reactions in which reactants and products are in their standard states, Equation 18.2 becomes

Equation 18.3	$\Delta G^{\circ} = -nFE_{cell}^{\circ}$

---

[1]Michael Faraday (1791–1867), an English chemist and physicist, is regarded by many as the greatest experimental scientist of the nineteenth century. He started as an apprentice to a bookbinder at the age of 13, but became interested in science after reading a book on chemistry. Faraday invented the electric motor and was the first person to demonstrate the principle governing electric generators. Besides making notable contributions to the fields of electricity and magnetism, Faraday also worked on optical activity and discovered and named benzene.

Equation 18.3 makes it possible to relate E_{cell}° to the equilibrium constant, K, of a redox reaction. In Section 15.4 we saw that the standard free-energy change, ΔG°, for a reaction is related to its equilibrium constant as follows [◄◄ Section 15.4, Equation 15.6]:

$$\Delta G^{\circ} = -RT \ln K$$

Therefore, if we combine Equations 15.6 and 18.3, we get

$$-nFE_{cell}^{\circ} = -RT \ln K$$

Solving for E_{cell}° gives

$$E_{cell}^{\circ} = \frac{RT}{nF} \ln K \qquad \textbf{Equation 18.4}$$

When $T = 298$ K and n moles of electrons are transferred per mole of reaction, Equation 18.4 can be simplified by inserting the values for R and F.

$$E_{cell}^{\circ} = \frac{\left(8.314 \dfrac{J}{K \cdot mol}\right)(298\ K)}{(n)\left(96,500 \dfrac{J}{V \cdot mol}\right)}$$

$$= \frac{0.0257\ V}{n} \ln K$$

And, by converting to the base-10 logarithm of K, we get

$$E_{cell}^{\circ} = \frac{0.0592\ V}{n} \log K \qquad (\text{at } 25^{\circ}C) \qquad \textbf{Equation 18.5}$$

Thus, if we know any one of the three quantities ΔG°, K, or E_{cell}°, we can convert to the others by using Equations 15.6, 18.3, and 18.5.

Worked Examples 18.4 and 18.5 demonstrate the interconversions among ΔG°, K, and E_{cell}°. For simplicity, the subscript "cell" is not shown.

Worked Example 18.4

Calculate the standard free-energy change for the following reaction at 25°C.

$$2Au(s) + 3Ca^{2+}(1.0\ M) \rightleftharpoons 2Au^{3+}(1.0\ M) + 3Ca(s)$$

Strategy Use E° values from Table 18.1 to calculate E° for the reaction, and then use Equation 18.3 to calculate the standard free-energy change.

Setup The half-cell reactions are

$$\text{Cathode (reduction):} \quad 3Ca^{2+}(aq) + 6e^{-} \longrightarrow 3Ca(s)$$

$$\text{Anode (oxidation):} \quad 2Au(s) \longrightarrow 2Au^{3+}(aq) + 6e^{-}$$

From Table 18.1, $E_{Ca^{2+}/Ca}^{\circ} = -2.87$ V and $E_{Au^{3+}/Au}^{\circ} = 1.50$ V.

Solution

$$E_{cell}^{\circ} = E_{cathode}^{\circ} - E_{anode}^{\circ}$$

$$= E_{Ca^{2+}/Ca}^{\circ} - E_{Au^{3+}/Au}^{\circ}$$

$$= -2.87\ V - 1.50\ V$$

$$= -4.37\ V$$

Next, substitute this value of E° into Equation 18.3 to obtain ΔG°.

$$\Delta G^{\circ} = -nFE^{\circ}$$

(Continued on next page)

The overall reaction shows that $n = 6$, so

$$\Delta G^\circ = -(6 \text{ mol } e^-)(96,500 \text{ J/V} \cdot \text{mol } e^-)(-4.37 \text{ V})$$

$$= 2.53 \times 10^6 \text{ J/mol}$$

$$= 2.53 \times 10^3 \text{ kJ/mol}$$

Think About It

The large positive value of ΔG° indicates that reactants are favored at equilibrium, which is consistent with the fact that E° for the reaction is negative.

Practice Problem **A**TTEMPT   Calculate ΔG° for the following reaction at 25°C.

$$3Mg(s) + 2Al^{3+}(aq) \rightleftharpoons 3Mg^{2+}(aq) + 2Al(s)$$

Practice Problem **B**UILD   Calculate ΔG° for the following reaction at 25°C.

$$Pb(s) + Ni^{2+}(aq) \rightleftharpoons Pb^{2+}(aq) + Ni(s)$$

Practice Problem **C**ONCEPTUALIZE   Which of the graphs [(i)–(iv)] best represents the relationship between ΔG° and E° for a chemical reaction?

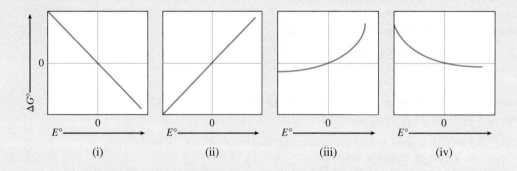

Worked Example 18.5

Calculate the equilibrium constant for the following reaction at 25°C.

$$Sn(s) + 2Cu^{2+}(aq) \rightleftharpoons Sn^{2+}(aq) + 2Cu^+(aq)$$

Strategy Use E° values from Table 18.1 to calculate E° for the reaction, and then calculate the equilibrium constant using Equation 18.5 (rearranged to solve for K).

Setup The half-cell reactions are

$$\textit{Cathode (reduction):} \quad 2Cu^{2+}(aq) + 2e^- \longrightarrow 2Cu^+(aq)$$

$$\textit{Anode (oxidation):} \quad Sn(s) \longrightarrow Sn^{2+}(aq) + 2e^-$$

From Table 18.1, $E^\circ_{Cu^{2+}/Cu^+} = 0.15$ V and $E^\circ_{Sn^{2+}/Sn} = -0.14$ V.

Solution

$$E^\circ_{cell} = E^\circ_{cathode} - E^\circ_{anode}$$

$$= E^\circ_{Cu^{2+}/Cu^+} - E^\circ_{Sn^{2+}/Sn}$$

$$= 0.15 \text{ V} - (-0.14 \text{ V})$$

$$= 0.29 \text{ V}$$

Solving Equation 18.5 for K gives

$$K = 10^{nE°/(0.0592\ V)}$$

$$= 10^{(2)(0.29\ V)/(0.0592\ V)}$$

$$= 6 \times 10^9$$

Think About It
A positive standard cell potential corresponds to a large equilibrium constant.

Practice Problem ATTEMPT Calculate the equilibrium constant for the following reaction at 25°C.

$$2Ag(s) + Fe^{2+}(aq) \rightleftharpoons 2Ag^+(aq) + Fe(s)$$

Practice Problem BUILD Like equilibrium constants, $E°_{cell}$ values are temperature dependent. At 80°C, $E°_{cell}$ for the cell diagram shown is 0.18 V.

$$Pt(s)\ |\ H_2(g)\ |\ HCl(aq)\ ||\ AgCl(s)\ |\ Ag(s)$$

The corresponding cell reaction is

$$H_2(g) + 2AgCl(s) \rightleftharpoons 2Ag(s) + 2H^+(aq) + 2Cl^-(aq)$$

Calculate the equilibrium constant for this reaction at 80°C.

Practice Problem CONCEPTUALIZE Which of the graphs [(i)–(iv)] best represents the relationship between K and $E°$ for a chemical reaction?

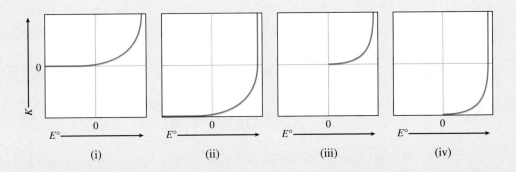

(i)　　　　(ii)　　　　(iii)　　　　(iv)

Section 18.4 Review

Spontaneity of Redox Reactions Under Standard State Conditions

18.4.1 Calculate K at 25°C for the following reaction.

$$Fe^{2+}(aq) + Ni(s) \rightleftharpoons Fe(s) + Ni^{2+}(aq)$$

(a) 6×10^{-4}　　　　(d) 1×10^{-13}
(b) 4×10^{-7}　　　　(e) 2×10^3
(c) 3×10^6

18.4.2 Calculate $\Delta G°$ at 25°C for the following reaction.

$$3Cu^{2+}(aq) + Cr(s) \rightleftharpoons 3Cu^+(aq) + Cr^{3+}(aq)$$

(a) 2.6×10^2 kJ/mol　　　　(d) -86 kJ/mol
(b) -2.6×10^2 kJ/mol　　　　(e) 86 kJ/mol
(c) 1×10^{45} kJ/mol

Thinking Outside the Box

Amalgam Fillings and Dental Pain

The pain caused by having aluminum foil contact an amalgam filling is the result of electrical stimulation of the nerve of a tooth caused by a current flowing between the aluminum foil and the metal in the filling. Historically, the material most commonly used to fill cavities is known as *dental amalgam*. (An *amalgam* is a substance made by combining mercury with one or more other metals.) Dental amalgam consists of liquid mercury mixed in roughly equal parts with an alloy powder containing silver, tin, copper, and sometimes smaller amounts of other metals such as zinc.

What happens is this: the aluminum and the dental filling act as *electrodes* in a galvanic cell. Aluminum acts as the anode, and the amalgam acts as the cathode. In effect, when the aluminum comes into contact with the amalgam, an electrochemical cell is established in the mouth, causing a current to flow. This current stimulates the nerve of the tooth, causing a *very* unpleasant sensation.

Another type of discomfort can result from the filling being made the anode in an electrochemical cell. This occurs when the filling touches a metal with a greater reduction potential than the components of the amalgam, such as gold. When an amalgam filling comes into contact with

a gold inlay, the tin in the filling (the most easily oxidized of the major amalgam components) is *oxidized*—creating an unpleasant metallic taste in the mouth. A simplified, unbalanced equation for the redox reaction that takes place is

$$Sn(s) + O_2(g) \longrightarrow Sn^{2+}(aq) + H_2O(l)$$

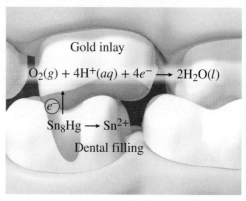

Gold inlay touching amalgam dental filling.

18.5 SPONTANEITY OF REDOX REACTIONS UNDER CONDITIONS OTHER THAN STANDARD STATE

So far we have focused on redox reactions in which the reactants and products are in their standard states. Standard-state conditions, though, are difficult to come by and usually impossible to maintain. Just as there is an equation that relates ΔG to $\Delta G°$ [◀◀ Section 15.4, Equation 15.5], there is an equation that relates E to $E°$. We will now derive this equation.

The Nernst Equation

Consider a redox reaction of the type

$$a\text{A} + b\text{B} \longrightarrow c\text{C} + d\text{D}$$

From Equation 15.5,

$$\Delta G = \Delta G° + RT \ln Q$$

Because $\Delta G = -nFE$ and $\Delta G° = -nFE°$, the equation can be expressed as

$$-nFE = -nFE° + RT \ln Q$$

Dividing the equation through by $-nF$, we get

Equation 18.6 $$E = E° - \frac{RT}{nF} \ln Q$$

where Q is the reaction quotient [◄◄ Section 15.2]. Equation 18.6 is known as the **Nernst[2] equation.** At 298 K, Equation 18.6 can be rewritten as

$$E = E° - \frac{0.0257 \text{ V}}{n} \ln Q$$

or using the base-10 logarithm of Q as

$$E = E° - \frac{0.0592 \text{ V}}{n} \log Q \qquad \textbf{Equation 18.7}$$

During the operation of a galvanic cell, electrons flow from the anode to the cathode, resulting in product formation and a decrease in reactant concentration. Thus, Q increases, which means that E decreases. Eventually, the cell reaches equilibrium. At equilibrium, there is no net transfer of electrons, so $E = 0$ and $Q = K$, where K is the equilibrium constant.

The Nernst equation enables us to calculate E as a function of reactant and product concentrations in a redox reaction. For example, for the cell pictured in Figure 18.1,

$$Zn(s) + Cu^{2+}(aq) \longrightarrow Zn^{2+}(aq) + Cu(s)$$

The Nernst equation for this cell at 25°C can be written as

$$E = 1.10 \text{ V} - \frac{0.0592 \text{ V}}{2} \log \frac{[Zn^{2+}]}{[Cu^{2+}]}$$

If the ratio $[Zn^{2+}]/[Cu^{2+}]$ is less than 1, $\log ([Zn^{2+}]/[Cu^{2+}])$ is a negative number, making the second term on the right-hand side of the preceding equation a positive quantity. Under this condition, E is *greater* than the standard potential $E°$. If the ratio is greater than 1, E is *smaller* than $E°$.

Worked Example 18.6 shows how to use the Nernst equation.

Student Hot Spot

Student data indicate you may struggle with using the Nernst equation. Access the SmartBook to view additional Learning Resources on this topic.

Worked Example 18.6

Predict whether the following reaction will occur spontaneously as written at 298 K:

$$Co(s) + Fe^{2+}(aq) \longrightarrow Co^{2+}(aq) + Fe(s)$$

assuming $[Co^{2+}] = 0.15 \ M$ and $[Fe^{2+}] = 0.68 \ M$.

Strategy Use $E°$ values from Table 18.1 to determine $E°$ for the reaction, and use Equation 18.7 to calculate E. If E is positive, the reaction will occur spontaneously.

Setup From Table 18.1,

Cathode (reduction): $Fe^{2+}(aq) + 2e^- \longrightarrow Fe(s)$

Anode (oxidation): $Co(s) \longrightarrow Co^{2+}(aq) + 2e^-$

$$E°_{cell} = E°_{cathode} - E°_{anode}$$

$$= E°_{Fe^{2+}/Fe} - E°_{Co^{2+}/Co}$$

$$= -0.44 \text{ V} - (-0.28 \text{ V})$$

$$= -0.16 \text{ V}$$

The reaction quotient, Q, for the reaction is $[Co^{2+}]/[Fe^{2+}]$. Therefore, $Q = (0.15/0.68) = 0.22$.

(Continued on next page)

---

[2]Walther Hermann Nernst (1864–1941), a German chemist and physicist, worked mainly on electrolyte solutions and thermodynamics. He also invented an electric piano. Nernst was awarded the Nobel Prize in Chemistry in 1920 for his contribution to thermodynamics.

Solution From Equation 18.7,

$$E = E° - \frac{0.0592 \text{ V}}{n} \log Q$$

$$= -0.16 \text{ V} - \frac{0.0592 \text{ V}}{2} \log 0.22$$

$$= -0.14 \text{ V}$$

The negative E value indicates that the reaction is *not* spontaneous as written under the conditions described.

Think About It

For this reaction to be spontaneous as written, the ratio of $[Fe^{2+}]$ to $[Co^{2+}]$ would have to be enormous. We can determine the required ratio by first setting E equal to zero.

$$0 \text{ V} = -0.16 \text{ V} - \frac{0.0592 \text{ V}}{2} \log Q$$

$$-\frac{(0.16 \text{ V})(2)}{0.0592 \text{ V}} = \log Q$$

$$\log Q = -5.4$$

$$Q = 10^{-5.4} = \frac{[Co^{2+}]}{[Fe^{2+}]} = 4 \times 10^{-6}$$

For E to be positive, therefore, the ratio of $[Fe^{2+}]$ to $[Co^{2+}]$, the reciprocal of Q, would have to be greater than 3×10^5 to 1.

Practice Problem **A**TTEMPT Will the following reaction occur spontaneously at 298 K if $[Fe^{2+}] = 0.60$ M and $[Cd^{2+}] = 0.010$ M?

$$Cd(s) + Fe^{2+}(aq) \longrightarrow Cd^{2+}(aq) + Fe(s)$$

Practice Problem **B**UILD Consider the electrochemical cell in Worked Example 18.5B (page 847), for which $E°_{cell} = 0.18$ V at 80°C.

$$H_2(g) + 2AgCl(s) \rightleftharpoons 2Ag(s) + 2H^+(aq) + 2Cl^-(aq)$$

If pH = 1.05 in the anode compartment, and $[Cl^-] = 2.5$ M in the cathode compartment, determine the partial pressure of H_2 necessary in the anode compartment for the cell potential to be 0.27 V at 80°C.

Practice Problem **C**ONCEPTUALIZE Consider a galvanic cell based on the reaction in Worked Example 18.5. Which of the graphs [(i)–(iv)] best represents what happens to the value of E as the Cu^{2+} ion concentration is increased?

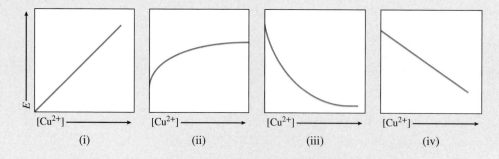

(i)   (ii)   (iii)   (iv)

Concentration Cells

Because electrode potential depends on ion concentrations, it is possible to construct a galvanic cell from two half-cells composed of the same material but differing in ion concentrations. Such a cell is called a ***concentration cell.***

Consider a galvanic cell consisting of a zinc electrode in 0.10 M zinc sulfate in one compartment and a zinc electrode in 1.0 M zinc sulfate in the other compartment (Figure 18.5). According to Le Châtelier's principle, the tendency for the reduction

$$Zn^{2+}(aq) + 2e^- \longrightarrow Zn(s)$$

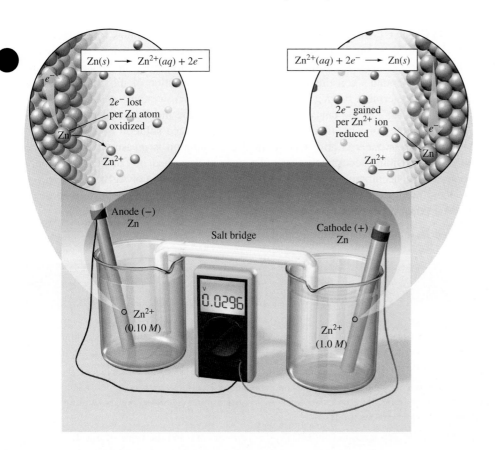

Figure 18.5 A concentration cell. Oxidation will occur in the container with the lower Zn^{2+} concentration. Reduction will occur in the container with the higher Zn^{2+} concentration.

to occur increases with increasing concentration of Zn^{2+} ions. Therefore, reduction should occur in the more concentrated compartment and oxidation should take place on the more dilute side. The cell diagram is

$$Zn(s) \mid Zn^{2+}(0.10\ M) \parallel Zn^{2+}(1.0\ M) \mid Zn(s)$$

and the half-cell reactions are

Oxidation: $\qquad\qquad Zn(s) \longrightarrow Zn^{2+}(0.10\ M) + 2e^{-}$

Reduction: $\quad Zn^{2+}(1.0\ M) + 2e^{-} \longrightarrow Zn(s)$

Overall: $\qquad\quad Zn^{2+}(1.0\ M) \longrightarrow Zn^{2+}(0.10\ M)$

The cell potential is

$$E = E^{\circ} - \frac{0.0592\ V}{2} \log \frac{[Zn^{2+}]_{dilute}}{[Zn^{2+}]_{concentrated}}$$

where the subscripts "dilute" and "concentrated" refer to the 0.10 M and 1.0 M concentrations, respectively. The E° for this cell is zero (because the same electrode-ion combination is used in both half-cells; i.e., $E^{\circ}_{cathode} = E^{\circ}_{anode}$), so

$$E = 0 - \frac{0.0592\ V}{2} \log \frac{0.10}{1.0}$$

$$= 0.030\ V$$

The cell potential for a concentration cell is typically *small* and decreases continually during the operation of the cell as the concentrations in the two compartments approach each other. When the concentrations of the ions in the two compartments are equal, E becomes zero and no further change occurs.

In Chapter 17, we studied solubility equilibria and learned how to use tabulated K_{sp} values to determine solubilities of ionic compounds that are only very slightly soluble. The tabulated K_{sp} values are determined by measuring the concentration of one of a compound's constituent ions in a saturated solution of the compound. For example, to determine the K_{sp} of silver bromide (AgBr), we must measure the concentration of Ag^+, or the concentration of Br^-. Because the K_{sp} of silver bromide is 7.7×10^{-13}, the concentration of either ion in a saturated solution is 8.8×10^{-7} M. You may have wondered how such a small concentration is measured. It cannot be done using visible spectrophotometry, the method described in Chapter 9; not only because the concentration is so low, but also because a solution of AgBr is colorless—meaning that it does not absorb visible light. In fact, concentrations of ions in these cases are measured using concentration cells.

Worked Example 18.7 illustrates this process.

Worked Example 18.7

An electrochemical cell is constructed for the purpose of determining the K_{sp} of silver cyanide (AgCN) at 25°C. One half-cell consists of a silver electrode in a 1.00-M solution of silver nitrate. The other half-cell consists of a silver electrode in a saturated solution of silver cyanide. The cell potential is measured and found to be 0.470 V. Determine the concentration of silver ion in the saturated silver cyanide solution and the value of K_{sp} for AgCN.

Strategy Use Equation 18.7 to solve for the unknown concentration of silver ion. The half-cell with the higher Ag^+ (1.0 M $AgNO_3$) concentration will be the cathode; the half-cell with the lower, unknown Ag^+ concentration (saturated AgCN solution) will be the anode. The overall reaction is $Ag^+(1.0\ M) \longrightarrow Ag^+(x\ M)$.

Setup Because this is a concentration cell, $E_{cell}^{\circ} = 0$ V. The reaction quotient, Q, is $(x\ M)/(1.00\ M)$; and the value of n is 1.

Solution

$$E_{cell} = E_{cell}^{\circ} - \frac{0.0592\ V}{n} \log Q$$

$$0.470\ V = 0 - \frac{0.0592\ V}{1} \log \frac{x}{1.00}$$

$$-7.939 = \log \frac{x}{1.00}$$

$$10^{-7.939} = 1.15 \times 10^{-8} = \log \frac{x}{1.00}$$

$$x = 1.15 \times 10^{-8}$$

Therefore, $[Ag^+] = 1.15 \times 10^{-8}$ M and K_{sp} for AgCN = $x^2 = 1.3 \times 10^{-16}$.

Think About It
Remember that in a saturated solution of a salt that dissociates into two ions, the ion concentrations are equal to each other and each ion concentration is equal to the square root of K_{sp} [◄◄ Section 17.4].

Practice Problem Ⓐ**TTEMPT** An electrochemical cell is constructed for the purpose of determining the K_{sp} of copper(I) chloride, CuCl, at 25°C. One half-cell consists of a copper electrode in a 1.00-M solution of copper(I) nitrate. The other half-cell consists of a copper electrode in a saturated solution of copper(I) chloride. The measured cell potential is 0.175 V. Determine the concentration of copper(I) ion in the saturated copper(I) chloride solution and the value of K_{sp} for CuCl.

Practice Problem Ⓑ**UILD** The K_{sp} of copper(II) ferrocyanide ($Cu_2[Fe(CN)_6]$) is 1.3×10^{-16} at 25°C. Determine the potential of a concentration cell in which one half-cell consists of a copper electrode in 1.00 M copper(II) nitrate, and the other consists of a copper electrode in a saturated solution of $Cu_2[Fe(CN)_6]$. Ferrocyanide, ($[Fe(CN)_6]^{4-}$), is a *complex ion* [◄◄ Section 17.5].

Practice Problem Ⓒ**ONCEPTUALIZE** When the circuit in a silver chloride concentration cell is first completed, the concentration of silver ion in the cathode compartment is significantly higher than it is in the anode compartment. During operation of the cell, the concentrations become closer as the concentration decreases in the cathode compartment and increases in the anode compartment. If the cell continues to operate for long enough, will the concentration of silver ion in the anode compartment ever be greater than that in the cathode compartment? Explain.

Section 18.5 Review

Spontaneity of Redox Reactions Under Conditions Other Than Standard State

18.5.1 Calculate E at 25°C for a galvanic cell based on the reaction

$$\text{Zn}(s) + \text{Cu}^{2+}(aq) \longrightarrow \text{Zn}^{2+}(aq) + \text{Cu}(s)$$

in which $[\text{Zn}^{2+}] = 0.55\ M$ and $[\text{Cu}^{2+}] = 1.02\ M$.
(a) 1.10 V   (d) 0.0118 V
(b) 1.11 V   (e) 1.03 V
(c) 1.09 V

18.5.2 Calculate the cell potential at 25°C of a galvanic cell consisting of an Ag electrode in 0.15 M AgNO$_3$ and an Ag electrode in 1.0 M AgNO$_3$.
(a) 0.0 V   (d) 0.024 V
(b) 0.049 V   (e) −0.024 V
(c) −0.049 V

18.5.3 Which of the following would cause an increase in the cell potential of the electrochemical cell in question 18.5.2?
(i) Adding AgNO$_3$ to the cathode compartment
(ii) Adding saturated AgCl solution to the anode compartment
(iii) Adding NaCl to the anode compartment

(a) i and iii   (d) i, ii, and iii
(b) ii and iii   (e) None of these
(c) iii only

18.5.4 Calculate E_{cell}° at 80°C for a galvanic cell based on the reaction

$$\text{H}_2(g) + 2\text{AgCl}(s) \rightleftharpoons 2\text{Ag}(s) + 2\text{H}^+(aq) + 2\text{Cl}^-(aq)$$

in which $[\text{H}^+] = 0.10\ M$, $[\text{Cl}^-] = 1.5\ M$, and $P = 1.25$ atm. ($E_{\text{cell}}^{\circ} = 0.18$ V at 80°C.)
(a) 0.12 V   (d) 0.22 V
(b) 0.16 V   (e) 0.24 V
(c) 0.18 V

18.6   BATTERIES

A *battery* is a galvanic cell, or a series of connected galvanic cells, that can be used as a portable, self-contained source of direct electric current. In this section we examine several types of batteries.

Dry Cells and Alkaline Batteries

The most common batteries, *dry cells* and *alkaline batteries,* are those used in flashlights, toys, and certain portable electronics such as MP3 players. The two are similar in appearance, but differ in the spontaneous chemical reaction responsible for producing a voltage. Although the reactions that take place in these batteries are somewhat complex, those shown here approximate the overall processes.

A dry cell, so named because it has no fluid component, consists of a zinc container (the anode) in contact with manganese dioxide and an electrolyte (Figure 18.6). The electrolyte consists of ammonium chloride and zinc chloride in water, to which starch is added to thicken the solution to a paste so that it is less likely to leak. A

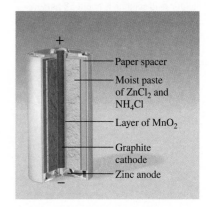

Figure 18.6 Interior view of the type of dry cell used in flashlights and other small devices.

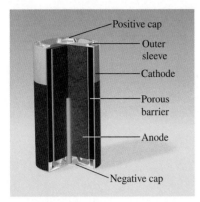

Figure 18.7 Interior view of an alkaline battery.

carbon rod, immersed in the electrolyte in the center of the cell, serves as the cathode. The cell reactions are

Anode: $$Zn(s) \longrightarrow Zn^{2+}(aq) + 2e^-$$

Cathode: $$2NH_4^+(aq) + 2MnO_2(s) + 2e^- \longrightarrow Mn_2O_3(s) + 2NH_3(aq) + H_2O(l)$$

Overall: $$Zn(s) + 2NH_4^+(aq) + 2MnO_2(s) \longrightarrow$$
$$Zn^{2+}(aq) + Mn_2O_3(s) + 2NH_3(aq) + H_2O(l)$$

The voltage produced by a dry cell is about 1.5 V.

An alkaline battery is also based on the reduction of manganese dioxide and the oxidation of zinc. However, the reactions take place in a *basic* medium, hence the name *alkaline* battery. The anode consists of powdered zinc suspended in a gel, which is in contact with a concentrated solution of KOH. The cathode is a mixture of manganese dioxide and graphite. The anode and cathode are separated by a porous barrier (Figure 18.7).

Anode: $$Zn(s) + 2OH^-(aq) \longrightarrow Zn(OH)_2(s) + 2e^-$$

Cathode: $$2MnO_2(s) + 2H_2O(l) + 2e^- \longrightarrow 2MnO(OH)(s) + 2OH^-(aq)$$

Overall: $$Zn(s) + 2MnO_2(s) + 2H_2O(l) \longrightarrow Zn(OH)_2(s) + 2MnO(OH)(s)$$

Alkaline batteries are more expensive than dry cells and offer superior performance and shelf life.

Lead Storage Batteries

The lead storage battery commonly used in automobiles consists of six identical cells joined together in series. Each cell has a lead anode and a cathode made of lead dioxide (PbO_2) packed on a metal plate (Figure 18.8). Both the cathode and the anode are immersed in an aqueous solution of sulfuric acid, which acts as the electrolyte. The cell reactions are

Anode: $$Pb(s) + SO_4^{2-}(aq) \longrightarrow PbSO_4(s) + 2e^-$$

Cathode: $$PbO_2(s) + 4H^+(aq) + SO_4^{2-}(aq) + 2e^- \longrightarrow PbSO_4(s) + 2H_2O(l)$$

Overall: $$Pb(s) + PbO_2(s) + 4H^+(aq) + 2SO_4^{2-}(aq) \longrightarrow 2PbSO_4(s) + 2H_2O(l)$$

Figure 18.8 Interior view of a lead storage battery. Under normal operating conditions, the concentration of the sulfuric acid solution is about 38 percent by mass.

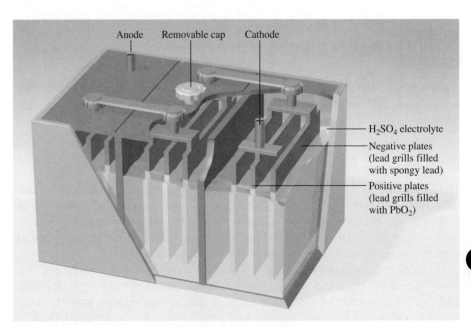

Under normal operating conditions, each cell produces 2 V. A total of 12 V from the six cells is used to power the ignition circuit of the automobile and its other electric systems. The lead storage battery can deliver large amounts of current for a short time, such as the time it takes to start the engine.

Unlike dry cells and alkaline batteries, the lead storage battery is rechargeable. Recharging the battery means reversing the normal electrochemical reaction by applying an external voltage at the cathode and the anode. (This kind of process is called *electrolysis,* which we discuss in Section 18.7.)

Lithium-Ion Batteries

Sometimes called "the battery of the future," lithium-ion batteries have several advantages over other battery types. The overall reaction that takes place in the lithium-ion battery is

$$\text{Anode:} \qquad \text{Li}(s) \longrightarrow \text{Li}^+ + e^-$$

$$\text{Cathode:} \quad \text{Li}^+ + \text{CoO}_2 + e^- \longrightarrow \text{LiCoO}_2(s)$$

$$\overline{\text{Overall:} \qquad \text{Li}(s) + \text{CoO}_2 \longrightarrow \text{LiCoO}_2(s)}$$

The overall cell potential is 3.4 V, which is relatively large. Lithium is also the lightest metal—only 6.941 g of Li (its molar mass) are needed to produce 1 mole of electrons. Furthermore, a lithium-ion battery can be recharged hundreds of times. These qualities make lithium batteries suitable for use in portable devices such as cell phones, digital cameras, and laptop computers.

Student Annotation: Lithium has the largest negative reduction potential of all the metals, making it a powerful reducing agent.

Fuel Cells

Fossil fuels are a major source of energy, but the conversion of fossil fuel into electric energy is a highly inefficient process. Consider the combustion of methane:

$$\text{CH}_4(g) + 2\text{O}_2(g) \longrightarrow \text{CO}_2(g) + 2\text{H}_2\text{O}(l) + \text{energy}$$

To generate electricity, heat produced by the reaction is first used to convert water to steam, which then drives a turbine, which then drives a generator. A significant fraction of the energy released in the form of heat is lost to the surroundings at each step (even the most efficient power plant converts only about 40 percent of the original chemical energy into electricity). Because combustion reactions are redox reactions, it is more desirable to carry them out directly by electrochemical means, thereby greatly increasing the efficiency of power production. This objective can be accomplished by a device known as a ***fuel cell,*** a galvanic cell that requires a continuous supply of reactants to keep functioning.

Student Annotation: Strictly speaking, a fuel cell is not a battery because it is not self-contained.

In its simplest form, a hydrogen-oxygen fuel cell consists of an electrolyte solution, such as a potassium hydroxide solution, and two inert electrodes. Hydrogen and oxygen gases are bubbled through the anode and cathode compartments (Figure 18.9), where the following reactions take place.

$$\text{Anode:} \qquad 2\text{H}_2(g) + 4\text{OH}^-(aq) \longrightarrow 4\text{H}_2\text{O}(l) + 4e^-$$

$$\text{Cathode:} \quad \text{O}_2(g) + 2\text{H}_2\text{O}(l) + 4e^- \longrightarrow 4\text{OH}^-(aq)$$

$$\overline{\text{Overall:} \qquad 2\text{H}_2(g) + \text{O}_2(g) \longrightarrow 2\text{H}_2\text{O}(l)}$$

Using $E°$ values from Table 18.1, the standard cell potential is calculated as follows:

$$E°_{\text{cell}} = E°_{\text{cathode}} - E°_{\text{anode}}$$

$$= 0.40 \text{ V} - (-0.83 \text{ V})$$

$$= 1.23 \text{ V}$$

856 **CHAPTER 18** Electrochemistry

Figure 18.9 A hydrogen-oxygen fuel cell. The Ni and NiO embedded in the porous carbon electrodes are catalysts.

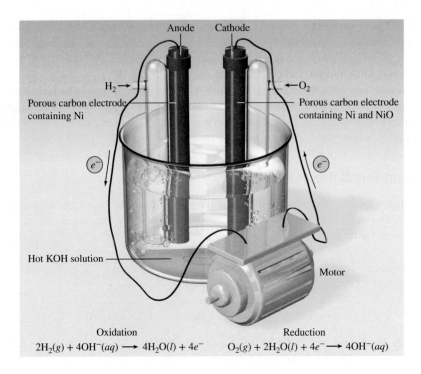

Oxidation
$$2H_2(g) + 4OH^-(aq) \longrightarrow 4H_2O(l) + 4e^-$$

Reduction
$$O_2(g) + 2H_2O(l) + 4e^- \longrightarrow 4OH^-(aq)$$

Thus, the cell reaction is spontaneous under standard state conditions. Note that the reaction is the same as the hydrogen combustion reaction, but the oxidation and reduction are carried out separately at the anode and the cathode. Like platinum in the standard hydrogen electrode, the electrodes serve two purposes. They serve as electrical conductors, and they provide the necessary surfaces for the initial decomposition of the molecules into atomic species, which must take place before electrons can be transferred. Electrodes that serve this particular purpose are called "electrocatalysts." Metals such as platinum, nickel, and rhodium also make good electrocatalysts.

In addition to the H_2-O_2 system, a number of other fuel cells have been developed. Among these is the propane-oxygen fuel cell. The corresponding half-cell reactions are

Anode: $\qquad$ $C_3H_8(g) + 6H_2O(l) \longrightarrow 3CO_2(g) + 20H^+(aq) + 20e^-$

Cathode: $\quad 5O_2(g) + 20H^+(aq) + 20e^- \longrightarrow 10H_2O(l)$

Overall: $\qquad$ $C_3H_8(g) + 5O_2(g) \longrightarrow 3CO_2(g) + 4H_2O(l)$

The overall reaction is identical to the burning of propane in oxygen.

Unlike batteries, fuel cells do not store chemical energy. Reactants must be constantly resupplied as they are consumed, and products must be removed as they form. However, properly designed fuel cells may be as much as 70 percent efficient, which is about twice as efficient as an internal combustion engine. In addition, fuel-cell generators are free of the noise, vibration, heat transfer, thermal pollution, and other problems normally associated with conventional power plants. Nevertheless, fuel cells are not yet in widespread use. One major problem is the expense of electrocatalysts that can function efficiently for long periods of time without contamination. One notable application of fuel cells is their use in space vehicles. Hydrogen-oxygen fuel cells provide electric power (and drinking water!) for space flight.

18.7 ELECTROLYSIS

In Section 18.6, we mentioned that lead storage batteries are rechargeable and that recharging means reversing the electrochemical processes by which the battery ordinarily operates through the application of an external voltage. This process, the use of electric energy to drive a nonspontaneous chemical reaction, is called *electrolysis.*

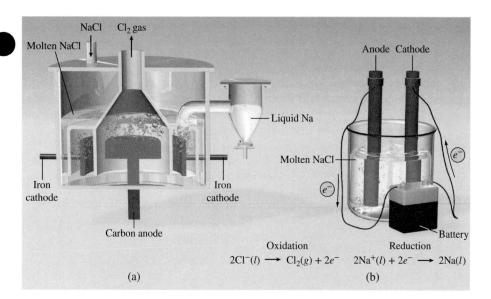

Figure 18.10 (a) A practical arrangement called a Downs cell for the electrolysis of molten NaCl (m.p. = 801°C). The sodium metal formed at the cathodes is in the liquid state. Because liquid sodium metal is lighter than molten NaCl, the sodium floats to the surface, as shown, and is collected. Chlorine gas forms at the anode and is collected at the top. (b) A simplified diagram showing the electrode reactions during the electrolysis of molten NaCl. The battery is needed to drive the nonspontaneous reaction.

An *electrolytic cell* is one used to carry out electrolysis. The same principles apply to the processes in both galvanic and electrolytic cells. In this section, we discuss three examples of electrolysis based on those principles. We then examine some of the quantitative aspects of electrolysis.

Electrolysis of Molten Sodium Chloride

In its molten (melted) state, sodium chloride, an ionic compound, can be electrolyzed to separate it into its constituent elements, sodium and chlorine. Figure 18.10(a) is a diagram of a Downs cell, which is used for the large-scale electrolysis of NaCl. In molten NaCl, the cations and anions are the Na^+ and Cl^- ions, respectively. Figure 18.10(b) is a simplified diagram showing the reactions that occur at the electrodes. The electrolytic cell contains a pair of electrodes connected to the battery. The battery serves to push electrons in the direction they would not flow spontaneously. The electrode toward which the electrons are pushed is the cathode, where reduction takes place. The electrode away from which electrons are drawn is the anode, where oxidation takes place. The reactions at the electrodes are

Anode (oxidation): $\qquad\qquad\qquad 2Cl^-(l) \longrightarrow Cl_2(g) + 2e^-$

Cathode (reduction): $\qquad 2Na^+(l) + 2e^- \longrightarrow 2Na(l)$

Overall: $\qquad\qquad 2Na^+(l) + 2Cl^-(l) \longrightarrow 2Na(l) + Cl_2(g)$

This process is a major industrial source of pure sodium metal and chlorine gas.

Using data from Table 18.1, we estimate E°_{cell} to be −4 V for this process. The negative standard reduction potential indicates that for the process to occur as written, a minimum of approximately 4 V must be supplied by the battery to drive the reaction in the desired direction. In practice, an even higher voltage is required because of inefficiencies in the electrolytic process and because of overvoltage, a phenomenon we will discuss later in this section.

Electrolysis of Water

Under ordinary atmospheric conditions (1 atm and 25°C), water will not spontaneously decompose to form hydrogen and oxygen gas, because the standard free-energy change for the reaction is a large positive quantity.

$$2H_2O(l) \longrightarrow 2H_2(g) + O_2(g) \qquad \Delta G^\circ = 474.4 \text{ kJ/mol}$$

Figure 18.11 Apparatus for small-scale electrolysis of water. The volume of hydrogen gas generated at the cathode is twice that of oxygen gas generated at the anode.
© McGraw-Hill Education/Stephen Frisch, photographer

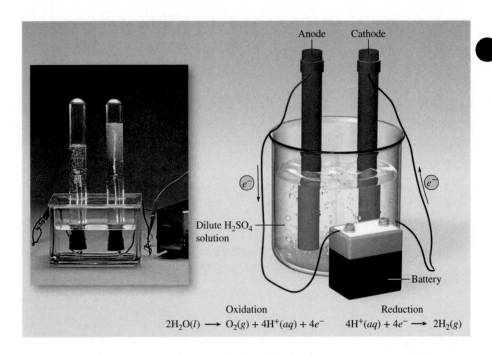

Anode   Cathode

e^-

Dilute H_2SO_4 solution

e^-

Battery

Oxidation   Reduction
$2H_2O(l) \longrightarrow O_2(g) + 4H^+(aq) + 4e^-$ $4H^+(aq) + 4e^- \longrightarrow 2H_2(g)$

However, this reaction can be made to occur in an electrolytic cell like the one shown in Figure 18.11. This cell consists of a pair of electrodes made of a nonreactive metal, such as platinum, immersed in water. When the electrodes are connected to the battery, nothing happens because there are not enough ions in pure water to carry much of an electric current. The reaction occurs readily in a 0.1 M H_2SO_4 solution, however, because there is a sufficient ion concentration to conduct electricity. Immediately, gas bubbles begin to appear at both electrodes.

Student Annotation: Remember that at 25°C pure water has only a very low concentration of ions [◄◄ Section 16.3]:

$$[H^+] = [OH^-] = 1 \times 10^{-7} \, M$$

The processes at the electrodes are

Anode: $2H_2O(l) \longrightarrow O_2(g) + 4H^+(aq) + 4e^-$

Cathode: $4H^+(aq) + 4e^- \longrightarrow 2H_2(g)$

Overall: $2H_2O(l) \longrightarrow O_2(g) + 2H_2(g)$

Note that there is no net consumption of the acid.

Electrolysis of an Aqueous Sodium Chloride Solution

An aqueous sodium chloride solution is the most complicated of the three examples of electrolysis considered here because NaCl(aq) contains several species that could be oxidized and reduced. The reductions that might occur at the cathode are

$$2H^+(aq) + 2e^- \longrightarrow H_2(g) \qquad\qquad E° = 0.00 \text{ V}$$

$$2H_2O(l) + 2e^- \longrightarrow H_2(g) + 2OH^-(aq) \qquad E° = -0.83 \text{ V}$$

or

$$Na^+(aq) + e^- \longrightarrow Na(s) \qquad\qquad E° = -2.71 \text{ V}$$

We can rule out the reduction of Na^+ ion because of the large negative $E°$ value. Under standard state conditions, the reduction of H^+ is more apt to occur than the reduction of H_2O; however, in a solution of NaCl, the H^+ concentration is very low, making the reduction of H_2O the more probable reaction at the cathode.

The oxidation reactions that might occur at the anode are

$$2Cl^-(aq) \longrightarrow Cl_2(g) + 2e^-$$

or

$$2H_2O(l) \longrightarrow O_2(g) + 4H^+(aq) + 4e^-$$

Referring to Table 18.1, we find

$$Cl_2(g) + 2e^- \longrightarrow 2Cl^-(aq) \qquad E° = 1.36 \text{ V}$$

$$O_2(g) + 4H^+(aq) + 4e^- \longrightarrow 2H_2O(l) \qquad E° = 1.23 \text{ V}$$

The standard reduction potentials of the two reactions are not very different, but the values do suggest that the oxidation of H_2O *should* occur more readily. However, by experiment we find that the gas produced at the anode is Cl_2, not O_2. In the study of electrolytic processes, we sometimes find that the voltage required for a reaction is considerably higher than the electrode potentials would indicate. The **overvoltage** is the difference between the calculated voltage and the actual voltage required to cause electrolysis. The overvoltage for O_2 formation is quite high. Under normal operating conditions, therefore, Cl_2 gas forms at the anode instead of O_2.

Thus, the half-cell reactions in the electrolysis of aqueous sodium chloride are

Anode (oxidation): $\qquad\qquad 2Cl^-(aq) \longrightarrow Cl_2(g) + 2e^-$

Cathode (reduction): $\qquad 2H_2O(l) + 2e^- \longrightarrow H_2(g) + 2OH^-(aq)$

Overall: $\qquad\qquad\quad 2H_2O(l) + 2Cl^-(aq) \longrightarrow H_2(g) + Cl_2(g) + 2OH^-(aq)$

Student Annotation: Keep in mind that in the electrolysis of aqueous solutions, the water itself may be oxidized and/or reduced.

As the overall reaction shows, the concentration of the Cl^- ions decreases during electrolysis and that of the OH^- ions increases. Therefore, in addition to H_2 and Cl_2, the useful by-product NaOH can be obtained by evaporating the aqueous solution at the end of the electrolysis.

Electrolysis has many important applications in industry, mainly in the extraction and purification of metals. We discuss some of these applications in Chapter 26.

Quantitative Applications of Electrolysis

The quantitative treatment of electrolysis was developed primarily by Faraday. He observed that the mass of product formed (or reactant consumed) at an electrode was proportional to both the amount of electricity transferred at the electrode and the molar mass of the substance being produced (or consumed). In the electrolysis of molten NaCl, for example, the cathode reaction tells us that one Na atom is produced when one Na^+ ion accepts an electron from the electrode. To reduce 1 mole of Na^+ ions, we must supply an Avogadro's number (6.02×10^{23}) of electrons to the cathode. On the other hand, stoichiometry tells us that it takes 2 moles of electrons to reduce 1 mole of Mg^{2+} ions and 3 moles of electrons to reduce 1 mole of Al^{3+} ions.

$$Na^+ + e^- \longrightarrow Na$$

$$Mg^{2+} + 2e^- \longrightarrow Mg$$

$$Al^{3+} + 3e^- \longrightarrow Al$$

In an electrolysis experiment, we generally measure the current (in amperes) that passes through an electrolytic cell in a given period of time. The relationship between charge (in coulombs) and the current is

$$1 \text{ C} = 1 \text{ A} \times 1 \text{ s}$$

That is, a coulomb is the quantity of electric charge passing any point in the circuit in 1 s when the current is 1 A. Therefore, if we know the current (in amperes) and how long it is applied (in seconds), we can calculate the charge (in coulombs). Knowing the charge enables us to determine the number of moles of electrons. And knowing

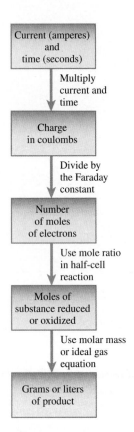

Current (amperes)
and
time (seconds)

↓ Multiply
current and
time

Charge
in coulombs

↓ Divide by
the Faraday
constant

Number
of moles
of electrons

↓ Use mole ratio
in half-cell
reaction

Moles of
substance reduced
or oxidized

↓ Use molar mass
or ideal gas
equation

Grams or liters
of product

Figure 18.12 Steps involved in calculating amounts of substances reduced or oxidized in electrolysis.

the number of moles of electrons allows us to use stoichiometry to determine the number of moles of product. Figure 18.12 shows the steps involved in calculating the quantities of substances produced in electrolysis.

To illustrate this approach, consider an electrolytic cell in which molten $CaCl_2$ is separated into its constituent elements, Ca and Cl_2. Suppose a current of 0.452 A is passed through the cell for 1.50 h. How much product will be formed at each electrode? The first step is to determine which species will be oxidized at the anode and which species will be reduced at the cathode. Here the choice is straightforward because we have only Ca^{2+} and Cl^- ions. The cell reactions are

Anode (oxidation): $2Cl^-(l) \longrightarrow Cl_2(g) + 2e^-$

Cathode (reduction): $Ca^{2+}(l) + 2e^- \longrightarrow Ca(l)$

Overall: $Ca^{2+}(l) + 2Cl^-(l) \longrightarrow Ca(l) + Cl_2(g)$

The quantities of calcium metal and chlorine gas formed depend on the number of electrons that pass through the electrolytic cell, which in turn depends on *charge,* or current × time.

$$\text{coulombs} = 0.452 \ \cancel{A} \times 1.50 \ \cancel{h} \times \frac{3600 \ \cancel{s}}{1 \ \cancel{h}} \times \frac{1 \ C}{1 \ \cancel{A} \cdot \cancel{s}} = 2.441 \times 10^3 \ C$$

Because 1 mol e^- = 96,500 C and 2 mol e^- are required to reduce 1 mole of Ca^{2+} ions, the mass of Ca metal formed at the cathode is calculated as follows:

$$\text{grams Ca} = (2.441 \times 10^3 \ \cancel{C})\left(\frac{1 \ \cancel{mol \ e^-}}{96,500 \ \cancel{C}}\right)\left(\frac{1 \ \cancel{mol \ Ca}}{2 \ \cancel{mol \ e^-}}\right)\left(\frac{40.08 \ g \ Ca}{1 \ \cancel{mol \ Ca}}\right) = 0.507 \ g \ Ca$$

The anode reaction indicates that 1 mole of chlorine is produced per 2 mol e^- of electricity. Hence, the mass of chlorine gas formed is

$$\text{grams Cl}_2 = (2.441 \times 10^3 \ \cancel{C})\left(\frac{1 \ \cancel{mol \ e^-}}{96,500 \ \cancel{C}}\right)\left(\frac{1 \ \cancel{mol \ Cl_2}}{2 \ \cancel{mol \ e^-}}\right)\left(\frac{70.90 \ g \ Cl_2}{1 \ \cancel{mol \ Cl_2}}\right) = 0.897 \ g \ Cl_2$$

Worked Example 18.8 applies this approach to electrolysis in an aqueous solution.

Worked Example 18.8

A current of 1.26 A is passed through an electrolytic cell containing a dilute sulfuric acid solution for 7.44 h. Write the half-cell reactions, and calculate the volume of gases generated at STP.

Strategy As shown in Figure 18.12, we can use current and time to determine charge. We can then convert charge to moles of electrons, and use the balanced half-reactions to determine how many moles of product form at each electrode. Finally, we can convert moles to volume.

Setup The half-cell reactions for the electrolysis of water are

Anode: $2H_2O(l) \longrightarrow O_2(g) + 4H^+(aq) + 4e^-$

Cathode: $4H^+(aq) + 4e^- \longrightarrow 2H_2(g)$

Overall: $2H_2O(l) \longrightarrow O_2(g) + 2H_2(g)$

Remember that STP for gases means 273 K and 1 atm.

Solution

$$\text{coulombs} = (1.26 \ A)(7.44 \ \cancel{h})\left(\frac{3600 \ \cancel{s}}{1 \ \cancel{h}}\right)\left(\frac{1 \ C}{1 \ \cancel{A} \cdot \cancel{s}}\right) = 3.375 \times 10^4 \ C$$

At the anode:

$$\text{moles O}_2 = (3.375 \times 10^4 \ \cancel{C})\left(\frac{1 \ \cancel{mol \ e^-}}{96,500 \ \cancel{C}}\right)\left(\frac{1 \ mol \ O_2}{4 \ \cancel{mol \ e^-}}\right) = 0.0874 \ mol \ O_2$$

The volume of 0.0874 mol O_2 at STP is given by

$$V = \frac{nRT}{P}$$

$$= \frac{(0.0874 \text{ mol})(0.08206 \text{ L} \cdot \text{atm/K} \cdot \text{mol})(273.15 \text{ K})}{1 \text{ atm}} = 1.96 \text{ L } O_2$$

Similarly, for hydrogen we write

$$\text{moles } H_2 = (3.375 \times 10^4 \text{ C}) \left(\frac{1 \text{ mol } e^-}{96,500 \text{ C}}\right)\left(\frac{1 \text{ mol } H_2}{2 \text{ mol } e^-}\right) = 0.175 \text{ mol } H_2$$

The volume of 0.175 mol H_2 at STP is given by

$$V = \frac{nRT}{P}$$

$$= \frac{(0.175 \text{ mol})(0.08206 \text{ L} \cdot \text{atm/K} \cdot \text{mol})(273.15 \text{ K})}{1 \text{ atm}} = 3.92 \text{ L } H_2$$

Think About It
The volume of H_2 is twice that of O_2 (see Figure 18.11), which is what we would expect based on Avogadro's law (at the same temperature and pressure, volume is directly proportional to the number of moles of gas: $V \propto n$) [◄◄ Section 11.4].

Practice Problem ATTEMPT A constant current of 0.912 A is passed through an electrolytic cell containing molten $MgCl_2$ for 18 h. What mass of Mg is produced?

Practice Problem BUILD A constant current is passed through an electrolytic cell containing molten $MgCl_2$ for 12 h. If 4.83 L of Cl_2 (at STP) is produced at the anode, what is the current in amperes?

Practice Problem CONCEPTUALIZE Diagram (i) shows the ions in an aqueous solution of sodium chloride. Which of the diagrams (ii) through (iv) could represent the system after electrolysis? (Water molecules are not shown.)

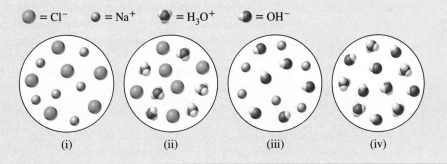

$\bullet = Cl^-$ $\bullet = Na^+$ $\bullet = H_3O^+$ $\bullet = OH^-$

(i)   (ii)   (iii)   (iv)

Section 18.7 Review

Electrolysis

18.7.1 In the electrolysis of molten $CaCl_2$, a current of 1.12 A is passed through the cell for 3.0 h. What is the mass of Ca produced at the cathode?
(a) 2.51 g   (d) 10.0 g
(b) 1.26 g   (e) 2.42×10^5 g
(c) 5.02 g

18.7.2 How long will a current of 0.995 A need to be passed through water (containing H_2SO_4) for 5.00 L of O_2 to be produced at STP?
(a) 6.0 h   (d) 3.0 h
(b) 8.2 h   (e) 24.0 h
(c) 1.5 h

18.7.3 The diagram shows an electrolytic cell being powered by a galvanic cell. Identify each of the electrodes from left to right as an anode or a cathode.

(a) Cathode, anode, anode, cathode   (d) Anode, anode, cathode, cathode

(b) Cathode, anode, cathode, anode   (e) Anode, cathode, cathode, anode

(c) Anode, cathode, anode, cathode

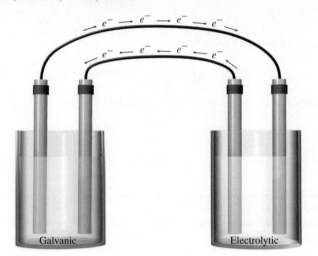

Galvanic   Electrolytic

18.8 CORROSION

The term ***corrosion*** generally refers to the deterioration of a metal by an electrochemical process. There are many examples of corrosion, including rust on iron, tarnish on silver, and the green layer that forms on copper and brass. In this section we discuss the processes involved in corrosion and some of the measures taken to prevent it.

The formation of rust on iron requires oxygen and water. Although the reactions involved are quite complex and not completely understood, the main steps are believed to be as follows. A region of the metal's surface serves as the anode, where the following oxidation occurs.

$$\text{Fe}(s) \longrightarrow \text{Fe}^{2+}(aq) + 2e^-$$

The electrons given up by iron reduce atmospheric oxygen to water at the cathode, which is another region of the same metal's surface.

$$\text{O}_2(g) + 4\text{H}^+(aq) + 4e^- \longrightarrow 2\text{H}_2\text{O}(l)$$

The overall redox reaction is

$$2\text{Fe}(s) + \text{O}_2(g) + 4\text{H}^+(aq) \longrightarrow 2\text{Fe}^{2+}(g) + 2\text{H}_2\text{O}(l)$$

With data from Table 18.1, the standard potential for this process can be calculated as follows:

$$E°_{\text{cell}} = E°_{\text{cathode}} - E°_{\text{anode}}$$

$$= 1.23 \text{ V} - (-0.44 \text{ V})$$

$$= 1.67 \text{ V}$$

Note that this reaction occurs in an *acidic* medium; the H$^+$ ions are supplied in part by the reaction of atmospheric carbon dioxide with water to form the weak acid, carbonic acid (H$_2$CO$_3$).

The Fe^{2+} ions formed at the anode are further oxidized by oxygen as follows:

$$4\text{Fe}^{2+}(aq) + \text{O}_2(g) + (4 + 2x)\text{H}_2\text{O}(l) \longrightarrow 2\text{Fe}_2\text{O}_3 \cdot x\text{H}_2\text{O}(s) + 8\text{H}^+(aq)$$

This hydrated form of iron(III) oxide is known as rust. The amount of water associated with the iron(III) oxide varies, so we represent the formula as Fe$_2$O$_3 \cdot x$H$_2$O.

Video 18.4
Electrochemistry—the production of aluminum.

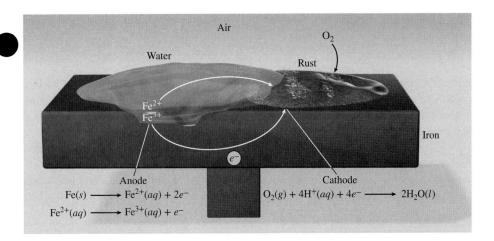

Figure 18.13 The electrochemical process involved in rust formation. The H^+ ions are supplied by H_2CO_3, which forms when CO_2 from air dissolves in water.

Figure 18.13 shows the mechanism of rust formation. The electric circuit is completed by the migration of electrons and ions; this is why rusting occurs so rapidly in saltwater. In cold climates, salts (NaCl or $CaCl_2$) spread on roadways to melt ice and snow are a major cause of rust formation on automobiles.

Other metals also undergo oxidation. Aluminum, for example, which is used to make airplanes, beverage cans, and aluminum foil, has a much greater tendency to oxidize than does iron. Unlike the corrosion of iron, though, corrosion of aluminum produces an insoluble layer of protective coating (Al_2O_3) that prevents the underlying metal from additional corrosion.

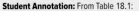

Student Annotation: From Table 18.1:

$$Fe^{2+}(aq) + 2e^- \longrightarrow Fe(s) \quad E° = -0.44 \text{ V}$$
$$Al^{3+}(aq) + 3e^- \longrightarrow Al(s) \quad E° = -1.66 \text{ V}$$

Al, with the more negative reduction potential, is less likely to be reduced (more likely to be oxidized) than Fe.

Coinage metals such as copper and silver also corrode, but much more slowly than either iron or aluminum.

$$Cu(s) \longrightarrow Cu^{2+}(aq) + 2e^-$$
$$Ag(s) \longrightarrow Ag^+(aq) + e^-$$

In ordinary atmospheric exposure, copper forms a layer of copper carbonate ($CuCO_3$), a green substance referred to as *patina,* that protects the metal underneath from further corrosion. Likewise, silverware that comes into contact with foodstuffs develops a layer of silver sulfide (Ag_2S).

Video 18.5
Electrochemistry—chrome plating.

A number of methods have been devised to protect metals from corrosion. Most of these methods are aimed at preventing rust formation. The most obvious approach is to coat the metal surface with paint to prevent exposure to the substances necessary for corrosion. If the paint is scratched or otherwise damaged, however, thus exposing even the smallest area of bare metal, rust will form under the paint layer. The surface of iron metal can be made inactive by a process called *passivation.* A thin oxide layer is formed when the metal is treated with a strong oxidizing agent such as concentrated nitric acid. A solution of sodium chromate is often added to cooling systems and radiators to prevent rust formation.

The tendency for iron to oxidize is greatly reduced when it is alloyed with certain other metals. For example, in stainless steel, an alloy of iron and chromium, a layer of chromium oxide forms that protects the iron from corrosion.

An iron container can be covered with a layer of another metal such as tin or zinc. A "tin" can is made by applying a thin layer of tin over iron. Rust formation is prevented as long as the tin layer remains intact. However, once the surface has been breached by a scratch or a dent, rusting occurs rapidly. If we look up the standard reduction potentials in Table 18.1, we find that when tin and iron are in contact with each other, tin, with its greater reduction potential, acts as a cathode. Iron acts as an anode and is therefore oxidized.

$$Sn^{2+}(aq) + 2e^- \longrightarrow Sn(s) \quad E° = -0.14 \text{ V}$$
$$Fe^{2+}(aq) + 2e^- \longrightarrow Fe(s) \quad E° = -0.44 \text{ V}$$

Zinc-plating, or **galvanization,** protects iron from corrosion by a different mechanism. According to Table 18.1,

$$Zn^{2+}(aq) + 2e^- \longrightarrow Zn(s) \qquad E^\circ = -0.76 \text{ V}$$

so zinc is *more* easily oxidized than iron. Like aluminum, zinc oxidizes to form a protective coating. Even when the zinc layer is compromised, though, and the underlying iron is exposed, zinc is still the more easily oxidized of the two metals and will act as the anode. Iron will be the cathode, thereby remaining reduced.

Galvanization is one example of *cathodic protection,* the process by which a metal is protected by being made the cathode in what amounts to a galvanic cell. Another example is the use of zinc or magnesium bars to protect underground storage tanks and ships. When a steel tank or hull is connected to a more easily oxidized metal, corrosion of the steel is prevented.

Learning Outcomes

- Use the half-reaction method to balance equations of oxidation-reduction reactions.
- Define key electrochemical terms: galvanic cell, anode, cathode, and salt bridge.
- Describe the conditions used to calculate the standard reduction potential for a reaction.
- Explain the importance of the standard hydrogen electrode.
- Determine the standard cell potential and the spontaneous cell reaction for a given chemical combination.
- Calculate E°_{cell} of a reaction given the value of ΔG and K and interconvert between these three.

- Use the Nernst equation to calculate the E_{cell} of a cell under nonstandard conditions.
- Define concentration cell.
- Distinguish between different types of batteries, including dry cell, alkaline, lead storage, lithium, and fuel cells.
- Describe in your own words the construction and operation of an electrolytic cell.
- Calculate the E° of a given electrolytic cell.
- Cite examples of electrolytic cells.
- Explain overvoltage.
- Describe in your own words the basis of corrosion and galvanization.

Chapter Summary

SECTION 18.1
- Redox reactions are those in which oxidation numbers change. **Half-reactions** are the separated oxidation and reduction reactions that make up the overall redox reaction.
- Redox equations can be balanced via the half-reaction method, which allows for the addition of H_2O to balance O, H^+ to balance H, and OH^- for reactions taking place in basic solution.

SECTION 18.2
- An electrochemical cell in which a spontaneous chemical reaction generates a flow of electrons through a wire is called a **galvanic cell.**
- Half-reactions in a galvanic cell take place in separate compartments called **half-cells.** Half-cells contain **electrodes** in solutions and are connected via an external wire and by a **salt bridge.**

- The electrode at which reduction occurs is called the **cathode;** the electrode at which oxidation occurs is called the **anode.**
- The difference in electric potential between the cathode and the anode is the **cell potential (E_{cell}).**

SECTION 18.3
- We use **standard reduction potentials (E°)** to calculate the standard cell voltage or *standard cell potential (E°_{cell}).*
- Half-cell potentials are measured relative to the **standard hydrogen electrode (SHE),** the half-reaction for which has an arbitrarily defined standard reduction potential of zero.

SECTION 18.4
- E°_{cell} is related to the standard free-energy change (ΔG°) and to the equilibrium constant, K. A positive E°_{cell} corresponds to a negative (ΔG°) value and a large K value.

SECTION 18.5

- E_{cell} under other than standard state conditions is determined from E°_{cell} and the reaction quotient, Q, using the **Nernst equation.**
- A **concentration cell** has the same type of electrode and the same ion in solution (at different concentrations) in the anode and cathode compartments.

SECTION 18.6

- **Batteries** are portable, self-contained sources of electric energy consisting of galvanic cells—or a series of galvanic cells.
- **Fuel cells** are not really batteries but also supply electric energy via a spontaneous redox reaction. Reactants must be supplied constantly for a fuel cell to operate.

SECTION 18.7

- **Electrolysis** is the use of electric energy to drive a nonspontaneous redox reaction. An electrochemical cell used for this purpose is called an **electrolytic cell.**

- The voltage that must actually be supplied to drive a nonspontaneous redox reaction is greater than the calculated amount because of **overvoltage.**
- Electrolysis is used to recharge lead storage batteries, separate compounds into their constituent elements, and separate and purify metals.
- We can calculate the amount of a substance produced in electrolysis if we know the current applied to the cell and the length of time for which it is applied.

SECTION 18.8

- **Corrosion** is the undesirable oxidation of metals.
- Corrosion can be prevented by coating the metal surface with paint, a less easily oxidized metal, or a more easily oxidized metal such as zinc.
- The use of a more easily oxidized metal is known as *cathodic protection,* wherein the metal being protected is made the cathode in a galvanic cell. *Galvanization* is the cathodic protection of iron or steel using zinc.

Key Words

Key Equations

18.1 $E^\circ_{cell} = E^\circ_{cathode} - E^\circ_{anode}$ Standard cell potential, E°_{cell}, is calculated by subtracting E°_{red} of the anode from E°_{red} of the cathode.

18.2 $\Delta G = -nFE_{cell}$ Free-energy change, ΔG, is calculated as the product of the number of electrons transferred in a redox reaction, n, the Faraday constant, F, and the cell potential, E_{cell}.

18.3 $\Delta G^\circ = -nFE^\circ_{cell}$ Standard free-energy change, ΔG°, is calculated as the product of the number of electrons transferred in a redox reaction, n, the Faraday constant, F, and the standard cell potential, E°_{cell}.

18.4 $E^\circ_{cell} = \dfrac{RT}{nF} \ln K$ Standard cell potential, E°_{cell}, is proportional to the natural log (ln) of the equilibrium constant, K.

18.5 $E^\circ_{cell} = \dfrac{0.0592\text{ V}}{n} \log K$ (at 25°C)   At 25°C, the relationship between E°_{cell} and K is simplified.

18.6 $E = E^\circ - \dfrac{RT}{nF} \ln Q$ At conditions other than standard, cell potential, E_{cell}, is proportional to the natural log (ln) of the reaction quotient, Q.

18.7 $E = E^\circ - \dfrac{0.0592\text{ V}}{n} \log Q$ At 25°C, the relationship between E_{cell} and Q is simplified.

Key Skills

Electrolysis of Metals

Electrolysis is used extensively in the processing and refining of metals. Converting between the current (and the time over which it is applied), and the amount of metal produced requires application of dimensional analysis [◀◀ Key Skills Chapter 1]. When we are given the current in amperes (A) and the time in seconds (s), we first convert to charge in coulombs (C).

$$\boxed{\text{Current (A)}} \times \boxed{\text{Time (s)}} = \boxed{\text{Charge (C)}}$$

We then divide by Faraday's constant to get the number of moles of electrons.

$$\boxed{\dfrac{\text{Charge (C)}}{96{,}500 \text{ C/mol } e^-}} = \boxed{\text{mol } e^-}$$

Using the reduction half-reaction, we can determine the number of moles of electrons needed to reduce a mole of metal.

$$M^{n+} + ne^- \longrightarrow M \qquad \boxed{\dfrac{1 \text{ mol metal}}{n \text{ mol } e^-}}$$

Finally, we convert moles electrons to moles metal.

$$\boxed{\text{mol } e^-} \times \boxed{\dfrac{1 \text{ mol metal}}{n \text{ mol } e^-}} = \boxed{\text{mol metal}}$$

Consider the following example. A current of 2.09 A is applied to a solution containing chromium(III) nitrate for 2.10 h. We determine the number of moles of Cr deposited as follows:

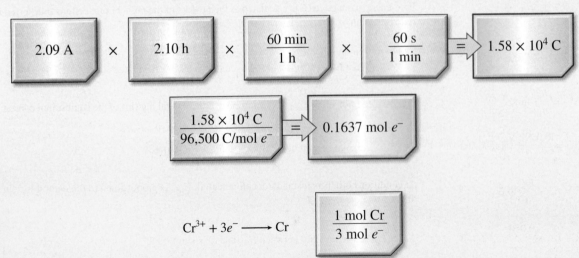

$$\boxed{2.09 \text{ A}} \times \boxed{2.10 \text{ h}} \times \boxed{\dfrac{60 \text{ min}}{1 \text{ h}}} \times \boxed{\dfrac{60 \text{ s}}{1 \text{ min}}} = \boxed{1.58 \times 10^4 \text{ C}}$$

$$\boxed{\dfrac{1.58 \times 10^4 \text{ C}}{96{,}500 \text{ C/mol } e^-}} = \boxed{0.1637 \text{ mol } e^-}$$

$$Cr^{3+} + 3e^- \longrightarrow Cr \qquad \boxed{\dfrac{1 \text{ mol Cr}}{3 \text{ mol } e^-}}$$

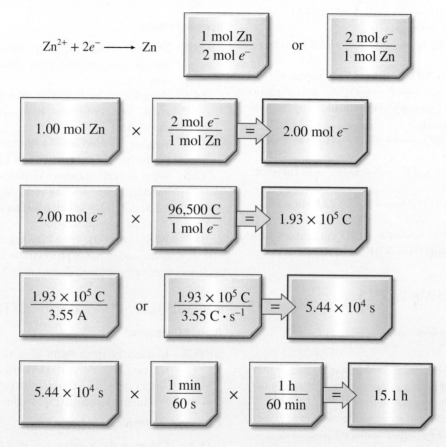

Conversely, we may know both the amount of metal to be produced and the current to be applied, and have to determine the amount of time required. If the goal is to produce 1.00 mole of zinc by electrolysis of a solution containing Zn^{2+} ions, and we will apply a current of 3.55 A, we determine the time over which the current must be applied as follows:

The reduction half-reaction indicates that 2 moles of electrons are needed to reduce a mole of zinc.

Key Skills Problems

18.1

How much copper metal can be produced by electrolysis of a solution containing Cu^{2+} ions by a current of 1.85 A applied for exactly 4 hours?
(a) 0.276 mol (b) 0.138 mol (c) 0.552 mol (d) 5.14×10^9 mol (e) 2.00 mol

18.2

What mass of cadmium will be produced by electrolysis of a solution of $Cd(NO_3)_2$ when a current of 4.83 A is applied for 6 hours and 15 minutes?
(a) 63.3 g (b) 127 g (c) 253 g (d) 31.6 g (e) 57.2 g

18.3

Of the following aqueous solutions, identify the one that would yield the *smallest* mass of metal and the one that would yield the *largest* mass of metal when it is electrolyzed with a current of 2.00 A for 25 minutes.

$$CuSO_4 \ AgNO_3 \ AuCl_3 \ ZnSO_4 \ Cr(NO_3)_3$$

(a) $ZnSO_4$, $CuSO_4$ (b) $CuSO_4$, $ZnSO_4$ (c) $AgNO_3$, $Cr(NO_3)_3$
(d) $Cr(NO_3)_3$, $AuCl_3$ (e) $AuCl_3$, $AgNO_3$

18.4

When a current of 5.22 A is applied over 3.50 hours to a solution containing metal ions, 20.00 grams of metal are produced. Which of the following could be the metal ion in the solution?
(a) Zn^{2+} (b) Au^{3+} (c) Ag^+ (d) Ni^{2+} (e) Cu^{2+}

Questions and Problems

SECTION 18.1: BALANCING REDOX REACTIONS

Conceptual Problems

18.1   Balance the following redox equations by the half-reaction method.
   (a) $H_2O_2 + Fe^{2+} \longrightarrow Fe^{3+} + H_2O$ (in acidic solution)
   (b) $Cu + HNO_3 \longrightarrow Cu^{2+} + NO + H_2O$ (in acidic solution)
   (c) $CN^- + MnO_4^- \longrightarrow CNO^- + MnO_2$ (in basic solution)
   (d) $Br_2 \longrightarrow BrO_3^- + Br^-$ (in basic solution)
   (e) $S_2O_3^{2-} + I_2 \longrightarrow I^- + S_4O_6^{2-}$ (in acidic solution)

18.2   Balance the following redox equations by the half-reaction method.
   (a) $Mn^{2+} + H_2O_2 \longrightarrow MnO_2 + H_2O$ (in basic solution)
   (b) $Bi(OH)_3 + SnO_2^{2-} \longrightarrow SnO_3^{2-} + Bi$ (in basic solution)
   (c) $Cr_2O_7^{2-} + C_2O_4^{2-} \longrightarrow Cr^{3+} + CO_2$ (in acidic solution)
   (d) $ClO_3^- + Cl^- \longrightarrow Cl_2 + ClO_2$ (in acidic solution)
   (e) $Mn^{2+} + BiO_3^- \longrightarrow Bi^{3+} + MnO_4^-$ (in acidic solution)

SECTION 18.2: GALVANIC CELLS

Visualizing Chemistry
Figure 18.1

VC 18.1   In the first scene of the animation, when a zinc bar is immersed in an aqueous copper sulfate solution, solid copper deposits on the bar. What reaction would take place if a *copper* bar were immersed in an aqueous *zinc* sulfate solution?
   (a) No reaction would take place.
   (b) Solid copper would still deposit on the bar.
   (c) Solid zinc would deposit on the bar.

VC 18.2   What causes the change in the potential of the galvanic cell in Figure 18.1 as the cell operates?
   (a) Changes in the sizes of the zinc and copper electrodes.
   (b) Changes in the concentrations of zinc and copper ions.
   (c) Changes in the volumes of solutions in the half-cells.

VC 18.3   Why does the color of the blue solution in the galvanic cell (Figure 18.1) fade as the cell operates?
   (a) Blue Cu^{2+} ions are replaced by colorless Zn^{2+} ions solution.
   (b) Blue Cu^{2+} ions are removed from solution by reduction.
   (c) Blue Cu^{2+} ions are removed from solution by oxidation.

VC 18.4   What happens to the mass of the copper electrode in the galvanic cell in Figure 18.1 as the cell operates?
   (a) It increases.
   (b) It decreases.
   (c) It does not change.

Review Questions

18.3   Define the following terms: *anode, cathode, cell voltage, electromotive force, standard reduction potential*.

18.4   Describe the basic features of a galvanic cell. Why are the two components of the cell separated from each other?

18.5   What is the function of a salt bridge? What kind of electrolyte should be used in a salt bridge?

18.6   What is a cell diagram? Write the cell diagram for a galvanic cell consisting of an Al electrode placed in a 1 M $Al(NO_3)_3$ solution and an Ag electrode placed in a 1 M $AgNO_3$ solution.

18.7   What is the difference between the half-reactions discussed in redox processes in Chapter 4 and the half-cell reactions discussed in Section 18.2?

SECTION 18.3: STANDARD REDUCTION POTENTIALS

Review Questions

18.8   Discuss the spontaneity of an electrochemical reaction in terms of its standard emf (E°_{cell}).

18.9   After operating a Daniell cell (see Figure 18.1) for a few minutes, the cell emf begins to drop. Explain.

Computational Problems

18.10   Calculate the standard emf of a cell that uses the Mg/Mg^{2+} and Cu/Cu^{2+} half-cell reactions at 25°C. Write the equation for the cell reaction that occurs under standard state conditions.

18.11   Calculate the standard emf of a cell that uses Ag/Ag^+ and Al/Al^{3+} half-cell reactions. Write the cell reaction that occurs under standard state conditions.

Conceptual Problems

18.12   Predict whether Fe^{3+} can oxidize I^- to I_2 under standard state conditions.

18.13   Which of the following reagents can oxidize H_2O to $O_2(g)$ under standard state conditions: $H^+(aq)$, $Cl^-(aq)$, $Cl_2(g)$, $Cu^{2+}(aq)$, $Pb^{2+}(aq)$, $MnO_4^-(aq)$ (in acid)?

18.14   Consider the following half-reactions.

$$MnO_4^-(aq) + 8H^+(aq) + 5e^- \longrightarrow Mn^{2+}(aq) + 4H_2O(l)$$

$$NO_3^-(aq) + 4H^+(aq) + 3e^- \longrightarrow NO(g) + 2H_2O(l)$$

Predict whether NO_3^- ions will oxidize Mn^{2+} to MnO_4^- under standard state conditions.

18.15 Predict whether the following reactions would occur spontaneously in aqueous solution at 25°C. Assume that the initial concentrations of dissolved species are all 1.0 M.
(a) $Ca(s) + Cd^{2+}(aq) \longrightarrow Ca^{2+}(aq) + Cd(s)$
(b) $2Br^-(aq) + Sn^{2+}(aq) \longrightarrow Br_2(l) + Sn(s)$
(c) $2Ag(s) + Ni^{2+}(aq) \longrightarrow 2Ag^+(aq) + Ni(s)$
(d) $Cu^+(aq) + Fe^{3+}(aq) \longrightarrow Cu^{2+}(aq) + Fe^{2+}(aq)$

18.16 Which species in each pair is a better oxidizing agent under standard state conditions: (a) Br_2 or Au^{3+}, (b) H_2 or Ag^+, (c) Cd^{2+} or Cr^{3+}, (d) O_2 in acidic media or O_2 in basic media?

18.17 Which species in each pair is a better reducing agent under standard state conditions: (a) Na or Li, (b) H_2 or I_2, (c) Fe^{2+} or Ag, (d) Br^- or Co^{2+}?

SECTION 18.4: SPONTANEITY OF REDOX REACTIONS UNDER STANDARD STATE CONDITIONS

Review Questions

18.18 Write the equations relating $\Delta G°$ and K to the standard emf of a cell. Define all the terms.

18.19 Compare the ease of measuring the equilibrium constant electrochemically with that by chemical means.

Computational Problems

18.20 Use the information in Table 2.1, and calculate the Faraday constant.

18.21 What is the equilibrium constant for the following reaction at 25°C?

$$Mg(s) + Zn^{2+}(aq) \rightleftharpoons Mg^{2+}(aq) + Zn(s)$$

18.22 The equilibrium constant for the reaction

$$Sr(s) + Mg^{2+}(aq) \rightleftharpoons Sr^{2+}(aq) + Mg(s)$$

is 2.69×10^{12} at 25°C. Calculate $E°$ for a cell made up of Sr/Sr^{2+} and Mg/Mg^{2+} half-cells.

18.23 Use the standard reduction potentials to find the equilibrium constant for each of the following reactions at 25°C.
(a) $Br_2(l) + 2I^-(aq) \rightleftharpoons 2Br^-(aq) + I_2(s)$
(b) $2Ce^{4+}(aq) + 2Cl^-(aq) \rightleftharpoons$
  $Cl_2(g) + 2Ce^{3+}(aq)$
(c) $5Fe^{2+}(aq) + MnO_4^-(aq) + 8H^+(aq) \rightleftharpoons$
  $Mn^{2+}(aq) + 4H_2O + 5Fe^{3+}(aq)$

18.24 Calculate $\Delta G°$ and K_c for the following reactions at 25°C.
(a) $Mg(s) + Pb^{2+}(aq) \rightleftharpoons Mg^{2+}(aq) + Pb(s)$
(b) $O_2(g) + 4H^+(aq) + 4Fe^{2+}(aq) \rightleftharpoons$
  $2H_2O(l) + 4Fe^{3+}(aq)$
(c) $2Al(s) + 3I_2(s) \rightleftharpoons 2Al^{3+}(aq) + 6I^-(aq)$

18.25 Under standard state conditions, what spontaneous reaction will occur in aqueous solution among the ions Ce^{4+}, Ce^{3+}, Fe^{3+}, and Fe^{2+}? Calculate $\Delta G°$ and K_c for the reaction.

18.26 Given that $E° = 0.52$ V for the reduction $Cu^+(aq) + e^- \longrightarrow Cu(s)$, calculate $E°$, $\Delta G°$, and K for the following reaction at 25°C.

$$2Cu^+(aq) \rightleftharpoons Cu^{2+}(aq) + Cu(s)$$

Conceptual Problems

18.27 Balance (in acidic medium) the equation for the oxidation of tin from an amalgam filling when it comes into contact with aluminum foil, and calculate the standard cell potential for the reaction.

18.28 Calculate the standard free-energy change and the equilibrium constant at 25°C for the reaction in Problem 18.27.

SECTION 18.5: SPONTANEITY OF REDOX REACTIONS UNDER CONDITIONS OTHER THAN STANDARD STATE

Review Questions

18.29 Write the Nernst equation, and explain all the terms.

Computational Problems

18.30 Write the Nernst equation for the following processes at some temperature T.
(a) $Mg(s) + Sn^{2+}(aq) \rightleftharpoons Mg^{2+}(aq) + Sn(s)$
(b) $2Cr(s) + 3Pb^{2+}(aq) \rightleftharpoons 2Cr^{3+}(aq) + 3Pb(s)$

18.31 What is the potential of a cell made up of Zn/Zn^{2+} and Cu/Cu^{2+} half-cells at 25°C if $[Zn^{2+}] = 0.25$ M and $[Cu^{2+}] = 0.15$ M?

18.32 Calculate $E°$, E, and ΔG for the following cell reactions.
(a) $Mg(s) + Sn^{2+}(aq) \rightleftharpoons Mg^{2+}(aq) + Sn(s)$
  $[Mg^{2+}] = 0.045$ M, $[Sn^{2+}] = 0.035$ M
(b) $3Zn(s) + 2Cr^{3+}(aq) \rightleftharpoons 3Zn^{2+}(aq) + 2Cr(s)$
  $[Cr^{3+}] = 0.010$ M, $[Zn^{2+}] = 0.0085$ M

18.33 Calculate the standard potential of the cell consisting of the Zn/Zn^{2+} half-cell and the SHE. What will the emf of the cell be if $[Zn^{2+}] = 0.45$ M, $P_{H_2} = 2.0$ atm, and $[H^+] = 1.8$ M?

18.34 What is the emf of a cell consisting of a Pb^{2+}/Pb half-cell and a $Pt/H^+/H_2$ half-cell if $[Pb^{2+}] = 0.10$ M, $[H] = 0.050$ M, and $P_{H_2} = 1.0$ atm?

18.35 Referring to the arrangement in Figure 18.1, calculate the $[Cu^{2+}]/[Zn^{2+}]$ ratio at which the following reaction is spontaneous at 25°C.

$$Cu(s) + Zn^{2+}(aq) \longrightarrow Cu^{2+}(aq) + Zn(s)$$

18.36 Calculate the emf of the following concentration cell.

$$Mg(s) \mid Mg^{2+}(0.24\ M) \parallel Mg^{2+}(0.53\ M) \mid Mg(s)$$

SECTION 18.6: BATTERIES

Review Questions

18.37 What is a battery? Describe several types of batteries.

18.38 Explain the differences between a primary galvanic cell—one that is not rechargeable—and a storage cell (e.g., the lead storage battery), which is rechargeable.

18.39   Discuss the advantages and disadvantages of fuel cells over conventional power plants in producing electricity.

Computational Problems

18.40   The hydrogen-oxygen fuel cell is described in Section 18.6. (a) What volume of $H_2(g)$, stored at 25°C at a pressure of 155 atm, would be needed to run an electric motor drawing a current of 8.5 A for 3.0 h? (b) What volume (in liters) of air at 25°C and 1.00 atm will have to pass into the cell per minute to run the motor? Assume that air is 20 percent O_2 by volume and that all the O_2 is consumed in the cell. The other components of air do not affect the fuel-cell reactions. Assume ideal gas behavior.

18.41   Calculate the standard emf of the propane fuel cell (discussed in Section 18.6) at 25°C, given that ΔG_f° for propane is −23.5 kJ/mol.

SECTION 18.7: ELECTROLYSIS

Review Questions

18.42   What is the difference between a galvanic cell (such as a Daniell cell) and an electrolytic cell?

18.43   Define the term *overvoltage*. How does overvoltage affect electrolytic processes?

Computational Problems

18.44   Calculate the number of grams of copper metal that can be produced by supplying 1.00 F to a solution of Cu^{2+} ions.

18.45   The half-reaction at an electrode is

$$Mg^{2+}(molten) + 2e^- \longrightarrow Mg(s)$$

Calculate the number of grams of magnesium that can be produced by supplying 1.00 F to the electrode.

18.46   Consider the electrolysis of molten barium chloride ($BaCl_2$). (a) Write the half-reactions. (b) How many grams of barium metal can be produced by supplying 0.50 A for 30 min?

18.47   Considering only the cost of electricity, would it be cheaper to produce a ton of sodium or a ton of aluminum by electrolysis?

18.48   If the cost of electricity to produce magnesium by the electrolysis of molten magnesium chloride is $155 per ton of metal, what is the cost (in dollars) of the electricity necessary to produce (a) 10.0 tons of aluminum, (b) 30.0 tons of sodium, and (c) 50.0 tons of calcium?

18.49   One of the half-reactions for the electrolysis of water is

$$2H_2O(l) \longrightarrow O_2(g) + 4H^+(aq) + 4e^-$$

If 0.076 L of O_2 is collected at 25°C and 755 mmHg, how many faradays of electricity had to pass through the solution?

18.50   How many faradays of electricity are required to produce (a) 0.84 L of O_2 at exactly 1 atm and 25°C from aqueous H_2SO_4 solution, (b) 1.50 L of Cl_2 at 750 mmHg and 20°C from molten NaCl, and (c) 6.0 g of Sn from molten $SnCl_2$?

18.51   Calculate the amounts of Cu and Br_2 produced in 1.0 h at inert electrodes in a solution of $CuBr_2$ by a current of 4.50 A.

18.52   In the electrolysis of an aqueous $AgNO_3$ solution, 0.67 g of Ag is deposited after a certain period of time. (a) Write the half-reaction for the reduction of Ag^+. (b) What is the probable oxidation half-reaction? (c) Calculate the quantity of electricity used (in coulombs).

18.53   A steady current was passed through molten $CoSO_4$ until 2.35 g of metallic cobalt was produced. Calculate the number of coulombs of electricity used.

18.54   A constant electric current flows for 3.75 h through two electrolytic cells connected in series. One contains a solution of $AgNO_3$ and the second a solution of $CuCl_2$. During this time, 2.00 g of silver is deposited in the first cell. (a) How many grams of copper are deposited in the second cell? (b) What is the current flowing (in amperes)?

18.55   What is the hourly production rate of chlorine gas (in kilograms) from an electrolytic cell using aqueous NaCl electrolyte and carrying a current of 1.500×10^3 A? The anode efficiency for the oxidation of Cl^- is 93.0%.

18.56   Chromium plating is applied by electrolysis to objects suspended in a dichromate solution, according to the following (unbalanced) half-reaction.

$$Cr_2O_7^{2-}(aq) + e^- + H^+(aq) \longrightarrow Cr(s) + H_2O(l)$$

How long (in hours) would it take to apply a chromium plating 1.0×10^{-2} mm thick to a car bumper with a surface area of 0.25 m^2 in an electrolytic cell carrying a current of 25.0 A? (The density of chromium is 7.19 g/cm^3.)

18.57   The passage of a current of 0.750 A for 25.0 min deposited 0.369 g of copper from a $CuSO_4$ solution. From this information, calculate the molar mass of copper.

18.58   A quantity of 0.300 g of copper was deposited from a $CuSO_4$ solution by passing a current of 3.00 A through the solution for 304 s. Calculate the value of the Faraday constant.

18.59   In a certain electrolysis experiment, 1.44 g of Ag were deposited in one cell (containing an aqueous $AgNO_3$ solution), while 0.120 g of an unknown metal X was deposited in another cell (containing an aqueous XCl_3 solution) in series with the $AgNO_3$ cell. Calculate the molar mass of X.

18.60   One of the half-reactions for the electrolysis of water is

$$2H^+(aq) + 2e^- \longrightarrow H_2(g)$$

If 0.845 L of H_2 is collected at 25°C and 782 mmHg, how many faradays of electricity had to pass through the solution?

SECTION 18.8: CORROSION

Review Questions

18.61 Steel hardware, including nuts and bolts, is often coated with a thin plating of cadmium. Explain the function of the cadmium layer.

18.62 "Galvanized iron" is steel sheet that has been coated with zinc; "tin" cans are made of steel sheet coated with tin. Discuss the functions of these coatings and the electrochemistry of the corrosion reactions that occur if an electrolyte contacts the scratched surface of a galvanized iron sheet or a tin can.

18.63 Tarnished silver contains Ag_2S. The tarnish can be removed by placing silverware in an aluminum pan containing an inert electrolyte solution, such as NaCl. Explain the electrochemical principle for this procedure. [The standard reduction potential for the half-cell reaction $Ag_2S(s) + 2e^- \longrightarrow 2Ag(s) + S^{2-}(aq)$ is -0.71 V.]

18.64 How does the tendency of iron to rust depend on the pH of the solution?

ADDITIONAL PROBLEMS

18.65 For each of the following redox reactions, (i) write the half-reactions, (ii) write a balanced equation for the whole reaction, (iii) determine in which direction the reaction will proceed spontaneously under standard-state conditions.
(a) $H_2(g) + Ni^{2+}(aq) \longrightarrow H^+(aq) + Ni(s)$
(b) $MnO_4^-(aq) + Cl^-(aq) \longrightarrow Mn^{2+}(aq) + Cl_2(g)$
(c) $Cr(s) + Zn^{2+}(aq) \longrightarrow Cr^{3+}(aq) + Zn(s)$

18.66 The oxidation of 25.0 mL of a solution containing Fe^{2+} requires 26.0 mL of 0.0250 M $K_2Cr_2O_7$ in acidic solution. Balance the following equation, and calculate the molar concentration of Fe^{2+}.
$$Cr_2O_7^{2-} + Fe^{2+} + H^+ \longrightarrow Cr^{3+} + Fe^{3+}$$

18.67 The SO_2 present in air is mainly responsible for the phenomenon of acid rain. The concentration of SO_2 can be determined by titrating against a standard permanganate solution as follows:
$$5SO_2 + 2MnO_4^- + 2H_2O \longrightarrow 5SO_4^{2-} + 2Mn^{2+} + 4H^+$$
Calculate the number of grams of SO_2 in a sample of air if 7.37 mL of 0.00800 M $KMnO_4$ solution is required for the titration.

18.68 A sample of iron ore weighing 0.2792 g was dissolved in an excess of a dilute acid solution. All the iron was first converted to Fe(II) ions. The solution then required 23.30 mL of 0.0194 M $KMnO_4$ for oxidation to Fe(III) ions. Calculate the percent by mass of iron in the ore.

18.69 The concentration of a hydrogen peroxide solution can be conveniently determined by titration against a standardized potassium permanganate solution in an acidic medium according to the following unbalanced equation.
$$MnO_4^- + H_2O_2 \longrightarrow O_2 + Mn^{2+}$$

(a) Balance this equation. (b) If 36.44 mL of a 0.01652 M $KMnO_4$ solution is required to completely oxidize 25.00 mL of an H_2O_2 solution, calculate the molarity of the H_2O_2 solution.

18.70 Oxalic acid ($H_2C_2O_4$) is present in many plants and vegetables. (a) Balance the following equation in acid solution.
$$MnO_4^- + C_2O_4^{2-} \longrightarrow Mn^{2+} + CO_2$$
(b) If a 1.00-g sample of plant matter requires 24.0 mL of 0.0100 M $KMnO_4$ solution to reach the equivalence point, what is the percent by mass of $H_2C_2O_4$ in the sample?

18.71 Calcium oxalate (CaC_2O_4) is insoluble in water. This property has been used to determine the amount of Ca^{2+} ions in blood. The calcium oxalate isolated from blood is dissolved in acid and titrated against a standardized $KMnO_4$ solution as described in Problem 18.70. In one test it is found that the calcium oxalate isolated from a 10.0-mL sample of blood requires 24.2 mL of 9.56×10^{-4} M $KMnO_4$ for titration. Calculate the number of milligrams of calcium per milliliter of blood.

18.72 Complete the following table. State whether the cell reaction is spontaneous, nonspontaneous, or at equilibrium.

E	ΔG	Cell Reaction
> 0		
	> 0	
$= 0$		

18.73 From the following information, calculate the solubility product of AgBr.
$$Ag^+(aq) + e^- \longrightarrow Ag(s) \qquad E° = 0.80 \text{ V}$$
$$AgBr(s) + e^- \longrightarrow Ag(s) + Br^-(aq) \qquad E° = 0.07 \text{ V}$$

18.74 Consider a galvanic cell composed of the SHE and a half-cell using the reaction $Ag^+(aq) + e^- \longrightarrow Ag(s)$. (a) Calculate the standard cell potential. (b) What is the spontaneous cell reaction under standard state conditions? (c) Calculate the cell potential when $[H^+]$ in the hydrogen electrode is changed to (i) 1.0×10^{-2} M and (ii) 1.0×10^{-5} M, all other reagents being held at standard state conditions. (d) Based on this cell arrangement, suggest a design for a pH meter.

18.75 A galvanic cell consists of a silver electrode in contact with 346 mL of 0.100 M $AgNO_3$ solution and a magnesium electrode in contact with 288 mL of 0.100 M $Mg(NO_3)_2$ solution. (a) Calculate E for the cell at 25°C. (b) A current is drawn from the cell until 1.20 g of silver has been deposited at the silver electrode. Calculate E for the cell at this stage of operation.

18.76 Explain why chlorine gas can be prepared by electrolyzing an aqueous solution of NaCl but fluorine gas cannot be prepared by electrolyzing an aqueous solution of NaF.

18.77 Calculate the emf of the following concentration cell at 25°C.

$$Cu(s) \mid Cu^{2+}(0.080\ M) \parallel Cu^{2+}(1.2\ M) \mid Cu(s)$$

18.78 The cathode reaction in the Leclanché cell is given by

$$2MnO_2(s) + Zn^{2+}(aq) + 2e^- \longrightarrow ZnMn_2O_4(s)$$

If a Leclanché cell produces a current of 0.0050 A, calculate how many hours this current supply will last if there is initially 4.0 g of MnO_2 present in the cell. Assume that there is an excess of Zn^{2+} ions.

18.79 For a number of years, it was not clear whether mercury(I) ions existed in solution as Hg^+ or as Hg_2^{2+}. To distinguish between these two possibilities, we could set up the following system:

$$Hg(l) \mid soln\ A \parallel soln\ B \mid Hg(l)$$

where soln A contained 0.263 g mercury(I) nitrate per liter and soln B contained 2.63 g mercury(I) nitrate per liter. If the measured emf of such a cell is 0.0289 V at 18°C, what can you deduce about the nature of the mercury(I) ions?

18.80 An aqueous KI solution to which a few drops of phenolphthalein have been added is electrolyzed using an apparatus like the one shown here.

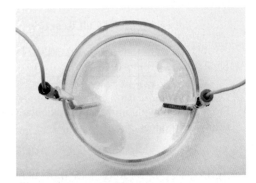

© David A. Tietz/Editorial Image, LLC

Describe what you would observe at the anode and the cathode. (*Hint:* Molecular iodine is only slightly soluble in water, but in the presence of I^- ions, it forms the brown color of I_3^- ions. See Problem 13.114 on page 614.)

18.81 A piece of magnesium metal weighing 1.56 g is placed in 100.0 mL of 0.100 *M* $AgNO_3$ at 25°C. Calculate $[Mg^{2+}]$ and $[Ag^+]$ in solution at equilibrium. What is the mass of the magnesium left? The volume remains constant.

18.82 Describe an experiment that would enable you to determine which is the cathode and which is the anode in a galvanic cell using copper and zinc electrodes.

18.83 An acidified solution was electrolyzed using copper electrodes. A constant current of 1.18 A caused the anode to lose 0.584 g after 1.52×10^3 s. (a) What is the gas produced at the cathode, and what is its volume at STP? (b) Given that the charge of an electron is 1.6022×10^{-19} C, calculate Avogadro's number. Assume that copper is oxidized to Cu^{2+} ions.

18.84 In a certain electrolysis experiment involving Al^{3+} ions, 60.2 g of Al is recovered when a current of 0.352 A is used. How many minutes did the electrolysis last?

18.85 Consider the oxidation of ammonia.

$$4NH_3(g) + 3O_2(g) \longrightarrow 2N_2(g) + 6H_2O(l)$$

(a) Calculate the $\Delta G°$ for the reaction. (b) If this reaction were used in a fuel cell, what would be the standard cell potential?

18.86 When an aqueous solution containing gold(III) salt is electrolyzed, metallic gold is deposited at the cathode and oxygen gas is generated at the anode. (a) If 9.26 g of Au is deposited at the cathode, calculate the volume (in liters) of O_2 generated at 23°C and 747 mmHg. (b) What is the current used if the electrolytic process took 2.00 h?

18.87 In an electrolysis experiment, a student passes the same quantity of electricity through two electrolytic cells, one containing a silver salt and the other a gold salt. Over a certain period of time, the student finds that 2.64 g of Ag and 1.61 g of Au are deposited at the cathodes. What is the oxidation state of gold in the gold salt?

18.88 Consider the electrochemical cell represented by the following diagram.

$$Pb(s) \mid Pb(NO_3)_2\ (0.60\ M) \parallel AgNO_3\ (0.40\ M) \mid Ag(s)$$

Determine the initial value of E_{cell} under the conditions shown in the cell diagram; and determine the initial value of E_{cell} if $[Ag^+]$ were increased by a factor of 4 at 25°C.

18.89 Given that

$$2Hg^{2+}(aq) + 2e^- \longrightarrow Hg_2^{2+}(aq) \qquad E° = 0.92\ V$$
$$Hg_2^{2+}(aq) + 2e^- \longrightarrow 2Hg(l) \qquad E° = 0.85\ V$$

calculate $\Delta G°$ and K for the following process at 25°C.

$$Hg_2^{2+}(aq) \longrightarrow Hg^{2+}(aq) + Hg(l)$$

(The preceding reaction is an example of a *disproportionation reaction* in which an element in one oxidation state is both oxidized and reduced.)

18.90 Fluorine (F_2) is obtained by the electrolysis of liquid hydrogen fluoride (HF) containing potassium fluoride (KF). (a) Write the half-cell reactions and the overall reaction for the process. (b) What is the purpose of KF? (c) Calculate the volume of F_2 (in liters) collected at 24.0°C and 1.2 atm after electrolyzing the solution for 15 h at a current of 502 A.

18.91 A 300-mL solution of NaCl was electrolyzed for 6.00 min. If the pH of the final solution was 12.24, calculate the average current used.

18.92 Industrially, copper is purified by electrolysis. The impure copper acts as the anode, and the cathode is made of pure copper. The electrodes are immersed in a $CuSO_4$ solution. During electrolysis, copper at the anode enters the solution as Cu^{2+} while Cu^{2+} ions are reduced at the cathode. (a) Write half-cell reactions and the overall reaction for the electrolytic process. (b) Suppose the anode was contaminated with Zn and Ag. Explain what happens to these impurities during electrolysis. (c) How many hours will it take to obtain 1.00 kg of Cu at a current of 18.9 A?

18.93 An aqueous solution of a platinum salt is electrolyzed at a current of 2.50 A for 2.00 h. As a result, 9.09 g of metallic Pt is formed at the cathode. Calculate the charge on the Pt ions in this solution.

18.94 Consider a galvanic cell consisting of a magnesium electrode in contact with 1.0 M $Mg(NO_3)_2$ and a cadmium electrode in contact with 1.0 M $Cd(NO_3)_2$. Calculate $E°$ for the cell, and draw a diagram showing the cathode, anode, and direction of electron flow.

18.95 A current of 6.00 A passes through an electrolytic cell containing dilute sulfuric acid for 3.40 h. If the volume of O_2 gas generated at the anode is 4.26 L (at STP), calculate the charge (in coulombs) on an electron.

18.96 Gold will not dissolve in either concentrated nitric acid or concentrated hydrochloric acid. However, the metal does dissolve in a mixture of the acids (one part HNO_3 and three parts HCl by volume), called *aqua regia*. (a) Write a balanced equation for this reaction. (*Hint:* Among the products are $HAuCl_4$ and NO_2.) (b) What is the function of HCl?

18.97 Explain why most useful galvanic cells give voltages of no more than 1.5 to 2.5 V. What are the prospects for developing practical galvanic cells with voltages of 5 V or more?

18.98 A silver rod and a SHE are dipped into a saturated aqueous solution of silver oxalate ($Ag_2C_2O_4$), at 25°C. The measured potential difference between the rod and the SHE is 0.589 V, the rod being positive. Calculate the solubility product constant for silver oxalate.

18.99 Zinc is an amphoteric metal; that is, it reacts with both acids and bases. The standard reduction potential is −1.36 V for the reaction

$$Zn(OH)_4^{2-}(aq) + 2e^- \longrightarrow Zn(s) + 4OH^-(aq)$$

Calculate the formation constant (K_f) for the reaction

$$Zn^{2+}(aq) + 4OH^-(aq) \rightleftharpoons Zn(OH)_4^{2-}(aq)$$

18.100 Use the data in Table 18.1 to determine whether hydrogen peroxide will undergo disproportionation in an acid medium: $2H_2O_2 \longrightarrow 2H_2O + O_2$.

18.101 The magnitudes (but *not* the signs) of the standard reduction potentials of two metals X and Y are

$$Y^{2+} + 2e^- \longrightarrow Y \quad |E°| = 0.34\ V$$
$$X^{2+} + 2e^- \longrightarrow X \quad |E°| = 0.25\ V$$

where the || notation denotes that only the magnitude (but not the sign) of the $E°$ value is shown. When the half-cells of X and Y are connected, electrons flow from X to Y. When X is connected to a SHE, electrons flow from X to SHE. (a) Are the $E°$ values of the half-reactions positive or negative? (b) What is the standard emf of a cell made up of X and Y?

18.102 A galvanic cell is constructed as follows. One half-cell consists of a platinum wire immersed in a solution containing 1.0 M Sn^{2+} and 1.0 M Sn^{4+}; the other half-cell has a thallium rod immersed in a solution of 1.0 M Tl^+. (a) Write the half-cell reactions and the overall reaction. (b) What is the equilibrium constant at 25°C? (c) What is the cell voltage if the Tl^+ concentration is increased 10-fold? ($E°_{Tl^+/Tl} = -0.34$ V.)

18.103 Given the standard reduction potential for Au^{3+} in Table 18.1 and

$$Au^+(aq) + e^- \longrightarrow Au(s) \quad E° = 1.69\ V$$

answer the following questions. (a) Why does gold not tarnish in air? (b) Will the following disproportionation occur spontaneously?

$$3Au^+(aq) \longrightarrow Au^{3+}(aq) + 2Au(s)$$

(c) Predict the reaction between gold and fluorine gas.

18.104 The ingestion of a very small quantity of mercury is not considered too harmful. Would this statement still hold if the gastric juice in your stomach were mostly nitric acid instead of hydrochloric acid?

18.105 When 25.0 mL of a solution containing both Fe^{2+} and Fe^{3+} ions is titrated with 23.0 mL of 0.0200 M $KMnO_4$ (in dilute sulfuric acid), all the Fe^{2+} ions are oxidized to Fe^{3+} ions. Next, the solution is treated with Zn metal to convert all the Fe^{3+} ions to Fe^{2+} ions. Finally, 40.0 mL of the same $KMnO_4$ solution is added to the solution to oxidize the Fe^{2+} ions to Fe^{3+}. Calculate the molar concentrations of Fe^{2+} and Fe^{3+} in the original solution.

18.106 Consider the Daniell cell in Figure 18.1. When viewed externally, the anode appears negative and the cathode positive (electrons are flowing from the anode to the cathode). Yet in solution anions are moving toward the anode, which means that it must appear positive to the anions. Because the anode cannot simultaneously be negative and positive, give an explanation for this apparently contradictory situation.

18.107 Use the data in Table 18.1 to show that the decomposition of H_2O_2 (a disproportionation reaction) is spontaneous at 25°C.

$$2H_2O_2(aq) \longrightarrow 2H_2O(l) + O_2(g)$$

18.108 The concentration of sulfuric acid in the lead-storage battery of an automobile over a period of time has decreased from 38.0% by mass (density = 1.29 g/mL) to 26.0% by mass (1.19 g/mL). Assume the volume of the acid remains constant at 724 mL. (a) Calculate the total charge in coulombs supplied by the battery. (b) How long (in hours) will it take to recharge the battery back to the original sulfuric acid concentration using a current of 22.4 A?

18.109 Consider a Daniell cell operating under nonstandard state conditions. Suppose that the cell's reaction is multiplied by 2. What effect does this have on each of the following quantities in the Nernst equation: (a) E, (b) $E°$, (c) Q, (d) ln Q, (e) n?

18.110 A spoon was silver-plated electrolytically in an $AgNO_3$ solution. (a) Sketch a diagram for the process. (b) If 0.884 g of Ag was deposited on the spoon at a constant current of 18.5 mA, how long (in min) did the electrolysis take?

18.111 Comment on whether F_2 will become a stronger oxidizing agent with increasing H^+ concentration.

18.112 In recent years there has been much interest in electric cars. List some advantages and disadvantages of electric cars compared to automobiles with internal combustion engines.

18.113 Calculate the pressure of H_2 (in atm) required to maintain equilibrium with respect to the following reaction at 25°C,

$$Pb(s) + 2H^+(aq) \rightleftharpoons Pb^{2+}(aq) + H_2(g)$$

given that $[Pb^{2+}] = 0.035\ M$ and the solution is buffered at pH 1.60.

18.114 A piece of magnesium ribbon and a copper wire are partially immersed in a 0.1 M HCl solution in a beaker. The metals are joined externally by another piece of metal wire. Bubbles are seen to evolve at both the Mg and Cu surfaces. (a) Write equations representing the reactions occurring at the metals. (b) What visual evidence would you seek to show that Cu is not oxidized to Cu^{2+}? (c) At some stage, NaOH solution is added to the beaker to neutralize the HCl acid. Upon further addition of NaOH, a white precipitate forms. What is it?

18.115 The zinc-air battery shows much promise for electric cars because it is lightweight and rechargeable.

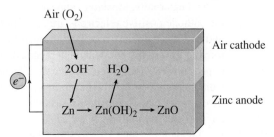

The net transformation is $Zn(s) + \frac{1}{2}O_2(g) \longrightarrow$ $ZnO(s)$. (a) Write the half-reactions at the zinc-air electrodes, and calculate the standard emf of the battery at 25°C. (b) Calculate the emf under actual operating conditions when the partial pressure of oxygen is 0.21 atm. (c) What is the energy density (measured as the energy in kilojoules that can be obtained from 1 kg of the metal) of the zinc electrode? (d) If a current of 2.1×10^5 A is to be drawn from a zinc-air battery system, what volume of air (in liters) would need to be supplied to the battery every second? Assume that the temperature is 25°C and the partial pressure of oxygen is 0.21 atm.

18.116 Calculate $E°$ for the reactions of mercury with (a) 1 M HCl and (b) 1 M HNO_3. Which acid will oxidize Hg to Hg_2^{2+} under standard state conditions? Can you identify which pictured test tube contains HNO_3 and Hg and which contains HCl and Hg?

© McGraw-Hill Education/Ken Karp, photographer.

18.117 Because all alkali metals react with water, it is not possible to measure the standard reduction potentials of these metals directly as in the case of, say, zinc. An indirect method is to consider the following hypothetical reaction.

$$Li^+(aq) + \tfrac{1}{2}H_2(g) \longrightarrow Li(s) + H^+(aq)$$

Using the appropriate equation presented in this chapter and the thermodynamic data in Appendix 2, calculate $E°$ for $Li^+(aq) + e^- \longrightarrow Li(s)$ at 298 K. Use 96,485.338 C/mol e^- for the Faraday constant. Compare your result with that listed in Table 18.1.

18.118 A galvanic cell using Mg/Mg^{2+} and Cu/Cu^{2+} half-cells operates under standard state conditions at 25°C, and each compartment has a volume of 218 mL. The cell delivers 0.22 A for 31.6 h. (a) How many grams of Cu are deposited? (b) What is the $[Cu^{2+}]$ remaining?

18.119 Given the following standard reduction potentials, calculate the ion product, K_w, for water at 25°C.

$$2H^+(aq) + 2e^- \longrightarrow H_2(g) \qquad\qquad E° = 0.00\ V$$
$$2H_2O(l) + 2e^- \longrightarrow H_2(g) + 2OH^-(aq) \quad E° = -0.83\ V$$

18.120 Compare the pros and cons of a fuel cell, such as the hydrogen-oxygen fuel cell, and a coal-fired power station for generating electricity.

18.121 Lead storage batteries are rated by ampere-hours, that is, the number of amperes they can deliver in an hour. (a) Show that 1 Ah = 3600 C. (b) The lead anodes of a certain lead storage battery have a total mass of 406 g. Calculate the maximum theoretical capacity of the battery in ampere-hours. Explain why in practice we can never extract this much energy from the battery. (*Hint:* Assume all the lead will be used up in the electrochemical reaction, and refer to the electrode reactions on page 854.) (c) Calculate $E°_{cell}$ and $\Delta G°$ for the battery.

18.122 Use Equations 14.10 and 18.3 to calculate the emf values of the Daniell cell at 25°C and 80°C. Comment on your results. What assumptions are used in the derivation? (*Hint:* You need the thermodynamic data in Appendix 2.)

18.123 A construction company is installing an iron culvert (a long cylindrical tube) that is 40.0 m long with a radius of 0.900 m. To prevent corrosion, the culvert must be galvanized. This process is carried out by first passing an iron sheet of appropriate dimensions through an electrolytic cell containing Zn^{2+} ions, using graphite as the anode and the iron sheet as the cathode. If the voltage is 3.26 V, what is the cost of electricity for depositing a layer 0.200 mm thick if the efficiency of the process is 95%? The electricity rate is $0.12 per kilowatt hour (kWh), where 1 W = 1 J/s and the density of Zn is 7.14 g/cm^3.

18.124   A 9.00×10^2 mL amount of 0.200 M MgI_2 solution was electrolyzed. As a result, hydrogen gas was generated at the cathode and iodine was formed at the anode. The volume of hydrogen collected at 26°C and 779 mmHg was 1.22×10^3 mL. (a) Calculate the charge in coulombs consumed in the process. (b) How long (in min) did the electrolysis last if a current of 7.55 A was used? (c) A white precipitate was formed in the process. What was it, and what was its mass in grams? Assume the volume of the solution was constant.

18.125   Based on the following standard reduction potentials,

$$Fe^{2+}(aq) + 2e^- \longrightarrow Fe(s) \qquad E_1^\circ = -0.44 \text{ V}$$
$$Fe^{3+}(aq) + e^- \longrightarrow Fe^{2+}(aq) \qquad E_2^\circ = 0.77 \text{ V}$$

calculate the standard reduction potential for the half-reaction

$$Fe^{3+}(aq) + 3e^- \longrightarrow Fe(s) \qquad E_3^\circ = ?$$

18.126   Which of the components of dental amalgam (mercury, silver, tin, copper, or zinc) would be oxidized when a filling is brought into contact with lead?

18.127   Calculate the equilibrium constant for the following reaction at 298 K.

$$Zn(s) + Cu^{2+}(aq) \rightleftharpoons Zn^{2+}(aq) + Cu(s)$$

18.128   Cytochrome c is a protein involved in biological electron transfer processes. The redox half-reaction is shown by the reduction of the Fe^{3+} ion to the Fe^{2+} ion.

$$cyt\ c(Fe^{3+}) + e^- \longrightarrow cyt\ c(Fe^{2+}) \qquad E^\circ = 0.254 \text{ V}$$

Calculate the number of moles of cyt $c(Fe^{3+})$ formed from cyt $c(Fe^{2+})$ with the Gibbs free energy derived from the oxidation of 1 mole of glucose.

18.129   The nitrite ion NO_2^- in soil is oxidized to the nitrate ion (NO_3^-) by the bacterium *Nitrobacter agilis* in the presence of oxygen. The half-reactions are

$$NO_3^- + 2H^+ + 2e^- \longrightarrow NO_2^- + H_2O \qquad E^\circ = 0.42 \text{ V}$$
$$O_2 + 4H^+ + 4e^- \longrightarrow 2H_2O \qquad E^\circ = 1.23 \text{ V}$$

Calculate the yield of ATP synthesis per mole of nitrite oxidized. (*Hint:* Refer to Section 14.6.)

18.130   Fluorine is a highly reactive gas that attacks water to form HF and other products. Follow the procedure in Problem 18.117 to show how you can determine indirectly the standard reduction potential for fluorine. Compare your result with the value in Table 18.1.

18.131   As discussed in Section 18.5, the potential of a concentration cell diminishes as the cell operates and the concentrations in the two compartments approach each other. When the concentrations in both compartments are the same, the cell ceases to operate. At this stage, is it possible to generate a cell potential by adjusting a parameter other than concentration? Explain.

Answers to In-Chapter Materials

PRACTICE PROBLEMS

18.1A $2MnO_4^- + H_2O + 3CN^- \longrightarrow 2MnO_2 + 2OH^- + 3CNO^-$.
18.1B $MnO_4^- + 5Fe^{2+} + 8H^+ \longrightarrow Mn^{2+} + 5Fe^{3+} + 4H_2O$.
18.2A $Cd + Pb^{2+} \longrightarrow Cd^{2+} + Pb$, $E_{cell}^\circ = 0.27$ V. **18.2B** In one cell, the iron electrode is the anode and the overall reaction is
$Fe(s) + Sn^{2+}(aq) \longrightarrow Fe^{2+}(aq) + Sn(s)$. $E_{cell}^\circ = E_{Sn^{2+}/Sn}^\circ - E_{Fe^{2+}/Fe}^\circ = (-0.14$ V$) - (-0.44$ V$) = 0.30$ V. In the other cell, the iron electrode is the cathode and the overall reaction is
$3Fe^{2+}(aq) + 2Cr(s) \longrightarrow 3Fe(s) + 2Cr^{3+}(aq)$. $E_{cell}^\circ = E_{Fe^{2+}/Fe}^\circ - E_{Cr^{3+}/Cr}^\circ = (-0.44$ V$) - (-0.74$ V$) = 0.30$ V. **18.3A** (a) No reaction, (b) $2H^+ + Pb \longrightarrow H_2 + Pb^{2+}$. **18.3B** Cobalt has a more positive reduction potential than iron. Cobalt ion would, therefore, be *reduced* in the presence of iron metal; and the iron in an iron container would oxidize to Fe^{2+}. Metal that is oxidized goes from the solid phase to the aqueous phase, meaning that the container would effectively *dissolve*. Cobalt has a less positive reduction potential than tin. Tin metal would not be oxidized by Co^{2+} ion, and the container would remain intact. A cobalt(II) chloride solution would be more safely stored in a tin container than in an iron container. **18.4A** −411 kJ/mol. **18.4B** 23.2 kJ/mol. **18.5A** 1×10^{-42}. **18.5B** 1.4×10^5. **18.6A** Yes, the reaction is spontaneous. **18.6B** 18 atm. **18.7A** $[Cu^+] = 1.11 \times 10^{-3}$ M, $K_{sp} = 1.2 \times 10^{-6}$. **18.7B** 0.16 V. **18.8A** 7.44 g Mg. **18.8B** 0.96 A.

SECTION REVIEW

18.1.1 a, c, e. **18.1.2** c. **18.3.1** d. **18.3.2** c. **18.3.3** b. **18.3.4** e. **18.4.1** b. **18.4.2** b. **18.5.1** b. **18.5.2** b. **18.5.3** d. **18.5.4** e. **18.7.1** a. **18.7.2** e. **18.7.3** c.

Chapter 19

Chemical Kinetics

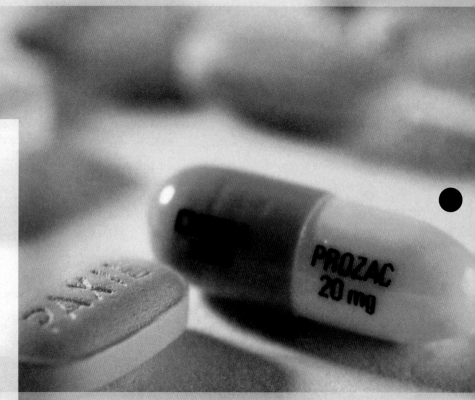

© Jonathan Nourok/Getty Images

WHEN DEVELOPING NEW DRUGS, it is critical to understand the fate of a new substance from the time it is administered to the point when it is eliminated from the body—and to understand the rate at which it is absorbed, metabolized, and eliminated. Pharmacokinetics, the study of the time course of drug absorption, distribution, metabolism, and excretion, has developed strong correlations between drug concentrations and their activity in physiological conditions. In particular, knowing a drug's bioavailability, the proportion of the drug that reaches the target area, allows clinicians to identify optimal dosages for maximum effect.

Before You Begin, Review These Skills

- The value of *R* expressed in J/K · mol [◄◄ Section 11.5]
- Use of logarithms [►►| Appendix 1]

19.1 REACTION RATES

Chemical kinetics is the study of how fast reactions take place. Many familiar reactions, such as the initial steps in vision and photosynthesis, happen almost instantaneously. Others, such as the rusting of iron or the conversion of diamond to graphite, take place on a timescale of days or even millions of years.

Knowledge of kinetics is important to many scientific endeavors, including drug design, pollution control, and food processing. The job of an industrial chemist often is to work on increasing the *rate* of a reaction rather than maximizing its yield or developing a new process. Among the factors that can increase reaction rate are increased reactant concentration, increased temperature, increased surface area of a solid reactant, and the presence of a catalyst. After examining how these things affect reaction rate at the molecular level, we will look at the quantitative impact of each factor in more detail.

19.2 COLLISION THEORY OF CHEMICAL REACTIONS

A chemical reaction can be represented by the general equation

$$\text{reactants} \longrightarrow \text{products}$$

This equation tells us that during the course of a reaction, reactants are consumed while products are formed. Chemical reactions generally occur as a result of collisions between reacting molecules. A greater frequency of collisions usually leads to a higher reaction rate. According to the **collision theory** of chemical kinetics, the reaction rate is directly proportional to the number of molecular collisions per second.

$$\text{rate} \propto \frac{\text{number of collisions}}{\text{s}}$$

Consider the reaction of A molecules with B molecules to form some product. Suppose that each product molecule is formed by the direct combination of an A molecule and a B molecule. If we doubled the concentration of A, then the number of A-B collisions would also double, because there would be twice as many A molecules that could collide with B molecules in any given volume. Consequently, the rate would increase by a factor of 2. Similarly, doubling the concentration of B molecules would increase the rate twofold.

This view of collision theory is something of a simplification, though, because not every collision between molecules results in a reaction. A collision that *does* result in a reaction is called an **effective collision.** A molecule in motion possesses kinetic energy; the faster it is moving, the greater its kinetic energy. When molecules collide, part of their kinetic energy is converted to vibrational energy. If the initial kinetic energies are large, then the colliding molecules will vibrate so strongly as to break

Student Annotation: Kinetic energy is the result of motion of the whole molecule, relative to its surroundings. Vibrational energy is the result of motion of the atoms in a molecule, relative to one another.

some of the chemical bonds. This bond breaking is the first step toward product formation. If the initial kinetic energies are small, the molecules will merely bounce off of each other intact. There is a minimum amount of energy, the ***activation energy*** (E_a) [◄◄ Section 19.6], required to initiate a chemical reaction. Without this minimum amount of energy at impact, a collision will be ineffective; that is, it will not result in a reaction.

When molecules react (as opposed to when *atoms* react), having sufficient kinetic energy is not the only requirement for a collision to be effective. Molecules must also be oriented in a way that favors reaction. The reaction between chlorine atoms and nitrosyl chloride (NOCl) illustrates this point.

$$Cl + NOCl \longrightarrow Cl_2 + NO$$

This reaction is most favorable when a free Cl atom collides directly with the Cl atom in the NOCl molecule [Figure 19.1(a)]. Otherwise, the reactants simply bounce off of each other and no reaction occurs [Figure 19.1(b)].

When an effective collision occurs between reactant molecules, they form an ***activated complex*** (a state also known as the ***transition state***), a temporary species formed by the reactant molecules as a result of the collision. Figure 19.2 shows a potential energy profile for the reaction between Cl and NOCl.

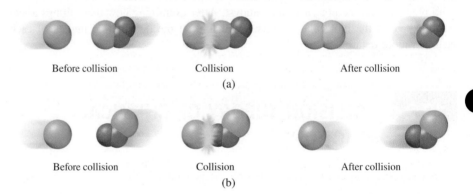

Before collision   Collision   After collision
(a)

Before collision   Collision   After collision
(b)

Figure 19.1 (a) For an effective collision to take place, the free Cl atom must collide directly with the Cl atom in NOCl. (b) Otherwise, the reactants bounce off of one another and the collision is ineffective—no reaction takes place.

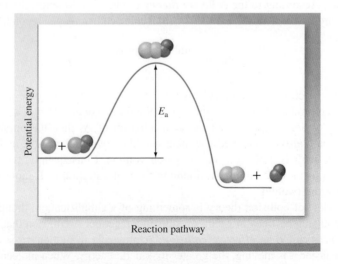

Figure 19.2 Energy profile for the reaction of Cl with NOCl. In addition to being oriented properly, reactant molecules must possess sufficient energy to overcome the activation energy.

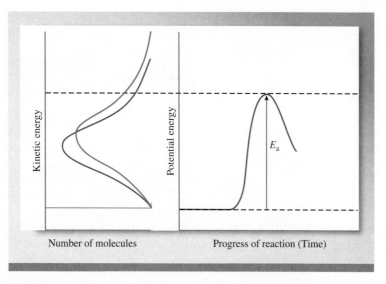

Figure 19.3 Kinetic molecular theory shows that the average speed and therefore average kinetic energy of a collection of molecules increases with increasing temperature. The blue line represents a collection of molecules at lower temperature; the red line represents the collection of molecules at higher temperature. At higher temperature, more molecules have sufficient kinetic energy to exceed the activation energy and undergo effective collision.

We can think of the activation energy as an energy *barrier* that prevents less energetic molecules from reacting. Because the number of reactant molecules in an ordinary reaction is very large, the speeds, and therefore also the kinetic energies of the molecules, vary greatly. Normally, only a small fraction of the colliding molecules are moving fast enough to have sufficient kinetic energy to exceed the activation energy. These molecules, the fastest-moving ones, can therefore take part in the reaction. This explains why reaction rates typically increase with increasing temperature. According to kinetic molecular theory, the average kinetic energy of a sample of molecules increases as the temperature increases [◄◄ Section 11.2, Figure 11.4]. Thus, at a higher temperature, more of the molecules in the sample have sufficient kinetic energy to exceed the activation energy (Figure 19.3), and the reaction rate increases. Figure 19.4 (pages 880–881) summarizes the factors that affect reaction rate.

19.3   MEASURING REACTION PROGRESS AND EXPRESSING REACTION RATE

We can follow the progress of a reaction by monitoring either the decrease in concentration of the reactants or the increase in the concentrations of the products. The method used to monitor changes in reactant or product concentrations depends on the specific reaction. In a reaction that either consumes or produces a colored species, we can measure the intensity of the color over time with a spectrometer. In a reaction that either consumes or produces a gas, we can measure the change in pressure over time with a manometer. Electrical conductance measurement can be used to monitor the progress if ionic species are consumed or produced.

Average Reaction Rate

Consider the hypothetical reaction represented by

$$A \longrightarrow B$$

Figure 19.4
Collision Theory

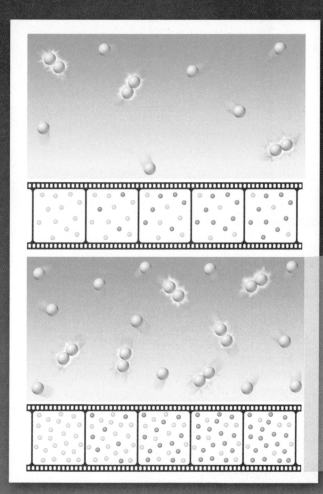

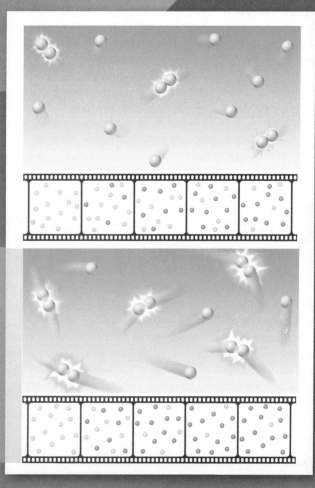

At higher concentration, reactant molecules collide more often, giving rise to a greater number of effective collisions—thereby increasing reaction rate.

At higher temperature, reactant molecules are moving faster, causing more frequent collisions and greater energy at impact. Both factors increase the number of effective collisions and rate increases.

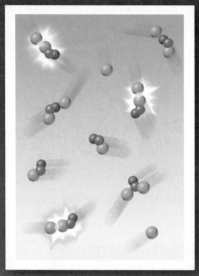

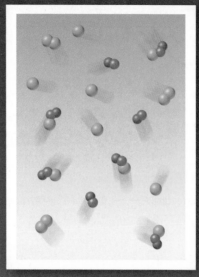

To be effective, a collision must occur between reactants with proper orientation. In the reaction of Cl and NOCl, collisions can be effective only when a Cl atom collides with the Cl atom in NOCl. (Effective collisions are identified by yellow.)

Product formation only occurs as the result of effective collision. Other collisions result in reactants simply bouncing off of one another. (Note that each effective collision resulted in the formation of one Cl_2 molecule and one NO molecule.)

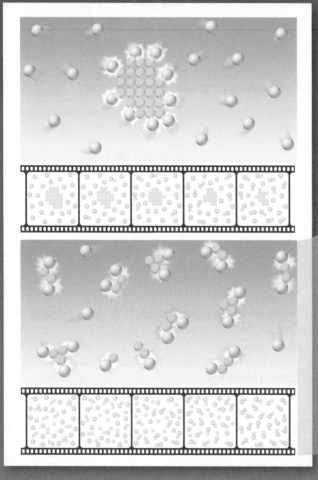

More finely divided solid reactant has more surface area where molecular collisions can occur. With greater surface area, reactant molecules can collide more frequently, giving rise to a greater number of effective collisions—thereby increasing reaction rate.

What's the point?

Several factors can influence the rate at which a chemical reaction ocurs including *reactant concentration, temperature, molecular orientation,* and *surface area.* In general:

- · Reaction rate increases as reactant concentration increases.
- · Reaction rate increases as temperature increases.
- · Reaction rate increases as the surface area of a solid reactant increases.
- · Only molecules oriented properly at collision will react.

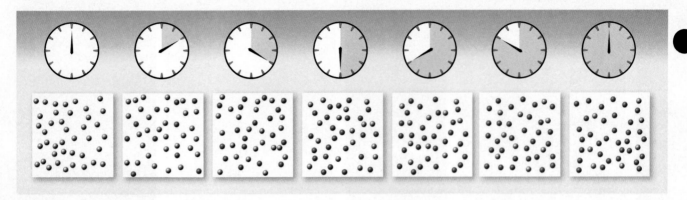

Figure 19.5 The progress of the reaction A ⟶ B. Initially, only A molecules (grey spheres) are present. As time progresses, there are more and more B molecules (red spheres).

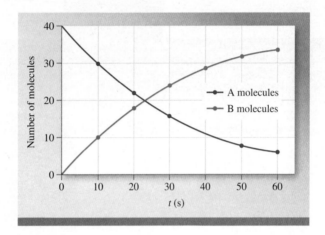

Figure 19.6 The rate of reaction A ⟶ B represented as the decrease of A molecules with time and as the increase of B molecules with time.

in which A molecules are converted to B molecules. Figure 19.5 shows the progress of this reaction as a function of time. The decrease in the number of A molecules and the increase in the number of B molecules with time are also shown graphically in Figure 19.6.

It is generally convenient to express the rate in terms of the change in concentration with time. Thus, for the reaction A ⟶ B, we can express the rate as

$$\text{rate} = -\frac{\Delta[\text{A}]}{\Delta t} \quad \text{or} \quad \text{rate} = \frac{\Delta[\text{B}]}{\Delta t}$$

where $\Delta[\text{A}]$ and $\Delta[\text{B}]$ are the changes in concentration (molarity) over a time period Δt. The rate expression containing $\Delta[\text{A}]$ has a minus sign because the concentration of A decreases during the time interval—that is, $\Delta[\text{A}]$ is a negative quantity. The rate expression containing $\Delta[\text{B}]$ does not have a minus sign because the concentration of B increases during the time interval. Rate is always a positive quantity, so when it is expressed in terms of the change in a reactant concentration, a minus sign is needed in the rate expression to make the rate positive. When the rate is expressed in terms of the change in a product concentration, no negative sign is needed to make the rate positive because the product concentration increases with time. Rates calculated in this way are average rates over the time period Δt.

To understand rates of chemical reactions and how they are determined, it is useful to consider some specific reactions. First, we consider the aqueous reaction of molecular bromine (Br_2) with formic acid (HCOOH).

$$Br_2(aq) + HCOOH(aq) \longrightarrow 2Br^-(aq) + 2H^+(aq) + CO_2(g)$$

Figure 19.7 From left to right: The decrease in bromine concentration as time elapses is indicated by the loss of color.

© McGraw-Hill Education/Ken Karp, photographer

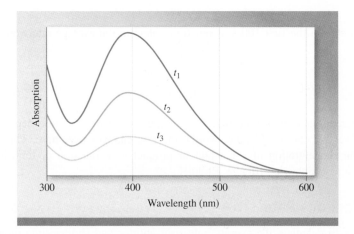

Figure 19.8 Plot of the absorption of bromine versus wavelength. The maximum absorption of visible light by bromine occurs at 393 nm. As the reaction progresses (t_1 to t_3), the absorption, which is proportional to [Br_2], decreases.

Molecular bromine is reddish brown, whereas all the other species in the reaction are colorless. As the reaction proceeds, the concentration of bromine decreases and its color fades (Figure 19.7). The decrease in intensity of the color (and, therefore in the concentration of bromine) can be monitored with a spectrometer, which registers the amount of visible light absorbed by bromine (Figure 19.8).

Measuring the bromine concentration at some initial time and then at some final time enables us to determine the average rate of the reaction during that time interval.

$$\text{average rate} = -\frac{\Delta[Br_2]}{\Delta t}$$

$$= -\frac{[Br_2]_{final} - [Br_2]_{initial}}{t_{final} - t_{initial}}$$

Using data from Table 19.1, we can calculate the average rate over the first 50-s time interval as follows:

$$\text{average rate} = -\frac{(0.0101 - 0.0120)\,M}{50.0} = 3.80 \times 10^{-5}\,M/s$$

	Rates of the Reaction of Molecular Bromine and Formic Acid at 25°C		
TABLE 19.1			
Time (s)	[Br$_2$] (M)	Rate (M/s)	
0.0	0.0120	4.20×10^{-5}	
50.0	0.0101	3.52×10^{-5}	
100.0	0.00846	2.96×10^{-5}	
150.0	0.00710	2.49×10^{-5}	
200.0	0.00596	2.09×10^{-5}	
250.0	0.00500	1.75×10^{-5}	
300.0	0.00420	1.48×10^{-5}	
350.0	0.00353	1.23×10^{-5}	
400.0	0.00296	1.04×10^{-5}	

If we had chosen the first 100 s as our time interval, the average rate would then be given by

$$\text{average rate} = -\frac{(0.00846 - 0.0120)\ M}{100.0} = 3.54 \times 10^{-5}\ M/s$$

These calculations demonstrate that the average rate of this reaction depends on the time interval we choose. In other words, the rate changes over time. This is why a plot of the concentration of a reactant or product as a function of time is a curve rather than a straight line [Figure 19.9(a)].

Instantaneous Rate

If we were to calculate the average rate over shorter and shorter time intervals, we could obtain the ***instantaneous rate,*** which is the rate for a specific instant in time. Figure 19.10 shows the plot of [Br$_2$] versus time based on the data from Table 19.1. The instantaneous rate is equal to the slope of a tangent to the curve at any particular time. Note that we can pick *any* two points along a tangent to calculate its slope. For a chemist, the instantaneous rate is generally a more useful quantity than the average rate. For the remainder of this chapter, therefore, the term *rate* will be used to mean "instantaneous rate" (unless otherwise stated).

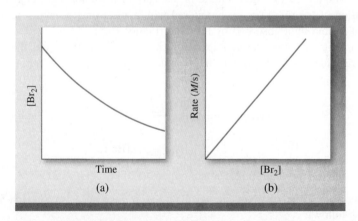

Figure 19.9 (a) The plot of [Br$_2$] against time is a curve because the reaction rate changes as [Br$_2$] changes with time. (b) The plot of the reaction rate against [Br$_2$] is a straight line because the rate is proportional to [Br$_2$].

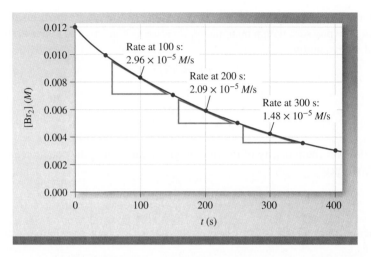

Figure 19.10 The instantaneous rates of the reaction between molecular bromine and formic acid at $t = 100$, 200, and 300 s are given by the slopes of the tangents at these times.

The slope of the tangent, and therefore the reaction *rate,* diminishes with time because the concentration of bromine decreases with time [Figure 19.9(a)]. The data in Table 19.1 show how the rate of this reaction depends on the concentration of bromine. At 50 s, for example, when the concentration of bromine is 0.0101 M, the rate is 3.52×10^{-5} M/s. When the concentration of bromine has been reduced by half (i.e., reduced to 0.00500 M), at 250 s, the rate is also reduced by half (i.e., reduced to 1.75×10^{-5} M/s).

$$\frac{[Br_2]_{50s}}{[Br_2]_{250s}} \approx 2 \quad \text{and} \quad \frac{\text{rate at 50.0 s}}{\text{rate at 250.0 s}} = \frac{3.52 \times 10^{-5} \ M/s}{1.75 \times 10^{-5} \ M/s} \approx 2$$

Thus, the rate is directly proportional to the concentration of bromine,

$$\text{rate} \propto [Br_2]$$

$$\text{rate} = k[Br_2]$$

where k, the proportionality constant, is called the **rate constant.**

Rearranging the preceding equation gives

$$k = \frac{\text{rate}}{[Br_2]}$$

We can use the concentration and rate data from Table 19.1 for any value of t to calculate the value of k for this reaction. For example, using the data for $t = 50.0$ s gives

$$k = \frac{3.52 \times 10^{-5} \ M/s}{0.0101 \ M} = 3.49 \times 10^{-3} \ s^{-1} \ (\text{at } t = 50.0 \text{ s})$$

Likewise, using the data for $t = 300.0$ s gives

$$k = \frac{1.48 \times 10^{-5} \ M/s}{0.00420 \ M} = 3.52 \times 10^{-3} \ s^{-1} \ (\text{at } t = 300.0 \text{ s})$$

Slight variations in the calculated values of k are due to experimental deviations in rate measurements. To two significant figures, we get $k = 3.5 \times 10^{-3} \ s^{-1}$ for this reaction, regardless of which line of data we chose from Table 19.1. It is important to note that the value of k does not depend on the concentration of bromine. The rate constant is *constant* at constant temperature. (In Section 19.6, we discuss how k depends on temperature.)

We now consider another specific reaction, the decomposition of hydrogen peroxide.

$$2H_2O_2(aq) \longrightarrow 2H_2O(l) + O_2(g)$$

Because one of the products is a gas, we can monitor the progress of this reaction by measuring the pressure with a manometer. Pressure is converted to concentration using the ideal gas equation:

$$PV = nRT$$

or

$$P_{O_2} = \frac{n}{V} RT = [O_2]RT$$

where n/V gives the molarity of the oxygen gas. Rearranging the equation and solving for $[O_2]$, we get

$$[O_2] = \frac{1}{RT} P_{O_2}$$

The reaction rate, which is expressed as the rate of oxygen production, can now be written as

$$rate = \frac{\Delta [O_2]}{\Delta t} = \frac{1}{RT} \frac{\Delta P_{O_2}}{\Delta t}$$

Determining the rate from pressure data is possible only if the temperature (in kelvins) at which the reaction is carried out is known. Figure 19.11(a) shows the apparatus used to monitor the pressure change in the decomposition of hydrogen peroxide. Figure 19.11(b) shows the plot of oxygen pressure, P_{O_2}, versus time. As we did in Figure 19.10, we can draw a tangent to the curve in Figure 19.11(b) to determine the instantaneous rate at any point.

Stoichiometry and Reaction Rate

For stoichiometrically simple reactions of the type A ⟶ B, the rate can be expressed either in terms of the decrease in reactant concentration with time, $-\Delta[A]/\Delta t$, or in terms of the increase in product concentration with time, $\Delta[B]/\Delta t$—both expressions give the same result. For reactions in which one or more of the stoichiometric coefficients is something other than 1, we must take extra care in expressing the rate. For example, consider again the reaction between molecular bromine and formic acid.

$$Br_2(aq) + HCOOH(aq) \longrightarrow 2Br^-(aq) + 2H^+(aq) + CO_2(g)$$

We have expressed the rate of this reaction in terms of the disappearance of bromine. But what if we chose instead to express the rate in terms of the appearance of bromide

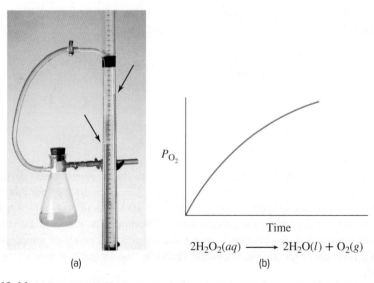

(a)   (b)

$$2H_2O_2(aq) \longrightarrow 2H_2O(l) + O_2(g)$$

Figure 19.11 (a) The rate of hydrogen peroxide decomposition can be measured with a manometer, which (b) shows the increase in the oxygen gas pressure with time.

(a) © McGraw-Hill Education/Ken Karp, photographer

ion? According to the balanced equation, 2 moles of Br^- are generated for each mole of Br_2 consumed. Thus, Br^- appears at twice the rate that Br_2 disappears. To avoid the potential ambiguity of reporting the rate of disappearance or appearance of a specific chemical *species,* we report the ***rate of reaction.*** We determine the rate of reaction such that the result is the same regardless of which species we monitor. For the hypothetical reaction,

$$A \longrightarrow 2B$$

the rate of reaction can be written as either

$$\text{rate} = -\frac{\Delta[A]}{\Delta t} \quad \text{or} \quad \text{rate} = \frac{1}{2}\frac{\Delta[B]}{\Delta t}$$

both of which give the same result. For the bromine and formic acid reaction, we can write the rate of reaction as either

$$\text{rate} = -\frac{\Delta[Br_2]}{\Delta t}$$

as we did earlier, or

$$\text{rate} = \frac{1}{2}\frac{\Delta[Br^-]}{\Delta t}$$

In general, for the reaction

$$aA + bB \longrightarrow cC + dD$$

the rate is given by

$$\text{rate} = -\frac{1}{a}\frac{\Delta[A]}{\Delta t} = -\frac{1}{b}\frac{\Delta[B]}{\Delta t} = \frac{1}{c}\frac{\Delta[C]}{\Delta t} = \frac{1}{d}\frac{\Delta[D]}{\Delta t} \qquad \textbf{Equation 19.1}$$

Expressing the rate in this fashion ensures that the rate of reaction is the same regardless of which species we measure to monitor the reaction's progress.

Worked Examples 19.1 and 19.2 show how to write expressions for reaction rates and how to take stoichiometry into account in rate expressions.

> **Student Annotation:** The rate of change in concentration of each species is divided by the coefficient of that species in the balanced equation.

Worked Example 19.1

Write the rate expressions for each of the following reactions.

(a) $I^-(aq) + OCl^-(aq) \longrightarrow Cl^-(aq) + OI^-(aq)$ (c) $4NH_3(g) + 5O_2(g) \longrightarrow 4NO(g) + 6H_2O(g)$

(b) $2O_3(g) \longrightarrow 3O_2(g)$

Strategy Use Equation 19.1 to write rate expressions for each of the reactions.

Setup For reactions containing gaseous species, progress is generally monitored by measuring pressure. Pressures are converted to molar concentrations using the ideal gas equation, and rate expressions are written in terms of molar concentrations.

Solution

(a) All the coefficients in this equation are 1. Therefore,

$$\text{rate} = -\frac{\Delta[I^-]}{\Delta t} = -\frac{\Delta[OCl^-]}{\Delta t} = \frac{\Delta[Cl^-]}{\Delta t} = \frac{\Delta[OI^-]}{\Delta t}$$

(b) $\text{rate} = -\frac{1}{2}\frac{\Delta[O_3]}{\Delta t} = \frac{1}{3}\frac{\Delta[O_2]}{\Delta t}$ (c) $\text{rate} = -\frac{1}{4}\frac{\Delta[NH_3]}{\Delta t} = -\frac{1}{5}\frac{\Delta[O_2]}{\Delta t} = \frac{1}{4}\frac{\Delta[NO]}{\Delta t} = \frac{1}{6}\frac{\Delta[H_2O]}{\Delta t}$

Think About It
Make sure that the change in concentration of each species is divided by the corresponding coefficient in the balanced equation. Also make sure that the rate expressions written in terms of reactant concentrations have a negative sign to make the resulting rate positive.

(Continued on next page)

Practice Problem Ⓐ**TTEMPT** Write the rate expressions for each of the following reactions.

(a) $CO_2(g) + 2H_2O(g) \longrightarrow CH_4(g) + 2O_2(g)$

(b) $3O_2(g) \longrightarrow 2O_3(g)$

(c) $2NO(g) + O_2(g) \longrightarrow 2NO_2(g)$

Practice Problem Ⓑ**UILD** Write the balanced equation corresponding to the following rate expressions.

(a) rate $= -\dfrac{1}{3}\dfrac{\Delta[CH_4]}{\Delta t} = -\dfrac{1}{2}\dfrac{\Delta[H_2O]}{\Delta t} = -\dfrac{\Delta[CO_2]}{\Delta t} = \dfrac{1}{4}\dfrac{\Delta[CH_3OH]}{\Delta t}$

(b) rate $= -\dfrac{1}{2}\dfrac{\Delta[N_2O_5]}{\Delta t} = \dfrac{1}{2}\dfrac{\Delta[N_2]}{\Delta t} = \dfrac{1}{5}\dfrac{\Delta[O_2]}{\Delta t}$

(c) rate $= -\dfrac{\Delta[H_2]}{\Delta t} = -\dfrac{\Delta[CO]}{\Delta t} = -\dfrac{\Delta[O_2]}{\Delta t} = \dfrac{\Delta[H_2CO_3]}{\Delta t}$

Practice Problem Ⓒ**ONCEPTUALIZE** The diagrams represent a system that initially consists of reactants A (red) and B (blue), which react to form product C (purple). Write the balanced chemical equation that corresponds to the reaction.

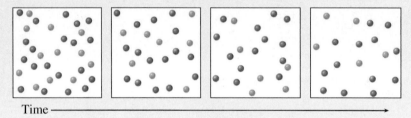

Time ────────────────────────▶

Worked Example 19.2

Consider the reaction

$$4NO_2(g) + O_2(g) \longrightarrow 2N_2O_5(g)$$

At a particular time during the reaction, nitrogen dioxide is being consumed at the rate of 0.00130 *M*/s. (a) At what rate is molecular oxygen being consumed? (b) At what rate is dinitrogen pentoxide being produced?

Strategy Determine the rate of reaction using Equation 19.1, and, using the stoichiometry of the reaction, convert to rates of change for the specified individual species.

Setup

$$\text{rate} = -\frac{1}{4}\frac{\Delta[NO_2]}{\Delta t} = -\frac{\Delta[O_2]}{\Delta t} = \frac{1}{2}\frac{\Delta[N_2O_5]}{\Delta t}$$

We are given

$$\frac{\Delta[NO_2]}{\Delta t} = -0.00130\ M/s$$

where the minus sign indicates that the concentration of NO_2 is decreasing with time. The rate of reaction, therefore, is

$$\text{rate} = -\frac{1}{4}\frac{\Delta[NO_2]}{\Delta t} = -\frac{1}{4}(-0.00130\ M/s)$$

$$= 3.25 \times 10^{-4}\ M/s$$

Solution

(a) $3.25 \times 10^{-4}\ M/s = -\dfrac{\Delta[O_2]}{\Delta t}$

$\dfrac{\Delta[O_2]}{\Delta t} = -3.25 \times 10^{-4}\ M/s$

Molecular oxygen is being consumed at a rate of $3.25 \times 10^{-4}\ M/s$.

(b) $3.25 \times 10^{-4} \ M/s = \dfrac{1}{2} \dfrac{\Delta[N_2O_5]}{\Delta t}$

$2(3.25 \times 10^{-4} \ M/s) = \dfrac{\Delta[N_2O_5]}{\Delta t}$

$\dfrac{\Delta[N_2O_5]}{\Delta t} = 6.50 \times 10^{-4} \ M/s$

Dinitrogen pentoxide is being produced at a rate of $6.50 \times 10^{-4} \ M/s$.

Think About It
Remember that the negative sign in a rate expression indicates that a species is being consumed rather than produced. Rates are always expressed as positive quantities.

Practice Problem A TTEMPT Consider the reaction

$$4PH_3(g) \longrightarrow P_4(g) + 6H_2(g)$$

At a particular point during the reaction, molecular hydrogen is being formed at the rate of 0.168 M/s. (a) At what rate is P_4 being produced? (b) At what rate is PH_3 being consumed?

Practice Problem B UILD Consider the following unbalanced equation.

$$A + B \longrightarrow C$$

When C is being formed at the rate of 0.086 M/s, A is being consumed at a rate of 0.172 M/s and B is being consumed at a rate of 0.258 M/s. Balance the equation based on the relative rates of formation and consumption of products and reactants.

Practice Problem C ONCEPTUALIZE Consider the reaction $2A + B \longrightarrow 2C$. Which graph could represent the concentrations of A, B, and C as the reaction progresses from $t = 0$ s?

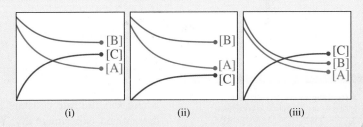

(i)   (ii)   (iii)

Section 19.3 Review

Measuring Reaction Progress and Expressing Reaction Rate

19.3.1 Write the rate expressions for the following reaction.

$$A + 2B \longrightarrow C + D$$

(a) rate $= -\dfrac{\Delta[A]}{\Delta t} = -\dfrac{1}{2}\dfrac{\Delta[B]}{\Delta t} = \dfrac{\Delta[C]}{\Delta t} = \dfrac{\Delta[D]}{\Delta t}$

(b) rate $= \dfrac{\Delta[A]}{\Delta t} = \dfrac{\Delta[B]}{\Delta t} = \dfrac{\Delta[C]}{\Delta t} = \dfrac{\Delta[D]}{\Delta t}$

(c) rate $= -\dfrac{\Delta[A]}{\Delta t} = -\dfrac{2\Delta[B]}{\Delta t} = \dfrac{\Delta[C]}{\Delta t} = \dfrac{\Delta[D]}{\Delta t}$

(d) rate $= -\dfrac{\Delta[A]}{\Delta t} = -\dfrac{1}{2}\dfrac{\Delta[B]}{\Delta t} = -\dfrac{\Delta[C]}{\Delta t} = -\dfrac{\Delta[D]}{\Delta t}$

(e) rate $= \dfrac{\Delta[A]}{\Delta t} = \dfrac{1}{2}\dfrac{\Delta[B]}{\Delta t} = -\dfrac{\Delta[C]}{\Delta t} = -\dfrac{\Delta[D]}{\Delta t}$

19.3.2 In the same reaction,

$$A + 2B \longrightarrow C + D$$

if the concentration of A is changing at a rate of 0.026 M/s, what is the rate of change in the concentration of B?

(a) 0.026 M/s (c) 0.013 M/s (e) 0.0004 M/s

(b) 0.052 M/s (d) 0.078 M/s

19.4 DEPENDENCE OF REACTION RATE ON REACTANT CONCENTRATION

We saw in Section 19.3 that the rate of reaction between bromine and formic acid is proportional to the concentration of bromine and that the proportionality constant k is the rate constant. We will now explore in more detail how the rate, the rate constant, and the reactant concentrations are related.

The Rate Law

The equation relating the rate of reaction to the concentration of molecular bromine,

$$\text{rate} = k[\text{Br}_2]$$

is an example of a ***rate law.*** The rate law is an equation that relates the rate of reaction to the concentrations of reactants. For the general reaction,

$$a\text{A} + b\text{B} \longrightarrow c\text{C} + d\text{D}$$

the rate law is

Equation 19.2	$\text{rate} = k[\text{A}]^x[\text{B}]^y$

Student Annotation: It is important to emphasize that the exponents in a rate law must be determined from experimental data and that, in general, they are *not* equal to the coefficients from the chemical equation.

where k is the rate constant and the exponents x and y are numbers that must be determined experimentally. When we know the values of k, x, and y, we can use Equation 19.2 to calculate the rate of the reaction, given the concentrations of A and B.

In the case of the reaction of molecular bromine and formic acid, the rate law is

$$\text{rate} = k[\text{Br}_2]^x[\text{HCOOH}]^y$$

where $x = 1$ and $y = 0$.

Student Hot Spot

Student data indicate you may struggle with rate laws. Access the SmartBook to view additional Learning Resources on this topic.

The values of the exponents in the rate law indicate the *order* of the reaction with respect to each reactant. In the reaction of bromine and formic acid, for example, the exponent for the bromine concentration, $x = 1$, means that the reaction is *first order* with respect to bromine. The exponent of 0 for the formic acid concentration indicates that the reaction is *zeroth order* with respect to formic acid. The sum of x and y is called the overall ***reaction order.*** Thus, the reaction of bromine and formic acid is first order in bromine, zeroth order in formic acid, and first order $(1 + 0 = 1)$ overall.

Experimental Determination of the Rate Law

To see how a rate law is determined from experimental data, consider the following reaction between fluorine and chlorine dioxide.

$$\text{F}_2(g) + 2\text{ClO}_2(g) \longrightarrow 2\text{FClO}_2(g)$$

Rate laws are commonly determined using a table of starting reactant concentrations and initial rates. The ***initial rate*** is the instantaneous rate at the beginning of the reaction. By varying the starting concentrations of reactants and observing the changes that result in the initial rate, we can determine how the rate depends on each reactant concentration.

TABLE 19.2	Initial Rate Data for the Reaction Between F_2 and ClO_2		
Experiment	[F_2] (*M*)	[ClO_2] (*M*)	Initial rate (*M*/s)
1	0.10	0.010	1.2×10^{-3}
2	0.10	0.040	4.8×10^{-3}
3	0.20	0.010	2.4×10^{-3}

Table 19.2 shows the initial rate data for the reaction of fluorine and chlorine dioxide being carried out three different times. Each time, the combination of reactant concentrations is different, and each time the initial rate is different.

To determine the values of the exponents *x* and *y* in the rate law,

$$\text{rate} = k[F_2]^x[ClO_2]^y$$

we must compare two experiments in which one reactant concentration *changes*, and the other remains constant. For example, we first compare the data from experiments 1 and 3.

Experiment	[F_2] (*M*)		[ClO_2] (*M*)		Initial rate (*M*/s)
1		0.10		0.010		1.2×10^{-3}
2	[F_2] doubles	0.10	[ClO_2] unchanged	0.040	Rate doubles	4.8×10^{-3}
3		0.20		0.010		2.4×10^{-3}

When the concentration of fluorine doubles, with the chlorine dioxide concentration held constant, the rate doubles.

$$\frac{[F_2]_3}{[F_2]_1} = \frac{0.20\ M}{0.10\ M} = 2 \qquad \frac{\text{rate}_3}{\text{rate}_1} = \frac{2.4 \times 10^{-3}\ M/s}{1.2 \times 10^{-3}\ M/s} = 2$$

This indicates that the rate is directly proportional to the concentration of fluorine and the value of *x* is 1.

$$\text{rate} = k[F_2][ClO_2]^y$$

Student Annotation: Recall that when an exponent is 1, it need not be shown [◀ Section 8.1].

Similarly, we can compare experiments 1 and 2.

Experiment	[F_2] (*M*)		[ClO_2] (*M*)		Initial rate (*M*/s)
1		0.10		0.010		1.2×10^{-3}
2	[F_2] unchanged	0.10	[ClO_2] quadruples	0.040	Rate quadruples	4.8×10^{-3}
3		0.20		0.010		2.4×10^{-3}

We find that the rate quadruples when the concentration of chlorine dioxide is quadrupled, but the fluorine concentration is held constant.

$$\frac{[ClO_2]_2}{[ClO_2]_1} = \frac{0.040\ M}{0.010\ M} = 4 \qquad \frac{\text{rate}_2}{\text{rate}_1} = \frac{4.8 \times 10^{-3}\ M/s}{1.2 \times 10^{-3}\ M/s} = 4$$

This indicates that the rate is also directly proportional to the concentration of chlorine dioxide, so the value of *y* is also 1. Thus, we can write the rate law as follows:

$$\text{rate} = k[F_2][ClO_2]$$

Because the concentrations of F_2 and ClO_2 are each raised to the first power, we say that the reaction is first order in F_2 and first order in ClO_2. The reaction is second order overall.

TABLE 19.3	Initial Rate Data for the Reaction Between A and B		
Experiment	**[A] (M)**	**[B] (M)**	**Initial rate (M/s)**
1	0.10	0.015	2.1×10^{-4}
2	0.20	0.015	4.2×10^{-4}
3	0.10	0.030	8.4×10^{-4}

Knowing the rate law, we can then use the data from any one of the experiments to calculate the rate constant. Using the data for the first experiment in Table 19.2, we can write

$$k = \frac{\text{rate}}{[F_2][ClO_2]} = \frac{1.2 \times 10^{-3}\ M/s}{(0.10\ M)(0.010\ M)} = 1.2\ M^{-1} \cdot s^{-1}$$

Table 19.3 contains initial rate data for the hypothetical reaction

$$a\text{A} + b\text{B} \longrightarrow c\text{C} + d\text{D}$$

which has the general rate law

$$\text{rate} = k[\text{A}]^x[\text{B}]^y$$

Comparing experiments 1 and 2, we see that when [A] doubles, with [B] unchanged, the rate also doubles.

Experiment	[A] (M)	[B] (M)	Initial rate (M/s)
1	[A] doubles { 0.10	[B] unchanged { 0.015	Rate doubles { 2.1×10^{-4}
2	0.20	0.015	4.2×10^{-4}
3	0.10	0.030	8.4×10^{-4}

Thus, $x = 1$.

Comparing experiments 1 and 3, when [B] doubles with [A] unchanged, the rate quadruples.

Experiment	[A] (M)	[B] (M)	Initial rate (M/s)
1	0.10	0.015	2.1×10^{-4}
2	[A] unchanged { 0.20	[B] doubles { 0.015	Rate quadruples { 4.2×10^{-4}
3	0.10	0.030	8.4×10^{-4}

Thus, the rate is *not* directly proportional to [B] to the first power, but rather it is directly proportional to [B] to the second power (i.e., $y = 2$).

$$\text{rate} \propto [\text{B}]^2$$

The overall rate law is

$$\text{rate} = k[\text{A}][\text{B}]^2$$

This reaction is therefore first order in A, second order in B, and third order overall.

Once again, knowing the rate law, we can use data from any of the experiments in the table to calculate the rate constant. Rearranging the rate law and using the data from experiment 1, we get

$$k = \frac{\text{rate}}{[\text{A}][\text{B}]^2} = \frac{2.1 \times 10^{-4}\ M/s}{(0.10\ M)(0.015\ M)^2} = 9.3\ M^{-2} \cdot s^{-1}$$

TABLE 19.4	Units of the Rate Constant k for Reactions of Various Overall Orders	
Overall reaction order	**Sample rate law**	**Units of k**
0	rate $= k$	$M \cdot s^{-1}$
1	rate $= k[A]$ or rate $= k[B]$	s^{-1}
2	rate $= k[A]^2$, rate $= k[B]^2$, or rate $= k[A][B]$	$M^{-1} \cdot s^{-1}$
3*	rate $= k[A]^2[B]$ or rate $= k[A][B]^2$	$M^{-2} \cdot s^{-1}$

*Another possibility for a third-order reaction is rate $= k[A][B][C]$, although such reactions are very rare.

Note that the units of this rate constant are different from those for the rate constant we calculated for the F_2-ClO_2 reaction and the bromine reaction. In fact, the units of a rate constant depend on the *overall* order of the reaction. Table 19.4 compares the units of the rate constant for reactions that are zeroth, first, second, and third order overall.

Following are three important things to remember about the rate law.

1. The exponents in a rate law must be determined from a table of experimental data—in general, they are not related to the stoichiometric coefficients in the balanced chemical equation.
2. Comparing changes in individual reactant concentrations with changes in rate shows how the rate depends on each reactant concentration.
3. Reaction order is always defined in terms of reactant concentrations, never product concentrations.

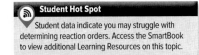

Student Hot Spot

Student data indicate you may struggle with determining reaction orders. Access the SmartBook to view additional Learning Resources on this topic.

Worked Example 19.3 shows how to use initial rate data to determine a rate law.

Worked Example 19.3

The gas-phase reaction of nitric oxide with hydrogen at 1280°C is

$$2NO(g) + 2H_2(g) \longrightarrow N_2(g) + 2H_2O(g)$$

From the following data collected at 1280°C, determine (a) the rate law, (b) the rate constant, including units, and (c) the rate of the reaction when $[NO] = 4.8 \times 10^{-3}$ M and $[H_2] = 6.2 \times 10^{-3}$ M.

Experiment	[NO] (M)	[H$_2$] (M)	Initial rate (M/s)
1	5.0×10^{-3}	2.0×10^{-3}	1.3×10^{-5}
2	1.0×10^{-2}	2.0×10^{-3}	5.0×10^{-5}
3	1.0×10^{-2}	4.0×10^{-3}	1.0×10^{-4}

Strategy Compare two experiments at a time to determine how the rate depends on the concentration of each reactant.

Setup The rate law is rate $= k[NO]^x[H_2]^y$. Comparing experiments 1 and 2, we see that the rate increases by approximately a factor of 4 when [NO] is doubled but [H$_2$] is held constant. Comparing experiments 2 and 3 shows that the rate doubles when [H$_2$] doubles but [NO] is held constant.

Solution (a) Dividing the rate from experiment 2 by the rate from experiment 1, we get

$$\frac{\text{rate}_2}{\text{rate}_1} = \frac{5.0 \times 10^{-5} \, M \cdot s^{-1}}{1.3 \times 10^{-5} \, M \cdot s^{-1}} \approx 4 = \frac{k(1.0 \times 10^{-2} \, M)^x(2.0 \times 10^{-3} \, M)^y}{k(5.0 \times 10^{-3} \, M)^x(2.0 \times 10^{-3} \, M)^y}$$

Canceling identical terms in the numerator and denominator gives

$$\frac{(1.0 \times 10^{-2} \, M)^x}{(5.0 \times 10^{-3} \, M)^x} = 2^x = 4$$

Student Annotation: A quotient of numbers, each raised to the same power, is equal to the quotient raised to that power: $x^n/y^n = (x/y)^n$.

Therefore, $x = 2$. The reaction is second order in NO.

(Continued on next page)

Dividing the rate from experiment 3 by the rate from experiment 2, we get

$$\frac{rate_3}{rate_2} = \frac{1.0 \times 10^{-4} \, M \cdot s^{-1}}{5.0 \times 10^{-5} \, M \cdot s^{-1}} = 2 = \frac{k(1.0 \times 10^{-2} \, M)^x(4.0 \times 10^{-3} \, M)^y}{k(1.0 \times 10^{-2} \, M)^x(2.0 \times 10^{-3} \, M)^y}$$

Canceling identical terms in the numerator and denominator gives

$$\frac{(4.0 \times 10^{-3} \, M)^y}{(2.0 \times 10^{-3} \, M)^y} = 2^y = 2$$

Therefore, $y = 1$. The reaction is first order in H_2. The overall rate law is

$$rate = k[NO]^2[H_2]$$

(b) We can use data from any of the experiments to calculate the value and units of k. Using the data from experiment 1 gives

$$k = \frac{rate}{[NO]^2[H_2]} = \frac{1.3 \times 10^{-5} \, M/s}{(5.0 \times 10^{-3} \, M)^2(2.0 \times 10^{-3} \, M)} = 2.6 \times 10^2 \, M^{-2} \cdot s^{-1}$$

(c) Using the rate constant determined in part (b) and the concentrations of NO and H_2 given in the problem statement, we can determine the reaction rate as follows:

$$rate = (2.6 \times 10^2 \, M^{-2} \cdot s^{-1})(4.8 \times 10^{-3} \, M)^2(6.2 \times 10^{-3} \, M)$$
$$= 3.7 \times 10^{-5} \, M/s$$

Think About It

The exponent for the concentration of H_2 in the rate law is 1, whereas the coefficient for H_2 in the balanced equation is 2. It is a common error to try to write a rate law using the stoichiometric coefficients as the exponents. Remember that, in general, the exponents in the rate law are not related to the coefficients in the balanced equation. Rate laws must be determined by examining a table of experimental data.

Practice Problem Ⓐ**TTEMPT** The reaction of peroxydisulfate ion ($S_2O_8^{2-}$) with iodide ion (I^-) is

$$S_2O_8^{2-}(aq) + 3I^-(aq) \longrightarrow 2SO_4^{2-}(aq) + I_3^-(aq)$$

From the following data collected at a certain temperature, determine the rate law and calculate the rate constant, including its units.

Experiment	$[S_2O_8^{2-}]$ (M)	$[I^-]$ (M)	Initial rate (M/s)
1	0.080	0.034	2.2×10^{-4}
2	0.080	0.017	1.1×10^{-4}
3	0.16	0.017	2.2×10^{-4}

Practice Problem Ⓑ**UILD** For the following general reaction, rate $= k[A]^2$ and $k = 1.3 \times 10^{-2} \, M^{-1} \cdot s^{-1}$.

$$A + B \longrightarrow 2C$$

Use this information to fill in the missing table entries.

Experiment	[A] (M)	[B] (M)	Initial rate (M/s)
1	0.013	0.250	2.20×10^{-6}
2	0.026	0.250	
3		0.500	2.20×10^{-6}

Practice Problem Ⓒ**ONCEPTUALIZE** Three initial-rate experiments are shown here depicting the reaction of X (red) and Y (yellow) to form Z (green). Using the diagrams, determine the rate law for the reaction X + Y $\longrightarrow$ Z.

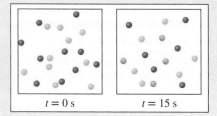

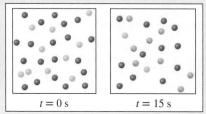

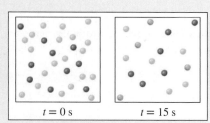

Section 19.4 Review

Dependence of Reaction Rate on Reactant Concentration

Answer questions 19.4.1 through 19.4.4 using the table of initial rate data for the reaction

$$A + 2B \longrightarrow 2C + D$$

Experiment	[A] (M)	[B] (M)	Initial rate (M/s)
1	0.12	0.010	2.2×10^{-2}
2	0.36	0.010	6.6×10^{-3}
3	0.12	0.020	2.2×10^{-3}

19.4.1 What is rate law for the reaction?
(a) rate = $k[A][B]^2$ (d) rate = $k[A]^2$
(b) rate = $k[A]^2[B]$ (e) rate = $k[A]$
(c) rate = $k[A]^3$

19.4.2 Calculate the rate constant.
(a) $0.15\ M^{-1} \cdot s^{-1}$ (d) $0.018\ s^{-1}$
(b) $0.15\ M \cdot s^{-1}$ (e) $0.018\ M^{-1} \cdot s^{-1}$
(c) $0.15\ s^{-1}$

19.4.3 What is the overall order of the reaction?
(a) 0   (b) 1   (c) 2   (d) 3   (e) 4

19.4.4 Determine the rate when [A] = 0.50 M and [B] = 0.25 M.
(a) $9.2 \times 10^{-3}\ M/s$ (d) $5.0 \times 10^{-3}\ M/s$
(b) $2.3 \times 10^{-3}\ M/s$ (e) $1.3 \times 10^{-2}\ M/s$
(c) $4.5 \times 10^{-3}\ M/s$

19.4.5 The diagrams represent three experiments in which the reaction A + B $\longrightarrow$ C is carried out with varied initial concentrations of A and B. Determine the rate law for the reaction. (A = green, B = yellow, C = red.)

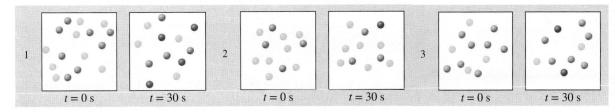

(a) rate = $k[A][B]$ (d) rate = $k[A]^2$
(b) rate = $k[A]^2[B]$ (e) rate = $k[B]$
(c) rate = $k[A][B]^2$

19.5   DEPENDENCE OF REACTANT CONCENTRATION ON TIME

We can use the rate law to determine the rate of a reaction using the rate constant and the reactant concentrations.

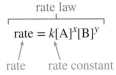

A rate law can also be used to determine the remaining concentration of a reactant at a specific time during a reaction. We will illustrate this use of rate laws using reactions that are first order overall and reactions that are second order overall.

First-Order Reactions

A *first-order reaction* is a reaction whose rate depends on the concentration of one of the reactants raised to the first power. Two examples are the decomposition of ethane (C_2H_6) into highly reactive fragments called methyl radicals ($\cdot CH_3$), and the decomposition of dinitrogen pentoxide (N_2O_5) into nitrogen dioxide (NO_2) and molecular oxygen (O_2).

$$C_2H_6 \longrightarrow 2 \cdot CH_3 \qquad\qquad \text{rate} = k[C_2H_6]$$

$$2N_2O_5(g) \longrightarrow 4NO_2(g) + O_2(g) \qquad\qquad \text{rate} = k[N_2O_5]$$

In a first-order reaction of the type

$$A \longrightarrow \text{products}$$

the rate can be expressed as the rate of change in reactant concentration,

$$\text{rate} = -\frac{\Delta[A]}{\Delta t}$$

as well as in the form of the rate law.

$$\text{rate} = k[A]$$

Setting these two expressions of the rate equal to each other we get

$$-\frac{\Delta[A]}{\Delta t} = k[A]$$

Applying calculus to the preceding equation, we can show that

Equation 19.3 $\qquad\qquad \ln\dfrac{[A]_t}{[A]_0} = -kt$

where ln is the natural logarithm, and $[A]_0$ and $[A]_t$ are the concentrations of A at times 0 and t, respectively. In general, time 0 refers to any specified time during a reaction—not necessarily the beginning of the reaction. Time t refers to any specified time *after* time 0. Equation 19.3 is sometimes called the *integrated rate law.*

In Worked Example 19.4 we apply Equation 19.3 to a specific reaction.

Student Annotation: In differential form, the preceding equation becomes

$$-\frac{d[A]}{dt} = k[A]$$

Rearranging, we get

$$\frac{d[A]}{[A]} = -kdt$$

Integrating between $t = 0$ and $t = t$ gives

$$\int_{[A]_0}^{[A]_t}\frac{d[A]}{[A]} = -k\int_0^t dt$$

$$\ln[A]_t - \ln[A]_0 = -kt$$

or

$$\ln\frac{[A]_t}{[A]_0} = -kt$$

Student Annotation: It is not necessary for you to be able to do the calculus required to arrive at Equation 19.3, but it is very important that you know how to *use* Equation 19.3.

Worked Example 19.4

The decomposition of hydrogen peroxide is first order in H_2O_2.

$$2H_2O_2(aq) \longrightarrow 2H_2O(l) + O_2(g)$$

The rate constant for this reaction at 20°C is 1.8×10^{-5} s^{-1}. If the starting concentration of H_2O_2 is 0.75 *M*, determine (a) the concentration of H_2O_2 remaining after 3 h and (b) how long it will take for the H_2O_2 concentration to drop to 0.10 *M*.

Strategy Use Equation 19.3 to find $[H_2O_2]_t$ where $t = 3$ h, and then solve Equation 19.3 for t to determine how much time must pass for $[H_2O_2]_t$ to equal 0.10 *M*.

Setup $[H_2O_2]_0 = 0.75$ *M*; time t for part (a) is (3 h)(60 min/h)(60 s/min) = 10,800 s.

Solution

(a) $\ln\dfrac{[H_2O_2]_t}{[H_2O_2]_0} = -kt$

$\ln\dfrac{[H_2O_2]_t}{0.75\ M} = -(1.8 \times 10^{-5}\ s^{-1})(10{,}800\ s) = -0.1944$

Take the inverse natural logarithm of both sides of the equation to get

Student Annotation: The inverse of ln x is e^x [▶◀ Appendix 1].

$$\dfrac{[H_2O_2]_t}{0.75\ M} = e^{-0.1944} = 0.823$$

$$[H_2O_2]_t = (0.823)(0.75\ M) = 0.62\ M$$

The concentration of H_2O_2 after 3 h is 0.62 M.

(b) $\ln\left(\dfrac{0.10\ M}{0.75\ M}\right) = -2.015 = -(1.8 \times 10^{-5}\ s^{-1})t$

$\dfrac{2.015}{1.8 \times 10^{-5}\ s^{-1}} = t = 1.12 \times 10^5\ s$

The time required for the peroxide concentration to drop to 0.10 M is 1.1×10^5 s or about 31 h.

Think About It

Don't forget the minus sign in Equation 19.3. If you calculate a concentration at time t that is greater than the concentration at time 0 (or if you get a negative time required for the concentration to drop to a specified level), check your solution for this common error.

Practice Problem ATTEMPT The rate constant for the reaction 2A ⟶ B is 7.5×10^{-3} s^{-1} at 110°C. The reaction is first order in A. How long (in seconds) will it take for [A] to decrease from 1.25 M to 0.71 M?

Practice Problem BUILD Refer again to the reaction 2A ⟶ B, for which $k = 7.5 \times 10^{-3}$ s^{-1} at 110°C. With a starting concentration of [A] = 2.25 M, what will [A] be after 2.0 min?

Practice Problem CONCEPTUALIZE The diagrams illustrate the first-order reaction of B (blue) to form C (yellow). Use the information given in the first set of diagrams to determine how much time is required for the change depicted in the second set of diagrams to take place.

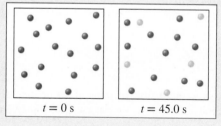
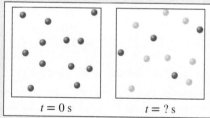

Equation 19.3 can be rearranged as follows:

$$\ln[A]_t = -kt + \ln[A]_0 \qquad\qquad \textbf{Equation 19.4}$$

Equation 19.4 has the form of the linear equation $y = mx + b$:

$$\ln[A]_t = (-k)(t) + \ln[A]_0$$
$$\ \ y \ \ = \ \ mx \ \ + \ \ b$$

Figure 19.12(a) shows the decrease in concentration of reactant A during the course of the reaction. As we saw in Section 19.2, the plot of reactant concentration as a function of time is not a straight line. For a first-order reaction, however, we do get a straight line if we plot the natural log of reactant concentration ($\ln[A]_t$) versus time (y versus x). The slope of the line is equal to $-k$ [Figure 19.12(b)], so we can determine the rate constant from the slope of this plot.

Student Annotation: Because pressure is proportional to concentration, for gaseous reactions [◀◀ Section 11.8] Equations 19.3 and 19.4 can be written as

$$\ln\dfrac{P_t}{P_0} = -kt$$

and

$$\ln P_t = -kt + \ln P_0$$

respectively, where P_0 and P_t are the pressures of reactant A at times 0 and t, respectively.

Student Annotation: This graphical determination is an alternative to using the method of initial rates to determine the value of k.

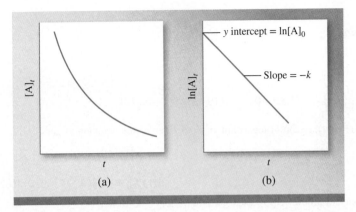

Figure 19.12 First-order reaction characteristics: (a) Decrease of reactant concentration with time. (b) A plot of $\ln[A]_t$ versus t. The slope of the line is equal to $-k$ and the y intercept is equal to $\ln[A]_0$.

Worked Example 19.5 shows how a rate constant can be determined from experimental data.

Worked Example 19.5

The rate of decomposition of azomethane is studied by monitoring the partial pressure of the reactant as a function of time.

$$CH_3-N=N-CH_3(g) \longrightarrow N_2(g) + C_2H_6(g)$$

The data obtained at 300°C are listed in the following table.

Time (s)	$P_{azomethane}$ (mmHg)
0	284
100	220
150	193
200	170
250	150
300	132

Determine the rate constant of the reaction at this temperature.

Strategy We can use Equation 19.3 only for first-order reactions, so we must first determine if the decomposition of azomethane is first order. We do this by plotting $\ln P$ against time. If the reaction is first order, we can use Equation 19.3 and the data at any two of the times in the table to determine the rate constant.

Setup The table expressed as $\ln P$ is

Time (s)	$\ln P$
0	5.649
100	5.394
150	5.263
200	5.136
250	5.011
300	4.883

Plotting these data gives a straight line, indicating that the reaction is indeed first order. Thus, we can use Equation 19.3 expressed in terms of pressure.

$$\ln \frac{P_t}{P_0} = -kt$$

P_t and P_0 can be pressures at any two times during the experiment. P_0 need not be the pressure at 0 s—it need only be at the earlier of the two times.

Solution Using data from times 100 s and 250 s of the original table ($P_{\text{azomethane}}$ versus t), we get

$$\ln\frac{150\ \text{mmHg}}{220\ \text{mmHg}} = -k(150\ \text{s})$$

$$\ln 0.682 = -k(150\ \text{s})$$

$$k = 2.55 \times 10^{-3}\ \text{s}^{-1}$$

Think About It

We could equally well have determined the rate constant by calculating the slope of the plot of ln P versus t. Using the two points labeled on the plot, we get

$$\text{slope} = \frac{5.011 - 5.394}{250 - 100}$$

$$= -2.55 \times 10^{-3}\ \text{s}^{-1}$$

Remember that slope = $-k$, so $k = 2.55 \times 10^{-3}\ \text{s}^{-1}$.

Practice Problem Ⓐ**TTEMPT** Ethyl iodide (C_2H_5I) decomposes at a certain temperature in the gas phase as follows:

$$C_2H_5I(g) \longrightarrow C_2H_4(g) + HI(g)$$

From the following data, determine the rate constant of this reaction. Begin by constructing a plot to verify that the reaction is first order.

Time (min)	[C$_2$H$_5$I] (M)
0	0.36
15	0.30
30	0.25
48	0.19
75	0.13

Practice Problem Ⓑ**UILD** Use the calculated k from Practice Problem A to fill in the missing values in the following table.

Time (min)	[C$_2$H$_5$I] (M)
0	0.45
10	—
20	—
30	—
40	—

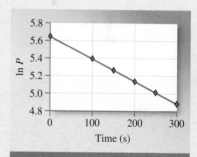

Practice Problem Ⓒ**ONCEPTUALIZE** Use the graph in the Setup section of Worked Example 19.5 to estimate the pressure of azomethane (in mmHg) at $t = 50$ s.

We often describe the rate of a reaction using the half-life. The **half-life ($t_{1/2}$)** is the time required for the reactant concentration to drop to *half* its original value. We obtain an expression for $t_{1/2}$ for a first-order reaction as follows:

$$t = \frac{1}{k}\ln\frac{[A]_0}{[A]_t}$$

According to the definition of half-life, $t = t_{1/2}$ when $[A]_t = \frac{1}{2}[A]_0$, so

$$t_{1/2} = \frac{1}{k}\ln\frac{[A]_0}{\frac{1}{2}[A]_0}$$

Because $\dfrac{[A]_0}{\frac{1}{2}[A]_0} = 2$, and ln 2 = 0.693, the expression for $t_{1/2}$ simplifies to

$$t_{1/2} = \frac{0.693}{k} \qquad \textbf{Equation 19.5}$$

Figure 19.13 A plot of [A] versus time for the first-order reaction A ⟶ products. The half-life of the reaction is 1 min. The concentration of A is halved every half-life.

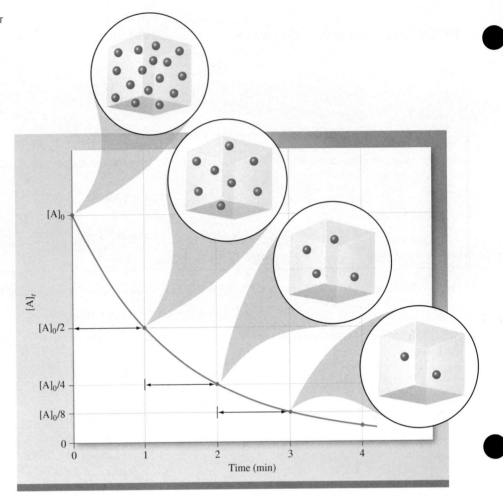

According to Equation 19.5, the half-life of a first-order reaction is independent of the initial concentration of the reactant. Thus, it takes the same time for the concentration of the reactant to decrease from 1.0 M to 0.50 M as it does for the concentration to decrease from 0.10 M to 0.050 M (Figure 19.13). Measuring the half-life of a reaction is one way to determine the rate constant of a first-order reaction.

The half-life of a first-order reaction is inversely proportional to its rate constant, so a short half-life corresponds to a large rate constant. Consider, for example, two radioactive isotopes used in nuclear medicine: ^{24}Na ($t_{1/2}$ = 19.7 h) and ^{60}Co ($t_{1/2}$ = 5.3 yr). Sodium-24, with the shorter half-life, decays faster. If we started with an equal number of moles of each isotope, most of the sodium-24 would be gone in a week whereas most of the cobalt-60 would remain unchanged.

Worked Example 19.6 shows how to calculate the half-life of a first-order reaction, given the rate constant.

Worked Example 19.6

The decomposition of ethane (C_2H_6) to methyl radicals (CH_3) is a first-order reaction with a rate constant of 5.36×10^{-4} s^{-1} at 700°C.

$$C_2H_6 \longrightarrow 2CH_3$$

Calculate the half-life of the reaction in minutes.

Strategy Use Equation 19.5 to calculate $t_{1/2}$ in seconds, and then convert to minutes.

Setup   seconds $\times \dfrac{1 \text{ minute}}{60 \text{ seconds}}$ = minutes

Solution

$$t_{1/2} = \frac{0.693}{k} = \frac{0.693}{5.36 \times 10^{-4} \text{ s}^{-1}} = 1293 \text{ s}$$

$$1293 \text{ s} \times \frac{1 \text{ min}}{60 \text{ s}} = 21.5 \text{ min}$$

The half-life of ethane decomposition at 700°C is 21.5 min.

> **Think About It**
> Half-lives and rate constants can be expressed using any units of time and reciprocal time, respectively. Track units carefully when you convert from one unit of time to another.

Practice Problem ATTEMPT Calculate the half-life of the decomposition of azomethane, discussed in Worked Example 19.5.

Practice Problem BUILD Calculate the rate constant for the first-order decay of ^{24}Na ($t_{1/2}$ = 19.7 h).

Practice Problem CONCEPTUALIZE The diagrams show a system in which A (red) is reacting to form B (blue) over the course of time. Use the information in the diagrams to determine the half-life of the reaction.

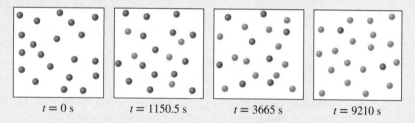

$t = 0$ s $\qquad$ $t = 1150.5$ s $\qquad$ $t = 3665$ s $\qquad$ $t = 9210$ s

Second-Order Reactions

A *second-order reaction* is a reaction whose rate depends on the concentration of one reactant raised to the second power or on the product of the concentrations of two different reactants (each raised to the first power). For simplicity, we will consider only the first type of reaction:

$$A \longrightarrow \text{product}$$

where the rate can be expressed as

$$\text{rate} = -\frac{\Delta[A]}{\Delta t}$$

or as

$$\text{rate} = k[A]^2$$

As before, we can combine the two expressions of the rate.

$$-\frac{\Delta[A]}{\Delta t} = k[A]^2$$

Again, using calculus, we obtain the following integrated rate law.

$$\frac{1}{[A]_t} = kt + \frac{1}{[A]_0} \qquad \textbf{Equation 19.6}$$

> **Student Annotation:** Equation 19.6 is the result of
>
> $$\int_{[A]_0}^{[A]_t} \frac{d[A]}{[A]^2} = -k \int_0^t dt$$
>
> **Again:** It is not necessary that you be able to derive Equation 19.6, only that you be able to use it to solve second-order kinetics problems.

Thus, for a second-order reaction, we obtain a straight line when we plot the reciprocal of concentration ($1/[A]_t$) against time (Figure 19.14), and the slope of the line is equal to the rate constant, k. As before, we can obtain the expression for the half-life by setting $[A]_t = \frac{1}{2}[A]_0$ in Equation 19.6.

$$\frac{1}{\frac{1}{2}[A]_0} = kt_{1/2} + \frac{1}{[A]_0}$$

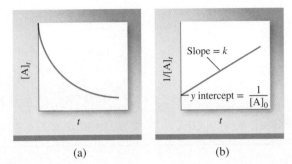

Figure 19.14 Second-order reaction characteristics: (a) Decrease of reactant concentration with time. (b) A plot of $\dfrac{1}{[A]_t}$. The slope of the line is equal to k, and the y intercept is equal to $\dfrac{1}{[A]_0}$.

Solving for $t_{1/2}$, we obtain

Equation 19.7	$t_{1/2} = \dfrac{1}{k[A]_0}$

Note that unlike the half-life of a first-order reaction, which is independent of the starting concentration, the half-life of a second-order reaction is inversely proportional to the initial reactant concentration. Determining the half-life at several different initial concentrations is one way to distinguish between first-order and second-order reactions.

Worked Example 19.7 shows how to use Equations 19.6 and 19.7 to calculate reactant concentrations and the half-life of a second-order reaction.

Worked Example 19.7

Iodine atoms combine to form molecular iodine in the gas phase.

$$I(g) + I(g) \longrightarrow I_2(g)$$

This reaction is second order and has a rate constant of $7.0 \times 10^9 \ M^{-1} \cdot s^{-1}$ at 23°C.
(a) If the initial concentration of I is 0.086 M, calculate the concentration after 2.0 min.
(b) Calculate the half-life of the reaction when the initial concentration of I is 0.60 M and when the initial concentration of I is 0.42 M.

Student Annotation: This equation can also be written as $2I(g) \longrightarrow I_2(g)$.

Strategy Use Equation 19.6 to determine $[I]_t$ at $t = 2.0$ min; use Equation 19.7 to determine $t_{1/2}$ when $[I]_0 = 0.60 \ M$ and when $[I]_0 = 0.42 \ M$.

Setup

$$t = (2.0 \text{ min})(60 \text{ s/min}) = 120 \text{ s}$$

Solution

(a) $\dfrac{1}{[A]_t} = kt + \dfrac{1}{[A]_0}$

$$= (7.0 \times 10^9 \ M^{-1} \cdot s^{-1})(120 \text{ s}) + \frac{1}{0.086 \ M}$$

$$= 8.4 \times 10^{11} \ M^{-1}$$

$$[A]_t = \frac{1}{8.4 \times 10^{11} \ M^{-1}} = 1.2 \times 10^{-12} \ M$$

The concentration of atomic iodine after 2 min is $1.2 \times 10^{-12} \ M$.

(b) When $[I]_0 = 0.60 \ M$,

$$t_{1/2} = \frac{1}{k[A]_0} = \frac{1}{(7.0 \times 10^9 \ M^{-1} \cdot s^{-1})(0.60 \ M)} = 2.4 \times 10^{-10} \text{ s}$$

When $[I]_0 = 0.42\ M$,

$$t_{1/2} = \frac{1}{k[A]_0} = \frac{1}{(7.0 \times 10^9\ M^{-1} \cdot s^{-1})(0.42\ M)} = 3.4 \times 10^{-10}\ s$$

Think About It

(a) Iodine, like the other halogens, exists as diatomic molecules at room temperature. It makes sense, therefore, that atomic iodine would react quickly, and essentially completely, to form I_2 at room temperature. The very low remaining concentration of I after 2 min makes sense. (b) As expected, the half-life of this second-order reaction is not constant. (A constant half-life is a characteristic of first-order reactions.)

Practice Problem **A**TTEMPT The reaction $2A \longrightarrow B$ is second order in A with a rate constant of $32\ M^{-1} \cdot s^{-1}$ at $25°C$. (a) Starting with $[A]_0 = 0.0075\ M$, how long will it take for the concentration of A to drop to $0.0018\ M$? (b) Calculate the half-life of the reaction for $[A]_0 = 0.0075\ M$ and for $[A]_0 = 0.0025\ M$.

Practice Problem **B**UILD Determine the initial concentration, $[A]_0$, for the reaction in Practice Problem A necessary for the half-life to be (a) 1.50 s, (b) 25.0 s, and (c) 175 s.

Practice Problem **C**ONCEPTUALIZE The diagrams show three different experiments with the reaction of A (red) and B (blue) to form C (purple). The reaction $A + B \longrightarrow C$ is second order in A. In the last experiment, the red spheres at $t = 0$ s are not shown. Determine how many red spheres must be included for the diagram to be correct.

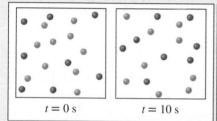
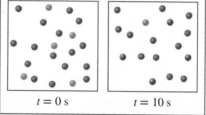
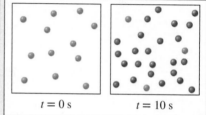

First- and second-order reactions are the most common reaction types. Reactions of overall order zero exist but are relatively rare. For a zeroth-order reaction

$$A \longrightarrow \text{products}$$

the rate law is given by

$$\text{rate} = k[A]^0 = k$$

Thus, the rate of a ***zeroth-order reaction*** is a constant, independent of reactant concentration. Third-order and higher-order reactions are quite rare and too complex to be covered in this book. Table 19.5 summarizes the kinetics for zeroth-order, first-order, and second-order reactions of the type $A \longrightarrow$ products.

TABLE 19.5	Summary of the Kinetics of Zeroth-Order, First-Order, and Second-Order Reactions		
Order	**Rate law**	**Integrated rate law**	**Half-Life**
0	rate $= k$	$[A]_t = -kt + [A]_0$	$\dfrac{[A]_0}{2k}$
1	rate $= k[A]$	$\ln\dfrac{[A]_t}{[A]_0} = -kt$	$\dfrac{0.693}{k}$
2	rate $= k[A]^2$	$\dfrac{1}{[A]_t} = kt + \dfrac{1}{[A]_0}$	$\dfrac{1}{k[A]_0}$

Section 19.5 Review

Dependence of Reactant Concentration on Time

The first-order decomposition of dinitrogen pentoxide (N_2O_5) is represented by

$$2N_2O_5(g) \longrightarrow 4NO_2(g) + O_2(g)$$

Use the table of data to answer questions 19.5.1 and 19.5.2.

t (s)	$[N_2O_5]$ (M)
0	0.91
300	0.75
600	0.64
1200	0.44
3000	0.16

19.5.1 What is the rate constant for the decomposition of N_2O_5?
(a) $9 \times 10^{-4}\,s^{-1}$ (d) $5 \times 10^{-3}\,s^{-1}$
(b) $4 \times 10^{-4}\,s^{-1}$ (e) $1 \times 10^{-3}\,s^{-1}$
(c) $6 \times 10^{-4}\,s^{-1}$

19.5.2 Approximately how long will it take for $[N_2O_5]$ to fall from 0.62 M to 0.10 M?
(a) 1000 s   (d) 4000 s
(b) 2000 s   (e) 5000 s
(c) 3000 s

19.5.3 Consider the first-order reaction A $\longrightarrow$ B in which A molecules (blue spheres) are converted to B molecules (yellow spheres). What is the half-life for the reaction?

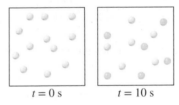

$t = 0$ s $t = 10$ s

(a) 5 s   (d) 20 s
(b) 10 s   (e) 30 s
(c) 15 s

19.5.4 A $\longrightarrow$ 2B is a second-order reaction for which $k = 5.3 \times 10^{-1}\,M^{-1} \cdot s^{-1}$. Calculate $t_{1/2}$ when $[A]_0 = 0.55\,M$.
(a) 1.3 s   (d) 1.8 s
(b) 0.52 s   (e) 1.9 s
(c) 3.4 s

19.6 DEPENDENCE OF REACTION RATE ON TEMPERATURE

Student Annotation: Cookbooks sometimes give alternate directions for cooking at high altitudes, where the lower atmospheric pressure results in water boiling at a lower temperature [◄◄ Section 12.5].

Nearly all reactions happen faster at higher temperatures. For example, the time required to cook food depends largely on the boiling point of water. The reaction involved in hard-boiling an egg happens faster at 100°C (about 10 min) than at 80°C (about 30 min). The dependence of reaction rate on temperature is the reason we keep food in a refrigerator—and why food keeps even longer in a freezer. The lower the temperature, the slower the processes that cause food to spoil.

The Arrhenius Equation

The dependence of the rate constant of a reaction on temperature can be expressed by the *Arrhenius equation,*

Equation 19.8 $$k = Ae^{-E_a/RT}$$

where E_a is the activation energy of the reaction (in kJ/mol), R is the gas constant (8.314 J/K · mol), T is the absolute temperature, and e is the base of the natural logarithm [▶▶ Appendix 1]. The quantity A represents the collision frequency and is called the frequency factor. It can be treated as a constant for a given reaction over a reasonably wide temperature range. Equation 19.8 shows that the rate constant decreases with increasing activation energy and increases with increasing temperature. This equation can be expressed in a more useful form by taking the natural logarithm of both sides:

$$\ln k = \ln Ae^{-E_a/RT}$$

or

$$\ln k = \ln A - \frac{E_a}{RT} \qquad \textbf{Equation 19.9}$$

which can be rearranged to give the following linear equation.

$$\ln k = \left(-\frac{E_a}{R}\right)\left(\frac{1}{T}\right) + \ln A$$
$$\quad y \;=\; \quad m \quad\; x \;\; + \;\; b \qquad \textbf{Equation 19.10}$$

Thus, a plot of $\ln k$ versus $1/T$ gives a straight line whose slope is equal to $-E_a/R$ and whose y intercept (b) is equal to $\ln A$.

Worked Example 19.8 demonstrates a graphical method for determining the activation energy of a reaction.

> **Student Annotation:** Absolute temperature is expressed in *kelvins* [◀◀ Section 1.2].

Worked Example 19.8

Rate constants for the reaction

$$CO(g) + NO_2(g) \longrightarrow CO_2(g) + NO(g)$$

were measured at four different temperatures. The data are shown in the table. Plot $\ln k$ versus $1/T$, and determine the activation energy (in kJ/mol) for the reaction.

k ($M^{-1} \cdot s^{-1}$)	T (K)
0.0521	288
0.101	298
0.184	308
0.332	318

Strategy Plot $\ln k$ versus $1/T$, and determine the slope of the resulting line. According to Equation 19.10, slope $= -E_a/R$.

Setup $R = 8.314$ J/mol · K. Taking the natural log of each value of k and the inverse of each value of T gives

$\ln k$	$1/T$ (K^{-1})
−2.95	3.47×10^{-3}
−2.29	3.36×10^{-3}
−1.69	3.25×10^{-3}
−1.10	3.14×10^{-3}

(Continued on next page)

Solution A plot of these data yields the following graph.

The slope is determined using the x and y coordinates of any two points on the line. Using the points that are labeled on the graph gives

$$\text{slope} = \frac{-1.4 - (-2.5)}{3.2 \times 10^{-3} \text{ K}^{-1} - 3.4 \times 10^{-3} \text{ K}^{-1}} = -5.5 \times 10^3 \text{ K}$$

The value of the slope is -5.5×10^3 K. Because slope $= -E_a/R$,

$$E_a = -(\text{slope})(R)$$

$$= -(-5.5 \times 10^3 \text{ K})(8.314 \text{ J/K} \cdot \text{mol})$$

$$= 4.6 \times 10^4 \text{ J/mol or 46 kJ/mol}$$

The activation energy is 46 kJ/mol.

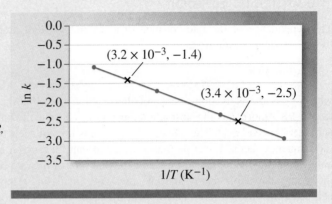

Think About It
Note that while k has units of $M^{-1} \cdot s^{-1}$, ln k has no units.

Practice Problem Ⓐ**TTEMPT** The second-order rate constant for the decomposition of nitrous oxide to nitrogen molecules and oxygen atoms has been determined at various temperatures.

k $(M^{-1} \cdot s^{-1})$	T (°C)
1.87×10^{-3}	600
0.0113	650
0.0569	700
0.244	750

Determine the activation energy graphically.

Practice Problem Ⓑ**UILD** Use the graph to determine the value of k at 475 K.

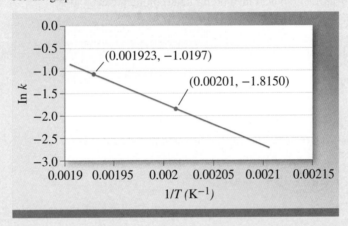

Practice Problem Ⓒ**ONCEPTUALIZE** Each line in the graph shown here represents a different reaction. List the reactions in order of increasing activation energy.

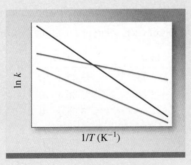

We can derive an even more useful form of the Arrhenius equation starting with Equation 19.9 written for two different temperatures, T_1 and T_2:

$$\ln k_1 = \ln A - \frac{E_a}{RT_1}$$

$$\ln k_2 = \ln A - \frac{E_a}{RT_2}$$

Student Annotation: Note that for two different temperatures, the only thing in Equation 19.9 that changes is k. The other variables, A and E_a (and of course R), are constant.

Subtracting $\ln k_2$ from $\ln k_1$ gives

$$\ln k_1 - \ln k_2 = \left(\ln A - \frac{E_a}{RT_1}\right) - \left(\ln A - \frac{E_a}{RT_2}\right)$$

$$\ln \frac{k_1}{k_2} = \frac{E_a}{R}\left(-\frac{1}{T_1} + \frac{1}{T_2}\right)$$

$$\ln \frac{k_1}{k_2} = \frac{E_a}{R}\left(\frac{1}{T_2} - \frac{1}{T_1}\right) \qquad \textbf{Equation 19.11}$$

Student Annotation: Equations 19.8 through 19.11 are all "Arrhenius equations," but Equation 19.11 is the form you will use most often to solve kinetics problems.

Equation 19.11 enables us to do two things:

1. If we know the rate constant at two different temperatures, we can calculate the activation energy.
2. If we know the activation energy and the rate constant at one temperature, we can determine the value of the rate constant at any other temperature.

Worked Examples 19.9 and 19.10 show how to use Equation 19.11.

Student Hot Spot

Student data indicate you may struggle with determining activation energy. Access the SmartBook to view additional Learning Resources on this topic.

Worked Example 19.9

The rate constant for a particular first-order reaction is given for three different temperatures.

T (K)	k (s^{-1})
400	2.9×10^{-3}
450	6.1×10^{-2}
500	7.0×10^{-1}

Using these data, calculate the activation energy of the reaction.

Strategy Use Equation 19.11 to solve for E_a.

Setup Solving Equation 19.11 for E_a gives

$$E_a = R\left(\frac{\ln \dfrac{k_1}{k_2}}{\dfrac{1}{T_2} - \dfrac{1}{T_1}}\right)$$

Solution Using the rate constants for 400 K (T_1) and 450 K (T_2), we get

$$E_a = \frac{8.314 \text{ J}}{\text{K} \cdot \text{mol}}\left(\frac{\ln \dfrac{2.9 \times 10^{-3} \text{ s}^{-1}}{6.1 \times 10^{-2} \text{ s}^{-1}}}{\dfrac{1}{450 \text{ K}} - \dfrac{1}{400 \text{ K}}}\right)$$

$$= 91{,}173 \text{ J/mol} = 91 \text{ kJ/mol}$$

The activation energy of the reaction is 91 kJ/mol.

(Continued on next page)

Think About It

A good way to check your work is to use the value of E_a that you calculated (and Equation 19.11) to determine the rate constant at 500 K. Make sure it agrees with the value in the table.

Practice Problem Ⓐ**TTEMPT** Use the data in the following table to determine the activation energy of the reaction.

T (K)	k (s^{-1})
625	1.1×10^{-4}
635	1.5×10^{-4}
645	2.0×10^{-4}

Practice Problem Ⓑ**UILD** Based on the data shown in Practice Problem A, what will be the value of k at 655 K?

Practice Problem Ⓒ**ONCEPTUALIZE** According to the Arrhenius equation, which graph [(i)–(iv)] best represents the relationship between temperature and rate constant?

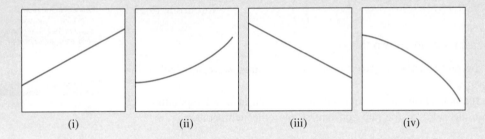

(i)   (ii)   (iii)   (iv)

Worked Example 19.10

A certain first-order reaction has an activation energy of 83 kJ/mol. If the rate constant for this reaction is 2.1×10^{-2} s^{-1} at 150°C, what is the rate constant at 300°C?

Strategy Use Equation 19.11 to solve for k_2. Pay particular attention to units in this type of problem.

Setup Solving Equation 19.11 for k_2 gives

$$k_2 = \frac{k}{e^{\left(\frac{E_a}{R}\right)\left(\frac{1}{T_2} - \frac{1}{T_1}\right)}}$$

$E_a = 8.3 \times 10^4$ J/mol, $T_1 = 423$ K, $T_2 = 573$ K, $R = 8.314$ J/K · mol, and $k_1 = 2.1 \times 10^{-2}$ s^{-1}.

Solution

$$k_2 = \frac{2.1 \times 10^{-2}\,\text{s}^{-1}}{e^{\left(\frac{8.3 \times 10^4\,\text{J/mol}}{8.314\,\text{J/K} \cdot \text{mol}}\right)\left(\frac{1}{573\,\text{K}} - \frac{1}{423\,\text{K}}\right)}}$$

$$= 1.0 \times 10^1\,\text{s}^{-1}$$

Student Annotation: Note that E_a is converted to joules so that the units will cancel properly. Alternatively, the R could be expressed as 0.008314 kJ/K · mol. Also note again that T must be expressed in kelvins.

The rate constant at 300°C is 10 s^{-1}.

Think About It

Make sure that the rate constant you calculate at a higher temperature is in fact higher than the original rate constant. According to the Arrhenius equation, the rate constant always increases with increasing temperature. If you get a smaller k at a higher temperature, check your solution for mathematical errors.

Practice Problem Ⓐ**TTEMPT** Calculate the rate constant at 200°C for a reaction that has a rate constant of 8.1×10^{-4} s^{-1} at 90°C and an activation energy of 99 kJ/mol.

Practice Problem Ⓑ**UILD** Calculate the rate constant at 200°C for a reaction that has a rate constant of 8.1×10^{-4} s^{-1} at 90°C and an activation energy of 59 kJ/mol.

Practice Problem Ⓒ**ONCEPTUALIZE** According to the Arrhenius equation, which graph [(i)–(iv)] best represents the relationship between temperature and activation energy?

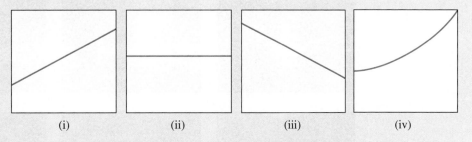

(i)   (ii)   (iii)   (iv)

Section 19.6 Review

Dependence of Reaction Rate on Temperature

Use the table of data collected for a first-order reaction to answer Questions 19.6.1 and 19.6.2.

T (K)	k (s^{-1})
300	3.9×10^{-2}
310	1.1×10^{-1}
320	2.8×10^{-1}

19.6.1 What is the activation energy of the reaction?
(a) 88 kJ/mol   (d) 8.8×10^{-3} kJ/mol
(b) 8.8×10^4 kJ/mol   (e) 80 kJ/mol
(c) 8.0×10^4 kJ/mol

19.6.2 What is the rate constant at 80°C?
(a) 5.8×10^{-40} s^{-1} (d) 5.2×10^{-3} s^{-1}
(b) 4.8 s^{-1} (e) 8.8×10^2 s^{-1}
(c) 2.8 s^{-1}

Thinking Outside the Box

Surface Area

As we saw in Section 19.2, one of the factors that can influence the rate of reaction is the surface area of a solid reactant. The reason is simple: aqueous or gaseous reactant molecules can only encounter and collide with solid reactant molecules at the surface of the solid. The more finely divided a solid reactant is, the more surface area is exposed, and the more collisions can take place. To illustrate this, we can compare the rate at which an Alka-Seltzer tablet reacts with water. In the first photo in each series, an Alka-Seltzer tablet is being added to a glass of lukewarm water. The top series of photos shows the reaction of a whole tablet; the bottom series shows the reaction of a tablet that has been ground to a powder. In the case of the powder, the reaction occurs much more rapidly.

(Continued on next page)

© David A. Tietz/Editorial Image, LLC

At the end of 30 seconds, the powder has reacted completely, while the *whole tablet* continues to react, as evidenced by its continued effervescence and the remaining tablet.

The effect of increased surface area on reaction rate can be dramatic and even dangerous. A tragic example of this occurred at the Imperial Sugar refinery (right) in Savanna, Georgia in 2008. Finely divided sugar dust in a storage silo was ignited by an unknown source, causing a series of violent explosions that destroyed the facility and killed more than a dozen workers.

© Stephen Morton/AP Images

19.7   REACTION MECHANISMS

A balanced chemical equation does not tell us much about *how* a reaction actually takes place. In many cases, the balanced equation is simply the sum of a series of steps. Consider the following hypothetical example. In the first step of a reaction, a molecule of reactant A combines with a molecule of reactant B to form a molecule of C.

$$A + B \longrightarrow C$$

In the second step, the molecule of C combines with another molecule of B to produce D.

$$C + B \longrightarrow D$$

The overall balanced equation is the sum of these two equations.

$$
\begin{array}{ll}
\textit{Step 1:} & A + B \longrightarrow \cancel{C} \\
\textit{Step 2:} & \underline{\cancel{C} + B \longrightarrow D} \\
& A + 2B \longrightarrow D
\end{array}
$$

The sequence of steps that sum to give the overall reaction is called the **reaction mechanism.** A reaction mechanism is comparable to the *route* traveled during a trip, whereas the overall balanced chemical equation specifies only the *origin* and the *destination.*

For a specific example of a reaction mechanism, we consider the reaction between nitric oxide and oxygen.

$$2NO(g) + O_2(g) \longrightarrow 2NO_2(g)$$

We know that NO_2 does not form as the result of a single collision between two NO molecules and one O_2 molecule because N_2O_2 is detected during the course of the reaction. We can envision the reaction taking place via the following two steps.

$$2NO(g) \longrightarrow N_2O_2(g)$$

$$N_2O_2(g) + O_2(g) \longrightarrow 2NO_2(g)$$

In the first step, two NO molecules collide to form an N_2O_2 molecule. This is followed by the step in which N_2O_2 and O_2 combine to give two molecules of NO_2. The net chemical equation, which represents the overall change, is given by the sum of the first and second steps.

$$
\begin{array}{ll}
\textit{Step 1:} & NO + NO \longrightarrow N_2O_2 \\
\textit{Step 2:} & \underline{N_2O_2 + O_2 \longrightarrow 2NO_2} \\
\textit{Overall reaction:} & 2NO(g) + O_2(g) \longrightarrow 2NO_2(g)
\end{array}
$$

Species such as N_2O_2 (and C in the hypothetical equation) are called **intermediates** because they appear in the mechanism of the reaction but not in the overall balanced equation. An intermediate is produced in an early step in the reaction and consumed in a later step.

Elementary Reactions

Each step in a reaction mechanism represents an **elementary reaction,** one that occurs in a single collision of the reactant molecules. The **molecularity** of an elementary reaction is essentially the *number* of reactant molecules involved in the collision. Elementary reactions may be **unimolecular** (one reactant molecule), **bimolecular** (two reactant molecules), or **termolecular** (three reactant molecules). These molecules may be of the same or different types. Each of the elementary steps in the formation of NO_2 from NO and O_2 is bimolecular, because there are *two* reactant molecules in each step. Likewise, each of the steps in the hypothetical A + 2B $\longrightarrow$ D reaction is bimolecular.

Knowing the steps of a reaction enables us to deduce the rate law. Suppose we have the following elementary reaction.

$$A \longrightarrow \text{products}$$

Because there is only one molecule present, this is a unimolecular reaction. It follows that the larger the number of A molecules present, the faster the rate of product

Student Annotation: Termolecular reactions actually are quite rare simply because the simultaneous encounter of three molecules is far less likely than that of two molecules.

formation. Thus, the rate of a unimolecular reaction is directly proportional to the concentration of A—the reaction is first order in A.

$$\text{rate} = k[A]$$

For a bimolecular elementary reaction involving A and B molecules,

$$A + B \longrightarrow \text{products}$$

the rate of product formation depends on how frequently A and B collide, which in turn depends on the concentrations of A and B. Thus, we can express the rate as

$$\text{rate} = k[A][B]$$

Therefore, this is a second-order reaction. Similarly, for a bimolecular elementary reaction of the type

$$A + A \longrightarrow \text{products}$$

or

$$2A \longrightarrow \text{products}$$

the rate becomes

$$\text{rate} = k[A]^2$$

Student Annotation: It is important to remember that this is not the case in general. It only applies to elementary reactions—and whether a reaction is elementary must be determined experimentally.

which is also a second-order reaction. The preceding examples show that the reaction order for each reactant in an elementary reaction is equal to its stoichiometric coefficient in the chemical equation for that step. In general, we cannot tell just by looking at the balanced equation whether the reaction occurs as shown or in a series of steps. This determination must be made using data obtained experimentally.

Rate-Determining Step

In a reaction mechanism consisting of more than one elementary step, the rate law for the overall process is given by the *rate-determining step,* which is the *slowest* step in the sequence. An analogy for a process in which there is a rate-determining step is the amount of time required to buy stamps at the post office when there is a long line of customers. The process consists of several steps: waiting in line, requesting the stamps, receiving the stamps, and paying for the stamps. At a time when the line is very long, the amount of time spent waiting in line (step 1) largely determines how much time the overall process takes.

To study reaction mechanisms, we first do a series of experiments to establish initial rates at various reactant concentrations. We then analyze the data to determine the rate constant and overall order of the reaction, and we write the rate law. Finally, we propose a plausible mechanism for the reaction in terms of logical elementary steps. The steps of the proposed mechanism must satisfy two requirements.

1. The sum of the elementary reactions must be the overall balanced equation for the reaction.
2. The rate-determining step must have the same rate law as that determined from the experimental data.

Let's consider the gas-phase reaction of hydrogen with iodine monochloride to form hydrogen chloride and iodine.

$$H_2 + 2ICl \longrightarrow 2HCl + I_2$$

The experimentally determined rate law for this reaction is rate $= k[H_2][ICl]$. Here are four different proposed mechanisms.

Mechanism 1

Step 1: $ICl + ICl \longrightarrow I_2 + Cl_2$ (slow)
Step 2: $Cl_2 + H_2 \longrightarrow 2HCl$

The steps in mechanism 1 sum to the correct overall reaction.

$$ICl + ICl \longrightarrow I_2 + \cancel{Cl_2} \text{ (slow)}$$
$$\cancel{Cl_2} + H_2 \longrightarrow 2HCl$$
$$\overline{\text{Sum: } H_2 + 2ICl \longrightarrow 2HCl + I_2}$$

However, the rate law of the rate-determining step is rate $= k[ICl]^2$. Therefore, mechanism 1 is not plausible. It meets the first requirement for a plausible mechanism, but it does not meet the second requirement.

Mechanism 2

Step 1: $\quad H_2 + ICl \longrightarrow HI + HCl$ (slow)
Step 2: $\quad ICl + HCl \longrightarrow HI + Cl_2$

The rate-determining step in mechanism 2 has the correct rate law.

$$\text{rate} = k[H_2][ICl]$$

However, the steps do not sum to the correct overall reaction.

$$H_2 + ICl \longrightarrow HI + \cancel{HCl} \text{ (slow)}$$
$$ICl + \cancel{HCl} \longrightarrow HI + Cl_2$$
$$\overline{\text{Sum: } H_2 + 2ICl \longrightarrow 2HI + Cl_2}$$

Therefore, mechanism 2 is not plausible. It meets the second requirement, but it does not meet the first requirement.

Mechanism 3

Step 1: $\quad H_2 \longrightarrow 2H$ (slow)
Step 2: $\quad ICl + H \longrightarrow HCl + I$
Step 3: $\quad H + I \longrightarrow HI$

The rate law for the rate-determining step in mechanism 3 (rate $= k[H_2]$) is not correct. Furthermore, the steps do not sum to the correct overall reaction.

$$H_2 \longrightarrow 2\cancel{H} \text{ (slow)}$$
$$ICl + \cancel{H} \longrightarrow HCl + \cancel{I}$$
$$\cancel{H} + \cancel{I} \longrightarrow HI$$
$$\overline{\text{Sum: } H_2 + ICl \longrightarrow HCl + HI}$$

Therefore, mechanism 3 is not plausible. It meets neither requirement 1 nor requirement 2.

Mechanism 4

Step 1: $\quad H_2 + ICl \longrightarrow HCl + HI$ (slow)
Step 2: $\quad HI + ICl \longrightarrow HCl + I_2$

The rate-determining step in mechanism 4 has the correct rate law.

$$\text{rate} = k[H_2][ICl]$$

Furthermore, the steps in mechanism 4 sum to the correct overall reaction.

$$H_2 + ICl \longrightarrow HCl + \cancel{HI} \text{ (slow)}$$
$$\cancel{HI} + ICl \longrightarrow HCl + I_2$$
$$\overline{\text{Sum: } H_2 + 2ICl \longrightarrow 2HCl + I_2}$$

Therefore, mechanism 4 is plausible. It meets both requirements.

Figure 19.15 The decomposition of hydrogen peroxide is catalyzed by the addition of an iodide salt. Some of the iodide ions are oxidized to molecular iodine, which then reacts with iodide ions to form the brown triiodide ion (I_3^-).

© McGraw-Hill Education/Charles D. Winters, photographer

The decomposition of hydrogen peroxide can be facilitated by iodide ions (Figure 19.15). The overall reaction is

$$2H_2O_2(aq) \longrightarrow 2H_2O(l) + O_2(g)$$

By experiment, we find the rate law to be

$$\text{rate} = k[H_2O_2][I^-]$$

Thus, the reaction is first order with respect to both H_2O_2 and I^-.

The decomposition of H_2O_2 is not an elementary reaction, because it does not occur in a single step. If it did, the reaction would be second order in H_2O_2 (as a result of the collision of two H_2O_2 molecules). What's more, the I^- ion, which is not even part of the overall equation, would not appear in the rate law expression. How can we reconcile these facts? First, we can account for the observed rate law by assuming that the reaction takes place in two separate elementary steps, each of which is bimolecular.

$$\textit{Step 1:}\quad H_2O_2 + I^- \xrightarrow{\;k_1\;} H_2O + IO^-$$
$$\textit{Step 2:}\quad H_2O_2 + IO^- \xrightarrow{\;k_2\;} H_2O + O_2 + I^-$$

If we further assume that step 1 is the rate-determining step, then the rate of the reaction can be determined from the first step alone:

$$\text{rate} = k_1[H_2O_2][I^-]$$

where $k_1 = k$. The IO^- ion is an intermediate because it is produced in the first step and consumed in the second step. It does not appear in the overall balanced equation. The I^- ion also does not appear in the overall equation, but it is consumed in the first step and then produced in the second step. In other words, it is present at the start of the reaction, and it is present at the end. The I^- ion is a *catalyst,* and its function is to *speed up* the reaction. Catalysts are discussed in greater detail in Section 19.8. Figure 19.16 shows the potential energy profile for a reaction like the decomposition of H_2O_2. The first step, which is the rate-determining step, has a larger activation energy than the second step. The intermediate, although stable enough to be observed, reacts quickly to form the products. Its existence is only fleeting.

Worked Example 19.11 lets you practice determining if a proposed reaction mechanism is plausible.

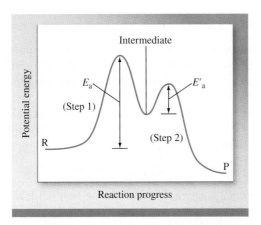

Figure 19.16 Potential energy profile for a two-step reaction in which the first step is rate determining. R and P represent reactants and products, respectively.

Worked Example 19.11

The gas-phase decomposition of nitrous oxide (N_2O) is believed to occur in two steps.

$$Step\ 1: \quad N_2O \xrightarrow{k_1} N_2 + O$$
$$Step\ 2: \quad N_2O + O \xrightarrow{k_2} N_2 + O_2$$

Experimentally the rate law is found to be rate = $k[N_2O]$. (a) Write the equation for the overall reaction. (b) Identify the intermediate(s). (c) Identify the rate-determining step.

Strategy Add the two equations, canceling identical terms on opposite sides of the arrow, to obtain the overall reaction. The canceled terms will be the intermediates if they were first generated and then consumed. Write rate laws for each elementary step; the one that matches the experimental rate law will be the rate-determining step.

Setup Intermediates are species that are generated in an earlier step and consumed in a later step. We can write rate laws for elementary reactions simply by using the stoichiometric coefficient for each species as its exponent in the rate law.

$$Step\ 1: \quad N_2O \xrightarrow{k_1} N_2 + O \qquad rate = k[N_2O]$$
$$Step\ 2: \quad N_2O + O \xrightarrow{k_2} N_2 + O_2 \qquad rate = k[N_2O][O]$$

Solution

(a) $2N_2O \longrightarrow 2N_2 + O_2$

(b) O (atomic oxygen) is the intermediate.

(c) Step 1 is the rate-determining step because its rate law is the same as the experimental rate law: rate = $k[N_2O]$.

> **Think About It**
> A species that gets canceled when steps are added may be an intermediate or a catalyst. In this case, the canceled species is an intermediate because it was first generated and then consumed. A species that is first consumed and then generated, but doesn't appear in the overall equation, is a catalyst.

Practice Problem Ⓐ**TTEMPT** The reaction between NO_2 and CO to produce NO and CO_2 is thought to occur in two steps.

$$Step\ 1: \quad NO_2 + NO_2 \xrightarrow{k_1} NO + NO_3$$
$$Step\ 2: \quad NO_3 + CO \xrightarrow{k_2} NO_2 + CO_2$$

The experimental rate law is rate = $k[NO_2]^2$. (a) Write the equation for the overall reaction. (b) Identify the intermediate(s). (c) Identify the rate-determining step.

Practice Problem Ⓑ**UILD** Propose a plausible mechanism for the reaction $A + 2B \longrightarrow C + D$ given that the rate law for the reaction is rate = $k[A][B]$.

Practice Problem Ⓒ**ONCEPTUALIZE** How many steps are there in the reaction represented by this potential-energy profile? Which step is the rate-determining step? How many intermediates, if any, are there in the reaction mechanism?

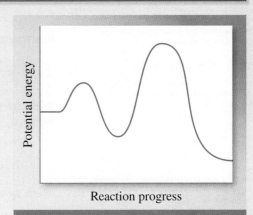

Mechanisms with a Fast First Step

For first- and second-order reactions, it is reasonably straightforward to propose a plausible mechanism. We simply use the experimentally determined rate law to write the rate-determining step as the first step, and then write one or more additional steps such that the appropriate species will cancel to give the correct overall equation. Sometimes, however, we will encounter a reaction with an experimentally determined rate law that suggests an unlikely scenario. For example, consider the gaseous reaction of nitric oxide with chlorine to produce nitrosyl chloride,

$$2NO(g) + Cl_2(g) \longrightarrow 2NOCl(g)$$

for which the experimentally determined rate law $= k[NO]^2[Cl_2]$. Note that the exponents in the rate law are the same as the coefficients in the balanced equation—making this reaction *third order* overall. Although one possibility is that the reaction is an elementary process, this would require the simultaneous collision of three molecules. As we have noted, such a termolecular reaction is unlikely. A more reasonable mechanism would be one in which a fast first step is followed by the slower, rate-determining step.

$$Step\ 1: \qquad NO(g) + Cl_2(g) \xrightarrow{k_1} NOCl_2(g)\ \text{(fast)}$$
$$Step\ 2: \quad NOCl_2(g) + NO(g) \xrightarrow{k_2} 2NOCl(g)\ \text{(slow)}$$

However, when we write the overall rate law using the equation for the rate-determining step, as we have done previously, the resulting rate law

$$\text{rate} = k_2[NOCl_2][NO]$$

includes the concentration of an *intermediate* ($NOCl_2$). Although this rate law is correct for the proposed mechanism, in this form it does not enable us to determine the plausibility of the mechanism by comparison to the experimentally determined rate law. To do this, we must derive a rate law in which only *reactants* from the overall equation appear.

When the intermediate ($NOCl_2$) is not consumed by step 2 as fast as it is produced by step 1, its concentration builds up, causing step 1 to happen in *reverse;* that is, $NOCl_2$ reacts to produce NO and Cl_2. This results in the establishment of a dynamic equilibrium [◄◄ Section 9.1], where the forward and reverse of step 1 are occurring at the same rate. We denote this by changing the single reaction arrow in step 1 to equilibrium arrows.

$$NO(g) + Cl_2(g) \underset{k_{-1}}{\overset{k_1}{\rightleftharpoons}} NOCl_2(g)$$

We can write the rate laws for the forward and reverse of step 1, both elementary processes, as

$$\text{rate}_{\text{forward}} = k_1[NO][Cl_2] \quad \text{and} \quad \text{rate}_{\text{reverse}} = k_{-1}[NOCl_2]$$

where k_1 and k_{-1} are the individual rate constants for the forward and reverse processes, respectively. Because the two rates are equal, we can write

$$k_1[NO][Cl_2] = k_{-1}[NOCl_2]$$

Rearranging to solve for the concentration of $NOCl_2$ gives

$$\frac{k_1}{k_{-1}}[NO][Cl_2] = [NOCl_2]$$

When we substitute the result into the original rate law for the concentration of $NOCl_2$, we get

$$\text{rate} = k_2[NOCl_2][NO] = k_2\frac{k_1}{k_{-1}}[NO]^2[Cl_2]$$

which agrees with the experimentally determined rate law, rate $= k[NO]^2[Cl_2]$, where k is equal to $\dfrac{k_2k_1}{k_{-1}}$.

Student Annotation: Interestingly, we can use this relationship to determine the equilibrium expression. At equilibrium, the forward and reverse reactions have the same rate. Rearranging this expression gives

$$\frac{k_1}{k_{-1}} = \frac{[NOCl_2]}{[NO][Cl_2]}$$

Because the ratio of the two rate constants is also a constant, we have

$$K_c = \frac{[NOCl_2]}{[NO][Cl_2]}$$

where K_c is the equilibrium constant. Compare this to the equilibrium expression written using the law of mass action [◄◄ Section 15.2].

Student Hot Spot

Student data indicate you may struggle with reaction mechanisms. Access the SmartBook to view additional Learning Resources on this topic.

Worked Example 19.12 lets you practice relating the experimentally determined rate law to a reaction mechanism in which a fast first step is followed by a slower, rate-determining step.

Worked Example 19.12

Consider the gas-phase reaction of nitric oxide and oxygen that was described at the beginning of Section 19.5.

$$2NO(g) + O_2(g) \longrightarrow 2NO_2(g)$$

Show that the following mechanism is plausible. The experimentally determined rate law is rate = $k[NO]^2[O_2]$.

$$\textit{Step 1:} \quad NO(g) + NO(g) \underset{k_{-1}}{\overset{k_1}{\rightleftharpoons}} N_2O_2(g) \text{ (fast)}$$

$$\textit{Step 2:} \quad N_2O_2(g) + O_2(g) \xrightarrow{k_2} 2NO_2(g) \text{ (slow)}$$

Strategy To establish the plausibility of a mechanism, we must compare the rate law of the rate-determining step to the experimentally determined rate law. In this case, the rate-determining step has an intermediate (N_2O_2) as one of its reactants, giving us a rate law of rate = $k_2[N_2O_2][O_2]$. Because we cannot compare this directly to the experimental rate law, we must solve for the intermediate concentration in terms of *reactant* concentrations.

Setup The first step is a rapidly established equilibrium. Both the forward and reverse of step 1 are elementary processes, which enables us to write their rate laws from the balanced equation.

$$\text{rate}_{\text{forward}} = k_1[NO]^2 \quad \text{and} \quad \text{rate}_{\text{reverse}} = k_{-1}[N_2O_2]$$

Solution Because at equilibrium the forward and reverse processes are occurring at the same rate, we can set their rates equal to each other and solve for the intermediate concentration.

$$k_1[NO]^2 = k_{-1}[N_2O_2]$$

$$[N_2O_2] = \frac{k_1[NO]^2}{k_{-1}}$$

Substituting the solution into the original rate law (rate = $k[N_2O_2][O_2]$) gives

$$\text{rate} = k_2\frac{k_1[NO]^2}{k_{-1}}[O_2] = k[NO]^2[O_2] \quad \text{where } k = \frac{k_2k_1}{k_{-1}}$$

Think About It
Not all reactions have a single rate-determining step. Analyzing the kinetics of reactions with two or more comparably slow steps is beyond the scope of this book.

Practice Problem Ⓐ**TTEMPT** Show that the following mechanism is consistent with the experimentally determined rate law of rate = $k[NO]^2[Br_2]$ for the reaction of nitric oxide and bromine: $2NO(g) + Br_2(g) \longrightarrow 2NOBr(g)$.

$$\textit{Step 1:} \quad NO(g) + Br_2(g) \underset{k_{-1}}{\overset{k_1}{\rightleftharpoons}} NOBr_2(g) \text{ (fast)}$$

$$\textit{Step 2:} \quad NOBr_2(g) + NO(g) \xrightarrow{k_2} 2NOBr(g) \text{ (slow)}$$

Practice Problem Ⓑ**UILD** The reaction $H_2(g) + I_2(g) \longrightarrow 2HI(g)$ proceeds via a two-step mechanism in which the rate law for the rate-determining step is rate = $k[H_2][I]^2$. Write the mechanism and rewrite the rate law using only reactant concentrations.

Practice Problem Ⓒ**ONCEPTUALIZE** The reaction of $A + B \longrightarrow C + D$ is believed to proceed via a two-step mechanism in which the second step is rate-limiting and the first step produces an intermediate (I). Which graph [(i)–(iii)] could represent the concentrations of the reactants and the intermediate as the reaction progresses?

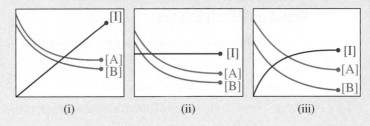

$H_2(g) + I_2(g) \longrightarrow 2HI(g)$

$rate = k[H_2][I_2]$

Proposed mechanism

Step 1: $I_2 \underset{k_{-1}}{\overset{k_1}{\rightleftharpoons}} 2I$

Step 2: $H_2 + 2I \xrightarrow{k_2} 2HI$

Experimental Support for Reaction Mechanisms

When you propose a mechanism based on experimental rate data, and your mechanism satisfies the two requirements listed on page 912, that the individual steps sum to the correct overall equation and that the rate-determining steps have the proper rate law, you can say that the proposed mechanism is *plausible*—not that it is necessarily *correct.* It is not possible to prove that a mechanism is correct using rate data alone. To determine whether a proposed reaction mechanism is actually correct, we must conduct other experiments. In the case of hydrogen peroxide decomposition, we might try to detect the presence of the IO^- ions. If we can detect them, it will support the proposed mechanism. Similarly, for the reaction of hydrogen and iodine to form hydrogen iodide, detection of iodine atoms would lend support to the proposed two-step mechanism shown in the margin. For example, I_2 dissociates into atoms when it is irradiated with visible light. Thus, we might predict that the formation of HI from H_2 and I_2 would speed up as the intensity of light is increased—because that would increase the concentration of I atoms. Indeed, this is just what is observed.

In one case, chemists wanted to know which C—O bond was broken in the reaction between methyl acetate and water to better understand the reaction

$$CH_3-\overset{\overset{\displaystyle O}{\|}}{C}-O-CH_3 \ + \ H_2O \longrightarrow CH_3-\overset{\overset{\displaystyle O}{\|}}{C}-OH \ + \ CH_3OH$$

The two possibilities of bond breaking are

$$CH_3-\overset{\overset{\displaystyle O}{\|}}{C}-\overset{\downarrow}{}\text{-}O-CH_3 \quad \text{and} \quad CH_3-\overset{\overset{\displaystyle O}{\|}}{C}-O-\overset{\downarrow}{}\text{-}CH_3$$

(a)   (b)

To distinguish between schemes (a) and (b), chemists used water containing the oxygen-18 isotope instead of ordinary water, which contains the oxygen-16 isotope. When the ^{18}O water was used, only the acetic acid formed contained ^{18}O.

$$CH_3-\overset{\overset{\displaystyle O}{\|}}{C}-{}^{18}O-H$$

Thus, the reaction must have occurred via bond-breaking scheme (a), because the product formed via scheme (b) would have retained both of its original oxygen atoms and would contain no ^{18}O.

Another example is photosynthesis, the process by which green plants produce glucose from carbon dioxide and water.

$$6CO_2 + 6H_2O \longrightarrow C_6H_{12}O_6 + 6O_2$$

A question that arose early in the studies of photosynthesis was whether the molecular oxygen produced came from the water, the carbon dioxide, or both. By using water containing only the oxygen-18 isotope, it was concluded that all the oxygen produced by photosynthesis came from the water and none came from the carbon dioxide, because the O_2 produced contained only ^{18}O.

These examples illustrate how creative chemists must be to study reaction mechanisms.

Section 19.7 Review

Reaction Mechanisms

Use the following information to answer questions 19.7.1 to 19.7.3. For the reaction

$$A + B \longrightarrow C + D$$

the experimental rate law is rate $= k[B]^2$.

19.7.1   What is the order of the reaction with respect to A and B, respectively?
(a) 0 and 1   (c) 0 and 2   (e) 2 and 0
(b) 1 and 0   (d) 2 and 1

19.7.2 What is the overall order of the reaction?

  (a) 0   (b) 1   (c) 2   (d) 3   (e) 4

19.7.3 Which of the following is a plausible mechanism for this reaction?

  (a) Step 1: A + B ⟶ C + E (slow)
  Step 2: E + A ⟶ D (fast)

  (b) Step 1: B + B ⟶ C + E (slow)
  Step 2: E + A ⟶ D + B (fast)

  (c) Step 1: A + A ⟶ B + D (slow)
  Step 2: B + B ⟶ A + C (fast)

  (d) Step 1: B + B ⟶ C + E (slow)
  Step 2: C + A ⟶ D + B (fast)

  (e) Step 1: A + B ⟶ D + E (slow)
  Step 2: E + A ⟶ C (fast)

19.7.4 A plausible mechanism for the reaction

$$H_2 + 2IBr \longrightarrow I_2 + 2HBr$$

is the following:

Step 1: $H_2 + IBr \xrightarrow{k_1} HI + HBr$ (slow)

Step 2: $HI + IBr \xrightarrow{k_2} I_2 + HBr$ (fast)

What is the rate law that will be determined experimentally?

  (a) rate = $k[H_2]^2$ (d) rate = $k[H_2][IBr]$
  (b) rate = $k[H_2][IBr]^2$ (e) rate = $k[HI][IBr]$
  (c) rate = $k[IBr]^2$

19.8 CATALYSIS

Recall from Section 19.7 that the reaction rate for the decomposition of hydrogen peroxide depends on the concentration of iodide ions, even though I^- does not appear in the overall equation. Instead, I^- acts as a catalyst for the reaction. A *catalyst* is a substance that increases the rate of a chemical reaction without itself being consumed. The catalyst may react to form an intermediate, but it is regenerated in a subsequent step of the reaction.

Video 19.1
Catalysis.

Molecular oxygen is prepared in the laboratory by heating potassium chlorate. The reaction is

$$2KClO_3(s) \longrightarrow 2KCl(s) + 3O_2(g)$$

However, this thermal decomposition process is very slow in the absence of a catalyst. The rate of decomposition can be increased dramatically by adding a small amount of manganese(IV) dioxide (MnO_2), a black powdery substance. All the MnO_2 can be recovered at the end of the reaction, just as all the I^- ions remain following the decomposition of H_2O_2.

A catalyst speeds up a reaction by providing a set of elementary steps with more favorable kinetics than those that exist in its absence. From Equation 19.8, we know that the rate constant k (and hence the rate) of a reaction depends on the frequency factor (A) and the activation energy (E_a)—the larger the value of A (or the smaller the value of E_a), the greater the rate. In many cases, a catalyst increases the rate by lowering the activation energy for the reaction.

Let's assume that the following reaction has a certain rate constant k and an activation energy E_a.

$$A + B \xrightarrow{k} C + D$$

In the presence of a catalyst, however, the rate constant is k_c, called the *catalytic rate constant.*

Figure 19.17 Comparison of the activation energy barriers of (a) an uncatalyzed reaction and (b) the same reaction with a catalyst. A catalyst lowers the energy barrier but does not affect the energies of the reactants or products. Although the reactants and products are the same in both cases, the reaction mechanisms and rate laws are different in (a) and (b).

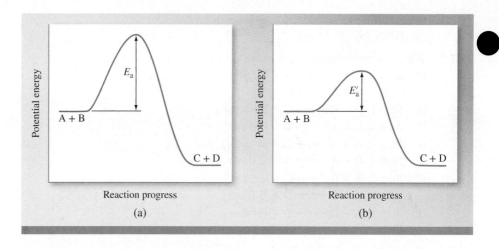

By the definition of a catalyst,

$$\text{rate}_{\text{catalyzed}} > \text{rate}_{\text{uncatalyzed}}$$

Figure 19.17 shows the potential energy profiles for both reactions. The total energies of the reactants (A and B) and those of the products (C and D) are unaffected by the catalyst; the only difference between the two is a lowering of the activation energy from E_a to E'_a. Because the activation energy for the reverse reaction is also lowered, a catalyst enhances the rates of the forward and reverse reactions equally.

The presence of a catalyst, however, does not alter the equilibrium constant, nor does it shift the position of an equilibrium system [◀◀ Section 15.6]. Adding a catalyst to a reaction mixture that is not at equilibrium will simply cause the mixture to reach equilibrium sooner. The same equilibrium mixture could be obtained without the catalyst, but it might take a much longer time.

There are three general types of catalysis, depending on the nature of the rate-increasing substance: heterogeneous catalysis, homogeneous catalysis, and enzyme catalysis.

Heterogeneous Catalysis

In *heterogeneous catalysis*, the reactants and the catalyst are in different phases. The catalyst is usually a solid, and the reactants are either gases or liquids. Heterogeneous catalysis is by far the most important type of catalysis in industrial chemistry, especially in the synthesis of many important chemicals. Heterogeneous catalysis is also used in the catalytic converters in automobiles.

At high temperatures inside a car's engine, nitrogen and oxygen gases react to form nitric oxide.

$$N_2(g) + O_2(g) \longrightarrow 2NO(g)$$

When released into the atmosphere, NO rapidly combines with O_2 to form NO_2. Nitrogen dioxide and other gases emitted by automobiles, such as carbon monoxide (CO) and various unburned hydrocarbons, make automobile exhaust a major source of air pollution.

Most new cars are equipped with catalytic converters [Figure 19.18(a)]. An efficient catalytic converter serves two purposes: It oxidizes CO and unburned hydrocarbons to CO_2 and H_2O, and it converts NO and NO_2 to N_2 and O_2. Hot exhaust gases into which air has been injected are passed through the first chamber of one converter to accelerate the complete burning of hydrocarbons and to decrease CO emissions. Because high temperatures increase NO production, however, a second chamber containing a different catalyst (a transition metal or a transition metal oxide such as CuO or Cr_2O_3) and operating at a lower temperature is required to dissociate NO into N_2 and O_2 before the exhaust is discharged through the tailpipe [Figure 19.18(b)].

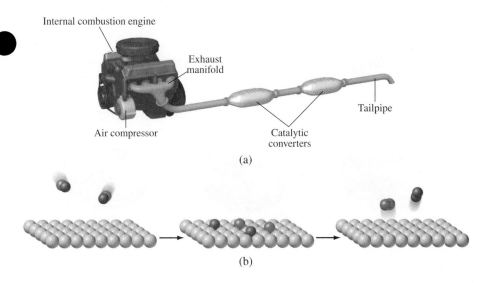

Internal combustion engine

Exhaust manifold

Tailpipe

Air compressor

Catalytic converters

(a)

(b)

Figure 19.18 (a) A two-stage catalytic converter for an automobile. (b) In the second stage, NO molecules bind to the surface of the catalyst. The N atoms bond to each other and the O atoms bond to each other, producing N_2 and O_2, respectively.

Homogeneous Catalysis

In *homogeneous catalysis,* the reactants and the catalyst are dispersed in a single phase, usually liquid. Acid and base catalyses are the most important types of homogeneous catalysis in liquid solution. For example, the reaction of ethyl acetate with water to form acetic acid and ethanol normally occurs too slowly to be measured.

$$CH_3COOC_2H_5 + H_2O \longrightarrow CH_3COOH + C_2H_5OH$$

In the absence of the catalyst, the rate law is given by

$$\text{rate} = k[CH_3COOC_2H_5]$$

The reaction, however, can be catalyzed by an acid. Often a catalyst is shown above the arrow in a chemical equation.

$$CH_3COOC_2H_5 + H_2O \xrightarrow{\text{H}^+} CH_3COOH + C_2H_5OH$$

In the presence of acid, the rate is faster and the rate law is given by

$$\text{rate} = k_c[CH_3COOC_2H_5][H^+]$$

Because $k_c > k$ in magnitude, the rate is determined solely by the catalyzed portion of the reaction.

Homogeneous catalysis has several advantages over heterogeneous catalysis. For one thing, the reactions can often be carried out under atmospheric conditions, thus reducing production costs and minimizing the decomposition of products at high temperatures. In addition, homogeneous catalysts can be designed to function selectively for particular types of reactions, and homogeneous catalysts cost less than the precious metals (e.g., platinum and gold) used in heterogeneous catalysis.

Enzymes: Biological Catalysts

Of all the intricate processes that have evolved in living systems, none is more striking or more essential than enzyme catalysis. *Enzymes* are *biological* catalysts. The amazing fact about enzymes is that not only can they increase the rate of biochemical reactions by factors ranging from 10^6 to 10^{18}, but they are also highly specific. An enzyme acts only on certain reactant molecules, called *substrates,* while leaving the rest of the system unaffected. It has been estimated that an average living cell may contain some 3000 different enzymes, each of them catalyzing a specific reaction in which a substrate is converted into the appropriate product(s). Enzyme catalysis is usually homogeneous because the substrate and enzyme are present in aqueous solution.

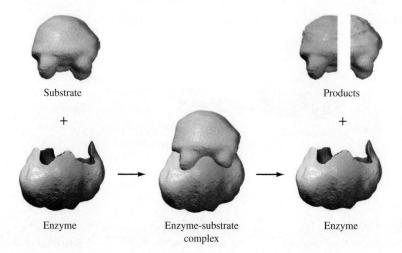

An enzyme is typically a large protein molecule that contains one or more active sites where interactions with substrates take place. These sites are structurally compatible with specific substrate molecules, in much the same way that a key fits a particular lock. In fact, the notion of a rigid enzyme structure that binds only to molecules whose shape exactly matches that of the active site was the basis of an early theory of enzyme catalysis, the so-called lock-and-key theory developed by Emil Fischer[1] in 1894 (Figure 19.19). Fischer's hypothesis accounts for the specificity of enzymes, but it contradicts research evidence that a single enzyme binds to substrates of different sizes and shapes. Chemists now know that an enzyme molecule (or at least its active site) has a fair amount of structural flexibility and can modify its shape to accommodate more than one type of substrate. Figure 19.20 shows a molecular model of an enzyme in action.

The mathematical treatment of enzyme kinetics is quite complex, even when we know the basic steps involved in the reaction. A simplified scheme is given by the following elementary steps:

$$E + S \underset{k_{-1}}{\overset{k_1}{\rightleftharpoons}} ES$$

$$ES \xrightarrow{k_2} E + P$$

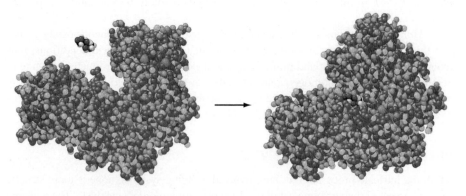

Figure 19.20 Left to right: The binding of glucose molecule (red) to hexokinase (an enzyme in the metabolic pathway). Note how the region at the active site closes around glucose after binding. Often, the geometries of both the substrate and the active site are altered to fit each other.

---

[1]Emil Fischer (1852–1919), a German chemist, is regarded by many as the greatest organic chemist of the nineteenth century. Fischer made many significant contributions in the synthesis of sugars and other important molecules. He was awarded the Nobel Prize in Chemistry in 1902.

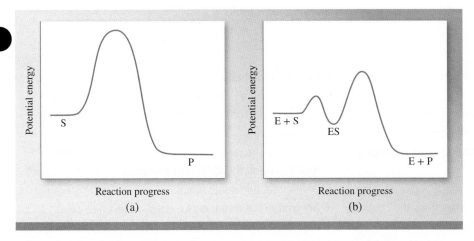

Figure 19.21 Comparison of (a) an uncatalyzed reaction and (b) the same reaction catalyzed by an enzyme. The plot in (b) assumes that the catalyzed reaction has a two-step mechanism, in which the second step (ES ⟶ E + P) is rate determining.

where E, S, and P represent enzyme, substrate, and product, respectively, and ES is the enzyme-substrate intermediate. It is often assumed that the formation of ES and its decomposition back to enzyme and substrate molecules occur rapidly and that the rate-determining step is the formation of product. Figure 19.21 shows the potential energy profile for the reaction.

In general, the rate of such a reaction is given by the equation

$$\text{rate} = \frac{\Delta[P]}{\Delta t}$$

$$= k[ES]$$

The concentration of the ES intermediate is itself proportional to the amount of the substrate present, and a plot of the rate versus the concentration of substrate typically yields a curve like that shown in Figure 19.22. Initially the rate rises rapidly with increasing substrate concentration.

Above a certain concentration, however, all the active sites are occupied, and the reaction becomes zeroth order in the substrate. In other words, the rate remains the same even though the substrate concentration increases. At and beyond this point, the rate of formation of product depends only on how fast the ES intermediate breaks down, not on the number of substrate molecules present.

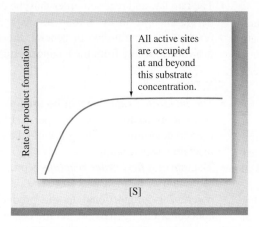

Figure 19.22 Plot of the rate of product formation versus substrate concentration in an enzyme-catalyzed reaction.

Learning Outcomes

- Describe the main factors that can increase the rate of a reaction.
- Describe the collision theory of chemical kinetics.
- Define effective collision and activation energy.
- Determine the average rate of a reaction given appropriate data.
- Determine the instantaneous rate of a reaction given a graph.
- Use the stoichiometry of a reaction to express the rate of a reaction in terms of a reactant or product.
- Define reaction order and provide examples of a zeroth-, first-, and second-order reaction rate law.
- Produce the rate law of a reaction given experimental data.
- Predict the units of the rate constant k for a reaction.
- Determine the order of a reaction from its rate law.
- Determine the concentration of reactant using the integrated rate law for the reaction.

- Calculate the half-life ($t_{1/2}$) of a reaction.
- Calculate the amount of reactant remaining after a certain period of time given the half-life and order of the reaction.
- Determine the rate constant or activation energy of a reaction using the Arrhenius equation.
- Define reaction mechanism.
- Determine the rate law of a reaction given its rate-determining step.
- Understand the criteria that must be met for a proposed mechanism to be plausible.
- Define elementary reaction.
- Define catalyst and intermediate.
- Describe the major types of catalysis: heterogeneous, homogeneous, and enzymatic.

Chapter Summary

SECTION 19.1

- Reaction rate refers to the speed at which a chemical reaction occurs. Factors that influence reaction rate are reactant concentration, temperature, surface area, and catalysis.

SECTION 19.2

- **Collision theory** explains why the rate constant, and therefore the reaction rate, increases with increasing temperature.
- The **activation energy (E_a)** is the minimum energy that colliding molecules must possess in order for the collision to be effective.
- Reactions occur when molecules of sufficient energy (and appropriate orientation) collide. **Effective collisions** are those that result in the formation of an **activated complex,** also called a **transition state.** Only *effective* collisions can result in product formation.

SECTION 19.3

- The rate of a chemical reaction is the change in concentration of reactants or products over time. Rates may be expressed as an *average* rate over a given time interval or as an **instantaneous rate.**
- The **rate constant (k)** is a proportionality constant that relates the **rate of reaction** with the concentration(s) of reactant(s). The rate constant k for a given reaction changes only with temperature.

SECTION 19.4

- The **rate law** is an equation that expresses the relationship between rate and reactant concentration(s). In general, the rate law for the reaction of A and B is rate = $k[A]^x[B]^y$.
- The **reaction order** is the power to which the concentration of a given reactant is raised in the rate law equation. The overall reaction order is the sum of the powers to which reactant concentrations are raised in the rate law.
- The **initial rate** is the instantaneous rate of reaction when the reactant concentrations are *starting* concentrations.
- The rate law and reaction order must be determined by comparing changes in the initial rate with changes in starting reactant concentrations. In general, the rate law cannot be determined solely from the balanced equation.

SECTION 19.5

- The **integrated rate law** can be used to determine reactant concentrations after a specified period of time. It can also be used to determine how long it will take to reach a specified reactant concentration.
- The rate of a **first-order reaction** is proportional to the concentration of a single reactant. The rate of a **second-order reaction** is proportional to the product of two reactant concentrations ([A][B]), or on the concentration of a single reactant squared ($[A]^2$ or $[B]^2$). The rate of a **zeroth-order reaction** does not depend on reactant concentration.

- The **half-life** ($t_{1/2}$) of a reaction is the time it takes for half of a reactant to be consumed. The half-life is *constant* for first-order reactions, and it can be used to determine the rate constant of the reaction.

SECTION 19.6

- The relationship between temperature and the rate constant is expressed by the **Arrhenius equation.**

SECTION 19.7

- A **reaction mechanism** may consist of a series of *steps,* called **elementary reactions.** Unlike rate laws in general, the rate law for an elementary reaction can be written from the balanced equation, using the stoichiometric coefficient for each reactant species as its exponent in the rate law.
- A species that is produced in one step of a reaction mechanism and subsequently consumed in another step is called an **intermediate.** A species that is first consumed and later regenerated is called a *catalyst.* Neither intermediates nor catalysts appear in the overall balanced equation.

- The rate law of each step in a reaction mechanism indicates the **molecularity** or overall *order* of the step. A **unimolecular** step is first order, involving just one molecule; a **bimolecular** step is second order, involving the collision of two molecules; and a **termolecular** step is third order, involving the collision of three molecules. Termolecular processes are relatively rare.
- If one step in a reaction is much slower than all the other steps, it is the **rate-determining step.** The rate-determining step has a rate law identical to the experimental rate law.

SECTION 19.8

- A **catalyst** speeds up a reaction, usually by lowering the value of the activation energy. *Catalysis* refers to the process by which a catalyst increases the reaction rate.
- Catalysis may be **heterogeneous,** in which the catalyst and reactants exist in different phases, or **homogeneous,** in which the catalyst and reactants exist in the same phase.
- **Enzymes** are biological catalysts with high specificity for the reactions that they catalyze.

Key Words

Activated complex, 878	Half-life ($t_{1/2}$), 899	Rate law, 890
Activation energy (E_a), 878	Heterogeneous catalysis, 920	Rate of reaction, 887
Arrhenius equation, 905	Homogeneous catalysis, 921	Reaction mechanism, 911
Bimolecular, 911	Initial rate, 890	Reaction order, 890
Catalyst, 919	Instantaneous rate, 884	Second-order reaction, 901
Collision theory, 877	Integrated rate law, 896	Termolecular, 911
Effective collision, 877	Intermediate, 911	Transition state, 878
Elementary reaction, 911	Molecularity, 911	Unimolecular, 911
Enzymes, 921	Rate constant (k), 885	Zeroth-order reaction, 903
First-order reaction, 896	Rate-determining step, 912	

Key Equations

19.1	$\text{rate} = -\dfrac{1}{a}\dfrac{\Delta[A]}{\Delta t} = -\dfrac{1}{b}\dfrac{\Delta[B]}{\Delta t} = \dfrac{1}{c}\dfrac{\Delta[C]}{\Delta t} = \dfrac{1}{d}\dfrac{\Delta[D]}{\Delta t}$	Rate can be expressed in terms of the concentration of any species in a reaction. The minus signs in the equation correct for the fact that although reactant concentrations decrease as a reaction proceeds, rate is always expressed as a positive quantity.
19.2	$\text{rate} = k[A]^x[B]^y$	Rate can also be expressed using the *rate law,* which relates the rate to reactant concentrations and a rate constant, k. Units of the rate constant depend on overall reaction order.
19.3	$\ln\dfrac{[A]_t}{[A]_0} = -kt$	The integrated rate law for a first-order reaction relates natural logarithm (ln) of the ratio of reactant concentration at time zero to reactant concentration at a subsequent time, $\ln\left(\frac{[A]_0}{[A]_t}\right)$, to the rate constant, k, and the time elapsed, t.
19.4	$\ln[A]_t = -kt + \ln[A]_0$	The integrated rate law for a first-order reaction can be written in the $y = mx + b$ form. This form of the equation indicates that when $\ln[A]_t$ is plotted against time, the slope of the resulting line is the negative of the rate constant, $-k$, and the intercept is $\ln[A]_0$.

19.5 $t_{1/2} = \dfrac{0.693}{k}$

The half-life, $t_{1/2}$, of a first-order reaction is constant and is calculated as the ratio of 0.693 (ln 2) to the rate constant of the reaction.

19.6 $\dfrac{1}{[A]_t} = kt + \dfrac{1}{[A]_0}$

The integrated rate law for a second-order reaction, written in $y = mx + b$ form, indicates that when $\frac{1}{[A]_t}$ is plotted against time, the slope of the resulting line is the rate constant and the intercept is $\frac{1}{[A]_0}$.

19.7 $t_{1/2} = \dfrac{1}{k[A]_0}$

The half-life of a second-order reaction is not constant, and is calculated as the reciprocal product of rate constant and reactant concentration at time 0.

19.8 $k = Ae^{-E_a/RT}$

The Arrhenius equation relates the rate constant to the frequency factor, A, the activation energy, E_a, and the absolute temperature. For units to cancel properly in this equation, R must be expressed using units of energy (J/K · mol or kJ/K · mol, depending on the units used for E_a).

19.9 $\ln k = \ln A - \dfrac{E_a}{RT}$

Another form of the Arrhenius equation.

19.10 $\ln k = \left(-\dfrac{E_a}{R}\right)\left(\dfrac{1}{T}\right) + \ln A$

The Arrhenius equation written in $y = mx + b$ form indicates that when $\ln k$ is plotted against $1/T$, the slope of the resulting line is $-E_a/R$ and the intercept is $\ln A$.

19.11 $\ln \dfrac{k_1}{k_2} = \dfrac{E_a}{R}\left(\dfrac{1}{T_2} - \dfrac{1}{T_1}\right)$

This is the most useful form of the Arrhenius equation. Subtracting Equation 19.10 at one temperature from Equation 19.10 at another temperature eliminates the frequency factor by cancellation and allows us to calculate the activation energy (E_a) for a reaction if we know the rate constants at two different temperatures. It also allows us to calculate rate constant at any other temperature, provided that we know E_a.

Key Skills

First-Order Kinetics

One of the important applications of kinetics is the analysis of radioactive decay, which is first order. Although the equations are the same as those we have already derived for first-order kinetics, the terms and symbols vary slightly from what we have seen to this point. For example, the amount of radioactive material is typically expressed using *activity* (A) in *disintegrations per second* (dps), or in numbers of nuclei (N), rather than in units of concentration. Thus, Equation 19.3 becomes

$$\ln \frac{A_t}{A_0} = -kt \qquad \text{or} \qquad \ln \frac{N_t}{N_0} = -kt$$

The use of activity is very similar to the use of concentration. As with other kinetics problems, we may be given the initial activity (A_0), the rate constant (k), and the time elapsed (t)—and be asked to determine the new activity (A_t). Or we may be asked to solve for one of the other parameters, which simply requires manipulation of the original equation. For example, we may be given A_0, A_t, and t and be asked to determine the rate constant—or the half-life. (Recall that for any first-order process, the half-life is constant, and is related to the rate constant by Equation 19.5.) In this case, we solve Equation 19.3 for k, or for $t_{1/2}$—or for whichever parameter we need to find.

When the missing parameter is either A_0 or A_t, we must pay special attention to the manipulation of logarithms [▶ Appendix 1]. Solving the activity version of Equation 19.3 for A_t, for example, gives

$$A_t = A_0(e^{-kt})$$

Student Annotation: Note that because mass is proportional to the number of nuclei for a given isotope, we can use masses in the number-of-nuclei version of Equation 19.3—as long as they are mass amounts of the same nucleus.

When radioactive decay is quantified using numbers of nuclei (N), some additional thought must to go into determining the "known" parameters. Rocks that contain uranium, for example, can be dated by measuring the amounts of uranium-238 and lead-206 they contain. ^{238}U is unstable and undergoes a series of radioactive decay steps [▶ Section 20.3], ultimately producing a ^{206}Pb nucleus for every ^{238}U nucleus that decays. Although rocks may contain other isotopes of lead, ^{206}Pb results strictly from the decay of ^{238}U. Therefore, we can assume that every ^{206}Pb nucleus was originally a ^{238}U nucleus. If we know the mass of ^{238}U (N_t) and the mass of ^{206}Pb in the rock, we can determine N_0 as follows:

$$\text{mass of } ^{206}\text{Pb} \times \frac{1}{\text{atomic mass of } ^{206}\text{Pb}} = \text{number of } ^{206}\text{Pb nuclei}$$

Because every ^{206}Pb nucleus was originally a ^{238}U nucleus,

$$\text{number of } ^{206}\text{Pb nuclei} = \text{number of } ^{238}\text{U nuclei}$$

$$\text{number of } ^{238}\text{U nuclei} \times \text{atomic mass of } ^{238}\text{U} = \text{mass of } ^{238}\text{U}$$

These operations condense to give

$$
\boxed{\text{mass of } ^{206}\text{Pb}} \times \boxed{\dfrac{\text{atomic mass of } ^{238}\text{U}}{\text{atomic mass of } ^{206}\text{Pb}}} = \boxed{\text{mass of } ^{238}\text{U}}
$$

This gives the mass of ^{238}U *not* accounted for in the mass of ^{238}U found in the rock. Adding this mass to the mass of ^{238}U in the rock gives the mass of ^{238}U *originally* present (N_0). When we have both N_t and N_0, given the rate constant for decay of ^{238}U (1.54×10^{-10} yr^{-1}), we can determine the age of the rock (t).

Consider the following example:

A rock is found to contain 23.17 g ^{238}U and 2.02 g ^{206}Pb. Its age is determined as follows:

$$
\boxed{2.02 \text{ g } ^{206}\text{Pb}} \times \boxed{\dfrac{238 \text{ g } ^{238}\text{U}}{206 \text{ g } ^{206}\text{Pb}}} = \boxed{2.334 \text{ g } ^{238}\text{U}}
$$

$$
\boxed{N_0 = (23.17 + 2.334) = 25.50 \text{ g}} \qquad \boxed{N_t = 23.17 \text{ g}}
$$

Solving Equation 19.3 for t gives

$$
\boxed{t = -\left(\dfrac{\ln \dfrac{N_t}{N_0}}{k} \right)}
$$

and

$$
\boxed{t = -\left(\dfrac{\ln \dfrac{23.17 \text{ g}}{25.50 \text{ g}}}{1.54 \times 10^{-10} \text{ yr}^{-1}} \right)} = \boxed{6.22 \times 10^8 \text{ yr}}
$$

Therefore, the rock is 622 million years old.

Key Skills Problems

19.1
It takes 218 hours for the activity of a certain radioactive isotope to fall to one-tenth of its original value. Calculate the half-life of the isotope.
(a) 3.18×10^{-4} h (b) 21.8 h (c) 0.0152 h (d) 65.6 h (e) 0.0106 h

19.2
^{61}Cu decays with a half-life of 3.35 h. Determine the original mass of a sample of ^{61}Cu if 612.8 mg remains after exactly 24 hours.
(a) 85.5 mg (b) 4.39×10^3 mg (c) 736 mg (d) 6.40×10^3 mg
(e) 8.78×10^4 mg

19.3
The rate constant for the radioactive decay of iodine-126 is 0.0533 d^{-1}. How much ^{126}I remains of a 2.55-g sample of ^{126}I after exactly 24 hours?
(a) 0.948 g (b) 2.42 g (c) 0.136 g (d) 0.710 g (e) 0.873 g

19.4
Determine the age of a rock that contains 45.7 mg ^{238}U and 1.02 mg ^{206}Pb.
(a) 165 million years (b) 2.50 billion years (c) 143 million years
(d) 6.49 billion years (e) 63.7 million years

Questions and Problems

SECTION 19.2: COLLISION THEORY OF CHEMICAL REACTIONS

Visualizing Chemistry
Figure 19.4

VC 19.1 The rate of a reaction in which the reactant concentration is reduced and the temperature is increased will
(a) increase.
(b) decrease.
(c) It is not possible to determine the effect on rate without additional information.

VC 19.2 The rate of a reaction in which the reactant concentration is reduced and the temperature is reduced will
(a) increase.
(b) decrease.
(c) It is not possible to determine the effect on rate without additional information.

VC 19.3 The rate of a reaction in which the reactant concentration is increased and the temperature is increased will
(a) increase.
(b) decrease.
(c) It is not possible to determine the effect on rate without additional information.

VC 19.4 Increasing the temperature of a reaction increases
(a) the number of collisions between reactant molecules.
(b) the kinetic energy of colliding molecules.
(c) both the number of collisions between reactant molecules and the kinetic energy of colliding molecules.

Review Questions

19.1 Define *activation energy*. What role does activation energy play in chemical kinetics?
19.2 Sketch a potential energy versus reaction progress plot for the following reactions.
(a) $S(s) + O_2(g) \longrightarrow SO_2(g)$ $\Delta H° = -296$ kJ/mol
(b) $Cl_2(g) \longrightarrow Cl(g) + Cl(g)$ $\Delta H° = 243$ kJ/mol
19.3 The reaction $H + H_2 \longrightarrow H_2 + H$ has been studied for many years. Sketch a potential energy versus reaction progress diagram for this reaction.

SECTION 19.3: MEASURING REACTION PROGRESS AND EXPRESSING REACTION RATE

Review Questions

19.4 What is meant by the *rate* of a chemical reaction? What are the units of the rate of a reaction?

19.5 Distinguish between average rate and instantaneous rate. Which of the two rates gives us an unambiguous measurement of reaction rate? Why?
19.6 What are the advantages of measuring the initial rate of a reaction?
19.7 Identify two reactions that are very slow (take days or longer to complete) and two reactions that are very fast (reactions that are over in minutes or seconds).

Computational Problems

19.8 Consider the reaction
$$N_2(g) + 3H_2(g) \longrightarrow 2NH_3(g)$$
Suppose that at a particular moment during the reaction molecular hydrogen is reacting at the rate of 0.082 *M*/s. (a) At what rate is ammonia being formed? (b) At what rate is molecular nitrogen reacting?
19.9 Consider the reaction
$$2NO(g) + O_2(g) \longrightarrow 2NO_2(g)$$
Suppose that at a particular moment during the reaction nitric oxide (NO) is reacting at the rate of 0.066 *M*/s. (a) At what rate is NO_2 being formed? (b) At what rate is molecular oxygen reacting?

Conceptual Problems

19.10 Write the reaction rate expressions for the following reactions in terms of the disappearance of the reactants and the appearance of products.
(a) $2H_2(g) + O_2(g) \longrightarrow 2H_2O(g)$
(b) $4NH_3(g) + 5O_2(g) \longrightarrow 4NO(g) + 6H_2O(g)$
19.11 Write the reaction rate expressions for the following reactions in terms of the disappearance of the reactants and the appearance of products.
(a) $H_2(g) + I_2(g) \longrightarrow 2HI(g)$
(b) $5Br^-(aq) + BrO_3^-(aq) + 6H^+(aq) \longrightarrow$
$$3Br_2(aq) + 3H_2O(l)$$

SECTION 19.4: DEPENDENCE OF REACTION RATE ON REACTANT CONCENTRATION

Review Questions

19.12 Explain what is meant by the *rate law* of a reaction.
19.13 Explain what is meant by the *order* of a reaction.
19.14 What are the units for the rate constants of first-order and second-order reactions?
19.15 Consider the zeroth-order reaction: A $\longrightarrow$ product. (a) Write the rate law for the reaction. (b) What are the units for the rate constant? (c) Plot the rate of the reaction versus [A].
19.16 The rate constant of a first-order reaction is 66 s^{-1}. What is the rate constant in units of minutes?
19.17 On which of the following properties does the rate constant of a reaction depend: (a) reactant concentrations, (b) nature of reactants, (c) temperature?

Computational Problems

19.18 Use the data in Table 19.2 to calculate the rate of the reaction at the time when $[F_2] = 0.040\ M$ and $[ClO_2] = 0.055\ M$.

19.19 The rate law for the reaction

$$NH_4^+(aq) + NO_2^-(aq) \longrightarrow N_2(g) + 2H_2O(l)$$

is given by rate $= k[NH_4^+][NO_2^-]$. At 25°C, the rate constant is $3.0 \times 10^{-4}/M \cdot s$. Calculate the rate of the reaction at this temperature if $[NH_4^+] = 0.036\ M$ and $[NO_2^-] = 0.065\ M$.

19.20 Consider the reaction

$$X + Y \longrightarrow Z$$

From the following data, obtained at 360 K,
(a) determine the order of the reaction, and
(b) determine the initial rate of disappearance of X when the concentration of X is 0.30 M and that of Y is 0.40 M.

Initial rate of disappearance of X (M/s)	[X] (M)	[Y] (M)
0.053	0.10	0.50
0.127	0.20	0.30
1.02	0.40	0.60
0.254	0.20	0.60
0.509	0.40	0.30

19.21 Consider the reaction

$$A + B \longrightarrow products$$

From the following data obtained at a certain temperature, determine the order of the reaction and calculate the rate constant.

[A] (M)	[B] (M)	Rate (M/s)
1.50	1.50	3.20×10^{-1}
1.50	2.50	3.20×10^{-1}
3.00	1.50	6.40×10^{-1}

19.22 Consider the reaction

$$A \longrightarrow B$$

The rate of the reaction is $1.6 \times 10^{-2}\ M$/s when the concentration of A is 0.15 M. Calculate the rate constant if the reaction is (a) first order in A and (b) second order in A.

19.23 Determine the overall orders of the reactions to which the following rate laws apply: (a) rate $= k\,[NO_2]^2$, (b) rate $= k$, (c) rate $= k\,[H_2]^2[Br_2]^{1/2}$, (d) rate $= k\,[NO]^2[O_2]$.

Conceptual Problems

19.24 The following gas-phase reaction was studied at 290°C by observing the change in pressure as a function of time in a constant-volume vessel.

$$ClCO_2CCl_3(g) \longrightarrow 2COCl_2(g)$$

Determine the order of the reaction and the rate constant based on the following data,

Time (s)	P (mmHg)
0	15.76
181	18.88
513	22.79
1164	27.08

where P is the total pressure.

19.25 Cyclobutane decomposes to ethylene according to the equation

$$C_4H_8(g) \longrightarrow 2C_2H_4(g)$$

Determine the order of the reaction and the rate constant based on the following pressures, which were recorded when the reaction was carried out at 430°C in a constant-volume vessel.

Time (s)	$P_{C_4H_8}$ (mmHg)
0	400
2,000	316
4,000	248
6,000	196
8,000	155
10,000	122

SECTION 19.5: DEPENDENCE OF REACTANT CONCENTRATION ON TIME

Review Questions

19.26 Write an equation relating the concentration of a reactant A at $t = 0$ to that at $t = t$ for a first-order reaction. Define all the terms, and give their units. Do the same for a second-order reaction.

19.27 Define *half-life*. Write the equation relating the half-life of a first-order reaction to the rate constant.

19.28 Write the equations relating the half-life of a second-order reaction to the rate constant. How does it differ from the equation for a first-order reaction?

19.29 For a first-order reaction, how long will it take for the concentration of reactant to fall to one-eighth its original value? Express your answer in terms of the half-life ($t_{1/2}$) and in terms of the rate constant k.

Computational Problems

19.30 The thermal decomposition of phosphine (PH_3) into phosphorus and molecular hydrogen is a first-order reaction.

$$4PH_3(g) \longrightarrow P_4(g) + 6H_2(g)$$

The half-life of the reaction is 35.0 s at 680°C. Calculate (a) the first-order rate constant for the reaction and (b) the time required for 15% of the phosphine to decompose.

19.31 What is the half-life of a compound if 42% of a given sample of the compound decomposes in 60 min? Assume first-order kinetics.

19.32 The rate constant for the second-order reaction

$$2NO_2(g) \longrightarrow 2NO(g) + O_2(g)$$

is 0.54/$M \cdot$ s at 300°C. How long (in seconds) would it take for the concentration of NO_2 to decrease from 0.25 M to 0.18 M?

19.33 The rate constant for the second-order reaction

$$2NOBr(g) \longrightarrow 2NO(g) + Br_2(g)$$

is 0.80/$M \cdot$ s at 10°C. (a) Starting with a concentration of 0.086 M, calculate the concentration of NOBr after 22 s. (b) Calculate the half-lives when $[NOBr]_0 = 0.072$ M and $[NOBr]_0 = 0.054$ M.

Conceptual Problems

19.34 Consider the first-order reaction X $\longrightarrow$ Y shown here. (a) What is the half-life of the reaction? (b) Draw pictures showing the number of X (red) and Y (blue) molecules at 20 s and at 30 s.

t = 0 s   t = 10 s

19.35 The reaction A $\longrightarrow$ B shown here follows first-order kinetics. Initially different amounts of A molecules are placed in three containers of equal volume at the same temperature. (a) What are the relative rates of the reaction in these three containers? (b) How would the relative rates be affected if the volume of each container were doubled? (c) What are the relative half-lives of the reactions in (i) to (iii)?

(i)   (ii)   (iii)

19.36 Consider the first-order reaction A $\longrightarrow$ B in which A molecules (blue spheres) are converted to B molecules (yellow spheres). The figure shows the progress of the reaction after 10 s.

t = 0 s   t = 10 s

Which figure represents the number of molecules present after 20 s for the previous reaction?

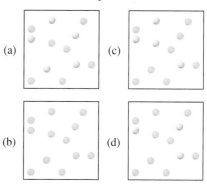

(a)   (c)

(b)   (d)

19.37 Considering the plots shown, which corresponds to a zeroth-order reaction, which corresponds to a first-order reaction, and which corresponds to a second-order reaction?

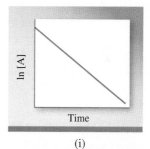

(i)

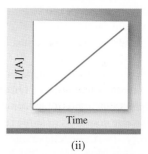

(ii)

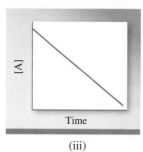

(iii)

SECTION 19.6: DEPENDENCE OF REACTION RATE ON TEMPERATURE

Review Questions

19.38 Write the Arrhenius equation, and define all terms.

19.39 Use the Arrhenius equation to show why the rate constant of a reaction (a) decreases with increasing activation energy and (b) increases with increasing temperature.

19.40 The burning of methane in oxygen is a highly exothermic reaction. Yet a mixture of methane and oxygen gas can be kept indefinitely without any apparent change. Explain.

19.41   Over the range of about ±3°C from normal body temperature, the metabolic rate, M_T, is given by $M_T = M_{37}(1.1)^{\Delta T}$, where M_{37} is the normal rate (at 37°C) and ΔT is the change in T. Discuss this equation in terms of a possible molecular interpretation.

Computational Problems

19.42   Given the same reactant concentrations, the reaction

$$CO(g) + Cl_2(g) \longrightarrow COCl_2(g)$$

at 250°C is 1.50×10^3 times as fast as the same reaction at 150°C. Calculate the activation energy for this reaction. Assume that the frequency factor is constant.

19.43   The rate of bacterial hydrolysis of fish muscle is twice as great at 2.2°C as at −1.1°C. Estimate an E_a value for this reaction. Is there any relation to the problem of storing fish for food?

19.44   The rate constant of a first-order reaction is 4.60×10^{-4} s^{-1} at 350°C. If the activation energy is 104 kJ/mol, calculate the temperature at which its rate constant is 8.80×10^{-4} s^{-1}.

19.45   For the reaction

$$NO(g) + O_3(g) \longrightarrow NO_2(g) + O_2(g)$$

the frequency factor A is 8.7×10^{12} s^{-1} and the activation energy is 63 kJ/mol. What is the rate constant for the reaction at 75°C?

19.46   The rate at which tree crickets chirp is 2.0×10^2 per minute at 27°C but only 39.6 per minute at 5°C. From these data, calculate the "activation energy" for the chirping process. (*Hint:* The ratio of rates is equal to the ratio of rate constants.)

19.47   The rate constants of some reactions double with every 10° rise in temperature. Assume that a reaction takes place at 295 K and 305 K. What must the activation energy be for the rate constant to double as described?

19.48   The activation energy for the denaturation of a protein is 196 kJ/mol. At what temperature will the rate of denaturation be 50% greater than its rate at 25.0°C?

Conceptual Problems

19.49   Variation of the rate constant with temperature for the first-order reaction

$$2N_2O_5(g) \longrightarrow 2N_2O_4(g) + O_2(g)$$

is given in the following table. Determine graphically the activation energy for the reaction.

T (K)	k (s^{-1})
298	1.74×10^{-5}
308	6.61×10^{-5}
318	2.51×10^{-4}
328	7.59×10^{-4}
338	2.40×10^{-3}

19.50   Diagram A describes the initial state of reaction

$$H_2 + Cl_2 \longrightarrow 2HCl$$

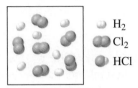

Diagram A

Suppose the reaction is carried out at two different temperatures as shown in diagram B. Which picture represents the result at the higher temperature? (The reaction proceeds for the same amount of time at both temperatures.)

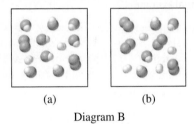

(a)   (b)

Diagram B

19.51   The directions for using Alka-Seltzer call for dropping two tablets into 6 ounces of lukewarm water, and drinking the mixture once the effervescence stops. Compare the time required for the effervescence to stop for the following scenarios: (a) Two whole tablets are dropped into lukewarm water, (b) two tablets each broken into two pieces are dropped into lukewarm water, (c) two whole tablets are dropped into cold water, (d) two crushed tablets are dropped into lukewarm water, (e) two crushed tablets are dropped into very warm water.

19.52   A chemist orders three different samples of zinc metal from a chemical supply company to try in an experiment where zinc metal is allowed to react with hydrochloric acid. The three types of zinc ordered are (a) zinc dust (particle size with a diameter of 0.044 mm), (b) zinc shot (particle size with a diameter of 1.40 mm), and (c) zinc powder (particle size with a diameter of 0.422 mm). Rank in order of increasing rate of reaction the various zinc samples used in the experiment.

SECTION 19.7: REACTION MECHANISMS

Review Questions

19.53   What do we mean by the *mechanism* of a reaction?

19.54   What is an elementary step? What is the molecularity of a reaction?

19.55   Reactions can be classified as unimolecular, bimolecular, and so on. Why are there no zero-molecular reactions? Explain why termolecular reactions are rare.

19.56 Determine the molecularity, and write the rate law for each of the following elementary steps.
(a) $X \longrightarrow$ products
(b) $X + Y \longrightarrow$ products
(c) $X + Y + Z \longrightarrow$ products
(d) $X + X \longrightarrow$ products
(e) $X + 2Y \longrightarrow$ products

19.57 What is the rate-determining step of a reaction? Give an everyday analogy to illustrate the meaning of *rate determining*.

19.58 The equation for the combustion of ethane (C_2H_6) is

$$2C_2H_6(g) + 7O_2(g) \longrightarrow 4CO_2(g) + 6H_2O(l)$$

Explain why it is unlikely that this equation also represents the elementary step for the reaction.

19.59 Specify which of the following species cannot be isolated in a reaction: activated complex, product, intermediate.

Conceptual Problems

19.60 Classify each of the following elementary steps as unimolecular, bimolecular, or termolecular.

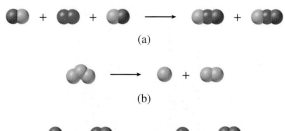

(a)

(b)

(c)

19.61 The rate law for the reaction

$$2NO(g) + Cl_2(g) \longrightarrow 2NOCl(g)$$

is given by rate = $k[NO][Cl_2]$. (a) What is the order of the reaction? (b) A mechanism involving the following steps has been proposed for the reaction

$$NO(g) + Cl_2(g) \longrightarrow NOCl_2(g)$$
$$NOCl_2(g) + NO(g) + NO(g) \longrightarrow 2NOCl(g)$$

If this mechanism is correct, what does it imply about the relative rates of these two steps?

19.62 For the reaction $X_2 + Y + Z \longrightarrow XY + XZ$, it is found that doubling the concentration of X_2 doubles the reaction rate, tripling the concentration of Y triples the rate, and doubling the concentration of Z has no effect. (a) What is the rate law for this reaction? (b) Why is it that the change in the concentration of Z has no effect on the rate? (c) Suggest a mechanism for the reaction that is consistent with the rate law.

19.63 The rate law for the reaction

$$2H_2(g) + 2NO(g) \longrightarrow N_2(g) + 2H_2O(g)$$

is rate = $k[H_2][NO]^2$. Which of the following mechanisms can be ruled out on the basis of the observed rate expression?

Mechanism I
$$H_2 + NO \longrightarrow H_2O + N \quad \text{(slow)}$$
$$N + NO \longrightarrow N_2 + O \quad \text{(fast)}$$
$$O + H_2 \longrightarrow H_2O \quad \text{(fast)}$$

Mechanism II
$$H_2 + 2NO \longrightarrow N_2O + H_2O \quad \text{(slow)}$$
$$N_2O + H_2 \longrightarrow N_2 + H_2O \quad \text{(fast)}$$

Mechanism III
$$2NO \rightleftharpoons N_2O_2 \quad \text{(fast equilibrium)}$$
$$N_2O_2 + H_2 \longrightarrow N_2O + H_2O \quad \text{(slow)}$$
$$N_2O + H_2 \longrightarrow N_2 + H_2O \quad \text{(fast)}$$

19.64 The rate law for the decomposition of ozone to molecular oxygen

$$2O_3(g) \longrightarrow 3O_2(g)$$

is

$$\text{rate} = k\frac{[O_3]^2}{[O_2]}$$

The mechanism proposed for this process is

$$O_3 \underset{k_{-1}}{\overset{k_1}{\rightleftharpoons}} O + O_2$$

$$O + O_3 \overset{k_2}{\longrightarrow} 2O_2$$

Derive the rate law from these elementary steps. Clearly state the assumptions you use in the derivation. Explain why the rate decreases with increasing O_2 concentration.

SECTION 19.8: CATALYSIS

Review Questions

19.65 How does a catalyst increase the rate of a reaction?

19.66 What are the characteristics of a catalyst?

19.67 Distinguish between homogeneous catalysis and heterogeneous catalysis.

19.68 Are enzyme-catalyzed reactions examples of homogeneous or heterogeneous catalysis? Explain.

19.69 The concentrations of enzymes in cells are usually quite small. What is the biological significance of this fact?

19.70 The first-order rate constant for the dehydration of carbonic acid

$$H_2CO_3 \longrightarrow CO_2 + H_2O$$

is about $1 \times 10^2 \text{ s}^{-1}$. In view of this rather high rate constant, explain why it is necessary to have the enzyme carbonic anhydrase to enhance the rate of dehydration in the lungs.

Conceptual Problems

19.71 Most reactions, including enzyme-catalyzed reactions, proceed faster at higher temperatures. However, for a given enzyme, the rate drops off abruptly at a certain temperature. Account for this behavior.

19.72 Consider the following mechanism for the enzyme-catalyzed reaction

$$E + S \underset{k_{-1}}{\overset{k_1}{\rightleftharpoons}} ES \quad \text{(fast equilibrium)}$$

$$ES \xrightarrow{k_2} E + P \quad \text{(slow)}$$

Derive an expression for the rate law of the reaction in terms of the concentrations of E and S. (*Hint:* To solve for [ES], make use of the fact that, at equilibrium, the rate of the forward reaction is equal to the rate of the reverse reaction.)

ADDITIONAL PROBLEMS

19.73 Classify the following elementary reactions as unimolecular, bimolecular, or termolecular.
(a) $2NO + Br_2 \longrightarrow 2NOBr$
(b) $CH_3NC \longrightarrow CH_3CN$
(c) $SO + O_2 \longrightarrow SO_2 + O$

19.74 Suggest experimental means by which the rates of the following reactions could be followed.
(a) $CaCO_3(s) \longrightarrow CaO(s) + CO_2(g)$
(b) $Cl_2(g) + 2Br^-(aq) \longrightarrow Br_2(aq) + 2Cl^-(aq)$
(c) $C_2H_6(g) \longrightarrow C_2H_4(g) + H_2(g)$
(d) $C_2H_5I(g) + H_2O(l) \longrightarrow$
$$C_2H_5OH(aq) + H^+(aq) + I^-(aq)$$

19.75 "The rate constant for the reaction

$$NO_2(g) + CO(g) \longrightarrow NO(g) + CO_2(g)$$

is $1.64 \times 10^{-6}/M \cdot s$." What is incomplete about this statement?

19.76 In a certain industrial process involving a heterogeneous catalyst, the volume of the catalyst (in the shape of a sphere) is 10.0 cm^3. Calculate the surface area of the catalyst. If the sphere is broken down into eight smaller spheres, each having a volume of 1.25 cm^3, what is the total surface area of the spheres? Which of the two geometric configurations of the catalyst is more effective? (The surface area of a sphere is $4\pi r^2$, where r is the radius of the sphere.) Based on your analysis here, explain why it is sometimes dangerous to work in grain elevators.

19.77 The following pictures represent the progress of the reaction $A \longrightarrow B$, where the red spheres represent A molecules and the green spheres represent B molecules. Calculate the rate constant of the reaction.

  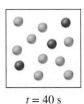

| $t = 0 \text{ s}$ | $t = 20 \text{ s}$ | $t = 40 \text{ s}$ |

19.78 The following pictures show the progress of the reaction $2A \longrightarrow A_2$. Determine whether the reaction is first order or second order, and calculate the rate constant.

| $t = 0 \text{ min}$ | $t = 15 \text{ min}$ | $t = 30 \text{ min}$ |

19.79 Explain why most metals used in catalysis are transition metals.

19.80 The following data were collected for the reaction between hydrogen and nitric oxide at 700°C.

$$2H_2(g) + 2NO(g) \longrightarrow 2H_2O(g) + N_2(g)$$

Experiment	$[H_2]$ (M)	$[NO]$ (M)	Initial rate (M/s)
1	0.010	0.025	2.4×10^{-6}
2	0.0050	0.025	1.2×10^{-6}
3	0.010	0.0125	0.60×10^{-6}

(a) Determine the order of the reaction. (b) Calculate the rate constant. (c) Suggest a plausible mechanism that is consistent with the rate law. (*Hint:* Assume that the oxygen atom is the intermediate.)

19.81 When methyl phosphate is heated in acid solution, it reacts with water.

$$CH_3OPO_3H_2 + H_2O \longrightarrow CH_3OH + H_3PO_4$$

If the reaction is carried out in water enriched with ^{18}O, the oxygen-18 isotope is found in the phosphoric acid product but not in the methanol. What does this tell us about the mechanism of the reaction?

19.82 The rate of the reaction

$$CH_3COOC_2H_5(aq) + H_2O(l) \longrightarrow$$
$$CH_3COOH(aq) + C_2H_5OH(aq)$$

shows first-order characteristics—that is, rate = $k[CH_3COOC_2H_5]$—even though this is a second-order reaction (first order in $CH_3COOC_2H_5$ and first order in H_2O). Explain.

19.83 List four factors that influence the rate of a reaction.

19.84 The reaction $2A + 3B \longrightarrow C$ is first order with respect to A and B. When the initial concentrations are $[A] = 1.6 \times 10^{-2} M$ and $[B] = 2.4 \times 10^{-3} M$, the rate is $4.1 \times 10^{-4} M/s$. Calculate the rate constant of the reaction.

19.85 The bromination of acetone is acid-catalyzed.

$$CH_3COCH_3 + Br_2 \xrightarrow[H^+]{\text{catalyst}} CH_3COCH_2Br + H^+ + Br^-$$

The rate of disappearance of bromine was measured for several different concentrations of acetone, bromine, and H^+ ions at a certain temperature.

	[CH₃COCH₃] (M)	[Br₂] (M)	[H⁺] (M)	Rate of disappearance of Br₂ (M/s)
(1)	0.30	0.050	0.050	5.7×10^{-5}
(2)	0.30	0.10	0.050	5.7×10^{-5}
(3)	0.30	0.050	0.10	1.2×10^{-4}
(4)	0.40	0.050	0.20	3.1×10^{-4}
(5)	0.40	0.050	0.050	7.6×10^{-5}

(a) What is the rate law for the reaction?
(b) Determine the rate constant. (c) The following mechanism has been proposed for the reaction.

$$CH_3{-}\overset{\overset{O}{||}}{C}{-}CH_3 + H_3O^+ \rightleftharpoons CH_3{-}\overset{\overset{+OH}{||}}{C}{-}CH_3 + H_2O \quad \text{(fast equilibrium)}$$

$$CH_3{-}\overset{\overset{+OH}{||}}{C}{-}CH_3 + H_2O \longrightarrow CH_3{-}\overset{\overset{OH}{|}}{C}{=}CH_2 + H_3O^+ \quad \text{(slow)}$$

$$CH_3{-}\overset{\overset{OH}{|}}{C}{=}CH_2 + Br_2 \longrightarrow CH_3{-}\overset{\overset{O}{||}}{C}{-}CH_2Br + HBr \text{ (fast)}$$

Show that the rate law deduced from the mechanism is consistent with that shown in part (a).

19.86 The decomposition of N_2O to N_2 and O_2 is a first-order reaction. At 730°C, the half-life of the reaction is 3.58×10^3 min. If the initial pressure of N_2O is 2.10 atm at 730°C, calculate the total gas pressure after one half-life. Assume that the volume remains constant.

19.87 The reaction $S_2O_8^{2-} + 2I^- \longrightarrow 2SO_4^{2-} + I_2$ proceeds slowly in aqueous solution, but it can be catalyzed by the Fe^{3+} ion. Given that Fe^{3+} can oxidize I^- and Fe^{2+} can reduce $S_2O_8^{2-}$, write a plausible two-step mechanism for this reaction. Explain why the uncatalyzed reaction is slow.

19.88 What are the units of the rate constant for a third-order reaction?

19.89 The integrated rate law for the zeroth-order reaction $A \longrightarrow B$ is $[A]_t = [A]_0 - kt$. (a) Sketch the following plots: (i) rate versus $[A]_t$ and (ii) $[A]_t$ versus t. (b) Derive an expression for the half-life of the reaction. (c) Calculate the time in half-lives when the integrated rate law is no longer valid, that is, when $[A]_t = 0$.

19.90 A flask contains a mixture of compounds A and B. Both compounds decompose by first-order kinetics. The half-lives are 50.0 min for A and 18.0 min for B. If the concentrations of A and B are equal initially, how long will it take for the concentration of A to be four times that of B?

19.91 Referring to Worked Example 19.5, explain how you would measure the partial pressure of azomethane experimentally as a function of time.

19.92 The rate law for the reaction $2NO_2(g) \longrightarrow N_2O_4(g)$ is rate $= k[NO_2]^2$. Which of the following changes will change the value of k? (a) The pressure of NO_2 is doubled. (b) The reaction is run in an organic solvent. (c) The volume of the container is doubled. (d) The temperature is decreased. (e) A catalyst is added to the container.

19.93 The reaction of G_2 with E_2 to form 2EG is exothermic, and the reaction of G_2 with X_2 to form 2XG is endothermic. The activation energy of the exothermic reaction is greater than that of the endothermic reaction. Sketch the potential energy profile diagrams for these two reactions on the same graph.

19.94 In the nuclear industry, workers use a rule of thumb that the radioactivity from any sample will be relatively harmless after 10 half-lives. Calculate the fraction of a radioactive sample that remains after this time period. (*Hint:* Radioactive decays obey first-order kinetics.)

19.95 Briefly comment on the effect of a catalyst on each of the following: (a) activation energy, (b) reaction mechanism, (c) enthalpy of reaction, (d) rate of forward reaction, (e) rate of reverse reaction.

19.96 When 6 g of granulated Zn is added to a solution of 2 M HCl in a beaker at room temperature, hydrogen gas is generated. For each of the following changes (at constant volume of the acid), state whether the rate of hydrogen gas evolution will be increased, decreased, or unchanged: (a) 6 g of powdered Zn is used, (b) 4 g of granulated Zn is used, (c) 2 M acetic acid is used instead of 2 M HCl, (d) temperature is raised to 40°C.

19.97 Strictly speaking, the rate law derived for the reaction in Problem 19.80 applies only to certain concentrations of H_2. The general rate law for the reaction takes the form

$$\text{rate} = \frac{k_1[NO]^2[H_2]}{1 + k_2[H_2]}$$

where k_1 and k_2 are constants. Derive rate law expressions under the conditions of very high and very low hydrogen concentrations. Does the result from Problem 19.80 agree with one of the rate expressions here?

19.98 A certain first-order reaction is 35.5% complete in 4.90 min at 25°C. What is its rate constant?

19.99 The decomposition of dinitrogen pentoxide has been studied in carbon tetrachloride solvent (CCl_4) at a certain temperature.

$$2N_2O_2 \longrightarrow 4NO_2 + O_2$$

[N₂O₅] (M)	Initial rate (M/s)
0.92	0.95×10^{-5}
1.23	1.20×10^{-5}
1.79	1.93×10^{-5}
2.00	2.10×10^{-5}
2.21	2.26×10^{-5}

Determine graphically the rate law for the reaction, and calculate the rate constant.

19.100 The thermal decomposition of N_2O_5 obeys first-order kinetics. At 45°C, a plot of $\ln [N_2O_5]$ versus t gives a slope of -6.18×10^{-4} min^{-1}. What is the half-life of the reaction?

19.101 When a mixture of methane and bromine is exposed to light, the following reaction occurs slowly.

$$CH_4(g) + Br_2(g) \longrightarrow CH_3Br(g) + HBr(g)$$

Suggest a reasonable mechanism for this reaction. (*Hint:* Bromine vapor is deep red; methane is colorless.)

19.102 The rate of the reaction between H_2 and I_2 to form HI increases with the intensity of visible light. (a) Explain why this fact supports the two-step mechanism given. (I_2 vapor is purple.) (b) Explain why the visible light has no effect on the formation of H atoms.

19.103 To prevent brain damage, a standard procedure is to lower the body temperature of someone who has been resuscitated after suffering cardiac arrest. What is the physiochemical basis for this procedure?

19.104 The second-order rate constant for the dimerization of a protein (P)

$$P + P \longrightarrow P_2$$

is $6.2 \times 10^{-3}/M \cdot$ s at 25°C. If the concentration of the protein is 2.7×10^{-4} M, calculate the initial rate (M/s) of formation of P_2. How long (in seconds) will it take to decrease the concentration of P to 2.7×10^{-5} M?

19.105 Consider the following elementary step.

$$X + 2Y \longrightarrow XY_2$$

(a) Write a rate law for this reaction. (b) If the initial rate of formation of XY_2 is 3.8×10^{-3} M/s and the initial concentrations of X and Y are 0.26 M and 0.88 M, respectively, what is the rate constant of the reaction?

19.106 In recent years, ozone in the stratosphere has been depleted at an alarmingly fast rate by chlorofluorocarbons (CFCs). A CFC molecule such as $CFCl_3$ is first decomposed by UV radiation.

$$CFCl_3 \longrightarrow CFCl_2 + Cl$$

The chlorine radical then reacts with ozone as follows:

$$Cl + O_3 \longrightarrow ClO + O_2$$
$$ClO + O \longrightarrow Cl + O_2$$

(a) Write the overall reaction for the last two steps. (b) What are the roles of Cl and ClO? (c) Why is the fluorine radical not important in this mechanism? (d) One suggestion to reduce the concentration of chlorine radicals is to add hydrocarbons such as ethane (C_2H_6) to the stratosphere. How will this work? (e) Draw potential energy versus reaction progress diagrams for the uncatalyzed and catalyzed (by Cl) destruction of ozone: $O_3 + O \longrightarrow 2O_2$. Use the thermodynamic data in Appendix 2 to determine whether the reaction is exothermic or endothermic.

19.107 Chlorine oxide (ClO), which plays an important role in the depletion of ozone (see Problem 19.106), decays rapidly at room temperature according to the equation

$$2ClO(g) \longrightarrow Cl_2(g) + O_2(g)$$

From the following data, determine the reaction order and calculate the rate constant of the reaction.

Time (s)	[ClO] (M)
0.12×10^{-3}	8.49×10^{-6}
0.96×10^{-3}	7.10×10^{-6}
2.24×10^{-3}	5.79×10^{-6}
3.20×10^{-3}	5.20×10^{-6}
4.00×10^{-3}	4.77×10^{-6}

19.108 A compound X undergoes two *simultaneous* first-order reactions as follows: X $\longrightarrow$ Y with rate constant k_1 and X $\longrightarrow$ Z with rate constant k_2. The ratio of k_1/k_2 at 40°C is 8.0. What is the ratio at 300°C? Assume that the frequency factors of the two reactions are the same.

19.109 Consider a car fitted with a catalytic converter. The first 5 min or so after it is started are the most polluting. Why?

19.110 The following scheme in which A is converted to B, which is then converted to C, is known as a consecutive reaction.

$$A \longrightarrow B \longrightarrow C$$

Assuming that both steps are first order, sketch on the same graph the variations of [A], [B], and [C] with time.

19.111 (a) What can you deduce about the activation energy of a reaction if its rate constant changes significantly with a small change in temperature? (b) If a bimolecular reaction occurs every time an A and a B molecule collide, what can you say about the orientation factor and activation energy of the reaction?

19.112 The rate law for the following reaction

$$CO(g) + NO_2(g) \longrightarrow CO_2(g) + NO(g)$$

is rate = $k[NO_2]^2$. Suggest a plausible mechanism for the reaction, given that the unstable species NO_3 is an intermediate.

19.113 Radioactive plutonium-239 ($t_{1/2} = 2.44 \times 10^5$ yr) is used in nuclear reactors and atomic bombs. If there are 5.0×10^2 g of the isotope in a small atomic bomb, how long will it take for the substance to decay to 1.0×10^2 g, too small an amount for an effective bomb?

19.114 Many reactions involving heterogeneous catalysts are zeroth order; that is, rate = k. An example is the decomposition of phosphine (PH_3) over tungsten (W):

$$4PH_3(g) \longrightarrow P_4(g) + 6H_2(g)$$

It is found that the reaction is independent of [PH_3] as long as phosphine's pressure is sufficiently high (≥ 1 atm). Explain.

19.115 Thallium(I) is oxidized by cerium(IV) as follows:

$$Tl^+ + 2Ce^{4+} \longrightarrow Tl^{3+} + 2Ce^{3+}$$

The elementary steps, in the presence of Mn(II), are as follows:

$$Ce^{4+} + Mn^{2+} \longrightarrow Ce^{3+} + Mn^{3+}$$
$$Ce^{4+} + Mn^{3+} \longrightarrow Ce^{3+} + Mn^{4+}$$
$$Tl^+ + Mn^{4+} \longrightarrow Tl^{3+} + Mn^{2+}$$

(a) Identify the catalyst, intermediates, and the rate-determining step if the rate law is rate = $k[Ce^{4+}]$ $[Mn^{2+}]$. (b) Explain why the reaction is slow without the catalyst. (c) Classify the type of catalysis (homogeneous or heterogeneous).

19.116 Sucrose ($C_{12}H_{22}O_{11}$), commonly called table sugar, undergoes hydrolysis (reaction with water) to produce fructose ($C_6H_{12}O_6$) and glucose ($C_6H_{12}O_6$).

$$\underset{\text{fructose} \quad \text{glucose}}{C_{12}H_{22}O_{11} + H_2O \longrightarrow C_6H_{12}O_6 + C_6H_{12}O_6}$$

This reaction is of considerable importance in the candy industry. First, fructose is sweeter than sucrose. Second, a mixture of fructose and glucose, called *invert sugar,* does not crystallize, so the candy containing this sugar would be chewy rather than brittle as candy containing sucrose crystals would be. (a) From the following data, determine the order of the reaction. (b) How long does it take for a sample of sucrose to be 95% hydrolyzed? (c) Explain why the rate law does not include [H_2O] even though water is a reactant.

Time (min)	[$C_{12}H_{22}O_{11}$] (M)
0	0.500
60.0	0.400
96.4	0.350
157.5	0.280

19.117 The first-order rate constant for the decomposition of dimethyl ether

$$(CH_3)_2O(g) \longrightarrow CH_4(g) + H_2(g) + CO(g)$$

is 3.2×10^{-4} s^{-1} at 450°C. The reaction is carried out in a constant-volume flask. Initially only dimethyl ether is present and the pressure is 0.350 atm. What is the pressure of the system after 8.0 min? Assume ideal behavior.

19.118 At 25°C, the rate constant for the ozone-depleting reaction

$$O(g) + O_3(g) \longrightarrow 2O_2(g)$$

is 7.9×10^{-15} cm³/molecule · s. Express the rate constant in units of $1/M \cdot$ s.

19.119 Consider the following elementary steps for a consecutive reaction.

$$A \xrightarrow{k_1} B \xrightarrow{k_2} C$$

(a) Write an expression for the rate of change of B. (b) Derive an expression for the concentration of B

under "steady-state" conditions; that is, when B is decomposing to C at the same rate as it is formed from A.

19.120 Ethanol is a toxic substance that, when consumed in excess, can impair respiratory and cardiac functions by interference with the neurotransmitters of the nervous system. In the human body, ethanol is metabolized by the enzyme alcohol dehydrogenase to acetaldehyde, which causes hangovers. Based on your knowledge of enzyme kinetics, explain why binge drinking (that is, consuming too much alcohol too fast) can prove fatal.

19.121 Strontium-90, a radioactive isotope, is a major product of an atomic bomb explosion. It has a half-life of 28.1 yr. (a) Calculate the first-order rate constant for the nuclear decay. (b) Calculate the fraction of ^{90}Sr that remains after 10 half-lives. (c) Calculate the number of years required for 99.0% of ^{90}Sr to disappear.

19.122 Consider the potential energy profiles for the following three reactions (from left to right). (1) Rank the rates (slowest to fastest) of the reactions. (2) Calculate ΔH for each reaction, and determine which reaction(s) are exothermic and which reaction(s) are endothermic. Assume the reactions have roughly the same frequency factors.

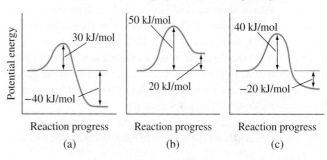

| (a) | (b) | (c) |

19.123 Consider the following potential energy profile for the A $\longrightarrow$ D reaction. (a) How many elementary steps are there? (b) How many intermediates are formed? (c) Which step is rate determining? (d) Is the overall reaction exothermic or endothermic?

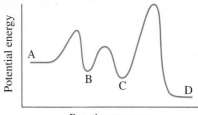

19.124 A factory that specializes in the refinement of transition metals such as titanium was on fire. The firefighters were advised not to douse the fire with water. Why?

19.125 The activation energy for the decomposition of hydrogen peroxide

$$2H_2O_2(aq) \longrightarrow 2H_2O(l) + O_2(g)$$

is 42 kJ/mol, whereas when the reaction is catalyzed by the enzyme catalase, it is 7.0 kJ/mol. Calculate the temperature that would cause the uncatalyzed decomposition to proceed as rapidly as the enzyme-catalyzed decomposition at 20°C. Assume the frequency factor A to be the same in both cases.

19.126   The *activity* of a radioactive sample is the number of nuclear disintegrations per second, which is equal to the first-order rate constant times the number of radioactive nuclei present. The fundamental unit of radioactivity is the *curie* (Ci), where 1 Ci corresponds to exactly 3.70×10^{10} disintegrations per second. This decay rate is equivalent to that of 1 g of radium-226. Calculate the rate constant and half-life for the radium decay. Starting with 1.0 g of the radium sample, what is the activity after 500 yr? The molar mass of Ra-226 is 226.03 g/mol.

19.127   To carry out metabolism, oxygen is taken up by hemoglobin (Hb) to form oxyhemoglobin (HbO_2) according to the simplified equation

$$Hb(aq) + O_2(aq) \xrightarrow{k} HbO_2(aq)$$

where the second-order rate constant is $2.1 \times 10^6/M \cdot s$ at 37°C. For an average adult, the concentrations of Hb and O_2 in the blood at the lungs are 8.0×10^{-6} M and 1.5×10^{-6} M, respectively. (a) Calculate the rate of formation of HbO_2. (b) Calculate the rate of consumption of O_2. (c) The rate of formation of HbO_2 increases to 1.4×10^{-4} M/s during exercise to meet the demand of the increased metabolism rate. Assuming the Hb concentration to remain the same, what must the oxygen concentration be to sustain this rate of HbO_2 formation?

19.128   At a certain elevated temperature, ammonia de-composes on the surface of tungsten metal as follows:

$$2NH_3 \longrightarrow N_2 + 3H_2$$

From the following plot of the rate of the reaction versus the pressure of NH_3, describe the mechanism of the reaction.

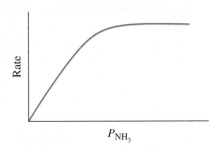

19.129   The following expression shows the dependence of the half-life of a reaction ($t_{1/2}$) on the initial reactant concentration $[A]_0$,

$$t_{1/2} \propto \frac{1}{[A]_0^{n-1}}$$

where n is the order of the reaction. Verify this dependence for zeroth-, first-, and second-order reactions.

19.130   Polyethylene is used in many items, including water pipes, bottles, electrical insulation, toys, and mailer

envelopes. It is a *polymer,* a molecule with a very high molar mass made by joining many ethylene molecules together. (Ethylene is the basic unit, or monomer for polyethylene.) The initiation step is

$$R_2 \xrightarrow{k_1} 2R\cdot \text{ (initiation)}$$

The R· species (called a radical) reacts with an ethylene molecule (M) to generate another radical.

$$R\cdot + M \longrightarrow M_1\cdot$$

The reaction of $M_1\cdot$ with another monomer leads to the growth or propagation of the polymer chain

$$M_1\cdot + M \xrightarrow{k_p} M_2\cdot \quad \text{(propagation)}$$

This step can be repeated with hundreds of monomer units. The propagation terminates when two radicals combine.

$$M'\cdot + M''\cdot \xrightarrow{k_t} M'{-}M'' \quad \text{(termination)}$$

The initiator frequently used in the polymerization of ethylene is benzoyl peroxide $[(C_6H_5COO)_2]$.

$$(C_6H_5COO)_2 \longrightarrow 2C_6H_5COO$$

This is a first-order reaction. The half-life of benzoyl peroxide at 100°C is 19.8 min. (a) Calculate the rate constant (in min^{-1}) of the reaction. (b) If the half-life of benzoyl peroxide is 7.30 h, or 438 min, at 70°C, what is the activation energy (in kJ/mol) for the decomposition of benzoyl peroxide? (c) Write the rate laws for the elementary steps in the preceding polymerization process, and identify the reactant, product, and intermediates. (d) What condition would favor the growth of long, high-molar-mass polyethylenes?

19.131   The rate constant for the gaseous reaction

$$H_2(g) + I_2(g) \longrightarrow 2HI(g)$$

is $2.42 \times 10^{-2}/M \cdot s$ at 400°C. Initially an equimolar sample of H_2 and I_2 is placed in a vessel at 400°C, and the total pressure is 1658 mmHg. (a) What is the initial rate (M/min) of formation of HI? (b) What are the rate of formation of HI and the concentration of HI (in molarity) after 10.0 min?

19.132   A protein molecule P of molar mass $\mathcal{M}$ dimerizes when it is allowed to stand in solution at room temperature. A plausible mechanism is that the protein molecule is first denatured (that is, loses its activity due to a change in overall structure) before it dimerizes,

$$P \xrightarrow{k} P^* \text{ (denatured)} \quad \text{slow}$$
$$2P^* \longrightarrow P_2 \quad \text{fast}$$

where the asterisk denotes a denatured protein molecule. Derive an expression for the average molar mass (of P and P_2), $\overline{\mathcal{M}}$, in terms of the initial protein concentration $[P]_0$ and the concentration at time t, $[P]_t$, and $\mathcal{M}$. Describe how you would determine k from molar mass measurements.

19.133 When the concentration of A in the reaction
A $\longrightarrow$ B was changed from 1.20 M to 0.60 M, the
half-life increased from 2.0 min to 4.0 min at 25°C.
Calculate the order of the reaction and the rate
constant. (*Hint:* Use the equation in Problem 19.129.)

19.134 At a certain elevated temperature, ammonia decomposes on the surface of tungsten metal as follows:

$$NH_3 \longrightarrow \frac{1}{2}N_2 + \frac{3}{2}H_2$$

The kinetic data are expressed as the variation of
the half-life with the initial pressure of NH_3.

P (mmHg)	$t_{1/2}$ (s)
264	456
130	228
59	102
16	60

(a) Determine the order of the reaction. (b) How
does the order depend on the initial pressure?
(c) How does the mechanism of the reaction vary
with pressure? (*Hint:* You need to use the equation
in Problem 19.129 and plot log $t_{1/2}$ versus log P.)

19.135 The activation energy for the reaction

$$N_2O(g) \longrightarrow N_2(g) + O(g)$$

is 2.4×10^2 kJ/mol at 600 K. Calculate the percentage
of the increase in rate from 600 K to 606 K. Comment
on your results.

19.136 The rate of a reaction was followed by the
absorption of light by the reactants and products as
a function of wavelengths (λ_1, λ_2, λ_3) as time
progresses. Which of the following mechanisms is
consistent with the experimental data?
(a) A $\longrightarrow$ B, A $\longrightarrow$ C
(b) A $\longrightarrow$ B + C
(c) A $\longrightarrow$ B, B $\longrightarrow$ C + D
(d) A $\longrightarrow$ B, B $\longrightarrow$ C

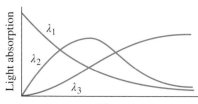

19.137 Magnesium reacts with strong acid to form Mg^{2+} ions
and hydrogen gas. Which of the following forms of
solid magnesium would react fastest with acid?

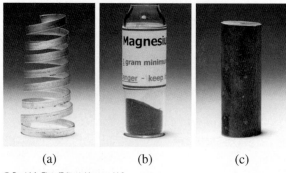

(a)   (b)   (c)

© David A. Tietz/Editorial Image, LLC.

19.138 Use the data in Worked Example 19.5 to determine
graphically the half-life of the reaction.

Answers to In-Chapter Materials

PRACTICE PROBLEMS

19.1A (a) rate $= -\dfrac{\Delta[CO_2]}{\Delta t} = -\dfrac{1}{2}\dfrac{\Delta[H_2O]}{\Delta t} = \dfrac{\Delta[CH_4]}{\Delta t} = \dfrac{1}{2}\dfrac{\Delta[O_2]}{\Delta t}$,

(b) rate $= -\dfrac{1}{3}\dfrac{\Delta[O_2]}{\Delta t} = \dfrac{1}{2}\dfrac{\Delta[O_3]}{\Delta t}$, (c) rate $= -\dfrac{1}{2}\dfrac{\Delta[NO]}{\Delta t}$

$= -\dfrac{\Delta[O_2]}{\Delta t} = \dfrac{1}{2}\dfrac{\Delta[NO_2]}{\Delta t}$. **19.1B** (a) $3CH_4 + 2H_2O + CO_2 \longrightarrow$
$4CH_3OH$, (b) $2N_2O_5 \longrightarrow 2N_2 + 5O_2$, (c) $H_2 + CO + O_2 \longrightarrow$
H_2CO_3. **19.2A** (a) 0.0280 M/s, (b) 0.112 M/s. **19.2B** $2A + 3B \longrightarrow C$.
19.3A rate $= k$ $[S_2O_8^{2-}]$ $[I^-]$, $k = 8.1 \times 10^{-2}/M \cdot$ s. **19.3B** 0.013 M,
8.8×10^{-6} M/s. **19.4A** 75 s. **19.4B** 0.91 M.
19.5A

(graph: y-axis labeled $\ln [C_2H_5I]$ ranging from -2.00 to -1.00; x-axis labeled Time (min) ranging from 0 to 60, showing a downward-sloping straight line through plotted data points)

$k = 1.4 \times 10^{-2}$/min. **19.5B** 0.39 M, 0.34 M, 0.30 M, 0.26 M.
19.6A $t_{1/2} = 272$ s. **19.6B** $k = 4.71 \times 10^{-2}$/h. **19.7A** (a) 13.2 s,
(b) $t_{1/2} = 4.2$ s, 13 s. **19.7B** (a) 0.0208 M, (b) 1.25×10^{-3} M,
(c) 1.79×10^{-4} M. **19.8A** 241 kJ/mol. **19.8B** 0.0655 s^{-1}. **19.9A** $1.0 \times$
10^2 kJ/mol. **19.9B** 2.7×10^{-4}/s. **19.10A** 1.7/s. **19.10B** 7.6×10^{-2}/s.
19.11A (a) $NO_2 + CO \longrightarrow NO + CO_2$, (b) NO_3, (c) step 1 is rate
determining. **19.11B** A $+$ B $\longrightarrow 2C$ (slow), C $+$ B $\longrightarrow$ D.
19.12A The first step is a rapidly established equilibrium. Setting
the rates of forward and reverse reactions equal to each other gives
$k_1[NO][Br_2] = k_{-1}[NOBr_2]$. Solving for $[NOBr_2]$ gives $k_1[NO][Br_2]/k_{-1}$.
Substituting this into the rate law for the rate-determining step, rate $=$
$k_2[NOBr_2][NO]$, gives rate $= (k_1k_2/k_{-1})[NO]^2[Br_2]$ or $k[NO]^2[Br_2]$.
19.12B *Step 1:* $I_2(g) \underset{k_{-1}}{\overset{k_1}{\rightleftharpoons}} 2I(g)$. *Step 2:* $H_2(g) + 2I(g) \longrightarrow 2HI(g)$,
rate $= k[H_2][I_2]$.

SECTION REVIEW

19.3.1 a. **19.3.2** b. **19.4.1** e. **19.4.2** d. **19.4.3** b. **19.4.4** a. **19.4.5** b.
19.5.1 c. **19.5.2** c. **19.5.3** b. **19.5.4** c. **19.6.1** e. **19.6.2** b. **19.7.1** c.
19.7.2 c. **19.7.3** b. **19.7.4** d.

Nuclear Chemistry

© Pallava Bagla/Corbis

NUCLEAR REACTORS have been employed for the production of electricity for a number of decades but there are significant issues facing their continued use. These include nuclear waste production, handling, and storage; as well as fears of limited worldwide uranium sources. Thus, nuclear chemists are now reinvestigating the extended use of thorium as reactor fuel. Shown here are thorium pellets which can be used in a nuclear reactor to produce fissile materials. Some of the advantages of a thorium fuel cycle include safer reactor conditions, better waste management options, and prevention of nuclear weapons proliferation.

Before You Begin, Review These Skills

- Radioactivity [◄◄ Section 2.2]
- Atomic number, mass number, and isotopes [◄◄ Section 2.3]
- Nuclear stability [◄◄ Section 2.4]
- First-order kinetics [◄◄ Section 19.5]

20.1 NUCLEI AND NUCLEAR REACTIONS

With the exception of hydrogen ($_1^1H$), all nuclei contain protons and neutrons. Some nuclei are unstable and undergo *radioactive decay,* emitting particles and/or electromagnetic radiation [◄◄ Section 2.2]. Spontaneous emission of particles or electromagnetic radiation is known as **radioactivity.** All elements having an atomic number greater than 83 are unstable and are therefore radioactive. Polonium-210 ($_{84}^{210}Po$), for example, decays spontaneously to Pb by emitting an α particle.

Another type of nuclear process, known as **nuclear transmutation,** results from the bombardment of nuclei by neutrons, protons, or other nuclei. An example of a nuclear transmutation is the conversion of atmospheric $_7^{14}N$ to $_6^{14}C$ and $_1^1H$, which results when the nitrogen isotope is bombarded by neutrons (from the sun). In some cases, heavier elements are synthesized from lighter elements. This type of transmutation occurs naturally in outer space, but it can also be achieved artificially, as we will see in Section 20.4.

Radioactive decay and nuclear transmutation are *nuclear reactions,* which differ significantly from ordinary chemical reactions. Table 20.1 summarizes the differences.

To discuss nuclear reactions in any depth, we must understand how to write and balance nuclear equations. Writing a nuclear equation differs somewhat from writing equations for chemical reactions. In addition to writing the symbols for the various chemical elements, we must also explicitly indicate the number of subatomic particles in *every* species involved in the reaction.

The symbols for subatomic particles are as follows:

$$\underset{\text{proton}}{_1^1H \text{ or } _1^1p} \quad \underset{\text{neutron}}{_0^1n} \quad \underset{\text{electron}}{_{-1}^0e \text{ or } _{-1}^0\beta} \quad \underset{\text{positron}}{_{+1}^0e \text{ or } _{+1}^0\beta} \quad \underset{\alpha \text{ particle}}{_2^4\alpha \text{ or } _2^4He}$$

In accordance with the notation introduced in Section 2.3, the superscript in each case denotes the mass number (the total number of neutrons and protons present) and the subscript is the atomic number (the number of protons). Thus, the "atomic number" of a proton is 1, because there is one proton present, and the "mass number" is also 1, because there is one proton but no neutrons present. On the other hand, the mass number

TABLE 20.1	Comparison of Chemical Reactions and Nuclear Reactions
Chemical reactions	**Nuclear reactions**
1. Atoms are rearranged by the breaking and forming of chemical bonds.	1. Elements are converted to other elements (or isotopes).
2. Only electrons in atomic or molecular orbitals are involved in the reaction.	2. Protons, neutrons, electrons, and other subatomic particles such as α particles may be involved.
3. Reactions are accompanied by the absorption or release of relatively small amounts of energy.	3. Reactions are accompanied by the absorption or release of tremendous amounts of energy.
4. Rates of reaction are influenced by temperature, pressure, concentration, and catalysts.	4. Rates of reaction normally are not affected by temperature, pressure, or catalysts.

of a neutron is 1, but its atomic number is 0, because there are no protons present. For the electron, the mass number is 0 (there are neither protons nor neutrons present), but the atomic number is −1, because the electron possesses a unit negative charge.

The symbol $_{-1}^{0}e$ represents an electron in or from an atomic orbital. The symbol $_{-1}^{0}\beta$, on the other hand, represents an electron that, although physically identical to any other electron, comes from a nucleus (in a decay process in which a neutron is converted to a proton and an electron) and not from an atomic orbital. The ***positron*** has the same mass as the electron, but bears a charge of +1. The α particle has two protons and two neutrons, so its atomic number is 2 and its mass number is 4.

In balancing any nuclear equation, we must balance the total of all atomic numbers and the total of all mass numbers for the products and reactants. If we know the atomic numbers and mass numbers of all but one of the species in a nuclear equation, we can identify the *unknown* species by applying these rules, as shown in Worked Example 20.1.

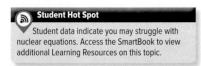

Student Hot Spot

Student data indicate you may struggle with nuclear equations. Access the SmartBook to view additional Learning Resources on this topic.

Student Annotation: An α particle is identical to a helium-4 nucleus and can be represented either as $_{2}^{4}\alpha$ or $_{2}^{4}$He.

Worked Example 20.1

Identify the missing species X in each of the following nuclear equations.

(a) $_{84}^{212}$Po $\longrightarrow$ $_{82}^{208}$Pb + X   (b) $_{38}^{90}$Sr $\longrightarrow$ X + $_{-1}^{0}\beta$ (c) X $\longrightarrow$ $_{8}^{18}$O + $_{+1}^{0}\beta$

Strategy Determine the mass number for the unknown species, X, by summing the mass numbers on both sides of the equation.

$$\Sigma \text{ reactant mass numbers} = \Sigma \text{ product mass numbers}$$

Similarly, determine the atomic number for the unknown species

$$\Sigma \text{ reactant atomic numbers} = \Sigma \text{ product atomic numbers}$$

Use the mass number and atomic number to determine the identity of the unknown species.

Setup (a) 212 = (208 + mass number of X); mass number of X = 4. 84 = (82 + atomic number of X); atomic number of X = 2.

(b) 90 = (mass number of X + 0); mass number of X = 90. 38 = [atomic number of X + (−1)]; atomic number of X = 39.

(c) Mass number of X = (18 + 0); mass number of X = 18. Atomic number of X = (8 + 1); atomic number of X = 9.

Solution (a) X = $_{2}^{4}\alpha$: $_{84}^{212}$Po $\longrightarrow$ $_{82}^{208}$Pb + $_{2}^{4}\alpha$ (b) X = $_{39}^{90}$Y: $_{38}^{90}$Sr $\longrightarrow$ $_{39}^{90}$Y + $_{-1}^{0}\beta$ (c) X = $_{9}^{18}$F: $_{9}^{18}$F $\longrightarrow$ $_{8}^{18}$O + $_{+1}^{0}\beta$

Think About It

The rules of summation that we apply to balance nuclear equations can be thought of as the *conservation of mass number* and the *conservation of atomic number.*

Practice Problem Ⓐ**TTEMPT** Identify X in each of the following nuclear equations.

(a) $_{33}^{78}$As $\longrightarrow$ X + $_{-1}^{0}\beta$ (b) $_{1}^{1}$H + $_{2}^{4}$He $\longrightarrow$ X   (c) $_{100}^{258}$Fm $\longrightarrow$ $_{100}^{257}$Fm + X

Practice Problem Ⓑ**UILD** Identify X in each of the following nuclear equations.

(a) X + $_{-1}^{0}\beta$ $\longrightarrow$ $_{94}^{244}$Pu   (b) $_{92}^{238}$U $\longrightarrow$ X + $_{2}^{4}$He   (c) X $\longrightarrow$ $_{7}^{14}$N + $_{-1}^{0}e$

Practice Problem Ⓒ**ONCEPTUALIZE** For each process, specify the identity of the product.

$$\xleftarrow{\text{electron capture}} \quad ^{222}\text{Rn} \quad \xrightarrow{\text{alpha emission}} \qquad \xleftarrow{\text{electron capture}} \quad ^{132}\text{Cs} \quad \xrightarrow{\text{beta emission}}$$

Section 20.1 Review

Nuclei and Nuclear Reactions

20.1.1 Identify the species X in the following nuclear equation.

$$_{86}^{222}\text{Rn} \longrightarrow X + _{2}^{4}\alpha$$

(a) $_{84}^{226}$Po   (b) $_{88}^{226}$Ra   (c) $_{84}^{212}$Po   (d) $_{84}^{218}$Po   (e) $_{88}^{218}$Ra

20.1.2 Identify the species X in the following nuclear equation.

$$_{8}^{15}\text{O} \longrightarrow X + _{-1}^{0}\beta$$

(a) $_{9}^{15}$F   (b) $_{8}^{14}$O   (c) $_{9}^{16}$F   (d) $_{7}^{15}$N   (e) $_{7}^{14}$N

20.2   NUCLEAR STABILITY

Recall from Chapter 2 that there are certain numbers of nucleons associated with nuclear stability [◄◄ Section 2.4]. Even numbers of protons or neutrons, for example, are associated with greater numbers of stable nuclei than odd numbers. The belt of stability (see Figure 2.9) also shows that the neutron-to-proton ratio is an important factor in nuclear stability. A nucleus that lies outside the belt will increase its stability by undergoing nuclear decay.

Types of Nuclear Decay

Above the belt of stability, the nuclei have higher neutron-to-proton ratios than those within the belt (for the same number of protons). To lower this ratio (and hence move down toward the belt of stability), these nuclei undergo the following process, called *β-particle emission.*

$$_0^1 n \longrightarrow {}_1^1 p + {}_{-1}^0 \beta$$

Beta-particle emission leads to an increase in the number of protons in the nucleus and a simultaneous decrease in the number of neutrons. Some examples are

$$_6^{14} C \longrightarrow {}_7^{14} N + {}_{-1}^0 \beta$$

$$_{19}^{40} K \longrightarrow {}_{20}^{40} Ca + {}_{-1}^0 \beta$$

$$_{40}^{97} Zr \longrightarrow {}_{41}^{97} Nb + {}_{-1}^0 \beta$$

Below the belt of stability, the nuclei have lower neutron-to-proton ratios than those in the belt (for the same number of protons). To increase this ratio (and hence move up toward the belt of stability), these nuclei may emit a positron.

$$_1^1 p \longrightarrow {}_0^1 n + {}_{+1}^0 \beta$$

An example of positron emission is

$$_{19}^{38} K \longrightarrow {}_{18}^{38} Ar + {}_{+1}^0 \beta$$

Alternatively, a nucleus may undergo electron capture.

$$_1^1 p + {}_{-1}^0 e \longrightarrow {}_0^1 n$$

Electron capture is the capture of an electron—usually a 1*s* electron—by the nucleus. The captured electron combines with a proton in the nucleus to form a *neutron* so that the atomic number decreases by 1 while the mass number remains the same. Electron capture has the same net effect on the nucleus as positron emission. Examples of electron capture are

$$_{18}^{37} Ar + {}_{-1}^0 e \longrightarrow {}_{17}^{37} Cl$$

$$_{26}^{55} Fe + {}_{-1}^0 e \longrightarrow {}_{25}^{55} Mn$$

Nuclear Binding Energy

A quantitative measure of nuclear stability is the ***nuclear binding energy,*** which is the energy required to break up a nucleus into its component protons and neutrons. This quantity represents the conversion of mass to energy that occurs during an exothermic nuclear reaction.

The concept of nuclear binding energy evolved from studies of nuclear properties showing that the masses of nuclei are always less than the sum of the masses of the *nucleons,* which is a general term for the protons and neutrons in a nucleus. For example, the $_9^{19}F$ isotope has an atomic mass of 18.99840 amu. The nucleus has 9 protons and 10 neutrons and therefore a total of 19 nucleons. Using the known masses of the proton

(1.00728 amu), the neutron (1.008665 amu), and the electron (5.4858×10^{-4} amu), we can carry out the following analysis. The mass of 9 protons is

$$9 \times 1.00728 \text{ amu} = 9.06552 \text{ amu}$$

and the mass of 9 electrons is

$$9 \times 5.4858 \times 10^{-4} \text{ amu} = 0.0049372 \text{ amu}$$

and the mass of 10 neutrons is

$$10 \times 1.008665 \text{ amu} = 10.08665 \text{ amu}$$

Therefore the atomic mass of an $^{19}_{9}\text{F}$ atom calculated from the known numbers of electrons, protons, and neutrons is

$$9.06552 \text{ amu} + 0.0049372 \text{ amu} + 10.08665 \text{ amu} = 19.15711 \text{ amu}$$

This value is larger than 18.99840 amu (the measured mass of $^{19}_{9}\text{F}$) by 0.15871 amu.

The difference between the mass of an atom and the sum of the masses of its protons, neutrons, and electrons is called the **mass defect.** According to relativity theory, the loss in mass shows up as energy (heat) given off to the surroundings. Thus, the formation of $^{19}_{9}\text{F}$ is *exothermic.* According to *Einstein's mass-energy equivalence relationship* ($E = mc^2$, where E is energy, m is mass, and c is the velocity of light), we can calculate the amount of energy released. We start by writing

Equation 20.1	$\Delta E = (\Delta m)c^2$

where ΔE and Δm are defined as follows:

$$\Delta E = \text{energy of product} - \text{energy of reactants}$$

$$\Delta m = \text{mass of product} - \text{mass of reactants}$$

Thus, the change in mass is

$$\Delta m = 18.99480 \text{ amu} - 19.15711 \text{ amu}$$

$$= -0.15871 \text{ amu}$$

or

$$\Delta m = (-0.15871 \text{ amu})\left(\frac{1 \text{ kg}}{6.0221418 \times 10^{26} \text{ amu}}\right)$$

$$= -2.6354 \times 10^{-28} \text{ kg}$$

Student Annotation: When you apply Einstein's equation, $E = mc^2$, it is important to remember that mass defect must be expressed in kilograms for the units to cancel properly.

$$1 \text{ kg} = 6.0221418 \times 10^{26} \text{ amu}$$

Student Annotation: Remember that joule is a *derived* unit:

$$1 \text{ J} = 1 \text{ kg} \cdot \text{m}^2/\text{s}^2$$

[◄◄ Section 3.1].

Because $^{19}_{9}\text{F}$ has a mass that is less than the mass calculated from the number of electrons and nucleons present, Δm is a negative quantity. Consequently, ΔE is also a negative quantity; that is, energy is released to the surroundings as a result of the formation of the fluorine-19 nucleus. We calculate ΔE as follows:

$$\Delta E = (-2.6354 \times 10^{-28} \text{ kg})(2.99792458 \times 10^8 \text{ m/s})^2$$

$$= -2.3686 \times 10^{-11} \text{ kg} \cdot \text{m}^2/\text{s}^2$$

$$= -2.3686 \times 10^{-11} \text{ J}$$

This is the amount of energy released when one fluorine-19 nucleus is formed from 9 protons and 10 neutrons. The nuclear binding energy of the nucleus is 2.3686×10^{-11} J, which is the amount of energy needed to decompose the nucleus into separate protons and neutrons. In the formation of 1 mole of fluorine nuclei, for instance, the energy released is

$$\Delta E = (-2.3686 \times 10^{-11} \text{ J})(6.0221418 \times 10^{23}/\text{mol})$$

$$= -1.4264 \times 10^{13} \text{ J/mol}$$

$$= -1.4264 \times 10^{10} \text{ kJ/mol}$$

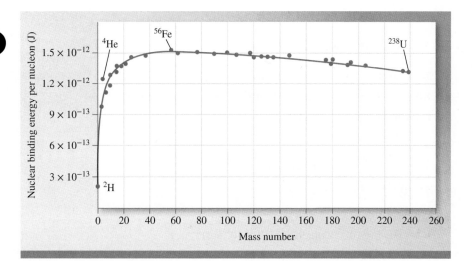

Figure 20.1 Plot of nuclear binding energy per nucleon versus mass number.

The nuclear binding energy, therefore, is 1.4264×10^{10} kJ for 1 mole of fluorine-19 nuclei, which is a tremendously large quantity when we consider that the enthalpies of ordinary chemical reactions are on the order of only 200 kJ. The procedure we have followed can be used to calculate the nuclear binding energy of any nucleus.

Student Annotation: Note that when we report a nuclear binding energy per mole, we give just the magnitude without the negative sign.

As we have noted, nuclear binding energy is an indication of the stability of a nucleus. When comparing the stability of any two nuclei, however, we must account for the fact that they have different numbers of nucleons. It makes more sense, therefore, to compare nuclei using the nuclear binding energy per nucleon:

Student Annotation: In general, the greater the nuclear binding energy per nucleon, the more stable the nucleus.

$$\text{nuclear binding energy per nucleon} = \frac{\text{nuclear binding energy}}{\text{number of nucleons}}$$

For the fluorine-19 nucleus,

$$\text{nuclear binding energy per nucleon} = \frac{2.3686 \times 10^{-11} \text{ J}}{19 \text{ nucleons}}$$

$$= 1.2466 \times 10^{-12} \text{ J/nucleon}$$

The nuclear binding energy per nucleon makes it possible to compare the stability of all nuclei on a common basis. Figure 20.1 shows the variation of nuclear binding energy per nucleon plotted against mass number. As you can see, the curve rises rather steeply. The highest binding energies per nucleon belong to elements with intermediate mass numbers—between 40 and 100—and are greatest for elements in the iron, cobalt, and nickel region (the Group 8B elements) of the periodic table. This means that the net attractive forces among the particles (protons and neutrons) are greatest for the nuclei of these elements.

Nuclear binding energy and nuclear binding energy per nucleon are calculated for an iodine nucleus in Worked Example 20.2.

Worked Example 20.2

The atomic mass of $^{127}_{53}\text{I}$ is 126.904473 amu. Calculate the nuclear binding energy of this nucleus and the corresponding nuclear binding energy per nucleon.

Strategy To calculate the nuclear binding energy, we first determine the difference between the mass of the nucleus and the mass of all the protons and neutrons, which yields the mass defect. Next, we must apply Einstein's mass-energy relationship [$\Delta E = (\Delta m)c^2$].

Solution There are 53 protons and 74 neutrons in the iodine nucleus. The mass of 53 protons is

$$53 \times 1.00728 \text{ amu} = 53.38584 \text{ amu}$$

(Continued on next page)

and the mass of 53 electrons is

$$53 \times 5.4858 \times 10^{-4} \text{ amu} = 0.029075 \text{ amu}$$

and the mass of 74 neutrons is

$$74 \times 1.008665 \text{ amu} = 74.64121 \text{ amu}$$

Therefore, the predicted mass for $^{127}_{53}\text{I}$ is $53.38584 + 0.029075 + 74.64121 = 128.056125$ amu, and the mass defect is

$$\Delta m = 126.904473 \text{ amu} - 128.056125 \text{ amu}$$
$$= -1.1517 \text{ amu}$$
$$= (-1.1517 \text{ amu})\left(\frac{1 \text{ kg}}{6.0221418 \times 10^{26} \text{ amu}}\right)$$
$$= -1.9124 \times 10^{-27} \text{ kg}$$

The energy released is

$$\Delta E = (\Delta m)c^2$$
$$= (-1.9124 \times 10^{-27} \text{ kg})(2.99792458 \times 10^8 \text{ m/s})^2$$
$$= 1.7188 \times 10^{-10} \text{ kg} \cdot \text{m}^2/\text{s}^2$$
$$= 1.7188 \times 10^{-10} \text{ J}$$

Thus the nuclear binding energy is 1.7188×10^{-10} J. The nuclear binding energy per nucleon is obtained as follows:

$$\frac{1.7188 \times 10^{-10} \text{ J}}{127 \text{ nucleons}} = 1.3533 \times 10^{-12} \text{ J/nucleon}$$

Think About It
To minimize rounding error, it is important in calculations such as these to use the *unrounded* values of Avogadro's constant and the speed of light: 6.0221418×10^{26} amu/1 kg and 2.99792458×10^8 m/s, respectively.

Practice Problem Ⓐ**TTEMPT** Calculate the nuclear binding energy (in joules) and the nuclear binding energy per nucleon of $^{209}_{83}\text{Bi}$ (208.980374 amu).

Practice Problem Ⓑ**UILD** The nuclear binding energy for $^{197}_{79}\text{Au}$ is 1.2683×10^{-12} J/nucleon. Determine the mass of a $^{197}_{79}\text{Au}$ atom.

Practice Problem Ⓒ**ONCEPTUALIZE** Which of the graphs [(i)–(iv)] best represents the relationship between energy released (ΔE) and mass defect (Δm)?

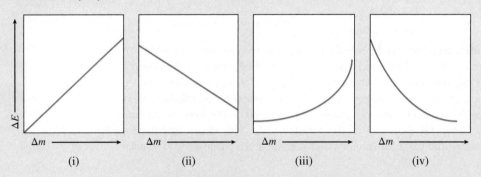

(i)   (ii)   (iii)   (iv)

Section 20.2 Review

Nuclear Stability

20.2.1 Determine the binding energy per nucleon in a ^{238}U nucleus (238.0507847 amu).
(a) 2.891×10^{-10} J/nucleon
(d) 2.823×10^{-10} J/nucleon
(b) 1.215×10^{-12} J/nucleon
(e) 3.212×10^{-24} J/nucleon
(c) 1.186×10^{-12} J/nucleon

20.2.2 What is the energy associated with a mass defect of 2.000 amu?
(a) 3.321×10^{-27} J
(d) 2.989×10^{-10} J
(b) 2.989×10^{-7} J
(e) 3.309×10^{-45} J
(c) 1.800×10^{17} J

20.2.3 What is the change in mass (in kg) for the following reaction? $\Delta H°$ for the reaction is −890.4 kJ/mol.

(a) 1.479×10^{-21} kg   (d) 4.932×10^{-33} kg
(b) 1.479×10^{-24} kg   (e) 9.907×10^{-15} kg
(c) 1.645×10^{-35} kg

20.2.4 What type of radioactive decay will the isotopes ^{13}B and ^{188}Au most likely undergo?
(a) Beta emission, positron emission
(b) Beta emission, beta emission
(c) Positron emission, beta emission
(d) Positron emission, positron emission

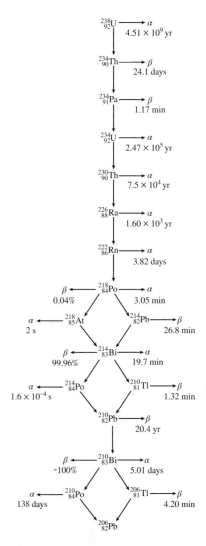

Figure 20.2 Decay series for uranium-238. (Times shown are half-lives.)

20.3 NATURAL RADIOACTIVITY

Nuclei that do not lie within the belt of stability, as well as nuclei with more than 83 protons, tend to be unstable. The spontaneous emission by unstable nuclei of particles or electromagnetic radiation, or both, is known as radioactivity. The main types of radioactivity are the emission of α particles (doubly charged helium nuclei, He^{2+}); the emission of β particles (electrons of nuclear origin); the emission of γ rays, which are very-short-wavelength (0.1 to 104 nm) electromagnetic waves; the emission of positrons; and electron capture.

The disintegration of a radioactive nucleus often is the beginning of a ***radioactive decay series,*** which is a *sequence* of nuclear reactions that ultimately result in the formation of a stable isotope. Figure 20.2 shows the decay series of naturally occurring uranium-238, which involves 14 steps. This decay scheme, known as the *uranium decay series,* also shows the half-lives of all the nuclei involved.

It is important to be able to balance the nuclear reaction for each of the steps in a radioactive decay series. For example, the first step in the uranium decay series is the decay of uranium-238 to thorium-234, with the emission of an α particle. Hence, the reaction is represented by

$$^{238}_{92}U \longrightarrow ^{234}_{90}Th + ^{4}_{2}\alpha$$

The next step is represented by

$$^{234}_{90}Th \longrightarrow ^{234}_{91}Pa + ^{0}_{-1}\beta$$

and so on. In a discussion of radioactive decay steps, the beginning radioactive isotope is called the *parent* and the product isotope is called the *daughter.* Thus, $^{238}_{92}U$ is the parent in the first step of the uranium decay series, and $^{234}_{90}Th$ is the daughter.

Kinetics of Radioactive Decay

All radioactive decays obey first-order kinetics. Therefore, the rate of radioactive decay at any time t is given by

$$\text{rate of decay at time } t = kN$$

where k is the first-order rate constant and N is the number of radioactive nuclei present at time t. According to Equation 19.3, the number of radioactive nuclei at time zero (N_0) and time t (N_t) is

$$\ln \frac{N_t}{N_0} = -kt$$

Student Annotation: Most nuclear scientists (and some general chemistry books) use the symbol λ instead of k for the rate constant of nuclear reactions.

Student Annotation: Although Equation 19.3 uses concentrations of a reactant at times t and 0, it is the *ratio* of the two that is important, so we can also use the *number* of radioactive nuclei in this equation [◀◀ Key Skills Chapter 19].

and the corresponding half-life of the reaction is given by Equation 19.5.

$$t_{1/2} = \frac{0.693}{k}$$

The half-lives and, therefore, the rate constants of radioactive isotopes vary greatly from nucleus to nucleus. Looking at Figure 20.2, for example, we find that $^{238}_{92}U$ and $^{214}_{84}Po$ are two extreme cases.

$$^{238}_{92}U \longrightarrow {}^{234}_{90}Th + {}^{4}_{2}\alpha \qquad t_{1/2} = 4.51 \times 10^9 \text{ yr}$$

$$^{214}_{84}Po \longrightarrow {}^{210}_{82}Pb + {}^{4}_{2}\alpha \qquad t_{1/2} = 1.6 \times 10^{-4} \text{ s}$$

These two rate constants, after conversion to the same time unit, differ by many orders of magnitude. Furthermore, the rate constants are unaffected by changes in environmental conditions such as temperature and pressure. These highly unusual features are not seen in ordinary chemical reactions (see Table 20.1).

Dating Based on Radioactive Decay

The half-lives of radioactive isotopes have been used as "atomic clocks" to determine the ages of certain objects. Some examples of dating by radioactive decay measurements will be described here.

The carbon-14 isotope is produced when atmospheric nitrogen is bombarded by cosmic rays.

$$^{14}_{7}N + {}^{1}_{0}n \longrightarrow {}^{14}_{6}C + {}^{1}_{1}H$$

The radioactive carbon-14 isotope decays according to the equation

$$^{14}_{6}C \longrightarrow {}^{14}_{7}N + {}^{0}_{-1}\beta$$

This reaction is the basis of radiocarbon or "carbon-14" dating. To determine the age of an object, we measure the *activity* (disintegrations per second) of ^{14}C and compare it to the activity of ^{14}C in living matter.

Worked Example 20.3 shows how to use radiocarbon dating to determine the age of an artifact.

Worked Example 20.3

A wooden artifact is found to have a ^{14}C activity of 9.1 disintegrations per second. Given that the ^{14}C activity of an equal mass of fresh-cut wood has a constant value of 15.2 disintegrations per second, determine the age of the artifact. The half-life of carbon-14 is 5715 years.

Strategy The activity of a radioactive sample is proportional to the number of radioactive nuclei. Thus, we can use Equation 19.3 with activity in place of concentration.

$$\ln \frac{{}^{14}C \text{ activity in artifact}}{{}^{14}C \text{ activity in fresh-cut wood}} = -kt$$

To determine k, though, we must solve Equation 19.5, using the value of $t_{1/2}$ for carbon-14 (5715 years) given in the problem statement.

Setup Solving Equation 19.5 for k gives

$$k = \frac{0.693}{5715 \text{ yr}} = 1.21 \times 10^{-4} \text{ yr}^{-1}$$

Solution

$$\ln \frac{9.1 \text{ distintegrations per second}}{15.2 \text{ disintegrations per second}} = -1.21 \times 10^{-4} \text{ yr}^{-1} (t)$$

$$t = \frac{-0.513}{-1.21 \times 10^{-4} \text{ yr}^{-1}} = 4240 \text{ yr}$$

Therefore, the age of the artifact is 4.2×10^3 years.

Think About It

Carbon dating cannot be used for objects older than about 60,000 years (about 10 half-lives). After that much time has passed, the activity of carbon-14 has fallen to a level too low to be measured reliably.

Practice Problem **A**TTEMPT A piece of linen cloth found at an ancient burial site is found to have a ^{14}C activity of 4.8 disintegrations per minute. Determine the age of the cloth. Assume that the carbon-14 activity of an equal mass of living flax (the plant from which linen is made) is 14.8 disintegrations per minute.

Practice Problem **B**UILD What would be the ^{14}C activity in a 2500-year-old wooden object? Assume that the ^{14}C activity of an equal mass of fresh-cut wood is 13.9 disintegrations per second.

Practice Problem **C**ONCEPTUALIZE The Think About It box in Worked Example 20.3 explains why carbon dating cannot be used to date objects that are older than 60,000 years. Explain why it also cannot be used to date objects that are only a few years old.

Because some of the intermediate products in the uranium decay series have very long half-lives (see Figure 20.2), this series is particularly suitable for estimating the age of rocks found on Earth and of extraterrestrial objects. The half-life for the first step ($^{238}_{92}$U to $^{234}_{90}$Th) is 4.51×10^9 years. This is about 20,000 times the second largest value (i.e., 2.47×10^5 years), which is the half-life for $^{234}_{92}$U to $^{230}_{90}$Th. As a good approximation, therefore, we can assume that the half-life for the *overall* process (i.e., from $^{238}_{92}$U to $^{206}_{82}$Pb) is equal to the half-life of the first step.

$$^{238}_{92}\text{U} \longrightarrow {}^{206}_{82}\text{Pb} + 8\,{}^4_2\alpha + 6\,{}^{\,0}_{-1}\beta \qquad t_{1/2} = 4.51 \times 10^9 \text{ yr}$$

In naturally occurring uranium minerals, we should and do find some lead-206 formed by radioactive decay. Assuming that no lead was present when the mineral was formed and that the mineral has not undergone chemical changes that would allow the lead-206 isotope to be separated from the parent uranium-238, it is possible to estimate the age of the rocks from the mass ratio of $^{206}_{82}$Pb to $^{238}_{92}$U. According to the preceding nuclear equation, 1 mol (206 g) of lead is formed for every 1 mol (238 g) of uranium that undergoes complete decay. If only half a mole of uranium-238 has undergone decay, the mass ratio ^{206}Pb/^{238}U becomes

$$\frac{206 \text{ g}/2}{238 \text{ g}/2} = 0.866$$

and the process would have taken a half-life of 4.51×10^9 years to complete (Figure 20.3). Ratios lower than 0.866 mean that the rocks are less than 4.51×10^9 years old, and higher ratios suggest a greater age. Interestingly, studies based on the uranium series, as well as other decay series, put the age of the oldest rocks and, therefore, probably the age of Earth itself, at 4.5×10^9, or 4.5 billion years.

One of the most important dating techniques in geochemistry is based on the radioactive decay of potassium-40. Radioactive potassium-40 decays by several different modes, but the one relevant for dating is that of electron capture.

$$^{40}_{19}\text{K} + {}^{\,0}_{-1}e \longrightarrow {}^{40}_{18}\text{Ar} \qquad t_{1/2} = 1.2 \times 10^9 \text{ yr}$$

The accumulation of gaseous argon-40 is used to gauge the age of a specimen. When a potassium-40 atom in a mineral decays, argon-40 is trapped in the lattice of the mineral and can escape only if the material is melted. Melting, therefore, is the procedure for analyzing a mineral sample in the laboratory. The amount of argon-40 present can be conveniently measured with a mass spectrometer. Knowing the ratio of argon-40 to potassium-40 in the mineral and the half-life of decay makes it possible to establish the ages of rocks ranging from millions to billions of years old.

Worked Example 20.4 shows how to use radioisotopes to determine the age of a specimen.

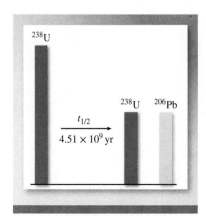

Figure 20.3 After one half-life, half of the original uranium-238 has been converted to lead-206.

Worked Example 20.4

A rock is found to contain 5.51 mg of ^{238}U and 1.63 mg of ^{206}Pb. Determine the age of the rock ($t_{1/2}$ of ^{238}U = 4.51 × 10^9 yr).

Strategy We must first determine what mass of ^{238}U decayed to produce the measured amount of ^{206}Pb and then use it to determine the original mass of ^{238}U. Knowing the initial and final masses of ^{238}U, we can use Equation 19.3 to solve for t.

Setup

$$1.63 \text{ mg } ^{206}\text{Pb} \times \frac{238 \text{ mg } ^{238}\text{U}}{206 \text{ mg } ^{206}\text{Pb}} = 1.88 \text{ mg } ^{238}\text{U}$$

Thus, the original mass of ^{238}U was 5.51 mg + 1.88 mg = 7.39 mg. The rate constant, k, is determined using Equation 19.5 and $t_{1/2}$ for ^{238}U.

$$k = \frac{0.693}{4.51 \times 10^9} = 1.54 \times 10^{-10} \text{ yr}^{-1}$$

Solution

$$\ln \frac{5.51 \text{ mg}}{7.39 \text{ mg}} = -1.54 \times 10^{-10} \text{ yr}^{-1} \, (t)$$

$$t = \frac{-0.294}{-1.54 \times 10^{-10} \text{ yr}^{-1}} = 1.91 \times 10^9 \text{ yr}$$

The rock is 1.9 billion years old.

Think About It

This is slightly more complicated than the radiocarbon problem. We cannot use the measured masses of the two isotopes in Equation 19.3 because they are masses of different elements.

Practice Problem **A**TTEMPT Determine the age of a rock that contains 12.75 mg of ^{238}U and 1.19 mg of ^{206}Pb.

Practice Problem **B**UILD How much ^{206}Pb will be in a rock sample that is 1.3 × 10^8 years old and that contains 3.25 mg of ^{238}U?

Practice Problem **C**ONCEPTUALIZE Isotope X decays to isotope Y with a half-life of 45 days. Which diagram most closely represents the sample of X after 105 days?

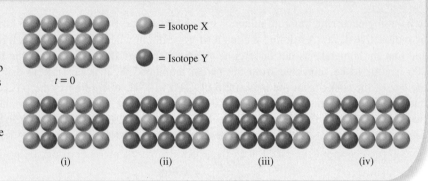

(i)   (ii)   (iii)   (iv)

Section 20.3 Review

Natural Radioactivity

20.3.1 Determine the age of a rock found to contain 4.31 mg ^{238}U and 2.47 mg of ^{206}Pb.
(a) 7.19 × 10^9 years   (c) 1.18 × 10^8 years   (e) 1.78 × 10^8 years
(b) 3.62 × 10^9 years   (d) 3.30 × 10^9 years

20.3.2 Determine the ^{14}C activity in disintegrations per second (dps) of a wooden artifact that is 23,000 years old. (Assume that fresh-cut wood of equal mass has an activity of 15.2 dps.)
(a) 16.3 dps   (c) 0.935 dps   (e) 4.31 dps
(b) 15.2 dps   (d) 11.5 dps

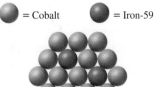

= Cobalt   = Iron-59

20.3.3 Iron-59 decays to cobalt via beta emission with a half-life of 45.1 days. From the diagram, determine how many half-lives have elapsed since the sample was pure iron-59.
(a) 2   (c) 0.25   (e) 14
(b) 3   (d) 8

20.4 NUCLEAR TRANSMUTATION

The scope of nuclear chemistry would be rather narrow if study were limited to natural radioactive elements. An experiment performed by Rutherford in 1919, however, suggested the possibility of producing radioactivity artificially. When he bombarded a sample of nitrogen with α particles, the following reaction took place.

$$^{14}_{7}\text{N} + ^{4}_{2}\alpha \longrightarrow ^{17}_{8}\text{O} + ^{1}_{1}\text{p}$$

An oxygen-17 isotope was produced with the emission of a proton. This reaction demonstrated for the first time the feasibility of converting one element into another, by the process of nuclear transmutation. Nuclear transmutation differs from radioactive decay in that transmutation is brought about by the *collision* of two particles.

The preceding reaction can be abbreviated as $^{14}_{7}\text{N}(\alpha,\text{p})^{17}_{8}\text{O}$. In the parentheses the bombarding particle is written first, followed by the emitted particle.

Worked Example 20.5 shows how to use this notation to represent nuclear transmutations.

Worked Example 20.5

Write the balanced nuclear equation for the reaction represented by $^{56}_{26}\text{Fe}(\text{d},\alpha)^{54}_{25}\text{Mn}$, where d represents a deuterium nucleus.

Strategy The species written first is a reactant. The species written last is a product. Within the parentheses, the bombarding particle (a reactant) is written first, followed by the emitted particle (a product).

Setup The bombarding and emitted particles are represented by $^{2}_{1}\text{H}$ and $^{4}_{2}\alpha$, respectively.

Solution

$$^{56}_{26}\text{Fe} + ^{2}_{1}\text{H} \longrightarrow ^{54}_{25}\text{Mn} + ^{4}_{2}\alpha$$

Think About It
Check your work by summing the mass numbers and the atomic numbers on both sides of the equation.

Practice Problem (A)TTEMPT Write an equation for the process represented by $^{106}_{46}\text{Pd}(\alpha,\text{p})^{109}_{47}\text{Ag}$.

Practice Problem (B)UILD Write the abbreviated form of the following process.

$$^{33}_{17}\text{Cl} + ^{1}_{0}\text{n} \longrightarrow ^{31}_{15}\text{P} + ^{3}_{2}\text{He}$$

Practice Problem (C)ONCEPTUALIZE One of the major aspirations of alchemy, the historical precursor to chemistry, was the transmutation of common metals into gold—although transmutation as we know it was not a process that was known to alchemists. Explain why we still mine gold today, despite the fact that it can be produced via transmutation of mercury.

Particle accelerators made it possible to synthesize the so-called *transuranium elements,* with atomic numbers greater than 92. Neptunium ($Z = 93$) was first prepared in 1940. Since then, 26 other transuranium elements have been synthesized. All isotopes of these elements are radioactive. Table 20.2 lists the transuranium elements that have been reported and some of the reactions by which they have been produced.

Although light elements are generally not radioactive, they can be made so by bombarding their nuclei with appropriate particles. As we saw in the previous section, the radioactive carbon-14 isotope can be prepared by bombarding nitrogen-14 with neutrons. Tritium ($^{3}_{1}\text{H}$) is prepared according to the following bombardment.

$$^{6}_{3}\text{Li} + ^{1}_{0}\text{n} \longrightarrow ^{3}_{1}\text{H} + ^{4}_{2}\alpha$$

Tritium decays with the emission of β particles.

$$^{3}_{1}\text{H} \longrightarrow ^{3}_{2}\text{He} + ^{0}_{-1}\beta \qquad t_{1/2} = 12.5 \text{ yr}$$

TABLE 20.2	Preparation of the Transuranium Elements		
Atomic number	**Name**	**Symbol**	**Preparation***
93	Neptunium	Np	$^{238}_{92}U + ^{1}_{0}n \longrightarrow ^{239}_{93}Np + ^{0}_{-1}\beta$
94	Plutonium	Pu	$^{239}_{93}Np \longrightarrow ^{239}_{94}Pu + ^{0}_{-1}\beta$
95	Americium	Am	$^{239}_{94}Pu + ^{1}_{0}n \longrightarrow ^{240}_{95}Am + ^{0}_{-1}\beta$
96	Curium	Cm	$^{239}_{94}Pu + ^{4}_{2}\alpha \longrightarrow ^{242}_{96}Cm + ^{1}_{0}n$
97	Berkelium	Bk	$^{241}_{95}Am + ^{4}_{2}\alpha \longrightarrow ^{243}_{97}Bk + 2^{1}_{0}n$
98	Californium	Cf	$^{242}_{96}Cm + ^{4}_{2}\alpha \longrightarrow ^{245}_{98}Cf + ^{1}_{0}n$
99	Einsteinium	Es	$^{238}_{92}U + 15^{1}_{0}n \longrightarrow ^{253}_{99}Es + 7^{0}_{-1}\beta$
100	Fermium	Fm	$^{238}_{92}U + 17^{1}_{0}n \longrightarrow ^{255}_{100}Fm + 8^{0}_{-1}\beta$
101	Mendelevium	Md	$^{253}_{99}Es + ^{4}_{2}\alpha \longrightarrow ^{256}_{101}Md + ^{1}_{0}n$
102	Nobelium	No	$^{246}_{96}Cm + ^{12}_{6}C \longrightarrow ^{254}_{102}No + 4^{1}_{0}n$
103	Lawrencium	Lr	$^{252}_{98}Cf + ^{10}_{5}B \longrightarrow ^{257}_{103}Lr + 5^{1}_{0}n$
104	Rutherfordium	Rf	$^{249}_{98}Cf + ^{12}_{6}C \longrightarrow ^{257}_{104}Rf + 4^{1}_{0}n$
105	Dubnium	Db	$^{249}_{98}Cf + ^{15}_{7}N \longrightarrow ^{260}_{105}Db + 4^{1}_{0}n$
106	Seaborgium	Sg	$^{249}_{98}Cf + ^{18}_{8}O \longrightarrow ^{263}_{106}Sg + 4^{1}_{0}n$
107	Bohrium	Bh	$^{209}_{83}Bi + ^{54}_{24}Cr \longrightarrow ^{262}_{107}Bh + ^{1}_{0}n$
108	Hassium	Hs	$^{208}_{82}Pb + ^{58}_{26}Fe \longrightarrow ^{265}_{108}Hs + ^{1}_{0}n$
109	Meitnerium	Mt	$^{209}_{83}Bi + ^{58}_{26}Fe \longrightarrow ^{266}_{109}Mt + ^{1}_{0}n$
110	Darmstadtium	Ds	$^{208}_{82}Pb + ^{62}_{28}Ni \longrightarrow ^{269}_{110}Ds + ^{1}_{0}n$
111	Roentgenium	Rg	$^{209}_{83}Bi + ^{64}_{28}Ni \longrightarrow ^{272}_{111}Rg + ^{1}_{0}n$
112	Copernicium	Cn	$^{208}_{82}Pb + ^{70}_{30}Zn \longrightarrow ^{277}_{112}Uub + ^{1}_{0}n$
113	Nihonium	Nh	$^{288}_{115}Mc \longrightarrow ^{284}_{113}Nh + ^{4}_{2}\alpha$
114	Flerovium	Fl	$^{244}_{94}Pu + ^{48}_{20}Ca \longrightarrow ^{289}_{114}Fl + 3^{1}_{0}n$
115	Moscovium	Mc	$^{243}_{95}Am + ^{48}_{20}Ca \longrightarrow ^{288}_{115}Mc + 3^{1}_{0}n$
116	Livermorium	Lv	$^{248}_{96}Cm + ^{48}_{20}Ca \longrightarrow ^{292}_{116}Lv + 4^{1}_{0}n$
117	Tennessine	Ts	$^{48}_{20}Ca + ^{249}_{97}Bk \longrightarrow ^{297}_{117}Ts + 4^{1}_{0}n$
118	Oganesson	Og	$^{249}_{98}Cf + ^{48}_{20}Ca \longrightarrow ^{294}_{118}Og + 3^{1}_{0}n$

*Some of the transuranium elements have been prepared by more than one method.

Many synthetic isotopes are prepared by using neutrons as projectiles. This approach is particularly convenient because neutrons carry no charges and therefore are not repelled by the targets—the nuclei. In contrast, when the projectiles are positively charged particles (e.g., protons or α particles), they must have considerable kinetic energy to overcome the electrostatic repulsion between themselves and the target nuclei. The synthesis of phosphorus from aluminum is one example.

$$^{27}_{13}Al + ^{4}_{2}\alpha \longrightarrow ^{30}_{15}P + ^{1}_{0}n$$

A *particle accelerator* uses electric and magnetic fields to increase the kinetic energy of charged species so that a reaction will occur (Figure 20.4). Alternating the polarity (i.e., + and −) on specially constructed plates causes the particles to accelerate along a spiral path. When they have the energy necessary to initiate the desired nuclear reaction, they are guided out of the accelerator into a collision with a target substance.

Various designs have been developed for particle accelerators, one of which accelerates particles along a linear path of about 3 km (Figure 20.5). It is now possible to accelerate particles to a speed well above 90 percent of the speed of light. (According to Einstein's theory of relativity, it is impossible for a particle to move *at* the speed of light. The only exception is the photon, which has a zero rest mass.) The

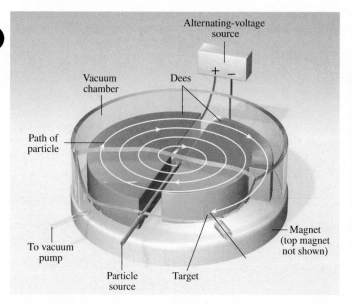

Figure 20.4 Schematic diagram of a cyclotron particle accelerator. The particle (an ion) to be accelerated starts at the center and is forced to move in a spiral path through the influence of electric and magnetic fields until it emerges at a high velocity. The magnetic fields are perpendicular to the plane of the dees (so-called because of their shape), which are hollow and serve as electrodes.

Figure 20.5 Section of a particle accelerator.
© Nick Wall/Science Source

extremely energetic particles produced in accelerators are employed by physicists to smash atomic nuclei to fragments. Studying the debris from such disintegrations provides valuable information about nuclear structure and binding forces.

Section 20.4 Review

Nuclear Transmutation

20.4.1 Identify the balanced nuclear equation for the reaction represented by $^{98}_{42}\text{Mo}(d,n)^{99}_{43}\text{Tc}$.

(a) $^{98}_{42}\text{Mo} + ^{1}_{1}\text{H} \longrightarrow ^{99}_{43}\text{Tc}$

(b) $^{98}_{42}\text{Mo} \longrightarrow ^{99}_{43}\text{Tc} + ^{2}_{1}\text{H} + ^{1}_{0}\text{n}$

(c) $^{98}_{42}\text{Mo} + ^{1}_{0}\text{n} \longrightarrow ^{99}_{43}\text{Tc} + ^{2}_{1}\text{H}$

(d) $^{98}_{42}\text{Mo} + ^{2}_{1}\text{H} + ^{1}_{0}\text{n} \longrightarrow ^{99}_{43}\text{Tc}$

(e) $^{98}_{42}\text{Mo} + ^{2}_{1}\text{H} \longrightarrow ^{99}_{43}\text{Tc} + ^{1}_{0}\text{n}$

20.4.2 Write the correct abbreviated form of the equation

$$^{125}_{53}\text{I} + ^{0}_{-1}\beta \longrightarrow ^{125}_{52}\text{Te} + \gamma$$

(a) $^{125}_{53}\text{I}(\gamma, \beta)^{125}_{52}\text{Te}$

(b) $^{125}_{53}\text{I}(\beta, \gamma)^{125}_{52}\text{Te}$

(c) $^{125}_{52}\text{Te}(\beta, \gamma)^{125}_{53}\text{I}$

(d) $^{125}_{52}\text{I}(\beta, \gamma)^{125}_{53}\text{Te}$

(e) $^{125}_{52}\text{I}(\gamma, \beta)^{125}_{53}\text{Te}$

20.5 NUCLEAR FISSION

Nuclear fission is the process in which a heavy nucleus (mass number > 200) divides to form smaller nuclei of intermediate mass and one or more neutrons. Because the heavy nucleus is less stable than its products (see Figure 20.1), this process releases a large amount of energy.

The first nuclear fission reaction to be studied was that of uranium-235 bombarded with slow neutrons, whose speed is comparable to that of air molecules at room temperature. Under these conditions, uranium-235 undergoes fission, as shown in Figure 20.6. Actually, this reaction is very complex: more than 30 different

Figure 20.6

Nuclear Fission and Fusion

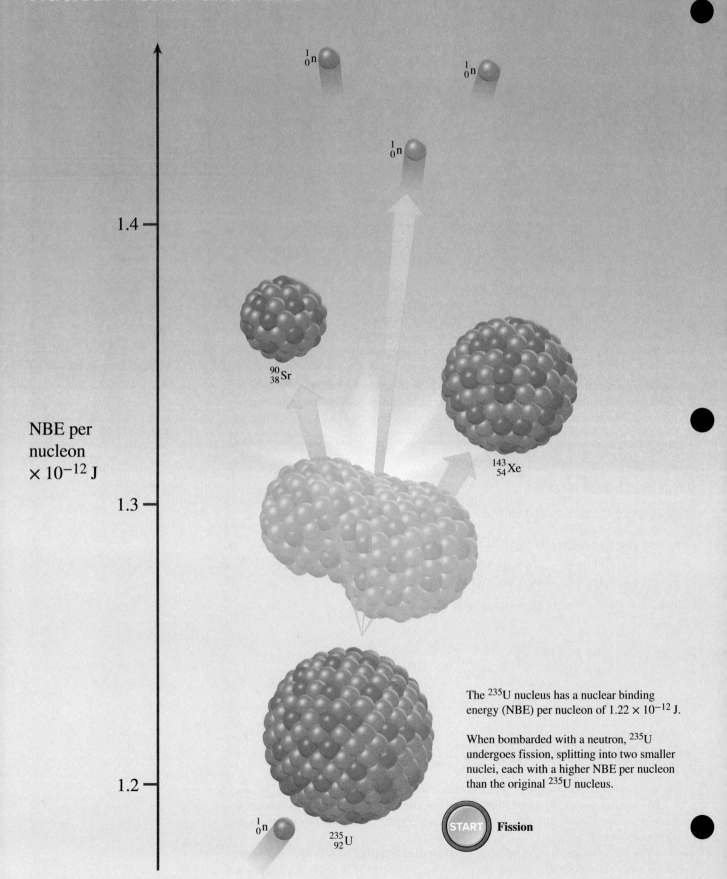

NBE per nucleon $\times 10^{-12}$ J

$^{1}_{0}n$

$^{1}_{0}n$

$^{1}_{0}n$

$^{90}_{38}Sr$

$^{143}_{54}Xe$

1.4

1.3

1.2

$^{1}_{0}n$

$^{235}_{92}U$

The ^{235}U nucleus has a nuclear binding energy (NBE) per nucleon of 1.22×10^{-12} J.

When bombarded with a neutron, ^{235}U undergoes fission, splitting into two smaller nuclei, each with a higher NBE per nucleon than the original ^{235}U nucleus.

START **Fission**

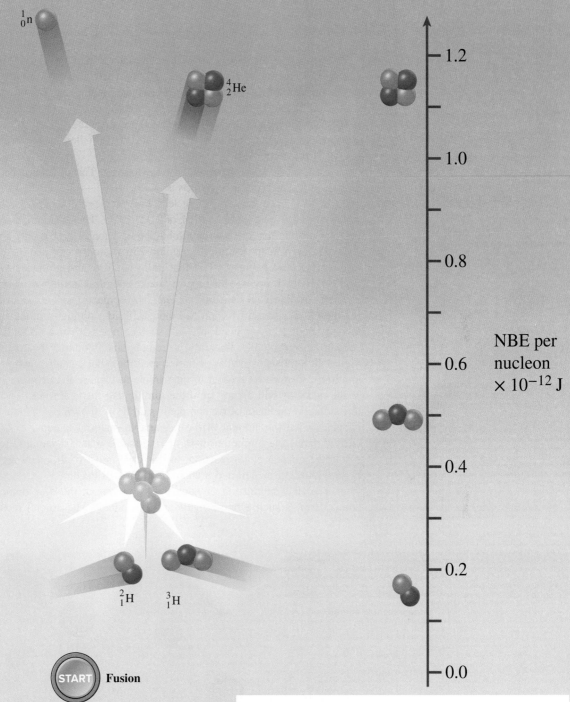

$^{1}_{0}n$

$^{4}_{2}He$

NBE per
nucleon
$\times 10^{-12}$ J

1.2

1.0

0.8

0.6

0.4

0.2

0.0

$^{2}_{1}H$ $^{3}_{1}H$

START **Fusion**

The ^{2}H and ^{3}H nuclei have the following
nuclear binding energies per nucleon:

^{2}H: 0.185×10^{-12} J
^{3}H: 0.451×10^{-12} J

At very high temperatures, the ^{2}H and ^{3}H nuclei
undergo fusion to produce a ^{4}He nucleus and a
neutron. The ^{4}He nucleus has a significantly
higher NBE per nucleon: 1.13×10^{-12} J.

What's the point?

Large nuclei, such as ^{235}U, can achieve greater nuclear stability
by splitting into smaller nuclei with greater NBE per nucleon.
Small nuclei achieve stability by undergoing fusion to produce
a larger nucleus with a greater NBE per nucleon. Note that
different scales are used to show the change in NBE per nucleon
for the two processes. There is a much greater change in NBE
per nucleon in the fusion process than in the fission process. As
with chemical reactions, nuclear reactions are favored when the
products are more stable than the reactants.

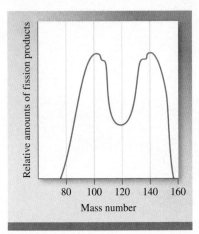

Figure 20.7 Relative yields of the products resulting from the fission of ^{235}U as a function of mass number.

Video 20.2
Nuclear chemistry—nuclear chain reaction.

elements have been found among the fission products (Figure 20.7). A representative reaction is

$$^{235}_{92}U + ^{1}_{0}n \longrightarrow ^{90}_{38}Sr + ^{143}_{54}Xe + 3^{1}_{0}n$$

Although many heavy nuclei can be made to undergo fission, only the fission of naturally occurring uranium-235 and of the artificial isotope plutonium-239 has any practical importance. Table 20.3 lists the nuclear binding energies of uranium-235 and its fission products. As the table shows, the binding energy per nucleon for uranium-235 is less than the sum of the binding energies for strontium-90 and xenon-143. Therefore, when a uranium-235 nucleus is split into two smaller nuclei, a certain amount of energy is released.

Let's estimate the magnitude of this energy. The difference between the binding energies of the reactants and products is $(1.23 \times 10^{-10} + 1.92 \times 10^{-10})$ J $- (2.82 \times 10^{-10})$ J, or 3.3×10^{-11} J per uranium-235 nucleus. For 1 mole of uranium-235, the energy released would be $(3.3 \times 10^{-11})(6.02 \times 10^{23})$, or 2.0×10^{13} J. This is an *extremely* exothermic reaction, considering that the heat of combustion of 1 ton of coal is only about 5×10^{7} J.

The significant feature of uranium-235 fission is not just the enormous amount of energy released, but the fact that more neutrons are produced than are originally captured in the process. This property makes possible a ***nuclear chain reaction,*** which is a self-sustaining sequence of nuclear fission reactions. The neutrons generated during the initial stages of fission can induce fission in other uranium-235 nuclei, which in turn produce more neutrons, and so on. In less than a second, the reaction can become uncontrollable, liberating a tremendous amount of heat to the surroundings. Figure 20.8 shows two types of fission reactions. For a chain reaction to occur, enough uranium-235 must be present in the sample to capture the neutrons. Otherwise, many of the neutrons will escape from the sample and the chain reaction will not occur. In this situation the mass of the sample is said to be *subcritical*. Figure 20.8 shows what happens when the amount of the fissionable material is equal to or greater than the ***critical mass,*** the minimum mass of fissionable material required to generate a self-sustaining nuclear chain reaction. In this case, most of the neutrons will be captured by uranium-235 nuclei, and a chain reaction will occur.

The first application of nuclear fission was in the development of the atomic bomb. How is such a bomb made and detonated? The crucial factor in the bomb's

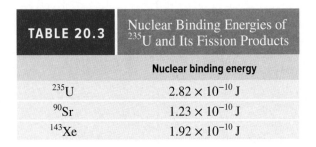

TABLE 20.3	Nuclear Binding Energies of ^{235}U and Its Fission Products
	Nuclear binding energy
^{235}U	2.82×10^{-10} J
^{90}Sr	1.23×10^{-10} J
^{143}Xe	1.92×10^{-10} J

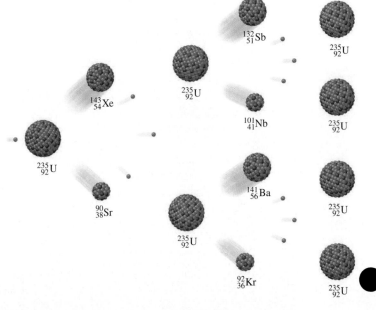

Figure 20.8 If a critical mass is present, many of the neutrons emitted during the fission process will be captured by other ^{235}U nuclei and a chain reaction will occur.

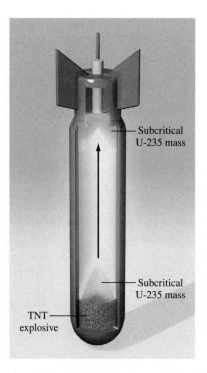

Figure 20.9 Schematic diagram of an atomic bomb. The TNT explosives are set off first. The explosion forces the sections of fissionable material together to form an amount considerably larger than the critical mass.

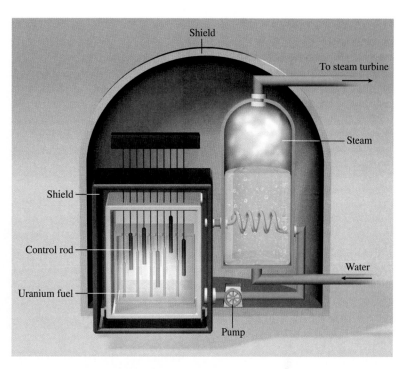

Figure 20.10 Schematic diagram of a nuclear fission reactor. The fission process is controlled by cadmium or boron rods. The heat generated by the process is used to produce steam for the generation of electricity via a heat exchange system.

design is the determination of the critical mass for the bomb. A small atomic bomb is equivalent to 20,000 tons of TNT (trinitrotoluene). Because 1 ton of TNT releases about 4×10^9 J of energy, 20,000 tons would produce 8×10^{13} J. Recall that 1 mole, or 235 g, of uranium-235 liberates 2.0×10^{13} J of energy when it undergoes fission. Thus, the mass of the isotope present in a small bomb must be at least

$$(235 \text{ g}) \left(\frac{8 \times 10^{13} \text{ J}}{2.0 \times 10^{13} \text{ J}} \right) \approx 1 \text{ kg}$$

An atomic bomb is never assembled with the critical mass already present. Instead, the critical mass is formed by using a conventional explosive, such as TNT, to force the fissionable sections together, as shown in Figure 20.9. Neutrons from a source at the center of the device trigger the nuclear chain reaction. Uranium-235 was the fissionable material in the bomb dropped on Hiroshima, Japan, on August 6, 1945. Plutonium-239 was used in the bomb exploded over Nagasaki three days later. The fission reactions generated were similar in these two cases, as was the extent of the destruction.

A peaceful but controversial application of nuclear fission is the generation of electricity using heat from a controlled chain reaction in a nuclear reactor. Currently, nuclear reactors provide about 20 percent of the electric energy in the United States. This is a small but by no means negligible contribution to the nation's energy production. Several different types of nuclear reactors are in operation; we will briefly discuss the main features of three of them, along with their advantages and disadvantages.

Most of the nuclear reactors in the United States are *light water reactors.* Figure 20.10 is a schematic diagram of such a reactor, and Figure 20.11 shows the refueling process in the core of a nuclear reactor.

An important aspect of the fission process is the speed of the neutrons. Slow neutrons split uranium-235 nuclei more efficiently than do fast ones. Because fission reactions are highly exothermic, the neutrons produced usually move at high velocities. For greater efficiency, they must be slowed down before they can be used to induce nuclear disintegration. To accomplish this goal, scientists use **moderators,** which are substances that can reduce the kinetic energy of neutrons. A good moderator

Video 20.3
Nuclear chemistry—nuclear fission.

Figure 20.11 Refueling the core of a nuclear reactor.
© Vanderlei Almeida/AFP/Getty

must satisfy several requirements: It should be nontoxic and inexpensive (as very large quantities of it are necessary), and it should resist conversion into a radioactive substance by neutron bombardment. Furthermore, it is advantageous for the moderator to be a fluid so that it can also be used as a coolant. No substance fulfills all these requirements, although water comes closer than many others that have been considered. Nuclear reactors that use light water (H_2O) as a moderator are called *light* water reactors because $_1^1H$ is the lightest isotope of the element hydrogen.

The nuclear fuel consists of uranium, usually in the form of its oxide, U_3O_8. Naturally occurring uranium contains about 0.7 percent of the uranium-235 isotope, which is too low a concentration to sustain a small-scale chain reaction. For effective operation of a light water reactor, uranium-235 must be enriched to a concentration of 3 or 4 percent. In principle, the main difference between an atomic bomb and a nuclear reactor is that the chain reaction that takes place in a nuclear reactor is kept under control at all times. The factor limiting the rate of the reaction is the number of neutrons present. This can be controlled by lowering cadmium or boron *control rods* between the fuel elements. These rods capture neutrons according to the equations

$$_{48}^{113}\text{Cd} + _0^1\text{n} \longrightarrow _{48}^{114}\text{Cd} + \gamma$$

$$_5^{10}\text{B} + _0^1\text{n} \longrightarrow _3^7\text{Li} + _2^4\alpha$$

where γ denotes gamma rays. Without the control rods, the reactor core would melt from the heat generated and release radioactive materials into the environment. Nuclear reactors have rather elaborate cooling systems that absorb the heat given off by the nuclear reaction and transfer it outside the reactor core, where it is used to produce enough steam to drive an electric generator. In this respect, a nuclear power plant is similar to a conventional power plant that burns fossil fuel. In both cases, large quantities of cooling water are needed to condense steam for reuse. Thus, most nuclear power plants are built near a river or a lake. Unfortunately, this method of cooling causes thermal pollution.

Another type of nuclear reactor uses D_2O, or *heavy* water, as the moderator, rather than H_2O. Deuterium absorbs neutrons much less efficiently than does ordinary hydrogen. Because fewer neutrons are absorbed, the reactor is more efficient and does not require enriched uranium. More neutrons leak out of the reactor, too, though this is not a serious disadvantage.

The main advantage of a heavy water reactor is that it eliminates the need for building expensive uranium enrichment facilities. However, D_2O must be prepared by either fractional distillation or electrolysis of ordinary water, which can be very expensive considering the amount of water used in a nuclear reactor. In countries where hydroelectric power is abundant, the cost of producing D_2O by electrolysis can be reasonably low. At present, Canada is the only nation successfully using heavy water nuclear reactors. The fact that no enriched uranium is required in a heavy water reactor allows a country to enjoy the benefits of nuclear power without undertaking work that is closely associated with weapons technology.

A ***breeder reactor*** uses uranium fuel, but unlike a conventional nuclear reactor, it produces more fissionable materials than it uses.

When uranium-238 is bombarded with fast neutrons, the following reactions take place.

$$^{238}_{92}U + {}^1_0n \longrightarrow {}^{239}_{92}U$$

$$^{239}_{92}U \longrightarrow {}^{239}_{93}Np + {}^0_{-1}\beta \qquad t_{1/2} = 23.4 \text{ min}$$

$$^{239}_{93}Np \longrightarrow {}^{239}_{94}Pu + {}^0_{-1}\beta \qquad t_{1/2} = 2.35 \text{ days}$$

In this manner, the nonfissionable uranium-238 is transmuted into the fissionable isotope plutonium-239 (Figure 20.12).

In a typical breeder reactor, nuclear fuel containing uranium-235 or plutonium-239 is mixed with uranium-238 so that breeding takes place within the core. For every uranium-235 (or plutonium-239) nucleus undergoing fission, more than one neutron is captured by uranium-238 to generate plutonium-239. Thus, the stockpile of fissionable material can be steadily increased as the starting nuclear fuels are consumed. It takes about 7 to 10 years to regenerate the sizable amount of material needed to refuel the original reactor and to fuel another reactor of comparable size. This interval is called the *doubling time*.

Another *fertile* isotope is $^{232}_{90}Th$. Upon capturing slow neutrons, thorium is transmuted to uranium-233, which, like uranium-235, is a fissionable isotope.

$$^{232}_{90}Th + {}^1_0n \longrightarrow {}^{233}_{90}Th$$

$$^{233}_{90}Th \longrightarrow {}^{233}_{91}Pa + {}^0_{-1}\beta \qquad t_{1/2} = 22 \text{ min}$$

$$^{233}_{91}Pa \longrightarrow {}^{233}_{92}U + {}^0_{-1}\beta \qquad t_{1/2} = 27.4 \text{ days}$$

Uranium-233 ($t_{1/2} = 1.6 \times 10^5$ years) is stable enough for long-term storage.

Although the amounts of uranium-238 and thorium-232 in Earth's crust are relatively plentiful (4 ppm and 12 ppm by mass, respectively), the development of breeder reactors has been very slow. To date, the United States does not have a single operating breeder reactor, and only a few have been built in other countries, such as France and Russia. One problem is economics; breeder reactors are more expensive to build than conventional reactors. There are also more technical difficulties associated with the construction of such reactors. As a result, the future of breeder reactors, in the United States at least, is rather uncertain.

Many people, including environmentalists, regard nuclear fission as a highly undesirable method of energy production. Many fission products such as strontium-90 are dangerous radioactive isotopes with long half-lives. Plutonium-239, used as a nuclear fuel and produced in breeder reactors, is one of the most toxic substances known. It is an α-emitter with a half-life of 24,400 years.

Accidents, too, present many dangers. An accident at the Three Mile Island reactor in Pennsylvania in 1979 first brought the potential hazards of nuclear plants to public attention. In this instance, very little radiation escaped the reactor, but the plant remained closed for more than a decade while repairs were made and safety issues addressed. Only a few years later, on April 26, 1986, a reactor at the Chernobyl nuclear plant in Ukraine surged out of control. The fire and explosion that followed released much radioactive material into the environment. People working near the plant died within weeks as a result of the exposure to the intense radiation. The long-term effect

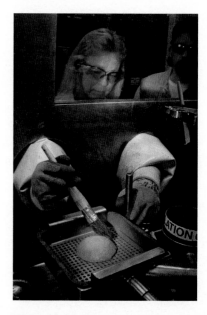

Figure 20.12 Radioactive plutonium oxide (PuO_2) has a red glow.
© Peter Essick/Aurora

of the radioactive fallout from this incident has not yet been clearly assessed, although agriculture and dairy farming were affected by the fallout. The number of potential cancer deaths attributable to the radiation contamination is estimated to be between a few thousand and more than 100,000.

In addition to the risk of accidents, the problem of radioactive waste disposal has not been satisfactorily resolved even for safely operated nuclear plants. Many suggestions have been made as to where to store or dispose of nuclear waste, including burial underground, burial beneath the ocean floor, and storage in deep geologic formations. But none of these sites has proved absolutely safe in the long run. Leakage of radioactive wastes into underground water, for example, can endanger nearby communities. The ideal disposal site would seem to be the sun, where a bit more radiation would make little difference, but this kind of operation requires space technology that is 100 percent reliable.

Because of the hazards, the future of nuclear reactors is clouded. What was once hailed as the ultimate solution to our energy needs in the twenty-first century is now being debated and questioned by both the scientific community and the general public. It seems likely that the controversy will continue for some time.

20.6   NUCLEAR FUSION

In contrast to the nuclear fission process, **nuclear fusion,** the combining of small nuclei into larger ones, is largely exempt from the waste disposal problem.

Figure 20.1 showed that for the lightest elements, nuclear stability increases with increasing mass number. This behavior suggests that if two light nuclei combine or fuse together to form a larger, more stable nucleus, an appreciable amount of energy will be released in the process. This is the basis for ongoing research into the harnessing of nuclear fusion for the production of energy.

Nuclear fusion occurs constantly in the sun. The sun is made up mostly of hydrogen and helium. In its interior, where temperatures reach about 15 million degrees Celsius, the following fusion reactions are believed to take place.

$$\mathrm{^{1}_{1}H + {}^{2}_{1}H \longrightarrow {}^{3}_{2}He}$$

$$\mathrm{^{3}_{2}He + {}^{3}_{2}He \longrightarrow {}^{4}_{2}He + 2{}^{1}_{1}H}$$

$$\mathrm{^{1}_{1}H + {}^{1}_{1}H \longrightarrow {}^{2}_{1}H + {}^{0}_{-1}\beta}$$

Because fusion reactions take place only at very high temperatures, they are often called **thermonuclear reactions.**

A major concern in choosing the proper nuclear fusion process for energy production is the temperature necessary to carry out the process. Some promising reactions are listed here.

Reaction	Energy released
$\mathrm{^{2}_{1}H + {}^{2}_{1}H \longrightarrow {}^{3}_{1}H + {}^{1}_{1}H}$	6.3×10^{-13} J
$\mathrm{^{2}_{1}H + {}^{3}_{1}H \longrightarrow {}^{3}_{2}He + 2{}^{1}_{0}n}$	2.8×10^{-12} J
$\mathrm{^{6}_{3}Li + {}^{2}_{1}H \longrightarrow 2{}^{4}_{2}He}$	3.6×10^{-12} J

These reactions must take place at extremely high temperatures, on the order of 100 million degrees Celsius, to overcome the repulsive forces between the nuclei. The first reaction is particularly attractive because the world's supply of deuterium is virtually inexhaustible. The total volume of water on Earth is about 1.5×10^{21} L. Because the natural abundance of deuterium is 0.015 percent, the total amount of deuterium present is roughly 4.5×10^{21} g, or 5.0×10^{15} tons. Although it is expensive to prepare deuterium, the cost is minimal compared to the value of the energy released by the reaction.

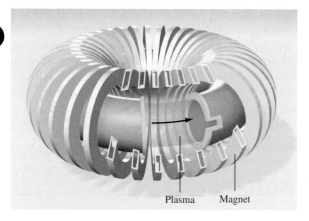

Figure 20.13 A magnetic plasma confinement design called a tokamak.

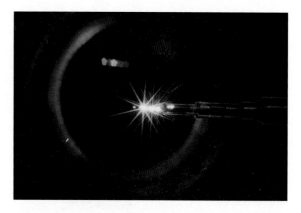

Figure 20.14 This small-scale fusion reaction was carried out at the Lawrence Livermore National Laboratory using one of the world's most powerful lasers, Nova.

© Science Source

In contrast to the fission process, nuclear fusion looks like a very promising energy source, at least on paper. Although thermal pollution would be a problem, fusion has the following advantages: (1) the fuels are cheap and almost inexhaustible and (2) the process produces little radioactive waste. If a fusion machine were turned off, it would shut down completely and instantly, without any danger of a meltdown.

If nuclear fusion is so great, why isn't there even one fusion reactor producing energy? Although we possess the scientific knowledge to design such a reactor, the technical difficulties have not yet been solved. The basic problem is finding a way to hold the nuclei together long enough, and at the appropriate temperature, for fusion to occur. At temperatures of about 100 million degrees Celsius, molecules cannot exist, and most or all of the atoms are stripped of their electrons. This state of matter, a gaseous mixture of positive ions and electrons, is called *plasma.* The problem of containing this plasma is a formidable one. No solid container can exist at such temperatures, unless the amount of plasma is small, but then the solid surface would immediately cool the sample and quench the fusion reaction. One approach to solving this problem is to use *magnetic confinement.* Because plasma consists of charged particles moving at high speeds, a magnetic field will exert a force on it. As Figure 20.13 shows, the plasma moves through a doughnut-shaped tunnel, confined by a complex magnetic field. Thus, the plasma never comes in contact with the walls of the container.

Another promising design employs high-power lasers to initiate the fusion reaction. In test runs, a number of laser beams transfer energy to a small fuel pellet, heating it and causing it to *implode*—that is, to collapse inward from all sides and compress into a small volume (Figure 20.14). Consequently, fusion occurs. Like the magnetic confinement approach, laser fusion presents a number of technical difficulties that still need to be overcome before it can be put to practical use on a large scale.

The technical problems inherent in the design of a nuclear fusion reactor do not affect the production of a *hydrogen bomb,* also called a *thermonuclear* bomb. In this case, the objective is all power and no control. Hydrogen bombs do not contain gaseous hydrogen or gaseous deuterium; they contain solid lithium deuteride (LiD), which can be packed very tightly. The detonation of a hydrogen bomb occurs in two stages—first a fission reaction and then a fusion reaction. The required temperature for fusion is achieved with an atomic bomb. Immediately after the atomic bomb explodes, the following fusion reactions occur, releasing vast amounts of energy (Figure 20.15).

$$^6_3\text{Li} + {}^2_1\text{H} \longrightarrow 2{}^4_2\alpha$$

$$^2_1\text{H} + {}^2_1\text{H} \longrightarrow {}^3_1\text{H} + {}^1_1\text{H}$$

Figure 20.15 Explosion of a thermonuclear bomb.

© Corbis

There is no critical mass in a fusion bomb, and the force of the explosion is limited only by the quantity of reactants present. Thermonuclear bombs are described as being "cleaner" than atomic bombs because the only radioactive isotopes they produce are tritium, which is a weak β-particle emitter ($t_{1/2} = 12.5$ years), and the products of the fission starter. Their damaging effects on the environment can be aggravated, however, by incorporating in the construction some nonfissionable material such as cobalt. Upon bombardment by neutrons, cobalt-59 is converted to cobalt-60, which is a very strong γ-ray emitter with a half-life of 5.2 years. The presence of radioactive cobalt isotopes in the debris or fallout from a thermonuclear explosion would be fatal to those who survived the initial blast.

20.7 USES OF ISOTOPES

Radioactive and stable isotopes alike have many applications in science and medicine. We have previously described the use of isotopes in the study of reaction mechanisms [◄◄ Section 19.7] and in determining the age of rocks [◄◄ Key Skills Chapter 19]. In this section we will discuss a few more examples.

Chemical Analysis

The formula of the thiosulfate ion is $S_2O_3^{2-}$. For some years, chemists were uncertain as to whether the two sulfur atoms occupied equivalent positions in the ion. The thiosulfate ion is prepared by treating the sulfite ion with elemental sulfur:

$$SO_3^{2-}(aq) + S(s) \longrightarrow S_2O_3^{2-}(aq)$$

Video 20.4
Nuclear chemistry—nuclear medical techniques.

When thiosulfate is treated with dilute acid, the reaction is reversed. The sulfite ion is re-formed, and elemental sulfur precipitates.

$$S_2O_3^{2-}(aq) \xrightarrow{\text{H}^+} SO_3^{2-}(aq) + S(s)$$

If this sequence is started with elemental sulfur enriched with the radioactive sulfur-35 isotope, the isotope acts as a "label" for S atoms. All the labels are found in the sulfur precipitate; none of them appears in the final sulfite ions. As a result, the two atoms of sulfur in $S_2O_3^{2-}$ are not structurally equivalent, as would be the case if the structure were

$$\left[\ddot{\text{O}}-\ddot{\text{S}}-\ddot{\text{O}}-\ddot{\text{S}}-\ddot{\text{O}}\right]^{2-}$$

If the sulfur atoms were equivalent, the radioactive isotope would be present in both the elemental sulfur precipitate and the sulfite ion. Based on spectroscopic studies, we now know that the structure of the thiosulfate ion is

$$\left[\begin{matrix} & \overset{\ddot{\text{S}}}{\|} & \\ :\ddot{\text{O}}-&\text{S}&-\ddot{\text{O}}: \\ & \| & \\ & \ddot{\text{O}} & \end{matrix}\right]^{2-}$$

The study of photosynthesis is also rich with isotope applications. The overall photosynthesis reaction can be represented as

$$6CO_2 + 6H_2O \longrightarrow C_6H_{12}O_6 + 6O_2$$

In Section 19.5 we learned that the ^{18}O isotope was used to determine the source of O_2. The radioactive ^{14}C isotope helped to determine the path of carbon in photosynthesis. Starting with $^{14}CO_2$, it was possible to isolate the intermediate products during photosynthesis and measure the amount of radioactivity of each carbon-containing compound. In this manner the path from CO_2 through various intermediate compounds to carbohydrate could be clearly charted. Isotopes, especially radioactive isotopes that are used to trace the path of the atoms of an element in a chemical or biological process, are called *tracers*.

Thinking Outside the Box

Nuclear Medicine

Brain tumors are some of the most difficult cancers to treat because the site of the malignant growth makes surgical excision difficult or impossible. Likewise, conventional radiation therapy using X rays or γ rays from outside the skull is usually not effective. An ingenious approach to this problem is *boron neutron capture therapy* (BNCT). This technique involves first administering a boron-10 compound that is selectively taken up by tumor cells and then applying a beam of low-energy neutrons to the tumor site. ^{10}B captures a neutron to produce ^{11}B, which disintegrates via the following nuclear reaction.

$$^{10}_{5}\text{B} + ^{1}_{0}\text{n} \longrightarrow ^{7}_{3}\text{Li} + ^{4}_{2}\alpha$$

The highly energetic particles produced by this reaction destroy the tumor cells in which the ^{10}B is concentrated. Because the particles are confined to just a few micrometers, they preferentially destroy tumor cells without damaging neighboring normal cells.

BNCT is a highly promising treatment and is an active area of research. One of the major goals of the research is to develop suitable compounds to deliver ^{10}B to the desired site. For such a compound to be effective, it must meet several criteria. It must have a high affinity for tumor cells, be able to pass through membrane barriers to reach the tumor site, and have minimal toxic effects on the human body.

In addition to BNCT, another promising treatment for brain tumors is brachytherapy using iodine-125. In brachytherapy, "seeds" containing ^{125}I are implanted directly into the tumor. As the radioisotope decays, γ rays destroy the tumor cells. Careful implantation prevents the radiation from harming nearby healthy cells.

Brachytherapy seeds (shown with a penny to illustrate their size).
© David A. Tietz/Editorial Image, LLC

Isotopes in Medicine

Tracers are also used for diagnosis in medicine. Sodium-24 (a β-emitter with a half-life of 14.8 h) injected into the bloodstream as a salt solution can be monitored to trace the flow of blood and detect possible constrictions or obstructions in the circulatory system. Iodine-131 (a β-emitter with a half-life of eight days) has been used to test the activity of the thyroid gland. A malfunctioning thyroid can be detected by giving the patient a drink of a solution containing a known amount of Na^{131}I and measuring the radioactivity just above the thyroid to see if the iodine is absorbed at the normal rate. Another radioactive isotope of iodine, iodine-123 (a γ-ray emitter), is used to image the brain (Figure 20.16). In each of these cases, though, the amount

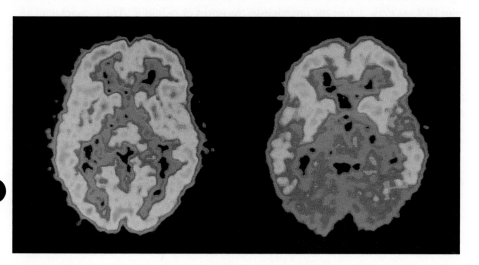

Figure 20.16 ^{123}I image of a normal brain (left) and the brain of an Alzheimer's victim (right).
© Mediscan/Corbis

Figure 20.17 Schematic diagram of a Geiger counter. Radiation (α, β, or γ rays) entering through the window ionizes the argon gas to generate a small current flow between the electrodes. This current is amplified and is used to flash a light or operate a counter with a clicking sound.

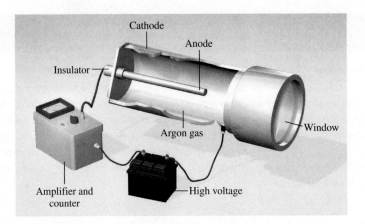

of radioisotope used must be kept small to prevent the patient from suffering permanent damage from the high-energy radiation.

Technetium, the first artificially prepared element, is one of the most useful elements in nuclear medicine. Although technetium is a transition metal, all its isotopes are radioactive. In the laboratory it is prepared by the nuclear reactions

$$^{98}_{42}\text{Mo} + ^{1}_{0}\text{n} \longrightarrow ^{99}_{42}\text{Mo}$$

$$^{99}_{42}\text{Mo} \longrightarrow ^{99m}_{43}\text{Tc} + ^{0}_{-1}\beta$$

where the superscript "m" denotes that the technetium-99 isotope is produced in its *excited* nuclear state. This isotope has a half-life of about six hours, decaying by γ radiation to technetium-99 in its nuclear *ground* state. Thus, it is a valuable diagnostic tool. The patient either drinks or is injected with a solution containing ^{99m}Tc. By detecting the γ rays emitted by ^{99m}Tc, doctors can obtain images of organs such as the heart, liver, and lungs.

A major advantage of using radioactive isotopes as tracers is that they are easy to detect. Their presence even in very small amounts can be detected by photographic techniques or by devices known as counters. Figure 20.17 is a diagram of a Geiger counter, an instrument widely used in scientific work and medical laboratories to detect radiation.

20.8   BIOLOGICAL EFFECTS OF RADIATION

In this section we will examine briefly the effects of radiation on biological systems. But first we must define the quantitative measures of radiation. The fundamental unit of radioactivity is the *curie* (Ci); 1 Ci corresponds to exactly 3.70×10^{10} nuclear disintegrations per second. This decay rate is equivalent to that of 1 g of radium. A *millicurie* (mCi) is one-thousandth of a curie. Thus, 10 mCi of a carbon-14 sample is the quantity that undergoes $(10 \times 10^{-3})(3.70 \times 10^{10}) = 3.70 \times 10^{8}$ disintegrations per second.

The intensity of radiation depends on the number of disintegrations as well as on the energy and type of radiation emitted. One common unit for the absorbed dose of radiation is the *rad* (radiation *a*bsorbed *d*ose), which is the amount of radiation that results in the absorption of 1×10^{-5} J per gram of irradiated material. The biological effect of radiation depends on the part of the body irradiated and the type of radiation. For this reason, the rad is often multiplied by a factor called the *RBE* (*r*elative *b*iological *e*ffectiveness). The product is called a *rem* (*r*oentgen *e*quivalent for *m*an).

$$\text{number of rems} = \text{number of rads} \times 1 \text{ RBE}$$

TABLE 20.4	Average Yearly Radiation Doses for Americans
Source	**Dose (mrem/yr)***
Cosmic rays	20–50
Ground and surroundings	25
Human body†	26
Medical and dental X rays	50–75
Air travel	5
Fallout from weapons tests	5
Nuclear waste	2
Total	133–188

*1 mrem = millirem = 1×10^{-3} rem.
†The radioactivity in the body comes from food and air.

Of the three types of nuclear radiation, α particles usually have the least-penetrating power. Beta particles are more penetrating than α particles, but less so than γ rays.

Gamma rays have very short wavelengths and high energies. Furthermore, because they carry no charge, they cannot be stopped by shielding materials as easily as α and β particles. If α- or β-emitters are ingested or inhaled, however, their damaging effects are greatly aggravated because the organs will be constantly subject to damaging radiation at close range. For example, strontium-90, a β-emitter, can replace calcium in bones, where it does the greatest damage.

Table 20.4 lists the average amounts of radiation an American receives every year. For short-term exposures to radiation, a dosage of 50 to 200 rems will cause a decrease in white blood cell counts and other complications, while a dosage of 500 rems or greater may result in death within weeks. Current safety standards permit nuclear workers to be exposed to no more than 5 rems per year and specify a maximum of 0.5 rem of human-made radiation per year for the general public.

The chemical basis of radiation damage is that of ionizing radiation. Radiation (of either particles or γ rays) can remove electrons from atoms and molecules in its path, leading to the formation of ions and radicals. *Radicals* (also called *free radicals*) are molecular fragments having one or more unpaired electrons; they are usually short lived and highly reactive. When water is irradiated with γ rays, for example, the following reactions take place.

$$H_2O \xrightarrow{\text{radiation}} H_2O^+ + e^-$$

$$H_2O^+ + H_2O \longrightarrow H_3O^+ + \cdot OH$$
$$\text{hydroxyl radical}$$

The electron (in the hydrated form) can subsequently react with water or with a hydrogen ion to form atomic hydrogen, and with oxygen to produce the superoxide ion (O_2^-) (a radical).

$$e^- + O_2 \longrightarrow \cdot O_2^-$$

In the tissues, the superoxide ions and other free radicals attack cell membranes and a host of organic compounds, such as enzymes and DNA molecules. Organic compounds can themselves be directly ionized and destroyed by high-energy radiation.

It has long been known that exposure to high-energy radiation can induce cancer in humans and other animals. Cancer is characterized by uncontrolled cellular growth. On the other hand, it is also well established that cancer cells can be destroyed by proper radiation treatment. In radiation therapy, a compromise is sought. The radiation to which the patient is exposed must be sufficient to destroy cancer cells without killing too many normal cells and, it is hoped, without inducing another form of cancer.

Radiation damage to living systems is generally classified as *somatic* or *genetic.* Somatic injuries are those that affect the organism during its own lifetime. Sunburn, skin rash, cancer, and cataracts are examples of somatic damage. Genetic damage means inheritable changes or gene mutations. For example, a person whose chromosomes have been damaged or altered by radiation may have deformed offspring.

Learning Outcomes

- List the particles or types of radiation that an unstable nucleus can produce.
- Identify a subatomic particle in a nuclear equation.
- Distinguish between chemical and nuclear reactions.
- Demonstrate use of subatomic particles in balancing nuclear reactions.
- List the rules for nuclear stability and use them to predict whether a particular nucleus is stable.
- Define nuclear binding energy.
- Calculate the nuclear binding energy of a nucleus.

- Use the half-life of a radioactive decay in calculations.
- Produce the balanced nuclear reaction for a nuclear transmutation.
- Distinguish between fusion and fission.
- Define critical mass and nuclear chain reaction.
- Describe the basis of a nuclear reactor.
- Provide examples of the uses of isotopes in science and medicine.
- Explain why high-energy radiation is biologically harmful.

Chapter Summary

SECTION 20.1

- Spontaneous emission of particles or radiation from unstable nuclei is known as **radioactivity.** Unstable nuclei emit α particles, β particles, **positrons,** or γ rays.
- **Nuclear transmutation** is the conversion of one nucleus to another. Nuclear reactions are balanced by summing the mass numbers and the atomic numbers.

SECTION 20.2

- Stable nuclei with low atomic numbers have neutron-to-proton ratios close to 1. Heavier stable nuclei have higher ratios. Nuclear stability is favored by certain numbers of nucleons including even numbers and "magic" numbers.
- The difference between the actual mass of a nucleus and the mass calculated by summing the masses of the individual nucleons is the **mass defect.**
- **Nuclear binding energy,** determined by using Einstein's equation $E = mc^2$, is a measure of nuclear stability.

SECTION 20.3

- Uranium-238 is the parent of a natural **radioactive decay series** that can be used to determine the ages of rocks. Radiocarbon dating is done using carbon-14.

SECTION 20.4

- **Transuranium elements** are created by bombarding other elements with accelerated neutrons, protons, α particles, or other nuclei.

SECTION 20.5

- **Nuclear fission** is the splitting of a large nucleus into two smaller nuclei and one or more neutrons. When the free neutrons are captured efficiently by other nuclei, a **nuclear chain reaction** can occur in which the fission process is sustained. The minimum amount of fissionable material required to sustain the reaction is known as the **critical mass.**

- Nuclear reactors use the heat from a controlled nuclear fission reaction to produce power. Fission is controlled, in part, by *moderators*—materials that limit the speed of liberated neutrons but that do not themselves undergo fission when bombarded with neutrons. The three important types of reactors are light water reactors, heavy water reactors, and *breeder reactors.* Breeder reactors produce more fissionable material than they consume.

SECTION 20.6
- *Nuclear fusion,* the type of reaction that occurs in the sun, is the combination of two light nuclei to form one heavier nucleus. Fusion reactions are sometimes referred to as *thermonuclear reactions* because they take place only at very high temperatures.

SECTION 20.7
- Radioactive isotopes are easy to detect and thus make excellent *tracers* in chemical reactions and in medical procedures.

SECTION 20.8
- High-energy radiation damages living systems by causing ionization and the formation of *radicals,* or *free radicals,* which are chemical species with unpaired electrons.

Key Words

Breeder reactor, 959	Nuclear fission, 953	Radioactive decay series, 947
Critical mass, 956	Nuclear fusion, 960	Radioactivity, 941
Mass defect, 944	Nuclear transmutation, 941	Thermonuclear reaction, 960
Moderator, 957	Positron, 942	Tracers, 962
Nuclear binding energy, 943	Radical, 965	Transuranium elements, 951
Nuclear chain reaction, 956		

Key Equation

20.1	$\Delta E = (\Delta m)c^2$	Einstein's equation allows us to calculate the energy change associated with the loss of a given mass (Δm) in a nuclear process. Mass must be expressed in kilograms for units to cancel properly.

Questions and Problems

SECTION 20.1: NUCLEI AND NUCLEAR REACTIONS

Review Questions

20.1 How do nuclear reactions differ from ordinary chemical reactions?

20.2 What are the steps in balancing nuclear equations?

20.3 What is the difference between $_{-1}^{0}e$ and $_{-1}^{0}\beta$?

20.4 What is the difference between an electron and a positron?

Conceptual Problems

20.5 Complete the following nuclear equations, and identify X in each case.
(a) $_{12}^{26}\text{Mg} + _{1}^{1}\text{p} \longrightarrow \alpha + \text{X}$
(b) $_{27}^{59}\text{Co} + _{1}^{2}\text{H} \longrightarrow _{27}^{60}\text{Co} + \text{X}$
(c) $_{92}^{235}\text{U} + _{0}^{1}\text{n} \longrightarrow _{36}^{94}\text{Kr} + _{56}^{139}\text{Ba} + 3\text{X}$
(d) $_{24}^{53}\text{Cr} + _{2}^{4}\alpha \longrightarrow _{0}^{1}\text{n} + \text{X}$
(e) $_{8}^{20}\text{O} \longrightarrow _{9}^{20}\text{F} + \text{X}$

20.6 Complete the following nuclear equations, and identify X in each case.
(a) $_{53}^{135}\text{I} \longrightarrow _{54}^{135}\text{Xe} + \text{X}$
(b) $_{19}^{40}\text{K} \longrightarrow _{-1}^{0}\beta + \text{X}$
(c) $_{27}^{59}\text{Co} + _{0}^{1}\text{n} \longrightarrow _{25}^{56}\text{Mn} + \text{X}$
(d) $_{92}^{235}\text{U} + _{0}^{1}\text{n} \longrightarrow _{40}^{99}\text{Zr} + _{52}^{135}\text{Te} + 2\text{X}$

SECTION 20.2: NUCLEAR STABILITY

Review Questions

20.7 What type of emission occurs when a nucleus is above the belt of stability? What type of emission occurs when a nucleus is below the belt of stability?

20.8 How does β-particle emission alter the nucleus? How does positron emission alter the nucleus?

20.9 What nuclear transformation occurs when a nucleus captures an electron?

20.10 Define *nuclear binding energy, mass defect,* and *nucleon.*

20.11 How does Einstein's equation, $E = mc^2$, enable us to calculate nuclear binding energy?

20.12 Why is it preferable to use nuclear binding energy per nucleon for a comparison of the stabilities of different nuclei?

Computational Problems

20.13 Given that

$$H(g) + H(g) \longrightarrow H_2(g) \qquad \Delta H° = -436.4 \text{ kJ/mol}$$

calculate the change in mass (in kilograms) per mole of H_2 formed.

20.14 Estimates show that the total energy output of the sun is 5×10^{26} J/s. What is the corresponding mass loss in kg/s of the sun?

20.15 Calculate the nuclear binding energy (in joules) and the binding energy per nucleon of the following isotopes: (a) $^{7}_{3}\text{Li}$ (7.01600 amu) and (b) $^{35}_{17}\text{Cl}$ (34.96885 amu).

20.16 Calculate the nuclear binding energy (in joules) and the binding energy per nucleon of the following isotopes: (a) $^{4}_{2}\text{He}$ (4.002603 amu) and (b) $^{184}_{74}\text{W}$ (183.950928 amu).

20.17 Given that the nuclear binding energy of ^{48}Cr is 1.37340×10^{-12} J/nucleon, calculate the mass of a single ^{48}Cr atom.

20.18 Given that the nuclear binding energy of ^{192}Ir is 1.27198×10^{-12} J/nucleon, calculate the mass of a single ^{192}Ir atom.

SECTION 20.3: NATURAL RADIOACTIVITY

Review Questions

20.19 Discuss factors that lead to nuclear decay.

20.20 Outline the principle for dating materials using radioactive isotopes.

Computational Problems

20.21 Fill in the blanks in the following radioactive decay series.

(a) $^{232}\text{Th} \xrightarrow{\alpha} \underline{\quad} \xrightarrow{\beta} \underline{\quad} \xrightarrow{\beta} {}^{228}\text{Th}$

(b) $^{235}\text{U} \xrightarrow{\alpha} \underline{\quad} \xrightarrow{\beta} \underline{\quad} \xrightarrow{\alpha} {}^{227}\text{Ac}$

(c) $\underline{\quad} \xrightarrow{\alpha} {}^{233}\text{Pa} \xrightarrow{\beta} \underline{\quad} \xrightarrow{\alpha} \underline{\quad}$

20.22 A radioactive substance undergoes decay as follows:

Time (days)	Mass (g)
0	500
1	438
2	383
3	335
4	294
5	257
6	225

Calculate the first-order decay constant and the half-life of the reaction.

20.23 The radioactive decay of Tl-206 to Pb-206 has a half-life of 4.20 min. Starting with 5.00×10^{22} atoms of Tl-206, calculate the number of such atoms left after 42.0 min.

20.24 A freshly isolated sample of ^{90}Y was found to have an activity of 9.8×10^5 disintegrations per minute at 1:00 P.M. on December 3, 2010. At 2:15 P.M. on December 17, 2010, its activity was measured again and found to be 2.6×10^4 disintegrations per minute. Calculate the half-life of ^{90}Y.

20.25 A wooden artifact has a ^{14}C activity of 18.9 disintegrations per minute, compared to 27.5 disintegrations per minute for live wood. Given that the half-life of ^{14}C is 5715 years, determine the age of the artifact.

20.26 In the thorium decay series, thorium-232 loses a total of six α particles and four β particles in a 10-stage process. What is the final isotope produced?

20.27 Consider the decay series $A \longrightarrow B \longrightarrow C \longrightarrow D$ where A, B, and C are radioactive isotopes with half-lives of 4.50 s, 15.0 days, and 1.00 s, respectively, and D is nonradioactive. Starting with 1.00 mole of A, and none of B, C, or D, calculate the number of moles of A, B, C, and D left after 30 days.

20.28 The activity of radioactive carbon-14 decay of a piece of charcoal found at a volcanic site is 11.2 disintegrations per second. If the activity of carbon-14 decay in an equal mass of living matter is 18.3 disintegrations per second, what is the age of the charcoal? (See Problem 20.25 for the half-life of carbon-14.)

20.29 The age of some animal bones was determined by carbon-14 dating to be 8.4×10^3 years old. Calculate the activity of the carbon-14 in the bones in disintegrations per minute per gram, given that the original activity was 15.3 disintegrations per minute per gram. (See Problem 20.25 for the half-life of carbon-14.)

20.30 Given that the half-life of ^{238}U is 4.51×10^9 years, determine (a) the age of a rock found to contain 1.09 mg ^{238}U and 0.08 mg ^{206}Pb, and (b) the ratio of ^{238}U to ^{206}Pb in a rock that is 1.7×10^8 years old.

20.31 Until relatively recently, ^{209}Bi was believed to be the heaviest stable isotope. In 2003, though, scientists determined that it decays by alpha emission and has the extraordinarily long half-life of 1.9×10^{19} years—more than a billion times the age of the universe. Determine the first-order rate constant for the decay of ^{209}Bi.

SECTION 20.4: NUCLEAR TRANSMUTATION

Review Questions

20.32 What is the difference between radioactive decay and nuclear transmutation?

20.33 How is nuclear transmutation achieved in practice?

Conceptual Problems

20.34 Write balanced nuclear equations for the following reactions, and identify X: (a) $X(p,\alpha)^{12}_{6}C$, (b) $^{27}_{13}Al(d,\alpha)X$, (c) $^{55}_{25}Mn(n,\gamma)X$.

20.35 Write the abbreviated forms for the following reactions.
(a) $^{9}_{4}Be + ^{4}_{2}\alpha \longrightarrow ^{12}_{6}C + ^{1}_{0}n$
(b) $^{14}_{7}N + ^{4}_{2}\alpha \longrightarrow ^{17}_{8}O + ^{1}_{1}p$
(c) $^{238}_{92}U + ^{2}_{1}H \longrightarrow ^{238}_{93}Np + 2^{1}_{0}n$

20.36 Write balanced nuclear equations for the following reactions, and identify X: (a) $^{80}_{34}Se(d,p)X$, (b) $X(d,2p)^{9}_{3}Li$, (c) $^{10}_{5}B(n,\alpha)X$.

20.37 Write the abbreviated forms for the following reactions.
(a) $^{40}_{20}Ca + ^{2}_{1}H \longrightarrow ^{41}_{20}Ca + ^{1}_{1}p$
(b) $^{32}_{16}S + ^{1}_{0}n \longrightarrow ^{32}_{15}P + ^{1}_{1}p$
(c) $^{239}_{94}Pu + ^{4}_{2}\alpha \longrightarrow ^{242}_{96}Cm + ^{1}_{0}n$

20.38 Two isotopes of element 115 (proposed IUPAC name moscovium, Mc) have been produced by bombardment of americium-243 with calcium-48. Four neutrons are produced with one isotope, and three neutrons are produced with the other isotope. Determine the mass numbers of these two isotopes of element 115.

20.39 A long-cherished dream of alchemists was to produce gold from cheaper and more abundant elements. This dream was finally realized when $^{198}_{80}Hg$ was converted into gold by neutron bombardment. Write a balanced equation for this reaction.

SECTION 20.5: NUCLEAR FISSION

Visualizing Chemistry
Figure 20.6

VC 20.1 The fission of ^{235}U can result in a variety of products, including those shown in Figure 20.6. Which of the following equations does *not* represent another possible fission process?
(a) $^{1}_{0}n + ^{235}_{92}U \longrightarrow ^{137}_{52}Te + ^{97}_{40}Zr + 2^{1}_{0}n$
(b) $^{1}_{0}n + ^{235}_{92}U \longrightarrow ^{142}_{56}Ba + ^{91}_{36}Kr + 3^{1}_{0}n$
(c) $^{1}_{0}n + ^{235}_{92}U \longrightarrow ^{137}_{55}Cs + ^{90}_{37}Rb + 3^{1}_{0}n$

VC 20.2 How many neutrons are produced in the fission reaction shown?
$$^{1}_{0}n + ^{239}_{94}Pu \longrightarrow ^{109}_{44}Ru + ^{129}_{50}Sn + \underline{\quad}^{1}_{0}n$$
(a) 1 (b) 2 (c) 3

VC 20.3 The fission of ^{235}U shown in Figure 20.6 is represented by the equation
$$^{1}_{0}n + ^{235}_{92}U \longrightarrow ^{143}_{54}Xe + ^{90}_{38}Sr + 3^{1}_{0}n$$

How does the combined mass of products compare to the combined mass of reactants for this process?
(a) The combined mass of products is smaller than the combined mass of reactants.
(b) The combined mass of products is larger than the combined mass of reactants.
(c) The combined mass of products is equal to the combined mass of reactants.

VC 20.4 The fusion of $^{2}_{1}H$ and $^{3}_{1}H$ shown in Figure 20.6 is represented by the equation
$$^{2}_{1}H + ^{3}_{1}H \longrightarrow ^{4}_{2}He + ^{1}_{0}n$$

How does the combined mass of products compare to the combined mass of reactants for this process?
(a) The combined mass of products is smaller than the combined mass of reactants.
(b) The combined mass of products is larger than the combined mass of reactants.
(c) The combined mass of products is equal to the combined mass of reactants.

Review Questions

20.40 Define *nuclear fission, nuclear chain reaction,* and *critical mass.*
20.41 Which isotopes can undergo nuclear fission?
20.42 Explain how an atomic bomb works.
20.43 Explain the functions of a moderator and a control rod in a nuclear reactor.
20.44 Discuss the differences between a light water and a heavy water nuclear fission reactor. What are the advantages of a breeder reactor over a conventional nuclear fission reactor?
20.45 No form of energy production is without risk. Make a list of the risks to society involved in fueling and operating a conventional coal-fired electric power plant, and compare them with the risks of fueling and operating a nuclear fission-powered electric plant.

SECTION 20.6: NUCLEAR FUSION

Review Questions

20.46 Define *nuclear fusion, thermonuclear reaction,* and *plasma.*
20.47 Why do heavy elements such as uranium undergo fission, whereas light elements such as hydrogen and lithium undergo fusion?
20.48 How does a hydrogen bomb work?
20.49 What are the advantages of a fusion reactor over a fission reactor? What are the practical difficulties in operating a large-scale fusion reactor?

SECTION 20.7: USES OF ISOTOPES

Computational Problems

20.50 ^{125}I is produced by a two-step process in which ^{124}Xe nuclei are bombarded with neutrons to produce ^{125}Xe—a process called *neutron activation*. ^{125}Xe then decays by electron capture to produce ^{125}I, which also decays by electron capture. Write nuclear equations for the two steps that produce ^{125}I from ^{124}Xe, and identify the product of the electron capture decay of ^{125}I.

20.51 The half-life of ^{125}I is 59.4 days. How long will it take for the activity of implanted ^{125}I seeds to fall to 5.00% of their original value?

Conceptual Problems

20.52 Describe how you would use a radioactive iodine isotope to demonstrate that the following process is in dynamic equilibrium.

$$PbI_2(s) \rightleftharpoons Pb^{2+}(aq) + 2I^-(aq)$$

20.53 Consider the following redox reaction.

$$IO_4^-(aq) + 2I^-(aq) + H_2O(l) \longrightarrow$$
$$I_2(s) + IO_3^-(aq) + 2OH^-(aq)$$

When KIO_4 is added to a solution containing iodide ions labeled with radioactive iodine-128, all the radioactivity appears in I_2 and none in the IO_3^- ion. What can you deduce about the mechanism for the redox process?

20.54 Explain how you might use a radioactive tracer to show that ions are not completely motionless in crystals.

20.55 Each molecule of hemoglobin, the oxygen carrier in blood, contains four Fe atoms. Explain how you would use the radioactive $^{59}_{26}Fe$ ($t_{1/2}$ = 46 days) to show that the iron in a certain food is converted into hemoglobin.

SECTION 20.8: BIOLOGICAL EFFECTS OF RADIATION

Review Questions

20.56 List the factors that affect the intensity of radiation from a radioactive element.

20.57 What are *rad* and *rem*, and how are they related?

20.58 Explain, with examples, the difference between somatic and genetic radiation damage.

20.59 Compare the extent of radiation damage done by α, β, and γ sources.

ADDITIONAL PROBLEMS

20.60 How does a Geiger counter work?

20.61 Strontium-90 is one of the products of the fission of uranium-235. This strontium isotope is radioactive,

with a half-life of 28.1 years. Calculate how long (in years) it will take for 1.00 g of the isotope to be reduced to 0.200 g by decay.

20.62 Nuclei with an even number of protons and an even number of neutrons are more stable than those with an odd number of protons and/or an odd number of neutrons. What is the significance of the even numbers of protons and neutrons in this case?

20.63 Tritium (^{3}H) is radioactive and decays by electron emission. Its half-life is 12.5 years. In ordinary water the ratio of ^{1}H to ^{3}H atoms is 1.0×10^{17} to 1. (a) Write a balanced nuclear equation for tritium decay. (b) How many disintegrations will be observed per minute in a 1.00-kg sample of water?

20.64 (a) What is the activity, in millicuries, of a 0.500-g sample of $^{237}_{93}NP$? (This isotope decays by α-particle emission and has a half-life of 2.20×10^6 years.) (b) Write a balanced nuclear equation for the decay of $^{237}_{93}NP$.

20.65 The following equations are for nuclear reactions that are known to occur in the explosion of an atomic bomb. Identify X.
(a) $^{235}_{92}U + ^1_0n \longrightarrow ^{140}_{56}Ba + 3^1_0n + X$
(b) $^{235}_{92}U + ^1_0n \longrightarrow ^{144}_{55}Cs + ^{90}_{37}Rb + 2X$
(c) $^{235}_{92}U + ^1_0n \longrightarrow ^{87}_{35}Br + 3^1_0n + X$
(d) $^{235}_{92}U + ^1_0n \longrightarrow ^{160}_{62}Sm + ^{72}_{30}Zn + 4X$

20.66 Calculate the nuclear binding energies (in J/nucleon) for the following species: (a) ^{10}B (10.0129 amu), (b) ^{11}B (11.009305 amu), (c) ^{14}N (14.003074 amu), (d) ^{56}Fe (55.93494 amu).

20.67 Write complete nuclear equations for the following processes: (a) tritium (^{3}H) undergoes β decay, (b) ^{242}Pu undergoes α-particle emission, (c) ^{131}I undergoes β decay, (d) ^{251}Cf emits an α particle.

20.68 The nucleus of nitrogen-18 lies above the stability belt. Write the equation for a nuclear reaction by which nitrogen-18 can achieve stability.

20.69 Why is strontium-90 a particularly dangerous isotope for humans? The half-life of strontium-90 is 29.1 years. Calculate the radioactivity in millicuries of 15.6 mg of ^{90}Sr.

20.70 How are scientists able to tell the age of a fossil?

20.71 After the Chernobyl accident, people living close to the nuclear reactor site were urged to take large amounts of potassium iodide as a safety precaution. What is the chemical basis for this action?

20.72 To detect bombs that may be smuggled onto airplanes, the Federal Aviation Administration (FAA) will soon require all major airports in the United States to install thermal neutron analyzers. The thermal neutron analyzer will bombard baggage with low-energy neutrons, converting some of the nitrogen-14 nuclei to nitrogen-15, with simultaneous emission of γ rays. Because nitrogen content is usually high in explosives, detection of a high dosage of γ rays will suggest that a bomb may be present. (a) Write an equation for the nuclear process. (b) Compare this technique with the conventional X-ray detection method.

20.73 Astatine, the last member of Group 7A, can be prepared by bombarding bismuth-209 with α particles. (a) Write an equation for the reaction. (b) Represent the equation in the abbreviated form as discussed in Section 20.4.

20.74 Explain why achievement of nuclear fusion in the laboratory requires a temperature of about 100 million degrees Celsius, which is much higher than that in the interior of the sun (15 million degrees Celsius).

20.75 The carbon-14 decay rate of a sample obtained from a young tree is 0.260 disintegration per second per gram of the sample. Another wood sample prepared from an object recovered at an archaeological excavation gives a decay rate of 0.186 disintegration per second per gram of the sample. What is the age of the object?

20.76 Tritium contains one proton and two neutrons. There is no significant proton-proton repulsion present in the nucleus. Why, then, is tritium radioactive?

20.77 The usefulness of radiocarbon dating is limited to objects no older than 60,000 years. What percent of the carbon-14, originally present in the sample, remains after this period of time?

20.78 The radioactive potassium-40 isotope decays to argon-40 with a half-life of 1.2×10^9 years. (a) Write a balanced equation for the reaction. (b) A sample of moon rock is found to contain 18% potassium-40 and 82% argon by mass. Calculate the age of the rock in years. (Assume that all the argon in the sample is the result of potassium decay.)

20.79 Nuclear waste disposal is one of the major concerns of the nuclear industry. In choosing a safe and stable environment to store nuclear wastes, consideration must be given to the heat released during nuclear decay. As an example, consider the β decay of ^{90}Sr (89.907738 amu).

$$^{90}_{38}Sr \longrightarrow ^{90}_{39}Y + ^{0}_{-1}\beta \qquad t_{1/2} = 28.1 \text{ yr}$$

The ^{90}Y (89.907152 amu) further decays as follows:

$$^{90}_{39}Y \longrightarrow ^{90}_{40}Zr + ^{0}_{-1}\beta \qquad t_{1/2} = 64 \text{ h}$$

Zirconium-90 (89.904703 amu) is a stable isotope. (a) Use the mass defect to calculate the energy released (in joules) in each of the preceding two decays. (The mass of the electron is 5.4857×10^{-4} amu.)

(b) Starting with 1 mole of ^{90}Sr, calculate the number of moles of ^{90}Sr that will decay in a year. (c) Calculate the amount of heat released (in kilojoules) corresponding to the number of moles of ^{90}Sr decayed to ^{90}Zr in part (b).

20.80 Which of the following poses a greater health hazard: a radioactive isotope with a short half-life or a radioactive isotope with a long half-life? Explain. [Assume the same type of radiation (α or β) and comparable energetics per particle emitted.]

20.81 From the definition of curie, calculate Avogadro's number, given that the molar mass of ^{226}Ra is 226.03 g/mol and that it decays with a half-life of 1.6×10^3 years.

20.82 As a result of being exposed to the radiation released during the Chernobyl nuclear accident, the dose of iodine-131 in a person's body is 7.4 mC ($1 \text{ mC} = 1 \times 10^{-3}$ Ci). Use the relationship rate = κN to calculate the number of atoms of iodine-131 to which this radioactivity corresponds. (The half-life of ^{131}I is 8.1 days.)

20.83 (a) Calculate the energy released when a U-238 isotope decays to Th-234. The atomic masses are as follows: U-238: 238.05078 amu; Th-234: 234.03596 amu; and He-4: 4.002603 amu. (b) The energy released in part (a) is transformed into the kinetic energy of the recoiling Th-234 nucleus and the α particle. Which of the two will move away faster? Explain.

20.84 A person received an anonymous gift of a decorative cube, which he placed on his desk. A few months later he became ill and died shortly afterward. After investigation, the cause of his death was linked to the box. The box was airtight and had no toxic chemicals on it. What might have killed the man?

20.85 Identify two of the most abundant radioactive elements that exist on Earth. Explain why they are still present. (You may wish to consult a website such as that of the University of Sheffield and WebElements Ltd, UK, webelements.com.)

20.86 Sources of energy on Earth include fossil fuels, geothermal power, gravity, hydroelectric power, nuclear fission, nuclear fusion, the sun, and wind. Which of these have a "nuclear origin," either directly or indirectly?

20.87 Cobalt-60 is an isotope used in diagnostic medicine and cancer treatment. It decays with γ-ray emission. Calculate the wavelength of the radiation in nanometers if the energy of the γ ray is 2.4×10^{-13} J/photon.

20.88 Americium-241 is used in smoke detectors because it has a long half-life (458 years) and its emitted

α particles are energetic enough to ionize air molecules. Using the given schematic diagram of a smoke detector, explain how it works.

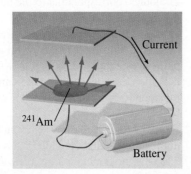

20.89 The constituents of wine contain, among others, carbon, hydrogen, and oxygen atoms. A bottle of wine was sealed about six years ago. To confirm its age, which of the isotopes would you choose in a radioactive dating study? The half-lives of the isotopes are ^{13}C: 5715 years; ^{15}O: 124 s; ^{3}H: 12.5 years. Assume that the activities of the isotopes were known at the time the bottle was sealed.

20.90 Name two advantages of a nuclear-powered submarine over a conventional submarine.

20.91 In 1997 a scientist at a nuclear research center in Russia placed a thin shell of copper on a sphere of highly enriched uranium-235. Suddenly, there was a huge burst of radiation, which turned the air blue. Three days later, the scientist died of radiation exposure. Explain what caused the accident. (*Hint:* Copper is an effective metal for reflecting neutrons.)

20.92 A radioactive isotope of copper decays as follows:

$$^{64}\text{Cu} \longrightarrow {}^{64}\text{Zn} + {}_{-1}^{0}\beta \qquad t_{\frac{1}{2}} = 12.8 \text{ h}$$

Starting with 84.0 g of ^{64}Cu, calculate the quantity of ^{64}Zn produced after 18.4 h.

20.93 A 0.0100-g sample of a radioactive isotope with a half-life of 1.3×10^9 years decays at the rate of 2.9×10^4 disintegrations per minute. Calculate the molar mass of the isotope.

20.94 The half-life of ^{27}Mg is 9.50 min. (a) Initially there were 4.20×10^{12} ^{27}Mg nuclei present. How many ^{27}Mg nuclei are left 30.0 min later? (b) Calculate the ^{27}Mg activities (in Ci) at $t = 0$ and $t = 30.0$ min. (c) What is the probability that any one ^{27}Mg nucleus decays during a 1-s interval? What assumption is made in this calculation?

20.95 (a) Assuming nuclei are spherical in shape, show that the radius (r) of a nucleus is proportional to the cube root of mass number (A). (b) In general, the radius of a nucleus is given by $r = r_0 A^{1/3}$, where r_0, the proportionality constant, is given by 1.2×10^{-15} m. Calculate the volume of the ^{238}U nucleus.

20.96 Modern designs of atomic bombs contain, in addition to uranium or plutonium, small amounts of tritium and deuterium to boost the power of explosion. What is the role of tritium and deuterium in these bombs?

20.97 Isotope X decays to isotope Y with a half-life of 45 days. Which diagram most closely represents the sample of X after 20 days?

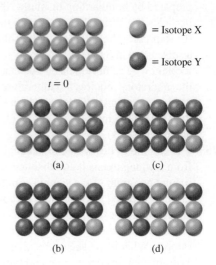

20.98 In 2006, an ex-KGB agent was murdered in London. The investigation following the agent's death revealed that he was poisoned with the radioactive isotope ^{210}Po, which had apparently been added to his food. (a) ^{210}Po is prepared by bombarding ^{209}Bi with neutrons. Write an equation for the reaction. (b) The half-life of ^{210}Po is 138 days. It decays by α-particle emission. Write the equation for the decay process. (c) Calculate the energy of an emitted α particle. Assume both the parent and daughter nuclei have zero kinetic energy. The atomic masses of ^{210}Po, ^{206}Pb, and $_{2}^{4}\alpha$ are 209.98286, 205.97444, and 4.00150 amu, respectively. (d) Ingestion of 1 μg of ^{210}Po could prove fatal. What is the total energy released by this quantity of ^{210}Po over the course of 138 days?

20.99 Alpha particles produced by radioactive decay eventually pick up electrons from their surroundings to form helium atoms. Calculate the volume (in milliliters) of He collected at STP when 1.00 g of pure ^{226}Ra is stored in a closed container for 125 years. (Assume that there are five α particles generated per ^{226}Ra as it decays to ^{206}Pb.)

20.100 An electron and a positron are accelerated to nearly the speed of light before colliding in a particle accelerator. The resulting collision produces an exotic particle having a mass many times that of a proton. Does this result violate the law of conservation of mass? Explain.

20.101 Iridium-192 can be used in brachytherapy. It is produced by a nuclear transmutation. (a) Identify the target nucleus X, and write the balanced nuclear equation for the reaction represented by $^{191}X(n,\gamma)^{192}Ir$. (b) The mass of an ^{125}I atom is 124.904624 amu. Calculate the nuclear binding energy and the nuclear binding energy per nucleon.

Answers to In-Chapter Materials

PRACTICE PROBLEMS

20.1A (a) $^{78}_{34}Se$, (b) $^{5}_{3}Li$, (c) $^{1}_{0}n$. **20.1B** (a) $^{244}_{95}Am$, (b) $^{234}_{90}Th$, (c) $^{14}_{6}C$. **20.2A** 2.6284×10^{-10} J; 1.2576×10^{-12} J/nucleon. **20.2B** 196.9665 amu. **20.3A** 9.3×10^3 yr. **20.3B** 1.0×10^1 dps. **20.4A** 6.6×10^8 yr. **20.4B** 5.7×10^{-2} mg. **20.5A** $^{106}_{46}Pd + ^{4}_{2}He \longrightarrow ^{1}_{1}H + ^{109}_{47}Ag$. **20.5B** $^{33}_{17}Cl(n,^{3}_{2}He)^{31}_{15}P$.

SECTION REVIEW

20.1.1 d. **20.1.2** a. **20.2.1** b. **20.2.2** d. **20.2.3** c. **20.2.4** a. **20.3.1** d. **20.3.2** c. **20.3.3** b. **20.4.1** e. **20.4.2** b.

Chapter **21**

Environmental Chemistry

© Digital Vision/Getty Images.

STRATOSPHERIC OZONE is responsible for the absorption of light that is known to cause cancer, genetic mutations, and the destruction of plant life. The balance of ozone destruction and regeneration can be disrupted, however, by the presence of substances not found naturally in the atmosphere. In 1973, F. Sherwood "Sherry" Rowland and Mario Molina, chemistry professors at the University of California–Irvine, discovered that although chlorofluorocarbon (CFCs) molecules were extraordinarily stable in the troposphere, the very stability that made them attractive as coolants and propellants also allowed them to survive the gradual diffusion into the stratosphere where they would ultimately be broken down by high-energy ultraviolet radiation. Rowland and Molina proposed that chlorine atoms liberated in the breakdown of CFCs could potentially catalyze the destruction of large amounts of ozone in the stratosphere. The work of Rowland and Molina, along with other atmospheric scientists, provoked a debate among the scientific and international communities regarding the fate of the ozone layer—and the planet. In 1995, Rowland and Molina, along with Dutch atmospheric chemist Paul Crutzen, were awarded the Nobel Prize in Chemistry for their elucidation of the role of human-made chemicals in the catalytic destruction of stratospheric ozone.

Before You Begin, Review These Skills

- Molecular polarity [◄◄ Section 7.2]
- Bond enthalpy [◄◄ Section 10.7]
- Catalysis [◄◄ Section 19.8]

21.1 EARTH'S ATMOSPHERE

Earth is unique among the planets of our solar system in having an atmosphere that is chemically active and rich in oxygen. Mars, for example, has a much thinner atmosphere that is about 90 percent carbon dioxide. Jupiter has no solid surface; it is made up, instead, of 90 percent hydrogen, 9 percent helium, and 1 percent other substances.

It is generally believed that 3 or 4 billion years ago, Earth's atmosphere consisted mainly of ammonia, methane, and water. There was little, if any, free oxygen present. Ultraviolet (UV) radiation from the sun probably penetrated the atmosphere, rendering the surface of Earth sterile. However, the same UV radiation may have triggered the chemical reactions (perhaps beneath the surface) that eventually led to life on Earth. Primitive organisms used energy from the sun to break down carbon dioxide (produced by volcanic activity) to obtain carbon, which they incorporated in their own cells. The major by-product of this process, called *photosynthesis,* is oxygen. Another important source of oxygen is the *photodecomposition* of water vapor by UV light. Over time, the more reactive gases such as ammonia and methane have largely disappeared, and today our atmosphere consists mainly of oxygen and nitrogen gases. Biological processes determine to a great extent the atmospheric concentrations of these gases, one of which is reactive (oxygen) and the other unreactive (nitrogen).

Table 21.1 shows the composition of dry air at sea level. The total mass of the atmosphere is about 5.3×10^{18} kg. Water is excluded from this table because its concentration in air can vary drastically from location to location.

Figure 21.1 shows the major processes involved in the cycle of nitrogen in nature. Molecular nitrogen, with its triple bond, is a very stable molecule. However, through biological and industrial **nitrogen fixation** (the conversion of molecular nitrogen into nitrogen compounds), atmospheric nitrogen gas is converted into nitrates and other compounds suitable for assimilation by algae and plants. Another important mechanism for producing nitrates from nitrogen gas is lightning. The steps are

$$N_2(g) + O_2(g) \xrightarrow{\text{electric energy}} 2NO(g)$$

$$2NO(g) + O_2(g) \longrightarrow 2NO_2(g)$$

$$2NO_2(g) + H_2O(l) \longrightarrow HNO_2(aq) + HNO_3(aq)$$

About 30 million tons of HNO_3 are produced this way annually. Nitric acid is converted to nitrate salts in the soil. These nutrients are taken up by plants, which in turn are ingested by animals. Animals use the nutrients from plants to make proteins and other essential biomolecules. Denitrification reverses nitrogen fixation to complete the cycle. For example, certain anaerobic organisms decompose animal wastes as well as dead plants and animals to produce free molecular nitrogen from nitrates.

The main processes of the global oxygen cycle are shown in Figure 21.2. This cycle is complicated by the fact that oxygen takes so many different chemical forms. Atmospheric oxygen is removed through respiration and various industrial processes (mostly combustion), which produce carbon dioxide. Photosynthesis is the major mechanism by which molecular oxygen is regenerated from carbon dioxide and water.

TABLE 21.1	Composition of Dry Air at Sea Level
Gas	**Composition (% by Volume)**
N_2	78.03
O_2	20.99
Ar	0.94
CO_2	0.033
Ne	0.0015
He	0.000524
Kr	0.00014
Xe	0.000006

Figure 21.1 The nitrogen cycle. Although the supply of nitrogen in the atmosphere is virtually inexhaustible, it must be combined with hydrogen or oxygen before it can be assimilated by higher plants, which in turn are consumed by animals. Juvenile nitrogen is nitrogen that has not previously participated in the nitrogen cycle.

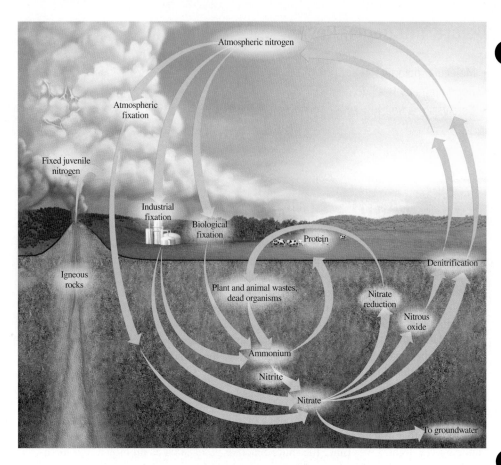

Figure 21.2 The oxygen cycle. The cycle is complicated because oxygen appears in so many chemical forms and combinations, primarily as molecular oxygen, in water, and in organic and inorganic compounds.

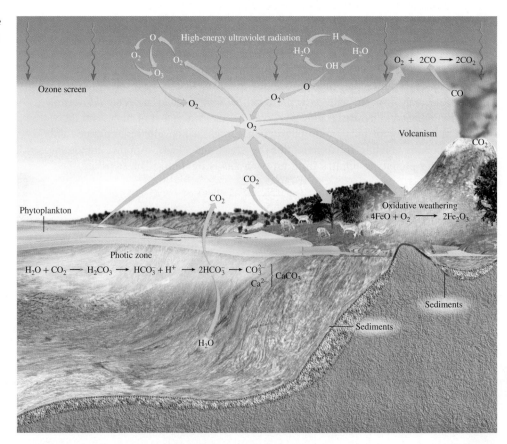

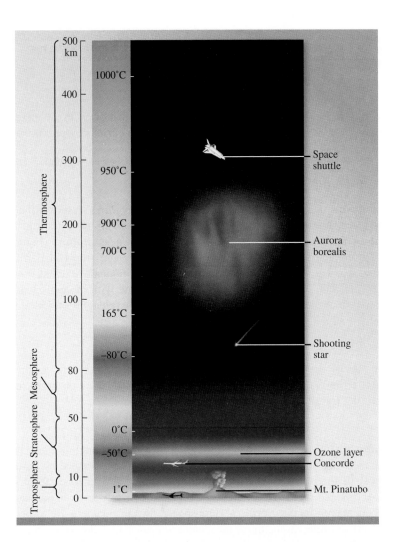

Scientists divide the atmosphere into several different layers according to temperature variation and composition (Figure 21.3). As far as visible events are concerned, the most active region is the *troposphere,* the layer of the atmosphere closest to Earth's surface. The troposphere contains about 80 percent of the total mass of air and practically all the atmosphere's water vapor. The troposphere is the thinnest layer of the atmosphere (10 km), but it is where all the dramatic events of weather—rain, lightning, hurricanes—occur. Temperature decreases almost linearly with increasing altitude in this region.

Above the troposphere is the *stratosphere,* which consists of nitrogen, oxygen, and ozone. In the stratosphere, the air temperature *increases* with altitude. This warming effect is the result of exothermic reactions triggered by UV radiation from the sun (see Section 21.3). One of the products of this reaction sequence is ozone (O_3), which serves to prevent harmful UV rays from reaching Earth's surface.

In the *mesosphere,* which is above the stratosphere, the concentration of ozone and other gases is low, and the temperature decreases again with increasing altitude. The *thermosphere,* or *ionosphere,* is the uppermost layer of the atmosphere. The increase in temperature in this region is the result of the bombardment of molecular oxygen and nitrogen and atomic species by energetic particles, such as electrons and protons, from the sun. Typical reactions are

$$N_2 \longrightarrow 2N \qquad \Delta H° = 941.4 \text{ kJ/mol}$$
$$N \longrightarrow N^+ + e^- \qquad \Delta H° = 1400 \text{ kJ/mol}$$
$$O_2 \longrightarrow O_2^+ + e^- \qquad \Delta H° = 1176 \text{ kJ/mol}$$

In reverse, these processes liberate the equivalent amount of energy, mostly as heat. Ionized particles are responsible for the reflection of radio waves back toward Earth.

Figure 21.4 Aurora borealis, commonly referred to as the northern lights.

© Dennis Fast/VisualWritten/The Image Works.

21.2 PHENOMENA IN THE OUTER LAYERS OF THE ATMOSPHERE

In this section we will discuss two dazzling phenomena that occur in the outer regions of the atmosphere. One is a natural event. The other is a curious by-product of human space travel.

Aurora Borealis and Aurora Australis

Violent eruptions on the surface of the sun, called *solar flares,* result in the ejection of myriad electrons and protons into space, where they disrupt radio transmission and provide us with spectacular celestial light shows known as *auroras* (Figure 21.4). These electrons and protons collide with the molecules and atoms in Earth's upper atmosphere, causing them to become ionized and electronically excited. Eventually, the excited molecules and ions return to the ground state with the emission of light. For example, an excited oxygen atom emits photons at wavelengths of 558 nm (green) and between 630 and 636 nm (red),

$$O^* \longrightarrow O + h\nu$$

where the asterisk denotes an electronically excited species and $h\nu$ denotes the emitted photon [◄◄ Section 3.3]. Similarly, the blue and violet colors often observed in auroras result from the transition in the ionized nitrogen molecule.

$$N_2^{+*} \longrightarrow N_2^+ + h\nu$$

The wavelengths for this transition fall between 391 and 470 nm.

The incoming streams of solar protons and electrons are oriented by Earth's magnetic field so that most auroral displays occur in doughnut-shaped zones about 2000 km in diameter centered on the North and South poles. *Aurora borealis* is the name given to this phenomenon in the Northern Hemisphere. In the Southern Hemisphere, it is called *aurora australis.* Sometimes, the number of solar particles is so immense that auroras are also visible from locations as far south as Olympia, Washington.

Worked Example 21.1 shows how to determine the maximum wavelength capable of breaking a chemical bond.

Worked Example 21.1

The bond enthalpy of O_2 is 498.7 kJ/mol. Calculate the maximum wavelength (in nanometers) of a photon that can cause the dissociation of an O_2 molecule.

Strategy We want to calculate the wavelength of a photon that will break an O=O bond. Therefore, we need the amount of energy in one bond. The bond energy of O_2 is given in units of kJ/mol. The units needed for the energy of one bond are J/molecule. Once we know the energy in one bond, we can calculate the minimum frequency and maximum wavelength needed to dissociate one O_2 molecule.

Setup The conversion steps are

$$\text{kJ/mol} \longrightarrow \text{J/molecule} \longrightarrow \text{frequency of photon} \longrightarrow \text{wavelength of photon}$$

Solution First we calculate the energy required to dissociate one O_2 molecule.

$$\text{energy per molecule} = \frac{498.7 \times 10^3 \text{ J}}{1 \text{ mol}} \times \frac{1 \text{ mol}}{6.022 \times 10^{23} \text{ molecules}} = 8.281 \times 10^{-19} \frac{\text{J}}{\text{molecule}}$$

The energy of the photon is given by $E = h\nu$ (Equation 3.4). Therefore,

$$\nu = \frac{E}{h} = \frac{8.281 \times 10^{-19} \text{ J}}{6.63 \times 10^{-34} \text{ J} \cdot \text{s}} = 1.25 \times 10^{15} \text{ s}^{-1}$$

Finally, we calculate the wavelength of the photon, given by $\lambda = c/v$ (Equation 3.3), as follows:

$$\lambda = \frac{3.00 \times 10^8 \text{ m/s}}{1.25 \times 10^{15} \text{ s}^{-1}} = 2.40 \times 10^{-7} \text{ m} = 240 \text{ nm}$$

Think About It
In principle, any photon with a wavelength of 240 nm or *shorter* can dissociate an O_2 molecule.

Practice Problem **A**TTEMPT   Calculate the wavelength (in nanometers) of a photon needed to dissociate an O_3 molecule.

$$O_3 \longrightarrow O + O_2 \qquad \Delta H° = 107.2 \text{ kJ/mol}$$

Practice Problem **B**UILD   Which of the following gaseous species is dissociated by visible light: CS_2, F_2, HI, ClF, HCN? (See Table 10.4.)

Practice Problem **C**ONCEPTUALIZE   Which of the graphs [(i)–(iv)] best represents the relationship between the wavelength of light and the maximum bond enthalpy it can dissociate?

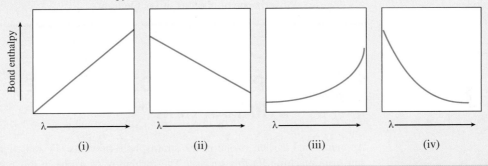

The Mystery Glow of Space Shuttles

A human-made light show that baffled scientists for several years was produced by space shuttles orbiting Earth. In 1983, astronauts first noticed an eerie orange glow on the outside surface of their spacecraft at an altitude about 300 km above Earth (Figure 21.5).

The light, which usually extends about 10 cm away from the protective silica heat tiles and other surface materials, is most pronounced on the parts of the shuttle facing its direction of travel. This fact led scientists to postulate that collision between oxygen atoms in the atmosphere and the fast-moving shuttle somehow produced the orange light. Spectroscopic measurements of the glow, as well as laboratory tests, strongly suggested that nitric oxide (NO) and nitrogen dioxide (NO_2) also played a part. It is believed that oxygen atoms interact with nitric oxide adsorbed on (i.e., bound to) the shuttle's surface to form electronically excited nitrogen dioxide.

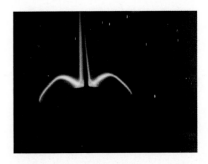

Figure 21.5 The glowing tail section of the space shuttle viewed from inside the vehicle.

NASA

$$O + NO \longrightarrow NO_2^*$$

As the NO_2^* leaves the shell of the spacecraft, it emits photons at a wavelength of 680 nm (orange).

$$NO_2^* \longrightarrow NO_2 + h\nu$$

Support for this explanation came inadvertently in 1991, when astronauts aboard *Discovery* released various gases, including carbon dioxide, neon, xenon, and nitric oxide, from the cargo bay in the course of an unrelated experiment. Expelled one at a time, these gases scattered onto the surface of the shuttle's tail. The nitric oxide caused the normal shuttle glow to intensify markedly, but the other gases had no effect on it.

What is the source of the nitric oxide on the outside of the spacecraft? Scientists believe that some of it may come from the exhaust gases emitted by the shuttle's rockets and that some of it is present in the surrounding atmosphere. The shuttle glow does not harm the vehicle, but it does interfere with spectroscopic measurements on distant objects made from the spacecraft.

Section 21.2 Review

Phenomena in the Outer Layers of the Atmosphere

21.2.1 What maximum wavelength (in nanometers) of light is necessary to break the bond in a nitrogen molecule (N_2)? (The bond energy of N_2 is 941.4 kJ/mol.)

(a) 127 nm   (d) 941 nm

(b) 211 nm   (e) 942 nm

(c) 236 nm

21.2.2 What process gives rise to the aurora borealis and the aurora australis?

(a) Formation of oxygen and nitrogen molecules

(b) Collisions between oxygen molecules

(c) Collisions between nitrogen molecules

(d) Emissions of photons of visible light

(e) Absorbance of photons of visible light

21.3 DEPLETION OF OZONE IN THE STRATOSPHERE

As mentioned earlier, ozone in the stratosphere prevents UV radiation emitted by the sun from reaching Earth's surface. The formation of ozone in this region begins with the *photodissociation* of oxygen molecules by solar radiation at wavelengths below 240 nm.

Equation 21.1

$$O_2 \xrightarrow{\text{UV} <240 \text{ nm}} O + O$$

The highly reactive O atoms combine with oxygen molecules to form ozone as follows:

Equation 21.2

$$O + O_2 + M \longrightarrow O_3 + M$$

where M is some inert substance such as N_2. The role of M in this exothermic reaction is to absorb some of the excess energy released and prevent the spontaneous decomposition of the O_3 molecule. The energy that is not absorbed by M is given off as heat. (As the M molecules themselves become de-excited, they release more heat to the surroundings.) In addition, ozone itself absorbs UV light between 200 and 300 nm.

Equation 21.3

$$O_3 \xrightarrow{\text{UV}} O + O_2$$

The process continues when O and O_2 recombine to form O_3 as shown in Equation 21.2, further warming the stratosphere.

If all the stratospheric ozone were compressed into a single layer at STP on Earth, that layer would be only about 3 mm thick! Although the concentration of ozone in the stratosphere is very low, it is sufficient to filter out (i.e., absorb) solar radiation in the 200- to 300-nm range (see Equation 21.3). In the stratosphere, it acts as our protective shield against UV radiation, which can induce skin cancer, cause genetic mutations, and destroy crops and other forms of vegetation.

The formation and destruction of ozone by natural processes is a dynamic equilibrium that maintains a constant concentration of ozone in the stratosphere. Since the mid-1970s scientists have been concerned about the harmful effects of CFCs on the ozone layer. Generally known by the trade name Freons, CFCs were first synthesized in the 1930s. Some of the common ones are $CFCl_3$ (Freon 11), CF_2Cl_2 (Freon 12), $C_2F_3Cl_3$ (Freon 113), and $C_2F_4Cl_2$ (Freon 114). Because these compounds are readily liquefied, relatively inert, nontoxic, noncombustible, and volatile, they have been used as coolants in refrigerators and air conditioners, in place of highly toxic liquid sulfur dioxide (SO_2) and ammonia (NH_3). Large quantities of CFCs are also used in the manufacture of disposable foam products such as cups and plates, as aerosol propellants in spray cans, and as solvents to clean newly soldered electronic circuit boards (Figure 21.6). In 1977, the peak year of production, nearly 1.5×10^6 tons of CFCs were produced in the United States. Most of the CFCs produced for commercial and industrial use are eventually discharged into the atmosphere.

Because of their relative inertness, the CFCs slowly diffuse unchanged up to the stratosphere, where UV radiation of wavelengths between 175 and 220 nm causes them to decompose.

$$CFCl_3 \longrightarrow CFCl_2 + Cl$$

$$CF_2Cl_2 \longrightarrow CF_2Cl + Cl$$

The reactive chlorine atoms then undergo the following reactions.

$$Cl + O_3 \longrightarrow ClO + O_2 \qquad \textbf{Equation 21.4}$$

$$ClO + O \longrightarrow Cl + O_2 \qquad \textbf{Equation 21.5}$$

The overall result (the sum of Equations 21.4 and 21.5) is the net removal of an O_3 molecule from the stratosphere.

$$O_3 + O \longrightarrow 2O_2 \qquad \textbf{Equation 21.6}$$

The oxygen atoms in Equation 21.5 are supplied by the photochemical decomposition of molecular oxygen and ozone described earlier. The Cl atom plays the role of a catalyst in the reaction mechanism scheme represented by Equations 21.4 and 21.5 because it is not used up and therefore can take part in many such reactions. In fact, one Cl atom can destroy up to 100,000 O_3 molecules before it is removed by some other reaction. The ClO (chlorine monoxide) species is an intermediate because it is produced in the first elementary step (Equation 21.4) and consumed in the second step (Equation 21.5). The preceding mechanism for the destruction of ozone has been supported by the detection of ClO in the stratosphere in recent years (Figure 21.7). As can be seen, the concentration of O_3 decreases in regions that have high amounts of ClO.

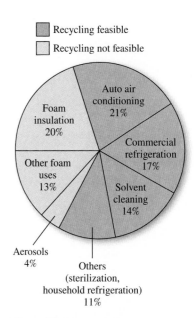

Recycling feasible
Recycling not feasible

Auto air conditioning 21%
Foam insulation 20%
Commercial refrigeration 17%
Other foam uses 13%
Solvent cleaning 14%
Aerosols 4%
Others (sterilization, household refrigeration) 11%

Figure 21.6 Uses of CFCs. Since 1978, the use of aerosol propellants has been banned in the United States.

Figure 21.7 The variations in concentrations of ClO and O_3 with latitude.

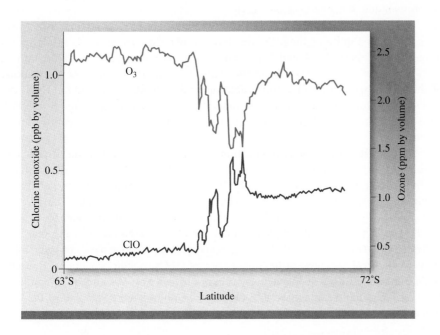

Another group of compounds that can destroy stratospheric ozone are the nitrogen oxides, generally denoted as NO_x. (Examples of NO_x are NO and NO_2.) These compounds come from the exhausts of high-altitude supersonic aircraft and from human and natural activities on Earth. Solar radiation decomposes a substantial amount of the other nitrogen oxides to nitric oxide (NO), which participates in the destruction of ozone as follows:

$$O_3 \longrightarrow O_2 + O$$
$$NO + O_3 \longrightarrow NO_2 + O_2$$
$$NO_2 + O \longrightarrow NO + O_2$$

Overall:
$$2O_3 \longrightarrow 3O_2$$

In this case, NO is the catalyst and NO_2 is the intermediate. Nitrogen dioxide also reacts with chlorine monoxide to form chlorine nitrate.

$$ClO + NO_2 \longrightarrow ClONO_2$$

Chlorine nitrate is relatively stable and serves as a "chlorine reservoir," which plays a role in the depletion of the stratospheric ozone over the North and South Poles.

Polar Ozone Holes

In the mid-1980s, evidence began to accumulate that an "Antarctic ozone hole" developed in late winter, depleting the stratospheric ozone over Antarctica by as much as 50 percent (Figure 21.8). In the stratosphere, a stream of air known as the "polar vortex" circles Antarctica in winter. Air trapped within this vortex becomes extremely cold during the polar night. This condition leads to the formation of ice particles known as polar stratospheric clouds (PSCs) (Figure 21.9). Acting as a heterogeneous catalyst, these PSCs provide a surface for reactions converting HCl (emitted from Earth) and chlorine nitrate to more reactive chlorine molecules.

$$HCl + ClONO_2 \longrightarrow Cl_2 + HNO_3$$

By early spring, the sunlight splits molecular chlorine into chlorine atoms,

$$Cl_2 + h\nu \longrightarrow 2Cl$$

which then attack ozone as shown earlier.

The situation is not as severe in the warmer Arctic region, where the vortex does not persist quite as long. Studies have shown that ozone levels in this region have declined between 4 and 8 percent in the past decade. Volcanic eruptions, such as that of Mount Pinatubo in the Philippines in 1991, inject large quantities of dust-sized particles and sulfuric acid aerosols into the atmosphere. These particles can perform the same catalytic function as the ice crystals at the South Pole. As a result, the Arctic hole is expected to grow larger for several years following an eruption.

Recognizing the serious implications of the loss of ozone in the stratosphere, nations throughout the world have acknowledged the need to drastically curtail or totally stop the production of CFCs. In 1978, the United States was one of the few countries to ban the use of CFCs in hair sprays and other aerosols. The Montreal protocol was signed by most industrialized nations in 1987, setting targets for cutbacks in CFC production and the complete elimination of these substances by the year 2000. Although progress has been made in this respect, many nations have not been able to abide by the treaty because of the importance of CFCs to their economies. Recycling could play a significant supplementary role in preventing CFCs already in appliances from escaping into the atmosphere. As Figure 21.6 shows, more than half of the CFCs in use are recoverable.

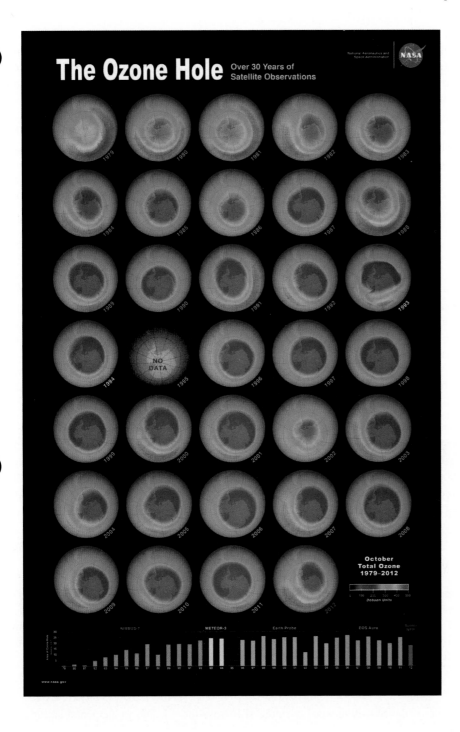

Figure 21.8 In recent decades, scientists have found that the ozone layer in the stratosphere over the South Pole has become thinner. This map, based on data collected over a number of years, shows the depletion of ozone in blue.
NASA

Figure 21.9 Polar stratospheric clouds containing ice particles can catalyze the formation of Cl atoms and lead to the destruction of ozone.

© *Time Life Pictures/NASA/Time Life Pictures/Getty Images*

An intense effort is underway to find CFC substitutes that are not harmful to the ozone layer. One of the promising candidates is called hydrochlorofluorocarbon-123, or HCFC-123 (CF_3CHCl_2). The presence of the hydrogen atom makes the compound more susceptible to oxidation in the lower atmosphere, so it never reaches the stratosphere. Specifically, it is attacked by the hydroxyl radical in the troposphere.

$$CF_3CHCl_2 + OH \longrightarrow CF_3CCl_2 + H_2O$$

The CF_3CCl_2 fragment reacts with oxygen, eventually decomposing to CO_2, water, and hydrogen halides that are removed by rainwater. Unfortunately, the same hydrogen atom also makes the compound more active biologically than the CFCs. Laboratory tests have shown that HCFC-123 can cause tumors in rats, although its toxic effect on humans is not known. Another promising group of compounds that can substitute for

CFCs are the hydrofluorocarbons (HFCs). Because they do not contain chlorine, HFCs will not promote the destruction of ozone even if they diffuse to the stratosphere. Examples of these compounds are CF_3CFH_2, CF_3CF_2H, CF_3CH_3, and CF_2HCH_3. In particular, CF_3CFH_2 is already widely used in place of CFCs in air-conditioning and refrigeration applications.

Although it is unclear whether the CFCs already released to the atmosphere will eventually result in catastrophic damage to life on Earth, it is conceivable that the depletion of ozone can be slowed by reducing the availability of Cl atoms. Indeed, some chemists have suggested sending a fleet of planes to spray 50,000 tons of ethane (C_2H_6) or propane (C_3H_8) high over the South Pole in an attempt to heal the hole in the ozone layer. Being a reactive species, the chlorine atom would react with the hydrocarbons as follows:

$$Cl + C_2H_6 \longrightarrow HCl + C_2H_5$$

$$Cl + C_3H_8 \longrightarrow HCl + C_3H_7$$

The products of these reactions would not affect the ozone concentration. A less realistic plan is to rejuvenate the ozone layer by producing large quantities of ozone and releasing it into the stratosphere from airplanes. Technically this solution is feasible, but it would be enormously costly and it would require the collaboration of many nations.

Having discussed the chemistry in the outer regions of Earth's atmosphere, we will focus in Sections 21.4 through 21.8 on events closer to us—that is, in the troposphere.

21.4 VOLCANOES

Volcanic eruptions, Earth's most spectacular natural displays of energy, are instrumental in forming large parts of Earth's crust. The upper mantle, immediately under the crust, is nearly molten. A slight increase in heat, such as that generated by the movement of one crustal plate under another, melts the rock. The molten rock, called *magma,* rises to the surface and generates some types of volcanic eruptions (Figure 21.10).

An active volcano emits gases, liquids, and solids. The gases spewed into the atmosphere include primarily N_2, CO_2, HCl, HF, H_2S, and water vapor. It is estimated that volcanoes are the source of about two-thirds of the sulfur in the air. On the slopes of Mount St. Helens, which last erupted in 1980, deposits of elemental sulfur are visible near the eruption site. At high temperatures, the hydrogen sulfide gas given off by a volcano is oxidized by air.

$$2H_2S(g) + 3O_2(g) \longrightarrow 2SO_2(g) + 2H_2O(g)$$

Some of the SO_2 is reduced by more H_2S from the volcano to elemental sulfur and water.

$$2H_2S(g) + SO_2(g) \longrightarrow 3S(s) + 2H_2O(g)$$

The rest of the SO_2 is released into the atmosphere, where it reacts with water to form acid rain (see Section 21.6).

The tremendous force of a volcanic eruption carries a sizable amount of gas into the stratosphere. There SO_2 is oxidized to SO_3, which is eventually converted to sulfuric acid aerosols in a series of complex reactions. In addition to destroying ozone in the stratosphere (see Section 21.3), these aerosols can also affect climate. Because the stratosphere is above the atmospheric weather patterns, the aerosol clouds often persist for more than a year. They absorb solar radiation and thereby cause a drop in temperature at Earth's surface. However, this cooling effect is local rather than global, because it depends on the site and frequency of volcanic eruptions.

Figure 21.10 A volcanic eruption.
© Jim Sugar/Corbis

21.5 THE GREENHOUSE EFFECT

Although carbon dioxide is only a trace gas in Earth's atmosphere, with a concentration of about 0.033 percent by volume (see Table 21.1), it plays a critical role in controlling our climate. The so-called **greenhouse effect** describes the trapping of heat near Earth's surface by gases in the atmosphere, particularly carbon dioxide. The glass roof of a greenhouse transmits visible sunlight and absorbs some of the outgoing infrared (IR) radiation, thereby trapping the heat. Carbon dioxide acts somewhat like a glass roof, except that the temperature rise in the greenhouse is due mainly to the restricted air circulation inside. Calculations show that if the atmosphere did not contain carbon dioxide, Earth would be 30°C cooler!

Figure 21.11 shows the carbon cycle in our global ecosystem. The transfer of carbon dioxide to and from the atmosphere is an essential part of the carbon cycle. Carbon dioxide is produced when any form of carbon or a carbon-containing compound is burned in an excess of oxygen. Many carbonates give off CO_2 when heated, and all give off CO_2 when treated with acid.

$$CaCO_3(s) \longrightarrow CaO(s) + CO_2(g)$$
$$CaCO_3(s) + 2HCl(aq) \longrightarrow CaCl_2(aq) + H_2O(l) + CO_2(g)$$

Carbon dioxide is also a by-product of the fermentation of sugar.

$$\underset{\text{glucose}}{C_6H_{12}O_6(aq)} \xrightarrow{\text{yeast}} \underset{\text{ethanol}}{2C_2H_5OH(aq)} + 2CO_2(g)$$

Figure 21.11 The carbon cycle.

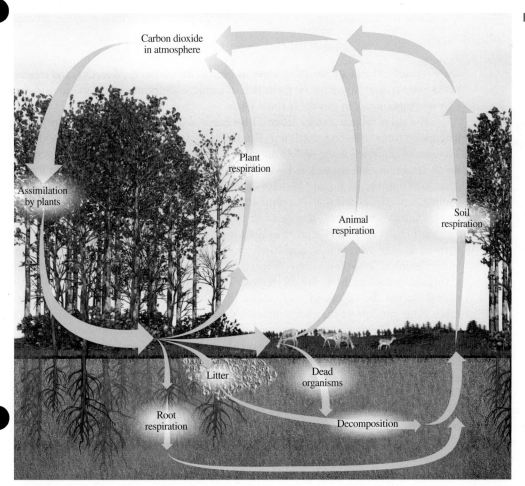

Figure 21.12 The incoming radiation from the sun and the outgoing radiation from Earth's surface.

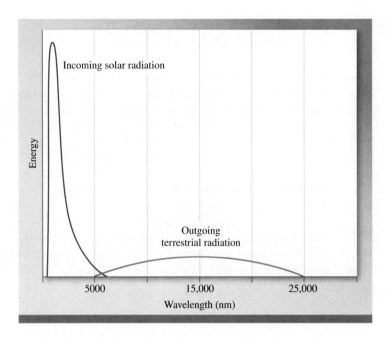

Carbohydrates and other complex carbon-containing molecules are consumed by animals, which respire and release CO_2 as an end product of metabolism.

$$C_6H_{12}O_6(aq) + 6O_2(g) \longrightarrow 6CO_2(g) + 6H_2O(l)$$

As mentioned earlier, another major source of CO_2 is volcanic activity. Carbon dioxide is removed from the atmosphere by photosynthetic plants and certain microorganisms.

$$6CO_2(g) + 6H_2O(l) \longrightarrow C_6H_{12}O_6(aq) + 6O_2(g)$$

After plants and animals die, the carbon in their tissues is oxidized to CO_2 and returns to the atmosphere. In addition, there is a dynamic equilibrium between atmospheric CO_2 and carbonates in the oceans and lakes.

Stable form

Stretched

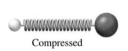

Compressed

Figure 21.13 Vibrational motion of a diatomic molecule. Chemical bonds can be stretched and compressed like a spring.

The solar radiant energy received by Earth is distributed over a band of wavelengths between 100 and 5000 nm, but much of it is concentrated in the 400- to 700-nm range, which is the visible region of the spectrum (Figure 21.12). By contrast, the thermal radiation emitted by Earth's surface is characterized by wavelengths longer than 4000 nm (the IR region) because of the much lower average surface temperature compared to that of the sun. The outgoing IR radiation can be absorbed by water and carbon dioxide, but not by nitrogen and oxygen.

All molecules vibrate, even at the lowest temperatures. The energy associated with molecular vibration is quantized, much like the electronic energies of atoms and molecules. To vibrate more energetically, a molecule must absorb a photon of a specific wavelength in the IR region. First, however, its dipole moment *must* change during the course of a vibration. [Recall that the dipole moment of a molecule is the product of the charge and the distance between charges (Equation 6.1).] Figure 21.13 shows how a diatomic molecule can vibrate. If the molecule is homonuclear like N_2 and O_2, there can be no change in the dipole moment; the molecule has a zero dipole moment no matter how far apart or close together the two atoms are. We call such molecules IR-inactive because they *cannot* absorb IR radiation. On the other hand, all heteronuclear diatomic molecules are IR-active; that is, they all can absorb IR radiation because their dipole moments constantly change as the bond lengths change.

A *polyatomic* molecule can vibrate in more than one way. Water, for example, can vibrate in three different ways as shown in Figure 21.14. Because water is a polar molecule, any of these vibrations results in a change in dipole moment because there

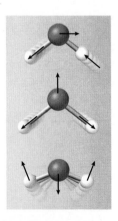

Figure 21.14 The three different modes of vibration of a water molecule. Each mode of vibration can be imagined by moving the atoms along the arrows and then reversing the direction of motion.

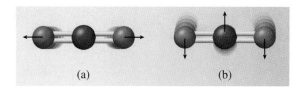

(a)
(b)

Figure 21.15 Two of the four ways a carbon dioxide molecule can vibrate. The vibration in (a) does not result in a change in dipole moment, but the vibration in (b) renders the molecule IR-active.

is a change in bond length. Therefore, an H_2O molecule is IR-active. Carbon dioxide has a linear geometry and is nonpolar. Figure 21.15 shows two of the four ways a CO_2 molecule can vibrate. One of them [Figure 21.15(a)] symmetrically displaces atoms from the center of gravity and will not create a dipole moment, but the other vibration [Figure 21.15(b)] is IR-active because the dipole moment changes from zero to a maximum value in one direction and then reaches the same maximum value when it changes to the other extreme position.

Upon receiving a photon in the IR region, a molecule of H_2O or CO_2 is promoted to a higher vibrational energy level (the asterisk denotes a vibrationally excited molecule).

$$H_2O + h\nu \longrightarrow H_2O^*$$

$$CO_2 + h\nu \longrightarrow CO_2^*$$

These energetically excited molecules soon lose their excess energy either by collision with other molecules or by spontaneous emission of radiation. Part of this radiation is emitted to outer space and part returns to Earth's surface.

Although the total amount of water vapor in our atmosphere has not altered noticeably over the years, the concentration of CO_2 has been rising steadily since the turn of the century as a result of the burning of fossil fuels (petroleum, natural gas, and coal). Figure 21.16 shows the percentages of CO_2 emitted due to human activities in the United States from 1990 to 2010, and Figure 21.17 shows the variation of carbon dioxide concentration over a period of years, as measured in Hawaii. In the Northern Hemisphere, the seasonal oscillations are caused by the removal of carbon dioxide by photosynthesis during the growing season and its buildup during the fall and winter months. The trend is toward an increase in CO_2. The current rate of increase is more than 1 ppm (>1 part CO_2 per million parts air) by volume per year, which is equivalent to roughly 10^{10} tons of CO_2! According to the website co2now.org, the CO_2 level exceeded 400 ppm in June 2015.

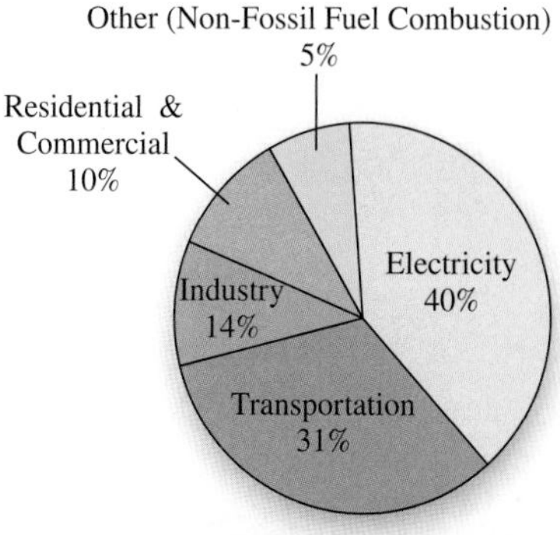

Other (Non-Fossil Fuel Combustion)
5%
Residential &
Commercial
10%
Industry
14%
Electricity
40%
Transportation
31%

Figure 21.16 Sources of carbon dioxide emission in the United States. Note that not all the emitted CO_2 enters the atmosphere. Some of it is taken up by carbon dioxide "sinks," such as the ocean.

380
360
340
320
CO_2 concentration (ppm by volume)
1960
1970
1980
1990
2000
2010
Year

Figure 21.17 Yearly variation of carbon dioxide concentration at Mauna Loa, Hawaii, the source of the longest running record of atmospheric CO_2. The general trend clearly points to an increase of carbon dioxide in the atmosphere.

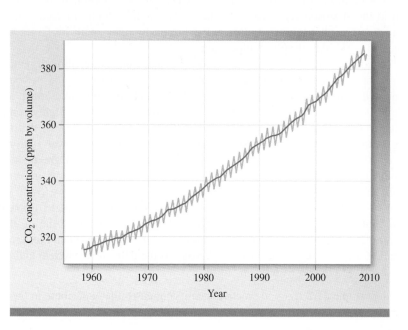

Figure 21.18 Temperature rise on Earth's surface from 1900 to 2002.

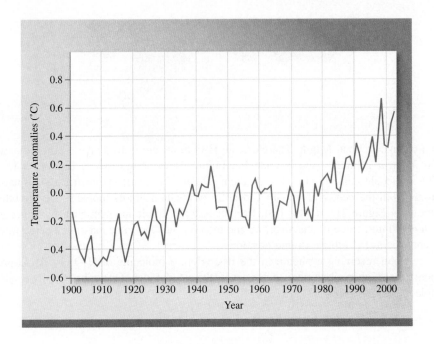

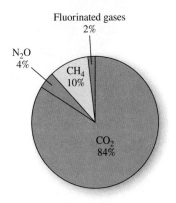

Figure 21.19 Contribution to global warming by various greenhouse gases. The concentrations of fluorinated gases and methane are much lower than that of carbon dioxide. However, because they can absorb IR radiation much more effectively than CO_2, they make significant contributions to the overall warming effect.

In addition to CO_2 and H_2O, other greenhouse gases, such as the CFCs, CH_4, NO_x, and N_2O, also contribute appreciably to the warming of the atmosphere. Figure 21.18 shows the gradual increase in temperature over the years, and Figure 21.19 shows the relative contributions of the greenhouse gases to global warming.

It is predicted by some meteorologists that should the buildup of greenhouse gases continue at its current rate, Earth's average temperature will increase by about $1°$ to $3°C$ in the twenty-first century. Although a temperature increase of a few degrees may seem insignificant, it is actually large enough to disrupt the delicate thermal balance on Earth and could cause glaciers and ice caps to melt. Consequently, the sea level would rise and coastal areas would be flooded. Predicting weather trends is extremely difficult, though, and there are other potentially moderating factors to take into account before concluding that global warming is inevitable and irreversible. For example, the ash from volcanic eruptions diffuses upward and can stay in the atmosphere for years. By reflecting incoming sunlight, volcanic ash can cause a cooling effect. Furthermore, the warming effect of CFCs in the troposphere is offset by its action in the stratosphere. Because ozone is a polar polyatomic molecule, it is also an effective greenhouse gas. A decrease in ozone brought about by CFCs actually produces a noticeable drop in temperature.

To combat the greenhouse effect, we must lower carbon dioxide emissions. This can be done by improving energy efficiency in automobiles and in household heating and lighting, and by developing nonfossil fuel energy sources, such as photovoltaic cells. Nuclear energy is a viable alternative, but its use is highly controversial due to the difficulty of disposing of radioactive waste and the fact that nuclear power stations are more prone to accidents than conventional power stations (see Chapter 20). The proposed phasing out of CFCs, the most potent greenhouse gas, will help to slow the warming trend. The recovery of methane gas generated at landfills and the reduction of natural gas leakages are other steps we could take to control CO_2 emissions. Finally, the preservation of the Amazon jungle, tropical forests in Southeast Asia, and other large forests is vital to maintaining the steady-state concentration of CO_2 in the atmosphere. Converting forests to farmland for crops and grassland for cattle may do irreparable damage to the delicate ecosystem and permanently alter the climate pattern on Earth.

Worked Example 21.2 lets you practice identifying gases that contribute to the greenhouse effect.

Student Note: Global climate change is the subject of the Academy Award–winning 2006 documentary, *An Inconvenient Truth*, presented by former vice president Al Gore.

Worked Example 21.2

Which of the following qualify as greenhouse gases: CO, NO, NO₂, Cl₂, H₂, Ne?

Strategy To behave as a greenhouse gas, either the molecule must possess a dipole moment or some of its vibrational motions must generate a temporary dipole moment.

Setup The necessary conditions immediately rule out homonuclear diatomic molecules and atomic species.

Solution Only CO, NO, and NO₂, which are all polar molecules, qualify as greenhouse gases. Both Cl₂ and H₂ are homonuclear diatomic molecules, and Ne is atomic. These three species are all IR-inactive.

> **Think About It**
> CO_2, the best-known greenhouse gas, is *nonpolar*. It is only necessary for at least one of a molecule's vibrational modes to induce a *temporary* dipole for it to act as a greenhouse gas.

Practice Problem Ⓐ**TTEMPT** Which of the following is a more effective greenhouse gas: CO or H_2O? Explain.

Practice Problem Ⓑ**UILD** Both O_2 and O_3 exhibit molecular vibration. Despite this, only O_3 acts as a greenhouse gas. Explain.

Practice Problem Ⓒ**ONCEPTUALIZE** Which of these molecules qualifies as a greenhouse gas?

$$Cl_2 \qquad SO_2 \qquad SO_3 \qquad CH_4 \qquad N_2 \qquad CS_2$$

Section 21.5 Review

The Greenhouse Effect

21.5.1 Which of the following can act as a greenhouse gas? (Select all that apply.)
(a) CH_4 (c) Rn   (e) Xe
(b) N_2 (d) O_3

21.5.2 The greenhouse effect is
(a) caused by depletion of stratospheric ozone.
(b) entirely the result of human activity.
(c) a natural phenomenon that has been enhanced by human activity.
(d) the absorption of the sun's energy by molecules and atoms in the upper atmosphere.
(e) responsible for the aurora borealis and the aurora australis.

21.6 ACID RAIN

Every year acid rain causes hundreds of millions of dollars' worth of damage to stone buildings and statues throughout the world. The term *stone leprosy* is used by some environmental chemists to describe the corrosion of stone by acid rain (Figure 21.20). Acid rain is also toxic to vegetation and aquatic life. Many well-documented cases show dramatically how acid rain has destroyed agricultural and forest lands and killed aquatic organisms.

Precipitation in the northeastern United States has an average pH of about 4.3 (Figure 21.21). Because atmospheric CO_2 in equilibrium with rainwater would not be expected to result in a pH less than 5.5, sulfur dioxide (SO_2) and, to a lesser extent, nitrogen oxides from auto emissions are believed to be responsible for the high acidity of rainwater. Acidic oxides, such as SO_2, react with water to give the corresponding acids. There are several sources of atmospheric SO_2. Nature itself contributes much

Figure 21.20 The effect of acid rain on a marble statue. The photos were taken just a few decades apart.

(first): © NYC Parks Photo Archive/Fundamental Photographs; (second): © Kristen Brochmann/ Fundamental Photographs

Figure 21.21 Mean precipitation pH in the United States in 1994. Most SO_2 comes from the midwestern states. Prevailing winds carry the acid droplets formed over the Northeast. Nitrogen oxides also contribute to acid rain formation.

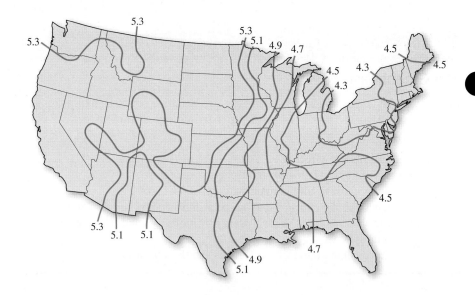

Figure 21.22 Sulfur dioxide and other air pollutants being released into the atmosphere from a coal-burning power plant.

© Larry Lee Photography/Corbis

SO_2 in the form of volcanic eruptions. Also, many metals exist combined with sulfur in nature. Extracting the metals often entails *smelting,* or *roasting,* the ores—that is, heating the metal sulfide in air to form the metal oxide and SO_2. For example:

$$2ZnS(s) + 3O_2(g) \longrightarrow 2ZnO(s) + 2SO_2(g)$$

The metal oxide can be reduced more easily than the sulfide (by a more reactive metal or in some cases by carbon) to the free metal.

Although smelting is a major source of SO_2, the burning of fossil fuels in industry, in power plants, and in homes accounts for most of the SO_2 emitted to the atmosphere (Figure 21.22). The sulfur content of coal ranges from 0.5 to 5 percent by mass, depending on the source of the coal. The sulfur content of other fossil fuels is similarly variable. Oil from the Middle East, for instance, is low in sulfur, whereas that from Venezuela has a high sulfur content. To a lesser extent, the nitrogen-containing compounds in oil and coal are converted to nitrogen oxides, which can also acidify rainwater.

All in all, some 50 to 60 million tons of SO_2 are released into the atmosphere each year! In the troposphere, SO_2 is almost all oxidized to H_2SO_4 in the form of aerosol, which ends up in wet precipitation or acid rain. The mechanism for the conversion of SO_2 to H_2SO_4 is quite complex and not fully understood. The reaction is believed to be initiated by the hydroxyl radical (OH).

$$OH + SO_2 \longrightarrow HOSO_2$$

The $HOSO_2$ radical is further oxidized to SO_3.

$$HOSO_2 + O_2 \longrightarrow HO_2 + SO_3$$

The sulfur trioxide formed would then rapidly react with water to form sulfuric acid.

$$SO_3 + H_2O \longrightarrow H_2SO_4$$

SO_2 can also be oxidized to SO_3 and then converted to H_2SO_4 on particles by heterogeneous catalysis. Eventually, the acid rain can corrode limestone and marble ($CaCO_3$). A typical reaction is

$$CaCO_3(s) + H_2SO_4(aq) \longrightarrow CaSO_4(s) + H_2O(l) + CO_2(g)$$

Sulfur dioxide can also attack calcium carbonate directly.

$$2CaCO_3(s) + 2SO_2(g) + O_2(g) \longrightarrow 2CaSO_4(s) + 2CO_2(g)$$

There are two ways to minimize the effects of SO_2 pollution. The most direct approach is to remove sulfur from fossil fuels before combustion, but this is technologically difficult to accomplish. A cheaper but less efficient way is to remove SO_2 as it is formed. For example, in one process powdered limestone is injected into the power plant boiler or furnace along with the coal (Figure 21.23). At high temperatures, the following decomposition occurs.

$$\underset{\text{limestone}}{CaCO_3(s)} \longrightarrow \underset{\text{quicklime}}{CaO(s)} + CO_2(g)$$

The quicklime reacts with SO_2 to form calcium sulfite and some calcium sulfate.

$$CaO(s) + SO_2(g) \longrightarrow CaSO_3(s)$$

$$2CaO(s) + 2SO_2(g) + O_2(g) \longrightarrow 2CaSO_4(s)$$

To remove any remaining SO_2, an aqueous suspension of quicklime is injected into a purification chamber prior to the gases' escape through the smokestack.

Video 21.1
Oil refining process.

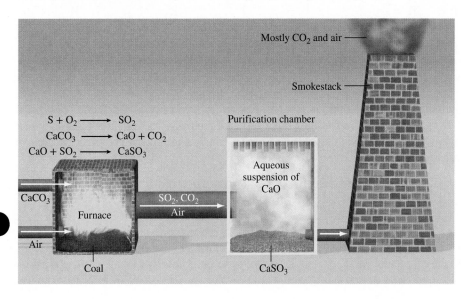

Figure 21.23 Common procedure for removing SO_2 from burning fossil fuel. Powdered limestone decomposes into CaO, which reacts with SO_2 to form $CaSO_3$. The remaining SO_2 is combined with an aqueous suspension of CaO to form $CaSO_3$.

Figure 21.24 Spreading calcium oxide (CaO) over acidified soil. This process is called liming.

© Wayne Hutchinson/Alamy

Quicklime is also added to lakes and soils in a process called *liming* to reduce their acidity (Figure 21.24). Installing a sulfuric acid plant near a metal ore refining site is also an effective way to cut SO_2 emission, because the SO_2 produced by roasting metal sulfides can be captured for use in the synthesis of sulfuric acid. This is a sensible way to turn what is a pollutant in one process into a starting material for another process!

21.7   PHOTOCHEMICAL SMOG

The word *smog* was coined to describe the combination of smoke and fog that shrouded London during the 1950s. The primary cause of this noxious cloud was sulfur dioxide. Today, however, ***photochemical smog,*** which is formed by the reactions of automobile exhaust in the presence of sunlight, is much more common.

Automobile exhaust consists mainly of NO, CO, and various unburned hydrocarbons. These gases are called *primary pollutants* because they set in motion a series of photochemical reactions that produce *secondary pollutants.* It is the secondary pollutants—chiefly NO_2 and O_3—that are responsible for the buildup of smog.

Nitric oxide is the product of the reaction between atmospheric nitrogen and oxygen at high temperatures inside an automobile engine.

$$N_2(g) + O_2(g) \longrightarrow 2NO(g)$$

Once released into the atmosphere, nitric oxide is oxidized to nitrogen dioxide.

$$2NO(g) + O_2(g) \longrightarrow 2NO_2(g)$$

Sunlight causes the photochemical decomposition of NO_2 (at a wavelength shorter than 400 nm) into NO and O.

$$NO_2(g) + hv \longrightarrow NO(g) + O(g)$$

Atomic oxygen is a highly reactive species that can initiate a number of important reactions, one of which is the formation of ozone,

$$O(g) + O_2(g) + M \longrightarrow O_3(g) + M$$

where M is some inert substance such as N_2. Ozone attacks the C=C linkage in rubber,

$$\underset{R}{\overset{R}{\diagdown}}C=C\underset{R}{\overset{R}{\diagup}} + O_3 \longrightarrow \underset{R}{\overset{R}{\diagdown}}C\underset{O-O}{\overset{O}{\diagup}}C\underset{R}{\overset{R}{\diagup}} \xrightarrow{H_2O} \underset{R}{\overset{R}{\diagdown}}C=O + O=C\underset{R}{\overset{R}{\diagup}} + H_2O_2$$

where R represents groups of C and H atoms. In smog-ridden areas, this reaction can cause automobile tires to crack. Similar reactions are also damaging to lung tissues and other biological substances.

Ozone can be formed also by a series of complex reactions involving unburned hydrocarbons, nitrogen oxides, and oxygen. One of the products of these reactions is peroxyacetyl nitrate (PAN).

$$CH_3-\underset{\underset{O}{\|}}{C}-O-O-NO_2$$

PAN is a powerful lachrymator, or tear producer, and causes breathing difficulties.

Figure 21.25 shows typical variations with time of primary and secondary pollutants. Initially, the concentration of NO_2 is quite low. As soon as solar radiation penetrates the atmosphere, though, more NO_2 is formed from NO and O_2. The

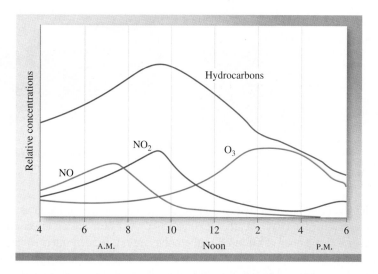

Figure 21.25 Typical variations with time in concentration of air pollutants on a smoggy day.

concentration of ozone remains fairly constant at a low level in the early morning hours. As the concentration of unburned hydrocarbons and aldehydes increases in the air, the concentrations of NO_2 and O_3 also rise rapidly. The actual amounts depend on the location, traffic, and weather conditions, but their presence is always accompanied by haze (Figure 21.26). The oxidation of hydrocarbons produces various organic intermediates, such as alcohols and carboxylic acids, which are all less volatile than the hydrocarbons themselves. These substances eventually condense into small droplets of liquid. The dispersion of these droplets in air, called an *aerosol,* scatters sunlight and reduces visibility. This interaction also makes the air look hazy.

As the mechanism of photochemical smog formation has become better understood, major efforts have been made to reduce the buildup of primary pollutants. Most automobiles now are equipped with catalytic converters designed to oxidize CO and unburned hydrocarbons to CO_2 and H_2O and to reduce NO and NO_2 to N_2 and O_2 [◄◄ Section 19.8]. More efficient automobile engines and better public transportation systems would also help to decrease air pollution in urban areas. A recent technological innovation to combat photochemical smog is to coat automobile radiators and air conditioner compressors with a platinum catalyst. So equipped, a running car can purify the air that flows under the hood by converting ozone and carbon monoxide to oxygen and carbon dioxide.

Figure 21.26 A smoggy day in a big city.
© Kent Knudson/PhotoLink/Getty Images

$$O_3(g) + CO(g) \xrightarrow{\text{Pt}} O_2(g) + CO_2(g)$$

In a city like Los Angeles, where the number of miles driven in one day equals nearly 300 million, this approach would significantly improve the air quality and reduce the "high-ozone level" warnings frequently issued to its residents.

21.8 INDOOR POLLUTION

Difficult as it is to avoid air pollution outdoors, it is no easier to avoid pollution indoors. The air quality in homes and in the workplace is affected by human activities, by construction materials, and by other factors in our immediate environment. The common indoor pollutants are radon, carbon monoxide, carbon dioxide, and formaldehyde.

The Risk from Radon

In a highly publicized case in 1984, an employee reporting for work at a nuclear power plant in Pennsylvania set off the plant's radiation monitor. Astonishingly, the source of his contamination turned out not to be the plant, but radon in his home!

Figure 21.27 (a) Sources of background radiation. (b) Radon occurrence in the United States.

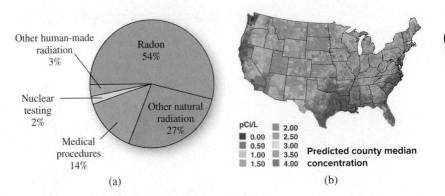

(a)   (b)

A lot has been said and written about the potential dangers of radon as an air pollutant. Just what is radon? Where does it come from? And how does it affect our health?

Radon is a member of Group 8A (the noble gases). It is an intermediate product of the radioactive decay of uranium-238 (see Figure 20.3). All isotopes of radon are radioactive, but radon-222 is the most hazardous because it has the longest half-life—3.8 days. Radon, which accounts for slightly over half the background radioactivity on Earth, is generated mostly from the phosphate minerals of uranium (Figure 21.27).

Since the 1970s, high levels of radon have been detected in homes built on reclaimed land above uranium mill tailing deposits. The colorless, odorless, and tasteless radon gas enters a building through tiny cracks in the basement floor. It is slightly soluble in water, so it can spread in different media. Radon-222 is an α-emitter. When it decays, it produces radioactive polonium-214 and polonium-218, which can build up to high levels in an enclosed space. These solid radioactive particles can adhere to airborne dust and smoke, which are inhaled into the lungs and deposited in the respiratory tract. Over a long period of time, the α particles emitted by polonium and its decay products, which are also radioactive, can cause lung cancer.

What can be done to combat radon pollution indoors? The first step is to measure the radon level in the basement with a reliable test kit. Short-term and long-term kits are available (Figure 21.28). The short-term tests use activated charcoal to collect the decay products of radon over a period of several days. The container is sent to a laboratory where a technician measures the radioactivity (γ rays) from radon-decay products lead-214 and bismuth-214. Knowing the length of exposure, the lab technician back-calculates to determine radon concentration. The long-term test kits use a piece of special polymer film on which an α particle will leave a "track." After several months' exposure, the film is etched with a sodium hydroxide solution and the number of tracks counted. Knowing the length of exposure enables the technician to calculate the radon concentration. If the radon level is unacceptably high, then the house must be regularly ventilated. This precaution is particularly important in recently built houses, which are well insulated. A more effective way to prevent radon pollution is to reroute the gas before it gets into the house (e.g., by installing a ventilation duct to draw air from beneath the basement floor to the outside).

Currently there is considerable controversy regarding the health effects of radon. The first detailed studies of the effects of radon on human health were carried out in the 1950s when it was recognized that uranium miners suffered from an abnormally high incidence of lung cancer. Some scientists have challenged the validity of these studies because the miners were also smokers. It seems quite likely that there is a synergistic effect between radon and smoking on the development of lung cancer. Radon decay products will adhere not only to tobacco tar deposits in the lungs, but also to the solid particles in cigarette smoke, which can be inhaled by smokers and nonsmokers. More systematic studies are needed to evaluate the environmental impact of radon. In the meantime, the Environmental Protection Agency (EPA) has recommended remedial action where the radioactivity level due to radon exceeds 4 picocuries (pCi) per liter of air. [A curie corresponds to 3.70×10^{10} disintegrations of radioactive nuclei per second; a picocurie is a trillionth of a curie, or 3.70×10^{-2} disintegrations per second (dps).]

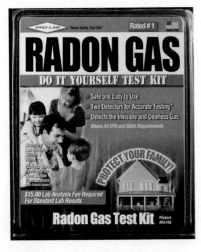

Figure 21.28 Home radon detector.

Worked Example 21.3 explores the first-order kinetics of nuclear decay.

Worked Example 21.3

The half-life of Rn-222 is 3.8 days. Starting with 1.0 g of Rn-222, how much will be left after 10 half-lives?

Strategy All radioactive decays obey first-order kinetics, making the half-life independent of the initial concentration.

Setup Because the question involves an integral number of half-lives, we can deduce the amount of Rn-222 remaining without using Equation 19.3.

Solution After one half-life, the amount of Rn left is 0.5×1.0 g, or 0.5 g. After two half-lives, only 0.25 g of Rn remains. Generalizing the fraction of the isotope left after n half-lives as $(1/2)^n$, where $n = 10$, we write

$$\text{quantity of Rn-222 left} = 1.0 \text{ g} \times (1/2)^{10}$$
$$= 9.8 \times 10^{-4} \text{ g}$$

Think About It

An alternative solution is to calculate the first-order rate constant from the half-life and use Equation 19.3:

$$\ln \frac{N_t}{N_0} = -kt$$

where N is the mass of Rn-222. Try this and verify that your answers are the same. (Since most of the kinetics problems we encounter do not involve an integral number of half-lives, we generally use Equation 19.3 as the *first* approach to solving them.)

Practice Problem **A**TTEMPT The concentration of Rn-222 in the basement of a house is 1.8×10^{-6} mol/L. Assume the air remains static, and calculate the concentration of the radon after 2.4 days.

Practice Problem **B**UILD How long will it take for the radioactivity due to radon to fall to a level considered acceptable by the EPA if the starting activity is 2.25×10^3 dps/L?

Practice Problem **C**ONCEPTUALIZE The first diagram represents a sample of radon gas. Which of the diagrams [(i)–(iii)] best represents the system after about 10 days?

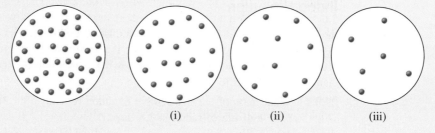

(i)   (ii)   (iii)

Carbon Dioxide and Carbon Monoxide

Both carbon dioxide (CO_2) and carbon monoxide (CO) are products of combustion.

In the presence of an abundant supply of oxygen, CO_2 is formed; in a limited supply of oxygen, both CO and CO_2 are formed. The indoor sources of these gases are gas cooking ranges, woodstoves, space heaters, tobacco smoke, human respiration, and exhaust fumes from cars (in garages). Carbon dioxide is not a toxic gas, but it does have an asphyxiating effect. In airtight buildings, the concentration of CO_2 can reach as high as 2000 ppm by volume (compared with 3 ppm outdoors). Workers exposed to high concentrations of CO_2 in skyscrapers and other sealed environments become fatigued more easily and have difficulty concentrating. Adequate ventilation is the solution to CO_2 pollution.

Like CO_2, CO is a colorless and odorless gas, but it differs from CO_2 in that it is highly poisonous. The toxicity of CO lies in its unusual ability to bind very strongly to hemoglobin, the oxygen carrier in blood. Both O_2 and CO bind to the Fe(II) ion in hemoglobin, but the affinity of hemoglobin for CO is about 200 times greater than it is for O_2. Hemoglobin molecules with tightly bound CO (called carboxyhemoglobin) cannot carry the oxygen needed for metabolic processes. At a concentration of 70 ppm, CO can cause drowsiness and headache; at higher concentrations, death may result when about half the hemoglobin molecules become complexed with CO. The best first-aid response to CO poisoning is to remove the victim immediately to an area with a plentiful oxygen supply or to give mouth-to-mouth resuscitation.

Formaldehyde

Formaldehyde (CH_2O) is a rather disagreeable-smelling liquid used as a preservative for laboratory specimens. Industrially, formaldehyde resins are used as bonding agents in building and furniture construction materials such as plywood and particle board. In addition, urea-formaldehyde insulation foams are used to fill wall cavities. The resins and foams slowly break down to release free formaldehyde, especially under acidic and humid conditions. Low concentrations of formaldehyde in the air can cause drowsiness, nausea, headaches, and other respiratory ailments. Laboratory tests show that breathing high concentrations of formaldehyde can induce cancers in animals, but whether it has a similar effect in humans is unclear. The safe standard of formaldehyde in indoor air has been set at 0.1 ppm by volume.

Because formaldehyde is a reducing agent, devices have been constructed to remove it by means of a redox reaction. Indoor air is circulated through an air purifier containing an oxidant such as $Al_2O_3/KMnO_4$, which converts formaldehyde to the less harmful and less volatile formic acid (HCOOH). Proper ventilation is the best way to remove formaldehyde. However, care should be taken not to remove the air from a room too quickly without replenishment, because a reduced pressure would cause the formaldehyde resins to decompose faster, resulting in the release of *more* formaldehyde.

Section 21.8 Review

Indoor Pollution

21.8.1   What is the risk posed by radon gas in homes?
    (a) respiratory distress   (d) coma
    (b) nausea   (e) lung cancer
    (c) dizziness

21.8.2   What mass of a 1.0-g sample of radon remains after 30 days? (The half-life of radon is 3.8 days.)
    (a) 0.030 g   (d) 0.97 g
    (b) 0.0040 g   (e) 0.0010 g
    (c) 0.99 g

Learning Outcomes

- Understand the process of nitrogen fixation.
- Name the regions of Earth's atmosphere.
- Know the process that gives rise to the aurora borealis and aurora australis.
- Recall the role of ozone in the stratosphere and how chlorofluorocarbons have affected the levels of ozone in the atmosphere.
- Identify the effects of volcanic eruptions.
- Understand the greenhouse effect and how it influences global warming.
- Recognize how sulfur dioxide causes acid rain.
- Understand how photochemical smog forms.
- Describe the major indoor air pollutants.

Chapter Summary

SECTION 21.1

- Earth's atmosphere is made up mainly of nitrogen and oxygen, plus a number of other trace gases. Molecular nitrogen in the atmosphere is incorporated into other compounds via **nitrogen fixation.**
- The regions of the atmosphere, from Earth's surface outward, are the **troposphere,** the **stratosphere,** the **mesosphere,** the **thermosphere,** and the **ionosphere.**
- The chemical processes that go on in the atmosphere are influenced by solar radiation, volcanic eruption, and human activities.

SECTION 21.2

- In the outer regions of the atmosphere, the bombardment of molecules and atoms by solar particles gives rise to the *aurora borealis* in the Northern Hemisphere and the *aurora australis* in the Southern Hemisphere. The glow on a space shuttle is caused by excitation of molecules adsorbed on the shuttle's surface.

SECTION 21.3

- Ozone in the stratosphere absorbs harmful UV radiation in the 200- to 300-nm range and protects life underneath. For many years, chlorofluorocarbons have been destroying the ozone layer.

SECTION 21.4

- Volcanic eruptions can lead to air pollution, deplete ozone in the stratosphere, and affect climate.

SECTION 21.5

- Carbon dioxide's ability to absorb infrared radiation enables it to trap some of the outgoing heat from Earth, warming its surface—a phenomenon known as the **greenhouse effect.** Other gases such as the CFCs and methane also contribute to the greenhouse effect. Global warming refers to the result of the *enhanced* greenhouse effect caused by human activities.

SECTION 21.6

- Sulfur dioxide, and to a lesser extent nitrogen oxides, generated mainly from the burning of fossil fuels and from the roasting of metal sulfides, causes acid rain.

SECTION 21.7

- **Photochemical smog** is formed by the photochemical reaction of automobile exhaust in the presence of sunlight. It is a complex reaction involving nitrogen oxides, ozone, and hydrocarbons.

SECTION 21.8

- Indoor air pollution is caused by radon, a radioactive gas formed during uranium decay; carbon monoxide and carbon dioxide, products of combustion; and formaldehyde, a volatile organic substance released from resins used in construction materials.

Key Words

Greenhouse effect, 985
Ionosphere, 977
Mesosphere, 977

Nitrogen fixation, 975
Photochemical smog, 992
Stratosphere, 977

Thermosphere, 977
Troposphere, 977

Questions and Problems

SECTION 21.1: EARTH'S ATMOSPHERE

Review Questions

21.1 Describe the regions of Earth's atmosphere.
21.2 Briefly outline the main processes of the nitrogen and oxygen cycles.
21.3 Explain why, for maximum performance, supersonic airplanes need to fly at a high altitude (in the stratosphere).
21.4 Jupiter's atmosphere consists mainly of hydrogen (90%) and helium (9%). How does this mixture of gases contrast with the composition of Earth's atmosphere? Why does the composition differ?

21.5 Describe the processes that result in the warming of the stratosphere.

Computational Problems

21.6 Calculate the partial pressure of CO_2 (in atm) in dry air when the atmospheric pressure is 754 mmHg.
21.7 Referring to Table 21.1, calculate the mole fraction of CO_2 and its concentration in parts per million by volume.
21.8 Calculate the mass (in kilograms) of nitrogen, oxygen, and carbon dioxide gases in the atmosphere. Assume that the total mass of air in the atmosphere is 5.25×10^{21} g.

SECTION 21.2: PHENOMENA IN THE OUTER LAYERS OF THE ATMOSPHERE

Review Questions

21.9   What process gives rise to the aurora borealis and aurora australis?

21.10   Why can astronauts not release oxygen atoms to test the mechanism of shuttle glow?

Computational Problems

21.11   The highly reactive OH radical (a species with an unpaired electron) is believed to be involved in some atmospheric processes. Table 10.6 lists the bond enthalpy for the oxygen-to-hydrogen bond in OH as 460 kJ/mol. What is the longest wavelength (in nanometers) of radiation that can bring about the following reaction?

$$OH(g) \longrightarrow O(g) + H(g)$$

21.12   The green color observed in the aurora borealis is produced by the emission of a photon by an electronically excited oxygen atom at 558 nm. Calculate the energy difference between the two levels involved in the emission process.

SECTION 21.3: DEPLETION OF OZONE IN THE STRATOSPHERE

Review Questions

21.13   Briefly describe the absorption of solar radiation in the stratosphere by O_2 and O_3 molecules.

21.14   Explain the processes that have a warming effect on the stratosphere.

21.15   List the properties of CFCs, and name four major uses of these compounds.

21.16   How do CFCs and nitrogen oxides destroy ozone in the stratosphere?

21.17   What causes the polar ozone holes?

21.18   How do volcanic eruptions contribute to ozone destruction?

21.19   Describe ways to curb the destruction of ozone in the stratosphere.

21.20   Discuss the effectiveness of some of the CFC substitutes.

Computational Problems

21.21   Given that the quantity of ozone in the stratosphere is equivalent to a 3.0-mm-thick layer of ozone on Earth at STP, calculate the number of ozone molecules in the stratosphere and their mass in kilograms. (*Hint:* The radius of Earth is 6371 km and the surface area of a sphere is $4\pi r^2$, where r is the radius.)

21.22   Referring to the answer in Problem 21.21, and assuming that the level of ozone in the stratosphere has already fallen 6.0%, calculate the number of kilograms of ozone that would have to be manufactured on a daily basis so that we could restore the ozone to the original level in 100 years. If ozone is made according to the process $3O_2(g) \longrightarrow 2O_3(g)$, how many kilojoules of energy would be required?

Conceptual Problems

21.23   Both Freon-11 and Freon-12 are made by the reaction of carbon tetrachloride (CCl_4) with hydrogen fluoride. Write equations for these reactions.

21.24   Why are CFCs not decomposed by UV radiation in the troposphere?

21.25   The average bond enthalpies of the C—Cl and C—F bonds are 340 and 485 kJ/mol, respectively. Based on this information, explain why the C—Cl bond in a CFC molecule is preferentially broken by solar radiation at 250 nm.

21.26   Like CFCs, certain bromine-containing compounds such as CF_3Br can also participate in the destruction of ozone by a similar mechanism starting with the Br atom.

$$CF_3Br \longrightarrow CF_3 + Br$$

Given that the average C—Br bond energy is 276 kJ/mol, estimate the longest wavelength required to break this bond. Will this compound be decomposed in the troposphere only or in both the troposphere and stratosphere?

21.27   Draw Lewis structures for chlorine nitrate ($ClONO_2$) and chlorine monoxide (ClO).

21.28   Draw Lewis structures for HCFC-123 (CF_3CHCl_2) and CF_3CFH_2.

SECTION 21.4: VOLCANOES

Review Questions

21.29   What are the effects of volcanic eruptions on climate?

21.30   Classify the reaction between H_2S and SO_2 that leads to the formation of sulfur at the site of a volcanic eruption.

SECTION 21.5: THE GREENHOUSE EFFECT

Review Questions

21.31   What is the greenhouse effect? What is the criterion for classifying a gas as a greenhouse gas?

21.32   Why is more emphasis placed on the role of carbon dioxide in the greenhouse effect than on that of water?

21.33   Describe three human activities that generate carbon dioxide. List two major mechanisms for the uptake of carbon dioxide.

21.34   Deforestation contributes to the greenhouse effect in two ways. What are they?

21.35   How does an increase in world population enhance the greenhouse effect?

21.36   Is ozone a greenhouse gas? If so, sketch three ways an ozone molecule can vibrate.

21.37 What effects do CFCs and their substitutes have on Earth's temperature?

21.38 Why are CFCs more effective greenhouse gases than methane and carbon dioxide?

Computational Problems

21.39 The annual production of zinc sulfide (ZnS) is 4.0×10^4 tons. Estimate the number of tons of SO_2 produced by roasting it to extract zinc metal.

21.40 Calcium oxide or quicklime (CaO) is used in steelmaking, cement manufacture, and pollution control. It is prepared by the thermal decomposition of calcium carbonate.

$$CaCO_3(s) \longrightarrow CaO(s) + CO_2(g)$$

Calculate the yearly release of CO_2 (in kilograms) to the atmosphere if the annual production of CaO in the United States is 1.7×10^{10} kg.

SECTION 21.6: ACID RAIN

Review Questions

21.41 Name the gas that is largely responsible for the acid rain phenomenon.

21.42 List three detrimental effects of acid rain.

21.43 Briefly discuss two industrial processes that lead to acid rain.

21.44 Discuss ways to curb acid rain.

21.45 Water and sulfur dioxide are both polar molecules, and their geometry is similar. Why is SO_2 not considered a major greenhouse gas?

SECTION 21.7: PHOTOCHEMICAL SMOG

Review Questions

21.46 What is photochemical smog? List the factors that favor the formation of photochemical smog.

21.47 What are primary and secondary pollutants?

21.48 Identify the gas that is responsible for the brown color of photochemical smog.

21.49 The safety limits of ozone and carbon monoxide are 120 ppb by volume and 9 ppm by volume, respectively. Why does ozone have a lower limit?

21.50 Suggest ways to minimize the formation of photochemical smog.

21.51 In which region of the atmosphere is ozone beneficial? In which region is it detrimental?

Computational Problems

21.52 The gas-phase decomposition of peroxyacetyl nitrate (PAN) obeys first-order kinetics,

$$CH_3COOONO_2 \longrightarrow CH_3COOO + NO_2$$

with a rate constant of 4.9×10^{-4} s^{-1}. Calculate the rate of decomposition (in M/s) if the concentration of PAN is 0.55 ppm by volume. Assume STP conditions.

21.53 On a smoggy day in a certain city, the ozone concentration was 0.42 ppm by volume. Calculate the partial pressure of ozone (in atm) and the number of ozone molecules per liter of air if the temperature and pressure were 20.0°C and 748 mmHg, respectively.

SECTION 21.8: INDOOR POLLUTION

Review Questions

21.54 List the major indoor pollutants and their sources.

21.55 What is the best way to deal with indoor pollution?

21.56 Why is it dangerous to idle a car's engine in a poorly ventilated place, such as the garage?

21.57 Describe the properties that make radon an indoor pollutant. Would radon be more hazardous if ^{222}Rn had a longer half-life?

Computational Problems

21.58 A volume of 5.0 L of polluted air at 18.0°C and 747 mmHg is passed through lime water [an aqueous suspension of $Ca(OH)_2$], so that all the carbon dioxide present is precipitated as $CaCO_3$. If the mass of the $CaCO_3$ precipitate is 0.026 g, calculate the percentage by volume of CO_2 in the air sample.

21.59 A concentration of 8.00×10^2 ppm by volume of CO is considered lethal to humans. Calculate the minimum mass of CO (in grams) that would become a lethal concentration in a closed room 17.6 m long, 8.80 m wide, and 2.64 m high. The temperature and pressure are 20.0°C and 756 mmHg, respectively.

ADDITIONAL PROBLEMS

21.60 As mentioned in the chapter, spraying the stratosphere with hydrocarbons such as ethane and propane should eliminate Cl atoms. What is the drawback of this procedure if used on a large scale for an extended period of time?

21.61 Briefly describe the harmful effects of the following substances: O_3, SO_2, NO_2, CO, $CH_3COOONO_2$ (PAN), Rn.

21.62 The equilibrium constant (K_P) for the reaction

$$N_2(g) + O_2(g) \rightleftharpoons 2NO(g)$$

is 4.0×10^{-31} at 25°C and 2.6×10^{-6} at 1100°C, the temperature of a running car's engine. Is this an endothermic or exothermic reaction?

21.63 Although the hydroxyl radical (OH) is present only in a trace amount in the troposphere, it plays a central role in its chemistry because it is a strong oxidizing agent and can react with many pollutants as well as some CFC substitutes. The hydroxyl radical is formed by the following reactions:

$$O_3 \xrightarrow{\lambda\,=\,320\ nm} O^* + O_2$$
$$O + H_2O \longrightarrow 2OH$$

where O* denotes an electronically excited atom. (a) Explain why the concentration of OH is so small even though the concentrations of O_3 and H_2O are quite large in the troposphere. (b) What property makes OH a strong oxidizing agent? (c) The reaction between OH and NO_2 contributes to acid rain. Write an equation for this process. (d) The hydroxyl radical can oxidize SO_2 to H_2SO_4. The first step is the formation of a neutral HSO_3 species, followed by its reaction with O_2 and H_2O to form H_2SO_4 and the hydroperoxyl radical (HO_2). Write equations for these processes.

21.64 The equilibrium constant (K_P) for the reaction $2CO(g) + O_2(g) \rightleftharpoons 2CO_2(g)$ is 1.4×10^{90} at 25°C. Given this enormous value, why doesn't CO convert totally to CO_2 in the troposphere?

21.65 How are past temperatures determined from ice cores obtained from the Arctic or Antarctica? (*Hint:* Look up the stable isotopes of hydrogen and oxygen. How does energy required for vaporization depend on the masses of H_2O molecules containing different isotopes? How would you determine the age of an ice core?)

21.66 The balance between SO_2 and SO_3 is important in understanding acid rain formation in the troposphere. From the following information at 25°C,

$$S(s) + O_2(g) \rightleftharpoons SO_2(g) \qquad K_1 = 4.2 \times 10^{52}$$

$$2S(s) + 3O_2(g) \rightleftharpoons 2SO_3(g) \qquad K_2 = 9.8 \times 10^{128}$$

calculate the equilibrium constant for the reaction

$$2SO_2(g) + O_2(g) \rightleftharpoons 2SO_3(g)$$

21.67 The effective incoming solar radiation per unit area on Earth is 342 W/m². Of this radiation, 6.7 W/m² is absorbed by CO_2 at 14,993 nm in the atmosphere. How many photons at this wavelength are absorbed per second in 1 m² by CO_2? (1 W = 1 J/s.)

21.68 A glass of water initially at pH 7.0 is exposed to dry air at sea level at 20°C. Calculate the pH of the water when equilibrium is reached between atmospheric CO_2 and CO_2 dissolved in the water, given that Henry's law constant for CO_2 at 20°C is 0.032 mol/L · atm. (*Hint:* Assume no loss of water due to evaporation, and use Table 21.1 to calculate the partial pressure of CO_2. Your answer should correspond roughly to the pH of rainwater.)

21.69 Ozone in the troposphere is formed by the following steps.

$$NO_2 \longrightarrow NO + O \qquad (1)$$

$$O + O_2 \longrightarrow O_3 \qquad (2)$$

The first step is initiated by the absorption of visible light (NO_2 is a brown gas). Calculate the longest wavelength required for step 1 at 25°C. (*Hint:* You need to first calculate ΔH and hence ΔE for step 1. Next, determine the wavelength for decomposing NO_2 from ΔE.)

21.70 Instead of monitoring carbon dioxide, suggest another gas that scientists could study to substantiate the fact that CO_2 concentration is steadily increasing in the atmosphere.

21.71 Describe the removal of SO_2 by CaO (to form $CaSO_3$) in terms of a Lewis acid-base reaction.

21.72 Which of the following settings is the most suitable for photochemical smog formation: (a) Gobi Desert at noon in June, (b) New York City at 1 P.M. in July, (c) Boston at noon in January? Explain your choice.

21.73 As stated in the chapter, about 50 million tons of sulfur dioxide is released into the atmosphere every year. (a) If 20% of the SO_2 is eventually converted to H_2SO_4, calculate the number of 1000-lb marble statues the resulting acid rain can damage. As an estimate, assume that the acid rain only destroys the surface layer of each statue, which is made up of 5% of its total mass. (b) What is the other undesirable result of the acid rain damage?

21.74 Peroxyacetyl nitrate (PAN) undergoes thermal decomposition as follows:

$$CH_3(CO)OONO_2 \longrightarrow CH_3(CO)OO + NO_2$$

The rate constant is 3.0×10^{-4} s^{-1} at 25°C. At the boundary between the troposphere and stratosphere, where the temperature is about −40°C, the rate constant is reduced to 2.6×10^{-7} s^{-1}. (a) Calculate the activation energy for the decomposition of PAN. (b) What is the half-life of the reaction (in min) at 25°C?

21.75 What is ironic about the following cartoon?

21.76 Calculate the standard enthalpy of formation (ΔH_f°) of ClO from the following bond energies: Cl_2: 242.7 kJ/mol; O_2: 498.7 kJ/mol; ClO: 206 kJ/mol.

21.77 The carbon dioxide level in the atmosphere today is often compared with that in preindustrial days. Explain how scientists use tree rings and air trapped in polar ice to arrive at the comparison.

21.78 A 14-m by 10-m by 3.0-m basement had a high radon content. On the day the basement was sealed off from its surroundings so that no exchange of air could take place, the partial pressure of ^{222}Rn was 1.2×10^{-6} mmHg. Calculate the number of ^{222}Rn isotopes ($t_{1/2} = 3.8$ days) at the beginning and end of 31 days. Assume STP conditions.

21.79 In 1991, it was discovered that nitrous oxide (N_2O) is produced in the synthesis of nylon. This compound, which is released into the atmosphere, contributes *both* to the depletion of ozone in the stratosphere and to the greenhouse effect. (a) Write equations representing the reactions between N_2O and oxygen atoms in the stratosphere to produce nitric oxide (NO), which is then oxidized by ozone to form nitrogen dioxide. (b) Is N_2O a more effective greenhouse gas than carbon dioxide? Explain. (c) One of the intermediates in nylon manufacture is adipic acid [$HOOC(CH_2)_4COOH$]. About 2.2×10^9 kg of adipic acid is consumed every year. It is estimated that for every mole of adipic acid produced, 1 mole of N_2O is generated. What is the maximum number of moles of O_3 that can be destroyed as a result of this process per year?

21.80 A person was found dead of carbon monoxide poisoning in a well-insulated cabin. Investigation showed that he had used a blackened bucket to heat water on a butane burner. The burner was found to function properly with no leakage. Explain, with an appropriate equation, the cause of his death.

21.81 Methyl bromide (CH_3Br, b.p. = 3.6°C) is used as a soil fumigant to control insects and weeds. It is also a marine by-product. Photodissociation of the C–Br bond produces Br atoms that can react with ozone similar to Cl, except more effectively. Do you expect CH_3Br to be photolyzed in the troposphere? The bond enthalpy of the C–Br bond is about 293 kJ/mol.

21.82 As stated in the chapter, carbon monoxide has a much higher affinity for hemoglobin than oxygen does. (a) Write the equilibrium constant expression (K_c) for the process

$$CO(g) + HbO_2(aq) \rightleftharpoons O_2(g) + HbCO(aq)$$

where HbO_2 and $HbCO$ are oxygenated hemoglobin and carboxyhemoglobin, respectively. (b) The

composition of a breath of air inhaled by a person smoking a cigarette is 1.9×10^{-6} mol/L CO and 8.6×10^{-3} mol/L O_2. Calculate the ratio of [HbCO] to [HbO_2], given that K_c is 212 at 37°C.

21.83 The molar heat capacity of a diatomic molecule is 29.1 J/K · mol. Assuming the atmosphere contains only nitrogen gas and there is no heat loss, calculate the total heat intake (in kilojoules) if the atmosphere warms up by 3°C during the next 50 years. Given that there are 1.8×10^{20} moles of diatomic molecules present, how many kilograms of ice (at the North and South poles) will this quantity of heat melt at 0°C? (The molar heat of fusion of ice is 6.01 kJ/mol.)

21.84 Assume that the formation of nitrogen dioxide,

$$2NO(g) + O_2(g) \longrightarrow 2NO_2(g)$$

is an elementary reaction. (a) Write the rate law for this reaction. (b) A sample of air at a certain temperature is contaminated with 2.0 ppm of NO by volume. Under these conditions, can the rate law be simplified? If so, write the simplified rate law. (c) Under the conditions described in part (b), the half-life of the reaction has been estimated to be 6.4×10^3 min. What would the half-life be if the initial concentration of NO were 10 ppm?

21.85 An electric power station annually burns 3.1×10^7 kg of coal containing 2.4% sulfur by mass. Calculate the volume of SO_2 emitted at STP.

21.86 The concentration of SO_2 in the troposphere over a certain region is 0.16 ppm by volume. The gas dissolves in rainwater as follows:

$$SO_2(g) + H_2O(l) \rightleftharpoons H^+(aq) + HSO_3^-(aq)$$

Given that the equilibrium constant for the preceding reaction is 1.3×10^{-2}, calculate the pH of the rainwater. Assume that the reaction does not affect the partial pressure of SO_2.

Answers to In-Chapter Materials

PRACTICE PROBLEMS

21.1A 1120 nm. **21.1B** HI, F_2, ClF. **21.2A** H_2O. **21.2B** To act as a greenhouse gas, a molecule must be IR-active. To be IR-active, a molecule must undergo a change in dipole moment as the result of one or more of its vibrations. **21.3A** 1.2×10^{-6} mol/L. **21.3B** 52.8 days.

SECTION REVIEW

21.2.1 a. **21.2.2** d. **21.5.1** a, d. **21.5.2** c. **21.8.1** e. **21.8.2** b.

Chapter 22

Coordination Chemistry

© Imaginechina via AP Images

BECAUSE THEY contain unpaired electrons, transition metals and their compounds
are attractive for use in magnetic applications. Chemists can manipulate the magnetic
states in transition-metal compounds and alloys, tailoring them for a variety of
specific uses. These powerful magnets can be incorporated into appplications such
as electric motors and generators, magnetic resonance imaging (MRI) instruments,
and magnetic levitation transportation (maglev) (shown here) which operates by
suspending, guiding, and propelling vehicles using electromagnetic forces.

Before You Begin, Review These Skills

- Lewis acids and bases [◄◄ Section 16.12]
- The shapes of *d* orbitals [◄◄ Section 3.8]

22.1 COORDINATION COMPOUNDS

Coordination compounds contain *coordinate covalent* bonds [◄◄ Section 6.6] formed by the reactions of metal ions with groups of *anions* or *polar molecules*. The metal ion in these kinds of reactions acts as a Lewis acid, accepting electrons, whereas the anions or polar molecules act as Lewis bases, donating pairs of electrons to form bonds to the metal ion. Thus, a coordinate covalent bond is a covalent bond in which one of the atoms donates *both* of the electrons that constitute the bond. Often a coordination compound consists of a *complex* ion and one or more *counter* ions. In writing formulas for such coordination compounds, we use square brackets to separate the complex ion from the counter ion.

Student Annotation: A complex ion is one in which a metal cation is covalently bound to one or more molecules or ions [◄◄ Section 17.5].

$$K_2[PtCl_6]$$

This compound consists of the complex ion $PtCl_6^{2-}$ and two K^+ counter ions.

Some coordination compounds, such as $Fe(CO)_5$, do not contain complex ions. Most but not all of the metals in coordination compounds are transition metals. Our understanding of the nature of coordination compounds stems from the classic work of Alfred Werner,[1] who prepared and characterized many coordination compounds. In 1893, at the age of 26, Werner proposed what is now commonly referred to as Werner's coordination theory.

Student Annotation: We can use the term *coordination complex* to refer to a compound, such as $Fe(CO)_5$, or to a complex ion.

Nineteenth-century chemists were puzzled by a certain class of reactions that seemed to violate valence theory. For example, the valences of the elements in cobalt(III) chloride and those in ammonia seem to be completely satisfied, and yet these two substances react to form a stable compound having the formula $CoCl_3 \cdot 6NH_3$. To explain this behavior, Werner postulated that most elements exhibit two types of valence: *primary valence* and *secondary valence*. In modern terminology, primary valence corresponds to the oxidation number and secondary valence to the coordination number of the element. In $CoCl_3 \cdot 6NH_3$, according to Werner, cobalt has a primary valence of 3 and a secondary valence of 6.

Today we use the formula $[Co(NH_3)_6]Cl_3$ to indicate that the ammonia molecules and the cobalt atom form a complex ion; the chloride ions are not part of the complex but are counter ions, held to the complex ion by Coulombic attraction.

Properties of Transition Metals

Transition metals are those that either have incompletely filled *d* subshells or form ions with incompletely filled *d* subshells (Figure 22.1). Incompletely filled *d* subshells give rise to several notable properties, including distinctive colors, the formation of paramagnetic compounds, catalytic activity, and the tendency to form complex ions. The most common transition metals are scandium through copper, which occupy the fourth row of the periodic table. Table 22.1 lists the electron configurations and some of the properties of these metals.

Student Annotation: The Group 2B metals—Zn, Cd, and Hg—do not fit either of these criteria. As a result, they are *d*-block metals, but they are not actually transition metals.

---

[1] Alfred Werner (1866–1919). Swiss chemist. Werner started as an organic chemist but did his most notable work in coordination chemistry. For his theory of coordination compounds, Werner was awarded the Nobel Prize in Chemistry in 1913.

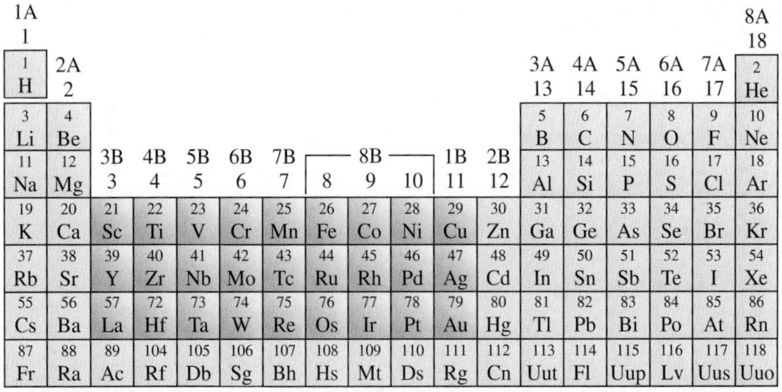

Figure 22.1 The transition metals (shown in green). Note that although the Group 2B elements (Zn, Cd, Hg) are described as transition metals by some chemists, neither the metals nor their ions possess incompletely filled *d* subshells.

1A 1																	8A 18
1 H	2A 2											3A 13	4A 14	5A 15	6A 16	7A 17	2 He
3 Li	4 Be	3B 3	4B 4	5B 5	6B 6	7B 7	8	—8B— 9	10	1B 11	2B 12	5 B	6 C	7 N	8 O	9 F	10 Ne
11 Na	12 Mg											13 Al	14 Si	15 P	16 S	17 Cl	18 Ar
19 K	20 Ca	21 Sc	22 Ti	23 V	24 Cr	25 Mn	26 Fe	27 Co	28 Ni	29 Cu	30 Zn	31 Ga	32 Ge	33 As	34 Se	35 Br	36 Kr
37 Rb	38 Sr	39 Y	40 Zr	41 Nb	42 Mo	43 Tc	44 Ru	45 Rh	46 Pd	47 Ag	48 Cd	49 In	50 Sn	51 Sb	52 Te	53 I	54 Xe
55 Cs	56 Ba	57 La	72 Hf	73 Ta	74 W	75 Re	76 Os	77 Ir	78 Pt	79 Au	80 Hg	81 Tl	82 Pb	83 Bi	84 Po	85 At	86 Rn
87 Fr	88 Ra	89 Ac	104 Rf	105 Db	106 Sg	107 Bh	108 Hs	109 Mt	110 Ds	111 Rg	112 Cn	113 Uut	114 Fl	115 Uup	116 Lv	117 Uus	118 Uuo

Student Annotation: The trends in radii, electronegativity, and ionization energy for the first row of transition metals are somewhat different from those for main group elements. This is largely the effect of *shielding* of the 4*s* electrons by the 3*d* electrons.

Student Annotation: The electron configurations of the first row transition metals and their ions were discussed in Section 3.10 and Section 4.5, respectively.

TABLE 22.1	Electron Configurations and Other Properties of the Fourth Period Transition Metals								
	Sc	Ti	V	Cr	Mn	Fe	Co	Ni	Cu
Electron configuration									
M	$4s^2 3d^1$	$4s^2 3d^2$	$4s^2 3d^3$	$4s^1 3d^5$	$4s^2 3d^5$	$4s^2 3d^6$	$4s^2 3d^7$	$4s^2 3d^8$	$4s^1 3d^{10}$
M^{2+}	—	$3d^2$	$3d^3$	$3d^4$	$3d^5$	$3d^6$	$3d^7$	$3d^8$	$3d^9$
M^{3+}	[Ar]	$3d^1$	$3d^2$	$3d^3$	$3d^4$	$3d^5$	$3d^6$	$3d^7$	$3d^8$
Electronegativity									
	1.3	1.5	1.6	1.6	1.5	1.8	1.9	1.9	1.9
Ionization energy (kJ/mol)									
First	631	658	650	652	717	759	760	736	745
Second	1235	1309	1413	1591	1509	1561	1645	1751	1958
Third	2389	2650	2828	2986	3250	2956	3231	3393	3578
Radius (pm)									
M	162	147	134	130	135	126	125	124	128
M^{2+}	—	90	88	85	80	77	75	69	72
M^{3+}	81	77	74	64	66	60	64		

Most of the transition metals exhibit a close-packed structure in which each atom has a coordination number of 12. Furthermore, these elements have relatively small atomic radii. The combined effect of closest packing and small atomic size results in strong metallic bonds. Therefore, transition metals have higher densities, higher melting points and boiling points, and higher heats of fusion and vaporization than the main group and Group 2B metals (Table 22.2).

TABLE 22.2	Physical Properties of Elements K to Zn										
	1A	2A				Transition metals					2B
	K	Ca	Sc	Ti	V	Cr	Mn	Fe	Co	Ni	Cu	Zn
Atomic radius (pm)	235	197	162	147	134	130	135	126	125	124	128	138
Melting point (°C)	63.7	838	1539	1668	1900	1875	1245	1536	1495	1453	1083	419.5
Boiling point (°C)	760	1440	2730	3260	3450	2665	2150	3000	2900	2730	2595	906
Density (g/cm³)	0.86	4.51	3.0	4.51	6.1	7.19	7.43	7.86	8.9	8.9	8.96	7.14

Transition metals exhibit variable oxidation states in their compounds. Figure 22.2 shows the oxidation states of the first row of transition metals. Note that all these metals can exhibit the oxidation state +3 and *nearly* all can exhibit the oxidation state +2. Of these two, the +2 oxidation state is somewhat more common for the heavier elements. The highest oxidation state for a transition metal is +7, exhibited by manganese $(4s^2 3d^5)$. Transition metals exhibit their highest oxidation states in compounds that contain highly electronegative elements such as oxygen and fluorine—for example, V_2O_5, CrO_3, and Mn_2O_7.

Student Annotation: The oxidation state of O in each of these compounds is −2, making those of V, Cr, and Mn +5, +6, and +7, respectively [◄◄ Section 9.4].

Ligands

The molecules or ions that surround the metal in a complex ion are called *ligands* (Table 22.3). The formation of covalent bonds between ligands and a metal can be thought of as a Lewis acid-base reaction. (Recall that a Lewis base is a species that donates a pair of electrons [◄◄ Section 16.12].) To be a ligand, a molecule or ion must have at least one unshared pair of valence electrons, as these examples illustrate:

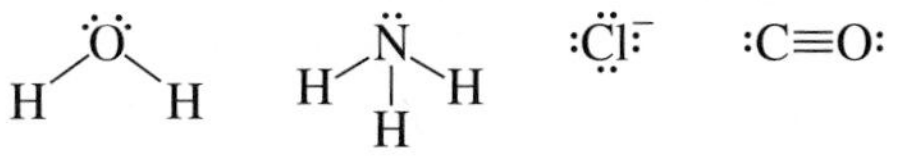

$$\overset{\displaystyle \ddot{O}}{\underset{\displaystyle H \quad\quad H}{}} \qquad \overset{\displaystyle \ddot{N}}{\underset{\displaystyle H \quad | \quad H}{} \atop H} \qquad :\ddot{\underset{..}{Cl}}:^{-} \qquad :C{\equiv}O:$$

Therefore, ligands play the role of Lewis bases. The transition metal, on the other hand, acts as a Lewis acid, accepting (and sharing) pairs of electrons from the Lewis bases.

				+7				
			+6	+6	+6			
		+5	+5	+5	+5			
	+4	+4	+4	+4	+4	+4		
+3	+3	+3	+3	+3	+3	+3	+3	+3
	+2	+2	+2	+2	+2	+2	+2	+2
								+1
Sc	Ti	V	Cr	Mn	Fe	Co	Ni	Cu

Figure 22.2 Oxidation states of the first-row transition metals. The most stable oxidation numbers are shown in red. The zero oxidation state is encountered in some compounds, such as $Ni(CO)_4$ and $Fe(CO)_5$.

TABLE 22.3	Common Ligands

Name	Structure
Monodentate	
Ammonia	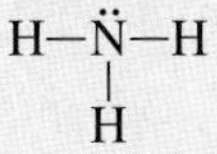 $H{-}\ddot{N}{-}H$ with H below N
Carbon monoxide	$:C{\equiv}O:$
Chloride ion	$:\ddot{\underset{..}{Cl}}:^{-}$
Cyanide ion	$[:C{\equiv}N:]^{-}$
Thiocyanate ion	$[:\ddot{\underset{..}{S}}{-}C{\equiv}N:]^{-}$
Water	$H{-}\ddot{O}{-}H$
Bidentate	
Ethylenediamine	$H_2\ddot{N}{-}CH_2{-}CH_2{-}\ddot{N}H_2$
Oxalate ion	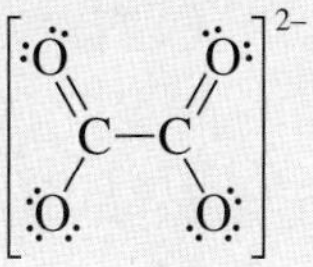 $\left[\begin{array}{c} :\ddot{O}: \quad\quad :\ddot{O}: \\ \backslash/ \\ C{-}C \\ /\backslash \\ :\ddot{O}. \quad\quad .\ddot{O}: \end{array} \right]^{2-}$
Polydentate	
Ethylenediaminetetraacetate ion (EDTA)	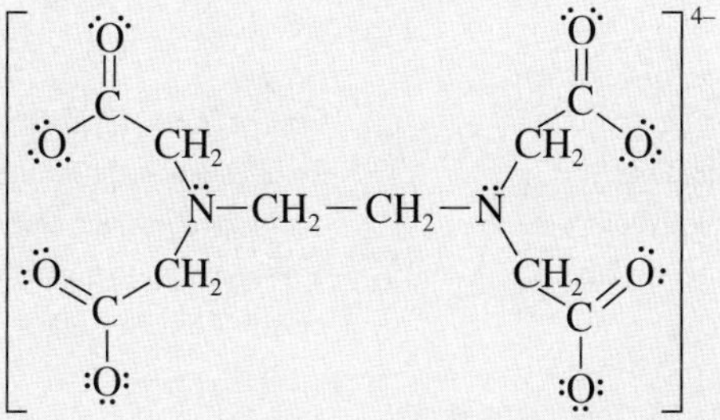 (structure) $^{4-}$

The atom in a ligand that is bound directly to the metal atom is known as the **donor atom.** For example, *nitrogen* is the donor atom in the $[Cu(NH_3)_4]^{2+}$ complex ion.

$$
\begin{bmatrix}
\begin{array}{c}
H \\
H-N-H \\
H \quad\quad H \\
H-N-Cu-N-H \\
H \quad\quad H \\
H-N-H \\
H
\end{array}
\end{bmatrix}^{2+}
$$

The **coordination number** in a coordination compound refers to the number of donor atoms surrounding the central metal atom in a complex ion. The coordination number of Cu^{2+} in $[Cu(NH_3)_4]^{2+}$ is 4. The most common coordination numbers are 4 and 6, although coordination numbers of 2 and 5 are also known.

Depending on the number of donor atoms a ligand possesses, it is classified as monodentate (1 donor atom), bidentate (2 donor atoms), or polydentate (> 2 donor atoms). Table 22.3 lists some common ligands. Figure 22.3 shows how ethylenediamine, sometimes abbreviated "en," forms two bonds to a metal atom.

Bidentate and polydentate ligands are also called **chelating agents** because of their ability to hold the metal atom like a claw (from the Greek *chele,* meaning "claw"). One example is EDTA (Figure 22.4), a polydentate ligand used to treat metal poisoning. Six donor atoms enable EDTA to form a stable complex ion with lead. This stable complex enables the body to remove lead from the blood.

The oxidation state of a transition metal in a complex ion is determined using the known charges of the *ligands* and the known *overall* charge of the complex ion. In the complex ion $[PtCl_6]^{2-}$, for example, each chloride ion ligand has an oxidation number of -1. For the overall charge of the ion to be -2, the Pt must have an oxidation number of $+4$.

Student Annotation: Remember that the oxidation number of a monoatomic ion is equal to the charge [◄◄ Section 9.4].

Worked Example 22.1 shows how to determine transition metal oxidation states in coordination compounds.

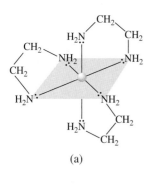

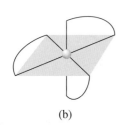

Figure 22.3 (a) Structure of a metal-ethylenediamine complex cation, such as $[Co(en)_3]^{2+}$. Each ethylenediamine molecule provides two N donor atoms and is therefore a bidentate ligand. (b) Simplified structure of the same complex cation.

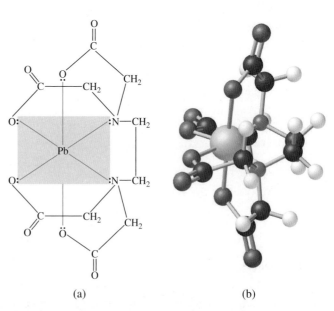

Figure 22.4 (a) EDTA complex of lead. The complex bears a net charge of 2– because each O donor atom has one negative charge and the lead ion carries two positive charges. Only the lone pairs that participate in bonding are shown. Note the octahedral geometry about the Pb^{2+} ion. (b) Molecular model of the Pb^{2+}-EDTA complex. The light green sphere is the Pb^{2+} ion.

Worked Example 22.1

Determine the oxidation state of the central metal atom in each of the following compounds: (a) $[Ru(NH_3)_5(H_2O)]Cl_2$, (b) $[Cr(NH_3)_6](NO_3)_3$, and (c) $Fe(CO)_5$.

Strategy Identify the components of each compound, and use known oxidation states and charges to determine the oxidation state of the metal.

Setup (a) $[Ru(NH_3)_5(H_2O)]Cl_2$ consists of a complex ion (the part of the formula enclosed in square brackets) and two Cl^- counter ions. Because the overall charge on the compound is zero, the complex ion is $[Ru(NH_3)_5(H_2O)]^{2+}$. There are six ligands: five ammonia molecules and one water molecule. Each molecule has a zero charge (i.e., each ligand is neutral), so the charge on the metal is equal to the overall charge on the complex ion.

(b) $[Cr(NH_3)_6](NO_3)_3$ consists of a complex ion and three NO_3^- ions, making the complex ion $[Cr(NH_3)_6]^{3+}$. Each of the six ammonia molecule ligands is neutral (i.e., each has a zero charge), making the charge on the metal equal to the overall charge on the complex ion.

(c) $Fe(CO)_5$ does not contain a complex ion. The ligands are CO molecules, which have a zero charge, so the central metal also has a zero charge.

Solution (a) +2 (b) +3 (c) 0

Think About It
To solve a problem like this, you must be able to recognize the common polyatomic ions and you must know their charges.

Practice Problem **A**TTEMPT Give oxidation numbers for the metals in (a) $K[Au(OH)_4]$ and (b) $K_4[Fe(CN)_6]$.

Practice Problem **B**UILD The oxidation state of cobalt in each of the following species is +3. Determine the overall charge on each complex: $[Co(NH_3)_4Cl_2]^?$, $[Co(CN)_6]^?$, $[Co(NH_3)_3Br_3]^?$, $[Co(en)_2Cl_2]^?$, $[Co(en)_3]^?$, $[Co(NH_3)_6]^?$, $[Co(NH_3)_5Cl]^?$.

Practice Problem **C**ONCEPTUALIZE Explain why the bonds between transition metals in very high oxidation states (>4) and nonmetals are covalent rather than ionic.

Nomenclature of Coordination Compounds

Now that we have discussed the various types of ligands and the oxidation numbers of metals, our next step is to learn how to name coordination compounds. The rules for naming ionic coordination compounds are as follows:

1. The cation is named before the anion, as in other ionic compounds. The rule holds regardless of whether the complex ion bears a net positive or a net negative charge. In the compounds $K_2[Fe(CN)_6]$ and $[Co(NH_3)_4]Cl$, for example, we name the K^+ and $[Co(NH_3)_4]^+$ cations first, respectively.
2. Within a complex ion, the ligands are named first, in alphabetical order, and the metal ion is named last.
3. The names of anionic ligands end with the letter *o*, whereas neutral ligands are usually called by the names of the molecules. The exceptions are H_2O (aqua), CO (carbonyl), and NH_3 (ammine). Table 22.4 lists some common ligands and their nomenclature.
4. When two or more of the same ligand are present, use Greek prefixes *di, tri, tetra, penta,* and *hexa,* to specify their number. Thus, the ligands in the cation $[Co(NH_3)_4Cl_2]^+$ are "tetraamminedichloro." (Note that prefixes are *not* used for the purpose of *alphabetizing* the ligands.)
5. The oxidation number of the metal is indicated in Roman numerals immediately following the name of the metal. For example, the Roman numeral III is used to indicate the +3 oxidation state of chromium in $[Cr(NH_3)_4Cl_2]^+$, which is called tetraamminedichlorochromium(III) ion.
6. If the complex is an anion, its name ends in *-ate.* In $K_4[Fe(CN)_6]$, for example, the anion $[Fe(CN)_6]^{4-}$ is called hexacyanoferrate(II) ion. Note that the Roman numeral indicating the oxidation state of the metal *follows* the suffix *-ate.* Table 22.5 lists the names of anions containing metal atoms.

TABLE 22.4	Names of Common Ligands in Coordination Compounds
Ligand	Name of ligand in coordination compound
Bromide (Br^-)	Bromo
Chloride (Cl^-)	Chloro
Cyanide (CN^-)	Cyano
Hydroxide (OH^-)	Hydroxo
Oxide (O^{2-})	Oxo
Carbonate (CO_3^{2-})	Carbonato
Nitrite (NO_2^-)	Nitro
Oxalate ($C_2O_4^{2-}$)	Oxalato
Ammonia (NH_3)	Ammine
Carbon monoxide (CO)	Carbonyl
Water (H_2O)	Aqua
Ethylenediamine	Ethylenediamine
Ethylenediaminetetraacetate	Ethylenediaminetetraacetate

TABLE 22.5	Names of Anions Containing Metal Atoms
Metal	Name of metal in anionic complex
Aluminum	Aluminate
Chromium	Chromate
Cobalt	Cobaltate
Copper	Cuprate
Gold	Aurate
Iron	Ferrate
Lead	Plumbate
Manganese	Manganate
Molybdenum	Molybdate
Nickel	Nickelate
Silver	Argentate
Tin	Stannate
Tungsten	Tungstate
Zinc	Zincate

Worked Examples 22.2 and 22.3 apply these rules to the nomenclature of coordination compounds.

Worked Example 22.2

Write the names of the following coordination compounds: (a) $[Co(NH_3)_4Cl_2]Cl$ and (b) $K_3[Fe(CN)_6]$.

Strategy For each compound, name the cation first and the anion second. Refer to Tables 22.4 and 22.5 for the names of ligands and anions containing metal atoms.

Setup (a) The cation is a complex ion containing four ammonia molecules and two chloride ions. The counter ion is chloride (Cl^-), so the charge on the complex cation is +1, making the oxidation state of cobalt +3.

(b) The cation is K^+, and the anion is a complex ion containing six cyanide ions. The charge on the complex ion is −3, making the oxidation state of iron +3.

Solution (a) Tetraamminedichlorocobalt(III) chloride (b) Potassium hexacyanoferrate(III)

Think About It
When the anion is a *complex* ion, its name must end in *-ate,* followed by the metal's oxidation state in Roman numerals. Also, do not use prefixes to denote numbers of counter ions.

Practice Problem (A)**TTEMPT** Give the correct name for (a) $[Co(NH_3)_4Br_2]Cl$, (b) $[Cr(H_2O)_4Cl_2]Cl$, and (c) $K_2[CuCl_4]$.

Practice Problem (B)**UILD** Give the correct name for (a) $Na_3[Fe(CN)_6]$, (b) $[Cr(en)_2Cl_2]Br$, and (c) $[Co(en)_3]Cl_3$. [*Hint:* For parts (b) and (c), read the Student Annotation in this chapter regarding the use of prefixes.]

Practice Problem (C)**ONCEPTUALIZE** Draw the structure of the en ligand.

Worked Example 22.3

Write formulas for the following compounds: (a) pentaamminechlorocobalt(III) chloride, and (b) dichloro*bis*(ethylenediamine) platinum(IV) nitrate.

Strategy If you can't remember them yet, refer to Tables 22.4 and 22.5 for the names of ligands and anions containing metal atoms.

Setup (a) There are six ligands: five NH_3 molecules and one Cl^- ion. The oxidation state of cobalt is +3, making the overall charge on the complex ion +2. Therefore, there are two chloride ions as counter ions.

(b) There are four ligands: two bidentate ethylenediamines and two Cl^- ions. The oxidation state of platinum is +4, making the overall charge on the complex ion +2. Therefore, there are two nitrate ions as counter ions.

Solution (a) $[Co(NH_3)_5Cl]Cl_2$ (b) $[Pt(en)_2Cl_2](NO_3)_2$

> **Think About It**
> Although ligands are alphabetized in a compound's name, they do not necessarily appear in alphabetical order in the compound's formula. In the formula of a coordination compound, neutral ligands are placed *before* anionic ligands.

Practice Problem (A)**TTEMPT** Write the formulas for (a) pentaaquabromoruthenium(II) nitrate, (b) potassium tetrabromodichloroplatinate(IV), and (c) sodium hexanitrocobaltate(III).

Practice Problem (B)**UILD** Write the formulas for (a) *bis*(ethylenediamine)oxalatovanadium(IV) chloride, (b) dibromo*bis*(ethylenediamine)chromium(III) nitrate, and (c) *tris*(ethylenediamine)platinum(IV) sulfate.

Practice Problem (C)**ONCEPTUALIZE** Explain why coordination-compound nomenclature uses the prefixes *bis, tris,* and *tetrakis,* instead of simply using *di, tri,* and *tetra.*

Thinking Outside the Box

Chelation Therapy

Although the Consumer Product Safety Commission (CPSC) banned the residential use of lead-based paint in 1978, millions of children remain at risk for exposure to lead from deteriorating paint in older homes. Lead poisoning is especially harmful to children under the age of 5 years because it interferes with growth and development and it has been shown to lower IQ. Symptoms of chronic exposure to lead include diminished appetite, nausea, malaise, and convulsions. *Blood lead level* (BLL), expressed as micrograms per deciliter (μg/dL), is used to monitor the effect of chronic exposure. A BLL $\leq 10\ \mu$g/dL is considered normal; a BLL $> 45\ \mu$g/dL requires medical and environmental intervention. At high levels ($>70\ \mu$g/dL), lead can cause seizures, coma, and death.

Treatment for lead poisoning involves *chelation therapy,* in which a chelating agent is administered orally, intravenously, or intramuscularly. Chelating agents form strong coordinate-covalent bonds to metal ions, forming stable, water-soluble complex ions that are easily removed from the body via the urine.

Elevated BLL and other heavy metal poisoning can be treated with one of several chelating agents, including DMSA and EDTA. EDTA is administered intravenously as either the sodium salt (Endrate) or as the calcium disodium salt (Versenate). Endrate is not approved for the treatment of lead poisoning because of its high affinity for calcium. It is approved, however, for treating *hypercalcemia,* a condition in which there is excess calcium in the blood—usually as a result of bone cancer. The accidental use of Endrate during treatment for lead poisoning resulted in the death of a 2-year-old girl in February 2005. The girl's death was attributed to sudden cardiac arrest caused by the removal of too much calcium from her blood.

© H.S. Photos/Alamy Stock Photo

Coordination Compounds

22.1.1 Write the correct name for the compound $[Cu(NH_3)_4]Cl_2$.
  (a) Coppertetraammine dichloride
  (b) Tetraammincopper(II) chloride
  (c) Tetraamminedichlorocuprate(II)
  (d) Dichlorotetraamminecopper(II)
  (e) Tetraamminedichlorocopper(II)

22.1.2 Write the correct name for the compound $K_3[FeF_6]$.
  (a) Tripotassiumironhexafluoride
  (b) Hexafluorotripotassiumferrate(III)
  (c) Hexafluoroiron(III) potassium
  (d) Potassium hexafluoroferrate(III)
  (e) Potassium ironhexafluorate

22.1.3 Write the correct formula for pentaamminenitrocobalt(III).
  (a) $[Co(NH_3)_5NO_2]^{3+}$ (d) $[Co(NH_3)_5](NO_2)$
  (b) $[Co(NH_3)_5NO_2]^{2+}$ (e) $[Co(NH_3)_5](NO_2)_2$
  (c) $Co(NH_3)_5NO_2$

22.1.4 Write the correct formula for tetraaquadichlorochromium(III) chloride.
  (a) $[Cr(H_2O)_4Cl_2]Cl_3$ (d) $[Cr(H_2O)_4]Cl_3$
  (b) $[Cr(H_2O)_4Cl_2]Cl_2$ (e) $[Cr(H_2O)_4]Cl_2$
  (c) $[Cr(H_2O)_4Cl_2]Cl$

22.2 STRUCTURE OF COORDINATION COMPOUNDS

The geometry of a coordination compound often plays a significant role in determining its properties. Figure 22.5 shows four different geometric arrangements for metal atoms with monodentate ligands. In these diagrams we see that structure and the coordination number of the metal relate to each other as follows:

Coordination number	Structure
2	Linear
4	Tetrahedral or square planar
6	Octahedral

In studying the geometry of coordination compounds, we sometimes find that there is more than one way to arrange the ligands around the central atom. Such compounds in which ligands are arranged differently, known as ***stereoisomers,*** have distinctly

Student Annotation: In general, stereoisomers are compounds that are made up of the same types and numbers of atoms, bonded together in the same sequence, but with different spatial arrangements.

Figure 22.5 Common geometries of complex ions. In each case M is a metal and L is a monodentate ligand.

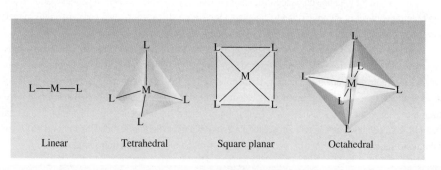

Linear   Tetrahedral   Square planar   Octahedral

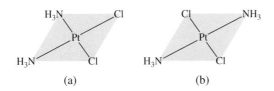

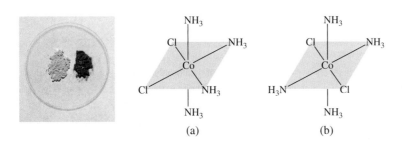

Figure 22.6 The (a) *cis* and (b) *trans* isomers of diamminedichloroplatinum(II). Note that the two Cl atoms are adjacent to each other in the *cis* isomer and diagonally across from each other in the *trans* isomer.

Figure 22.7 The (a) *cis* and (b) *trans* isomers of tetraamminedichlorocobalt(III) ion, $[Co(NH_3)_4Cl_2]^+$. The ion has only two geometric isomers.

© McGraw-Hill Education/Ken Karp, Photographer

different physical and chemical properties. Coordination compounds may exhibit two types of stereoisomerism: *geometric* and *optical.*

Geometrical isomers are stereoisomers that cannot be interconverted without breaking chemical bonds. Geometric isomers come in pairs. We use the terms *cis* and *trans* to distinguish one geometric isomer of a compound from the other. *Cis* means that two particular atoms (or groups of atoms) are adjacent to each other, whereas *trans* means that the atoms (or groups of atoms) are on opposite sides in the structural formula. The *cis* and *trans* isomers of coordination compounds generally have quite different colors, melting points, dipole moments, and chemical reactivities. Figure 22.6 shows the *cis* and *trans* isomers of diamminedichloroplatinum(II). Note that although the types of bonds are the same in both isomers (two Pt—N and two Pt—Cl bonds), the spatial arrangements are different. Another example is the tetraamminedichlorocobalt(III) ion, shown in Figure 22.7.

Optical isomers are nonsuperimposable mirror images. (*Superimposable* means that if one structure is laid over the other, the positions of all the atoms will match.) Like geometric isomers, optical isomers come in pairs. However, the optical isomers of a compound have *identical* physical and chemical properties, such as melting point, boiling point, dipole moment, and chemical reactivity toward molecules that are not *themselves* optical isomers. Optical isomers differ from each other, though, in their interactions with plane-polarized light, as we will see.

The structural relationship between two optical isomers is analogous to the relationship between your left and right hands. If you place your left hand in front of a mirror, the image you see will look like your right hand (Figure 22.8). Your left hand and right hand are mirror images of each other. They are nonsuperimposable, however, because when you place your left hand over your right hand (with both palms facing down), they do not match. This is why a right-handed glove will not fit comfortably on your left hand.

Figure 22.9 shows the *cis* and *trans* isomers of dichloro*bis*(ethylenediamine) cobalt(III) ion and the mirror image of each. Careful examination reveals that the *trans* isomer and its mirror image are superimposable, but the *cis* isomer and its mirror image are not. Thus, the *cis* isomer and its mirror image are *optical isomers.*

Optical isomers are described as *chiral* (from the Greek word for "hand") because, like your left and right hands, chiral molecules are nonsuperimposable. Isomers that are superimposable with their mirror images are said to be *achiral.* Chiral molecules play a vital role in enzyme reactions in biological systems. Many drug molecules are chiral, although only one of a pair of chiral isomers is biologically effective.

Chiral molecules are said to be optically active because of their ability to rotate the plane of polarization of polarized light as it passes through them. Unlike ordinary

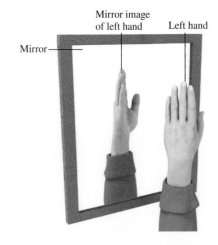

Figure 22.8 A left hand and its mirror image.

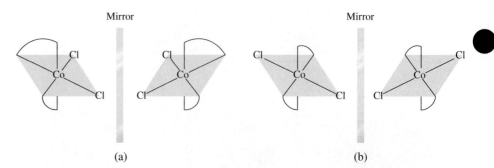

(a)   (b)

Figure 22.9 The (a) *cis* and (b) *trans* isomers of dichlorobis(ethylenediamine)cobalt(III) ion and their mirror images. If you could rotate the mirror image in (b) 90° clockwise about the vertical position and place the ion over the *trans* isomer, you would find that the two are superimposable. No matter how you rotate the *cis* isomer and its mirror image in (a), however, you cannot superimpose one on the other.

light, which vibrates in all directions, plane-polarized light vibrates only in a single plane. We use a ***polarimeter*** to measure the rotation of polarized light by optical isomers (Figure 22.10). A beam of unpolarized light first passes through a Polaroid sheet, called the polarizer, and then through a sample tube containing a solution of an optically active, chiral compound. As the polarized light passes through the sample tube, its plane of polarization is rotated either to the right (clockwise) or to the left (counterclockwise). This rotation can be measured directly by turning the analyzer in the appropriate direction until minimal light transmission is achieved (Figure 22.11). If the plane of polarization is rotated to the right, the isomer is said to be ***dextrorotatory*** and the isomer is labeled *d;* if the rotation is to the left, the isomer is ***levorotatory*** and the isomer is labeled *l.* The *d* and *l* isomers of a chiral substance, called ***enantiomers,*** always rotate the plane of polarization by the same amount, but in opposite directions. Thus, in an equimolar mixture of two enantiomers, called a ***racemic mixture,*** the net rotation is zero.

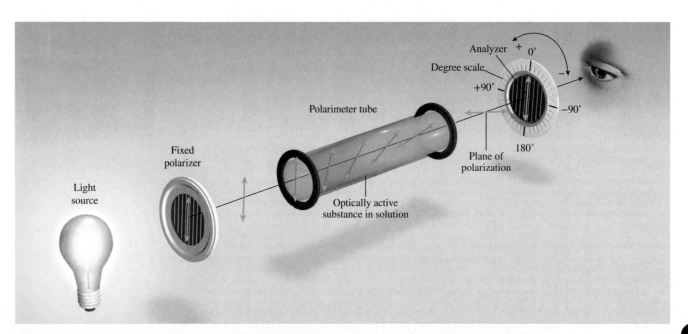

Figure 22.10 Operation of a polarimeter. Initially, the tube is filled with an achiral compound. The analyzer is rotated so that its plane of polarization is perpendicular to that of the polarizer. Under this condition, no light reaches the observer. Next, a chiral compound is placed in the tube as shown. The plane of polarization of the polarized light is rotated as it travels through the tube so that some light reaches the observer. Rotating the analyzer (either to the left or to the right) until no light reaches the observer again allows the angle of optical rotation to be measured.

22.3 BONDING IN COORDINATION COMPOUNDS: CRYSTAL FIELD THEORY

A satisfactory theory of bonding in coordination compounds must account for properties such as color and magnetism, as well as stereochemistry and bond strength. No single theory as yet does all this for us. Rather, several different approaches have been applied to transition metal complexes. We will consider only one of them here—crystal field theory—because it accounts for both the color and magnetic properties of many coordination compounds.

We will begin our discussion of crystal field theory with the most straightforward case—namely, complex ions with octahedral geometry. Then we will see how it is applied to tetrahedral and square-planar complexes.

Crystal Field Splitting in Octahedral Complexes

Crystal field theory explains the bonding in complex ions purely in terms of electrostatic forces. In a complex ion, two types of electrostatic interaction come into play. One is the attraction between the positive metal ion and the negatively charged ligand or the negatively charged end of a polar ligand. This is the force that binds the ligands to the metal. The second type of interaction is the electrostatic repulsion between the lone pairs on the ligands and the electrons in the d orbitals of the metals.

The d orbitals have different orientations [◀◀ Section 3.8], but in the absence of an external disturbance, they all have the same energy. In an octahedral complex, a central metal atom is surrounded by six lone pairs of electrons (on the six ligands), so all five d orbitals experience electrostatic repulsion. The magnitude of this repulsion depends on the orientation of the d orbital that is involved. Take the $d_{x^2-y^2}$ orbital as an example. In Figure 22.12, we see that the lobes of this orbital point toward the corners of the octahedron along the x and y axes, where the lone-pair electrons are positioned. Thus, an electron residing in this orbital would experience a greater repulsion from the ligands than an electron would in the d_{xy}, d_{yz}, or d_{xz} orbitals. For this reason, the energy of the $d_{x^2-y^2}$ orbital is increased relative to the d_{xy}, d_{yz}, and d_{xz} orbitals. The d_{z^2} orbital's energy is also greater, because its lobes are pointed at the

Figure 22.12 The five d orbitals in an octahedral environment. The metal atom (or ion) is at the center of the octahedron, and the six lone pairs on the donor atoms of the ligands are at the corners.

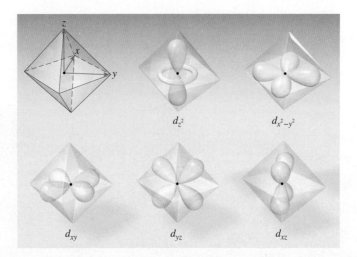

d_{z^2} $d_{x^2-y^2}$

d_{xy} d_{yz} d_{xz}

ligands along the z axis. As a result of these metal-ligand interactions, the five d orbitals in an octahedral complex are split between two sets of energy levels: a higher level with two orbitals ($d_{x^2-y^2}$ and d_{z^2}) having the same energy, and a lower level with three equal-energy orbitals (d_{xy}, d_{yz}, and d_{xz}), as shown in Figure 22.13. The ***crystal field splitting (Δ)*** is the energy difference between two sets of d orbitals in a metal atom when ligands are present. The magnitude of Δ depends on the metal and the nature of the ligands; it has a direct effect on the color and magnetic properties of complex ions.

Color

In Chapter 3 we learned that white light, such as sunlight, is a combination of all colors. A substance appears black if it absorbs all the visible light that strikes it. If it absorbs no visible light, it is white or colorless. An object appears green if it absorbs all light but reflects the green component. An object also looks green if it reflects all colors except red, the *complementary* color of green (Figure 22.14).

What has been said of reflected light also applies to *transmitted* light (i.e., the light that passes *through* the medium, such as a solution). Consider the hydrated cupric ion ($[Cu(H_2O)_6]^{2+}$); it absorbs light in the orange region of the spectrum, so a solution of $CuSO_4$ appears blue to us. Recall from Chapter 3 that when the energy of a photon is equal to the difference between the ground state and an excited state, absorption occurs as the photon strikes the atom (or ion or compound), and an electron is promoted to a higher level. Using these concepts, we can calculate the energy change involved in the electron transition. The energy of a photon is given by

$$E = h\nu$$

where h represents Planck's constant (6.63×10^{-34} J · s) and ν is the frequency of the

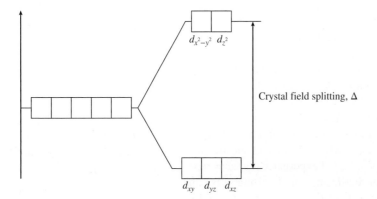

$d_{x^2-y^2}$ d_{z^2}

Crystal field splitting, Δ

d_{xy} d_{yz} d_{xz}

Figure 22.13 Crystal field splitting between d orbitals in an octahedral complex.

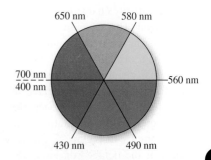

650 nm   580 nm

700 nm
400 nm   560 nm

430 nm   490 nm

Figure 22.14 A color wheel with appropriate wavelengths. Complementary colors, such as red and green, are on opposite sides of the wheel.

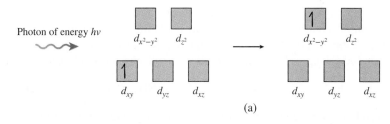

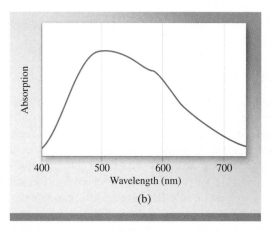

radiation, which is $5.00 \times 10^{14} \ s^{-1}$ for a wavelength of 600 nm. Here $E = \Delta$, so we have

$$\Delta = hv$$

$$= (6.63 \times 10^{-34} \ J \cdot s)(5.00 \times 10^{14} \ s^{-1})$$

$$= 3.32 \times 10^{-19} \ J$$

This value is very small, but it is the energy absorbed by only *one* ion. If the wavelength of the photon absorbed by an ion lies outside the visible region, then the transmitted light looks the same (to us) as the incident light—white—and the ion appears colorless.

The best way to measure crystal field splitting is to use spectroscopy to determine the wavelength at which light is absorbed. The $[Ti(H_2O)_6]^{3+}$ ion provides a straightforward example, because Ti^{3+} has only one $3d$ electron (Figure 22.15). The $[Ti(H_2O)_6]^{3+}$ ion absorbs light in the visible region of the spectrum (Figure 22.16). The wavelength corresponding to maximum absorption is 498 nm [Figure 22.15(b)]. To calculate the crystal field splitting energy, we start by writing

$$\Delta = hv$$

Next, recall that

$$v = \frac{c}{\lambda}$$

where c is the speed of light and λ is the wavelength. Therefore,

$$\Delta = \frac{hc}{\lambda} = \frac{(6.63 \times 10^{-34} \ J \cdot s)(3.00 \times 10^8 \ m/s)}{(498 \ nm)(1 \times 10^{-9} \ m/nm)} = 3.99 \times 10^{-19} \ J$$

This is the energy required to excite *one* $[Ti(H_2O)_6]^{3+}$ ion. To express this energy difference in the more convenient units of kJ/mol, we write

$$\Delta = (3.99 \times 10^{-19} \ J/ion)(6.02 \times 10^{23} \ ions/mol)$$

$$= 240,000 \ J/mol$$

$$= 240 \ kJ/mol$$

Figure 22.16 Colors of some of the first-row transition metal ions in solution. From left to right: Ti^{3+}, Cr^{3+}, Mn^{2+}, Fe^{3+}, Co^{2+}, Ni^{2+}, Cu^{2+}. The Sc^{3+} and V^{5+} ions are colorless.
© McGraw-Hill Education/Charles D. Winters, photographer

Aided by spectroscopic data for a number of complexes, all having the same metal ion but different ligands, chemists calculated the crystal field splitting for each ligand and established the following *spectrochemical series,* which is a list of ligands arranged in increasing order of their abilities to split the d orbital energy levels.

$$I^- < Br^- < Cl^- < OH^- < F^- < H_2O < NH_3 < en < CN^- < CO$$

These ligands are arranged in the order of increasing value of Δ. CO and CN^- are called *strong-field ligands,* because they cause a large splitting of the d orbital energy levels. The halide ions and hydroxide ion are *weak-field ligands,* because they split the d orbitals to a lesser extent.

Magnetic Properties

The magnitude of the crystal field splitting also determines the magnetic properties of a complex ion. The $[Ti(H_2O)_6]^{3+}$ ion, having only one d electron, is always paramagnetic. However, for an ion with several d electrons, the situation is less immediately clear. Consider, for example, the octahedral complexes $[FeF_6]^{3-}$ and $[Fe(CN)_6]^{3-}$ (Figure 22.17). The electron configuration of Fe^{3+} is $[Ar]3d^5$, and there are two possible ways to distribute the five d electrons among the d orbitals. According to Hund's rule [◄◄ Section 3.9], maximum stability is reached when the electrons are placed in five separate orbitals with parallel spins. This arrangement can be achieved only at a cost, however, because two of the five electrons must be promoted to the higher-energy $d_{x^2-y^2}$ and d_{z^2} orbitals. No such energy investment is needed if all five electrons enter the d_{xy}, d_{yz}, and d_{xz} orbitals. According to Pauli's exclusion principle [◄◄ Section 3.9], there will be only one unpaired electron present in this case.

Figure 22.18 shows the distribution of electrons among d orbitals that results in low- and high-spin complexes. The actual arrangement of the electrons is determined by the amount of stability gained by having maximum parallel spins versus the investment in energy required to promote electrons to higher d orbitals. Because F^- is a weak-field ligand, the five d electrons enter five separate d orbitals with parallel spins to create a high-spin complex. The cyanide ion is a strong-field ligand, though, so it is energetically preferable for all five electrons to be in the lower orbitals, thus forming a low-spin complex. High-spin complexes are more paramagnetic than low-spin complexes.

The actual number of unpaired electrons (or spins) in a complex ion can be found by magnetic measurements, and in general, experimental findings support predictions based on crystal field splitting. However, a distinction between low- and high-spin complexes can be made only if the metal ion contains more than three and fewer than eight d electrons, as shown in Figure 22.18.

Worked Example 22.4 shows how to determine the number of spins in an octahedral complex.

Figure 22.17 Energy-level diagrams for the Fe^{3+} ion and for the $[FeF_6]^{3-}$ and $[Fe(CN)_6]^{3-}$ complex ions.

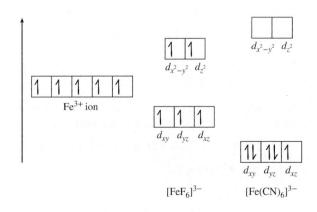

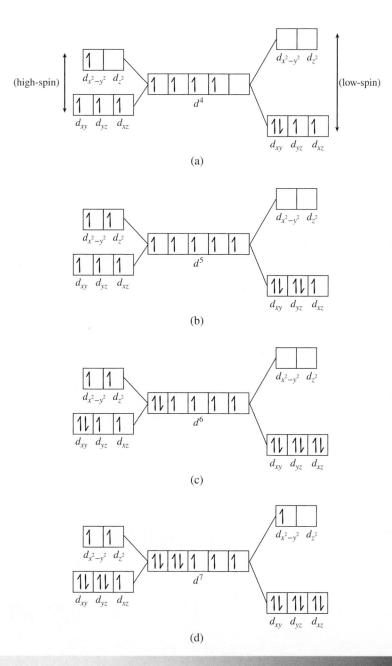

Figure 22.18 Orbital diagrams for the high-spin (left) and low-spin (right) octahedral complexes corresponding to the electron configurations of (a) d^4, (b) d^5, (c) d^6, and (d) d^7.

Worked Example 22.4

Predict the number of unpaired spins in the $[Cr(en)_3]^{2+}$ ion.

Strategy The magnetic properties of a complex ion depend on the strength of the ligands. Strong-field ligands, which cause a high degree of splitting among the d orbital energy levels, result in low-spin complexes. Weak-field ligands, which cause only a small degree of splitting among the d orbital energy levels, result in high-spin complexes.

Setup The electron configuration of Cr^{2+} is $[Ar]3d^4$; and en is a strong-field ligand.

Solution Because en is a *strong*-field ligand, we expect $[Cr(en)_3]^{2+}$ to be a low-spin complex. According to Figure 22.18, all four electrons will be placed in the lower-energy d orbitals (d_{xy}, d_{yz}, and d_{xz}) and there will be a total of two unpaired spins.

> **Think About It**
> It is easy to draw the wrong conclusion regarding high- and low-spin complexes. Remember that the term *high spin* refers to the number of spins (*unpaired electrons*), not to the energy levels of the d orbitals. The greater the energy gap between the lower-energy and higher-energy d orbitals, the greater the chance that the complex will be *low* spin.

(Continued on next page)

Practice Problem **A**TTEMPT How many unpaired spins are in $[Mn(H_2O)_6]^{2+}$? (*Hint:* H_2O is a weak-field ligand.)

Practice Problem **B**UILD Visible transitions can occur in metal ions having as few as one *d* electron or as many as nine *d* electrons. Which numbers of *d* electrons (1–9) would result in the same number of unpaired spins in both the high-spin and low-spin states?

Practice Problem **C**ONCEPTUALIZE Transition metal complexes containing CN^- ligands are often yellow in color, whereas those containing H_2O ligands are often green or blue. Explain.

Tetrahedral and Square-Planar Complexes

So far we have concentrated on octahedral complexes. The splitting of the *d* orbital energy levels in tetrahedral and square-planar complexes, though, can also be accounted for satisfactorily by the crystal field theory. In fact, the splitting pattern for a tetrahedral ion is just the reverse of that for octahedral complexes. In this case, the d_{xy}, d_{yz}, and d_{xz} orbitals are more closely directed at the ligands and therefore have more energy than the $d_{x^2-y^2}$ and d_{z^2} orbitals (Figure 22.19). Most tetrahedral complexes are high-spin complexes. Presumably, the tetrahedral arrangement reduces the magnitude of the metal-ligand interactions, resulting in a smaller Δ value. This is a reasonable assumption because the number of ligands is smaller in a tetrahedral complex.

As Figure 22.20 shows, the splitting pattern for square-planar complexes is the most complicated. The $d_{x^2-y^2}$ orbital possesses the highest energy (as in the octahedral case), and the d_{xy} orbital is the next highest. However, the relative placement of the d_{z^2} and the d_{xz} and d_{yz} orbitals cannot be determined simply by inspection and must be calculated.

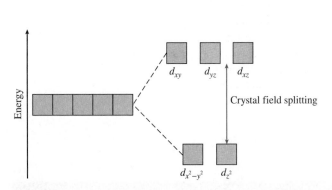

Figure 22.19 Crystal field splitting between *d* orbitals in a tetrahedral complex.

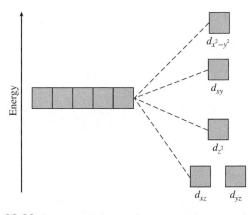

Figure 22.20 Energy-level diagram for a square-planar complex. Because there are more than two energy levels, we cannot define crystal field splitting as we can for octahedral and tetrahedral complexes.

Section 22.3 Review

Bonding in Coordination Compounds: Crystal Field Theory

22.3.1 How many unpaired spins would you expect the $[Mn(CO)_6]^{2+}$ ion to have?
    (a) zero   (d) three
    (b) one   (e) five
    (c) two

22.3.2 Which of the following metal ions can potentially form both low-spin and high-spin complexes? (Select all that apply.)
    (a) Ti^{2+} (d) Ni^{2+}
    (b) Cu^+ (e) Cr^{3+}
    (c) Fe^{2+}

22.4 REACTIONS OF COORDINATION COMPOUNDS

Complex ions undergo *ligand exchange* (or *substitution*) reactions in solution. The rates of these reactions vary widely, depending on the nature of the metal ion and the ligands.

In studying ligand exchange reactions, it is often useful to distinguish between the stability of a complex ion and its tendency to react, which we call *kinetic lability.* Stability in this context is a thermodynamic property, which is measured in terms of the species' formation constant K_f [◀ Section 17.5]. For example, we say that the complex ion tetracyanonickelate(II) is *stable* because it has a large formation constant ($K_f = 1 \times 10^{30}$).

$$Ni^{2+} + 4CN^- \rightleftharpoons [Ni(CN)_4]^{2-}$$

By using cyanide ions labeled with the radioactive isotope carbon-14, chemists have shown that $[Ni(CN)_4]^{2-}$ undergoes ligand exchange very rapidly in solution. The following equilibrium is established almost as soon as the species are mixed:

$$[Ni(CN)_4]^{2-} + 4*CN^- \rightleftharpoons [Ni(*CN)_4]^{2-} + 4CN^-$$

where the asterisk denotes a ^{14}C atom. Complexes like the tetracyanonickelate(II) ion are termed *labile complexes* because they undergo *rapid* ligand exchange reactions. Thus, a thermodynamically *stable* species (i.e., one that has a *large* formation constant) is not necessarily unreactive.

A complex that is thermodynamically *unstable* in acidic solution is $[Co(NH_3)_6]^{3+}$. The equilibrium constant for the following reaction is about 1×10^{20}.

$$[Co(NH_3)_6]^{3+} + 6H^+ + 6H_2O \rightleftharpoons [Co(H_2O)_6]^{3+} + 6NH_4^+$$

When equilibrium is reached, the concentration of the $[Co(NH_3)_6]^{3+}$ ion is very low. This reaction requires several days to complete, however, because the $[Co(NH_3)_6]^{3+}$ ion is so inert. This is an example of an *inert complex*—a complex ion that undergoes very slow exchange reactions (on the order of hours or even days). It shows that a thermodynamically unstable species is not necessarily chemically reactive. The rate of reaction is determined by the energy of activation, which is high in this case.

Most complex ions containing Co^{3+}, Cr^{3+}, and Pt^{2+} are kinetically inert. Because they exchange ligands very slowly, they are easy to study in solution. As a result, our knowledge of the bonding, structure, and isomerism of coordination compounds has come largely from studies of these compounds.

22.5 APPLICATIONS OF COORDINATION COMPOUNDS

Coordination compounds are found in living systems and have many uses in the home, in industry, and in medicine. We briefly describe a few examples in this section.

Metallurgy

The extraction of silver and gold by the formation of cyanide complexes and the purification of nickel by converting the metal to the gaseous compound $Ni(CO)_4$ are typical examples of the use of coordination compounds in metallurgical processes.

Chelation Therapy

Earlier we mentioned that chelation therapy is used in the treatment of lead poisoning. Other metals, such as arsenic and mercury, can also be removed using chelating agents.

Chemotherapy

Several platinum-containing coordination compounds, including cisplatin $[Pt(NH_3)_2Cl_2]$ and carboplatin $[Pt(NH_3)_2(OCO)_2C_4H_6]$, can effectively inhibit the growth of cancerous cells. The mechanism for the action of cisplatin is the *chelation* of DNA, the molecule that contains the genetic code. During cell division, the double-stranded DNA unwinds into two single strands, which must be accurately copied for the new cells to be identical to their parent cell. X-ray studies show that cisplatin binds to DNA by forming cross-links in which the two chlorides on cisplatin are replaced by nitrogen atoms in the adjacent guanine bases on the same strand of the DNA. (Guanine is one of the four bases in DNA.) This causes a bend in the double-stranded structure at the binding site. It is believed that this structural distortion is a key factor in inhibiting replication. The damaged cell is then destroyed by the body's immune system. Because the binding of cisplatin to DNA requires both Cl atoms to be on the same side of the complex, the *trans* isomer of the compound is totally ineffective as an anticancer drug.

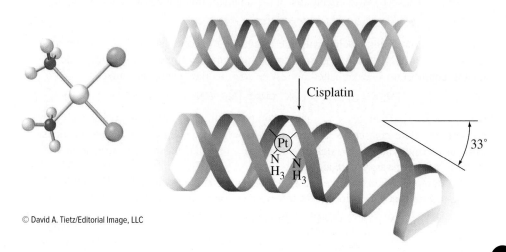

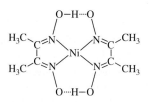

Figure 22.21 Structure of nickel dimethylglyoxime. Note that the overall structure is stabilized by hydrogen bonds.

Chemical Analysis

Although EDTA has a great affinity for a large number of metal ions (especially 2+ and 3+ ions), other chelates are more selective in binding. Dimethylglyoxime, for example, forms an insoluble brick-red solid with Ni^{2+} and an insoluble bright-yellow solid with Pd^{2+}. These characteristic colors are used in qualitative analysis to identify nickel and palladium. Furthermore, the quantities of ions present can be determined by gravimetric analysis [◀◀ Section 9.6] as follows: To a solution containing Ni^{2+} ions, say, we add an excess of dimethylglyoxime reagent, and a brick-red precipitate forms. The precipitate is then filtered, dried, and weighed. Knowing the formula of the complex (Figure 22.21), we can readily calculate the amount of nickel in the original solution.

Detergents

The cleansing action of soap in hard water is hampered by the reaction of the Ca^{2+} ions in the water with the soap molecules to form insoluble salts or curds. In the late 1940s the detergent industry introduced a "builder," usually sodium tripolyphosphate, to circumvent this problem. The tripolyphosphate ion is an effective chelating agent that forms stable, soluble complexes with Ca^{2+} ions. Sodium tripolyphosphate revolutionized the detergent industry. Because phosphates are plant nutrients, however, wastewater containing phosphates discharged into rivers and lakes causes algae to grow, resulting in oxygen depletion. Under these conditions, most or all aquatic life eventually succumbs. This process is called *eutrophication*. Consequently, many states have banned phosphate detergents since the 1970s, and manufacturers have reformulated their products to eliminate phosphates.

Sequestrants

In addition to its use in medicine and chemical analysis, EDTA is used as a food additive to sequester metal ions. EDTA sequesters copper, iron, and nickel ions that would otherwise catalyze the oxidation reactions that cause food to spoil. EDTA is a common preservative in a wide variety of consumer products.

Learning Outcomes

- Define coordination compound.
- Describe some of the properties of transition metals.
- List some common ligands found in transition metal complexes.
- Define coordination number.
- Explain what a chelating agent does.
- Name a coordination compound using nomenclature rules.
- Determine the structure of a coordination compound.
- Distinguish between geometric and optical isomers.

- Define the following stereochemical terms: *chiral, dextrorotatory, levorotatory,* and *enantiomer.*
- Use concepts of crystal field theory to explain properties of coordination compounds (e.g., color and magnetic properties).
- Describe the nature of ligand exchange reactions including an explanation of kinetic lability.
- Provide examples of common applications of coordination compounds.

Chapter Summary

SECTION 22.1

- ***Coordination compounds*** contain coordinate covalent bonds between a metal ion (often a transition metal ion) and two or more polar molecules or ions.
- The molecules or anions that surround a metal in a coordination complex are called ***ligands.***
- Many coordination compounds consist of a *complex ion* and a *counter* ion.
- Transition metals are those that have incompletely filled *d* subshells—or that *give rise* to ions with incompletely filled *d* subshells.
- Transition metals exhibit variable oxidation states ranging from +1 to +7.
- To act as a ligand, a molecule or ion must have at least one unshared pair of electrons. The atom that bears the unshared pair of electrons is the ***donor atom.***
- Ligands are classified as *monodentate, bidentate,* or *polydentate,* based on the number of donor atoms they contain. Bidentate and polydentate ions are also known as ***chelating agents.***
- The ***coordination number*** is the number of donor atoms surrounding a metal in a complex.
- Ionic coordination compounds are named by first naming the cation and then the anion. Complex ions are named by listing the ligands in alphabetical order, followed by the metal and its oxidation state (as a Roman numeral). When the complex ion is the anion, the anion's name ends in -*ate.*

SECTION 22.2

- The coordination number largely determines the geometry of a coordination complex.
- Coordination compounds containing different arrangements of the same ligands are ***stereoisomers.*** The two types of stereoisomerism are geometric and optical.
- ***Geometrical isomers*** contain the same atoms and bonds arranged differently in space.
- ***Optical isomers*** are nonsuperimposable mirror images. We call a *pair* of optical isomers ***enantiomers.*** The rotation of polarized light is measured with a ***polarimeter.***

- Enantiomers rotate the plane of plane-polarized light in opposite directions. The enantiomer that rotates it to the right is called ***dextrorotatory*** and is labeled *d.* The enantiomer that rotates it to the left is called ***levorotatory*** and is labeled *l.* An equal mixture of a pair of enantiomers, called a ***racemic mixture,*** does not cause any net rotation of plane-polarized light.

SECTION 22.3

- Ligands in a coordination complex cause the energy levels of the *d* orbitals on a metal to split. The difference in energy between the lower and higher *d* orbital energy levels is called the ***crystal field splitting (Δ).***
- The magnitude of Δ depends on the nature of the ligands in the complex. The ***spectrochemical series*** orders some common ligands in order of increasing *field strength.*
- *Strong*-field ligands give rise to a larger Δ value; *weak*-field ligands yield a smaller Δ value.
- Crystal field splitting sometimes changes the number of unpaired electrons, and therefore the magnetic properties, of a metal.
- Complexes containing transition metals with d^4, d^5, d^6, or d^7 configurations may be *high spin* or *low spin.* In high-spin complexes, the number of unpaired electrons is maximized because Δ is small; in low-spin complexes, the number of unpaired electrons is minimized because Δ is large.

SECTION 22.4

- Complex ions undergo ligand exchange in solution. The rate at which ligand exchange occurs is a measure of a complex's *kinetic lability* and does not necessarily correspond directly to the complex's *thermodynamic stability.*

SECTION 22.5

- Coordination chemistry is important in many biological, medical, and industrial processes.

Key Words

Chelating agent, 1006
Coordination compound, 1003
Coordination number, 1006
Crystal field splitting (Δ), 1014
Dextrorotatory, 1012

Donor atom, 1006
Enantiomers, 1012
Geometrical isomers, 1011
Levorotatory, 1012
Ligand, 1005

Optical isomers, 1011
Polarimeter, 1012
Racemic mixture, 1012
Spectrochemical series, 1016
Stereoisomers, 1010

Questions and Problems

SECTION 22.1: COORDINATION COMPOUNDS

Review Questions

22.1   What distinguishes a transition metal from a main group metal?

22.2   Why is zinc not considered a transition metal?

22.3   Explain why atomic radii decrease very gradually from scandium to copper.

22.4   Without referring to the text, write the ground-state electron configurations of the first-row transition metals. Explain any irregularities.

22.5   Write the electron configurations of the following ions: V^{5+}, Cr^{3+}, Mn^{2+}, Fe^{3+}, Cu^{2+}, Sc^{3+}, Ti^{4+}.

22.6   Why do transition metals have more oxidation states than other elements?

22.7   Give the highest oxidation states for scandium to copper.

22.8   Define the following terms: *coordination compound, ligand, donor atom, coordination number, chelating agent.*

22.9   Describe the interaction between a donor atom and a metal atom in terms of a Lewis acid-base reaction.

Conceptual Problems

22.10   Complete the following statements for the complex ion $[Co(en)_2(H_2O)CN]^{2+}$. (a) en is the abbreviation for _____. (b) The oxidation number of Co is _____. (c) The coordination number of Co is _____. (d) _____ is a bidentate ligand.

22.11   Complete the following statements for the complex ion $[Cr(C_2O_4)_2(H_2O)_2]^-$. (a) The oxidation number of Cr is _____. (b) The coordination number of Cr is _____. (c) _____ is a bidentate ligand.

22.12   Give the oxidation numbers of the metals in the following species: (a) $K_3[Fe(CN)_6]$, (b) $K_3[Cr(C_2O_4)_3]$, (c) $[Ni(CN)_4]^{2-}$.

22.13   Give the oxidation numbers of the metals in the following species: (a) Na_2MoO_4, (b) $MgWO_4$, (c) $Fe(CO)_5$.

22.14   What are the systematic names for the following ions and compounds?
(a) $[Co(NH_3)_4Cl_2]^+$ (c) $[Co(en)_2Br_2]^+$
(b) $[Cr(NH_3)_3Cl_3]$ (d) $[Co(NH_3)_6]Cl_3$

22.15   What are the systematic names for the following ions and compounds?
(a) $[cis\text{-}Co(en)_2Cl_2]^+$ (c) $[Co(NH_3)_5Cl]Cl_2$
(b) $[Pt(NH_3)_5Cl]Cl_3$

22.16   Write the formulas for each of the following ions and compounds: (a) tetrahydroxozincate(II), (b) penta-aquachlorochromium(III) chloride, (c) tetrabromo-cuprate(II), (d) ethylenediaminetetraacetatoferrate(II).

22.17   One of the drugs commonly used for chelation therapy is dimercaptosuccinic acid (DMSA), marketed under the name Chemet. How many donor atoms does DMSA have?

Dimercaptosuccinic acid (DMSA)

SECTION 22.2: STRUCTURE OF COORDINATION COMPOUNDS

Review Questions

22.18   Define the following terms: *stereoisomers, geometric isomers, optical isomers, plane-polarized light.*

22.19   Specify which of the following structures can exhibit geometric isomerism: (a) linear, (b) square planar, (c) tetrahedral, (d) octahedral.

22.20   What determines whether a molecule is chiral? How does a polarimeter measure the chirality of a molecule?

22.21   Explain the following terms: *enantiomers, racemic mixtures.*

Conceptual Problems

22.22   The complex ion $[Ni(CN)_2Br_2]^{2-}$ has a square-planar geometry. Draw the structures of the geometric isomers of this complex.

22.23 How many geometric isomers are in the following species: (a) $[Co(NH_3)_2Cl_4]^-$, (b) $[Co(NH_3)_3Cl_3]$?

22.24 Draw structures of all the geometric and optical isomers of each of the following cobalt complexes.
(a) $[Co(NH_3)_6]^{3+}$ (c) $[Co(C_2O_4)_3]^{3-}$
(b) $[Co(NH_3)_5Cl]^{2+}$

22.25 Draw structures of all the geometric and optical isomers of each of the following cobalt complexes:
(a) $[Co(NH_3)_4Cl_2]^+$, (b) $[Co(en)_3]^{3+}$.

SECTION 22.3: BONDING IN COORDINATION COMPOUNDS: CRYSTAL FIELD THEORY

Review Questions

22.26 Briefly describe crystal field theory.

22.27 Define the following terms: *crystal field splitting, high-spin complex, low-spin complex, spectrochemical series.*

22.28 What is the origin of color in a coordination compound?

22.29 Compounds containing the Sc^{3+} ion are colorless, whereas those containing the Ti^{3+} ion are colored. Explain.

22.30 For the same type of ligands, explain why the crystal field splitting for an octahedral complex is always greater than that for a tetrahedral complex.

Conceptual Problems

22.31 Transition metal complexes containing CN^- ligands are often yellow in color, whereas those containing H_2O ligands are often green or blue. Explain.

22.32 Predict the number of unpaired electrons in the following complex ions: (a) $[Cr(CN)_6]^{4-}$, (b) $[Cr(H_2O)_6]^{2+}$.

22.33 The absorption maximum for the complex ion $[Co(NH_3)_6]^{3+}$ occurs at 470 nm. (a) Predict the color of the complex, and (b) calculate the crystal field splitting in kJ/mol.

22.34 From each of the following pairs, choose the complex that absorbs light at a longer wavelength: (a) $[Co(NH_3)_6]^{2+}$, $[Co(H_2O)_6]^{2+}$; (b) $[FeF_6]^{3-}$, $[Fe(CN)_6]^{3-}$; (c) $[Cu(NH_3)_4]^{2+}$, $[CuCl_4]^{2-}$.

22.35 A solution made by dissolving 0.875 g of $Co(NH_3)_4Cl_3$ in 25.0 g of water freezes at $-0.56°C$. Calculate the number of moles of ions produced when 1 mole of $Co(NH_3)_4Cl_3$ is dissolved in water, and suggest a structure for the complex ion present in this compound.

22.36 In biological systems, the Cu^{2+} ions coordinated with S atoms tend to form tetrahedral complexes, whereas those coordinated with N atoms tend to form octahedral complexes. Explain.

22.37 Plastocyanin, a copper-containing protein found in photosynthetic systems, is involved in electron transport, with the copper ion switching between the +1 and +2 oxidation states. The copper ion is coordinated with two histidine residues, a cysteine residue, and a methionine residue in a tetrahedral configuration. How does the crystal field splitting (Δ) change between these two oxidation states?

SECTION 22.4: REACTIONS OF COORDINATION COMPOUNDS

Review Questions

22.38 Define the terms *labile complex* and *inert complex.*

22.39 Explain why a thermodynamically stable species may be chemically reactive and a thermodynamically unstable species may be unreactive.

Conceptual Problems

22.40 Oxalic acid ($H_2C_2O_4$) is sometimes used to clean rust stains from sinks and bathtubs. Explain the chemistry underlying this cleaning action.

22.41 The $[Fe(CN)_6]^{3-}$ complex is more labile than the $[Fe(CN)_6]^{4-}$ complex. Suggest an experiment that would prove that $[Fe(CN)_6]^{3-}$ is a labile complex.

22.42 Aqueous copper(II) sulfate solution is blue in color. When aqueous potassium fluoride is added, a green precipitate is formed. When aqueous potassium chloride is added instead, a bright-green solution is formed. Explain what is happening in these two cases.

22.43 When aqueous potassium cyanide is added to a solution of copper(II) sulfate, a white precipitate, soluble in an excess of potassium cyanide, is formed. No precipitate is formed when hydrogen sulfide is bubbled through the solution at this point. Explain.

22.44 A concentrated aqueous copper(II) chloride solution is bright green in color. When diluted with water, the solution becomes light blue. Explain.

22.45 In a dilute nitric acid solution, Fe^{3+} reacts with thiocyanate ion (SCN^-) to form a dark-red complex.

$$[Fe(H_2O)_6]^{3+} + SCN^- \rightleftharpoons H_2O + [Fe(H_2O)_5NCS]^{2+}$$

The equilibrium concentration of $[Fe(H_2O)_5NCS]^{2+}$ may be determined by how darkly colored the solution is (measured by a spectrometer). In one such experiment, 1.0 mL of 0.20 M $Fe(NO_3)_3$ was mixed with 1.0 mL of 1.0×10^{-3} M KSCN and 8.0 mL of dilute HNO_3. The color of the solution quantitatively indicated that the $[Fe(H_2O)_5NCS]^{2+}$ concentration was 7.3×10^{-5} M. Calculate the formation constant for $[Fe(H_2O)_5NCS]^{2+}$.

SECTION 22.5: APPLICATIONS OF COORDINATION COMPOUNDS

Review Questions

22.46 Describe and give examples of the applications of coordination compounds.

ADDITIONAL PROBLEMS

22.47 Suffocation victims usually look purple, but a person poisoned by carbon monoxide often has rosy cheeks. Explain.

22.48 As we read across the first-row transition metals from left to right, the +2 oxidation state becomes more stable in comparison with the +3 state. Why is this so?

22.49 Which is a stronger oxidizing agent in aqueous solution, Mn^{3+} or Cr^{3+}? Explain your choice.

22.50 Carbon monoxide binds to Fe in hemoglobin some 200 times more strongly than oxygen. This is the reason why CO is a toxic substance. The metal-to-ligand sigma bond is formed by donating a lone pair from the donor atom to an empty sp^3d^2 orbital on Fe. (a) On the basis of electronegativities, would you expect the C or O atom to form the bond to Fe? (b) Draw a diagram illustrating the overlap of the orbitals involved in the bonding.

22.51 What are the oxidation states of Fe and Ti in the ore ilmenite ($FeTiO_3$)? (*Hint:* Look up the ionization energies of Fe and Ti in Table 22.1; the fourth ionization energy of Ti is 4180 kJ/mol.)

22.52 A student has prepared a cobalt complex that has one of the following three structures: $[Co(NH_3)_6]Cl_3$, $[Co(NH_3)_5Cl]Cl_2$, or $[Co(NH_3)_4Cl_2]Cl$. Explain how the student would distinguish between these possibilities by an electrical conductance experiment. At the student's disposal are three strong electrolytes—$NaCl$, $MgCl_2$, and $FeCl_3$—which may be used for comparison purposes.

22.53 Chemical analysis shows that hemoglobin contains 0.34% of Fe by mass. What is the minimum possible molar mass of hemoglobin? The actual molar mass of hemoglobin is about 65,000 g. How do you account for the discrepancy between your minimum value and the actual value?

22.54 Explain the following facts: (a) Copper and iron have several oxidation states, whereas zinc has only one. (b) Copper and iron form colored ions, whereas zinc does not.

22.55 A student in 1895 prepared three coordination compounds containing chromium, with the following properties.

Formula	Color	Cl^- ions in solution per formula unit
(a) $CrCl_3 \cdot 6H_2O$	Violet	3
(b) $CrCl_3 \cdot 6H_2O$	Light green	2
(c) $CrCl_3 \cdot 6H_2O$	Dark green	1

Write modern formulas for these compounds, and suggest a method for confirming the number of Cl^- ions present in solution in each case. (*Hint:* Some of the compounds may exist as hydrates.)

22.56 The formation constant for the reaction $Ag^+ + 2NH_3 \rightleftharpoons [Ag(NH_3)_2]^+$ is 1.5×10^7, and that for the reaction $Ag^+ + 2CN^- \rightleftharpoons [Ag(CN)_2]^-$ is 1.0×10^{21} at 25°C (see Table 17.5). Calculate the equilibrium constant and $\Delta G°$ at 25°C for the reaction

$$[Ag(NH_3)_2]^+ + 2CN^- \rightleftharpoons [Ag(CN)_2]^- + 2NH_3$$

22.57 From the standard reduction potentials listed in Table 18.1 for Zn/Zn^{2+} and Cu/Cu^{2+}, calculate $\Delta G°$ and the equilibrium constant for the reaction

$$Zn(s) + 2Cu^{2+}(aq) \longrightarrow Zn^{2+}(aq) + 2Cu^+(aq)$$

22.58 Using the standard reduction potentials listed in Table 18.1 and the *Handbook of Chemistry and Physics,* show that the following reaction is favorable under standard-state conditions.

$$2Ag(s) + Pt^{2+}(aq) \longrightarrow 2Ag^+(aq) + Pt(s)$$

What is the equilibrium constant of this reaction at 25°C?

22.59 The Co^{2+}-porphyrin complex is more stable than the Fe^{2+}-porphyrin complex. Why, then, is iron the metal ion in hemoglobin (and other heme-containing proteins)?

22.60 What are the differences between geometric isomers and optical isomers?

22.61 Oxyhemoglobin is bright red, whereas deoxyhemoglobin is purple. Show that the difference in color can be accounted for qualitatively on the basis of high-spin and low-spin complexes. (*Hint:* O_2 is a strong-field ligand.)

22.62 Hydrated Mn^{2+} ions are practically colorless (see Figure 22.16) even though they possess five $3d$ electrons. Explain. (*Hint:* Electronic transitions in which there is a change in the number of unpaired electrons do not occur readily.)

22.63 Which of the following hydrated cations are colorless: $Fe^{2+}(aq)$, $Zn^{2+}(aq)$, $Cu^+(aq)$, $Cu^{2+}(aq)$, $V^{5+}(aq)$, $Ca^{2+}(aq)$, $Co^{2+}(aq)$, $Sc^{3+}(aq)$, $Pb^{2+}(aq)$? Explain your choice.

22.64 Aqueous solutions of $CoCl_2$ are generally either light pink or blue. Low concentrations and low temperatures favor the pink form, whereas high concentrations and high temperatures favor the blue form. Adding hydrochloric acid to a pink solution of $CoCl_2$ causes the solution to turn blue; the pink color is restored by the addition of $HgCl_2$. Account for these observations.

22.65 Suggest a method that would allow you to distinguish between *cis*-$Pt(NH_3)_2Cl_2$ and *trans*-$Pt(NH_3)_2Cl_2$.

22.66 You are given two solutions containing $FeCl_2$ and $FeCl_3$ at the same concentration. One solution is light yellow, and the other one is brown. Identify these solutions based only on color.

22.67 The label of a certain brand of mayonnaise lists EDTA as a food preservative. How does EDTA prevent the spoilage of mayonnaise?

22.68 The compound 1,1,1-trifluoroacetylacetone (tfa) is a bidentate ligand.

$$\overset{\displaystyle O \quad\;\; O}{\underset{\displaystyle CF_3CCH_2CCH_3}{\| \quad\;\; \|}}$$

It forms a tetrahedral complex with Be^{2+} and a square-planar complex with Cu^{2+}. Draw structures of these complex ions, and identify the type of isomerism exhibited by these ions.

22.69 How many geometric isomers can the following square-planar complex have?

a \ / b
   Pt
d / \ c

22.70 $[Pt(NH_3)_2Cl_2]$ is found to exist in two geometric isomers designated I and II, which react with oxalic acid as follows:

$$I + H_2C_2O_4 \longrightarrow [Pt(NH_3)_2C_2O_4]$$
$$II + H_2C_2O_4 \longrightarrow [Pt(NH_3)_2(HC_2O_4)_2]$$

Comment on the structures of I and II.

22.71 Commercial silver-plating operations frequently use a solution containing the complex $Ag(CN)_2^-$ ion. Because the formation constant (K_f) is quite large, this procedure ensures that the free Ag^+ concentration in solution is low for uniform electrodeposition. In one process, a chemist added 9.0 L of 5.0 M NaCN to 90.0 L of 0.20 M $AgNO_3$. Calculate the concentration of free Ag^+ ions at equilibrium. See Table 17.5 for K_f value.

22.72 Draw qualitative diagrams for the crystal field splittings in (a) a linear complex ion ML_2, (b) a trigonal-planar complex ion ML_3, and (c) a trigonal-bipyramidal complex ion ML_5.

22.73 (a) The free Cu(I) ion is unstable in solution and has a tendency to disproportionate.

$$2Cu^+(aq) \rightleftharpoons Cu^{2+}(aq) + Cu(s)$$

Use the information in Table 18.1 to calculate the equilibrium constant for the reaction. (b) Based on your results in part (a), explain why most Cu(I) compounds are insoluble.

22.74 Consider the following two ligand exchange reactions.

$$[Co(H_2O)_6]^{3+} + 6NH_3 \rightleftharpoons [Co(NH_3)_6]^{3+} + 6H_2O$$
$$[Co(H_2O)_6]^{3+} + 3en \rightleftharpoons [Co(en)_3]^{3+} + 6H_2O$$

(a) Which of the reactions should have a larger $\Delta S°$?
(b) Given that the Co—N bond strength is approximately the same in both complexes, which

reaction will have a larger equilibrium constant? Explain your choices.

22.75 The Cr^{3+} ion forms octahedral complexes with two neutral ligands X and Y. The color of CrX_6^{3+} is blue while that of CrY_6^{3+} is yellow. Which is a stronger field ligand, X or Y?

22.76 Explain the difference between these two compounds: $CrCl_3 \cdot 6H_2O$ and $[Cr(H_2O)_6]Cl_3$.

22.77 The K_f for the formation of the complex ion between Pb^{2+} and $EDTA^{4-}$,

$$Pb^{2+} + EDTA^{4-} \rightleftharpoons [Pb(EDTA)]^{2-}$$

is 1.0×10^{18} at 25°C. Calculate $[Pb^{2+}]$ at equilibrium in a solution containing 1.0×10^{-3} M Pb^{2+} and 2.0×10^{-3} M $EDTA^{4-}$.

22.78 Locate the transition metal atoms and ions in the periodic table shown here. Atoms: (a) $[Kr]5s^24d^5$. (b) $[Xe]6s^24f^{14}5d^4$. Ions: (c) $[Ar]3d^3$ (ion with charge 4+). (d) $[Xe]4f^{14}5d^8$ (ion with charge 3+).

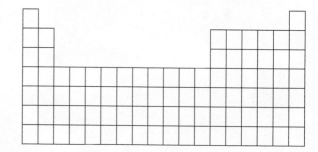

22.79 Manganese forms three low-spin complex ions with the cyanide ion with the formulas $[Mn(CN)_6]^{5-}$, $[Mn(CN)_6]^{4-}$, and $[Mn(CN)_6]^{3-}$. For each complex ion, determine the oxidation number of Mn and the number of unpaired d electrons.

22.80 Copper is known to exist in the +3 oxidation state, which is believed to be involved in some biological electron-transfer reactions. (a) Would you expect this oxidation state of copper to be stable? Explain. (b) Name the compound K_3CuF_6 and predict the geometry and magnetic properties of the complex ion. (c) Most of the known Cu(III) compounds have square-planar geometry. Are these compounds diamagnetic or paramagnetic?

Answers to In-Chapter Materials

PRACTICE PROBLEMS

22.1A (a) K: +1, Au: +3, (b) K: +1, Fe: +2. **22.1B** (a) +1, −3, 0, +1, +3, +3, +2. **22.2A** (a) tetraamminedibromocobalt(III) chloride, (b) tetraaquadichlorochromium(III) chloride, (c) potassium tetrachlorocuprate(II). **22.2B** (a) sodium hexacyanoferrate(III), (b) dichlorobis(ethylenediamine)chromium(III) chloride,

(c) tris(ethylenediamine)cobalt(III) chloride. **22.3A** (a) $[Ru(H_2O)_5Br]$ NO_3, (b) $K_2[PtBr_4Cl_2]$, (c) $Na_3[Co(NO_2)_6]$. **22.3B** (a) $[V(en)_2C_2O_4]Cl_2$, (b) $[Cr(en)_2Br_2]NO_3$, (c) $[Pt(en)_3](SO_4)_2$. **22.4A** 5. **22.4B** 1, 2, 3, 8, and 9.

SECTION REVIEW

22.1.1 b. **22.1.2** d. **22.1.3** b. **22.1.4** c. **22.3.1** b. **22.3.2** c.

Chapter 23

Organic Chemistry

© Digital Vision/Getty

EARTH'S OCEANS, which cover 79 percent of our planet, are a rich and fertile source for bioprospecting—the effort to find new and useful natural products. Corals, sponges, fish, and even marine microorganisms produce a bounty of biologically potent organic compounds with significant anti-inflammatory, antiviral, and anticancer properties. A natural product found to be of major importance is known as a *lead* (lēd) compound. Chemists may devise a way to synthesize a lead compound, or to use it as a precursor for synthesis or other useful compounds. This work can provide chemists a template for synthesizing structurally different, yet chemically related compounds to develop new drugs.

Before You Begin, Review These Skills

- Lewis structures and formal charge [◄◄ Sections 6.3 and 6.4]
- Resonance [◄◄ Section 6.5]
- Molecular geometry and polarity [◄◄ Sections 7.1 and 7.2]

23.1 WHY CARBON IS DIFFERENT

Organic chemistry is usually defined as the study of compounds that contain carbon. This definition is not entirely satisfactory, though, because it would include such things as cyanide and cyanate complexes and metal carbonates, which are considered to be *inorganic*. A somewhat more useful definition of organic chemistry is the study of compounds that contain carbon and hydrogen, although many organic compounds also contain other elements, such as oxygen, sulfur, nitrogen, phosphorus, or the halogens, and many do not contain hydrogen. Examples of organic compounds include the following:

Student Annotation: Carbon-containing compounds that are considered *inorganic* are generally those that are obtained from minerals.

$$CH_4 \quad C_2H_5OH \quad H_2C_5H_6O_6 \quad CH_3NH_2 \quad CCl_4$$

Methane   Ethanol   Ascorbic acid   Methylamine   Carbon tetrachloride

Early in the study of organic chemistry there was thought to be some fundamental difference between compounds that came from living things, such as plants and animals, and those that came from nonliving things, such as rocks. Compounds obtained from plants or animals were called *organic,* whereas compounds obtained from nonliving sources were called *inorganic.* In fact, until early in the nineteenth century, scientists believed that only nature could produce organic compounds. In 1829, however, Friedrich Wöhler prepared urea, a well-known organic compound, by combining the inorganic substances lead cyanate and aqueous ammonia:

$$Pb(OCN)_2 + 2NH_3 + 2H_2O \xrightarrow{\Delta} 2NH_2CONH_2 + Pb(OH)_2$$

Urea

Wöhler's synthesis of urea dispelled the notion that organic compounds were fundamentally different from inorganic compounds—and that they could only be produced by nature. We now know that it is possible to synthesize a wide variety of organic compounds in the laboratory; in fact, many thousands of new organic compounds are produced in research laboratories each year. In this chapter, we will consider several types of organic compounds that are important biologically.

Because of its unique nature, carbon is capable of forming millions of different compounds. Carbon's position in the periodic table (Group 4A, period 2) gives it the following set of unique characteristics.

Student Annotation: By contrast, there are only about 100,000 known *inorganic* compounds.

- The electron configuration of carbon ($[He]2s^22p^2$) effectively prohibits ion formation. This and carbon's electronegativity, which is intermediate between those of metals and nonmetals, cause carbon to complete its octet by sharing electrons. In nearly all its compounds, carbon forms four covalent bonds, which can be oriented in as many as four different directions. Boron and nitrogen, carbon's neighbors in Groups 3A and 5A, respectively, usually form covalent compounds, too, but B and N form ions more readily than C.

Student Annotation: To form an ion that is isoelectronic with a noble gas, a C atom would have to either gain or lose *four* electrons [◄◄ Section 4.5]—something that is energetically impossible under ordinary conditions. This is not to say that carbon *cannot* form ions. But in the vast majority of its compounds, carbon acquires a complete octet by *sharing* electrons—rather than by gaining or losing them.

Methane
(4 σ bonds)

Formaldehyde
(3 σ bonds & 1 π bond)

Carbon dioxide
(2 σ bonds & 2 π bonds)

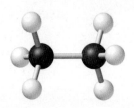

Ethane

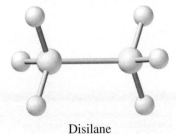

Disilane

- Carbon's small atomic radius allows the atoms to approach one another closely, giving rise to short, *strong,* carbon-carbon bonds and *stable* carbon compounds. In addition, carbon atoms that are *sp*- or *sp*^{2}-hybridized approach one another closely enough for their singly occupied, unhybridized *p* orbitals to overlap effectively—giving rise to relatively strong π bonds [◀◀ Section 7.6]. Recall that elements in the same group generally exhibit similar chemical behavior. Silicon atoms, however, are bigger than carbon atoms, so silicon atoms generally cannot approach one another closely enough for their unhybridized *p* orbitals to overlap significantly. As a result, very few compounds exhibit significant π bonding between Si atoms.

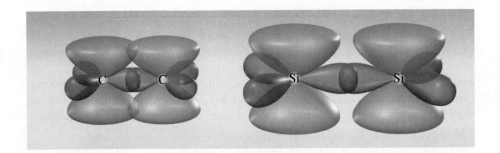

- Carbon's valence electrons are in the second shell ($n = 2$), where there are no *d* orbitals. The valence electrons of silicon, on the other hand, are in the third shell ($n = 3$) where there are *d* orbitals, which can be occupied or attacked by lone pairs on another substance—resulting in a reaction. This reactivity makes silicon compounds far less stable than the analogous carbon compounds. Ethane (CH_3-CH_3), for example, is stable in both water and air, whereas disilane (SiH_3-SiH_3) is unstable—breaking down in water and combusting spontaneously in air.

Student Annotation: Carbon's formation of chains is called *catenation.*

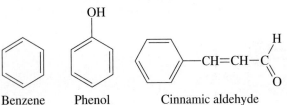

These attributes enable carbon to form chains (straight, branched, and cyclic) containing single, double, and triple carbon-carbon bonds, This, in turn, results in an endless array of organic compounds containing any number and arrangement of carbon atoms. Each carbon atom in a compound can be classified by the number of other carbon atoms to which it is bonded. A carbon atom that is bonded to just one other carbon atom is called a *primary* carbon; one that is bonded to two other carbon atoms is called a *secondary* carbon; one that is bonded to three other carbon atoms is called a *tertiary* carbon; and one that is bonded to four other carbon atoms is called a *quaternary* carbon. These four types of carbon atoms are identified with the labels 1°, 2°, 3°, and 4°, respectively. Each of the four carbon types is labeled in the structure on the left.

One of the important organic molecules that we encountered in Chapter 7 is benzene (C_6H_6). Organic compounds that are related to benzene, or that contain one or more benzene rings, are called *aromatic* compounds. Organic molecules that do not contain the benzene ring are called *aliphatic* compounds.

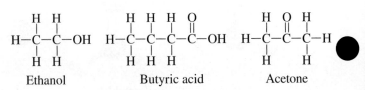

Benzene   Phenol   Cinnamic aldehyde

Aromatic compounds

Ethanol   Butyric acid   Acetone

Aliphatic compounds

23.2 CLASSES OF ORGANIC COMPOUNDS

Organic molecules occur in seemingly limitless variety. Different types of organic compounds, each with their own characteristic properties, result from the following:

1. Carbon's ability to form chains by bonding with itself
2. The presence of elements other than carbon and hydrogen
3. Functional groups
4. Multiple bonds

In this section, we will discuss several types of organic compounds and how we represent them. Consider two isomers of C_3H_6O:

$$\underset{\text{Propanaldehyde}}{\overset{\overset{\displaystyle O}{\|}}{CH_3CH_2CH}} \qquad \underset{\text{Acetone}}{\overset{\overset{\displaystyle O}{\|}}{CH_3CCH_3}}$$

> **Student Annotation:** A functional group is a group of atoms that determines many of a molecule's properties.

> **Student Annotation:** *Isomers* are different compounds with the same chemical formula.

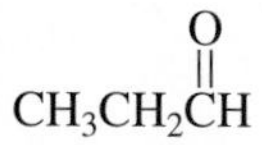

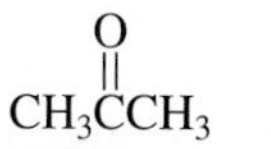

Although the two isomers contain exactly the same atoms, their different arrangements of atoms result in two very different compounds. The first is an *aldehyde* called propanal. The second is a *ketone* called acetone. Aldehydes and ketones are two classes of organic compounds. The classes of organic compounds that we will discuss in this chapter are alcohols, carboxylic acids, aldehydes, ketones, esters, amines, and amides.

A class of organic compounds often is represented with a general formula that shows the atoms of the functional group(s) explicitly, and the remainder of the molecule using one or more Rs, where R represents an alkyl group. An *alkyl group* is a portion of a molecule that resembles an alkane. In fact, an alkyl group is formed by removing one hydrogen atom from the corresponding alkane. The *methyl group* ($-CH_3$), for example, is formed by removing a hydrogen atom from methane (CH_4), the simplest alkane. Methyl groups are found in many organic molecules. Table 23.1 lists some of the simplest alkyl groups. Table 23.2 gives the general formula for each of the classes of organic compounds discussed in this chapter and the Lewis structure of each functional group.

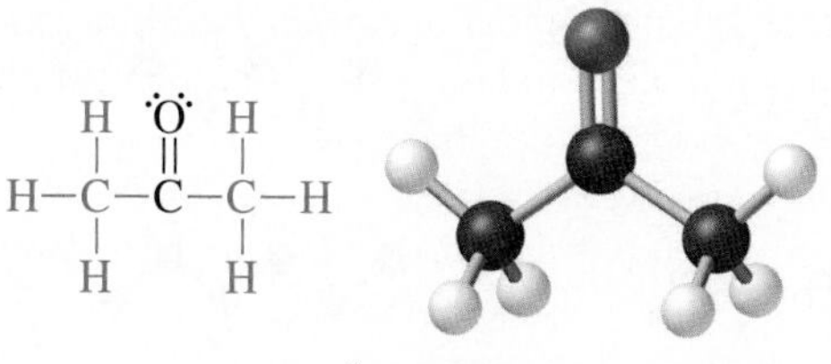

Acetone

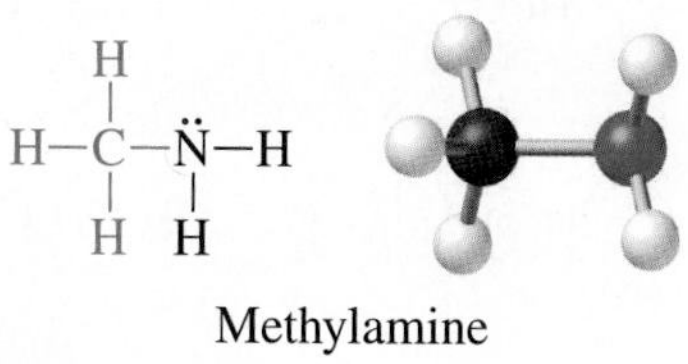

Methylamine

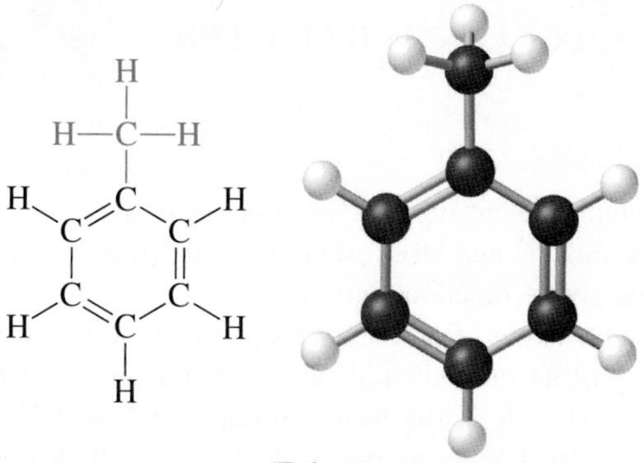

Toluene

The functional groups in the types of compounds shown in Table 23.2 are the *hydroxy* group (in *alcohols*), the *carboxy* group (in *carboxylic acids*), the $-COOR$ group (in *esters*), the *carbonyl* group (in *aldehydes* and *ketones*), the *amino* group (in *amines*), and the *amide* group (in *amides*). Functional groups determine many of the properties

TABLE 23.1	Alkyl Groups	
Name	**Formula**	**Model**
Methyl	$-CH_3$	
Ethyl	$-CH_2CH_3$	
Propyl	$-CH_2CH_2CH_3$	
Isopropyl	$-CH(CH_3)_2$	
Butyl	$-CH_2CH_2CH_2CH_3$	
tert-Butyl	$-C(CH_3)_3$	
Pentyl	$-CH_2CH_2CH_2CH_2CH_3$	
Isopentyl	$-CH_2CH_2CH(CH_3)_2$	
Hexyl	$-CH_2CH_2CH_2CH_2CH_2CH_3$	
Heptyl	$-CH_2CH_2CH_2CH_2CH_2CH_2CH_3$	
Octyl	$-CH_2CH_2CH_2CH_2CH_2CH_2CH_2CH_3$	

of a compound, including what types of reactions it is likely to undergo. Figure 23.1 shows ball-and-stick models and electrostatic potential maps of the hydroxy, carboxy, carbonyl, amino, and amide functional groups.

A compound consisting of an alkyl group and the functional group $-OH$ is an *alcohol*. The identity of an individual alcohol depends on the identity of R, the alkyl group. For example, when R is the *methyl* group, we have CH_3OH. This is methyl alcohol or methanol, also known as wood alcohol. It is highly toxic and can cause blindness or even death in relatively small doses. When R is the *ethyl* group, we have CH_3CH_2OH. This is ethyl alcohol or ethanol. Ethanol is the alcohol in alcoholic beverages. When R is the *isopropyl* group, we have $(CH_3)_2CHOH$. This is isopropyl alcohol. Isopropyl alcohol, what we commonly call "rubbing alcohol," is widely used as a disinfectant.

TABLE 23.2	General Formulas for Select Classes of Organic Compounds		
Class	**General formula**	**Structure**	**Functional group**
Alcohol	ROH	—Ö—H	Hydroxy group
Carboxylic acid	RCOOH	$\overset{\displaystyle\overset{..}{\underset{..}{O}}}{\underset{\displaystyle\ }{\|}}$ —C—Ö—H	Carboxy group
Ester*	RCOOR′	—C—Ö—R′	Ester group
Aldehyde	RCHO	—C—H	Carbonyl group
Ketone	RCOR′	—C—R′	Carbonyl group
Amine†	RNH₂	—N̈—H | H	Amino group (primary, 1°)
Amine	RNR′H	—N̈—R′ | H	Amino group (secondary, 2°)
Amine	RNR′R″	—N̈—R′ | R″	Amino group (tertiary, 3°)
Amide	RCONH₂	—C—N̈—H | H	Amide group (primary, 1°)
Amide	RCONR′H	—C—N̈—R′ | H	Amide group (secondary, 2°)
Amide	RCONR′R″	—C—N̈—R′ | R″	Amide group (tertiary, 3°)

*R′ represents a second alkyl group that may or may not be identical to the first alkyl group R. Likewise, R″ represents a third alkyl group that may or may not be identical to R or R′.

†The designations 1°, 2°, and 3° refer to how many R groups are bonded to the N atom.

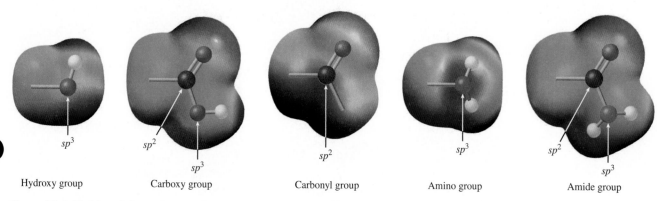

Student Annotation: Some functional groups have special names and some do not.

Figure 23.1 Models and electrostatic potential maps of the hydroxy, carboxy, carbonyl, amino, and amide functional groups.

Methanol

© David A. Tietz/Editorial Image, LLC

Ethanol

© John A. Rizzo/Getty

Isopropyl alcohol

© David A. Tietz/Editorial Image, LLC

Many compounds contain more than one functional group. An *amino acid,* for example, contains both the amine group and the carboxy group.

$$
\begin{array}{ccc}
\text{H} & \text{H} & \text{O} \\
| & | & || \\
\text{H}-\text{C}-\text{C}-\text{C}-\text{OH} \\
| & | \\
\text{H} & \text{NH}_2
\end{array}
$$

Alanine

Worked Example 23.1 lets you practice identifying functional groups in molecules.

Worked Example 23.1

Many familiar substances are organic compounds. Some examples include aspartame, the artificial sweetener in the sugar substitute Equal and in many diet sodas; salicylic acid, found in some acne medicines and wart-removal treatments; and amphetamine, a stimulant used to treat narcolepsy, attention-deficit hyperactivity disorder (ADHD), and obesity. Identify the functional group(s) in each molecule.

Aspartame
(a)

Salicylic acid
(b)

Amphetamine
(c)

Strategy Look for and identify the combinations of atoms shown in Table 23.2.

Setup (a) From right to left, aspartame contains a $-$COOH group, an $-$NH$_2$ group, a $-$CONHR group, and a $-$COOR group.

(b) Salicylic acid contains an $-$OH group and a $-$COOH group.

(c) Amphetamine contains an —NH₂ group.

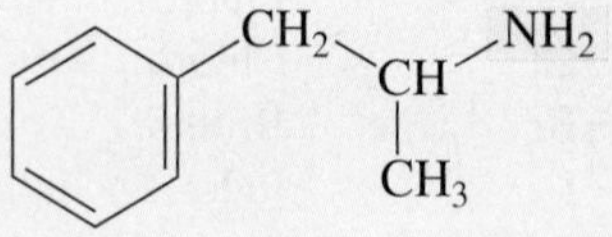

Solution (a) From right to left, aspartame contains a *carboxy* group, an *amino* group, an *amide* group, and a —COOR (ester) group.

(b) Salicylic acid contains a *hydroxy* group and a *carboxy* group.

(c) Amphetamine contains an *amino* group.

Think About It

The salicylic acid molecule [part (b)] contains a benzene ring and is therefore *aromatic*. When the hydroxy group is attached to a benzene ring, the resulting aromatic compound is a *phenol*, not an alcohol. Amphetamine [part (c)] has several legitimate medicinal uses, but it is also one of the most commonly misused drugs in the United States. Because it frequently is prescribed to adolescents for ADHD, much of it finds its way into high schools, where its misuse is a serious problem. A closely related compound that frequently makes headlines is *methamphetamine*.

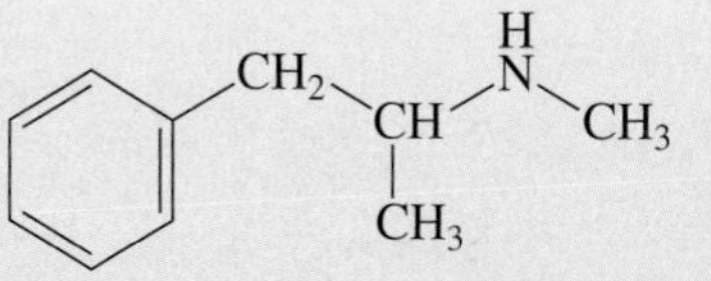

In August 2005, an issue of *Newsweek* magazine devoted a cover story to methamphetamine and its abuse.

Practice Problem Ⓐ**TTEMPT** Identify the functional groups in each of the following molecules.

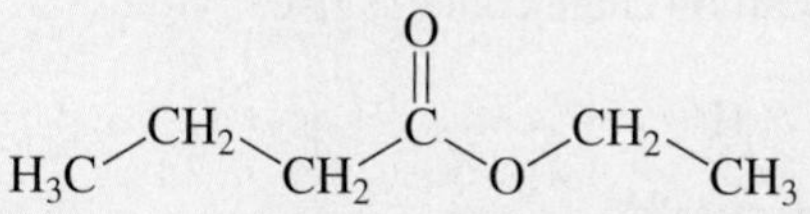

| Ethyl butyrate | Aspirin | Tartaric acid |
| (a) | (b) | (c) |

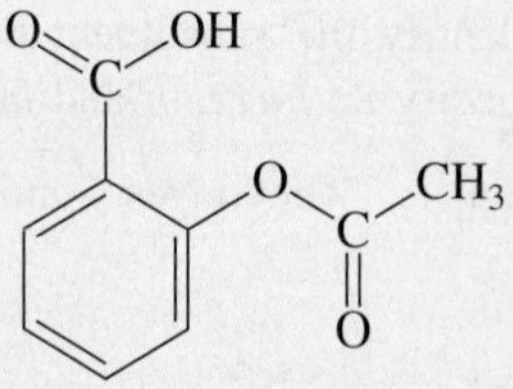

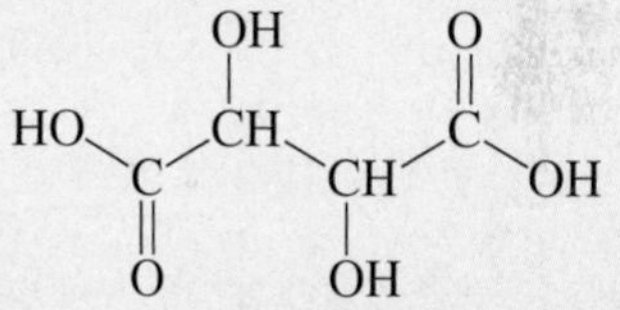

Practice Problem Ⓑ**UILD** Identify the functional groups in each of the following cyclic compounds.

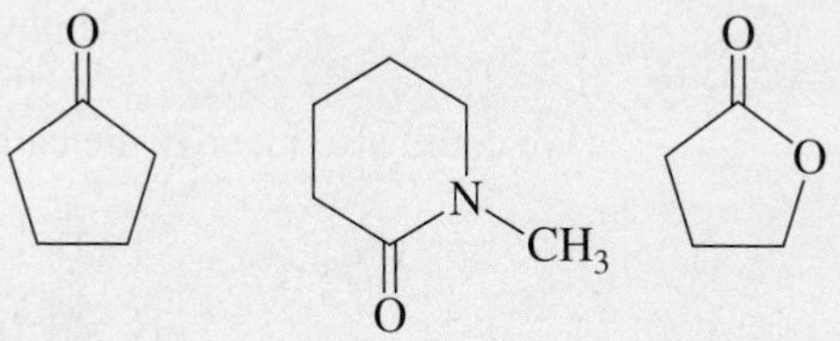

Practice Problem Ⓒ**ONCEPTUALIZE** Which of the molecules in Practice Problem 23.1B can form hydrogen bonds with itself?

Basic Nomenclature

Organic compounds are named systematically using International Union of Pure and Applied Chemistry (IUPAC) rules.

Root Names for Alkanes and Alkyl Groups

Number of carbons	Name	Number of carbons	Name
1	Meth-	6	Hex-
2	Eth-	7	Hept-
3	Prop-	8	Oct-
4	But-	9	Non-
5	Pent-	10	Dec-

Prefixes for Halogen Substituents

−F	Fluoro
−Cl	Chloro
−Br	Bromo
−I	Iodo

Alkanes

In Chapter 5, we encountered the names of some simple, straight-chain alkanes such as pentane.

$$
\begin{array}{c}
\quad\; H \;\; H \;\; H \;\; H \;\; H \\
\quad\; | \;\;\; | \;\;\; | \;\;\; | \;\;\; | \\
H-C-C-C-C-C-H \\
\quad\; | \;\;\; | \;\;\; | \;\;\; | \;\;\; | \\
\quad\; H \;\; H \;\; H \;\; H \;\; H
\end{array}
$$

To name substituted alkanes (i.e., those that have **substituents,** which are groups other than −H bonded to the carbons of the chain), we follow a series of steps.

1. Identify the longest continuous carbon chain to get the *parent name.*
2. Number the carbons in the continuous chain, beginning at the end closest to the substituent. Commonly encountered substituents include alkyl groups and halogens.
3. Identify the substituent and use a *number* followed by a dash and a *prefix* to specify its *location* and *identity,* respectively.

Step 1:   The longest continuous carbon chain contains five C atoms.

2-Methylpentane

(We could also identify the carbon chain as

but the number of C atoms in the chain is the same either way.) The parent name of a five-carbon chain is *pentane.* (See root names.)

Step 2:   We number the carbon atoms beginning at the end nearest the substituent (shaded in green).

Step 3:   The substituent is a methyl group, −CH$_3$. It is attached to carbon 2. Therefore, the name is 2-methylpentane.

Worked Example 23.2 lets you practice naming some simple organic compounds.

Worked Example 23.2

Give names for the following compounds.

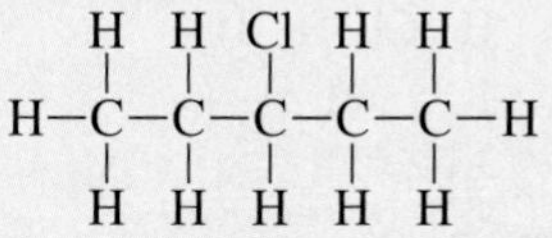

(a) 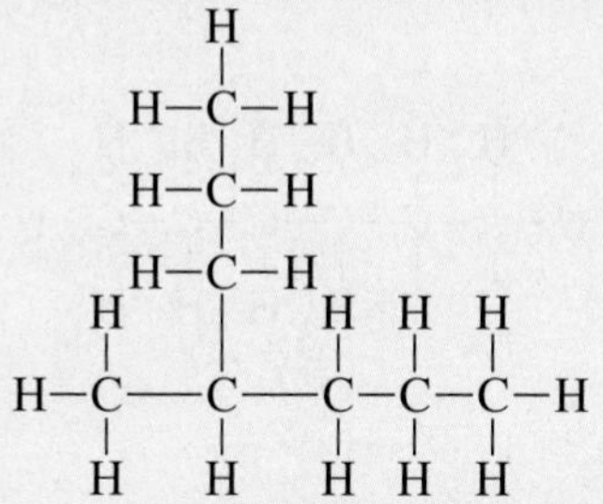 (b) 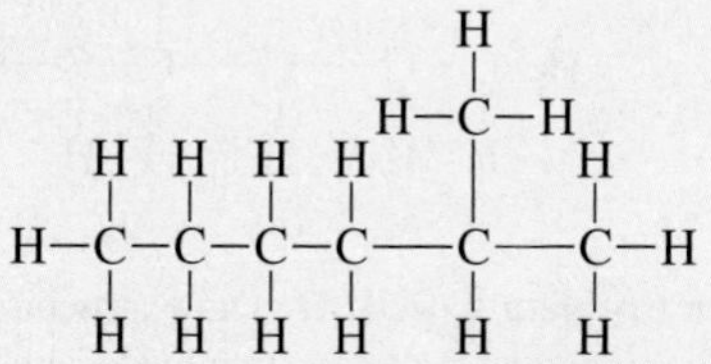 (c)

Strategy Use the three-step procedure for naming substituted alkanes: (1) name the parent alkane, (2) number the carbons, and (3) name and number the substituent. (Consult Table 5.8 for parent alkane names.)

Setup (a) This is a five-carbon chain. We can number the carbons starting at either end because the Cl substituent will be located on carbon 3 either way.

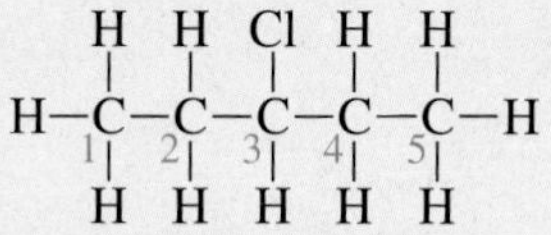

(b) This may look like a substituted pentane, too, but the longest carbon chain in this molecule is seven carbons long.

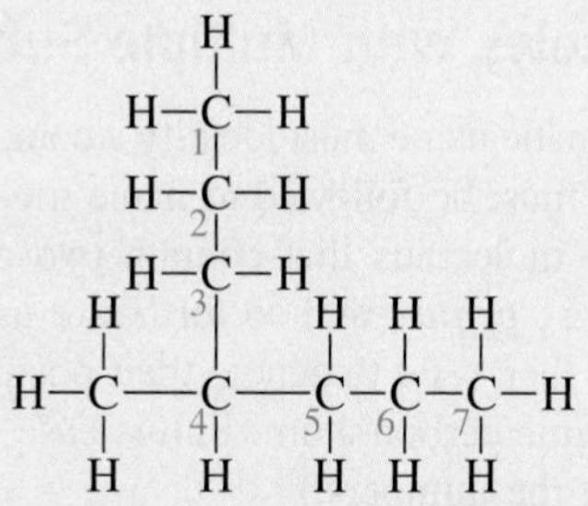

Although Lewis structures appear to be flat and to contain 90° angles, the C atoms in this molecule are all *sp*³-hybridized (four electron domains around each) and there is free rotation about the C—C bonds [◄◄ Section 7.6]. Thus, the molecule can also be drawn as

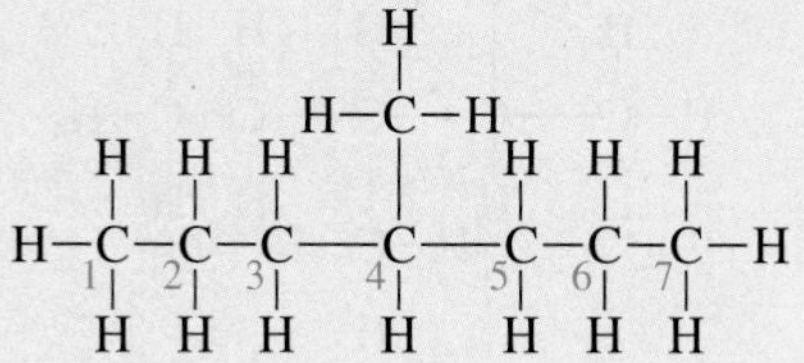

The substituent is a methyl group on carbon 4.

(c) This is a substituted *hexane*.

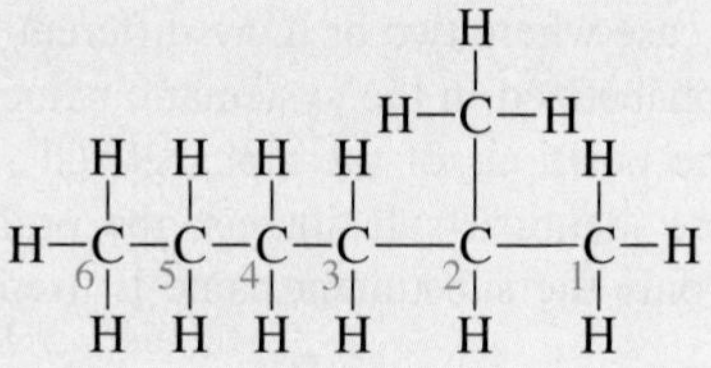

Solution (a) 3-chloropentane, (b) 4-methylheptane, (c) 2-methylhexane

---

Think About It

A common error is to misidentify the parent alkane. Double-check to be sure you have identified the *longest* continuous carbon chain in the molecule. Also be sure to number the carbon atoms so as to give the substituent the *lowest* possible number.

(Continued on next page)

Practice Problem **A**TTEMPT Give the systematic IUPAC name for each of the following:

(a)

(b)

(c)

Practice Problem **B**UILD Draw structures for (a) 4-ethyloctane, (b) 2-fluoropentane, and (c) 3-methyldecane.

Practice Problem **C**ONCEPTUALIZE How many carbons are there in the longest carbon chain in this molecule?

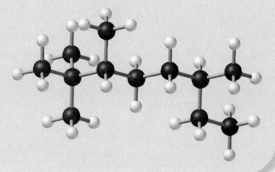

Molecules with Multiple Substituents

A systematic name must identify a compound unambiguously. Therefore, a special system of rules must be followed to name molecules that contain more than one substituent.

In molecules that contain two or more identical substituents, the prefixes *di-, tri-, tetra-, penta-,* and so forth, are used to denote the number of substituents. Numbers are then used to denote their positions. (Note that two substituents may be bonded to the same carbon atom. In this case, the carbon's number is repeated with a comma between the numbers.)

2,3-Dimethylpentane   2,2-Dichlorohexanoic acid

In the case where two or more different substituents are present, the substituent names are alphabetized in the systematic name of the compound. Numbers are used to indicate the positions of the alphabetized substituents. If a prefix is used to denote two or more identical substituents, the prefix is *not* used to determine the alphabetization—only the substituent name is used.

4-Ethyl-2-methylhexane   4-Ethyl-2,2-dimethylhexane

Molecules with Specific Functional Groups

Alcohols

Identify the longest continuous carbon chain that includes the carbon to which the —OH group is attached. This is the parent alkyl group. Name it according to the number of carbons it contains, and change the *-e* ending to *-ol*. Number the C atoms such that the —OH group has the lowest possible number, and, when necessary, use a number to indicate the position of the —OH group.

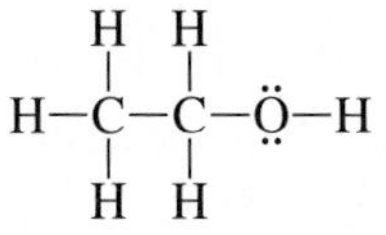

Ethanol 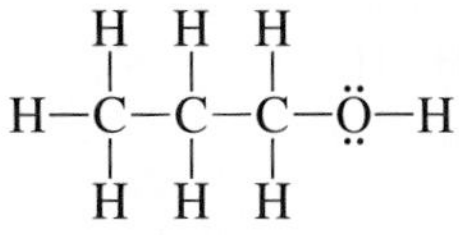   1-Propanol 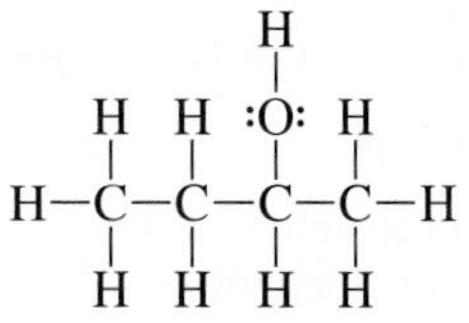   2-Butanol

When the chain that bears the —OH group also bears an alkyl substituent, the chain is numbered in the direction that gives the lowest possible number to the carbon attached to —OH.

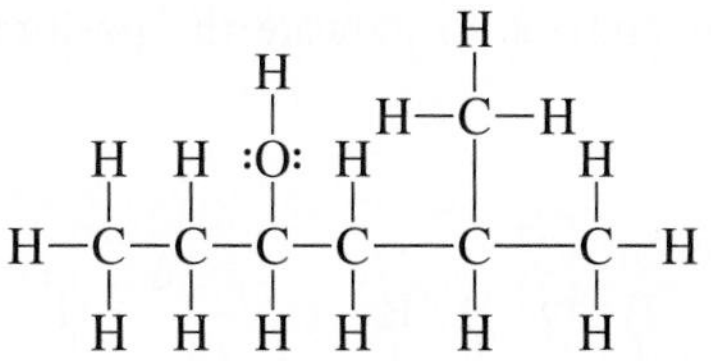

5-Methyl-3-hexanol

Carboxylic Acids

Identify the longest continuous carbon chain that includes the carboxy group. Name it according to the number of carbons it contains, and change the *-e* ending to *-oic* acid. Number the C atoms starting with the *carbonyl carbon.* Use numbers and prefixes to indicate the position and the identity of any substituents.

Many organic compounds have common names in addition to their systematic names. Common names for some of the carboxylic acids shown here are given in parentheses.

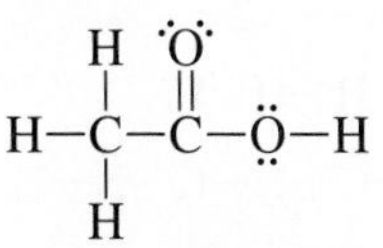

Student Annotation: The carbonyl carbon is the one that is doubly bonded to oxygen.

Ethanoic acid (acetic acid) 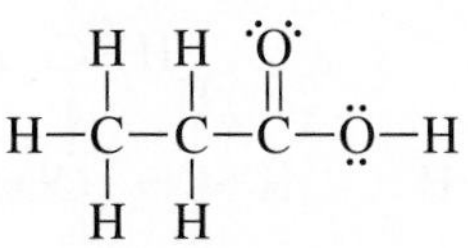   Propanoic acid (propionic acid) 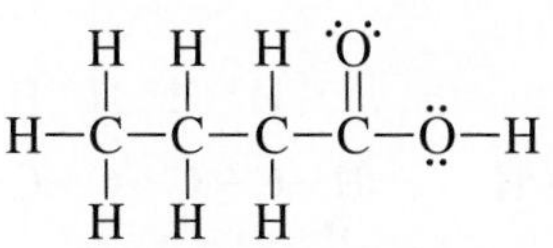   Butanoic acid (butyric acid) 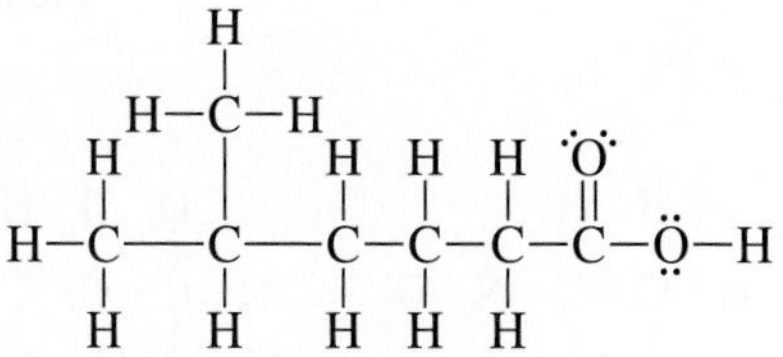   5-Methylhexanoic acid

Esters

Name esters as derivatives of carboxylic acids by replacing the *-ic* acid ending with *-ate.*

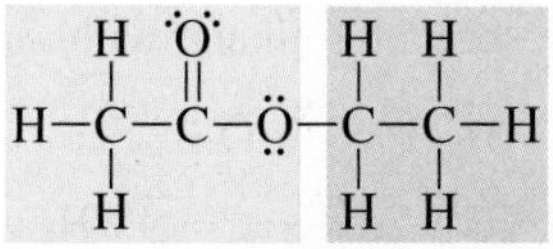

Ethyl ethanoate
Based on acetic acid

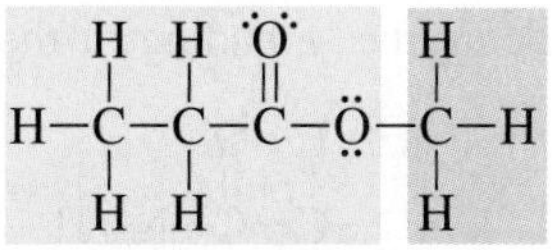

Methyl propanoate
Based on propionic acid

(The first part of an ester's name specifies the substituent that replaces the ionizable hydrogen of the corresponding carboxylic acid.)

Aldehydes

Identify the longest continuous carbon chain that includes the carbonyl group. Name it according to the number of carbons it contains, and change the -e ending to -al. Number the C atoms starting with the carbonyl carbon. Use numbers and prefixes to indicate the position and the identity of any substituents.

Ethanal
(acetaldehyde)

Propanal
(propionaldehyde)

Butanal

5-Methylhexanal

Ketones

Identify the longest continuous carbon chain that includes the carbonyl group. Name it according to the number of carbons it contains, and change the -e ending to -one. If necessary, number the C atoms to give the carbonyl carbon the lowest possible number. Use numbers and prefixes to indicate the position and the identity of any substituents.

Propanone
(acetone)

2-Butanone
(ethyl methyl ketone)

5-Methyl-3-hexanone

Primary Amines

Identify the longest continuous carbon chain that includes the carbon to which the $-NH_2$ group is bonded. Name it according to the number of carbons it contains, and change the -e ending to -amine. Number the C atoms starting with the carbon to which the $-NH_2$ group is bonded. Use numbers and prefixes to indicate the position and the identity of any substituents.

Ethanamine
(ethylamine)

1-Propanamine
(propylamine)

1-Butanamine
(butylamine)

5-Methyl-1-hexanamine
(5-methylhexylamine)

Primary Amides

Primary amides are named as derivatives of carboxylic acids, but they can also be named by replacing the -e ending of the corresponding alkane with -amide.

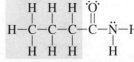

Propanamide
(propionamide)

Butanamide
(butyramide)

Section 23.2 Review

Classes of Organic Compounds

23.2.1 Write the name of the following compound.

(a) 2-Ethylpropane	(d) 2-Methylpentane
(b) 2-Ethylbutane	(e) 3-Methylbutane
(c) 2-Methylbutane	

23.2.2 Write the name of the following compound.

(a) 1-Chloroethane	(d) 1-Chloropropane
(b) 2-Chloropropane	(e) 3-Chloroethane
(c) 2-Chloromethane	

23.2.3 Write the name of the following compound.

(a) 4-Pentanone	(d) Propylpentanal
(b) 2-Pentanone	(e) Propylethanal
(c) Ethylpentanal	

23.2.4 Write the name of the following compound.

(a) 3-Methyl-1-ethanamine	(d) 2-Methyl-1-butanamine
(b) 2-Methyl-1-ethanamine	(e) 3-Methyl-1-butanamine
(c) 1-Pentanamine	

23.2.5 Identify the functional group(s) in the following molecule.

(Select all that apply.)

(a) Hydroxy	(d) Amino
(b) Carboxy	(e) Amide
(c) Carbonyl	

23.2.6   Identify the functional group(s) in the following molecule.

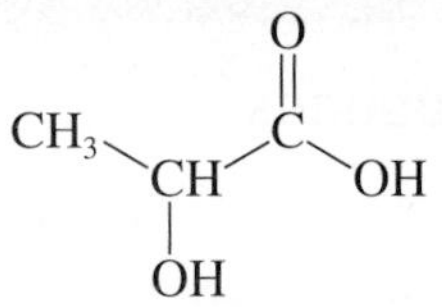

$$CH_3 - CH - C \overset{\displaystyle O}{\underset{\displaystyle OH}{\parallel}} OH$$

(Select all that apply.)

(a) Hydroxy   (d) Amino
(b) Carboxy   (e) Amide
(c) Carbonyl

23.3   REPRESENTING ORGANIC MOLECULES

You've learned previously how to represent molecules using *molecular* and *structural formulas* [◄◄ Section 5.5], as well as using Lewis structures [◄◄ Section 6.3]. In this section, we will learn several additional ways to represent molecules—ways that are particularly useful in the study of organic chemistry.

The representation of organic molecules is especially important because the atoms in an organic molecule, unlike those in inorganic compounds, may be arranged in an enormous variety of different ways. For example, there are literally dozens of different ways that a compound containing five carbon atoms, one oxygen atom, and the necessary number of hydrogen atoms can be arranged, with each arrangement representing a unique organic compound. Here are 10 possibilities:

Student Annotation: The "necessary" number of H atoms is the number necessary to *complete the octet* of each C and O atom [◄◄ Section 6.3].

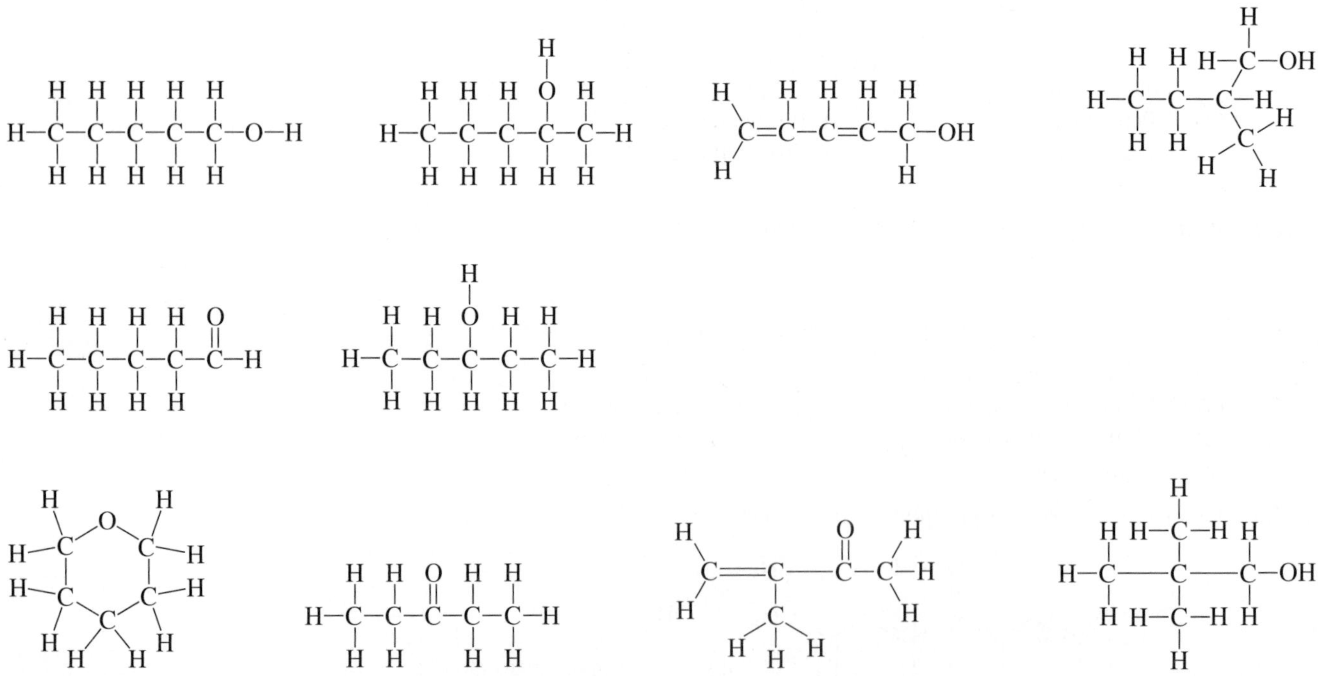

Condensed Structural Formulas

A *condensed structural formula,* or simply a *condensed structure,* shows the same information as a *structural formula,* but in a *condensed* form. For instance, the molecular formula, structural formula, and condensed structural formula of octane are as follows:

C_8H_{18} $CH_3CH_2CH_2CH_2CH_2CH_2CH_2CH_3$ $CH_3(CH_2)_6CH_3$
Molecular formula   Structural formula   Condensed structural formula

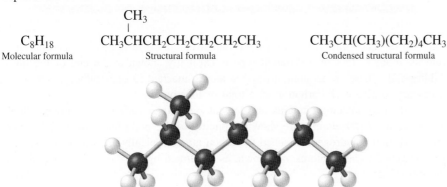

Octane

In the condensed structural formula, the identical adjacent groups of atoms (in this case $-CH_2-$ groups) are enclosed in parentheses and subscripted to denote their number.

In molecules where the carbon atoms do not form a single, unbranched chain, branches are indicated using additional parentheses in the condensed structural formula. For example, 2-methylheptane has the same molecular formula (C_8H_{18}) as octane. In effect, the chain is one C atom *shorter* than in octane and one of the hydrogen atoms on the second C atom has been replaced by a methyl group. The molecular formula, structural formula, and condensed structural formula of 2-methylheptane are as follows:

$$CH_3$$
$$|$$

C_8H_{18} $CH_3CHCH_2CH_2CH_2CH_2CH_3$ $CH_3CH(CH_3)(CH_2)_4CH_3$
Molecular formula   Structural formula   Condensed structural formula

2-Methylheptane

Kekulé Structures

Kekulé structures are similar to Lewis structures except that they do not show lone pairs. For molecules that contain no lone pairs, such as octane, the Lewis and Kekulé structures are identical.

$$
\begin{array}{c}
\text{H}\ \ \text{H}\ \ \text{H}\ \ \text{H}\ \ \text{H}\ \ \text{H}\ \ \text{H}\ \ \text{H} \\
|\ \ \ |\ \ \ |\ \ \ |\ \ \ |\ \ \ |\ \ \ |\ \ \ | \\
\text{H}-\text{C}-\text{C}-\text{C}-\text{C}-\text{C}-\text{C}-\text{C}-\text{C}-\text{H} \\
|\ \ \ |\ \ \ |\ \ \ |\ \ \ |\ \ \ |\ \ \ |\ \ \ | \\
\text{H}\ \ \text{H}\ \ \text{H}\ \ \text{H}\ \ \text{H}\ \ \text{H}\ \ \text{H}\ \ \text{H}
\end{array}
$$

Octane

(Many of the structures already shown in this chapter are Kekulé structures.) The Kekulé structures for several familiar organic molecules are as follows:

$$
\begin{array}{ccc}
\begin{array}{c}
\text{H}\ \ \text{H} \\
|\ \ \ | \\
\text{H}-\text{C}-\text{C}-\text{O}-\text{H} \\
|\ \ \ | \\
\text{H}\ \ \text{H}
\end{array}
&
\begin{array}{c}
\text{O} \\
|| \\
\text{H}-\text{C}-\text{H}
\end{array}
&
\begin{array}{c}
\text{H}\ \ \text{O} \\
|\ \ \ || \\
\text{H}-\text{C}-\text{C}-\text{O}-\text{H} \\
| \\
\text{H}
\end{array} \\
\text{Ethanol} & \text{Formaldehyde} & \text{Acetic acid}
\end{array}
$$

Bond-Line Structures

Bond-line structures are especially useful for representing complex organic molecules. A *bond-line structure* consists of straight lines that represent carbon-carbon bonds. The carbon atoms themselves (and the attached hydrogen atoms) are not shown, but you need to know that they are there. The structural formulas and bond-line structures for several hydrocarbons [◄◄ Section 5.6] are as follows:

Name	Structural formula	Bond-line structure
Pentane	$CH_3CH_2CH_2CH_2CH_3$	
Isopentane	$\overset{\displaystyle CH_3}{\underset{\displaystyle |}{CH_3CHCH_2CH_3}}$	
Neopentane	$\overset{\displaystyle CH_3}{\underset{\displaystyle CH_3}{\overset{|}{\underset{|}{CH_3CCH_3}}}}$	

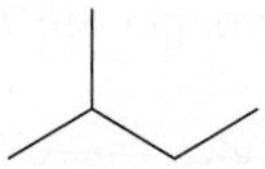

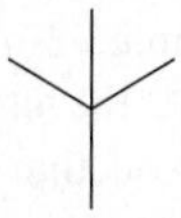

The end of each straight line in a bond-line structure corresponds to a carbon atom (unless another atom of a different type is explicitly shown at the end of the line). Additionally, there are as many hydrogen atoms attached to each carbon atom as are necessary to give each carbon atom a total of four bonds.

When a molecule contains an element *other* than carbon or hydrogen, those atoms, called *heteroatoms,* are shown explicitly in the bond-line structure. Furthermore, while the hydrogens attached to carbon atoms typically are not shown, hydrogens attached to heteroatoms *are* shown, as illustrated by the molecule ethylamine in the following:

Student Annotation: Study these structural formulas and bond-line structures, and those shown earlier, and make sure you understand how to interpret the bond-line structures.

Name	Structural formula	Bond-line structure
Propanone (acetone)	$CH_3\overset{\displaystyle O}{\overset{\|}{C}}CH_3$	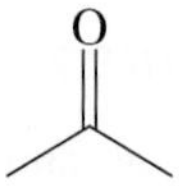
Ethylamine	$CH_3CH_2NH_2$	 NH_2
Tetrahydrofuran	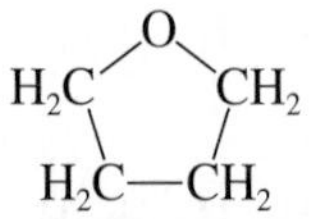 $\begin{array}{c} H_2C{\overset{O}{\diagup}}\diagdown CH_2 \\ H_2C{-}CH_2 \end{array}$	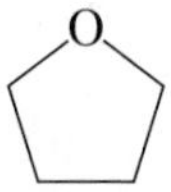

Although carbon and hydrogen atoms need not be shown in a bond-line structure, some of the C and H atoms *can* be shown for the purpose of emphasizing a particular part of a molecule. Often when we show a molecule, we choose to emphasize the *functional group(s),* which are largely responsible for the properties and reactivity of the compound [◄◄ Section 5.6].

Worked Example 23.3 shows how to interpret bond-line structures.

Worked Example 23.3

Write a molecular formula and a structural formula (or *condensed* structural formula) for the following:

(a)  [bond-line structure]   (b) 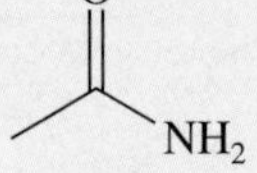 [structure with C=O and NH₂]

Strategy Count the C atoms represented and the heteroatoms shown. Determine how many H atoms are present using the octet rule.

Setup Each line represents a bond. (Double lines represent double bonds.) Count one C atom at the end of each line unless another atom is shown there. Count the number of H atoms necessary to complete the octet of each C atom.

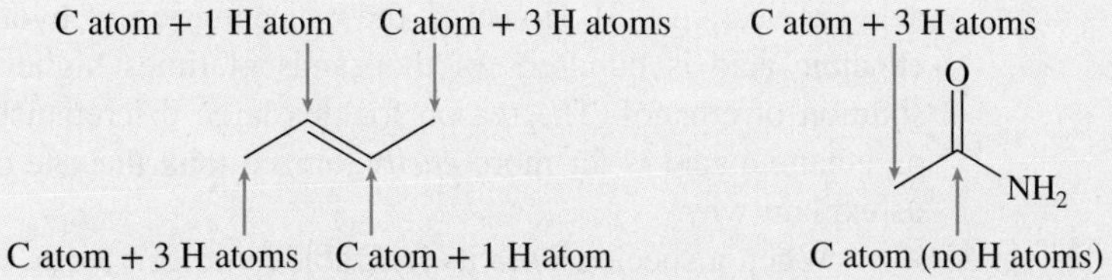

C atom + 1 H atom   C atom + 3 H atoms   C atom + 3 H atoms

[structures with labels]

C atom + 3 H atoms   C atom + 1 H atom   C atom (no H atoms)

Solution (a) Molecular formula: C_4H_8; structural formula: $CH_3(CH)_2CH_3$

(b) Molecular formula: C_2H_5NO; structural formula: CH_3CONH_2

> **Think About It**
> Make sure that each C atom is surrounded by four electron pairs: four single bonds, two single bonds and a double bond, or two double bonds. Remember that the single bonds to H typically are not shown in a bond-line structure—you have to remember that they are there and account for the H atoms when you deduce the formula.

Practice Problem ATTEMPT Write a structural formula for the compound represented by the following bond-line structure.

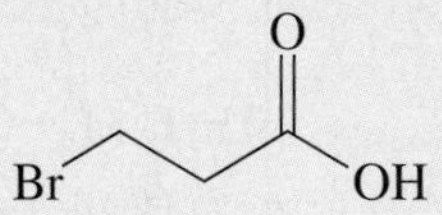

[bond-line structure with Br and COOH]

Practice Problem BUILD Draw the bond-line structure for $(CH_3)_2C=CHNH_2$.

Practice Problem CONCEPTUALIZE How many hydrogen atoms are there in the molecule represented here?

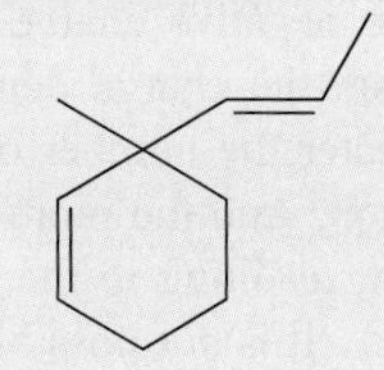

[bond-line structure]

Resonance

Recall from Chapter 6 that many molecules and ions can be represented by more than one Lewis structure [◄◄ Section 6.5]. Furthermore, two or more equally valid Lewis structures that differ only in the position of their electrons are called *resonance structures*. For example, SO_3 can be represented by three different Lewis structures [◄◄ Worked Example 6.7].

None of these Lewis structures represents the SO_3 molecule accurately. The bonds in SO_3 are actually *equivalent*—equal in length and strength, which we would not expect if two were single bonds and one were a double bond. Each individual resonance structure implies that electron pairs are *localized* in specific bonds or on specific atoms. The concept of resonance allows us to envision certain electron pairs as *delocalized* over several atoms [◄◄ Section 7.8]. Delocalization of electron pairs imparts additional stability to a molecule (or polyatomic ion), and a species that can be represented by two or more resonance structures is said to be *resonance stabilized*.

Chemists sometimes use curved arrows to specify the differences in positions of electrons in resonance structures. In SO_3, for example, the "repositioning" of electrons in the resonance structures can be indicated as follows:

Student Annotation: Remember that the real structure of the molecule is neither the first structure nor the second or third, but rather something in between that cannot be represented by a single structure [⏮ Section 6.5].

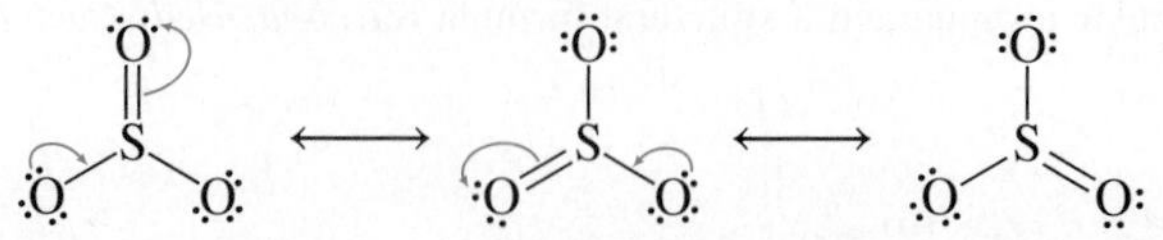

Resonance stabilization is observed in many organic species as well and affects chemical properties such as in the acidic behavior of ethanol (CH_3CH_2OH) and ethanoic acid (CH_3COOH), also commonly known as *acetic* acid. Each of these molecules has one *ionizable hydrogen atom* [⏮ Section 5.6], enabling it to behave as a Brønsted acid [⏮ Section 9.3]. However, the concentration of hydronium ions in a solution of ethanoic acid is hundreds of thousands of times higher than that in a comparable solution of ethanol. The reason for this large discrepancy is that the hydrogen atom on ethanoic acid is far more *easily* ionized than the one on ethanol. *Resonance* helps us explain why.

When a species loses its ionizable hydrogen atom, what remains is an anion. In the case of ethanoic acid and ethanol, the anions are

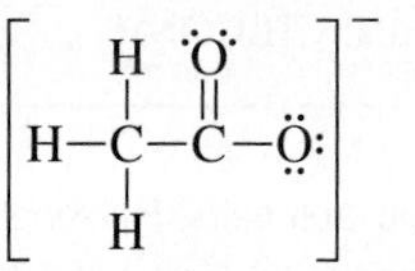
and

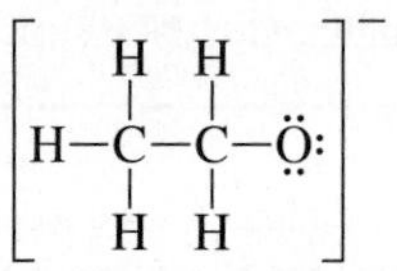

We can draw a second resonance structure for the anion produced by the ionization of ethanoic acid by repositioning the electron pairs as follows:

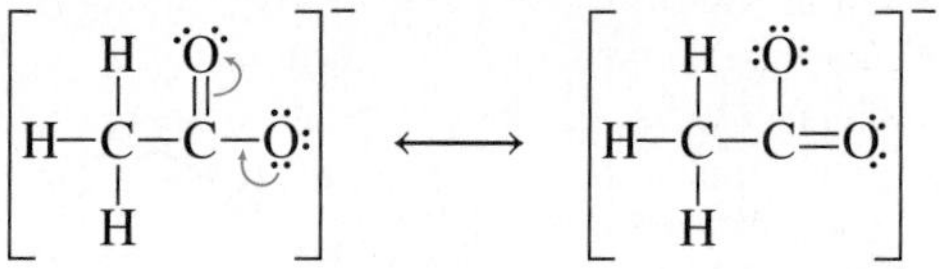

The negative charge on the anion resides on an oxygen atom. Which oxygen atom bears the charge depends on which resonance structure we look at. In essence, the greater the number of possible locations for the negative charge, the more stable the anion. And the more stable the anion, the more easily the ionizable hydrogen atom is lost, resulting in the production of more hydronium ions in solution.

Student Annotation: We determine where a charge resides in a polyatomic anion by calculating the *formal charge* on each atom [⏮ Section 6.4].

It is not possible to draw additional resonance structures for the anion produced by the ionization of ethanol because there is nowhere else to put the lone pairs that reside on the O atom. They cannot be moved in between the O and C atoms because C can only have four electron pairs around it. (We were able to do this with CH_3COO^- because we could also move one of the pairs already around C to the other O atom. See the curved arrows in the preceding resonance structure.)

Worked Example 23.4 illustrates the use of curved arrow notation and the determination of resonance structures.

Worked Example 23.4

Adenosine triphosphate (ATP) is sometimes called the "universal energy carrier" or "molecular energy currency." It contains two high-energy bonds (shown in red) that, when *hydrolyzed* (broken by the addition of water), release the energy necessary for cell function. Resonance stabilization of the hydrogen phosphate ion is one of the reasons the breakdown of ATP releases energy.

Adenosine triphosphate + H₂O ⟶ Adenosine diphosphate + Inorganic phosphate

Draw all the possible resonance structures for the hydrogen phosphate ion (HPO_4^{2-}). Use curved arrows to indicate how electrons are repositioned, and determine the position(s) of the negative charges.

Strategy Draw a valid Lewis structure for HPO_4^{2-}, and determine whether and where electrons can be repositioned to produce one or more additional structures. Indicate the movement of electrons with curved arrows, and draw all possible resonance structures. Calculate the formal charge on each atom to determine the placement of charges.

Setup A valid Lewis structure for the hydrogen phosphate ion is

For the purpose of determining formal charges, P and O have five and six valence electrons, respectively.

Solution A lone pair can be moved from one of the oxygen atoms to create a double bond to the phosphorus, and a pair of electrons from the *original* double bond can be moved onto *that* oxygen atom. The net result is simply a repositioning of the double bond by moving two electron pairs. This can be done once more, giving a total of three resonance structures for HPO_4^{2-}. In each of the resonance structures, the formal charge on phosphorus is $[5 - (5)] = 0$. The formal charge on each singly bonded oxygen is $[6 - (1 + 6)] = -1$, and the formal charge on the doubly bonded oxygen is $[6 - (2 + 4)] = 0$.

Think About It
ATP can also be hydrolyzed to give AMP (adenosine *mono*phosphate) and *pyrophosphate* ($P_2O_7^{4-}$). Pyrophosphate hydrolyzes, in turn, to give two hydrogen phosphate ions. The oxygen atoms that can help delocalize the negative charges are highlighted.

These structures can also be drawn with one double bond to each phosphorus atom, to minimize formal charges [◀◀ Section 6.4].

Practice Problem **A**TTEMPT Follow the curved arrows to draw a second resonance structure for the HCOO⁻ ion.

Practice Problem **B**UILD Given the following two resonance structures, draw the curved arrows on the first structure that will give rise to the second structure.

(Continued on next page)

Practice Problem **C**ONCEPTUALIZE In which of the following examples do the curved arrows *not* correspond correctly to the repositioning of electrons needed to arrive at the resonance structure shown?

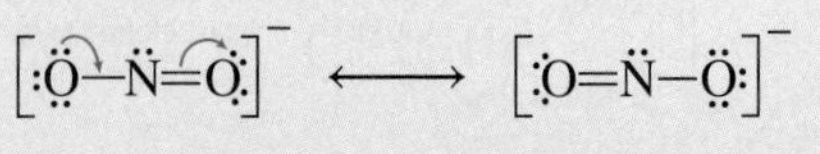

$$[:\overset{..}{O}-\overset{..}{N}=\overset{..}{O}:]^{-} \longleftrightarrow [:O=\overset{..}{N}-\overset{..}{O}:]^{-} \qquad [:\overset{..}{O}-\overset{..}{N}=\overset{..}{O}:]^{-} \longleftrightarrow [:O=\overset{..}{N}-\overset{..}{O}:]^{-} \qquad [:\overset{..}{O}-\overset{..}{N}=\overset{..}{O}:]^{-} \longleftrightarrow [:O=\overset{..}{N}-\overset{..}{O}:]^{-}$$

(i) 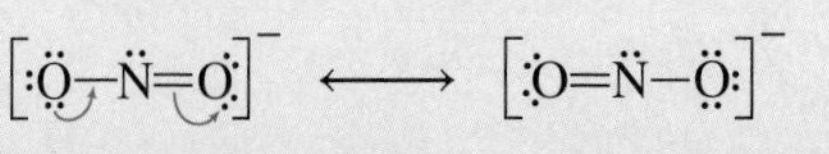   (ii) 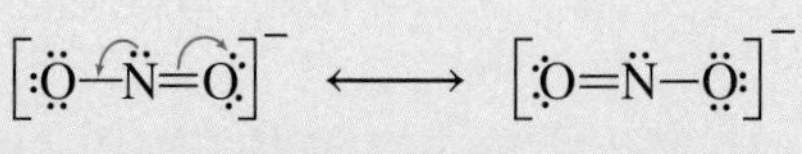   (iii)

Section 23.3 Review

Representing Organic Molecules

23.3.1 Give the molecular formula of the compound represented by

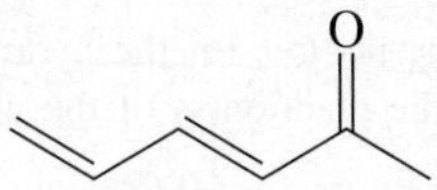

(a) $C_5H_{12}O$ (c) $C_6H_{12}O$ (e) $C_6H_{13}O$
(b) C_6H_8O (d) C_5H_6O

23.3.2 Give the molecular formula of the compound represented by

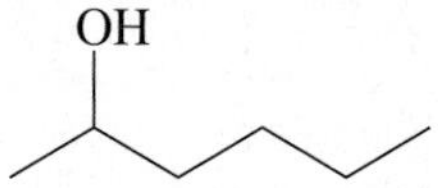

(a) $C_5H_{12}O$ (c) $C_6H_{12}O$ (e) $C_4H_{13}O$
(b) $C_6H_{13}O$ (d) $C_6H_{14}O$

23.3.3 Which of the following pairs of species are resonance structures?

(a)

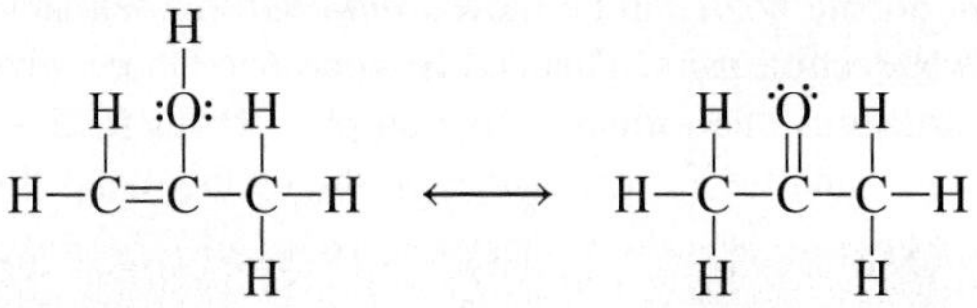

(b)

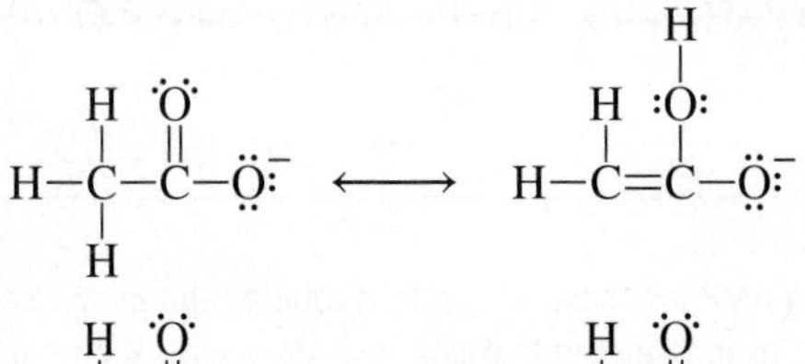

(c)

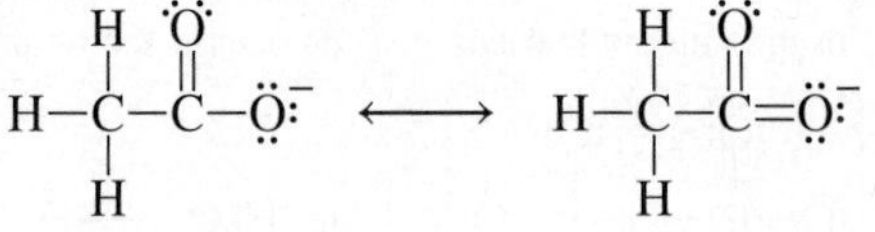

(d)

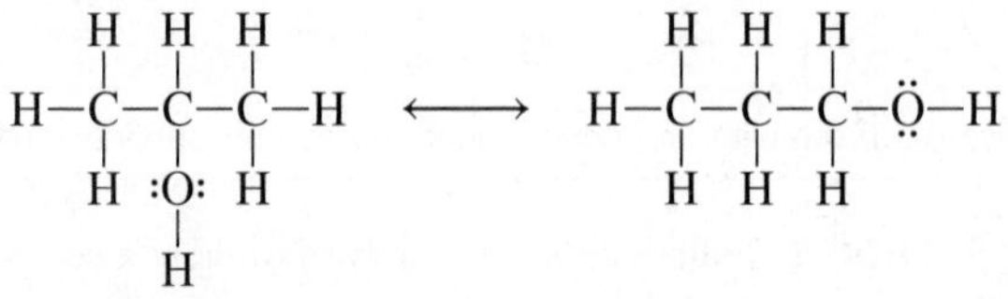

(e)

23.3.4 Which of the following structural formulas represents a species that has two or more resonance structures? (Select all that apply.)
(a) HCOOH   (c) CH_2CH_2 (e) CO_2
(b) $HCCO^-$ (d) O_3

23.4 ISOMERISM

We first encountered the term *isomer* in the context of molecular geometry and molecular polarity [◄◄ Section 7.2]. Isomers are different compounds that have the same chemical formula. In this section, we will discuss the different types of isomerism (namely, constitutional isomerism and stereoisomerism) and their importance to organic chemistry.

Constitutional Isomerism

Also known as *structural* isomers, **constitutional isomers** occur when the same atoms can be connected in two or more different ways. For example, there are three different ways to arrange the atoms in a compound with the chemical formula C_5H_{12}. Constitutional isomers have distinct names and generally have different physical and chemical properties. Table 23.3 lists the three constitutional isomers of C_5H_{12} along with their boiling points for comparison.

Stereoisomerism

Stereoisomers are those that contain identical bonds but differ in the orientation of those bonds in space. Two types of stereoisomers exist: geometrical isomers and optical isomers. **Geometrical isomers** occur in compounds that have restricted rotation around a bond. For instance, compounds that contain carbon-carbon double bonds can form geometrical isomers. Individual geometrical isomers have the same names but are distinguished by a prefix such as *cis-* or *trans-*. Dichloroethylene, ethylene in which two of the H atoms (one on each C atom) have been replaced by Cl atoms, exists as two geometrical isomers. The isomer in which the Cl atoms both lie on the same side (above or below, in this example) of the double bond is called the *cis* isomer. The isomer in which the Cl atoms lie on opposite sides of the double bond is

Video 23.1
Organic and biochemistry—structural isomers of hexane.

TABLE 23.3	**Constitutional Isomers of C_5H_{12}**			
Name	**Structural formula**	**Bond-line structure**	**Ball-and-stick model**	**BP (°C)**
Pentane (*n*-pentane)				36.1
Methylbutane (isopentane)				27.8
2,2-Dimethylpropane (neopentane)				9.5

Figure 23.2 Three isomers of dichloro-ethylene.

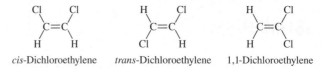

cis-Dichloroethylene   trans-Dichloroethylene   1,1-Dichloroethylene

called the **trans** isomer. (The compound in which both Cl atoms are attached to the same C atom is a *constitutional* isomer rather than a stereoisomer.) Figure 23.2 depicts the isomers of $C_2H_2Cl_2$.

Geometrical isomers usually have different physical and chemical properties. *Trans* isomers tend to be more stable and are generally easier to synthesize than their *cis* counterparts. The existence of *cis* isomers in living systems is a testament to how much better nature is at chemical synthesis than we are. Geometrical isomerism is sometimes of tremendous biological significance.

Stereoisomers that are mirror images of each other, but are not superimposable, are called **optical isomers.** Consider the hypothetical organic molecule shown in Figure 23.3. It consists of an sp^3-hybridized carbon atom that is bonded to four different groups. Its mirror image appears identical to it, just as your right and left hands appear identical to each other. But, if you have ever tried to put a right-handed glove on your left hand, or vice versa, you know that your hands are not identical. Imagine rotating the molecule on the right so that its green and red spheres coincide with those of the molecule on the left. Doing so results in the yellow sphere of one molecule coinciding with the blue sphere on the other. These two molecules are mirror images of each other, but they are not identical.

Molecules with nonsuperimposable mirror images are called **chiral;** and a pair of such mirror-image molecules are called **enantiomers.** Most of the chemical properties of enantiomers and all their physical properties (with the exception of the direction of rotation of plane-polarized light) are identical. Their chemical properties differ only in reactions that involve another chiral species, such as a chiral molecule or a receptor site that is shaped to fit only one enantiomer. Most biochemical processes consist of a series of chemically specific reactions that use chiral receptor sites to facilitate reaction by allowing only the specific reactants to fit (and thus react).

In organic chemistry, it often is necessary to represent tetrahedral molecules (three-dimensional objects) on paper (a two-dimensional surface). By convention, this is done using solid lines to represent bonds that lie in the plane of the page, dashes

Video 23.2
Organic and biochemistry—chiral molecules.

Figure 23.3 (a) Nonsuperimposable mirror images. (b) Despite being mirror images of each other, enantiomers are different compounds.

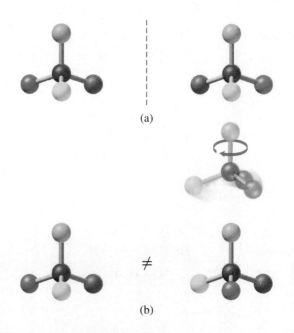

(a)

(b)

to represent bonds that point behind the page, and wedges to represent bonds that point in front of the page.

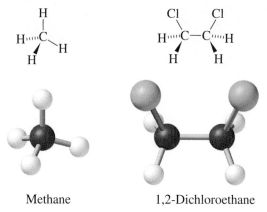

Methane   1,2-Dichloroethane

One property of chiral molecules is that the two enantiomers rotate the plane of plane-polarized light in opposite directions; that is, they are *optically active.* Unlike ordinary light, which oscillates in all directions, plane-polarized light oscillates only in a single plane. We use a polarimeter, shown schematically in Figure 23.4, to measure the rotation of polarized light by optical isomers. A beam of unpolarized light first passes through a Polaroid sheet, called the *polarizer,* and then through a sample tube containing a solution of an optically active, chiral species. As the polarized light passes through the sample tube, its plane of polarization is rotated either to the right or to the left. The amount of rotation can be measured by turning the analyzer in the appropriate direction until minimal light transmission is achieved.

Minimal transmission occurs when the plane of polarization of the light is perpendicular to that of the analyzer through which it is viewed. This effect can be demonstrated using two pairs of polarized sunglasses as shown in Figure 23.5. If the plane of polarization is rotated to the right, the isomer is said to be *dextrorotatory* and is labeled *d*; if it is rotated to the left, the isomer is called *levorotatory* and labeled *l*. Enantiomers always rotate the light by the same amount, but in opposite directions. Thus, in an equimolar mixture of both enantiomers, called a ***racemic mixture,*** the net rotation is zero.

A fascinating use of plane-polarized light is in 3-D movies. We see in three dimensions because our eyes view the world from slightly different positions. Our brains synthesize a three-dimensional (3-D) picture based on the two different pictures

Student Annotation: There are several conventions used to designate specific enantiomers. *Dextro-* and *levo-* prefixes refer to the direction of rotation of polarized light. *R* and *S*, the most commonly used designations, are assigned based on the "priority" assigned to each of the four groups attached to the chiral carbon—something that is beyond the scope of this text.

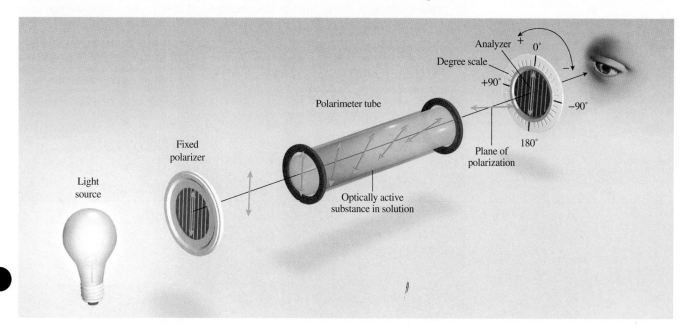

Figure 23.4 Schematic of a polarimeter.

Figure 23.5 Two pairs of polarized sunglasses. (a) When two polarized lenses overlap with their planes of polarization parallel to each other, light is transmitted. (b) When one pair is rotated so that its plane of polarization is perpendicular to the other, no light is transmitted through the overlapped lenses.

Source: © McGraw-Hill Education/Charles D. Winters, photographer.

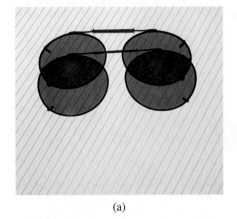 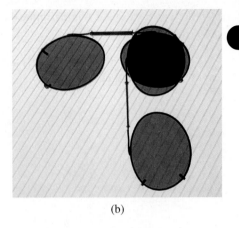

(a)   (b)

sent to it by our eyes. Modern 3-D movies make use of this phenomenon to make it seem as though objects on the screen are actually moving toward the viewer.

Three-dimensional movies are filmed using two different cameras at slightly different angles to the action. Thus, there are actually *two* movies that must be shown to us simultaneously. To make sure that our two eyes receive two different perspectives, each movie is projected through a polarizer, which polarizes the two projections in directions perpendicular to each other (Figure 23.6).

If we were to watch the movie without the special glasses provided by the movie house, we would see the blurry combination of the two movies. However, the 3-D glasses consist of polarized lenses, with planes of polarization that are mutually perpendicular. The left lens, polarized in one direction, blocks the image that is polarized perpendicular to it. The right lens, polarized in the other direction, blocks the other image. Our eyes are "tricked" into seeing two different movies, which our brain combines to form one 3-D image. The results of this process can be quite impressive!

Numerous processes important to biological function involve one enantiomer of a chiral compound. Many drugs, including thalidomide (see Thinking Outside the Box in this chapter), are chiral with only one enantiomer having the desired properties. Some such drugs have been manufactured and marketed as racemic mixtures. It has become common, though, for drug companies to invest in *chiral switching*, the preparation of single-isomer versions of drugs originally marketed as racemic mixtures, in an effort to improve on existing therapies and to combat the revenue losses caused by generic drugs. A fairly high-profile example of this is the single-isomer drug

Student Annotation: In some cases, the other enantiomer has no biological activity; in other cases, the other enantiomer has a different type of activity, making it detrimental or even deadly—as in the case of thalidomide.

Figure 23.6 Two different versions of the same movie are projected onto the screen. Polarized lenses make it so that each eye sees only one version. The result is the perception of three-dimensional action.

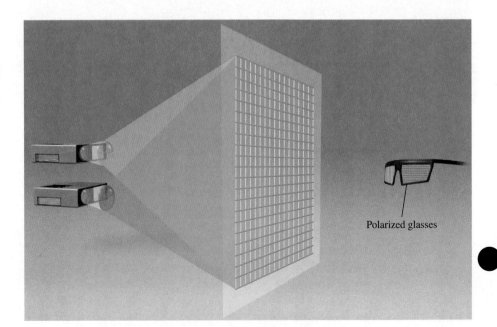

Polarized glasses

Nexium, the so-called purple pill. AstraZeneca, makers of Nexium, held a patent for the drug Prilosec, originally a prescription heartburn medication that is now available over the counter. Prilosec is a racemic mixture of the chiral compound omeprazole. Prior to the 2002 expiration of its patent on Prilosec, AstraZeneca began producing and marketing Nexium, which contains only the therapeutically effective enantiomer (*S*)-omeprazole or *esomeprazole.*

Thinking Outside the Box

Thalidomide Analogues

Beginning in 1957, the drug thalidomide was marketed in 48 countries around the world as a sleeping pill and as an antinausea medicine for pregnant women suffering from morning sickness. By 1962, the drug was shown to have caused horrific birth defects and an untold number of fetal deaths. Thalidomide interferes with spinal cord and limb development, and more than 10,000 babies had been born with severe spinal cord abnormalities and malformed or absent limbs. Many of the victims were born to mothers who reportedly had taken just *one* thalidomide pill early in their pregnancies. At the time, thalidomide was not approved for use in the United States, but the otherwise worldwide tragedy did prompt the U.S. Congress to enact a new law to give the FDA more control over the testing and approval of new drugs.

In August 1998, the FDA approved thalidomide for the treatment of *erythema nodosum leprosum* (ENL), a painful inflammatory skin condition associated with leprosy. This approval is controversial because of the drug's infamous history, but thalidomide has shown tremendous promise in the treatment of a wide variety of painful and debilitating conditions, including complications from certain cancers, AIDS, and some autoimmune disorders such as lupus and rheumatoid arthritis. Because of the dangers known to be associated with thalidomide, researchers are working on developing *analogues*—drugs that are chemically similar enough to have the same therapeutic benefits but chemically *different* enough *not* to have the undesirable and/or dangerous properties. Two such analogues that are currently being investigated are lenalidomide and CC-4047, shown here:

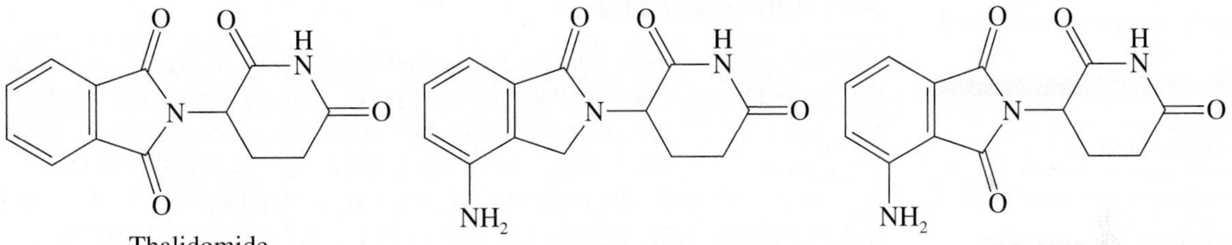

Thalidomide   Lenalidomide   CC-4047

Although it was approved by the FDA in 1998, thalidomide is the most regulated prescription drug in history, because it is known to harm developing fetuses. The drug's manufacturer, Celgene Corporation, has developed the *System for Thalidomide Education and Prescribing Safety* (STEPS) program. In order for physicians to prescribe thalidomide to their patients, they must be registered in the STEPS program. Female patients must have a negative pregnancy test within 24 hours of beginning treatment and must undergo periodic pregnancy testing throughout treatment. Both female and male patients must comply with mandatory contraceptive measures, patient registration, and patient surveys. Moreover, new patients must view an informational video in which a thalidomide victim explains the potential dangers of the drug.

Another example of chiral switching is that of the selective serotonin reuptake inhibitor (SSRI) antidepressant Celexa, which was introduced to the market in 1998 by Forest Laboratories. Celexa is a racemic mixture of (*R*)-citalopram oxalate and (*S*)-citalopram oxalate. While only the (*S*) enantiomer has therapeutic antidepressant properties, both enantiomers contribute to the side effects of the drug and therefore limit effectiveness and patient tolerance. In 2002, the FDA approved Lexapro, a new antidepressant derived from Celexa but from which the therapeutically ineffective (*R*) enantiomer has been removed. The benefits of isolating the active isomer include smaller required dosages, reduced side effects, and a faster and better patient response to the drug.

Although the intellectual property laws regarding single-isomer drug patents are somewhat ambiguous, chiral switching has enabled some pharmaceutical companies to extend the time that they are able to market their popular prescriptions exclusively. Strictly speaking, the FDA does not consider a single enantiomer of an already

Student Annotation: Thalidomide was not approved for use in the United States thanks in large part to the vigilance of one doctor at the FDA. She was troubled by inadequate research into the safety of the drug and steadfastly refused to approve the drug-maker's application.

approved chiral drug to be a "new chemical entity," which is a requirement for obtaining a patent on a compound. Early in the history of chiral switching, however, there was some disagreement among patent examiners regarding what constituted a new chemical entity, and patents were granted on single-isomer drugs that might not be granted today.

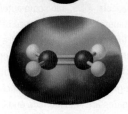

Figure 23.7 The partial positive charge on H in HCl is attracted to the region of electron density in ethylene's double bond.

Student Annotation: Although it ionizes completely in aqueous solution [◄◄ Section 9.1], HCl exists as *molecules* in the gas phase.

Student Annotation: When an electrophile approaches another species and accepts electrons from it to form a bond, this is called *electrophilic attack.*

Student Annotation: Curved arrows are used to illustrate the mechanism by which an organic reaction occurs. Unlike their use in resonance structures, curved arrows in a reaction mechanism correspond to the actual *movement* of electrons.

23.5   ORGANIC REACTIONS

Coulomb's law, which we first encountered in Section 4.4, measures the force of the attraction between opposite charges. The attraction between regions of opposite charge on neighboring species and the resulting movement of electrons are the basis for our understanding of many organic reactions. We begin by defining what *electrophiles* and *nucleophiles* are, two terms used frequently in organic chemistry.

An **electrophile** is a species with a positive or partial positive charge. Literally, an electrophile "loves electrons." Thus, an electrophile is attracted to a region of negative or partial negative charge. An electrophile may be a cation, such as H^+, or the positive portion of a polar molecule, such as the H atom in HCl. Electrophiles are *electron poor.*

A **nucleophile** is a species with a negative or partial negative charge. Literally, a nucleophile "loves a nucleus." A nucleophile is attracted to a region of positive or partial positive charge (i.e., an electrophile). A nucleophile may be an anion, such as Cl^-, or the negative portion of a polar molecule, such as the Cl atom in HCl. Nucleophiles are *electron rich.* Electron-rich sites and electron-poor sites are attracted to one another.

Addition Reactions

The electrostatic potential maps of HCl and C_2H_4 shown in Figure 23.7 demonstrate that both molecules have regions of partial positive charge and partial negative charge. For example, HCl is a polar molecule, with H bearing a partial positive *charge,* due to the large difference in electronegativity between hydrogen and chlorine. Moreover, the carbon-carbon double bond in C_2H_4 consists of two pairs of shared electrons, one pair in a sigma bond and one pair in a pi bond [◄◄ Section 7.6], making the double bond a region of partial negative charge. The partial positive charge on the H in the HCl molecule is an electrophile. The double bond in ethylene, a region of relatively high electron density, is a nucleophile.

A reaction takes place when the positive end of the HCl molecule approaches the double bond in ethylene. The pi bond *breaks,* and the electrons it contained move as indicated by the *curved arrows* shown in the following equation, forming a sigma bond between the H atom of the HCl molecule and one of the C atoms. As this new bond forms, two things happen.

1. Because there cannot be more than one bond to the H atom, the original bond between H and Cl breaks. Both of the electrons originally shared by H and Cl go with the Cl atom. The resulting intermediate species are shown in square brackets in the following equation (dashed lines represent bonds that are being formed):

The C atom on the right bears a positive charge after the valence electron it originally shared (in the pi bond with the other C atom) is removed from it completely. A species such as this, in which one of the carbons is surrounded by only six electrons, is called a **carbocation.** Although carbon must obey the octet in any stable compound, some reactions involve transient, *intermediate* species in which a carbon atom may be electron deficient—having only three electron pairs around it.

2. The C atom forming the new sigma bond to the H atom changes from sp^2-hybridized to sp^3-hybridized.

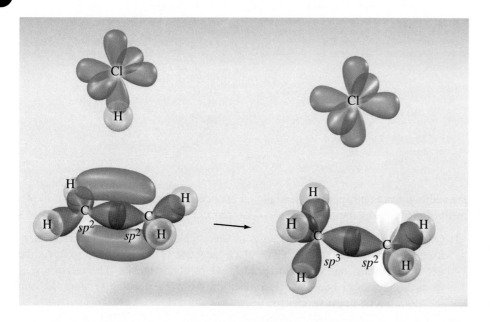

The carbon atom bearing the positive charge is still sp^2-hybridized at this point.

The chloride ion produced when both of the electrons originally shared by H and Cl go with Cl is a *nucleophile*. It is attracted to the newly formed positive charge and two of its electrons form a bond to the positively charged C atom as shown by the curved arrows. The formation of a new sigma bond between the Cl and C atoms causes the hybridization of the second C to change from sp^2 to sp^3.

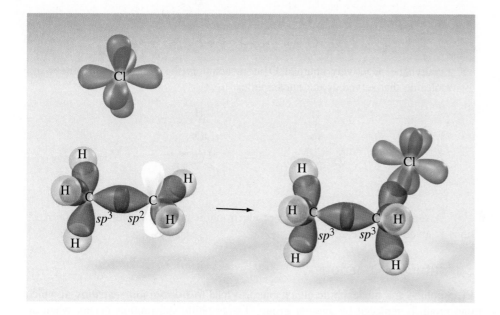

The overall reaction is called an *addition reaction*. Specifically, this is an **electrophilic addition** reaction because it begins with the attack on HCl by the region of electron density in the double bond.

Addition reactions can also begin with nucleophilic attack, in which case the reaction is called a **nucleophilic addition.** In a nucleophilic addition reaction, a bond forms when a nucleophile donates a pair of electrons to an electron-deficient atom. Water, for example, reacts with carbon dioxide to produce carbonic acid. Although

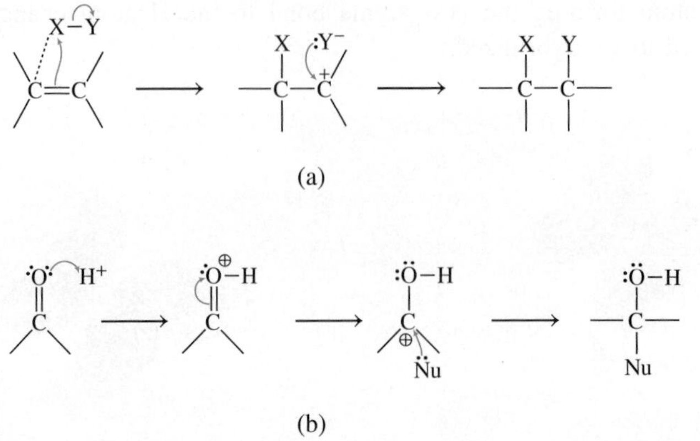

Figure 23.8 (a) Electrophilic addition reaction. (b) Nucleophilic addition reaction. Curved arrows indicate movement of electrons. (Nu represents a nucleophile.)

(a)

(b)

the addition reaction happens essentially all at once, it is helpful to think of the movement of electrons as the following stepwise process.

Student Annotation: Nucleophilic attack in which only one electron is donated by the attacking species can also occur. These are called *radical* reactions.

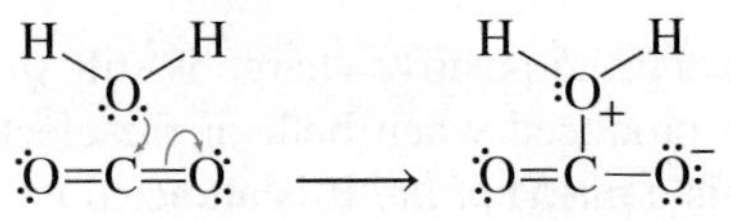

1. One of the lone pairs on the O atom in water attacks the C atom, which is electron deficient because of the highly electronegative O atoms bonded to it. As the new bond forms, one of the original C—O pi bonds breaks. The electron pair from the broken pi bond is repositioned on the corresponding O atom, leaving the atom with a negative charge.

2. One of the O—H bonds in water breaks, with both electrons remaining with the O atom. The H atom separates from the water molecule as a proton.

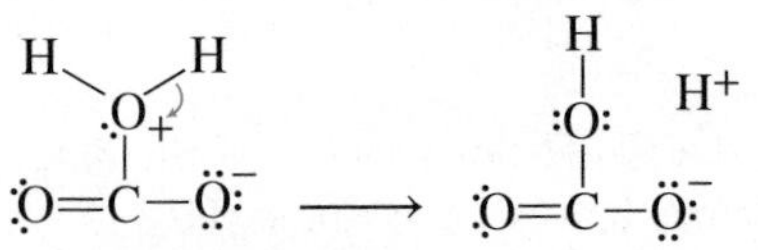

3. Finally, the negatively charged O atom acquires the proton lost by the water molecule that served as the nucleophile.

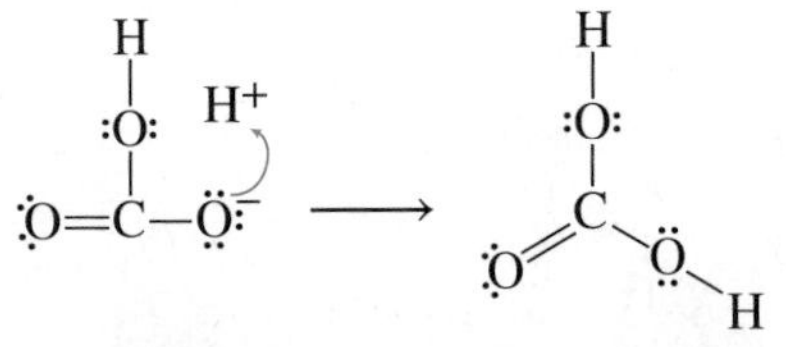

A summary of the mechanisms of electrophilic and nucleophilic addition reactions is given in Figure 23.8.

Substitution Reactions

Student Annotation: Specifically, this type of reaction is called an *electrophilic aromatic substitution* reaction.

Electrophilic and nucleophilic attack can also lead to *substitution reactions* in which one group is replaced by another group. Electrophilic substitution occurs when an electrophile attacks an aromatic molecule and replaces a hydrogen atom. Nucleophilic substitution occurs when a nucleophile replaces another group on a carbon atom. Figure 23.9 shows the general mechanisms for substitution reactions.

The nitration of benzene is an example of an electrophilic substitution reaction.

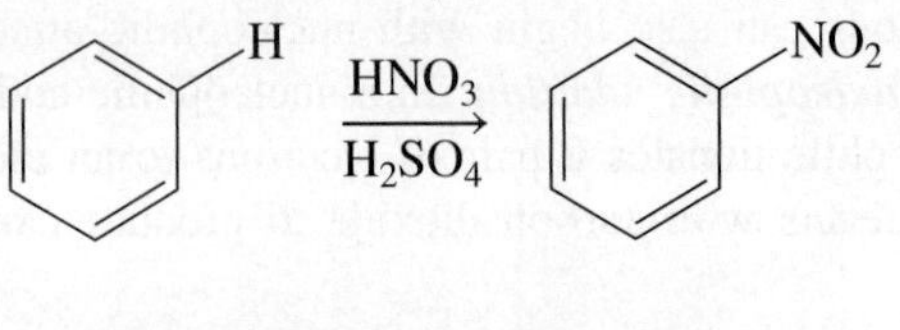

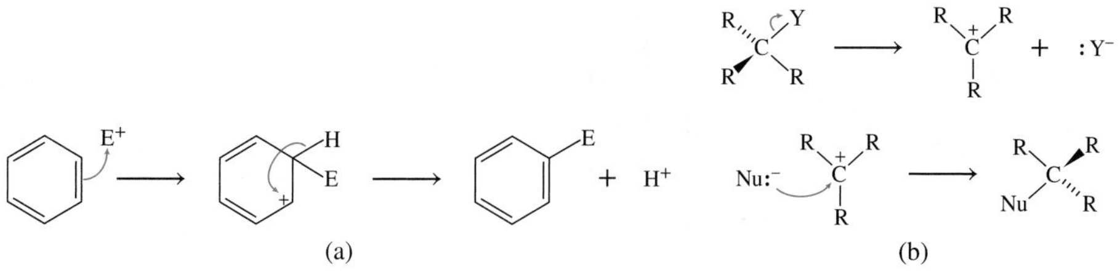

Figure 23.9 (a) Electrophilic substitution. Benzene attacks an electrophile (E^+). (b) Nucleophilic substitution reaction.

Nitric acid and sulfuric acid react to produce the nitronium ion ($^+NO_2$), which acts as the electrophile.

$$H{-}\ddot{O}{-}NO_2 + H{-}OSO_3H \longrightarrow H{-}\overset{H}{\overset{|}{O^+}}{-}NO_2 \longrightarrow H_2\ddot{O}: + \overset{+}{N}O_2 = :\ddot{O}{=}\overset{+}{N}{=}\ddot{O}:$$

$$+ \; HSO_4^- \qquad\qquad \text{Electrophile}$$

The positively charged nitronium ion is attracted to the electron-rich pi bonds of the benzene ring. A bond forms between one of the carbon atoms and the nitronium ion, breaking one of benzene's pi bonds.

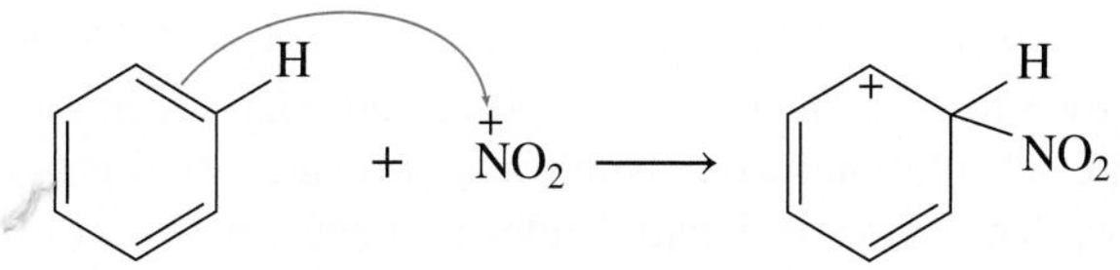

The resulting carbocation is stabilized by resonance.

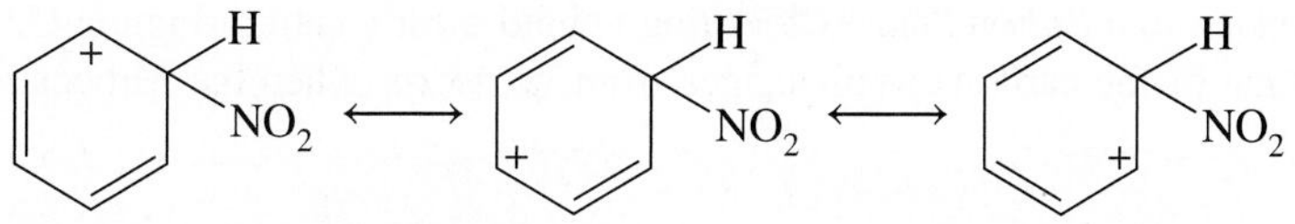

Finally, the electrons in the C–H bond move to the ring, restoring the original pi bond, and the hydrogen atom leaves as a proton, H^+.

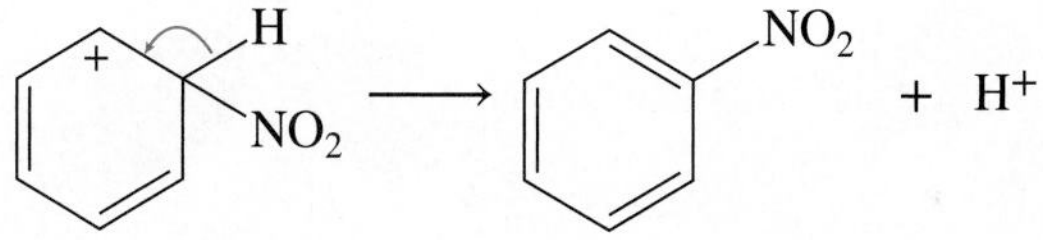

When necessary, one additional step can convert the $-NO_2$ group into the $-NH_2$ group.

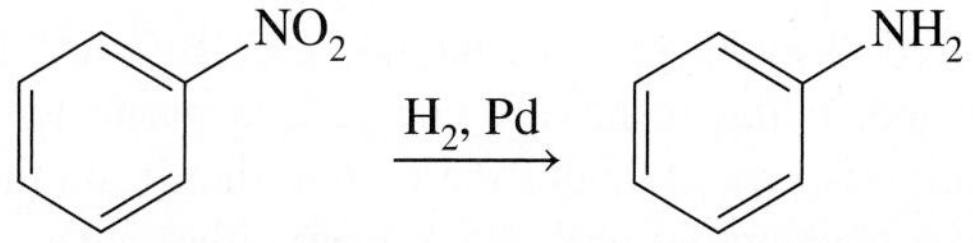

A simple example of nucleophilic substitution is the reaction of an alkyl halide such as methyl bromide with a nucleophile such as the chloride ion. The chloride ion is attracted to the partial positive charge on the C atom in methyl bromide. A lone pair on the chloride ion moves to form a sigma bond between the Cl atom and the C atom. And, because there can be no more than four electron pairs around the C atom, the original C–Br bond breaks, with both of the electrons originally shared by C and Br going to the Br atom. The result is a methyl chloride molecule and a bromide ion.

Student Annotation: The C atom in methyl bromide bears a partial positive charge because it is bonded to the somewhat more electronegative bromine atom [◀◀ Section 6.2].

$$:\ddot{\underset{..}{Cl}}:^- + H{-}\underset{H}{\overset{H}{C}}{-}\ddot{\underset{..}{Br}}: \longrightarrow :\ddot{\underset{..}{Cl}}{-}\underset{H}{\overset{H}{C}}{-}H + :\ddot{\underset{..}{Br}}:^-$$

Nucleophilic substitutions are particularly important in living systems. The hydrolysis of ATP (adenosine triphosphate), which was described in Section 23.3, is

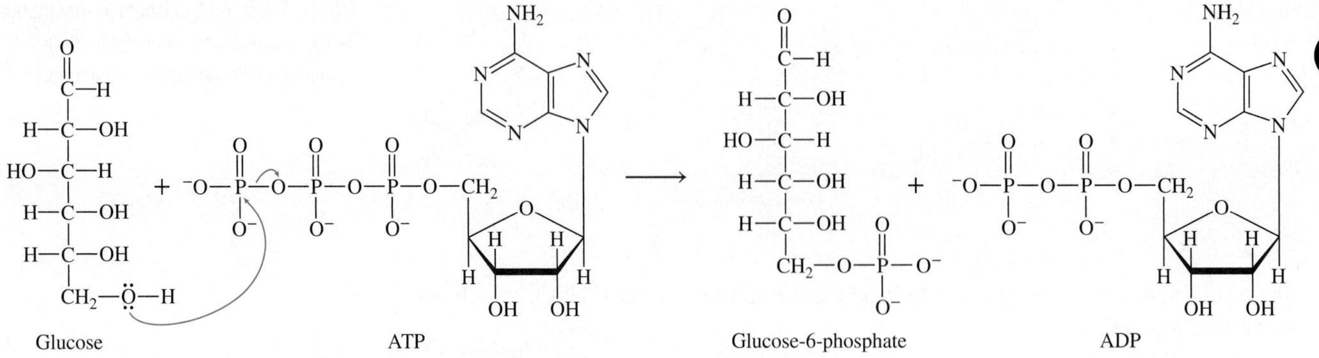

Figure 23.10 Reaction of glucose and ATP to produce glucose-6-phosphate and ADP.

an example of nucleophilic substitution. In hydrolysis, the oxygen atom in water acts as the nucleophile, attacking the electron-deficient phosphorus atom. A similar reaction happens between glucose and ATP, as shown in Figure 23.10. The P atom that is attacked is electron deficient because of the four highly electronegative O atoms bonded to it. As a bond forms between the attacking O and the P, one of the original P—O bonds breaks. The net result is the replacement of the original —H group on glucose with a —PO$_4^{3-}$ group to give glucose-6-phosphate and ADP (adenosine diphosphate).

Nucleophilic substitution reactions fall into two categories, called S$_N$1 reactions and S$_N$2 reactions. (The numbers 1 and 2 refer to a specific aspect of the *kinetics* of the reactions [◄◄ Chapter 19].) The nucleophilic substitution that converts one enantiomer to a mixture of both is an S$_N$1 reaction. An S$_N$1 reaction begins when one of the groups bonded to a carbon "leaves," leaving behind a carbocation [Figure 23.9(b)]. The hybridization of the carbon atom changes from sp^3 to sp^2 when the carbocation forms.

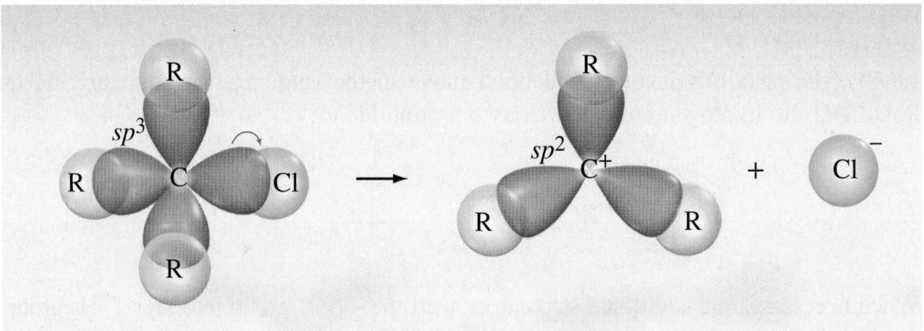

> **Student Annotation:** Digestion of proteins also begins with a nucleophilic substitution reaction.

Commonly encountered "leaving groups" include Cl$^-$, Br$^-$, and I$^-$. A carbocation is an unstable species and, being positively charged, is prone to nucleophilic attack. Because the carbocation is *planar* about the carbon that bears the positive charge, it is equally likely that a nucleophile will attack from either side.

> **Student Annotation:** Recall that sp^2-hybrid orbitals form a trigonal *plane* [◄◄ Table 7.2].

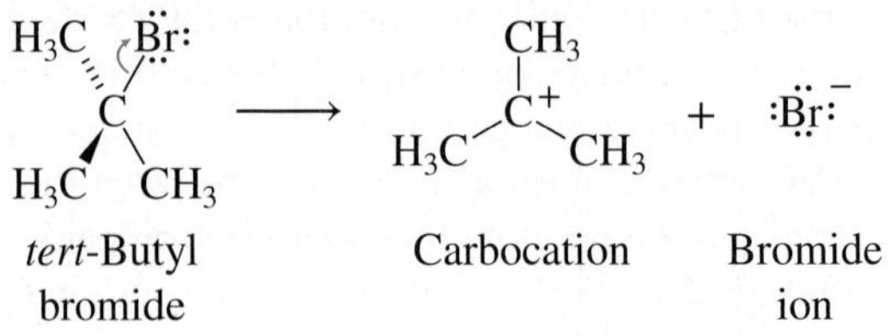

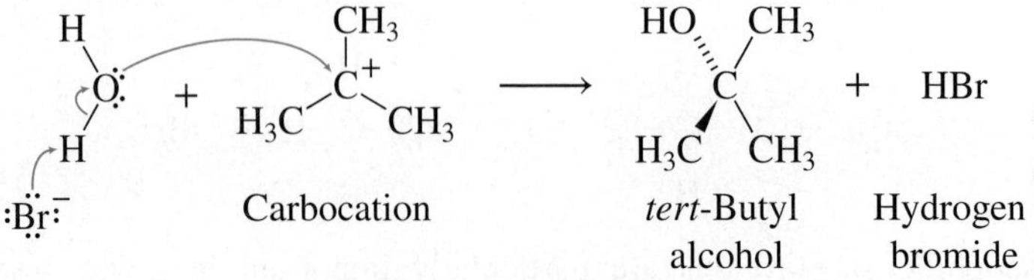

In the reaction of (CH$_3$)$_3$CBr with H$_2$O, the product is the same regardless of whether the nucleophile (water) attacks from the front or from the back. When a molecule loses a leaving group from a chiral carbon, however, as would be the case with the thalidomide molecule, attack from one side of the carbocation will yield one enantiomer whereas attack from the other side will yield the other enantiomer. The result is a racemic mixture. The conversion of a single enantiomer to a racemic mixture of both enantiomers is called *racemization.*

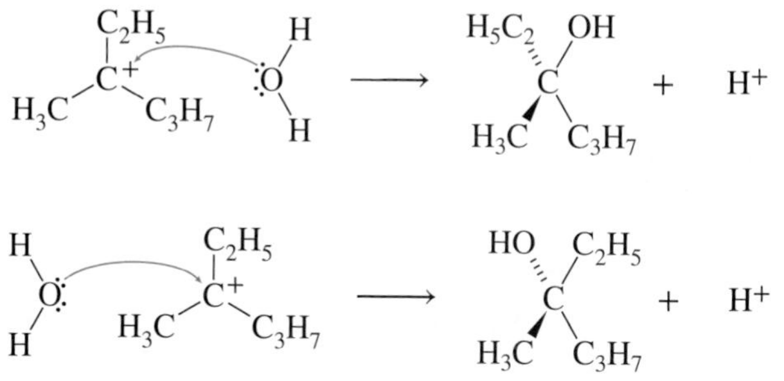

Nucleophilic attack on opposite sides of the carbocation leads to two different products, which are mirror images of each other.

Worked Example 23.5 shows how to draw mechanisms for addition and substitution reactions.

Worked Example 23.5

Using curved arrows to indicate the movement of electrons, draw the mechanism for each of the following reactions: (a) nucleophilic addition of CN^- to CH_3CHO and (b) electrophilic substitution of benzene with $^+SO_3H$. (Draw all resonance structures for the carbocation intermediate.)

Strategy For nucleophilic addition, draw Lewis structures with the nucleophile close to the electron-poor atom where attack will occur (in this case, the carbonyl carbon). For electrophilic substitution, draw Lewis structures with the electrophile close to the site of attack on the benzene ring. Remember that nucleophiles are attracted to and react with electron-poor atoms whereas electrophiles are attracted to and react with electron-rich areas of the molecule. Using this information, and the octet rule, determine which electrons are likely to be involved in the reaction and indicate their repositioning with curved arrows.

Setup (a) The nucleophile is CN^-. The site of attack is the carbonyl C in CH_3CHO, which is electron poor because it is bonded to the more electronegative O atom.

(b) The electrophile is $^+SO_3H$. The site of attack is the electron-rich, delocalized pi bonds of the benzene ring.

Solution

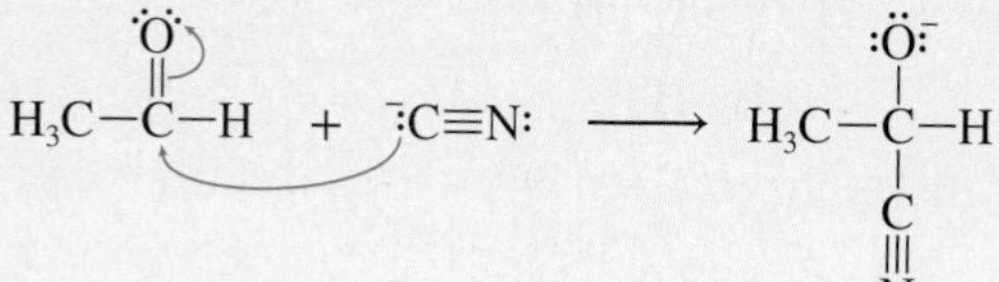

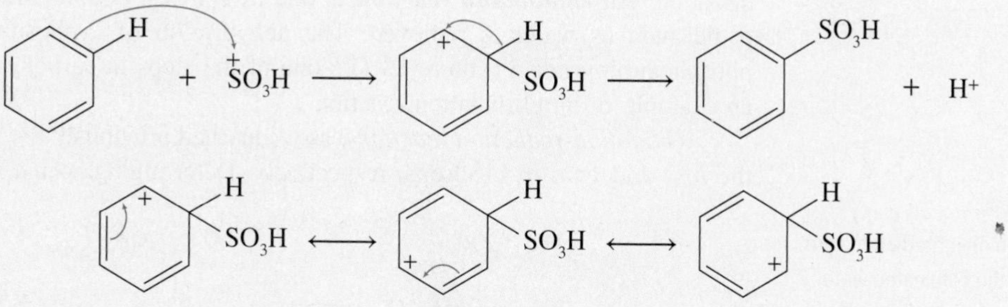

(Continued on next page)

Think About It

The C atoms in benzene are all equivalent, so the choice of which C will bear the substituent in part (b) is arbitrary. All the following represent the same product.

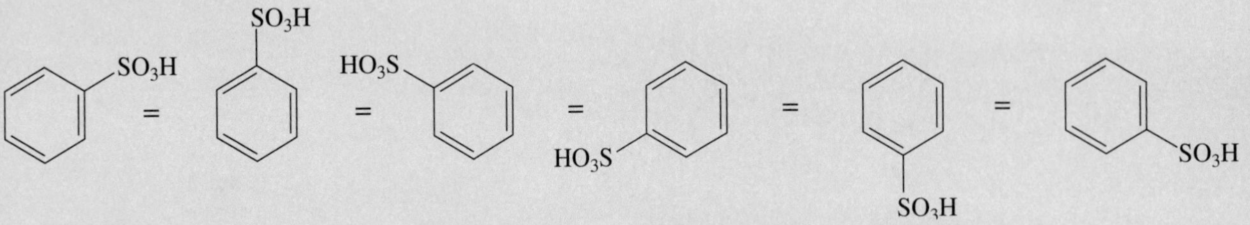

Practice Problem **A**TTEMPT Draw mechanisms for (a) nucleophilic addition of H^- to CH_3COCH_3 and (b) electrophilic substitution of benzene with Cl^+.

Practice Problem **B**UILD Draw all the possible products that could result from the following electrophilic substitution reaction.

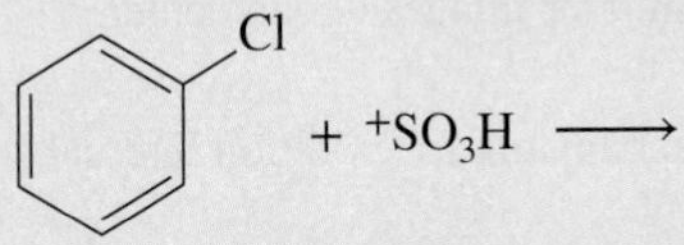

Practice Problem **C**ONCEPTUALIZE In which of the following examples do the curved arrows correspond correctly to the proper movement of electrons in the mechanism of electrophilic addition?

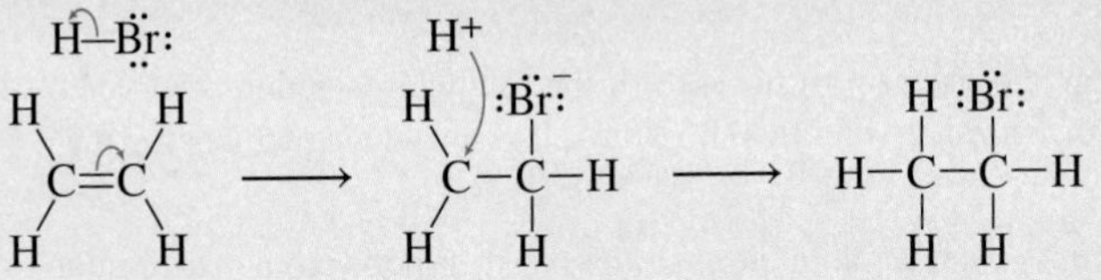

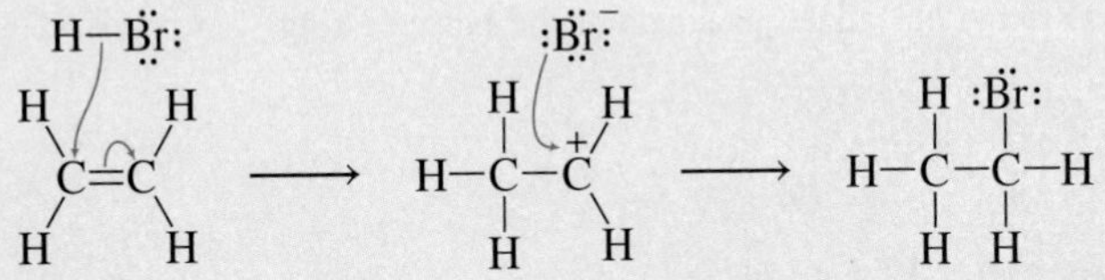

Other Types of Organic Reactions

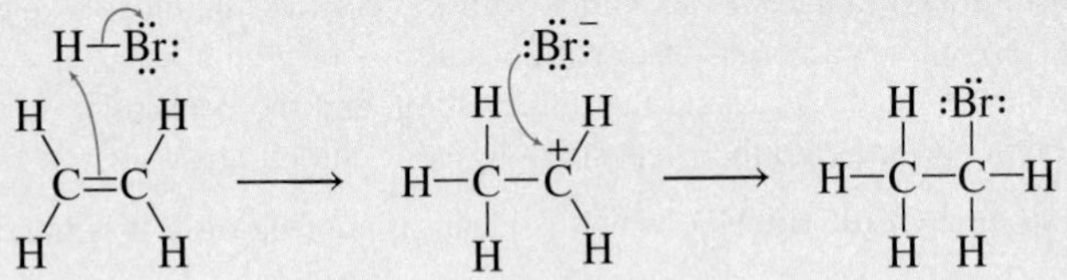

Other important organic reaction types are elimination, oxidation reduction, and isomerization. An *elimination reaction* is one in which a double bond forms and a molecule such as water is removed. The dehydration of 2-phosphoglycerate to form phosphoenolpyruvate (Figure 23.11), one of the steps in carbohydrate metabolism, is an example of an elimination reaction.

Oxidation-reduction reactions, as we learned previously [◀◀ Section 9.4], involve the *loss* and *gain* of electrons, respectively. Determining which species have lost or

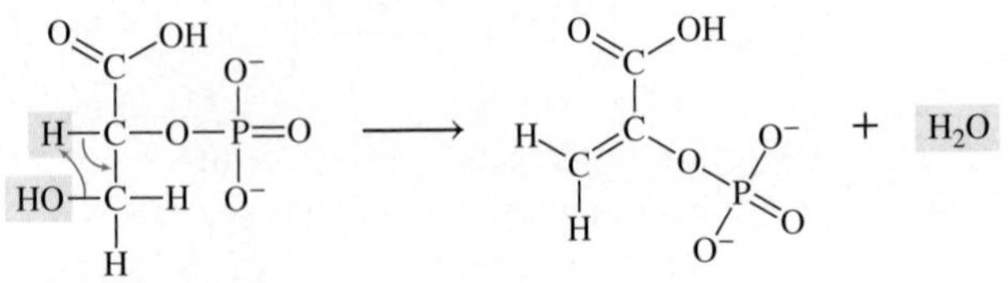

Figure 23.11 The highlighted atoms are those that constitute the eliminated water molecule.

gained electrons may seem less straightforward with organic reactions than with the inorganic reactions we have encountered, but the following guidelines can help you decide when an organic molecule has been oxidized or reduced.

1. When a molecule gains O or loses H, it is *oxidized.*
2. When a molecule loses O or gains H, it is *reduced.*

An important biological example of oxidation is the enzyme-catalyzed reaction by which ethanol is converted to acetaldehyde in the liver.

$$CH_3CH_2OH \longrightarrow CH_3CHO$$

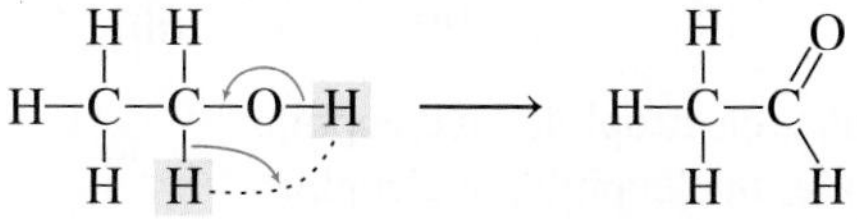

The ethanol molecule loses two H atoms in this reaction, so it is *oxidized.*

Isomerization reactions are those in which one isomer is converted to another. The interconversion between the sugars aldose and ketose is an example.

Aldose   Ketose

Student Annotation: Many of the organic reactions in living systems require *enzymes,* natural *catalysts,* in order to occur rapidly enough to be useful [◄◄ Chapter 19].

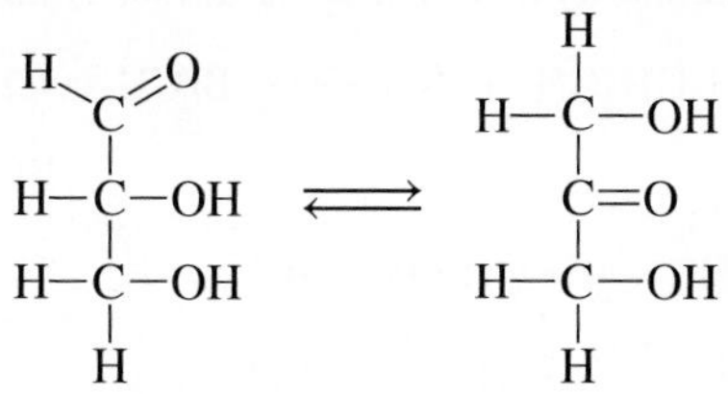

Vision, our ability to perceive light, is the result of an isomerization reaction. Our eyes contain millions of cells called rods that are packed with *rhodopsin,* an 11-*cis*-retinal molecule

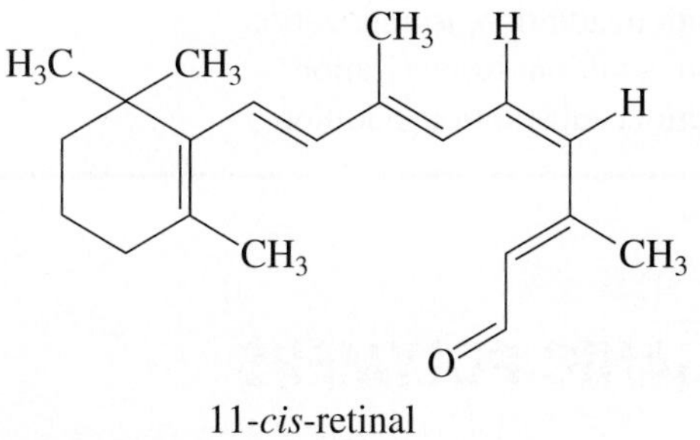

11-*cis*-retinal

bonded to a large protein. When visible light strikes rhodopsin, the retinal molecule isomerizes to the all-*trans* isomer.

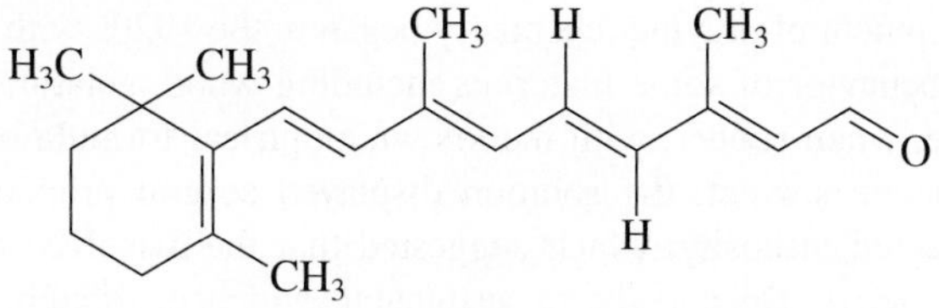

All-*trans*-retinal

This isomerization causes such a significant change in the structure of retinal that it separates from the protein. These events trigger the electrical impulses that stimulate the optic nerve and result in the brain receiving the signals that we know as vision. The all-*trans*-retinal diffuses away from the protein and is converted back to 11-*cis*-retinal. The regenerated 11-*cis*-retinal can then rebind with the protein. Interestingly, the conversion of the all-*trans* isomer back to the 11-*cis* isomer, which is necessary for us to perceive light, is much slower than the light-induced conversion from *cis* to *trans.* This is why after looking at a bright light, you have a "blind spot" for a period of time.

Section 23.5 Review

Organic Reactions

23.5.1   Identify each species as a nucleophile or an electrophile.

$$CH_3-\overset{..}{\underset{..}{O}}{:}^-\qquad H-\overset{..}{\underset{..}{S}}{:}^-\qquad H_3C-\overset{\overset{\displaystyle CH_3}{|}}{\underset{\underset{\displaystyle CH_3}{|}}{C}}{}^+$$

(i)   (ii)   (iii)

(a) Nucleophile, electrophile, nucleophile
(b) Nucleophile, nucleophile, nucleophile
(c) Electrophile, nucleophile, electrophile
(d) Nucleophile, nucleophile, electrophile
(e) Electrophile, electrophile, electrophile

23.5.2   Identify each reaction as addition, substitution, elimination, or isomerization.

(i) $H_2C=CHCH_2CH_2CH_3 + Br_2 \longrightarrow BrCH_2-\underset{\underset{\displaystyle Br}{|}}{C}HCH_2CH_2CH_3$

(ii) $CH_3Br + OH^- \longrightarrow CH_3OH + Br^-$

(iii)
$$\underset{\underset{\displaystyle H}{}}{\overset{\overset{\displaystyle H_3C}{}}{C}}=\underset{\underset{\displaystyle CH_3}{}}{\overset{\overset{\displaystyle H}{}}{C} \longrightarrow \underset{\underset{\displaystyle H}{}}{\overset{\overset{\displaystyle H_3C}{}}{C}}=\underset{\underset{\displaystyle H}{}}{\overset{\overset{\displaystyle CH_3}{}}{C}}$$

(a) Addition, substitution, isomerization
(b) Isomerization, substitution, addition
(c) Substitution, addition, isomerization
(d) Addition, addition, isomerization
(e) Substitution, substitution, addition

23.6   ORGANIC POLYMERS

Student Annotation: Polymers typically have very high molar masses—sometimes thousands or even millions of grams.

Student Annotation: Some of the other properties that suggested a high molar mass solute were low osmotic pressure and negligible freezing-point depression. These are known as *colligative* properties [◄◄ Chapter 13].

Polymers are molecular compounds, either natural or synthetic, that are made up of many *repeating units* called *monomers*. The physical properties of these so-called macromolecules differ greatly from those of small, ordinary molecules.

The development of polymer chemistry began in the 1920s with the investigation of the puzzling behavior of some materials including wood, gelatin, cotton, and rubber. For example, when rubber, with the known empirical formula of C_5H_8, was dissolved in an organic solvent, the solution displayed several properties, including a higher than expected viscosity, which suggested that the dissolved compound had a very high molar *mass*. Despite the experimental evidence, though, scientists at the time were not ready to accept the idea that such giant molecules could exist. Instead, they postulated that materials such as rubber consisted of aggregates of small molecular units, like C_5H_8 or $C_{10}H_{16}$, held together by intermolecular forces. This misconception persisted for a number of years, until Hermann Staudinger[1] clearly showed that

---

[1]Hermann Staudinger (1881–1963), a German chemist, was one of the pioneers in polymer chemistry. Staudinger was awarded the Nobel Prize in Chemistry in 1953.

these so-called aggregates were, in fact, enormously large molecules, each of which contained many thousands of atoms held together by covalent bonds.

Once the *structures* of these macromolecules were understood, the way was open for the synthesis of polymers, which now pervade almost every aspect of our daily lives. About 90 percent of today's chemists, including biochemists, work with polymers. In this section, we will discuss the reactions that result in polymer formation and some of the natural polymers that are important to biology.

Video 23.3
Organic and biochemistry—natural and synthetic polymers.

Addition Polymers

Addition polymers form when monomers such as ethylene join end to end to make polyethylene. Reactions of this type can be initiated by a *radical*—a species that contains an unpaired electron [◄◄ Section 6.6]. The mechanism of addition polymerization is as follows:

1. The radical, which is unstable because of its unpaired electron, attacks a carbon atom on an ethylene molecule. This is the *initiation* of the reaction.
2. This attack would result in the carbon atom in question having more than eight electrons around it. To keep the carbon atom from having too many electrons around it, the double bond breaks.
3. One of the electrons in the double bond, together with the electron from the radical, becomes a new bond between the ethylene molecule and the radical.
4. The other electron remains with the other carbon atom, generating a new radical species.
5. The new radical species, also unstable, attacks another ethylene molecule, causing the same sequence of events to happen again. The generation of a new, highly reactive radical species at each step is known as *propagation* of the reaction.
6. Each step lengthens the chain of carbon atoms and results in the formation of a new radical, continuing the propagation of the reaction. Reactions such as this are known as *chain* reactions. They continue until the system runs out of ethylene molecules or until the radical species encounters another radical species—resulting in *termination* of the reaction.

Initiation step

Propagation step

Termination step

Some familiar and important addition polymers are listed in Table 23.4.

TABLE 23.4	Addition Polymers		
Name	**Monomer unit**	**Structure**	**Uses**
Polytetrafluoroethylene (Teflon)	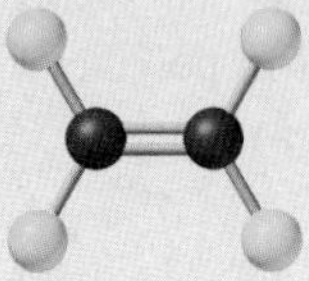	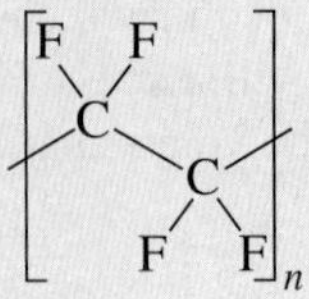 $\left[\begin{array}{c} F\ F \\ C \\ C \\ F\ F \end{array}\right]_n$	Nonstick coatings
Polyethylene	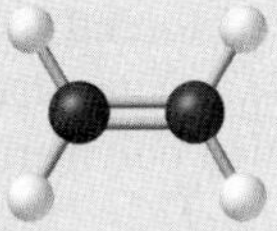	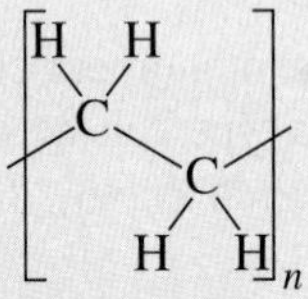 $\left[\begin{array}{c} H\ H \\ C \\ C \\ H\ H \end{array}\right]_n$	Plastic bags, bottles, toys
Polypropylene	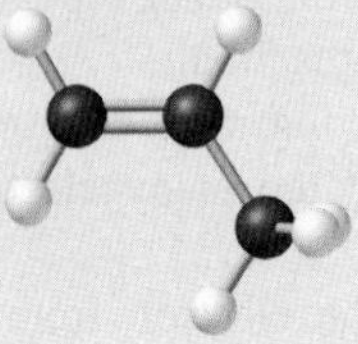	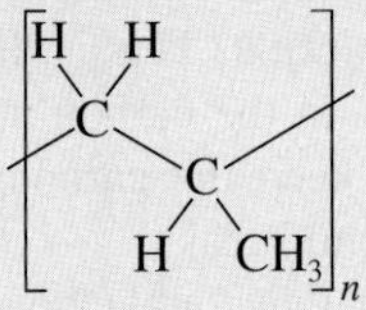 $\left[\begin{array}{c} H\ H \\ C \\ C \\ H\ CH_3 \end{array}\right]_n$	Carpeting, bottles
Polyvinylchloride (PVC)	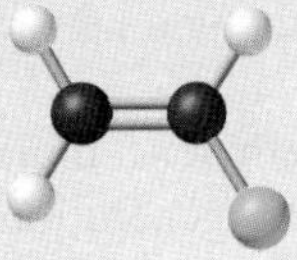	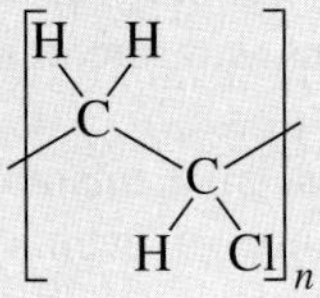 $\left[\begin{array}{c} H\ H \\ C \\ C \\ H\ Cl \end{array}\right]_n$	Water pipes, garden hoses, plastic wrap
Polystyrene	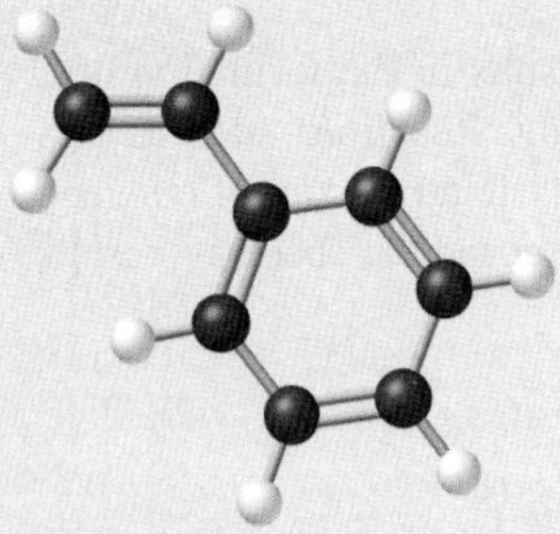	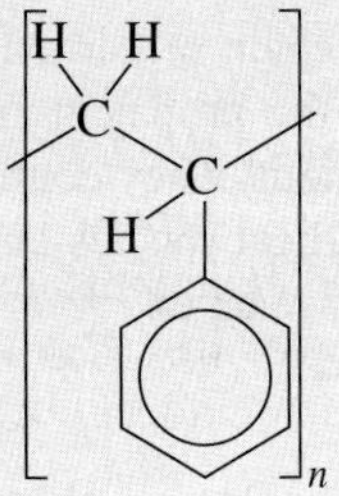 $\left[\begin{array}{c} H\ H \\ C \\ C \\ H \end{array}\right]_n$	Packing material, insulation, furniture

Condensation Polymers

Reactions in which two or more molecules become connected with the elimination of a small molecule, often water, are called *condensation reactions* (Figure 23.12). *Condensation polymers* form when molecules with two different functional groups combine, with the elimination of a small molecule, often water. Many condensation polymers are *copolymers,* meaning that they are made up of two or more *different* monomers.

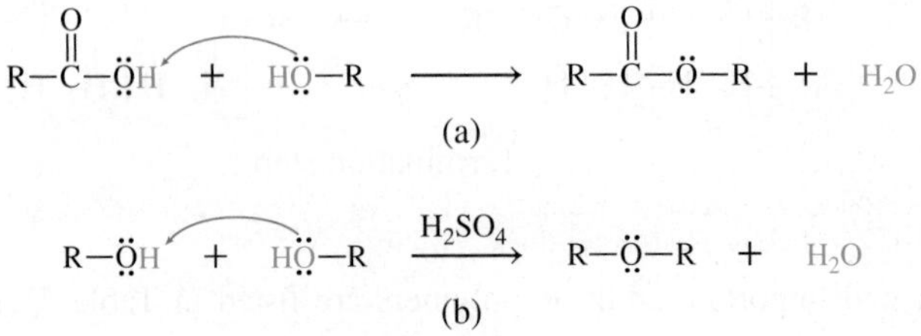

Figure 23.12 (a) Condensation reaction between an alcohol and a carboxylic acid to form an ester. (b) Condensation of two alcohol molecules in the presence of sulfuric acid to form an ether.

$$R-\overset{\overset{O}{\|}}{C}-\ddot{\underset{}{O}}H \;+\; H\ddot{\underset{}{O}}-R \;\longrightarrow\; R-\overset{\overset{O}{\|}}{C}-\ddot{\underset{}{O}}-R \;+\; H_2O$$
(a)

$$R-\ddot{\underset{}{O}}H \;+\; H\ddot{\underset{}{O}}-R \;\xrightarrow{H_2SO_4}\; R-\ddot{\underset{}{O}}-R \;+\; H_2O$$
(b)

The first synthetic fiber, nylon 66, is a condensation copolymer of two molecules: one with carboxy groups at each end (adipic acid) and one with amine groups at both ends (hexamethylenediamine).

$$H_2N-(CH_2)_6-NH_2 \quad + \quad HOOC-(CH_2)_4-COOH$$
Hexamethylenediamine $\qquad\qquad$ Adipic acid

$$\downarrow \text{Condensation}$$

$$H_2N-(CH_2)_6-\underset{H}{\overset{\overset{\displaystyle O}{\|}}{N}}-C-(CH_2)_4-COOH \quad + \quad H_2O$$

$$\downarrow \text{Further condensation reactions}$$

$$-(CH_2)_4-\overset{\overset{\displaystyle O}{\|}}{C}-\underset{H}{N}-(CH_2)_6-\underset{H}{N}-\overset{\overset{\displaystyle O}{\|}}{C}-(CH_2)_4-\overset{\overset{\displaystyle O}{\|}}{C}-\underset{H}{N}-(CH_2)_6-$$

Student Annotation: The number 66 refers to the fact that there are six C atoms in each monomer. Other nylons, such as nylon 610, are made from various combinations of molecules similar to adipic acid and hexamethylenediamine.

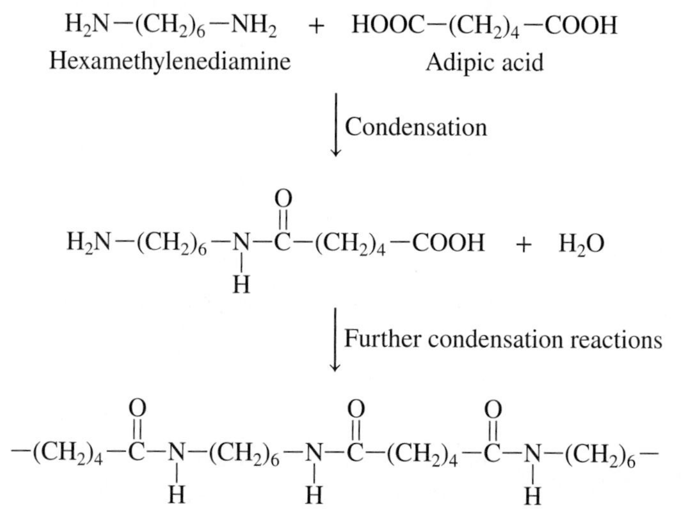

Nylon was first made by Wallace Carothers[2] at DuPont in 1931. The versatility of nylons is so great that the annual production of nylons and related substances now amounts to several billion pounds.

Biological Polymers

Naturally occurring polymers include *proteins, polysaccharides,* and *nucleic acids.*

Proteins, polymers of amino acids, play an important role in nearly all biological processes. The human body contains an estimated 100,000 different kinds of proteins, each of which has a specific physiological function. An amino acid has both the carboxylic acid functional group and the amino functional group. Amino acids are joined together into chains when a condensation reaction occurs between a carboxy group on one molecule and an amino group on another molecule (Figure 23.13).

The bonds that form between amino acids are called **peptide bonds.** Very long chains of amino acids assembled in this way are called *proteins,* while shorter chains are called *polypeptides.*

Amino acids consist of a central carbon atom bonded to four different groups: an amino group, a carboxy group, a hydrogen atom, and an additional group (highlighted in Figure 23.14) consisting of carbon, hydrogen, and sometimes other elements such as nitrogen or sulfur. Proteins are made essentially from

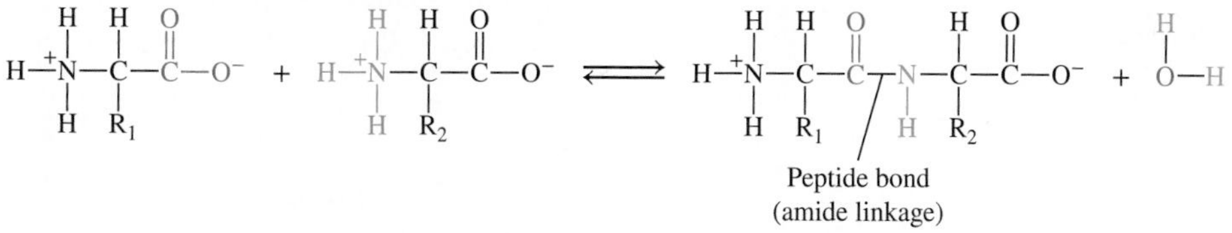

Student Annotation: Peptide bonds are also called *amide bonds* or *amide linkages* because they contain the amide functional group.

$$H-\underset{H}{\overset{+}{N}}-\underset{R_1}{C}-\overset{\overset{\displaystyle O}{\|}}{C}-O^- \quad + \quad H-\underset{H}{\overset{+}{N}}-\underset{R_2}{C}-\overset{\overset{\displaystyle O}{\|}}{C}-O^- \quad \rightleftharpoons \quad H-\underset{H}{\overset{+}{N}}-\underset{R_1}{C}-\underset{\substack{\\ \text{Peptide bond} \\ \text{(amide linkage)}}}{C}-\underset{H}{N}-\underset{R_2}{C}-\overset{\overset{\displaystyle O}{\|}}{C}-O^- \quad + \quad \underset{H}{O}-H$$

Figure 23.13 Formation of a peptide bond with elimination of water.

[2] Wallace H. Carothers (1896–1937) was an American chemist. Besides its enormous commercial success, Carothers's work on nylon is ranked with that of Staudinger in clearly elucidating macromolecular structure and properties. Depressed by the death of his sister and wrongly believing that his life's work had been a failure, Carothers committed suicide at the age of 41.

Name	Abbreviation	Structure
Alanine	Ala	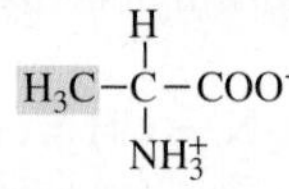
Arginine	Arg	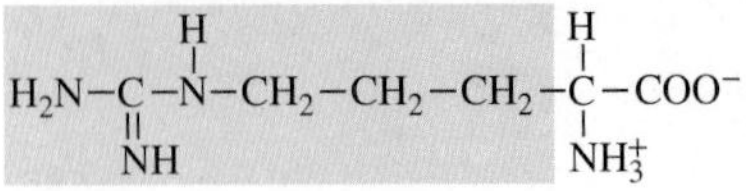
Asparagine	Asn	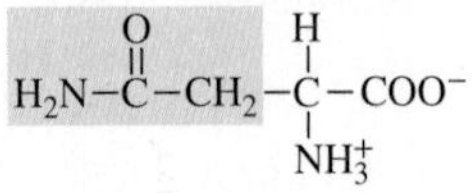
Aspartic acid	Asp	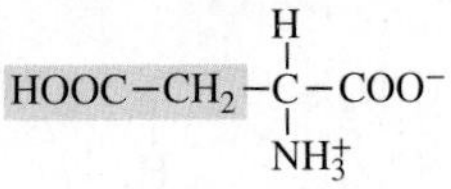
Cysteine	Cys	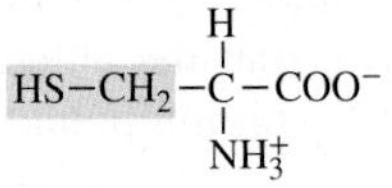
Glutamic acid	Glu	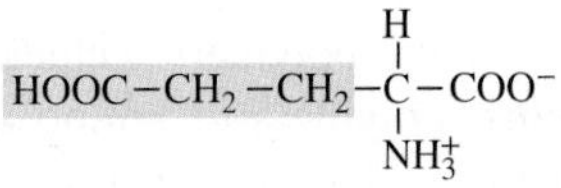
Glutamine	Gln	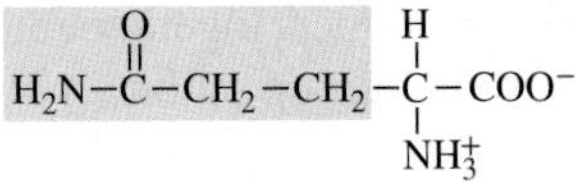
Glycine	Gly	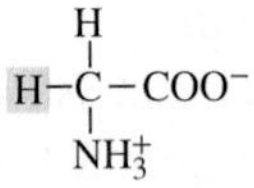
Histidine	His	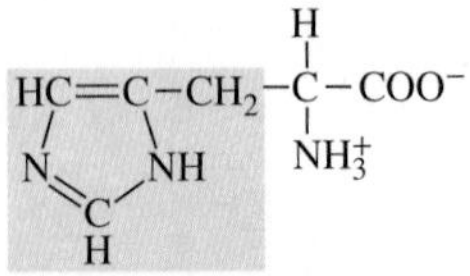
Isoleucine	Ile	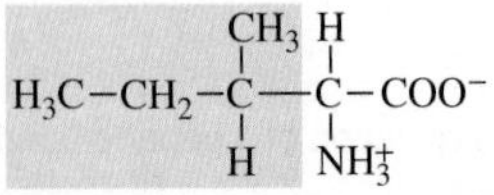

Figure 23.14 The 20 amino acids essential to living organisms. The shaded area represents the R group.

Name	Abbreviation	Structure
Leucine	Leu	H_3C–CH–CH$_2$–C(H)(NH$_3^+$)–COO$^-$ with H_3C
Lysine	Lys	H_2N–CH$_2$–CH$_2$–CH$_2$–CH$_2$–C(H)(NH$_3^+$)–COO$^-$
Methionine	Met	H_3C–S–CH$_2$–CH$_2$–C(H)(NH$_3^+$)–COO$^-$
Phenylalanine	Phe	C$_6$H$_5$–CH$_2$–C(H)(NH$_3^+$)–COO$^-$
Proline	Pro	H_2N^+—C(H)—COO$^-$ ring with H_2C–CH$_2$–CH$_2$
Serine	Ser	HO–CH$_2$–C(H)(NH$_3^+$)–COO$^-$
Threonine	Thr	H_3C–C(OH)(H)–C(H)(NH$_3^+$)–COO$^-$
Tryptophan	Trp	indole–CH$_2$–C(H)(NH$_3^+$)–COO$^-$
Tyrosine	Tyr	HO–C$_6$H$_4$–CH$_2$–C(H)(NH$_3^+$)–COO$^-$
Valine	Val	H_3C–CH–C(H)(NH$_3^+$)–COO$^-$ with H_3C

the 20 different amino acids shown in Figure 23.14. The identity of a protein depends on which of the 20 amino acids it contains and on the order in which the amino acids are assembled.

Polysaccharides are polymers of sugars such as glucose and fructose. Starch and cellulose are two polymers of glucose with slightly different linkages—and very different properties. In starch, glucose molecules are connected by what biochemists call α linkages. This enables animals, including humans, to digest the starch in such foods as corn, wheat, potatoes, and rice.

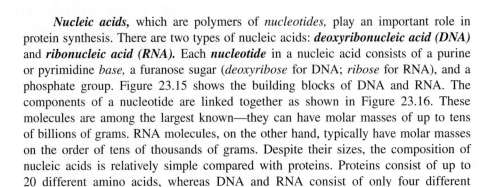

Starch

In cellulose, the glucose molecules are connected by β linkages. Digestion of cellulose requires enzymes that most animals do not have. Species that *do* digest cellulose, such as termites and ruminants (including cattle, sheep, and llamas), do so with the help of enzyme-producing symbiotic bacteria in the gut.

Cellulose

Video 23.4
Organic and biochemistry—molecular structure in DNA.

Nucleic acids, which are polymers of *nucleotides,* play an important role in protein synthesis. There are two types of nucleic acids: *deoxyribonucleic acid (DNA)* and *ribonucleic acid (RNA).* Each *nucleotide* in a nucleic acid consists of a purine or pyrimidine *base,* a furanose sugar (*deoxyribose* for DNA; *ribose* for RNA), and a phosphate group. Figure 23.15 shows the building blocks of DNA and RNA. The components of a nucleotide are linked together as shown in Figure 23.16. These molecules are among the largest known—they can have molar masses of up to tens of billions of grams. RNA molecules, on the other hand, typically have molar masses on the order of tens of thousands of grams. Despite their sizes, the composition of nucleic acids is relatively simple compared with proteins. Proteins consist of up to 20 different amino acids, whereas DNA and RNA consist of only four different nucleotides each.

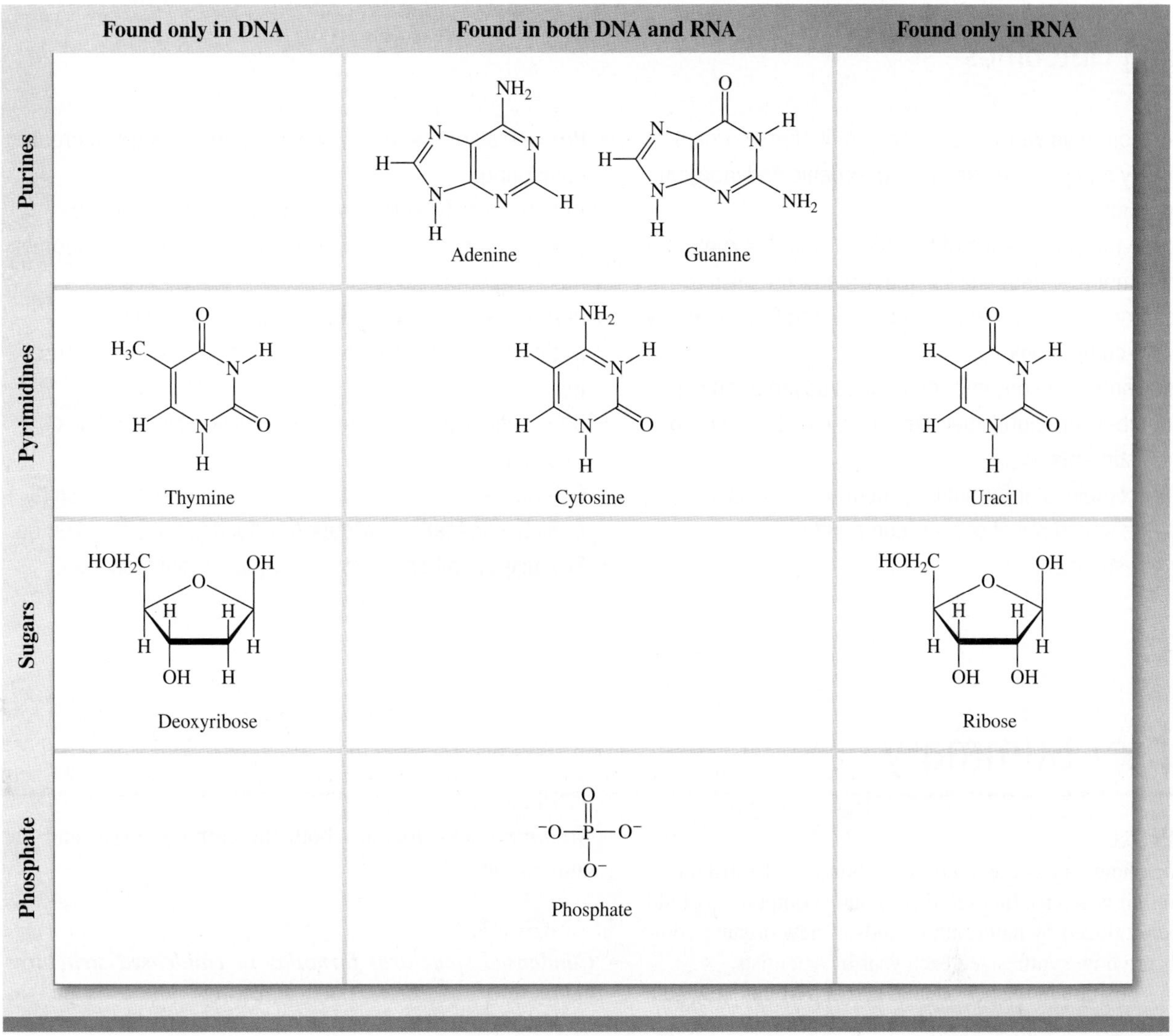

	Found only in DNA	Found in both DNA and RNA	Found only in RNA
Purines		Adenine Guanine	
Pyrimidines	Thymine	Cytosine	Uracil
Sugars	Deoxyribose		Ribose
Phosphate		Phosphate	

Figure 23.15 Components of the nucleic acids DNA and RNA.

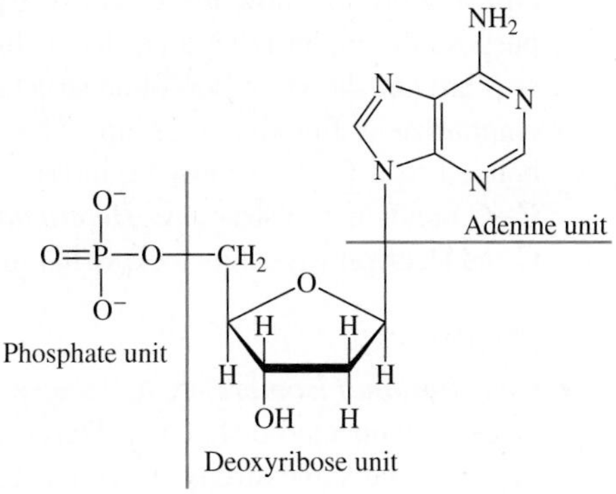

Figure 23.16 Structure of a nucleotide, one of the repeating units in DNA.

Learning Outcomes

- Give in your own words a workable definition of organic chemistry and provide examples of organic and inorganic compounds.
- Describe the significance of Friedrich Wöhler's experiment to contemporary understanding of organic chemistry.
- List characteristics of carbon that allow it to form so many different compounds.
- Classify and name organic compounds based upon carbon number, structural characteristics, functional group, and constituents.
- Construct molecular formulas, structural formulas, and bond-line structures of organic compounds.
- Define resonance.

- Produce all of the resonance structures of a given organic compound.
- Define, identify, and provide examples of the various types of isomers, including constitutional, stereo, geometrical, and optical isomers.
- Produce all of the isomers of a given compound.
- Define chiral and identify chiral carbons in a given structure.
- Differentiate between nucleophiles and electrophiles.
- Describe a carbocation.
- Categorize organic reactions as addition, substitution (S_N1 or S_N2), elimination, oxidation-reduction, or isomerization.
- Produce the mechanism of a simple organic reaction.

Chapter Summary

SECTION 23.1

- Organic chemistry is the study of carbon-based substances. Although it was once thought that organic compounds could only be produced by nature, thousands of new organic compounds are now synthesized each year by scientists.
- Carbon's position in the periodic table makes it uniquely able to form long, stable chains—a process called *catenation.*
- *Aromatic* compounds contain one or more benzene rings. *Aliphatic* compounds are organic compounds that do not contain benzene rings.

SECTION 23.2

- *Alkyl groups* consist of just carbon and hydrogen. They are derived from the corresponding alkane by removing one hydrogen atom and are represented generically in the formulas of organic compounds with the letter R.
- Functional groups are specific arrangements of atoms that are responsible for the properties and reactivity of organic compounds. Common functional groups and their formulas include

Alcohol	ROH
Carboxylic acid	RCOOH
Aldehyde	RCHO
Ketone	RCOR′
Ester	RCOOR′
Amine	RNH₂, RNHR′, or RNR′R″
Amide	RCONH₂, RCONHR′, or RCONR′R″

- Many organic compounds contain more than one functional group. A *substituent* in an alkane is a group other than hydrogen that is bonded to the carbon chain.

- An *amino acid* contains both the carboxy group and the amino group.

SECTION 23.3

- *Condensed structural formulas* or *condensed structures* abbreviate a series of repeating units, such as $-CH_2CH_2CH_2CH_2-$, into a more compact form, such as $-(CH_2)_4-$. *Kekulé structures* are similar to Lewis structures but do not show the lone pairs on a molecule. *Bond-line structures* use straight lines to represent C—C bonds. They typically do not show the C atoms explicitly except for the purpose of emphasizing a particular functional group. H atoms are not shown in bond-line structures—again, except to emphasize a functional group. The number of H atoms bonded to a C atom must be inferred from the number of C—C bonds in the structure. *Heteroatoms,* atoms other than C and H, are always shown explicitly in a bond-line structure.

SECTION 23.4

- *Constitutional isomers* are molecules in which the same atoms are connected differently. *Stereoisomers* are molecules in which the same atoms are connected by the same bonds but the bonds are oriented differently. *Geometrical isomers* arise due to restricted rotation about a carbon-carbon double bond. *Cis* and *trans* isomers are geometrical isomers.
- Molecules that are nonsuperimposable mirror images of each other are *optical isomers.* They are also referred to as *chiral,* and each of the mirror images is called an *enantiomer.* Optical isomers are so called because they rotate the plane of plane-polarized light. The degree of rotation is the same for

both enantiomers, but the directions of rotation are opposite each other. An equal mixture of both enantiomers is called a *racemic mixture.* A racemic mixture does not rotate the plane of plane-polarized light.

SECTION 23.5

- An *electrophile* generally is a positively charged ion that is attracted to electrons. A *nucleophile* is a negatively charged ion or a partially negatively charged atom in a polar molecule. Nucleophiles and electrophiles are attracted to each other.
- A *carbocation* is an intermediate species in which one of the carbon atoms is surrounded by only six electrons and bears a positive charge.
- *Electrophilic addition* reactions and *nucleophilic addition* reactions involve the addition of a molecule or an ion to another molecule.
- *Substitution reactions* occur when an electrophile replaces a hydrogen on an aromatic ring or when a nucleophile replaces a leaving group on a carbon atom.
- Carbocations are the intermediate species in *racemization,* a nucleophilic substitution reaction in which a single enantiomer is converted into a racemic mixture.
- An *elimination reaction* is one in which a double bond forms and a small molecule, such as water, is eliminated.
- *Isomerization reactions* convert one isomer into another.

SECTION 23.6

- *Polymers* are long chains of repeating molecular units called *monomers. Polysaccharides* are polymers of sugars. *Addition polymers* form when a radical species attacks a double bond, forming a new, longer radical species that attacks another double bond, and so on.
- An elimination reaction that joins two molecules is a *condensation reaction. Condensation polymers* form when molecules with two different functional groups undergo a condensation reaction.
- *Copolymers* are polymers that contain more than one type of monomer.
- *Proteins* and *polypeptides* are biological polymers in which the monomers are amino acids. Amino acids are joined by *peptide bonds,* which result from the condensation reaction between the amino group of one amino acid and the carboxyl group of another amino acid.
- *Nucleic acids* are polymers of *nucleotides.* The two types of nucleic acid are *deoxyribonucleic acid (DNA)* and *ribonucleic acid (RNA).* Each nucleotide in a nucleic acid consists of a purine or pyrimidine base, a furanose sugar (deoxyribose for DNA; ribose for RNA), and a phosphate group linked together.

Key Words

Addition polymer, 1061
Alcohol, 1029
Aldehyde, 1029
Aliphatic, 1028
Alkyl group, 1029
Amide, 1029
Amine, 1029
Amino acid, 1032
Aromatic, 1028
Bond-line structure, 1042
Carbocation, 1052
Carboxylic acid, 1029
Catenation, 1068
Chiral, 1048
Cis, 1047
Condensation polymer, 1062
Condensation reaction, 1062

Condensed structural formula, 1040
Condensed structure, 1040
Constitutional isomers, 1047
Copolymer, 1062
Deoxyribonucleic acid (DNA), 1066
Electrophile, 1052
Electrophilic addition, 1053
Elimination reaction, 1058
Enantiomer, 1048
Ester, 1029
Geometrical isomer, 1047
Heteroatom, 1042
Isomerization reaction, 1059
Kekulé structure, 1041
Ketone, 1029
Monomer, 1060
Nucleic acid, 1066

Nucleophile, 1052
Nucleophilic addition, 1053
Nucleotide, 1066
Optical isomers, 1048
Peptide bond, 1063
Polymer, 1060
Polypeptide, 1063
Polysaccharides, 1066
Protein, 1063
Racemic mixture, 1049
Racemization, 1057
Ribonucleic acid (RNA), 1066
Stereoisomers, 1047
Substituent, 1034
Substitution reaction, 1054
Trans, 1048

Questions and Problems

SECTION 23.1: WHY CARBON IS DIFFERENT

Review Questions

23.1 Explain why carbon is able to form so many more compounds than any other element.

23.2 Why was Wöhler's synthesis of urea so important for the development of organic chemistry?

23.3 What are aromatic organic compounds? What are aliphatic organic compounds?

SECTION 23.2: CLASSES OF ORGANIC COMPOUNDS

Review Questions

23.4   What are functional groups? Why is it logical and useful to classify organic compounds according to their functional groups?

23.5   Draw the Lewis structure for each of the following functional groups: alcohol, aldehyde, ketone, carboxylic acid, amine.

23.6   Name the classes to which the following compounds belong.
(a) C_4H_9OH
(b) C_2H_5CHO
(c) C_6H_5COOH
(d) CH_3NH_2

Conceptual Problems

23.7   Classify each of the following molecules as alcohol, aldehyde, ketone, carboxylic acid, or amine.
(a) $CH_3-CH_2-NH_2$

(b) $CH_3-CH_2-C\overset{\displaystyle O}{\underset{\displaystyle H}{}}$

(c) $CH_3-\overset{}{\underset{\displaystyle \overset{\|}{O}}{C}}-CH_2-CH_3$

(d) $H-\overset{\displaystyle \overset{O}{\|}}{C}-OH$

(e) $CH_3-CH_2CH_2-OH$

23.8   Draw structures for molecules with the following formulas.
(a) CH_4O
(b) C_2H_6O
(c) $C_3H_6O_2$
(d) C_3H_8O

23.9   Name each of the following compounds.

(a)

(b) $CH_3\overset{\displaystyle \overset{CH_3}{|}}{\underset{\displaystyle \underset{CH_3}{|}}{C}}CH_2CH_2CH_2\overset{\displaystyle \overset{OH}{|}}{C}HCH_3$

(c) $Cl\overset{}{\underset{\displaystyle \underset{CH_2CH_3}{|}}{C}}HCH_2CH_2\overset{\displaystyle \overset{O}{\|}}{C}H$

23.10   Name each of the following compounds.

(a) $\underset{\displaystyle NH_2}{}$

(c)

(b) $\overset{\displaystyle O}{\underset{\displaystyle NH_2}{}}$

23.11   Give the name of the alkane represented by the model shown.

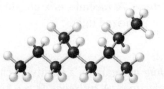

23.12   The molecular formula corresponding to the model is $C_5H_9ClO_2$. What is the name of the compound?

23.13   Write structural formulas for each of the following based on their systematic names. The names quoted in parentheses are so-called common names. Common names are not systematic and are more difficult, sometimes impossible, to connect with a unique structure.
(a) 2,2,4-trimethylpentane ("isooctane")
(b) 3-methyl-1-butanol ("isoamyl alcohol")
(c) hexanamide ("caproamide")
(d) 2,2,2-trichloroethanal ("chloral")

23.14   Write structural formulas for each of the following:
(a) 3,3-dimethyl-2-butanone ("pinacolone")
(b) 3-hydroxybutanal ("acetaldol")
(c) ethyl pentanoate ("ethyl valerate")
(d) 6-methyl-2-heptanamine ("isooctylamine")

23.15   Classify the oxygen-containing groups in the plant hormone abscisic acid.

23.16   Identify the functional groups in the antipsychotic drug haloperidol.

23.17   PABA was the active UV-absorbing compound in earlier versions of sunblock creams. What functional groups are present in PABA?

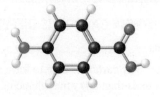

23.18   Lidocaine ($C_{14}H_{22}N_2O$) is a widely used local anesthetic. Classify its nitrogen-containing functional groups.

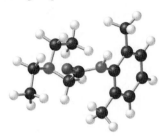

SECTION 23.3: REPRESENTING ORGANIC MOLECULES

Conceptual Problems

23.19   Write structural formulas for the following organic compounds: (a) 3-methylhexane, (b) 2,3-dimethylpentane, (c) 2-bromo-4-phenylpentane, (d) 3,4,5-trimethyloctane.

23.20   Write structural formulas for the following compounds: (a) 1,1,3-trichloro-2-propanol, (b) 3-methyl-3-pentanamine, (c) 3-bromo-1-chloro-2-butanone, (d) propyl-4-bromobutanoate.

23.21   (a) Convert $CH_3(CH_2)_4C(O)CH_2CO_2H$ to a Kekulé structure and to a bond-line structure.
(b) Convert the following to a condensed structure and to a bond-line structure.

$$CH_3CH_2CHCH_2COCCH_3$$

(with O double bond above the second C from right, CH₃CH₂ below the CH, and CH₃ below and CH₃ below the final C)

(c) Convert the following to a condensed structure and to a Kekulé structure.

(isobutyl-N(H)-isopropyl amine structure)

23.22   (a) Convert $(CH_3)_2C{=}CHCO_2H$ to a Kekulé structure and to a bond-line structure.
(b) Convert the following to a condensed structure and to a bond-line structure.

$$CH_3CH_2NCH_2CH_3$$
$$\quad\quad\quad | $$
$$\quad\quad CH_3$$

(c) Convert the bond-line structure of the general anesthetic isoflurane to a condensed structure and to a Kekulé structure.

(isoflurane structure: CF₃ with F, O, CHF₂, Cl, F)

23.23   Convert each of the molecular models to a condensed structural formula, a Kekulé structure, and a bond-line structure.

(a)  C_3H_7NO: DMF, a widely used organic solvent.

(b)  $C_6H_8O_7$: Citric acid, which contributes to the tart taste of citrus fruits.

(c) 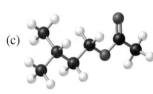 $C_7H_{14}O_2$: Commonly known as "isoamyl acetate," this ester is largely responsible for the characteristic odor of bananas.

23.24   Convert each of the molecular models to a condensed structural formula, a Kekulé structure, and a bond-line structure.

(a)  $C_3H_6O_3$: Dihydroxyacetone, an intermediate in glycolysis, the process by which glucose is converted to energy.

(b) 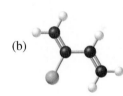 C_4H_5Cl: Chloroprene, which gives neoprene on polymerization.

(c) 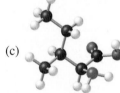 $C_6H_{13}NO_2$: Isoleucine, one of the 20 amino acids that occur in proteins.

23.25   Given a structural formula or bond-line structure, rewrite it in the other style.

(a) $CH_3CH_2CHCH_2CH_2CH_2CH_3$
$$\quad\quad\quad\quad\quad |$$
$$\quad\quad\quad\quad OH$$

(b) (bond-line structure with tert-butyl group, Br, and aldehyde O=C–H)

(c) (ring structure with $CH(CH_3)_2$, H_2C, H_2C, O, and C=O)

23.26 Given a structural formula or bond-line structure, rewrite it in the other style.
(a) $ClCH_2CH_2CH_2CH{=}CH_2$

(b) 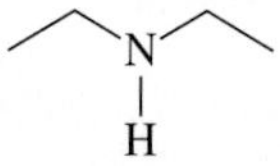 [bond-line structure of diethylamine, N–H]

(c) 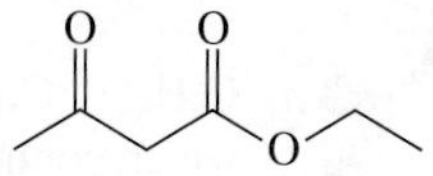 [bond-line structure with two C=O groups and an O-ethyl ester]

23.27 Using the curved arrows as a guide to placing the electrons, write a resonance structure for each of the compounds shown. The resonance structure should include formal charges where appropriate.

(a) $CH_3{-}C{\equiv}N\!:$

(b) 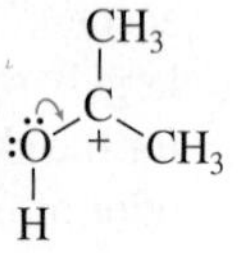 [structure with CH_3, C, :O–H, +CH_3]

(c) [structure with CH_3, C, $^-$:O, CH_2]

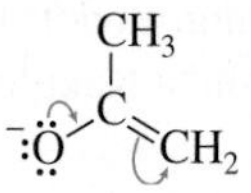

23.28 Using the curved arrows as a guide to placing the electrons, write a resonance structure for each of the compounds shown. The resonance structure should include formal charges where appropriate.

(a) 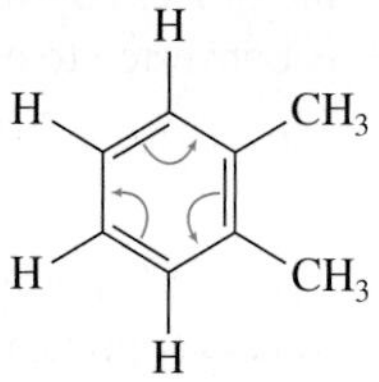 [benzene ring structure with H's and two CH_3 groups]

(b) 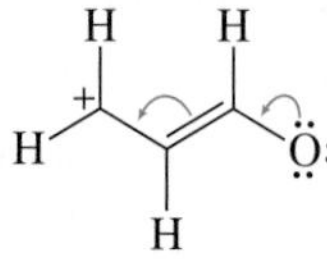 [structure H, +, H, C=C, H, :O:$^-$, H]

(c) [cyclopentanone-type structure with :O: and $^-$:O:, C, H]

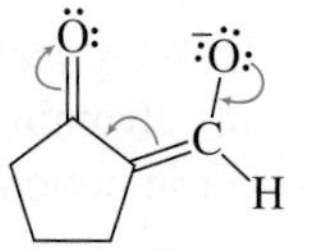

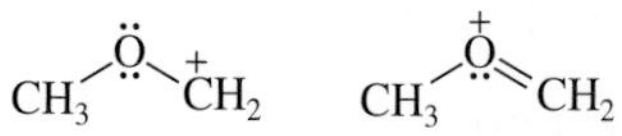

23.29 Use curved arrows to show how the resonance structure on the left can be transformed to the one on the right.

(a) 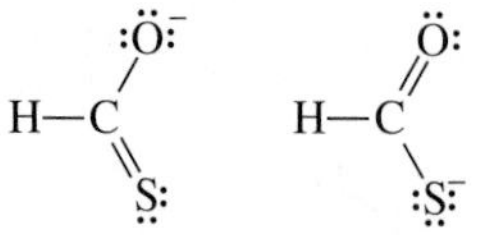 $H{-}C$ with :O:$^-$ and :S: → $H{-}C$ with O: and :S:$^-$

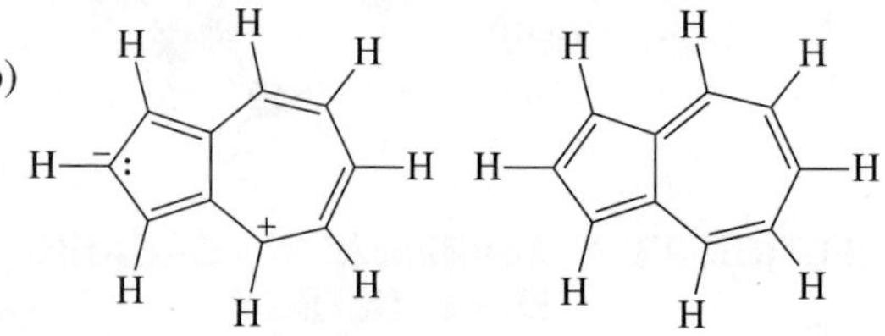

(b) CH_3, :O:, $^+CH_2$ → CH_3, $^+$O:, CH_2

(c) [benzene-type ring H–, $^+CH_2$ → H–, +, =CH_2]

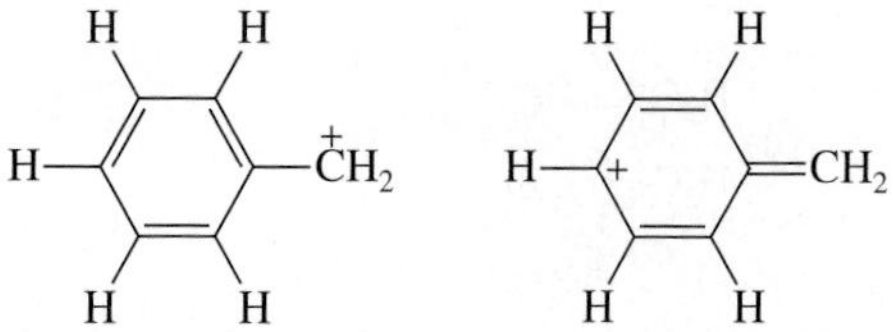

23.30 Use curved arrows to show how the resonance structure on the left can be transformed to the one on the right.

(a) 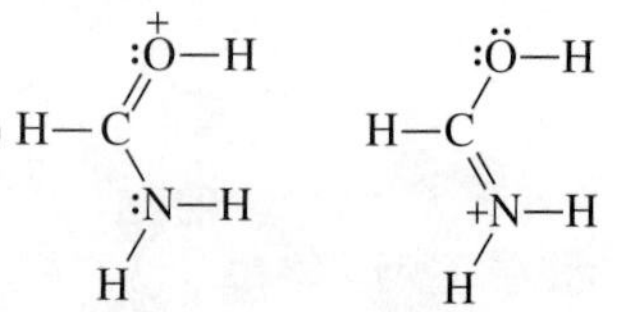 $H{-}C$ with $^+$:O–H, :N–H, H → $H{-}C$ with :O–H, $^+$N–H, H

(b) [azulene-type bicyclic structures with H's, H–:, +, and H–]

(c) [two bicyclic aromatic structures with H's]

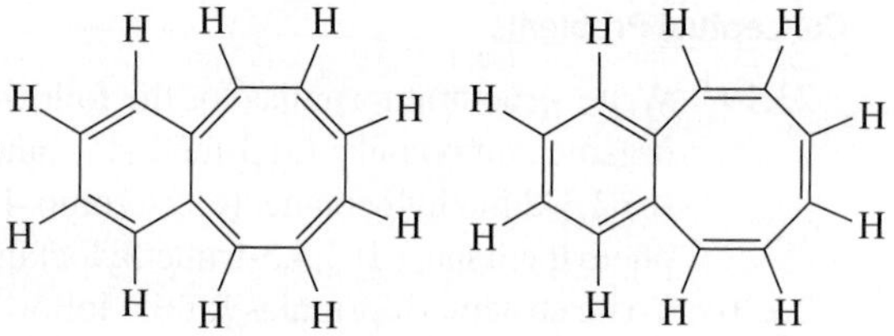

SECTION 23.4: ISOMERISM

Review Questions

23.31 Alkenes exhibit geometrical isomerism because rotation about the C=C bond is restricted. Explain.

23.32 Why is it that alkanes and alkynes, unlike alkenes, have no geometrical isomers?

23.33 Define the term *chiral*. What are enantiomers?

23.34 What factor determines whether a carbon atom in a compound is chiral?

23.35 Fill in the blanks in the given paragraph with the most appropriate term from the following: chiral, *cis*, constitutional isomers, enantiomers, resonance structures, stereoisomers, *trans*.

Isomers are different compounds that have the *same molecular formula*. Isomers that have their atoms connected in a different order (branched versus unbranched chain, for example) or a different sequence of bond types (C=CCC versus CC=CC, for example) are termed _____.
Isomers with the same order of connections and sequence of bond types, but which differ in the spatial arrangement of the atoms are called _____. This is often seen in compounds where substituents may be on the same or opposite sides of a carbon-carbon double bond. Substituents on the same side are described as _____; those on opposite sides are _____. A different kind of isomerism characterizes a _____ molecule, that is, a molecule with a structure that allows for two nonsuperimposable mirror-image forms. Two nonsuperimposable mirror images are _____ of the other.

Conceptual Problems

23.36 Write structural formulas for all the C_5H_{10} alkenes, and identify the relationship (constitutional isomer or stereoisomer) of each one to the others. Are any chiral?

23.37 Draw all possible structural isomers for the following alkane: C_7H_{16}.

23.38 Draw all possible isomers for the molecule C_4H_8.

23.39 Draw all possible isomers for the molecule C_3H_5Br.

23.40 Which of the following amino acids are chiral: (a) $CH_3CH(NH_2)COOH$, (b) $CH_2(NH_2)COOH$, (c) $CH_2(OH)CH(NH_2)COOH$?

23.41 Draw all the possible structural isomers for the molecule having the formula C_7H_7Cl. All isomers contain one benzene ring.

23.42 Draw all the structural isomers of compounds with the formula $C_4H_8Cl_2$. Indicate which isomers are chiral, and give them systematic names.

23.43 Indicate the asymmetric carbon atoms in the following compounds.

(a) $CH_3-CH_2-\overset{\overset{\displaystyle CH_3}{|}}{CH}-\overset{\overset{\displaystyle O}{\|}}{\underset{\underset{\displaystyle NH_2}{|}}{CH}}-C-NH_2$

(b)
```
       H
      /
  H  / Br  H
   \/      |
   /\      |
  H        |
     H     Br
```

23.44 Suppose benzene contained three distinct single bonds and three distinct double bonds. How many different isomers would there be for dichlorobenzene ($C_6H_4Cl_2$)? Draw all your proposed structures.

23.45 Write the structural formula of an aldehyde that is a structural isomer of acetone.

23.46 How many asymmetric carbon atoms are present in each of the following compounds?

(a) $H-\overset{\overset{\displaystyle H}{|}}{\underset{\underset{\displaystyle H}{|}}{C}}-\overset{\overset{\displaystyle H}{|}}{\underset{\underset{\displaystyle Cl}{|}}{C}}-\overset{\overset{\displaystyle H}{|}}{\underset{\underset{\displaystyle H}{|}}{C}}-Cl$

(b) $H_3C-\overset{\overset{\displaystyle OH}{|}}{\underset{\underset{\displaystyle H}{|}}{C}}-\overset{\overset{\displaystyle CH_3}{|}}{\underset{\underset{\displaystyle H}{|}}{C}}-CH_2OH$

(c)
```
            CH2OH
             |
        H    C---O    OH
         \   |H       /
          C     OH  H C
          |  OH    H  |
        HO C-------C  H
             |     |
             H     OH
```

23.47 Draw structures of each of the compounds shown, using wedges and dashes to show stereochemistry.

(a)  (C_4H_9Br)

(b) 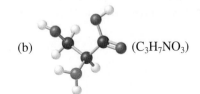 ($C_3H_7NO_3$)

(c) 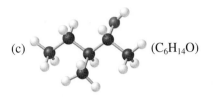 ($C_6H_{14}O$)

(d) 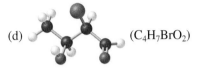 ($C_4H_7BrO_2$)

23.48 Write the structural formulas of the alcohols with the formula $C_6H_{14}O$ and indicate those that are chiral. Show only the C atoms and the $-OH$ groups.

SECTION 23.5: ORGANIC REACTIONS

Review Questions

23.49 What property distinguishes an electrophile from a nucleophile? Of the two, which would you expect to be more reactive toward a cation? Toward an anion? Give a specific example of an electrophile and a nucleophile.

23.50 Classify each of the following according to whether you think it reacts as an electrophile or a nucleophile.

(a) $CH_3-\overset{\overset{\displaystyle CH_3}{|}}{\underset{\underset{\displaystyle CH_3}{|}}{C}}-\ddot{\overset{..}{O}}{:}^{\bar{}}$

(b) $CH_3CH_2\ddot{N}H_2$

(c) $\overset{\overset{\displaystyle CH_3}{|}}{\underset{\underset{\displaystyle H_3C \quad {}^{+} \quad CH_3}{}}{C}}$

(d) $CH_3-C\equiv\overset{+}{\overset{..}{O}}{:}$

23.51 (a) The compound 2-bromopropane [$(CH_3)_2CHBr$] can undergo both substitution and elimination when treated with $CH_3CH_2O^-$, which is a strong base. Predict the organic product in each case, and write a separate chemical equation for each reaction. (b) The compound 1,2-dibromoethane ($BrCH_2CH_2Br$) was formerly used in large amounts as an agricultural chemical. Write a chemical equation showing how this compound could be prepared from ethylene by an addition reaction. (c) Certain reactions of aldehydes and ketones begin with isomerization of the aldehyde or ketone to an *enol* isomer. Enols contain an $-OH$ group attached to a carbon-carbon double bond. Write a chemical equation for the isomerization of acetone [$(CH_3)_2C=O$] to its enol isomer.

23.52 Classify the following reactions according to whether they are addition, substitution, elimination, or isomerization.

(a)
```
                O  O
                ‖  ‖
  ═══\/\OPOPOH      Enzyme      \/\═OPOPOH
              |  |      ──────►           |  |
              HO OH                       HO OH
   /                                  /
```

(b) $CH_3CH=CH_2 + Cl_2 \xrightarrow{heat} ClCH_2CH=CH_2 + HCl$

(c) $CH_3CH=CH_2 + Cl_2 \longrightarrow CH_3\underset{\underset{Cl}{|}}{C}HCH_2Cl$

(d) $CH_3\underset{\underset{Cl}{|}}{C}HCH_2Cl + NaSH \longrightarrow CH_3\underset{\underset{SH}{|}}{C}HCH_2SH + NaCl$

(e) $=O + H-C\equiv N \longrightarrow$

(f) $\xrightarrow[heat]{H_2SO_4}$ $-C\equiv N + H_2O$

23.53 (a) Define *carbocation*.
(b) Which of the following are carbocations?

$$CH_3CH_2-\overset{H}{\underset{H}{\overset{|}{\underset{|}{O:}}}}{}^+ \qquad CH_3-\overset{+}{\underset{..}{O}}-CH_2 \qquad H_2\overset{+}{C}-CH=CH_2$$

(i)   (ii)   (iii)

$$CH_3CH_2\overset{+}{C}H_2 \qquad CH_3-\overset{\overset{CH_3}{|}}{\underset{\underset{CH_3}{|}}{\overset{+}{N}}}-CH_3$$

(iv)   (v)

Conceptual Problems

23.54 Halogenated hydrocarbons are biodegraded in the natural environment by reactions catalyzed by dehalogenase enzymes. The reaction that takes place with 1,2-dichloroethane begins with nucleophilic substitution involving a carboxylate site of the enzyme.

$$Enzyme-\overset{\overset{\ddot{O}:}{||}}{C}_{\overset{|}{\ddot{O}:^-}} + :\ddot{C}l-CH_2CH_2-\ddot{C}l: \longrightarrow$$

$$Enzyme-\overset{\overset{\ddot{O}:}{||}}{C}_{\overset{|}{:\ddot{O}-CH_2CH_2-\ddot{C}l:}} + :\ddot{C}l^-$$

Expand this equation by adding curved arrows to show the movement of electrons.

23.55 A common reaction in carbohydrate biochemistry is the conversion of an aldose to a ketose. The glucose to fructose isomerization is a specific example; the equation illustrates the general case.

The first stage in the reaction is shown in the following equation. Enzymes facilitate the

reaction, but for simplicity the overall change can be approximated with water molecules. Use curved arrows to show the flow of electrons in the equation.

23.56 (a) Benzene reacts with *tert*-butyl cation $[(CH_3)_3C^+]$ by a two-step electrophilic aromatic substitution mechanism to yield *tert*-butylbenzene $[C_6H_5C(CH_3)_3]$. Write a chemical equation for each step in the mechanism and use curved arrows to track electron flow.
(b) Electrophilic addition of hydrogen chloride to styrene gives the product shown. Write the mechanism for this reaction, including curved arrows.

$-CH=CH_2 + HCl \longrightarrow$ $-\underset{\underset{Cl}{|}}{C}HCH_3$

23.57 (a) Acetylide ion undergoes nucleophilic addition to aldehydes and ketones to give the species shown. Subsequent addition of water yields an acetylenic alcohol. Add curved arrows to the equations to show how the reaction occurs.

$$CH_3CH_2\overset{\overset{\ddot{O}}{||}}{C}H + {}^-:C\equiv CH \longrightarrow CH_3CH_2\underset{\underset{:\ddot{O}:^-}{|}}{C}HC\equiv CH \xrightarrow{H-\overset{H}{\underset{|}{\ddot{O}:}}}$$

$$CH_3CH_2\underset{\underset{:\ddot{O}H}{|}}{C}HC\equiv CH + {}^-:\ddot{O}-H$$

(b) A nucleophilic site on a molecule can substitute for a halogen elsewhere in the same molecule to form a ring. Use curved arrows to show how the following *cyclization* is related to a conventional nucleophilic substitution.

$$^-:\ddot{S}-CH_2CH_2CH_2CH_2-\ddot{B}r: \longrightarrow$$ $+ :\ddot{B}r^-$

23.58 Complete the following reaction by including essential unshared electron pairs and formal charges. Use curved arrows to show electron flow.

$$(CH_3)_3B + $$ $\longrightarrow (CH_3)_3B-O$

23.59 The product of the reaction of chlorobenzene with NaNH$_2$ is a very reactive species called *benzyne*. Add necessary electron pairs and formal charges to the net ionic equation, and use curved arrows to show how benzyne is formed. To what general type of reaction does this belong?

23.60 Consider the following reactions of butanal.

(i) $CH_3CH_2CH_2CH \xrightarrow[H_2O]{NaBH_4} CH_3CH_2CH_2CH_2OH$ (with O double bonded to the CH)

(ii) $CH_3CH_2CH_2CH \xrightarrow[H_2O]{H_2CrO_4} CH_3CH_2CH_2COH$ (with O double bonded)

In which reaction is butanal oxidized? In which reaction is it reduced?

23.61 Esters can be prepared by the acid-catalyzed condensation of a carboxylic acid and an alcohol.

$CH_3CH_2COH + CH_3OH \longrightarrow CH_3CH_2COCH_3 + H_2O$

Is this an oxidation-reduction reaction? If so, identify the species being oxidized and the species being reduced.

23.62 A compound has the empirical formula $C_5H_{12}O$. Upon controlled oxidation, it is converted into a compound of empirical formula $C_5H_{10}O$, which behaves as a ketone. Draw possible structures for the original compound and the final compound.

23.63 Isopropanol is prepared by reacting propylene (CH_3CHCH_2) with sulfuric acid, followed by treatment with water. (a) Show the sequence of steps leading to the product. What is the role of sulfuric acid? (b) Draw the structure of an alcohol that is an isomer of isopropanol. (c) Is isopropanol a chiral molecule?

23.64 Identify the functional groups in thalidomide.

23.65 Write molecular formulas for thalidomide, lenalidomide, and CC-4047.

SECTION 23.6: ORGANIC POLYMERS

Review Questions

23.66 Define the following terms: *monomer, polymer, copolymer.*

23.67 Name 10 objects that contain synthetic organic polymers.

23.68 Calculate the molar mass of a particular polyethylene sample, $+CH_2-CH_2+_n$, where $n = 4600$.

23.69 Describe the two major mechanisms of organic polymer synthesis.

23.70 What are the steps involved in polymer formation by chain reaction?

23.71 Polysaccharides, proteins, and nucleic acids comprise the three main classes of biopolymers. Compare and contrast them with respect to structure and function. What are the building block units for each? What are the key functional groups involved in linking the units together? In which biopolymer is there the greatest variety of building block structure? In which is there the least?

Conceptual Problems

23.72 Teflon is formed by a radical addition reaction involving the monomer tetrafluoroethylene. Show the mechanism for this reaction.

23.73 Vinyl chloride ($H_2C=CHCl$) undergoes copolymerization with 1,1-dichloroethylene ($H_2C=CCl_2$) to form a polymer commercially known as Saran. Draw the structure of the polymer, showing the repeating monomer units.

23.74 Deduce plausible monomers for polymers with the following repeating units.
(a) $+CH_2-CF_2+_n$

(b)

23.75 Deduce plausible monomers for polymers with the following repeating units.
(a) $+CH_2-CH=CH-CH_2+_n$
(b) $+CO+CH_2+_6NH+_n$

23.76 Draw the structures of the dipeptides that can be formed from the reaction between the amino acids glycine and alanine.

23.77 Draw the structures of the dipeptides that can be formed from the reaction between the amino acids glycine and lysine.

23.78 From among the given nucleotides, identify those that occur naturally in RNA and those that occur in DNA. Do any not occur in either RNA or DNA?

(a)

(b)

(c)

(d)

ADDITIONAL PROBLEMS

23.79 Write structural formulas for all the constitutionally isomeric C_4H_9 alkyl groups. Check your answers with Table 23.1, and note the names of these groups.

23.80 *Ethers* are compounds (excluding esters) that contain the C—O—C functional group. An acceptable way to name them is to list the two groups attached to oxygen in alphabetical order as separate words, followed by the word *ether*. If the two groups are the same, add the prefix *di-* to the name of the alkyl group. Thus, $CH_3OCH_2CH_3$ is "ethyl methyl ether" and $CH_3CH_2OCH_2CH_3$ is "diethyl ether."

Write structural formulas and provide names for all the constitutionally isomeric ethers in which

only the C_4H_9 alkyl groups from Problem 23.79 are attached to oxygen. Which of these ethers is potentially chiral?

23.81 The α-amino acids found in proteins are based on the structural formula

$$RCHCO_2H$$
$$|$$
$$NH_2$$

From the C_4H_9 alkyl groups in Problem 23.79, find the ones that correspond to "R" among the amino acids listed in Figure 23.14. Match the name of the amino acid with the name of the group.

23.82 *Alkynes* are hydrocarbons that contain a carbon-carbon triple bond.
(a) Write structural formulas for all the isomeric alkynes of molecular formula C_5H_8.
(b) Are any of the alkynes chiral?
(c) Are any of the alkynes stereoisomeric?

23.83 Among the many alkenes of molecular formula C_6H_{12}, only one is chiral.
(a) Write a structural formula for this alkene.
(b) Place substituents on the tetrahedral carbons so as to represent the two enantiomers of this alkene.

23.84 (a) Use VSEPR to predict the geometry at the carbon shown in bold in the carbocation, radical, and anion derived from the *tert*-butyl group. Is the arrangement of bonds to this carbon linear, tetrahedral, trigonal planar, or trigonal pyramidal?

$^+\mathbf{C}(CH_3)_3$	$\cdot\mathbf{C}(CH_3)_3$	$:\bar{\mathbf{C}}(CH_3)_3$
tert-Butyl	Cation	Radical	Anion

(b) Which of these species is the most nucleophilic? Which is the most electrophilic?
(c) Predict which species reacts with water to give $HC(CH_3)_3$.
(d) Write a chemical equation for the reaction in part (c), and use curved arrows to show the flow of electrons.

23.85 In each of the following pairs, specify whether the two structural formulas represent constitutional isomers, *cis/trans*-stereoisomers, enantiomers, resonance structures, or are simply a different representation of the same structure.

(a)

(b)

(c)

(d)

23.86 In each of the following pairs, specify whether the two structural formulas represent constitutional isomers, *cis/trans*-stereoisomers, enantiomers, resonance structures, or are simply a different representation of the same structure.

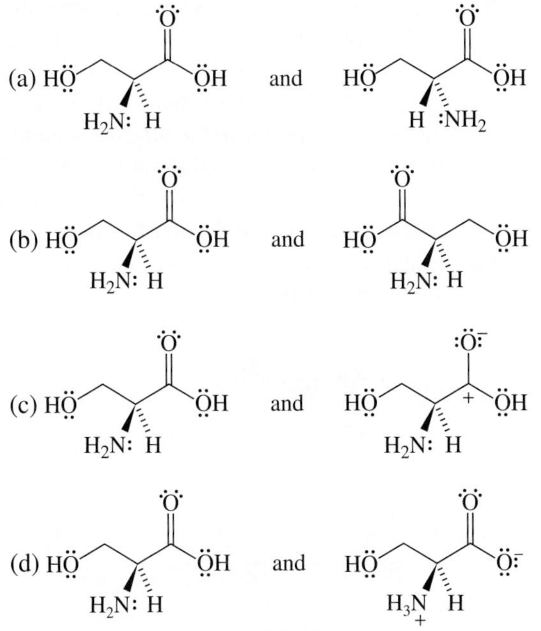

(a) and

(b) and

(c) and

(d) and

23.87 The electrophile in the following aromatic substitution is $(CH_3)_2CH^+$, and the mechanism involves the intermediate shown. Write two other resonance structures for this intermediate.

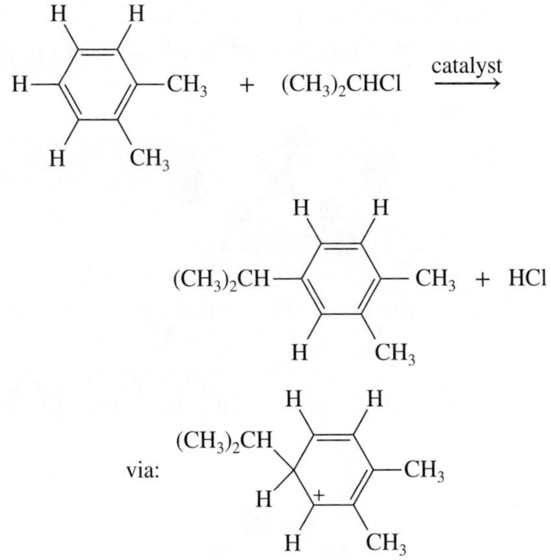

23.88 Two isomeric alkenes are formed in the dehydration of 2-methyl-2-butanol. What are these two alkenes? Are they constitutional isomers or stereoisomers?

23.89 Electrophilic addition of HCl to *cis*-2-butene gave 2-chlorobutane, which was determined *not* to be optically active when examined with a polarimeter.

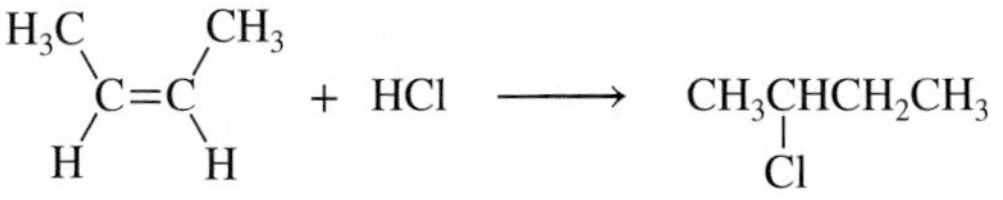

Which of the following is the better explanation for the lack of optical activity in the 2-chlorobutane formed in this reaction?
(a) 2-Chlorobutane is not chiral.
(b) Two enantiomers of 2-chlorobutane were formed in equal amounts.

23.90 Reactions such as the following have been used in carbohydrate synthesis since the nineteenth century.

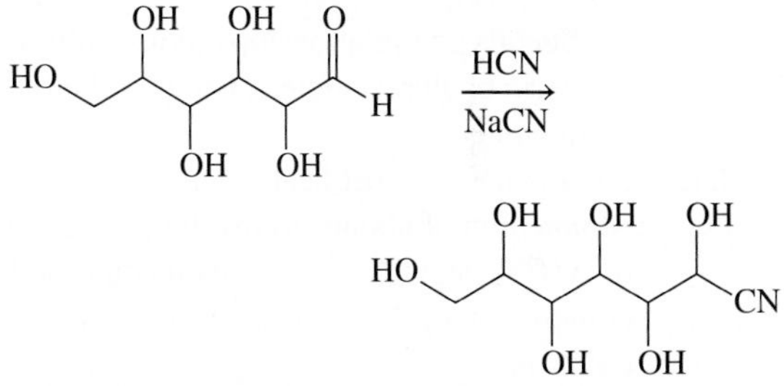

Classify this procedure according to reaction type.
(a) electrophilic addition
(b) electrophilic substitution
(c) nucleophilic addition
(d) nucleophilic substitution

23.91 Excluding compounds that have rings, there are three hydrocarbons that have the molecular formula C_4H_6. Write their structural formulas, and specify the hybridization of each carbon in these isomers.

23.92 Give the structures of the two tertiary amines that are isomers of $CH_3CH_2CH_2CH_2CH_2NH_2$.

23.93 In which class of compounds, esters or amides, do you think electron donation into the carbonyl group is more pronounced? Explain.

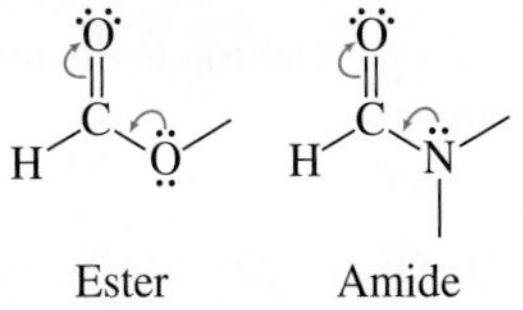

23.94 (a) A plane of symmetry in a molecule is a plane passing through the middle of the molecule that bisects the molecule into two mirror-image halves. If a molecule has a plane of symmetry, it cannot be chiral and cannot be optically active. Which of the following have at least one plane of symmetry? Do any have more than one?

Cl
Cl

1,1-Dichlorocyclopropane

Cl   Cl

H   H

cis-1,2-Dichlorocyclopropane

Cl   H

H   Cl

trans-1,2-Dichlorocyclopropane

(b) Specify the relationships (constitutionally isomeric or stereoisomeric) among these compounds.

23.95 All alkanes give off heat when burned in air. Such *combustion* of alkanes is exothermic, and the sign of $\Delta H°$ is negative. The general equation for the combustion of the alkanes of molecular formula C_5H_{12} is

$$C_5H_{12} + 8O_2 \longrightarrow 5CO_2 + 6H_2O$$

The values of $\Delta H°$ for the combustion of the three pentane isomers are

$CH_3CH_2CH_2CH_2CH_3 = -3536$ kJ/mol
$(CH_3)_2CHCH_2CH_3 = -3529$ kJ/mol
$(CH_3)_4C = -3515$ kJ/mol

(a) What do these data tell you about the effect of chain branching on the relative potential energies and stabilities of these isomers?
(b) Assume this effect is general to predict which one of the 18 C_8H_{18} isomers should be the most stable.

23.96 Carbon dioxide reacts with sodium hydroxide according to the following equation.

$$CO_2 + 2NaOH \longrightarrow Na_2CO_3 + H_2O$$

The overall reaction is the result of two separate reactions.

Reaction I: H—Ö: + :O=C=O: ⟶ H—O—C—O:⁻ (with =O above C)

Reaction II:

H—Ö:⁻ + H—O—C—O:⁻ (with =O) ⟶ :O—C—O:⁻ (with =O) + H—O—H

(a) Use curved arrows to track the flow of electrons in reaction I.
(b) Use curved arrows to track the flow of electrons in reaction II.

(c) Classify reaction I as electrophilic addition, nucleophilic addition, electrophilic substitution, or acid-base.
(d) Classify reaction II according to the choices in part (c).

23.97 The combustion of 20.63 mg of compound Y, which contains only C, H, and O, with excess oxygen gave 57.94 mg of CO_2 and 11.85 mg of H_2O. (a) Calculate how many milligrams of C, H, and O were present in the original sample of Y. (b) Derive the empirical formula of Y. (c) Suggest a plausible structure for Y if the empirical formula is the same as the molecular formula.

23.98 Match each molecular model with the correct line-wedge-dash structure.

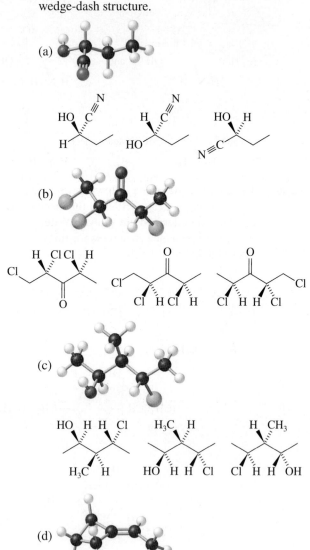

(a)

(b)

(c)

(d)

23.99 Kevlar is a copolymer used in bulletproof vests. It is formed in a condensation reaction between the following two monomers.

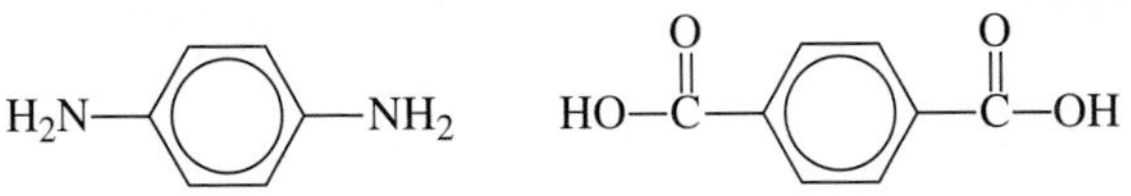

Sketch a portion of the polymer chain showing several monomer units. Write the overall equation for the condensation reaction.

23.100 Describe the formation of polystyrene.

23.101 The amino acid glycine can be condensed to form a polymer called polyglycine. Draw the repeating monomer unit.

23.102 Nylon can be destroyed easily by strong acids. Explain the chemical basis for the destruction. (*Hint:* The products are the starting materials of the polymerization reaction.)

23.103 How many different tripeptides can be formed by lysine and alanine?

23.104 Nylon was designed to be a synthetic silk. (a) The average molar mass of a batch of nylon 66 is 12,000 g/mol. How many monomer units are there in this sample? (b) Which part of nylon's structure is similar to a polypeptide's structure? (c) How many different tripeptides (made up of three amino acids) can be formed from the amino acids alanine (Ala), glycine (Gly), and serine (Ser), which account for most of the amino acids in silk?

23.105 Thalidomide is converted to the drug CC-4047 by substitution of an amino group for one of the H atoms on the aromatic portion of the molecule. Using curved arrows, draw the mechanism for this reaction and all the resonance structures for the carbocation intermediate.

Answers to In-Chapter Materials

PRACTICE PROBLEMS

23.1A (a) 3-ethylpentane, (b) 2-bromohexane, (c) 2-chlorobutane.
23.1B (a) $CH_3CH_2CH_2CHCH_2CH_2CH_2CH_3$, (b) $CH_3CHCH_2CH_2CH_3$,

(c) $CH_3CH_2CHCH_2CH_2CH_2CH_2CH_2CH_3$. **23.2A** (a) ester,

(b) carboxy group; ester, (c) 2 hydroxy groups, 2 carboxy groups.
23.2B (a) ketone, (b) 3° amide, (c) ester.

23.3A CH_2BrCH_2COOH. **23.3B**

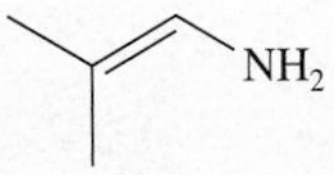

23.4A H—C≡O:

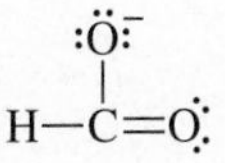

23.4B

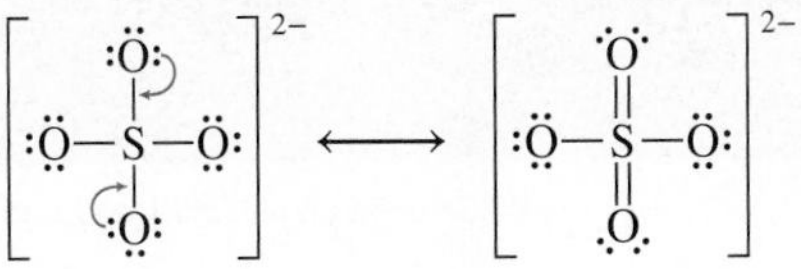

23.5A (a)

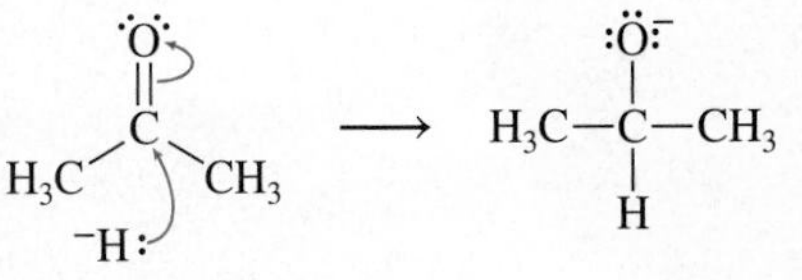

(b)

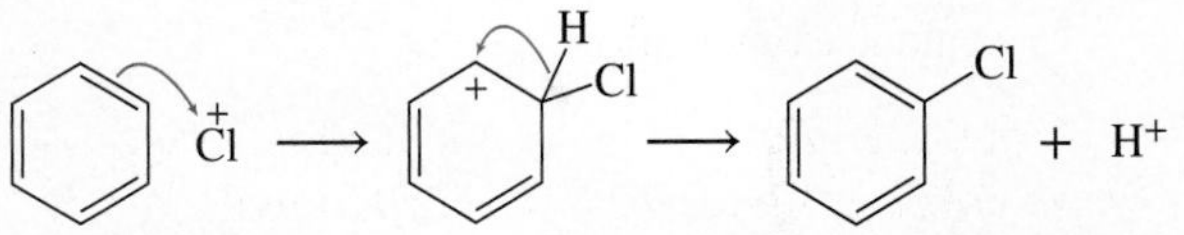

23.5B

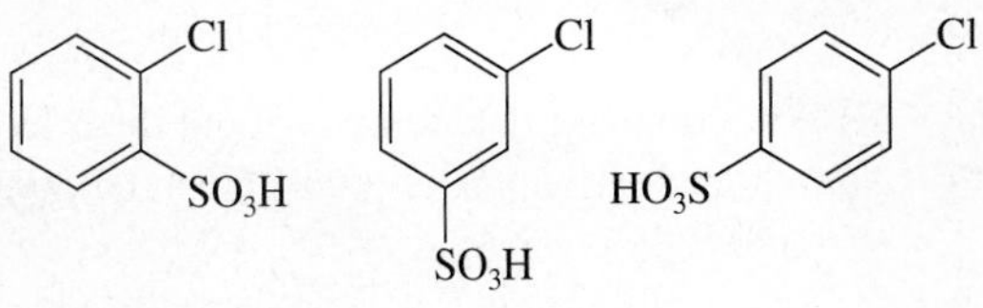

SECTION REVIEW

23.2.1 c. **23.2.2** b. **23.2.3** b. **23.2.4** e. **23.2.5** b,c. **23.2.6** a,b. **23.3.1** b. **23.3.2** d. **23.3.3** e. **23.3.4** b,d. **23.5.1** d. **23.5.2** a.

Chapter 24

Modern Materials

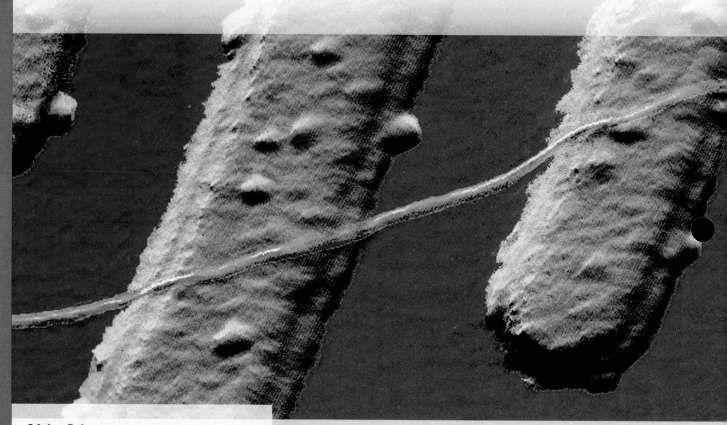

© Delft University of Technology/Science Source

MODERN ELECTRONIC devices heavily rely on semiconducting materials but heretofore existing engineering technologies have imposed a limit on how small such devices can be built. With nanotechnology, chemists and materials scientists are able to build working devices that are just *tens* of atoms wide. Shown here is a nanowire, the world's smallest electrical wire, a mere 10 atoms wide (1.5 nanometers across). Devices fabricated at this tiny scale possess properties so different from their predecessors that they may ultimately lead to an entirely new era in electronics.

Before You Begin, Review These Skills

- Functional groups [◄◄ Section 23.2]
- Organic polymers [◄◄ Section 23.6]

24.1 POLYMERS

Many biological molecules, such as DNA, starch, and proteins, have very large molecular masses. These molecules are **polymers** because they are made up of many smaller parts linked together [◄◄ Section 23.6]. The prefix *poly-* comes from a Greek root meaning "many," and the small molecules that make up the individual building blocks of polymers are called **monomers.** Many natural polymers, such as silk and rubber, now have synthetic analogs that are manufactured by the chemical industry.

Polymers are used in many medical applications, such as flexible wound dressings, surgical implants, and prosthetics. Some of these polymers are **thermoplastic,** which means that they can be melted and reshaped, or heated and bent. Others are **thermosetting,** which means that their shape is determined as part of the chemical process that formed the polymer. Thermosetting polymers cannot be reshaped easily and are not easily recycled, whereas thermoplastic polymers can be melted down and cast into new shapes in different products.

Bottles that can be recycled are made of thermoplastic polymers.

© Tracy Montana/PhotoLink/Getty

Addition Polymers

The simplest type of polymerization reaction involves the bonding of monomer molecules by movement of electrons from a multiple bond into new single bonds between molecules. This type of polymerization is called **addition polymerization** [◄◄ Section 23.6]. A molecule with a carbon-carbon double bond is called an *alkene,* and a molecule with a carbon-carbon triple bond is called an *alkyne.* The synthesis of polyethylene from ethylene (C_2H_4) illustrates an addition polymerization. The arrows with half-pointed heads show movement of a single electron. The process of "splitting open" the double bond is initiated by a molecule with a single $O-O$ bond that can be broken in half by heating. The free radical initiator molecule attaches to one of the ethylene carbon atoms and breaks the pi bond between the two carbon atoms in ethylene. Figure 24.1 shows the scheme of addition polymerization with respect to the double bonds opening up to form single bonds between monomer units to form polyethylene from ethylene.

The structure of polyethylene can be represented as $+CH_2-CH_2+_n$, where n is the number of ethylene monomer molecules that reacted (which could be hundreds, thousands, or more). The end of the chain contains a carbon atom with an unpaired electron (i.e., $-CH_2CH_2\cdot$), though, so it must be closed by bonding with another atom. This might be a hydrogen atom (e.g., $-CH_2CH_2-H$), but it might also be part of the initiator molecule (e.g., $-CH_2CH_2-OR$). Although the residue

Student Annotation: The systematic name of C_2H_4 is *ethene,* but the common name *ethylene* is still widely used.

Student Annotation: Recall that species with an unpaired electron are called *free radicals* [◄◄ Section 6.6].

Video 24.1
Organic and biochemistry—natural and synthetic polymers.

Figure 24.1 Addition polymerization to form polyethylene from ethylene. Each curved, single-headed arrow represents the movement of one electron.

Figure 24.2 Linear polyethylene is used to make plastic milk bottles. Plastic bags and wrap like Glad Cling Wrap are made from branched polyethylene.
© David A. Tietz/Editorial Image, LLC

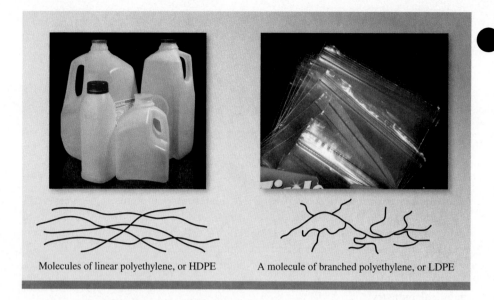

Molecules of linear polyethylene, or HDPE   A molecule of branched polyethylene, or LDPE

of the initiator molecule has very different chemical properties than the main polymer chain, it will not significantly affect the polymer's properties because the polyethylene chain is so long.

Although the structure of a polyethylene chain is drawn in Figure 24.1 as a straight line, the bond angles around a central carbon atom with four bonds are approximately 109.5° each [◄◄ Section 7.1]. Thus, the "straight" chain actually zigzags. The polymer chain bends, too, because there is free rotation about carbon-carbon single bonds. These molecular-level features manifest themselves in such macroscopic properties as flexibility.

It is possible, depending on the reaction conditions, for branches to form off the main polyethylene chain during polymerization. Polymer chains with many branches cannot pack together as efficiently as "straight" or unbranched chains, so branched-chain polyethylene is more flexible and lower in density than straight-chain polyethylene. The difference between branched and unbranched polyethylene is shown schematically in Figure 24.2.

Polyethylene consisting primarily of unbranched chains is known as high-density polyethylene (HDPE), whereas polyethylene that consists primarily of branched chains is known as low-density polyethylene (LDPE). HDPE is used in plastic bottles and containers where it is important to maintain shape, whereas LDPE is used in plastic food bags and wraps.

Natural rubber formed from latex is an addition polymer. The monomer unit is isoprene, a molecule with two C=C double bonds. In the reaction that forms *polyisoprene,* a pair of electrons moves from one double bond to form a new double bond between the middle two carbon atoms, and the other double bond opens up to form a carbon-carbon single bond to the next monomer unit.

$$H_2C \qquad CH_2$$
$$C — C$$
$$H_3C \qquad H$$

Isoprene

$$\cdots H_2C \qquad CH_2 \cdots\cdots H_2C \qquad CH_2 \cdots\cdots H_2C \qquad CH_2 \cdots$$
$$C=C \qquad C=C \qquad C=C$$
$$H_3C \qquad H \qquad H_3C \qquad H \qquad H_3C \qquad H$$

cis-Polyisoprene

Figure 24.3 Vulcanization.

cis-Polyisoprene + Sulfur → Cross-linked polyisoprene

AP Photo/Robert E. Klein

Natural rubber is not very durable—it only retains its useful properties in a relatively narrow temperature range. To address this problem, Charles Goodyear[1] developed industrial *vulcanization* in 1839, a process in which sulfur is used to create linkages between individual polymer chains in the rubber. By heating the rubber with sulfur (S_8) under carefully controlled conditions, some of the carbon-hydrogen bonds are broken and replaced with carbon-sulfur bonds. The sulfur atoms are bonded to one or more additional sulfur atoms and then to another polymer chain through another sulfur-carbon bond (Figure 24.3). These linkages, called **cross-links,** make the rubber much stronger and more rigid than natural polyisoprene. Rubber is an **elastomer,** a material that can stretch or bend and then return to its original shape as long as the limits of its elasticity are not exceeded. The cross-linking of the polyisoprene chains plays an important role in maintaining rubber's elastomer properties.

Goodyear, the current supplier of tires to NASCAR teams, varies the composition of the rubber in its tires for each race depending on the expected tire wear and banking of the track. The left- and right-side tires can even be of different compositions depending on the track layout and conditions. Softer tire compounds offer more grip but wear out faster, whereas harder compounds last longer but grip less.

Many other molecules containing carbon-carbon double bonds can form addition polymers. Styrene [Figure 24.4(a)] is an analogue of ethylene in which a hydrogen atom has been replaced with a C_6H_5 (a *phenyl*) group. The polymerization of styrene forms *polystyrene* (Figure 24.5). As a foam (with air or some other gas blown into the solid), polystyrene is used in disposable cups and plates as well as Styrofoam insulation and craft materials. Vinyl chloride [Figure 24.4(b)], an analogue of ethylene in which one of the hydrogen atoms has been replaced by a chlorine atom, is another monomer that can be used to form a polymer with many common uses. Poly(vinyl chloride) (PVC) is used in pipes and fittings, some plastic wraps, and the classic "vinyl" records that were the predecessors to musical recordings on cassettes and CDs.

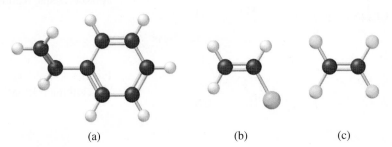

(a)   (b)   (c)

Figure 24.4 (a) Styrene, (b) vinyl chloride, and (c) tetrafluoroethylene (the monomer of Teflon).

© Comstock Images/Alamy

[1]Charles Goodyear (1800–1860), an American chemist, was the first person to realize the potential of natural rubber. His vulcanization process made the use of rubber practical, contributing greatly to the development of the automobile industry.

Polytetrafluoroethylene, more commonly known as the trademarked brand name Teflon, is formed from the addition polymerization of tetrafluoroethylene. Tetrafluoroethylene, shown in Figure 24.4(c), is an analogue of ethylene in which fluorine atoms have replaced all four of the hydrogen atoms. Teflon is a very good electrical insulator, so it is commonly used to coat wires. It is probably best known, though, as a nonstick substance used to coat bakeware, frying pans, and pots. It is also used in films that can be inserted into threaded joints between metal pipes to make it easier to unscrew the connection when necessary. Because Teflon is chemically inert, it is not possible to cross-link the chains like an elastomer. The structures of some addition polymers, including the structures of their respective monomers and what they are typically used for, are summarized in Table 24.1.

The diversity of polymers increases greatly when different monomers are used in the same polymer chain. **Copolymers** are polymers made from two or more different monomers [◄◄ Section 23.6]. Different combinations of the two or more monomers give rise to different types of copolymers. These different copolymers are summarized in Table 24.2.

Figure 24.5 Structure of polystyrene.

TABLE 24.1	Some Common Addition Polymers		
Polymer	**Structure**	**Monomer**	**Uses**
Polyethylene	$-(CH_2-CH_2)_n-$	$H_2C=CH_2$	Plastic bags and wraps, bottles, toys, Tyvek wrap
Polystyrene			Disposable cups and plates, insulation, packing materials
Poly(vinyl chloride)			Pipes and fittings, plastic wraps, records
Polyacrylonitrile			Fibers for carpeting and clothing (Orlon)
Polybutadiene			Synthetic rubber

Copolymer type	Representation*
Random	... −A−A−B−A−B−B−B−A−A−B−A−B−A−B−A−B−B−A− ...
Alternating	... −A−B−A−B−A−B−A−B−A−B−A−B−A−B−A−B−A−B− ...
Block	... −A−A−A−A−A−A−B−B−B−B−B−B−A−A−A−A−A−A−A− ...
Graft	... −A−A−A−A−A−A−A−A−A−A−A−A−A−A−A−A−A−A− ... 　　　　| 　　B−B−B−B−B−B−B−B−B ...

TABLE 24.2 Types of Copolymers

*A and B represent different monomers (e.g., ethylene and styrene).

Worked Example 24.1 lets you practice deducing the structure of an addition polymer based on the structure of its monomer.

Worked Example 24.1

Polypropylene is used in applications that range from indoor-outdoor carpeting to soda bottles to bank notes. It is produced by the addition polymerization of propylene, $H_2C=CH-CH_3$. Draw the structure of polypropylene, showing at least two repeating monomer units, and write its general formula.

Student Annotation: An Australian bank note issued in 1988 is printed on polypropylene for increased durability and security.

Strategy Addition polymers form via a free-radical reaction in which the pi electrons in the carbon-carbon double bond of a monomer molecule are used to form carbon-carbon single bonds to other monomer molecules. Draw the structural formula of propylene such that the double bond can be "opened up" to form single bonds between consecutive monomer units.

Setup The structural formula of propylene is

By drawing three adjacent propylene molecules, we can rearrange the bonds to show three repeating units.

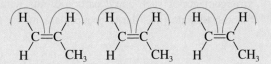

Solution The structure of polypropylene showing three monomer units is

$$
\begin{array}{c}
\quad\; H \;\; H \;\; H \;\; H \;\; H \;\; H \\
\;\;|\;\;\;\;|\;\;\;\;|\;\;\;\;|\;\;\;\;|\;\;\;\;| \\
-C-C-C-C-C-C- \\
\;\;|\;\;\;\;|\;\;\;\;|\;\;\;\;|\;\;\;\;|\;\;\;\;| \\
\; H \; CH_3 \; H \; CH_3 \; H \; CH_3
\end{array}
$$

Thus, the general formula of polypropylene is

$$
\left(\!\!
\begin{array}{c}
H \;\; H \\
|\;\;\;\; | \\
C-C \\
|\;\;\;\; | \\
H \;\; CH_3
\end{array}
\!\!\right)_{\!n}
$$

Think About It

The general formula looks like the structure of the monomer, except that the carbon-carbon double bond has been opened up to form bonds to monomer units on each side. This means the answer is reasonable. In fact, the opposite process (i.e., restoring the carbon-carbon double bond) can be used to determine the structure of the monomer used to create an addition polymer.

(Continued on next page)

Practice Problem A TTEMPT Saran Wrap, the original plastic film, is polyvinylidene chloride, which is formed by the addition polymerization of vinylidene chloride, $H_2C=CCl_2$. Draw the structure of polyvinylidene chloride, showing at least two repeating units, and write its general structure.

Practice Problem B UILD The general structure of polymethyl methacrylate, an addition polymer used in Plexiglas and Lucite, is as follows:

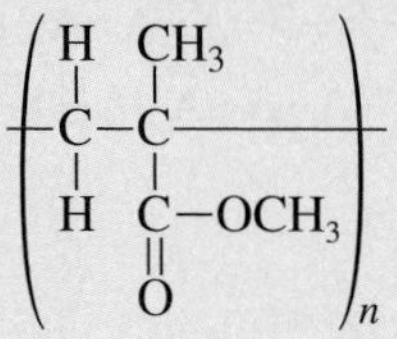

Draw the structure of the monomer used to make this polymer.

Practice Problem C ONCEPTUALIZE Which diagram best represents the movement of electrons in the formation of an addition polymer? (Electrons in carbon-carbon double bonds are color coded for identification.)

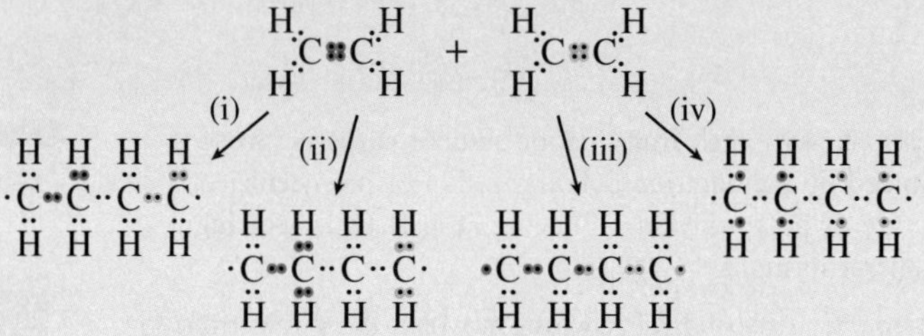

Butadiene ($CH_2=CH-CH=CH_2$) and styrene, two monomers listed in Table 24.1, can form a block copolymer called poly(styrene-butadiene-styrene) or SBS rubber [Figure 24.6(a)], in which there are alternating blocks of polybutadiene and polystyrene. They also form a graft polymer known as high-impact polystyrene (HIPS) [Figure 24.6(b)], in which chains of polybutadiene branch off of the main polystyrene chain.

Polymers containing a vinyl group ($-CH_2-CHR-$) can exist as optical isomers because the carbon atom bonded to the R group is tetrahedral and bonded to four *different* groups [◄◄ Section 23.4]. *Tacticity* is the term that describes the relative arrangements of such chiral carbon atoms within a polymer. Polymers in which all the substituents (i.e., the R groups) are in the same relative orientation (i.e., on the

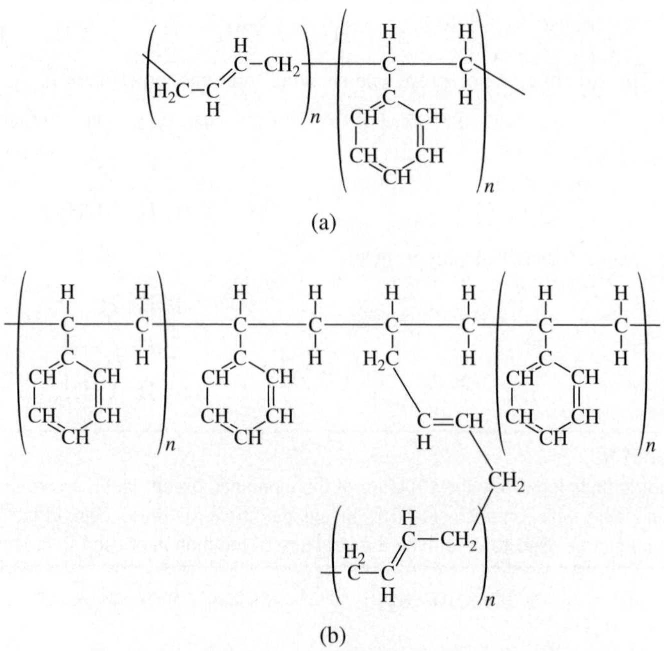

Figure 24.6 Structures of (a) SBS rubber and (b) HIPS. SBS consists of alternating blocks of polybutadiene and polystyrene, whereas HIPS consists of a polystyrene backbone with polybutadiene branches.

(a)

(b)

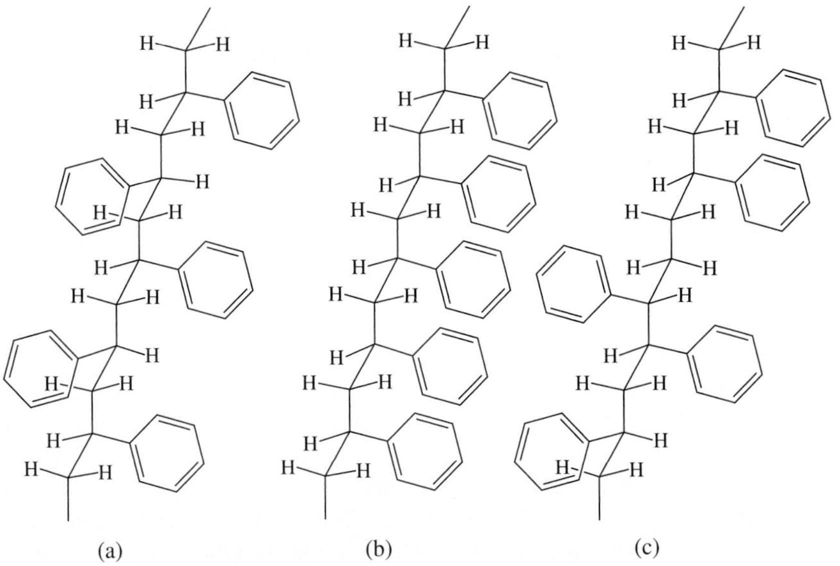

(a)   (b)   (c)

same side of the polymer chain) are said to be *isotactic.* Polymers in which the substituents alternate positions along the polymer chain are said to be *syndiotactic.* Finally, polymers in which the substituents are oriented randomly along the polymer chain are said to be *atactic.* The tacticity of polystyrene is shown in Figure 24.7. Polymers that are formed by free-radical polymerization processes, such as polystyrene and PVC, are usually atactic. Special catalysts have been developed, though, that can be used to synthesize isotactic and syndiotactic polymers.

Condensation Polymers

Condensation polymers form when monomers with two different functional groups combine, resulting in the elimination of a small molecule (usually water) [◀◀ Section 23.6]. Proteins are biological condensation polymers that can have molecular weights in the thousands or millions of grams per mole. Proteins form between amino acid monomers. Each amino acid contains an amino group ($-NH_2$) and a carboxy group ($-COOH$), and the condensation reaction occurs between the amino group of one amino acid and the carboxy group of another. (See Figure 23.13.) One water molecule is produced as each new C–N bond is formed.

The C–N bond formed between an amino group and a carboxy group is called an amide bond because $RCONR'R''$ is the general form of an amide [◀◀ Section 23.2, Table 23.2]. Polymers in which the monomers are connected by amide linkages are called *polyamides.* The amide bond between amino acids is more specifically called a peptide bond, and proteins are routinely referred to as polypeptides. Although polypeptides can form from a single amino acid (e.g., polyvaline), most proteins in organisms consist of chains of different amino acids. As a result, most proteins are random copolymers (see Table 24.2).

Nylon 66 is a synthetic condensation polymer formed from hexamethylenediamine, a molecule with two amino groups, and adipic acid, which has two carboxy groups. Water is the small molecule eliminated in this condensation reaction.

$$n\text{H}_2\text{N}\left(\text{CH}_2\right)_6\text{NH}_2 + n\text{OH}-\overset{\text{O}}{\overset{\|}{\text{C}}}\left(\text{CH}_2\right)_4\overset{\text{O}}{\overset{\|}{\text{C}}}-\text{OH} \longrightarrow \left[\text{NH}\left(\text{CH}_2\right)_6\text{NH}-\overset{\text{O}}{\overset{\|}{\text{C}}}\left(\text{CH}_2\right)_4\overset{\text{O}}{\overset{\|}{\text{C}}}\right]_n + 2n\text{H}_2\text{O}$$

Both hexamethylenediamine and adipic acid contain six carbon atoms, which is why the resulting polymer is called nylon 66. Like proteins, nylon 66 contains amide linkages between monomers, so nylon 66 is a polyamide, too. Also like proteins, nylon 66 is a copolymer, although nylon 66 is an *alternating* copolymer (Table 24.2), not a *random* copolymer.

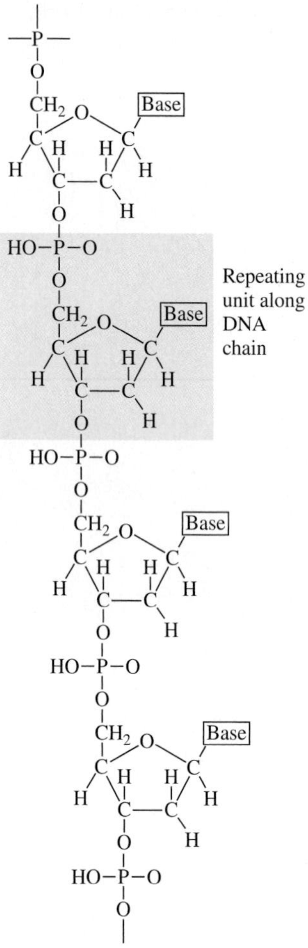

Figure 24.8 Structure of DNA.

Dacron is the trade name for the condensation copolymer formed from ethylene glycol (a diol—a molecule with two alcohol groups) and *para*-terephthalic acid (a diacid—a molecule with two carboxy groups). The condensation reaction removes the hydrogen atom from the alcohol on one monomer and the OH group from the carboxy group on a second monomer, producing water.

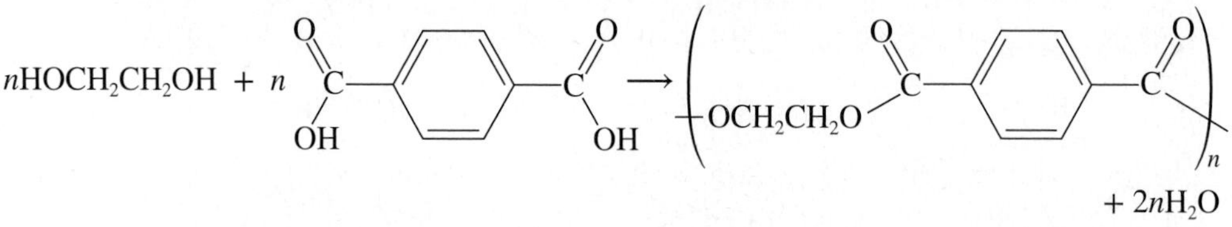

The bond that forms between each ethylene glycol monomer and each *para*-terephthalic acid monomer is called an ester linkage because RCOOR′ is the general form of an ester [◄◄ Section 23.2, Table 23.2]. Polymers in which the monomers are connected by ester linkages are called **polyesters.**

Deoxyribonucleic acid (DNA; see Figure 24.8), which stores the genetic information for every known organism and has a molecular weight as large as billions of grams per mole, is a biological condensation copolymer. If your body contained anything but a tiny fraction of a mole of DNA, you would be extremely heavy! DNA is a copolymer of a five-carbon sugar called deoxyribose, four different molecular bases (adenine, thymine, guanine, and cytosine), and phosphoric acid (H_3PO_4). Ribonucleic acid (RNA) is a biological condensation copolymer analogous to DNA, except that RNA is made from a five-carbon sugar called ribose and the four different molecular bases are adenine, uracil, guanine, and cytosine. RNA molecules vary greatly in size. They are large molecules, but still much smaller than DNA molecules.

Worked Example 24.2 shows how to determine the monomers in a copolymer.

Worked Example 24.2

Kevlar, a condensation polymer used in bulletproof vests, has the following general structure.

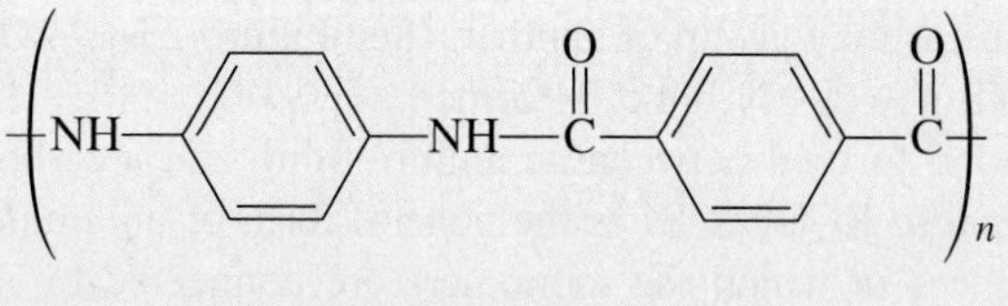

Write the structure of the two monomers that form Kevlar.

Strategy Identify the condensation linkages, split them apart, and then add H to one of the resulting exposed bonds and OH to the other.

Setup Kevlar contains an *amide* (linkage between $\overset{\overset{\text{O}}{\|}}{\text{C}}$ and N). We split the C—N bond, adding an H to the exposed N and an OH to the exposed C to produce an amine and a carboxylic acid, respectively.

Solution To determine what the monomers are, first remove the C—N bond.

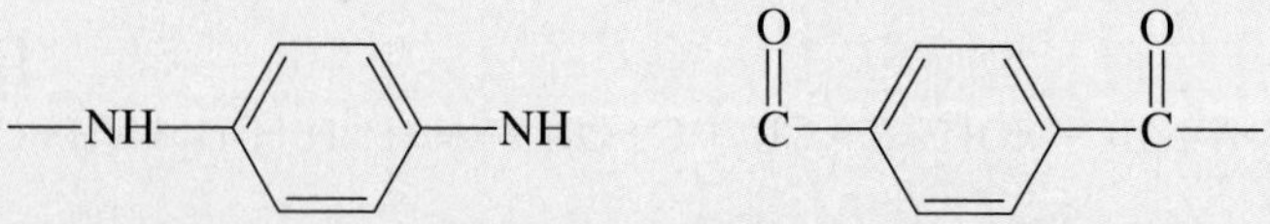

Next, add H_2O as H to the N and OH to the C.

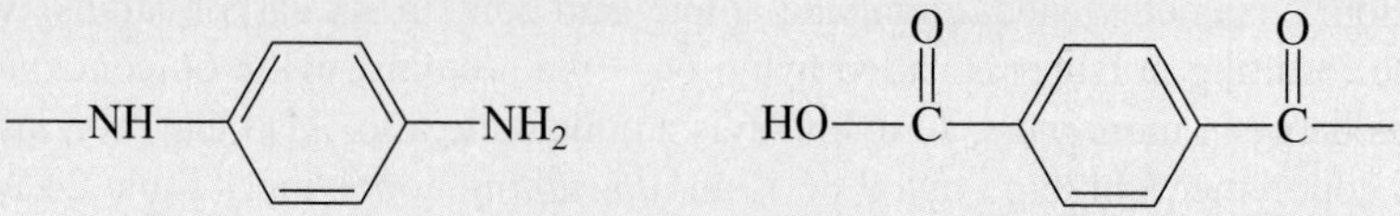

Finally, add the components of another water molecule to the open ends of the molecules—H to N and OH to C, as before.

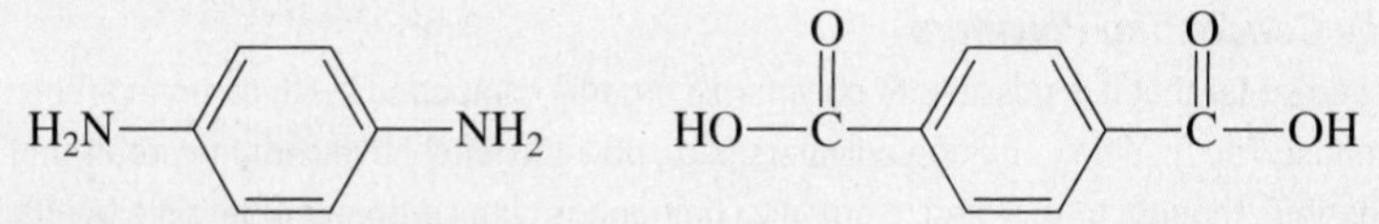

$$H_2N-\bigcirc-NH_2 \qquad HO-\overset{\overset{\displaystyle O}{\|}}{C}-\bigcirc-\overset{\overset{\displaystyle O}{\|}}{C}-OH$$

Think About It

To determine whether the monomers in the final answer are reasonable, recombine them in a condensation reaction to see if they re-form Kevlar, the polymer in the problem.

Practice Problem **A**TTEMPT Nylon 6, like nylon 66, is a condensation polymer (a polyamide). Unlike nylon 66, however, nylon 6 is *not* a copolymer because it is synthesized from a single monomer called 6-aminohexanoic acid.

$$H_2NCH_2(CH_2)_3CH_2\overset{\overset{\displaystyle O}{\|}}{C}OH$$

Draw the structure of at least three repeating units of nylon 6.

Practice Problem **B**UILD Kodel is a polymer used to make stain-resistant carpets.

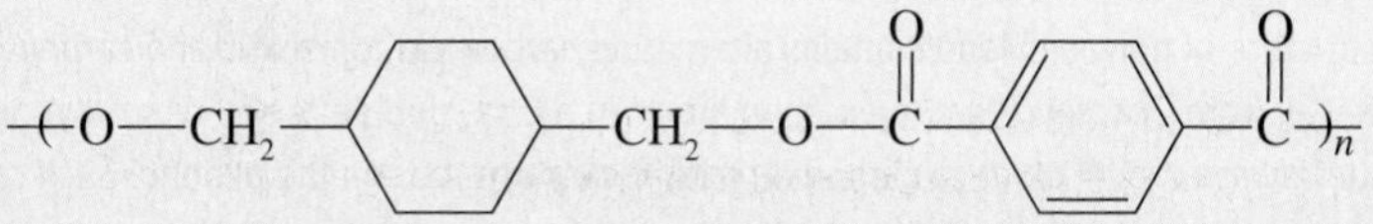

$$-\!\!\left(\!O-CH_2-\bigcirc-CH_2-O-\overset{\overset{\displaystyle O}{\|}}{C}-\bigcirc-\overset{\overset{\displaystyle O}{\|}}{C}\!\right)_{\!\!n}$$

Draw the structure of the monomer. Is this an additional polymer or a condensation polymer?

Practice Problem **C**ONCEPTUALIZE Draw the structure of the substance that would be eliminated if the molecule shown here were to form a condensation polymer.

$$H-O-\underset{\underset{\displaystyle H}{|}}{\overset{\overset{\displaystyle H}{|}}{C}}=\underset{\underset{\displaystyle H}{|}}{\overset{\overset{\displaystyle H}{|}}{C}}-\underset{\underset{\displaystyle H}{|}}{\overset{}{C}}=C-\overset{\overset{\displaystyle O}{\|}}{C}-O-\underset{\underset{\displaystyle H}{|}}{\overset{\overset{\displaystyle H}{|}}{C}}-H$$

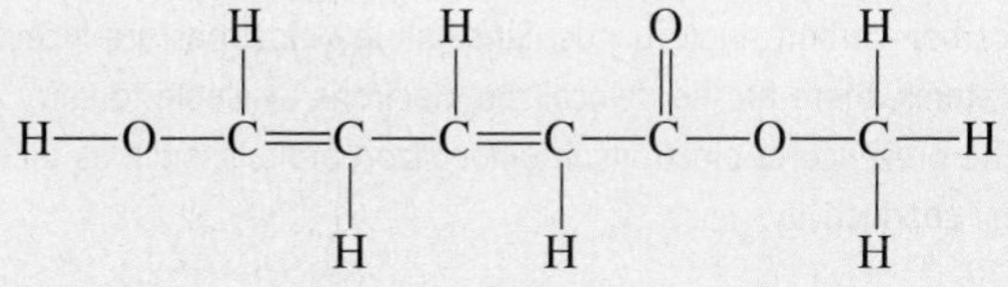

Section 24.1 Review

Polymers

24.1.1 Classify the following copolymer. (Select all that apply.)

$$-B-B-B-B-B-A-A-A-A-A-B-B-B-B-B-A-A-A-A-A-$$

(a) Block
(b) Random
(c) Graft
(d) Alternating
(e) All of the above

24.1.2 What feature is common to molecules that can undergo polymerization?
(a) Fluorine
(b) Hydrogen bonds
(c) Sulfur
(d) Multiple bonds
(e) Lone pairs

Thinking Outside the Box

Electrically Conducting Polymers

You have seen so far that it is possible to polymerize organic compounds with carbon-carbon double bonds (alkenes). The resulting addition polymers (e.g., polyethylene) have carbon-carbon single bonds. It is also possible, though, to polymerize organic compounds with carbon-carbon triple bonds. The simplest of these compounds (known collectively as *alkynes*) is **acetylene** (C_2H_2), which is commonly used in welding torches. The polymerization of acetylene is very similar to the addition polymerization of ethylene. That is, a free-radical initiator molecule attaches to one of the carbon atoms and breaks one of the pi bonds between the two carbon atoms in acetylene. The resulting polymer, polyacetylene, thus retains a carbon-carbon double bond:

$$H-C\equiv C-H \quad H-C\equiv C-H \quad H-C\equiv C-H \longrightarrow -\overset{H}{\underset{H}{C}}=\overset{H}{C}-\overset{H}{\underset{H}{C}}=\overset{}{C}-\overset{H}{\underset{H}{C}}=\overset{}{C}-$$

The generalized form of the reaction can be represented as follows:

$$n\,H-C\equiv C-H \longrightarrow \left(\!\!\begin{array}{c} H \\ | \\ C=C \\ | \\ H \end{array}\!\!\right)_{n}$$

The structure of polyacetylene contains alternating carbon-carbon single and carbon-carbon double bonds. Recall from Chapter 7 that the carbon atoms in a C=C double bond are sp^2-hybridized and that it is the leftover p orbital on each carbon atom that overlaps to form the pi bond. Each carbon atom in polyacetylene is identical (structurally and electronically), so there is extended overlap of the p orbitals throughout the polymer chain. The p electrons are *delocalized*—that is, they can move throughout the network of overlapping orbitals that extend the length of the polymer chain. By being able to move electrons from one end of the polymer chain to the other, polyacetylene can conduct electricity much like a wire. That is, polyacetylene is a plastic that conducts electricity!

For comparison, consider polyethylene and Teflon. Both are polyalkenes in which the polymer chain contains only carbon-carbon single bonds. Since all the electrons are localized in sigma bonds between the carbon atoms, there are no delocalized electrons available to carry a charge throughout the polymer chain. The presence of electrons in *delocalized* orbitals, such as those in polyacetylene, is the key to electrical conductivity.

24.2   CERAMICS AND COMPOSITE MATERIALS

Ceramics

The use of ceramics in the form of pottery dates back to antiquity. Along with common ceramic substances like brick and cement, modern advanced ceramics are found in electronic devices and on the exterior of spaceships. All these *ceramics* are polymeric inorganic compounds that share the properties of hardness, strength, and high melting points. Ceramics are usually formed by melting and then solidifying inorganic substances (including clays).

Most ceramic materials are electrical insulators, but some conduct electricity well at very cold temperatures. Ceramics are also good heat insulators, which is why the outer layer of the space shuttle (shown in Figure 24.9) was made of ceramic tiles. These ceramic tiles could withstand the very hot temperatures of reentry into the atmosphere (where temperatures can reach 1650°C), while keeping the underlying metal structural body of the orbiter (and the astronauts within) relatively cool.

Figure 24.9 Captain Wendy Lawrence inspects the ceramic tiles on the exterior of the space shuttle.

NASA Kennedy Space Center (NASA-KSC)

Ceramics can be prepared by heating a slurry of a powder of the inorganic substance in water to a very high temperature under high pressure. This process, called *sintering,* bonds the particles to each other, thus producing the finished ceramic. Sintering is a relatively easy process to perform, but the resulting particle size is irregular, so the solid may contain cracks, spaces, and imperfections. Imperfections in the interior of a solid ceramic can be impossible to detect, but they make weak spots in the substance that can cause the ceramic piece to fail. To avoid these problems, the *sol-gel process* is frequently employed for modern ceramics that are used in structural applications. The sol-gel process produces particles of nearly uniform size that are much more likely to produce a solid ceramic without gaps or cracks.

The first step in the sol-gel process is the preparation of an alkoxide of the metal or metalloid that is going to be made into the ceramic. This can be illustrated with the yttrium(III) ion, which is used to prepare a yttrium-oxygen ceramic.

$$Y^{3+} + 3CH_3CH_2OH \longrightarrow Y(OCH_3CH_2)_3 + 3H^+$$
$$\text{yttrium alkoxide}$$

The resulting metal alkoxide is dissolved in alcohol. Water is then added to generate the metal hydroxide and regenerate the alcohol.

$$Y(OCH_3CH_2)_3 + 3H_2O \longrightarrow Y(OH)_3 + 3CH_3CH_2OH$$
$$\text{yttrium alkoxide} \qquad\qquad \text{yttrium hydroxide} \quad \text{ethanol}$$

The metal hydroxide, once formed, undergoes condensation polymerization to form a chain with bridging oxygen atoms between the metal atoms.

$$\begin{array}{cccc} \text{OH} & \text{OH} & \text{OH} & \text{OH} \\ | & | & | & | \\ \text{HO}-\text{Y}-\text{OH} & \text{HO}-\text{Y}-\text{OH} \longrightarrow & \text{HO}-\text{Y}-\text{O}-\text{Y}-\text{OH} & + \; H_2O \; \text{(water)} \end{array}$$

Just as in condensation polymerization between organic molecules, a small molecule (water, in this case) is produced by the combination of atoms from the two monomers. Repeated condensation produces an insoluble chain of metal atoms connected by oxygen atoms. The polymer chain may contain branches (i.e., it may be three dimensional) because all three of the hydroxide groups attached to the central metal ion are identical (and are thus equally likely to react). The suspension of the metal oxide-hydroxide polymer is called a *gel*. The gel is carefully heated to remove the liquid, and what remains is a collection of tiny, remarkably uniformly sized particles. Sintering of material produced by the sol-gel process produces a ceramic with relatively few imperfections.

Student Annotation: The sintering of bronze in the manufacture of bearings takes advantage of the inherent imperfections produced by the process. Porous bearings are sometimes preferred because the porosity allows a lubricant to flow throughout the material.

Student Annotation: The "sol" in sol-gel refers to a colloidal suspension of individual particles. The "gel" refers to the suspension of the resulting polymer.

Composite Materials

A *composite material* is made from two or more substances with different properties that remain separate in the bulk material. Each contributes properties to the overall material, though, such that the composite exhibits the best properties of each of its components. One of the oldest examples of a human-made composite material is the formation of bricks from mud and straw, a process that dates back to biblical times.

Modern composite materials commonly include reinforcing fibers, analogous to the straw in mud bricks, in a polymer matrix called a resin. Fiberglass, Kevlar, or carbon fibers can be used to give strength to the composite, whereas polyester, polyamide, or epoxy forms the matrix that holds the fibers together. These types of composites are called *polymer matrix composites.* Composites made from a metal and a ceramic, organic polymer, or another metal are called *metal matrix composites.* Carbide drill tips are made with a combination of softer cobalt and tougher tungsten carbide. Toyota has used a metal matrix composite in the engine block of some of its cars, and some bicycles are made with aluminum metal matrix composites. Composites made of a ceramic as the primary reinforcing material accompanied by organic polymers are called *ceramic matrix composites.* A common biological ceramic matrix composite is bone. Bones consist of reinforcing fibers made of collagen (a protein) surrounded by a matrix of hydroxyapatite [$Ca_5(PO_4)_3OH$].

Another type of composite material that does not fit neatly into one of these categories is a reinforced carbon-carbon composite (RCC), which consists of carbon fibers in a graphite matrix. To prevent oxidation, a silicon carbide coating is applied. RCC is used on the nose cones of missiles, but it is most commonly known as the coating on the nose of the space shuttle. It is believed that the Space Shuttle *Columbia* disaster in 2003 occurred because the ceramic tiles and RCC panel material were both damaged on liftoff by the impact of falling foam insulation. The extreme heat of reentry into the atmosphere damaged the orbiter's structure through the breaches in the ceramic tiles and the RCC panel, and the orbiter and all seven astronauts on board were lost.

Carbon fibers can be woven into fabrics and threads to be used as the structural components of vehicles and for sporting equipment such as bicycles, tennis rackets, and skateboards.

24.3 LIQUID CRYSTALS

Liquid crystals are substances that exhibit properties of both liquids, such as the ability to flow and to take on the shape of a container, and those of crystals, such as a regular arrangement of particles in a lattice. Some substances exhibit liquid crystal behavior when they are melted from a solid. As the temperature increases, the solid liquefies but retains some order in one or two dimensions. At a higher temperature, the ordered liquid becomes a more conventional liquid in which there is no consistent orientation of the molecules.

Liquids are *isotropic,* because their properties are independent of the direction of testing. Because particles in a liquid are free to rotate, there are molecules present in every possible orientation, so the bulk sample gives the same measurements regardless of the direction in which the measurements are taken or observations are made. Liquid crystals are *anisotropic* because the properties they display depend on the direction (orientation) of the measurement. A box full of pencils, all neatly aligned, is an analogy for anisotropy, because the pencils are long and narrow. Looking at the pencils end on is different than looking at them from the side. What you see depends on the direction from which you view the pencils. Similarly, what you see when you look at anisotropic molecules depends on the direction from which you view them.

Frederick Reinitzer discovered the first compound to exhibit liquid crystal behavior in 1888. Cholesteryl benzoate was observed to form an ordered liquid crystalline phase when melted, and this liquid crystalline phase became an ordinary liquid at higher temperatures.

Cholesteryl benzoate

The structure of cholesteryl benzoate is fairly rigid due to the presence of the fused rings and sp^2-hybridized carbon atoms. The molecule is relatively long compared to its width, too. Rigidity and this particular shape makes it possible for the cholesteryl benzoate molecules to arrange themselves in an orderly manner, in much the way that pencils, chopsticks, or tongue depressors can be arranged.

There are three different types of ordering (called nematic, smectic, and cholesteric) that a liquid crystal can adopt (see Figure 24.10). A **nematic** liquid crystal contains molecules ordered in one dimension only. Nematic molecules are all aligned parallel to one another, but there is no organization into layers or rows. A **smectic** liquid crystal is ordered in two dimensions. Smectic molecules are aligned parallel to one another and are further arranged in layers that are parallel to one another. **Cholesteric** liquid crystal molecules are parallel to each other within each layer, but each layer is rotated with respect to the layers above and below it. The layers are rotated because there are repulsive intermolecular forces between layers.

Liquid crystalline substances were curiosities for many years but have found applications in many fields. Liquid crystal displays (LCDs) in calculators, watches, and laptop computers are possible, for example, because polarized light can be transmitted through liquid crystals in one phase but not transmitted through another (Figure 24.11). Polarized light is produced by passing ordinary light (which oscillates in all directions perpendicular to the beam) through a filter that allows only the light waves oscillating in one particular plane to pass through. This plane-polarized light can be rotated by a twisted liquid crystal such that it is allowed to pass through another filter when a voltage is applied to the liquid. If the light passes through the liquid when the voltage is off, though, the light will *not* pass through. This principle can be used in calculator and watch displays because each digit is made up of no more than the seven segments shown in Figure 24.11, and the specific digit depends on which segments are bright and which are dark. LCDs in laptop computers and televisions are more sophisticated, but still operate on the same basic principle.

Liquid crystals can also be used in thermometers (Figure 24.12). The spacing between crystal layers depends on temperature, and the wavelength of light reflected by the crystal depends on this spacing. Thus, the color reflected by the liquid crystal will change with temperature, and this color can be used to indicate the temperature to which the liquid crystal is exposed.

Worked Example 24.3 lets you practice determining whether a molecule might exhibit liquid crystal properties.

Student Annotation: Recall from Chapter 7 that there is free rotation about carbon-carbon single bonds (i.e., bonds between sp^3-hybridized carbon atoms), but not about carbon-carbon double bonds (i.e., bonds between sp^2-hybridized carbon atoms).

Student Annotation: A material that changes color as the temperature changes is called *thermochromic*.

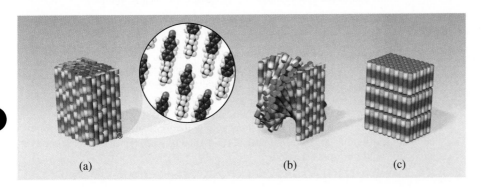

Figure 24.10 (a) Nematic, (b) cholesteric, and (c) smectic liquid crystals.

(a)   (b)   (c)

Figure 24.11 Liquid crystal display. When the current is on, the liquid crystal molecules are aligned and polarized light cannot pass through. When the current is off, the molecules are not aligned and light can pass through.

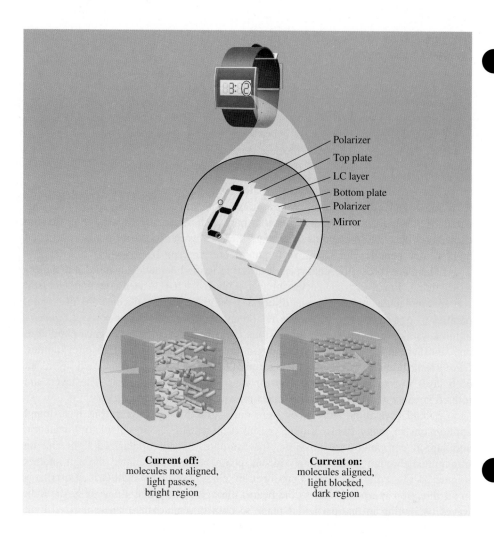

Current off:
molecules not aligned,
light passes,
bright region

Current on:
molecules aligned,
light blocked,
dark region

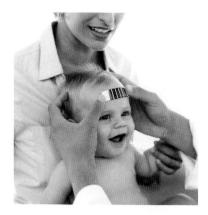

Figure 24.12 LCD thermometer.
© Stockbyte/Getty

Worked Example 24.3

Would the following molecule make a good liquid crystal? Why or why not?

$$CH_3-\overset{\overset{\displaystyle O}{\|}}{C}-\bigcirc-\overset{\overset{\displaystyle O}{\|}}{C}-\bigcirc-CH_2-CH_2-CH_2-CH_2-CH_2-CH_3$$

Strategy Cholesteryl benzoate exhibits liquid crystal properties because it has structurally rigid regions (fused rings and sp^2-hybridized carbon atoms) and because it is relatively long compared to its width. Examine the structure in question to see if it has rigid regions and to see if it is longer in one dimension than in another.

Setup Carbon atoms that are sp^2-hybridized contribute to the rigidity of a molecule's shape.

Solution The left-hand portion of the molecule contains an sp^2-hybridized carbon atom, a benzene ring, another sp^2-hybridized carbon atom, and another benzene ring. These features are relatively rigid and should allow the molecule to maintain its shape when heated. The chain of CH_2 groups ending in a CH_3 group is not rigid due to free rotation about the carbon-carbon single bonds (i.e., sp^3-hybridized carbon atoms). The overall shape of the molecule is longer than it is wide, though, which combined with a large portion that is rigid should make the molecule a good candidate for liquid crystal behavior.

Think About It
Double bonds in a molecular structure indicate the presence of pi bonds. Recall that it is the pi bonds that restrict rotation about bonds in a molecule [◄◄ Section 7.6] and lend rigidity to the structure.

Practice Problem **A**TTEMPT Based on its structure, explain why you would expect this molecule to exhibit liquid crystal properties.

Practice Problem **B**UILD Would you expect this molecule to exhibit liquid crystal properties? Explain your answer.

Practice Problem **C**ONCEPTUALIZE This compound does not possess liquid crystal properties. What changes are needed (e.g., addition of a group) to convert it to a liquid crystal?

Section 24.3 Review

Liquid Crystals

24.3.1 Which of the following is a good analogy for anisotropy? (Select all that apply.)
  (a) Glass of water
  (b) Bucket of ice cubes
  (c) Box of spaghetti
  (d) Box of rice
  (e) Box of macaroni

24.3.2 What characteristics make a molecule likely to exhibit liquid crystal properties? (Select all that apply.)
  (a) Long narrow shape
  (b) Flexibility
  (c) Rigidity
  (d) Low molar mass
  (e) High molar mass

24.4   BIOMEDICAL MATERIALS

Many modern materials are finding uses in medical applications. Replacement joints, dental implants, and artificial organs all contain modern polymers, composites, and ceramics. To function successfully in a biomedical application, though, a material must first be compatible with the living system. The human body very easily recognizes foreign substances and attacks them in an effort to rid them from the body. Thus, a biomaterial must be designed with enough similarity to the body's own systems that the body will accept the material as its own. Additionally, if the polymer, ceramic, or composite contains leftover chemicals from its manufacture, these contaminants may lead to adverse reactions with the body over the lifetime of the medical implant. It is important, therefore, that the substance be as pure as possible.

The physical properties of a biomedical material are important, too, because the longer the material lasts, the less often the medical device has to be replaced during the lifetime of the patient. Most biomedical materials must possess great strength and flexibility to perform in the body. For example, the materials in artificial joints and heart valves must be able to flex many times without breaking. Materials used in dental implants must show great hardness, moreover, so as not to crack during biting and chewing. It takes years of research to develop successful biomedical materials that meet these needs.

Dental Implants

Dental implant materials have been used for many years. The oldest dental fillings were made of various materials including lead (which fell out of favor due to its softness—the danger of lead poisoning was as yet unknown), tin, platinum, gold, and aluminum. The remains of Confederate soldiers have been shown to include fillings made of a lead-tungsten mixture, probably from shotgun pellets; tin-iron; a mercury amalgam; gold; and even radioactive thorium. (The dentist probably thought he was using tin, which is similar in appearance.) Many modern fillings are made of dental *amalgam* [◄◄ Chapter 18], a solution of several metals in mercury. Modern dental amalgam consists of 50 percent mercury and 50 percent of an alloy powder that usually contains (in order of abundance) silver, tin, copper, and zinc. These metals tend to expand slightly with age, causing fissures and cracks in the tooth that may require further intervention (e.g., crowns, root canals, or tooth replacement). Some people consider amalgam fillings to be unsightly, too.

Dental composite materials are now used that have several advantages over traditional amalgams. The composites can be made in a wide range of colors, for example, to match the color of the other teeth. The existing healthy portion of the tooth can be etched with acid, moreover, to create pores into which the composite material can bond. With traditional dental amalgam, the dentist must create indentations in the healthy tooth to hold the amalgam in place. Destroying a portion of the healthy tooth is undesirable and can be avoided with the use of the composite.

The dental composite consists of a matrix (made from a methacrylate resin) and a silica filler. The composite material is applied to the cavity in a puttylike consistency and then is dried and polymerized by a photochemical reaction initiated by light of a particular wavelength. Because the light does not penetrate very far into the composite, the thickness of the applied composite is critical. If the layer of composite is too thick, some of the composite will remain soft. A properly constructed dental composite filling will last more than 10 years. Over time, though, composite fillings tend to shrink, leaving breaches that can cause leakage, a situation that must be addressed to prevent further tooth decay.

Porcelain ceramic fillings and crowns are very common, but the ceramic has two disadvantages: It is very hard and can wear on the opposing teeth, and it is brittle and may crack if subjected to great force. Most dentists, therefore, do not recommend porcelain ceramic crowns and fillings for the molar teeth, which do the bulk of the grinding work during chewing.

A material used in dentistry must have properties that maximize both patient comfort and the lifetime of the implant. Dental fillings and tooth replacements must be resistant to acids, for example, because many foods (such as citrus fruits and soft drinks) contain acids directly. Any food containing carbohydrates can produce acids, though, if traces are left in the mouth because the bacteria that reside there consume carbohydrates, producing acids in the process. These acids eat away at the natural hydroxyapatite in teeth, creating caries (cavities). Materials like dental amalgam, gold, and dental composite resist attack by acids in the mouth.

A dental material should also have low thermal conductivity; that is, it should conduct heat poorly. This is especially important in applications where the implant material is in contact with the nerve inside the tooth. If the implant material conducts heat well, then whenever hot or cold foods come in contact with the implant, the hot or cold will be transmitted through the material to the nerve, causing pain. Metals are good thermal conductors, so dental amalgams must not touch nerves.

Finally, dental materials should resist wear (so they last a long time) and they should resist expansion and contraction as the temperature fluctuates. The close spacing and precision fits involved in dental fillings and implants would result in discomfort and possibly failure of the implant if the material expanded or contracted appreciably with changes in temperature.

Soft Tissue Materials

Burn victims who have lost large amounts of skin do not have enough cells to grow new skin, so a synthetic substitute must be used. The most promising artificial skin material is based on a polymer of lactic acid and glycolic acid. Both of these compounds contain an alcohol group ($-OH$) and a carboxy group ($-COOH$), so they can form a polyester copolymer in a condensation reaction that mirrors the formation of Dacron polyester (Section 24.1).

$$HOCH_2\overset{\overset{O}{\|}}{C}\boxed{-OH\quad H-}OCH\overset{\overset{O}{\|}}{C}-OH \longrightarrow \left(-OCH_2\overset{\overset{O}{\|}}{C}-OCH\overset{\overset{O}{\|}}{C}-\right)_n + H_2O$$
$$\qquad\qquad\qquad\qquad\quad CH_3 \qquad\qquad\qquad\qquad\qquad CH_3$$

glycolic acid   lactic acid

This copolymer forms the structural mesh that supports the growth of skin tissue cells from a source other than the patient. Once the skin cells grow on the structural mesh, the artificial skin is applied to the patient, and the copolymer mesh eventually disappears as the ester linkages are hydrolyzed. During this treatment, the patients must take drugs that will suppress their immune systems, which ordinarily would attack these substances as foreign to prevent their bodies from rejecting the new skin.

Student Annotation: Hydrolysis is essentially the opposite of a condensation reaction.

Sutures, commonly known as stitches, are now made of the same lactic acid–glycolic acid copolymer as the structural mesh of artificial skin. Before, sutures were made of materials that did not degrade so they eventually had to be removed again. This type of suture is still used in cases where the degrading suture is deemed a risk or there might be a need to reopen the incision at a later time. Biodegradable sutures are routinely used during operations on animals, too, which decreases the number of trips to the veterinarian, thereby reducing the stress on the pet and its owner.

When portions of blood vessels need to be replaced or circumvented in the body, it may or may not be possible to obtain a suitable vascular graft from another part of the patient's body. Dacron polyester (Section 24.1) can be woven into a tubular shape for this purpose, and pores can be worked into the material so that the graft can integrate itself into the tissue around it, as well as to allow capillaries to connect. Unfortunately, the body's immune system recognizes this as a foreign substance and blood platelets stick to the walls of the Dacron graft, causing clotting.

Artificial hearts and heart valves are potential solutions to the shortage of donor organs. Artificial hearts have not yet been perfected to the point that they can work on a permanent basis, but valve replacement is a common procedure that can successfully prolong a patient's life. The most common replacement heart valve is fixed in position by surrounding the valve with lactic acid–glycolic acid copolymer, the same polyester that is used in skin grafts and biodegradable sutures. The polymer is woven into a mesh ring that allows the body tissue to grow into the ring and hold it in place.

Researchers today are using *electrospinning* to place nanofibers directly onto wounds. (Nanofibers are products of nanotechnology; see Section 24.5.) Electrospinning uses a high-voltage electric field to create a jet of polymer solution that transforms into a spiraling dry fiber only nanometers in diameter. The wound dressings applied by direct electrospinning have high rates of moisture transmission and bacterial resistance. Enzymes can be added to the solution containing the nanofiber polymer, and these enzymes can be released into the skin after the dressing is applied. Proteins can also be cross-linked onto the nanofiber, making possible many applications of the nanofiber wound dressing.

Artificial Joints

Joint replacements, such as artificial knees, elbows, and hips, require major surgery but have become relatively commonplace because suitable materials have become available for not only the bone replacement but also the contact surfaces. In knee replacement, for example, the goal is usually to replace worn-out cartilage surfaces (which may have resulted from arthritis or injury). It is possible to replace shattered bones that cannot heal back to their original state as well.

The materials originally used in joint replacements were ivory, metals, and glass, but they are fast being replaced by polymeric plastics and ceramics. Polymethyl methacrylate (PMMA), which was the subject of Practice Problem 24.1B in Section 24.1, and polyethylene have been used in many total joint replacements. However, these joints tend to fail over time when they are implanted in younger, more active patients, so efforts are under way to improve the wear quality of the PMMA and polyethylene parts.

Although metallurgical research provided new substances that were much stronger and less likely to break, including titanium-based alloys, total joint replacements were predicted to last no more than 20 years in a patient before needing to be replaced again. Ultrahigh-molecular-weight polyethylene, treated with radiation to initiate cross-linking of the polymer chains and stored under inert nitrogen gas to stop free-radical damage to the chain from the radiation, is now commonly used in many replacement joints, with the expectation of improved wear. It remains to be seen whether ultrahigh-molecular-weight polyethylene will prolong the life of these devices, since many of the joints have been in service in patients for less than a decade.

Other surfaces have also been researched to reduce wear. Metal-on-metal bearing surfaces suffered early setbacks due to the manufacturing tolerances of the cobalt-based alloy used. As the bearing surfaces wore, they produced metal sludge, and high friction between components caused loosening of the joint. Modern metal-on-metal bearing surfaces have improved as the manufacturing has become more consistent and produced components lower in friction.

Ceramic-on-ceramic bearing surfaces have also been developed. Some ceramics have been prone to cracking, but many have shown great promise for long-lasting use. When a ceramic component in a replacement joint does crack, however, it can be disastrous because the entire joint may need to be replaced and further healthy bone may need to be sacrificed to install the new replacement joint.

Despite the progress made so far on developing safe, durable, and useful artificial joints, much work remains to be done. One long-term issue that has yet to be researched extensively, for example, is the effect of the release of metal cations from the joint into the body over the lifetime of the joint replacement. This may become a more urgent health issue as the lifetime of the artificial joints increases and the age at which patients get the replacements decreases.

24.5   NANOTECHNOLOGY

Atoms and molecules have sizes on the order of a tenth of a nanometer (where 1 nm = 10^{-9} m) [◄◄ Section 1.2]. **Nanotechnology** is the development and study of such small-scale materials and objects. An atomic force microscope [Figure 24.13(a)] is one way to image the surface of a material at the atomic or molecular level.

A sharp-pointed stylus is moved over the surface of the substance under study, and the intermolecular forces between the material and the stylus force the stylus up

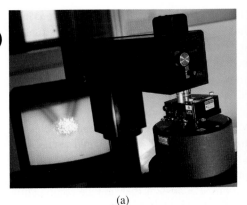
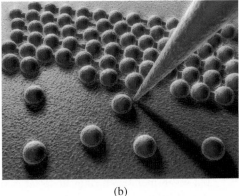
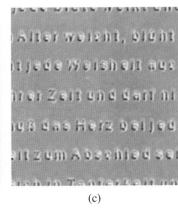

(a)	(b)	(c)

Figure 24.13 (a) Atomic force microscope (AFM). (b) Atomic force micrograph. (c) Poem etched on a polymer surface using AFM and dynamic plowing lithography (DPL).

(a): © Tek Image/Science Source; (b): © Flirt/Alamy Stock Photo; (c): © Digital Instruments/Veeco/Science Source

and down. The stylus is mounted on a probe that reflects a laser beam into a detector. From the differences in deflection of the laser beam, the electronics attached to the probe can create an image of the surface of the object [Figure 24.13(b)].

The scanning tunneling microscope (STM) works on a similar principle but only for samples that conduct electricity. The STM measures the peak and valley heights of the sample from the flow of electric current. The tip of the STM must move up and down to keep the current constant, and the up-and-down movements are translated into an image of the atomic and molecular structure of the substance [Figure 24.13(c)].

Graphite, Buckyballs, and Nanotubes

Carbon exists as several allotropes, one of which is graphite. Graphite consists of sheets of carbon atoms that are all sp^2-hybridized [◄◄ Section 7.5].

> **Student Annotation:** Recall that *allotropes* are different forms of the same element [◄◄ Section 5.5].

Intermolecular forces [◄◄ Chapter 7] hold the sheets together. Because of the delocalization of the electrons in the extended network of p orbitals perpendicular to the graphite sheet, graphite is an electrical conductor within its sheets, but not between them. Electrical conductivity is much more common for metals than it is for nonmetals like graphite.

Graphite and diamond were long thought to be the only two allotropes of carbon. In 1985, though, a compound consisting of 60 carbon atoms was isolated from graphite rods that had been vaporized with an intense pulse of laser light. The structure of the compound (C_{60}) was identical to that of a soccer ball, with 12 pentagonal and 20 hexagonal faces and carbon atoms at the corners of each face:

(left): © Science Picture Co/Science Source; (right): © Getty Images

This pattern also resembles an architectural structure called a geodesic dome, so the C_{60} molecule was named *buckminsterfullerene* in honor of R. Buckminster Fuller, the designer of the geodesic dome. The nickname for C_{60} is the *buckyball*.

Since the discovery of C_{60}, elongated and elliptical cages of 70 and 80 carbon atoms have also been discovered. These molecules, called *fullerenes* as a group, have been proposed to exist in such exotic places as stars and interstellar media and have been observed in such mundane places as deposits of chimney soot.

A single sheet of graphite is known as *graphene.* A graphene is essentially a two-dimensional carbon crystal consisting of an array of six-membered rings. Graphenes are extraordinarily strong (hundreds of times stronger than steel), they conduct electricity, and their shapes can be customized by the introduction of five-membered and seven-membered rings at specific locations. Because of their unique combination of properties, graphenes currently are the subject of a great deal of research. Proposed uses of graphenes include replacement of copper interconnects in integrated circuits, production of transparent conduction electrodes—useful for touchscreens and liquid crystal displays—and ultrahigh-sensitivity detection devices.

Student Annotation: The 2010 Nobel Prize in Physics was awarded to Andre Geim and Konstantin Novoselov for their work with graphene.

Unless otherwise constrained, graphene sheets will wrap around on themselves forming cylinders with diameters on the scale of nanometers. Such cylinders are called *carbon nanotubes.* Two types of nanotube structures are known, as shown in Figure 24.14. The differences in arrangement of hexagons give the two kinds of nanotubes different electrical conductivities.

Student Annotation: An analogy is to take a piece of paper and roll it so that you have a tube.

Carbon nanotubes of different diameters can nest inside each other, producing multiple-walled nanotubes. Most nanotubes are single-walled, though, and they can be as narrow as 1 nm in diameter. These nanotubes are remarkably strong—stronger, in fact, than a comparably sized piece of *steel*. Because nanotubes consist only of carbon atoms, they can come into close contact, thereby allowing intermolecular forces to hold large numbers of tubes together. This, in turn, makes it possible to align groups of tubes, potentially forming long strands and fibers. Lightweight fibers and strands with tremendous strength have many applications in such things as auto parts, medical implants, and sports equipment.

Besides carbon, other elements have been fashioned into nanotubes, too. Molybdenum sulfide nanotubes have been synthesized, for example, with openings between atoms that are just the right size to permit some gas molecules to pass through, but not others. This may have potential for the nanoscale storage of gases for fuel cells.

Many universities have developed large-scale facilities for nanotechnology research in collaboration with industrial partners. Since the lower limits of size have been reached in the production of silicon chip-based electronic circuits, nanotechnology may provide the pathway to circuits that are much smaller than have ever been imagined.

Figure 24.14 Carbon nanotubes.

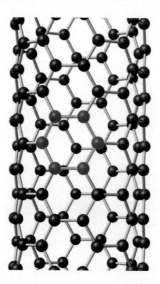 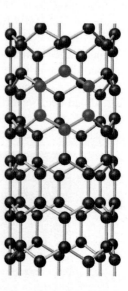

24.6 SEMICONDUCTORS

Recall from Section 7.7 that bonds between atoms can be described as resulting from the combination of *atomic* orbitals to form *molecular* orbitals. In a bulk sample of metal containing many, many atoms, there are equally as many molecular orbitals. These molecular orbitals are so close in energy that instead of forming individual bonding and antibonding orbitals for each pair of atomic orbitals that combine, they form a band of bonding levels and a band of antibonding levels (Figure 24.15.).

The energies of the bonding and antibonding bands depend on the energies of the atomic orbitals that combined to form them in the first place. As a result, the band structure of a bulk sample depends on the original atoms' energy levels. The band energies and the gaps (or lack of gaps) between the bonding band (called the **valence band**) and the antibonding band (called the **conduction band**) make it possible to classify a substance as an electrical conductor, a semiconductor, or an electrical insulator based on the behavior of the electrons in the bands (Figure 24.16).

The valence band is filled with the valence (bonding) electrons, whereas the conduction band is empty or only partially filled with electrons. It is the conduction band that allows electrons to move between atoms. The gap between the valence band and the conduction band can vary from nothing (for an electrical conductor) to a small amount (for a semiconductor) to a very large amount (for an electrical insulator). Thus, the size of the band gap determines the conduction behavior of the material.

Metals have no band gap, so they are good conductors of electricity. The valence band and conduction band actually overlap or are one and the same in metallic conductors, so there is no energy barrier to the movement of electrons from one atom to another in a metal. **Semiconductors** have a band gap, but it is relatively small, so there is limited movement of electrons from the valence band to the conduction band.

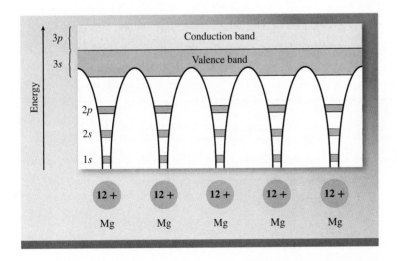

Figure 24.15 Formation of conduction bands in magnesium. The electrons in the 1s, 2s, and 2p orbitals are localized on each Mg atom. However, the 3s and 3p orbitals overlap to form delocalized molecular orbitals. Electrons in these orbitals can travel throughout the metal, and this accounts for the electrical conductivity of the metal.

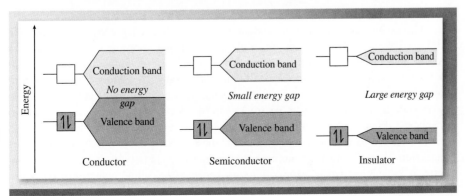

Figure 24.16 Energy bands in metals (conductors), semiconductors, and insulators.

Nonconductors (electrical insulators) have large band gaps, so it is nearly impossible to promote electrons from the valence band to the conduction band.

Silicon, germanium, and carbon in the form of graphite are the only elemental semiconductors at room temperature. All three of these elements are in Group 4A and have *four* valence electrons. (Tin and lead are also in Group 4A, but they are metals, not semiconductors.) The use of semiconductors in transistors and solar cells, to name two applications, has revolutionized the electronic industry in recent decades, leading to the increased miniaturization of electronic equipment.

Other semiconductors consist of combinations of elements whose valence electron count totals eight. For example, gallium (Group 3A) and phosphorus (Group 5A) form a semiconductor because gallium contributes three valence electrons and phosphorus contributes five, giving a total of eight valence electrons. Semiconductors have also been formed between Group 2B elements (particularly Zn and Cd) and Group 6A elements.

Worked Example 24.4 lets you practice identifying combinations of elements that can exhibit semiconductor properties.

> **Student Annotation:** The group number for a main group element corresponds to the number of valence electrons that the atom has.

Worked Example 24.4

State whether each of the following combinations of elements could form a semiconductor: (a) Ga-Se, (b) In-P, (c) Cd-Te.

Strategy Count the valence electrons in each type of atom. If they total eight for the two elements, then the combination will probably form a semiconductor.

Setup Locate each element in the periodic table. All of them are main group elements, so we can use their group numbers to directly determine the number of valence electrons each element contributes: (a) Ga is in Group 3A, and Se is in Group 6A. (b) In is in Group 3A, and P is in Group 5A. (c) Cd is in Group 2B, and Te is in Group 6A.

Solution

(a) Ga has three valence electrons, and Se has six. This gives a total of nine, which is too many to form a semiconductor.

(b) In (three valence electrons) and P (five valence electrons) combine for a total of eight, so In-P should be a semiconductor.

(c) Cd (two valence electrons) and Te (six valence electrons) also combine for a total of eight, so Cd-Te should be a semiconductor, too.

> **Think About It**
> Semiconductors consisting of Group 3 and Group 5 elements are used in the contact layers of light-emitting diodes (LEDs).

Practice Problem ATTEMPT Name an element in each case that could be combined with each of the following elements to form a semiconductor: (a) O, (b) Sb, (c) Zn.

Practice Problem BUILD State whether each combination of elements would be expected to produce a semiconductor: (a) In and P, (b) P and As, (c) Ga and N.

Practice Problem CONCEPTUALIZE For two elements to combine to produce a semiconductor, their valence electrons must sum to eight. Explain how pure members of Group 4A can be semiconductors.

The conductivity of a semiconductor can be enhanced greatly by **doping,** the addition of very small quantities of an element with one more or one fewer valence electron than the natural semiconductor. For the purpose of simplicity, we will consider a pure silicon semiconductor. Silicon is a Group 4A element, so it has four valence electrons per atom ($[Ne]3s^2 3p^2$). A small (parts per million) amount of phosphorus (Group 5A, five valence electrons) can be added, thus doping the silicon with phosphorus (Figure 24.17). Since each phosphorus atom ($[Ne]3s^2 3p^3$) has an "extra" electron relative to the pure semiconductor, these extra electrons must reside in the conduction band, where they increase its conductivity. This type of doped semiconductor is called an **n-type semiconductor** because the semiconductivity has been enhanced by the addition of negative particles, the extra electrons.

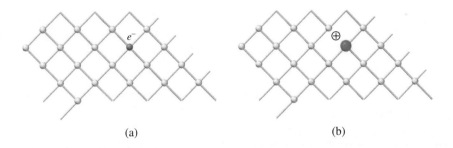

(a)　　　　　　　　　　(b)

Figure 24.17 (a) Silicon crystal doped with phosphorus. (b) Silicon crystal doped with gallium. Note the formation of a negative center in (a) and a positive center in (b).

It is also possible to dope a semiconductor with an element that has fewer valence electrons than the semiconductor itself. For example, a silicon semiconductor could be doped with small amounts (again, parts per million) of gallium (Group 3A, three valence electrons, ([Ar]$4s^2 3d^{10} 4p^1$)). Now, instead of an "extra" electron, there is a "hole" (one less electron) in every place that a gallium atom has replaced a silicon atom (Figure 24.17). These holes are effectively positive charges (because each is the absence of an electron), so this type of material is called a **_p-type semiconductor._** A _p_-type semiconductor has increased conductivity because the holes (which are in the valence band) move through the silicon rather than electrons. The energy required to move an electron from the valence band into a hole (also in the valence band) is considerably less than the energy needed to promote an electron from the valence band to the conduction band of a semiconductor (Figure 24.18). Thus, the movement of the holes results from the movement of charge, and conductivity is the movement of charge.

In both the _p_-type and _n_-type semiconductors, the energy gap between the valence band and the conduction band is effectively reduced, so only a small amount of energy is needed to excite the electrons. Typically, the conductivity of a semiconductor is increased by a factor of 100,000 or so by the presence of impurity atoms.

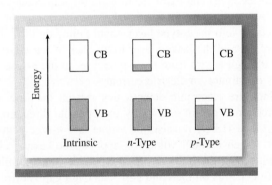

Figure 24.18 The effects of doping on the conduction of semiconductors.

Worked Example 24.5 lets you practice identifying _n_-type and _p_-type semiconductors.

Worked Example 24.5

State whether silicon would form _n_-type or _p_-type semiconductors with the following elements: (a) Ga, (b) Sb.

Strategy Determine the number of valence electrons in each element. If the element has more valence electrons than silicon, it will form an _n_-type semiconductor with silicon. If the element has fewer valence electrons than silicon, it will form a _p_-type semiconductor.

Setup Locate each element in the periodic table. All of them are main group elements, so we can use their group numbers to directly determine the number of valence electrons each element contributes: (a) Ga is in Group 3A and has three valence electrons. (b) Sb is in Group 5A and five valence electrons.

(_Continued on next page_)

Solution

(a) Ga has three valence electrons, and Si has four valence electrons. This combination will form a *p*-type semiconductor.

(b) Sb (five valence electrons) and Si (four valence electrons) will combine to make an *n*-type semiconductor.

> **Think About It**
> Gallium (Ga) is to the left of silicon (Si) on the periodic table and therefore has fewer valence electrons than silicon. Antimony (Sb) is to the right of silicon on the periodic table and has more valence electrons than silicon.

Practice Problem ATTEMPT Determine whether an *n*-type semiconductor or a *p*-type semiconductor is produced when silicon is doped with (a) Al and (b) As.

Practice Problem BUILD For each of the elements C, As, and Ga, indicate which can be combined with silicon to produce a semiconductor, which can be added to silicon as a dopant to produce a *p*-type semiconductor, and which can be added to silicon as a dopant to produce an *n*-type semiconductor.

Practice Problem CONCEPTUALIZE Titanium dioxide (TiO_2) can be doped with small amounts of Ti^{3+} ion to produce an *n*-type semiconductor. Explain how the presence of Ti^{3+} ions leads to an *n*-type semiconductor.

Figure 24.19 LED arrays in traffic signals. Many communities have replaced the red and green signals with LED arrays that are much more energy efficient than the incandescent bulbs used previously. The yellow lights are being replaced more slowly, however, because LED arrays are expensive and the short duration of yellow lights would not necessarily allow an overall cost saving when factoring in energy efficiency.

© David A. Tietz/Editorial Image, LLC

The growth of the semiconductor industry since the early 1960s has been truly remarkable. Today semiconductors are essential components of nearly all electronic equipment, ranging from radios and television sets to pocket calculators and computers. One of the main advantages of solid-state devices over vacuum-tube electronics is that the former can be made on a single "chip" of silicon no larger than the cross section of a pencil eraser. Consequently, much more equipment can be packed into a small volume—a point of particular importance in space travel, as well as in handheld calculators and microprocessors (computers-on-a-chip).

A combination of *p*-type and *n*-type semiconductors can be used to create a solar cell in which the holes and extra electrons in the semiconductors are at equilibrium in the dark. At equilibrium, the positive holes in the *n*-type layer have offset the movement of electrons into the *p*-type layer. Light falling on the solar cell causes the electrons that have entered the *p*-type layer to go back to the *n*-type layer, and other electrons to pass through a wire from the *n*-type layer to the *p*-type layer. This movement of electrons constitutes an electric current.

Diodes, which are electronic devices that restrict the flow of electrons in a circuit to one direction, work in essentially the opposite direction from solar cells. A light-emitting diode (LED) consists of *n*-type and *p*-type semiconductor layers placed in contact. When a small voltage is applied, the "extra" electrons from the *n*-type side combine with the holes in the *p*-type side, thus emitting energy (light) whose wavelength depends on the band gap. The band gap is different for different semiconductors (e.g., Ga-As versus Ga-P), so the specific combination of semiconductor materials can be chosen to make the band gap correspond to a desired color of light. LEDs are finding increasing use in applications previously dominated by traditional incandescent lightbulbs such as emergency exit signs, car tail lights, and traffic signals (Figure 24.19).

Section 24.6 Review

Semiconductors

24.6.1 Which group of the periodic table contains elements that are semiconductors at room temperature?

(a) 2A   (d) 4A

(b) 2B   (e) 5A

(c) 3A

24.7 SUPERCONDUCTORS

No ordinary electrical conductor is perfect. Even metals have some resistance to the flow of electrons, which wastes energy in the form of heat. *Superconductors* have no resistance to the flow of electrons and thus could be very useful for the transmission of electricity over the long distances between power plants and cities and towns. The first superconductor was discovered in 1911 by the Dutch physicist H. Kamerlingh-Onnes. He received the 1913 Nobel Prize in Physics for showing that mercury was a superconductor at 4 K, the temperature of liquid helium. Liquid helium is very expensive, however, so there were few feasible applications of a mercury superconductor.

Since that time, more superconductors have been discovered, including a lanthanum-, barium-, copper-, and oxygen-containing ceramic compound, and a series of copper oxides ("cuprates"), all of which become superconducting below 77 K, the temperature of liquid nitrogen (a much more common and less expensive refrigerant than liquid helium). The temperature below which an element, compound, or material becomes superconducting is called the *superconducting transition temperature, T_c.* The higher the T_c, the more useful the superconductor.

By 1987, superconductivity was observed in another ceramic, yttrium barium copper oxide (YBCO) ($YBa_2Cu_3O_7$). Commonly called a "1-2-3 compound" because of the 1:2:3 ratio of yttrium to barium to copper, YBCO has a T_c of 93 K (above the temperature of liquid nitrogen). Although 93 K is still cold, it qualifies as quite warm in the world of superconductors, so YBCO is considered to be the first *high-temperature superconductor.* YBCO was first prepared by combining the three metal carbonates at high temperature (1000 to 1300 K). The copper ions are present in a mixture of +2 and +3 oxidation states.

$$4BaCO_3 + Y_2(CO_3)_3 + 4CuCO_3 + Cu_2(CO_3)_3 \longrightarrow 2YBa_2Cu_3O_7 + 14CO_2$$

YBCO must be sintered carefully to obtain the optimal superconductivity.

Superconducting wires were first made from $Bi_2Sr_2CaCu_2O_8$, known as BSCCO-2212, a high-temperature superconductor with $T_c = 95$ K. It can be prepared in wire forms because it has layers of bismuth and oxygen atoms that YBCO does not have. BSCCO-2212 is so named because its formula consists of two bismuth, two strontium, one calcium, and two copper atoms. Other BSCCO compounds exist, too, and they all have a general formula of $Bi_2Sr_2Ca_nCu_{n+1}O_{2n+6}$. BSCCO-2223, for example, which has the formula $Bi_2Sr_2Ca_2Cu_3O_{10}$ (where $n = 2$), is a superconductor with $T_c = 107$ K.

Superconductors at or below T_c exhibit the *Meissner effect,* the exclusion of magnetic fields. In Figure 24.20, for example, a magnet is shown levitating above a YBCO pellet, a superconductor at or below −180°C, its T_c.

Magnetic levitation is being researched for use in trains, because the amount of friction between the train and the tracks would be drastically reduced in a magnetically levitated ("maglev") train versus one running on ordinary tracks and wheels. The engineering challenge to a maglev train is that the superconductor must be kept at or below its T_c for the magnetic levitation to occur, and the highest temperatures obtained for superconducting substances are approximately 140 K—warm on the superconductivity scale but still very cold by ordinary standards.

Superconductivity in metals can be explained satisfactorily by BCS theory, first proposed by John Bardeen, Leon Neil Cooper, and John Robert Schrieffer in 1957. They received the Nobel Prize in Physics in 1972 for their work. BCS theory treats superconductivity using quantum mechanical effects, proposing that electrons with opposite spin can pair due to fundamental attractive forces between the electrons. At temperatures below T_c, the paired electrons resist energetic interference from other atoms and experience no resistance to flow. Superconductivity in ceramics has yet to be satisfactorily explained.

Student Annotation: Allex Müller and Georg Bednorz were awarded the 1987 Nobel Prize in Physics "for their important breakthrough in the discovery of superconductivity in ceramic materials."

Student Annotation: The decomposition of a metal carbonate to the metal oxide and carbon dioxide gas is a very common reaction.

Student Annotation: Many a Nobel Prize has been won by scientists who research and develop modern materials!

Student Annotation: You might be skeptical that two negatively charged electrons would attract each other, but there is evidence that all subatomic particles do exert some attractive forces on each other.

Figure 24.20 The Meissner effect. A magnet levitates above a YBCO pellet at the temperature of liquid nitrogen (−196°C).

© Bill Pierce//Time Life Pictures/Getty Images

Learning Outcomes

- Define polymer and monomer.
- Describe the mechanism of addition polymerization.
- Differentiate between thermoplastic and thermosetting polymers.
- Distinguish between addition and condensation polymers.
- Cite examples of polymers: addition, condensation, biological (nucleic acid, protein, polysaccharide), and copolymer.
- List some of the common characteristics of ceramics.
- Describe how ceramics are produced.

- Give examples of composite materials.
- Describe the characteristics of a liquid crystal.
- Define isotropic and anisotropic.
- Discriminate between the different types of liquid crystals.
- List biomedical applications associated with modern materials.
- Differentiate between graphite, buckyballs, and nanotubes.
- Define semiconductor.
- Define superconductor.

Chapter Summary

SECTION 24.1

- **Polymers** are made up of many repeating units, called **monomers,** that are linked together by covalent bonds.
- **Thermoplastic** polymers can be melted and reshaped, whereas **thermosetting** polymers assume their final shape as part of the chemical reaction that forms them in the first place.
- **Addition polymers** are formed via a radical reaction in which a pi bond in an alkene or alkyne is opened up to form new bonds to adjacent monomers.

- Polymer chains can be **cross-linked** to provide added durability and flexibility.
- **Elastomers,** such as natural rubber, can stretch and bend and then return to their original shape.
- **Copolymers** contain more than one type of monomer. Copolymers can be random, alternating, block, or graft, depending on the sequence of the different monomers and the way that the monomers are attached in the polymer.
- **Tacticity** describes the relative geometric arrangements of the groups attached to the chiral carbon atoms in vinyl

polymers. In *isotactic* polymers, all the groups attached to the chiral carbon atoms have identical arrangements. In *syndiotactic* polymers, they alternate positions along the polymer chain. In *atactic* polymers, they have random arrangements.

- *Condensation polymers* form when two different groups on monomers react, thus producing a covalent bond between monomers and a small molecule (usually water) that is expelled. The two reacting groups may be on the same monomer (as in the amino acids that form polypeptides and proteins), or they may be on two different monomers (as in hexamethylenediamine and adipic acid, which form nylon 66). Most condensation polymers are actually copolymers.

- Amines and organic acids form condensation polymers known as *polyamides.* Alcohols and organic acids form condensation polymers known as *polyesters.* Amino acids form polyamide polymers known as polypeptides. Long, straight-chain polypeptides are known as proteins.

- DNA and RNA are condensation polymers consisting of sugars, organic bases, and phosphate ions.

SECTION 24.2

- *Ceramics* are polymeric inorganic compounds that are hard and strong and have high melting points.

- Ceramics are *sintered* to bond the particles to each other and to close any gaps that may exist in the material.

- The sol-gel process is used to prepare ceramics for structural applications because it produces particles of nearly uniform size that are much more likely to produce a solid ceramic without gaps or cracks.

- *Composite materials* contain several components each of which contributes properties to the overall material. Modern composites include reinforcing fibers and a surrounding matrix. The reinforcing fibers may be made of polymers, metals, or ceramics.

SECTION 24.3

- *Liquid crystals* are substances that maintain an ordered arrangement in the liquid phase. Molecules containing structurally rigid regions (e.g., rings and multiple bonds) and molecules that are longer in one dimension than in another seem to form the best liquid crystals. Although ordinary liquids are *isotropic,* liquid crystals are *anisotropic,* meaning that their apparent properties depend on the direction from which we view them.

- Liquid crystals may be *nematic, smectic,* or *cholesteric,* depending on the type of ordering the molecules can adopt. Nematic liquid crystals contain order in only one dimension (parallel molecules), smectic liquid crystals contain order in two dimensions (parallel molecules in parallel layers), and cholesteric liquid crystals contain layers that are rotated with respect to one another. Within each cholesteric layer, the molecules are parallel.

SECTION 24.4

- Dental amalgam consists of several metals dissolved in an alloy with mercury. Dental composite materials consist of a methacrylate polymer in a silica matrix. The composite can be varied in color to match existing teeth, making the repair more aesthetically pleasing than metallic amalgam.

- Biodegradable *sutures* and artificial skin are made of synthetic polyesters that support the growth of tissue around the polymer and then hydrolyze into the body.

- Artificial joint replacements can be made of metal, ceramics, and organic polymers.

SECTION 24.5

- *Nanotechnology* is the study and development of objects with sizes on the order of a nanometer (one billionth of a meter).

- Pure carbon exists as diamond, graphite, *fullerenes,* and *carbon nanotubes.*

- Two-dimensional sheets of sp^2-hybridized carbon atoms are called *graphenes.*

SECTION 24.6

- Electrical conductors (e.g., metals) have no gap between their *valence band* and *conduction band,* so electrons are easily promoted from the valence to the conduction band. Semiconductors, which conduct electricity more than nonconductors (insulators) but less than metallic conductors, have a small gap. The gap in *nonconductors* is too large for electrons to be promoted from the valence to the conduction band.

- *Semiconductors* generally have a total of eight valence electrons per formula unit. Thus, they typically form between two Group 4A elements, one Group 3A element and one Group 5A element, or one Group 2B element and one Group 6A element.

- Semiconductors can be *doped* with an element with fewer valence electrons to produce a *p-type semiconductor,* or with an element with more valence electrons to produce an *n-type semiconductor.*

- *p*-Type and *n*-type semiconductors can be used together in solar cells and *diodes.*

SECTION 24.7

- *Superconductors* offer no resistance to the flow of electrons. The earliest known superconductors only exhibited the behavior at liquid helium temperature (4 K), but modern *high-temperature superconductors* conduct at temperatures as high as 140 K.

- The highest temperature at which a superconductor exhibits superconductivity is called the *superconducting transition temperature* (T_c).

- Superconducting substances exhibit the *Meissner effect,* the exclusion of magnetic fields in the substance's volume. This can be demonstrated by levitating a magnet above a superconductor.

Key Words

Questions and Problems

SECTION 24.1: POLYMERS

Review Questions

24.1 Bakelite, the first commercially produced polymer, contains monomer units of phenol and formaldehyde. If an item made of Bakelite were broken, it could not be melted down and re-formed. Is Bakelite a thermoplastic or thermosetting polymer? Can Bakelite be recycled?

24.2 How are SBS rubber [Figure 24.6(a)] and nylon 66 similar? How are they different?

Conceptual Problems

24.3 What type of monomer unit is needed to make an electrically conducting organic polymer?

24.4 Draw the structure of the addition polymer formed by 1-butene.

$$H_2C\!=\!\overset{\textstyle H}{\underset{\textstyle H_2}{C}}\!-\!\overset{}{\underset{}{C}}\!-\!CH_3$$

Show at least two repeating units.

24.5 Draw the structure of the alternating copolymer of these two compounds, showing at least two repeating units.

$$\underset{H_3C}{\overset{H_3C}{>}}\!\!\!\overset{CH_2}{\underset{}{C}}\!=\!\overset{}{\underset{CH_3}{C}}\!\!\!\!<\qquad H\!-\!C\!\equiv\!C\!-\!H$$

24.6 Draw the structure of an alternating copolymer of these two amino acids, showing at least two repeating units.

$$\underset{SH}{\overset{O}{\underset{CH_2}{H_2N\!-\!CH\!-\!C\!-\!OH}}}\qquad \underset{CH_3}{\overset{O}{\underset{CH\!-\!OH}{H_2N\!-\!CH\!-\!C\!-\!OH}}}$$

SECTION 24.2: CERAMICS AND COMPOSITE MATERIALS

Review Questions

24.7 What types of atoms and bonds are present in the polymer "backbone" in a ceramic containing a transition metal?

24.8 Describe two natural types of composite materials and the substances that they contain.

Conceptual Problems

24.9 What steps would be used to form a ceramic material from scandium and ethanol (CH_3CH_2OH) using the sol-gel process?

24.10 Amorphous silica (SiO_2) can be formed in uniform spheres from a sol-gel process using methanol (CH_3OH) solvent. List the chemical reactions involved in this process.

24.11 Bakelite, described in Review Question 24.1, is manufactured using materials like cellulose paper or glass fibers that are compressed and heated among the monomer units. Is Bakelite a simple polymer? If not, what classification best fits Bakelite?

SECTION 24.3: LIQUID CRYSTALS

Review Questions

24.12 Is a normal liquid isotropic or anisotropic? How is an anisotropic material different from an isotropic material?

24.13 Describe how nematic, smectic, and cholesteric liquid crystals are similar and different.

Conceptual Problems

24.14 Would each of these molecules be likely to form a liquid crystal? If not, why not?

(a)

(b)

(c) H₂C=C

24.15 Would long-chain hydrocarbons or addition polymers of alkene molecules form liquid crystals? Why or why not?

24.16 Would an ionic compound form a liquid crystal? Why or why not?

SECTION 24.4: BIOMEDICAL MATERIALS

Review Questions

24.17 Name three major concerns that must be addressed when proposing any material to be used in medical applications inside the body.

24.18 Name a necessary chemical property and a necessary physical property of dental materials.

Conceptual Problems

24.19 Is the polymer commonly used in artificial skin material (shown here) an addition polymer or a condensation polymer? Is it a block, alternating, graft, or random copolymer?

$$\left(OCH_2\overset{\overset{\textstyle O}{\|}}{C}-O\overset{\overset{\textstyle O}{\|}}{C}HC \right)_n$$
$$\underset{CH_3}{|}$$

24.20 What are some advantages and disadvantages of the use of ceramics in replacement joint bearing surfaces relative to metals?

24.21 How do amalgam and composite tooth fillings differ in the ways they fail over time?

SECTION 24.5: NANOTECHNOLOGY

Review Questions

24.22 How does an STM measure the peak and valley heights along the surface of a sample?

24.23 Name four allotropic forms of carbon.

Conceptual Problems

24.24 How does an AFM differ from an STM? Why can graphite possibly be investigated with either instrument whereas diamond can only be investigated with an AFM?

24.25 What is the hybridization of the carbon atoms in a single-walled nanotube?

24.26 What structural or bonding aspect of graphite and carbon nanotubes allows their use as semiconductors, whereas diamond is an electrical insulator?

24.27 What type of intermolecular forces holds the sheets of carbon atoms together in graphite? What type of intermolecular forces hold carbon nanotubes together in strands and fibers?

SECTION 24.6: SEMICONDUCTORS

Review Questions

24.28 How do the band gaps differ among conductors, semiconductors, and insulators?

24.29 The only elemental semiconductors (Si, Ge, and graphite) are members of Group 4A. Why are Sn and Pb, which are also in Group 4A, not classified as semiconductors?

Conceptual Problems

24.30 State whether each combination of elements would be expected to produce a semiconductor: (a) Ga and Sb, (b) As and N, (c) B and P, (d) Zn and Sb, (e) Cd and S.

24.31 What type of semiconductor would be formed (a) if Si were doped with Sb and (b) if Ge were doped with B?

24.32 Describe the movement of electrons in a solar battery that uses p-type and n-type semiconductors when light shines on the solar cell.

SECTION 24.7: SUPERCONDUCTORS

Review Questions

24.33 Why is YBCO called a "high-temperature" semiconductor when it exhibits superconductivity only below −180°C, which is very cold?

24.34 How might the Meissner effect be used in a real-world application? What are the engineering challenges to this application with the current superconductors?

Conceptual Problems

24.35   What is the molecular formula of BSCCO-2201?

24.36   Superconductivity at room temperature is sometimes a subject of science fiction or fantasy. For example, in the movie *Terminator 2: Judgment Day,* a cyborg has a CPU that contains a room-temperature superconductor. How might a computer with such a CPU differ in design from current computers?

24.37   What are the electron configurations of the copper ions in YBCO? Are both ions expected oxidation states for copper?

Additional Problems

24.38   Draw representations of block copolymers and graft copolymers using the symbols A and B for two types of monomer units. Are the categories exclusive? Using a third monomer type, see if you can draw a representation of a polymer that is both a block and a graft copolymer.

24.39   Draw the structures of the two monomer units in the polymer

24.40   Is the polymer in Problem 24.39 a condensation polymer or an addition polymer? Is it a polyester, a polyamide, or something else?

24.41   What types of bonding (covalent, ionic, network, metallic) are present in a plastic polymer? In a ceramic material?

24.42   Draw representations of isotactic, syndiotactic, and atactic polyacrylonitrile (for the monomer unit, see Table 24.1).

24.43   Fluoride ion is commonly used in drinking water supplies and in toothpaste (in the form of sodium

fluoride) to prevent tooth decay. The fluoride ions replace hydroxide ions in the enamel mineral hydroxyapatite [$Ca_5(PO_4)_3OH$, the same mineral as in bone], forming fluoroapatite [$Ca_5(PO_4)_3F$]. Why does replacing the hydroxide ion with fluoride ion prevent tooth decay? How does this relate to the properties of materials used in artificial fillings?

24.44   Would the metal alkoxide in the sol-gel process be acidic or basic?

24.45   Would the compound shown form a liquid crystal? Why or why not?

24.46   The semiconductor band gap energy determines the color of an LED. Knowing what you learned in Chapter 3 about the energies of photons of different colors of light, which semiconductor would have the larger band gap: the one in a green exit sign's LEDs, or the one in a red traffic signal?

24.47   Draw the structures of poly(1-butyne) and poly(1-butene), showing three or more repeating units. How are they different? How are they similar? Which of them, if any, do you think will conduct electricity?

$$H-C{\equiv}C-CH_2CH_3 \qquad H_2C{=}CH-CH_2CH_3$$
$$\text{1-butyne} \qquad\qquad \text{1-butene}$$

24.48   Propyne ($HC{\equiv}CCH_3$) can be used to form an electrically conducting polymer. Draw the structure of polypropyne, showing at least three repeating units, and write the general formula for the polymer.

Answers to In-Chapter Materials

PRACTICE PROBLEMS

24.1B

$-(CH_2CCl_2)-$

24.2A

24.2B condensation.

24.3A The molecule is long, thin, and has conjugated rings that limit free rotation. **24.3B** No. This molecule lacks the rigidity needed for liquid crystal behavior. **24.4A** (a) elements with two valence electrons, e.g., Be, Sr, Cd, (b) elements with three valence electrons, e.g., B, Ga, In, (c) elements with six valence electrons, e.g., O, Se, Te. **24.4B** Carbon can be combined with silicon to produce a semiconductor. Arsenic can be used to dope silicon to produce an *n*-type semiconductor; and gallium can be used to dope silicon to produce a *p*-type semiconductor.

SECTION REVIEW

24.1.1 a. **24.1.2** d. **24.3.1** c. **24.3.2** a, c. **24.6.1** d.

Appendix 1

Scientific Notation

Chemists often deal with numbers that are either extremely large or extremely small. For example, in 1 g of the element hydrogen there are roughly

$$602,200,000,000,000,000,000,000$$

hydrogen atoms. Each hydrogen atom has a mass of only

$$0.00000000000000000000000166 \text{ g}$$

These numbers are cumbersome to handle, and it is easy to make mistakes when using them in arithmetic computations. Consider the following multiplication.

$$0.0000000056 \times 0.00000000048 = 0.000000000000000002688$$

It would be easy for us to miss one zero or add one more zero after the decimal point. Consequently, when working with very large and very small numbers, we use a system called *scientific notation*. Regardless of their magnitude, all numbers can be expressed in the form

$$N \times 10^n$$

where N is a number between 1 and 10 and n, the exponent, is a positive or negative integer (whole number). Any number expressed in this way is said to be written in scientific notation.

Suppose that we are given a certain number and asked to express it in scientific notation. Basically, this assignment calls for us to find n. We count the number of places that the decimal point must be moved to give the number N (which is between 1 and 10). If the decimal point has to be moved to the left, then n is a positive integer; if it has to be moved to the right, n is a negative integer. The following examples illustrate the use of scientific notation.

1. Express 568.762 in scientific notation.

$$568.762 = 5.68762 \times 10^2$$

Note that the decimal point is moved to the left by two places and $n = 2$.
2. Express 0.00000772 in scientific notation.

$$0.00000772 = 7.72 \times 10^{-6}$$

Here the decimal point is moved to the right by six places and $n = -6$.

Keep in mind the following two points. First, $n = 0$ is used for numbers that are not expressed in scientific notation. For example, 74.6×10^0 ($n = 0$) is equivalent to 74.6. Second, the usual practice is to omit the superscript when $n = 1$. Thus the scientific notation for 74.6 is 7.46×10 and not 7.46×10^1.

Next, we consider how scientific notation is handled in arithmetic operations.

Addition and Subtraction

To add or subtract using scientific notation, we first write each quantity—say N_1 and N_2—with the same exponent n. Then we combine N_1 and N_2; the exponents remain the same. Consider the following examples.

$$(7.4 \times 10^3) + (2.1 \times 10^3) = 9.5 \times 10^3$$

$$(4.31 \times 10^4) + (3.9 \times 10^3) = (4.31 \times 10^4) + (0.39 \times 10^4)$$

$$= 4.70 \times 10^4$$

$$(2.22 \times 10^{-2}) - (4.10 \times 10^{-3}) = (2.22 \times 10^{-2}) - (0.41 \times 10^{-2})$$

$$= 1.81 \times 10^{-2}$$

Multiplication and Division

To multiply numbers expressed in scientific notation, we multiply N_1 and N_2 in the usual way, but *add* the exponents together. To divide using scientific notation, we divide N_1 and N_2 as usual and subtract the exponents. The following examples show how these operations are performed.

$$(8.0 \times 10^4) \times (5.0 \times 10^2) = (8.0 \times 5.0)(10^{4+2})$$

$$= 40 \times 10^6$$

$$= 4.0 \times 10^7$$

$$(4.0 \times 10^{-5}) \times (7.0 \times 10^3) = (4.0 \times 7.0)(10^{-5+3})$$

$$= 28 \times 10^{-2}$$

$$= 2.8 \times 10^{-1}$$

$$\frac{6.9 \times 10^7}{3.0 \times 10^{-5}} = \frac{6.9}{3.0} \times 10^{7-(-5)}$$

$$= 2.3 \times 10^{12}$$

$$\frac{8.5 \times 10^4}{5.0 \times 10^9} = \frac{8.5}{5.0} \times 10^{4-9}$$

$$= 1.7 \times 10^{-5}$$

Basic Trigonometry

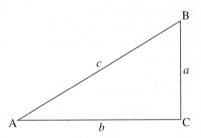

In the triangle shown, A, B, and C are angles (C = 90°) and a, b, and c are side lengths. To calculate unknown angles or sides, we use the following relationships.

$$a^2 + b^2 = c^2$$

$$\sin A = \frac{a}{c}$$

$$\cos A = \frac{b}{c}$$

$$\tan A = \frac{a}{b}$$

Logarithms

Common Logarithms

The concept of logarithms is an extension of the concept of exponents, which is discussed on page A-1. The common, or base-10, logarithm of any number is the power to which 10 must be raised to equal the number. The following examples illustrate this relationship.

Logarithm	Exponent
$\log 1 = 0$	$10^0 = 1$
$\log 10 = 1$	$10^1 = 10$
$\log 100 = 2$	$10^2 = 100$
$\log 10^{-1} = -1$	$10^{-1} = 0.1$
$\log 10^{-2} = -2$	$10^{-2} = 0.01$

In each case the logarithm of the number can be obtained by inspection.

Because the logarithms of numbers are exponents, they have the same properties as exponents. Thus, we have

Logarithm	Exponent
$\log AB = \log A + \log B$	$10^A \times 10^B = 10^{A+B}$
$\log \dfrac{A}{B} = \log A - \log B$	$\dfrac{10^A}{10^B} = 10^{A-B}$

Furthermore, $\log A^n = n \log A$.

Now suppose we want to find the common logarithm of 6.7×10^{-4}. On most electronic calculators, the number is entered first and then the log key is pressed. This operation gives us

$$\log 6.7 \times 10^{-4} = -3.17$$

Note that there are as many digits *after* the decimal point as there are significant figures in the original number. The original number has two significant figures, and the "17" in −3.17 tells us that the log has two significant figures. The "3" in −3.17 serves only to locate the decimal point in the number 6.7×10^{-4}. Other examples are

Number	Common Logarithm
62	1.79
0.872	−0.0595
1.0×10^{-7}	−7.00

Sometimes (as in the case of pH calculations) it is necessary to obtain the number whose logarithm is known. This procedure is known as taking the antilogarithm; it is simply the reverse of taking the logarithm of a number. Suppose in a certain calculation we have pH = 1.46 and are asked to calculate $[H^+]$. From the definition of pH (pH = $-\log [H^+]$) we can write

$$[H^+] = 10^{-1.46}$$

Many calculators have a key labeled $\log^{-1}$ or INV log to obtain antilogs. Other calculators have a 10^x or y^x key (where x corresponds to −1.46 in our example and y is 10 for base-10 logarithm). Therefore, we find that $[H^+] = 0.035$ M.

Natural Logarithms

Logarithms taken to the base e instead of 10 are known as natural logarithms (denoted by ln or $\log_e$); e is equal to 2.7183. The relationship between common logarithms and natural logarithms is as follows:

$$\log 10 = 1 \qquad\qquad 10^1 = 10$$

$$\ln 10 = 2.303 \qquad\qquad e^{2.303} = 10$$

Thus,

$$\ln x = 2.303 \log x$$

To find the natural logarithm of 2.27, say, we first enter the number on the electronic calculator and then press the ln key to get

$$\ln 2.27 = 0.820$$

If no ln key is provided, we can proceed as follows:

$$2.303 \log 2.27 = 2.303 \times 0.356$$

$$= 0.820$$

Sometimes we may be given the natural logarithm and asked to find the number it represents. For example,

$$\ln x = 59.7$$

On many calculators, we simply enter the number and press the e key.

$$e^{59.7} = 8 \times 10^{25}$$

The Quadratic Equation

A quadratic equation takes the form

$$ax^2 + bx + c = 0$$

If coefficients a, b, and c are known, then x is given by

$$x = \frac{-b \pm \sqrt{b^2 - 4ac}}{2a}$$

Suppose we have the following quadratic equation.

$$2x^2 + 5x - 12 = 0$$

Solving for x, we write

$$x = \frac{-5 \pm \sqrt{(5)^2 - 4(2)(-12)}}{2(2)}$$

$$= \frac{-5 \pm \sqrt{25 + 96}}{4}$$

Therefore,

$$x = \frac{-5 + 11}{4} = \frac{3}{2}$$

and

$$x = \frac{-5 - 11}{4} = -4$$

Successive Approximation

In the determination of hydrogen ion concentration in a weak acid solution, use of the quadratic equation can sometimes be avoided using a method known as *successive approximation*. Consider the example of a 0.0150-*M* solution of hydrofluoric acid (HF). The K_a for HF is 7.10×10^{-4}. To determine the hydrogen ion concentration in this solution, we construct an equilibrium table and enter the initial concentrations, the expected change in concentrations, and the equilibrium concentrations of all species.

$$HF(aq) \rightleftharpoons H^+(aq) + F^-(aq)$$

Initial concentration (M):	0.0150	0	0
Change in concentration (M):	$-x$	$+x$	$+x$
Equilibrium concentration (M):	$0.0150 - x$	x	x

Using the rule that x can be neglected if the initial acid concentration divided by the K_a is greater than 100, we find that in this case x cannot be neglected ($0.0150/7.1 \times 10^{-4} \approx 21$). Successive approximation involves first neglecting x with respect to initial acid concentration

$$\frac{x^2}{0.0150 - x} \approx \frac{x^2}{0.0150} = 7.10 \times 10^{-4}$$

and solving for x.

$$x^2 = (0.0150)(7.10 \times 10^{-4}) = 1.07 \times 10^{-5}$$

$$x = \sqrt{1.07 \times 10^{-5}} = 0.00326 \, M$$

We then solve for x again, this time using the calculated value of x on the bottom of the fraction.

$$\frac{x^2}{0.0150 - x} = \frac{x^2}{0.0150 - 0.00326} = 7.10 \times 10^{-4}$$

$$x^2 = (0.0150 - 0.00326)(7.10 \times 10^{-4}) = 8.33 \times 10^{-6}$$

$$x = \sqrt{8.33 \times 10^{-6}} = 0.00289 \, M$$

Note that the calculated value of x decreased from 0.00326 to 0.00289. We now use the new calculated value of x on the bottom of the fraction and solve for x again.

$$\frac{x^2}{0.0150 - x} = \frac{x^2}{0.0150 - 0.00289} = 7.10 \times 10^{-4}$$

$$x^2 = (0.0150 - 0.00289)(7.10 \times 10^{-4}) = 8.60 \times 10^{-6}$$

$$x = \sqrt{8.60 \times 10^{-6}} = 0.00293 \, M$$

This time the value of x increased slightly. We use the new calculated value and solve for x again.

$$\frac{x^2}{0.0150 - x} = \frac{x^2}{0.0150 - 0.00293} = 7.10 \times 10^{-4}$$

$$x^2 = (0.0150 - 0.00293)(7.10 \times 10^{-4}) = 8.57 \times 10^{-6}$$

$$x = \sqrt{8.57 \times 10^{-6}} = 0.00293 \, M$$

This time we find that the answer is still 0.00293 *M,* so there is no need to repeat the process. In general, we apply the method of successive approximation until the value of x obtained does not differ from the value obtained in the previous step. The value of x determined using successive approximation is the same value we would get if we were to use the quadratic equation.

Appendix 2

THERMODYNAMIC DATA AT 1 ATM AND 25°C*

Inorganic Substances

Substance	ΔH_f° (kJ/mol)	ΔG_f° (kJ/mol)	S° (J/K · mol)
Ag(s)	0	0	42.7
Ag$^+$(aq)	105.9	77.1	73.9
AgCl(s)	−127.0	−109.7	96.1
AgBr(s)	−99.5	−95.9	107.1
AgI(s)	−62.4	−66.3	114.2
AgNO$_3$(s)	−123.1	−32.2	140.9
Al(s)	0	0	28.3
Al^{3+}(aq)	−524.7	−481.2	−313.38
Al$_2$O$_3$(s)	−1669.8	−1576.4	50.99
As(s)	0	0	35.15
AsO$_4^{3-}$(aq)	−870.3	−635.97	−144.77
AsH$_3$(g)	171.5		
H$_3$AsO$_4$(s)	−900.4		
Au(s)	0	0	47.7
Au$_2$O$_3$(s)	80.8	163.2	125.5
AuCl(s)	−35.2		
AuCl$_3$(s)	−118.4		
B(s)	0	0	6.5
B$_2$O$_3$(s)	−1263.6	−1184.1	54.0
H$_3$BO$_3$(s)	−1087.9	−963.16	89.58
H$_3$BO$_3$(aq)	−1067.8	−963.3	159.8
Ba(s)	0	0	66.9
Ba^{2+}(aq)	−538.4	−560.66	12.55
BaO(s)	−558.2	−528.4	70.3
BaCl$_2$(s)	−860.1	−810.66	125.5

*The thermodynamic quantities of ions are based on the reference states that $\Delta H_f^\circ[H^+(aq)] = 0$, $\Delta G_f^\circ[H^+(aq)] = 0$, and $S^\circ[H^+(aq)] = 0$.

Substance	ΔH_f° (kJ/mol)	ΔG_f° (kJ/mol)	S° (J/K · mol)
$BaSO_4(s)$	−1464.4	−1353.1	132.2
$BaCO_3(s)$	−1218.8	−1138.9	112.1
$Be(s)$	0	0	9.5
$BeO(s)$	−610.9	−581.58	14.1
$Br_2(l)$	0	0	152.3
$Br_2(g)$	30.7	3.14	245.13
$Br^-(aq)$	−120.9	−102.8	80.7
$HBr(g)$	−36.2	−53.2	198.48
$C(graphite)$	0	0	5.69
$C(diamond)$	1.90	2.87	2.4
$CCl_4(g)$	−95.7	−62.3	309.7
$CCl_4(l)$	−128.2	−66.4	216.4
$CO(g)$	−110.5	−137.3	197.9
$CO_2(g)$	−393.5	−394.4	213.6
$CO_2(aq)$	−412.9	−386.2	121.3
$CO_3^{2-}(aq)$	−676.3	−528.1	−53.1
$HCO_3^-(aq)$	−691.1	−587.1	94.98
$H_2CO_3(aq)$	−699.7	−623.2	187.4
$CS_2(g)$	115.3	65.1	237.8
$CS_2(l)$	87.3	63.6	151.0
$HCN(aq)$	105.4	112.1	128.9
$CN^-(aq)$	151.0	165.69	117.99
$(NH_2)_2CO(s)$	−333.19	−197.15	104.6
$(NH_2)_2CO(aq)$	−319.2	−203.84	173.85
$Ca(s)$	0	0	41.6
$Ca(g)$	179.3	145.5	154.8
$Ca^{2+}(aq)$	−542.96	−553.0	−55.2
$CaO(s)$	−635.6	−604.2	39.8
$Ca(OH)_2(s)$	−986.6	−896.8	83.4
$CaF_2(s)$	−1214.6	−1161.9	68.87
$CaCl_2(s)$	−794.96	−750.19	113.8
$CaSO_4(s)$	−1432.69	−1320.3	106.69
$CaCO_3(s)$	−1206.9	−1128.8	92.9
$Cd(s)$	0	0	51.46
$Cd^{2+}(aq)$	−72.38	−77.7	−61.09
$CdO(s)$	−254.6	−225.06	54.8

(Continued)

Substance	ΔH_f° (kJ/mol)	ΔG_f° (kJ/mol)	S° (J/K · mol)
$CdCl_2(s)$	−389.1	−342.59	118.4
$CdSO_4(s)$	−926.17	−820.2	137.2
$Cl_2(g)$	0	0	223.0
$Cl(g)$	121.7	105.7	165.2
$Cl^-(aq)$	−167.2	−131.2	56.5
$HCl(g)$	−92.3	−95.27	187.0
$Co(s)$	0	0	28.45
$Co^{2+}(aq)$	−67.36	−51.46	155.2
$CoO(s)$	−239.3	−213.38	43.9
$Cr(s)$	0	0	23.77
$Cr^{2+}(aq)$	−138.9		
$Cr_2O_3(s)$	−1128.4	−1046.8	81.17
$CrO_4^{2-}(aq)$	−863.16	−706.26	38.49
$Cr_2O_7^{2-}(aq)$	−1460.6	−1257.29	213.8
$Cs(s)$	0	0	82.8
$Cs(g)$	76.50	49.53	175.6
$Cs^+(aq)$	−247.69	−282.0	133.05
$CsCl(s)$	−442.8	−414.4	101.2
$Cu(s)$	0	0	33.3
$Cu^+(aq)$	51.88	50.2	−26.4
$Cu^{2+}(aq)$	64.39	64.98	−99.6
$CuO(s)$	−155.2	−127.2	43.5
$Cu_2O(s)$	−166.69	−146.36	100.8
$CuCl(s)$	−134.7	−118.8	91.6
$CuCl_2(s)$	−205.85	?	?
$CuS(s)$	−48.5	−49.0	66.5
$CuSO_4(s)$	−769.86	−661.9	113.39
$F_2(g)$	0	0	203.34
$F(g)$	80.0	61.9	158.7
$F^-(aq)$	−329.1	−276.48	−9.6
$HF(g)$	−271.6	−270.7	173.5
$Fe(s)$	0	0	27.2
$Fe^{2+}(aq)$	−87.86	−84.9	−113.39
$Fe^{3+}(aq)$	−47.7	−10.5	−293.3
$FeO(s)$	−272.0	−255.2	60.8
$Fe_2O_3(s)$	−822.2	−741.0	90.0
$Fe(OH)_2(s)$	−568.19	−483.55	79.5
$Fe(OH)_3(s)$	−824.25	?	?
$H(g)$	218.2	203.2	114.6
$H_2(g)$	0	0	131.0

Substance	ΔH_f° (kJ/mol)	ΔG_f° (kJ/mol)	S° (J/K · mol)
$H^+(aq)$	0	0	0
$OH^-(aq)$	−229.94	−157.30	−10.5
$H_2O(g)$	−241.8	−228.6	188.7
$H_2O(l)$	−285.8	−237.2	69.9
$H_2O_2(g)$	−136.1	−105.5	232.9
$H_2O_2(l)$	−187.6	−118.1	109.6
$Hg(l)$	0	0	77.4
$Hg^{2+}(aq)$		−164.38	
$HgO(s)$	−90.7	−58.5	72.0
$HgCl_2(s)$	−230.1		
$Hg_2Cl_2(s)$	−264.9	−210.66	196.2
$HgS(s)$	−58.16	−48.8	77.8
$HgSO_4(s)$	−704.17		
$Hg_2SO_4(s)$	−741.99	−623.92	200.75
$I_2(g)$	62.25	19.37	260.57
$I_2(s)$	0	0	116.7
$I(g)$	106.8	70.21	180.67
$I^-(aq)$	−55.9	−51.67	109.37
$HI(g)$	25.9	1.30	206.3
$K(s)$	0	0	63.6
$K^+(aq)$	−251.2	−282.28	102.5
$KOH(s)$	−425.85		
$KCl(s)$	−435.87	−408.3	82.68
$KClO_3(s)$	−391.20	−289.9	142.97
$KClO_4(s)$	−433.46	−304.18	151.0
$KBr(s)$	−392.17	−379.2	96.4
$KI(s)$	−327.65	−322.29	104.35
$KNO_3(s)$	−492.7	−393.1	132.9
$Li(s)$	0	0	28.0
$Li(g)$	159.3	126.6	138.8
$Li^+(aq)$	−278.46	−293.8	14.2
$LiCl(s)$	−408.3	−384.0	59.30
$Li_2O(s)$	−595.8	?	?
$LiOH(s)$	−487.2	−443.9	50.2
$Mg(s)$	0	0	32.5
$Mg(g)$	150	115	148.55
$Mg^{2+}(aq)$	−461.96	−456.0	−117.99
$MgO(s)$	−601.8	−569.6	26.78
$Mg(OH)_2(s)$	−924.66	−833.75	63.1
$MgCl_2(s)$	−641.8	−592.3	89.5

(Continued)

Substance	ΔH_f° (kJ/mol)	ΔG_f° (kJ/mol)	S° (J/K · mol)
$MgSO_4(s)$	−1278.2	−1173.6	91.6
$MgCO_3(s)$	−1112.9	−1029.3	65.69
$Mn(s)$	0	0	31.76
$Mn^{2+}(aq)$	−218.8	−223.4	−83.68
$MnO_2(s)$	−520.9	−466.1	53.1
$N(g)$	470.4	455.5	153.3
$N_2(g)$	0	0	191.5
$N_3^-(aq)$	245.18	?	?
$NH_3(g)$	−46.3	−16.6	193.0
$NH_4^+(aq)$	−132.80	−79.5	112.8
$NH_4Cl(s)$	−315.39	−203.89	94.56
$NH_3(aq)$	−80.3	−26.5	111.3
$N_2H_4(l)$	50.4		
$NO(g)$	90.4	86.7	210.6
$NO_2(g)$	33.85	51.8	240.46
$N_2O_4(g)$	9.66	98.29	304.3
$N_2O(g)$	81.56	103.6	219.99
$HNO_2(aq)$	−118.8	−53.6	
$HNO_3(l)$	−173.2	−79.9	155.6
$NO_3^-(aq)$	−206.57	−110.5	146.4
$Na(s)$	0	0	51.05
$Na(l)$	2.41	0.50	57.56
$Na(g)$	107.7	77.3	153.7
$Na^+(aq)$	−239.66	−261.87	60.25
$NaOH(aq)$	−469.6	−419.2	49.8
$Na_2O(s)$	−415.9	−376.56	72.8
$NaCl(s)$	−410.9	−384.0	72.38
$NaI(s)$	−288.0		
$Na_2SO_4(s)$	−1384.49	−1266.8	149.49
$NaNO_3(s)$	−466.68	−365.89	116.3
$Na_2CO_3(s)$	−1130.9	−1047.67	135.98
$NaHCO_3(s)$	−947.68	−851.86	102.09
$Ni(s)$	0	0	30.1
$Ni^{2+}(aq)$	−64.0	−46.4	−159.4
$NiO(s)$	−244.35	−216.3	38.58
$Ni(OH)_2(s)$	−538.06	−453.1	79.5
$O(g)$	249.4	230.1	160.95
$O_2(g)$	0	0	205.0
$O_3(aq)$	−12.09	16.3	110.88
$O_3(g)$	142.2	163.4	237.6

Substance	ΔH_f° (kJ/mol)	ΔG_f° (kJ/mol)	S° (J/K · mol)
P(white)	0	0	44.0
P(red)	−18.4	13.8	29.3
$PCl_3(l)$	−319.7	−272.3	217.1
$PCl_3(g)$	−288.07	−269.6	311.7
$PCl_5(g)$	−374.9	−305.0	364.5
$PO_4^{3-}(aq)$	−1284.07	−1025.59	−217.57
$P_4O_{10}(s)$	−3012.48		
$PH_3(g)$	9.25	18.2	210.0
$HPO_4^{2-}(aq)$	−1298.7	−1094.1	−35.98
$H_2PO_4^-(aq)$	−1302.48	−1135.1	89.1
Pb(s)	0	0	64.89
$Pb^{2+}(aq)$	1.6	−24.3	21.3
PbO(s)	−217.86	−188.49	69.45
$PbO_2(s)$	−276.65	−218.99	76.57
$PbCl_2(s)$	−359.2	−313.97	136.4
PbS(s)	−94.3	−92.68	91.2
$PbSO_4(s)$	−918.4	−811.2	147.28
Pt(s)	0	0	41.84
$PtCl_4^{2-}(aq)$	−516.3	−384.5	175.7
Rb(s)	0	0	69.45
Rb(g)	85.8	55.8	170.0
$Rb^+(aq)$	−246.4	−282.2	124.27
RbBr(s)	−389.2	−378.1	108.3
RbCl(s)	−435.35	−407.8	95.90
RbI(s)	−328	−326	118.0
S(rhombic)	0	0	31.88
S(monoclinic)	0.30	0.10	32.55
SO(g)	5.01	−19.9	221.8
$SO_2(g)$	−296.4	−300.4	248.5
$SO_3(g)$	−395.2	−370.4	256.2
$SO_3^{2-}(aq)$	−624.25	−497.06	43.5
$SO_4^{2-}(aq)$	−907.5	−741.99	17.15
$H_2S(g)$	−20.15	−33.0	205.64
$HSO_3^-(aq)$	−627.98	−527.3	132.38
$HSO_4^-(aq)$	−885.75	−752.87	126.86
$H_2SO_4(l)$	−811.3	?	?
$SF_6(g)$	−1096.2	?	?
Si(s)	0	0	18.70
$SiO_2(s)$	−859.3	−805.0	41.84
Sr(s)	0	0	54.39

(Continued)

Substance	ΔH_f° (kJ/mol)	ΔG_f° (kJ/mol)	S° (J/K · mol)
$Sr^{2+}(aq)$	−545.5	−557.3	−39.33
$SrCl_2(s)$	−828.4	−781.15	117.15
$SrSO_4(s)$	−1444.74	−1334.28	121.75
$SrCO_3(s)$	−1218.38	−1137.6	97.07
$U(s)$	0	0	50.21
$UF_6(g)$	−2147	−2064	378
$Zn(s)$	0	0	41.6
$Zn^{2+}(aq)$	−152.4	−147.2	−106.48
$ZnO(s)$	−348.0	−318.2	43.9
$ZnCl_2(s)$	−415.89	−369.26	108.37
$ZnS(s)$	−202.9	−198.3	57.7
$ZnSO_4(s)$	−978.6	−871.6	124.7

Organic Substances

Substance	Formula	ΔH_f° (kJ/mol)	ΔG_f° (kJ/mol)	S° (J/K · mol)
Acetic acid(l)	CH_3COOH	−484.2	−389.45	159.8
Acetaldehyde(g)	CH_3CHO	−166.35	−139.08	264.2
Acetone(l)	CH_3COCH_3	−246.8	−153.55	198.7
Acetylene(g)	C_2H_2	226.6	209.2	200.8
Benzene(l)	C_6H_6	49.04	124.5	172.8
Butane(g)	C_4H_{10}	−124.7	−15.7	310.0
Ethanol(l)	C_2H_5OH	−276.98	−174.18	161.0
Ethane(g)	C_2H_6	−84.7	−32.89	229.5
Ethylene(g)	C_2H_4	52.3	68.1	219.5
Formic acid(l)	$HCOOH$	−409.2	−346.0	129.0
Glucose(s)	$C_6H_{12}O_6$	−1274.5	−910.56	212.1
Methane(g)	CH_4	−74.85	−50.8	186.2
Methanol(l)	CH_3OH	−238.7	−166.3	126.8
Propane(g)	C_3H_8	−103.9	−23.5	269.9
Sucrose(s)	$C_{12}H_{22}O_{11}$	−2221.7	−1544.3	360.2

Appendix 3

Compound	Dissolution equilibrium	K_{sp}
Bromides		
Copper(I) bromide	$CuBr(s) \rightleftharpoons Cu^+(aq) + Br^-(aq)$	4.2×10^{-8}
Lead(II) bromide	$PbBr_2(s) \rightleftharpoons Pb^{2+}(aq) + 2Br^-(aq)$	6.6×10^{-6}
Mercury(I) bromide	$Hg_2Br_2(s) \rightleftharpoons Hg_2^{2+}(aq) + 2Br^-(aq)$	6.4×10^{-23}
Silver bromide	$AgBr(s) \rightleftharpoons Ag^+(aq) + Br^-(aq)$	7.7×10^{-13}
Carbonates		
Barium carbonate	$BaCO_3(s) \rightleftharpoons Ba^{2+}(aq) + CO_3^{2-}(aq)$	8.1×10^{-9}
Calcium carbonate	$CaCO_3(s) \rightleftharpoons Ca^{2+}(aq) + CO_3^{2-}(aq)$	8.7×10^{-9}
Lead(II) carbonate	$PbCO_3(s) \rightleftharpoons Pb^{2+}(aq) + CO_3^{2-}(aq)$	3.3×10^{-14}
Magnesium carbonate	$MgCO_3(s) \rightleftharpoons Mg^{2+}(aq) + CO_3^{2-}(aq)$	4.0×10^{-5}
Silver carbonate	$Ag_2CO_3(s) \rightleftharpoons 2Ag^+(aq) + CO_3^{2-}(aq)$	8.1×10^{-12}
Strontium carbonate	$SrCO_3(s) \rightleftharpoons Sr^{2+}(aq) + CO_3^{2-}(aq)$	1.6×10^{-9}
Chlorides		
Lead(II) chloride	$PbCl_2(s) \rightleftharpoons Pb^{2+}(aq) + 2Cl^-(aq)$	2.4×10^{-4}
Mercury(I) chloride	$Hg_2Cl_2(s) \rightleftharpoons Hg_2^{2+}(aq) + 2Cl^-(aq)$	3.5×10^{-18}
Silver chloride	$AgCl(s) \rightleftharpoons Ag^+(aq) + Cl^-(aq)$	1.6×10^{-10}
Chromates		
Lead(II) chromate	$PbCrO_4(s) \rightleftharpoons Pb^{2+}(aq) + CrO_4^{2-}(aq)$	2.0×10^{-14}
Silver chromate	$Ag_2CrO_4(s) \rightleftharpoons 2Ag^+(aq) + CrO_4^{2-}(aq)$	1.2×10^{-12}
Fluorides		
Barium fluoride	$BaF_2(s) \rightleftharpoons Ba^{2+}(aq) + 2F^-(aq)$	1.7×10^{-6}
Calcium fluoride	$CaF_2(s) \rightleftharpoons Ca^{2+}(aq) + 2F^-(aq)$	4.0×10^{-11}
Lead(II) fluoride	$PbF_2(s) \rightleftharpoons Pb^{2+}(aq) + 2F^-(aq)$	4.0×10^{-8}
Hydroxides		
Aluminum hydroxide	$Al(OH)_3(s) \rightleftharpoons Al^{3+}(aq) + 3OH^-(aq)$	1.8×10^{-33}
Calcium hydroxide	$Ca(OH)_2(s) \rightleftharpoons Ca^{2+}(aq) + 2OH^-(aq)$	8.0×10^{-6}
Chromium(III) hydroxide	$Cr(OH)_3(s) \rightleftharpoons Cr^{3+}(aq) + 3OH^-(aq)$	3.0×10^{-29}
Copper(II) hydroxide	$Cu(OH)_2(s) \rightleftharpoons Cu^{2+}(aq) + 2OH^-(aq)$	2.2×10^{-20}
Iron(II) hydroxide	$Fe(OH)_2(s) \rightleftharpoons Fe^{2+}(aq) + 2OH^-(aq)$	1.6×10^{-14}
Iron(III) hydroxide	$Fe(OH)_3(s) \rightleftharpoons Fe^{3+}(aq) + 3OH^-(aq)$	1.1×10^{-36}
Magnesium hydroxide	$Mg(OH)_2(s) \rightleftharpoons Mg^{2+}(aq) + 2OH^-(aq)$	1.2×10^{-11}
Strontium hydroxide	$Sr(OH)_2(s) \rightleftharpoons Sr^{2+}(aq) + 2OH^-(aq)$	3.2×10^{-4}
Zinc hydroxide	$Zn(OH)_2(s) \rightleftharpoons Zn^{2+}(aq) + 2OH^-(aq)$	1.8×10^{-14}

Compound	Dissolution equilibrium	K_{sp}
Iodides		
Copper(I) iodide	$CuI(s) \rightleftharpoons Cu^+(aq) + I^-(aq)$	5.1×10^{-12}
Lead(II) iodide	$PbI_2(s) \rightleftharpoons Pb^{2+}(aq) + 2I^-(aq)$	1.4×10^{-8}
Silver iodide	$AgI(s) \rightleftharpoons Ag^+(aq) + I^-(aq)$	8.3×10^{-17}
Phosphates		
Calcium phosphate	$Ca_3(PO_4)_2(s) \rightleftharpoons 3Ca^{2+}(aq) + 2PO_4^{3-}(aq)$	1.2×10^{-26}
Iron(III) phosphate	$FePO_4(s) \rightleftharpoons Fe^{3+}(aq) + PO_4^{3-}(aq)$	1.3×10^{-22}
Sulfates		
Barium sulfate	$BaSO_4(s) \rightleftharpoons Ba^{2+}(aq) + SO_4^{2-}(aq)$	1.1×10^{-10}
Calcium sulfate	$CaSO_4(s) \rightleftharpoons Ca^{2+}(aq) + SO_4^{2-}(aq)$	2.4×10^{-5}
Lead(II) sulfate	$PbSO_4(s) \rightleftharpoons Pb^{2+}(aq) + SO_4^{2-}(aq)$	1.8×10^{-8}
Mercury(I) sulfate	$Hg_2SO_4(s) \rightleftharpoons Hg_2^{2+}(aq) + SO_4^{2-}(aq)$	6.5×10^{-7}
Silver sulfate	$Ag_2SO_4(s) \rightleftharpoons 2Ag^+(aq) + SO_4^{2-}(aq)$	1.5×10^{-5}
Strontium sulfate	$SrSO_4(s) \rightleftharpoons Sr^{2+}(aq) + SO_4^{2-}(aq)$	3.8×10^{-7}
Sulfides		
Bismuth sulfide	$Bi_2S_3(s) \rightleftharpoons 2Bi^{3+}(aq) + 3S^{2-}(aq)$	1.6×10^{-72}
Cadmium sulfide	$CdS(s) \rightleftharpoons Cd^{2+}(aq) + S^{2-}(aq)$	8.0×10^{-28}
Cobalt(II) sulfide	$CoS(s) \rightleftharpoons Co^{2+}(aq) + S^{2-}(aq)$	4.0×10^{-21}
Copper(II) sulfide	$CuS(s) \rightleftharpoons Cu^{2+}(aq) + S^{2-}(aq)$	6.0×10^{-37}
Iron(II) sulfide	$FeS(s) \rightleftharpoons Fe^{2+}(aq) + S^{2-}(aq)$	6.0×10^{-19}
Lead(II) sulfide	$PbS(s) \rightleftharpoons Pb^{2+}(aq) + S^{2-}(aq)$	3.4×10^{-28}
Manganese(II) sulfide	$MnS(s) \rightleftharpoons Mn^{2+}(aq) + S^{2-}(aq)$	3.0×10^{-14}
Mercury(II) sulfide	$HgS(s) \rightleftharpoons Hg^{2+}(aq) + S^{2-}(aq)$	4.0×10^{-54}
Nickel(II) sulfide	$NiS(s) \rightleftharpoons Ni^{2+}(aq) + S^{2-}(aq)$	1.4×10^{-24}
Silver sulfide	$Ag_2S(s) \rightleftharpoons 2Ag^+(aq) + S^{2-}(aq)$	6.0×10^{-51}
Tin(II) sulfide	$SnS(s) \rightleftharpoons Sn^{2+}(aq) + S^{2-}(aq)$	1.0×10^{-26}
Zinc sulfide	$ZnS(s) \rightleftharpoons Zn^{2+}(aq) + S^{2-}(aq)$	3.0×10^{-23}

Appendix 4

DISSOCIATION CONSTANTS FOR WEAK ACIDS AND BASES AT 25°C

Weak Acids

Name	Formula	K_{a_1}	K_{a_2}	K_{a_3}
Acetic	$HC_2H_3O_2$ (CH_3COOH)	1.8×10^{-5}		
Acetylsalicylic	$HC_9H_7O_4$	3.0×10^{-4}		
Ascorbic	$H_2C_6H_6O_6$	8.0×10^{-5}	1.6×10^{-12}	
Benzoic	$HC_7H_5O_2$ (C_6H_5COOH)	6.5×10^{-5}		
Carbonic	H_2CO_3	4.2×10^{-7}	4.8×10^{-11}	
Chloroacetic	$HC_2H_2O_2Cl$ ($CH_2ClCOOH$)	1.4×10^{-3}		
Chlorous	$HClO_2$	1.1×10^{-2}		
Citric	$H_3C_6H_5O_7$	7.4×10^{-4}	1.7×10^{-5}	4.0×10^{-7}
Dichloroacetic	$HC_2HO_2Cl_2$ ($CHCl_2COOH$)	5.5×10^{-2}		
Formic	$HCHO_2$ ($HCOOH$)	1.7×10^{-4}		
Hydrocyanic	HCN	4.9×10^{-10}		
Hydrofluoric	HF	7.1×10^{-4}		
Hydrosulfuric	H_2S	9.5×10^{-8}	$\sim 1 \times 10^{-19}$	
Nitrous	HNO_2	4.5×10^{-4}		
Oxalic	$H_2C_2O_4$	6.5×10^{-2}	6.1×10^{-5}	
Phenol	C_6H_5OH	1.3×10^{-10}		
Phosphoric	H_3PO_4	7.5×10^{-3}	6.2×10^{-8}	4.8×10^{-13}
Phosphorous	H_3PO_3	5×10^{-2}	2×10^{-7}	
Sulfuric	H_2SO_4	very large	1.3×10^{-2}	
Sulfurous	H_2SO_3	1.3×10^{-2}	6.3×10^{-8}	
Trichloroacetic	$HC_2O_2Cl_3$ (CCl_3COOH)	2.2×10^{-1}		
Trifluoroacetic	$HC_2O_2F_3$ (CF_3COOH)	3.0×10^{-1}		

Weak Bases		
Name	**Formula**	**K_b**
Ammonia	NH_3	1.8×10^{-5}
Aniline	$C_6H_5NH_2$	3.8×10^{-10}
Ethylamine	$C_2H_5NH_2$	5.6×10^{-4}
Methylamine	CH_3NH_2	4.4×10^{-4}
Pyridine	C_5H_5N	1.7×10^{-9}
Urea	H_2NCONH_2	1.5×10^{-14}

Answers
TO ODD-NUMBERED PROBLEMS

Chapter 1

1.5 (a) law (b) theory (c) hypothesis **1.11** 3.12 g/mL **1.13** (a) 35°C
(b) −11°C (c) 39°C (d) 1011°C (e) −459.67°F **1.15** 2.52 mL
1.17 (a) 386 K (b) 310 K (c) 630 K **1.19** 64.3 g; 20°C; 11.35 g/cm^3
1.25 (a) 0.0152 (b) 0.0000000778 (c) 0.00000329 (d) 0.841
1.27 (a) 1.8×10^{-2} (b) 1.14×10^{10} (c) -5×10^4 (d) 1.3×10^3
1.29 (a) one (b) three (c) three (d) four (e) three (f) one (g) one or
two (h) three **1.31** (a) 1.28 (b) 3.18×10^{-3} mg (c) 8.14×10^7 dm
1.33 Carpenter Z: most accurate measurement; Carpenter Y: least
accurate measurement; Carpenter X: most precise measurement;
Carpenter Y: least precise measurement **1.35** upper thermometer:
41.85°; lower thermometer: 41.85° **1.37** 1.1 g/L **1.39** (a) 1.10×10^8 mg
(b) 6.83×10^{-5} m^3 (c) 7.2×10^3 L (d) 6.24×10^{-8} lb **1.41** (a) 66 amu
(b) 6.50×10^{-3} amu (c) 83 Å (d) 1.32 Å **1.43** 3.1557×10^7 s
1.45 (a) 81 in/s (b) 1.2×10^2 m/min (c) 7.4 km/h **1.47** 32 km/h
1.49 3.7×10^{-3} g Pb **1.51** (a) 1.85×10^{-7} m (b) 1.4×10^{17} s (c) 7.12×10^{-5} m^3 (d) 8.86×10^4 L **1.53** 6.25×10^{-4} g/cm^3 **1.57** (a) homogeneous
mixture (b) pure substance (c) pure substance (d) homogeneous
mixture (e) homogeneous mixture (f) pure substance (g) heterogeneous
mixture (h) pure substance (i) pure substance **1.63** (a) Quantitative
(b) Qualitative (c) Quantitative (d) Qualitative (e) Qualitative
1.65 (a) physical change (b) chemical change (c) physical change
(d) chemical change (e) physical change **1.67** (a) upper ruler: 2.5 cm
(b) lower ruler: 2.55 cm **1.69** (a) chemical (b) chemical (c) physical
(d) physical (e) chemical **1.71** 4.75×10^7 tons of sulfuric acid
1.73 (a) 8.08×10^4 g (b) 1.4×10^{-6} g (c) 39.9 g **1.75** 31.35 cm^3
1.77 21.5 g/cm^3 **1.79** 8.96 g/cm^3 **1.81** Fahrenheit thermometer: 0.1%;
Celsius thermometer: 0.3% **1.83** 72.8°S **1.85** 5×10^2 mL/breath
1.87 4.8×10^{19} kg NaCl; 5.3×10^{16} tons NaCl **1.89** (a) 75.0 g Au
(b) A troy ounce is heavier than an ounce. **1.91** (a) 0.5% (b) 3.1%
1.93 1.77×10^6 g Cu **1.95** 9.5×10^{10} kg CO$_2$ **1.97** 1.1×10^2 yr
1.99 4×10^{-19} g/L **1.101** density: 7.20 g/cm^3; radius: 0.853 cm
1.103 It would be more difficult to prove that the unknown substance
is a pure substance. Most mixtures can be separated into two or more
different substances. **1.105** Gently heat the liquid to see if any solid
remains after the liquid evaporates. Also, collect the vapor and then
compare the densities of the condensed liquid with the original liquid.
The composition of a mixed liquid frequently changes with evaporation
along with its density. **1.107** The volume occupied by the ice is larger
than the volume of the glass bottle. The glass bottle would break.
1.109 (a) yes (b) no (c) no **1.111** (a) in °C: 527°C and 1227°C; in
kelvins: 800 K and 1500 K; range span: 700°C and 700 K (b) 5.833
$\times 10^{-2}$F/s; 3.241×10^{-2}°C/s; 3.241×10^{-2}°K/s **1.113** (a) − 40°C
(b) 574.59 K or 574.59°F (c) No. Since the value of Kelvin is always
equal to a value that is 273.15 greater than the Celsius value, there
is no temperature at which the values can be numerically equal.
1.115 (a) 20°C to 25°C (b) 1.18 g/mL (c) 1180 g/L; 1180 kg/m^3
(d) 420 mg

Key Skills
1.1 (c) **1.2** (c) **1.3** (e) **1.4** (e)

Chapter 2

2.9 0.16 mi; 2.5×10^2 m **2.15** 149 **2.17** $^{15}_{7}$N: protons = 7, neutrons = 8,
electrons = 7; $^{55}_{25}$Mn: protons = 25, neutrons = 30, electrons = 25;

$^{81}_{35}$Br: protons = 35, neutrons = 46, electrons = 35; $^{207}_{82}$Pb: protons =
82, neutrons = 125, electrons = 82; $^{115}_{49}$In: protons = 49, neutrons =
66, electrons = 49; $^{94}_{40}$Zr: protons = 40; neutrons = 54, electrons = 40.
2.19 (a) $^{187}_{75}$Re (b) $^{209}_{83}$Bi (c) $^{75}_{33}$As (d) $^{236}_{93}$Np **2.21** (a) 35 (b) 32 (c) 121
(d) 105 **2.27** (a) lithium-9 (b) sodium-25 (c) scandium-48
2.29 (a) neon-17 (b) calcium-45 (c) technetium-92 (All technetium
isotopes are radioactive.) **2.35** 20.18 amu **2.37** ^{6}Li: 7.5%; ^{7}Li: 92.5%
2.39 23.98 amu **2.45** (a) Metallic character increases down a group.
(b) Metallic character decreases from left to right. **2.47** Na and K;
N and P; F and Cl **2.49** iron: Fe, Period 4, upper-left square of Group
8B; iodine: I, Period 5, Group 7A; sodium: Na, Period 3, Group 1A;
phosphorus: P, Period 3, Group 5A; sulfur: S, Period 3, Group 6A;
magnesium: Mg, Period 3, Group 2A

Atomic number 26, iron, Fe, (present in hemoglobin for
transporting oxygen)
Atomic number 53, iodine, I, (present in the thyroid gland)
Atomic number 11, sodium, Na, (present in intra- and extracellular
fluids)
Atomic number 15, phosphorus, P, (present in bones and teeth)
Atomic number 16, sulfur, S, (present in proteins)
Atomic number 12, magnesium, Mg, (present in chlorophyll
molecules)
2.53 2.10×10^5 atoms **2.55** 1.49×10^{-14} mol Ni **2.57** 5.15×10^3 g Pt
2.59 (a) 2.023×10^{-22} g/Sb atom (b) 1.767×10^{-22} g/Pd atom
2.61 5.48×10^{22} Sc atoms **2.63** 7.5×10^{-22} mole of silver
2.65 8.1×10^{22} Fr atoms **2.67** $^{131}_{54}$Xe **2.69** A: $^{22}_{10}$Ne; B: $^{185}_{75}$Re; C: $^{42}_{21}$Sc;
To write a correct symbol for atom D, the number of neutrons
would need to be known. **2.71** It establishes a standard mass unit
that permits the measurement of masses of all other isotopes relative
to carbon-12. **2.73** (a) $^4_2$He: protons = 2, neutrons = 2; $^{20}_{10}$Ne:
protons = 10, neutrons = 10; $^{40}_{18}$Ar: protons = 18, neutrons = 22;
$^{84}_{36}$Kr: protons = 36, neutrons = 48; $^{132}_{54}$Xe: protons = 54, neutrons = 78
(b) The neutron/proton ratio increases with increasing atomic
number. **2.75** 192 amu; this is the atomic mass that appears in the
periodic table. **2.77** (a) iodine (b) radon (c) selenium (d) sodium
(e) lead **2.79** (a) ^{40}Mg: 28 neutrons (b) ^{44}Si: 30 neutrons (c) ^{48}Ca: 28
neutrons (d) ^{43}Al: 30 neutrons **2.81** Symbol: ^{101}Ru, ^{181}Ta, ^{150}Sm;
Protons: 44, 73, 62; Neutrons: 57, 108, 88; Electrons: 44, 73, 62
2.83 (a) 6.6×10^{22} Pt atoms (b) 1.4×10^2 pm

Key Skills
2.1 (e) **2.2** (c) **2.3** (a) **2.4** (c)

Chapter 3

3.5 (a) 20.7 J (b) 1.8×10^5 J (c) 1.03×10^{-25} J (d) 1.75×10^{-21} J **3.7** (a) 1.05×10^3 m/s (b) 328 m/s (c) 39.1 amu; potassium **3.9** (a) 1.5 times as large (b) $\frac{2}{3}d$ **3.15** (a) 1.58×10^4 nm (b) 1.28×10^{15} Hz **3.17** 3.26×10^7 nm; microwave region **3.19** 2.5 min **3.25** 2.82×10^{-19} J **3.27** (a) 4.6×10^7 nm; no (b) 4.3×10^{-24} J/photon (c) 2.6 J/mol **3.29** 3.56×10^{-20} J **3.31** 1.29×10^{-15} J **3.33** Infrared photons have insufficient energy to cause the chemical changes. **3.35** 9.29×10^{-13} m; 3.23×10^{20} s^{-1} **3.37** 7.40×10^{-19} J **3.39** Yellow light would eject more electrons. Blue light will eject electrons with greater kinetic energy.

Visualizing Chemistry
VC 3.1 b **VC 3.2** a **VC 3.3** c **VC 3.4** b

3.45 3.027×10^{-19} J **3.47** 1.60×10^{14} Hz; 1.88×10^3 nm **3.49** 5 **3.51** Analyze the emitted light by passing it through a prism. **3.53** Excited atoms of the chemical elements emit the same characteristic frequencies or lines in a terrestrial laboratory, in the sun, or in a star many light-years distant from earth. **3.57** 0.565 nm **3.59** 4.14×10^{-30} cm **3.61** 377 m/s **3.69** $\Delta u \geq 2.0 \times 10^{-5}$ m/s **3.71** $\Delta u \geq 4.38 \times 10^{-26}$ m/s; This uncertainty is far smaller than what can be measured. **3.75** $\ell = 0$ or 1; $m_\ell = -1$, 0, or 1 **3.77** (b) is wrong: ℓ cannot be equal to n; and (d) is wrong: When ℓ is 1, m_ℓ can only be equal to -1, 0, or $+1$. **3.83** (a) 2p: $n = 2$, $\ell = 1$, $m_\ell = 1$, 0, or -1 (b) 3s: $n = 3$, $\ell = 0$, $m_\ell = 0$ (c) 5d: $n = 5$, $\ell = 2$, $m_\ell = 2$, 1, 0, -1, or -2 **3.85** A 2s orbital is larger than a 1s orbital and exhibits a node. Both have the same spherical shape. The 1s orbital is lower in energy than the 2s. **3.87** In H, energy depends only on n, but for all other atoms, energy depends on n and ℓ. **3.89** (a) 2s (b) 3p (c) equal (d) equal (e) 5s **3.91** (a) orbital (b); (b) orbitals (a) and (d); (c) none **3.95** (a) three (b) six (c) none **3.97** (a) 2 (b) 8 (c) 18 (d) 32 **3.99** Al: two 2p electrons missing, $1s^2 2s^2 2p^6 3s^2 3p^1$; B: too many 2p electrons, $1s^2 2s^2 2p^1$; F: too many 2p electrons, $1s^2 2s^2 2p^5$ **3.101** B = F < C = O < N **3.103** (a) 1; paramagnetic (b) 3; paramagnetic (c) 1; paramagnetic (d) 6; paramagnetic (e) zero; diamagnetic **3.105** (a) Wrong, because m_ℓ can have only whole number values. (b) Wrong, because m_ℓ can have only the value 0 when ℓ is 0. (c) Wrong, because m_ℓ can have only the value 0 when ℓ is 0. (d) acceptable (e) Wrong, because m_s can have only half-integral values. **3.115** [Xe]$6s^2 4f^{14} 5d^7$ **3.117** Ge: [Ar]$4s^2 3d^{10} 4p^2$; Fe: [Ar]$4s^2 3d^6$; Zn: [Ar]$4s^2 3d^{10}$; Ni: [Ar]$4s^2 3d^8$; W: [Xe]$6s^2 4f^{14} 5d^4$; Tl: [Xe]$6s^2 4f^{14} 5d^{10} 6p^1$ **3.119** (a) Ga (b) Y (c) Cr (d) Ac **3.121** Part (b) is correct in the view of contemporary quantum theory. Bohr's explanation of emission and absorption line spectra appears to have universal validity. Parts (a) and (c) are artifacts of Bohr's early planetary model of the hydrogen atom and are *not* considered to be valid today. **3.123** (a) 4; 2s and 2p. (b) 6; 4p, 4d, and 4f. (c) 10; 3d. (d) 1; 2s. (e) 2; 4f. **3.125** 0.381 pm; 7.86×10^{20} s^{-1} **3.127** (a) 1.20×10^{18} photons (b) 3.76×10^8 W **3.129** krypton **3.131** 3.0×10^{19} photons/s **3.133** He$^+$: $n = 3 \longrightarrow 2$: $\lambda = 164$ nm; $n = 4 \longrightarrow 2$: $\lambda = 121$ nm; $n = 5 \longrightarrow 2$: $\lambda = 108$ nm; $n = 6 \longrightarrow 2$: $\lambda = 103$ nm. H: $n = 3 \longrightarrow 2$: $\lambda = 656$ nm; $n = 4 \longrightarrow 2$: $\lambda = 486$ nm; $n = 5 \longrightarrow 2$: $\lambda = 434$ nm; $n = 6 \longrightarrow 2$: $\lambda = 410$ nm. All the Balmer transitions for He$^+$ are in the ultraviolet region, whereas the transitions for H are all in the visible region. **3.135** $\frac{1}{\lambda_1} = \frac{1}{\lambda_2} + \frac{1}{\lambda_3}$ **3.137** (a) He, $1s^2$ (b) N, $1s^2 2s^2 2p^3$ (c) Na, $1s^2 2s^2 2p^6 3s^1$ (d) As, [Ar]$4s^2 3d^{10} 4p^3$ (e) Cl, [Ne]$3s^2 3p^5$

3.139 (a) (b) [Ne]

(c) [Ar]

3.141 0.596 m; microwave/radio **3.143** (a) false (b) false (c) true (d) false (e) true **3.145** blue light: 42 photons; green light: 53 photons; red light: 56 photons **3.147** 2.2×10^5 J **3.149** Only (b) and (d) are allowed transitions. Any of the transitions in Figure 3.11 is possible as long as ℓ for the final state differs from ℓ of the initial state by 1. **3.151** $\Delta u \geq 2.79 \times 10^{-26}$ m/s; This uncertainty is far smaller than can be measured. **3.153** (a) 2.29×10^{-6} nm (b) 6.0×10^{-2} kg; the equation is important only for speeds close to that of light.

Key Skills
3.1 (e) **3.2** (d) **3.3** (b) **3.4** (d)

Chapter 4

4.15 sulfur (S); The electron configuration is $1s^2 2s^2 2p^6 3s^2 3p^4$ or [Ne]$3s^2 3p^4$. **4.17** (a) and (d); (b) and (e); (c) and (f) **4.19** (a) Group 1A or 1 (b) Group 5A or 15 (c) Group 5A or 15 (d) Group 8B or 8 **4.21** (a) Sn (b) V (c) Cl (d) Ba **4.25** (a) $\sigma = 2$ and $Z_{eff} = +4$ (b) for 2s electrons: $Z_{eff} = +3.22$, for 2p electrons: $Z_{eff} = +3.14$; These values are lower than those in part (a) because the 2s and 2p electrons actually do shield one another to some extent. **4.39** 8.40×10^6 kJ/mol **4.41** (a) four times greater (b) six times greater (c) reduce the force by one-half (d) eight times greater (e) no change **4.43** ii < iii < iv < i **4.45** Cl < P < Si < Al < Mg **4.47** oxygen **4.49** left to right: S, Se, Ca, and K **4.51** K < Ca < P < F < Ne **4.53** The Group 3A elements (such as Al) all have a single electron in the outermost p subshell, which is well shielded from the nuclear charge by the inner electrons and the ns^2 electrons. Therefore, less energy is needed to remove a single p electron than to remove a paired s electron from the same principal energy level (such as for Mg). **4.55** 496 kJ/mol is paired with $1s^2 2s^2 2p^6 3s^1$. 2080 kJ/mol is paired with $1s^2 2s^2 2p^6$, a very stable noble gas configuration. **4.57** A: Group 3A or 3; B: Group 2A or 2; C: Group 1A or 1 **4.59** Cl **4.61** Alkali metals have a valence electron configuration of ns^1, so they can accept another electron in the ns orbital. On the other hand, alkaline earth metals have a valence electron configuration of ns^2, so they have little tendency to accept another electron unless it goes into a higher energy p orbital. **4.63** The calculated number for magnesium is about 78 percent larger than that for sodium. **4.69** (a) [Kr] (b) [Kr] (c) [Kr]$5s^2 4d^{10}$ (d) [Kr]$5s^2 4d^{10} 5p^6$ (e) [Xe] (f) [Kr]$4d^{10}$ (g) [Xe]$6s^2 4f^{14} 5d^{10}$ (h) [Xe]$4f^{14} 5d^{10}$ **4.71** (a) [Ar]$3d^6$ (b) [Ar]$3d^9$ (c) [Ar]$3d^7$ (d) [Ar]$3d^5$ **4.73** (a) [Ar]$3d^8$ (b) [Ar]$3d^{10}$ (c) [Kr]$4d^{10}$ (d) [Xe]$4f^{14} 5d^{10}$ (e) [Xe]$4f^{14} 5d^8$ **4.75** (a) Cr^{3+} (b) Sc^{3+} (c) Rh^{3+} (d) Ir^{3+} **4.77** Be^{2+} and He (2 e^-); N^{3-} and F$^-$ (10 e^-); Fe^{2+} and Co^{3+} (24 e^-); S^{2-} and Ar (18 e^-) **4.79** Cu$^+$ < Ti^{3+} < V^{3+} < Mn^{4+} < Fe^{2+} **4.81** selenium, Se **4.83** iron, Fe **4.85** (a) Fe^{2+} or Co^{3+} (b) Ag$^+$ or Cd^{2+} (c) Cu^{2+} (d) Cu$^+$ or Zn^{2+} (e) V^{2+} or Cr^{3+} (f) Co^{2+} or Ni^{3+} **4.89** (a) Cl (b) Na$^+$ (c) O^{2-} (d) Al^{3+} (e) Au^{3+} **4.91** The Cu$^+$ ion is larger than the Cu^{2+} ion because it has one more electron. **4.93** (d) **4.95** In^{3+} **4.97** (a) bromine, Br (b) nitrogen, N (c) rubidium, Rb (d) magnesium, Mg **4.99** O^{2-} < F$^-$ < Na$^+$ < Mg^{2+} **4.101** O$^+$ and N; S^{2-} and Ar; N^{3-} and Ne; As^{3+} and Zn; Cs$^+$ and Xe **4.103** (a) and (d) **4.105** The binding of a cation to an anion results from electrostatic attraction. As the +2 cation gets smaller (from Ba^{2+} to Mg^{2+}), the distance between the opposite charges decreases and the electrostatic attraction increases. **4.107** (a) B < D < A < C (b) C < A < D < B **4.109** This anomaly is a result of the lanthanide contraction. The lanthanide contraction is a term used to describe a decrease in atomic radius of the lanthanide elements. The lanthanide contraction also decreases the radius of the elements following the lanthanides in the periodic table, including Hf. The lanthanide contraction is caused by poor shielding of the nuclear charge from the partially filled 4f electron orbitals.

4.111 H$^-$; Since H$^-$ has only one proton compared to two protons for He, the nucleus of H$^-$ will attract the two electrons less strongly compared to He. Therefore, H$^-$ is larger. **4.113** 0.66 **4.115** The effective nuclear charge changes less than the nuclear charge as we move down a column of the periodic table. Gallium has an additional ten *core* electrons in its 3d shell to shield valence electrons. Therefore, the effective nuclear charge is not as large as might be expected. **4.117** 77.5% **4.119** 6.94 × 10^{-19} J/electron; If there are no other electrons with lower kinetic energy, then this is the electron from the valence shell. To ensure that the ejected electron is the valence electron, UV light of the *longest* wavelength (lowest energy) that can still eject electrons should be used. **4.121** A: Rb; B: F; C: Mg; D: Rb$^+$; E: F$^-$ **4.123** (a) $IE_1 = 3s^1$ electron; $IE_2 = 2p^6$ electron; $IE_3 = 2p^5$ electron; $IE_4 = 2p^4$ electron; $IE_5 = 2p^3$ electron; $IE_6 = 2p^2$ electron; $IE_7 = 2p^1$ electron; $IE_8 = 2s^2$ electron; $IE_9 = 2s^1$ electron; $IE_{10} = 1s^2$ electron; $IE_{11} = 1s^1$ electron. (b) Each break ($IE_1 \longrightarrow IE_2$ and $IE_9 \longrightarrow IE_{10}$) represents the transition to another shell ($n = 3 \longrightarrow 2$ and $n = 2 \longrightarrow 1$). **4.125** Considering electron configurations, Fe^{2+}[Ar]3$d^6 \longrightarrow$ Fe^{3+}[Ar]3d^5, Mn^{2+}[Ar]3$d^5 \longrightarrow$ Mn^{3+}[Ar]3d^4, a half-filled shell has extra stability. In oxidizing Fe^{2+}, the product is a half-filled d^5 shell. In oxidizing Mn^{2+}, a half-filled d^5 shell electron is being lost, which requires more energy. **4.127** Z_{eff} increases from left to right across the table, so electrons are held more tightly. (This explains the electron affinity values of C and O.) Nitrogen has a zero value of electron affinity because of the stability of the half-filled 2p subshell (that is, N has little tendency to accept another electron). **4.129** Once an atom gains an electron forming a negative ion, adding additional electrons is typically an unfavorable process due to electron-electron repulsions. Second and third electron affinities do not occur spontaneously and are therefore difficult to measure. **4.131** (a) It was determined that the periodic table was based on atomic number, not atomic mass. (b) argon: 39.95 amu, potassium: 39.10 amu **4.133** element 119; electron configuration: [Rn]7$s^2$5f^{14}6d^{10}7$p^6$8s^1. **4.135** Group 2A; There is a large jump from the second to the third ionization energy, indicating a change in the principal quantum number, n. **4.137** Group 2A or 2 **4.139** Sn^{4+}

Key Skills
4.1 (e) **4.2** (d) **4.3** (b) **4.4** (a)

Chapter 5

5.7 (a) $\ddot{\text{B}}\text{r}\cdot$ (b) $\dot{\ddot{\text{N}}}\cdot$ (c) $\ddot{\ddot{\text{I}}}\cdot$ (d) $\dot{\text{A}}\ddot{\text{s}}\cdot$ (e) $\ddot{\text{F}}\cdot$ (f) $\left[\ddot{\ddot{\text{P}}}\ddot{\cdot}\right]^{3-}$ (g) Na$^+$ (h) Mg^{2+} (i) $\left[\cdot\ddot{\text{As}}\cdot\right]^{3+}$ (j) $\left[\cdot\text{Pb}\cdot\right]^{2+}$ **5.9** (a) Group 4A or 14 (b) Group 6A or 16 (c) Group 7A or 17 **5.15** (a) decreases the ionic bond energy (b) triples the ionic bond energy (c) increases the bond energy by a factor of 4 (d) increases the bond energy by a factor of 2

5.17 (a) $\dot{\text{B}}\text{a} + \cdot\ddot{\text{O}}\cdot \longrightarrow \text{Ba}^{2+}\left[\ddot{\ddot{\text{O}}}\ddot{\cdot}\right]^{2-}$

(b) $\dot{\text{A}}\dot{\text{l}}\cdot + \cdot\ddot{\text{N}}\cdot \longrightarrow \text{Al}^{3+}\left[\ddot{\ddot{\text{N}}}\ddot{\cdot}\right]^{3-}$

5.23 (a) K$_3$P, potassium phosphide (b) Fe$_2$Se$_3$, iron(III) selenide (c) Ba$_3$N$_2$, barium nitride (d) Ga$_2$S$_3$, gallium sulfide **5.25** (a) Na$_2$O (b) FeS (c) Co$_2$Te$_3$ (d) BaF$_2$ **5.27** (a) barium hydride (b) potassium nitride (c) zinc oxide (d) barium peroxide

5.29 (a) [image] (b) [image] (c) [image] (d) [image] **5.41** (a) element (b) compound (c) compound **5.43** elements: N$_2$, S$_8$, H$_2$; compounds: NH$_3$, NO, CO, CO$_2$, and SO$_2$ **5.45** 2:3 **5.47** (a) CH$_3$ (b) CH$_2$ (c) CH$_2$ (d) P$_2$S$_3$ (e) C$_2$H$_6$O **5.49** C$_3$H$_7$NO$_2$ **5.55** (a) disulfur dibromide (b) iodine pentafluoride (c) tetraphosphorus heptoxide (d) dibromine trioxide **5.57** (a) NF$_3$: nitrogen trifluoride (b) PBr$_5$: phosphorus

pentabromide (c) SCl$_2$: sulfur dichloride **5.59** alcohol and aldehyde functional groups **5.61** (a) potassium hypochlorite (b) silver carbonate (c) hydrogen nitrite (d) potassium permanganate (e) cesium chlorate (f) potassium ammonium sulfate (g) iron(II) perbromate (h) potassium hydrogen phosphate **5.63** (a) CuCN (b) Sr(ClO$_2$)$_2$ (c) HBrO$_4$ (d) HI (e) Na$_2$NH$_4$PO$_4$ (f) PbCO$_3$ (g) SnSO$_3$ (h) Cd(SCN)$_2$ **5.67** (a) 50.48 amu (b) 92.02 amu (c) 64.07 amu (d) 84.16 amu (e) 34.02 amu (f) 342.3 amu (g) 17.03 amu **5.69** (a) 50.48 amu (b) 44.02 amu (c) 64.07 amu (d) 84.16 amu (e) 102.89 amu (f) 361.87 amu (g) 601.8 amu **5.71** (a) 120.92 amu (b) 216.3 amu (c) 128.4 amu **5.75** %Sn = 78.77%; %O = 21.23% **5.77** %Ca = 39.89%; %P = 18.50%; %O = 41.41%; %H = 0.2007% **5.79** ammonia, NH$_3$ **5.83** 409 g/mol **5.85** 3.01 × 10^{22} C atoms; 6.02 × 10^{22} H atoms; 3.01 × 10^{22} O atoms **5.87** 4.0 × 10^9 molecules **5.89** (a) %C = 80.56%; %H = 7.51%; %O = 11.93% (b) 4.614 × 10^{21} molecules **5.91** 1.72 g F **5.93** 5.28 × 10^{21} molecules **5.95** C$_{18}$H$_{20}$O$_2$ **5.97** 981 g **5.99** Na$_2$O and SrO **5.101** NaCl is an ionic compound; it does not consist of molecules. **5.103** (a) molecule and compound (b) element and molecule (c) element (d) molecule and compound (e) element (f) element and molecule (g) element and molecule (h) molecule and compound (i) compound, not molecule (j) element (k) element and molecule (l) compound, not molecule **5.105** (a) Yes; This is consistent with the Law of Multiple Proportions which states that if two elements combine to form more than one compound, the masses of one element that combine with a fixed mass of the other element are in ratios of small whole numbers. In this case, the ratio of the masses of hydrogen in the two compounds is 3:1. (b) ethane: any formula with C:H = 1:3 (CH$_3$, C$_2$H$_6$, etc.); acetylene: any formula with C:H = 1:1 (CH, C$_2$H$_2$, etc.) **5.107** ionic: LiF, BaCl$_2$, and KCl; molecular: SiCl$_4$, B$_2$H$_6$, and C$_2$H$_4$ **5.109** ionic: NaF, MgF$_2$, and AlF$_3$; covalent: SiF$_4$, PF$_5$, SF$_6$, and ClF$_3$ **5.111** cation: Mg^{2+}, Sr^{2+}, Fe^{3+}, Mn^{2+}, Sn^{4+}; anion: HCO$_3^-$, Cl$^-$, NO$_2^-$, ClO$_3^-$, Br$^-$; formula: Mg(HCO$_3$)$_2$, SrCl$_2$, Fe(NO$_2$)$_3$, Mn(ClO$_3$)$_2$, SnBr$_4$; name: magnesium bicarbonate, strontium chloride, iron(III) nitrite, manganese(II) chlorate, tin(IV) bromide **5.113** C$_2$H$_6$ < C$_3$H$_8$ < C$_2$H$_2$ = C$_6$H$_6$ **5.115** (a) covalent; BF$_3$; boron trifluoride (b) ionic; KBr; potassium bromide **5.117** C$_3$H$_2$OF$_5$Cl; 184.50 g/mol; 0.0207 mol **5.119** (a) 0.212 mol O (b) 0.424 mol O **5.121** ionic: RbCl and KO$_2$; covalent: PF$_5$, BrF$_3$, and CI$_4$ **5.123** MgMoO$_4$ **5.125** 0.0011 mol chlorophyll **5.127** HClO$_4$ **5.129** %C = 30.20%; %H = 5.069%; %Cl = 44.57%; %S = 20.16% **5.131** (a) 6.532 × 10^4 g (b) 7.6 × 10^{-2} g HG **5.133** (a) 4.24 × 10^{22} K$^+$ ions; 4.24 × 10^{22} Br$^-$ ions (b) 4.58 × 10^{22} Na$^+$ ions; 2.29 × 10^{22} SO$_4^{2-}$ ions (c) 4.34 × 10^{22} Ca^{2+} ions; 2.89 × 10^{22} PO$_4^{3-}$ ions **5.135** (e) 0.50 mol Cl$_2$ **5.137** 28.97 g/mol

Key Skills
5.1 (d) **5.2** (c) **5.3** (a) **5.4** (c)

Chapter 6

6.9 (a) ionic; BaF$_2$; barium fluoride (b) covalent; CBr$_4$; carbon tetrabromide **6.13** 0.151 D **6.15** 15.9 D **6.17** C—H (ΔEN = 0.4) < Br—H (ΔEN = 0.7) < F—H (ΔEN = 1.9) < Li—Cl (ΔEN = 2.0) < Na—Cl (ΔEN = 2.1) < K—F (ΔEN = 3.2) **6.19** (a) polar covalent; The electronegativity difference is 0.9. (b) nonpolar covalent; The electronegativities of C and S are identical. (c) nonpolar covalent; The electronegativity difference is 0.4. (d) ionic; The electronegativity difference is 2.5. **6.21** C—O < K—I < B—F < Li—F **6.23**

(a) $\ddot{\ddot{\text{F}}}-\ddot{\text{O}}-\ddot{\ddot{\text{F}}}$

(b) $\ddot{\ddot{\text{F}}}-\ddot{\text{N}}=\ddot{\text{N}}-\ddot{\ddot{\text{F}}}$

(c)
$$\begin{array}{ccc} \text{H} & & \text{H} \\ | & & | \\ \text{H}-\text{Si}- & \text{Si}-\text{H} \\ | & & | \\ \text{H} & & \text{H} \end{array}$$

(d) $\ddot{\text{O}}$—H or $\left[:\ddot{\text{O}}-\text{H}\right]^-$ (e) H—C—C—$\ddot{\text{O}}$: or $\left[\text{H}-\text{C}-\text{C}-\ddot{\text{O}}:\right]^-$ (with :O: double bond and :Cl: substituents)

(f) H—C—$\overset{+}{\text{N}}$—H or $\left[\text{H}-\text{C}-\text{N}-\text{H}\right]^+$ (with H's around C and N)

6.25

(a) $:\ddot{\text{F}}-\ddot{\text{C}}\text{l}-\ddot{\text{F}}:$ with $:\ddot{\text{F}}:$ below   (b) H—$\ddot{\text{S}}$e—H   (c) H—$\ddot{\text{N}}$—H with $:\ddot{\text{O}}$—H below   (d) $:\ddot{\text{C}}\text{l}-\text{P}-\ddot{\text{C}}\text{l}:$ with :O: above and :Cl: below

(e) H—C—C—$\ddot{\text{B}}$r: (with H's)   (f) $:\ddot{\text{C}}\text{l}-\ddot{\text{N}}-\ddot{\text{C}}\text{l}:$ with $:\ddot{\text{C}}\text{l}:$ below   (g) H—C—N: (with H's)

6.29

(a) $\ddot{\text{O}}=\overset{+}{\text{N}}=\ddot{\text{O}}$ or $\left[\ddot{\text{O}}=\text{N}=\ddot{\text{O}}\right]^+$ (b) $\ddot{\text{S}}-\text{C}\equiv\text{N}:$ or $\left[:\ddot{\text{S}}-\text{C}\equiv\text{N}:\right]^-$

(c) $\ddot{\text{S}}-\ddot{\text{S}}:$ or $\left[:\ddot{\text{S}}-\ddot{\text{S}}:\right]^{2-}$ (d) $:\ddot{\text{F}}-\overset{+}{\ddot{\text{C}}\text{l}}-\ddot{\text{F}}:$ or $\left[:\ddot{\text{F}}-\ddot{\text{C}}\text{l}-\ddot{\text{F}}:\right]^+$

6.31 (a) Neither oxygen atom has a complete octet, and the leftmost hydrogen atom shows two bonds (4 electrons). Hydrogen can hold only two electrons in its valence shell.

(b) H—C—C—$\ddot{\text{O}}$—H (with H and :O: on C) **6.35** (a) H—C $\longleftrightarrow$ H—C (with :O: and :O: resonance)

(b) resonance structures:
$\overset{\text{H}}{\underset{\text{H}}{}}$C=$\overset{+}{\text{N}}$ $\longleftrightarrow$ C—$\overset{+}{\text{N}}$ $\longleftrightarrow$ C—$\overset{+}{\text{N}}$ (with :O:− and :O: groups)

6.37 H—$\ddot{\text{N}}$=$\overset{+}{\text{N}}$=$\ddot{\text{N}}$: $\longleftrightarrow$ H—$\ddot{\text{N}}$—$\overset{+}{\text{N}}$≡N: $\longleftrightarrow$ H—$\overset{+}{\text{N}}$≡$\overset{+}{\text{N}}$—$\ddot{\text{N}}$:$^{-2}$

6.39 $\ddot{\text{O}}$=C=$\ddot{\text{N}}^-$ $\longleftrightarrow$ $:\ddot{\text{O}}$—C≡N: $\longleftrightarrow$ $\overset{+}{\text{O}}$≡C—$\ddot{\text{N}}$:$^{-2}$

6.41 (b) and (e)

6.43 (purine ring structures with resonance, NH$_2$, O and O$^-$, N—H)

6.49 No. An octet of electrons on Be can only be formed by making two double bonds as shown: $\overset{+}{\ddot{\text{C}}\text{l}}$=$\overset{-2}{\text{Be}}$=$\overset{+}{\ddot{\text{C}}\text{l}}$. It is not a plausible Lewis structure because this places a high negative formal charge on Be and positive formal charges on the Cl atoms. This structure distributes the formal charges counter to the electronegativities of the elements.

6.51 $:\ddot{\text{C}}\text{l}-\text{Al}-\ddot{\text{C}}\text{l}:$ + $:\ddot{\text{C}}\text{l}:^-$ $\longrightarrow$ $:\ddot{\text{C}}\text{l}-\overset{-}{\text{Al}}-\ddot{\text{C}}\text{l}:$ (with :Cl: groups); coordinate covalent bond.

6.53 $:\ddot{\text{C}}\text{l}$—Sb—$\ddot{\text{C}}\text{l}:$ with Cl groups; no.

6.55 H—$\ddot{\text{O}}$—$\ddot{\text{O}}$:; $:\ddot{\text{C}}\text{l}-\ddot{\text{O}}$:

6.57 completed octet on S: $\left[:\ddot{\text{O}}-\ddot{\text{S}}-\ddot{\text{O}}:\right]^{2-}$ with :O: below;

zero formal charge on S: $\left[:\ddot{\text{O}}-\ddot{\text{S}}-\ddot{\text{O}}:\right]^{2-}$ with :O: below

6.59 $\ddot{\text{N}}$=$\overset{+}{\text{N}}$=$\ddot{\text{N}}$$^-$ $\longleftrightarrow$ $:\text{N}\equiv\overset{+}{\text{N}}-\ddot{\text{N}}:^{-2}$ $\longleftrightarrow$ $^{-2}\ddot{\text{N}}-\overset{+}{\text{N}}\equiv\text{N}:$

6.61 (a) $AlCl_4^-$ (b) AlF_6^{3-} (c) $AlCl_3$ **6.63** CF_2 would be very unstable because carbon does not have an octet; LiO_2 would not be stable because the lattice energy between Li^+ and superoxide O_2^- would be too low to stabilize the solid; $CsCl_2$ requires a Cs^{2+} cation. The second ionization energy is too large to be compensated by the increase in lattice energy. PI_5 appears to be a reasonable species. However, the iodine atoms are too large to have five of them "fit" around a single P atom. **6.65** (a) False (b) False (c) False **6.67** Only N_2O_4 has a nitrogen-nitrogen single bond. Therefore, it has the longest bond length. **6.69** CH_4 and NH_4^+; N_2 and CO; C_6H_6 and $B_3N_3H_6$

6.71 (large fused-ring organic structure with C, N—H, O and —O—H groups)

6.73 $\left[:\ddot{\text{O}}-\ddot{\text{S}}=\ddot{\text{O}}\right]^-$ $\left[:\ddot{\text{O}}-\ddot{\text{S}}=\ddot{\text{O}}\right]^-$ (with :O: below)

6.75 The central iodine atom in I_3^- has *ten* electrons surrounding it: two bonding pairs and three lone pairs. The central iodine has an expanded octet. Elements in the second period such as fluorine cannot have an expanded octet as would be required for F_3^-.

6.77 H—C—$\ddot{\text{N}}$=C=$\ddot{\text{O}}$ $\longleftrightarrow$ H—C—$\overset{+}{\text{N}}$≡C—$\ddot{\text{O}}$: (with H's on C)

6.79 Form (a) is the most important resonance structure with no formal charges and all satisfied octets. Form (b) is likely not as important as (a) because of the positive formal charge on O. Forms (c) and (d) do not satisfy the octet rule for all atoms and are likely not important.

6.81 (a) $:\ddot{\text{F}}$—C—$\ddot{\text{C}}\text{l}$: (b) $:\ddot{\text{F}}$—C—$\ddot{\text{C}}\text{l}$: (c) H—C—$\ddot{\text{C}}\text{l}$: (d) $:\ddot{\text{F}}$—C—C—H (with :Cl:, :F:, :F: groups)

6.83 (a) $\text{C}\equiv\overset{+}{\text{O}}$: (b) $\text{N}=\overset{+}{\text{O}}$: (c) $\text{C}\equiv\text{N}:$ (d) $\text{N}\equiv\text{N}:$

6.85 The noble gases violate the octet rule. This is because each noble gas atom already has completely filled ns and np subshells.

6.87 (a) $:\text{N}=\ddot{\text{O}}$ $\longleftrightarrow$ $\ddot{\text{N}}=\overset{+}{\ddot{\text{O}}}$; The first structure is the most important. (b) No. **6.89** (Al$_2$Cl$_6$ dimer structure with bridging Cl atoms); The arrows indicate coordinate covalent bonds. This dimer does not possess a dipole moment. **6.91** RbF: 62.5%; RbI: 64.25%; We would expect the percent ionic character to be higher for RbF than it is for RbI, given that fluorine is more electronegative. **6.93** compound: B, C, A; dipole moment: 1.94 D, 0.524 D, 10.3 D.

6.95

The C—N bond length will be shorter than that of a single bond (143 pm) and longer than that of a double bond (138 pm), approximately 141 pm.

6.97

$$:N≡\overset{+}{N}-\overset{+}{N}=\overset{..}{\underset{..}{N}}{}^{-} \longleftrightarrow {}^{-}\overset{..}{\underset{..}{N}}=\overset{+}{N}=\overset{+}{N}=N: \longleftrightarrow :N≡\overset{+}{N}-\overset{..}{\underset{..}{N}}{}^{-}\overset{+}{N}≡N:$$

6.99 Using Mulliken's definition, EN(O) = 727.5, EN(F) = 1004.5, and EN(Cl) = 802.5. When converted to the Pauling scale, EN(O) = 3.16 or 3.2, EN(F) = 4.37 or 4.4, and EN(Cl) = 3.49 or 3.5. These values compare to the Pauling values for oxygen of 3.5, fluorine of 4.0, and chlorine of 3.0.

Key Skills
6.1 (a, c) **6.2** (b, e) **6.3** (c) **6.4** (a)

Chapter 7

7.7 (a) tetrahedral (b) trigonal planar (c) trigonal pyramidal (d) bent (e) bent **7.9** (a) trigonal pyramidal (b) tetrahedral (c) tetrahedral (d) seesaw **7.11** (a) linear (b) linear (c) linear **7.13** carbon at the center of H_3C: electron-domain geometry = tetrahedral, molecular geometry = tetrahedral; carbon at center of COOH: electron-domain geometry = trigonal planar, molecular geometry = trigonal planar; oxygen in OH: electron-domain geometry = tetrahedral, molecular geometry = bent **7.15** (a) T-shaped: Angles of <90° are possible. (b) Seesaw-shaped: It is not possible for this angle to be >120°. (c) Trigonal pyramidal: Angles of <109.5° are possible. (d) Square pyramidal: It is not possible for this angle to be >90°. **7.19** (a) polar (b) polar (c) nonpolar **7.21** Only (c) is polar. **7.29** Butane would be a liquid in winter (boiling point −44.5°C), and on the coldest days even propane would become a liquid (boiling point −0.5°C). Only methane would remain gaseous (boiling point −161.6°C). **7.31** (a) dispersion (b) dispersion and dipole-dipole (c) dispersion and dipole-dipole (d) dispersion and ionic (e) dispersion **7.33** (e) CH_3COOH (acetic acid) **7.35** 1-Butanol has greater intermolecular forces because it can form hydrogen bonds. Therefore, it has a higher boiling point. **7.37** (a) Xe: it is larger and therefore has stronger dispersion forces. (b) CS_2: it is larger (both molecules are nonpolar) and therefore has stronger dispersion forces. (c) Cl_2: it is larger (both the molecules are nonpolar) and therefore has stronger dispersion forces. (d) LiF: it is an ionic compound, and the ion-ion attractions are much stronger than the dispersion forces between F_2 molecules. (e) NH_3: it can form hydrogen bonds and PH_3 cannot. **7.39** (a) dispersion and dipole-dipole, including hydrogen bonding (b) dispersion only (c) dispersion only (d) covalent bonds **7.41** PH_3 will condense to a liquid at a higher temperature because it is polar and has stronger intermolecular forces. **7.43** The compound with −NO_2 and −OH groups on adjacent carbons can form hydrogen bonds with itself (*intra*molecular hydrogen bonds). Such bonds do not contribute to *inter*molecular attraction and do not help raise the melting point of the compound. The other compound, with the −NO_2 and −OH groups on the opposite sides of the ring, can form only *inter*molecular hydrogen bonds; therefore, it will take a higher temperature to escape into the gas phase. **7.49** BF_3; Valence bond theory fails to explain the bonding in molecules in which the central atom in its ground state does not have enough unpaired electrons to form the

observed number of bonds. **7.55** SiH_4: sp^3; H_3Si—SiH_3: sp^3 **7.57** Since there are five electron domains around P, the electron-domain geometry is trigonal bipyramidal (AB_5 with no lone pairs) and we conclude that P is sp^3d hybridized. **7.59** Boron is sp^2 hybridized in BF_3 and nitrogen is sp^3 hybridized in NH_3. After the reaction, boron and nitrogen are both sp^3 hybridized.

Visualizing Chemistry
VC 7.1 c **VC 7.2** b **VC 7.3** b **VC 7.4** b

7.63 (a) sp (b) sp (c) sp **7.65** sp **7.67** nine σ bonds and nine π bonds **7.69** 36 σ bonds and 10 π bonds **7.75** In order for the two hydrogen atoms to combine to form an H_2 molecule, the electrons must have opposite spins. If two H atoms collide and their electron spins are parallel, no bond will form. **7.77** order of increasing stability: $Li_2^- = Li_2^+ < Li_2$; Li_2: $(\sigma_{1s})^2(\sigma_{1s}^*)^2(\sigma_{2s})^2$: bond order = 1; Li_2^+: $(\sigma_{1s})^2(\sigma_{1s}^*)^2(\sigma_{2s})^1$: bond order = $\frac{1}{2}$; Li_2^-: $(\sigma_{1s})^2(\sigma_{1s}^*)^2(\sigma_{2s})^2(\sigma_{2s}^*)^1$: bond order = $\frac{1}{2}$ **7.79** B_2^+, with a bond order of $\frac{1}{2}$. (B_2 has a bond order of 1.) **7.81** The Lewis diagram has all electrons paired (incorrect) and a double bond (correct). The MO diagram has two unpaired electrons (correct), and a bond order of 2 (correct). **7.83** O_2: bond order = 2, paramagnetic; O_2^+: bond order = 2.5, paramagnetic; O_2^-: bond order = 1.5, paramagnetic; O_2^{2-}: bond order = 1, diamagnetic **7.85** The two shared electrons that make up the single bond in B_2 both lie in pi molecular orbitals and constitute a pi bond. The four shared electrons that make up the double bond in C_2 all lie in pi molecular orbitals and constitute two pi bonds. **7.87** Be_2 **7.91** The left symbol shows three delocalized double bonds (correct). The right symbol shows three localized double bonds and three single bonds (incorrect).

7.93 (a) $\overset{..}{O}=N-\overset{..}{\underset{..}{O}}\overset{..}{}{}^- \longleftrightarrow {}^-\overset{..}{\underset{..}{O}}-N=\overset{..}{O}$
         $\quad\quad\;\;\underset{+}{}\quad\quad\quad\quad\quad\underset{+}{}$

(b) sp^2 (c) Sigma bonds join the nitrogen atom to the fluorine and oxygen atoms. There is a pi molecular orbital delocalized over the N and O atoms. **7.95** The central oxygen atom is sp^2 hybridized. The unhybridized $2p_z$ orbital on the central oxygen overlaps with the $2p_z$ orbitals on the two terminal atoms. **7.97** $:\overset{..}{\underset{..}{Br}}-Hg-\overset{..}{\underset{..}{Br}}:$; You could establish the geometry of $HgBr_2$ by measuring its dipole moment. **7.99** bent; sp^3 **7.101** (a) $C_6H_8O_6$ (b) Five central O atoms sp^3; hybridization of the sixth (double bonded) peripheral O atom is sp^2. Three C atoms sp^2; three C atoms sp^3. (c) Five central O atoms bent; three sp^2 C atoms are trigonal planar; and the three sp^3 C atoms are tetrahedral. **7.103** (a) 180° (b) 120° (c) 109.5° (d) 109.5° (e) 180° (f) 120° (g) 109.5° (h) 109.5°

7.105 (a) ; planar (b) $\left[\overset{..}{\underset{..}{O}}-\overset{|}{\underset{\underset{..}{O}}{Cl}}-\overset{..}{\underset{..}{O}}\right]^-$; nonplanar

(c) H—C≡N:; polar (d) ; polar (e) ; greater than 120° **7.107** CCl_4 **7.109** Only ICl_2^- and $CdBr_2$ will be linear. **7.111** (a) sp^2 (b) The structure on the right is polar. Molecule on the left will be nonpolar. **7.113** In 1,2-dichloroethane, the two C atoms are joined by a sigma bond. Rotation about a sigma bond does not destroy the bond, and the bond is therefore free (or relatively free) to rotate. Thus, all angles are permitted and the molecule is nonpolar because the C—Cl bond moments cancel each other because of the averaging effect brought about by rotation. In *cis*-dichloroethylene, the two C—Cl bonds are locked in position. The π bond between the C atoms prevents rotation (in order to rotate, the π bond must be broken, using an energy source such as light or heat). Therefore, there is no rotation about the C=C in *cis*-dichloroethylene, and the molecule is polar.

7.115 (a) K_2S: Ionic forces are much stronger than the dipole-dipole forces in $(CH_3)_3N$. (b) Br_2: Both the molecules are nonpolar; but Br_2 has a larger mass.

7.117

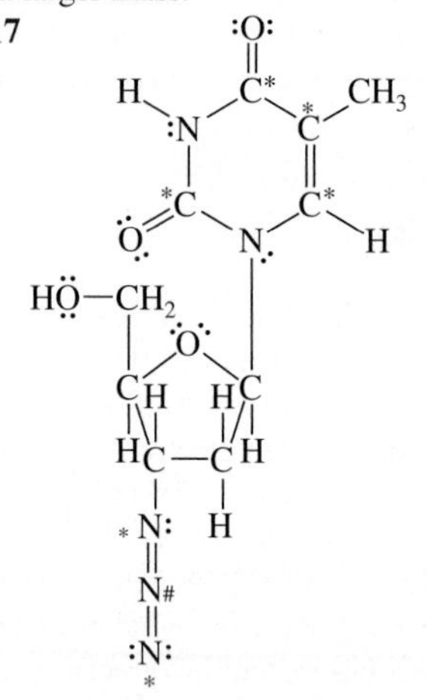

The carbon atoms and nitrogen atoms marked with an asterisk (C* and N*) are sp^2 hybridized; unmarked carbon atoms and nitrogen atoms are sp^3 hybridized; and the nitrogen atom marked with (#) is sp hybridized. **7.119** (a) no sp hybrid orbitals; (b) two sp^2 hybrid orbitals; (c) five sp^3 orbitals; (d) no sp^3d orbitals **7.121** The carbons are all sp^2 hybridized. The nitrogen double bonded to the carbon in the ring is sp^2 hybridized. The other nitrogens are sp^3 hybridized. **7.123** The molecules are all polar. The F atoms can form H-bonds with water and other −OH and −NH groups in the membrane, so water solubility plus easy attachment to the membrane would allow these molecules to pass the blood-brain barrier. **7.125** All except homonuclear diatomic molecules N_2 and O_2. **7.127** The S−S bond is a normal two-electron shared pair covalent bond. Each S is sp^3 hybridized, so the X−S−S angle is about 109°. **7.129** (a) O_2 (b) 2 (c) two **7.131** figure (c) **7.133** The second ("asymmetric stretching mode") and third ("bending mode").

7.135

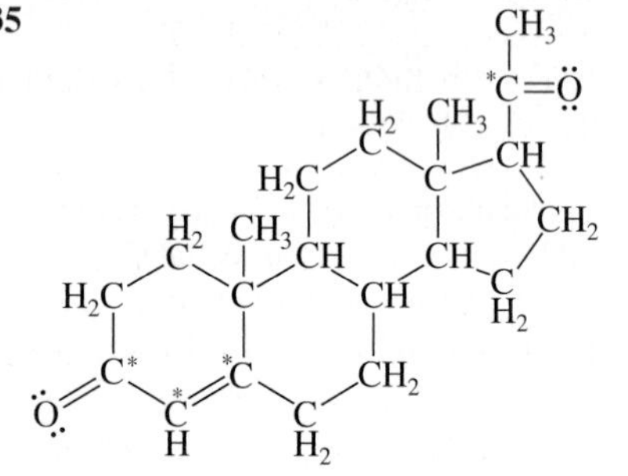

The four carbons marked with an asterisk are sp^2 hybridized. The remaining carbons are sp^3 hybridized. **7.137** (a) $[Ne_2](\sigma_{3s})^2(\sigma_{3s}^*)^2$ $(\pi_{3p_z})^2(\pi_{3p_z}^*)^2(\sigma_{3p_x})^2$ (b) 3 (c) diamagnetic **7.139** The Lewis structure of O_2 shows four pairs of electrons on the two oxygen atoms. From Figure 7.23 of the text, we see that these eight valence electrons are placed in the σ_{2p_x}, π_{2p_y}, π_{2p_z}, $\pi_{2p_y}^*$, and $\pi_{2p_z}^*$ orbitals. For all the electrons to be paired, energy is needed to flip the spin in one of the antibonding molecular orbitals ($\pi_{2p_y}^*$ or $\pi_{2p_z}^*$). According to Hund's rule, this arrangement is less stable than the ground-state configuration shown in Figure 7.23, and hence the Lewis structure shown actually corresponds to an excited state of the oxygen molecule. **7.141** (a) Although the O atoms are sp^3 hybridized, they are locked in a planar structure by the benzene rings. The molecule is symmetrical and therefore is not polar. (b) 20 σ bonds and 6 π bonds. **7.143** (a) $:\overset{-}{C}\equiv\overset{+}{O}:$ The electronegativity difference between O and C suggests that electron density should concentrate on the

O atom, but assigning formal charges places a negative charge on the C atom. Therefore, we expect CO to have a small dipole moment. (b) CO is isoelectronic with N_2 and has a bond order of 3. This agrees with the triple bond in the Lewis structure. (c) Since C has a negative formal charge, it is more likely to form bonds with Fe^{2+} ($OC-Fe^{2+}$ rather than $CO-Fe^{2+}$). **7.145** tetrahedral **7.147** We can assume that the electron is removed from an antibonding orbital. This will increase the ratio of bonding to antibonding electrons, increase the bond order, and shorten the bond length.

Key Skills
7.1 (c) **7.2** (d) **7.3** (b) **7.4** (e)

Chapter 8

8.5 (a) $KOH + H_3PO_4 \longrightarrow K_3PO_4 + H_2O$ (b) $Zn + AgCl \longrightarrow ZnCl_2 + Ag$ (c) $NaHCO_3 \longrightarrow Na_2CO_3 + H_2O + CO_2$ (d) $NH_4NO_2 \longrightarrow N_2 + H_2O$ (e) $CO_2 + KOH \longrightarrow K_2CO_3 + H_2O$
8.7 (a) Potassium and water react to form potassium hydroxide and hydrogen. (b) Barium hydroxide and hydrochloric acid react to form barium chloride and water. (c) Copper and nitric acid react to form copper nitrate, nitrogen monoxide, and water. (d) Aluminum and sulfuric acid react to form aluminum sulfate and hydrogen. (e) Hydrogen iodide reacts to form hydrogen and iodine.
8.9 (a) $2N_2O_5 \longrightarrow 2N_2O_4 + O_2$ (b) $2KNO_3 \longrightarrow 2KNO_2 + O_2$ (c) $NH_4NO_3 \longrightarrow N_2O + 2H_2O$ (d) $NH_4NO_2 \longrightarrow N_2 + 2H_2O$ (e) $2NaHCO_3 \longrightarrow Na_2CO_3 + H_2O + CO_2$ (f) $P_4O_{10} + 6H_2O \longrightarrow 4H_3PO_4$ (g) $2HCl + CaCO_3 \longrightarrow CaCl_2 + H_2O + CO_2$ (h) $2Al + 3H_2SO_4 \longrightarrow Al_2(SO_4)_3 + 3H_2$ (i) $CO_2 + 2KOH \longrightarrow K_2CO_3 + H_2O$ (j) $CH_4 + 2O_2 \longrightarrow CO_2 + 2H_2O$ (k) $Be_2C + 4H_2O \longrightarrow 2Be(OH)_2 + CH_4$ (l) $3Cu + 8HNO_3 \longrightarrow 3Cu(NO_3)_2 + 2NO + 4H_2O$ (m) $S + 6HNO_3 \longrightarrow H_2SO_4 + 6NO_2 + 2H_2O$ (n) $2NH_3 + 3CuO \longrightarrow 3Cu + N_2 + 3H_2O$ **8.11** (d)
8.13 (a) combustion (b) combination (c) decomposition
8.17 (a) diagram (b); (b) diagram (a) **8.19** $C_{10}H_{20}O$ **8.21** CH_2Cl
8.23 empirical formula: $C_2H_3O_2$; molecular formula: $C_4H_6O_4$
8.27 1.01 mol Cl_2 **8.29** 2.0×10^1 mol CO_2 **8.31** (a) $2NaHCO_3 \longrightarrow Na_2CO_3 + CO_2 + H_2O$ (b) 78.3 g $NaHCO_3$ **8.33** 255.9 g C_2H_5OH; 0.324 L **8.35** 19.1 g KCN **8.37** (a) $NH_4NO_3(s) \longrightarrow N_2O(g) + 2H_2O(g)$ (b) 2.0×10^1 g N_2O **8.39** 18.0 g O_2

Visualizing Chemistry
VC 8.1 a **VC 8.2** b **VC 8.3** c **VC 8.4** b

8.45 6 mol NH_3 produced and 1 mol H_2 left **8.47** $4NO_2 + O_2 \longrightarrow 2N_2O_5$; The limiting reactant is NO_2. **8.49** 0.709 g NO_2 produced; O_3 is the limiting reactant; 0.0069 mol NO remaining. **8.51** HCl is the limiting reactant; 23.4 g Cl_2 **8.53** (a) 7.05 g O_2 (b) 92.9% **8.55** 3.48×10^3 g C_6H_{14} **8.57** 8.55 g S_2Cl_2; 76.6% **8.59** 100% **8.63** The electron configuration of helium is $1s^2$ and that of the other noble gases is ns^2np^6. The completely filled subshell of noble gases represents great stability. Consequently, these elements are chemically unreactive.
8.65 diagram (b) **8.67** Cl_2O_7 **8.69** (a) 0.212 mol O (b) 0.424 mol O **8.71** 39.64% NO **8.73** 700 g **8.75** (a) $Zn(s) + H_2SO_4(aq) \longrightarrow ZnSO_4(aq) + H_2(g)$ (b) 64.2% (c) We assume that the impurities are inert and do not react with the sulfuric acid to produce hydrogen.
8.77 (a) 0.307 g salicylic acid (b) 0.410 g salicylic acid (c) 12.08 g aspirin; 90.2% **8.79** 65.4 amu; Zn **8.81** Mg_3N_2; magnesium nitride **8.83** PbC_8H_{20} **8.85** (a) $C_3H_8(g) + 3H_2O(g) \longrightarrow 3CO(g) + 7H_2(g)$ (b) 909 kg H_2 **8.87** (a) \$0.47/kg (b) 0.631 kg K_2O **8.89** (a) NH_4NO_2 is the *only* reactant. When a reaction involves only a single reactant, we usually do not describe that reactant as the "limiting reactant"

because it is understood that this is the case. (b) In principle, two or more reactants can be limiting reactants if they are exhausted simultaneously. In practice, it is virtually impossible to measure the reactants so precisely that two or more run out at the same time. **8.91** 2.60 g H_2O **8.93** $2NO(g) + SO_2(g) \longrightarrow SO_3(g) + N_2O(g)$. **8.95** 0.56 mol $NH_4V_3O_8$ **8.97** The atom economy for the recent synthesis, 100%, is nearly three times higher than it is for the traditional preparation, 25.46%. **8.99** (a) compound X: MnO_2; compound Y: Mn_3O_4 (b) $3MnO_2(s) \longrightarrow Mn_3O_4(s) + O_2(g)$

Key Skills
8.1 b **8.2** c **8.3** e **8.4** a

Chapter 9

9.9 $A_2F < A_2G < A_2E$ **9.11** diagram (c) **9.13** (a) strong electrolyte (b) nonelectrolyte (c) weak electrolyte (d) strong electrolyte (e) weak electrolyte **9.17** diagram (c) **9.19** reaction (d) **9.21** (a) insoluble (b) insoluble (c) soluble (d) soluble (e) insoluble **9.23** (a) ionic: $2Ag^+(aq) + 2NO_3^-(aq) + 2Na^+(aq) + SO_4^{2-}(aq) \longrightarrow Ag_2SO_4(s) + 2Na^+(aq) + 2NO_3^-(aq)$; net ionic: $2Ag^+(aq) + SO_4^{2-}(aq) \longrightarrow Ag_2SO_4(s)$ (b) ionic: $Ba^{2+}(aq) + 2Cl^-(aq) + Zn^{2+}(aq) + SO_4^{2-}(aq) \longrightarrow BaSO_4(s) + Zn^{2+}(aq) + 2Cl^-(aq)$; net ionic: $Ba^{2+}(aq) + SO_4^{2-}(aq) \longrightarrow BaSO_4(s)$ (c) ionic: $2NH_4^+(aq) + CO_3^{2-}(aq) + Ca^{2+}(aq) + 2Cl^-(aq) \longrightarrow CaCO_3(s) + 2NH_4^+(aq) + 2Cl^-(aq)$; net ionic: $Ca^{2+}(aq) + CO_3^{2-}(aq) \longrightarrow CaCO_3(s)$ **9.25** A precipitation reaction will likely result from (b) mixing a $BaCl_2$ solution with a K_2SO_4 solution; (a) no net ionic equation (b) $Ba^{2+}(aq) + SO_4^{2-}(aq) \longrightarrow BaSO_4(s)$ (c) no net ionic equation **9.33** (a) Brønsted base (b) Brønsted base (c) Brønsted acid (d) either Brønsted acid or Brønsted base **9.35** (a) balanced: $HC_2H_3O_2(aq) + KOH(aq) \longrightarrow KC_2H_3O_2(aq) + H_2O(l)$; ionic: $HC_2H_3O_2(aq) + K^+(aq) + OH^-(aq) \longrightarrow C_2H_3O_2^-(aq) + K^+(aq) + H_2O(l)$; net ionic: $HC_2H_3O_2(aq) + OH^-(aq) \longrightarrow C_2H_3O_2^-(aq) + H_2O(l)$ (b) balanced: $H_2CO_3(aq) + 2NaOH(aq) \longrightarrow Na_2CO_3(aq) + 2H_2O(l)$; ionic: $H_2CO_3(aq) + 2Na^+(aq) + 2OH^-(aq) \longrightarrow 2Na^+(aq) + CO_3^{2-}(aq) + 2H_2O(l)$; net ionic: $H_2CO_3(aq) + 2OH^-(aq) \longrightarrow CO_3^{2-}(aq) + 2H_2O(l)$ (c) balanced: $2HNO_3(aq) + Ba(OH)_2(aq) \longrightarrow Ba(NO_3)_2(aq) + 2H_2O(l)$; ionic: $2H^+(aq) + 2NO_3^-(aq) + Ba^{2+}(aq) + 2OH^-(aq) \longrightarrow Ba^{2+}(aq) + 2NO_3^-(aq) + 2H_2O(l)$; net ionic: $2H^+(aq) + 2OH^-(aq) \longrightarrow 2H_2O(l)$ or $H^+(aq) + OH^-(aq) \longrightarrow H_2O(l)$ **9.43** (a) half-reactions: $2Sr \longrightarrow 2Sr^{2+} + 4e^-$ and $O_2 + 4e^- \longrightarrow 2O^{2-}$; oxidizing agent: O_2; reducing agent: Sr (b) half-reactions: $2Li \longrightarrow 2Li^+ + 2e^-$ and $H_2 + 2e^- \longrightarrow 2H^-$; oxidizing agent: H_2; reducing agent: Li (c) half-reactions: $2Cs \longrightarrow 2Cs^+ + 2e^-$ and $Br_2 + 2e^- \longrightarrow 2Br^-$; oxidizing agent: Br_2; reducing agent: Cs (d) half-reactions: $3Mg \longrightarrow 3Mg^{2+} + 6e^-$ and $N_2 + 6e^- \longrightarrow 2N^{3-}$; oxidizing agent: N_2; reducing agent: Mg **9.45** H_2S (−2), S^{2-}(−2), HS^-(−2) < S_8(0) < SO_2(+4) < SO_3(+6), H_2SO_4(+6); The number in parentheses denotes the oxidation number of sulfur. **9.47** (a) +1 (b) +7 (c) −4 (d) −1 (e) −2 (f) +6 (g) +6 (h) +7 (i) +4 (j) 0 (k) +5 (l) −1/2 (m) +5 (n) +3 **9.49** (a) +1 (b) −1 (c) +3 (d) +3 (e) +4 (f) +6 (g) +2 (h) +4 (i) +2 (j) +3 (k) +5 **9.51** If nitric acid is a strong oxidizing agent and zinc is a strong reducing agent, then zinc metal will probably reduce nitric acid when the two react; that is, N will gain electrons and the oxidation number of N must decrease. Since the oxidation number of nitrogen in nitric acid is +5, then the nitrogen-containing product must have a smaller oxidation number for nitrogen. The only compound in the list that doesn't have a nitrogen oxidation number less than +5 is N_2O_5 (what is the oxidation number of N in N_2O_5?) This is never a product of the reduction of nitric acid. **9.53** Molecular oxygen is a powerful oxidizing agent. In SO_3, the oxidation number of the element bound to oxygen (S) is at its maximum value (+6); the sulfur cannot be oxidized further. The other elements bound to oxygen in this problem have less than their maximum oxidation number and can undergo further oxidation. Only SO_3 does not react with molecular oxygen. **9.55** (a) decomposition/disproportionation (b) displacement (c) decomposition (d) combination

Visualizing Chemistry
VC 9.1 b **VC 9.2** c **VC 9.3** c **VC 9.4** a

9.63 411 g CaI_2 **9.65** 0.571 g $MgCl_2$ **9.67** (a) 1.16 M (b) 0.608 M (c) 1.78 M **9.69** (a) 136 mL soln (b) 62.2 mL soln (c) 47 mL soln **9.71** Dilute 323 mL of the 2.00 M HCl solution to a final volume of 1.00 L. **9.73** Dilute 3.00 mL of the 4.00 M HNO_3 solution to a final volume of 60.0 mL. **9.75** 1.41 M $KMnO_4$ **9.77** 0.00699 M $C_6H_{12}O_6$ **9.79** (a) 2.57 (b) 8.72 (c) 1.25 **9.81** (a) 3.8×10^{-3} M (b) 6.2×10^{-12} M (c) 1.1×10^{-7} M (d) 1×10^{-15} M **9.83** A solution of pH < 7 and $[H^+]$ > 1.0×10^{-7} M will be acidic at 25°C. A solution of pH > 7 and $[H^+]$ < 1.0×10^{-7} M will be basic at 25°C. A solution of pH = 7 and $[H^+]$ = 1.0×10^{-7} M will be neutral at 25°C. **9.85** (a) $BaCl_2$: 0.300 M Cl^-; NaCl: 0.566 M Cl^-; $AlCl_3$: 3.606 M Cl^- (b) 1.28 M $Sr(NO_3)_2$ **9.87** 2.325 M **9.89** (a) 0.26; 0.40; 0.53; 0.73 (b) 0.659 L/mol · cm **9.91** diagram (e) **9.99** 3.58 g AgCl **9.101** 0.165 g NaCl; $Ag^+(aq) + Cl^-(aq) \longrightarrow AgCl(s)$ **9.103** 0.1106 M **9.105** (a) 59.85 mL (b) 221.7 mL (c) 110.8 mL **9.107** diagram (c) **9.109** Diagram (d) corresponds to the neutralization of NaOH. Diagram (c) corresponds to the neutralization of $Ba(OH)_2$. **9.111** Choice (d), 0.20 M $Mg(NO_3)_2$, has the greatest concentration of ions. **9.113** 773 mL **9.115** (a) weak electrolyte (b) strong electrolyte (c) strong electrolyte (d) nonelectrolyte **9.117** (a) $C_2H_5ONH_2$ molecules (b) K^+ and F^- ions (c) NH_4^+ and NO_3^- ions (d) C_3H_7OH molecules **9.119** 1146 g/mol **9.121** 1.28 M **9.123** 43.4 g $BaSO_4$ **9.125** (a) redox (b) precipitation (c) acid-base (d) combination (e) redox (f) redox (g) precipitation (h) redox (i) redox (j) redox **9.127** (a) Magnesium hydroxide is insoluble in water. It can be prepared by mixing a solution containing Mg^{2+} ions such as $MgCl_2(aq)$ or $Mg(NO_3)_2(aq)$ with a solution containing hydroxide ions such as $NaOH(aq)$. $Mg(OH)_2$ will precipitate and can be collected by filtration. (b) 0.78 L **9.129** 1.72 M **9.131** Electrolysis to ascertain if hydrogen and oxygen were produced; the reaction with an alkali metal to see if a base and hydrogen gas were produced; the dissolution of a metal oxide to see if a base was produced (or a nonmetal oxide to see if an acid was produced). **9.133** diagram (a); $AgOH(aq) + HNO_3(aq) \longrightarrow H_2O(l) + AgNO_3(aq)$ **9.135** (a) Check with litmus paper; combine with carbonate or bicarbonate to see if CO_2 gas is produced; combine with a base and check for neutralization with an indicator. (b) Titrate a known quantity of acid with a standard NaOH solution. (c) Visually compare the conductivity of the acid with a standard NaCl solution of the same molar concentration. **9.137** (a) $Pb(NO_3)_2(aq) + Na_2SO_4(aq) \longrightarrow PbSO_4(s) + 2NaNO_3(aq)$; $Pb^{2+}(aq) + SO_4^{2-}(aq) \longrightarrow PbSO_4(s)$ (b) 6.34×10^{-5} M **9.139** (a) $HI(aq) + KOH(aq) \longrightarrow KI(aq) + H_2O(l)$; The resulting solution could be evaporated to dryness to isolate the KI. (b) $2HI(aq) + K_2CO_3(aq) \longrightarrow 2KI(aq) + CO_2(g) + H_2O(l)$; The resulting solution could be evaporated to dryness to isolate the KI. **9.141** (a) Combine any soluble magnesium salt with a soluble hydroxide; filter the precipitate. (b) Combine any soluble silver salt with any soluble iodide salt; filter the precipitate. (c) Combine any soluble barium salt with any soluble phosphate salt; filter the precipitate. **9.143** (a) Add Na_2SO_4. (b) Add KOH. (c) Add $AgNO_3$. (d) Add $Ca(NO_3)_2$. (e) Add $Mg(NO_3)_2$. **9.145** reaction 1: $SO_3^{2-}(aq) + H_2O_2(aq) \longrightarrow SO_4^{2-}(aq) + H_2O(l)$; reaction 2: $SO_4^{2-}(aq) + Ba^{2+}(aq) \longrightarrow BaSO_4(s)$ **9.147** Cl_2O (+1), Cl_2O_3 (+3), ClO_2 (+4), Cl_2O_6 (+6), Cl_2O_7 (+7)

9.149 4.99 grains of aspirin in one tablet **9.151** (a) $CaF_2(s)$ + $H_2SO_4(aq) \longrightarrow CaSO_4(s) + 2HF(g); 2NaCl(s) + H_2SO_4(aq) \longrightarrow Na_2SO_4(aq) + 2HCl(g)$ (b) The sulfuric acid would oxidize Br^- and I^- ions to Br_2 and I_2. (c) $PBr_3(l) + 3H_2O(l) \longrightarrow 3HBr(g) + H_3PO_3(aq)$ **9.153** electric furnace method: $P_4(s) + 5O_2(g) \longrightarrow P_4O_{10}(s)$ (redox); $P_4O_{10}(s) + 6H_2O(l) \longrightarrow 4H_3PO_4(aq)$ (acid-base); wet process: $Ca_5(PO_4)_3F(s) + 5H_2SO_4(aq) \longrightarrow 3H_3PO_4(aq) + HF(aq) + 5CaSO_4(s)$ (acid-base and precipitation) **9.155** (a) $4KO_2(s) + 2CO_2(g) \longrightarrow 2K_2CO_3(s) + 3O_2(g)$ (b) $-1/2$ (c) 34 L air **9.157** No. In both O_2 (molecular oxygen) and O_3 (ozone), the oxidation number of oxygen is zero. **9.159** (a) acid: H_3O^+; base: OH^-

OH⁻   H₃O⁺   H₂O   H₂O

(b) acid: NH_4^+; base: NH_2^-

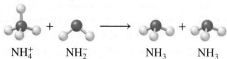

NH₄⁺   NH₂⁻   NH₃   NH₃

9.161 When a solid dissolves in solution, the volume of the solution usually changes. **9.163** (a) The precipitate $CaSO_4$ formed over Ca preventing the Ca from reacting with the sulfuric acid. (b) Aluminum is protected by a tenacious oxide layer with the composition Al_2O_3. (c) These metals react more readily with water: $2Na(s) + 2H_2O(l) \longrightarrow 2NaOH(aq) + H_2(g)$ (d) The metal should be placed below Fe and above H. (e) Any metal above Al in the activity series will react with Al^{3+}. Metals from Mg to Li will work. **9.165** (a) (1) $Cu(s) + 4HNO_3(aq) \longrightarrow Cu(NO_3)_2(aq) + 2NO_2(g) + 2H_2O(l)$ (redox) (2) $Cu(NO_3)_2(aq) + 2NaOH(aq) \longrightarrow Cu(OH)_2(s) + 2NaNO_3(aq)$ (precipitation) (3) $Cu(OH)_2(s) \longrightarrow CuO(s) + H_2O(g)$ (decomposition) (4) $CuO(s) + H_2SO_4(aq) \longrightarrow CuSO_4(aq) + H_2O(l)$ (acid-base) (5) $CuSO_4(aq) + Zn(s) \longrightarrow Cu(s) + ZnSO_4(aq)$ (redox) (6) $Zn(s) + 2HCl(aq) \longrightarrow ZnCl_2(aq) + H_2(g)$ (redox) (b) (1) 194 g $Cu(NO_3)_2$, (2) 101 g $Cu(OH)_2$, (3) 82.1 g CuO, (4) 165 g $CuSO_4$, (5) 65.6 g Cu (6) The amount of copper recovered in Step 6 will vary with the washing technique used. (c) All the reaction steps are clean and almost quantitative; therefore, the recovery yield should be high. **9.167** [Na^+]: 0.5295 M; [NO_3^-]: 0.4298 M; [OH^-]: 0.09968 M; [Mg^{2+}]: 0 M

Key Skills
9.1 b **9.2** c **9.3** c **9.4** d

Chapter 10

10.5 law of conservation of energy **10.7** Energy is needed to break chemical bonds, whereas energy is released when bonds are formed. **10.11** 61 J **10.13** -177 J (work done by the system) **10.15** (a) diagram (iii) (b) diagram (iii) (c) diagrams (i), (ii), and (iii) **10.21** (a) 0 J (b) -33 J (c) -94.7 J **10.23** 4.51 kJ/g **10.25** 4.80×10^2 kJ **10.27** 595.8 kJ/mol

Visualizing Chemistry
VC 10.1 c **VC 10.2** a **VC 10.3** b **VC 10.4** c **VC 10.5** c **VC 10.6** a **VC 10.7** b **VC 10.8** c

10.31 90.1 kJ **10.33** 50.7°C **10.35** 26.3°C **10.37** 48 °C/day; 4.1 kg water/day; A person would have to drink approximately this much water each day to maintain normal body temperature, so the human body is not an isolated system. Other external factors, such as the choice of clothing that allows evaporation, must also play a role in maintaining proper body temperature. **10.39** 2.36 J/g · °C

10.41 choice (d) **10.45** -238.7 kJ/mol **10.47** 0.30 kJ/mol **10.49** -180 kJ/mol **10.55** O_2 **10.57** 177.8 kJ/mol **10.59** (a) -571.6 kJ/mol (b) -2599 kJ/mol **10.61** (a) -724 kJ/mol (b) -1.37×10^3 kJ/mol (c) -2.01×10^3 kJ/mol **10.63** -3924 kJ/mol **10.65** -3.43×10^4 kJ **10.67** $\Delta H_f^\circ[H_2O(l)]$ **10.69** $\Delta H_f^\circ[H_2O(l)]$ is even more negative, meaning that $2H_2O(l) + O_2(g)$ is more stable than $H_2O_2(l)$. **10.71** -746.8 kJ/mol **10.75** 303.0 kJ/mol **10.77** (a) -2759 kJ/mol (b) -2855.4 kJ/mol **10.79** -651 kJ/mol

Visualizing Chemistry
VC 10.9 b **VC 10.10** c **VC 10.11** c **VC 10.12** b

10.83 2255 kJ/mol **10.85** -112.7 kJ/mol **10.87** -44.35 kJ/mol **10.89** -175.3 kJ **10.91** -67 kJ/mol **10.93** (a) -10.5 kJ/mol (b) -9.2 kJ/mol **10.95** -350.7 kJ/mol **10.97** 0.492 J/g · °C **10.99** The first reaction is exothermic and produces the energy required by the second reaction, which is endothermic. **10.101** 446 mol H_2 **10.103** 0.167 mol ethane **10.105** 5.60 kJ/mol **10.107** reaction (a) **10.109** (a) No work is done by a gas expanding in a vacuum. (b) -9.1 J **10.111** -8.68×10^4 kJ **10.113** -1.84×10^3 kJ **10.115** 3.0 billion atomic bombs **10.117** (a) $2LiOH(aq) + CO_2(g) \longrightarrow Li_2CO_3(aq) + H_2O(l)$ (b) 1.1 kg CO_2, 1.2 kg LiOH **10.119** -5.2×10^6 kJ **10.121** -3.60×10^2 kJ/mol Zn (or -3.60×10^2 kJ/2 mol Ag^+) **10.123** (a) and (b) The ΔH_{rxn}° is not the ΔH_f° value because more than 1 mole of a substance is formed. (c) The ΔH_{rxn}° is the ΔH_f° value because 1 mole of a substance is formed from its constituent elements in their standard states. (d) The ΔH_{rxn}° is not the ΔH_f° value because $Cu^{2+}(aq)$ is not copper's standard state and CO_3^{2-} is not an element. (e) The ΔH_{rxn}° is the ΔH_f° value because 1 mole of a substance is formed from its constituent elements in their standard states. **10.125** Water has a larger specific heat than air. Thus cold, damp air can extract more heat from the body than cold, dry air. By the same token, hot, humid air can deliver more heat to the body. **10.127** 5.8×10^2 m **10.129** This law does not hold for Hg because it is a liquid. **10.131** $\Delta H_f^\circ(C_2H_2) = 226.6$ kJ/mol; $\Delta H_f^\circ(C_6H_6) = 49.0$ kJ/mol; $\Delta H_{rxn} = -630.8$ kJ/mol. **10.133** (a) -46 kJ/mol (b) 260 nm **10.135** (a) 114 kJ/mol (b) The bond in F_2^- is weaker. **10.137** (a) glucose: 31 kJ; sucrose: 33 kJ (b) glucose: 15 m; sucrose: 16 m **10.139** 347 kJ/mol **10.141** -564.2 kJ/mol **10.143** 96.21% **10.145** (a) We could easily measure ΔH_{rxn}° for this combustion by burning a mole of glucose in a bomb calorimeter. (b) 1.1×10^{19} kJ **10.147** (a) $CaC_2(s) + 2H_2O(l) \longrightarrow Ca(OH)_2(s) + C_2H_2(g)$ (b) 1.51×10^6 J **10.149** $q = \frac{1}{2}mu^2$; The amount of heat generated must be proportional to the braking distance, d: $d \propto q$, $d \propto u^2$; Therefore, as u increases to $2u$, d increases to $(2u)^2 = 4u^2$ which is proportional to $4d$. **10.151** -413 kJ/mol **10.153** -5153 kJ/mol; 0 **10.155** 4.1 cents **10.157** C—C: 347 kJ/mol; N—N: 193 kJ/mol; O—O: 142 kJ/mol; Lone pairs appear to weaken the bond.

Key Skills
10.1 e **10.2** a **10.3** c **10.4** b

Chapter 11

11.9 u_{rms} (N_2) = 490 m/s; u_{rms} (O_2) = 459 m/s; u_{rms} (O_3) = 375 m/s **11.11** RMS speed = 2.8 m/s; average speed = 2.7 m/s; The root-mean-square value is always greater than the average value, because squaring favors the larger values compared to just taking the average value. **11.13** 43.8 g/mol; CO_2 **11.15** (a) Since more of the yellow gas was able to escape the container in the same amount of time, the rate of effusion for the yellow gas is greater. Therefore, the molar mass of the yellow gas is lower. (b) More of the red gas escaped; hence, the molar mass of the blue gas is greater.

11.23 0.493 atm; 0.500 bar; 375 torr; 5.00×10^4 Pa **11.25** 11.9 m **11.27** 5.7 atm; 5.8 bar **11.31** 6.10 L **11.33** 587 mmHg **11.35** 31.8 L **11.37** one volume **11.39** (a) diagram (d); (b) diagram (b) **11.43** 6.6 atm **11.45** 1.8 atm **11.47** 0.70 L **11.49** 63.31 L **11.51** 6.1×10^{-3} atm **11.53** 35.1 g/mol **11.55** 2.1×10^{22} N_2 molecules; 5.6×10^{22} O_2 molecules; 2.7×10^{20} Ar atoms **11.57** 2.98 g/L **11.59** SF_4 **11.61** 1590°C **11.67** 18.0 atm **11.71** (a) 0.89 atm (b) 1.4 L at STP **11.73** $P_{Ne} = 349$ mmHg **11.75** $P_{N_2} = 217$ mmHg; $P_{H_2} = 650$ mmHg **11.77** 13 atm **11.79** (a) box on the right (b) box on the right

Visualizing Chemistry
VC 11.1 b **VC 11.2** a **VC 11.3** a **VC 11.4** b

11.81 3.70×10^2 L **11.83** 88.9% **11.85** M(s) + 3HCl(aq) ⟶ $1.5H_2(g) + MCl_3(aq)$; M_2O_3; $M_2(SO_4)_3$ **11.87** 94.7%; The impurity (or impurities) must not react with HCl to produce CO_2 gas. **11.89** $C_2H_5OH(l) + 3O_2(g) \longrightarrow 2CO_2(g) + 3H_2O(l)$; 1.44×10^3 L air **11.91** 19.8 g Zn **11.93** 1.7×10^2 L; $P_{CO_2} = 0.49$ atm, $P_{H_2O} = 0.41$ atm, $P_{N_2} = 0.25$ atm, and $P_{O_2} = 0.041$ atm **11.95** (a) $NH_4NO_2(s) \longrightarrow N_2(g) + 2H_2O(l)$ (b) 0.273 g **11.97** (a) $P_{ii} = 4.0$ atm; $P_{iii} = 2.7$ atm (b) total pressure: 2.7 atm; $P_A = 1.3$ atm, $P_B = 1.3$ atm **11.99** (a) 9.54 atm (b) If the tetracarbonylnickel gas starts to decompose significantly above 43°C, then for every mole of tetracarbonylnickel gas decomposed, four moles of carbon monoxide would be produced. **11.101** 1.3×10^{22} molecules; CO_2, O_2, N_2, and H_2O **11.103** (a) $2KClO_3(s) \longrightarrow 2KCl(s) + 3O_2(g)$ (b) 6.21 L **11.105** 0.0701 M **11.107** $P_{He} = 0.16$ atm, $P_{Ne} = 2.0$ atm **11.109** When the water enters the flask from the dropper, some hydrogen chloride dissolves, creating a partial vacuum. Pressure from the atmosphere forces more water up the vertical tube. **11.111** 7 **11.113** (a) 61.2 m/s (b) 4.58×10^{-4} s (c) 328 m/s, $u_{rms} = 366$ m/s; The magnitudes of the speeds are comparable, but not identical. The rms speed is derived for three-dimensional motion and is not applicable to the speed distributions in directed molecular beams. Also, the experiment is measuring a most-probable speed, not an rms speed. **11.115** 2.09×10^4 g O_2; 1.58×10^4 L O_2 **11.117** The fruit ripens more rapidly because the quantity (partial pressure) of ethylene gas inside the bag increases. **11.119** As the pen is used the amount of ink decreases, increasing the volume inside the pen. As the volume increases, the pressure inside the pen decreases. The hole is needed to equalize the pressure as the volume inside the pen increases. **11.121** (a) $NH_4NO_3(s) \longrightarrow N_2O(g) + 2H_2O(l)$ (b) 0.0821 L · atm/K · mol **11.123** C_6H_6 **11.125** The gases inside the mine were a mixture of carbon dioxide, carbon monoxide, methane, and other harmful compounds. The low atmospheric pressure caused the gases to flow out of the mine (the gases in the mine were at a higher pressure), and the man suffocated. **11.127** (a) 4.90 L (b) 6.0 atm (c) 1 atm **11.129** (a) 4×10^{-22} atm (b) 6×10^{20} L/g of H **11.131** 91% **11.133** 1.7×10^{12} O_2 molecules **11.135** NO_2 **11.137** 1.4 g LiH **11.139** P_{CO_2}: 8.7×10^6 Pa; P_{N_2}: 3.2×10^5 Pa; P_{SO_2}: 1.4×10^3 Pa **11.141** (a) (i) He atoms in both flasks have the same rms speed and average kinetic energy. (ii) The atoms in the larger volume strike the walls of their container with the same average force as those in the smaller volume, but pressure and frequency of collisions in the larger volume will be less. (b) (i) u_2 is greater than u_1 by a factor of $\sqrt{T_2/T_1}$. (ii) Both the frequency of collision and the average force of collision of the atoms at T_2 will be greater than those at T_1. (c) (i) False. The rms speed of He is greater than that of Ne because He is less massive. (ii) True. The average kinetic energies of the two gases are equal. (iii) The rms speed of the whole collection of atoms is 1.47×10^3 m/s, but the speeds of the individual atoms are random and constantly changing. **11.143** vapor = 3.7 nm; liquid = 0.31 nm; Water molecules are packed very closely together in the liquid, but much farther apart in the steam.

11.145 9.98 atm **11.147** 1.41 atm **11.149** methane = 0.789; ethane = 0.211 **11.151** (a) $8(\frac{4}{3}\pi r^3)$. (b) $4N_A(\frac{4}{3}\pi r^3)/1$ mole; The excluded volume per mole is four times the volume actually occupied by a mole of molecules. **11.153** NO **11.155** Reaction (b) **11.157** 1.80×10^{-4} g/mL **11.159** The kinetic energy of a particle depends on its mass and its speed ($E_k = \frac{1}{2}mv^2$). At a given temperature, the more massive SF_6 molecules move more slowly than do the Ne molecules. **11.161** Intermolecular attractions tend to lower Z at low pressures and excluded volume tends to increase Z at high pressures.

Key Skills
11.1 d **11.2** c **11.3** b **11.4** e

Chapter 12

12.13 29.9 kJ/mol **12.15** Ethylene glycol has two —OH groups, allowing it to exert strong intermolecular forces through hydrogen bonding. Its viscosity should fall between ethanol (one —OH group) and glycerol (three —OH groups). **12.17** Liquid X has a larger ΔH_{vap} than does liquid Y. **12.27** Simple cubic cell: one sphere; body-centered cubic cell: two spheres; face-centered cubic cell: four spheres **12.29** 6.20×10^{23} atoms/mol **12.31** 458 pm **12.33**: XY_3 **12.35** 0.220 nm **12.37** ZnO **12.41** molecular solid **12.43** molecular: Se_8, HBr, CO_2, P_4O_6, and SiH_4; covalent: Si and C **12.45** Diamond: Each carbon atom is covalently bonded to four other carbon atoms. Because these bonds are strong and uniform, diamond is a very hard substance. Graphite: The carbon atoms in each layer are linked by strong bonds, but the layers are bound by weak dispersion forces. As a result, graphite may be cleaved easily between layers and is not hard. In graphite, all the atoms are sp^2 hybridized; each atom is covalently bonded to three other atoms. The remaining unhybridized $2p$ orbital is used in pi bonding forming a delocalized molecular orbital. The electrons are free to move around in this extensively delocalized molecular orbital, making graphite a good conductor of electricity in directions along the planes of carbon atoms. **12.65** 664 kJ **12.67** Two phase changes occur in this process. First, the liquid is turned to solid (freezing), then the solid ice is turned to gas (sublimation). **12.69** When steam condenses to liquid water at 100°C, it releases a large amount of heat equal to the enthalpy of vaporization. Thus steam at 100°C exposes one to more heat than an equal amount of water at 100°C. **12.71** The solid ice turns to vapor (sublimation). The temperature is too low for melting to occur. **12.75** Initially, the ice melts because of the increase in pressure. As the wire sinks into the ice, the water above the wire refreezes. Eventually the wire moves completely through the ice block without cutting it in half. **12.77** (a) Ice would melt. (If heating continues, the liquid water would eventually boil and become a vapor.) (b) Liquid water would vaporize.(c) Water vapor would solidify without becoming a liquid. **12.79** covalent **12.81** (d) very low vapor pressure **12.83** 0.26 cm; 7.69×10^6 base pairs **12.85** 760 mmHg **12.87** It has reached the critical point; the point of critical temperature (T_c) and critical pressure (P_c). **12.89** crystalline SiO_2; Its regular structure results in a more efficient packing. **12.91** (a) and (b) **12.93** 8.3×10^{-3} atm **12.95** 62.4 kJ/mol **12.97** SO_2 will behave less ideally because it is polar and has greater intermolecular forces. **12.99** Smaller ions can approach water molecules more closely, resulting in larger ion-dipole interactions. The greater the ion-dipole interaction, the larger is the heat of hydration. **12.101** (a) 30.7 kJ/mol (b) 192.5 kJ/mol; It requires more energy to break the bond than to vaporize the molecule. **12.103** (a) decreases (b) no change (c) no change **12.105** $CaCO_3(s) \longrightarrow CaO(s) + CO_2(g)$; initial state: one solid phase; final state: two solid phase components and one gas phase component.

12.107 The time required to cook food depends on the boiling point of the water in which it is cooked. The boiling point of water increases when the pressure inside the cooker increases. **12.109** Extra heat is produced when steam condenses at 100°C. Steaming avoids the extraction of ingredients caused by boiling in water. **12.111** (a) ~2.3 K (b) ~10 atm (c) ~5 K (d) no **12.113** (a) Pumping allows Ar atoms to escape, thus removing heat from the liquid phase. Eventually the liquid freezes. (b) The slope of the solid-liquid line of cyclohexane is positive. Therefore, its melting point increases with pressure. (c) These droplets are supercooled liquids. (d) When the dry ice is added to water, it sublimes. The cold CO_2 gas generated causes nearby water vapor to condense, hence the appearance of fog. **12.115** (a) two triple points: diamond/graphite/liquid and graphite/liquid/vapor. (b) diamond (c) Apply high pressure at high temperature. **12.117** Ethanol mixes well with water. The mixture has a lower surface tension and readily flows out of the ear channel. **12.119** When water freezes it releases heat, helping keep the fruit warm enough not to freeze. Also, a layer of ice is a thermal insulator. **12.121** The fuel source for the Bunsen burner is most likely methane gas. When methane burns in air, carbon dioxide and water are produced. The water vapor produced during the combustion condenses to liquid water when it comes in contact with the outside of the cold beaker. **12.123** 6.019×10^{23} Fe atoms/mol **12.125** 127 mmHg **12.127** 55°C **12.129** 0.833 g/L; The hydrogen-bonding interactions in HF are relatively strong, and since the ideal gas equation ignores intermolecular forces, it underestimates significantly the density of HF gas near its boiling point. **12.131** 14.3° **12.133** 3.2×10^2 kJ; 1.1×10^3 kJ

Key Skills

12.1 (a, d, e) **12.2** (a, d) **12.3** (b) **12.4** (d)

Chapter 13

13.9 "Like dissolves like." Naphthalene and benzene are nonpolar, whereas CsF is ionic. **13.11** $O_2 < Br_2 < LiCl < CH_3OH$ **13.17** (a) 0.310 m (b) 4.38 m **13.19** (a) 1.31 m (b) 1.08 m (c) 3.60 m **13.21** 14 g **13.23** 1.1×10^{-3} M **13.31** 45.9 g KNO_3 **13.33** 1.0×10^{-5} mol/L **13.35** $\chi(O_2)$: 0.32; $\chi(N_2)$: 0.68 Because of the greater solubility of oxygen, it has a larger mole fraction in solution than it does in the air. **13.37** Vitamin C: water soluble; Vitamin E: fat soluble **13.39** $P = 3.3$ atm; This pressure is only an estimate since we ignored the amount of CO_2 that was present in the unopened container in the gas phase. **13.41** red solute: exothermic; green solute: endothermic; ΔH_{soln} is greater for the red solute, since changing the temperature produces a greater difference in solubility. The formation of chemical bonds of the precipitate in the red solute releases energy, whereas the breaking of chemical bonds in the dissolution in the green solute absorbs energy. **13.57** 1.3×10^3 g sucrose **13.59** $P_{ethanol}$: 30.0 mmHg; $P_{1\text{-propanol}}$: 26.3 mmHg **13.61** 187 g urea **13.63** 0.59 m **13.65** −5.4°C **13.67** (a) freezing point: −10.0°C; boiling point: 102.8°C (b) freezing point: −7.14°C; boiling point: 102.0°C **13.69** Both NaCl and $CaCl_2$ are strong electrolytes. Urea and sucrose are nonelectrolytes. NaCl or $CaCl_2$ will yield more particles per mole of the solid dissolved, resulting in greater freezing point depression. Also, sucrose and urea would make a mess when the ice melts. **13.71** 2.47 **13.73** 9.16 atm **13.75** (a) $CaCl_2$ (b) urea (c) $CaCl_2$; $CaCl_2$ is an ionic compound and is therefore an electrolyte in water. Assuming that $CaCl_2$ is a strong electrolyte and completely dissociates (no ion pairs, van't Hoff factor $i = 3$), the total ion concentration will be $3 \times 0.35 = 1.05$ m, which is larger than the urea (nonelectrolyte) concentration of 0.90 m.

13.77 0.15 m $C_6H_{12}O_6$ > 0.15 m CH_3COOH > 0.10 m Na_3PO_4 > 0.20 m $MgCl_2$ > 0.35 m NaCl **13.79** (a) NaCl (b) $MgCl_2$ (c) KBr **13.83** 4.3×10^2 g/mol; $C_{24}H_{20}P_4$ **13.85** 1.75×10^4 g/mol **13.87** 342 g/mol **13.89** 15.7% **13.95** levels before: 7.77×10^{-6} mol glucose/mL blood, 3.9×10^{-2} mol glucose, 7.0 g glucose; levels after: 1.33×10^{-5} mol glucose/mL blood, 6.7×10^{-2} mol glucose, 12 g glucose **13.97** Water migrates through the semipermeable cell walls of the cucumber into the concentrated salt solution. **13.99** 3.5 **13.101** X: water; Y: NaCl; Z: urea **13.103** The pill is in a hypotonic solution. Consequently, by osmosis, water moves across the semipermeable membrane into the pill. The increase in pressure pushes the elastic membrane to the right, causing the drug to exit through the small holes at a constant rate. **13.105** Reverse osmosis involves no phase changes and is usually cheaper than distillation or freezing; 34 atm. **13.107** 0.295 M; −0.55°C **13.109** (a) 2.14×10^3 g/mol (b) 4.50×10^4 g/mol **13.111** (a) 0.099 L (b) 9.9 **13.113** (a) At reduced pressure, the solution is supersaturated with CO_2. (b) As the escaping CO_2 expands, it cools, condensing water vapor in the air to form fog. **13.115** (a) Runoff of the salt solution into the soil increases the salinity of the soil. If the soil becomes hypertonic relative to the tree cells, osmosis would reverse, and the tree would lose water to the soil and eventually die of dehydration. (b) Assuming the collecting duct acts as a semipermeable membrane, water would flow from the urine into the hypertonic fluid, thus returning water to the body. **13.117** 33 mL and 67 mL **13.119** 12.3 M **13.121** 14.2% **13.123** (a) Solubility decreases with increasing lattice energy. (b) Ionic compounds are more soluble in a polar solvent. (c) Solubility increases with enthalpy of hydration of the cation and anion. **13.125** 1.43 g/mL; 37.0 m **13.127** NH_3 can form hydrogen bonds with water; NCl_3 cannot. (Like dissolves like.) **13.129** 3% **13.131** NaBr: 8.47%; KCl: 17.7%; toluene: 11% **13.133** Yes **13.135** As the water freezes, dissolved minerals in the water precipitate from the solution. The minerals refract light and create an opaque appearance. **13.137** 1.8×10^2 g/mol **13.139** 1.9 m **13.141** P_A°: 1.9×10^2 mmHg; P_B°: 4.0×10^2 mmHg **13.143** −0.737°C **13.145** about 0.4 molal **13.147** 1.0×10^{-4} % F^-; 5.3×10^{-5} m NaF **13.149** (a) χ_A: 0.52; χ_B: 0.48 (b) P_A: 51 mmHg; P_B: 20 mmHg (c) χ_A: 0.72; χ_B: 0.28 (d) P_A: 71 mmHg; P_B: 12 mmHg

Key Skills

13.1 (a, c) **13.2** (d) **13.3** (e) **13.4** (a)

Chapter 14

14.7 (a) with barrier: 16; without barrier: 64 (b) 16; 16; 32; both particles on one side: $S = 3.83 \times 10^{-23}$ J/K; particles on opposite sides: $S = 4.78 \times 10^{-23}$ J/K; The most probable state is the one with the larger entropy (particles on opposite sides).

Visualizing Chemistry

VC 14.1 b **VC 14.2** a **VC 14.3** c **VC 14.4** b

14.13 (a) −0.048 J/K (b) −0.86 J/K (c) 12.6 J/K **14.15** (a) 47.5 J/K · mol (b) −12.5 J/K · mol (c) −242.8 J/K · mol **14.17** (c) < (d) < (e) < (a) < (b) (a) Ne(g): a monatomic gas of higher molar mass than H_2. (For gas phase molecules, increasing molar mass tends to increase the molar entropy. While Ne(g) is structurally less complex than $H_2(g)$, its much larger molar mass more than offsets the complexity difference.) (b) $SO_2(g)$: a polyatomic gas of higher complexity and higher molar mass than Ne(g) (c) Na(s): highly ordered, crystalline material (d) NaCl(s): highly ordered crystalline material, but with more particles per mole than Na(s) (e) H_2: a diatomic gas, hence of higher entropy than a solid **14.21** (a) 291 J/K · mol; spontaneous

(b) 2.10×10^3 J/K · mol; spontaneous (c) 2.99×10^3 J/K · mol; spontaneous **14.23** (a) $\Delta S_{rxn} = -75.6$ J/K · mol; $\Delta S_{surr} = 185$ J/K · mol; spontaneous (b) $\Delta S_{rxn} = 215.8$ J/K · mol; $\Delta S_{surr} = -304$ J/K · mol; not spontaneous (c) $\Delta S_{rxn} = 98.2$ J/K · mol; $\Delta S_{surr} = -1.46 \times 10^3$ J/K · mol; not spontaneous (d) $\Delta S_{rxn} = -282$ J/K · mol; $\Delta S_{surr} = 7.20 \times 10^3$ J/K · mol; spontaneous **14.29** 64 kJ/mol
14.31 (a) -1139 kJ/mol (b) -140.0 kJ/mol (c) -2935 kJ/mol
14.33 (a) all temperatures (b) below 111 K **14.35** $\Delta S_{fus} = 99.9$ J/K · mol; $\Delta S_{vap} = 93.7$ J/K · mol **14.37** 75.9 kJ **14.41** $\Delta H_{fus} > 0$; $\Delta S_{fus} > 0$;
(a) $\Delta G_{fus} < 0$ (b) $\Delta G_{fus} = 0$ (c) $\Delta G_{fus} > 0$ **14.43** 42°C **14.45** ΔS must be positive ($\Delta S > 0$). **14.47** (a) benzene: $\Delta S_{vap} = 87.8$ J/K · mol; hexane: $\Delta S_{vap} = 90.1$ J/K · mol; mercury: $\Delta S_{vap} = 93.7$ J/K · mol; toluene: $\Delta S_{vap} = 91.8$ J/K · mol; Trouton's rule is a statement about ΔS_{vap}°. In most substances, the molecules are in constant and random motion in both the liquid and gas phases, so $\Delta S_{vap}^{\circ} = 90$ J/K · mol.
(b) ethanol: $\Delta S_{vap} = 112$ J/K · mol; water: $\Delta S_{vap} = 109$ J/K · mol; In ethanol and water, there are fewer possible arrangements of the molecules due to the network of H-bonds, so ΔS_{vap}° is greater.
14.49 (a) positive (b) negative (c) positive (d) positive **14.51** 625 K; ΔH° and ΔS° do not depend on temperature. **14.53** No. A negative ΔG° tells us that a reaction has the potential to happen but gives no indication of the rate. **14.55** ΔH and ΔS are both positive, indicating this is an endothermic process with high entropy; 320 K = 47°C
14.57 $\Delta S_{sys} = 91.1$ J/K; $\Delta S_{surr} = -91.1$ J/K; $\Delta S_{univ} = 0$; The system is at equilibrium. **14.59** 249 J/K **14.61** $\Delta S_{sys} = -327.2$ J/K · mol; $\Delta S_{surr} = 1918$ J/K · mol; $\Delta S_{univ} = 1591$ J/K · mol **14.63** q and w are not state functions. **14.65** ΔG must be negative; ΔS must be negative; ΔH must be negative. **14.67** U and H

Key Skills
14.1 (e) **14.2** (c) **14.3** (b) **14.4** (b)

Chapter 15
15.9 (a) $\dfrac{[N_2][H_2O]^2}{[NO]^2[H_2]^2}$ (b) 32 **15.11** 1.08×10^7 **15.13** six

15.23 2.40×10^{33} **15.25** 3.5×10^{-7} **15.27** (a) 0.082 (b) 0.29
15.29 $K_P = 0.105$; $K_c = 2.05 \times 10^{-3}$ **15.31** 7.09×10^{-3} **15.33** $K_P = 9.6 \times 10^{-3}$; $K_c = 3.9 \times 10^{-4}$ **15.35** 4.0×10^{-6} **15.37** 5.6×10^{23}
15.39 (1) diagram (a); (2) diagram (d); Volume will cancel from the K_c expression. **15.43** The forward reaction will not occur.
15.45 79 kJ/mol **15.47** (a) $\Delta G_{rxn}^{\circ} = 35.4$ kJ/mol; $K_P = 6.2 \times 10^{-7}$
(b) $\Delta G^{\circ} = 44.6$ kJ/mol **15.49** (a) 1.6×10^{-23} atm (b) 0.535 atm
15.51 3.1×10^{-2} atm or 23.6 mmHg **15.53** 0.35 **15.55** diagrams (a), (b), and (c)

Visualizing Chemistry
VC 15.1 a **VC 15.2** c **VC 15.3** c **VC 15.4** c

15.57 0.173 mol H_2 **15.59** $[H_2] = [Br_2] = 1.80 \times 10^{-4}$ M; $[HBr] = 0.267$ M **15.61** $P_{COCl_2} = 0.408$ atm; $P_{CO} = P_{Cl_2} = 0.352$ atm
15.63 $P_{CO_2} = 2.54$ atm; $P_{CO} = 1.96$ atm

Visualizing Chemistry
VC 15.5 c **VC 15.6** b **VC 15.7** a and c **VC 15.8** a, b, and c **15.9** c **15.10** a **15.11** b **15.12** c

15.69 reaction (b) **15.71** (b), (d), and (e) **15.73** (a) The equilibrium would shift to the right. (b) The equilibrium would be unaffected. (c) The equilibrium would be unaffected. **15.75** (a) no effect (b) no effect (c) shift to the left (d) no effect (e) shift to the left **15.77** (a) shifts to the right (b) shifts to the left (c) shifts to the right (d) shifts to the left (e) A catalyst has no effect on equilibrium position.

15.79 (a) shifts to the right (b) no effect (c) no effect (d) shifts to the left (e) shifts to the right (f) shifts to the left (g) shifts to the right
15.81 (a) $2O_3(g) \rightleftharpoons 3O_2(g)$; $\Delta H_{rxn}^{\circ} = -284.4$ kJ/mol (b) The number of O_3 molecules would increase and the number of O_2 molecules would decrease. **15.83** The concentration of oxyhemoglobin will decrease as acidosis occurs. **15.85** diagrams (b) and (d) **15.87** (a) 8×10^{-44}
(b) A mixture of H_2 and O_2 can be kept at room temperature because of the large amount of energy necessary to get the reaction started. This quantity of energy, known as the *activation energy*, is discussed in detail in Chapter 19. **15.89** (a) $2CO + 2NO \longrightarrow 2CO_2 + N_2$
(b) The oxidizing agent is NO; the reducing agent is CO. (c) 3.29×10^{120} (d) 1.2×10^{14}; to the right (e) no **15.91** (a) 1.7 (b) $P_A = 0.69$ atm; $P_B = 0.81$ atm **15.93** 1.5×10^5 **15.95** (a) 86.7 kJ/mol (b) 4.0×10^{-31}
(c) 2.62×10^{-6} (d) Lightning supplies the energy necessary to drive this reaction, converting the two most abundant gases in the atmosphere into $NO(g)$. The NO gas dissolves in the rain, which carries it into the soil where it is converted into nitrate and nitrite by bacterial action. This "fixed" nitrogen is a necessary nutrient for plants. **15.97** $P_{H_2} = 0.28$ atm; $P_{Cl_2} = 0.0507$ atm; $P_{HCl} = 1.67$ atm
15.99 1.78×10^{70}; Because of the large magnitude of K, you would expect this equilibrium to lie very far to the right. However, some reactions do not occur spontaneously despite large K values because of the large amount of energy required to get them started. This quantity of energy, referred to as the *activation energy*, is discussed in detail in Chapter 19. **15.101** 0.0384 **15.103** 5.0×10^1 atm
15.105 6.28×10^{-4} **15.107** 3.3×10^2 atm **15.109** (a) disproportionation redox reaction (b) 8.25×10^{15}; This method is feasible for removing SO_2. (c) less effective **15.111** (a) $\Delta G^{\circ} = -106.4$ kJ/mol; $K_P = 3.87 \times 10^{18}$
(b) $\Delta G^{\circ} = -53.2$ kJ/mol; $K_P = 2.17 \times 10^9$; The K_P in (a) is the square of the K_P in (b). Both ΔG° and K_P depend on the number of moles of reactants and products specified in the balanced equation.
15.113 $P_{N_2} = 0.860$ atm; $P_{H_2} = 0.366$ atm; $P_{NH_3} = 4.40 \times 10^{-3}$ atm
15.115 (a) 1.16 (b) 53.7% **15.117** 0.97 **15.119** (a) 0.49 atm (b) 0.23 (c) 0.037 (d) The smallest number of moles of CuO needed to establish equilibrium must be slightly greater than 0.037 mol.
15.121 $[H_2] = 0.070$ M; $[I_2] = 0.182$ M; $[HI] = 0.825$ M **15.123** gas (c); N_2O_4(colorless) $\longrightarrow$ $2NO_2$(brown) is consistent with the observations. The reaction is endothermic, so heating darkens the color. Above 150°C, the NO_2 breaks up into colorless NO and $2NO_2(g) \longrightarrow 2NO(g) + O_2(g)$. An increase in pressure shifts the equilibrium back to the left, restoring the color by producing NO_2.
15.125 (a) 4.2×10^{-4} (b) 0.83 (c) 1.1 (d) K_P in part (b): 2.3×10^3; K_P in part (c): 2.1×10^{-2} **15.127** (a) color deepens (b) increases (c) decreases (d) increases (e) unchanged **15.129** Potassium is more volatile than sodium. Therefore, its removal shifts the equilibrium from left to right. **15.131** 3.5×10^{-2} **15.133** (a) shifts to right (b) shifts to right (c) no change (d) no change (e) no change (f) shifts to left **15.135** $P_{N_2O_4} = 0.09$ atm; $P_{NO_2} = \approx 0.10$ atm **15.137** (a) 1.03 atm (b) 0.39 atm (c) 1.67 atm (d) 0.620 **15.139** (a) $K_c = 1.1 \times 10^{-7}$; $K_P = 2.6 \times 10^{-6}$ (b) 2.2 g; 22 mg/m³; yes **15.141** (a) 7.62×10^{14}
(b) 4.07×10^{-12}; The activity series is correct. The very large value of K for reaction (a) indicates that *products* are highly favored, whereas the very small value of K for reaction (b) indicates that *reactants* are highly favored. **15.143** There is a temporary dynamic equilibrium between the melting ice cubes and the freezing of water between the ice cubes. **15.145** $[NH_3] = 0.042$ M; $[N_2] = 0.086$ M; $[H_2] = 0.26$ M

15.147 (a) $K_P = \dfrac{\left(\dfrac{4x^2}{1+x}\right)P}{1-x} = \dfrac{4x^2}{1-x^2}P$ (b) K_P is a constant

(at constant temperature). If P increases, the fraction $\dfrac{4x^2}{1-x^2}$

(and therefore x) must decrease. Equilibrium shifts to the left to produce less NO_2 and more N_2O_4 as predicted. **15.149** $P_{SO_2Cl_2}$ = 3.58 atm; P_{SO_2} = P_{Cl_2} = 2.71 atm **15.151** 4.0 **15.153** −3 **15.155** reactions (d) and (g); reactions (c) and (f) **15.157** (a) Assume an endothermic reaction, $\Delta H° > 0$ and $T_2 > T_1$. Then, $\dfrac{\Delta H°}{R}\left(\dfrac{1}{T_2} - \dfrac{1}{T_1}\right) < 0$,

meaning that $\ln \dfrac{K_1}{K_2} < 0$ or $K_1 < K_2$. A larger K_2 indicates that there are more products at equilibrium as the temperature is raised. This agrees with Le Châtelier's principle that an increase in temperature favors the forward endothermic reaction. The opposite of the above discussion holds for an exothermic reaction. (b) 434 kJ/mol

Key Skills
15.1 (e) **15.2** (a) **15.3** (b) **15.4** (d)

Chapter 16

16.3 (a) both (b) base (c) acid (d) base (e) acid (f) base (g) base (h) base (i) acid (j) acid **16.5** (a) NO_2^- (b) HSO_4^- (c) HS^- (d) CN^- (e) $HCOO^-$ **16.7** (a) CH_2ClCOO^- (b) IO_4^- (c) $H_2PO_4^-$ (d) HPO_4^{2-} (e) PO_4^{3-} (f) HSO_4^- (g) SO_4^{2-} (h) IO_3^- (i) SO_3^{2-} (j) NH_3 (k) HS^- (l) S^{2-} (m) OCl^- **16.11** (a) $H_2SO_4 > H_2SeO_4$ (b) $H_3PO_4 > H_3AsO_4$ **16.13** The conjugate bases are $C_6H_5O^-$ from phenol and CH_3O^- from methanol. The $C_6H_5O^-$ is stabilized by resonance:

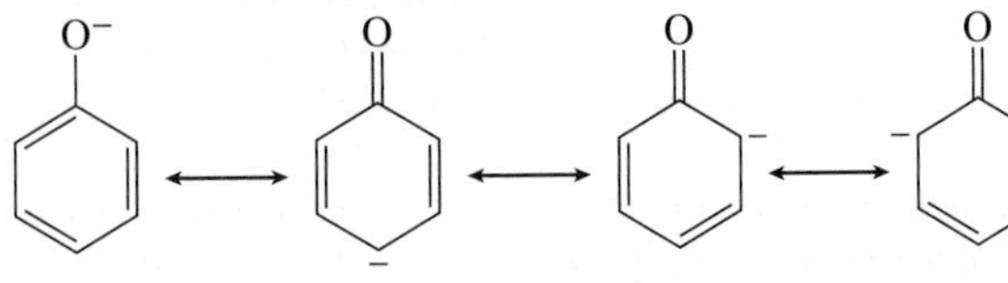

The CH_3O^- ion has no such resonance stabilization. A more stable conjugate base means an increase in the strength of the acid.
16.19 (a) 4.0×10^{-13} M (b) 6.0×10^{-10} M (c) 1.2×10^{-12} M (d) 5.7×10^{-3} M **16.21** (a) 2.05×10^{-11} M (b) 3.07×10^{-8} M (c) 5.95×10^{-11} M (d) 2.93×10^{-1} M **16.25** 7.1×10^{-12} M **16.27** (a) 3.00 (b) 13.89 **16.29** 2.5×10^{-5} M **16.31** 2.2×10^{-3} g **16.35** (a) 1.12 (b) 5.19 (c) 3.82 **16.37** (a) 1.2×10^{-3} M (b) 4.7×10^{-6} M (c) 1.5 M **16.39** (a) pOH = 1.146; pH = 12.854 (b) pOH = 1.356; pH = 12.644 (c) pOH = 0.77; pH = 13.23 **16.41** (a) 8.5×10^{-6} M (b) 4.3×10^{-6} M **16.43** 2.24×10^{-5} M

Visualizing Chemistry
VC 16.1 b **VC 16.2** a **VC 16.3** a **VC 16.4** b

16.51 2.59 **16.53** 5.17 **16.55** (a) 10% (b) 42% (c) 0.97% **16.57** 1.3×10^{-6} **16.59** 4.8×10^{-9} **16.61** 2.3×10^{-3} M **16.63** 6.80 **16.65** (a) strong base (b) weak base (c) weak base (d) weak base (e) strong base **16.69** 1.3×10^{-7} **16.71** 11.98 **16.73** 8.64 **16.77** $K_b(CN^-) = 2.0 \times 10^{-5}$; $K_b(F^-) = 1.4 \times 10^{-11}$; $K_b(CH_3COO^-)$ = 5.6×10^{-10}; $K_b(HCO_3^-) = 2.4 \times 10^{-8}$ **16.79** (a) A^- (b) B^- **16.83** pH (0.040 M HCl) = 1.40; pH (0.040 M H_2SO_4) = 1.31 **16.85** $[H^+] = [HCO_3^-] = 1.0 \times 10^{-4}$ M; $[CO_3^{2-}] = 4.8 \times 10^{-11}$ M **16.87** 1.00 **16.89** (1) diagram (c); (2) diagrams (b) and (d) **16.95** 4.82 **16.97** 5.39 **16.99** (a) neutral (b) basic (c) acidic (d) acidic **16.101** HZ < HY < HX **16.103** pH > 7 **16.107** (a) $Al_2O_3 < BaO < K_2O$ (b) $CrO_3 < Cr_2O_3 < CrO$ **16.109** $Al(OH)_3(s) + OH^-(aq) \longrightarrow Al(OH)_4^-(aq)$; This is a Lewis acid-base reaction. **16.111** (a) $Li_2O(s) + H_2O(l) \longrightarrow 2LiOH(aq)$ (b) $CaO(s) + H_2O(l) \longrightarrow Ca(OH)_2(aq)$ (c) $SO_3(g) + H_2O(l) \longrightarrow H_2SO_4(aq)$ **16.115** $AlCl_3$ is a Lewis acid with an incomplete octet of electrons and Cl^- is the Lewis base donating a pair of electrons. **16.117** CO_2, SO_2, and BCl_3

16.119 (a) $AlBr_3$ is the Lewis acid; Br^- is the Lewis base. (b) Cr is the Lewis acid; CO is the Lewis base. (c) Cu^{2+} is the Lewis acid; CN^- is the Lewis base. **16.121** strong acid: diagram (b); weak acid: diagram (c); very weak acid: diagram (d) **16.123** The products will be $CH_3COO^-(aq)$ and $HCl(aq)$, but this reaction will *not* occur to any measurable extent. **16.125** 1.70; 2.3% **16.127** choice (c) **16.129** (a) For the forward reaction NH_4^+ and NH_3 are the conjugate acid and base pair, respectively. For the reverse reaction NH_3 and NH_2^- are the conjugate acid and base pair, respectively. (b) H^+ corresponds to NH_4^+; OH^- corresponds to NH_2^-. For the neutral solution, $[NH_4^+] = [NH_2^-]$. **16.131** $K_a = \dfrac{[H^+][A^-]}{[HA]}$; $[HA] \approx 0.1$ M, $[A^-] \approx 0.1$ M; therefore, $K_a = [H^+] = \dfrac{K_w}{[OH^-]}$ and $[OH^-] = \dfrac{K_w}{K_a}$
16.133 1.7×10^{10} **16.135** (a) $H^-(base_1) + H_2O(acid_2) \longrightarrow OH^-(base_2) + H_2(acid_1)$ (b) H^- is the reducing agent, and H_2O is the oxidizing agent. **16.137** 2.8×10^{-2} **16.139** PH_3 is a weaker base than NH_3. **16.141** (a) HNO_2 (b) HF (c) BF_3 (d) NH_3 (e) H_2SO_3 (f) HCO_3^- and CO_3^{2-} **16.143** (a) trigonal pyramidal (b) H_4O^{2+} does *not* exist because the positively charged H_3O^+ has no affinity to accept the positive H^+ ion. If H_4O^{2+} existed, it would have a tetrahedral geometry. **16.145** $Cl_2(g) + H_2O(l) \rightleftharpoons HCl(aq) + HClO(aq)$; $HCl(aq) + AgNO_3(aq) \rightleftharpoons AgCl(s) + HNO_3(aq)$; In the presence of OH^- ions, the first equation is shifted to the right: H^+ (from HCl) $+ OH^- \longrightarrow H_2O$. Therefore, the concentration of HClO increases. (The "bleaching action" is due to ClO^- ions.) **16.147** 11.80 **16.149** (a) 18.1% (b) 4.63×10^{-3} **16.151** 4.26 **16.153** 7.2×10^{-3} g NaCN **16.155** 1.000 **16.157** (a) The pH of the solution of HA would be lower. (b) The electrical conductance of the HA solution would be greater. (c) The rate of hydrogen evolution from the HA solution would be greater. **16.159** 1.4×10^{-4} **16.161** 2.7×10^{-3} g **16.163** (a) NH_2^- (base) $+ H_2O$ (acid) $\longrightarrow NH_3 + OH^-$; N^{3-} (base) $+ 3H_2O$ (acid) $\longrightarrow NH_3 + 3OH^-$ (b) N^{3-} **16.165** When smelling salt is inhaled, some of the powder dissolves in the basic solution. The ammonium ions react with the base as follows: $NH_4^+(aq) + OH^-(aq) \longrightarrow NH_3(aq) + H_2O(l)$; It is the pungent odor of ammonia that prevents a person from fainting. **16.167** reaction (c) **16.169** 21 mL **16.171** magnesium **16.173** Both NaF and SnF_2 provide F^- ions in solution. The F^- ions replace OH^- ions during the remineralization process. $5Ca^{2+} + 3PO_4^{3-} + F^- \longrightarrow Ca_5(PO_4)_3F$ (fluorapatite). Because F^- is a weaker base than OH^-, fluorapatite is more resistant to attacks by acids compared to hydroxyapatite. **16.175** Li_2O, lithium oxide, basic; BeO, beryllium oxide, amphoteric; B_2O_3, diboron trioxide, acidic; CO_2, carbon dioxide, acidic; N_2O_5, dinitrogen pentoxide, acidic **16.177** $[SO_4^{2-}] = 4.0 \times 10^{-5}$ M; $[H^+] = 8.0 \times 10^{-5}$ M; $[HSO_4^-] = 0$ M; $[H_2SO_4] = 0$ M

Key Skills
16.1 (b) **16.2** (d) **16.3** (d) **16.4** (b)

Chapter 17

17.5 (a) 2.57 (b) 4.44

Visualizing Chemistry
VC 17.1 b **VC 17.2** c **VC 17.3** a **VC 17.4** a

17.9 8.89 **17.11** 0.024 **17.13** 0.58 **17.15** 9.25; 9.18 **17.17** solutions (c) and (d) **17.19** $NaHA/H_2A$ **17.21** (1) solutions (a), (b), and (c); (2) solution (a) **17.27** 177 g/mol **17.29** 0.550 M **17.31** (a) 1.10×10^2 g/mol (b) 1.6×10^{-6} **17.33** 5.82 **17.35** (a) 2.87 (b) 4.56 (c) 5.34 (d) 8.78 (e) 12.10 **17.37** (a) cresol red and phenolphthalein (b) most of the indicators in Table 17.3, except thymol blue and, to a lesser extent, bromophenol blue and methyl orange (c) bromophenol blue, methyl orange, methyl red, and chlorophenol blue **17.39** red **17.41** (1) diagram (c); (2) diagram (b); (3) diagram (d); (4) diagram (a); The pH at the equivalence point is below 7 (acidic). **17.49** (a) $[I^-]$ = 1.9×10^{-10} M (b) $[Al^{3+}]$ = 2.1×10^{-12} M **17.51** 1.8×10^{-11} **17.53** 2.8×10^{-92} **17.55** 12.94 **17.57** yes **17.59** (c) AlA_3

Visualizing Chemistry
VC 17.5 c **VC 17.6** a **VC 17.7** c **VC 17.8** a

17.65 (a) 0.013 M or 1.3×10^{-2} M (b) 2.2×10^{-4} M (c) 3.3×10^{-3} M **17.67** (a) 1.0×10^{-5} M (b) 1.1×10^{-10} M **17.69** (a) 0.016 M or 1.6×10^{-2} M (b) 1.6×10^{-6} M **17.71** A precipitate of $Fe(OH)_2$ will form. **17.73** $[Cd^{2+}]$ = 1.1×10^{-18} M; $[Cd(CN)_4^{2-}]$ = 4.2×10^{-3} M; $[CN^-]$ = 0.48 M **17.75** 3.5×10^{-5} M **17.77** (a) 3×10^{-60} (b) 1.2×10^{-8} M **17.79** (b), (c), (d), and (e) **17.83** The pH must be *greater than* 2.68 but *less than* 8.11. **17.85** 0.011 M **17.87** Chloride ion will precipitate Ag^+ but not Cu^{2+}. Hence, dissolve some solid in H_2O and add HCl. If a precipitate forms, the salt is $AgNO_3$. A flame test will also work. Cu^{2+} gives a green flame test. **17.89** 1.3 M **17.91** $[H^+]$ = 3.0×10^{-13} M; $[HCOOH]$ = 8.8×10^{-11} M; $[HCOO^-]$ = 0.0500 M; $[OH^-]$ = 0.0335 M; $[Na^+]$ = 0.0835 M **17.93** $Cd(OH)_2(s) + 2OH^-(aq) \rightleftharpoons$ $Cd(OH)_4^{2-}(aq)$; This is a Lewis acid-base reaction. **17.95** reaction (d) **17.97** $[Ag^+]$ = 2.0×10^{-9} M; $[Cl^-]$ = 0.080 M; $[Zn^{2+}]$ = 0.070 M; $[NO_3^-]$ = 0.060 M **17.99** 0.035 g/L **17.101** 2.4×10^{-13} **17.103** $[Ba^{2+}]$ = 1.0×10^{-5} M; $Ba(NO_3)_2$ is too soluble to be used for this purpose. **17.105** (a) AgBr (b) 1.8×10^{-7} M (c) 0.0018% or 1.8×10^{-3}% **17.107** 3.0×10^{-8} **17.109** (a) 1.0×10^{14} (b) 1.8×10^9 (c) 1.8×10^9 (d) 3.2×10^4 **17.111** (a) Mix 500 mL of 0.40 M CH_3COOH with 500 mL of 0.40 M CH_3COONa. (b) Mix 500 mL of 0.80 M CH_3COOH with 500 mL of 0.40 M NaOH. (c) Mix 500 mL of 0.80 M CH_3COONa with 500 mL of 0.40 M HCl. **17.113** (a) figure (b); (b) figure (a) **17.115** (a) Add Na_2CO_3. (b) Add HCl. (c) Add H_2S. **17.117** They are insoluble in water. **17.119** The ionized polyphenols have a dark color. In the presence of citric acid from lemon juice, the anions are converted to the lighter-colored acids. **17.121** yes **17.123** (c) **17.125** Decreasing the pH would increase the solubility of calcium oxalate and should help minimize the formation of calcium oxalate kidney stones. **17.127** At pH = 1.0: $^+NH_3-CH_2-COOH$; at pH = 7.0: $^+NH_3-CH_2-COO^-$; at pH = 12.0: $NH_2-CH_2-COO^-$ **17.129** (a) before dilution: pH = 4.74; after a 10-fold dilution: pH = 4.74 (b) before dilution: pH = 2.52; after dilution, pH = 3.02 **17.131** AgBr(s) will precipitate first. $Ag_2SO_4(s)$ will precipitate last. **17.133** compound (c) **17.135** 8.8×10^{-12} **17.137** 4.75 **17.139** (a) diagram (b); (b) diagram (e); (c) diagram (d); (d) diagram (a) **17.141** 2.51 to 4.41

Key Skills
17.1 (e) **17.2** (d) **17.3** (e) **17.4** (a)

Chapter 18

18.1 (a) $2H^+ + H_2O_2 + 2Fe^{2+} \longrightarrow 2Fe^{3+} + 2H_2O$ (b) $6H^+ + 2HNO_3 + 3Cu \longrightarrow 3Cu^{2+} + 2NO + 4H_2O$ (c) $3CN^- + 2MnO_4^- + H_2O \longrightarrow 3CNO^- + 2MnO_2 + 2OH^-$ (d) $6OH^- + 3Br_2 \longrightarrow BrO_3^- + 3H_2O + 5Br^-$ (e) $2S_2O_3^{2-} + I_2 \longrightarrow S_4O_6^{2-} + 2I^-$

Visualizing Chemistry
VC 18.1 a **VC 18.2** b **VC 18.3** b **VC 18.4** a

18.11 $Al(s) + 3Ag^+(1.0\,M) \longrightarrow Al^{3+}(1.0\,M) + 3Ag(s)$; E_{cell}° = 2.46 V **18.13** $Cl_2(g)$ and $MnO_4^-(aq)$ **18.15** (a) spontaneous (b) not spontaneous (c) not spontaneous (d) spontaneous **18.17** (a) Li (b) H_2 (c) Fe^{2+} (d) Br^- **18.21** 3×10^{54} **18.23** (a) 2×10^{18} (b) 3×10^8

(c) 3×10^{62} **18.25** $Ce^{4+}(aq) + Fe^{2+}(aq) \longrightarrow Ce^{3+}(aq) + Fe^{3+}(aq)$; ΔG° = −81 kJ/mol; $K = 2 \times 10^{14}$ **18.27** $2Sn + 4H^+ + O_2 \longrightarrow 2Sn^{2+} + 2H_2O$; 1.37 V **18.31** 1.09 V **18.33** E_{cell}° = 0.76 V; E = 0.78 V **18.35** 6×10^{-38} **18.41** 1.09 V **18.45** 12.2 g Mg **18.47** It is cheaper to prepare 1 ton of sodium by electrolysis. **18.49** 0.012 F **18.51** 5.33 g Cu; 13.4 g Br_2 **18.53** 7.70×10^3 C **18.55** 1.84 kg Cl_2/h **18.57** 63.3 g/mol **18.59** 27.0 g/mol **18.65** (a) half-reactions: $H_2(g) \longrightarrow 2H^+(aq) + 2e^-$, $Ni^{2+}(aq) + 2e^- \longrightarrow Ni(s)$; balanced equation: $H_2(g) + Ni^{2+}(aq) \longrightarrow 2H^+(aq) + Ni(s)$; The reaction will proceed to the left. (b) half-reactions: $5e^- + 8H^+(aq) + MnO_4^-(aq) \longrightarrow Mn^{2+}(aq) + 4H_2O(l)$, $2Cl^-(aq) \longrightarrow Cl_2(g) + 2e^-$; balanced equation: $16H^+(aq) + 2MnO_4^-(aq) + 10Cl^-(aq) \longrightarrow 2Mn^{2+}(aq) + 8H_2O(l) + 5Cl_2(g)$; The reaction will proceed to the right. (c) half-reactions: $Cr(s) \longrightarrow Cr^{3+}(aq) + 3e^-$, $Zn^{2+}(aq) + 2e^- \longrightarrow Zn(s)$; balanced equation: $2Cr(s) + 3Zn^{2+}(aq) \longrightarrow 2Cr^{3+}(aq) + 3Zn(s)$; The reaction will proceed to the left. **18.67** 0.00944 g **18.69** (a) $2MnO_4^- + 6H^+ + 5H_2O_2 \longrightarrow 2Mn^{2+} + 8H_2O + 5O_2$ (b) 0.06020 M **18.71** 0.232 mg Ca^{2+}/mL blood **18.73** 5×10^{-13} **18.75** (a) 3.14 V (b) 3.13 V **18.77** 0.035 V **18.79** Mercury(I) is Hg_2^{2+}. **18.81** 1.44 g Mg; $[Ag^+]$ = 7×10^{-55} M; $[Mg^{2+}]$ = 0.0500 M **18.83** (a) H_2; 0.206 L (b) 6.09×10^{23} e^-/mol e^- **18.85** (a) −1356.8 kJ/mol (b) 1.17 V **18.87** +3 **18.89** ΔG° = 6.8 kJ/mol; K = 0.064 **18.91** 1.4 A **18.93** +4 **18.95** 1.60×10^{-19} C/e^- **18.97** Cells of higher voltage require very reactive oxidizing and reducing agents, which are difficult to handle. (From Table 18.1 of the text, we see that 5.92 V is the theoretical limit of a cell made up of Li^+/Li and F_2/F^- electrodes under standard-state conditions.) Batteries made up of several cells in series are easier to use. **18.99** 2×10^{20} **18.101** (a) E_{red}° for X is negative. Thus E_{red}° for Y is positive. (b) 0.59 V **18.103** (a) Gold does not tarnish in air because the reduction potential for oxygen is not sufficiently positive to result in the oxidation of gold. (b) yes (c) $2Au + 3F_2 \longrightarrow 2AuF_3$ **18.105** $[Fe^{2+}]$ = 0.0920 M; $[Fe^{3+}]$ = 0.0680 M **18.107** $H_2O_2(aq) + 2H^+(aq) + 2e^- \longrightarrow 2H_2O(l)$, $E_{cathode}^\circ$ = 1.77 V; $H_2O_2(aq) \longrightarrow O_2(g) + 2H^+(aq) + 2e^-$, E_{anode}° = 0.68 V; $E_{cell}^\circ = E_{cathode}^\circ - E_{anode}^\circ$ = 1.09 V; Because E° is positive, the decomposition is spontaneous. **18.109** (a) unchanged (b) unchanged (c) squared (d) doubled (e) doubled **18.111** As $[H^+]$ increases, F_2 does become a stronger oxidizing agent. **18.113** 4.4×10^2 atm **18.115** (a) 1.65 V (b) 1.63 V (c) 4.86×10^3 kJ/kg Zn (d) 64 L air **18.117** −3.05 V **18.119** 1×10^{-14} **18.121** (a) 1 A · h = 1 A × 3600 s = 3600 C (b) 105 A · h; This ampere · hour cannot be fully realized because the concentration of H_2SO_4 keeps decreasing. (c) E_{cell}° = 2.01 V; ΔG° = −388 kJ/mol **18.123** $220 **18.125** −0.037 V **18.127** 2×10^{37} **18.129** 5.0 mol ATP/mol NO_2^- **18.131** A small non-zero emf will appear if the temperatures of the two half-cells are different.

Key Skills
18.1 (b) **18.2** (a) **18.3** (d) **18.4** (d)

Chapter 19

Visualizing Chemistry
VC 19.1 c **VC 19.2** b **VC 19.3** a **VC 19.4** c

19.9 (a) 0.066 M/s (b) 0.033 M/s

19.11 (a) rate = $-\dfrac{\Delta[H_2]}{\Delta t} = -\dfrac{\Delta[I_2]}{\Delta t} = \dfrac{1}{2}\dfrac{\Delta[HI]}{\Delta t}$

(b) rate = $-\dfrac{1}{5}\dfrac{\Delta[Br^-]}{\Delta t} = -\dfrac{\Delta[BrO_3^-]}{\Delta t} = -\dfrac{1}{6}\dfrac{\Delta[H^+]}{\Delta t} = \dfrac{1}{3}\dfrac{\Delta[Br_2]}{\Delta t}$

19.19 7.0×10^{-7} M/s **19.21** The reaction is first-order in A and first-order overall; k = 0.213 s^{-1} **19.23** (a) 2 (b) 0 (c) 2.5 (d) 3

19.25 first-order; $k = 1.19 \times 10^{-4} \text{ s}^{-1}$ **19.31** 76 min **19.33** (a) 0.034 M (b) 17 s; 23 s **19.35** (a) 4:3:6 (b) The relative rates would be unaffected; each absolute rate would decrease by 50%. (c) 1:1:1 **19.37** (a) first-order (b) second-order (c) zeroth-order **19.43** 1.3×10^2 kJ/mol; For maximum freshness, fish should be frozen immediately after capture and kept frozen until cooked. **19.45** $3.0 \times 10^3 \text{ s}^{-1}$ **19.47** 51.8 kJ/mol **19.49** 103 kJ/mol **19.51** in order of increasing time required for the effervescence to stop: e < d < b < a < c **19.61** (a) second-order (b) The first step is the slower (rate-determining) step. **19.63** Mechanism I can be discarded. Mechanisms II and III are possible. **19.71** At higher temperatures, the enzyme becomes denatured; that is, it loses its activity due to a change in the overall structure. **19.73** (a) termolecular (b) unimolecular (c) bimolecular **19.75** The temperature must be specified. **19.77** 0.035 s^{-1} **19.79** Most transition metals have several stable oxidation states. This allows the metal atoms to act as either a source or a receptor of electrons in a broad range of reactions. **19.81** Since the methanol contains no oxygen-18, the oxygen atom must come from the phosphate group and not the water. The mechanism must involve a bond-breaking process like:

$$CH_3-O \overset{O}{\underset{O-H}{\overset{\|}{-}P-O-H}}$$

19.83 temperature, energy of activation, concentration of reactants, and a catalyst **19.85** (a) rate = $k[CH_3COCH_3][H^+]$ (b) $3.8 \times 10^{-3}/Ms$ (c) $k = k_1 k_2/k_{-1}$ **19.87** (I) Fe^{3+} oxidizes I^-: $2Fe^{3+} + 2I^- \longrightarrow 2Fe^{2+} + I_2$; (II) Fe^{2+} reduces $S_2O_8^{2-}$: $2Fe^{2+} + S_2O_8^{2-} \longrightarrow 2Fe^{3+} + 2SO_4^{2-}$; The uncatalyzed reaction is slow because both I^- and $S_2O_8^{2-}$ are negatively charged, which makes their mutual approach unfavorable. **19.89** (a) (i) rate = $k[A]^0 = k$,

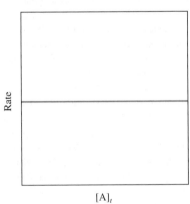

(ii) $[A] = -kt + [A]_0$,

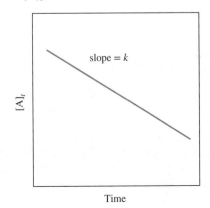

(b) $t_{1/2} = \dfrac{[A]_0}{2k}$ (c) $t = 2t_{1/2}$ **19.91** There are three gases present and we can measure only the total pressure of the gases. To measure the partial pressure of azomethane at a particular time, we must withdraw a sample of the mixture, analyze and determine the mole fractions. Then, $P_{\text{azomethane}} = P_T X_{\text{azomethane}}$

19.93

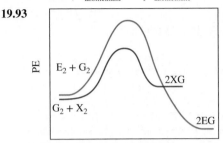

Reaction progress

19.95 (a) A catalyst works by changing the reaction mechanism, thus lowering the activation energy. (b) A catalyst changes the reaction mechanism. (c) A catalyst does not change the enthalpy of reaction. (d) A catalyst increases the forward rate of reaction. (e) A catalyst increases the reverse rate of reaction.

19.97 at very high $[H_2]$: $k_2[H_2] \gg 1$, rate = $\dfrac{k_1[NO]^2[H_2]}{k_2[H_2]} = \dfrac{k_1}{k_2}[NO]^2$;

at very low $[H_2]$: $k_2[H_2] \ll 1$, rate = $\dfrac{k_1[NO]^2[H_2]}{1} = k_1[NO]^2[H_2]$;

The result from Problem 19.80 agrees with the rate law determined for low $[H_2]$. **19.99** rate = $k[N_2O_5]$; $k = 1.0 \times 10^{-5} \text{ s}^{-1}$ **19.101** The red bromine vapor absorbs photons of blue light and dissociates to form bromine atoms: $Br_2 \longrightarrow 2Br\cdot$. The bromine atoms collide with methane molecules and abstract hydrogen atoms: $Br\cdot + CH_4 \longrightarrow HBr + \cdot CH_3$. The methyl radical then reacts with Br_2, giving the observed product and regenerating a bromine atom to start the process over again: $\cdot CH_3 + Br_2 \longrightarrow CH_3Br + Br\cdot$, $Br\cdot + CH_4 \longrightarrow HBr + \cdot CH_3$ and so on. **19.103** Lowering the body temperature would slow down all chemical reactions, which would be especially important for those that might damage the brain. **19.105** (a) rate = $k[X][Y]^2$ (b) $0.019 \text{ } M^{-2}\text{s}^{-1}$ **19.107** second-order; $k = 2.4 \times 10^7 \text{ } M^{-1}\text{s}^{-1}$ **19.109** During the first five minutes or so the engine is relatively cold, so the exhaust gases will not fully react with the components of the catalytic converter. Remember, for almost all reactions, the rate of reaction increases with temperature. **19.111** (a) The activation energy must be high. (b) The molecules must have the appropriate orientation, and the energy that the colliding particles possess must be greater than the activation energy. **19.113** 5.7×10^5 yr **19.115** (a) catalyst: Mn^{2+}; intermediates: Mn^{3+}, Mn^{4+}; First step is rate-determining. (b) Without the catalyst, the reaction would be a termolecular one involving three cations! (one Tl^+ and two Ce^{4+}). The reaction would be slow. (c) The catalysis is homogeneous because the product has the same phase (aqueous) as the reactants. **19.117** 0.45 atm **19.119** (a) $\dfrac{\Delta[B]}{\Delta t} = k_1[A] - k_2[B]$

(b) $[B] = \dfrac{k_1}{k_2}[A]$ **19.121** (a) $k = 0.0247 \text{ yr}^{-1}$ (b) 9.8×10^{-4} (c) 187 yr

19.123 (a) three (b) two (c) the third step (d) exothermic **19.125** 1.8×10^3 K **19.127** (a) 2.5×10^{-5} M/s (b) 2.5×10^{-5} M/s (c) 8.3×10^{-6} M

19.129 $n = 0$, $t_{1/2} = C\dfrac{1}{[A]_0^{-1}} = C[A]_0$; $n = 1$, $t_{1/2} = C\dfrac{1}{[A]_0^0} = C$;

$n = 2$, $t_{1/2} = C\dfrac{1}{[A]_0}$

19.131 (a) 1.13×10^{-3} M/min (b) 6.83×10^{-4} M/min; 8.80×10^{-3} M
19.133 second-order; $k = 0.42/M$ min **19.135** 60% increase; The result shows the profound effect of an exponential dependence.
19.137 form (b)

Key Skills
19.1 (d) **19.2** (e) **19.3** (b) **19.4** (a)

Chapter 20

20.5 (a) $^{23}_{11}$Na (b) $^1_1$p or $^1_1$H (c) $^1_0$n (d) $^{56}_{26}$Fe (e) $^0_{-1}\beta$ **20.13** -4.85×10^{-12} kg/mol H_2 **20.15** (a) Li-7: 6.290×10^{-12} J; 8.986×10^{-13} J/nucleon
(b) Cl-35: 4.779×10^{-11} J; 1.365×10^{-12} J/nucleon **20.17** 47.9541 amu

20.21 (a) $^{232}_{90}$Th $\xrightarrow{\alpha}$ $^{228}_{88}$Ra $\xrightarrow{\beta}$ $^{228}_{89}$Ac $\xrightarrow{\beta}$ $^{228}_{90}$Th

(b) $^{235}_{92}$U $\xrightarrow{\alpha}$ $^{231}_{90}$Th $\xrightarrow{\beta}$ $^{231}_{91}$Pa $\xrightarrow{\alpha}$ $^{227}_{89}$Ac

(c) $^{237}_{93}$Np $\xrightarrow{\alpha}$ $^{233}_{91}$Pa $\xrightarrow{\beta}$ $^{233}_{92}$U $\xrightarrow{\alpha}$ $^{229}_{90}$Th

20.23 4.88×10^{19} atoms **20.25** 3.09×10^3 yr **20.27** no A remains; 0.25 mole of B; no C is left; 0.75 mole of D **20.29** 5.5 dpm
20.31 3.7×10^{-20} yr^{-1} **20.35** (a) ^{14}N$(\alpha,p)^{17}$O (b) ^{9}Be$(\alpha,n)^{12}$C
(c) ^{238}U$(d,2n)^{238}$Np **20.37** (a) ^{40}Ca$(d,p)^{41}$Ca (b) ^{32}S$(n,p)^{32}$P
(c) ^{239}Pu$(\alpha,n)^{242}$Cm **20.39** $^{198}_{80}$Hg $+ ^1_0$n $\longrightarrow$ $^{199}_{80}$Hg $\longrightarrow$ $^{198}_{79}$Au $+ ^1_1$p

Visualizing Chemistry
VC 20.1 c **VC 20.2** b **VC 20.3** a **VC 20.4** a

20.51 257 days **20.53** The fact that the radioisotope appears only in the I_2 shows that the IO_3^- is formed only from the IO_4^-. **20.55** Add iron-59 to the person's diet, and allow a few days for the iron-59 isotope to be incorporated into the person's body. Isolate red blood cells from a blood sample and monitor radioactivity from the hemoglobin molecules present in the red blood cells. **20.61** 65.3 yr
20.63 (a) $^1_1$H $\longrightarrow$ $^3_2$He $+ ^0_{-1}\beta$ (b) 70.5 dpm **20.65** (a) $^{93}_{36}$Kr (b) $^1_0$n
(c) $^{146}_{57}$La (d) $^1_0$n **20.67** (a) $^1_1$H $\longrightarrow$ $^3_2$He $+ ^0_{-1}\beta$ (b) $^{242}_{94}$Pu $\longrightarrow$ $^4_2\alpha +$
$^{238}_{92}$U (c) $^{131}_{53}$I $\longrightarrow$ $^{131}_{54}$Xe $+ ^0_{-1}\beta$ (d) $^{251}_{98}$Cf $\longrightarrow$ $^{247}_{96}$Cm $+ ^4_2\alpha$
20.69 Because both Ca and Sr belong to Group 2A, radioactive strontium that has been ingested into the human body becomes concentrated in bones (replacing Ca) and can damage blood cell production; 2.12×10^3 mCi **20.71** Normally the human body concentrates iodine in the thyroid gland. The purpose of the large doses of KI is to displace radioactive iodine from the thyroid and allow its excretion from the body. **20.73** (a) $^{209}_{83}$Bi $+ ^4_2\alpha \longrightarrow$
$^{211}_{85}$At $+ 2^1_0$n (b) $^{209}_{83}$Bi$(\alpha, 2n)^{211}_{85}$At **20.75** 2.77×10^3 yr **20.77** 0.070%
20.79 (a) 5.59×10^{-15} J; 2.84×10^{-13} J (b) 0.0244 mol (c) 4.26×10^6 kJ **20.81** 6.1×10^{23} atoms/mol **20.83** (a) 1.83×10^{-12} J (b) The α particle will move away faster because it is smaller. **20.85** U-238, $t_{1/2} = 4.5 \times 10^9$ yr and Th-232, $t_{1/2} = 1.4 \times 10^{10}$ yr; They are still present because of their long half-lives. **20.87** 8.3×10^{-4} nm
20.89 Only ^{3}H has a suitable half-life. The other half-lives are either too long or too short to determine the time span of 6 years accurately.
20.91 A small-scale chain reaction (fission of ^{235}U) took place. Copper played the crucial role of reflecting neutrons from the splitting uranium-235 atoms back into the uranium sphere to trigger the chain reaction. Note that a sphere has the most appropriate geometry for such a chain reaction. In fact, during the implosion process prior to an atomic explosion, fragments of uranium-235 are pressed roughly into a sphere for the chain reaction to occur (see Section 20.5 of the text). **20.93** 2.1×10^2 g/mol **20.95** (a) $r = r_0A^{1/3}$, where r_0 is a proportionality constant (b) 1.7×10^{-42} m^3
20.97 diagram (d) **20.99** 3.4 mL **20.101** (a) X $= ^{191}_{77}$Ir;
$^{191}_{77}$Ir $+ ^1_0$n $\longrightarrow$ $^{192}_{77}$Ir $+ \gamma$ (b) 1.6927×10^{-10} J;
1.3541×10^{-12} J/nucleon

Chapter 21

21.5 3.3×10^{-4}; 330 ppm. **21.7** In the stratosphere, the air temperature rises with altitude. This warming effect is the result of exothermic reactions triggered by UV radiation from the sun.
21.11 260 nm. **21.21** 4.0×10^{37} molecules; 3.2×10^{12} kg O_3.
21.23 $CCl_4 + HF \longrightarrow HCl + CFCl_3$ (Freon-11); $CFCl_3 + HF$
$\longrightarrow HCl + CF_2Cl_2$ (Freon-12). **21.25** $E = 479$ kJ/mol. Solar radiation preferentially breaks the C—Cl bond. There is not enough energy to break the C—F bond.

21.27 :Cl̈—Ö—N—Ö̈ (structure) :Ö: ; :Cl̈—Ö̈·. **21.39** 2.6×10^4 tons.

21.47 Primary pollutants, such as automobile exhaust consisting mainly of NO, CO, and various unburned hydrocarbons, set in motion a series of photochemical reactions that produce secondary pollutants. It is the secondary pollutants, chiefly NO_2 and O_3, that are responsible for the buildup of smog. **21.49** While carbon monoxide is a primary pollutant, it is the secondary pollutants, such as ozone, that are responsible for the buildup of smog. **21.51** Most automobiles now are equipped with catalytic converters designed to oxidize CO and unburned hydrocarbons to CO_2 and H_2O and to reduce NO and NO_2 to N_2 and O_2. More efficient automobile engines and better public transportation systems would help to decrease air pollution in urban areas. A recent technological innovation to combat photochemical smog is to coat automobile radiators and air conditioner compressors with a platinum catalyst. So equipped, a running car can purify the air that flows under the hood by converting ozone and carbon monoxide to oxygen and carbon dioxide. **21.53** 4.1×10^{-7} atm; 1×10^{16} molecules/L. **21.59** 378 g.
21.61 O_3: greenhouse gas, toxic to humans, attacks rubber; SO_2: toxic to humans, forms acid rain; NO_2: forms acid rain, destroys ozone; CO: toxic to humans; PAN: a powerful lachrymator, causes breathing difficulties; Rn: causes lung cancer. **21.63** (a) Its small concentration is the result of the high reactivity of the OH radical. (b) OH has an unpaired electron; free radicals are always good oxidizing agents. (c) OH $+ NO_2 \longrightarrow HNO_3$. (d) OH $+ SO_2 \longrightarrow$ HSO_3; $HSO_3 + O_2 + H_2O \longrightarrow H_2SO_4 + HO_2$. **21.65** Most water molecules contain oxygen-16, but a small percentage of water molecules contain oxygen-18. The ratio of the two isotopes in the ocean is essentially constant, but the ratio in the water vapor evaporated from the oceans is temperature-dependent, with the vapor becoming slightly enriched with oxygen-18 as temperature increases. The water locked up in ice cores provides a historical record of this oxygen-18 enrichment, and thus ice cores contain information about past global temperatures. **21.67** 5.1×10^{20} photons. **21.69** 394 nm. **21.71** The lone pair on the S in SO_2 functions as the Lewis base and the Ca in CaO functions as a Lewis acid. **21.73** (a) 6.2×10^8. (b) The CO_2 liberated from limestone contributes to global warming. **21.75** The use of the aerosol liberates CFC's that destroy the ozone layer. **21.77** The size of tree rings can be related to CO_2 content, where the number of rings indicates the age of the tree. The amount of CO_2 in ice can be directly measured from portions of polar ice in different layers obtained by drilling. The "age" of CO_2 can be determined by radiocarbon dating and other methods. **21.79** (a) $N_2O + O \longrightarrow$ $2NO$; $2NO + 2O_3 \longrightarrow 2NO_2 + 2O_2$; overall: $N_2O + O + 2O_3$ $\longrightarrow 2NO_2 + 2O_2$. (b) N_2O is a more effective greenhouse gas than CO_2 because it has a permanent dipole. (c) 3.0×10^{10} mol.
21.81 Yes. Light of wavelength 409 nm (visible) or shorter will break the C—Br bond. **21.83** 1.6×10^{19} kJ; 4.8×10^{16} kg.
21.85 5.2×10^8 L.

Chapter 22

22.11 (a) +3 (b) 6 (c) oxalate ion ($C_2O_4^{2-}$) **22.13** (a) Na: +1; Mo: +6 (b) Mg: +2; W: +6 (c) Fe: 0 **22.15** (a) *cis*-dichlorobis(ethylenediamine) cobalt(III) (b) pentaamminechloroplatinum(IV) chloride (c) pentaamminechlorocobalt(III) chloride **22.17** six
22.23 (a) two (b) two

22.25 (a)

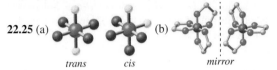

trans   *cis*   *mirror*

(b)

22.31 Cyanide ion is a very strong field ligand. It causes a larger crystal field splitting than water, resulting in the absorption of higher energy (shorter wavelength) radiation when a *d* electron is excited to a higher energy *d* orbital. **22.33** (a) orange (b) 255 kJ/mol **22.35** two moles; [Co(NH₃)₄Cl₂]Cl; Refer to Problem 22.25(a) for a diagram of the structure. **22.37** Δ would be greater for the higher oxidation state. **22.41** Use a radioactive label such as $^{14}CN^-$ (in NaCN). Add NaCN to a solution of $K_3Fe(CN)_6$. Isolate some of the $K_3Fe(CN)_6$ and check its radioactivity. If the complex shows radioactivity, then it must mean that the CN^- ion has participated in the exchange reaction. **22.43** $Cu(CN)_2$ is the white precipitate. It is soluble in KCN(*aq*), due to formation of $[Cu(CN)_4]^{2-}$, so the concentration of Cu^{2+} is too small for Cu^{2+} ions to precipitate with sulfide. **22.45** 1.4×10^2 **22.47** The purple color is caused by the buildup of deoxyhemoglobin. When either oxyhemoglobin or deoxyhemoglobin takes up CO, the carbonylhemoglobin takes on a red color, the same as oxyhemoglobin. **22.49** Mn^{3+} is $3d^4$ and Cr^{3+} is $3d^5$. Therefore, Mn^{3+} has a greater tendency to accept an electron and is a stronger oxidizing agent. The $3d^5$ electron configuration of Cr^{3+} is a stable configuration. **22.51** Ti^{3+}; Fe^{3+} **22.53** 1.6×10^4 g hemoglobin/mol Fe. The discrepancy between our minimum value and the actual value can be explained by realizing that there are four iron atoms per mole of hemoglobin. **22.55** (a) $[Cr(H_2O)_6]Cl_3$, number of ions: 4 (b) $[Cr(H_2O)_5Cl]Cl_2 \cdot H_2O$, number of ions: 3 (c) $[Cr(H_2O)_4Cl_2]Cl \cdot 2H_2O$, number of ions: 2; Compare the compounds with equal molar amounts of NaCl, MgCl₂, and FeCl₃ in an electrical conductance experiment. The solution that has similar conductance to the NaCl solution contains(c); the solution with the conductance similar to MgCl₂ contains (b); and the solution with conductance similar to FeCl₃ contains (a). **22.57** $\Delta G° = -1.8 \times 10^2$ kJ/mol; $K = 6 \times 10^{30}$ **22.59** Iron is much more abundant than cobalt. **22.61** Oxyhemoglobin absorbs higher energy light than deoxyhemoglobin. Oxyhemoglobin is diamagnetic (low spin), while deoxyhemoglobin is paramagnetic (high spin). These differences occur because oxygen (O_2) is a strong-field ligand. **22.63** Complexes are expected to be colored when the highest occupied orbitals have between one and nine *d* electrons. Such complexes can therefore have $d \longrightarrow d$ transitions (that are usually in the visible part of the electromagnetic radiation spectrum). Zn^{2+}, Cu^+, and Pb^{2+} are d^{10} ions. V^{5+}, Ca^{2+}, and Sc^{3+} are d^0 ions. **22.65** Dipole moment measurement. Only the *cis* isomer has a dipole moment. **22.67** EDTA sequesters metal ions (like Ca^{2+} and Mg^{2+}) which are essential for the growth and function of bacteria. **22.69** three **22.71** 2.2×10^{-20} *M* **22.73** (a) 2.7×10^6 (b) Free Cu^+ ions are unstable in solution [as shown in part (a)]. Therefore, the only stable compounds containing Cu^+ ions are insoluble. **22.75** Y **22.77** 0.0 *M* **22.79** $[Mn(CN)_6]^{5-}$: Mn is +1, one unpaired *d* electron; $[Mn(CN)_6]^{4-}$: Mn is +2, one unpaired *d* electron; $[Mn(CN)_6]^{3-}$: Mn is +3, two unpaired *d* electrons.

Chapter 23

23.7 (a) amine (b) aldehyde (c) ketone (d) carboxylic acid (e) alcohol **23.9** (a) 3-ethyl-2,4,4-trimethylhexane (b) 6,6-dimethyl-2-heptanol (c) 4-chlorohexanal **23.11** 3,5-dimethyloctane
23.13 (a) $(CH_3)_3CCH_2CH(CH_3)_2$ (b) $HO(CH_2)_2CH(CH_3)_2$ (c) $CH_3(CH_2)_4C(O)NH_2$ (d) Cl_3CCHO

23.15

A = Carbonyl (Ketone)
B = Carboxy (Carboxylic acid)
C = Hydroxy (Alcohol)

23.17 Primary amino / Carboxy

23.19 (a) $CH_3CH_2CHCH_2CH_2CH_3$ with CH_3 branch (b) $CH_3CHCHCH_2CH_3$ with CH_3 branch
(c) $CH_3CHCH_2CHCH_3$ with Br and C_6H_5 branches (d) $CH_3CH_2CHCHCH_2CH_2CH_3$ with CH_3 and CH_3 branches

23.21 (a) Kekulé:

bond-line:

(b) condensed: $(C_2H_5)_2CHCH_2CO_2C(CH_3)_3$;

bond-line:

(c) condensed: $(CH_3)_2CHCH_2NHCH(CH_3)_2$;

Kekulé:

23.23 (a) condensed structural: $(CH_3)_2NCHO$

Kekulé: ; bond-line:

(b) condensed structural: $(CH_2COOH)_2C(OH)COOH$;

Kekulé: ; bond-line:

(c) condensed structural: $(CH_3)_2CH(CH_2)_2OC(O)CH_3$

Kekulé:

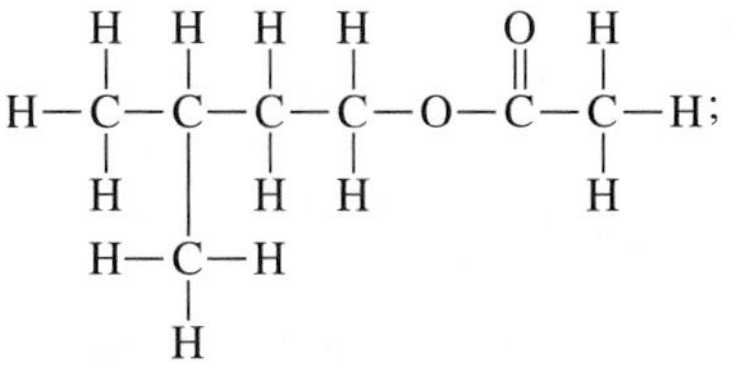

$$H-\overset{\overset{H}{|}}{\underset{\underset{H}{|}}{C}}-\overset{\overset{H}{|}}{\underset{\underset{H-\overset{\overset{H}{|}}{\underset{\underset{H}{|}}{C}}-H}{|}}{C}}-\overset{\overset{H}{|}}{\underset{\underset{H}{|}}{C}}-\overset{\overset{H}{|}}{\underset{\underset{H}{|}}{C}}-O-\overset{\overset{O}{||}}{C}-\overset{\overset{H}{|}}{\underset{\underset{H}{|}}{C}}-H;$$

bond-line:

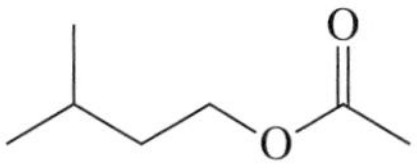

23.25 (a)

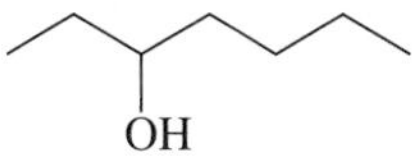

(b) 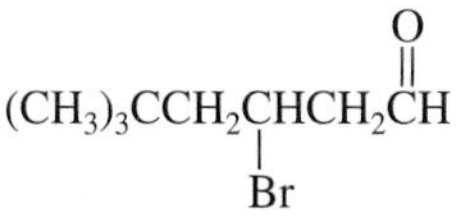 $(CH_3)_3CCH_2CHCH_2CH$
        $\underset{Br}{|}$ $\overset{O}{||}$

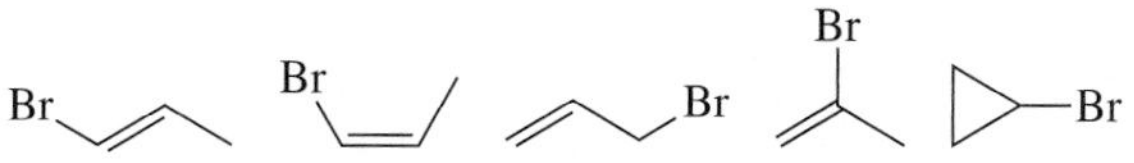

(c)

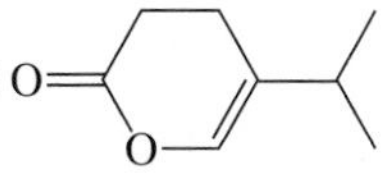

23.27 (a)  $CH_3-C\equiv N: \longleftrightarrow CH_3-C\overset{+}{\equiv}\overset{..}{N}:^-$

(b)

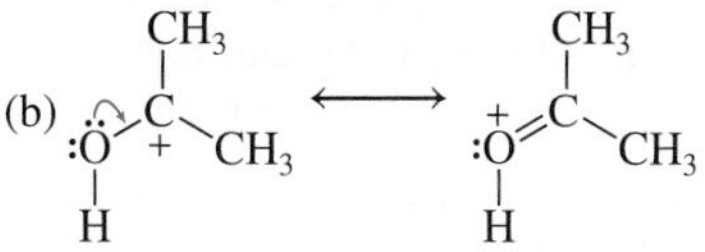

(c)

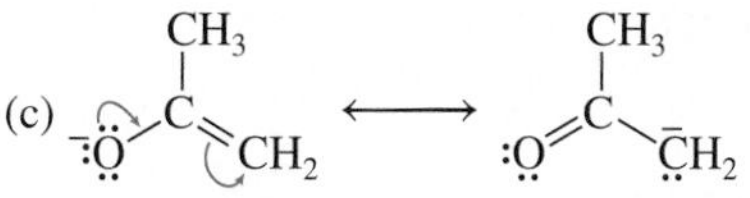

23.29 (a)

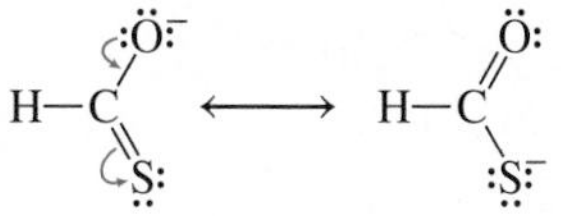

(b) 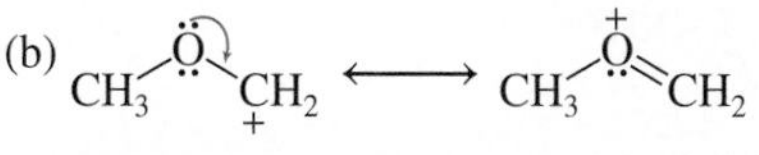 $CH_3\overset{\overset{..}{O}}{\underset{+}{C}}CH_2 \longleftrightarrow CH_3\overset{\overset{+}{O}}{C}CH_2$

(c)

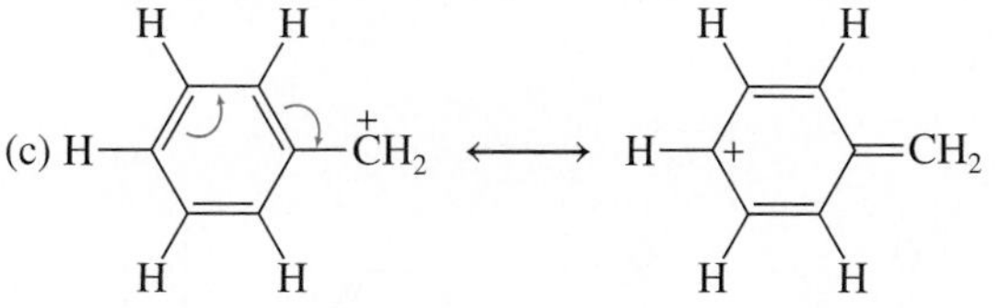

23.37

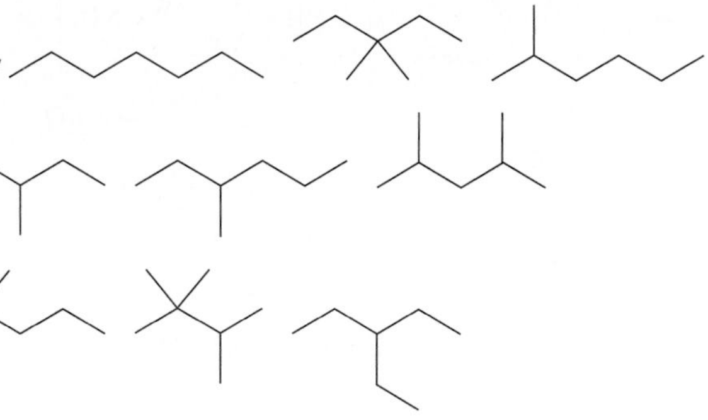

23.39 Br

23.41 Cl

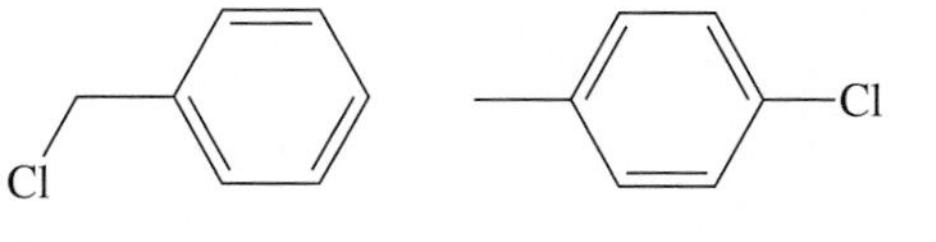

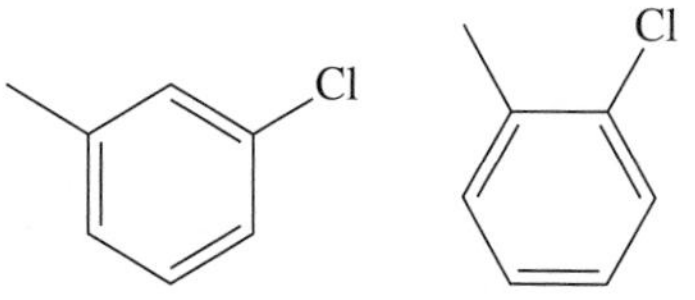

23.43 (a) 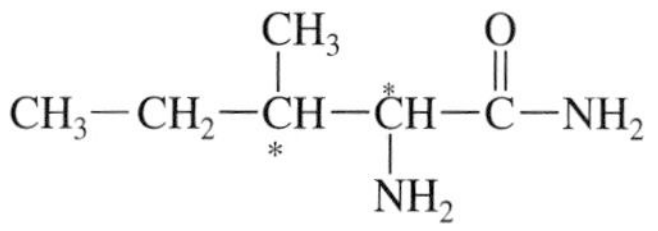 $CH_3-CH_2-\overset{*}{\underset{\underset{NH_2}{|}}{CH}}-\overset{\overset{CH_3}{|}}{\overset{*}{CH}}-\overset{\overset{O}{||}}{C}-NH_2$ (b)

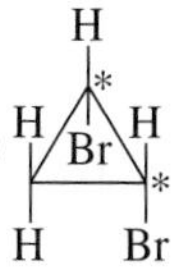

23.45 $CH_3CH_2\overset{\overset{O}{||}}{C}-H$

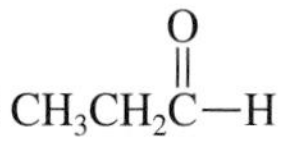

23.47 (a)

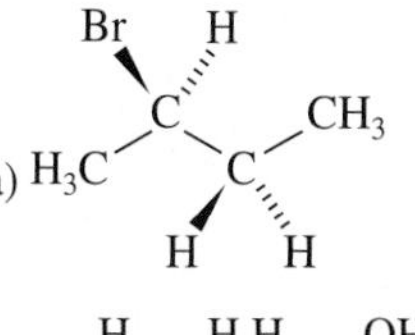

(b)

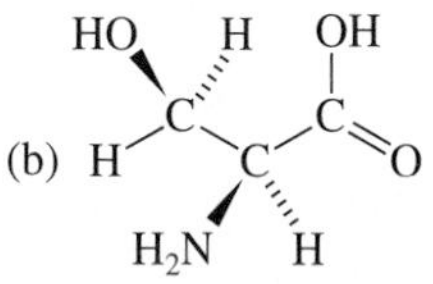

(c)

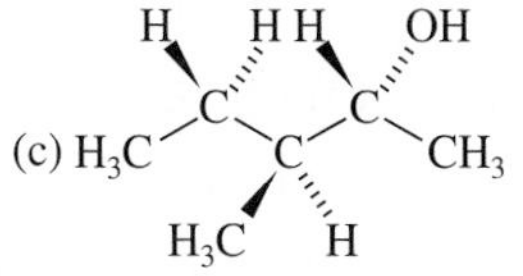

(d)

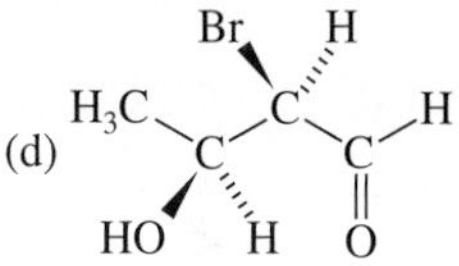

23.55

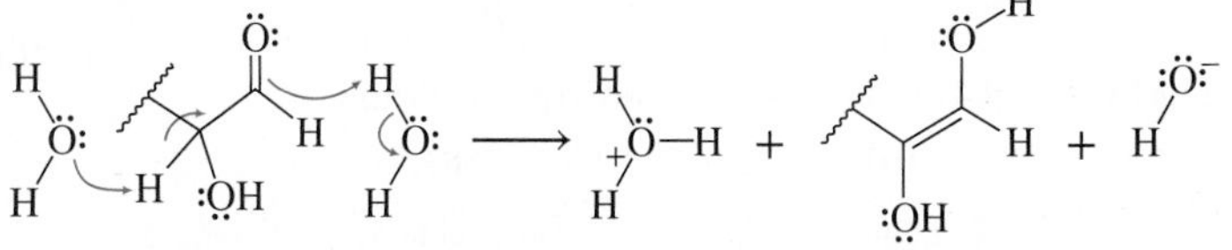

23.57 (a)

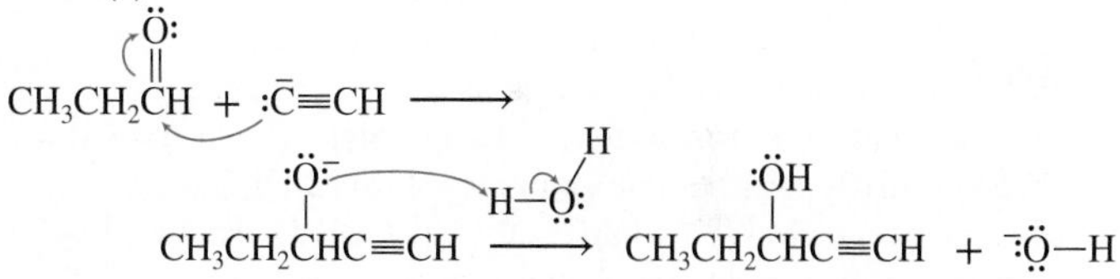

(b)

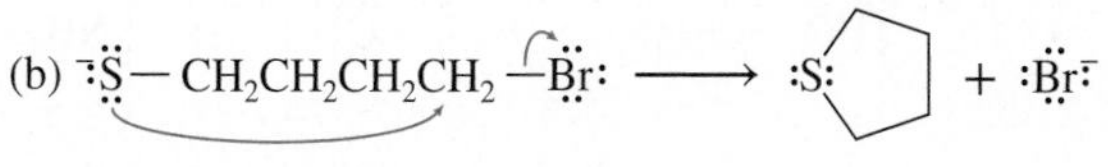

23.59

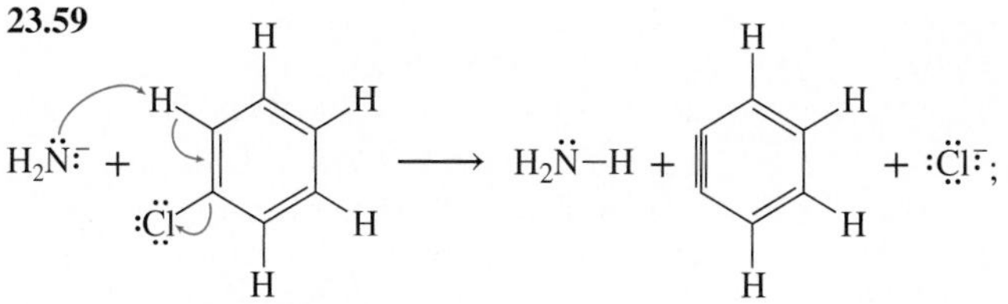

elimination reaction

23.61 No. This is not an oxidation-reduction reaction. There are no changes in oxidation states for any of the atoms. **23.63** (a) The sulfuric acid releases protons, which then protonate the terminal CH_2 group. This creates a cation; nucleophilic attack by water produces the alcohol with release of another proton. Sulfuric acid is thus a catalyst;

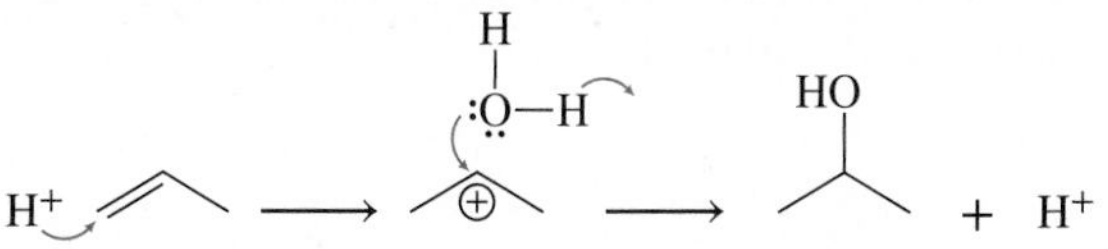

(b) *n*-propanol: OH

(c) no **23.65** thalidomide:
$C_{13}H_{10}N_2O_4$; lenalidomide: $C_{13}H_{13}N_3O_3$; CC-4047: $C_{13}H_{11}N_3O_4$

23.73

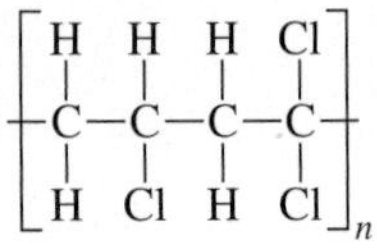

$$\left[\begin{array}{cccc} \overset{H}{\underset{H}{C}} - \overset{H}{\underset{Cl}{C}} - \overset{H}{\underset{H}{C}} - \overset{Cl}{\underset{Cl}{C}} \end{array} \right]_n$$

23.75 (a) $CH_2=CHCH=CH_2$

(b)

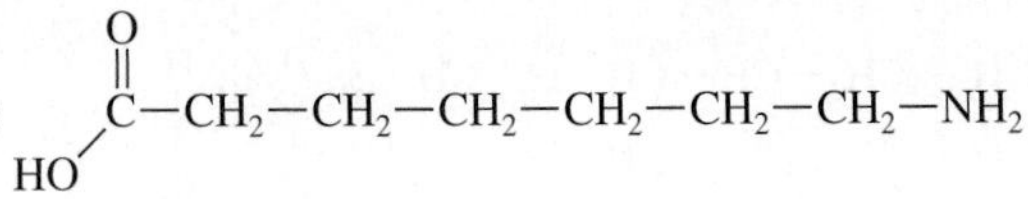

23.77

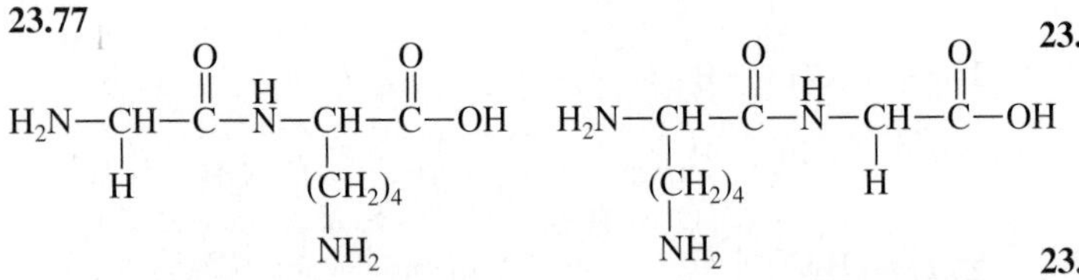

GlyLys   LysGly

23.79

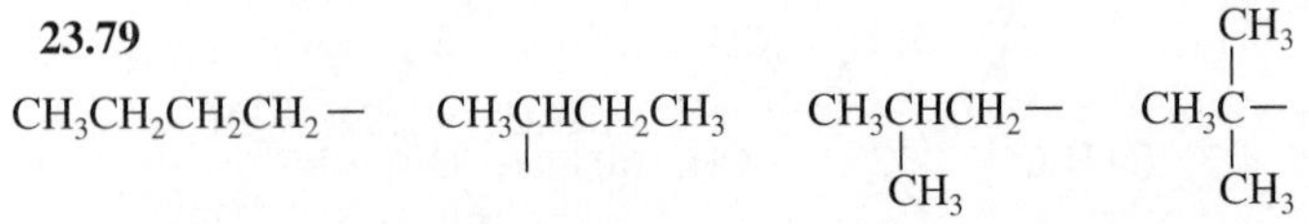

$CH_3CH_2CH_2CH_2-$ $CH_3CHCH_2CH_3$ CH_3CHCH_2- CH_3C-

Butyl   *sec*-Butyl   Isobutyl   *tert*-Butyl

23.81 $CH_3CHCH_2CHCO_2H$ $CH_3CH_2CHCHCO_2H$

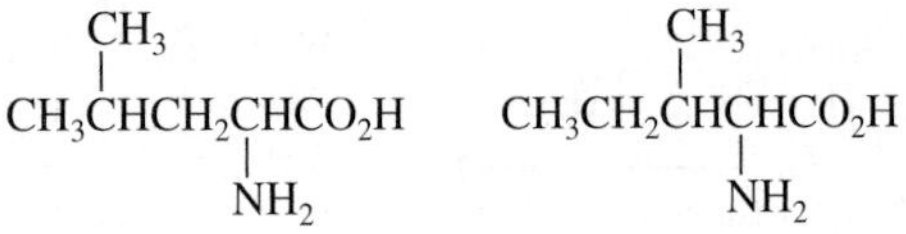

Leucine   Isoleucine

23.83 (a) 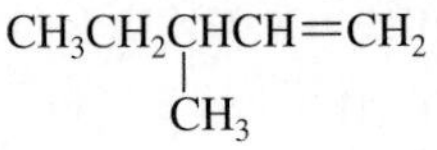 $CH_3CH_2CHCH=CH_2$

(b)

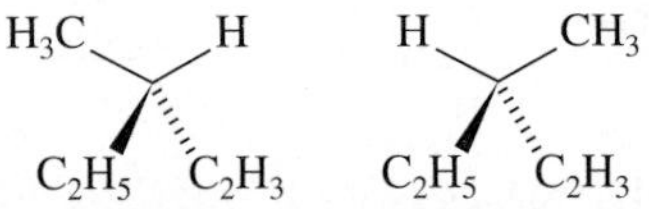

23.85 (a) *cis/trans* stereoisomers (b) constitutional isomers
(c) resonance structures (d) different representations of the same
structure

23.87

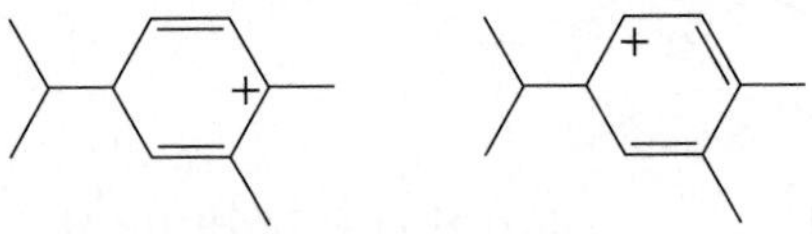

23.89 (b)

23.91 $CH_3CH_2C{\equiv}CH$ $CH_3C{\equiv}CCH_3$ $H_2C=CHCH=CH_2$
all sp^2

23.93 Since N is less electronegative than O, electron donation in the
amide would be more pronounced. **23.95** (a) The more negative $\Delta H°$
implies stronger alkane bonds; branching decreases the total bond
enthalpy (and overall stability) of the alkane. (b) the least highly
branched isomer (*n*-octane) **23.97** (a) 15.81 mg C; 1.32 mg H;
3.49 mg O (b) C_6H_6O

(c) 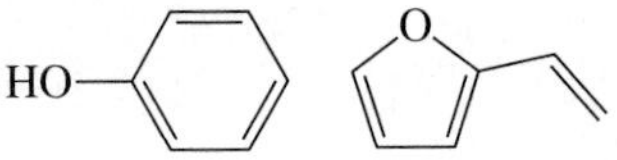 HO

23.99

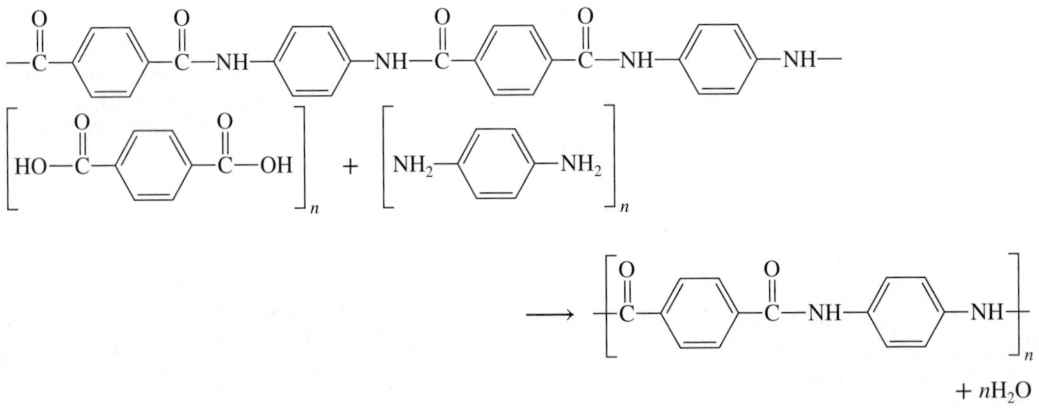

23.101

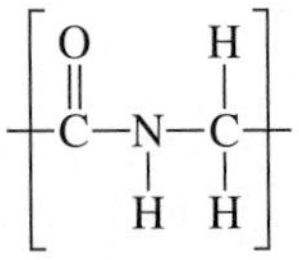

23.103 eight different tripeptides: Lys−Lys−Lys, Lys−Lys−Ala,
Lys−Ala−Lys, Ala−Lys−Lys, Lys−Ala−Ala, Ala−Lys−Ala,
Ala−Ala−Lys, Ala−Ala−Ala

23.105

so it should form a liquid crystal. **24.47** $-CH=C(CH_2CH_3)-CH=$ $C(CH_2CH_3)-CH=C(CH_2CH_3)-$; $-CH_2-CH(CH_2CH_3)-CH_2-$ $CH(CH_2CH_3)-CH_2-CH(CH_2CH_3)-$; The two structures are similar in that they both have ethyl groups on alternating carbons. They are different in that poly(1-butyne) is a conjugated system and poly(1-butene) is not. The first structure is more likely to be an electrically conducting polymer.

Chapter 25

25.11 $CaH_2(s) + 2H_2O(l) \longrightarrow Ca(OH)_2(aq) + 2H_2(g)$; 22.7 g CaH_2. **25.13** HCl; CaH_2; A water solution of HCl is called hydrochloric acid. Calcium hydride will react according to the equation $CaH_2(s) + 2H_2O(l) \longrightarrow Ca(OH)_2(aq) + 2H_2(g)$. **25.15** NaH: ionic compound, reacts with water as follows: $NaH(s) + H_2O(l) \longrightarrow NaOH(aq) + H_2(g)$; CaH_2: ionic compound, reacts with water as follows: $CaH_2(s) + 2H_2O(l) \longrightarrow Ca(OH)_2(aq) + 2H_2(g)$; CH_4: covalent compound, unreactive, burns in air or oxygen: $CH_4(g) + 2O_2(g) \longrightarrow CO_2(g) + 2H_2O(l)$; NH_3: covalent compound, weak base in water: $NH_3(aq) + H_2O(l) \rightleftharpoons NH_4^+(aq) + OH^-(aq)$; H_2O: covalent compound, forms strong intermolecular hydrogen bonds, good solvent for both ionic compounds and substances capable of forming hydrogen bonds; HCl: covalent compound (polar), acts as a strong acid in water: $HCl(g) + H_2O(l) \longrightarrow H_3O^+(aq) + Cl^-(aq)$. **25.17** (a) H_2 can reduce Cu^{2+}: $CuO(s) + H_2(g) \longrightarrow Cu(s) + H_2O(l)$, (b) H_2 cannot reduce Na^+. There is no reaction. **25.25** $[:C\equiv C:]^{2-}$. **25.27** (a) $2NaHCO_3(s) \longrightarrow Na_2CO_3(s) + H_2O(g) + CO_2(g)$, (b) $Ca(OH)_2(aq) + CO_2(g) \longrightarrow CaCO_3(s) + H_2O(l)$. **25.29** Heat causes bicarbonates to decompose according to the reaction: $2HCO_3^- \longrightarrow CO_3^{2-} + H_2O + CO_2$. Generation of carbonate ion causes precipitation of the insoluble $MgCO_3$. **25.31** $NaHCO_3$ plus some Na_2CO_3. **25.33** yes. **25.37** $KNO_3(s) + C(s) \longrightarrow KNO_2(s) + CO(g)$; 48.0 g. **25.39** (a) 86.7 kJ/mol, (b) 4×10^{-31}, (c) 4×10^{-31}. **25.41** 125 g/mol; P_4. **25.43** $4HNO_3(aq) + P_4O_{10}(s) \longrightarrow 2N_2O_5(g) + 4HPO_3(l)$; 60.4 g. **25.45** The P_4 molecule is unstable due to the structural strain of having the P—P bonds compressed in a tetrahedron. **25.47** (a) $2NaNO_3(s) \longrightarrow 2NaNO_2(s) + O_2(g)$, (b) $NaNO_3(s) + C(s) \longrightarrow NaNO_2(s) + CO(g)$. **25.49** $2NH_3(g) + CO_2(g) \longrightarrow (NH_2)_2CO(s) + H_2O(l)$; The reaction should be run at high pressure. **25.51** The oxidation state of N in nitric acid is +5, the highest oxidation state for N. N can be reduced to an oxidation state of -3. **25.53** sp^3. **25.55** (a) $NH_4NO_3(s) \longrightarrow N_2O(g) + 2H_2O(l)$, (b) $2KNO_3(s) \longrightarrow 2KNO_2(s) + O_2(g)$, (c) $Pb(NO_3)_2(s) \longrightarrow PbO(s) + 2NO_2(g) + O_2(g)$. **25.63** $\Delta G° = -198.3$ kJ/mol; $K_P = K_c = 6 \times 10^{34}$. **25.65** 35 g. **25.67** (a) to exclude light, (b) 0.37 L O_2. **25.69** To form OF_6, there would have to be six bonds (12 electrons) around the oxygen atom. This would violate the octet rule. Since oxygen does not have accessible d orbitals, it cannot have an expanded octet. **25.71** (a) $HCOOH(l) \rightleftharpoons CO(g) + H_2O(l)$, (b) $4H_3PO_4(l) \rightleftharpoons P_4O_{10}(s) + 6H_2O(l)$, (c) $2HNO_3(l) \rightleftharpoons N_2O_5(g) + H_2O(l)$, (d) $2HClO_3(l) \rightleftharpoons Cl_2O_5(l) + H_2O(l)$. **25.73** F: -1; O: 0. **25.75** $9H_2SO_4(aq) + 8NaI(aq) \longrightarrow 4I_2(s) + H_2S(g) + 4H_2O(l) + 8NaHSO_4(aq)$. **25.79** 25.3 L. **25.81** (a) $H-\ddot{F}-H-\ddot{F}:$, (b) $[:\ddot{F}-H-\ddot{F}:]^-$. **25.83** (a) linear, (b) tetrahedral, (c) trigonal bipyramidal, (d) see-saw. **25.85** $I_2O_5(s) + 5CO(g) \longrightarrow 5CO_2(g) + I_2(s)$; Iodine is reduced; carbon is oxidized. **25.87** As with iodide salts, a redox reaction occurs between sulfuric acid and sodium bromide. $2H_2SO_4(aq) + 2NaBr(aq) \longrightarrow SO_2(g) + Br_2(l) + 2H_2O(l) + Na_2SO_4(aq)$. **25.89** (a) $2H_3PO_3(aq) \longrightarrow H_3PO_4(aq) + PH_3(g) + O_2(g)$, (b) $Li_4C(s) + 4HCl(aq) \longrightarrow 4LiCl(aq) + CH_4(g)$, (c) $2HI(g) + 2HNO_2(aq) \longrightarrow I_2(s) + 2NO(g) + 2H_2O(l)$, (d) $H_2S(g) + 2Cl_2(g) \longrightarrow 2HCl(g) + SCl_2(l)$.

Chapter 24

24.3 The monomer must have a triple bond.

24.5

24.9 (1) Produce the alkoxide: $Sc(s) + 3C_2H_5OH(l) \longrightarrow Sc(OC_2H_5)_3(alc) + 3H^+(alc)$ ("alc" indicates a solution in alcohol); (2) hydrolyze to produce hydroxide pellets: $Sc(OC_2H_5)_3(alc) + 3H_2O(l) \longrightarrow Sc(OH)_3(s) + 3C_2H_5OH(alc)$; (3) sinter pellets to produce ceramic: $2Sc(OH)_3(s) \longrightarrow Sc_2O_3(s) + 3H_2O(g)$. **24.11** No. Bakelite is not a simple polymer; its permanence is due to cross-linking of the polymer chains. Bakelite is best described as a thermosetting composite polymer. **24.15** No. These molecules are too flexible, and liquid crystals require long, relatively rigid molecules. **24.19** As shown, it is an alternating condensation copolymer of the polyester class. **24.21** Metal amalgams expand with age; composite fillings tend to shrink. **24.25** sp^2 **24.27** dispersion forces; dispersion forces **24.31** (a) n-type (b) p-type **24.35** $Bi_2Sr_2CuO_6$ **24.37** Two are $+2$ ([Ar]$3d^9$), one is $+3$ ([Ar]$3d^8$). The $+3$ oxidation state is unusual for copper.

24.39 $H_2N-(CH_2)_8-NH_2$ $\quad$ $HO-\overset{O}{\underset{}{C}}-(CH_2)_4-\overset{O}{\underset{}{C}}-OH$

24.41 plastic (organic) polymer: covalent bonds, disulfide (covalent) bonds, hydrogen bonds, and dispersion forces; ceramics: ionic and network covalent bonds. **24.43** Fluoroapatite is less soluble than hydroxyapatite, particularly in acidic solutions. Dental fillings must also be insoluble. **24.45** The molecule is long, flat and rigid,

25.91 (a) $SiCl_4$, (b) F^-, (c) F, (d) CO_2. **25.93** There is no change in oxidation number; it is zero for both compounds.

25.95 PCl_4^+: $\begin{bmatrix} & Cl & \\ & | & \\ Cl-P-Cl \\ & | & \\ & Cl & \end{bmatrix}^+$, sp^3; PCl_6^-: $\begin{bmatrix} Cl & & Cl \\ & \diagdown P \diagup & \\ Cl & & Cl \\ & \diagup \quad \diagdown & \\ Cl & & Cl \end{bmatrix}^-$, sp^3d^2.

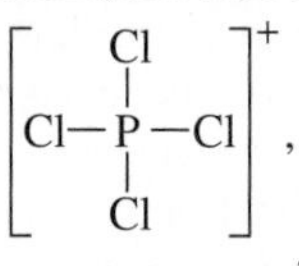

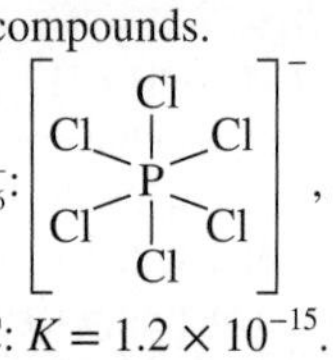

25.97 25°C: $K = 9.61 \times 10^{-22}$; 100°C: $K = 1.2 \times 10^{-15}$. **25.99** The glass is etched (dissolved) by the reaction: $6HF(aq) + SiO_2(s) \longrightarrow H_2SiF_6(aq) + 2H_2O(l)$. This process gives the glass a frosted appearance. **25.101** 1.18. **25.103** 0.833 g/L; The molar mass derived from the observed density is 74.41, which suggests that the molecules are associated to some extent in the gas phase. This makes sense due to strong hydrogen bonding in HF.

Chapter 26

26.3 (a) $CaCO_3$ (b) $CaCO_3 \cdot MgCO_3$ (c) CaF_2 (d) NaCl (e) Al_2O_3 (f) Fe_3O_4 (g) $Be_3Al_2Si_6O_{18}$ (h) PbS (i) $MgSO_4 \cdot 7H_2O$ (j) $CaSO_4$ **26.13** $K_P = 4.5 \times 10^5$ **26.15** (a) 8.9×10^{12} cm^3 (b) 4.0×10^8 kg SO_2 **26.17** Ag, Pt, and Au will not be oxidized, but the other metals will. **26.19** Al, Na, and Ca **26.33** (a) $2Na(s) + 2H_2O(l) \longrightarrow 2NaOH(aq) + H_2(g)$ (b) $2NaOH(aq) + CO_2(g) \longrightarrow Na_2CO_3(aq) + H_2O(l)$ (c) $Na_2CO_3(s) + 2HCl(aq) \longrightarrow 2NaCl(aq) + CO_2(g) + H_2O(l)$ (d) $NaHCO_3(aq) + HCl(aq) \longrightarrow NaCl(aq) + CO_2(g) + H_2O(l)$ (e) $2NaHCO_3(s) \longrightarrow Na_2CO_3(s) + CO_2(g) + H_2O(g)$ (f) no reaction **26.35** (a) $2K(s) + 2H_2O(l) \longrightarrow 2KOH(aq) + H_2(g)$ (b) $NaH(s) + H_2O(l) \longrightarrow NaOH(aq) + H_2(g)$ (c) $2Na(s) + O_2(g) \longrightarrow Na_2O_2(s)$ (d) $K(s) + O_2(g) \longrightarrow KO_2(s)$ **26.39** $3Mg(s) + 8HNO_3(aq) \longrightarrow 3Mg(NO_3)_2(aq) + 4H_2O(l) + 2NO(g)$; The magnesium nitrate is recovered from solution by evaporation, dried, and heated in air to obtain magnesium oxide: $2Mg(NO_3)_2(s) \longrightarrow 2MgO(s) + 4NO_2(g) + O_2(g)$. **26.41** The electron configuration of magnesium is $[Ne]3s^2$. The $3s$ electrons are outside the neon core (shielded), so they have relatively low ionization energies. Removing the third electron means separating an electron from the neon (closed shell) core, which requires a great deal more energy. **26.43** Even though helium and the Group 2A metals have ns^2 outer electron configurations, helium has a closed shell noble gas configuration and the Group 2A metals do not. The electrons in He are much closer to and more strongly attracted by the nucleus. Hence, the electrons in He are not easily removed. Helium is inert. **26.45** (a) CaO(s) (b) $Ca(OH)_2(s)$ **26.49** (a) 1.03 V (b) 3.32×10^4 kJ/mol **26.51** tetrahedral and octahedral; The accepted explanation for the nonexistence of $AlCl_6^{3-}$ is that the chloride ion is too big to form an octahedral cluster around a very small Al^{3+} ion. **26.53** $4Al(NO_3)_3(s) \longrightarrow 2Al_2O_3(s) + 12NO_2(g) + 3O_2(g)$. **26.55** The "bridge" bonds in Al_2Cl_6 break at high temperature: $Al_2Cl_6(g) \rightleftharpoons 2AlCl_3(g)$. This increases the number of molecules in the gas phase and causes the pressure to be higher than expected for pure Al_2Cl_6. **26.57** In Al_2Cl_6, each aluminum atom is surrounded by four bonding pairs of electrons (AB_4-type molecule), and therefore each aluminum atom is sp^3 hybridized. VSEPR analysis shows $AlCl_3$ to be an AB_3-type molecule (no lone pairs on the central atom). The geometry should be trigonal planar, and the aluminum atom should therefore be sp^2 hybridized. **26.59** 65.4 g/mol **26.61** Copper(II) ion is more easily reduced than either water or hydrogen ion, so there should be no reduction of water or hydrogen ion at the cathode. Copper metal is more easily oxidized than water, so water should not be affected by the copper purification process under standard conditions and it should not be oxidized at the anode. **26.63** (a) 1482 kJ/mol (b) 3152.8 kJ/mol **26.65** Mg(s) reacts with $N_2(g)$ at high temperatures to produce $Mg_3N_2(s)$. Ti(s) also reacts with $N_2(g)$ at high temperatures to produce TiN(s). **26.67** (a) In water, the aluminum(III) ion causes an increase in the concentration of hydrogen ion (lower pH). This results from the effect of the small diameter and high charge (3+) of the aluminum ion on surrounding water molecules. The aluminum ion draws electrons in the O—H bonds to itself, thus allowing easy formation of H^+ ions. (b) $Al(OH)_3$ is an amphoteric hydroxide. It will dissolve in strong base with the formation of a complex ion. $Al(OH)_3(s) + OH^-(aq) \longrightarrow Al(OH)_4^-(aq)$. The concentration of OH^- in aqueous ammonia is too low for this reaction to occur. **26.69** $CaO(s) + 2HCl(aq) \longrightarrow CaCl_2(aq) + H_2O(l)$ **26.71** Metals have closely spaced energy levels and (referring to Figure 26.10 of the text) a very small energy gap between filled and empty levels. Consequently, many electronic transitions can take place with absorption and subsequent emission of light continually occurring. Some of these transitions fall in the visible region of the spectrum and give rise to the flickering appearance. **26.73** NaF: cavity prevention; Li_2CO_3: antidepressant; $Mg(OH)_2$: laxative; $CaCO_3$: calcium supplement, antacid; $BaSO_4$: radiocontrast agent. **26.75** Both Li and Mg form oxides (Li_2O and MgO). Other Group 1A metals (Na, K, etc.) also form peroxides and superoxides. In Group 1A, only Li forms nitride (Li_3N), like Mg (Mg_3N_2). Li resembles Mg in that its carbonate, fluoride, and phosphate have low solubilities. **26.77** Zn **26.79** 87.66% Na_2O; 12.34% Na_2O_2 **26.81** 727 atm

Glossary

A

absolute temperature scale A scale based on −273.15°C (absolute zero) being the lowest point. (11.4)

absolute zero Theoretically the lowest obtainable temperature: −273.15°C or 0 K. (11.4)

absorbance Negative base ten logarithm of transmittance. (9.5)

absorption spectrum Plot of absorbance as a function of wavelength of incident light. (9.5)

accuracy The closeness of a measurement to the true or accepted value. (1.3)

acid See *Arrhenius acid, Brønsted acid,* and *Lewis acid.*

acid ionization constant (K_a) The equilibrium constant that indicates to what extent a weak acid ionizes. (16.6)

actinide series Series of elements that has partially filled $5f$ and/or $6d$ subshells. (3.10)

activated complex A transient species that forms when molecules collide in an effective collision. (19.2)

activation energy (E_a) The minimum amount of energy to begin a chemical reaction. (19.2)

activity series A list of metals arranged from top to bottom in order of decreasing ease of oxidation. (9.4)

actual yield The amount of product actually obtained from a reaction. (8.4)

addition polymer A large molecule that forms when small molecules known as monomers join together. (23.6)

addition polymerization Process by which monomers combine to form polymers without the elimination of small molecules, such as water. (24.1)

adhesion The attractions between unlike molecules. (12.2)

alcohol A compound consisting of an alkyl group and the functional group —OH. (23.2)

aldehyde A compound containing a hydrogen atom bonded to a carbonyl group. (23.2)

aliphatic Describes organic molecules that do not contain the benzene ring. (23.1)

alkali metal An element from Group 1A, with the exception of H (i.e., Li, Na, K, Rb, Cs, and Fr). (2.6)

alkaline earth metal An element from Group 2A (Be, Mg, Ca, Sr, Ba, and Ra). (2.6)

alkane Hydrocarbons having the general formula C_nH_{2n+2}, where $n = 1, 2, \ldots$ (5.6)

alkyl group A portion of a molecule that resembles an alkane. (23.2)

allotrope One of two or more distinct forms of an element. (5.5)

alloy Homogeneous mixture of two or more metals. (26.2)

alpha particle A helium ion with a positive charge of +2. (2.2)

alpha ray See *alpha particle.*

amalgam A substance made by combining mercury with one or more other metals. (26.2)

amide An organic molecule that contains an amide group. (23.2)

amine An organic molecule that contains an amino group. (23.2)

amino acid A compound that contains both an amino group and a carboxy group. (23.2)

amorphous solid A solid that lacks a regular three-dimensional arrangement of atoms. (12.3)

amphoteric Describes an oxide that displays both acidic and basic properties. (16.3)

amplitude The vertical distance from the midline of a wave to the top of the peak or the bottom of the trough. (3.2)

angstrom A measure of length that is equal to 1×10^{-10} m. (1.2)

angular momentum quantum number (ℓ) Describes the shape of the atomic orbital. (3.7)

anion An ion with a negative charge. (4.4)

anisotropic Dependent upon the axis of measurement. (24.3)

anode The electrode at which oxidation occurs. (18.2)

antibonding molecular orbital A molecular orbital that is higher in energy than the atomic orbitals that combined to produce it. (7.7)

aqueous Dissolved in water. (8.1)

aromatic Describes organic compounds that are related to benzene or that contain one or more benzene rings. (23.1)

Arrhenius acid Substance that increases H^+ concentration when added to water. (9.3)

Arrhenius base Substance that increases OH^- concentration when added to water. (9.3)

Arrhenius equation An equation that gives the dependence of the rate constant of a reaction on temperature: $k = Ae^{-E_a/RT}$. (19.6)

atactic Describes polymers in which the substituents are oriented randomly along the polymer chain. (24.1)

atom The basic unit of an element that can enter into chemical combination. (2.1)

atomic ion Atom that has lost or gained one or more electrons, giving it a positive or negative charge. (see *monatomic ion*)

atomic mass The mass of the atom given in atomic mass units (amu). (2.5)

atomic mass unit (amu) A mass exactly equal to one-twelfth the mass of one carbon-12 atom. (1.2)

atomic number (Z) The number of protons in the nucleus of each atom of an element. (2.3)

atomic orbital The wave function of an electron in an atom. (3.6)

atomic radius Metallic: One-half the distance between the nuclei in the two adjacent atoms of the same element in a metal. Covalent: One-half the distance between the nuclei of the two identical atoms in a diatomic molecule. (4.4)

Aufbau principle The process by which the periodic table can be built up by successively adding one proton to the nucleus and one electron to the appropriate atomic orbital. (3.9)

autoionization of water Ionization of water molecules to give H^+ and OH^- ions. (16.3)

average atomic mass The weighted average of the atomic masses of all natural isotopes of an element. (2.5)

Avogadro's law The volume of a sample of gas (V) is directly proportional to the number of moles (n) in the sample at constant temperature and pressure: $V \propto n$. (11.4)

Avogadro's number (N_A) The number of atoms in exactly 12 g of carbon-12: 6.022×10^{23}. (2.7)

axial Describes the two bonds that form an axis perpendicular to the trigonal plane in trigonal bipyramidal geometry. (7.1)

B

band theory A theory wherein atomic orbitals merge to form energy bands. (26.3)

barometer An instrument used to measure atmospheric pressure. (11.3)

base See *Arrhenius base, Brønsted base,* and *Lewis base.*

base ionization constant (K_b) The equilibrium constant that indicates to what extent a weak base ionizes. (16.7)

battery A portable, self-contained source of electric energy consisting of galvanic cells or a series of galvanic cells. (18.6)

Beer-Lambert law Equation relating absorbance to molar absorptivity, solution concentration, and distance light travels through a solution. (9.5)

beta particle An electron. (2.2)

beta ray See *beta particle.*

bimolecular Describes a reaction in which two reactant molecules collide. (19.7)

binary compound A substance that consists of just two different elements. (5.3)

blackbody radiation The electromagnetic radiation emitted from a heated solid. (3.3)

body-centered cubic cell A unit cell with one atom at the center of the cube and one atom at each of the eight corners. (12.3)

boiling point The temperature at which vapor pressure equals atmospheric pressure. (12.2)

bond angle The angle between two adjacent A—B bonds. (7.1)

bond enthalpy The enthalpy change associated with breaking a particular bond in 1 mole of gaseous molecules. (10.7)

bond length Distance between two nuclei connected by a covalent bond. (6.1)

bond order A number based on the number of electrons in bonding and antibonding molecular orbitals that indicates, qualitatively, how stable a bond is. (7.7)

bond-line structure A structure in which straight lines represent carbon-carbon bonds. (23.3)

bonding molecular orbital A molecular orbital that is lower in energy than the atomic orbitals that combined to produce it. (7.7)

Born-Haber cycle The cycle that relates the lattice energy of an ionic compound to quantities that can be measured. (10.8)

Boyle's law The pressure of a fixed amount of gas at a constant temperature is inversely proportional to the volume of the gas: $P \propto 1/V$. (11.4)

Bragg equation An equation relating the wavelength of X rays, the angle of diffraction, and the spacing between atoms in a lattice. (12.3)

breeder reactor A nuclear reactor that produces more fissionable material than it consumes. (20.5)

Brønsted acid A substance that donates a proton (H^+). (9.3, 16.1)

Brønsted base A substance that accepts a proton (H^+). (9.3, 16.1)

buffer A solution that contains significant concentrations of both members of a conjugate pair (weak acid/conjugate base or weak base/conjugate acid). (17.2)

C

calorimetry The measurement of heat changes. (10.4)

capillary action The movement of liquid up a narrow tube as the result of adhesive forces. (12.2)

carbocation A species in which one of the carbons is surrounded by only six electrons. (23.5)

carbon nanotube A tube made of carbon atoms with dimensions on the order of nanometers. (24.5)

carboxylic acid An organic acid that contains a carboxy group. (23.2)

catalyst A substance that increases the rate of a chemical reaction without itself being consumed. (19.8)

catenation The formation of long carbon chains. (23.1)

cathode The electrode at which reduction occurs. (18.2)

cathode rays Radiation (electron beams) emitted from the negatively charged plate in an evacuated tube when a high voltage is applied. (2.2)

cation An ion with a positive charge. (4.4)

cell potential (E_{cell}) The difference in electric potential between the cathode and the anode. (18.2)

Celsius Temperature scale originally based on the freezing point (0°C) and boiling point (100°C) of water. (1.2)

ceramics Polymeric inorganic compounds that share the properties of hardness, strength, and high melting points. (24.2)

chalcogens Elements in Group 6A (O, S, Se, Te, and Po). (2.6)

Charles and Gay-Lussac's law See *Charles's law*.

Charles's law The volume of a fixed amount of gas (V) maintained at constant pressure is directly proportional to absolute temperature (T): $V \propto T$. Also known as *Charles's and Gay-Lussac's law*. (11.4)

chelating agent A polydentate ligand that forms complex ions with metal ions in solution. (22.1)

chemical change A process in which one or more substances are changed into one or more new substances. (1.6)

chemical energy Energy stored within the structural units (molecules or polyatomic ions) of chemical substances. (3.1)

chemical equation Chemical symbols used to represent a chemical reaction. (8.1)

chemical formula Chemical symbols and numerical subscripts used to denote the composition of the substance. (5.3)

chemical process A process in which the identity of matter changes. (1.6, 4.4)

chemical property Any property of a substance that cannot be studied without converting the substance into some other substance. (1.6)

chemical reaction A process that neither creates nor destroys atoms but rearranges atoms in elements and/or chemical compounds. (8.1)

chemistry The study of matter and the changes it undergoes. (1.1)

chiral Describes molecules with nonsuperimposable mirror images. (23.4)

cholesteric Describes molecules that are parallel to each other within each layer, but where each layer is rotated with respect to the layers above and below it. (24.3)

cis Describes the isomer in which two substituents both lie on the same side of a double bond. (23.4)

Clausius-Clapeyron equation A linear relationship that exists between the natural log of vapor pressure and the reciprocal of absolute temperature. (12.2)

closed system A system that can exchange energy (but not mass) with the surroundings. (10.2)

coefficients See *stoichiometric coefficients*.

cohesion The attraction between like molecules. (12.2)

colligative properties Properties that depend on the number of solute particles in solution but do not depend on the nature of the solute particles. (13.5)

collision theory The reaction rate is directly proportional to the number of molecular collisions per second. (19.2)

colloid A dispersion of particles of one substance throughout another substance. (13.7)

combination reaction A reaction in which two or more reactants combine to form a single product. (8.1)

combined gas law The mathematical relationship that correlates pressure (P), volume (V), moles (n), and temperature (T) as these quantities change values. (11.4)

combustion analysis An experimental determination of an empirical formula by a reaction with oxygen to produce carbon dioxide and water. (8.2)

combustion Burning in air. (8.1)

combustion reaction A reaction in which a substance burns in the presence of oxygen. (8.1)

common ion effect The presence of a common ion suppresses the ionization of a weak acid or weak base. (17.1)

complex ion Charged species consisting of a central metal cation bonded to two or more anions or polar molecules. (17.5)

composite material A material made from two or more substances with different properties that remain separate in the bulk material. (24.2)

compound A substance composed of atoms of two or more elements chemically united in fixed proportions. (5.1)

concentration Amount of solute relative to the volume of a solution or to the amount of solvent in a solution. (9.5)

concentration cell A cell that has the same type of electrode and the same ion in solution (at different concentrations) in the anode and cathode compartments. (18.5)

condensation The phase transition from gas to liquid. (12.2)

condensation polymer A large molecule that forms when small molecules undergo condensation reactions. (23.6, 24.1)

condensation reaction An elimination reaction in which two or more molecules become connected with the elimination of a small molecule, often water. (23.6)

condensed structural formula Shows the same information as a structural formula, but in a condensed form. (23.3)

condensed structure Chemical structure simplified by using abbreviations for repeating structural units. (23.3)

conduction band The antibonding band. (24.6)

conductor A substance through which electrons move freely. (26.3)

conjugate acid The cation that results when a Brønsted base accepts a proton. (16.1)

conjugate base The anion that remains when a Brønsted acid donates a proton. (16.1)

conjugate pair The combination of a Brønsted acid and its conjugate base (or the combination of a Brønsted base and its conjugate acid). (16.1)

constitutional isomers Compounds with the same chemical formula but different structures. (23.4)

conversion factor A fraction in which the same quantity is expressed one way in the numerator and another way in the denominator. (1.4)

coordinate covalent bond A covalent bond in which one of the atoms donates both electrons. (6.6)

coordination compound A compound that contains coordinate covalent bonds between a metal ion (often a transition metal ion) and two or more polar molecules or ions. (12.3, 22.1)

coordination number The number of atoms surrounding an atom in a crystal lattice. (12.3) The number of donor atoms surrounding a metal in a complex. (22.1)

copolymer A polymer made of two or more different monomers. (23.6, 24.1)

corrosion The undesirable oxidation of metals. (18.8)

Coulomb's law The force (F) between two charged objects (Q_1 and Q_2) is directly proportional to the product of the two charges and inversely proportional to the distance (d) between the objects squared. (4.4)

covalent bond A shared pair of electrons. (5.5)

covalent bonding Two atoms sharing a pair of electrons. (5.5)

covalent radius Half the distance between adjacent, identical nuclei in a molecule. (4.4)

critical mass The minimum amount of fissionable material required to sustain a reaction. (20.5)

critical pressure (P_c) The minimum pressure that must be applied to liquefy a substance at its critical temperature. (12.5)

critical temperature (T_c) The temperature at which the gas phase cannot be liquefied, no matter how great the applied pressure. (12.5)

cross-link A bond that forms between a functional group off the backbone of the polymer chain that interacts with another functional group of a second polymer strand creating a new covalent bond and causing the polymer to be stronger and more rigid. (24.1)

crystal field splitting (Δ) The difference in energy between the lower and higher d-orbital energy levels. (22.3)

crystalline solid A solid that possesses rigid and long-range order; its atoms, molecules, or ions occupy specific positions. (12.3)

D

d orbital Atomic orbital in which the angular momentum quantum number (ℓ) is 2. (3.8, 6.6)

Dalton's law of partial pressures The total pressure exerted by a gas mixture is the sum of the partial pressures exerted by each component of the mixture. (11.7)

dative bond A covalent bond in which one of the atoms donates both electrons. (see *coordinate covalent bond*)

de Broglie wavelength A wavelength calculated using the following equation: $\lambda = h/mu$. (3.5)

decomposition reaction A reaction in which one reactant forms two or more products. (8.1)

degenerate Having an equal energy. (3.9)

delocalized Describes bonding electron density that is spread out over the molecule or part of the molecule, rather than confined between two specific atoms. (7.8)

density The ratio of mass to volume. (1.2)

deoxyribonucleic acid (DNA) Biological polymer arranged in a double-helix shape, consisting of two long chains of nucleotides in which the sugar is deoxyribose. (23.6)

deposition The phase change from gas to solid. (12.5)

dextrorotatory The term used to describe the enantiomer that rotates the plane-polarized light to the right. (22.2)

diagonal relationships Similarities in chemical properties of elements that are in different groups but that are positioned diagonally to one another in the periodic table. (4.4)

diamagnetic A species without unpaired electrons that is weakly repelled by magnetic fields. (3.9)

diatomic molecule A molecule that contains two atoms. (5.5)

dibasic A substance that produces two moles of hydroxide ion per mole of compound dissolved in aqueous solution. (9.3)

diffusion The mixing of gases. (11.2)

dilution The process of preparing a less concentrated solution from a more concentrated one. (9.5)

dimensional analysis The use of conversion factors in problem solving. (1.4)

diode An electronic device that restricts the flow of electrons in a circuit to one direction. (24.6)

dipole moment (μ) A quantitative measure of the polarity of a bond. (6.2)

dipole-dipole interactions Attractive forces that act between polar molecules. (7.3)

diprotic acid An acid with two ionizable protons. (9.3)

dispersion forces See *London dispersion forces*.

displacement reaction Reaction in which two reactants trade components (double displacement) or where a component of a reactant is removed (single displacement). (9.4)

disproportionation reaction When an element undergoes both oxidation and reduction in the same reaction. (9.4)

dissociation The process by which an ionic compound, upon dissolution, breaks apart into its constituent ions. (9.1)

donor atom The atom that bears the unshared pair of electrons. (22.1)

doping The addition of very small quantities of an element with one more or one fewer valence electron than the natural semiconductor. (24.6)

double bond A multiple bond in which the atoms share two pairs of electrons. (6.1)

double replacement Reaction in which compounds exchange ions. (9.2)

dynamic equilibrium Occurs when a forward process and reverse process are occurring at the same rate. (12.2)

E

effective collision A collision that results in a reaction. (19.2)

effective nuclear charge (Z_{eff}) The actual magnitude of positive charge that is "experienced" by an electron in the atom. (4.3)

effusion The escape of a gas from a container into a vacuum. (11.2)

elastomer A material that can stretch or bend and then return to its original shape as long as the limits of its elasticity are not exceeded. (24.1)

electricity The movement of electrons. (9.1)

electrode A piece of conducting metal in an electrochemical cell at which either oxidation or reduction takes place. (18.2)

electrolysis The use of electric energy to drive a nonspontaneous redox reaction. (18.7)

electrolyte A substance that dissolves in water to yield a solution that conducts electricity. (9.1)

electrolytic cell An electrochemical cell used for electrolysis. (18.7)

electromagnetic spectrum Consists of radio waves, microwave radiation, infrared radiation, visible light, ultraviolet radiation, X rays, and gamma rays. (3.2)

electromagnetic wave A wave that has an electric field component and a magnetic field component. (3.2)

electron A negatively charged subatomic particle found outside the nucleus of all atoms. (2.2)

electron affinity (EA) The energy released (the negative of the enthalpy change, $-\Delta H$) when an atom in the gas phase accepts an electron. (4.4)

electron configuration The distribution of electrons in the atomic orbitals of an atom. (3.9)

electron density The probability that an electron will be found in a particular region of an atom. (3.6)

electron domain A lone pair or a bond, regardless of whether the bond is single, double, or triple. (7.1)

electron spin quantum number (m_s) The fourth quantum number that differentiates two electrons in the same orbital. (3.7)

electron-domain geometry The arrangement of electron domains (bonds and lone pairs) around a central atom. (7.1)

electronegativity The ability of an atom in a compound to draw electrons to itself. (6.2)

electrophile A region of positive or partial positive charge. (23.5)

electrophilic addition An addition reaction that begins when an electrophile approaches a region of electron density. (23.5)

electrostatic energy The potential energy that results from the interaction of charged particles. (3.1)

element A substance that cannot be separated into simpler substances by chemical means. (2.1)

elementary reaction A reaction that occurs in a single collision of the reactant molecules. (19.7)

elimination reaction A reaction in which a double bond forms and a molecule such as water is removed. (23.5)

emission spectrum The light emitted, either as a continuum or in discrete lines, by a substance in an excited electronic state. (3.4)

empirical formula The chemical formula that conveys with the smallest possible whole numbers the ratio of combination of elements in a compound. (5.5)

empirical formula mass The mass in amu of one empirical formula of a compound. (5.9)

enantiomers Molecules that are mirror images of each other but cannot be superimposed. (22.2, 23.4)

endothermic process A process that absorbs heat. (10.1)

endpoint The point at which the color of the indicator changes. (9.6, 17.3)

energy The capacity to do work or transfer heat. (3.1)

enthalpy (H) A thermodynamic quantity defined by the equation $H = U + PV$. The change in enthalpy, ΔH, is equal to the heat exchanged between the system and surroundings at constant pressure q_P. (10.3)

enthalpy of reaction (ΔH_{rxn}) The difference between the enthalpy of the products and the enthalpy of the reactants. (10.3)

entropy (S) A thermodynamic state function that describes how dispersed a system's energy is. (13.2, 14.2)

enzymes Biological catalysts. (19.8)

equatorial Describes the three bonds that are arranged in a trigonal plane in a trigonal bipyramidal geometry. (7.1)

equilibrium A state in which forward and reverse processes are occurring at the same rate. (15.1)

equilibrium constant (K_c) A number equal to the ratio of the equilibrium concentrations of products to the equilibrium concentrations of reactants, with each concentration raised to the power of its stoichiometric coefficient. (15.2)

equilibrium expression The quotient of product concentrations and reactant concentrations, each raised to the power of its stoichiometric coefficient. (15.2)

equilibrium process A process that can be made to occur by the addition or removal of energy but does not happen on its own. (14.4)

equilibrium vapor pressure The pressure exerted by the molecules that have escaped to the gas phase, once the pressure has stopped increasing. (12.2)

equivalence point The point in a titration where the reaction (e.g., neutralization) is complete. (9.6)

ester An organic molecule containing a —COOR group. (23.2)

evaporation The phase change from liquid to gas at a temperature below the boiling point. (12.2)

excess reactant The reactant present in a greater amount than necessary to react with all the limiting reactant. (8.4)

excited state A state that is higher in energy than the ground state. (3.4)

exothermic process A process that gives off heat. (10.1)

extensive property A property that depends on the amount of matter involved. (1.6)

F

f orbital Atomic orbital in which the angular momentum quantum number (ℓ) is 3. (3.8)

face-centered cubic cell A cubic unit cell with one atom on each of the six faces and one atom at each of the eight corners. (12.3)

family The elements in a vertical column of the periodic table. (2.5)

first law of thermodynamics Energy can be converted from one form to another, but cannot be created or destroyed. (10.2)

first-order reaction A reaction whose rate depends on the reactant concentration raised to the first power. (19.5)

formal charge Method of electron "bookkeeping" in which shared electrons are divided equally between the atoms that share them. (6.4)

formation constant (K_f) The equilibrium constant that indicates to what extent complex-ion formation reactions occur. (17.5)

formula mass The mass of a formula unit. (5.8)

formula weight See *formula mass*.

fractional precipitation The separation of a mixture based upon the components' solubilities. (17.6)

free energy The energy available to do work. Also known as *Gibbs free energy*. (14.5)

free radical A molecule with an odd number of electrons. (6.6)

freezing point The temperature at which a liquid undergoes the phase transition to solid. (12.5)

frequency (ν) The number of waves that pass through a particular point in 1 s. (3.2)

fuel cell A voltaic cell in which reactants must be continually supplied. (18.6)

fullerenes Molecules with elongated and elliptical cages of 70 and 80 carbon atoms. (24.5)

functional group The part of a molecule characterized by a special arrangement of atoms that is largely responsible for the chemical behavior of the parent molecule. (5.6)

fusion The phase transition from solid to liquid (melting). (12.5) The combination of light nuclei to form heavier nuclei. (20.5)

G

galvanic cell An electrochemical cell in which a spontaneous chemical reaction generates a flow of electrons. (18.2)

galvanization The cathodic protection of iron or steel using zinc. (18.8)

gamma ray High-energy radiation. (2.2)

gas constant (R) The proportionality constant that appears in the ideal gas equation. (11.2)

gas laws Equations that relate the volume of a gas sample to its other parameters: temperature (T), pressure (P), and number of moles (n). (11.4)

geometrical isomers Molecules that contain the same atoms and bonds arranged differently in space. (22.2, 23.4)

Gibbs free energy The energy available to do work. (14.5)

glass Commonly refers to an optically transparent fusion product of inorganic materials that has cooled to a rigid state without crystallizing. (12.3)

Graham's law The rates of diffusion and effusion are inversely proportional to the square root of the molar mass of the gas. (11.2)

graphene Two-dimensional sheet of sp^2-hybridized carbon atoms. (24.5)

gravimetric analysis An analytical technique based on the measurement of mass. (9.6)

greenhouse effect Describes the trapping of heat near Earth's surface by gases in the atmosphere, particularly carbon dioxide. (21.5)

ground state The lowest energy state of an atom. (3.4)

group The elements in a vertical column of the periodic table. (2.6)

H

half-cell One compartment of an electrochemical cell containing an electrode immersed in a solution. (18.2)

half-life ($t_{1/2}$) The time required for the reactant concentration to drop to half its original value. (19.5)

half-reaction method Balancing an oxidation-reduction equation by separating the oxidation and the reduction, balancing them separately, and adding them back together. (9.4)

half-reactions The separated oxidation and reduction reactions that make up the overall redox reaction. (9.4, 18.1)

Hall process Electrolytic reduction of aluminum from anhydrous aluminum oxide (corundum). (26.7)

halogens The elements in Group 7A (F, Cl, Br, I, and At). (2.6)

heat The transfer of thermal energy between two bodies that are at different temperatures. (10.1)

heat capacity (C) The amount of heat required to raise the temperature of an object by 1°C. (10.4)

Heisenberg uncertainty principle It is impossible to know simultaneously both the momentum (p) (defined as mass times velocity, $m \times u$) and the position (x) of a particle with certainty. (3.6)

Henderson-Hasselbalch equation $pH = pK_a + \log$ ([conjugate base]/[weak acid]). (17.2)

Henry's law The solubility of a gas in a liquid is proportional to the pressure of the gas over the solution. (13.4)

Henry's law constant (k) The proportionality constant that is specific to the gas-solvent combination and varies with temperature. (13.4)

Hess's law The change in enthalpy that occurs when reactants are converted to products in a reaction is the same whether the reaction takes place in one step or in a series of steps. (10.5)

heteroatom Any atom in an organic molecule other than carbon or hydrogen. (23.3)

heterogeneous catalysis A catalysis in which the reactants and the catalyst are in different phases. (19.8)

heterogeneous mixture A mixture in which the composition varies. (1.5)

heteronuclear Containing two or more different elements. (5.5)

high-temperature superconductor A material that shows no resistance to the flow of electrons at an unusually high temperature. (24.7)

homogeneous catalysis A catalysis in which the reactants and the catalyst are in the same phase. (19.8)

homogeneous mixture A mixture in which the composition is uniform. Also called a solution. (1.5)

homonuclear Containing atoms of only one element. (5.5)

Hund's rule The most stable arrangement of electrons in orbitals of equal energy is the one in which the number of electrons with the same spin is maximized. (3.9)

hybrid orbitals Orbitals formed by hybridization of some combination of s, p, or d atomic orbitals. (7.5)

hybridization The mixing of atomic orbitals. (7.5)

hydrate A compound with a specific number of water molecules within its solid structure. (5.7)

hydration The process by which water molecules surround solute particles in an aqueous solution. (9.2)

hydrocarbon A compound containing only carbon and hydrogen. (5.6)

hydrogen bonding A special type of dipole-dipole interaction that occurs only in molecules that contain H bonded to a small, highly electronegative atom, such as N, O, or F. (7.3)

hydrogen displacement Reaction in which hydrogen ion is reduced to hydrogen gas. (9.4)

hydronium ion A hydrated proton (H_3O^+). (9.3)

hydrophilic Water loving. (13.7)

hydrophobic Water fearing. (13.7)

hypertonic Describes a solution with a relatively higher concentration of dissolved substances. (13.5)

hypothesis A tentative explanation for a set of observations. (1.1)

hypotonic Describes a solution that has a relatively lower concentration of dissolved substances. (13.5)

I

ideal gas A hypothetical sample of gas whose pressure-volume-temperature behavior is predicted accurately by the ideal gas equation. (11.5)

ideal gas equation An equation that describes the relationship among the four variables P, V, n, and T. (11.5)

ideal solution A solution that obeys Raoult's law. (13.5)

indicator A substance that has a distinctly different color in acidic and basic media. (9.6)

initial rate The instantaneous rate at the beginning of a reaction. (19.4)

inorganic compounds Compounds that do not contain carbon or that are derived from nonliving sources. (5.6)

instantaneous dipole A fleeting nonuniform distribution of electron density in a molecule without a permanent dipole. (7.3)

instantaneous rate The reaction rate at a specific time. (19.3)

insulator A substance that does not conduct electricity. (26.3)

integrated rate law $\ln([A]_t/[A]_0) = -kt$. (19.5)

intensive property A property that does not depend on the amount of matter involved. (1.6)

intermediate A chemical species that is produced in one step of a reaction mechanism and consumed in a subsequent step. (19.7)

intermolecular forces The attractive forces that hold particles together in the condensed phases. (7.3)

internal energy The sum of energy attributed to a system, by virtue of the particles it contains, and their relative positions. (10.2)

International System of Units See *SI units*. (1.2)

ion Atom or molecule that has lost or gained one or more electrons giving it a positive or negative charge. (4.4)

ion pair Ions in solution that are held together by electrostatic forces. (13.5)

ion-dipole interactions Coulombic attractions between ions and polar molecules. (7.3)

ion-product constant Another name for K_w, the equilibrium constant for the autoionization of water. (16.3)

ionic bonding An electrostatic attraction that holds oppositely charged ions together in an ionic compound. (5.3)

ionic compound Substance consisting of ions held together by electrostatic attraction. (5.3)

ionic equation Chemical equation in which all strong electrolytes are shown as ions. Also known as *complete ionic equation*. (9.2)

ionic radius The radius of a cation or an anion. (4.6)

ionizable hydrogen atom A hydrogen atom that can be lost as a hydrogen ion, (H^+). (5.6)

ionization The process by which a molecular compound forms ions when it dissolves. (9.1)

ionization energy (IE) The minimum energy required to remove an electron from an atom in the gas phase. (4.4)

ionosphere The outermost layer of the atmosphere—also known as the thermosphere. (21.1)

isoelectronic Describes two or more species with identical electron configurations. (4.5)

isoelectronic series A series of two or more species that have identical electron configurations but different nuclear charges. (4.6)

isolated system A system that can exchange neither energy nor mass with the surroundings. (10.2)

isomerization reaction A reaction in which one isomer is converted to another. (23.5)

isotactic Polymers in which all the substituents (i.e., the R groups) are in the same relative orientation (i.e., on the same side of the polymer chain). (24.1)

isotonic Equal in concentration and osmotic pressure. (13.5)

isotopes Atoms that have the same atomic number (Z) but different mass numbers (A). (2.3)

isotropic Independent of the axis of measurement. (24.3)

J

joule (J) SI unit of energy, 1 kg · m²/s². (3.1)

K

Kekulé structure A structure similar to a Lewis structure in which the lone pairs may not be shown. (23.3)

kelvin The SI base unit of temperature. (1.2)

Kelvin temperature scale A temperature scale offset from the Celsius scale by 273.15. One kelvin (1 K) is equal in magnitude to one degree Celsius (1°C). (1.4, 11.4)

ketone An organic compound consisting of two R groups bonded to a carbonyl. (23.2)

kinetic energy The energy that results from motion. (3.1)

kinetic molecular theory A theory that explains how the molecular nature of gases gives rise to their macroscopic properties. (11.2)

L

lanthanide (rare earth) series A series of 14 elements that have incompletely filled 4f subshells or that readily give rise to cations that have incompletely filled 4f subshells. (3.10)

lattice A three-dimensional array of cations and anions. (5.3)

lattice energy The amount of energy required to convert a mole of ionic solid to its constituent ions in the gas phase. (5.3, 10.8)

lattice points Positions occupied by atoms, ions, or molecules in a unit cell. (12.3)

lattice structure The arrangement of the particles in a crystalline solid. (12.3)

law A concise verbal or mathematical statement of a reliable relationship between phenomena. (1.1)

law of conservation of energy First law of thermodynamics stating that energy can be neither created nor destroyed. (3.1)

law of conservation of mass An alternate statement of the first law of thermodynamics stating that matter can be neither created nor destroyed. (8.1)

law of definite proportions Different samples of a given compound always contain the same elements in the same mass ratio. (5.5)

law of mass action For a reversible reaction at equilibrium and a constant temperature, the reaction quotient, Q, has a constant value, K (the equilibrium constant). (15.2)

law of multiple proportions Different compounds made up of the same elements differ in the number of atoms of each kind that combine. (5.5)

Le Châtelier's principle When a stress is applied to a system at equilibrium, the system will respond by shifting in the direction that minimizes the effect of the stress. (15.6)

levorotatory The term used to describe the enantiomer that rotates the plane-polarized light to the left. (22.2)

Lewis acid A species that can accept a pair of electrons. (6.6, 16.12)

Lewis base A species that can donate a pair of electrons. (6.6, 16.12)

Lewis dot symbol An elemental symbol surrounded by dots, where each dot represents a valence electron. (5.2)

Lewis structure A representation of covalent bonding in which shared electron pairs are shown either as dashes or as pairs of dots between two atoms, and lone pairs are shown as pairs of dots on individual atoms. (6.1)

Lewis theory of bonding A chemical bond involves atoms sharing electrons. (5.5)

ligand A molecule or anion that can form coordinate bonds to a metal to form a coordination complex. (22.1)

limiting reactant The reactant that is completely consumed and determines the amount of product formed. (8.4)

line spectra The emission or absorption of light only at discrete wavelengths. (3.4)

liquid crystal A substance that exhibits properties of both a liquid, such as the ability to flow and to take on the shape of a container, and those of a crystal, such as a regular arrangement of particles in a lattice. (24.3)

localized Describes electrons that are shared between two specific atoms and that cannot be repositioned to generate additional resonance structures. (7.8)

London dispersion forces Attractive forces that act between all molecules, including nonpolar molecules, resulting from the formation of instantaneous dipoles and induced dipoles. Also known simply as *dispersion forces*. (7.3)

lone pair A pair of valence electrons that are not involved in covalent bond formation. (6.1)

M

magnetic quantum number (m_ℓ) Describes the orientation of an orbital in space. (3.7)

main group elements Elements in the s and p blocks of the periodic table. (4.2)

manometer A device used to measure the pressure of gases relative to atmospheric pressure. (11.3)

mass A measure of the amount of matter in an object or sample. (1.2)

mass defect The difference between the actual mass of a nucleus and the mass calculated by summing the masses of the individual nucleons. (20.2)

mass number (A) The number of neutrons and protons present in the nucleus of an atom of an element. (2.3)

matter Anything that occupies space and has mass. (1.1)

Meissner effect The exclusion of magnetic fields. (24.7)

melting point The temperature at which solid and liquid phases are in equilibrium. (12.3)

mesosphere The region located above the stratosphere in which the concentration of ozone and other gases is low and temperature decreases again with increasing altitude. (21.1)

metal Element with a tendency to lose electrons, located left of the zigzag line on the periodic table. (2.6)

metallic radius Half the distance between the nuclei of two adjacent, identical metal atoms. (4.4)

metalloid Elements with properties intermediate between metals and nonmetals. (2.6, 4.4)

metallurgy The preparation, separation, and purification of metals. (26.2)

metathesis Another name for a double replacement reaction in which compounds exchange ions. (9.2)

microstate A specific microscopic configuration of a system. (14.2)

mineral A naturally occurring substance with a characteristic chemical composition and specific physical properties. (26.1)

miscible Mutually soluble in any proportions. (13.2)

mixture A combination of two or more substances in which the substances retain their distinct identities. (1.5)

moderator A material that limits the speed of liberated neutrons but does not itself undergo fission when bombarded with neutrons. (20.5)

molality (m) The number of moles of solute dissolved in 1 kg (1000 g) of solvent. (13.3)

molar absorptivity Proportionality constant relating absorbance to solution concentration and distance that light travels through a solution. (9.5)

molar concentration See *molarity (M)*.

molar heat (enthalpy) of fusion (ΔH_{fus}) The energy, usually expressed in kJ/mol, required to melt 1 mole of a solid. (12.5)

molar heat (enthalpy) of sublimation (ΔH_{sub}) The energy, usually expressed in kilojoules, required to sublime 1 mole of a solid. (12.5)

molar heat of vaporization (ΔH_{vap}) The amount of heat required to vaporize a mole of substance at its boiling point. (12.5)

molar mass ($\mathcal{M}$) The mass in grams of 1 mole of a substance. (2.7, 5.10)

molar solubility The number of moles of solute in 1 liter of saturated solution (mol/L). (17.4)

molarity (M) The number of moles of solute per liter of solution. Also known as *molar concentration*. (9.5)

mole (mol) The amount of a substance that contains as many elementary entities (atoms, molecules, formula units, etc.) as there are atoms in exactly 0.012 kg (12 g) of carbon-12. (2.7)

mole fraction (χ_i) The number of moles of a component divided by the total number of moles in a mixture. (11.7)

molecular equation A chemical equation written with all compounds represented by their chemical formulas. (9.2)

molecular formula A chemical formula that gives the number of atoms of each element in a molecule. (5.5)

molecular geometry The arrangement of bonded atoms. (7.1)

molecular mass The sum of the atomic masses (in amu) of the atoms that make up a molecule. (5.8)

molecular orbital theory A theory that describes the orbitals in a molecule as bonding and antibonding combinations of atomic orbitals. (7.7)

molecular orbitals Orbitals that result from the interaction of atomic orbitals of bonding atoms. (7.7)

molecular weight Average molecular mass. (5.8)

molecularity The number of molecules involved in a specific step in a reaction mechanism. (19.7)

molecule A combination of two or more atoms in a specific arrangement held together by chemical bonds. (5.5)

monatomic ion An ion that contains only one atom. (4.5)

monobasic A substance that produces one mole of hydroxide ion per mole of compound dissolved in aqueous solution. (9.3)

monomer A small molecule that can be linked in large numbers to form a large molecule (polymer). (23.6, 24.1)

monoprotic acid An acid with one ionizable hydrogen atom. (5.7, 9.3)

multiple bond A chemical bond in which two atoms share two or more pairs of electrons. (6.1)

N

n-type semiconductor A semiconductor in which an electron-rich impurity is added to enhance conduction. (24.6, 26.3)

nanotechnology The development and study of extremely small-scale materials and objects. (24.5)

nematic Describes an arrangement in which molecules are all aligned parallel to one another but with no organization into layers or rows. (24.3)

Nernst equation An equation relating the emf of a galvanic cell with the standard emf and the concentrations of reactants and products. (18.5)

net ionic equation Chemical equation from which spectator ions have been removed. (9.2)

neutralization reaction A reaction between an acid and a base. (9.3)

neutron An electrically neutral subatomic particle with a mass slightly greater than that of a proton. (2.2)

newton (N) The SI unit of force. (11.3)

nitrogen fixation The conversion of molecular nitrogen into nitrogen compounds. (21.1)

noble gas core A representation in an electron configuration that shows in brackets the most recently completed noble gas. (3.10)

noble gases Elements in Group 8A (He, Ne, Ar, Kr, Xe, and Rn). (2.6)

nodes Points at which electron density in an atom is zero. (3.5)

nonconductor A substance that does not conduct electricity. (24.6)

nonelectrolyte A substance that dissolves in water to yield a solution that does not conduct electricity. (9.1)

nonmetal Element with a tendency to gain electrons, located in the upper right portion of the periodic table. (2.6)

nonpolar Having a uniform distribution of electron density. (6.2)

nonspontaneous process A process that does not occur under a specified set of conditions. (14.1)

nonvolatile Having no measurable vapor pressure. (13.5)

normal boiling point Temperature at which a substance boils at 1 atm pressure. (12.2)

normal freezing point Temperature at which a substance freezes at 1 atm pressure. Also known as the *normal melting point*. (12.5)

normal melting point Temperature at which a substance melts at 1 atm pressure. Also known as the *normal freezing point*. (12.5)

nuclear binding energy The energy required to separate the nucleons in a nucleus. (20.2)

nuclear chain reaction A self-sustaining reaction sequence of fission reactions. (20.5)

nuclear fission The splitting of a large nucleus into smaller nuclei and one or more neutrons. (20.5)

nuclear fusion The combination of two light nuclei to form one heavier nucleus. (20.6)

nuclear transmutation The conversion of one nucleus to another. (20.1)

nucleic acid Macromolecule formed by polymerization of nucleotides. (23.6)

nucleons Protons and neutrons. (2.3)

nucleophile A region of negative or partial negative charge. (23.5)

nucleophilic addition An addition reaction that begins when a nucleophile donates a pair of electrons to an electron-deficient atom. (23.5)

nucleotide A structural unit consisting of a sugar (ribose or deoxyribose) bonded to both a cyclic-amine base and a phosphate group. (23.6)

nucleus The central core of the atom that contains the protons and neutrons. (2.2)

O

octet rule Rule stating that atoms will lose, gain, or share electrons to achieve a noble gas electron configuration. (6.1)

open system A system that can exchange mass and energy with its surroundings. (10.2)

optical isomers Nonsuperimposable mirror images. (22.2, 23.4)

ore A mineral deposit concentrated enough to allow economical recovery of a desired metal. (26.1)

organic compounds Compounds containing carbon and hydrogen, sometimes in combination with other elements such as oxygen, nitrogen, sulfur, and the halogens. (5.6)

osmosis The selective passage of solvent molecules through a porous membrane from a more dilute solution to a more concentrated one. (13.5)

osmotic pressure (π) The pressure required to stop osmosis. (13.5)

overvoltage The difference between the electrode potential and the actual voltage required to cause electrolysis. (18.7)

oxidation Loss of electrons. (9.4)

oxidation number See *oxidation state*.

oxidation state The charge an atom would have if electrons were transferred completely. Also known as *oxidation number*. (9.4)

oxidation-reduction reaction See *redox reaction*.

oxidizing agent A species that accepts electrons. (9.4)

oxoacid An acid consisting of one or more ionizable protons and an oxoanion. (5.7)

oxoanion A polyatomic anion that contains one or more oxygen atoms bonded to a central atom. (5.7)

P

p orbital Atomic orbital in which the angular momentum quantum number (ℓ) is 1. (3.8)

p-type semiconductor A semiconductor in which an electron-poor impurity is added to enhance conduction. (24.6, 26.3)

paramagnetic A species with unpaired electrons that are attracted by magnetic fields. (3.9)

partial pressure (P_i) The pressure exerted by a component in a gas mixture. (11.7)

pascal (Pa) The SI unit of pressure. (11.3)

Pauli exclusion principle No two electrons in an atom can have the same four quantum numbers. (3.9)

peptide bond The bond that forms between amino acids. (23.6)

percent by mass The ratio of the mass of an individual component to the total mass, multiplied by 100 percent. (13.3)

percent composition by mass The percent of the total mass contributed by each element in a compound. (5.9)

percent dissociation The percentage of dissolved molecules (or formula units, in the case of an ionic compound) that separate into ions in solution. (13.6)

percent ionic character The ratio of the observed dipole moment (μ) to the calculated dipole moment (μ), multiplied by 100. (6.2)

percent ionization The percentage of dissolved molecules that separate into ions in solution. (13.6)

percent yield The ratio of actual yield to theoretical yield, multiplied by 100 percent. (8.4)

period A horizontal row of the periodic table. (2.6)

periodic table A chart in which elements having similar chemical and physical properties are grouped together. (2.6)

pH A scale used to measure acidity. $pH = -\log[H^+]$. (9.5)

phase The sign of a wave's amplitude. (7.7)

phase change When a substance goes from one phase to another phase. (12.5)

phase diagram Summarizes the conditions (temperature and pressure) at which a substance exists as a solid, a liquid, or a gas. (12.6)

photochemical smog Air pollution resulting from the interaction of sunlight with nitrogen monoxide and carbon monoxide from automobile exhaust. (21.7)

photoelectric effect A phenomenon in which electrons are ejected from the surface of a metal exposed to light of at least a certain minimum frequency. (3.3)

photon A particle of light. (3.3)

physical change A process in which the state of matter changes but the identity of the matter does not change. (1.6)

physical process A process in which the state of matter changes but the identity of the matter does not change. (1.5)

physical property A property that can be observed and measured without changing the identity of a substance. (1.6)

pi (π) bond A bond that forms from the interaction of parallel p orbitals. (7.6)

pOH A scale used to measure basicity. $pOH = -\log[OH^-]$. (16.4)

polar Having a nonuniform electron density. (6.2)

polar covalent bonds Bonds in which electrons are unequally shared. (6.2)

polarimeter Device used to measure the angle of rotation of plane-polarized light caused by an optically active compound. (22.2)

polarized Describes a molecule in which a dipole moment has been induced. (7.3)

polyamide Polymers in which the monomers are connected by amide linkages. (24.1)

polyatomic ion Molecule that has lost or gained one or more electrons giving it a positive or negative charge. (5.7)

polyatomic molecules Molecules containing more than two atoms. (5.5)

polyester Polymers in which the monomers are connected by ester linkages. (24.1)

polymer Molecular compounds, either natural or synthetic, that are made up of many repeating units called monomers. (23.6, 24.1)

polypeptide Short chains of amino acids. (23.6)

polyprotic Refers to an acid with more than two ionizable protons. (5.7, 9.3)

polysaccharides Biological polymer consisting of sugars. (23.6)

positron A subatomic particle with the same mass as an electron, but with a positive charge. (20.1)

potential energy The energy possessed by an object by virtue of its position. (3.1)

precipitate An insoluble solid product that separates from a solution. (9.2)

precipitation reaction A chemical reaction in which a precipitate forms. (9.2)

precision The closeness of agreement of two or more measurements of the same quantity. (1.3)

pressure The force applied per unit area. (11.3)

principal quantum number (n) Designates the size of the orbital. (3.7)

product A substance that forms in a chemical reaction. (8.1)

protein A polymer of amino acids. (23.6)

proton A positively charged particle in the nucleus of an atom. (2.2)

pyrometallurgy Metallurgic processes carried out at high temperatures. (26.2)

Q

qualitative analysis The determination of the types of ions present in a solution. (17.6)

qualitative property A property of a system that can be determined by general observation. (1.6)

quantitative property A property of a system that can be measured and expressed with a number. (1.6)

quantum The smallest quantity of energy that can be emitted (or absorbed) in the form of electromagnetic radiation. (3.3)

quantum numbers Numbers required to describe the arrangement of electrons in an atom. (3.7)

R

racemic mixture An equimolar mixture of enantiomers that does not rotate the plane of plane-polarized light. (22.2, 23.4)

racemization The conversion of a single enantiomer to a racemic mixture of both enantiomers. (23.5)

radiation The emission and transmission of energy through space in the form of waves. (2.2)

radical A chemical species with an odd number of electrons. (8.8, 20.8)

radioactive decay series A sequence of nuclear reactions that ultimately result in the formation of a stable isotope. (20.3)

radioactivity The spontaneous emission of particles or radiation from unstable nuclei. (2.2, 20.1)

Raoult's law The partial pressure of a solvent over a solution, P_i, is given by the vapor pressure of the pure solvent, $P°$, times the mole fraction of the solvent in the solution, χ_i. (13.5)

rate constant (k) The proportionality constant in a rate law. (19.3)

rate law An equation relating the rate of reaction to the concentrations of reactants. (19.4)

rate of reaction The change in concentration of reactants or products per unit time. (19.3)

rate-determining step The slowest step in a reaction mechanism. (19.7)

reactant A substance that is consumed in a chemical reaction. (8.1)

reaction mechanism Series of steps by which a chemical reaction occurs. (19.7)

reaction order The sum of the powers to which all reactant concentrations appearing in the rate law are raised. (19.4)

reaction quotient (Q_c) A fraction with product concentrations in the numerator and reactant concentrations in the denominator, each raised to its stoichiometric coefficient. (15.2)

redox reaction A chemical reaction in which electrons are transferred from one reactant to another. Also known as an *oxidation-reduction reaction*. (9.4)

reducing agent A species that can donate electrons. (9.4)

reduction A gain of electrons. (9.4)

resonance structures Two or more equally valid Lewis structures for a single molecule that differ only in the positions of electrons. (6.5)

reversible process A process in which the products can react to form reactants. (15.1)

ribonucleic acid (RNA) Biological polymer consisting of nucleotides in which the sugar is ribose. (23.6)

root-mean-square speed (u_{rms}) The molecular speed that is inversely proportional to the molecular mass. (11.2)

S

s orbital Atomic orbital in which the angular momentum quantum number (ℓ) is 0. (3.8)

salt An ionic compound made up of the cation from a base and the anion from an acid. (9.3)

salt bridge An inverted U tube containing an inert electrolyte solution, such as KCl or NH_4NO_3, that maintains electrical neutrality in an electrochemical cell. (18.2)

salt hydrolysis The reaction of a salt's constituent ions with water to produce either hydroxide ions or hydronium ions. (16.10)

saturated solution A solution that contains the maximum amount of a solute that will dissolve in a solvent at a specific temperature. (13.1)

scientific method A systematic approach to experimentation. (1.1)

second law of thermodynamics The entropy of the universe increases in a spontaneous process and remains unchanged in an equilibrium process. (14.4)

second-order reaction A reaction whose rate depends on the concentration of one reactant raised to the second power or on the product of the concentrations of two different reactants each raised to the first power. (19.5)

semiconductor A substance that normally does not conduct electricity but that will conduct at elevated temperatures or when combined with a small amount of certain other elements. (24.6, 26.3)

semipermeable membrane A membrane that allows the passage of solvent molecules but blocks the passage of solute molecules. (13.5)

shielding The partial obstruction of nuclear charge by core electrons. (4.3)

SI units International System of Units. A system of units based on metric units. (1.2)

sigma (σ) bond A bond in which the shared electron density is concentrated directly along the internuclear axis. (7.6)

significant figures Meaningful digits in a measured or calculated value. (1.3)

simple cubic cell The basic repeating unit in which there is one atom at each of the eight corners in a cube. (12.3)

single bond A pair of electrons shared by two atoms. (6.1)

sintering A method used to form objects by heating a finely divided substance. (24.2)

smectic Containing molecules ordered in two dimensions. The molecules are aligned parallel to one other and are further arranged in layers that are parallel to one another. (24.3)

solubility The maximum amount of solute that will dissolve in a given quantity of solvent at a specific temperature. (9.2, 13.1, 17.4)

solubility product constant (K_{sp}) The equilibrium constant that indicates to what extent a slightly soluble ionic compound dissolves in water. (17.4)

solute The dissolved substance in a solution. (9.1)

solution A homogeneous mixture consisting of a solvent and one or more solutes. (9.1)

solvation Process by which solute particles become surrounded by solvent molecules. (13.2)

solvent A substance in a solution that is present in the largest amount. (9.1)

specific heat (s) The amount of heat required to raise the temperature of 1 g of a substance by 1°C. (10.4)

spectator ion An ion that does not participate in the reaction and appears on both the reactant and product side in the ionic equation. (9.2)

spectrochemical series A list of ligands arranged in increasing order of their abilities to split the d orbital energy levels. (22.3)

spontaneous process A process that occurs under a specified set of conditions. (14.1)

standard atmospheric pressure The pressure that would support a column of mercury exactly 760 mm high at 0°C. (11.3)

standard enthalpy of formation ($\Delta H_f°$) The heat change that results when 1 mole of a compound is formed from its constituent elements in their standard states. (10.6)

standard enthalpy of reaction ($\Delta H_{rxn}°$) The enthalpy of a reaction carried out under standard conditions. (10.6)

standard entropy ($S°$) The absolute entropy of a substance at 1 atm. (14.3)

standard free energy of formation ($\Delta G_f°$) The free-energy change that occurs when 1 mole of the compound forms from its constituent elements in their standard states. (14.5)

standard free energy of reaction (ΔG°_{rxn}) The free-energy change for a reaction when it occurs under standard-state conditions. (14.5)

standard hydrogen electrode (SHE) A half-cell based on the half-reaction $2H^+(1\,M) + 2e^- \rightarrow H_2(1\text{ atm})$, which has an arbitrarily defined standard reduction potential of zero. (18.3)

standard reduction potential (E°) The potential associated with a reduction half-reaction at an electrode when the ion concentration is $1\,M$ and the gas pressure is 1 atm. (18.3)

standard solution A solution of precisely known concentration. (9.6)

standard temperature and pressure (STP) 0°C and 1 atm. (11.5)

state function Properties that are determined by the state of the system, independent of how the state was achieved. (10.2)

state of a system The values of all relevant macroscopic properties, such as composition, energy, temperature, pressure, and volume. (10.2)

stereoisomers Molecules that contain identical bonds but differ in the orientation of those bonds in space. (22.2, 23.4)

stoichiometric amount Quantity of reactant in the same relative amount as that represented in the balanced chemical equation. (8.3)

stoichiometric coefficients The numeric values written to the left of each species in a chemical equation to balance the equation. Also known as *coefficients*. (8.1)

stratosphere The region of the atmosphere located above the troposphere and consisting of nitrogen, oxygen, and ozone. (21.1)

strong conjugate acid The conjugate acid of a weak base. Acts as a weak Brønsted acid in water. (16.8)

strong conjugate base The conjugate base of a weak acid. Acts as a weak Brønsted base in water. (16.8)

strong electrolyte An electrolyte that ionizes or dissociates completely. (9.1)

structural formula A chemical formula that shows the general arrangement of atoms within the molecule. (5.5)

structural isomers Molecules that have the same chemical formula but different arrangements of atoms. (7.2)

sublimation The phase change from solid to gas. (12.5)

substance Matter with a definite (constant) composition and distinct properties. (1.5)

substituent A group other than —H bonded to the carbons of an organic molecule. (23.2)

substitution reaction One group is replaced by another group by electrophilic or nucleophilic attack. (23.5)

superconducting transition temperature (T_c) The temperature below which an element, compound, or material becomes superconducting. (24.7)

superconductor A substance with no resistance to the flow of electrons. (24.7)

supercooling A phenomenon in which a liquid can be temporarily cooled to below its freezing point. (12.5)

supercritical fluid A fluid at a temperature and pressure that exceed T_c and P_c. (12.5)

supersaturated solution A solution that contains more dissolved solute than is present in a saturated solution. (13.1)

surface tension The amount of energy required to stretch or increase the surface of a liquid by a unit area. (12.2)

surroundings The part of the universe not included in the system. (10.1)

suture Thread used by physicians to close wounds and/or incisions. (24.4)

syndiotactic Polymers in which the substituents alternate positions along the polymer chain. (24.1)

system The specific part of the universe that is of interest to us. (10.1)

T

tacticity Describes the relative arrangements of chiral carbon atoms within a polymer. (24.1)

termolecular Involving three reactant molecules. (19.7)

theoretical yield The maximum amount of product that can be obtained from a reaction. (8.4)

theory A unifying principle that explains a body of experimental observations and the laws that are based on them. (1.1)

thermal energy The energy associated with the random motion of atoms and molecules. (3.1)

thermochemical equation A chemical equation that includes the enthalpy change. (10.3)

thermochemistry The study of the heat associated with chemical reactions and physical processes. (10.1)

thermodynamics The scientific study of the interconversion of heat and other kinds of energy. (10.2)

thermonuclear reaction Generally refers to a fusion reaction. (20.6)

thermoplastic Polymers that can be melted and reshaped. (24.1)

thermosetting Polymers that assume their final shape as part of the chemical reaction that forms them. (24.1)

thermosphere The outermost layer of the atmosphere—also known as the ionosphere. (21.1)

third law of thermodynamics The entropy of a pure crystalline solid is zero at absolute zero (0 K). (14.4)

threshold frequency The minimum frequency of light required to eject an electron from the surface of a metal in the photoelectric effect. (3.3)

titration The gradual addition of a solution of known concentration to another solution of unknown concentration until the chemical reaction between the two solutions is complete. (9.6)

tracers Radioactive isotopes that are used to trace the path of the atoms of an element in a chemical or biological process. (20.7)

trans The isomer in which two substituents lie on opposite sides of a double bond. (23.4)

transition element An element that has—or readily forms one or more ions that have—an incompletely filled *d* subshell (Groups 3B to 8B or Group 1B). (2.6)

transition metals The elements in Group 1B and Groups 3B to 8B. (2.6)

transition state See *activated complex.*

transmittance Ratio of the intensity of light transmitted through a sample (I) to the intensity of the incident light (I_0). (9.5)

transuranium elements Elements with atomic numbers greater than 92, created by bombarding other elements with accelerated neutrons, protons, alpha particles, or other nuclei. (20.4)

triple bond A multiple bond in which the atoms share three pairs of electrons. (6.1)

triple point The point at which all three phase boundary lines meet. (12.6)

triprotic acid An acid molecule with three ionizable protons. (9.3)

troposphere The layer of the atmosphere closest to Earth's surface. (21.1)

Tyndall effect The scattering of visible light by colloidal particles. (13.7)

U

unimolecular Describes a reaction involving one reactant molecule. (19.7)

unit cell The basic repeating structural unit of a crystalline solid. (12.3)

unsaturated solution A solution that contains less solute than it has the capacity to dissolve. (13.1)

V

valence band The bonding band. (24.6)

valence bond theory Atoms share electrons when an atomic orbital on one atom overlaps with an atomic orbital on the other. (7.4)

valence electrons The outermost electrons of an atom. (4.2)

valence-shell electron-pair repulsion (VSEPR) A model that accounts for electron pairs in the valence shell of an atom repelling one another. (7.1)

van der Waals equation An equation relating the volume of a real gas to the other parameters, P, T, and n. (11.6)

van der Waals forces The attractive forces that hold particles together in the condensed phases that include dipole-dipole interactions (including hydrogen bonding) and dispersion forces. (7.3)

van't Hoff factor (*i*) The ratio of the actual number of particles in solution after dissociation to the number of formula units initially dissolved in solution. (13.5)

vaporization The phase change from liquid to gas at the boiling point. (12.2)

viscosity A measure of a fluid's resistance to flow. (12.2)

visible spectrophotometry Measurement of absorption of visible light for qualitative and quantitative analysis. (9.5)

volatile Describes a substance that has a high vapor pressure. (12.2, 13.5)

W

wavelength (λ) The distance between identical points on successive waves. (3.2)

weak acid An acid that ionizes only partially. (16.6)

weak base A base that ionizes only partially. (16.7)

weak conjugate acid A conjugate acid of a strong base. Does not react with water. (16.8)

weak conjugate base A conjugate base of a strong acid. Does not react with water. (16.8)

weak electrolyte A molecular compound that produces ions upon dissolving but exists in solution predominantly as molecules that are not ionized. (9.1)

X

X rays Electromagnetic radiation emitted when cathode rays are directed to glass or metal. (2.2)

X-ray diffraction A method of using X rays to bombard a crystalline sample to determine the structure of the crystal. (12.3)

Z

zeroth-order reaction A constant rate, independent of reactant concentration. (19.5)

Index